2006

To many, many years of appreciating "birdies".
Keep up your wonderful passion, it's part
makes you who you are, and I love who you
My love to you, Evelyne

The Howard and Moore
COMPLETE CHECKLIST OF THE
BIRDS OF THE WORLD

Revised and Enlarged Third Edition

The Howard and Moore
COMPLETE CHECKLIST OF THE BIRDS OF THE WORLD

Revised and Enlarged Third Edition

Edited by Edward C. Dickinson

Research Associate, Natural History Museum of the Netherlands, Leiden

Regional Consultants
Edward Dickinson (Asia)
David Pearson (Africa)
Van Remsen (The Americas)
Kees Roselaar (The Palaearctic)
Richard Schodde (Australasia)

Special Adviser
Joel Cracraft

Compilers
Norbert Bahr
Nigel Cleere

PRINCETON UNIVERSITY PRESS
PRINCETON AND OXFORD

Published in the United States, Canada, and the Philippine Islands
by Princeton University Press, 41 William Street, Princeton, New Jersey 08540

In the United Kingdom and European Union, published by Christopher Helm,
an imprint of A & C Black Publishers Ltd., 37 Soho Square, London W1D 3QZ

ISBN 0-691-11701-2

Library of Congress Control Number 2003101580

This book has been composed in New Baskerville

www.birds.princeton.edu

Printed in Hong Kong

10 9 8 7 6 5 4 3 2 1

Recommended citation:

Dickinson, E.C. (Editor) 2003. *The Howard & Moore Complete Checklist of the Birds of the World*. 3rd Edition. Princeton University Press, Princeton, New Jersey.

Citation to the chapter by Cracraft *et al.* is recommended in the following format:

Cracraft, J., Barker, F.K. & Cibois, A. 2003. Avian higher-level phylogenetics and the Howard & Moore Checklist of Birds. Pp. 16-21 in: Dickinson, E.C. (Editor) 2003. *The Howard & Moore Complete Checklist of the Birds of the World*. 3rd Edition. Princeton University Press, Princeton, New Jersey.

Contents

Foreword

by
Richard Howard

It is some twenty-five years since work began on the first edition of this book. The late Alick Moore deserves much of the credit for that, and more than half the credit for the second edition. He and I were perhaps lucky to recognise the appeal of a compact list of the world's birds that included subspecies. Our effort was the first such book, and for twenty years the only such list.

It would never have been possible without the tremendously valuable and important groundwork that created a *Check-list of Birds of the World* by J.L. Peters and others. The first edition of 'Howard & Moore' (1980) appeared before volume 11 of 'Peters' in 1986, and by 1979 some volumes of 'Peters' had been in print for over 40 years. In consequence, Alick's work on the passerines involved areas that had not yet been covered and my work on the non-passerines required the updating of much that had been done many years earlier. In our second edition (1991), complemented in 1994 by an Appendix, the focus was on bringing in the additions and changes from the recent literature. Alick happily oversaw this process.

Based on a wealth of feedback, this third edition is essentially a new work; made easier to achieve perhaps by the existence of our foundation work, but nonetheless entirely revised throughout. The main driver of change is that birdwatchers have become serious about their birds. A list that does not explain in depth will not begin to satisfy the raft of intelligent questions that readers ask. Also, as the years pass, Peters Check-list is more and more in need of a supplement: one that treats essentially every new name proposed since each of its volumes, and provides a full reference to both these and to most of the many recent revisions. This revised edition sets out to achieve these objectives.

Some five years ago I asked Edward Dickinson for his advice. Realising that the recommendations that he made corresponded to what users of 'Howard & Moore' would now expect, I was delighted to accept his proposal that the revision be far-reaching and organised on a collaborative basis with five regional consultants: one each for the Americas, Africa, Asia, Australasia and the Palaearctic. This has led to far wider consultation and has resulted in a very thorough revision that goes far beyond simple updating and the correction of the errors and omissions of the last edition. Incidentally, our readers pointed out a good few of these errors and I am most grateful to them.

The introductory section will offer a perspective on the reasons for these changes and explain in detail the revised structure of the book, the extra content and the new features.

I hope that birdwatchers and ornithologists will continue to demand regular updates of this book. Part of the royalties from this and future editions will not be distributed, but used to help finance the vital literature surveillance as an on-going process rather than a search over a number of past years. We also intend to keep a team in place that will contribute on a regular basis and then later engage on a focused revision.

Introduction

OUR OBJECTIVES

First, as in previous editions, we set out to provide **a comprehensive list**.

Not only have our regional consultants helped catch omissions and other mistakes, but we have brought together what we believe to be a *nearly* complete list of all the taxa described since Peters Checklist. We have used this to strengthen our approach. In recent years it has been necessary for someone reviewing the taxonomy of a species to take the subspecies given by Peters (1937) (1823) and to try to find all subsequent relevant literature – both newly proposed subspecies and other related papers – in order to ensure a thorough review. The opportunity to search on-line databases makes researching the last 20 years or so simpler. The years prior to that, however, depend on a painstaking examination of the Zoological Record; sadly, however, that does not suffice, as its coverage – although better and better over time — is less than complete.

We believe that 98% of all names that are new since Peters Checklist are included here: some recognised, some in synonymy. Space does not permit a complete synonymy: ours includes only names that are new *and* in synonymy, and names that we listed as recognised taxa in the last edition but that we now have in synonymy. In 95% of cases we cite an authority for the placement there.

Second, we try to present **a conservative list** that is essentially compiled from the judgements of published authorities. However, we have taken our conservatism a step farther. We have chosen to consider Peters Checklist as something of a watershed. Despite some unevenness in the treatment, which we discuss further below, it is a good discipline to look for persuasive published reasons for changes. We have done so and we have deferred acceptance of many proposals because sufficient detail has not yet appeared. We should like to stress that in many of the cases where we defer the acceptance of a 'split' we are confident of the abilities of the observers who have recommended such splits; all that remains is to do the paperwork thoroughly. We expand on this topic later.

Third, our aim is **a list with a high standard of nomenclatural accuracy** — one that we shall strive to improve each time. In this context we focus on three things: a) correctly formed scientific names, drawing on the expert advice of Normand David; b) correct attribution of names to authors and c) correct dates. Some two hundred scientific names are spelled differently in this edition.

As experienced users of Peters Checklist we believe that the section authors of that work verified most citations to the originals, but we also know that some were not checked, and that mistakes slipped in.

Since the initiation of that Checklist in 1931 the *International Code of Zoological Nomenclature* has been introduced and has reached its fourth edition (I.C.Z.N., 1999) (1178). There are now agreed rules explaining how to attribute names in circumstances where more than one author may be involved. We have begun to make the changes needed to apply these rules (see Appendix I). Furthermore bibliographic studies have not stood still and in some cases we can date work more accurately today than could the authors of Peters. Late in the compilation of this edition we began to collaborate with various correspondents including Murray Bruce, Rémy Bruckert, Normand David, Steven Gregory, Alan Peterson (whose website www.zoonomen.net we recommend), Aasheesh Pittie (whose CD-ROM (1901A) is very useful), and Michael Walters. Between us we will gradually check out citations where there are doubts about the authorship or the dates. A first few steps in this direction are taken herein; for details see Appendix II.

METHODOLOGY

As we had to add an author and a date to every taxon name we have taken the opportunity to review every entry; however with over 25,000 names to check it is likely that some candidates for change or correction have slipped past us. Improved range statements are offered for most forms (but the range statements for most New World taxa will be dealt with mainly in the next edition). Our expert advisers on certain groups where field identification is particularly difficult have persuaded us to limit range statements to exclude sight records that were doubtful and potentially misleading. We do not normally include vagrancy details. Introduced species create a problem for a list like this and we have largely ignored introduced ranges, except in well-established cases.

In a first stage of our revision a compiler entered most of the authors' names and dates and, based on a literature search, proposed various revisions. A primary objective of this exercise was to provide the regional consultants with a more manageable task.

The listing has not been a Committee process; for any given region the regional consultant has worked with the Editor. In general the views of the regional consultant have been followed for range statements and vernacular names. This is only very slightly less true in the context of taxonomic judgements; here, in some cases, there has been a broader approach with special advice being sought from experts on the family.

In our two previous editions we provided an initial list of our main references. The extent to which particular decisions cited one or other of these works as our 'authority' was limited.

In this edition we place more reliance on major works that are strong on detail, for example, *The Birds of the Western Palearctic*, the *Handbook of Australian, New Zealand and Antarctic Birds* and the *Birds of Africa*. For the New World we have drawn heavily on the A.O.U. Checklist and on members of the Committee working on a Checklist of the Birds of South America. Other major works such as the *Handbook of Birds of the World* have also been consulted. In this some of the authors have worked hard on the taxonomy of their groups, but there has been too little space for them to spell out the detailed reasons for novel treatments. In some cases such treatments are promised us in the future. Where we are able we have preferred to cite the works in which such details appear. Numerous field-guides and monographs are available that have helped with one group or another, but taxonomic revision should normally be found in the primary literature, the refereed journals.

However, whatever we cite is gathered in one place: the list of references. That list is arranged in alphabetical order, yet every item is numbered and to save space it is just this number that is cited, in rounded brackets, throughout the text and the footnotes. The result of citing authorities for most decisions and of giving full references for all new taxa is a very substantial reference list. We hope and believe this will be helpful to all readers.

We have been privileged to draw on knowledgeable advisers on generic names and their genders.

RELATED TOPICS

List sequence

The Howard and Moore Checklist has been closely based on the sequence adopted for Peters Checklist since 1980. Our 1991 edition followed too closely on the heels of Sibley & Monroe's proposals (2262) to consider changing.

For this edition the idea of change could not be ignored. We therefore invited Joel Cracraft to provide us with a framework, setting the sequence of families to use (and in so doing advising on issues relating to familial ranking). He describes this in the following chapter. We have sought to follow his views precisely and misinterpretations, if any, must be blamed on the editor alone. Below family level we have usually mentioned subfamilies; most being those that have been traditionally accepted, but some we use with less certain grounding and we introduce these to help put across our understanding of the structure and more as suggestions for others to consider than as matters of fact. In several cases genera stand out that seem, in the light of molecular evidence, to deserve familial status. Where formal proposals to this effect have not been published we have preferred to await such publication. In such cases we place these genera, sometimes side by side (e.g. *Melampitta* and *Ifrita*) under the heading Genera Incertae Sedis. Clearly, however, not all genera placed in these holding patterns deserve family status.

The list itself disregards all suprafamilial rankings. Reasons to do this can be found in the following chapter by Cracraft *et al.* As in previous editions we do not include the names of tribes.

Our initial intention was to limit changes to those already proposed but to achieve a reasonably coherent list we have needed to include work that is in press. The sequence of families that we use appears on p.22.

We fully expect future editions to require further revision to this list sequence as more evidence accumulates, and corroboration of molecular evidence by other evidence strengthens due to more focussed studies.

Species concepts

The introduction to the 1991 edition made several points that remain fundamental to the role of a list like this. It is as true now as it was then that the established order and relationships of birds are questioned more often and more deeply than they are in other animal groups, and that most issues of many leading ornithological journals contain new ideas or discoveries affecting such lists. It is true too that new techniques, including molecular methods, are playing a significant role. However, increasingly the results of molecular studies are being interpreted in the light of all the other evidence available. Corroborative evidence is important and probably no single technique will suffice to sufficiently inform us.

The last decade has witnessed the debate between the protagonists of two different species concepts, the traditional Biological Species Concept (BSC) and the newer Phylogenetic Species Concept (PSC). See for example: Haffer, 1992, 1997 (1053a, 1054a); Snow, 1997 (2294a); Sangster *et al.*, 1999 (2165a); Schodde & Mason, 1999 (2189); and Collinson, 2001 (549a).

The debate continues, and this book is not the place to summarise it. However, the implications for a list of this kind are important and some comments are needed. The shared goal is to achieve a broad consensus as to what should be called a species. This may, or may not, come down to which taxa are truly distinct and which are not. Any detailed examination of a polytypic species is likely to reveal that just how distinct each taxon is from each of the others varies. There are a number of reasons for this. Perhaps the most important is that evolutionary change seems to happen more or less quickly according to a range of influences that will have differed both in time and in space from one population to another. Another important factor regarding observable differences in distinctness is human judgement, which varies from individual to individual as well as over time as both the state of scientific knowledge, the views of the individual, and the fashions in taxonomy change.

What does "truly distinct" mean? Clearly not simply that one can see a difference, for it is now apparent that birds can see differences that we cannot. Acoustic evidence is of increasing value, and nowhere more usefully than in the context of "look-alikes" or sibling species. For believers in the BSC the ability of two forms to interbreed is the evidence needed to show that they are conspecific. Proof of such ability requires geographic contact and an opportunity to interbreed. Where two *species* interbreed we talk not of intergradation but of hybridisation. When the resultant gene flow seems to create a seamless whole we no longer talk of hybridisation. Here we speak, or should speak, of intergradation. Where it seems abruptly broken we suspect secondary contact and hybridisation. Here then is a process of interpretation. Very limited zones of more or less constant hybridisation do sometimes occur between two good species, apparently without the gene pool more distant from the mixing being affected. This biologists take to support the notion that disadvantages inherent in the process cause the two species to maintain their integrity and that the restricted zone of hybridisation helps to do this. It easy to understand that there is room here for reasonable people to take opposing views at least before such a zone stabilizes.

Forms that we consider to be related subspecies are often separated by large geographical gaps. These birds are so treated because it has been inferred that they are alike enough that they would be expected to interbreed if the two populations came into contact. Again this is a subjective judgement, but one tries to use as objective a methodology of evaluation as possible. Once based on anatomy and morphology, the decision is now increasingly supported by evidence from voice and "genetic distance" as evidenced by DNA studies. Now that we know that many birds have the ability to see ultraviolet light we know that what they see is not what we see. It is very possible therefore that some pairs of species that to us look very closely alike look much more distinct to the birds themselves.

The PSC claims to be focused on the degree of perceptible difference (without abstraction to breeding biology), and so to be able to dispense with the problems that arise in trying to judge situations in which disjunct ranges prevent any test of whether interbreeding might occur. Disjunct taxa thought related under the BSC are very often "diagnosably distinct" and, when this is so, can quite easily be judged species under the PSC. This does not make the degree of difference a significantly less subjective judgement, but to empower the process a wide range of traits can be compared and again increasing use is made of different kinds of supporting evidence. However there are problems with this approach (Johnson *et al.*, 1999) (1216a). Because the PSC leads to the recognition of more species it is likely to find favour in a number of contexts and it has been suggested that there is reason to expect the two concepts to continue to find supporters and to co-exist. Time will tell.

For a book of this kind a choice has to be made. Here we believe it helpful to adhere, as best we can, to the Biological Species Concept. Our reasons are 1) that much more work has yet to be done to offer a world list based on the PSC, and 2) that we are not yet convinced that the PSC is inherently any less subjective.

Our second point touches on whether we *need* to go beyond the BSC. Here we think it useful to instance the numerous changes to Australian avian taxonomy, to be found in Schodde & Mason (1999) (2189), not based on acceptance of the PSC. What has made these changes possible is the collation of specimens that much better represent the picture of variation that is actually found on the ground. For many areas of the world such samples cannot be assembled due to gaps in collecting, the quantity and quality of available material and museum loan rules.

Regarding both the first point and the second see Johnson *et al.* (1999) (1216a), who rejected the PSC and broadened the BSC by introducing a more comprehensive version, which seems to us to be helpful. In some ways parallel to that is the Evolutionary Species Concept (ESC), of which we were reminded by Peterson (1998) (1834a).

For most of us the matter is still *sub judice* and the easy option is to hold to the tried and tested older concept until we are convincingly shown that we must adapt to the pressures of change.

Species and subspecies

It is important to realise that there is no one correct list. Every compilation is built on a pyramid of judgements, each with potential for reasonable people to disagree. We have already stated that our list is intended to be conservative, and that we have excluded some species that common sense tells us should be recognised. We contend only that there is a burden of proof required before conviction is general.

Our arrangement of genera is inconsistent. Sometimes, where substantial molecular studies have sampled enough taxa, we have tried to follow published phylogenies, in these cases listing from oldest to youngest, i.e. from deepest branch to least deep. Such arrangements are footnoted and references cited. We are aware that the sequencing of just one unsampled genus could change the construction of the tree. Otherwise we have arranged genera within families or subfamilies in a traditional arrangement, generally that used in Peters Checklist.

Species are again arranged by default as Peters treated them, but where there are persuasive recent reviews we follow them with a citation. We do however have two guiding principles that we have tried to follow, first we keep together those species that comprise a superspecies, second in such a context we list species as we do subspecies, i.e. from northwest to southeast. We have not been wholly consistent here. For example, in the case of subspecies, we have made exceptions where there are evident groups of subspecies that form incipient (or actual but not yet documented) species. We have also sometimes left subspecies in a previous order where they were previously arranged zoogeographically. Most typically this has been our usual practice for the Philippines, where, normally, the logic applied by Dickinson *et al.* (1991) (745) has been retained.

Taxon 'recognition'

We would not advocate the view that our list *recognises* taxa. It is very largely a compilation of other works that do that and thus the credit for recognition does not belong to us. By contrast it can be argued that we have not recognised particular taxa and indeed here our biases will show.

Earlier we stated that we considered Peters Checklist to be a watershed. The editors of Peters's Checklist did their best to impose on each author a methodology that required a re-examination of a range of study skins that would be sufficient to support the conclusions. Even if many authors failed to publish preliminary papers, it may be assumed that a robust editorial process kept the work close to the standard set by Peters himself. Admittedly the ethic of the Checklist may now be thought to have favoured a process of lumping. Such lumping may from case to case now seem to have been arbitrary. To replace the lumping by an arbitrary split (or re-elevation to species status) does no more than continue an anarchic process. Two wrongs do not make a right. The same must, of course, be said in cases where there were splits that we now see being lumped again. The minimum requirement should be a reference to whose treatment is followed.

There are many reversals of the treatment in Peters that have now stood the test of time for years. Clearly we do not seek to go back and re-examine the level of proof in such cases; we do however presume that major regional handbooks or checklists will examine proposals for rearrangement. By and large our readers will find that we base our conclusions for all earlier proposals on these sources and that our footnotes reference them. In essence, the point is that we regard the editors of such works as being the arbiters and our role as being that of the reporter. Only to provide a more comprehensive treatment do we render a *provisional* judgement in later cases.

If, ten years ago, the introduction to the last edition could refer to change flowing from the pages of most issues of the journals, this introduction must mention the extent to which change is now proposed elsewhere. Books that we would traditionally have seen as secondary references, such as monographs and handbooks, now include innovation. It is very possible that this could, for a while, be defended on the rational grounds that the evolution of the practice of ornithology had orphaned the taxonomic review. Few professional ornithologists, outside the museum community, contributed to this work until molecular technology and other recent developments stimulated a new round of work. Monographs, handbooks, and even field guides, may well be as thoroughly reviewed by the writer's peers as are the best ornithological journals. But, even if this is so, the process is not transparent. The space made available in such books tends to contain the conclusions with little of the supporting detail (details of the quantum of evidence examined, the statistical tables and the graphs for example). There is, in fact, an insufficient 'audit trail' for subsequent reviewers to judge.

A list like this faces the danger of either being uncritical or of being biased in its decisions over which changes to admit. We have explained above that the process needs the fuel of evidence to run on. In general, where we have found no regional authority to follow, we have been readier to accept changes proposed in refereed journals than those proposed elsewhere. We have not knowingly accepted the elevation of any taxon to species level unless it was originally named binomially or has been fully reviewed. If it was originally named binomially then we apply the concept of prevailing usage in judging whether detailed reasoning should have been published.

Occasionally we accept a view rejected by a Checklist Committee. We have tried to footnote a reference to the alternate view in such cases. Again, just occasionally, personal experience has led us to move names into or out of synonymy and in these cases the initials of the Editor or of the relevant regional consultant will appear in the footnote.

Our conservative bias still allowed some proposals for elevation to specific rank to qualify for full inclusion. Some regional works, for example Ridgely & Tudor (1989) (2018), included small print rationales for their taxonomic decisions, usually, but not always, referring to prior publications. We like this approach and we encourage others to use it; the onus of proof is on the writer, but presumably the space must be negotiated with the publishers. In some cases we have not accepted such a proposal, preferring to wait for fuller 'primary' publication in an appropriate refereed journal or for recognition by an appropriate authority. Our approach to this may look arbitrary as the regional consultants came to such situations with their personal views and sought to apply broad guidelines, not a set of strict rules. We apologise to any who think that discrimination has worked against them.

The Editor has found the periodic papers entitled "New species of birds described from ..... to ......" very useful. They were begun by Meise, but continued by Mayr and others (1479, 1483, 1496, 2551, 2552). There are two problems, however, apart from the fact that the views in these papers are not always those that come to be accepted. These are, first, that a list like this cannot wait for such reviews to appear, and, second, that no attempt is made to review species that reach recognition through elevation from prior subspecies level treatment. Despite these drawbacks we hope the series continues and, as explained in 'Notes on the list presentation', we flag taxa that we believe will be treated in the next such list.

At the level of new subspecies it should be noted that those who seek to judge them cannot do so without access to type specimens and if these are deposited in small museums every effort should be made to place paratypes in one or two major museums.

Scientific names

The rules concerning scientific names are governed by the International Code of Zoological Nomenclature (I.C.Z.N., 1999) (1178). We have mentioned the correctness of spelling earlier. Here we wish to touch on the concept of 'original spelling'.

Over that last forty years there has been considerable discussion amongst the specialists about the idea that the original spelling should always be maintained. Against this view could be found those arguing that names which had been emended many decades ago should not now revert to such spellings as stability was preferable to exactitude. Other related subjects surfaced; among them the idea that corrections to spelling that could be shown to reflect reality should be permitted. The 3rd edition of the Code (1985) left some of these matters for an eventual 4th edition.

Unfortunately when the list of Sibley & Monroe (1990) (2262) was published it still seemed probable that the vote would favour original spellings over stability and the acceptance of genuine corrections (such as the emendation of eponyms — names honouring people — to ensure their names were spelled right). It was in anticipation of this vote that Monroe included changes that 'jumped the gun' (A.P. Peterson *in litt.*).

Finally the 4th edition of the Code was not what Monroe had expected: stability was provided for, such that names that had truly been out of use for many years could no longer be brought into use just because they had priority. Equally, *clearer* rules were set out governing emendations. The result is that changes, and even formal proposals, made in the period from 1985 to 1999 that ran counter to prevailing practice should probably be ignored, although in rare cases they may have quickly established prevalence. There are, in fact, no clear guidelines on judging 'prevailing usage'; despite this however it seems clear that consistent usage from 1900 to 1985 is likely to outweigh usage introduced post 1985. We footnote those cases that have attracted our attention, explaining the decision made in this list.

English names

English names used in this edition have not been a main focus of revision.

It is not our intention that our list should be seen as a 'standard list' of vernacular names. We have made changes to bring the list into greater conformity with regional lists for those species that are endemic to those regions, and we have considered the English names chosen for the published volumes of the *Handbook of the Birds of the World*.

In a few cases we have included two names or two alternatives (placing our preference first). We have also applied certain ground rules to hyphenation and to the capitalisation of group-names (which, for the sake of grammatical consistency, may run counter to locally established practice).

We urge our readers, especially those in the Americas and Australasia, to reconsider the tendency to use hyphens so widely. Not only does this make heavy weather of the hunt through the index, but there are some 'rules of grammar' worth considering (see Inskipp *et al.*, 1996: 7-10) (1182). We have bent those rules here! We do this in the case of group-names comprised of two group-names. For example we use Hawk-Eagle rather than Hawk Eagle (which would be grammatically better). This exception is intended to help the Americans and Australians, in particular, to come to terms with 'losing' many of their hyphens. We have also tried to be consistent over the adjectival form of geographic names (e.g. using Madagascan in place of Madagascar).

Otherwise our changes are few, since in our opinion there are too many different alternative English names already.

Yet there is a paucity of good 'generic names' and too often the same one, e.g. Robin, is used for two or more unrelated groups that simply include one member with a passing resemblance to the species first so named. Although this is a product of history, and has acquired local sentimental value, it does not help us to explain relationships in a world where printed matter gains far wider currency than in the 19th century. Yet we feel it best to resist making changes until we have extracted most of what molecular studies can tell us about these relationships and until there is more consensus on the value of 'generic vernaculars'.

We understand why some people would like to see a standard list of English names. We also understand why others oppose this. No doubt any consensus list published will be of 'Recommended Names'. When this comes out, as one will under the auspices of the 2002 International Ornithological Congress, we hope that there will first be a consultation phase and thereafter to be in sufficient agreement with its style and content that we can adopt it whole-heartedly.

Extinct taxa

Our one objective has been to include these as a service to certain users. We make no claim that our information about whether or not each is actually extinct is up to date. Nor do we seek to include threat categories for endangered species.

Range statements

We have tried to maintain brief statements, but to improve them. Most New World range statements need further revision and more consistency in the indication of which taxa are montane and which are not; this will be a focus of work on the next edition.

We have updated names that have changed, using the current names of Chinese provinces and of regions in and around Russia. Except where more precision is needed Russia can usefully be seen as four regions: a) European Russia lying west of the Urals, b) Siberia stretching across the north from there to the Bering Sea, and south of this c) Transbaikalia, essentially from Lake Baikal to Mongolia, and d) the Russian Far East.

We use Bioko in place of Fernando Po, Bacan in place of Batjan, and Kiritimati for the "Christmas Island" in the Pacific — this leaves the name Christmas Island to the island in the Indian Ocean. We have also made several other similar changes, mostly minor, but we have retained Burma in preference to Myanmar.

We have also allowed rather more repetition of the names of island groups to help users quickly relate to the geography. Maps are under consideration for a future edition. Here we have used spellings from the Times Atlas and we would expect readers to find modern geographic names without difficulty; future maps are likely to focus on helping with the historical names and boundaries of older states and provinces.

To state the wintering ranges of migrants birds represent a challenge. As before we have defined the breeding ranges and then after >> indicated the main wintering grounds. It is implied that birds will be found on passage in the area between the two. It is also well known that there are often problems identifying wintering individuals to subspecies, and that much work remains to be done in the field to understand the extent of overlap of winter ranges. We have tried to avoid including accidental occurrences.

The List of References

We have used EndNotes software to maintain our bibliographic database. The result is a list that includes only works that we have cited and not all that we have consulted. We should have liked our list to give the page numbers where each new description would be found. Some papers cited, e.g. Koelz (1954) (1284) contain dozens of new names, and we decided that this would have required too many pages.

Cut off date

We have scanned the journals to 31 December 2000. Over the following 9 months or so we have included such material as has come to our notice. Essentially the same dates apply to books. Our search for next edition will begin with all 2001 publications.

Our efforts to achieve comprehensive coverage were not always up to the task of obtaining papers published in small local journals, even national ones. This is especially true when such journals deal with natural history in general, and when the editorial staff of the Zoological Record does not scan the journal. We sincerely apologise to any author whose proposal may have been overlooked and would ask that a copy be sent to us care of our publisher.

Avian Higher-level Phylogenetics and the Howard and Moore Checklist of Birds

by

Joel Cracraft, F. Keith Barker, and Alice Cibois

Department of Ornithology, American Museum of Natural History
Central Park West at 79th Street, New York, New York 10024

As phylogenetic knowledge has increased over the past 40 years following the introduction of cladistic theory and methodology, systematists have increasingly attempted to make classifications reflect phylogenetic relationships. Given an hypothesis of relationships it is reasonably straight-forward to express those relationships within the context of hierarchical classifications in a number of different ways. Ornithologists have addressed this issue for over 20 years (e.g., Cracraft 1981) (578)[1]. Without question the best-known attempt at constructing a phylogenetic classification for avian higher taxa was that derived from DNA hybridization distances (Sibley *et al.* 1988; Sibley & Ahlquist 1990; Sibley & Monroe 1990) (2264, 2261, 2262). However there are many other examples at lower hierarchical levels within birds (e.g., Livezey's 1998 classification for gruiforms) (1368).

The present volume — a list of avian species — does not arrange these within a series of hierarchies but instead uses, aside from the Linnean binomial of genus and species, two hierarchical levels, the family and the subfamily. Because knowledge of avian phylogenetic relationships is still clouded with uncertainties, any decision to recognize a complex classificatory hierarchy would have resulted in numerous arbitrary choices. We have attempted to reflect currently accepted hypotheses of relationships to the extent possible within a linear sequence. We have followed a simple set of conventions. First, the list sequentially clusters families in a way that represents what are thought to be groupings of related families; many of these follow traditional orders when compelling alternative evidence is not available. Second, within each of these clusters, subgroups of related families are listed near one another. Finally, within all such groups and subgroups, the presumed basal families are listed first.

This volume is not the place to present an extended discussion about the higher-level relationships of birds. Nevertheless, the present list attempts to incorporate recent evidence on relationships, with the caveat that often the best evidence presently available is only of a single kind — for example, either morphology or molecular. Ideally we would like to have congruence across data sets before concluding that a satisfactory understanding of those relationships has been attained. This is a time of unprecedented activity in avian systematics, with more investigators collecting data than at any time in history. Consequently, understanding of avian relationships is changing rapidly. In discussing the evidence, only that work which is published, or in press, is mentioned. The list errs on the side of conservatism because a list of species, designed in part for a popular audience, is not an appropriate venue for making radical changes. Over the years, and continuing to the present, there have been many calls to dismember traditional orders, but most of these opinions are based on evidence that is still not widely accepted within the systematic community, where ideas about relationships remain contentious.

In the following remarks, not all groupings (orders) will be discussed, either because there is relatively little dispute over relationships or because there has been no new information about relationships published in recent years.

Basal lineages: Palaeognathae, Neognathae, and Galloanserae

The identity of, and relationships among, the three major basal lineages of living birds (Neornithes) have gained increasing support. See literature reviewed in Cracraft & Clarke 2001 (583); see also Cracraft 1988, 2001 (580, 582); Groth & Barrowclough 1999 (1023); van Tuinen *et al.* 2000 (2488); Livezey & Zusi 2001 (1369); Paton *et al.*, in press (1794). These morphological and molecular data support the division of the neornithines into the Palaeognathae (tinamous and ratites) and the Neognathae (all other groups). Within the latter lineage, the Galloanserae (Galliformes and Anseriformes) are the sister-group of all other birds (the Neoaves).

---

[1] Numbers in brackets such as this refer to the main List of References for the book.

Relationships within the palaeognaths are still very uncertain, as there is conflict among morphological and molecular data, and within the latter the results can differ depending on the data set and the method of analysis (see Lee *et al.* 1997; Haddrath & Baker 2001; Cooper *et al.* 2001) (1346, 1050, 569). Relationships within the anseriforms, at least at higher taxonomic levels, do not appear to be too controversial (Livezey 1986, 1997) (1364, 1367), and both morphological and molecular data (Cracraft 1981, 1988; Livezey & Zusi 2001; Dimcheff *et al.* 2000) (578, 580, 1369, 751) support at least a tripartite pattern of relationships for galliforms: (Megapodiidae (Cracidae + phasianoids)). Within the phasianoids, the numidids appear to be the sister-group to all others (Cracraft 1981; Dimcheff *et al.* 2000) (578, 751).

Neoaves

Although the interrelationships of the basal lineages of the neornithes have been clarified in recent years, we have no firm idea how the basal relationships of the Neoaves might be resolved and this constitutes a major unsolved problem for avian systematics. Sibley & Ahlquist (1990) (2261) placed the Pici at the base of their DNA hybridization tree, but their evidence was hardly persuasive and that result has not emerged from any other study. Resolving the basal lineages of the Neoaves is likely to be a very difficult problem that probably will require considerable data to solve.

Sphenisciformes, Gaviiformes, Procellariiformes

The interrelationships of penguins, loons, and procellariiform seabirds are still not established convincingly. However, various authors using morphological (Cracraft 1988; Livezey & Zusi 2001) (580, 1369) or molecular data (Sibley & Ahlquist 1990; van Tuinen *et al.* 2001) (2261, 2488) have them clustering together or near one another, with penguins and procellariiforms generally being sister-taxa.

Podicipediformes, Phoenicopteriformes, Ciconiiformes

It has become fashionable in recent years to dismember the traditional Ciconiiformes (e.g., Sibley & Ahlquist 1990) (2261). It now seems there may be at least a core group of taxa that are related, including ciconiids, threskiornithids, and ardeids (van Tuinen *et al.* 2001; Livezey & Zusi 2001) (2488, 1369). However, other former ciconiiforms (*Balaeniceps* and *Scopus*) appear closer to pelecaniforms (see below). Surprising recent molecular analyses place flamingos with grebes (van Tuinen *et al.* 2001) (2488), and a preliminary analysis of cranial characters only (Livezey & Zusi 2001) (1369) put these two in the vicinity of one another.

Pelecaniformes

Like the ciconiiforms, the pelecaniforms have also been the subject of exuberant claims of paraphyly (Sibley & Ahlquist 1990; Hedges & Sibley 1994) (2261, 1093), yet the idea that they are not related was never adequately supported by the data. Morphological (Cracraft 1985; Livezey & Zusi 2001) (579, 1369) and behavioral (van Tets 1965; Kennedy *et al.* 1996) (2487, 1255) analyses found characters uniting them, but this was often dismissed as convergence (Hedges & Sibley 1994) (1093). These conflicts and confusion over pelecaniform relationships may have been magnified by the traditional exclusion of the Shoebill (*Balaeniceps*) and the Hamerkop (*Scopus*) from the pelecaniforms (e.g., Cracraft 1988) (580) and their placement in ciconiiforms (but see Cottam 1957 (575), for an exception). And yet mounting evidence suggests they belong near pelecanids (e.g., van Tuinen *et al.* 2001; see also Livezey & Zusi 2001) (2488, 1369). Phaethontids and fregatids have been problem groups for molecular analysis because both, especially tropicbirds, are relatively long-branch taxa; see, for example, the data of van Tuinen *et al.* 2001 (2488), for *Phaethon*. This, combined with the fact they are fairly ancient lineages, may be the reason for their questionable position on some trees (Sibley & Ahlquist 1990; Hedges & Sibley 1994) (2261, 1093). Although additional study is clearly needed, it would not be surprising to see most, perhaps all, traditional pelecaniforms reunited in an expanded clade that includes the Shoebill and Hamerkop.

Falconiformes

Much ado has also been made of the nonmonophyly of the falconiforms (Ligon 1967; Sibley & Ahlquist 1990; Avise *et al.* 1994) (1360, 2261, 71). This has mostly concerned the placement of the cathartids with respect to storks, although the evidence supporting the various alternative hypotheses has not been very compelling. The weight of the morphological evidence, at least, argues for falconiform monophyly (Cracraft 1981; Griffiths 1994; Livezey & Zusi 2001) (578, 1012, 1369). Previous ideas (Cracraft 1988) (580) that owls and falconiforms might be related seem incorrect.

Gruiformes

This list follows the detailed morphological analysis of Livezey (1998) (1368), although a broader comparison of cranial characters alone (Livezey & Zusi 2001) (1369) did not result in gruiform monophyly. The placement of the otidids is particularly uncertain.

Charadriiformes

Current evidence supports the hypothesis that virtually all the groups traditionally included in the charadriiforms comprise a monophyletic lineage (Sibley & Ahlquist 1990; Livezey & Zusi 2001) (2261, 1369), the major uncertainty being the turnicids (see for example Rotthowe & Starck 1998) (2109A). Moreover, charadriforms do not represent the primitive neornithine morphotype (see Paton *et al.* in press) (1794).

Musophagiformes, Cuculiformes, *Opisthocomus*

These three groups are listed together merely for want of a better alternative. They may very well prove unrelated. Hughes & Baker (1999) (1162) proposed that *Opisthocomus* is closer to musophagids than cuculids, but their study effectively included only these three taxa and so is inconclusive. Livezey & Zusi (2001) (1369), using cranial characters, placed *Opisthocomus* with core gruiform taxa, thus suggesting a novel arrangement. Based on a small amount of nuclear data, Johansson *et al.* (2001) (1210) found that cuculids and musophagids were distant from one another. Veron & Winney (2000) (2523) examined relationships within turacos using partial cytochrome *b* sequences, but their comparisons to cuckoos and the Hoatzin were also inconclusive because of restricted taxon sampling.

Caprimulgiformes, Apodiformes, Strigiformes

DNA hybridization distances placed these three groups in a clade (Sibley & Ahlquist 1990; Bleiweiss *et al.* 1994) (2261, 170), and that arrangement is also supported by cranial morphological characters (Livezey & Zusi 2001) (1369). The monophyly of swifts and hummingbirds is strongly supported (Cracraft 1988) (580); however, given that caprimulgiform taxa do not often group together on molecular trees (Johansson *et al.* 2001) (1210), and are not clustered by Livezey & Zusi's data (2001) (1369), questions remain about their monophyly.

Coraciiformes, Coliiformes, Trogoniformes, Galbulae

Most recent work suggests that these four groups are related in some way or another, often in association with the Pici and/or the Passeriformes (Espinosa de los Monteros 2000; Johansson *et al.* 2001) (819, 1210). However the data are insufficient to resolve their relationships clearly. The coraciiforms, as traditionally constituted, are apparently separable into at least two major groups that may or may not be related (Johansson *et al.* 2001) (1210). One includes the alcedinids, momotids, todids, and the three groups of rollers, the leptosomatids, coraciids, and brachypteraciids (Kirchman *et al.* 2001) (1271); the second, bucerotids and upupids. Groups such as trogonids, coliids, Galbulae, Pici, and possibly passeriforms are probably related to these coraciiform lineages, but the topology is unclear. At present, it is difficult to say what the molecular data mean, since most studies have had restricted taxon and character samples. Finally, even morphology breaks up the coraciiforms (Livezey & Zusi 2001) (1369). Clearly, much work is needed.

Pici

Relationships among the Pici now seem moderately well established. Picids and indicatorids are sister-taxa, and they, in turn, are related to barbets and toucans (Simpson & Cracraft 1981; Swiercewski & Raikow 1981) (2274, 2380). It is now apparent, however, that barbets are seriously paraphyletic. New World taxa are nonmonophyletic and some, but not all, are related to toucans, and to make matters more complicated, Old World barbets are also paraphyletic, with some probably being more closely related to the New World taxa (Prum 1988; Barker & Lanyon 2000) (1949, 97). The position of the Galbulae is still a question, but molecular data, none of which is entirely adequate at present, associate them most often with various coraciiform taxa (see above).

Passeriformes: basal relationships and major lineages

The monophyly of passeriforms is firmly established (Raikow 1982) (1959A) and considerable evidence supports a basal division into two large monophyletic groups, the suboscines and oscines (e.g., see review in Sibley & Ahlquist 1990; Barker *et al.* 2002) (2261, 98). A major question has always been the position of the New Zealand wrens (Acanthisittidae). Using sequences from the nuclear genes RAG-1 and *c-mos*, Barker *et al.* (2002) (98) found strong support for a sister-group relationship between acanthisittids and all other passeriforms; using a smaller data set of partial RAG-1 and *c-myc* sequences, a similar conclusion was reached by Ericson *et al.* (2002) (817).

Suboscines

The suboscines are divisible into two monophyletic groups: (a) the Old World pittids and eurylaimids-philepittids (Prum 1993, Barker *et al.* 2002) (1952, 981), and possibly the South American genus *Sapayoa* (Sibley & Ahlquist 1990) (2261), and (b) the New World Tyranni. The latter clade is a diverse assemblage whose systematic relationships are not well resolved, but molecular evidence (Sibley & Ahlquist 1990; Irestedt *et al.* 2001; Barker *et al.* 2002) (2261, 1184, 98) supports of a sister-group relationship of piprids, cotingids, and tyrannids, on the one hand, and furnariids, dendrocolaptids, formicariids, thamnophilids, rhinocryptids, and conopophagids, on the other.

Basal lineages of the oscines

Based on DNA hybridization evidence (Sibley & Ahlquist 1985, 1990) (2260A, 2261) it has become widely accepted that oscines can be divided into two monophyletic groups, the Australasian Corvida and the worldwide Passerida. Recent and ongoing studies using nuclear genes are proving this incorrect. The corvidans are paraphyletic: some are at the base of the oscines and others may be more closely related to passeridans than to core corvidans (Barker *et al.* (2002) (98) using RAG-1 and *c-mos* nuclear sequences; Ericson *et al.* 2002 (817), using fragments of RAG-1 and *c-myc*).. Barker *et al.* (2002) (98) placed *Menura* as the sister-group of the remaining oscines, and within the latter group relationships have the structure: ((climacterids + ptilo-norhynchids) (meliphagoids + all other oscines)). The core "corvidans" and passeridans are monophyletic, but several other groups, including pomatostomids, orthonychids, and petroicids have ambiguous relationships.

"Corvida"

With the exception of the DNA hybridization data (Sibley & Ahlquist 1990) (2261), taxon sampling across the core corvidans has been insufficient to specify a clear set of relationships among this large assemblage. Currently available sequence data (Cracraft & Feinstein 2000; Barker *et al.* 2002) (584, 98) suggest major conflicts with the DNA hybridization results, which is not surprising given the very short internodes indicated by the DNA hybridization phenograms. A major finding of the mitochondrial sequences (and morphology) is that cnemophilines are not paradisaeids but near the base of corvidans (Cracraft & Feinstein 2000) (2920), whereas RAG-1 and *c-mos* sequences remove the melanocharitids from the passeridans and move them to a basal position within corvoids (Barker *et al.* 2002) (98). Two clades strongly supported by the sequence data include (many other groups were not sampled in these studies): (a) *Aegithina*, vangids, artamids, cracticids, and (b) paradisaeids, corvids, laniids, monarchids, dicrurids. Groups such as oriolids, vireonids, campephagids, and pachycephalids appear to be more basal within corvoids than do these two clades.

Passerida

The passeridans are a very large monophyletic group whose relationships, at least at the higher taxonomic levels, are becoming better understood as DNA sequences accumulate. At the same time, these new studies make it clear that many traditional families are not monophyletic, and that a fuller understanding of passeridan phylogeny will only unfold as more and more of its diversity is sampled genetically.

  Both DNA hybridization (Sibley & Ahlquist 1990) (2261) and nuclear gene sequences (Barker *et al.* 2002) (98) delineate three major clades of passeridans — conveniently termed passeroids, muscicapoids, and sylviioids — yet the placement of a number of "families" in these three groups differs between the two data sets. Many, but not all, of the differences are due to very short internodal distances, especially on the DNA hybridization phenogram, and poor internodal support on the nuclear gene tree because of insufficient data (it will take additional sequence information to provide satisfactory resolution).

  The nuclear gene data of Barker *et al.* (2002) (98) suggest the following sequence of families, beginning with those most basal within the three major clades (again, keeping in mind that not all these family-group names represent monophyletic groups):

Sylvioidea: Bombycillidae + Paridae, Hirundinidae, (Pycnonotidae, Sylviidae, Zosteropidae, Timaliidae), (Aegithalidae, Alaudidae, Cisticolidae)

Muscicapoidea: (Troglodytidae, Sittidae, Certhiidae), ((Mimidae+Sturnidae), (Turdidae, Muscicapidae), Cinclidae)

Passeroidea: ((Chloropseidae, Dicaeidae, Nectariniidae), ((Prunellidae, Ploceidae), (Passeridae, (Motacillidae, (Fringillidae, ((Icteridae, Parulidae), (Emberizidae, (Cardinalidae, Thraupidae)))))))

Additionally, the Irenidae and Regulidae represent ancient lineages the relationships of which remain uncertain. Although a number of nodes in the nuclear gene passeridan tree are not well supported — especially within the muscicapoids and passeroids — these results are more consistent with the DNA hybridization experiments of Sheldon & Gill (1996) (2235), which were undertaken with stringent analytical procedures, than with those of Sibley & Ahlquist (1990) (2261). Thus, Sheldon & Gill (1996) (2235), unlike Sibley & Ahlquist (1990) (2261), found the alaudids were not passeroids but sylvioids, and troglodytes, sittids, and certhiids went with muscicapoids rather than sylvioids.

Literature Cited

Avise, J. C., W. S. Nelson, & C. G. Sibley. 1994. DNA sequence support for a close phylogenetic relationship between some storks and New World vultures. *Proc. Natl. Acad. Sci.* 91: 5173-5177.

Barker, F. K., G. F. Barrowclough, & J. G. Groth. 2002. A phylogenetic hypothesis for passerine birds: taxonomic and biogeographic implications of an analysis of nuclear DNA sequence data. *Proc. Roy. Soc. Lond.* 269: 295-308.

Barker, F. K. & S. M. Lanyon. 2000. The impact of parsimony weighting schemes on inferred relationships among toucans and Neotropical barbets (Aves: Piciformes). *Mol. Phylogen. Evol.* 15: 215-234.

Bleiweiss, R. J. A. W. Kirsch, F.-J. Lapointe. 1994. DNA-DNA hybridization-based phylogeny for "higher" nonpasserines: reevaluating a key portion of the avian family tree. *Mol. Phylogen. Evol.* 3: 248-255.

Cooper, A., C. Lalueza-Fox, S. Anderson, A. Rambaut, J. Austin, & R. Ward. 2001. Complete mitochondrial genome sequences of two extinct moas clarify ratite evolution. *Nature* 409: 704-707.

Cottam, P. A. 1957. The pelecaniform characters of the skeleton of the Shoebill Stork, *Balaeniceps rex. Bull. Brit. Mus. (Nat. Hist.) Zool.* 5:49-72.

Cracraft, J. 1981. Toward a phylogenetic classification of the Recent birds of the world (Class Aves). *Auk* 98: 681-714.

Cracraft, J. 1985. Monophyly and phylogenetic relationships of the Pelecaniformes: a numerical cladistic analysis. *Auk* 102: 834-853.

Cracraft, J. 1988 The major clades of birds. Pp. 339-361 in *The phylogeny and classification of the tetrapods* (ed. M. J. Benton), vol. 1. Oxford: Clarendon Press.

Cracraft, J. 2001. Avian evolution, Gondwana biogeography, and the Cretaceous-Tertiary mass extinction event. *Proc. Roy. Soc. Lond.* 268B: 459-469.

Cracraft, J. & J. Clarke. 2001. The basal clades of modern birds. Pp. 143-156 in *New perspectives on the origin and early evolution of birds: proceedings of the International Symposium in Honor of John H. Ostrom* (J. Gauthier & L. F. Gall, eds.). New Haven: Yale Univ. Peabody Mus. Nat. Hist.

Cracraft, J. & J. Feinstein. 2000. What is not a bird of paradise? Molecular and morphological evidence places *Macgregoria* in the Meliphagidae and the Cnemophilinae near the base of the corvoid tree. *Proc. Roy. Soc. Lond.* 267: 233-241.

Dimcheff, D. E., S. V. Drovetski, M. Krishnan, & D. P. Mindell. 2000. Cospeciation and horizontal transmission of avian sarcoma and leukosis virus *gag* genes in galliform birds. *J. Virology* 74: 3984-3995.

Ericson, P. G. P., L. Christidis, A. Cooper, M. Irestedt, J. Jackson, U. S. Johansson, & J. A. Norman. 2002. A Gondwanan origin of passerine birds supported by DNA sequences of the endemic New Zealand wrens. *Proc. Roy. Soc. Lond.*, Ser. B. 269: 235-241.

Espinosa de los Monteros, A. 2000. Higher-level phylogeny of Trogoniformes. *Mol. Phylogen. Evol.* 14: 20-34.

Griffiths, C. S. 1994. Monophyly of the Falconiformes based on syringeal morphology. *Auk* 111: 787-805.

Groth, J. G. & G. F. Barrowclough. 1999. Basal divergences in birds and the phylogenetic utility of the nuclear RAG-1 gene. *Mol. Phylogen. Evol.* 12: 115-123.

Haddrath, O. & A. J. Baker. 2001. Complete mitochondrial DNA genome sequences of extinct birds: ratite phylogenetics and the vicariance biogeography hypothesis. *Proc. Roy. Soc. Lond.* 268: 939-945.

Hedges, S. B. & C. G. Sibley. 1994. Molecules vs. morphology in avian systematics: the case of the "pelecaniform" birds. *Proc. Natl. Acad. Sci.* 91: 9861-9865.

Hughes, J. M. & A. J. Baker. 1999. Phylogenetic relationships of the enigmatic hoatzin (*Opisthocomus hoazin*) resolved using mitochondrial and nuclear gene sequences. *Mol. Biol. Evol.* 16(9): 1300-1307.

Irestedt, M., U. S. Johansson, T. J. Parsons, & P. G. P. Ericson. 2001. Phylogeny of major lineages of suboscines (Passeriformes) analysed by nuclear DNA sequence data. *J. Avian Biol.* 32: 15-25.

Johansson, U. S., T. J. Parsons, M. Irestedt, & P. G. P. Ericson. 2001. Clades within the 'higher land birds,' evaluated by nuclear DNA sequences. *J. Zool. Syst. Evol. Research* 39: 37-51.

Kennedy, M., H. G. Spencer, & R. D. Gray. 1996. Hop, step and gape: do the social displays of the Pelecaniformes reflect phylogeny? *Anim. Behav.* 51: 273-291.

Kirschman, J. J., S. J. Hackett, S. M. Goodman, & J. M. Bates. 2001. Phylogeny and systematics of ground rollers (Brachypteracidae) of Madagascar. *Auk* 118: 849-863.

Lee, K., J. Feinstein, & J. Cracraft. 1997. Phylogenetic relationships of the ratite birds: resolving conflicts between molecular and morphological data sets. Pp. 173-211 in *Avian Molecular Evolution and Systematics* (D. P. Mindell, ed.). Academic Press, New York.

Ligon, J. D. 1967. Relationships of the cathartid vultures. *Occ. Pap. Univ. Michigan Mus. Zool.* 651:1-26

Livezey, B. C. 1986. A phylogenetic analysis of Recent anseriform genera using morphological characters. *Auk* 105: 681-698.

Livezey, B. C. 1997. A phylogenetic analysis of basal Anseriformes, the fossil *Presbyornis*, and the interordinal relationships of waterfowl. *Zool. J. Linnean Soc.* 121:361-428.

Livezey, B. C. 1998. A phylogenetic analysis of the Gruiformes (Aves) based on morphological characters, with an emphasis on the rails (Rallidae). *Phil. Trans. Roy. Soc. Lond.* 353B: 2077-2151.

Livezey, B. C. & R. L. Zusi. 2001. Higher-order phylogenetics of modern Aves based on comparative anatomy. *Netherlands J. Zool.* 51:179-205.

Paton, T., O. Haddrath, & A. J. Baker. 2002. Complete mtDNA genome sequences show that modern birds are not descended from transitional shorebirds. *Proc. Roy. Soc. Lond.* Ser. B. 269: 839-846.

Prum, R. O. 1988. Phylogenetic relationships of the barbets (Aves: Capitonidae) and toucans (Aves: Ramphastidae) based on morphology with comparisons to DNA-DNA hybridization. *Zool. J. Linn. Soc.* 92: 313-343.

Prum, R. O. 1993. Phylogeny, biogeography, and evolution of the broadbills (Eurylaimidae) and asities (Philepittidae) based on morphology. *Auk* 110: 304-324.

Raikow, R. J. 1982. Monophyly of the Passeriformes: test of a phylogenetic hypothesis. *Auk* 99: 431-445.

Rotthowe, K. & J. M. Starck. 1998. Evidence for a phylogenetic position of button quails (Turnicidae: Aves) among the gruiformes. *J. Zool. Syst. Evol. Research* 36: 39-51.

Sheldon, F. H. & F. B. Gill. 1996. A reconsideration of songbird phylogeny, with emphasis on the evolution of titmice and their sylvioid relatives. *Syst. Biol.* 45: 473-495.

Sibley, C. G. & Ahlquist, J. E. & B.L. Monroe. 1988. A classification of the living birds of the world based on DNA-DNA hybridization studies. *Auk* 105: 409-423.

Sibley, C. G. & J. E. Ahlquist. 1985. The phylogeny and classification of the Australo-Papuan passerine birds. *Emu* 85: 1-14.

Sibley, C. G. & Ahlquist, J. E. 1990. *Phylogeny and classification of birds: a study in molecular evolution.* Yale University Press, New Haven.

Sibley, C. G. & B. L. Monroe, Jr. 1990. *Distribution and taxonomy of birds of the world.* Yale University Press, New Haven.

Simpson, S. F. & J. Cracraft. 1981. The phylogenetic relationships of the Piciformes (Class Aves). *Auk* 98: 481-494.

Swierczewski, E. V. & R. J. Raikow. 1981. Hind limb morphology, phylogeny, and classification of the Piciformes. *Auk* 98: 466-480.

van Tets, G. F. 1965. A comparative study of some social communication patterns in the Pelecaniformes. *Amer. Ornithol. Union, Ornithological Monographs* No. 2.

van Tuinen, M., D. B. Butvill, J. A. W. Kirsch, & S. B. Hedges. 2001. Convergence and divergence in the evolution of aquatic birds. *Proc. Roy. Soc. Lond.* 268B: 1-6.

van Tuinen, M., Sibley, C. G. & S. B. Hedges. 2000 The early history of modern birds inferred from DNA sequences of nuclear and mitochondrial ribosomal genes. *Mol. Biol. Evol.* 17: 451-457.

The Sequence of Families of Birds

Family Name	Peters Checklist Vol. No.	Number of genera	Number of species	No. of extinct species	Page
TINAMIDAE (TINAMOUS)	I: 12	9	47	1	31
STRUTHIONIDAE (OSTRICHES)	I: 3	1	1		34
RHEIDAE (RHEAS)	I: 5	2	2		35
CASUARIIDAE (CASSOWARIES)	I: 7	1	3		35
DROMAIIDAE (EMUS)	I: 9	1	3	2	35
APTERYGIDAE (KIWIS)	I: 10	1	3		35
MEGAPODIIDAE (MEGAPODES)	II: 3	7	22		35
CRACIDAE (CHACHALACAS, CURASSOWS, GUANS)	II: 9	11	50	1	37
NUMIDIDAE (GUINEAFOWL)	II: 133	4	6		40
ODONTOPHORIDAE (NEW WORLD QUAILS)	II: 42	9	32		41
PHASIANIDAE (TURKEYS, GROUSE, PHEASANTS AND PARTRIDGES)	II: 24, 58	49	180	2	44
ANHIMIDAE (SCREAMERS)	I: 505	2	3		61
ANSERANATIDAE (MAGPIE-GOOSE)	I: 426	1	1		61
ANATIDAE (DUCKS, GEESE, SWANS)	I: 427	49	158	5	61
SPHENISCIDAE (PENGUINS)	I: 121	6	17		70
GAVIIDAE (DIVERS OR LOONS)	I: 135	1	5		71
DIOMEDEIDAE (ALBATROSSES)	I: 48	4	13		72
PROCELLARIIDAE (PETRELS, SHEARWATERS)	I: 58	14	74	4	73
HYDROBATIDAE (STORM PETRELS)	I: 102	7	21	1	77
PELECANOIDIDAE (DIVING PETRELS)	I: 108	1	4		78
PODICIPEDIDAE (GREBES)	I: 140	6	22	3	79
PHOENICOPTERIDAE (FLAMINGOS)	I: 269	3	5		80
CICONIIDAE (STORKS)	I: 245	6	19		81
THRESKIORNITHIDAE (IBISES AND SPOONBILLS)	I: 253	14	32		82
ARDEIDAE (HERONS, BITTERNS)	I: 193	19	65	1	84
PHAETHONTIDAE (TROPICBIRDS)	I: 155	1	3		88
FREGATIDAE (FRIGATEBIRDS)	I: 159	1	5		89
SCOPIDAE (HAMMERKOP)	I: 244	1	1		89
BALAENICIPITIDAE (SHOEBILL)	I: 252	1	1		89
PELECANIDAE (PELICANS)	I: 188	1	7		89
SULIDAE (GANNETS, BOOBIES)	I: 181	3	10		90
PHALACROCORACIDAE (CORMORANTS)	I: 163	1	36	1	91
ANHINGIDAE (ANHINGAS)	I: 179	1	2		92
CATHARTIDAE (NEW WORLD VULTURES)	I: 274	5	7		93
FALCONIDAE (FALCONS, CARACARAS)	I: 390	11	64	1	93
ACCIPITRIDAE (SECRETARY BIRD, OSPREY, KITES, HAWKS AND EAGLES)	I: 278	67	233		98
OTIDAE (BUSTARDS)	II: 217	11	26		114
MESITORNITHIDAE (MESITES)	II: 141	2	3		116
CARIAMIDAE (SERIEMAS)	II: 216	2	2		116
RHYNOCHETIDAE (KAGU)	II: 215	1	1		116

Family Name	Peters Checklist Vol. No.	Number of genera	Number of species	No. of extinct species	Page
— EURYPYGIDAE (SUNBITTERN)	II: 215	1	1		117
— RALLIDAE (RAILS, WATERHENS, COOTS)	II: 157	33	141	13	117
HELIORNITHIDAE (FINFOOTS)	II: 213	3	3		127
PSOPHIIDAE (TRUMPETERS)	II: 155	1	3		127
— GRUIDAE (CRANES)	II: 150	4	15		128
— ARAMIDAE (LIMPKIN)	II: 155	1	1		129
TURNICIDAE (BUTTONQUAILS)	II: 142	2	16		129
BURHINIDAE (THICK-KNEES)	II: 293	2	9		130
CHIONIDAE (SHEATHBILLS AND ALLIES)	II: 308	2	3		131
— HAEMATOPODIDAE (OYSTERCATCHERS)	II: 231	1	11	1	131
DROMADIDAE (CRAB PLOVER)	II: 293	1	1		132
IBIDORHYNCHIDAE (IBISBILL)	II: 288	1	1		132
— RECURVIROSTRIDAE (STILTS, AVOCETS)	II: 289	3	7		132
— CHARADRIIDAE (PLOVERS)	II: 234	10	66	1	133
ROSTRATULIDAE (PAINTED-SNIPE)	II: 230	2	2		137
— JACANIDAE (JACANAS)	II: 226	6	8		137
PEDIONOMIDAE (PLAINS WANDERER)	II: 150	1	1		138
— THINOCORIDAE (SEEDSNIPES)	II: 306	2	4		138
— SCOLOPACIDAE (SANDPIPERS, SNIPE)	II: 258	23	92	4	139
GLAREOLIDAE (COURSERS, PRATINCOLES)	II: 298	5	18		145
— LARIDAE (GULLS, TERNS AND SKIMMERS)	II: 312	15	97		146
— STERCORARIIDAE (SKUAS)	II: 309	1	7		153
— ALCIDAE (AUKS)	II: 350	11	24	1	154
PTEROCLIDIDAE (SANDGROUSE)	III: 3	2	16		156
RAPHIDAE (DODOS, SOLITAIRES)	III: 10	2	3	3	157
— COLUMBIDAE (DOVES, PIGEONS)	III: 11	42	308	10	157
— PSITTACIDAE (COCKATOOS AND PARROTS)	III: 141	85	364	12	181
OPISTHOCOMIDAE (HOATZIN)	II: 141	1	1		205
MUSOPHAGIDAE (TURACOS)	IV: 3	6	23		205
— CUCULIDAE (CUCKOOS)	IV: 12	35	138	1	207
— TYTONIDAE (BARN OWLS)	IV: 77	2	15		218
— STRIGIDAE (OWLS)	IV: 86	27	180	1	220
PODARGIDAE (FROGMOUTHS)	IV: 175	2	12		237
STEATORNITHIDAE (OILBIRD)	IV: 174	1	1		238
NYCTIBIIDAE (POTOOS)	IV: 179	1	7		238
— CAPRIMULGIDAE (NIGHTJARS)	IV: 184	16	89	2	238
AEGOTHELIDAE (OWLET-NIGHTJARS)	IV: 181	2	9	1	245
— APODIDAE (SWIFTS)	IV: 220	19	94		246
HEMIPROCNIDAE (TREE SWIFTS)	IV: 257	1	4		254
— TROCHILIDAE (HUMMINGBIRDS)	V: 3	104	331	4	255
COLIIDAE (MOUSEBIRDS)	V: 143	2	6		278
— TROGONIDAE (TROGONS)	V: 148	6	39		279
CORACIIDAE (ROLLERS)	V: 241	2	12		282
BRACHYPTERACIIDAE (GROUND ROLLERS)	V: 240	4	5		283
LEPTOSOMATIDAE (CUCKOO-ROLLERS)	V: 239	1	1		283

Family Name	Peters Checklist Vol. No.	Number of genera	Number of species	No. of extinct species	Page
— ALCEDINIDAE (KINGFISHERS)	V: 165	17	91	1	284
TODIDAE (TODIES)	V: 220	1	5		293
— MOMOTIDAE (MOTMOTS)	V: 221	6	10		293
MEROPIDAE (BEE-EATERS)	V: 229	3	25		295
— UPUPIDAE (HOOPOES)	V: 247	1	1		297
PHOENICULIDAE (WOOD HOOPOES)	V: 250	2	8		297
BUCEROTIDAE (HORNBILLS)	V: 254	13	49		298
BUCORVIDAE (GROUND HORNBILLS)	V: 272	1	2		301
— RAMPHASTIDAE (BARBETS AND TOUCANS)	VI: 24, 70	20	120		301
INDICATORIDAE (HONEYGUIDES)	VI: 63	4	17		310
— PICIDAE (WOODPECKERS)	VI: 86	29	210	2	311
— GALBULIDAE (JACAMARS)	VI: 3	5	18		331
BUCCONIDAE (PUFFBIRDS)	VI: 10	10	33		332
— ACANTHISITTIDAE (NEW ZEALAND WRENS)	VIII: 331	2	4	2	335
EURYLAIMIDAE (BROADBILLS)	VII: 3	9	14		335
PHILEPITTIDAE (ASITIES)	VIII: 330	2	4		337
SAPAYAOIDAE (SAPAYAO)	VIII: 249	1	1		337
PITTIDAE (PITTAS)	VIII: 310	1	30		337
— PIPRIDAE (MANAKINS)	VIII: 245	13	48		339
— COTINGIDAE (COTINGAS)	VIII: 281	33	96	1	343
GENERA INCERTAE SEDIS [Piprites and Calyptura]	VIII: 249, 293	2	4		349
— TYRANNIDAE (TYRANT-FLYCATCHERS)	VIII: 1	98	400		349
— THAMNOPHILIDAE (ANTBIRDS)	VII: 162	46	206	1	379
CONOPOPHAGIDAE (GNATEATERS)	VII: 273	1	8		394
— RHINOCRYPTIDAE (TAPACULOS)	VII: 278	12	55	1	395
FORMICARIIDAE (ANT-THRUSHES AND ANTPITTAS)	VII: 153	7	62		397
— FURNARIIDAE (OVENBIRDS)	VII: 58	55	236		402
— DENDROCOLAPTIDAE (WOODCREEPERS)	VII: 13	13	50		419
— MENURIDAE (LYREBIRDS)	VIII: 333	1	2		426
ATRICHORNITHIDAE (SCRUBBIRDS)	VIII: 335	1	2		426
— PTILONORHYNCHIDAE (BOWERBIRDS)	XV: 172	8	18		426
— CLIMACTERIDAE (AUSTRALIAN TREECREEPERS)	XII: 162	2	7		428
— MALURIDAE (AUSTRALASIAN WRENS)	XI: 390	5	28		429
— MELIPHAGIDAE (HONEYEATERS)	XII: 338	44	174	5	431
DASYORNITHIDAE (BRISTLEBIRDS)	XI: 409	1	3		443
— PARDALOTIDAE (PARDALOTES)	XII: 202	1	4		444
— ACANTHIZIDAE (THORNBILLS, GERYGONES)	XI: 409	14	60		444
POMATOSTOMIDAE (AUSTRALASIAN BABBLERS)	X: 279	2	5		450
ORTHONYCHIDAE (LOGRUNNERS)	X: 228	1	3		450
CNEMOPHILIDAE (SATIN BIRDS)	XV: 181	2	3		450
MELANOCHARITIDAE (BERRYPECKERS)	XII: 167	6	12	1	451
CALLAEATIDAE (WATTLED CROWS)	XV: 157	3	3	1	452
— EUPETIDAE (WHIPBIRDS, WEDGEBILLS AND JEWELBABBLERS)	X: 238	4	10		452
CINCLOSOMATIDAE (QUAIL-THRUSHES)	X: 231	1	5		453

Family Name	Peters Checklist Vol. No.	Number of genera	Number of species	No. of extinct species	Page
PLATYSTEIRIDAE (SHRIKE-FLYCATCHERS, WATTLE-EYES AND BATISES)	XI: 376	6	28		454
GENERA INCERTAE SEDIS [Tephrodornis and Philentoma]	IX: 219, XI: 471	2	4		456
MALACONOTIDAE (HELMET-SHRIKES, BUSH SHRIKES AND PUFFBACKS)	IX: 314	10	52		456
MACHAERIRHYNCHIDAE (BOATBILL)	XI: 527	1	2		461
VANGIDAE (VANGAS)	IX: 365, XII: 124	15	22	1	461
CRACTICIDAE (BUTCHER BIRDS)	XV: 166	4	13		463
ARTAMIDAE (WOODSWALLOWS)	XV: 160	1	10		464
AEGITHINIDAE (IORAS)	IX: 300	1	4		465
PITYRIASIDAE (BRISTLEHEAD)	IX: 364	1	1		465
CAMPEPHAGIDAE (CUCKOO-SHRIKES)	IX: 167	7	81		465
NEOSITTIDAE (SITELLAS)	XII: 145	1	3		473
FALCUNCULIDAE (SHRIKE-TITS)	XII: 3	2	4		474
GENUS INCERTAE SEDIS [Turnagra]	XII: 52	1	1	1	474
PACHYCEPHALIDAE (WHISTLERS)	XII: 5	6	41		474
LANIIDAE (SHRIKES)	IX: 341	4	30		478
VIREONIDAE (VIREOS AND ALLIES)	XIV: 103	4	52		481
ORIOLIDAE (ORIOLES AND FIGBIRDS)	XV: 122	2	29		486
COLLURICINCLIDAE (SHRIKE-THRUSHES AND ALLIES)	XII: 36	3	14		489
DICRURIDAE (DRONGOS)	XV: 137	2	22		491
RHIPIDURIDAE (FANTAILS)	XI: 530	1	43		493
MONARCHIDAE (MONARCHS)	XI: 472	15	87		497
CORVIDAE (CROWS, JAYS)	XV: 204	24	117	1	504
CORCORACIDAE (AUSTRALIAN MUDNESTERS)	XV: 160	2	2		515
GENERA INCERTAE SEDIS [Melampitta and Ifrita]	X: 239	2	3		515
PARADISAEIDAE (MANUCODES AND BIRDS OF PARADISE)	XV: 183	16	40		515
PETROICIDAE (AUSTRALASIAN ROBINS)	XI: 556	13	45		518
PICATHARTIDAE (BALD CROWS)	X: 442	1	2		522
BOMBYCILLIDAE (WAXWINGS AND ALLIES)	IX: 369	5	8		522
DULIDAE (PALMCHAT)	IX: 373	1	1		523
PARIDAE (TITS, CHICKADEES)	XII: 70	3	54		523
REMIZIDAE (PENDULINE TITS)	XII: 62	5	10		530
HIRUNDINIDAE (SWALLOWS, MARTINS)	IX: 80	20	84		531
AEGITHALIDAE (LONG-TAILED TITS)	XII: 52	4	11		538
ALAUDIDAE (LARKS)	IX: 3	19	92		540
CISTICOLIDAE (CISTICOLAS AND ALLIES)	XI: 84	21	110		551
GENERA SEDIS INCERTAE [Neomixis, Orthotomus, Artisornis, Poliolais]	XI: 3, 173, 195	4	17		563
PYCNONOTIDAE (BULBULS)	IX: 221	22	118		565
GENERA SEDIS INCERTAE [Neolestes, Nicator, Erythrocercus and Elminia]	IX: 274, 299; XI: 465	4	12		576
SYLVIIDAE (OLD WORLD WARBLERS)	XI: 4	48	265	2	576
TIMALIIDAE (BABBLERS AND PARROTBILLS)	X: 240	50	273	1	599

Family Name	Peters Checklist Vol. No.	Number of genera	Number of species	No. of extinct species	Page
GENERA INCERTAE SEDIS [Myzornis,Malia, Horizornis, Chaetops, Modulatrix]	X: 427	5	7		624
ZOSTEROPIDAE (WHITE EYES)	XII: 289	14	95		625
IRENIDAE (FAIRY BLUEBIRDS)	IX: 307	1	2		632
REGULIIDAE (GOLDCRESTS/KINGLETS)	XI: 286	1	5		632
TROGLODYTIDAE (WRENS)	IX: 379	16	76		633
GENUS INCERTAE SEDIS [Donacobius]	IX: 456	1	1		642
POLIOPTILIDAE (GNATCATCHERS)	X: 443	3	14		643
SITTIDAE (NUTHATCHES AND WALLCREEPER)	XII: 125	2	25		644
CERTHIIDAE (TREECREEPERS)	XII: 150	2	8		647
MIMIDAE (MOCKINGBIRDS, THRASHERS)	IX: 440	12	34		648
RHABDORNITHIDAE (PHILIPPINE CREEPERS)	XII: 161	1	2		651
STURNIDAE (STARLINGS)	XV: 75	25	115	3	651
TURDIDAE (THRUSHES)	X: 14 etc	24	165	5	659
MUSCICAPIDAE (CHATS AND OLD WORLD FLYCATCHERS)	X: 14 etc. XI: 295	48	275		674
CINCLIDAE (DIPPERS)	IX: 374	1	5		698
CHLOROPSEIDAE (LEAFBIRDS)	IX: 303	1	8		699
DICAEIDAE (FLOWERPECKERS)	XII: 171	2	44		699
NECTARINIIDAE (SUNBIRDS) (16: 126)	XII: 208	16	127		703
PROMEROPIDAE (SUGARBIRDS) (1: 2)	XII: 409	1	2		715
PASSERIDAE (SPARROWS, SNOWFINCHES AND ALLIES)	XV: 5	11	40		715
PLOCEIDAE (WEAVERS, SPARROWS)	XV: 30	11	108		719
ESTRILDIDAE (WAXBILLS, GRASS FINCHES, MUNIAS AND ALLIES)	XIV: 306	26	130		726
VIDUIDAE (INDIGOBIRDS AND ALLIES)	XIV: 390	2	20		737
PRUNELLIDAE (ACCENTORS)	X: 3	1	13		737
PEUCEDRAMIDAE (OLIVE WARBLER)	XIV: 77	1	1		739
MOTACILLIDAE (WAGTAILS, PIPITS)	IX: 129	5	64		739
FRINGILLIDAE (FINCHES AND HAWAIIAN HONEYCREEPERS)	XIV: 93, 202	42	168	17	745
PARULIDAE (NEW WORLD WARBLERS)	XIV: 3	24	112	2	759
GENERA INCERTAE SEDIS [Granatellus and Xenoligea]	XIV: 79	2	4		768
ICTERIDAE (NEW WORLD BLACKBIRDS)	XIV: 138	26	98	1	768
COEREBIDAE (BANANAQUIT)	XIV: 87	1	1		776
EMBERIZIDAE (BUNTINGS, AMERICAN SPARROWS AND ALLIES)	XIII: 3	73	308	1	776
THRAUPIDAE (TANAGERS)	XIII: 246	50	202	2	802
GENERA INCERTAE SEDIS [Chlorospingus, Piranga, Habia, Chlorothraupis, Nesospingus, Phaenicophilus, Calyptophilus, Spindalis, Rhodinocichla, Euphonia, Chlorophonia, Catamblyrhynchus]	XIII: var.	12	69		817
CARDINALIDAE (CARDINAL, GROSBEAKS, SALTATORS AND ALLIES	XIII: 216	11	42		823
		2161	9721	128	

Presentation of the List

Below are some selected, and cut-down, non-consecutive portions of the list:

○ †*Xenicus lyalli* (Rothschild, 1894) Stephens Island Wren   Stephens I. (New Zealand)

COTINGIDAE COTINGAS[1] (33: 96)
TITYRINAE

TITYRA Vieillot, 1816 F
○ *Tityra inquisitor* Black-crowned Tityra
   — *T.i.albitorques* Du Bus, 1847[2]   E Panama to Peru, N Brazil
   — *T.i.buckleyi* Salvin & Godman, 1890   SE Colombia, E Ecuador
   — *T.i.erythrogenys* (Selby, 1826)   E Colombia, Venezuela, the Guianas
   — *T.i.inquisitor* (M.H.K. Lichtenstein, 1823)   E Brazil, Paraguay, N Argentina

DOLIORNIS Taczanowski, 1874[3] M
○ *Doliornis remseni* Robbins, Rosenberg & Molina, 1994 Chestnut-bellied Cotinga #
                        Andes of Ecuador to extreme N Peru (2084)

○ *Phyllomyias burmeisteri* Rough-legged Tyrannulet[4]
   — *P.b.burmeisteri* Cabanis & Heine, 1859   SE Brazil, E Bolivia, Paraguay, N Argentina
   — *P.b.zeledoni* (Lawrence, 1869)   Costa Rica, W Panama
   — *P.b.wetmorei* (Aveledo & Pons, 1953)   NW Venezuela
   — *P.b.leucogonys* (Sclater & Salvin, 1871)   Colombia, Ecuador, Peru

○ *Phyllomyias virescens* (Temminck, 1824) Greenish Tyrannulet[5]   SE Brazil, Paraguay, NE Argentina

GENERA INCERTAE SEDIS (2: 4)

PIPRITES Cabanis, 1847[6] F[7]
○ *Piprites griseiceps* Salvin, 1865 Grey-headed Piprites   Guatemala to Costa Rica

---

[1] Our generic sequence is based on the branching shown by Prum *et al.* (2000) (1957) and Prum (2001) (1955).
[2] Du Bus is an abbreviation for Du Bus de Gisignies.
[3] For reasons to recognise this genus see Robbins *et al.* (1994) (2084).
[4] We follow A.O.U. (1998: 380) (3) in retaining a single species here.
[5] For reasons to separate this and the next species see Stotz (1990)(2354).
[6] It has been suggested that this genus be moved to the Tyrannidae (Prum & Lanyon, 1989) (1956); however A.O.U. (1998: 416) (3) preferred to treat the genus as Incertae Sedis following Prum (1990) (1950). Prum (2001) (1955) placed it outside the cluster of subfamilies of the Cotingidae in his phylogenetic tree.
[7] Gender addressed by David & Gosselin (2002) (614), multiple changes occur in specific names.

The selected elements above allow a fairly full explanation of the presentation of the list. Each family name, in bold and in scientific form, is followed by the English group name or a number of group names that figure in the family. The figures in brackets, in the example, report that the family, as we treat it, comprises 96 species and is divided into 33 genera. A footnote attached to the family name explains our sequence of genera. Where no such explanation is given we have probably retained that found in Peters Checklist (and in our previous editions). For some subfamily names we give a vernacular name for the group.

The generic name (in italics) comes with its author and date. The citation will be found in Peters Checklist. However generic names that are new since Peters Checklist are linked to the List of References (as discussed below). Following the author and date is one of the three letters M, F and N. These indicate the gender (masculine, feminine or neuter) of the generic name. For more information in this connection see Appendix I.

Of the first two species set out above, the first is monotypic and the second polytypic. In the second case you will know that the author of the species is the author of the nominate (or oldest) subspecies. Every subspecies is provided with an author and date.

Before *Xenicus lyalli* the symbol † implying that this is generally thought to be extinct. When †? is used we believe that there is some doubt about extinction. This edition includes all extinct taxa that have been based on birds described from feathered remains, or from a picture of the feathered bird. Names based on bones (sub-fossils) and fossils are not included.

The third species selected here (*Doliornis remseni*) has been described since Peters Checklist. After the distribution statement is a bracketed number. This may be looked up in the List of References. All taxa that are listed and are new since Peters Checklist are dealt with in this way. Over the same period names have also been proposed that have been found not to deserve recognition. Such names go into synonymy. We give these synonyms in footnotes showing where in synonymy each name is thought to lie. The footnotes again have bracketed numbers after the author and date, these are links to the List of References. Following the English name of the species the sign # may be found. If it is, it means that specific status was proposed by the describer(s) and that the series "New species of birds described from .... to ...." (begun by Meise and extended by Mayr and others) has not yet reviewed the proposal. Our judgement of it is provisional. In due course consideration will be given to what the reviewers write. One further symbol used, but not in the examples, is ¶ —this implies that we are unsure that anyone has re-evaluated the name since its proposal. This is often because the material to do this is not widely available.

Further down is 'GENERA INCERTAE SEDIS', meaning genera of uncertain origins. Some species are judged to be close to, but not members of, a particular family. Usually we place such a group after that family but in some cases, where the following family is perhaps an equal candidate for the closest family, we may place it just before its likely relatives.

It will be observed that some authors have their initials given. For some explanations of the conventions observed here see Appendix II.

Finally, the other footnotes will give a flavour of the content of some 4,000 of these, many of which are also linked to the List of References.

Acknowledgements

First and foremost the Editor is profoundly grateful to the Regional Consultants (David Pearson for Africa, Van Remsen for the Americas, Cees Roselaar for the Palearctic and Dick Schodde for Australasia). Each one worked hard to provide us with guidance in respect of his region and offered a number of more general ideas that we have adopted.

Next we should like to thank Joel Cracraft and Mary LeCroy for providing a significant foundation to our list of taxa newly described since Peters Checklist. And also Norbert Bahr, who complemented that list with a major one that he had developed.

Our thanks are also due to Joel Cracraft and his colleagues Keith Barker and Alice Cibois for their combined work on the sequence, including their patient responses to numerous questions.

We are extremely grateful to Nigel Cleere for his hard work as our initial compiler. His role was to comb the literature for proposals for taxonomic changes that needed our attention. He also traced many of the entries needed for the list of references, including many relating to names already in previous editions of Howard & Moore and nothing to do with the new decision process. Most of the inclusion of authors' names and dates was also his work. Norbert Bahr played a major and growing role in completing the search for names and citations and has also researched and revised our listing of extinct taxa. The responsibility for all decisions, including those deferring recognition of proposed taxa, lies with us and not with them.

Numerous people have assisted us with matters great and small, ranging from the review of the whole texts of major sections to resolving minor questions of spelling and bibliographic issues. We first intended to include next to each name a brief indication of the subject on which help was received. Later we decided that there were better reasons not to do this. We are truly grateful to every one of these people:

Salman Abdulali, Alexandre Aleixo, Per Alström, Richard Banks, Keith Barker, Eustace Barnes, Rauri Bowen, Axel Braunlich, Murray Bruce, Rémy Bruckert, Kevin J. Burns, Don Causey, Roger and Liz Charlwood, Robert Cheke, Alice Cibois, Carla Cicero, Mario Cohn-Haft, Joel Cracraft, Adrian Craig, Ann Datta, Normand David, Geoffrey Davison, Richard Dean, René Dekker, Ding Chang-qing, Bob Dowsett, Jonathan Eames, Siegfried Eck, Per Ericson, Richard Fairbank, Chris Feare, Anita Gamauf, Margaret Gardner, Frank Gill, Oleg Goroshko, Gary Graves, Stephen Gregory, Graeme Green, Alison Harding, Frau Dr. A. Hausinger, Frank Hawkins, Cees Hazevoet, Alain Hennache, Christoph Hinkelmann, Roman Holynski, Dominique Homberger, Jon Hornbuckle, Jesper Hornskov, Mort and Phyllis Isler, Helen James, James Jobling, Leo Joseph, Mikhail Kalyakin, Janet Kear, Robert S. Kennedy, Peter Kennerley, Niels Krabbe, Daniel F. Lane, Mary LeCroy, Brad Livezey, Vladimir Loskot, Clive Mann, Curtis Marantz, Manuel Marin, Jochen Martens, Philip McGowan, Bob McGowan, Chris Milensky, Hiroyuki Morioka, Robert G. Moyle, Manuel Nores, Robin Panza, Teresa Pardo, Stephen Parry, Eric Pasquet, Robert Payne, John Penhallurick, Chris Perrins, A. Alan Peterson, A.Townsend Peterson, Aasheesh Pittie, Manuel Plenge, Robert Prys-Jones, Michael Rank, Pamela Rasmussen, Ronald de Ruiter, Robin Restall, Clemencia Rodner, Michael Reiser, Peter Ryan, Roger Safford, Karl-Ludwig Schuchmann, Thomas Schulenberg, Lucia Severinghaus, Frederick H. Sheldon, S. Somadikarta, Frank Steinheimer, Leo Stepanyan, Douglas F. Stotz, Pavel Tomkovich, Philip Tubbs, Angela Turner, Renate van den Elzen, Paulo Vanzolini, Jean-François and Claire Voisin, Andrew Wakeham-Dawson, Michael Walters, F.E. (Effie) Warr, Jason D. Weckstein, Andre Weller, David Wells, Roger Wilkinson, Chris Witt, Martin Woodcock and Kevin Zimmer.

Finally, we are particularly grateful to Andy Richford for all his early guidance and facilities from Academic Press, to Nigel Redman at A&C Black for taking over the production of the book, and to Julie Dando (Fluke Art) for design and layout and the substantial work in the creation of the indexes. We are also grateful to Erica Moore who made available Alick Moore's database and agreed that this work should be undertaken. We are very pleased to dedicate this edition to Alick.

TINAMIDAE TINAMOUS (9: 47)

TINAMINAE

TINAMUS Hermann, 1783 M
○ *Tinamus tao* GREY TINAMOU
  — *T.t.larensis* Phelps & Phelps, Jr., 1949   C Colombia, W Venezuela
  — *T.t.septentrionalis* Brabourne & Chubb, 1913   NE Venezuela, Guyana
  — *T.t.tao* Temminck, 1815   NC Brazil
  — *T.t.kleei* (Tschudi, 1843)   SC Colombia, E Ecuador, E Bolivia, W Brazil

○ *Tinamus solitarius* SOLITARY TINAMOU
  — *T.s.pernambucensis* Berla, 1946   EC Brazil
  — *T.s.solitarius* (Vieillot, 1819)   E Brazil, Paraguay, NE Argentina

○ *Tinamus osgoodi* BLACK TINAMOU
  — *T.o.hershkovitzi* Blake, 1953   S Colombia
  — *T.o.osgoodi* Conover, 1949   S Peru

● *Tinamus major* GREAT TINAMOU
  — *T.m.robustus* P.L.Sclater & Salvin, 1868   SE Mexico, Honduras
  — *T.m.percautus* van Tyne, 1935   S Mexico, Guatemala, Belize
  — *T.m.fuscipennis* Salvador, 1895   N Nicaragua to Panama
  — *T.m.brunneiventris* Aldrich, 1937   SC Panama
  ✓ *T.m.castaneiceps* Salvadori, 1895   SW Costa Rica, W Panama
  — *T.m.saturatus* Griscom, 1929   E Panama, N Colombia
  — *T.m.zuliensis* Osgood & Conover, 1922   NE Colombia, N Venezuela
  — *T.m.latifrons* Salvadori, 1895   W Colombia, W Ecuador
  — *T.m.major* (J.F. Gmelin, 1789)   E Venezuela to NE Brazil
  — *T.m.serratus* (Spix, 1825)   NW Brazil
  — *T.m.olivascens* Conover, 1937   Amazonian Brazil
  — *T.m.peruvianus* Bonaparte, 1856   SE Colombia, E Ecuador to W Brazil

○ *Tinamus guttatus* Pelzeln, 1863 WHITE-THROATED TINAMOU   Amazonia

NOTHOCERCUS Bonaparte, 1856 M
○ *Nothocercus bonapartei* HIGHLAND TINAMOU
  — *N.b.frantzii* (Lawrence, 1868)   Costa Rica, W Panama
  — *N.b.bonapartei* (G.R. Gray, 1867)   NC Colombia, W Venezuela
  — *N.b.intercedens* Salvadori, 1895   W Colombia
  — *N.b.discrepans* Friedmann, 1947   E Colombia
  — *N.b.plumbeiceps* Lönnberg & Rendahl, 1922   E Ecuador, N Peru

○ *Nothocercus julius* (Bonaparte, 1854) TAWNY-BREASTED TINAMOU   W Venezuela and Colombia to S Peru

○ *Nothocercus nigrocapillus* HOODED TINAMOU
  — *N.n.cadwaladeri* Carriker, 1933   NW Peru
  — *N.n.nigrocapillus* (G.R. Gray, 1867)   C Peru to N Bolivia

CRYPTURELLUS Brabourne & Chubb, 1914 M
○ *Crypturellus berlepschi* (Rothschild, 1897) BERLEPSCH'S TINAMOU   W Colombia to NW Ecuador

○ *Crypturellus cinereus* (J.F. Gmelin, 1789) CINEREOUS TINAMOU   The Guianas to SE Colombia, N Bolivia, E Peru

● *Crypturellus soui* LITTLE TINAMOU
  — *C.s.meserythrus* (P.L. Sclater, 1859)   S Mexico to SE Nicaragua
  ✓ *C.s.modestus* (Cabanis, 1869)   Costa Rica, W Panama
  — *C.s.capnodes* Wetmore, 1963   NW Panama
  — *C.s.poliocephalus* (Aldrich, 1937)   C and E Panama
  — *C.s.panamensis* (Carriker, 1910)   Pearl Is (Panama)
  — *C.s.mustelinus* (Bangs, 1905)   NE Colombia, NW Venezuela
  — *C.s.soui* (Hermann, 1783)   E Colombia, Venezuela, the Guianas, NE Brazil
  — *C.s.andrei* (Brabourne & Chubb, 1914)   Trinidad, NE Venezuela

    __ *C.s.caucae* (Chapman, 1912)   NC Colombia

    __ *C.s.harterti* (Brabourne & Chubb, 1914)   W Colombia, W Ecuador

    __ *C.s.caquetae* (Chapman, 1915)   SE Colombia

    __ *C.s.nigriceps* (Chapman, 1923)   E Ecuador, NE Peru

    __ *C.s.albigularis* (Brabourne & Chubb, 1914)   N and E Brazil

    __ *C.s.inconspicuus* Carriker, 1935   C and E Peru, E Bolivia

○ *Crypturellus ptaritepui* Zimmer & Phelps, 1945 TEPUI TINAMOU   SE Venezuela

○ *Crypturellus obsoletus* BROWN TINAMOU

    __ *C.o.cerviniventris* (Sclater & Salvin, 1873)   N Venezuela

    __ *C.o.knoxi* W.H. Phelps, Jr., 1976[1]   NW Venezuela

    __ *C.o.castaneus* (P.L. Sclater, 1858)   E Colombia, E Ecuador, N Peru

    __ *C.o.ochraceiventris* (Stolzmann, 1926)   C Peru

    __ *C.o.traylori* Blake, 1961   E Peru

    __ *C.o.punensis* (Chubb, 1917)   S Peru, N Bolivia

    __ *C.o.griseiventris* (Salvadori, 1895)   NC Brazil

    __ *C.o.hypochraceus* (Miranda-Ribeiro, 1938)[2]   SW Brazil

    __ *C.o.obsoletus* (Temminck, 1815)   S Brazil, Paraguay, NE Argentina

○ *Crypturellus undulatus* UNDULATED TINAMOU

    __ *C.u.manapiare* Phelps & Phelps, Jr., 1952 S Venezuela

    __ *C.u.simplex* (Salvadori, 1895)   S Guyana, NC Brazil north of the Amazon

    __ *C.u.yapura* (Spix, 1825)   W Amazonia

    __ *C.u.adspersus* (Temminck, 1815)   C Brazil south of the Amazon

    __ *C.u.vermiculatus* (Temminck, 1825)   E Brazil south of the Amazon

    __ *C.u.undulatus* (Temminck, 1815)   SE Peru to N Argentina

○ *Crypturellus transfasciatus* (Sclater & Salvin, 1878) PALE-BROWED TINAMOU

                                        W Ecuador to NW Peru

○ *Crypturellus strigulosus* (Temminck, 1815) BRAZILIAN TINAMOU   S Amazonian Brazil, E Peru, N Bolivia,

○ *Crypturellus duidae* J.T. Zimmer, 1938 GREY-LEGGED TINAMOU   E Columbia, SE Venezuela

○ *Crypturellus erythropus* RED-LEGGED TINAMOU

    __ *C.e.cursitans* Wetmore & Phelps, 1956   N Colombia, NW Venezuela

    __ *C.e.idoneus* (Todd, 1919)   NE Colombia, W Venezuela

    __ *C.e.columbianus* (Salvadori, 1895)   NC Colombia

    __ †?*C.e.saltuarius* Wetmore, 1950   S Magdalena Prov. (NC Colombia)

    __ *C.e.spencei* (Brabourne & Chubb, 1914) N Venezuela

    __ *C.e.margaritae* Phelps & Phelps, Jr., 1948 Margarita I.

    __ *C.e.erythropus* (Pelzeln, 1863)   E Venezuela, Guyana, Surinam, N Brazil

○ *Crypturellus noctivagus* YELLOW-LEGGED TINAMOU

    __ *C.n.zabele* (Spix, 1825)   NE Brazil

    __ *C.n.noctivagus* (Wied, 1820)   SE Brazil

○ *Crypturellus atrocapillus* BLACK-CAPPED TINAMOU

    __ *C.a.atrocapillus* (Tschudi, 1844)   SE Peru

    __ *C.a.garleppi* (Berlepsch, 1892)   N Bolivia

○ *Crypturellus cinnamomeus* THICKET TINAMOU

    __ *C.c.occidentalis* (Salvadori, 1895)   WC Mexico

    __ *C.c.mexicanus* (Salvadori, 1895)   EC Mexico

    __ *C.c.soconuscensis* Brodkorb, 1939   C Chiapas (SW Mexico)

    __ *C.c.vicinior* Conover, 1933   Chiapas to C Honduras

    __ *C.c.sallaei* (Bonaparte, 1856)   SE Mexico

    __ *C.c.goldmani* (Nelson, 1901)   SE Mexico, N Guatemala, N Belize

    __ *C.c.cinnamomeus* (Lesson, 1842)   S Guatemala, El Salvador, S Honduras

    __ *C.c.delattrei* (Bonaparte, 1854)   W Nicaragua

    __ *C.c.praepes* (Bangs & Peters, 1927)   NW Costa Rica

---

[1] Omitted in our last Edition; see Blake (1979: 25) (165).
[2] For correction of spelling see David & Gosselin (2002) (613).

○ *Crypturellus boucardi* SLATY-BREASTED TINAMOU
   — *C.b.boucardi* (P.L. Sclater, 1859)   S Mexico to NW Honduras
   — *C.b.costaricensis* (Dwight & Griscom, 1924)   N and E Honduras to Costa Rica

○ *Crypturellus kerriae* (Chapman, 1915) CHOCO TINAMOU   SE Panama, NW Colombia

○ *Crypturellus variegatus* (J.F. Gmelin, 1789) VARIEGATED TINAMOU   The Guianas, Amazonia, E Brazil

○ *Crypturellus brevirostris* (Pelzeln, 1863) RUSTY TINAMOU   E Peru, N Brazil, French Guiana

○ *Crypturellus bartletti* (Sclater & Salvin, 1873) BARTLETT'S TINAMOU   E Peru, N Bolivia, W Brazil

○ *Crypturellus parvirostris* (Wagler, 1827) SMALL-BILLED TINAMOU   SE Peru, C and S Brazil to NE Argentina

○ *Crypturellus casiquiare* (Chapman, 1929) BARRED TINAMOU   E Colombia, S Venezuela

○ *Crypturellus tataupa* TATAUPA TINAMOU
   — *C.t.inops* Bangs & Noble, 1918   SW Ecuador, NW Peru
   — *C.t.peruvianus* (Cory, 1915)[1]   C Peru
   — *C.t.lepidotus* (Swainson, 1837)   NE Brazil south of the Amazon
   — *C.t.tataupa* (Temminck, 1815)   S Brazil, N and E Bolivia, Paraguay, N Argentina

RHYNCHOTINAE

RHYNCHOTUS Spix, 1825   M
○ *Rhynchotus rufescens* RED-WINGED TINAMOU
   — *R.r.catingae* Reiser, 1905   C and NE Brazil
   — *R.r.rufescens* (Temminck, 1815)   SE Peru, N Bolivia east to NE Brazil and south to
                                         Uruguay
   — *R.r.maculicollis* G.R. Gray, 1867   NW Bolivia to NW Argentina
   — *R.r.pallescens* Kothe, 1907   Lowlands of NE and C Argentina

NOTHOPROCTA Sclater & Salvin, 1873   F
○ *Nothoprocta taczanowskii* Sclater & Salvin, 1875 TACZANOWSKI'S TINAMOU
                                        Andes of C and S Peru

○ †?*Nothoprocta kalinowskii* Berlepsch & Stolzmann, 1901 KALINOWSKI'S TINAMOU
                                        Andes of N and S Peru

○ *Nothoprocta ornata* ORNATE TINAMOU
   — *N.o.branickii* Taczanowski, 1875   C Peru
   — *N.o.ornata* (G.R. Gray, 1867)   S Peru, NE Bolivia
   — ¶*N.o.jimenezi* Cabot, 1997   NW Altiplano of Bolivia (287)
   — ¶*N.o.labradori* Cabot, 1997   Alpine E Andes of Bolivia (287)
   — *N.o.rostrata* Berlepsch, 1907   SE Bolivia, N Chile, NW Argentina

◑ *Nothoprocta perdicaria* CHILEAN TINAMOU
   ↙ *N.p.perdicaria* (Kittlitz, 1830)   NC Chile
   — *N.p.sanborni* Conover, 1924   S Chile

○ *Nothoprocta cinerascens* BRUSHLAND TINAMOU
   — *N.c.cinerascens* (Burmeister, 1860)   NW and C Argentina, NW Paraguay, SE Bolivia
   — *N.c.parvimaculata* Olrog, 1959   NW Argentina

○ *Nothoprocta pentlandii* ANDEAN TINAMOU
   — *N.p.ambigua* Cory, 1915   S Ecuador, NW Peru
   — *N.p.niethammeri* Koepcke, 1968   WC Peru
   — *N.p.oustaleti* Berlepsch & Stolzmann, 1901   C and S Peru
   — *N.p.fulvescens* Berlepsch, 1902   SE Peru
   — *N.p.pentlandii* (G.R. Gray, 1867)   Bolivia to NW Argentina
   — *N.p.mendozae* Banks & Bohl, 1968   WC Argentina
   — *N.p.doeringi* Cabanis, 1878   C Argentina

---

[1] For correction of spelling see David & Gosselin (2002) (613).

○ *Nothoprocta curvirostris* Curve-billed Tinamou
   — *N.c.curvirostris* Sclater & Salvin, 1873   C Ecuador to N Peru
   — *N.c.peruviana* Taczanowski, 1886   N and C Peru

NOTHURA Wagler, 1827 F
○ *Nothura boraquira* (Spix, 1825) White-bellied Nothura   NE Brazil, E Bolivia, NE Paraguay

○ *Nothura minor* (Spix, 1825) Lesser Nothura   S Brazil

○ *Nothura darwinii* Darwin's Nothura
   — *N.d.peruviana* Berlepsch & Stolzmann, 1906   S Peru
   — *N.d.agassizii* Bangs, 1910   SE Peru, W Bolivia
   — *N.d.boliviana* Salvadori, 1895   W Bolivia
   — *N.d.salvadorii* E. Hartert, 1909   W Argentina
   — *N.d.darwinii* G.R. Gray, 1867   SC Argentina

○ *Nothura maculosa* Spotted Nothura
   — *N.m.cearensis* Naumburg, 1932   S Ceará (Brazil)
   — *N.m.major* (Spix, 1825)   EC Brazil
   — *N.m.paludivaga* Conover, 1950   C Paraguay, NC Argentina
   — *N.m.maculosa* (Temminck, 1815)   SE Brazil, E Paraguay, Uruguay, NE Argentina
   — *N.m.pallida* Olrog, 1959   NW Argentina
   — *N.m.annectens* Conover, 1950   E Argentina
   — *N.m.submontana* Conover, 1950   SW Argentina
   — *N.m.nigroguttata* Salvadori, 1895   SC Argentina

○ *Nothura chacoensis* Conover, 1937 Chaco Nothura[1]   NW Paraguay, NC Argentina

TAONISCUS Gloger, 1842 M
○ *Taoniscus nanus* (Temminck, 1815) Dwarf Tinamou   SE Brazil, NE Argentina

EUDROMIA I. Geoffroy Saint-Hilaire, 1832 F
○ *Eudromia elegans* Elegant Crested Tinamou
   — *E.e.intermedia* (Dabbene & Lillo, 1913)   N Andes of NW Argentina
   — *E.e.magnistriata* Olrog, 1959   Hills of NC Argentina
   — *E.e.riojana* Olrog, 1959   Andes of WC Argentina
   — *E.e.albida* (Wetmore, 1921)   Lowland WC Argentina
   — *E.e.wetmorei* Banks, 1977   Foothills W Argentina
   — *E.e.numida* Banks, 1977   C Argentina
   — *E.e.elegans* I. Geoffroy Saint-Hilaire, 1832   SC Argentina
   — *E.e.multiguttata* Conover, 1950   EC Argentina
   — *E.e.devia* Conover, 1950   Foothills SW Argentina
   — *E.e.patagonica* Conover, 1950   Andes of S Chile, S Argentina

○ *Eudromia formosa* (Lillo, 1905) Quebracho Crested Tinamou   NC Argentina, WC Paraguay

TINAMOTIS Vigors, 1837 F
○ *Tinamotis pentlandii* Vigors, 1837 Puna Tinamou   Andes of C Peru, W Bolivia, W Argentina, N Chile

○ *Tinamotis ingoufi* Oustalet, 1890 Patagonian Tinamou   S Chile, SW Argentina

STRUTHIONIDAE OSTRICHES (1: 1)

STRUTHIO Linnaeus, 1758 M
○ *Struthio camelus* Ostrich[2]
   — †?*S.c.syriacus* Rothschild, 1919   Syrian and Arabian deserts
   — *S.c.camelus* Linnaeus, 1758   S Morocco and Mauritania to Sudan, N and W Ethiopia, N Uganda
   — *S.c.molybdophanes* Reichenow, 1883   Somalia, SE Ethiopia, N and E Kenya

---

[1] Treated as a race of *N. maculosa* by Blake (1979: 43) (165) but see Cabot (1992: 137) (286).
[2] Freitag & Robinson (1993) (873) reviewed molecular evidence for treatment of this as one species or two. We interpret the authors to be supportive of single species treatment.

— *S.c.massaicus* Neumann, 1898   S Kenya to C Tanzania
— *S.c.australis* J.H. Gurney, 1868   Southern Africa

RHEIDAE RHEAS (2: 2)

RHEA Brisson, 1760 F
○ *Rhea americana* GREATER RHEA
   — *R.a.americana* (Linnaeus, 1758)   N and E Brazil
   — *R.a intermedia* Rothschild & Chubb, 1914   SE Brazil, Uruguay
   — *R.a.nobilis* Brodkorb, 1939   E Paraguay
   — *R.a.araneipes* Brodkorb, 1938   W Paraguay, E Bolivia, SW Brazil
   — *R.a.albescens* Lynch & Holmberg, 1878   NE and E Argentina

PTEROCNEMIA G.R. Gray, 1871 F
○ *Pterocnemia pennata* LESSER RHEA
   — *P.p.garleppi* Chubb, 1913   SE Peru, Bolivia, NW Argentina
   — *P.p.tarapacensis* Chubb, 1913   N Chile
   — *P.p.pennata* (d'Orbigny, 1834)   S Chile, WC and S Argentina

CASUARIIDAE CASSOWARIES (1: 3)

CASUARIUS Brisson, 1760 M
○ *Casuarius casuarius* (Linnaeus, 1758) SOUTHERN CASSOWARY[1]   Seram, New Guinea, Cape York Pen. and NE Queensland

○ *Casuarius bennetti* Gould, 1857 DWARF CASSOWARY[2]   Montane New Guinea, Yapen I. and New Britain

○ *Casuarius unappendiculatus* Blyth, 1860 NORTHERN CASSOWARY[3]   Salawati, lowland New Guinea and Yapen I.

DROMAIIDAE EMUS (1: 3)

DROMAIUS Vieillot, 1816 M
● *Dromaius novaehollandiae* EMU
   ✓ *D.n.rothschildi* (Mathews, 1912)   SW Australia
   — *D.n.novaehollandiae* (Latham, 1790)[4]   Australia (except SW corner)
   — †*D.n.diemenensis* Le Souef, 1907   Tasmania

○ †*Dromaius ater* Vieillot, 1817 KING ISLAND EMU[5]   King I. (Australia)

○ †*Dromaius baudinianus* S. Parker, 1984 KANGAROO ISLAND EMU   Kangaroo I. (Australia) (1746)

APTERYGIDAE KIWIS (1: 3)

APTERYX Shaw, 1813 F
○ *Apteryx australis* BROWN KIWI
   — *A.a.mantelli* A.D. Bartlett, 1852[6][7]   North Island (New Zealand)
   — *A.a.australis* Shaw, 1813   South Island (New Zealand)
   — *A.a.lawryi* Rothschild, 1893   Stewart I.

○ *Apteryx owenii* Gould, 1847 LITTLE SPOTTED KIWI   New Zealand[8]

○ *Apteryx haastii* Potts, 1872 GREAT SPOTTED KIWI   W South Island (New Zealand)

MEGAPODIIDAE[9] MEGAPODES (7: 22)

ALECTURA Latham, 1824 F
○ *Alectura lathami* BRUSH TURKEY
   — *A.l.purpureicollis* (Le Souef, 1898)   N Queensland
   — *A.l.lathami* J.E. Gray, 1831   C and S Queensland, New South Wales

---

[1] We follow Mayr (1979) (1484) in monotypic treatment. See also Mayr (1940) (1469) for some of the reasons for this.
[2] We follow Mayr (1979) (1484) in monotypic treatment; *shawmeyeri* Rothschild, 1937 (2109) has been described since Peters (1931) (1821).
[3] Again monotypic treatment follows Mayr (1979) (1484).
[4] Implicitly includes *woodwardi* see Marchant & Higgins (1990: 58) (1426).
[5] For comments on the name *diemenianus* see Jouanin (1959) (1228).
[6] Implicitly includes *novaezelandiae* see Marchant & Higgins (1990: 80) (1426).
[7] For suggestions that this warrants treatment as a species see Baker et al. (1995) (78).
[8] Extinct North Island form, not usually recognised, called *iredalei* (Mathews, 1935) (1456).
[9] Unless otherwise stated changes in species limits follow Roselaar (1994) (2106) and Jones et al. (1995) (1223).

AEPYPODIUS Oustalet, 1880 M
○ *Aepypodius arfakianus* WATTLED BRUSH TURKEY
    __ *A.a.arfakianus* (Salvadori, 1877)   Mts. of New Guinea
    __ *A.a.misoliensis* Ripley, 1957   Misool I. (2039)

○ *Aepypodius bruijnii* (Oustalet, 1880) BRUIJN'S BRUSH TURKEY   Waigeo I.

TALEGALLA Lesson, 1828 F
○ *Talegalla cuvieri* RED-BILLED BRUSH TURKEY
    __ *T.c. cuvieri* Lesson, 1828   Vogelkop Pen. (NW New Guinea), Misool I.
    __ *T.c.granti* Roselaar, 1994   S foothills of Snow Mts. (WC New Guinea) (2106)

○ *Talegalla fuscirostris* BLACK-BILLED BRUSH TURKEY
    __ *T.f.fuscirostris* Salvadori, 1877   SE New Guinea
    __ *T.f.occidentis* C.M.N. White, 1938   SW New Guinea (Triton Bay to upper Lorentz R.) (2623)
    __ *T.f.aruensis* Roselaar, 1994   Aru Is., S New Guinea (Kolepom I to Fly R.) (2106)
    __ *T.f.meyeri* Roselaar, 1994   Head of Geelvink Bay (N New Guinea) (2106)

○ *Talegalla jobiensis* BROWN-COLLARED BRUSH TURKEY
    __ *T.j.jobiensis* A.B. Meyer, 1874   Yapen I.
    __ *T.j.longicaudus* A.B. Meyer, 1891   N and NE New Guinea

LEIPOA Gould, 1840 F
○ *Leipoa ocellata* Gould, 1840 MALLEEFOWL   Southern Australia

MACROCEPHALON S. Müller, 1846 N
○ *Macrocephalon maleo* S. Müller, 1846 MALEO   Sulawesi, Butong Is.

EULIPOA Ogilvie-Grant, 1893 F
○ *Eulipoa wallacei* (G.R. Gray, 1860) MOLUCCAN MEGAPODE   Moluccas, Misool I.

MEGAPODIUS Gaimard, 1823 M
○ *Megapodius pritchardii* G.R. Gray, 1864 POLYNESIAN MEGAPODE   Niuafo'ou I. (N Tonga)

○ *Megapodius laperouse* MICRONESIAN MEGAPODE
    __ *M.l.senex* Hartlaub, 1867   Palau Is.
    __ *M.l.laperouse* Gaimard, 1823   Mariana Is.

○ *Megapodius nicobariensis* NICOBAR MEGAPODE
    __ *M.n.nicobariensis* Blyth, 1846   C. Nicobar Is.
    __ *M.n.abbotti* Oberholser, 1919   Gt. and Little Nicobar

○ *Megapodius cumingii* PHILIPPINE MEGAPODE[1]
    __ *M.c.cumingii* Dillwyn, 1853   Islets off E Borneo, Balabac, Palawan, the Calamianes
                           and Sulu Is.
    __ *M.c.dillwyni* Tweeddale, 1877   Luzon, Mindoro, Marinduque and nearby islands
    __ *M.c.pusillus* Tweeddale, 1877   W and E Visayas, W Mindanao and Basilan
    __ *M.c.tabon* Hachisuka, 1931   E Mindanao
    __ *M.c.gilbertii* G.R. Gray, 1861   Sulawesi and Togian Is.
    __ *M.c.sanghirensis* Schlegel, 1880   Sangihe Is.
    __ *M.c.talautensis* Roselaar, 1994   Talaud Is. (2106)

○ *Megapodius bernsteinii* Schlegel, 1866 SULA MEGAPODE[2]   Sula and Banggai Is.

○ *Megapodius tenimberensis* P.L. Sclater, 1883 TANIMBAR MEGAPODE   Tanimbar Is.

○ *Megapodius freycinet* DUSKY MEGAPODE
    __ *M.f.quoyii* G.R. Gray, 1861   N Moluccas (997)
    __ *M.f.freycinet* Gaimard, 1823   Gebe, Waigeo, Misool and nearby islets
    __ *M.f.oustaleti* Roselaar, 1994   Batanta, Salawati, small islets off NW New Guinea (2106)

○ *Megapodius geelvinkianus* A.B. Meyer, 1874 BIAK MEGAPODE   Biak, Numfor and Mios Num Is. (N New Guinea)

---

[1] We follow Roselaar (1994) (2106) as his review updates that in Dickinson *et al.* (1991) (745).
[2] Roselaar in Jones *et al.* (1995) (1223) considered *perrufus* Neumann, 1939 (1646), a synonym of this species.

○ *Megapodius forstenii* Forster's Megapode
 — *M.f.forstenii* G.R. Gray, 1847   Seram and islands nearby
 — *M.f.buruensis* Stresemann, 1914[1]   Buru I.

○ *Megapodius eremita* Hartlaub, 1867 Melanesian Megapode   S Melanesia, Admiralty Is., Bismarck Arch., Solomon Is.

○ *Megapodius layardi* Tristram, 1879 Vanuatu Megapode   Vanuatu

○ *Megapodius decollatus* Oustalet, 1878 New Guinea Megapode[2]   N New Guinea, Yapen I. and offshore islands east to Manam I.

○ *Megapodius reinwardt* Orange-footed Scrubfowl[3]
 — *M.r.reinwardt* Dumont, 1823[4]   Lesser Sundas (except Tanimbar), islands in Java, Banda and Flores Seas, Watubela, Kai and Aru Is., NW, S and SE New Guinea
 — *M.r.macgillivrayi* G.R. Gray, 1861   Louisiade Arch., D'Entrecasteaux Arch., Trobriand Is.
 — *M.r.tumulus* Gould, 1842   NW Australia east to Groote Eylandt
 — *M.r.yorki* Mathews, 1929   N Queensland south to Cooktown area
 — *M.r.castanonotus* Mayr, 1938   C Queensland (from about Cairns to Yeppoon) (1466)

CRACIDAE CHACHALACAS, CURASSOWS, GUANS[5] (11: 50)

PENELOPINAE

ORTALIS Merrem, 1786 F
○ *Ortalis vetula* Plain Chachalaca[6]
 — *O.v.mccallii* S.F. Baird, 1858   S Texas, Mexico south to Vera Cruz
 — *O.v.vetula* (Wagler, 1830)[7]   E Mexico to NW Costa Rica
 — *O.v.pallidiventris* Ridgway, 1887   N Yucatán
 — *O.v.intermedia* J.L. Peters, 1913   S Yucatán to Belize and Guatemala
 — *O.v.deschauenseei* Bond, 1936   Utila I. (Honduras) (181)

● *Ortalis cinereiceps* (G.R. Gray, 1867) Grey-headed Chachalaca[8]   SE Honduras to NW Colombia

○ *Ortalis garrula* (Humboldt, 1805) Chestnut-winged Chachalaca   NW Colombia

○ *Ortalis ruficauda* Rufous-vented Chachalaca
 — *O.r.ruficrissa* (Sclater & Salvin, 1870)[9]   N Colombia, NW Venezuela, Venezuela
 — *O.r.ruficauda* (Jardine, 1847)   Venezuela, Tobago I., Lesser Antilles

○ *Ortalis erythroptera* (Sclater & Salvin, 1870) Rufous-headed Chachalaca   W Ecuador, NW Peru

○ *Ortalis wagleri* (G.R. Gray, 1867) Rufous-bellied Chachalaca[10]   N and W Mexico

○ *Ortalis poliocephala* (Wagler, 1830) West Mexican Chachalaca[11]   W Mexico

○ *Ortalis canicollis* Chaco Chachalaca
 — *O.c.canicollis* (Wagler, 1830)   E Bolivia, W Paraguay, N Argentina
 — *O.c.pantanalensis* Cherrie & Reichenberger, 1921[12]   E Paraguay, SW Brazil

○ *Ortalis leucogastra* (Gould, 1843) White-bellied Chachalaca[13]   SW Mexico to NW Nicaragua

○ *Ortalis guttata* Speckled Chachalaca
 — *O.g.araucuan* (Spix, 1825)   E Brazil
 — *O.g.squamata* Lesson, 1829[14]   SE Brazil

---

[1] Dated 1912 in "Peters", a *lapsus*.
[2] Called *affinis* in last Edition; however the surviving type of *affinis* is identifiable as *M. reinwardt* (see Roselaar, 1994; Jones *et al.*, 1995) (1223, 2106).
[3] In last Edition the species concept included *tenimberensis* and *forstenii*.
[4] Includes *aruensis* Mayr, 1938 (1466).
[5] Generic limits in this group, here taken from del Hoyo (1994) (673), require molecular study.
[6] The race *intermedia* was omitted in last Edition.
[7] Includes *vallicola* Brodkorb, 1942 (223); see Blake (1977) (164).
[8] Includes *olivacea* Aldrich & Bole, 1937 (25) and *chocoensis* Meyer de Schauensee, 1950 (1566); see Blake (1977) (164).
[9] Includes *baliola* and the more recent *lamprophonia* Wetmore, 1953 (2596); see Blake (1977) (164).
[10] Includes *griseiceps* van Rossem, 1934 (2470); see Banks (1990) (88).
[11] This species and the next split by Banks (1990) (88); *lajuelae* Moore & Medina, 1957 (1622), is a synonym of *poliocephala*.
[12] Includes *ungeri* J. Steinbacher, 1962 (2314); see Blake (1977) (164).
[13] Dated 1844 by A.O.U. (1998) (3); a *lapsus*, we follow Duncan (1937) (770).
[14] Includes *remota* Pinto, 1960 (1891); see Blake (1977) (164).

— *O.g.colombiana* Hellmayr, 1906[1]   Colombia, NW Upper Amazonia
— *O.g.guttata* (Spix, 1825)   W Upper Amazonia
— *O.g.subaffinis* Todd, 1932   NE and E Bolivia, W Brazil

○ *Ortalis motmot* LITTLE CHACHALACA[2]
— *O.m.motmot* (Linnaeus, 1766)   S Venezuela, the Guianas, N Brazil
— *O.m.ruficeps* (Wagler, 1830)   N C Brazil

○ *Ortalis superciliaris* G.R. Gray, 1867 BUFF-BROWED CHACHALACA NE Brazil

PENELOPE Merrem, 1786 F
○ *Penelope argyrotis* BAND-TAILED GUAN
— *P.a.colombiana* Todd, 1912   Santa Marta Mts. (N Colombia)
— *P.a.argyrotis* (Bonaparte, 1856)[3] Andes of N Colombia, W and N Venezuela,

○ *Penelope barbata* Chapman, 1921 BEARDED GUAN[4] S Ecuador, NW Peru

○ *Penelope ortoni* Salvin, 1874 BAUDO GUAN W Colombia, W Ecuador

○ *Penelope montagnii* ANDEAN GUAN
— *P.m.montagnii* (Bonaparte, 1856) N and C Colombia
— *P.m.atrogularis* Hellmayr & Conover, 1932 S Colombia, WC Ecuador
— *P.m.brooki* Chubb, 1917   S Colombia, EC Ecuador
— *P.m.plumosa* Berlepsch & Stolzmann, 1902[5] C Peru (E slope of Andes)
— *P.m.sclateri* G.R. Gray, 1860   C Bolivia

○ *Penelope marail* MARAIL GUAN
— *P.m.jacupeba* Spix, 1825   SE Venezuela, N Brazil
— *P.m.marail* (Statius Müller, 1776) E Venezuela, the Guianas

○ *Penelope superciliaris* RUSTY-MARGINED GUAN[6]
— *P.s.superciliaris* Temminck, 1815[7] NC and E Brazil
— *P.s.jacupemba* Spix, 1825[8]   C and S Brazil
— *P.s.major* Bertoni, 1901   S Brazil, E Paraguay, NE Argentina

○ *Penelope dabbenei* Hellmayr & Conover, 1942 RED-FACED GUAN S Bolivia, NW Argentina (1104)

○ *Penelope purpurascens* CRESTED GUAN
— *P.p.purpurascens* Wagler, 1830   NE and NW Mexico to Honduras, Nicaragua
— *P.p.aequatorialis* Salvadori & Festa, 1900 S Honduras to NW Colombia
— *P.p.brunnescens* Hellmayr & Conover, 1932 N Colombia, E Venezuela

○ *Penelope perspicax* Bangs, 1911 CAUCA GUAN[9] W and C Colombia

○ *Penelope albipennis* Taczanowski, 1878 WHITE-WINGED GUAN NW Peru

○ *Penelope jacquacu* SPIX'S GUAN
— *P.j.granti* Berlepsch, 1908   E Venezuela, Guyana
— *P.j.orienticola* Todd, 1932   NW Brazil, SE Venezuela
— *P.j.jacquacu* Spix, 1825   Upper Amazonia
— *P.j.speciosa* Todd, 1915   C and E Bolivia

○ *Penelope obscura* DUSKY-LEGGED GUAN
— *P.o.bronzina* Hellmayr, 1914   E Brazil
— *P.o.obscura* Temminck, 1815   S Brazil, E Paraguay, Uruguay, NE Argentina
— *P.o.bridgesi* G.R. Gray, 1860   C Bolivia, NW Argentina

---

[1] Includes *caucae* see Vaurie (1965) (2512).
[2] In both our previous Editions a broad species *motmot* was adopted that included *superciliaris* and *guttata*; here we revert to the treatment of Vaurie (1965) (2512).
[3] Includes *olivaceiceps* and *albicauda* Phelps & Gilliard, 1940 (1842), and *mesaeus* Conover, 1945 (563); see Vaurie (1966) (2514). *P. barbata* was treated therein as conspecific, but see Hilty & Brown (1986: 124) (1116).
[4] Includes *inexpectata* Carriker, 1934 (301); see Vaurie (1966) (2514).
[5] Includes *marcapatensis* Blake, 1963 (161); see Blake (1977) (164).
[6] The recently named races ¶*alagoensis* Nardelli, 1993 (1635), and ¶*cyanosparius* Nardelli, 1993 (1635), await review.
[7] Includes *pseudonyma* Neumann, 1933 (1642); see Vaurie (1966) (2514).
[8] Includes *argyromitra* Neumann, 1933 (1642), and *ochromitra* Neumann, 1933 (1642); see Vaurie (1966) (2514).
[9] Treated as a distinct endemic Colombian species by Hilty & Brown (1986: 125) (1116), but treated as a form of *P. jacquacu* by Vaurie (1966) (2513) and detailed case for split apparently not published.

○ *Penelope pileata* Wagler, 1830 WHITE-CRESTED GUAN   NC Brazil

○ *Penelope ochrogaster* Pelzeln, 1870 CHESTNUT-BELLIED GUAN   SC Brazil

○ *Penelope jacucaca* Spix, 1825 WHITE-BROWED GUAN   NE Brazil

PIPILE Bonaparte, 1856 M
○ *Pipile pipile* (Jacquin, 1784) TRINIDAD PIPING GUAN[1]   Trinidad

○ *Pipile cumanensis* BLUE-THROATED PIPING GUAN
    — *P.c.cumanensis* (Jacquin, 1784)   The Guianas to C Colombia, NW Peru, W Brazil
    — *P.c.grayi* (Pelzeln, 1870)   E Bolivia, NE Paraguay, SW Brazil

○ *Pipile cujubi* RED-THROATED PIPING GUAN
    — *P.c.cujubi* (Pelzeln, 1858)   NC Brazil
    — *P.c.nattereri* Reichenbach, 1862   S and W Amazonia

○ *Pipile jacutinga* (Spix, 1825) BLACK-FRONTED PIPING GUAN   SE Brazil, SE Paraguay

ABURRIA Reichenbach, 1853 F
○ *Aburria aburri* (Lesson, 1828) WATTLED GUAN   N Colombia, E Venezuela to SC Peru

CHAMAEPETES Wagler, 1832 M
● *Chamaepetes unicolor* Salvin, 1867 BLACK GUAN   Costa Rica, Panama

○ *Chamaepetes goudotii* SICKLE-WINGED GUAN
    — *C.g.goudotii* (Lesson, 1828)   C and W Colombia
    — *C.g.sanctaemarthae* Chapman, 1912   Santa Marta Mts.
    — *C.g.fagani* Chubb, 1917   SW Colombia, W Ecuador
    — *C.g.tschudii* Taczanowski, 1886   EC Ecuador, N Peru
    — *C.g.rufiventris* (Tschudi, 1843)   EC Peru
    — *C.g.*subsp ?   N Bolivia

PENELOPINA Reichenbach, 1862 F
○ *Penelopina nigra* (Fraser, 1852) HIGHLAND GUAN[2]   S Mexico to N Nicaragua

OREOPHASIS G.R. Gray, 1844 M
○ *Oreophasis derbianus* G.R. Gray, 1844 HORNED GUAN   SE Mexico, SW Guatemala

CRACINAE

NOTHOCRAX Burmeister, 1856 M
○ *Nothocrax urumutum* (Spix, 1825) NOCTURNAL CURASSOW   Upper Amazonia

MITU Lesson, 1831 N[3]
○ *Mitu tomentosum* (Spix, 1825) CRESTLESS CURASSOW   SE Colombia, S Venezuela, Guyana, NW Brazil

○ *Mitu salvini* (Reinhardt, 1879) SALVIN'S CURASSOW   SW Colombia, E Ecuador, NE Peru

○ *Mitu tuberosum* Spix, 1825 RAZOR-BILLED CURASSOW[4]   S Amazonia

○ †?*Mitu mitu* (Linnaeus, 1766) ALAGOAS CURASSOW   NE Brazil[5]

PAUXI Temminck, 1813 F
○ *Pauxi pauxi* NORTHERN HELMETED CURASSOW
    — *P.p.pauxi* (Linnaeus, 1766)   NC to W Venezuela
    — *P.p.gilliardi* Wetmore & Phelps, 1943   Venezuela-Colombia border (2611)

---

[1] Treated as a separate species from *cumanensis* in del Hoyo (1994) (673), apparently based mainly on disjunct insular range. The two are apparently closely related.
[2] Includes *dickeyi* van Rossem, 1934 (2472), and *rufescens* van Rossem, 1934 (2472) (see del Hoyo, 1994) (673).
[3] Gender addressed by David & Gosselin (2002) (614): the following taxa are now spelled differently: *tomentosum* and *tuberosum*.
[4] This name *may* more correctly apply to the E Brazilian form; its separation as a species distinct from *mitu* requires more evidence and an investigation of the nomenclature, which is confused (*apud* Blake, 1977: 426) (164).
[5] Survives in captivity at least.

○ *Pauxi unicornis* Southern Helmeted Curassow
    — *P.u.unicornis* Bond & Meyer de Schauensee, 1939   NE Bolivia (189)
    — *P.u.koepckeae* Weske & Terborgh, 1971   E Peru (2585)

CRAX Linnaeus, 1758[1] F
○ *Crax rubra* Great Curassow
    — *C.r.rubra* Linnaeus, 1758   E Mexico to W Colombia and W Ecuador
    — *C.r.griscomi* Nelson, 1926   Cozumel I (off Yucatán)

○ *Crax alberti* Fraser, 1852   Blue-billed Curassow   N Colombia

○ *Crax daubentoni* G.R. Gray, 1867   Yellow-knobbed Curassow   N Venezuela

○ *Crax alector* Linnaeus, 1766   Black Curassow   E Colombia to the Guianas, N Brazil

○ *Crax globulosa* Spix, 1825   Wattled Curassow   W Amazonia

○ *Crax fasciolata* Bare-faced Curassow[2]
    — *C.f.fasciolata* Spix, 1825   C and SW Brazil, Paraguay
    — *C.f.pinima* Pelzeln, 1870   NE Brazil
    — *C.f.grayi* Ogilvie-Grant, 1893   E Bolivia

○ *Crax blumenbachii* Spix, 1825   Red-billed Curassow   SE Brazil

NUMIDIDAE GUINEAFOWL (4: 6)

AGELASTES Bonaparte, 1850 M
○ *Agelastes meleagrides* Bonaparte, 1850   White-breasted Guineafowl   Liberia to Ghana

○ *Agelastes niger* (Cassin, 1857)   Black Guineafowl[3]   S Cameroon to C Zaire

NUMIDA Linnaeus, 1766 M[4]
○ *Numida meleagris* Helmeted Guineafowl
    — *N.m.sabyi* E. Hartert, 1919   NW Morocco
    — *N.m.galeatus* Pallas, 1767[5]   W Africa to S Chad, S to N Angola. Intr. Cape Verde Is.
    — *N.m.meleagris* (Linnaeus, 1758)[6]   E Chad to W Ethiopia, S to N Zaire, Uganda and N Kenya
    — *N.m.somaliensis* Neumann, 1899   NE Ethiopia, Somalia, NE Kenya
    — *N.m.reichenowi* Ogilvie-Grant, 1894   S Kenya to C Tanzania
    — *N.m.mitratus* Pallas, 1767   E Kenya, E and S Tanzania to Zimbabwe, W Transvaal
    — *N.m.marungensis* Schalow, 1884[7]   S Zaire to W Angola and Zambia
    — *N.m.papillosus* Reichenow, 1894[8]   Botswana, N and S Namibia
    — *N.m.coronatus* J.H. Gurney, 1868   E Sth. Africa

GUTTERA Wagler, 1832 F
○ *Guttera plumifera* Plumed Guineafowl
    — *G.p.plumifera* (Cassin, 1857)   Cameroon, Gabon, N Angola
    — *G.p.schubotzi* Reichenow, 1912   N Zaire

○ *Guttera pucherani* Crested Guineafowl[9]
    — *G.p.pucherani* (Hartlaub, 1860)   SW Somalia, E Kenya, NE and E Tanzania, Zanzibar and Tumbatu Is.
    — *G.p.verreauxi* (Elliot, 1870)[10]   Guinea to W Cameroon, E Congo to W Kenya and C Tanzania south to Angola and W Zambia
    — *G.p.sclateri* Reichenow, 1898   NW Cameroon

[1] *Crax estudilloi* Allen, Allen & Allen, 1977 (29), has not been satisfactorily substantiated. See also Joseph (1999) (1225).
[2] The proposed race ¶ *xavieri* Nardelli, 1993 (1635), has not been reviewed.
[3] Treated in the genus *Phasidus* in last Edition but see Crowe *et al.* (1986: 3) (599).
[4] Gender addressed by David & Gosselin (2002) (614), multiple changes occur in specific names.
[5] Includes *marchei, callewaerti* and *blancoui* Grote, 1936 (1019) see Martínez (1994: 565) (1446), and presumably *strasseni*. Also includes *bannermani* Frade, 1976 (869), a name apparently assigned to stock introduced to the Cape Verde Is.
[6] Includes *major, macroceras, toruensis, intermedius* and *uhehensis* see Martínez (1994: 565) (1446).
[7] Includes *maximus, rikwae,* and *bodalyae* Boulton, 1934 (200), see Martínez (1994: 565) (1446).
[8] Includes *mulondensis* Monard, 1934 (1601) see White (1965: 96) (2649). Treated as *damarensis* in Crowe *et al.* (1986: 9) (599) and in Martínez (1994: 565) (1446) with *papillosa* in synonymy — but the latter is the prior name. Reasons to avoid its use may exist, but have not been traced.
[9] In last Edition we erred in placing *pucherani* in a species *edouardi* as the former has 7 years priority.
[10] Includes *schoutedeni, chapini, sethsmithi* and *kathleenae* White, 1943 (2625) see Martínez (1994: 566) (1446).

— *G.p.barbata* Ghigi, 1905[1]   SE Tanzania, N Mozambique, Malawi
— *G.p.edouardi* (Hartlaub, 1867)[2]   E Zambia to S Mozambique and Natal

ACRYLLIUM G.R. Gray, 1840 N

⃝ *Acryllium vulturinum* (Hardwicke, 1834) Vulturine Guineafowl   E Uganda, S Somalia, E Kenya, NE Tanzania

ODONTOPHORIDAE NEW WORLD QUAILS (9: 32)

DENDRORTYX Gould, 1844 M
⃝ ***Dendrortyx barbatus*** Gould, 1846 Bearded Wood Partridge[3]   Vera Cruz (Mexico)

⃝ ***Dendrortyx macroura*** Long-tailed Wood Partridge
     — *D.m.macroura* (Jardine & Selby, 1828)   EC Mexico
     — *D.m.griseipectus* Nelson, 1897   C Mexico
     — *D.m.diversus* Friedmann, 1943   NW Jalisco (Mexico) (875)
     — *D.m.striatus* Nelson, 1897   WC Mexico
     — *D.m.inesperatus* A.R. Phillips, 1966[4]   S Mexico (1869)
     — *D.m.oaxacae* Nelson, 1897   W Oaxaca (Mexico)

⃝ ***Dendrortyx leucophrys*** Buffy-crowned Wood Partridge
     — *D.l.leucophrys* (Gould, 1844)[5]   SE Mexico, Guatemala, Honduras, Nicaragua
     — *D.l.hypospodius* Salvin, 1896   Costa Rica

OREORTYX S.F. Baird, 1858 M[6]
⬤ ***Oreortyx pictus*** Mountain Quail
     ⤊ *O.p.pictus* (Douglas, 1829)[7]   SW Washington to N. California W of Cascades
     — *O.p.plumifer* (Gould, 1837)   Oregon, NE California, Nevada, E of Cascades
     — *O.p.russelli* A.H. Miller, 1946   Little San Bernardino Mts. (1580)
     — *O.p.eremophilus* van Rossem, 1937   Southern California (2473)
     — *O.p.confinis* Anthony, 1889   Northern Baja California

CALLIPEPLA Wagler, 1832[8] F
⬤ ***Callipepla squamata*** Scaled Quail
     ⤊ *C.s.pallida* Brewster, 1881   SW USA, NW Mexico
     — ¶*C.s.hargravei* Rea, 1973   New Mexico (1997)
     — *C.s.castanogastris* Brewster, 1883   S Texas, NE Mexico
     — *C.s.squamata* (Vigors, 1830)   NC and C Mexico

⃝ ***Callipepla douglasii*** Elegant Quail
     — *C.d.bensoni* Ridgway, 1887[9]   Sonora, C Chihuahua
     — *C.d.douglasii* (Vigors, 1829)   Sinaloa, NW Durango
     — *C.d.teres* (Friedmann, 1943)[10]   Nayartit and Jalisco (876)

⬤ ***Callipepla californica*** Californian Quail
     ⤊ *C.c.brunnescens* (Ridgway, 1884)   Coast from SW Oregon to C California
     ⤊ *C.c.canfieldae* (van Rossem, 1939)   EC California (2475)
     — *C.c.californica* (Shaw, 1798)   E Oregon to Baja California and NW Mexico
     — *C.c.catalinensis* (Grinnell, 1906)   Santa Catalina I. (Los Coronados Is.)
     — *C.c.achrustera* (J.L. Peters, 1923)[11]   S Baja California

⬤ ***Callipepla gambelii*** Gambel's Quail
     ⤊ *C.g.gambelii* (Gambel, 1843)[12]   SW USA, NW Mexico incl. Tiburon I.
     — *C.g.fulvipectus* Nelson, 1899[13]   Rest of W Mexico S to S Sonora (NW to SW Sonora)

[1] Includes *suahelica* see White (1965: 98) (2649).
[2] Includes *symonsi* see White (1965: 98) (2649).
[3] Birds of this genus are called Tree Quail in del Hoyo *et al.* (1994) (674).
[4] Recognised by Binford (1989) (150).
[5] Includes *nicaraguae*; see Monroe (1968) (1604).
[6] Gender addressed by David & Gosselin (2002) (614), multiple changes occur in specific names.
[7] Includes *palmeri*; see Browning (1977) (245).
[8] Subsumes the genus *Lophortyx*; see A.O.U. (1983) (2). R. Banks (*in litt.*) anticipates reductions in the numbers of races recognised in these species.
[9] Includes *languens* Friedmann, 1943 (876) (Banks, *in litt.*).
[10] Includes *impedita* Friedmann, 1943 (876) (Banks, *in litt.*).
[11] Includes *decolorata* van Rossem, 1946 (2479) (Banks *in litt.*).
[12] Includes *sana, pembertoni* and the more recent *ignoscens* Friedmann, 1943 (876) (Banks, *in litt.*).
[13] Includes *friedmanni* R.T. Moore, 1947 (1615) and *stephensi* A.R. Phillips, 1959 (1867) (Banks, *in litt.*).

PHILORTYX Gould, 1846 M
○ *Philortyx fasciatus* (Gould, 1844) BANDED QUAIL[1] Colima, Guerrero, Puebla (SW Mexico)

COLINUS Goldfuss, 1820 M
● *Colinus virginianus* NORTHERN BOBWHITE[2]
   ✓ *C.v.virginianus* (Linnaeus, 1758) C and E USA
   — *C.v.floridanus* (Coues, 1872) Florida, Bahama Is.
   — †*C.v.insulanus* Howe, 1904 Key West, Florida
   — *C.v.cubanensis* (G.R. Gray, 1846) Cuba
   — *C.v.taylori* Lincoln, 1915 South Dakota to N Texas and W Missouri
   — *C.v.ridgwayi* Brewster, 1885 N and SC Sonora
   — *C.v.texanus* (Lawrence, 1853) NE Mexico
   — *C.v.maculatus* Nelson, 1899 C Mexico
   — *C.v.aridus* Aldrich, 1942 NC Mexico (22)
   — *C.v.graysoni* (Lawrence, 1867) WC Mexico
   — *C.v.nigripectus* Nelson, 1897 SC Mexico
   — *C.v.pectoralis* (Gould, 1843) C Vera Cruz
   — *C.v.godmani* Nelson, 1897 E Vera Cruz
   — *C.v.minor* Nelson, 1901 NE Chiapas
   — *C.v.insignis* Nelson, 1897 NW Guatemala, W and C Chiapas
   — *C.v.salvini* Nelson, 1897 S Chiapas
   — *C.v.coyolcos* (Statius Müller, 1776) S Mexico
   — *C.v.thayeri* Bangs & Peters, 1928 NE Oaxaca
   — *C.v.harrisoni* Orr & Webster, 1968 SW Oaxaca (1728)
   — *C.v.atriceps* (Ogilvie-Grant, 1893) W Oaxaca

● *Colinus nigrogularis* BLACK-THROATED BOBWHITE
   — *C.n.caboti* van Tyne & Trautman, 1941 Campeche (Mexico) (2491)
   — *C.n.persiccus* van Tyne & Trautman, 1941 N Yucatán (2491)
   — *C.n.nigrogularis* (Gould, 1843) Yucatán
   — *C.n.segoviensis* Ridgway, 1888 Guatemala, Honduras

● *Colinus leucopogon* SPOT-BELLIED BOBWHITE[3]
   — *C.l.incanus* Friedman, 1944 S Guatemala (877)
   — *C.l.hypoleucus* (Gould, 1860) W Guatemala, W El Salvador
   — *C.l.leucopogon* (Lesson, 1842) El Salvador, W Honduras
   — *C.l.leylandi* (T.J. Moore, 1859) NW Honduras
   — *C.l.sclateri* (Bonaparte, 1856) C and SW Honduras, NW Nicaragua (179)
   ✓ *C.l.dickeyi* Conover, 1932 NW and C Costa Rica

○ *Colinus cristatus* CRESTED BOBWHITE
   — *C.c.mariae* Wetmore, 1962 Chiriqui area, NW Panama (2600)
   — *C.c.panamensis* Dickey & van Rossem, 1930 SW Panama
   — *C.c.decoratus* (Todd, 1917) Coastal N Colombia
   — *C.c.littoralis* (Todd, 1917) N Colombia (Santa Marta foothills)
   — *C.c.cristatus* (Linnaeus, 1766)[4] NE Colombia, NW Venezuela, Aruba, Curacao
   — *C.c.horvathi* (Madarász, 1904) Merida Mts. (NW Venezuela)
   — *C.c.barnesi* Gilliard, 1940 WC Venezuela (937)
   — *C.c.sonnini* (Temminck, 1815) NC Venezuela, the Guianas, N Brazil
   — *C.c.mocquerysi* (E. Hartert, 1894) NE Venezuela, Margarita I.
   — *C.c.leucotis* (Gould, 1844) NC Colombia (Magdalena Valley)
   — *C.c.badius* Conover, 1938 C Colombia (W Andes) (562)
   — *C.c.bogotensis* Dugand, 1943 C Colombia (E Andes) (766)
   — *C.c.parvicristatus* (Gould, 1843) EC Colombia

---

[1] Not 1843, as sometimes given, see Duncan (1937) (770).
[2] Although some additional races are listed for this species by Carroll (1994) (303), we follow Banks (*in litt.*) in placing *nelsoni* Brodkorb, 1942 (224), in the synonymy of *insignis*, and in recognising *taylori* and *harrisoni*.
[3] A.O.U. (1983, 1998: 126) (2, 3) treated this as conspecific with *C. cristatus*; we prefer to follow Blake (1977) (164) as we did in our first Edition.
[4] Including *continentis* recognised in last Edition, but see Carroll (1994) (303).

ODONTOPHORUS Vieillot, 1816 M
○ *Odontophorus gujanensis* MARBLED WOOD QUAIL

    — *O.g.castigatus* Bangs, 1901   SW Costa Rica (and NW Panama ?)
    — *O.g.marmoratus* (Gould, 1843)[1]   E Panama, N Colombia, NW Venezuela
    — *O.g.medius* Chapman, 1929   Mt. Duida (SW Venezuela), NW Brazil
    — *O.g.gujanensis* (J.F. Gmelin, 1789)[2]   SE Venezuela, the Guianas, Brazil, N Paraguay
    — *O.g.buckleyi* Chubb, 1919   SE Colombia, E Ecuador
    — *O.g.rufogularis* Blake, 1959   NE Peru (159)
    — *O.g.pachyrhynchus* Tschudi, 1844   E Peru
    — *O.g.simonsi* Chubb, 1919   La Paz Prov., Bolivia

○ *Odontophorus capueira* (Spix, 1825) SPOT-WINGED WOOD QUAIL[3]   E Brazil, E Paraguay, NE Argentina

○ *Odontophorus melanotis* BLACK-EARED WOOD QUAIL[4]
    — *O.m.verecundus* J.L. Peters, 1929   Honduras
    — *O.m.melanotis* Salvin, 1865[5]   Nicaragua, Costa Rica, Panama

○ *Odontophorus erythrops* RUFOUS-FRONTED WOOD QUAIL
    — *O.e.parambae* Rothschild, 1897   W Colombia, NW Ecuador
    — *O.e.erythrops* Gould, 1859   E Ecuador

○ *Odontophorus atrifrons* BLACK-FRONTED WOOD QUAIL
    — *O.a.atrifrons* J.A. Allen, 1900   Santa Marta Mts. (N Colombia)
    — *O.a.variegatus* Todd, 1919   Andes of NE Colombia
    — *O.a.navai* Aveledo & Pons, 1952   W Venezuela (70)

○ *Odontophorus hyperythrus* Gould, 1858 CHESTNUT WOOD QUAIL   Andes of Colombia

○ *Odontophorus melanonotus* Gould, 1860 BLACK-BACKED WOOD QUAIL   SW Colombia, NW Ecuador

○ *Odontophorus speciosus* RUFOUS-BREASTED WOOD QUAIL
    — *O.s.soderstromii* Lönnberg & Rendahl, 1922[6]   C Ecuador
    — *O.s.speciosus* Tschudi, 1843   E Ecuador, C Peru
    — *O.s.loricatus* Todd, 1932   S Peru to C Bolivia

○ *Odontophorus dialeucos* Wetmore, 1963 TACARCUNA WOOD QUAIL   Cerros Tacarcuna and Malí, E Panama (2601)

○ *Odontophorus strophium* (Gould, 1844) GORGETED WOOD QUAIL   C Colombia

○ *Odontophorus columbianus* Gould, 1850 VENEZUELAN WOOD QUAIL Andes of N Venezuela

○ *Odontophorus leucolaemus* Salvin, 1867 BLACK-BREASTED WOOD QUAIL Costa Rica, W Panama

○ *Odontophorus balliviani* Gould, 1846 STRIPE-FACED WOOD QUAIL   Andes of S Peru and N Bolivia

○ *Odontophorus stellatus* (Gould, 1843) STARRED WOOD QUAIL   W Amazonia

○ *Odontophorus guttatus* (Gould, 1838) SPOTTED WOOD QUAIL[7]   S Mexico to W Panama

DACTYLORTYX Ogilvie-Grant, 1893 M
○ *Dactylortyx thoracicus* SINGING QUAIL
    — *D.t.pettingilli* Warner & Harrell, 1957   NE Mexico (2561)
    — *D.t.thoracicus* (Gambel, 1848)[8]   SE Mexico
    — *D.t.sharpei* Nelson, 1903   N and W of Yucatán Pen. (Mexico)
    — *D.t.paynteri* Warner & Harrell, 1955   S Quintana Roo (Mexico) (2560)
    — *D.t.devius* Nelson, 1898   Jalisco to Colima (W Mexico)
    — *D.t.melodus* Warner & Harrell, 1957   Guerrero (Mexico) (2561)
    — *D.t.chiapensis* Nelson, 1898[9]   SE Oaxaca, C Chiapas (Mexico)

---

[1] Including *polionotus* recognised in last Edition, but see Carroll (1994) (303).
[2] Including *snethlagei* Pinto, 1938 (1887) (*fide* Hellmayr & Conover, 1942) (1104).
[3] Including *plumbeicollis* based on a unique type.
[4] For recognition as separate from the next species see A.O.U. (1998: 127) (3).
[5] Includes *coloratus* see Wetmore (1965: 326) (2603).
[6] The emendation to oe (i.e. *soederstromi*) that has been suggested, based on the ö in the name Söderstrom, is not correct when applied to a Swedish name; see Art. 32.5.2.1 (I.C.Z.N., 1999) (1178).
[7] Thought to include *matudae* Brodkorb, 1939 (219), but see Storer (1988) (2351).
[8] Includes *lineolatus*; see Banks (1987) (86).
[9] Includes 3 names proposed by Warner & Harrell (1957) (2561): *ginetensis*, *edwardsi* and *moorei*; see Banks (1987) (86).

___ *D.t.dolichonyx* Warner & Harrell, 1957[1]	S Chiapas (Mexico), W Guatemala (2561)
___ *D.t.salvadoranus* Dickey & van Rossem, 1928[2]	El Salvador
___ *D.t.fuscus* Conover, 1937[3]	C Honduras (560)
___ *D.t.conoveri* Warner & Harrell, 1957	E Honduras (2561)

CYRTONYX Gould, 1844 M
○ *Cyrtonyx montezumae* MONTEZUMA QUAIL[4]

___ *C.m.mearnsi* Nelson, 1900	S USA, NW Mexico
___ *C.m.montezumae* (Vigors, 1830)[5]	N and C Mexico
___ *C.m.merriami* Nelson, 1897	Mt Orizaba (Vera Cruz, Mexico)
___ *C.m.rowleyi* A.R. Phillips, 1966[6]	Oaxaca (Mexico) (1869)
___ *C.m.sallei* J. Verreaux, 1859	SW Mexico

○ *Cyrtonyx ocellatus* (Gould, 1837) OCELLATED QUAIL[7]   SW Mexico, W Honduras, N Nicaragua

RHYNCHORTYX Ogilvie-Grant, 1893 M
○ *Rhynchortyx cinctus* TAWNY-FACED QUAIL

___ *R.c.pudibundus* J.L. Peters, 1929	Honduras and N Nicaragua
___ *R.c.cinctus* (Salvin, 1876)[8]	S Nicaragua to Panama
___ *R.c.australis* Chapman, 1915	W Colombia, NW Ecuador

PHASIANIDAE TURKEYS, GROUSE, PHEASANTS, PARTRIDGES (49: 180)

MELEAGRIDINAE

MELEAGRIS Linnaeus, 1758 F
◑ *Meleagris gallopavo* WILD TURKEY

✓ *M.g.silvestris* Vieillot, 1817	SE USA
___ *M.g.osceola* W.E.D. Scott, 1890	S Florida
✓ *M.g.intermedia* Sennett, 1879	N Texas to NE Mexico
___ *M.g.mexicana* Gould, 1856[9]	NW and NC Mexico (966)
✓ *M.g.merriami* Nelson, 1900	SW USA, NW Mexico
___ *M.g.gallopavo* Linnaeus, 1758	WC Mexico

● *Meleagris ocellata* Cuvier, 1820 OCELLATED TURKEY[10]   SE Mexico, Belize, Guatemala

TETRAONINAE[11]

BONASA Stephens, 1819 F
● *Bonasa umbellus* RUFFED GROUSE

___ *B.u.yukonensis* Grinnell, 1916	Alaska, NW Canada
✓ *B.u.umbelloides* (Douglas, 1829)	S British Columbia to Manitoba and N Colorado
___ *B.u.labradorensis* Ouellet, 1990	Labrador Pen. (1732)
___ *B.u.castanea* Aldrich & Friedmann, 1943[12]	Mt. Olympus, Washington (26)
___ *B.u.affinis* Aldrich & Friedmann, 1943	British Columbia, Oregon (26)
___ *B.u.obscura* Todd, 1947[13]	N Ontario (2407)
___ *B.u.sabini* (Douglas, 1829)	Coast of British Columbia to California
___ *B.u.brunnescens* Conover, 1935	Vancouver I. (559)
___ *B.u.togata* (Linnaeus, 1766)[14]	NC and NE USA and Nova Scotia

---

[1] Includes *calophonus* Warner & Harrell, 1957 (2561); see Banks (1987) (86).
[2] Includes *taylori*; see Banks (1987) (86).
[3] Includes *rufescens* Warner & Harrell, 1957 (2561); see Banks (1987) (86).
[4] In last Edition *C. sallei* was treated at specific level; we now follow A.O.U. (1998) (3).
[5] Includes ¶ *morio* van Rossem, 1942 (2478).
[6] For the recognition of this see Binford (1989) (150).
[7] Treated as monotypic following Carroll (1994) (303).
[8] Includes *hypopius* see Wetmore (1965: 333) (2603).
[9] Includes *onusta* R.T. Moore, 1938 (1612); see A.O.U. (1957) (1).
[10] We follow A.O.U. (1998) (3) in submerging the genus *Agriocharis*.
[11] Sequence and genera recognised based on Gutiérrez *et al.* (2000) (1026).
[12] For correction of spelling see David & Gosselin (2002) (613).
[13] A new name for *canescens* Todd, 1940 (2406).
[14] Includes *thayeri* see Godfrey (1986) (948); and *helmei* Bailey, 1941 (76).

— *B.u.mediana* Todd, 1940[1]   Minnesota (2406)

— *B.u.phaia* Aldrich & Friedmann, 1943   Idaho (26)

— *B.u.incana* Aldrich & Friedmann, 1943   Utah (26)

— *B.u.monticola* Todd, 1940   W Virginia (2406)

— *B.u.umbellus* (Linnaeus, 1766)   EC USA

TETRASTES Keyserling & Blasius, 1840   M

○ ***Tetrastes bonasia*** HAZEL GROUSE[2]

— *T.b.rhenana* (O. Kleinschmidt, 1917)   W Germany and Belgium to NE France

— *T.b.styriacus* (von Jordans & Schiebel, 1944) [3]   Jura, Alps and Carpathian Mts. north to E Poland (2541)

— *T.b.schiebeli* (A. Kleinschmidt, 1943)   S Slovenia to N Serbia, Greece and Bulgaria (1275)

— *T.b.rupestris* (C.L. Brehm, 1831)   C and S Germany and W Poland to N Austria

— *T.b.bonasia* (Linnaeus, 1758)[4]   W and S Scandinavia, S Finland, C Poland to C Ukraine and C European Russia

— *T.b.griseonotus* Salomonsen, 1947   N Scandinavia to NW European Russia (2152)

— *T.b.sibiricus* Buturlin, 1916   NE European Russia, N and C Siberia south to Transbaikalia and N Mongolia

— *T.b.kolymensis* Buturlin, 1916   E Yakutia to Sea of Okhotsk (E Siberia)

— *T.b.amurensis* Riley, 1916[5]   NE China, Amurland to Korea

— *T.b.yamashinai* Momiyama, 1928   Sakhalin

— *T.b.vicinitas* Riley, 1915   Hokkaido

○ ***Tetrastes sewerzowi*** SEVERTSOV'S HAZEL GROUSE

— *T.s.sewerzowi* Przevalski, 1876   Gansu, NE Qinghai, N Sichuan

— *T.s.secunda* Riley, 1925   SE Qinghai, E Xizang, W Sichuan

FALCIPENNIS Elliot, 1864[6]   M

● ***Falcipennis falcipennis*** (Hartlaub, 1855)   SIBERIAN GROUSE   NE Asia, Sakhalin

CANACHITES Stejneger, 1885   M

● ***Canchites canadensis*** SPRUCE GROUSE

— *C.c.osgoodi* Bishop, 1900   N Alaska

— *C.c.atratus* Grinnell, 1910   S Alaska

— *C.c.isleibi* (Dickerman & Gustafson, 1996)   SE Alaska (727)

— *C.c.canadensis* (Linnaeus, 1758)[7]   C Alberta to Labrador and Nova Scotia

— *C.c.canace* (Linnaeus, 1766)   SE Canada, N and NE USA

— *C.c.franklinii* (Douglas, 1829)   SE Alaska to NW Wyoming

TETRAO Linnaeus, 1758   M

● ***Tetrao urogallus*** WESTERN CAPERCAILLIE[8]

— *T.u.cantabricus* Castroviejo, 1967   NW Spain (306)

— *T.u.aquitanicus* Ingram, 1915   Pyrenees

— *T.u.major* C.L. Brehm, 1831   Alps to W and S Balkans and N Greece and to Estonia and N Belarus

— *T.u.rudolfi* Dombrowski, 1912   S and E Carpathian Mts

— *T.u.volgensis* Buturlin, 1907   S and C Belarus to C European Russia

✔ *T.u.urogallus* Linnaeus, 1758[9]   Scotland, Scandinavia to White Sea and Lake Onega

— *T.u.kureikensis* Buturlin, 1927[10]   NE European Russia to W Putorana Plateau

— *T.u.uralensis* Menzbier, 1887[11]   SE European Russia, S Urals, SW Siberia to NW Altai

— *T.u.taczanowskii* (Stejneger, 1885)   NE Altai to Lake Baikal and Olekma R.

---

[1] For correction of spelling see David & Gosselin (2002) (613).
[2] Revision based on Bauer (1960) (107) and Stepanyan (1990) (2326).
[3] Includes *horicei* Hachler, 1950 (1046) see Stepanyan (1990: 137) (2326).
[4] Includes *volgensis* see Stepanyan (1990: 138) (2326).
[5] Includes *gilacorum* Buturlin, 1935 (285) see Stepanyan (1990: 139) (2326).
[6] In separating this we follow A.O.U. (1998) (3) and de Juana (1994) (623).
[7] Includes *torridus* Uttal, 1939 (2456) following Godfrey (1986) (948).
[8] We have been unable to trace the name *hiomanus* Loudon, 1951, to its source.
[9] Includes *lugens, pleskei* (a population intermediate with *volgensis*), *karelicus* and, more recently, *lonnbergi* Johansen, 1957 (1208), see Stepanyan (1990: 133) (2326).
[10] For recognition see Stepanyan (1990: 133) (2326). Includes *obsoletus* Snigerewski, 1937 (2290).
[11] Includes *grisescens* see Stepanyan (1990: 135) (2326).

○ *Tetrao parvirostris* Black-billed Capercaillie[1]

 __ *T.p.parvirostris* Bonaparte, 1856[2]   E Asia from C Siberia and N Mongolia to Anadyr and Russian Far East

 __ *T.p kamschaticus* Kittlitz, 1858   Kamchatka

LYRURUS Swainson, 1832 M

● *Lyrurus tetrix* Eurasian Black Grouse[3]

 __ *L.t.britannicus* Witherby & Lönnberg, 1913   Scotland, Wales, N England

 ⤵ *L.t.tetrix* (Linnaeus, 1758)[4]   Norway, France and Balkans to W and N European Russia and N Siberia

 __ *L.t.viridanus* (Lorenz, 1891)   SE European Russia, SW Siberia and N Kazakhstan east to N Altai and Tuva Rep.

 __ *L.t.baikalensis* (Lorenz, 1911)   Lake Biakal and N Mongolia to W Amurland and N Nei Mongol

 __ *L.t.mongolicus* (Lönnberg, 1904)   Tien Shan to Russian and Mongolian Altai

 __ *L.t.ussuriensis* (Kohts, 1911)   C Amurland and NE Manchuria to Russian Far East and N Korea

○ *Lyrurus mlokosiewiczi* (Taczanowski, 1875) Caucasian Black Grouse   S and E Turkey, Caucasus, Transcaucasia

CENTROCERCUS Swainson, 1832 M

○ *Centrocercus urophasianus* (Bonaparte, 1827) Greater Sage Grouse[5]   S British Columbia to E California, N Colorado and Nebraska

○ *Centrocercus minimus* Young, Braun, Oyler-McCance, Quinn & Hupp, 2001 Gunnison Sage Grouse[6]
                  SW Colorado, E Utah (2702)

DENDRAGAPUS Elliot, 1864 M

● *Dendragapus obscurus* Blue Grouse

 __ *D.o.sitkensis* Swarth, 1921   SE Alaska

 __ *D.o.fuliginosus* (Ridgway, 1873)   S Yukon to NW California

 __ *D.o.richardsonii* (Douglas, 1829)   N British Columbia to SE Idaho

 __ *D.o.pallidus* Swarth, 1931   SE British Columbia to NE Oregon

 __ *D.o.sierrae* Chapman, 1904   Oregon to NC California

 __ *D.o.howardi* Dickey & van Rossem, 1923   C California

 __ *D.o.oreinus* Behle & Selander, 1951   E Nevada, W Utah (113)

 __ *D.o.obscurus* (Say, 1823)   S Utah to New Mexico

TYMPANUCHUS Gloger, 1841 M

○ *Tympanuchus pallidicinctus* (Ridgway, 1873) Lesser Prairie Chicken   Kansas to New Mexico

● *Tympanuchus phasianellus* Sharp-tailed Grouse

 __ *T.p.caurus* (Friedmann, 1943)   Alaska to S Yukon, N British Colombia and N Alberta (874)

 __ *T.p.kennicottii* (Suckley, 1862)   Mackenzie

 __ *T.p.phasianellus* (Linnaeus, 1758)   N Manitoba, N Ontario, WC Quebec

 __ *T.p columbianus* (Ord, 1815)   British Columbia to W Colorado

 ⤵ *T.p.jamesi* (Lincoln, 1917)   NC Alberta to Wyoming and Nebraska

 __ *T.p.campestris* (Ridgway, 1884)   C Canada to Wisconsin

 __ †*T.p.hueyi* Dickermann & Hubbard, 1994   New Mexico (728)

○ *Tympanuchus cupido* Greater Prairie Chicken

 __ *T.c.pinnatus* (Brewster, 1885)   SC Canada to NE Texas

 __ *T.c.attwateri* Bendire, 1893   Coast of Texas and SW Louisiana

 __ †*T.c.cupido* (Linnaeus, 1758)   S New England

---

[1] Most Russian literature, at least prior to Stepanyan (1990) (2326), treated this species under the name *urogalloides* Middendorff, 1851, but this name is preoccupied see Ogilvie-Grant (1893: 64, 66) (1675).

[2] Synonyms include *janensis*, *kolymensis* and *macrurus*. More recently includes *turensis* Buturlin, 1933 (281) and *stegmanni* Potapov, 1985 (1912), see Stepanyan (1990: 136).

[3] The name *tschusii* relates to intergrades.

[4] Includes *fedjuschini* Charlemagne, 1934 (318) see Stepanyan (1990: 129) (2326).

[5] Includes *phaios* Aldrich, 1946 (23).

[6] For recognition see A.O.U. (2000: 850) (4); however the authors mentioned therein (Bradbury & Vehrencamp, 1998) (208) did not write for the purpose defined in Art. 8.1.1. (I.C.Z.N., 1999) (1178).

LAGOPUS Brisson, 1760 F[1]

● ***Lagopus leucura*** White-tailed Ptarmigan

 ☑ *L.l.peninsularis* Chapman, 1902   C Alaska, Yukon

 — *L.l.leucura* (Richardson, 1831)   N British Columbia

 — *L.l.rainierensis* Taylor, 1920   Mt. Rainier, C and S Washington

 — *L.l.saxatilis* Cowan, 1939   Vancouver I. (576)

 — *L.l.altipetens* Osgood, 1901   Rocky Mts. from Montana to New Mexico

● ***Lagopus muta*** Rock Ptarmigan

 ☑ *L.m.muta* (Montin, 1781)   N Europe

 — *L.m.millaisi* E. Hartert, 1923   Scotland

 — *L.m.helvetica* (Thienemann, 1829)   Alps to the mountains of the Balkans

 — *L.m.pyrenaica* E. Hartert, 1921   C and E Pyrenees

 — *L.m.pleskei* Serebrovsky, 1926[2]   Siberia (from the Ural Mts. to Kamchatka)

 — *L.m.nadezdae* Serebrovsky, 1926[3]   Tarbagatay and Altai Mts. to south end of Lake Baikal
                 and N Mongolia; Tajikistan (this race?)

 — *L.m.ridgwayi* Stejneger, 1884   Commander Is.

 — *L.m.kurilensis* N. Kuroda, Sr., 1924   N and C Kuril Is.

 — *L.m.japonica* Clark, 1907   Hondo, Japan

 — *L.m.evermanni* Elliot, 1896   Attu I. (W Aleutians)

 — *L.m.townsendi* Elliot, 1896   Kiska I. (W Aleutians)

 — *L.m.gabrielsoni* Murie, 1944   Amchitka, Little Sitkin and Rat Is. (W Aleutians) (1633)

 — *L.m.sanfordi* Bent, 1912   Tanaga I. (WC Aleutians)

 — *L.m.chamberlaini* Clark, 1907   Adak I. (WC Aleutians)

 — *L.m.atkhensis* Turner, 1882   Atka I. (WC Aleutians)

 — *L.m.yunaskensis* Gabrielson & Lincoln, 1951   Yunaska I. (C Aleutians) (903)

 — *L.m.nelsoni* Stejneger, 1884[4]   Unimak I., Unalaska I. (E Aleutians) to S Alaska

 — *L.m.dixoni* Grinnell, 1909   Glacier Bay islands

 — *L.m.rupestris* (J.F. Gmelin, 1789)[5]   N Nth. America, S Greenland

 — *L.m.welchi* Brewster, 1885   Newfoundland

 — *L.m.saturata* Salomonsen, 1950   NW Greenland (2153)

 — *L.m.capta* J.L. Peters, 1934   NE Greenland

 — *L.m.hyperborea* Sundevall, 1845   Svalbard, Franz Josef Land

 — *L.m.islandorum* (Faber, 1822)   Iceland

● ***Lagopus lagopus*** Willow Grouse/Red Grouse[6]

 — *L.l.scotica* (Latham, 1787)[7]   Britain and Ireland

 — *L.l.variegata* Salomonsen, 1936   Islets in Trondheim Fjord (W Norway) (2150)

 ☑ *L.l.lagopus* (Linnaeus, 1758)   N Europe

 — *L.l.rossica* Serebrovsky, 1926[8]   Belarus and C European Russia to WC Urals

 — *L.l.koreni* Thayer & Bangs, 1914[9]   N Siberia E to Chukotskiy Pen., Kamchatka and Kuril Is.

 — *L.l.maior* Lorenz, 1904   S Ural steppe, SW Siberia, N Kazakhstan, W Altai

 — *L.l.brevirostris* Hesse, 1912   Altai to W Mongolia and W Tuva Republic

 — *L.l.kozlowae* Portenko, 1931   Khangai (N Mongolia) and E Tuva Republic

 — *L.l.sserebrowskii* Domaniewski, 1933[10]   NE Mongolia and Lake Baikal area to SE Siberia and
                 NE China (753)

 — *L.l.okadai* Momiyama, 1928[11]   Sakhalin I.

 ☑ *L.l.alascensis* Swarth, 1926   N and W (most of) Alaska

 — *L.l.muriei* Gabrielson & Lincoln, 1949   Aleutian, Shumagin and Kodiak Is. (902)

 — *L.l.alexandrae* Grinnell, 1909   S and SE Alaska islands, NW British Columbia

---

[1] Gender addressed by David & Gosselin (2002) (614), multiple changes occur in specific names.
[2] Including *komensis*, *transbaicalica* and, more recently, *barguzinensis* Buturlin, 1935 (285), and *krascheninnikowi* Potapov, 1985 (1912); see Stepanyan (1990: 128) (2326).
[3] Includes *macrorhyncha* see Stepanyan (1990: 129) (2326).
[4] Implicitly includes *kelloggae* see A.O.U. (1957: 133-134) (1).
[5] Includes *reinhardi* (see Vaurie, 1965: 245) (2510) and *carpathicus* Keve, 1948 (1261A).
[6] The name Red Grouse is used for *scotica* which some, e.g. Gutiérrez *et al.* (2000) (1026), consider a separate species.
[7] Peters (1934) (1822) dated the name from 1789, for use of 1787 see Vaurie (1965: 240) (2510). Includes *hibernica* see Vaurie (1965: 240) (2510).
[8] Includes *pallasi* nom. nov. Portenko, 1972 (1910) see Stepanyan (1990: 125) (2326).
[9] Including *birulai* and more recently *septentrionalis* Buturlin, 1935 (284) see Stepanyan (1990: 125) (2326).
[10] Spelling of trinomial not verified. Including *dybowskii* Stegmann, 1934 (2310) see Stepanyan (1990: 126) (2326).
[11] Recognised by Stepanyan (1990) (2326), who included *kamtschatkensis* which we place in *koreni* following the range statement of Vaurie (1965: 241) (2510).

__ *L.l.leucoptera* Taverner, 1932	Northern Canadian archipelago
__ *L.l.alba* (J.F. Gmelin, 1789)	Mainland of northern Canada (except NE)
__ *L.l.ungavus* Riley, 1911	N Canada (E of Hudson's Bay)
__ *L.l.alleni* Stejneger, 1884	Newfoundland

PERDICINAE

LERWA Hodgson, 1837 F
○ *Lerwa lerwa* SNOW PARTRIDGE

__ *L.l.lerwa* (Hodgson, 1833)	Himalayas and S Xizang
__ *L.l.major* R. Meinertzhagen, 1927	N Yunnan and W Sichuan
__ *L.l.callipygia* Stegmann, 1938	N Sichuan and Gansu (2312)

TETRAOPHASIS Elliot, 1871 M
○ *Tetraophasis obscurus* (J. Verreaux, 1869) VERREAUX'S MONAL PARTRIDGE

E Xizang, W Sichuan, E Qinghai, SW Gansu

○ *Tetraophasis szechenyii* Madarász, 1885 SZECHENYI'S MONAL PARTRIDGE SE Xizang, N Yunnan, SE Qinghai, W Sichuan

TETRAOGALLUS J.E. Gray, 1832 M
○ *Tetraogallus caucasicus* (Pallas, 1811) CAUCASIAN SNOWCOCK Caucasus

○ *Tetraogallus caspius* CASPIAN SNOWCOCK

__ *T.c.tauricus* Dresser, 1876[1]	S and E Turkey to W Armenia
__ *T.c.caspius* (S.G. Gmelin, 1784)	C Armenia, SW and SE Azerbaijan, N Iran, SW Turkmenistan
__ *T.c.semenowtianschanskii* Zarudny, 1908	Zagros Mts. (SW Iran)

○ *Tetraogallus himalayensis* HIMALAYAN SNOWCOCK

__ *T.h.saurensis* Potapov, 1993[2]	Tarbagatai and Saur Mts. (1913)
__ *T.h.sewerzowi* Zarudny, 1910	W Tien Shan to Dzhungarian Alatau and Karlik Shan
__ *T.h.incognitus* Zarudny, 1911[3]	Mts. of S Tajikistan, N Afghanistan
__ *T.h.himalayensis* (G.R. Gray, 1843)	E Afghanistan and W Himalayas to W Nepal
__ *T.h.grombczewskii* Bianchi, 1898	W Kun Lun (SW Xinjiang)
__ *T.h.koslowi* Bianchi, 1898	SE Xinjiang, N Qinghai, SW Gansu

○ *Tetraogallus tibetanus* TIBETAN SNOWCOCK

__ *T.t.tschimenensis* Sushkin, 1926	Kun Lun and Altun Shan Mts.
__ *T.t.tibetanus* Gould, 1854	E Pamir and E Afghanistan to W Tibetan Plateau and Ladakh
__ *T.t.aquilonifer* R. & A. Meinertzhagen, 1926	N Nepal, Sikkim, N Bhutan, N Assam, S Xizang
__ *T.t.yunnanensis* Yang & Xu, 1987	N Yunnan (2698)
__ *T.t.henrici* Oustalet, 1891	E Xizang, W Sichuan
__ *T.t.przewalskii* Bianchi, 1907[4]	Qinghai, N Sichuan, W Gansu

○ *Tetraogallus altaicus* (Gebler, 1836) ALTAI SNOWCOCK[5] Mongolia to C Russian Altai and Tuva Rep.

ALECTORIS Kaup, 1829 F
○ *Alectoris graeca* ROCK PARTRIDGE[6]

__ *A.g.saxatilis* (Bechstein, 1805)[7]	Alps, mainland Italy, Croatia, Montenegro
__ *A.g.graeca* (Mesiner, 1804)	Serbia and Albania to Bulgaria and Greece
__ *A.g.whitakeri* Schiebel, 1934	Sicily (2172)

---

[1] Includes *tauricus* Buturlin, 1933 (281) (a different taxon proposed in its own right), see Stepanyan (1990: 140) (2326).
[2] In the title of the paper, but not in the description, called *sauricus*.
[3] Includes *bendi* Koelz, 1951 (1281), see Stepanyan (1990: 140) (2326).
[4] Includes *centralis* see Cheng (1987: 131) (333).
[5] Treated as monotypic by Stepanyan (1990: 141) (2326).
[6] In previous Editions we followed Watson (1962) (2563) in splitting these first three species; here we correct the error we made in attributing the Central Asian races to *graeca*.
[7] Includes *orlandoi* Priolo, 1984 (1948) see McGowan (1994: 458) (1502).

○ *Alectoris chukar* CHUKAR PARTRIDGE

— *A.c.kleini* E. Hartert, 1925[1]   SE Bulgaria, NE Greece, NW and N Turkey, Caucasus

— *A.c.cypriotes* E. Hartert, 1917[2]   Crete, Rhodes, Cyprus, SW and SC Turkey

— *A.c.kurdestanica* R. Meinertzhagen, 1923[3]   SE Turkey, N Syria, N Iraq, Transcaucasia, N Iran

— *A.c.sinaica* (Bonaparte, 1858)   Syria to Sinai

— *A.c.werae* (Zarudny & Loudon, 1904)[4]   E Iraq, SW Iran

— *A.c.koroviakovi* (Zarudny, 1914)[5]   NE and E Iran, SW and S Turkmenistan, W and S Afghanistan, W Pakistan

— *A.c.subpallida* (Zarudny, 1914)   C Turkmenistan to C Uzbekistan and N Afghanistan

— *A.c.falki* E. Hartert, 1917[6]   Pamir-Alai and W and C Tien Shan Mts.

— *A.c.dzungarica* Sushkin, 1927   E Tien Shan and Zaisan basin to Altai, Tuva Rep., and NW Mongolia

— *A.c.pallida* (Hume, 1873)[7]   Tarim basin east to W Nan Shan (Xinjiang)

— *A.c.pallescens* (Hume, 1873)   S Pamir to W Tibetan Plateau

— *A.c.chukar* (J.E. Gray, 1830)   W Himalayan foothills from E Afghanistan to Nepal

— *A.c.potanini* Sushkin, 1927   NE Xinjiang and SW Mongolia to N Ningxia

— *A.c.pubescens* (Swinhoe, 1871)[8]   E Qinghai and Gansu to E China

○ *Alectoris magna* (Przevalski, 1876) PRZEVALSKI'S PARTRIDGE   N and NE Qinghai and SW Gansu

○ *Alectoris philbyi* P.R. Lowe, 1934 PHILBY'S PARTRIDGE   SW Arabia (1390)

○ *Alectoris barbara* BARBARY PARTRIDGE

— *A.b.koenigi* (Reichenow, 1899)   Canary Is., NW Morocco

— *A.b.spatzi* (Reichenow, 1895)[9]   S Morocco, C and S Algeria, S Tunisia

— *A.b.barbara* (Bonnaterre, 1792)   NE Morocco, N Algeria, N Tunisia

— *A.b.barbata* (Reichenow, 1896)   NE Libya, NW Egypt

● *Alectoris rufa* RED-LEGGED PARTRIDGE

✓ *A.r.rufa* (Linnaeus, 1758)[10]   Britain to Pyrenees, Corsica and N Italy

— *A.r.hispanica* (Seoane, 1894)   N and NW Spain, N Portugal

— *A.r.intercedens* (A.E. Brehm, 1857)[11]   E and S Iberia, Balearic Is.

— *A.r.corsa* (Parrot, 1910)   Corsica

○ *Alectoris melanocephala* ARABIAN PARTRIDGE

— *A.m.melanocephala* (Rüppell, 1835)   SW Arabia to SW Yemen

— *A.m.guichardi* R. Meinertzhagen, 1951   E Hadhramaut (E Yemen) (1550)

AMMOPERDIX Gould, 1851 M[12]

○ *Ammoperdix griseogularis* (J.F. Brandt, 1843) SEE-SEE PARTRIDGE[13]   SE Turkey to Pakistan

○ *Ammoperdix heyi* SAND PARTRIDGE

— *A.h.heyi* (Temminck, 1825)   Sinai, Israel and Jordan to NW and inland Arabian Peninsula

— *A.h.nicolli* E. Hartert, 1919   NE Egypt (Nile valley to Gulf of Suez)

— *A.h.cholmleyi* Ogilvie-Grant, 1897   SE Egypt, NE Sudan

— *A.h.intermedius* E. Hartert, 1917   SW Arabian penin. and NE Oman

PTILOPACHUS Swainson, 1837 M

○ *Ptilopachus petrosus* STONE PARTRIDGE

— *P.p.petrosus* (J.F. Gmelin, 1789)[14]   Gambia to Cameroon

---

[1] Includes *caucasica fide* Roselaar (*in litt.*); Stepanyan (1990: 142) (2326) puts it in *kurdestanica*.
[2] Includes *scotti* White, 1937 (2622) see Vaurie (1965: 270) (2510).
[3] Includes *daghestanica* and *armenica* Buturlin, 1933 (281), see Stepanyan (1990: 142) (2326).
[4] Includes *farsiana* Koelz, 1954 (1284) see Vaurie (1965: 271) (2510).
[5] Includes *shestoperovi* and both *laptevi* Dementiev, 1945 (688), and *dementievi* Rustamov, 1948 (2142) see Vaurie (1965: 271) (2510).
[6] The name *falki* was provided to replace *kakelik* Falk, 1786, which was considered indeterminate by Hartert (1921: 1907) (1083), but this name was long retained for the species by Russian workers. Includes *obscurata fide* Roselaar (*in litt.*); Vaurie (1965: 273) (2510) and Stepanyan (1990: 143) (2326), put it in *dzungarica*.
[7] Includes *fallax* see Cheng (1987: 135) (333).
[8] Tentatively includes ¶*ordoscensis* Zhang, Yin-sun *et al.*, 1989 (2705); map published clearly implies more work is needed.
[9] Includes *duprezi* Lavauden, 1930 (1336), and *theresae* R. Meinertzhagen, 1939 (1547), unless overlooked by Etchécopar & Hue (1964) (821).
[10] Includes *corsa* see Cramp *et al.* (1980: 463) (589).
[11] Includes *australis* (population considered introduced).
[12] Gender addressed by David & Gosselin (2002) (614), multiple changes occur in specific names.
[13] Treated as monotypic following Cramp *et al.* (1980: 473) (589); includes *peraticus* Koelz, 1950 (1280).
[14] Includes *saturatior* see White (1965: 92) (2649).

— *P.p.brehmi* Neumann, 1908   Lake Chad to C Sudan

— *P.p.major* Neumann, 1908   W Eritrea, N Ethiopia

— *P.p.florentiae* Ogilvie-Grant, 1900   S Sudan and S Ethiopia to NE Zaire, N Uganda and C Kenya

FRANCOLINUS Stephens, 1819[1] M

● *Francolinus francolinus* BLACK FRANCOLIN

— *F.f.francolinus* (Linnaeus, 1766)[2]   Cyprus, Turkey and Levant to Transcaucasia and SW Turkmenistan

— *F.f.arabistanicus* Zarudny & Härms, 1913   C and S Iraq, SW Iran

— *F.f.bogdanovi* Zarudny, 1906[3]   Afghanistan, SW Pakistan

— *F.f.henrici* Bonaparte, 1856   C and S Pakistan, SE Afghanistan

— *F.f.asiae* Bonaparte, 1856[4]   W and C India to C Nepal

— *F.f.melanonotus* Hume, 1888   E Nepal and E Bangladesh to Assam

○ *Francolinus pictus* PAINTED FRANCOLIN

— *F.p.pallidus* (J.E. Gray, 1831)[5]   W and NW India

— *F.p.pictus* (Jardine & Selby, 1828)   C and S India

— *F.p.watsoni* Legge, 1880   Sri Lanka (1348)

○ *Francolinus pintadeanus* CHINESE FRANCOLIN

— *F.p.phayrei* (Blyth, 1843)   NE India to S China, Indochina

— *F.p.pintadeanus* (Scopoli, 1786)   SE China

● *Francolinus pondicerianus* GREY FRANCOLIN

— *F.p.mecranensis* Zarudny & Härms, 1913   S Iran to W Pakistan

— *F.p.interpositus* E. Hartert, 1917[6]   N India

— *F.p.pondicerianus* (J.F. Gmelin, 1789)[7]   S India and Sri Lanka

○ *Francolinus gularis* (Temminck, 1815) SWAMP FRANCOLIN[8]   N India and Nepal to Assam and Bangladesh

○ *Francolinus lathami* LATHAM'S FOREST FRANCOLIN

— *F.l.lathami* Hartlaub, 1854   Sierra Leone to W Zaire and Angola

— *F.l.schubotzi* Reichenow, 1912   N and NE Zaire, Uganda, SW Sudan, NW Tanzania

○ *Francolinus coqui* COQUI FRANCOLIN

— *F.c.spinetorum* Bates, 1928   S Mauritania to N Nigeria

— *F.c.maharao* W.L. Sclater, 1927[9]   S Ethiopia to C and E Kenya

— *F.c.hubbardi* Ogilvie-Grant, 1895   W and S Kenya, N Tanzania

— *F.c.coqui* (A. Smith, 1836)[10]   S Zaire, Uganda, NW and NE Tanzania and coastal Kenya south to South Africa.

○ *Francolinus albogularis* WHITE-THROATED FRANCOLIN

— *F.a.albogularis* Hartlaub, 1854   Senegal and Gambia

— *F.a.buckleyi* Ogilvie-Grant, 1892   E Ivory Coast to N Cameroon

— *F.a.dewittei* Chapin, 1937[11]   SE Zaire, NW Zambia, E Angola (311)

○ *Francolinus schlegelii* Heuglin, 1863 SCHLEGEL'S FRANCOLIN[12]   WC Cameroon to SW Sudan

○ *Francolinus streptophorus* Ogilvie-Grant, 1891 RING-NECKED FRANCOLIN

       N Cameroon, W Kenya, N and C Uganda to NW Tanzania

○ *Francolinus africanus* Stephens, 1819 GREY-WING FRANCOLIN[13]   S Africa

---

[1] Species sequence follows Crowe *et al.* (1986) (599) for African species. We defer recognition of several genera until more have been sequenced in molecular studies.

[2] Includes *billypayni* R. Meinertzhagen, 1933 (1545), see Vaurie (1965: 282) (2510).

[3] Includes *festinus* Koelz, 1954 (1284), see Vaurie (1965: 283) (2510).

[4] Includes *parkerae* Van Tyne & Koelz, 1936 (2490), see Ripley (1982: 70) (2055).

[5] See Wheeler (1998) (2615) for date of publication.

[6] Includes *prepositus* Koelz, 1954 (1284), *paganus* Koelz, 1954 (1284), and *titar* Koelz, 1954 (1284) see Ripley (1982: 71) (2055).

[7] Includes *ceylonensis* Whistler, 1941 (2618) see Ripley (1982: 72) (2055).

[8] Includes *ridibundus* Koelz, 1954 (1284) see Ripley (1982: 72) (2055).

[9] Includes *thikae* Grant & Mackworth-Praed, 1934 (968) see Crowe *et al.* (1986: 30) (599).

[10] Includes *angolensis, ruahdae, campbelli, lynesi, vernayi, bourquii* Monard, 1934 (1601), *hoeschianus* Stresemann, 1937 (2360), *kasaicus* White, 1945 (2630) see White (1965: 69-70) (2649) and Crowe *et al.* (1986: 29) (599).

[11] Includes *meinertzhageni* White, 1944 (2627) see Crowe *et al.* (1986: 31) (599).

[12] Includes *confusus* Neumann, 1933 (1643) see White (1965: 69) (2649).

[13] Includes *proximus* Clancey, 1957 (374) see White (1965: 73) (2649). The resolution of this species complex, not reflected in previous Editions, involved moving several forms we had attached to a species *africanus*; *gutturalis* and *lorti* are now considered part of *F. levaillantoides*, *uluensis* is seen as a race of *F. shelleyi* and *ellenbecki* as a synonym of *F. p. psilolaemus* (see Crowe *et al.*, 1986) (599).

○ *Francolinus levaillantii* RED-WING FRANCOLIN
— *F.l.kikuyuensis* Ogilvie-Grant, 1897[1] E Zaire east to WC Kenya, south to Angola and Zambia
— *F.l.crawshayi* Ogilvie-Grant, 1896 Highlands of N Malawi
— *F.l.levaillanti* (Valenciennes, 1825) Zimbabwe, Transvaal, Natal, NE Cape Province

○ *Francolinus finschi* Bocage, 1881 FINSCH'S FRANCOLIN SW Zaire, W Angola

○ *Francolinus shelleyi* SHELLEY'S FRANCOLIN
— *F.s.uluensis* Ogilvie-Grant, 1892 C and S Kenya, N Tanzania
— *F.s.macarthuri* van Someren, 1938 Chyulu hills (SE Kenya) (2485)
— *F.s.shelleyi* Ogilvie-Grant, 1890[2] S Uganda and C Tanzania to E Transvaal and Natal
— *F.s.whytei* Neumann, 1908 SE Zaire, Zambia, N Malawi
— *F.s.canidorsalis* Lawson, 1963 SE Zimbabwe, N Transvaal, W Mozambique (1339)

○ *Francolinus psilolaemus* MOORLAND FRANCOLIN
— *F.p.psilolaemus* G.R. Gray, 1867 C Ethiopia
— *F.p.elgonensis* Ogilvie-Grant, 1891[3] E Uganda to C Kenya

○ *Francolinus levaillantoides* ORANGE RIVER FRANCOLIN
— *F.l.gutturalis* (Rüppell, 1835)[4] Eritrea, N Ethiopia
— *F.l.archeri* W.L. Sclater, 1927[5] S Ethiopia, SE Sudan, N Uganda, NW Kenya
— *F.l.lorti* Sharpe, 1897 N Somalia
— *F.l.jugularis* Büttikofer, 1889[6] SW Angola
— *F.l.pallidior* Neumann, 1908[7] N Namibia
— *F.l.levaillantoides* (A. Smith, 1836)[8] E Namibia, S Botswana, N and W Sth. Africa, Lesotho (2280)

○ *Francolinus sephaena* CRESTED FRANCOLIN
— *F.s.spilogaster* Salvadori, 1888[9] E Ethiopia, Somalia, NE Kenya
— *F.s.grantii* Hartlaub, 1866 S Sudan, W Ethiopia to NC Tanzania
— *F.s.rovuma* G.R. Gray, 1867[10] Coastal Kenya to N Mozambique
— *F.s.zambesiae* Mackworth-Praed, 1920[11] WC Mozambique to Namibia and S Angola
— *F.s.sephaena* A. Smith, 1836[12] SE Botswana, SE Zimbabwe to Mozambique and NE Sth. Africa

○ *Francolinus squamatus* SCALY FRANCOLIN
— *F.s.squamatus* Cassin, 1857[13] SE Nigeria to N, C and W Zaire
— *F.s.schuetti* Cabanis, 1880[14] E Zaire to Uganda, W and C Kenya, Ethiopia
— *F.s.maranensis* Mearns, 1910[15] S and SE Kenya to N Tanzania
— *F.s.usambarae* Conover, 1928 NE Tanzania
— *F.s.uzungwensis* Bangs & Loveridge, 1931 SE Tanzania
— *F.s.doni* Benson, 1939 Vipya Plateau (Malawi) (115)

○ *Francolinus ahantensis* AHANTA FRANCOLIN[16]
— *F.a.hopkinsoni* Bannerman, 1934 W Senegal and W Gambia to Guinea Bissau (93)
— *F.a.ahantensis* Temminck, 1851 S Guinea to SW Nigeria

○ *Francolinus griseostriatus* Ogilvie-Grant, 1890 GREY-STRIPED FRANCOLIN
W Angola

○ *Francolinus nahani* A.J.C. Dubois, 1905 NAHAN'S FRANCOLIN NE Zaire, W Uganda

---

[1] Includes *benguellensis* (renamed *momboloensis* White, 1952 (2634)) and *clayi* White, 1944 (2626) see White (1965: 76) (2649).
[2] Includes *trothae,* and *sequestris* Clancey, 1960 (385) see White (1965: 73) (2649).
[3] Moved from the next species to this by White (1965: 74) (2649); followed by Crowe *et al.* (1986: 41) (599); *theresae* R. Meinertzhagen, 1937 (1546), is a synonym see Crowe *et al.* (1986: 41) (599).
[4] Includes *eritreae* see White (1965: 76) (2649).
[5] Includes *friedmanni* Grant & Mackworth-Praed, 1934 (969), *stantoni* Cave, 1940 (307) see White (1965: 75) (2649).
[6] White (1965: 75) (2649) placed *stresemanni* Hoesch & Niethammer, 1940 (1126) in *cunenensis* which is now included herein see Crowe *et al.* (1986: 42) (599).
[7] Includes *wattii* Macdonald, 1953 (1408) see White (1965: 75) (2649).
[8] The belated uptake of this name is discussed in Macdonald (1953) (1408). Includes *kalaharicus* see Crowe *et al.* (1986: 41) (599).
[9] Includes *somaliensis* Grant & Mackworth-Praed, 1934 (967) see White (1965: 72) (2649).
[10] Treated at specific level in last Edition, but see White (1965: 72) (2649) and Crowe *et al.* (1986: 43) (599).
[11] Includes *thompsoni* see White (1965: 72) (2649).
[12] Includes *zuluensis* see White (1965: 71) (2649).
[13] Includes *whitei* Schouteden, 1954 (2200) a new name for *confusus* Schouteden, 1954 (2199A), and see White (1965: 89) (2649).
[14] Includes *zappeyi* and *tetraoninus* see White (1965: 89) (2649).
[15] Includes *chyuluensis* van Someren, 1939 (2486) see White (1965: 89) (2649).
[16] We follow White (1965: 88) (2649) in recognizing *hopkinsoni* (93).

◯ *Francolinus hartlaubi* Hartlaub's Francolin
   — *F.h.hartlaubi* Bocage, 1869   SW Angola
   — *F.h.bradfieldi* (Roberts, 1928)   C and N Namibia
   — *F.h.crypticus* Stresemann, 1939   NW Namibia (2362)

◯ *Francolinus hildebrandti* Hildebrandt's Francolin
   — *F.h.altumi* Fischer & Reichenow, 1884   W Kenya
   — *F.h.hildebrandti* Cabanis, 1878[1]   E and C Kenya, N and W Tanzania, SE Zaire, NE Zambia
   — *F.h.johnstoni* Shelley, 1894[2]   SE Tanzania, N Mozambique, Malawi.

◯ *Francolinus natalensis* Natal Francolin
   — *F.n.neavei* Mackworth-Praed, 1920   S Zambia, W Mozambique
   — *F.n.natalensis* A. Smith, 1834[3]   Zimbabwe to Transvaal and Natal

◯ *Francolinus bicalcaratus* Double-spurred Francolin
   — *F.b.ayesha* E. Hartert, 1917   W Morocco
   — *F.b.bicalcaratus* (Linnaeus, 1766)[4]   Senegal across sub-Saharan W Africa to SW Chad
   — *F.b.thornei* Ogilvie-Grant, 1902   Sierra Leone to Ivory Coast
   — *F.b.ogilviegranti* Bannerman, 1922[5]   S Nigeria, S Cameroon

◯ *Francolinus icterorhynchus* Heuglin, 1863 Heuglin's/Yellow-billed Francolin[6]
                                C African Republic, SW Sudan, N Zaire, W Uganda

◯ *Francolinus clappertoni* Clapperton's Francolin
   — *F.c.clappertoni* Children, 1826   Mali to W Sudan
   — *F.c.konigseggi* Madarász, 1914   E Sudan
   — *F.c.heuglini* Neumann, 1907   SW and SC Sudan
   — *F.c.sharpii* Ogilvie-Grant, 1892   Eritrea, N and C Ethiopia
   — *F.c.nigrosquamatus* Neumann, 1902[7]   E Sudan, W Ethiopia
   — *F.c.gedgii* Ogilvie-Grant, 1891[8]   SE Sudan to NE Uganda

◯ *Francolinus harwoodi* Blundell & Lovat, 1899 Harwood's Francolin   NW Ethiopia

◯ *Francolinus capensis* (J.F. Gmelin, 1789) Cape Francolin   W Cape Province

◯ *Francolinus adspersus* Red-billed Francolin
   — *F.a.adspersus* Waterhouse, 1838[9]   SW Angola and NW Namibia east to N Botswana
   — *F.a.mesicus* (Clancey, 1996)   NC highlands of Namibia; and NW Zimbabwe, SW
                                Zambia (527)

◯ *Francolinus swierstrai* (Roberts, 1929) Swierstra's Francolin[10]   W Angola

◯ *Francolinus camerunensis* Alexander, 1909 Mount Cameroon Francolin
                                Mt. Cameroon

◯ *Francolinus nobilis* Reichenow, 1908 Handsome Francolin[11]   E Zaire, W and SW Uganda

◯ *Francolinus jacksoni* Ogilvie-Grant, 1891 Jackson's Francolin[12]   Mts. of WC Kenya

◯ *Francolinus castaneicollis* Chestnut-naped Francolin
   — *F.c.ogoensis* Mackworth-Praed, 1920   N Somalia
   — *F.c.castaneicollis* Salvadori, 1888[13]   WC to E Ethiopia
   — *F.c.kaffanus* Grant & Mackworth-Praed, 1934[14]   SW Ethiopia (968)
   — *F.c.atrifrons* Conover, 1930   S Ethiopia

◯ *Francolinus ochropectus* Dorst & Jouanin, 1952 Djibouti Francolin   Goda and Mabla Mts. (Djibouti) (754)

---

[1] Includes *helleri* and *fischeri* see White (1965: 78) (2649).
[2] Includes *grotei* see White (1965: 78) (2649).
[3] Includes *thamnobium* Clancey, 1953 (370) see White (1965: 77) (2649).
[4] Includes *adamauae* (see White (1965: 79) (2649).
[5] Includes *molunduensis* Grote, 1948 (1020) see White (1965: 79) (2649).
[6] Now generally treated as monotypic Crowe *et al.* (1986: 55) (599); McGowan, 1994: 497) (1502).
[7] Includes *testis* thought, based on the type locality, to be within the range of this race, although placed in *sharpii* by Mackworth-Praed & Grant, (1957: 280) (1411).
[8] Includes *cavei* Macdonald, 1940 (1406) see White (1965: 80) (2649).
[9] Includes *kalahari* see Clancey (1996) (527).
[10] Includes *cruzi* Themido, 1937 (2396) see White (1965: 87) (2649).
[11] Monotypic, see for example White (1965: 86) (2649); includes *chapini* Mackworth-Praed & Grant, 1934 (970) and *ruandae* van Saceghem, 1942 (2484).
[12] Now generally treated as monotypic (e.g. White, 1965: 86) (2649); includes *pollenorum* R. Meinertzhagen, 1937 (1546), and presumably ¶*patriciae* Ripley & Bond, 1971 (2064).
[13] Includes *bottegi* and *gofanus* see White (1965: 87) (2649).
[14] Includes *patrizii* Toschi, 1958 (2420) see White (1965: 87) (2649).

● *Francolinus erckelii* (Rüppell, 1835) ERCKEL'S FRANCOLIN   NE Sudan, Eritrea, Ethiopia

○ *Francolinus swainsonii* SWAINSON'S FRANCOLIN
    — *F.s.lundazi* (C.M.N. White, 1947)   NE Zambia, N and W Zimbabwe, S Mozambique (2631)
    — *F.s.swainsonii* (A. Smith, 1836)[1]   SW Angola, N Namibia to S and NE Sth. Africa

○ *Francolinus afer*[2] RED-NECKED FRANCOLIN[3]
    — *F.a.harterti* (Reichenow, 1909)   E Zaire, Rwanda, Burundi, NW Tanzania
    — *F.a.cranchii* (Leach, 1818)[4]   S Zaire and N and C Angola to Uganda, W Kenya, W and C Tanzania, N Malawi, N and W Zambia
    — *F.a.leucoparaeus* Fischer & Reichenow, 1884   SE Kenya, N Tanzania
    — *F.a.melanogaster* (Neumann, 1898)[5]   E Tanzania to N Mozambique and NE Zambia
    — *F.a.afer* (Statius Müller, 1776)[6]   W Angola, NW Namibia
    — *F.a.swynnertoni* (W.L. Sclater, 1921)   E Zimbabwe, S Mozambique
    — *F.a.castaneiventer* (Gunning & Roberts, 1911)[7]   S and E Cape Province

○ *Francolinus rufopictus* (Reichenow, 1887) GREY-BREASTED FRANCOLIN   SE Lake Victoria

○ *Francolinus leucoscepus* G.R. Gray, 1867 YELLOW-NECKED FRANCOLIN[8]   SE Sudan, Eritrea, S and E Ethiopia, Somalia and Kenya to NE Tanzania

PERDIX Brisson, 1760 F
● *Perdix perdix* GREY PARTRIDGE
    — *P.p.perdix* (Linnaeus, 1758)[9]   Norway, Britain and N France to C Europe, W Bulgaria and Greece
    — *P.p.armoricana* E. Hartert, 1917   W and C France
    — *P.p.sphagnetorum* (Altum, 1894)   NE Netherlands, NW Germany
    — *P.p.hispaniensis* Reichenow, 1892   N Spain, NE Portugal
    — *P.p.italica* E. Hartert, 1917   Italy
    — *P.p.lucida* (Altum, 1894)   Finland to NE Bulgaria to Urals and N Caucasus
    — *P.p.canescens* Buturlin, 1906[10]   Turkey to Caucasus area and NW Iran
    — *P.p.robusta* Homeyer & Tancré, 1883[11]   SW Siberia and Kazakhstan to NW China and Tuva Rep.

○ *Perdix dauurica* DAURIAN PARTRIDGE[12]
    — *P.d.dauurica* (Pallas, 1811)[13]   Tien Shan and Altai to Transbaikalia and Mongolia, N China
    — *P.d.suschkini* Poliakov, 1915[14]   NE China and Russian Far East

○ *Perdix hodgsoniae* TIBETAN PARTRIDGE
    — *P.h.caraganae* R. & A. Meinertzhagen, 1926   Edge of SW Tibetan Plateau (E Kashmir to N Kumaon)
    — *P.h.hodgsoniae* (Hodgson, 1857)   Edge of SE Tibetan Plateau (N Kumaon to SE Xizang)
    — *P.h.sifanica* Przevalski, 1876   Edge of E Tibetan Plateau (N Yunnan to Gansu)

RHIZOTHERA G.R. Gray, 1841 F
○ *Rhizothera longirostris* (Temminck, 1815) LONG-BILLED PARTRIDGE   Malay Pen., Sumatra, W Borneo

○ *Rhizothera dulitensis* Ogilvie-Grant, 1895 HOSE'S PARTRIDGE[15]   Mts. of N Borneo

MARGAROPERDIX Reichenbach, 1853 M
○ *Margaroperdix madagarensis*[16] (Scopoli, 1786) MADAGASCAN PARTRIDGE
    Madagascar (not extreme S)

[1] Includes *damarensis* and *gilli* (as in Peters, 1934) (1822). The name *cooperi* Roberts, 1947 (2090) was applied to a hybrid of this species and *F. afer* see Irwin (1981: 87) (1189).
[2] Type species of the genus *Pternistis* here submerged.
[3] We keep the name spurfowl for the genus *Galloperdix*. The name *humboldtii* is considered to relate to an unstable intergrading population involving *loangwae* (now in the synonymy of *melanogaster*) and *swynnertoni* see White (1965: 83) (2649).
[4] Includes *punctulatus, boehmi, benguellensis, intercedens, nyanzae* and *itigi*, and the more recent names *tornowi* Meise, 1933 (1551), *mackenziei* White, 1945 (2628), *manueli* White, 1945 (2628), and *camabatelae* Meise, 1958 (1553) following White (1965: 84) (2649).
[5] Includes *tertius* Meise, 1934 (1551), *loangwae* Grant & Mackworth-Praed, 1934 (969), and *aylwinae* White, 1947 (2631) following Crowe *et al.* (1986: 71) (599) and White (1965: 82) (2649).
[6] Includes *chio* Meise, 1958 (1553), and *palliditectus* White, 1958 (2636) see White (1965: 82) (2649).
[7] Includes *notatus* Roberts, 1924 (2087) and *lehmanni* following Crowe *et al.* (1986: 70) (599).
[8] For monotypic treatment see McGowan (1994: 499) (1502); includes *oldowai* van Someren, 1939 (2486) placed in *infuscatus* by White (1965: 85) (2649).
[9] Includes *borkumensis* Harrison, 1952 (1076) (nom. nov. for *pallida* Harrison, 1952 (1075)) see Vaurie (1965: 285) (2510).
[10] Includes *furvescens* see Stepanyan (1990: 146) (2326).
[11] Includes *arenicola* see Stepanyan (1990: 146) (2326).
[12] Forms now in synonymy as given by Vaurie (1965: 288-289) (2510).
[13] Spelling as used by McGowan (1994: 505) (1502) is correct; the original spelling *dauuricae* (see Vaurie, 1965: 288) (2510) is in the genitive case and must be corrected.
[14] Includes *castaneothorax* Hachisuka & Taka-Tsukasa, 1939 (1045) see Vaurie (1965: 289) (2510).
[15] For reasons to treat this as distinct from *R. longirostris* see Davison (2000) (622).
[16] The original spelling; *madagascarensis* is an emendation.

MELANOPERDIX Jerdon, 1864 M
○ ***Melanoperdix niger*** BLACK PARTRIDGE[1]
    __ *M.n.niger* (Vigors, 1829)   Malay Pen., Sumatra
    __ *M.n.borneensis* (Rothschild, 1917)   Borneo

COTURNIX Bonnaterre, 1791 F
○ ***Coturnix coturnix*** COMMON QUAIL
    __ *C.c.coturnix* (Linnaeus, 1758)[2]   Canary Is., Madeira and NW Africa, and Europe east to Lena Valley, Mongolia and N India >> Africa, C and S India

    __ *C.c.conturbans* E. Hartert, 1917   Azores
    __ *C.c.inopinata* E. Hartert, 1917   Cape Verde Is.
    __ *C.c.africana* Temminck & Schlegel, 1849   Sthn. Africa, Mauritius, Madagascar, Comoros Is.
    __ *C.c.erlangeri* Zedlitz, 1912   E and NE Africa

○ ***Coturnix japonica*** Temminck & Schlegel, 1849 JAPANESE QUAIL[3]   Transbaikalia and Mongolia to Sakhalin, Japan and Korea >> NE India and China to N Indochina

○ ***Coturnix coromandelica*** (J.F. Gmelin, 1789) RAIN QUAIL   Pakistan to Burma and Sri Lanka

○ ***Coturnix delegorguei*** HARLEQUIN QUAIL
    __ *C.d.delegorguei* Delegorgue, 1847   Ivory Coast to Ethiopia, south to N Namibia and E Cape Province

    __ *C.d.histrionica* Hartlaub, 1849   São Tomé
    __ *C.d.arabica* Bannermann, 1929   S Arabia

○ ***Coturnix pectoralis*** Gould, 1837 STUBBLE QUAIL   Australia, Tasmania

○ †***Coturnix novaezelandiae*** Quoy & Gaimard, 1830 NEW ZEALAND QUAIL
                                  New Zealand

○ ***Coturnix ypsilophora*** BROWN QUAIL
    __ *C.y.raaltenii* (S. Müller, 1842)[4]   Lesser Sundas (except Sumba)
    __ *C.y.pallidior* (E. Hartert, 1897)   Sumba, Savu
    __ *C.y.saturatior* (E. Hartert, 1930)   N New Guinea lowlands
    __ *C.y.dogwa* Mayr & Rand, 1935   S New Guinea lowlands (1492)
    __ *C.y.plumbea* (Salvadori, 1894)[5]   NE New Guinea lowlands
    __ *C.y.monticola* Mayr & Rand, 1935[6]   Alpine New Guinea (1492)
    __ *C.y.mafulu* Mayr & Rand, 1935   Montane New Guinea (1492)
    __ *C.y.australis* (Latham, 1802)[7]   Australia
    __ *C.y.ypsilophora* Bosc, 1792   Tasmania

○ ***Coturnix chinensis*** BLUE-BREASTED QUAIL/KING QUAIL
    __ *C.c.chinensis* (Linnaeus, 1766)   India to Sri Lanka, Taiwan, Indochina and Malay Pen.
    __ *C.c.trinkutensis* (Richmond, 1902)   Nicobar Is
    __ *C.c.palmeri* (Riley, 1919)   Sumatra, Java
    __ *C.c.lineata* (Scopoli, 1786)[8]   Philippines, Borneo, Lesser Sundas, Sulawesi, Moluccas
    __ *C.c.novaeguineae* Rand, 1941   Montane New Guinea (1965)
    __ *C.c.papuensis* Mayr & Rand, 1936   SE New Guinea (1494)
    __ *C.c.lepida* (Hartlaub, 1879)   Bismarck Arch.
    __ *C.c.victoriae* (Mathews, 1912)[9]   E Australia
    __ *C.c.colletti* (Mathews, 1912)   N Australia

○ ***Coturnix adansonii*** J. & E. Verreaux, 1851 AFRICAN BLUE QUAIL[10]   Sub-Saharan Africa

---

[1] For correction of spelling see David & Gosselin (2002) (614).
[2] Includes *confisa* see Cramp *et al.* (1980: 496) (589). Also includes *ragonierii* Trischitta, 1939 (2439), and *parisii* Trischitta, 1939 (2440), see Vaurie (1965: 292) (2510).
[3] Includes *ussuriensis* (placed in wrong species in last Ed.) see Stepanyan (1990: 148) (2326).
[4] Includes *castaneus* Mayr, 1944 (1474) see White & Bruce (1986: 145) (2652).
[5] For correction of spelling see David & Gosselin (2002) (613).
[6] Includes *lamonti* Mayr & Gilliard, 1951 (1489) (R. Schodde *in litt.*). Whether the alpine and montane forms are conspecific with lowland forms requires review.
[7] For reasons to use 1802 not 1801 see Browing & Monroe (1991) (258). Includes *cervinus* and *queenslandicus* see Schodde & Tidemann (1986) (2196).
[8] Includes *lineatula* (see White & Bruce, 1986: 145) (2652). For placement of Bornean birds here (*contra* Peters, 1934) (1822) see Chasen (1935: 5fn) (319) and Smythies (1960) (2287).
[9] Mistakenly listed as *australis* in last Edition, but this name is preoccupied.
[10] Treated as a species by McGowan (1994: 511) (1502).

ANUROPHASIS van Oort, 1910 M
○ *Anurophasis monorthonyx* van Oort, 1910 SNOW MOUNTAIN QUAIL   Kemabu Plateau and Snow Mts. (New Guinea)

PERDICULA Hodgson, 1837 F
○ *Perdicula asiatica* JUNGLE BUSH QUAIL
    — *P.a.punjaubi* Whistler, 1939   NW India (2617)
    — *P.a.asiatica* (Latham, 1790)   N and C India
    — *P.a.vidali* Whistler & Kinnear, 1936   SW India (2620)
    — *P.a.vellorei* Abdulali & Reuben, 1964[1]   S India (17)
    — *P.a.ceylonensis* Whistler & Kinnear, 1936   Sri Lanka (2620)

○ *Perdicula argoondah* ROCK BUSH QUAIL[2]
    — *P.a.meinertzhageni* Whistler, 1937   W and NW India (2616)
    — *P.a.argoondah* (Sykes, 1832)   C and SE India
    — *P.a.salimali* Whistler, 1943   S India (27)

○ *Perdicula erythrorhyncha* PAINTED BUSH QUAIL
    — *P.e.blewitti* (Hume, 1874)   C and SE India
    — *P.e.erythrorhyncha* (Sykes, 1832)   SW India

○ *Perdicula manipurensis* MANIPUR BUSH QUAIL
    — *P.m.inglisi* (Ogilvie-Grant, 1909)   N Bengal to N Assam
    — *P.m.manipurensis* Hume, 1880   E Bangladesh and S Assam

OPHRYSIA Bonaparte, 1856 F
○ †?*Ophrysia superciliosa* (J.E. Gray, 1846) HIMALAYAN QUAIL   Kumaon (NW Himalayas)

XENOPERDIX Dinesen, Lehmberg, Svendsen, Hansen & Fjeldså, 1994 M
○ *Xenoperdix udzungwensis* Dinesen, Lehmberg, Svendsen, Hansen & Fjeldså, 1994 UDZUNGWA FOREST PARTRIDGE[3] #
                                   Udzungwa Mts. (Tanzania) (752)

ARBOROPHILA Hodgson, 1837[4] F
○ *Arborophila torqueola* COMMON HILL PARTRIDGE
    — *A.t.millardi* (E.C.S. Baker, 1921)   W Himalayas to W Nepal
    — *A.t.torqueola* (Valenciennes, 1825)   C and E Himalayas to N Burma, N Yunnan and SE
                                   Xizang
    — *A.t.interstincta* Ripley, 1951   S and SE Assam (2035)
    — *A.t.batemani* (Ogilvie-Grant, 1906)   W and NW Burma
    — *A.t.griseata* Delacour & Jabouille, 1930   NW Vietnam

○ *Arborophila rufogularis* RUFOUS-THROATED HILL PARTRIDGE
    — *A.r.rufogularis* (Blyth, 1850)   Foothills of C and E Himalayas
    — *A.r.intermedia* (Blyth, 1856)[5]   S and E Assam to W and NW Burma
    — *A.r.tickelli* (Hume, 1880)   E and SE Burma to SW Laos
    — *A.r.euroa* (Bangs & Phillips, 1914)   SE Yunnan to N Laos
    — *A.r.guttata* Delcour & Jabouille, 1928   C Vietnam
    — *A.r.annamensis* (Robinson & Kloss, 1919)   SC Vietnam

○ *Arborophila atrogularis* (Blyth, 1850) WHITE-CHEEKED HILL PARTRIDGE[6]
                                   Assam to W and NW Burma

○ *Arborophila crudigularis* (Swinhoe, 1864) TAIWAN HILL PARTRIDGE   Taiwan

○ *Arborophila mandellii* Hume, 1874 CHESTNUT-BREASTED HILL PARTRIDGE
                                   Sikkim to NE Assam

○ *Arborophila brunneopectus* BAR-BACKED HILL PARTRIDGE
    — *A.b.brunneopectus* (Blyth, 1855)   E Assam, W and S Yunnan to W Thailand and E and
                                 S Burma

---

[1] Accepted by Ripley (1982: 74) (2055).
[2] Polytypic treatment follows Ripley (1982: 75) (2055).
[3] We note that within this article (but not in conjunction with the specific name where proposed) the authors name the type species of the genus and cite the authors in a different sequence. We take this to be a *lapsus*.
[4] Now includes *Tropicoperdix* see Davison (1982) (621).
[6] Includes *tenebrarum* Koelz, 1954 (1284) see Ripley (1982: 77) (2055).
[6] Includes *rupchandi* Koelz, 1953 (1283) see Ripley (1982: 77) (2055).

— *A.b.henrici* (Oustalet, 1896)   NW and C Vietnam

— *A.b.albigula* (Robinson & Kloss, 1919)   SC Vietnam

○ *Arborophila rufipectus* Boulton, 1932 Sıchuan Hıll Partridge   W Sichuan

○ *Arborophila gingica* (J.F. Gmelin, 1789) Collared Hıll Partridge   Guanxi and Hunan to Fujian hills

○ *Arborophila davidi* Delacour, 1927 Orange-necked Hıll Partridge   SC Vietnam

○ *Arborophila cambodiana* Chestnut-headed Hıll Partridge

— *A.c.cambodiana* Delacour & Jabouille, 1928   Cambodia

— *A.c.diversa* Riley, 1930   SE Thailand

○ *Arborophila campbelli* (Robinson, 1904) Campbell's Hıll Partridge[1]

     W Malaysia

○ *Arborophila rolli* (Rothschild, 1909) Roll's Hıll Partridge   NW Sumatra

○ *Arborophila sumatrana* Ogilvie-Grant 1891 Sumatran Hıll Partridge   C Sumatra

○ *Arborophila orientalis* (Horsfield, 1821) Horsfield's Hıll Partridge   E Java

○ *Arborophila javanica* Chestnut-bellied Hıll Partridge

— *A.j.javanica* (J.F. Gmelin, 1789)[2]   W and WC Java

— *A.j.lawuana* M. Bartels, Jr., 1938   EC Java (100)

○ *Arborophila rubrirostris* (Salvadori, 1879) Red-billed Hıll Partridge   Sumatra

○ *Arborophila hyperythra* (Sharpe, 1879) Red-breasted Hıll Partridge   NW Borneo

○ *Arborophila ardens* (Styan, 1892) Hainan Hıll Partridge   Hainan I.

○ *Arborophila charltonii* Chestnut-necklaced Hıll Partridge

— *A.c.charltonii* (Eyton, 1845)   Malay Pen.

— *A.c.atjenensis* Meyer de Schauensee & Ripley, 1940[3]   N Sumatra (1578)

— *A.c.tonkinensis* (Delacour, 1927)   N and NC Vietnam

— *A.c.graydoni* (Sharpe & Chubb, 1906)   Borneo

○ *Arborophila chloropus* Green-legged Hıll Partridge[4]

— *A.c.chloropus* (Blyth, 1859)   N and E Burma to W Thailand

— *A.c.cognacqi* (Delacour & Jabouille, 1924)   SC Vietnam

— *A.c.merlini* (Delacour & Jabouille, 1924)   WC Vietnam

— *A.c.vivida* (Delacour, 1926)   EC Vietnam

— *A.c.peninsularis* Meyer de Schauensee, 1940   SW Thailand (1559)

— *A.c.olivacea* (Delacour & Jabouille, 1928)   Laos, Cambodia

CALOPERDIX Blyth, 1861 M[5]

○ *Caloperdix oculeus* Ferruginous Partridge

— *C.o.oculeus* (Temminck, 1815)   SE Burma, SW Thailand, W Malaysia

— *C.o.ocellatus* (Vigors, 1829)   Sumatra

— *C.o.borneensis* Ogilvie-Grant, 1892   Borneo

HAEMATORTYX Sharpe, 1879 M

○ *Haematortyx sanguiniceps* Sharpe, 1879 Crimson-headed Partridge   N Borneo

ROLLULUS Bonnaterre, 1791 M

○ *Rollulus rouloul* (Scopoli, 1786) Roulroul[6]   S Thailand, W Malaysia, Sumatra, Borneo

BAMBUSICOLA Gould, 1863 M

○ *Bambusicola fytchii* Mountain Bamboo Partridge

— *B.f.fytchii* Anderson, 1871   E Burma and NW Vietnam to Yunnan and W Sichuan

— *B.f.hopkinsoni* Godwin-Austen, 1874[7]   S and E Assam, W Burma

---

[1] This and the next three species previously treated as conspecific, but see Mees (1996) (1543).
[2] Includes *bartelsi* see Mees (1996) (1543).
[3] Recently, and perhaps correctly, placed in synonymy by Wells (1999: 7) (2579) *contra* van Marle & Voous (1988) (2468).
[4] Here we follow Davison (1982) (621) in lumping *merlini* with *chloropus*.
[5] Gender addressed by David & Gosselin (2002) (614), multiple changes occur in specific names.
[6] Original spelling as given, but frequently spelled *roulroul* (as used for the vernacular name).
[7] Includes *rincheni* Koelz, 1954 (1284) see Ripley (1982: 78) (2055).

○ *Bambusicola thoracicus* Chinese Bamboo Partridge[1]
   — *B.t.thoracicus* (Temminck, 1815)   S and C China. Introd. Japan
   — *B.t.sonorivox* Gould, 1863   Taiwan

GALLOPERDIX Blyth, 1844 F
○ *Galloperdix spadicea* Red Spurfowl
   — *G.s.spadicea* (J.F. Gmelin, 1789)   N and C India
   — *G.s.caurina* Blanford, 1898   S Rajasthan
   — *G.s.stewarti* E.C.S. Baker, 1919   C and S Kerala

○ *Galloperdix lunulata* (Valenciennes, 1825) Painted Spurfowl   NC, C and SE India

○ *Galloperdix bicalcarata* (J.R. Forster, 1781) Ceylon Spurfowl   Sri Lanka

PHASIANINAE[2]

ITHAGINIS Wagler, 1832 M
○ *Ithaginis cruentus* Blood Pheasant[3]
   — *I.c.cruentus* (Hardwicke, 1821)   N Nepal
   — *I.c.affinis* Beebe, 1912   Sikkim and W Bhutan
   — *I.c.tibetanus* E.C.S. Baker, 1914   E Bhutan and neighbouring Xizang
   — *I.c.kuseri* Beebe, 1912   N and E Assam, SE Xixang and NW Yunnan
   — *I.c.marionae* Mayr, 1941   NE Burma and Shweli/Salween watershed (W
                                      Yunnan) (2306)

   — *I.c.rocki* Riley, 1925[4]   W Yunnan (Salween/Yangtze watershed)
   — *I.c.clarkei* Rothschild, 1920   N Yunnan (Lijiang Mts)
   — *I.c.geoffroyi* J. Verreaux, 1867   E Xizang and SW Sichuan
   — *I.c.berezowskii* Bianchi, 1903[5]   W and N Sichuan to S Gansu
   — *I.c.beicki* Mayr & Birckhead, 1937   NE Qinghai (152)
   — *I.c.michaelis* Bianchi, 1903   N Qinghai and nearby Gansu
   — *I.c.sinensis* David, 1873   Shensi

TRAGOPAN Cuvier, 1829 M
○ *Tragopan melanocephalus* (J.E. Gray, 1829) Western Tragopan   NW Pakistan to N Uttar Pradesh (NW Himalayas)

○ *Tragopan satyra* (Linnaeus, 1758) Satyr Tragopan   Garhwal to N Assam (C and E Himalayas)

○ *Tragopan blythii* Blyth's Tragopan
   — *T.b.molesworthi* E.C.S. Baker, 1914   E Bhutan and SE Xizang to N and NE Assam
   — *T.b.blythii* (Jerdon, 1870)[6]   S and SE Assam, W and NW Burma

○ *Tragopan temminckii* (J.E. Gray, 1831) Temminck's Tragopan NE Assam to NW Vietnam, N to N Sichuan and S Shaanxi

○ *Tragopan caboti* Cabot's Tragopan
   — *T.c.caboti* (Gould, 1857)   Hills of N Guangdong and Fujian (SE China)
   — *T.c.guangxiensis* Cheng & Wu, 1979   NE Guangxi (SC China) (337)

PUCRASIA G.R. Gray, 1841 F
○ *Pucrasia macrolopha* Koklass Pheasant
   — *P.m.castanea* Gould, 1855   E Afghanistan, extr. NW Pakistan
   — *P.m.biddulphi* G.F.L. Marshall, 1879[7]   N Kashmir, Ladakh, N Himachal Pradesh
   — *P.m.macrolopha* (Lesson, 1829)   S Himachal Pradesh to N Uttar Pradesh (W Himalayas)
   — *P.m.nipalensis* Gould, 1855   W Nepal
   — *P.m.meyeri* Madarász, 1886   E Xizang, N Yunnan, W Sichuan
   — *P.m.ruficollis* David & Oustalet, 1877   WC Sichuan to SW Gansu and S Shaanxi
   — *P.m.xanthospila* G.R. Gray, 1864   N Shaanxi and Shanxi to W Hebei
   — *P.m.joretiana* Heude, 1883   SW Anhui
   — *P.m.darwini* Swinhoe, 1872   E Sichuan and Guizhou to Fujian and Zhejiang

[1] For correction of spelling see David & Gosselin (2002) (614).
[2] *Sensu stricto* i.e. excluding the Perdicinae.
[3] Treatment follows Vaurie (1965) (2510) except that *affinis* is retained, see Cheng (1987: 149) (333). However see also Yang *et al.* (1994) (2699).
[4] Includes *holoptilus* Greenway, 1933 (1003) see Cheng (1987: 150) (333).
[5] Includes *annae* Mayr & Birckhead, 1937 (152) see Cheng (1987: 148) (333).
[3] Includes *rupchandi* Koelz, 1954 (1284), see Ripley (1982: 82) (2055).
[7] Includes *bethelae* Fleming, 1947 (853), see Vaurie (1965: 302) (2510) and Ripley (1982: 86 fn) (2055).

LOPHOPHORUS Temminck, 1813 M
○ *Lophophorus impejanus* (Latham, 1790) HIMALAYAN MONAL

E Afghanistan and Himalayas west to Bhutan

○ *Lophophorus sclateri* SCLATER'S MONAL
— *L.s.sclateri* Jerdon, 1870
— *L.s.orientalis* G.W.H. Davison, 1974

SE Xixang, N Assam, N Burma, N Yunnan
NE Burma and NW Yunnan (620)

○ *Lophophorus lhuysii* A. Geoffroy Saint-Hilaire, 1866 CHINESE MONAL

E Qinghai, W and N Sichuan, SW Gansu

GALLUS Brisson, 1766 M
◉ *Gallus gallus* RED JUNGLEFOWL[1]
— *G.g.murghi* Robinson & Kloss, 1920[2]
— *G.g.spadiceus* (Bonnaterre, 1791)
— *G.g.jabouillei* Delacour & Kinnear, 1928
— *G.g.gallus* (Linnaeus, 1758)
— *G.g.bankiva* Temminck, 1813

N Pakistan to N Assam and C India
E Assam, N Burma, SW Yunnan, N Indochina
N and NC Vietnam
C Burma to S Indochina and Sumatra
Java

○ *Gallus sonneratii* Temminck, 1813 GREY JUNGLEFOWL[3]

WC and S India

○ *Gallus lafayettii* Lesson, 1831 CEYLON JUNGLEFOWL[4]

Sri Lanka

○ *Gallus varius* (Shaw, 1798) GREEN JUNGLEFOWL

Java east to Sumba and Alor

LOPHURA J. Fleming, 1822 F
◉ *Lophura leucomelanos* KALIJ PHEASANT[5]
— *L.l.hamiltonii* (J.E. Gray, 1829)
— *L.l.leucomelanos* (Latham, 1790)
— *L.l.melanota* (Hutton, 1848)
— *L.l.moffitti* (Hachisuka, 1938)
— *L.l.lathami* (J.E. Gray, 1829)
— *L.l.williamsi* (Oates, 1898)
— *L.l.oatesi* (Ogilvie-Grant, 1893)[6]

W Himalayas (N Pakistan to WC Nepal)
C Nepal
E Nepal, Sikkim, W Bhutan
C Bhutan (1036)
E Bhutan, Assam, and N Burma to Xizang
NW and W Burma
WC Burma (Arakan Yomas)

○ *Lophura nycthemera* SILVER PHEASANT[7]
— *L.n.omeiensis* Cheng, Chang & Tang, 1964
— *L.n.rongjiangensis* Tan & Wu, 1981
— *L.n.nycthemera* (Linnaeus, 1758)
— *L.n.fokiensis* Delacour, 1948
— *L.n.whiteheadi* (Ogilvie-Grant, 1899)
— *L.n.occidentalis* Delacour, 1948
— *L.n.rufipes* (Oates, 1898)
— *L.n.jonesi* (Oates, 1903)

— *L.n.lineata* (Vigors, 1831)
— *L.n.ripponi* (Sharpe, 1902)
— *L.n.crawfurdii* (J.E. Gray, 1829)[8]
— *L.n.beaulieui* Delacour, 1948
— *L.n.berliozi* (Delacour & Jabouille, 1928)
— *L.n.beli* (Oustalet, 1898)
— *L.n.annamensis* (Ogilvie-Grant, 1906)
— *L.n.lewisi* (Delacour & Jabouille, 1928)
— *L.n.engelbachi* Delacour, 1948

WC Sichuan (340)
S and W Guizhou (2384)
Guangxi and Guangzdong (S China), NE Vietnam
NW Fokien (678)
Hainan I.
Extreme NW Yunnan, NE Burma (678)
SW Yunnan, NC Burma
Extreme E Burma (Kengtung), S Yunnan and north-
    ernmost Thailand (south along Dong Phaya Fai
    range to NE Thailand)
EC Burma (Karen Hills), extreme W Thailand
EC Burma (E of R. Salween)
SE Burma and SW Thailand (998)
SE Yunnan, N Laos, NW Vietnam (678)
EC Laos and NC Vietnam
WC Vietnam
SC Vietnam
Cambodia
S Laos (678)

○ *Lophura imperialis* (Delacour & Jabouille, 1924) IMPERIAL PHEASANT[9] C Vietnam

[1] Introduced in the Philippines (sometimes recognised as *philippensis* Hachisuka, 1939 (1039)) and Micronesia (named *micronesiae* Hachisuka, 1939 (1039)).
[2] Includes *gallina* Koelz, 1954 (1284), see Ripley (1982: 85) (2055).
[3] Includes *wangyeli* Koelz, 1954 (1284), see Ripley (1982: 85) (2055).
[4] Includes *xanthimaculatus* Deraniyagala, 1955 (693), see Ripley (1982: 86) (2055). The original spelling of the specific name was *lafayetii*, as given by Peters (1934: 119) (1822), but the vernacular name correctly spelled Lafayette clearly implied a *lapsus*.
[5] The original spelling. For many years spelled *leucomelana*. For a review of this species and the next see McGowan & Panchen (1994) (1503); we follow them in their main proposal (with a slight exception), but refrain from lumping subspecies as DNA studies are ongoing.
[6] Tentatively attached here, but review in process.
[7] As suggested by McGowan & Panchen (1994) (1503) this species is probably oversplit; DNA studies may soon throw light on this.
[8] This name, not in Peters, reintroduced by Deignan (1943) (636).
[9] Evidence is accumulating that this name has been given to a hybrid between *L. edwardsi* and *L. nycthemera*; once confirmation is published this name will disappear from future Editions.

○ *Lophura edwardsi* (Oustalet, 1896) EDWARDS'S PHEASANT   C Vietnam

○ *Lophura hatinhensis* Vo Quy, 1975 VIETNAMESE PHEASANT[1] #   NC Vietnam (2531)

○ *Lophura swinhoii* (Gould, 1863) SWINHOE'S PHEASANT   Taiwan

○ *Lophura inornata* SALVADORI'S PHEASANT
    — *L.i.hoogerwerfi* (Chasen, 1939)   N Sumatra (322)[2]
    — *L.i.inornata* (Salvadori, 1879)   C and S Sumatra

○ *Lophura erythrophthalma* CRESTLESS FIREBACK
    — *L.e.erythrophthalma* (Raffles, 1822)   Malay Pen., NE Sumatra
    — *L.e.pyronota* (G.R. Gray, 1841)   S and W Borneo

○ *Lophura ignita* CRESTED FIREBACK
    — *L.i.rufa* (Raffles, 1822)   Malay Pen., N and C Sumatra
    — *L.i.macartneyi* (Temminck, 1813)   SE Sumatra
    — *L.i.nobilis* (P.L. Sclater, 1863)   N Borneo
    — *L.i.ignita* (Shaw, 1797)   S Borneo

○ *Lophura diardi* (Bonaparte, 1856) SIAMESE FIREBACK   W Burma, Thailand, C and S Indochina

○ *Lophura bulweri* (Sharpe, 1874) BULWER'S PHEASANT   Borneo

CROSSOPTILON Hodgson, 1838 N
○ *Crossoptilon crossoptilon* WHITE EARED PHEASANT
    — *C.c.dolani* Meyer de Schauensee, 1937   SE Qinghai (1558)
    — *C.c.crossoptilon* (Hodgson, 1838)   W and N Sichuan to SE Xizang, Yunnan
    — *C.c.lichiangense* Delacour, 1945   N Yunnan to SW Sichuan (676)
    — *C.c.drouynii* J. Verreaux, 1868   SW Qinghai to E and SE Xizang and NW Sichuan
    — *C.c.harmani* Elwes, 1881[3]   S Xizang and N Himalayas to N Assam

○ *Crossoptilon mantchuricum* Swinhoe, 1863 BROWN EARED PHEASANT   Shanxi to Hebei and neighbouring Nei Mongol

○ *Crossoptilon auritum* (Pallas, 1811) BLUE EARED PHEASANT   N Sichuan, N and E Qinghai to Gansu and Ningxia

CATREUS Cabanis, 1851 M
○ *Catreus wallichii* (Hardwicke, 1827) CHEER PHEASANT   Himalayas (S Kashmir to Nepal)

SYRMATICUS Wagler, 1832 M
○ *Syrmaticus ellioti* (Swinhoe, 1872) ELLIOT'S PHEASANT   SE China (Fujian to S Anhui and Zhejiang)

○ *Syrmaticus humiae* MRS. HUME'S PHEASANT
    — *S.h.humiae* (Hume, 1881)   E Assam and NW Burma
    — *S.h.burmanicus* (Oates, 1898)   NE Burma, SW Yunnan and northernmost Thailand

○ *Syrmaticus mikado* (Ogilvie-Grant, 1906) MIKADO PHEASANT   Taiwan

○ *Syrmaticus soemmerringi* COPPER PHEASANT
    — *S.s.scintillans* (Gould, 1866)   N and W Honshu
    — *S.s.subrufus* (N. Kuroda, Sr., 1919)   Pacific side of C and SW Honshu and Shikoku
    — *S.s.intermedius* (N. Kuroda, Sr., 1919)   Northern side of SW Honshu and Shikoku
    — *S.s.soemmerringi* (Temminck, 1830)   N and C Kyushu
    — *S.s.ijimae* (Dresser, 1902)   S Kyushu

○ *Syrmaticus reevesii* (J.E. Gray, 1829) REEVES'S PHEASANT   NE Sichuan, Hubei and Anhui to EC Nei Mongol and
                                                         Hebei

PHASIANUS Linnaeus, 1758 M
● *Phasianus colchicus* COMMON PHEASANT[4]
    — *P.c.septentrionalis* Lorenz, 1888   N Caucasus, W Caspian Sea to Volga Delta

---

[1] Misspelled *haitiensis* in last Edition. Treated as a race of *L. edwardsi* by McGowan (1994: 535) (1502). This is also expected to prove to be a hybrid.
[2] Separated from the preceding species in van Marle & Voous (1988) (2468). The male, closely similar to that of *inornata*, has since been discovered (Garson, 2000) (928), and separation from *inornata* may have been premature. Females reportedly show "very slight differences" (McGowan, 1994) (1502) or are "noticeably different" (Garson, 2000) (928).
[3] May deserve specific rank, see Ludlow (1951) (1402) and Cheng (1997) (335).
[4] The area of introduction is excluded; the name *europaeus* Hachisuka, 1937 (1035), probably relates to some such stock.

— *P.c.colchicus* Linnaeus, 1758	Transcaucasia, east to W and N Azerbaijan
— *P.c.talischensis* Lorenz, 1888	E and SE Azerbaijan to NC Iran
— *P.c.persicus* Severtsov, 1875	SW Turkmenistan and NE Iran
— *P.c.principalis* P.L. Sclater, 1885	SE Turkmenistan, NW Afghanistan
— *P.c.chrysomelas* Severtsov, 1875	W Uzbekistan and N Turkmenistan
— *P.c.zarudnyi* Buturlin, 1904	Amu Dar'ya valley of E Turkmenistan
— *P.c.bianchii* Buturlin, 1904	SE Uzbekistan, SW Tajikistan, NE Afghanistan
— *P.c.zerafschanicus* Tarnovski, 1893	Regions of Bukhara and Samarkand
— *P.c.turcestanicus* Lorenz, 1896[1]	Syr Dar'ya valley (S Kazakhstan) to Fergana Basin
— *P.c.mongolicus* J.F. Brandt, 1844	SE Kazakhstan and N Kyrgyzstan
— *P.c.shawii* Elliot, 1870	W Tarim Basin (Xinjiang, W China)
— *P.c.tarimensis* Pleske, 1888	E Tarim Basin east to Lop Nur
— *P.c.vlangalii* Przevalski, 1876	Qaidam Basin (N Qinghai, China)
— *P.c.strauchi* Przevalski, 1876	Xining area (NE Qinghai)
— *P.c.sohokotensis* Buturlin, 1908	N foothills of Nan Shan in E Gansu
— *P.c.satscheuensis* Pleske, 1892	N foothills of Nan Shan in W Gansu
— *P.c.hagenbecki* Rothschild, 1901	W Mongolia
— *P.c.edzinensis* Sushkin, 1926	Lower Ruo Shui R. (W Nei Mongol)
— *P.c.alaschanicus* Alphéraky & Bianchi, 1908	Ningxia and SC Nei Mongol
— *P.c.kiangsuensis* Buturlin, 1904	N Shaanxi and Shanxi to W Hebei and SE Nei Mongol
— *P.c.karpowi* Buturlin, 1904	S Manchuria, C and S Korea, Cheju-do and Tsu-shima Is.
— *P.c.pallasi* Rothschild, 1901	SE Siberia, N and E Manchuria, and NE Korea
— *P.c.suehschanensis* Bianchi, 1906	N Sichuan and S Gansu
— *P.c.elegans* Elliot, 1870	W Sichuan to NW Yunnan and NE Burma
— *P.c.decollatus* Swinhoe, 1870	NE Yunnan, E Sichuan, Guizhou and W Hubei
— *P.c.rothschildi* La Touche, 1922	SE Yunnan, NW Vietnam
— *P.c.takatsukasae* Delacour, 1927	SW Guandong and NE Vietnam
— *P.c.torquatus* J.F. Gmelin, 1789	Anhui and Shandong to N Guangxi and N Guangdong
— *P.c.formosanus* Elliot, 1870	Taiwan

◯ *Phasianus versicolor* GREEN PHEASANT[2]

— *P.v.robustipes* N. Kuroda, Sr., 1919[3]	NW Honshu and Sado
— *P.v.tohkaidi* Momiyama, 1922	C and W Honshu, Shikoku
— *P.v.tanensis* N. Kuroda, Sr., 1919	S peninsulas and islands S of Honshu and Kyushu
— *P.v.versicolor* Vieillot, 1825	Extreme W Honshu, Kyushu and islands W of Kyushu

CHRYSOLOPHUS J.E. Gray, 1834 M

◯ *Chrysolophus pictus* (Linnaeus, 1758) GOLDEN PHEASANT — N Guangxi and N Guangdong to S Gansu and S Shaanxi

◯ *Chrysolophus amherstiae* (Leadbeater, 1829) LADY AMHERST'S PHEASANT

N and NE Burma, N Yunnan to W Sichuan and W Guizhou

POLYPLECTRON Temminck, 1813 N

◯ *Polyplectron chalcurum* BRONZE-TAILED PEACOCK-PHEASANT

— *P.c.scutulatum* Chasen, 1941	N Sumatra (324)
— *P.c.chalcurum* (Lesson, 1831)	S Sumatra

◯ *Polyplectron inopinatum* (Rothschild, 1903) MOUNTAIN PEACOCK-PHEASANT

C Malay Pen.

◯ *Polyplectron germaini* Elliot, 1866 GERMAIN'S PEACOCK-PHEASANT — S Vietnam

◯ *Polyplectron bicalcaratum* GREY PEACOCK-PHEASANT

— *P.b.bakeri* P.R. Lowe, 1925	Sikkim to N and E Assam
— *P.b.bicalcaratum* (Linnaeus, 1758)[4]	E Bangladesh to Burma and Laos
— *P.b.ghigii* (Delacour & Jabouille, 1924)	N and C Vietnam
— *P.b.katsumatae* Rothschild, 1906	Hainan I.

◯ *Polyplectron malacense* (Scopoli, 1786) MALAYAN PEACOCK-PHEASANT — SW Thailand, W Malaysia, Sumatra

---

[1] Includes *bergii* see Vaurie (1965: 319) (2510).
[2] Sometimes treated as conspecific with the previous species, see for example Orn. Soc. Japan (2000) (1727) from which we take our subspecific treatment.
[3] Includes *kigis* Taka-Tsukasa, 1944 (2383).
[4] Includes *bailyi* see Deignan (1946) (638).

○ *Polyplectron schleiermacheri* Brüggemann, 1877 Bornean Peacock-Pheasant[1]

Borneo

○ *Polyplectron napoleonis* (Lesson, 1831)[2] Palawan Peacock-Pheasant Palawan

RHEINARDIA Maingounat, 1882[3] F
○ *Rheinardia ocellata* Crested Argus
  — *R.o.ocellata* (Elliot, 1871) C Vietnam
  — *R.o.nigrescens* (Rothschild, 1902) C Malay Pen.

ARGUSIANUS Rafinesque, 1815 M
○ *Argusianus argus* Great Argus
  — *A.a.grayi* (Elliot, 1865) Borneo
  — *A.a.argus* (Linnaeus, 1766) Malay Pen., Sumatra

PAVO Linnaeus, 1758 M
○ *Pavo cristatus* Linnaeus, 1758 Indian Peafowl NE Pakistan east to C Assam and south to Sri Lanka

○ *Pavo muticus* Green Peafowl
  — *P.m.spicifer* Shaw & Nodder, 1804 SE Assam and E Bangladesh to W Burma
  — *P.m.imperator* Delacour, 1951 E Burma, Thailand, Indochina (679)
  — *P.m.muticus* Linnaeus, 1766 W Malaysia, Java

AFROPAVO Chapin, 1936 M
○ *Afropavo congensis* Chapin, 1936 Congo Peafowl Ituri Forest (EC Zaire) (310)

ANHIMIDAE SCREAMERS (2: 3)

ANHIMA Brisson, 1760 F
○ *Anhima cornuta* (Linnaeus, 1766) Horned Screamer Colombia, Venezuela, Guianas to S Bolivia and S Brazil

CHAUNA Illiger, 1811 F
○ *Chauna chavaria* (Linnaeus, 1766) Northern Screamer N Colombia, NW Venezuela

○ *Chauna torquata* (Oken, 1816) Southern Screamer E Bolivia, S Brazil to Paraguay, N Argentina

ANSERANATIDAE MAGPIE-GOOSE (1: 1)

ANSERANAS Lesson, 1828 F
○ *Anseranas semipalmata* (Latham, 1798) Magpie-Goose S New Guinea, N and E Australia

ANATIDAE DUCKS, GEESE, SWANS (49: 158)

DENDROCYGNINAE

DENDROCYGNA Swainson, 1837 F
○ *Dendrocygna viduata* (Linnaeus, 1766) White-faced Whistling Duck Costa Rica and Trinidad to N Argentine, sub-Saharan Africa, Madagascar, Comoro Is.

● *Dendrocygna autumnalis* Black-bellied Whistling Duck
  ⤴ *D.a.autumnalis* (Linnaeus, 1758) SE Texas to Panama
  — *D.a.discolor* Sclater & Salvin, 1873 E Panama to Ecuador and E of Andes to N Argentina

○ *Dendrocygna guttata* Schlegel, 1866 Spotted Whistling Duck Mindanao (Philippines) to Sulawesi, Moluccas, New Guinea and Bismarck Archipelago

○ *Dendrocygna arborea* (Linnaeus, 1758) West Indian Whistling Duck West Indies

○ *Dendrocygna bicolor* (Vieillot, 1816) Fulvous Whistling Duck[4] S USA to N and C Sth. America, C and E Africa, Madagascar, S Asia east to Burma

[1] In previous Editions we followed Delacour (1947) (677) and Smythies (1957, 1960) (2286, 2287) in lumping this with the preceding species. See however Davison in Smythies (2000) (2289).
[2] Recently known as *Polyplectron emphanum* Temminck, but this name has been misdated and does not have priority (Dickinson, 2001) (738).
[3] For correct spelling (not *Rheinartia*) see Peters (1934: 132 fn.) (1822) and Banks (1993) (89).
[4] Includes *helva* see Johnsgard (1979: 428) (1211).

○ *Dendrocygna eytoni* (Eyton, 1838) Plumed Whistling Duck | N and E Australia

○ *Dendrocygna arcuata* Wandering Whistling Duck
— *D.a.arcuata* (Horsfield, 1824) | Sundas, Sulawesi, Philippines, Moluccas
— *D.a.pygmaea* Mayr, 1945 | N New Guinea, New Britain; formerly Fiji Is.
— *D.a australis* Reichenbach, 1850 | Tropical Australia, S New Guinea

○ *Dendrocygna javanica* (Horsfield, 1821) Lesser Whistling Duck | Pakistan east to Taiwan and Ryukyu Is. and south to Sri Lanka and Gtr. Sundas

THALASSORNIS Eyton, 1838 M
○ *Thalassornis leuconotos* White-backed Duck
— *T.l.leuconotos* Eyton, 1838 | E Cameroon to S Ethiopia and south to Cape Province
— *T.l.insularis* Richmond, 1897 | Madagascar

ANSERINAE

CEREOPSIS Latham, 1802[1] F
○ *Cereopsis novaehollandiae* Cape Barren Goose
— *C.n.novaehollandiae* Latham, 1802 | S Australia, Tasmania
— *C.n.grisea* (Vieillot, 1818)[2] | Recherche Arch. (W Australia) (2524)

ANSER Brisson, 1760 M
○ *Anser cygnoides* (Linnaeus, 1758) Swan Goose | C Asia to SE Siberia and Mongolia >> China

○ *Anser fabalis* Bean Goose
— *A.f.fabalis* (Latham, 1787) | N Europe to NC Siberia >> W, C and SE Europe, SW Asia
— *A.f.johanseni* Delacour, 1951 | W Siberia >> Iran, Turkestan, W China
— *A.f.middendorffi* Severtsov, 1873 | E Siberia, Transbaikalia >> E China and Japan
— *A.f.rossicus* Buturlin, 1933 | N Russia, NW Siberia >> W and C Europe, SW Asia
— *A.f.serrirostris* Swinhoe, 1871 | NE Siberia >> China, Korea, Japan

○ *Anser brachyrhynchus* Baillon, 1834 Pink-footed Goose | Greenland, Iceland, Svalbard >> W Europe

● *Anser anser* Greylag Goose
⤶ *A.a.anser* (Linnaeus, 1758) | N, W, WC, and SE Europe >> W and S Europe, N Africa and Middle East
— *A.a.rubrirostris* Swinhoe, 1871 | EC Europe and Iraq to China >> SE Europe and S and E Asia

● *Anser albifrons* Greater White-fronted Goose
⤶ *A.a.elgasi* Delacour & Ripley, 1975 | C Alaska >> C California
— *A.a.gambeli* Hartlaub, 1852[3] | NW Canada >> S USA and Mexico
— *A.a.flavirostris* Dalgety & Scott, 1948 | W Greenland >> British Isles
— *A.a.albifrons* (Scopoli, 1769) | N Eurasia >> W and S Europe, Middle East, Transcaspia, N India, China
— *A.a.frontalis* S.F. Baird, 1858 | NE Siberia, W and N Alaska >> W USA, China, Japan

○ *Anser erythropus* (Linnaeus, 1758) Lesser White-fronted Goose | N Eurasia >> C and SE Europe and China

○ *Anser indicus* (Latham, 1790) Bar-headed Goose | Montane C Asia >> N India, N Burma

○ *Anser canagicus* (Sevastianov, 1802) Emperor Goose | NE Siberia, NW Alaska >> Kamchatka, Aleutian Is., S Alaska

● *Anser caerulescens* Snow Goose[4]
⤶ *A.c.caerulescens* (Linnaeus, 1758) | NE Siberia, N Alaska, NW Canada >> S USA, N Mexico, Japan
— *A.c.atlanticus* (Kennard, 1927) | NE Canada, NW Greenland >> NE USA

● *Anser rossii* Cassin, 1861 Ross's Goose | N Canada >> W and C USA

---

[1] For reasons to use 1802 not 1801 see Browing & Monroe (1991) (258).
[2] For recognition see Storr (1980) (2353) (J. Kear *in litt.*).
[3] The emended name used here is judged to be in prevailing usage; see Art. 33.2 of the Code (I.C.Z.N., 1999) (1178) and p. 000 herein.
[4] The generic name *Chen* Boie, 1822, was retained for this and the next two species by A.O.U. (1998: 58) (3).

BRANTA Scopoli, 1769 F

● *Branta canadensis* CANADA GOOSE[1]

    __ †*B.c.asiatica* Aldrich, 1946   Bering I., N Kuril Is.

    __ *B.c.leucopareia* (J.F. Brandt, 1836)   Aleutian Is. >> Japan and W USA

    ✓ *B.c.minima* Ridgway, 1885   W Alaska >> SW Canada to W USA

    __ *B.c.occidentalis* (S.F. Baird, 1858)   Gulf of Alaska

    ✓ *B.c.fulva* Delacour, 1951   S Alaska, W British Columbia

    __ *B.c.maxima* Delacour, 1951   SC Canada (reintroduced)

    __ *B.c.taverneri* Delacour, 1951   NE Alaska, N Canada >> SW USA, Mexico

    __ *B.c.parvipes* (Cassin, 1852)   C Alaska,C Canada >> SC USA

    __ *B.c.moffitti* Aldrich, 1946   SW Canada, NW USA >> NW USA to N Mexico

    __ *B.c.hutchinsii* (Richardson, 1832)   Arctic C Canada >> Texas, Mexico

    __ *B.c.interior* Todd, 1938   SC Canada >> E USA

    ✓ *B.c.canadensis* (Linnaeus, 1758)   E Canada >> E USA

● *Branta sandvicensis* (Vigors, 1833) HAWAIIAN GOOSE   Hawaii

● *Branta bernicla* BRENT GOOSE/BRANT

    __ *B.b.bernicla* (Linnaeus, 1758)   NW and NC Russia >> NW Europe

    __ *B.b.hrota* (O.F. Müller, 1776)   NE Canada, Greenland, Svalbard, Franz Josef Land >> NE USA, NW Europe (Ireland)

    ✓ *B.b.nigricans* (Lawrence, 1846)[2]   NE Siberia, Alaska, NW Canada >> E Asia and W Nth. America

● *Branta leucopsis* (Bechstein, 1803) BARNACLE GOOSE   NE Greenland, Svalbard, NW Russia, Baltic area >> NW Europe

○ *Branta ruficollis* (Pallas, 1769) RED-BREASTED GOOSE   NW Siberia >> E Balkan area, S Russia, Middle East

COSCOROBA Reichenbach, 1853 F

● *Coscoroba coscoroba* (Molina, 1782) COSCOROBA SWAN   S South America >> north to about 25°S.

CYGNUS Bechstein, 1803 M

● *Cygnus atratus* (Latham, 1790) BLACK SWAN   S Australia, Tasmania; introduced New Zealand

● *Cygnus melancoryphus* (Molina, 1782)[3] BLACK-NECKED SWAN   S South America, Falkland Is. >> north to Tropic of Capricorn

● *Cygnus olor* (J.F. Gmelin, 1789) MUTE SWAN   Europe to C Asia >> N Africa, India

● *Cygnus buccinator* Richardson, 1832 TRUMPETER SWAN   Alaska, W Canada >> USA

● *Cygnus columbianus* TUNDRA SWAN

    ✓ *C.c.columbianus* (Ord, 1815)   Alaska, N Canada >> coasts of USA

    __ *C.c.bewickii* Yarrell, 1830[4]   N Eurasia >> W and C Europe, C and E Asia to Japan

● *Cygnus cygnus* (Linnaeus, 1758) WHOOPER SWAN[5]   N Eurasia >> W and C Europe, C Asia, China

STICTONETTINAE

STICTONETTA Reichenbach, 1853 F

○ *Stictonetta naevosa* (Gould, 1841) FRECKLED DUCK   Inland S Australia

TADORNINAE

HYMENOLAIMUS G.R. Gray, 1843 M

○ *Hymenolaimus malacorhynchos* (J.F. Gmelin, 1789) BLUE DUCK[6]   Alpine W South Island and C North Island

---

[1] Some authors accept two species; but more research is needed. Populations in the U.K. and New Zealand were introduced.
[2] Includes *orientalis* see Stepanyan (1990: 45) (2326).
[3] For correction of spelling see David & Gosselin (2002) (613).
[4] Includes *jankowskyi* see Johnsgard (1979: 433) (1211); if two species are recognised then *jankowskii* merits recognition as a race of *bewickii*.
[5] Includes *islandicus* see Johnsgard (1979: 433) (1211).
[6] We follow Kear (in press) (1235) in accepting a monotypic species.

TACHYERES Owen, 1875 M
○ *Tachyeres patachonicus* (P.P. King, 1831)[1] FLYING STEAMER DUCK   S Chile, S Argentina, Falkland Is.

○ *Tachyeres pteneres* (J.R. Forster, 1844)[2] MAGELLANIC STEAMER DUCK   S Chile, S Argentina

○ *Tachyeres brachypterus* (Latham, 1790) FALKLAND ISLAND STEAMER DUCK Falkland Is.

○ *Tachyeres leucocephalus* Humphrey & Thompson, 1981 WHITE-HEADED STEAMER DUCK
S Argentina (1164)

MERGANETTA Gould, 1842 F
○ *Merganetta armata* TORRENT DUCK
— *M.a.colombiana* Des Murs, 1845   Andes W Venezuela, Colombia to C Ecuador
— *M.a.leucogenis* (Tschudi, 1843)   Andes of N Peru
— *M.a.turneri* Sclater & Salvin, 1869   Andes of S Peru
— *M.a.garleppi* Berlepsch, 1894   Andes of N Bolivia
— *M.a.berlepschi* E. Hartert, 1909   Andes of S Bolivia, NW Argentina
— *M.a.armata* Gould, 1842[3]   Andes of C Chile, W Argentina

PLECTROPTERUS Stephens, 1824 M
○ *Plectropterus gambensis* SPUR-WINGED GOOSE
— *P.g.gambensis* (Linnaeus, 1766)   Gambia to Sudan and C Africa
— *P.g.niger* P.L. Sclater, 1877   Southern Africa

SARKIDIORNIS Eyton, 1838 M
○ *Sarkidiornis melanotos* COMB DUCK
— *S.m.melanotos* (Pennant, 1769)   Sub-Saharan Africa, Madagascar, Pakistan to SE China
south to Sri Lanka and Indochina
— *S.m.sylvicola* H. & R. Ihering, 1907[4] E Panama, N Colombia, Venezuela and Brazil to

CYANOCHEN Bonaparte, 1856 F
○ *Cyanochen cyanoptera* (Rüppell, 1845) BLUE-WINGED GOOSE[5]   Highlands of Eritrea, Ethiopia

ALOPOCHEN Stejneger, 1885 F
○ *Alopochen aegyptiaca* (Linnaeus, 1766) EGYPTIAN GOOSE[6]   Sub-Saharan Africa, S Egypt; introduced U.K.

NEOCHEN Oberholser, 1918 F
○ *Neochen jubata* (Spix, 1825) ORINOCO GOOSE[7]   N Colombia, Venezuela, the Guianas south to Paraguay,

CHLOEPHAGA Eyton, 1838 F
○ *Chloephaga melanoptera* (Eyton, 1838) ANDEAN GOOSE   Andes of C Peru to Tierra del Fuego >> lower elevations

○ *Chloephaga picta* UPLAND GOOSE/MAGELLAN GOOSE
— *C.p.picta* (J.F. Gmelin, 1789)   S Chile, S Argentina
— *C.p.leucoptera* (J.F. Gmelin, 1789)   Falkland Is.

○ *Chloephaga hybrida* KELP GOOSE
— *C.h.hybrida* (Molina, 1782)   Coastal S Chile
— *C.h.malvinarum* J.C. Phillips, 1916   Falkland Is.

○ *Chloephaga poliocephala* P.L. Sclater, 1857 ASHY-HEADED GOOSE   S Chile, Argentina

○ *Chloephaga rubidiceps* P.L. Sclater, 1861 RUDDY-HEADED GOOSE   Tierra del Fuego, Falkland Is.

TADORNA Boie, 1822[8] F
● *Tadorna tadorna* (Linnaeus, 1758) COMMON SHELDUCK   W Europe to E Asia >> N Africa, India, S China

---

[1] The prior name *Oidemia patachonica* King, 1828, has been suppressed, it seems to have been used for a different species. See Opinion No. 1648 (I.C.Z.N. (1991) (1177).
[2] Drawings with Forster's proposed names were published many years after his death (d. 1798).
[3] Includes *fraenata* see Johnsgard (1979: 460) (1211).
[4] Includes *carunculata* see Johnsgard (1979: 455) (1211).
[5] For correction of spelling see David & Gosselin (2002) (614).
[6] For correction of spelling see David & Gosselin (2002) (614).
[7] For correction of spelling see David & Gosselin (2002) (614).
[8] Cited as Fleming, 1822, by Peters (1931: 157) (1821) and as Boie, 1822, by Johnsgard (1979: 449) (1211), but with an erroneous 'label' of Fleming still present (although the details give the reference to Boie and imply his was earlier than Fleming's usage). Recent references to the name in Oken, 1817, seem to refer to use in an invalid context.

○ *Tadorna radjah* RADJAH SHELDUCK
   — *T.r.radjah* (Lesson, 1828)   S Moluccas, Aru Is., New Guinea
   — *T.r.rufitergum* E. Hartert, 1905   Tropical N and E Australia

○ *Tadorna ferruginea* (Pallas, 1764) RUDDY SHELDUCK   S Europe, NW and NE Africa and SW and C Asia >> S Europe, N Africa, S and E Asia

○ *Tadorna cana* (J.F. Gmelin, 1789) SOUTH AFRICAN SHELDUCK   Sth. Africa, Namibia

○ *Tadorna tadornoides* (Jardine & Selby, 1828) AUSTRALIAN SHELDUCK   S Australia, Tasmania

◑ *Tadorna variegata* (J.F. Gmelin, 1789) PARADISE SHELDUCK   New Zealand

○ †?*Tadorna cristata* (N. Kuroda, Sr., 1917) CRESTED SHELDUCK   NE Asia >> Russian Far East, Korea, S Japan, NE China

ANATINAE

MALACORHYNCHUS Swainson, 1831 M
○ *Malacorhynchus membranaceus* (Latham, 1802)[1] PINK-EARED DUCK   Australia (inland waters)

SALVADORINA Rothschild & Hartert, 1894[2] F
○ *Salvadorina waigiuensis* Rothschild & Hartert, 1894 SALVADORI'S TEAL
                                       Montane New Guinea[3]

CAIRINA J. Fleming, 1822 F
● *Cairina moschata* (Linnaeus, 1758) MUSCOVY DUCK   S Mexico to Peru and N Argentina

ASACORNIS Salvadori, 1895 F
○ *Asarcornis scutulata* (S. Müller, 1842) WHITE-WINGED DUCK   Assam and Thailand south to Sumatra and Java[4]

PTERONETTA Salvadori, 1895 F
○ *Pteronetta hartlaubii* (Cassin, 1859) HARTLAUB'S DUCK[5]   Equatorial W Africa, Zaire

AIX Boie, 1828 F
● *Aix sponsa* (Linnaeus, 1758) WOOD DUCK   S Canada, USA, Cuba >> S USA

○ *Aix galericulata* (Linnaeus, 1758) MANDARIN DUCK   SE Siberia, Korea, E China, Japan >> south of 40°N. Introduced U.K., California etc.

CHENONETTA J.F. Brandt, 1836 F
● *Chenonetta jubata* (Latham, 1802)[6] MANED DUCK   Australia, Tasmania

NETTAPUS J.F. Brandt, 1836 M
○ *Nettapus auritus* (Boddaert, 1783) AFRICAN PYGMY-GOOSE   Sub-Saharan Africa (mainly coastal belt), Madagascar

○ *Nettapus coromandelianus* COTTON TEAL/INDIAN PYGMY-GOOSE
   — *N.c.coromandelianus* (J.F. Gmelin, 1789)   India to S China, Philippines, N Indonesia and N New Guinea
   — *N.c.albipennis* Gould, 1842   NE Australia (E Queensland)

○ *Nettapus pulchellus* Gould, 1842 GREEN PYGMY-GOOSE   Sulawesi, S Moluccas, S New Guinea, N Australia

AMAZONETTA Boetticher, 1929 F
○ *Amazonetta brasiliensis* BRAZILIAN TEAL
   — *A.b.brasiliensis* (J.F. Gmelin, 1789)   E Colombia, Venezuela, Guyana, E Brazil
   — *A.b.ipecutiri* (Vieillot, 1816)   E Bolivia, S Brazil to Paraguay, N Argentina

CALLONETTA Delacour, 1936 F
○ *Callonetta leucophrys* (Vieillot, 1816) RINGED TEAL   S Bolivia, S Brazil to Paraguay, N Argentina

---

[1] For reasons to use 1802 not 1801 see Browing & Monroe (1991) (258).
[2] For reasons to retain this genus see Mlikovsky (1989) (1595).
[3] Not known to occur on Waigeo.
[4] Birds of Sumatra and Java may deserve subspecific recognition, see Green (1992, 1993) (999, 1000) (J. Kear *in litt.*).
[5] Last Edition listed *albifrons* probably based on Scott (1972) (2220); but see Peters (1931) and Johnsgard (1979: 454) (1211).
[6] For reasons to use 1802 not 1801 see Browing & Monroe (1991) (258).

LOPHONETTA Riley, 1914 F
○ *Lophonetta specularioides* CRESTED DUCK
  — *L.s.alticola* (Ménégaux, 1909)
  — *L.s.specularioides* (P.P. King, 1828)

Andean lakes of C Peru, Bolivia, C Chile, NW Argentina
S Chile, WC and S Argentina, Falkland Is.

SPECULANAS Boettischer, 1929 F
● *Speculanas specularis* (P.P. King, 1828) SPECTACLED DUCK

S Chile, S Argentina >> somewhat further north

ANAS Linnaeus, 1758 F
○ *Anas capensis* J.F. Gmelin, 1789 CAPE TEAL

Chad to Ethiopia, E Africa, Angola, Botswana and
    Cape Province

● *Anas strepera* GADWALL
  ☑ *A.s.strepera* Linnaeus, 1758

N and C Eurasia, Nth America >> SE and SC Eurasia,
    C Nth America, Africa

  — †*A.s.couesi* (Streets, 1876)

Tabuaeran I. (Kiribati Is.)

○ *Anas falcata* Georgi, 1775 FALCATED TEAL

NE Asia >> NE China, Korea, Japan to N Indochina

● *Anas sibilatrix* Poeppig, 1829 CHILOE WIGEON

S South America >> northern part of range

● *Anas penelope* Linnaeus, 1758 EURASIAN WIGEON

N Europe, N Asia >> N, NE Africa, India, Japan

● *Anas americana* J.F. Gmelin, 1789 AMERICAN WIGEON

Alaska, Canada, N USA >> coasts of Nth America to
    Costa Rica, W Indies, Hawaii

○ *Anas sparsa* AFRICAN BLACK DUCK
  — *A.s.leucostigma* Rüppell, 1845[1]
  — *A.s.sparsa* Eyton, 1838

Gabon; Sudan and Ethiopia to Tanzania and Angola
Namibia and Zimbabwe to Cape Province

● *Anas rubripes* Brewster, 1902 AMERICAN BLACK DUCK

NE North America >> SE USA

● *Anas platyrhynchos* MALLARD[2]
  ☑ *A.p.platyrhynchos* Linnaeus, 1758[3]

Europe, Asia, N America >> S to N Africa, India,
    Mexico

  — *A.p.conboschas* C.L. Brehm, 1831
  — *A.p.diazi* Ridgway, 1886

Greenland
Highlands of N and C Mexico

● *Anas fulvigula* MOTTLED DUCK
  ☑ *A.f.fulvigula* Ridgway, 1874
  — *A.f.maculosa* Sennett, 1889

Florida (SE USA)
SC USA, NE Mexico

● *Anas wyvilliana* P.L. Sclater, 1878 HAWAIIAN DUCK

Hawaiian Is.

○ *Anas laysanensis* Rothschild, 1892 LAYSAN DUCK

Laysan I. (NW Hawaiian Is.)

○ *Anas luzonica* Fraser, 1839 PHILIPPINE DUCK

Philippines

◐ *Anas superciliosa* PACIFIC BLACK DUCK
  — *A.s.rogersi* Mathews, 1912
  ☑ *A.s.superciliosa* J.F. Gmelin, 1789
  — *A.s.pelewensis* Hartlaub & Finsch, 1872

Sumatra, Java, Lesser Sundas, Sulawesi
S New Guinea, Australia, Tasmania, New Zealand
Micronesia[4] to N New Guinea, Bismarck Archipelago,
    Solomon Is, C and E Melanesia and W Polynesia to
    Iles Australes

○ *Anas poecilorhyncha* SPOT-BILLED DUCK
  — *A.p.poecilorhyncha* J.R. Forster, 1781
  — *A.p.haringtoni* (Oates, 1907)
  — *A.p.zonorhyncha* Swinhoe, 1866

India, Sri Lanka
Burma, SW China and N Indochina
SE Siberia, NE and E China >> SE Asia

○ *Anas undulata* YELLOW-BILLED DUCK
  — *A.u.ruppelli* Blyth, 1855
  — *A.u.undulata* C.F. Dubois, 1839

Ethiopia, E Sudan, N Kenya
S Kenya to Angola and Cape Province

○ *Anas melleri* P.L. Sclater, 1865 MELLER'S DUCK

E Madagascar

[1] Includes *maclatchyi* see Johnsgard (1979: 472) (1211).
[2] For comments on the relationships of this and the next three species see A.O.U. (1998: 69) (3). Some authors treat *diazi* as a species.
[3] Includes *neoboria* Oberholser, 1974 (1672), see Browning (1978) (246).
[4] The name *rukensis* N. Kuroda, Sr., 1939 (1309), was given to Micronesian birds, but see Baker (1951) (82).

● *Anas discors* Linnaeus, 1766 BLUE-WINGED TEAL[1] S Canada and USA >> C America, West Indies, N Sth. America

● *Anas cyanoptera* CINNAMON TEAL
   ✓ *A.c.septentrionalium* Snyder & Lumsden, 1951 W North America >> N W South America
   — *A.c.tropicus* Snyder & Lumsden, 1951 Cauca and Magdalena valleys of NW Colombia
   — *A.c.borreroi* Snyder & Lumsden, 1951 E Andes of Colombia
   — *A.c.orinomus* (Oberholser, 1906) Andes of Peru, Bolivia, N Chile, NW Argentina
   — *A.c.cyanoptera* Vieillot, 1816 SW Peru, N Chile; SE Bolivia, Paraguay, SE Brazil to Tierra del Fuego

○ *Anas smithii* (E. Hartert, 1891) CAPE SHOVELER S Angola to W Zimbabwe and Cape Province

◑ *Anas platalea* Vieillot, 1816 RED SHOVELER Peru, S Brazil to Tierra del Fuego

○ *Anas rhynchotis* AUSTRALASIAN SHOVELER
   — *A.r.rhynchotis* Latham, 1802[2] S Australia, Tasmania
   — *A.r.variegata* (Gould, 1856) New Zealand

● *Anas clypeata* Linnaeus, 1758 NORTHERN SHOVELER Europe, Asia, N America >> N and E Africa, India, China, Mexico

○ *Anas bernieri* (Hartlaub, 1860) MADAGASCAN TEAL W Madagascar

○ *Anas gibberifrons* SUNDA TEAL
   — *A.g.albogularis* (Hume, 1873)[3] Andaman Is., Cocos Is.
   — *A.g.gibberifrons* S. Müller, 1842 Java to Sulawesi and Lesser Sundas to Timor I., Wetar I.

○ *Anas gracilis* Buller, 1869 GREY TEAL[4] Australia, New Guinea, New Zealand, vagrant to Banda Sea and SW Pacific Is.

◑ *Anas castanea* (Eyton, 1838) CHESTNUT-BREASTED TEAL S Australia, Tasmania

○ *Anas chlorotis* G.R. Gray, 1845 BROWN TEAL[5] New Zealand. Formerly Chatham Is.

○ *Anas aucklandica* (G.R. Gray, 1844) AUCKLAND ISLANDS TEAL[6] Auckland Is.

○ *Anas nesiotis* (J.H. Fleming, 1935) CAMPBELL ISLAND TEAL Campbell I.

○ *Anas bahamensis* WHITE-CHEEKED PINTAIL
   — *A.b.bahamensis* Linnaeus, 1758 Bahamas, Gtr. Antilles, N South America
   — *A.b.rubrirostris* Vieillot, 1816 S Ecuador and S Brazil to C Peru and N Argentina
   — *A.b.galapagensis* (Ridgway, 1889) Galapagos Is.

○ *Anas erythrorhyncha* J.F. Gmelin, 1789 RED-BILLED DUCK Ethiopia to E and S Africa, Madagascar

● *Anas flavirostris* SPECKLED TEAL
   — *A.f.altipetens* (Conover, 1941) Andes of E Colombia, W Venezuela
   — *A.f.andium* (Sclater & Salvin, 1873) Andes of Ecuador, C and S Colombia
   — *A.f.oxyptera* Meyen, 1834 Andes of N Peru to N Chile, N Argentina
   ✓ *A.f.flavirostris* Vieillot, 1816 C Chile, S Argentina to Tierra del Fuego >> Paraguay and S Brazil

◑ *Anas georgica* YELLOW-BILLED PINTAIL
   — †?*A.g.niceforoi* Wetmore & Borrero, 1946 EC Colombia
   ✓ *A.g.spinicauda* Vieillot, 1816 S Colombia to Tierra del Fuego, Falkland Is.
   — *A.g.georgica* J.F. Gmelin, 1789 South Georgia I.

● *Anas acuta* Linnaeus, 1758 NORTHERN PINTAIL N Europe, Asia, N America >> N tropical and E Africa, China, C America

○ *Anas eatoni* KERGUELEN PINTAIL[7]
   — *A.e.eatoni* (Sharpe, 1875) Kerguelen
   — *A.e.drygalskii* Reichenow, 1904 Crozet Is.

● *Anas querquedula* Linnaeus, 1758 GARGANEY W Europe to Japan >> Africa, India, SE Asia, Indonesia, N Australia, central Pacific Ocean

[1] Includes *orphna* see Johnsgard (1979: 477) (1211).
[2] For reasons to use 1802 not 1801 see Browing & Monroe (1991) (258).
[3] This taxon requires review. Although geographically closest to *gibberifrons* it may not belong with it. Not sufficiently addressed by studies to date.
[4] For comments on relationship to *castanea* see Kennedy & Spencer (2000) (1254). Includes *remissa* from Rennell Island (J. Kear *in litt.*), now thought extinct.
[5] The English name New Zealand Teal is left unused here (to be used by those who prefer the broad species).
[6] For reasons to split *nesiotis* and *chlorotis* from *A. aucklandica* see Daugherty *et al.* (1999) (611) and Kennedy & Spencer (2000) (1254).
[7] For recognition of this at specific level see Livezey (1991) (1365).

○ *Anas formosa* Georgi, 1775 BAIKAL TEAL

NE Asia >> E China, Korea, Japan

● *Anas crecca* COMMON TEAL/GREEN-WINGED TEAL[1]

    __ *A.c.crecca* Linnaeus, 1758

N and C Europe and Asia >> C and S Europe, N Africa, C, S and E Asia

    __ *A.c.nimia* Friedmann, 1948

Aleutian Is.

    ✔ *A.c.carolinensis* J.F. Gmelin, 1789

N Nth. America >> W and S Nth. America, C America, W Indies

○ *Anas versicolor* SILVER TEAL

    __ *A.v.puna* Tschudi, 1844[2]

Andes of C Peru to N Chile and NW Argentina

    __ *A.v.versicolor* Vieillot, 1816

C Chile; SE Bolivia, Paraguay, S Brazil to S Argentina

    __ *A.v.fretensis* P.P. King, 1831

S Chile, S Argentina, Falkland Is. >> C Argentina, Uruguay

○ *Anas hottentota* (Eyton, 1838) HOTTENTOT TEAL

N Nigeria, Eritrea to E and S Africa, Madagascar

MARMARONETTA Reichenbach, 1853 F

○ *Marmaronetta angustirostris* (Ménétriés, 1832) MARBLED TEAL

S Spain and NW Africa to Kazakhstan and NW India

RHODONESSA Reichenbach, 1853 F

○ †?*Rhodonessa caryophyllacea* (Latham, 1790) PINK-HEADED DUCK

NE and E India

NETTA Kaup, 1829 F

● *Netta rufina* (Pallas, 1773) RED-CRESTED POCHARD

C and S Europe, SW and C Asia >> S Europe, N and NE Africa, S Asia

○ *Netta peposaca* (Vieillot, 1816) ROSYBILL

C Chile, N and C Argentina >> Paraguay, E Bolivia, S Brazil

○ *Netta erythrophthalma* SOUTHERN POCHARD

    __ *N.e.brunnea* (Eyton, 1838)

Ethiopia, E and S Africa and Cape Province

    __ *N.e.erythrophthalma* (Wied, 1832)

S Venezuela through E Brazil to N Argentina

AYTHYA Boie, 1822 F

● *Aythya valisineria* (A. Wilson, 1814) CANVASBACK

W and C Canada, NW USA >> S USA, Mexico

● *Aythya americana* (Eyton, 1838) REDHEAD

C Alaska, Canada, N USA >> NW Mexico

● *Aythya ferina* (Linnaeus, 1758) COMMON POCHARD

W Europe to C Asia >> N and E Africa, India, S China

● *Aythya australis* HARDHEAD

    ✔ *A.a.australis* (Eyton, 1838)[3]

Australia, Tasmania, New Zealand

    __ *A.a.extima* Mayr, 1940

Vanuatu, New Caledonia

○ †?*Aythya innotata* (Salvadori, 1894) MADAGASCAN POCHARD

N and E Madagascar

○ *Aythya baeri* (Radde, 1863) BAER'S POCHARD

SE Siberia and NE China >> SE and E Asia

○ *Aythya nyroca* (Güldenstädt, 1770) FERRUGINOUS DUCK

C and S Europe and NW Africa to C Asia >> N and NE Africa and S Asia

● *Aythya novaeseelandiae* (J.F. Gmelin, 1789) NEW ZEALAND SCAUP

New Zealand

● *Aythya collaris* (Donovan, 1809) RING-NECKED DUCK

E Alaska, Canada, N USA >> S USA, Central America, W Indies

● *Aythya fuligula* (Linnaeus, 1758) TUFTED DUCK

N and C Eurasia >> W Europe and N Africa to E Asia and SW Pacific

● *Aythya marila* GREATER SCAUP

    __ *A.m.marila* (Linnaeus, 1761)

N Europe and NW Asia >> W and S Europe, NW India

    ✔ *A.m.nearctica* Stejneger, 1885[4]

NE Asia and N and W Canada >> E Asia; W and E USA, W Indies

● *Aythya affinis* (Eyton, 1838) LESSER SCAUP

W and C Canada, W USA >> S USA to Panama, W Indies

---

[1] We follow A.O.U. (1998: 73) in retaining one species in spite of contrary views espoused in Europe (Sangster *et al.,* 2001, 2002) (2166, 2167). If treated as two species then *carolinensis* is the Green-winged Teal (and *nimia* attaches to *crecca*).

[2] Since treatment at species level by Blake (1977) (164) now more usually treated here see Kear (in press) (1235).

[3] Includes *ledeboeri* and *papuana* based on apparent colonists of Java and on vagrants to New Guinea respectively. See Johnsgard (1979: 485) (1211).

[4] The name *mariloides* (Vigors, 1839), which has sometimes been applied to all or part of this population, is unavailable as it was attached to specimens of the Lesser Scaup see Banks (1986) (85). Its type locality has been corrected to San Francisco Bay.

POLYSTICTA Eyton, 1836 F

○ *Polysticta stelleri* (Pallas, 1769) STELLER'S EIDER | NE Europe and N Siberia to W and N Alaska >> Baltic Sea, N Norway, Bering Sea shores, S Alaska

SOMATERIA Leach, 1819 F

○ *Somateria fischeri* (J.F. Brandt, 1847) SPECTACLED EIDER | NE Siberia, Alaska >> offshore in Bering Sea

○ *Somateria spectabilis* (Linnaeus, 1758) KING EIDER | N Europe, N Asia, N Canada >> Arctic and subarctic ocean coasts

◑ *Somateria mollissima* COMMON EIDER
    ✍ *S.m.v-nigra* G.R. Gray, 1856 | Alaska, NE Asia >> Bering Sea and Aleutian Is
    — *S.m.borealis* (C.L. Brehm, 1824) | NE Canada, Greenland, Iceland, Svalbard, Franz Josef Land >> N Atlantic

    — *S.m.sedentaria* Snyder, 1941 | Hudson Bay
    ✍ *S.m.dresseri* Sharpe, 1871 | SE Canada and NE USA >> NW Atlantic coasts
    — *S.m.faroeensis* C.L. Brehm, 1831 | Faroe Is., Shetlands, Orkneys and Outer Hebrides
    ✓ *S.m.mollissima* (Linnaeus, 1758) | Mainland Scotland and Ireland to Baltic Sea, Norway and NW Russia and S to France >> NW and C Europe

HISTRIONICUS Lesson, 1828 M

◉ *Histrionicus histrionicus* (Linnaeus, 1758) HARLEQUIN DUCK[1] | E Canada to Greenland and Iceland; NE Siberia and NW Nth. America >> N Pacific and NW Atlantic coasts

CAMPTORHYNCHUS Bonaparte, 1838 M

○ †*Camptorhynchus labradorius* (J.F. Gmelin, 1789) LABRADOR DUCK | Labrador >> NW Atlantic shore

MELANITTA Boie, 1822 F

◉ *Melanitta perspicillata* (Linnaeus, 1758) SURF SCOTER | Alaska, N Canada >> Aleutian Is. to Baja California, SE Canada to NE USA

◉ *Melanitta fusca* WHITE-WINGED SCOTER/VELVET SCOTER[2]
    — *M.f.fusca* (Linnaeus, 1758) | N Europe, NW Siberia, E Turkey and Transcaucasia >> W and S Europe and SW Asia

    — *M.f.stejnegeri* (Ridgway, 1887) | C and E Siberia >> coasts of NE Pacific
    ✍ *M.f.deglandi* (Bonaparte, 1850)[3] | Alaska and Canada >> shores W and E Nth. America

◉ *Melanitta nigra* BLACK SCOTER[4]
    — *M.n.nigra* (Linnaeus, 1758) | N Europe, NW and NC Siberia >> W and S Europe and NW Africa

    ✓ *M.n.americana* (Swainson, 1832) | NE Siberia, W Alaska, Canada >> N Pacific, Great Lakes, and NW Atlantic

CLANGULA Leach, 1819 F

◉ *Clangula hyemalis* (Linnaeus, 1758) LONG-TAILED DUCK/OLDSQUAW | N Europe, N Asia, N America >> W and N Europe, E Asia, W and E Nth. America

BUCEPHALA S.F. Baird, 1858 F

◉ *Bucephala albeola* (Linnaeus, 1758) BUFFLEHEAD | Canada, NW and NC USA >> Aleutian Is. to S USA, N Mexico

◉ *Bucephala clangula* COMMON GOLDENEYE
    ✓ *B.c.clangula* (Linnaeus, 1758) | C and N Europe, N Asia >> W and S Europe, C, S, and E Asia

    ✓ *B.c.americana* (Bonaparte, 1838) | Alaska, Canada, N USA >> coastal Nth. America

◉ *Bucephala islandica* (J.F. Gmelin, 1789) BARROW'S GOLDENEYE | Alaska, Canada, W USA, Greenland and Iceland >> W and E Nth. America

MERGELLUS Selby, 1840[5] M

○ *Mergellus albellus* (Linnaeus, 1758) SMEW | N Europe, N Asia >> W and S Europe, C and E Asia

LOPHODYTES Reichenbach, 1853[6] M

◉ *Lophodytes cucullatus* (Linnaeus, 1758) HOODED MERGANSER | SE Alaska, C Canada, USA >> W, S, and E USA, N Mexico

---

[1] Includes *pacificus* see Johnsgard (1979: 491) (1211).
[2] We also await acoustic and behavioural evidence to split these.
[3] Includes *dixoni* see Johnsgard (1979: 494) (1211).
[4] Some authors split this, but behavioural and vocal evidence is awaited.
[5] We follow A.O.U. (1998: 83) (3) in using this generic name.
[6] We follow A.O.U. (1998: 83) (3) in using this generic name.

MERGUS Linnaeus, 1758 M
○ †*Mergus australis* Hombron & Jacquinot, 1841 AUCKLAND ISLAND MERGANSER
                                               Auckland Is.

○ *Mergus octosetaceus* Vieillot, 1817 BRAZILIAN MERGANSER   S Brazil, E Paraguay, NE Argentina

● *Mergus merganser* GOOSANDER/COMMON MERGANSER
    ⚏ *M.m.merganser* Linnaeus, 1758   Iceland to NE Asia >> Mediterranean and China
    — *M.m.orientalis* Gould, 1845[1]   C Asia to Himalayas >> Assam and Japan
    ⚏ *M.m.americanus* Cassin, 1852   Canada, W USA >> E and S USA

● *Mergus serrator* Linnaeus, 1758 RED-BREASTED MERGANSER[2] Greenland, N Europe, Asia, N America >> China, Mexico

○ *Mergus squamatus* Gould, 1864 SCALY-SIDED MERGANSER NE China, Korea and SE Siberia >> China, Taiwan

HETERONETTA Salvadori, 1865[3] F
○ *Heteronetta atricapilla* (Merrem, 1841) BLACK-HEADED DUCK C Chile, E Bolivia, Paraguay, N and C Argentina

NOMONYX Ridgway, 1880[4] M
○ *Nomonyx dominicus* (Linnaeus, 1766) MASKED DUCK S USA and West Indies to N Argentina

OXYURA Bonaparte, 1828 F
● *Oxyura jamaicensis* RUDDY DUCK
    ⚏ *O.j.rubida* (A. Wilson, 1814)   S Canada, USA >> USA to N Mexico
    — *O.j.jamaicensis* (J.F. Gmelin, 1789)   W Indies

○ *Oxyura ferruginea* (Eyton, 1838) ANDEAN DUCK[5] Andes of Colombia to Tierra del Fuego

● *Oxyura vittata* (Philippi, 1860) LAKE DUCK C Chile N Argentina to Tierra del Fuego >> S Brazil

○ *Oxyura australis* Gould, 1836 BLUE-BILLED DUCK S Australia, Tasmania

○ *Oxyura maccoa* (Eyton, 1838) MACCOA DUCK Ethiopia to N Tanzania, Zimbabwe and Namibia to Cape Province

○ *Oxyura leucocephala* (Scopoli, 1769) WHITE-HEADED DUCK Mediterranean to C Asia >> N Africa, N India

BIZIURA Stephens, 1824 F
○ *Biziura lobata* (Shaw, 1796) MUSK DUCK S Australia, Tasmania >> C Australia

SPHENISCIDAE PENGUINS (6: 17)

APTENODYTES J.F. Miller, 1778 M
○ *Aptenodytes patagonicus* J.F. Miller, 1778 KING PENGUIN[6] Prince Edward, South Georgia, Falkland, Macquarie, Kerguelen, Heard, Crozet, Marion Is.

○ *Aptenodytes forsteri* G.R. Gray, 1844 EMPEROR PENGUIN Antarctica

PYGOSCELIS Wagler, 1832 M
○ *Pygoscelis papua* GENTOO PENGUIN
    — *P.p.papua* (J.R. Forster, 1781)[7]   Subantarctica to about 60°S
    — *P.p.ellsworthi* Murphy, 1947   Antarctic Pen. to S Sandwich Is.

○ *Pygoscelis adeliae* (Hombron & Jacquinot, 1841) ADELIE PENGUIN Antarctica, S Orkney Is., S Shetland Is.

○ *Pygoscelis antarcticus* (J.R. Forster, 1781) CHINSTRAP PENGUIN[8] Antarctic Ocean, S Atlantic

---

[1] This name may well have been attached to a bird (types in BMNH) representative of the nominate form. If so the distinct population of C Asia must be called *comatus* Salvadori, 1895 (CSR).
[2] Treated as monotypic following Roselaar in Cramp *et al.* (1977: 680) (588).
[3] In the tribe of stifftails (here the genera *Heteronetta*, *Nomonyx* and *Oxyura*) we follow McCracken *et al.* (1999) (1501).
[4] We follow Livezey (1995) (1366) and A.O.U. (1998: 85) (3) in recognising this genus.
[5] We tentatively follow A.O.U. (1998: 86) (3) in separating this from *O. jamaicensis* and Fjeldså (1986) (844), as updated in Fjeldså & Krabbe (1990) (850), in considering *andina* as a name applicable to hybrids. But see also McCracken *et al.* (1999) (1501).
[6] Treated as essentially monotypic following Marchant & Higgins (1990: 135) (1426).
[7] Includes *taeniata* see Stonehouse (1970) (2346) = *taeniatus* see David & Gosselin (2002) (614).
[8] For correction of spelling see David & Gosselin (2002) (614).

EUDYPTES Vieillot, 1816 M

○ *Eudyptes pachyrhynchus* G.R. Gray, 1845 Fiordland Penguin   South Island and offshore islands (New Zealand)

○ *Eudyptes robustus* Oliver, 1953 Snares Penguin[1]   Snares I.

○ *Eudyptes sclateri* Buller, 1888 Erect-crested Penguin   Auckland, Bounty and Antipodes Is.

○ *Eudyptes chrysocome* Rockhopper Penguin
  — *E.c.chrysocome* (J.R. Forster, 1781)   Tierra del Fuego, Falkland Is.
  — *E.c.filholi* Hutton, 1879   Kerguelen, Prince Edward, Marion, Crozet, Heard, Macquarie Is. and islands south of New Zealand
  — *E.c.moseleyi* Mathews & Iredale, 1921   Tristan da Cunha, Gough, St. Paul, Amsterdam Is.

○ *Eudyptes schlegeli* Finsch, 1876 Royal Penguin   Macquarie I.

○ *Eudyptes chrysolophus* (J.F. Brandt, 1837) Macaroni Penguin   Antarctic Pen. to S Sandwich Is., S Georgia, Kerguelen, Falkland, Prince Edward, Marion, Crozet Is.

MEGADYPTES A. Milne-Edwards, 1880 M

● *Megadyptes antipodes* (Hombron & Jacquinot, 1841) Yellow-eyed Penguin   South Island (New Zealand) and adjacent southern islands

EUDYPTULA Bonaparte, 1856 F

● *Eudyptula minor* Little Penguin[2]
  — *E.m.novaehollandiae* (Stephens, 1826)   Tasmania, S Australia
  — *E.m.iredalei* Mathews, 1911   N of North Island (New Zealand)
  — *E.m.variabilis* Kinsky & Falla, 1976   S of North Island (New Zealand), Cook Strait
  ✓ *E.m.minor* (J.R. Forster, 1781)   W and SE of South Island (New Zealand), Stewart I.
  — *E.m.albosignata* Finsch, 1874   E of South Island (New Zealand)
  — *E.m.chathamensis* Kinsky & Falla, 1976   Chatham Is.

SPHENISCUS Brisson, 1760 M

○ *Spheniscus demersus* (Linnaeus, 1758) Jackass Penguin   S and SW coast of S Africa and islands

○ *Spheniscus magellanicus* (J.R. Forster, 1781) Magellanic Penguin[3]   S South America and islands

○ *Spheniscus humboldti* Meyen, 1834 Humboldt Penguin   W coast of S America and islands

○ *Spheniscus mendiculus* Sundevall, 1871 Galapagos Penguin   Galapagos Is.

GAVIIDAE DIVERS OR LOONS (1: 5)

GAVIA J.R. Forster, 1788 F

● *Gavia stellata* (Pontoppidan, 1763) Red-throated Diver/Red-throated Loon   Holarctic, circumpolar >> mainly near coasts of S Europe, S China, Florida

● *Gavia arctica* Black-throated Diver/Arctic Loon
  — *G.a.arctica* (Linnaeus, 1758)[4]   N Eurasia east to R Lena >> near coasts of Europe to SW and C Asia
  ✓ *G.a.viridigularis* Dwight, 1918   Lena Basin (NE Siberia) to W Alaska >> near coasts of E Asia and W North America

● *Gavia pacifica* (Lawrence, 1858) Pacific Loon/Pacific Diver[5]   NE Siberia, Alaska, N Canada >> near coasts of NE Asia and W North America

● *Gavia immer* (Brünnich, 1764) Common Loon/Great Northern Diver[6]   N North America, Iceland >> near coasts of SW and E Canada, USA, seas round Britain and off Brittany

○ *Gavia adamsii* (G.R. Gray, 1859) White-billed Diver/Yellow-billed Loon   N Eurasia, Alaska and NW Canada >> near coasts of W and N Europe, N Pacific south to Japan and NW USA

[1] The name *Eudyptes atratus* Finsch, 1875, has been suppressed as regards Priority but not as regards Homonymy see Opinion 1056 (I.C.Z.N., 1976) (1174).
[2] Treatment follows Falla & Mougin (1979: 132) (824).
[3] This species is closer to *S. demersus* than is *S. humboldti* see Thumser & Karron (1994) (2402).
[4] Includes *suschkini* see Storer (1979: 138) (2349).
[5] Treated as a separate species by A.O.U. (1998: 4) (3). See also Flint & Kistchinski (1982) (855).
[6] Includes *elasson* see Storer (1979: 139) (2349).

DIOMEDEIDAE ALBATROSSES[1] (4: 13)

PHOEBASTRIA Reichenbach, 1853 F

○ *Phoebastria immutabilis* (Rothschild, 1893) LAYSAN ALBATROSS — Ogasawara-shoto (Bonin) Is. (Japan), W Hawaii, islands off W Mexico >> N Pacific

● *Phoebastria nigripes* (Audubon, 1849) BLACK-FOOTED ALBATROSS — Islands off S Japan, W Hawaii Is. >> N Pacific

○ *Phoebastria irrorata* (Salvin, 1883) WAVED ALBATROSS — Hood I. (Galapagos) and La Plata I. (off Ecuador) >> EC Pacific Ocean

○ *Phoebastria albatrus* (Pallas, 1769) SHORT-TAILED ALBATROSS — Tori-shima I. (S Japan) >> N Pacific

DIOMEDEA Linnaeus, 1758 F

○ *Diomedea epomophora* ROYAL ALBATROSS
— *D.e.sanfordi* Murphy, 1917 — Otago Pen. (New Zealand), Chatham Is. >> SW Pacific
— *D.e.epomophora* Lesson, 1825 — Campbell and Auckland Is. >> S Australia, S Sth. America

○ *Diomedea exulans* WANDERING ALBATROSS[2]
— *D.e.exulans* Linnaeus, 1758[3] — Tristan da Cunha, Gough I. >> S Atlantic
— *D.e.chionoptera* Salvin, 1896 — South Georgia I. east to Kerguelen and Macquarie Is. >> S Oceans
— *D.e.amsterdamensis* Roux, Jouventin, Mougin, Stahl & Weimerskirch, 1983[4] — Amsterdam I. (2112) >> S Ocean
— *D.e.antipodensis* Robertson & Warham, 1992 — Antipodes and Campbell Is. >> Australasian seas, S Pacific (2093)
— *D.e.gibsoni* Robertson & Warham, 1992 — Auckland Is. >> SW Pacfic Ocean (2093)

PHOEBETRIA Reichenbach, 1853 F

○ *Phoebetria fusca* (Hilsenberg, 1822) SOOTY ALBATROSS — Southern Ocean, Tristan da Cunha, Gough I. east to St Paul I. >> S Ocean

○ *Phoebetria palpebrata* (J.R. Forster, 1785) LIGHT-MANTLED ALBATROSS[5] — Southern Ocean, South Georgia east to Antipodes Is. >> S Ocean

THALASSARCHE Reichenbach, 1853 F

○ *Thalassarche chlororhynchos* (J.F. Gmelin, 1789) YELLOW-NOSED ALBATROSS[6] — Tristan da Cunha, Gough Is., S Indian Ocean >> S Atlantic and Indian Oceans

○ *Thalassarche chrysostoma* (J.R. Forster, 1785) GREY-HEADED ALBATROSS — Southern Ocean from Cape Horn to Campbell I. >> Southern Ocean

○ *Thalassarche melanophrys* BLACK-BROWED ALBATROSS[7]
— *T.m.impavida* Mathews, 1912 — Islands south of New Zealand (Campbell I.) >> S Pacific
— *T.m.melanophrys* (Temminck, 1828) — S Sth. America to Antipodes Is. >> S Ocean

○ *Thalassarche bulleri* BULLER'S ALBATROSS
— *T.b. bulleri* (Rothschild, 1893)[8] — Solander and Snares Is. >> S Pacific
— *T.b.*subsp. ? — Three Kings and Chatham Is. >> S Pacific

○ *Thalassarche cauta* SHY ALBATROSS[9]
— *T.c.cauta* (Gould, 1841) — Tasmania >> Sthn. Australasian waters)
— *T.c.steadi* Falla, 1933[10] — Auckland Is. >> Sthn. Australasian waters
— *T.c.eremita* Murphy, 1930 — Chatham Is. >> New Zealand waters
— *T.c.salvini* (Rothschild, 1893) — S Sth. America, Snares I.,Bounty I. >> S Pacific

---

[1] The genera used and their sequence and species essentially follow Nunn *et al.* (1996) (1666) and Robertson & Nunn (1998) (2092).
[2] Situation confusing, but Mathews considered Tristan da Cunha birds provided the type for Linnaeus, see Jouanin & Mougin (1979: 51 fn) (1229). *Contra* Carboneras (1992: 211) (291) we revert to prior treatment and recognise *chionoptera* as the name for the larger populations of Kerguelen etc., as apparently do Robertson & Nunn (1998) (2092).
[3] Includes *dabbenena* see Jouanin & Mougin (1979: 51) (1229).
[4] Molecular studies place this as a race of *D. exulans* see Robertson & Nunn (1998) (2092).
[5] It is now unclear upon what authority our previous Editions recognized *huttoni*.
[6] Two subspecies listed by Carboneras (1992: 214) (291) however we have not resolved the correct nomenclature of such forms.
[7] The emended name used here is judged to be in prevailing usage; see Art. 33.2 of the Code (I.C.Z.N., 1999) (1178) and p. 000 herein.
[8] Although recognised by Jouanin & Mougin (1979: 57) (1229) apparently this name was attached to a juvenile plumage phase of nominate *bulleri* see Robertson & Nunn (1998) (2092).
[9] Oceanic ranges by subspecies very tentative.
[10] For recognition of this see Hillcoat (1997) (1115).

PROCELLARIIDAE PETRELS, SHEARWATERS[1] (14: 74)

MACRONECTES Richmond, 1905 M

● *Macronectes giganteus* (J.F. Gmelin, 1789) Southern Giant Petrel — Antarctic coast and islands north to 40°S >> S Oceans

○ *Macronectes halli* Mathews, 1912 Northern Giant Petrel — Falkland, South Georgia, Prince Edward, Crozet, Kerguelen, Macquarie Is. and islands off New Zealand >> S Oceans

FULMARUS Stephens, 1826 M

◉ *Fulmarus glacialis* Northern Fulmar

— *F.g.glacialis* (Linnaeus, 1761)[2] — NE Canada, N and NE Greenland, Svalbard, Bear I., and Franz Josef Land >> N Atlantic Ocean

— *F.g.auduboni* Bonaparte, 1857 — SE Canada, W and SE Greenland, and W and N Europe north to Iceland, Jan Mayen, Norway and Novaya Zemlya >> N Atlantic

✔ *F.g.rodgersii* Cassin, 1862 — N Pacific

○ *Fulmarus glacialoides* (A. Smith, 1840) Southern Fulmar — Antarctica, Southern Ocean

THALASSOICA Reichenbach, 1853 F

○ *Thalassoica antarctica* (J.F. Gmelin, 1789) Antarctic Petrel — Antarctic and surrounding seas

DAPTION Stephens, 1826 N

● *Daption capense* Cape Petrel

— *D.c.capense* (Linnaeus, 1758) — Antarctica and subantarctic islands
✔ *D.c.australe* Mathews, 1913 — Seas and islands round S New Zealand

PAGODROMA Bonaparte, 1856 F

○ *Pagodroma nivea* Snow Petrel

— *P.n.nivea* (G. Forster, 1777) — Islands in the Scotia Arc and surrounding seas
— *P.n.confusa* Mathews, 1912[3] — Adélie Land (Antarctica) and nearby islands and seas

HALOBAENA Bonaparte, 1856 F

○ *Halobaena caerulea* (J.F. Gmelin, 1789) Blue Petrel — Subantarctic, Ramírez Is. and Cape Horn, east to Macquarie I.

PACHYPTILA Illiger, 1811 F

○ *Pachyptila vittata* (G. Forster, 1777) Broad-billed Prion — Tristan da Cunha, South Georgia, Chatham Is., SW New Zealand >> nearby seas

○ *Pachyptila salvini* Salvin's Prion[4]

— *P.s.salvini* (Mathews, 1912) — Marion, Prince Edward, Crozet Is. >> S Indian Ocean
— *P.s.macgillivrayi* (Mathews, 1912) — Amsterdam, St. Paul Is. >> S Indian Ocean

○ *Pachyptila desolata* (J.F. Gmelin, 1789) Antarctic Prion[5] — Circumpolar, from Scotia Arc to Macquarie and Auckland Is.

○ *Pachyptila belcheri* (Mathews, 1912) Slender-billed Prion — Kerguelen, Crozet, Falkland Is., Noir I. (Chile) >> southern seas

○ *Pachyptila turtur* Fairy Prion

— *P.t.turtur* (Kuhl, 1820) — Poor Knights, Chatham, Snares Is. (New Zealand), Bass Strait Is. (Australia) >> nearby seas
— *P.t.subantarctica* Oliver, 1955[6] — Marion, Falkland, Crozet, Macquarie, Antipodes Is. >> nearby seas

○ *Pachyptila crassirostris* Fulmar-Prion

— *P.c.eatoni* (Mathews, 1912) — Heard, Auckland Is. >> nearby seas
— *P.c.crassirostris* (Mathews, 1912)[7] — Snares, Bounty, Chatham Is. >> nearby seas

---

[1] Based on Jouanin & Mougin (1979) (1229), with modifications derived from Imber (1985) (1179) and Bretagnolle *et al.* (1998) (212).
[2] Includes *minor* see Jouanin & Mougin (1979: 62) (1229). Here we recognise subspecies based on colour accepting that size varies clinally (van Franeker, 1995; van Franeker & Wattel, 1982) (2465, 2466).
[3] In last Edition we used the name *major* for this, but our source for that is now unclear. Range not well understood.
[4] For reasons to transfer *macgillivrayi* from *P. vittata* see Roux *et al.* (1986) (2113).
[5] Treated as monotypic following Harper (1980) (1069).
[6] For reasons to recognise *subantarctica* see Harper (1980) (1069).
[7] Includes *pyramidalis* see Harper (1980) (1069).

APHRODROMA Olson, 2000 (1706)[1] F
○ *Aphrodroma brevirostris* (Lesson, 1831) KERGUELEN PETREL — Gough, Kerguelen, Crozet, Marion Is. >> S Oceans

PTERODROMA Bonaparte, 1856[2] F
○ *Pterodroma macroptera* GREAT-WINGED PETREL
— *P.m.macroptera* (A. Smith, 1840) — SW Australia, Tristan da Cunha, Gough, Crozet, Marion, Kerguelen Is., SW Australia >> southern seas

— *P.m.gouldi* (Hutton, 1869) — North Island (New Zealand) >> southern seas

○ *Pterodroma lessonii* (Garnot, 1826) WHITE-HEADED PETREL — Crozet, Kerguelen, Macquarie, Antipodes, Auckland Is. >> southern seas

○ *Pterodroma incerta* (Schlegel, 1863) ATLANTIC PETREL — Gough I., Tristan da Cunha >> S Atlantic

○ *Pterodroma solandri* (Gould, 1844) PROVIDENCE PETREL — Lord Howe I., Philip I. >> W Pacific Ocean

○ *Pterodroma magentae* (Giglioli & Salvadori, 1869) MAGENTA PETREL — Chatham Is. >> nearby seas

○ *Pterodroma ultima* Murphy, 1949 MURPHY'S PETREL — Tuamotu, Pitcairn, Iles Australes >> SC Pacific

○ *Pterodroma mollis* (Gould, 1844) SOFT-PLUMAGED PETREL[3] — Tristan da Cunha, Gough, Antipodes, Crozet, Kerguelen, Marion, Prince Edward, Macquarie Is. >> S oceans (mainly Indian and Atlantic)

○ *Pterodroma madeira* Mathews, 1934 MADEIRA PETREL — Madeira >> nearby seas

○ *Pterodroma feae* (Salvadori, 1899) CAPE VERDE PETREL — Cape Verde Is., Bugio I. (off Madeira) >> nearby seas

○ *Pterodroma cahow* (Nichols & Mowbray, 1916) BERMUDA PETREL — Bermuda >> nearby seas

○ *Pterodroma hasitata* BLACK-CAPPED PETREL
— *P.h.hasitata* (Kuhl, 1820) — Cuba and Hispaniola to Martinique >> nearby seas
— †?*P.h.caribbaea* Carte, 1866 — Jamaica

○ *Pterodroma externa* (Salvin, 1875) JUAN FERNANDEZ PETREL[4] — Robinson Crusoe I. (Juan Fernandez Is., Chile) >> central Pacific Ocean

○ *Pterodroma occulta* Imber & Tennyson, 2001 VANUATU PETREL — Mera Lava, Banks Is. (Vanuatu) >> SW Pacific waters (1180)

○ *Pterodroma neglecta* KERMADEC PETREL
— *P.n.neglecta* (Schlegel, 1863) — Lord Howe, Kermadec, Tuamotu, Austral, Pitcairn Is. >> Pacific Ocean

— *P.n.juana* Mathews, 1935 — Juan Fernandez, San Ambrosio, San Felix Is. >> Pacific Ocean

○ †?*Pterodroma phillipii* G.R. Gray, 1862 MOUNT PITT PETREL — Kermadec, Lord Howe, Austral and Tuamoto Is.; Juan Fernandez Is.

○ *Pterodroma arminjoniana* HERALD PETREL
— *P.a.arminjoniana* (Giglioli & Salvadori, 1869) — S Trinidade I. (Atlantic Oc.), Round I. (Mauritius) >> nearby seas

— *P.a.heraldica* (Salvin, 1888)[5] — Chesterfield, Raine, Tonga, Tuamotu, Pitcairn, Marquesas, Easter Is. >> the tropical Pacific Ocean

○ *Pterodroma atrata* (Mathews, 1912) HENDERSON PETREL[6] — Henderson I. (SE Pacific) >> ?

○ *Pterodroma alba* (J.F. Gmelin, 1789) PHOENIX PETREL — Kiritimati, Phoenix, Tonga, Marquesas, Pitcairn, Tuamotu, Kermadec Is. >> WC Pacific Ocean

○ *Pterodroma baraui* (Jouanin, 1964) BARAU'S PETREL — Mascarene Is. (Indian Oc.) >> nearby seas

○ *Pterodroma phaeopygia* DARK-RUMPED PETREL
— *P.p.phaeopygia* (Salvin, 1876) — Galapagos Is. >> C Pacific
— *P.p.sandwichensis* (Ridgway, 1884) — Hawaiian Is. >> C Pacific

○ *Pterodroma inexpectata* (J.R. Forster, 1844)[7] MOTTLED PETREL — S New Zealand, Stewart I., Snares Is. >> Pacific Ocean

[1] This name was proposed to replace *Lugensa* see Olson (2000) (1706), but see Bourne (2001) (204).
[2] We have grouped species following Imber (1985) (1179), but we sequence the species as close to the traditional sequence as his groups allow. This contradicts his arguments in some measure, and is here purely pragmatic.
[3] Clancey *et al.* (1981) (531) and Carboneras (1992: 242) (292) used the name *dubius* Mathews, 1924. This and the new name *fusca* Imber, 1985 (1179) are included here following Bourne (1987) (203). See also Sangster *et al.* (2002) (2167).
[4] In last Edition we listed *tristrani*, see the synonymy in Jouanin & Mougin (1979: 74) (1229) the location Tristan da Cunha is not now accepted for this.
[5] Treated as a species by Imber (1985) (1179), but see Bourne (1987) (203) and A.O.U. (1998: 13) (3). Includes *paschae* see Jouanin & Mougin (1979: 71) (1229). For further support for two species see Brooke & Rowe (1996) (225).
[6] For recognition of this see Brooke & Rowe (1996) (225).
[7] Drawings with Forster's proposed names were published many years after his death (d. 1798).

○ *Pterodroma cervicalis* (Salvin, 1891) WHITE-NECKED PETREL

Kermadec Is. >> W Pacific

○ *Pterodroma nigripennis* (Rothschild, 1893) BLACK-WINGED PETREL

Lord Howe, Norfolk, Kermadec Is., north to New Caledonia, south to New Zealand, Austral and Chatham Is. >> tropical Pacific Ocean

○ *Pterodroma axillaris* (Salvin, 1893) CHATHAM ISLANDS PETREL

Chatham Is. >> nearby seas

○ *Pterodroma hypoleuca* (Salvin, 1888) BONIN ISLANDS PETREL

Ogasawara-shoto (Bonin) Is., Iwo (Volcano) Is., W Hawaiian Is. >> NW Pacific

○ *Pterodroma leucoptera* GOULD'S PETREL
   — *P.l.leucoptera* (Gould, 1844)

Cabbage Tree and Boondelbah Is. (New South Wales) >> SW and C Pacific

   — *P.l.caledonica* de Naurois, 1978

New Caledonia (627) >> ??

   — *P.l.brevipes* (Peale, 1848)[1]

Fiji and Cook Is. (? Solomon Is.) >> SW and C Pacific

○ *Pterodroma cookii* (G.R. Gray, 1843) COOK'S PETREL[2]

Great and Little Barrier Is., Codfish I., N and S New Zealand >> N and C Pacific

○ *Pterodroma defilippiana* (Giglioli & Salvadori, 1869) MASATIERRA[3] PETREL

Juan Fernandez and Los Desventurados Is. (Chile) >> EC Pacific

○ *Pterodroma longirostris* (Stejneger, 1893) STEJNEGER'S PETREL[4]

Robinson Crusoe I. (Juan Fernandez Is.) >> ?NC Pacific

○ *Pterodroma pycrofti* Falla, 1933 PYCROFT'S PETREL

NE North Island (New Zealand) >> nearby seas

PSEUDOBULWERIA Mathews, 1936[5] F
○ †?, *Pterodroma aterrima* (Bonaparte, 1857) MASCARENE PETREL

SW Mascarene Is. >> nearby seas

○ *Pseudobulweria rostrata* TAHITI PETREL
   — *P.r.rostrata* (Peale, 1848)

Marquesas Is., Society Is. >> WC Pacific

   — *P.r.trouessarti* (Brasil, 1917)

New Caledonia >> WC Pacific

○ †?*Pseudobulweria becki* (Murphy, 1928) BECK'S PETREL[6]

? >> Bismarck Arch., Solomon Is.

○ †?*Pseduobulweria macgillivrayi* (G.R. Gray, 1860) FIJI PETREL

Gau I. (Fiji Is.) >> ?

PROCELLARIA Linnaeus, 1758 F
○ *Procellaria cinerea* J.F. Gmelin, 1789 GREY PETREL

Tristan da Cunha, Gough east to Antipodes Is. >> southern oceans

○ *Procellaria aequinoctialis* WHITE-CHINNED PETREL[7]
   — *P.a.aequinoctialis* Linnaeus, 1758[8]

South Georgia, Falkland east to Antipodes Is. >> southern oceans

   — *P.a.conspicillata* Gould, 1844

Inaccessible I., Tristan da Cunha >> southern oceans

○ *Procellaria parkinsoni* G.R. Gray, 1862 PARKINSON'S/BLACK PETREL

North Island (New Zealand) >> C Pacific

○ *Procellaria westlandica* Falla, 1946 WESTLAND PETREL

NW South Island (New Zealand) >> nearby seas

CALONECTRIS Mathews & Iredale, 1915 F
○ *Calonectris leucomelas* (Temminck, 1836[9]) STREAKED SHEARWATER

NW Pacific >> SW Pacific Ocean, New Guinea

○ *Calonectris diomedea* CORY'S SHEARWATER[10]
   — *C.d.diomedea* (Scopoli, 1769)

Mediterranean Islands >> S Atlantic Ocean

   — *C.d.borealis* (Cory, 1881)

Portugal, Canaries, Madeira, Azores Is. >> W Atlantic

   — *C.d.edwardsii* (Oustalet, 1883)

Cape Verde Is. >> ?

---

[1] Sometimes treated as a species, e.g. by Imber (1985) (1179).
[2] Includes *orientalis* see Jouanin & Mougin (1979: 77) (1229).
[3] Recently renamed Alejandro Selkirk Island!
[4] Includes *masafuerae* Jouanin & Mougin (1979: 77) (1229); described as a form of *leucoptera* and still so attached in our last Edition. See also A.O.U. (1998: 16) (3).
[5] For recognition of this genus see Imber (1985) (1179) and Bretagnolle *et al.* (1998) (212).
[6] Imber (1985) (1179) recognised this and we do so tentatively; its breeding grounds remain to be discovered and the range given relates to specimens taken at sea.
[7] We retain the conspecific treatment of Jouanin & Mougin (1979: 86) (1229), but for a contrary opinion see Ryan (1998) (2143).
[8] Includes *steadi* see Jouanin & Mougin (1979: 86) (1229).
[9] The source ("Planches Coloriées", Livraison 99) must date from 1836 not 1835 (Dickinson, 2001) (739). In footnotes about the "Planches" later in this book we just cite "Livr.".
[10] Now often treated as three species, but we believe these views need to gain acceptance. The name *flavirostris* is considered indeterminate, see Bourne (1955) (202).

PUFFINUS Brisson, 1760[1] M
○ *Puffinus nativitatis* Streets, 1877 Kiritimati Shearwater[2]   Central Pacific (Hawaii, Kiritimati, Phoenix, Tuamotu, Pitcairn, Austral, Marquesas Is.

● *Puffinus pacificus* (J.F. Gmelin, 1789) Wedge-tailed Shearwater   Tropical and subtropical Indian and Pacific Oceans

○ *Puffinus bulleri* Salvin, 1888 Buller's Shearwater   Islands off N New Zealand >> N Pacific

○ *Puffinus puffinus* (Brünnich, 1764) Manx Shearwater[3]   North Atlantic >> SW Atlantic

○ *Puffinus yelkouan* (Acerbi, 1827) Yelkouan Shearwater   Mediterranean >> local seas

○ *Puffinus mauretanicus* P.R. Lowe, 1921 Balearic Shearwater[4]   Balearic Is. >> E Atlantic

○ *Puffinus opisthomelas* Coues, 1864 Black-vented Shearwater   W coast of North America

○ *Puffinus auricularis* Townsend's Shearwater
   — *P.a.auricularis* C.H. Townsend, 1890   Revillagigedo Is. (W Mexico) >> local seas
   — *P.a.newelli* Henshaw, 1900   Hawaiian Is. >> local seas

○ *Puffinus gavia* (J.R. Forster, 1844)[5] Fluttering Shearwater   Islands off N New Zealand >> Tasman Sea

○ *Puffinus huttoni* Mathews, 1912 Hutton's Shearwater   NE South Island (New Zealand) >> Australian waters

○ *Puffinus lherminieri* Audubon's Shearwater[6]
   — *P.l.persicus* Hume, 1873   Kuriamuria Is. (Oman)
   — *P.l.nicolae* Jouanin, 1971   Seychelles, Amirantes, Maldives, Chagos
   — *P.l.colstoni* Shirihai & Christie, 1996[7]   Aldabra I. (2245)
   — *P.l.temptator* Louette & Herremans, 1985   Comoro Is. (1382)
   — *P.l.bailloni* (Bonaparte, 1857)[8]   Mauritius, Reunion, Europa Is.
   — *P.l.bannermani* Mathews & Iredale, 1915   Ogasawara-shoto (Bonin), Iwo (Volcano) Is.
   — *P.l.dichrous* Finsch & Hartlaub, 1867[9]   Central Pacific (Palau, Phoenix, Kiritimati Is.)
   — *P.l.heinrothi* Reichenow, 1919   New Britain
   — *P.l.gunax* Mathews, 1930[10]   Vanuatu
   — *P.l.subalaris* Ridgway, 1897   Galapagos Is.
   — *P.l.lherminieri* Lesson, 1839   West Indies incl. islets off Venezuela
   — *P.l.loyemilleri* Wetmore, 1959   Islands off E Panama

○ *Puffinus assimilis* Little Shearwater[11]
   — *P.a.baroli* (Bonaparte, 1857)   Azores, Madeira, Canary Is.
   — *P.a.boydi* Mathews, 1912[12]   Cape Verde Is.
   — *P.a.elegans* Giglioli & Salvadori, 1869[13]   Tristan da Cunha, Gough, Chatham, Antipodes Is.
   — *P.a.tunneyi* Mathews, 1912   Islands off SW Australia
   — *P.a.assimilis* Gould, 1838   Lord Howe, Norfolk Is.
   — *P.a.haurakiensis* Fleming & Serventy, 1943   NE North Island (New Zealand)
   — *P.a.myrtae* Bourne, 1959[14]   Iles Australes (central Pacific)
   — *P.a.kermadecensis* Murphy, 1927   Kermadec Is., ? Chiloe I. (off Chile)

● *Puffinus griseus* (J.F. Gmelin, 1789) Sooty Shearwater   SE Australia, Tasmania, New Zealand to Chile, Falkland Is. ranging widely in the Pacific and Atlantic Oceans

○ *Puffinus tenuirostris* (Temminck, 1836)[15] Short-tailed Shearwater   SE Australia, Tasmania >> waters of W Pacific

● *Puffinus creatopus* Coues, 1864 Pink-footed Shearwater   Juan Fernandez, Mocha Is. >> E Pacific

[1] Some forms in some of the polytypic "species" will no doubt prove to deserving splitting.
[2] Previously Christmas Island Shearwater, based on Kiritimati's old name; that name is now misleading as Christmas Island of the Indian Ocean retains that name.
[3] For comments on reasons to treat this and the next two taxa as species see Sangster *et al.* (2002) (2167).
[4] Separated tentatively from *P. yelkouan*, but see also Bourne (1988) (206).
[5] Drawings with Forster's proposed names were published many years after his death (d. 1798).
[6] There remains much to learn of this group; treatment here very conservative and tentative.
[7] Perhaps within the range of variation of *nicolae*, see Bretagnolle *et al.* (2000) (213).
[8] The name *atrodorsalis* Shirihai *et al.*, 1995 (2245) is tentatively placed here following comments by Bretagnolle *et al.* (2000) (213).
[9] Includes *polynesiae* see Jouanin & Mougin (1979: 98) (1229).
[10] We used the name *nugax* in last Edition (overlooking the new name found to be needed).
[11] May well be more than one species. See, for example, Bourne (1955: 536) (201).
[12] Placed in this species by Carboneras (1992: 256) (292), but in *P. lherminieri* in Jouanin & Mougin (1979: 99) (1229).
[13] The names *Procellaria munda* Kuhl, 1820, and *Nectris munda* Kuhl, 1820, have been suppressed as regards Priority but not as regards Homonymy see Opinion 497 I.C.Z.N. (1957) (1170).
[14] Omitted in previous Editions.
[15] Livr. 99 must date from 1836 not 1835 (Dickinson, 2001) (739).

○ *Puffinus carneipes* Gould, 1844 Flesh-footed Shearwater[1]   SW Australia, Lord Howe I., New Zealand >> W Pacific
and Indian Oceans

○ *Puffinus gravis* (O'Reilly, 1818) Great Shearwater   Tristan da Cunha, Gough, Falkland Is. >> Atlantic
Ocean

BULWERIA Bonaparte, 1843 F
○ *Bulweria bulwerii* (Jardine & Selby, 1828) Bulwer's Petrel   Tropical and temperate waters of the three oceans

○ *Bulweria fallax* Jouanin, 1955 Jouanin's Petrel   Tropical waters, NW Indian Ocean

HYDROBATIDAE STORM PETRELS (7: 21)

OCEANITINAE

OCEANITES Keyserling & Blasius, 1840 M
○ *Oceanites oceanicus* Wilson's Storm Petrel
— *O.o.oceanicus* (Kuhl, 1820)   Falkland, South Georgia, Crozet, Kerguelen Is. >> all
main oceans

— *O.o.exasperatus* Mathews, 1912[2]   Antarctica and closer islands >> all main oceans

○ *Oceanites gracilis* White-vented Storm Petrel
— *O.g.gracilis* (Elliot, 1859)   W coast of South America >> Humboldt current seas
— *O.g.galapagoensis* P.R. Lowe, 1921   Galapagos Is. >> Humboldt current seas

GARRODIA Forbes, 1881 F
○ *Garrodia nereis* (Gould, 1841) Grey-backed Storm Petrel   South Georgia, Falkland, Crozet, Kerguelen, Macquarie
and islands neighbouring New Zealand >> ? local seas

PELAGODROMA Reichenbach, 1853 F
○ *Pelagodroma marina* White-faced Storm Petrel
— *P.m.hypoleuca* (Moquin-Tandon, 1841)   Salvage Is. (Madeira), NE Canary Is. >> N Atlantic
— *P.m.eadesi* Bourne, 1953   Cape Verde Is. >> N Atlantic
— *P.m.marina* (Latham, 1790)   Tristan da Cunha, Gough Is. >> S Atlantic
— *P.m.dulciae* Mathews, 1912   SW to SE Australia >> Indian and S Pacific Oceans
— *P.m.maoriana* Mathews, 1912   New Zealand islands, Auckland, Chatham Is. >> ?
— *P.m.albiclunis* Murphy & Irving, 1951   Kermadec Is. >> ?

FREGETTA Bonaparte, 1855 F
○ *Fregetta grallaria* White-bellied Storm Petrel
— *F.g.grallaria* (Vieillot, 1817)[3]   Admiralty, Lord Howe, Kermadec Is.
— *F.g.leucogaster* (Gould, 1844)[4]   Tristan da Cunha group, Gough, St. Paul, (?) Amster-
dam Is.
— *F.g.segethi* (Philippi & Landbeck, 1860)[5]   Juan Fernandez Is. (Chile)
— *F.g.titan* Murphy, 1928   Iles Australes

○ *Fregetta tropica* Black-bellied Storm Petrel
— *F.t.tropica* (Gould, 1844)[6]   Circumpolar. Scotia Arc to Antipodes Is. >> tropical
and subtropical parts of the three oceans
— *F.t.melanoleuca* Salvadori, 1908   Tristan da Cunha, Gough Is.

NESOFREGETTA Mathews, 1912 F
○ *Nesofregetta fuliginosa* (J.F. Gmelin, 1789) Polynesian Storm Petrel[7]   South-central equatorial Pacific Ocean from Vanuatu
east to Marquesas and Gambier Is.

HYDROBATINAE

HYDROBATES Boie, 1822 M
○ *Hydrobates pelagicus* (Linnaeus, 1758) European Storm Petrel   E North Atlantic, Mediterranean >> E South Atlantic

---

[1] Includes *hullianus* see Peters (1931) (1821).
[2] Includes *maoriana* see Jouanin & Mougin (1979: 104) (1229); considered a morph and thus not eligible for possible listing as an extinct race.
[3] Includes *royana* see Jouanin & Mougin (1979: 109) (1229).
[4] Omitted in error from last Edition.
[5] Submerged in *grallaria* in last Edition.
[6] Includes *lineata* see Jouanin & Mougin (1979: 108) (1229).
[7] Includes *moestissima* see Jouanin & Mougin (1979: 110) (1229).

OCEANODROMA Reichenbach, 1853 F

○ *Oceanodroma microsoma* (Coues, 1864) LEAST STORM PETREL[1] — Baja California >> local waters south to Ecuador

○ *Oceanodroma tethys* WEDGE-RUMPED STORM PETREL

     — *O.t.tethys* (Bonaparte, 1852) — Galapagos Is. >> Humboldt Current seas

     — *O.t.kelsalli* (P.R. Lowe, 1925) — Coast of Peru >> Humboldt Current seas

○ *Oceanodroma castro* (Harcourt, 1851) BAND-RUMPED STORM PETREL — E Atlantic Is. (Azores and Madeira to St Helena I.);
           NW to SE Pacific (Japan, Hawaii to Galapagos Is.)

○ *Oceanodroma monorhis* (Swinhoe, 1867) SWINHOE'S STORM PETREL — Japan to Yellow Sea, N to SE USSR >> N Indian Ocean

○ *Oceanodroma leucorhoa* LEACH'S STORM PETREL

     — *O.l.leucorhoa* (Vieillot, 1818)[2] — N Pacific and N Atlantic coasts >> tropical waters

     — *O.l.chapmani* Berlepsch, 1906[3] — San Benito and Coronados Is. (W Mexico)

     — *O.l.cheimomnestes* Ainley, 1980 — Guadalupe I. (Baja California) (winter breeder) (19)

     — *O.l.socorroensis* C.H. Townsend, 1890[4] — Islets off Guadalupe I. (summer breeder)

○ †?*Oceanodroma macrodactyla* W.E. Bryant, 1887 GUADALUPE STORM PETREL
         Guadalupe I. (Baja California)

○ *Oceanodroma markhami* (Salvin, 1883) MARKHAM'S STORM PETREL — Central W coast Sth. America >> WC Pacific

○ *Oceanodroma tristrami* Salvin, 1896 TRISTRAM'S STORM PETREL[5] — Japan, Laysan I., Midway I. >> local seas

○ *Oceanodroma melania* (Bonaparte, 1854) BLACK STORM PETREL — Baja California >> tropical W Pacific

○ *Oceanodroma matsudairae* N. Kuroda, Sr., 1922 MATSUDAIRA'S STORM PETREL
         Ogasawara-shoto (Bonin) Is. (Japan) >> Timor Sea
           and tropical waters Indian Ocean

○ *Oceanodroma homochroa* (Coues, 1864) ASHY STORM PETREL — Islands off California to Coronados Is. >> nearby seas

○ *Oceanodroma hornbyi* (G.R. Gray, 1854) HORNBY'S STORM PETREL / RINGED STORM PETREL
         Humboldt Current (W Sth. America)

○ *Oceanodroma furcata* FORK-TAILED STORM PETREL[6]

     — *O.f.furcata* (J.F. Gmelin, 1789) — N Kuril, Commander and Aleutian Is. >> N Pacific

     — *O.f.plumbea* (Peale, 1848) — Alaska to N California >> N Pacific Ocean

PELECANOIDIDAE DIVING PETRELS (1: 4)

PELECANOIDES Lacépède, 1799 M

○ *Pelecanoides garnotii* (Lesson, 1828) PERUVIAN DIVING PETREL — Coasts of Peru and Chile

○ *Pelecanoides magellani* (Mathews, 1912) MAGELLANIC DIVING PETREL — S Chile to Cape Horn

○ *Pelecanoides georgicus* Murphy & Harper, 1916[7] SOUTH GEORGIA DIVING PETREL
         Antarctic waters near colonies: South Georgia I.,
           S Indian Ocean Is., islands off S New Zealand

○ *Pelecanoides urinatrix* COMMON DIVING PETREL[8]

     — *P.u.dacunhae* Nicholl, 1906[9] — Tristan da Cunha, Gough I.

     — *P.u.berard* (Gaimard, 1823) — Falkland Is.

     — *P.u.urinatrix* (J.F. Gmelin, 1789) — SE Australia, Tasmania, N New Zealand

     — *P.u.chathamensis* Murphy & Harper, 1916 — Chatham, Snares, Stewart Is.

     — *P.u.exsul* Salvin, 1896 — South Georgia, Crozet, Marion, Heard, Kerguelen,
           Macquarie, Antipodes, Auckland Is.

     — *P.u.coppingeri* Mathews, 1912 — S Chile

---

[1] Treated in the genus *Halocyptena* in last Edition, but see A.O.U. (1998: 26) (3).
[2] Includes *beali* see Ainley (1980) (19).
[3] Ainley (1980) (19) placed this in the synonymy of the nominate form. Distinction between the two is not reliable for all birds, see Ainley (1983) (20) and Bourne & Jehl (1982) (205). Includes *willetti* see Bourne & Jehl (1982) (205).
[4] Includes *keadingi* see Jouanin & Mougin (1979: 115) (1229).
[5] Includes *owstoni* see Jouanin & Mougin (1979: 116) (1229).
[6] Treated as monotypic in last Edition.
[7] In last Edition a race *novus* was included in error.
[8] In Jouanin & Mougin (1979: 120) (1229) named *urinator*, but see David & Gosselin (2002) (613).
[9] Includes *elizabethae* see Jouanin & Mougin (1979: 120) (1229).

PODICIPEDIDAE GREBES (6: 22)

TACHYBAPTUS Reichenbach, 1853 M

○ †?*Tachybaptus rufolavatus* Delacour, 1932 ALAOTRA GREBE Lake Alaotra (Madagascar)

● *Tachybaptus ruficollis* LITTLE GREBE

  ✔ *T.r.ruficollis* (Pallas, 1764) Europe to Urals and N Caucasus, NW Africa

  — *T.r.iraquensis* (Ticehurst, 1923) Iraq, SW Iran

  — *T.r.capensis* (Salvadori, 1884) Caucasus area, SW and C Asia to Tien Shan, S Asia to Burma, Ghana to Ethiopia and Cape Province, Madagascar

  — *T.r.poggei* (Reichenow, 1902)[1] Ladakh to C and NE China and Japan, south to Malay Pen., Indochina and E China

  — *T.r.philippensis* (Bonnaterre, 1791) Taiwan, Philippines (excl. Mindanao)

  — *T.r.cotabato* (Rand, 1948) Mindanao

  — *T.r.tricolor* (G.R. Gray, 1861)[2] Java, the Lesser Sundas and Sulawesi to New Guinea

  — *T.r.collaris* (Mayr, 1945)[3] NE New Guinea to Bougainville I. (Solomons)

○ *Tachybaptus novaehollandiae* AUSTRALASIAN GREBE[4]

  — *T.n.javanicus* (Mayr, 1943) Java

  — *T.n.fumosus* (Mayr, 1943) Talaud Is., Sangihe Is.?

  — *T.n.incola* (Mayr, 1943) N New Guinea

  — *T.n.novaehollandiae* (Stephens, 1826)[5] Australia, New Zealand, S New Guinea, vagrant (?) to Timor

  — *T.n.leucosternos* (Mayr, 1931) Vanuatu, New Caledonia

  — *T.n.rennellianus* (Mayr, 1931) Rennell I.

○ *Tachybaptus pelzelnii* Hartlaub, 1861 MADAGASCAN GREBE Madagascar

● *Tachybaptus dominicus* LEAST GREBE[6]

  ✔ *T.d.brachypterus* (Chapman, 1899) S Texas to Panama

  — *T.d.bangsi* (van Rossem & Hachisuka, 1937) S Baja California

  — *T.d.dominicus* (Linnaeus, 1766) Gtr. Antilles

  — *T.d.speciosus* (Lynch Arribálzaga, 1877) South America (except Ecuador) S to N Argentina

  — *T.d.eisenmanni* Storer & Getty, 1985 W Ecuador (2352)

PODILYMBUS Lesson, 1831 M

● *Podilymbus podiceps* PIED-BILLED GREBE

  ✔ *P.p.podiceps* (Linnaeus, 1758) Canada and USA to W Panama >> C America and West Indies

  — *P.p.antillarum* Bangs, 1913 Gtr. and Lesser Antilles

  ✔ *P.p.antarcticus* (Lesson, 1842) E Panama, Colombia and Venezuela to S Argentina

○ †?*Podilymbus gigas* Griscom, 1929 ATITLAN GREBE Lake Atitlan (Guatemala)

ROLLANDIA Bonaparte, 1856 F

○ *Rollandia rolland* WHITE-TUFTED GREBE

  — *R.r.morrisoni* (Simmons, 1962) C Peru

  — *R.r.chilensis* (Lesson, 1828) S Brazil and NW Peru to Tierra del Fuego

  — *R.r.rolland* (Quoy & Gaimard, 1824) Falkland Is.

○ *Rollandia microptera* (Gould, 1868) SHORT-WINGED GREBE SE Peru and W Bolivia

POLIOCEPHALUS Selby, 1840[7] M

● *Poliocephalus poliocephalus* (Jardine & Selby, 1827) HOARY-HEADED GREBE Australia, Tasmania

○ *Poliocephalus rufopectus* G.R. Gray, 1843 NEW ZEALAND GREBE New Zealand (mainly North Island)

[1] Includes *kunikyonis* see Vaurie (1965: 11) (2510).
[2] Includes *vulcanorum* see White & Bruce (1986: 89) (2652).
[3] This race accidentally omitted in last Edition.
[4] In our last Edition the separation of this species from *T. ruficollis* was mishandled, limiting it to a single race, see Mayr (1943) (1473)
[5] Includes *timorensis* see White & Bruce (1986: 89) (2652).
[6] In last Edition listed in the genus *Podiceps*, but see Storer (1979: 145) (2350).
[7] We follow Storer (1979: 147) (2350) in the generic name used.

PODICEPS Latham, 1787 M

◐ *Podiceps major* GREAT GREBE[1]

    ☑ *P.m.major* (Boddaert, 1783)   Paraguay, SE Brazil to S Argentina and C Chile >> coasts of W Peru, Chile, Argentina

    __ *P.m.navasi* Manghi, 1984   S Chile (1417)

◐ *Podiceps grisegena* RED-NECKED GREBE

    __ *P.g.grisegena* (Boddaert, 1783)   N and C Europe to SW Siberia and Middle East >> N Africa, W and S Europe, SW Asia

    ☑ *P.g.holboellii* Reinhardt, 1854   N America, NE Asia >> China, Japan, S USA

◐ *Podiceps cristatus* GREAT CRESTED GREBE

    ☑ *P.c.cristatus* (Linnaeus, 1758)   Europe to Japan, south to N Africa, Middle East and N India >> southern parts of range

    __ *P.c.infuscatus* Salvadori, 1884   Ethiopia, E and S Africa

    ☑ *P.c.australis* Gould, 1844[2]   S Australia, Tasmania, South Island (New Zealand)

◐ *Podiceps auritus* SLAVONIAN GREBE/HORNED GREBE

    ☑ *P.a.auritus* (Linnaeus, 1758)   Iceland and N Scotland to Kamchatka >> W Europe and C and E Asia

    ☑ *P.a.cornutus* (J.F. Gmelin, 1789)[3]   Canada, N USA >> North America south to Texas

◐ *Podiceps nigricollis* BLACK-NECKED GREBE/EARED GREBE

    ☑ *P.n.nigricollis* C.L. Brehm, 1831   Europe to C Asia >> W and SW Palaearctic, C and E Asia

    __ *P.n.gurneyi* (Roberts, 1919)   Ethiopia, to N Tanzania, Namibia to Transvaal and Cape Province

    ☑ *P.n.californicus* Heermann, 1854   S Canada, W and C USA, and N Mexico >> USA and C America

○ †?*Podiceps andinus* Meyer de Schauensee, 1959 COLOMBIAN GREBE[4]   C Colombia

◐ *Podiceps occipitalis* SILVERY GREBE

    __ *P.o.juninensis* Berlepsch & Stolzmann, 1894   Andes from Colombia to NW Argentina

    ☑ *P.o.occipitalis* Garnot, 1826   C and S Chile, C and S Argentina, Falkland Is.

○ *Podiceps taczanowskii* Berlepsch & Stolzmann, 1894 JUNIN FLIGHTLESS GREBE

                                        Lake Junin (Peru)

○ *Podiceps gallardoi* Rumboll, 1974 HOODED GREBE   SW Argentina

AECHMOPHORUS Coues, 1862 M

◐ *Aechmophorus occidentalis* WESTERN GREBE

    ☑ *A.o.occidentalis* (Lawrence, 1858)   SW Canada, NW USA >> Pacific coast
    __ *A.o.ephemeralis* Dickerman, 1986   SC and WC Mexico (715)

◐ *Aechmophorus clarkii* CLARK'S GREBE

    __ *A.c.clarkii* (Lawrence, 1858)   N and C Mexico
    ☑ *A.c.transitionalis* Dickerman, 1986   W USA (715)

PHOENICOPTERIDAE FLAMINGOS (3: 5)

PHOENICOPTERUS Linnaeus, 1758 M

◐ *Phoenicopterus ruber* GREATER FLAMINGO

    ☑ *P.r.ruber* Linnaeus, 1758   Caribbean coasts of Central and South America, West Indies

    __ *P.r.roseus* Pallas, 1811   S Europe, C Asia, NW India, N, E and S Africa

○ *Phoenicopterus chilensis* Molina, 1782 CHILEAN FLAMINGO   Peru and Uruguay to Tierra del Fuego

PHOENICONAIAS G.R. Gray, 1869 F

○ *Phoeniconaias minor* (E. Geoffroy Saint-Hilaire, 1798) LESSER FLAMINGO

                                        Mauritania, S and E Africa, NW India >> sub-Saharan Africa, Madagascar and S Asia

[1] Storer (1979: 149) (2350) footnoted the distinctness of this species. See also Bochenski (1994) (172).
[2] Includes *christiani* see Vaurie (1965) (2510).
[3] For recognition of this see Storer (1979: 150) (2350).
[4] For elevation to specific rank see Fjeldså (1982) (841) and A.O.U. (1998: 8) (3).

PHOENICOPARRUS Bonaparte, 1856 M
○ *Phoenicoparrus andinus* (Philippi, 1854) ANDEAN FLAMINGO | Andes of Bolivia, Chile and NW Argentina

○ *Phoenicoparrus jamesi* (P.L. Sclater, 1886) JAMES'S FLAMINGO | Andes of S Peru, Bolivia, N Chile, NW Argentina

CICONIIDAE STORKS (6: 19)

MYCTERIA Linnaeus, 1758 F
● *Mycteria americana* Linnaeus, 1758 WOOD STORK | SE USA to C South America

○ *Mycteria cinerea* (Raffles, 1822) MILKY STORK | Indochina, W Malaysia, Sumatra, Java and Bali

○ *Mycteria ibis* (Linnaeus, 1766) YELLOW-BILLED STORK | Senegal to Eritrea and Cape Province

○ *Mycteria leucocephala* (Pennant, 1769) PAINTED STORK | India and Sri Lanka to SE China, Indochina

ANASTOMUS Bonnaterre, 1791 M
○ *Anastomus oscitans* (Boddaert, 1783) ASIAN OPENBILL | India and Sri Lanka to Indochina

○ *Anastomus lamelligerus* AFRICAN OPENBILL
    — *A.l.lamelligerus* Temminck, 1823 | Mali to Ethiopia, Botswana and Natal
    — *A.l.madagascariensis* A. Milne-Edwards, 1880 | W Madagascar

CICONIA Brisson, 1760 F
● *Ciconia nigra* (Linnaeus, 1758) BLACK STORK | a) Europe to N China >> N, E Africa, S Asia east to Indochina
    | b) S Africa >> E Africa

○ *Ciconia abdimii* M.H.K. Lichtenstein, 1823 ABDIM'S STORK | Ethiopia to Angola >> south to S Africa

○ *Ciconia episcopus* WOOLLY-NECKED STORK
    — *C.e.microscelis* G.R. Gray, 1848 | Gambia to Eritrea south to N Botswana and Natal
    — *C.e.episcopus* (Boddaert, 1783) | India and Sri Lanka to SE Asia, Sumatra and the Philippines
    — *C.e.neglecta* (Finsch, 1904) | Java, Lesser Sundas, Sulawesi

○ *Ciconia stormi* (W.H. Blasius, 1896) STORM'S STORK | SW Thailand, Malay Pen., Borneo

○ *Ciconia maguari* (J.F. Gmelin, 1789) MAGUARI STORK | E Colombia, Venezuela, Guianas; C Brazil and E Bolivia to Paraguay and N Argentina

● *Ciconia ciconia* EUROPEAN WHITE STORK
    ✓ *C.c.ciconia* (Linnaeus, 1758) | a) Europe, Middle East and N Africa >> E, S Africa
    | b) S Africa >> E Africa
    — *C.c.asiatica* Severtsov, 1873 | C Asia >> India

○ *Ciconia boyciana* Swinhoe, 1873 ORIENTAL WHITE STORK | SE Siberia, NE China, Korea, formerly Japan >> S China

EPHIPPIORHYNCHUS Bonaparte, 1855 M
○ *Ephippiorhynchus asiaticus* BLACK-NECKED STORK
    — *E.a.asiaticus* (Latham, 1790) | India to Malay Pen. and Vietnam
    — *E.a.australis* (Shaw, 1800) | New Guinea, N and E Australia

○ *Ephippiorhynchus senegalensis* (Shaw, 1800) SADDLE-BILL STORK | Senegal to Eritrea south to N Botswana and N Natal

JABIRU Hellmayr, 1906 F
○ *Jabiru mycteria* (M.H.K. Lichtenstein, 1819) JABIRU | S Mexico to C Argentina

LEPTOPTILOS Lesson, 1831 M
○ *Leptoptilos javanicus* (Horsfield, 1821) LESSER ADJUTANT | C India to S China, Gtr. Sundas

○ *Leptoptilos dubius* (J.F. Gmelin, 1789) GREATER ADJUTANT | India (formerly east to Indochina and Gtr. Sundas)

○ *Leptoptilos crumeniferus* (Lesson, 1831) MARABOU | Senegal to Eritrea south to Botswana and Natal

81

THRESKIORNITHIDAE IBISES AND SPOONBILLS (14: 32)

THRESKIORNITHINAE

THRESKIORNIS G.R. Gray, 1842 M
O ***Threskiornis aethiopicus*** SACRED IBIS
    — *T.a.aethiopicus* (Latham, 1790)   Sub-Saharan Africa, Iraq
    — *T.a.abbotti* (Ridgway, 1893)   Aldabra I.
    — *T.a.bernieri* (Bonaparte, 1855)[1]   W Madagascar

O ***Threskiornis melanocephalus*** (Latham, 1790) BLACK-HEADED IBIS   India to China, SE Asia and Java >> S Asia and Philippines

● ***Threskiornis molucca*** AUSTRALIAN IBIS
    ✓ *T.m.molucca* (Cuvier, 1829)[2]   E Lesser Sundas, Moluccas, Kai Is., S New Guinea,
                                          Australia
    — *T.m.pygmaeus* Mayr, 1931[3]   Rennell and Bellona Is. (Solomons)

O ***Threskiornis spinicollis*** (Jameson, 1835) STRAW-NECKED IBIS   S New Guinea, Australia, Tasmania

PSEUDIBIS Hodgson, 1844 F
O ***Pseudibis papillosa*** (Temminck, 1824) INDIAN BLACK IBIS   N India, Nepal, Pakistan

O ***Pseudibis davisoni*** (Hume, 1875) WHITE-SHOULDERED IBIS[4]   S Vietnam, Borneo

O ***Pseudibis gigantea*** (Oustalet, 1877) GIANT IBIS   S Thailand, S Indochina

GERONTICUS Wagler, 1832 M
O ***Geronticus eremita*** (Linnaeus, 1758) NORTHERN BALD IBIS   Morocco, Turkey >> NE Africa

O ***Geronticus calvus*** (Boddaert, 1783) SOUTHERN BALD IBIS   S Africa

NIPPONIA Reichenbach, 1850[5] F
O ***Nipponia nippon*** (Temminck, 1835) CRESTED IBIS   E China, formerly Japan

BOSTRYCHIA Reichenbach, 1853[6] F
O ***Bostrychia olivacea*** OLIVE IBIS
    — *B.o.olivacea* (Du Bus, 1838)[7]   Sierra Leone to Ivory Coast
    — *B.o.cupreipennis* (Reichenow, 1903)   S Cameroon to Zaire
    — †?*B.o.rothschildi* (Bannerman, 1919)   Principe I.
    — *B.o.bocagei* (Chapin, 1923)   São Tomé I.
    — *B.o.akleyorum* (Chapman, 1912)   Mts. of Kenya and Tanzania

O ***Bostrychia rara*** (Rothschild, Hartert & Kleinschmidt, 1897) SPOT-BREASTED IBIS
                                          Liberia to E Zaire and Angola

O ***Bostrychia hagedash*** HADADA IBIS
    — *B.h.nilotica* Neumann, 1909   Sudan and Ethiopia to Uganda and NW Tanzania
    — *B.h.brevirostris* (Reichenow, 1907)[8]   Gambia to Zaire, Kenya, Somalia, E and S Tanzania
                                          and Zambia
    — *B.h.hagedash* (Latham, 1790)   Southern Africa

O ***Bostrychia carunculata*** (Rüppell, 1837) WATTLED IBIS   Ethiopia

THERISTICUS Wagler, 1832[9] M
O ***Theristicus caerulescens*** Vieillot, 1817 PLUMBEOUS IBIS   C Brazil and E Bolivia to N Argentina

---

[1] Sometimes treated as a separate species, e.g. by Roselaar in Cramp *et al.* (1977: 351) (588), but no detailed review published.
[2] Includes *strictipennis* see Mees (1982) (1533).
[3] For recognition see Payne (1979: 264) (1797).
[4] For arguments supporting separation from *P. papillosa* see Sözer *et al.* (1997) (2300).
[5] Dated 1853 by Steinbacher (1979: 266) (2315) but see Dekker *et al.* (2001) (672); many other names – including *Bostrychia* — may be affected and need review.
[6] The expansion of this genus follows Payne (1979) (1797).
[7] Du Bus is here abbreviated from Du Bus de Gisignies.
[8] Includes *erlangeri* see Payne (1979: 260) (1797).
[9] The expansion of this genus follows Payne (1979) (1797).

○ *Theristicus caudatus* BUFF-NECKED IBIS
  — *T.c.caudatus* (Boddaert, 1783)   E Colombia, Venezuela, Guianas, N Brazil
  — *T.c.hyperorius* Todd, 1948   E Bolivia, S Brazil, Paraguay, N Argentina

◐ *Theristicus melanopis* BLACK-FACED IBIS
  — *T.m.branickii* Berlepsch & Stolzmann, 1894   Andes of Ecuador, Peru, N Bolivia and N Chile
  ✓ *T.m.melanopis* (J.F. Gmelin, 1789)   SW Peru, W Chile, S Argentina

CERCIBIS Wagler, 1832 F
○ *Cercibis oxycerca* (Spix, 1825) SHARP-TAILED IBIS   SE Colombia to Surinam, N Brazil

MESEMBRINIBIS J.L. Peters, 1930 F
○ *Mesembrinibis cayennensis* (J.F. Gmelin, 1789) GREEN IBIS   Costa Rica to Paraguay, NE Argentina

PHIMOSUS Wagler, 1832 M
○ *Phimosus infuscatus* BARE-FACED IBIS
  — *P.i.berlepschi* Hellmayr, 1903   N Colombia, Venezuela, Guianas, N Brazil
  — *P.i.nudifrons* (Spix, 1825)   C and S Brazil
  — *P.i.infuscatus* (M.H.K. Lichtenstein, 1823)   E Bolivia, Paraguay, Uruguay, C Argentina

EUDOCIMUS Wagler, 1832 M
◉ *Eudocimus albus* (Linnaeus, 1758) WHITE IBIS   SE USA to N Sth. America

○ *Eudocimus ruber* (Linnaeus, 1758) SCARLET IBIS   N South America, Trinidad

PLEGADIS Kaup, 1829 M
◉ *Plegadis falcinellus* (Linnaeus, 1766) GLOSSY IBIS[1]   S Europe, Africa, SW and S Asia to Java, Sulawesi,
                  Philippines and Australia; Central America, N Sth.
                  America

◉ *Plegadis chihi* (Vieillot, 1817) WHITE-FACED IBIS   W and S USA to SC South America

○ *Plegadis ridgwayi* (J.A. Allen, 1876) PUNA IBIS   Andes of Peru, Bolivia, N Chile, NW Argentina

LOPHOTIBIS Reichenbach, 1853 F
○ *Lophotibis cristata* MADAGASCAN CRESTED IBIS
  — *L.c.cristata* (Boddaert, 1783)   N and E Madagascar
  — *L.c.urschi* Lavauden, 1929   W and S Madagascar

PLATALEINAE

PLATALEA Linnaeus, 1758 F
◉ *Platalea leucorodia* EURASIAN SPOONBILL
  ✓ *P.l.leucorodia* Linnaeus, 1758[2]   Europe to N China and Sri Lanka >> Africa (mainly
                 north of the Equator) and S Asia

  — *P.l.balsaci* Naurois & Roux, 1974   W Mauritania
  — *P.l.archeri* Neumann, 1928   Red Sea coasts, Somalia

○ *Platalea minor* Temminck & Schlegel, 1849[3] BLACK-FACED SPOONBILL   E and NE China, Korea >> S China, Taiwan, Indochina
                 and formerly Philippines

○ *Platalea alba* Scopoli, 1786 AFRICAN SPOONBILL   Senegal to Eritrea and south to Cape Province

◉ *Platalea regia* Gould, 1838 ROYAL SPOONBILL   Java, Lesser Sundas, Sulawesi, Moluccas, S New
                 Guinea, Australia, New Zealand to Solomons

○ *Platalea flavipes* Gould, 1838 YELLOW-BILLED SPOONBILL   Australia

AJAIA Reichenbach, 1853[4] F
◉ *Ajaia ajaja* (Linnaeus, 1758) ROSEATE SPOONBILL   S USA to C Argentina and Chile

---

[1] For monotypic treatment see Amadon & Woolfenden (1952) (49).
[2] Includes *major* see Payne (1979: 267) (1797).
[3] Should probably be dated 1850 see Holthuis & Sakai (1970) (1128).
[4] Although placed in *Platalea* by Payne (1979: 268) (1797), A.O.U. (1998: 50) (3) retained the genus *Ajaia*.

ARDEIDAE HERONS, BITTERNS AND EGRETS[1] (19: 65)

TIGRISOMATINAE

ZONERODIUS Salvadori, 1882 M
○ **Zonerodius heliosylus** (Lesson, 1828) FOREST BITTERN — New Guinea, Aru Is.

TIGRIORNIS Sharpe, 1895 F
○ **Tigriornis leucolopha** (Jardine, 1846) WHITE-CRESTED TIGER HERON[2] — Sierra Leone to SW and NE Zaire

TIGRISOMA Swainson, 1827 N
○ **Tigrisoma lineatum** RUFESCENT TIGER HERON
    — T.l.lineatum (Boddaert, 1783) — Honduras south through C America and S America to N Bolivia and Amazonian Brazil
    — T.l.marmoratum (Vieillot, 1817) — E and S Brazil, Paraguay, Uruguay, N Argentina

○ **Tigrisoma fasciatum** FASCIATED TIGER HERON
    — T.f.salmoni Sclater & Salvin, 1875 — Costa Rica, Panama, Colombia to Bolivia
    — T.f.fasciatum (Such, 1825) — SE Brazil, N Argentina
    — T.f.pallescens Olrog, 1950 — NW Argentina

● **Tigrisoma mexicanum** Swainson, 1834 BARE-THROATED TIGER HERON — W Mexico to E Panama, NW Colombia

AGAMIA Reichenbach, 1853 F
○ **Agamia agami** (J.F. Gmelin, 1789) AGAMI HERON — E Mexico to Brazil and N Bolivia

COCHLEARIINAE

COCHLEARIUS Brisson, 1760 M
● **Cochlearius cochlearius** BOAT-BILLED HERON
    — C.c.zeledoni (Ridgway, 1885) — W Mexico
    — C.c.phillipsi Dickerman, 1973 — S Mexico, Belize, N Guatemala
    — C.c.ridgwayi Dickerman, 1973 — Guatemala, Honduras
    ↙ C.c.panamensis Griscom, 1926 — S Costa Rica, Panama
    — C.c.cochlearius (Linnaeus, 1766) — N and C South America

BOTAURINAE

ZEBRILUS Bonaparte, 1855 M
○ **Zebrilus undulatus** (J.F. Gmelin, 1789) ZIGZAG HERON — Colombia, Venezuela, the Guianas south to N Bolivia and C Brazil

BOTAURUS Stephens, 1819 M
○ **Botaurus stellaris** EURASIAN BITTERN
    — B.s.stellaris (Linnaeus, 1758) — Europe to E Asia >> N and C Africa
    — B.s.capensis (Schlegel, 1863) — S Tanzania to E Angola and Cape Province

○ **Botaurus poiciloptilus** (Wagler, 1827) AUSTRALASIAN BITTERN — S Australia, Tasmania, New Zealand

● **Botaurus lentiginosus** (Rackett, 1813) AMERICAN BITTERN — Alaska, Canada to SW and NE USA >> C America

○ **Botaurus pinnatus** PINNATED BITTERN
    — B.p.caribaeus Dickerman, 1961 — SE Mexico, Belize
    — B.p.pinnatus (Wagler, 1829) — El Salvador, Nicaragua to S Brazil, Paraguay and N Argentina

IXOBRYCHUS Billberg, 1828 M
○ **Ixobrychus involucris** (Vieillot, 1823) STRIPE-BACKED BITTERN — E Columbia, Guianas, Trinidad, S Brazil to Patagonia

● **Ixobrychus exilis** LEAST BITTERN
    ↙ I.e.exilis (J.F. Gmelin, 1789) — Nth. America >> C America, West Indies

---

[1] Generic sequence based on Sheldon *et al.* (1995) (2237), McCracken & Sheldon (1998) (1500), and Sheldon *et al.* (2000) (2239). Species arrangements therein largely follow Martínez-Vilalta & Motis (1992) (1447).
[2] For correction of spelling see David & Gosselin (2002) (614).

— *I.e.pullus* van Rossem, 1930 — S Sonora (Mexico)

— *I.e.erythromelas* (Vieillot, 1817) — E Panama, Colombia, Venezuela, Trinidad, the Guianas to N Bolivia, Paraguay, and N Argentina

— *I.e.limoncochae* Norton, 1965[1] — E Ecuador

— *I.e.bogotensis* Chapman, 1914 — C Colombia

— *I.e.peruvianus* Bond, 1955 — Coast of Peru

○ *Ixobrychus minutus* LITTLE BITTERN

— *I.m.minutus* (Linnaeus, 1766) — S and C Europe to C Asia and NW India >>Africa

— *I.m.payesii* (Hartlaub, 1858) — Senegal to Yemen and Cape Province

— *I.m.podiceps* (Bonaparte, 1855) — Madagascar

— *I.m.dubius* Mathews, 1912 — E and SW Australia

○ †*Ixobrychus novaezelandiae* (Potts, 1871) NEW ZEALAND BITTERN[2] — New Zealand

○ *Ixobrychus sinensis* (J.F. Gmelin, 1789) YELLOW BITTERN — Manchuria, Japan >> India, Sundas, Philippines

○ *Ixobrychus eurhythmus* (Swinhoe, 1873) SCHRENK'S BITTERN — E Asia >> Sundas and Philippines

○ *Ixobrychus cinnamomeus* (J.F. Gmelin, 1789) CINNAMON BITTERN — S and E Asia to Philippines and Sulawesi

○ *Ixobrychus sturmii* (Wagler, 1827) DWARF BITTERN — Senegal to Ethiopia and Cape Province

DUPETOR Heine & Reichenow, 1890 M
○ *Dupetor flavicollis* BLACK BITTERN

— *D.f.flavicollis* (Latham, 1790) — S Asia to Gtr. Sundas, Philippines and Sulawesi

— *D.f.australis* (Lesson, 1831)[3] — Moluccas, Timor, Kai Is., Australia, New Guinea, Bismarck Arch.

— *D.f.woodfordi* (Ogilvie-Grant, 1888) — Solomon Is.

ARDEINAE

GORSACHIUS Bonaparte, 1855 M
○ *Gorsachius magnificus* (Ogilvie-Grant, 1899) WHITE-EARED NIGHT HERON — SE China, Hainan I.

○ *Gorsachius goisagi* (Temminck, 1836[4]) JAPANESE NIGHT HERON — Japan >> E China, Japan

○ *Gorsachius melanolophus* (Raffles, 1822) MALAYSIAN NIGHT HERON — India to S Ryukyu Is., SE Asia, Gtr. Sundas, Philippines, Christmas I. (Indian Ocean)

○ *Gorsachius leuconotus* (Wagler, 1827) WHITE-BACKED NIGHT HERON — Senegal to W Ethiopia and Cape Province

NYCTICORAX T. Forster, 1817 M
● *Nycticorax nycticorax* BLACK-CROWNED NIGHT HERON

— *N.n.nycticorax* (Linnaeus, 1758) — C and S Europe east to Kazakhstan and through S and SE Asia to E China and Japan, Taiwan, the Sundas and Philippines, N Africa >> C Africa and SE Asia

✓ *N.n.hoactli* (J.F. Gmelin, 1789) — SE Canada to Argentina, Hawaii Is.

✓ *N.n.obscurus* Bonaparte, 1855 — SW Peru, Chile, SW Argentina

— *N.n.falklandicus* E. Hartert, 1914 — Falkland Is.

○ *Nycticorax caledonicus* RUFOUS NIGHT HERON

— †*N.c.crassirostris* Vigors, 1839 — Ogasawara-shoto (Bonin) Is. (Japan)

— *N.c.manillensis* Vigors, 1831 — Java, N Borneo, Sulawesi, Philippines

— *N.c.hilli* Mathews, 1912 — Wallacea, New Guinea, Australia, New Zealand

— *N.c.pelewensis* Mathews, 1926 — Palau Is., Caroline Is.

— *N.c.caledonicus* (J.F. Gmelin, 1789) — New Caledonia

— *N.c.mandibularis* Ogilvie-Grant, 1888 — Bismarck Arch., Solomon Is.

---

[1] Omitted from last Edition.
[2] For reasons to separate this distinct form see Marchant & Higgins (1990: 1045) (1427).
[3] White & Bruce (1986: 109) (2652) recognised *gouldi* restricting *australis* to Timor; unresolved.
[4] Livr. 98 must date from 1836 not 1835 (Dickinson, 2001) (739).

NYCTANASSA Stejneger, 1887[1] F
● *Nyctanassa violacea* YELLOW-CROWNED NIGHT HERON

✔ *N.v.violacea* (Linnaeus, 1758)	EC USA, E Central America
— *N.v.bancrofti* Huey, 1927[2]	W Baja California, Socorro I., West Indies
— *N.v.caliginis* Wetmore, 1946	Panama and Columbia on Pacific coast to NW Peru
— *N.v.cayennensis* (J.F. Gmelin, 1789)	Panama and Columbia on Atlantic coast east to N Brazil
— *N.v.pauper* (Sclater & Salvin, 1870)	Galapagos Is.

BUTORIDES Blyth, 1852 F[3]
● *Butorides virescens* GREEN HERON

✔ *B.v.anthonyi* (Mearns, 1895)	SW USA, NW Mexico incl. N Baja California
— *B.v.frazari* (Brewster, 1888)	S Baja California
✔ *B.v.virescens* (Linnaeus, 1758)	E Nth. America and C America >> Panama
— *B.v.maculata* (Boddaert, 1783)[4]	Bahama Is., the Antilles, Cayman Is. and Tobago

○ *Butorides sundevalli* (Reichenow, 1877) LAVA HERON[5] · Galapagos Is.

○ *Butorides striata* STRIATED HERON[6]

— *B.s.striata* (Linnaeus, 1758)	E Panama to N Argentina, Bolivia
— *B.s.atricapilla* (Afzelius, 1804)	Senegal to Sudan and Cape Province
— *B.s.brevipes* (Hemprich & Ehrenberg, 1833)	Somalia, Red Sea coast
— *B.s.rutenbergi* (Hartlaub, 1880)	Madagascar, Réunion
— *B.s.rhizophorae* Salomonsen, 1934	Comoro Is.
— *B.s.crawfordi* Nicoll, 1906	Amirantes, Aldabra Is.
— *B.s.degens* E. Hartert, 1920	Seychelles
— *B.s.albolimbata* Reichenow, 1900	Chagos and Maldive Is.
— *B.s.chloriceps* (Bonaparte, 1855)	Pakistan, India, Sri Lanka, Laccadive Is.
— *B.s.spodiogaster* Sharpe, 1894	Andamans, Nicobars, W Sumatran Is.
— *B.s.amurensis* (von Schrenk, 1860)	SE Siberia, NE China, Japan >> Philippines, Sulawesi, Gtr. Sundas
— *B.s.actophila* Oberholser, 1912	S China, N Indochina >> SE Asia to Gtr. Sundas
— *B.s.javanica* (Horsfield, 1821)[7]	Burma, Thailand, S Indochina, Hainan, Taiwan, Malay Pen., Gtr. and Lesser Sundas, Sulawesi, Philippines
— *B.s.moluccarum* E. Hartert, 1920	Moluccas and Kai Is.
— *B.s.papuensis* Mayr, 1940	NW New Guinea
— *B.s.idenburgi* Rand, 1941	N, E and SE New Guinea
— *B.s.macrorhyncha* (Gould, 1848)[8]	E and NE Australia
— *B.s.stagnatilis* (Gould, 1848)[9]	NW and NC Australia
— *B.s.patruelis* (Peale, 1848)	Tahiti Is.
— *B.s.solomonensis* Mayr, 1940	Solomon Is.

ARDEOLA Boie, 1822 F

○ *Ardeola ralloides* (Scopoli, 1769) SQUACCO HERON	S Europe to Kazakhstan; C Africa and Madagascar
○ *Ardeola grayii* (Sykes, 1832) INDIAN POND HERON	Iran to India, Sri Lanka, Burma and islands in N Indian Ocean
○ *Ardeola bacchus* (Bonaparte, 1855) CHINESE POND HERON	Inner Mongolia through China to E India and SE Asia to Sulawesi

○ *Ardeola speciosa* JAVAN POND HERON

— *A.s.continentalis* Salomonsen, 1933	SE Burma, C Thailand, Cambodia, S Vietnam
— *A.s.speciosa* (Horsfield, 1821)	Borneo, Sumatra, Java, E Lesser Sundas, Sulawesi, Mindanao

○ *Ardeola idae* (Hartlaub, 1860) MADAGASCAN POND HERON	Aldabra, Madagascar >> C and E Africa
○ *Ardeola rufiventris* (Sundevall, 1851) RUFOUS-BELLIED HERON	S Uganda, Tanzania to N Botswana and Natal

---

[1] For reasons to use *Nyctanassa* see A.O.U. (1998: 47) (3).
[2] Includes *gravirostris* see Payne (1979: 228) (1797).
[3] Gender addressed by David & Gosselin (2002) (614), multiple changes occur in specific names.
[4] Includes *bahamensis* (C.S.R.).
[5] Implicitly recognised by A.O.U. (1998: 45) (3).
[6] For reasons to separate *B. virescens* and *B. striata* see Monroe & Browning (1992) (1605).
[7] Includes *steini* and *carcinophila* see Parkes (1971) (1763), White & Bruce (1986) (2652), and Cheng (1987) (333).
[8] Includes *littleri* see Schodde *et al.* (1980) (2191).
[9] Includes *cinerea* and *rogersi* see Schodde *et al.* (1980) (2191).

BUBULCUS Bonaparte, 1855 M
● ***Bubulcus ibis*** Cattle Egret
    ✓ *B.i.ibis* (Linnaeus, 1758)[1]   Spain to Iran, N and C Africa to Mascarene and
                                          Seychelles Is.; E Nth. America, C and N Sth. America

    — *B.i.coromandus* (Boddaert, 1783)   India and Pakistan to S Japan, Philippines, Indonesia,

ARDEA Linnaeus, 1758 F
● ***Ardea cinerea*** Grey Heron
    ✓ *A.c.cinerea* Linnaeus, 1758   Europe to E Asia, W China, Africa
    — *A.c.jouyi* Clark, 1907   Korea, Japan to Sumatra and Java
    — *A.c.monicae* Jouanin & Roux, 1963   Banc d'Arguin, Mauretania
    — *A.c.firasa* E. Hartert, 1917   Madagascar, Aldabra, Comoro Is.

● ***Ardea herodias*** Great Blue Heron
    ✓ *A.h.fannini* Chapman, 1901   W Nth. America
    ✓ *A.h.herodias* Linnaeus, 1758[2]   NE Nth. America >> C America
    ✓ *A.h.occidentalis* Audubon, 1835   SE USA >> Cuba, Jamaica, West Indies
    — *A.h.cognata* Bangs, 1903   Galapagos Is.

● ***Ardea cocoi*** Linnaeus, 1766 Cocoi Heron   S. America

● ***Ardea pacifica*** Latham, 1802[3] White-necked Heron   Australia, Tasmania, S New Guinea

○ ***Ardea melanocephala*** Vigors & Children, 1826 Black-headed Heron   Senegal to Ethiopia and Cape Province

○ ***Ardea humbloti*** Milne-Edwards & Grandidier, 1885 Humblot's Heron
                                        W Madagascar, Aldabra I.

○ ***Ardea insignis*** Hume, 1878 White-bellied Heron[4]   Sikkim to C Burma

○ ***Ardea sumatrana*** Raffles, 1822 Great-billed Heron[5]   SE Asia, the Sundas, Philippines, Wallacea, New
                                        Guinea and N Australia

○ ***Ardea goliath*** Cretzschmar, 1827 Goliath Heron   Senegal to Ethiopia and Cape Province (vagrant S Asia)

● ***Ardea purpurea*** Purple Heron
    ✓ *A.p.purpurea* Linnaeus, 1766[6]   S and C Europe to Kazakhstan and Middle East, Cape
                                        Verde Is., Africa
    — *A.p.madagascariensis* van Oort, 1910   Madagascar
    — *A.p.manilensis* Meyen, 1834   S and E Asia east to SE Siberia through SE Asia to E
                                        Lesser Sundas, Sulawesi, Philippines

● ***Ardea alba*** Great Egret
    ✓ *A.a.alba* Linnaeus, 1758   C and SE Europe, N Asia to N Japan and NE China >>
                                        C and S Europe, Africa, S and E Asia
    — *A.a.modesta* J.E. Gray, 1831   S and E Asia to S Japan and Australasia
    — *A.a.melanorhynchos* Wagler, 1827   Senegal to Ethiopia and Cape Province
    ✓ *A.a.egretta* J.F. Gmelin, 1789   S Canada to Patagonia

PILHERODIUS Reichenbach, 1850[7] M
○ ***Pilherodius pileatus*** (Boddaert, 1783) Capped Heron   E Panama to Bolivia, Paraguay, and S Brazil

SYRIGMA Ridgway, 1878 F
○ ***Syrigma sibilatrix*** Whistling Heron
    — *S.s.fostersmithi* Friedmann, 1949   NW Venezuela, NE Colombia
    — *S.s.sibilatrix* (Temminck, 1824)   S Brazil, E Bolivia, Paraguay, Uruguay, NE Argentina

EGRETTA T. Forster, 1817 F
○ ***Egretta intermedia*** Intermediate Egret[8]
    — *E.i.brachyrhyncha* (A.E. Brehm, 1854)   Sudan to Cape Province

[1] Includes *seychellarum* see Payne (1979: 210) (1797).
[2] Includes *wardi* see Payne (1979: 198) (1797).
[3] For reasons to use 1802 not 1801 see Browing & Monroe (1991) (258).
[4] Payne (1979: 201) (1797) used the name *imperialis*, but see Walters (2001) (2558).
[5] Includes *mathewsae* see Payne (1979: 202) (1797).
[6] Includes *bournei* see Peters (1979: 202) (1797).
[7] Usually attributed to Bonaparte, 1855, but see Reichenbach (1850: xvi) (2001).
[8] We follow Sheldon *et al.* (2000) (2239) in placing this species in *Egretta* rather than retain a genus *Mesophoyx*.

○ *Egretta picata* (Gould, 1845) PIED HERON | Sulawesi to New Guinea, N Australia

— *E.i.intermedia* (Wagler, 1829)[1] | S India to Japan, Philippines, Gtr. Sundas, S Moluccas, New Guinea, Australia

○ *Egretta picata* (Gould, 1845) PIED HERON | Sulawesi to New Guinea, N Australia

● *Egretta novaehollandiae* (Latham, 1790) WHITE-FACED HERON[2] | Lombok to Timor, Australia, Loyalty Is., New Zealand >> Sulawesi, Moluccas

● *Egretta rufescens* REDDISH EGRET

   ⊥ *E.r.rufescens* (J.F. Gmelin, 1789) | S USA, Mexico, Cuba, Jamaica, Hispaniola
   — *E.r.dickeyi* (van Rossem, 1926) | NW Mexico, Baja California

○ *Egretta ardesiaca* (Wagler, 1827) BLACK HERON | Senegal to Ethiopia and Cape Province

○ *Egretta vinaceigula* (Sharpe, 1895) SLATY EGRET | Zambia to N Botswana

● *Egretta tricolor* TRICOLORED HERON/LOUISIANA HERON

   — *E.t.ruficollis* Gosse, 1847 | E and SE USA to NW South America
   — *E.t.occidentalis* (Huey, 1927)[3] | Baja California >> SW USA, NW Mexico
   — *E.t.tricolor* (Statius Müller, 1776) | N Sth. America (except extreme NW)

● *Egretta caerulea* (Linnaeus, 1758) LITTLE BLUE HERON | S USA to C Sth. America

● *Egretta thula* SNOWY EGRET

   ⊻ *E.t.brewsteri* Thayer & Bangs, 1909 | W and SE USA to N Chile, N Argentina
   ⊻ *E.t.thula* (Molina, 1782) | E and S USA to C Chile and N Argentina

● *Egretta garzetta* LITTLE EGRET[4]

   ⊻ *E.g.garzetta* (Linnaeus, 1766) | Europe to Japan, India, Philippines and Africa
   ⊻ *E.g.nigripes* (Temminck, 1840)[5] | Sundas to New Guinea, Solomons, Australia, New Zealand

○ *Egretta gularis* WESTERN REEF EGRET

   — *E.g.gularis* (Bosc, 1792) | Coast tropical W Africa
   — *E.g.schistacea* (Hemprich & Ehrenberg, 1828) | Coastal NE Africa to W India, Sri Lanka

○ *Egretta dimorpha* E. Hartert, 1914 DIMORPHIC EGRET | Coastal S Kenya and N Tanzania, Pemba I., Madagascar to Seychelles

● *Egretta sacra* PACIFIC REEF EGRET

   ⊻ *E.s.sacra* (J.F. Gmelin, 1789)[6] | Coastal E and SE Asia to Australia, New Zealand, Micronesia and Polynesia
   — *E.s.albolineata* (G.R. Gray, 1859) | New Caledonia, Loyalty Is.

○ *Egretta eulophotes* (Swinhoe, 1860) CHINESE EGRET | E China and Korea >> SE China and Taiwan to Philippines and Indonesia

PHAETHONTIDAE TROPICBIRDS (1: 3)

PHAETHON Linnaeus, 1758 M

○ *Phaethon aethereus* RED-BILLED TROPICBIRD

   — *P.a.mesonauta* J.L. Peters, 1930[7] | E Pacific (California to Galapagos and coast of Peru), Caribbean, E Atlantic
   — *P.a.aethereus* Linnaeus, 1758 | S Atlantic
   — *P.a.indicus* Hume, 1876 | Persian Gulf, Gulf of Aden, Red Sea

● *Phaethon rubricauda* RED-TAILED TROPICBIRD

   — *P.r.rubricauda* Boddaert, 1783 | W Indian Ocean (Mauritius to Aldabra Is.)
   — *P.r.westralis* Mathews, 1912 | Indian Ocean (Christmas I., Cocos Keeling Is., islands off mid W Australia)
   — *P.r.roseotinctus* (Mathews, 1926)[8] | Raine to Lord Howe, Norfolk, Kermadec Is.

---

[1] Includes *plumifera* see Mees (1975) (1527) and Payne (1979: 209) (1797).
[2] Includes *parryi* see Peters (1931: 102) (1821).
[3] For recognition see Payne (1979: 208) (1797).
[4] We separate *gularis* and *dimorpha* following Payne (1979: 213) (1797).
[5] Includes *immaculata* see Payne (1979: 213) (1797).
[6] Apparently includes *micronesiae* Momiyama, 1926 (1599); omitted by Peters (1931) (1821).
[7] Includes *limatus* see Dorst & Mougin (1979: 156) (755).
[8] For correction of spelling see David & Gosselin (2002) (613).

√ *P.r.melanorhynchos* J.F. Gmelin, 1789[1]   Ogasawara-shoto (Bonin) and Iwo (Volcano) Is. through Micronesia to Vanuatu, New Caledonia and Fiji and central S Pacific Is. and Hawaii

● *Phaethon lepturus* WHITE-TAILED TROPICBIRD
- — *P.l.catesbyi* J.F. Brandt, 1838   West Indies, Bahama, Bermuda Is.
- — *P.l.ascensionis* (Mathews, 1915)   Tropical S Atlantic Ocean
- — *P.l.lepturus* Daudin, 1802   Mascarene, Seychelles, Andaman Is.
- — *P.l.europae* Le Corre & Jouventin, 1999   Europa I. (Mozambique Channel) (1342)
- — *P.l.fulvus* J.F. Brandt, 1838   Christmas I. (Indian Ocean)
- √ *P.l.dorotheae* Mathews, 1913   Micronesia to New Caledonia, islands of central south Pacific and Hawaii

FREGATIDAE FRIGATEBIRDS (1: 5)

FREGATA Lacépède, 1799 F

○ *Fregata aquila* (Linnaeus, 1758) ASCENSION FRIGATEBIRD   Ascension I. (Atlantic Ocean)

○ *Fregata andrewsi* Mathews, 1914 CHRISTMAS ISLAND FRIGATEBIRD   Christmas I. (Indian Ocean)

● *Fregata magnificens* Mathews, 1914 MAGNIFICENT FRIGATEBIRD[2]   Coasts of the Americas (mainly intertropical), West Indies, Galapagos, Cape Verde Is.

● *Fregata minor* GREAT FRIGATEBIRD
- — *F.m.nicolli* Mathews, 1914   S Atlantic Ocean
- — *F.m.aldabrensis* Mathews, 1914   W Indian Ocean
- — *F.m.minor* (J.F. Gmelin, 1789)   C and E Indian Ocean to South China Sea
- √ *F.m.palmerstoni* (J.F. Gmelin, 1789)[3]   SW and C Pacific Ocean to Hawaii
- — *F.m.ridgwayi* Mathews, 1914   E Pacific Ocean

○ *Fregata ariel* LESSER FRIGATEBIRD
- — *F.a.trinitatis* Miranda-Ribeiro, 1919   S Trindade, Martin Vaz Is. (Brazil)
- — *F.a.iredalei* Mathews, 1914   W Indian Ocean
- — *F.a.ariel* (G.R. Gray, 1845)   E Indian Ocean to Philippines, N Australia and C Pacific Ocean

SCOPIDAE HAMMERKOP (1: 1)

SCOPUS Brisson, 1760 M
○ *Scopus umbretta* HAMMERKOP
- — *S.u.umbretta* J.F. Gmelin, 1789[4]   Africa (except coastal W Africa), SW Arabia; Madagascar[5]
- — *S.u.minor* Bates, 1931   Coast from Sierra Leone to E Nigeria

BALAENICIPITIDAE SHOEBILL (1: 1)

BALAENICEPS Gould, 1850 M
○ *Balaeniceps rex* Gould, 1850 SHOEBILL   Sudan to Zambia

PELECANIDAE PELICANS (1: 7)

PELECANUS Linnaeus, 1758 M
○ *Pelecanus onocrotalus* Linnaeus, 1758 GREAT WHITE PELICAN   S Europe, Africa, C Asia

○ *Pelecanus rufescens* J.F. Gmelin, 1789 PINK-BACKED PELICAN   Sub-Saharan Africa

○ *Pelecanus philippensis* J.F. Gmelin, 1789 SPOT-BILLED PELICAN   S Asia, Iran to Philippines

○ *Pelecanus crispus* Bruch, 1832 DALMATIAN PELICAN   SE Europe to China

---

[1] Includes *rothschildi* see Dorst & Mougin (1979: 158) (755).
[2] Includes *lowei* and *rothschildi* see Dorst & Mougin (1979: 160) (755).
[3] Includes *strumosa* see Peters (1931: 96) (1821) and *peninsulae* see Dorst & Mougin (1979: 161) (755).
[4] Includes *bannermanni* see Payne (1979: 245) (1797).
[5] If a Madagascan form is to be recognised the name *tenuirostris* is available.

○ *Pelecanus conspicillatus* Temminck, 1824 AUSTRALIAN PELICAN[1]   Australia, New Guinea; vagrant to Lesser Sundas, Banda Sea, Solomons, Vanuatu, Fiji Is.

● *Pelecanus erythrorhynchos* J.F. Gmelin, 1789 AMERICAN WHITE PELICAN   N and C America, West Indies

● *Pelecanus occidentalis* BROWN PELICAN
      — *P.o.occidentalis* Linnaeus, 1766   West Indies
      ✓ *P.o.carolinensis* J.F. Gmelin, 1789   Coasts of S Carolina to Venezuela
      ✓ *P.o.californicus* Ridgway, 1884   Coast of California, W Mexico
      — *P.o.murphyi* Wetmore, 1945   Coasts of W Colombia and Ecuador
      — *P.o.urinator* Wetmore, 1945   Galapagos Is.
      ✓ *P.o.thagus* Molina, 1782   Coasts of Peru and Chile

SULIDAE GANNETS, BOOBIES (3: 10)

MORUS Vieillot, 1816 M
● *Morus bassanus* (Linnaeus, 1758) NORTHERN GANNET   E Canada, Iceland, British Isles, W France, Heligoland, Norway

○ *Morus capensis* (M.H.K. Lichtenstein, 1823) CAPE GANNET   Southern Africa, St Paul (Indian Ocean)

● *Morus serrator* (G.R. Gray, 1843) AUSTRALASIAN GANNET[2]   St. Paul (Indian Ocean), S Australian coast, Tasmania, North Island (New Zealand)

PAPASULA Olson & Warheit, 1988 (1710) F
○ *Papasula abbotti* (Ridgway, 1893) ABBOTT'S BOOBY   Christmas I. (Indian Ocean)

SULA Brisson, 1760 F
○ *Sula nebouxii* BLUE-FOOTED BOOBY
      — *S.n.nebouxii* A. Milne-Edwards, 1882   California to Peru
      — *S.n.excisa* Todd, 1948   Galapagos Is.

● *Sula variegata* (Tschudi, 1843) PERUVIAN BOOBY   Coasts of Peru and Chile

○ *Sula dactylatra* MASKED BOOBY
      — *S.d.dactylatra* Lesson, 1831   Caribbean, Bahama Is., Ascension I.
      — *S.d.melanops* Heuglin, 1859   Red Sea to NW Indian Ocean
      — *S.d.bedouti* Mathews, 1913   S and E Indian Ocean (Christmas I., Lesser Sundas, central N coast W Australia)
      — *S.d.fullagari* O'Brien & Davies, 1990   N Tasman Sea (Norfolk, Lord Howe, Kermadec Is.) (1674)
      — *S.d.personata* Gould, 1846   Coral Sea (NE Australia) to Philippines, Micronesia and Polynesia (W and C Pacific Ocean)
      — *S.d.californica* Rothschild, 1915   Islands off W Mexico

○ *Sula granti* Rothschild, 1902 NAZCA BOOBY[3]   Galapagos Is., Cocos Is., Malpelos Is., La Plata Is.

◉ *Sula sula* RED-FOOTED BOOBY
      — *S.s.sula* (Linnaeus, 1766)   Caribbean Is. and S Trinidade I. (Brazil)
      ✓ *S.s.rubripes* Gould, 1838   Tropical Indian and Pacific Oceans
      — *S.s.websteri* Rothschild, 1898   Coast of W Mexico to Galapagos Is.

● *Sula leucogaster* BROWN BOOBY[4]
      — *S.l.leucogaster* (Boddaert, 1783)   Caribbean and Atlantic Is.
      — *S.l.brewsteri* Goss, 1888[5]   Coast of California and W Mexico
      — *S.l.etesiaca* Thayer & Bangs, 1905   Pacific coast of C America and Colombia
      ✓ *S.l.plotus* (J.R. Forster, 1844)[6]   Red and Arabian Seas through tropical Indian Ocean and Banda Sea to Micronesia and Polynesia (W and C Pacific Ocean)

[1] Includes *novaezealandiae* Scarlett, 1966 (2170).
[2] Includes *rex* see Dorst & Mougin (1979: 183) (755).
[3] For reasons to treat as a separate species see Pitman & Jehl (1998) (1901); see A.O.U. (2000: 848) (4). The population on Los Desventurados Is. may be this species.
[4] Variation may be polymorphic and irrespective of geography.
[5] Includes *nesiotes* see Dorst & Mougin (1979: 187) (755).
[6] Drawings with Forster's proposed names were published many years after his death (d. 1798).

PHALACROCORACIDAE CORMORANTS[1] (1: 36)

PHALACROCORAX Brisson, 1760 M

● *Phalacrocorax melanoleucos* LITTLE PIED CORMORANT
- ☑ *P.m.melanoleucos* (Vieillot, 1817)[2] — E Java to W Melanesia, Australia, Tasmania
- — *P.m.brevicauda* Mayr, 1931 — Rennell I. (Solomon Is.)
- ☑ *P.m.brevirostris* Gould, 1837 — New Zealand and islands in region

○ *Phalacrocorax africanus* LONG-TAILED CORMORANT/REED CORMORANT
- — *P.a.africanus* (J.F. Gmelin, 1789) — Sub-Saharan Africa
- — *P.a.pictilis* Bangs, 1918 — Madagascar

○ *Phalacrocorax coronatus* (Wahlberg, 1855) CROWNED CORMORANT — Coast of SW Africa

○ *Phalacrocorax niger* (Vieillot, 1817) LITTLE CORMORANT — India and Sri Lanka to Greater Sundas

○ *Phalacrocorax pygmeus*[3] (Pallas, 1773) PYGMY CORMORANT — C Europe to C Asia[4]

○ †*Phalacrocorax perspicillatus* Pallas, 1811 SPECTACLED CORMORANT — Bering Island, Commander Is.

● *Phalacrocorax penicillatus* (J.F. Brandt, 1837) BRANDT'S CORMORANT — S Alaska to NW Mexico

○ *Phalacrocorax harrisi* Rothschild, 1898 FLIGHTLESS CORMORANT[5] — Galapagos Is.

○ *Phalacrocorax neglectus* (Wahlberg, 1855) BANK CORMORANT — Coasts of S Africa

○ *Phalacrocorax fuscescens* (Vieillot, 1817) BLACK-FACED CORMORANT — S Australia, Tasmania

● *Phalacrocorax brasilianus* NEOTROPICAL CORMORANT[6]
- ☑ *P.b.mexicanus* (J.F. Brandt, 1837)[7] — S USA to Nicaragua and Cuba
- ☑ *P.b.brasilianus* (J.F. Gmelin, 1789)[8] — Panama to Tierra del Fuego

● *Phalacrocorax auritus* DOUBLE-CRESTED CORMORANT
- — *P.a.cincinatus* (J.F. Brandt, 1837) — S Alaska to SW Canada
- ☑ *P.a.albociliatus* Ridgway, 1884 — SW Canada to California and Revillagigedo Is.
- — *P.a.auritus* (Lesson, 1831) — C and E Canada, NE USA >> W Atlantic and N Caribbean
- ☑ *P.a.floridanus* (Audubon, 1835) — S and SE USA >> N and W Caribbean
- — *P.a.heuretus* Watson, Olson & Miller, 1991 — Bahama Is., ? Cuba (2567)

○ *Phalacrocorax fuscicollis* Stephens, 1826 INDIAN CORMORANT — India, Sri Lanka, Burma, C Thailand

● *Phalacrocorax sulcirostris* (J.F. Brandt, 1837) LITTLE BLACK CORMORANT[9]
— Java and S Borneo to New Guinea, Australia, Tasmania and New Zealand

● *Phalacrocorax varius* PIED CORMORANT
- — *P.v.hypoleucos* (J.F. Brandt, 1837) — Australia, Tasmania
- ☑ *P.v.varius* (J.F. Gmelin, 1789) — New Zealand

● *Phalacrocorax carbo* GREAT CORMORANT
- — *P.c.carbo* (Linnaeus, 1758) — E Canada to British Isles
- — *P.c.sinensis* (Blumenbach, 1798) — C Europe to India and China
- — *P.c.hanedae* N. Kuroda, Sr., 1925 — Japan
- — *P.c.maroccanus* E. Hartert, 1906 — NW Africa
- — *P.c.lugubris* Rüppell, 1845 — NE Africa
- — *P.c.lucidus* (M.H.K. Lichtenstein, 1823) — Cape Verde Is., Senegal, Lake Chad, E and S Africa
- ☑ *P.c.novaehollandiae* Stephens, 1826[10] — Australia, Tasmania, New Zealand, Chatham Is.

---

[1] See Siegel-Causey (1988) (2272); we have broadly arranged the species in his groups, although his treatment at generic level is disputed. For a proposal to use just two genera see Christidis & Bowles (1994) (351); however, we have felt it better, in the present state of our knowledge, to use just one; the splits among Australasian taxa are accepted. For others see below.
[2] Includes *melvillensis* see Dorst & Mougin (1979: 178) (755).
[3] The original spelling repeated by Peters (1931) (1821) often emended to *pygmaeus*.
[4] Previously in N Africa too.
[5] Type species of the genus *Nannopterum* treated in a monotypic genus by some recent authors.
[6] Previously called *P. olivaceus* (Humboldt, 1805), but see A.O.U. (1998: 32) (3).
[7] Includes *chanco* see Dorst & Mougin (1979: 169) (755).
[8] Includes *hornensis* see Orta (1992: 343) (1729).
[9] Treated as monotypic following Dorst & Mougin (1979: 170) (755).
[10] Includes *steadi* see Dorst & Mougin (1979: 167) (755).

○ *Phalacrocorax capillatus* (Temminck & Schlegel, 1848[1]) JAPANESE CORMORANT
NE Asia to China, Japan

○ *Phalacrocorax nigrogularis* Ogilvie-Grant & Forbes, 1899 SOCOTRA CORMORANT
S Red Sea, Persian Gulf

○ *Phalacrocorax capensis* (Sparrman, 1788) CAPE CORMORANT — Coasts of W South Africa

● *Phalacrocorax bougainvillii* (Lesson, 1837) GUANAY CORMORANT — Coasts of Peru, Chile and S Argentina

○ *Phalacrocorax atriceps* IMPERIAL SHAG[2]
— *P.a.atriceps* P.P. King, 1828[3] — S Sth. America, Antarctic Pen. and islands
— *P.a.nivalis* Falla, 1937 — Heard I.
— *P.a.georgianus* Lönnberg, 1906 — South Georgia, South Orkney and South Sandwich Is., Shag Rocks

○ *Phalacrocorax albiventer* KING CORMORANT
— *P.a.albiventer* (Lesson, 1831) — S Sth. America, Falkland Is.
— *P.a.melanogenis* (Blyth, 1860) — Marion, Crozet, Prince Edward, Kerguelen Is.
— *P.a.purpurascens* (J.F. Brandt, 1837) — Macquarie I.

○ *Phalacrocorax verrucosus* (Cabanis, 1875) KERGUELEN SHAG — Kerguelen Is.

○ *Phalacrocorax carunculatus* (J.F. Gmelin, 1789) ROUGH-FACED SHAG — South Island (New Zealand)

● *Phalacrocorax chalconotus* (G.R. Gray, 1845) STEWART SHAG — Stewart I.

○ *Phalacrocorax onslowi* Forbes, 1893 CHATHAM SHAG — Chatham Is.

○ *Phalacrocorax campbelli* (Filhol, 1878) CAMPBELL SHAG — Campbell I.

○ *Phalacrocorax colensoi* Buller, 1888 AUCKLAND SHAG — Auckland Is.

○ *Phalacrocorax ranfurlyi* Ogilvie-Grant, 1901 BOUNTY SHAG — Bounty I.

○ *Phalacrocorax magellanicus* (J.F. Gmelin, 1789) ROCK SHAG — Tierra del Fuego, Falkland Is.

● *Phalacrocorax pelagicus* PELAGIC CORMORANT[4]
— *P.p.pelagicus* Pallas, 1811 — N Pacific islands
✓ *P.p.resplendens* Audubon, 1838 — British Columbia to Mexico

○ *Phalacrocorax urile* (J.F. Gmelin, 1789) RED-FACED SHAG — Bering Sea to Japan.

● *Phalacrocorax aristotelis* EUROPEAN SHAG
— *P.a.aristotelis* (Linnaeus, 1761) — Iceland and W European coast from Lapland to Portugal
— *P.a.riggenbachi* E. Hartert, 1923 — Coast of Morocco
✓ *P.a.desmarestii* (Payraudeau, 1826) — C Mediterranean

○ *Phalacrocorax gaimardi* (Lesson & Garnot, 1828) RED-LEGGED CORMORANT[5]
Coast of S South America

● *Phalacrocorax punctatus* SPOTTED SHAG
✓ *P.p.punctatus* (Sparrman, 1786) — New Zealand
✓ *P.p.oliveri* Mathews, 1930 — Stewart I. off South Island (New Zealand)

○ *Phalacrocorax featherstoni* Buller, 1873 PITT SHAG — Chatham Is.

ANHINGIDAE ANHINGAS (1: 2)

ANHINGA Brisson, 1760[6] F
○ *Anhinga melanogaster* DARTER
— *A.m.rufa* (Daudin, 1802) — Africa south of the Sahara; Iraq
— *A.m.vulsini* Bangs, 1918 — Madagascar (mainly W)
— *A.m.melanogaster* Pennant, 1769 — India through Greater Sundas to Philippines, Sulawesi
— *A.m.novaehollandiae* (Gould, 1847) — New Guinea, Australia

[1] Not 1850, see Holthuis & Sakai (1970) (1128) and Dekker *et al.* (2001) (672).
[2] We have retained our previous treatment of this species and the next as the relationship of the two birds in southernmost South America is still unresolved (P. Rasmussen *in litt.*).
[3] Includes *bransfieldensis* see Dorst & Mougin (1979: 175) (755).
[4] Includes *kenyoni* Siegel-Causey, 1991 (2273) see Rohwer *et al.* (2000) (2098).
[5] Includes *cirriger* see Peters (1931: 91) (1821).
[6] The arrangement used is that of Orta (1992) (1730).

◑ *Anhinga anhinga* ANHINGA
    ✓ *A.a.leucogaster* (Vieillot, 1816)   SE USA to Colombia
    — *A.a.anhinga* (Linnaeus, 1766)   Brazil, Argentina, E Bolivia

CATHARTIDAE NEW WORLD VULTURES (5: 7)

CATHARTES Illiger, 1811 M
◑ **Cathartes aura** TURKEY-VULTURE
    ✓ *C.a.aura* (Linnaeus, 1758)   S Canada to Costa Rica and Cuba
    ✓ *C.a.septentrionalis* Wied, 1839   E North America
    — *C.a.ruficollis* Spix, 1824   Panama to N Argentina, Trinidad
    ✓ *C.a.jota* (Molina, 1782)   Colombia to Patagonia, Falkland Is.

○ **Cathartes burrovianus** Cassin, 1845 LESSER YELLOW-HEADED VULTURE   E Mexico to N Argentina

○ **Cathartes melambrotus** Wetmore, 1964 GREATER YELLOW-HEADED VULTURE
                                       C South America

CORAGYPS Le Maout, 1853[1] M
◑ **Coragyps atratus** (Bechstein, 1793) BLACK VULTURE[2]   W and S USA and Mexico to Chile

SARCORAMPHUS Duméril, 1806 M
○ **Sarcoramphus papa** (Linnaeus, 1758) KING VULTURE   C Mexico to N Argentina, Trinidad

GYMNOGYPS Lesson, 1842 M
◕ **Gymnogyps californianus** (Shaw, 1798) CALIFORNIAN CONDOR   S California

VULTUR Linnaeus, 1758 M
◐ **Vultur gryphus** Linnaeus, 1758 ANDEAN CONDOR   Andes from W Venezuela to Tierra del Fuego

FALCONIDAE FALCONS, CARACARAS (11: 64)

POLYBORINAE

DAPTRIUS Vieillot, 1816 M
○ **Daptrius ater** Vieillot, 1816 BLACK CARACARA   Amazonia

IBYCTER Vieillot, 1816[3] M
○ **Ibycter americanus** (Boddaert, 1783) RED-THROATED CARACARA   S Mexico to C Peru and S Brazil, Bolivia

PHALCOBOENUS Orbigny, 1834 M
○ **Phalcoboenus carunculatus** Des Murs, 1853 CARUNCULATED CARACARA   Andes of SW Colombia and Ecuador

◑ **Phalcoboenus megalopterus** (Meyen, 1834) MOUNTAIN CARACARA   Andes from N Peru to C Chile, NW Argentina

○ **Phalcoboenus albogularis** (Gould, 1837) WHITE-THROATED CARACARA   S Chile, S Argentina

○ **Phalcoboenus australis** (J.F. Gmelin, 1788) STRIATED CARACARA   Falkland Is., Cape Horn islands

CARACARA Merrem, 1826[4] F
◑ **Caracara cheriway** (Jacquin, 1784) CRESTED CARACARA[5]   S USA to Sth. America (N of R. Amazon)

○ †**Caracara lutosa** (Ridgway, 1876) GUADALUPE CARACARA   Guadalupe I. (Mexico)

◑ **Caracara plancus** (J.F. Miller, 1777) SOUTHERN CARACARA   Sth. America (south of R. Amazon), Falkland Is.

MILVAGO Spix, 1824 F
◑ **Milvago chimachima** YELLOW-HEADED CARACARA
    ✓ *M.c.cordata* Bangs & Penard, 1918[6]   S Costa Rica and Panama through northern Sth.
                                           America to the Amazon

    — *M.c.chimachima* (Vieillot, 1816)   E Bolivia and S Brazil south to N Argentina

---

[1] For this correction see Gregory, 1998 (1011).
[2] Includes *brasiliensis* and *foetens* see Stresemann & Amadon (1979: 275) (2368).
[3] For use of this genus see A.O.U. (2000: 849) (4).
[4] For use of this generic name in place of *Polyborus* see A.O.U. (1998: 106) (3).
[5] For reasons to treat *C. plancus* as three species see Dove & Banks (1999) (757).
[6] For correction of spelling see David & Gosselin (2002) (614).

◉ *Milvago chimango* CHIMANGO CARACARA
    — *M.c.chimango* (Vieillot, 1816)   Paraguay and Uruguay to S Argentina
    ⤓ *M.c.temucoensis* W.L. Sclater, 1918   S Chile, Tierra del Fuego

HERPETOTHERES Vieillot, 1817 M
● *Herpetotheres cachinnans* LAUGHING FALCON
    — *H.c.chapmani* Bangs & Penard, 1918   Mexico to Honduras
    ⤓ *H.c.cachinnans* (Linnaeus, 1758)[1]   Nicaragua to Peru and C Brazil
    — *H.c.queribundus* Bangs & Penard, 1919   E Brazil to Paraguay, E Bolivia and N Argentina

MICRASTUR G.R. Gray, 1841 M
○ *Micrastur ruficollis* BARRED FOREST FALCON
    — *M.r.guerilla* Cassin, 1848   Mexico to Nicaragua
    — *M.r.interstes* Bangs, 1907   Costa Rica to W Colombia and Ecuador
    — *M.r.zonothorax* (Cabanis, 1865)   N Venezuela, E Colombia
    — *M.r.pelzelni* Ridgway, 1876   W Brazil, E Peru
    — *M.r ruficollis* (Vieillot, 1817)   E Brazil, Paraguay, N Argentina
    — *M.r.olrogi* Amadon, 1964   NW Argentina

○ *Micrastur plumbeus* W.L. Sclater, 1918 PLUMBEOUS FOREST FALCON[2]   W Colombia, NW Ecuador

○ *Micrastur gilvicollis* (Vieillot, 1817) LINED FOREST FALCON   S Venezuela, Guianas, Amazonia

○ *Micrastur mirandollei* (Schlegel, 1862) SLATY-BACKED FOREST FALCON   E Costa Rica to C and E Brazil

○ *Micrastur semitorquatus* COLLARED FOREST FALCON
    — *M.s.naso* (Lesson, 1842)   Mexico to NW Peru
    — *M.s.semitorquatus* (Vieillot, 1817)   E Colombia and N Peru to Brazil and N Argentina

○ *Micrastur buckleyi* Swann, 1919 BUCKLEY'S FOREST FALCON   E Ecuador, NE Peru

FALCONINAE

SPIZIAPTERYX Kaup, 1852 F
○ *Spiziapteryx circumcincta* (Kaup, 1852) SPOT-WINGED FALCONET[3]   S Bolivia, Paraguay, N and W Argentina

POLIHIERAX Kaup, 1847 M
○ *Polihierax semitorquatus* AFRICAN PYGMY FALCON
    — *P.s.castanotus* (Heuglin, 1860)   S Sudan and Ethiopia to C Tanzania
    — *P.s.semitorquatus* (A. Smith, 1836)   S Angola, Namibia, Botswana, NW Sth. Africa

○ *Polihierax insignis* WHITE-RUMPED PYGMY FALCON[4]
    — *P.i.insignis* Walden, 1872   W and C Burma
    — *P.i.cinereiceps* E.C.S. Baker, 1927   S Burma, NW Thailand, N Indochina
    — *P.i.harmandi* (Oustalet, 1876)   S and C Laos, S Vietnam, Cambodia

MICROHIERAX Sharpe, 1874 M
○ *Microhierax caerulescens* COLLARED FALCONET
    — *M.c.caerulescens* (Linnaeus, 1758)   Himalayas, N India
    — *M.c.burmanicus* Swann, 1920   Burma to Indochina

○ *Microhierax fringillarius* (Drapiez, 1824) BLACK-THIGHED FALCONET   Malay Pen., Gtr. Sundas (except NW Borneo) and Bali

○ *Microhierax latifrons* Sharpe, 1879 WHITE-FRONTED FALCONET   NW Borneo

○ *Microhierax erythrogenys* PHILIPPINE FALCONET
    — *M.e.erythrogenys* (Vigors, 1831)   Bohol, Catanduanes, Luzon, Mindoro and Negros
    — *M.e.meridionalis* Ogilvie-Grant, 1897[5]   Cebu, Leyte, Mindanao and Samar

○ *Microhierax melanoleucos* (Blyth, 1843)[6] PIED FALCONET   Assam, SE China, N Indochina

---

[1] Includes *fulvescens* see Peters (1931: 276) (1821).
[2] For recognition as a species see Hilty & Brown (1986: 117) (1116), but see Bierregaard (1994: 253) (147).
[3] For correction of spelling see David & Gosselin (2002) (614).
[4] Type species of the genus *Neohierax* here submerged.
[5] Omitted in last Edition.
[6] For correction of spelling see David & Gosselin (2002) (613).

FALCO Linnaeus, 1758 M

○ *Falco naumanni* Fleischer, 1818 LESSER KESTREL   S Europe to China >> Africa

● *Falco tinnunculus* COMMON KESTREL

    ↙ *F.t.tinnunculus* Linnaeus, 1758   Europe and NW Africa to Lena basin (EC Siberia) >> south to E Africa and S Asia

    — *F.t.perpallidus* (Clark, 1907)[1]   NE Siberia to NE China and Korea >> E and SE Asia
    — *F.t.interstinctus* McClelland, 1840   Himalayas and Indochina to Japan >> E and S Asia and Philippines

    — *F.t.objurgatus* (E.C.S. Baker, 1927)   S India
    — *F.t.canariensis* (Koenig, 1890)   Madeira, W Canary Is.
    — *F.t.dacotiae* E. Hartert, 1913   E Canary Is.
    — *F.t.neglectus* Schlegel, 1873   N Cape Verde Is.
    — *F.t.alexandri* Bourne, 1955   S Cape Verde Is.
    — *F.t.rupicolaeformis* (C.L. Brehm, 1855)   Egypt, S Yemen
    — *F.t.archeri* Hartert & Neumann, 1932   Socotra I., Somalia, NE Kenya
    — *F.t.rufescens* Swainson, 1837   Guinea and N Angola to Ethiopia and Tanzania
    — *F.t.rupicolus* Daudin, 1800   C Angola and S Tanzania to Cape Province

○ *Falco newtoni* MADAGASCAN KESTREL
    — *F.n.newtoni* (J.H. Gurney, 1863)   Madagascar
    — *F.n.aldabranus* Grote, 1928   Aldabra, Anjouan Is.

○ *Falco punctatus* Temminck, 1821 MAURITIUS KESTREL   Mauritius

○ *Falco araea* (Oberholser, 1917) SEYCHELLES KESTREL   Seychelles

○ *Falco moluccensis* SPOTTED KESTREL
    — *F.m.moluccensis* (Bonaparte, 1850)   Morotai, Halmahera, Bacan, Obi, Buru, Ambon, Seram
    — *F.m.microbalius* (Oberholser, 1919)[2][3]   Java, Bali, Sulawesi, Lesser Sundas to Kai Is.

● *Falco cenchroides* Vigors & Horsfield, 1827 AUSTRALIAN KESTREL[4]   Australia, Tasmania >> Christmas I., New Guinea

● *Falco sparverius* AMERICAN KESTREL
    ↙ *F.s.sparverius* Linnaeus, 1758   Alaska and Canada to W Mexico >> Panama
    — *F.s.paulus* (Howe & King, 1902)   SE USA
    — *F.s.peninsularis* Mearns, 1892   Baja California, NW Mexico
    — *F.s.tropicalis* (Griscom, 1930)   S Mexico, Guatemala, N Honduras
    — *F.s.nicaraguensis* T.R. Howell, 1965   NW Honduras, Nicaragua
    — *F.s.sparverioides* Vigors, 1827   S Bahamas Is., Cuba
    — *F.s.dominicensis* J.F. Gmelin, 1788   Hispaniola
    — *F.s.caribaearum* J.F. Gmelin, 1788   Puerto Rico, Virgin Is., Lesser Antilles
    — *F.s.brevipennis* (Berlepsch, 1892)   Aruba, Curaçao, Bonaire
    — *F.s.isabellinus* Swainson, 1837   The Guianas, E Venezuela, N Brazil
    — *F.s.ochraceus* (Cory, 1915)   Andes of E Colombia, NW Venezuela
    — *F.s.caucae* (Chapman, 1915)[5]   Andes of W Colombia
    — *F.s.aequatorialis* Mearns, 1892   NW Colombia, N Ecuador
    — *F.s.peruvianus* (Cory, 1915)   SW Ecuador, Peru, N Chile
    ↙ *F.s.cinnamominus* Swainson, 1837   SE Peru to Paraguay and Tierra del Fuego
    — *F.s.fernandensis* (Chapman, 1915)   Alejandro Selkirk I., Juan Fernandez Is. (Chile)
    — *F.s.cearae* (Cory, 1915)   S Brazil

○ *Falco rupicoloides* GREATER KESTREL
    — *F.r.fieldi* (Elliot, 1897)   Eritrea, N Somalia, NE Ethiopia, N Kenya
    — *F.r.arthuri* (J.H. Gurney, 1884)   C and S Kenya, N Tanzania
    — *F.r.rupicoloides* A. Smith, 1829   Cape Province to S and E Angola , SW Zambia

○ *Falco alopex* (Heuglin, 1861) FOX KESTREL   Mali to Sudan and N and W Kenya

○ *Falco ardosiaceus* Vieillot, 1823 GREY KESTREL   Senegal to W Ethiopia, W Tanzania and Angola

---

[1] For recognition see Stepanyan (1990) (2326).
[2] For correction of spelling see David & Gosselin (2002) (613).
[3] Includes *jungei* Mees, 1961 (1510), see White & Bruce (1986: 132) (2652).
[4] Includes *baru* (New Guinea birds are winter visitors).
[5] Omitted in last Edition.

○ *Falco dickinsoni* P.L. Sclater, 1864 DICKINSON'S KESTREL Angola to C Tanzania and S Mozambique

○ *Falco zoniventris* W.K.H. Peters, 1854 BANDED KESTREL Madagascar

○ *Falco chicquera* RED-NECKED FALCON
— *F.c.chicquera* Daudin, 1800 SE Iran, Pakistan, India, Bangladesh
— *F.c.ruficollis* Swainson, 1837 Gambia to W Ethiopia, E Africa to Zambia and N Mozambique
— *F.c.horsbrughi* Gunning & Roberts, 1911 Namibia to S Mozambique and Cape Province

○ *Falco vespertinus* Linnaeus, 1766 RED-FOOTED FALCON C Europe to C Asia >> C and S Africa

○ *Falco amurensis* Radde, 1863 AMUR FALCON E Siberia, N China >> E and S Asia, SE Africa

○ *Falco eleonorae* Géné, 1839 ELEONORA'S FALCON Canary Is., Mediterranean basin >> Madagascar

○ *Falco concolor* Temminck, 1825 SOOTY FALCON NE Africa to Israel and Arabia >> Madagascar and SE Africa

○ *Falco femoralis* APLOMADO FALCON
— *F.f.septentrionalis* Todd, 1916 SW USA and Mexico
— *F.f.femoralis* Temminck, 1822 C and S America
— *F.f.pichinchae* Chapman, 1925 Andes from Colombia to Chile

● *Falco columbarius* MERLIN
— *F.c.subaesalon* C.L. Brehm, 1827 Iceland >> W Europe
— *F.c.aesalon* Tunstall, 1771 Europe to N Russia and NW Siberia
— *F.c.insignis* (Clark, 1907) NC Siberia >> NE Africa and S and E Asia
— *F.c.pacificus* (Stegmann, 1929)[1] NE Siberia and Sakhalin >> Japan and E China
— *F.c.pallidus* (Sushkin, 1900) S Urals and N and NE Kazakhstan
— *F.c.lymani* Bangs, 1913 Pamir-Alai and Altai Mts to Transbaikalia and Mongolia
— *F.c.columbarius* Linnaeus, 1758 Alaska to Newfoundland and N USA >> USA to N Sth. America
— *F.c.suckleyi* Ridgway, 1873 SE Alaska to NW USA
— *F.c.richardsoni* Ridgway, 1871 Great Plains >> C USA to N Mexico

● *Falco rufigularis* BAT FALCON
⊥ *F.r.petoensis* Chubb, 1918[2] Mexico to W Ecuador (W of the Andes)
— *F.r.rufigularis* Daudin, 1800 Northern Sth. America E of the Andes through Amazonia to S Brazil and NE Argentina
— *F.r.ophryophanes* (Salvadori, 1895) E Bolivia, S Brazil (tableland), Paraguay, NW Argentina

○ *Falco deiroleucus* Temminck, 1825 ORANGE-BREASTED FALCON C America to N Argentina

○ *Falco subbuteo* EURASIAN HOBBY
— *F.s.subbuteo* Linnaeus, 1758[3] Europe to Japan, N and W China and N India >> Africa and S Asia
— *F.s.streichi* Hartert & Neumann, 1907 Burma, N Indochina and S China

○ *Falco cuvieri* A. Smith, 1830 AFRICAN HOBBY Guinea to Ethiopia, E Africa to Namibia and Transvaal

○ *Falco severus* Horsfield, 1821 ORIENTAL HOBBY[4]
— *F.s.severus* Horsfield, 1821 E Himalayas to Philippines and Gtr. Sundas
— *F.s.papuanus* Meyer & Wiglesworth, 1894 Sulawesi, Moluccas, Papuasia to Solomons

○ *Falco longipennis* AUSTRALIAN HOBBY
— *F.l.hanieli* Hellmayr, 1914 Lesser Sundas
— *F.l.longipennis* Swainson, 1837 SW and SE Australia, Tasmania >> to New Guinea and Moluccas
— *F.l.murchisonianus* Mathews, 1912 Australia except SE, SW; also S Papuasia to Moluccas and New Britain

○ *Falco novaezeelandiae* J.F. Gmelin, 1788 NEW ZEALAND FALCON New Zealand, Auckland Is.

---

[1] Omitted in last Edition.
[2] Includes *petrophilus* see Stresemann & Amadon (1979: 414) (2368).
[3] Includes *jugurtha* see Stresemann & Amadon (1979: 415) (2368).
[4] For treatment as polytypic see Stresemann & Amadon (1979: 416) (2368).

○ *Falco berigora* Brown Falcon
     __ *F.b.novaeguineae* (A.B. Meyer, 1894)   C and E New Guinea, coastal N Australia
     __ *F.b.berigora* Vigors & Horsfield, 1827[1]   E, C and N Australia
     __ *F.b.occidentalis* (Gould, 1844)   SW and WC Australia
     __ *F.b.tasmanicus* (Mathews, 1916)[2]   Tasmania

○ *Falco hypoleucos* Gould, 1841 Grey Falcon   W, N and C Australia

○ *Falco subniger* G.R. Gray, 1843 Black Falcon   Australia

○ *Falco biarmicus* Lanner Falcon
     __ *F.b.feldeggii* Schlegel, 1843   Italy to Transcaucasia and NW Levant
     __ *F.b.erlangeri* O. Kleinschidt, 1901   NW Africa
     __ *F.b.tanypterus* Schlegel, 1843   NE Africa, Israel, Arabia and Iraq
     __ *F.b.abyssinicus* Neumann, 1904   N tropical Africa
     __ *F.b.biarmicus* Temminck, 1825   S Uganda and Kenya to Angola and Cape Province

○ *Falco jugger* J.E. Gray, 1834 Laggar Falcon   Baluchistan, Himalayas, N and C India

○ *Falco cherrug* Saker Falcon[3]
     __ *F.c.cherrug* J.E. Gray, 1834[4]   C Europe to SC Siberia and N Kazakhstan >> south to
                                           NE Africa and SW Asia
     __ *F.c.milvipes* Jerdon, 1871[5]   C Tien Shan and Altai to N China and S Transbaikalia
     __ *F.c.coatsi* Dementiev, 1945   Transcaspian plains to E Uzbekistan and S Kazakhstan
     __ *F.c.hendersoni* Hume, 1871   Pamir Mts. to Tibetan Plateau

● *Falco rusticolus* Linnaeus, 1758 Gyrfalcon   Arctic Europe, Asia, N America

● *Falco mexicanus* Schlegel, 1851 Prairie Falcon   SW Canada, W USA, NW Mexico >> NC Mexico

● *Falco peregrinus* Peregrine Falcon
     ✔ *F.p.tundrius* C.M. White, 1968   N Alaska to Greenland south to Sth. America
     __ *F.p.pealei* Ridgway, 1873   Commander Is., Aleutian Is., S Alaska, SW Canada >>
                                           south to Japan and W USA
     __ *F.p.anatum* Bonaparte, 1838   Aleutians through coastal Alaska to Washington State
                                           (USA) >> south to C America
     __ *F.p.cassini* Sharpe, 1873[6]   Southern Sth. America, Falkland Is.
     ✔ *F.p.peregrinus* Tunstall, 1771   W, C and SE Europe to SE Siberia and NE China
     __ *F.p.calidus* Latham, 1790   N European Russia and N Siberia east to Yena R. >>
                                           S Europe, Africa, S and E Asia to New Guinea
     __ *F.p.japonensis* J.F. Gmelin, 1788   NE Siberia to Japan >> Japan, China, Borneo and
                                           Philippines
     __ *F.p.brookei* Sharpe, 1873   Mediterranean to Crimea, Caucasus area and N Iran
     __ *F.p.peregrinator* Sundevall, 1837   Pakistan, India, Sri Lanka to S and E China
     __ *F.p.fruitii* Momiyama, 1927   Ogasawara-shoto (Bonin) Is. (Japan)
     __ *F.p.madens* Ripley & Watson, 1963   Cape Verde Is.
     __ *F.p.minor* Bonaparte, 1850   Gambia to Somalia and Cape Province
     __ *F.p.radama* Hartlaub, 1861   Madagascar, Comoro Is.
     __ *F.p.ernesti* Sharpe, 1894   Indonesia, Philippines, New Guinea, Bismarck Arch.
     ✔ *F.p.macropus* Swainson, 1837[7]   Australia, Tasmania
     __ *F.p.nesiotes* Mayr, 1941   Solomons, Vanuatu, Loyalty Is., New Caledonia, Fiji

○ *Falco pelegrinoides* Barbary Falcon[8]
     __ *F.p.babylonicus* P.L. Sclater, 1861   E Iran to Mongolia and Pakistan
     __ *F.p.pelegrinoides* Temminck, 1829   Canary Is, Saharan hills to Somalia, Arabia, and Israel
                                           to SW Iran >> N tropical Africa

○ *Falco fasciinucha* Reichenow & Neumann, 1895 Taita Falcon   S Ethiopia to Zimbabwe

[1] Includes *centralia* (R.S.)
[2] For correction of spelling see David & Gosselin (2002) (613).
[3] Races recognised follow Stepanyan (1990) (2326).
[4] Incudes *cyanopus* see Stresemann & Amadon (1979: 419) (2368).
[5] Includes *altaicus* a name applied to a morph of this race (Stepanyan, 1990) (2326).
[6] Includes *kreyenborgi* a colour morph, see Ellis & Peres (1983) (799).
[7] Includes *submelanogenys* see Marchant & Higgins (1993: 319) (1428).
[8] Here tentatively treated as a separate species due to distinctiveness, following Vaurie, (1961) (2507), but although *babylonicus* is thought to breed more or less sympatrically with nominate *peregrinus* in NE Iran this needs substantiation.

ACCIPITRIDAE OSPREY, KITES, HAWKS AND EAGLES (67: 233)

SAGITTARIINAE

SAGITTARIUS Hermann, 1783 M
○ *Sagittarius serpentarius* (J.F. Miller, 1779) SECRETARY BIRD Senegal to Somalia and Cape Province

PANDIONINAE

PANDION Savigny, 1809 M
◕ *Pandion haliaetus* OSPREY
    __ *P.h.haliaetus* (Linnaeus, 1758) Europe, Asia >> Africa, India, Sundas
    ☑ *P.h.carolinensis* (J.F. Gmelin, 1788) N America >> C South America
    __ *P.h.ridgwayi* Maynard, 1887 Bahama Is., E Belize, Cuba
    ☑ *P.h.cristatus* (Vieillot, 1816)[1] Australia to New Caledonia, Papuasia, Indonesia, Philippines, Palau Is.

ACCIPITRINAE[2]

AVICEDA Swainson, 1836 F
○ *Aviceda cuculoides* AFRICAN CUCKOO-HAWK
    __ *A.c.cuculoides* Swainson, 1837 Gambia to N Nigeria and N Zaire
    __ *A.c.batesi* (Swann, 1920) Guinea to Cameroon and Congo basin
    __ *A.c.verreauxi* Lafresnaye, 1846[3] Angola to Uganda and E Cape Province

○ *Aviceda madagascariensis* (A. Smith, 1834) MADAGASCAN CUCKOO-HAWK
    Madagascar

○ *Aviceda jerdoni* JERDON'S BAZA
    __ *A.j.jerdoni* (Blyth, 1842) E Himalayas, N India to Indochina, Sumatra and Philippines
    __ *A.j.ceylonensis* (Legge, 1876) SW India, Sri Lanka
    __ *A.j.borneensis* (Brüggemann, 1876) Borneo
    __ *A.j.magnirostris* (Kaup, 1847) Philippines
    __ *A.j.celebensis* (Schlegel, 1873) Sulawesi, Banggai and Sula Is.

○ *Aviceda subcristata* PACIFIC BAZA
    __ *A.s.timorlaoensis* (A.B. Meyer, 1894) Lesser Sundas east to Tanimbar
    __ *A.s.rufa* (Schlegel, 1866) Obi (C Moluccas)
    __ *A.s.stresemanni* (Siebers, 1930) Buru (S Moluccas)
    __ *A.s.reinwardtii* (Schlegel & Müller, 1841) Seram and Ambon (S Moluccas)
    __ *A.s.pallida* (Stresemann, 1913) Seramlaut Is., Kai Is. (S Moluccas)
    __ *A.s.waigeuensis* Mayr, 1940 Waigeo I.
    __ *A.s.obscura* Junge, 1956 Biak I.
    __ *A.s.stenozona* (G.R. Gray, 1858) W New Guinea, Aru Is., Misool I.
    __ *A.s.megala* (Stresemann, 1913) E New Guinea, Yapen, Fergusson and Goodenough Is.
    __ *A.s.coultasi* Mayr, 1945 Admiralty Is.
    __ *A.s.bismarckii* (Sharpe, 1888) E Bismarck Arch.
    __ *A.s.gurneyi* (E.P. Ramsay, 1882)[4] Solomon Is.
    __ *A.s.subcristata* (Gould, 1838)[5] N and E Australia

○ *Aviceda leuphotes* BLACK BAZA
    __ *A.l.syama* (Hodgson, 1837) C and E Himalayas, N Burma and S China >> Indochina
    __ *A.l.wolfei* Deignan, 1948 Sichuan >> ??
    __ *A.l.leuphotes* (Dumont, 1820)[6] Indian plains to Indochina and Hainan
    __ *A.l.andamanica* Abdulali, 1970 Andaman Is.

---

[1] Includes *melvillensis* and *microhaliaetus* see Prevost (1983) (1926). Some authors use *leucocephalus* Gould, 1838, for the Australasian form.
[2] With rare exceptions the sequence of genera and species used here is that of Thiollay (1994) (2398).
[3] Includes *emini* see Stresemann & Amadon (1979: 280) (2368).
[4] Includes *robusta* and *proxima* see Debus (1994: 108) (631).
[5] Includes *njikena* see Marchant & Higgins (1993: 34) (1428).
[6] Includes *burmana* see Stresemann & Amadon (1979: 284) (2368).

LEPTODON Sundevall, 1836 M
- ● **Leptodon cayanensis** GREY-HEADED KITE
  - ☑ *L.c.cayanensis* (Latham, 1790)[1]　　　　　EC Mexico to NW Sth. America and Trinidad
  - — *L.c.monachus* (Vieillot, 1817)　　　　　C Brazil to N Argentina

CHONDROHIERAX Lesson, 1843 M
- ○ **Chondrohierax uncinatus** HOOK-BILLED KITE
  - — *C.u.uncinatus* (Temminck, 1822)[2]　　　　Mexico to N Argentina, Trinidad
  - — *C.u.mirus* Friedmann, 1934　　　　　Grenada
  - — *C.u.wilsonii* (Cassin, 1847)　　　　　E Cuba

HENICOPERNIS G.R. Gray, 1859 M[3]
- ○ **Henicopernis longicauda** (Garnot, 1828) LONG-TAILED BUZZARD[4]　New Guinea, Geelvink Bay, W Papua Is, Aru Is.

- ○ **Henicopernis infuscatus** J.H. Gurney, 1882 NEW BRITAIN HONEY BUZZARD
  　　　　　New Britain

PERNIS Cuvier, 1816[5] M
- ● **Pernis apivorus** (Linnaeus, 1758) EUROPEAN HONEY BUZZARD　Europe, N Asia >> Africa

- ○ **Pernis ptilorhynchus** ORIENTAL HONEY BUZZARD
  - — *P.p.orientalis* Taczanowski, 1891　　　　SE Siberia to NE China >> SE Asia to Sundas and
    　　　　　Philippines
  - — *P.p.ruficollis* Lesson, 1830　　　　　India, Sri Lanka, Burma, SW China
  - — *P.p.torquatus* Lesson, 1830　　　　　Malay Pen., Thailand, Sumatra, Borneo
  - — *P.p.ptilorhynchus* (Temminck, 1821)[6]　　Java
  - — *P.p.palawanensis* Stresemann, 1940　　　Palawan I.
  - — *P.p.philippensis* Mayr, 1939　　　　　Philippines (except Palawan)

- ○ **Pernis celebensis** BARRED HONEY BUZZARD
  - — *P.c.celebensis* Wallace, 1868　　　　　Sulawesi
  - — *P.c.winkleri* Gamauf & Preleuthner, 1998　Luzon I. (908)
  - — *P.c.steerei* W.L. Sclater, 1919　　　　Philippines (main islands except Luzon and Palawan)

LOPHOICTINIA Kaup, 1847 F
- ○ **Lophoictinia isura** (Gould, 1838) SQUARE-TAILED KITE　Australia

HAMIROSTRA T. Brown, 1846 F
- ○ **Hamirostra melanosternon** (Gould, 1841) BLACK-BREASTED BUZZARD　N and C Australia

ELANOIDES Vieillot, 1818 M
- ● **Elanoides forficatus** SWALLOW-TAILED KITE
  - ☑ *E.f.forficatus* (Linnaeus, 1758)　　　　S USA, N Mexico >> S America
  - — *E.f.yetapa* (Vieillot, 1818)　　　　　S Mexico to N Argentina >> S America

MACHEIRAMPHUS Bonaparte, 1850[7] M
- ○ **Macheiramphus alcinus** BAT HAWK
  - — *M.a.anderssoni* (J.H. Gurney, 1866)　　Gambia to Somalia and Natal; Madagascar
  - — *M.a.alcinus* Bonaparte, 1850　　　　Malay Pen., Sumatra, Borneo
  - — *M.a.papuanus* Mayr, 1940　　　　　Lowland E New Guinea

GAMPSONYX Vigors, 1825 M
- ○ **Gampsonyx swainsonii** PEARL KITE
  - — *G.s.leonae* Chubb, 1918　　　　　W Nicaragua, N South America
  - — *G.s.magnus* Chubb, 1918　　　　　S Ecuador, N Peru
  - — *G.s.swainsonii* Vigors, 1825　　　　C Brazil to N Argentina

---

[1]　The validity of *Leptodon forbesi* Swann, 1922, seems not to have been established, but see Teixeira *et al.* (1987) (2393) and Bierregaard (1994: 108) (147).
[2]　Includes *aquilonis* see Bierregaard (1994: 109) (147).
[3]　Gender addressed by David & Gosselin (in press) (614), multiple changes occur in specific names.
[4]　Includes *minimus* and *fraterculus* see Debus (1994: 111) (631).
[5]　Dated 1817 by Stresemann & Amadon (1979: 287) (2368), but see Browning & Monroe (1991) (258) and www.zoonomen.net (A.P. Peterson *in litt.*).
[6]　Original spelling on the wrapper see Froriep (1821) (888), not an emendation *contra* Orta (1994: 111) (1731); the spelling *ptilorhyncus* reported by
　　Stresemann & Amadon (1979) (2368) comes from the text issued some time later.
[7]　See Deignan (1960) (661) and Brooke & Clancey (1981) (238).

ELANUS Savigny, 1809 M

● *Elanus caeruleus* Black-winged Kite[1]

    — *E.c.caeruleus* (Desfontaines, 1789)   SW Iberian Pen., Africa, SW Arabia

    — *E.c.vociferus* (Latham, 1790)   Pakistan to Yunnan (China), Indochina, Malay Pen.

    — *E.c.hypoleucus* Gould, 1859[2]   Sumatra and Java to Philippines, Sulawesi and New Guinea

○ *Elanus axillaris* (Latham, 1802)[3] Black-shouldered Kite[4]   Australia

● *Elanus leucurus* White-tailed Kite

    ✔ *E.l.majusculus* Bangs & Penard, 1920   S USA and E Mexico

    ✔ *E.l.leucurus* (Vieillot, 1818)   N South America to C Chile

○ *Elanus scriptus* Gould, 1842 Letter-winged Kite   C Australia

CHELICTINIA Lesson, 1843 F

○ *Chelictinia riocourii* (Vieillot, 1822) Scissor-tailed Kite   Senegal to Eritrea and N Kenya

ROSTRHAMUS Lesson, 1830 M

● *Rostrhamus sociabilis* Snail Kite

    ✔ *R.s.plumbeus* Ridgway, 1874   Florida

    — *R.s.levis* Friedmann, 1933   Cuba, Isla de la Juventud

    — *R.s.major* Nelson & Goldman, 1933   E Mexico, Guatemala, Belize

    — *R.s.sociabilis* (Vieillot, 1817)   Honduras and Nicaragua to C Argentina

○ *Rostrhamus hamatus* (Temminck, 1821) Slender-billed Kite   E Panama to N Bolivia and E Brazil

HARPAGUS Vigors, 1824 M

○ *Harpagus bidentatus* Double-toothed Kite

    — *H.b.fasciatus* Lawrence, 1869   S Mexico to W Colombia and W Ecuador

    — *H.b.bidentatus* (Latham, 1790)   E Bolivia to E Colombia and E Brazil

○ *Harpagus diodon* (Temminck, 1823) Rufous-thighed Kite   The Guianas, E and C Brazil to Paraguay and N Argentina

ICTINIA Vieillot, 1816 F

◑ *Ictinia mississippiensis* (A. Wilson, 1811) Mississippi Kite   S USA >> C South America

○ *Ictinia plumbea* (J.F. Gmelin, 1788) Plumbeous Kite   EC Mexico to Paraguay and N Argentina >> South America

MILVUS Lacépède, 1799 M

◑ *Milvus milvus* Red Kite

    ✔ *M.m.milvus* (Linnaeus, 1758)   Europe and NW Africa to Turkey and N Iran >> S of range

    — *M.m.fasciicauda* E. Hartert, 1914   Cape Verde Is.

● *Milvus migrans* Black Kite

    ✔ *M.m.migrans* (Boddaert, 1783)[5]   S and C Europe, Cape Verde Is., Madeira, N Africa, SW Asia to W Pakistan >> Africa

    — *M.m.lineatus* (J.E. Gray, 1831)   Asia from Ural Mts to Japan, south to Himalayas and N Indochina >> S Asia

    — *M.m.govinda* Sykes, 1832   Pakistan to S Indochina, Sri Lanka and Malay Pen.

    — *M.m.formosanus* N. Kuroda, Sr., 1920   Hainan I., Taiwan

    — *M.m.affinis* Gould, 1838   Lesser Sundas, Sulawesi, New Guinea, Bismarck Arch., Australia

    — *M.m.aegyptius* (J.F. Gmelin, 1788)[6]   Egypt and Red Sea >> Kenya

    — *M.m.parasitus* (Daudin, 1800)   Senegal to Ethiopia and Cape Province, Madagascar

---

[1] Mees (1982) (1534) argued for recognition of a separate species *hypoleucus* and reviewed alternative treatments; we do not follow him in this or in sinking *vociferus* in nominate *caeruleus*.

[2] Includes *sumatranus* and *wahgiensis* see Mees (1982) (1534).

[3] For reasons to use 1802 not 1801 see Browing & Monroe (1991) (258).

[4] Previously *E. notatus* Gould, 1838.

[5] Includes *tenebrosus* see Hazevoet (1995) (1092).

[6] Includes *arabicus* see Stresemann & Amadon (1979: 298) (2368).

HALIASTUR Selby, 1840 M
○ *Haliastur sphenurus* (Vieillot, 1818) WHISTLING KITE   E New Guinea, New Caledonia, Australia

○ *Haliastur indus* BRAHMINY KITE
     — *H.i.indus* (Boddaert, 1783)   India, Sri Lanka to S China, Burma, Thailand, Vietnam
     — *H.i.intermedius* Blyth, 1865   Malay Pen., Gtr. and Lesser Sundas, Philippines, Sulawesi
     — *H.i.girrenera* (Vieillot, 1822)   Moluccas, Australia, New Guinea, Bismarck Arch.
     — *H.i.flavirostris* Condon & Amadon, 1954   Solomon Is.

HALIAEETUS Savigny, 1809 M
● *Haliaeetus leucogaster* (J.F. Gmelin, 1788) WHITE-BELLIED SEA EAGLE   India to SE China, through New Guinea to Australia

○ *Haliaeetus sanfordi* Mayr, 1935 SANFORD'S SEA EAGLE   Solomon Is.

○ *Haliaeetus vocifer* (Daudin, 1800) AFRICAN FISH EAGLE   Sub-Saharan Africa

○ *Haliaeetus vociferoides* Des Murs, 1845 MADAGASCAN FISH EAGLE   W Madagascar

○ *Haliaeetus leucoryphus* (Pallas, 1771) PALLAS'S FISH EAGLE   C Asia to Pakistan, N India and Burma

○ *Haliaeetus albicilla* WHITE-TAILED SEA EAGLE
     — *H.a.albicilla* (Linnaeus, 1758)   Europe, N Asia, Japan to India, China
     — *H.a.groenlandicus* C.L. Brehm, 1831   Greenland

● *Haliaeetus leucocephalus* BALD EAGLE
     ☑ *H.l.washingtoniensis* (Audubon, 1827)[1]   Aleutian Is., Alaska, Canada, N USA
     — *H.l.leucocephalus* (Linnaeus, 1766)   S USA to NW Mexico

○ *Haliaeetus pelagicus* STELLER'S SEA EAGLE
     — *H.p.pelagicus* (Pallas, 1811)   NE Asia >> NE China, Japan
     — *H.p.niger* Heude, 1887   Korea

ICHTHYOPHAGA[2] Lesson, 1843 F
○ *Ichthyophaga humilis* LESSER FISH EAGLE
     — *I.h.plumbea* (Jerdon, 1871)   Himalayas from Kashmir E and through N Thailand and N Indochina to Hainan
     — *I.h.humilis* (Müller & Schlegel, 1841)   Malay Pen., Borneo, Sumatra to Sulawesi

○ *Ichthyophaga ichthyaetus* (Horsfield, 1821) GREY-HEADED FISH EAGLE   India and Sri Lanka east through mainland SE Asia, the Sundas and Philippines

GYPOHIERAX Rüppell, 1836 M
○ *Gypohierax angolensis* (J.F. Gmelin, 1788) PALM-NUT VULTURE   Gambia to S Sudan and Kenya, south to Cape Province

GYPAETUS Storr, 1784 M
○ *Gypaetus barbatus* BEARDED VULTURE / LAMMERGEIER
     — *G.b.aureus* (Hablizl, 1783)[3]   S Europe, Tibetan Plateau north to Pamir, Tien Shan and Altai Mts., SW and S Asia, east to W Pakistan through the Himalayas to Nepal
     — *G.b.barbatus* (Linnaeus, 1758)   NW Africa
     — *G.b.meridionalis* Keyserling & Blasius, 1840   Yemen, NE and E Africa, Sth. Africa

NEOPHRON Savigny, 1809 M
○ *Neophron percnopterus* EGYPTIAN VULTURE
     — *N.p.percnopterus* (Linnaeus, 1758)   S Europe, Middle East, Africa
     — ¶*N.p.majorensis* Donázar *et al.*, 2002   Canary Islands (753A)
     — *N.p.ginginianus* (Latham, 1790)   Himalayas to S India

NECROSYRTES Gloger, 1841 M
○ *Necrosyrtes monachus* (Temminck, 1823[4]) HOODED VULTURE[5]   Mali to Eritrea, south to South Africa

---

[1] Replaces *alascanus* see Stresemann & Amadon (1979: 301) (2368).
[2] There is still some dispute about the original spelling of this name.
[3] Includes *haemachalanus* see Stresemann & Amadon (1979: 303) (2368). A Central Asian race *altaicus* may prove separable (Roselaar, *in litt.*)
[4] Where plate 222 not 22 — a *lapsus* in Stresemann & Amadon (1979: 305) (2368).
[5] Treated as monotypic by White (1965) (2649).

GYPS Savigny, 1809 M

○ *Gyps africanus* Salvadori, 1865 AFRICAN WHITE-BACKED VULTURE — Senegal to Eritrea and E Africa, south to Botswana and N Transvaal

○ *Gyps bengalensis* (J.F. Gmelin, 1788) INDIAN WHITE-BACKED VULTURE — S Asia from SE Iran to S Indochina

○ *Gyps tenuirostris* G.R. Gray, 1844[1] SLENDER-BILLED VULTURE — Himalayan foothills from Kashmir to Assam and the Gangetic Plain

○ *Gyps indicus* (Scopoli, 1786) INDIAN VULTURE[2] — Plains of Pakistan and India (south of the Gangetic Plain); formerly east to Indochina

○ *Gyps rueppelli* RÜPPELL'S GRIFFON
    — *G.r.rueppelli* (A.E. Brehm, 1852) — Senegal to Sudan, Uganda Kenya and N Tanzania
    — *G.r.erlangeri* Salvadori, 1908 — Ethiopia, Somalia

○ *Gyps himalayensis* Hume, 1869 HIMALAYAN GRIFFON — C Asia to N India

● *Gyps fulvus* EURASIAN GRIFFON
    ⟟ *G.f.fulvus* (Hablizl, 1783) — S Europe, N and NE Africa to C Asia
    — *G.f.fulvescens* Hume, 1869 — Afghanistan to N India

○ *Gyps coprotheres* (J.R. Forster, 1798) CAPE GRIFFON — Namibia, Botswana and SW Zimbabwe to Cape Province

SARCOGYPS Lesson, 1842[3] M

○ *Sarcogyps calvus* (Scopoli, 1786) RED-HEADED VULTURE — Pakistan to Yunnan, Indochina and Malay Pen.

TRIGONOCEPS Lesson, 1842[4] M

○ *Trigonoceps occipitalis* (Burchell, 1824) WHITE-HEADED VULTURE — N tropical, E and S Africa

AEGYPIUS Savigny, 1809 M

○ *Aegypius monachus* (Linnaeus, 1766) CINEREOUS VULTURE — S Europe to SW and C Asia >> N Africa, India and China

TORGOS Kaup, 1828[5] M

○ *Torgus tracheliotus* LAPPET-FACED VULTURE
    — *T.t.negevensis* Bruun, Mendelssohn & Bull, 1981 — Arabia north to S Israel (264)
    — *T.t.nubicus* (H. Smith, 1829) — Mauritania to Egypt and Kenya
    — *T.t.tracheliotus* (J.R. Forster, 1791) — Angola and C Kenya to S Africa

CIRCAETUS Vieillot, 1816 M

○ *Circaetus gallicus* (J.F. Gmelin, 1788) SHORT-TOED EAGLE[6] — S Europe, N Africa to India and China

○ *Circaetus beaudouini* Verreaux & Des Murs, 1862 BEAUDOUIN'S SNAKE EAGLE[7] — Senegal to Sudan

○ *Circaetus pectoralis* A. Smith, 1829 BLACK-BREASTED SNAKE EAGLE — E and S Africa

○ *Circaetus cinereus* Vieillot, 1818 BROWN SNAKE EAGLE — N tropical, E and S Africa

○ *Circaetus fasciolatus* J.H. Gurney, 1861 SOUTHERN BANDED SNAKE EAGLE — E Kenya to Natal

○ *Circaetus cinerascens* J.W. von Müller, 1851 WESTERN BANDED SNAKE EAGLE — Guinea to W Ethiopia, south to Angola and N Zimbabwe

TERATHOPIUS Lesson, 1830 M

○ *Terathopius ecaudatus* (Daudin, 1800) BATELEUR — Senegal to Eritrea, south to Botswana and E Transvaal

SPILORNIS G.R. Gray, 1840 M

○ *Spilornis cheela* CRESTED SERPENT EAGLE[8]
    — *S.c.cheela* (Latham, 1790) — N India, Nepal

---

[1] Previously called *nudiceps* but see Stresemann & Amadon (1979: 306) (2368). For separation of this from the next taxon see Rasmussen & Parry (2001) (1992) and Rasmussen *et al.* (2001) (1996).

[2] This scientific name may require change (P. Rasmussen, *in litt.*).

[3] Treated in the genus *Aegypius* in last Edition.

[4] Treated in the genus *Aegypius* in last Edition.

[5] Treated in the genus *Aegypius* in last Edition.

[6] Includes *heptneri* see Stresemann & Amadon (1979: 309) (2368).

[7] For a defence of the separation of this species and the next see Clark (1999) (534).

[8] Treatment of this has changed often, see Amadon (1974) (46). Stepanyan (1993) (2328) has suggested that there are two species, one small and one large, with both represented on mainland Asia. We await more evidence of this. The Philippine populations is allopatric and we keep *holospilus* in this species, see Dickinson *et al.* (1991) (745).

— *S.c.melanotis* (Jerdon, 1844)	S India
— *S.c.spilogaster* (Blyth, 1852)	Sri Lanka
— *S.c.burmanicus* Swann, 1920	Burma to C and S Indochina
— *S.c.ricketti* W.L. Sclater, 1919[1]	SC and SE China, N Vietnam
— *S.c.malayensis* Swann, 1920	Malay Pen., N Sumatra
— *S.c.davisoni* Hume, 1873	Andaman Is.
— *S.c.minimus* Hume, 1873[2]	C Nicobar Is.
— *S.c.perplexus* Swann, 1922	S Ryukyu Is.
— *S.c.hoya* Swinhoe, 1866	Taiwan
— *S.c.rutherfordi* Swinhoe, 1870	Hainan I.
— *S.c.pallidus* Walden, 1872	N Borneo
— *S.c.richmondi* Swann, 1922[3]	S Borneo
— *S.c.natunensis* Chasen, 1934	Bunguran I., Belitung I.
— *S.c.sipora* Chasen & Kloss, 1926	Mentawei Is., Sipora I.
— *S.c.batu* Meyer de Schauensee & Ripley, 1939	Batu I., S Sumatra
— *S.c.asturinus* A.B. Meyer, 1884	Nias I.
— *S.c.abbotti* Richmond, 1903	Simeuluë I.
— *S.c.bido* (Horsfield, 1821)	Java, Bali
— *S.c.baweanus* Oberholser, 1917	Bawean I.
— *S.c.palawanensis* W.L. Sclater, 1919	Palawan, Balabac, Calamian Is.
— *S.c.holospilus* (Vigors, 1831)[4]	Philippines (except Palawan group)

○ ***Spilornis klossi*** Richmond, 1902 NICOBAR SERPENT EAGLE | Gt. Nicobar I.

○ ***Spilornis kinabaluensis*** W.L. Sclater, 1919 KINABALU SERPENT EAGLE | NE Borneo

○ ***Spilornis rufipectus*** SULAWESI SERPENT EAGLE[5]

— *S.r.rufipectus* Gould, 1858	Sulawesi
— *S.r.sulaensis* (Schlegel, 1866)	Sula, Banggai Is.

○ ***Spilornis elgini*** (Blyth, 1863) ANDAMAN SERPENT EAGLE[6] | Andaman Is.

DRYOTRIORCHIS Shelley, 1874 M
○ ***Dryotriorchis spectabilis*** CONGO SERPENT EAGLE

— *D.s.spectabilis* (Schlegel, 1863)	Liberia to N Cameroon
— *D.s.batesi* Sharpe, 1904	S Cameroon and Gabon to C Zaire

EUTRIORCHIS Sharpe, 1875 M
○ ***Eutriorchis astur*** Sharpe, 1875 MADAGASCAN SERPENT EAGLE | Madagascar (E rainforest)

CIRCUS Lacépède, 1799 M
◑ ***Circus aeruginosus*** WESTERN MARSH HARRIER

⚔ *C.a.aeruginosus* (Linnaeus, 1758)	Europe and Middle East to C Asia >> S Europe, Africa, S Asia
— *C.a.harterti* Zedlitz, 1914	NW Africa

○ ***Circus spilonotus*** EASTERN MARSH HARRIER

— *C.s.spilonotus* Kaup, 1847	E Asia to Japan >> SE Asia to Gtr. Sundas and Philippines
— *C.s.spilothorax* Salvadori & D'Albertis, 1875	New Guinea

◑ ***Circus approximans*** Peale, 1848 SWAMP HARRIER | S New Guinea, Australia[7], New Zealand, SW Pacific islands

○ ***Circus ranivorus*** (Daudin, 1800) AFRICAN MARSH HARRIER | S Sudan and Kenya to Cape Province

○ ***Circus maillardi*** RÉUNION HARRIER

— *C.m.maillardi* J. Verreaux, 1862	Reunion I.
— *C.m.macrosceles* A. Newton, 1863[8]	Madagascar, Comoro Is.

[1] Tentatively includes ¶ *tonkinensis* Stepanyan, 1992 (2327).
[2] In placing this taxon here and separating *klossi* we tentatively follow Clark (1994: 133) (533); we are unsure there is sympatry in the Nicobars.
[3] This was lumped in *pallidus* in Smythies (1981) (2288) and in our last Edition, but recognised in Stresemann & Amadon (1979: 313) (2368) and in Clark (1994: 133) (533).
[4] Includes *panayensis* see Dickinson *et al.* (1991: 125) (745).
[5] May be conspecific with *cheela*; no sympatry involved.
[6] For reasons to recognise this as a second species in the Andamans see Amadon (1974) (46).
[7] Tasmanian birds migrate to Australia and perhaps S New Guinea.
[8] For recognition see Stresemann & Amadon (1979: 321) (2368).

○ *Circus buffoni* (J.F. Gmelin, 1788) LONG-WINGED HARRIER   Colombia, Trinidad, Guianas to C Argentina

○ *Circus assimilis* Jardine & Selby, 1828 SPOTTED HARRIER   Sulawesi, Timor, Sumba, New Guinea, W, N and E
                                                                       Australia

○ *Circus maurus* (Temminck, 1828) BLACK HARRIER   Cape Province >> S Namibia and Transvaal

● *Circus cyaneus* NORTHERN HARRIER/HEN HARRIER
  ✓ *C.c.cyaneus* (Linnaeus, 1766)   Europe, C and N Asia >> south to N Africa, S Asia
  ✓ *C.c.hudsonius* (Linnaeus, 1766)   N and C America >> S to C America and West Indies

● *Circus cinereus* Vieillot, 1816 CINEREOUS HARRIER   W and S Sth. America, Falkland Is.

○ *Circus macrourus* (S.G. Gmelin, 1770) PALLID HARRIER   E Europe and C Asia >> Africa and India

○ *Circus melanoleucos* (Pennant, 1769) PIED HARRIER   EC Asia >> S and mainland SE Asia and Philippines

○ *Circus pygargus* (Linnaeus, 1758) MONTAGU'S HARRIER   W Europe to EC Asia >> sub-Sharan Africa and S Asia

POLYBOROIDES A. Smith, 1829 M
○ *Polyboroides typus* AFRICAN HARRIER-HAWK
  — *P.t.typus* A. Smith, 1829   Sudan and E Africa to N Botswana and E Cape Province
  — *P.t.pectoralis* Sharpe, 1903   Gambia to Gabon and Zaire

○ *Polyboroides radiatus* (Scopoli, 1786) MADAGASCAN HARRIER-HAWK   Madagascar

MELIERAX G.R. Gray, 1840 M
○ *Melierax metabates* DARK CHANTING GOSHAWK
  — *M.m.theresae* R. Meinertzhagen, 1939   SW Morocco
  — *M.m.ignoscens* Friedmann, 1928   Yemen
  — *M.m.neumanni* E. Hartert, 1914   Mali east to N Sudan
  — *M.m.metabates* Heuglin, 1861   Senegal to Eritrea and N Tanzania
  — *M.m.mechowi* Cabanis, 1882   S Tanzania to N Namibia and E Transvaal

○ *Melierax poliopterus* Cabanis, 1869 EASTERN CHANTING GOSHAWK   Ethiopia and Somalia to N Tanzania

○ *Melierax canorus* (Rislachi, 1799)[1] PALE CHANTING GOSHAWK   Namibia, Botswana, South Africa

MICRONISUS G.R. Gray, 1840 M
○ *Micronisus gabar* GABAR GOSHAWK[2]
  — *M.g.niger* (Vieillot, 1823)   N Nigeria to S Yemen south to
  — *M.g.aequatorius* Clancey, 1987   C Africa (509)
  — *M.g.gabar* (Daudin, 1800)   Cape Province to Angola and Malawi

ACCIPITER Brisson, 1760 M
○ *Accipiter poliogaster* (Temminck, 1824) GREY-BELLIED HAWK   N and C Sth. America

○ *Accipiter trivirgatus* CRESTED GOSHAWK
  — *A.t.indicus* (Hodgson, 1836)   NE India to S China, Thailand, Indochina and the
                                                                       Malay Pen.

  — *A.t.formosae* Mayr, 1949[3]   Taiwan
  — *A.t.peninsulae* Koelz, 1949   SW India
  — *A.t.layardi* (Whistler, 1936)   Sri Lanka
  — *A.t.trivirgatus* (Temminck, 1824)   Sumatra
  — *A.t.niasensis* Mayr, 1949   Nias I.
  — *A.t.javanicus* Mayr, 1949   Java
  — *A.t.microstictus* Mayr, 1949   Borneo
  — *A.t.palawanus* Mayr, 1949   Palawan, Calamian Is.
  — *A.t.castroi* Manuel & Gilliard, 1952   Polillo Is.
  — *A.t.extimus* Mayr, 1945   Negros, Samar, Leyte, Mindanao

○ *Accipiter griseiceps* (Schlegel, 1862) SULAWESI GOSHAWK   Sulawesi, Muna, Buton

---

[1] Was called *M. musicus* Daudin, 1800, but this name preoccupied, not due to Thunberg, 1801 (*contra* White, 1965 (2649)), but to Rislachi, 1799, see
   Stresemann & Amadon (1979: 323 fn) (2368).
[2] For recognition of these subspecies, but not *defensorum*, see Clancey (1987) (509).
[3] For recognition see Stresemann & Amadon (1979: 324) (2368); omitted in last Edition.

○ *Accipiter toussenelii* RED-CHESTED GOSHAWK[1]
— *A.t.macroscelides* (Hartlaub, 1855)   Sierra Leone to W Cameroon
— *A.t.lopezi* (Alexander, 1903)   Bioko
— *A.t.toussenelii* (J. & E. Verreaux, 1855)   Cameroon, Gabon, N and W Zaire

○ *Accipiter tachiro* AFRICAN GOSHAWK
— *A.t.canescens* (Chapin, 1921)   E Zaire to W Uganda
— *A.t.unduliventer* (Rüppell, 1836)   Eritrea, N and E Ethiopia
— *A.t.croizati* Desfayes, 1974   SW Ethiopia
— *A.t.sparsimfasciatus* (Reichenow, 1895)   Angola to N Mozambique, E Africa and Somalia
— *A.t.pembaensis* Benson & Elliott, 1975[2]   Pemba I. (Tanzania)
— *A.t.tachiro* (Daudin, 1800)   S Angola and Mozambique to Transvaal and Cape Province

○ *Accipiter castanilius* CHESTNUT-FLANKED SPARROWHAWK
— *A.c.castanilius* Bonaparte, 1853   S Nigeria to Gabon; W Zaire
— *A.c.beniensis* Lönnberg, 1917   E Zaire

○ *Accipiter badius* SHIKRA
— *A.b.sphenurus* (Rüppell), 1836)   Gambia to Eritrea, south to N Tanzania
— *A.b.polyzonoides* A. Smith, 1838   S Tanzania and S Zaire to Cape Province
— *A.b.cenchroides* (Severtsov, 1873)   Transcaucasia, N Iran to C Asia >> NW India
— *A.b.dussumieri* (Temminck, 1824)   Himalayas, N India
— *A.b.badius* (J.F. Gmelin, 1788)   S India, Sri Lanka
— *A.b.poliopsis* (Hume, 1874)   Assam to S China, Indochina, Taiwan

○ *Accipiter butleri* NICOBAR SPARROWHAWK[3]
— *A.b.butleri* (J.J. Gurney, 1898)   Car Nicobar I.
— *A.b.obsoletus* (Richmond, 1902)   Katchal, Camorta (C Nicobars)

○ *Accipiter brevipes* (Severtsov, 1850) LEVANT SPARROWHAWK   Balkans to S Russia >> NE Africa

○ *Accipiter soloensis* (Horsfield, 1821) CHINESE GOSHAWK   S and E China to Korea >> E China to Indonesia and Philippines

○ *Accipiter francesii* FRANCES'S SPARROWHAWK
— *A.f.francesii* A. Smith, 1834   Madagascar
— *A.f.griveaudi* Benson, 1960   Grande Comore I.
— *A.f.pusillus* (J.H. Gurney, 1875)   Anjouan
— *A.f.brutus* (Schlegel, 1866)   Mayotte

○ *Accipiter trinotatus* Bonaparte, 1850 SPOT-TAILED GOSHAWK   Sulawesi, Muna and Buton Is.

○ *Accipiter novaehollandiae* VARIABLE GOSHAWK/GREY GOSHAWK
— *A.n.sylvestris* Wallace, 1864   Sumbawa, Flores, Pantar, Alor Is. (Lesser Sundas)
— *A.n.polionotus* (Salvadori, 1890)   Babar, Damar, Banda, Tanimbar Is. (Banda Sea)
— *A.n.mortyi* E. Hartert, 1925   Morotai I. (Moluccas)
— *A.n.griseogularis* (G.R. Gray, 1860)   Halmahera and Bacan (N Moluccas)
— *A.n.obiensis* (E. Hartert, 1903)   Obi I. (Moluccas)
— *A.n.hiogaster* (S. Müller, 1841)   Seram, Ambon (Moluccas)
— *A.n.pallidiceps* (Salvadori, 1879)   Buru (Moluccas)
— *A.n.albiventris* (Salvadori, 1875)   Kai Is.
— *A.n.leucosomus* (Sharpe, 1874)   New Guinea and Is.
— *A.n.misoriensis* (Salvadori, 1875)   Biak I. (Geelvink Bay)
— *A.n.pallidimas* Mayr, 1940   D'Entrecasteaux Arch. (New Guinea)
— *A.n.misulae* Mayr, 1940   Louisiade Arch. (New Guinea)
— *A.n.lavongai* Mayr, 1945   New Hanover (Bismarck Arch.)
— *A.n.matthiae* Mayr, 1945   St. Matthias I. (Bismarck Arch.)
— *A.n.manusi* Mayr, 1945   Admiralty Is.
— *A.n.dampieri* (J.H. Gurney, 1882)   New Britain, Umboi I.
— *A.n.lihirensis* Stresemann, 1933   Lihir Is., Tanga Is.

[1] For separation of this species see Prigogine (1980) (1937).
[2] Not recognised in last Edition; see Stresemann & Amadon (1979: 326) (2368).
[3] See Rasmussen (2000) (1987) on the sole Great Nicobar record.

— *A.n.bougainvillei* (Rothschild & Hartert, 1905)   Bougainville I., Shortland Is. (Solomons)
— *A.n.rufoschistaceus* (Rothschild & Hartert, 1902)   Ysabel I., Choiseul I. (Solomons)
— *A.n.rubianae* (Rothschild & Hartert, 1905)   New Georgia, Rendova, Vella Lavella I. (Solomons)
— *A.n.malaitae* Mayr, 1931   Malaita I. (Solomons)
— *A.n.pulchellus* (E.P. Ramsay, 1881)   Guadalcanal I. (Solomons)
— *A.n.novaehollandiae* (J.F. Gmelin, 1788)   N and E Australia, Tasmania

○ *Accipiter fasciatus* BROWN GOSHAWK
— *A.f.natalis* (Lister, 1889)   Christmas I. (Indian Ocean)
— *A.f.wallacii* (Sharpe, 1874)   Lombok to Wetar, Damar, Moa I. (Lesser Sundas)
— *A.f.tjendanae* Stresemann, 1925   Sumba I.
— *A.f.stresemanni* Rensch, 1931   Djampea I., Tukangbesi Is.
— *A.f.savu* Mayr, 1941   Savu I. (Lesser Sundas)
— *A.f.hellmayri* Stresemann, 1922[1]   Alor, Samao, Timor (Lesser Sundas)
— *A.f.buruensis* Stresemann, 1914[2]   Buru (Moluccas)
— *A.f.dogwa* Rand, 1941   S New Guinea
— *A.f.polycryptus* Rothschild & Hartert, 1915   E New Guinea
— *A.f.vigilax* (Wetmore, 1926)   New Caledonia, Vanuatu
— *A.f.didimus* (Mathews, 1912)   N Australia
— *A.f.fasciatus* (Vigors & Horsfield, 1827)   Australia, E Solomons

○ *Accipiter melanochlamys* BLACK-MANTLED GOSHAWK
— *A.m.melanochlamys* (Salvadori, 1875)   Montane NW New Guinea
— *A.m.schisacinus* (Rothschild & Hartert, 1903)[3]   Montane C and E New Guinea

○ *Accipiter albogularis* PIED GOSHAWK
— *A.a.eichhorni* E. Hartert, 1926   Feni I. (Bismarck Arch.)
— *A.a.woodfordi* (Sharpe, 1888)   Bougainville, Guadalcanal, Malaita, Choiseul Is.
— *A.a.gilvus* Mayr, 1945   New Georgia, Rendova, Vella Lavella, Kolombangara Is.
— *A.a.albogularis* G.R. Gray, 1870   San Cristobal, Ugi, Santa Ana Is. (Solomons)
— *A.a.sharpei* (Oustalet, 1875)   Vanikoro I., Utupua I. (Santa Cruz Is., Solomons)

○ *Accipiter haplochrous* P.L. Sclater, 1859 WHITE-BELLIED GOSHAWK   New Caledonia

○ *Accipiter rufitorques* (Peale, 1848) FIJI GOSHAWK   Fiji Is.

○ *Accipiter henicogrammus* (G.R. Gray, 1860) MOLUCCAN GOSHAWK   Bacan, Halmahera, Morotai Is. (N Moluccas)

○ *Accipiter luteoschistaceus* Rothschild & Hartert, 1926 SLATY-BACKED SPARROWHAWK
New Britain

○ *Accipiter imitator* E. Hartert, 1926 IMITATOR SPARROWHAWK   Bougainville, Choiseul, Guadalcanal and Santa
Ysabel Island

○ *Accipiter poliocephalus* G.R. Gray, 1858 GREY-HEADED GOSHAWK   New Guinea and islands

○ *Accipiter princeps* Mayr, 1934 NEW BRITAIN GOSHAWK   Montane New Britain

○ *Accipiter superciliosus* TINY HAWK
— *A.s.fontanieri* Bonaparte, 1853   SE Nicaragua to W Colombia, W Ecuador
— *A.s.superciliosus* (Linnaeus, 1766)   E Peru to Venezuela, Guianas, N Argentina

○ *Accipiter collaris* P.L. Sclater, 1860 SEMI-COLLARED HAWK   W Venezuela, E and S Colombia, Ecuador

○ *Accipiter erythropus* RED-THIGHED SPARROWHAWK
— *A.e.erythropus* (Hartlaub, 1855)   Gambia to Nigeria
— *A.e.zenkeri* Reichenow, 1894   Cameroon to W Uganda and N Angola

○ *Accipiter minullus* AFRICAN LITTLE SPARROWHAWK[4]
— *A.m.tropicalis* Reichenow, 1898   S Somalia and coastal Kenya to E Mozambique
— *A.m.minullus* (Daudin, 1800)[5]   Ethiopia and inland E Africa to Angola and Cape
Province

---

[1] Retained following White & Bruce (1986: 122) (2652).
[2] Recognition tentative as status uncertain, see White & Bruce (1986: 123) (2652).
[3] For recognition see Stresemann & Amadon (1979: 333) (2368).
[4] Treated as polytypic by White (1965: 49) (2649).
[5] Tentatively includes ¶*infrequens* Clancey, 1982 (488).

○ *Accipiter gularis* Japanese Sparrowhawk[1][2]

   — *A.g.sibiricus* Stepanyan, 1959   Altai and Mongolia to SE Yakutia and Sea of Okhotsk >> SE Asia to Greater Sundas

   — *A.g.gularis* (Temminck & Schlegel, 1844[3])   Russian Far East, NE China, Sakhalin, Japan >> SE China, Java, Borneo, Philippines and Lesser Sundas

   — *A.g.iwasakii* Mishima, 1962   Iriomote I., Ishigaki I. (S Ryukyu Is.)

○ *Accipiter virgatus* Besra

   — *A.v.affinis* Hodgson, 1836   W Himalayas to N Burma, S and C China, Thailand and N Indochina >> Sthn. mainland SE Asia

   — *A.v.fuscipectus* Mees, 1970   Taiwan
   — *A.v.besra* Jerdon, 1839   S India, Sri Lanka
   — *A.v.vanbemmelli* Voous, 1950   Sumatra
   — *A.v.rufotibialis* Sharpe, 1887   N Borneo
   — *A.v.virgatus* (Temminck, 1822)   Java
   — *A.v.quinquefasciatus* Mees, 1984   Flores (1535)
   — *A.v.abdulalii* Mees, 1981   Andaman Is. (1532)
   — *A.v.confusus* E. Hartert, 1910   Catanduanes Luzon, Mindoro, Negros
   — *A.v.quagga* Parkes, 1973   Bohol, Cebu, Leyte, Mindanao, Samar

○ *Accipiter nanus* (W.H. Blasius, 1897) Dwarf Sparrowhawk   Sulawesi

○ *Accipiter erythrauchen* Rufous-necked Sparrowhawk

   — *A.e.erythrauchen* G.R. Gray, 1861   Bacan, Halmahera, Morotai, Obi Is. (N and C Moluccas)
   — *A.e.ceramensis* (Schlegel, 1862)   Seram, Buru Is. (S Moluccas)

○ *Accipiter cirrhocephalus* Collared Sparrowhawk

   — *A.c.papuanus* (Rothschild & Hartert, 1913)   New Guinea and nearby islands
   — *A.c.rosselianus* Mayr, 1940   Rossel I., Louisiade Arch.
   — *A.c.cirrhocephalus* (Vieillot, 1817)[4]   Australia, Tasmania

○ *Accipiter brachyurus* (E.P. Ramsay, 1879) New Britain Sparrowhawk   New Britain

○ *Accipiter rhodogaster* Vinous-breasted Sparrowhawk

   — *A.r.rhodogaster* (Schlegel, 1862)   Sulawesi, Buton, Muna Is.
   — *A.r.sulaensis* (Schlegel, 1866)   Peleng, Banggai, Sula Is.

○ *Accipiter madagascariensis* A. Smith, 1834 Madagascan Sparrowhawk   Madagascar

○ *Accipiter ovampensis* J.H. Gurney, 1875 Ovambo Sparrowhawk   Ghana to Ethiopia south to N Namibia and E Transvaal

◕ *Accipiter nisus* Eurasian Sparrowhawk

   ✓ *A.n.nisus* (Linnaeus, 1758)   Europe to SW Siberia, N Iran and SW Turkmenistan >> south to NE Africa and SW Asia

   — *A.n.nisosimilis* (Tickell, 1833)   NW and C Siberia to Altai, N China and Japan >> SW, C and E Asia

   — *A.n.dementjevi* Stepanyan, 1958[5]   Pamir-Alai to Tien Shan Mts.
   — *A.n.melaschistos* Hume, 1869   E Afghanistan and Himalayas to W Sichuan
   — *A.n.wolterstorffi* O. Kleinschmidt, 1901   Corsica, Sardinia
   — *A.n.punicus* Erlanger, 1897   NW Africa
   — *A.n.granti* Sharpe, 1890   Madeira, Canary Is.

○ *Accipiter rufiventris* Rufous-breasted Sparrowhawk

   — *A.r.perspicillaris* (Rüppell, 1836)   Ethiopia
   — *A.r.rufiventris* A. Smith, 1830   Zaire and Kenya to Cape Province

◕ *Accipiter striatus* Sharp-shinned Hawk[6]

   — *A.s.perobscurus* Snyder, 1938   Queen Charlotte Is. >> south to Oregon
   ✓ *A.s.velox* (A. Wilson, 1812)   Canada, USA >> Middle America

---

[1] Subspecies recognised as in Stresemann & Amadon (1979: 337) (2368). Treated as monotypic in last Edition.
[2] Tentatively includes *nisoides* see Mees (1981) (1532).
[3] Should perhaps date from 1845 see Holthuis & Sakai (1970) (1128).
[4] The emended name used here is judged to be in prevailing usage; see Art. 33.2 of the Code (I.C.Z.N., 1999) (1178) and p. 000 herein. Includes *quaesitandus* see Debus (1994:157) (631).
[5] For recognition see Stresemann & Amadon (1979: 341) (2368); omitted in last Edition.
[6] We follow Stresemann & Amadon (1979: 344) (2368) and A.O.U. (1998: 93) (3) in retaining S American forms herein.

__ *A.s.suttoni* van Rossem, 1939	N Mexico
__ *A.s.madrensis* Storer, 1952	SW Mexico
__ *A.s.fringilloides* Vigors, 1827	Cuba
__ *A.s.striatus* Vieillot, 1808[1]	Hispaniola
__ *A.s.venator* Wetmore, 1914	Puerto Rico
__ *A.s.chionogaster* (Kaup, 1852)	S Mexico, Guatemala to Nicaragua
__ *A.s.ventralis* P.L. Sclater, 1866	Andes of W Venezuela and Colombia to C Bolivia
__ *A.s.erythronemius* (Kaup, 1850)	E Bolivia and S Brazil to Uruguay and N Argentina

◍ *Accipiter cooperii* (Bonaparte, 1828) Cooper's Hawk   S Canada to N Central America

○ *Accipiter gundlachii* Gundlach's Hawk

__ *A.g.gundlachii* Lawrence, 1860	W and C Cuba
__ *A.g.wileyi* Wotzkow, 1991	E Cuba (2691)

○ *Accipiter bicolor* Bicoloured Hawk

__ *A.b.fidens* Bangs & Noble, 1918	SW Mexico
__ *A.b.bicolor* (Vieillot, 1817)	SE Mexico to E Bolivia
__ *A.b.pileatus* (Temminck, 1823)	S Brazil
__ *A.b.guttifer* Hellmayr, 1917	S Bolivia, W Paraguay, N Argentina
__ *A.b.chilensis* Philippi & Landbeck, 1864[2]	Andes of Chile, Argentina to Tierra del Fuego

○ *Accipiter melanoleucus* Black Sparrowhawk

__ *A.m.temminckii* (Hartlaub, 1855)	Liberia to W Zaire and N Angola
__ *A.m.melanoleucus* A. Smith, 1830	Sudan, Ethiopia and E Africa to E Cape Province

○ *Accipiter henstii* (Schlegel, 1873) Henst's Goshawk   W, N and E Madagascar

◐ *Accipiter gentilis* Northern Goshawk

__ *A.g.buteoides* (Menzbier, 1882)	NE Europe, NW and CN Siberia to Lena R. >> C Europe to C Asia
__ *A.g.albidus* (Menzbier, 1882)	NE Siberia (Lena R. to Kamchatka)
__ *A.g.schvedowi* (Menzbier, 1882)	S Ural region and SW Siberia to Kuril Is., Sakhalin and N China >> C and E Asia
__ *A.g.fujiyamae* (Swann & Hartert, 1923)	Japan
__ *A.g.gentilis* (Linnaeus, 1758)	C and N Europe east to S and C European Russia
__ *A.g.marginatus* (Piller & Mitterpacher, 1783)[3]	Italy and Balkans to Caucasus and N Iran
__ *A.g.arrigonii* (O. Kleinschmidt, 1903)	N Morocco, Iberia, Corsica, Sardinia
__ *A.g.atricapillus* (A. Wilson, 1812)	N America (except as below) >> Sthn. N America
__ *A.g.laingi* (Taverner, 1940)	SW Canada (islands of Brit. Columbia)
✓ *A.g.apache* van Rossem, 1938	SW USA, NW Mexico

○ *Accipiter meyerianus* (Sharpe, 1878) Meyer's Goshawk   Moluccas, Yapen I., New Guinea, New Britain, Solomon Is.

ERYTHROTRIORCHIS Sharpe, 1875[4] M
○ *Erythrotriorchis buergersi* (Reichenow, 1914) Chestnut-shouldered Goshawk
                             Montane E New Guinea
○ *Erythrotriorchis radiatus* (Latham, 1802)[5] Red Goshawk   N and E Australia

MEGATRIORCHIS Salvadori & D'Albertis, 1875 M
○ *Megatriorchis doriae* Salvadori & D'Albertis, 1875 Doria's Goshawk   New Guinea

UROTRIORCHIS Sharpe, 1874 M
○ *Urotriorchis macrourus* (Hartlaub, 1855) Long-tailed Hawk[6]   Liberia to E Zaire

KAUPIFALCO Bonaparte, 1854 M
○ *Kaupifalco monogrammicus* Lizard Buzzard

__ *K.m.monogrammicus* (Temminck, 1824)	Senegal to Ethiopia and Kenya
__ *K.m.meridionalis* (Hartlaub, 1860)	Tanzania to Angola and Natal

[1] Dated 1807 by Stresemann & Amadon (1979: 344) (2368), but see Browning & Monroe (1991) (258).
[2] We follow Stresemann & Amadon (1979: 345) (2368) and A.O.U. (1998: 94) (3) in retaining *chilensis*.
[3] For recognition see Stresemann & Amadon (1979: 346) (2368); omitted in last Edition.
[4] We follow Thiollay (1994) (2398) in using the genera *Erythrotriorchis* and *Megatriorchis*.
[5] For reasons to use 1802 not 1801 see Browing & Monroe (1991) (258).
[6] If treated within the genus *Accipiter* must be called *amadoni* Wolters, 1978,

BUTASTUR Hodgson, 1843 M
○ *Butastur rufipennis* (Sundevall, 1851) GRASSHOPPER BUZZARD   Senegal to Somalia >> N Zaire and E Africa

○ *Butastur teesa* (Franklin, 1831) WHITE-EYED BUZZARD   SE Iran to India and Burma

○ *Butastur liventer* (Temminck, 1827) RUFOUS-WINGED BUZZARD   S Burma to S China, south to Sulawesi and Greater and Lesser Sundas

○ *Butastur indicus* (J.F. Gmelin, 1788) GREY-FACED BUZZARD   E Asia to Japan >> Indochina to Indonesia and Philippines

GERANOSPIZA Kaup, 1844 F
○ *Geranospiza caerulescens* CRANE-HAWK
    __ *G.c.livens* Bangs & Penard, 1921   NW Mexico
    __ *G.c.nigra* (Du Bus, 1847)[1]   Mexico to C Panama
    __ *G.c.balzarensis* W.L. Sclater, 1918   E Panama to NW Peru
    __ *G.c.caerulescens* (Vieillot, 1817)   E Ecuador and Colombia to Guianas, N Brazil
    __ *G.c.gracilis* (Temminck, 1821)   NE Brazil
    __ *G.c.flexipes* J.L. Peters, 1935   S Brazil and Bolivia to N Argentina

LEUCOPTERNIS Kaup, 1847 M[2]
○ *Leucopternis plumbeus* Salvin, 1872 PLUMBEOUS HAWK   E Panama to NW Peru

○ *Leucopternis schistaceus* (Sundevall, 1851) SLATE-COLOURED HAWK   Amazonia

○ *Leucopternis princeps* P.L. Sclater, 1865 BARRED HAWK   Costa Rica to NW Ecuador

○ *Leucopternis melanops* (Latham, 1790) BLACK-FACED HAWK   N Amazonia

○ *Leucopternis kuhli* Bonaparte, 1850 WHITE-BROWED HAWK   S Amazonia

○ *Leucopternis lacernulatus* (Temminck, 1827) WHITE-NECKED HAWK   SE Brazil

◐ *Leucopternis semiplumbeus* Lawrence, 1861 SEMIPLUMBEOUS HAWK   Honduras to NW Ecuador

○ *Leucopternis albicollis* WHITE HAWK
    __ *L.a.ghiesbreghti* (Du Bus, 1845)[3]   S Mexico to Nicaragua
    __ *L.a.costaricensis* W.L. Sclater, 1919   Honduras to NW Colombia
    __ *L.a.williaminae* Meyer de Schauensee, 1950   NW Colombia to W Venezuela
    __ *L.a.albicollis* (Latham, 1790)   Trinidad, C Venezuela and the Guianas to Bolivia

○ *Leucopternis occidentalis* Salvin, 1876 GREY-BACKED HAWK   W Ecuador, NW Peru

○ *Leucopternis polionotus* (Kaup, 1847) MANTLED HAWK   SE Brazil, E Paraguay, E Uruguay

BUTEOGALLUS Lesson, 1830 M
○ *Buteogallus aequinoctialis* (J.F. Gmelin, 1788) RUFOUS CRAB HAWK   Coasts from Venezuela to SE Brazil

◐ *Buteogallus anthracinus* COMMON BLACK HAWK
    ☑ *B.a.anthracinus* (Deppe, 1830)   SW USA to NW Guyana, Trinidad, St Vincent I.
    __ *B.a.utilensis* Twomey, 1956   Bay I., Honduras
    __ *B.a.gundlachii* (Cabanis, 1855)   Cuba, Isla de la Juventud

○ *Buteogallus subtilis* MANGROVE BLACK HAWK[4]
    __ *B.s.rhizophorae* Monroe, 1963   Pacific coast of El Salvador and Honduras
    __ *B.s.bangsi* (Swann, 1922)   Pacific coast of Costa Rica and Panama
    __ *B.s.subtilis* (Thayer & Bangs, 1905)   Pacific coast of Colombia, Ecuador, N Peru

◐ *Buteogallus urubitinga* GREAT BLACK HAWK
    ☑ *B.u.ridgwayi* (J.H. Gurney, 1884)   N Mexico to Panama
    __ *B.u.urubitinga* (J.F. Gmelin, 1788)   N and C Sth. America, Trinidad

○ *Buteogallus meridionalis* (Latham, 1790) SAVANNA HAWK   E Panama and Trinidad south to C Argentina

---

[1] Du Bus is abbreviated from Du Bus de Gisignies.
[2] Gender addressed by David & Gosselin (2002) (614), multiple changes occur in specific names.
[3] Du Bus is abbreviated from Du Bus de Gisignies.
[4] Treated as monotypic in last Edition, but see Stresemann & Amadon (1979: 356) (2368).

PARABUTEO Ridgway, 1874 M
● **Parabuteo unicinctus** HARRIS's HAWK
    — *P.u.harrisi* (Audubon, 1837)[1]   SE California and S Texas to N Peru
    ↓ *P.u.unicinctus* (Temminck, 1824)   S America (except NW)

BUSARELLUS Lesson, 1843 M
○ **Busarellus nigricollis** BLACK-COLLARED HAWK
    — *B.n.nigricollis* (Latham, 1790)   Mexico to N Brazil
    — *B.n.leucocephalus* (Vieillot, 1816)   Paraguay, Uruguay, N Argentina

GERANOAETUS Kaup, 1844 M
○ **Geranoaetus melanoleucus** BLACK-CHESTED BUZZARD-EAGLE
    — *G.m.australis* Swann, 1922   Venezuela and western Sth. America, Tierra del Fuego
    — *G.m.melanoleucus* (Vieillot, 1819)   S Brazil, Uruguay, Paraguay, NE Argentina

HARPYHALIAETUS Lafresnaye, 1842 M
○ **Harpyhaliaetus solitarius** SOLITARY EAGLE
    — *H.s.sheffleri* (van Rossem, 1948)[2]   W Mexico to Panama
    — *H.s.solitarius* (Tschudi, 1844)   Colombia to the Guianas, south to NW Argentina

○ **Harpyhaliaetus coronatus** (Vieillot, 1817) CROWNED EAGLE   S Brazil to SE Bolivia and C Argentina

ASTURINA Vieillot, 1816[3] F
○ **Asturina nitida** GREY HAWK
    — *A.n.plagiata* Schlegel, 1862   S USA to NW Costa Rica
    — *A.n.costaricensis* Swann, 1922   SW Costa Rica to W Ecuador
    — *A.n.nitida* (Latham, 1790)   Trinidad, N Amazonia
    — *A.n.pallida* Todd, 1915   S Brazil, E Bolivia to NC Argentina

BUTEO Lacépède, 1799 M
● **Buteo magnirostris** ROADSIDE HAWK
    ↓ *B.m.griseocauda* Ridgway, 1873   Mexico to W Peru
    — *B.m.conspectus* (J.L. Peters, 1913)   SE Mexico, N Belize
    — *B.m.gracilis* (Ridgway, 1885)   Cozumel, Holbox Is.
    — *B.m.sinushonduri* Bond, 1936   Bonacca I., Ruatan I. (Honduras)
    — *B.m.petulans* van Rossem, 1935   SW Costa Rica, SW Panama
    — *B.m.alius* (Peters & Griscom, 1929)   Pearl Is. (Panama)
    — *B.m.magnirostris* (J.F. Gmelin, 1788) S America (N of R. Amazon)
    — *B.m.occiduus* (Bangs, 1911)   E Peru, W Brazil
    — *B.m.saturatus* (Sclater & Salvin, 1876) Bolivia to W Argentina
    — *B.m.nattereri* (Sclater & Salvin, 1869) NE Brazil
    — *B.m.magniplumis* (Bertoni, 1901)   S Brazil, NW Argentina
    — *B.m.pucherani* (J. & E. Verreaux, 1855) E Argentina, Paraguay

● **Buteo lineatus** RED-SHOULDERED HAWK
    ↓ *B.l.lineatus* (J.F. Gmelin, 1788)   E Nth America
    — *B.l.alleni* Ridgway, 1885   Florida to E Texas
    — *B.l.extimus* Bangs, 1920   S Florida and Keys
    — *B.l.texanus* Bishop, 1912   SC Texas to C Mexico
    ↓ *B.l.elegans* Cassin, 1856   S Oregon to Baja California

○ **Buteo ridgwayi** (Cory, 1883) RIDGWAY's HAWK   Hispaniola

● **Buteo platypterus** BROAD-WINGED HAWK
    ↓ *B.p.platypterus* (Vieillot, 1823)   SE Canada, E USA >> Peru and Brazil
    — *B.p.cubanensis* Burns, 1911   Cuba
    — *B.p.brunnescens* Danforth & Smyth, 1935 Puerto Rico
    — *B.p.insulicola* Riley, 1908   Antigua I.

---

[1] Includes *superior* see Stresemann & Amadon (1979: 358) (2368).
[2] For recognition see Stresemann & Amadon (1979: 360) (2368); treated as a synonym in last Edition.
[3] Genus retained following A.O.U. (1998: 97) (3), but placed next to *Buteo* as in Thiollay (1994) (2398).

◯ — *B.p.rivierei* Verrill, 1905   Dominica, Martinique, St Lucia Is.

— *B.p.antillarum* Clark, 1905   Barbados, St Vincent, Grenada, Tobago Is

◯ ***Buteo leucorrhous*** (Quoy & Gaimard, 1824)   WHITE-RUMPED HAWK   Andes from Venezuela to Bolivia; also NE Argentina, Paraguay, SE Brazil

◯ ***Buteo brachyurus*** SHORT-TAILED HAWK

— *B.b.fuliginosus* P.L. Sclater, 1858   S Florida to Panama

— *B.b.brachyurus* Vieillot, 1816   S America (Trinidad to N Argentina)

◯ ***Buteo albigula*** Philippi, 1899   WHITE-THROATED HAWK   Andes from Colombia to C Chile and NW Argentina

◑ ***Buteo swainsoni*** Bonaparte, 1838   SWAINSON'S HAWK   W Canada, W USA, NW Mexico >> South America

◯ ***Buteo albicaudatus*** WHITE-TAILED HAWK

— *B.a.hypospodius* J.H. Gurney, 1876   Texas to N Colombia, W Venezuela

— *B.a.colonus* Berlepsch, 1892   E Colombia to Surinam

— *B.a.albicaudatus* Vieillot, 1816   S Brazil to C Argentina

◯ ***Buteo galapagoensis*** (Gould, 1837)   GALAPAGOS HAWK   Galapagos Is.

● ***Buteo polyosoma*** RED-BACKED HAWK

✓ *B.p.polyosoma* (Quoy & Gaimard, 1824)   NW Colombia to Tierra del Fuego, Falkland Is.

— *B.p.exsul* Salvin, 1875   Robinson Crusoe I. (Juan Fernandez Is. (Chile)[1]

◯ ***Buteo poecilochrous*** J.H. Gurney, 1879   PUNA HAWK/VARIABLE HAWK

     Andes from S Colombia to N Chile, NW Argentina

◑ ***Buteo albonotatus*** Kaup, 1847   ZONE-TAILED HAWK   SW USA to NC Sth. America

◑ ***Buteo solitarius*** Peale, 1848   HAWAIIAN HAWK   Hawaii Is.

◑ ***Buteo jamaicensis*** RED-TAILED HAWK

— *B.j.alascensis* Grinnell, 1909   SE Alaska, SW Canada

✓ *B.j.harlani* (Audubon, 1830)   N Brit. Columbia, N Alberta >> S USA

✓ *B.j.calurus* Cassin, 1856   W Nth America >> C America

✓ *B.j.borealis* (J.F. Gmelin, 1788)   E Nth America >> N Mexico

✓ *B.j.kriderii* Hoopes, 1873[2]   SC Canada, NC USA >> S USA

— *B.j.socorroensis* Nelson, 1898   Socorro I. (W of Mexico)

— *B.j.suttoni* Dickermann, 1993   S Baja California (723)

✓ *B.j.fuertesi* Sutton & Van Tyne, 1935   SW USA, NW Mexico

— *B.j.fumosus* Nelson, 1898   Tres Marias Is. (off WC Mexico)

— *B.j.hadropus* Storer, 1962   north to SC Mexico

— *B.j.kemsiesi* Oberholser, 1959   S Mexico to Nicaragua

— *B.j.costaricensis* Ridgway, 1874   Costa Rica, W Panama

— *B.j.umbrinus* Bangs, 1901   S Florida, Bahama Is.

— *B.j.solitudinis* Barbour, 1935   Cuba, Isla de la Juventud, Hispaniola, Puerto Rico, N Lesser Antilles

— *B.j.jamaicensis* (J.F. Gmelin, 1788)   Jamaica

◯ ***Buteo ventralis*** Gould, 1837   RUFOUS-TAILED HAWK   S Chile, S Argentina

◑ ***Buteo buteo*** EURASIAN BUZZARD[3]

— *B.b.bannermani* Swann, 1919   Cape Verde Is.

— *B.b.insularum* Floericke, 1903[4]   Canary Is. and Azores

— *B.b.harterti* Swann, 1919   Madeira

✓ *B.b.pojana* (Savi, 1822)   Corsica, Sardinia, Sicily, C and S Italy

✓ *B.b.buteo* (Linnaeus, 1758)   Europe (except SC) east to S Finland, Rumania and W Turkey

— *B.b.vulpinus* (Gloger, 1833)   N and E Europe, C Asia >> E and S Africa and S Asia

— *B.b.menetriesi* Bogdanov, 1879   Crimea, C and E Turkey and Caucasus to N Iran

— *B.b.japonicus* Temminck & Schlegel, 1844[5]   NE China and E Siberia to to Sakhalin and Japan >> S and E Asia

[1] Bierregaard (1994: 183) (147) wrote Alejandro Selkirk I., which is Masatierra; Másafuera is Robinson Crusoe I.
[2] 'Probably not a valid subspecies' (Preston & Beane, 1993) (1925).
[3] Subspecies recognised following Stresemann & Amadon (1979: 371) (2368).
[4] Includes *rothschildi* but this may be recognisable.
[5] Should perhaps date from 1845 see Holthuis & Sakai (1970) (1128).

○ — *B.b.refectus* Portenko, 1929   Himalayas and E Tibetan Plateau

— *B.b.toyoshimai* Momiyama, 1927   Ogasawara-shoto (Bonin) Is., Izu Is.

— *B.b.oshiroi* N. Kuroda, Jr., 1971   Daito Is.

○ ***Buteo oreophilus*** Mountain Buzzard

— *B.o.oreophilus* Hartert & Neumann, 1914   Ethiopia to Malawi

— *B.o.trizonatus* Rudebeck, 1957   Transvaal to S Cape Province

○ ***Buteo brachypterus*** Hartlaub, 1860 Madagascan Buzzard   Madagascar

○ ***Buteo rufinus*** Long-legged Buzzard

— *B.r.rufinus* (Cretzschmar, 1827)   C Europe to C Asia >> N tropical Africa

— *B.r.cirtensis* (Levaillant, 1850)   N Africa

○ ***Buteo hemilasius*** Temminck & Schlegel, 1844[1] Upland Buzzard   C and E Asia

● ***Buteo regalis*** (G.R. Gray, 1844) Ferruginous Hawk   SW Canada, WC USA >> N Mexico

● ***Buteo lagopus*** Rough-legged Buzzard

— *B.l.lagopus* (Pontoppidan, 1763)   N Europe, NW Siberia >> S to C and E Europe, SW Asia

— *B.l.menzbieri* Dementiev, 1951   NC and NE Siberia >> S to SE Europe, C and E Asia

— *B.l.kamchatkensis* Dementiev, 1931   Shores of Sea of Okhotsk to Kamchatka, N Kuril Is.

✔ *B.l.sanctijohannis* (J.F. Gmelin, 1788)   Alaska, Canada >> N USA

○ ***Buteo auguralis*** Salvadori, 1865 Red-necked Buzzard   Sierra Leone to Ethiopia, W Uganda and Angola

○ ***Buteo augur*** Augur Buzzard[2]

— *B.a.archeri* W.L. Sclater 1918   N Somalia

— *B.a.augur* Rüppell, 1836   E Sudan and Ethiopia to Zimbabwe, C Namibia and W Angola

○ ***Buteo rufofuscus*** (J.R. Forster, 1798) Jackal Buzzard   S Namibia, South Africa

MORPHNUS Dumont, 1816 M

○ *Morphnus guianensis* (Daudin, 1800) Crested Eagle   Honduras to N Paraguay and N Argentina

HARPIA Vieillot, 1816 F

○ *Harpia harpyja* (Linnaeus, 1758) Harpy Eagle   S Mexico to E Bolivia and N Argentina

HARPYOPSIS Salvadori, 1875 F

○ *Harpyopsis novaeguineae* Salvadori, 1875 Papuan Harpy Eagle   New Guinea

PITHECOPHAGA Ogilvie-Grant, 1896 F

○ *Pithecophaga jefferyi* Ogilvie-Grant, 1896 Philippine Eagle   Luzon, Samar, Mindanao

ICTINAETUS Blyth, 1843 M

○ *Ictinaetus malayensis* Indian Black Eagle

— *I.m.perniger* (Hodgson, 1836)   India, Sri Lanka

— *I.m.malayensis* (Temminck, 1822)   Burma to S China and Moluccas

AQUILA Brisson, 1760 F

○ ***Aquila pomarina*** Lesser Spotted Eagle

— *A.p.pomarina* C.L. Brehm, 1831   C Europe to Ukraine, Caucasus area and N Iran >> SW Asia and E and S Africa

— *A.p.hastata* (Lesson, 1831)[3]   India, N Burma

○ ***Aquila clanga*** Pallas, 1811 Greater Spotted Eagle   E Europe to E Asia >> SE Europe, NE Africa, S Asia to Malay Pen.

○ ***Aquila rapax*** Tawny Eagle

— *A.r.vindhiana* Franklin, 1831   Pakistan, India, N Burma

— *A.r.belisarius* (Levaillant, 1850)   Morocco to Nigeria and Ethiopia, SW Arabia

— *A.r.rapax* (Temminck, 1828)   Angola, Uganda and Kenya to Sth. Africa

---

[1] Should perhaps date from 1845 see Holthuis & Sakai (1970) (1128).
[2] Treated as two species by Thiollay (1994) (2398); we await detailed studies.
[3] Given as 1834 in Stresemann & Amadon (1979) (2368) but see Browning & Monroe (1991) (258). The priority of this name compared to Brehm's *pomarina* has not been explored.

○ *Aquila nipalensis* STEPPE EAGLE[1]
   — *A.n.orientalis* Cabanis, 1854   E Europe to C Kazakhstan >> SW Asia, Middle East
                                    and E Africa

   — *A.n.nipalensis* Hodgson, 1833   E Kazakhstan to Mongolia and N China >> S Asia

○ *Aquila adalberti* C.L. Brehm, 1861 SPANISH IMPERIAL EAGLE[2]   Iberia

○ *Aquila heliaca* Savigny, 1809 EASTERN IMPERIAL EAGLE   SE Europe to C Siberia >> NE Africa, S Asia

○ *Aquila gurneyi* G.R. Gray, 1860 GURNEY'S EAGLE   Moluccas, New Guinea and nearby islands

◉ *Aquila chrysaetos* GOLDEN EAGLE
   ✓ *A.c.chrysaetos* (Linnaeus, 1758)   Europe (except Iberia), NW and CN Asia to Yenisey R.
   — *A.c.kamtschatica* Severtsov, 1888[3]   NE Asia (Yenisey and Mongolia to Kamchatka and NE
                                    China)

   — *A.c.japonica* Severtsov, 1888   Korea, Japan
   — *A.c.daphanea* Severtsov, 1873   Mts. of C Asia from Pamir to C China
   — *A.c.homeyeri* Severtsov, 1888   Iberia, N Africa, Middle East and Arabia to Caucasus
                                    area, Iran and E Uzbekistan

   — *A.c.*subsp. ?   Ethiopia
   ✓ *A.c.canadensis* (Linnaeus, 1758)   Canada, W USA

○ *Aquila audax* WEDGE-TAILED EAGLE
   — *A.a.audax* (Latham, 1802)[4]   S New Guinea, Australia
   — *A.a.fleayi* Condon & Amadon, 1954   Tasmania

○ *Aquila verreauxii* Lesson, 1830 VERREAUX'S EAGLE   Israel, Arabia, Ethiopia, Sudan to Cape Province

HIERAAETUS Kaup, 1844 M
○ *Hieraaetus wahlbergi* (Sundevall, 1851) WAHLBERG'S EAGLE[5]   Gambia to Eritrea south to Transvaal and Natal

○ *Hieraaetus fasciatus* BONELLI'S EAGLE
   — *H.f.fasciatus* (Vieillot, 1822)   N Africa and S Europe to S and C Asia and along Red
                                    Sea

   — *H.f.renschi* Stresemann, 1932   Lesser Sundas

○ *Hieraaetus spilogaster* (Bonaparte, 1850) AFRICAN HAWK-EAGLE   Senegal to Eritrea south to Cape Province

◉ *Hieraaetus pennatus* (J.F. Gmelin, 1788) BOOTED EAGLE[6]   S Europe and N Africa to NW India and C Asia, Sth.
                                    Africa >> E and S Africa

○ *Hieraaetus morphnoides* LITTLE EAGLE
   — *H.m.weiskei* (Reichenow, 1900)   Montane E and C New Guinea
   — *H.m.morphnoides* (Gould, 1841)   Australia

○ *Hieraaetus ayresii* (J.H. Gurney, 1862) AYRES'S HAWK-EAGLE[7]   Nigeria to Ethiopia and Cape Province

○ *Hieraaetus kienerii* RUFOUS-BELLIED HAWK-EAGLE
   — *H.k.kienerii* (I. Geoffroy Saint-Hilaire, 1835)   S Himalayas, W India, Sri Lanka
   — *H.k.formosus* Stresemann, 1924   Burma through mainland SE Asia to Sundas,
                                    Philippines and Sulawesi

POLEMAETUS Heine, 1890[8] M
○ *Polemaetus bellicosus* (Daudin, 1800) MARTIAL EAGLE   Senegal to Somalia and Cape Province

SPIZASTUR G.R. Gray, 1841 M
○ *Spizastur melanoleucus* (Vieillot, 1816) BLACK-AND-WHITE HAWK-EAGLE   E and S Mexico to Paraguay and NE Argentina

LOPHAETUS Kaup, 1847[9] M
○ *Lophaetus occipitalis* (Daudin, 1800) LONG-CRESTED EAGLE   Senegal to Ethiopia, S to Angola and E Cape Province

---

[1] For reasons to split this from *A. rapax* see Clark (1992) (532) and Sangster *et al.* (2002) (2167).
[2] For reasons to separate this from the next species see Hiraldo *et al.* (1976) (1123) and Sangster *et al.* (2002) (2167).
[3] Recognised by Stepanyan (1990) (2326).
[4] For reasons to use 1802 not 1801 see Browing & Monroe (1991) (258).
[5] For treatment in this genus we follow Smeenk (1974) (2279), as in last Edition, but confirmation needed.
[6] Includes *milvoides* and *harterti* as we now prefer to treat this as monotypic, see Kemp (1994: 199) (1250). However ¶*minisculus* Yosef *et al.*, 2000 (2701), needs evaluation.
[7] Previously called *H. dubius* (e.g. in Stresemann & Amadon, 1979) (2368), but Kemp (1994: 200) (1250) confirmed that that name related to *H. pennatus*.
[8] This genus is employed following Stresemann & Amadon (1979: 390) (2368).
[9] This genus is used as in Stresemann & Amadon (1979: 385) (2368).

SPIZAETUS Vieillot, 1816 M
○ *Spizaetus africanus* (Cassin, 1865) CASSIN'S HAWK-EAGLE | Togo to Zaire and W Uganda

○ *Spizaetus cirrhatus* CHANGEABLE HAWK-EAGLE
— *S.c.limnaeetus* (Horsfield, 1821) | Himalaya foothills E mainland SE Asia, and south to Gtr. Sundas and Philippines.

— *S.c.cirrhatus* (J.F. Gmelin, 1788) | S India
— *S.c.ceylanensis* (J.F. Gmelin, 1788) | Sri Lanka
— *S.c.andamanensis* Tytler, 1865 | Andaman Is.
— *S.c.vanheurni* Junge, 1936 | Simeuluě I. (W Sumatran Is.)
— *S.c.floris* (E. Hartert, 1898) | Sumbawa, Flores Is. (Lesser Sundas)

○ *Spizaetus nipalensis* MOUNTAIN HAWK-EAGLE
— *S.n.nipalensis* (Hodgson, 1836) | Himalayas east to China and south to Malay Pen.
— *S.n.kelaarti* Legge, 1878 | Sri Lanka
— *S.n.orientalis* Temminck & Schlegel, 1844[1] | Japan >> NE China

○ *Spizaetus alboniger* (Blyth, 1845) BLYTH'S HAWK-EAGLE | S Burma to Sumatra and Borneo

○ *Spizaetus bartelsi* Stresemann, 1924 JAVAN HAWK-EAGLE | Java

○ *Spizaetus lanceolatus* Temminck & Schlegel, 1844[2] SULAWESI HAWK-EAGLE
| Sulawesi, Banggai and Sula Is.

○ *Spizaetus philippensis* PHILIPPINE HAWK-EAGLE
— *S.p.philippensis* Gould, 1863 | Luzon
— *S.p.pinskeri* Preleuthner & Gamauf, 1998 | Basilan, Leyte, Mindanao, Mindoro, Negros, Samar, Siquijor (1923)

○ *Spizaetus nanus* WALLACE'S HAWK-EAGLE
— *S.n.nanus* Wallace, 1868 | Malay Pen., Sumatra and Borneo
— *S.n.stresemanni* Amadon, 1953 | Nias I. (off W Sumatra)

○ *Spizaetus tyrannus* BLACK HAWK-EAGLE
— *S.t.serus* Friedmann, 1950 | C Mexico to E Peru
— *S.t.tyrannus* (Wied, 1820) | E and S Brazil

○ *Spizaetus ornatus* ORNATE HAWK-EAGLE
— *S.o.vicarius* Friedmann, 1935 | Mexico to Colombia, W Ecuador
— *S.o.ornatus* (Daudin, 1800) | C Colombia to Guianas and N Argentina

STEPHANOAETUS W.L. Sclater, 1922[3] M
○ *Stephanoaetus coronatus* (Linnaeus, 1766) CROWNED HAWK-EAGLE | Guinea to W Ethiopia; Zaire, E Africa and E Cape Province

OROAETUS Ridgway, 1920[4] M
○ *Oroaetus isidori* (Des Murs, 1845) BLACK-AND-CHESTNUT EAGLE | W Venezuela to Bolivia and NW Argentina

OTIDAE BUSTARDS (11: 26)

OTIS Linnaeus, 1758 F
○ *Otis tarda* GREAT BUSTARD
— *O.t.tarda* Linnaeus, 1758[5] | SW and C Europe and NW Africa to Kazakhstan and W Xinjiang

— *O.t.dybowskii* Taczanowski, 1874 | Altai and Mongolia to Russian Far East and NE China

ARDEOTIS Le Mahout, 1853 F
○ *Ardeotis arabs* ARABIAN BUSTARD
— †?*A.a.lynesi* (Bannerman, 1930) | W Morocco
— *A.a.stieberi* (Neumann, 1907)[6] | Senegal and Gambia to E Sudan

---

[1] Should perhaps date from 1845 see Holthuis & Sakai (1970) (1128).
[2] Should perhaps date from 1845 see Holthuis & Sakai (1970) (1128).
[3] This genus is used as in Stresemann & Amadon (1979: 389) (2368).
[4] This genus is used as in Stresemann & Amadon (1979: 389) (2368).
[5] Includes *korejewi* (see Cramp *et al.*, 1980) (589).
[6] Includes *geyri* Niethammer, 1954 (1651).

— *A.a.arabs* (Linnaeus, 1758) NE Sudan, Eritrea, Ethiopia, NW Somalia, Djibouti, SW Arabia

— *A.a.butleri* (Bannerman, 1930) S Sudan, NW Kenya

○ *Ardeotis kori* Kori Bustard

    — *A.k.struthiunculus* (Neumann, 1907) N Somalia and Ethiopia to Kenya and N Tanzania
    — *A.k.kori* (Burchell, 1822) S Angola to Zimbabwe and Sth. Africa

○ *Ardeotis nigriceps* (Vigors, 1831) Great Indian Bustard W and C India

○ *Ardeotis australis* (J.E. Gray, 1829) Australian Bustard S New Guinea, Australia

CHLAMYDOTIS Lesson, 1839 F

○ *Chlamydotis undulata* Houbara Bustard[1]

    — *C.u.fuertaventurae* (Rothschild & Hartert, 1894) E Canary Is.
    — *C.u.undulata* (Jacquin, 1784) N Africa east to Nile valley

○ *Chlamydotis macqueenii* (J.E. Gray, 1832) Macqueen's Bustard

    Arabia, Middle East and lower Ural R. to W Pakistan, N China and Mongolia >> N Africa to NW India, C China

NEOTIS Sharpe, 1893 F

○ *Neotis ludwigii* (Rüppell, 1837) Ludwig's Bustard SW Angola, W Namibia to S Sth. Africa

○ *Neotis denhami* Denham's Bustard

    — *N.d.denhami* (Children, 1826) Senegal, Gambia and SW Mauritania to W Ethiopia and N Uganda

    — *N.d.jacksoni* Bannerman, 1930 Kenya to N Zimbabwe, S Congo and W Zaire, SW Angola
    — *N.d.stanleyi* (J.E. Gray, 1831) Sth. Africa

○ *Neotis heuglinii* (Hartlaub, 1859) Heuglin's Bustard Eritrea, E and S Ethiopia, Djibouti, N and SC Somalia, N Kenya

○ *Neotis nuba* (Cretzschmar, 1826) Nubian Bustard[2] W Mauritania to E Sudan

EUPODOTIS Lesson, 1839 F

○ *Eupodotis senegalensis* White-bellied Bustard

    — *E.s.senegalensis* (Vieillot, 1820) SW Mauritania and Guinea to C Sudan and Eritrea
    — *E.s.canicollis* (Reichenow, 1881)[3] Ethiopia and Somalia to Uganda, N and E Kenya and NE Tanzania

    — *E.s.erlangeri* (Reichenow, 1905) SW Kenya, W Tanzania
    — *E.s.mackenziei* C.M.N. White, 1945 E Gabon, S Zaire to W Zambia, C and S Angola (2629)
    — *E.s.barrowii* (J.E. Gray, 1829) C and SE Botswana, E Sth. Africa, Swaziland

○ *Eupodotis caerulescens* (Vieillot, 1820) Blue Bustard E Cape Province to S Transvaal, Lesotho

○ *Eupodotis vigorsii* Vigors's Bustard

    — *E.v.namaqua* (Roberts, 1932)[4] S Namibia, NW Cape Province
    — *E.v.vigorsii* (A. Smith, 1831)[5] Orange Free State, S Cape Province

○ *Eupodotis rueppellii* Rüppell's Bustard

    — *E.r.rueppellii* (Wahlberg, 1856) NW Namibia, S Angola
    — *E.r.fitzsimonsi* (Roberts, 1937) WC Namibia (2089)

○ *Eupodotis humilis* (Blyth, 1856) Little Brown Bustard E Ethiopia, N Somalia

LOPHOTIS Reichenbach, 1848 F

○ *Lophotis savilei* Lynes, 1920 Savile's Bustard SW Mauritania and Senegal to C Sudan

○ *Lophotis gindiana* (Oustalet, 1881) Buff-crested Bustard Ethiopia and Somalia to NE Tanzania

---

[1] We provisionally accept the suggestion of Gaucher *et al.* (1996) (930) that *macqueeni* should be separated.
[2] Includes *agaze* Vaurie, 1961 (2506) (see Collar, 1996) (544).
[3] Includes *parva* Moltoni, 1935 (1598) see White (1965: 115) (2649).
[4] Includes *harei* Roberts, 1937 (2089), see White (1965: 114) (2649), and apparently *barlowi* (Roberts, 1937) (2089) still recognised by White.
[5] Includes *karrooensis* Vincent, 1949 (2530), see White (1965: 114) (2649).

○ *Lophotis ruficrista* (A. Smith, 1836) Red-crested Bustard[1]
    S Angola, S Zambia and NE Namibia to S
      Mozambique, N Sth. Africa, Swaziland

AFROTIS G.R. Gray, 1855 F
○ *Afrotis afra* (Linnaeus, 1758) Black Bustard
    W and S Cape Province

○ *Afrotis afraoides* White-quilled Bustard
   — *A.a.etoschae* (Grote, 1922)
    NW Namibia, NE Botswana
   — *A.a.damarensis* Roberts, 1926[2]
    Namibia, W and C Botswana
   — *A.a.afraoides* (A. Smith, 1831)[3]
    SE Botswana and N and NE Sth. Africa to Lesotho

LISSOTIS Reichenbach, 1848 F
○ *Lissotis melanogaster* Black-bellied Bustard
   — *L.m.melanogaster* (Rüppell, 1835)
    Senegal to Ethiopia, south to S Angola and Zambia
   — *L.m.notophila* Oberholser, 1905
    Zimbabwe and S Mozambique to E Cape Province

○ *Lissotis hartlaubii* (Heuglin, 1863) Hartlaub's Bustard
    E and S Sudan and E Ethiopia to Kenya and NE
      Tanzania

HOUBAROPSIS Sharpe, 1893 F
○ *Houbaropsis bengalensis* Bengal Florican
   — *H.b.bengalensis* (J.F. Gmelin, 1789)
    S Nepal to NE India
   — *H.b.blandini* Delacour, 1928
    S Cambodia, S Vietnam

SYPHEOTIDES Lesson, 1839 M
○ *Sypheotides indicus* (J.F. Miller, 1782) Lesser Florican[4]
    Pakistan, India, S Nepal

TETRAX T. Forster, 1817 M
○ *Tetrax tetrax* (Linnaeus, 1758) Little Bustard[5]
    SW Europe and NW Africa to NW China >>
      Mediterranean to S Asia

MESITORNITHIDAE MESITES (2: 3)

MESITORNIS Bonaparte, 1855 M
○ *Mesitornis variegatus* (I. Geoffroy Saint-Hilaire, 1838) White-breasted Mesite[6]
    W and N Madagascar

○ *Mesitornis unicolor* (Des Murs, 1845) Brown Mesite
    E Madagascar

MONIAS Oustalet & Grandidier, 1903 M
○ *Monias benschi* Oustalet & Grandidier, 1903 Subdesert Mesite
    SW Madagascar

CARIAMIDAE SERIEMAS (2: 2)

CARIAMA Brisson, 1760 F
○ *Cariama cristata* (Linnaeus, 1766) Red-legged Seriema[7]
    C and E Brazil, E Bolivia and Paraguay to Uruguay and
      C Argentina

CHUNGA Hartlaub, 1860 F
○ *Chunga burmeisteri* (Hartlaub, 1860) Black-legged Seriema
    S Bolivia and W Paraguay to C Argentina

RHYNOCHETIDAE KAGU (1: 1)

RHYNOCHETOS Verreaux & Des Murs, 1860 M
○ *Rhynochetos jubatus* Verreaux & Des Murs, 1860 Kagu
    New Caledonia

---

[1] Tentatively includes *ochrofacies* Clancey, 1977 (461).
[2] Includes *kalaharica* Roberts, 1932, see Clancey (1980) (482). Others perceive different bounds to these subspecies and place *damarensis* in *afraoides*, using *kalaharica* for Botswana birds alone.
[3] Includes *chiversi* Roberts, 1933 (2088), and *boehmeri* Hoesch & Niethammer, 1940 (1126) see White (1965: 113) (2649).
[4] For correction of spelling see David & Gosselin (2002) (614).
[5] We now follow Cramp *et al.* (1980) (589) in treating this as monotypic.
[6] For correction of spelling see David & Gosselin (2002) (614).
[7] The Miranda-Ribeiro (1938) (1588) names *azarae*, *bicincta*, *leucofimbria* and *schistofimbria* are said to be based on individual variation (Gonzaga, 1996) (950).

EURYPYGIDAE SUNBITTERN (1: 1)

EURYPYGA Illiger, 1811 F
● *Eurypyga helias* Sunbittern
    √ *E.h.major* Hartlaub, 1844   Guatemala to W Ecuador
    __ *E.h.meridionalis* Berlepsch & Stolzmann, 1902   SC Peru
    __ *E.h.helias* (Pallas, 1781)   Colombia to the Guianas, south to C Brazil and
                                             E Bolivia

RALLIDAE RAILS, WATERHENS, COOTS[1] (33: 141)

SAROTHRURA Heine, 1890 F
○ *Sarothrura pulchra* White-spotted Flufftail
    __ *S.p.pulchra* (J.E. Gray, 1829)   Senegal and Gambia to N and C Cameroon
    __ *S.p.zenkeri* Neumann, 1908   SE Nigeria, coastal Cameroon, Gabon
    __ *S.p.batesi* Bannerman, 1922   S Cameroon
    __ *S.p.centralis* Neumann, 1908   Congo and N Angola to S Sudan, W Kenya, NW
                                             Tanzania and NW Zambia

○ *Sarothrura elegans* Buff-spotted Flufftail
    __ *S.e.reichenovi* (Sharpe, 1894)   Sierra Leone to Uganda south to N Angola
    __ *S.e.elegans* (A. Smith, 1839)   S Ethiopia and Kenya to Zimbabwe and E Sth. Africa

○ *Sarothrura rufa* Red-chested Flufftail
    __ *S.r.bonapartii* (Bonaparte, 1856)   Sierra Leone to Cameroon and Gabon
    __ *S.r.elizabethae* van Someren, 1919   Ethiopia, Central African Republic and N Zaire to
                                          Uganda and W Kenya
    __ *S.r.rufa* (Vieillot, 1819)   C Kenya to NE Tanzania, Angola and S Zaire to Sth.
                                           Africa

○ *Sarothrura lugens* Chestnut-headed Flufftail
    __ *S.l.lugens* (Böhm, 1884)[2]   Cameroon to Zaire, Rwanda, W Tanzania
    __ *S.l.lynesi* (Grant & Mackworth-Praed, 1934)   C Angola, Zambia and Zimbabwe (969)

○ *Sarothrura boehmi* Reichenow, 1900 Streaky-breasted Flufftail   Nigeria to Kenya, south to Zimbabwe

○ *Sarothrura affinis* Striped Flufftail
    __ *S.a.antonii* Madaràsz & Neumann, 1911   S Sudan to E Zimbabwe
    __ *S.a.affinis* (A. Smith, 1828)   Sth. Africa

○ *Sarothrura insularis* (Sharpe, 1870) Madagascan Flufftail   E Madagascar

○ *Sarothrura ayresi* (J.H. Gurney, 1877) White-winged Flufftail   Ethiopia, Zimbabwe and Sth. Africa

○ *Sarothrura watersi* (E. Bartlett, 1880) Slender-billed Flufftail   E Madagascar

HIMANTORNIS Hartlaub, 1855 M
○ *Himantornis haematopus* Hartlaub, 1855 Nkulengu Rail   Sierra Leone to W Uganda, south to coastal and C Zaire

CANIRALLUS Bonaparte, 1856 M
○ *Canirallus oculeus* (Hartlaub, 1855) Grey-throated Rail   Sierra Leone to W Uganda

○ *Canirallus kioloides* Madagascan Forest Rail[3]
    __ *C.k.berliozi* (Salomonsen, 1934)   NW Madagascar (2149)
    __ *C.k.kioloides* (Pucheran, 1845)   E Madagascar

COTURNICOPS G.R. Gray, 1855 M
○ *Coturnicops exquisitus* (Swinhoe, 1873) Swinhoe's Rail   Baikal area to Russian Far East and NE China >>
                                             Japan, Korea to S China

[1] Unlike last Edition here we broadly follow the systematic treatment of Olson in Ripley (1977) (2050), as modified in Taylor (1996) (2387). Comparison with last Edition would show numerous changes in generic assignment.
[2] Includes *modesta* Monard, 1949 (1602), see Mayr (1957) (1479) and Keith el al. (1970) (1243A).
[3] We have elected to reserve the generic vernacular Wood Rail for members of the genus *Aramides*.

○ *Coturnicops noveboracensis* YELLOW RAIL[1]
   — *C.n.noveboracensis* (J.F. Gmelin, 1789)[2]   SC and SE Canada to NC and NE USA >> S and SE USA

   — *C.n.goldmani* (Nelson, 1904)   River Lerma (C Mexico)

○ *Coturnicops notatus* (Gould, 1841) SPECKLED RAIL   Colombia, Venezuela, Guyana, S Brazil to N Argentina

MICROPYGIA Bonaparte, 1856 F
○ *Micropygia schomburgkii* OCELLATED CRAKE
   — *M.s.schomburgkii* (Schomburgk, 1848)   Costa Rica, Colombia, SE Peru, Bolivia, Venezuela, the Guianas

   — *M.s.chapmani* (Naumburg, 1930)   C to SE Brazil

RALLINA G.R. Gray, 1846 F
○ *Rallina rubra* CHESTNUT FOREST RAIL
   — *R.r.rubra* (Schlegel, 1871)   W New Guinea
   — *R.r.klossi* (Ogilvie-Grant, 1913)[3]   WC New Guinea
   — *R.r.telefolminensis* (Gilliard, 1961)   C New Guinea (940)

○ *Rallina leucospila* (Salvadori, 1875) WHITE-STRIPED FOREST RAIL   NW New Guinea

○ *Rallina forbesi* FORBES'S FOREST RAIL
   — *R.f.steini* (Rothschild, 1934)   C New Guinea (2369)
   — *R.f.parva* (Pratt, 1982)   Adelbert Mts. (NE New Guinea) (1921)
   — *R.f.dryas* (Mayr, 1931)   Huon Pen. (NE New Guinea)
   — *R.f.forbesi* (Sharpe, 1887)   SE New Guinea

○ *Rallina mayri* MAYR'S FOREST RAIL
   — *R.m.mayri* (E. Hartert, 1930)   Cyclops Mts. (N New Guinea)
   — *R.m.carmichaeli* (Diamond, 1969)   Torricelli and Bewani Mts. (N New Guinea) (698)

○ *Rallina tricolor* G.R. Gray, 1858 RED-NECKED CRAKE[4]   N and E Queensland, New Guinea, Bismarck Arch., S Moluccas, E Lesser Sundas

○ *Rallina canningi* (Blyth, 1863) ANDAMAN CRAKE   Andaman Is.

○ *Rallina fasciata* (Raffles, 1822) RED-LEGGED CRAKE   NE India to Philippines, south to Flores I., Moluccas, Greater and Lesser Sundas

○ *Rallina eurizonoides* SLATY-LEGGED CRAKE
   — *R.e.amauroptera* (Jerdon, 1864)   NW Pakistan, India >> Sri Lanka
   — *R.e.telmatophila* Hume, 1878   Burma to Indochina and SE China >>S Thailand to Java
   — *R.e.sepiaria* (Stejneger, 1887)   Ryukyu Is.
   — *R.e.formosana* Seebohm, 1894   Taiwan and nearby Lanyu I.
   — *R.e.alvarezi* Kennedy & Ross, 1987   Batan Is. (1257)
   — *R.e.eurizonoides* (Lafresnaye, 1845)   Philippines (except Batan), Palau Is.
   — *R.e.minahasa* Wallace, 1863   Sulawesi, Peleng, Sula Is.

ANUROLIMNAS Sharpe, 1893 M
○ *Anurolimnas castaneiceps* CHESTNUT-HEADED CRAKE
   — *A.c.coccineipes* Olson, 1973   S Colombia, NE Ecuador (1690)
   — *A.c.castaneiceps* (Sclater & Salvin, 1868)   E Ecuador, N Peru, NW Bolivia

○ *Anurolimnas viridis* RUSSET-CROWNED CRAKE
   — *A.v.brunnescens* (Todd, 1932)   E Colombia
   — *A.v.viridis* (Statius Müller, 1776)   S Venezuela, the Guianas, Brazil, south to N Bolivia

○ *Anurolimnas fasciatus* (Sclater & Salvin, 1867) BLACK-BANDED CRAKE   S and SE Colombia to NE Peru, NW Brazil

---

[1] Treated as conspecific with *exquisitus* in last Edition, but see A.O.U. (1983:149) (2).
[2] A.O.U. (1957: 157) (1) implicitly included *emersoni* Bailey, 1935 (75), and *richii* Bailey, 1935 (75).
[3] Tentatively includes *subrubra* Rand, 1940 (1963), but see Mayr (1944) (1474).
[4] Treatment of this as monotypic may be inappropriate; it presently includes recent names *laeta* Mayr, 1949 (1475), and *maxima* Mayr, 1949 (1475). A review is needed; some migration occurs.

LATERALLUS G.R. Gray, 1855 M
○ *Laterallus melanophaius* RUFOUS-SIDED CRAKE
    __ *L.m.oenops* (Sclater & Salvin, 1880)   SE Colombia, E Ecuador, E Peru, W Brazil
    __ *L.m.melanophaius* (Vieillot, 1819)   N Venezuela, east to Surinam, south to N Argentina

○ *Laterallus levraudi* (Sclater & Salvin, 1868) RUSTY-FLANKED CRAKE   N Venezuela

○ *Laterallus ruber* (Sclater & Salvin, 1860) RUDDY CRAKE   S and E Mexico to NW Costa Rica

○ *Laterallus albigularis* WHITE-THROATED CRAKE
    __ *L.a.cinereiceps* (Lawrence, 1875)   SE Honduras and Nicaragua to NW Panama
    __ *L.a.albigularis* (Lawrence, 1861)   SW Costa Rica to W Ecuador
    __ *L.a.cerdaleus* Wetmore, 1958   E Colombia (2599)

○ *Laterallus exilis* (Temminck, 1831) GREY-BREASTED CRAKE   Guatemala south to N Ecuador, Venezuela and the
                                                      Guianas south to Bolivia, Paraguay and SE Brazil

○ *Laterallus jamaicensis* BLACK RAIL
    __ *L.j.coturniculus* (Ridgway, 1874)   California, Baja California
    __ *L.j.jamaicensis* (J.F. Gmelin, 1789)   E USA, E Central America >> S USA to Guatemala and
                                                      Greater Antilles

    __ *L.j.murivagans* (Riley, 1916)   W Peru
    __ *L.j.tuerosi* Fjeldså, 1983   C Peru (842)
    __ *L.j.salinasi* (Philippi, 1857)   C Chile, W Argentina

○ *Laterallus spilonotus* (Gould, 1841) GALAPAGOS RAIL   Galapagos Is.

○ *Laterallus leucopyrrhus* (Vieillot, 1819) RED-AND-WHITE CRAKE   SE Brazil, Paraguay, Uruguay, NE Argentina

○ *Laterallus xenopterus* Conover, 1934 RUFOUS-FACED CRAKE   SC Brazil, C Paraguay (557)

NESOCLOPEUS J.L. Peters, 1932[1] M
○ *Nesoclopeus poecilopterus* BAR-WINGED RAIL[2]
    __ *N.p.woodfordi* (Ogilvie-Grant, 1889)[3]   Guadalcanal and Santa Isabel Is.
    __ *N.p.tertius* Mayr, 1949   Bougainville I. (1475)
    __ †?*N.p.poecilopterus* (Hartlaub, 1866)   Viti Levu I., Ovalau (S Fiji Is.)

GALLIRALLUS Lafresnaye, 1841 M
● *Gallirallus australis* WEKA
    __ *G.a.greyi* (Buller, 1888)   North Island (New Zealand)
    __ *G.a.australis* (Sparrman, 1786)   N and W South Island (New Zealand)
    __ *G.a.hectori* (Hutton, 1874)   E South Island, Chatham Is. (New Zealand)
    ✓ *G.a.scotti* (Ogilvie-Grant, 1905)   Stewart I.

○ †?*Gallirallus lafresnayanus* Verreaux & Des Murs, 1860 NEW CALEDONIAN RAIL
                                          New Caledonia

○ *Gallirallus sylvestris* (P.L. Sclater, 1870[4]) WOODHEN/LORD HOWE RAIL
                                          Lord Howe I.

○ *Gallirallus okinawae* (Yamashina & Mano, 1981) OKINAWA RAIL   N Okinawa I. (2696)

○ *Gallirallus torquatus* BARRED RAIL
    __ *G.t.torquatus* (Linnaeus, 1766)[5]   Philippines
    __ *G.t.celebensis* (Quoy & Gaimard, 1830)[6]   Sulawesi
    __ *G.t.sulcirostris* (Wallace, 1863)[7]   Peling, Sula Is.
    __ *G.t.kuehni* (Rothschild, 1902)   Tukangbesi Is.
    __ *G.t.limarius* (J.L. Peters, 1934)   Salawati I., NW New Guinea

○ *Gallirallus insignis* (P.L. Sclater, 1880) NEW BRITAIN RAIL   New Britain

[1]   Treated in the genus *Rallus* in last Edition.
[2]   Treated as one species by Pratt *et al.* (1987) (1920) and in last Edition, but *poecilopterus* sometimes separated as a monotypic species, see Taylor (1996: 161)
    (2387). No detailed review traced.
[3]   Includes *immaculatus* Mayr, 1949 (1475) see Kratter *et al.* (2001) (1305A).
[4]   For correct date see Duncan (1937) (770).
[5]   Parkes (1971) (1763) placed *maxwelli* Lowe, 1944 (1391), *quisumbingi* Gilliard, 1949 (938), and *sanfordi* Gilliard, 1949 (938), in synonymy.
[6]   Includes *remigialis* Stresemann, 1936 (2359), see White & Bruce (1986: 151) (2652).
[7]   Includes *simillimus* Neumann, 1939 (1646), see White & Bruce (1986: 151) (2652).

○ †*Gallirallus pacificus* (J.F. Gmelin, 1789)[1] TAHITIAN RAIL   Tahiti

◉ *Gallirallus philippensis* BUFF-BANDED RAIL
    — *G.p.andrewsi* (Mathews, 1911)   Cocos Keeling Is.
    — *G.p.xerophilus* (van Bemmel & Hoogerwerf, 1940)   Gunung Api (Lesser Sundas) (2461)
    — *G.p.wilkinsoni* (Mathews, 1911)   Flores
    — *G.p.philippensis* (Linnaeus, 1766)   Philippines, Sulawesi, Sumba to Timor
    — *G.p.pelewensis* (Mayr, 1933)   Palau Is. (1464)
    — *G.p.anachoretae* (Mayr, 1949)   Kaniet Is. (1475)
    — *G.p.admiralitatis* (Stresemann, 1929)   Admiralty Is.
    — *G.p.praedo* Mayr, 1949   Skoki I., Sabben group (1475)
    — *G.p.lesouefi* (Mathews, 1911)   New Hanover, Tabar and Tanga Is.
    — *G.p.meyeri* (E. Hartert, 1930)   Witu Is., New Britain
    — *G.p.christophori* (Mayr, 1938)   Solomon Is. (1467)
    — *G.p.sethsmithi* (Mathews, 1911)   Vanuatu, Fiji
    — *G.p.swindellsi* (Mathews, 1911)   New Caledonia, Loyalty Is.
    ✓ *G.p.goodsoni* (Mathews, 1911)   Samoa and Niue Is.
    — *G.p.ecaudatus* (J.F. Miller, 1783)   Tonga
    — *G.p.assimilis* (G.R. Gray, 1843)   New Zealand and Auckland Is.
    — †*G.p.macquariensis* (Hutton, 1879)   Macquarie Is.
    — *G.p.lacustris* (Mayr, 1938)   N New Guinea (1467)
    — *G.p.reductus* (Mayr, 1938)   NE and C New Guinea, Long I. (1467)
    — *G.p.randi* (Mayr & Gilliard, 1951)   W montane New Guinea (1489)
    — *G.p.wahgiensis* (Mayr & Gilliard, 1951)   E montane New Guinea (1489)
    — *G.p.tounelieri* Schodde & de Naurois, 1982   Coral Sea and islands in Gt. Barrier Reef (2182)
    — *G.p.mellori* (Mathews, 1912)[2]   Moluccas[3], S and SW New Guinea, Australia, Tasmania, Norfolk I.

○ *Gallirallus owstoni* (Rothschild, 1895) GUAM RAIL   Guam I., Rota I.

○ †*Gallirallus wakensis* (Rothschild, 1903) WAKE ISLAND RAIL   Wake I. (tropical N Pacific)

○ *Gallirallus rovianae* Diamond, 1991 ROVIANA RAIL #   C Solomon Is. (703)

○ †*Gallirallus dieffenbachii* (G.R. Gray, 1843) DIEFFENBACH'S RAIL   Chatham Is.

○ †*Gallirallus modestus* (Hutton, 1872) CHATHAM ISLANDS RAIL   Chatham Is.

○ *Gallirallus striatus* SLATY-BREASTED RAIL
    — *G.s.albiventer* (Swainson, 1838)[4]   India and Sri Lanka eastward through Burma and Thailand to S China
    — *G.s.obscurior* (Hume, 1874)[5]   Andaman and Nicobar Is.
    — *G.s.jouyi* (Stejneger, 1886)   SE China, Hainan
    — *G.s.taiwanus* (Yamashina, 1932)   Taiwan
    — *G.s.gularis* (Horsfield, 1821)   Vietnam and Cambodia to Sumatra, Java, S Borneo
    — *G.s.striatus* (Linnaeus, 1766)[6]   Borneo, Philippines (incl. Sulu Is.), Sulawesi

○ †*Gallirallus sharpii* (Büttikofer, 1893) SHARPE'S RAIL   Unknown[7]

RALLUS Linnaeus, 1758 M
○ *Rallus longirostris* CLAPPER RAIL[8]
    — *R.l.obsoletus* Ridgway, 1874   C California
    — *R.l.levipes* Bangs, 1899   SC California to Baja California
    — *R.l.yumanensis* Dickey, 1923   SE California, S Arizona, NW Mexico
    — *R.l.beldingi* Ridgway, 1882[9]   S Baja California

[1] Treated as *Rallus ecaudata* J.F. Miller by Peters (1934: 166) (1822), but this name is now attributed to the Tongan form of *G. pectoralis* see Lysaght (1953) (1405).
[2] Includes *norfolkensis*; birds occurring on Norfolk I. are now thought to be Australian visitors.
[3] White & Bruce (1986) (2652) stated that Moluccan birds (including those from Buru) were not safely distinguishable from *mellori*. However, they tentatively recognised them as *yorki*; in addition they listed two forms from Buru (the second being the nominate). A review is needed.
[4] Some authors (e.g. Cheng, 1987) (333) use the name *gularis* for Chinese birds; fresh comparison with Indian birds seems indicated and Indochinese birds need review.
[5] Includes *nicobarensis* Abdulali, 1967 (8) see Ripley (1982: 93) (2055).
[6] Includes *paratermus* (see Dickinson *et al.* 1991) (745) and *deignani* Ripley, 1970 (2049), see Ripley & Olson (1973) (2070).
[7] See Olson (1987) (1699).
[8] The salt-water counterpart of the next species, see Olson (1997) (1704).
[9] Includes *magdalenae* van Rossem, 1947 (2480) see Ripley (1977) (2050).

— *R.l.crepitans* J.F. Gmelin, 1789   Connecticut to NE North Carolina
— *R.l.waynei* Brewster, 1899   SE North Carolina to Florida
— *R.l.saturatus* Ridgway, 1880   SW Alabama to NE Mexico
— *R.l.scotti* Sennett, 1888   Coastal Florida
— *R.l.insularum* W.S. Brooks, 1920   Florida Keys
— *R.l.coryi* Maynard, 1887   Bahama Is.
— *R.l.leucophaeus* Todd, 1913   Isla de la Juventud
— *R.l.caribaeus* Ridgway, 1880[1]   Cuba to Puerto Rico, Lesser Antilles to Antigua, Guadeloupe
— *R.l.pallidus* Nelson, 1905   Yucatán (Mexico)
— *R.l.grossi* Paynter, 1950   Quintana Roo (Mexico) (1805)
— *R.l.belizensis* Oberholser, 1937   Ycacos Lagoon (Belize) (1671)
— *R.l.phelpsi* Wetmore, 1941[2]   NE Colombia, NW Venezuela (2590)
— *R.l.margaritae* Zimmer & Phelps, 1944   Margarita I. (Venezuela) (2722)
— *R.l.pelodramus* Oberholser, 1937   Trinidad (1671)
— *R.l.longirostris* Boddaert, 1783   Coast of the Guianas
— *R.l.crassirostris* Lawrence, 1871   Coast of Brazil
— *R.l.cypereti* Taczanowski, 1877   Coastal SW Colombia to NW Peru

○ *Rallus elegans* KING RAIL[3]
— *R.e.elegans* Audubon, 1834   E Canada, NE USA
— *R.e.tenuirostris* Ridgway, 1874   C Mexico
— *R.e.ramsdeni* Riley, 1913   Cuba, Isla de la Juventud

○ *Rallus wetmorei* Zimmer & Phelps, 1944 PLAIN-FLANKED RAIL   N Venezuela (2722)

◑ *Rallus limicola* VIRGINIA RAIL
✓ *R.l.limicola* Vieillot, 1819   S Canada, W, C and NE USA >> SW Canada to Guatemala
— *R.l.friedmanni* Dickerman, 1966   SC Mexico to Guatemala (705)
— *R.l.aequatorialis* Sharpe, 1894   SW Colombia, Ecuador
— *R.l.meyerdeschauenseei* Fjeldså, 1990   Coastal Peru (south to Arequipa) (845).

○ *Rallus semiplumbeus* BOGOTA RAIL
— *R.s.semiplumbeus* P.L. Sclater, 1856   E Colombia
— †?*R.s.peruvianus* Taczanowski, 1886[4]   Peru

○ *Rallus antarcticus* P.P. King, 1828 AUSTRAL RAIL   C Chile, N Argentina

○ *Rallus aquaticus* WATER RAIL
— *R.a.hibernans* Salomonsen, 1931   Iceland
— *R.a.aquaticus* Linnaeus, 1758   Europe, N Africa, W Asia
— *R.a.korejewi* Zarudny, 1905[5]   Aral Sea to Iran, Kashmir and WC China
— *R.a.indicus* Blyth, 1849   N Mongolia, Tuva Rep. and Baikal area to SE Siberia and Japan >> E Bengal to Indochina and S Japan

○ *Rallus caerulescens* J.F. Gmelin, 1789 AFRICAN RAIL   Ethiopia, Uganda and S Kenya south to N Namibia and Sth. Africa

○ *Rallus madagascariensis* J. Verreaux, 1833 MADAGASCAN RAIL   E Madagascar

LEWINIA G.R. Gray, 1855 F[6][7]
○ *Lewinia mirifica* (Parkes & Amadon, 1959) BROWN-BANDED RAIL[8]   Luzon (1785)

○ *Lewinia pectoralis* LEWIN'S RAIL
— *L.p.exsul* (E. Hartert, 1898)   Flores I.
— *L.p.mayri* (E. Hartert, 1930)   NW New Guinea
— *L.p.capta* Mayr & Gilliard, 1951[9]   Montane C New Guinea (1489)

---

[1] Includes *manglecola* Danforth, 1934 (609), and *limnetis* Oberholser, 1937 (1671) see Ripley (1977) (2050).
[2] Includes ¶ *dillonripleyi* Phelps & Aveledo, 1987 (1839), which requires evaluation.
[3] In last Edition treated as part of a broad species *R. longirostris* following Ripley (1977) (2050), but see A.O.U. (1998: 131) (3).
[4] Taxon virtually unknown and not recognised in last Edition, but may not be extirpated, see Fjeldså (1990) (845).
[5] Includes *deserticolor* Sudilovskaya, 1934 (2373), *tsaidamensis* Buturlin, 1935 (285), and *arjanicus* Koelz, 1954 (1284), see Stepanyan (1990: 158) (2326).
[6] These species were treated in the genus *Rallus* in last Edition, but see Taylor (1996: 170-171) (2387).
[7] Gender addressed by David & Gosselin (2002) (614), multiple changes occur in specific names.
[8] May be conspecific with *pectoralis*; breeding range still unknown.
[9] Includes *connectens* Junge, 1952 (1232), see comments by Rand & Gilliard (1966: 109) (1975).

— *L.p.alberti* (Rothschild & Hartert, 1907)[1]   E New Guinea

— †?*L.p.clelandi* (Mathews, 1911)   SW Australia

— *L.p.pectoralis* (Temminck, 1831)   SE and E Australia

— *L.p.brachipus* (Swainson, 1838)   Tasmania

○ *Lewinia muelleri* (Rothschild, 1893) Auckland Rail   Auckland Is. (S New Zealand)

DRYOLIMNAS Sharpe, 1893 M

○ *Dryolimnas cuvieri* White-throated Rail

— *D.c.cuvieri* (Pucheran, 1845)   Madagascar, Mauritius

— *D.c.aldabranus* (Günther, 1879)   Aldabra I.

— †*D.c.abbotti* (Ridgway, 1894)   Assumption I. (Indian Ocean)

CREX Bechstein, 1803 F

○ *Crex egregia* (W.K.H. Peters, 1854) African Crake[2]   Senegal and Gambia to S Sudan and Kenya, south to N Namibia and Sth. Africa

○ *Crex crex* (Linnaeus, 1758) Corncrake   Europe to C Siberia and NW China >> E Zaire and S Tanzania to E Sth. Africa

ROUGETIUS Bonaparte, 1856 M

○ *Rougetius rougetii* (Guérin-Méneville, 1843) Rouget's Rail   Ethiopia and Eritrea

ARAMIDOPSIS Sharpe, 1893 F

○ *Aramidopsis plateni* (W.H. Blasius, 1886) Snoring Rail   N, NC and SE Sulawesi

ATLANTISIA P.R. Lowe, 1923 F

○ *Atlantisia rogersi* P.R. Lowe, 1923 Inaccessible Rail   Inaccessible I.

ARAMIDES Pucheran, 1845 M[3]

○ *Aramides mangle* (Spix, 1825) Little Wood Rail   E Brazil

● *Aramides axillaris* Lawrence, 1863 Rufous-necked Wood Rail   S Mexico to Panama, NW Colombia to NW Peru, east to Guyana

● *Aramides cajanea* Grey-necked Wood Rail

— *A.c.mexicanus* Bangs, 1907   S Mexico

— *A.c.vanrossemi* Dickey, 1929   S Mexico to S Guatemala and El Salvador

— *A.c.albiventris* Lawrence, 1867   Yucatán to Belize and N Guatemala

— *A.c.pacificus* Miller & Griscom, 1921   Honduras, Nicaragua

— *A.c.plumbeicollis* Zeledon, 1888   NE Costa Rica

— *A.c.latens* Bangs & Penard, 1918   San Miguel and Viveros (Pearl Is., Panama)

— *A.c.morrisoni* Wetmore, 1946   San José and Pedro González (Pearl Is., Panama) (2592)

— *A.c.cajanea* (Statius Müller, 1776)[4]   Costa Rica to Paraguay and N Argentina

— *A.c.avicenniae* Stotz, 1992   SE Brazil (São Paulo coast)[5] (2355)

○ *Aramides wolfi* Berlepsch & Taczanowski, 1884 Brown Wood Rail   Colombia to SW Ecuador and NW Peru

○ *Aramides ypecaha* (Vieillot, 1819) Giant Wood Rail   E and SE Brazil to NE Argentina

○ *Aramides saracura* (Spix, 1825) Slaty-breasted Wood Rail   SE Brazil, Paraguay, NE Argentina

○ *Aramides calopterus* Sclater & Salvin, 1878 Red-winged Wood Rail   E Ecuador, NE Peru, W Brazil

AMAUROLIMNAS Sharpe, 1893 M

○ *Amaurolimnas concolor* Uniform Crake

— *A.c.guatemalensis* (Lawrence, 1863)   S Mexico to Ecuador

— *A.c.castaneus* (Pucheran, 1851)   Venezuela, the Guianas, Brazil, E Peru, E Bolivia

— †*A.c.concolor* (Gosse, 1847)   Jamaica

---

[1] Includes *insulsa* Greenway, 1935 (1004), see comments by Rand & Gilliard (1966: 109) (1975).
[2] Treated in the genus *Porzana* in last Edition. We follow Keith & Taylor (1986: 98) (1242) in placing it here.
[3] Gender addressed by David & Gosselin (2002) (614), multiple changes occur in specific names.
[4] Includes *gutturalis* see Taylor (1996: 176) (2387).
[5] Not included in Taylor (1996) (2387); see also Bornschein & Reinert (1996) (194).

GYMNOCREX Salvadori, 1875 F
○ *Gymnocrex rosenbergii* (Schlegel, 1866) Bald-faced Rail — N and NC Sulawesi, Peleng I.

○ *Gymnocrex talaudensis* Lambert, 1998 Talaud Rail # — Karakelong I. (Talaud Is.) (1317)

○ *Gymnocrex plumbeiventris* Bare-eyed Rail
— *G.p.plumbeiventris* (G.R. Gray, 1862) — N Moluccas, Misool I., N New Guinea, Karkar, New Ireland

— *G.p.hoeveni* (von Rosenberg, 1866) — S New Guinea, Aru Is.

AMAURORNIS Reichenbach, 1853 F[1]
○ *Amaurornis akool* Brown Crake
— *A.a.akool* (Sykes, 1832) — India, Bangladesh, W Burma
— *A.a.coccineipes* (Slater, 1891) — SE China to NE Vietnam

○ *Amaurornis isabellina* (Schlegel, 1865) Isabelline Bush-hen — Sulawesi

○ *Amaurornis olivacea* (Meyen, 1834) Plain Bush-hen[2]
— *A.o.olivacea* (Meyen, 1834) — Philippines
— *A.o.magnirostris* Lambert, 1998[3] # — Karakelong I. (Talaud Is.) (1316)
— *A.o.moluccana* (Wallace, 1865) — N and W New Guinea, Misool I, Moluccas, Sangihe Is.
— *A.o.nigrifrons* (E. Hartert, 1926) — Bismarck Arch., W Solomon Is.
— *A.o.ultima* Mayr, 1949 — E Solomon Is. (1475)
— *A.o.ruficrissa* (Gould, 1869)[4] — S and E New Guinea, N and NE Australia

○ *Amaurornis phoenicurus* White-breasted Waterhen[5]
— *A.p.phoenicurus* (Pennant, 1769)[6] — Pakistan to E China and Philippines, south to Maldives, Sri Lanka, Java, Borneo and Sangihe Is.
— *A.p.insularis* Sharpe, 1894[7] — Andaman and Nicobar Is.
— *A.p.leucomelana* (S. Müller, 1842)[8] — Lesser Sundas, W Moluccas, Sulawesi

○ *Amaurornis flavirostra* (Swainson, 1837) Black Crake[9] — Senegal to Ethiopia and Sth. Africa

○ †?*Amaurornis olivieri* (Grandidier & Berlioz, 1929) Sakalava Rail — WC Madagascar

PORZANA Vieillot, 1816 F

○ *Porzana bicolor* Walden, 1872 Black-tailed Crake[10] — NE India east to Yunnan, W Sichuan, NW Thailand and N Indochina

○ *Porzana parva* (Scopoli, 1769) Little Crake — S, C and E Europe to NW China >> Mediterranean, W and NE Africa, Arabia to NW India

○ *Porzana pusilla* Baillon's Crake
— *P.p.intermedia* (Hermann, 1804) — N Africa and Europe to Asia Minor >> N sub-Saharan Africa
— *P.p.obcura* Neumann, 1897 — S and E Africa, Madagascar
— *P.p.pusilla* (Pallas, 1776)[11] — Ukraine, S European Russia and Iran to NW India, China and Japan >> S and SE Asia
— *P.p.mira* Riley, 1938[12] — Borneo (2023)
— *P.p.mayri* Junge, 1952 — New Guinea (1232)
— *P.p.palustris* Gould, 1843 — E New Guinea, Australia, Tasmania
— *P.p.affinis* (J.E. Gray, 1846) — New Zealand, Chatham Is.

○ *Porzana porzana* (Linnaeus, 1766) Spotted Crake — W Europe to C Asia >>Africa, Pakistan, India, W Burma

[1] Gender addressed by David & Gosselin (2002) (614), multiple changes occur in specific names.
[2] The case for splitting this species needs to be made in more detail, especially after the description of *magnirostris*.
[3] We see this as a link form between *olivaceus* and *moluccanus* and closer to the former.
[4] For correction of spelling see David & Gosselin (2002) (613).
[5] Possibly overlumped, a thorough review is needed; some authors on Asia (e.g. White & Bruce, 1986) (2652) follow Ripley (1977) (2050), but others retain *chinensis* (e.g. Cheng, 1987) (333) and *javanica* (e.g. Mees, 1986; Dickinson *et al.*, 1991) (1539, 745).
[6] Includes *maldivus* Phillips & Sims, 1958 (1882), see Ripley (1982: 97) (2055).
[7] Includes *leucocephalus* Abdulali, 1964 (6), see Ripley (1982: 97) (2055); and tentatively *midnicobarica* Abdulali, 1979 (13), not listed by Ripley (1982: 97) (2055). But see Ripley & Beehler (1989) (2061A).
[8] Includes *variabilis* Stresemann, 1936 (2359) (see White & Bruce, 1986) (2652).
[9] For correction of spelling see David & Gosselin (2002) (613).
[10] For a review see Inskipp & Round (1989) (1181).
[11] Includes *bareji* Dunajewski, 1937 (767) (see Stepanyan, 1990: 160) (2326).
[12] The validity of this needs review.

○ *Porzana fluminea* Gould, 1843 AUSTRALIAN CRAKE | Australia, Tasmania

◐ *Porzana carolina* (Linnaeus, 1758) SORA RAIL | SE Alaska, Canada, USA >> S USA, West Indies, Central America to C Peru and the Guianas

○ *Porzana spiloptera* Durnford, 1877 DOT-WINGED CRAKE | S Uruguay, N Argentina

○ *Porzana albicollis* ASH-THROATED CRAKE
— *P.a.olivacea* (Vieillot, 1819) | Colombia, Venezuela and Trinidad to the Guianas, N Brazil
— *P.a.albicollis* (Vieillot, 1819) | E and S Brazil, N and E Bolivia, Paraguay, N Argentina

○ *Porzana fusca* RUDDY-BREASTED CRAKE[1]
— *P.f.erythrothorax* (Temminck & Schlegel, 1849) | NE China to Japan >> mainland SE Asia
— *P.f.phaeopyga* Stejneger, 1887 | Ryukyu Is.
— *P.f.bakeri* E. Hartert, 1917 | Pakistan, N India, Burma to SE China and Indochina
— *P.f.fusca* (Linnaeus, 1766) | SW India, Sri Lanka, Malay Pen. and Philippines to Sulawesi and Lesser Sundas

○ *Porzana paykullii* (Ljungh, 1813) BAND-BELLIED CRAKE[2] | SE Siberia, N Korea, NE China >> E Asia to Gtr. Sundas

○ *Porzana tabuensis* SPOTLESS CRAKE
— *P.t.tabuensis* (J.F. Gmelin, 1789) | Philippines to lowland New Guinea, Australia, Tasmania, New Zealand, Chatham Is., New Caledonia, SW Pacific Is., Micronesia, Polynesia
— *P.t.richardsoni* Rand, 1940 | NW New Guinea (Mts. in the Vogelkop) (1963)
— *P.t.edwardi* Gyldenstolpe, 1955 | WC New Guinea (Snow Mts.) (1032)

○ †?*Porzana monasa* (Kittlitz, 1858) KOSRAE CRAKE | Kosrae I. (E Caroline Is.)

○ *Porzana atra* North, 1908 HENDERSON CRAKE | Henderson I. (C Pitcairn Is.)

○ †*Porzana sandwichensis* (J.F. Gmelin, 1789) HAWAIIAN CRAKE[3] | Hawaii (Hawaiian Is.)

○ *Porzana flaviventer* YELLOW-BREASTED CRAKE
— *P.f.gossii* (Bonaparte, 1856) | Cuba, Jamaica
— *P.f.hendersoni* Bartsch, 1917 | Hispaniola, Puerto Rico
— *P.f.woodi* van Rossem, 1934 | C Mexico to NW Costa Rica (2471)
— *P.f.bangsi* Darlington, 1931 | N Colombia
— *P.f.flaviventer* (Boddaert, 1783) | Panama, Colombia to the Guianas, N Brazil south to N Argentina

○ *Porzana cinerea* (Vieillot, 1819) WHITE-BROWED CRAKE[4] | Malay Pen. through the Sundas, Philippines and Moluccas to New Guinea, N Australia, Micronesia, Melanesia, WC Polynesia

○ †*Porzana palmeri* (Frohawk, 1892) LAYSAN CRAKE | Laysan I. (Hawaiian Is.)

AENIGMATOLIMNAS J.L. Peters, 1932 M
○ *Aenigmatolimnas marginalis* (Hartlaub, 1857) STRIPED CRAKE | N Ivory Coast to Cameroon and coastal Congo, E Zaire to Sth. Africa

CYANOLIMNAS Barbour & Peters, 1927 M
○ *Cyanolimnas cerverai* Barbour & Peters, 1927 ZAPATA RAIL | Zapata Swamp (WC Cuba)

NEOCREX Sclater & Salvin, 1868[5] F
○ *Neocrex columbiana* COLOMBIAN CRAKE
— *N.c.ripleyi* Wetmore, 1967 | C Panama, NW Colombia (2604)
— *N.c.columbiana* Bangs, 1898 | N and W Colombia, W Ecuador

○ *Neocrex erythrops* PAINT-BILLED CRAKE
— *N.e.olivascens* Chubb, 1917 | E Colombia, Venezuela, the Guianas to NW Argentina
— *N.e.erythrops* (P.L. Sclater, 1867) | Peru, Galapagos Is.

---

[1] After a review by CSR we follow Whistler (1944) (2619) in uniting southern birds from S India and Sri Lanka to the Philippines.
[2] Includes *Amaurornis boineti* Bourret, 1944 (207) see Stepanyan (1990: 161) (2326).
[3] *Pennula millsi* is a synonym see A.O.U. (1983: 156) (2).
[4] We follow Taylor (1996: 190) (2387) in treating this as monotypic (but see Mees, 1982) (1534). Includes *micronesiae* Hachisuka, 1939 (1038). Also includes †*brevipes* from the Ogasawara-shoto (Bonin) Is.
[5] This genus was submerged in *Porzana* in last Edition.

PARDIRALLUS Bonaparte, 1856 M
○ *Pardirallus maculatus* SPOTTED RAIL
    — *P.m.insolitus* (Bangs & Peck, 1908)   Mexico to Costa Rica
    — *P.m.maculatus* (Boddaert, 1783)   Colombia to E Brazil, south to Argentina, Trinidad
                                and Tobago, W Cuba, Hispaniola, Jamaica

○ *Pardirallus nigricans* BLACKISH RAIL
    — *P.n.caucae* (Conover, 1949)   SW Colombia (564)
    — *P.n.nigricans* (Vieillot, 1819)   E Ecuador to E Brazil, south to NE Argentina

○ *Pardirallus sanguinolentus* PLUMBEOUS RAIL
    — *P.s.simonsi* Chubb, 1917   W Peru, N Chile
    — *P.s.tschudii* Chubb, 1919   Peru, C and SE Bolivia
    — *P.s.zelebori* (Pelzeln, 1865)   SE Brazil
    — *P.s.sanguinolentus* (Swainson, 1838)   SE Brazil, Paraguay, Uruguay, W Argentina
    — *P.s.landbecki* (Hellmayr, 1932)   C Chile, SW Argentina
    — *P.s.luridus* (Peale, 1848)   S Chile, S Argentina

EULABEORNIS Gould, 1844 M
○ *Eulabeornis castaneoventris* CHESTNUT-BELLIED RAIL
    — *E.c.sharpei* Rothschild, 1906   Aru Is.
    — *E.c.castaneoventris* Gould, 1844   Coast of N Australia

HABROPTILA G.R. Gray, 1860 F
○ *Habroptila wallacii* G.R. Gray, 1860 INVISIBLE RAIL   Halmahera I. (Moluccas)

MEGACREX D'Albertis & Salvadori, 1879 F
○ *Megacrex inepta* NEW GUINEA FLIGHTLESS RAIL
    — *M.i.pallida* Rand, 1938   NC New Guinea (1962)
    — *M.i.inepta* D'Albertis & Salvadori, 1879   SC New Guinea

GALLICREX Blyth, 1852 F
○ *Gallicrex cinerea* (J.F. Gmelin, 1789) WATERCOCK   Pakistan to Japan, south to Sri Lanka, Indochina and
                                Philippines >> S and SE Asia to W and C Indonesia

PORPHYRIO Brisson, 1760 M
● *Porphyrio porphyrio* PURPLE SWAMPHEN[1]
    — *P.p.porphyrio* (Linnaeus, 1758)   SW Europe, NW Africa
    — *P.p.madagascariensis* (Latham, 1802)[2]   Egypt, sub-Saharan Africa, Madagascar
    — *P.p.caspius* E. Hartert, 1917   Caspian Sea, NW Iran, Turkey
    — *P.p.seistanicus* Zarudny & Härms, 1911   Iraq to NW India
    — *P.p.poliocephalus* (Latham, 1802)[3]   India, Sri Lanka, Andaman and Nicobar Is. to Yunnan,
                                Burma, W Thailand, Malay Pen. and N Sumatra
    — *P.p.viridis* Begbie, 1834   C Thailand and S and C Indochina to the Malay Pen.
    — *P.p.indicus* Horsfield, 1821   Sumatra, Java, SE Borneo
    — *P.p.melanopterus* Bonaparte, 1856   Sulawesi, Moluccas and Lesser Sundas
    ✓ *P.p.melanotus* Temminck, 1820   Kai Is., Aru Is., New Guinea, Trobriand Is., Woodlark I.,
                                Misima I., N and E Australia, Tasmania, Lord Howe I.,
                                Norfolk I., Kermadec Is., New Zealand and
                                Chatham Is.
    — *P.p.bellus* Gould, 1841   SW Australia
    — *P.p.pulverulentus* Temminck, 1826   Philippines (and Talaud Is.?)
    — *P.p.pelewensis* Hartlaub & Finsch, 1872   Palau Is.
    ✓ *P.p.samoensis* Peale, 1848   Admiralty Is and Bismarck Arch. to New Caledonia, E
                                through SW Pacific to Samoa

○ †*Porphyrio albus* (Shaw, 1790)[4] LORD HOWE SWAMPHEN   Lord Howe I. (Tasman Sea)

---

[1] Molecular studies by Trewick (1997) (2437), covering part of this assemblage, suggest that this species as treated here is polyphyletic. Although some
    populations are reasonably well understood (see Roselaar in Cramp *et al.*, 1980) (589), others, especially in SE Asia, are not. Colour morphs seem to be
    present (see Mayr, 1938, 1949) (1467, 1475), but may represent sympatry, and await adequate studies. Pending these we retain one species.
[2] For reasons to use 1802 not 1801 see Browing & Monroe (1991) (258).
[3] For reasons to use 1802 not 1801 see Browing & Monroe (1991) (258).
[4] Scientific names in White's 'Journal of a Voyage to New South Wales' were not his but Shaw's see Schodde & Mason (1997: 127) (2188).

○ *Porphyrio hochstetteri* (A.B. Meyer, 1883) SOUTHERN TAKAHE[1]   SW South Island (New Zealand)

○ *Porphyrio alleni* Thomson, 1842 ALLEN's GALLINULE[2]   Senegal and Gambia to Eritrea south to N Namibia and E Sth. Africa

◑ *Porphyrio martinica* (Linnaeus, 1766) PURPLE GALLINULE[3]   SE USA to N Argentina, West Indies

○ *Porphyrio flavirostris* (J.F. Gmelin, 1789) AZURE GALLINULE   E Colombia to the Guianas, south to N Bolivia, N, W and C Brazil, Paraguay, N Argentina

GALLINULA Brisson, 1760 F

○ †?*Gallinula silvestris* (Mayr, 1933) SAN CRISTOBAL MOORHEN   San Cristobal I. (Solomon Is.) (1463)

○ *Gallinula nesiotis* TRISTAN DA CUNHA MOORHEN[4]
    — †*G.n.nesiotis* P.L. Sclater, 1861   Tristan da Cunha
    — *G.n.comeri* (J.A. Allen, 1892)   Gough Is.

● *Gallinula chloropus* COMMON MOORHEN
    ⌦ *G.c.chloropus* (Linnaeus, 1758)[5]   Europe, N Africa, Azores and Canary Is., east to Japan, south to Sri Lanka and mainland SE Asia (some birds winter further south)

    — *G.c.meridionalis* (C.L. Brehm, 1855)   Sub-Saharan Africa, St Helena
    — *G.c.pyrrhorrhoa* A. Newton, 1861   Madagascar, Reunion, Mauritius, Comoros Is.
    — *G.c.orientalis* Horsfield, 1821[6]   Seychelles, Andaman Is., Malay Pen., Greater and Lesser Sundas to Philippines and Palau Is.

    — *G.c.guami* E. Hartert, 1917   N Mariana Is.
    ✓ *G.c.sandvicensis* Streets, 1877   Hawaiian Group
    ✓ *G.c.cachinnans* Bangs, 1915   SE Canada, USA to W Panama, Bermuda, Galapagos Is.
    — *G.c.cerceris* Bangs, 1910   Greater and Lesser Antilles
    — *G.c.barbadensis* Bond, 1954   Barbados (184)
    — *G.c.pauxilla* Bangs, 1915   E Panama, N and W Colombia, W Ecuador, NW Peru
    — *G.c.garmani* J.A. Allen, 1876[7]   Peru, Bolivia, Chile, NW Argentina
    — *G.c.galeata* (M.H.K. Lichtenstein, 1818)   Trinidad, the Guianas S to Uruguay and N Argentina

● *Gallinula tenebrosa* DUSKY MOORHEN
    — *G.t.frontata* Wallace, 1863   Sulawesi, Lesser Sundas and S Moluccas to W and SE New Guinea

    — *G.t.neumanni* E. Hartert, 1930   NC New Guinea
    ⌦ *G.t.tenebrosa* Gould, 1846   Australia

○ *Gallinula angulata* Sundevall, 1851 LESSER MOORHEN   Senegal and Gambia to Ethiopia, S to E Sth. Africa

○ *Gallinula melanops* SPOT-FLANKED GALLINULE
    — *G.m.bogotensis* (Chapman, 1914)   C Colombia
    — *G.m.melanops* (Vieillot, 1819)   E and S Brazil, Uruguay, Paraguay, NE Argentina, E Bolivia

    — *G.m.crassirostris* (J.E. Gray, 1829)   Chile, Argentina

○ *Gallinula ventralis* Gould, 1837 BLACK-TAILED NATIVE-HEN   Australia

○ *Gallinula mortierii* (Du Bus, 1840)[8] TASMANIAN NATIVE-HEN   Tasmania

○ †*Gallinula pacifica* (Hartlaub & Finsch, 1871) SAMOAN WOODHEN   Samoa

FULICA Linnaeus, 1758 F

○ *Fulica cristata* J.F. Gmelin, 1789 RED-KNOBBED COOT   S Spain and N Morocco, Ethiopia to Sth. Africa, Madagascar

---

[1] For evidence of separation from North Island *P. mantelli* (Owens, 1848) – known only from subfossil remains — see Trewick (1997) (2437).
[2] Type species of the genus *Porphyrula* submerged here.
[3] Includes *georgica* Pereyra, 1944 (1819) see Mayr (1957) (1479). Based on a vagrant juvenile.
[4] For an account of the questioned relationships of these two populations see Beintema (1972) (114). A remnant population on SE Tristan da Cunha seems to resemble *comeri* and is believed to have been introduced (see Taylor, 1996: 200) (2387) in 1956 (though this was unknown to Beintema).
[5] Tentatively includes *indica* and *correiana*, and the more recently named *lucida* Dunajewski, 1938 (768), and *vestigialis* Clancey, 1939 (361) (see Stepanyan, 1990: 163) (2326); some confusion is engendered by migrant specimens and by southward expansion by the Eurasian population(s).
[6] Tentatively includes the intermediate Philippine population called *lozanoi* by Dickinson *et al.* (1991) (745).
[7] Includes *hypomelaena* Todd, 1954 (2415) see Krabbe & Fjeldså (1990) (850).
[8] An abbreviated name; in full Du Bus de Gisignies.

● *Fulica atra* Common Coot
 ↙ *F.a.atra* Linnaeus, 1758   Europe and N Africa to Japan, S to India and Sri Lanka
 — *F.a.lugubris* S. Müller, 1847[1]   Java, Bali and upland New Guinea
 ✔ *F.a.australis* Gould, 1845   Australia, Tasmania, New Zealand

● *Fulica alai* Peale, 1848 Hawaiian Coot[2]   Hawaiian Group

● *Fulica americana* American Coot
 ↙ *F.a.americana* J.F. Gmelin, 1789   SE Alaska, Canada to Costa Rica, W Indies
 — *F.a.colombiana* Chapman, 1914   Colombia, N Ecuador
 — ¶*F.a.peruviana* Morrison, 1939   Peru (1628)

○ *Fulica caribaea* Ridgway, 1884 Caribbean Coot   S Bahamas to Trinidad, Curaçao, NW Venezuela

◐ *Fulica leucoptera* Vieillot, 1817 White-winged Coot   Bolivia and S Brazil to Cape Horn

○ *Fulica ardesiaca* Slate-coloured Coot[3]
 — *F.a.atrura* Fjeldså, 1983   S Colombia, Ecuador, NW Peru (843)
 — *F.a.ardesiaca* Tschudi, 1843   C Peru to N Chile and NW Argentina

◐ *Fulica armillata* Vieillot, 1817 Red-gartered Coot   Paraguay and S Brazil to Cape Horn, Falkland Is.

○ *Fulica rufifrons* Philippi & Landbeck, 1861 Red-fronted Coot   S Peru, SE Brazil, C Chile, Paraguay, Uruguay to
                 S Argentina

○ *Fulica gigantea* Eydoux & Souleyet, 1841 Giant Coot   S Peru, W Bolivia, N Chile, NW Argentina

○ *Fulica cornuta* Bonaparte, 1853 Horned Coot   SW Bolivia, N Chile, NW Argentina

HELIORNITHIDAE FINFOOTS (3: 3)

PODICA Lesson, 1831 F
○ *Podica senegalensis* African Finfoot
 — *P.s.senegalensis* (Vieillot, 1817)   Senegal to Nigeria, N and NE Zaire, Uganda and NW
                Tanzania, Ethiopia
 — *P.s.camerunensis* Sjöstedt, 1893   S Cameroon, Gabon, C Congo basin
 — *P.s.somereni* Chapin, 1954   C and SE Kenya, NE Tanzania (313)
 — *P.s.petersii* Hartlaub, 1852[4]   Angola to Mozambique, south to E Sth. Africa

HELIOPAIS Sharpe, 1893 M
○ *Heliopais personatus* (G.R. Gray, 1849) Masked Finfoot[5]   Bangladesh and NE India to Indochina, Sumatra

HELIORNIS Bonnaterre, 1791 M
○ *Heliornis fulica* (Boddaert, 1783) Sungrebe   SE Mexico south to Bolivia and NE Argentina

PSOPHIIDAE TRUMPETERS (1: 3)

PSOPHIA Linnaeus, 1758 F
○ *Psophia crepitans* Grey-winged Trumpeter
 — *P.c.crepitans* Linnaeus, 1758   SE Colombia to the Guianas, N Brazil
 — *P.c.napensis* Sclater & Salvin, 1873   SE Colombia to NE Peru, NW Brazil

○ *Psophia leucoptera* Pale-winged Trumpeter
 — *P.l.leucoptera* Spix, 1825   E Peru, NE Bolivia, WC Brazil
 — *P.l.ochroptera* Pelzeln, 1857   NW Brazil

○ *Psophia viridis* Dark-winged Trumpeter[6]
 — *P.v.viridis* Spix, 1825   C Brazil
 — *P.v.dextralis* Conover, 1934   EC Brazil (558)
 — *P.v.obscura* Pelzeln, 1857   NE Brazil

[1] Includes *novaeguineae* Rand, 1940 (1963) (though *australis* occurs in New Guinea as a lowland vagrant) and *anggiensis* Thompson & Temple, 1964 (2400) (R.S.).
[2] For separation from *americana* see Pratt (1987) (1920).
[3] For the latest treatment of this see Fjeldså (1983) (843).
[4] Includes *albipectus*, which White (1965: 107) (2649) treated as a synonym of the nominate form.
[5] For correction of spelling see David & Gosselin (2002) (614).
[6] The name *interjecta* Griscom & Greenway, 1937 (1017), is considered applicable to intergrades, see Sherman (1996: 107) (2243).

GRUIDAE CRANES (4: 15)

BALEARICINAE

BALEARICA Brisson, 1760 F
O ***Balearica regulorum*** GREY CROWNED CRANE
   — *B.r.gibbericeps* Reichenow, 1892[1]   S Uganda and C and S Kenya to N Zimbabwe and N Mozambique

   — *B.r.regulorum* (E.T. Bennett, 1834)   S Angola and N Namibia to E Sth. Africa

O ***Balearica pavonina*** BLACK CROWNED CRANE
   — *B.p.pavonina* (Linnaeus, 1758)   Senegal and Gambia to Chad
   — *B.p.ceciliae* Mitchell, 1904   Sudan to W Ethiopia, N Uganda and NW Kenya

GRUINAE

ANTHROPOIDES Vieillot, 1816 M
O ***Anthropoides virgo*** (Linnaeus, 1758) DEMOISELLE CRANE   SE European Russia and E Turkey to Inner Mongolia >> NE Africa and S Asia. Vagr. to Burma

O ***Anthropoides paradiseus*** (A.A.H. Lichtenstein, 1793) BLUE CRANE[2]   E and S Sth. Africa, Etosha Pan (Namibia)

BUGERANUS Gloger, 1842 M
O ***Bugeranus carunculatus*** (J.F. Gmelin, 1789) WATTLED CRANE   Ethiopia, SE Zaire and SW Tanzania to N Botswana, Zimbabwe and C Mozambique, Angola, SE South Africa

GRUS Pallas, 1766 F
O ***Grus leucogeranus*** Pallas, 1773 SIBERIAN CRANE   NW and EC Siberia >> Iran, NW India and China

● ***Grus canadensis*** SANDHILL CRANE
   ✔ *G.c.canadensis* (Linnaeus, 1758)   NE Siberia, Alaska and Canada to Baffin I >> S USA and N Mexico

   — *G.c.rowani* Walkinshaw, 1965   British Colombia to N Ontario (2553)
   ✔ *G.c.tabida* (J.L. Peters, 1925)   SW Canada, W and C USA >> S USA, NC Mexico
   — *G.c.pratensis* F.A.A. Meyer, 1794   Georgia, Florida
   — *G.c.pulla* Aldrich, 1972   Mississippi (24)
   — *G.c.nesiotes* Bangs & Zappey, 1905   Isla de la Juventud, Cuba

O ***Grus antigone*** SARUS CRANE
   — *G.a.antigone* (Linnaeus, 1758)   N India, Nepal, Bangladesh
   — *G.a.sharpii* Blanford, 1895[3]   Cambodia, S Laos (in past Luzon, Philippines)
   — *G.a.gillae* Schodde, Blackman & Haffenden, 1989[4]   NE Australia (2177)

O ***Grus rubicunda*** (Perry, 1810) BROLGA   S New Guinea, N and E Australia

O ***Grus vipio*** Pallas, 1811 WHITE-NAPED CRANE   N Mongolia, S and E Transbaikalia, NE China and Russian Far East >> Korea, S Japan, EC China

● ***Grus grus*** COMMON CRANE
   ✔ *G.g.grus* (Linnaeus, 1758)   C, N, and E Europe to the Urals >> S Europe, N Africa
   — *G.g.lilfordi* Sharpe, 1894   Turkey, lower Volga river, and Urals to EC Asia >> NE Africa and S Asia east to SE China

O ***Grus monacha*** Temminck, 1835 HOODED CRANE   S Siberia, N China >> Japan, South Korea, C and E China

O ***Grus americana*** (Linnaeus, 1758) WHOOPING CRANE   N Saskatchewan >> SE Texas

O ***Grus nigricollis*** Przevalski, 1876 BLACK-NECKED CRANE   Tibetan Plateau >> Bhutan, NE India to SC China

O ***Grus japonensis*** (Statius Müller, 1776) RED-CROWNED CRANE   N Japan, NE China, SE Russia >> Korea, EC China

---

[1] Mistakenly assigned to *pavonina* in previous Editions (the two species were once considered conspecific).
[2] For correction of spelling see David & Gosselin (2002) (614).
[3] Includes *luzonica* Hachisuka, 1941 (1040), see Dickinson *et al.* (1991) (745).
[4] The spelling *gilliae* as used in Archibald & Meine, (1996: 87) (51) is not correct. The name *fordi* Bruce & McAllan, 1989 (260), appeared later and is a synonym.

ARAMIDAE LIMPKIN (1: 1)

ARAMUS Vieillot, 1816 M
● *Aramus guarauna* LIMPKIN

— *A.g.pictus* (F.A.A. Meyer, 1794)	Florida, Cuba, Jamaica
— *A.g.elucus* J.L. Peters, 1925	Hispaniola, Puerto Rico
⩗ *A.g.dolosus* J.L. Peters, 1925	SE Mexico to Panama
— *A.g.guarauna* (Linnaeus, 1766)	N Sth. America to Paraguay and Argentina

TURNICIDAE BUTTONQUAILS (2: 16)

TURNIX Bonnaterre, 1791 M[1]
○ *Turnix sylvaticus* COMMON BUTTONQUAIL

— *T.s.sylvaticus* (Desfontaines, 1787)	S Spain, NW Africa
— *T.s.lepuranus* (A. Smith, 1836)[2]	Sub-Saharan Africa
— *T.s.dussumier* (Temminck, 1828)	Pakistan to Burma
— *T.s.davidi* Delacour & Jabouille, 1930[3]	C Thailand, Indochina, S China, Taiwan
— *T.s.bartelsorum* Neumann, 1929	Java, Bali
— *T.s.whiteheadi* Ogilvie-Grant, 1897	Luzon
— *T.s.celestinoi* McGregor, 1907[4]	Bohol, Mindanao
— *T.s.nigrorum* duPont, 1976	Negros (775)
— *T.s.suluensis* Mearns, 1905[5]	Sulu Is.

○ *Turnix maculosus* RED-BACKED BUTTONQUAIL[6]

— *T.m.kinneari* Neumann, 1939[7]	Peleng I. (1645)
— *T.m.beccarii* Salvadori, 1875	Sulawesi, Muna and Tukangbesi Is.
— *T.m.obiensis* Sutter, 1955	Obi and Kai Is., Babar I. (2375)
— *T.m.sumbanus* Sutter, 1955	Sumba I. (2375)
— *T.m.floresianus* Sutter, 1955	Sumbawa, Komodo, Padar, Flores, Alor Is. (2375)
— *T.m.maculosus* (Temminck, 1815)	Semau, Roti, Timor, Wetar, Kisar, Moa Is.
— *T.m.savuensis* Sutter, 1955	Sawu I. (2375)
— *T.m.saturatus* Forbes, 1882	New Britain, Duke of York I
— *T.m.furvus* Parkes, 1949	NE New Guinea (1752)
— *T.m.giluwensis* Sims, 1954	Montane C New Guinea (2275)
— *T.m.horsbrughi* Ingram, 1909	S New Guinea
— *T.m.mayri* Sutter, 1955	Louisiade Arch. (New Guinea) (2375)
— *T.m.salomonis* Mayr, 1938	Guadalcanal (1467)
— *T.m.melanotus* (Gould, 1837)[8]	N and E Australia

○ *Turnix hottentottus* BLACK-RUMPED BUTTONQUAIL

— *T.h.nanus* (Sundevall, 1851)[9]	Senegal to W Kenya, south to N Angola, Zimbabwe and E Sth. Africa
— *T.h.hottentottus* Temminck, 1815	S Sth. Africa

○ *Turnix tanki* YELLOW-LEGGED BUTTONQUAIL

— *T.t.tanki* Blyth, 1843	C Pakistan, India, Nicobar and Andaman Is.
— *T.t.blanfordii* Blyth, 1863	Russian Far East, Korea and NE China S to Burma, SE Thailand and Indochina >> NE India and SE Asia

○ *Turnix ocellatus* SPOTTED BUTTONQUAIL

— *T.o.benguetensis* Parkes, 1968	N Luzon (1760)
— *T.o.ocellatus* (Scopoli, 1786)	C Luzon

[1] Gender addressed by David & Gosselin (2002) (614), multiple changes occur in specific names.
[2] Includes *alleni* and *arenarius* Stresemann, 1938 (2361), see Debus (1996: 53) (632).
[3] Includes *mikado* see Sutter (1955) (2375); but retained by Deignan (1963) (663) and needs review.
[4] Placed here not in *maculosus* and includes *masaaki*, see Dickinson, *et al.* (1991) (745), *contra* Sutter (1955) (2375).
[5] Misplaced in *maculosus* in last Edition, see Sutter (1955) (2375) and Dickinson *et al.* (1991) (745).
[6] Based on the revision of Sutter (1955) (2375), but *celestinoi* moved (see above).
[7] Described as a race of *sylvaticus*, but see Sutter (1955) (2375).
[8] All Australian birds are now considered to be *melanotus* see Marchant & Higgins (1993: 432) (1428), hence *pseutes* listed in last Edition is now a synonym, as is *yorki*.
[9] Includes *luciana* and *insolata* Ripley & Heinrich, 1960 (2066) (see White, 1965: 151) (2649).

○ *Turnix suscitator* Barred Buttonquail
  — *T.s.plumbipes* (Hodgson, 1837)   Nepal and NE India to NE Burma
  — *T.s.bengalensis* Blyth, 1852   C and S of W Bengal (India)
  — *T.s.taigoor* (Sykes, 1832)   Rest of India
  — *T.s.leggei* E.C.S. Baker, 1920   Sri Lanka
  — *T.s.okinavensis* A.R. Phillips, 1947   S Kyushu (Japan) and west to Ryukyu Is. (1866)
  — *T.s.rostratus* Swinhoe, 1865   Taiwan
  — *T.s.blakistoni* (Swinhoe, 1871)   E Burma to S China and N Vietnam
  — *T.s.pallescens* Robinson & Baker, 1928   SC Burma
  — *T.s.thai* Deignan, 1946   NW and C Thailand (640)
  — *T.s.interrumpens* Robinson & Baker, 1928   S Burma, Thailand
  — *T.s.atrogularis* (Eyton, 1839)   Malay Pen.
  — *T.s.suscitator* (J.F. Gmelin, 1789)[1]   Sumatra, Java, Bawean, Belitung, Bangka and Bali
  — *T.s.powelli* Guillemard, 1885   Lombok to Alor (Lesser Sundas)
  — *T.s.rufilatus* Wallace, 1865   Sulawesi
  — *T.s.haynaldi* W.H. Blasius, 1888   Palawan and nearby islands
  — *T.s.fasciatus* (Temminck, 1815)   Luzon, Mindoro, Masbate, Sibuyan
  — *T.s.nigrescens* Tweeddale, 1878   Cebu, Guimaras, Negros, Panay

○ *Turnix nigricollis* (J.F. Gmelin, 1789) Madagascan Buttonquail   Madagascar

○ *Turnix melanogaster* (Gould, 1837) Black-breasted Buttonquail   SE Queensland, NE New South Wales, Fraser I.

○ *Turnix castanotus* (Gould, 1840) Chestnut-backed Buttonquail[2]   NW and NC Australia and offshore islands

○ *Turnix olivii* Robinson, 1900 Buff-breasted Buttonquail   Cape York Pen. (Australia)

○ *Turnix varius* Painted Buttonquail
  — †? *T.v.novaecaledoniae* Ogilvie-Grant, 1893   New Caledonia
  — *T.v.scintillans* (Gould, 1845)   Abrolhos Is. (SW Australia)
  — *T.v.varius* (Latham, 1802)[3]   SW, E and SE Australia

○ *Turnix worcesteri* McGregor, 1904 Worcester's Buttonquail   Luzon

○ *Turnix everetti* E. Hartert, 1898 Sumba Buttonquail   Sumba I.

○ *Turnix pyrrhothorax* (Gould, 1841) Red-chested Buttonquail   N and E Australia

○ *Turnix velox* (Gould, 1841) Little Buttonquail   Arid parts of Australia (not Tasmania)

ORTYXELOS Vieillot, 1825 M
○ *Ortyxelos meiffrenii* (Vieillot, 1819) Quail-plover   Senegal to C Sudan, S Ethiopia to NW and E Kenya

BURHINIDAE THICK-KNEES (2: 9)

BURHINUS Illiger, 1811 M
○ *Burhinus oedicnemus* Eurasian Stone Curlew
  — *B.o.distinctus* (Bannerman, 1914)   W Canary Is.
  — *B.o.insularum* (Sassi, 1908)   E Canary Is.
  — *B.o.oedicnemus* (Linnaeus, 1758)   W and S Europe to N Balkans, Ukraine and Caucasus area
  — *B.o.saharae* (Reichenow, 1894)[4]   N Africa, Balearic Is., Malta, Cyprus, Levant to SC Turkey and SW Iran >> S Europe, N Africa, and Arabia
  — *B.o.harterti* Vaurie, 1963   Volga Delta, Transcaspia and NE Iran to W Pakistan and E Kazakhstan >> NE Africa and SW Asia (2509)
  — *B.o.indicus* (Salvadori, 1865)[5]   C Pakistan, India and Sri Lanka to east to mainland SE Asia

---

[1] Includes *kuiperi* Chasen, 1937 (320), and *baweanus* Hoogerwerf, 1962 (1141), see Mees (1986) (1539). Also includes *machetes* Deignan, 1946 (640), although this seems to have been overlooked, e.g. by van Marle & Voous (1986) (2468), and may need review.
[2] In last Edition treated as polytypic, but see Condon (1975) (556); thus includes *olivii*.
[3] For reasons to use 1802 not 1801 see Browing & Monroe (1991) (258).
[4] Includes *jordansi*, *astutus* and *theresae* R. Meinertzhagen, 1948 (1548) see Vaurie (1965: 447) (2510).
[5] Includes *mayri* Koelz, 1939 (1279), see Ripley (1982: 106) (2055).

○ *Burhinus senegalensis* (Swainson, 1837) Senegal Thick-knee[1]   Senegal to Sudan, Egypt, Ethiopia, N Uganda and NW Kenya

○ *Burhinus vermiculatus* Water Dikkop
   — *B.v.buettikoferi* (Reichenow, 1898)   Coastal Nigeria and Gabon
   — *B.v.vermiculatus* (Cabanis, 1868)   Zaire, S Somalia, and S Kenya to Angola, N Botswana and E and S Cape Province

○ *Burhinus capensis* Spotted Dikkop
   — *B.c.maculosus* (Temminck, 1824)[2]   Senegal to Ethiopia and Somalia, south to N Uganda and N Kenya
   — *B.c.dodsoni* (Ogilvie-Grant, 1899)[3]   S Arabia, Eritrea and N Somalia
   — *B.c.capensis* (M.H.K. Lichtenstein, 1823)   Sth. Africa and E Botswana to E and NW Angola, Tanzania and S Kenya
   — *B.c.damarensis* (Reichenow, 1905)   Namibia, Botswana, SW Angola

○ *Burhinus bistriatus* Double-striped Thick-knee
   — *B.b.bistriatus* (Wagler, 1829)[4]   S Mexico to NW Costa Rica
   — *B.b.vocifer* (L'Herminier, 1837)   N Colombia to Guyana and N Brazil
   — *B.b.pediacus* Wetmore & Borrero, 1964   N Colombia (2609)
   — *B.b.dominicensis* (Cory, 1883)   Hispaniola

○ *Burhinus superciliaris* (Tschudi, 1843) Peruvian Thick-knee   SW Ecuador to SW Peru

○ *Burhinus grallarius* (Latham, 1802)[5] Bush Stone Curlew[6]   Australia, SC New Guinea

ESACUS Lesson, 1831 M
○ *Esacus recurvirostris* (Cuvier, 1829) Great Stone Curlew   SE Iran to India, Sri Lanka, SE Asia and Hainan
○ *Esacus magnirostris* (Vieillot, 1818) Beach Stone Curlew[7]   Coasts of Andaman Is. and islets off Malay Pen. to New Guinea and Australia

CHIONIDAE SHEATHBILLS AND ALLIES (2: 3)

CHIONINAE

CHIONIS J.R. Forster, 1788 M
○ *Chionis albus* (J.F. Gmelin, 1789) Snowy Sheathbill[8]   Antarctic Pen., S Shetland Is., Elephant I., S Georgia, S Orkneys >> Falkland Is., Tierra del Fuego and Patagonia

○ *Chionis minor* Black-faced Sheathbill
   — *C.m.marionensis* Reichenow, 1908   Prince Edward and Marion Is.
   — *C.m.crozettensis* (Sharpe, 1896)   Crozet Is.
   — *C.m.minor* Hartlaub, 1841[9]   Kerguelen Is.
   — *C.m.nasicornis* Reichenow, 1904   Heard and McDonald Is.

PLUVIANELLINAE[10]

PLUVIANELLUS G.R. Gray, 1846 M
○ *Pluvianellus socialis* G.R. Gray, 1846 Magellanic Plover   S Chile, S Argentina

HAEMATOPODIDAE OYSTERCATCHERS (1: 11)

HAEMATOPUS Linnaeus, 1758 M
○ *Haematopus leucopodus* Garnot, 1826 Magellanic Oystercatcher   SC Chile and SC Argentina S to Cape Horn, Falkland Is.

[1] Includes *inornatus* see White (1965: 118) (2649).
[2] Includes *affinis* implicitly placed here by White (1965: 119) (2649).
[3] Includes *ehrenbergi* see White (1965: 119) (2649).
[4] Includes *vigilans* van Rossem, 1934 (2469) see Hume (1996: 362) (1163).
[5] For reasons to use 1802 not 1801 see Browing & Monroe (1991) (258).
[6] Treated as monotypic by Schodde & Mason (1980) (2186). For some years known as *B. magnirostris* (including in our last Edition). The name *grallarius*, however, appears to have been properly selected by Sharpe (1896) (2233) following Gould as "First Reviser" and was then in general use for over 40 years.
[7] Two alternative names have been proposed for this (*giganteus* and *neglectus*). We believe application to the ICZN is needed to resolve this dispute and for now follow Hume (1996: 363) (1163) pending a ruling.
[8] For correction of spelling see David & Gosselin (2002) (614).
[9] Includes *nivea* Mathews, 1933 (1455).
[10] For reasons to treat this here see Wolters (1983) (2686) and Chu (1995) (354).

○ *Haematopus ater* Vieillot & Oudart, 1825 Blackish Oystercatcher   N Peru to Tierra del Fuego, north to SC Argentina, Falkland Is.

● *Haematopus bachmani* Audubon, 1838 American Black Oystercatcher
                                                 W coast of N America

● *Haematopus palliatus* American Oystercatcher
     ⊥ *H.p.palliatus* Temminck, 1820[1]   Coasts of N and S America, W Indies
     — *H.p.galapagensis* Ridgway, 1886   Galapagos Is.

○ †?*Haematopus meadewaldoi* Bannerman, 1913 Canary Island Oystercatcher
                                                 E Canary Is.

○ *Haematopus moquini* Bonaparte, 1856 African Black Oystercatcher   N Namibia to Natal

● *Haematopus ostralegus* Eurasian Oystercatcher
     ⊥ *H.o.ostralegus* Linnaeus, 1758[2]   Iceland to NW European Russia, south to W Mediterranean >> W and S Europe and W Africa

     — *H.o.longipes* Buturlin, 1910   Balkan, Ukraine, and Turkey to S and C European Russia and W Siberia >> E Mediterranean, coastal E Africa, Arabia

     — *H.o.buturlini* Dementiev, 1941[3]   Lower Ural R. basin and Transcaspia to E Kazakhstan and NW China>> Red Sea and SW Asia to India (687)

     — *H.o.osculans* Swinhoe, 1871   Shores Sea of Okhotsk and Kamchatka to Korea and NE China >> E China and Korea

● *Haematopus finschi* G.H. Martens, 1897 South Island Pied Oystercatcher[4]
                                Inland South Island >> North Island (New Zealand)

● *Haematopus longirostris* Vieillot, 1817 Australian Pied Oystercatcher
                                Kai Is., Aru Is., S New Guinea, Australia, Tasmania

● *Haematopus unicolor* Variable Oystercatcher[5]
     ⊥ *H.u.unicolor* J.R. Forster, 1844[6]   Coasts of New Zealand
     — *H.u.chathamensis* E. Hartert, 1827   Chatham Is. (New Zealand)

○ *Haematopus fuliginosus* Sooty Oystercatcher
     — *H.f.fuliginosus* Gould, 1845   Coast of S Australia, Tasmania
     — *H.f.opthalmicus* Castelnau & Ramsay, 1877   Coast of N Australia

DROMADIDAE CRAB PLOVER (1: 1)

DROMAS Paykull, 1805 M
○ *Dromas ardeola* Paykull, 1805 Crab Plover   Coasts (mainly west) of Indian Ocean >> south to SE Africa and Madagascar, east to Andaman Is.

IBIDORHYNCHIDAE IBISBILL (1: 1)

IBIDORHYNCHA Vigors, 1832 F
○ *Ibidorhyncha struthersii* Vigors, 1832 Ibisbill   Tien Shan, Pamir, and Tibetan Plateau to Shaanxi, Shanxi and W and N Hebei >> S to nearby southern foothill rivers beds from N India to NE Burma

RECURVIROSTRIDAE STILTS, AVOCETS (3: 7)

HIMANTOPUS Brisson, 1760 M
● *Himantopus himantopus* Black-winged Stilt[7]
     — *H.h.himantopus* (Linnaeus, 1758)[8]   S Europe to C Asia, Africa, India to Sri Lanka, Indochina and Taiwan >> S Europe, Africa, S and E Asia to Gtr. Sundas and Philippines

---

[1] Includes *prattii*, *pitanay* and *durnfordi*, but not *frazari* thought to have been used to name a hybrid swarm, see Hockey (1996: 322) (1125).
[2] Includes *malacophaga* (which, contrary to its original spelling, should be rendered as *malacophagus*) and *occidentalis* see Stepanyan (1990: 184) (2326).
[3] For recognition of this see Stepanyan (1990: 185) (2326).
[4] For reasons to separate this from *longirostris* see Marchant & Higgins (1993: 715) (1428).
[5] We prefer to follow Sibley & Monroe (1990) (2262) and treat *chathamensis* as conspecific
[6] Drawings with Forster's proposed names were published many years after his death (d. 1798).
[7] Mayr & Short (1970) (1495) recommended splitting, which we did in last Edition. Despite the reservations of CSR we now follow Pierce (1996: 345) (1883), finding each of the alternative groupings of subspecies unsatisfactory.
[8] Includes *ceylonensis* Whistler, 1944 (2619), although recognised by Ripley (1982: 105) (2055).

___ *H.h.leucocephalus* Gould, 1837   Java east to New Guinea, Australia, New Zealand
            (migrant >> Philippines ?)

✓ *H.h.knudseni* Stejneger, 1887   Hawaiian Is.

✓ *H.h.mexicanus* (Statius Müller, 1776)   W and S USA to N and C Sth. America, W Indies,
            Galapagos

✓ *H.h.melanurus* Vieillot, 1817   C and S Andes to Sthn. S America

○ *Himantopus novaezelandiae* Gould, 1841   Black Stilt   New Zealand

CLADORHYNCHUS G.R. Gray, 1840   M
○ *Cladorhynchus leucocephalus* (Vieillot, 1816)   Banded Stilt   C and S Australia

RECURVIROSTRA Linnaeus, 1758   F
◗ *Recurvirostra avosetta* Linnaeus, 1758   Pied Avocet   C and S Europe to SW and C Asia, N, E, and S Africa
            >> W Europe to NW India and SE China

◗ *Recurvirostra americana* J.F. Gmelin, 1789   American Avocet   C and W North America, E USA, C Mexico >> S and
            SE USA to N Honduras, Bahamas to Cuba

○ *Recurvirostra novaehollandiae* Vieillot, 1816   Red-necked Avocet   Australia

○ *Recurvirostra andina* Philippi & Landbeck, 1861   Andean Avocet   C Peru to N Chile, NW Argentina

CHARADRIIDAE PLOVERS (10: 66)

VANELLINAE

VANELLUS Brisson, 1760   M
◑ *Vanellus vanellus* (Linnaeus, 1758)   Northern Lapwing   W Europe to China and Japan

○ *Vanellus crassirostris*   Long-toed Lapwing
    ___ *V.c.crassirostris* (Hartlaub, 1855)[1]   N Nigeria, S Sudan to E Zaire, Kenya and N Tanzania
    ___ *V.c.leucopterus* Reichenow, 1889   S Zaire to Malawi, N Botswana and Natal

○ *Vanellus armatus* (Burchell, 1822)   Blacksmith Plover[2]   Angola, Tanzania and S Kenya to Cape Province

○ *Vanellus spinosus* (Linnaeus, 1758)   Spur-winged Plover   SE Europe, Middle East, sub-Saharan Africa south to
            Ghana, Central African Republic and Kenya

○ *Vanellus duvaucelii* (Lesson, 1826)   River Lapwing   N and C India to Indochina

○ *Vanellus tectus*   Black-headed Lapwing
    ___ *V.t.tectus* (Boddaert, 1783)   Senegal and S Mauritania to Ethiopia, Uganda and
            NW Kenya
    ___ *V.t.latifrons* (Reichenow, 1881)   S Somalia to E Kenya

○ *Vanellus malabaricus* (Boddaert, 1783)   Yellow-wattled Lapwing   India, Sri Lanka

○ *Vanellus albiceps* Gould, 1834   White-headed Lapwing   Senegal and Gambia to SW Sudan and N Angola, S
            Tanzania, Zambia to N Transvaal and S Mozambique

○ *Vanellus lugubris* (Lesson, 1826)   Lesser Black-winged Lapwing   S Mali and Sierra Leone to SW Nigeria, Gabon and
            Zaire to E Africa, Mozambique and Natal

○ *Vanellus melanopterus*   Black-winged Lapwing
    ___ *V.m.melanopterus* (Cretzschmar, 1829)   Ethiopia and Eritrea to SW Arabia
    ___ *V.m.minor* (Zedlitz, 1908)   Kenya and N Tanzania, Transvaal to E Cape Province

○ *Vanellus coronatus*   Crowned Plover
    ___ *V.c.demissus* (Friedmann, 1928)   N Somalia
    ___ *V.c.coronatus* (Boddaert, 1783)   Ethiopia and E Africa to Zambia and Sth. Africa
    ___ *V.c.xerophilus* Clancey, 1960[3]   SW Angola, Namibia, Botswana, W Zimbabwe, W
            Transvaal (387)

○ *Vanellus senegallus*   African Wattled Lapwing
    ___ *V.s.senegallus* (Linnaeus, 1766)   Senegal and Gambia to Sudan, NE Zaire and N Uganda

---

[1] Includes *hybrida* see White (1965: 120) (2649).
[2] The generic name *Antibyx* was retained for this in last Edition.
[3] For acceptance of this race see White (1965: 122) (2649).

— *V.s.major* (Neumann, 1914) — W and C Ethiopia, Eritrea

— *V.s.lateralis* A. Smith, 1839[1] — S and E Zaire to S Uganda, and W Kenya, south to N Namibia and E Sth. Africa

○ *Vanellus melanocephalus* (Rüppell, 1845) Spot-breasted Lapwing — Ethiopian highlands

○ *Vanellus superciliosus* (Reichenow, 1886) Brown-chested Lapwing — Togo to NE Zaire >> Uganda, NW Tanzania, SE Zaire

○ *Vanellus cinereus* (Blyth, 1842) Grey-headed Lapwing — NE and E China, S Russia Far East, Japan >> Nepal to N Indochina, S China, Taiwan and S Japan

○ *Vanellus indicus* Red-wattled Lapwing

— *V.i.aigneri* (Laubman, 1913) — Turkey to W Pakistan and NE Arabia

— *V.i.indicus* (Boddaert, 1783) — C Pakistan, India, and Nepal to N Assam and Bangladash

— *V.i.lankae* (Koelz, 1939)[2] — Sri Lanka (1279)

— *V.i.atronuchalis* (Jerdon, 1864) — S Assam east to W Yunnan and S through mainland SE Asia to the Malay Pen. and N Sumatra

○ †?*Vanellus macropterus* (Wagler, 1827) Javan Wattled Lapwing — Java

○ *Vanellus tricolor* (Vieillot, 1818) Banded Lapwing — C and S Australia, Tasmania

● *Vanellus miles* Masked Lapwing

— *V.m.miles* (Boddaert, 1783) — Coastal New Guinea, Aru Is., N Australia >> SE Wallacea

⤋ *V.m.novaehollandiae* Stephens, 1819 — S Australia, Tasmania, New Zealand

○ *Vanellus gregarius* (Pallas, 1771) Sociable Plover — SC Russia, Kazakhstan >> NE Africa, Arabia, N India

○ *Vanellus leucurus* (M.H.K. Lichtenstein, 1823) White-tailed Plover — Middle East to Iran and Kazakhstan>> NE Africa to NW India

○ *Vanellus cayanus* (Latham, 1790) Pied Lapwing — N and C Sth. America

◑ *Vanellus chilensis* Southern Lapwing

— *V.c.cayennensis* (J.F. Gmelin, 1789) — N Sth. America

— *V.c.lampronotus* (Wagler, 1827) — C, E and S Brazil to N Chile and N Argentina

⤋ *V.c.chilensis* (Molina, 1782) — C Chile, SW Argentina

— *V.c.fretensis* (Brodkorb, 1934)[3] — S Chile, S Argentina (217)

○ *Vanellus resplendens* (Tschudi, 1843) Andean Lapwing — SW Colombia to N Chile, NW Argentina

ERYTHROGONYS Gould, 1838[4] M

○ *Erythrogonys cinctus* Gould, 1838 Red-kneed Dotterel — Australia

PLUVIALINAE

PLUVIALIS Brisson, 1760 F

○ *Pluvialis apricaria* Eurasian Golden Plover

— *P.a.altifrons* (C.L. Brehm, 1831) — EC Greenland, Iceland, Faroe Is., Norway, and W and N Sweden to Taymyr >> W and S Europe, N Africa, and Middle East

— *P.a.apricaria* (Linnaeus, 1758)[5] — British Isles to S Sweden and Baltic States >> W and S Europe

● *Pluvialis fulva* (J.F. Gmelin, 1789) Pacific Golden Plover[6] — NC and NE Russia, N Siberia, W Alaska >> E Africa, India, SE Asia to Oceania, Australia and New Zealand

○ *Pluvialis dominica* (Statius Müller, 1776) American Golden Plover

Alaska and N Canada >> C Sth. America

◑ *Pluvialis squatarola* Grey Plover/Black-bellied Plover[7]

⤋ *P.s.squatarola* (Linnaeus, 1758) — N Eurasia (except Wrangel I.) and Alaska >> W and S Europe, Africa, S and E Asia, Australasia, W America

— *P.s.tomkovichi* Engelmoer & Roselaar, 1998 — Wrangel I. (NE Siberia) >> ??

— *P.s.cynosurae* (Thayer & Bangs, 1914) — N Canada >> coastal N and S America

[1] Apparently includes *solitaneus* Clancey, 1979 (466).
[2] For recognition of this see Ripley (1982: 112) (2055).
[3] For recognition of this see Wiersma (1996: 420) (2662).
[4] The need to recognise this genus and the next was discussed by Christian *et al.* (1992) (348).
[5] Includes *oreophilos*. The ranges used in last Edition for this and for *apricaria* were those used by Peters (1934) (1822) See also Cramp *et al.* (1983: 215) (590).
[6] For details of separation from *P. dominica* see references in Sangster *et al.* (2002) (2167).
[7] For polytypic treatment see Englemoer & Rosendaal (1998) (800).

CHARADRIINAE

CHARADRIUS Linnaeus, 1758 M

● *Charadrius obscurus* J.F. Gmelin, 1789 Red-breasted Plover[1] New Zealand

● *Charadrius hiaticula* Common Ringed Plover[2]

    __ *C.h.psammodroma* Salomonsen, 1930   NE Canada, Greenland, Iceland, Svalbard >> SW
            Europe and W Africa

    ✓ *C.h.hiaticula* Linnaeus, 1758[3]   British Isles and W France to S Norway, S Sweden, and
            Baltic States >> W Europe and NW Africa

    __ *C.h.tundrae* (P.R. Lowe, 1915)[4]   N Europe and N Eurasia >> S Europe, Africa and SW
            Asia

● *Charadrius semipalmatus* Bonaparte, 1825 Semipalmated Plover   Alaska, N Canada >> coastal N, C and S America,
            Bermuda, West Indies, Galapagos Is.

○ *Charadrius placidus* J.E. & G.R. Gray, 1863 Long-billed Plover[5]   Russian Far East to NE and E China and Japan >> E
            Nepal to S China and Japan

● *Charadrius dubius* Little Ringed Plover

    ✓ *C.d.curonicus* J.F. Gmelin, 1789   N Africa, Europe, Asia (except SC and SE) >> Africa
            and S Asia to Indonesia and Philippines

    __ *C.d.jerdoni* (Legge, 1880)   India to S China and Indochina

    __ *C.d.dubius* Scopoli, 1786[6]   Philippines to New Guinea and Bismarck Arch. >> SE
            Asia to Australasia

○ *Charadrius wilsonia* Wilson's Plover

    __ *C.w.wilsonia* Ord, 1814   E USA to E Mexico and Belize

    __ *C.w.rufinucha* (Ridgway, 1874)   Bahama Is., Gtr. and N Lesser Antilles

    __ *C.w.beldingi* (Ridgway, 1919)   Baja California to C Peru

    __ *C.w.cinnamominus* (Ridgway, 1919)   Colombia to French Guiana, Dutch Antilles, Trinidad,
            Grenada

● *Charadrius vociferus* Killdeer

    ✓ *C.v.vociferus* Linnaeus, 1758   Canada, USA, Mexico >> south to NW Sth. America

    __ *C.v.ternominatus* Bangs & Kennard, 1920   Bahamas, Gtr. Antilles

    __ *C.v.peruvianus* (Chapman, 1920)   Peru, NW Chile

● *Charadrius melodus* Ord, 1824 Piping Plover   C and E North America >> S USA, W Mexico, Bahama
            Is., Gtr. Antilles

○ *Charadrius thoracicus* (Richmond, 1896) Black-banded Plover/Madagascan Plover
            W and SW Madagascar

○ *Charadrius pecuarius* Temminck, 1823 Kittlitz's Sand Plover[7]   Egypt, Senegal to Sudan and Cape Province, Madagascar

○ *Charadrius sanctaehelenae* (Harting, 1873) St. Helena Plover   St. Helena I.

○ *Charadrius tricollaris* Three-banded Plover

    __ *C.t.tricollaris* Vieillot, 1818[8]   Ethiopia and E Africa to Gabon, S Zaire and Sth. Africa
    __ *C.t.bifrontatus* Cabanis, 1882   Madagascar

○ *Charadrius forbesi* (Shelley, 1883) Forbes's Plover[9]   S Senegal to SW Sudan and S to N Angola and N Zambia

○ *Charadrius marginatus* White-fronted Plover

    __ *C.m.mechowi* (Cabanis, 1884)[10]   Sub-Saharan Africa to inland N Namibia, N Botswana
            and Transvaal

    __ *C.m.marginatus* Vieillot, 1818   Coastal SW Angola to SW Cape Province

    __ *C.m.arenaceus* Clancey, 1971[11]   Coastal S Mozambique to S Cape Province (432)

    __ *C.m.tenellus* Hartlaub, 1861   Madagascar

[1] Treated in *Pluvialis* in last Edition, but see Christian *et al.* (1992) (348). Tentatively includes *aquilonius* Dowding, 1994 (758), but this awaits evaluation.
[2] We follow Englemoer & Rosendaal (1998) (800) in recognising *psammodroma*.
[3] Includes *harrisoni* Clancey, 1949 (362) see Ibis, 93: 298.
[4] Includes *kolymensis* Buturlin, 1934 (283), see Stepanyan (1990: 173) (2326).
[5] Includes *japonicus* Mishima, 1956 (1589), see Orn. Soc. Japan (1974) (1726).
[6] Includes *papuanus* Mayr, 1938 (1467), see Mees (1982) (1534).
[7] Includes *allenbyi* and *tephricolor* Clancey, 1971 (431), see Cramp *et al.* (1983: 146) (590).
[8] Thought to include *pelodromus* Clancey, 1981 (484).
[9] Already treated as a species by White (1965: 125) (2649).
[10] Includes *pons* see White (1965: 125) (2649).
[11] For recognition see Urban (1986: 237) (2451).

◑ *Charadrius alexandrinus* KENTISH PLOVER
    __ *C.a.alexandrinus* Linnaeus, 1758[1]
    W Europe, Atlantic islands and N Africa to NE China >> N tropical Africa, S Asia, W Indonesia

    __ *C.a.dealbatus* (Swinhoe, 1870)[2]
    S Japan, E and SE China >> S to Philippines, and Borneo
    __ *C.a.seebohmi* Hartert & Jackson, 1915[3]
    S India, Sri Lanka
    ✓ *C.a.nivosus* (Cassin, 1858)[4]
    USA, Mexico, Bahamas, Greater and Dutch Antilles >> south to Panama

    __ *C.a.occidentalis* (Cabanis, 1872)
    Coastal Peru to SC Chile

○ *Charadrius javanicus* Chasen, 1938 JAVAN PLOVER[5]
    Java (321)

○ *Charadrius ruficapillus* Temminck, 1821[6] RED-CAPPED PLOVER
    Australia, Tasmania

○ *Charadrius peronii* (Schlegel, 1865) MALAYSIAN PLOVER[7]
    S Thailand, Malay Pen. and S Vietnam to Philippines, Sulawesi, Borneo, Sumatra, Bali to Timor

○ *Charadrius pallidus* CHESTNUT-BANDED PLOVER[8]
    __ *C.p.venustus* Fischer & Reichenow, 1884
    S Kenya, N Tanzania
    __ *C.p.pallidus* Strickland, 1852
    SW Angola to Mozambique, south to NC Sth Africa

○ *Charadrius collaris* Vieillot, 1818 COLLARED PLOVER
    WC Mexico to N Argentina and C Chile

○ *Charadrius alticola* (Berlepsch & Stolzmann, 1902) PUNA PLOVER
    C Peru to W Bolivia and NW Argentina

○ *Charadrius falklandicus* Latham, 1790 TWO-BANDED PLOVER
    S Sth. America, Falkland Is.

◑ *Charadrius bicinctus* DOUBLE-BANDED PLOVER
    ✓ *C.b.bicinctus* Jardine & Selby, 1827
    New Zealand, Chatham Is. >> S and E Australia, Tasmania, New Caledonia, Vanuatu and Fiji
    __ *C.b.exilis* Falla, 1978
    Auckland Is. (823)

○ *Charadrius mongolus* LESSER SAND PLOVER[9]
    __ *C.m.pamirensis* (Richmond, 1896)
    W Tien Shan, Pamir and NW Himalaya to W Kunlun Shan >> E and S Africa to India (2015)

    __ *C.m.atrifrons* Wagler, 1829
    C and E Himalayas and S Tibetan Plateau>> S China and Gtr. Sundas
    __ *C.m.schaeferi* Meyer de Schauensee, 1937[10]
    E Tibetan Plateau to S Mongolia >> S China to Gtr. Sundas (1558)

    __ *C.m.mongolus* Pallas, 1776
    Lake Baikal to Sea of Okhotsk >> SE China and Taiwan to Australia
    __ *C.m.stegmanni* Portenko, 1939[11]
    NE Siberia, Kamchatka and Commander Is. >> Taiwan, Philippines, New Guinea, E Australia and W and SW Pacific Is. (1904)

○ *Charadrius leschenaultii* GREATER SAND PLOVER[12]
    __ *C.l.columbinus* Wagler, 1829
    Turkey and Syria to SW Iran >> SE Mediterranean, Red Sea, Persian Gulf
    __ *C.l.crassirostris* (Severtsov, 1873)
    E Transcaucasia and Transcaspia to Syr Dar'ya basin and Karatau (WC Kazakhstan) >> shores NE and E Africa to W India

    __ *C.l.leschenaultii* Lesson, 1826
    Chu basin (EC Kazakhstan) and W China east to Mongolia and Nei Mongol >> shores of Indian Ocean (incl. E Africa), Indonesia, Philippines and Australasia

[1] Includes *spatzi* see Stepanyan (1990: 178) (2326) and implictly *hesperius* see White (1965: 125) (2649).
[2] Includes *nihonensis* Deignan, 1941 (635), see Stepanyan (1990: 178) (2326).
[3] Includes *leggei* Whistler & Kinnear, 1937 (2621), see Ripley (1982: 116) (2055).
[4] We follow A.O.U. (1998: 146) (3) in including *occidentalis* and *nivosus*, with *tenuirostris* as a synonym, see Wiersma (1996: 432) (2662), in this species.
[5] This split, suggested by Roselaar in Cramp *et al.* (1983: 165) (590), has not yet been properly documented, partly due to the relative inaccessibility of the type, and the taxon might be better attached to *C. ruficapillus*.
[6] Dated 1822 in Peters (1934: 250) (1822) a *lapsus*.
[7] Van Marle & Voous (1988: 87) (2468) implicitly placed *chaseni* Junge, 1939 (1230), in synonymy.
[8] In last Edition we named this *C. venustus* believing that *pallidus* applied to a synonym of *C. alexandrinus* see Peters (1934: 250) (1822). This was wrong and we now follow White (1965: 126) (2649).
[9] We now follow the treatment in Cramp *et al.* (1983: 166) (590).
[10] Sometimes cited as 1938. This date is not that given by Schäfer (1939) (2171).
[11] A new name for *litoralis* Stegmann, 1937 (2311).
[12] In last Edition treated as monotypic. We now follow Cramp *et al.* (1983: 170) (590); see also Hirschfeld *et al.* (2000) (1124) and works in Russian and German cited therein.

○ *Charadrius asiaticus* Pallas, 1773 Caspian Plover — Caspian Sea to E Kazakhstan and NW China >> E and S Africa

○ *Charadrius veredus* Gould, 1848 Oriental Plover — Tuva Republic (Russia) and Mongolia to SE Transbaikalia and N Nei Mongol >> to Australia

○ *Charadrius morinellus* Linnaeus, 1758 Eurasian Dotterel[1] — N Europe and N Asia to E Russia >> Mediterranean to W Iran

○ *Charadrius modestus* M.H.K. Lichtenstein, 1823 Rufous-chested Dotterel — SC and S Chile, WC and S Argentina, Falkland Is. >> north to N Chile and SE Brazil

○ *Charadrius montanus* J.K. Townsend, 1837 Mountain Plover — W USA >> SW USA to NE Mexico

PELTOHYAS Sharpe, 1896[2] M
○ *Peltohyas australis* (Gould, 1841) Inland Dotterel — SW, SC and EC Australia

THINORNIS G.R. Gray, 1845[3] M
○ *Thinornis rubricollis* (J.F. Gmelin, 1789) Hooded Dotterel — S Australia, Tasmania

○ *Thinornis novaeseelandiae* (J.F. Gmelin, 1789) Shore Dotterel — Rangatira I. (Chatham Is.)

ELSEYORNIS Mathews, 1914 M
○ *Elseyornis melanops* (Vieillot, 1818) Black-fronted Dotterel — Australia, Tasmania, New Zealand

OREOPHOLUS Jardine & Selby, 1835[4] M
○ *Oreopholus ruficollis* Tawny-throated Dotterel
— *O.r.pallidus* Carriker, 1935 — N Peru (302)
— *O.r.ruficollis* (Wagler, 1829) — C Peru to Tierra del Fuego

ANARHYNCHUS Quoy & Gaimard, 1830 M
● *Anarhynchus frontalis* Quoy & Gaimard, 1830 Wry-bill — New Zealand

PHEGORNIS G.R. Gray, 1846 M
● *Phegornis mitchellii* (Fraser, 1845) Diademed Plover — NC Peru to SC Chile and SC Argentina

ROSTRATULIDAE PAINTED-SNIPE (2: 2)

ROSTRATULA Vieillot, 1816 F
○ *Rostratula benghalensis* Greater Painted-snipe
— *R.b.benghalensis* (Linnaeus, 1758) — Africa, Madagascar, Asia Minor through S Asia to Japan and Philippines, south east to Gtr. and Lesser Sundas (east to Flores)

— *R.b.australis* (Gould, 1838) — N and E Australia

NYCTICRYPHES Wetmore & Peters, 1923 M
○ *Nycticryphes semicollaris* (Vieillot, 1816) South American Painted-snipe — E and SE Brazil, Paraguay, Uruguay, C Chile to C Argentina

JACANIDAE JACANAS (6: 8)

MICROPARRA Cabanis, 1877 F
○ *Microparra capensis* (A. Smith, 1839) Lesser Jacana — Mali to Sudan and Ethiopia, south to E Sth. Africa

ACTOPHILORNIS Oberholser, 1899 M
○ *Actophilornis africanus* (J.F. Gmelin, 1789) African Jacana — Sub-Saharan Africa

---

[1] Treated in the genus *Eudromias* in last Edition, but see Cramp *et al.* (1983: 114) (590).
[2] The species included is close to Charadrius see Christian *et al.* (348).
[3] We suspect this genus and the next will prove to be close to the *Vanellinae*.
[4] Treated in the genus *Eudromias* in last Edition, but see Piersma (1996: 386) (1884). Position in generic sequence doubtful.

○ *Actophilornis albinucha* (I. Geoffroy Saint-Hilaire, 1832) MADAGASCAN JACANA

W Madagascar

IREDIPARRA Mathews, 1911 F
○ *Irediparra gallinacea* (Temminck, 1828) COMB-CRESTED JACANA[1]

Borneo and Mindanao to S Moluccas and Lesser Sundas to New Guinea and N and E Australia

HYDROPHASIANUS Wagler, 1832 M
○ *Hydrophasianus chirurgus* (Scopoli, 1786) PHEASANT-TAILED JACANA

Pakistan east to SE and E China and south to Sri Lanka, Philippines and Gtr. Sundas >> (some) E Arabia, S Asia

METOPIDIUS Wagler, 1832 M
○ *Metopidius indicus* (Latham, 1790) BRONZE-WINGED JACANA

India E to mainland SE Asia S to Sumatra and Java

JACANA Brisson, 1760 F
● *Jacana spinosa* NORTHERN JACANA
　— *J.s.gymnostoma* (Wagler, 1831)[2]

Mexico and Cozumel I.

　— *J.s.violacea* (Cory, 1881)

Cuba, Isla de la Juventud, Jamaica, Hispaniola

　— *J.s.spinosa* (Linnaeus, 1758)[3]

Belize and Guatemala to W Panama

○ *Jacana jacana* WATTLED JACANA
　— *J.j.hypomelaena* (G.R. Gray, 1846)

WC Panama, N Colombia

　— *J.j.melanopygia* (P.L. Sclater, 1857)

W Colombia, W Venezuela

　— *J.j.intermedia* (P.L. Sclater, 1857)

N and C Venezuela

　— *J.j.jacana* (Linnaeus, 1766)

SE Colombia, S Venezuela, Trinidad, the Guianas to E Bolivia, N Argentina and Uruguay

　— *J.j.scapularis* Chapman, 1922

W Ecuador, NW Peru

　— *J.j.peruviana* J.T. Zimmer, 1930

NE Peru, NW Brazil

PEDIONOMIDAE PLAINS WANDERER[4] (1: 1)

PEDIONOMUS Gould, 1841 M
○ *Pedionomus torquatus* Gould, 1841 PLAINS WANDERER

Inland E Australia

THINOCORIDAE SEEDSNIPES (2: 4)

ATTAGIS Geoffroy Saint-Hilaire & Lesson, 1831 M
○ *Attagis gayi* RUFOUS-BELLIED SEEDSNIPE
　— *A.g.latreillii* Lesson, 1831

N Ecuador

　— *A.g.simonsi* Chubb, 1918

C Peru to NW Argentina

　— *A.g.gayi* Geoffroy Saint-Hilaire & Lesson, 1831

Chile, Argentina

○ *Attagis malouinus* (Boddaert, 1783) WHITE-BELLIED SEEDSNIPE[5]

S Chile and S Argentina

THINOCORUS Eschscholtz, 1829 M
● *Thinocorus orbignyianus* GREY-BREASTED SEEDSNIPE
　— *T.o.ingae* Tschudi, 1843

N Peru to N Chile, NW Argentina

　— *T.o.orbignyianus* Geoffroy Saint-Hilaire & Lesson, 1831

NC and S Chile and Argentina

○ *Thinocorus rumicivorus* LEAST SEEDSNIPE
　— *T.r.pallidus* Salvadori & Festa, 1910

SW Ecuador, NW Peru

　— *T.r.cuneicauda* (Peale, 1848)

Peruvian desert

　— *T.r.bolivianus* P.R. Lowe, 1921

S Peru, N Chile and W Bolivia to NW Argentina

　— *T.r.rumicivorus* Eschscholtz, 1829[6]

Patagonia >> C Chile, NE Argentina and Uruguay

---

[1] Sometimes treated as polytypic see Mees (1982) (1534).
[2] Implicitly includes *lowi* van Rossem, 1938 (2474) see Jenni (1996: 291) (1205).
[3] Includes *dorsalis* Brodkorb, 1939 (220) see Blake (1977: 534) (164).
[4] For a review of this and its placement see Olson & Steadman (1981) (1709).
[5] Includes *cheeputi* see Fjeldså (1996: 545) (849).
[6] Includes *patagonicus* see Fjeldså (1996: 545) (849).

SCOLOPACIDAE SANDPIPERS, SNIPE (23: 92)

SCOLOPACINAE

SCOLOPAX Linnaeus, 1758 F

○ *Scolopax rusticola* Linnaeus, 1758 EURASIAN WOODCOCK[1]   Palaearctic >> W and S Europe and N Africa to SE Asia

○ *Scolopax mira* E. Hartert, 1916 AMAMI WOODCOCK   C Ryukyu Is.

○ *Scolopax saturata* Horsfield, 1821 JAVAN WOODCOCK   N and SC Sumatra, W Java

○ *Scolopax rosenbergii* Schlegel, 1871 NEW GUINEA WOODCOCK[2]   New Guinea

○ *Scolopax bukidnonensis* Kennedy, Fisher, Harrap, Diesmos & Manamtam, 2001 BUKIDNON WOODCOCK #
Mts. of N and C Luzon and Mindanao (1260)

○ *Scolopax celebensis* Riley, 1921 SULAWESI WOODCOCK
— *S.c.heinrichi* Stresemann, 1932[3]   NE Sulawesi
— *S.c.celebensis* Riley, 1921   C Sulawesi

○ *Scolopax rochussenii* Schlegel, 1866 MOLUCCAN WOODCOCK   Obi and Bacan Is. (N Moluccas)

◉ *Scolopax minor* J.F. Gmelin, 1789 AMERICAN WOODCOCK   SC and SE Canada, E USA >> SE USA

GALLINAGININAE

COENOCORYPHA G.R. Gray, 1855 F

○ *Coenocorypha pusilla* (Buller, 1869) CHATHAM SNIPE[4]   Chatham Is.

○ *Coenocorypha aucklandica* NEW ZEALAND SNIPE[5]
— *C.a.huegeli* (Tristram, 1893)   Snares Is.
— †? *C.a.iredalei* Rothschild, 1921   Islands off Stewart I.
— † *C.a.barrierensis* Oliver, 1955   Little Barrier I. (Hauraki Gulf), New Zealand (1676)
— *C.a.meinertzhagenae* Rothschild, 1927   Antipodes Is.
— *C.a.aucklandica* (G.R. Gray, 1845)   Auckland Is.

LYMNOCRYPTES Boie, 1826 M

○ *Lymnocryptes minimus* (Brünnich, 1764) JACK SNIPE[6] [7]   N Europe and Baltic to Siberia E to Kolyma basin >>
W Europe S to W and E Africa, E to Indochina

GALLINAGO Brisson, 1760 F

○ *Gallinago solitaria* SOLITARY SNIPE
— *G.s.solitaria* Hodgson, 1831   Mts. of S Siberia and perhaps N Mongolia >> S, C and
E Asia south to E Pakistan to Burma

— *G.s.japonica* (Bonaparte, 1856)   NE Mongolia to NE China and Kamchatka >>
Amurland to Kamchatka, Korea, Japan, E China

○ *Gallinago hardwickii* (J.E. Gray, 1831) LATHAM'S SNIPE   Russian Far East, S Sakhalin and S Kurils to N Japan >>
E Australia, Tasmania

○ *Gallinago nemoricola* Hodgson, 1836 WOOD SNIPE   Himalayas to Burma and SC China >> India,
Bangladesh, Burma, N Indochina

○ *Gallinago stenura* (Bonaparte, 1830) PINTAIL SNIPE   NE European Russia to NE Siberia, S to N Mongolia
and NE China >> Saudi Arabia, E Africa and
Aldabra Is. (W Indian Ocean), India to SE China
and Taiwan, S to Philippines and W Indonesia

○ *Gallinago megala* Swinhoe, 1861 SWINHOE'S SNIPE   C and SE Siberia, south to N Mongolia >> India to
N Australia

○ *Gallinago nigripennis* AFRICAN SNIPE
— *G.n.aequatorialis* Rüppell, 1845[8]   Ethiopia to E Zaire, Malawi, E Zimbabwe and Mozambique
— *G.n.angolensis* Bocage, 1868   Angola and N Namibia to Zambia and W Zimbabwe
— *G.n.nigripennis* Bonaparte, 1839   Sth. Africa

---

[1] Includes *ultimus* Koelz, 1954 (1284), see Ripley (1982: 125) (2055).
[2] For treatment of this at specific level see Kennedy *et al.* (2001) (1260).
[3] Considered distinct by Kennedy *et al.* (2001) (1260) *contra* White & Bruce (1986: 175) (2652).
[4] We follow Higgins & Davies (1996: 47-53) (1113) in separating *C. pusilla* from *C. aucklandica*.
[5] Higgins & Davies (1996: 47-53) (1113) included, but did not employ, a recommended treatment splitting this into a monotypic *huegeli* and two other species — each with a second race attached.
[6] For correction of spelling see David & Gosselin (2002) (614).
[7] Includes *nipponensis* Momiyama, 1939 (1600), see Stepanyan (1990: 203) (2326).
[8] For recognition see Clancey (1974) (442).

○ *Gallinago macrodactyla* Bonaparte, 1839 MADAGASCAN SNIPE | N and E Madagascar

○ *Gallinago media* (Latham, 1787) GREAT SNIPE | N and E Europe to W and C Siberia >> sub-Saharan Africa and Iran

● *Gallinago gallinago* COMMON SNIPE
  ✔ *G.g.delicata* (Ord, 1825) | N and C North America >> NW and C USA to N Sth. America
  — *G.g.faeroeensis* (C.L. Brehm, 1831) | Iceland, Faroe Is., Orkney and Shetland Is. >> British Is.
  — *G.g.gallinago* (Linnaeus, 1758) | N and C Palaearctic >> W Europe and N, W and E Africa to S Japan and W Indonesia

○ *Gallinago paraguaiae* SOUTH AMERICAN SNIPE
  — *G.p.paraguaiae* (Vieillot, 1816) | Trinidad, Colombia to the Guianas S to N Argentina and Uruguay
  — *G.p.magellanica* (P.P. King, 1828) | C Chile and C Argentina to Tierra del Fuego, Falkland Is.

○ *Gallinago andina* Taczanowski, 1875 PUNA SNIPE[1] | S Peru, W Bolivia, N Chile, NW Argentina

○ *Gallinago nobilis* P.L. Sclater, 1856 NOBLE SNIPE | NW Venezuela to Ecuador

○ *Gallinago undulata* GIANT SNIPE
  — *G.u.undulata* (Boddaert, 1783) | W and E Colombia to the Guianas
  — *G.u.gigantea* (Temminck, 1826) | E Bolivia, Paraguay, SE Brazil, NE Argentina

○ *Gallinago stricklandii* (G.R. Gray, 1845) CORDILLERAN SNIPE | S Chile, S Argentina, Tierra del Fuego

○ *Gallinago jamesoni* (Bonaparte, 1855) JAMESON'S SNIPE[2] | N Colombia and W Venezuela to WC Bolivia

○ *Gallinago imperialis* Sclater & Salvin, 1869 IMPERIAL SNIPE | Colombia, Peru

LIMNODROMUS Wied, 1833 M
● *Limnodromus griseus* SHORT-BILLED DOWITCHER[3]
  ✔ *L.g.caurinus* Pitelka, 1950 | S Alaska and S Yukon >> C USA to Peru (1900)
  — *L.g.hendersoni* Rowan, 1932 | C Canada >> SE USA to Panama
  — *L.g.griseus* (J.F. Gmelin, 1789) | NE Canada >> S USA to Brazil

● *Limnodromus scolopaceus* (Say, 1823) LONG-BILLED DOWITCHER[4] | Taymyr (NC Siberia) east to Chukotkiy and Koryakskiy Mts. (NE Siberia), W Alaska, N Yukon >> >> W and S USA to Panama and French Guiana

○ *Limnodromus semipalmatus* (Blyth, 1848) ASIAN DOWITCHER | S Siberia (patchy), N Mongolia and NE China >> E India to Indochina, south to N Australia

TRINGINAE

LIMOSA Brisson, 1760 F
● *Limosa limosa* BLACK-TAILED GODWIT
  — *L.l.islandica* C.L. Brehm, 1831[5] | Iceland to Shetland Is. and W and N Norway >> British Is. and SW Netherlands to Iberia
  — *L.l.limosa* (Linnaeus, 1758) | W and C Europe to C Asia >> Mediterranean and Africa to W India
  — *L.l.melanuroides* Gould, 1846 | Lena Basin and Lake Baikal to Kamchatka >> India to Indochina, south to Australia

○ *Limosa haemastica* (Linnaeus, 1758) HUDSONIAN GODWIT | NW Alaska to Hudson Bay >> SE Sth. America and SC Chile

● *Limosa lapponica* BAR-TAILED GODWIT
  ✔ *L.l.lapponica* (Linnaeus, 1758) | N Europe to NC European Russia >> North Sea coasts south to Sth. Africa
  — *L.l.taymyrensis* Engelmoer & Roselaar, 1998 | NW and NC Siberia (Yamal Pen. to Anabar basin) >> Africa to W India (800)
  — *L.l.menzbieri* Portenko, 1936[6] | Lena delta to Chaunskaya Bay (NE Siberia) >> SE Asia to NW Australia (1902)
  ✔ *L.l.baueri* J.F. Naumann, 1836[7] | E Anadyrland (NE Siberia) to N and W Alaska >> China to Australia, New Zealand and Pacific Is.

---

[1] For reasons to separate this from *G. paraguaiae* see Fjeldså & Krabbe (1990) (850). Includes *innotata* see van Gils & Wiersma (1996: 496) (2467).
[2] Thought to include *chapmani* R.T. Moore, 1937 (1610), but may need re-evaluation.
[3] For recognition of three races see A.O.U. (1957: 201) (1).
[4] Includes *fasciatus* Brodkorb, 1933 (216), see Pitelka (1950: 68) (1900).
[5] For recognition of this see Vaurie (1965: 420) (2510) and Engelmoer & Roselaar (1998) (800).
[6] In synonymy in last Edition following Vaurie (1965) (2510) but recognised by Stepanyan (1990: 212) (2326).
[7] Includes *anadyrensis* Engelmoer & Roselaar, 1998 (800), see Tomkovich & Serra (1999) (2419).

● *Limosa fedoa* MARBLED GODWIT
    — *L.f.fedoa* (Linnaeus, 1758)   C Canada, NW USA >> S USA to Panama and
                                                    NW Sth. America
    ✓ *L.f.beringiae* Gibson & Kessel, 1989   Alaska >> W USA (936)

NUMENIUS Brisson, 1760 M
○ *Numenius minutus* Gould, 1841 LITTLE CURLEW   NC and NE Siberia >> New Guinea, Australia

○ †?*Numenius borealis* (J.R. Forster, 1772) ESKIMO CURLEW   NC Canada >> Uruguay, Argentina

● *Numenius phaeopus* WHIMBREL[1]
    — *N.p.islandicus* C.L. Brehm, 1831   Iceland and British Is. >> W Africa
    — *N.p.phaeopus* (Linnaeus, 1758)   Norway and Estonia to Evenkia (NC Siberia) >> Africa
                                        and S Asia east to W Burma and south to Sumatra
    — *N.p.alboaxillaris* P.R. Lowe, 1921   Lower Volga R. to N Kazakhstan, SW Siberia >> Persian
                                        Gulf, Red Sea, W Indian Ocean islands and SE Africa
    — *N.p.variegatus* (Scopoli, 1786)   NE Siberia (E from Yana basin) >> E India to Australasia
    ✓ *N.p.rufiventris* Vigors, 1829   Alaska and NW Canada >> W Nth. America, C and
                                        S America
    ✓ *N.p.hudsonicus* Latham, 1790   Hudson Bay area to NE Canada >> Caribbean and
                                        Sth. America, Australasia

○ *Numenius tahitiensis* (J.F. Gmelin, 1789) BRISTLE-THIGHED CURLEW   W Alaska >> Marshall and Hawaiian Is. to Society Is.

○ *Numenius tenuirostris* Vieillot, 1817 SLENDER-BILLED CURLEW   SW Siberia and N Kazakhstan >> NW Africa and Iraq
                                        and Persian Gulf

● *Numenius arquata* EURASIAN CURLEW[2]
    ✓ *N.a.arquata* (Linnaeus, 1758)   W, N and C Europe to N European Russia >> W and
                                        S Europe and W Africa
    — *N.a.orientalis* C.L.Brehm, 1831   W and C Siberia to Transbaikalia >> coasts of
                                        E Mediterranean, Africa, and S and E Asia south to
                                        Philippines and Gtr. Sundas
    — *N.a.suschkini* Neumann, 1929   Lower Volga R. to SW Siberia and N Kazakhstan >>
                                        coasts of sub-Saharan Africa and SW Asia

○ *Numenius madagascariensis* (Linnaeus, 1766) FAR EASTERN CURLEW   E Siberia and Russian Far East >> Taiwan and Indonesia
                                        to Australia and New Zealand

● *Numenius americanus* LONG-BILLED CURLEW
    — *N.a.parvus* Bishop, 1910[3]   SW and SC Canada to SW USA >> SW USA to Mexico
    — *N.a.americanus* Bechstein, 1812   WC USA >> SW USA to Guatemala

BARTRAMIA Lesson, 1831 F
○ *Bartramia longicauda* (Bechstein, 1812) UPLAND SANDPIPER   C Alaska, C and S Canada, NC USA >> Surinam,
                                        Paraguay and S Brazil to C Argentina

TRINGA Linnaeus, 1758 F
○ *Tringa erythropus* (Pallas, 1764) SPOTTED REDSHANK   N Europe, N and NE Siberia >> W and S Europe, N,
                                        W and E Africa to SE Asia, SE China and Taiwan

● *Tringa totanus* COMMON REDSHANK[4]
    — *T.t.robusta* (Schiøler, 1919)   Iceland and Faroe Is. >> British Is. and W Europe
    ✓ *T.t.totanus* (Linnaeus, 1758)[5]   N and W Europe from the British Isles and Spain to
                                      W Siberia >> W Africa and the Mediterranean east
                                      to India, Indonesia
    — *T.t.ussuriensis* Buturlin, 1934   S Siberia and Mongolia to E Asia >> E Mediterranean
                                        and E Africa to India and SE Asia (283)
    — *T.t.terrignotae* R. & A. Meinertzhagen, 1926   S Manchuria, E China >> SE and E Asia, ? Australia
    — *T.t.craggi* Hale, 1971   Xinjiang (NW China) >> S Asia, SE Asia (to Australia ?) (1057)
    — *T.t.eurhinus* (Oberholser, 1900)[6]   Pamir Mts. (Tajikhstan), N India, C and S Tibet >>
                                        India to the Malay Pen.

---

[1] Based on review by Engelmoer & Roselaar (1998) (800).
[2] For recognition of *suschkini* see Engelmoer & Roselaar (1998) (800).
[3] Van Gils & Wiersma (1996: 505) (2467) used the name *parvus* as did A.O.U. (1957: 181) (1); it is apparently unknown whether the earlier name *occidentalis* Woodhouse, 1852, would apply or not. It was based on a bird taken in New Mexico.
[4] We follow Hale (1971, 1973) (1057, 1058); a more recent review, that of Engelmoer & Roselaar (1998) (800), is apparently to be further revised and we defer changes until then.
[5] Including *britannica* Mathews, 1935.
[6] Includes *meinertzhageni* Thiede, 1963 (2397), see Hale (1971: 251) (1057).

○ *Tringa stagnatilis* (Bechstein, 1803) MARSH SANDPIPER — E Baltic and Ukraine to E Siberia and NE China >> Mediterranean and sub-Saharan Africa to Indonesia and Australasia

● *Tringa nebularia* (Gunnerus, 1767) COMMON GREENSHANK — N Palaearctic >> W Europe, Africa and Middle East to Australasia

○ *Tringa guttifer* (Nordmann, 1835) NORDMANN'S GREENSHANK — Russian Far East (Sea of Okhotsk) >> NE India to mainland SE Asia

● *Tringa melanoleuca* (J.F. Gmelin, 1789) GREATER YELLOWLEGS — S Alaska and E across C Canada to NE Nova Scotia >> coastal and S USA, C and Sth. America

● *Tringa flavipes* (J.F. Gmelin, 1789) LESSER YELLOWLEGS — Alaska to C Canada >> S USA, West Indies, C and Sth. America

● *Tringa ochropus* Linnaeus, 1758 GREEN SANDPIPER — N Palaearctic >> W Europe and Africa to Philippines and N Borneo

● *Tringa solitaria* SOLITARY SANDPIPER

    ⊥ *T.s.cinnamomea* (Brewster, 1890) — C and S Alaska to NE Manitoba >> Sth. America

    — *T.s.solitaria* A. Wilson, 1813 — E British Columbia to Quebec and Labrador >> W Indies, C and Sth. America

○ *Tringa glareola* Linnaeus, 1758 WOOD SANDPIPER — N Palaearctic >> Africa, S Asia, Philippines, Indonesia, Australia

XENUS Kaup, 1829 M

○ *Xenus cinereus* (Güldenstädt, 1774) TEREK SANDPIPER — E Baltic area, Belarus and N Ukraine to NE Siberia >> S and E Africa, Middle East, S Asia, Indonesia to Australasia

ACTITIS Illiger, 1811 M

● *Actitis hypoleucos* (Linnaeus, 1758) COMMON SANDPIPER — Europe and N and C Asia >> S Europe, Africa, and S and E Asia to Australasia

● *Actitis macularius* (Linnaeus, 1766) SPOTTED SANDPIPER[1][2] — Canada and N USA (but not Arctic) >> W Indies, S USA to C Sth. America

HETEROSCELUS S.F. Baird, 1858 M

○ *Heteroscelus brevipes* (Vieillot, 1816) GREY-TAILED TATTLER — NC and NE Siberia >> Taiwan and Malay Pen. to Australia and New Zealand

● *Heteroscelus incanus* (J.F. Gmelin, 1789) WANDERING TATTLER — NE Siberia, S Alaska to NW British Columbia >> SW USA and W Mexico, Ecuador and Galapagos Is., Pacific islands to E Australia and New Zealand

CATOPTROPHORUS Bonaparte, 1827 M

● *Catoptrophorus semipalmatus* WILLET

    — *C.s.inornatus* (Brewster, 1887) — SC Canada and NC USA >> S USA to N Sth. America

    √ *C.s.semipalmatus* (J.F. Gmelin, 1789) — SE Canada, E and S USA, West Indies >> S USA, C America, West Indies to S Brazil

AECHMORHYNCHUS Coues, 1874[3] M

○ †*Aechmorhynchus cancellatus* (J.F. Gmelin, 1789) KIRITIMATI SANDPIPER[4] — ? Kiritimati I. (Pacific Ocean)

○ *Aechmorhynchus parvirostris* (Peale, 1848) TUAMOTU SANDPIPER — Tuamotu Is.

PROSOBONIA Bonaparte, 1850 F

○ †*Prosobonia leucoptera* (J.F. Gmelin, 1789) TAHITI SANDPIPER — Tahiti (Society Is.)

○ †*Prosobonia ellisi* Sharpe, 1906 MOOREAN SANDPIPER — Moorea (Society Is.)

ARENARIINAE

ARENARIA Brisson, 1760 F

● *Arenaria interpres* RUDDY TURNSTONE

    √ *A.i.interpres* (Linnaeus, 1758)[5] — NE Canada, Greenland and N Europe to NE Siberia and W Alaska >> W and S Europe, Africa and S and E Asia, Australasia, Pacific Is., and W USA to W Mexico

    √ *A.i.morinella* (Linnaeus, 1766) — NE Alaska and N Canada >> S USA to Sth. America

---

[1] For correction of spelling see David & Gosselin (2002) (614).
[2] Includes *rava* Burleigh, 1960 (272), see van Gils & Wiersma (1996: 513) (2467).
[3] Lumped with *Prosobonia* by Zusi & Jehl (1970) (2738), but better retained (C.S.R.).
[4] For the retention of this taxon, not lumping it with Tuamotu birds, see Walters (1993) (2554).
[5] Birds from Siberia east to W Alaska may be recognised as *oahuensis* (Bloxham, 1826).

◕ *Arenaria melanocephala* (Vigors, 1829) Black Turnstone   W and S Alaska >> SE Alaska to NW Mexico

CALIDRIDINAE

APHRIZA Audubon, 1839[1] F
◕ *Aphriza virgata* (J.F. Gmelin, 1789) Surfbird   S and C Alaska and C Yukon >> Pacific coast of N, C and Sth. America

CALIDRIS Merrem, 1804 F
○ *Calidris tenuirostris* (Horsfield, 1821) Great Knot   NE Siberia >> E Arabia to Indochina, New Guinea and Australia

○ *Calidris canutus* Red Knot
  — *C.c.canutus* (Linnaeus, 1758)   NC Siberia >> W and S Africa
  — *C.c.piersmai* Tomkovich, 2001   New Siberian Is. >> NW Australia (2418)
  — *C.c.rogersi* Mathews, 1913   Mountains of Chukotka and Chukotskiy Pen. >> Australasia
  — *C.c.roselaari* Tomkovich, 1990   Wrangel I. (NE Siberia) and NW Alaska >> shores of Gulf of Mexico (2417)
  — *C.c.rufa* (A. Wilson, 1813)   N Canada >> NE and S Sth. America
  — *C.c.islandica* (Linnaeus, 1767)[2]   Islands off N Canada, N Greenland >> W Europe

● *Calidris alba* Sanderling[3]
  ✔ *C.a.alba* (Pallas, 1764)   Ellesmere I., N and E Greenland, Svalbard, Franz Josef Land and Taymyr >> coastal W Europe, Africa, Asia and Australasia
  ✔ *C.a.rubida* (J.F. Gmelin, 1789)   NE Siberia, Alaska and Canada >> coastal E Asia and W and S America

◕ *Calidris pusilla* (Linnaeus, 1766) Semipalmated Sandpiper   N Chukotskiy Pen., W and N Alaska, N Canada >> C and S America, W Indies

◕ *Calidris mauri* (Cabanis, 1857) Western Sandpiper   N and E Chukotskiy Pen., W and N Alaska >> coastal USA to N Sth. America, West Indies

○ *Calidris ruficollis* (Pallas, 1776) Red-necked Stint   NC Taymyr NE Siberia, W and N Alaska >> SE Asia to Australasia

○ *Calidris minuta* (Leisler, 1812) Little Stint   N Europe and N Siberia >> SW Europe and Africa to India

○ *Calidris temminckii* (Leisler, 1812) Temminck's Stint   N Europe, N Siberia >> W and E Africa, S Asia to S China, south through mainland SE Asia to the Malay Pen. and Borneo

○ *Calidris subminuta* (Middendorff, 1853) Long-toed Stint   SW to E Siberia >> India and Indochina to Australasia

◕ *Calidris minutilla* (Vieillot, 1819) Least Sandpiper   Alaska and N Canada >> S USA to N Sth. America, West Indies

○ *Calidris fuscicollis* (Vieillot, 1819) White-rumped Sandpiper   NE Alaska and N Canada to S Baffin I. >> S Sth. America

◕ *Calidris bairdii* (Coues, 1861) Baird's Sandpiper   Extreme NE Siberia, Alaska and NW Canada to NW Greenland >> W and S Sth. America

◕ *Calidris melanotos* (Vieillot, 1819) Pectoral Sandpiper   N Siberia, W and N Alaska, and N Canada >> SE Australia, New Zealand, S Sth. America

○ *Calidris acuminata* (Horsfield, 1821) Sharp-tailed Sandpiper   NC and NE Siberia >> New Guinea to Australia and New Zealand

○ *Calidris ferruginea* (Pontoppidan, 1763) Curlew Sandpiper   N Siberia (rarely N Alaska) >> Africa to Australasia

○ *Calidris maritima* Purple Sandpiper[4]
  — *C.m.belcheri* Engelmoer & Roselaar, 1998   E Hudson Bay and James Bay (Canada) >> E Nth. America (800)
  — *C.m.maritima* (Brünnich, 1764)[5]   N Canada, Greenland, Svalbard and Scandinavia to NC Siberia >> E Nth. America and W and N Europe
  — *C.m.littoralis* (C.L. Brehm, 1831)   Iceland

---

[1] For placement next to the knots see Jehl (1968) (1203); we retain the genus *Aphriza* with reservations, submergence in *Calidris* may be more appropriate.
[2] For recognition of this see Roselaar (1983) (2104).
[3] For acceptance of *rubida* see Engelmoer & Roselaar (1998) (800).
[4] Revision based on Engelmoer & Roselaar (1998) (800).
[5] Includes *groenlandica* Løvenskiold, 1950 (1388), see Vaurie (1965: 397) (2510) and Engelmoer & Roselaar (1998) (800).

○ *Calidris ptilocnemis* ROCK SANDPIPER

— *C.p.tschuktschorum* (Portenko, 1937)[1] E Chukotskiy Pen., St. Lawrence and Nunivak Is., W Alaska >> NW North America (1903)

— *C.p.ptilocnemis* (Coues, 1873) St. Mathew, Hall and Pribilof Is. >>Alaska Pen.
— *C.p.couesi* (Ridgway, 1880) Aleutian Is., Alaska Pen.
— *C.p.quarta* (E. Hartert, 1920)[2] Kuril Is., S Kamchatka and Commander Is. >> E Japan

● *Calidris alpina* DUNLIN[3]

— *C.a.arctica* (Schiøler, 1922) NE Greenland >> NW Africa
— *C.a.schinzii* (C.L. Brehm, 1822) SE Greenland, Iceland and British Is. to S Scandinavia and Baltic >> SW Europe and NW Africa

— *C.a.alpina* (Linnaeus, 1758) N Europe and NW Siberia to Yenisey R >> W Europe, Mediterranean, Africa and ? SW Asia to India

— *C.a.centralis* (Buturlin, 1932) Taymyr to Kolyma delta (NE Siberia) >> E Mediterranean, Red Sea and SW and S Asia

— *C.a.sakhalina* (Vieillot, 1816) Extreme NE Asia >> E China, Korea, Japan, Taiwan
— *C.a.kistchinskii* Tomkovich, 1986[4] N Sea of Okhotsk and S Koryakland to N Kuril Is. and Kamchatka >> E Asia (2416)

— *C.a.actites* Nechaev & Tomkovich, 1988[5] N Sakhalin >> E Asia (1641)
— *C.a.arcticola* (Todd, 1953) NW and N Alaska and NW Canada >> E China, Korea, Japan (2414)

✓ *C.a.pacifica* (Coues, 1861) W and S Alaska >> W USA and W Mexico
— *C.a.hudsonia* (Todd, 1953) NC Canada east to Hudson Bay >> SE USA (2414)

○ *Calidris himantopus* (Bonaparte, 1826) STILT-SANDPIPER[6] N Alaska and N Canada >> S USA, C Sth. America

EURYNORHYNCHUS Nilsson, 1821 M
○ *Eurynorhynchus pygmeus* (Linnaeus, 1758) SPOON-BILLED SANDPIPER Chukotskiy Pen. to N Kamchatka >> SE India, Bangladesh, Vietnam, Thailand and the Malay Pen. S to Singapore

LIMICOLA Koch, 1816 F
○ *Limicola falcinellus* BROAD-BILLED SANDPIPER
— *L.f.falcinellus* (Pontoppidan, 1763) N Europe, NW Siberia >> E and S Africa to W and S India and Sri Lanka

— *L.f.sibirica* Dresser, 1876 NE Siberia >> NE India to Australia

TRYNGITES Cabanis, 1857 M
○ *Tryngites subruficollis* (Vieillot, 1819) BUFF-BREASTED SANDPIPER Extreme NE Siberia, N Alaska, N Canada >> SC Sth. America

PHILOMACHUS Merrem, 1804 M
○ *Philomachus pugnax* (Linnaeus, 1758) RUFF[7] N and C Europe and across N Siberia to Sakhalin >> W and S Europe, Africa and S Asia east to India, more rarely SE Asia and Australia

PHALAROPODINAE

PHALAROPUS Brisson, 1760 M
● *Phalaropus tricolor* (Vieillot, 1819) WILSON'S PHALAROPE[8] S Canada, NW and CN USA >> W and S Sth. America and subantarctic islands

● *Phalaropus lobatus* (Linnaeus, 1758) RED-NECKED PHALAROPE N Holarctic >> pelagic areas off WC Sth. America, Arabian Sea, Philippines, Wallacea to W Melanesia and Australasia

● *Phalaropus fulicarius* (Linnaeus, 1758)[9] GREY PHALAROPE/RED PHALAROPE

N Holarctic >> pelagic areas off W Sth. America and W and SW Africa and subantarctic

[1] For recognition see Stepanyan (1990: 199) (2326).
[2] Includes *kurilensis* see Vaurie (1965: 399) (2510), but perhaps valid (P. Tomkovich *in litt.*).
[3] For a partial review, recognising *actites*, *sakhalina* and *arcticola* see Browning (1991) (252). Engelmoer & Roselaar (1998) (800) validated *centralis* and *hudsonia*.
[4] For evaluation and acceptance of this race and the next see Browning (1991) (252).
[5] A new name for *litoralis* Nechaev & Tomkovich, 1987 (1640), preoccupied.
[6] van Gils & Wiersma (1996: 530) (2467) retained the genus *Micropalama* but A.O.U. (1998: 173) (3) did not.
[7] Includes *occidentalis* Verhayen, 1941 (2519), see Vaurie (1965: 405) (2510).
[8] A.O.U. (1998: 180) (3) acknowledged work showing the next two species to be closer to each other than to this, but keep this in *Phalaropus*. We accept this and submerge *Steganopus*.
[9] For an explanation of the spelling *fulicarius* see David & Gosselin (2000) (612).

GLAREOLIDAE COURSERS, PRATINCOLES (5: 18)

CURSORIINAE

PLUVIANUS Vieillot, 1816 M
○ *Pluvianus aegyptius* (Linnaeus, 1758) EGYPTIAN PLOVER[1]
    Senegal to Nigeria, Sudan and W Ethiopia, south to N Angola, W and N Zaire, N Uganda

CURSORIUS Latham, 1790 M
○ *Cursorius cursor* CREAM-COLOURED COURSER
    — *C.c.cursor* (Latham, 1787)[2]   Canary Is. and Sahara to Arabia and Iraq
    — *C.c.bogolubovi* Zarudny, 1885   SE Turkey to W Pakistan
    — *C.c.exsul* E. Hartert, 1920   Cape Verde Is.

○ *Cursorius somalensis* SOMALI COURSER[3]
    — *C.s.somalensis* Shelley, 1885   Eritrea, E Ethiopia, N Somalia
    — *C.s.littoralis* Erlanger, 1905   S Somalia, N and E Kenya, SE Sudan

○ *Cursorius rufus* Gould, 1837 BURCHELL'S COURSER[4]   SW Angola to Sth. Africa

○ *Cursorius temminckii* Swainson, 1822 TEMMINCK'S COURSER[5]   Sub-Saharan Africa

○ *Cursorius coromandelicus* (J.F. Gmelin, 1789) INDIAN COURSER   Pakistan, Nepal, India, N Sri Lanka

RHINOPTILUS Strickland, 1852 M
○ *Rhinoptilus africanus* DOUBLE-BANDED COURSER[6]
    — *R.a.raffertyi* Mearns, 1915   E Ethiopia, Eritrea, Djibouti
    — *R.a.hartingi* Sharpe, 1893   SE Ethiopia, Somalia
    — *R.a.gracilis* (Fischer & Reichenow, 1884)[7]   SW Kenya to C Tanzania
    — *R.a.bisignatus* (Hartlaub, 1865)   Coastal SW Angola
    — *R.a.erlangeri* Niethammer & Wolters, 1966[8]   NW Namibia (1654)
    — *R.a.traylori* Irwin, 1963   NE Botswana (1187)
    — *R.a.africanus* (Temminck, 1807)[9]   C and S Namibia, W and S Botswana, N Cape Province, SW and C Kalahari
    — *R.a.granti* W.L. Sclater, 1921   Karoo and W Cape Province

○ *Rhinoptilus cinctus* HEUGLIN'S COURSER
    — *R.c.mayaudi* Érard, Hémery & Pasquet, 1993   Ethiopia and N Somalia (815)
    — *R.c.balsaci* Érard, Hémery & Pasquet, 1993   S Somalia, NE Kenya (815)
    — *R.c.cinctus* (Heuglin, 1863)   SE Sudan, NW Kenya
    — *R.c.emini* Zedlitz, 1914   S Kenya, Tanzania, N Zambia
    — *R.c.seebohmi* Sharpe, 1893   S Angola and N Namibia to N Botswana, Zimbabwe and N Transvaal

○ *Rhinoptilus chalcopterus* (Temminck, 1824) BRONZE-WINGED COURSER[10]
    Senegal to W Ethiopia, Uganda and S Kenya, south to Angola, Namibia and N and E Sth. Africa

○ *Rhinoptilus bitorquatus* (Blyth, 1848) JERDON'S COURSER   EC India

GLAREOLINAE

STILTIA G.R. Gray, 1855 F
○ *Stiltia isabella* (Vieillot, 1816) AUSTRALIAN PRATINCOLE   Australia >> N Australia and S New Guinea to S Borneo

---

[1] Includes *angolae* see White (1965: 134) (2649).
[2] Includes *dahlakensis* see White (1965: 134) (2649) and implicitly *bannermani* see Cramp et. al (1983: 91) (590).
[3] For reasons to separate *somaliensis* and *rufus* from *cursor* see Pearson & Ash (1996) (1815).
[4] Includes *theresae* R. Meinerthagen, 1949 (1549), see White (1965: 135) (2649).
[5] Includes *damarensis* see White (1965: 135) (2649), a name applied to birds in winter quarters and thus a synonym for the nominate form (Clancey, 1989), and *aridus* Clancey, 1989 (513), proposed for SW African residents previously thought to be *damarensis*.
[6] Type species of the genus *Smutsornis*, here submerged.
[7] Includes *illustris* see White (1965: 136) (2649).
[8] For recognition see Clancey (1980) (482).
[9] Irwin (1963) (1187) and White (1965: 136) (2649) recognised *sharpei* but Niethammer & Wolters (1966) (1654) selected a lectotype making it a synonym of nominate *africanus*.
[10] Includes *albofasciatus* see White (1965: 137) (2649).

GLAREOLA Brisson, 1760 F

○ *Glareola pratincola* COLLARED PRATINCOLE
   — *G.p.pratincola* (Linnaeus, 1766)[1]   S Europe, N Africa to Pakistan >> N tropical Africa
   — *G.p.erlangeri* Neumann, 1920   Coastal S Somalia, N Kenya
   — *G.p.fuelleborni* Neumann, 1910[2]   Senegal to S Kenya south to E Zambia, Zimbabwe and
                                               E Sth. Africa
   — ¶*G.p.riparia* Clancey, 1979   Angola, SW Zambia, NE Namibia, NW Botswana >>
                                                 coastal Mozambique

○ *Glareola maldivarum* J.R. Forster, 1795 ORIENTAL PRATINCOLE   C, E and SE Asia >> south to Australia

○ *Glareola nordmanni* J.G. Fischer, 1842[3] BLACK-WINGED PRATINCOLE   SE Europe and C Asia >> S and SW Africa

○ *Glareola ocularis* J. Verreaux, 1833 MADAGASCAN PRATINCOLE   Madagascar (except extreme S) >> E Africa

○ *Glareola nuchalis* ROCK PRATINCOLE
   — *G.n.liberiae* Schlegel, 1881   Sierra Leone to W Cameroon
   — *G.n.nuchalis* G.R. Gray, 1849   Chad to Ethiopia, south to Mozambique
   — *G.n.torrens* Clancey, 1981   S Angola, NE Namibia to W Zambia >> E Zaire,
                                               Uganda (483)

○ *Glareola cinerea* Fraser, 1843 GREY PRATINCOLE[4]   Ghana to Central African Republic, W Zaire, NW Angola

○ *Glareola lactea* Temminck, 1820 LITTLE PRATINCOLE   India, Sri Lanka to S Indochina

LARIDAE SKUAS, GULLS AND TERNS (15: 97)

LARINAE[5]

LEUCOPHAEUS Bruch, 1853[6] M
○ *Leucophaeus scoresbii* (Traill, 1823) DOLPHIN GULL   S coast Sth. America, Falkland Is.

LARUS Linnaeus, 1758[7] M
○ *Larus pacificus* PACIFIC GULL
   — *L.p.georgii* P.P. King, 1826[8]   W, SW and SC Australia
   — *L.p.pacificus* Latham, 1802[9]   SE Australia, Tasmania

○ *Larus belcheri* Vigors, 1829 BELCHER'S GULL   Coastal N Peru to N Chile >> to N Ecuador and C Chile

○ *Larus atlanticus* Olrog, 1958 OLROG'S GULL[10]   SE Sth. America (1678)

○ *Larus crassirostris* Vieillot, 1818 BLACK-TAILED GULL   SE Russia, Japan, Korea, E China

○ *Larus modestus* Tschudi, 1843 GREY GULL   S Peru and N Chile

● *Larus heermanni* Cassin, 1852 HEERMANN'S GULL   NW Mexico >> Pacific coast of USA

○ *Larus leucophthalmus* Temminck, 1825 WHITE-EYED GULL   Red Sea, Gulf of Aden

○ *Larus hemprichii* Bruch, 1853 SOOTY GULL   Red Sea, Gulf of Aden, Persian Gulf to S Pakistan and
                                               N Kenya >> south to Tanzania

● *Larus canus* MEW GULL[11]
   — *L.c.canus* Linnaeus, 1758   Iceland and British Is. to White Sea >> Europe and N
                                               Africa to Persian Gulf
   — *L.c.heinei* Homeyer, 1853[12]   NC and C European Russia and W and C Siberia >> C
                                               and S Europe, Middle East and Caspian Sea
   — *L.c.kamtschatschensis* (Bonaparte, 1857)   NE Siberia >> SE Asia
   ⤢ *L.c.brachyrhynchus* Richardson, 1831   N Alaska, W Canada >> W USA

---

[1] Includes *limbata*, based at least in part on juvenile nominate birds, see Cramp *et al.* (1983: 107) (590).
[2] Includes *boweni* see White (1965: 138) (2649).
[3] Fischer is an abbreviation of Fischer von Waldheim.
[4] Includes *colorata* see White (1965: 139) (2649).
[5] Generic treatments traditional. Likely to change when molecular work is completed.
[6] In last Edition this species and the next were placed in the genus *Gabianus*; however *pacificus* is type species of that genus and as *pacificus* is now placed in *Larus* the applicable and available generic name is *Leucophaeus*.
[7] The systematics of many species are still subject to much debate; here, except where explained, we follow Burger & Gochfeld (1996) (270).
[8] For recognition of this see Higgins & Davies (1996: 471) (1113).
[9] For reasons to use 1802 not 1801 see Browing & Monroe (1991) (258).
[10] For reasons to treat this as a species separate from *L. belcheri* see Devillers (1977) (696) and A.O.U. (1998: 187) (3).
[11] For recognition of *heini* and *kamtschatkensis* see Vaurie (1965: 477) (2510).
[12] Includes *stegmanni* Brodkorb, 1936 (218), an unnecessary new name for *major* Middendorff, see Vaurie (1965: 477) (2510).

⦿ *Larus audouinii* Payraudeau, 1826 Audouin's Gull

Mediterranean >> south to Senegal and Gambia

⦿ *Larus delawarensis* Ord, 1815 Ring-billed Gull

Canada, N USA >> south to S USA, C America, Greater Antilles

⦿ *Larus californicus* Californian Gull
  — *L.c.albertaensis* Jehl, 1987[1]

C Canada, NW USA (east of the Rocky Mts.) >> SW Canada to SW Mexico (1204)

  ☑ *L.c.californicus* Lawrence, 1854

NW USA (S of *albertaensis*) >> W USA to SW Mexico

⦿ *Larus marinus* Linnaeus, 1758 Great Black-backed Gull

E North America, S Greenland, Iceland, W and N Europe to White Sea >> south to West Indies and SW Europe

⦿ *Larus dominicanus* Kelp Gull
  ☑ *L.d.dominicanus* M.H.K. Lichtenstein, 1823

W and SE Sth. America, New Zealand, S Australia, subantarctic islands

  — *L.d.vetula* (Bruch, 1853)[2]

Madagascar, S Africa

⦿ *Larus glaucescens* J.F.Naumann, 1840 Glaucous-winged Gull

St. Lawrence and Commander Is. to S Alaska and N Oregon >> Bering Sea to N Japan and Baja California

⦿ *Larus occidentalis* Western Gull
  ☑ *L.o.occidentalis* Audubon, 1839

W USA

  ☑ *L.o.wymani* Dickey & Van Rossem, 1925

C California to C Baja California

○ *Larus livens* Dwight, 1919 Yellow-footed Gull

Gulf of California

⦿ *Larus hyperboreus* Glaucous Gull
  — *L.h.hyperboreus* Gunnerus, 1767

N Europe and NW Siberia

  — *L.h.pallidissimus* Portenko, 1939

Taymyr Pen. to Bering Sea (1905)

  ☑ *L.h.barrovianus* Ridgway, 1886

Alaska to W Canada

  — *L.h.leuceretes* Schleep, 1819[3]

N Canada, Greenland, Iceland

○ *Larus glaucoides* Iceland Gull
  — *L.g.kumlieni* Brewster, 1883[4]

NE Canada >> SE Canada, NE USA

  — *L.g.glaucoides* B. Meyer, 1822

S and W Greenland and Novaya Zemlya >> south to N Europe (1557)

⦿ *Larus thayeri* W.S. Brooks, 1915 Thayer's Gull

Arctic Canada, W Greenland >> W coast North America

⦿ *Larus argentatus* Herring Gull[5]
  ☑ *L.a.smithsonianus* Coues, 1862

North America >> south to C America

  ☑ *L.a.argenteus* C.L. Brehm, 1822

Iceland, Faroe Is., W Europe >> south to N Iberia

  — *L.a.argentatus* Pontoppidan, 1763[6]

Denmark to E Kola Pen. >> N and W Europe

  — *L.a.vegae* Palmén, 1887[7]

NE Siberia >> south to China

○ *Larus cachinnans* Yellow-legged Gull[8]
  — *L.c.atlantis* Dwight, 1922

Azores, Madeira, Canary Is.

  — *L.c.michahellis* J.F. Naumann, 1840[9]

W and S Europe, NW Africa, Mediterranean

  — *L.c.cachinnans* Pallas, 1811[10]

Black Sea to C Kazakhstan and Caspian Sea >> south to Middle East and SW Asia

  — *L.c.barabensis* H. Johansen, 1960[11]

C Asia >> SW Asia (1209)

  — *L.c.mongolicus* Sushkin, 1925[12]

SE Altai (incl. E Kazakhstan) and Lake Baikal to Mongolia >> S Asia

○ *Larus armenicus* Buturlin, 1934 Armenian Gull

E Turkey and NW Iran to Caucasus >> N Red Sea, E Mediterranean and SE Black Sea (282)

---

[1] For recognition see Browning (1990) (251).
[2] For recognition of *vetula* see Brooke & Cooper (1979) (239).
[3] This race requires re-evaluation.
[4] For recognition of *kumlieni* see, for example, A.O.U. (1998: 191) (3).
[5] Compared to last Edition we have accepted the split of *L. cachinnans* see Marion *et al.* (1985) (1434) and A.O.U. (1998: 190) (3).
[6] Including *omissus* see Vaurie (1965: 470) (2510) and Burger & Gochfeld (1996: 609) (270), which may deserve recognition.
[7] Includes *birulae* see Vaurie (1965: 470) (2510); but this may deserve recognition.
[8] Some, including Roselaar (*in litt.*), would treat this as three species: *michahellis* (with a race *atlantis*), *cachinnans* (with a race *barabensis*) and *mongolicus*. We prefer to wait for adoption of this by a broader consensus before we follow.
[9] Includes *lusitianus* Joiris, 1978 (1222), but this may belong in synonymy of *L. c. atlantis*.
[10] Includes *ponticus* Stegmann, 1934 (2309) see Stepanyan (1990: 222) (2326).
[11] Treated as a synonym of *cachinnans* in last Edition as in Vaurie (1965) (2510) and Stepanyan (1990: 222) (2326), but it was recognised by Cramp *et al.* (1983) (590) and by Burger & Gochfeld (1996: 610) (270).
[12] Not listed in last Edition, but see Vaurie (1965: 472) (2510).

○ *Larus schistisagus* Stejneger, 1884 SLATY-BACKED GULL[1]   NE Siberia to Japan and E Russia >> south to Taiwan

◉ *Larus fuscus* LESSER BLACK-BACKED GULL
    ✓ *L.f.graellsii* A.E. Brehm, 1857   Iceland, Faroe Is., W Europe >> SW Europe, E coast USA, W Africa

    __ *L.f.intermedius* Schiøler, 1922   Netherlands, Germany, Denmark, SW Sweden and coastal Norway >> W Europe, W Africa

    __ *L.f.fuscus* Linnaeus, 1758   Baltic Sea, Finland, and Karelia (inland NW European Russia) >> NE and E Africa, SW Asia

    __ *L.f.heuglini* Bree, 1876[2]   N European Russia and NW Siberia east to Taymyr Pen. >> Middle East to E Africa, SW Asia and NW India

○ *Larus ichthyaetus* Pallas, 1773 GREAT BLACK-HEADED GULL   S Russia to NW Mongolia and N China >> E Mediterranean to Burma and NE Africa

○ *Larus brunnicephalus* Jerdon, 1840 BROWN-HEADED GULL   SC Asia >> E Arabia to SE Asia

○ *Larus cirrocephalus* GREY-HOODED GULL
    __ *L.c.cirrocephalus* Vieillot, 1818   Coastal Ecuador, Peru and C Brazil to Argentina
    __ *L.c.poiocephalus* Swainson, 1837   W, C, E and S Africa, Madagascar

○ *Larus hartlaubii* Bruch, 1853 HARTLAUB'S GULL   Coastal C Namibia to SW Cape Province

◉ *Larus novaehollandiae* SILVER GULL[3]
    __ *L.n.forsteri* (Mathews, 1912)   New Caledonia, Loyalty Is., SW Pacific
    ✓ *L.n.novaehollandiae* Stephens, 1826   Australia
    __ *L.n.gunni* Mathews, 1912   Tasmania
    ✓ *L.n.scopulinus* J.R. Forster, 1844[4]   New Zealand, Stewart, Chatham, Bounties, Snares, Auckland and Campbell Is.

◉ *Larus bulleri* Hutton, 1871 BLACK-BILLED GULL   New Zealand

◉ *Larus maculipennis* M.H.K. Lichtenstein, 1823 BROWN-HOODED GULL   S Sth. America, Falkland Is.

◉ *Larus ridibundus* Linnaeus, 1766 COMMON BLACK-HEADED GULL   NE North America, S Greenland, Iceland, Europe, Asia >> south to W and E Africa, India, SE Asia, Philippines

○ *Larus genei* Brème, 1839 SLENDER-BILLED GULL   Senegal, Gambia and Mauritania, Mediterranean, Black Sea, and Asia Minor to C Asia, and NW India >> coastal regions to Sthn. Africa

◉ *Larus philadelphia* (Ord, 1815) BONAPARTE'S GULL   W Alaska, W, C and E Canada >> coastal USA

○ *Larus saundersi* (Swinhoe, 1871) SAUNDERS'S GULL   Coastal NE China >> S Japan and South Korea to N Vietnam

○ *Larus serranus* Tschudi, 1844 ANDEAN GULL   Andean lakes from N Ecuador to N Argentina >> coastal W Sth. America

◉ *Larus melanocephalus* Temminck, 1820 MEDITERRANEAN GULL   Europe to Black Sea and Caucasus >> W Europe, NW Africa, Mediterranean

○ *Larus relictus* Lönnberg, 1931 RELICT GULL   SE Kazkhstan to NW China >> Korea to E China

○ *Larus fuliginosus* Gould, 1841 LAVA GULL   Galapagos Is.

◉ *Larus atricilla* LAUGHING GULL
    ✓ *L.a.megalopterus* Bruch, 1855[5]   Coastal SE Canada, E, S and SW USA, W Mexico and E Central America >> south to S Peru

    __ *L.a.atricilla* Linnaeus, 1758   West Indies to N Venezuela Is >> south to N Brazil

◉ *Larus pipixcan* Wagler, 1831 FRANKLIN'S GULL   C and S Canada, C USA >> coastal W Central and Sth. America

◉ *Larus minutus* Pallas, 1776 LITTLE GULL   Baltic area to Lena basin and Transbaikalia and Nei Mongol (China) >> E North America, W Europe, Mediterranean, Black Sea, China (E coast)

---

[1] Implicitly includes *ochotensis* Portenko, 1963 (1909), see Stepanyan (1990: 223) (2326) for monotypic treatment.
[2] In last Edition treated as part of *L. argentatus* and perhaps closely related to *L.a.vegae*, and both *heuglini* and *vegae* require further study if either is to be recognised, see Burger & Gochfeld (1996: 611) (270), who include *antelius* and *taimyrensis* as synonyms.
[3] We follow Higgins & Davies (1996: 517) (1113) in retaining *scopulinus* within this species and recognising the Tasmanian form.
[4] Drawings with Forster's proposed names were published many years after his death (d. 1798).
[5] For recognition of this see Burger & Gochfeld (1996: 618) (270).

PAGOPHILA Kaup, 1829 F
O *Pagophila eburnea* (Phipps, 1774) IVORY GULL   Circumpolar from N Canada to NC Siberia >> Arctic

RHODOSTETHIA W. Macgillivray, 1842 F
O *Rhodostethia rosea* W. Macgillivray, 1842 ROSS's GULL   N Siberia, N Canada, SW Greenland >> Arctic

XEMA Leach, 1819 F
◐ *Xema sabini* (Sabine, 1819) SABINE's GULL[1]   Arctic N America and N Eurasia >> S America and
                                                                    S Africa

CREAGRUS Bonaparte, 1854 M
O *Creagrus furcatus* (Néboux, 1846) SWALLOW-TAILED GULL   Galapagos Is., Malpelo Is. (E Pacific Ocean) >> Pacific
                                                                    coast Sth. America

RISSA Stephens, 1826 F
◐ *Rissa tridactyla* (Linnaeus, 1758) BLACK-LEGGED KITTIWAKE[2]   N Atlantic and Pacific Oceans

O *Rissa brevirostris* (Bruch, 1853) RED-LEGGED KITTIWAKE   Pribilof Is., Commander Is., Aleutians >> N Pacific Ocean

STERNINAE[3]

STERNA Linnaeus, 1758[4] F
O *Sterna nilotica* GULL-BILLED TERN
 — *S.n.nilotica* J.F. Gmelin, 1789   Europe to Mauritania, W India, and NE China >>
                                                                    Africa to SE Asia

 — *S.n.affinis* Horsfield, 1821[5]   E India to SE and E China >> mainland and
                                                                    archipelagic SE Asia to Australia

 — *S.n.macrotarsa* Gould, 1837   Australia >> N Australia and S New Guinea
 — *S.n.aranea* A. Wilson, 1814   E and S USA, Bahamas, Puerto Rico, Cuba >> coastal
                                                                    C America south to Peru and Brazil

 — *S.n.vanrossemi* (Bancroft, 1929)   S California to Sinaloa (Mexico) >> south to Ecuador
 — *S.n.groenvoldi* (Mathews, 1912)   French Guiana to NE Argentina

◐ *Sterna caspia* Pallas, 1770 CASPIAN TERN   N America, N Europe, Africa, Madagascar, C and S
                                                                    Asia, Australia, New Zealand

◐ *Sterna elegans* Gambel, 1849 ELEGANT TERN[6]   S California to C Baja California and Gulf of California
                                                                    >> W USA to W Sth. America

O *Sterna bengalensis* LESSER CRESTED TERN
 — *S.b.emigrata* Neumann, 1934   Libya and Persian Gulf >> W and NE Africa and W
                                                                    India (1644)

 — *S.b.bengalensis* Lesson, 1831[7]   Red Sea and Indian Ocean >> E and S Africa
 — *S.b.torresii* (Gould, 1843)   Australasia >> Gulf of Bengal and Philippines to
                                                                    Bismarck Arch.

◐ *Sterna sandvicensis* SANDWICH TERN
 — *S.s.sandvicensis* Latham, 1787   Europe to Caspian Sea >> Mediterranean and Africa
                                                                    to NW India and Sri Lanka

 ✓ *S.s.acuflavida* Cabot, 1848   E North America to S Caribbean >> Caribbean to S
                                                                    Peru and Uruguay

 — *S.s.eurygnatha* Saunders, 1876   Islands off Venezuela and French Guiana, N and E
                                                                    Sth. America

O *Sterna bernsteini* Schlegel, 1863 CHINESE CRESTED TERN   Coastal E China, Taiwan >> E China to Philippines and
                                                                    Indonesia

◐ *Sterna maxima* ROYAL TERN
 ✓ *S.m.maxima* Boddaert, 1783   S California to Peru, Florida to Argentina, W Indies
 — *S.m.albididorsalis* E. Hartert, 1921   Coastal W Africa >> Angola

---

[1] Includes *palaearctica* Stegmann, 1934 (2308), *tschuktschorum* Portenko, 1939 (1905), and *woznesenskii* Portenko, 1939 (1905), see Stepanyan (1990: 226) (2326).
[2] Includes *pollicaris*; we follow Vaurie (1965: 480) (2510) and Stepanyan (1990: 227) (2326) in treating this as monotypic.
[3] In contradiction to our treatment of the gulls we here take a broad view of the genus *Sterna*.
[4] With some reservations we employ a broad genus engulfing *Gelochelidon*, *Hydroprogne* and *Thalasseus*.
[5] This name, based on a specimen taken in Java, presumably a migrant, has priority over *addenda* and is used by most recent authors on Asian birds, e.g.
    Cheng, 1987 (333), Deignan, 1963 (663), Dickinson *et al.*, 1991 (745), Ripley, 1982 (2055), van Marle & Voous, 1988 (2468), Wells, 1999 (2579), and White &
    Bruce, 1986 (2652).
[6] The name *ichla* Mathews, 1937 (1457), a *nomen novum*, is apparently not needed.
[7] Includes *par* based on Red Sea birds. Ranges misunderstood in last Edition.

● *Sterna bergii* GREATER CRESTED TERN
   — *S.b.bergii* M.H.K. Lichtenstein, 1823   Namibia and Sth. Africa
   — *S.b.enigma* Clancey, 1979[1]   Zambezi delta (467)
   — *S.b.thalassina* Stresemann, 1914   Tanzania and Madagascar to Mascarenes and Chagos Arch. >> S Somalia
   — *S.b.velox* Cretzschmar, 1827   Red Sea and NW Somalia to Burma and W Sumatra, Maldives and Sri Lanka >> Kenya
   ✓ *S.b.cristata* Stephens, 1826[2]   Coasts of E and S China and Java Seas to W, N and E Australia, east to Ryukyu and Tuamotu Is.

○ *Sterna aurantia* J.E. Gray, 1831 RIVER TERN   Rivers systems of E Pakistan to SW China

○ *Sterna dougallii* ROSEATE TERN
   — *S.d.dougallii* Montagu, 1813   E USA to N Venezuela, NW Europe, Azores, E and S Africa >> mid-Atlantic, W Africa
   — *S.d.arideensis* Mathews, 1912   Seychelles and ? Madagascar and Rodrigues I.
   — *S.d.korustes* (Hume, 1874)   Sri Lanka, Andaman Is., Mergui Arch. (SW Burma)
   — *S.d.bangsi* Mathews, 1912   Ryukyu Is. and S China to Java Seas and east to N and NE Australia and SW Pacific Is.
   — *S.d.gracilis* Gould, 1845   W coasts of Australia >> NW Australia and Wallacea

● *Sterna striata* J.F. Gmelin, 1789 WHITE-FRONTED TERN[3]   Tasmania, islands in Bass Strait, New Zealand, Auckland, Chatham and Snares Is.

○ *Sterna sumatrana* BLACK-NAPED TERN
   — *S.s.mathewsi* Stresemann, 1914   Islands in W Indian Ocean
   — *S.s.sumatrana* Raffles, 1822   Islands in E Indian and W Pacific Oceans, N Australia

● *Sterna hirundinacea* Lesson, 1831 SOUTH AMERICAN TERN   S Peru and EC Brazil to Tierra del Fuego

● *Sterna hirundo* COMMON TERN
   ✓ *S.h.hirundo* Linnaeus, 1758   N America to N Sth. America, Europe, N and W Africa, W Asia >> south of Tropic of Capricorn
   — *S.h.tibetana* Saunders, 1876   Tibetan Plateau to Pamir, Tien Shan and Ningxia >> E Africa, India and SE Asia
   — *S.h.minussensis* Sushkin, 1925   Mongolia and upper Yenisey R. to Lake Baikal >> N Indian Ocean
   ✓ *S.h.longipennis* Nordmann, 1835   NE Siberia to NE China >> SE Asia to Australia

○ *Sterna repressa* E. Hartert, 1916 WHITE-CHEEKED TERN   Red Sea to Kenya and W India

● *Sterna paradisaea* Pontoppidan, 1763 ARCTIC TERN   Arctic regions south to NE USA and W Europe >> Antarctica

○ *Sterna vittata* ANTARCTIC TERN
   — *S.v.tristanensis* Murphy, 1938   Tristan da Cunha, Gough, St. Paul, Amsterdam Is. (1634)
   — *S.v.georgiae* Reichenow, 1904   S Georgia, S Orkneys, S Sandwich Is.
   — *S.v.gaini* Murphy, 1938   S Shetland Is., Antarctic Pen. (1634)
   — *S.v.vittata* J.F. Gmelin, 1789   Prince Edward, Marion, Crozet, Kerguelen, Bouvet and Heard Is.
   — *S.v.bethunei* Buller, 1896[4]   Stewart, Snares, Auckland, Bounty, Antipodes, Campbell and Macquarie Is.

○ *Sterna virgata* KERGUELEN TERN
   — *S.v.mercuri* Voisin, 1971   Crozet and Prince Edward Is. (2535)
   — *S.v.virgata* Cabanis, 1875   Kerguelen Is. (S Indian Ocean)

● *Sterna forsteri* Nuttall, 1834 FORSTER'S TERN   S Canada, USA, NE Mexico >> S USA, C America

○ *Sterna trudeaui* Audubon, 1838 TRUDEAU'S TERN   S Sth. America

○ *Sterna albifrons* LITTLE TERN
   — *S.a.albifrons* Pallas, 1764   Europe and N Africa to C Asia, Persian Gulf, Indian Ocean (Seychelles and Comoros), India >> Africa

[1] For recognition of this see Britton (1986: 381) (215).
[2] Includes *gwendolenae* see Mees (1982) (1534).
[3] Higgins & Davies (1996: 632) (1113) stated "no races described"; their monotypic treatment nonetheless apparently included two synonyms *incerta* and *aucklandorna*.
[4] Includes *macquariensis* Falla, 1937 (822), see Higgins & Davies (1996: 677, 691) (1113).

○ _S.a.guineae_ Bannerman, 1931   Mauritania to Gabon

— _S.a.sinensis_ J.F. Gmelin, 1789[1]
a) E and SE Asia (incl. Philippines) >> SE Asia and Philippines to N Australia
b) N and E Australia and Tasmania >> Indonesia and tropical SW Pacific

○ **_Sterna saundersi_** Hume, 1877   SAUNDERS'S TERN   Red Sea and S Arabian Gulf to S Somalia, coastal Pakistan and islands off NW Sri Lanka >> Arabian Sea to E Africa[2]

● **_Sterna antillarum_** LEAST TERN

— _S.a.athalassos_ Burleigh & Lowery, 1942[3]   C North America >> south to N Brazil (277)

— _S.a.antillarum_ (Lesson, 1847)[4]   E and S USA and Bermuda to N Venezuela >> south to N Brazil

— _S.a.browni_ Mearns, 1916[5]   C California to W Mexico >> C America

○ **_Sterna superciliaris_** Vieillot, 1819   YELLOW-BILLED TERN   N, C and E Sth. America

○ **_Sterna lorata_** Philippi & Landbeck, 1861   PERUVIAN TERN   Coastal C Ecuador to N Chile

○ **_Sterna nereis_**   FAIRY TERN

— _S.n.exsul_ (Mathews, 1912)   New Caledonia, Coral Sea

— _S.n.nereis_ (Gould, 1843)[6]   W and S Australia, Victoria, Tasmania

— _S.n.davisae_ (Mathews & Iredale, 1913)   N North Island (New Zealand)

○ **_Sterna balaenarum_** (Strickland, 1852)   DAMARA TERN   Namibia to Cape Province >> WC African coast

○ **_Sterna acuticauda_** J.E. Gray, 1832   BLACK-BELLIED TERN[7]   Pakistan, Nepal, India, Burma, SW China, Indochina

● **_Sterna aleutica_** S.F. Baird, 1869   ALEUTIAN TERN   Sakhalin, Sea of Okhotsk and Bering Sea to S Alaska >> Hong Kong, Philippines, South China Sea south to waters near Singapore

○ **_Sterna lunata_** Peale, 1848   GREY-BACKED TERN   N Mariana to Hawaiian Is., Fiji, Austral, Tuamotu Is. >> some to N Moluccas

○ **_Sterna anaethetus_** BRIDLED TERN

— _S.a.melanoptera_ Swainson, 1837[8]   Caribbean Sea and W Africa

— _S.a.antarctica_ Lesson, 1831[9]   Red Sea, Persian Gulf, and W Indian Ocean

— _S.a.anaethetus_ Scopoli, 1786[10]   E Indian Ocean and W Pacific from S Japan to Australasia

— _S.a.nelsoni_ Ridgway, 1919[11]   W coast of Mexico and C America

○ **_Sterna fuscata_** SOOTY TERN

— _S.f.fuscata_ Linnaeus, 1766   Gulf of Mexico, W Indies, Gulf of Guinea, S Atlantic

— _S.f.nubilosa_ Sparrman, 1788[12]   S Red Sea, Gulf of Aden and Indian Ocean to Ryukyu Is., Gtr. Sundas and Philippines

— _S.f.serrata_ Wagler, 1830   W and N Australia, New Guinea through S Pacific Ocean to Easter I.

— _S.f.kermadeci_ (Mathews, 1916)   Kermadec, Lord Howe and Norfolk Is.

— _S.f.oahuensis_ Bloxham, 1826   Hawaii, Ogasawara-shoto (Bonin) Is. and tropical N Pacific Ocean south to Kiritimati I.

— _S.f.crissalis_ (Lawrence, 1872)   Islands off W Mexico and C America to Galapagos Is.

— _S.f.luctuosa_ Philippi & Landbeck, 1866[13]   Juan Fernández Is. (Chile)

---

[1] Includes _pusilla_ see Junge (1948) (1231); and _placens_ see Vaurie (1965: 501) (2510) — it would appear that breeding birds in E China are not distinguishable from those of Australia.
[2] For reasons to omit Malay Pen. from range see Wells (1999: 319) (2579).
[3] For a validation of this form see Johnson _et al._ (1998) (1216).
[4] Thought to include _staebleri_ Brodkorb, 1940 (221), but may need re-evaluation.
[5] Includes _mexicanus_ Van Rossem & Hachisuka, 1937 (2481), see Gochfeld & Burger (1996: 657) (946).
[6] Includes _horni_ see Higgins & Davies (1996: 736) (1113).
[7] This name has been used almost universally since Deignan (1945: 134) (637) and replaced _S. melanogaster_ which was used by Peters (1934: 336) (1822). For a dissenting view see Mees (1989: 370) (1540).
[8] Includes _recognita_ see Higgins & Davies (1996: 738) (1113).
[9] Includes _fuligula_ see Higgins & Davies (1996: 738) (1113).
[10] Includes _rogersi_ see Higgins & Davies (1996: 738) (1113).
[11] Perhaps not distinct from _melanoptera_ see Olsen & Larsson (1995) (1689).
[12] Includes _somaliensis_ (listed in last Edition). See also Gochfeld & Burger (1996: 661) (946).
[13] Omitted in last Edition.

◕ *Sterna albistriata* (G.R. Gray, 1845) Black-fronted Tern[1]   New Zealand

CHLIDONIAS Rafinesque, 1822 M
○ *Chlidonias hybrida* Whiskered Tern[2] [3]
   — *C.h.hybrida* (Pallas, 1811)[4]   N Africa and SW and C Europe, C Asia, NC India
         (Gangetic plain), SE Siberia (Lake Khanka), China
         >> Africa, S Asia, mainland SE Asia south to Malay
         Pen., Sulawesi and Philippines

   — *C.h.delalandei* (Mathews, 1912)[5]   S and E Africa, Madagascar
   — *C.h.javanicus* (Horsfield, 1821)[6]   Australia >> Philippines, Sundas, Wallacea, N Australia
         and New Guinea

○ *Chlidonias leucopterus* (Temminck, 1815) White-winged Black Tern   C Europe to E Asia >> Africa, S and SE Asia and
         Australasia

◕ *Chlidonias niger* Black Tern
   — *C.n.niger* (Linnaeus, 1758)   Europe to C Asia >> Africa
   ✓ *C.n.surinamensis* (J.F. Gmelin, 1789)   Canada, N USA >> C America, N Sth. America

PHAETUSA Wagler, 1832 F
○ *Phaetusa simplex* Large-billed Tern
   — *P.s.simplex* (J.F. Gmelin, 1789)   N, C and E Sth. America
   — *P.s.chloropoda* (Vieillot, 1819)   SC and SE Sth. America

ANOUS Stephens, 1826 M
◕ *Anous stolidus* Brown Noddy
   ✓ *A.s.pileatus* (Scopoli, 1786)[7]   Red Sea, Indian Ocean, and Pacific Ocean east to
         Marquesas Is., Easter I., and Los Desventurados Is.
   — *A.s.galapagensis* Sharpe, 1879   Galapagos Is
   — *A.s.ridgwayi* Anthony, 1898   Islands off W Mexico and W Central America
   — *A.s.stolidus* (Linnaeus, 1758)   Caribbean and tropical Atlantic to Tristan da Cunha Is.

○ *Anous tenuirostris* Lesser Noddy
   — *A.t.tenuirostris* (Temminck, 1823)   Seychelles, Madagascar, Mascarene Is., Maldive and
         Chagos Arch.
   — *A.t.melanops* Gould, 1846   Houtman Abrolhos Is. and Ashmore Reef (W
         Australia)

◕ *Anous minutus* Black Noddy
   — *A.m.worcesteri* (McGregor, 1911)   Cavilli I. and Tubbataha Reef (Sulu Sea)
   — *A.m.minutus* Boie, 1844   NE Australia and New Guinea to Tuamotu and
         Kermadec Is. and Samoa
   — *A.m.marcusi* (Bryan, 1903)   Marcus and Wake Is. to Caroline Is.
   ✓ *A.m.melanogenys* G.R. Gray, 1846   Hawaiian Is.
   — *A.m.diamesus* (Heller & Snodgrass, 1901)   Clipperton Is., Cocos I.
   — *A.m.americanus* (Mathews, 1912)   Islands in the Caribbean
   — *A.m.atlanticus* (Mathews, 1912)   Tropical S Atlantic islands

PROCELSTERNA Lafresnaye, 1842 F
○ *Procelsterna cerulea* Blue Noddy[8]
   — *P.c.saxatilis* W.K. Fisher, 1903   NW Hawaiian Is. to N Marshall Is. and Marcus I.
   — *P.c.cerulea* (F.D. Bennett, 1840)   Kiritimati I. (Pacific Ocean)
   — *P.c.nebouxi* Mathews, 1912   Phoenix, Ellice, Samoa Is.
   — *P.c.teretirostris* (Lafresnaye, 1841)   Tuamotu, Cook, Austral, Society and Marquesas Is.
   — *P.c.murphyi* Mougin & de Naurois, 1981   Gambier Is. (1629)

---

[1] Originally spelled *albistriata* in the text and on the plate *contra* Gochfeld & Burger (1996) (946) who used albostriata which is an emendation.
[2] For note on spelling see David & Gosselin (2002) (613).
[3] Our arrangement follows Mees (1977) (1530) except that *hybrida* reaches Sulawesi (White & Bruce, 1986) (2652) and Luzon in good numbers (Dickinson *et al.*, 1991) (745).
[4] Includes *indicus* and *swinhoei* see Mees (1977: 47) (1530).
[5] Includes *sclateri* see Mees (1977: 47) (1530).
[6] Includes *fluviatilis* see Mees (1977: 47) (1530).
[7] Includes *plumbeigularis* see Cramp *et al.* (1985: 163) (591).
[8] In separating *P. albivitta* we follow Gochfeld & Burger (1996: 666) (946).

○ *Procelsterna albivitta* GREY NODDY
    — *P.a.albivitta* Bonaparte, 1856   Lord Howe, Norfolk, Kermadec Is., New Zealand (Bay of Plenty)

    — *P.a.skottsbergii* Lönnberg, 1921   Henderson I., Easter I., Sala y Gómez I.
    — *P.a.imitatrix* Mathews, 1912   San Ambrosio and San Félix Is. (Chile)

GYGIS Wagler, 1832 F
◑ *Gygis alba* WHITE TERN[1]
    — *G.a.alba* (Sparrman, 1786)   Tropical S Atlantic Ocean
    ✔ *G.a.candida* (J.F. Gmelin, 1789)   Seychelles and Mascarenes to tropical Pacific Ocean (except for ranges of next two subspecies)

    — *G.a.microrhyncha* Saunders, 1876   Marquesas Is. and Kiribati Is.
    — *G.a.leucopes* Holyoak & Thibault, 1976   Henderson I., Pitcairn I. (1132)

LAROSTERNA Blyth, 1852 F
○ *Larosterna inca* (Lesson, 1827) INCA TERN   Coastal N Peru to C Chile

RYNCHOPINAE

RYNCHOPS Linnaeus, 1758 M
● *Rynchops niger* BLACK SKIMMER
    ✔ *R.n.niger* Linnaeus, 1758[2]   Coastal USA and Mexico >> south to Panama
    — *R.n.cinerascens* Spix, 1825   N Sth. America including Orinoco and Amazon rivers
    — *R.n.intercedens* Saunders, 1895   E and SE Sth. America (south of the Amazon)

○ *Rynchops flavirostris* Vieillot, 1816 AFRICAN SKIMMER   Senegal to Sudan and SW Ethiopia, south to N Namibia, N Botswana and S Mozambique

○ *Rynchops albicollis* Swainson, 1838 INDIAN SKIMMER   E Pakistan, N and EC India to Burma, Cambodia

STERCORARIIDAE SKUAS OR JAEGERS (1: 7)

STERCORARIUS Brisson, 1760[3] M
○ *Stercorarius chilensis* Bonaparte, 1857 CHILEAN SKUA   S Sth. America >> Chile and Argentina to S Peru and Falkland Is.

○ *Stercorarius maccormicki* Saunders, 1893 SOUTH POLAR SKUA   Antarctica >> N Pacific and NW Atlantic

○ *Stercorarius antarcticus* BROWN SKUA/SOUTHERN SKUA
    — *S.a.antarcticus* (Lesson, 1831)   Falkland Is., S Argentina
    — *S.a.lonnbergi* (Mathews, 1912)   Circumpolar on subantarctic islands and Antarctic Pen. >> Tropic of Capricorn

    — *S.a.hamiltoni* (Hagen, 1952)   Gough I., Tristan da Cunha (1055)

○ *Stercorarius skua* (Brünnich, 1764) GREAT SKUA   N Atlantic >> C Atlantic

○ *Stercorarius pomarinus* (Temminck, 1815) POMARINE SKUA   N Holarctic >> Peru, S Africa, India, Australia, N New Zealand

○ *Stercorarius parasiticus* (Linnaeus, 1758) PARASITIC JAEGER/ARCTIC SKUA[4]
                                        N Holarctic >> S Sth. America, S Africa, India, Australia, New Zealand

● *Stercorarius longicaudus* LONG-TAILED JAEGER
    — *S.l.longicaudus* Vieillot, 1819   Scandinavia, N Russia >> subantarctic, S Sth. America and S Africa

    ✔ *S.l.pallescens* Løppenthin, 1932[5]   Greenland, Nearctic, E Siberia >> subantarctic, S Sth. America, S Africa, temperate Australasia

---

[1] This arrangement follows A.O.U. (1998: 207) (3), but *leucopes* is tentatively recognised.
[2] Tentatively includes *oblita* Griscom, 1935 (1016), but this may require re-evaluation as A.O.U. (1957: 244) (1) did not treat the species in a way that contradicted Griscom's review.
[3] A.O.U. (2000: 850) (4) recommended placing *Catharacta* in *Stercorarius*. We follow.
[4] Includes *parallelus* Dementiev, 1940 (686), see Stepanyan (1990: 216) (2326).
[5] For recognition see Cramp *et al.* (1983: 675) (590).

ALCIDAE AUKS[1] (11: 24)

ALLE Link, 1806 F
○ *Alle alle* LITTLE AUK/DOVEKIE

— *A.a.alle* (Linnaeus, 1758)	N Atlantic Ocean to Svalbard >> New Jersey and W Europe
— *A.a.polaris* Stenhouse, 1930	Franz Josef Land and Novaya Zemlya >> Barents Sea

URIA Brisson, 1760 F
○ *Uria lomvia* THICK-BILLED MURRE/BRUNNICH'S GUILLEMOT[2]

— *U.l.lomvia* (Linnaeus, 1758)[3]	N Atlantic (NE Canada east to Franz Josef Land and Novaya Zemlya)
— *U.l.eleonorae* Portenko, 1937	E Taymyr Pen. to New Siberian Is. and De Long I. (1903)
— *U.l.heckeri* Portenko, 1944	Wrangel I., Herald I., N Chukotskiy Pen. (1906)
— *U.l.arra* (Pallas, 1811)	N Pacific (Sakhalin to S and E Chukotka, Aleutians and NW and S Alaska)

● *Uria aalge* GUILLEMOT/COMMON MURRE

— *U.a.aalge* (Pontoppidan, 1763)[4]	E Canada, SW Greenland, Iceland, Faroe Is., Scotland, C and S Norway and Baltic Sea
— *U.a.hyperborea* Salomonsen, 1932	Jan Mayen I., N Norway, Svalbard and N European Russia
— *U.a.albionis* Witherby, 1923	Ireland and Britain to Helgoland and W Portugal
— *U.a.inornata* Salomonsen, 1932	E Korea, Japan and Bering Sea to W Alaska and SW Canada
✓ *U.a.californica* (H. Bryant, 1861)	N Washington to California (USA)

ALCA Linnaeus, 1758 F
○ *Alca torda* RAZORBILL

— *A.t.torda* Linnaeus, 1758	E North America, Greenland to N Scandinavia and White Sea
— *A.t.islandica* C.L. Brehm, 1831[5]	Iceland, Faroe Is., British Isles to Heligoland and NW France

PINGUINUS Bonnaterre, 1790 M
○ †*Pinguinus impennis* (Linnaeus, 1758) GREAT AUK — N Atlantic islands >> offshore waters to south

CEPPHUS Pallas, 1769 M
○ *Cepphus grylle* BLACK GUILLEMOT

— *C.g.mandtii* (Mandt, 1822)[6]	NE Canada, N Greenland, Svalbard, Franz Josef Land, N Siberia and N Alaska
— *C.g.arcticus* (C.L..Brehm, 1824)[7]	SE Canada, NE USA, S Greenland, British Isles, Denmark, SW Sweden and Norway to White Sea
— *C.g.islandicus* Hørring, 1937[8]	Iceland (1148)
— *C.g.faeroeensis* (C.L. Brehm, 1831)[9]	Faroe Is.
— *C.g.grylle* (Linnaeus, 1758)	Baltic Sea

● *Cepphus columba* PIGEON GUILLEMOT

✓ *C.c.columba* Pallas, 1811[10]	Kamchatka to Chukotka, Bering Sea Is., EC and E Aleutians, W and S Alaska and W USA >> Bering Sea to Japan and NW Mexico
— *C.c.snowi* Stejneger, 1897	N and C Kuril Is. (Paramushir to Iturup) >> Hokkaido
— *C.c.kaiurka* Portenko, 1937	Commander Is. to WC Aleutian Is. >> Hokkaido (1903)

[1] Generic sequence as in A.O.U. (1998) (3).
[2] This arrangement, recognising *eleonorae* and *heckeri*, follows Vaurie (1965: 511) (2510) and Cramp *et al.* (1985: 184) (591).
[3] Includes *arroides* Portenko, 1937 (1903), see Stepanyan (1990: 239) (2326).
[4] Includes *intermedia* and *spiloptera* see Vaurie (1965: 509) (2510).
[5] Includes *britannica* Ticehurst, 1936 (2403), see Vaurie (1965: 506) (2510).
[6] Includes *ultimus* Salomonsen, 1944 (2151), and *tajani* Portenko, 1944 (1906), see Vaurie (1965: 514) (2510).
[7] For recognition see Cramp *et al.* (1985: 208) (591) who apparently included *atlantis* Salomonsen, 1944 (2151) herein (Roselaar *in litt.*).
[8] For recognition see Vaurie (1965: 515) (2510).
[9] For recognition see Cramp *et al.* (1985: 208) (591).
[10] This arrangement, with *adianta* Storer, 1950 (2347), correctly *adiantus* see David & Gosselin (2002) (613), and *eureka* Storer, 1950 (2347), in synonymy, follows Stepanyan (1990: 241) (2326).

○ *Cepphus carbo* Pallas, 1811 SPECTACLED GUILLEMOT — Russian Far East north to Sea of Okhotsk and S Kuril Is., coasts of N and C Japan

BRACHYRAMPHUS J.F. Brandt, 1837[1] M
○ *Brachyramphus perdix* (Pallas, 1811) LONG-BILLED MURRELET[2] — Shores of Sea of Okhotsk and Kamchatka to Hokkaido (N Japan) >> Honshu

● *Brachyramphus marmoratus* (J.F. Gmelin, 1789) MARBLED MURRELET — W Aleutians to SE Alaska and C California

○ *Brachyramphus brevirostris* (Vigors, 1829) KITTLITZ'S MURRELET — Bering Sea to Chukchi Sea and Gulf of Alaska

SYNTHLIBORAMPHUS J.F. Brandt, 1837 M
○ *Synthliboramphus hypoleucus* XÁNTUS'S MURRELET[3]
 — *S.h.scrippsi* Green & Arnold, 1939[4] — S Californian Channel Is., W Baja California Is. (1001)
 — *S.h.hypoleucus* (Xántus de Vesey, 1860) — NW Baja California

○ *Synthliboramphus craveri* (Salvadori, 1865) CRAVERI'S MURRELET — Gulf of California, W Baja Califonia

○ *Synthliboramphus antiquus* ANCIENT MURRELET
 — *S.a.antiquus* (J.F. Gmelin, 1789) — Yellow Sea and Sea of Japan to S Alaska and Queen Charlotte Is.
 — *S.a.microrhynchos* Stepanyan, 1972[5] — Commander Is. (2318)

○ *Synthliboramphus wumizusume* (Temminck, 1836[6]) JAPANESE MURRELET — Islands off Korea, Japan (Kyushu and the Izu Is.)

PTYCHORAMPHUS J.F. Brandt, 1837 M
○ *Ptychoramphus aleuticus* CASSIN'S AUKLET
 — *P.a.aleuticus* (Pallas, 1811) — Aleutians to SW and SE Alaska, S to N Baja California
 — *P.a.australis* van Rossem, 1939[7] — W Baja Califonia (2476)

AETHIA Merrem, 1788[8] F
○ *Aethia psittacula* (Pallas, 1769) PARAKEET AUKLET — Coasts and islands of Bering Sea south to Sea of Okhotsk and Kodiak I. >> N Pacific Ocean

○ *Aethia pusilla* (Pallas, 1811) LEAST AUKLET — Bering Sea islands and coasts >> N Pacific Ocean

○ *Aethia pygmaea* (J.F. Gmelin, 1789) WHISKERED AUKLET — NE Sea of Okhotsk to C Kuril Is. and Aleutian Is.

○ *Aethia cristatella* (Pallas, 1769) CRESTED AUKLET — Bering Sea south to C Kuril Is. and Kodiak I. >> N Pacific Ocean

CERORHINCA Bonaparte, 1828 F
● *Cerorhinca monocerata* (Pallas, 1811) RHINOCEROS AUKLET — Korea to S Sea of Okhotsk and S Kuril Is., and SE Alaska to W California >> N Pacific Ocean

FRATERCULA Brisson, 1760 F
○ *Fratercula arctica* ATLANTIC PUFFIN
 — *F.a.naumanni* Norton, 1901 — NW and E Greenland to Svalbard and N Novaya Zemlya >> subarctic Atlantic Ocean
 — *F.a.arctica* (Linnaeus, 1758) — E Canada, NE USA, SW Greenland, Iceland, C and N Norway to Bear I., Kola Pen. and S Novaya Zemlya >> N Atlantic and W Mediterranean
 — *F.a.grabae* (C.L. Brehm, 1831) — Faroe Is., S Norway, British Isles, NW France >> N Atlantic and W Mediterranean

● *Fratercula corniculata* (J.F. Naumann, 1821) HORNED PUFFIN — Sakhalin to Chukotka, Bering Sea Is., Aleutians and W and S Alaska >> N Pacific Ocean

● *Fratercula cirrhata* (Pallas, 1769) TUFTED PUFFIN[9] — Russian Far East and Hokkaido to Chukotka, Bering Sea Is., Aleutians, W and S Alaska to W California >> N Pacific Ocean

[1] Often incorrectly cited as "M. Brandt" but his papers in French published in Russia used "M" meaning Monsieur!
[2] For split from *B. marmoratus* see A.O.U. (1998: 212) (3).
[3] This species, and the next, once in the genus *Endomychura* were returned to *Brachyramphus* by Peters (1934: 355) (1822), but later moved to this genus see A.O.U. (1983: 245) (2).
[4] For recognition of this see A.O.U. (1957; 1998) (1, 3). Includes *pontilis* van Rossem, 1939 (2476).
[5] Omitted in previous Edition.
[6] Livr. 98 must date from 1836 not 1835 (Dickinson, 2001) (739).
[7] Recognised by A.O.U. (1957) (1).
[8] We follow A.O.U. (1998: 214) (3) in submerging *Cyclorrhynchus* in *Aethia*.
[9] Previously treated in a genus *Lunda* now submerged.

PTEROCLIDIDAE SANDGROUSE (2: 16)

SYRRHAPTES Illiger, 1811 M

○ *Syrrhaptes tibetanus* Gould, 1850 TIBETAN SANDGROUSE[1]   Tibetan Plateau to E Tajikistan, Qinghai, W Sichuan and W Gansu

○ *Syrrhaptes paradoxus* (Pallas, 1773) PALLAS'S SANDGROUSE   S Urals and Transcaspia to Mongolia and Nei Mongol

PTEROCLES Temminck, 1815 M

○ *Pterocles alchata* PIN-TAILED SANDGROUSE
— *P.a.alchata* (Linnaeus, 1766)   Iberian Pen., SE France
— *P.a.caudacutus* (S.G. Gmelin, 1774)   N Africa and Middle East to Uzbekistan and S Kazakhstan >> Arabia and Pakistan

○ *Pterocles namaqua* (J.F. Gmelin, 1789) NAMAQUA SANDGROUSE[2]   SW Angola and Namibia to Transvaal and S Cape Province

○ *Pterocles exustus* CHESTNUT-BELLIED SANDGROUSE[3]
— *P.e.exustus* Temminck, 1825   Senegal, Gambia and Mauritania to C Sudan
— †*P.e.floweri* Nicoll, 1921   Nile Valley (Egypt)
— *P.e.ellioti* Bogdanov, 1881   SE Sudan, Eritrea, N Ethiopia, Somalia
— *P.e.olivascens* (E. Hartert, 1909)   SW Ethiopia, Kenya, N Tanzania
— *P.e.erlangeri* (Neumann, 1909)   S and W Arabia
— *P.e.hindustan* R. Meinertzhagen, 1923   SE Iran, Pakistan, India

○ *Pterocles senegallus* (Linnaeus, 1771) SPOTTED SANDGROUSE   N Africa, Middle East, Arabia, Pakistan, NW India

○ *Pterocles orientalis* BLACK-BELLIED SANDGROUSE
— *P.o.orientalis* (Linnaeus, 1758)[4]   E Canary Is., N Africa, Iberian Pen., Cyprus and Israel to Turkey and Transcaucasia
— *P.o.arenarius* (Pallas, 1775)   Lower Volga R., Transcaspia and Iran to E Kazakhstan, NW China and SW Pakistan

○ *Pterocles gutturalis* YELLOW-THROATED SANDGROUSE
— *P.g.saturatior* E. Hartert, 1900   Ethiopia to N Zambia
— *P.g.gutturalis* A. Smith, 1836   S Zambia to W Transvaal and N Cape Province

○ *Pterocles coronatus* CROWNED SANDGROUSE
— *P.c.coronatus* M.H.K. Lichtenstein, 1823   W and C Sahara to W Egypt and W Sudan
— *P.c.vastitas* R. Meinertzhagen, 1928   NE Egypt, Sinai, Israel, Jordan
— *P.c.saturatus* Kinnear, 1927   Hills of N Oman
— *P.c.atratus* E. Hartert, 1902   W and S Arabia, Iran and S Afghanistan to SW Pakistan
— *P.c.ladas* Koelz, 1954   Pakistan (except SW) (1284)

○ *Pterocles decoratus* BLACK-FACED SANDGROUSE
— *P.d.ellenbecki* Erlanger, 1905   S Somalia, N Kenya, S Ethiopia, NE Uganda
— *P.d.decoratus* Cabanis, 1868   SE Kenya, E Tanzania
— *P.d.loveridgei* (Friedmann, 1928)   SW Kenya, C Tanzania

○ *Pterocles personatus* Gould, 1843 MADAGASCAN SANDGROUSE   N, W and S Madagascar

○ *Pterocles lichtensteinii* LICHTENSTEIN'S SANDGROUSE
— *P.l.targius* Geyr von Schweppenburg, 1916   W, C and S Sahara east to Chad
— *P.l.lichtensteinii* Temminck, 1825   SE Egypt and Sudan to E Kenya and Socotra, Arabia (except SE Yemen), Israel, Jordan and S Iran to W Pakistan
— *P.l.sukensis* Neumann, 1909   SE Sudan and S Ethiopia to C Kenya
— *P.l.arabicus* Neumann, 1909   S Arabia to S Iran, S Afghanistan and Pakistan
— *P.l.ingramsi* Bates & Kinnear, 1937   Hadhramaut (SE Yemen) (104)

○ *Pterocles indicus* (J.F. Gmelin, 1789) PAINTED SANDGROUSE   E Pakistan and India to Bengal

---

[1] Includes *pamirensis* Koelz, 1939 (1279), see Vaurie (1965) (2510), Stepanyan (1990) (2326).
[2] Includes *furva* Clancey, 1959 (383), see White (1965: 147) (2649).
[3] The revised subspecific ranges follow de Juana (1997) (624). Peters (1937: 5) (1823) suggested that the introduction of the name *ellioti* would involve the submergence of several names here recognized.
[4] Includes *aragonica* Latham, 1790 (1334A) and *bangsi* Koelz, 1939 (1279), see Vaurie (1965) (2510).

○ *Pterocles quadricinctus* Temminck, 1815 Four-banded Sandgrouse | Senegal and Gambia to Ethiopia, W Eritrea, N Uganda and NW Kenya

○ *Pterocles bicinctus* Double-banded Sandgrouse
— *P.b.ansorgei* Benson, 1947 | SW Angola (118)
— *P.b.multicolor* E. Hartert, 1908[1] | S and E Zambia, S Malawi and W Mozambique to Transvaal
— *P.b.bicinctus* Temminck, 1815[2] | Namibia, Botswana and NW Cape Province

○ *Pterocles burchelli* W.L. Sclater, 1922 Burchell's Sandgrouse[3] | SE Angola, Namibia and Botswana to W Transvaal, N Cape Province and W Orange Free State

RAPHIDAE DODOS, SOLITAIRES (2: 3)

RAPHUS Brisson, 1760 M
○ †*Raphus cucullatus* (Linnaeus, 1758) Dodo | Mauritius

○ †*Raphus solitarius* (Sélys-Longchamps, 1848) Réunion Solitaire Réunion I.

PEZOPHAPS Strickland, 1848 F
○ †*Pezophaps solitaria* (J.F. Gmelin, 1789) Rodriguez Solitaire | Rodriguez I. (Mauritius)

COLUMBIDAE DOVES, PIGEONS (42: 308)

COLUMBINAE

COLUMBA Linnaeus, 1758 F
◉ *Columba livia* Rock Dove[4]
↙ *C.l.livia* J.F. Gmelin, 1789 | W and S Europe to S Urals and SW Siberia, W and N Turkey, Cyprus, N and W Caucasus, W Transcaucasia, Atlantic Is., N Africa to NW Egypt
— *C.l.gymnocycla* G.R. Gray, 1856[5][6] | Senegal and Gambia to S Mali and Nigeria
— *C.l.targia* Geyr von Schweppenburg, 1916 | Mts. of C Sahara (N Mali and S Algeria to C Sudan
— *C.l.dakhlae* R. Meinertzhagen, 1928 | Oases of W Egypt
— *C.l.schimperi* Bonaparte, 1854[7] | Nile Valley and E Egypt to E Sudan and N Eritrea
— *C.l.palestinae* Zedlitz, 1912 | Sinai, W and S Arabia, Israel to Syria
— *C.l.gaddi* Zarundy & Loudon, 1906 | E Turkey, Armenia, Azerbaijan and Iran to Uzbekistan and W and N Afghanistan
— *C.l.neglecta* Hume, 1873 | W Pakistan, E Afghanistan and Tajikistan to Himalayas and Tien Shan
— *C.l.intermedia* Strickland, 1844 | S India, Sri Lanka

○ *Columba rupestris* Hill Pigeon
— *C.r.turkestanica* Buturlin, 1908 | Mts. of C Asia, from Pamir to Himalayas, W Sichuan and Gansu, and from Alai to Tien Shan and Altai
— *C.r.rupestris* Pallas, 1811 | W Mongolia and SE Siberia to E Sichuan, Shandong, Korea, and Russian Far East

○ *Columba leuconota* Snow Pigeon
— *C.l.leuconota* Vigors, 1831 | Himalayas
— *C.l.gradaria* E. Hartert, 1916 | E Tibet, E Nan Shan to Yunnan and N Burma

○ *Columba guinea* Speckled Pigeon
— *C.g.guinea* Linnaeus, 1758 | Senegal and Gambia to Ethiopia and N Malawi
— *C.g.phaeonota* G.R. Gray, 1856 | SW Angola to Zimbabwe and Cape Province

---

[1] Includes *usheri* Benson, 1947 (117), see White (1965: 150) (2649).
[2] Includes *elizabethae* Macdonald, 1954 (1409), see White (1965: 150) (2649).
[3] Thought to include *delabati* Winterbottom, 1964 (2680).
[4] Treatment based on Cramp *et al.* (1985: 285) (591). The populations named *atlantis*, *canariensis* and *nigricans* (listed in last Edition) are here considered feral populations. Such populations occur in many towns and cities in and outside the range of the wild species and interbreeding confuses our understanding of wild populations.
[5] For correction of spelling see David & Gosselin (2002) (613).
[6] Includes *lividior* see White (1965: 153) (2649).
[7] Includes *butleri* see Vaurie (1965: 546) (2510); but White (1965: 152) (2649) placed *butleri* in *gaddi*.

○ *Columba albitorques* Rüppell, 1837 WHITE-COLLARED PIGEON — Ethiopian and Eritrean highlands

○ *Columba oenas* STOCK DOVE
   — *C.o.oenas* Linnaeus, 1758[1] — NW Africa and Europe to N Iran, N Kazakhstan and SW Siberia

   — *C.o.yarkandensis* Buturlin, 1908 — SW Turkmenistan to Tien Shan and fringes of Tarim Basin (W China) to Lop Nor Lake

○ *Columba eversmanni* Bonaparte, 1856 PALE-BACKED PIGEON/YELLOW-EYED PIGEON
   NE Iran and N Afghanistan to S Kazakhstan and NW China >> Iran to NW India

○ *Columba oliviae* Stephenson Clarke, 1918 SOMALI PIGEON — N and NE Somalia

● *Columba palumbus* WOODPIGEON
   ⌐ *C.p.palumbus* Linnaeus, 1758[2] — NW Africa and Europe to Iraq, Caucasus area and Urals
   — †*C.p.maderensis* Tschusi, 1904 — Madeira
   — *C.p.azorica* E. Hartert, 1905 — E and C Azores
   — *C.p.excelsa* (Bonaparte, 1856) — NW Africa
   — *C.p.iranica* (Zarudny, 1910) — SW and N Iran to SW Turkmenistan
   — *C.p.casiotis* (Bonaparte, 1854)[3] — N Oman, SE Iran and Afghanistan to NW and C Himalayas and Tien Shan

○ *Columba trocaz* Heineken, 1829 MADEIRA LAUREL PIGEON — Madeira

○ *Columba bollii* Godman, 1872 DARK-TAILED LAUREL PIGEON — La Palma, Gomera and Tenerife (W Canary Is.)

○ *Columba junoniae* E. Hartert, 1916 WHITE-TAILED LAUREL PIGEON — La Palma, Gomera and Tenerife (W Canary Is.)

○ *Columba unicincta* Cassin, 1860 AFEP PIGEON — Liberia to NW Angola east to Uganda

○ *Columba arquatrix* Temminck, 1809 AFRICAN OLIVE PIGEON — Ethiopia and E Africa to Angola and Sth. Africa

○ *Columba sjostedti* Reichenow, 1898 CAMEROON OLIVE PIGEON — E Nigeria, Cameroon, Bioko I.

○ *Columba thomensis* Bocage, 1888 SÃO TOMÉ OLIVE PIGEON — São Tomé I.

○ *Columba pollenii* Schlegel, 1866 COMORO OLIVE PIGEON — Comoro Is.

○ *Columba hodgsonii* Vigors, 1832 SPECKLED WOODPIGEON — W Himalayas to Burma, NW Thailand, NW Laos, Yunnan, E Qinghai and W Gansu

○ *Columba albinucha* Sassi, 1911 WHITE-NAPED PIGEON — E Zaire to W Uganda, W Cameroon

○ *Columba pulchricollis* Blyth, 1846 ASHY WOODPIGEON — C Himalayas and Assam hills through most of upland Burma to NW Thailand, N Laos, S China and Taiwan

○ *Columba elphinstonii* (Sykes, 1833) NILGIRI WOODPIGEON — Western Ghats (SW India)

○ *Columba torringtoni* (Bonaparte, 1854) SRI LANKA WOODPIGEON — Sri Lanka

○ *Columba punicea* Blyth, 1842 PALE-CAPPED PIGEON — S Tibet and E India through much of Burma and Thailand to Indochina and Hainan

○ *Columba argentina* Bonaparte, 1855 SILVERY PIGEON — Islands E and W of Sumatra and W of Borneo

○ *Columba palumboides* (Hume, 1873) ANDAMAN WOODPIGEON — Andaman and Nicobar Is.

○ *Columba janthina* JAPANESE WOODPIGEON
   — *C.j.janthina* Temminck, 1830[4] — Tsushima and S Japan to Izu Is. and Ryukyu Is.
   — †?*C.j.nitens* (Stejneger, 1887) — Ogasawara and Iwo Is.
   — *C.j.stejnegeri* (N. Kuroda, Sr., 1923)[5] — Yaeyama Is. (Ryukyu Is.)

○ †*Columba versicolor* Kittlitz, 1832 BONIN WOODPIGEON — Parry Group, Ogasawara-shoto (Bonin) Is.

○ †*Columba jouyi* (Stejneger, 1887) RYUKYU WOODPIGEON — C Ryukyu Is. and Borodino Is.

○ *Columba vitiensis* WHITE-THROATED PIGEON
   — *C.v.griseogularis* (Walden & Layard, 1872) — Philippines, Sulu Arch., N Bornean Is.

---

[1] Includes *hyrcana* see Vaurie (1965: 541) (2510).
[2] Includes *excelsa* see Vaurie (1965: 538) (2510); also includes *ghigii* Trischitta, 1939 (2440), see Vaurie (1965: 538) (2510), and *kleinschmidti* Clancey, 1950 (363), see Cramp *et al.* (1985: 328) (591).
[3] Includes *kirmanica* Koelz, 1954 (1284), see Vaurie (1965: 539) (2510).
[4] Date correct but Livr. 85 not 86 (Dickinson, 2001) (739).
[5] For recognition see Orn. Soc. Japan (2000: 146) (1727).

— *C.v.anthracina* (Hachisuka, 1939)	Palawan (1038)
— *C.v.metallica* Temminck, 1835	Lesser Sundas (Lombok east to Damar)
— *C.v.halmaheira* (Bonaparte, 1855)[1]	Banggai and Sula Is., Moluccas to New Guinea, Louisiade Arch. and Solomon Is.
— *C.v.leopoldi* (Tristram, 1879)	Vanuatu
— *C.v.hypoenochroa* (Gould, 1856)	New Caledonia, Isle of Pines, Loyalty Is.
— *C.v.vitiensis* Quoy & Gaimard, 1830	Fiji
— *C.v.castaneiceps* Peale, 1848	Samoa
○ *Columba leucomela* Temminck, 1821 WHITE-HEADED PIGEON	E Australia
○ *Columba pallidiceps* (E.P. Ramsay, 1877) YELLOW-LEGGED PIGEON	Solomon Is., Bismarck Arch.
○ *Columba leucocephala* Linnaeus, 1758 WHITE-CROWNED PIGEON	Florida Keys, Bahamas, Greater and Lesser Antilles, Caribbean islands off C America, N Columbia
○ *Columba squamosa* Bonnaterre, 1792 SCALY-NAPED PIGEON	Greater and Lesser Antilles
○ *Columba speciosa* J.F. Gmelin, 1789 SCALED PIGEON	S Mexico to E Peru, N and E Bolivia and S Brazil, Trinidad
○ *Columba picazuro* PICAZURO PIGEON	
— *C.p.marginalis* Naumburg, 1932	NE Brazil
— *C.p.picazuro* Temminck, 1813	E and S Brazil and Bolivia to SC Argentina
○ *Columba corensis* Jacquin, 1784 BARE-EYED PIGEON	Coastal N Colombia and N Venezuela, Aruba, Curaçao, Bonaire, Margarita, Blanquilla
○ *Columba maculosa* SPOT-WINGED PIGEON	
— *C.m.albipennis* Sclater & Salvin, 1876	C Peru to NW Argentina
— *C.m.maculosa* Temminck, 1813	S Bolivia to Uruguay and SC Argentina
● *Columba fasciata* BAND-TAILED PIGEON	
✓ *C.f.monilis* Vigors, 1839[2]	Pacific NW USA to California
— *C.f.fasciata* Say, 1823	Rocky Mts. (W USA) to Guatemala, Honduras and NC Nicaragua
— *C.f.vioscae* Brewster, 1888	S Baja California
— *C.f.letonai* Dickey & van Rossem, 1926	Border of Honduras and El Salvador
— *C.f.parva* Griscom, 1935	N Nicaragua
— *C.f.crissalis* Salvadori, 1893	Costa Rica, W Panama
— *C.f.albilinea* Bonaparte, 1854	Colombia to NW Argentina
— *C.f.roraimae* Chapman, 1929	Mt. Roraima (Venezuela)
● *Columba araucana* Lesson, 1827 CHILEAN PIGEON	C and S Chile, Mocha I., SC and S Argentina
○ *Columba caribaea* Jacquin, 1784 RING-TAILED PIGEON	Jamaica
○ *Columba cayennensis* PALE-VENTED PIGEON	
— *C.c.pallidicrissa* Chubb, 1910[3]	S Mexico to N Colombia
— *C.c.occidentalis* Stolzmann, 1926	W Ecuador
— *C.c.andersoni* Cory, 1915[4]	SE Colombia and E Ecuador east to Venezuela and N Brazil
— *C.c.tobagensis* Cory, 1915	Trinidad and Tobago Is.
— *C.c.cayennensis* Bonnaterre, 1792	The Guianas
— *C.c.sylvestris* Vieillot, 1818	E Peru to N Argentina and Uruguay
● *Columba flavirostris* RED-BILLED PIGEON	
— *C.f.flavirostris* Wagler, 1831	Texas to E Costa Rica
— *C.f.madrensis* Nelson, 1898	Tres Marias Is.
— *C.f.restricta* van Rossem, 1930	W Mexico
✓ *C.f.minima* Carriker, 1910	W Costa Rica
○ *Columba oenops* Salvin, 1895 PERUVIAN PIGEON	SE Ecuador, N Peru

---

[1] Includes *mendeni* Neumann, 1939 (1645), see White & Bruce (1986) (2652).
[2] For recognition see A.O.U. (1957: 259) (1).
[3] Apparently includes *tamboensis* Conover, 1938 (561); may need re-evaluation.
[4] Apparently includes *obsoleta* Todd, 1947 (2410); may need re-evaluation.

○ *Columba inornata* PLAIN PIGEON[1]
    — *C.i.inornata* Vigors, 1827   Cuba and Isla de la Juventud, Hispaniola
    — *C.i.exigua* (Ridgway, 1915)   Jamaica
    — *C.i.wetmorei* J.L. Peters, 1937   Puerto Rico

○ *Columba plumbea* PLUMBEOUS PIGEON[2]
    — *C.p.delicata* Berlepsch & Stolzmann, 1902   E Colombia to N Brazil and N and E Bolivia
    — *C.p.chapmani* (Ridgway, 1916)   NW Ecuador
    — *C.p.pallescens* E. Snethlage, 1908   Sthn. tributaries of the Amazon
    — *C.p.baeri* Hellmayr, 1908   C Brazil
    — *C.p.plumbea* Vieillot, 1818   SE Brazil, NE and E Paraguay

○ *Columba subvinacea* RUDDY PIGEON[3]
    — *C.s.subvinacea* (Lawrence, 1868)   Costa Rica, Panama
    — *C.s.berlepschi* E. Hartert, 1898   SE Panama to SW Ecuador
    — *C.s.peninsularis* Chapman, 1915   NE Venezuela
    — *C.s.zuliae* Cory, 1915   NE Colombia, W Venezuela
    — *C.s.purpureotincta* Ridgway, 1888   SE Colombia, S Venezuela, the Guianas
    — *C.s.bogotensis* (Berlepsch & Leverkühn, 1890)   W Colombia to C Brazil and N and E Bolivia

◉ *Columba nigrirostris* P.L. Sclater, 1860 SHORT-BILLED PIGEON   SE Mexico to NW Colombia

○ *Columba goodsoni* E. Hartert, 1902 DUSKY PIGEON   W Colombia, NW Ecuador

○ *Columba delegorguei* EASTERN BRONZE-NAPED PIGEON
    — *C.d.sharpei* (Salvadori, 1893)   SE Sudan, Uganda, Kenya, Tanzania, Zanzibar I.
    — *C.d.delegorguei* Delegorgue, 1847   Malawi to E Sth. Africa

○ *Columba iriditorques* Cassin, 1856 WESTERN BRONZE-NAPED PIGEON   Sierra Leone to NW Angola west to W Uganda

○ *Columba malherbii* J. & E. Verreaux, 1851 SÃO TOMÉ BRONZE-NAPED PIGEON
                                     São Tomé, Príncipe and Pagalu Is.

○ *Columba larvata* AFRICAN LEMON DOVE[4]
    — *C.l.hypoleuca* (Salvadori, 1903)   Sierra Leone to NW Cameroon, Bioko I., Pagalu I.
    — *C.l.principalis* (Hartlaub, 1866)   Príncipe I.
    — *C.l.simplex* (Hartlaub, 1849)   São Tomé
    — *C.l.bronzina* (Rüppell, 1837)   SE Sudan, Ethiopia
    — *C.l.larvata* Temminck, 1810   S Sudan to Cape Province
    — *C.l.jacksoni* (Sharpe, 1904)   E Zaire and SW Uganda to W Tanzania and NC Zambia
    — *C.l.samaliyae* (C.M.N. White, 1948)   Angola, extreme NW Zambia (2633)

NESOENAS Salvadori, 1893[5] F
○ *Nesoenas mayeri* (Prévost, 1843) PINK PIGEON   Mauritius

STREPTOPELIA Bonaparte, 1855 F
◉ *Streptopelia turtur* EUROPEAN TURTLE DOVE
    ⚓ *S.t.turtur* (Linnaeus, 1758)[6]   Europe to W Siberia, Madeira, Canary Is. >> Senegal east to Ethiopia

    — *S.t.arenicola* (E. Hartert, 1894)   N Africa, Iraq and Iran to NW China >> Senegal east to Ethiopia

    — *S.t.hoggara* (Geyr von Schweppenburg, 1916)   Hoggar and Aïr Massifs (S Sahara)
    — *S.t.rufescens* (C.L. Brehm, 1855)[7]   Nile Delta and Dakhla and Kharga Oases (Egypt) >> C Sudan and NW Ethiopia

○ *Streptopelia lugens* (Rüppell, 1837) DUSKY TURTLE DOVE[8]   SW Arabia, SE Sudan to Somalia and Malawi

---

[1] For tentative recognition of subspecies see Baptista *et al.* (1997: 127) (94).
[2] Treatment of this and the next species follows Baptista *et al.* (1997: 128) (94) and is close to that of Peters (1937) (1823). In our last Edition we listed *wallacei* (now a synonym of *pallescens*) and mistakenly associated *bogotensis* with this species, displacing the name *delicata*
[3] Baptista *et al.* (1997) (94) appear to have placed *anolaimae* in *zuliae*, *ogilviegranti* in *purpureotincta*, and *ruberrima* Meyer de Schauensee, 1950 (1566), in *bogotensis* where we believe *recondita* Todd, 1937 (2405), also belongs.
[4] In last Edition we retained the monotypic genus *Aplopelia* but see Fry *et al.*, 1985 (898). Baptista *et al.* (1997) (94) separated *simplex* on acoustic evidence, but admitted other insular forms might then not belong with *larvata*. For an early account of the complexity of this group see Bannerman (1916) (92); we look forward to a detailed revision of these birds. We use the name *hypoleuca* because the name *inornata* Reichenow, 1892, is preoccupied in the genus *Columba* by a Caribbean species (see Plain Pigeon).
[5] For use of this generic name see Baptista *et al.* (1997: 132) (94).
[6] Includes *moltonii* Trischitta, 1939 (2440), see Vaurie (1965: 553) (2510).
[7] The name *isabellina* was used in our last Edition, but is one year younger.
[8] Treated as monotypic by Fry *et al.* (1984) (898); includes *funebrea* and *bishaensis* Goodwin, 1963 (958) — originally named *arabica* Goodwin, 1963 (957A).

○ *Streptopelia hypopyrrha* (Reichenow, 1910) ADAMAWA TURTLE DOVE NC and E Nigeria, N Cameroon, SW Chad[1]

○ *Streptopelia orientalis* ORIENTAL TURTLE DOVE
    — *S.o.meena* (Sykes, 1832)   W Siberia to Nepal and Himalayas >> India
    — *S.o.orientalis* (Latham, 1790)   C Siberia to Japan and Himalayas >> S and E Asia
    — *S.o.stimpsoni* (Stejneger, 1887)   Ryukyu Is.
    — *S.o.orii* Yamashina, 1932   Taiwan
    — *S.o.erythrocephala* (Bonaparte, 1855)[2]   S India
    — *S.o.agricola* (Tickell, 1833)[3]   NE India to SC China

○ *Streptopelia bitorquata* ISLAND COLLARED DOVE
    — *S.b.dusumieri* (Temminck, 1823)   Philippines, N Borneo
    — *S.b.bitorquata* (Temminck, 1810)   Java to Timor

◉ *Streptopelia decaocto* EURASIAN COLLARED DOVE
    ☑ *S.d.decaocto* (Frivaldszky, 1838)[4]   Europe to W China
    — *S.d.xanthocycla* (Newman, 1906)[5]   Burma to E China

○ *Streptopelia roseogrisea* AFRICAN COLLARED DOVE
    — *S.r.roseogrisea* (Sundevall, 1857)   Senegal, Gambia and Mauritania to N Sudan and NW Ethiopia

    — *S.r.arabica* (Neumann, 1904)   NE Sudan, Eritrea, NE Ethiopia and N Somalia, SW Arabia

○ *Streptopelia reichenowi* (Erlanger, 1901) WHITE-WINGED COLLARED DOVE
                                    S Somalia, SE Ethiopia, NE Kenya

○ *Streptopelia decipiens* MOURNING COLLARED DOVE[6]
    — *S.d.shelleyi* (Salvadori, 1893)   Senegal, Gambia and Mauritania to C Nigeria
    — *S.d.logonensis* (Reichenow, 1921)   L Chad to S Sudan, N and E Zaire and N Uganda
    — *S.d.decipiens* (Hartlaub & Finsch, 1870)   W and C Sudan, N and C Ethiopia, NW Somalia
    — *S.d.elegans* (Zedlitz, 1913)   S Ethiopia, S Somalia, N and E Kenya
    — *S.d.perspicillata* (Fischer & Reichenow, 1884)   W and S Kenya, C Tanzania
    — *S.d.ambigua* (Bocage, 1877)   E Angola to Malawi and Limpopo and Zambezi Rivers

○ *Streptopelia semitorquata* (Rüppell, 1837) RED-EYED TURTLE DOVE[7]   SW Arabia, sub-Saharan Africa

○ *Streptopelia capicola* RING-NECKED DOVE
    — *S.c.electa* (Madarász, 1913)   W Ethiopia
    — *S.c.somalica* (Erlanger, 1905)   E Ethiopia, Somalia, N Kenya
    — *S.c.tropica* (Reichenow, 1902)[8]   C Kenya to Angola and Sth. Africa
    — *S.c.onguati* Macdonald, 1957   N Namibia, SW Angola (1410)
    — *S.c.damarensis* (Hartlaub & Finsch, 1870)   Namibia, Botswana, SW Zimbabwe
    — *S.c.capicola* (Sundevall, 1857)   W Cape Province

○ *Streptopelia vinacea* (J.F. Gmelin, 1789) VINACEOUS DOVE   Senegal, Gambia and Mauritania to Eritrea, C Ethiopia and C Uganda

○ *Streptopelia tranquebarica* RED TURTLE DOVE
    — *S.t.humilis* (Temminck, 1824)   NE Tibet to N China and N Philippines
    — *S.t.tranquebarica* (Hermann, 1804)   India

○ *Streptopelia picturata* MADAGASCAN TURTLE DOVE[9]
    — *S.p.rostrata* (Bonaparte, 1855)   Seychelles
    — *S.p.aldabrana* (P.L. Sclater, 1872)   Amirantes Is.
    — *S.p.coppingeri* (Sharpe, 1884)   Aldabra (South I.), Glorieuse Is.
    — *S.p.comorensis* (E. Newton, 1877)   Comoro Is.
    — *S.p.picturata* (Temminck, 1813)   Madagascar

---

[1] For an account of the discovery of a disjunct taxon thought to be related to this see Baillon (1992) (77).
[2] Includes *sylvicola* Koelz, 1939 (1279), see Ripley (1982: 155) (2055).
[3] Includes *khasiana* Koelz, 1954 (1284), and *meridionalis* Koelz, 1954 (1284), see Ripley (1982: 155) (2055).
[4] Includes *stoliczkae* see Vaurie (1965: 551) (2510).
[5] For correction of spelling see David & Gosselin (2002) (613).
[6] Name from Goodwin (1967) (959).
[7] Includes *chersophila* Clancey, 1986 (500). We follow White (1965: 156) (2649) in considering monotypic treatment appropriate for this species.
[8] Includes *abunda* Clancey, 1960 (385), see White (1965: 158) (2649).
[9] We tentatively follow the revised attribution of subspecific names in Baptista *et al.* (1997: 140) (94).

● *Streptopelia chinensis* SPOTTED-NECKED DOVE[1]

    — *S.c.suratensis* (J.F. Gmelin, 1789)[2]   Pakistan, Nepal, Bhutan and N Assam S through India to Sri Lanka

    — *S.c.tigrina* (Temminck, 1810)[3]   Bangladesh, S Assam through W, S Burma and mainland SE Asia to W Yunnan. Introd. Philippines and the Sundas east to Timor

    — *S.c.chinensis* (Scopoli, 1786)[4]   NE Burma east to C and E China and Taiwan

    — *S.c.hainana* E. Hartert, 1910   Hainan

○ *Streptopelia senegalensis* LAUGHING DOVE

    — *S.s.phoenicophila* E. Hartert, 1916   S Morocco, Algeria, Tunisia, NW Libya

    — *S.s.aegyptiaca* (Latham, 1790)   Nile valley (Egypt)

    — *S.s.sokotrae* C.H.B. Grant, 1914   Socotra I.

    — *S.s.senegalensis* (Linnaeus, 1766)[5]   W Arabia, sub-Saharan Africa

    — *S.s.cambayensis* (J.F. Gmelin, 1789)[6]   E Arabia and E Iran to Pakistan, India and Bangladesh

    — *S.s.ermanni* (Bonaparte, 1856)   Transcaspia east to Kazakhstan, Xinjiang and N Afghanistan

MACROPYGIA Swainson, 1837 F

○ *Macropygia unchall* BARRED CUCKOO-DOVE

    — *M.u.tusalia* (Blyth, 1843)   Himalayas to Sichuan, SW Yunnan and Shan States

    — *M.u.minor* Swinhoe, 1870   SE China, N Indochina, Hainan I.

    — *M.u.unchall* (Wagler, 1827)   Malay Pen., Sumatra and Java to Flores I.

○ *Macropygia amboinensis* BROWN CUCKOO-DOVE[7]

    — *M.a.sanghirensis* Salvadori, 1878   Sangihe Is., Talaud Is.

    — *M.a.atrata* Ripley, 1941   Togian I. (Sulawesi) (2024)

    — *M.a.albicapilla* Bonaparte, 1854[8]   Sulawesi, Banggai Is., Sula Is.

    — *M.a.albiceps* Bonaparte, 1856[9]   N Moluccas

    — *M.a.amboinensis* (Linnaeus, 1766)   S Moluccas

    — *M.a.keyensis* Salvadori, 1876   Kai Is.

    — *M.a.doreya* Bonaparte, 1854   NW New Guinea, W Papuan islands

    — *M.a.maforensis* Salvadori, 1878   Numfor I.

    — *M.a.griseinucha* Salvadori, 1876   Mios Num I.

    — *M.a.cinereiceps* Tristram, 1889[10]   Yapen I., New Guinea (except NW and perhaps NC) and D'Entrecasteaux Arch.

    — *M.a.balim* Rand, 1941   Montane NC New Guinea (1964)

    — *M.a.meeki* Rothschild & Hartert, 1915   Manam I.

    — *M.a.admiralitatis* Mayr, 1937   Admiralty Is. (1465)

    — *M.a.carteretia* Bonaparte, 1854   New Britain, New Ireland, Lihir Is.

    — *M.a.hueskeri* Neumann, 1922   New Hanover

    — *M.a.cunctata* E. Hartert, 1899   Louisiade Arch.

    — *M.a.quinkan* Schodde, 1989   Cape York Pen. (2177)

    — *M.a.robinsoni* Mathews, 1912   N and C E Queensland

    — *M.a.phasianella* (Temminck, 1821)   S and C E Australia

○ *Macropygia magna* DUSKY CUCKOO-DOVE

    — *M.m.macassariensis* Wallace, 1865   SW Sulawesi, Saleyar I., Tana Keke I.

    — *M.m.longa* Meise, 1930   Tanahjampea I., Kalaotoa I.

---

[1] The treatment in Baptista *et al.* (1997: 140) (94) is odd as it suggests that *suratensis* alone has been considered distinct and potentially a monotypic species. This is not our understanding. Vaurie (1965: 561 fn) (2510) suggested that a species *suratensis* might need recognition and that it would include *tigrina, ceylonensis, forresti* and *vacillans*. Here, drawing on Gibbs *et al.* (2001) (935), and lumping those that differ but slightly, we recognise the same three races with revised ranges, plus the Hainan form. The first two make up the *suratensis* group (*sensu* Vaurie). Gibbs (2001: 251) agreed with Baptista *et al.* (op. cit.) and considered that the *suratensis* group did not include *tigrina*. In the light of these contrasting opinions a detailed review dealing fully with distinctions and synonymy is needed.

[2] Includes *ceylonensis* and *edwardi* Ripley, 1948 (2031).

[3] Includes *forresti* "doubtfully distinct" in Peters (1937: 97) (1823) and *vacillans* see Gibbs (2001: 251).

[4] Includes *setzeri* Deignan, 1955 (655) see Cheng (1987: 294) (333) and *formosa* see Gibbs (2001: 251).

[5] Includes *divergens* Clancey, 1970 (423), see Irwin (1981) (1189).

[6] Includes *kirmanensis* Koelz, 1950 (1280), see Vaurie (1965: 560) (2510).

[7] Baptista *et al.* (1997:143) (94) treated *amboinensis* and *phasianella* as separate species, but see Schodde & Mason (1997) (2188). The many subspecies between the Moluccas and Australia still need review and may be oversplit.

[8] Here we tentatively include *sedecima* Neumann, 1939 (1645); we have retained *atrata* based on the comments of White & Bruce (1986) (2652), they also retained *sedecima* but with some doubt.

[9] For the use of this name, not *batchianensis*, see Mees (1982) (1533).

[10] Includes *goldiei* see Mayr (1941) (1472) and *kerstingi* see Mees (1982) (1534). The montane form *balim* may also belong herein.

— *M.m.magna* Wallace, 1864   E Lesser Sundas

— *M.m.timorlaoensis* A.B. Meyer, 1884   Tanimbar Is.

○ *Macropygia rufipennis* Blyth, 1846 ANDAMAN CUCKOO-DOVE[1]   Andaman and Nicobar Is.

○ *Macropygia tenuirostris* PHILIPPINE CUCKOO-DOVE[2]

— *M.t.phaea* McGregor, 1904[3]   Lanyu I. (off S Taiwan), Batan and Calayan (Philippines)

— *M.t.tenuirostris* Bonaparte, 1854   Philippines incl. Sulu Arch.

○ *Macropygia emiliana* RUDDY CUCKOO-DOVE[4]

— *M.e.hypopercna* Oberholser, 1912   Simeuluë I.

— *M.e.modiglianii* Salvadori, 1887   Nias I.

— *M.e.elassa* Oberholser, 1912   Mentawai Is., Pagi Is.

— *M.e.cinnamomea* Salvadori, 1892   Enggano I.

— *M.e.emiliana* Bonaparte, 1854   Java to W Lesser Sundas (Flores and Paloe I.)

— *M.e.megala* Siebers, 1929   Kangean Is.

— *M.e.borneensis* Robinson & Kloss, 1921   Borneo

○ *Macropygia nigrirostris* Salvadori, 1875 BLACK-BILLED CUCKOO-DOVE   New Guinea, Yapen and Karkar Is., Bismarck Arch., D'Entrecasteaux Arch.

○ *Macropygia mackinlayi* MACKINLAY'S CUCKOO-DOVE

— *M.m.goodsoni* E. Hartert, 1924   Admiralty and St Matthias Is., Squally, Watom and Witu Is., WC New Britain

— *M.m.krakari* Rothschild & Hartert, 1915   Karkar I.

— *M.m.arossi* Tristram, 1879   Solomon Is.

— *M.m.mackinlayi* E.P. Ramsay, 1878[5]   Santa Cruz Is., Banks Is., Vanuatu

○ *Macropygia ruficeps* LITTLE CUCKOO-DOVE

— *M.r.assimilis* Hume 1874   S Burma and SC China to NW, NE and W Thailand

— *M.r.engelbachi* Delacour, 1928   NW Vietnam, N Laos

— *M.r.malayana* Chasen & Kloss, 1931   Mts. of W Malaysia

— *M.r.simalurensis* Richmond, 1902   Simeuluë I.

— *M.r.sumatrana* Robinson & Kloss, 1919   Sumatra

— *M.r.nana* Stresemann, 1913   Borneo

— *M.r.ruficeps* (Temminck, 1835[6])   Java, Bali I.

— *M.r.orientalis* E. Hartert, 1896   Lesser Sundas (Lombok to Timor)

REINWARDTOENA Bonaparte, 1854 F

○ *Reinwardtoena reinwardtii* GREAT CUCKOO-DOVE

— *R.r.reinwardtii* (Temminck, 1824)[7]   Moluccas

— *R.r.griseotincta* E. Hartert, 1896   New Guinea and islands from Misool and Waigeo (except Biak) round N coast to D'Entrecasteaux Arch.

— *R.r.brevis* J.L. Peters, 1937   Biak I. (New Guinea)

○ *Reinwardtoena browni* (P.L. Sclater, 1877) PIED CUCKOO-DOVE[8]   Admiralty Is., Bismarck Arch.

○ *Reinwardtoena crassirostris* (Gould, 1856) CRESTED CUCKOO-DOVE   Solomon Is.

TURACOENA Bonaparte, 1854 F

○ *Turacoena manadensis* (Quoy & Gaimard, 1830) WHITE-FACED CUCKOO-DOVE

Sulawesi, Togian, Banggai and Sula Is.

○ *Turacoena modesta* (Temminck, 1835) BLACK CUCKOO-DOVE   Timor and Wetar Is. (Lesser Sundas)

---

[1] Tentatively includes *andamanica* Abdulali, 1967 (8), and *tiwarii* Abdulali, 1979 (13), but comparative material may only be available in India. See also Ripley & Beehler (1989) (2061A).

[2] Specific limits in this genus are still relatively obscure, see Kennedy *et al.* (2000: 147) (1259). The separation of this species and of *emiliana* (see Sibley & Monroe, 1990) (2262) from *phasianella* is less comfortable in light of the apparent conspecificity of *amboinensis* and *phasianella*. See also Mayr (1944: 148, 191) (1474).

[3] Tentatively includes *septentrionalis*; this was recognised in Dickinson *et al.* (1991) (745) but the promised review of northern forms has not appeared.

[4] Forms recognised largely follow van Marle & Voous (1988) (2468).

[5] Presumably including *troughtoni* Kinghorn, 1937 (1269).

[6] Not 1834 as in Peters (1937: 78) (1823) see Dickinson (2001) (739).

[7] The emended name used here is judged to be in prevailing usage; see Art. 33.2 of the Code (I.C.Z.N., 1999) (1178) and p. 000 herein.

[8] Presumably includes *solitaria* Salomonsen, 1972 (2160), but see Greenway (1978: 55) (1010).

TURTUR Boddaert, 1783 M

○ *Turtur chalcospilos* (Wagler, 1827) EMERALD-SPOTTED WOOD DOVE   Coastal Gabon to SW Angola. Ethiopia, S Somalia and Kenya to N Namibia and E Cape Province

○ *Turtur abyssinicus* (Sharpe, 1902) BLACK-BILLED WOOD DOVE   Senegal, Gambia and S Mauritania to Eritrea

○ *Turtur afer* (Linnaeus, 1766) BLUE-SPOTTED WOOD DOVE[1]   Senegal and Gambia to Ethiopia, south to Angola and N Transvaal

○ *Turtur tympanistria* (Temminck, 1810) TAMBOURINE DOVE   S Senegal to S Nigeria, Zaire and Ethiopia, south to N Angola and E and S Sth. Africa, Bioko I., Comoro Is.

○ *Turtur brehmeri* BLUE-HEADED WOOD DOVE

    ___ *T.b.infelix* J.L. Peters, 1937   Coastal Guinea and Sierra Leone to coastal Cameroon

    ___ *T.b.brehmeri* (Hartlaub, 1865)   S Cameroon to E Zaire and Cabinda

OENA Swainson, 1837 F

○ *Oena capensis* NAMAQUA DOVE

    ___ *O.c.capensis* (Linnaeus, 1766)   Sub-Saharan Africa, Arabia, Socotra I., S Israel

    ___ *O.c.aliena* Bangs, 1918   Madagascar

CHALCOPHAPS Gould, 1843 F

○ *Chalcophaps indica* EMERALD DOVE

    ___ *C.i.indica* (Linnaeus, 1758)[2]   India, S China, Taiwan, the Yaeyama Is. (Japan) south through SE Asia and the Philippines to Gtr. and W Lesser Sundas, Sulawesi, Moluccas and W Papuan Is.

    ___ *C.i.robinsoni* E.C.S. Baker, 1928   Sri Lanka
    ___ *C.i.maxima* E. Hartert, 1931   Andaman Is.
    ___ *C.i.augusta* Bonaparte, 1855   Nicobar Is.
    ___ *C.i.natalis* Lister, 1889   Christmas I. (Indian Ocean)
    ___ *C.i.timorensis* Bonaparte, 1856   E Lesser Sundas
    ___ *C.i.minima* E. Hartert, 1931   Numfoor, Biak and Mios Num Is., Irian Jaya
    ___ *C.i.longirostris* Gould, 1848[3]   N Northern Territory, N Western Australia
    ___ *C.i.rogersi* Mathews, 1912   E Australia, Lord Howe I., Norfolk I., New Guinea (incl. eastern arch.)
    ___ *C.i.sandwichensis* E.P. Ramsay, 1878   Santa Cruz Is., Banks Is., Vanuatu, New Caledonia

○ *Chalcophaps stephani* STEPHAN'S DOVE

    ___ *C.s.wallacei* Brüggemann, 1877   Sulawesi, Sula Is.
    ___ *C.s.stephani* Pucheran, 1853   Kai, Aru and W Papuan Is., New Guinea to Bismarck Arch. and D'Entrecasteaux Arch.
    ___ *C.s.mortoni* E.P. Ramsay, 1882   Solomon Is.

HENICOPHAPS G.R. Gray, 1862 F

○ *Henicophaps albifrons* NEW GUINEA BRONZEWING

    ___ *H.a.albifrons* G.R. Gray, 1862   W Papuan Is., New Guinea, Yapen I.
    ___ *H.a.schlegeli* (Rosenberg, 1866)   Aru Is.

○ *Henicophaps foersteri* Rothschild & Hartert, 1906 NEW BRITAIN BRONZEWING   S Bismarck Arch.

PHAPS Selby, 1835 F

○ *Phaps chalcoptera* (Latham, 1790) COMMON BRONZEWING[4]   Australia, Tasmania

○ *Phaps elegans* BRUSH BRONZEWING

    ___ *P.e.occidentalis* Schodde, 1989   SW Western Australia (2177)
    ___ *P.e.elegans* (Temminck, 1809)[5]   SE Queensland to SC South Australia, Tasmania

○ *Phaps histrionica* (Gould, 1841) HARLEQUIN BRONZEWING/FLOCK BRONZEWING[6]   Inland N, E and C Australia

---

[1] The proposed race *liberiensis* Rand, 1949 (1968), was not mentioned by Baptista *et al.* (1997: 149) (94) and may need review.
[2] Includes *formosanus* see Cheng (1987: 298) (333); *yamashinai* Hachisuka, 1939 (1037) see Vaurie (1965: 563), but recognised by Orn. Soc. Japan (2000) (1727); *nana* Neumann, 1941 (1647), see White & Bruce (1986) (2652), and *salimalii* Mukherjee, 1960 (1632) see Ripley (1982: 157) (2055).
[3] Includes *melvillensis*. Regarding *chrysochlora* see Schodde & Mason (1997: 26) (2188).
[4] For treatment as monotypic see Schodde & Mason (1997) (2188).
[5] Includes *neglecta* see Schodde & Mason (1997) (2188). For the correct date see Stresemann (1953) (2367).
[6] Treated as monotypic following Schodde & Mason (1997) (2188).

OCYPHAPS G.R. Gray, 1842 F
● *Ocyphaps lophotes* Crested Pigeon
    — *O.l.whitlocki* Mathews, 1912   NW Australia
    ⤳ *O.l.lophotes* (Temminck, 1822)   NE, C, S and E Australia

GEOPHAPS G.R. Gray, 1842[1] F
○ *Geophaps plumifera* Spinifex Pigeon[2]
    — *G.p.ferruginea* (Gould, 1865)   WC Western Australia
    — *G.p.plumifera* Gould, 1842[3]   N Western Australia, W Northern Territory
    — *G.p.leucogaster* (Gould, 1867)   C to inland NE Australia

○ *Geophaps scripta* Squatter Pigeon
    — *G.s.peninsulae* H.L. White, 1922   NE Queensland
    — *G.s.scripta* (Temminck, 1821)   C and S Queensland, NC New South Wales

○ *Geophaps smithii* Partridge-Pigeon[4]
    — *G.s.blaauwi* Mathews, 1912   Kimberley region (NE Western Australia)
    — *G.s.smithii* (Jardine & Selby, 1830)   NE Western Australia, N Northern Territory

TRUGON G.R. Gray, 1849[5] F
○ *Trugon terrestris* Thick-billed Ground Pigeon
    — *T.t.terrestris* G.R. Gray, 1849   NW New Guinea, Salawati I.
    — *T.t.mayri* Rothschild, 1931   NC New Guinea
    — *T.t.leucopareia* (A.B. Meyer, 1886)   S New Guinea

LEUCOSARCIA Gould, 1843 F
○ *Leucosarcia melanoleuca* (Latham, 1802)[6] Wonga Pigeon   EC Queensland to SC Victoria

PETROPHASSA Gould, 1841 F
○ *Petrophassa rufipennis* Collett, 1898 Chestnut-quilled Rock Pigeon   W Arnhem Land (Australia)

○ *Petrophassa albipennis* White-quilled Rock Pigeon
    — *P.a.albipennis* Gould, 1841   N and NE Western Australia, N NW Northern Territory
    — *P.a.boothi* Goodwin, 1969   Central NW Northern Territory (961)

GEOPELIA Swainson, 1837 F
○ *Geopelia cuneata* (Latham, 1802)[7] Diamond Dove   Arid interior of Australia

● *Geopelia striata* (Linnaeus, 1766) Zebra Dove[8]   S Burma to Sumatra, Borneo, Philippines, Java and
                                          Lombok I.

○ *Geopelia placida* Peaceful Dove
    — *G.p.papua* Rand, 1938   S New Guinea (1962)
    — *G.p.placida* Gould, 1844[9]   N and E Australia
    — *G.p.clelandi* Mathews, 1912   WC Western Australia

○ *Geopelia maugei* (Temminck, 1809)[10] Barred Dove   Sumbawa to Tanimbar and Kai Is.

○ *Geopelia humeralis* Bar-shouldered Dove[11]
    — *G.h.gregalis* Bangs & Peters, 1926   S New Guinea
    — *G.h.headlandi* Mathews, 1913   NC Western Australia
    — *G.h.inexpectata* Mathews, 1912   N Australia, Cape York Pen.
    — *G.h.humeralis* (Temminck, 1821)   E Australia

---

[1] Lumped in the genus *Petrophassa* in our last Edition, but see Frith (1982) (887) and Schodde & Mason (1997) (2188).
[2] For inclusion of *ferruginea* in this species see Crome *et al.* (1980) (597) and Schodde & Mason (1997) (2188). The name *mungi* was given to an intergrading population.
[3] Includes *proxima* Mayr, 1951 (1476), see Schodde & Mason (1997) (2188).
[4] Treated as monotypic in last Edition, but see Johnstone (1981) (1218) and Schodde & Mason (1997) (2188).
[5] This genus and the next best placed close to *Geophaps* (R. Schodde *in litt.*).
[6] For reasons to use 1802 not 1801 see Browning & Monroe (1991) (258).
[7] For reasons to use 1802 not 1801 see Browning & Monroe (1991) (258).
[8] For information on the revision of this and the next species see Schodde & Mason (1997) (2188).
[9] Including *tranquilla*.
[10] For the date of this name see Stresemann (1953) (2367). The emended name used here is judged to be in prevailing usage; see Art. 33.2 of the Code (I.C.Z.N., 1999) (1178) and p. 000 herein. Treated as monotypic, see White & Bruce (1986: 192) (2652).
[11] For revision see Schodde & Mason (1997) (2188).

ECTOPISTES Swainson, 1827 M
○ †*Ectopistes migratorius* (Linnaeus, 1766) Passenger Pigeon   Nthn. N America >> Sthn. N America

ZENAIDA Bonaparte, 1838 F
◉ *Zenaida macroura* Mourning Dove
     ✓ *Z.m.marginella* (Woodhouse, 1852)[1]   W Nth. America to SC Mexico
     ✓ *Z.m.carolinensis* (Linnaeus, 1766)   E Nth. America, Bahama Is., Bermuda
     __ *Z.m.macroura* (Linnaeus, 1758)   Cuba, Isla de la Juventud, Hispaniola, Puerto Rico, Jamaica
     __ *Z.m.clarionensis* (C.H. Townsend, 1890)   Clarion I.
     __ *Z.m.turturilla* Wetmore, 1956   Costa Rica, W Panama (2597)

○ †*Zenaida graysoni* (Lawrence, 1871) Socorro Dove[2]   Socorro I. (W Mexico)

◉ *Zenaida auriculata* Eared Dove[3]
     __ *Z.a.caucae* Chapman, 1922   W Colombia
     __ *Z.a.antioquiae* Chapman, 1917   NC Colombia
     __ *Z.a.ruficauda* Bonaparte, 1855   E Colombia, W Venezuela
     __ *Z.a.vinaceorufa* Ridgway, 1884   Curaçao, Aruba and Bonaire Is.
     __ *Z.a.rubripes* Lawrence, 1885[4]   Lesser Antilles, Trinidad, C Colombia through S Venezuela to Guyana and N Brazil
     __ *Z.a.jessieae* Ridgway, 1888   Lower Amazon (Brazil)
     __ *Z.a.marajoensis* Berlepsch, 1913   River Amazon estuary
     __ *Z.a.noronha* Sharpe, 1890   NE Brazil and Fernando de Noronha Is.
     __ *Z.a.hypoleuca* Bonaparte, 1855   W Ecuador, Peru
     __ *Z.a.virgata* Bertoni, 1901[5]   Bolivia, Uruguay, NW Argentina, C Brazil
     ✓ *Z.a.auriculata* (Des Murs, 1847)   C Chile, WC Argentina

○ *Zenaida aurita* Zenaida Dove
     __ *Z.a.salvadorii* Ridgway, 1916   N Yucatán Pen., Cozumel, Holbox and Mujeres Is.
     __ *Z.a.zenaida* (Bonaparte, 1825)   Bahama, Cuba, Isla de la Juventud, Cayman Is., Jamaica and Puerto Rico to Virgin Is.
     __ *Z.a.aurita* (Temminck, 1810)   Lesser Antilles

○ *Zenaida galapagoensis* Galapagos Dove
     __ *Z.g.galapagoensis* Gould, 1839   Galapagos Is.
     __ *Z.g.exsul* (Rothschild & Hartert, 1899)   Culpepper and Wenman Is. (N Galapagos Is.)

◉ *Zenaida asiatica* White-winged Dove[6]
     __ *Z.a.mearnsi* (Ridgway, 1915)   SW USA, W Mexico, Tres Marias Is.
     ✓ *Z.a.asiatica* (Linnaeus, 1758)[7]   S USA to Nicaragua, S Bahamas to Puerto Rico and Cayman Is.
     ✓ *Z.a.australis* (J.L. Peters, 1913)   W Costa Rica, W Panama

○ *Zenaida meloda* (Tschudi, 1843) West Peruvian Dove[8]   SW Ecuador to N Chile

COLUMBINA Spix, 1825 F
◉ *Columbina inca* (Lesson, 1847) Inca Dove[9]   SW and SC USA to NW Costa Rica

○ *Columbina squammata* Scaled Dove
     __ *C.s.ridgwayi* (Richmond, 1896)   Coastal Colombia to French Guiana
     __ *C.s.squammata* (Lesson, 1831)   C and E Brazil to Paraguay and NE Argentina

◉ *Columbina passerina* Common Ground Dove
     ✓ *C.p.passerina* (Linnaeus, 1758)   South Carolina to SE Texas

[1] Includes *tresmariae* (recognised in last Edition) see Gibbs *et al.* (2001) (935).
[2] Not included in last Edition due to presumed extinction. A captive breeding population survives (N. Bahr).
[3] We follow Baptista *et al.* (1997: 159) (94); *stenura* and *pentheria* (recognised in our last Edition) are therein submerged.
[4] Includes *vulcania* Meyer de Schauensee, 1944 (1561), of last Edition. See Gibbs *et al.* (2001) (935).
[5] Includes *chrysauchenia* of last Edition, see Gibbs *et al.* (2001) (935).
[6] Numerous forms (*collina, grandis, insularis, monticola, palustris, panama* and *peninsulae*) proposed by Saunders (1968) (2169) and *clara* van Rossem, 1947 (2480), are submerged in the simplified arrangement of Baptista *et al.* (1997) (94), who also treated *meloda* as a separate species.
[7] Includes *alticola* Saunders, 1951 (2168).
[8] The English vernacular name Pacific Dove is awkward as it implies an insular, not coastal, provenance.
[9] This and the next species treated in the genus *Scardafella* in last Edition; we follow A.O.U. (1998: 225-226) (3) in placing them in *Columbina*.

_ *C.p.pallescens* (S.F. Baird, 1860) — SW USA to Guatemala and Belize
_ *C.p.socorroensis* (Ridgway, 1887) — Socorro I. (W Mexico)
↙ *C.p.neglecta* (Carriker, 1910) — Honduras to Panama
_ *C.p.bahamensis* (Maynard, 1887) — Bahamas, Bermuda
_ *C.p.exigua* (Riley, 1905)[1] — Great Inagua and Mona Is.
_ *C.p.insularis* (Ridgway, 1888)[2] — Cuba, Isla de la Juventud, Cayman Is., Hispaniola and nearby islands
_ *C.p.jamaicensis* (Maynard, 1899) — Jamaica
_ *C.p.navassae* (Wetmore, 1930) — Navassa I. (SW Hispaniola)
_ *C.p.portoricensis* (P.R. Lowe, 1908) — Puerto Rico, Virgin Is., Culebras and Vieques Is.
_ *C.p.nigrirostris* (Danforth, 1935) — St Croix I., N Lesser Antilles
_ *C.p.trochila* (Bonaparte, 1855) — Martinique
_ *C.p.antillarum* (P.R. Lowe, 1908) — S Lesser Antilles
_ *C.p.albivitta* (Bonaparte, 1855)[3] — N Colombia, N Venezuela, Aruba I. to Trinidad
_ *C.p.parvula* (Todd, 1913) — C Colombia
_ *C.p.nana* (Todd, 1913) — W Colombia
_ *C.p.quitensis* (Todd, 1913) — C Ecuador
_ *C.p.griseola* (Spix, 1825) — S Venezuela, the Guianas, N and NE Brazil

○ *Columbina minuta* PLAIN-BREASTED GROUND DOVE
_ *C.m.interrupta* (Griscom, 1929) — S Mexico, S Guatemala, Belize to Nicaragua
_ *C.m.elaeodes* (Todd, 1913) — Costa Rica to WC Colombia
_ *C.m.minuta* (Linnaeus, 1766) — Venezuela to the Guianas and NE Argentina
_ *C.m.amazilia* (Bonaparte, 1855)[4] — SW Ecuador, W Peru

○ *Columbina buckleyi* ECUADORIAN GROUND DOVE
_ *C.b.buckleyi* (Sclater & Salvin, 1877) — NW Ecuador to NW Peru
_ *C.b.dorsti* (Koepcke, 1962)[5] — NW Peru (Marañon Valley) (1289)

◐ *Columbina talpacoti* RUDDY GROUND DOVE
_ *C.t.eluta* (Bangs, 1901) — W Mexico
_ *C.t.rufipennis* (Bonaparte, 1855) — EC Mexico to N Colombia, N Venezuela, Trinidad, Tobago
_ *C.t.caucae* (Chapman, 1915) — W Colombia
_ *C.t.talpacoti* (Temminck, 1811) — E Ecuador E to the Guianas and S to N Argentina

◐ *Columbina picui* PICUI GROUND DOVE
_ *C.p.strepitans* Spix, 1825 — NE Brazil
↙ *C.p.picui* (Temminck, 1813) — Bolivia and S Brazil to C Chile and S Argentina

○ *Columbina cruziana* Prévost, 1842 CROAKING GROUND DOVE — N Ecuador to N Chile

○ *Columbina cyanopis* (Pelzeln, 1870) BLUE-EYED GROUND DOVE — SC Brazil

CLARAVIS Oberholser, 1899 F
◐ *Claravis pretiosa* (Ferrari-Pérez, 1886) BLUE GROUND DOVE — E Mexico to Paraguay and N Argentina

○ *Claravis godefrida* (Temminck, 1811) PURPLE-WINGED GROUND DOVE
— SE Brazil, E Paraguay, NE Argentina

○ *Claravis mondetoura* MAROON-CHESTED GROUND DOVE[6]
_ *C.m.ochoterena* van Rossem, 1934 — SE Mexico
_ *C.m.salvini* Griscom, 1930 — Guatemala, El Salvador, Honduras
_ *C.m.umbrina* Griscom, 1930 — Costa Rica
_ *C.m.pulchra* Griscom, 1930 — W Panama
_ *C.m.mondetoura* (Bonaparte, 1856) — N and NW Venezuela, Colombia, E Ecuador
_ *C.m.inca* van Rossem, 1934 — Peru, WC Bolivia

[1] Presumably includes ¶*volitans* Bond, 1945 (182).
[2] Presumably includes *umbrina* Buden, 1985 (265).
[3] Presumably includes *tortugensis* Fernandez-Yepez, 1945 (832).
[4] Omitted from last Edition.
[5] For recognition see Gibbs (2001) (935).
[6] Baptista *et al.* (1997) (94) followed Peters (1937) (1823); we tentatively follow, having treated this as monotypic in last Edition.

METRIOPELIA Bonaparte, 1855 F
○ *Metriopelia ceciliae* Bare-faced Ground Dove
    __ *M.c.ceciliae* (Lesson, 1845)   W Peru
    __ *M.c.obsoleta* (J.T. Zimmer, 1924)   E Peru
    __ *M.c.zimmeri* J.L. Peters, 1937[1]   S Peru to N Chile

○ *Metriopelia morenoi* (Sharpe, 1902) Moreno's Ground Dove   NW Argentina

○ *Metriopelia melanoptera* Black-winged Ground Dove
    __ *M.m.saturatior* Chubb, 1917   SW Colombia, Ecuador
    __ *M.m.melanoptera* (Molina, 1782)   Peru to Chile and W Argentina

○ *Metriopelia aymara* (Prévost, 1840) Golden-spotted Ground Dove   S Peru to N Chile and NW Argentina

UROPELIA Bonaparte, 1855 F
○ *Uropelia campestris* Long-tailed Ground Dove[2]
    __ *U.c.campestris* (Spix, 1825)   NC and C Brazil
    __ *U.c.figginsi* Oberholser, 1931   WC Brazil, N and E Bolivia

LEPTOTILA Swainson, 1837 F
● *Leptotila verreauxi* White-tipped Dove
    __ *L.v.capitalis* Nelson, 1898   Tres Marias Is.
    __ *L.v.angelica* Bangs & Penard, 1922[3]   S Texas, N and C Mexico
    __ *L.v.fulviventris* (Lawrence, 1882)   SE Mexico, NE Guatemala, Belize
    __ *L.v.bangsi* Dickey & van Rossem, 1926   W Guatemala to W Nicaragua and W Honduras
    __ *L.v.nuttingi* Ridgway, 1915   Lake Nicaragua
    ↙ *L.v.verreauxi* (Bonaparte, 1855)[4]   SW Nicaragua and Costa Rica to N Venezuela and the
                                               Dutch Antilles
    __ *L.v.zapluta* J.L. Peters, 1937   Trinidad
    __ *L.v.tobagensis* Hellmayr & Seilern, 1915   Tobago I.
    __ ¶*L.v.hernandezi* Romero & Morales, 1981   SW Colombia (2101)
    __ *L.v.decolor* (Salvin, 1895)   W Colombia to W and N Peru
    __ *L.v.brasiliensis* (Bonaparte, 1856)   The Guianas, N Brazil
    __ *L.v.approximans* (Cory, 1917)   NE Brazil
    __ *L.v.decipiens* (Salvadori, 1871)   E Peru, E Bolivia, W Brazil
    __ *L.v.chalcauchenia* (Sclater & Salvin, 1870)   S Bolivia to Uruguay and NC Argentina

○ *Leptotila megalura* (Sclater & Salvin, 1879) Large-tailed Dove[5]   N, E and SE Bolivia, NW Argentina

○ *Leptotila rufaxilla* Grey-fronted Dove
    __ *L.r.pallidipectus* Chapman, 1915   E Colombia, W Venezuela
    __ *L.r.dubusi* (Bonaparte, 1855)   SE Colombia to E Peru and SC Venezuela
    __ *L.r.rufaxilla* (Richard & Bernard, 1792)[6]   E Venezuela, the Guianas, N Brazil
    __ *L.r.hellmayri* Chapman, 1915   NE Venezuela, Trinidad
    __ *L.r.bahiae* (Berlepsch, 1885)   C Brazil
    __ *L.r.reichenbachii* (Pelzeln, 1870)   C and S Brazil to Paraguay, NE Argentina and Uruguay

○ *Leptotila plumbeiceps* Grey-headed Dove[7]
    __ *L.p.plumbeiceps* (Sclater & Salvin, 1868)   E Mexico to W Costa Rica and W Colombia
    __ *L.p.notia* J.L. Peters, 1931[8]   W Panama

○ *Leptotila pallida* (Berlepsch & Taczanowski, 1884) Pallid Dove   W Colombia, W Ecuador

○ *Leptotila battyi* Brown-backed Dove
    __ *L.b.malae* Griscom, 1927   S Veraguas, W Herrera and Cebaco I (W Panama)
    __ *L.b.battyi* (Rothschild, 1901)   Coiba I. (SC Panama)

---

[1] We used the name *gymnops* in the last Edition for reasons not now clear.
[2] Treated as monotypic in last Edition, but see Baptista *et al.* (1997) (94).
[3] Presumably includes *santiago* van Rossem & Hachisuka, 1937 (2482).
[4] Includes *insularis* of Margarita I. not Trinidad (see last Edition). Baptista *et al.* (1997) (94) listed *riottei* although doubtfully distinct (Peters, 1937) (1823); as in last Edition we prefer to place it herein.
[5] Treated as monotypic following Baptista *et al.* (1997) (94).
[6] Presumably includes *hypochroos* Griscom & Greenway, 1937 (1017).
[7] In last Edition we treated *pallida* as one species and *plumbeiceps* including *battyi* as another. We here separate *battyi* and sequence these species as we do tentatively following Baptista *et al.* (1997) (94).
[8] For correction of spelling see David & Gosselin (2002) (613).

○ *Leptotila wellsi* (Lawrence, 1884) GRENADA DOVE   Grenada I.

○ *Leptotila jamaicensis* CARIBBEAN DOVE
    — *L.j.gaumeri* (Lawrence, 1885)   N Yucatán Pen. and islands, NE Belize, Honduran Is.
    — *L.j.collaris* (Cory, 1886)   Cayman Is.
    — *L.j.jamaicensis* (Linnaeus, 1766)   Jamaica
    — *L.j.neoxena* (Cory, 1887)   San Andrés I. (off EC Nicaragua)

○ *Leptotila cassini* GREY-CHESTED DOVE
    — *L.c.cerviniventris* (Sclater & Salvin, 1868)   E Guatemala to W Panama
    — *L.c.rufinucha* (Sclater & Salvin, 1873)   SW Costa Rica, NW Panama
    — *L.c.cassini* (Lawrence, 1867)   E Panama, N Colombia

○ *Leptotila ochraceiventris* Chapman, 1914 OCHRE-BELLIED DOVE   SW Ecuador, NW Peru

○ *Leptotila conoveri* Bond & Meyer de Schauensee, 1943 TOLIMA DOVE C Colombia (192)

GEOTRYGON Gosse, 1847 F
○ *Geotrygon lawrencii* Salvin, 1874 PURPLISH-BACKED QUAIL-DOVE   Costa Rica, Panama

○ *Geotrygon carrikeri* (Wetmore, 1941) TUXTLA QUAIL-DOVE[1]   SE Vera Cruz (SE Mexico) (2590)

○ *Geotrygon costaricensis* Lawrence, 1868 BUFF-FRONTED QUAIL-DOVE   Costa Rica, W Panama

○ *Geotrygon goldmani* RUSSET-CROWNED QUAIL-DOVE
    — *G.g.oreas* (Wetmore, 1950)   Cerro Chucantí (E Panama) (2593)
    — *G.g.goldmani* Nelson, 1912   E Panama, NW Colombia

○ *Geotrygon saphirina* SAPPHIRE QUAIL-DOVE
    — *G.s.purpurata* (Salvin, 1878)   NW Colombia to NW Ecuador
    — *G.s.saphirina* Bonaparte, 1855   E Ecuador to SE Peru and W Amazonian Brazil, N Bolivia
    — *G.s.rothschildi* (Stolzmann, 1926)   Marcapata valley (SE Peru)

○ *Geotrygon veraguensis* (Lawrence, 1867) OLIVE-BACKED QUAIL-DOVE   E Costa Rica to NW Ecuador

○ *Geotrygon caniceps* GREY-HEADED QUAIL-DOVE
    — *G.c.caniceps* (Gundlach, 1852)   Cuba
    — *G.c.leucometopia* (Chapman, 1917)[2]   Dominican Republic

○ *Geotrygon versicolor* (Lafresnaye, 1846) CRESTED QUAIL-DOVE   Jamaica

○ *Geotrygon albifacies* P.L. Sclater, 1858 WHITE-FACED QUAIL-DOVE[3]   SE Mexico to NW Nicaragua

● *Geotrygon chiriquensis* P.L. Sclater, 1856 RUFOUS-BREASTED QUAIL-DOVE
                    Costa Rica, W Panama

○ *Geotrygon linearis* (Prévost, 1843) LINED QUAIL-DOVE[4]   NE and C Colombia to N and W Venezuela, Trinidad,
                    Tobago I.

○ *Geotrygon frenata* WHITE-THROATED QUAIL-DOVE[5]
    — *G.f.bourcieri* Bonaparte, 1855   W Colombia, W Ecuador
    — *G.f.erythropareia* Salvadori, 1893   E Ecuador
    — *G.f.frenata* (Tschudi, 1843)[6]   N Peru to C Bolivia
    — *G.f.margaritae* Blake, Hoy & Cantino, 1961   S Bolivia to NW Argentina (167)

○ *Geotrygon chrysia* Bonaparte, 1855 KEY WEST QUAIL-DOVE   Bahama Is., Cuba, Isla de la Juventud, Hispaniola,
                    Puerto Rico

○ *Geotrygon mystacea* (Temminck, 1811) BRIDLED QUAIL-DOVE[7]   Puerto Rico, Virgin Is., Lesser Antilles to St Lucia

○ *Geotrygon violacea* VIOLACEOUS QUAIL-DOVE
    — *G.v.albiventer* Lawrence, 1865   Nicaragua to NE Colombia and W, N and SC Venezuela
    — *G.v.violacea* (Temminck, 1810)   Surinam to Paraguay and NE Argentina

---

[1] For separation of this see Peterson (1993) (1834).
[2] For correction of spelling see David & Gosselin (2002) (614).
[3] Includes *anthonyi, rubida* and *silvestris* from our last Edition.
[4] Includes *infusca* and *trinitatis* from last Edition; the treatment reverts to that of Peters (1937) (1823) following Baptista *et al.* (1997) (94), but *veraguensis*
seems unrelated and we place it separately like A.O.U. (1998) (3).
[5] See Baptista *et al.* (1997) (94) on morphological problems affecting subspecific evaluation.
[6] Presumably includes *loricata* Todd, 1947 (2410).
[7] Includes *beattyi* Danforth, 1938 (610).

○ **Geotrygon montana** RUDDY QUAIL-DOVE
    — *G.m.martinica* (Linnaeus, 1766)   Lesser Antilles Is.
    — *G.m.montana* (Linnaeus, 1758)   Mexico to NE Argentina, Greater Antilles, Trinidad

STARNOENAS Bonaparte, 1838 F
○ **Starnoenas cyanocephala** (Linnaeus, 1758) BLUE-HEADED QUAIL-DOVE
                                       Cuba, Isla de la Juventud

CALOENAS G.R. Gray, 1840 F
○ **Caloenas nicobarica** NICOBAR PIGEON
    — *C.n.nicobarica* (Linnaeus, 1758)   Andaman and Nicobar Is. along coasts and on islets
                                       from the Malay Pen. through the Sundas and
                                       Wallacea to islands off New Guinea and Solomon Is.
    — *C.n.pelewensis* Finsch, 1875   Palau Is.

GALLICOLUMBA Heck, 1849 F
○ **Gallicolumba luzonica** LUZON BLEEDING-HEART
    — *G.l.griseolateralis* Parkes, 1962   N Luzon (1755)
    — *G.l.luzonica* (Scopoli, 1786)   C and S Luzon
    — *G.l.rubiventris* Gonzales, 1983   Catanduanes (953)

○ **Gallicolumba crinigera** MINDANAO BLEEDING-HEART[1]
    — *G.c.leytensis* (E. Hartert, 1918)   Samar, Leyte, Bohol
    — *G.c.crinigera* (Pucheran, 1853)   Mindanao, Dinagat
    — *G.c.bartletti* (P.L. Sclater, 1863)   Basilan (2217)

○ **Gallicolumba platenae** (Salvadori, 1893) MINDORO BLEEDING-HEART   Mindoro

○ **Gallicolumba keayi** (W.E. Clarke, 1900) NEGROS BLEEDING-HEART   Negros and Panay

○ **Gallicolumba menagei** (Bourns & Worcester, 1894) SULU BLEEDING-HEART
                                       Tawitawi (S Sulu Is.)

○ **Gallicolumba rufigula** CINNAMON GROUND DOVE
    — *G.r.rufigula* (Pucheran, 1853)   W Papuan islands and Vogelkop
    — *G.r.helviventris* (Rosenberg, 1867)   Aru Is.
    — *G.r.septentrionalis* (Rand, 1941)   N New Guinea east to Huon Gulf (1964)
    — *G.r.orientalis* (Rand, 1941)[2]   SE New Guinea (1964)
    — *G.r.alaris* (Rand, 1941)   S New Guinea

○ **Gallicolumba tristigmata** SULAWESI GROUND DOVE
    — *G.t.tristigmata* (Bonaparte, 1855)   N and NC Sulawesi
    — *G.t.auripectus* Stresemann, 1941[3]   SC and SE Sulawesi (2364)
    — *G.t.bimaculata* (Salvadori, 1892)   S Sulawesi

○ **Gallicolumba hoedtii** (Schlegel, 1871) WETAR GROUND DOVE   Wetar and Timor Is.

○ **Gallicolumba jobiensis** WHITE-BIBBED GROUND DOVE
    — *G.j.jobiensis* (A.B. Meyer, 1875)   New Guinea, Bismarck Arch., D'Entrecasteaux Arch.
    — *G.j.chalconota* Mayr, 1935   Vella Lavella and Guadalcanal Is. (C and S Solomon Is.)

○ **Gallicolumba kubaryi** (Finsch, 1880) CAROLINES GROUND DOVE   Chuuk and Pohnpei (C and E Caroline Is.)

○ †?**Gallicolumba erythroptera** (J.F. Gmelin, 1789) POLYNESIAN GROUND DOVE
                                       Tuamotu Arch.

○ **Gallicolumba xanthonura** (Temminck, 1823) WHITE-THROATED GROUND DOVE
                                       Mariana Is., Yap I.

○ **Gallicolumba stairi** (G.R. Gray, 1856) SHY GROUND DOVE[4]   Fiji, Samoa, Tonga, Wallis and Futuna Is.

○ **Gallicolumba sanctaecrucis** Mayr, 1935 SANTA CRUZ GROUND DOVE   Santa Cruz Is.

○ †?**Gallicolumba salamonis** (E.P. Ramsay, 1882) THICK-BILLED GROUND DOVE
                                       San Cristobal and Ramos Is. (SE Solomon Is.)

---

[1] For correction of spelling see David & Gosselin (2002) (613).
[2] An intergrade between *septentrionalis* and *alaris*. Perhaps better submerged.
[3] This form requires confirmation, see White & Bruce (1986: 193) (2652).
[4] Includes *vitiensis* see Watling (1982) (2562). For use of this vernacular name see Pratt (1987: 194) (1920).

○ *Gallicolumba rubescens* (Vieillot, 1818) MARQUESAN GROUND DOVE   Fatu Huku and Hatuta Is. (Marquesas Is.)

○ *Gallicolumba beccarii* BRONZE GROUND DOVE
  — *G.b.beccarii* (Salvadori, 1876)   Montane New Guinea
  — *G.b.johannae* (P.L. Sclater, 1877)[1]   Karkar I., Bismarck Arch., Nissan I.
  — *G.b.eichhorni* E. Hartert, 1924   St. Matthias Group
  — *G.b.admiralitatis* (Rothschild & Hartert, 1914) Admiralty Is.
  — *G.b.intermedia* (Rothschild & Hartert, 1905) W Solomon Is.
  — *G.b.solomonensis* (Ogilvie-Grant, 1888) E Solomon Is.

○ *Gallicolumba canifrons* (Hartlaub & Finsch, 1872) PALAU GROUND DOVE
                     Palau Is.

MICROGOURA Rothschild, 1904 F
○ †*Microgoura meeki* Rothschild, 1904 CHOISEUL PIGEON   Choiseul I. (NW Solomon Is.)

OTIDIPHABINAE

OTIDIPHAPS Gould, 1870 F
○ *Otidiphaps nobilis* PHEASANT PIGEON
  — *O.n.aruensis* Rothschild, 1928   Aru Is.
  — *O.n.nobilis* Gould, 1870   Batanta and Waigeo Is., W New Guinea
  — *O.n.cervicalis* E.P. Ramsay, 1880   E and SE New Guinea
  — *O.n.insularis* Salvin & Godman, 1883 Fergusson I.

GOURINAE

GOURA Stephens, 1819 F
○ *Goura cristata* WESTERN CROWNED PIGEON
  — *G.c.cristata* (Pallas, 1764)   NW New Guinea (Vogelkop), Salwati I.
  — *G.c.minor* Schlegel, 1864   W Papuan islands (except Salwati and Misool)
  — *G.c.pygmaea* Mees, 1965   Misool (1515)

○ *Goura scheepmakeri* SOUTHERN CROWNED PIGEON
  — *G.s.sclaterii* Salvadori, 1876[2]   S New Guinea
  — *G.s.scheepmakeri* Finsch, 1876   SE New Guinea

○ *Goura victoria* VICTORIA CROWNED PIGEON
  — *G.v.victoria* (Fraser, 1844)   Yapen and Biak Is.
  — *G.v.beccarii* Salvadori, 1876   N New Guinea

DIDUNCULINAE

DIDUNCULUS Peale, 1848 M
○ *Didunculus strigirostris* (Jardine, 1845) TOOTH-BILLED PIGEON   Upolu and Savaii Is. (Western Samoa)

TRERONINAE

PHAPITRERON Bonaparte, 1854 M[3]
○ *Phapitreron leucotis* WHITE-EARED BROWN DOVE
  — *P.l.leucotis* (Temminck, 1823)   Catanduanes, Luzon, Mindoro
  — *P.l.nigrorum* (Sharpe, 1877)   W Visayas (Philippines)
  — *P.l.brevirostris* (Tweeddale, 1877)[4]   E Visayas, Mindanao
  — *P.l.occipitalis* (Salvadori, 1893)   Basilan, Sulu Is.

○ *Phapitreron amethystinus* AMETHYST BROWN DOVE
  — *P.a.amethystinus* Bonaparte, 1855[5]   N, E and SE Philippines
  — *P.a.imeldae* de la Paz, 1976   Marinduque (625)

---

[1] Probably includes *masculina* Salomonsen, 1972 (2160) (but see Greenway, 1978: 66-67) (1010).
[2] Includes *wadai* Yamashina, 1944 (2695).
[3] Gender addressed by David & Gosselin (2002) (614), multiple changes occur in specific names.
[4] Includes *albifrons* see Dickinson *et al.* (1991) (745).
[5] Includes *celestinoi* Manuel, 1936 (1418), and *mindanaoensis* Manuel, 1936 (1418), see Dickinson *et al.* (1991) (745).

___ *P.a.maculipectus* (Bourns & Worcester, 1894)   Negros
___ †*P.a.frontalis* (Bourns & Worcester, 1894)   Cebu

○ *Phapitreron cinereiceps* DARK-EARED BROWN DOVE[1]
___ *P.c.brunneiceps* (Bourns & Worcester, 1894)   Mindanao, Basilan
___ *P.c.cinereiceps* (Bourns & Worcester, 1894)   Tawitawi I.

TRERON Vieillot, 1816 M[2]
○ *Treron fulvicollis* CINNAMON-HEADED GREEN PIGEON
    ___ *T.f.fulvicollis* (Wagler, 1827)   Malay Pen. to Sumatra
    ___ *T.f.melopogenys* (Oberholser, 1912)   Nias and Siberut I.
    ___ *T.f.oberholseri* Chasen, 1934   Natuna Is.
    ___ *T.f.baramensis* A.B. Meyer, 1891   Borneo and islands off N Borneo

○ *Treron olax* (Temminck, 1823) LITTLE GREEN PIGEON   Malay Pen. to Sumatra, Java and Borneo

○ *Treron vernans* (Linnaeus, 1771) PINK-NECKED GREEN PIGEON[3]   Mainland SE Asia to Gtr. and W Lesser Sundas, Sulawesi and the Philippines.

○ *Treron bicinctus* ORANGE-BREASTED GREEN PIGEON
    ___ *T.b.leggei* E. Hartert, 1910   Sri Lanka
    ___ *T.b.bicinctus* (Jerdon, 1840)   India to mainland SE Asia
    ___ *T.b.domvilii* (Swinhoe, 1870)   Hainan I.
    ___ *T.b.javanus* Robinson & Kloss, 1923   Java, Bali

○ *Treron pompadora* POMPADOUR GREEN PIGEON[4]
    ___ *T.p.conoveri* Rand & Fleming, 1953   Nepal (1973)
    ___ *T.p.phayrei* (Blyth, 1862)   W Bengal through Burma and Thailand to SW China and S Indochina
    ___ *T.p.affinis* (Jerdon, 1840)   SW India
    ___ *T.p.pompadora* (J.F. Gmelin, 1789)   Sri Lanka
    ___ *T.p.chloropterus* Blyth, 1846   Andaman and Nicobar Is.
    ___ *T.p.amadoni* Parkes, 1965   N Luzon (1757)
    ___ *T.p.axillaris* (Bonaparte, 1855)   Catanduanes, Luzon, Mindoro, Polillo
    ___ *T.p.canescens* Parkes, 1965   W and E Visayas, Mindanao, Basilan (1757)
    ___ *T.p.everetti* (Rothschild, 1894)   Sulu Arch.
    ___ *T.p.aromaticus* (J.F. Gmelin, 1789)   Islands in Flores Sea and Buru I. (S Moluccas)

○ *Treron curvirostra* THICK-BILLED GREEN PIGEON[5]
    ___ *T.c.nipalensis* (Hodgson, 1836)   W Nepal through Assam and Burma to Thailand and Indochina
    ___ *T.c.hainanus* Hartert & Goodson, 1918   Hainan I.
    ___ *T.c.curvirostra* (J.F. Gmelin, 1789)[6]   Malay Pen., Sumatra
    ___ *T.c.haliplous* Oberholser, 1912   Simeuluë I.
    ___ *T.c.pegus* Oberholser, 1912   Nias I.
    ___ *T.c.smicrus* Oberholser, 1912   Sipora, Siberut, Batu Is.
    ___ *T.c.hypothapsinus* Oberholser, 1912   Enggano I.
    ___ *T.c.nasica* Schlegel, 1863   Borneo
    ___ *T.c.erimacrus* Oberholser, 1924   Balabac, Palawan, Mindoro

○ *Treron griseicauda* GRAY-CHEEKED GREEN PIGEON
    ___ *T.g.sangirensis* Brüggemann, 1876   Sangihe Is.
    ___ *T.g.wallacei* (Salvadori, 1893)[7]   Sulawesi, Salayar I., Muna I., Buton I., Tukangbesi I., Banggai Is., Sula Is.

[1] Occurs in sympatry with *amethystina* and recognised at specific level, see Dickinson *et al.* (1991) (745).
[2] Gender addressed by David & Gosselin (2002) (614), multiple changes occur in specific names.
[3] Mees (1996: 20-22) (1543) suggested that the birds from the whole of this region vary individually such that recognition of races is unsafe on present evidence (except perhaps for larger insular forms). We tentatively accept this view. In last Edition we listed *griseicapilla, parvus, miza, mesochlous, adinus* and *purpureus*. Recent names not mentioned in that Edition are *kangeanus* Mayr, 1938 (1468), and *karimuniensis* Hoogerwerf, 1962 (1141).
[4] Our treatment follows Ripley (1982) (2055) as regards populations in the Indian subcontinent, and Dickinson *et al.* (1991) (745) for the Philippines; *pallidior* and *ada* (see last Edition) move to *T. griseicauda* see White & Bruce (1986: 194-197) (2652).
[5] Baptista *et al.* (1997: 197) (94) considered this species monotypic.
[6] Including *harterti* see van Marle & Voous (1988: 101) (2468) and *chaseni* Stresemann, 1950 (2365), see Hoogerwerf (1962) (1139) and White & Bruce (1986: 195) (2652).
[7] The name *griseicauda* Wallace, 1863, is a synonym used for Sulawesi birds from 1863-1893 and then by Peters (1937) (1823) and until recently, including our last Edition. However, it is preoccupied by *griseicauda* Bonaparte, 1854. See below. Includes *dehaani* Voous *in* van Bemmel & Voous, 1951 (2462) see White & Bruce (1986: 197) (2652).

— *T.g.vordermani* Finsch, 1900	Kangean Is.
— *T.g.pallidior* (E. Hartert, 1896)[1]	Kalao, Kalaotoa, Tanahjampea
— *T.g.griseicauda* Bonaparte, 1854[2]	Java, Bali[3]

○ **Treron teysmannii** Schlegel, 1879 Sumba Green Pigeon — Sumba I.

○ **Treron floris** Wallace, 1864 Flores Green Pigeon — Lombok to Alor (WC Lesser Sundas)

○ **Treron psittaceus** (Temminck, 1808) Timor Green Pigeon — Roti, Timor and Semau Is. (EC Lesser Sundas)

○ **Treron capellei** Large Green Pigeon

— *T.c.magnirostris* Strickland, 1844[4]	Malay Pen., Sumatra, Borneo (171)
— *T.c.capellei* (Temminck, 1822[5])	Java

○ **Treron phoenicopterus** Yellow-legged Green Pigeon

— *T.p.phoenicopterus* (Latham, 1790)	E Pakistan and N India to Bangladesh and Assam
— *T.p.chlorigaster* (Blyth, 1843)	C and S India
— *T.p.phillipsi* Ripley, 1949	Sri Lanka (2033)
— *T.p.viridifrons* Blyth, 1846	Burma, Yunnan and NW Thailand
— *T.p.annamensis* (Ogilvie-Grant, 1909)	E Thailand, S Laos, S Vietnam

○ **Treron waalia** (F.A.A. Meyer, 1793) Bruce's Green Pigeon[6] — Senegal to Eritrea, N Somalia and N Uganda, Socotra I., SW Arabia

○ **Treron australis** Madagascan Green Pigeon

— *T.a.griveaudi* Benson, 1960	Mwali (Moheli) I. (Comoro Is.) (122)
— *T.a.xenius* Salomonsen, 1934	W Madagascar
— *T.a.australis* (Linnaeus, 1771)	E Madagascar

○ **Treron calvus** African Green Pigeon[7]

— *T.c.nudirostris* (Swainson, 1837)	Senegal, Gambia, Guinea
— *T.c.sharpei* (Reichenow, 1902)	Sierra Leone to C Nigeria
— *T.c.calvus* (Temminck, 1808)	E Nigeria to C Angola and east to E and S Zaire
— *T.c.poensis* Hartert & Goodson, 1918	Bioko I.
— *T.c.virescens* Amadon, 1953	Principe I. (42)
— *T.c.uellensis* (Reichenow, 1912)	N Zaire to S Sudan, NW Uganda and SW Ethiopia
— *T.c.gibberifrons* (Madarász, 1915)[8]	S Sudan to Lake Victoria basin
— *T.c.brevicera* Hartert & Goodson, 1918	S Ethiopia, E Kenya, N and C Tanzania
— *T.c.wakefieldii* (Sharpe, 1874)	Coastal Kenya and NE Tanzania
— *T.c.granti* (van Someren, 1919)	E Tanzania
— *T.c.salvadorii* (A.J.C. Dubois, 1897)[9]	SE Zaire, W Tanzania, Zambia, N Malawi, NW Mozambique
— *T.c.ansorgei* Hartert & Goodson, 1918	W Angola
— *T.c.schalowi* Reichenow, 1880[10]	E Angola, NE Namibia, N Botswana, W Zimbabwe
— *T.c.vylderi* Glydenstolpe, 1924	NW Namibia
— *T.c.delalandii* (Bonaparte, 1854)[11]	SE Tanzania to Sth. Africa

○ **Treron pembaensis** Pakenham, 1940 Pemba Green Pigeon — Pemba I. (NE Tanzania) (1740)

○ **Treron sanctithomae** (J.F. Gmelin, 1789) São Tomé Green Pigeon — São Tomé I.

○ **Treron apicauda** Pin-tailed Green Pigeon

— *T.a.apicauda* Blyth, 1846	C and E Himalayas to Assam, W Yunnan, SW Sichuan, N and W Thailand and Burma
— *T.a.laotinus* (Delacour, 1926)	SE Yunnan, N Laos, N Vietnam
— *T.a.lowei* (Delacour & Jabouille, 1924)	W, NW and NE Thailand, C Laos, C Vietnam

---

[1] Includes *ada* see White & Bruce (1986: 197) (2652).
[2] Baptista *et al.* (1997: 197) (94) used the name *griseicauda* G. R. Gray, 1856, but this is preoccupied by Bonaparte's use of the name. The merits of this have been argued by Sims & Warren (1955) (2276); Mayr (1956) (1478), and Mees (1973) (1524). In last Edition we used *pulverulentus* Wallace, 1863, the synonym defended by Mayr. Includes *goodsoni* see White & Bruce (1986) (2652).
[3] Records from Sumatra of the Javan form were not accepted by van Marle & Voous (1988: 101) (2468).
[4] For recognition of this, named in a paper by Blyth, see Chasen (1940) (323).
[5] Livr. 24 dates from 1822; Peters (1937: 13) (1823) gave 1823 *contra* Sherborn (1898) (2242).
[6] Includes *jubaensis* Benson, 1942 (116).
[7] The treatment here follows Baptista *et al.* (1997: 201) (94).
[8] Includes *granviki* see White (1965: 165) (2649).
[9] Includes *clayi* White, 1943 (2624).
[10] Includes *chobiensis* and *damarensis* see White (1965: 166) (2649).
[11] Includes *orientalis* see White (1965: 167) (2649) and *glaucus* Clancey, 1967 (410), see Baptista *et al.* (1997: 201) (94).

○ *Treron oxyurus* (Temminck, 1823) Sᴜᴍᴀᴛʀᴀɴ Gʀᴇᴇɴ Pɪɢᴇᴏɴ Sumatra, W Java

○ *Treron seimundi* Yᴇʟʟᴏᴡ-ᴠᴇɴᴛᴇᴅ Gʀᴇᴇɴ Pɪɢᴇᴏɴ
___ *T.s.modestus* (Delacour, 1926) W and SE Thailand, N and C Laos, C and S Vietnam
___ *T.s.seimundi* (Robinson, 1910) SC Malay Pen.

○ *Treron sphenurus* Wᴇᴅɢᴇ-ᴛᴀɪʟᴇᴅ Gʀᴇᴇɴ Pɪɢᴇᴏɴ
___ *T.s.sphenurus* (Vigors, 1832)[1] Himalayas to Assam, Burma, Yunnan and Sichuan
___ *T.s.delacouri* Biswas, 1950[2] C Vietnam (153)
___ *T.s.robinsoni* (Ogilvie-Grant, 1906) Mts. of W Malaysia
___ *T.s.etorques* (Salvadori, 1879) Sumatra
___ *T.s.korthalsi* (Bonaparte, 1855) Java, Bali, Lombok

○ *Treron sieboldii* Wʜɪᴛᴇ-ʙᴇʟʟɪᴇᴅ Gʀᴇᴇɴ Pɪɢᴇᴏɴ
___ *T.s.sieboldii* (Temminck, 1835) Japan, E China
___ *T.s.fopingensis* Cheng, Tan & Sung, 1973 E Sichuan and S Shaanxi (342)
___ *T.s.sororius* (Swinhoe, 1866)[3] Taiwan
___ *T.s.murielae* (Delacour, 1927)[4] SC China, N and C Vietnam, N Thailand

○ *Treron formosae* Wʜɪsᴛʟɪɴɢ Gʀᴇᴇɴ Pɪɢᴇᴏɴ
___ *T.f.permagnus* Stejneger, 1887 N Ryukyu Is. (Tanegashima to Okinawa)
___ *T.f.medioximus* (Bangs, 1901) S Ryukyu Is.
___ *T.f.formosae* Swinhoe, 1863 Taiwan and nearby Lanyu I.
___ *T.f.filipinus* Hachisuka, 1952[5] Islands to N of Luzon (1043)

PTILINOPUS Swainson, 1825 M
○ *Ptilinopus cinctus* Bᴀɴᴅᴇᴅ Fʀᴜɪᴛ Dᴏᴠᴇ[6]
___ *P.c.baliensis* E. Hartert, 1896 Bali
___ *P.c.albocinctus* Wallace, 1864 Lombok, Sumbawa and Flores
___ *P.c.everetti* Rothschild, 1898 Pantar and Alor
___ *P.c.cinctus* (Temminck, 1810) Timor, Wetar and Romang
___ *P.c.lettiensis* (Schlegel, 1873) Leti, Moa, Luang, Sermata and Teun Is.
___ *P.c.ottonis* E. Hartert, 1904 Damar, Babar and Nila Is.
___ *P.c.alligator* (Collett, 1898) W Arnhem Land (N Australia)

○ *Ptilinopus dohertyi* (Rothschild, 1896) Rᴇᴅ-ɴᴀᴘᴇᴅ Fʀᴜɪᴛ Dᴏᴠᴇ Sumba I.

○ *Ptilinopus porphyreus* (Temminck, 1822[7]) Pɪɴᴋ-ʜᴇᴀᴅᴇᴅ Fʀᴜɪᴛ Dᴏᴠᴇ Sumatra, Java, Bali

○ *Ptilinopus marchei* (Oustalet, 1880) Fʟᴀᴍᴇ-ʙʀᴇᴀsᴛᴇᴅ Fʀᴜɪᴛ Dᴏᴠᴇ Luzon

○ *Ptilinopus merrilli* Cʀᴇᴀᴍ-ʙᴇʟʟɪᴇᴅ Fʀᴜɪᴛ Dᴏᴠᴇ
___ *P.m.faustinoi* (Manuel, 1936) Mt. Tabuan (N Luzon) (1419)
___ *P.m.merrilli* (McGregor, 1916) E and S Luzon, Catanduanes and Polillo

○ *Ptilinopus occipitalis* Yᴇʟʟᴏᴡ-ʙʀᴇᴀsᴛᴇᴅ Fʀᴜɪᴛ Dᴏᴠᴇ
___ *P.o.occipitalis* Gray & Mitchell, 1844 N and C Philippines
___ *P.o.incognitus* (Tweeddale, 1877) Mindanao

○ *Ptilinopus fischeri* Rᴇᴅ-ᴇᴀʀᴇᴅ Fʀᴜɪᴛ Dᴏᴠᴇ
___ *P.f.fischeri* (Brüggemann, 1876) N Sulawesi
___ *P.f.centralis* (A.B. Meyer, 1903) C and SE Sulawesi
___ *P.f.meridionalis* (Meyer & Wiglesworth, 1893) SW Sulawesi

○ *Ptilinopus jambu* (J.F. Gmelin, 1789) Jᴀᴍʙᴜ Fʀᴜɪᴛ Dᴏᴠᴇ Malay Pen., Sumatra, Borneo, Java

○ *Ptilinopus subgularis* Mᴀʀᴏᴏɴ-ᴄʜɪɴɴᴇᴅ Fʀᴜɪᴛ Dᴏᴠᴇ
___ *P.s.epia* (Oberholser, 1918)[8] Sulawesi

---

[1] Includes *yunnanensis* although this is recognised by Cheng (1987) (333).
[2] We recognise this because Wells (1999) (2579) considered *robinsoni* endemic to the Malay Peninsula.
[3] Recognised by Cheng (1987) (333).
[4] Includes *oblitus* (treated as a race of *T. sphenurus* in last Edition, but see Cheng, 1987) (333). A recent synonym is *Sphenocercus sphenurus lungchowensis* Yen & Chong, 1937 (2700), see Cheng *et al.* (1973).
[5] The name *mcgregorii* Husain, 1958 (1166), is a synonym see Dickinson *et al.* (1991) (745).
[6] In last Edition we treated *alligator* as a separate species, but see Hartert (1904) (1081) and the corroborative views of Schodde & Mason (1997) (2188).
[7] Livr. 18 dates from 1822; Peters (1937: 27) (1823) gave 1823 *contra* Sherborn (1898) (2242).
[8] Includes *restrictus* Ripley, 1941 (2024), see White & Bruce (1986) (2652).

○ — *P.s.subgularis* (Meyer & Wiglesworth, 1896) Banggai Is.
— *P.s.mangoliensis* Rothschild, 1898 Sula Is.

○ *Ptilinopus leclancheri* BLACK-CHINNED FRUIT DOVE
— *P.l.taiwanus* Ripley, 1962[1] Taiwan (2043)
— *P.l.longialis* (Manuel, 1936) Lanyu I. (off S Taiwan), islands N of Luzon
  (Philippines) (1419)
— *P.l.leclancheri* (Bonaparte, 1855) Philippines (except northern isles and Palawan group)
— *P.l.gironieri* (Verreaux & Des Murs, 1862) Palawan and nearby islands

○ *Ptilinopus bernsteinii* SCARLET-BREASTED FRUIT DOVE[2]
— *P.b.micrus* (Jany, 1955) Obi I. (1200)
— *P.b.bernsteinii* (Schlegel, 1863) Halmahera, Ternate and Bacan

○ *Ptilinopus magnificus* WOMPOO FRUIT DOVE
— *P.m.puella* (Lesson, 1827)[3] W Papuan Is., Vogelkop (NW New Guinea)
— *P.m.interpositus* (E. Hartert, 1930)[4][5] W New Guinea
— *P.m.septentrionalis* (A.B. Meyer, 1893) N New Guinea and islands
— *P.m.poliurus* (Salvadori, 1878)[6] SE New Guinea
— *P.m.assimilis* (Gould, 1850) E Cape York Pen.
— *P.m.keri* (Mathews, 1912) NE Queensland
— *P.m.magnificus* (Temminck, 1821) S Queensland, E New South Wales

○ *Ptilinopus perlatus* PINK-SPOTTED FRUIT DOVE
— *P.p.perlatus* (Temminck, 1835) NW New Guinea and W Papuan Is.
— *P.p.plumbeicollis* (A.B. Meyer, 1890) NC and NE New Guinea
— *P.p.zonurus* (Salvadori, 1876) S New Guinea, Aru Is., D'Entrecasteaux Arch.

○ *Ptilinopus ornatus* ORNATE FRUIT DOVE
— *P.o.ornatus* (Schlegel, 1871)[7] NW New Guinea
— *P.o.gestroi* (D'Albertis & Salvadori, 1875) C and E New Guinea

○ *Ptilinopus tannensis* (Latham, 1790) TANNA FRUIT DOVE Vanuatu, Banks Is.

○ *Ptilinopus aurantiifrons* (G.R. Gray, 1858) ORANGE-FRONTED FRUIT DOVE
  W Papuan Is., New Guinea, Aru Is.

○ *Ptilinopus wallacii* (G.R. Gray, 1858) WALLACE'S FRUIT DOVE S Moluccas, E Lesser Sundas, Aru Is., SW New Guinea

○ *Ptilinopus superbus* SUPERB FRUIT DOVE
— *P.s.temminckii* (Des Murs & Prévost, 1849) Sulawesi; Sulu Is. (once)
— *P.s.superbus* (Temminck, 1809)[8] Moluccas and New Guinea to Bismarck Arch.,
  Solomon Is. and E Australia

◉ *Ptilinopus perousii* MANY-COLOURED FRUIT DOVE
✓ *P.p.perousii* Peale, 1848 Samoa
— *P.p.mariae* (Jacquinot & Pucheran, 1853) Tonga, Fiji

◉ *Ptilinopus porphyraceus* PURPLE-CAPPED FRUIT DOVE[9]
— *P.p.ponapensis* (Finsch, 1878) Chuuk and Pohnpei (Caroline Is.)
— *P.p.hernsheimi* Finsch, 1880 Kosrae (Caroline Is.)
— *P.p.porphyraceus* (Temminck, 1821) Niue, Tonga, Fiji Is.
✓ *P.p.fasciatus* Peale, 1848 Samoa

○ *Ptilinopus pelewensis* Hartlaub & Finsch, 1868 PALAU FRUIT DOVE Palau Is.

○ *Ptilinopus rarotongensis* RAROTONGAN FRUIT DOVE
— *P.r.rarotongensis* Hartlaub & Finsch, 1871 Rarotonga I. (Cook Is.)
— *P.r.goodwini* Holyoak, 1974 Atiu I. (Cook Is.) (1129)

[1] Accepted by Cheng (1987: 282) (333) and by Gibbs *et al.* (2001) (935).
[2] Used to share a genus with the next species and be called *Megaloprepia formosus* but this binomen unavailable in *Ptilinopus* see Mees (1973, 1982) (1524, 1534).
[3] Includes *alaris* see Mayr (1941) (1472).
[4] For correction of spelling see David & Gosselin (2002) (613).
[5] Mees (1982) (1534) suggested that *interposita* should be placed in the synonymy of *puella* but this does not appear to have been accepted or perhaps re-evaluated by later writers.
[6] For correction of spelling see David & Gosselin (2002) (613).
[7] For the submergence of *kaporensis* – still listed in last Edition — see Peters (1937: 39) (1823) and Rand & Gilliard (1967) (1975).
[8] For correction of date to 1809 see Schodde & Mason (1997) (2188).
[9] For reasons to suppress *graeffi* see Baptista *et al.* (1997: 205) (94).

○ *Ptilinopus roseicapilla* (Lesson, 1831) MARIANAS FRUIT DOVE — Marianas Is.

○ *Ptilinopus regina* ROSE-CROWNED FRUIT DOVE
— *P.r.flavicollis* Bonaparte, 1855 — Flores, Roti, Sawu, Semau and W Timor
— *P.r.roseipileum* E. Hartert, 1904 — E Timor, Wetar, Romang, Kisar, Moa and Leti
— *P.r.xanthogaster* (Wagler, 1827) — Banda, Kai, Damar, Babar, Teun, Nila, Tanimbar, Aru Is.
— *P.r.ewingii* Gould, 1842 — N Australia, Melville I.
— *P.r.regina* Swainson, 1825 — Torres Straits Is., E Australia

○ *Ptilinopus richardsii* SILVER-CAPPED FRUIT DOVE
— *P.r.richardsii* (E.P. Ramsay, 1882) — E Solomon Is.
— *P.r.cyanopterus* Mayr, 1931 — Rennell and Bellona Is. (SE Solomon Is.)

○ *Ptilinopus purpuratus* GREY-GREEN FRUIT DOVE[1]
— *P.p.chrysogaster* (G.R. Gray, 1854) — W Society Is.
— *P.p.frater* Ripley & Birckhead, 1942 — Moorea I. (E Society, Is.) (2062)
— *P.p.purpuratus* (J.F. Gmelin, 1789) — Tahiti (E Society Is.)

○ *Ptilinopus chalcurus* (G.R. Gray, 1859) MAKATEA FRUIT DOVE — Makatea I. (Tuamotu Arch.)

○ *Ptilinopus coralensis* Peale, 1848 ATOLL FRUIT DOVE — Tuamotu Arch.

○ *Ptilinopus greyii* Bonaparte, 1857 RED-BELLIED FRUIT DOVE — Ndai I., Santa Cruz Is., Banks Group, Vanuatu to Loyalty Is., New Caledonia and Isle of Pines

○ *Ptilinopus huttoni* (Finsch, 1874) RAPA FRUIT DOVE — Rapa I.

○ *Ptilinopus dupetithouarsii* WHITE-CAPPED FRUIT DOVE
— *P.d.viridior* (Murphy, 1924) — N Marquesas Is.
— *P.d.dupetithouarsii* (Neboux, 1840) — S Marquesas Is.

○ *Ptilinopus mercierii* RED-MOUSTACHED FRUIT DOVE
— †*P.m.mercierii* (Des Murs & Prévost, 1849) — Nukuhiva I. (N Marquesas Is.)
— †?*P.m.tristrami* (Salvadori, 1892) — Hivaoa I. (S Marquesas Is.)

○ *Ptilinopus insularis* (North, 1908) HENDERSON FRUIT DOVE — Henderson I. (Pitcairn Is.)

○ *Ptilinopus coronulatus* CORONETED FRUIT DOVE
— *P.c.trigeminus* (Salvadori, 1875) — W Vogelkop (NW New Guinea), Salawati I
— *P.c.geminus* (Salvadori, 1875) — Western N New Guinea (west to E Vogelkop), Yapen I.
— *P.c.quadrigeminus* (A.B. Meyer, 1890) — Eastern N New Guinea, Manam and Kairiru Is.
— *P.c.huonensis* (A.B. Meyer, 1892) — N coast of SE New Guinea
— *P.c.coronulatus* (G.R. Gray, 1858) — S New Guinea, Aru Is.

○ *Ptilinopus pulchellus* BEAUTIFUL FRUIT DOVE
— *P.p.decorus* (Madarász, 1910) — NC New Guinea
— *P.p.pulchellus* (Temminck, 1835) — W, S and E New Guinea and W Papuan Is.

○ *Ptilinopus monacha* (Temminck, 1824) BLUE-CAPPED FRUIT DOVE — N Moluccas

○ *Ptilinopus rivoli* WHITE-BIBBED FRUIT DOVE
— *P.r.prasinorrhous* (G.R. Gray, 1858)[2] — Moluccas, islands off W Papua, Aru Is., Kai Is., islands of Geelvink Bay
— *P.r.bellus* (P.L. Sclater, 1874) — New Guinea, Karkar I., Goodenough I.
— *P.r.miquelii* (Schlegel, 1871) — Yapen, Meos Num Is.
— *P.r.rivoli* (Prévost, 1843) — Bismarck Arch.
— *P.r.strophium* Gould, 1850 — Louisiade Arch., Egum Atoll (Trobriand Is.)

○ *Ptilinopus solomonensis* YELLOW-BIBBED FRUIT DOVE
— *P.s.speciosus* (Schlegel, 1871) — Islands in Geelvink Bay (NW New Guinea)
— *P.s.johannis* (P.L. Sclater, 1877) — Admiralty Is., St. Matthias Group, New Hanover (Bismarck Arch.)
— *P.s.meyeri* E. Hartert, 1926 — New Britain and Is. (Bismarck Arch.)
— *P.s.neumanni* E. Hartert, 1926 — Nissan I.
— *P.s.bistictus* Mayr, 1931 — Bougainville and Buka Is.

---

[1] We follow Pratt *et al.* (1987) (1920) in separating the species *chalcurus* and *coralensis*.
[2] Includes *buruensis* see White & Bruce (1986) (2652).

— *P.s.vulcanorum* Mayr, 1931 — C Solomon Is.

— *P.s.ocularis* Mayr, 1931 — Guadalcanal I.

— *P.s.ambiguus* Mayr, 1931 — Malaita I.

— *P.s.solomonensis* (G.R. Gray, 1870) — Makira and Uki Is.

○ *Ptilinopus viridis* CLARET-BREASTED FRUIT DOVE

— *P.v.viridis* (Linnaeus, 1766) — S Moluccas

— *P.v.pectoralis* (Wagler, 1829) — W Papuan Is., Vogelkop (NW New Guinea)

— *P.v.geelvinkianus* (Schlegel, 1871)[1][2] — Geelvink Bay Is.

— *P.v.salvadorii* (Rothschild, 1892) — N New Guinea, Yapen I.

— *P.v.vicinus* (E. Hartert, 1895) — Trobriand Is., D'Entrecasteaux Arch.

— *P.v.lewisii* (E.P. Ramsay, 1882) — Hibernian and all Solomon Is. east to Guadalcanal, Malaita Is.

○ *Ptilinopus eugeniae* (Gould, 1856) WHITE-HEADED FRUIT DOVE — Makira, Malaupaina and Uki Is. (E Solomon Is.)

○ *Ptilinopus iozonus* ORANGE-BELLIED FRUIT DOVE[3]

— *P.i.iozonus* (G.R. Gray, 1858) — Aru Is.

— *P.i.humeralis* (Wallace, 1862) — W Papuan Is., NW New Guinea

— *P.i.jobiensis* (Schlegel, 1873) — Yapen I., N New Guinea

— *P.i.pseudohumeralis* Rand, 1938 — SW and SC New Guinea (1962)

— *P.i.finschi* Mayr, 1931 — SE New Guinea

○ *Ptilinopus insolitus* KNOB-BILLED FRUIT DOVE

— *P.i.insolitus* (Schlegel, 1863) — Bismarck Arch. (excl. St. Matthias group)

— *P.i.inferior* E. Hartert, 1924 — St. Matthias group (N Bismarck Arch.)

○ *Ptilinopus hyogastrus* (Temminck, 1824)[4] GREY-HEADED FRUIT DOVE — N Molucas

○ *Ptilinopus granulifrons* E. Hartert, 1898 CARUNCULATED FRUIT DOVE — Obi I. (C Moluccas)

○ *Ptilinopus melanospilus* BLACK-NAPED FRUIT DOVE[5]

— *P.m.bangueyensis* (A.B. Meyer, 1891) — N Bornean Is., Palawan, Mindanao, Basilan and Sulu Arch.

— *P.m.xanthorrhous* (Salvadori, 1875)[6][7] — Sangihe, Talaud and Doi Is.

— *P.m.melanospilus* (Salvadori, 1875) — Sulawesi and islands to NE

— *P.m.chrysorrhous* (Salvadori, 1875)[8] — Sula, Peleng and Banggai Is., Obi and Seram Is.

— *P.m.melanauchen* (Salvadori, 1875)[9] — Java and Bali to Alor, also Matasiri and Kangean Is. and S Sulawesi Is.

○ *Ptilinopus nainus* DWARF FRUIT DOVE

— *P.n.minimus* Stresemann & Paludan, 1932 — W Papuan islands

— *P.n.nainus* (Temminck, 1835)[10] — W, S and SE New Guinea

○ †?*Ptilinopus arcanus* Ripley & Rabor, 1955 NEGROS FRUIT DOVE # — Mt. Canlaon (NC Negros, Philippines) (2071)

○ *Ptilinopus victor* ORANGE DOVE

— *P.v.victor* (Gould, 1872) — Vanua Levu, Rabi, Kioa, Taveuni (N Fiji)

— *P.v.aureus* Amadon, 1943 — Qamea, Laucala (NE Fiji) (41)

○ *Ptilinopus luteovirens* (Hombron & Jacquinot, 1841) GOLDEN DOVE — Waya Group, Vitu Levu, Beqa, Ovalau, Gau (WC Fiji)

○ *Ptilinopus layardi* (Elliot, 1878) WHISTLING DOVE — Kadavu and Ono (SW Fiji)

DREPANOPTILA Bonaparte, 1855 F

○ *Drepanoptila holosericea* (Temminck, 1810) CLOVEN-FEATHERED DOVE — New Caledonia and nearby Isle of Pines

---

[1] For correction of spelling see David & Gosselin (2002) (613).

[2] Includes *pseudogeelvinkianus* Junge, 1952 (1232), see Rand & Gilliard (1967) (1975).

[3] Mees (1982: 70) (1534) believed that nominate *iozonus* occupied S New Guinea east to near Merauke (but did not discuss *pseudohumeralis* now thought to extend westward).

[4] Spelled *hyogastra* in the original text; for correction of spelling see David & Gosselin (2002) (613). Temminck (1839) (2395) emended it to *hyogaster*.

[5] For correction of spelling see David & Gosselin (2002) (613).

[6] For correction of spelling see David & Gosselin (2002) (613).

[7] Includes *loloda* Jany, 1955 (1200), see White & Bruce (1986) (2652).

[8] For correction of spelling see David & Gosselin (2002) (613).

[9] Includes *massoptera* see White & Bruce (1986) (2652).

[10] The original name used here is judged to be in prevailing usage; see Art. 33.2 of the Code (I.C.Z.N., 1999) (1178) and p. 000 herein.

ALECTROENAS G.R. Gray, 1840 F

○ †*Alectroenas nitidissima* (Scopoli, 1786) Mauritius Blue Pigeon   Mauritius

○ *Alectroenas madagascariensis* (Linnaeus, 1766) Madagascan Blue Pigeon
                                                                  N and E Madagascar

○ *Alectroenas sganzini* Comoro Blue Pigeon
    — *A.s.minor* Berlepsch, 1898   Aldabra Is.
    — *A.s.sganzini* (Bonaparte, 1854)   Comoro Is.

○ *Alectroenas pulcherrima* (Scopoli, 1786) Seychelles Blue Pigeon   Seychelles

DUCULA Hodgson, 1836 F

○ *Ducula poliocephala* (Gray & Mitchell, 1844) Pink-bellied Imperial Pigeon
                                                                  Philippine Is.

○ *Ducula forsteni* (Bonaparte, 1854) White-bellied Imperial Pigeon   Sulawesi, Sula Is.

○ *Ducula mindorensis* (J. Whitehead, 1896) Mindoro Imperial Pigeon   Mindoro I.

○ *Ducula radiata* (Quoy & Gaimard, 1830) Grey-headed Imperial Pigeon
                                                                  Mts. of Sulawesi

○ *Ducula carola* Spotted Imperial Pigeon
    — *D.c.carola* (Bonaparte, 1854)   Luzon, Mindoro and Sibuyan
    — *D.c.nigrorum* (J. Whitehead, 1897)   Negros and Siquijor
    — *D.c.mindanensis* (Ogilvie-Grant, 1905)   Mindanao

○ *Ducula aenea* Green Imperial Pigeon[1]
    — *D.a.sylvatica* (Tickell, 1833)[2]   C and E Himalayas, Assam, S Yunnan, and Burma,
                                                                      Thailand (south to mid Malay Pen.), east to
                                                                      Indochina and Hainan

    — *D.a.pusilla* (Blyth, 1849)   S India, Sri Lanka
    — *D.a.andamanica* Abdulali, 1964   Andaman Is. (6)
    — *D.a.nicobarica* (Pelzeln, 1865)   Nicobar Is.
    — *D.a.consobrina* (Salvadori, 1887)[3]   Islands of western Sumatra (except Enggano)
    — *D.a.oenothorax* (Salvadori, 1892)   Enggano I.
    — *D.a.polia* Oberholser, 1917   S Malay Pen., Gtr. and Lesser Sundas, Anambas and
                                                                      Natunas

    — *D.a.palawanensis* (W.H. Blasius, 1888)   Banggi I. (Sabah) to Palawan and nearby islands
    — *D.a.fugaenis* (Hachisuka, 1930)   Islands to N of N Luzon
    — *D.a.nuchalis* (Cabanis, 1882)   N Luzon (Philippines)
    — *D.a.aenea* (Linnaeus, 1766)[4]   Rest of the Philippines
    — *D.a.intermedia* (Meyer & Wiglesworth, 1894)   Talaud Is., Sangihe Is.
    — *D.a.paulina* Bonaparte, 1854[5]   Sulawesi, Togian, Banggai and Sula Is.

○ *Ducula perspicillata* White-eyed Imperial Pigeon
    — *D.p.perspicillata* (Temminck, 1824)   N Moluccas and Buru
    — *D.p.neglecta* (Schlegel, 1865)   Seram, Ambon and nearby small islands

○ *Ducula concinna* (Wallace, 1865) Blue-tailed Imperial Pigeon[6]   Talaud and Sangihe Is. to S Moluccas, E Lesser Sundas
                                                                      and Aru Is.

● *Ducula pacifica* Pacific Imperial Pigeon[7]
    — *D.p.sejuncta* Amadon, 1943[8]   NC New Guinea Is. to W Bismarck Arch. (41)
    ✓ *D.p.pacifica* (J.F. Gmelin, 1789)[9]   Louisiade Arch. to Solomons, Santa Cruz, Vanuatu,
                                                                      Ellice, Phoenix and Cook Is., Samoa, Tonga and
                                                                      Niue

---

[1] In last Edition we did not accept the views of Stresemann (1952) (2366); here we do, nominate *aenea* is thus a Philippine bird; for the races we follow
     Ripley (1982) (2055), White & Bruce (1986) (2652), van Marle & Voous (1988) (2468), Dickinson *et al.* (1991) (745) and Wells (1999) (2579).
[2] Includes *kwangtungensis* Chou, 1955 (347), which was accepted with reservations by Cheng (1987: 283 fn.) (333).
[3] Includes *mista, babiensis* and *vicina* see van Marle & Voous (1986) (2468).
[4] Includes *glaucocauda* Manuel, 1936 (1420).
[5] Includes *pallidinucha* Mayr, 1944 (1474), and *intercedens* Eck, 1976 (788).
[6] Includes *aru* see Schodde & Mathews (1977) (2193).
[7] We follow Baptista *et al.* (1997: 230) (94).
[8] It is quite probable that the prior name *farquhari* Sharpe, 1900, should be applied.
[9] Includes *tarrali, intensitincta* and *microcera*.

O *Ducula oceanica* MICRONESIAN IMPERIAL PIGEON
    — *D.o.monacha* (Momiyama, 1922)   Yap I., Palau Is.
    — *D.o.teraokai* (Momiyama, 1922)   Chuuk I. (C Caroline Is.)
    — *D.o.townsendi* (Wetmore, 1919)   Pohnpei I. (E Caroline Is.)
    — *D.o.oceanica* (Lesson & Garnot, 1826)   Kosrae I. (E Caroline Is.)
    — *D.o.ratakensis* (Taka-Tsukasa & Yamashina, 1932)   Marshall Is.

O *Ducula aurorae* POLYNESIAN IMPERIAL PIGEON
    — *D.a.aurorae* (Peale, 1848)   Makatea (Tuamotu Arch.)
    — *D.a.wilkesii* (Peale, 1848)   Tahiti (Society Is.)

O *Ducula galeata* (Bonaparte, 1855) MARQUESAN IMPERIAL PIGEON   Nukuhiva I. (Marquesas Is.)

O *Ducula rubricera* RED-KNOBBED IMPERIAL PIGEON
    — *D.r.rubricera* (Bonaparte, 1854)   Bismarck Arch., Hibernian Is.
    — *D.r.rufigula* (Salvadori, 1878)   Solomon Is.

O *Ducula myristicivora* SPICE IMPERIAL PIGEON
    — *D.m.myristicivora* (Scopoli, 1786)   Widi Is. (NE Moluccas), W Papuan islands
    — *D.m.geelvinkiana* (Schlegel, 1873)   Islands in Geelvink Bay (W New Guinea)

O *Ducula rufigaster* PURPLE-TAILED IMPERIAL PIGEON
    — *D.r.rufigaster* (Quoy & Gaimard, 1830)[1]   W Papuan Is., NW and S New Guinea
    — *D.r.uropygialis* Streseman & Paludan, 1932   N New Guinea, Yapen I.

O *Ducula basilica* CINNAMON-BELLIED IMPERIAL PIGEON
    — *D.b.basilica* Bonaparte, 1854   N Moluccas (except Obi)
    — *D.b.obiensis* (E. Hartert, 1898)   Obi I.

O *Ducula finschii* (E.P. Ramsay, 1882) FINSCH'S IMPERIAL PIGEON   Bismarck Arch.

O *Ducula chalconota* SHINING IMPERIAL PIGEON
    — *D.c.chalconota* (Salvadori, 1874)   Mts. of Vogelkop (NW New Guinea)
    — *D.c.smaragdina* Mayr, 1931   Mts. of W, C and E New Guinea

O *Ducula pistrinaria* ISLAND IMPERIAL PIGEON
    — *D.p.rhodinolaema* (P.L.Sclater, 1877)   N New Guinea Is., Admiralty Is., St Matthias Group,
                                              New Hanover
    — *D.p.vanwyckii* (Cassin, 1862)   Main islands of E and C Bismarck Arch.
    — *D.p.postrema* E. Hartert, 1926   D'Entrecasteaux Arch., Woodlark Group, Louisiade
                                              Arch.
    — *D.p.pistrinaria* Bonaparte, 1855   Lihir, Feni and Green Is. to Solomon Is.

O *Ducula rosacea* (Temminck, 1836[2]) PINK-HEADED IMPERIAL PIGEON   Islands in Java, Flores and S Banda Seas, all Lesser
                                              Sundas east to Kai Is.

O *Ducula whartoni* (Sharpe, 1887) CHRISTMAS IMPERIAL PIGEON   Christmas I. (Indian Ocean)

O *Ducula pickeringii* (Cassin, 1854) GREY IMPERIAL PIGEON[3]   N Bornean islets to Sulu Is., Miangas and Talaud Is.

O *Ducula latrans* (Peale, 1848) PEALE'S IMPERIAL PIGEON   Fiji Is.

O *Ducula brenchleyi* (G.R. Gray, 1870) CHESTNUT-BELLIED IMPERIAL PIGEON
                                              E Solomon Is.

O *Ducula bakeri* (Kinnear, 1928) VANUATU IMPERIAL PIGEON   N Vanuatu

O *Ducula goliath* (G.R. Gray, 1859) NEW CALEDONIAN IMPERIAL PIGEON   New Caledonia and nearby Isle of Pines

O *Ducula pinon* PINON'S IMPERIAL PIGEON
    — *D.p.pinon* (Quoy & Gaimard, 1824)   W Papuan islands, NW to all S New Guinea and N
                                              coast of SE New Guinea, Aru Is.
    — *D.p.jobiensis* (Schlegel, 1871) [4]   N New Guinea E to Huon Gulf, Yapen I., Dampier Is.
    — *D.p.salvadorii* (Tristram, 1882)   D'Entrecasteaux Arch., Louisiade Arch.

---

[1] Includes *pallida* Junge, 1952 (1232); but see doubts of Diamond (1972) (699).
[2] Livr. 98 must date from 1836 not 1835 see Dickinson (2001) (739).
[3] Treated as monotypic as in last Edition, *contra* Dickinson *et al.* (1991) (745).
[4] The name *rubiensis* was applied to an intergrading population.

○ *Ducula melanochroa* (P.L. Sclater, 1878) BISMARCK IMPERIAL PIGEON   Bismarck Arch.

○ *Ducula mullerii* COLLARED IMPERIAL PIGEON
    — *D.m.mullerii* (Temminck, 1835)   N New Guinea
    — *D.m.aurantia* (A.B. Meyer, 1893)   S New Guinea, Aru Is.

○ *Ducula zoeae* (Lesson, 1826) BANDED IMPERIAL PIGEON/ZOE'S IMPERIAL PIGEON
                       New Guinea, Salawati, Yapen I., Aru Is. to Fergusson I., Louisiade Arch.

○ *Ducula badia* MOUNTAIN IMPERIAL PIGEON
    — *D.b.cuprea* (Jerdon, 1840)   SW India
    — *D.b.insignis* Hodgson, 1836   C and E Himalayas and hills of N Assam
    — *D.b.griseicapilla* Walden, 1875[1]   S Assam, Burma and Thailand to W and S Yunnan, Indochina and Hainan
    — *D.b.badia* (Raffles, 1822)[2]   Malay Pen., Sumatra, Borneo, W Java

○ *Ducula lacernulata* DARK-BACKED IMPERIAL PIGEON
    — *D.l.lacernulata* (Temminck, 1822[3])   W and C Java
    — *D.l.williami* (E. Hartert, 1896)   E Java, Bali
    — *D.l.sasakensis* (E. Hartert, 1896)   Lombok and Flores

○ *Ducula cineracea* (Temminck, 1835) TIMOR IMPERIAL PIGEON[4]   Timor and Wetar

○ *Ducula bicolor* PIED IMPERIAL PIGEON[5]
    — *D.b.bicolor* (Scopoli, 1786)[6]   Andaman and Nicobar Is, W Burma and Malay Pen. to Philippines, Gtr. and Lesser Sundas, Moluccas, W Papuan Is.
    — *D.b.luctuosa* (Temminck, 1824)   Sulawesi to Banggai and Sula Is.
    — *D.b.spilorrhoa* (G.R. Gray, 1858)[7]   Aru Is., coastal New Guinea and islands to N and NE Australia
    — *D.b.subflavescens* (Finsch, 1886)   Bismarck Arch., Admiralty Is.

LOPHOLAIMUS Gould, 1841 M
○ *Lopholaimus antarcticus* (Shaw, 1794) TOPKNOT PIGEON   E Australia

HEMIPHAGA Bonaparte, 1854 F
◉ *Hemiphaga novaeseelandiae* NEW ZEALAND PIGEON
    ↙ *H.n.novaeseelandiae* (J.F. Gmelin, 1789)   New Zealand
    — *H.n.chathamensis* (Rothschild, 1891)   Chatham Is.
    — †*H.n.spadicea* (Latham, 1802)[8]   Norfolk I.

CRYPTOPHAPS Salvadori, 1893 F
○ *Cryptophaps poecilorrhoa* (Brüggemann, 1876) SOMBRE PIGEON   Mts. of N, S and SE Sulawesi

GYMNOPHAPS Salvadori, 1874 F
○ *Gymnophaps albertisii* PAPUAN MOUNTAIN PIGEON
    — *G.a.exsul* (E. Hartert, 1903)   Bacan I. (NC Moluccas)
    — *G.a.albertisii* Salvadori, 1874   Yapen I., New Guinea, New Britain, New Ireland, Goodenough I.

○ *Gymnophaps mada* LONG-TAILED MOUNTAIN PIGEON
    — *G.m.mada* (E. Hartert, 1899)   Buru I.
    — *G.m.stalkeri* (Ogilvie-Grant, 1911)   Seram I.

○ *Gymnophaps solomonensis* Mayr, 1931 PALE MOUNTAIN PIGEON   Solomon Is.

---

[1] Includes *carolinae* Ripley, 1953 (2037), see Ripley (1982) (2055) and *obscurata* see Baptista *et al.* (1997: 239) (94).
[2] Includes *capistrata* see Baptista *et al.* (1997: 239) (94).
[3] Livr. 28 dates from 1822; Peters (1937: 51) (1823) gave 1823 *contra* Sherborn (1898) (2242).
[4] Includes *schistacea* Mayr, 1944 (1474), see White & Bruce (1986) (2652).
[5] Our treatment follows that of Johnstone (1981) (1217).
[6] Includes *siebersi* van Bemmel, 1940 (2460), see White & Bruce (1986) (2652).
[7] Includes *melvillensis, tarara* Rand, 1941 (1964) and *constans* Bruce, 1989 (259) see Schodde & Mason (1997) (2188).
[8] For reasons to use 1802 not 1801 see Browning & Monroe (1991) (258).

PSITTACIDAE[1] COCKATOOS AND PARROTS (85: 364)

NESTORINAE[2]

NESTOR Lesson, 1830 M

◉ *Nestor notabilis* Gould, 1856 Kea — South Island (New Zealand)

○ †*Nestor productus* (Gould, 1836) Norfolk Island Kaka — Norfolk I.

○ *Nestor meridionalis* Kaka[3]
    — *N.m.meridionalis* (J.F. Gmelin, 1788) — South Island and offshore islands, Stewart I.
    — *N.m.septentrionalis* Lorenz, 1896 — North Island (New Zealand) and offshore islands

STRIGOPINAE[4]

STRIGOPS G.R. Gray, 1845 F
○ *Strigops habroptila* G.R. Gray, 1845 Kakapo[5] — Introduced to Little Barrier, Maud, Codfish and Pearl Is.

PSITTRICHADINAE[6]

PSITTRICHAS Lesson, 1831 M
○ *Psittrichas fulgidus* (Lesson, 1830) Pesquet's Parrot — Midmontane New Guinea

LORICULINAE[7]

LORICULUS Blyth, 1850 M
○ *Loriculus vernalis* Vernal Hanging Parrot
    — *L.v.vernalis* (Sparrman, 1787) — SW, S and E India to Indochina
    — *L.v.phileticus* Deignan, 1956[8] — Pen. Thailand (south to Trang) (657)

○ *Loriculus beryllinus* (J.R. Forster, 1781) Sri Lankan Hanging Parrot — Sri Lanka

○ *Loriculus philippensis* Colasisi/Philippine Hanging Parrot[9]
    — *L.p.philippensis* (Statius Müller, 1776) — Luzon and adjacent islands
    — *L.p.mindorensis* Steere, 1890 — Mindoro
    — *L.p.bournsi* McGregor, 1905 — Sibuyan
    — *L.p.regulus* Souancé, 1856 — Guimaras, Negros, Ticao, Tablas, Masbate, Panay
    — †?*L.p.chrysonotus* P.L. Sclater, 1872 — Cebu
    — *L.p.worcesteri* Steere, 1890 — Samar, Leyte, Bohol
    — †?*L.p.siquijorensis* Steere, 1890 — Siquijor
    — *L.p.apicalis* Souancé, 1856 — Mindanao, Dinagat, Siargao and Camiguin Sur I.
    — *L.p.dohertyi* E. Hartert, 1906 — Basilan
    — *L.p.bonapartei* Souancé, 1856 — Sulu Arch.

○ *Loriculus galgulus* (Linnaeus, 1758) Blue-crowned Hanging Parrot — Malay Pen., Borneo, Sumatra and islands

○ *Loriculus stigmatus* (S. Müller, 1843) Sulawesi Hanging Parrot[10] — Sulawesi and islands nearby

○ *Loriculus amabilis* Moluccan Hanging Parrot[11]
    — *L.a.amabilis* Wallace, 1862 — Halmahera, Bacan
    — *L.a.sclateri* Wallace, 1863 — Sula I.
    — *L.a.ruber* Meyer & Wiglesworth, 1896 — Peleng, Banggai Is.

○ *Loriculus catamene* Schlegel, 1873 Sangihe Hanging Parrot — Sangihe I.

---

[1] For the whole of the Psittaciformes we have been guided by Homberger (1980a, b; 1985, 1991, 2002) (1134, 1135, 1136, 1137, 1138) and D. Homberger (*in litt.*).
[2] Homberger (op. cit.) would prefer to treat this as a family.
[3] For recognition of subspecies see Forshaw & Cooper (1989: 156) (868). The vernacular name is retained without a qualifier. It is the only extant Kaka.
[4] Homberger (op. cit.) would prefer to treat this as a family.
[5] For correction of spelling see David & Gosselin (2002) (614).
[6] Homberger (op. cit.) would prefer to treat this as a family.
[7] Homberger (op. cit.) would prefer to treat this as a family.
[8] For recent recognition see Wells (1999: 365) (2579).
[9] Following Dickinson *et al.* (1991) (745). *L. salvadori* was probably based on an aberrant specimen; see also Dickinson *et al.* (op. cit. p. 209).
[10] Includes *croconotus* Jany, 1955 (1201), see White & Bruce (1986) (2652).
[11] Collar (1997) (545) treated this as two species with *ruber* in the synonymy of *sclateri*. Detailed evidence not seen. We prefer to follow White & Bruce (1986: 220) (2652).

○ *Loriculus aurantiifrons* Orange-fronted Hanging Parrot
    — *L.a.aurantiifrons* Schlegel, 1871   Misool I.
    — *L.a.batavorum* Stresemann, 1913   Waigeo I., NW and W New Guinea
    — *L.a.meeki* E. Hartert, 1895   E New Guinea, Fergusson, Goodenough and Karkar Is.
                                                (D'Entrecasteaux Arch.)
    — *L.a.tener* P.L. Sclater, 1877   Bismarck Arch.[1]

○ *Loriculus exilis* Schlegel, 1866 Pygmy Hanging Parrot   Sulawesi

○ *Loriculus pusillus* G.R. Gray, 1859 Yellow-throated Hanging Parrot Java, Bali

○ *Loriculus flosculus* Wallace, 1864 Wallace's Hanging Parrot   Flores

MICROPSITTINAE[2]

MICROPSITTA Lesson, 1831 F
○ *Micropsitta keiensis* Yellow-capped Pygmy Parrot
    — *M.k.keiensis* (Salvadori, 1875)   Kai Is., Aru Is.
    — *M.k.chloroxantha* Oberholser, 1917[3]   W Papuan Is., NW New Guinea
    — *M.k.viridipectus* (Rothschild, 1911)   SW New Guinea

○ *Micropsitta geelvinkiana* Geelvink Pygmy Parrot
    — *M.g.geelvinkiana* (Schlegel, 1871)   Numfor I. (Geelvink Bay)
    — *M.g.misoriensis* (Salvadori, 1875)   Biak I. (Geelvink Bay)

○ *Micropsitta pusio* Buff-faced Pygmy Parrot
    — *M.p.beccarii* (Salvadori, 1876)   N and NE New Guinea
    — *M.p.pusio* (P.L. Sclater, 1866)   SE New Guinea, Bismarck Arch.
    — *M.p.harterti* Mayr, 1940   Fergusson I. (D'Entrecasteaux Arch.) (1470)
    — *M.p.stresemanni* E. Hartert, 1926   Misima and Tagula Is. (Louisiade Arch.)

○ *Micropsitta meeki* Yellow-breasted Pygmy Parrot
    — *M.m.meeki* Rothschild & Hartert, 1914   Admiralty Is.
    — *M.m.proxima* Rothschild & Hartert, 1924   St. Matthias and Squally Is.

○ *Micropsitta finschii* Green Pygmy Parrot
    — *M.f.viridifrons* (Rothschild & Hartert, 1899)   New Hanover, New Ireland, Lihir Is.
    — *M.f.nanina* (Tristram, 1891)   N Solomon Is.
    — *M.f.tristrami* (Rothschild & Hartert, 1902)   C Solomon Is.
    — *M.f.aolae* (Ogilvie-Grant, 1888)   EC Solomon Is.
    — *M.f.finschii* (E.P. Ramsay, 1881)   SE Solomon Is.

○ *Micropsitta bruijnii* Red-breasted Pygmy Parrot
    — *M.b.pileata* Mayr, 1940   Buru and Seram Is. (S Moluccas) (1470)
    — *M.b.bruijnii* (Salvadori, 1875)   Mts. of New Guinea
    — *M.b.necopinata* E. Hartert, 1925   New Britain, New Ireland
    — *M.b.rosea* Mayr, 1940[4]   Bougainville, Guadalcanal and Kolambangra Is. (1470)

CACATUINAE[5]

PROBOSCIGER Kuhl, 1820 M
○ *Probosciger aterrimus* Palm Cockatoo[6]
    — *P.a.stenolophus* (van Oort, 1911)   Yapen I., N and E New Guinea
    — *P.a.goliath* (Kuhl, 1820)   W Papuan Is., W, C and SE New Guinea
    — *P.a.aterrimus* (J.F. Gmelin, 1788)[7]   Aru and Misool Is.
    — *P.a.macgillivrayi* (Mathews, 1912)   S New Guinea, Cape York Pen.

---

[1] Collar (1997) (545) treated *tener* as a separate species (it lacks the orange front).
[2] Homberger (op. cit.) would prefer to treat this as a family.
[3] Tentatively includes *sociabilis* Greenway, 1966 (1007), but see Greenway (1978) (1010).
[4] Including *brevis* Mayr, 1940 (1470).
[5] Homberger (op. cit.) would prefer to treat this as a family with three subfamilies (Calyptorynchinae, Cacatuinae and Nymphicinae).
[6] Revised following Schodde & Mason (1997) (2188).
[7] Peters (1937: 171) reported a restricted type locality in Australia, but this has been corrected to the Aru Islands by Mees (1957) (1508).

CALYPTORHYNCHUS Desmarest, 1826 M
○ *Calyptorhynchus banksii* Red-tailed Black Cockatoo[1]
　　— *C.b.macrorhynchus* Gould, 1843　　　　　NW Australia to E Gulf of Carpentaria
　　— *C.b.banksii* (Latham, 1790)　　　　　　　E Australia (Cape York Pen. to NE New South Wales)
　　— *C.b.graptogyne* Schodde, Saunders & Homberger, 1989[2]　SE South Australia, SW Victoria (2177)
　　— *C.b.samueli* Mathews, 1917　　　　　　　WC to EC Australia (arid zone river systems)
　　— *C.b.naso* Gould, 1837　　　　　　　　　　SW Australia (forest belt)

○ *Calyptorhynchus lathami* Glossy Black Cockatoo
　　— *C.l.erebus* Schodde & Mason, 1993　　　Coastal EC Queensland (2192)
　　— *C.l.lathami* (Temminck, 1807)　　　　　　E Australia
　　— *C.l.halmaturinus* Mathews, 1912　　　　　Kangaroo I.[3]

◉ *Calyptorhynchus funereus* Yellow-tailed Black Cockatoo[4]
　　— *C.f.funereus* (Shaw, 1794)　　　　　　　EC Queensland to E and SC Victoria
　　— *C.f.whiteae* Mathews, 1912　　　　　　　SW Victoria, SC South Australia, Kangaroo I.
　　— *C.f.xanthanotus* Gould, 1838[5]　　　　　Tasmania, Flinders I., King I.

○ *Calyptorhynchus latirostris* Carnaby, 1948 Short-billed Black Cockatoo
　　　　　　　　　　　　　　　　　　　　　　　　Forested extreme SW Western Australia (300)

○ *Calyptorhynchus baudinii* Lear, 1832 Long-billed Black Cockatoo
　　　　　　　　　　　　　　　　　　　　　　　　SW Western Australia (wheat belt and mallee)

CALLOCEPHALON Lesson, 1837 N
◉ *Callocephalon fimbriatum* (J. Grant, 1803) Gang-gang Cockatoo　　SE Australia

EOLOPHUS Bonaparte, 1854 M
◉ *Eolophus roseicapilla* Galah[6] [7]
　　— *E.r.kuhli* (Mathews, 1912)　　　　　　　N Australia
　　— *E.r.roseicapilla* (Vieillot, 1817)　　　　　W and WC Australia
　　⊿ *E.r.albiceps* Schodde, 1989　　　　　　　EC and E Australia to Tasmania (2177)

CACATUA Vieillot, 1817[8] F
○ *Cacatua tenuirostris* (Kuhl, 1820) Long-billed Corella[9]　　SE Australia

○ *Cacatua pastinator* Western Corella
　　— *C.p.pastinator* (Gould, 1841)　　　　　　Extreme SW Western Australia
　　— *C.p.derbyi* (Mathews, 1916)[10]　　　　　SW to WC Western Australia

○ *Cacatua sanguinea* Little Corella
　　— *C.s.transfreta* Mees, 1982　　　　　　　S New Guinea (1534)
　　— *C.s.sanguinea* Gould, 1843　　　　　　　NW Australia, Northern Australia
　　— *C.s.westralensis* (Mathews, 1917)　　　　WC and C Western Australia
　　— *C.s.gymnopis* P.L. Sclater, 1871　　　　　C and inland E Australia
　　— *C.s.normantoni* (Mathews, 1917)　　　　　W Cape York Pen.

○ *Cacatua goffini* (Finsch, 1863) Tanimbar Corella[11]　　Tanimbar Is., Tula (Kai Is.)[12]

○ *Cacatua ducorpsii* Pucheran, 1853 Solomon Islands Corella　　Bougainville to Guadalcanal (Solomon Is.)

○ *Cacatua haematuropygia* (Statius Müller, 1776) Philippine Cockatoo
　　　　　　　　　　　　　　　　　　　　　　　　Philippines incl. Palawan

---

[1] Called *Calyptorhynchus magnificus* in Peters (1937) (1823) and in our last Edition.
[2] The name *Calyptorhynchus magnificus*, which might have been thought available, was based on a specimen of another species (Schodde & Mason, 1997) (2188).
[3] For recognition see Schodde, Mason & Wood (1993) (2192).
[4] Revised, including splitting the species, and using the name *latirostris* for one taxon (the name *tenuirostris* being a junior synonym of *baudinii*), following Schodde & Mason (1997) (2188).
[5] Nec *xanthonotus*.
[6] Revision based on Schodde & Mason (1997) (2188).
[7] For correction of spelling see David & Gosselin (2002) (613).
[8] Most revision in this genus is based on Schodde & Mason (1997) (2188).
[9] Thought to include *mcallani* Wells & Wellington, 1992 (2583).
[10] For recognition see Schodde & Mason (1997) (2188), and for treatment of the name *butleri* Ford, 1987 (864).
[11] A recent proposal to rename this *tanimberensis* Roselaar & Prins, 2000 (2108), because *goffini* is a junior synonym of *ducorpsii*, will be considered in our next Edition.
[12] Feral in Singapore.

○ *Cacatua leadbeateri* Major Mitchell's Cockatoo[1]

    — *C.l.mollis* (Mathews, 1912)   SC to NC Western Australia to C Australia and W South Australia (Eyre Pen.)

    — *C.l.leadbeateri* (Vigors, 1831)   EC Australia (Murray-Darling to E Lake Eyre basins)

○ *Cacatua sulphurea* Yellow-crested Cockatoo[2]

    — *C.s.sulphurea* (J.F. Gmelin, 1788)   Sulawesi and Is.

    — *C.s.abbotti* (Oberholser, 1917)   Masalembu Besar I. (Java Sea)

    — *C.s.parvula* (Bonaparte, 1850)   Main chain of Lesser Sundas (Lombok to Alor and Timor)

    — *C.s.citrinocristata* (Fraser, 1844)   Sumba I.

● *Cacatua galerita* Sulphur-crested Cockatoo

    — *C.g.triton* Temminck, 1849   New Guinea and Is.

    — *C.g.eleonora* (Finsch, 1867)   Aru Is.

    — *C.g.fitzroyi* (Mathews, 1912)   N Australia

    ↙ *C.g.galerita* (Latham, 1790)   E and SE Australia to Tasmania

○ *Cacatua ophthalmica* P.L. Sclater, 1864 Blue-eyed Cockatoo   New Britain

○ *Cacatua alba* (Statius Müller, 1776) White Cockatoo   N and C Moluccas

○ *Cacatua moluccensis* (J.F. Gmelin, 1788) Salmon-crested Cockatoo   S Moluccas

NYMPHICUS Wagler, 1832 M
○ *Nymphicus hollandicus* (Kerr, 1792) Cockatiel   Interior of Australia

LORIINAE[3] [4]

CHALCOPSITTA Bonaparte, 1850 F
○ *Chalcopsitta atra* Black Lory[5]

    — *C.a.bernsteini* Rosenberg, 1861   Misool I.

    — *C.a.atra* (Scopoli, 1786)   Batanta and Salawati Is., NW New Guinea (W Vogelkop)

    — *C.a.insignis* Oustalet, 1878   NW New Guinea (S Vogelkop), Amberpon I.

○ *Chalcopsitta duivenbodei* Brown Lory

    — *C.d.duivenbodei* (A.J.C. Dubois, 1884)   Western NC New Guinea

    — *C.d.syringanuchalis* (Neumann, 1915)   Eastern NC New Guinea

○ *Chalcopsitta sintillata* Yellow-streaked Lory

    — *C.s.rubrifrons* G.R. Gray, 1858   Aru Is.

    — *C.s.sintillata* (Temminck, 1835)   SW New Guinea

    — *C.s.chloroptera* (Salvadori, 1876)   SE New Guinea

○ *Chalcopsitta cardinalis* (G.R. Gray, 1849) Cardinal Lory   Hibernian Is. (E of New Ireland) to all Solomon Is.

EOS Wagler, 1832 F
○ *Eos histrio* Red-and-blue Lory

    — *E.h.challengeri* Salvadori, 1891   Miangas I. (Philippines)

    — *E.h.talautensis* Meyer & Wiglesworth, 1894   Talaud Is. (N of the Moluccas)

    — *E.h.histrio* (Statius Müller, 1776)   Sangihe, Siau and Ruang Is. (near Sulawesi)

○ *Eos squamata* Violet-necked Lory

    — *E.s.riciniata* (Bechstein, 1811)[6]   N Moluccas

    — *E.s.obiensis* Rothschild, 1899   Obi and Bisa Is. (NC Moluccas)

    — *E.s.squamata* (Boddaert, 1783)[7]   W Papuan Is., Schildpad Is.

---

[1] Revision based on Schodde (1993) (2179).
[2] Feral in Singapore.
[3] Homberger (op. cit.) would prefer to treat this as a family).
[4] For a discussion see Schodde & Mason (1997: 110-111) (2188).
[5] In last Edition we listed *spectabilis* but Mayr (1941: 52) (1472) suggested hybrid origin and this is now considered the case.
[6] Tentatively including *atrocaerulea* Jany, 1955 (1200) see Mees (1965) (1515).
[7] Includes *attenua* Ripley, 1957 (2039), not *attenuata* as given by Mees (1965) (1515).

○ *Eos rubra* RED LORY[1]
   — *E.r.cyanonotha* (Vieillot, 1818)[2]   Buru I.
   — *E.r.rubra* (J.F. Gmelin, 1788)   Ambon and Seram to Kai Is.

○ *Eos reticulata* (S. Müller, 1841) BLUE-STREAKED LORY   Tanimbar, Kai and Damar Is.

🔍 ○ *Eos cyanogenia* Bonaparte, 1850 BLACK-WINGED LORY   Islands in Geelvink Bay (New Guinea)

○ *Eos semilarvata* Bonaparte, 1850 BLUE-EARED LORY   Seram (Moluccas)

PSEUDOS J.L. Peters, 1935 F
○ *Pseudeos fuscata* (Blyth, 1858) DUSKY LORY   New Guinea, Salawati and Yapen Is.

TRICHOGLOSSUS Vigors & Horsfield, 1827 M
○ *Trichoglossus ornatus* (Linnaeus, 1758) ORNATE LORIKEET   Sulawesi and nearby islands

○ *Trichoglossus forsteni* SCARLET-BREASTED LORIKEET
   — *T.f.djampeanus* E. Hartert, 1897   Tanahjampea I.
   — *T.f.stresemanni* Meise, 1929   Kalaotoa I.
   — *T.f.mitchellii* G.R. Gray, 1859   Bali and Lombok Is.
   — *T.f.forsteni* Bonaparte, 1850   Sumbawa I.

○ *Trichoglossus weberi* (Büttikofer, 1894) FLORES LORIKEET   Flores I.

○ *Trichoglossus capistratus* MARIGOLD LORIKEET
   — *T.c.fortis* E. Hartert, 1898   Sumba I.
   — *T.c.capistratus* (Bechstein, 1811)   Timor I.
   — *T.c.flavotectus* Hellmayr, 1914   Wetar and Romang Is.

◉ *Trichoglossus haematodus* RAINBOW LORIKEET[3]
   — *T.h.haematodus* (Linnaeus, 1771)[4]   S Moluccas, W Papuan Is., NW and SW New Guinea
   — *T.h.rosenbergii* Schlegel, 1871   Biak I.
   — *T.h.intermedius* Rothschild & Hartert, 1901   NC New Guinea
   — *T.h.nigrogularis* G.R. Gray, 1858[5]   Kai Is., Aru Is., S New Guinea
   — *T.h.caeruleiceps* D'Albertis & Salvadori, 1879   SC New Guinea
   — *T.h.massena* Bonaparte, 1854[6]   SE, NE New Guinea, Louisiade Arch., Karkar I,
                                      Bismarck Arch., Solomon Is, Vanuatu
   — *T.h.nesophilus* Neumann, 1929   Ninigo Is., Hermit Is.
   — *T.h.flavicans* Cabanis & Reichenow, 1876   New Hanover, St. Mathias I., Admiralty Is.
   — *T.h.deplanchii* Verreaux & Des Murs, 1860   New Caledonia, Loyalty Is.
   — *T.h.septentrionalis* Robinson, 1900   Cape York Pen.
   — *T.h.moluccanus* (J.F. Gmelin, 1788)   NE and SE Australia, Tasmania
   — *T.h.eyrei* Mathews, 1912   SE and SC South Australia, W Victoria

○ *Trichoglossus rubritorquis* Vigors & Horsfield, 1827 RED-COLLARED LORIKEET
                                      Easternmost Lesser Sundas, NW Australia east to S
                                        Gulf of Carpentaria

○ *Trichoglossus euteles* (Temminck, 1835) OLIVE-HEADED LORIKEET   Timor, Lomblen Is. to Babar I. (S Banda Sea)

○ *Trichoglossus flavoviridis* YELLOW-AND-GREEN LORIKEET
   — *T.f.meyeri* Walden, 1871   Sulawesi
   — *T.f.flavoviridis* Wallace, 1863   Sula Is.

○ *Trichoglossus johnstoniae* E. Hartert, 1903 MINDANAO LORIKEET[7]   W and SC Mindanao

○ *Trichoglossus rubiginosus* (Bonaparte, 1850) POHNPEI LORIKEET   Pohnpei I. (E Caroline Is.)

○ *Trichoglossus chlorolepidotus* (Kuhl, 1820) SCALY-BREASTED LORIKEET   E Australia

---

[1] For reasons to use this name in place of *E. bornea* see Walters (1998) (2557). It has been suggested that *E. goodfellowi* was a name applied to the immature see Collar (1997: 342) (545).
[2] Not *cyanothorus* as given in Collar (1997) (545). Although the title page says 1817 this volume is usually and we believe correctly dated 1818.
[3] Revised in accordance with Schodde & Mason (1997: 114-115) (2188) with 5 species resulting.
[4] Including *berauensis* Cain, 1955 (288) (see Mees, 1980) (1531).
[5] Including *brooki* see Mayr (1941: 55) (1472). The next taxon is but slightly distinct.
[6] Tentatively includes *micropteryx* (R. Schodde, *in litt.*).
[7] Including *pistra* Rand & Rabor, 1959 (1976) see Collar (1997) (545).

PSITTEUTELES Bonaparte, 1854[1] M
○ *Psitteuteles versicolor* (Lear, 1831) VARIED LORIKEET — N Australia

○ *Psitteuteles iris* IRIS LORIKEET
   — *P.i.iris* (Temminck, 1835)[2] — Timor
   — *P.i.wetterensis* (Hellmayr, 1912) — Wetar I.

○ *Psitteuteles goldiei* (Sharpe, 1882) GOLDIE'S LORIKEET — Montane New Guinea

LORIUS Vigors, 1825 M
○ *Lorius garrulus* CHATTERING LORY
   — *L.g.morotaianus* (van Bemmel, 1940) — Morotai and Rau Is. (N Moluccas) (2460)
   — *L.g.garrulus* (Linnaeus, 1758) — Halmahera and adjacent Is.
   — *L.g.flavopalliatus* Salvadori, 1877 — Bacan, Obi, Kasiruta and Mandiole Is.

○ *Lorius domicella* (Linnaeus, 1758)[3] PURPLE-NAPED LORY[4] — Seram and Ambon Is. (S Moluccas)

○ *Lorius lory* BLACK-CAPPED LORY
   — *L.l.lory* (Linnaeus, 1758) — W Papuan Is., Vogelkop (NW New Guinea)
   — *L.l.cyanauchen* (S. Müller, 1841)[5] — Biak I. (Geelvink Bay)
   — *L.l.jobiensis* (A.B. Meyer, 1874) — Yapen and Mios Num Is. (Geelvink Bay)
   — *L.l.viridicrissalis* de Beaufort, 1909 — Western N New Guinea
   — *L.l.salvadorii* A.B. Meyer, 1891 — Eastern N New Guinea
   — *L.l.erythrothorax* Salvadori, 1877 — SW to SE New Guinea
   — *L.l.somu* (Diamond, 1967)[6] — Inland SC New Guinea (697)

○ *Lorius hypoinochrous* PURPLE-BELLIED LORY
   — *L.h.devittatus* E. Hartert, 1898[7] — Bismarck Arch., SE New Guinea, Trobriand Is., D'Entrecasteaux Arch., Woodlark I.
   — *L.h.hypoinochrous* G.R. Gray, 1859 — Misima and Tagula Is. (C Louisiade Arch.)
   — *L.h.rosselianus* Rothschild & Hartert, 1918 — Rossel I. (E Louisiade Arch.)

○ *Lorius albidinucha* (Rothschild & Hartert, 1924) WHITE-NAPED LORY[8] — SE New Ireland

○ *Lorius chlorocercus* Gould, 1856 YELLOW-BIBBED LORY — E Solomon Is.

PHIGYS G.R. Gray, 1870 M
○ *Phigys solitarius* (Suckow, 1800) COLLARED LORY — Fiji Is.

VINI Lesson, 1831 F
○ *Vini australis* (J.F. Gmelin, 1788) BLUE-CROWNED LORIKEET — Niue, Samoa and Tonga Is. to Lau Arch.
○ *Vini kuhlii* (Vigors, 1824) KUHL'S LORIKEET — Rimitara, Kiritimati, Tabuaeran Teraina and Kiribati Is.
○ *Vini stepheni* (North, 1908) HENDERSON LORIKEET — Henderson I. (Pitcairn Is.)
○ *Vini peruviana* (Statius Müller, 1776) BLUE LORIKEET — Cook Is., Society Is., Tuamotu Arch.
○ *Vini ultramarina* (Kuhl, 1820) ULTRAMARINE LORIKEET — Marquesas Is.

GLOSSOPSITTA Bonaparte, 1854 F
○ *Glossopsitta concinna* MUSK LORIKEET
   — *G.c.concinna* (Shaw, 1791) — E and SE Australia, Kangaroo I.
   — *G.c.didimus* Mathews, 1915[9] — Tasmania

○ *Glossopsitta pusilla* (Shaw, 1790) LITTLE LORIKEET — E and SE Australia

○ *Glossopsitta porphyrocephala* (Dietrichsen, 1837) PURPLE-CROWNED LORIKEET — SW to inland SE Australia, Kangaroo I.

[1] These three species were treated within the genus *Trichoglossus* in our last Edition. In using this generic name we follow Collar (1997) (545).
[2] For reasons to treat *rubripileum* as a synonym see White & Bruce (1986: 216) (2652).
[3] Not *domicellus* (as in last Edition).
[4] Collar (1997: 349) (545) associated the name *L. tibialis* with this.
[5] Although recently queried, "1841" as used by Peters (1937) was sustained by Husson & Holthuis (1955) (1169).
[6] This form requires more investigation due to its apparently highly local range.
[7] Collar (1997: 349) (545) associated the name *L. amabilis* with this.
[8] For correction of spelling see David & Gosselin (2002) (613).
[9] For recognition see Schodde & Mason (1997) (2188).

CHARMOSYNA Wagler, 1832 F
○ *Charmosyna palmarum* (J.F. Gmelin, 1788) PALM LORIKEET   Vanuatu, Banks Is., Santa Cruz Is., Duff Is.

○ *Charmosyna rubrigularis* (P.L. Sclater, 1881) RED-CHINNED LORIKEET[1]
                                                                   New Britain, New Ireland, Karkar I.

○ *Charmosyna meeki* (Rothschild & Hartert, 1901) MEEK'S LORIKEET Solomon Is.

○ †?*Charmosyna toxopei* (Siebers, 1930) BLUE-FRONTED LORIKEET   Buru (S Moluccas)

○ *Charmosyna multistriata* (Rothschild, 1911) STRIATED LORIKEET Central SW New Guinea

○ *Charmosyna wilhelminae* (A.B. Meyer, 1874) PYGMY LORIKEET   Montane W, C and E New Guinea

○ *Charmosyna rubronotata* RED-FRONTED LORIKEET
     — *C.r.rubronotata* (Wallace, 1862)   Salawati I., NW and NC New Guinea
     — *C.r.kordoana* (A.B. Meyer, 1874)   Biak I.

○ *Charmosyna placentis* RED-FLANKED LORIKEET
     — *C.p.intensior* (Kinnear, 1928)[2]   N Moluccas, Gebe I. (W Papuan Is.)
     — *C.p.placentis* (Temminck, 1835)[3][4]   S Moluccas, Aru Is., SW to SC New Guinea
     — *C.p.ornata* Mayr, 1940   W Papuan Is., NW New Guinea (1470)
     — *C.p.subplacens* (P.L. Sclater, 1876)   NE and SE New Guinea
     — *C.p.pallidior* (Rothschild & Hartert, 1905)   Woodlark I., Bismarck Arch. to W Solomon Is.

○ †?*Charmosyna diadema* (Verreaux & Des Murs, 1860) NEW CALEDONIAN LORIKEET
                                                                   New Caledonia

○ *Charmosyna amabilis* (E.P. Ramsay, 1875) RED-THROATED LORIKEET Viti Levu, Vanu Levu, Ovalau, Taveuni Is. (Fiji Is.)

○ *Charmosyna margarethae* Tristram, 1879 DUCHESS LORIKEET   Solomon Is.

○ *Charmosyna pulchella* FAIRY LORIKEET
     — *C.p.pulchella* G.R. Gray, 1859[5]   Mts. of Vogelkop and central cordillera, New Guinea
     — *C.p.rothschildi* (E. Hartert, 1930)   Cyclops Mts. of NC New Guinea

○ *Charmosyna josefinae* JOSEPHINE'S LORIKEET
     — *C.j.josefinae* (Finsch, 1873)   Montane W New Guinea
     — *C.j.cyclopum* E. Hartert, 1930   Cyclops Mts. (NC New Guinea)
     — *C.j.sepikiana* Neumann, 1922   Montane C New Guinea

○ *Charmosyna papou* PAPUAN LORIKEET
     — *C.p.papou* (Scopoli, 1786)   Mts. of the Vogelkop (NW New Guinea)
     — *C.p.goliathina* Rothschild & Hartert, 1911   Mts. of W and C New Guinea
     — *C.p.wahnesi* Rothschild, 1906   Mts. of Huon Pen. (New Guinea)
     — *C.p.stellae* A.B. Meyer, 1886   Mts. of SE New Guinea

OREOPSITTACUS Salvadori, 1877 M
○ *Oreopsittacus arfaki* PLUM-FACED LORIKEET
     — *O.a.arfaki* (A.B. Meyer, 1874)   Mts. of the Vogelkop (NW New Guinea)
     — *O.a.major* Ogilvie-Grant, 1914   Mts. of WC New Guinea
     — *O.a.grandis* Ogilvie-Grant, 1895   Mts. of C and E New Guinea

NEOPSITTACUS Salvadori, 1875 M
○ *Neopsittacus musschenbroekii* YELLOW-BILLED LORIKEET
     — *N.m.musschenbroekii* (Schlegel, 1871)   Mts. of the Vogelkop (NW New Guinea)
     — *N.m.major* Neumann, 1924[6]   Mts. of NC to SE New Guinea

○ *Neopsittacus pullicauda* ORANGE-BILLED LORIKEET
     — *N.p.alpinus* Ogilvie-Grant, 1914   Mts. of WC New Guinea
     — *N.p.socialis* Mayr, 1931   Mts. of NE New Guinea
     — *N.p.pullicauda* E. Hartert, 1896   Mts. of EC and SE New Guinea

---

[1] In last Edition we listed a subspecies *krakari*, but the species is now treated as monotypic see Forshaw (1989: 98) (868).
[2] Reviewed by Mees (1965) (1515), but no material of *brevipennis* Jany, 1955 (1200), was available for study; White & Bruce (1986) (2652) placed it here.
[3] Livr. 93 dates from 1835; Peters (1937: 159) (1823) gave 1834 *contra* Sherborn (1898) (2242).
[4] Includes *occidentalis* Mayr, 1940 (1470), see White & Bruce (1986) (2652).
[5] Includes *bella* see Forshaw (1978) (867).
[6] Includes *medius* Stresemann, 1936 (1085), see Collar (1997) (545).

PSITTACINAE[1]

PROSOPEIA Bonaparte, 1854[2] F

O *Prosopeia splendens* (Peale, 1848) CRIMSON SHINING PARROT

Kandavu, Ono and Viti Levu Is. (Fiji Is.)

O *Prosopeia personata* (G.R. Gray, 1848) MASKED SHINING PARROT

Viti Levu I. (Fiji Is.)

O *Prosopeia tabuensis* RED SHINING PARROT
    — *P.t.taviunensis* (E.L. Layard, 1876)

Taveuni and Ngamea Is. (Fiji Is.)

    — *P.t.tabuensis* (J.F. Gmelin, 1788)

Vanua Levu, Kioa, Koro and Gau Is. (Fiji Is.), Eua I. (Tonga)

EUNYMPHICUS J.L. Peters, 1937[3] M

O *Eunymphicus cornutus* HORNED PARAKEET
    — *E.c.cornutus* (J.F. Gmelin, 1788)

New Caledonia

    — *E.c.uvaeensis* (E.L. & E.L.C. Layard, 1882)

Ouvea I. (Loyalty Is.)

CYANORAMPHUS Bonaparte, 1854[4] M

O †*Cyanoramphus zealandicus* (Latham, 1790) BLACK-FRONTED PARAKEET

Tahiti (Society Is.)

O †*Cyanoramphus ulietanus* (J.F. Gmelin, 1788) RAIATEA PARAKEET

Raiatea (Society Is.)

O *Cyanoramphus saisseti* Verreaux & Des Murs, 1860 RED-CROWNED PARAKEET[5]

New Caledonia

O *Cyanoramphus forbesi* Rothschild, 1893 CHATHAM PARAKEET[6]

Chatham Is.

O *Cyanoramphus cookii* (G.R. Gray, 1859) NORFOLK ISLAND PARAKEET[7]

Norfolk I.

O *Cyanoramphus unicolor* (Lear, 1831) ANTIPODES PARAKEET

Antipodes Is.

O *Cyanoramphus auriceps* (Kuhl, 1820) YELLOW-FRONTED PARAKEET

New Zealand and nearby islands, Stewart I., Auckland Is.

O *Cyanoramphus malherbi* Souancé, 1857 MALHERBE'S PARAKEET[8]

N South Island (New Zealand)

O *Cyanoramphus novaezelandiae* RED-FRONTED PARAKEET
    — †*C.n.subflavescens* Salvadori, 1891

Lord Howe I.

    — *C.n.cyanurus* Salvadori, 1891

Kermadec Is.

    — *C.n.novaezelandiae* (Sparrman, 1787)

New Zealand and nearby islands, Auckland Is., Stewart I.

    — *C.n.chathamensis* Oliver, 1930

Chatham Is.

    — *C.n.hochstetteri* (Reischek, 1889)

Antipodes Is.

    — †*C.n.erythrotis* (Wagler, 1832)

Macquarie I.

PURPUREICEPHALUS Bonaparte, 1854 M

O *Purpureicephalus spurius* (Kuhl, 1820) RED-CAPPED PARROT

SW Western Australia

BARNARDIUS Bonaparte, 1854 M

O *Barnardius zonarius* AUSTRALIAN RINGNECK[9]
    — *B.z.macgillivrayi* (North, 1900)

NW Queensland to E Northern Territory

    — *B.z.barnardi* (Vigors & Horsfield, 1827)

Interior of SE Australia

    — *B.z.semitorquatus* (Quoy & Gaimard, 1830)

SW Western Australia

    — *B.z.zonarius* (Shaw, 1805)

WC and SC Western Australia to W and C South Australia

PLATYCERCUS Vigors, 1825 M

O *Platycercus caledonicus* GREEN ROSELLA[10]
    — *P.c.caledonicus* (J.F. Gmelin, 1788)

Tasmania, Flinders I.

    — *P.c.brownii* (Kuhl, 1820)

King I. (Bass Strait)

---

[1] Homberger (op. cit.) would prefer to treat this as a family (with five subfamilies: Platycercinae, Psittaculinae, Psittacinae, Arinae and Cyclopsittacinae).
[2] This is the first genus in the Platycercini.
[3] This genus not recognised in last Edition, but see, for example, Forshaw (1978) (867) whom we also follow in accepting just one species.
[4] Species sequence based on Wee *et al.* (2001) (2569).
[5] For reasons to elevate this to specific status see Wee *et al.* (2001) (2569).
[6] For treatment at species level see Triggs & Daugherty (1996) (2438).
[7] For reasons to elevate this to specific status see Wee *et al.* (2001) (2569).
[8] Omitted from our last Edition, thought to be a colour morph (and may well be one); for conservation reasons to treat it as a species see Triggs & Daugherty (1996) (2438). Tentatively accepted here following Higgins (1999: 492) (1112).
[9] We follow Schodde & Mason (1997) (2188) in lumping *barnardi* and *zonarius* in one species and reducing the recognised forms to four.
[10] For recognition of subspecies see Schodde & Mason (1997) (2188).

● *Platycercus elegans* CRIMSON ROSELLA[1]
    ⊽ *P.e.nigrescens* E.P. Ramsay, 1888   NE Queensland
    ✓ *P.e.elegans* (J.F. Gmelin, 1788)[2]   EC Queensland to SE South Australia
    — *P.e.melanopterus* North, 1906[3]   Kangaroo I.
    — *P.e.fleurieuensis* Ashby, 1917   Mt. Lofty range (South Australia)
    — *P.e.subadelaidae* Mathews, 1912   S Flinders range (South Australia)
    — *P.e.flaveolus* Gould, 1837   Murray River (interior of SE Australia

○ *Platycercus venustus* NORTHERN ROSELLA[4]
    — *P.v.venustus* (Kuhl, 1820)   N Northern Territory and extreme NW Queensland
    — *P.v.hilli* Mathews, 1910   NE Western Australia, NW Northern Territory

○ *Platycercus adscitus* PALE-HEADED ROSELLA[5]
    — *P.a.adscitus* (Latham, 1790)   Cape York Pen.
    — *P.a.palliceps* Lear, 1832[6]   NE Queensland to N New South Wales

● *Platycercus eximius* EASTERN ROSELLA
    — *P.e.elecica* Schodde & Short, 1989[7]   Extreme SE Queensland, NE New South Wales (2177)
    — *P.e.eximius* (Shaw, 1792)   SE to SC Australia
    — *P.e.diemenensis* North, 1911   Tasmania

○ *Platycercus icterotis* WESTERN ROSELLA
    — *P.i.icterotis* (Kuhl, 1820)   Coastal SW Western Australia
    — *P.i.xanthogenys* Salvadori, 1891   Inland SW Western Australia

NORTHIELLA Mathews, 1912 F
○ *Northiella haematogaster* BLUEBONNET
    — *N.h.haematorrhous* (Gould, 1865)[8]   SE Queensland, NE and EC New South Wales
    — *N.h.haematogaster* (Gould, 1838)   SC Queensland, W and C New South Wales, NW
                                             Victoria, EC South Australia
    — *N.h.pallescens* (Salvadori, 1891)   NE South Australia
    — *N.h.narethae* (H.L. White, 1921)   Extreme SW South Australia, SE Western Australia

PSEPHOTUS Gould, 1845 M
○ *Psephotus haematonotus* RED-RUMPED PARROT
    — *P.h.caeruleus* Condon, 1941   NE South Australia, SW Queensland (555)
    — *P.h.haematonotus* (Gould, 1838)   Inland SE Australia

○ *Psephotus varius* Clark, 1910 MULGA PARROT[9]   WC, SC, C and inland S and EC Australia

○ *Psephotus dissimilis* Collett, 1898 HOODED PARROT[10]   C Arnhem Land (Northern Territory)

○ *Psephotus chrysopterygius* Gould, 1858 GOLDEN-SHOULDERED PARROT   S Cape York Pen.

○ †*Psephotus pulcherrimus* (Gould, 1845) PARADISE PARROT   C and S Queensland, N New South Wales

NEOPSEPHOTUS Mathews, 1912 M
○ *Neopsephotus bourkii* (Gould, 1841) BOURKE'S PARROT   W, SC and EC Australia

NEOPHEMA Salvadori, 1891 F
○ *Neophema chrysostoma* (Kuhl, 1820) BLUE-WINGED PARROT   SE Australia, Tasmania

○ *Neophema elegans* ELEGANT PARROT
    — *N.e.elegans* (Gould, 1837)   SC South Australia, Kangaroo I.
    — *N.e.carteri* (Mathews, 1912)[11]   SW Western Australia

---

[1] This treatment, which includes lumping *P. flaveolus* and the placement of *adelaidae* in synonymy, follows Schodde & Mason (1997) (2188).
[2] Including *filewoodi* McAllan & Bruce, 1989 (1499) see Schodde & Mason (1997) (2188).
[3] For correction of spelling see David & Gosselin (2002) (613).
[4] Two forms recognised following Schodde & Mason (1997) (2188).
[5] There is a broad area of introgression between this species and the next and the two may be conspecific (Schodde & Mason, 1997) (2188). In the context of evaluating this complex the name *aureodorsalis* McAllan & Bruce, 1989 (1499), a synonym of *elecica*, was also proposed.
[6] Includes *amathusia* and *mackaiensis* Cain, 1955 (288) see Schodde & Mason (1997) (2188).
[7] The description attached to *P. splendidus* for which the name *cecilae* Mathews was proposed as a nomen novum is now considered insufficient to identify the taxon.
[8] Spelling as advised by N. David (*in litt.*) see also David & Gosselin (2002) (614).
[9] For submergence of *orientalis* see Schodde & Mason (1997) (2188). The name *dulciei* Mathews, 1911 (1453), is a synonym which is not in Peters (1937) (1823).
[10] For separation of this from *chrysopterygius* see Schodde & Mason (1997) (2188).
[11] For the resurrection of this form see Schodde & Mason (1997) (2188).

○ *Neophema petrophila* ROCK PARROT
  — *N.p.petrophila* (Gould, 1841)   Coastal SW Western Australia
  — *N.p.zietzi* (Mathews, 1912)   Coastal C and W South Australia

○ *Neophema chrysogaster* (Latham, 1790) ORANGE-BELLIED PARROT   W Tasmania, coastal S Victoria and SE South Australia

○ *Neophema pulchella* (Shaw, 1792) TURQUOISE PARROT   Inland SE Queensland to NC Victoria

○ *Neophema splendida* (Gould, 1841) SCARLET-CHESTED PARROT   SC Australia

LATHAMUS Lesson, 1830 M
○ *Lathamus discolor* (Shaw, 1790)[1] SWIFT PARROT   Tasmania >> SE and E Australia

MELOPSITTACUS Gould, 1840 M
○ *Melopsittacus undulatus* (Shaw, 1805) BUDGERIGAR   Inland Australia

PEZOPORUS Illiger, 1811 M
○ *Pezoporus wallicus* GROUND PARROT[2]
  — *P.w.flaviventris* North, 1911   Coastal extreme SW Western Australia
  — *P.w.wallicus* (Kerr, 1792)   Coastal SE and E Australia
  — *P.w.leachi* Mathews, 1912   W Tasmania

○ *Pezoporus occidentalis* (Gould, 1861) NIGHT PARROT[3]   Inland W and C Australia

PSITTACELLA Schlegel, 1873 F
○ *Psittacella brehmii* BREHM'S TIGER PARROT
  — *P.b.brehmii* Schlegel, 1871   Mts. of the Vogelkop (NW New Guinea)
  — *P.b.intermixta* E. Hartert, 1930   Mts. of WC New Guinea
  — *P.b.harterti* Mayr, 1931   Mts. of Huon Pen. (E New Guinea)
  — *P.b.pallida* A.B. Meyer, 1886   Mts. of EC and SE New Guinea

○ *Psittacella picta* PAINTED TIGER PARROT
  — *P.p.lorentzi* van Oort, 1910   Alpine WC New Guinea
  — *P.p.excelsa* Mayr & Gilliard, 1951   Alpine C New Guinea (1489)
  — *P.p.picta* Rothschild, 1896   Alpine SE New Guinea

○ *Psittacella modesta* MODEST TIGER PARROT
  — *P.m.modesta* Schlegel, 1871   Mts. of the Vogelkop (NW New Guinea)
  — *P.m.subcollaris* Rand, 1941   Nthn. mts. of C New Guinea (1964)
  — *P.m.collaris* Ogilvie-Grant, 1914   Sthn. mts. of WC New Guinea

○ *Psittacella madaraszi* MADARÁSZ'S TIGER PARROT
  — *P.m.major* Rothschild, 1936   Mts. of WC New Guinea (1085)
  — *P.m.hallstromi* Mayr & Gilliard, 1951   Mts. of C New Guinea (1489)
  — *P.m.huonensis* Mayr & Rand, 1935   Mts. of Huon Pen. (E New Guinea)
  — *P.m.madaraszi* A.B. Meyer, 1886   Mts. of SE New Guinea

PSITTINUS Blyth, 1842[4] M
○ *Psittinus cyanurus* BLUE-RUMPED PARROT
  — *P.c.cyanurus* (J.R. Forster, 1795)   S Burma and S Thailand to Sumatra and Borneo
  — *P.c.abbotti* Richmond, 1902   Simeuluë and Siumat Is. (NW Sumatra)
  — *P.c.pontius* Oberholser, 1912   Mentawai Is. (NW Sumatra)

GEOFFROYUS Bonaparte, 1850 M
○ *Geoffroyus geoffroyi* RED-CHEEKED PARROT
  — *G.g.cyanicollis* (S. Müller, 1841)   N Moluccas
  — *G.g.obiensis* (Finsch, 1868)   Obi (NC Moluccas)
  — *G.g.rhodops* (Schlegel, 1864)[5]   S Moluccas

---

[1] For the correction of the author's name, given as White by Peters (1937) (1823), see Schodde & Mason (1997) (2188).
[2] For recognition of *leachi* see Schodde & Mason (1997) (2188).
[3] Reasons for treatment in this genus rather than in *Geopsittacus* can be found in Schodde & Mason (1997) (2188).
[4] This is the first genus in the tribe Psittaculini.
[5] Includes *explorator* see White & Bruce (1986: 223) (2652).

— *G.g.keyensis* Finsch, 1868   Kai Is.

— *G.g.floresianus* Salvadori, 1891   W Lesser Sundas

— *G.g.geoffroyi* (Bechstein, 1811)   E Lesser Sundas

— *G.g.timorlaoensis* A.B. Meyer, 1884   Tanimbar Is.

— *G.g.pucherani* Souancé, 1856   W Papuan Is., Vogelkop (NW New Guinea)

— *G.g.minor* Neumann, 1922   N New Guinea

— *G.g.jobiensis* (A.B. Meyer, 1874)   Yapen and Meos Num Is. (Geelvink Bay)

— *G.g.mysoriensis* (A.B. Meyer, 1874)   Biak and Numfor Is. (Geelvink Bay)

— *G.g.orientalis* A.B. Meyer, 1891   Huon Pen. (E New Guinea)

— *G.g.sudestiensis* De Vis, 1890   Misima and Tagula Is. (Louisiade Arch.)

— *G.g.cyanicarpus* E. Hartert, 1899   Rossel I. (Louisiade Arch.)

— *G.g.aruensis* (G.R. Gray, 1858)   Aru Is., SW to SE New Guinea

— *G.g.maclennani* (W.D.K. Macgillivray, 1913)[1]   Cape York Pen.

○ *Geoffroyus simplex* Blue-collared Parrot

— *G.s.simplex* (A.B. Meyer, 1874)   Vogelkop (NW New Guinea)

— *G.s.buergersi* Neumann, 1922   WC to SE New Guinea

○ *Geoffroyus heteroclitus* Singing Parrot

— *G.h.heteroclitus* (Hombron & Jacquinot, 1841)   Bismarck Arch. to Solomon Is. (except Rennell I.).

— *G.h.hyacinthinus* Mayr, 1931   Rennell I.

PRIONITURUS Wagler, 1832 M

○ *Prioniturus montanus* Montane Racquet-tail[2]

— *P.m.montanus* Ogilvie-Grant, 1895   Luzon

— *P.m.malindangensis* Mearns, 1909   W Mindanao

— *P.m.waterstradti* Rothschild, 1904   SE Mindanao

○ *Prioniturus platenae* W.H. Blasius, 1888 Blue-headed Racquet-tail   Palawan and nearby islands

○ *Prioniturus luconensis* Steere, 1890 Green Racquet-tail   Luzon, Marinduque

○ *Prioniturus discurus* Blue-crowned Racquet-tail

— *P.d.whiteheadi* Salomonsen, 1953[3]   Luzon to Leyte and Bohol (2155)

— *P.d.mindorensis* Steere, 1890   Mindoro

— *P.d.discurus* (Vieillot, 1822)   Mindanao, Basilan; Jolo and Olutanga (Sulu Is.)

○ *Prioniturus verticalis* Sharpe, 1893 Blue-winged Racquet-tail   Sulu Arch.

○ *Prioniturus flavicans* Cassin, 1853 Yellowish-breasted Racquet-tail   N and C Sulawesi

○ *Prioniturus platurus* Golden-mantled Racquet-tail

— *P.p.talautensis* E. Hartert, 1898   Talaud Is.

— *P.p.platurus* (Vieillot, 1818)   Sulawesi and nearby islands

— *P.p.sinerubris* Forshaw, 1971   Taliabu I. (866)

○ *Prioniturus mada* E. Hartert, 1900 Buru Racquet-tail   Buru (S Moluccas)

TANYGNATHUS Wagler, 1832 M

○ *Tanygnathus megalorhynchos* Great-billed Parrot

— *T.m.megalorhynchos* (Boddaert, 1783)[4]   Talaud and Sangihe Is. to Lesser Sundas, N and C Moluccas and W Papuan Is.

— *T.m.affinis* Wallace, 1863   S Moluccas

— *T.m.sumbensis* A.B. Meyer, 1881   Sumba I.

— *T.m.hellmayri* Mayr, 1944   W Timor, Semao I. (1474)

— *T.m.subaffinis* P.L. Sclater, 1883   Babar and Tanimbar Is.

○ *Tanygnathus lucionensis* Blue-naped Parrot

— *T.l.lucionensis* (Linnaeus, 1766)   Luzon, Mindoro

— *T.l.hybridus* Salomonsen, 1952   Polillo I. (2154)

---

[1] For the resurrection of this race see Schodde & Mason (1997) (2188).
[2] We follow the arrangement of Dickinson *et al.* (1991) (745).
[3] Includes *nesophilus* Salomonsen, 1953 (2155), see Dickinson *et al.* (1991) (745).
[4] Includes *insularum* Jany, 1955 (1200), see White & Bruce (1986) (2652).

___ *T.l.salvadorii* Ogilvie-Grant, 1896[1]   C Philippines to NE Borneo Is.
___ *T.l.talautensis* Meyer & Wiglesworth, 1895   Talaud Is.

○ *Tanygnathus sumatranus* Blue-backed Parrot[2]
  ___ *T.s.duponti* Parkes, 1971   Luzon (1764)
  ___ *T.s.freeri* McGregor, 1910   Polillo I.
  ___ *T.s.everetti* Tweeddale, 1877   Visayan Is., Mindanao
  ___ *T.s.burbidgii* Sharpe, 1879   Sulu Is.
  ___ *T.s.sumatranus* (Raffles, 1822)[3]   Sulawesi and nearby islands to the north and west including Sangihe, Talaud, Peleng and Banggai

○ *Tanygnathus gramineus* (J.F. Gmelin, 1788) Black-lored Parrot   Buru I. (S Moluccas)

ECLECTUS Wagler, 1832 M
○ *Eclectus roratus* Eclectus Parrot
  ___ *E.r.vosmaeri* (Rothschild, 1922)   N and C Moluccas
  ___ *E.r.roratus* (Statius Müller, 1776)   S Moluccas
  ___ *E.r.cornelia* Bonaparte, 1850   Sumba I.
  ___ *E.r.riedeli* A.B. Meyer, 1882   Tanimbar Is.
  ___ *E.r.aruensis* G.R. Gray, 1858   Aru Is.
  ___ *E.r.biaki* (E. Hartert, 1932)   Biak I.
  ___ *E.r.polychloros* (Scopoli, 1786)   New Guinea and Is.
  ___ *E.r.solomonensis* Rothschild & Hartert, 1901   Admiralty Is., Bismarck Arch., Solomon Is.
  ___ *E.r.macgillivrayi* Mathews, 1913   Cape York Pen. (Australia)

ALISTERUS Mathews, 1911 M
○ *Alisterus amboinensis* Moluccan King Parrot
  ___ *A.a.hypophonius* (S. Müller, 1843)   Halmahera
  ___ *A.a.sulaensis* (Reichenow, 1881)   Sula Is.
  ___ *A.a.versicolor* Neumann, 1939[4]   Peleng I. (1646)
  ___ *A.a.buruensis* (Salvadori, 1876)   Buru
  ___ *A.a.amboinensis* (Linnaeus, 1766)   Ambon and Seram Is.
  ___ *A.a.dorsalis* (Quoy & Gaimard, 1830)   W Papuan Is., Vogelkop (NW New Guinea)

○ *Alisterus chloropterus* Papuan King Parrot
  ___ *A.c.moszkowskii* (Reichenow, 1911)   NC New Guinea
  ___ *A.c.callopterus* (D'Albertis & Salvadori, 1879)   SC New Guinea
  ___ *A.c.chloropterus* (E.P. Ramsay, 1879)   E New Guinea

◉ *Alisterus scapularis* Australian King Parrot
  ___ *A.s.minor* Mathews, 1911   NE Queensland
  ✓ *A.s.scapularis* (M.H.K. Lichtenstein, 1816[5])   EC Queensland to S Victoria

APROSMICTUS Gould, 1843 M
○ *Aprosmictus jonquillaceus* Jonquil Parrot
  ___ *A.j.jonquillaceus* (Vieillot, 1818)   Timor and Roti Is.
  ___ *A.j.wetterensis* (Salvadori, 1891)   Wetar I.

○ *Aprosmictus erythropterus* Red-winged Parrot
  ___ *A.e.papua* Mayr & Rand, 1936   S New Guinea (1493)
  ___ *A.e.coccineopterus* (Gould, 1865)   N Australia
  ___ *A.e.erythropterus* (J.F. Gmelin, 1788)   EC Australia

POLYTELIS Wagler, 1832 M
○ *Polytelis swainsonii* (Desmarest, 1826) Superb Parrot   Inland SE Australia

---

[1] Dickinson *et al.* (1991) (745) recognised *salvadori* (and the Talaud form as an endemic) and included *nigrorum* Salomonsen, 1953 (2155), and *siquijorensis* Salomonsen, 1953 (2155) in their synonymy.
[2] Still widely called Müller's Parrot in avicultural circles. Collar (1997: 393) (545) suggested that *T. heterurus* applied to an aberrant specimen of this.
[3] Including *incognitus* Eck, 1976 (788), an aberrant bird named *heterurus* that we listed in last Edition, but see White & Bruce (1986) (2652) and above. Also includes *sangirensis* see White & Bruce (1986: 229) (2652).
[4] Doubtfully distinct from *sulaensis* (R. Schodde, *in litt.*)
[5] 1818, and correctly page 26, in Peters (1937) (1823), however dates from the 1816 edition of this rare book where the description is on page 29.

○ *Polytelis anthopeplus* REGENT PARROT
    — *P.a.anthopeplus* (Lear, 1831)[1]   SW Western Australia
    — *P.a.monarchoides* Schodde, 1993   SW New South Wales, NW Victoria, EC South
                                              Australia (2180)

○ *Polytelis alexandrae* Gould, 1863 PRINCESS PARROT   Interior C and W Australia

MASCARINUS Lesson, 1830 M
○ †*Mascarinus mascarin* (Linnaeus, 1771) MASCARENE PARROT   Réunion (Indian Ocean)

PSITTACULA Cuvier, 1800[2] F
○ *Psittacula eupatria* ALEXANDRINE PARAKEET
    — *P.e.nipalensis* (Hodgson, 1836)   E Afghanistan to Bangladesh
    — *P.e.eupatria* (Linnaeus, 1766)   Sri Lanka, S India
    — *P.e.magnirostris* (Ball, 1872)   Andaman Is.
    — *P.e.avensis* (Kloss, 1917)   E Assam, N Burma
    — *P.e.siamensis* (Kloss, 1917)   N and W Thailand, Indochina

○ †*Psittacula wardi* (E. Newton, 1867) SEYCHELLES PARAKEET   Seychelles

○ *Psittacula krameri* ROSE-RINGED PARAKEET
    — *P.k.krameri* (Scopoli, 1769)   S Mauritania and Senegal to S Sudan
    — *P.k.parvirostris* (Souancé, 1856)   EC Sudan to NW Somalia
    — *P.k.borealis* (Neumann, 1915)[3]   NW Pakistan and N India to SE China
    — *P.k.manillensis* (Bechstein, 1800)   S India, Sri Lanka

○ *Psittacula echo* (A. & E. Newton, 1876) MAURITIUS PARAKEET   Mauritius

○ *Psittacula himalayana* (Lesson, 1831)[4] SLATY-HEADED PARAKEET   E Afghanistan to W Assam

○ *Psittacula finschii* (Hume, 1874) GREY-HEADED PARAKEET[5]   W Bengal to S China and N Indochina

○ *Psittacula cyanocephala* (Linnaeus, 1766) PLUM-HEADED PARAKEET   India, Sri Lanka

○ *Psittacula roseata* BLOSSOM-HEADED PARAKEET
    — *P.r.roseata* Biswas, 1951   W Bengal to N Burma (155)
    — *P.r.juneae* Biswas, 1951   S Assam to Indochina (155)

○ *Psittacula columboides* (Vigors, 1830) MALABAR PARAKEET   Western Ghats (SW India)

○ *Psittacula calthorpae* (Blyth, 1849) LAYARD'S PARAKEET[6]   Sri Lanka

○ *Psittacula derbiana* (Fraser, 1852) LORD DERBY'S PARAKEET   NE Assam, SE Tibet, SW China

○ *Psittacula alexandri* RED-BREASTED PARAKEET
    — *P.a.fasciata* (Statius Müller, 1776)   N India to S China, Hainan and Indochina
    — *P.a.abbotti* (Oberholser, 1919)   Andaman Is.
    — *P.a.cala* (Oberholser, 1912)   Simeuluë I.
    — *P.a.major* (Richmond, 1902)   Lasia and Babi Is.
    — *P.a.perionca* (Oberholser, 1912)   Nias I.
    — *P.a.alexandri* (Linnaeus, 1758)   Java, Bali, S Borneo
    — *P.a.dammermani* Chasen & Kloss, 1932   Karimundjawa Is.
    — *P.a.kangeanensis* Hoogerwerf, 1962   Kangean I. (1141)

○ *Psittacula caniceps* (Blyth, 1846) NICOBAR PARAKEET   Nicobar Is.

○ †*Psittacula exsul* (A. Newton, 1872) RODRIGUEZ PARAKEET   Rodriguez I. (Mauritius)

○ *Psittacula longicauda* LONG-TAILED PARAKEET
    — *P.l.tytleri* (Hume, 1874)   Andaman Is., Cocos Is.
    — *P.l.nicobarica* (Gould, 1857)   Nicobar Is.
    — *P.l.longicauda* (Boddaert, 1783)   S Malay Pen., Borneo, Sumatra

---

[1] Historically had its type locality wrongly restricted (Schodde, 1993) (2180); when corrected the name has priority over *westralis* Mathews, 1915.
[2] The name *intermedia* (a species was listed under this name in previous Editions) has been shown to relate to hybrids (Rasmussen & Collar, 1999) (1991).
[3] Includes *fragosa* Koelz, 1954 (1284), see Ripley (1982: 159) (2055).
[4] Given as 1832 in Peters (1937) (1823), but see Browning & Monroe (1991) (258).
[5] This species was split from *himalayana* by Husain (1959) (1167), and is only accepted tentatively see Collar (1997: 400) (545).
[6] The emended name used here is judged to be in prevailing usage; see Art. 33.2 of the Code (I.C.Z.N., 1999) (1178) and p. 000 herein.

— *P.l.modesta* (Fraser, 1845)	Enggano I.
— *P.l.defontainei* Chasen, 1934	Natuna Is.

AGAPORNIS Selby, 1836 M[1]

○ ***Agapornis canus*** GREY-HEADED LOVEBIRD

— *A.c.canus* (J.F. Gmelin, 1788)	Madagascar (except S)
— *A.c.ablectaneus* Bangs, 1918	S Madagascar

○ ***Agapornis pullarius*** RED-HEADED LOVEBIRD

— *A.p.pullarius* (Linnaeus, 1758)[2]	S Guinea to Sudan, W Zaire and NW Angola
— *A.p.ugandae* Neumann, 1908	W Ethiopia to NW Tanzania and E Zaire

○ ***Agapornis taranta*** (Stanley, 1814) BLACK-WINGED LOVEBIRD — Eritrea and Ethiopia

○ ***Agapornis swindernianus*** BLACK-COLLARED LOVEBIRD

— *A.s.swindernianus* (Kuhl, 1820)	Liberia, Ivory Coast, Ghana
— *A.s.zenkeri* Reichenow, 1895	Cameroon and Gabon to SW Central African Republic and C Congo
— *A.s.emini* Neumann, 1908	N, C and E Zaire, W Uganda

○ ***Agapornis roseicollis*** ROSY-FACED LOVEBIRD

— *A.r.catumbella* B.P. Hall, 1952	SW Angola (1059)
— *A.r.roseicollis* (Vieillot, 1817)[3]	Namibia, NW Cape Province

○ ***Agapornis fischeri*** Reichenow, 1887 FISCHER'S LOVEBIRD — Rwanda and Burundi to NW Tanzania

○ ***Agapornis personatus*** Reichenow, 1887 YELLOW-COLLARED LOVEBIRD — NE and C Tanzania

○ ***Agapornis lilianae*** Shelley, 1894 NYASA LOVEBIRD — S Tanzania, NW Mozambique, S Malawi, E Zambia, N Zimbabwe

○ ***Agapornis nigrigenis*** W.L. Sclater, 1906 BLACK-CHEEKED LOVEBIRD — SW Zambia to NW Zimbabwe

CORACOPSIS Wagler, 1832[4] F

○ ***Coracopsis vasa*** GREATER VASA PARROT

— *C.v.comorensis* (W.K.H. Peters, 1854)[5]	Grande Comore, Mwali (Moheli) and Anjouan Is.
— *C.v.drouhardi* Lavauden, 1929	W and S Madagascar
— *C.v.vasa* (Shaw, 1812)	E Madagascar

○ ***Coracopsis nigra*** LESSER VASA PARROT

— *C.n.sibilans* Milne-Edwards & Oustalet, 1885[6]	Grande Comore and Anjouan Is., Praslin I.
— *C.n.libs* Bangs, 1927	W and S Madagascar
— *C.n.nigra* (Linnaeus, 1758)	E Madagascar. Intr. Mauritius

PSITTACUS Linnaeus, 1758 M

○ ***Psittacus erithacus*** GREY PARROT

— *P.e.timneh* Fraser, 1844	Guinea-Bissau, S Mali, Sierra Leone to SW Ivory Coast
— *P.e.erithacus* Linnaeus, 1758[7]	SE Ivory Coast to W Kenya and N Angola, Príncipe and São Tomé Is.

POICEPHALUS Swainson, 1837 M

○ ***Poicephalus robustus*** BROWN-NECKED PARROT

— *P.r.fuscicollis* (Kuhl, 1820)	Gambia to EC Nigeria and N Angola
— *P.r.suahelicus* Reichenow, 1898	Rwanda and C Tanzania to NE Transvaal and SC Angola
— *P.r.robustus* (J.F. Gmelin, 1788)	E Cape Province and Natal to E Transvaal

○ ***Poicephalus gulielmi*** RED-FRONTED PARROT

— *P.g.fantiensis* Neumann, 1908	Liberia to Ghana
— *P.g.gulielmi* (Jardine, 1849)	S Cameroon to N Angola, E Zaire and SW Uganda
— *P.g.massaicus* (Fischer & Reichenow, 1884)[8]	W and C Kenya, N Tanzania

---

[1] Gender addressed by David & Gosselin (2002) (614), multiple changes occur in specific names.
[2] Implicitly includes *guineensis* Statius Müller, 1776 (2307A), see Fry & Harwin (1988: 11) (896).
[3] The volume (25) containing this name is dated 1817, but as volumes 20 onwards seem to have appeared in 1818 it is thought that this did too.
[4] This is the first genus of those making up the tribe Psittacini.
[5] Tentatively includes *makawa* Benson, 1960 (122).
[6] Includes *barklyi* (see Collar, 1997) (545).
[7] Includes *princeps* (see Collar, 1997) (545).
[8] Includes *permistus* see White (1965: 168) (2649).

○ *Poicephalus meyeri* BROWN PARROT
    — *P.m.meyeri* (Cretzschmar, 1827)   S Chad and NE Cameroon to C and S Sudan, Eritrea
                                                 and W Ethiopia

    — *P.m.saturatus* (Sharpe, 1901)   Uganda and W Kenya to E Zaire and NW Tanzania
    — *P.m.matschiei* Neumann, 1898   SE Zaire and W and C Tanzania to E Angola, N
                                                 Zambia and N Malawi

    — *P.m.reichenowi* Neumann, 1898   W Angola
    — *P.m.damarensis* Neumann, 1898   S Angola, N Namibia, NW Botswana
    — *P.m.transvaalensis* Neumann, 1899 S Zambia and W Mozambique to E Botswana and
                                                 NW Transvaal

○ *Poicephalus rueppellii* (G.R. Gray, 1849) RÜPPELL'S PARROT   SW Angola, N and C Namibia

○ *Poicephalus cryptoxanthus* BROWN-HEADED PARROT
    — *P.c.tanganyikae* Bowen, 1930[1]   Coastal Kenya, E Tanzania, Zanzibar and Pemba Is. to
                                               N and C Mozambique

    — *P.c.cryptoxanthus* (W.K.H. Peters, 1854) SE Zimbabwe and S Mozambique to NE Sth Africa

○ *Poicephalus crassus* (Sharpe, 1884) NIAM-NIAM PARROT   SW Chad to SW Sudan and N Zaire

○ *Poicephalus rufiventris* RED-BELLIED PARROT
    — *P.r.pallidus* van Someren, 1922   Somalia, E Ethiopia
    — *P.r.rufiventris* (Rüppell, 1845)   C Ethiopia to NE Tanzania

○ *Poicephalus senegalus* SENEGAL PARROT
    — *P.s.senegalus* (Linnaeus, 1766)   Gambia and Guinea-Bissau to S Niger and NW Nigeria
    — *P.s.mesotypus* Reichenow, 1910   NE Nigeria, N Cameroon and SW Chad
    — *P.s.versteri* Finsch, 1863   NW Ivory Coast to SW Nigeria

○ *Poicephalus flavifrons* (Rüppell, 1845) YELLOW-FRONTED PARROT[2] Ethiopia

ANODORHYNCHUS Spix, 1824[3] M
○ *Anodorhynchus hyacinthinus* (Latham, 1790) HYACINTH MACAW   NC to SC Brazil, E Bolivia

○ *Anodorhynchus leari* Bonaparte, 1856 INDIGO MACAW   NE Brazil

○ †*Anodorhynchus glaucus* (Vieillot, 1816) GLAUCOUS MACAW   SE Brazil, E Paraguay, W Uruguay, NE Argentina

CYANOPSITTA Bonaparte, 1854 F
○ *Cyanopsitta spixii* (Wagler, 1832) SPIX'S MACAW   N Bahia (E Brazil)

ARA Lacépède, 1799[4] M[5]
○ *Ara ararauna* (Linnaeus, 1758) BLUE-AND-YELLOW MACAW   E Panama south to Paraguay, S Brazil and N Argentina

○ *Ara glaucogularis* Dabbene, 1921 BLUE-THROATED MACAW[6]   N Bolivia

○ *Ara militaris* MILITARY MACAW
    — *A.m.mexicanus* Ridgway, 1915[7]   N and C Mexico
    — *A.m.militaris* (Linnaeus, 1766)   NW Venezuela, W Colombia, E Ecuador, N Peru
    — *A.m.bolivianus* Reichenow, 1908   Bolivia, NW Argentina

○ *Ara ambiguus* GREAT GREEN MACAW[8]
    — *A.a.ambiguus* (Bechstein, 1811)   E Honduras and Nicaragua to NW Colombia
    — *A.a.guayaquilensis* Chapman, 1925 W Ecuador

● *Ara macao* SCARLET MACAW
    — *A.m.cyanopterus* Wiedenfeld, 1995   SE Mexico to Nicaragua (2661)
    √ *A.m.macao* (Linnaeus, 1758)   Costa Rica to Bolivia and C Brazil

---

[1] For recognition see Forshaw & Cooper (1989: 320) (868); implicitly includes *zanzibaricus* see Fry & Harwin (1988: 11) (896).
[2] Includes *aurantiiceps* see Fry & Harwin (1988: 11) (896).
[3] This the first genus in the tribe Arini. Genera employed follow Collar (1997) (545) as does their sequence; except for later information. Treated as a subfamily as in A.O.U. (1998: 233) (3). Homberger (*in litt.*) considers this preferable, but only when eight families of parrots are recognised.
[4] Reasons for restoring three small genera out of *Ara*, and which here follow it, are given by Collar (1977: 285) (545).
[5] Gender addressed by David & Gosselin (2002) (614), multiple changes occur in specific names.
[6] Called *Ara caninde* in last Edition and in Peters (1937) (1823).
[7] Presumed to include *sheffleri* van Rossem & Hachisuka, 1939 (2483).
[8] Treated as monotypic in last Edition; for recognition of two subspecies see Collar (1997) (545).

○ *Ara chloropterus* G.R. Gray, 1859 RED-AND-GREEN MACAW   E Panama to S Brazil and Paraguay

○ †*Ara tricolor* Bechstein, 1811 HISPANIOLAN MACAW[1]   Hispaniola

○ *Ara rubrogenys* Lafresnaye, 1847 RED-FRONTED MACAW   C Bolivia

○ *Ara severus* CHESTNUT-FRONTED MACAW[2]
— *A.s.castaneifrons* Lafresnaye, 1847   E Panama to N Bolivia and C Brazil
— *A.s.severus* (Linnaeus, 1758)   E Venezuela, the Guianas, N Brazil

ORTHOPSITTACA Ridgway, 1912 F
○ *Orthopsittaca manilata* (Boddaert, 1783) RED-BELLIED MACAW   SE Colombia south to N Bolivia and east to NC and NE Brazil and Venezuela and the Guianas

PRIMOLIUS Bonaparte, 1857[3] M
○ *Primolius couloni* (P.L. Sclater, 1876) BLUE-HEADED MACAW   E Peru, SW Brazil, N Bolivia

○ *Primolius maracana* (Vieillot, 1816) BLUE-WINGED MACAW   NE, C and SE Brazil, Paraguay, NE Argentina

○ *Primolius auricollis* (Cassin, 1853) YELLOW-COLLARED MACAW   NE Bolivia, EC and SW Brazil, N Paraguay, N Argentina

DIOPSITTACA Ridgway, 1912 F
○ *Diopsittaca nobilis* RED-SHOULDERED MACAW
— *D.n.nobilis* (Linnaeus, 1758)   E Venezuela, the Guianas, N Brazil
— *D.n.cumanensis* (M.H.K. Lichtenstein, 1823)   NC Brazil (S of the Amazon)
— *D.n.longipennis* (Neumann, 1931)   C and S Brazil, NE Bolivia, SE Peru

RHYNCHOPSITTA Bonaparte, 1854 F
○ *Rhynchopsitta pachyrhyncha* (Swainson, 1827) THICK-BILLED PARROT   Sierra Madre Occid. (W Mexico)

○ *Rhynchopsitta terrisi* R.T. Moore, 1947 MAROON-FRONTED PARROT   Sierra Madre Oriental (NE Mexico) (1615)

OGNORHYNCHUS Bonaparte, 1857 M
○ *Ognorhynchus icterotis* (Massena &Souancé, 1854) YELLOW-EARED PARROT
Andean S Colombia, N Ecuador

GUAROUBA Lesson, 1831[4] F
○ *Guarouba guarouba* (J.F. Gmelin, 1788) GOLDEN PARAKEET   NE Brazil

ARATINGA Spix, 1824 F[5]
○ *Aratinga acuticaudata* BLUE-CROWNED PARAKEET
— *A.a.koenigi* Arndt, 1996[6]   N Venezuela, NE Colombia (54)
— *A.a.neoxena* (Cory, 1909)   Margarita I. (Venezuela)
— *A.a.haemorrhous* Spix, 1824   Inland NE Brazil
— *A.a.neumanni* Blake & Traylor, 1947   Upland E Bolivia (168)
— *A.a.acuticaudata* (Vieillot, 1818)   E Bolivia and S Brazil to N Argentina and W Uruguay

○ *Aratinga holochlora* GREEN PARAKEET[7]
— *A.h.brewsteri* Nelson, 1928   NW Mexico
— *A.h.holochlora* (P.L. Sclater, 1859)   NE to S Mexico
— *A.h.brevipes* (Lawrence, 1871)   Socorro I. (SW Mexico)
— *A.h.rubritorquis* (P.L. Sclater, 1887)   E Guatemala to N Nicaragua

○ *Aratinga strenua* (Ridgway, 1915) PACIFIC PARAKEET[8]   Pacific slope of S Mexico to NW Nicaragua

○ *Aratinga wagleri* SCARLET-FRONTED PARAKEET
— *A.w.wagleri* (G.R. Gray, 1845)   W and N Colombia, NW Venezuela
— *A.w.transilis* J.L. Peters, 1927   E Colombia, N Venezuela

[1] Wetherbee (1985) (2589) described *cubensis* an extinct subfossil form from Cuba and concluded that the origins of *tricolor* were Hispaniola not Cuba.
[2] Collar (1997) (545) presumed there would be intergradation and treated this as monotypic.
[3] Penhallurick (2001) (1818) has shown this name to have priority over *Propyrrhura* Mirando-Ribiero, 1920.
[4] Spelling verified. There is a conflicting error in usage by Lesson, but he was simply creating a genus on the back of Gmelin's specific epithet.
[5] Gender addressed by David & Gosselin (2002) (614), multiple changes occur in specific names.
[6] Arndt (1996) (55) provided the name *nigrirostris* in the German edition of his book (the two names are thought to refer to the same population; we believe the English edition appeared first and that *koenigi* must be used).
[7] Although split by some authors, including Collar (1997) (545), A.O.U. (1998) (3) refrained from doing so.
[8] For recognition see A.O.U. (1998: 234) (3).

○   — *A.w.frontata* (Cabanis, 1846)   W Ecuador, W Peru
     — *A.w.minor* Carriker, 1933   C and S Peru

○ ***Aratinga mitrata*** MITRED PARAKEET
     — *A.m.mitrata* (Tschudi, 1844)   C Peru to NW Argentina
     — *A.m.alticola* Chapman, 1921   Montane C Peru

○ ***Aratinga erythrogenys*** (Lesson, 1844)   RED-MASKED PARAKEET   W Ecuador, NW Peru

○ ***Aratinga finschi*** (Salvin, 1871) CRIMSON-FRONTED PARAKEET   S Nicaragua to W Panama

○ ***Aratinga leucophthalma*** WHITE-EYED PARAKEET
     — *A.l.nicefori* Meyer de Schauensee, 1946   Meta (E Colombia) (1563)
     — *A.l.callogenys* (Salvadori, 1891)   W Amazonia
     — *A.l.leucophthalma* (Statius Müller, 1776)[1]   E Venezuela and the Guianas through Brazil and E
                                    Bolivia to Paraguay and N Argentina

○ ***Aratinga euops*** (Wagler, 1832) CUBAN PARAKEET   Cuba

○ ***Aratinga chloroptera*** HISPANIOLAN PARAKEET
      A.c.chloroptera (Souancé, 1856)   Hispaniola
      †*A.c.maugei* (Souancé, 1856)   Mona I.

○ ***Aratinga solstitialis*** (Linnaeus, 1758) SUN PARAKEET   Guyana, Surinam, N Brazil (N of the Amazon)

○ ***Aratinga jandaya*** (J.F. Gmelin, 1788) JANDAYA PARAKEET   NE Brazil (S of the Amazon)

○ ***Aratinga auricapillus*** GOLDEN-CAPPED PARAKEET[2]
     — *A.a.auricapillus* (Kuhl, 1820)   EC Brazil
     — *A.a.aurifrons* Spix, 1824   SE Brazil

○ ***Aratinga weddellii*** (Deville, 1851) DUSKY-HEADED PARAKEET   SE Colombia to NE Bolivia

● ***Aratinga nana*** OLIVE-THROATED PARAKEET
     — *A.n.vicinalis* (Bangs & Penard, 1919)   NE Mexico
     √ *A.n.astec* (Souancé, 1857)[3]   SE Mexico to W Panama
     — *A.n.nana* (Vigors, 1830)   Jamaica

● ***Aratinga canicularis*** ORANGE-FRONTED PARAKEET
     — *A.c.clarae* R.T. Moore, 1937   Pacific slope WC Mexico (1609)
     — *A.c.eburnirostrum* (Lesson, 1842)   Pacific slope SW Mexico
     √ *A.c.canicularis* (Linnaeus, 1758)   Pacific slope S Mexico to W Costa Rica

○ ***Aratinga aurea*** (J.F. Gmelin, 1788) PEACH-FRONTED PARAKEET[4]   S Surinam through Brazil to SE Peru, E Bolivia,
                                    Paraguay and N Argentina

○ ***Aratinga pertinax*** BROWN-THROATED PARAKEET
     — *A.p.ocularis* (Sclater & Salvin, 1864)   Panama
     — *A.p.aeruginosa* (Linnaeus, 1758)   N Colombia, NW Venezuela
     — *A.p.griseipecta* Meyer de Schauensee, 1950   NE Colombia (1566)
     — *A.p.lehmanni* Dugand, 1943   E Colombia (766)
     — *A.p.arubensis* (E. Hartert, 1892)   Aruba I.
     — *A.p.pertinax* (Linnaeus, 1758)   Curaçao I.
     — *A.p.xanthogenia* (Bonaparte, 1850)   Bonaire I.
     — *A.p.tortugensis* (Cory, 1909)   Tortuga I. (Venezuela)
     — *A.p.margaritensis* (Cory, 1918)   Margarita I. (Venezuela)
     — *A.p.venezuelae* Zimmer & Phelps, 1951   Venezuela (2727)
     — *A.p.surinama* Zimmer & Phelps, 1951   NE Venezuela, the Guianas (2727)
     — *A.p.chrysophrys* (Swainson, 1838)   SE Venezuela, NW Brazil
     — *A.p.chrysogenys* (Massena & Souancé, 1854)   NW Brazil
     — *A.p.paraensis* Sick, 1959[5]   NC Brazil (2267)

---

[1] Includes *propinqua* Sclater, 1862 (2216A) of SE Brazil see Collar (1997: 431) (545).
[2] For correction of spelling see David & Gosselin (2002) (613).
[3] Including *melloni* Twomey, 1950 (2443), see Parkes (1976) (1769).
[4] Treated as monotypic following Collar (1997) (545).
[5] Described as a race of *cactorum*.

○ *Aratinga cactorum* Cactus Parakeet
    — *A.c.caixana* Spix, 1824   NE Brazil
    — *A.c.cactorum* (Kuhl, 1820)   EC Brazil

NANDAYUS Bonaparte, 1854 M
○ *Nandayus nenday* (Vieillot, 1823) Nanday Parakeet   SE Bolivia and SW Brazil to C Paraguay and N Argentina

LEPTOSITTACA Berlepsch & Stolzmann, 1894 F
○ *Leptosittaca branickii* Berlepsch & Stolzmann, 1894 Golden-plumed Parakeet
                                      Montane S Colombia to C Peru

CONUROPSIS Salvadori, 1891 F
○ *Conuropsis carolinensis* Carolina Parakeet
    — †*C.c.carolinensis* (Linnaeus, 1758)   SE USA
    — †*C.c.ludoviciana* (J.F. Gmelin, 1788)   C USA

CYANOLISEUS Bonaparte, 1854 M
○ *Cyanoliseus patagonus* Burrowing Parakeet
    — *C.p.andinus* Dabbene & Lillo, 1913   NW Argentina
    — *C.p.conlara* Nores & Yzurieta, 1983   San Luis and Cordoba (WC Argentina) (1660)
    — *C.p.patagonus* (Vieillot, 1818)   C to SE Argentina
    — *C.p.bloxami* Olson, 1995[1]   C Chile (1702)

PYRRHURA Bonaparte, 1856 F
○ *Pyrrhura cruentata* (Wied, 1820) Blue-throated Parakeet   E Brazil

○ *Pyrrhura devillei* (Massena & Souancé, 1854) Blaze-winged Parakeet
                                        N Paraguay, S Brazil

○ *Pyrrhura frontalis* Maroon-bellied Parakeet
    — *P.f.frontalis* (Vieillot, 1818)[2]   E Brazil
    — *P.f.chiripepe* (Vieillot, 1818)   SE Brazil, SE Paraguay, N Argentina

○ *Pyrrhura perlata* (Spix, 1824) Crimson-bellied Parakeet[3]   C Brazil to N Bolivia

○ *Pyrrhura lepida* Pearly Parakeet[4]
    — *P.l.lepida* (Wagler, 1832)   NE Para, NW Maranhao (NE Brazil)
    — *P.l.anerythra* Neumann, 1927   E Para (NE Brazil)
    — *P.l.coerulescens* Neumann, 1927   W and C Maranhao (NC Brazil)

○ *Pyrrhura molinae* Green-cheeked Parakeet[5]
    — *P.m.flavoptera* Maijer, Herzog, Kessler, Friggens & Fjeldså, 1998
                                        W Bolivia (1415)
    — *P.m.molinae* (Massena & Souancé, 1854)   E Bolivia (highlands)
    — *P.m.phoenicura* (Schlegel, 1864)   NE Bolivia, W Brazil
    — *P.m.sordida* Todd, 1947   Extreme E Bolivia, SW Brazil (2408)
    — *P.m.restricta* Todd, 1947   E Bolivia (lowlands) (2408)
    — *P.m.australis* Todd, 1915   S Bolivia to NW Argentina

○ *Pyrrhura eisenmanni* Delgado, 1985 Azuero Parakeet[6]   Azuero Pen. (Panama) (683)

○ *Pyrrhura caeruleiceps* Todd, 1947 Todd's Parakeet[7]   N Colombia, NW Venezuela (2408)

○ *Pyrrhura subandina* Todd, 1917 Sinú Parakeet   Sinú valley (NW Colombia)

○ *Pyrrhura picta* Painted Parakeet
    — *P.p.picta* (Statius Müller, 1776)[8]   Venezuela, the Guianas and N Brazil
    — *P.p.amazonum* Hellmayr, 1906[9]   SE Amazonia Brazil

---

[1] Listed as *byroni* in last Edition, but see Olson (1995) (1702).
[2] Includes *kriegi* based on the ranges given in Collar (1997) (545).
[3] Listed as *P. rhodogaster* in last Edition, but see note on next species.
[4] This name was used by Collar (1997: 440) (545) and is employed pending further investigation. The facts are complex and under review.
[5] The name *hypoxantha*, used in last Edition, has been shown to belong to a colour morph of the nominate race (Collar, 1997) (545).
[6] Treated as a separate species on the advice of L. Joseph (*in litt.*).
[7] Includes *pantchenkoi* W.H. Phelps, Jr., 1977 (1836), see Joseph (2000) (1226).
[8] The name *orinocoensis* Todd, 1947 (2408), and *cuchivera* Phelps & Phelps, 1949 (1845), are synonyms see Phelps & Phelps (1958) (1860).
[9] Includes *microptera* Todd, 1947 (2408).

_ *P.p.lucianii* (Deville, 1851)	SE Ecuador, NE Peru, N Bolivia, NW Brazil
_ *P.p.roseifrons* (G.R. Gray, 1859)	W Brazil, E Peru

○ **Pyrrhura leucotis** MAROON-FACED PARAKEET
_ *P.l.emma* Salvadori, 1891[1]	N Venezuela
_ *P.l.griseipectus* Salvadori, 1900	NE Brazil
_ *P.l.leucotis* (Kuhl, 1820)	E and SE Brazil

○ **Pyrrhura pfrimeri** Miranda-Ribeiro, 1920 PFRIMER'S PARAKEET[2] C Brazil

○ **Pyrrhura viridicata** Todd, 1913 SANTA MARTA PARAKEET Santa Marta Mts. (N Colombia)

○ **Pyrrhura egregia** FIERY-SHOULDERED PARAKEET
_ *P.e.egregia* (P.L. Sclater, 1881)	SE Venezuela, SW Guyana
_ *P.e.obscura* Zimmer & Phelps, 1946	S Venezuela, NE Brazil (2723)

○ **Pyrrhura melanura** MAROON-TAILED PARAKEET
_ *P.m.pacifica* Chapman, 1915	SW Colombia, NW Ecuador
_ *P.m.chapmani* Bond & Meyer de Schauensee, 1940	Andes of Colombia (190)
_ *P.m.melanura* (Spix, 1824)	S Venezuela, SE Colombia, NW Brazil, E Ecuador, NE Peru
_ *P.m.souancei* (J. Verreaux, 1858)	SC Colombia
_ *P.m.berlepschi* Salvadori, 1891	SE Ecuador, N Peru

○ **Pyrrhura orcesi** Ridgely & Robbins, 1988 EL ORO PARAKEET SW Ecuador (2016)

○ **Pyrrhura rupicola** BLACK-CAPPED PARAKEET
_ *P.r.rupicola* (Tschudi, 1844)	C Peru
_ *P.r.sandiae* Bond & Meyer de Schauensee, 1944	SE Peru, W Brazil, N Bolivia (193)

○ **Pyrrhura albipectus** Chapman, 1914 WHITE-NECKED PARAKEET SE Ecuador

○ **Pyrrhura calliptera** (Massena & Souancé, 1854) BROWN-BREASTED PARAKEET E Andes of Colombia

○ **Pyrrhura hoematotis** RED-EARED PARAKEET
_ *P.h.immarginata* Zimmer & Phelps, 1944	NW Venezuela (2722)
_ *P.h.hoematotis* Souancé, 1857	N Venezuela

○ **Pyrrhura rhodocephala** (Sclater & Salvin, 1871) ROSE-HEADED PARAKEET W Venezuela

○ **Pyrrhura hoffmanni** SULPHUR-WINGED PARAKEET
_ *P.h.hoffmanni* (Cabanis, 1861)	S Costa Rica
_ *P.h.gaudens* Bangs, 1906	W Panama

ENICOGNATHUS G.R. Gray, 1840 M
● **Enicognathus ferrugineus** AUSTRAL PARAKEET
✓ *E.f.minor* (Chapman, 1919)	S Chile, SW Argentina
_ *E.f.ferrugineus* (Statius Müller, 1776)	Extreme S of Chile and Argentina

○ **Enicognathus leptorhynchus** (P.P. King, 1831) SLENDER-BILLED PARAKEET C Chile

MYIOPSITTA Bonaparte, 1854 F
○ **Myiopsitta monachus** MONK PARAKEET[3]
_ *M.m.luschi* (Finsch, 1868)	C Bolivia
_ *M.m.cotorra* (Vieillot, 1818)	S Bolivia, S Brazil, Paraguay, NW Argentina
_ *M.m.monachus* (Boddaert, 1783)	SE Brazil, Uruguay, NE Argentina
_ *M.m.calita* (Jardine & Selby, 1830)	W Argentina

PSILOPSIAGON Ridgway, 1912 F
○ **Psilopsiagon aymara** (d'Orbigny, 1839) GREY-HOODED PARAKEET NC Bolivia to NW Argentina

[1] Includes *auricularis* Zimmer & Phelps, 1949 (2725) see Joseph (2000) (1226).
[2] For reasons to treat as a species see Olmos *et al.* (1997) (1677).
[3] Treated as two species by Collar (1997) (545).

○ *Psilopsiagon aurifrons* MOUNTAIN PARAKEET
— *P.a.robertsi* Carriker, 1933   NW Peru
— *P.a.aurifrons* (Lesson, 1831)   C Peru
— *P.a.margaritae* (Berlioz & Dorst, 1956)   S Peru and WC Bolivia to N Chile and NW Argentina (143)
— *P.a.rubrirostrus* (Burmeister, 1860)   NW Argentina to C Chile

BOLBORHYNCHUS Bonaparte, 1857 M
○ *Bolborhynchus lineola* BARRED PARAKEET
— *B.l.lineola* (Cassin, 1853)   S Mexico to W Panama
— *B.l.tigrinus* (Souancé, 1856)   N and W Venezuela to C Peru

○ *Bolborhynchus orbygnesius* (Souancé, 1856) ANDEAN PARAKEET   Upper Andes of Peru and NC Bolivia

○ *Bolborhynchus ferrugineifrons* (Lawrence, 1880) RUFOUS-FRONTED PARAKEET
     C Andes of WC Colombia

FORPUS Boie, 1858 M
○ *Forpus cyanopygius* MEXICAN PARROTLET
— *F.c.insularis* (Ridgway, 1888)   Tres Marias Is.
— *F.c.cyanopygius* (Souancé, 1856)[1]   W Mexico

○ *Forpus passerinus* GREEN-RUMPED PARROTLET
— *F.p.cyanophanes* (Todd, 1915)   N Colombia
— *F.p.viridissimus* (Lafresnaye, 1848)   N Venezuela, Trinidad
— *F.p.passerinus* (Linnaeus, 1758)   The Guianas
— *F.p.cyanochlorus* (Schlegel, 1864)   Roraima (NW Brazil)
— *F.p.deliciosus* (Ridgway, 1888)   Lower Amazon (Brazil)

○ *Forpus xanthopterygius* BLUE-WINGED PARROTLET[2]
— *F.x.spengeli* (Hartlaub, 1885)   N Colombia (isolate)
— *F.x.crassirostris* (Taczanowski, 1883)[3]   SE Colombia to N Peru and W Brazil
— *F.x.flavescens* (Salvadori, 1891)   SE Peru, E Bolivia
— *F.x.flavissimus* Hellmayr, 1929   NE Brazil
— *F.x.xanthopterygius* (Spix, 1824)[4]   E and SE Brazil, Paraguay, N Argentina

○ *Forpus conspicillatus* SPECTACLED PARROTLET
— *F.c.conspicillatus* (Lafresnaye, 1848)   E Panama, NC Colombia
— *F.c.metae* Borrero & Camacho, 1961   C Colombia to W Venezuela (198)
— *F.c.caucae* (Chapman, 1915)   SW Colombia

○ *Forpus sclateri* DUSKY-BILLED PARROTLET
— *F.s.eidos* J.L. Peters, 1937   E Colombia and S Venezuela to the Guianas and N Brazil
— *F.s.sclateri* (G.R. Gray, 1859)   SE Colombia to N Bolivia and across the Amazon basin S of the river

○ *Forpus coelestis* (Lesson, 1847) PACIFIC PARROTLET   W Ecuador, NW Peru

○ *Forpus xanthops* (Salvin, 1895) YELLOW-FACED PARROTLET   Upper Marañón valley (NC Peru)

BROTOGERIS Vigors, 1825 F[5]
○ *Brotogeris tirica* (J.F. Gmelin, 1788) PLAIN PARAKEET   E Brazil from EC to SE

○ *Brotogeris versicolurus* (Statius Müller, 1776) CANARY-WINGED PARAKEET[6]
     SE Colombia and NE Peru to S French Guiana and CN Brazil

○ *Brotogeris chiriri* YELLOW-CHEVRONED PARAKEET
— *B.c.chiriri* (Vieillot, 1818)   N Bolivia, Paraguay, N Argentina to NE, C and SE Brazil
— *B.c.behni* Neumann, 1931   C and S Bolivia

---

[1] Including *pallidus* see Collar (1997: 449) (545).
[2] Treated as *F. crassirostris* by Collar (1997) (545), but see Whitney & Pacheco (1999) (2656) for correction. See also the note attached to the name *xanthopterygius*.
[3] The name *olallae* Gyldenstolpe, 1945 (1030), is apparently a synonym, see Collar (1997: 449) (545).
[4] The name *vividus* in Collar (1997) (545) is a junior synonym, see Whitney & Pacheco (1999) (2656).
[5] Gender addressed by David & Gosselin (2002) (614), multiple changes occur in specific names.
[6] This form and the next are reportedly sympatric and were split by Collar (1997) (545).

○ *Brotogeris pyrrhoptera* (Latham, 1802)[1] GREY-CHEEKED PARAKEET   W and SW Ecuador, NW Peru

○ *Brotogeris jugularis* ORANGE-CHINNED PARAKEET
— *B.j.jugularis* (Statius Müller, 1776)   SW Mexico to N Colombia and NW Venezuela
— *B.j.exsul* Todd, 1917   E Colombia, W Venezuela

○ *Brotogeris cyanoptera* COBALT-WINGED PARAKEET
— *B.c.cyanoptera* (Salvadori, 1891)   SE Colombia and S Venezuela to E Peru and WC Brazil
— *B.c.gustavi* Berlepsch, 1889   N Peru
— *B.c.beniensis* Gyldenstolpe, 1941   N Bolivia (1029)

○ *Brotogeris chrysoptera* GOLDEN-WINGED PARAKEET
— *B.c.chrysoptera* (Linnaeus, 1766)   NE Venezuela, CN Brazil, the Guianas
— *B.c.tenuifrons* Friedmann, 1945   NW Brazil (878)
— *B.c.solimoensis* Gyldenstolpe, 1941   Amazonas, NC Brazil (1028)
— *B.c.tuipara* (J.F. Gmelin, 1788)   Para and Maranhão, C Brazil
— *B.c.chrysosema* P.L. Sclater, 1864   W Brazil

○ *Brotogeris sanctithomae* TUI PARAKEET
— *B.s.sanctithomae* (Statius Müller, 1776)   W Amazonia
— *B.s.takatsukasae* Neumann, 1931   Lower Amazon basin, Brazil

NANNOPSITTACA Ridgway, 1912 F
○ *Nannopsittaca panychlora* (Salvin & Godman, 1883) TEPUI PARROTLET
   N, S and E Venezuela, S Guyana, N Brazil

○ *Nannopsittaca dachilleae* O'Neill, Munn & Franke, 1991 AMAZONIAN PARROTLET #
   SE Peru, NE Bolivia (1718)

TOUIT G.R. Gray, 1855 M[2]
○ *Touit batavicus* (Boddaert, 1783) LILAC-TAILED PARROTLET   N Venezuela, the Guianas, Trinidad, Tobago I.

○ *Touit huetii* (Temminck, 1830) SCARLET-SHOULDERED PARROTLET   C Colombia, NE and S Venezuela, N Guyana, NE
   Brazil, E Ecuador to N Bolivia

○ *Touit costaricensis* (Cory, 1913) RED-FRONTED PARROTLET[3]   SE Costa Rica, W Panama

○ *Touit dilectissimus* (Sclater & Salvin, 1871) BLUE-FRONTED PARROTLET   E Panama, N Colombia to W Venezuela and south to
   NW Ecuador

○ *Touit purpuratus* SAPPHIRE-RUMPED PARROTLET
— *T.p.purpuratus* (J.F. Gmelin, 1788)   S Venezuela east to the Guianas and NE Brazil
— *T.p.viridiceps* Chapman, 1929   S Venezuela west to SE Colombia, NW and NC Brazil

○ *Touit melanonotus* (Wied, 1818) BROWN-BACKED PARROTLET   SE Brazil

○ *Touit surdus* GOLDEN-TAILED PARROTLET
— *T.s.chryseura* (Swainson, 1820)[4]   Recife, NE Brazil
— *T.s.surdus* (Kuhl, 1820)   E Brazil

○ *Touit stictopterus* (P.L. Sclater, 1862) SPOT-WINGED PARROTLET   S Colombia, Ecuador, N Peru

PIONITES Heine, 1890 M[5]
○ *Pionites melanocephalus* BLACK-HEADED PARROT
— *P.m.melanocephalus* (Linnaeus, 1758)   SE Colombia to the Guianas and N Brazil
— *P.m.pallidus* (Berlepsch, 1889)   S Colombia to NE Peru

○ *Pionites leucogaster* WHITE-BELLIED PARROT
— *P.l.xanthomerius* (P.L. Sclater, 1858)   E Peru, N Bolivia, W Brazil
— *P.l.xanthurus* Todd, 1925   R. Purús and R. Juruá to R. Madeira (Amazonian Brazil)
— *P.l.leucogaster* (Kuhl, 1820)   R. Madeira to R. Maranhão (N Brazil)

PIONOPSITTA Bonaparte, 1854 F
○ *Pionopsitta vulturina* (Kuhl, 1820) VULTURINE PARROT   R. Madeira to R. Maranhão (N Brazil)

---

[1] For reasons to use 1802 not 1801 see Browing & Monroe (1991) (258).
[2] Gender addressed by David & Gosselin (2002) (614), multiple changes occur in specific names.
[3] Previously listed as a race of the next species but see A.O.U. (1998: 240) (3).
[4] This name has priority over *ruficauda* Berla, 1954 (136) (used in last Edition), see Collar (1997) (545).
[5] Gender addressed by David & Gosselin (2002) (614), multiple changes occur in specific names.

● *Pionopsitta haematotis* Brown-hooded Parrot
    ⏚ *P.h.haematotis* (Sclater & Salvin, 1860)   SE Mexico to W Panama
    — *P.h.coccinicollaris* (Lawrence, 1862)   E Panama, NW Colombia

○ *Pionopsitta pulchra* Berlepsch, 1897 Rose-faced Parrot   Pacific lowlands W Colombia, NW Ecuador

○ *Pionopsitta barrabandi* Orange-cheeked Parrot
    — *P.b.barrabandi* (Kuhl, 1820)   SW Colombia, S Venezuela, NW Brazil (N of Amazon)
    — *P.b.aurantiigena* Gyldenstolpe, 1951   E Ecuador to N Bolivia, NW Brazil (S of Amazon) (1031)

○ *Pionopsitta pyrilia* (Bonaparte, 1853) Saffron-headed Parrot   E Panama, N and C Colombia, NW Venezuela

○ *Pionopsitta caica* (Latham, 1790) Caica Parrot   E Venezuela, the Guianas, N Brazil (south to Amazon)

○ *Pionopsitta pileata* (Scopoli, 1769) Pileated Parrot   SE Brazil, E Paraguay, NE Argentina

HAPALOPSITTACA Ridgway, 1912 F
○ *Hapalopsittaca melanotis* Black-winged Parrot
    — *H.m.peruviana* (Carriker, 1932)   C and S Peru
    — *H.m.melanotis* (Lafresnaye, 1847)   WC Bolivia

○ *Hapalopsittaca amazonina* Rusty-faced Parrot[1]
    — *H.a.theresae* (Hellmayr, 1915)   NE Colombia, NW Venezuela
    — *H.a.velezi* Graves & Uribe Restrepo, 1989   C Andes of Colombia (992)
    — *H.a.amazonina* (Des Murs, 1845)   E Andes of Colombia

○ *Hapalopsittaca fuertesi* (Chapman, 1912) Indigo-winged Parrot   WC Andes of Colombia

○ *Hapalopsittaca pyrrhops* (Salvin, 1876) Red-faced Parrot   SW Ecuador, NW Peru

GRAYDIDASCALUS Bonaparte, 1854 M
○ *Graydidascalus brachyurus* (Kuhl, 1820) Short-tailed Parrot[2]   E Ecuador and E Peru through Amazon basin to
                                          French Guiana

PIONUS Wagler, 1832 M
○ *Pionus menstruus* Blue-headed Parrot
    — *P.m.rubrigularis* Cabanis, 1881   NE Costa Rica to W Ecuador
    — *P.m.menstruus* (Linnaeus, 1766)   E Colombia to N Bolivia, the Guianas and NE Brazil
    — *P.m.reichenowi* (Heine, 1884)[3]   E Brazil

○ *Pionus sordidus* Red-billed Parrot
    — *P.s.saturatus* Todd, 1915   Santa Marta massif (N Colombia)
    — *P.s.ponsi* Aveledo & Ginés, 1950   N Colombia, NW Venezuela (64)
    — *P.s.sordidus* (Linnaeus, 1758)   N Venezuela
    — *P.s.antelius* Todd, 1947   NE Venezuela (2408)
    — *P.s.corallinus* Bonaparte, 1854   E Andes of Colombia south to N Bolivia
    — *P.s.mindoensis* Chapman, 1925   W Ecuador

○ *Pionus maximiliani* Scaly-headed Parrot
    — *P.m.maximiliani* (Kuhl, 1820)   NE and EC Brazil
    — *P.m.siy* Souancé, 1856   SE Bolivia, Paraguay, WC and SC Brazil, N Argentina
    — *P.m.lacerus* (Heine, 1884)   NW Argentina
    — *P.m.melanoblepharus* Miranda-Ribeiro, 1920   SE Brazil, E Paraguay, NE Argentina

○ *Pionus tumultuosus* Speckle-faced Parrot[4]
    — *P.t.seniloides* (Massena & Souancé, 1854)   W Venezuela to NW Peru
    — *P.t.tumultuosus* (Tschudi, 1844)   C and S Peru, W Bolivia

● *Pionus senilis* (Spix, 1824) White-crowned Parrot   NE Mexico to W Panama

○ *Pionus chalcopterus* (Fraser, 1841) Bronze-winged Parrot[5]   W Venezuela and NE Colombia to NW Peru

○ *Pionus fuscus* (Statius Müller, 1776) Dusky Parrot   NE Colombia, NE Venezuela through the Guianas to
                                          N Brazil

[1] Split by Graves & Restrepo (1989) (992) such that the next two species listed are spin-offs.
[2] The name *insulsus* Griscom & Greenway, 1937 (1017), is a synonym.
[3] The name *chlorocyanescens* Pinto, 1962 (1893), was proposed as a new name — probably for *cyanescens* Pinto, 1960 (1891) – and is thought to be a synonym.
[4] For reasons to unite *seniloides* with this see O'Neill & Parker (1977) (1715).
[5] Includes *cyanescens* Meyer de Schauensee, 1944 (1560).

AMAZONA Lesson, 1830[1] F

○ *Amazona leucocephala* CUBAN PARROT
  — *A.l.bahamensis* (H. Bryant, 1867)   Great Inagua and Abaco (Bahamas)
  — *A.l.leucocephala* (Linnaeus, 1758)[2]   Cuba and Isla de la Juventud
  — *A.l.caymanensis* (Cory, 1886)   Grand Cayman I.
  — *A.l.hesterna* Bangs, 1916   Cayman Brac I.

○ *Amazona collaria* (Linnaeus, 1758) YELLOW-BILLED PARROT   Jamaica

○ *Amazona ventralis* (Statius Müller, 1776) HISPANIOLAN PARROT   Hispaniola and satellite islands

● *Amazona albifrons* WHITE-FRONTED PARROT
  — *A.a.saltuensis* Nelson, 1899   NW Mexico
  — *A.a.albifrons* (Sparrman, 1788)   S Mexico to SW Guatemala
  ☑ *A.a.nana* W. de W. Miller, 1905   E Mexico, Belize and N Guatemala to NW Costa Rica

○ *Amazona xantholora* (G.R. Gray, 1859) YELLOW-LORED PARROT   Yucatán Peninsula and Cozumel I. (Mexico), N Belize

○ *Amazona agilis* (Linnaeus, 1758) BLACK-BILLED PARROT   Jamaica

○ *Amazona vittata* PUERTO RICAN PARROT
  — *A.v.vittata* (Boddaert, 1783)   Puerto Rico
  — †*A.v.gracilipes* Ridgway, 1915   Culebra I. (off Puerto Rico)

○ *Amazona tucumana* (Cabanis, 1885) TUCUMAN PARROT   SE Bolivia (?), NW Argentina

○ *Amazona pretrei* (Temminck, 1830) RED-SPECTACLED PARROT   SE Brazil (mainly Rio Grande do Sul), rarely Paraguay
                                                         and Argentina

○ *Amazona viridigenalis* (Cassin, 1853) RED-CROWNED PARROT   NE Mexico

○ *Amazona finschi* (P.L. Sclater, 1864) LILAC-CROWNED PARROT[3]   W Mexico

● *Amazona autumnalis* RED-LORED PARROT
  — *A.a.autumnalis* (Linnaeus, 1758)   Caribbean slope from E Mexico to N Nicaragua
  ☑ *A.a.salvini* (Salvadori, 1891)   N Nicaragua to SW Colombia and NW Venezuela
  — *A.a.lilacina* Lesson, 1844   W Ecuador
  — *A.a.diadema* (Spix, 1824)   Lower R Negro and N bank of Amazon (NW Brazil)

○ *Amazona dufresniana* (Shaw, 1812) BLUE-CHEEKED PARROT   SE Venezuela, the Guianas

○ *Amazona rhodocorytha* (Salvadori, 1890) RED-BROWED PARROT[4]   E Brazil

○ *Amazona brasiliensis* (Linnaeus, 1758) RED-TAILED PARROT   SE Brazil

○ *Amazona festiva* FESTIVE PARROT
  — *A.f.bodini* (Finsch, 1873)   E Colombia to E Venezuela
  — *A.f.festiva* (Linnaeus, 1758)   W Amazonia

○ *Amazona xanthops* (Spix, 1824) YELLOW-FACED PARROT   Interior EC Brazil to N Paraguay and E Bolivia

○ *Amazona barbadensis* (J.F. Gmelin, 1788) YELLOW-SHOULDERED PARROT[5]
                                                         N Venezuela and offshore islands of La Blanquilla,
                                                            Bonaire and Margarita

○ *Amazona aestiva* BLUE-FRONTED PARROT
  — *A.a.aestiva* (Linnaeus, 1758)   E Brazil
  — *A.a.xanthopteryx* (Berlepsch, 1896)   N and E Bolivia, SW Brazil, Paraguay, N Argentina

○ *Amazona oratrix* YELLOW-HEADED PARROT[6]
  — *A.o.tresmariae* Nelson, 1900   Tres Marias Is (off WC Mexico)
  — *A.o.oratrix* Ridgway, 1887   SW and S Mexico
  — *A.o.belizensis* Monroe & Howell, 1966   Belize (1606)

---

[1] The vernacular name Amazon is often used for this genus; we believe New World usage favours Parrot instead.
[2] Includes *palmarum* see Collar (1997: 467) (545).
[3] Includes *woodi* R.T. Moore, 1937 (1611) see Collar (1997: 469) (545).
[4] Split here from *dufresniana* tentatively following Collar (1997) (545) but, as he said, one or both may be conspecific with *brasiliensis*.
[5] Includes *rothschildi* see Collar 91997: 473) (545).
[6] The relationships of *oratrix*, *auropalliata* and *ochrocephala* are not understood. Collar (1997) (545) treated them as one species *ochrocephala*. We tentatively
     follow A.O.U. (1998: 244-245) (3), which equates with our treatment last Edition; however the description of *hondurensis* Lousada & Howell, 1997 (1387),
     was perhaps not yet taken into account, and we defer assigning it. It has been suggested that these birds may be recent introductions to the Sula valley,
     where they *may* replace an original population of the same superspecies (A.O.U., 1998) (3).

● *Amazona auropalliata* YELLOW-NAPED PARROT
   ⊿ *A.a.auropalliata* (Lesson, 1842)   S Mexico to NW Costa Rica
   — *A.o.caribaea* Lousada, 1989   Bay Is. (off Honduras)[1] (1386)
   — *A.a.parvipes* Monroe & Howell, 1966   N Honduras to NE Nicaragua (1606)

○ *Amazona ochrocephala* YELLOW-CROWNED PARROT
   — *A.o.magna* Monroe & Howell, 1966   Tamaulipas to Tabasco (E Mexico) (1606)
   — *A.o.panamensis* (Cabanis, 1874)   W Panama, NW Colombia, Pearl Is
   — *A.o.nattereri* (Finsch, 1865)   S Colombia to E Peru, N Bolivia and W Brazil
   — *A.o.xantholaema* Berlepsch, 1913   Marajó I (Amazon Delta, N Brazil)
   — *A.o.ochrocephala* (J.F. Gmelin, 1788)   E Colombia to Trinidad, the Guianas and N Brazil

○ *Amazona amazonica* (Linnaeus, 1766) ORANGE-WINGED PARROT[2]   N and E Colombia south to N Bolivia, east through most of nthn. Sth. America

○ *Amazona mercenaria* SCALY-NAPED PARROT
   — *A.m.canipalliata* (Cabanis, 1874)   Andes of Venezuela, Colombia and Ecuador
   — *A.m.mercenaria* (Tschudi, 1844)   Andes of Peru and Bolivia

○ *Amazona kawalli* Grantsau & Camargo, 1989 KAWALL'S PARROT[3] (976) #
     Amazon basin (Brazil)

● *Amazona farinosa* MEALY PARROT
   — *A.f.guatemalae* (P.L. Sclater, 1860)   SE Mexico to NW Honduras
   ⊿ *A.f.virenticeps* (Salvadori, 1891)   Sula Valley (N Honduras) to W Panama
   — *A.f.farinosa* (Boddaert, 1783)[4]   E Panama to the Amazon basin; EC and SE Brazil; W Columbia and NW Ecuador.

○ *Amazona vinacea* (Kuhl, 1820) VINACEOUS PARROT   E Brazil, E Paraguay, NE Argentina

○ *Amazona versicolor* (Statius Müller, 1776) ST. LUCIA PARROT   St. Lucia (Lesser Antilles)

○ *Amazona arausiaca* (Statius Müller, 1776) RED-NECKED PARROT   Dominica (Lesser Antilles)

○ *Amazona guildingii* (Vigors, 1837) ST. VINCENT PARROT   St. Vincent (Lesser Antilles)

○ *Amazona imperialis* Richmond, 1899 IMPERIAL PARROT   Dominica (Lesser Antilles)

DEROPTYUS Wagler, 1832 M
○ *Deroptyus accipitrinus* RED-FAN PARROT
   — *D.a.accipitrinus* (Linnaeus, 1758)   SE Colombia to NE Peru, the Guianas and N Brazil
   — *D.a.fuscifrons* Hellmayr, 1905   C Brazil (S of Amazon)

TRICLARIA Wagler, 1832 F
○ *Triclaria malachitacea* (Spix, 1824) BLUE-BELLIED PARROT   SE Brazil

CYCLOPSITTA Reichenbach, 1850[5] [6] F
○ *Cyclopsitta gulielmitertii* ORANGE-BREASTED FIG PARROT
   — *C.g.melanogenia* (Schlegel, 1866)   Aru Is.
   — *C.g.gulielmitertii* (Schlegel, 1866)   Salawati I., Vogelkop (NW New Guinea)
   — *C.g.nigrifrons* (Reichenow, 1891)   N New Guinea
   — *C.g.ramuensis* (Neumann, 1915)   Ramu R. (NE New Guinea)
   — *C.g.amabilis* (Reichenow, 1891)   E New Guinea
   — *C.g.suavissima* P.L. Sclater, 1876   SE New Guinea
   — *C.g.fuscifrons* (Salvadori, 1876)   S New Guinea

○ *Cyclopsitta diophthalma* DOUBLE-EYED FIG PARROT
   — *C.d.diophthalma* (Hombron & Jacquinot, 1841)   W Papuan Is., W and N New Guinea
   — *C.d.aruensis* (Schlegel, 1874)   Aru Is., SC New Guinea
   — *C.d.coccineifrons* (Sharpe, 1882)   E and EC New Guinea
   — *C.d.virago* (E. Hartert, 1895)   Goodenough and Fergusson Is. (D'Entrecasteaux Arch.)

---

[1] Validity questioned by Parkes (1990) (1779), but reaffirmed by Lousada & Howell (1997) (1387) and references therein.
[2] Treated as monotypic following Collar (1997) (545); includes *tobagensis* and *micra* Griscom & Greenway, 1937 (1017).
[3] See Collar & Pittman (1996) (548).
[4] Includes *inornata* and *chapmani* Traylor, 1948 (2421) see Collar (1997: 475) (545).
[5] This is the first genus in the tribe Cyclopsittini.
[6] Includes *Opopsitta* see Schodde (1978) (2176).

— *C.d.inseparabilis* (E. Hartert, 1898)	Tagula I. (Louisiade Arch.)
— *C.d.marshalli* (Iredale, 1946)	Cape York Pen. (1183)
— *C.d.macleayana* E.P. Ramsay, 1874	NE Queensland
— *C.d.coxeni* Gould, 1867	SE Queensland, NE New South Wales

PSITTACULIROSTRIS J.E. & G.R. Gray, 1859 M
○ *Psittaculirostris desmarestii* LARGE FIG PARROT

— *P.d.blythii* (Wallace, 1864)[1]	Misool I. (W Papuan Is.)
— *P.d.occidentalis* (Salvadori, 1875)	Salawati and Batanta Is., West Vogelkop (NW New Guinea)
— *P.d.desmarestii* (Desmarest, 1826)	E Vogelkop (NW New Guinea)
— *P.d.intermedius* (van Oort, 1909)	S Vogelkop (NW New Guinea)
— *P.d.godmani* (Ogilvie-Grant, 1911)	SW New Guinea
— *P.d.cervicalis* (Salvadori & D'Albertis, 1875)	SE to NE New Guinea

○ *Psittaculirostris edwardsii* (Oustalet, 1885) EDWARDS'S FIG PARROT   EC northern New Guinea

○ *Psittaculirostris salvadorii* (Oustalet, 1880) SALVADORI'S FIG PARROT   WC northern New Guinea

BOLBOPSITTACUS Salvadori, 1891 M
○ *Bolbopsittacus lunulatus* GUAIABERO

— *B.l.lunulatus* (Scopoli, 1786)	Luzon
— *B.l.callainipictus* Parkes, 1971	Samar (1764)
— *B.l.intermedius* Salvadori, 1891	Leyte and Panaon Is.
— *B.l.mindanensis* (Steere, 1890)	Mindanao

OPISTHOCOMIDAE HOATZIN[2] (1: 1)

OPISTHOCOMUS Illiger, 1811 M
○ *Opisthocomus hoazin* (Statius Müller, 1776) HOATZIN   E Colombia to the Guianas, south to Bolivia

MUSOPHAGIDAE TURACOS[3] (6: 23)

CORYTHAEOLINAE

CORYTHAEOLA Heine, 1860 F
○ *Corythaeola cristata* (Vieillot, 1816) GREAT BLUE TURACO   Guinea-Bissau to W Kenya, NW Tanzania and N Angola, Bioko I.

MUSOPHAGINAE[4]

TAURACO Kluk, 1779 M
○ *Tauraco persa* GREEN TURACO[5]

— *T.p.buffoni* (Vieillot, 1819)	Senegal and Gambia to Liberia
— *T.p.persa* (Linnaeus, 1758)	Ivory Coast to Cameroon
— *T.p.zenkeri* (Reichenow, 1896)	S Cameroon to NW Zaire, Congo and N Angola

○ *Tauraco livingstonii* LIVINGSTONE'S TURACO[6]

— *T.l.reichenowi* (G.A. Fischer, 1880)	Nguru and Uluguru Mts. to Njombe highlands (Tanzania)
— *T.l.cabanisi* (Reichenow, 1883)	Coastal Tanzania to Mozambique and NE Zululand
— *T.l.livingstonii* (G.R. Gray, 1864)	Malawi and N Mozambique south to E Zimbabwe

○ *Tauraco schalowi* (Reichenow, 1891) SCHALOW'S TURACO[7]   SW Kenya and N Tanzania, W Tanzania and Malawi to W, C and E Angola

---

[1] Erroneously spelled *blythi* by Peters (1937: 163) (1823).
[2] In last Edition treated as a family within the Cuculiformes, but see also Thomas (1996) (2399).
[3] Three subfamilies accepted following Turner (1997: 480) (2441), and Veron & Winney (2000) (2523).
[4] Sequence of genera follows Turner (1997) (2441).
[5] In last edition treated as a broader species following Brosset & Fry (1988: 30) (243), but see Turner (1997: 499) (2441), whose subspecific treatment we follow.
[6] In recognising subspecies we follow Turner (1997: 499) (2441), who also mentioned a population in E Burundi and W Tanzania not assigned to a subspecies.
[7] Includes *chalcolophus* see Turner (1997: 500) (2441).

○ *Tauraco corythaix* KNYSNA TURACO
— *T.c.phoebus* (Neumann, 1907)   E and N Transvaal, NW Swaziland
— *T.c.corythaix* (Wagler, 1827)   Natal, W Zululand, S Swaziland, S and E Cape Province

○ *Tauraco schuetti* BLACK-BILLED TURACO
— *T.s.emini* (Reichenow, 1893)   E Zaire to Uganda, W Kenya and SE Sudan
— *T.s.schuetti* (Cabanis, 1879)   W and N Zaire to N Angola

○ *Tauraco fischeri* FISCHER'S TURACO
— *T.f.fischeri* (Reichenow, 1878)   Coastal S Somalia, Kenya, NE Tanzania
— *T.f.zanzibaricus* (Pakenham, 1937)   Zanzibar

○ *Tauraco macrorhynchus* YELLOW-BILLED TURACO
— *T.m.macrorhynchus* (Fraser, 1839)   Sierra Leone to Ghana
— *T.m.verreauxi* (Schlegel, 1854)   S Nigeria to W Zaire and Cabinda (N Angola), Bioko I.

○ *Tauraco leucolophus* (Heuglin, 1855) WHITE-CRESTED TURACO   SE Nigeria and N Cameroon to S Sudan, NE Zaire, N and C Uganda and W Kenya

○ *Tauraco bannermani* (Bates, 1923) BANNERMAN'S TURACO   Bamenda-Banso highlands (NW Cameroon)

○ *Tauraco erythrolophus* (Vieillot, 1819) RED-CRESTED TURACO   W and C Angola

○ *Tauraco hartlaubi* (Fischer & Reichenow, 1884) HARTLAUB'S TURACO
     Kenyan highlands to E Uganda and N Tanzania

○ *Tauraco leucotis* WHITE-CHEEKED TURACO
— *T.l.leucotis* (Rüppell, 1835)   W Ethiopian highlands, SE Sudan, Eritrea
— *T.l.donaldsoni* (Sharpe, 1895)   EC Ethiopian highlands

○ *Tauraco ruspolii* (Salvadori, 1896) RUSPOLI'S TURACO   S Ethiopia

○ *Tauraco porphyreolophus* PURPLE-CRESTED TURACO
— *T.p.chlorochlamys* (Shelley, 1881)   S Kenya to Malawi and N Mozambique
— *T.p.porphyreolophus* (Vigors, 1831)   Zimbabwe and C and S Mozambique to E Transvaal and Natal

RUWENZORORNIS Neumann, 1903[1] M
○ *Ruwenzorornis johnstoni* RUWENZORI TURACO
— *R.j.johnstoni* (Sharpe, 1901)   W Uganda (Ruwenzori Mts.), NE and E Zaire
— *R.j.kivuensis* Neumann, 1908   Highlands of E Zaire, Rwanda, Burundi and SW Uganda
— *R.j.bredoi* Verheyen, 1947[2]   Mt. Kabobo, E Zaire (2521)

MUSOPHAGA Isert, 1789 F
○ *Musophaga violacea* Isert, 1789 VIOLET TURACO   Gambia and Guinea to N Nigeria and NW Cameroon, S Chad and N Central African Republic

○ *Musophaga rossae* Gould, 1852 ROSS'S TURACO   Cameroon to S Sudan and W Kenya, south to NW Botswana and N and E Angola

CRINIFERINAE

CORYTHAIXOIDES A. Smith, 1833 M
○ *Corythaixoides concolor* GREY GO-AWAY BIRD[3]
— *C.c.molybdophanes* Clancey, 1964   NE Angola to S Tanzania, N Malawi and N Mozambique (396)

— *C.c.pallidiceps* Neumann, 1899   W Angola to C Namibia
— *C.c.bechuanae* Roberts, 1932   S Angola, NE Namibia, SW Zambia, Botswana, Zimbabwe and N and W Transvaal

— *C.c.concolor* (A. Smith, 1833)   S Malawi, C and S? Mozambique to E Transvaal, E Swaziland and E Zululand

○ *Corythaixoides personatus* BARE-FACED GO-AWAY BIRD
— *C.p.personatus* (Rüppell, 1842)   Ethiopian Rift Valley
— *C.p.leopoldi* (Shelley, 1881)   S Uganda, SW Kenya and Tanzania to SE Zaire, NE Zambia, N Malawi

---

[1] Placed in *Musophaga* in Brosset & Fry (1988: 45) (243) and in our last Edition, but see Turner (1997) (2441).
[2] For recognition of *bredoi* see Louette *et al.* (2000) (1384).
[3] Races follow Turner (1997: 505) (2441); in this arrangement *chobiensis* and *cuanhamae* da Rosa Pinto, 1962 (604) become synonyms of *bechuanae*.

○ *Corythaixoides leucogaster* (Rüppell, 1842) White-bellied Go-away Bird[1]

NW, C and S Somalia, E and S Ethiopia and SE Sudan to NE Uganda, Kenya and N Tanzania

CRINIFER Jarocki, 1821 M
○ *Crinifer piscator* (Boddaert, 1783) Western Grey Plantain-eater

S Senegal and Gambia to Central African Republic; W Zaire and Congo

○ *Crinifer zonurus* (Rüppell, 1835) Eastern Grey Plantain-eater

Eritrea, Ethiopia and S and W Sudan to NE and E Zaire, Uganda, W Kenya and NW Tanzania; SE Chad and W Sudan (Darfur)

CUCULIDAE CUCKOOS[2] (35: 138)

CUCULINAE

CLAMATOR Kaup, 1829[3] M
○ *Clamator jacobinus* Jacobin Cuckoo

— *O.j.serratus* (Sparrman, 1786) — Sth. Africa to S Zambia >> Uganda and Ethiopia

— *O.j.pica* (Hemprich & Ehrenberg, 1833) — Sub-Saharan Africa south to N Zambia and Malawi; NW India to Nepal and Burma >> Africa south to N Namibia and N Botswana

— *O.j.jacobinus* (Boddaert, 1783) — S and E India, C and S Burma, Sri Lanka >> SE Africa

○ *Clamator levaillantii* (Swainson, 1829) Levaillant's Cuckoo[4]

N tropical to E and S Africa >> coastal W Africa and C Africa

○ *Clamator coromandus* (Linnaeus, 1766) Chestnut-winged Cuckoo

N India and Nepal to S and E China and Indochina >> S to Gtr. Sundas, Sulawesi and the Philippines

○ *Clamator glandarius* (Linnaeus, 1758) Great Spotted Cuckoo

a) S Europe and Middle East south to W Africa, Zaire and N Tanzania >> N, C and E Africa

b) Sthn. Africa N to Zambia and S Tanzania[5] >> C Africa

PACHYCOCCYX Cabanis, 1882 M
○ *Pachycoccyx audeberti* Thick-billed Cuckoo

— *P.a.brazzae* (Oustalet, 1886) — Guinea to Cameroon, Congo and W Zaire

— *P.a.validus* (Reichenow, 1879)[6] — E Zaire, SE Kenya and Tanzania to Mozambique and E Transvaal

— *P.a.audeberti* (Schlegel, 1879) — Madagascar

CUCULUS Linnaeus, 1758 M
○ *Cuculus crassirostris* (Walden, 1872) Sulawesi Hawk-Cuckoo — Montane Sulawesi

○ *Cuculus sparverioides* Large Hawk-Cuckoo

— *C.s.sparverioides* Vigors, 1832 — Himalayas to N, C and E China and SE Asia >> E India to Greater Sundas and Philippines

— *C.s.bocki* (Wardlaw Ramsay, 1886) — Malay Pen., Sumatra, Borneo

○ *Cuculus varius* Common Hawk-Cuckoo

— *C.v.varius* Vahl, 1797 — India, Nepal, Bangladesh, Burma >> Sri Lanka

— *C.v.ciceliae* (W.W.A. Phillips, 1949) — Sri Lanka (1880)

○ *Cuculus vagans* S. Müller, 1845 Moustached Hawk-Cuckoo

S Burma, S Thailand, S Laos and Malay Pen. to Sumatra and Borneo. Java as a vagrant only?

○ *Cuculus fugax* Hodgson's Hawk-Cuckoo

— *C.f.nisicolor* Blyth, 1843 — C and E Himalayas and mainland SE Asia north to S and C China, Hainan >> SE Asia to Greater Sundas

— *C.f.fugax* Horsfield, 1821 — Malay Pen., Sumatra, Java, Borneo

[1] Treated as the sole species in the genus *Criniferoides* in last Edition.
[2] Treatment based on Payne (1997) (1800) and R.B. Payne (*in litt.*). The changes from Payne (1997) (1800) include the reduction of the Coccyzinae to a group within the Phaenicophaeinae (where most old generic names are restored and the sequence of genera and species has yet to be refined).
[3] Including *Oxylophus* see Payne (1997) (1800). Type species *levaillantii*.
[4] The older name *cafer* once used for this species was a misidentification. See Payne (1997: 547) (1800).
[5] If recognised as a distinct population, as in Cramp *et al.* (1985) (591), the name *choragium* Clancey, 1951 (365) applies.
[6] Recognised by Irwin (1988: 66) (1190) and by Payne (1997) (1800).

○ *Cuculus hyperythrus* Gould, 1856 NORTHERN HAWK-CUCKOO[1]   NE China, Korea, Russian Far East and Japan >> S China, SE Asia, Philippines, Borneo and Sulawesi

○ *Cuculus pectoralis* (Cabanis & Heine, 1863) PHILIPPINE HAWK-CUCKOO[2]
Palawan, Luzon, Cebu, Mindoro, Sibuyan, Negros, Leyte and Mindanao

○ *Cuculus solitarius* Stephens, 1815 RED-CHESTED CUCKOO[3]   Senegal to Ethiopia and S Africa, Bioko I

○ *Cuculus clamosus* BLACK CUCKOO
— *C.c.gabonensis* Lafresnaye, 1853   Liberia to Zaire, NW Angola, SW Ethiopia and W Kenya
— *C.c.clamosus* Latham, 1802[4]   N and C Ethiopia, E and C Kenya to E and S Angola and Sth. Africa

○ *Cuculus micropterus* INDIAN CUCKOO
— *C.m.micropterus* Gould, 1838[5] [6]   India and Sri Lanka to SE Asia, Russian Far East and NE China >> S and SE Asia to Philippines and Greater Sundas

— *C.m.concretus* S. Müller, 1845   Malay Pen., Sumatra, Java, Borneo

● *Cuculus canorus* COMMON CUCKOO
↓ *C.c.canorus* Linnaeus, 1758[7]   Europe and Middle East to Kamchatka, Japan and NC and NE China >> Africa and S Asia

— *C.c.bangsi* Oberholser, 1919   Iberia, Balearic Is., NW Africa >> Africa
— *C.c.subtelephonus* Zarundy, 1914   Turkmenistan and lower Urals to Afghanistan, NW China and SE Kazakhstan >> S Asia and Africa

— *C.c.bakeri* E. Hartert, 1912[8]   Himalayan foothills, Assam, N and E Burma and NW Thailand >> NE India and mainland SE Asia

○ *Cuculus gularis* Stephens, 1815 AFRICAN CUCKOO   Senegal and Gambia to Eritrea and E Sth. Africa

○ *Cuculus saturatus* HIMALAYAN CUCKOO
— *C.s.saturatus* Blyth, 1843[9]   N Pakistan and Himalayas to S China and Taiwan >> mainland SE Asia and Philippines to New Guinea and N and E Australia

— *C.s.optatus* Gould, 1845[10]   E European Russia to Kuril Is., Japan, Korea and N China >> SE Asia to N and E Australia

— *C.s.lepidus* S. Müller, 1845[11]   Submontane Malay Pen., Greater and Lesser Sundas

○ *Cuculus poliocephalus* Latham, 1790 LESSER CUCKOO[12]   N Afghanistan and Himalayas to Burma and China, north to Russian Far East and Japan >> SE Africa and S and SE Asia

○ *Cuculus rochii* Hartlaub, 1863 MADAGASCAN CUCKOO   Madagascar >> E and C Africa

○ *Cuculus pallidus* (Latham, 1802)[13] PALLID CUCKOO[14]   Australia, Tasmania >> north to Wallacea

CERCOCOCCYX Cabanis, 1882 M
○ *Cercococcyx mechowi* Cabanis, 1882 DUSKY LONG-TAILED CUCKOO   Sierra Leone to Uganda and NW Angola

○ *Cercococcyx olivinus* Sassi, 1912 OLIVE LONG-TAILED CUCKOO   Guinea and Ghana to W Uganda, NW Angola and SE Zaire

○ *Cercococcyx montanus* BARRED LONG-TAILED CUCKOO
— *C.m.montanus* Chapin, 1928   Ruwenzori Mts. to Lake Tanganyika (EC Africa)
— *C.m.patulus* Friedmann, 1928   Kenya to Mozambique and E Zimbabwe

[1] Separation advised by R.B. Payne (*in litt.*).
[2] We follow Payne (1997: 512) (1800) in treating this as an endemic Philippine species.
[3] Includes *magnirostris* Amadon, 1953 (42), see Payne (1997: 553) (1800).
[4] For reasons to use 1802 not 1801 see Browing & Monroe (1991) (258).
[5] The relevant page in the 1837 volume of the PZS appeared in June 1838 see Duncan (1937) (770).
[6] Includes *ognevi* Vorobyev, 1951 (2549) see Vaurie (1965) (2510) and Stepanyan (1990) (2326) and *fatidicus* Koelz, 1954 (1284), see Ripley (1982: 166) (2055) and Stepanyan (1990) (2326).
[7] Includes *sardus* Trischitta, 1939 (2439) see Vaurie (1965: 570) (2510).
[8] Includes *fallax* see Payne (1997: 554) (1800) and R.B. Payne (*in litt.*).
[9] Schodde & Mason (1997) (2188) argued that this name should be attributed to Hodgson as Blyth believed the specimens to be age-related variants of *micropterus*. However, the only description given there, "dark ash-coloured specimens", would appear to have been written by Blyth, and it is not clear that Hodgson did more than offer a MS name.
[10] This name antedates *horsfieldi* see Schodde & Mason (1997) (2188). It was treated as a separate species by Payne (1997: 512) (1800), but apparently is now considered conspecific (R.B. Payne *in litt.*).
[11] Includes *insulindae* see Payne (1997: 554) (1800); for attachment to this species see Becking (1975) (109) and Wells & Becking (1975) (2581).
[12] Includes *assamicus* Koelz, 1952 (1282), see Ripley (1982: 167) (2055).
[13] For reasons to use 1802 not 1801 see Browing & Monroe (1991) (258).
[14] Includes *occidentalis* see Payne (1997) (1800) and Schodde & Mason (1997) (2188).

CACOMANTIS S. Müller, 1843 M

○ *Cacomantis sonneratii* BANDED BAY CUCKOO[1]

 — *C.s.sonneratii* (Latham, 1790)[2]   India and Nepal to Yunnan, Thailand and Indochina
                         and the Malay Pen.

 — *C.s.waiti* (E.C.S. Baker, 1919)   Sri Lanka

 — *C.s.schlegeli* (Junge, 1948)[3]   Sumatra, Borneo, Palawan (1231)

 — *C.s.musicus* (Ljungh, 1804)   Java

○ *Cacomantis passerinus* (Vahl, 1797) GREY-BELLIED CUCKOO   Himalayas, India >> south to Sri Lanka and Maldive Is.

○ *Cacomantis merulinus* PLAINTIVE CUCKOO

 — *C.m.querulus* Heine, 1863   E Himalayas to S China and mainland SE Asia

 — *C.m.threnodes* Cabanis & Heine, 1863[4]   Malay Pen., Sumatra, Borneo

 — *C.m.lanceolatus* (S. Müller, 1843)   Java, Sulawesi

 — *C.m.merulinus* (Scopoli, 1786)   Philippines

○ *Cacomantis variolosus* BRUSH CUCKOO[5]

 — *C.v.sepulcralis* (S. Müller, 1843)   Malay Pen., Greater and Lesser Sundas and
                         Philippines (except as below)

 — *C.v.everetti* E. Hartert, 1925   Basilan, Sulu Is.

 — *C.v.virescens* (Brüggeman, 1876)[6]   Sulawesi, Tukangbesi, Butung I., Banggai I.

 — *C.v.aeruginosus* Salvadori, 1878   Sula, Buru, Ambon and Seram Is. (W and C Moluccas)

 — *C.v.infaustus* Cabanis & Heine, 1863[7]   N and E Moluccas east to lowland New Guinea

 — *C.v.oreophilus* E. Hartert, 1925[8]   Highlands of E and S New Guinea and
                         D'Entrecasteaux Arch.

 — *C.v.blandus* Rothschild & Hartert, 1914   Admiralty Is.

 — *C.v.macrocercus* Stresemann, 1921[9]   New Britain, New Ireland and Tabar I. (Bismarck Arch.)

 — *C.v.websteri* E. Hartert, 1898   New Hanover (Bismarck Arch.)

 — *C.v.addendus* Rothschild & Hartert, 1901   Solomon Is.

 — *C.v.dumetorum* (Gould, 1845)[10]   N Australia (Kimberley to Cape York Pen. and NE
                         Queensland) >>S New Guinea, Moluccas, ?E Lesser
                         Sundas

 — *C.v.variolosus* (Vigors & Horsfield, 1827)[11]   EC and SE Australia >> New Guinea, Moluccas, Timor
                         and E Lesser Sundas and Philippines

○ *Cacomantis castaneiventris* CHESTNUT-BREASTED CUCKOO

 — *C.c.arfakianus* Salvadori, 1889   W Papuan Is., montane NW New Guinea

 — *C.c.weiskei* Reichenow, 1900   Montane EC and E New Guinea

 — *C.c.castaneiventris* (Gould, 1867)   Aru Is., E Cape York Pen.

○ *Cacomantis flabelliformis* FAN-TAILED CUCKOO

 — *C.f.excitus* Rothschild & Hartert, 1907   Mountains of New Guinea

 — *C.f.flabelliformis* (Latham, 1802)[12]   EC to E Australia, Tasmania, SW Western Australia

 — *C.f.pyrrophanus* (Vieillot, 1817)[13]   New Caledonia, Loyalty Is. >> Solomon Is.

 — *C.f.schistaceigularis* Sharpe, 1900   Vanuatu

 — *C.f.simus* (Peale, 1848)   Fiji Is.

CHRYSOCOCCYX Boie, 1826[14] M

○ *Chrysococcyx osculans* (Gould, 1847) BLACK-EARED CUCKOO   Australia >> Aru and Kai Is., Moluccas, Banda Sea Is

---

[1] Listed in a monotypic genus *Penthoceryx* in last Edition, but see Payne (1997) (1800).
[2] Includes *malayanus* (R.B. Payne *in litt.*)
[3] In previous Editions we used the name *fasciolatus*. Junge (1948) (1231) considered this a synonym of *musicus* and Payne (1997: 559) (1800) adopted
  schlegeli. Now R.B. Payne (*in litt.*) believes this needs revisiting.
[4] Includes *subpallidus* see Van Marle & Voous (1988) (2468).
[5] Treatment here is tentative, some Australian birds migrate north, and much has yet to be learned. Further taxa may require recognition and two species
  may be involved.
[6] Includes *fistulator* Stresemann, 1940 (2363).
[7] Includes *obscuratus*, *chivae* Mayr & Meyer de Schauensee, 1939 (1491), and *obiensis* Jany, 1955 (1200) and *heinrichi* (R.B. Payne, *in litt.*) although this was
  treated as a separate species by Payne (1997: 561) (1800).
[8] Includes *fortior*.
[9] Includes *tabarensis* Amadon, 1942 (40).
[10] This may be better submerged in nominate *variolosus* (R.B. Payne *in litt.*).
[11] Includes *whitei* Bruce, 1986 (2652), from Timor see Schodde & Mason (1997) (2188), where *sepulcralis* and *variolosus* are treated as separate species (with
  tymbonomus associated to the latter). Also includes *tymbonomus* (R.B. Payne *in litt.*).
[12] The identity of the bird so named has been discussed by Mason (1982) (1448); it proves to be a prior name for what was called *C. pyrrophanus prionurus* in
  Peters (1940) (1825) and in our previous Editions. For reasons to use 1802 not 1801 see Browing & Monroe (1991) (258).
[13] Includes *meeki* based on a migrant or vagrant to the Solomons.
[14] Payne (1997) (1800) accepted a broad genus *Chrysococcyx* (see Friedmann, 1968) (882), but this is not always accepted, see Marchant (1972) (1425) and
  Schodde & Mason (1997) (2188).

○ *Chrysococcyx basalis* (Horsfield, 1821) HORSFIELD'S BRONZE CUCKOO   Australia, Tasmania >> S New Guinea, Moluccas, Lesser Sundas, Java, Sumatra and Belitung

○ *Chrysococcyx minutillus* LITTLE BRONZE CUCKOO[1]
___ *C.m.peninsularis* S. Parker, 1981   S Thailand, Malay Pen. (1744)
___ *C.m.albifrons* (Junge, 1938)   Sumatra, W Java
___ *C.m.aheneus* (Junge, 1938)[2]   Borneo, S Philippines
___ *C.m.jungei* (Stresemann, 1938)   Sulawesi, Madu, Flores
___ *C.m.subsp.?*   Timor[3]
___ *C.m.rufomerus* E. Hartert, 1900   Romang to Wetar (E Lesser Sundas)
___ *C.m.misoriensis* (Salvadori, 1875)   Biak (Geelvink Bay)
___ *C.m.poecilurus* G.R. Gray, 1862[4]   Seram, coastal lowlands of New Guinea and adjacent islands inc. Misool and D'Entecasteaux Arch. Cape York Pen. to C Queensland (NE Australia)
___ *C.m.minutillus* Gould, 1859[5]   N Australia (incl. Melville I.)
___ *C.m.barnardi* Mathews, 1912   SE Queensland, NE New South Wales >> NE Queensland, Cape York Pen. >> New Guinea

○ *Chryscoccyx crassirostris* PIED BRONZE CUCKOO[6]
___ *C.c.crassirostris* (Salvadori, 1878)   Tayandu and Kai Is. (Moluccas), Yamdena and Lartat (Tanimbar Is.)[7]
___ *C.c.salvadorii* (Hartert & Stresemann, 1925)   Tepa (Babar Is., Lesser Sundas)

○ *Chrysococcyx lucidus* SHINING BRONZE CUCKOO
___ *C.l.harterti* (Mayr, 1932)   Rennell and Bellona Is. (S Solomons)
___ *C.l.layardi* Mathews, 1912[8]   New Caledonia, Loyalty Is., Vanuatu, Banks Is., Santa Cruz Is.
___ *C.l.plagosus* (Latham, 1802)[9]   E, SW Australia, Tasmania >> Lesser Sundas, New Guinea, N. Melanesia
___ *C.l.lucidus* (J.F. Gmelin, 1788)   New Zealand, Chatham Is., Lord Howe I., Norfolk I. >> N Melanesia to Solomons

○ *Chrysococcyx ruficollis* (Salvadori, 1875) RUFOUS-THROATED BRONZE CUCKOO   Mountains of New Guinea

○ *Chrysococcyx meyeri* Salvadori, 1874 WHITE-EARED BRONZE CUCKOO[10]   Mountains of New Guinea and Batanta I.

○ *Chrysococcyx maculatus* (J.F. Gmelin, 1788) ASIAN EMERALD CUCKOO   Himalayas to SC China and Indochina, Andaman and Nicobar Is. >> south to N Vietnam, Malay Pen. and Sumatra

○ *Chrysococcyx xanthorhynchus* VIOLET CUCKOO
___ *C.x.xanthorhynchus* (Horsfield, 1821)[11]   NE India and Bangladesh to mainland SE Asia, Greater Sundas and Palawan
___ *C.x.amethystinus* (Vigors, 1831)   Philippines

○ *Chrysococcyx flavigularis* Shelley, 1880 YELLOW-THROATED CUCKOO[12]   Sierra Leone to Gabon, east to W Uganda

○ *Chrysococcyx klaas* (Stephens, 1815) KLAAS's CUCKOO[13]   Senegal to Ethiopia south to N Namibia and E Sth. Africa, Bioko I.

○ *Chrysococcyx cupreus* (Shaw, 1792) AFRICAN EMERALD CUCKOO[14]   Senegal to S Sudan and Ethiopia, S to N Angola and E Sth. Africa, Bioko, São Tomé, Príncipé and Pagalu Is.

---

[1] This arrangement follows Payne (1997) (1800). In last Edition we followed Parker (1981) (1744) and this assemblage was split between three monotypic species (*poecilurus*, *rufomerus* and *crassirostris*) and polytypic *russatus* and *minutillus*. Questions remain and previous reports of sympatry should be kept in mind as migrants may not be involved. See also Ford (1981) (858) and Schodde & Mason (1997) (2188).
[2] Includes *cleis* S. Parker, 1981 (1744).
[3] For the suggestion that there may be an undescribed form see Bruce *in* White & Bruce (1986: 238) (2652). The one Timor specimen (AMNH 627051) seems to match *poecilurus* (R.B. Payne *in litt.*); perhaps there is no breeding population in Timor and this is a vagrant.
[4] Includes *russatus* (R.B. Payne *in litt.*).
[5] Includes *nieuwenhuysi* see White & Bruce (1986) (2652).
[6] This species has a distinctive red eye-ring.
[7] Specimens from Sorong in the Vogelkop (New Guinea) are of disputed provenance, as to a lesser extent are those from the N Moluccas, see Parker (1981) (1744) (R.B. Payne *in litt.*).
[8] Apparently includes *aeneus* Warner, 1951 (2559) of Vanuatu.
[9] For recognition of this form as distinct from more eastern birds see Schodde & Mason (1997) (2188). For reasons to use 1802 not 1801 see Browing & Monroe (1991) (258).
[10] This species has a distinctive red eye-ring.
[11] Includes *bangueyensis* and *limborgi* Tweeddale, 1877 (2442A) (for the latter name see Chasen, 1935: 127) (319). *C. malayanus* was based on a female specimen of this species (Parker, 1981) (1744).
[12] Includes *parkesi* Dickerman, 1994 (724), see Payne (1997: 565) (1800).
[13] Apparently includes *somereni* Chapin, 1954 (312).
[14] Includes *insularum* Moreau & Chapin, 1951 (1624), see Payne (1997: 565) (1800).

○ *Chrysococcyx caprius* (Boddaert, 1783) Dideric Cuckoo   Sub-Saharan Africa, S Arabia

RHAMPHOMANTIS Salvadori, 1878 M
○ *Rhamphomantis megarhynchus* (G.R. Gray, 1858) Long-billed Cuckoo[1]
    New Guinea, Waigeo and Aru Is.

SURNICULUS Lesson, 1830[2] M
○ *Surniculus lugubris* Asian Drongo-Cuckoo[3]
  — *S.l.dicruroides* (Hodgson, 1839)[4]   N India to W Sichuan, S China (incl. Hainan),
        N Thailand and Vietnam >> S to Greater Sundas

  — *S.l.lugubris* (Horsfield, 1821)[5]   SW India and Sri Lanka, Malay Pen., Sumatra, Java,
        Bali, Bangka, Borneo and the Palawan group

  — *S.l.musschenbroeki* A.B. Meyer, 1878   Sulawesi and Butung to Bacan and Halmahera

○ *Surniculus velutinus* Philippine Drongo-Cuckoo
  — *S.v.chalybaeus* Salomonsen, 1953[6]   Luzon, Mindoro and Negros (2155)
  — *S.v.velutinus* Sharpe, 1877[7]   Bohol, Leyte, Samar, Mindanao, Basilan and the Sulu Is.

CALIECHTHRUS Cabanis & Heine, 1863 M
○ *Caliechthrus leucolophus* (S. Müller, 1840) White-crowned Koel   New Guinea and Salawati I.

MICRODYNAMIS Salvadori, 1878 F
○ *Microdynamis parva* Dwarf Koel
  — *M.p.grisecens* Mayr & Rand, 1936   Humboldt Bay to Kumusi R (NC and NE New Guinea)
  — *M.p.parva* (Salvadori, 1875)   Lowland forests of New Guinea, D'Entrecasteaux Arch.

EUDYNAMYS Vigors & Horsfield, 1827 M[8]
○ *Eudynamys scolopaceus* Common Koel[9]
  — *E.s.scolopaceus* (Linnaeus, 1758)   Nepal, Pakistan, India and Sri Lanka to Laccadives
        and Maldives

  — *E.s.dolosus* Ripley, 1946   Andamans, Nicobar Is. (2029)
  — *E.s.chinensis* Cabanis & Heine, 1863   S and C China, Indochina >> south to Borneo
  — *E.s.harterti* Ingram, 1912   Hainan
  — *E.s.malayanus* Cabanis & Heine, 1863   NE India and Bangladesh to Sumatra, Borneo and W
        Lesser Sundas

  — *E.s.simalurensis* Junge, 1936   Simeuluë and Kokos and Babi Is. (off NW Borneo)
  — *E.s.melanorhynchus* S. Müller, 1843[10]   Sulawesi, Sula Is.
  — *E.s.frater* McGregor, 1904   Calayan and Fuga (NE Philippines)
  — *E.s.mindanensis* (Linnaeus, 1766)[11]   Palawan and the Philippines to Sangihe Is., Talaud Is.,
        Halmahera, Bacan, Ternate (N Moluccas)

  — *E.s.orientalis* (Linnaeus, 1766)   S Moluccas
  — *E.s.picatus* S. Müller, 1843   E Lesser Sundas, Kai Is.
  — *E.s.rufiventer* (Lesson, 1830)   New Guinea
  — *E.s.hybrida* J.M. Diamond, 2002   Long I. off NE New Guinea (703A)
  — *E.s.salvadorii* E. Hartert, 1900   Bismarck Arch.
  — *E.s.alberti* Rothschild & Hartert, 1907   Solomon Is.
  — *E.s.subcyanocephalus* Mathews, 1912   N Australia (Kimberley to Cape York Pen. and NE
        Queensland) >> New Guinea

  — *E.s.cyanocephalus* (Latham, 1802)[12]   EC Queensland to all E New South Wales >> Arafura
        and Banda Sea Is., Moluccas, E Lesser Sundas

[1] Includes *sanfordi* (R.B. Payne, *in litt.*).
[2] This genus is probably genetically closest to *Cuculus* and *Cacomantis* (R.B. Payne, *in litt.*).
[3] We follow Payne (1997: 569) (1800); and for further consolidation, R.B. Payne (*in litt.*).
[4] The name *barussarum* has been shown to apply to migrants (Mees, 1986) (1539).
[5] Includes *stewarti* see Payne (1997: 569) (1800), and *minimus* and *brachyurus* (R.B. Payne *in litt.*).
[6] Includes *mindorensis* Ripley & Rabor, 1958 (2072), see Dickinson *et al.* (1991) (745).
[7] Includes *suluensis* Salomonsen, 1953 (2155), see Dickinson *et al.* (1991) (745).
[8] Gender addressed by David & Gosselin (2002) (614), multiple changes occur in specific names.
[9] Schodde & Mason (1997) (2188) treated a group of forms from the Moluccas to Australasia as an allospecies *orientalis*.
[10] Black-billed and treated as a species by White & Bruce (1986) (2652), but not by Payne (1997: 570) (1800). Tentatively lumped here. Includes *facialis* see
    White & Bruce (1986: 240) (2652).
[11] Including *paraguena* see Dickinson *et al.* (1991) (745), and *corvinus* (R.B. Payne *in litt.*).
[12] For reasons to use 1802 not 1801 see Browing & Monroe (1991) (258).

URODYNAMIS Salvadori, 1880 F
○ *Urodynamis taitensis* (Sparrman, 1787)[1] LONG-TAILED KOEL   New Zealand, Great and Little Barrier, Kapiti, Stewart

SCYTHROPS Latham, 1790 M
○ *Scythrops novaehollandiae* CHANNEL-BILLED CUCKOO
    __ *S.n.novaehollandiae* Latham, 1790   N and E Australia >> New Guinea and nearby islands and Moluccas, E Lesser Sundas

    __ *S.n.fordi* Mason, 1996   Sulawesi, Banggai and Tukangbesi Is. (1450)
    __ *S.n.schoddei* Mason, 1996   Bismarck Arch. (1450)

PHAENICOPHAEINAE

CEUTHMOCHARES Cabanis & Heine, 1863 M
○ *Ceuthmochares aereus* YELLOWBILL
    __ *C.a.flavirostris* (Swainson, 1837)   Gambia to SW Nigeria
    __ *C.a.aereus* (Vieillot, 1817)[2]   SE Nigeria and Bioko to W Kenya south to N Angola and N Zambia
    __ *C.a.australis* Sharpe, 1873[3]   SW Ethiopia, SE Somalia south to Mozambique, coastal Natal and E Cape

RHOPODYTES Cabanis & Heine, 1863 M
○ *Rhopodytes diardi* BLACK-BELLIED MALKOHA
    __ *R.d.diardi* (Lesson, 1830)   Malay Pen. and Sumatra
    __ *R.d.borneensis* Salvadori, 1874   Borneo

○ *Rhopodytes sumatranus* (Raffles, 1822) CHESTNUT-BELLIED MALKOHA[4]   Malay Pen., Sumatra, Borneo

○ *Rhopodytes tristis* GREEN-BILLED MALKOHA
    __ *R.t.tristis* (Lesson, 1830)   W Himalayas to Assam and Bangladesh
    __ *R.t.saliens* Mayr, 1938   N Burma, N Indochina, SW China
    __ *R.t.hainanus* E. Hartert, 1910   Hainan and S Guangdong
    __ *R.t.longicaudatus* (Blyth, 1842)   Most of Burma and Thailand, W Malaysia, S Indochina
    __ *R.t.elongatus* (S. Müller, 1835)   Sumatra
    __ *R.t.kangeangensis* Vorderman, 1893   Kangean Is.

○ *Rhopodytes viridirostris* (Jerdon, 1840) BLUE-FACED MALKOHA   Peninsular India, Sri Lanka

TACCOCUA Lesson, 1830 F
○ *Taccocua leschenaultii* SIRKEER MALKOHA
    __ *T.l.sirkee* (J.E. Gray, 1831)[5]   Pakistan, NW India
    __ *T.l.infuscata* Blyth, 1845[6]   C and E Himalayan foothills to W Assam
    __ *T.l.leschenaultii* Lesson, 1830   S India, Sri Lanka

RHINORTHA Vigors, 1830 F
○ *Rhinortha chlorophaea* RAFFLES'S MALKOHA
    __ *R.c.chlorophaea* (Raffles, 1822)[7]   S Burma and Malaysia to Sumatra, Bangka and N, E and S Borneo

    __ *R.c.facta* Ripley, 1942   Batu Is. (off W Sumatra) (2025)
    __ *R.c.fuscigularis* E.C.S. Baker, 1919   Sarawak and NW Kalimantan (NW Borneo)

ZANCLOSTOMUS Swainson, 1837 M
○ *Zanclostomus javanicus* RED-BILLED MALKOHA[8]
    __ *Z.j.pallidus* Robinson & Kloss, 1921   Malay Pen., Sumatra and Borneo

---

[1] Treated in the genus *Urodynamis* in last Edition. Payne (1997) (1800) disagreed, but has now recommended it be retained (R.B. Payne *in litt.*).
[2] Includes *extensicaudus* Clancey, 1962 (392), see White (1965: 187) (2649).
[3] Includes *dendrobates* Clancey, 1962 (392), see White (1965: 188) (2649) who erroneously called it *dendrobiastes*.
[4] Includes *minor* from Borneo, see Mees (1986) (1539), although Wells (1999) (2579) thought it might prove distinct, and *rodolphi* Ripley, 1942 (2025).
[5] Includes *vantynei* Koelz, 1954 (1284), see Ripley (1982: 172) (2055).
[6] Includes *affinis* see Ripley (1982: 172) (2055).
[7] Includes *mayri* Delacour, 1947 (677), see Payne (1997) (1800), and *bangkanus* Meyer de Schauensee, 1958 (1570), see Mees (1986: 49) (1539), although Wells (1999) (2579) excluded Borneo from the range of nominate *chlorophaeus*, presumably placing all Bornean birds in *fuscigularis*. R.B. Payne (*in litt.*) now favours treatment as a monotypic species.
[8] Payne (1997) (1800) treated it as monotypic. In last Edition we erroneously listed *factus* as a fourth race. Wells (1999) (2579) recognised *pallidus* as distinct and specimens in ZMA support this.

— *Z.j.natunensis* Chasen, 1934   Natuna Is.

— *Z.j.javanicus* (Horsfield, 1821)   Java

○ ***Zanclostomus calyorhynchus*** Yellow-billed Malkoha

— *Z.c.calyorhynchus* (Temminck, 1825)   N, E and SE Sulawesi, Togian Is.

— *Z.c.meridionalis* (Meyer & Wiglesworth, 1896)   C and S Sulawesi

— *Z.c.rufiloris* (E. Hartert, 1903)   Butung I.

○ ***Zanclostomus curvirostris*** Chestnut-breasted Malkoha

— *Z.c.singularis* (Parrot, 1907)[1]   S Burma to Sumatra

— *Z.c.oeneicaudus* (J. & E. Verreaux, 1855)   Mentawai Is. (off W Sumatra)

— *Z.c.curvirostris* (Shaw, 1810)   W and C Java

— *Z.c.deningeri* (Stresemann, 1913)   E Java, Bali

— *Z.c.borneensis* (Blasius & Nehrkorn, 1881)[2]   Borneo, Bangka I.

— *Z.c.harringtoni* (Sharpe, 1877)   Calamianes, Palawan and Balabac (SW Philippines)

PHAENICOPHAEUS Stephens, 1815   M

○ ***Phaenicophaeus pyrrhocephalus*** (Pennant, 1769)   Red-faced Malkoha
     Sri Lanka

DASYLOPHUS Swainson, 1837   M

○ ***Dasylophus superciliosus*** Rough-crested Malkoha

— *D.s.cagayanensis* (Rand & Rabor, 1967)   Cagayan Province (NE Luzon) (1978)

— *D.s.superciliosus* Dumont, 1823   Luzon

LEPIDOGRAMMUS Reichenbach, 1849   M

○ ***Lepidogrammus cumingi*** (Fraser, 1839)   Scale-feathered Malkoha   Luzon, Marinduque and Catanduanes (N Philippines)

COCCYZUS Vieillot, 1816[3] M

○ ***Coccyzus pumilus*** Strickland, 1852   Dwarf Cuckoo   N Colombia, NE Venezuela

○ ***Coccyzus cinereus*** Vieillot, 1817   Ash-coloured Cuckoo   N and E Bolivia, Paraguay, Uruguay, C and S Brazil, N Argentina (migration and winter range unclear)

○ ***Coccyzus erythopthalmus*** (A. Wilson, 1811)   Black-billed Cuckoo   SC and SE Canada to SC and SE USA >> NW and WC Sth. America

● ***Coccyzus americanus*** (Linnaeus, 1758)   Yellow-billed Cuckoo[4]   SE Canada and USA to Mexico and N West Indies >> Sth. America

○ ***Coccyzus euleri*** Cabanis, 1873   Pearly-breasted Cuckoo   NE Colombia, the Guianas and N Brazil to SE Brazil and NE Argentina; perhaps only a winter visitor in the north

○ ***Coccyzus minor*** (J.F. Gmelin, 1788)   Mangrove Cuckoo[5]   Florida to Greater and Lesser Antilles, S Mexico to Colombia, Trinidad and coastal N Sth. America

○ ***Coccyzus ferrugineus*** Gould, 1843   Cocos Island Cuckoo   Cocos I. (off Costa Rica)

○ ***Coccyzus melacoryphus*** Vieillot, 1817   Dark-billed Cuckoo   W Colombia to E and S Brazil and C Argentina >> Venezuela, Guianas and Trinidad; Galapagos Is. (status?)

○ ***Coccyzus lansbergi*** Bonaparte, 1850   Grey-capped Cuckoo   N Colombia and N Venezuela >> SW Ecuador and W Peru

SAUROTHERA Vieillot, 1816   F

○ ***Saurothera merlini*** Great Lizard Cuckoo

— *S.m.bahamensis* H. Bryant, 1864[6]   Andros, New Providence and Eleuthera Is. (Bahamas)

— *S.m.santamariae* Garrido, 1971   Islands off NC Cuba (919)

— *S.m.merlini* d'Orbigny, 1839   Cuba

— *S.m.decolor* Bangs & Zappey, 1905   Isla de la Juventud

---

[1] Called *erythrognathus* in previous Editions, but see Deignan (1952) (650).
[2] Within a broad genus *Phaenicophaeus* the name *borneensis* is preoccupied and the name *microrhinus* von Berlepsch, 1895, must be used, see Mees (1986: 49) (1539).
[3] This and the next three genera (the American cuckoos) should be treated within the Phaenicophaeinae (R.B. Payne, *in litt.*).
[4] Treated as monotypic, see A.O.U. (1983, 1998) (2, 3).
[5] Reasons to treat this as monotypic are given by Payne (1997: 596) (1800).
[6] Includes *andria* see Payne (1997: 597) (1800), there spelled *andrina*.

○ *Saurothera vielloti* Bonaparte, 1850 Puerto Rican Lizard Cuckoo   Puerto Rico

○ *Saurothera vetula* (Linnaeus, 1758) Jamaican Lizard Cuckoo   Jamaica

○ *Saurothera longirostris* Hispaniolan Lizard Cuckoo
     __ *S.l.petersi* Richmond & Swales, 1924   La Mahotiere and Gonâve I
     __ *S.l.longirostris* (Hermann, 1783)[1]   Hispaniola, Sãona I.

HYETORNIS P.L. Sclater, 1862 M
○ *Hyetornis rufigularis* (Hartlaub, 1852) Bay-breasted Cuckoo[2]   Hispaniola and Gonâve I.

○ *Hyetornis pluvialis* (J.F. Gmelin, 1788) Chestnut-bellied Cuckoo   Jamaica

PIAYA Lesson, 1830 F
● *Piaya cayana* Squirrel Cuckoo
     __ *P.c.mexicana* (Swainson, 1827)   W Mexico
     ⊿ *P.c.thermophila* P.L. Sclater, 1860[3]   E Mexico to Panama and NW Colombia
     __ *P.c.nigricrissa* (Cabanis, 1862)[4]   W Colombia, W Ecuador, NW and EC Peru, N Bolivia
     __ *P.c.mehleri* Bonaparte, 1850   NE Colombia, N Venezuela
     __ *P.c.mesura* (Cabanis & Heine, 1863)   E Colombia, E Ecuador
     __ *P.c.circe* Bonaparte, 1850   W Venezuela
     __ *P.c.cayana* (Linnaeus, 1766)[5]   E and S Venezuela, the Guianas, N Brazil
     __ *P.c.insulana* Hellmayr, 1906   Trinidad
     __ *P.c.obscura* E. Snethlage, 1908   C Brazil
     __ *P.c.hellmayri* Pinto, 1938[6]   C and E Brazil
     __ *P.c.pallescens* (Cabanis & Heine, 1863)   E Brazil
     __ *P.c.cabanisi* J.A. Allen, 1893   SC Brazil
     __ *P.c.macroura* Gambel, 1849   SE Brazil, NE Argentina, Paraguay, Uruguay
     __ *P.c.mogenseni* J.L. Peters, 1926   S Bolivia to NW Argentina

○ *Piaya melanogaster* (Vieillot, 1817) Black-bellied Cuckoo[7]   SE Colombia to the Guianas, NW and WC Brazil and N Bolivia

○ *Piaya minuta* Little Cuckoo
     __ *P.m.gracilis* (Heine, 1863)[8]   E Panama, Colombia (N and W of the E Andes), W Ecuador, Bolivia
     __ *P.m.minuta* (Vieillot, 1817)[9]   E Colombia to the Guianas, Trinidad S to Amazonian Peru and Brazil

CARPOCOCCYX G.R. Gray, 1840 M
○ *Carpococcyx radiceus* (Temminck, 1832)[10] Bornean Ground Cuckoo   Borneo

○ *Carpococcyx viridis* Salvadori, 1879 Sumatran Ground Cuckoo[11]   Lower hills of C and S Sumatra

○ *Carpococcyx renauldi* Oustalet, 1896 Coral-billed Ground Cuckoo   SE Thailand, Indochina

COUA Schinz, 1821 F
○ †*Coua delalandei* (Temminck, 1827) Snail-eating Coua[12]   Madagascar (NE?)

○ *Coua gigas* (Boddaert, 1783) Giant Coua   Coastal W and S Madagascar

○ *Coua coquereli* A. Grandidier, 1867 Coquerel's Coua   W Madagascar

○ *Coua serriana* Pucheran, 1845 Red-breasted Coua   NE Madagascar

○ *Coua reynaudii* Pucheran, 1845 Red-fronted Coua   N and E Madagascar

---

[1] Includes *saonae* see Payne (1997) (1800).
[2] The A.O.U. vernacular name.
[3] Includes *stirtoni*.
[4] Includes *boliviana*.
[5] Thought to include *inexpectata* Todd, 1947 (2409).
[6] The name *cearae* relates to birds intermediate between this and the next form.
[7] Includes *ochracea*.
[8] Includes *panamensis* and *chaparensis* (R.B. Payne *in litt.*).
[9] Includes *barinensis* Aveledo & Perez, 1994 (69) (R.B. Payne, *in litt.*).
[10] The name *radiatus* cannot be construed as an emendation, it has a different root meaning. As such *radiatus* is a junior synonym of *radiceus*. As *radiceus* has been in use since 1899 and has priority its use must be maintained, *contra* Collar & Long (1996) (547). The two conditions that would allow Art. 23.9. of the Code (I.C.Z.N., 1999) (1178) to apply are not met. See also p. 000.
[11] For separation of this from *C. radiceus* [nec *radiatus*] see Collar & Long (1996) (547).
[12] Extinct taxa were omitted in last Edition.

○ *Coua cursor* A. Grandidier, 1867 RUNNING COUA — SW and S Madagascar

○ *Coua ruficeps* RED-CAPPED COUA
— *C.r.ruficeps* G.R. Gray, 1846 — W and SW Madagascar
— *C.r.olivaceiceps* (Sharpe, 1873) — S Madagascar

○ *Coua cristata* CRESTED COUA
— *C.c.cristata* (Linnaeus, 1766) — NW, NE and E Madagascar
— *C.c.dumonti* Delacour, 1931 — WC Madagascar
— *C.c.pyropyga* A. Grandidier, 1867 — S and W Madagascar
— *C.c.maxima* Milon, 1950 — SE Madagascar (1585)

○ *Coua verreauxi* A. Grandidier, 1867 VERREAUX'S COUA — SW Madagascar

○ *Coua caerulea* (Linnaeus, 1766) BLUE COUA — NW, NE and E Madagascar

CENTROPODINAE

CENTROPUS Illiger, 1811 M
○ *Centropus celebensis* BAY COUCAL
— *C.c.celebensis* Quoy & Gaimard, 1830 — N Sulawesi, Togian Is.
— *C.c.rufescens* (Meyer & Wiglesworth, 1896) — C, E, S and SE Sulawesi, and islands off SE coast

○ *Centropus unirufus* (Cabanis & Heine, 1863) RUFOUS COUCAL — Luzon, Polillo and Catanduanes (N Philippines)

○ *Centropus melanops* BLACK-FACED COUCAL
— *C.m.melanops* Lesson, 1830 — Dinagat, Siargao, Mindanao and Basilan (S Philippines)
— *C.m.banken* Hachisuka, 1934 — Bohol, Leyte, and Samar (EC Philippines)

○ *Centropus nigrorufus* (Cuvier, 1816)[1] SUNDA COUCAL — Java

○ *Centropus milo* BUFF-HEADED COUCAL
— *C.m.albidiventris* Rothschild, 1904 — Vella Lavella, Kolombangara, New Georgia, Gizo and Rendova (C Solomons)
— *C.m.milo* Gould, 1856 — Florida and Guadalcanal Is. (S Solomons)

○ *Centropus goliath* Bonaparte, 1850 GOLIATH COUCAL — N Moluccas

○ *Centropus violaceus* Quoy & Gaimard, 1830 VIOLACEOUS COUCAL — New Ireland, New Britain (Bismarck Arch.)

○ *Centropus menbeki* GREATER BLACK COUCAL
— *C.m.menbeki* Lesson & Garnot, 1828[2] — W Papuan Is., New Guinea, Yapen and Numfor Is.
— *C.m.aruensis* (Salvadori, 1878) — Aru Is.

○ *Centropus ateralbus* Lesson, 1826 PIED COUCAL — New Britain, New Ireland (Bismarck Arch.)

○ *Centropus phasianinus* PHEASANT-COUCAL
— *C.p.mui* Mason & McKean, 1984 — Timor (1452)
— *C.p.propinquus* Mayr, 1937 — N New Guinea
— *C.p.nigricans* (Salvadori, 1876)[3] — SE New Guinea and nearby islands
— *C.p.thierfelderi* Stresemann, 1927 — S New Guinea, NW Torres Straits Is.
— *C.p.melanurus* Gould, 1847[4] — N and NW Australia
— *C.p.phasianinus* (Latham, 1802)[5] — E Australia

○ *Centropus spilopterus* G.R. Gray, 1858 KAI COUCAL — Kai Is. (SE Moluccas)

○ *Centropus bernsteini* LESSER BLACK COUCAL
— *C.b.bernsteini* Schlegel, 1866 — West to SC and NC New Guinea
— *C.b.manam* Mayr, 1937 — Manam I. (NE New Guinea)

○ *Centropus chalybeus* (Salvadori, 1875) BIAK COUCAL — Biak I. (Geelvink Bay, N New Guinea)

○ *Centropus rectunguis* Strickland, 1847 SHORT-TOED COUCAL — Malay Peninsula, Sumatra, Borneo

[1] Given as 1817 by Peters (1940: 71) (1825) but see Browning & Monroe (1991) (258).
[2] Includes *jobiensis* (Payne, *in litt.*).
[3] We believe this includes *obscuratus* of the D'Entrecasteaux Is.
[4] The names *melanurus* and *macrourus* (see our last Edition) are of identical date. Mathews (1927) (1454) must be followed as First Reviser in the use of *melanurus*, see Art. 24.2.1. of the Code (I.C.Z.N., 1999) (1178). This contradicts usage in Peters (1940: 69) (1825) see Schodde & Mason (1997: 185) (2188).
[5] For reasons to use 1802 not 1801 see Browing & Monroe (1991) (258).

○ *Centropus steerii* Bourns & Worcester, 1894 BLACK-HOODED COUCAL | Mindoro (Philippines)

○ *Centropus sinensis* GREATER COUCAL
 __ *C.s.sinensis* (Stephens, 1815) | C Pakistan and N India to N Assam, E and SE China
 __ *C.s.parroti* Stresemann, 1913 | C and S India, Sri Lanka
 __ *C.s.intermedius* (Hume, 1873) | Bangladesh and S Assam to Yunnan, Indochina, S Thailand and SE Burma
 __ *C.s.bubutus* Horsfield, 1821 | Malay Pen., Sumatra, islands off W Sumatra, Natuna Is. to Borneo, Balabac and Palawan, and Java and Bali
 __ *C.s.anonymus* Stresemann, 1913 | Basilan and Sulu Is. (SW Philippines)
 __ *C.s.kangeanensis* Vorderman, 1893 | Kangean Is. (Java Sea)

○ *Centropus andamanensis* Beavan, 1867 ANDAMAN COUCAL[1] | Andaman Is., Cocos Is.

○ *Centropus viridis* PHILIPPINE COUCAL
 __ *C.v.major* Parkes & Niles, 1988 | Babuyanes Is. (N Philippines) (1786)
 __ *C.v.viridis* (Scopoli, 1786) | N and E Philippines
 __ *C.v.mindorensis* (Steere, 1890) | Mindoro and Semirara Is. (NC Philippines)
 __ *C.v.carpenteri* Mearns, 1907 | Batan group (E Philippines)

○ *Centropus toulou* MADAGASCAN COUCAL
 __ *C.t.toulou* (Statius Müller, 1776) | Madagascar
 __ *C.t.insularis* Ridgway, 1894 | Aldabra I.
 __ †*C.t.assumptionis* Nicoll, 1906 | Assumption I.

○ *Centropus grillii* Hartlaub, 1861 AFRICAN BLACK COUCAL | Senegal and Gambia to W Ethiopia, Angola and NE Natal

○ *Centropus bengalensis* LESSER COUCAL[2]
 __ *C.b.bengalensis* (J.F. Gmelin, 1788)[3] | India and Nepal to Burma, Thailand and Indochina
 __ *C.b.lignator* Swinhoe, 1861 | S and SE China, Hainan, Taiwan
 __ *C.b.javanensis* (Dumont, 1818) | Malay Pen. to Sumatra, Java, Borneo and the Palawan group and the Sulu Is. (Philippines)
 __ *C.b.philippinensis* Mees, 1971 | Philippines (except Palawan and the Sulus) (1522)
 __ *C.b.sarasinorum* Stresemann, 1912[4] | Sulawesi, Lesser Sundas
 __ *C.b.medius* Bonaparte, 1850 | Moluccas

○ *Centropus chlororhynchos* Blyth, 1849 GREEN-BILLED COUCAL[5] | Sri Lanka

○ *Centropus leucogaster* BLACK-THROATED COUCAL
 __ *C.l.leucogaster* (Leach, 1814) | S Senegal and Guinea-Bissau to SE Nigeria
 __ *C.l.efulensis* Sharpe, 1904 | SW Cameroon, Gabon
 __ *C.l.neumanni* Alexander, 1908[6] | NE Zaire

○ *Centropus anselli* Sharpe, 1874 GABON COUCAL | S Cameroon to NW Angola and C Zaire

○ *Centropus monachus* BLUE-HEADED COUCAL
 __ *C.m.fischeri* Reichenow, 1887[7] | Ivory Coast and Ghana to S Sudan, W Kenya and N Angola
 __ *C.m.monachus* Rüppell, 1837 | Ethiopia to C Kenya

○ *Centropus cupreicaudus* Reichenow, 1896 COPPERY-TAILED COUCAL[8] | SE Zaire and SW Tanzania to Malawi, N Botswana and Angola

○ *Centropus senegalensis* SENEGAL COUCAL
 __ *C.s.aegyptius* (J.F. Gmelin, 1788) | R Nile to El Minya (Egypt)
 __ *C.s.senegalensis* (Linnaeus, 1766) | Senegal and Gambia to Eritrea, south to NW Angola, SC Zaire and Uganda
 __ *C.s.flecki* Reichenow, 1893 | E Angola and NE Namibia to S Tanzania, Malawi and Zimbabwe

---

[1] Ripley (1982: 173) (2055) was in doubt whether to treat this as a separate species.
[2] Subspecies recognised follow Payne (1997) (1800) except for the acceptance of *philippinensis* (see Dickinson *et al.*, 1991: 221) (745).
[3] Includes *chamnongi* Deignan, 1955 (655) see Payne (1997: 590) (1800).
[4] Omitted in our last Edition.
[5] Blyth's spelling was *chlororhynchos* as given by Peters (1940: 69) (1825).
[6] In last Edition we treated this as a species following Louette (1986) (1378); however see Dowsett & Dowsett-Lemaire (1993: 335) (761).
[7] Includes *verheyeni* Louette, 1986 (1378).
[8] Includes *songweensis* Benson, 1948 (119), described as a race of *monachus*, see Payne (1997) (1800).

○ *Centropus superciliosus* WHITE-BROWED COUCAL
  — *C.s.sokotrae* C.H.B. Grant, 1915   Tihama (SW Arabia), Socotra I.
  — *C.s.superciliosus* Hemprich & Ehrenberg, 1833   E Sudan to W Somalia, NE Uganda, N, C and E Kenya and NE Tanzania
  — *C.s.loandae* C.H.B. Grant, 1915   W and S Uganda and SW Kenya to N Malawi, N Zimbabwe and Angola
  — *C.s.burchellii* Swainson, 1838[1]   SE Tanzania to E Botswana and Sth. Africa

CROTOPHAGINAE

CROTOPHAGA Linaeus, 1758 F
○ *Crotophaga major* J.F. Gmelin, 1788 GREATER ANI   E Panama to N Argentina, Trinidad

○ *Crotophaga ani* Linnaeus, 1758 SMOOTH-BILLED ANI   C and S Florida, Bahamas, Caribbean Is., Trinidad, Tobago, Mexico to N Argentina

● *Crotophaga sulcirostris* GROOVE-BILLED ANI
  — †*C.s.pallidula* Bangs & Penard, 1921   S Baja California
  ⦓ *C.s.sulcirostris* Swainson, 1827   S Texas and Mexico to N Chile and NW Argentina, Netherlands Antilles

GUIRA Lesson, 1830 F
○ *Guira guira* (J.F. Gmelin, 1788) GUIRA CUCKOO   E and S Brazil to Bolivia, Paraguay, N and E Argentina and Uruguay

NEOMORPHINAE

TAPERA Thunberg, 1819 F
○ *Tapera naevia* STRIPED CUCKOO
  — *T.n.excellens* (P.L. Sclater, 1858)[2]   S Mexico to E Panama
  — *T.n.naevia* (Linnaeus, 1766)[3]   Colombia to Brazil, Bolivia, Paraguay and N Argentina, Trinidad, Margarita I (Venezuela)

DROMOCOCCYX Wied, 1832 M
○ *Dromococcyx phasianellus* (Spix, 1824) PHEASANT-CUCKOO[4]   S Mexico to Brazil, N Bolivia, Paraguay and N Argentina

○ *Dromococcyx pavoninus* Pelzeln, 1870 PAVONINE CUCKOO[5]   Venezuela and Guyana to N Bolivia, Brazil, Paraguay and N Argentina

MOROCOCCYX P.L. Sclater, 1862 M
○ *Morococcyx erythropygus* LESSER GROUND CUCKOO[6]
  — *M.e.mexicanus* Ridgway, 1915[7]   Pacific slope of SW and S Mexico
  — *M.e.erythropygus* (Lesson, 1842)[8]   S Mexico to NW Costa Rica

GEOCOCCYX Wagler, 1831 M
● *Geococcyx californianus* (Lesson, 1829) GREATER ROADRUNNER[9]   SW USA to SC Mexico

○ *Geococcyx velox* (Wagner, 1836) LESSER ROADRUNNER[10]   W Mexico, Yucatán Peninsula, Guatemala to C Nicaragua

NEOMORPHUS Gloger, 1827 M
○ *Neomorphus geoffroyi* RUFOUS-VENTED GROUND CUCKOO
  — *N.g.salvini* P.L. Sclater, 1866   Nicaragua to W Colombia
  — *N.g.aequatorialis* Chapman, 1923   SE Colombia to N Peru
  — *N.g.australis* Carriker, 1935   Peru, NW Bolivia
  — *N.g.geoffroyi* (Temminck, 1820)   Para (S of Amazon, Brazil)
  — *N.g.maximiliani* Pinto, 1962   Bahia (E Brazil) (1892)
  — *N.g.dulcis* E. Snethlage, 1927   SE Brazil

[1] Includes *fasciipygialis* see Payne (1997: 592) (1800).
[2] Apparently includes *major* Brodkorb, 1940 (221).
[3] Includes *chochi* see Payne (1997) (1800).
[4] For reasons to treat this as monotypic see Payne (1997: 605) (1800).
[5] Includes *perijanus* Aveledo & Ginés, 1950 (64).
[6] We follow Payne (1997: 605) (1800) in listing two forms.
[7] Payne (1997: 606) (1800) implicitly includes *dilutus* and *simulans*.
[8] Payne (1997: 606) (1800) implicitly includes *macrourus*.
[9] Includes *dromicus* Oberholser, 1974 (1672), see Browning (1978, 1990) (246, 251); the name *conklingi* listed in last Edition was given to a fossil (Hughes, 1996) (1161).
[10] Includes *melanchima, affinis, pallidus* and *longisignum* see Payne (1997) (1800).

○ *Neomorphus squamiger* Todd, 1925 Scaled Ground Cuckoo[1] Lower Tapajóz region, (S of Amazon, Brazil)

○ *Neomorphus radiolosus* Sclater & Salvin, 1878 Banded Ground Cuckoo

Foothills of SW Colombia and NW Ecuador

○ *Neomorphus rufipennis* (G.R. Gray, 1849) Rufous-winged Ground Cuckoo[2]

S Venezuela, Guyana, N Brazil

○ *Neomorphus pucheranii* Red-billed Ground Cuckoo
  — *N.p.pucheranii* (Deville, 1851) NE Peru, W Brazil (N of R. Amazon)
  — *N.p.lepidophanes* Todd, 1925 E Peru, W Brazil (S of R. Amazon)

TYTONIDAE BARN OWLS (2: 15)

TYTONINAE

TYTO Billberg, 1828 F
○ *Tyto tenebricosa* Greater Sooty Owl
  — *T.t.arfaki* (Schlegel, 1879) New Guinea, Yapen I.
  — *T.t.tenebricosa* (Gould, 1845) EC to SE Australia

○ *Tyto multipunctata* Mathews, 1912 Lesser Sooty Owl NE Queensland

○ *Tyto novaehollandiae* Australian Masked Owl[3]
  — *T.n.calabyi* Mason, 1983 S New Guinea (1449)
  — *T.n.kimberli* Mathews, 1912 NW Australia
  — *T.n.melvillensis* Mathews, 1912 Melville and Bathurst Is. (Northern Territory)
  — *T.n.galei* Mathews, 1914 Cape York Pen.
  — *T.n.novaehollandiae* (Stephens, 1826) Sthn. coastal Australia to NE Queensland
  — *T.n.castanops* (Gould, 1837) Tasmania

○ *Tyto aurantia* (Salvadori, 1881) Golden Masked Owl New Britain

○ *Tyto manusi* Rothschild & Hartert, 1914 Manus Masked Owl[4] Manus I. (Admiralty Is.)

○ *Tyto sororcula* Lesser Masked Owl
  — *T.s.sororcula* (P.L. Sclater, 1883) Tanimbar Is.
  — *T.s.cayelii* (E. Hartert, 1900) Buru I. (Moluccas)

○ *Tyto nigrobrunnea* Neumann, 1939 Taliabu Masked Owl Sula Is. (1645)

○ *Tyto inexspectata* (Schlegel, 1879) Minahassa Masked Owl N Sulawesi

○ *Tyto rosenbergii* Sulawesi Masked Owl
  — *T.r.pelengensis* Neumann, 1939 Peleng I. (1645)
  — *T.r.rosenbergii* (Schlegel, 1866) Sulawesi, Sangihe Is.

● *Tyto alba* Barn Owl[5]
  ✓ *T.a.alba* (Scopoli, 1769) W and S Europe east to W Balkans
  — *T.a.guttata* (C.L. Brehm, 1831) C Europe to Ukraine and E Balkans
  — *T.a.ernesti* (O. Kleinschmidt, 1901) Corsica, Sardinia
  — *T.a.erlangeri* W.L. Sclater, 1921[6] N Africa, Arabia, Crete and Middle East east to S Iran
  — *T.a.affinis* (Blyth, 1862) S Mauritania to Sudan and Cape Province
  — *T.a.schmitzi* (E. Hartert, 1900) Madeira
  — *T.a.gracilirostris* (E. Hartert, 1905) E Canary Is.
  — *T.a.detorta* E. Hartert, 1913 Cape Verde Is.
  — *T.a.poensis* (Fraser, 1842) Bioko I.
  — *T.a.thomensis* (Hartlaub, 1852) São Tomé I.
  — *T.a.hypermetra* Grote, 1928 Comoro Is., Madagascar

---

[1] Includes *iungens* Griscom & Greenway, 1945 (1018). Treated as a race of the preceding species by Payne (1997) (1800), but his comments seem doubtful and we defer lumping.
[2] Treated as monotypic by Payne (1997) (1800); includes *nigrogularis*.
[3] For a review see Mason (1983) (1449). Re-examined by Schodde & Mason (1997) (2188).
[4] For reasons to separate from *novaehollandiae* see McAllan & Bruce (1988) (1499).
[5] The Australasian treatment follows Schodde & Mason (1980) (2186).
[6] Includes *microsticta* Koelz, 1950 (1280), see Vaurie (1965: 632) (2510).

___ *T.a.stertens* E. Hartert, 1929[1] Pakistan, India, Sri Lanka E through C Burma and Thailand (except Pen.) to Indochina

___ *T.a.javanica* (J.F. Gmelin, 1788) C and S Malay Pen. to Sumatra, Java to Lesser Sundas and islands S of Sulawesi (? Borneo)

___ *T.a.deroepstorffi* (Hume, 1875) S Andaman Is.
___ *T.a.sumbaensis* (E. Hartert, 1897) Sumba I.
___ *T.a.meeki* (Rothschild & Hartert, 1907) SE New Guinea, Vulcan, Dampier Is.
✓ *T.a.delicatula* (Gould, 1837)[2] Australia >> E Lesser Sundas, SW Pacific Is. (Solomon, Santa Cruz, Banks Is., Vanuatu, New Caledonia, Fiji, Tonga, Samoa Is.)

___ *T.a.crassirostris* Mayr, 1935 Hibernian Is. (NE Bismarck Arch. to Solomons)
___ *T.a.pratincola* (Bonaparte, 1838) C and NE USA to E Nicaragua, Hispaniola
___ *T.a.guatemalae* (Ridgway, 1874) W Guatemala to Panama
___ *T.a.bondi* Parkes & Phillips, 1978 Bay Islands (off N Honduras) (1788)
___ *T.a.lucayana* Riley, 1913 Bahama Is.
___ *T.a.niveicauda* Parkes & Phillips, 1978 Isla de la Juventud (Cuba) (1788)
___ *T.a.furcata* (Temminck, 1827) Cuba, Cayman Is., Jamaica
___ *T.a.nigrescens* (Lawrence, 1878) Dominica I.
___ *T.a.insularis* (Pelzeln, 1872) S Lesser Antilles
___ *T.a.bargei* (E. Hartert, 1892) Curacao I.
___ *T.a.subandeana* L. Kelso, 1938 Submontane Colombia, Ecuador
___ *T.a.contempta* (E. Hartert, 1898) Andes from Venezuela to Peru
___ *T.a.hellmayri* Griscom & Greenway, 1937 The Guianas, N Brazil
___ *T.a.tuidara* (J.E. Gray, 1829) C Brazil to Chile, Argentina
___ *T.a.punctatissima* (G.R. Gray, 1838) Galapagos Is.

○ **Tyto glaucops** (Kaup, 1852) ASHY-FACED OWL Hispaniola and Tortuga Is.

○ **Tyto soumagnei** (A. Milne-Edwards, 1878) MADAGASCAN RED OWL E Madagascar

○ **Tyto capensis** GRASS OWL[3]

___ *T.c.capensis* (A. Smith, 1834)[4] [5] Cameroon; Kenya, W Tanzania and SE Zaire south to S Angola, N Namibia and Cape Province

___ *T.c.longimembris* (Jerdon, 1839)[6] India to Indochina and Lesser Sunda Is., N and E Australia, Sulawesi ?, Fiji Is.

___ *T.c.chinensis* E. Hartert, 1929[7] Fujian, Guanxi, Guangdong (China)
___ *T.c.pithecops* (Swinhoe, 1866) Taiwan
___ *T.c.amauronota* (Cabanis, 1872) Philippine Is.
___ *T.c.papuensis* E. Hartert, 1929[8] Midmontane grasslands of New Guinea

PHODILINAE

PHODILUS I. Geoffroy Saint-Hilaire, 1830 M
○ **Phodilus prigoginei** Schouteden, 1952 CONGO BAY OWL[9] E Zaire, NW Tanzania (2199)

○ **Phodilus badius** ORIENTAL BAY OWL
___ *P.b.saturatus* Robinson, 1927 C and E Himalayas, Assam hills and Burma east to S Yunnan and Indochina

___ *P.b.ripleyi* Hussain & Reza Khan, 1978 SW India (1168)
___ *P.b.assimilis* Hume, 1877 Sri Lanka
___ *P.b.badius* (Horsfield, 1821) S Burma, Thailand, Malay Pen., Sumatra, Java, Borneo
___ *P.b.parvus* Chasen, 1937 Belitung I.
___ *P.b.arixuthus* Oberholser, 1932 Bunguran I. (Natuna Is.)

---

[1] Includes *crypta* Koelz, 1939 (1279) see Ripley (1982: 175) (2055).
[2] Includes *lulu*, *interposita* and also *bellonae* Bradley, 1962 (209), see Schodde & Mason (1980) (2186).
[3] See Amadon (1959) (44) and Mees (1964) (1514).
[4] Mees (1964) (1514) suggested that the name *capensis* A. Smith was preoccupied. However, *capensis* Daudin, 1800, has been suppressed for the purpose of priority and Smith's name validated, see Opinion 895, I.C.Z.N. (1970) (1173).
[5] Includes *damarensis*, *liberatus* — correctly spelled *liberata* see David & Gosselin (2002) (613) — and *cameroonensis* Serle, 1949 (2223); treated as a monotypic species by Konig *et al.* (1999) (1297).
[6] Includes *walleri* and *maculosa* Glauert, 1945 (945), see Mees (1964) (1514).
[7] Includes *melli* see Amadon & Jewett (1946) (48).
[8] Includes *baliem* Ripley, 1964 (2044), see Konig *et al.* (1999) (1297).
[9] Placed in *Tyto* by König *et al.* We await detailed molecular evidence.

STRIGIDAE OWLS[1] (27: 180)

STRIGINAE

OTUS Pennant, 1769[2] M

○ *Otus sagittatus* (Cassin, 1848) White-fronted Scops Owl   Malay Pen.

○ *Otus rufescens* Reddish Scops Owl[3]
  — *O.r.malayensis* Hachisuka, 1934   S Malay Pen.
  — *O.r.rufescens* (Horsfield, 1821)   Sumatra, Bangka, Java, Borneo

○ *Otus icterorhynchus* Sandy Scops Owl
  — *O.i.icterorhynchus* (Shelley, 1873)   Ghana
  — *O.i.holerythrus* (Sharpe, 1901)   S Cameroon to N Zaire

○ *Otus ireneae* Ripley, 1966 Sokoke Scops Owl   SE Kenya and NE Tanzania (2047)

○ *Otus balli* (Hume, 1873) Andaman Scops Owl   Andaman Is.

○ *Otus alfredi* (E. Hartert, 1897) Flores Scops Owl   Flores I.

○ *Otus spilocephalus* Mountain Scops Owl
  — *O.s.huttoni* (Hume, 1870)   W Himalayas
  — *O.s.spilocephalus* (Blyth, 1846)[4]   E Himalayas to Burma
  — *O.s.latouchi* (Rickett, 1900)   SE China, N Indochina
  — *O.s.hambroecki* (Swinhoe, 1870)   Taiwan
  — *O.s.siamensis* Robinson & Kloss, 1922   Thailand (south to C Malay Pen.), S Indochina
  — *O.s.vulpes* (Ogilvie-Grant, 1906)   S Malay Pen.
  — *O.s.vandewateri* (Robinson & Kloss, 1916)   Sumatra
  — *O.s.luciae* (Sharpe, 1888)   Borneo

○ *Otus brookii* Rajah Scops Owl
  — *O.b.solokensis* (E. Hartert, 1893)   Mts. of Sumatra
  — *O.b.brookii* (Sharpe, 1892)   Mts. of Borneo, Java

○ *Otus angelinae* (Finsch, 1912) Javan Scops Owl   W Java

○ *Otus mentawi* Chasen & Kloss, 1926 Mentawai Scops Owl[5]   Siberut, Sipora, Pagi Is.

○ *Otus bakkamoena* Collared Scops Owl[6]
  — *O.b.deserticolor* Ticehurst, 1922   SE Saudi Arabia, S Iran, S Pakistan, W India
  — *O.b.plumipes* (Hume, 1870)   N Pakistan and NW Himalayas
  — *O.b.lettia* (Hodgson, 1836)[7]   Submontane C and E Himalayas to Assam, Burma, Thailand (except Pen.) and C Indochina
  — *O.b.gangeticus* Ticehurst, 1922[8]   Rajasthan and C Himalayan foothills
  — *O.b.marathae* Ticehurst, 1922   C India
  — *O.b.bakkamoena* Pennant, 1769   S India, Sri Lanka
  — *O.b.ussuriensis* (Buturlin, 1910)[9]   E and NE China to SE Siberia and Korea
  — *O.b.semitorques* Temminck & Schlegel, 1844[10]   Sakhalin, Kuril Is., Japan, Cheju-do (Quelpart Is.)
  — *O.b.pryeri* (J.H. Gurney, 1889)   Hachijo (Izu Is.), Okinawa (Ryukyu Is.)
  — *O.b.erythrocampe* (Swinhoe, 1874)   C and S China, NW Vietnam
  — *O.b.glabripes* (Swinhoe, 1870)   Taiwan
  — *O.b.umbratilis* (Swinhoe, 1870)   Hainan I.
  — *O.b.condorensis* Kloss, 1930   Pulo Condor I.[11]

---

[1] The sequence of genera and species follows Marks *et al.* (1999) (1436), except that *Ciccaba* is maintained (limited to two species).
[2] Hekstra's new names are cited from his formal publication and not his thesis, which may have been earlier (those only in the thesis are not included).
[3] For reasons to omit *burbidgei* see Dickinson *et al.* (1991) (745).
[4] Includes *rupchandi* Koelz, 1952 (1282), see Ripley (1982: 176) (2055).
[5] Specific treatment, based strongly on acoustic evidence (Marshall, 1978) (1439), needs re-evaluation along with the *bakkamoena* complex.
[6] The case, based on acoustic evidence, to split this species (here containing forms with feathered toes and with bare toes) has yet to be convincingly tied to morphology, and vocal repertoire is apparently not yet understood (Wells, 1999) (2579). It may even be premature to have separated two species from the Philippines (tentatively recognised here).
[7] Includes *manipurensis* Roonwal & Nath, 1949 (2102), and *alboniger* Koelz, 1952 (1282), see Ripley (1982: 179) (2055).
[8] Includes *stewarti* Koelz, 1939 (1279), see Ripley (1982: 178) (2055).
[9] Includes *aurorae* Allison, 1946 (30), see Deignan (1950) (646).
[10] This name should *perhaps* date from 1845 see Holthuis & Sakai (1970) (1128).
[11] We agree with the view that birds from the north of the Malay Peninsula called *condorensis* by Deignan (1950, 1963) (645, 663), based on Robinson & Kloss (1930) (2094), should not be given that name without stronger evidence. Perhaps topotypical specimens should be compared with "small island" species!

___ *O.b.lempiji* (Horsfield, 1821)[1]   Malay Pen., Sumatra, Bankga, Belitung, Java, Bali,
                                      Borneo, Natuna Is.

___ *O.b.kangeanus* Mayr, 1938   Kangean I. (Java Sea)

○ *Otus fuliginosus* (Sharpe, 1888) PALAWAN SCOPS OWL   Palawan

○ *Otus megalotis* PHILIPPINE SCOPS OWL
   ___ *O.m.megalotis* (Walden, 1875)   Luzon, Catanduanes, Marinduque
   ___ *O.m.nigrorum* Rand, 1950   Negros, Panay (1969)
   ___ *O.m.everetti* (Tweeddale, 1879)[2]   Bohol, Leyte, Samar, Mindanao, Basilan

○ *Otus silvicola* (Wallace, 1864) WALLACE'S SCOPS OWL   Flores, Sumbawa

○ *Otus mirus* Ripley & Rabor, 1968 MINDANAO SCOPS OWL   Mindanao I. (2075)

○ *Otus longicornis* (Ogilvie-Grant, 1894) LUZON SCOPS OWL   Luzon I.

○ *Otus mindorensis* (J. Whitehead, 1899) MINDORO SCOPS OWL   Mindoro I.

○ *Otus brucei* PALLID SCOPS OWL
   ___ *O.b.exiguus* Mukherjee, 1958   Israel and S Iraq to W Pakistan and NE Arabia (1631)
   ___ *O.b.obsoletus* (Cabanis, 1875)   S Turkey and Syria to S Uzbekistan and N Afghanistan
   ___ *O.b.brucei* (Hume, 1873)   Aral Sea and E Uzbekistan to Tajikistan and Kyrgyzstan
                                        >> Pakistan and N India
   ___ *O.b.semenowi* (Zarudny & Härms, 1902)   Tarim Basin (W China) >> N Pakistan and NW India

○ *Otus senegalensis* AFRICAN SCOPS OWL[3]
   ___ *O.s.pamelae* Bates, 1937   W and S Saudi Arabia north to Israel
   ___ *O.s.socotranus* (Ogilvie-Grant & Forbes, 1899)   Socotra I.
   ___ *O.s.senegalensis* (Swainson, 1837)[4]   Senegal to Somalia, south to N Namibia and E and S
                                        South Africa
   ___ *O.s.feae* (Salvadori, 1903)   Annobon I.

◉ *Otus scops* EURASIAN SCOPS OWL

HEARD   ___ *O.s.pulchellus* (Pallas, 1771)   Kazakhstan to Altai and Tien Shan >> NE Africa and
                                        SW Asia
   ✓ *O.s.scops* (Linnaeus, 1758)[5]   France and Italy to Caucasus area >> W and C Africa
   ___ *O.s.mallorcae* von Jordans, 1924[6]   Iberia, Balearic Is., NW Africa >> W Africa
   ___ *O.s.cycladum* (Tschusi, 1904)   S Greece, Cyclades Is., Crete, SW Turkey, Levant
   ___ *O.s.cyprius* (Madarász, 1901)   Cyprus
   ___ *O.s.turanicus* (Loudon, 1905)   Iraq and SW Turkmenistan to Afghanistan >> NE
                                        Africa and SW Asia

○ *Otus sunia* ORIENTAL SCOPS OWL[7]
   ___ *O.s.japonicus* Temminck & Schlegel, 1844[8]   Hokkaido and N and C Honshu (Japan) >> S Japan
   ___ *O.s.stictonotus* (Sharpe, 1875)   SE Siberia, E and NE China, Taiwan, N Korea >> S
                                        China and Indochina to Pen. Thailand
   ___ *O.s.malayanus* (Hay, 1845)[9]   S China to Indochina >> SE Asia south to Sumatra
   ___ *O.s.sunia* (Hodgson, 1836)[10]   N Pakistan, N and C India to Burma >> lower and to S
                                        part of range
   ___ *O.s.distans* Friedmann & Deignan, 1939[11]   N and E Thailand (883)
   ___ *O.s.rufipennis* (Sharpe, 1875)   S India
   ___ *O.s.leggei* Ticehurst, 1923   Sri Lanka

[1] Includes *hypnodes* Deignan, 1950 (645), and *cnephaeus* Deignan, 1950 (645), see Wells (1999) (2579). However Wells (1999) (2579) erred in stating that Mees (1986) (1539) had submerged *hypnodes* in *lempiji*, so Wells, himself, took that step. Also includes *lemurum* Deignan, 1957 (658), see Konig *et al.* (1999) (1297).
[2] Includes *boholensis* see Dickinson *et al.* (1991) (745).
[3] The many races listed in last Edition could usefully have been reduced to the African forms of *O. scops* recognised by White (1965: 193) (2649); of the five he listed all but *hartlaubi* appear here.
[4] Includes *huszari* Keve, 1959 (1262), and *nivosus* Keith & Twomey, 1968 (1243) which require re-evaluation.
[5] Includes *vincii* Trischitta, 1939 (2439), see Vaurie (1965: 599) (2510).
[6] For recognition see Cramp *et al.* (1985: 454) (591).
[7] Separation of *sunia* from *scops* is founded on two distinct morphological *groups* (Vaurie, 1965: 597) (2510), this one, which is dimorphic, extending beyond that treated here to include Philippine forms, and later acoustic evidence (Marshall, 1978) (1439). The specific separation (Roberts & King, 1986) (2091) now widely accepted, although some populations insufficiently studied, draws support from range disjunction.
[8] This name should *perhaps* date from 1845 see Holthuis & Sakai (1970) (1128).
[9] Given as 1847 in Peters (1940: 91) (1825), but other taxa described in this paper are given as 1844 or 1845!
[10] Includes *khasiensis* Koelz, 1954 (1284), see Ripley (1982: 177) (2055).
[11] Requires re-evaluation; the Feb. type is reportedly unlike either *sunia* or northern migrants (Marshall, 1978: 10) (1439). It is sometimes treated as a synonym within a broad ranging *modestus* (e.g. by Holt *et al.*, 1999: 164) (1127).

— *O.s.modestus* (Walden, 1874)   Andaman Is.

— *O.s.nicobaricus* (Hume, 1876)[1]   Nicobar Is.

○ *Otus flammeolus* (Kaup, 1853) FLAMMULATED OWL[2]   SW Canada to Mexico and Guatemala

○ *Otus magicus* MOLUCCAN SCOPS OWL[3]

  — *O.m.morotensis* (Sharpe, 1875)   Morotai, Ternate Is.

  — *O.m.leucospilus* (G.R. Gray, 1860)   Halmahera, Kasiruta, Bacan

  — *O.m.obira* Jany, 1955   Obi Is. (1200)

  — *O.m.magicus* (S. Müller, 1841)[4]   Buru, Seram, Ambon

  — *O.m.mendeni* Neumann, 1939   Banggai Is. (1646)

  — *O.m.siaoensis* (Schlegel, 1873)   Siao I.

  — *O.m.sulaensis* (E. Hartert, 1898)   Sula Is.

  — *O.m.kalidupae* (E. Hartert, 1903)   Kalidupa I.

  — *O.m.albiventris* (Sharpe, 1875)   Lombok, Sumbawa, Flores, Lomblen (Lesser Sundas)

  — *O.m.tempestatis* (E. Hartert, 1904)   Wetar I. (Lesser Sundas)

○ *Otus mantananensis* MANTANANI SCOPS OWL

  — *O.m.mantananensis* (Sharpe, 1892)   Mantanani I. (near Borneo)

  — *O.m.cuyensis* McGregor, 1904   Cuyo I. (WC Philippines)

  — *O.m.romblonis* McGregor, 1905   Romblon I. (C Philippines)

  — *O.m.sibutuensis* (Sharpe, 1893)[5]   Sibutu I., Tumindao I. (Sulu Is.)

○ *Otus elegans* ELEGANT SCOPS OWL

  — *O.e.interpositus* N. Kuroda, Sr., 1923   Daito Is. (Borodinos)

  — *O.e.elegans* (Cassin, 1852)   Ryukyu Is.

  — *O.e.botelensis* N. Kuroda, Sr., 1928   Lanyu I. (off S Taiwan)

  — *O.e.calayensis* McGregor, 1904[6]   Calayan I. (off N Luzon)

○ *Otus manadensis* (Quoy & Gaimard, 1830) SULAWESI SCOPS OWL[7]   Sulawesi

○ *Otus collari* Lambert & Rasmussen, 1998 SANGIHE SCOPS OWL #   Sangihe I. (1318)

○ *Otus beccarii* (Salvadori, 1876) BIAK SCOPS OWL   Biak I. (Geelvink Bay)

○ *Otus insularis* (Tristram, 1880) SEYCHELLES SCOPS OWL   Mahe I. (Seychelles)

○ *Otus umbra* SIMEULUË SCOPS OWL[8]

  — *O.u.umbra* (Richmond, 1903)   Simeuluë I. (W of Sumatra)

  — *O.u.enganensis* Riley, 1927   Enggano I. (W of Sumatra)

○ *Otus alius* Rasmussen, 1998 NICOBAR SCOPS OWL #   Great Nicobar I (1985)

○ *Otus pembaensis* Pakenham, 1937 PEMBA SCOPS OWL   Pemba I. (off Tanzania)

○ *Otus pauliani* Benson, 1960 GRAND COMORO SCOPS OWL   Grand Comore I. (Comoro Is.) (122)

○ *Otus capnodes* (J.H. Gurney, 1889) ANJOUAN SCOPS OWL   Anjouan I. (Comoro Is.)

○ *Otus moheliensis* Lafontaine & Moulaert, 1998 MOHELI SCOPS OWL #   Moheli (Comoro Is.) (1314)

○ *Otus mayottensis* Benson, 1960 MAYOTTE SCOPS OWL   Mayotte I. (Indian Ocean) (122)

○ *Otus madagascariensis* (A. Grandidier, 1867) TOROTOROKA SCOPS OWL[9]

                          S, W and central high plateau of Madagascar

○ *Otus rutilus* (Pucheran, 1849) RAINFOREST SCOPS OWL   Madagascar (mainly E)

○ *Otus hartlaubi* (Giebel, 1872) SÃO TOMÉ SCOPS OWL   São Tomé I. (WC Africa)

---

[1] Omitted from last Edition, but see Ripley (1982: 178) (2055).

[2] For reasons to treat this as monotypic see Marshall (1967) (1438). Browning (1990) (251) reviewed *borealis* Hekstra, 1982 (1097), and *frontalis* Hekstra, 1982 (1097), accepting the later as valid but not in the context of Marshall (1967) (1438) for *idahoensis* is recognised; *meridionalis* Hekstra, 1982, awaits re-evaluation see Browning (1989) (248).

[3] We again follow the tentative treatment of White & Bruce (1986: 251) (2652) in placing forms from islands close to Sulawesi in *magicus*. There seems to be no valid basis to move them to *manadensis*, see Lambert & Rasmussen (1998) (1318). However they are probably not part of this species either.

[4] Includes *bouruensis* see White & Bruce (1986: 252) (2652).

[5] Includes *steerei* see Dickinson *et al.* (1991) (745).

[6] Includes *batanensis* Manuel & Gilliard, 1952 (1422), see Dickinson *et al.* (1991) (745).

[7] The name *obsti* Eck, 1973 (786), has been associated with this species (793); *obsti* is from Java, which is unlikely to hold *manadensis*, it apparently awaits resolution.

[8] No satisfactory evidence has been published for treatment of *enganensis* as a separate species.

[9] For a revision of this and the next few species see Rasmussen (2000) (1995).

○ **Otus asio** EASTERN SCREECH OWL[1]
— *O.a.maxwelliae* (Ridgway, 1877)[2] SC Canada, NC USA
— *O.a.asio* (Linnaeus, 1758)[3] NE and C USA south to E Virginia to Kansas
— *O.a.hasbroucki* Ridgway, 1914 C Oklahoma to N Texas
— *O.a.mccallii* (Cassin, 1854)[4] S Texas to C Mexico
— *O.a.floridanus* (Ridgway, 1873) S Louisiana, Florida

○ **Otus kennicotti** WESTERN SCREECH OWL
— *O.k.kennicotti* (Elliot, 1867)[5] S Alaska to coastal Oregon
— *O.k.bendirei* (Brewster, 1882)[6] British Columbia and Idaho to SE California
— *O.k.aikeni* (Brewster, 1891)[7] Nevada, E California, Utah, Arizona, SE Colorado, W Oklahoma and Sonora (N Mexico)
— *O.k.cardonensis* Huey, 1926[8] S California, N Baja California
— *O.k.xantusi* (Brewster, 1902) S Baja California
— *O.k.yumanensis* A.H. & L. Miller, 1951 Colorado desert, lower Colorado R. and NW Sonora (Mexico) (1583)
— *O.k.vinaceus* (Brewster, 1888)[9] C Sonora to N Sinaloa
— *O.k.suttoni* R.T. Moore, 1941[10] SW Texas to Mexican Plateau (1613)

○ **Otus seductus** R.T. Moore, 1941 BALSAS SCREECH OWL[11] Michoacan and Guerrero (WC Mexico) (1613)

○ **Otus cooperi** PACIFIC SCREECH OWL
— *O.c.lambi* Moore & Marshall, 1959 C Oaxaca (Mexico) (1621)
— *O.c.cooperi* (Ridgway, 1878)[12] S Oaxaca along W coast to Costa Rica

○ **Otus trichopsis** WHISKERED SCREECH OWL[13]
— *O.t.aspersus* (Brewster, 1888) SE Arizona, N Mexico
— *O.t.trichopsis* (Wagler, 1832)[14] C Mexico
— *O.t.mesamericanus* van Rossem, 1932[15] SE Mexico to W Panama

○ **Otus choliba** TROPICAL SCREECH OWL[16]
— *O.c.luctisonus* Bangs & Penard, 1921 Costa Rica to NW Colombia
— *O.c.margaritae* Cory, 1915 Margarita I. (off N Venezuela)
— *O.c.duidae* Chapman, 1929 S Venezuela
— *O.c.crucigerus* (Spix, 1824)[17] E Colombia, E Peru to Venezuela, Trinidad, the Guianas and NE Brazil
— *O.c.suturutus* L. Kelso, 1941 Bolivia (1246)
— *O.c.decussatus* (M.H.K. Lichtenstein, 1823)[18] C and E Brazil
— *O.c.choliba* (Vieillot, 1817)[19] S Brazil, E Paraguay
— *O.c.wetmorei* Brodkorb, 1937[20] W Paraguay, N Argentina
— *O.c.uruguaiensis* Hekstra, 1982[21] NE Argentina, SE Brazil, Uruguay (1097)

○ **Otus roboratus** PERUVIAN SCREECH OWL
— *O.r.pacificus* Hekstra, 1982[22] SW Ecuador, NW Peru (south to Lambayeque) (1097)
— *O.r.roboratus* Bangs & Noble, 1918 Extreme S Ecuador, NW Peru (between W and C Andes)

[1] Treatment here and in the next three species follows Marshall (1967) (1438), with his four groups elevated to species (A.O.U., 1998) (3). The subspecific name *ocreatus* Hekstra, 1982 (published only within his dissertation) has not been assigned.
[2] Includes *swenki* see Marshall (1967) (1438).
[3] Includes *naevius* see Marshall (1967) (1438).
[4] Includes *semplei* Sutton & Burleigh, 1939 (2378), see Marshall (1967) (1438).
[5] Includes *brewsteri* and *saturatus* see Marshall (1967) (1438)
[6] Includes *macfarlanei* and *quercinus* see Marshall (1967) (1438).
[7] Includes *inyoensis, cineraceus, mychophilus* and *gilmani* see Marshall (1967) (1438).
[8] Includes *clazus* see Marshall (1967) (1438).
[9] Includes *sinaloensis* see Marshall (1967) (1438).
[10] Includes *sortilegus* R.T. Moore, 1941 (1613), see Marshall (1967) (1438).
[11] Includes *colimensis* Hekstra, 1982 (1097) see Holt et. al. (1999: 173) (1127).
[12] Includes *chiapensis* R.T. Moore, 1947 (1616), see Marshall (1967) (1438).
[13] Treatment based on Marshall (1967) (1438).
[14] Includes *ridgwayi* and *guerrensis* see Marshall (1967) (1438).
[15] Includes *pumilis* see Marshall (1967) (1438).
[16] Treatment follows Holt *et al.* (1999: 175) (1127), with reservations.
[17] Includes *alticola, montanus* Hekstra, 1982 (1097), *kelsoi* Hekstra, 1982 (1097), *caucae* Hekstra, 1982 (1097), and *guyanensis* Hekstra, 1982 (1097), see Holt *et al.* (1999: 175) (1127). Also includes *portoricensis* Lesson, see Ridgway (1914) (2022), mistakenly ascribed to Kelso (1942) (1247) by Hekstra (1982) (1097). Hekstra's freshly described bird from Trinidad is tentatively included here, but requires re-evaluation.
[18] Includes *caatingensis* Hekstra, 1982 (1097), see Holt *et al.* (1999: 175) (1127).
[19] Includes *chapadensis* Hekstra, 1982 (1097), see Holt *et al.* (1999: 175) (1127).
[20] Includes *alilicuco* Hekstra, 1982 (1097), see Holt *et al.* (1999: 175) (1127).
[21] For selection of this spelling over *urugaii* see Browning (1989) (248).
[22] For recognition see Holt *et al.* (1999: 175) (1127).

○ *Otus koepckeae* Hekstra, 1982 KOEPCKE'S SCREECH OWL[1]   W Peru, NW Bolivia (1097)

○ *Otus clarkii* L. & E.H. Kelso, 1935 BARE-SHANKED SCREECH OWL   Costa Rica to extreme NW Colombia

○ *Otus barbarus* (Sclater & Salvin, 1868) BEARDED SCREECH OWL   S Mexico, N Guatemala

○ *Otus ingens* RUFESCENT SCREECH OWL
    — *O.i.venezuelanus* Phelps & Phelps, Jr., 1954   NW Venezuela, N Colombia (1854)
    — *O.i.ingens* (Salvin, 1897)[2]   NE Ecuador to WC Bolivia

○ *Otus colombianus* Traylor, 1952 COLOMBIAN SCREECH OWL   W Colombia, N Ecuador, N Peru (2423)

○ *Otus petersoni* Fitzpatrick & O'Neill, 1986[3] CINNAMON SCREECH OWL

                                                                     N Peru (836)

○ *Otus marshalli* Weske & Terborgh, 1981 CLOUD-FOREST SCREECH OWL[4]

                                                                     SC Peru (2588)

○ *Otus watsonii* TAWNY-BELLIED SCREECH OWL
    — *O.w.watsonii* (Cassin, 1848)[5]   E Colombia to Surinam, NE Ecuador, Amazonian
                                                                        Brazil (North of Amazon) and NE Peru
    — *O.w.usta* (P.L. Sclater, 1859)[6]   Peru, C Brazil, to N Bolivia

○ *Otus guatemalae* MIDDLE AMERICAN SCREECH OWL[7]
    — *O.g.tomlini* R.T. Moore, 1937   NW Mexico
    — *O.g.hastatus* (Ridgway, 1887)[8]   SW Sinaloa to Oaxaca
    — *O.g.cassini* (Ridgway, 1878)   NE Mexico
    — *O.g.fuscus* Moore & Peters, 1939   C Vera Cruz
    — *O.g.thompsoni* Cole, 1906   Yucatán Pen., Cozumel I.
    — *O.g.guatemalae* (Sharpe, 1875)[9]   SE Vera Cruz, Guatemala, Honduras
    — *O.g.dacrysistactus* Moore & Peters, 1939   N Nicaragua

○ *Otus vermiculatus* VERMICULATED SCREECH OWL[10]
    — *O.v.vermiculatus* (Ridgway, 1887)[11] [12]   Costa Rica, Panama, NW Colombia, N Venezuela
    — *O.v.napensis* Chapman, 1928[13]   E Ecuador and E Colombia to Peru and N Bolivia
    — *O.v.roraimae* (Salvin, 1897)   S Venezuela, N Brazil (Tepui region)

○ *Otus hoyi* Koenig & Straneck, 1989 MONTANE FOREST SCREECH OWL[14] S Bolivia, NW Argentina (1295)

○ *Otus atricapilla* (Temminck, 1822) BLACK-CAPPED SCREECH OWL[15][16] C and S Brazil

○ *Otus sanctaecatarinae* (Salvin, 1897) LONG-TUFTED SCREECH OWL   SE Brazil, NE Argentina, Uruguay

○ *Otus nudipes* PUERTO RICAN SCREECH OWL
    — *O.n.nudipes* (Daudin, 1800)[17]   Puerto Rico
    — *O.n.newtoni* (Lawrence, 1860)   St. Thomas, St. John, St. Croix Is.

○ *Otus albogularis* WHITE-THROATED SCREECH OWL[18]
    — *O.a.obscurus* Phelps & Phelps, Jr., 1953   NW Venezuela (1852)
    — *O.a.meridensis* (Chapman, 1923)   W Venezuela
    — *O.a.macabrum* (Bonaparte, 1850)   Colombia to N Peru

---

[1] Original spelling *koepckei*, emended by Greenway (1978) (1010) see Browning (1989) (248). Considered a distinct species by Marshall & King *in* Amadon & Bull (1988) (47).
[2] Includes *minimus* see Holt *et al.* (1999: 179) (1127).
[3] Treated as a species by Holt *et al.* (1999: 179) (1127), there is a substantial size difference compared to *colombianus*. See also Vuilleumier *et al.* (1992) (2552).
[4] Status as a species endorsed by Vuilleumier *et al.* (1992) (2552).
[5] Apparently includes *ater* Hekstra, 1982 (1097), and *moreliensis* Hekstra, 1982 (1097), originally *morelius* but emended, see Browning, 1989 (248). See also Konig *et al.* (1999) (1297).
[6] Apparently includes *inambarii* Hekstra, 1982 (1097), emended to *inambariensis* by Browning (1989) (248), and *fulvescens* Hekstra, 1982 (1097), see Konig *et al.* (1999) (1297).
[7] A.O.U. (1998: 256) (3) continued to treat a broader species including the *versicolor* group.
[8] Includes *pettingilli* Hekstra, 1982 (1097), see Holt *et al.* (1999: 180) (1127).
[9] Includes *petenensis* Hekstra, 1982 (1097) (originally *peteni* but emended by Browning (1989) (248), see Holt *et al.* (1999: 180) (1127).
[10] Treatment follows Holt *et al.* (1999: 180) (1127); information to date suggests the three taxa may deserve specific status, but further evidence needed.
[11] Thought to include *centralis* Hekstra, 1982 (1097), and *pallidus* Hekstra, 1982 (1097).
[12] Includes *inexpectus* Hekstra, 1982 (1097), see Browning (1989) (248).
[13] Includes *helleri* L. Kelso, 1940 (1245), and *bolivianus* Bond & Meyer de Schauensee, 1941 (191), see Holt *et al.* (1999: 180) (1127), and thought to include *rufus* Hekstra, 1982 (1097).
[14] Tentatively treated as a species following Holt *et al.* (1999: 180) (1127), but see Vuilleumier *et al.* (1992) (2552).
[15] For correction of spelling see David & Gosselin (2002) (613).
[16] Includes *argentinus* see Holt *et al.* (1999: 181) (1127). Several forms named by Hekstra (1982) (1097) are in the synonymy of the species *watsoni* (see above).
[17] Thought to include *krugii* Gundlach, 1874 (1025).
[18] Treatment, including recognition of two extra forms (compared to last Edition), follows Holt *et al.* (1999: 181) (1127), with the exception that we follow Traylor (1952) (2423) and Konig *et al.* (1297) in considering *aequatorialis* a name applying to variants of nominate *O. ingens*.

— *O.a.albogularis* (Cassin, 1849)   Colombia, N Ecuador

— *O.a.remotus* Bond & Meyer de Schauensee, 1941   Peru to C Bolivia (191)

PYRROGLAUX Yamashina, 1938[1] F

○ ***Pyrroglaux podargina*** (Hartlaub & Finsch, 1872) PALAU SCOPS OWL   Palau Is.

GYMNOGLAUX Cabanis, 1855 F

○ ***Gynoglaux lawrencii*** CUBAN SCREECH OWL/BARE-LEGGED OWL

— *G.l.lawrencii* Sclater & Salvin, 1868   C and E Cuba

— *G.l.exsul* (Bangs, 1913)   W Cuba, Isla de la Juventud

PTILOPSIS Kaup, 1851[2] F

○ ***Ptilopsis leucotis*** (Temminck, 1820) NORTHERN WHITE-FACED OWL[3]   Senegal to Ethiopia and Kenya

○ ***Ptilopsis granti*** (Kollibay, 1910)[4] SOUTHERN WHITE-FACED OWL   S Zaire and Tanzania to Cape Province

MIMIZUKU Hachisuka, 1934 N

○ ***Mimizuku gurneyi*** (Tweeddale, 1879) GIANT SCOPS OWL[5]   Mindanao (Philippines)

NYCTEA Stephens, 1826 F

○ ***Nyctea scandiaca*** (Linnaeus, 1758) SNOWY OWL   N Asia, N Canada, Holarctic region

BUBO Duméril, 1806 M

● ***Bubo virginianus*** GREAT HORNED OWL[6] [7]

— *B.v.lagophonus* (Oberholser, 1904)	C Alaska to NE Oregon and Idaho
✓ *B.v.saturatus* Ridgway, 1877	SE Alaska to California
— *B.v.pacificus* Cassin, 1854	Coastal California to NW Baja California
— *B.v.subarcticus* Hoy, 1853[8]	WC Canada to WC USA
✓ *B.v.pallescens* Stone, 1897	SW USA to Mexico
— *B.v.heterocnemis* (Oberholser, 1904)	NE Canada south to Great Lakes
— *B.v.virginianus* (J.F. Gmelin, 1788)	SE Canada, C and E USA
— *B.v.elachistus* Brewster, 1902	S Baja California
— *B.v.mayensis* Nelson, 1901	Yucatán Pen.
— *B.v.mesembrinus* (Oberholser, 1904)	Isthmus of Tehuantepec to W Panama
— *B.v.nigrescens* Berlepsch, 1884	Colombia to NW Peru
— *B.v.nacurutu* (Vieillot, 1817)	E Colombia E to the Guianas and S to EC Argentina
— *B.v.magellanicus* (J.F. Gmelin, 1788)[9] [10]	C Peru, W Bolivia, W Argentina

○ ***Bubo bubo*** EURASIAN EAGLE-OWL[11]

— *B.b.hispanus* Rothschild & Hartert, 1910	Iberian Pen.
— *B.b.bubo* (Linnaeus, 1758)[12]	Scandinavia to France, Sicily, W and N Romania and C and NW European Russia
— *B.b.interpositus* Rothschild & Hartert, 1910	S Ukraine to N Greece east to Turkey, NW Iran, Caucasus area, Mangyshlak and lower Don R.
— *B.b.nikolskii* Zarudny, 1905	Iraq to C and S Afghanistan and W and N Pakistan
— *B.b.ruthenus* Buturlin & Zhitkov, 1906	E and S European Russia south to lower Volga and foot of N Caucasus
— *B.b.sibiricus* (Gloger, 1833)	SW and C Ural region and W Siberia to Ob' R.

---

[1] In last Edition we placed this in *Otus*; we did the same for the following species for which see also A.O.U. (1998: 257) (3).
[2] Treated in the genus *Otus* in last Edition, but see Wink & Heidrich in König *et al.* (1999: 42) (1297).
[3] Includes *margarethae* see White (1965: 194) (2649).
[4] This name was proposed as a new name for *Scops erlangeri* Ogilvie-Grant, 1906, preoccupied in *Otus*. The latter seems unlikely to be preoccupied in *Ptilopsis* and should probably be restored (Art. 59.3., I.C.Z.N., 1998) (1178), but has not been here pending confirmation. For specific separation see Holt *et al.* (1999: 183) (1127).
[5] Preliminary molecular information seems to suggest that this is closer to *Otus* than to *Bubo* (Miranda *et al.*, 1997) (1587).
[6] Treatment follows Holt *et al.* (1999: 185) (1127): all other forms listed in last Edition, including *colombianus* Lehmann, 1946 (1349), as well as *scalariventris* Snyder, 1961 (2296), are treated as exemplars of individual variation and, with one exception, not assigned to synonymy by them. They also note the replacement of the name *wapacuthu* J.F. Gmelin, 1788 (see Peters, 1940, p. 111 fn) by *subarcticus*. Gmelin's name was retained by König *et al.* (1994), but see Browning (1990) (250).
[7] For a rather different treatment see Weick (2001) (2570).
[8] Includes *occidentalis* see Holt *et al.* (1999: 185) (1127).
[9] Treated as a species by Weick (2001) (2570), but substantiation needed.
[10] Includes *andicolus* L. Kelso, 1941 (1246) see Weick (2001) (2570).
[11] Treatment based on Cramp *et al.* (1985) (591), Cheng (1987) (333) and Stepanyan (1990: 263) (2326).
[12] Includes *ognevi* Dementiev, 1952 (690), and *meridionalis* Orlando, 1957 (1724), see Stepanyan (1990: 264) (2326).

— *B.b.yenisseensis* Buturlin, 1911[1]   WC and SC Siberia from Ob' R. to Lake Baikal, south to extreme N Xinjiang and N Mongolia

— *B.b.jakutensis* Buturlin, 1908   NC and NE Siberia, south to Stanovoi Mts.

— *B.b.turcomanus* (Eversmann, 1835)[2]   N Kazakhstan and SW Siberia from lower Ural R. to Irtysh R. and Lake Zaisan

— *B.b.omissus* Dementiev, 1932   Turkmenistan to S Uzbekistan, SW Tajikistan and N Afghanistan

— *B.b.bengalensis* (Franklin, 1831)[3]   Pakistan, India, Nepal and SW Burma

— *B.b.hemachalanus* Hume, 1873[4] [5]   Pamir and Tien Shan Mts. to SE Kazakhstan, N and W Xinjiang, W Xizang and W Himalaya

— *B.b.tibetanus* Bianchi, 1906   Tibetan Plateau from C Xizang to SW Gansu, W Sichuan and NW Yunnan

— *B.b.tarimensis* Buturlin, 1928   E Tarim Basin (E from Hami and Lop Nor) to SW and S Mongolia

— *B.b.kiautschensis* Reichenow, 1903[6]   Korea and E China from S Gansu and Shandong to Yunnan and Guangdong

— *B.b.ussuriensis* Poliakov, 1915[7]   SE Siberia from Lake Baikal and E Mongolia to NE China and Russian Far East

— *B.b.borrisowi* Hesse, 1915   Sakhalin and S Kuril Is. north to Urup I.

○ *Bubo ascalaphus* Savigny, 1809 DESERT EAGLE-OWL[8]   Sahara, and from Sinai and Syria to SW Iran and Arabia

○ *Bubo capensis* CAPE EAGLE-OWL
— *B.c.dillonii* Des Murs & Prévost, 1846   Ethiopia
— *B.c.mackinderi* Sharpe, 1899   Kenya, Tanzania, Malawi, Zimbabwe and W Mozambique
— *B.c.capensis* A. Smith, 1834   South Africa to S Namibia

○ *Bubo africanus* SPOTTED EAGLE-OWL
— *B.a.africanus* (Temminck, 1821)[9]   S Zaire, S Uganda and W, C and S Kenya to Cape Province

— *B.a.milesi* Sharpe, 1886   S Saudi Arabia

○ *Bubo cinerascens* Guérin-Méneville, 1843 GREYISH EAGLE-OWL[10]   Senegal and Gambia to Eritrea and W Somalia, south to Cameroon, N Uganda and N Kenya

○ *Bubo poensis* Fraser, 1853 FRASER'S EAGLE-OWL   Liberia to N Angola (Cabinda), N and C Zaire and W Uganda

○ *Bubo vosseleri* Reichenow, 1908 USAMBARA EAGLE-OWL[11]   Usambara Mts. (NE Tanzania)

○ *Bubo nipalensis* SPOT-BELLIED EAGLE-OWL
— *B.n.nipalensis* Hodgson, 1836   Himalayas south to S India and east to C Burma, N and E Thailand, Laos and Vietnam

— *B.n.blighi* Legge, 1878   Sri Lanka

○ *Bubo sumatranus* BARRED EAGLE-OWL[12]
— *B.s.sumatranus* (Raffles, 1822)   S Burma, SW Thailand, Malay Pen., Sumatra, Bangka
— *B.s.strepitans* (Temminck, 1823)   Java, Bali
— *B.s.tenuifasciatus* Mees, 1964   Borneo (1512)

○ *Bubo shelleyi* (Sharpe & Ussher, 1872) SHELLEY'S EAGLE-OWL   Liberia to S Cameroon and N Zaire

○ *Bubo lacteus* (Temminck, 1820) VERREAUX'S EAGLE-OWL   Senegal and C Mali to Eritrea and Somalia south to Cape Province

---

[1] Includes *zaissanensis* see Stepanyan (1990: 265) (2326).
[2] Includes *gladkovi* Zaletaev, 1962 (2703), see Stepanyan (1990: 267) (2326), but the name attaches to intermediates between this and *omissus*.
[3] Ripley (1982: 180) (2055) retained this in *B. bubo*. It is perhaps premature to separate this as not all subspecies have known DNA profiles.
[4] We follow Stepanyan (1990: 267) (2326) who included *auspicabilis*, which was recognised by Cheng (1987: 325) (333).
[5] For correction of spelling see David & Gosselin (2002) (613).
[6] Includes *swinhoei* and *jarlandi* see Cheng (1987: 326) (333) and *tenuipes* see Orn. Soc. Japan (2000: 154) (1727).
[7] Includes *dauricus* and *inexpectatus* see Stepanyan (1990: 266) (2326).
[8] Separation from *B. bubo* tentative as reported to intergrade with *B.b.nikolskii*. Seen as more distinct than *bengalensis* by Vaurie (1965: 580) (2510), but see Dowsett & Dowsett-Lemaire (1993: 337) (761).
[9] Includes *tanae* Keith & Twomey, 1968 (1243), see König *et al.* (1999) (1297).
[10] Separation from *B. africanus* tentative, but based on agreement between Holt *et al.* (1999: 188) (1127) and König *et al.* (1999) (1297). Includes *kollmannspergeri* Niethammer, 1957 (1653), see White (1965: 195) (2649).
[11] For separation from *B. poensis* see Turner *et al.* (1991) (2442).
[12] For correction of spelling see David & Gosselin (2002) (613).

○ *Bubo coromandus* DUSKY EAGLE-OWL
— *B.c.coromandus* (Latham, 1790) N and C India
— *B.c.klossii* Robinson, 1911 S Burma, Malay Pen.

○ *Bubo leucostictus* Hartlaub, 1855 AKUN EAGLE-OWL Sierra Leone to C and E Zaire

○ *Bubo philippensis* PHILIPPINE EAGLE-OWL
— *B.p.philippensis* (Kaup, 1851) Luzon, Catanduanes
— *B.p.mindanensis* (Ogilvie-Grant, 1906) Leyte, Samar, Mindanao

○ *Bubo blakistoni* BLAKISTON'S EAGLE-OWL[1]
— *B.b.doerriesi* Seebohm, 1895[2] NE China and the Russian Far East north to the W and N shore of the Sea of Okhotsk
— *B.b.blakistoni* Seebohm, 1884[3] Sakhalin, S Kuril Is., Hokkaido

KETUPA Lesson, 1830[4] F[5]
○ *Ketupa zeylonensis* BROWN FISH OWL
— *K.z.semenowi* Zarudny, 1905 Middle East to W and NW Pakistan
— *K.z.leschenaultii* (Temminck, 1820)[6] NE Pakistan, India, Burma, Thailand and N tip of W Malaysia
— *K.z.zeylonensis* (J.F. Gmelin, 1788) Sri Lanka
— *K.z.orientalis* Delacour, 1926 NE Burma to S China, Hainan and Indochina

○ *Ketupa flavipes* (Hodgson, 1836) TAWNY FISH OWL Himalayas to W China, Indochina

○ *Ketupa ketupu* BUFFY FISH OWL[7]
— *K.k.ketupu* (Horsfield, 1821) S Assam east to Vietnam and south to the Malay Pen., Sumatra, Java, Borneo, Bangka, Belitung
— *K.k.minor* Büttikofer, 1896[8] Nias

SCOTOPELIA Bonaparte, 1850 F
○ *Scotopelia peli* (Bonaparte, 1850) PEL'S FISHING OWL Senegal to Central African Republic and Ethiopia, south to N Botswana and E Cape Province

○ *Scotopelia ussheri* Sharpe, 1871 RUFOUS FISHING OWL Sierra Leone to Ghana

○ *Scotopelia bouvieri* Sharpe, 1875 VERMICULATED FISHING OWL S Cameroon to N Angola and W and N Zaire

STRIX Linnaeus, 1758 F
○ *Strix seloputo* SPOTTED WOOD OWL
— *S.s.seloputo* Horsfield, 1821 S Burma east to S Indochina and south to Malay. Pen., Java (? Sumatra)
— *S.s.baweana* Oberholser, 1917 Bawean I. (Java Sea)
— *S.s.wiepkeni* (W.H. Blasius, 1888) Palawan group (Philippines)

○ *Strix ocellata* MOTTLED WOOD OWL
— *S.o.grisescens* Koelz, 1950 NW and NC India (1280)
— *S.o.grandis* Koelz, 1950 Kathiawar Pen., Gujarat (India) (1280)
— *S.o.ocellata* (Lesson, 1839) C and S India[9]

○ *Strix leptogrammica* BROWN WOOD OWL[10]
— *S.l.ticehursti* Delacour, 1930 SE China, N Indochina
— *S.l.caligata* (Swinhoe, 1863) Taiwan, Hainan
— *S.l.laotiana* Delacour, 1926 N Thailand, S Indochina
— *S.l.newarensis* (Hodgson, 1836) Himalayas, Assam hills, N Burma, W Yunnan

[1] We follow König *et al.* (1999) (1297) in placing this in *Bubo* rather than in *Ketupa*.
[2] Includes *piscivorus* see Stepanyan (1990: 268) (2326) and Cheng (1987) (333).
[3] Includes *karafutonis* see Stepanyan (1990: 268) (2326) and Orn. Soc. Japan (2000: 154) (1727).
[4] Treated within a broader genus *Bubo* in last Edition.
[5] For correct gender see David & Gosselin (2002) (614).
[6] The wrapper apparently carried the name spelled thus, see Froriep (1821) (888) (Dickinson (2001) (739)).
[7] Includes *pageli* see Mayr (1938) (1468) and *aagaardi* see Wells (1999: 423) (2579).
[8] The name *minor* is not available for this taxon if *Ketupu* is placed in the genus *Bubo*. See Chasen (1935: 84) (319) who proposed the substitute name *buettikoferi*.
[9] Range quite wrong in last Edition!
[10] Lumped by Ripley (1977) (2051). Given that two species *may* be involved the decision of Holt *et al.* (1999: 197) (1127) to go back to basics seems preferable, but a fresh published review is needed.

— *S.l.indranee* Sykes, 1832[1]	C and S India
— *S.l.ochrogenys* (Hume, 1873)	Sri Lanka
— *S.l.maingayi* (Hume, 1878)	Malay Pen.
— *S.l.myrtha* (Bonaparte, 1850)	Sumatra
— *S.l.nyctiphasma* Oberholser, 1924	Banyak I. (W Sumatra)
— *S.l.niasensis* (Salvadori, 1887)	Nias I. (W Sumatra)
— *S.l.chaseni* Hoogerwerf & deBoer, 1947[2]	Belitung I. (SE Sumatra) (1147)
— *S.l.leptogrammica* Temminck, 1832[3]	S and C Borneo,
— *S.l.vaga* Mayr, 1938	N Borneo
— *S.l.bartelsi* (Finsch, 1906)	Java[4]

○ **Strix aluco** TAWNY OWL

— *S.a.aluco* Linnaeus, 1758[5]	N, C and SE Europe east to W foot of Ural Mts. and south to N Caucasus foothills
— *S.a.sylvatica* Shaw, 1809	Britain, W France and Iberian Pen.; W and C Turkey and Levant
— *S.a.mauritanica* (Witherby, 1905)	NW Africa
— *S.a.siberiae* Dementiev, 1933	Ural Mts. and W Siberia
— *S.a.willkonskii* (Menzbier, 1896)[6]	Caucasus area, NE Turkey, N Iran, SW Turkmenistan
— *S.a.sanctinicolai* (Zarudny, 1905)	Iraq, W and SW Iran
— *S.a.harmsi* (Zarudny, 1911)	W foothills of Tienshan and Pamir Mts.
— *S.a.biddulphi* Scully, 1881	N and NE Afghansitan, N Pakistan, Kashmir
— *S.a.nivicola* (Blyth, 1845)[7]	Himalayas, Assam hills, N and E Burma to C China
— *S.a.yamadae* Yamashina, 1936	S Taiwan
— *S.a.ma* (Clark, 1907)	NE China, Korea

○ **Strix butleri** (Hume, 1878) HUME'S OWL

SE Egypt, Sinai, Arabia north to SE Israel and SW Jordan, S Iran

● **Strix occidentalis** SPOTTED OWL

⌇ *S.o.caurina* (Merriam, 1898)	S Brit. Columbia to N California
— *S.o.occidentalis* (Xántus de Vesey, 1860)	S California
— *S.o.lucida* (Nelson, 1903)	SW USA to C Mexico
— *S.o.juanaphillipsae* Dickerman, 1997	C Mexico (726)

● **Strix varia** BARRED OWL

⌇ *S.v.varia* Barton, 1799	S Canada, EC USA
— *S.v.georgica* Latham, 1802[8]	S and SE USA
— *S.v.helveola* (Bangs, 1899)	SC Texas
— *S.v.sartorii* (Ridgway, 1873)	N and C Mexico

○ **Strix fulvescens** (Sclater & Salvin, 1868) FULVOUS OWL

S Mexico, W Guatemala, Honduras

○ **Strix hylophila** Temminck, 1825 RUSTY-BARRED OWL

Brazil, Paraguay, N Argentina

○ **Strix chacoensis** Cherrie & Reichenberger, 1921 CHACO OWL[9]

S Bolivia, Paraguay, N Argentina

○ **Strix rufipes** RUFOUS-LEGGED OWL

— *S.r.sanborni* Wheeler, 1938	Chiloe I. (Chile)
— *S.r.rufipes* P.P. King, 1828	S Chile, S Argentina

○ **Strix uralensis** URAL OWL

— *S.u.macroura* Wolf, 1810[10]	C and SE Europe
— *S.u.liturata* Lindroth, 1788	N Scandinavia and NE Poland to C European Russia and N Belarus

---

[1] Includes *connectens* Koelz, 1950 (1280), see Ripley (1982: 187) (2055).
[2] Mees (1986) (1539) rejected this, but he did so saying "if Ripley is right that the populations of …. Sumatra and Borneo are consubspecific" and as van Marle & Voous (1988) (2468) did not agree with that premise they quoted Mees (op. cit.) out of context.
[3] Pl. 525 was in Livr. 89 (not 88 as given by Peters, 1940: 159) and dates from 1832 (not 1831) see Dickinson (2001) (739).
[4] Specimen said to be from Sumatra likely to be mistakenly labelled.
[5] Includes *volhyniae* Dunajewski, 1948 (769), see Stepanyan (1990: 282) (2326).
[6] Includes *obscurata* see Stepanyan (1990: 283) (2326).
[7] Includes *obrieni* Koelz, 1954 (1284), see Ripley (1982: 187) (2055).
[8] For reasons to use 1802 not 1801 see Browing & Monroe (1991) (258).
[9] Reasons to accept this as a species are given by Straneck *et al.* (1995) (2357).
[10] Includes *carpathica* Dunajewski, 1940 (768A), see Stepanyan (1990: 285) (2326).

— *S.u.uralensis* Pallas, 1771[1]	E European Russia to W Siberia
— *S.u.yenisseensis* Buturlin, 1915	C and NE Siberia to NW Mongolia
— *S.u.daurica* Stegmann, 1929[2]	Lake Baikal and NE Mongolia to W and N Amurland and W and N Manchuria
— *S.u.nikolskii* (Buturlin, 1907)[3]	E Amurland, Sakhalin south to NE China and Korea
— *S.u.japonica* Clark, 1907	S Kuril Is., Hokkaido
— *S.u.hondoensis* (Clark, 1907)	N Honshu
— *S.u.momiyamae* Taka-Tsukasa, 1931	C Honshu
— *S.u.fuscescens* Temminck & Schlegel, 1850[4]	S Honshu, Kyushu and Shikoku Is.
— *S.u.davidi* (Sharpe, 1875)[5]	W and N Sichuan, NE Qinghai

● *Strix nebulosa* GREAT GREY OWL

✓ *S.n.nebulosa* J.R. Forster, 1772	N Nth. America
— *S.n.lapponica* Thunberg, 1798[6]	N Eurasia east to Kolyma Mts. and Sakhalin, south to N Mongolia

○ *Strix woodfordi* AFRICAN WOOD OWL

— *S.w.umbrina* (Heuglin, 1863)	Ethiopia and SE Sudan
— *S.w.nigricantior* (Sharpe, 1897)[7]	S Somalia, Kenya, Tanzania, E Zaire
— *S.w.nuchalis* (Sharpe, 1870)	Sierra Leone to S Sudan, Uganda, N and W Zaire and NC Angola
— *S.w.woodfordi* (A. Smith, 1834)	S Angola, SE Zaire and Malawi to N Botswana and E and S Sth. Africa

○ *Strix virgata* MOTTLED OWL

— *S.v.squamulata* (Bonaparte, 1850)	Sonora to Guerrero (NW Mexico)
— *S.v.tamaulipensis* J.C. Phillips, 1911	Nuevo Leon and S Tamaulipas (NE Mexico)
— *S.v.centralis* (Griscom, 1929)	S Mexico to W Panama
— *S.v.virgata* (Cassin, 1849)[8]	E Panama east to Venezuela and Trinidad, and south to Ecuador
— *S.v.macconnelli* (Chubb, 1916)	The Guianas
— *S.v.superciliaris* (Pelzeln, 1863)	NC and NE Brazil
— *S.v.borelliana* (W. Bertoni, 1901)	S Brazil, Paraguay, NE Argentina

○ *Strix albitarsis* RUFOUS-BANDED OWL[9]

— *S.a.albitarsis* (Bonaparte, 1850)	Colombia, Ecuador, Venezuela
— *S.a.opaca* J.L. Peters, 1943	C Peru (1831)
— *S.a.tertia* Todd, 1947	Bolivia (2411)

CICCABA Wagler, 1832[10] F
○ *Ciccaba nigrolineata* P.L. Sclater, 1859 BLACK-AND-WHITE OWL   S Mexico to W Ecuador

○ *Ciccaba huhula* BLACK-BANDED OWL

— *C.h.huhula* (Daudin, 1800)	N Sth. America (east of the Andes)
— *C.h.albomarginata* (Spix, 1824)	NE Argentina, Paraguay, SE Brazil

JUBULA Bates, 1929 F
○ *Jubula lettii* (Büttikofer, 1889) MANED OWL   Liberia to N Zaire

LOPHOSTRIX Lesson, 1836 F
○ *Lophostrix cristata* CRESTED OWL

— *L.c.stricklandi* Sclater & Salvin, 1859	S Mexico to W Panama, W Colombia
— *L.c.wedeli* Griscom, 1932	E Panama, N Colombia, NW Venezuela
— *L.c.cristata* (Daudin, 1800)[11]	Amazonia

[1] Includes *buturlini* Dementiev, 1951 (689), see Stepanyan (1990: 284) (2326).
[2] For recognition see Stepanyan (1990: 285) (2326).
[3] Includes *tatibanai* see Stepanyan (1990: 285) (2326), and *coreensis* see Holt *et al.* (1999: 203).
[4] For reasons to use this date see Dekker *et al.* (2001) (672).
[5] May be a distinct species. We agree with Holt *et al.* (1999: 203) (1127) that more study is needed.
[6] Includes *elisabethae* see Stepanyan (1990: 286) (2326).
[7] Includes *sokokensis* Ripley & Bond, 1971 (2064), see Holt *et al.* (1999: 204) (1127).
[8] Presumably includes *minuscula* L. Kelso, 1940 (1245), and *occidentalis* von Sneidern, 1957 (2544); the latter name is preoccupied in *Strix* and Eck (1971) (785) renamed it *sneiderni*.
[9] For correction of spelling see David & Gosselin (2002) (613).
[10] This narrow use of *Ciccaba* was an option suggested by Voous (1964) (2546).
[11] Thought to includes *amazonica* L. Kelso, 1940 (1245).

PULSATRIX Kaup, 1848 F
○ *Pulsatrix perspicillata* SPECTACLED OWL
  __ *P.p.saturata* Ridgway, 1914   S Mexico to W Panama
  __ *P.p.chapmani* Griscom, 1932   E Costa Rica to W Ecuador, NW Peru
  __ *P.p.perspicillata* (Latham, 1790)   Venezuela, the Guianas, Amazonia
  __ *P.p.trinitatis* Bangs & Penard, 1918   Trinidad
  __ *P.p.boliviana* L. Kelso, 1933   S Bolivia to N Argentina
  __ *P.p.pulsatrix* (Wied, 1820)   E Brazil, Paraguay

○ *Pulsatrix koeniswaldiana* (M. & W. Bertoni, 1901) TAWNY-BROWED OWL
     S Brazil, NE Argentina

○ *Pulsatrix melanota* BAND-BELLIED OWL
  __ *P.m.melanota* (Tschudi, 1844)   E Ecuador, E Peru
  __ *P.m.philoscia* Todd, 1947   W Bolivia (2411)

SURNIINAE

SURNIA Duméril, 1806 F
◉ *Surnia ulula* NORTHERN HAWK-OWL
  __ *S.u.ulula* (Linnaeus, 1758)   N Eurasia
  __ *S.u.tianschanica* Smallbones, 1906   Tien Shan to N China (S of *ulula*)
  ✓ *S.u.caparoch* (Statius Müller, 1776)   Alaska and Canada >> south to N USA

GLAUCIDIUM Boie, 1826 N
○ *Glaucidium passerinum* EURASIAN PYGMY OWL
  __ *G.p.passerinum* (Linnaeus, 1758)   C and N Europe and SW Siberia
  __ *G.p.orientale* Taczanowski, 1891   NW and C Siberia to Sakhalin and NE China

○ *Glaucidium brodiei* COLLARED OWLET
  __ *G.b.brodiei* (E. Burton, 1836)[1]   Himalayas east to S China and S through mainland SE Asia to the Malay Pen.
  __ *G.b.sylvaticum* Bonaparte, 1850[2]   Sumatra
  __ *G.b.borneense* Sharpe, 1893   Borneo
  __ *G.b.pardalotum* (Swinhoe, 1863)   Taiwan

○ *Glaucidium perlatum* PEARL-SPOTTED OWLET
  __ *G.p.perlatum* (Vieillot, 1817)   Senegal to W Sudan
  __ *G.p.licua* (M.H.K. Lichtenstein, 1842)[3]   E, NE and Sthn. Africa

○ *Glaucidium gnoma* NORTHERN PYGMY OWL[4]
  __ *G.g.grinnelli* Ridgway, 1914   SE Alaska to N California
  __ *G.g.swarthi* Grinnell, 1913   Vancouver I.
  __ *G.g.californicum* P.L. Sclater, 1857   C Brit. Columbia to S California
  __ *G.g.pinicola* Nelson, 1910   WC USA
  __ *G.g.hoskinsii* Brewster, 1888   Baja California
  __ *G.g.gnoma* Wagler, 1832   N and C Mexico
  __ *G.g.cobanense* Sharpe, 1875   Guatemala

○ *Glaucidium costaricanum* L. Kelso, 1937 COSTA RICAN PYGMY OWL[5] Costa Rica, Panama

○ *Glaucidium nubicola* Robbins & Stiles, 1999 CLOUD-FOREST PYGMY OWL #
     W Colombia, N Ecuador (2086)

○ *Glaucidium jardinii* (Bonaparte, 1855) ANDEAN PYGMY OWL   Andes from Venezuela and Colombia, to C Peru

○ *Glaucidium bolivianum* König, 1991 YUNGAS PYGMY OWL #   S Peru, Bolivia, NW Argentina (1294)

[1] Includes *garoense* Koelz, 1952 (1282), see Ripley (1982: 182) (2055).
[2] For the use of this name in place of *peritum* see Mees (1967) (1518).
[3] Includes *kilimense* Reichenow, 1893 (2002) of last Edition, not mentioned in Peters as previously synonymised by Reichenow (1900: 674) (2003), and *diurnum* Clancey, 1968 (415), see Holt *et al.* (1999: 210) (1127).
[4] Although a split may be anticipated, the evidence is not all in print yet.
[5] For reasons to separate from *jardinii* see Heidrich *et al.* (1995) (1094), A.O.U. (1998: 259) (3).

○ *Glaucidium palmarum* Colima Pygmy Owl[1]
  — *G.p.oberholseri* R.T. Moore, 1937   C and S Sinaloa (NW Mexico)
  — *G.p.palmarum* Nelson, 1901   W Mexico
  — *G.p.griscomi* R.T. Moore, 1947   SW Morelos, NE Guerrero (S Mexico) [1616]

○ *Glaucidium sanchezi* Lowery & Newman, 1949   Tamaulipas Pygmy Owl
     NE Mexico [1393]

○ *Glaucidium griseiceps* Central American Pygmy Owl
  — *G.g.occultum* R.T. Moore, 1947   E Oaxaca, Chiapas (Mexico) [1617]
  — *G.g.griseiceps* Sharpe, 1875   E Guatemala, Belize, E Honduras
  — *G.g.rarum* Griscom, 1931   Costa Rica, Panama

○ *Glaucidium parkeri* Robbins & Howell, 1995   Subtropical Pygmy Owl #
     E Ecuador, E Peru [2079]

○ *Glaucidium hardyi* Vielliard, 1989   Amazonian Pygmy Owl #   Amazonia [2527]

○ *Glaucidium minutissimum* (Wied, 1830)   Least Pygmy Owl   SE Brazil, Paraguay

○ *Glaucidium brasilianum* Ferruginous Pygmy Owl[2]
  — *G.b.cactorum* van Rossem, 1937   S Arizona to W Mexico
  — *G.b.ridgwayi* Sharpe, 1875[3]   S Texas to C Panama and NW Colombia
  — *G.b.medianum* Todd, 1916   N Colombia
  — *G.b.margaritae* Phelps & Phelps, Jr., 1951   Margarita I. (off Venezuela) [1849]
  — *G.b.phalaenoides* (Daudin, 1800)   N Venezuela, Trinidad
  — *G.b.duidae* Chapman, 1929   Mt. Duida (Venezuela)
  — *G.b.olivaceum* Chapman, 1939   Mt. Auyun-tepui (Venezuela) [316]
  — *G.b.ucayalae* Chapman, 1929   SE Colombia to Peru
  — *G.b.brasilianum* (J.F. Gmelin, 1788)   W and S Amazonia, NE Argentina, N Uruguay
  — *G.b.pallens* Brodkorb, 1938   E Bolivia, W Paraguay, N Argentina
  — *G.b.stranecki* König & Wink, 1995   C Argentina to S Uruguay [1296]
  — *G.b.tucumanum* Chapman, 1922   W Argentina

○ *Glaucidium peruanum* König, 1991   Peruvian Pygmy Owl #   W Ecuador, Peru [1294]

○ *Glaucidium nanum* (P.P. King, 1828)   Austral Pygmy Owl   S Chile, S Argentina

○ *Glaucidium siju* Cuban Pygmy Owl
  — *G.s.siju* (d'Orbigny, 1839)   Cuba
  — *G.s.vittatum* Ridgway, 1914   Isla de la Juventud

○ *Glaucidium tephronotum* Red-chested Owlet
  — *G.t.tephronotum* Sharpe, 1875   Sierra Leone, Liberia, Ivory Coast, Ghana
  — *G.t.pycrafti* Bates, 1911   Cameroon
  — *G.t.medje* Chapin, 1932[4]   Zaire, Rwanda, SW Uganda
  — *G.t.elgonense* Granvik, 1934   E Uganda, W Kenya

○ *Glaucidium sjostedti* Reichenow, 1893   Sjostedt's Owlet   Cameroon to Gabon and C Zaire

○ *Glaucidium cuculoides* Asian Barred Owlet
  — *G.c.cuculoides* (Vigors, 1831)   W and C Himalayas
  — *G.c.austerum* Ripley, 1948   NE Assam [2032]
  — *G.c.rufescens* E.C.S. Baker, 1926   E Himalayas, Assam hills, W and N Burma to SW Yunnan
  — *G.c.whitelyi* (Blyth, 1867)   C Yunnan, Sichuan and S Gansu to E China, NE Vietnam
  — *G.c.persimile* E. Hartert, 1910   Hainan
  — *G.c.delacouri* Ripley, 1948   NE Laos, NW and C Vietnam [2032]
  — *G.c.deignani* Ripley, 1948   SE Thailand, S Vietnam, ? Cambodia [2032]
  — *G.c.bruegeli* (Parrot, 1908)   S Burma, S Thailand

○ *Glaucidium castanopterum* (Horsfield, 1821)   Javan Owlet   Java, Bali

---

[1] In last Edition a broad concept *G. minutissimum* began here, but see A.O.U. (1998: 259) (3) for sources of accepted treatment as seven species.
[2] A split is to be expected when more evidence is published.
[3] Here treated as including *saturatum* Brodkorb, 1941 (222) and *intermedium* A.R. Phillips, 1966 (1869); but both require re-evaluation.
[4] Includes *lukolelae* and *kivuense* Verheyen, 1946 (2520), see White (1965: 198) (2649).

○ *Glaucidium radiatum* Jungle Owlet
 — *G.r.radiatum* (Tickell, 1833)[1]   India, Sri Lanka
 — *G.r.malabaricum* (Blyth, 1846)   SW India

○ *Glaucidium castanonotum* (Blyth, 1852) Chestnut-backed Owlet[2]   Sri Lanka

○ *Glaucidium capense* African Barred Owlet[3]
 — *G.c.etchecopari* Érard & Roux, 1983   Liberia, Ivory Coast (814)
 — *G.c.castaneum* Neumann, 1893   NE Zaire, SW Uganda
 — *G.c.albertinum* Prigogine, 1983[4]   NE Zaire and N Rwanda (1939)
 — *G.c.scheffleri* Neumann, 1911   SE Kenya, NE Tanzania
 — *G.c.ngamiense* (Roberts, 1932)[5]   SE Zaire, NE Angola, S and W Tanzania, W
                  Mozambique to N Botswana and E Transvaal
 — *G.c.capense* (A. Smith, 1834)   S Mozambique to Natal and E Cape Province

XENOGLAUX O'Neill & Graves, 1977 F
○ *Xenoglaux loweryi* O'Neill & Graves, 1977 Long-whiskered Owlet   N Peru (1713)

MICRATHENE Coues, 1866 F
● *Micrathene whitneyi* Elf Owl
 — *M.w.whitneyi* (J.G. Cooper, 1861)   SW USA, NW Mexico >> C Mexico
 ⚲ *M.w.idonea* (Ridgway, 1914)   Texas, C Mexico
 — *M.w.sanfordi* (Ridgway, 1914)   S Baja California and W Mexico
 — *M.w.graysoni* Ridgway, 1886   Socorro I.

ATHENE Boie, 1822 F
○ *Athene noctua* Little Owl
 — *A.n.vidalii* A.E. Brehm, 1857[6]   W and C Europe east to Baltic Rep. and Belarus
 — *A.n.noctua* (Scopoli, 1769)[7]   Corsica, Sardinia and Italy to Slovakia, NW Romania
                  and Croatia
 — *A.n.indigena* C.L. Brehm, 1855   Albania to S European Russia, south to Greece,
                  Turkey, N Levant and Transcaucasia
 — *A.n.lilith* E. Hartert, 1913   Cyprus, Sinai, N Arabia, Jordan and C Syria to SE
                  Turkey and Iraq
 — *A.n.bactriana* Blyth, 1847   Lower Ural R., Transcaspia and Iran to E Kazakhstan
                  and W Pakistan
 — *A.n.orientalis* Severtsov, 1873   Pamir and Tien Shan Mts. to Qinghai (W China)
 — *A.n.ludlowi* E.C.S. Baker, 1926   S and E Tibetan Plateau from Ladakh to W Sichuan
 — *A.n.impasta* Bangs & Peters, 1928   NE Qinghai, SW Gansu
 — *A.n.plumipes* Swinhoe, 1870   S Altai and Mongolia to S Transbaikalia and NC and
                  NE China
 — *A.n.glaux* (Savigny, 1809)   Coastal N Africa, Nile Valley, SW Israel
 — *A.n.saharae* (O. Kleinschmidt, 1909)[8]   Sahara E to Red Sea (except Nile Valley) and C Arabia
 — *A.n.spilogastra* (Heuglin, 1869)   E Sudan, Eritrea, NE Ethiopia
 — *A.n.somaliensis* Reichenow, 1905   E Ethiopia, Somalia

○ *Athene brama* Spotted Owlet[9]
 — *A.b.indica* (Franklin, 1831)[10]   S Iran to N and C India and Bangladesh
 — *A.b.ultra* Ripley, 1948   E Assam (2031)
 — *A.b.brama* (Temminck, 1821)   S India
 — *A.b.pulchra* Hume, 1873   Burma
 — *A.b.mayri* Deignan, 1941   Thailand (except SW and S), S Laos, S Vietnam,
                  Cambodia (634)

---

[1] Includes *principum* Koelz, 1950 (1280), see Greenway (1978) (1010).
[2] In last Edition treated as a race of *cuculoides* in contradiction to Ripley (1948, 1982) (2032, 2055); for recognition see also Konig *et al.* (1999) (1297).
[3] We treat this as a broad species, including *albertinum*, following Dowsett & Dowsett-Lemaire (1993) (762). For an alternative treatment see Prigogine (1985) (1942).
[4] Treated as a species by Kemp (1988: 143) (1249); the acoustic evidence for lumping is perhaps somewhat ambivalent.
[5] Includes *clanceyi* Prigogine, 1985 (1942) see Kemp (1988: 141) (1249).
[6] Includes *grueni* von Jordans & Steinbacher, 1942 (2542), and *cantabriensis* Harrison, 1957 (1077), see Vaurie (1965: 610) (2510).
[7] Includes *sarda* see Vaurie (1965: 610) (2510), and *salentina* Trischitta, 1939 (2439), and *daciae* Keve & Kohl, 1961 (1266), see Stepanyan (1990: 276) (2326).
[8] Includes *solitudinis* see Vaurie (1965: 609) (2510).
[9] Holt et al (1999: 226) (1127) reported that *A. b. poikila* Yang & Li, 1989 (2697), has been reidentified as *Aegolius funereus*.
[10] Includes *albida* Koelz, 1950 (1280), see Vaurie (1965: 613) (2510).

⬧ *Athene cunicularia* BURROWING OWL
     ☑ *A.c.hypugaea* (Bonaparte, 1825)   SW Canada to W Mexico
     __ *A.c.rostrata* (C.H. Townsend, 1890)   Clarion I.
     __ *A.c.floridana* (Ridgway, 1874)   C and S Florida, Bahama Is.
     __ †*A.c.amaura* Lawrence, 1878   Antigua, W Indies
     __ †*A.c.guadeloupensis* Ridgway, 1874   Guadeloupe (Lesser Antilles)
     __ ¶*A.c.guantanamensis* Garrido, 2001   Cuba (924)
     __ *A.c.troglodytes* (Wetmore & Swales, 1931)   Hispaniola, Gonave I.
     __ *A.c.arubensis* (Cory, 1915)   Aruba I.
     __ *A.c.brachyptera* (Richmond, 1896)   Margarita I. and N Venezuela
     __ *A.c.apurensis* (Gilliard, 1940)   NC Venezuela (937)
     __ *A.c.minor* (Cory, 1918)   S Guyana, S Surinam, NE Brazil
     __ *A.c.carrikeri* (Stone, 1922)   E Colombia
     __ *A.c.tolimae* (Stone, 1899)   W Colombia
     __ *A.c.pichinchae* (von Boetticher, 1929)   W Ecuador
     __ *A.c.punensis* (Chapman, 1914)   SW Ecuador, NW Peru
     __ *A.c.intermedia* (Cory, 1915)   W Peru
     __ *A.c.nanodes* (Berlepsch & Stolzmann, 1892)   SW Peru
     __ *A.c.juninensis* (Berlepsch & Stolzmann, 1902)   C Peru, W Bolivia
     __ *A.c.boliviana* (L. Kelso, 1939)   Bolivia (1244)
     __ *A.c.grallaria* (Temminck, 1822)   E and C Brazil
     __ *A.c.cunicularia* (Molina, 1782)   S Bolivia and S Brazil to Tierra del Fuego
     __ *A.c.partridgei* Olrog, 1976   Corrientes (Argentina) (1685)

HETEROGLAUX Hume, 1873[1] F
○ *Heteroglaux blewitti* Hume, 1873 FOREST OWLET   C India

AEGOLIUS Kaup, 1829 M
○ *Aegolius funereus* TENGMALM'S OWL/BOREAL OWL
     __ *A.f.richardsoni* (Bonaparte, 1838)   Alaska, N Canada and N USA
     __ *A.f.funereus* (Linnaeus, 1758)   N, C and SE Europe to Lena basin and Lake Baikal,
                                               south to Tarbagatay and N Mongolia
     __ *A.f.magnus* (Buturlin, 1907)[2]   NE Siberia east to Anadyr and Kamchatka
     __ *A.f.sibiricus* (Buturlin, 1910)   Lake Baikal to W shore of Sea of Okhotsk, Sakhalin
                                               and NE China
     __ *A.f.pallens* (Schalow, 1908)   Russian and Chinese Tien Shan
     __ *A.f.caucasicus* (Buturlin, 1907)   N Caucasus
     __ *A.f.beickianus* Stresemann, 1928[3]   NW India and NE Qinghai (China)

● *Aegolius acadicus* NORTHERN SAW-WHET OWL
     ☑ *A.a.acadicus* (J.F. Gmelin, 1788)[4]   Canada, N and SW USA south to N Mexico
     __ *A.a.brooksi* (J.H. Fleming, 1916)   Queen Charlotte Is.

○ *Aegolius ridgwayi* UNSPOTTED SAW-WHET OWL[5]
     __ *A.r.tacanensis* R.T. Moore, 1947   S Mexico (1617)
     __ *A.r.rostratus* (Griscom, 1930)   Guatemala
     __ *A.r.ridgwayi* (Alfaro, 1905)   Costa Rica

○ *Aegolius harrisii* BUFF-FRONTED OWL
     __ *A.h.harrisii* (Cassin, 1849)   Colombia, Ecuador, Venezuela
     __ *A.h.iheringi* (Sharpe, 1899)   SE Brazil, Paraguay, NE Argentina
     __ *A.h.dabbenei* Olrog, 1979[6]   NW Argentina (1686)

NINOX Hodgson, 1837 F
○ *Ninox rufa* RUFOUS OWL
     __ *N.r.humeralis* (Bonaparte, 1850)   New Guinea, Waigeo I.

---

[1] For reasons to use this generic name see Rasmussen & Collar (1999) (1990).
[2] Includes *jakutorum* see Stepanyan (1990: 274) (2326), but an intergrade with *pallidus*.
[3] Includes *juniperi* Koelz, 1939 (1279), placed in *caucasicus* by Ripley (1982: 189) (2055), but NW Indian birds reassigned by Roselaar in Cramp *et al.* (1985: 606) (591).
[4] Probably includes *brodkorbi* Briggs, 1954 (214), based on a juvenile (Holt *et al.*, 1999: 228) (1127).
[5] The two races tentatively recognised may prove to be hybrids (Holt *et al.*, 1999: 229) (1127).
[6] Recognition tentative; requires evaluation.

__ *N.r.aruensis* (Schlegel, 1866)   Aru Is.

__ *N.r.rufa* (Gould, 1846)   NW Australia (Kimberley to Arnhem Land)

__ *N.r.meesi* Mason & Schodde, 1980   Cape York Pen. (1451)

__ *N.r.queenslandica* Mathews, 1911[1]   NE to EC Queensland

○ *Ninox strenua* (Gould, 1838) POWERFUL OWL   EC to SE Australia

○ *Ninox connivens* BARKING OWL

__ *N.c.rufostrigata* (G.R. Gray, 1860)   N Moluccas

__ *N.c.assimilis* Salvadori & D'Albertis, 1875   E New Guinea, Vulcan, Dampier Is.

__ *N.c.peninsularis* Salvadori, 1876[2]   N Australia

__ *N.c.connivens* (Latham, 1802)[3]   SW, E and SE Australia

○ *Ninox rudolfi* A.B. Meyer, 1882 SUMBA BOOBOOK   Sumba I.

○ *Ninox boobook* SOUTHERN BOOBOOK[4]

__ *N.b.rotiensis* Johnstone & Darnell, 1997   Roti I. (1221)

__ *N.b.fusca* (Vieillot, 1817)   Timor I.

__ *N.b.plesseni* Stresemann, 1929   Alor I.

__ *N.b.moae* Mayr, 1943   Moa, Leti, Roma Is. (1473)

__ *N.b.cinnamomina* E. Hartert, 1906   Babar I.

__ *N.b.remigialis* Stresemann, 1930   Kai Is.

__ *N.b.pusilla* Mayr & Rand, 1935   S New Guinea

__ *N.b.ocellata* (Bonaparte, 1850)[5]   Australia W of Murray-Darling basin and South Australian gulfs, Sawu I.

__ *N.b.halmaturina* Mathews, 1912[6]   Kangaroo I. (South Australia)

__ *N.b.lurida* De Vis, 1889   NE Queensland

__ *N.b.boobook* (Latham, 1802)[7]   E to SC Australia

○ *Ninox novaeseelandiae* MOREPORK

__ *N.n.leucopsis* (Gould, 1838)[8]   Tasmania

__ †*N.n.albaria* E.P. Ramsay, 1888   Lord Howe I.

__ †?*N.n.undulata* (Latham, 1802)[9]   Norfolk I.

__ *N.n.novaeseelandiae* (J.F. Gmelin, 1788)   New Zealand

○ *Ninox scutulata* BROWN HAWK-OWL

__ *N.s.florensis* (Wallace, 1864)[10]   Russian Far East and N Korea to C and NE China >> main winter quarters?[11]

__ *N.s.japonica* (Temminck & Schlegel, 1845[12])   S Korea, Japan >> Philippines, Sundas, Wallacea

__ *N.s.lugubris* (Tickell, 1833)   NE Pakistan, N and C India to W Assam

__ *N.s.burmanica* Hume, 1876   S Assam east to Thailand (not Pen.), Indochina and S China

__ *N.s.hirsuta* (Temminck, 1824)   S India, Sri Lanka

__ *N.s.obscura* Hume, 1873   Andaman, Nicobar Is.

__ *N.s.totogo* Momiyama, 1931[13]   Ryukyu Is., Taiwan and nearby Lanyu I.

__ *N.s.scutulata* (Raffles, 1822)[14]   Malay Pen., Riau Arch, Sumatra, Bangka

__ *N.s.javanensis* Stresemann, 1928   Java (? Bali)

__ *N.s.borneensis* (Bonaparte, 1850)   Borneo, N Natuna Is.

---

[1] Includes *marginata* Mees, 1964 (1514), see Mason & Schodde (1980) (2186).

[2] Includes *connivens* see Schodde & Mason (1980) (2186). The name *enigma* Mathews & Neumann, 1939 (1458), is thought to relate to an aberrant specimen, perhaps of this form (Mees, 1964) (1514).

[3] For reasons to use 1802 not 1801 see Browing & Monroe (1991) (258). Includes *addenda* see Mees (1964) (1514).

[4] For reasons to separate *boobook* from *novaeseelandiae* see Schodde & Mason (1980) (2186).

[5] Includes *arida* Mayr, 1943 (1473), see Mees (1964) (1514) who also put *marmorata* in nominate *boobook* but Schodde & Mason (1980) (2186) reassessed the distribution patterns and placed the name in *ocellata*. Also includes *rufigaster* Mees, 1961 (1510), see Schodde & Mason (1980) (2186).

[6] For recognition of this see Mees (1961, 1964) (1510, 1514).

[7] For reasons to use 1802 not 1801 see Browing & Monroe (1991) (258).

[8] This taxon appears to be genetically closer to *N. novaeseelandiae* see Christidis & Norman (1998) (1661), but its morphological similarity to *N. boobook* suggests that it may be best treated as a distinct species.

[9] For reasons to use 1802 not 1801 see Browing & Monroe (1991) (258).

[10] This name must replace *ussuriensis* see Dickinson *et al.* (1991: 229) (745).

[11] Scattered specimens thought to be this form recorded from Thailand to Philippines and Lesser Sundas; all older records need review against voucher specimens selected to take account of changes in nomenclature (see other footnotes to this species).

[12] This name must date from 1845 see Holthuis & Sakai (1970) (1128).

[13] Mees (1970) (1521) considered this a synonym of *japonica*; this has not been accepted by Cheng (1987: 334) (333) or Orn. Soc. Japan (2000: 159) (1727) and *totogo* is usefully maintained for this non-migratory population. Includes *yamashinae* Ripley, 1953 (2036), see Orn. Soc. Japan (1958) (1725).

[14] Includes *malaccensis* see Dickinson (1975) (735).

— *N.s.palawanensis* Ripley & Rabor, 1962[1]   Palawan (2074)

— *N.s.randi* Deignan, 1951   Philippine Is. (except Palawan group) (649)

○ *Ninox affinis* ANDAMAN HAWK-OWL

— *N.a.affinis* Beavan, 1867   Andaman Is.

— *N.a.isolata* E.C.S. Baker, 1926   Car Nicobar I.

— *N.a.rexpimenti* Abdulali, 1979   Gt. Nicobar I. (13)

○ *Ninox superciliaris* (Vieillot, 1817) WHITE-BROWED HAWK-OWL   SW and S Madagascar (rarely NE)

○ *Ninox sumbensis* Olsen, Wink, Sauer-Gürth & Trost, 2002 LITTLE SUMBA HAWK-OWL #

Sumba (Lesser Sundas) (1688A)

○ *Ninox philippensis* PHILIPPINE HAWK-OWL[2]

— *N.p.philippensis* Bonaparte, 1855   Luzon, Polillo, Catanduanes, Marinduque, Leyte, Samar

— *N.p.mindorensis* Ogilvie-Grant, 1896   Mindoro

— *N.p.proxima* Mayr, 1945   Masbate (682)

— *N.p.ticaoensis* duPont, 1972   Ticao (773)

— *N.p.centralis* Mayr, 1945   Panay, Guimaras, Negros, Bohol (682)

— *N.p.spilonota* Bourns & Worcester, 1894   Tablas, Sibuyan, Cebu, Camiguin Sur

— *N.p.spilocephala* Tweeddale, 1879   Mindanao, Dinagat, Siargao, Basilan

— *N.p.reyi* Oustalet, 1880[3]   Sulu Arch.

○ *Ninox ochracea* (Schlegel, 1866) OCHRE-BELLIED HAWK-OWL   Sulawesi

○ *Ninox ios* Rasmussen, 1999 CINNABAR HAWK-OWL #   N Sulawesi (1986)

○ *Ninox squamipila* MOLUCCAN HAWK-OWL

— *N.s.hypogramma* (G.R. Gray, 1860)   N Moluccas

— *N.s.hantu* (Wallace, 1863)   Buru I.

— *N.s.squamipila* (Bonaparte, 1850)   Seram I.

— *N.s.forbesi* P.L. Sclater, 1883   Tanimbar Is.

○ *Ninox natalis* Lister, 1889 CHRISTMAS ISLAND HAWK-OWL[4]   Christmas I. (Indian Ocean)

○ *Ninox theomacha* JUNGLE HAWK-OWL

— *N.t.hoedtii* (Schlegel, 1871)   Waigeo, Misool Is.

— *N.t.goldii* J.H. Gurney, 1883   D'Entrecasteaux Arch.

— *N.t.theomacha* (Bonaparte, 1855)   New Guinea

— *N.t.rosseliana* Tristram, 1889   Louisiade Arch.

○ *Ninox meeki* Rothschild & Hartert, 1914 MANUS HAWK-OWL   Admiralty Is.

○ *Ninox punctulata* (Quoy & Gaimard, 1830) SPECKLED HAWK-OWL   Sulawesi

○ *Ninox variegata* BISMARCK HAWK-OWL

— *N.v.superior* E. Hartert, 1925   New Hanover (Bismarck Arch.)

— *N.v.variegata* (Quoy & Gaimard, 1830)   New Ireland, New Britain

○ *Ninox odiosa* P.L. Sclater, 1877 RUSSET HAWK-OWL   New Britain (Bismarck Arch.)

○ *Ninox jacquinoti* SOLOMON ISLANDS HAWK-OWL

— *N.j.eichhorni* (E. Hartert, 1927)   Bougainville, Choiseul Is.

— *N.j.mono* Mayr, 1935   Mono I.

— *N.j.jacquinoti* (Bonaparte, 1850)   Sta. Isabel, St George Is.

— *N.j.floridae* Mayr, 1935   Florida I.

— *N.j.granti* Sharpe, 1888   Guadalcanal I.

— *N.j.malaitae* Mayr, 1931   Malaita I.

— *N.j.roseoaxillaris* (E. Hartert, 1929)   San Cristobal I.

UROGLAUX Mayr, 1937 F

○ *Uroglaux dimorpha* (Salvadori, 1874) PAPUAN HAWK-OWL   New Guinea, Yapen I.

[1] For recognition of this see Dickinson *et al.* (1991) (745).
[2] Arrangement follows Dickinson *et al.* (1991) (745); as presently arranged probably comprises two or three species but problems of separation unresolved.
[3] Includes *everetti*.
[4] For reasons to separate from distant *N. squampila* see Norman *et al.* (1998) (1661).

SCELOGLAUX Kaup, 1848 F

○ *Sceloglaux albifacies* LAUGHING OWL
  — †?*S.a.rufifacies* Buller, 1904   North Island (New Zealand)
  — †?*S.a.albifacies* (G.R. Gray, 1844)   South Island (New Zealand)

ASIONINAE[1]

PSEUDOSCOPS Kaup, 1848 M

○ *Pseudoscops grammicus* (Gosse, 1847) JAMAICAN OWL   Jamaica

○ *Pseudoscops clamator* STRIPED OWL[2]
  — *P.c.forbesi* (Lowery & Dalquest, 1951)   S Mexico to Panama (1392)
  — *P.c.clamator* (Vieillot, 1808)[3]   N Sth. America south to E Peru and C Brazil
  — *P.c.oberi* (E.H. Kelso, 1936)   Trinidad, Tobago Is.
  — *P.c.midas* (Schlegel, 1862)   E Bolivia, S Brazil, Paraguay, Uruguay, N Argentina

NESASIO J.L. Peters, 1937 M

○ *Nesasio solomonensis* (E. Hartert, 1901) FEARFUL OWL   Bougainville, Choiseul, Sta. Isabel Is. (Solomons)

ASIO Brisson, 1760 M

○ *Asio stygius* STYGIAN OWL
  — *A.s.lambi* R.T. Moore, 1937   NW Mexico
  — *A.s.robustus* L. Kelso, 1934   E Mexico, Guatemala, Nicaragua
  — *A.s.siguapa* (d'Orbigny, 1839)   Cuba, Isla de la Juventud
  — *A.s.noctipetens* Riley, 1916   Hispaniola, Gonave I.
  — *A.s.stygius* (Wagler, 1832)   C and S Brazil
  — *A.s.barberoi* W. Bertoni, 1930   Paraguay, N Argentina

○ *Asio otus* LONG-EARED OWL
  — *A.o.otus* (Linnaeus, 1758)   Eurasia to NW Xinjiang and NE China and Japan;
                                                    south to Mediterranean, Middle East, N Pakistan,
                                                    Azores, NW Africa,

  — *A.o.canariensis* Madarász, 1901   Canary Is.
  — *A.o.tuftsi* Godfrey, 1948   W Canada south to NW Mexico (947)
  — *A.o.wilsonianus* (Lesson, 1830)   Canada and USA (east of *tuftsi*)

○ *Asio abyssinicus* AFRICAN LONG-EARED OWL
  — *A.a.abyssinicus* (Guérin-Méneville, 1843)   Highlands of Eritrea to C Ethiopia
  — *A.a.graueri* Sassi, 1912   Mts. of E Zaire, W Uganda and Mt. Kenya

○ *Asio madagascariensis* (A. Smith, 1834) MADAGASCAN LONG-EARED OWL
                                               Madagascar

● *Asio flammeus* SHORT-EARED OWL
  ✓ *A.f.flammeus* (Pontoppidan, 1763)   Europe, N Asia, N Africa, Nth. America
  — *A.f.portoricensis* Ridgway, 1882   Puerto Rico
  — *A.f.domingensis* (Statius Müller, 1776)   Hispaniola
  — *A.f.galapagoensis* (Gould, 1837)   Galapagos Is.
  — *A.f.bogotensis* Chapman, 1915   Colombia, Ecuador
  — *A.f.pallidicaudus* Friedmann, 1949   Venezuela (880)
  — *A.f.suinda* (Vieillot, 1817)   S Peru, S Brazil to Tierra del Fuego
  — *A.f.sanfordi* Bangs, 1919   Falkland Is.
  ✓ *A.f.sandwichensis* (Bloxham, 1826)   Hawaiian Is.
  — *A.f.ponapensis* Mayr, 1933   Mariana and Caroline Is. (W Pacific)

○ *Asio capensis* MARSH OWL
  — *A.c.tingitanus* (Loche, 1867)   NW Africa,
  — *A.c.capensis* (A. Smith, 1834)   Senegal to Cameroon; Ethiopia to Angola and Cape
                                                    Province

  — *A.c.hova* (Stresemann, 1922)   Madagascar

---

[1] The generic sequence is based on Olson (1995) (1701).
[2] Treated in the genus *Asio* in last Edition; for reasons to place this in *Pseudoscops* see Olson (1995) (1701).
[3] Dated 1807 by Peters (1940: 166) (1825), but see Browning & Monroe (1991) (258).

PODARGIDAE FROGMOUTHS (2: 12)

PODARGINAE

PODARGUS Vieillot, 1818 M
○ ***Podargus ocellatus*** MARBLED FROGMOUTH

— *P.o.ocellatus* Quoy & Gaimard, 1830	New Guinea and islands
— *P.o.marmoratus* Gould, 1855	Cape York Pen.
— *P.o.plumiferus* Gould, 1846[1]	EC Australia
— *P.o.intermedius* E. Hartert, 1895	Triobriand Is., D'Entrecasteaux Arch.
— *P.o.meeki* E. Hartert, 1898	Louisiade Arch. (Sudest I.)
— *P.o.inexpectatus* E. Hartert, 1901	Solomon Is.

○ ***Podargus papuensis*** PAPUAN FROGMOUTH[2]

— *P.p.papuensis* Quoy & Gaimard, 1830	New Guinea and islands, Aru Is.
— *P.p.rogersi* Mathews, 1912	Cape York Pen.
— *P.p.baileyi* Mathews, 1912	Humid NE Queensland

○ ***Podargus strigoides*** TAWNY FROGMOUTH[3]

— *P.s.strigoides* (Latham, 1802)[4]	E Australia west to E Murray-Darling and E Lake Eyre basins, Tasmania
— *P.s.brachypterus* Gould, 1841	W and C Australia east to W Lake Eyre and Murray-Darling basins
— *P.s.phalaenoides* Gould, 1840[5]	N Australia (Kimberley to Cape York Pen.)

BATRACHOSTOMINAE

BATRACHOSTOMUS Gould, 1838 M

○ ***Batrachostomus auritus*** (J.E. Gray, 1829) LARGE FROGMOUTH	Malay Pen., Sumatra, Borneo, Natuna Is.
○ ***Batrachostomus harterti*** Sharpe, 1892 DULIT FROGMOUTH	Mts. of N and C Borneo

○ ***Batrachostomus septimus*** PHILIPPINE FROGMOUTH

— *B.s.microrhynchus* Ogilvie-Grant, 1895	Luzon, Catanduanes (N Philippines)
— *B.s.menagei* Bourns & Worcester, 1894	Panay, Negros (WC Philippines
— *B.s.septimus* Tweeddale, 1877	Samar, Leyte, Bohol, Mindanao, Basilan (EC and S Philippines)

○ ***Batrachostomus stellatus*** (Gould, 1837) GOULD'S FROGMOUTH	C and S Malay Pen., Sumatra, Borneo, Bangka, Bunguran (Natuna Is.)
○ ***Batrachostomus moniliger*** Blyth, 1849 SRI LANKAN FROGMOUTH	SW India, Sri Lanka

○ ***Batrachostomus hodgsoni*** HODGSON'S FROGMOUTH

— *B.h.hodgsoni* (G.R. Gray, 1859)[6]	NE India, E Bangladesh, W and N Burma
— *B.h.indochinae* Stresemann, 1937	E Burma, N Thailand, S Yunnan, NW Laos, C Vietnam

○ ***Batrachostomus poliolophus*** SHORT-TAILED FROGMOUTH[7]

— *B.p.poliolophus* E. Hartert, 1892 (Apr.)	Mts. of Sumatra
— *B.p.mixtus* Sharpe, 1892 (Nov.)	Mts. of Borneo

○ ***Batrachostomus javensis*** HORSFIELD'S FROGMOUTH[8]

— *B.j.continentalis* Stresemann, 1937	SE Burma, NW and S Thailand (south to N Malay Pen.), S Laos, C Vietnam
— *B.j.affinis* Blyth, 1847	SE Thailand, S Malay Pen., Sumatra, Borneo
— *B.j.javensis* (Horsfield, 1821)	W and C Java
— *B.j.chaseni* Stresemann, 1937	Palawan, Banggi and Calamianes Is[9].

---

[1] For recognition see Schodde & Mason (1980) (2186).
[2] We follow Schodde & Mason (1980) (2186); another opinion is offered by Holyoak (2001) (1131A).
[3] This treatment follows Ford (1986) (862) and Schodde & Mason (1997) (2188).
[4] For reasons to use 1802 not 1801 see Browing & Monroe (1991) (258).
[5] Includes *lilae* Deignan, 1951 (648), see Schodde & Mason (1997) (2188).
[6] Includes *rupchandi* Koelz, 1954 (1284), see Ripley (1982: 289) (2055). Holyoak (2001) (1131A) treated the species as monotypic.
[7] May represent two species see Peters (1940) (1825). So treated by Marshall (1978) (1439), but no acoustic evidence then available.
[8] For comments on why *affinis* was at times treated as a separate species see Marshall (1978) (1439); but see also Wells (1999: 442) (2579).
[9] Sometimes treated as *javensis*, e.g. by Holyoak (2001) (1131A), but Marshall (1978) (1439) thought these birds matched *affinis*. *Batrachostomus pygmaeus* Alviola, 1997 (39), proposed as a new species #, is here considered a synonym of *B.j.chaseni* see Holyoak (1999: 287) (1131).

○ *Batrachostomus cornutus* Sundan Frogmouth
  — *B.c.cornutus* (Temminck, 1822)   Sumatra, Borneo, Bangka, Belitung Is.
  — *B.c.longicaudatus* Hoogerwerf, 1962   Kangean I. (Java Sea) (1141)

STEATORNITHIDAE OILBIRD (1: 1)

STEATORNIS Humboldt, 1814 M
○ *Steatornis caripensis* Humboldt, 1817 Oilbird   Panama, Colombia, Venezuela, the Guianas, Trinidad
                                                  and N Brazil; in Andes south to Bolivia

NYCTIBIIDAE POTOOS (1: 7)

NYCTIBIUS Vieillot, 1816 M
○ *Nyctibius grandis* Great Potoo
  — *N.g.guatemalensis* Land & Schultz, 1963[1]   S Mexico, Guatemala (1320)
  — *N.g.grandis* (J.F. Gmelin, 1789)   Nicaragua to Panama, Sth. America south to C Bolivia
                                             and S Brazil

○ *Nyctibius aethereus* Long-tailed Potoo
  — *N.a.chocoensis* Chapman, 1921   W Colombia
  — *N.a.longicaudatus* (Spix, 1825)   Amazonia
  — *N.a.aethereus* (Wied, 1820)   SE Brazil, SE Paraguay, NE Argentina

○ *Nyctibius jamaicensis* Northern Potoo[2]
  — *N.j.lambi* J. Davis, 1959[3]   WC Mexico (616)
  — *N.j.mexicanus* Nelson, 1900   E and S Mexico to Honduras
  — *N.j.costaricensis* Ridgway, 1912[4]   NW Costa Rica
  — *N.j.jamaicensis* (J.F. Gmelin, 1789)   Jamaica
  — *N.j.abbotti* Richmond, 1917   Hispaniola, Gonave I.

○ *Nyctibius griseus* Common Potoo
  — *N.g.panamensis* Ridgway, 1912   E Nicaragua to W Panama and Colombia and Ecuador
                                             (W of the Andes)
  — *N.g.griseus* (J.F. Gmelin, 1789)[5]   N and C Sth. America (E of the Andes)

○ *Nyctibius maculosus* Ridgway, 1912 Andean Potoo[6]   Andes from Venezuela to Bolivia

○ *Nyctibius leucopterus* (Wied, 1821) White-winged Potoo   The Guianas and Amazonia

○ *Nyctibius bracteatus* Gould, 1846 Rufous Potoo   The Guianas and Amazonia

CAPRIMULGIDAE NIGHTJARS (16: 89)[7]

CHORDEILINAE

LUROCALIS Cassin, 1851 M
○ *Lurocalis semitorquatus* Short-tailed Nighthawk/Semi-collared Nighthawk
  — *L.s.stonei* Huber, 1923   SE Mexico, NE Guatemala, N Honduras, NE Nicaragua
  — *L.s.noctivagus* Griswold, 1936   Costa Rica, Panama, W Colombia, NW Ecuador
  — *L.s.semitorquatus* (J.F. Gmelin, 1789)   N Colombia to the Guianas, N Brazil
  — *L.s.schaeferi* Phelps & Phelps, Jr., 1952   N Venezuela (Aragua) (1850)
  — *L.s.nattereri* (Temminck, 1822)   C Ecuador, N and E Peru, C and S Brazil, Bolivia,
                                            Paraguay, N Argentina

○ *Lurocalis rufiventris* Taczanowski, 1884 Rufous-bellied Nighthawk[8]

                                               Andes from Venezuela to Bolivia

---

[1] For recognition see Cohn-Haft (1999) (299) (542).
[2] For separation from *N. griseus* see A.O.U. (1998: 274) (3).
[3] For recognition see Cleere (1998: 139) (535), but see also Cohn-Haft (1999: 299) (542).
[4] Treated as a race of *griseus* by Cleere (1998: 141) (535), but moved to this species by Cohn-Haft (1999: 299) (542), who considered it doubtfully valid.
[5] Includes *cornutus* see Cohn-Haft (1999: 300) (542).
[6] For reasons to separate from *N. leucopterus* see Schulenberg *et al.* (1984) (2214).
[7] We have followed Cleere (1999) (536) as regards subfamilies and generic placements. Holyoak (2001) (1131A) differs in several respects; his sequences within genera also differ.
[8] For reasons to separate from *L. semitorquatus* see Parker *et al.* (1991) (1750).

CHORDEILES Swainson, 1832 M

○ ***Chordeiles pusillus*** Least Nighthawk

   — *C.p.septentrionalis* (Hellmayr, 1908)   C and E Venezuela, NE Colombia, Guyana, W Surinam

   — *C.p.esmeraldae* Zimmer & Phelps, 1947   E Colombia, S Venezuela, NW Brazil (2724)

   — *C.p.xerophilus* Dickerman, 1988   NE Brazil (719)

   — *C.p.novaesi* Dickerman, 1988   NE Brazil (719)

   — *C.p.pusillus* Gould, 1861   E Brazil

   — *C.p.saturatus* Pinto & Camargo, 1957   WC Brazil, E Bolivia and (?) NE Argentina (1898)

○ ***Chordeiles vielliardi*** (Lencioni-Neto, 1994) Bahian Nighthawk #   E Brazil (1351)

○ ***Chordeiles rupestris*** Sand-coloured Nighthawk[1]

   — *C.r.xyostictus* Oberholser, 1914   C Colombia (Cundinamarca)

   — *C.r.rupestris* (Spix, 1825)   Amazonia

● ***Chordeiles acutipennis*** Lesser Nighthawk

   — *C.a.texensis* Lawrence, 1856   SW USA, N and C Mexico >> C America, W Colombia

   — *C.a.littoralis* Brodkorb, 1940   S Mexico to Costa Rica (221)

   — *C.a.micromeris* Oberholser, 1914   N Yucatán Pen. (Mexico) >>Costa Rica, Panama

   — *C.a.acutipennis* (Hermann, 1783)   N and C Sth. America

   — *C.a.crissalis* A.H. Miller, 1959   SW Colombia (1582)

   — *C.a.aequatorialis* Chapman, 1923   W Colombia, W Ecuador

   — *C.a.exilis* (Lesson, 1839)   W Peru

● ***Chordeiles minor*** Common Nighthawk

   ✓ *C.m.minor* (J.R. Forster, 1771)   Canada, C and E USA >> S Sth. America

   — *C.m.hesperis* Grinnell, 1905[2] [3]   SW Canada, W USA >> C Sth. America

   — *C.m.sennetti* Coues, 1888   NW USA >> C Sth. America

   — *C.m.howelli* Oberholser, 1914   WC USA >> C Sth. America

   — *C.m.henryi* Cassin, 1855   SW USA >> C Sth. America

   ✓ *C.m.aserriensis* Cherrie, 1896   SE Texas, NE Mexico >> C Sth. America

   — *C.m.chapmani* Coues, 1888   SE USA >> C Sth. America

   — *C.m.neotropicalis* Selander & Alvarez del Toro, 1955[4]   E and S Mexico >> C Sth. America (2221)

   — *C.m.panamensis* Eisenmann, 1962   S Central America >> Sth. America (795)

○ ***Chordeiles gundlachii*** Antillean Nighthawk

   — *C.g.vicinus* Riley, 1903   SE USA, Bahama Is.

   — *C.g.gundlachii* Lawrence, 1856   Cuba, Jamaica, Hispaniola, Puerto Rico

NYCTIPROGNE Bonaparte, 1857 F

○ ***Nyctiprogne leucopyga*** Band-tailed Nighthawk

   — *N.l.exigua* Friedmann, 1945   E Colombia, S Venezuela (879)

   — *N.l.pallida* Phelps & Phelps, Jr., 1952   W and C Venezuela (1850)

   — *N.l.leucopyga* (Spix, 1825)   E Venezuela, the Guianas, N Brazil

   — *N.l.latifascia* Friedmann, 1945   S Venezuela (879)

   — *N.l.majuscula* Pinto & Camargo, 1952   NE Peru, E Bolivia, W Brazil, N Paraguay (1896)

PODAGER Wagler, 1832 M

○ ***Podager nacunda*** Nacunda Nighthawk

   — *P.n.minor* Cory, 1915   N Sth. America

   — *P.n.nacunda* (Vieillot, 1817)   E Peru and C Brazil to C Argentina

EUROSTOPODINAE

EUROSTOPODUS Gould, 1838 M

○ ***Eurostopodus argus*** E. Hartert, 1892 Spotted Nightjar[5]   Australia >> islands in Arafura Sea

---

[1] Holyoak (2001) (1131A) treated this species as monotypic.
[2] Includes *divisus* Oberholser, 1974 (1672), see Browning (1990) (251).
[3] Presumably includes *twomeyi* Hawkins, 1948 (1089).
[4] For recognition see Cleere (1998: 167) (535).
[5] Includes *harterti*, *gilberti* Deignan, 1950 (647), and *insulanus* Deignan, 1950 (647), see Schodde & Mason (1980) (2186).

○ *Eurostopodus mystacalis* WHITE-THROATED NIGHTJAR[1]
  — *E.m.mystacalis* (Temminck, 1826)   E Australia >> New Guinea
  — *E.m.nigripennis* E.P. Ramsay, 1881   Solomon Is.
  — *E.m.exul* Mayr, 1941   New Caledonia (1471)

○ *Eurostopodus diabolicus* Stresemann, 1931 HEINRICH'S NIGHTJAR   N and C Sulawesi

○ *Eurostopodus papuensis* (Schlegel, 1866) PAPUAN NIGHTJAR[2]   Salawati I., lowland New Guinea

○ *Eurostopodus archboldi* (Mayr & Rand, 1935) ARCHBOLD'S NIGHTJAR   Upper montane New Guinea

○ *Eurostopodus temminckii* (Gould, 1838) MALAYSIAN EARED NIGHTJAR   S Malay Pen., Sumatra, Nias, Bangka, Belitung, Borneo

○ *Eurostopodus macrotis* GREAT EARED NIGHTJAR
  — *E.m.cerviniceps* (Gould, 1838)   S and E Assam east to Burma, Yunnan and mainland
                                                        SE Asia south to Kedah (W Malaysia)
  — *E.m.bourdilloni* (Hume, 1875)   SW India
  — *E.m.macrotis* (Vigors 1831)   Philippines
  — *E.m.jacobsoni* (Junge, 1936)[3]   Simeuluĕ I. (W Sumatra)
  — *E.m.macropterus* (Bonaparte, 1850)   Sulawesi, Sangihe and Peleng Is.

CAPRIMULGINAE

VELES Bangs, 1918 M
○ *Veles binotatus* (Bonaparte, 1850) BROWN NIGHTJAR[4]   Ghana to S Cameroon, N Gabon and C Zaire

NYCTIDROMUS Gould, 1838 M
◉ *Nyctidromus albicollis* COMMON PAURAQUE
  — *N.a.insularis* Nelson, 1898   Tres Marias Is. (W Mexico)
  ⏌ *N.a.merrilli* Sennett, 1888   S Texas, NE Mexico
  — *N.a.yucatanensis* Nelson, 1901   W and E Mexico to C Guatemala
  — *N.a.intercedens* Griscom, 1929[5]   S Guatemala to W Panama
  — *N.a.gilvus* Bangs, 1902   C and E Panama, N Colombia
  — *N.a.albicollis* (J.F. Gmelin, 1789)   E and S Venezuela to Peru and E Brazil
  — *N.a.derbyanus* Gould, 1838   C and S Brazil, Paraguay, NE Argentina, N and E Bolivia

PHALAENOPTILUS Ridgway, 1880 M
○ *Phalaenoptilus nuttallii* COMMON POORWILL
  — *P.n.nuttallii* (Audubon, 1844)   W and WC USA >> C Mexico
  — *P.n.californicus* Ridgway, 1887   W California, NW Baja California
  — *P.n.hueyi* Dickey, 1928   SE California, SW Arizona
  — *P.n.dickeyi* Grinnell, 1928   S Baja California
  — *P.n.adustus* van Rossem, 1941   S Arizona to C Sonora (2477)
  — *P.n.centralis* R.T. Moore, 1947   C Mexico (1617)

SIPHONORHIS P.L. Sclater, 1861 F
○ †*Siphonorhis americana* (Linnaeus, 1758) JAMAICAN PAURAQUE[6]   Jamaica

○ *Siphonorhis brewsteri* (Chapman, 1917) LEAST PAURAQUE   Hispaniola, Gonave I.

NYCTIPHRYNUS Bonaparte, 1857[7] M
○ *Nyctiphrynus mcleodii* EARED POORWILL
  — *N.m.mcleodii* (Brewster, 1888)   Chihuahua, Sinaloa, Jalisco, Colima (Mexico)
  — *N.m.rayi* (A.H. Miller, 1948)   Guerrero (Mexico) (1581)[8]

○ *Nyctiphrynus yucatanicus* (E. Hartert, 1892) YUCATÁN POORWILL   SE Mexico, N Guatemala, N Belize

---

[1] The prior name *guttatus* Vigors & Horsfield, 1827 – not 1826, see Raphael (1970) (1981) — applied by Mathews to specimens thought to have been named *argus* has been shown by Schodde & Mason (1980) (2186) to be founded on *E. mystacalis* (Temminck, 1826).
[2] Includes *astrolabae* see Peters (1940: 191) (1825). The subspecies is recognised by Holyoak (2001) (1331A).
[3] Acoustic evidence may show this isolate to belong, as zoogeographic logic suggests, or be closest to *E. temmincki* (Wells, 1999) (2579).
[4] For a defence of the validity of this genus see Cleere (2001) (537).
[5] Tentatively listed by Cleere (1998: 186) (535). Placed in synonymy of *albicollis* by Holyoak (2001) (1131A).
[6] For correction of spelling see David & Gosselin (2002) (614).
[7] Includes *Otophanes* see A.O.U. (1998: 270) (3).
[8] Not recognised by Holyoak (2001) (1131A).

○ *Nyctiphrynus ocellatus* OCELLATED POORWILL
　— *N.o.lautus* Miller & Griscom, 1925　　　　　E Honduras, NE Nicaragua, NW Costa Rica
　— *N.o.ocellatus* (Tschudi, 1844)　　　　　　　SE Colombia to E Brazil, south to NE Argentina

○ *Nyctiphrynus rosenbergi* (E. Hartert, 1895) CHOCO POORWILL[1]　W Colombia, NW Ecuador

CAPRIMULGUS Linnaeus, 1758 M
○ *Caprimulgus carolinensis* J.F. Gmelin, 1789 CHUCK-WILL'S WIDOW　　EC and S USA >> E Mexico, C and N Sth. America

○ *Caprimulgus rufus* RUFOUS NIGHTJAR
　— *C.r.minimus* Griscom & Greenway, 1937[2]　　S Costa Rica to Venezuela and N Colombia, Trinidad
　— *C.r.rufus* Boddaert, 1783　　　　　　　　　S Venezuela, the Guianas, N Brazil
　— *C.r.otiosus* (Bangs, 1911)[3]　　　　　　　　NE St Lucia
　— *C.r.rutilus* (Burmeister, 1856)　　　　　　　C and S Brazil, Bolivia, Paraguay, N Argentina
　— *C.r.saltarius* Olrog, 1979[4]　　　　　　　　NW Argentina (1686)

○ *Caprimulgus cubanensis* GREATER ANTILLEAN NIGHTJAR[5]
　— *C.c.cubanensis* (Lawrence, 1860)　　　　　　Cuba
　— *C.c.insulaepinorun* Garrido, 1983[6]　　　　　Isla de la Juventud (922)
　— *C.c.ekmani* (Lönnberg, 1929)　　　　　　　　Hispaniola

○ *Caprimulgus salvini* E. Hartert, 1892 TAWNY-COLLARED NIGHTJAR　　E Mexico

○ *Caprimulgus badius* (Bangs & Peck, 1908) YUCATÁN NIGHTJAR　　Yucatán Pen. (Mexico)

○ *Caprimulgus sericocaudatus* SILKY-TAILED NIGHTJAR
　— *C.s.mengeli* Dickerman, 1975　　　　　　　　N Brazil, N Peru, N Bolivia (709)
　— *C.s.sericocaudatus* (Cassin, 1849)　　　　　　SE Brazil, E Paraguay, NE Argentina

○ *Caprimulgus ridgwayi* BUFF-COLLARED NIGHTJAR
　— *C.r.ridgwayi* (Nelson, 1897)　　　　　　　　W Mexico
　— *C.r.troglodytes* Griscom, 1930　　　　　　　C Guatemala, C Honduras, C Nicaragua

○ *Caprimulgus vociferus* WHIP-POOR-WILL
　— *C.v.vociferus* A. Wilson, 1812　　　　　　　　S Canada, E USA >> C America
　— *C.v.arizonae* (Brewster, 1881)　　　　　　　SW USA, N and C Mexico >> C Mexico
　— *C.v.setosus* van Rossem, 1934　　　　　　　E Mexico
　— *C.v.oaxacae* (Nelson, 1900)　　　　　　　　SW Mexico
　— *C.v.chiapensis* (Nelson, 1900)　　　　　　　SE Mexico, Guatemala
　— *C.v.vermiculatus* (Dickey & van Rossem, 1928)　Honduras, El Salvador

○ *Caprimulgus noctitherus* (Wetmore, 1919) PUERTO RICAN NIGHTJAR　　SW Puerto Rico

○ *Caprimulgus saturatus* (Salvin, 1870) DUSKY NIGHTJAR　　Costa Rica, W Panama

○ *Caprimulgus longirostris* BAND-WINGED NIGHTJAR
　— *C.l.ruficervix* (P.L. Sclater, 1866)　　　　　Colombia, Venezuela, Ecuador
　— *C.l.roraimae* (Chapman, 1929)　　　　　　　The Pantepui (S Venezuela)
　— *C.l.decussatus* Tschudi, 1844　　　　　　　SW Peru, N Chile
　— *C.l.atripunctatus* (Chapman, 1923)　　　　　Peru, Bolivia, N Chile, NW Argentina
　— *C.l.bifasciatus* Gould, 1837　　　　　　　　Chile, W Argentina
　— *C.l.longirostris* Bonaparte, 1825　　　　　　SE Brazil, NW and C Paraguay, NE Argentina, Uruguay
　— *C.l.patagonicus* Olrog, 1962[7]　　　　　　　C and S Argentina (1680)

○ *Caprimulgus cayennensis* WHITE-TAILED NIGHTJAR
　— *C.c.albicauda* (Lawrence, 1875)　　　　　　Costa Rica to N Colombia
　— *C.c.apertus* J.L. Peters, 1940　　　　　　　W Colombia
　— *C.c.cayennensis* J.F. Gmelin, 1789　　　　　C and NE Colombia, Venezuela, the Guianas, N Brazil

[1] For reasons to separate this from *N. ocellatus* see Robbins & Ridgely (1992) (2082).
[2] Thought to include *maximus* Bond & Meyer de Schauensee, 1940 (190), and *noctivigulus* Wetmore & Phelps, 1953 (2613), but both may require re-evaluation. They are recognised by Holyoak (2001) (1131A).
[3] Placed here following A.O.U. (1998: 271) (3).
[4] We note that Cleere (1998: 199) (535) moves this taxon here, from *C. sericocaudatus*, but with reservations and we share them (we omitted it last Edition). Holyoak (2001) (1131A) made this a synonym of *C.r.rutilus*.
[5] We defer splitting *ekmani* as did A.O.U. (1998: 271) (3).
[6] Tentatively accepted; requires evaluation, measurements overlap those of nominate. See Cleere (1998: 200) (535).
[7] Omitted from last Edition.

___ *C.c.insularis* (Richmond, 1902) Curacao, Bonaire, Margarita Is., N Venezuela, NE Colombia

___ *C.c.leopetes* Jardine & Selby, 1830 Trinidad, Tobago I.[1]
___ *C.c.manati* (Pinchon, 1963)[2] Martinique (1886)

○ *Caprimulgus candicans* (Pelzeln, 1867) WHITE-WINGED NIGHTJAR C and SW Brazil, E Paraguay, E Bolivia

○ *Caprimulgus maculicaudus* (Lawrence, 1862) SPOT-TAILED NIGHTJAR SE Mexico to S Brazil and E Paraguay

○ *Caprimulgus parvulus* LITTLE NIGHTJAR
___ *C.p.heterurus* (Todd, 1915) N Colombia, N and C Venezuela
___ *C.p.parvulus* Gould, 1837 E Peru to E Brazil and C Argentina

○ *Caprimulgus anthonyi* (Chapman, 1923) SCRUB NIGHTJAR[3] W Ecuador, NW Peru

○ †?*Caprimulgus maculosus* (Todd, 1920) CAYENNE NIGHTJAR Mana River, French Guiana

○ *Caprimulgus nigrescens* Cabanis, 1848 BLACKISH NIGHTJAR[4] Amazonian Sth. America

○ *Caprimulgus whitelyi* (Salvin, 1885) RORAIMAN NIGHTJAR Mt. Roraima and tepuis of S Venezuela

○ *Caprimulgus hirundinaceus* PYGMY NIGHTJAR
___ *C.h.cearae* (Cory, 1917) NE Brazil
___ *C.h.hirundinaceus* Spix, 1825 E Brazil
___ *C.h.vielliardi* Ribon, 1995 EC Brazil (2014)

○ *Caprimulgus ruficollis* RED-NECKED NIGHTJAR
___ *C.r.ruficollis* Temminck, 1820 Iberia and W Morocco >> W Africa
___ *C.r.desertorum* Erlanger, 1899 NE Morocco, Algeria, Tunisia >> W Africa

○ *Caprimulgus indicus* GREY NIGHTJAR
___ *C.i.jotaka* Temminck & Schlegel, 1845[5] Transbaikalia and Japan S to C and SE China >> S China to mainland SE Asia, Indonesia and Philippines
___ *C.i.hazarae* Whistler & Kinnear, 1935[6] NW Pakistan, Himalayas, S Assam, Burma, SW China (E Xizang, NW Yunnan) >> mainland SE Asia south to Malay Pen.
___ *C.i.indicus* Latham, 1790 C and S India
___ *C.i.kelaarti* Blyth, 1852 Sri Lanka[7]
___ *C.i.phalaena* Hartlaub & Finsch, 1872 Palau Is. (WC Pacific)

○ *Caprimulgus europaeus* EUROPEAN NIGHTJAR[8]
___ *C.e.europaeus* Linnaeus, 1758 C and N Europe, C and N Siberia and N Kazakhstan to Lake Baikal >> C and S Africa
___ *C.e.meridionalis* E. Hartert, 1896 N Africa and S Europe to Crimea and Caucasus area >> W, C and S Africa
___ *C.e.sarudnyi* E. Hartert, 1912 W Siberia, C Asia >> E and SE Africa
___ *C.e.unwini* Hume, 1871 Iraq and Transcaspia to W Tien Shan, W Tarim Basin and W and N Pakistan >> SE Africa and NW India
___ *C.e.plumipes* Przevalski, 1876 E Tien Shan to SW and S Mongolia >> SE Africa
___ *C.e.dementievi* Stegmann, 1949 NE Mongolia, S Transbaikalia >> SE Africa (2313)

○ *Caprimulgus fraenatus* Salvadori, 1884 SOMBRE NIGHTJAR[9] Eritrea and Ethiopia to C and S Kenya and NE Tanzania

○ *Caprimulgus rufigena* RUFOUS-CHEEKED NIGHTJAR
___ *C.r.damarensis* Strickland, 1852 N Cape Province, Namibia, Botswana and coastal W Angola >> W Africa
___ *C.r.rufigena* A. Smith, 1845 W and S Cape to Transvaal, Zimbabwe, SW Zambia and S Angola >> W Africa

---

[1] Placed in the synonymy of the nominate form by Holyoak (2001) (1131A).
[2] Omitted from last Edition.
[3] For reasons to separate from *C. parvulus* see Robbins *et al.*, (1994) (2085).
[4] Includes *australis* Gyldenstolpe, 1941 (1029).
[5] This name dates from 1844 or 1845 see Holthuis & Sakai (1970) (1128); not from 1847 as given by Peters (1940: 204) (1825).
[6] Includes *memnon* Koelz, 1954 (1284), see Ripley (1982: 191). Both *memnon* and *hazarae* as placed in the nominate form by Holyoak (2001) (1131A).
[7] Treated as a synonym of the nominate form by Holyoak (2001) (1131A).
[8] We follow Cramp *et al.* (1985: 620) (591).
[9] Includes *keneiensis* see Cleere (1999: 361) (536).

○ *Caprimulgus aegyptius* EGYPTIAN NIGHTJAR
  — *C.a.saharae* Erlanger, 1899   NW Africa east to W Egypt >> Senegal to Nigeria
  — *C.a.aegyptius* M.H.K. Lichtenstein, 1823[1]   Nile Valley and Delta of Egypt to Middle East and WC
                                       Asia >> E Sahel to Sudan

○ *Caprimulgus mahrattensis* Sykes, 1832 SYKES'S NIGHTJAR   SE Iran and S Afghanistan to NW India >> N and C India

○ *Caprimulgus centralasicus* Vaurie, 1960 VAURIE'S NIGHTJAR[2]   SW Xinjiang (W China) (2505)

○ *Caprimulgus nubicus* NUBIAN NIGHTJAR
  — *C.n.tamaricis* Tristram, 1864   Dead Sea to Yemen >> NE Africa
  — *C.n.nubicus* M.H.K. Lichtenstein, 1823   W and C Sudan
  — *C.n.torridus* Lort Phillips, 1898[3]   Somalia to N Tanzania
  — *C.n.jonesi* Ogilvie-Grant & Forbes, 1899   Socotra I.

○ *Caprimulgus eximius* GOLDEN NIGHTJAR
  — *C.e.simplicior* E. Hartert, 1921   S Mauritania to C Chad
  — *C.e.eximius* Temminck, 1826   W and C Sudan

○ *Caprimulgus atripennis* JERDON'S NIGHTJAR
  — *C.a.atripennis* Jerdon, 1845   S India
  — *C.a.aequabilis* Ripley, 1945[4]   Sri Lanka (2028)

○ *Caprimulgus macrurus* LARGE-TAILED NIGHTJAR
  — *C.m.albonotatus* Tickell, 1833[5]   NE Pakistan, N and NE India
  — *C.m.bimaculatus* Peale, 1848[6]   S Assam east to S Yunnan and Hainan and S through
                                       SE Asia to Malay Pen. and Sumatra
  — *C.m.andamanicus* Hume, 1873   Andaman Is.
  — *C.m.macrurus* Horsfield, 1821   Java, Bali
  — *C.m.salvadori* Sharpe, 1875[7]   Borneo, Banggi and S Sulu Is.
  — *C.m.johnsoni* Deignan, 1955[8]   Palawan (656)
  — *C.m.schlegelii* A.B. Meyer, 1874[9]   Lesser Sundas north to Salayar and Tanahjampea Is.,
                                       Moluccas to New Guinea, Bismarck Arch. and
                                       Lihir I., N and NE Australia

○ *Caprimulgus manillensis* Walden, 1875 PHILIPPINE NIGHTJAR[10]   Philippines (except Palawan and Sulu groups)

○ *Caprimulgus celebensis* SULAWESI NIGHTJAR[11]
  — *C.c.celebensis* Ogilvie-Grant, 1894   Sulawesi
  — *C.c.jungei* Neumann, 1939   Sula Is. (1645)

○ *Caprimulgus donaldsoni* Sharpe, 1895 DONALDSON-SMITH'S NIGHTJAR
                                       Somalia and E Ethiopia to E Kenya and NE Tanzania

○ *Caprimulgus nigriscapularis* Reichenow, 1893 BLACK-SHOULDERED NIGHTJAR
                                       Guinea to S Sudan, NE and E Zaire and Uganda

○ *Caprimulgus pectoralis* FIERY-NECKED NIGHTJAR[12]
  — *C.p.shelleyi* Bocage, 1879   Angola and S Zaire to N Malawi, C and NE Tanzania
                                       and SE Kenya
  — *C.p.fervidus* Sharpe, 1875   S Angola, and N Namibia to NE Sth. Africa to Zimbabwe
  — *C.p.crepusculans* Clancey, 1994   Mozambique to Natal (525)
  — *C.p.pectoralis* Cuvier, 1816[13]   Cape Province

○ *Caprimulgus poliocephalus* Rüppell, 1840 ABYSSINIAN NIGHTJAR[14]   Saudi Arabia, Ethiopia to N Tanzania

---

[1] Includes *arenicolor* Severtsov, 1875 (2225A), see Cramp *et al.* (1985: 646) (591).
[2] For reservations as to its validity see Cleere (1999: 362) (536).
[3] Includes *taruensis* see Zimmermann *et al.* (1996) (2735); but accepted by Holyoak (2001) (1131A).
[4] Listed as a race of *C. macrurus* in last Edition (with *albonotatus* treated as a race of *C. atripennis*), but see Ripley & Beehler (1987) (2061). Holyoak (2001) (1131A) placed it in the synonymy of the nominate form making *C. atripennis* a monotypic species.
[5] Includes *noctuvigilis* Koelz, 1954 (1284), see Ripley (1982: 192) (2055).
[6] Includes *silvanus* Koelz, 1954 (1284), see Ripley (1982: 193) (2055) and *hainanus* see Mees (1977) (1529).
[7] Omitted last Edition, but see Mees (1977) (1529).
[8] For recognition see Mees (1977) (1529).
[9] Includes *obiensis* Jany, 1955 (1200), see Mees (1977) (1529).
[10] For reasons to separate this (including *celebensis*) from *macrurus* see Mees (1986) (1537).
[11] For reasons to separate from *manillensis*, here accepted tentatively, see Rozendaal (1990) (2121).
[12] Polytypic treatment follows Cleere (1998) (535).
[13] Given as 1817 by Peters (1940: 208) (1825) but see Browning & Monroe (1991) (258).
[14] Treated as polytypic in last Edition; the races then attached to it have been transferred to *C. ruwenzori* by Cleere (1998: 260) (535), except that he kept *koesteri* in synonymy; it was recognised by White (1965: 204) (2649) and is apparently very isolated. Holyoak (2001) (1131A) agreed with Cleere.

○ *Caprimulgus ruwenzorii* MONTANE NIGHTJAR
— *C.r.ruwenzorii* Ogilvie-Grant, 1908 — SW Uganda, E Zaire, Burundi, Rwanda
— *C.r.guttifer* Grote, 1921 — NE and SW Tanzania, N Malawi, NE Zambia
— *C.r.koesteri* Neumann, 1931 — Highlands of C Angola

○ *Caprimulgus asiaticus* INDIAN NIGHTJAR
— *C.a.asiaticus* Latham, 1790[1] — S Pakistan, India, Burma, Thailand (except S), C and S Laos, C and S Vietnam, Cambodia
— *C.a.eidos* J.L. Peters, 1940 — Sri Lanka

○ *Caprimulgus madagascariensis* MADAGASCAN NIGHTJAR
— *C.m.aldabrensis* Ridgway, 1894 — Aldabra
— *C.m.madagascariensis* Sganzin, 1840 — Madagascar

○ *Caprimulgus natalensis* SWAMP NIGHTJAR
— *C.n.accrae* Shelley, 1875 — Liberia to W Cameroon
— *C.n.natalensis* A. Smith, 1845[2] — Mali and Sudan to Central Africa, Gabon and E Sth. Africa

○ *Caprimulgus solala* Safford, Ash, Duckworth, Telfer & Zewdie, 1995 NECHISAR NIGHTJAR[3] #
Nechisar Plains (?) (S Ethiopia) (2147)

○ *Caprimulgus inornatus* Heuglin, 1869 PLAIN NIGHTJAR[4] — S Mauritania and N Senegal to Ethiopia, NW Somalia and SW Arabia >> Liberia, N Zaire and C Tanzania

○ *Caprimulgus stellatus* Blundell & Lovat, 1899 STAR-SPOTTED NIGHTJAR — Djibouti and Ethiopia, to N Kenya and SE Sudan

○ *Caprimulgus affinis* SAVANNA NIGHTJAR[5]
— *C.a.monticolus* Franklin, 1831[6] — NE Pakistan, India, SE Xizang, Burma, Thailand (not S)
— *C.a.amoyensis* E.C.S. Baker, 1931 — S and SE China
— *C.a.stictomus* Swinhoe, 1863 — Indochina, Taiwan
— *C.a.affinis* Horsfield, 1821 — S tip of Malay Pen., Sumatra, Nias, Bangka, Belitung, Borneo, Java, Bali, Lombok
— *C.a.undulatus* Mayr, 1944 — Flores, Sumbawa (1474)
— *C.a.kasuidori* Hachisuka, 1932 — Savu, Sumba Is.
— *C.a.timorensis* Mayr, 1944 — Alor, Timor, Roti and Kisar Is. (1474)
— *C.a.griseatus* Walden, 1875 — N and C Philippines
— *C.a.mindanensis* Mearns, 1905 — Mindanao
— *C.a.propinquus* Riley, 1918 — C and S Sulawesi

○ *Caprimulgus tristigma* FRECKLED NIGHTJAR
— *C.t.sharpei* Alexander, 1901 — Senegal to C Central African Republic
— *C.t.pallidogriseus* Parker & Benson, 1971 — Nigeria (1743)
— *C.t.tristigma* Rüppell, 1840 — S Sudan and Ethiopia to Burundi and N Tanzania
— *C.t.granosus* Clancey, 1965[7] — S Tanzania and SE Zaire to NE and E Sth. Africa (403)
— *C.t.lentiginosus* A. Smith, 1845 — Angola to W Sth. Africa

○ *Caprimulgus concretus* Bonaparte, 1850 BONAPARTE'S NIGHTJAR — Sumatra, Borneo, Belitung I.

○ *Caprimulgus pulchellus* SALVADORI'S NIGHTJAR
— *C.p.pulchellus* Salvadori, 1879 — Mts. of Sumatra
— *C.p.bartelsi* Finsch, 1902 — Mts. of Java

○ *Caprimulgus prigoginei* Louette, 1990 PRIGOGINE'S NIGHTJAR — Lake Kivu dist., (E Zaire) (1379)

○ *Caprimulgus enarratus* G.R. Gray, 1871 COLLARED NIGHTJAR — NW, N and E Madagascar

○ *Caprimulgus batesi* Sharpe, 1906 BATES'S NIGHTJAR — S Cameroon to C and E Zaire

[1] Includes *gurgaoni* Koelz, 1939 (1279), see Ripley (1982: 193) (2055), and *siamensis* see Deignan (1963) (663). Holyoak (2001) (1131A) also included *eidos* making a monotypic species.
[2] Includes *carpi* Smithers, 1954 (2284), and *mpasa* Smithers, 1954 (2283), recognised by White (1965) (2649), but see Harwin (1983) (1088). Detailed review required.
[3] Described, and still only known, from a wing. Requires substantiation.
[4] Includes *malbranti* Niethammer, 1957 (1653), see White (1965: 206) (2649). Also includes *ludovicianus* see Cleere (1999: 371) (536).
[5] Two species in last Edition, but see Mayr (1944: 153) (1474).
[6] Includes *burmanicus* see Ticehurst (1939) (2307).
[7] For recognition of this see Fry & Harwin (1988: 181) (896).

○ *Caprimulgus climacurus* Long-tailed Nightjar[1]
  — *C.c.climacurus* Vieillot, 1825 — Senegal to NW Ethiopia
  — *C.c.nigricans* (Salvadori, 1868) — Upper Nile (C Sudan)
  — *C.c.sclateri* (Bates, 1927) — Sierra Leone to N Zaire

○ *Caprimulgus clarus* Reichenow, 1892 Slender-tailed Nightjar[2] — Djibouti to C Tanzania

○ *Caprimulgus fossii* Mozambique Nightjar
  — *C.f.fossii* Hartlaub, 1857 — Gabon
  — *C.f.welwitschii* Bocage, 1867 — S Zaire and S Tanzania to Natal >> N Zaire to S Kenya[3]
  — *C.f.griseoplurus* Clancey, 1965 — W Botswana (404) >> ?

MACRODIPTERYX Swainson, 1837 F
○ *Macrodipteryx longipennis* (Shaw, 1796) Standard-winged Nightjar — S Senegal to Uganda >> Sahel belt of W Africa, Sudan and Ethiopia

○ *Macrodipteryx vexillarius* (Gould, 1838) Pennant-winged Nightjar[4] — Angola and S Tanzania to Transvaal >> Nigeria to Zaire, Uganda and W Kenya

HYDROPSALIS Wagler, 1832 F
○ *Hydropsalis climacocerca* Ladder-tailed Nightjar
  — *H.c.schomburgki* P.L. Sclater, 1866 — E Venezuela, the Guianas
  — *H.c.climacocerca* (Tschudi, 1844) — W Amazonia
  — *H.c.pallidior* Todd, 1937 — NC Brazil
  — *H.c.intercedens* Todd, 1937 — C Brazil (Obidos, W Para)
  — *H.c.canescens* Griscom & Greenway, 1937 — E Brazil

○ *Hydropsalis torquata* Scissor-tailed Nightjar
  — *H.t.torquata* (J.F. Gmelin, 1789)[5] — S Surinam, Brazil, EC Peru
  — *H.t.furcifer* (Vieillot, 1817)[6] — E Bolivia and S Brazil to and C Argentina

UROPSALIS W. de W. Miller, 1915[7] F
○ *Uropsalis segmentata* Swallow-tailed Nightjar
  — *U.s.segmentata* (Cassin, 1849) — Andes from Colombia to Ecuador
  — *U.s.kalinowskii* (Berlepsch & Stolzmann, 1894) — Andes from Peru to C Bolivia

○ *Uropsalis lyra* Lyre-tailed Nightjar
  — *U.l.lyra* (Bonaparte, 1850) — Andes from Venezuela to Ecuador
  — *U.l.peruana* (Berlepsch & Stolzmann, 1906) — Andes from Peru to C Bolivia
  — *U.l.argentina* Olrog, 1975[8] — Andes of N Argentina (1684)

MACROPSALIS P.L. Sclater, 1866 F
○ *Macropsalis creagra* (Bonaparte, 1850) Long-trained Nightjar[9] — SE Brazil, NE Argentina

ELEOTHREPTUS G.R. Gray, 1840 M
○ *Eleothreptus anomalus* (Gould, 1837) Sickle-winged Nightjar — S Brazil, S Paraguay, N and NE Argentina

AEGOTHELIDAE OWLET-NIGHTJARS (2: 9)

EUAEGOTHELES Mathews, 1918[10] M
○ *Euaegotheles insignis* (Salvadori, 1875) Feline Owlet-Nightjar[11] — Mts. of New Guinea

○ *Euaegotheles tatei* (Rand, 1941) Spangled Owlet-Nightjar[12] — Lowland E and C New Guinea (1964)

○ *Euaegotheles crinifrons* (Bonaparte, 1850) Moluccan Owlet-Nightjar — Halmahera, Bacan Is. (Moluccas)

[1] This species, *climacurus*, and the next *fossii*, were listed in the genus *Scotornis* in last Edition.
[2] A note on the recognition of this appears in Safford *et al.* (1995) (2147).
[3] Holyoak (2001) (1131A) split this and recognised a further subspecies.
[4] For comment on spelling see David & Gosselin (2002) (614).
[5] For reasons to use this name in place of *H. brasiliana* Gmelin, 1789, see Pacheco & Whitney (1998) (1737).
[6] For correction of spelling see David & Gosselin (2002) (613).
[7] Holyoak (2001) (1131A) submerges this genus in *Macropsalis*.
[8] Omitted from last Edition.
[9] Cleere (1999: 385) (536) used the name *forcipata* Nitzsch, 1840 (1656A), but used *creagra* in Cleere (1998) (535). This change, about which see Pacheco & Whitney (1998) (1737), is associated with unresolved issues and, although we accept that *forcipata* is not a *nomen oblitum*, we defer its adoption.
[10] The recognition of this genus was proposed at the 2nd Southern Hemisphere Ornithological Congress in Brisbane. Not employed by Holyoak (2001) (1131A).
[11] Includes *pulcher* see Rand (1942) (1966).
[12] For recognition at specific rank see Pratt (2000) (1922); but this is controversial (N. Cleere *in litt.*).

AEGOTHELES Vigors & Horsfield, 1827 M

○ *Aegotheles wallacii* WALLACE'S OWLET-NIGHTJAR
     — *A.w.wallacii* G.R. Gray, 1859   NW to SC New Guinea, Aru Is.
     — *A.w.gigas* Rothschild, 1931   WC New Guinea
     — *A.w.manni* Diamond, 1969   N New Guinea (698)

○ *Aegotheles archboldi* Rand, 1941 ARCHBOLD'S OWLET-NIGHTJAR   Montane C New Guinea (1964)

○ *Aegotheles albertisi* MOUNTAIN OWLET-NIGHTJAR
     — *A.a.albertisi* P.L. Sclater, 1874   NW New Guinea (Mts. of Vogelkop)
     — *A.a.wondiwoi* Mayr & Rand, 1936   NW New Guinea (Mts. at head of Geelvink Bay)
     — *A.a.salvadorii* E. Hartert, 1892   Mts. of C and E New Guinea

○ †?*Aegotheles savesi* E.L. & E.L.C. Layard, 1881 NEW CALEDONIAN OWLET-NIGHTJAR
                                       New Caledonia[1]

○ *Aegotheles bennettii* BARRED OWLET-NIGHTJAR
     — *A.b.affinis* Salvadori, 1875   NW New Guinea (Vokelkop Pen.)
     — *A.b.wiedenfeldi* Laubmann, 1914   N New Guinea
     — *A.b.terborghi* Diamond, 1967   Montane SC New Guinea (697)
     — *A.b.bennettii* Salvadori & D'Albertis, 1875   Aru Is., Lowland SE to SC New Guinea
     — *A.b.plumiferus* E.P. Ramsay, 1883   D'Entrecasteaux Arch.

○ *Aegotheles cristatus* AUSTRALIAN OWLET-NIGHTJAR
     — *A.c.cristatus* (Shaw, 1790)[2]   Australia, S New Guinea
     — *A.c.tasmanicus* Mathews, 1918   Tasmania

APODIDAE SWIFTS (19: 94)

CYPSELOIDINAE

CYPSELOIDES Streubel, 1848 M

○ *Cypseloides cherriei* Ridgway, 1893 SPOT-FRONTED SWIFT   Costa Rica, N Venezuela, NC and SW Colombia, N Ecuador

○ *Cypseloides cryptus* J.T. Zimmer, 1945 WHITE-CHINNED SWIFT   Belize to Panama, Colombia to Peru, S Venezuela, Guyana (2709)

○ *Cypseloides storeri* Navarro, Peterson, Escalante & Benitez, 1992 WHITE-FRONTED SWIFT #
                                         SW Mexico (1637)

○ *Cypseloides fumigatus* (Streubel, 1848) SOOTY SWIFT   E Bolivia, E Paraguay, NW Argentina, SE Brazil

○ *Cypseloides rothschildi* J.T. Zimmer, 1945 ROTHSCHILD'S SWIFT[3]   NW Argentina (2709)

● *Cypseloides niger* BLACK SWIFT[4]
     ✓ *C.n.borealis* (Kennerly, 1858)   SE Alaska to SW USA >> Sth. America
     — *C.n.costaricensis* Ridgway, 1910   C Mexico to Costa Rica
     — *C.n.niger* (J.F. Gmelin, 1789)   W Indies, Trinidad

○ *Cypseloides lemosi* Eisenmann & Lehmann, 1962 WHITE-CHESTED SWIFT
                                       SW Colombia (and perhaps NC Colombia, NE Ecuador, NW Peru) (796)[5]

○ *Cypseloides senex* (Temminck, 1826) GREAT DUSKY SWIFT[6]   C and S Brazil, E Paraguay, N Argentina

STREPTOPROCNE Oberholser, 1906 F

○ *Streptoprocne phelpsi* (Collins, 1972) TEPUI SWIFT   S Venezuela, NW Guyana (549)

○ *Streptoprocne rutila* CHESTNUT-COLLARED SWIFT[7]
     — *S.r.griseifrons* (Nelson, 1900)   W Mexico

---

[1] Recent reports suggest this *is* still extant.
[2] Includes *major* and *leucogaster* see Schodde & Mason (1980) (2186).
[3] Called *Cypseloides major* in last Edition. With the genus *Aerornis* submerged, this 1931 name became preoccupied by the name *Chaetura major* Bertoni, 1900, a synonym of *C. senex*. Under the current Code (I.C.Z.N., 1999) (1178) the name *major* might not require replacement.
[4] The genus *Nephoectes*, used for this species in last Edition, is submerged following A.O.U. (1998: 275) (3).
[5] The bracketed records are sight records (M. Marin *in litt.* 23 Jul. 2001).
[6] In last Edition treated in the genus *Aerornis*.
[7] For reasons to transfer *rutilus* from *Cypseloides* see Marin & Stiles (1992) (1432) and A.O.U. (1998: 276) (3). The sister species *phelpsi* moves with it.

— _S.r.brunnitorques_ (Lafresnaye, 1844)[1] SE Mexico to Bolivia
— _S.r.rutila_ (Vieillot, 1817) Venezuela to Guyana, Trinidad

● _Streptoprocne zonaris_ WHITE-COLLARED SWIFT[2]
    — _S.z.mexicana_ Ridgway, 1910 S Mexico to Belize
    — _S.z.bouchellii_ Huber, 1923 Nicaragua to Panama
    — _S.z.pallidifrons_ (E. Hartert, 1896) Greater and Lesser Antilles
    — _S.z.minor_ (Lawrence, 1882) Coastal N Venezuela, Trinidad
    — _S.z.albicincta_ (Cabinis, 1862) Guyana, Venezuela
    — _S.z.subtropicalis_ Parkes, 1994 Colombia south to Peru (1782)
    — _S.z.altissima_ Chapman, 1914 Colombia, Ecuador
    — _S.z.kuenzeli_ Niethammer, 1953 Bolivia, NW Argentina (1650)
    — _S.z.zonaris_ (Shaw, 1796) S Brazil, Bolivia, C and E Paraguay

○ _Streptoprocne biscutata_ BISCUTATE SWIFT
    — _S.b.seridoensis_ Sick, 1991 NE Brazil (2271)
    — _S.b.biscutata_ (P.L. Sclater, 1865) E to SE Brazil, Paraguay, N Argentina

○ _Streptoprocne semicollaris_ (DeSaussure, 1859) WHITE-NAPED SWIFT W Mexico

APODINAE

HYDROCHOUS Brooke, 1970 M
○ _Hydrochous gigas_ (Hartert & Butler, 1901) WATERFALL SWIFT Malaysia, Sumatra, Borneo, W Java

COLLOCALIA G.R. Gray, 1840 F
○ _Collocalia esculenta_ GLOSSY SWIFTLET[3]
    — _C.e.affinis_ Beavan, 1867 Andaman, Nicobar Is.
    — _C.e.elachyptera_ Oberholser, 1906 Mergui Arch. (S Burma)
    — _C.e.cyanoptila_ Oberholser, 1906 Malay Pen., Sumatra, Belitung I., Borneo, (?) **Palawan group (Philippines)**
    — _C.e.vanderbilti_ Meyer de Schauensee & Ripley, 1940 Nias I. (1579)
    — _C.e.oberholseri_ Stresemann, 1912 Mentawi Is., Batu Is.
    — _C.e.natalis_ Lister, 1889 Christmas I. (Indian Ocean)
    — _C.e.septentrionalis_ Mayr, 1945 Babuyan, Calayan, N Camiguin Is. (off N Luzon) (682)
    — _C.e.isonata_ Oberholser, 1906 N Luzon
    — _C.e.marginata_ Salvadori, 1882 C and S Luzon and E and W Visayas (C Philippines)
    — _C.e.bagobo_ Hachisuka, 1930[4] Mindoro, Mindanao, Bongao (Sulu Arch.)
    — _C.e.manadensis_ Salomonsen, 1983 N Sulawesi to Sangihe, (?) Talaud Is. (2161)
    — _C.e.esculenta_ (Linnaeus, 1758) C and S Sulawesi; Buru, Seram, Obi (C and S Moluccas), Kai Is.
    — _C.e.minuta_ Stresemann, 1925 Tanahdjampea, Kalao Is. (off Sulawesi)
    — _C.e.spilura_ G.R. Gray, 1866[5] Bacan, Halmahera, Morotai, Ternate (N Moluccas)
    — _C.e.sumbawae_ Stresemann, 1925 Sumbawa, Flores, Sumba Is. (W Lesser Sundas)
    — _C.e.neglecta_ G.R. Gray, 1866 Timor, Roti (S Lesser Sundas)
    — _C.e.perneglecta_ Mayr, 1944 Alor to Damar Is. (EC Lesser Sundas) (1474)
    — _C.e.amethystina_ Salomonsen, 1983 Waigeo I. (2161)
    — _C.e.numforensis_ Salomonsen, 1983 Numfor I., (?) Biak I. (2161)
    — _C.e.nitens_ Ogilvie-Grant, 1914 New Guinea (except alpine W), Misoöl, Batanta, Yapen, Karkar Is.
    — _C.e.erwini_ Collin & Hartert, 1927 Alpine W New Guinea
    — _C.e.stresemanni_ Rothschild & Hartert, 1914 Admiralty Is.
    — _C.e.kalili_ Salomonsen, 1983 New Hanover, New Ireland, Djaul I. (2161)
    — _C.e.spilogaster_ Salomonsen, 1983 Tabar Is., Lihir group (2161)
    — _C.e.hypogrammica_ Salomonsen, 1983 Nissan I. (Green Is.) (2161)
    — _C.e.tametamele_ Stresemann, 1921 New Britain, Witu Is., Bougainville I.

[1] Includes _nubicola_ see Chantler (1999: 419) (308).
[2] For recognition of some taxa see Parkes (1994) (1782).
[3] The inclusion of birds with pale scalloped rumps (_marginata_ – treated as a separate species in last Edition) remains tentative, see Dickinson (1990) (737). We have tentatively accepted the forms proposed by Salomonsen (1983) (2161), but re-evaluation may suggest some consolidation.
[4] Includes _mindanensis_ Hachisuka, 1941 (1041), see Dickinson _et al._ (1991) (745).
[5] Includes _nubila_ Ripley, 1959 (2040), see White & Bruce (1986: 264) (2652).

    — *C.e.misimae* Salomonsen, 1983   Louisiade Arch. (2161)

    — *C.e.becki* Mayr, 1931   NE and C Solomon Is.

    — *C.e.makirensis* Mayr, 1931   San Cristobal I.

    — *C.e.desiderata* Mayr, 1931   Rennell I.

    — *C.e.uropygialis* G.R. Gray, 1866   Santa Cruz Is., Vanuatu

    — *C.e.albidior* Salomonsen, 1983   Loyalty Is., New Caledonia (2161)

○ *Collocalia linchi* Cave Swiftlet

    — *C.l.ripleyi* Somadikarta, 1986   Mts. of Sumatra (2298)

    — *C.l.linchi* Horsfield & Moore, 1854[1]   Java, Bawean I.

    — *C.l.dedii* Somadikarta, 1986[2]   Bali, Lombok I. (2298)

    — *C.l.dodgei* Richmond, 1905   Mt. Kinabalu (N Borneo)

○ *Collocalia troglodytes* Gray & Mitchell, 1845 Pygmy Swiftlet   Philippine Is.

AERODRAMUS Oberholser, 1906 M

○ *Aerodramus elaphrus* (Oberholser, 1906) Seychelles Swiftlet   Seychelles

○ *Aerodramus francicus* (J.F. Gmelin, 1789) Mascarene Swiftlet   Mauritius, Reunion

○ *Aerodramus unicolor* (Jerdon, 1840) Indian Swiftlet   SW India, Sri Lanka

○ *Aerodramus mearnsi* (Oberholser, 1912) Philippine Swiftlet[3]   Luzon, Mindoro, Palawan, Negros, Bohol, Mindanao Is.

○ *Aerodramus infuscatus* Moluccan Swiftlet[4]

    — *A.i.sororum* (Stresemann, 1931)   C, S and SE Sulawesi

    — *A.i.infuscatus* (Salvadori, 1880)   N Moluccas, Sangihe, Siau

    — *A.i.ceramensis* (van Oort, 1911)   S Moluccas

○ *Aerodramus hirundinaceus* Mountain Swiftlet

    — *A.h.baru* (Stresemann & Paludan, 1932)   Yapen I.

    — *A.h.hirundinaceus* (Stresemann, 1914)   Montane New Guinea, Dampier I.

    — *A.h.excelsus* (Ogilvie-Grant, 1914)   Snow Mts. (SW New Guinea)

◉ *Aerodramus spodiopygius* White-rumped Swiftlet[5]

    — *A.s.delichon* (Salomonsen, 1983)   Admiralty Is. (2161)

    — *A.s.eichhorni* (E. Hartert, 1924)   NC Bismarck Arch.

    — *A.s.noonaedanae* (Salomonsen, 1983)   New Ireland, New Britain (2161)

    — *A.s.reichenowi* (Stresemann, 1912)   S and E Solomon Is.

    — *A.s.desolatus* (Salomonsen, 1983)   Duff Is., Swallow Is., Santa Cruz Is. (2161)

    — *A.s.epiensis* (Salomonsen, 1983)   N and C Vanuatu (2161)

    — *A.s.ingens* (Salomonsen, 1983)   S Vanuatu (2161)

    — *A.s.leucopygius* (Wallace, 1864)   Loyalty Is., New Caledonia

    — *A.s.assimilis* (Stresemann, 1912)   Fiji

    — *A.s.townsendi* (Oberholser, 1906)   Tonga

    ☑ *A.s.spodiopygius* (Peale, 1848)   Samoa

○ *Aerodramus terraereginae* Australian Swiftlet

    — *A.t.terraereginae* (E.P. Ramsay, 1875)   E Queensland

    — *A.t.chillagoensis* Pecotich, 1982   NE Queensland (1817)

○ *Aerodramus brevirostris* Himalayan Swiftlet

    — *A.b.brevirostris* (Horsfield, 1840)   Himalayas, NE India, W Yunnan, N Burma

    — *A.b.innominatus* (Hume, 1873)   Sichuan and Hubei (C China) to N Vietnam >> SE Asia incl. Andamans

    — *A.b.rogersi* (Deignan, 1955)[6]   E Burma, N and W Thailand, S Yunnan, N Laos, NW Vietnam >> Malay Pen. (654)

    — *A.b.vulcanorum* (Stresemann, 1926)[7]   W Java

[1] Includes *plesseni* Meise, 1941 (1552) according to Somadikarta (1986) (2298).

[2] Somadikarta (1986) (2298) thought birds from Nusa Penida (type not then traced) to be nominate *linchi*. However, Penida lies between Bali and Lombok and the proposed distribution is unlikely; we omit Nusa Penda from the two range statements, but we suspect that *plesseni* will take priority over *dedii*.

[3] Includes *apoensis* see Dickinson (1989) (736); wrongly treated as a form of *whiteheadi* in last Edition.

[4] Bruce in White & Bruce (1986) (2652) noted that Salomonsen (1983) (2161) had observed that groups of forms comprising the 'species' *spodiopygius* might be treated as species within a superspecies; this is now generally accepted for *infuscatus, hirundinaceus* and *terraereginae*. The last of these was not yet accepted in last Edition. *A. mearnsi* has also been suggested as part of the superspecies.

[5] We tentatively accept the new forms of Salomonsen (1983) (2161), but re-evaluation is needed.

[6] An enigmatic and unresolved form (or forms) see Wells (1999: 455) (2579).

[7] Now that breeding records of the species from Sumatra have been discarded (Wells, 1975) (2578), and the purported Philippine subspecies detached (Dickinson, 1989) (736), this taxon will probably be found to belong elsewhere (*rogersi* may be attached to it).

○ *Aerodramus whiteheadi* WHITEHEAD'S SWIFTLET[1]
   — *A.w.whiteheadi* (Ogilvie-Grant, 1895)   Mt. Data (Luzon)
   — *A.w.origenis* (Oberholser, 1906)   Mindanao, Panay

○ *Aerodramus nuditarsus* (Salomonsen, 1962) BARE-LEGGED SWIFTLET   E New Guinea (2156)

○ *Aerodramus orientalis* MAYR'S SWIFTLET
   — *A.o.leletensis* (Salomonsen, 1962)   C New Ireland (2156)
   — *A.o.orientalis* (Mayr, 1935)   Guadalcanal (Solomon Is.)

○ *Aerodramus salangana* MOSSY-NEST SWIFTLET
   — *A.s.* subsp.?   Sumatra[2]
   — *A.s.aerophilus* (Oberholser, 1912)   W Sumatran Is.
   — *A.s.natunae* (Stresemann, 1930)   N Borneo, Natuna Is.
   — *A.s.maratua* (Riley, 1927)   Maratua Is. (Borneo), Basilan (SW Philippines)
   — *A.s.salangana* (Streubel, 1848)   Java

○ *Aerodramus vanikorensis* UNIFORM SWIFTLET
   — *A.v.amelis* (Oberholser, 1906)[3]   Philippines (excl. Palawan)
   — *A.v.palawanensis* (Stresemann, 1914)[4]   Palawan
   — *A.v.aenigma* (Riley, 1918)   N and C Sulawesi
   — *A.v.heinrichi* (Stresemann, 1932)   S Sulawesi
   — *A.v.moluccarum* (Stresemann, 1914)   S Moluccas
   — *A.v.waigeuensis* (Stresemann & Paludan, 1932)   N Moluccas, W Papuan Is.
   — *A.v.steini* (Stresemann & Paludan, 1932)   Numfor I., Biak I.
   — *A.v.yorki* (Mathews, 1916)[5]   New Guinea, Aru Is., D'Entrecasteaux Is.
   — *A.v.tagulae* (Mayr, 1937)   Louisiade Arch., Trobriand Is., Woodlark I.
   — *A.v.coultasi* (Mayr, 1937)   Admiralty Is., St Matthias Group
   — *A.v.pallens* (Salomonsen, 1983)   New Hanover, New Ireland, Djaul I., New Britain (2161)
   — *A.v.lihirensis* (Mayr, 1937)   Lihir, Nuguria and Hibernian Is., Feni Is., Tabar Is.
   — *A.v.lugubris* (Salomonsen, 1983)   Solomon Is. (2161)
   — *A.v.vanikorensis* (Quoy & Gaimard, 1830)   Santa Cruz, Vanuatu

○ *Aerodramus pelewensis* (Mayr, 1935) PALAU SWIFTLET[6]   Palau Is.

○ *Aerodramus bartschi* (Mearns, 1909) MARIANAS SWIFTLET   S Mariana Is.

○ *Aerodramus inquietus* CAROLINE ISLANDS SWIFTLET
   — *A.i.rukensis* (N. Kuroda, Sr., 1915)   Chuuk (C Caroline Is.)
   — *A.i.ponapensis* (Mayr, 1935)   Pohnpei (E Caroline Is.)
   — *A.i.inquietus* (Kittlitz, 1858)   Kusaie (E Caroline Is.)

○ *Aerodramus leucophaeus* POLYNESIAN SWIFTLET
   — *A.l.leucophaeus* (Peale, 1848)   Society Islands
   — *A.l.sawtelli* (Holyoak, 1974)[7]   Atiu I. (S Cook Is.) (1129)

○ *Aerodramus ocistus* MARQUESAN SWIFTLET[8]
   — *A.o.ocistus* (Oberholser, 1906)   N Marquesas Is.
   — *A.o.gilliardi* (Somadikarta, 1994)   S Marquesas Is. (2299)

○ *Aerodramus maximus* BLACK-NEST SWIFTLET[9]
   — *A.m.maximus* (Hume, 1878)   S Burma, Malaysia, SE Vietnam[10]
   — *A.m.lowi* (Sharpe, 1879)   Sumatra, N and W Borneo, Java
   — *A.m.tichelmani* (Stresemann, 1926)   SE Borneo

[1] The two other forms listed in last Edition do not belong to this species (Dickinson, 1989) (736) and are mentioned in other footnotes.
[2] Wells (1975) (2578) using the specific name *vanikorensis* described breeding. Some name in synonymy may prove applicable. As now grouped this species may require splitting.
[3] Treated as a species in last Edition but see Dickinson, 1989 (736).
[4] Attached to *maximus* in last Edition but see Dickinson (1989) (736), who included *tsubame* — treated as a form of *whiteheadi* in last Edition.
[5] Apparently has priority over the name *granti*. However, the two types differ significantly (Somadikarta, *in litt.*).
[6] There is limited published evidence as a basis for separation of this species and the next from *vanikorensis*.
[7] Placement as advised by Somadikarta (*in litt.*).
[8] For reasons to separate this from *A. leucophaeus* see Somadikarta (1994) (2299).
[9] Tacit acceptance has been given almost universally to use of this name (in place of the well-established *lowii*) following Deignan (1955) (653), but see Salomonsen (1983: 6) (2161). In view of the quantum of publication since 1955, during which progress has been made on the taxonomy of swiftlets, stability is probably best served by accepting the *fait accompli* however this is under fresh review.
[10] A report by Medway (1962) (1504) that both *brevirostris* and *maximus* specimens are known from Bhutan and S Tibet was contradicted by Grimmett *et al.* (1998: 424) (1014).

○ *Aerodramus fuciphagus* EDIBLE-NEST SWIFTLET[1]

— *A.f.inexpectatus* (Hume, 1873)   Andaman and Nicobar Is.
— *A.f.germani* (Oustalet, 1878)   SE Asia from SW Burma to N Vietnam S to C Malay
                                                      Pen., Borneo, Palawan and scattered islands of W
                                                      Philippines and off Borneo

— *A.f.amechanus* (Oberholser, 1912)   S Malay Pen., Singapore, Anamba Is.
— *A.f.vestitus* (Lesson, 1843)   Sumatra, Belitung, Borneo
— *A.f.perplexus* (Riley, 1927)[2]   Maratua Is. (E of N Borneo)
— *A.f.fuciphagus* (Thunberg, 1812)[3]   Java, Kangean I., W Lesser Sundas
— *A.f.dammermani* (Rensch, 1931)   Flores (NC Lesser Sundas)
— *A.f.micans* (Stresemann, 1914)   Sumba, Sawu, Timor (SC Lesser Sundas)

○ *Aerodramus papuensis* (Rand, 1941) PAPUAN SWIFTLET/THREE-TOED SWIFTLET
                                                      Montane WC to E New Guinea (1964)

SCHOUTEDENAPUS de Roo, 1968 M
○ *Schoutedenapus myoptilus* SCARCE SWIFT
— *S.m.myoptilus* (Salvadori, 1888)   Ethiopia south to Zimbabwe
— *S.m.poensis* (Alexander, 1903)   Bioko I.
— *S.m.chapini* (Prigogine, 1957)   E Zaire, Rwanda, SW Uganda (1927)

○ *Schoutedenapus schoutedeni* (Prigogine, 1960) SCHOUTEDEN'S SWIFT   E Zaire (1928)

MEARNSIA Ridgway, 1911 F
○ *Mearnsia picina* (Tweeddale, 1879) PHILIPPINE SPINETAILED SWIFT   SC and S Philippines

○ *Mearnsia novaeguineae* PAPUAN SPINETAILED SWIFT
— *M.n.buergersi* (Reichenow, 1917)   N New Guinea
— *M.n.novaeguineae* (D'Albertis & Salvadori, 1879) S and SE New Guinea

ZOONAVENA Mathews, 1918 F
○ *Zoonavena grandidieri* MADAGASCAN SPINETAILED SWIFT
— *Z.g.grandidieri* (J. Verreaux, 1867)   Madagascar
— *Z.g.mariae* (Benson, 1960)[4]   Grande Comore (122)

○ *Zoonavena thomensis* (E. Hartert, 1900) SÃO TOMÉ SPINETAILED SWIFT   São Tomé and Príncipe

○ *Zoonavena sylvatica* (Tickell, 1846) WHITE-RUMPED SPINETAILED SWIFT   India, Burma

TELACANTHURA Mathews, 1918 F
○ *Telacanthura ussheri* MOTTLED SPINETAILED SWIFT
— *T.u.ussheri* (Sharpe, 1870)   Senegal and Gambia to N Nigeria
— *T.u.sharpei* (Neumann, 1908)   S Cameroon, Gabon, Zaire and Uganda
— *T.u.stictilaema* (Reichenow, 1879)[5]   Kenya, NE and C Tanzania
— *T.u.benguellensis* (Neumann, 1908)   W Angola, SE Africa

○ *Telacanthura melanopygia* (Chapin, 1915) BLACK SPINETAILED SWIFT   Sierra Leone to Cameroon and Gabon, SW Central
                                                      African Republic, NE Zaire, NE Angola

RHAPHIDURA Oates, 1883 F
○ *Rhaphidura leucopygialis* (Blyth, 1849) SILVER-RUMPED SPINETAILED SWIFT
                                                      S Burma to Sumatra, Java, Borneo

○ *Raphidura sabini* (J.E. Gray, 1829) SABINE'S SPINETAILED SWIFT   Sierra Leone to Uganda and W Kenya, Bioko

NEAFRAPUS Mathews, 1918 M
○ *Neafrapus cassini* (P.L. Sclater, 1863) CASSIN'S SPINETAILED SWIFT   S Ivory Coast to Gabon and Zaire and W Uganda

○ *Neafrapus boehmi* BÖHM'S SPINETAILED SWIFT
— *N.b.boehmi* (Schalow, 1882)   W Angola to W Tanzania
— *N.b.sheppardi* (Roberts, 1922)[6]   SE Kenya, E Tanzania, E and S Zambia to Mozambique

---

[1] Treatment here follows Medway (1966) (1506); the arrangement in last Edition (with two forms included in a species *germani*, one in *amelis*, and *vestitus*, *dammermani* and *micans* in synonymy) demonstrated considerable confusion.
[2] Attached to *amelis* in last Edition, but see Medway (1966) (1506).
[3] As the synonymy is unusually complex, note that this includes *bartelsi* and *javensis*.
[4] Omitted in last Edition.
[5] Includes *marwitzi* see Chantler (1999: 438) (308).
[6] In synonymy in last Edition but Fry (1988: 205) (889). Presumably includes *madaraszi* Keve, 1959 (1262), but may need re-evaluation.

HIRUNDAPUS Hodgson, 1837 M
● ***Hirundapus caudacutus*** White-throated Needletail[1]

   ↙ *H.c.caudacutus* (Latham, 1802)[2]   C and E Siberia to Japan, Korea, NE China and
                                                Mongolia >> E Australia

   — *H.c.nudipes* (Hodgson, 1837)   Himalayas and Assam east to Sichuan and Yunnan >>
                                                Malay Pen., Java

○ ***Hirundapus cochinchinensis*** (Oustalet, 1878) Silver-backed Needletail[3] [4]
                                                C Himalayas and Assam to S China and Taiwan >>
                                                mainland SE Asia and Gtr. Sundas

○ ***Hirundapus giganteus*** Brown-backed Needletail
   — *H.g.indicus* (Hume, 1873)   SW India and Sri Lanka; Andaman Is.; E India east to
                                                mainland SE Asia south to SW Thailand

   — *H.g.giganteus* (Temminck, 1825)   Malay Pen., Gtr. Sundas, Palawan I.

○ ***Hirundapus celebensis*** (P.L. Sclater, 1865) Purple Needletail[5]   N Sulawesi, Philippines (excl. Palawan)

CHAETURA Stephens, 1826 F
○ ***Chaetura martinica*** (Hermann, 1783) Lesser Antillean Swift   Lesser Antilles

○ ***Chaetura spinicaudus*** Band-rumped Swift[6]
   — *C.s.aetherodroma* Wetmore, 1951   C and E Panama, W Colombia, W Ecuador (2594)
   — *C.s.spinicaudus* (Temminck, 1839)[7]   E Venezuela, the Guianas, N and C Brazil

○ ***Chaetura fumosa*** Salvin, 1870 Costa Rican Swift[8]   S Costa Rica to Chiriqui (Panama)

○ ***Chaetura egregia*** Todd, 1916 Pale-rumped Swift[9]   SW Amazonia

○ ***Chaetura cinereiventris*** Grey-rumped Swift
   — *C.c.phaeopygos* Hellmayr, 1906   E Nicaragua to Panama
   — *C.c.lawrencei* Ridgway, 1893   N Venezuela, Grenada, Trinidad, Tobago Is.
   — *C.c.schistacea* Todd, 1937   E Colombia, W Venezuela
   — *C.c.guianensis* E. Hartert, 1892   W Guyana, E Venezuela
   — *C.c.occidentalis* Berlepsch & Taczanowski, 1884   W Colombia, W Ecuador
   — *C.c.sclateri* Pelzeln, 1868   S Venezuela to N Brazil, E Peru to NW Bolivia
   — *C.c.cinereiventris* P.L. Sclater, 1862   E Brazil, Paraguay, NE Argentina

● ***Chaetura vauxi*** Vaux's Swift[10]
   ↙ *C.v.vauxi* (J.K. Townsend, 1839)   SW Canada to SW USA >> C America
   — *C.v.tamaulipensis* Sutton, 1941   E Mexico (2376)
   — *C.v.gaumeri* Lawrence, 1882   Yucatán Penin, Cozumel I.
   — *C.v.richmondi* Ridgway, 1910[11]   S Mexico to Costa Rica
   — *C.v.ochropygia* Aldrich, 1937   E Panama
   — *C.v.aphanes* Wetmore & Phelps, 1956   N Venezuela (2614)
   — *C.v.andrei* (Berlepsch & Hartert, 1902)   E Venezuela

○ ***Chaetura meridionalis*** Hellmayr, 1907 Sick's Swift[12]   C Sth. America >> Colombia

● ***Chaetura pelagica*** (Linnaeus, 1758) Chimney Swift   S Canada to S USA >> WC Sth. America

○ ***Chaetura chapmani*** Hellmayr, 1907 Chapman's Swift   French Guiana, Trinidad

○ ***Chaetura viridipennis*** Cherrie, 1916 Amazonian Swift[13]   C Brazil

○ ***Chaetura brachyura*** Short-tailed Swift
   — *C.b.praevelox* Bangs & Penard, 1918   Grenada, St. Vincent, Tobago

---

[1] In last Edition *formosanus* was attached here. It was transferred to *cochinchinensis* by Mees (1973) (1525) and is now in its synonymy (Mees, 1977) (1528).
[2] For reasons to use 1802 not 1801 see Browing & Monroe (1991) (258). Includes *bourreti* David-Beaulieu, 1944 (615), see Dickinson (1970) (734).
[3] Mees (1973) (1525) listed *ernsti* as a synonym — overlooked by van den Hoek Ostende *et al.* (1997) (2464).
[4] Includes *formosanus* and *rupchandi* Biswas, 1951 (154), see Mees (1973, 1977) (1525, 1528).
[5] Includes *manobo* Salomonsen, 1953 (2155), see Mees (1985) (1538) and Dickinson *et al.* (1991) (745).
[6] For correction of spelling see David & Gosselin (2002) (613).
[7] Includes *aethalea* and *latirostris* Zimmer & Phelps, 1952 (2728) see Marin (2000) (1431).
[8] For reasons to treat as a species distinct from *spinicauda* see Marin (2000) (1431).
[9] Includes *pachiteae* Meise, 1964, see Marin (2000) (1431).
[10] The form *andrei* was treated as the nominate race of a polytypic species in last Edition, but see Marin (1997) (1430).
[11] Presumably includes *warneri* A.R. Phillips, 1966 (1869).
[12] For reasons to separate from a "species" *andrei* (as in last Edition and in "HBW") see Marin (1997) (1430).
[13] For reasons to separate from *chapmani* see Marin (1997) (1430).

___ *C.b.brachyura* (Jardine, 1846)   The Guianas south to N Bolivia

___ *C.b.ocypetes* J.T. Zimmer, 1953   SW Ecuador, NW Peru (2720)

___ *C.b.cinereocauda* (Cassin, 1850)   NC Brazil

AERONAUTES E. Hartert, 1892 M

● *Aeronautes saxatalis* WHITE-THROATED SWIFT[1]

    ✓ *A.s.saxatalis* (Woodhouse, 1853)[2]   SW Canada to SW USA >> Mexico

    ___ *A.s.nigrior* Dickey & van Rossem, 1928   S Mexico to C Honduras

○ *Aeronautes montivagus* WHITE-TIPPED SWIFT

    ___ *A.m.montivagus* (d'Orbigny & Lafresnaye, 1837)   Andes from N Venezuela south to NW Argentina

    ___ *A.m.tatei* (Chapman, 1929)   S Venezuela, N Brazil

○ *Aeronautes andecolus* ANDEAN SWIFT

    ___ *A.a.parvulus* (Berlepsch & Stolzmann, 1892)   W Peru, N Chile

    ___ *A.a.peruvianus* (Chapman, 1919)   SE Peru

    ___ *A.a.andecolus* (d'Orbigny & Lafresnaye, 1837)   C Bolivia, W Argentina

TACHORNIS Gosse, 1847 F

○ *Tachornis phoenicobia* ANTILLEAN PALM SWIFT

    ___ *T.p.iradii* (Lembeye, 1850)   Cuba, Isla de la Juventud

    ___ *T.p.phoenicobia* Gosse, 1847   Hispaniola, Jamaica

○ *Tachornis furcata* PYGMY SWIFT

    ___ *T.f.furcata* (Sutton, 1928)   NE Colombia, NW Venezuela

    ___ *T.f.nigrodorsalis* (Aveledo & Pons, 1952)   W Venezuela (70)

○ *Tachornis squamata* FORK-TAILED PALM SWIFT

    ___ *T.s.squamata* (Cassin, 1853)[3]   Trinidad, the Guianas, C and E Brazil

    ___ *T.s.semota* (Riley, 1933)   E Ecuador and NE Peru east to S and E Venezuela, NW Brazil

PANYPTILA Cabanis, 1847 F

○ *Panyptila sanctihieronymi* Salvin, 1863 GREAT SWALLOW-TAILED SWIFT   SW Mexico, C Guatemala, S Honduras

○ *Panyptila cayennensis* LESSER SWALLOW-TAILED SWIFT

    ___ *P.c.veraecrucis* R.T. Moore, 1947   E Mexico to N Honduras (1617)

    ___ *P.c.cayennensis* (J.F. Gmelin, 1789)   S Honduras to N and C Sth. America

CYPSIURUS Lesson, 1843 M

○ *Cypsiurus parvus* AFRICAN PALM SWIFT

    ___ *C.p.parvus* (M.H.K. Lichtenstein, 1823)   Senegal and Gambia to N Ethiopia and SW Arabia

    ___ *C.p.brachypterus* (Reichenow, 1903)   Sierra Leone to NE and S Zaire and Angola, Bioko

    ___ *C.p.myochrous* (Reichenow, 1886)   S Sudan through E Zaire, W Tanzania,and Zimbabwe to NE South Africa,

    ___ *C.p.laemostigma* (Reichenow, 1905)[4]   S Somalia to C Mozambique

    ___ *C.p.griveaudi* Benson, 1960[5]   Comoro Is. (122)

    ___ *C.p.gracilis* (Sharpe, 1871)   Madagascar

    ___ *C.p.hyphaenes* Brooke, 1972   N Namibia, N Botswana (233)

    ___ *C.p.celer* Clancey, 1983   S Mozambique to Natal (491)

○ *Cypsiurus balasiensis* ASIAN PALM SWIFT[6]

    ___ *C.b.balasiensis* (J.E. Gray, 1829)   Pakistan, India (except Assam), Sri Lanka

    ___ *C.b.infumatus* (P.L. Sclater, 1865)   Assam east to Hainan, south through mainland SE Asia to Sumatra, Borneo

    ___ *C.b.bartelsorum* Brooke, 1972   Java, Bali (233)

    ___ *C.b.pallidior* McGregor, 1905   Philippine Is

---

[1] Wrongly spelled *saxatilis* in last Edition.
[2] Apparently includes *sclateri* Rogers, 1939 (2097), but recognised by A.O.U. (1957: 300) (1).
[3] Presumably includes *orientalis* Pinto & de Camargo, 1961 (1899), but may need evaluation.
[4] For recognition of this see Clancey (1983) (491).
[5] Omitted in error in last Edition.
[6] Spelling corrected by Medway (1965) (1505).

TACHYMARPTIS Roberts, 1922 M
- ◐ *Tachymarptis melba* ALPINE SWIFT
  - — *T.m.melba* (Linnaeus, 1758)[1]

  - — *T.m.tuneti* (Tschudi, 1904)
  - — *T.m.archeri* (E. Hartert, 1928)
  - — *T.m.maximus* (Ogilvie-Grant, 1907)
  - — *T.m.africanus* (Temminck, 1815)
  - — *T.m.marjoriae* (Bradfield, 1935)
  - — *T.m.willsi* (E. Hartert, 1896)
  - — *T.m.nubifuga* (Koelz, 1954)[2]
  - — *T.m.bakeri* (E. Hartert, 1928)

- ○ *Tachymarptis aequatorialis* MOTTLED SWIFT
  - — *T.a.lowei* (Bannerman, 1920)
  - — *T.a.bamendae* (Serle, 1949)[3]
  - — *T.a.furensis* (Lynes, 1920)
  - — *T.a.aequatorialis* (J.W. von Müller, 1851)

  - — *T.a.gelidus* (Brooke, 1967)

APUS Scopoli, 1777 M
- ○ *Apus alexandri* E. Hartert, 1901 ALEXANDER'S SWIFT

- ● *Apus apus* COMMON SWIFT
  - ⚓ *A.a.apus* (Linnaeus, 1758)
  - — *A.a.pekinensis* (Swinhoe, 1870)

- ○ *Apus unicolor* (Jardine, 1830) PLAIN SWIFT

- ○ *Apus niansae* NYANZA SWIFT
  - — *A.n.niansae* (Reichenow, 1887)[4]
  - — *A.n.somalicus* (Stephenson Clarke, 1919)

- ○ *Apus pallidus* PALLID SWIFT
  - — *A.p.brehmorum* E. Hartert, 1901

  - — *A.p.illyricus* Tschusi, 1907
  - — *A.p.pallidus* (Shelley, 1870)

- ○ *Apus barbatus* AFRICAN SWIFT
  - — *A.b.barbatus* (P.L. Sclater, 1865)
  - — *A.b.hollidayi* Benson & Irwin, 1960
  - — *A.b.oreobates* Brooke, 1970
  - — *A.b.balstoni* (E. Bartlett, 1880)
  - — *A.b.mayottensis* (Nicoll, 1906)
  - — *A.b.sladeniae* (Ogilvie-Grant, 1904)
  - — *A.b.serlei* de Roo, 1970
  - — *A.b.roehli* Reichenow, 1906
  - — *A.b.glanvillei* Benson, 1967

- ○ *Apus berliozi* FORBES-WATSON'S SWIFT
  - — *A.b.berliozi* Ripley, 1965
  - — *A.b.bensoni* Brooke, 1969

- ○ *Apus bradfieldi* BRADFIELD'S SWIFT
  - — *A.b.bradfieldi* (Roberts, 1926)
  - — *A.b.deserticola* Brooke, 1970

Distributions (right column):

N Morocco, S Europe, Turkey, NW Iran >> W, C and NE Africa
C Morocco to N Libya, Levant, Iran to W Pakistan
N Somalia, SW Arabia to Dead Sea
Mt. Ruwenzori (Zaire/Uganda)
Ethiopia, E Africa to N Namibia and Cape Province
C Namibia to NW Cape
Madagascar
NW Himalayas >> C India (1284)
Sri Lanka

Sierra Leone to Nigeria
NW Cameroon
W Sudan
Eritrea, Ethiopia and E Africa south to S Mozambique and E Zimbabwe, WC Angola
W Zimbabwe (226)

Cape Verde Is.

N Africa, W Europe to Lake Baikal >> Africa
Iran and Transcaspia to N China and Mongolia >> E and Sthn. Africa
Madeira, Canary Is. >> Morocco, Mauritania

Mts. of Eritrea and Ethiopia to N Tanzania
N Somalia

Madeira, Canary Is., N Africa to NW Egypt, S Europe, Cyprus, Turkey >> Sahel and Uganda
E coast Adriatic Sea >> E Africa
W Mauritania, Saharan hills, Egypt, Levant to W Pakistan >> Sahel to S Pakistan

S Africa
Victoria Falls (Zambia/Zimbabwe) (129)
Zimbabwe to Mozambique (230)
Madagascar
Comoro Is
SE Nigeria, W Cameroon, Bioko, W Angola
W Cameroon (629)
Kenya to E Zaire, south to Malawi, NE Angola
Sierra Leone (126)

Socotra I. (2046)
Somalia >> coastal Kenya (227)

SW Angola, Namibia
N Cape Prov. (S Africa) (230)

---

[1] Includes *obscurus* Trischitta, 1939 (2440), see Vaurie (1965: 655) (2510).
[2] For recognition see Ripley (1982: 198) (2055), who included *dorabtatai* Abdulali, 1965 (7).
[3] Not recognised by Chantler (1995: 191) (309); but recognised by White (1965: 210) (2649) and listed on advice of R. Dowsett (*in litt.*).
[4] Includes *kapnodes* Friedmann, 1964 (881), see White (1967) (2650).

○ *Apus pacificus* Fork-tailed Swift/Pacific Swift
    __ *A.p.leuconyx* (Blyth, 1845)
    __ *A.p.kanoi* Yamashina, 1942[1]

    __ *A.p.pacificus* (Latham, 1802)[2]

    __ *A.p.cooki* (Harington, 1913)

○ *Apus acuticauda* (Jerdon, 1864) Dark-rumped Swift[3]

○ *Apus affinis* Little Swift[4]
    __ *A.a.galilejensis* (Antinori, 1855)

    __ *A.a.bannermani* E. Hartert, 1928
    __ *A.a.aerobates* Brooke, 1969[5]
    __ *A.a.theresae* R. Meinertzhagen, 1949

    __ *A.a.affinis* (J.E. Gray, 1830)

    __ *A.a.singalensis* Madarász, 1911

○ *Apus nipalensis* House Swift
    __ *A.n.nipalensis* (Hodgson, 1837)
    __ *A.n.subfurcatus* (Blyth, 1849)

    __ *A.n.furcatus* Brooke, 1971
    __ *A.n.kuntzi* Deignan, 1958

○ *Apus horus* Horus Swift
    __ *A.h.horus* (Heuglin, 1869)[6]

    __ *A.h.fuscobrunneus* Brooke, 1971

○ *Apus caffer* (M.H.K. Lichtenstein, 1823) White-rumped Swift

○ *Apus batesi* (Sharpe, 1904) Bates's Swift

Himalayas and Assam hills >> S India
SE Tibet to S China, Taiwan, Lanyu I., Batan Is. (off N Luzon) >> Assam, mainland SE Asia (2693)
Altai to Kamchatka, Japan and NC China >> Indonesia, Melanesia, Australia
E Burma, N Thailand, Laos, C Vietnam >> Malay Pen.

Nepal, S Assam (hills)

NW Africa, N Sahel to N Somalia, Middle East to Uzbekistan and W Pakistan
São Tomé, Príncipe, Bioko
Mauritania to Ethiopia, S to C Angola and Natal (228)
W and S Angola and S Zambia S to Cape Province (1549)
E Somalia to N Mozambique; Indian plains to Assam >> S India
S India, Sri Lanka

Himalayas >> ? Malay Pen.
Assam east to S China south through mainland SE Asia and Philippines to Sumatra and Borneo
Java, Bali (232)
Taiwan (659)

Ethiopia to Sth. Africa, Nigeria, N Cameroon, W Chad, W and EC Sudan, W Zaire, NW Angola
SW Angola (231)

Iberian Pen., NW Africa, Senegal and Gambia east to Somalia, south to Sth. Africa (sub-Saharan Africa)
Guinea to Cameroon, E Zaire

HEMIPROCNIDAE TREESWIFTS (1: 4)

HEMIPROCNE Nitzsch, 1829 F
○ *Hemiprocne coronata* (Tickell, 1833) Crested Treeswift[7]

○ *Hemiprocne longipennis* Grey-rumped Treeswift
    __ *H.l.harterti* Stresemann, 1913

    __ *H.l.perlonga* (Richmond, 1903)[8]
    __ *H.l.longipennis* (Rafinesque, 1802)
    __ *H.l.wallacii* (Gould, 1859)[9]

○ *Hemiprocne comata* Whiskered Treeswift
    __ *H.c.comata* (Temminck, 1824)[10]
    __ *H.c.major* (E. Hartert, 1895)[11]

India and Sri Lanka to Yunnan and SE Asia south to SW Thailand

S Burma, Malaysia, Sumatra, Borneo, Tawitawi (Sulu Is., SW Philippines)
W Sumatran Is.
Java, Lombok and Kangean Is.
Sula Is., Sulawesi

Malaysia, Sumatra and W Sumatran Is., Borneo
Philippines incl. Sulu Is.

---

[1] Includes *salimalii* Lack, 1958 (1311A) see Ripley (1982: 200) (2055).
[2] For reasons to use 1802 not 1801 see Browing & Monroe (1991) (258). Includes *shiraii* Mishima, 1960 (1592) see Stepanyan (1990: 294) (2326).
[3] Includes *rupchandi* Koelz, 1954 (1284), see Ripley (1982: 199) (2055).
[4] Two groups, this and *nipalensis*, identified by Brooke (1971) (232), the later since usually treated as a species (but evidence not yet wholly convincing).
[5] Includes *gyrator* Clancey, 1980 (480), see Clements & Driessens (1995: 211) (309).
[6] The name *toulsoni* attaches to what is thought to be a local colour morph (Brooke, 1971) (231).
[7] For reasons to separate this see Brooke (1969) (229). Includes *dryas* Ripley, 1950 (2034), see Ripley (1982: 202) (2055).
[8] Includes *ocyptera* and *thoa* see Wells (1999: 465) (2580).
[9] Includes *dehaani* Somadikarta, 1975 (2297), and *mendeni* Somadikarta, 1975 (2297), see Wells (1999: 465) (2580).
[10] Includes *stresemanni* see Wells (1999: 465) (2580).
[11] Includes *nakamurai* and *barbarae* Peters, 1939 (1824), see Dickinson *et al.* (1991: 235) (745).

○ *Hemiprocne mystacea* MOUSTACHED TREESWIFT[1]

— *H.m.confirmata* Stresemann, 1914   Moluccas, Aru and Boano Is.
— *H.m.mystacea* (Lesson, 1827)   New Guinea, W Papuan and Geelvink Bay islands
— *H.m.aeroplanes* Stresemann, 1921   Bismarck Arch.
— *H.m.macrura* Salomonsen, 1983   Admiralty Is. (2161)
— *H.m.woodfordiana* (E. Hartert, 1896)   Solomon Is. (except San Cristobal and Sta. Cruz Is.)
— *H.m.carbonaria* Salomonsen, 1983   San Cristobal I. (SC Solomons) (2161)

TROCHILIDAE HUMMINGBIRDS[2] (1O4: 331)

PHAETHORNITHINAE

RAMPHODON Lesson, 1830 M
○ *Ramphodon naevius* (Dumont, 1818) SAW-BILLED HERMIT[3]   SE Brazil

EUTOXERES Reichenbach, 1849 M[4]
○ *Eutoxeres aquila* WHITE-TIPPED SICKLEBILL
— *E.a.salvini* Gould, 1868[5]   Costa Rica to W Colombia
— *E.a.heterurus* Gould, 1868   SW Colombia, W Ecuador
— *E.a.aquila* (Bourcier, 1847)   E Colombia to N Peru

○ *Eutoxeres condamini* BUFF-TAILED SICKLEBILL
— *E.c.condamini* (Bourcier, 1851)   SE Colombia to N Peru
— *E.c.gracilis* Berlepsch & Stolzmann, 1902 C Peru to N Bolivia

GLAUCIS Boie, 1831 M[6]
○ *Glaucis dohrnii* (Bourcier & Mulsant, 1852) HOOK-BILLED HERMIT SE Brazil

○ *Glaucis hirsutus* RUFOUS-BREASTED HERMIT
— *G.h.insularum* Hellmayr & von Seilern-Aspang, 1913   Grenada, Trinidad, Tobago Is.
— *G.h.hirsutus* (J.F. Gmelin, 1788)[7] [8]   Panama, Colombia to Bolivia, Venezuela, the Guianas, Brazil

○ *Glaucis aeneus* Lawrence, 1867 BRONZY HERMIT   E Honduras and E Nicaragua to W Panama, W Colombia, W Ecuador

THRENETES Gould, 1852 M
○ *Threnetes ruckeri* BAND-TAILED BARBTHROAT[9]
— *T.r.ventosus* Bangs & Penard, 1924   E Guatemala to Panama
— *T.r.ruckeri* (Bourcier, 1847)   N and W Colombia, W Ecuador
— *T.r.venezuelensis* Cory, 1913   NW Venezuela

○ *Threnetes niger* PALE-TAILED BARBTHROAT[10]
— *T.n.cervinicauda* Gould, 1855   E Colombia, E Ecuador, NE Peru, W Brazil
— *T.n.rufigastra* Cory, 1915   C Peru to NC Bolivia
— *T.n.leucurus* (Linnaeus, 1766)   The Orinoco and Amazon basins (Guianas, Venezuela, Brazil and NE Bolivia)
— *T.n.niger* (Linnaeus, 1758)   French Guiana, N Brazil
— *T.n.loehkeni* Grantsau, 1969[11] [12]   NE Brazil (N of R. Amazon) (974)
— *T.n.medianus* Hellmayr, 1929   NE Brazil (S of R. Amazon)

---

[1] The two new races are accepted tentatively.
[2] Gerwin & Zink (1998) (934) supported the view that two broad groups (here subfamilies) should be recognised. Names of hybrids are omitted (or footnoted if recently declared to be hybrids); two probable hybrids described since Peters (1945) (1826) are: *Augasma cyaneo-beryllina* Berlioz, 1965 (140), and *Phasmornis mystica* Oberholser, 1974 (1672). Mayr (1971) (1483) discussed the former and Browning (1978) (246) the latter.
[3] The name *freitasi* Ruschi, 1978 (2139), is a synonym see Hinkelmann (1999: 537) (1121).
[4] Gender addressed by David & Gosselin (2002) (614), multiple changes occur in specific names.
[5] Includes *mundus* see Hinkelmann (1999: 537) (1121) although an intergrade.
[6] Gender addressed by David & Gosselin (2002) (614), multiple changes occur in specific names.
[7] Includes *affinis*, *mazeppa* and *abrawayae* Ruschi, 1973 (2135), see Hinkelmann (1999: 538) (1121).
[8] The name *Threnetes grzimeki* Ruschi, 1973 (2134), is thought to have been applied to immature specimens of this taxon, see Hinkelmann (1988) (1117).
[9] The name *darienensis* is thought to relate to a population where two or more forms meet, see Hinkelmann (1999: 538) (1121).
[10] Hinkelmann (1999: 539) (1121) explained the lumping of *niger* and *leucurus*.
[11] Includes *freirei* Ruschi, 1976 (2138), see Hinkelmann (1999: 539) (1121).
[12] Includes *cristinae* Ruschi, 1975 (2137), see Hinkelmann (1988) (1117).

ANOPETIA Simon, 1918[1] F
○ *Anopetia gounellei* (Boucard, 1891) Broad-tipped Hermit E Brazil

PHAETHORNIS Swainson, 1827 M
○ *Phaethornis yaruqui* (Bourcier, 1851) White-whiskered Hermit[2] W Colombia, W Ecuador

○ *Phaethornis guy* Green Hermit
  __ *P.g.coruscus* Bangs, 1902 Costa Rica, Panama, NW Colombia
  __ *P.g.emiliae* (Bourcier & Mulsant, 1846) WC Colombia
  __ *P.g.apicalis* (Tschudi, 1844) N Colombia and NW Venezuela to SE Peru
  __ *P.g.guy* (Lesson, 1832) NE Venezuela, Trinidad

○ *Phaethornis hispidus* (Gould, 1846) White-bearded Hermit E Venezuela, E Colombia, E Ecuador, E Peru, Bolivia, W Brazil

○ *Phaethornis longirostris* Western Long-tailed Hermit
  __ *P.l.griseoventer* A.R. Phillips, 1962 W Mexico (1868)
  __ *P.l.mexicanus* E. Hartert, 1897 SW Mexico
  __ *P.l.longirostris* (DeLattre, 1843)[3] S Mexico to N Honduras
  __ *P.l.cephalus* (Bourcier & Mulsant, 1848)[4] E Honduras to NW Colombia
  __ *P.l.sussurus* Bangs, 1901 N Colombia
  __ *P.l.baroni* E. Hartert, 1897 W Ecuador, NW Peru

◕ *Phaethornis superciliosus* Eastern Long-tailed Hermit[5]
  __ *P.s.superciliosus* (Linnaeus, 1766)[6] S Venezuela, the Guianas, N Brazil
  __ *P.s.muelleri* Hellmayr, 1911 N Brazil

○ *Phaethornis malaris* Great-billed Hermit
  __ *P.m.malaris* (Nordmann, 1835) The Guianas, N Brazil
  __ *P.m.insolitus* J.T. Zimmer, 1950 E Colombia, S Venezuela, N Brazil (2713)
  __ *P.m.moorei* Lawrence, 1858[7] E Colombia, E Ecuador, N Peru
  __ *P.m.ochraceiventris* Hellmayr, 1907 NE Peru, W Brazil
  __ *P.m.bolivianus* Gould, 1861[8] SE Peru, Bolivia, W Brazil
  __ *P.m.margarettae* Ruschi, 1972[9] E Brazil (2132)

○ *Phaethornis syrmatophorus* Tawny-bellied Hermit
  __ *P.s.syrmatophorus* Gould, 1851 W Colombia to SW Ecuador
  __ *P.s.columbianus* Boucard, 1891[10] E Colombia, E Ecuador, N Peru

○ *Phaethornis koepckeae* Weske & Terborgh, 1977 Koepcke's Hermit Peru (2587)

○ *Phaethornis philippii* (Bourcier, 1847) Needle-billed Hermit E Peru, N Bolivia, W Brazil

○ *Phaethornis bourcieri* Straight-billed Hermit
  __ *P.b.bourcieri* (Lesson, 1832)[11] E Colombia, NE Peru to Brazil, S Venezuela and Guianas
  __ *P.b.major* Hinkelmann, 1989 Brazil (1119)

○ *Phaethornis anthophilus* Pale-bellied Hermit[12]
  __ *P.a.hyalinus* Bangs, 1901 Pearl Is. (Panama)
  __ *P.a.anthophilus* (Bourcier, 1843) C Panama, N Colombia, N Venezuela

○ *Phaethornis eurynome* Scale-throated Hermit
  __ *P.e.eurynome* (Lesson, 1832)[13] SE Brazil
  __ *P.e.paraguayensis* M. & W. Bertoni, 1901[14] E Paraguay, NE Argentina

[1] This generic name was reintroduced in Hinkelmann (1999: 539) (1121). In last Edition we placed this in the genus *Phaethornis* following Peters (1945) (1826), see Peters p. 7 for a relevant footnote.
[2] Includes *sanctijohannis*, a name given to immature birds, see Hinkelmann (1999: 539) (1121).
[3] Includes *veraecrucis* see Hinkelmann (1996) (1120).
[4] Includes *cassinii* see Hinkelmann (1996) (1120).
[5] The broad species concept used in our last Edition we replace with the species concept of Hinkelmann (1996, 1999) (1120, 1121); some forms are attached to a species *malaris*, and *longirostris* is treated as a good polytypic species.
[6] Includes *saturatior* see Hinkelmann (1996) (1120).
[7] The name *ucayali* J.T. Zimmer, 1950 (2713) relates to intergrades between *moorei* and *bolivianus* see Hinkelmann (1999: 543) (1121).
[8] Includes *insignis* see Hinkelmann (1996) (1120).
[9] For attachment of this taxon here see Hinkelmann (1988) (1117). Includes *camargoi* Grantsau, 1988 (975).
[10] Includes *huallagae* see Hinkelmann (1999: 543) (1121).
[11] Includes *whitelyi* see Hinkelmann (1999: 544) (1121).
[12] The name *fuliginosus*, treated by Peters (1945) (1826) as a subspecies, attached to a melanistic bird of doubtful origin, see Hinkelmann (1999: 544) (1121).
[13] The name *Phaethornis nigrirostris* Ruschi, 1973 (2133), has been given to aberrant black-billed individuals, see Hinkelmann (1988) (1117) and *pinheiroi* Ruschi, 1965 (2131) is a synonym (Hinkelmann *in litt.*).
[14] In our last Edition the species *eurynome* was treated as monotypic, but see Hinkelmann (1999: 544) (1121).

○ *Phaethornis pretrei* (Lesson & DeLattre, 1839) PLANALTO HERMIT[1] Bolivia, N Argentina, Paraguay, SE and E Brazil

○ *Phaethornis augusti* SOOTY-CAPPED HERMIT
— *P.a.curiosus* Wetmore, 1956 N Colombia (2597)
— *P.a.augusti* (Bourcier, 1847)[2] E Colombia, N Venezuela
— *P.a.incanescens* (Simon, 1921) S Venezuela, W Guyana

○ *Phaethornis subochraceus* Todd, 1915 BUFF-BELLIED HERMIT N Bolivia, S Brazil

○ *Phaethornis squalidus* (Temminck, 1822) DUSKY-THROATED HERMIT SE Brazil

○ *Phaethornis rupurumii* STREAK-THROATED HERMIT[3]
— *P.r.rupurumii* Boucard, 1892 E Colombia, C and E Venezuela, W Guyana, N Brazil
— *P.r.amazonicus* Hellmayr, 1906 NC Brazil

● *Phaethornis longuemareus* (Lesson, 1832) LITTLE HERMIT[4] [5] NE Venezuela to French Guiana, Trinidad

○ *Phaethornis idaliae* (Bourcier & Mulsant, 1856) MINUTE HERMIT SE Brazil

○ *Phaethornis nattereri* Berlepsch, 1887 CINNAMON-THROATED HERMIT[6] NE and SW Brazil, E Bolivia

○ *Phaethornis ruber* REDDISH HERMIT
— *P.r.episcopus* Gould, 1857 C and E Venezuela, Guyana, N Brazil
— *P.r.ruber* (Linnaeus, 1758) Surinam, French Guiana, Brazil, SE Peru, N Bolivia
— *P.r.nigricinctus* Lawrence, 1858 SW Venezuela, E and S Colombia, to N Peru
— *P.r.longipennis* Berlepsch & Stolzmann, 1902 S Peru

○ *Phaethornis stuarti* E. Hartert, 1897 WHITE-BROWED HERMIT SE Peru to C Bolivia

○ *Phaethornis atrimentalis* BLACK-THROATED HERMIT
— *P.a.atrimentalis* Lawrence, 1858 Colombia, Ecuador, N Peru
— *P.a.riojae* Berlepsch, 1889[7] C Peru

○ *Phaethornis striigularis* STRIPE-THROATED HERMIT
— *P.s.saturatus* Ridgway, 1910[8] [9] S Mexico to NW Colombia
— *P.s.subrufescens* Chapman, 1917 W Colombia, W Ecuador
— *P.s.striigularis* Gould, 1854 N Colombia, W Venezuela
— *P.s.ignobilis* Todd, 1913 N Venezuela

○ *Phaethornis griseogularis* GREY-CHINNED HERMIT
— *P.g.griseogularis* Gould, 1851 SE Venezuela, N Brazil, E Colombia, E Ecuador, E Peru
— *P.g.zonura* Gould, 1860 E Andean slope NW Peru
— *P.g.porcullae* Carriker, 1935 W Andean NW Peru

TROCHILINAE

ANDRODON Gould, 1863 M
○ *Androdon aequatorialis* Gould, 1863 TOOTH-BILLED HUMMINGBIRD E Panama, W Colombia, NW Ecuador

DORYFERA Gould, 1847 F
○ *Doryfera ludovicae* GREEN-FRONTED LANCEBILL
— *D.l.veraguensis* Salvin, 1867 NC Costa Rica, W Panama
— *D.l.ludovicae* (Bourcier & Mulsant, 1847)[10] E Panama, Colombia, W Venezuela to NW Bolivia

○ *Doryfera johannae* BLUE-FRONTED LANCEBILL
— *D.j.johannae* (Bourcier, 1847) EC Colombia to NE Peru
— *D.j.guianensis* (Boucard, 1893) SE Venezuela, S Guyana, N Brazil

[1] The names *schwarti* Ruschi, 1975 (2136), and *minor* Grantsau, 1988 (975), relate to individual variation (Hinkelmann, 1999: 544) (1121).
[2] Includes *vicarius* see Peters (1945) (1826).
[3] Separated from the species *squalidus* by Hinkelmann & Schuchmann (1997) (1122).
[4] This species previously treated as conspecific with *striigularis* and *atrigularis* but see Hinkelmann (1999: 545) (1121).
[5] The name *imatacae* Phelps & Phelps, Jr., 1952 (1850) relates to a hybrid, so too does *aethopyga* J.T. Zimmer, 1950 (2713) see Hinkelmann (1996) (1120).
[6] The name *maranhaoensis* Grantsau, 1968 (973), was given to the previously undescribed male, see Hinkelmann (1988) (1118).
[7] Submerged in nominate *atrimentalis* in our last Edition.
[8] Includes *adolphi* Gould, 1857, preoccupied; a substitute name *cordobae* J.T. Zimmer, 1950 (2713), yields priority to *saturatus* see Hinkelmann (1999: 547) (1121).
[9] Includes *nelsoni* see Hinkelmann (1999: 547) (1121).
[10] Includes *rectirostris* and *grisea*, see Stiles (1999: 549) (2344).

CAMPYLOPTERUS Swainson, 1827[1] M
○ *Campylopterus cuvierii* SCALY-BREASTED HUMMINGBIRD
  __ *C.c.roberti* (Salvin, 1861)   SE Mexico, N Guatemala to E NE Costa Rica
  __ *C.c.maculicauda* (Griscom, 1932)   Pacific slope Costa Rica
  __ *C.c.furvescens* (Wetmore, 1967)[2]   Pacific slope W Panama (2604)
  __ *C.c.saturatior* (E. Hartert, 1901   Coiba I. (Panama)
  __ *C.c.cuvierii* (DeLattre & Bourcier, 1846) E and C Panama
  __ *C.c.berlepschi* (Hellmayr, 1915)   N Colombia

○ *Campylopterus curvipennis* WEDGE-TAILED SABREWING
  __ *C.c.curvipennis* (M.H.K. Lichtenstein, 1830) SE Mexico
  __ *C.c.excellens* (Wetmore, 1941)[3]   S Mexico
  __ *C.c.yucatanensis* (Simon, 1921)   Yucatán Pen. (Mexico)
  __ *C.c.pampa* (Lesson, 1832)[4]   N Guatemala, Belize, Honduras

○ *Campylopterus largipennis* GREY-BREASTED SABREWING
  __ *C.l.largipennis* (Boddaert, 1783)   E Venezuela, the Guianas, NW Brazil
  __ *C.l.obscurus* Gould, 1848   NE Brazil
  __ *C.l.aequatorialis* Gould, 1861   E Colombia and NW Brazil to N Bolivia
  __ *C.l.diamantinensis* Ruschi, 1963[5]   SE Brazil (2130)

○ *Campylopterus rufus* Lesson, 1840 RUFOUS SABREWING S Mexico, W Guatemala, El Salvador

○ *Campylopterus hyperythrus* Cabanis, 1848 RUFOUS-BREASTED SABREWING
                   SE Venezuela, NW Brazil

● *Campylopterus hemileucurus* VIOLET SABREWING
  __ *C.h.hemileucurus* (M.H.K. Lichtenstein, 1830) S Mexico to SC Nicaragua
  ✓ *C.h.mellitus* Bangs, 1902   Costa Rica, W Panama

○ *Campylopterus ensipennis* (Swainson, 1822) WHITE-TAILED SABREWING NE Venezuela, Tobago I.

○ *Campylopterus falcatus* (Swainson, 1821) LAZULINE SABREWING NC Venezuela to NE Ecuador

○ *Campylopterus phainopeplus* Salvin & Godman, 1879 SANTA MARTA SABREWING
                   NE Colombia

○ *Campylopterus villaviscensio* (Bourcier, 1851) NAPO SABREWING S Colombia, E Ecuador, NE Peru

○ *Campylopterus duidae* BUFF-BREASTED SABREWING
  __ *C.d.duidae* Chapman, 1929[6]   S Venezuela, N Brazil
  __ *C.d.guayquinimae* Zimmer & Phelps, 1946 Mt. Guaiquinima (S Venezuela) (2723)

APHANTOCHROA Gould, 1853 F
○ *Aphantochroa cirrochloris* (Vieillot, 1818) SOMBRE HUMMINGBIRD[7] E Brazil

EUPETOMENA Gould, 1853 F
○ *Eupetomena macroura* SWALLOW-TAILED HUMMINGBIRD[8]
  __ *E.m.macroura* (J.F. Gmelin, 1788)   The Guianas, Brazil, Paraguay
  __ *E.m.simoni* Hellmayr, 1929   NE Brazil
  __ *E.m.cyanoviridis* Grantsau, 1988   SE Brazil (975)
  __ *E.m.hirundo* Gould, 1875   E Peru
  __ *E.m.boliviana* J.T. Zimmer, 1950   Beni (NW Bolivia) (2714)

FLORISUGA Bonaparte, 1850 F
● *Florisuga mellivora* WHITE-NECKED JACOBIN
  ✓ *F.m.mellivora* (Linnaeus, 1758)   C America, N Sth. America
  __ *F.m.flabellifera* (Gould, 1846)   Tobago I.

○ *Florisuga fusca* (Vieillot, 1817) BLACK JACOBIN[9] E Brazil to SE Paraguay, NE Argentina, N Uruguay

---

[1] Including the genus *Phaeochroa* see Schuchmann (1999: 550) (2203).
[2] For recognition of this see Stiles (1999: 550) (2344).
[3] Treated as a separate monotypic species in our last Edition, but see Zûchner (1999: 550) (2737).
[4] Omitted in our last Edition.
[5] Omitted in our last Edition.
[6] Includes *zuloagae* Phelps & Phelps, 1948 (1844), see Sánchez Osés (1999: 554) (2163).
[7] Placed in the genus *Campylopterus* by Zûchner (1999: 554) (2737); we prefer to await more detailed evidence, but we have listed the species next to that genus.
[8] Placed in the genus *Campylopterus* by Schuchmann (1999: 554) (2203); we prefer to await more detailed evidence.
[9] Previously treated in a monotypic genus *Melanotrochilus*.

COLIBRI Spix, 1824 M

○ *Colibri delphinae* (Lesson, 1839) BROWN VIOLET-EAR[1] Guatemala to Panama, N and W Sth. America

● *Colibri thalassinus* GREEN VIOLET-EAR

— *C.t.thalassinus* (Swainson, 1827)[2] C and S Mexico to NC Nicaragua

✓ *C.t.cabanidis* (Heine, 1863) Costa Rica, W Panama

— *C.t.cyanotus* (Bourcier, 1843) Mts. of Colombia, Ecuador and NW Venezuela

— *C.t.kerdeli* Aveledo & Perez, 1991 Cordillera Oriental (NE Venezuela) (68)

— *C.t.crissalis* Todd, 1942 Andes of Peru, Bolivia and NW Argentina

○ *Colibri coruscans* SPARKLING VIOLET-EAR

— *C.c.coruscans* (Gould, 1846) Andes of NW Venezuela and Colombia to NW Argentina

— *C.c.rostratus* Phelps & Phelps, Jr., 1952 Mts. of Amazonas (S Venezuela) (1850)

— *C.c.germanus* (Salvin & Godman, 1884) Tepuis of SE Venezuela, E Guyana, N Brazil

○ *Colibri serrirostris* (Vieillot, 1816) WHITE-VENTED VIOLET-EAR E Bolivia, C Paraguay, S Brazil, N Argentina

ANTHRACOTHORAX Boie, 1831 M

○ *Anthracothorax viridigula* (Boddaert, 1783) GREEN-THROATED MANGO

NE Venezuela, Trinidad, the Guianas, N Brazil

● *Anthracothorax prevostii* GREEN-BREASTED MANGO

— *A.p.prevostii* (Lesson, 1832) E and S Mexico to Guatemala, Belize and El Salvador

✓ *A.p.gracilirostris* Ridgway, 1910 El Salvador to C Costa Rica

— *A.p.hendersoni* (Cory, 1887)[3] San Andrés I., Providencia I.

— *A.p.viridicordatus* Cory, 1913 NE Colombia, N Venezuela

— *A.p.iridescens* (Gould, 1861)[4] W Colombia, SW Ecuador, NW Peru

○ *Anthracothorax nigricollis* (Vieillot, 1817) BLACK-THROATED MANGO W Panama, Sth. America south to NE Argentina

○ *Anthracothorax veraguensis* Reichenbach, 1855 VERAGUAS MANGO Panama

○ *Anthracothorax dominicus* ANTILLEAN MANGO

— *A.d.dominicus* (Linnaeus, 1766) Hispaniola and nearby islands

— *A.d.aurulentus* (Audebert & Vieillot, 1801) Puerto Rico and nearby islands

○ *Anthracothorax viridis* (Audebert & Vieillot, 1801) GREEN MANGO Puerto Rico

○ *Anthracothorax mango* (Linnaeus, 1758) JAMAICAN MANGO Jamaica

○ *Anthracothorax recurvirostris* (Swainson, 1822) FIERY-TAILED AWLBILL[5] SE Venezuela, the Guianas, NE Brazil, E Ecuador

TOPAZA G.R. Gray, 1840 F

○ *Topaza pella* CRIMSON TOPAZ

— *T.p.pella* (Linnaeus, 1758)[6] E Venezuela, the Guianas

— *T.p.smaragdula* (Bosc, 1792) Surinam and NE Brazil (north of the Amazon)

— *T.p.microrhyncha* Butler, 1926 S bank of Amazon (NC Brazil)

○ *Topaza pyra* FIERY TOPAZ

— *T.p.pyra* (Gould, 1846) E Colombia, S Venezuela, NW Brazil

— *T.p.amaruni* Hu, Joseph & Agro, 2000 E Ecuador, NW Peru (1156)

— *T.p.*subsp. ? WC Brazil[7]

EULAMPIS Boie, 1831 M

○ *Eulampis jugularis* (Linnaeus, 1766) PURPLE-THROATED CARIB Montane Lesser Antilles

○ *Eulampis holosericeus* GREEN-THROATED CARIB[8]

— *E.h.holosericeus* (Linnaeus, 1758) E Puerto Rico, most of Lesser Antilles

— *E.h.chlorolaemus* Gould, 1857 Grenada I.

---

[1] Includes *greenwalti* Ruschi, 1962 (2128), see Stiles (1999: 557) (2344).
[2] Includes *minor* see Stiles (1999: 557) (2344).
[3] Includes *pinchoti* see Stiles (1999: 558) (2344).
[4] Previously attached to *A. nigricollis* and still a matter of opinion see Stiles (1999: 558) (2344).
[5] Previously treated in the monotypic genus *Avocettula*, but see Schuchmann (1999: 560) (2203).
[6] Includes *pampreta* see Hu *et al.* (2000) (1156).
[7] See map in Hu *et al.* (2000) (1156).
[8] Previously treated in the monotypic genus *Sericotes*, but see A.O.U. (1998: 289) (3).

CHRYSOLAMPIS Boie, 1831 F
○ *Chrysolampis mosquitus* (Linnaeus, 1758) RUBY TOPAZ

E Panama, Colombia E to NE Brazil, S to E Bolivia

ORTHORHYNCUS Lacépède, 1799 M
○ *Orthorhyncus cristatus* ANTILLEAN CRESTED HUMMINGBIRD
    __ *O.c.exilis* (J.F. Gmelin, 1788)
    __ *O.c.ornatus* Gould, 1861
    __ *O.c.cristatus* (Linnaeus, 1758)
    __ *O.c.emigrans* Lawrence, 1877

E Puerto Rico to Lesser Antilles and St. Lucia I.
St Vincent I.
Barbados I.
Grenada and Grenadines Is.

KLAIS Reichenbach, 1854 F
● *Klais guimeti* VIOLET-HEADED HUMMINGBIRD
    ↲ *K.g.merrittii* (Lawrence, 1860)[1]
    __ *K.g.guimeti* (Bourcier, 1843)
    __ *K.g.pallidiventris* Stolzmann, 1926

E Honduras to E Panama
N and W Venezuela to N Peru, N and NW Brazil
E Peru, WC Bolivia

STEPHANOXIS Simon, 1897 M
○ *Stephanoxis lalandi* PLOVERCREST
    __ *S.l.lalandi* (Vieillot, 1818)
    __ *S.l.loddigesii* (Gould, 1831)

E Brazil
S Brazil, E Paraguay, NE Argentina

ABEILLIA Bonaparte, 1850 F
○ *Abeillia abeillei* EMERALD-CHINNED HUMMINGBIRD
    __ *A.a.abeillei* (Lesson & DeLattre, 1839)
    __ *A.a.aurea* Miller & Griscom, 1925

Mts. of SE Mexico to N Honduras
Mts. of S Honduras, N Nicaragua

LOPHORNIS Lesson, 1829[2] M[3]
○ *Lophornis ornatus* (Boddaert, 1783) TUFTED COQUETTE

E Venezuela, the Guianas, Trinidad, N Brazil

○ *Lophornis gouldii* (Lesson, 1833) DOT-EARED COQUETTE

N and C Brazil

○ *Lophornis magnificus* (Vieillot, 1817) FRILLED COQUETTE

EC Brazil

○ *Lophornis brachylophus* R.T. Moore, 1949 SHORT-CRESTED COCQUETTE[4]

SW Mexico (1618)

○ *Lophornis delattrei* RUFOUS-CRESTED COQUETTE[5]
    __ *L.d.lessoni* Simon, 1921
    __ *L.d.delattrei* (Lesson, 1839)

SW Costa Rica, Panama, C Colombia
E and C Peru, N Bolivia

○ *Lophornis stictolophus* Salvin & Elliot, 1873 SPANGLED COQUETTE

Andes of W Venezuela and E Colombia S to N Peru

○ *Lophornis chalybeus* FESTIVE COQUETTE
    __ *L.c.verreauxii* J. & E. Verreaux, 1853
    __ *L.c.klagesi* Berlepsch & Hartert, 1902
    __ *L.c.chalybeus* (Vieillot, 1823)

E Colombia to C Bolivia, NW Brazil
SE Venezuela
SE Brazil

○ *Lophornis pavoninus* PEACOCK COQUETTE
    __ *L.p.pavoninus* Salvin & Godman, 1882[6]

    __ *L.p.duidae* Chapman, 1929

Cerro Ptaretepui, Mt. Roraima and the Merumé Mts.
   (Venezuela/Guyana)
Mt. Duida (SE Venezuela)

○ *Lophornis helenae* (DeLattre, 1843) BLACK-CRESTED COQUETTE

S Mexico to E Costa Rica

○ *Lophornis adorabilis* Salvin, 1870 WHITE-CRESTED COQUETTE

C Costa Rica to W Panama

DISCOSURA Bonaparte, 1850[7] F
○ *Discosura popelairii* (Du Bus, 1846)[8] WIRE-CRESTED THORNTAIL

E Colombia, E Ecuador, NE Peru

---

[1] For recognition see Stiles (1999: 564) (2344).
[2] Includes the two species (*helenae* and *adorabilis*) treated in the genus *Paphosia* in last Edition. See Schuchmann (1999: 569) (2203).
[3] Gender addressed by David & Gosselin (2002) (614), multiple changes occur in specific names.
[4] For reasons to separate this from the next species see Banks (1990) (87).
[5] Walters (1997) (2556) has discussed *Lophornis melania* concluding that the name was probably applied to aberrant specimens or faded skins of *L. d. delattrei*.
[6] Includes *punctigula* Zimmer & Phelps, 1946 (2723), see Zūchner (1999: 568) (2737).
[7] Includes the four species (*popelairii, langsdorffi, letitiae* and *conversii*) treated in the genus *Popelairia* in last Edition. See Schuchmann (1999: 571) (2203).
[8] Du Bus is an abbreviation for Du Bus de Gisignies.

○ *Discosura langsdorffi* BLACK-BELLIED THORNTAIL
    — *D.l.melanosternon* (Gould, 1868)   SE Colombia, S Venezuela, E Ecuador, E Peru, W Brazil
    — *D.l.langsdorffi* (Temminck, 1821)   E Brazil

○ *Discosura letitiae* (Bourcier & Mulsant, 1852) COPPERY THORNTAIL   NE Bolivia

○ *Discosura conversii* (Bourcier & Mulsant, 1846) GREEN THORNTAIL   Costa Rica to W Ecuador

○ *Discosura longicaudus* (J.F. Gmelin, 1788) RACQUET-TAILED COQUETTE[1]
                                        S Venezuela, the Guianas, NC and NE Brazil

TROCHILUS Linnaeus, 1758 M
○ *Trochilus polytmus* Linnaeus, 1758 RED-BILLED STREAMERTAIL   Jamaica

○ *Trochilus scitulus* (Brewster & Bangs, 1901) BLACK-BILLED STREAMERTAIL[2]
                                          E Jamaica

CHLOROSTILBON Gould, 1853 M
○ *Chlorostilbon notatus* BLUE-CHINNED SAPPHIRE[3] [4]
    — *C.n.notatus* (Reich, 1795)   NE Colombia to Trinidad and Tobago, the Guianas,
                                          E Brazil (2000)
    — *C.n.puruensis* Riley, 1913   NW Brazil to SE Colombia and NE Peru
    — *C.n.obsoletus* (J.T. Zimmer, 1950)   NE Peru (2715)

○ *Chlorostilbon auriceps* (Gould, 1852) GOLDEN-CROWNED EMERALD   W and C Mexico

○ *Chlorostilbon forficatus* Ridgway, 1885 COZUMEL EMERALD   Cozumel Is.

○ *Chlorostilbon canivetii* CANIVET'S EMERALD
    — *C.c.canivetii* (Lesson, 1832)   SE Mexico, Belize, N Guatemala, Bay Is.
    — *C.c.osberti* Gould, 1860   SE Mexico, C and W Guatemala, El Salvador,
                                          Honduras, Nicaragua
    — *C.c.salvini* Cabanis & Heine, 1860   W Costa Rica

○ *Chlorostilbon assimilis* Lawrence, 1861 GARDEN EMERALD   SW Costa Rica, SW Panama, Pearl Is.

○ *Chlorostilbon mellisugus* BLUE-TAILED EMERALD[5]
    — *C.m.pumilus* Gould, 1872   W Colombia, W Ecuador
    — *C.m.melanorhynchus* Gould, 1860   Andes of W Colombia, W and C Ecuador
    — *C.m.gibsoni* (Fraser, 1840)   C Colombia
    — *C.m.chrysogaster* (Bourcier, 1843)   N Colombia
    — *C.m.nitens* Lawrence, 1861   NE Colombia, NW Venezuela
    — *C.m.caribaeus* Lawrence, 1871   Netherlands Antilles, Trinidad, NE Venezuela
    — *C.m.duidae* Zimmer & Phelps, 1952   Mt. Duida (S Venezuela) (2728)
    — *C.m.subfurcatus* Berlepsch, 1887   E and S Venezuela, Guyana, NW Brazil
    — *C.m.mellisugus* (Linnaeus, 1758)   Surinam, French Guiana, NE Brazil
    — *C.m.phoeopygus* (Tschudi, 1844)[6]   Upper Amazonia
    — *C.m.peruanus* Gould, 1861   Peru, E Bolivia

○ *Chlorostilbon olivaresi* Stiles, 1996 CHIRIBIQUETE EMERALD #   SE Colombia (2341)

○ *Chlorostilbon aureoventris* GLITTERING-BELLIED EMERALD[7]
    — *C.a.pucherani* (Bourcier & Mulsant, 1848)   E Brazil
    — *C.a.aureoventris* (d'Orbigny & Lafresnaye, 1838)   Bolivia, Paraguay, WC Brazil, NW Argentina
    — *C.a.berlepschi* Pinto, 1938   S Brazil, Uruguay, NE Argentina

○ †*Chlorostilbon bracei* (Lawrence, 1877) BRACE'S EMERALD[8]   Bahamas

○ *Chlorostilbon ricordii* (Gervais, 1835) CUBAN EMERALD   Cuba, Isla de la Juventud, Bahamas

○ *Chlorostilbon swainsonii* (Lesson, 1829) HISPANIOLAN EMERALD   Hispaniola

---

[1] For correction of spelling see David & Gosselin (2002) (613).
[2] Treated as separate from *T. polytmus* by Schuchmann (1978; 1999: 572) (2201, 2203); not recognised as separate by A.O.U. (1998: 297) (3).
[3] Treated in a monotypic genus *Chlorestes* in last Edition. For correction of the attribution of this name (not of Reichenbach, 1795) see Walters (1996) (2555).
[4] See Bündgen (1999: 572) (269) for comments on synonymy and some doubtful associated names.
[5] Bündgen (1999: 573) (269) treated this species more broadly and included the four preceding ones. We choose to follow A.O.U. (1998: 292) (3).
[6] This, not *phaeopygus*, is Tschudi's original spelling. Includes *vitticeps* see Bündgen (1999: 573) (269); previously listed as a separate species.
[7] The name *minutus* Berlioz, 1949 (138) is thought to have been attached to a hybrid.
[8] For treatment at specific level see Graves & Olson (1987) (991).

○ *Chlorostilbon maugaeus* (Audebert & Vieillot, 1801) Puerto Rican Emerald
Puerto Rico

○ *Chlorostilbon russatus* (Salvin & Godman, 1881) Coppery Emerald   NE Colombia, NW Venezuela

○ *Chlorostilbon stenurus* Narrow-tailed Emerald[1]
  __ *C.s.stenurus* (Cabanis & Heine, 1860)   NE Colombia, NW Venezuela, NE Ecuador
  __ *C.s.ignotus* Todd, 1942[2]   NW Venezuela

○ *Chlorostilbon alice* (Bourcier & Mulsant, 1848) Green-tailed Emerald
N Venezuela

○ *Chlorostilbon poortmani* Short-tailed Emerald
  __ *C.p.poortmani* (Bourcier, 1843)   E Colombia, NW Venezuela
  __ *C.p.euchloris* (Reichenbach, 1854)   C Colombia

○ †?*Chlorostilbon elegans* (Gould, 1860) Caribbean Emerald[3]   ? Caribbean

PANTERPE Cabanis & Heine, 1860 F
○ *Panterpe insignis* Fiery-throated Hummingbird
  __ *P.i.eisenmanni* Stiles, 1985   N Costa Rica (2338)
  __ *P.i.insignis* Cabanis & Heine, 1860   NC Costa Rica to W Panama

ELVIRA Mulsant and J. & E. Verreaux, 1866 F
○ *Elvira chionura* (Gould, 1851) White-tailed Emerald   S Costa Rica to C Panama
● *Elvira cupreiceps* (Lawrence, 1866) Coppery-headed Emerald   N and C Costa Rica

EUPHERUSA Gould, 1857 F
○ *Eupherusa cyanophrys* Rowley & Orr, 1964 Oaxaca Hummingbird   S Oaxaca (Mexico) (2117)

○ *Eupherusa poliocerca* Elliot, 1871 White-tailed Hummingbird   S Mexico

◉ *Eupherusa eximia* Stripe-tailed Hummingbird
  __ *E.e.nelsoni* Ridgway, 1910   E and SE Mexico
  __ *E.e.eximia* (DeLattre, 1843)   E Mexico to C Nicaragua
  ◿ *E.e.egregia* Sclater & Salvin, 1868   Costa Rica, W Panama

○ *Eupherusa nigriventris* Lawrence, 1867 Black-bellied Hummingbird   C Costa Rica, W Panama

GOETHALSIA Nelson, 1912 F
○ *Goethalsia bella* Nelson, 1912 Pirre Hummingbird   E Panama, NW Colombia

GOLDMANIA Nelson, 1911 F
○ *Goldmania violiceps* Nelson, 1911 Violet-capped Hummingbird   E Panama

CYNANTHUS Swainson, 1827 M
○ *Cynanthus sordidus* (Gould, 1859) Dusky Hummingbird   To north of Sierra Madre del Sur (S Mexico)

○ *Cynanthus latirostris* Broad-billed Hummingbird[4]
  __ *C.l.magicus* (Mulsant & Verreaux, 1872)   Arizona south to Nayarit and Durango
  __ *C.l.propinquus* R.T. Moore, 1939   NE Jalico, Guanajuaoto (C Mexico)
  __ *C.l.latirostris* Swainson, 1827   C and E Mexico
  __ *C.l.toroi* Berlioz, 1937   Colima and Michoacán (SC Mexico)
  __ *C.l.lawrencei* (Berlepsch, 1887)   Tres Marias Arch.
  __ *C.l.doubledayi* (Bourcier, 1847)   Coastal plain of SW Mexico

CYANOPHAIA Reichenbach, 1854 F
○ *Cyanophaia bicolor* (J.F. Gmelin, 1788) Blue-headed Hummingbird   Dominica and Martinique Is.

THALURANIA Gould, 1848[5] F
○ *Thalurania ridgwayi* Nelson, 1900 Mexican Woodnymph   W Mexico

---

[1] For correction of spelling see David & Gosselin (2002) (613).
[2] For correction of spelling see David & Gosselin (2002) (613).
[3] For a discussion see Weller (1999) (2572).
[4] For alternative views to those of Powers (1999: 581) (1914), supporting some specific separation see Navarro & Peterson (1999) (1638).
[5] The arrangement of the first four species follows Schuchmann (1999: 585-586) (2203), although tentatively; in last Edition these four species were treated as two.

● *Thalurania colombica* Purple-crowned Woodnymph
   — *T.c.townsendi* Ridgway, 1888   E Guatemala to SE Honduras
   ↙ *T.c.venusta* (Gould, 1851)   Nicaragua to C Panama
   — *T.c.colombica* (Bourcier, 1843)   N Colombia, NW Venezuela
   — *T.c.rostrifera* Phelps & Phelps, Jr., 1956   NW Venezuela (1858)

○ *Thalurania fannyi* Green-crowned Woodnymph
   — *T.f.fannyi* (DeLattre & Bourcier, 1846)   E Panama, NW to SW Colombia
   — *T.f.subtropicalis* Griscom, 1932   WC Colombia
   — *T.f.verticeps* (Gould, 1851)   SW Colombia, NW Ecuador
   — *T.f.hypochlora* Gould, 1871   W Ecuador

○ *Thalurania furcata* Fork-tailed Woodnymph
   — *T.f.refulgens* Gould, 1853   NE Venezuela
   — *T.f.furcata* (J.F. Gmelin, 1788)   E Venezuela, the Guianas, NE Brazil
   — *T.f.fissilis* Berlepsch & Hartert, 1902   E Venezuela, W Guyana, NE Brazil
   — *T.f.orenocensis* Hellmayr 1921   Upper Orinoco (S Venezuela)
   — *T.f.nigrofasciata* (Gould, 1846)   SE Colombia, S Venezuela, NW Brazil
   — *T.f.viridipectus* Gould, 1848[1]   E Colombia, E Ecuador, NE Peru
   — *T.f.jelskii* Taczanowski, 1874   E Peru, W Brazil
   — *T.f.simoni* Hellmayr, 1906   SE Peru, SW Brazil
   — *T.f.balzani* Simon, 1896   NC Brazil
   — *T.f.furcatoides* Gould, 1861   E Brazil
   — *T.f.boliviana* Boucard, 1894   SE Peru, NE Bolivia
   — *T.f.baeri* Hellmayr, 1907   NE and C Brazil to SE Bolivia and NW and NC
                           Argentina
   — *T.f.eriphile* (Lesson, 1832)   SE Brazil, Paraguay, NE Argentina

○ *Thalurania watertonii* (Bourcier, 1847) Long-tailed Woodnymph   Guyana, E Brazil

○ *Thalurania glaucopis* (J.F. Gmelin, 1788) Violet-capped Woodnymph
                          E and S Brazil, E Paraguay, NE Argentina

DAMOPHILA Reichenbach, 1854 F
○ *Damophila julie* Violet-bellied Hummingbird
   — *D.j.panamensis* Berlepsch, 1884   C Panama
   — *D.j.julie* (Bourcier, 1842)   N and C Colombia
   — *D.j.feliciana* (Lesson, 1844)   SW Colombia to NW Peru

LEPIDOPYGA Reichenbach, 1855 F
○ *Lepidopyga coeruleogularis* Sapphire-throated Hummingbird
   — *L.c.coeruleogularis* (Gould, 1851)   W Panama
   — *L.c.confinis* Griscom, 1932   E Panama, NW Colombia
   — *L.c.coelina* (Bourcier, 1856)   N Colombia

○ *Lepidopyga lilliae* Stone, 1917 Sapphire-bellied Hummingbird   Coastal NC Colombia

○ *Lepidopyga goudoti* Shining-green Hummingbird
   — *L.g.luminosa* (Lawrence, 1862)   N Colombia
   — *L.g.goudoti* (Bourcier, 1843)   Magdalena valley (C Colombia)
   — *L.g.zuliae* Cory, 1918   NE Colombia, NW Venezuela
   — *L.g.phaeochroa* Todd, 1942   NW Venezuela

HYLOCHARIS Boie, 1831 F
● *Hylocharis eliciae* Blue-throated Goldentail
   ↙ *H.e.eliciae* (Bourcier & Mulsant, 1846)   SE Mexico to S Costa Rica
   — *H.e.earina* Wetmore, 1967[2]   W Panama to NW Colombia (2604)

○ *Hylocharis sapphirina* (J.F. Gmelin, 1788) Rufous-throated Sapphire
                          E Colombia, S Venezuela and the Guianas to NE Peru
                          and C Brazil, NE Bolivia, Paraguay, NE Argentina

---

[1] Includes *taczanowksii* see Stiles (1999: 586) (2344).
[2] Not recognised in our last Edition, for recognition see Stiles (1999: 589) (2344).

○ *Hylocharis cyanus* WHITE-CHINNED SAPPHIRE[1]
   — *H.c.viridiventris* Berlepsch, 1880
                   N and E Colombia, W and S Venezuela, the Guianas, N Brazil

   — *H.c.rostrata* Boucard, 1895   E Peru, NE Bolivia, W Brazil
   — *H.c.conversa* J.T. Zimmer, 1950   E Bolivia, N Paraguay, SW Brazil (2716)
   — *H.c.cyanus* (Vieillot, 1818)   Coastal E Brazil
   — *H.c.griseiventris* Grantsau, 1988 SE Brazil, NE Argentina (975)

○ *Hylocharis chrysura* (Shaw, 1812) GILDED HUMMINGBIRD[2]
                   NE and SC Bolivia, Paraguay, C and SE Brazil, Uruguay, N Argentina

○ *Hylocharis grayi* BLUE-HEADED SAPPHIRE
   — *H.g.humboldtii* (Bourcier & Mulsant, 1852) SE Panama to NW Ecuador
   — *H.g.grayi* (DeLattre & Bourcier, 1846)   W Colombia, N Ecuador

CHRYSURONIA Bonaparte, 1850 F
○ *Chrysuronia oenone* GOLDEN-TAILED SAPPHIRE
   — *C.o.oenone* (Lesson, 1832)[3]
                   N and E Venezuela, E Colombia, E Ecuador, NE Peru, W Brazil

   — *C.o.josephinae* (Bourcier & Mulsant, 1848) E Peru
   — *C.o.alleni* (Elliot, 1888)[4]   N Bolivia

LEUCOCHLORIS Reichenbach, 1854 M
○ *Leucochloris albicollis* (Vieillot, 1818) WHITE-THROATED HUMMINGBIRD
                   E Bolivia, SE Brazil, E Paraguay, N Argentina

POLYTMUS Brisson, 1760 M
○ *Polytmus guainumbi* WHITE-TAILED GOLDENTHROAT
   — *P.g.andinus* Simon, 1921[5]   E Colombia
   — *P.g.guainumbi* (Pallas, 1764)   Venezuela, the Guianas, Trinidad, N Brazil
   — *P.g.thaumantias* (Linnaeus, 1766) E and C Brazil, E Paraguay, E Bolivia, NE Argentina

○ *Polytmus milleri* (Chapman, 1929) TEPUI GOLDENTHROAT   SE Venezuela and Mt. Roraima (N Brazil)

○ *Polytmus theresiae* GREEN-TAILED GOLDENTHROAT
   — *P.t.theresiae* (Da Silva Maia, 1843)   The Guianas, NC Brazil
   — *P.t.leucorrhous* Sclater & Salvin, 1867 E Colombia and S Venezuela to NE Peru and NW Brazil

LEUCIPPUS Bonaparte, 1850[6] M
○ *Leucippus fallax* (Bourcier, 1843) BUFFY HUMMINGBIRD[7] Coastal Colombia and Venezuela, and offshore islands

○ *Leucippus baeri* Simon, 1901 TUMBES HUMMINGBIRD   SW Ecuador and NW Peru

○ *Leucippus taczanowskii* (P.L. Sclater, 1879) SPOT-THROATED HUMMINGBIRD[8]
                   N and C Peru

○ *Leucippus chlorocercus* Gould, 1866 OLIVE-SPOTTED HUMMINGBIRD E Peru

TAPHROSPILUS Simon, 1910[9] M
○ *Taphrospilus hypostictus* (Gould, 1862) MANY-SPOTTED HUMMINGBIRD[10]
                   E Ecuador, E Peru, C and SE Bolivia to SW Brazil

AMAZILIA Lesson, 1843[11] F
○ *Amazilia chionogaster* WHITE-BELLIED HUMMINGBIRD
   — *A.c.chionogaster* (Tschudi, 1844)   N and C Peru
   — *A.c.hypoleuca* (Gould, 1846)   SE Peru, Bolivia, NW Argentina

---

[1] In last Edition we listed a species *H. pyropygia* now thought to be a hybrid, see Weller (1999: 589) (2573).
[2] Includes ¶ *lessoni* Pinto & Camargo, 1955 (1897); this needs evaluation.
[3] Includes *longirostris* and *azurea*, while *intermedia* thought to be a hybrid, see Stiles (1999: 590) (2344).
[4] For recognition of this taxon see Stiles (1999: 590) (2344).
[5] Schuchmann (1999: 591) (2203) explained that the name *andinus* is not preoccupied. The name *doctus* is thus a synonym.
[6] A broader genus is employed in Schuchmann (1999: 592-593) (2203), but we defer adopting this until more detailed reasons are published.
[7] Proposed races, including *occidentalis* Phelps & Phelps, 1949 (1845), now thought to relate to individual, not geographic variation see Zúchner (1999: 592) (2737).
[8] Includes *fractus* see Weller (1999: 592) (2573).
[9] Schuchmann (1999: 593) (2203) placed this in the genus *Leucippus*.
[10] Includes *peruvianus* see Schuchmann (1999: 593) (2203).
[11] In Schuchmann (1999: 595-605) (2203) this genus is split. We defer this pending publication of a detailed analysis. Two of his three genera were treated as subgenera in Peters (1945) (1826); the third *Agyrtria* Reichenbach, 1854, was in the synonymy of the subgenus *Polyerata* Heine, 1863 (apparently a junior name).

○ *Amazilia viridicauda* (Berlepsch, 1883) Green-and-white Hummingbird
                                                  C Peru

● *Amazilia rutila* Cinnamon Hummingbird
    — *A.r.diluta* van Rossem, 1938   NW Mexico
    — *A.r.graysoni* Lawrence, 1866   Tres Marías Is. (off W Mexico)
    — *A.r.rutila* (Lesson, 1842)   W and SW Mexico
    ⌀ *A.r.corallirostris* (Bourcier & Mulsant, 1846)   S and SE Mexico to Costa Rica

● *Amazilia yucatanensis* Buff-bellied Hummingbird
    ⌀ *A.y.chalconota* Oberholser, 1898   S Texas, NE Mexico
    — *A.y.cerviniventris* (Gould, 1856)   E Mexico
    — *A.y.yucatanensis* (Cabot, 1845)   SE Mexico, NW Guatemala, N Belize

● *Amazilia tzacatl* Rufous-tailed Hummingbird
    ⌀ *A.t.tzacatl* (De la Llave, 1833)   EC Mexico to C Panama
    — *A.t.handleyi* Wetmore, 1963[1]   I. Escudo de Veraguas (NW Panama) (2601)
    — *A.t.fuscicaudata* (Fraser, 1840)   N Colombia, W Venezuela (872)
    — *A.t.brehmi* Weller & Schuchmann 1999   Nariño (SW Colombia) (2577)
    — *A.t.jucunda* (Heine, 1863)   W Colombia, W Ecuador

○ *Amazilia castaneiventris* (Gould, 1856) Chestnut-bellied Hummingbird
                                                  NC Colombia

○ *Amazilia amazilia* Amazilia Hummingbird
    — *A.a.dumerilii* (Lesson, 1832)   W and SE Ecuador, N Peru
    — *A.a.leucophoea* Reichenbach, 1854[2]   NW Peru
    — *A.a.amazilia* (Lesson, 1828)   W Peru
    — *A.a.caeruleigularis* Carriker, 1933   SW Peru

○ *Amazilia alticola* Gould, 1860 Loja Hummingbird[3]   S Ecuador

○ *Amazilia leucogaster* Plain-bellied Emerald
    — *A.l.leucogaster* (J.F. Gmelin, 1788)   E Venezuela, the Guianas, NE Brazil
    — *A.l.bahiae* (E. Hartert, 1899)   E Brazil

○ *Amazilia versicolor* Versicoloured Emerald
    — *A.v.millerii* (Bourcier, 1847)   E Colombia to E Peru, N Brazil
    — *A.v.hollandi* (Todd, 1913)   SE Venezuela
    — *A.v.nitidifrons* (Gould, 1860)   NE Brazil
    — *A.v.versicolor* (Vieillot, 1818)   SE Brazil
    — *A.v.kubtchecki* Ruschi, 1959[4]   NE Bolivia, E Paraguay, NE Argentina, SW Brazil (2127)

○ *Amazilia rondoniae* Ruschi, 1982 Blue-green Emerald #   WC Brazil, N Bolivia (2140)

○ *Amazilia brevirostris* White-chested Emerald[5]
    — *A.b.chionopectus* (Gould, 1859)   Trinidad
    — *A.b.brevirostris* (Lesson, 1829)[6]   E Venezuela and Guyana to NC Brazil
    — *A.b.orienticola* Todd, 1942   Coastal French Guiana

○ *Amazilia franciae* Andean Emerald
    — *A.f.franciae* (Bourcier & Mulsant, 1846)   NW and C Colombia
    — *A.f.viridiceps* (Gould, 1860)   SW Colombia, W Ecuador
    — *A.f.cyanocollis* (Gould, 1854)   N Peru

○ *Amazilia candida* White-bellied Emerald
    — *A.c.genini* (Meise, 1938)[7]   E and S Mexico
    — *A.c.candida* (Bourcier & Mulsant, 1846)   SE Mexico (Chiapas) to Nicaragua
    — *A.c.pacifica* (Griscom, 1929)   SE Mexico (Yucatán Pen.), S Guatemala

---

[1] Omitted in last Edition but see Weller & Schuchmann (1999) (2577).
[2] Spelled *leucophaea* in Weller (1999: 595) (2573), but Peters (1945) (1826) used *leucophoea*. Original spelling verified. ECD.
[3] For reasons to recognise this see Weller (2000) (2574).
[4] For recognition of this see Weller (1999: 597) (2573).
[5] In last Edition we listed this as *A. chionopectus*, however *Ornismya brevirostris* Lesson, treated by Peters (1945: 64) (1826) as a synonym of *A. versicolor*, has been found to be a senior synonym of *whitelyi* as used for this population. See Weller (1999: 598) (2573).
[6] Includes *whitelyi* see Weller (1999: 598) (2573).
[7] If placed in a genus *Argyrtria* Reichenbach the authors' names should remain bracketed (*contra* Weller (1999: 598) (2573), Meise described it as *Argyrtrina v. genini* using the generic name of Chubb, 1910, not the slightly different one of Reichenbach.

○ *Amazilia cyanocephala* Azure-crowned Hummingbird
    — *A.c.cyanocephala* (Lesson, 1829)[1]   E and S Mexico to E Honduras and NC Nicaragua
    — *A.c.chlorostephana* (T.R. Howell, 1965)[2]   E Honduras to NE Nicaragua (1151)

○ *Amazilia violiceps* Violet-crowned Hummingbird
    — *A.v.ellioti* (Berlepsch, 1889)   SW USA to NW and C Mexico
    — *A.v.violiceps* (Gould, 1859)   SW Mexico

○ *Amazilia viridifrons* Green-fronted Hummingbird[3]
    — *A.v.viridifrons* (Elliot, 1871)   Guerrero (S Mexico)
    — *A.v.wagneri* A.R. Phillips, 1966[4]   E and SC Oaxaca and Chiapas (S Mexico) (1872)

○ *Amazilia fimbriata* Glittering-throated Emerald[5]
    — *A.f.elegantissima* Todd, 1942[6]   NE Colombia, N and NW Venezuela
    — *A.f.fimbriata* (J.F. Gmelin, 1788)[7]   NE Venezuela, the Guianas, N Brazil
    — *A.f.apicalis* (Gould, 1861)   E Colombia
    — *A.f.fluviatilis* (Gould, 1861)   SE Colombia, E Ecuador
    — *A.f.laeta* (E. Hartert, 1900)   NE Peru
    — *A.f.nigricauda* (Elliot, 1878)   E Bolivia to C Brazil
    — *A.f.tephrocephala* (Vieillot, 1818)   SE Brazil

○ *Amazilia lactea* Sapphire-spangled Emerald
    — *A.l.zimmeri* (Gilliard, 1941)   SE Venezuela
    — *A.l.lactea* (Lesson, 1832)   C and S Brazil
    — *A.l.bartletti* (Gould, 1866)   E and SE Peru, N Bolivia

● *Amazilia amabilis* Blue-chested Hummingbird[8]
    ⚖ *A.a.amabilis* (Gould, 1853)   NE Nicaragua to NC Colombia, W Ecuador[9]
    — *A.a.decora* (Salvin, 1891)   SW Costa Rica, W Panama

○ *Amazilia rosenbergi* (Boucard, 1895) Purple-chested Hummingbird   W Colombia, NW Ecuador

○ *Amazilia boucardi* (Mulsant, 1877) Mangrove Hummingbird   W Costa Rica

○ *Amazilia luciae* (Lawrence, 1867) Honduran Emerald   Honduras

● *Amazilia saucerrottei* Steely-vented Hummingbird[10]
    ⚖ *A.s.hoffmanni* (Cabanis & Heine, 1860)   SW Nicaragua to C Costa Rica
    — *A.s.warscewiczi* (Cabanis & Heine, 1860)   N and C Colombia
    — *A.s.saucerrottei* (DeLattre & Bourcier, 1846)[11]   Andean slopes and valleys of NW Colombia
    — *A.s.braccata* (Heine, 1863)   Andes of Merida, Venezuela

○ *Amazilia cyanifrons* (Bourcier, 1843) Indigo-capped Hummingbird   N and C Colombia

○ †*Amazilia alfaroana* Underwood, 1896 Miravalles Hummingbird   Volcán Miravalles (Costa Rica)[12]

○ *Amazilia edward* Snowy-breasted Hummingbird
    — *A.e.niveoventer* (Gould, 1851)[13]   SW Costa Rica, W and C Panama, Coiba I
    — *A.e.edward* (DeLattre & Bourcier, 1846)   WC Panama
    — *A.e.collata* Wetmore, 1952   EC Panama (2595)
    — *A.e.margaritarum* (Griscom, 1927)   E Panama and islands in N Gulf of Panama

---

[1] Includes *guatemalensis* see Weller (1999: 599) (2573).
[2] Omitted from last Edition.
[3] Includes *villadai* Peterson & Navarro, 2000 (1834B). These authors treated *wagneri* as a species. They considered *rowleyi* S.N.G. Howell, 1993 (1150) rightly placed as a synonym of *wagneri* by Weller (1999: 599) (2573). The AOU Checklist Committee apparently still considers *wagneri* conspecific with *viridifrons*. We follow with reservations; ¶*villadai* Peterson & Navarro-Siguenza (2000) (1834B) may deserve recognition.
[4] Given as 1964 in Weller (1999: 599) (2573), but actually published in 1966 see Dickerman & Parkes (1997: 226) (729).
[5] In last Edition we listed *Amazilia distans* Wetmore & Phelps, 1956 (2614), next after this. It is now thought this name attaches to a hybrid, see Weller (1999: 601) (2573).
[6] Includes *obscuricauda* Zimmer & Phelps, 1951 (2727), see Weller (1999: 601) (2573).
[7] Includes *maculicauda* and *alia* J.T. Zimmer, 1950 (2727), see Weller (1999: 601) (2573).
[8] Type species of the genus *Polyerata* still submerged here, pending further publication.
[9] It is now thought the range of *amabilis* excludes that of *decora*, which is found on the Pacific slope; hence Weller (1999: 601) (2573) united these two taxa as there is no sympatry.
[10] Type species of the genus *Saucerrottia* still submerged here, pending further publication.
[11] Includes *australis* Meyer de Schauensee, 1951 (1567), see Weller (1999: 603) (2573).
[12] The recognition of this was discussed by Weller (1999: 603; 2001) (2573, 2576).
[13] Includes the aberrant *ludibunda* Wetmore, 1952 (2595), see Weller (1999: 603) (2573).

○ *Amazilia cyanura* Blue-tailed Hummingbird
 — *A.c.guatemalae* (Dearborn, 1907)   SE Mexico to S Guatemala
 — *A.c.cyanura* Gould, 1859   S Honduras to NW Nicaragua
 — *A.c.impatiens* (Bangs, 1906)[1]   NW and C Costa Rica

○ *Amazilia beryllina* Berylline Hummingbird
 — *A.b.viola* (W. de W. Miller, 1905)   NW and W Mexico
 — *A.b.beryllina* (M.H.K. Lichtenstein, 1830) C Mexico
 — *A.b.lichtensteini* R.T. Moore, 1950   W Chiapas (S Mexico) (1619)
 — *A.b.sumichrasti* Salvin, 1891[2]   C and S Chiapas (Mexico)
 — *A.b.devillei* (Bourcier & Mulsant, 1848) S Guatemala to C Honduras

○ *Amazilia viridigaster* Green-bellied Hummingbird
 — *A.v.viridigaster* (Bourcier, 1843)   NC Colombia
 — *A.v.iodura* (Reichenbach, 1854)[3]   NE Colombia, W Venezuela

○ *Amazilia cupreicauda* Copper-tailed Hummingbird[4]
 — *A.c.duidae* (Chapman, 1929)   S and C Venezuela
 — *A.c.cupreicauda* Salvin & Godman, 1884 E Venezuela, W Guyana, N Brazil
 — *A.c.laireti* (Phelps & Aveledo, 1988) Extreme S Venezuela (1840)
 — *A.c.pacaraimae* (Weller, 2000)   SE Venezuela (Co. Urutani) (2575)

○ *Amazilia tobaci* Copper-rumped Hummingbird
 — *A.t.monticola* (Todd, 1913)   NW Venezuela
 — *A.t.feliciae* (Lesson, 1840)   NC Venezuela
 — *A.t.caudata* Zimmer & Phelps, 1949 NE Venezuela (2725)
 — *A.t.aliciae* Richmond, 1895   NE Venezuela, Margarita I.
 — *A.t.erythronotos* (Lesson, 1829)   Trinidad
 — *A.t.tobaci* (J.F. Gmelin, 1788)   Tobago I.
 — *A.t.caurensis* (Berlepsch & Hartert, 1902) E and SE Venezuela

MICROCHERA Gould, 1858 F
○ *Microchera albocoronata* Snowcap
 — *M.a.parvirostris* (Lawrence, 1865)   S Honduras, Nicaragua, Costa Rica
 — *M.a.albocoronata* (Lawrence, 1855) WC Panama

ANTHOCEPHALA Cabanis & Heine, 1860 F
○ *Anthocephala floriceps* Blossomcrown
 — *A.f.floriceps* (Gould, 1854)   NE Colombia
 — *A.f.berlepschi* Salvin, 1893   C Colombia

CHALYBURA Reichenbach, 1854 F
○ *Chalybura buffonii* White-vented Plumeleteer
 — *C.b.micans* Bangs & Barbour, 1922   C and E Panama, W Colombia
 — *C.b.buffonii* (Lesson, 1832)   C and NE Colombia, NW Venezuela
 — *C.b.aeneicauda* Lawrence, 1865   N Colombia, NC and W Venezuela
 — *C.b.caeruleogaster* (Gould, 1847)   N and C Colombia

● *Chalybura urochrysia* Bronze-tailed Plumeleteer
 ✓ *C.u.melanorrhoa* Salvin, 1865   Nicaragua, Costa Rica
 — *C.u.isaurae* (Gould, 1861)[5]   E Panama, NW Colombia
 — *C.u.urochrysia* (Gould, 1861)   SE Panama, NC and W Colombia, NW Ecuador
 — *C.u.intermedia* E. & C. Hartert, 1894[6] SW Ecuador

LAMPORNIS Swainson, 1827 M
○ *Lampornis clemenciae* Blue-throated Hummingbird
 — *L.c.phasmorus* Oberholser, 1974[7]   Chisos Mts. (Texas) (1672)

---

[1] For reasons for recognition see Weller (1997: 604) (2573).
[2] Recognised by Weller (1999: 604) (2573).
[3] Recognised by Weller (1999: 604) (2573).
[4] Separated from *viridigaster* by Weller (1999: 604; 2000) (2573, 1575).
[5] Includes *incognita* see Stiles (1999: 608) (2344).
[6] In last Edition treated as a race of *C. buffonii* but see Stiles (199: 608) (2344).
[7] For recognition of this see Browning (1978) (246).

— *L.c.bessophilus* (Oberholser, 1918) SW USA, NW Mexico
— *L.c.clemenciae* (Lesson, 1829) S USA, NE, S and C Mexico

○ *Lampornis amethystinus* AMETHYST-THROATED HUMMINGBIRD
 — *L.a.amethystinus* Swainson, 1827 W, C and E Mexico
 — *L.a.circumventris* (A.R. Phillips, 1966)[1] S Mexico (1870)
 — *L.a.margaritae* (Salvin & Godman, 1889)[2] SW Mexico
 — *L.a.salvini* (Ridgway, 1908) S Mexico, Guatemala, El Salvador
 — *L.a.nobilis* Griscom, 1932 Honduras

○ *Lampornis viridipallens* GREEN-THROATED MOUNTAIN-GEM
 — *L.v.amadoni* J.S. Rowley, 1968[3] S Mexico (2116)
 — *L.v.ovandensis* (Brodkorb, 1939) S Mexico, NW Guatemala
 — *L.v.viridipallens* (Bourcier & Mulsant, 1846)[4] Guatemala, N El Salvador, W Honduras
 — *L.v.nubivagus* Dickey & van Rossem, 1929 El Salvador

○ *Lampornis sybillae* (Salvin & Godman, 1892) GREEN-BREASTED MOUNTAIN-GEM
        E Honduras, NC Nicaragua

○ *Lampornis hemileucus* (Salvin, 1865) WHITE-BELLIED MOUNTAIN-GEM NC Costa Rica to W Panama

● *Lampornis calolaemus* PURPLE-THROATED MOUNTAIN-GEM
 — *L.c.pectoralis* (Salvin, 1891) S Nicaragua to NC Costa Rica
 — *L.c.calolaemus* (Salvin, 1865)[5] C Costa Rica
 — *L.c.homogenes* Wetmore, 1967[6] S Costa Rica, W Panama (2604)

○ *Lampornis castaneoventris* WHITE-THROATED MOUNTAIN-GEM[7]
 — *L.c.cinereicauda* (Lawrence, 1867) S Costa Rica
 — *L.c.castaneoventris* (Gould, 1851) W Panama

BASILINNA Boie, 1831[8] F
○ *Basilinna xantusii* (Lawrence, 1860) XANTUS'S HUMMINGBIRD S Baja California, Cerralvo I.

○ *Basilinna leucotis* WHITE-EARED HUMMINGBIRD
 — *B.l.borealis* (Griscom, 1929) SE Arizona, N Mexico
 — *B.l.leucotis* (Vieillot, 1818) C and S Mexico, Guatemala
 — *B.l.pygmaea* Simon & Hellmayr, 1908 El Salvador, Honduras, Nicaragua

LAMPROLAIMA Reichenbach, 1854 F
○ *Lamprolaima rhami* (Lesson, 1838) GARNET-THROATED HUMMINGBIRD[9]
        C and S Mexico south to Honduras and El Salvador

ADELOMYIA Bonaparte, 1854 F
○ *Adelomyia melanogenys* SPECKLED HUMMINGBIRD
 — *A.m.cervina* Gould, 1872 W and C Colombia
 — *A.m.connectens* Meyer de Schauensee, 1945 S Colombia (1562)
 — *A.m.melanogenys* (Fraser, 1840) E Colombia to SC Peru, W Venezuela
 — *A.m.debellardiana* Aveledo & Perez, 1994[10] W Venezuela (Perija Mts.) (69)
 — *A.m.aeneosticta* Simon, 1889 C and N Venezuela
 — *A.m.maculata* Gould, 1861 Ecuador to N Peru
 — *A.m.chlorospila* Gould, 1872 SE Peru
 — *A.m.inornata* (Gould, 1846) Bolivia, NW Argentina

PHLOGOPHILUS Gould, 1860 M
○ *Phlogophilus hemileucurus* Gould, 1860 ECUADORIAN PIEDTAIL S Colombia to NE Peru

○ *Phlogophilus harterti* Berlepsch & Stolzmann, 1901 PERUVIAN PIEDTAIL
        C and SE Peru

[1] For recognition see Züchner (1999: 608) (2737).
[2] Includes *brevirostris* see Züchner (1999: 608) (2737).
[3] Züchner (1999: 609) (2737) considered this a more distinct race than *ovandensis* and *nubivagus*.
[4] Includes *connectens* see Züchner (1999: 609) (2737).
[5] For correction of spelling see David & Gosselin (2002) (614).
[6] For recognition see Stiles (1999: 611) (2344). Not all authors place this taxon in this species.
[7] For this and the next species we retain the treatment of A.O.U. (1998: 304-305) (3).
[8] We tentatively follow Schuchmann (1999) (2203) in reintroducing this generic name.
[9] Includes *saturatior* and *occidentalis* A.R. Phillips, 1966 (1870), see Schuchmann (1999: 612) (2203).
[10] The type has been examined by A. Weller (*in litt.*) who provisionally recognises this form.

CLYTOLAEMA Gould, 1853 F
○ ***Clytolaema rubricauda*** (Boddaert, 1783) Brazilian Ruby   SE Brazil

HELIODOXA Gould, 1849[1]
○ ***Heliodoxa xanthogonys*** Velvet-browed Brilliant
    — *H.x.willardi* Weller & Renner, 2001   Extreme S Venezuela (2576A)
    — *H.x. xanthogonys* Salvin & Godman, 1882   E Venezuela, Guyana, NC Brazil

○ ***Heliodoxa gularis*** (Gould, 1860) Pink-throated Brilliant   S Colombia, NE Ecuador, NE Peru, NW Brazil

○ ***Heliodoxa branickii*** (Taczanowski, 1874) Rufous-webbed Brilliant   SE Peru, NW Bolivia

○ ***Heliodoxa schreibersii*** Black-throated Brilliant
    — *H.s.schreibersii* (Bourcier, 1847)   SE Colombia, E Ecuador, NE Peru, NW Brazil
    — *H.s.whitelyana* (Gould, 1872)   E Peru

○ ***Heliodoxa aurescens*** (Gould, 1846) Gould's Jewelfront   Amazonia

○ ***Heliodoxa rubinoides*** Fawn-breasted Brilliant
    — *H.r.rubinoides* (Bourcier & Mulsant, 1846)   C and E Colombia
    — *H.r.aequatorialis* (Gould, 1860)   W Colombia, W Ecuador
    — *H.r.cervinigularis* (Salvin, 1892)   E Ecuador, E Peru, C Bolivia

● ***Heliodoxa jacula*** Green-crowned Brilliant
    ✔ *H.j.henryi* Lawrence, 1867   Costa Rica, W Panama
    — *H.j.jacula* Gould, 1850   E Panama, N and C Colombia
    — *H.j.jamesoni* (Bourcier, 1851)   SW Colombia, W Ecuador

○ ***Heliodoxa imperatrix*** (Gould, 1856) Empress Brilliant   WC Colombia, NW Ecuador

○ ***Heliodoxa leadbeateri*** Violet-fronted Brilliant
    — *H.l.leadbeateri* (Bourcier, 1843)   N Venezuela
    — *H.l.parvula* Berlepsch, 1887   NW Venezuela, N, C and S Colombia
    — *H.l.sagitta* (Reichenbach, 1854)   E Ecuador, N Peru
    — *H.l.otero* (Tschudi, 1844)[2]   C and S Peru, NW Bolivia

EUGENES Gould, 1856 M
○ ***Eugenes fulgens*** Magnificent Hummingbird
    — *E.f.fulgens* (Swainson, 1827)[3]   SW USA to Honduras and NE Nicaragua
    — *E.f.spectabilis* (Lawrence, 1867)   Costa Rica, W Panama

HYLONYMPHA Gould, 1873 F
○ ***Hylonympha macrocerca*** Gould, 1873 Scissor-tailed Hummingbird Paria Pen. (NE Venezuela)

STERNOCLYTA Gould, 1858 F
○ ***Sternoclyta cyanopectus*** (Gould, 1846) Violet-chested Hummingbird NW Venezuela

UROCHROA Gould, 1856 F
○ ***Urochroa bougueri*** White-tailed Hillstar
    — *U.b.bougueri* (Bourcier, 1851)   SW Colombia, NW Ecuador
    — *U.b.leucura* Lawrence, 1864[4]   S Colombia, E Ecuador, NE Peru

BOISSONNEAUA Reichenbach, 1854 F
○ ***Boissonneaua flavescens*** Buff-tailed Coronet
    — *B.f.flavescens* (Loddiges, 1832)   Colombia, NW Venezuela
    — *B.f.tinochlora* Oberholser, 1902   SW Colombia, W Ecuador

○ ***Boissonneaua matthewsii*** (Bourcier, 1847) Chestnut-breasted Coronet
              SE Colombia, Ecuador, Peru

○ ***Boissonneaua jardini*** (Bourcier, 1851) Velvet-purple Coronet SW Colombia, NW Ecuador

[1] The species treated here and their sequence follows Gerwin & Zink (1989) (933).
[2] Apparently omitted from our previous Edition.
[3] Includes *viridiceps* see Powers (1999: 619) (1914).
[4] The inclusion of a "race" *eulcura* in previous Editions seems to have been a *lapsus*, but perhaps *leucura* has been found to be preoccupied and *eulcura* proposed as a substitute!

AGLAEACTIS Gould, 1848 F[1]
○ *Aglaeactis cupripennis* Shining Sunbeam
　　— *A.c.cupripennis* (Bourcier, 1843)[2]　　　Colombia, Ecuador, N Peru
　　— *A.c.caumatonota* Gould, 1848　　　　　C and S Peru

○ *Aglaeactis aliciae* Salvin, 1896 Purple-backed Sunbeam　　Marañon Valley (N Peru)

○ *Aglaeactis castelnaudii* White-tufted Sunbeam
　　— *A.c.regalis* J.T. Zimmer, 1951[3]　　　C Peru (2718)
　　— *A.c.castelnaudii* (Bourcier & Mulsant, 1848)　　S Peru

○ *Aglaeactis pamela* (d'Orbigny, 1838) Black-hooded Sunbeam　　Andes of N and C Bolivia

OREOTROCHILUS Gould, 1847 M
○ *Oreotrochilus chimborazo* Ecuadorian Hillstar[4]
　　— *O.c.jamesonii* Jardine, 1849　　　　　N Ecuador, S Colombia
　　— *O.c.soderstromi* Lönnberg & Rendahl, 1922　　Mt. Quillotoa (C Ecuador)
　　— *O.c.chimborazo* (DeLattre & Bourcier, 1846)　　Mt. Chimborazo (C Ecuador)

○ *Oreotrochilus estella* Andean Hillstar
　　— *O.e.stolzmanni* Salvin, 1895　　　　　N and C Peru
　　— *O.e.estella* (d'Orbigny & Lafresnaye, 1838)　　SW Peru to N Chile and NW Argentina
　　— *O.e.bolivianus* Boucard, 1893　　　　C Bolivia

● *Oreotrochilus leucopleurus* Gould, 1847 White-sided Hillstar　　S Bolivia to SC Chile and S Argentina

○ *Oreotrochilus melanogaster* Gould, 1847 Black-breasted Hillstar　　C Peru

○ *Oreotrochilus adela* (d'Orbigny & Lafresnaye, 1838) Wedge-tailed Hillstar
　　　　　　　　　　　　　　　　　　　　　S Bolivia

LAFRESNAYA Bonaparte, 1850 F
○ *Lafresnaya lafresnayi* Mountain Velvetbreast
　　— *L.l.liriope* Bangs, 1910　　　　　　NE Colombia
　　— *L.l.greenewalti* Phelps & Phelps, Jr., 1961　　W Venezuela (1863)
　　— *L.l.lafresnayi* (Boissonneau, 1840)[5]　　　W Venezuela, C and E Colombia
　　— *L.l.saul* (DeLattre & Bourcier, 1846)　　SW Colombia, Ecuador, N Peru
　　— *L.l.rectirostris* Berlepsch & Stolzmann, 1902[6]　　N and C Peru

COELIGENA Lesson, 1832 F
○ *Coeligena coeligena* Bronzy Inca
　　— *C.c.ferruginea* (Chapman, 1917)　　　W and C Colombia
　　— *C.c.columbiana* (Elliot, 1876)　　　　E and C Colombia, NW Venezuela
　　— *C.c.coeligena* (Lesson, 1833)[7]　　　N Venezuela
　　— *C.c.zuliana* Phelps & Phelps, Jr., 1953　　W Venezuela, NE Colombia (1853)
　　— *C.c.obscura* (Berlepsch & Stolzmann, 1902)[8]　　S Colombia, Ecuador, Peru
　　— *C.c.boliviana* (Gould, 1861)　　　　C and SE Bolivia

○ *Coeligena wilsoni* (DeLattre & Bourcier, 1846) Brown Inca　　W and SW Colombia, W Ecuador

○ *Coeligena prunellei* (Bourcier, 1843) Black Inca　　NC Colombia

○ *Coeligena torquata* Collared Inca[9]
　　— *C.t.conradii* (Bourcier, 1847)　　　E Colombia, NW Venezuela
　　— *C.t.torquata* (Boissonneau, 1840)　　NW Venezuela, Colombia, E Ecuador, N Peru
　　— *C.t.fuligidigula* (Gould, 1854)　　　W Ecuador
　　— *C.t.margaretae* J.T. Zimmer, 1948　　N Peru (2712)

[1] Gender addressed by David & Gosselin (2002) (614), multiple changes occur in specific names.
[2] Includes *parvula, ruficauda* and *cajabambae* J.T. Zimmer, 1951 (2718), see Schuchmann (1999: 621) (2203).
[3] Recognised by Fjeldså (1999: 621) (848).
[4] In our last Edition this was treated as part of *O. estella*; for separation see Fjeldså (1999: 623) (848); we defer specific level treatment for *stolzmanni* pending further publication.
[5] Includes *tamae* Phelps & Aveledo, 1987 (1839) (nec *tamai* as given in last Edition), see Schuchmann, 1999: 624) (2203).
[6] Includes *orestes* J.T. Zimmer, 1951 (2718), see Schuchmann (1999: 624) (2203).
[7] Includes *zuloagae* Phelps & Phelps, Jr., 1959 (1861), see Züchner (1999: 627) (2737).
[8] Not recognised in last Edition but see Züchner (1999: 627) (2737).
[9] We have chosen not to follow Züchner (1999: 628) (2737) yet in the separation of a species *C. inca*. He was right to highlight the need for further study.

— *C.t.eisenmanni* Weske, 1985   S Peru (2584)

— *C.t.insectivora* (Tschudi, 1844)   C Peru

— *C.t.omissa* J.T. Zimmer, 1948   SE Peru (2712)

— *C.t.inca* (Gould, 1852)   Bolivia

○ *Coeligena phalerata* (Bangs, 1898) WHITE-TAILED STARFRONTLET   NE Colombia

○ *Coeligena bonapartei* GOLDEN-BELLIED STARFRONTLET[1]

— *C.b.consita* Wetmore & Phelps, 1952   NE Colombia, NW Venezuela (2612)

— *C.b.bonapartei* (Boissonneau, 1840)   E Colombia

— *C.b.orina* Wetmore, 1953   NC Colombia (2596)

— *C.b.eos* (Gould, 1848)   W Venezuela

○ *Coeligena helianthea* BLUE-THROATED STARFRONTLET

— *C.h.helianthea* (Lesson, 1838)   N and E Colombia

— *C.h.tamai* Berlioz & Phelps, 1953[2]   W Venezuela (144)

○ *Coeligena lutetiae* (DeLattre & Bourcier, 1846) BUFF-WINGED STARFRONTLET

                                                                 C Colombia, C Ecuador, N Peru

○ *Coeligena violifer* VIOLET-THROATED STARFRONTLET

— *C.v.dichroura* (Taczanowski, 1874)   S Ecuador, N and C Peru, W Peru

— *C.v.albicaudata* Schuchmann & Züchner, 1997   S Peru (2207)

— *C.v.osculans* (Gould, 1871)   SE Peru

— *C.v.violifer* (Gould, 1846)   NW Bolivia

○ *Coeligena iris* RAINBOW STARFRONTLET

— *C.i.hesperus* (Gould, 1865)   SW Ecuador

— *C.i.iris* (Gould, 1854)   S Ecuador, NW Peru (Piura)

— *C.i.aurora* (Gould, 1854)   NW Peru (Cutervo)

— *C.i.flagrans* J.T. Zimmer, 1951   W slope of W Andes of Peru (2717)

— *C.i.eva* (Salvin, 1897)   E slope of W Andes of N Peru

— *C.i.fulgidiceps* (Simon, 1921)[3]   E Andes of N Peru (E of the Marañon)

ENSIFERA Lesson, 1843 F
○ *Ensifera ensifera* (Boissonneau, 1840) SWORD-BILLED HUMMINGBIRD   W Venezuela, Colombia to N Bolivia

PTEROPHANES Gould, 1849 M
○ *Pterophanes cyanopterus* GREAT SAPPHIREWING

— *P.c.cyanopterus* (Fraser, 1839)   NC Colombia

— *P.c.caeruleus* J.T. Zimmer, 1951   S Colombia (2718)

— *P.c.peruvianus* Boucard, 1895   Ecuador, Peru, N Bolivia

PATAGONA G.R. Gray, 1840 F
◑ *Patagona gigas* GIANT HUMMINGBIRD

— *P.g.peruviana* Boucard, 1893   SW Colombia to N Chile, NW Argentina

✓ *P.g.gigas* (Vieillot, 1824)   C and S Chile, WC Argentina

SEPHANOIDES G.R. Gray, 1840 M
● *Sephanoides sephaniodes* (Lesson, 1827) GREEN-BACKED FIRECROWN[4]

                                                                 Chile, W Argentina

○ *Sephanoides fernandensis* JUAN FERNANDEZ FIRECROWN

— *S.f.fernandensis* (P.P. King, 1831)   Robinson Crusoe I. (Juan Fernandez Is.)

— †?*S.f.leyboldi* (Gould, 1870)   Alejandro Selkirk I. (Juan Fernandez Is.)

HELIANGELUS Gould, 1848 M
○ *Heliangelus mavors* Gould, 1848 ORANGE-THROATED SUNANGEL   E Colombia, NW Venezuela

---

[1] We have chosen not to follow Züchner (1999: 628) (2737) yet in the separation of a species *C. eos*.
[2] Erroneously spelled *tamae* in last Edition.
[3] Includes *hypocrita* see Züchner (1999: 631) (2737).
[4] Monotypic treatment of Roy (1999) (2118) presumably includes *albicans* A. Kleinschmidt, 1970 (1273), which seems to be an infrasubspecific name and unavailable, and *mariannae* A. Kleinschmidt, 1970 (1273).

○ *Heliangelus amethysticollis* Amethyst-throated Sunangel[1]

    — *H.a.violiceps* Phelps & Phelps, Jr., 1953   NE Colombia (1853)

    — *H.a.clarisse* (Longuemare, 1841)[2]   E Colombia, W Venezuela

    — *H.a.spencei* (Bourcier, 1847)   Mérida Andes (NW Venezuela)

    — *H.a.laticlavius* Salvin, 1891   S Ecuador, N Peru

    — *H.a.decolor* J.T. Zimmer, 1951   C Peru (2718)

    — *H.a.amethysticollis* (d'Orbigny & Lafresnaye, 1838)   S Peru, NW Bolivia

○ *Heliangelus strophianus* (Gould, 1846) Gorgeted Sunangel   SW Colombia, NW Ecuador

○ *Heliangelus exortis* (Fraser, 1840) Tourmaline Sunangel   Colombia, NW Ecuador

○ *Heliangelus micraster* Little Sunangel

    — *H.m.micraster* Gould, 1872   SE Ecuador, N Peru

    — *H.m.cutervensis* Simon, 1921   NW Peru

○ *Heliangelus viola* (Gould, 1853) Purple-throated Sunangel   S Ecuador, N Peru

○ †?*Heliangelus zusii* Graves, 1993 Bogotá Sunangel #   Bogota (Colombia) (988)

○ *Heliangelus regalis* Fitzpatrick, Willard & Terborgh, 1979 Royal Sunangel

                                     N Peru (840)

ERIOCNEMIS Reichenbach, 1849[3] F[4]

○ *Eriocnemis nigrivestis* (Bourcier & Mulsant, 1852) Black-breasted Puffleg

                                     NW Ecuador

○ *Eriocnemis vestita* Glowing Puffleg

    — *E.v.paramillo* (Chapman, 1917)   NW Colombia

    — *E.v.vestita* (Lesson, 1838)   E Colombia, NW Venezuela

    — *E.v.smaragdinipectus* Gould, 1868   SW Colombia, E Ecuador

    — *E.v.arcosi* Schuchmann, Weller & Heynen, 2001   S Ecuador, N Peru (2206)

○ *Eriocnemis derbyi* (DeLattre & Bourcier, 1846) Black-thighed Puffleg[5]

                                     C Colombia, NW Ecuador

○ *Eriocnemis godini* (Bourcier, 1851) Turquoise-throated Puffleg   NW Ecuador

○ *Eriocnemis cupreoventris* (Fraser, 1840) Coppery-bellied Puffleg   E Colombia, NW Venezuela

○ *Eriocnemis luciani* Sapphire-vented Puffleg[6]

    — *E.l.meridae* Schuchmann, Weller & Heynen, 2001   Merida Andes of NW Venezuela (2206)

    — *E.l.luciani* (Bourcier, 1847)   W Ecuador, SW Colombia

    — *E.l.baptistae* Schuchmann, Weller & Heynen, 2001   Andes of SC to S Ecuador (2206)

    — *E.l.catharina* Salvin, 1897   N Peru

    — *E.l.sapphiropygia* Taczanowski, 1874[7]   C and S Peru

○ *Eriocnemis mosquera* (DeLattre & Bourcier, 1846) Golden-breasted Puffleg

                                     SW and C Colombia, NW Ecuador

○ *Eriocnemis glaucopoides* (d'Orbigny & Lafresnaye, 1838) Blue-capped Puffleg

                                     C Bolivia to NW Argentina

○ *Eriocnemis mirabilis* Meyer de Schauensee, 1967 Colourful Puffleg

                                     W Colombia (1574)

○ *Eriocnemis alinae* Emerald-bellied Puffleg

    — *E.a.alinae* (Bourcier, 1842)   E Colombia, E Ecuador

    — *E.a.dybowskii* Taczanowski, 1882   N and C Peru

HAPLOPHAEDIA Simon, 1918[8] F

○ *Haplophaedia aureliae* Greenish Puffleg

    — *H.a.caucensis* (Simon, 1911)[9]   SE Panama, W and C Andes of Colombia

---

[1] We have chosen not to follow Heynen (1999: 636) (1111) in the separation of a species *H. clarisse* (now preferring to lump even *spencei* — which we treated as a separate species in last Edition), but the two groups seem to have widely disjunct ranges.

[2] Includes *verdiscutatus* Phelps & Phelps, Jr., 1955 (1855) (nec *viridiscutatus* as in our last Edition), see Heynen (1999: 636) (1111).

[3] The names *soederstroemi* and *isaacsonii* probably attach to hybrids, see Heynen (1999: 639-640) (1111).

[4] Gender addressed by David & Gosselin (2002) (614), multiple changes occur in specific names.

[5] Includes *longirostris* see Heynen (1999: 639) (1111).

[6] We have not followed Heynen (1999: 640) (1111) yet in the separation of a species *E. sapphiropygia*.

[7] Includes *marcapatae* J.T. Zimmer, 1951 (2718) see Heynen (1999: 640) (1111).

[8] For this genus we follow Schuchmann *et al.* (2000) (2205).

[9] Includes *galindoi* Wetmore, 1967 (2604) and *floccus*, see Heynen (1999) (1111).

    — *H.a.aureliae* (Bourcier & Mulsant, 1846)[1]   C and E Colombia

    — *H.a.russata* (Gould, 1871)   N and C Ecuador

    — *H.a.cutucuensis* Schuchmann, Weller & Heynen, 2000   S Ecuador (2205)

○ *Haplophaedia assimilis* Buff-thighed Puffleg[2]

    — *H.a.affinis* (Taczanowski, 1884)   N and C Peru

    — *H.a.assimilis* (Elliot, 1876)   S Peru, N Bolivia

○ *Haplophaedia lugens* (Gould, 1851) Hoary Puffleg   SW Colombia, NW Ecuador

UROSTICTE Gould, 1853 F

○ *Urosticte benjamini* (Bourcier, 1851) Whitetip[3]

    — *U.b.benjamini* (Bourcier, 1851)[4][5]   W Colombia to SW Ecuador and NE Peru

    — *U.b.ruficrissa* Lawrence, 1864   S Colombia, E Ecuador

OCREATUS Gould, 1846 M

○ *Ocreatus underwoodii* Racket-tailed Puffleg[6]

    — *O.u.polystictus* Todd, 1942   N Venezuela

    — *O.u.discifer* (Heine, 1863)   NW Venezuela, NE Colombia

    — *O.u.underwoodii* (Lesson, 1832)   E Colombia

    — *O.u.incommodus* (O. Kleinschmidt, 1943)[7]   W and C Colombia (1274)

    — *O.u.melanantherus* (Jardine, 1851)   E Ecuador

    — *O.u.peruanus* (Gould, 1849)   E Ecuador, NE Peru

    — *O.u.annae* (Berlepsch & Stolzmann, 1894)   C and S Peru

    — *O.u.addae* (Bourcier, 1846)   Bolivia

LESBIA Lesson, 1832 F

○ *Lesbia victoriae* Black-tailed Trainbearer

    — *L.v.victoriae* (Bourcier & Mulsant, 1846)[8]   NE and S Colombia, Ecuador

    — *L.v.juliae* (E. Hartert, 1899)   N and C Peru

    — *L.v.berlepschi* (Hellmayr, 1915)   SE Peru

○ *Lesbia nuna* Green-tailed Trainbearer

    — *L.n.gouldii* (Loddiges, 1832)   Colombia, W Venezuela

    — *L.n.gracilis* (Gould, 1846)   Ecuador

    — *L.n.pallidiventris* (Simon, 1902)   N Peru

    — *L.n.eucharis* (Bourcier & Mulsant, 1848)[9]   C Peru

    — *L.n.nuna* (Lesson, 1832)[10]   SW Peru, N Bolivia

SAPPHO Reichenbach, 1849 F

○ *Sappho sparganura* Red-tailed Comet

    — *S.s.sparganura* (Shaw, 1812)   N and C Bolivia

    — *S.s.sapho* (Lesson, 1829)   S Bolivia, N and W Argentina

POLYONYMUS Heine, 1863 M

○ *Polyonymus caroli* (Bourcier, 1847) Bronze-tailed Comet   C and S Peru

RAMPHOMICRON Bonaparte, 1850 N

○ *Ramphomicron dorsale* Salvin & Godman, 1880 Black-backed Thornbill

                               NE Colombia

○ *Ramphomicron microrhynchum* Purple-backed Thornbill

    — *R.m.andicola* (Simon, 1921)[11]   W Venezuela

---

[1] Includes *bernali* Romero & Hérnandez, 1979 (2100) see Heynen (1999) (1111).
[2] For reasons to split this from the previous species see Schuchmann *et al.* (2000) (2205).
[3] We have not followed Schuchmann (1999: 643-644) (2203) who treated this as two species.
[4] Includes *rostrata* see Schuchmann (1999: 643) (2203).
[5] Includes *intermedia* see Schuchmann (1999: 643) (2203).
[6] We have preferred not to follow Schuchmann (1999: 644) (2203) but to restrict the generic name "Racquet-tail" to the genus *Prioniturus*.
[7] Includes *ambiguus* J.T. Zimmer, 1951 (2718) see Schuchmann (1999: 644) (2203). In last Edition we listed that.
[8] Includes *aequatorialis* see Züchner (1999: 647) (2737).
[9] See comments in Züchner (1999: 647) (2737); the discovery that this came from C Peru requires *chlorura* to be treated as a synonym.
[10] Includes *boliviana* see Züchner (1999: 647) (2737).
[11] For correction of spelling see David & Gosselin (2002) (613).

__ *R.m.microrhynchum* (Boissonneau, 1839)   Colombia, Ecuador, NW Peru

__ *R.m.albiventre* Carriker, 1935   C Peru

__ *R.m.bolivianum* Schuchmann, 1984   N Bolivia (2202)

OREONYMPHA Gould, 1869 F

○ *Oreonympha nobilis* BEARDED MOUNTAINEER

__ *O.n.albolimbata* Berlioz, 1938   WC Peru

__ *O.n.nobilis* Gould, 1869   SC Peru

OXYPOGON Gould, 1848 M

○ *Oxypogon guerinii* BEARDED HELMETCREST

__ *O.g.cyanolaemus* Salvin & Godman, 1880   NE Colombia

__ *O.g.lindenii* (Parzudaki, 1845)   NW Venezuela

__ *O.g.stuebelii* A.B. Meyer, 1884   C Colombia

__ *O.g.guerinii* (Boissonneau, 1840)   E Colombia

METALLURA Gould, 1847 F

○ *Metallura tyrianthina* TYRIAN METALTAIL

__ *M.t.districta* Bangs, 1899   NE Colombia

__ *M.t.chloropogon* (Cabanis & Heine, 1860)   N Venezuela

__ *M.t.oreopola* Todd, 1913   W Venezuela

__ *M.t.tyrianthina* (Loddiges, 1832)   NW Venezuela, Colombia, E and S Ecuador, N Peru

__ *M.t.quitensis* Gould, 1861   NW Ecuador

__ *M.t.septentrionalis* E. Hartert, 1899   N Peru

__ *M.t.smaragdinicollis* (d'Orbigny & Lafresnaye, 1838)[1]   C and S Peru, N Bolivia

○ *Metallura iracunda* Wetmore, 1946 PERIJA METALTAIL   N Colombia, W Venezuela (2591)

○ *Metallura williami* VIRIDIAN METALTAIL

__ *M.w.recisa* Wetmore, 1970[2]   NC Colombia (2606)

__ *M.w.williami* (DeLattre & Bourcier, 1846)   C Colombia

__ *M.w.primolina* Bourcier, 1853   N Ecuador, S Colombia

__ *M.w.atrigularis* Salvin, 1893   S Ecuador

○ *Metallura baroni* Salvin, 1893 VIOLET-THROATED METALTAIL   SC Ecuador

○ *Metallura odomae* Graves, 1980 NEBLINA METALTAIL   S Ecuador, N Peru (977)

○ *Metallura theresiae* COPPERY METALTAIL

__ *M.t.parkeri* Graves, 1981   N Peru (979)

__ *M.t.theresiae* Simon, 1902   NE Peru

○ *Metallura eupogon* (Cabanis, 1874) FIRE-THROATED METALTAIL   C Peru

○ *Metallura aeneocauda* SCALED METALTAIL

__ *M.a.aeneocauda* (Gould, 1846)   S Peru, N Bolivia

__ *M.a.malagae* Berlepsch, 1897   C Bolivia

○ *Metallura phoebe* (Lesson & DeLattre, 1839) BLACK METALTAIL   Peru, Bolivia, N Chile

CHALCOSTIGMA Reichenbach, 1854 N

○ *Chalcostigma ruficeps* (Gould, 1846) RUFOUS-CAPPED THORNBILL[3]   SE Ecuador, Peru to C Bolivia

○ *Chalcostigma olivaceum* OLIVACEOUS THORNBILL

__ *C.o.pallens* Carriker, 1935   C Peru

__ *C.o.olivaceum* (Lawrence, 1864)   S Peru to C Bolivia

○ *Chalcostigma stanleyi* BLUE-MANTLED THORNBILL

__ *C.s.stanleyi* (Bourcier, 1851)   Ecuador

__ *C.s.versigulare* J.T. Zimmer, 1924[4]   N Peru

__ *C.s.vulcani* (Gould, 1852)   S Peru to WC Bolivia

[1] Includes *peruviana* see Heindl & Schuchmann (1998) (1096).
[2] For recognition see Heindl & Schuchmann (1998) (1096).
[3] Includes *aureofastigata* see Heindl (1999: 655) (1095).
[4] For correction of spelling see David & Gosselin (2002) (613).

○ *Chalcostigma heteropogon* (Boissonneau, 1839) Bronze-tailed Thornbill

E Colombia, W Venezuela

○ *Chalcostigma herrani* Rainbow-bearded Thornbill
— *C.h.tolimae* O. Kleinschmidt, 1927
W Colombia
— *C.h.herrani* (DeLattre & Bourcier, 1846)
S Colombia to N Peru

OPISTHOPRORA Cabanis & Heine, 1860 F
○ *Opisthoprora euryptera* (Loddiges, 1832) Mountain Avocetbill   S Colombia, NE Ecuador

TAPHROLESBIA Simon, 1918 F
○ *Taphrolesbia griseiventris* (Taczanowski, 1883) Grey-bellied Comet   NW Peru (Marañon Valley)

AGLAIOCERCUS J.T. Zimmer, 1930 M
○ *Aglaiocercus kingi* Long-tailed Sylph
— *A.k.berlepschi* (E. Hartert, 1898)[1]
NE Venezuela
— *A.k.margarethae* (Heine, 1863)
NC and N Venezuela
— *A.k.caudatus* (Berlepsch, 1892)[2]
N Colombia, W Venezuela
— *A.k.emmae* (Berlepsch, 1892)
Colombia, NW Ecuador
— *A.k.kingi* (Lesson, 1832)
E Colombia
— *A.k.mocoa* (DeLattre & Bourcier, 1846)
S Colombia to N Peru
— *A.k.smaragdinus* (Gould, 1846)
C Peru to C Bolivia

○ *Aglaiocercus coelestis* Violet-tailed Sylph
— *A.c.coelestis* (Gould, 1861)[3]
W Colombia, N and C Ecuador
— *A.c.aethereus* (Chapman, 1925)
SW Ecuador

AUGASTES Gould, 1849 M
○ *Augastes scutatus* Hyacinth Visorbearer[4]
— *A.s.scutatus* (Temminck, 1824)
C and E Minas Gerais (Brazil)
— *A.s.ilseae* Grantsau, 1967
C and E Minas Gerais (Brazil) (972)
— *A.s.soaresi* Ruschi, 1963
SC Minas Gerais (Brazil) (2129)

○ *Augastes lumachella* (Lesson, 1838) Hooded Visorbearer[5]   E Brazil

SCHISTES Gould, 1851[6] M
○ *Schistes geoffroyi* Wedge-billed Hummingbird
— *S.g.albogularis* (Gould, 1851)
W Colombia, W Ecuador
— *S.g.geoffroyi* (Bourcier, 1843)
N Venezuela, E Colombia to S Peru
— *S.g.chapmani* (Berlioz, 1941)[7]
C Bolivia (137)

HELIOTHRYX Boie, 1831 M[8]
● *Heliothryx barroti* (Bourcier, 1843) Purple-crowned Fairy[9]   SE Mexico to W Ecuador

○ *Heliothryx auritus* Black-eared Fairy
— *H.a.auritus* (J.F. Gmelin, 1788)
SE Colombia and E Ecuador to the Guianas, N Brazil
— *H.a.phainolaemus* Gould, 1855
NC Brazil
— *H.a.auriculatus* (Nordmann, 1835)
E Peru, C Bolivia, C and E Brazil

HELIACTIN Boie, 1831 F
○ *Heliactin bilophus* (Temminck, 1821) Horned Sungem   S Surinam, C and E Brazil, E Bolivia

LODDIGESIA Bonaparte, 1850 F
○ *Loddigesia mirabilis* (Bourcier, 1847) Marvellous Spatuletail   N Peru

---

[1] Altshuler (1999: 659) (37) treated this as a separate monotypic species; we await a more detailed published study.
[2] For correction of spelling see David & Gosselin (2002) (613).
[3] Includes *pseudocoelestis* see Schuchmann (1999: 659) (2203).
[4] For acceptance of three races see Schuchmann (1999:660) (2203).
[5] For correction of spelling see David & Gosselin (2002) (613).
[6] Submerged in *Augastes* in Schuchmann (1999: 660) (2203).
[7] Omitted from last Edition.
[8] Gender addressed by David & Gosselin (2002) (614), multiple changes occur in specific names.
[9] In last Edition we listed *major* as a race of *H. auritus*. Schuchmann (1999:663) (2203) suggested this name may have been applied to immature males of *H. barroti*.

HELIOMASTER Bonaparte, 1850 M
○ *Heliomaster constantii* PLAIN-CAPPED STARTHROAT
    — *H.c.pinicola* (Gould, 1853)   NW Mexico
    — *H.c.leocadiae* (Bourcier & Mulsant, 1852)[1]   W and SW Mexico, W Guatemala
    — *H.c.constantii* (DeLattre, 1843)   El Salvador to Costa Rica

○ *Heliomaster longirostris* LONG-BILLED STARTHROAT
    — *H.l.pallidiceps* Gould, 1861[2]   S Mexico to Nicaragua
    — *H.l.longirostris* (Audebert & Vieillot, 1801)[3]   E and SW Costa Rica to N and C Sth. America, Trinidad
    — *H.l.albicrissa* Gould, 1871   W Ecuador, NW Peru

○ *Heliomaster squamosus* (Temminck, 1823) STRIPE-BREASTED STARTHROAT
                            E Brazil

○ *Heliomaster furcifer* (Shaw, 1812) BLUE-TUFTED STARTHROAT   C and S Brazil to Bolivia, Paraguay and N Argentina

RHODOPIS Reichenbach, 1854 F
○ *Rhodopis vesper* OASIS HUMMINGBIRD
    — *R.v.koepckeae* Berlioz, 1974   NW Peru (142)
    — *R.v.vesper* (Lesson, 1829)[4]   Peru, N Chile
    — *R.v.atacamensis* (Leybold, 1869)   N Chile

THAUMASTURA Bonaparte, 1850 F
○ *Thaumastura cora* (Lesson & Garnot, 1827) PERUVIAN SHEARTAIL   W Peru

TILMATURA Reichenbach, 1855 F
○ *Tilmatura dupontii* (Lesson, 1832) SPARKLING-TAILED WOODSTAR   C and S Mexico to N Nicaragua

DORICHA Reichenbach, 1854 F
○ *Doricha enicura* (Vieillot, 1818) SLENDER SHEARTAIL   S Mexico, Guatemala, El Salvador, Honduras

○ *Doricha eliza* (Lesson & DeLattre, 1839) MEXICAN SHEARTAIL   N Yucatán Pen. (SE Mexico)

CALLIPHLOX Boie, 1831[5] F
○ *Calliphlox amethystina* (Boddaert, 1783) AMETHYST WOODSTAR   E Colombia, S Venezuela, Ecuador and the Guianas
                                      south to N Bolivia, Paraguay, NE Argentina

○ *Calliphlox evelynae* BAHAMA WOODSTAR
    — *C.e.evelynae* (Bourcier, 1847)[6]   Bahama and Caicos Is.
    — *C.e.lyrura* (Gould, 1869)   Gt. Inagua I. (Bahamas)

● *Calliphlox bryantae* (Lawrence, 1867) MAGENTA-THROATED WOODSTAR   N Costa Rica, W Panama

○ *Calliphlox mitchellii* (Bourcier, 1847) PURPLE-THROATED WOODSTAR   E Panama, W Colombia, W Ecuador

MICROSTILBON Todd, 1913 M
○ *Microstilbon burmeisteri* (P.L. Sclater, 1887) SLENDER-TAILED WOODSTAR
                                      C Bolivia to NW Argentina

CALOTHORAX G.R. Gray, 1840 M
○ *Calothorax lucifer* (Swainson, 1827) LUCIFER HUMMINGBIRD   SC USA to SC Mexico. Winters in S of range.

○ *Calothorax pulcher* Gould, 1859 BEAUTIFUL HUMMINGBIRD   S Mexico

MELLISUGA Brisson, 1760 F
○ *Mellisuga minima* VERVAIN HUMMINGBIRD
    — *M.m.minima* (Linnaeus, 1758)   Jamaica
    — *M.m.vieilloti* (Shaw, 1812)   Hispaniola and nearby islands

○ *Mellisuga helenae* (Lembeye, 1850) BEE HUMMINGBIRD[7]   Cuba, Isla de la Juventud

[1] Includes *surdus* see Stiles (1999: 664) (2344).
[2] *Contra* Stiles (1999: 664) (2344) we think that *masculinus* A.R. Phillips, 1966 (1870) must, due to its type locality, be a synonym of this form and not of *longirostris*.
[3] Includes *stuartae* see Stiles (1999: 664) (2344).
[4] Includes *tertia* see Schuchmann (1999: 665) (2203).
[5] We follow Schuchmann (1999) (2203) in submerging *Philodice* herein.
[6] Includes *salita* see Schuchmann (1999: 667) (2203).
[7] In last Edition we treated this in the genus *Calypte* but see Schuchmann (1999: 671) (2203).

ARCHILOCHUS Reichenbach, 1854 M

● *Archilochus colubris* (Linnaeus, 1758) RUBY-THROATED HUMMINGBIRD S Canada, C and E USA >> Florida, C Mexico to W Panama

● *Archilochus alexandri* (Bourcier & Mulsant, 1846) BLACK-CHINNED HUMMINGBIRD

SW Canada, W USA, N Mexico >> W and SC Mexico

CALYPTE Gould, 1856 F

● *Calypte anna* (Lesson, 1829) ANNA'S HUMMINGBIRD SW Canada, W USA, NW Mexico >> SC USA

○ *Calypte costae* (Bourcier, 1839) COSTA'S HUMMINGBIRD SW and WC USA, NW Mexico >> W Mexico

ATTHIS Reichenbach, 1854 F

○ *Atthis heloisa* BUMBLEBEE HUMMINGBIRD
    — *A.h.margarethae* R.T. Moore, 1937 NW and W Mexico
    — *A.h.heloisa* (Lesson & DeLattre, 1839) NE, C and S Mexico

○ *Atthis ellioti* WINE-THROATED HUMMINGBIRD
    — *A.e.ellioti* Ridgway, 1878 S Mexico, Guatemala
    — *A.e.selasphoroides* Griscom, 1932 Honduras

MYRTIS Reichenbach, 1854 F

○ *Myrtis fanny* PURPLE-COLLARED WOODSTAR
    — *M.f.fanny* (Lesson, 1838) W and SE Ecuador, W Peru
    — *M.f.megalura* J.T. Zimmer, 1953[1] N Peru (2719)

EULIDIA Mulsant & Verreaux, 1877[2] F

○ *Eulidia yarrellii* (Bourcier, 1847) CHILEAN WOODSTAR S Peru, N Chile

MYRMIA Mulsant, 1876 F

○ *Myrmia micrura* (Gould, 1854) SHORT-TAILED WOODSTAR W Ecuador, NW Peru

CHAETOCERCUS G.R. Gray, 1855[3] M

○ *Chaetocercus mulsanti* (Bourcier, 1842)[4] WHITE-BELLIED WOODSTAR Colombia to C Bolivia

○ *Chaetocercus bombus* Gould, 1871 LITTLE WOODSTAR SW Colombia, W and E Ecuador, N and NC Peru

○ *Chaetocercus heliodor* GORGETED WOODSTAR
    — *C.h.heliodor* (Bourcier, 1840)[5] W Ecuador to Venezuela
    — *C.h.cleavesi* R.T. Moore, 1934 NE Ecuador

○ *Chaetocercus astreans* (Bangs, 1899) SANTA MARTA WOODSTAR NE Colombia

○ *Chaetocercus berlepschi* Simon, 1889 ESMERALDAS WOODSTAR W Ecuador

○ *Chaetocercus jourdanii* RUFOUS-SHAFTED WOODSTAR
    — *C.j.andinus* Phelps & Phelps, Jr., 1949 NE Colombia , W Venezuela (1846)
    — *C.j.rosae* (Bourcier & Mulsant, 1846) N Venezuela
    — *C.j.jourdanii* (Bourcier, 1839) NE Venezuela, Trinidad

SELASPHORUS Swainson, 1832 M

● *Selasphorus platycercus* (Swainson, 1827) BROAD-TAILED HUMMINGBIRD[6]

W and WC USA to Mexico, W Guatemala >> Mexico, C America

● *Selasphorus rufus* (J.F. Gmelin, 1788) RUFOUS HUMMINGBIRD SE Alaska to NW USA >> Mexico and Gulf coast of USA

● *Selasphorus sasin* ALLEN'S HUMMINGBIRD
    ✓ *S.s.sasin* (Lesson, 1829) Coastal S Oregon and California (USA) >> SC Mexico
    — *S.s.sedentarius* Grinnell, 1929 Offshore islands of S California (USA)

---

[1] Omitted from last Edition.
[2] Schuchmann (1999: 674) (2203) submerged *Eulidia* in the genus *Myrtis*.
[3] We follow Schuchmann (1999: 677) (2203) in submerging *Acestrura* in *Chaetocercus*.
[4] The emended name used here is judged to be in prevailing usage; see Art. 33.2 of the Code (I.C.Z.N., 1999) (1178) and p. 000 herein.
[5] Züchner (1999) (2737) made no mention of placing *meridae* Zimmer & Phelps, 1950 (2726) as being in synonymy; it was placed in synonymy by Graves (1986) (982).
[6] Includes *guatemalae* see Schuchmann (1999: 679) (2203).

○ *Selasphorus flammula* VOLCANO HUMMINGBIRD
    — *S.f.flammula* Salvin, 1865   Irazu and Turrialba Volc. (Costa Rica)
    — *S.f.simoni* Carriker, 1910   Poás and Barba Volc. (Costa Rica)
    — *S.f.torridus* Salvin, 1870   Talamanca range (Costa Rica), W Panama

○ *Selasphorus ardens* Salvin, 1870 GLOW-THROATED HUMMINGBIRD   WC Panama

○ *Selasphorus scintilla* (Gould, 1851) SCINTILLANT HUMMINGBIRD   Costa Rica, W Panama

STELLULA Gould, 1861 F
● *Stellula calliope* (Gould, 1847) CALLIOPE HUMMINGBIRD[1]   SW Canada, W USA >> WC and SC Mexico

COLIIDAE MOUSEBIRDS (2: 6)

COLIUS Brisson, 1760 M
○ *Colius striatus* SPECKLED MOUSEBIRD[2]
    — *C.s.nigricollis* Vieillot, 1817   E Nigeria to W Zaire
    — *C.s.leucophthalmus* Chapin, 1921   N Zaire, SE Central African Republic, SW Sudan
    — *C.s.leucotis* Rüppell, 1839[3]   E Sudan, W and C Ethiopia, Eritrea
    — *C.s.hilgerti* Zedlitz, 1910   NE Ethiopia, NW Somalia
    — *C.s.jebelensis* Mearns, 1915   S Sudan, N Uganda
    — *C.s.mombassicus* van Someren, 1919   S Somalia to NE Tanzania
    — *C.s.kikuyensis* van Someren, 1919   W and C Kenya
    — *C.s.cinerascens* Neumann, 1900   N, C and W Tanzania to NE Zambia
    — *C.s.affinis* Shelley, 1885   E Tanzania to N Mozambique, S Malawi and NE
                                                   Zimbabwe
    — *C.s.berlepschi* E. Hartert, 1899   S Tanzania, Malawi, NE Zambia, NW Mozambique
    — *C.s.kiwuensis* Reichenow, 1908[4]   C and S Uganda, NW Tanzania, Rwanda, Burundi,
                                                   E Zaire
    — *C.s.congicus* Reichenow, 1923[5]   S and SE Zaire, W Zambia, E Angola
    — *C.s.simulans* Clancey, 1979   C Mozambique to SE Malawi (468)
    — *C.s.integralis* Clancey, 1957   S Mozambique, SE Zimbabwe, E Transvaal (375)
    — *C.s.rhodesiae* Grant & Mackworth-Praed, 1938   E Zimbabwe to W Mozambique
    — *C.s.minor* Cabanis, 1876   N Transvaal to Natal, E Cape Province and Swaziland
    — *C.s.striatus* J.F. Gmelin, 1789   SW and S Cape Province

○ *Colius leucocephalus* WHITE-HEADED MOUSEBIRD
    — *C.l.turneri* van Someren, 1919   N and NE Kenya
    — *C.l.leucocephalus* Reichenow, 1879   S Somalia, SE Kenya to NE Tanzania

○ *Colius castanotus* J. & E. Verreaux, 1855 RED-BACKED MOUSEBIRD   W Angola

○ *Colius colius* (Linnaeus, 1766) WHITE-BACKED MOUSEBIRD   Namibia, S Botswana, W Sth. Africa

UROCOLIUS Bonaparte, 1854 M
○ *Urocolius macrourus* BLUE-NAPED MOUSEBIRD
    — *U.m.macrourus* (Linnaeus, 1766)[6]   Senegal and SW Mauritania to Eritrea
    — *U.m.laeneni* (Niethammer, 1955)   Niger (Aïr Mountains) (1652)
    — *U.m.abyssinicus* Schifter, 1975   C and S Ethiopia, N Kenya, NW Somalia (2174)
    — *U.m.pulcher* Neumann, 1900   SE Sudan, NE Uganda, Kenya, S Somalia, N Tanzania
    — *U.m.griseogularis* (van Someren, 1919)   S Sudan, W and S Uganda to E Zaire, NW Tanzania
    — *U.m.massaicus* Schifter, 1975   C to coastal Tanzania (2174)

○ *Urocolius indicus* RED-FACED MOUSEBIRD
    — *U.i.mossambicus* Reichenow, 1896[7]   E Angola, Zambia, SE Zaire, Malawi, SW Tanzania

---

[1] Includes *lowei* see Calder (1999: 674) (289).
[2] Treatment follows that of Schifter (1972) (2173).
[3] The range in Decoux (1988: 250) (633) implicitly includes *erlangeri*.
[4] Includes *ugandensis* see White (1965: 219) (2649).
[5] Includes *lungae* White, 1947 (2632), see White (1965: 219) (2649).
[6] We follow Decoux (1988: 247) (633) who implicitly treated *syntactus* as a synonym. However Schifter (1975) (2174) recognised this form and his lectotype probably needs re-examination.
[7] Presumably includes *lualabae* Verheyen, 1951 (2522); White (1965: 222) (2649) thought this 'doubtfully distinct' from *pallidus* (but White did not recognise *mossambicus*).

— *U.i.lacteifrons* Sharpe, 1892   W Angola to N Namibia
— *U.i.pallidus* Reichenow, 1896   SE Tanzania, NE Mozambique
— *U.i.transvaalensis* Roberts, 1922   SW Zambia, Botswana, Zimbabwe C and S
         Mozambique to N and E Cape Province
— *U.i.indicus* Latham, 1790   S Cape Province

TROGONIDAE TROGONS[1] (6: 39)

APALODERMA Swainson, 1833 N
○ *Apaloderma narina* NARINA'S TROGON
     — *A.n.constantia* (Sharpe & Ussher, 1872)   Sierra Leone to Ghana
     — *A.n.brachyurum* Chapin, 1923   S Cameroon to SW Sudan, N, C and S Zaire
     — *A.n.narina* (Stephens, 1815)[2]   Ethiopia to inland E Africa, E Zaire, Angola and
         E Cape Province
     — *A.n.littorale* van Someren, 1931[3]   S Somalia to E Tanzania and Mozambique

○ *Apaloderma aequatoriale* (Sharpe, 1901) BARE-CHEEKED TROGON   S Cameroon to Gabon, NE and S Zaire

○ *Apaloderma vittatum* (Shelley, 1882) BAR-TAILED TROGON   W Cameroon, Angola, E Zaire to Kenya, Malawi and
     N Mozambique

HARPACTES Swainson, 1833 M
○ *Harpactes reinwardtii* RED-BILLED TROGON[4]
     — *H.r.mackloti* (S. Müller, 1835)   Mts. of Sumatra
     — *H.r.reinwardtii* (Temminck, 1822)   Mts. of Java

○ *Harpactes fasciatus* MALABAR TROGON
     — *H.f.malabaricus* (Gould, 1834)   SW and S India
     — *H.f.legerli* Koelz, 1939   W and E Ghats (C India)
     — *H.f.fasciatus* (Pennant, 1769)[5]   Sri Lanka

○ *Harpactes kasumba* RED-NAPED TROGON
     — *H.k.kasumba* (Raffles, 1822)   SC and S Malay Pen., Sumatra
     — *H.k.impavidus* (Chasen & Kloss, 1931)   Borneo

○ *Harpactes diardii* DIARD'S TROGON
     — *H.d.sumatranus* W.H. Blasius, 1896   C and S Malay Pen., Sumatra
     — *H.d.diardii* (Temminck, 1832)   Borneo, Bangka I.

○ *Harpactes ardens* PHILIPPINE TROGON[6]
     — *H.a.herberti* Parkes, 1970   NE Luzon (1762)
     — *H.a.luzoniensis* Rand & Rabor, 1952   Luzon, Marinduque, Catanduanes (1762)
     — *H.a.minor* Manuel, 1958   Polillo (1421)
     — *H.a.linae* Rand & Rabor, 1959   Bohol, Leyte and Samar (1976)
     — *H.a.ardens* (Temminck, 1826)   Basilan, Dinagat and Mindanao

○ *Harpactes whiteheadi* Sharpe, 1888 WHITEHEAD'S TROGON   Mt. Kinabalu (N Borneo)

○ *Harpactes orrhophaeus* CINNAMON-RUMPED TROGON
     — *H.o.orrhophaeus* (Cabanis & Heine, 1863)   SC and S Malay Pen., Sumatra
     — *H.o.vidua* Ogilvie-Grant, 1892   N Borneo

○ *Harpactes duvaucelii* (Temminck, 1824) SCARLET-RUMPED TROGON   C and S Malay Pen., Sumatra, Borneo

○ *Harpactes oreskios* ORANGE-BREASTED TROGON
     — *H.o.stellae* Deignan, 1941   Kengtung (E Burma), Thailand (except Pen.),
         S Yunnan south to S Indochina
     — *H.o.uniformis* (Robinson, 1917)   Malay Pen., Sumatra

[1] For recent views on phylogeny see Espinosa de los Monteros (2000) (819).
[2] Includes *arcanum* Clancey, 1959 (380) see White (1965: 222) (2649).
[3] For correction of spelling see David & Gosselin (2002) (613).
[4] This is treated as two species by Collar (2001: 106) (546) with mention of unpublished research (which, if now out, escaped our attention). Awaiting this research, we prefer to retain one species, and to defer the use of a separate generic name. Espinosa de los Monteros (1998) (818) did not suggest any split in *Harpactes*.
[5] Includes *parvus* Deraniyagala, 1955 (694), see Ripley (1981: 202) (2055).
[6] Revised from last Edition; now follows Dickinson *et al.* (1991) (745).

___ *H.o.oreskios* (Temminck, 1823)   Java
___ *H.o.dulitensis* Ogilvie-Grant, 1892   Mts. of NW Borneo
___ *H.o.nias* Meyer de Schauensee & Ripley, 1939 Nias I.

○ *Harpactes erythrocephalus*   Red-headed Trogon
 ___ *H.e.erythrocephalus* (Gould, 1834)   C Himalayas, S and E Assam, W and SE Burma,
                   NW Thailand
 ___ *H.e.helenae* Mayr, 1941   Arunachal Pradesh, W Yunnan, N and NE Burma
 ___ *H.e.yamakenensis* Rickett, 1899[1]   SC and SE China west to Guizhou and S Sichuan
 ___ *H.e.intermedius* (Kinnear, 1925)   N Laos, N Vietnam
 ___ *H.e.annamensis* (Robinson & Kloss, 1919) NE Thailand, S Indochina
 ___ *H.e.klossi* (Robinson, 1915)   W Cambodia
 ___ *H.e.chaseni* Riley, 1934   C and S Malay Pen.
 ___ *H.e.hainanus* Ogilvie-Grant, 1900   Hainan I.
 ___ *H.e.flagrans* (S. Müller, 1835)   Sumatra

○ *Harpactes wardi* (Kinnear, 1927)   Ward's Trogon Bhutan to N Burma, NE Vietnam

PRIOTELUS G.R. Gray, 1840[2]   M
○ *Priotelus temnurus*   Cuban Trogon
 ___ *P.t.temnurus* (Temminck, 1825)   Cuba
 ___ *P.t.vescus* Bangs & Zappey, 1905   Isla de la Juventud

○ *Priotelus roseigaster* (Vieillot, 1817)   Hispaniolan Trogon[3] Hispaniola

TROGON Brisson, 1760   M
● *Trogon melanocephalus* Gould, 1835   Black-headed Trogon[4] E Mexico to N Costa Rica

○ *Trogon citreolus*   Citreoline Trogon
 ___ *T.c.citreolus* Gould, 1835   W Mexico
 ___ *T.c.sumichrasti* Brodkorb, 1942   S Mexico

○ *Trogon viridis*   White-tailed Trogon
 ___ *T.v.chionurus* Sclater & Salvin, 1871 E Panama, W Colombia, W Ecuador
 ___ *T.v.viridis* Linnaeus, 1766   E Colombia to Peru, Brazil and Trinidad
 ___ *T.v.melanopterus* Swainson, 1838[5]   SE Brazil

○ *Trogon bairdii* Lawrence, 1868   Baird's Trogon SW Costa Rica, W Panama

○ *Trogon surrucura*   Surucua Trogon
 ___ *T.s.aurantius* Spix, 1824   E Brazil
 ___ *T.s.surrucura* Vieillot, 1817   S Brazil, Paraguay, Uruguay, N Argentina

○ *Trogon curucui*   Blue-crowned Trogon
 ___ *T.c.peruvianus* Gould, 1858[6]   S Colombia to Bolivia and W Brazil
 ___ *T.c.curucui* Linnaeus, 1766   E Brazil
 ___ *T.c.behni* Gould, 1875   E Bolivia, S Brazil, Paraguay, N Argentina

● *Trogon violaceus*   Violaceous Trogon
 ___ *T.v.braccatus* (Cabanis & Heine, 1856) C Mexico to Nicaragua
 ⤷ *T.v.concinnus* Lawrence, 1862   Costa Rica to W Ecuador
 ___ *T.v.caligatus* Gould, 1838   N Colombia, W Venezuela
 ___ *T.v.violaceus* J.F. Gmelin, 1788   Venezuela, the Guianas, N Brazil, Trinidad
 ___ *T.v.ramonianus* Deville & Des Murs, 1849 Upper Amazonia
 ___ *T.v.crissalis* (Cabanis & Heine, 1863) E Brazil

○ *Trogon mexicanus*   Mountain Trogon
 ___ *T.m.clarus* Griscom, 1932   NW Mexico
 ___ *T.m.mexicanus* Swainson, 1827   C Mexico to W Guatemala
 ___ *T.m.lutescens* Griscom, 1932[7]   Honduras

[1] Includes *rosa* see Cheng (1987: 357) (333).
[2] Our sequence for New World trogons draws on Espinosa de los Monteiros (1998) (818).
[3] Placed in the genus *Temnotrogon* by Collar (2001: 123) (546).
[4] Includes *illaetabilis* see Collar (2001: 113) (546).
[5] Placed in synonymy by Collar (2001: 113) (546), but we note the range gap and await further details on this.
[6] Includes *bolivianus* see Collar (2001: 121) (546).
[7] Placed in synonymy by Collar (2001: 117) (546), but his text implies that he did not examine specimens.

○ *Trogon elegans* ELEGANT TROGON
— *T.e.canescens* van Rossem, 1934   S Arizona, NW Mexico
— *T.e.ambiguus* Gould, 1835   S Texas, E and C Mexico
— *T.e.elegans* Gould, 1834   Guatemala
— *T.e.lubricus* J.L. Peters, 1945   Nicaragua, Costa Rica
— *T.e.goldmani* Nelson, 1898   Tres Marias Is.

○ *Trogon collaris* COLLARED TROGON
— *T.c.puella* Gould, 1845   C Mexico to W Panama
— *T.c.extimus* Griscom, 1929   NE Panama
— *T.c.hoethinus* Wetmore, 1967   E Panama (2604)
— *T.c.virginalis* Cabanis & Heine, 1863   W Colombia, W Ecuador, NW Peru
— *T.c.subtropicalis* J.T. Zimmer, 1948   C Colombia (2711)
— *T.c.exoptatus* Cabanis & Heine, 1863   N Venezuela
— *T.c.collaris* Vieillot, 1817   Colombia to Bolivia, Trinidad, S Brazil
— *T.c.castaneus* Spix, 1824   SE Colombia, NW Brazil

○ *Trogon personatus* MASKED TROGON
— *T.p.sanctaemartae* J.T. Zimmer, 1948   N Colombia (2711)
— *T.p.ptaritepui* Zimmer & Phelps, 1946   Venezuela (2723)
— *T.p.personatus* Gould, 1842   W Venezuela, E Colombia, E Peru
— *T.p.assimilis* Gould, 1846   W Ecuador
— *T.p.temperatus* (Chapman, 1923)   C Colombia, Ecuador
— *T.p.heliothrix* Tschudi, 1844   Temperate Peru
— *T.p.submontanus* Todd, 1943   Bolivia
— *T.p.duidae* Chapman, 1929   Mt. Duida (S Venezuela)
— *T.p.roraimae* (Chapman, 1929)   SE Venezuela, S Guyana

● *Trogon aurantiiventris* ORANGE-BELLIED TROGON
↙ *T.a.underwoodi* Bangs, 1908   NW Costa Rica
— *T.a.aurantiiventris* Gould, 1856   C Costa Rica to W Panama
— *T.a.flavidior* (Griscom, 1925)   E Panama

● *Trogon rufus* BLACK-THROATED TROGON
↙ *T.r.tenellus* Cabanis, 1862   SE Honduras to NW Colombia
— *T.r.cupreicauda* (Chapman, 1914)   W Colombia, W Ecuador
— *T.r.rufus* J.F. Gmelin, 1788   E Venezuela, the Guianas, N Brazil
— *T.r.sulphureus* Spix, 1824   E Peru, W Brazil
— *T.r.amazonicus* Todd, 1943   NE Brazil
— *T.r.chrysochloros* Pelzeln, 1856   S Brazil, Paraguay, NE Argentina

● *Trogon massena* SLATY-TAILED TROGON
— *T.m.massena* Gould, 1838   S Mexico to Nicaragua
↙ *T.m.hoffmanni* (Cabanis & Heine, 1863)   Costa Rica, Panama
— *T.m.australis* (Chapman, 1915)   W Colombia

○ *Trogon clathratus* Salvin, 1866 LATTICE-TAILED TROGON   E Costa Rica, W Panama

○ *Trogon melanurus* BLACK-TAILED TROGON
— *T.m.macroura* Gould, 1838   E Panama, N Colombia
— *T.m.mesurus* (Cabanis & Heine, 1863)   W Ecuador, NW Peru
— *T.m.eumorphus* J.T. Zimmer, 1948   W Amazonia (2711)
— ¶ *T.m.occidentalis* Pinto, 1950   Brazil (1889)
— *T.m.melanurus* Swainson, 1838   Colombia to Bolivia and N Brazil

○ *Trogon comptus* J.T. Zimmer, 1948 BLUE-TAILED TROGON   C Colombia (2711)

EUPTILOTIS Gould, 1858 M
○ *Euptilotis neoxenus* (Gould, 1838) EARED QUETZAL   C Mexico

PHAROMACHRUS De La Llave, 1832 M
○ *Pharomachrus pavoninus* (Spix, 1824) PAVONINE QUETZAL[1]   Upper and central Amazonia

[1] Includes *viridiceps* see Collar (2001: 125) (546).

○ *Pharomachrus auriceps* Golden-headed Quetzal
   — *P.a.auriceps* (Gould, 1842)[1]
      E Panama and Andes from N Colombia to E Peru and N Bolivia

   — *P.a.hargitti* (Oustalet, 1891)
      Andes of NW Venezuela

○ *Pharomachrus fulgidus* White-tipped Quetzal
   — *P.f.festatus* Bangs, 1899
      N Colombia
   — *P.f.fulgidus* (Gould, 1838)
      NE and NC Venezuela

● *Pharomachrus mocinno* Resplendent Quetzal
   — *P.m.mocinno* de la Llave, 1832
      S Mexico to N Nicaragua
   ☑ *P.m.costaricensis* Cabanis, 1869
      Costa Rica to W Panama

○ *Pharomachrus antisianus* (d'Orbigny, 1837) Crested Quetzal
      W Venezuela, Colombia to Brazil

CORACIIDAE ROLLERS (2: 12)

CORACIAS Linnaeus, 1758 M[2]
○ *Coracias naevius* Rufous-crowned Roller
   — *C.n.naevius* Daudin, 1800
      Senegal to Ethiopia and N Tanzania
   — *C.n.mosambicus* Dresser, 1890
      Angola and S Zaire to Namibia and N Sth. Africa

○ *Coracias benghalensis* Indian Roller
   — *C.b.benghalensis* (Linnaeus, 1758)
      E Saudi Arabia to N India and Bangladesh
   — *C.b.indicus* Linnaeus, 1766
      C and S India, Sri Lanka, Laccadive and Maldive Is.
   — *C.b.affinis* McClelland, 1840
      Bhutan and Assam E to S Sichuan and W Yunnan and S through mainland SE Asia to NE of W Malaysia

○ *Coracias temminckii* (Vieillot, 1819) Purple-winged Roller/Sulawesi Roller
      Sulawesi

○ *Coracias spatulatus* Racquet-tailed Roller
   — *C.s.weigalli* Dresser, 1890[3]
      S Tanzania, Malawi and N Mozambique
   — *C.s.spatulatus* Trimen, 1880
      Angola to Zambia, Zimbabwe and S Mozambique

○ *Coracias caudatus* Lilac-breasted Roller
   — *C.c.lorti* Shelley, 1885
      Eritrea, Ethiopia, W Somalia, NE Kenya
   — *C.c.caudatus* Linnaeus, 1766
      Uganda, S and W Kenya to N Namibia and N Sth. Africa

○ *Coracias abyssinicus* Hermann, 1783 Abyssinian Roller
      Senegal to Eritrea south to N Zaire and NW Kenya

○ *Coracias garrulus* European Roller
   — *C.g.garrulus* Linnaeus, 1758
      N Africa and S and C Europe to Middle East, NW Himalayas and SW Siberia >> Africa
   — *C.g.semenowi* Loudon & Tschudi, 1902
      Iraq and Transcaspia to W Pakistan, Xinjiang and C Kazakhstan >> NE Africa and SW Asia

○ *Coracias cyanogaster* Cuvier, 1816[4] Blue-bellied Roller
      Senegal to C Nigeria, N Zaire and SW Sudan

EURYSTOMUS Vieillot, 1816 M
○ *Eurystomus gularis* Blue-throated Roller
   — *E.g.gularis* Vieillot, 1819
      Guinea to W Cameroon
   — *E.g.neglectus* Neumann, 1908
      S Cameroon to N Angola and Uganda

○ *Eurystomus glaucurus* Broad-billed Roller[5]
   — *E.g.afer* (Latham, 1790)[6]
      Senegal to Liberia, N Zaire, Sudan and Ethiopia
   — *E.g.suahelicus* Neumann, 1905
      S Somalia and Uganda to N Zambia and Tanzania
   — *E.g.pulcherrimus* Neumann, 1905
      Angola, S Zambia and Mozambique south to Natal
   — *E.g.glaucurus* (Statius Müller, 1776)
      Madagascar >> SE and C Africa

---

[1] Mistakenly attached to *P. pavoninus* in last Edition. Includes *heliactin* see Peters (1945: 149) (1826).
[2] Gender addressed by David & Gosselin (2002) (614), multiple changes occur in specific names.
[3] For recognition see Fry & Fry (1992: 293) (895).
[4] Given as 1817 by Peters (1945: 244) (1826) but see Browning & Monroe (1991) (258).
[5] Treatment follows Cramp *et al.* (1985: 783) (591).
[6] Implicitly includes *aethiopicus* — recognised by Fry (1988: 354) (889) and Fry & Fry (1992: 303) (895).

○ *Eurystomus orientalis* Dollarbird[1]

— *E.o.calonyx* Sharpe, 1890[2]   E Himalayas and N Assam east to C and NE China, Korea, SE Siberia and Japan >> S and SE Asia to Gtr. Sundas

— *E.o.orientalis* (Linnaeus, 1766)[3]   NE India (?), Burma, C and S Thailand, Indochina S to the Malay Pen., Sumatra, Java, Borneo and Philippines

— *E.o.laetior* Sharpe, 1890   SW India

— *E.o.gigas* Stresemann, 1913   S Andaman Is.

— †?*E.o.irisi* Deraniyagala, 1951[4]   Dry Zone of Sri Lanka (692)

— *E.o.oberholseri* Junge, 1936   Simaleue[5]

— *E.o.pacificus* (Latham, 1802)[6]   Lesser Sundas , N and E Australia >> Wallacea and New Guinea to Bismarck Arch.

— *E.o.waigiouensis* Elliot, 1871   W, S and E New Guinea and nearby islands

— *E.o.crassirostris* P.L. Sclater, 1869   Bismarck Arch.

— *E.o.solomonensis* Sharpe, 1890   Feni I. (Solomon Is.)

○ *Eurystomus azureus* G.R. Gray, 1860 Purple Roller   N Moluccas

BRACHYPTERACIIDAE GROUND ROLLERS (4: 5)

BRACHYPTERACIAS Lafresnaye, 1834 M

○ *Brachypteracias leptosomus* (Lesson, 1833) Short-legged Ground Roller

    NE and E Madagascar

GEOBIASTES Sharpe, 1871[7] M

○ *Geobiastes squamiger* (Lafresnaye, 1838) Scaly Ground Roller[8]   NE Madagascar

ATELORNIS Pucheran, 1846 M

○ *Atelornis pittoides* (Lafresnaye, 1834) Pitta-like Ground Roller   NE and E Madagascar

○ *Atelornis crossleyi* Sharpe, 1875 Rufous-headed Ground Roller   NE and E Madagascar

URATELORNIS Rothschild, 1895 M

○ *Uratelornis chimaera* Rothschild, 1895 Long-tailed Ground Roller   SW Madagascar

LEPTOSOMATIDAE CUCKOO-ROLLER (1: 1)

LEPTOSOMUS Vieillot, 1816 M

○ *Leptosomus discolor* Cuckoo-Roller

— *L.d.gracilis* Milne-Edwards & Oustalet, 1885   Grande Comore I.

— *L.d.intermedius* Hartert & Neumann, 1924   Anjouan I.

— *L.d.discolor* (Hermann, 1783)   Madagascar, Mayotte I.

---

[1] Differences in nomenclatural treatment can be traced to Stresemann (1952) (2366). Those who accepted his views, including Dickinson *et al.* (1991) (745) and Kennedy *et al.* (2000) (1259), called mainland Asian birds, stretching to and including those in the Philippines, *cyanocollis* Vieillot, 1819, not mentioned in Peters 1945 (1826). By implication they would treat nominate *orientalis* as the Australian form (first obtained as a migrant in Amboina). Mees (1965) (1515) rejected this view and a majority of writers has followed his opinion, including Fry (2001: 376) (894). Although we accept this majority view here, we note that this implies "cherry-picking" from the decisions proposed by Stresemann (*op. cit.*).

[2] Ripley (1942) (2026) argued that the name *calonyx* Sharpe, 1890, could not be used and introduced the name *abundus* as a *nomen novum*. His argument rested on the existence in synonymy at an earlier date of the *nomen nudum calonyx* Hodgson, 1844. Under the Code (I.C.Z.N., 1999) (1178) this is no longer an objection. We agree with Fry (2001: 376) (894) and follow him in the use of *calonyx* Sharpe, 1890. The name *latouchei* Allison, 1946 (30), is a synonym see Deignan (1950) (646), Stepanyan (1990) (2326).

[3] Implicitly includes *deignani* see Fry & Fry (1992: 306) (895).

[4] Reputedly a rare breeding form (Phillips, 1978) (1881); perhaps extinct (Ripley, 1982) (2055).

[5] This population has been minimally sampled and requires review; accepted by van Marle & Voous (1988: 133) (2468) who apparently had no specimens available.

[6] For reasons to use 1802 not 1801 see Browing & Monroe (1991) (258). Includes *connectens* see White & Bruce (1986: 286) (2652), although if proved to breed may warrant the use of this name.

[7] For the use of this generic name see Kirchman *et al.* (2001) (1271).

[8] For correction of spelling see David & Gosselin (2002) (614).

ALCEDINIDAE KINGFISHERS[1] (17: 91)

HALCYONINAE[2]

ACTENOIDES Hombron & Jacquinot, 1853 M
○ *Actenoides monachus* GREEN-BACKED KINGFISHER[3]
    __ *A.m.monachus* (Bonaparte, 1850)[4]   N and C Sulawesi
    __ *A.m.capucina* (Meyer & Wiglesworth, 1896)   E, SE and S Sulawesi

○ *Actenoides princeps* SCALY KINGFISHER[5]
    __ *A.p.princeps* (Reichenbach, 1851)   NE Sulawesi
    __ *A.p.erythrorhampha* (Stresemann, 1931)[6]   NW and C Sulawesi
    __ *A.p.regalis* (Stresemann, 1932)   SE Sulawesi

○ *Actenoides bougainvillei* MOUSTACHED KINGFISHER
    __ *A.b.bougainvillei* (Rothschild, 1904)   Bougainville I.
    __ *A.b.excelsus* (Mayr, 1941)   Guadalcanal I.

○ *Actenoides lindsayi* SPOTTED KINGFISHER
    __ *A.l.lindsayi* (Vigors, 1831)   Catanduanes, Marinduque, Luzon
    __ *A.l.moseleyi* Steere, 1890   Negros, Panay

○ *Actenoides hombroni* Bonaparte, 1850 BLUE-CAPPED KINGFISHER[7]   Mindanao (Philippines)

○ *Actenoides concretus* RUFOUS-COLLARED KINGFISHER
    __ *A.c.peristephes* (Deignan, 1946)   N Malay Pen. (641)
    __ *A.c.concretus* (Temminck, 1825)   S Malay Pen., Sumatra, Bangka, Belitung and
                                      Mentawai Is.
    __ *A.c.borneanus* (Chasen & Kloss, 1930   Borneo

MELIDORA Lesson, 1830 F
○ *Melidora macrorrhina* HOOK-BILLED KINGFISHER
    __ *M.m.waigiuensis* E. Hartert, 1930   Waigeo I.
    __ *M.m.macrorrhina* (Lesson, 1827)   Misool, Batanta Is., W, S and E lowland New Guinea
    __ *M.m.jobiensis* Salvadori, 1880   Yapen I., N New Guinea

LACEDO Reichenbach, 1851 F
○ *Lacedo pulchella* BANDED KINGFISHER
    __ *L.p.amabilis* (Hume, 1873)[8]   SC and SE Burma, Thailand (except S Pen.), S Vietnam
    __ *L.p.pulchella* (Horsfield, 1821)   C and S Malay Pen., Sumatra, Java, N Natuna Is.
    __ *L.p.melanops* (Bonaparte, 1850)   Borneo, Bangka I.

TANYSIPTERA Vigors, 1828 F
○ *Tanysiptera galatea* COMMON PARADISE KINGFISHER
    __ *T.g.emiliae* Sharpe, 1871   Rau I. (N Moluccas)
    __ *T.g.doris* Wallace, 1862   Morotai I. (N Moluccas)
    __ *T.g.browningi* Ripley, 1983   Halmahera (2056)
    __ *T.g.brunhildae* Jany, 1955   Doi I. (N Moluccas) (1200)
    __ *T.g.margarethae* Heine, 1859   Bacan I. (N Moluccas)
    __ *T.g.sabrina* G.R. Gray, 1860   Kayoa I. (N Moluccas)
    __ *T.g.obiensis* Salvadori, 1877   Obi, Oblilatu Is. (N Moluccas)
    __ *T.g.acis* Wallace, 1863   Buru I. (S Moluccas)
    __ *T.g.boanensis* Mees, 1964   Boano I. (S Moluccas) (1513)
    __ *T.g.nais* G.R. Gray, 1860   Seram, Ambon, Manipa, Seramlaut Is. (S Moluccas)

---

[1] With two exceptions (the placement of the genera *Melidora* and *Lacedo*) we follow the sequence used by Woodall (2001) (2689). In doing this we bring in
the generic name *Todiramphus* for part of *Halcyon* which we used in last Edition.
[2] We use this subfamily following Woodall (2001) (2689).
[3] For correction of spelling see David & Gosselin (2002) (613).
[4] Includes *intermedia* see White & Bruce (1986: 270) (2652).
[5] The name *princeps* is said to be preoccupied in *Halcyon* and *kleinschmidti* Laubmann, 1950 (1335), was proposed to replace nominate *princeps*; the
availability of *princeps* in *Todiramphus* is presumed here.
[6] For correction of spelling see David & Gosselin (2002) (613).
[7] Treated as part of *lindsayi* in last Edition, but see Dickinson *et al.* (1991) (745) where *burtoni* duPont, 1976 (775), is a synonym.
[8] Includes *deignani* Meyer de Schauensee, 1946 (1564), see Wells (1999: 505) (2579), type locality N of Trang.

— *T.g.galatea* G.R. Gray, 1859	NW New Guinea and W Papuan islands
— *T.g.meyeri* Salvadori, 1889	N New Guinea
— *T.g.minor* Salvadori & D'Albertis, 1875	S and SE New Guinea
— *T.g.vulcani* Rothschild & Hartert, 1915	Manam I. (NE New Guinea)
— *T.g.rosseliana* Tristram, 1889	Rossel I. (E Louisade Arch.)

○ *Tanysiptera ellioti* Sharpe, 1870 Kafiau Paradise Kingfisher — Kafiau I. (west of Misool)

○ *Tanysiptera riedelii* J. Verreaux, 1866 Biak Paradise Kingfisher — Biak I. (Geelvink Bay)

○ *Tanysiptera carolinae* Schlegel, 1871 Numfor Paradise Kingfisher — Numfor I. (Geelvink Bay)

○ *Tanysiptera hydrocharis* G.R. Gray, 1858 Little Paradise Kingfisher — S New Guinea, Aru Is.

○ *Tanysiptera sylvia* Buff-breasted Paradise Kingfisher

— *T.s.leucura* Neumann, 1915	Umboi I. (Bismarck Arch.)
— *T.s.nigriceps* P.L. Sclater, 1877	New Britain and Duke of York Is. (Bismarck Arch.)
— *T.s.salvadoriana* E.P. Ramsay, 1878	SE New Guinea
— *T.s.sylvia* Gould, 1850	N Queensland >> N and S New Guinea

○ *Tanysiptera nympha* G.R. Gray, 1840 Red-breasted Paradise Kingfisher

SC, W and north to NE New Guinea

○ *Tanysiptera danae* Sharpe, 1880 Brown-headed Paradise Kingfisher — SE New Guinea

CITTURA Kaup, 1848 F
○ *Cittura cyanotis* Lilac-cheeked Kingfisher

— *C.c.sanghirensis* Sharpe, 1868	Sangihe Is.
— *C.c.cyanotis* (Temminck, 1824)[1]	Sulawesi and nearby Lembeh I.

CLYTOCEYX Sharpe, 1880 M
○ *Clytoceyx rex* Shovel-billed Kookaburra

— *C.r.rex* Sharpe, 1880	Submontane New Guinea (except Hellwig Mts.)
— *C.r.imperator* van Oort, 1909	Hellwig Mts. (SW New Guinea)

DACELO Leach, 1815[2] F
● *Dacelo novaeguineae* Laughing Kookaburra

— *D.n.minor* Robinson, 1900	Cape York Pen.
✓ *D.n.novaeguineae* (Hermann, 1783)	NE Queensland to SC South Australia, SW Western Australia, Tasmania

○ *Dacelo leachii* Blue-winged Kookaburra

— *D.l.superflua* Mathews, 1918[3]	SW New Guinea
— *D.l.intermedia* Salvadori, 1876	SC to SE New Guinea
— *D.l.occidentalis* Gould, 1870[4]	WC Australia
— *D.l.leachii* Vigors & Horsfield, 1826[5]	Australia (except WC), Melville I.

○ *Dacelo tyro* Spangled Kookaburra

— *D.t.archboldi* (Rand, 1938)	S New Guinea
— *D.t.tyro* G.R. Gray, 1858	Aru Is.

○ *Dacelo gaudichaud* Quoy & Gaimard, 1824 Rufous-bellied Kookaburra

New Guinea and islands, Aru Is.

CARIDONAX Cabanis & Heine, 1860 M
○ *Caridonax fulgidus* White-rumped Kingfisher

— *C.f.fulgidus* (Gould, 1857)	Lombok, Sumbawa Is. (Lesser Sundas)
— *C.f.gracilirostris* (Rensch, 1928)	Flores I. (Lesser Sundas)

PELARGOPSIS Gloger, 1841 F
○ *Pelargopsis capensis* Stork-billed Kingfisher

— *P.c.capensis* (Linnaeus, 1766) — Nepal, India, Sri Lanka

[1] Includes *modesta* see White & Bruce (1986) (2652).
[2] Given that this genus comes next to *Clytoceyx* we have taken the largest species first here.
[3] For recognition of this see Mees (1982) (1534).
[4] The name *cliftoni* is a junior synonym.
[5] Tentatively includes *kempi* (see Schodde & Mason, 1997) (2188), but further revision expected.

__ *P.c.osmastoni* (E.C.S. Baker, 1934)[1]   Andaman Is. (80)
__ *P.c.intermedia* Hume, 1874   Nicobar Is.
__ *P.c.burmanica* Sharpe, 1870   Burma east to Indochina and south to N Malay Pen.
__ *P.c.malaccensis* Sharpe, 1870   C and S Malay Pen., Riau and Lingga Arch.
__ *P.c.cyanopteryx* (Oberholser, 1909)   Sumatra, Bangka, Belitung Is.
__ *P.c.simalurensis* Richmond, 1903   Simaleue I.
__ *P.c.sodalis* Richmond, 1903[2]   Banyak, Nias, Batu, Mentawai Is.
__ *P.c.innominata* (van Ooort, 1910)   Borneo
__ *P.c.javana* (Boddaert, 1783)[3]   Java
__ *P.c.floresiana* Sharpe, 1870   Bali to Flores I.
__ *P.c.gouldi* Sharpe, 1870   Lubang, Mindoro and Palawan (N Philippines)[4]
__ *P.c.gigantea* Walden, 1874[5]   C and S Philippines (incl. Sulu Is.)

◯ *Pelargopsis melanorhyncha* BLACK-BILLED KINGFISHER
__ *P.m.melanorhyncha* (Temminck, 1826)   Sulawesi, Muna, Buton, Salayar Is.
__ *P.m.dichrorhyncha* Meyer & Wiglesworth, 1896   Peleng, Banggai Is.
__ *P.m.eutreptorhyncha* E. Hartert, 1898   Sula Is.

◯ *Pelargopsis amauroptera* (J.T. Pearson, 1841) BROWN-WINGED KINGFISHER[6]
    Coasts of Orissa (E India) to Malay Pen. (W coast only) south to Langkawi I.

HALCYON Swainson, 1821 F
◯ *Halcyon coromanda* RUDDY KINGFISHER
__ *H.c.coromanda* (Latham, 1790)   Nepal to S Yunnan S through mainland SE Asia to N Malay Pen.

__ *H.c.major* (Temminck & Schlegel, 1848)[7]   SE Manchuria (incl. Jilin, China), Korea, Japan >> E China, Philippines and Sulawesi

__ *H.c.bangsi* (Oberholser, 1915)   Ryukyu Is., Taiwan and nearby Lanyu I. >> Taiwan, Philippines, Talaud Is.

__ *H.c.mizorhina* (Oberholser, 1915*)*   Andaman and Nicobar Is.
__ *H.c.minor* (Temminck & Schlegel, 1848)   S'most Malay Pen., Gtr. Sundas, W Sumatran Is., Bangka, Belitung

__ *H.c.linae* Hubbard & duPont, 1974   Palawan I. (1159)
__ *H.c.claudiae* Hubbard & duPont, 1974   Tawitawi group, Sulu Is. (1159)
__ *H.c.rufa* Wallace, 1863   Sulawesi, Sangihe, Talaud, Togian, Muna and Buton Is.
__ *H.c.pelingensis* Neumann, 1939   Peling and Banggai Is.
__ *H.c.sulana* Mees, 1970   Sula Is. (1521)

◯ *Halcyon smyrnensis* WHITE-THROATED KINGFISHER
__ *H.s.smyrnensis* (Linnaeus, 1758)   Asia Minor and Levant to NW India
__ *H.s.fusca* (Boddaert, 1783)   W India and Sri Lanka
__ *H.s.perpulchra* Madarász, 1904[8]   Bhutan south to E India and east to mainly SE Asia and W Java

__ *H.s.saturatior* Hume, 1874   Andaman Is.
__ *H.s.fokiensis* Laubmann & Götz, 1926   S and E China
__ *H.s.gularis* (Kuhl, 1820*)*   Philippines

◯ *Halcyon cyanoventris* (Vieillot, 1818) JAVAN KINGFISHER   Java, Bali

◯ *Halcyon badia* J. & E. Verreaux, 1851 CHOCOLATE-BACKED KINGFISHER[9]
    Liberia to Gabon and E Zaire, W Uganda

◯ *Halcyon pileata* (Boddaert, 1783) BLACK-CAPPED KINGFISHER   India and Sri Lanka east to N Indochina, China and Korea >> mainland SE Asia, Gtr. Sundas, Sulawesi and Philippines

---

[1] Includes *shekarii* Abdulali, 1964 (6), see Ripley (1982: 208) (2055).
[2] Includes *nesoeca* and *isoptera* see van Marle & Voous (1988) (2468).
[3] The names used in last Edition for the Javan (*fraseri*) and Bornean (*javana*) populations were taken from Peters (1945); but see Mees (1971) (1523) for the correction of the names and for reasons to recognise the Bornean population, followed by Smythies (1981) (2288).
[4] No recent records known from Luzon (and status of old ones doubtful).
[5] Includes *smithi* see Dickinson *et al.* (1991) (745).
[6] For spelling see David & Gosselin (2002) (613).
[7] Includes *ochrothorectis* see Dickinson *et al.* (1991) (745).
[8] Sometimes claimed to be identical with *fusca*; details of this claim possibly await publication.
[9] Includes *obscuridorsalis* Dickerman & Cane, 1994 (731), see Woodall (2001: 209) (2689).

○ *Halcyon leucocephala* GREY-HEADED KINGFISHER
    — *H.l.acteon* (Lesson, 1830)
    — *H.l.leucocephala* (Statius Müller, 1776)[1]
    — *H.l.semicaerulea* (J.F. Gmelin, 1788)
    — *H.l.hyacinthina* Reichenow, 1900
    — *H.l.pallidiventris* Cabanis, 1880

Cape Verde Is.
Senegal to Ethiopia, south to N Zaire and N Tanzania
Yemen
S Somalia to NE Tanzania
S Zaire and W Tanzania south to N Namibia and E
    Transvaal >> Kenya, Uganda and N Zaire

○ *Halcyon albiventris* BROWN-HOODED KINGFISHER
    — *H.a.orientalis* W.K.H. Peters, 1868[2]

    — *H.a.albiventris* (Scopoli, 1786)
    — *H.a.vociferans* Clancey, 1952
    — *H.a.prentissgrayi* Bowen, 1930[3]

Coasts of Somalia and Kenya to C Mozambique,
    Malawi, S Zambia and N Botswana, Gabon and
    Angola to S Somalia, N Kenya and Mozambique
Natal to S and E Cape Province
Zimbabwe and S Mozambique to NE Sth. Africa (368)
Congo, Angola, S Zaire, N Zambia, W and C Tanzania,
    S and C Kenya

○ *Halcyon chelicuti* STRIPED KINGFISHER
    — *H.c.eremogiton* E. Hartert, 1921
    — *H.c.chelicuti* (Stanley, 1814)[4]

C Mali to C Sudan
Senegal and Gambia to Nigeria, Ethiopia, and E and S
    Africa to Botswana and Natal

○ *Halcyon malimbica* BLUE-BREASTED KINGFISHER
    — *H.m.torquata* Swainson, 1837[5]
    — *H.m.forbesi* Sharpe, 1892
    — *H.m.dryas* Hartlaub, 1854
    — *H.m.malimbica* (Shaw, 1811)[6]

S Senegal, Gambia, Guinea-Bissau, W Mali
Sierra Leone to Mt. Cameroon and Bioko
Principe I.
Cameroon to Uganda and Zambia

○ *Halcyon senegalensis* WOODLAND KINGFISHER
    — *H.s.fuscopileus* Reichenow, 1906[7]
    — *H.s.senegalensis* (Linnaeus, 1766)
    — *H.s.cyanoleuca* (Vieillot, 1818)

Sierra Leone to Zaire and N Angola
Senegal to Ethiopia and W Kenya
S Angola to W Tanzania south to N Namibia and
    Transvaal >> S Sudan and Kenya

○ *Halcyon senegaloides* A. Smith, 1834 MANGROVE KINGFISHER[8]
Coast of E and SE Africa

TODIRAMPHUS Lesson, 1827[9] M[10]
○ *Todiramphus nigrocyaneus* BLUE-BLACK KINGFISHER
    — *T.n.nigrocyaneus* (Wallace, 1862)
    — *T.n.quadricolor* (Oustalet, 1880)
    — *T.n.stictolaemus* (Salvadori, 1876)

Lowland NW to SC New Guinea and islands
Yapen I., N New Guinea
SC to SE New Guinea

○ *Todiramphus winchelli* RUFOUS-LORED KINGFISHER[11]
    — *T.w.nigrorum* (Hachisuka, 1934)

    — *T.w.nesydrionetes* (Parkes, 1966)
    — *T.w.mindanensis* (Parkes, 1966)
    — *T.w.winchelli* (Sharpe, 1877)
    — *T.w.alfredi* (Oustalet, 1890)

Bohol, Cebu, Leyte, Negros and Samar (C and EC
    Philippines)
Romblon, Sibuyan and Tablas (NC Philippines) (1758)
Mindanao (1758)
Basilan
Sulu Is. (SW Philippines) (1735)

○ *Todiramphus diops* (Temminck, 1824) BLUE-AND-WHITE KINGFISHER
N Moluccas

○ *Todiramphus lazuli* (Temminck, 1830) LAZULI KINGFISHER
S Moluccas (except Buru)

○ *Todiramphus macleayii* FOREST KINGFISHER
    — *T.m.elizabeth* (Heine, 1883)
    — *T.m.macleayii* (Jardine & Selby, 1830)[12]
    — *T.m.incinctus* (Gould, 1838)

SE New Guinea
N Northern Territory, Arus Is., ? Sermata Is.
NE and EC Australia >> S, E and N New Guinea, Kai
    Is., New Britain

[1] Includes *centralis* see White (1965: 230) (2649).
[2] Includes *erlangeri* see White (1965: 229) (2649).
[3] Thought to include *hylophila* Clancey, 1986 (501), see Woodall (2001: 210) (2689).
[4] Includes *damarensis* see White (1965: 228) (2649).
[5] Includes *fortis* only doubtfully recognised by White (1965) (2649).
[6] Includes *prenticei* see White (1965: 228) (2649).
[7] For correction of spelling see David & Gosselin (2002) (613).
[8] Includes *ranivorus* see White (1965: 227) (2649).
[9] Treated in a broad genus *Halcyon* in last Edition.
[10] For correct gender and spellings see David & Gosselin (2002) (614).
[11] Just two forms recognised in last Edition, but see Dickinson *et al.* (1991) (745).
[12] Includes *insularis* see Mees (1982) (1534).

○ *Todiramphus albonotatus* (E.P. Ramsay, 1885) NEW BRITAIN KINGFISHER/WHITE-BACKED KINGFISHER
New Britain

○ *Todiramphus leucopygius* (J. Verreaux, 1858) ULTRAMARINE KINGFISHER
Solomon Is.

○ *Todiramphus farquhari* (Sharpe, 1899) CHESTNUT-BELLIED KINGFISHER   C Vanuatu

○ *Todiramphus funebris* Bonaparte, 1850 SOMBRE KINGFISHER   Halmahera (N Moluccas)

◉ *Todiramphus chloris* COLLARED KINGFISHER

—— *T.c.abyssinicus* (Pelzeln, 1856)	Coast of Red Sea and NW Somalia
—— *T.c.kalbaensis* (Cowles, 1980)	NE Arabia (577)
—— *T.c.vidali* (Sharpe, 1892)	SW India
—— *T.c.davisoni* (Sharpe, 1892)	Andaman Is., Cocos Is. (off Burma)
—— *T.c.occipitalis* Blyth, 1846	Nicobar Is.
—— *T.c.humii* (Sharpe, 1892)	Coastal W Bengal, Bangladesh, Burma and C and S Malay Pen., NE Sumatra
—— *T.c.armstrongi* (Sharpe, 1892)	Thailand (on E coast south to C Malay Penin) round to Vietnam
—— *T.c.laubmannianus* (Grote, 1933)	Sumatra, Borneo and islands nearby
—— *T.c.chloropterus* (Oberholser, 1919)	W Sumatran Is. (but not Enggano)
—— *T.c.azelus* (Oberholser, 1919)	Enggano I.
—— *T.c.palmeri* (Oberholser, 1919)	Java, Bali, Bawean and Kangean Is.
—— *T.c.collaris* (Scopoli, 1786)	Philippines incl. Palawan group
—— *T.c.chloris* (Boddaert, 1783)	Talaud Is., Sulawesi to NW New Guinea, and Lesser Sundas to Kai Is. and Tanimbar
—— *T.c.sordidus* (Gould, 1842)	N and NE Australia, Aru Is., S New Guinea
—— *T.c.pilbara* (Johnstone, 1983)	W and N Western Australia (1219)
—— *T.c.colcloughi* (Mathews, 1916)[1]	EC Australia
—— *T.c.colonus* (E. Hartert, 1896)[2]	Islands off SE New Guinea, Louisiade Arch.
—— *T.c.teraokai* (N. Kuroda, Sr., 1915)	Palau Is.
—— *T.c.owstoni* (Rothschild, 1904)	Ascuncion, Pagan, Alamagan Is. (N Marianas)
—— *T.c.albicilla* (Dumont, 1823)	Saipan, Tinan Is. (S Marianas)
—— *T.c.orii* (Taka-Tsukasa & Momiyama, 1931)	Rota I. (S Marianas)
—— *T.c.nusae* (Heinroth, 1902)	Bismarck Arch. (except as below)
—— *T.c.matthiae* (Heinroth, 1902)	St. Matthias Is. (Bismarck Arch.)
—— *T.c.stresemanni* (Laubmann, 1923)	French, Umboi Is. (Bismarck Arch.)
—— *T.c.novaehiberniae* (E. Hartert, 1925)	SW New Ireland (Bismarck Arch.)
—— *T.c.bennetti* (Ripley, 1947)	Nissan I. (Bismarck Arch.) (2030)
—— *T.c.tristrami* (E.L. Layard, 1880)	New Britain
—— *T.c.alberti* (Rothschild & Hartert, 1905)	N and C Solomons to Guadalcanal
—— *T.c.mala* (Mayr, 1935)	Malaita I. (NE Solomons)
—— *T.c.pavuvu* (Mayr, 1935)	Pavuvu I. (SC Solomons)
—— *T.c.solomonis* (E.P. Ramsay, 1882)	Ugi, San Cristobal, St. Ana Is. (SE Solomons)
—— *T.c.sororum* (I.C.J. & E.H. Galbraith, 1962)	Malau Is. (E Solomons) (907)
—— *T.c.amoenus* (Mayr, 1931)	Rennell, Bellona Is. (S Solomons)
—— *T.c.brachyurus* (Mayr, 1931)	Reef I. (N Sta. Cruz Is.)
—— *T.c.vicina* (Mayr, 1931)	Duff I. (N Sta. Cruz Is.)
—— *T.c.ornatus* (Mayr, 1931)	Ndeni, Tinakula Is. (N Santa Cruz Is.)
—— *T.c.utupuae* (Mayr, 1931)	Utupua I. (S Santa Cruz Is.)
—— *T.c.melanodera* (Mayr, 1931)	Vanikoro I. (S Santa Cruz Is.)
—— *T.c.torresianus* (Mayr, 1931)	Torres Is. (N Vanuatu)
—— *T.c.santoensis* (Mayr, 1931)	Espiritu Santo, Banks Is. (N Vanuatu)
—— *T.c.juliae* (Heine, 1860)	Aneitum, Efate Is. (C Vanuatu)
—— *T.c.erromangae* (Mayr, 1938)	Erromanga I. (S Vanuatu)
—— *T.c.tannensis* (Sharpe, 1892)	Tanna I. (S Vanuatu)
—— *T.c.vitiensis* (Peale, 1848)	Viti Levu, Vanua Levu, Taviuni Is. (Fiji)
—— *T.c.eximius* (Mayr, 1941)	Kandavu, Ono, Vanua Kula Is. (Fiji)

---

[1] For recognition of this form see Schodde & Mason (1997: 349) (2188).
[2] For correction of spelling see David & Gosselin (2002) (613).

— *T.c.marinus* (Mayr, 1941)　　　　　　　　　　Lau Arch. (Fiji)

— *T.c.sacer* (J.F. Gmelin, 1788)　　　　　　　　C and S Tonga Is.

— *T.c.regina* (Mayr, 1941)　　　　　　　　　　Futuna I. (Ilês de Horn)

✓ *T.c.pealei* (Finsch & Hartlaub, 1867)　　　　Tutuila, Samoa Is.

— *T.c.manuae* (Mayr, 1941)　　　　　　　　　　Ofu, Olosinga, Tau Is.

○ *Todiramphus enigma* E. Hartert, 1904 TALAUD KINGFISHER　　Talaud Is.

○ *Todiramphus cinnamominus* MICRONESIAN KINGFISHER

— *T.c.pelewensis* (Wiglesworth, 1891)　　　　Palau Is.

— *T.c.reichenbachii* Hartlaub, 1852　　　　　Pohnpei I. (E Carolines)

— *T.c.cinnamominus* (Swainson, 1821)　　　　Guam I. (S Marianas)

○ †*Todiramphus miyakoensis* (N. Kuroda, Sr., 1919) MIYAKO ISLAND KINGFISHER

　　　　　　　　　　　　　　　　　　　　　　Miyako I. (Ryukyu Is., Japan)

○ *Todiramphus saurophagus* BEACH KINGFISHER

— *T.s.saurophagus* (Gould, 1843)　　　　　　N Moluccas to Solomon Is.

— *T.s.anachoreta* (Reichenow, 1898)　　　　　Hermit, Ninigo Is. (off E New Guinea)

— *T.s.admiralitatis* (Sharpe, 1892)　　　　　Admiralty Is. (Bismarck Arch.)

● *Todiramphus sanctus* SACRED KINGFISHER

— *T.s.sanctus* (Vigors & Horsfield, 1827)　　Australia, E Solomon Is. >> New Guinea, Solomon Is.,
　　　　　　　　　　　　　　　　　　　　　　Moluccas, Sulawesi, Lesser and Gtr. Sundas

✓ *T.s.vagans* (Lesson, 1830)[1]　　　　　　　New Zealand, Lord Howe I., Kermadec Is

— *T.s.norkolkiensis* (Tristram, 1885)[2]　　　Norfolk I.

— *T.s.recurvirostris* Lafresnaye, 1842[3]　　　W Samoa

— *T.s.canacorum* (Brasil, 1916)　　　　　　New Caledonia

— *T.s.macmillani* (Mayr, 1940)　　　　　　　Loyalty Is.

○ *Todiramphus australasia* CINNAMON-BANDED KINGFISHER[4]

— *T.a.australasia* (Vieillot, 1818)　　　　　Lombok, Sumba, Timor, Wetar, Roma Is (W and C
　　　　　　　　　　　　　　　　　　　　　　Lesser Sundas)

— *T.a.dammerianus* (E. Hartert, 1900)[5]　　Damar, Babar, Leti and Moa Is. (E Lesser Sundas)

— *T.a.odites* (J.L. Peters, 1945)　　　　　　Tanimbar I.

○ *Todiramphus tutus* CHATTERING KINGFISHER[6]

— *T.t.tutus* (J.F. Gmelin, 1788)　　　　　　Society Is. (except Moorea)

— *T.t.atiu* (Holyoak, 1974)　　　　　　　　Atiu I. (Cook Is.) (1129)

— *T.t.mauke* (Holyoak, 1974)　　　　　　　Mauke I. (Cook Is.) (1129)

— *T.t.ruficollaris* (Holyoak, 1974)[7]　　　　Mangaia I. (Cook Is.) (1129)

○ *Todiramphus veneratus* TAHITIAN KINGFISHER

— *T.v.veneratus* (J.F. Gmelin, 1788)　　　　Tahiti I. (Society Is.)

— *T.v.youngi* Sharpe, 1892　　　　　　　　Moorea I. (Society Is.)

○ *Todiramphus gambieri* TUAMOTU KINGFISHER

— †*T.g.gambieri* (Oustalet, 1895)　　　　　Mangareva I. (Tuamoto Arch.)

— *T.g.gertrudae* Murphy, 1924　　　　　　Niau I. (Tuamoto Arch.)

○ *Todiramphus godeffroyi* (Finsch, 1877) MARQUESAN KINGFISHER　　S Marquesas Is.

○ *Todiramphus pyrrhopygius* (Gould, 1841) RED-BACKED KINGFISHER[8]　Australia

SYMA Lesson, 1827 F

○ *Syma torotoro* YELLOW-BILLED KINGFISHER

— *S.t.torotoro* Lesson, 1827[9]　　　　　　W Papuan Is., north to W and SC New Guinea, Aru Is.

— *S.t.meeki* Rothschild & Hartert, 1901　　SE New Guinea

[1] Includes *adamsi* of Lord Howe I. see Fry & Fry (1992: 185) (895).
[2] For recognition of this form see Schodde & Mason (1997: 349) (2188).
[3] Treated at specific level in last Edition, but see Fry & Fry (1992) (895); for reasons to include this here see Woodall (2001: 223) (2689). We follow, but with reservations.
[4] We continue to follow White & Bruce (1986) (2652).
[5] Includes *interpositus* see White & Bruce (1986: 279) (2652).
[6] For correction of spelling see David & Gosselin (2002) (613).
[7] A substitute name *mangaia* Holyoak, 1976 (1130), was proposed, but see Schodde & Holyoak (1977) (2183).
[8] For correction of spelling see David & Gosselin (2002) (613).
[9] Includes *brevirostris*, *pseutes* and *tentelare* see Mees (1982) (1534).

___ *S.t.ochracea* Rothschild & Hartert, 1901   D'Entrecasteaux Arch.
___ *S.t.flavirostris* (Gould, 1850)   N Cape York Pen. (Australia)

○ *Syma megarhyncha* MOUNTAIN KINGFISHER
   ___ *S.m.wellsi* Mathews, 1918   Mts. of WC New Guinea
   ___ *S.m.sellamontis* Reichenow, 1919   Huon Pen. (E New Guinea)
   ___ *S.m.megarhyncha* Salvadori, 1896   Mts. of SE to C New Guinea

ALCEDININAE

CEYX Lacépède, 1799 M
○ *Ceyx lecontei* (Cassin, 1856) AFRICAN DWARF KINGFISHER[1]   Sierra Leone to Liberia and S Uganda

○ *Ceyx pictus* AFRICAN PYGMY KINGFISHER[2]
   ___ *C.p.pictus* (Boddaert, 1783)   Senegal to Eritrea, south to Uganda and N Tanzania
   ___ *C.p.ferruginus* Clancey, 1984[3]   Guinea-Bissau to W Uganda south to NW Zambia (492)
   ___ *C.p.natalensis* (A. Smith, 1832)   S Angola, Zambia and S Tanzania to E Cape Province
            >> coastal Kenya

○ *Ceyx erithaca*[4] ORIENTAL DWARF KINGFISHER[5]
   ___ *C.e.erithaca* (Linnaeus, 1758)[6]   India to SE China S through mainland SE Asia to
            Malay Pen., Sumatra, Bangka, Belitung, and
            Simaleue and Batu Is. (off W Sumatra)
   ___ *C.e.macrocarus* Oberholser, 1917   Andaman, Nicobar Is.
   ___ *C.e.motleyi* Chasen & Kloss, 1929   N Bornean Is., Borneo, Philippines, Java, Sumbawa, Flores

○ *Ceyx melanurus* PHILIPPINE DWARF KINGFISHER
   ___ *C.m.melanurus* (Kaup, 1848)   Luzon, Polillo (N Philippines)
   ___ *C.m.samarensis* Steere, 1890   Samar, Leyte (EC Philippines)
   ___ *C.m.mindanensis* Steere, 1890   Mindanao, Basilan (S Philippines)

○ *Ceyx fallax* SULAWESI DWARF KINGFISHER
   ___ *C.f.sangirensis* (Meyer & Wiglesworth, 1898)   Sangihe Is.
   ___ *C.f.fallax* (Schlegel, 1866)   Sulawesi

○ *Ceyx madagascariensis* MADAGASCAN PYGMY KINGFISHER
   ___ *C.m.madagascariensis* (Linnaeus, 1766)   NW, N and E Madagascar
   ___ *C.m.dilutus* (Benson, 1974)   SW Madagascar (128)

○ *Ceyx lepidus* VARIABLE DWARF KINGFISHER
   ___ *C.l.margarethae* W.H. Blasius, 1890[7]   W Visayas (Panay, Negros), Mindanao, Basilan and the
            Sulu Is. (Philippines)[8]
   ___ *C.l.wallacii* Sharpe, 1868   Sula Is. (Indonesia)
   ___ *C.l.uropygialis* G.R. Gray, 1860   N and C Moluccas (996)
   ___ *C.l.lepidus* Temminck, 1836   S Moluccas (except Buru)
   ___ *C.l.cajeli* Wallace, 1863   Buru I.
   ___ *C.l.solitarius* Temminck, 1836   New Guinea and islands, Aru Is., D'Entrecasteaux Arch.
   ___ *C.l.dispar* Rothschild & Hartert, 1914   Admiralty Is.
   ___ *C.l.mulcatus* Rothschild & Hartert, 1914   New Hanover, New Ireland, Lihir Is.
   ___ *C.l.sacerdotis* E.P. Ramsay, 1882   New Britain, Umboi Is.
   ___ *C.l.meeki* Rothschild, 1901   Choiseul, Ysabel, Bougainville, Buka Is. (N Solomons)
   ___ *C.l.collectoris* Rothschild & Hartert, 1901   Vellalavella, New Georgia, Rendova Is. (C Solomons)
   ___ *C.l.malaitae* Mayr, 1935   Malaita I. (E Solomons)
   ___ *C.l.nigromaxilla* Rothschild & Hartert, 1905   Guadalcanal I. (S Solomons)
   ___ *C.l.gentiana* Tristram, 1879   San Cristobal (Makira) I. (SE Solomons)

---

[1] We choose to follow White (1965: 226) (2649) in treating this as monotypic.
[2] Type species of the genus *Ispidina* here submerged.
[3] Clancey spelled this name *ferrugina* not *ferruginea*. It should not be rendered as *ferrugineus* (N. David *in litt.*).
[4] For an explanation of the spelling *erithaca* see David & Gosselin (2000) (612).
[5] We prefer to interpret variation in this species as relating to primary differentiation with geographic separation and subsequent reoccupation of territory with introgression, showing full speciation was not achieved. In this arrangement the name *rufidorsa*, applied to rufous-backed individuals, is a synonym of *erithaca* the population of the Malay Pen. including such birds among the intergrades (Wells, 1999) (2579).
[6] Includes *jungei* see Fry & Fry (1992: 198) (895).
[7] An extremely variable population; in last Edition we listed a species *goodfellowi*; that name and associated ones (including *virgicapitus*) are all synonyms of this see Dickinson *et al.* (1991) (745).
[8] Apparently absent from the E Visayas (Samar and Leyte), Luzon and the Palawan group.

ALCEDO Linnaeus, 1758 F

○ *Alcedo leucogaster* WHITE-BELLIED KINGFISHER[1]
  — *A.l.bowdleri* Neumann, 1908   Guinea to Mali and Ghana
  — *A.l.leucogaster* (Fraser, 1843)[2]   Nigeria to NW Angola, Bioko I.
  — *A.l.nais* Kaup, 1848   Principe I.
  — *A.l.leopoldi* (A.J.C. Dubois, 1905) E Congo to S Uganda, south to NW Zambia

○ *Alcedo cristata* MALACHITE KINGFISHER
  — *A.c.galerita* Statius Müller, 1776   Senegal to Nigeria and E Africa S to N Angola and
                                           Malawi
  — *A.c.cristata* Pallas, 1764   S Angola, SW Zambia, Zimbabwe and S Mozambique
                                           to Transvaal and E Cape Province
  — *A.c.thomensis* (Salvadori, 1902)   São Tomé
  — *A.c.stuartkeithi* Dickerman, 1989 Sudan, Ethiopia (720)

○ *Alcedo vintsioides* MADAGASCAN KINGFISHER
  — *A.v.vintsioides* Eydoux & Gervais, 1836 Madagascar
  — *A.v.johannae* (R. Meinertzhagen, 1924) Comoro Is.

○ *Alcedo cyanopectus* INDIGO-BANDED KINGFISHER[3]
  — *A.c.cyanopectus* (Lafresnaye, 1840)   Luzon, Mindoro and smaller islands nearby (N
                                         Philippines
  — *A.c.nigrirostris* (Bourns & Worcester, 1894) Cebu, Negros, Panay (WC Philippines)

○ *Alcedo argentata* SILVERY KINGFISHER
  — *A.a.flumenicola* (Steere, 1890)   Bohol, Leyte, Samar Is. (EC Philippines)
  — *A.a.argentata* (Tweeddale, 1877)[4] Mindanao, Basilan, Dinagat, Siargao Is. (S Philippines)

○ *Alcedo coerulescens* Vieillot, 1818 CERULEAN KINGFISHER S Sumatra and Java east to Sumbawa I. and Kangean I.

○ *Alcedo euryzona* BLUE-BANDED KINGFISHER
  — *A.e.peninsulae* Laubmann, 1941   SE Burma, SW Thailand, Malay Pen., Sumatra,
                                         Bangka, Borneo[5]
  — *A.e.euryzona* Temminck, 1830   Java

○ *Alcedo quadribrachys* SHINING-BLUE KINGFISHER
  — *A.q.quadribrachys* Bonaparte, 1850 Gambia to WC Nigeria
  — *A.q.guentheri* Sharpe, 1892   S Nigeria to Uganda, N Angola and NW Zambia

○ *Alcedo azurea* AZURE KINGFISHER
  — *A.a.affinis* (G.R. Gray, 1860)   N Moluccas
  — *A.a.lessonii* (Cassin, 1850)[6]   W Papuan Is., W, S, and E New Guinea, Aru Is.
  — *A.a.ochrogaster* (Reichenow, 1903) N New Guinea, Geelvink Bay Is.
  — *A.a.ruficollaris* (Bankier, 1841)[7] N Australia, ? Tanimbar Is.
  — *A.a.azurea* Latham, 1802[8]   E and SE Australia
  — *A.a.diemenensis* (Gould, 1846)[9]   Tasmania

○ *Alcedo websteri* (E. Hartert, 1898) BISMARCK KINGFISHER Bismarck Arch.

○ *Alcedo pusilla* LITTLE KINGFISHER
  — *A.p.halmaherae* (Salomonsen, 1934) Bacan and Halmahera (N Moluccas)
  — *A.p.laetior* (Rand, 1941)   N New Guinea
  — *A.p.pusilla* (Temminck, 1836)[10]   W Papuan Is., W, S and E New Guinea, Kai Is., Aru Is.,
                                           D'Entrecasteaux Arch.

---

[1] Woodall (2001: 233) (2689) separated *nais* of Principe as a species, but considered the island form on Bioko essentially undifferentiated from birds of the nearby African coast. We tentatively retain the treatment of Fry & Fry (1992) (895).
[2] Includes *batesi* see White (1965: 225) (2649).
[3] For correct spelling see David & Gosselin (2002) (613).
[4] The range given in last Edition was a *lapsus* hiding the fact that this is an allospecies of *cyanopectus*.
[5] For attribution to this form see Mees (1971) (1523).
[6] Includes *wallaceanus* given as doubtfully distinct by Peters (1945) (1826).
[7] We follow White & Bruce (1986) (2652) in doubting the distinctness of *yamdenae*.
[8] For reasons to use 1802 not 1801 see Browing & Monroe (1991) (258). Includes *mixtus* see Schodde & Mason (1976) (2185).
[9] For recognition of this see Schodde & Mason (1976) (2185).
[10] In last Edition we had nominate *pusilla* extending to the Moluccas and we still accepted *laetior*. White & Bruce (1986: 281) (2652) seemed to suggest that both *halmaherae* and *laetior* should be submerged in nominate *pusilla*. However both are recognised by Woodall (2000: 238) (2689) and we tentatively follow him.

— *A.p.ramsayi* (North, 1912)   N coast of Northern Territory, Melville and Bathurst Is., Groote Eylandt, W Cape York Pen.

— *A.p.halli* (Mathews, 1912)   NE Queensland to E Cape York Pen.

— *A.p.masauji* (Mathews, 1927)   Bismarck Arch.

— *A.p.bougainvillei* (Ogilvie-Grant, 1914)   Bougainville, Ysabel, Choiseul Is. (N Solomons)

— *A.p.richardsi* (Tristram, 1882)   Vellalavella, Kolombangara Is. (C Solomons)

— *A.p.aolae* (Ogilvie-Grant, 1914)   Guadalcanal (SE Solomons)

○ *Alcedo meninting* BLUE-EARED KINGFISHER

— *A.m.coltarti* E.C.S. Baker, 1919[1]   Nepal and India (except SW) east to N and C Thailand and Indochina

— *A.m.phillipsi* E.C.S. Baker, 1927   SW India[2], Sri Lanka

— *A.m.scintillans* E.C.S. Baker, 1919   N Malay Pen.

— *A.m.rufigastra* Walden, 1873[3]   S and C Andaman Is.

— *A.m.meninting* Horsfield, 1821[4] [5]   Malay Pen., Borneo, the Palawan group and the Sulu Is. (Philippines), SW Sulawesi and Sula Is. (Indonesia), and Sumatra and W Sumatran Is., Java to Lombok

○ *Alcedo atthis* COMMON KINGFISHER

— *A.a.ispida* Linnaeus, 1758   NW and C Europe >> W, C and S Europe

— *A.a.atthis* (Linnaeus, 1758)   Mediterranean basin and E Europe to SW and C Asia >> N Africa and Sudan

— *A.a.bengalensis* J.F. Gmelin, 1788[6]   N India to Sakhalin and Japan >> S and E Asia south to Sulawesi and the Moluccas

— *A.a.taprobana* O. Kleinschmidt, 1894   S India, Sri Lanka

— *A.a.floresiana* Sharpe, 1892   Bali to Timor and Romang Is. (Lesser Sundas)

— *A.a.hispidoides* Lesson, 1837   Sangihe Is., Sulawesi to Moluccas, coastal New Guinea and islands, Bismarck Arch. and Louisade Arch.

— *A.a.salomonensis* Rothschild & Hartert, 1905   Solomon Is.

○ *Alcedo semitorquata* Swainson, 1823 HALF-COLLARED KINGFISHER[7]   Ethiopia and Tanzania S to Transvaal and E Cape Province

○ *Alcedo hercules* Laubmann, 1917 BLYTH'S KINGFISHER   E Himalayas east to N Vietnam

CERYLINAE

CHLOROCERYLE Kaup, 1848 F

○ *Chloroceryle aenea* AMERICAN PYGMY KINGFISHER

— *C.a.stictoptera* (Ridgway, 1884)   S Mexico to Nicaragua

— *C.a.aenea* (Pallas, 1764)   Costa Rica, Panama, N Sth. America, Trinidad

○ *Chloroceryle inda* (Linnaeus, 1766) GREEN-AND-RUFOUS KINGFISHER[8]   Panama and tropical Sth. America

● *Chloroceryle americana* GREEN KINGFISHER

— *C.a.hachisukai* Laubmann, 1942   SC USA, NW Mexico

√ *C.a.septentrionalis* (Sharpe, 1892)[9]   SE Texas and E Mexico south to N Columbia and W Venezuela

— *C.a.americana* (J.F. Gmelin, 1788)[10]   N Sth. America (east of Andes)

— *C.a.mathewsii* Laubmann, 1927   C and SC Sth. America

— *C.a.cabanisii* (Tschudi, 1846)[11]   W Colombia south to N Chile (west of the Andes)

● *Chloroceryle amazona* (Latham, 1790) AMAZON KINGFISHER[12]   S Mexico to N Sth. America

---

[1] Includes *laubmanni* see Ripley (1982: 206) (2055).
[2] Based on Woodall (2001: 238) (2689), not Ripley (1982) (2055), and tentatively accepted here on zoogeographical grounds.
[3] For correction of spelling see David & Gosselin (2002) (613).
[4] Includes *verreauxi* see Mees (1986) (1539), Wells (1999) (2579), and thus also *proxima*, *subviridis* and *callima* (see Woodall, 2001: 238) (2689).
[5] Includes *amadoni* duPont, 1971 (771). Dickinson *et al.* (1991) (745) relied on an evaluation prior to the review of Mees (1986) (1539) establishing the nature of variation; with the sinking of *verreauxi* it is now best to put both Philippine populations, thought the same by Woodall (2001: 238) (2689), in nominate *meninting*.
[6] Includes *japonica* see Orn. Soc. Japan (2000) (1727).
[7] Includes *heuglini* and *tephria* Clancey, 1951 (364), see White (1965: 224) (2649).
[8] Includes *chocoensis* see Woodall (2001: 243) (2689).
[9] Includes *isthmica* see Woodall (2001: 243) (2689).
[10] Includes *croteta* see Woodall (2001: 243) (2689); presumably also includes ¶*bottomeana* Aveledo & Perez, 1994 (69).
[11] Includes *hellmayri* and *ecuadorensis* Yamashina, 1943 (2694), see Woodall (2001: 248) (2689).
[12] Includes *mexicana* see monotypic treatment in Fry *et al.* (1992) (895).

MEGACERYLE Kaup, 1848 F
○ *Megaceryle lugubris* Crested Kingfisher
 — *M.l.continentalis* (E. Hartert, 1900)   W and C Himalayas
 — *M.l.guttulata* (Stejneger, 1892)   E Himalayas, Assam hills and Burma, N Thailand, S
              and C China and N Indochina
 — *M.l.pallida* (Momiyama, 1927)   Hokkaido I., S Kuril Is.
 — *M.l.lugubris* (Temminck, 1834)   C and S Japan, Korea

○ *Megaceryle maxima* Giant Kingfisher
 — *M.m.maxima* (Pallas, 1769)   Senegal to N Nigeria, Ethiopia and S Kenya to Cape
              Province
 — *M.m.gigantea* (Swainson, 1837)   Liberia to S Nigeria and N Angola E to W Tanzania; Bioko

● *Megaceryle torquata* Ringed Kingfisher
 — *M.t.stictipennis* (Lawrence, 1885)   Guadeloupe I., Dominica I.
 ✓ *M.t.torquata* (Linnaeus, 1766)   Mexico to N Argentina, Peru
 — *M.t.stellata* (Meyen, 1834)   Chile, Argentina

● *Megaceryle alcyon* (Linnaeus, 1758) Belted Kingfisher[1]   Alaska and C and E Canada to N Sth. America and W
              Indies

CERYLE Boie, 1828 M[2]
○ *Ceryle rudis* Pied Kingfisher
 — *C.r.syriacus* Roselaar, 1995[3]   Asia Minor and the Levant to SW Iran (2107)
 — *C.r.rudis* (Linnaeus, 1758)   Egypt and sub-Saharan Africa
 — *C.r.leucomelanurus* Reichenbach, 1851   E Afghanistan, S to C and SE India, E to W and S Yunnan,
              Thailand (except SW and Pen.) and Indochina
 — *C.r.travancoreensis* Whistler, 1935   SW India
 — *C.r.insignis* E. Hartert, 1910   SE China, Hainan I.

TODIDAE TODIES (1: 5)

TODUS Brisson, 1760 M
○ *Todus multicolor* Gould, 1837 Cuban Tody   Cuba, Isla de la Juventud

○ *Todus subulatus* G.R. Gray, 1847 Broad-billed Tody   Hispaniola, Gonave I.

○ *Todus angustirostris* Lafresnaye, 1851 Narrow-billed Tody   Hispaniola

○ *Todus todus* (Linnaeus, 1758) Jamaican Tody   Jamaica

○ *Todus mexicanus* Lesson, 1838 Puerto Rican Tody   Puerto Rico

MOMOTIDAE MOTMOTS[4] (6: 10)

HYLOMANES M.H.K. Lichtenstein, 1839 M
○ *Hylomanes momotula* Tody-Motmot[5]
 — *H.m.chiapensis* Brodkorb, 1938   S Mexico (Pacific)
 — *H.m.momotula* M.H.K. Lichtenstein, 1839   Caribbean slope of S Mexico to Honduras
 — *H.m.obscurus* Nelson, 1911   NW Costa Rica to NW Colombia

ASPATHA Sharpe, 1892 F
○ *Aspatha gularis* (Lafresnaye, 1840) Blue-throated Motmot   S Mexico to Honduras, El Salvador

MOMOTUS Brisson, 1760 M
○ *Momotus mexicanus* Russet-crowned Motmot
 — *M.m.vanrossemi* R.T. Moore, 1932   NW Mexico
 — *M.m.mexicanus* Swainson, 1827   NC and C Mexico
 — *M.m.saturatus* Nelson, 1897   SW Mexico
 — *M.m.castaneiceps* Gould, 1855   C Guatemala

[1] Includes *caurina* see Woodall (2001: 248) (2689).
[2] Gender addressed by David & Gosselin (2002) (614), multiple changes occur in specific names, but *Megaceryle* remains feminine and names therein do not change.
[3] For an opinion opposing recognition see Kasparek (1996) (1234).
[4] For convenience the sequence of genera is revised to follow A.O.U. (1998) (3).
[5] Although not normally hyphenated we prefer to hyphenate wherever two 'generic vernaculars' appear together and to use lower case for the second if that name, with a capital, would suggest a presumption of close relationship that is unwarranted.

● *Momotus momota* BLUE-CROWNED MOTMOT
  — *M.m.coeruliceps* (Gould, 1836)   NE and C Mexico
  — *M.m.goldmani* Nelson, 1900   SE Mexico, N Guatemala
  — *M.m.exiguus* Ridgway, 1912   S Mexico
  ⅃ *M.m.lessonii* Lesson, 1842   S Mexico to W Panama
  — *M.m.conexus* Thayer & Bangs, 1906[1]   C Panama, NW Colombia
  — *M.m.spatha* Wetmore, 1946   Guajira Pen. (Colombia) (2591)
  — *M.m.olivaresi* Hernández & Romero, 1978   NC Colombia (1107)
  — *M.m.subrufescens* P.L. Sclater, 1853   N Colombia, N Venezuela
  — *M.m.momota* (Linnaeus, 1766)   E Venezuela, the Guianas, N Brazil
  — *M.m.osgoodi* Cory, 1913   W Venezuela
  — *M.m.bahamensis* (Swainson, 1837)   Trinidad, Tobago I.
  — *M.m.microstephanus* P.L. Sclater, 1858   SE Colombia, E Ecuador, NW Brazil
  — *M.m.argenticinctus* Sharpe, 1892   W Ecuador, NW Peru
  — *M.m.ignobilis* Berlepsch, 1889   E Peru, W Brazil
  — *M.m.nattereri* P.L. Sclater, 1858   NE Bolivia
  — *M.m.simplex* Chapman, 1923   W Brazil
  — *M.m.cametensis* E. Snethlage, 1912   NC Brazil
  — *M.m.parensis* Sharpe, 1892   NE Brazil
  — *M.m.marcgraviniana* Pinto & Camargo, 1961[2]   Easternmost Brazil (1899)
  — *M.m.pilcomajensis* Reichenow, 1919   S Bolivia, S Brazil, NW Argentina

○ *Momotus aequatorialis* HIGHLAND MOTMOT[3]
  — *M.a.aequatorialis* Gould, 1857   WC Colombia, E Ecuador
  — *M.a.chlorolaemus* Berlepsch & Stolzmann, 1902   E Peru

BARYPHTHENGUS Cabanis & Heine, 1859 M
○ *Baryphthengus martii* RUFOUS MOTMOT[4]
  — *B.m.semirufus* (P.L. Sclater, 1853)[5]   E Honduras to NW Columbia and W Ecuador
  — *B.m.martii* (Spix, 1824)   W Amazonia

○ *Baryphthengus ruficapillus* (Vieillot, 1818) RUFOUS-CAPPED MOTMOT[6]   S and E Brazil, Paraguay, NE Argentina

ELECTRON Gistel, 1848 N
○ *Electron carinatum* (Du Bus, 1847)[7] KEEL-BILLED MOTMOT   S Mexico to NW Costa Rica

● *Electron platyrhynchum* BROAD-BILLED MOTMOT
  ⅃ *E.p.minus* (E. Hartert, 1898)[8]   E Honduras to C Colombia
  — *E.p.platyrhynchum* (Leadbeater, 1829)   NW Colombia to SW Ecuador
  — *E.p.colombianum* Meyer de Schauensee, 1950   Lowland N Colombia (1566)
  — *E.p.pyrrholaemum* (Berlepsch & Stolzmann, 1902)   E Colombia, E Ecuador, Peru, N Bolivia
  — *E.p.orienticola* Oberholser, 1920   W Amazonian Brazil
  — *E.p.chlorophrys* Miranda-Ribeiro, 1931   C Brazil (S of the Amazon)

EUMOMOTA P.L. Sclater, 1858 F
● *Eumomota superciliosa* TURQUOISE-BROWED MOTMOT
  — *E.s.bipartita* Ridgway, 1912   SW Mexico, W Guatemala
  ⅃ *E.s.superciliosa* (Sandbach, 1837)   SE Mexico
  — *E.s.vanrossemi* Griscom, 1929   C Guatemala
  — *E.s.sylvestris* Carriker & Meyer de Schauensee, 1935   E Guatemala
  — *E.s.apiaster* (Lesson, 1842)   El Salvador, W Honduras, NW Nicaragua
  — *E.s.euroaustris* Griscom, 1929   N Honduras
  ⅃ *E.s.australis* Bangs, 1906   SW Nicaragua, NW Costa Rica

---

[1] Includes *reconditus* see Snow (2001: 283) (2295).
[2] For recognition see Snow (2001: 283) (2295).
[3] See A.O.U. (1998: 321) (3) for a discussion of relationships, and the implicit recognition of this following Fjeldså & Krabbe (1990) (850).
[4] For reasons to separate this species from *ruficapillus* see A.O.U. (1998: 321) (3).
[5] Includes *costaricensis* see Snow (2001: 280) (2295).
[6] Straube & Bornschein (1991) (2358) considered the species monotypic. The name *berlai* Stager, 1959 (2302), is considered invalid by Snow (2001: 281) (2295).
[7] Du Bus is an abbreviation of Du Bus de Gisignies.
[8] For correction of spelling see David & Gosselin (2002) (613).

MEROPIDAE BEE-EATERS (3: 25)

NYCTYORNIS Jardine & Selby, 1830 M[1]
○ *Nyctyornis amictus* (Temminck, 1824) RED-BEARDED BEE-EATER   SE Burma, SW Thailand, Malay Pen., Sumatra, Borneo, Bangka

○ *Nyctyornis athertoni* BLUE-BEARDED BEE-EATER
— *N.a.athertoni* (Jardine & Selby, 1828)[2][3]   SW India and S Himalayas east to Indochina and south to N Malay Pen.

— *N.a.brevicaudatus* (Koelz, 1939)   Hainan I.

MEROPOGON Bonaparte, 1850 M
○ *Meropogon forsteni* Bonaparte, 1850 PURPLE-BEARDED BEE-EATER   Sulawesi

MEROPS Linnaeus, 1758 M
○ *Merops breweri* (Cassin, 1859) BLACK-HEADED BEE-EATER   Gabon to W and N Zaire, Central African Republic, SE Nigeria

○ *Merops muelleri* BLUE-HEADED BEE-EATER
— *M.m.mentalis* Cabanis, 1889   Sierra Leone to SW Cameroon
— *M.m.muelleri* (Cassin, 1857)   S Cameroon to E Zaire, W Kenya

○ *Merops gularis* BLACK BEE-EATER
— *M.g.gularis* Shaw, 1798   Sierra Leone to S Nigeria
— *M.g.australis* (Reichenow, 1885)   SE Nigeria to W Uganda and N Angola

○ *Merops hirundineus* SWALLOW-TAILED BEE-EATER
— *M.h.chrysolaimus* Jardine & Selby, 1830   Senegal to Ghana and Central African Republic
— *M.h.heuglini* (Neumann, 1906)   NE Zaire, N Uganda, S Sudan, SW Ethiopia
— *M.h.furcatus* Stanley, 1814   SE Zaire and Tanzania to Zimbabwe and Mozambique
— *M.h.hirundineus* A.A.H. Lichtenstein, 1793[4]   Angola to Namibia, Botswana and N Sth. Africa

○ *Merops pusillus* LITTLE BEE-EATER
— *M.p.pusillus* Statius Müller, 1776   Senegal to SW Sudan and N Zaire
— *M.p.ocularis* (Reichenow, 1900)   NE Zaire, N Uganda, S Sudan, W Ethiopia
— *M.p.cyanostictus* Cabanis, 1869   C and E Ethiopia, W Somalia, Kenya
— *M.p.meridionalis* (Sharpe, 1892)   S and E Zaire to Tanzania south to C Angola, Zimbabwe and Natal
— *M.p.argutus* Clancey, 1967[5]   SW Angola, Botswana, SW Zambia (412)

○ *Merops variegatus* BLUE-BREASTED BEE-EATER
— *M.v.lafresnayii* Guérin, 1843   Ethiopia
— *M.v.loringi* (Mearns, 1915)   SE Nigeria and Cameroon to Uganda and W Kenya
— *M.v.variegatus* Vieillot, 1817   Gabon to C Zaire, Rwanda, Burundi and N Angola
— *M.v.bangweoloensis* (C.H.B. Grant, 1915)   S Zaire, Zambia, W Tanzania

○ *Merops oreobates* (Sharpe, 1892) CINNAMON-CHESTED BEE-EATER   S Sudan and E Zaire to Kenya and N Tanzania

○ *Merops bulocki* RED-THROATED BEE-EATER
— *M.b.bulocki* Vieillot, 1817   Senegal to N Central African Republic
— *M.b.frenatus* Hartlaub, 1854   S Sudan to W Ethiopia, NE Zaire and N Uganda

○ *Merops bullockoides* A. Smith, 1834 WHITE-FRONTED BEE-EATER[6]   Gabon and Angola to Kenya, Tanzania, Malawi and Natal

○ *Merops revoilii* Oustalet, 1882 SOMALI BEE-EATER   SE Ethiopia, W Somalia, NE Kenya

○ *Merops albicollis* Vieillot, 1817 WHITE-THROATED BEE-EATER   Senegal to Eritrea and Kenya >> Congo, Rwanda and N Tanzania, Yemen

○ *Merops boehmi* Reichenow, 1882 BOEHM'S BEE-EATER   Tanzania, E Zambia, Malawi

---

[1] Gender addressed by David & Gosselin (2002) (614), multiple changes occur in specific names.
[2] Not 1830 as in Peters (1945) (1826); for the dates of Jardine & Selby's parts see Sherborn (1894) (2241).
[3] Includes *bartletti* Koelz, 1954 (1284), see Ripley (1982: 213) (2055).
[4] Includes *strenuus* Clancey, 1962 (393), see White (1965: 235) (2649).
[5] The name *landanae* da Rosa Pinto, 1972 (608), is thought to be a synonym, it may need evaluation.
[6] Includes *randorum* Clancey, 1953 (370) see White (1965: 234) (2649).

○ *Merops orientalis* GREEN BEE-EATER
    __ *M.o.viridissimus* Swainson, 1837[1]   Senegal to Eritrea, S Sudan
    __ *M.o.cleopatra* Nicoll, 1910   Nile valley, Egypt south to N Sudan
    __ *M.o.cyanophrys* (Cabanis & Heine, 1860)   Dead Sea depression to W and S Arabia
    __ *M.o.muscatensis* Sharpe, 1886[2]   C and E Arabia
    __ *M.o.beludschicus* Neumann, 1910   S Iran to NW India
    __ *M.o.orientalis* Latham, 1802[3]   India (except N and Assam)
    __ *M.o.ceylonicus* Whistler, 1944   Sri Lanka (2619)
    __ *M.o.ferrugeiceps* Anderson, 1878[4]   Assam, Burma and Yunnan to Indochina

○ *Merops persicus* BLUE-CHEEKED BEE-EATER
    __ *M.p.persicus* Pallas, 1773   N Egypt and Middle East to Kazakhstan and NW India >> E and S Africa
    __ *M.p.chrysocercus* Cabanis & Heine, 1860   W Sahara >> W Africa

○ *Merops superciliosus* OLIVE BEE-EATER[5]
    __ *M.s.superciliosus* Linnaeus, 1766   Madagascar; Somalia to Mozambique and Angola
    __ *M.s.alternans* Clancey, 1971   W Angola, NW Namibia (428)

○ *Merops philippinus* BLUE-TAILED BEE-EATER[6]
    __ *M.p.javanicus* Horsfield, 1821   India and Sri Lanka east to S China and south to Indochina and Penang >> S Asia and S mainland SE Asia to Gtr. Sundas
    __ *M.p.philippinus* Linnaeus, 1766   Philippines
    __ *M.p.celebensis* W.H. Blasius, 1885   C and S Sulawesi, ? Lesser Sundas
    __ *M.p.salvadorii* A.B. Meyer, 1891   New Guinea and New Britain

○ *Merops ornatus* Latham, 1802[7] RAINBOW BEE-EATER   Australia, SE New Guinea >> New Guinea, Bismarck Arch. and Solomon Is. and Moluccas to Lesser Sundas and Sulawesi

○ *Merops viridis* BLUE-THROATED BEE-EATER
    __ *M.v.viridis* Linnaeus, 1758   S and SE China through E Thailand and Indochina to Gtr. Sundas >> SE Asia to Indonesia
    __ *M.v.americanus* Statius Müller, 1776   Philippines

○ *Merops leschenaulti* CHESTNUT-HEADED BEE-EATER
    __ *M.l.leschenaulti* Vieillot, 1817   India and Sri Lanka east to W and S Yunnan and south to Malay Pen.
    __ *M.l.quinticolor* Vieillot, 1817   Sumatra, Java, Bali
    __ *M.l.andamanensis* Marien, 1950   Andaman Is. (1429)

○ *Merops apiaster* Linnaeus, 1758 EUROPEAN BEE-EATER   a) S Europe and N Africa to C Asia >> sub-Saharan W and E Africa
     b) Sthn. Africa >> SC Africa (N to S Zaire and Zambia)

○ *Merops malimbicus* Shaw, 1806 ROSY BEE-EATER   Ghana to Gabon and W Zaire

○ *Merops nubicus* CARMINE BEE-EATER
    __ *M.n.nubicus* J.F. Gmelin, 1788   Senegal to Somalia >> NE Tanzania
    __ *M.n.nubicoides* Des Murs & Pucheran, 1846   Angola to Natal >> S Zaire and S Tanzania

---

[1] The name *flavoviridis* Niethammer, 1955 (1652), was not accepted by White (1965: 231) (2649). Fry (2001: 334) (893) accepted it.
[2] Includes *najdanus* see Fry (2001: 334) (893).
[3] For reasons to use 1802 not 1801 see Browing & Monroe (1991) (258).
[4] The name *ferrugeiceps* was validly used by Anderson; its availability is not affected by *ferrugiceps* Hodgson, 1844, a *nomen nudum*. See *nomen nudum* in the Glossary of the Code (I.C.Z.N., 1999: 111) (1178). Anderson's type locality was restricted to Mandalay by Deignan (1963) (663). The name *birmanus* is a junior synonym.
[5] We agree with Fry (2001) (893) that it is best to separate this from *M. persicus*; these two were previously separated from *philippinus* by Vaurie (1965: 675) (2510).
[6] Monotypic in Fry (2001: 338) (893), but most writers on Asian birds recognise subspecies following Deignan (1955) (652).
[7] For reasons to use 1802 not 1801 see Browing & Monroe (1991) (258).

UPUPIDAE HOOPOES (1: 1)

UPUPA Linnaeus, 1758 F
● ***Upupa epops*** COMMON HOOPOE[1]
    ⅃ *U.e.epops* Linnaeus, 1758[2][3]

    — *U.e.ceylonensis* Reichenbach, 1853
    — *U.e.longirostris* Jerdon, 1862

    — *U.e.major* C.L. Brehm, 1855
    — *U.e.senegalensis* Swainson, 1837
    — *U.e.waibeli* Reichenow, 1913
    — *U.e.africana* Bechstein, 1811
    — *U.e.marginata* Cabanis & Heine, 1860[4]

NW Africa and S and C Europe E to E Siberia and N and
    E China S to Tibetan Plateau >> Africa, S and E Asia
Himalayan foothills to Sri Lanka and Bangladesh
Assam to Indochina east to S China and south to Pen.
    Thailand (formerly to W Malaysia)
Egypt
Senegal to Somalia
Cameroon to NW Kenya
S Zaire to Uganda, C Kenya and Cape Province
Madagascar

PHOENICULIDAE WOOD HOOPOES (2: 8)

PHOENICULUS Jarocki, 1821 M
○ ***Phoeniculus castaneiceps*** FOREST WOOD HOOPOE
    — *P.c.castaneiceps* (Sharpe, 1871)
    — *P.c.brunneiceps* (Sharpe, 1903)

Liberia to SW Nigeria
S Cameroon to NE Zaire, Uganda

○ ***Phoeniculus bollei*** WHITE-HEADED WOOD HOOPOE
    — *P.b.bollei* (Hartlaub, 1858)
    — *P.b.jacksoni* (Sharpe, 1890)
    — *P.b.okuensis* Serle, 1949

Liberia to NW Zaire
E Zaire to C Kenya
S Cameroon (2222)

○ ***Phoeniculus purpureus*** GREEN WOOD HOOPOE
    — *P.p.senegalensis* (Vieillot, 1822)
    — *P.p.guineensis* (Reichenow, 1902)
    — *P.p.niloticus* (Neumann, 1903)
    — *P.p.marwitzi* (Reichenow, 1906)
    — *P.p.angolensis* (Reichenow, 1902)
    — *P.p.purpureus* (J.F. Miller, 1784)

S Senegal to S Ghana
N Senegal, Mali to S Chad and Central African Republic
Sudan, W Ethiopia, E Zaire to NW Kenya
E Uganda, C and S Kenya, E Zaire to Natal, Zanzibar
Angola, NE Namibia to W Zambia, N Botswana
C and SW Sth. Africa

○ ***Phoeniculus somaliensis*** BLACK-BILLED WOOD HOOPOE
    — *P.s.abyssinicus* (Neumann, 1903)
    — *P.s.somaliensis* (Ogilvie-Grant, 1901)
    — *P.s.neglectus* (Neumann, 1905)

N Ethiopia and Eritrea
SE Ethiopia, SW Somalia, NE Kenya
C and SW Ethiopia

○ ***Phoeniculus damarensis*** VIOLET WOOD HOOPOE
    — *P.d.damarensis* (Ogilvie-Grant, 1901)
    — *P.d.granti* (Neumann, 1903)

W Angola, NW Namibia
S Ethiopia, C and SE Kenya

RHINOPOMASTUS Jardine, 1828 M
○ ***Rhinopomastus aterrimus*** BLACK SCIMITARBILL
    — *R.a.aterrimus* (Stephens, 1826)

    — *R.a.emini* (Neumann, 1905)[5]
    — *R.a.notatus* (Salvin, 1892)
    — *R.a.anchietae* (Bocage, 1892)[6]

Senegal and S Sahara to W Sudan south to Nigeria
    and N Zaire
C Sudan to NE Zaire, Uganda
Ethiopia
SW Zaire, Angola, W Zambia

○ ***Rhinopomastus cyanomelas*** COMMON SCIMITARBILL
    — *R.c.schalowi* Neumann, 1900

    — *R.c.cyanomelas* (Vieillot, 1819)

S Uganda to S Somalia south to Zambia, Mozambique
    and Natal
Angola to Namibia, Botswana and W Transvaal

[1] Both *africana* and *marginata* are treated as species by some authors.
[2] Includes *saturata* see Stepanyan (1990: 304) (2326).
[3] May also include *orientalis* which was listed as a synonym of *ceylonensis* by Ripley (1982: 216) (2055).
[4] Dowsett & Dowsett-Lemaire (1993) (762) were "fairly confident" that vocal differences were of a character justifying specific status for this.
[5] Includes *cavei* Macdonald, 1946 (1407), see White (1965: 242) (2649).
[6] Includes *anomalus* Traylor, 1964 (2426), see Ligon (2001: 433) (1361).

◯ *Rhinopomastus minor* ABYSSINIAN SCIMITARBILL
    — *R.m.minor* (Rüppell, 1845)   E and SE Ethiopia to Somalia
    — *R.m.cabanisi* (Defilippi, 1853)   SE Sudan, SW Ethiopia, Kenya to C Tanzania

BUCEROTIDAE HORNBILLS (13: 49)

ANORRHINUS Reichenbach, 1849 M
◯ *Anorrhinus tickelli* BROWN HORNBILL
    — *A.t.austeni* Jerdon, 1872[1]   S and E Assam east to Indochina
    — *A.t.tickelli* (Blyth, 1855)   S Burma, SW Thailand

◯ *Anorrhinus galeritus* (Temminck, 1831) BUSHY-CRESTED HORNBILL[2]   Malay Pen., Sumatra, Borneo, Natuna Is.

TOCKUS Lesson, 1830 M
◯ *Tockus alboterminatus* CROWNED HORNBILL[3]
    — *T.a.alboterminatus* (Büttikofer, 1889)   W Angola
    — *T.a.geloensis* (Neumann, 1905)   SW Ethiopia, Uganda and W Kenya to N Angola and N Zambia

    — *T.a.suahelicus* (Neumann, 1905)   S Somalia and E Kenya to Mozambique and coastal Cape Province

◯ *Tockus bradfieldi* (Roberts, 1930) BRADFIELD'S HORNBILL   S Angola, N Namibia, N Botswana, W Zimbabwe

◯ *Tockus fasciatus* AFRICAN PIED HORNBILL
    — *T.f.semifasciatus* (Hartlaub, 1855)   S Senegal to S Nigeria
    — *T.f.fasciatus* (Shaw, 1811)   SE Nigeria to N Zaire, Uganda and N Angola

◯ *Tockus hemprichii* (Ehrenberg, 1833) HEMPRICH'S HORNBILL   Ethiopia, Eritrea, NW Somalia, NW Kenya

◯ *Tockus pallidirostris* PALE-BILLED HORNBILL
    — *T.p.pallidirostris* (Hartlaub & Finsch, 1870)   Angola, S Zaire, W Zambia
    — *T.p.neumanni* (Reichenow, 1894)   SE Tanzania, Malawi, N Mozambique

◯ *Tockus nasutus* AFRICAN GREY HORNBILL
    — *T.n.nasutus* (Linnaeus, 1766)[4]   Senegal to Eritrea, N Uganda and W and C Kenya, SW Arabia

    — *T.n.epirhinus* (Sundevall, 1851)[5]   S Uganda, SE Kenya to Angola, Botswana, N Sth. Africa

◯ *Tockus monteiri* Hartlaub, 1865 MONTEIRO'S HORNBILL   SW Angola, N Namibia

◯ *Tockus erythrorhynchus* RED-BILLED HORNBILL
    — *T.e.kempi* Tréca & Érard, 2000   Senegal, Gambia, Guinea and W Mali (2436)
    — *T.e.erythrorhynchus* (Temminck, 1823)   Cameroon east to Somalia and Tanzania
    — *T.e.rufirostris* (Sundevall, 1851)[6]   Malawi and Zambia to SW Angola and Transvaal
    — *T.e.damarensis* (Shelley, 1888)   N Namibia

◯ *Tockus leucomelas* SOUTHERN YELLOW-BILLED HORNBILL
    — *T.l.elegans* Hartlaub, 1865   SW Angola
    — *T.l.leucomelas* (M.H.K. Lichtenstein, 1842)[7]   Namibia to W Mozambique and Natal

◯ *Tockus flavirostris* (Rüppell, 1835) EASTERN YELLOW-BILLED HORNBILL   Eritrea and Somalia to NE Tanzania

◯ *Tockus deckeni* (Cabanis, 1869) VON DER DECKEN'S HORNBILL   C Ethiopia, S Somalia, N and NE Kenya to C Tanzania

◯ *Tockus jacksoni* (Ogilvie-Grant, 1891) JACKSON'S HORNBILL[8]   NE Uganda, NW Kenya

◯ *Tockus hartlaubi* BLACK DWARF HORNBILL
    — *T.h.hartlaubi* Gould, 1861   Sierra Leone to S Cameroon
    — *T.h.granti* (E. Hartert, 1895)   WC to NE Zaire, W Uganda

---

[1] Includes *indochinensis* see Kemp (2001: 491) (1252) and no doubt earlier.
[2] For treatment as a monotypic species see Kemp (2001: 491) (1252)
[3] We prefer the arrangement of White (1965: 248) (2649) to monotypic treatment.
[4] Includes *forskali* which White (1965: 244) (2649) found only slightly darker.
[5] Includes *dorsalis* Sanft, 1954 (2164), see Kemp (1985: 399) (1249).
[6] Probably includes *degens* Clancey, 1964 (397); but may need re-evaluation.
[7] Includes *parvior* Clancey, 1960 (384), see White (1965: 246) (2649).
[8] We do not believe this should be lumped with the preceding species and follow Zimmerman, *et al.* (1996) (2735).

○ *Tockus camurus* Cassin, 1857 RED-BILLED DWARF HORNBILL   Sierra Leone to NE Zaire, W Uganda

TROPICRANUS W.L. Sclater, 1922 M
○ *Tropicranus albocristatus* LONG-TAILED HORNBILL[1]
   — *T.a.albocristatus* (Cassin, 1848)   Sierra Leone to Ivory Coast
   — *T.a.macrourus* (Bonaparte, 1850)   Ghana and Togo
   — *T.a.cassini* (Finsch, 1903)   W Nigeria to Gabon, Zaire and W Uganda

OCYCEROS Hume, 1873[2] M
○ *Ocyceros griseus* (Latham, 1790) MALABAR GREY HORNBILL   SW India

○ *Ocyceros gingalensis* (Shaw, 1811) SRI LANKAN GREY HORNBILL[3]   Sri Lanka

○ *Ocyceros birostris* (Scopoli, 1786) INDIAN GREY HORNBILL   NE Pakistan, N and C India, NW Bangladesh

ANTHRACOCEROS Reichenbach, 1849[4] M
○ *Anthracoceros coronatus* (Boddaert, 1783) MALABAR PIED HORNBILL[5]   C and S India, Sri Lanka

○ *Anthracoceros albirostris* ORIENTAL PIED HORNBILL
   — *A.a.albirostris* (Shaw & Nodder, 1808)   Himalayan foothills and W Bengal to S China, Indochina and N Malay Pen.
   — *A.a.convexus* (Temminck, 1832[6])   S Malay Pen. to Gtr. Sundas, W Sumatran Is., Bali and Natuna Is.

○ *Anthracoceros marchei* Oustalet, 1885 PALAWAN HORNBILL   Calamianes, Palawan, Balabac Is.

○ *Anthracoceros malayanus* (Raffles, 1822) BLACK HORNBILL[7]   Malay Pen., Sumatra, Borneo, Bangka, Belitung

○ *Anthracoceros montani* (Oustalet, 1880) SULU HORNBILL   Jolo, Tawitawi Is.

BUCEROS Linnaeus, 1758 M
○ *Buceros bicornis* Linnaeus, 1758 GREAT HORNBILL   SW India[8]; Himalayas east to Indochina and south to Malay Pen., Sumatra

○ *Buceros rhinoceros* RHINOCEROS HORNBILL
   — *B.r.rhinoceros* Linnaeus, 1758[9]   S Malay Pen., Sumatra
   — *B.r.silvestris* Vieillot, 1816   Java
   — *B.r.borneoensis* Schlegel & Müller, 1840   Borneo

○ *Buceros hydrocorax* RUFOUS HORNBILL
   — *B.h.hydrocorax* Linnaeus, 1766   Luzon, Marinduque
   — *B.h.semigaleatus* Tweeddale, 1878   Panay, Bohol, Leyte, Samar
   — *B.h.mindanensis* Tweeddale, 1877[10]   Basilan, Mindanao

RHINOPLAX Gloger, 1841 F
○ *Rhinoplax vigil* (J.R. Forster, 1781) HELMETED HORNBILL   C and S Malay Pen., Sumatra, Borneo

PENELOPIDES Reichenbach, 1849 M[11]
○ *Penelopides exarhatus* SULAWESI HORNBILL
   — *P.e.exarhatus* (Temminck, 1823)   N Sulawesi
   — *P.e.sanfordi* (Stresemann, 1932)   C, SE, E and S Sulawesi, Muna, Buton

○ *Penelopides panini* TARICTIC HORNBILL[12]
   — *P.p.manillae* (Boddaert, 1783)   Luzon, Marinduque

[1] This genus, used by White (1965: 249), is still preferred by East African ornithologists, although the species was placed in *Tockus* by Kemp (1988: 381) (1249).
[2] Treated within the genus *Tockus* in last Edition, but see Kemp (1995) (1251).
[3] For the separation of *gingalensis* from *griseus* see Kemp & Crowe (1985) (1253).
[4] Based on the revision of Frith & Frith (1983) (884); related nomenclatural issues require further verification (mainly because some types could not be examined).
[5] Synonyms presumably include both *malabaricus* and *leucogaster* but types probably not extant. Many records attached to these names belonged to *albirostris* not *coronatus* see Frith & Frith (1983) (884).
[6] Pl. 530 appears in Livr. 89 (not 88 as given by Peters, 1945: 267) (1826) and this dates from 1832 not 1831 (see Dickinson, 2001) (739).
[7] Includes *deminutus* Sanft, 1960 (2165), recognised by Davison in Smythies (2000) (2289) but not by Frith & Frith (1983) (884) or Wells (1999) (2579).
[8] This disjunct population was called *cavatus* in last Edition, a name that "cannot be used" (Ripley (1961: 231 fn) (2042).
[9] Includes *sumatranus* see Chasen (1935: 105) (319).
[10] Includes *basilanicus* see Dickinson et al. (1991) (745).
[11] Gender addressed by David & Gosselin (2002) (614), multiple changes occur in specific names.
[12] Although Kemp (1988, 1995) (1248, 1251) considered this several species Kennedy et al. (2000) (1259) have preferred to retain just one awaiting more extensive acoustic evidence, although there are recognisable groups of forms.

__ *P.p.subniger* McGregor, 1910	Polillo Is.
__ *P.p.mindorensis* Steere, 1890	Mindoro
__ *P.p.samarensis* Steere, 1890	Samar, Leyte, Bohol
__ *P.p.affinis* Tweeddale, 1877	Dinagat, Mindanao
__ *P.p.basilanicus* Steere, 1890	Basilan
__ *P.p.ticaensis* Hachisuka, 1930	Ticao I.
__ *P.p.panini* (Boddaert, 1783)	Masbate, Panay, Negros

BERENICORNIS Bonaparte, 1850[1] M
○ ***Berenicornis comatus*** (Raffles, 1822) White-crowned Hornbill — Malay Pen., Sumatra, Borneo

ACEROS J.E. Gray, 1844 M
○ ***Aceros nipalensis*** (Hodgson, 1829) Rufous-necked Hornbill[2] — Nepal E to S Yunnan, N Vietnam and S to SW Thailand

○ ***Aceros cassidix*** (Temminck, 1823) Knobbed Hornbill[3] — Sulawesi and nearby islands

○ ***Aceros corrugatus*** Wrinkled Hornbill
| __ *A.c.rugosus* Begbie, 1834[4] [5] | Malay Pen., Sumatra, Batu Is. (111) |
| __ *A.c.corrugatus* (Temminck, 1832) | Borneo |

○ ***Aceros waldeni*** (Sharpe, 1877) Walden's Hornbill[6] — Panay, Guimaras, Negros

○ ***Aceros leucocephalus*** (Vieillot, 1816) Writhed Hornbill — Mindanao, Dinagat, Camiguin (South) I.

RHYTICEROS Reichenbach, 1849 M
○ ***Rhyticeros plicatus*** Papuan Hornbill[7]
__ *R.p.plicatus* (J.R. Forster, 1781)	Seram, Ambon (S Moluccas)
__ *R.p.ruficollis* (Vieillot, 1816)	N Moluccas, W Papuan Is., W New Guinea
__ *R.p.jungei* Mayr, 1937	N, E and S New Guinea, Yapen I., Fergusson and Goodenough Is.
__ *R.p.dampieri* Mayr, 1934	Bismarck Arch.
__ *R.p.harterti* Mayr, 1934	Buka, Bougainville, Shortland (N Solomons) Is.
__ *R.p.mendanae* E. Hartert, 1924	Choiseul, Ysabel, Guadalcanal, Malaita Is.

○ ***Rhyticeros narcondami*** Hume, 1873 Narcondam Hornbill — Narcondam I. (Andaman Is.)

○ ***Rhyticeros subruficollis*** (Blyth, 1843) Plain-pouched Hornbill[8] — S Burma, SW Thailand, Malay Pen., (Sumatra?)

○ ***Rhyticeros undulatus*** Wreathed Hornbill
| __ *R.u.undulatus* (Shaw, 1811) | Bhutan, W Bengal and S Assam east to Indochina and south to Malay Pen., Sumatra, Java, Bali |
| __ *R.u.aequabilis* (Sanft, 1960)[9] | Borneo (2165) |

○ ***Rhyticeros everetti*** (Rothschild, 1897) Sumba Hornbill — Sumba I.

BYCANISTES Cabanis & Heine, 1860 M
○ ***Bycanistes fistulator*** Piping Hornbill
__ *B.f.fistulator* (Cassin, 1852)	Senegal to W Nigeria
__ *B.f.sharpii* (Elliot, 1873)	SE Nigeria to Gabon and NW Angola
__ *B.f.duboisi* W.L. Sclater, 1922	C and N Zaire to Central African Republic and W Uganda

○ ***Bycanistes bucinator*** (Temminck, 1824) Trumpeter Hornbill — SE Kenya, C Tanzania and S Zaire to S Mozambique and SE Cape Province

○ ***Bycanistes cylindricus*** Brown-cheeked Hornbill
| __ *B.c.cylindricus* (Temminck, 1831) | Sierra Leone to Ghana |
| __ *B.c.albotibialis* (Cabanis & Reichenow, 1877) | Benin and S Nigeria to W Uganda and NW Angola |

---

[1] Like Wells (1999) (2579) we see retain a monotypic genus for this distinctive species. Molecular evidence now supports this (Hübner et al., 2001) (1160).
[2] Includes ¶*yunnanensis* Guan, 1989 (1024).
[3] Includes *brevirostris* van Bemmel, 1951 (2462), see White & Bruce (1986: 288) (2652).
[4] Van Marle & Voous (1988: 134) (2468) and Wells (1999: 498) (2579) agree that Bornean birds differ from all others. For the resurrection of the name *rugosus* see Sanft (1960: 106) (2165).
[5] Includes *megistus* see Van Marle & Voous (1988: 134) (2468).
[6] Separated from *leucocephalus* by Kemp (1988) (1248) and by Kennedy *et al.* (2000) (1259).
[7] Kemp (2001: 515) (1252) treats this as monotypic; our polytypic treatment is tentative.
[8] For treatment as a species separate from *plicatus* see Kemp (1988) (1248). See also Rasmussen (2000) (1988).
[9] For recognition of this form see Wells (1999) (2579) and Davison in Smythies (2000) (2289).

○ *Bycanistes subcylindricus* GREY-CHEEKED HORNBILL
    — *B.s.subcylindricus* (P.L. Sclater, 1870)   Ghana to SW Nigeria
    — *B.s.subquadratus* Cabanis, 1880   SE Nigeria to N Zaire, W Kenya and Burundi, Angola

○ *Bycanistes brevis* Friedmann, 1929 SILVERY-CHEEKED HORNBILL   Ethiopia, Kenya to Malawi and Mozambique

CERATOGYMNA Bonaparte, 1854 F
○ *Ceratogymna atrata* (Temminck, 1835) BLACK-CASQUED HORNBILL   Liberia to Zaire, SW Sudan and NW Angola

○ *Ceratogymna elata* (Temminck, 1831) YELLOW-CASQUED HORNBILL   Guinea to W Cameroon

BUCORVIDAE GROUND HORNBILLS[1] (1: 2)

BUCORVUS Lesson, 1830 M
○ *Bucorvus abyssinicus* (Boddaert, 1783) NORTHERN GROUND HORNBILL   Senegal and Gambia to NW Kenya, Ethiopia

○ *Bucorvus leadbeateri* (Vigors, 1825)[2] SOUTHERN GROUND HORNBILL   S Kenya to Angola, W Zaire, N Namibia and E Cape
                                                        Province

RAMPHASTIDAE TOUCANS AND BARBETS[3] (20: 120)

RAMPHASTINAE

AULACORHYNCHUS Gould, 1835 M
● *Aulacorhynchus prasinus* EMERALD TOUCANET
    — *A.p.wagleri* (Sturm, 1841)   SW Mexico
    — *A.p.warneri* Winker, 2000   EC Mexico (2675)
    — *A.p.prasinus* (Gould, 1834)   SE Mexico
    — *A.p.virescens* Ridgway, 1912[4]   N Guatemala, Belize to N Nicaragua
    — *A.p.volcanius* Dickey & van Rossem, 1930   E El Salvador
    ↙ *A.p.caeruleogularis* (Gould, 1854)[5]   Costa Rica to W Panama
    — *A.p.cognatus* (Nelson, 1912)   E Panama
    — *A.p.griseigularis* Chapman, 1915   NW Colombia
    — *A.p.phaeolaemus* (Gould, 1874)   W Colombia
    — *A.p.lautus* (Bangs, 1898)   N Colombia
    — *A.p.albivitta* (Boissonneau, 1840)   E Colombia, E Ecuador, W Venezuela
    — *A.p.cyanolaemus* (Gould, 1866)   SE Ecuador, N Peru
    — *A.p.dimidiatus* (Ridgway, 1886)   N Peru
    — *A.p.atrogularis* (Sturm, 1841)   E Peru, N Bolivia

○ *Aulacorhynchus sulcatus* GROOVE-BILLED TOUCANET
    — *A.s.calorhynchus* (Gould, 1874)   N Colombia, NW Venezuela
    — *A.s.sulcatus* (Swainson, 1820)   N Venezuela
    — *A.s.erythrognathus* (Gould, 1874)   NE Venezuela

○ *Aulacorhynchus derbianus* CHESTNUT-TIPPED TOUCANET
    — *A.d.nigrirostris* Traylor, 1951   SE Colombia, E Ecuador, NE Peru (2422)
    — *A.d.derbianus* Gould, 1835   SE Peru, NE Bolivia
    — *A.d.duidae* Chapman, 1929   S Venezuela (Mt. Duida)
    — *A.d.whitelianus* (Salvin & Godman, 1882)   S Venezuela (Auyan-tepui and Mt. Roraima and on
                                            Guyana border)
    — *A.d.osgoodi* Blake, 1941   S Guyana (Acary Mts.)

○ *Aulacorhynchus haematopygus* CRIMSON-RUMPED TOUCANET
    — *A.h.haematopygus* (Gould, 1835)   Andes of Colombia and NW Venezuela
    — *A.h.sexnotatus* (Gould, 1868)   W Ecuador

---

[1] We treat the Ground Hornbills as a full family following Wolters (1976: 119).
[2] Kemp & Crowe (1985) (1253) and others used the name *cafer*, but see Browning (1992) (253).
[3] It is quite possible that further information will change our views that the barbets and toucans should be united at family level and that a suprafamilial linkage will be preferable.
[4] Includes *stenorhabdus* see Haffer (1974: 197) (1053).
[5] Includes *maxillaris* see Short & Horne (2001) (2255A).

○ *Aulacorhynchus huallagae* Carriker, 1933 YELLOW-BROWED TOUCANET   Huallaga Valley (E Peru)

○ *Aulacorhynchus coeruleicinctis* d'Orbigny, 1840 BLUE-BANDED TOUCANET
         E Andes of S Peru, N Bolivia

PTEROGLOSSUS Illiger, 1811 M

○ *Pteroglossus viridis* (Linnaeus, 1766) GREEN ARACARI   E Venezuela, the Guianas, N Brazil

○ *Pteroglossus inscriptus* LETTERED ARACARI[1]
   — *P.i.inscriptus* Swainson, 1822[2]   C and S Brazil
   — *P.i.humboldti* Wagler, 1827   SE Colombia to N Bolivia, W Brazil

○ *Pteroglossus bitorquatus* RED-NECKED ARACARI
   — *P.b.sturmii* Natterer, 1842   E Bolivia, WC Brazil
   — *P.b.reichenowi* E. Snethlage, 1907   NC Brazil
   — *P.b.bitorquatus* Vigors, 1826   NE Brazil

○ *Pteroglossus azara* IVORY-BILLED ARACARI[3]
   — *P.a.flavirostris* Fraser, 1841   N Upper Amazonia
   — *P.a.azara* (Vieillot, 1819)   NW Brazil
   — *P.a.mariae* Gould, 1854   NE Peru, W Brazil, N Bolivia

○ *Pteroglossus aracari* BLACK-NECKED ARACARI
   — *P.a.atricollis* (Statius Müller, 1776)[4]   Venezuela, The Guianas. , N Brazil
   — *P.a.aracari* (Linnaeus, 1758)   C and E Brazil
   — *P.a.vergens* Griscom & Greenway, 1937[5]   E Brazil

○ *Pteroglossus castanotis* CHESTNUT-EARED ARACARI
   — *P.c.castanotis* Gould, 1834   E Colombia, E Ecuador, E Peru, NW Brazil
   — *P.c.australis* Cassin, 1867   E Bolivia, W Brazil, NE Argentine

○ *Pteroglossus pluricinctus* Gould, 1836 MANY-BANDED ARACARI   E Peru and E Colombia to E Venezuela

● *Pteroglossus torquatus* COLLARED ARACARI
   ⩔ *P.t.torquatus* (J.F. Gmelin, 1788)   S Mexico to W Panama
   — *P.t.erythrozonus* Ridgway, 1912   SE Mexico, N Guatemala, Belize
   — *P.t.nuchalis* Cabanis, 1862[6]   N Colombia, N Venezuela

○ *Pteroglossus sanguineus* Gould, 1854 STRIPE-BILLED ARACARI[7]   W Colombia, NW Ecuador

○ *Pteroglossus erythropygius* Gould, 1843 PALE-MANDIBLED ARACARI[8]   W Ecuador

○ *Pteroglossus frantzii* Cabanis, 1861 FIERY-BILLED ARACARI   Costa Rica, W Panama

○ *Pteroglossus beauharnaesii* Wagler, 1832 CURL-CRESTED ARACARI   E Peru, N Bolivia, W Brazil

BAILLONIUS Cassin, 1867 M
○ *Baillonius bailloni* (Vieillot, 1819) SAFFRON TOUCANET   SE Brazil

SELENIDERA Gould, 1837 F
○ *Selenidera spectabilis* Cassin, 1857 YELLOW-EARED TOUCANET   Honduras to Panama, NC and NW Colombia

○ *Selenidera culik* (Wagler, 1827) GUIANAN TOUCANET   The Guianas, NE Brazil

○ *Selenidera reinwardtii* GOLDEN-COLLARED TOUCANET
   — *S.r.reinwardtii* (Wagler, 1827)   S Colombia, E Ecuador, NE Peru
   — *S.r.langsdorffii* (Wagler, 1827)   E Peru, W Brazil, N Bolivia

○ *Selenidera nattereri* (Gould, 1836) TAWNY-TUFTED TOUCANET   E Venezuela, NW Brazil

○ *Selenidera gouldii* (Natterer, 1837) GOULD'S TOUCANET[9]   C Brazil, E Bolivia

---

[1] For reasons to separate this species see Haffer (1974: 210) (1053).
[2] The name *didymus* attaches to subadult specimens of *inscriptus* see Haffer (1974: 212) (1053).
[3] Erroneously listed as *P. flavirostris* in last Edition, and in Peters (1948: 78) (1827), who also listed *Pteroglossus olallae*, probably a hybrid; for a discussion see Haffer (1974: 219) (1053).
[4] Includes *roraimae* see Haffer (1974: 228) (1053).
[5] Haffer (1974: 229) (1053) thought the name *vergens* probably related to hybrid birds.
[6] Includes *pectoralis* see Short & Horne (2001) (2255A).
[7] For treatment at specific level see Hilty & Brown (1986: 325) (1116) and Hackett & Lehn (1997) (1049).
[8] For treatment at specific level see Meyer de Schauensee (1970) (1575) and and Hackett & Lehn (1997) (1049).
[9] Includes *baturitensis* Pinto & Camargo, 1961 (1899), but see Novaes & Lima (1991) (1664). Also includes *hellmayri* see Short & Horne (2001) (2255A); this had previously been attached to *S. maculirostris*.

○ *Selenidera maculirostris* (M.H.K. Lichtenstein, 1823) Spot-billed Toucanet

E and SE Brazil, NE Argentina

ANDIGENA Gould, 1851 F

○ *Andigena hypoglauca* Grey-breasted Mountain Toucan

— *A.h.hypoglauca* (Gould, 1833) C Colombia
— *A.h.lateralis* Chapman, 1923 E Ecuador, E Peru

○ *Andigena laminirostris* Gould, 1851 Plate-billed Mountain Toucan SW Colombia, W Ecuador

○ *Andigena cucullata* (Gould, 1846) Hooded Mountain Toucan S Peru, N Bolivia

○ *Andigena nigrirostris* Black-billed Mountain Toucan

— *A.n.occidentalis* Chapman, 1915 W Colombia
— *A.n.spilorhynchus* Gould, 1858 S Colombia, NE Ecuador
— *A.n.nigrirostris* (Waterhouse, 1839) E Colombia, E Ecuador, W Venzuela

RAMPHASTOS Linnaeus, 1758 M

○ *Ramphastos dicolorus* Linnaeus, 1766 Red-breasted Toucan SE Brazil, Paraguay, NE Argentina

○ *Ramphastos vitellinus* Channel-billed Toucan

— *R.v.culminatus* Gould, 1833 Upper Amazonia
— *R.v.citreolaemus* Gould, 1844 C Colombia
— *R.v.vitellinus* M.H.K. Lichtenstein, 1823 Trinidad, Venezuela, the Guianas, N Brazil
— *R.v.ariel* Vigors, 1826[1] C and E Brazil

○ *Ramphastos brevis* Meyer de Schauensee, 1945 Choco Toucan W Colombia, W Ecuador

● *Ramphastos sulfuratus* Keel-billed Toucan

— *R.s.sulfuratus* Lesson, 1830 S Mexico, N Guatemala, Belize
— *R.s.brevicarinatus* Gould, 1854 SE Guatemala to N Colombia, NW Venezuela

○ *Ramphastos toco* Toco Toucan

— *R.t.toco* Statius Müller, 1776 The Guianas, N and E Brazil
— *R.t.albogularis* Cabanis, 1862 E and S Brazil, Paraguay, Bolivia, N Argentina

○ *Ramphastos tucanus* White-throated Toucan[2]

— *R.t.tucanus* Linnaeus, 1758 SE Venezuela, the Guianas, N Brazil
— *R.t.cuvieri* Wagler, 1827 Upper Amazonia
— *R.t.inca* Gould, 1846 E Bolivia

● *Ramphastos swainsonii* Gould, 1833 Chestnut-mandibled Toucan[3] SE Honduras to W Ecuador

○ *Ramphastos ambiguus* Black-mandibled Toucan

— *R.a.ambiguus* Swainson, 1823 N Colombia to C Peru
— *R.a.abbreviatus* Cabanis, 1862 W Venezuela, NE Colombia

CAPITONINAE (NEW WORLD BARBETS)

CAPITO Vieillot, 1816 M

○ *Capito aurovirens* (Cuvier, 1829) Scarlet-crowned Barbet SE Colombia, E Ecuador, E Peru, W Brazil

○ *Capito wallacei* O'Neill, Lane, Kratter, Caparella & Joo, 2000 Scarlet-banded Barbet #

NC Peru (1719)

○ *Capito maculicoronatus* Spot-crowned Barbet

— *C.m.maculicoronatus* Lawrence, 1861 W Panama
— *C.m.rubrilateralis* Chapman, 1912[4] E Panama, NW Colombia

○ *Capito squamatus* Salvin, 1876 Orange-fronted Barbet SW Colombia, W Ecuador

○ *Capito hypoleucus* White-mantled Barbet

— *C.h.hypoleucus* Salvin, 1897 N Colombia
— *C.h.carrikeri* Graves, 1986 NC Colombia (981)
— *C.h.extinctus* Graves, 1986 C Colombia (981)

---

[1] Includes *pintoi* see Short & Horne (2001) (2255A); this is intermediate between *culminatus* and *ariel* as is *theresae*.
[2] As regards the name *oblitus* see Haffer (1974: 307) (1053).
[3] Treated as a race of *R. ambiguus* in Short & Horne (2001) (2255A).
[4] Includes *pirrensis* and *melas* see Wetmore (1968) (2605).

○ *Capito dayi* Cherrie, 1916 BLACK-GIRDLED BARBET — E Bolivia, C Brazil

○ *Capito brunneipectus* Chapman, 1921 BROWN-CHESTED BARBET — N Brazil

○ *Capito niger* (P.L.Status Müller, 1776) BLACK-SPOTTED BARBET[1] — The Guianas, NE Brazil, E Venezuela

○ *Capito auratus* GILDED BARBET
    __ *C.a.punctatus* (Lesson, 1830) — E Colombia to E Peru
    __ *C.a.aurantiicinctus* Dalmas, 1900[2] — WC and E Venezuela
    __ *C.a.auratus* (Dumont, 1816) — NE Peru
    __ *C.a.orosae* Chapman, 1928 — EC Peru, NW Brazil
    __ *C.a.amazonicus* Deville & Des Murs, 1849[3] — NW and NC Brazil
    __ *C.a.nitidior* Chapman, 1928[4] — SW Venezuela, NW Brazil, SE Colombia
    __ *C.a.hypochondriacus* Chapman, 1928 — NC Brazil
    __ *C.a.bolivianus* Ridgway, 1912[5] — SE Peru, N and W Bolivia, W Brazil

○ *Capito quinticolor* Elliot, 1865 FIVE-COLOURED BARBET — W Colombia, NW Ecuador

EUBUCCO Bonaparte, 1850 M
○ *Eubucco richardsoni* LEMON-THROATED BARBET
    __ *E.r.richardsoni* (G.R. Gray, 1846) — SE Colombia, E Ecuador, E Peru
    __ *E.r.nigriceps* Chapman, 1928 — NE Peru
    __ *E.r.aurantiicollis* P.L. Sclater, 1858[6] — C and E Peru, N Bolivia
    __ *E.r.purusianus* Gyldenstolpe, 1951[7] — W Amazonian Brazil (1031)

○ *Eubucco bourcierii* RED-HEADED BARBET
    __ *E.b.salvini* (Shelley, 1891) — Costa Rica, W Panama
    __ *E.b.anomalus* Griscom, 1929 — E Panama
    __ *E.b.occidentalis* Chapman, 1914 — W Colombia
    __ *E.b.bourcierii* (Lafresnaye, 1845) — C and E Colombia
    __ *E.b.aequatorialis* (Salvadori & Festa, 1900) — W Ecuador
    __ *E.b.orientalis* Chapman, 1914 — E Ecuador, N Peru

○ *Eubucco tucinkae* (Seilern, 1913)[8] SCARLET-HOODED BARBET — SE Peru

○ *Eubucco versicolor* VERSICOLOURED BARBET
    __ *E.v.steerii* (P.L. Sclater, 1878) — N Peru
    __ *E.v.glaucogularis* (Tschudi, 1844) — C Peru
    __ *E.v.versicolor* (Statius Müller, 1776) — S Peru, N Bolivia

SEMNORNIS Richmond, 1900 M
● *Semnornis frantzii* (P.L. Sclater, 1864) PRONG-BILLED BARBET — Costa Rica, W Panama

○ *Semnornis ramphastinus* TOUCAN-BARBET
    __ *S.r.caucae* Gyldenstolpe, 1941 — Central Andes of Colombia
    __ *S.r.ramphastinus* (Jardine, 1855) — W Andes of Colombia and Ecuador

MEGALAIMINAE (ASIAN BARBETS)

PSILOPOGON S. Müller, 1835 M
○ *Psilopogon pyrolophus* S. Müller, 1835 FIRE-TUFTED BARBET — Mts. of W Malaysia, Sumatra

MEGALAIMA G.R. Gray, 1842 F
○ *Megalaima virens* GREAT BARBET
    __ *M.v.marshallorum* (Swinhoe, 1870) — NW Himalayas to W Nepal
    __ *M.v.magnifica* E.C.S. Baker, 1926 — C and E Himalayas, S Assam, NW Burma, SW Yunnan
    __ *M.v.mayri* Ripley, 1948 — Mishmi hills (NE Assam) (2031)

---

[1] Separation of this from *C. brunneipectus* and *C. auratus* follows Haffer (1997) (1054). Short & Horne (2001) (2255A) have placed *auratus* within the species *niger*.
[2] Includes *intermedius* see Short & Horne (2001) (2255A).
[3] Includes *novaolindae* and *arimae* see Short & Horne (2001) (2255A).
[4] Includes *transilens* see Short & Horne (2001) (2255A).
[5] Includes *insperatus* see Short & Horne (2001) (2255A).
[6] Includes *coccineus* see Short & Horne (2001) (2255A).
[7] For recognition of this see Short & Horne (2001) (2255A).
[8] The usual abbreviated form of von Seilern-Aspang.

— *M.v.clamator* Mayr, 1941[1] NE Burma and extreme W Yunnan
— *M.v.virens* (Boddaert, 1783) C and E Burma, N Thailand, N Laos to E China
— *M.v.indochinensis* Rand, 1953 N Vietnam (1970)

○ *Megalaima lagrandieri* RED-VENTED BARBET
— *M.l.rothschildi* (Delacour, 1927) N Laos, N Vietnam
— *M.l.lagrandieri* J. Verreaux, 1868 S Laos, C and S Vietnam

○ *Megalaima zeylanica* BROWN-HEADED BARBET
— *M.z.inornata* Walden, 1870 W India
— *M.z.caniceps* (Franklin, 1831)[2] NW Himalayas to C India
— *M.z.zeylanica* (J.F. Gmelin, 1788) SW India, Sri Lanka

○ *Megalaima lineata* LINEATED BARBET
— *M.l.hodgsoni* (Bonaparte, 1850)[3] WC Himalayas and Assam east to Indochina and south to Malay Pen.
— *M.l.lineata* (Vieillot, 1816) Java, Bali

○ *Megalaima viridis* (Boddaert, 1783) WHITE-CHEEKED BARBET SW and S India

○ *Megalaima faiostricta* GREEN-EARED BARBET
— *M.f.praetermissa* (Kloss, 1918) SC Guangdong (S China), N Vietnam
— *M.f.faiostricta* (Temminck, 1832)[4] Thailand (not Pen.), Laos, Cambodia, C and S Vietnam

○ *Megalaima corvina* (Temminck, 1831) BROWN-THROATED BARBET Java

○ *Megalaima chrysopogon* GOLD-WHISKERED BARBET
— *M.c.laeta* (Robinson & Kloss, 1918) Malay Pen.
— *M.c.chrysopogon* (Temminck, 1824) Sumatra
— *M.c.chrysopsis* Goffin, 1863 Borneo

○ *Megalaima rafflesii* (Lesson, 1839) RED-CROWNED BARBET[5] Malay Pen., Sumatra, Bangka and Belitung Is., Borneo

○ *Megalaima mystacophanos* RED-THROATED BARBET
— *M.m.mystacophanos* (Temminck, 1824)[6] Malay Pen., Sumatra, Borneo
— *M.m.ampala* (Oberholser, 1912) Batu Is.

○ *Megalaima javensis* (Horsfield, 1821) BLACK-BANDED BARBET Java

○ *Megalaima flavifrons* (Cuvier, 1816)[7] YELLOW-FRONTED BARBET Sri Lanka

○ *Megalaima franklinii* GOLDEN-THROATED BARBET
— *M.f.franklinii* (Blyth, 1842) E Himalayas and Assam hills to S China and N Vietnam
— *M.f.ramsayi* (Walden, 1875)[8] Hills of E and SE Burma, N and W Thailand
— *M.f.auricularis* (Robinson & Kloss, 1919) Mts. of S Laos, SC Vietnam
— *M.f.trangensis* (Riley, 1934) Mts. of S Thailand
— *M.f.minor* (Kloss & Chasen 1926) Mts. of W Malaysia

○ *Megalaima oorti* BLACK-BROWED BARBET
— *M.o.sini* (Stresemann, 1929) Dayao Shan (Guangxi, China)
— *M.o.nuchalis* (Gould, 1863) Taiwan
— *M.o.faber* (Swinhoe, 1870) Mts. of Hainan I.
— *M.o.annamensis* (Robinson & Kloss, 1919) Mts. S Laos and SC Vietnam
— *M.o.oorti* (S. Müller, 1835) Mts. of W Malaysia, Sumatra

○ *Megalaima asiatica* BLUE-THROATED BARBET
— *M.a.asiatica* (Latham, 1790)[9] Himalyan foothills and W Bengal to W, N, and NE Burma and SW Yunnan
— *M.a.davisoni* Hume, 1877 SE Burma to SE Yunnan and N Indochina
— *M.a.chersonesus* (Kloss & Chasen, 1927) Mts. of S Thailand

[1] Short & Horne (2001) (2255A) subsume *mayri* and *indochinensis*.
[2] Includes *kangrae* see Ripley (1982: 221) (2055).
[3] Includes *rana* Ripley, 1950 (2034), see Ali & Ripley (1983) (28), and *kutru* Mukherjee, 1952 (1630), see Ripley (1982: 222) (2055).
[4] *Contra* Peters (1948: 33) pl. 527 was in Livr. 89 and appeared in 1832 not 1831, see Dickinson (2001) (739).
[5] Includes *billitonis* see Voous (1961) (2545), as well as *borneensis* and *malayensis* see Mees (1986) (1539).
[6] Includes *humii* see Smythies (1960) (2287).
[7] Dated 1817 by Peters (1948: 35) (1827) but see Browning & Monroe (1991) (258) and www.zoonomen.net (A.P. Peterson *in litt.*).
[8] Short & Horne (2001) (2255A) subsume *trangensis* and *minor*.
[9] Includes erythristic *rubescens* see Ripley (1982: 223) (2055).

○ *Megalaima monticola* (Sharpe, 1889) MOUNTAIN BARBET[1] Mts. of N Borneo

○ *Megalaima incognita* MOUSTACHED BARBET
  __ *M.i.incognita* Hume, 1874 SE Burma
  __ *M.i.elbeli* Deignan, 1956 NE and E Thailand (657)
  __ *M.i.euroa* (Deignan, 1939) SE Thailand, Indochina

○ *Megalaima henricii* YELLOW-CROWNED BARBET
  __ *M.h.henricii* (Temminck, 1831) C and S Malay Pen., Sumatra
  __ *M.h.brachyrhyncha* (Neumann, 1908) Lowland Borneo

○ *Megalaima armillaris* FLAME-FRONTED BARBET[2]
  __ *M.a.armillaris* (Temminck, 1821) Java
  __ *M.a.baliensis* (Rensch, 1928) Bali

○ *Megalaima pulcherrima* (Sharpe, 1888) GOLDEN-NAPED BARBET Mts. of NW Borneo

○ *Megalaima australis* BLUE-EARED BARBET
  __ *M.a.cyanotis* (Blyth, 1847) E Himalayas and N Bengal to Yunnan, N Thailand
  __ *M.a.orientalis* (Robinson, 1915) E and SE Thailand, Indochina
  __ *M.a.stuarti* (Robinson & Kloss, 1919) S Burma, SW and Pen. Thailand
  __ *M.a.duvaucelii* (Lesson, 1830) W Malaysia, Sumatra, Borneo, Bangka
  __ *M.a.gigantorhina* (Oberholser, 1912)[3] Nias I.
  __ *M.a.tanamassae* (Meyer de Schauensee, 1929) Batu Is.
  __ *M.a.australis* (Horsfield, 1821)[4] Java, Bali

○ *Megalaima eximia* BORNEAN BARBET
  __ *M.e.cyanea* (Harrisson & Hartley, 1934) Mts. of Sabah (NE Borneo)
  __ *M.e.eximia* (Sharpe, 1892) Mts. of NC Borneo

○ *Megalaima rubricapillus* CRIMSON-FRONTED BARBET[5]
  __ *M.r.malabarica* (Blyth, 1847) SW India
  __ *M.r.rubricapillus* (J.F. Gmelin, 1788) Sri Lanka

○ *Megalaima haemacephala* COPPERSMITH BARBET[6]
  __ *M.h.indica* (Latham, 1790) Pakistan, India and Sri Lanka east to Yunnan, and mainland SE Asia

  __ *M.h.delica* (Parrot, 1907) Sumatra
  __ *M.h.rosea* (Dumont, 1816) Java, Bali
  __ *M.h.haemacephala* (Statius Müller, 1776) Luzon and Mindoro
  __ *M.h.celestinoi* Gilliard, 1949 Samar and Leyte (938)
  __ *M.h.mindanensis* Rand, 1948 Mindanao (1967)
  __ *M.h.intermedia* (Shelley, 1891) Guimaras, Negros and Panay
  __ *M.h.cebuensis* Dziadosz & Parkes, 1984 Cebu (780)
  __ *M.h.homochroa* Dziadosz & Parkes, 1984 Tablas, Romblon and Masbate (780)

CALORHAMPHUS Lesson, 1839 M
○ *Calorhamphus fuliginosus* BROWN BARBET
  __ *C.f.hayii* (J.E. Gray, 1831)[7] Malay Pen., Sumatra
  __ *C.f.fuliginosus* (Temminck, 1830) Borneo (except North)
  __ *C.f.tertius* Chasen & Kloss, 1929 N Borneo

LYBIINAE (AFRICAN BARBETS)[8]

GYMNOBUCCO Bonaparte, 1850 M
○ *Gymnobucco bonapartei* GREY-THROATED BARBET
  __ *G.b.bonapartei* Hartlaub, 1854 SW Cameroon and Gabon to C Zaire
  __ *G.b.cinereiceps* Sharpe, 1891 S Sudan, NE and E Zaire to Uganda and W Kenya

[1] Treated as a form of *asiatica* by Peters (1948: 37) (1827), but Smythies (1957) (2286) treated it as a distinct species. Needs re-evaluation as no detailed reasons given.
[2] For revision to ranges see Mees (1996: 25) (1543).
[3] For correction of spelling see David & Gosselin (2002) (613).
[4] Includes *hebereri* see Mees (1996: 27) (1543).
[5] For correction of spelling see David & Gosselin (2002) (613).
[6] Philippine subspecies recognised follow Dickinson *et al.* (1991) (745).
[7] Includes *detersus* Deignan, 1960 (660), see Wells (1999: 599) (2579). But see also Short & Horne (2001) (2255A).
[8] Here we follow Short & Horne (1988) (2255).

○ *Gymnobucco sladeni* Ogilvie-Grant, 1907 Sᴌᴀᴅᴇɴ's Bᴀʀʙᴇᴛ

N, C and E Zaire

○ *Gymnobucco peli* Hartlaub, 1857 Bʀɪsᴛᴌᴇ-ɴᴏsᴇᴅ Bᴀʀʙᴇᴛ

Ivory Coast to S Nigeria and W Zaire

○ *Gymnobucco calvus* Nᴀᴋᴇᴅ-ғᴀᴄᴇᴅ Bᴀʀʙᴇᴛ
   — *G.c.calvus* (Lafresnaye, 1841)
   — *G.c.congicus* Chapin, 1932
   — *G.c.vernayi* Boulton, 1931

Sierra Leone to S Nigeria, Gabon
W Zaire, NW Angola
WC Angola

STACTOLAEMA C.H.T. & G.F.L. Marshall, 1870 F
○ *Stactolaema leucotis* Wʜɪᴛᴇ-ᴇᴀʀᴇᴅ Bᴀʀʙᴇᴛ
   — *S.l.kilimensis* (Shelley, 1889)
   — *S.l.leucogrammica* (Reichenow, 1915)
   — *S.l.leucotis* (Sundevall, 1850)[1]

C and SE Kenya, NE Tanzania
EC Tanzania
Mozambique to S Malawi, E Zimbabwe and Natal

○ *Stactolaema whytii* Wʜʏᴛᴇ's Bᴀʀʙᴇᴛ
   — *S.w.buttoni* (C.M.N. White, 1945)
   — *S.w.stresemanni* (Grote, 1934)
   — *S.w.terminata* (Clancey, 1956)[2]
   — *S.w.angoniensis* (Benson, 1964)
   — *S.w.whytii* (Shelley, 1893)[3]
   — *S.w.sowerbyi* Sharpe, 1898[4]

NW Zambia
SW Tanzania, NE Zambia
SC Tanzania (Iringa) (372)
W Malawi, E Zambia (123)
S Tanzania, S Malawi
E and NE Zimbabwe, NW Mozambique (Tete)

○ *Stactolaema anchietae* Aɴᴄʜɪᴇᴛᴀ's Bᴀʀʙᴇᴛ
   — *S.a.katangae* (Vincent, 1934)
   — *S.a.anchietae* (Bocage, 1869)
   — *S.a.rex* (Neumann, 1908)

NE Angola, SE Zaire, N Zambia
SC Angola to W Zambia
WC Angola

○ *Stactolaema olivacea* Gʀᴇᴇɴ Bᴀʀʙᴇᴛ
   — *S.o.olivacea* (Shelley, 1880)
   — *S.o.howelli* Jensen & Stuart, 1982
   — *S.o.woodwardi* Shelley, 1895[5]
   — *S.o.rungweensis* Benson, 1948[6]
   — *S.o.belcheri* (W.L. Sclater, 1927)

SE Kenya, NE Tanzania
EC Tanzania (E Uzungwes, Mahenge) (1207)
SE Tanzania (Rondo), N Natal
SW Tanzania, N Malawi (120)
S Malawi, NW Mozambique

POGONIULUS Lafresnaye, 1844 M
○ *Pogoniulus scolopaceus* Sᴘᴇᴄᴋᴌᴇᴅ Tɪɴᴋᴇʀʙɪʀᴅ
   — *P.s.scolopaceus* (Bonaparte, 1850)
   — *P.s.stellatus* (Jardine & Fraser, 1852)
   — *P.s.flavisquamatus* (J. & E. Verreaux, 1855)

Sierra Leone and Liberia to SE Nigeria
Bioko
Cameroon to W Kenya to N Angola

○ *Pogoniulus simplex* (Fischer & Reichenow, 1884) Eᴀsᴛᴇʀɴ Gʀᴇᴇɴ Tɪɴᴋᴇʀʙɪʀᴅ[7]

Coastal Kenya, NE Tanzania, E Mozambique to SE Malawi

○ *Pogoniulus leucomystax* (Sharpe, 1892) Mᴏᴜsᴛᴀᴄʜᴇᴅ Gʀᴇᴇɴ Tɪɴᴋᴇʀʙɪʀᴅ[8]

W Kenya to W Malawi

○ *Pogoniulus coryphaeus* Wᴇsᴛᴇʀɴ Gʀᴇᴇɴ Tɪɴᴋᴇʀʙɪʀᴅ
   — *P.c.coryphaeus* (Reichenow, 1892)
   — *P.c.hildamariae* (W.L. Sclater, 1938)
   — *P.c.angolensis* (Boulton, 1931)

SE Nigeria, SW Cameroon
E Zaire, Rwanda, SW Uganda
W Angola

○ *Pogoniulus atroflavus* (Sparrman, 1798) Rᴇᴅ-ʀᴜᴍᴘᴇᴅ Tɪɴᴋᴇʀʙɪʀᴅ[9]

S Senegal to S Nigeria, Gabon, W, C and NE Zaire and W Uganda

○ *Pogoniulus subsulphureus* Yᴇᴌᴌᴏᴡ-ᴛʜʀᴏᴀᴛᴇᴅ Tɪɴᴋᴇʀʙɪʀᴅ
   — *P.s.chrysopygus* (Shelley, 1889)
   — *P.s.flavimentum* (J. & E. Verreaux, 1851)
   — *P.s.subsulphureus* (Fraser, 1843)

Guinea to Ghana
S Nigeria to Gabon, W, C and NE Zaire and Uganda
Bioko

---

[1] Includes *rufopectoralis* Irwin, 1957 (1185) see White (1965: 262) (2649).
[2] For correction of spelling see David & Gosselin (2002) (613).
[3] Includes *euroum* Clancey, 1956 (372) see White (1965: 265) (2649).
[4] Includes *irwini* Benson, 1956 (121) see White (1965: 265) (2649).
[5] Thought to include *hylophona* Clancey, 1979 (472); re-evaluation may be indicated.
[6] Thought to include *ulugurensis* Ripley & Heinrich, 1969 (2069); re-evaluation may be indicated.
[7] Includes *hylodytes* Clancey, 1984 (494). May need review.
[8] Includes *meridionalis* Ripley & Heinrich, 1969 (2069); may need re-evaluation.
[9] Includes *vividus* Clancey, 1984 (493). May need review.

○ *Pogoniulus bilineatus* YELLOW-RUMPED TINKERBIRD[1]
— *P.b.leucolaimus* (J. & E. Verreaux, 1851)[2]   Gambia to S Nigeria, S Sudan and C Uganda south to
                                                       N Angola and SE Zaire

— *P.b.poensis* (Alexander, 1908)   Bioko
— *P.b.mfumbiri* (Ogilvie-Grant, 1907)   SW Uganda, E Zaire, and SW Tanzania,to N Zambia
— *P.b.jacksoni* (Sharpe, 1897)   C and W Kenya, Rwanda, Burundi, NW Tanzania
— *P.b.fischeri* (Reichenow, 1880)[3]   Coastal Kenya, NE Tanzania, Zanzibar I.
— *P.b.bilineatus* (Sundevall, 1850)[4]   SE Tanzania to Malawi, Mozambique, NE Sth. Africa

○ *Pogoniulus pusillus* RED-FRONTED TINKERBIRD
— *P.p.uropygialis* (Heuglin, 1862)   C Ethiopia,to Eritrea and N Somalia
— *P.p.affinis* (Reichenow, 1879)   S Ethiopia and S Somalia to N Tanzania
— *P.p.pusillus* (Dumont, 1816)[5]   Mozambique to E Cape Province

○ *Pogoniulus chrysoconus* YELLOW-FRONTED TINKERBIRD
— *P.c.chrysoconus* (Temminck, 1832)   Senegal to Sudan, Uganda and W Kenya
— *P.c.xanthostictus* (Blundell & Lovat, 1899)   SW and C Ethiopia
— *P.c.extoni* (E.L. Layard, 1871)[6]   S Angola and N Namibia to S Tanzania, Mozambique
                                            and Transvaal

BUCCANODON Hartlaub, 1857 M
○ *Buccanodon duchaillui* (Cassin, 1856) YELLOW-SPOTTED BARBET[7]   Sierra Leone and Gabon to Uganda and W Kenya

TRICHOLAEMA J. & E. Verreaux, 1855 F[8]
○ *Tricholaema hirsuta* HAIRY-BREASTED BARBET
— *T.h.hirsuta* (Swainson, 1821)[9]   Sierra Leone to Togo
— *T.h.flavipunctata* J. & E. Verreaux, 1855   S Nigeria, Cameroon, N and C Gabon
— *T.h.angolensis* Neumann, 1908   S Gabon to W and S Zaire, N Angola
— *T.h.ansorgii* Shelley, 1895   E Cameroon to E Zaire, Uganda, W Kenya

○ *Tricholaema diademata* RED-FRONTED BARBET
— *T.d.diademata* (Heuglin, 1861)   S Sudan, Ethiopia, NW Somalia, to N Uganda, W and
                                        C Kenya
— *T.d.massaica* (Reichenow, 1887)   Kenya to C and SW Tanzania

○ *Tricholaema frontata* (Cabanis, 1880) MIOMBO PIED BARBET   Angola to Malawi

○ *Tricholaema leucomelas*[10] ACACIA PIED BARBET
— *T.l.centralis* (Roberts, 1932)   W and S Angola, Botswana and W Zimbabwe to N
                                        Cape, W Transvaal
— *T.l.affinis* (Shelley, 1880)   E Zimbabwe and E Transvaal to Natal
— *T.l.leucomelas* (Boddaert, 1783)   C and SW Cape Province

○ *Tricholaema lacrymosa* SPOT-FLANKED BARBET
— *T.l.lacrymosa* Cabanis, 1878   N and C Uganda to C and SE Kenya, NE Tanzania
— *T.l.radcliffei* Ogilvie-Grant, 1904   E Zaire, and S Uganda to Rwanda, W Tanzania, SW Kenya

○ *Tricholaema melanocephala* BLACK-THROATED BARBET
— *T.m.melanocephala* (Cretzschmar, 1826)   Eritrea to C Ethiopia, NW Somalia
— *T.m.blandi* Lort Phillips, 1897   N Somalia
— *T.m.stigmatothorax* Cabanis, 1878   S Ethiopia, S Somalia to NE Tanzania
— *T.m.flavibuccalis* Reichenow, 1893   N Tanzania

LYBIUS Hermann, 1783 M
○ *Lybius undatus* BANDED BARBET
— *L.u.thiogaster* Neumann, 1903   Eritrea and NE Ethiopia

[1] Includes *makawai* Benson & Irwin, 1965 (130) see Short & Horne (2002:184) (2255B).
[2] For correction of spelling see David & Gosselin (2002) (613).
[3] Includes *pallidus* Ripley & Bond, 1971 (2064) see Short & Horne (1988: 436) (2255).
[4] Includes *riparium* Clancey, 1952 (368), see White (1965: 269) (2649) and *oreonesus* Clancey, 1971 (429) and *deceptor* Clancey, 1972 (437), see Short &
    Horne (1988: 437) (2255).
[5] Includes *niethammeri* Clancey, 1952 (368) see White (1965: 267) (2649).
[6] Includes *dryas* Clancey & Lawson, 1961 (389), see White (1965: 269) (2649).
[7] Includes *bannermani* Serle, 1949 (2222) see White (1965: 264) (2649).
[8] Gender addressed by David & Gosselin (2002) (614), but was largely correctly applied in last Edition.
[9] Misspelled in last Edition.
[10] For the use of this spelling see Dowsett (1989) (759) and Art. 31.2.3. of the Code (I.C.Z.N., 1999) (1178).

— *L.u.undatus* (Rüppell, 1837) NW to C Ethiopia
— *L.u.leucogenys* (Blundell & Lovat, 1899) W and SW Ethiopia
— *L.u.salvadori* Neumann, 1903 SE Ethiopia

○ *Lybius vieilloti* VIEILLOT'S BARBET
— *L.v.buchanani* E. Hartert, 1924 S Sahara east to Chad
— *L.v.rubescens* (Temminck, 1823) Senegal to N Cameroon and N Zaire
— *L.v.vieilloti* (Leach, 1815) C and S Sudan to NE Zaire and N Eritrea

○ *Lybius leucocephalus* WHITE-HEADED BARBET
— *L.l.adamauae* Reichenow, 1921 N Nigeria to S Chad, W Central African Republic
— *L.l.leucocephalus* (Defilippi, 1853) S Sudan, Uganda, NE Zaire
— *L.l.senex* (Reichenow, 1887) C and SE Kenya
— *L.l.albicauda* (Shelley, 1881)[1] SW Kenya, N Tanzania
— *L.l.lynesi* Grant & Mackworth-Praed, 1938 C and SC Tanzania
— *L.l.leucogaster* (Bocage, 1877) SW Angola

○ *Lybius chaplini* Stevenson Clarke, 1920 CHAPLIN'S BARBET S Zambia

○ *Lybius rubrifacies* (Reichenow, 1892) RED-FACED BARBET S Uganda, NW Tanzania

○ *Lybius guifsobalito* Hermann, 1783 BLACK-BILLED BARBET Ethiopia, E and S Sudan, Uganda, NE Zaire, W Kenya

○ *Lybius torquatus* BLACK-COLLARED BARBET
— *L.t.zombae* (Shelley, 1893)[2] S Malawi, C and N Mozambique, S Tanzania
— *L.t.pumilio* Grote, 1927 E Zaire, to Burundi, W Tanzania, E Zambia, N Malawi
— *L.t.irroratus* (Cabanis, 1878) E Kenya, NE Tanzania
— *L.t.congicus* (Reichenow, 1898) SE Zaire, N Angola, N Zambia, SW Tanzania
— *L.t.vivacens* Clancey, 1977 SC and W Mozambique to E Zimbabwe and S Malawi (463)
— *L.t.bocagei* (de Sousa, 1886)[3] [4] SW Angola to N Botswana, SW Zambia, W Zimbabwe
— *L.t.torquatus* (Dumont, 1816) SE Botswana to Transvaal and E Cape Province

○ *Lybius melanopterus* (W.K.H. Peters, 1854) BROWN-BREASTED BARBET S Somalia to Malawi, N Mozambique

○ *Lybius minor* BLACK-BACKED BARBET
— *L.m.minor* (Cuvier, 1816)[5] S Gabon to SW Zaire, NW Angola
— *L.m.macclounii* (Shelley, 1899) S and SE Zaire, NE Angola, N Zambia, W Tanzania, NW Malawi

○ *Lybius bidentatus* DOUBLE-TOOTHED BARBET
— *L.b.bidentatus* (Shaw, 1798) Guinea to Cameroon and N Angola
— *L.b.aequatorialis* (Shelley, 1889) N and E Zaire to Uganda and W Kenya, C Ethiopia

○ *Lybius dubius* (J.F. Gmelin, 1788) BEARDED BARBET Senegal to N Cameroon and N Central African Republic

○ *Lybius rolleti* (Defilippi, 1853) BLACK-BREASTED BARBET S Chad, N Central African Republic to W and SW Sudan

TRACHYLAEMUS Reichenow, 1891[6] M
○ *Trachylaemus purpuratus* YELLOW-BILLED BARBET
— *T.p.goffinii* (Schlegel, 1863) Sierra Leone to Ghana
— *T.p.togoensis* Reichenow, 1891 Togo, SW Nigeria
— *T.p.purpuratus* (J. & E. Verreaux, 1851) SE Nigeria to N and C Zaire and NW Angola
— *T.p.elgonensis* (Sharpe, 1891) NE Zaire, Uganda and W Kenya

TRACHYPHONUS Ranzani, 1821 M
○ *Trachyphonus vaillantii* CRESTED BARBET
— *T.v.suahelicus* Reichenow, 1887 Angola, SE Zaire and Tanzania to C Zimbabwe
— *T.v.vaillantii* Ranzani, 1821 E Botswana, S Zimbabwe and C Mozambique to E Cape Province

---

[1] Thought to include *pareensis* Ripley & Heinrich, 1966 (2068).
[2] Includes *nampunju* Williams, 1966 (2666), a colour morph, see Short & Horne (1988: 467) (2255).
[3] See Peters (1948: 43) (1827) and White (1965: 256) (2649).
[4] Includes *lucidiventris* Clancey, 1956 (373), as the Matobo Hills are in S Zimbabwe; White (1965: 256) (2649) placed this in the synonymy of a broad subspecies *zombae*.
[5] Dated 1817 by Peters (1948: 59) (1827) but see Browning & Monroe (1991) (258) and www.zoonomen.net (A.P. Peterson *in litt.*).
[6] We use this genus following Zimmermann *et al.* (1996) (2735).

○ *Trachyphonus erythrocephalus* RED-AND-YELLOW BARBET
     __ *T.e.shelleyi* Hartlaub, 1886   E Ethiopia, N Somalia
     __ *T.e.versicolor* Hartlaub, 1883   NE Uganda, N Kenya, S Ethiopia
     __ *T.e.erythrocephalus* Cabanis, 1878   S Kenya, N Tanzania

○ *Trachyphonus margaritatus* YELLOW-BREASTED BARBET
     __ *T.m.margaritatus* (Cretzschmar, 1826)   C Mali to C Sudan, N Ethiopia and Eritrea
     __ *T.m.somalicus* Zedlitz, 1910   E Ethiopia to N Somalia and Gabon to C and E Zaire

○ *Trachyphonus darnaudii* D'ARNAUD'S BARBET
     __ *T.d.darnaudii* (Prévost & Des Murs, 1847)[1]   SE Sudan and SW Ethiopia to WC Kenya
     __ *T.d.boehmi* Fischer & Reichenow, 1884   E Ethiopia to S Somalia, E Kenya, NE Tanzania
     __ *T.d.usambiro* Neumann, 1908[2]   SW Kenya to NW Tanzania
     __ *T.d.emini* Reichenow, 1891   C and E Tanzania

INDICATORIDAE HONEYGUIDES (4:17)

PRODOTISCUS Sundevall, 1850 M
○ *Prodotiscus insignis* CASSIN'S HONEYBIRD
     __ *P.i.flavodorsalis* Bannerman, 1923   Sierra Leone to SW Nigeria
     __ *P.i.insignis* (Cassin, 1856)   SE Nigeria to N Angola, NE Zaire, Uganda, W Kenya

○ *Prodotiscus zambesiae* GREEN-BACKED HONEYBIRD
     __ *P.z.ellenbecki* Erlanger, 1901   S Ethiopia, C Kenya to NE Tanzania
     __ *P.z.zambesiae* Shelley, 1894[3]   Angola to SW Tanzania, Zimbabwe and Mozambique

○ *Prodotiscus regulus* WAHLBERG'S HONEYBIRD
     __ *P.r.camerunensis* Reichenow, 1921[4]   Nigeria and Cameroon
     __ *P.r.regulus* Sundevall, 1850[5]   Ethiopia to Angola and E Cape Province

MELIGNOMON Reichenow, 1898 M
○ *Melignomon zenkeri* Reichenow, 1898 ZENKER'S HONEYGUIDE   S Cameroon to W Uganda

○ *Melignomon eisentrauti* Louette, 1981 YELLOW-FOOTED HONEYGUIDE   Liberia; Mt. Cameroon (1377)

INDICATOR Stephens, 1815 M
○ *Indicator pumilio* Chapin, 1958 DWARF HONEYGUIDE   E Zaire, SW Uganda, Rwanda, Burundi (314)

○ *Indicator willcocksi* WILLCOCKS'S HONEYGUIDE
     __ *I.w.ansorgei* C.H.B. Grant, 1915   Guinea Bissau
     __ *I.w.willcocksi* Alexander, 1901   Liberia to S Nigeria, Zaire and W Uganda
     __ *I.w.hutsoni* Bannerman, 1928   C Nigeria, N Cameroon to SW Sudan

○ *Indicator meliphilus* PALLID HONEYGUIDE[6]
     __ *I.m.meliphilus* (Oberholser, 1905)   Kenya to C Tanzania
     __ *I.m.angolensis* Monard, 1934[7]   E Zaire to N Zambia, C Angola, Malawi and C
                                        Mozambique

○ *Indicator exilis* LEAST HONEYGUIDE
     __ *I.e.poensis* Alexander, 1903   Bioko
     __ *I.e.exilis* (Cassin, 1856)[8]   Senegal and Gambia to SW Sudan, south to Angola
                                          and NW Zambia
     __ *I.e.pachyrhynchus* (Heuglin, 1864)   E Zaire, S Uganda to W Kenya

[1] Peters (1948: 62) (1827) bracketed the date implying uncertainty. However, see Sherborn & Woodward (1902) (2242A).
[2] Treated at specific level in last Edition following Wickler (1973) (2660), but see Dowsett & Dowsett-Lemaire (1993: 346) (761).
[3] Includes *lathburyi* B.P. Hall, 1958 (1060), see White (1965: 278) (2649).
[4] Includes *caurinus* Clancey, 1975 (449) see Short & Horne (1988) (2255).
[5] Includes *adustoides* Clancey, 1952 (366), see White (1965: 278) (2649).
[6] Mayr (1971) (1483) questioned whether *narokensis* Jackson, 1906, belonged in the synonymy of this taxon, but it is the immature of this species see Short et al. (1987) (2256).
[7] For recognition see Short & Horne (2001) (2255A).
[8] Includes *propinquus* see Mayr (1957) (1479). Apparently includes ¶*cerophagus* Clancey, 1977 (459) of NW Zambia – not mentioned by Short & Horne (1988) (2255).

○ *Indicator conirostris* T<small>HICK-BILLED</small> H<small>ONEYGUIDE</small>
  — *I.c.ussheri* Sharpe, 1902   Liberia to Ghana
  — *I.c.conirostris* (Cassin, 1856)[1]   Nigeria to W Uganda, W Kenya

○ *Indicator minor* L<small>ESSER</small> H<small>ONEYGUIDE</small>
  — *I.m.senegalensis* Neumann, 1908   Senegal to N Cameroon, Chad, W Sudan
  — *I.m.riggenbachi* Zedlitz, 1915   Cameroon to SW Sudan, W Uganda
  — *I.m.diadematus* Rüppell, 1837   E Sudan, Ethiopia, N Somalia
  — *I.m.teitensis* Neumann, 1900[2]   E and S Uganda, Kenya to C Angola, Zimbabwe and
                    Mozambique
  — *I.m.damarensis* (Roberts, 1928)   S Angola to C Namibia
  — *I.m.minor* Stephens, 1815   South Africa

○ *Indicator maculatus* S<small>POTTED</small> H<small>ONEYGUIDE</small>
  — *I.m.maculatus* G.R. Gray, 1847   Gambia to S Nigeria
  — *I.m.stictithorax* Reichenow, 1877   Cameroon to E Zaire, SW Uganda

○ *Indicator variegatus* Lesson, 1830 S<small>CALY-THROATED</small> H<small>ONEYGUIDE</small>[3]   Ethiopia to Angola and E Cape Province

○ *Indicator xanthonotus* Y<small>ELLOW-RUMPED</small> H<small>ONEYGUIDE</small>
  — *I.x.xanthonotus* Blyth, 1842[4]   W Himalayas to Bhutan
  — *I.x.fulvus* Ripley, 1951   NE and SE Assam (2035)

○ *Indicator archipelagicus* Temminck, 1832 M<small>ALAYSIAN</small> H<small>ONEYGUIDE</small>   W and SW Thailand to Malay Pen., Sumatra, Borneo

○ *Indicator indicator* (Sparrman, 1777) G<small>REATER</small> H<small>ONEYGUIDE</small>[5]   Senegal to Ethiopia and Cape Province

MELICHNEUTES Reichenow, 1910 M
○ *Melichneutes robustus* (Bates, 1909) L<small>YRE-TAILED</small> H<small>ONEYGUIDE</small>   S Cameroon

PICIDAE WOODPECKERS[6] (29: 210)

JYGNINAE

JYNX Linnaeus, 1758 F
○ *Jynx torquilla* N<small>ORTHERN</small> W<small>RYNECK</small>
  — *J.t.torquilla* Linnaeus, 1758   Europe south to Alps, Bulgaria, Caucasus area >> sub-
                    Saharan Africa
  — *J.t.sarudnyi* Loudon, 1912   W Siberia to Yenisey >> NE Africa, SW and S Asia
  — *J.t.chinensis* Hesse, 1911   E Siberia to Kolyma, Sakhalin and NE and C China >>
                    S and SE Asia
  — *J.t.himalayana* Vaurie, 1959   N Pakistan to NW Himalayas (2504)
  — *J.t.tschusii* O. Kleinschmidt, 1907   Iberia (?), Italy, Sardinia, Corsica, Croatia to ? Greece
  — *J.t.mauretanica* Rothschild, 1909   NW Africa

○ *Jynx ruficollis* R<small>UFOUS-BREASTED</small> W<small>RYNECK</small>
  — *J.r.pulchricollis* Hartlaub, 1884   Cameroon and W Zaire to SW Sudan, W Uganda
  — *J.r.aequatorialis* Rüppell, 1842   Ethiopia
  — *J.r.ruficollis* Wagler, 1830[7]   Kenya to Angola and E Sth. Africa

PICUMNINAE

PICUMNUS Temminck, 1825 M
○ *Picumnus innominatus* S<small>PECKLED</small> P<small>ICULET</small>
  — *P.i.innominatus* E. Burton, 1836   W Himalayas to N Assam and SE Xizang

---

[1] Includes *pallidus* Marchant, 1950 (1424), see White (1965: 276) (2649).
[2] Includes *valens* Clancey, 1977 (456), see Short & Horne (1988: 504) (2255).
[3] The proposed races *pseudonymus* Clancey, 1977 (458), *dryas* Clancey, 1979 (469), and *satyrus* Clancey, 1979 (469), appear to be synonyms.
[4] Includes *radcliffi* recognised by Ripley (1982: 225) (2055), but see his footnote. See also Short & Horne (2001) (2255A).
[5] Includes *inquisitor* Clancey, 1970 (422).
[6] We took Short (1982) (2252) as our basis, but here we usually recognise more races of the Eurasian species.
[7] In last Edition we included *pectoralis* Sharpe & Bouvier, 1878, based on Clancey (1987) (508). However, the name he referred to is preoccupied by *pectoralis* Vigors, 1831. Both are synonyms herein; so too are *striaticula* Clancey, 1952 (367), *rougeoti* Berlioz, 1953 (139), and *diloloensis* da Rosa Pinto, 1962 (604), see White (1965: 279) (2649).

__ *P.i.malayorum* E. Hartert, 1912	S and E India to SW Yunnan, Indochina, Sumatra, Borneo
__ *P.i.chinensis* (Hargitt, 1881)	S and C China

○ *Picumnus aurifrons* Bar-breasted Piculet

__ *P.a.aurifrons* Pelzeln, 1870	C Brazil
__ *P.a.transfasciatus* Hellmayr & Gyldenstolpe, 1937	EC Brazil
__ *P.a.borbae* Pelzeln, 1870	C Brazil
__ *P.a.wallacii* Hargitt, 1889	WC Brazil
__ *P.a.purusianus* Todd, 1946	W Brazil
__ *P.a.flavifrons* Hargitt, 1889	NE Peru, W Brazil
__ *P.a.juruanus* Gyldenstolpe, 1941	E Peru, W Brazil

○ *Picumnus lafresnayi* Lafresnaye's Piculet

__ *P.l.lafresnayi* Malherbe, 1862	S Colombia, E Ecuador, S Peru
__ *P.l.punctifrons* Taczanowski, 1886	E Peru
__ *P.l.taczanowskii* Domaniewski, 1925	NE Peru
__ *P.l.pusillus* Pinto, 1936	W Brazil

○ *Picumnus pumilus* Cabanis & Heine, 1863 Orinoco Piculet[1]   N Brazil, S Colombia, S Venezuela

○ *Picumnus exilis* Golden-spangled Piculet

__ *P.e.salvini* Hargitt, 1893[2]	E Venezuela
__ *P.e.clarus* Zimmer & Phelps, 1946	EC Venezuela
__ *P.e.undulatus* Hargitt, 1889	SE Venezuela, S Guyana, N Brazil
__ *P.e.buffoni* Lafresnaye, 1845	Surinam, Cayenne, NE Brazil
__ *P.e.pernambucensis* J.T. Zimmer, 1947	E Brazil (2710)
__ *P.e.alegriae* Hellmayr, 1929	NE Brazil
__ *P.e.exilis* (M.H.K. Lichtenstein, 1823)	E Brazil

○ *Picumnus sclateri* Ecuadorean Piculet

__ *P.s.parvistriatus* Chapman, 1921	W Ecuador
__ *P.s.sclateri* Taczanowski, 1877	SW Ecuador, NW Peru
__ *P.s.porcullae* Bond, 1954	N Peru (185)

○ *Picumnus squamulatus* Scaled Piculet

__ *P.s.squamulatus* Lafresnaye, 1854	N and E Colombia
__ *P.s.roehli* Zimmer & Phelps, 1944	NE Colombia, N Venezuela
__ *P.s.obsoletus* J.A. Allen, 1892	NE Venezuela
__ *P.s.apurensis* Phelps & Aveledo, 1987	Venezuela (1839)
__ *P.s.lovejoyi* Phelps & Aveledo, 1987	Venezuela (1839)

○ *Picumnus spilogaster* White-bellied Piculet

__ *P.s.orinocensis* Zimmer & Phelps, 1950	E Venezuela (2726)
__ *P.s.spilogaster* Sundevall, 1866	NE Brazil, the Guianas
__ *P.s.pallidus* E. Snethlage, 1924	NE Brazil

○ *Picumnus minutissimus* (Pallas, 1782) Guianan Piculet   The Guianas

○ *Picumnus pygmaeus* (M.H.K. Lichtenstein, 1823) Spotted Piculet[3]   NE Brazil

○ *Picumnus steindachneri* Taczanowski, 1882 Speckle-chested Piculet   NE Peru

○ *Picumnus varzeae* E. Snethlage, 1912 Varzea Piculet   Islands in mid-Amazon

○ *Picumnus cirratus* White-barred Piculet

__ *P.c.macconnelli* Sharpe, 1901	NE Brazil
__ *P.c.confusus* Kinnear, 1927	Guyana, Cayenne
__ *P.c.cirratus* Temminck, 1825	S Brazil, E Paraguay
__ *P.c.pilcomayensis* Hargitt, 1891	SW Brazil, Paraguay, N Argentina
__ *P.c.tucumanus* E. Hartert, 1909	N Argentina
__ *P.c.thamnophiloides* Bond & Meyer de Schauensee, 1942	S Bolivia, NW Argentina

---

[1] This was separated from *P. lafresnayi* by Short (1982) (2252).
[2] Includes *nigropunctatus* Zimmer & Phelps, 1950 (2726) see Short (1982) (2252); but see also Rodner at al. (2000) (2096), who believe *salvini* is a junior synonym of *P. squamulatus obsoletus*. Publication of details awaited, but *nigropunctatus* may be shown to be valid.
[3] Includes *distinctus* Pinto & Camargo, 1961 (1899), which may need re-evaluation.

○ *Picumnus dorbygnianus* Ocellated Piculet
     — *P.d.jelskii* Taczanowski, 1882   E Peru
     — *P.d.dorbygnianus* Lafresnaye, 1845   N and C Bolivia

○ *Picumnus temminckii* Lafresnaye, 1845 Ochre-collared Piculet   E Paraguay, SE Brazil, NE Argentina

○ *Picumnus albosquamatus* White-wedged Piculet
     — *P.a.albosquamatus* d'Orbigny, 1840   Bolivia, W Brazil
     — *P.a.guttifer* Sundevall, 1866   C Brazil

○ *Picumnus fuscus* Pelzeln, 1870 Rusty-necked Piculet   SW Brazil, E Bolivia

○ *Picumnus rufiventris* Rufous-breasted Piculet
     — *P.r.rufiventris* (Bonaparte, 1838)   SW Colombia, E Ecuador
     — *P.r.grandis* Carriker, 1930   E Peru
     — *P.r.brunneifrons* Stager, 1968   NW Bolivia (2304)

○ *Picumnus limae* E. Snethlage, 1924 Ochraceous Piculet [1]   E Brazil

○ *Picumnus fulvescens* Stager, 1961 Tawny Piculet   NE Brazil (2303)

○ *Picumnus nebulosus* Sundevall, 1866 Mottled Piculet   S Brazil

○ *Picumnus castelnau* Malherbe, 1862 Plain-breasted Piculet   SE Colombia, E Ecuador, NE Peru

○ *Picumnus subtilis* Stager, 1968 Fine-barred Piculet   SE Peru (Cuzco) (2305)

● *Picumnus olivaceus* Olivaceous Piculet
     — *P.o.dimotus* Bangs, 1903   E Guatemala to Nicaragua
     ☑ *P.o.flavotinctus* Ridgway, 1889   Costa Rica to E Panama and NW Colombia
     — *P.o.malleolus* Wetmore, 1965[2]   N Colombia (2602)
     — *P.o.olivaceus* Lafresnaye, 1845   NC Colombia
     — *P.o.eisenmanni* Phelps & Aveledo, 1966[3]   NE Colombia, NW Venezuela (1837)
     — *P.o.tachirensis* Phelps & Gilliard, 1941   C Colombia, NE Venezuela
     — *P.o.harterti* Hellmayr, 1909   SW Colombia, W Ecuador

○ *Picumnus granadensis* Greyish Piculet
     — *P.g.antioquensis* Chapman, 1915   N Colombia
     — *P.g.granadensis* Lafresnaye, 1847   NC Colombia

○ *Picumnus cinnamomeus* Chestnut Piculet
     — *P.c.cinnamomeus* Wagler, 1829   N Colombia
     — *P.c.perijanus* Zimmer & Phelps, 1944[4]   NW Venezuela
     — *P.c.persaturatus* Haffer, 1961   C Colombia (1051)
     — *P.c.venezuelensis* Cory, 1913   WC Venezuela

SASIA Hodgson, 1836 F
○ *Sasia africana* J. & E. Verreaux, 1855 African Piculet   Nigeria to C Zaire

○ *Sasia abnormis* Rufous Piculet
     — *S.a.abnormis* (Temminck, 1825)   Malay Pen., Sumatra to Borneo
     — *S.a.magnirostris* E. Hartert, 1901   Nias I

○ *Sasia ochracea* White-browed Piculet
     — *S.o.ochracea* Hodgson, 1836   C and E Himalayas
     — *S.o.reichenowi* Hesse, 1911[5]   S and E Assam, Burma, W and S Yunnan, Thailand,
                                           C Indochina
     — *S.o.kinneari* Stresemann, 1929   SE China, N Vietnam

NESOCTITES Hargitt, 1890 M
○ *Nesoctites micromegas* Antillean Piculet
     — *N.m.micromegas* (Sundevall, 1866)   Hispaniola
     — *N.m.abbotti* Wetmore, 1928   Gonave I.

---

[1] Thought to include *saturatus* Pinto & Camargo, 1961 (1899) see Short (1982: 91) (2252), but Short examined no specimens. Re-evaluation needed; possibly a synonym of *P. fulvescens*.
[2] Recognised by Hilty & Brown (1986: 332) (1116).
[3] A new name for *perijanus* Phelps & Phelps, Jr., 1953 (1853), preoccupied.
[4] Thought to include ¶ *larensis* Aveledo, 1998 (63A).
[5] Includes *ferruginea* Koelz, 1952 (1282), see Ripley (1982) (2055).

PICINAE

MELANERPES Swainson, 1832 M

○ *Melanerpes candidus* (Otto, 1796) WHITE WOODPECKER N and E Brazil to C Argentina

● *Melanerpes lewis* (G.R. Gray, 1849) LEWIS'S WOODPECKER W Canada to NW Mexico

○ *Melanerpes herminieri* (Lesson, 1830) GUADELOUPE WOODPECKER Guadeloupe I.

○ *Melanerpes portoricensis* (Daudin, 1803) PUERTO RICAN WOODPECKER Puerto Rico

● *Melanerpes erythrocephalus* (Linnaeus, 1758) RED-HEADED WOODPECKER[1]
                                         S Canada, WC and E USA

● *Melanerpes formicivorus* ACORN WOODPECKER
    ✓ *M.f.bairdi* Ridgway, 1881 Oregon to N Baja California
    — *M.f.angustifrons* S.F. Baird, 1870 S Baja California
    ✓ *M.f.formicivorus* (Swainson, 1827)[2] SW USA to C Mexico
    — *M.f.albeolus* Todd, 1910 SE Mexico to Belize
    — *M.f.lineatus* (Dickey & van Rossem, 1927) S Mexico to N Nicaragua
    — *M.f.striatipectus* Ridgway, 1874 Nicaragua to W Panama
    — *M.f.flavigula* (Malherbe, 1849) Andes of Colombia

○ *Melanerpes cruentatus* (Boddaert, 1783) YELLOW-TUFTED WOODPECKER The Guianas to NE Bolivia

○ *Melanerpes flavifrons* (Vieillot, 1818) YELLOW-FRONTED WOODPECKER S Brazil, Paraguay, N Argentina

● *Melanerpes chrysauchen* GOLDEN-NAPED WOODPECKER
    ✓ *M.c.chrysauchen* Salvin, 1870 SW Costa Rica, W Panama
    — *M.c.pulcher* P.L. Sclater, 1870 N Colombia

● *Melanerpes pucherani* (Malherbe, 1849) BLACK-CHEEKED WOODPECKER S Mexico to W Ecuador

○ *Melanerpes cactorum* (d'Orbigny, 1840) WHITE-FRONTED WOODPECKER S Peru to C Argentina

○ *Melanerpes striatus* (Statius Müller, 1776) HISPANIOLAN WOODPECKER
                                           Hispaniola

○ *Melanerpes radiolatus* (Wagler, 1827) JAMAICAN WOODPECKER Jamaica

○ *Melanerpes chrysogenys* GOLDEN-CHEEKED WOODPECKER[3]
    — *M.c.chrysogenys* (Vigors, 1839) NW Mexico
    — *M.c.flavinuchus* (Ridgway, 1911) SW Mexico

○ *Melanerpes hypopolius* (Wagler, 1829) GREY-BREASTED WOODPECKER Mexico

● *Melanerpes pygmaeus* RED-VENTED WOODPECKER
    — *M.p.tysoni* (Bond, 1936) Guanaja I. (Bay Is., Honduras)
    ✓ *M.p.rubricomus* J.L. Peters, 1948 Yucatán (SE Mexico)
    — *M.p.pygmaeus* (Ridgway, 1885) Cozumel I. (off Yucatán)

○ *Melanerpes rubricapillus* RED-CROWNED WOODPECKER
    — *M.r.rubricapillus* (Cabanis, 1862) SW Costa Rica to Surinam and Tobago
    — *M.r.subfusculus* (Wetmore, 1957) Coiba I. (Panama) (2598)
    — *M.r.seductus* Bangs, 1901 San Miguel I. (Pearl Is., Panama)
    — *M.r.paraguanae* (Gilliard, 1940)[4] Paraguana Pen. (N Venezuela)

● *Melanerpes uropygialis* GILA WOODPECKER
    ✓ *M.u.uropygialis* (S.F. Baird, 1854) SW Nth. America
    — *M.u.cardonensis* (Grinnell, 1927) N Baja California
    — *M.u.brewsteri* (Ridgway, 1911) S Baja California
    — *M.u.fuscescens* (van Rossem, 1934) S Sonora (Mexico)

● *Melanerpes hoffmanni* (Cabanis, 1862) HOFFMANN'S WOODPECKER Nicaragua, W Costa Rica

[1] Includes *brodkorbi* and *caurinus* see Short (1982) (2252).
[2] Includes *phasma* Oberholser, 1974 (1672), see Browning (1978) (246).
[3] Includes *morelensis* R.T. Moore, 1951 (1620), from Central Mexico.
[4] Recognised by Meyer de Schauensee & Phelps (1978: 178) (1577).

● *Melanerpes aurifrons* Golden-fronted Woodpecker
  ↙ *M.a.aurifrons* (Wagler, 1829) — Texas to SC Mexico
  — *M.a.polygrammus* (Cabanis, 1862) — SW Mexico
  — *M.a.grateloupensis* (Lesson, 1839) — E Mexico
  ↙ *M.a.dubius* (Cabot, 1844) — SE Mexico, Belize
  — *M.a.leei* (Ridgway, 1885) — Cozumel I.
  — *M.a.santacruzi* (Bonaparte, 1838) — S Chiapas to N Nicaragua
  — *M.a.pauper* (Ridgway, 1888) — E Honduras
  — *M.a.hughlandi* Dickerman, 1987 — Guatemala (717)
  — *M.a.turneffensis* Russell, 1963[1] — Turneffe Is. (2141)
  — *M.a.insulanus* (Bond, 1936) — Utilla I. (Bay Is., Honduras)
  — *M.a.canescens* (Salvin, 1889) — Roatan, Barburat Is. (Bay Is., Honduras)

◉ *Melanerpes carolinus* (Linnaeus, 1758) Red-bellied Woodpecker[2] — C, E and S USA

○ *Melanerpes superciliaris* West Indian Woodpecker
  — *M.s.nyeanus* (Ridgway, 1886)[3] — Grand Bahama, San Salvador I.
  — *M.s.blakei* (Ridgway, 1886) — Abaco I. (Bahamas)
  — *M.s.superciliaris* (Temminck, 1827) — Cuba[4]
  — *M.s.murceus* (Bangs, 1910) — Isla de la Juventud
  — *M.s.caymanensis* (Cory, 1886) — Grand Cayman I.

SPHYRAPICUS S.F. Baird, 1858 M
● *Sphyrapicus thyroideus* Williamson's Sapsucker
  ↙ *S.t.thyroideus* (Cassin, 1851) — S Brit. Columbia to S California
  — *S.t.nataliae* (Malherbe, 1854) — SE Brit. Columbia, WC USA, C Mexico

◉ *Sphyrapicus varius* (Linnaeus, 1766) Yellow-bellied Sapsucker[5] — S Canada, NE USA >> C and S USA to C America and Antilles

◉ *Sphyrapicus nuchalis* S.F. Baird, 1858 Red-naped Sapsucker — W Canada, W USA, W Mexico

◉ *Sphyrapicus ruber* Red-breasted Sapsucker
  ↙ *S.r.ruber* (J.F. Gmelin, 1788) — SE Alaska to W Oregon
  ↙ *S.r.daggetti* Grinnell, 1901 — S Oregon to SC California

XIPHIDIOPICUS Bonaparte, 1854 M
○ *Xiphidiopicus percussus* Cuban Green Woodpecker[6]
  — *X.p.percussus* (Temminck, 1826) — Cuba
  — *X.p.insulaepinorum* Bangs, 1910 — Isla de la Juventud

CAMPETHERA G.R. Gray, 1841 F
○ *Campethera punctuligera* Fine-spotted Woodpecker
  — *C.p.punctuligera* (Wagler, 1827) — Senegal to Nigeria and Central African Republic
  — *C.p.balia* (Heuglin, 1871) — S Chad, S Sudan, N Zaire

○ *Campethera bennettii* Bennett's Woodpecker
  — *C.b.bennettii* (A. Smith, 1836) — W Tanzania and SE Zaire to Zimbabwe S Mozambique and Natal
  — *C.b.capricorni* Strickland, 1853[7] — S Angola, N Namibia to N Botswana, SW Zambia
  — *C.b.scriptoricauda* (Reichenow, 1896)[8] — C and E Tanzania, N Mozambique, Malawi

○ *Campethera nubica* Nubian Woodpecker
  — *C.n.nubica* (Boddaert, 1783) — Sudan, Ethiopia and Kenya to SW Tanzania
  — *C.n.pallida* (Sharpe, 1902) — S Somalia

---

[1] Recognised by Short (1982: 164) (2252).
[2] Includes *harpaceus* Koelz, 1954 (1285) see Short (1982: 170) (2252).
[3] In our last Edition we placed Watling I. (now San Salvador) in the range of the Bahamian form (*bahamensis*), but the reverse is necessary as the name *nyaneus* has priority.
[4] The proposed races *florentinoi* Garrido, 1966 (915), and *sanfelipensis* Garrido, 1973 (920), await re-evaluation.
[5] Includes *appalachiensis* Ganier, 1954 (909) see Short (1982: 176) (2252).
[6] The proposed races *gloriae* Garrido, 1971 (918), *monticola* Garrido, 1971 (918), *marthae* Regalado Ruiz, 1977 (1998), and *cocoensis* Garrido, 1978 (921), were reviewed by Kirkconnell (2000) (1269A).
[7] Includes *buysi* Winterbottom, 1966 (2681), *fide* R. Dowsett (*in litt.*).
[8] Includes *vincenti* Grant & Mackworth-Praed, 1953 (971), see White (1965: 282) (2649). As regards the place of *scriptoricauda* within this species see Dowsett & Dowsett-Lemaire (1993: 346) (761).

○ *Campethera abingoni* GOLDEN-TAILED WOODPECKER
    — *C.a.chrysura* (Swainson, 1837) — Senegal to Sudan and Malawi
    — *C.a.kavirondensis* van Someren, 1926 — SW Kenya to C Tanzania
    — *C.a.suahelica* (Reichenow, 1902)[1] — S Tanzania to Zimbabwe and Mozambique
    — *C.a.abingoni* (A. Smith, 1836) — Botswana to S Mozambique and Natal
    — *C.a.anderssoni* (Roberts, 1936) — Namibia
    — *C.a.constricta* Clancey, 1965 — Natal (406)

○ *Campethera mombassica* (Fischer & Reichenow, 1884) MOMBASA WOODPECKER[2]
    Coastal S Somalia to N Tanzania

○ *Campethera notata* (M.H.K. Lichtenstein, 1823) KNYSNA WOODPECKER[3]
    S Natal, S and E Cape Province

○ *Campethera cailliautii* GREEN-BACKED WOODPECKER
    — *C.c.permista* (Reichenow, 1876) — Ghana to SW Zaire, SW Uganda, C Zaire and NW Angola
    — *C.c.nyansae* (Neumann, 1900)[4] — SW Kenya, W Tanzania to E Zaire, NE Angola and N Zambia
    — *C.c.cailliautii* (Malherbe, 1849) — S Somalia to NE Tanzania
    — *C.c.loveridgei* E. Hartert, 1920[5] — EC Tanzania to Mozambique

○ *Campethera maculosa* (Valenciennes, 1826) LITTLE GREEN WOODPECKER
    Guinea to Ghana

○ *Campethera tullbergi* TULLBERG'S WOODPECKER
    — *C.t.tullbergi* (Sjöstedt, 1892)[6] — SE Nigeria, W Cameroon, Bioko
    — *C.t.taeniolaema* (Reichenow & Neumann, 1895) — W Kenya, W Uganda to SE Zaire
    — *C.t.hausburgi* (Sharpe, 1900) — E and C Kenya

○ *Campethera nivosa* BUFF-SPOTTED WOODPECKER
    — *C.n.nivosa* (Swainson, 1837)[7] — Guinea to W Zaire, N Angola
    — *C.n.poensis* Alexander, 1903 — Bioko
    — *C.n.herberti* (Alexander, 1908)[8] — N, C and E Zaire, Uganda, W Kenya

○ *Campethera caroli* BROWN-EARED WOODPECKER
    — *C.c.arizela* (Oberholser, 1899) — Sierra Leone, Liberia
    — *C.c.caroli* (Malherbe, 1852) — Cameroon to W Kenya, S Zaire, NW Angola

GEOCOLAPTES Swainson, 1832 M
○ *Geocolaptes olivaceus* (J.F. Gmelin, 1788) GROUND WOODPECKER[9] — Cape Province to Transvaal and Natal (369)

DENDROPICOS Malherbe, 1849 M
○ *Dendropicos elachus* Oberholser, 1919 LITTLE GREY WOODPECKER — Senegal to W Sudan

○ *Dendropicos poecilolaemus* Reichenow, 1893 SPECKLE-BREASTED WOODPECKER
    Cameroon to Uganda, W Kenya

○ *Dendropicos abyssinicus* (Stanley, 1814) GOLD-MANTLED WOODPECKER — Ethiopia

○ *Dendropicos fuscescens* CARDINAL-WOODPECKER
    — *D.f.lafresnayei* Malherbe, 1849 — Senegal to Nigeria
    — *D.f.sharpii* (Oustalet, 1879) — Cameroon to S Sudan and N Angola
    — *D.f.lepidus* (Cabanis & Heine, 1863) — E Zaire to SWEthiopia, C Kenya, NW Tanzania
    — *D.f.hemprichii* (Ehrenberg, 1833) — N and E Ethiopia, Somalia, N and E Kenya
    — *D.f.massaicus* (Neumann, 1900) — SC Ethiopia, W and C Kenya, to NC Tanzania
    — *D.f.centralis* (Neumann, 1900) — Angola to W Tanzania, Zimbabwe, N Namibia
    — *D.f.hartlaubii* Malherbe, 1849 — SE Kenya to Malawi and N Mozambique
    — *D.f.intermedius* Roberts, 1924[10] — E Transvaal, E Natal, S and C Mozambique
    — *D.f.fuscescens* (Vieillot, 1818) — C Namibia to Cape Province, S Transvaal, W Natal

[1] Includes *vibrator* Clancey, 1953 (371), see White (1965: 284) (2649).
[2] Separated from *C. abingoni* by Short & Horne (1988) (2254). See also Clancey (1988) (512).
[3] Includes *relicta* Clancey, 1958 (376), see White (1965: 284) (2649).
[4] Includes *fuelleborni* see Clancey (1970) (425), but this name has been widely misapplied.
[5] Includes *quadrosi* da Rosa Pinto, 1959 (603), see Clancey (1970) (425).
[6] Thought to include *poensis* Eisentraut, 1968 (797), but preoccupied by *C.n.poensis* Alexander, 1903.
[7] Includes *maxima* Traylor, 1970 (2429) see Short (1982: 204) (2252), who later considered it doubtfully distinct (Short, 1988: 530) (2253).
[8] Includes *canzelae* Meise, 1958 (1553), see White (1965: 286) (2649).
[9] Includes *theresae* R. Meinertzhagen, 1949 (1549), *petrobates* Clancey, 1952 (369) and *prometheus* Clancey, 1952, see Earlé (1986) (784A).
[10] Includes *xylobates* Clancey, 1980 (481) see Short (1988: 540) (2253).

○ *Dendropicos gabonensis* GABON WOODPECKER
   — *D.g.lugubris* Hartlaub, 1857[1]   Sierra Leone to SW Nigeria
   — *D.g.reichenowi* Sjöstedt, 1893   S Nigeria, SW Cameroon
   — *D.g.gabonensis* (J. & E. Verreaux, 1851)   Cameroon and Gabon, NE and C Zaire

○ *Dendropicos stierlingi* Reichenow, 1901 STIERLING'S WOODPECKER   S Tanzania, N Mozambique, S Malawi

○ *Dendropicos namaquus* BEARDED WOODPECKER
   — *D.n.schoensis* (Rüppell, 1842)   Ethiopia, Somalia, N Kenya
   — *D.n.namaquus* (A.A.H. Lichtenstein, 1793)   Central African Republic, C and S Kenya and Tanzania
                        to N Namibia, Botswana and Transvaal
   — *D.n.coalescens* Clancey, 1958   E Cape Province to S Mozambique (377)

○ *Dendropicos xantholophus* Hargitt, 1883 YELLOW-CRESTED WOODPECKER
                        S Cameroon to NW Angola and W Kenya

○ *Dendropicos pyrrhogaster* (Malherbe, 1845) FIRE-BELLIED WOODPECKER   Sierra Leone to SE Nigeria

○ *Dendropicos elliotii* ELLIOT'S WOODPECKER
   — *D.e.johnstoni* (Shelley, 1887)   E Nigeria, Mt. Cameroon, Bioko
   — *D.e.elliotii* (Cassin, 1863)   S Cameroon and Gabon to Uganda
   — *D.e.kupeensis* Serle, 1952   Mt. Kupe (Cameroon) (2224)
   — *D.e.gabela* Rand & Traylor, 1959   NW Angola (1980)

○ *Dendropicos goertae* GREY WOODPECKER
   — *D.g.koenigi* (Neumann, 1903)   Mali to WC Sudan
   — *D.g.abessinicus* (Reichenow, 1900)   E Sudan, N and W Ethiopia
   — *D.g.spodocephalus* (Bonaparte, 1850)[2]   High plateaux of C and S Ethiopia
   — *D.g.rhodeogaster* (Fischer & Reichenow, 1884)   Highland C and SE Kenya to NE Tanzania
   — *D.g.goertae* (Statius Müller, 1776)   Senegal to S Sudan, W Kenya, Uganda, NW Tanzania
   — *D.g.meridionalis* Louette & Prigogine, 1982   SE Zaire, NW Angola, S Gabon (1383)

○ *Dendropicos griseocephalus* OLIVE WOODPECKER
   — *D.g.ruwenzori* (Sharpe, 1902)   Angola, Zambia and SE Zaire to W Uganda and C
                        Tanzania
   — *D.g.kilimensis* (Neumann, 1926)   N and E Tanzania
   — *D.g.griseocephalus* (Boddaert, 1783)[3]   E Transvaal and Natal to S Cape Province

○ *Dendropicos obsoletus* BROWN-BACKED WOODPECKER[4]
   — *D.o.obsoletus* (Wagler, 1829)   Gambia to W and S Sudan and C Uganda
   — *D.o.heuglini* (Neumann, 1904)   E Sudan to N Ethiopia
   — *D.o.ingens* (E. Hartert, 1900)   S Ethiopia, NE Uganda, W and C Kenya
   — *D.o.crateri* (Sclater & Moreau, 1935)   N Tanzania

HYPOPICUS Bonaparte, 1854[5] M
○ *Hypopicus hyperythrus* RUFOUS-BELLIED WOODPECKER
   — *H.h.marshalli* (E. Hartert, 1912)   NW Himalayas
   — *H.h.hyperythrus* (Vigors, 1831)[6]   C and E Himalayas to W Sichuan, Yunnan and
                        N Thailand
   — *H.h.subrufinus* (Cabanis & Heine, 1863)   NE China, S Ussuriland, Korea >> S China and
                        N Indochina
   — *H.h.annamensis* (Kloss, 1925)   Mts. of S Laos, SC Vietnam

DENDROCOPOS Koch, 1816 M
○ *Dendrocopos temminckii* (Malherbe, 1849) SULAWESI PYGMY WOODPECKER
                        Sulawesi, Togian Is.

○ *Dendrocopos maculatus* PHILIPPINE PYGMY WOODPECKER
   — *D.m.validirostris* (Blyth, 1849)[7]   Luzon, Catanduanes and Mindoro

---

[1] For an opinion on whether this should be treated as a separate species see Dowsett & Dowsett-Lemaire (1993: 347) (761).
[2] This form and the next have been treated as a species *D. spodocephalus* by Prigogine & Louette (1983) (1946) and by Winkler *et al.* (1995) (2676). Their views were not accepted by Short (1992: 548) (2255) or Dowsett & Dowsett-Lemaire (1993) (761).
[3] Tentatively includes *aureovirens* Clancey, 1975 (447); re-evaluation needed.
[4] We follow Goodwin (1968) (960) in placing this species in *Dendropicos*.
[5] Zusi & Marshall (1970) (2739) demonstrated that this species should not be placed in the genus *Sphyrapicus*; however, in the context of the separation of *Dendrocopus* from *Picoides*, the behaviour and morphology seem sufficiently different from *Dendrocopos* to justify retention of the genus *Hypopicus*.
[6] Includes *haemorrhous* Koelz, 1952 (1282), and *henoticus* Koelz, 1954 (1284), see Ripley (1982: 235) (2055).
[7] Includes *igorotus* Salomonsen, 1953 (2155) see Dickinson *et al.* (1991) (745).

__ *D.m.leytensis* (Steere, 1890)	Bohol, Leyte and Samar
__ *D.m.fulvifasciatus* (Hargitt, 1881)	Basilan, Dinagat and Mindanao
__ *D.m.ramsayi* (Hargitt, 1881)	Bongao, Jolo and Tawitawi Is. (Sulu Is.)
__ *D.m.siasiensis* (Mearns, 1909)	Siasi I. (Sulu Is.)
__ *D.m.maculatus* (Scopoli, 1786)	Cebu, Guimaras, Negros and Panay
__ *D.m.menagei* (Bourns & Worcester, 1894)	Sibuyan I.

○ *Dendrocopos moluccensis* BROWN-CAPPED WOODPECKER[1]

__ *D.m.nanus* (Vigors, 1832)	NE Pakistan, W and N India, Bangladesh
__ *D.m.hardwickii* (Jerdon, 1844)[2]	C India
__ *D.m.cinereigula* (Malherbe, 1849)	SW India
__ *D.m.gymnophthalmus* (Blyth, 1849)	Sri Lanka
__ *D.m.moluccensis* (J.F. Gmelin, 1788)	W Malaysia, Sumatra, Java, Borneo, Bangka, Belitung
__ *D.m.grandis* (Hargitt, 1882)	Lombok to Alor (Lesser Sunda Is.)

○ *Dendrocopos kizuki* JAPANESE PYGMY WOODPECKER[3]

__ *D.k.permutatus* (Meise, 1934)	Liaoning, N Korea, SE Siberia
__ *D.k.seebohmi* (Hargitt, 1884)[4]	Sakhalin, S Kuril Is., Hokkaido
__ *D.k.nippon* (N. Kuroda, Sr., 1922)	Shandong, N Hebei, S Korea, Chejo-do, Honshu (except SW)
__ *D.k.shikokuensis* (N. Kuroda, Sr., 1922)	SW Honshu, Shikoku
__ *D.k.kizuki* (Temminck, 1836)	Kyushu,
__ *D.k.matsudairai* (N. Kuroda, Sr., 1921)	Yakushima, Izu Is.
__ *D.k.kotataki* (N. Kuroda, Sr., 1922)	Tsushima, Oki Is. (S Japan)
__ *D.k.amamii* (N. Kuroda, Sr., 1922)	Amami Oshima (N Ryukyu Is.)
__ *D.k.nigrescens* (Seebohm, 1887)	Okinawa (C Ryukyu Is.)
__ *D.k.orii* (N. Kuroda, Sr., 1923)	Iriomote (S Ryukyu Is.)

○ *Dendrocopos canicapillus* GREY-CAPPED PYGMY WOODPECKER[5]

__ *D.c.doerriesi* (Hargitt, 1881)	SE Siberia, E Manchuria, Korea
__ *D.c.scintilliceps* (Swinhoe, 1853)	E China north to Hebei
__ *D.c.szetschuanensis* (Rensch, 1924)	WC Sichuan to S Shaanxi
__ *D.c.omissus* (Rothschild, 1922)	N and W Yunnan and W Sichuan to E Qinghai and SW Gansu
__ *D.c.nagamichii* (La Touche, 1932)	S China (E Yunnan to Fujian)
__ *D.c.obscurus* (La Touche, 1921)	SW and S Yunnan
__ *D.c.kaleensis* (Swinhoe, 1863)	Taiwan
__ *D.c.swinhoei* (E. Hartert, 1910)	Hainan I.
__ *D.c.mitchelli* (Malherbe, 1849)	W Himalayas (N Pakistan to W Nepal)
__ *D.c.semicoronatus* (Malherbe, 1849)[6]	E Himalayas
__ *D.c.canicapillus* (Blyth, 1845)	SE Assam, C and S Burma, N Thailand, Laos
__ *D.c.delacouri* (Meyer de Schauensee, 1938)	E and SE Thailand, Cambodia, S Vietnam
__ *D.c.auritus* (Eyton, 1845)	SW Thailand and Malay Pen.
__ *D.c.volzi* (Stresemann, 1920)	Sumatra
__ *D.c.aurantiiventris* (Salvadori, 1868)	N and E Borneo

● *Dendrocopos minor* LESSER SPOTTED WOODPECKER

__ *D.m.comminutus* E. Hartert, 1907	England, Wales
__ *D.m.minor* (Linnaeus, 1758)	Scandinavia, NE Poland and Russia to Ural Mts.
__ *D.m.kamtschatkensis* (Malherbe, 1861)	Siberia from Ural Mts. to Sea of Okhotsk and N Mongolia
__ *D.m.immaculatus* Stejneger, 1884[7]	Kamchatka and Anadyr basin
__ *D.m.amurensis* (Buturlin, 1908)[8]	NE China, SE Siberia, Sakhalin and Korea
__ *D.m.hortorum* (C.L. Brehm, 1831)	C Europe (N France and N Alps to C Poland and Rumania)

---

[1] This species is often split. Ripley (1982: 240) (2055) used the name *nanus* implying a split, but gave the global range for the species as named here. We retain one species following Short (1982) (2252), but see Wells (1999: 543) (2579).
[2] For recognition see Ripley (1982: 240) (2055).
[3] Based on O.S.J. (2000) (1727), but as regards the names to apply to the populations of Hokkaido and Honshu it is necessary to revert to the treatment of Peters (1948) (1827) see Dickinson *et al.* (2001) (750).
[4] Includes *ijimae* see Dickinson *et al.* (2001) (750).
[5] We follow the treatment of Cheng (1987) (333) for China.
[6] Includes *gigantiusculus* Koelz, 1952 (1282) see Ripley (1982: 239) (2055).
[7] For recognition see Stepanyan (1990: 320) (2326).
[8] Includes *kemaensis* Won, 1962 (2688), see Stepanyan (1990) (2326).

— *D.m.buturlini* (E. Hartert, 1912)[1]   S Europe (Iberia to Bulgaria and NW Greece)

— *D.m.danfordi* (Hargitt, 1883)   C and E Greece, Turkey

— *D.m.colchicus* (Buturlin, 1908)   Caucasus and Transcaucasia

— *D.m.quadrifasciatus* (Radde, 1884)   SE Azerbaijan

— *D.m.hyrcanus* (Zarudny & Bilkevitch, 1913)   N Iran

— *D.m.morgani* Zarudny & Loudon, 1904   SW Iran

— *D.m.ledouci* (Malherbe, 1855)[2]   NW Africa

○ *Dendrocopos macei* Fulvous-breasted Woodpecker

— *D.m.westermanni* (Blyth, 1870)   W Himalayas

— *D.m.macei* (Vieillot, 1818)[3]   E Himalayas and W Bengal to N Burma

— *D.m.longipennis* Hesse, 1912   C and S Burma through Thailand to S Vietnam

— *D.m.andamanensis* (Blyth, 1859)   Andaman Is.

— *D.m.analis* (Bonaparte, 1850)   S Sumatra, Java, Bali

○ *Dendrocopos atratus* Stripe-breasted Woodpecker

— *D.a.atratus* (Blyth, 1849)   S and E Assam, Burma, N Thailand, Laos

— *D.a.vietnamensis* Stepanyan, 1988   Vietnam (2325)

○ *Dendrocopos auriceps* Brown-fronted Woodpecker

— *D.a.auriceps* (Vigors, 1831)   E Afghanistan, N Pakistan, W Himalayas

— *D.a.incognitus* (Scully, 1879)[4]   C Nepal

○ *Dendrocopos mahrattensis* Yellow-crowned Woodpecker

— *D.m.pallescens* Biswas, 1951   C Pakistan, NW India (155)

— *D.m.mahrattensis* (Latham, 1802)[5]   India, Sri Lanka, Burma

○ *Dendrocopos dorae* (Bates & Kinnear, 1935) Arabian Woodpecker   SW Arabia

○ *Dendrocopos cathpharius* Crimson-breasted Woodpecker

— *D.c.cathpharius* (Blyth, 1843)   E Himalayas

— *D.c.ludlowi* Vaurie, 1959   SE Xizang (China) (2503)

— *D.c.pyrrhothorax* (Hume, 1881)[6]   S and E Assam, W Burma

— *D.c.tenebrosus* (Rothschild, 1926)   N and E Burma to SW Yunnan and NW Vietnam

— *D.c.pernyii* (J. Verreaux, 1867)   N Yunnan, WC Sichuan, S Gansu (C China)

— *D.c.innixus* (Bangs & Peters, 1928)   NE Sichuan and SE Shaanxi to W Hubei (NC China)

○ *Dendrocopos darjellensis* Darjelling Woodpecker

— *D.d.darjellensis* (Blyth, 1845)[7]   C and E Himalayas and S Assam to Yunnan and NW Vietnam

— *D.d.desmursi* (J. Verreaux, 1870)[8]   W Sichuan

○ *Dendrocopos medius* Middle-spotted Woodpecker

— *D.m.medius* (Linnaeus, 1758)   Europe to NW Turkey

— *D.m.caucasicus* (Bianchi, 1904)   N Asia Minor, Caucasus, Transcaucasia

— *D.m.anatoliae* (E. Hartert, 1912)[9]   W and S Asia Minor to N Iraq

— *D.m.sanctijohannis* (Blanford, 1873)   Zagros Mts. (SW Iran)

○ *Dendrocopos leucotos* White-backed Woodpecker[10]

— *D.l.leucotos* (Bechstein, 1803)   C and N Europe to E Alps, Carpathians and EC Europe, S Russia to SE Siberia, Sakhalin and NE China

— *D.l.uralensis* (Malherbe, 1861)   W Ural Mts. to Lake Baikal area

— *D.l.lilfordi* (Sharpe & Dresser, 1871)   Pyrenees; Balkans, Turkey, Caucasus area

— *D.l.tangi* Cheng, Tso-hsin, 1956   W Sichuan (China) (328)

— *D.l.subcirris* (Stejneger, 1886)   S Kuril Is., Hokkaido

---

[1] Includes *heinrichi* von Jordans, 1940 (2539), see Vaurie (1965: 724) (2510).
[2] Omitted in last Edition.
[3] Includes *humei* Koelz, 1954 (1284) see Ripley (1982: 238) (2055).
[4] For recognition see Ripley (1982: 238) (2055), who placed *conoveri* Rand & Fleming, 1956 (1974), in the synonymy of this.
[5] For reasons to use 1802 not 1801 see Browing & Monroe (1991) (258). Includes *koelzi* Biswas, 1951 (155) see Ripley (1982: 239) (2055), who synonymised *pallescens* but apparently recognised *blanfordi* of Burma.
[6] Includes *cruentipectus* Koelz, 1954 (1284) see Ripley (1982: 237) (2055).
[7] Includes *fumidus* Ripley, 1951 (2035), and *diatropus* Koelz, 1954 (1284), see Ripley (1982: 237) (2055).
[8] Omitted in last Edition. For recognition see Vaurie (1965: 710) (2510).
[9] For recognition see Cramp *et al.* (1985: 882) (591).
[10] For this species we follow Vaurie (1965: 714) (2510).

__ *D.l.stejnegeri* (N. Kuroda, Sr., 1921)   N and C Honshu, Sado-shima

__ *D.l.namiyei* (Stejneger, 1886)   W Honshu, Shikoku, Kyushu, Oki Is.

__ *D.l.takahashii* (Kuroda & Mori, 1920)   Ullung-do (Dagelet Is., S Korea)

__ *D.l.owstoni* (Ogawa, 1905)   Amami-Oshima (Ryukyu Is.)

__ *D.l.quelpartensis* (Kuroda & Mori, 1918)   Cheju-do (Quelpart I., S Korea)

__ *D.l.fohkiensis* (Buturlin, 1908)   NW Fujian (SE China)

__ *D.l.insularis* (Gould, 1836)   Taiwan

○ ***Dendrocopos himalayensis*** HIMALAYAN WOODPECKER

__ *D.h.albescens* (E.C.S. Baker, 1926)   NW Himalayas east to Chamba

__ *D.h.himalayensis* (Jardine & Selby, 1831)[1]   WC Himalayas (Simla to W Nepal)

○ ***Dendrocopos assimilis*** (Blyth, 1849) SIND WOODPECKER   SE Iran, W and NW Pakistan

● ***Dendrocopos syriacus*** SYRIAN WOODPECKER[2]

✓ *D.s.syriacus* (Hemprich & Ehrenberg, 1833)   C Europe to Turkey, Levant and SW Iran

__ *D.s.transcaucasicus* Buturlin, 1910   Transcaucasia and N Iran

__ *D.s.milleri* Zarudny, 1909   Kuh-e Taftan Mts. (SE Iran)

○ ***Dendrocopos leucopterus*** (Salvadori, 1870) WHITE-WINGED WOODPECKER[3]
      E Turkmenistan, C Uzbekistan, SW Kazakhstan, Pamir and Tien Shan Mts. to SE Kazakstan and Xinjiang

● ***Dendrocopos major*** GREAT SPOTTED WOODPECKER[4]

__ *D.m.major* (Linnaeus, 1758)   Scandinavia and NE Poland to Ural Mts.

__ *D.m.brevirostris* (Reichenbach, 1854)[5]   Ural Mts. to W shore Sea of Okhotsk, NE China, N Mongolia and NE Tien Shan

__ *D.m.kamtschaticus* (Dybowskii, 1883)   Kamchatka and N shore Sea of Okhotsk

__ *D.m.anglicus* E. Hartert, 1900   Great Britain

✓ *D.m.pinetorum* (C.L. Brehm, 1831)   C Europe (Denmark and France to Alps and Carpathians)

✓ *D.m.parroti* E. Hartert, 1911[6]   Corsica

__ *D.m.harterti* Arrigoni, 1902   Sardinia

__ *D.m.italiae* (Stresemann, 1919)   Mainland Italy, Sicily, W Slovenia

__ *D.m.hispanus* (Schlüter, 1908)   Iberian Pen.

__ *D.m.canariensis* (König, 1889)   Tenerife I.

__ *D.m.thanneri* Le Roi, 1911   Gran Canaria I.

__ *D.m.mauritanus* (C.L. Brehm, 1855)   Morocco

__ *D.m.numidus* (Malherbe, 1843)   N Algeria, Tunisia

__ *D.m.candidus* (Stresemann, 1919)   Rumania and S Ukraine to Greece

__ *D.m.paphlagoniae* (Kumerlöwe & Niethammer, 1935)[7]   N Asia Minor

__ *D.m.tenuirostris* Buturlin, 1906[8]   Caucasus and Transcaucasia

__ *D.m.poelzami* (Bogdanov, 1879)   SE Azerbaijan, N Iran, SW Turkmenistan

__ *D.m.japonicus* (Seebohm, 1883)   SE Siberia and N and E Manchuria to S Kuril Is., N and C Japan and Korea

__ *D.m.wulashanicus* Cheng, Xian, Zhang & Jiang, 1975[9]   SC Nei Mongol (343)

__ *D.m.cabanisi* (Malherbe, 1854)   S Manchuria to N Anhui and N Jiangxu (E China)

__ *D.m.beicki* (Stresemann, 1927)   NE Sichuan and E Qinghai to Gansu and Shaanxi

__ *D.m.mandarinus* (Malherbe, 1857)   E Burma, S China, N Laos, N Vietnam

__ *D.m.stresemanni* (Rensch, 1924)   S and E Assam, NE Burma, Yunnan, W Sichuan

__ *D.m.hainanus* Hartert & Hesse, 1911   Hainan

PICOIDES Lacépède, 1799 M

○ ***Picoides mixtus*** CHECKERED WOODPECKER

__ *P.m.cancellatus* (Wagler, 1829)   SE Brazil

---

[1] Not 1835 as given by Peters (1948) (1827); see Sherborn (1894) (2241).
[2] In last Edition we followed Short (1982) (2252) and treated this as monotypic, but we now follow Cramp *et al.* (1985: 874) (591).
[3] Stepanyan (1990: 312) (2326) may advisedly treat this as polytypic. His synonymy includes *spangenbergi* Gladkov, 1951 (944).
[4] Except as indicated and for one new taxon we follow Vaurie (1965: 699) (2510).
[5] Cheng (1987: 393) (333) and Stepanyan (1990: 311) (2326) recognised *tianshanicus* and may be right to do so or it may be an unstable intergrade. In a broad genus *Picoides* this name may conflict with *tianschanicus* Buturlin, 1907, although Art. 58 of the Code (I.C.Z.N., 1999) (1178) seems not to apply.
[6] For recognition see Cramp *et al.* (1985: 856) (591).
[7] For recognition see Cramp *et al.* (1985: 856) (591).
[8] If treated in the genus *Picoides* this must be called *kitsutsuki* Hachisuka, 1952 (1042), as *tenuirostris* is preoccupied in *Picoides*.
[9] Tentatively recognised. Multiple original spellings; we treat Cheng (1987) (333) as First Reviser under Art. 24.2 of the Code (I.C.Z.N., 1999) (1178).

— *P.m.mixtus* (Boddaert, 1783)   E Paraguay, Uruguay, E Argentina

— *P.m.malleator* (Wetmore, 1922)   W Paraguay, SE Bolivia, N and W Argentina

— *P.m.berlepschi* (Hellmayr, 1915)   S and W Argentina

⬤ *Picoides lignarius* (Molina, 1782)   STRIPED WOODPECKER   C Bolivia, Chile, S and W Argentina

⬤ *Picoides scalaris*   LADDER-BACKED WOODPECKER

    ☑ *P.s.cactophilus* (Oberholser, 1911)   SW USA to C and E Mexico

    — *P.s.eremicus* (Oberholser, 1911)   N Baja California

    — *P.s.lucasanus* (S.F. Baird, 1859)   S Baja California

    — *P.s.soulei* Banks, 1963[1]   Cerralvo I. (84)

    — *P.s.graysoni* (S.F. Baird, 1874)   Tres Marias Is.

    — *P.s.sinaloensis* (Ridgway, 1887)   W Mexico

    — *P.s.scalaris* (Wagler, 1829)[2]   S Mexico

    — *P.s.parvus* (Cabot, 1845)   N Yucatán, Cozumel I.

    — *P.s.leucoptilurus* (Oberholser, 1911)   Belize to N Nicaragua

◐ *Picoides nuttallii* (Gambel, 1843)   NUTTALL'S WOODPECKER   W California, NW Baja California

⬤ *Picoides pubescens*   DOWNY WOODPECKER[3]

    — *P.p.glacialis* (Grinnell, 1910)   SE Alaska

    — *P.p.medianus* (Swainson, 1832)   C Alaska to SC and E USA

    — *P.p.fumidus* (Maynard, 1889)   SW Canada, W Washington (1462)

    ☑ *P.p.gairdnerii* (Audubon, 1839)   W Oregon to NW California

    — *P.p.turati* (Malherbe, 1861)   Interior Washington and Oregon to California

    — *P.p.leucurus* (Hartlaub, 1852)[4]   E British Columbia, W USA

    ☑ *P.p.pubescens* (Linnaeus, 1766)   SE USA

⬤ *Picoides villosus*   HAIRY WOODPECKER

    ☑ *P.v.septentrionalis* (Nuttall, 1840)   SC Alaska, S Canada, NW USA

    — *P.v.picoideus* (Osgood, 1901)   Queen Charlotte Is.

    — *P.v.sitkensis* (Swarth, 1911)   SE Alaska, N Brit. Columbia

    ☑ *P.v.harrisi* (Audubon, 1838)   S Brit. Columbia to N California

    — *P.v.terraenovae* (Batchelder, 1908)   Newfoundland

    — *P.v.villosus* (Linnaeus, 1766)   SE Canada, NC and E USA

    ☑ *P.v.orius* (Oberholser, 1911)   SC Washington to W Texas

    ☑ *P.v.audubonii* (Swainson, 1832)   SE USA

    — *P.v.hyloscopus* (Cabanis & Heine, 1863)   W and S California

    — *P.v.icastus* (Oberholser, 1911)   SE Arizona, NW Mexico

    — *P.v.jardinii* (Malherbe, 1845)   E and S Mexico

    — *P.v.sanctorum* (Nelson, 1897)   S Mexico to N Nicaragua

    — *P.v.piger* (G.M. Allen, 1905)   Grand Bahama, Mores, Abaco Is.

    — *P.v.maynardi* (Ridgway, 1887)   New Providence, Andros Is. (Bahamas)

◯ *Picoides arizonae*   ARIZONA WOODPECKER[5]

    — *P.a.arizonae* (Hargitt, 1886)   Arizona, NW Mexico

    — *P.a.fraterculus* (Ridgway, 1887)[6]   W Mexico

◯ *Picoides stricklandi* (Malherbe, 1845)   STRICKLAND'S WOODPECKER[7]   C and E Mexico

⬤ *Picoides borealis* (Vieillot, 1809)[8]   RED-COCKADED WOODPECKER   SC and SE USA

⬤ *Picoides albolarvatus*   WHITE-HEADED WOODPECKER

    ☑ *P.a.albolarvatus* (Cassin, 1850)   W USA

    — *P.a.gravirostris* (Grinnell, 1902)   S California

---

[1] Omitted in last Edition.
[2] Includes *lambi* A.R. Phillips, 1996 (1870) see Short (1982) (2252).
[3] We follow Browning (1997) (257).
[4] Includes *parvirostris* Burleigh, 1961 (275), see Browning (1990) (251).
[5] For restoration of recognition to this species see A.O.U. (2000: 851) (4). Treated within *P. stricklandi* in last Edition.
[6] Includes *websteri* A.R. Phillips, 1962 (1868), see Short (1982) (2252).
[7] Includes *aztecus* R.T. Moore, 1946 (1614), see Short (1982) (2252). May deserve recognition.
[8] Dated 1807 by Peters (1948: 211) (1827), but see Browning & Monroe (1991) (258).

● *Picoides tridactylus* THREE-TOED WOODPECKER[1]
— *P.t.tridactylus* (Linnaeus, 1758) — N Europe to Urals; S Urals through S Siberia and N Mongolia to Sakhalin and N Manchuria to NE Asia
— *P.t.alpinus* C.L. Brehm, 1831 — C and SE Europe to the Carpathian Mts.
— *P.t.crissoleucus* (Reichenbach, 1854) — C and N Urals to Sea of Okhotsk and Anadyr
— *P.t.albidior* Stejneger, 1885 — Kamchatka
— *P.t.tianschanicus* Buturlin, 1907 — C and E Tien Shan and Dzhungarian Alatau
— *P.t.kurodai* Yamashina, 1930 — SE Manchuria and NE Korea
— *P.t.inouyei* Yamashina, 1943[2] — Hokkaido
— *P.t.funebris* J. Verreaux, 1870 — N Yunnan and W Sichuan to E Qinghai and W Gansu
— *P.t.fasciatus* S.F. Baird, 1870 — W Nth. America south to NW USA
— *P.t.dorsalis* S.F. Baird, 1858 — N Montana to C New Mexico and C Arizona
— *P.t.bacatus* Bangs, 1900 — C and E Canada, NE USA

● *Picoides arcticus* (Swainson, 1832) BLACK-BACKED WOODPECKER — Canada, N and W USA

VENILIORNIS Bonaparte, 1854 M
○ *Veniliornis callonotus* SCARLET-BACKED WOODPECKER
— *V.c.callonotus* (Waterhouse, 1841) — W Ecuador
— *V.c.major* (Berlepsch & Taczanowski, 1884) — SW Ecuador, N Peru

○ *Veniliornis dignus* YELLOW-VENTED WOODPECKER
— *V.d.dignus* (Sclater & Salvin, 1877)[3] — W Colombia, SW Venezuela
— *V.d.baezae* Chapman, 1923 — E Ecuador
— *V.d.valdizani* (Berlepsch & Taczanowski, 1894) — Peru

○ *Veniliornis nigriceps* BAR-BELLIED WOODPECKER
— *V.n.equifasciatus* Chapman, 1912 — WC Colombia, N Ecuador
— *V.n.pectoralis* (Berlepsch & Taczanowski, 1902) — C Peru
— *V.n.nigriceps* (d'Orbigny, 1840) — N Bolivia

○ *Veniliornis fumigatus* SMOKY-BROWN WOODPECKER
— *V.f.oleagineus* (Reichenbach, 1854) — E Mexico
— *V.f.sanguinolentus* (P.L. Sclater, 1859) — C and S Mexico to W Panama
— *V.f.reichenbachi* (Cabanis & Heine, 1863)[4] — E Panama, N Venezuela, Colombia, E Ecuador
— *V.f.fumigatus* (d'Orbigny, 1840) — Upper Amazonia
— *V.f.obscuratus* Chapman, 1927 — NW Peru to NW Argentina

○ *Veniliornis passerinus* LITTLE WOODPECKER
— *V.p.fidelis* (Hargitt, 1889) — E Colombia, W Venezuela
— *V.p.modestus* J.T. Zimmer, 1942 — NE Venezuela
— *V.p.diversus* J.T. Zimmer, 1942 — N Brazil
— *V.p.agilis* (Cabanis & Heine, 1863) — E Ecuador to N Bolivia, W Brazil
— *V.p.insignis* J.T. Zimmer, 1942 — WC Brazil
— *V.p.tapajozensis* Gyldenstolpe, 1941 — C Brazil
— *V.p.passerinus* (Linnaeus, 1766) — The Guianas, NE Brazil
— *V.p.taenionotus* (Reichenbach, 1854) — E Brazil
— *V.p.olivinus* (Natterer & Malherbe, 1845) — S Bolivia, S Brazil, Paraguay, N Argentina

○ *Veniliornis frontalis* (Cabanis, 1883) DOT-FRONTED WOODPECKER — Andes from C Bolivia to NW Argentina

○ *Veniliornis spilogaster* (Wagler, 1827) WHITE-SPOTTED WOODPECKER — SE Brazil, Paraguay, NE Argentina

○ *Veniliornis sanguineus* (A.A.H. Lichtenstein, 1793)[5] BLOOD-COLOURED WOODPECKER — The Guianas

○ *Veniliornis kirkii* RED-RUMPED WOODPECKER
— *V.k.neglectus* Bangs, 1901 — SW Costa Rica, W Panama
— *V.k.cecilii* (Malherbe, 1849) — E Panama, W Colombia, W Ecuador
— *V.k.continentalis* Hellmayr, 1906 — N and W Venezuela
— *V.k.monticola* Hellmayr, 1918 — Mt. Roraima (Venezuela)
— *V.k.kirkii* (Malherbe, 1845) — Trinidad, Tobago I.

[1] Eurasian races now follow Vaurie (1965: 730) (2510).
[2] Despite the doubts of Vaurie (1965) (2510) recognised by Orn. Soc. Japan (2000: 176) (1727).
[3] Includes *abdominalis* Phelps & Phelps, Jr., 1956 (1857), see Short (1982: 345) (2252).
[4] Implicitly includes *exsul* and *tectricalis* see Winkler *et al.* (1995) (2676).
[5] Misdated 1783 in Peters (1948: 175) (1827), a *lapsus*.

○ *Veniliornis affinis* Red-stained Woodpecker
— *V.a.chocoensis* Todd, 1919[1]   NW Columbia
— *V.a.orenocensis* Berlepsch & Hartert, 1902[2]   E Colombia, E Venezuela, N Brazil
— *V.a.hilaris* (Cabanis & Heine, 1863)   E Ecuador, E Peru, N Bolivia, W Brazil
— *V.a.ruficeps* (Spix, 1824)   C and E Brazil
— *V.a.affinis* (Swainson, 1821)   E Brazil

○ *Veniliornis cassini* (Malherbe, 1862) Golden-collared Woodpecker   Venezuela, the Guianas, N Brazil

○ *Veniliornis maculifrons* (Spix, 1824) Yellow-eared Woodpecker   SE Brazil

PICULUS Spix, 1824 M

○ *Piculus simplex* (Salvin, 1870) Rufous-winged Woodpecker[3]   Honduras to W Panama

○ *Piculus callopterus* (Lawrence, 1862) Stripe-cheeked Woodpecker   E Panama

○ *Piculus leucolaemus* White-throated Woodpecker
— *P.l.leucolaemus* (Natterer & Malherbe, 1845)   E Colombia to Bolivia, W Brazil
— *P.l.litae* (Rothschild, 1901)   W Colombia, NW Ecuador

○ *Piculus flavigula* Yellow-throated Woodpecker
— *P.f.flavigula* (Boddaert, 1783)   N Amazonia
— *P.f.magnus* (Cherrie & Reichenberger, 1921)   SE Colombia, NW Brazil
— *P.f.erythrops* (Vieillot, 1818)   E and SE Brazil

○ *Piculus chrysochloros* Golden-green Woodpecker
— *P.c.aurosus* (Nelson, 1912)   Panama
— *P.c.xanthochlorus* (Sclater & Salvin, 1875)   NE Colombia, NW Venezuela
— *P.c.capistratus* (Malherbe, 1862)   C Colombia to Guyana, NW Brazil
— *P.c.guianensis* Todd, 1937   Cayenne
— *P.c.laemostictus* Todd, 1937   W Brazil
— *P.c.hypochryseus* Todd, 1937   W Brazil, N Bolivia
— *P.c.paraensis* (E. Snethlage, 1907)   NE Brazil
— *P.c.polyzonus* (Valenciennes, 1826)   SE Brazil
— *P.c.chrysochlorus* (Vieillot, 1818)   C and S Brazil, Bolivia, N Argentina

○ *Piculus aurulentus* (Temminck, 1821[4]) White-browed Woodpecker   SE Brazil, N Argentina

○ *Piculus rubiginosus* Golden-olive Woodpecker
— *P.r.aeruginosus* (Malherbe, 1862)   NE Mexico
— *P.r.yucatanensis* (Cabot, 1844)   C Mexico to W Panama
— *P.r.alleni* (Bangs, 1902)   N Colombia
— *P.r.buenavistae* (Chapman, 1915)   E Colombia, E Ecuador
— *P.r.meridensis* (Ridgway, 1911)   W Venezuela
— *P.r.rubiginosus* (Swainson, 1820)   N Venezuela
— *P.r.deltanus* Aveledo & Ginés, 1953   NE Venezuela (66)
— *P.r.paraquensis* Phelps & Phelps, Jr., 1948   E Venezuela (1844)
— *P.r.guianae* (Hellmayr, 1918)   SE Venezuela, S Guyana
— *P.r.viridissimus* Chapman, 1939   S Venezuela
— *P.r.nigriceps* Blake, 1941[5]   Guyana, Surinam
— *P.r.trinitatis* (Ridgway, 1911)   Trinidad
— *P.r.tobagensis* (Ridgway, 1911)   Tobago I
— *P.r.gularis* (Hargitt, 1889)   C and SW Colombia
— *P.r.rubripileus* (Salvadori & Festa, 1900)   SW Colombia, NW Peru
— *P.r.coloratus* (Chapman, 1923)   NC Peru
— *P.r.chrysogaster* (Berlepsch & Stolzmann, 1902)   C Peru
— *P.r.canipileus* (d'Orbigny, 1840)   N Bolivia
— *P.r.tucumanus* (Cabanis, 1883)   C Bolivia to NW Argentina

---

[1] Some recent authors treat as a species, but more detailed reasons awaited.
[2] Includes *caquetanus* Meyer de Schauensee, 1949 (1565), see Short (1974) (2251).
[3] For the separation of this and *callopterus* from *P. leucolaemus* see A.O.U. (1998: 324) (3).
[4] Dated 1823 in Peters (1948: 114) (1827) *contra* Sherborn (1898) (2242); see also Dickinson (2001) (739).
[5] Includes *fortirostris* Mees, 1974 (1526) and *poliocephalus* Mees, 1974 (1526), see Short (1982: 366) (2252).

○ *Piculus auricularis* (Salvin & Godman, 1889) Grey-crowned Woodpecker

W and NW Mexico

○ *Piculus rivolii* Crimson-mantled Woodpecker
  __ *P.r.quindiuna* (Chapman, 1923)
  __ *P.r.rivolii* (Boissonneau, 1840)[1]
  __ *P.r.meridae* (Chapman, 1923)
  __ *P.r.brevirostris* (Taczanowski, 1875)
  __ *P.r.atriceps* (Sclater & Salvin, 1876)

NC Colombia
W Venezuela, EC Colombia
W Venezuela
SW Colombia to C Peru
SE Peru, Bolivia

COLAPTES Vigors, 1826 M

○ *Colaptes atricollis* Black-necked Woodpecker
  __ *C.a.atricollis* (Malherbe, 1850)
  __ *C.a.peruvianus* (Reichenbach, 1854)

W Peru
E Peru

○ *Colaptes punctigula* Spot-breasted Woodpecker
  __ *C.p.ujhelyii* (Madarász, 1912)
  __ *C.p.striatigularis* (Chapman, 1914)
  __ *C.p.punctipectus* (Cabanis & Heine, 1863)
  __ *C.p.zuliae* (Cory, 1915)
  __ *C.p.punctigula* (Boddaert, 1783)
  __ *C.p.guttatus* (Spix, 1824)

N Colombia
E Panama to WC Colombia
E Colombia, Venezuela
NW Venezuela
Surinam, Cayenne
Upper Amazonia

○ *Colaptes melanochloros* Green-barred Woodpecker
  __ *C.m.nattereri* (Malherbe, 1845)
  __ *C.m.melanochloros* (J.F. Gmelin, 1788)
  __ *C.m.melanolaimus* (Malherbe, 1857)
  __ *C.m.nigroviridis* (C.H.B. Grant, 1911)
  __ *C.m.leucofrenatus* Leybold, 1873

E Bolivia, S Brazil
C Brazil
Bolivia, W Argentina
S Bolivia, Paraguay
C and S Argentina

● *Colaptes auratus* Northern Flicker[2]
  ✓ *C.a.cafer* (J.F. Gmelin, 1788)
  ✓ *C.a.collaris* Vigors, 1829
  __ †*C.a.rufipileus* Ridgway, 1876
  __ *C.a.nanus* Griscom, 1934
  __ *C.a.mexicanus* Swainson, 1827
  __ *C.a.mexicanoides* Lafresnaye, 1844
  ✓ *C.a.luteus* Bangs, 1898
  ✓ *C.a.auratus* (Linnaeus, 1758)
  __ *C.a.chrysocaulosus* Gundlach, 1858
  __ *C.a.gundlachi* Cory, 1886

S Alaska to N California
SE Brit. Columbia to NW Mexico
Guadalupe I. (Mexico)
NE Mexico
C Mexico
Highland S Mexico to N Nicaragua
Canada, E, S and C USA
SE USA
Cuba
Grand Cayman I.

● *Colaptes chrysoides* Gilded Flicker[3]
  __ *C.c.mearnsi* Ridgway, 1911
  ✓ *C.c.tenebrosus* van Rossem, 1930
  __ *C.c.brunnescens* Anthony, 1895
  __ *C.c.chrysoides* (Malherbe, 1852)

SE California, NW Baja California
NW Mexico
C Baja California
S and W Baja California

○ *Colaptes fernandinae* Vigors, 1827 Fernandina's Flicker

Cuba

● *Colaptes pitius* (Molina, 1782) Chilean Flicker

C and S Chile, S Argentina

○ *Colaptes rupicola* Andean Flicker
  __ *C.r.cinereicapillus* Reichenbach, 1854[4]
  __ *C.r.puna* Cabanis, 1883
  __ *C.r.rupicola* d'Orbigny, 1840

N Peru
S and C Peru
Bolivia, N Chile, NW Argentina

○ *Colaptes campestris* Campo Flicker[5]
  __ *C.c.campestris* (Vieillot, 1818)
  __ *C.c.campestroides* (Malherbe, 1849)

E Bolivia, C and E Brazil
SC Sth. America

---

[1] Tentatively includes *zuliensis* Aveledo & Perez, 1989 (67), but requires evaluation.
[2] We recognise races following Short (1982: 380) (2252).
[3] For reasons to treat at specific level see A.O.U. (1998: 345) (3).
[4] For correction of spelling see David & Gosselin (2002) (613).
[5] The two races herein treated as species in last Edition, but see Short (1982: 389) (2252).

CELEUS Boie, 1831 M

○ ***Celeus brachyurus*** RUFOUS WOODPECKER[1]

   — *C.b.jerdonii* (Malherbe, 1849)[2]   W and SW India, Sri Lanka

   — *C.b.humei* (Kloss, 1918)   WC Himalayas (N Uttar Pradesh to W Nepal)

   — *C.b.phaioceps* (Blyth, 1845)   C and E Himalayas and Assam hills to W and

                                              S Yunnan, Burma and N, E and SE Thailand

   — *C.b.annamensis* (Delacour & Jabouille, 1924)   Laos, Cambodia, C and S Vietnam

   — *C.b.fokiensis* (Swinhoe, 1863)   SE China, NE Vietnam

   — *C.b.holroydi* (Swinhoe, 1870)   Hainan I.

   — *C.b.williamsoni* (Kloss, 1918)[3]   SW and C Thailand, N Malay Pen.

   — *C.b.squamigularis* (Sundevall, 1866)   S Malay Pen., Sumatra, Bangka, Belitung, Nias Is.

   — *C.b.brachyurus* (Vieillot, 1818)   Java

   — *C.b.badiosus* (Bonaparte, 1850)   Borneo, N Natuna Is.

○ ***Celeus loricatus*** CINNAMON WOODPECKER

   — *C.l.diversus* Ridgway, 1914   SE Costa Rica, W Panama

   — *C.l.mentalis* Cassin, 1860   E Panama, NW Colombia

   — *C.l.innotatus* Todd, 1917   N and NC Colombia

   — *C.l.loricatus* (Reichenbach, 1854)   W Colombia, W Ecuador

○ ***Celeus undatus*** WAVED WOODPECKER

   — *C.u.amarcurensis* Phelps & Phelps, Jr., 1950   NE Venezuela (1848)

   — *C.u.undatus* (Linnaeus, 1766)   The Guianas, N Brazil

   — *C.u.multifasciatus* (Natterer & Malherbe, 1845)   NE Brazil

○ ***Celeus grammicus*** SCALY-BREASTED WOODPECKER

   — *C.g.verreauxii* (Malherbe, 1858)   SE Colombia, E Ecuador

   — *C.g.grammicus* (Natterer & Malherbe, 1845)   S Venezuela, NE Peru, W Brazil

   — *C.g.subcervinus* Todd, 1937   NC Brazil

   — *C.g.latifasciatus* Seilern, 1936[4]   SE Peru, NE Bolivia

○ ***Celeus castaneus*** (Wagler, 1829) CHESTNUT-COLOURED WOODPECKER   SE Mexico to C Panama

○ ***Celeus elegans*** CHESTNUT WOODPECKER

   — *C.e.hellmayri* Berlepsch, 1908   E Venezuela, NE Brazil, Guyana, Surinam

   — *C.e.deltanus* Phelps & Phelps, Jr., 1950   NE Venezuela (1848)

   — *C.e.leotaudi* Hellmayr, 1906   Trinidad

   — *C.e.elegans* (Statius Müller, 1776)   NE Brazil, Cayenne

   — *C.e.citreopygius* Sclater & Salvin, 1867   SE Colombia, E Ecuador, Peru

   — *C.e.jumanus* (Spix, 1824)[5]   E Colombia to N Bolivia

○ ***Celeus lugubris*** PALE-CRESTED WOODPECKER

   — ¶ *C.l."castaneus"* Olrog, 1963[6]   NE Bolivia (1681)

   — *C.l.lugubris* (Malherbe, 1851)   WC Brazil, E Bolivia

   — *C.l.kerri* Hargitt, 1891   N Argentina, Paraguay, S Brazil

○ ***Celeus flavescens*** BLOND-CRESTED WOODPECKER

   — *C.f.intercedens* Hellmayr, 1908   NE Brazil

   — *C.f.ochraceus* (Spix, 1824)   E Brazil

   — *C.f.flavescens* (J.F. Gmelin, 1788)   S Brazil, SE Bolivia, E Paraguay

○ ***Celeus flavus*** CREAM-COLOURED WOODPECKER

   — *C.f.flavus* (Statius Müller, 1776)   C Colombia to the Guianas, N Brazil

   — *C.f.peruvianus* (Cory, 1919)   W Brazil, E Peru

   — *C.f.tectricialis* (Hellmayr, 1922)   NE Brazil

   — *C.f.subflavus* Sclater & Salvin, 1877   E Brazil

---

[1] For reasons to consider retaining the monotypic genus *Micropternus*, see Goodwin (1968) (960).

[2] Includes *kanarae* Koelz, 1950 (1280), see Ripley (1982: 228) (2055).

[3] For recognition see Wells (1999: 547) (2579).

[4] The usual abbreviated form of von Seilern-Aspang.

[5] For correction of spelling see David & Gosselin (2002) (613).

[6] Duly proposed as a new race of *lugubris* and apparently overlooked, see Short (1972) (2249); the oversight has no doubt been due to the fact that the name is preoccupied by *Celeus castaneus* (Wagler, 1829) and, if validated, this form will require a new name.

○ *Celeus spectabilis* RUFOUS-HEADED WOODPECKER
  — *C.s.spectabilis* Sclater & Salvin, 1880
  — *C.s.obrieni* Short, 1973
  — *C.s.exsul* Bond & Meyer de Schauensee, 1941

E Ecuador, NE Peru
Piauhy (Brazil) (2250)
SE Peru, N Bolivia

○ *Celeus torquatus* RINGED WOODPECKER[1]
  — *C.t.torquatus* (Boddaert, 1783)
  — *C.t.occidentalis* (Hargitt, 1889)
  — *C.t.tinnunculus* (Wagler, 1829)

E Venezuela, the Guianas, NE Brazil
W Amazonia
E Brazil

DRYOCOPUS Boie, 1826 M
○ *Dryocopus galeatus* (Temminck, 1822) HELMETED WOODPECKER

SE Brazil, Paraguay, NE Argentina

○ *Dryocopus schulzi* (Cabanis, 1883) BLACK-BODIED WOODPECKER

SC Bolivia, W Paraguay, N Argentina

● *Dryocopus lineatus* LINEATED WOODPECKER
  — *D.l.scapularis* (Vigors, 1829)
  √ *D.l.similis* (Lesson, 1847)
  — *D.l.lineatus* (Linnaeus, 1766)
  — *D.l.fuscipennis* P.L. Sclater, 1860
  — *D.l.erythrops* (Valenciennes, 1826)

W Mexico
NE Mexico to NW Costa Rica
E Costa Rica to C Brazil
W Ecuador, NW Peru
SE and S Brazil

● *Dryocopus pileatus* PILEATED WOODPECKER
  √ *D.p.abieticola* (Bangs, 1898)
  √ *D.p.pileatus* (Linnaeus, 1758)

Canada, west to NE USA
C and E USA

○ *Dryocopus javensis* WHITE-BELLIED WOODPECKER
  — *D.j.hodgsonii* (Jerdon, 1840)
  — *D.j.richardsi* Tristram, 1879
  — *D.j.forresti* Rothschild, 1922
  — *D.j.feddeni* (Blyth, 1863)
  — *D.j.javensis* (Horsfield, 1821)
  — *D.j.parvus* (Richmond, 1902)
  — *D.j.hargitti* (Sharpe, 1884)
  — *D.j.esthloterus* Parkes, 1971
  — *D.j.confusus* (Stresemann, 1913)
  — *D.j.pectoralis* (Tweeddale, 1878)[2]
  — *D.j.multilunatus* (McGregor, 1907)
  — *D.j.suluensis* (W.H. Blasius, 1890)
  — *D.j.philippinensis* (Steere, 1890)
  — †?*D.j.cebuensis* Kennedy, 1987
  — *D.j.mindorensis* (Steere, 1890)

W, SW, and C India
C and S Korea
N Burma to S Sichuan and N Vietnam
Burma, N, E and SW Thailand, S Indochina
C and S Malay Pen., Gtr. Sundas, Bangka, Nias I.
Simeuluë I.
Palawan
N Luzon (1763)
C and S Luzon
Bohol, Leyte and Samar
Basilan, Dinagat and Mindanao
Sulu Arch.
Guimaras, Masbate, Negros and Panay
Cebu (1256)
Mindoro

○ *Dryocopus hodgei* (Blyth, 1860) ANDAMAN WOODPECKER[3]

Andaman Is.

○ *Dryocopus martius* BLACK WOODPECKER
  — *D.m.martius* (Linnaeus, 1758)
  — *D.m.khamensis* (Buturlin, 1908)

W Europe and Caucasus area E to Kamchatka and Japan
Fringe of SE and E Tibetan Plateau (W China)

CAMPEPHILUS G.R. Gray, 1840 M
○ *Campephilus pollens* POWERFUL WOODPECKER
  — *C.p.pollens* (Bonaparte, 1845)
  — *C.p.peruvianus* (Cory, 1915)

Andes of Colombia, Ecuador
Andes of N and C Peru

○ *Campephilus haematogaster* CRIMSON-BELLIED WOODPECKER
  — *C.h.splendens* (Hargitt, 1889)
  — *C.h.haematogaster* (Tschudi, 1844)

Panama to NW Ecuador
N Colombia to C Peru

○ *Campephilus rubricollis* RED-NECKED WOODPECKER
  — *C.r.rubricollis* (Boddaert, 1783)
  — *C.r.trachelopyrus* (Malherbe, 1857)
  — *C.r.olallae* (Gyldenstolpe, 1945)

C Colombia to the Guianas, N Brazil
E Peru, N Bolivia, W Brazil
W Bolivia to C and E Brazil

[1] Includes *pieteroyensi* Oren, 1993 (1721), awaiting evaluation.
[2] Includes *samarensis* Parkes, 1960 (1754), see Dickinson *et al.* (1991) (745).
[3] For reasons to separate this from *D. javensis* see Winkler *et al.* (1995) (2676).

○ *Campephilus robustus* (M.H.K. Lichtenstein, 1819) Robust Woodpecker

SE Brazil, Paraguay, NE Argentina

○ *Campephilus melanoleucos* Crimson-crested Woodpecker
— *C.m.malherbii* G.R. Gray, 1845 E Panama to W Venezuela
— *C.m.melanoleucos* (J.F. Gmelin, 1788) C and W Amazonia
— *C.m.cearae* (Cory, 1915) NE Brazil

● *Campephilus guatemalensis* Pale-billed Woodpecker
— *C.g.regius* Reichenbach, 1854 E and C Mexico
— *C.g.nelsoni* (Ridgway, 1911) N and W Mexico
— *C.g.guatemalensis* (Hartlaub, 1844) S Mexico to W Panama

○ *Campephilus gayaquilensis* (Lesson, 1845) Guayaquil Woodpecker W Ecuador, NW Peru

○ *Campephilus leucopogon* (Valenciennes, 1826) Cream-backed Woodpecker[1]

C Bolivia to Uruguay, NE Argentina

○ *Campephilus magellanicus* (P.P. King, 1828) Magellanic Woodpecker Chile, SW Argentina

○ *Campephilus principalis* Ivory-billed Woodpecker
— †*C.p.principalis* (Linnaeus, 1758) SE USA
— †?*C.p.bairdii* Cassin, 1863 Cuba

○ †?*Campephilus imperialis* (Gould, 1832) Imperial Woodpecker NW and W Mexico

PICUS Linnaeus, 1758 M

○ *Picus miniaceus* Banded Woodpecker
— *P.m.perlutus* (Kloss, 1918) S Burma, SW Thailand
— *P.m.malaccensis* Latham, 1790 Malay Pen., Sumatra, Bangka, Belitung, Borneo
— *P.m.niasensis* (Büttikofer, 1897) Nias I.
— *P.m.miniaceus* Pennant, 1769 W and C Java

○ *Picus puniceus* Crimson-winged Woodpecker
— *P.p.observandus* (E. Hartert, 1896) S Burma, Malay Pen., Sumatra, Borneo, Bangka
— *P.p.soligae* Meyer de Schauensee & Ripley, 1940 Nias I.
— *P.p.puniceus* Horsfield, 1821 Java

○ *Picus chlorolophus* Lesser Yellow-naped Woodpecker
— *P.c.simlae* R. Meinertzhagen, 1924 NW Himalayas (Dharmsala to EC Nepal)
— *P.c.chlorolophus* Vieillot, 1818 E Himalayas and S Assam to SW Yunnan and W and N Thailand
— *P.c.laotianus* Delacour, 1926[2] N Laos and NE Thailand (S Yunnan ?)
— *P.c.chlorigaster* Jerdon, 1844 C and S India
— *P.c.wellsi* R. Meinertzhagen, 1924 Sri Lanka
— *P.c.citrinocristatus* (Rickett, 1901) Fujian and N Vietnam
— *P.c.longipennis* (E. Hartert, 1910) Hainan I.
— *P.c.annamensis* R. Meinertzhagen, 1924 E and SE Thailand, S Laos, C and S Vietnam
— *P.c.rodgeri* (Hartert & Butler, 1898) Mts. of W Malaysia
— *P.c.vanheysti* (Robinson & Kloss, 1919) Mts. of Sumatra

○ *Picus mentalis* Chequer-throated Woodpecker
— *P.m.humii* (Hargitt, 1889) S Burma, Malay Pen., Sumatra, Borneo, Bangka
— *P.m.mentalis* Temminck, 1826[3] Java

○ *Picus flavinucha* Greater Yellow-naped Woodpecker
— *P.f.kumaonensis* (Koelz, 1950) Central Himalayas (1280)
— *P.f.flavinucha* (Gould, 1834)[4] E Himalayas and W Bengal to N and NE Burma
— *P.f.ricketti* (Styan, 1898) C Fujian
— *P.f.styani* (Ogilvie-Grant, 1899) S Guangxi, NE Vietnam, Hainan I.
— *P.f.lylei* (Kloss, 1918)[5] W and N Thailand, E Burma, S Yunnan, N Laos, NW Vietnam

[1] Includes *major* Olrog, 1958 (1678), see Short (1982: 446) (2252).
[2] We prefer to retain this race as did Deignan (1963) (663).
[3] Dated 1825 in Peters (1948: 142) (1827) *contra* Sherborn (1898) (2242); see also Dickinson (2001) (739), although erroneously omitted from Appendix III therein.
[4] Includes *marianae* Biswas, 1952 (156), see Ripley (1982: 230) (2055).
[5] We prefer to recognise this race as did Deignan (1963) (663).

— *P.f.pierrei* Oustalet, 1889   S and E Thailand, S Indochina

— *P.f.wrayi* (Sharpe, 1888)   Mts. of W Malaysia

— *P.f.mystacalis* (Salvadori, 1879)   Mts. of N Sumatra

— *P.f.korinchi* (Chasen, 1940)   Mts. of S Sumatra

○ *Picus viridanus* STREAK-BREASTED WOODPECKER[1]

— *P.v.viridanus* Blyth, 1843   Burma (except NW and E), forested SW Thailand[2]

— *P.v.weberi* (A. Müller, 1882)[3]   N and C Malay Pen.

○ *Picus vittatus* LACED WOODPECKER

— *P.v.eisenhoferi* Gyldenstolpe, 1916   E Burma, SW Yunnan, W, N and EC Thailand, Laos, Vietnam

— *P.v.eurous* Deignan, 1955   SE Thailand, Cambodia (651)

— *P.v.connectens* (Robinson & Kloss, 1919)   C and coastal SW Thailand; Langkawi I.[4]

— *P.v.vittatus* Vieillot, 1818[5]   S Malay Pen., Sumatra, Java, Bali and Kangean

○ *Picus xanthopygaeus* (J.E. & G.R. Gray, 1846) STREAK-THROATED WOODPECKER[6]
    C Himalayas to Sri Lanka, Burma, W Yunnan and S Indochina

○ *Picus squamatus* SCALY-BELLIED WOODPECKER

— *P.s.flavirostris* (Menzbier, 1886)   E Iran and S Turkmenistan to W Pakistan

— *P.s.squamatus* Vigors, 1831   W and C Himalayas (N Pakistan to Sikkim)

○ *Picus awokera* [7]JAPANESE WOODPECKER

— *P.a.awokera* Temminck, 1836   Honshu

— *P.a.horii* Taka-Tsukasa, 1918   Kyushu, Shikoku

— *P.a.takatsukasae* N. Kuroda, Sr., 1921[8]   Tanegashima, Yakushima

● *Picus viridis* GREEN WOODPECKER

— *P.v.viridis* Linnaeus, 1758   N and C Europe

— *P.v.karelini* J.F. Brandt, 1841[9]   Italy to Bulgaria, Turkey, N Iran and SW Turkmenistan

— *P.v.sharpei* (Saunders, 1872)   Iberia

— *P.v.innominatus* (Zarudny & Loudon, 1905)   SW and S Iran

○ *Picus vaillantii* (Malherbe, 1847) LEVAILLANT'S WOODPECKER[10]   Morocco to Tunisia

○ *Picus rabieri* (Oustalet, 1898) RED-COLLARED WOODPECKER   Laos and NW and C Vietnam

○ *Picus erythropygius* BLACK-HEADED WOODPECKER

— *P.e.nigrigenis* (Hume, 1874)   SE Burma, SW Thailand

— *P.e.erythropygius* (Elliot, 1865)   SE Thailand, S Indochina

○ *Picus canus* GREY-HEADED WOODPECKER

— *P.c.canus* J.F. Gmelin, 1788   N and C Europe to Turkey and C Siberia

— *P.c.jessoensis* Stejneger, 1886[11]   Tarbagatai and Altai Mts. and N Mongolia to Sakhalin, Hokkaido and NE China

— *P.c.kogo* (Bianchi, 1906)   E Qinghai, Gansu, W Sichuan

— *P.c.guerini* (Malherbe, 1849)[12]   E China (C Sichuan and Shaanxi to Shandong and Zhejiang)

— *P.c.sobrinus* J.L. Peters, 1948   SE China, NE Vietnam

— *P.c.tancolo* (Gould, 1862)   Taiwan, Hainan

— *P.c.sordidior* (Rippon, 1906)   NE Burma and Yunnan to SW Sichuan

— *P.c.sanguiniceps* E.C.S. Baker, 1926   NW Himalayas

[1] The separation of this species and the very similar next one remains a matter of opinion. Deignan (1955) (651) united the two, arguing "that habitat partitioning does not constitute true sympatry" (Rasmussen, 2000) (1989). However, since then the two have usually been treated as (allopatric) species. Most recently Wells (1999: 558) (2579) recognised both, but noted that Short (1982) (2252) admitted no subspecies of *viridanus*. See footnote associated with *weberi*.

[2] A vagrant to S Bangladesh (Rasmussen, 2000) (1989).

[3] This includes some birds Deignan (1963) (663) put in *connectens*.

[4] Note range gap between Thai range and Langkawi; nominate *vittatus* is further south yet, see map in Wells (1999: 558) (2579).

[5] Includes *limitans* see Mees (1996) (1543).

[6] Called *P. myrmecophoneus* Stresemann, 1920, by Deignan (1963) (663) and Ripley (1982) (2055), but see Art. 11.6 (I.C.Z.N., 1999) (1178).

[7] Dated 1826 in Peters (1948: 133) (1827) a *lapsus*.

[8] Recognised by O.S.J. (2000: 170) (1727).

[9] Recognised by Cramp *et al.* (1985: 824) (591); includes *bampurensis*.

[10] Treated as a separate species by Voous (1973) (2547).

[11] Includes *biedermanni* see Stepanyan (1990: 308) (2326).

[12] Includes *brunneatus* Allison, 1946 (30), see Vaurie (1965: 694) (2510).

— *P.c.hessei* Gyldenstolpe, 1916   E Himalayas and Assam Hills to Burma and Indochina
— *P.c.robinsoni* (Ogilvie-Grant, 1906)   Mts. of W Malaysia
— *P.c.dedemi* (van Oort, 1911)   Sumatra

DINOPIUM Rafinesque, 1814 N
◯ ***Dinopium rafflesii*** OLIVE-BACKED WOODPECKER
    — *D.r.rafflesii* (Vigors & Horsfield, 1830)   S Burma, SW Thailand, Malay Pen., Sumatra, Bangka
    — *D.r.dulitense* Delacour, 1946   Borneo

◯ ***Dinopium shorii*** HIMALAYAN FLAME-BACKED WOODPECKER
    — *D.s.shorii* (Vigors, 1832)   Himalayas
    — *D.s.anguste* Ripley, 1952   W and N Burma (NE Assam?) (2034)

◯ ***Dinopium javanense*** COMMON FLAME-BACKED WOODPECKER
    — *D.j.malabaricum* Whistler & Kinnear, 1934   SW India
    — *D.j.intermedium* (Blyth, 1845)   NE India to E Qinghai, SW Gansu, Indochina and Thailand (except Pen.)
    — *D.j.javanense* (Ljungh, 1797)   Malay Pen., Sumatra, Java
    — *D.j.exsul* (E. Hartert, 1901)   Bali
    — *D.j.borneonense* (A.J.C. Dubois, 1897)[1]   Borneo
    — *D.j.everetti* (Tweeddale, 1878)   Balabac, Palawan, Calamian Is.

◯ ***Dinopium benghalense*** BLACK-RUMPED WOODPECKER
    — *D.b.benghalense* (Linnaeus, 1758)   N and C India to Assam
    — *D.b.dilutum* (Blyth, 1849)[2]   Pakistan to W India
    — *D.b.tehminae* (Whistler & Kinnear, 1934)   SW India
    — *D.b.puncticolle* (Malherbe, 1845)   E India
    — *D.b.jaffnense* (Whistler, 1944)[3]   N Sri Lanka
    — *D.b.psarodes* (A.A.H. Lichtenstein, 1793)   C and S Sri Lanka

CHRYSOCOLAPTES Blyth, 1843 M
◯ ***Chrysocolaptes lucidus*** GREATER FLAME-BACKED WOODPECKER[4]
    — *C.l.sultaneus* (Hodgson, 1837)   NW Himalayas
    — *C.l.guttacristatus* (Tickell, 1833)   E Himalayas and SE India to Burma, W and S Yunnan, Thailand (except penin.) and S Indochina
    — *C.l.socialis* Koelz, 1939[5]   SW India
    — *C.l.indomalayicus* Hesse, 1911[6]   Malay Pen., Sumatra, Bangka, W and C Java
    — *C.l.stricklandi* (E.L. Layard, 1854)   Sri Lanka
    — *C.l.strictus* (Horsfield, 1821)   E Java, Kangean and Bali
    — *C.l.kangeanensis* Hoogerwerf, 1963[7]   Kangean I. (1144)
    — *C.l.andrewsi* Amadon, 1943   NE Sabah (Borneo)
    — *C.l.erythrocephalus* Sharpe, 1877   Palawan, Balabac, Busuanga Is.
    — *C.l.haematribon* (Wagler, 1827)[8]   Catanduanes, Luzon, Marinduque
    — *C.l.grandis* Hachisuka, 1930[9]   Polillo Is.
    — *C.l.rufopunctatus* Hargitt, 1889   Samar, Leyte, Bohol
    — *C.l.montanus* Ogilvie-Grant, 1905   Mindanao I. (except Zamboanga Pen.)
    — *C.l.lucidus* (Scopoli, 1786)   Mindanao (Zamboanga Pen.), Basilan I.
    — *C.l.xanthocephalus* Walden & Layard, 1872   Guimaras, Masbate, Panay, Negros and Ticao.

◯ ***Chrysocolaptes festivus*** WHITE-NAPED WOODPECKER
    — *C.f.festivus* (Boddaert, 1783)   C and S India
    — *C.f.tantus* Ripley, 1946   Sri Lanka

[1] Includes *raveni* see Smythies (1981) (2288).
[2] Includes *girensis* Koelz, 1954 (1284), see Ripley (1982: 231) (2055).
[3] For recognition see Ripley (1982: 232) (2055).
[4] See Mees (1986, 1996) (1539, 1543). His proposals included splitting the species in three and treating both *socialis* and *indomalayicus* as synonyms of *guttacristatus*. Although distinction may be difficult we feel it better illustrates the facts to list these forms.
[5] Tentatively recognised. In his review Mees (1986) (1539) did not explore the distinctness of SW Indian birds but gave measurements for *indomalayicus* that are larger than those given by Abdulali (1975) (10).
[6] We accept the view of Mees (1986) (1539) that this name antedates *chersonesus* Kloss, 1918.
[7] Although Mees (1996) seemed to treat *kangeanensis* as a synonym of *strictus* he made no explicit statement on this, and earlier had not seen specimens from Kangean (Mees, 1986) (1539).
[8] Includes *ramosi* Gilliard, 1949 (938), and *montium* Salomonsen, 1952 (2154), see Dickinson *et al.* (1991: 267) (745).
[9] For recognition see Dickinson *et al.* (1991: 267) (745).

GECINULUS Blyth, 1845 M
○ *Gecinulus grantia* BAMBOO WOODPECKER[1]
    — *G.g.grantia* (McClelland, 1840)[2]   E Nepal and Assam to W Burma
    — *G.g.indochinensis* Delacour, 1927   W and S Yunnan and NE Burma to N Vietnam
    — *G.g.viridanus* Slater, 1897   SE China
    — *G.g.viridis* Blyth, 1862   C and E Burma, SW, C, and E Thailand
    — *G.g.poilanei* Deignan, 1950[3]   S Vietnam (644)
    — *G.g.robinsoni* Kloss, 1918   Malay Pen.

SAPHEOPIPO Hargitt, 1890 F
○ *Sapheopipo noguchii* (Seebohm, 1887) OKINAWAN WOODPECKER   Okinawa I.

BLYTHIPICUS Bonaparte, 1854 M
○ *Blythipicus rubiginosus* (Swainson, 1837) MAROON WOODPECKER   Malay Pen., Sumatra, Borneo

○ *Blythipicus pyrrhotis* BAY WOODPECKER
    — *B.p.pyrrhotis* (Hodgson, 1837)[4]   NE Himalayas and Assam to SW Sichuan and south to N and NE Thailand and N Indochina
    — *B.p.sinensis* (Rickett, 1897)   S and SE China
    — *B.p.annamensis* Kinnear, 1926   SW and S Yunnan to S Indochina
    — *B.p.hainanus* (Ogilvie-Grant, 1899)   Hainan I.
    — *B.p.cameroni* Robinson, 1928   Mts. of W Malaysia

REINWARDTIPICUS Bonaparte, 1854 M
○ *Reinwardtipicus validus* ORANGE-BACKED WOODPECKER
    — *R.v.xanthopygius* (Finsch, 1905)   Malay Pen., Sumatra, Bangka, Borneo, N Natuna I.
    — *R.v.validus* (Temminck, 1825)   W and C Java

MEIGLYPTES Swainson, 1837 M
○ *Meiglyptes tristis* BUFF-RUMPED WOODPECKER
    — *M.t.grammithorax* (Malherbe, 1862)   Malay Pen., Sumatra, Bangka, Nias, Borneo, N Natuna I.
    — *M.t.tristis* (Horsfield, 1821)   Java

○ *Meiglyptes jugularis* (Blyth, 1845) BLACK-AND-BUFF WOODPECKER   Burma through Thailand (not penin.) to S Indochina

○ *Meiglyptes tukki* BUFF-NECKED WOODPECKER
    — *M.t.tukki* (Lesson, 1839)   Malay Pen., Sumatra, Belitung, Banyak Is, N Natuna I, Borneo
    — *M.t.pulonis* Chasen & Kloss, 1929   Banggi I. (off N Sabah)
    — *M.t.percnerpes* Oberholser, 1924   S Borneo
    — *M.t.infuscatus* Salvadori, 1887   Nias I.
    — *M.t.batu* Meyer de Schauensee & Ripley, 1940   Batu I.

HEMICIRCUS Swainson, 1837 M
○ *Hemicircus concretus* GREY-AND-BUFF WOODPECKER
    — *H.c.sordidus* (Eyton, 1845)   Malay Pen., Sumatra[5], Bangka, Mentawai Is., Borneo
    — *H.c.concretus* (Temminck, 1821)   W and C Java

○ *Hemicircus canente* (Lesson, 1830) HEART-SPOTTED WOODPECKER   W India east to S Indochina and south to N Malay Pen.

MULLERIPICUS Bonaparte, 1854 M
○ *Mulleripicus fulvus* ASHY WOODPECKER
    — *M.f.fulvus* (Quoy & Gaimard, 1830)   N Sulawesi, Togian Is.
    — *M.f.wallacei* Tweeddale, 1877   C, E, S and SE Sulawesi

○ *Mulleripicus funebris* SOOTY WOODPECKER
    — *M.f.funebris* (Valenciennes, 1826)[6]   Catanduanes, Luzon, Marinduque and Polillo
    — *M.f.fuliginosus* Tweeddale, 1877   Samar, Leyte, Mindanao Is.

[1] We prefer to follow Short (1982) (2252) and await further evidence before splitting this.
[2] Includes *aristus* Koelz, 1954 (1284), see Ripley (1982: 233) (2055).
[3] Accepted by Delacour (1951) (680), but perhaps not reviewed since.
[4] Includes *porphyreus* Koelz, 1954, and *pyrrhopipra* Koelz, 1954 (1284), see Ripley (1982: 241) (2055).
[5] Van Marle & Voous (1988: 142) (2468) recognised *coccometopus* for Sumatran birds.
[6] Includes *mayri* Gilliard, 1949 (938), and *parkesi* Manuel, 1957 (1421), see Dickinson *et al.* (1991: 262) (745).

○ *Mulleripicus pulverulentus* GREAT SLATY WOODPECKER
    — *M.p.mohun* Ripley, 1950   C Himalayas to N Assam (2034)
    — *M.p.harterti* Hesse, 1911[1]   S and E Assam east to C and S Indochina and south to
                                         N Malay Pen. (Trang)

    — *M.p.pulverulentus* (Temminck, 1826*)*   S Malay Pen., Sumatra, Borneo, N Natuna, Palawan,
                                       Balabac Is.

GALBULIDAE JACAMARS[2] (5:18)

GALBALCYRHYNCHUS Des Murs, 1845 M
○ *Galbalcyrhynchus leucotis* Des Murs, 1845 WHITE-EARED JACAMAR   N Upper Amazonia

○ *Galbalcyrhynchus purusianus* Goeldi, 1904 CHESTNUT JACAMAR   E and S Peru, W Brazil, N Bolivia

BRACHYGALBA Bonaparte, 1854 F
○ *Brachygalba salmoni* Sclater & Salvin, 1879 DUSKY-BACKED JACAMAR[3]   E Panama and extreme NW Andes of Colombia

○ *Brachygalba goeringi* Sclater & Salvin, 1869 PALE-HEADED JACAMAR   E Colombia, N Venezuela

○ *Brachygalba lugubris* BROWN JACAMAR
    — *B.l.fulviventris* P.L. Sclater, 1891   E Colombia
    — *B.l.caquetae* Chapman, 1917   SE Colombia to E Peru
    — *B.l.lugubris* (Swainson, 1838)   E and S Venezuela, the Guianas, N Brazil
    — *B.l.obscuriceps* Zimmer & Phelps, 1947   S Venezuela, NW Brazil
    — *B.l.naumbergi* Chapman, 1931   NE Brazil
    — *B.l.phaeonota* Todd, 1943[4]   WC Brazil (N bank Rio Solimoës)
    — *B.l.melanosterna* P.L. Sclater, 1855   E Bolivia, C and SW Brazil

○ *Brachygalba albogularis* (Spix, 1824) WHITE-THROATED JACAMAR   E Peru

JACAMARALCYON Lesson, 1830 F
○ *Jacamaralcyon tridactyla* (Vieillot, 1817) THREE-TOED JACAMAR   SE Brazil

GALBULA Brisson, 1760 F
○ *Galbula albirostris* YELLOW-BILLED JACAMAR
    — *G.a.chalcocephala* Deville, 1849   S Colombia, Ecuador, NW Peru, W Brazil
    — *G.a.albirostris* Latham, 1790   E Venezuela, the Guianas, N Brazil

○ *Galbula cyanicollis* Cassin, 1851 BLUE-CHEEKED JACAMAR   C Brazil

● *Galbula ruficauda* RUFOUS-TAILED JACAMAR
    ↙ *G.r.melanogenia* P.L. Sclater, 1852   S Mexico to W Ecuador
    — *G.r.ruficauda* Cuvier, 1816[5]   C Colombia, Venezuela, the Guianas, N Brazil
    — *G.r.pallens* Bangs, 1898   N Colombia
    — *G.r.brevirostris* Cory, 1913   NW Venezuela
    — *G.r.rufoviridis* Cabanis, 1851   S Brazil, N Bolivia, Paraguay, N Argentina
    — *G.r.heterogyna* Todd, 1932   E Bolivia, SW Brazil

○ *Galbula galbula* (Linnaeus, 1766) GREEN-TAILED JACAMAR   E and S Venezuela, the Guianas, N and C Brazil

○ *Galbula pastazae* Taczanowski & Berlepsch, 1885 COPPERY-CHESTED JACAMAR
                                         E Ecuador

○ *Galbula tombacea* WHITE-CHINNED JACAMAR
    — *G.t.tombacea* Spix, 1824   S Colombia, Ecuador, Peru, E Brazil
    — *G.t.mentalis* Todd, 1943   C and WC Brazil

○ *Galbula cyanescens* Deville, 1849 BLUISH-FRONTED JACAMAR   W Brazil, E Peru, N Bolivia

○ *Galbula chalcothorax* P.L. Sclater, 1855 PURPLISH JACAMAR[6]   W Brazil, E Ecuador, E Peru

---

[1] The name *celadinus* Deignan, 1955 (655), applies to intergrades with the nominate form (Wells, 1999).
[2] For reasons to consider a level above that of a family for this group see Harshman (1994) (1078).
[3] Includes *carmenensis* Haffer, 1962 (1052) see Haffer (1974) (1053).
[4] Treated at specific level by Peters (1948: 4) (1827), but see Haffer (1974: 322) (1053).
[5] Dated 1817 by Peters (1948: 7) (1827) but see Browning & Monroe (1991) (258) and www.zoonomen.net (A.P. Peterson *in litt.*).
[6] For the separation of this species from *G. leucogastra* see Parker & Remsen (1987: 98) (1748).

○ *Galbula leucogastra* BRONZY JACAMAR
   — *G.l.leucogastra* Vieillot, 1817   W Venezuela, the Guianas, W Brazil
   — *G.l.viridissima* Griscom & Greenway, 1937   C Brazil

○ *Galbula dea* PARADISE JACAMAR
   — *G.d.dea* (Linnaeus, 1758)   Venezuela, the Guianas, N Brazil
   — *G.d.amazonum* (P.L. Sclater, 1855)   N Bolivia, SW Brazil
   — *G.d.brunneiceps* (Todd, 1943)   E Colombia, E Peru, W Brazil
   — *G.d.phainopepla* (Todd, 1943)   W C Brazil

JACAMEROPS Lesson, 1830 M[1]
○ *Jacamerops aureus* GREAT JACAMAR
   — *J.a.penardi* Bangs & Barbour, 1922   Costa Rica to W Colombia
   — *J.a.aureus* (Statius Müller, 1776)   E Colombia, the Guianas, Venezuela
   — *J.a.ridgwayi* Todd, 1943   NE and C Brazil
   — *J.a.isidori* Deville, 1849   E Peru, N Bolivia, W Brazil

BUCCONIDAE PUFFBIRDS (10: 33)

NOTHARCHUS Cabanis & Heine, 1863 M
○ *Notharchus macrorhynchos* WHITE-NECKED PUFFBIRD
   — *N.m.cryptoleucus* van Rossem, 1934   El Salvador, NW Nicaragua
   — *N.m.hyperrynchus* (P.L. Sclater, 1856)   S Mexico to NW Sth. America
   — *N.m.macrorhynchos* (J.F. Gmelin, 1788)   The Guianas, N Brazil
   — *N.m.paraensis* Sassi, 1932   E Brazil
   — *N.m.swainsoni* (G.R. Gray, 1846)   SE Brazil, E Paraguay, NE Argentina

○ *Notharchus pectoralis* (G.R. Gray, 1846) BLACK-BREASTED PUFFBIRD   E Panama to NW Ecuador

○ *Notharchus ordii* (Cassin, 1851) BROWN-BANDED PUFFBIRD   S Venezuela, NW Brazil

○ *Notharchus tectus* PIED PUFFBIRD
   — *N.t.subtectus* (P.L. Sclater, 1860)   E Panama to C Colombia and SW Ecuador
   — *N.t.picatus* (P.L. Sclater, 1856)   E Ecuador, E Peru
   — *N.t.tectus* (Boddaert, 1783)   S Venezuela, the Guianas, N Brazil

BUCCO Brisson, 1760 M
○ *Bucco macrodactylus* (Spix, 1824) CHESTNUT-CAPPED PUFFBIRD[2]   N and W Amazonia

○ *Bucco tamatia* SPOTTED PUFFBIRD
   — *B.t.pulmentum* P.L. Sclater, 1856   S Colombia, E Ecuador, E Peru, W Brazil
   — *B.t.tamatia* J.F. Gmelin, 1788[3]   E Colombia, Venezuela, the Guianas, N Brazil
   — *B.t.inexpectatus* (Todd, 1943)   NC Brazil
   — *B.t.punctuliger* (Todd, 1943)   C Brazil, E Bolivia
   — *B.t.hypneleus* (Cabanis & Heine, 1863)   EC Brazil
   — *B.t.interior* (Cherrie & Reichenberger, 1921)   SW Brazil

○ *Bucco noanamae* Hellmayr, 1909 SOOTY-CAPPED PUFFBIRD   W Colombia

○ *Bucco capensis* COLLARED PUFFBIRD
   — *B.c.dugandi* Gilliard, 1949   SE Colombia, Ecuador, C Peru (939)
   — *B.c.capensis* Linnaeus, 1766   The Guianas, Brazil, E Peru

NYSTALUS Cabanis & Heine, 1863 M
○ *Nystalus radiatus* (P.L. Sclater, 1854) BARRED PUFFBIRD   W Panama to W Ecuador

○ *Nystalus chacuru* WHITE-EARED PUFFBIRD
   — *N.c.uncirostris* (Stolzmann, 1926)   E Peru, E Bolivia
   — *N.c.chacuru* (Vieillot, 1816)   S Brazil, Paraguay, NE Argentina

[1] Gender addressed by David & Gosselin (2002) (614).
[2] Includes *caurensis* see Peters (1948: 12) (1827); recognised by Meyer de Schauensee & Phelps (1978: 165) (1577) and in our last Edition, but needs re-evaluation.
[3] Includes *cuyunii* Phelps & Phelps, Jr., 1949 (1846) see Meyer de Schauensee & Phelps (1978: 165) (1577).

○ *Nystalus striolatus* Sᴛʀɪᴏʟᴀᴛᴇᴅ Pᴜꜰꜰʙɪʀᴅ
— *N.s.striolatus* (Pelzeln, 1856)   W Amazonia
— *N.s.torridus* Bond & Meyer de Schauensee, 1940   S Brazil

○ *Nystalus maculatus* Sᴘᴏᴛ-ʙᴀᴄᴋᴇᴅ Pᴜꜰꜰʙɪʀᴅ
— *N.m.maculatus* (J.F. Gmelin, 1788)   E Brazil
— *N.m.parvirostris* (Hellmayr, 1908)   C Brazil
— *N.m.pallidigula* Cherrie & Reichenberger, 1923   SW Brazil
— *N.m.striatipectus* (P.L. Sclater, 1854)   E and S Bolivia, N Argentina

HYPNELUS Cabanis & Heine, 1863 M
○ *Hypnelus ruficollis* Rᴜssᴇᴛ-ᴛʜʀᴏᴀᴛᴇᴅ Pᴜꜰꜰʙɪʀᴅ
— *H.r.ruficollis* (Wagler, 1829)   N Colombia, W Venezuela
— *H.r.decolor* Todd & Carriker, 1922   NE Colombia, NW Venezuela
— *H.r.striaticollis* Phelps & Phelps, Jr., 1958   NW Venezuela (1859)
— *H.r.coloratus* Ridgway, 1914   W Venezuela
— *H.r.bicinctus* (Gould, 1837)   N Venezuela
— *H.r.stoicus* Wetmore, 1939   Margarita I.

MALACOPTILA G.R. Gray, 1841 F
○ *Malacoptila striata* Cʀᴇsᴄᴇɴᴛ-ᴄʜᴇsᴛᴇᴅ Pᴜꜰꜰʙɪʀᴅ
— *M.s.minor* Sassi, 1911   E Brazil
— *M.s.striata* (Spix, 1824)   SE Brazil

○ *Malacoptila fusca* Wʜɪᴛᴇ-ᴄʜᴇsᴛᴇᴅ Pᴜꜰꜰʙɪʀᴅ
— *M.f.fusca* (J.F. Gmelin, 1788)   N and NW Amazonia
— *M.f.venezuelae* Phelps & Phelps, Jr., 1947   S Venezuela (1843)

○ *Malacoptila semicincta* Todd, 1925 Sᴇᴍɪᴄᴏʟʟᴀʀᴇᴅ Pᴜꜰꜰʙɪʀᴅ   S Peru, N Bolivia, W Brazil

○ *Malacoptila fulvogularis* Bʟᴀᴄᴋ-sᴛʀᴇᴀᴋᴇᴅ Pᴜꜰꜰʙɪʀᴅ
— *M.f.substriata* P.L. Sclater, 1854   Colombia
— *M.f.huilae* Meyer de Schauensee, 1946   N Colombia
— *M.f.fulvogularis* P.L. Sclater, 1854   E Ecuador to Bolivia

○ *Malacoptila rufa* Rᴜꜰᴏᴜs-ɴᴇᴄᴋᴇᴅ Pᴜꜰꜰʙɪʀᴅ
— *M.r.rufa* (Spix, 1824)   E Ecuador, E Peru, W Brazil, E Bolivia
— *M.r.brunnescens* J.T. Zimmer, 1931   C Brazil

○ *Malacoptila panamensis* Wʜɪᴛᴇ-ᴡʜɪsᴋᴇʀᴇᴅ Pᴜꜰꜰʙɪʀᴅ
— *M.p.inornata* (Du Bus, 1847)[1]   S Mexico to N Nicaragua
— *M.p.fuliginosa* Richmond, 1893   SE Nicaragua to W Panama
— *M.p.panamensis* Lafresnaye, 1847   SW Costa Rica to NW Colombia
— *M.p.chocoana* Meyer de Schauensee, 1950   Pacific coast Choco (W Colombia) (1566)
— *M.p.poliopis* P.L. Sclater, 1862   SW Colombia, W Ecuador
— *M.p.magdalenae* Todd, 1943   N Colombia

○ *Malacoptila mystacalis* (Lafresnaye, 1850) Mᴏᴜsᴛᴀᴄʜᴇᴅ Pᴜꜰꜰʙɪʀᴅ[2]   Colombia, NW Venezuela

MICROMONACHA P.L. Sclater, 1881 F
○ *Micromonacha lanceolata* Lᴀɴᴄᴇᴏʟᴀᴛᴇᴅ Mᴏɴᴋʟᴇᴛ
— *M.l.austinsmithi* Dwight & Griscom, 1924   E Costa Rica, W Panama
— *M.l.lanceolata* (Deville, 1849)   W Colombia, E Ecuador, E Peru, W Brazil

NONNULA P.L. Sclater, 1854 F
○ *Nonnula rubecula* Rᴜsᴛʏ-ʙʀᴇᴀsᴛᴇᴅ Nᴜɴʟᴇᴛ[3]
— *N.r.tapanahoniensis* Mees, 1968   Surinam, extreme NE Brazil (1519)
— *N.r.duidae* Chapman, 1914   Cerro Duida and C Yapacana (S Venezuela)
— *N.r.interfluvialis* Parkes, 1970   Watershed between the Orinoco and the R Negro (S Venezuela, N Brazil) (1761)

---

[1] Du Bus is an abbreviation for Du Bus de Gisignies.
[2] Includes *pacifica* Meyer de Schauensee, 1949 (1565), see Hilty & Brown (1986: 317) (1116).
[3] Arranged according to Parkes (1970) (1761).

— *N.r.simulatrix* Parkes, 1970	Watershed between the R. Negro and R. Solimões (Brazil N of the Amazon) (1761)
— *N.r.cineracea* P.L. Sclater, 1881	NE Peru, W Brazil
— *N.r.simplex* Todd, 1937	NE Brazil (south bank of the Amazon)
— *N.r.rubecula* (Spix, 1824)	SE Brazil, Paraguay, NE Argentina

○ *Nonnula sclateri* Hellmayr, 1907 FULVOUS-CHINNED NUNLET   N Bolivia, W Brazil

○ *Nonnula brunnea* P.L. Sclater, 1881 BROWN NUNLET   S Colombia, E Ecuador, E Peru

○ *Nonnula frontalis* GREY-CHEEKED NUNLET[1]

— *N.f.frontalis* (P.L. Sclater, 1854)	Nthn. E Panama, NW Colombia
— *N.f.stulta* Wetmore, 1953	Sthn. E Panama, W Colombia (2596)
— *N.f.pallescens* Todd, 1919	NE Colombia

○ *Nonnula ruficapilla* RUFOUS-CAPPED NUNLET

— *N.r.rufipectus* Chapman, 1928	NE Peru
— *N.r.ruficapilla* (Tschudi, 1844)	E Peru, W Brazil
— *N.r.inundata* Novaes, 1991	Rio Tocantins near Jacundá, Pará (E Brazil) (1663)
— *N.r.nattereri* Hellmayr, 1921	SW Brazil

○ *Nonnula amaurocephala* Chapman, 1921 CHESTNUT-HEADED NUNLET   W Brazil

HAPALOPTILA P.L. Sclater, 1881 F

○ *Hapaloptila castanea* (J. Verreaux, 1866) WHITE-FACED NUNBIRD   W Colombia, W Ecuador, N Peru

MONASA Vieillot, 1816 F

○ *Monasa atra* (Boddaert, 1783) BLACK NUNBIRD   S Venezuela, the Guianas, N Brazil

○ *Monasa nigrifrons* BLACK-FRONTED NUNBIRD

— *M.n.nigrifrons* (Spix, 1824)	W and NW Amazonia
— *M.n.canescens* Todd, 1937	E Bolivia

○ *Monasa morphoeus* WHITE-FRONTED NUNBIRD

— *M.m.grandior* Sclater & Salvin, 1868	Nicaragua to NW Panama
— *M.m.fidelis* Nelson, 1912	E Panama
— *M.m.pallescens* Cassin, 1860	E Panama, NW Colombia
— *M.m.sclateri* Ridgway, 1912	N Colombia
— *M.m.peruana* P.L. Sclater, 1856	SE Colombia, E Peru, NW Brazil
— *M.m.morphoeus* (Hahn & Küster, 1823)	S and E Brazil
— *M.m.boliviana* Carriker, 1935	NE Bolivia

○ *Monasa flavirostris* Strickland, 1850 YELLOW-BILLED NUNBIRD   E Colombia, E Ecuador, E Peru, W Brazil, NE Bolivia

CHELIDOPTERA Gould, 1837 F

○ *Chelidoptera tenebrosa* SWALLOW-WING PUFFBIRD

— *C.t.tenebrosa* (Pallas, 1782)[2]	E Colombia, E Ecuador, E Peru and N Bolivia east to the Guianas and south to C Brazil
— *C.t.brasiliensis* P.L. Sclater, 1862	E and SE Brazil

---

[1] For treatment as a species see Hilty & Brown (1986: 318) (1116).
[2] Probably includes *pallida*, recognised in last Edition *contra* Peters (1948: 23) (1827)

ACANTHISITTIDAE NEW ZEALAND WRENS[1] (2: 4)

ACANTHISITTA Lafresnaye, 1842 F
● *Acanthisitta chloris* Rifleman
   — *A.c.granti* Mathews & Iredale, 1913   North Island (New Zealand)
   — *A.c.chloris* (Sparrman, 1787)[2]   South Island (New Zealand)

XENICUS G.R. Gray, 1855 M
○ *Xenicus longipes* Bush Wren
   — †?*X.l.stokesii* G.R. Gray, 1862   North Island (New Zealand)
   — †?*X.l.longipes* (J.F. Gmelin, 1789)   South Island (New Zealand)
   — †?*X.l.variabilis* Stead, 1936   Islands off Stewart I. (New Zealand)

○ *Xenicus gilviventris* Pelzeln, 1867 Alpine Rock Wren   Alpine South Island (New Zealand)

○ †*Xenicus lyalli* (Rothschild, 1894) Stephens Island Wren   Stephens I. (New Zealand)

EURYLAIMIDAE BROADBILLS[3] (9: 14)

SMITHORNITHINAE

SMITHORNIS Bonaparte, 1850 M
○ *Smithornis capensis* African Broadbill[4]
   — *S.c.delacouri* Bannerman, 1923   Sierra Leone, Liberia, Ghana, Ivory Coast
   — *S.c.camerunensis* Sharpe, 1905   S Cameroon, Gabon, Central African Republic
   — *S.c.albigularis* E. Hartert, 1904   N Malawi, N Zambia, Zaire, W Tanzania
   — *S.c.meinertzhageni* van Someren, 1919   NE Zaire, Rwanda, Uganda, W Kenya
   — *S.c.medianus* Hartert & van Someren, 1916   C Kenya, NE Tanzania
   — *S.c.suahelicus* Grote, 1926   Coastal Kenya to E Tanzania
   — *S.c.cryptoleucus* Clancey, 1963   E Zimbabwe, NE Transvaal, and Mozambique to S
                                    Malawi, SE Tanzania (394)
   — *S.c.conjunctus* Clancey, 1963   S Angola, N Botswana, S and SE Zambia, N Zimbabwe,
                                    NW Mozambique (394)
   — *S.c.capensis* (A. Smith, 1840)   Natal, S Zululand

○ *Smithornis sharpei* Grey-headed Broadbill
   — *S.s.sharpei* Alexander, 1903   Bioko
   — *S.s.zenkeri* Reichenow, 1903   S Cameroon, N Gabon
   — *S.s.eurylaemus* Neumann, 1923   E Zaire

○ *Smithornis rufolateralis* Rufous-sided Broadbill
   — *S.r.rufolateralis* G.R. Gray, 1864   Liberia to W Zaire
   — *S.r.budongoensis* van Someren, 1921   C and NE Zaire, W Uganda

CALYPTOMENINAE

CALYPTOMENA Raffles, 1822 F
○ *Calyptomena viridis* Green Broadbill
   — *C.v.caudacuta* Swainson, 1838[5]   Malay Pen. (south to and incl. Singapore)
   — *C.v.viridis* Raffles, 1822   Borneo, N Natuna Is., Sumatra and Nias I.
   — *C.v.siberu* Chasen & Kloss, 1926   Siberut, Pagi Is.

○ *Calyptomena hosii* Sharpe, 1892 Hose's Broadbill   Mts. of N Borneo

○ *Calyptomena whiteheadi* Sharpe, 1888 Whitehead's Broadbill   Mt. Kinabalu (N Borneo)

---

[1] For the use of Acanthisittidae see Bock (1994) (173). For a molecular review see Lovette *et al.* (2000) (1389).
[2] Includes *citrina* see Higgins *et al.* (2001: 75) (1114).
[3] See Prum (1993) (1952), Lambert & Woodcock (1996) (1319). As regards Asian species see Dekker & Dickinson (2000) (670) and Dekker *et al.*
(2000) (671).
[4] For a review recognising *meinertzhageni* and *suahelicus* see Clancey (1970) (421).
[5] For reasons to use this name see Dekker & Dickinson (2000) (670).

EURYLAIMINAE

CYMBIRHYNCHUS Vigors & Horsfield, 1830 M
○ ***Cymbirhynchus macrorhynchos*** BLACK-AND-RED BROADBILL
   — *C.m.affinis* Blyth, 1846   W and C Burma
   — *C.m.malaccensis* Salvadori, 1874[1]   SE Burma, SW and SE Thailand, S Laos, S Vietnam, Cambodia, Malay Pen.
   — *C.m.macrorhynchos* (J.F. Gmelin, 1788)[2]   Sumatra, Bangka, Belitung, Borneo

PSARISOMUS Swainson, 1837 M
○ ***Psarisomus dalhousiae*** LONG-TAILED BROADBILL
   — *P.d.dalhousiae* (Jameson, 1835)   Himalayas, Burma, N and W Thailand, Laos, N Vietnam
   — *P.d.divinus* Deignan, 1947   SC Vietnam (Langbian Plateau)
   — *P.d.cyanicauda* Riley, 1935   SE Thailand, Cambodia
   — *P.d.psittacinus* (S. Müller, 1835)   W Malaysia, Sumatra
   — *P.d.borneensis* E. Hartert, 1905   Mts. of NW Borneo

SERILOPHUS Swainson, 1837 M
○ ***Serilophus lunatus*** SILVER-BREASTED BROADBILL
   — *S.l.rubropygius* (Hodgson, 1839)   Nepal, NW Burma (west of the Irrawaddy)
   — *S.l.elisabethae* La Touche, 1921[3]   SW Yunnan, NE Burma (E of the Irrawaddy), E Burma, N Laos, E Thailand and NW and C Vietnam
   — *S.l.polionotus* Rothschild, 1903   Hainan I.
   — *S.l.lunatus* (Gould, 1834)[4]   E Burma and NW Thailand
   — *S.l.impavidus* Deignan, 1948   S Laos (Bolovens Plateau)
   — *S.l.stolidus* Robinson & Kloss, 1919   Malay Pen. (excl. south)
   — *S.l.rothschildi* Hartert & Butler, 1898 S Malay Pen.
   — *S.l.intensus* Robinson & Kloss, 1916[5] Sumatra

EURYLAIMUS Horsfield, 1821 M
○ ***Eurylaimus javanicus*** BANDED BROADBILL
   — *E.j.pallidus* Chasen, 1935   SE Burma, Thailand, Laos, S Vietnam and Malay Pen.
   — *E.j.harterti* van Oort, 1909[6]   Sumatra, Bangka and Belitung
   — *E.j.javanicus* Horsfield, 1821   Java
   — *E.j.brookei* Robinson & Kloss, 1919   Borneo, N Natuna Is.

○ ***Eurylaimus ochromalus*** Raffles, 1822 BLACK-AND-YELLOW BROADBILL[7]   SE Burma, Malay Pen., Sumatra, Batu and Banjak Is., Borneo and N Natuna Is.

SARCOPHANOPS Sharpe, 1879[8] M
○ ***Sarcophanops steerii*** WATTLED BROADBILL
   — *S.s.steerii* (Sharpe, 1876)[9]   Mindanao, Basilan, Dinagat, Siargao
   — *S.s.samarensis* Steere, 1890   Samar, Leyte

CORYDON Lesson, 1828 M
○ ***Corydon sumatranus*** DUSKY BROADBILL
   — *C.s.laoensis* Meyer de Schauensee, 1929[10] S Burma, Thailand, S Laos, Vietnam and Cambodia
   — *C.s.sumatranus* (Raffles, 1822)[11]   Malay Pen., Sumatra
   — *C.s.brunnescens* E. Hartert, 1916   Sarawak and N Natuna Is.
   — *C.s.orientalis* Mayr, 1938   Borneo (excl. Sarawak)

---

[1] Includes *siamensis* see Deignan (1963: 95) (663).
[2] Includes *lemniscatus* and *tenebrosus* see Mees (1986) (1539) and Dekker *et al.* (2000) (671).
[3] Includes *atrestus* see Dekker & Dickinson (2000) (670).
[4] Includes *intrepidus* see Peters (1951: 10) (1828).
[5] Includes *moderatus* see Deignan (1948) (643).
[6] Includes *billitonis* see Mees (1986) (1539).
[7] For monotypic treatment see Mees (1986) (1539) and Dekker *et al.* (2000) (671).
[8] For reasons to use this generic name see Lambert & Woodcock (1996: 227) (1319).
[9] Includes *mayri* Salomonsen, 1953 (2155) see Dekker *et al.* (2000) (671).
[10] Includes *morator* and *ardescens* see Deignan (1963: 94) (663).
[11] Includes *pallescens* see Deignan (1963: 95) (663) for implicit submergence.

PSEUDOCALYPTOMENINAE

PSEUDOCALYPTOMENA Rothschild, 1909 F
○ *Pseudocalyptomena graueri* Rothschild, 1909 GRAUER'S BROADBILL   E Zaire, SW Uganda

PHILEPITTIDAE ASITIES (2: 4)

PHILEPITTA I. Geoffroy Saint-Hilaire, 1838 F
○ *Philepitta castanea* (Statius Müller, 1776) VELVET ASITY   N and E Madagascar

○ *Philepitta schlegeli* Schlegel, 1867 SCHLEGEL'S ASITY   W Madagascar

NEODREPANIS Sharpe, 1875 F
○ *Neodrepanis coruscans* Sharpe, 1875 COMMON SUNBIRD-ASITY   Rain forests of E Madagascar

○ *Neodrepanis hypoxantha* Salomonsen, 1933 YELLOW-BELLIED SUNBIRD-ASITY
                                                                    Montane forest of E Madagascar

SAPAYOAIDAE SAPAYOA[1] (1: 1)

SAPAYOA E. Hartert, 1903[2] F
○ *Sapayoa aenigma* E. Hartert, 1903 SAPAYOA   E Panama to NW Ecuador

PITTIDAE PITTAS[3] (1: 30)

PITTA Vieillot, 1816 F
○ *Pitta phayrei* (Blyth, 1863) EARED PITTA   Burma, Thailand, Laos, Vietnam, Cambodia; vagrant
                                                                    (?) Bangladesh

○ *Pitta nipalensis* BLUE-NAPED PITTA
— *P.n.nipalensis* (Hodgson, 1837)   E Himalayas to Bangladesh and NE Burma
— *P.n.hendeei* Bangs & Van Tyne, 1931   E Burma, N Thailand, Laos, N Vietnam

○ *Pitta soror* BLUE-RUMPED PITTA
— *P.s.tonkinensis* Delacour, 1927   C China to N Vietnam
— *P.s.douglasi* Ogilvie-Grant, 1910   Hainan
— *P.s.petersi* Delacour, 1934   NC Vietnam
— *P.s.soror* Wardlaw Ramsay, 1881   S and C Vietnam, S Laos
— *P.s.flynnstonei* Rozendaal, 1993   S Cambodia, SE Thailand (2122)

○ *Pitta oatesi* RUSTY-NAPED PITTA
— *P.o.oatesi* (Hume, 1873)   E Burma, W Thailand, NW Laos
— *P.o.castaneiceps* Delacour & Jabouille, 1930   SE China, N Laos, N Vietnam
— *P.o.bolovenensis* Delacour, 1932   S Laos
— *P.o.deborah* B.F. King, 1978   Mts. of W Malaysia

○ *Pitta schneideri* E. Hartert, 1909 SCHNEIDER'S PITTA   Batak and Kerinji Mts. (Sumatra)

○ *Pitta caerulea* GIANT PITTA
— *P.c.caerulea* (Raffles, 1822)   Malay Pen., Sumatra
— *P.c.hosei* E.C.S. Baker, 1918   N and C Borneo

○ *Pitta cyanea* BLUE PITTA
— *P.c.cyanea* Blyth, 1843   Himalayas east to Yunnan, Burma, N Laos, N, W and
                                                                    SW Thailand
— *P.c.aurantiaca* Delacour & Jabouille, 1928   SE Thailand, Cambodia
— *P.c.willoughbyi* Delacour, 1926   C Laos to S Vietnam

○ *Pitta elliotii* Oustalet, 1874 ELLIOT'S/BAR-BELLIED PITTA   Vietnam, Laos and Cambodia.

---

[1] Family status is derived from Lanyon (1985) (1321) and Sibley & Monroe (1990) (2262).
[2] It has been suggested that this genus be moved to the Tyrannidae (Prum & Lanyon, 1989; Prum, 1990) (1950, 1956), however A.O.U. (1998: 416) (3)
    preferred to treat the genus as Incertae Sedis, given the genetic data of Lanyon (1985) (1321).
[3] See Lambert & Woodcock (1996) (1319), Dickinson & Dekker (2000) (741) and Dickinson *et al.* (2000) (747).

○ *Pitta guajana* BANDED PITTA

   — *P.g.ripleyi* Deignan, 1946   NC Malay Pen. (Thailand)

   — *P.g.irena* Temminck, 1836   S Malay Pen., Sumatra

   — *P.g.guajana* (Statius Müller, 1776)[1]   Java, Bali

   — *P.g.schwaneri* Bonaparte, 1850   Borneo

○ *Pitta gurneyi* Hume, 1875 GURNEY'S PITTA   N Malay Pen. (Thailand)

○ *Pitta kochi* Brüggemann, 1876 WHISKERED PITTA   Luzon (montane)

○ *Pitta erythrogaster* RED-BELLIED PITTA

   — *P.e.erythrogaster* Temminck, 1823   Philippines (except Palawan group)

   — *P.e.thompsoni* Ripley & Rabor, 1962   Culion I.

   — *P.e.propinqua* (Sharpe, 1877)   Balabac, Palawan

   — *P.e.inspeculata* Meyer & Wiglesworth, 1894   Talaut Is. (Indonesia)

   — *P.e.caeruleitorques* Salvadori, 1876   Gtr. Sangihe

   — *P.e.palliceps* Brüggemann, 1876   Siao and Tagulendang (Sangihe Is.)

   — *P.e.celebensis* Müller & Schlegel, 1845   Sulawesi

   — *P.e.dohertyi* Rothschild, 1898[2]   Sula Is.

   — *P.e.rufiventris* (Heine, 1859)   Bacan, Halmahera, Morotai (N Moluccas)

   — *P.e.cyanonota* G.R. Gray, 1860   Ternate (N Moluccas)

   — *P.e.obiensis* Hachisuka, 1935   Obi (C Moluccas)

   — *P.e.bernsteini* Junge, 1958   Gebe I. (W Papuan Is.)

   — *P.e.rubrinucha* Wallace, 1862   Buru (S Moluccas)

   — *P.e.piroensis* Muir & Kershaw, 1910   Seram (S Moluccas)

   — *P.e.kuehni* Rothschild, 1899   Kai Is.

   — *P.e.aruensis* Rothschild & Hartert, 1901   Aru Is.

   — *P.e.digglesi* Krefft, 1869[3]   Cape York Pen.

   — *P.e.macklotii* Temminck, 1834[4]   W and S New Guinea, N Queensland

   — *P.e.loriae* Salvadori, 1890   N of far E New Guinea

   — *P.e.habenichti* Finsch, 1912   N New Guinea

   — *P.e.finschii* E.P. Ramsay, 1884   D'Entrecasteaux Arch.

   — *P.e.meeki* Rothschild, 1898   Rossel I.

   — *P.e.gazellae* Neumann, 1908   New Britain (Bismarck Arch.)

   — *P.e.novaehibernicae* E.P. Ramsay, 1878   New Ireland

   — *P.e.extima* Mayr, 1955   New Hanover

   — *P.e.splendida* Mayr, 1955   Tabar I.

○ *Pitta arquata* Gould, 1871 BLUE-BANDED PITTA[5]   Lower montane Borneo

○ *Pitta granatina* GARNET PITTA

   — *P.g.coccinea* Eyton, 1839   Malay Pen., lowland Sumatra

   — *P.g.ussheri* Gould, 1877[6]   N Borneo

   — *P.g.granatina* Temminck, 1830   S Borneo

○ *Pitta venusta* S. Müller, 1835 GRACEFUL PITTA   Midmontane Sumatra

○ *Pitta baudii* Müller & Schlegel, 1839[7] BLUE-HEADED PITTA   Lowland Borneo

○ *Pitta sordida* HOODED PITTA

   — *P.s.cucullata* Hartlaub, 1843   Himalayas to Thailand, Laos, N Vietnam >> Malay
                                              Pen., Sumatra, Java and Bangka

   — *P.s.abbotti* Richmond, 1902   Nicobar Is.

   — *P.s.mulleri* Bonaparte, 1850   Malay Pen., Sumatra, Java, Borneo, Sibutu I.

   — *P.s.bangkana* Schlegel, 1863   Bangka, Belitung

   — *P.s.sordida* (Statius Müller, 1776)   Philippines (excl. Balabac and Palawan)

---

[1] Includes *affinis* see Mees (1996) (1543) and *bangkae* see Mees (1986) (1539).

[2] This is a well-marked form and the whole species needs revision but we prefer not to single this race out to treat as a species as it is entirely surrounded by associated forms.

[3] For recognition of this see Schodde & Mason (1999) (2189).

[4] Includes *oblita* (R. Schodde, *in litt.*).

[5] For a discussion on the spelling see Dickinson & Dekker (2000) (741); we believe that since 1981 the original name has regained prevailing usage and we maintain it, see Art. 33.2.3.1. of the Code (I.C.Z.N., 1999) (1178). See also p. 000.

[6] Treated as a race of *venusta* in last Edition, but see Rozendaal (1994) (2123).

[7] For an explanation of this date correction see Dickinson *et al.* (2000: 108) (747).

— *P.s.palawanensis* Parkes, 1960   Balabac, Palawan

— *P.s.sanghirana* Schlegel, 1866   Sangihe Is.

— *P.s.forsteni* Bonaparte, 1850   N Sulawesi

— *P.s.goodfellowi* C.M.N. White, 1937   Aru Is.

— *P.s.mefoorana* Schlegel, 1874   Numfor I. (Geelvink Bay)

— *P.s.rosenbergii* Schlegel, 1871   Biak I. (Geelvink Bay)

— *P.s.novaeguineae* Müller & Schlegel, 1845[1]   W Papuan Is., New Guinea, Karkar I.

○ *Pitta maxima* IVORY-BREASTED PITTA

— *P.m.maxima* Müller & Schlegel, 1845   Bacan, Halmahera (N Moluccas)

— *P.m.morotaiensis* van Bemmel, 1939   Morotai I. (N Moluccas)

○ *Pitta steerei* AZURE-BREASTED PITTA

— *P.s.coelestis* Parkes, 1971   Bohol, Leyte, Samar (EC Philippines)

— *P.s.steerei* (Sharpe, 1876)   Mindanao (S Philippines)

○ *Pitta superba* Rothschild & Hartert, 1914 SUPERB PITTA   Admiralty Is.

○ *Pitta angolensis* AFRICAN PITTA

— *P.a.pulih* Fraser, 1843   Sierra Leone to S Cameroon

— *P.a.angolensis* Vieillot, 1816   Cameroon to N Angola

— *P.a.longipennis* Reichenow, 1901   Uganda, E Zaire to Transvaal

○ *Pitta reichenowi* Madarász, 1901 GREEN-BREASTED PITTA   Cameroon to Uganda

○ *Pitta brachyura* (Linnaeus, 1766) INDIAN PITTA   S Himalayas, C India >> S India, Sri Lanka

○ *Pitta nympha* Temminck & Schlegel, 1850 FAIRY PITTA   E China, Korea, S Japan and Taiwan >> S China, Vietnam, Laos and Borneo

○ *Pitta moluccensis* (Statius Müller, 1766) BLUE-WINGED PITTA   S China, Burma, Thailand, N Malay Pen. >> S Malay Pen., Sumatra, Borneo and nearby islands

○ *Pitta megarhyncha* Schlegel, 1863 MANGROVE PITTA   Coasts of Bangladesh, Burma, W Malay Pen. to Sumatra and islands nearby

○ *Pitta elegans* ELEGANT PITTA

— *P.e.virginalis* E. Hartert, 1896   Djampea, Kalaotua, Kalao

— *P.e.vigorsii* Gould, 1838   Tukangbesi, Banda, Tanimbar, Kai Is.

— *P.e.hutzi* Meise, 1941   Nusa Penida (Lombok Str.)

— *P.e.concinna* Gould, 1857   Lombok, Flores, Sumbawa, Alor

— *P.e.maria* E. Hartert, 1896   Sumba

— *P.e.elegans* Temminck, 1836   Timor, Sula Is., S Moluccas

○ *Pitta iris* RAINBOW PITTA

— *P.i.iris* Gould, 1842   N Northern Territory (Arnhem Land)

— *P.i.johnstoneiana* Schodde & Mason, 1999   N Western Australia (Kimberley) (2189)

○ *Pitta versicolor* NOISY PITTA

— *P.v.simillima* Gould, 1868   S New Guinea, Cape York Pen.

— *P.v.intermedia* Mathews, 1912   Coastal NE and C Queensland

— *P.v.versicolor* Swainson, 1825   SE Queensland, NE New South Wales

○ *Pitta anerythra* BLACK-FACED PITTA

— *P.a.pallida* Rothschild, 1904   Bougainville I.

— *P.a.anerythra* Rothschild, 1901   Ysabel I. (Solomon Is.)

— *P.a.nigrifrons* Mayr, 1935   Choiseul I. (Solomon Is.)

PIPRIDAE MANAKINS (13: 48)

NEOPELMA P.L. Sclater, 1860 N

○ *Neopelma pallescens* (Lafresnaye, 1853) PALE-BELLIED TYRANT-MANAKIN

S Brazil, extreme E Bolivia

○ *Neopelma chrysocephalum* (Pelzeln, 1868) SAFFRON-CRESTED TYRANT-MANAKIN

S Venezuela, E Colombia, the Guianas, N Brazil

[1] Includes *hebetior* see Diamond & LeCroy (1979) (704).

○ *Neopelma aurifrons* (Wied, 1831) Wied's Tyrant-Manakin   Coastal E Brazil

○ *Neopelma chrysolophum* Pinto, 1944 Sierra do Mar Tyrant-Manakin[1]   SE Brazil

○ *Neopelma sulphureiventer* (Hellmayr, 1903) Sulphur-bellied Tyrant-Manakin
                                                                    E Peru, N Bolivia, W Brazil

TYRANNEUTES Sclater & Salvin, 1881 M
○ *Tyranneutes stolzmanni* (Hellmayr, 1906) Dwarf Tyrant-Manakin   W and S Amazonia

○ *Tyranneutes virescens* (Pelzeln, 1868) Tiny Tyrant-Manakin   E Venezuela, the Guaianas, NE Brazil

ILICURA Reichenbach, 1850 F
○ *Ilicura militaris* (Shaw & Nodder, 1808) Pin-tailed Manakin   SE Brazil

MASIUS Bonaparte, 1850 M
○ *Masius chrysopterus* Golden-winged Manakin
    __ *M.c.bellus* Hartert & Hellmayr, 1903   W Colombia
    __ *M.c.pax* Meyer de Schauensee, 1952   SE Colombia
    __ *M.c.coronulatus* P.L. Sclater, 1860   SW Colombia, W Ecuador
    __ *M.c.chrysopterus* (Lafresnaye, 1843)   E Colombia, NW Venezuela, E Ecuador
    __ *M.c.peruvianus* Carriker, 1934   N Peru

CORAPIPO Bonaparte, 1854 F
○ *Corapipo altera* White-ruffed Manakin[2]
    __ *C.l.altera* Hellmayr, 1906   E Nicaragua, NW Costa Rica, Panama
    __ *C.l.heteroleuca* Hellmayr, 1910   SW Costa Rica, W Panama

○ *Corapipo leucorrhoa* (P.L. Sclater, 1863) White-bibbed Manakin N Colombia, NW Venezuela

○ *Corapipo gutturalis* (Linnaeus, 1766) White-throated Manakin   E Venezuela, the Guinas, NE Brazil

MACHAEROPTERUS Bonaparte, 1854 M
○ *Machaeropterus deliciosus* (P.L. Sclater, 1860) Club-winged Manakin W Colombia, W Ecuador

○ *Machaeropterus regulus* Striped Manakin
    __ *M.r.antioquiae* Chapman, 1924   W and C Colombia
    __ *M.r.striolatus* (Bonaparte, 1838)   E Colombia, E Ecuador, NE Peru
    __ *M.r.obscurostriatus* Phelps & Gilliard, 1941   NW Venezuela
    __ *M.r.zulianus* Phelps & Phelps, Jr., 1952   W Venezuela
    __ *M.r.aureopectus* Phelps & Gilliard, 1941   SE Venezuela
    __ *M.r.regulus* (Hahn, 1819)   SE Brazil

○ *Machaeropterus pyrocephalus* Fiery-capped Manakin
    __ *M.p.pallidiceps* J.T. Zimmer, 1936   E Venezuela, ? N Brazil
    __ *M.p.pyrocephalus* (P.L. Sclater, 1852)   Amazonian Brazil, E Peru, N Bolivia

LEPIDOTHRIX Bonaparte, 1854[3] F
○ *Lepidothrix coronata* Blue-crowned Manakin
    __ *L.c.velutina* (Berlepsch, 1883)   SW Costa Rica, W Panama
    __ *L.c.minuscula* (Todd, 1919)   E Panama to NW Ecuador
    __ *L.c.caquetae* (Meyer de Schauensee, 1953)   C Colombia
    __ *L.c.carbonata* (Todd, 1925)   SE Colombia, NE Ecuador, NW Brazil
    __ *L.c.coronata* (Spix, 1825)   NW Brazil
    __ *L.c.caelestipileata* (Goeldi, 1905)   W Brazil, SE Peru
    __ *L.c.exquisita* (Hellmayr, 1905)   C Peru
    __ *L.c.regalis* (Bond & Meyer de Schauensee, 1940)   N Bolivia

○ *Lepidothrix isidorei* Blue-rumped Manakin[4]
    __ *L.i.isidorei* (P.L. Sclater, 1852)   Andes of E Colombia, E Ecuador
    __ *L.i.leucopygia* (Hellmayr, 1903)   Andes of N Peru

---

[1] For reasons to treat this as a species separate from *N. aurifrons* see Whitney *et al.* (1995) (2658).
[2] For reasons to separate this from *C. leucorrhoa* see Wetmore (1972) (2608) and A.O.U. (1998: 424) (3).
[3] For recognition of this genus see Prum (1992) (1951).
[4] In last Edition this and the next six species were attached to the genus *Pipra*, but see Prum (1992, 1994) (1951, 1953).

○ *Lepidothrix coeruleocapilla* (Tschudi, 1844) Cerulean-capped Manakin
Andes of C and SE Peru

○ *Lepidothrix nattereri* (P.L. Sclater, 1865) Snow-capped Manakin   SW Brazil

○ *Lepidothrix vilasboasi* (Sick, 1959) Golden-crowned Manakin   SC Brazil

○ *Lepidothrix iris* Opal-crowned Manakin
— *L.i.eucephala* (Todd, 1928)   C Brazil
— *L.i.iris* (Schinz, 1851)   EC Brazil

○ *Lepidothrix serena* (Linnaeus, 1766) White-fronted Manakin   The Guianas, NE Brazil

○ *Lepidothrix suavissima* (Salvin & Godman, 1882) Orange-bellied Manakin[1]
SE Venezuela, NC Brazil, N Guyana

MANACUS Brisson, 1760 M
● *Manacus manacus* Bearded Manakin[2]
✓ *M.m.candei* (Parzudaki, 1841)   SE Mexico to Costa Rica
— *M.m.vitellinus* (Gould, 1843)   E Panama
— *M.m.milleri* Chapman, 1915   N Colombia
— *M.m.viridiventris* Griscom, 1929   W Colombia
— *M.m.cerritus* J.L. Peters, 1927   NW Panama
— *M.m.aurantiacus* (Salvin, 1870)   SW Costa Rica, W Panama
— *M.m.trinitatis* (E. Hartert, 1912)   Trinidad, NE Venezuela
— *M.m.abditivus* Bangs, 1899   N Colombia
— *M.m.flaveolus* Cassin, 1851   E Colombia
— *M.m.bangsi* Chapman, 1914   SW Colombia
— *M.m.interior* Chapman, 1914   S Colombia, E Ecuador, N Peru, NW Brazil
— *M.m.umbrosus* Friedmann, 1944   S Venezuela
— *M.m.manacus* (Linnaeus, 1766)   The Guianas, N Brazil
— *M.m.leucochlamys* Chapman, 1914   W Ecuador
— *M.m.maximus* Chapman, 1924   SW Ecuador
— *M.m.expectatus* Gyldenstolpe, 1941   W Brazil, NE Peru
— *M.m.longibarbatus* J.T. Zimmer, 1936   C Brazil
— *M.m.purissimus* Todd, 1928   E Brazil
— *M.m.gutturosus* (Desmarest, 1806)   SE Brazil, E Paraguay, NE Argentina
— *M.m.purus* Bangs, 1899   N Brazil
— *M.m.subpurus* Cherrie & Reichenberger, 1923   C Brazil

ANTILOPHIA Reichenbach, 1850 F
○ *Antilophia bokermanni* Coelho & Silva, 1998 Araripe Manakin #   NE Brazil (540)

○ *Antilophia galeata* (M.H.K. Lichtenstein, 1823) Helmeted Manakin   E and S Brazil, extreme E Bolivia, NE Paraguay

CHIROXIPHIA Cabanis, 1847 F
● *Chiroxiphia linearis* Long-tailed Manakin
— *C.l.linearis* (Bonaparte, 1838)   S Mexico, Guatemala
✓ *C.l.fastuosa* (Lesson, 1842)   El Salvador, W Nicaragua, Costa Rica

○ *Chiroxiphia lanceolata* (Wagler, 1830) Lance-tailed Manakin   SW Costa Rica, Panama to NW Venezuela, N Colombia

○ *Chiroxiphia pareola* Blue-backed Manakin
— *C.p.atlantica* Dalmas, 1900   Tobago I.
— *C.p.pareola* (Linnaeus, 1766)   The Guianas, N and E Brazil
— *C.p.regina* P.L. Sclater, 1856   N Brazil
— *C.p.napensis* W. de W. Miller, 1908   SE Colombia, E Ecuador, NE Peru

○ *Chiroxiphia boliviana* J.A. Allen, 1889 Yungas Manakin[3]   SE Peru, N Bolivia

○ *Chiroxiphia caudata* (Shaw & Nodder, 1793) Blue Manakin   SE Brazil, E Paraguay, NE Argentina

[1] For separation from *L. serena* see Prum (1990, 1994) (1950, 1954).
[2] Species limits highly uncertain; recent genetic data indicate that narrowly defined *Manacus manacus* is paraphyletic with respect to non-white-bellied taxa. We therefore return to a broadly defined *M. manacus* for now. See Brumfield & Braun (2001) (261) and Brumfield *et al.* (2001) (263).
[3] For reasons to separate this from *C. pareola* see Parker & Remsen (1987: 103) (1748).

XENOPIPO Cabanis, 1847 F

○ **Xenopipo holochlora** GREEN MANAKIN[1]
- — *X.h.suffusa* (Griscom, 1932)   E Panama
- — *X.h.litae* (Hellmayr, 1906)   E Panama to NW Ecuador
- — *X.h.holochlora* (P.L. Sclater, 1888)   E Colombia, E Ecuador, N Peru
- — *X.h.viridior* (Chapman, 1924)   S Peru

○ **Xenopipo flavicapilla** (P.L. Sclater, 1852) YELLOW-HEADED MANAKIN   Andes of S Colombia, N Ecuador

○ **Xenopipo unicolor** (Taczanowski, 1884) JET MANAKIN   Andes of Ecuador and Peru

○ **Xenopipo uniformis** OLIVE MANAKIN
- — *X.u.duidae* (Chapman, 1929)   S Venezuela
- — *X.u.uniformis* (Salvin & Godman, 1884)   Guyana, N Brazil

○ **Xenopipo atronitens** Cabanis, 1847 BLACK MANAKIN   E Colombia, S Venezuela, The Guianas, NE and EC Brazil, E Bolivia, SE Peru

HETEROCERCUS P.L. Sclater, 1862 M

○ **Heterocercus aurantiivertex** Sclater & Salvin, 1880 ORANGE-CROWNED MANAKIN   E Ecuador, ? NE Peru[2]

○ **Heterocercus flavivertex** Pelzeln, 1868 YELLOW-CROWNED MANAKIN   E Colombia, S Venezuela, N Brazil

○ **Heterocercus linteatus** (Strickland, 1850) FLAME-CROWNED MANAKIN   NE Peru, WC Brazil

PIPRA Linnaeus, 1764 F

○ **Pipra pipra** WHITE-CROWNED MANAKIN[3]
- — *P.p.anthracina* Ridgway, 1906   Costa Rica, W Panama
- — *P.p.bolivari* Meyer de Schauensee, 1950   NW Colombia
- — *P.p.minima* Chapman, 1917[4]   W Colombia
- — *P.p.unica* Meyer de Schauensee, 1945   NC Colombia
- — *P.p.pipra* (Linnaeus, 1758)   The Guianas, E Venezuela, N Brazil
- — *P.p.coracina* P.L. Sclater, 1856   E Colombia, E Ecuador, NW Peru
- — *P.p.discolor* J.T. Zimmer, 1936   Amazonian N Peru
- — *P.p.pygmaea* J.T. Zimmer, 1936   Amazonian NC Peru
- — *P.p.occulta* J.T. Zimmer, 1936   Andean slope NE Peru
- — *P.p.comata* Berlepsch & Stolzmann, 1894   SC Peru
- — *P.p.microlopha* J.T. Zimmer, 1929   Amazonian far E Peru, W Brazil
- — *P.p.separabilis* J.T. Zimmer, 1936   C Brazil
- — *P.p.cephaleucos* Thunberg, 1822   E Brazil

○ **Pipra aureola** CRIMSON-HOODED MANAKIN
- — *P.a.aureola* (Linnaeus, 1758)   The Guianas, NE Venezuela, NE Brazil
- — *P.a.borbae* J.T. Zimmer, 1936   C Brazil (S of the Amazon)
- — *P.a.aurantiicollis* Todd, 1925   EC Brazil (NW Para)
- — *P.a.flavicollis* P.L. Sclater, 1851   EC Brazil (WC Para)

○ **Pipra filicauda** WIRE-TAILED MANAKIN
- — *P.f.subpallida* (Todd, 1928)   E Colombia, NW Venezuela
- — *P.f.filicauda* Spix, 1825   NE Peru to SC Venezuela, WC Brazil

○ **Pipra fasciicauda** BAND-TAILED MANAKIN
- — *P.f.calamae* Hellmayr, 1910   C and W Brazil
- — *P.f.saturata* J.T. Zimmer, 1936   N Peru
- — *P.f.purusiana* E. Snethlage, 1907   W Brazil, E Peru
- — *P.f.fasciicauda* Hellmayr, 1906   E Bolivia, SE Peru
- — *P.f.scarlatina* Hellmayr, 1915   E Paraguay, S Brazil

○ **Pipra cornuta** Spix, 1825 SCARLET-HORNED MANAKIN   S Venezuela, NC Brazil, W Guyana

---

[1] This species and the next three were treated in last Edition as comprising the genus *Chloropipo* which is here submerged, see Prum (1992) (1951).
[2] See Bond (1951) (183) for a view that an intermediate form may link *H. aurantiivertex* and *H. flavivertex*.
[3] May comprise more than one species, see A.O.U. (1998: 425) (3).
[4] For correction of spelling see David & Gosselin (2002) (613).

○ *Pipra mentalis* Red-capped Manakin
    — *P.m.mentalis* P.L. Sclater, 1857   SE Mexico to E Costa Rica
    — *P.m.ignifera* Bangs, 1901   W Costa Rica, W Panama
    — *P.m.minor* E. Hartert, 1898   E Panama to W Ecuador

○ *Pipra chloromeros* Tschudi, 1844 Round-tailed Manakin   E Peru, N Bolivia

○ *Pipra erythrocephala* Golden-headed Manakin
    — *P.e.erythrocephala* (Linnaeus, 1758)   E Panama, N Colombia, Venezuela, the Guianas, N Brazil
    — *P.e.flammiceps* Todd, 1919   E Colombia
    — *P.e.berlepschi* Ridgway, 1906   W Amazonia

○ *Pipra rubrocapilla* Temminck, 1821 Red-headed Manakin   E Peru, W Brazil, N Bolivia

COTINGIDAE COTINGAS[1] (33: 96)

TITYRINAE

TITYRA Vieillot, 1816 F
● *Tityra inquisitor* Black-crowned Tityra
    ☑ *T.i.fraserii* (Kaup, 1852)   SE Mexico to W Panama
    — *T.i.albitorques* Du Bus, 1847[2]   E Panama to Peru, N Brazil
    — *T.i.buckleyi* Salvin & Godman, 1890   SE Colombia, E Ecuador
    — *T.i.erythrogenys* (Selby, 1826)   E Colombia, Venezuela, the Guianas
    — *T.i.pelzelni* Salvin & Godman, 1890   E Bolivia, W Brazil
    — *T.i.inquisitor* (M.H.K. Lichtenstein, 1823)   E Brazil, Paraguay, N Argentina

○ *Tityra cayana* Black-tailed Tityra
    — *T.c.cayana* (Linnaeus, 1766)   N Sth. America, Trinidad
    — *T.c.braziliensis* (Swainson, 1837)   E and S Brazil, E Paraguay, NE Argentina

● *Tityra semifasciata* Masked Tityra
    — *T.s.hannumi* van Rossem & Hachisuka, 1937   NW Mexico
    — *T.s.griseiceps* Ridgway, 1888   W Mexico
    — *T.s.personata* Jardine & Selby, 1827   E Mexico to El Salvador
    ☑ *T.s.costaricensis* Ridgway, 1906   S Honduras to W Panama
    — *T.s.columbiana* Ridgway, 1906   E Panama, Colombia, W Venezuela
    — *T.s.nigriceps* J.A. Allen, 1888   SW Colombia, W Ecuador
    — *T.s.semifasciata* (Spix, 1825)   N and NW Amazonia
    — *T.s.fortis* Berlepsch & Stolzmann, 1896   C and SE Peru, N and E Bolivia, C Brazil

○ *Tityra leucura* Pelzeln, 1868 White-tailed Tityra   Rio Madeira, Amazonian Brazil (once)[3]

SCHIFFORNIS Bonaparte, 1854[4] F[5]
○ *Schiffornis major* Varzea Schiffornis
    — *S.m.major* Des Murs, 1856   W Brazil, E Peru, E Ecuador
    — *S.m.duidae* J.T. Zimmer, 1936   SE Venezuela

○ *Schiffornis turdina* Thrush-like Schiffornis
    — *S.t.veraepacis* (Sclater & Salvin, 1860)   SE Mexico to Costa Rica
    — *S.t.dumicola* (Bangs, 1903)   W Panama
    — *S.t.panamensis* Hellmayr, 1929   E Panama, NW Colombia
    — *S.t.acrolophites* Wetmore, 1972   E of E Panama, N Choco (Colombia)
    — *S.t.rosenbergi* (E. Hartert, 1898)   WC Colombia south to W Ecuador
    — *S.t.stenorhyncha* (Sclater & Salvin, 1869)   NE Colombia, N Venezuela
    — *S.t.olivacea* (Ridgway, 1906)   E Venezuela, Guyana
    — *S.t.aenea* J.T. Zimmer, 1936   E Andes of Ecuador to N Peru

---

[1] Our generic sequence is based on the branching shown by Prum *et al.* (2000) (1957) and Prum (2001) (1955).
[2] Du Bus is an abbreviation for Du Bus de Gisignies.
[3] This enigmatic bird seems unlikely to represent a valid wild taxon (a color morph or hybrid seems likely).
[4] A.O.U. (1998: 416) (3) preferred to treat this genus as Incertae Sedis, but see Prum (2001) (1955) for close association with *Tityra*.
[5] Gender addressed by David & Gosselin (2002) (614), multiple changes occur in specific names.

—— *S.t.amazona* (P.L. Sclater, 1860)	S Venezuela to E Peru, W Brazil
—— *S.t.wallacii* (Sclater & Salvin, 1867)	N Brazil, French Guiana, Surinam
—— *S.t.steinbachi* Todd, 1928	N Bolivia, SE Peru
—— *S.t.intermedia* Pinto, 1954	E Brazil
—— *S.t.turdina* (Wied, 1831)	SE Brazil

○ *Schiffornis virescens* (Lafresnaye, 1838) GREENISH SCHIFFORNIS   SE Brazil, E Paraguay, NE Argentina

LANIOCERA Lesson, 1840 F
○ *Laniocera rufescens* SPECKLED MOURNER

—— *L.r.rufescens* (P.L. Sclater, 1858)	Guatemala to Panama, W Colombia
—— *L.r.tertia* (E. Hartert, 1902)	NW Ecuador, SW Colombia
—— *L.r.griseigula* Meyer de Schauensee, 1950	NC Colombia

○ *Laniocera hypopyrra* (Vieillot, 1817) CINEREOUS MOURNER   The Guianas, S Venezuela, E Colombia, Amazonia, SE Brazil

IODOPLEURA Lesson, 1839 F
○ *Iodopleura pipra* BUFF-THROATED PURPLETUFT

—— †?*I.p.leucopygia* Salvin, 1885	The Guianas (?)
—— *I.p.pipra* (Lesson, 1831)	SE Brazil

○ *Iodopleura fusca* (Vieillot, 1817) DUSKY PURPLETUFT   E Venezuela, NC Brazil, the Guianas

○ *Iodopleura isabellae* WHITE-BROWED PURPLETUFT

—— *I.i.isabellae* Parzudaki, 1847	S Venezuela, E Colombia, E Ecuador, E Peru, N and W Brazil, N Bolivia
—— *I.i.paraensis* Todd, 1950	E Brazil

LANIISOMA Swainson, 1832 N
○ *Laniisoma elegans* SHRIKE-LIKE COTINGA

—— *L.e.venezuelense* Phelps & Gilliard, 1941[1]	NE Colombia, NW Venezuela
—— *L.e.buckleyi* (Sclater & Salvin, 1880)	E Peru, E Ecuador
—— *L.e.cadwaladeri* Carriker, 1935	Bolivia
—— *L.e.elegans* (Thunberg, 1823)	SE Brazil

XENOPSARIS Ridgway, 1891 M
○ *Xenopsaris albinucha* WHITE-NAPED XENOPSARIS

—— *X.a.minor* Hellmayr, 1920	Venezuela, N Brazil
—— *X.a.albinucha* (Burmeister, 1869)	S Brazil, Paraguay, N Argentina

PACHYRAMPHUS G.R. Gray, 1840 M
○ *Pachyramphus viridis* GREEN-BACKED BECARD[2]

—— *P.v.griseigularis* Salvin & Godman, 1883	SE Venezuela, E Brazil
—— *P.v.viridis* (Vieillot, 1816)	E Bolivia, Paraguay, N Argentina to E and S Brazil
—— *P.v.xanthogenys* Salvadori & Festa, 1898	E Ecuador
—— *P.v.peruanus* Hartert & Goodson, 1917	C Peru

○ *Pachyramphus versicolor* BARRED BECARD

—— *P.v.costaricensis* Bangs, 1908	Costa Rica
—— *P.v.versicolor* (Hartlaub, 1843)	W Venezuela, Colombia, Ecuador, N Peru
—— *P.v.meridionalis* Carriker, 1934	Peru, N Bolivia

○ *Pachyramphus spodiurus* P.L. Sclater, 1860 SLATY BECARD   W Ecuador, NW Peru

○ *Pachyramphus rufus* CINEREOUS BECARD

—— *P.r.rufus* (Boddaert, 1783)	Panama, N Sth. America
—— *P.r.juruanus* Gyldenstolpe, 1951	NE Peru, W Brazil

○ *Pachyramphus castaneus* CHESTNUT-CROWNED BECARD

—— *P.c.saturatus* Chapman, 1914	SE Colombia, E Ecuador, N Peru, NW Brazil

---

[1] For correction of spelling see David & Gosselin (2002) (614).
[2] We defer splitting this until more data are published.

— *P.c.intermedius* Berlepsch, 1879   N Venezuela
— *P.c.parui* Phelps & Phelps, Jr., 1949   S Venezuela
— *P.c.amazonus* J.T. Zimmer, 1936   NE Brazil
— *P.c.castaneus* (Jardine & Selby, 1827)   E Brazil, E Paraguay, NE Argentina

● *Pachyramphus cinnamomeus* Cinnamon Becard
    ✓ *P.c.fulvidior* Griscom, 1932   SE Mexico to W Panama
    — *P.c.cinnamomeus* Lawrence, 1861   E Panama, W Colombia, W Ecuador
    — *P.c.magdalenae* Chapman, 1914   N and E Colombia, W Venezuela
    — *P.c.badius* Phelps & Phelps, Jr., 1955   W Venezuela

○ *Pachyramphus polychopterus* White-winged Becard
    — *P.p.similis* Cherrie, 1891   Guatemala to Panama
    — *P.p.cinereiventris* P.L. Sclater, 1862   N Colombia
    — *P.p.dorsalis* P.L. Sclater, 1862   W Colombia, NW Ecuador
    — *P.p.tenebrosus* J.T. Zimmer, 1936   S Colombia
    — *P.p.tristis* (Kaup, 1852)   N and NE Sth. America
    — *P.p.nigriventris* P.L. Sclater, 1857   W Amazonia
    — *P.p.polychopterus* (Vieillot, 1818)   E Brazil
    — *P.p.spixii* (Swainson, 1837)   S Brazil, E Bolivia to N Argentina >> SW Amazonia

○ *Pachyramphus marginatus* Black-capped Becard
    — *P.m.nanus* Bangs & Penard, 1921   S Venezuela, the Guianas, Amazonia
    — *P.m.marginatus* (M.H.K. Lichtenstein, 1823)   E Brazil

○ *Pachyramphus albogriseus* Black-and-white Becard
    — *P.a.ornatus* Cherrie, 1891   W Nicaragua to W Panama
    — *P.a.coronatus* Phelps & Phelps, Jr., 1953   N Colombia, NW Venezuela
    — *P.a.albogriseus* P.L. Sclater, 1857   E Colombia, N Venezuela
    — *P.a.guayaquilensis* J.T. Zimmer, 1936   E Ecuador
    — *P.a.salvini* Richmond, 1899   C Peru, E Ecuador

○ *Pachyramphus major* Grey-collared Becard
    — *P.m.uropygialis* Nelson, 1899   W Mexico
    — *P.m.major* (Cabanis, 1847)   E Mexico
    — *P.m.matudai* A.R. Phillips, 1966   S Mexico, N Guatemala
    — *P.m.itzensis* Nelson, 1901   SE Mexico, Belize
    — *P.m.australis* Miller & Griscom, 1925   Guatemala to E Nicaragua

○ *Pachyramphus surinamus* (Linnaeus, 1766) Glossy-backed Becard   Surinam, E Brazil, French Guiana

○ *Pachyramphus aglaiae* Rose-throated Becard
    — *P.a.albiventris* (Lawrence, 1867)   S Arizona, W Mexico
    — *P.a.gravis* (van Rossem, 1938)   NE Mexico
    — *P.a.yucatanensis* (Ridgway, 1906)   E Mexico
    — *P.a.insularis* (Ridgway, 1887)   Tres Marias Is.
    — *P.s.aglaiae* (Lafresnaye, 1839)   NE Mexico
    — *P.a.sumichrasti* (Nelson, 1897)   SE Mexico to El Salvador
    — *P.a.hypophaeus* (Ridgway, 1891)   Honduras to NE Costa Rica
    — *P.a.latirostris* Bonaparte, 1854   W Nicaragua, W Costa Rica

○ *Pachyramphus homochrous* One-coloured Becard
    — *P.h.homochrous* P.L. Sclater, 1859   E Panama to N Peru
    — *P.h.quimarinus* (Meyer de Schauensee, 1950)   NW Colombia
    — *P.h.canescens* (Chapman, 1912)   N Colombia, NW Venezuela

○ *Pachyramphus minor* (Lesson, 1830) Pink-throated Becard   S Venezuela, The Guianas, Amazonia

○ *Pachyramphus validus* Crested Becard
    — *P.v.audax* (Cabanis, 1873)   S Peru, C Bolivia, NW Argentina
    — *P.v.validus* (M.H.K. Lichtenstein, 1823)   E Bolivia to C Brazil, N Argentina

○ *Pachyramphus niger* (J.F. Gmelin, 1788) Jamaican Becard   Jamaica

PHIBALURA Vieillot, 1816 F
○ *Phibalura flavirostris* SWALLOW-TAILED COTINGA
    — *P.f.flavirostris* Vieillot, 1816   E Paraguay, SE Brazil, NE Argentina
    — *P.f.boliviana* Chapman, 1930   N Bolivia (La Paz)

PHYTOTOMINAE

AMPELION Tschudi, 1845 M
○ *Ampelion rubrocristatus* (d'Orbigny & Lafresnaye, 1837) RED-CRESTED COTINGA[1]
                                       Andes from W Venezuela, Colombia to C Bolivia

○ *Ampelion rufaxilla* CHESTNUT-CRESTED COTINGA
    — *A.r.antioquiae* (Chapman, 1924)   W and C Andes of Colombia and Ecuador
    — *A.r.rufaxilla* (Tschudi, 1844)   Andes from N Peru to N Bolivia

ZARATORNIS Koepcke, 1954[2] M
○ *Zaratornis stresemanni* Koepcke, 1954 WHITE-CHEEKED COTINGA   W Peru

DOLIORNIS Taczanowski, 1874[3] M
○ *Doliornis remseni* Robbins, Rosenberg & Molina, 1994 CHESTNUT-BELLIED COTINGA #
                                         Andes of Ecuador to extreme N Peru (2084)

○ *Doliornis sclateri* Taczanowski, 1874 BAY-VENTED COTINGA   Andes of N and C Peru

PHYTOTOMA Molina, 1782[4] F
○ *Phytotoma raimondii* Taczanowski, 1883 PERUVIAN PLANTCUTTER   W Peru

○ *Phytotoma rutila* WHITE-TIPPED PLANTCUTTER
    — *P.r.angustirostris* d'Orbigny & Lafresnaye, 1837   SE Bolivia, NW Argentina
    — *P.r.rutila* Vieillot, 1818   N Argentina, W Uruguay, W Paraguay

● *Phytotoma rara* Molina, 1782 RUFOUS-TAILED PLANTCUTTER   S Chile, S Argentina, Falkland Is.

RUPICOLINAE

CARPORNIS G.R. Gray, 1846 F[5]
○ *Carpornis cucullata* (Swainson, 1821) HOODED BERRYEATER   SE Brazil

○ *Carpornis melanocephala* (Wied, 1820) BLACK-HEADED BERRYEATER   SE Brazil

PIPREOLA Swainson, 1837 F
○ *Pipreola riefferii* GREEN-AND-BLACK FRUITEATER
    — *P.r.occidentalis* (Chapman, 1914)   W Colombia, W Ecuador
    — *P.r.riefferii* (Boissonneau, 1840)   E Colombia, E Ecuador
    — *P.r.melanolaema* P.L. Sclater, 1856   NW Venezuela
    — *P.r.confusa* J.T. Zimmer, 1936   Peru, N Bolivia
    — *P.r.chachapoyas* (Hellmayr, 1915)   N Peru
    — *P.r.tallmanorum* O'Neill & Parker, 1981   C Peru (1716)

○ *Pipreola intermedia* BAND-TAILED FRUITEATER
    — *P.i.intermedia* Taczanowski, 1884   N and C Peru
    — *P.i.signata* (Hellmayr, 1917)   S Peru, N Bolivia

○ *Pipreola arcuata* BARRED FRUITEATER
    — *P.a.arcuata* (Lafresnaye, 1843)   Andes from N Colombia, W Venezuela to N Peru
    — *P.a.viridicauda* Meyer de Schauensee, 1953   Andes from C Peru to N Bolivia

○ *Pipreola aureopectus* GOLDEN-BREASTED FRUITEATER[6]
    — *P.a.decora* Bangs, 1899   N Colombia

[1] For correction of spelling see David & Gosselin (2002) (614).
[2] Treated within the genus *Ampelion* in last Edition, but see Ridgely & Tudor (1994: 742) (2019).
[3] For reasons to recognise this genus see Robbins *et al.* (1994) (2084).
[4] For reasons to treat this genus in the Cotingidae see Lanyon & Lanyon (1989) (1324).
[5] Gender addressed by David & Gosselin (2002) (614), multiple changes occur in specific names.
[6] The species *aureopectus* in last Edition embodied *jucunda*, *lubomirskii* and *pulchra*, but see Ridgely & Tudor (1994: 747) (2019).

_ *P.a.festiva* (Todd, 1912)   NW Venezuela
_ *P.a.aureopectus* (Lafresnaye, 1843)   E Colombia, W Venezuela

○ *Pipreola jucunda* P.L. Sclater, 1860   ORANGE-BREASTED FRUITEATER   W Colombia, W Ecuador

○ *Pipreola lubomirskii* Taczanowski, 1879   BLACK-CHESTED FRUITEATER   Andes of S Colombia, E Ecuador, N Peru

○ *Pipreola pulchra* (Hellmayr, 1917)   MASKED FRUITEATER   Andes of N and C Peru

○ *Pipreola frontalis*   SCARLET-BREASTED FRUITEATER
_ *P.f.squamipectus* (Chapman, 1925)   Andes of Ecuador to N Peru
_ *P.f.frontalis* (P.L. Sclater, 1858)   Andes of S Peru to N Bolivia

○ *Pipreola chlorolepidota* Swainson, 1837   FIERY-THROATED FRUITEATER   Andes of Colombia to C Peru

○ *Pipreola formosa*   HANDSOME FRUITEATER
_ *P.f.formosa* (Hartlaub, 1849)   N Venezuela
_ *P.f.rubidior* (Chapman, 1925)   NE Venezuela
_ *P.f.pariae* Phelps & Phelps, Jr., 1949   NE Venezuela

○ *Pipreola whitelyi*   RED-BANDED FRUITEATER
_ *P.w.kathleenae* Zimmer & Phelps, 1944   SE Venezuela
_ *P.w.whitelyi* Salvin & Godman, 1884   W Guyana

OXYRUNCUS Temminck, 1820   M
○ *Oxyruncus cristatus*   SHARPBILL
_ *O.c.frater* (Sclater & Salvin, 1868)   Costa Rica, W Panama
_ *O.c.brooksi* Bangs & Barbour, 1922   E Panama
_ *O.c.hypoglaucus* (Salvin & Godman, 1883)   The Guianas, SE Venezuela, NE Brazil
_ *O.c.tocantinsi* Chapman, 1939   C Brazil
_ *O.c.cristatus* Swainson, 1821   SE Brazil, Paraguay
_ *O.c.*subsp?   E Peru, N Bolivia

AMPELIOIDES J. Verreaux, 1867   M
○ *Ampelioides tschudii* (G.R. Gray, 1846)   SCALED FRUITEATER   Andes from NW Venezuela to N Bolivia

RUPICOLA Brisson, 1760   M[1]
○ *Rupicola rupicola* (Linnaeus, 1766)   GUIANAN COCK-OF-THE-ROCK   E Colombia to the Guianas, N Brazil

○ *Rupicola peruvianus*   ANDEAN COCK-OF-THE-ROCK
_ *R.p.sanguinolentus* Gould, 1859   W Colombia, W Ecuador
_ *R.p.aequatorialis* Taczanowski, 1889   W Venezuela, Colombia, E Ecuador, N Peru
_ *R.p.peruvianus* (Latham, 1790)   C Peru
_ *R.p.saturatus* Cabanis & Heine, 1859   S Peru, N Bolivia

PHOENICIRCUS Swainson, 1832   M
○ *Phoenicircus carnifex* (Linnaeus, 1758)   GUIANAN RED COTINGA   E Venezuela, the Guianas, NE Brazil

○ *Phoenicircus nigricollis* Swainson, 1832   BLACK-NECKED RED COTINGA   SE Colombia, WC Brazil, E Ecuador, NE Peru

COTINGINAE

COTINGA Brisson, 1760   F
○ *Cotinga amabilis* Gould, 1857   LOVELY COTINGA   SE Mexico to Costa Rica

○ *Cotinga ridgwayi* Ridgway, 1887   TURQUOISE COTINGA   SW Costa Rica, Panama

○ *Cotinga nattererii* (Boissonneau, 1840)   BLUE COTINGA   NW Venezuela, N and W Colombia, NW Ecuador

○ *Cotinga maynana* (Linnaeus, 1766)   PLUM-THROATED COTINGA   W Amazonia

○ *Cotinga cotinga* (Linnaeus, 1766)   PURPLE-BREASTED COTINGA   E Colombia to the Guianas, N Brazil

○ *Cotinga maculata* (Statius Müller, 1776)   BANDED COTINGA   SE Brazil

○ *Cotinga cayana* (Linnaeus, 1766)   SPANGLED COTINGA   S Venezuela, the Guianas, Amazonia

[1] Gender addressed by David & Gosselin (2002) (614), multiple changes occur in specific names.

PROCNIAS Illiger, 1811 M[1]
○ *Procnias tricarunculatus* (J. & E. Verreaux, 1853) THREE-WATTLED BELLBIRD
Nicaragua to W Panama

○ *Procnias albus* WHITE BELLBIRD
— *P.a.albus* (Hermann, 1783) E Venezuela, the Guyanas, NE Brazil
— *P.a.wallacei* Oren & Novaes, 1985 Belem (Brazil) (1722)

○ *Procnias averano* BEARDED BELLBIRD
— *P.a.carnobarba* (Cuvier, 1816)[2] N and E Venezuela, Trinidad, W Guyana, NC Brazil
— *P.a.averano* (Hermann, 1783) NE Brazil

○ *Procnias nudicollis* (Vieillot, 1817) BARE-THROATED BELLBIRD E and SE Brazil, E Paraguay, NE Argentina

TIJUCA Férussac, 1829 F
○ *Tijuca atra* Férussac, 1829 BLACK-AND-GOLD COTINGA SE Brazil

○ *Tijuca condita* Snow, 1980 GREY-WINGED COTINGA SE Brazil (2294)

LIPAUGUS Boie, 1828[3] M
○ *Lipaugus weberi* Cuervo, Salaman, Donegan & Ochoa, 2001 CHESTNUT-CAPPED COTINGA #
Central cordillera of the Andes of Colombia (600)

○ *Lipaugus fuscocinereus* (Lafresnaye, 1843) DUSKY PIHA Andes from N Colombia to extreme N Peru

○ *Lipaugus uropygialis* (Sclater & Salvin, 1876) SCIMITAR-WINGED PIHA[4] Extreme S Peru, N Bolivia

○ *Lipaugus unirufus* RUFOUS PIHA
— *L.u.unirufus* P.L. Sclater, 1859 SE Mexico to N Colombia
— *L.u.castaneotinctus* (E. Hartert, 1902) SW Colombia to NW Ecuador

○ *Lipaugus vociferans* (Wied, 1820) SCREAMING PIHA S Venezuela, the Guianas, Amazonia and E Brazil

○ *Lipaugus lanioides* (Lesson, 1844) CINNAMON-VENTED PIHA SE Brazil

○ *Lipaugus streptophorus* (Salvin & Godman, 1884) ROSE-COLLARED PIHA
W Guyana, SE Venezuela, NC Brazil

CONIOPTILON Lowery & O'Neill, 1966 N
○ *Conioptilon mcilhennyi* Lowery & O'Neill, 1966 BLACK-FACED COTINGA
SE Peru

SNOWORNIS Prum, 2001 (1955) M
○ *Snowornis subalaris* (P.L. Sclater, 1861) GREY-TAILED PIHA Andes from S Colombia to E C Peru

○ *Snowornis cryptolophus* OLIVACEOUS PIHA
— *S.c.mindoensis* (Hellmayr & Seilern, 1914) Andes of W Ecuador, SW Colombia
— *S.c.cryptolophus* (Sclater & Salvin, 1877) Andes from SE Colombia to C Peru

PORPHYROLAEMA Bonaparte, 1854 F
○ *Porphyrolaema porphyrolaema* (Deville & Sclater, 1852) PURPLE-THROATED COTINGA
W Amazonia

XIPHOLENA Gloger, 1842 F
○ *Xipholena punicea* (Pallas, 1764) POMPADOUR COTINGA S Venezuela, the Guianas, C Brazil, extreme E Bolivia

○ *Xipholena lamellipennis* (Lafresnaye, 1839) WHITE-TAILED COTINGA E Brazil

○ *Xipholena atropurpurea* (Wied, 1820) WHITE-WINGED COTINGA SE Brazil

CARPODECTES Salvin, 1865 M
○ *Carpodectes hopkei* Berlepsch, 1897 BLACK-TIPPED COTINGA E Panama to NW Ecuador

● *Carpodectes nitidus* Salvin, 1865 SNOWY COTINGA E Honduras to E Panama

○ *Carpodectes antoniae* Ridgway, 1884 YELLOW-BILLED COTINGA W Costa Rica, W Panama

[1] Gender addressed by David & Gosselin (in press) (614), multiple changes occur in specific names.
[2] Given as 1817 by Snow (1979: 306) (2293) but see Browning & Monroe (1991) (258).
[3] A.O.U. (1998: 416) (3) preferred to treat this genus as Incertae Sedis. Prum (2000) (1955) considered it a member of the subfamily Cotinginae.
[4] Treated in a separate genus *Chirocylla* in last Edition, as in Snow (1979) (2293), but see Prum (2000) (1955) for the submergence of that genus.

GYMNODERUS E. Geoffroy Saint-Hilaire, 1809 M
○ *Gymnoderus foetidus* (Linnaeus, 1758) BARE-NECKED FRUITCROW   S Venezuela, the Guianas, Amazonia

QUERULA Vieillot, 1816 F
○ *Querula purpurata* (Statius Müller, 1776) PURPLE-THROATED FRUITCROW

Costa Rica, Panama, SE Venezuela, the Guianas,
Amazonia

HAEMATODERUS Bonaparte, 1854 M
○ *Haematoderus militaris* (Shaw, 1792) CRIMSON FRUITCROW   The Guianas, NE Brazil

PYRODERUS G.R. Gray, 1840 M
○ *Pyroderus scutatus* RED-RUFFED FRUITCROW
  ___ *P.s.occidentalis* Chapman, 1914   W Colombia
  ___ *P.s.granadensis* (Lafresnaye, 1846)   E Colombia, W Venezuela
  ___ *P.s.orenocensis* (Lafresnaye, 1846)   NE Venezuela, W Guyana
  ___ *P.s.masoni* Ridgway, 1886   E Peru
  ___ *P.s.scutatus* (Shaw, 1792)   SE Brazil, E Paraguay, NE Argentina

PERISSOCEPHALUS Oberholser, 1899 M
○ *Perissocephalus tricolor* (Statius Müller, 1776) CAPUCHINBIRD   E Venezuela, the Guianas, N Brazil

CEPHALOPTERUS E. Geoffroy Saint-Hilaire, 1809 M
○ *Cephalopterus glabricollis* Gould, 1851 BARE-NECKED UMBRELLABIRD   Costa Rica, W Panama

○ *Cephalopterus ornatus* E. Geoffroy Saint-Hilaire, 1809 AMAZONIAN UMBRELLABIRD
S Venezuela, SW Guyana, W and C Amazonia

○ *Cephalopterus penduliger* P.L. Sclater, 1859 LONG-WATTLED UMBRELLABIRD
Andes of SW Colombia, Ecuador

GENERA INCERTAE SEDIS (2: 4)

PIPRITES Cabanis, 1847[1] F[2]
○ *Piprites griseiceps* Salvin, 1865 GREY-HEADED PIPRITES   Guatemala to Costa Rica

○ *Piprites chloris* WING-BARRED PIPRITES
  ___ *P.c.antioquiae* Chapman, 1924   NC Colombia
  ___ *P.c.perijana* Phelps & Phelps, Jr., 1949   N Colombia, W Venezuela
  ___ *P.c.tschudii* (Cabanis, 1874)   S Colombia to C Peru, NW Brazil
  ___ *P.c.chlorion* (Cabanis, 1847)   The Guianas, N Brazil
  ___ *P.c.grisescens* Novaes, 1964   Para (Brazil)
  ___ *P.c.boliviana* Chapman, 1924   N Bolivia, C Brazil
  ___ *P.c.chloris* (Temminck, 1822)   SE Brazil, Paraguay, NE Argentina

○ *Piprites pileata* (Temminck, 1822) BLACK-CAPPED PIPRITES   SE Brazil, NE Argentina

CALYPTURA Swainson, 1832 F
○ *Calyptura cristata* (Vieillot, 1818) KINGLET CALYPTURA   SE Brazil

TYRANNIDAE TYRANT-FLYCATCHERS (98: 400)

PHYLLOMYIAS Cabanis & Heine, 1859 M
○ *Phyllomyias fasciatus* PLANALTO TYRANNULET
  ___ *P.f.cearae* Hellmayr, 1927   NE Brazil
  ___ *P.f.fasciatus* (Thunberg, 1822)   C and E Brazil, E Bolivia
  ___ *P.f.brevirostris* (Spix, 1825)   S Brazil, Paraguay, NE Argentina

---

[1] It has been suggested that this genus be moved to the Tyrannidae (Prum & Lanyon, 1989) (1956), however A.O.U. (1998: 416) (3) preferred to treat the genus as Incertae Sedis following Prum (1990) (1950). Prum (2001) (1955) placed it outside the cluster of subfamilies of the Cotingidae in his phylogenetic tree.
[2] Gender addressed by David & Gosselin (2002) (614), multiple changes occur in specific names.

○ *Phyllomyias burmeisteri* ROUGH-LEGGED TYRANNULET[1]
— *P.b.burmeisteri* Cabanis & Heine, 1859   SE Brazil, E Bolivia, Paraguay, N Argentina
— *P.b.zeledoni* (Lawrence, 1869)   Costa Rica, W Panama
— *P.b.viridiceps* (Zimmer & Phelps, 1944)   N Venezuela
— *P.b.wetmorei* (Aveledo & Pons, 1953)   NW Venezuela
— *P.b.bunites* (Wetmore & Phelps, 1956)   SC Venezuela
— *P.b.leucogonys* (Sclater & Salvin, 1871)   Colombia, Ecuador, Peru

○ *Phyllomyias virescens* (Temminck, 1824) GREENISH TYRANNULET[2]   SE Brazil, Paraguay, NE Argentina

○ *Phyllomyias reiseri* Hellmayr, 1905 REISER'S TYRANNULET   E Brazil

○ *Phyllomyias urichi* (Chapman, 1899) URICH'S TYRANNULET[3]   NE Venezuela

○ *Phyllomyias sclateri* SCLATER'S TYRANNULET
— *P.s.subtropicalis* (Chapman, 1919)   SE Peru
— *P.s.sclateri* Berlepsch, 1901   Bolivia, NW Argentina

○ *Phyllomyias griseocapilla* P.L. Sclater, 1861 GREY-CAPPED TYRANNULET   SE Brazil

○ *Phyllomyias griseiceps* SOOTY-HEADED TYRANNULET
— *P.g.cristatus* Berlepsch, 1884   E Panama, N Colombia, N Venezuela
— *P.g.caucae* Chapman, 1915   WC Colombia
— *P.g.griseiceps* (Sclater & Salvin, 1871)   Ecuador
— *P.g.pallidiceps* J.T. Zimmer, 1941   NE Brazil, Peru, SE Venezuela

○ *Phyllomyias plumbeiceps* (Lawrence, 1869) PLUMBEOUS-CROWNED TYRANNULET
    Colombia, Ecuador, Peru

○ *Phyllomyias nigrocapillus* BLACK-CAPPED TYRANNULET
— *P.n.aureus* (J.T. Zimmer, 1941)   W Venezuela
— *P.n.flavimentum* (Chapman, 1912)   N Colombia
— *P.n.nigrocapillus* (Lafresnaye, 1845)   W Colombia, Ecuador, C Peru

○ *Phyllomyias cinereiceps* (P.L. Sclater, 1860) ASHY-HEADED TYRANNULET   Colombia, Ecuador, Peru

○ *Phyllomyias uropygialis* (Lawrence, 1869) TAWNY-RUMPED TYRANNULET
    Venezuela, Colombia to S Bolivia

TYRANNULUS Vieillot, 1816 M
○ *Tyrannulus elatus* (Latham, 1790) YELLOW-CROWNED TYRANNULET   Panama to N Bolivia and C Brazil

MYIOPAGIS Salvin & Godman, 1888 F
○ *Myiopagis gaimardii* FOREST ELAENIA
— *M.g.macilvainii* (Lawrence, 1871)   E Panama, N Colombia
— *M.g.trinitatis* (Hartert & Goodson, 1917)   Trinidad
— *M.g.bogotensis* (Berlepsch, 1907)   E Colombia, N Venezuela
— *M.g.guianensis* (Berlepsch, 1907)   The Guianas, N Brazil
— *M.g.gaimardii* (d'Orbigny, 1840)   S Venezuela, Brazil, E Peru, N Bolivia
— *M.g.subcinerea* J.T. Zimmer, 1941[4]   EC Brazil

○ *Myiopagis caniceps* GREY ELAENIA
— *M.c.parambae* (Hellmayr, 1904)   W Colombia, NW Ecuador
— *M.c.absita* (Wetmore, 1963)   Panama
— *M.c.cinerea* (Pelzeln, 1868)   S Venezuela, E Colombia, E Ecuador, Peru, NW Brazil
— *M.c.caniceps* (Swainson, 1836)   E and S Brazil, Paraguay, N Argentina

○ *Myiopagis olallai* Coopmans & Krabbe, 2000 FOOTHILL ELAENIA #   Andean foothills of Ecuador and Peru (570)

○ *Myiopagis subplacens* (P.L. Sclater, 1861) PACIFIC ELAENIA   SW Ecuador, NW Peru

○ *Myiopagis flavivertex* (P.L. Sclater, 1887) YELLOW-CROWNED ELAENIA   French Guiana, Surinam, S Venezuela, N Peru, N Brazil

○ *Myiopagis viridicata* GREENISH ELAENIA
— *M.v.jaliscensis* Nelson, 1900   SW Mexico

---

[1] We follow A.O.U. (1998: 380) (3) in retaining a single species here.
[2] For reasons to separate this and the next species see Stotz (1990) (2354).
[3] Treated as a separate species following Cardoso da Silva (1996) (294).
[4] For correction of spelling see David & Gosselin (2002) (614).

___ *M.v.minima* Nelson, 1898   Tres Marias Is.

___ *M.v.placens* (P.L. Sclater, 1859)   S Mexico to Honduras

___ *M.v.pacifica* (Brodkorb, 1943)   SE Chiapas (Mexico)

___ *M.v.accola* Bangs, 1902   Nicaragua to Panama, NW Colombia

___ *M.v.pallens* Bangs, 1902   NW Colombia, N Venezuela

___ *M.v.restricta* Todd, 1952   N coast of Venezuela

___ *M.v.zuliae* Zimmer & Phelps, 1955   N Venezuela

___ *M.v.implacens* (P.L. Sclater, 1861)   S Colombia, W Ecuador

___ *M.v.viridicata* (Vieillot, 1817)   SE Peru, E Bolivia, Brazil, Paraguay, N Argentina

○ *Myiopagis cotta* (Gosse, 1849) JAMAICAN ELAENIA Jamaica

ELAENIA Sundevall, 1836 F

○ *Elaenia flavogaster* YELLOW-BELLIED ELAENIA

___ *E.f.subpagana* P.L. Sclater, 1860   Pacific coast of S Mexico to Panama

___ *E.f.flavogaster* (Thunberg, 1822)   Trinidad, N Sth. America

___ *E.f.semipagana* P.L. Sclater, 1861   W Ecuador, NW Peru

___ *E.f.pallididorsalis* Aldrich, 1937   Panama, Pearl Is.

○ *Elaenia martinica* CARIBBEAN ELAENIA

___ *E.m.martinica* (Linnaeus, 1766)   Lesser Antilles

___ *E.m.chinchorrensis* Griscom, 1926   Great Key I., E Mexico

___ *E.m.barbadensis* Cory, 1888   Barbados

___ *E.m.riisii* P.L. Sclater, 1860   Virgin Is., Antigua, Curacao

___ *E.m.caymanensis* Berlepsch, 1907   Cayman Is.

___ *E.m.cinerescens* Ridgway, 1884   Isla Andrès and Isla Providencia (W Caribbean)

___ *E.m.remota* Berlepsch, 1907   Yucatán coast islands

○ *Elaenia spectabilis* Pelzeln, 1868 LARGE ELAENIA NE Peru, N and C Brazil, W Bolivia, N Argentina

○ *Elaenia ridleyana* Sharpe, 1888 NORONHA ELAENIA[1] Fernando de Noronha I.

● *Elaenia albiceps* WHITE-CRESTED ELAENIA

___ *E.a.griseigularis* P.L. Sclater, 1858   Ecuador

___ *E.a.diversa* J.T. Zimmer, 1941   NC Peru

___ *E.a.urubambae* J.T. Zimmer, 1941   SE Peru

___ *E.a.albiceps* (d'Orbigny & Lafresnaye, 1837) Bolivia, Brazil

___ *E.a.modesta* Tschudi, 1844   Peru, NW Chile

✓ *E.a.chilensis* Hellmayr, 1927   S Chile, SW Argentina

○ *Elaenia parvirostris* Pelzeln, 1868 SMALL-BILLED ELAENIA SE Bolivia, Paraguay, N Argentina, S Brazil >> Amazonia

○ *Elaenia mesoleuca* (M.H.K. Lichtenstein, 1830) OLIVACEOUS ELAENIA SE Brazil, Paraguay, NE Argentina

○ *Elaenia strepera* Cabanis, 1883 SLATY ELAENIA C Bolivia, NW Argentina >> Venezuela

○ *Elaenia gigas* P.L. Sclater, 1871 MOTTLE-BACKED ELAENIA Andes from C Colombia to C Bolivia

○ *Elaenia pelzelni* Berlepsch, 1907 BROWNISH ELAENIA W Brazil, NE Peru, NE Bolivia

○ *Elaenia cristata* PLAIN-CRESTED ELAENIA

___ *E.c.alticola* Zimmer & Phelps, 1946   S Venezuela, N Brazil

___ *E.c.cristata* Pelzeln, 1868   French Guiana and Venezuela to E Peru and E Bolivia

○ *Elaenia chiriquensis* LESSER ELAENIA

___ *E.c.chiriquensis* Lawrence, 1865   SW Costa Rica, Panama

___ *E.c.brachyptera* Berlepsch, 1907   SW Colombia, NW Ecuador

___ *E.c.albivertex* Pelzeln, 1868   NW and NC South America, Trinidad

○ *Elaenia ruficeps* Pelzeln, 1868 RUFOUS-CROWNED ELAENIA The Guianas, N Brazil, S Venezuela

○ *Elaenia frantzii* MOUNTAIN ELAENIA

___ *E.f.ultima* Griscom, 1935   Guatemala to Nicaragua

___ *E.f.frantzii* Lawrence, 1865   Nicaragua to Panama

___ *E.f.browni* Bangs, 1898   N Colombia, N Venezuela

___ *E.f.pudica* P.L. Sclater, 1871   E Colombia, W Venezuela

---

[1] For separation of this see Sick (1985) (2270).

○ *Elaenia obscura* HIGHLAND ELAENIA
　　__ *E.o.sordida* J.T. Zimmer, 1941　　　　　　　　　　SE Brazil, E Paraguay, NE Argentina
　　__ *E.o.obscura* (d'Orbigny & Lafresnaye, 1837)　　　S Brazil, Peru, Bolivia, Paraguay, N Argentina

○ *Elaenia dayi* GREAT ELAENIA
　　__ *E.d.dayi* Chapman, 1929　　　　　　　　　　　　Mt. Roraima (S Venezuela)
　　__ *E.d.auyantepui* Zimmer & Phelps, 1952　　　　　SE Venezuela
　　__ *E.d.tyleri* Chapman, 1929　　　　　　　　　　　S Venezuela

○ *Elaenia pallatangae* SIERRAN ELAENIA
　　__ *E.p.pallatangae* P.L. Sclater, 1861　　　　　　　Andes of W Colombia, Ecuador
　　__ *E.p.davidwillardi* Dickerman & Phelps, 1987　　Tepuis of Venezuela (730)
　　__ *E.p.olivina* Salvin & Godman, 1884　　　　　　　Tepuis of S Guyana, S Venezuela
　　__ *E.p.intensa* J.T. Zimmer, 1941　　　　　　　　　Peruvian Andes
　　__ *E.p.exsul* Todd, 1952　　　　　　　　　　　　　Bolivian Andes

○ *Elaenia fallax* GREATER ANTILLEAN ELAENIA
　　__ *E.f.fallax* P.L. Sclater, 1861　　　　　　　　　　Jamaica
　　__ *E.f.cherriei* Cory, 1895　　　　　　　　　　　　Hispaniola

ORNITHION Hartlaub, 1853　N
○ *Ornithion semiflavum* (Sclater & Salvin, 1860) YELLOW-BELLIED TYRANNULET
　　　　　　　　　　　　　　　　　　　　　　　　　　　　S Mexico to Costa Rica

○ *Ornithion brunneicapillus* BROWN-CAPPED TYRANNULET[1]
　　__ *O.b.brunneicapillus* (Lawrence, 1862)　　　　　Costa Rica, Panama, W Colombia, NW Ecuador
　　__ *O.b.dilutum* (Todd, 1913)　　　　　　　　　　　NW Venezuela, N Colombia

○ *Ornithion inerme* Hartlaub, 1853　WHITE-LORED TYRANNULET　　S Venezuela, the Guianas, E Colombia, E Ecuador, E
　　　　　　　　　　　　　　　　　　　　　　　　　　　　　　Peru, N Bolivia, Amazonian and SE Brazil

CAMPTOSTOMA P.L. Sclater, 1857　N
○ *Camptostoma imberbe* NORTHERN BEARDLESS TYRANNULET
　　__ *C.i.imberbe* P.L. Sclater, 1857　　　　　　　　　S USA, Mexico to NW Costa Rica
　　__ *C.i.ridgwayi* (Brewster, 1882)　　　　　　　　　SW USA, W Mexico
　　__ *C.i.thyellophilum* Parkes & Phillips, 1999　　　Cozumel I. (1789)

○ *Camptostoma obsoletum* SOUTHERN BEARDLESS TYRANNULET
　　__ *C.o.flaviventre* Sclater & Salvin, 1864　　　　　W Costa Rica, Panama
　　__ *C.o.orphnum* Wetmore, 1957　　　　　　　　　　Coiba I., Afuerita I.
　　__ *C.o.majus* Griscom, 1932[2]　　　　　　　　　　Pearl Is.
　　__ *C.o.caucae* Chapman, 1914　　　　　　　　　　W Colombia
　　__ *C.o.bogotense* J.T. Zimmer, 1941　　　　　　　C Colombia
　　__ *C.o.pusillum* (Cabanis & Heine, 1859)　　　　　N Colombia, NW Venezuela
　　__ *C.o.napaeum* (Ridgway, 1888)　　　　　　　　　N Venezuela, the Guianas, N Brazil
　　__ *C.o.venezuelae* J.T. Zimmer, 1941　　　　　　　N and C Venezuela, Trinidad
　　__ *C.o.maranonicum* Carriker, 1933　　　　　　　　N Peru
　　__ *C.o.olivaceum* (Berlepsch, 1889)　　　　　　　　S Colombia, E Ecuador, NE Peru, W Brazil
　　__ *C.o.sclateri* (Berlepsch & Taczanowski, 1883)　W Ecuador, W Peru
　　__ *C.o.griseum* Carriker, 1933　　　　　　　　　　W Peru
　　__ *C.o.bolivianum* J.T. Zimmer, 1941　　　　　　　C Bolivia, NW Argentina
　　__ *C.o.cinerascens* (Wied, 1831)　　　　　　　　　E Bolivia, E and C Brazil
　　__ *C.o.obsoletum* (Temminck, 1824)　　　　　　　　SE Brazil, Uruguay, N Argentina

SUIRIRI d'Orbigny, 1840　F
○ *Suiriri suiriri* SUIRIRI FLYCATCHER
　　__ *S.s.affinis* (Burmeister, 1856)　　　　　　　　　C Brazil, Surinam, NW Bolivia
　　__ *S.s.bahiae* (Berlepsch, 1893)[3]　　　　　　　　E Brazil
　　__ *S.s.suiriri* (Vieillot, 1818)　　　　　　　　　　E Bolivia, S Brazil, Uruguay, Paraguay, N Argentina

○ *Suiriri islerorum* Zimmer, Whittaker & Oren, 2001　CHAPADA FLYCATCHER #
　　　　　　　　　　　　　　　　　　　　　　　　　　　　SC Brazil, E Bolivia (2734)

---

[1] For correction of spelling see David & Gosselin (2002) (613).
[2] The name *major* is rendered as *majus* in the neuter form.
[3] Hayes (2001) (1090A) suspected this form may be of hybrid origin.

MECOCERCULUS P.L. Sclater, 1862 M

○ **Mecocerculus leucophrys** White-throated Tyrannulet

— *M.l.montensis* (Bangs, 1899)	N Colombia
— *M.l.chapmani* Dickerman, 1985	S Venezuela [714]
— *M.l.nigriceps* Chapman, 1899	N Venezuela
— *M.l.notatus* Todd, 1919	W Colombia
— *M.l.setophagoides* (Bonaparte, 1845)	E Colombia
— *M.l.palliditergum* Phelps & Phelps, Jr., 1947	Coast of N Venezuela
— *M.l.gularis* (Madarász, 1903)	E Venezuela
— *M.l.parui* Phelps & Phelps, Jr., 1950	S Venezuela
— *M.l.rufomarginatus* (Lawrence, 1869)	S Colombia, Ecuador, NW Peru
— *M.l.roraimae* Hellmayr, 1921	C Venezuela
— *M.l.brunneomarginatus* Chapman, 1924	C Peru
— *M.l.pallidior* Carriker, 1933	W Peru
— *M.l.leucophrys* (d'Orbigny & Lafresnaye, 1837)	SE Peru, Bolivia, NW Argentina

○ **Mecocerculus poecilocercus** (Sclater & Salvin, 1873) White-tailed Tyrannulet

Colombia, Ecuador, Peru

○ **Mecocerculus hellmayri** Berlepsch, 1907 Buff-banded Tyrannulet[1] SE Peru, Bolivia, Argentina

○ **Mecocerculus calopterus** (P.L. Sclater, 1859) Rufous-winged Tyrannulet

W Ecuador, NW Peru

○ **Mecocerculus minor** (Taczanowski, 1879) Sulphur-bellied Tyrannulet

E Colombia, Venezuela, N Peru

○ **Mecocerculus stictopterus** White-banded Tyrannulet

— *M.s.albocaudatus* Phelps & Gilliard, 1941	NW Venezuela
— *M.s.stictopterus* (P.L. Sclater, 1858)	Colombia, Ecuador, N Peru
— *M.s.taeniopterus* Cabanis, 1874	SE Peru, N Bolivia

ANAIRETES Reichenbach, 1850[2] M

○ **Anairetes nigrocristatus** Taczanowski, 1884 Black-crested Tit-Tyrant[3]

S Ecuador, N Peru

○ **Anairetes reguloides** Pied-crested Tit-Tyrant

— *A.r.albiventris* (Chapman, 1924)	W Peru
— *A.r.reguloides* (d'Orbigny & Lafresnaye, 1837)	SW Peru, NW Chile

○ **Anairetes alpinus** Ash-breasted Tit-Tyrant

— *A.a.alpinus* (Carriker, 1933)	C Peru
— *A.a.bolivianus* (Carriker, 1935)	N Bolivia

○ **Anairetes flavirostris** Yellow-billed Tit-Tyrant

— *A.f.huancabambae* (Chapman, 1924)	NW Peru
— *A.f.arequipae* (Chapman, 1926)	SW Peru, NW Chile
— *A.f.cuzcoensis* (Chapman, 1924)	SE Peru
— *A.f.flavirostris* Sclater & Salvin, 1876	W Peru, Bolivia, N Chile, W Argentina

● **Anairetes parulus** Tufted Tit-Tyrant

— *A.p.aequatorialis* Berlepsch & Taczanowski, 1884	S Colombia, Ecuador, Peru, Bolivia
— *A.p.patagonicus* (Hellmayr, 1920)	C Argentina
✓ *A.p.parulus* (Kittlitz, 1830)	S and C Chile, W Argentina

○ **Anairetes fernandezianus** (Philippi, 1857) Juan Fernandez Tit-Tyrant Robinson Crusoe I. (Juan Fernandez Is.)

○ **Anairetes agilis** (P.L. Sclater, 1856) Agile Tit-Tyrant[4]

E Colombia, NW Venezuela, N Ecuador

○ **Anairetes agraphia** Unstreaked Tit-Tyrant

— *A.a.agraphia* Chapman, 1919	SE Peru
— *A.a.plengei* (Schulenberg & Graham, 1981)	C Peru [2210]
— *A.a.squamigera* (O'Neill & Parker, 1976)	E Peru

---

[1] Including ¶ *argentinus* Olrog, 1972 [1682], which may need re-evaluation.
[2] Including *Uromyias* as per Roy *et al.* (1999) [2119].
[3] We follow Fjeldså & Krabbe (1990) [850] in treating this as a species distinct from *A. reguloides*.
[4] Type species of the genus *Uromyias*.

SERPOPHAGA Gould, 1839 F

○ *Serpophaga cinerea* TORRENT TYRANNULET
    __ *S.c.grisea* Lawrence, 1871   Costa Rica, W Panama
    __ *S.c.cinerea* (Tschudi, 1844)   Andes from Colombia and Venezuela to C Bolivia

○ *Serpophaga hypoleuca* RIVER TYRANNULET
    __ *S.h.venezuelana* J.T. Zimmer, 1940   NE Venezuela
    __ *S.h.hypoleuca* Sclater & Salvin, 1866   SE Colombia, NW Venezuela, E Peru, NE Bolivia
    __ *S.h.pallida* E. Snethlage, 1907   EC Brazil

○ *Serpophaga nigricans* (Vieillot, 1817) SOOTY TYRANNULET   SE Brazil, Uruguay, Paraguay, N Argentina, S Bolivia

○ *Serpophaga subcristata* WHITE-CRESTED TYRANNULET
    __ *S.s.straminea* (Temminck, 1822)   SC Brazil, Uruguay
    __ *S.s.subcristata* (Vieillot, 1817)   Bolivia to E Brazil, Argentina

○ *Serpophaga munda* Berlepsch, 1893 WHITE-BELLIED TYRANNULET[1]   E Bolivia, N Argentina

PHAEOMYIAS Berlepsch, 1902[2] F

○ *Phaeomyias murina* MOUSE-COLOURED TYRANNULET
    __ *P.m.eremonoma* Wetmore, 1953   Panama
    __ *P.m.incomta* (Cabanis & Heine, 1859)   Trinidad, the Guianas, Colombia, Venezuela, N Brazil
    __ *P.m.tumbezana* (Taczanowski, 1877)   SW Ecuador, N Peru
    __ *P.m.inflava* Chapman, 1924   W Peru
    __ *P.m.maranonica* J.T. Zimmer, 1941   NC Peru
    __ *P.m.wagae* (Taczanowski, 1884)   E Peru, W Bolivia
    __ *P.m.ignobilis* Bond & Meyer de Schauensee, 1941   S Bolivia, Paraguay, NW Argentina
    __ *P.m.murina* (Spix, 1825)   S Brazil

CAPSIEMPIS Cabanis & Heine, 1869[3] F

○ *Capsiempis flaveola* YELLOW TYRANNULET
    __ *C.f.semiflava* (Lawrence, 1865)   Nicaragua to Panama
    __ *C.f.cerulua* Wetmore, 1939   Venezuela, Colombia, Ecuador
    __ *C.f.amazona* J.T. Zimmer, 1955   French Guiana, N Brazil
    __ *C.f.leucophrys* Berlepsch, 1907   Colombia, W Venezuela
    __ *C.f.magnirostris* E. Hartert, 1898   SW Ecuador
    __ *C.f.flaveola* (M.H.K. Lichtenstein, 1823) S Colombia, SE Brazil, E Bolivia, Paraguay

POLYSTICTUS Reichenbach, 1850 M

○ *Polystictus pectoralis* BEARDED TACHURI
    __ *P.p.bogotensis* (Chapman, 1915)   Colombia
    __ *P.p.brevipennis* (Berlepsch & Hartert, 1902) Guyana, S Venezuela, N Brazil
    __ *P.p.pectoralis* (Vieillot, 1817)   E Bolivia, SW Brazil, Uruguay, Paraguay, N Argentina

○ *Polystictus superciliaris* (Wied, 1831) GREY-BACKED TACHURI   SE Brazil

NESOTRICCUS C.H. Townsend, 1895 M

○ *Nesotriccus ridgwayi* C.H. Townsend, 1895 COCOS ISLAND FLYCATCHER[4]
                                            Cocos I. (off Costa Rica)

PSEUDOCOLOPTERYX Lillo, 1905 F

○ *Pseudocolopteryx dinelliana* Lillo, 1905 DINELLI'S DORADITO[5]   NW Argentina >> SE Bolivia, W Paraguay

○ *Pseudocolopteryx sclateri* (Oustalet, 1892) CRESTED DORADITO   Trinidad, Guyana, S Brazil, E Bolivia, Paraguay, E
                                              Argentina

○ *Pseudocolopteryx acutipennis* (Sclater & Salvin, 1873) SUBTROPICAL DORADITO
                                          Colombia, Ecuador, Peru, Bolivia, W Argentina

○ *Pseudocolopteryx flaviventris* (d'Orbigny & Lafresnaye, 1837) WARBLING DORADITO
                                          Uruguay, C Chile, N Argentina, S Brazil >> Paraguay

---

[1] For reasons to keep this separate from *S. subcristata* see Herzog (2001) (1110).
[2] Spelled *Phaiomyias* in last Edition, but see Traylor (1979) (2431).
[3] Treated within *Phylloscartes* in last Edition, but the genus *Casiempis* was recognised by A.O.U. (1998: 375) (3).
[4] See Lanyon (1984) (1327).
[5] For correction of spelling see David & Gosselin (2002) (614).

PSEUDOTRICCUS Taczanowski & Berlepsch, 1885 M

○ ***Pseudotriccus pelzelni*** Bronze-olive Pygmy Tyrant

— *P.p.berlepschi* Nelson, 1913   E Panama

— *P.p.annectens* (Salvadori & Festa, 1899)   W Colombia, W Ecuador

— *P.p.pelzelni* Taczanowski & Berlepsch, 1885   E Colombia, E Ecuador

— *P.p.peruvianus* Bond, 1947   E Peru

○ ***Pseudotriccus simplex*** (Berlepsch, 1901) Hazel-fronted Pygmy Tyrant

Bolivia, SE Peru

○ ***Pseudotriccus ruficeps*** (Lafresnaye, 1843) Rufous-headed Pygmy Tyrant

Colombia to NW Bolivia

CORYTHOPIS Sundevall, 1836 M[1]

○ ***Corythopis torquata*** Ringed Antpipit

— *C.t.sarayacuensis* Chubb, 1918   SE Colombia, E Ecuador, NE Peru

— *C.t.anthoides* (Pucheran, 1855)   S Venezuela, the Guianas, N and E Brazil

— *C.t.torquatus* Tschudi, 1844   C Peru, W Brazil

— *C.t.subtorquatus* Todd, 1927   NE Bolivia

○ ***Corythopis delalandi*** (Lesson, 1830-31) Southern Antpipit   E Bolivia, S Brazil, E Paraguay, NE Argentina

EUSCARTHMUS Wied, 1831 M

○ ***Euscarthmus meloryphus*** Tawny-crowned Pygmy Tyrant

— *E.m.paulus* (Bangs, 1899)   N Venezuela, N Colombia

— *E.m.fulviceps* P.L. Sclater, 1871   W Ecuador, W Peru

— *E.m.meloryphus* Wied, 1831   Brazil, E Bolivia, Paraguay, N Argentina

○ ***Euscarthmus rufomarginatus*** Rufous-sided Pygmy Tyrant

— *E.r.rufomarginatus* (Pelzeln, 1868)   SE Brazil, E Bolivia

— *E.r.savannophilus* Mees, 1968   Surinam

PSEUDELAENIA W.E. Lanyon, 1988 (1330) F

○ ***Pseudelaenia leucospodia*** Grey-and-white Tyrannulet

— *P.l.cinereifrons* (Salvadori & Festa, 1899)   SW Ecuador

— *P.l.leucospodia* (Taczanowski, 1877)   NW Peru

STIGMATURA Sclater & Salvin, 1866 F

○ ***Stigmatura napensis*** Lesser Wagtail-Tyrant

— *S.n.napensis* Chapman, 1926   SE Colombia, NE Peru, W Brazil

— *S.n.bahiae* Chapman, 1926   NE Brazil

○ ***Stigmatura budytoides*** Greater Wagtail-Tyrant

— *S.b.budytoides* (d'Orbigny & Lafresnaye, 1837)   NC Bolivia

— *S.b.inzonata* Wetmore & Peters, 1923   SE Bolivia, N Argentina

— *S.b.flavocinerea* (Burmeister, 1861)   C Argentina

— *S.b.gracilis* J.T. Zimmer, 1955   Brazil

ZIMMERIUS Traylor, 1977 M

○ ***Zimmerius vilissimus*** Paltry Tyrannulet[2]

— *Z.v.vilissimus* (Sclater & Salvin, 1859)   S Mexico to Honduras

— *Z.v.parvus* (Lawrence, 1862)   Nicaragua to Panama

— *Z.v.tamae* (Phelps & Phelps, Jr., 1954)   N Colombia

— *Z.v.improbus* (Sclater & Salvin, 1871)   N Colombia, W Venezuela

— *Z.v.petersi* (Berlepsch, 1907)   N Venezuela

○ ***Zimmerius bolivianus*** Bolivian Tyrannulet

— *Z.b.viridissimus* (P.L. Sclater, 1874)   SE Peru

— *Z.b.bolivianus* (d'Orbigny, 1840)   N Bolivia

○ ***Zimmerius cinereicapilla*** (Cabanis, 1873) Red-billed Tyrannulet[3]   Ecuador to S Peru

---

[1] Gender addressed by David & Gosselin (2002) (614), multiple changes occur in specific names.
[2] See comments on potential for separation of a species *parvus* in A.O.U. (1998: 380) (3).
[3] For correction of spelling see David & Gosselin (2002) (613).

○ *Zimmerius villarejoi* Alvarez & Whitney, 2001 MISHIMA TYRANNULET # S part of Rio Nanay drainage, Depto. Loreto, Peru (38)

○ *Zimmerius gracilipes* SLENDER-FOOTED TYRANNULET
    __ *Z.g.acer* (Salvin & Godman, 1883)   S Venezuela, the Guianas, N Brazil
    __ *Z.g.gracilipes* (Sclater & Salvin, 1867)   S Venezuela, S Guyana, N Brazil, N Bolivia, E Peru
    __ *Z.g.gilvus* (J.T. Zimmer, 1941)   W Brazil, NW Bolivia

○ *Zimmerius chrysops* GOLDEN-FACED TYRANNULET
    __ *Z.c.minimus* (Chapman, 1912)   N Colombia
    __ *Z.c.cumanensis* (J.T. Zimmer, 1941)   NE Venezuela
    __ *Z.c.albigularis* (Chapman, 1924)   W Ecuador, SW Colombia
    __ *Z.c.flavidifrons* (P.L. Sclater, 1860)   SW Ecuador
    __ *Z.c.chrysops* (P.L. Sclater, 1859)   N Peru, Ecuador, S Colombia, W Venezuela

○ *Zimmerius viridiflavus* (Tschudi, 1844) PERUVIAN TYRANNULET[1]   C Peru

PHYLLOSCARTES Cabanis & Heine, 1859 M
○ *Phylloscartes poecilotis* VARIEGATED BRISTLE TYRANT
    __ *P.p.pifanoi* (Aveledo & Pons, 1952)   NE Colombia, NW Venezuela
    __ *P.p.poecilotis* (P.L. Sclater, 1862)   N Colombia to S Peru

○ *Phylloscartes chapmani* CHAPMAN'S BRISTLE TYRANT
    __ *P.c.chapmani* Gilliard, 1940   S Venezuela
    __ *P.c.duidae* Phelps & Phelps, Jr., 1951   SE Venezuela

○ *Phylloscartes ophthalmicus* MARBLE-FACED BRISTLE TYRANT
    __ *P.o.purus* (Todd, 1952)   Coast of Venezuela
    __ *P.o.ophthalmicus* (Taczanowski, 1874)   NW Venezuela, Colombia, Ecuador to C Peru
    __ *P.o.ottonis* (Berlepsch, 1901)   S Peru, N Bolivia

○ *Phylloscartes orbitalis* (Cabanis, 1873) SPECTACLED BRISTLE TYRANT   Andes from C Colombia to C Bolivia

○ *Phylloscartes venezuelanus* (Berlepsch, 1907) VENEZUELAN BRISTLE TYRANT
    N Venezuela

○ *Phylloscartes lanyoni* Graves, 1988 ANTIOQUIA BRISTLE TYRANT   N Colombia (984)

○ *Phylloscartes eximius* (Temminck, 1822) SOUTHERN BRISTLE TYRANT   SE Brazil, E Paraguay, N Argentina

○ *Phylloscartes ventralis* MOTTLE-CHEEKED TYRANNULET
    __ *P.v.angustirostris* (d'Orbigny & Lafresnaye, 1837)   Peru, Bolivia, NW Argentina
    __ *P.v.tucumanus* J.T. Zimmer, 1940   NW Argentina
    __ *P.v.ventralis* (Temminck, 1824)   SE Brazil, Uruguay, Paraguay, NE Argentina

○ *Phylloscartes ceciliae* Teixeira, 1987 ALAGOAS TYRANNULET #   NE Brazil (2388)

○ *Phylloscartes kronei* Willis & Oniki, 1992 RESTINGA TYRANNULET #   SE Brazil (2671)

○ *Phylloscartes beckeri* Gonzaga & Pacheco, 1995 BAHIA TYRANNULET #   S Bahia (Brazil) (952)

○ *Phylloscartes flavovirens* (Lawrence, 1862) YELLOW-GREEN TYRANNULET   Panama

○ *Phylloscartes virescens* Todd, 1925 OLIVE-GREEN TYRANNULET   The Guianas

○ *Phylloscartes gualaquizae* (P.L. Sclater, 1887) ECUADOREAN TYRANNULET
    E Ecuador, N Peru

○ *Phylloscartes nigrifrons* (Salvin & Godman, 1884) BLACK-FRONTED TYRANNULET
    S Venezuela, NC Brazil

○ *Phylloscartes superciliaris* RUFOUS-BROWED TYRANNULET
    __ *P.s.superciliaris* (Sclater & Salvin, 1868)   Costa Rica to W Panama
    __ *P.s.palloris* (Griscom, 1935)   E Panama
    __ *P.s.griseocapillus* Phelps & Phelps, Jr., 1952   NW Venezuela

○ *Phylloscartes flaviventris* (E. Hartert, 1897) RUFOUS-LORED TYRANNULET
    NW Venezuela

---

[1] This species separated from *Z. chrysops* by Zimmer (1941) (2708).

○ *Phylloscartes parkeri* Fitzpatrick & Stotz, 1997 Cinnamon-faced Tyrannulet #

SE Peru, N Bolivia (837)

○ *Phylloscartes roquettei* E. Snethlage, 1928[1] Minas Gerais Tyrannulet   Minas Gerais (Brazil)

○ *Phylloscartes paulistus* H. & R. von Ihering, 1907 São Paulo Tyrannulet

SE Brazil, Paraguay

○ *Phylloscartes oustaleti* (P.L. Sclater, 1887) Oustalet's Tyrannulet   SE Brazil

○ *Phylloscartes difficilis* (H. & R. von Ihering, 1907) Sierra do Mar Tyrannulet

SE Brazil

○ *Phylloscartes sylviolus* (Cabanis & Heine, 1859) Bay-ringed Tyrannulet

SE Brazil, Paraguay

MIONECTES Cabanis, 1844 M

○ *Mionectes striaticollis* Streak-necked Flycatcher

__ *M.s.columbianus* Chapman, 1919	Colombia, E Ecuador
__ *M.s.selvae* Meyer de Schauensee, 1952	WC Colombia
__ *M.s.viridiceps* Chapman, 1924	W Ecuador
__ *M.s.palamblae* Chapman, 1927	N Peru
__ *M.s.poliocephalus* Tschudi, 1844	N and C Peru
__ *M.s.striaticollis* (d'Orbigny & Lafresnaye, 1837)	S Peru to C Bolivia

○ *Mionectes olivaceus* Olive-striped Flycatcher

__ *M.o.olivaceus* Lawrence, 1868	Costa Rica, W Panama
__ *M.o.hederaceus* Bangs, 1910	E Panama, W Colombia, W Ecuador
__ *M.o.galbinus* Bangs, 1902	N Colombia
__ *M.o.pallidus* Chapman, 1914	E Colombia
__ *M.o.venezuelanus* Ridgway, 1906	N Venezuela, Trinidad
__ *M.o.fasciaticollis* Chapman, 1923	E Ecuador, E Peru, N Bolivia
__ *M.o.meridae* J.T. Zimmer, 1941	NW Venezuela, NE Colombia

○ *Mionectes oleagineus* Ochre-bellied Flycatcher

__ *M.o.assimilis* P.L. Sclater, 1859	S Mexico to Costa Rica
__ *M.o.obscurus* (Dickey & van Rossem, 1925)	El Salvador
__ *M.o.dyscolus* Bangs, 1901	W Costa Rica, W Panama
__ *M.o.lutescens* (Griscom, 1927)	W Panama
__ *M.o.parcus* Bangs, 1900	E Panama, N Colombia, NW Venezuela
__ *M.o.chloronotus* (d'Orbigny & Lafresnaye, 1837)	W Amazonia
__ *M.o.abdominalis* (Phelps & Phelps, Jr., 1955)	N Venezuela
__ *M.o.pallidiventris* Hellmayr, 1906	NE Venezuela, Trinidad
__ *M.o.intensus* (Zimmer & Phelps, 1946)	E Venezuela, Guyana
__ *M.o.dorsalis* (Phelps & Phelps, Jr., 1952)	SE Venezuela
__ *M.o.pacificus* (Todd, 1921)	W Colombia, W Ecuador
__ *M.o.hauxwelli* (Chubb, 1919)	E Ecuador, NE Peru
__ *M.o.wallacei* (Chubb, 1919)	N Brazil, the Guianas
__ *M.o.maynanus* (Stolzmann, 1926)[2]	E Peru
__ *M.o.oleagineus* (M.H.K. Lichtenstein, 1823)	SE Brazil

○ *Mionectes macconnelli* McConnell's Flycatcher

__ *M.m.mercedesfosterae* Dickerman & Phelps, 1987	Venezuela (730)
__ *M.m.macconnelli* (Chubb, 1919)	Guyana, French Guiana, N Brazil, E Venezuela
__ *M.m.roraimae* (Chubb, 1919)	S Guyana, S Venezuela
__ *M.m.peruanus* (Carriker, 1930)	C Peru
__ *M.m.amazonus* (Todd, 1921)	C Brazil, E Bolivia

○ *Mionectes rufiventris* Cabanis, 1846 Grey-hooded Flycatcher   SE Brazil, Paraguay, N Argentina

LEPTOPOGON Cabanis, 1844[3] M

○ *Leptogogon amaurocephalus* Sepia-capped Flycatcher

__ *L.a.pileatus* Cabanis, 1865	S Mexico, Guatemala, Honduras

[1] Probably 1927 see J.f. Orn. (1930: 134).
[2] For correction of spelling see David & Gosselin (2002) (613).
[3] For species sequencing we drew on Bates & Zink (1994) (105).

— *L.a.faustus* Bangs, 1907   Costa Rica, Panama
— *L.a.idius* Wetmore, 1957   Coiba I.
— *L.a.diversus* Todd, 1913   N Colombia
— *L.a.orinocensis* Zimmer & Phelps, 1946   W and SW Venezuela
— *L.a.obscuritergum* Zimmer & Phelps, 1946   S Venezuela
— *L.a.peruvianus* Sclater & Salvin, 1867   N and W Amazonia
— *L.a.amaurocephalus* Tschudi, 1846   SE Brazil, E Bolivia, N Argentina, Paraguay

○ *Leptopogon superciliaris* SLATY-CAPPED FLYCATCHER
— *L.s.superciliaris* Tschudi, 1844   SE Colombia, N Peru
— *L.s.poliocephalus* Cabanis & Heine, 1859   N and C Colombia
— *L.s.hellmayri* Griscom, 1929   Costa Rica to W Panama
— *L.s.venezuelensis* Hartert & Goodson, 1917   N Venezuela, N Brazil
— *L.s.pariae* Phelps & Phelps, Jr., 1949   NE Venezuela, Trinidad
— *L.s.transandinus* Berlepsch & Taczanowski, 1883   SW Colombia, W Ecuador
— *L.s.albidiventer* Hellmayr, 1918   Bolivia, SE Peru

○ *Leptopogon rufipectus* RUFOUS-BREASTED FLYCATCHER
— *L.r.venezuelanus* Phelps & Phelps, Jr., 1957   NE Colombia, NW Venezuela
— *L.r.rufipectus* (Lafresnaye, 1846)   E Ecuador, E Colombia, N Peru

○ *Leptopogon taczanowskii* Hellmayr, 1917 INCA FLYCATCHER   E Peru

SUBLEGATUS Sclater & Salvin, 1868 M
○ *Sublegatus arenarum* NORTHERN SCRUB FLYCATCHER[1]
— *S.a.arenarum* (Salvin, 1863)   SW Costa Rica
— *S.a.atrirostris* (Lawrence, 1871)   Panama to NE Colombia
— *S.a.glaber* Sclater & Salvin, 1868   N Venezuela, Trinidad
— *S.a.tortugensis* Phelps & Phelps, Jr., 1946   Tortuga
— *S.a.pallens* J.T. Zimmer, 1941   Aruba, Curacao, Bonaire
— *S.a.orinocensis* J.T. Zimmer, 1941   C Venezuela

○ *Sublegatus obscurior* Todd, 1920 AMAZONIAN SCRUB FLYCATCHER   W Amazonia

○ *Sublegatus modestus* SOUTHERN SCRUB FLYCATCHER
— *S.m.modestus* (Wied, 1831)   E Peru to S Brazil
— *S.m.brevirostris* (d'Orbigny & Lafresnaye, 1837)   E Bolivia to W Argentina

INEZIA Cherrie, 1909 F
○ *Inezia tenuirostris* (Cory, 1913) SLENDER-BILLED TYRANNULET   N Colombia, NW Venezuela

○ *Inezia inornata* (Salvadori, 1897) PLAIN TYRANNULET   SW Brazil, E Bolivia, W Paraguay, S Peru

○ *Inezia subflava* AMAZONIAN TYRANNULET
— *I.s.obscura* J.T. Zimmer, 1939   S Venezuela, NW Brazil
— *I.s.subflava* (Sclater & Salvin, 1873)   NC Brazil

○ *Inezia caudata* PALE-TIPPED TYRANNULET[2]
— *I.c.intermedia* Cory, 1913   NW and NC Venezuela, N Colombia
— *I.c.caudata* (Salvin, 1897)   C and NE Venezuela, the Guianas, N Brazil

MYIOPHOBUS Reichenbach, 1850[3] M
○ *Myiophobus flavicans* FLAVESCENT FLYCATCHER
— *M.f.flavicans* (P.L. Sclater, 1860)   Colombia, Ecuador, N Peru
— *M.f.perijanus* Phelps & Phelps, Jr., 1957   NW Venezuela
— *M.f.venezuelanus* (Hellmayr, 1920)   N Venezuela
— *M.f.caripensis* Zimmer & Phelps, 1954   NE Venezuela
— *M.f.superciliosus* (Taczanowski, 1874)   Peru

○ *Myiophobus phoenicomitra* ORANGE-CRESTED FLYCATCHER
— *M.p.litae* (E. Hartert, 1900)   NW Ecuador, W Colombia
— *M.p.phoenicomitra* (Taczanowski & Berlepsch, 1885)   E Ecuador, N Peru

---

[1] This species was treated as part of *S. modestus* in last Edition, but see Traylor (1982) (2434), and A.O.U. (1998: 381) (3).
[2] Separated from *I. subflava* by Zimmer & Whittaker, 2000 (2733).
[3] We expect to split this genus when better data are available.

○ *Myiophobus inornatus* Carriker, 1932 Unadorned Flycatcher S Peru, N Bolivia

○ *Myiophobus roraimae* Roraiman Flycatcher
— *M.r.sadiecoatsae* (Dickerman & Phelps, 1987) Venezuela (730)
— *M.r.roraimae* (Salvin & Godman, 1883) S Venezuela, SE Colombia, W Guyana
— *M.r.rufipennis* Carriker, 1932 SE Peru

○ *Myiophobus pulcher* Handsome Flycatcher
— *M.p.pulcher* (P.L. Sclater, 1860) W Colombia, W Ecuador
— *M.p.bellus* (P.L. Sclater, 1862) E Colombia, E Ecuador
— *M.p.oblitus* Bond, 1943 SE Peru

○ *Myiophobus lintoni* Meyer de Schauensee, 1951 Orange-banded Flycatcher
S Ecuador, N Peru

○ *Myiophobus ochraceiventris* (Cabanis, 1873) Ochraceous-breasted Flycatcher
N Peru to N Bolivia

○ *Myiophobus cryptoxanthus* (P.L. Sclater, 1860) Olive-chested Flycatcher
S Ecuador, N Peru

○ *Myiophobus fasciatus* Bran-coloured Flycatcher
— *M.f.furfurosus* (Thayer & Bangs, 1905) SW Costa Rica, Panama
— *M.f.fasciatus* (Statius Müller, 1776) The Guianas, Trinidad, N Venezuela, Colombia
— *M.f.crypterythrus* (P.L. Sclater, 1860) S Colombia, W Ecuador, N Peru
— *M.f.saturatus* (Berlepsch & Stolzmann, 1906) E Peru
— *M.f.rufescens* (Salvadori, 1864) W Peru, N Chile
— *M.f.auriceps* (Gould, 1839) SE Peru, N and E Bolivia, N Argentina, W Paraguay
— *M.f.flammiceps* (Temminck, 1822) Brazil, Bolivia, Uruguay, Paraguay, N Argentina

MYIOTRICCUS Ridgway, 1905 M
○ *Myiotriccus ornatus* Ornate Flycatcher
— *M.o.ornatus* (Lafresnaye, 1853) C Colombia
— *M.o.stellatus* (Cabanis, 1873) W Colombia, W Ecuador
— *M.o.phoenicurus* (P.L. Sclater, 1855) SE Colombia, E Ecuador, N Peru
— *M.o.aureiventris* (P.L. Sclater, 1874) S Peru

TACHURIS Lafresnaye, 1836 F
○ *Tachuris rubrigastra* Many-coloured Rush Tyrant
— *T.r.libertatis* Hellmayr, 1920 W Peru
— *T.r.alticola* (Berlepsch & Stolzmann, 1896) Peru, W Bolivia, NW Argentina
— *T.r.rubrigastra* (Vieillot, 1817) SE Brazil, Paraguay, Uruguay, Chile, N Argentina
— *T.r.loaensis* Philippi & Johnson, 1946 N Chile

CULICIVORA Swainson, 1827 F
○ *Culicivora caudacuta* (Vieillot, 1818) Sharp-tailed Tyrant S Brazil, E Bolivia, Paraguay, NE Argentina

HEMITRICCUS Cabanis & Heine, 1859 M
○ *Hemitriccus diops* (Temminck, 1822) Drab-breasted Pygmy Tyrant SE Brazil, Paraguay, NE Argentina

○ *Hemitriccus obsoletus* Brown-breasted Pygmy Tyrant
— *H.o.obsoletus* (Miranda-Ribeiro, 1906) SE Brazil
— *H.o.zimmeri* Traylor, 1979 Southernmost Brazil

○ *Hemitriccus flammulatus* Flammulated Pygmy Tyrant
— *H.f.flammulatus* Berlepsch, 1901 Peru, N Bolivia, W Brazil
— *H.f.olivascens* (Todd, 1915) E Bolivia

○ *Hemitriccus minor* Snethlage's Tody-Tyrant
— *H.m.minor* (E. Snethlage, 1907) SE Amazonian Brazil
— *H.m.snethlageae* (E.H. Snethlage, 1937)[1] SC Amazonian Brazil
— *H.m.pallens* (Todd, 1925) C Amazonian Brazil, E Bolivia

○ *Hemitriccus spodiops* (Berlepsch, 1901) Yungas Tody-Tyrant[2] NW Bolivia[3]

[1] Replaces *minimus* Todd, 1925 (see Stotz, 1992) (2356).
[2] For placement next to *H. minor* see Cohn-Haft (1996) (541).
[3] Erroneously given a Brazilian range in last Edition.

○ *Hemitriccus josephinae* (Chubb, 1914) Boat-billed Tody-Tyrant   Guyana, W Surinam, NE Brazil

○ *Hemitriccus zosterops* White-eyed Tody-Tyrant
  — *H.z.zosterops* (Pelzeln, 1868)   S Venezuela, S Colombia, the Guianas, NW Brazil
  — *H.z.flaviviridis* (J.T. Zimmer, 1940)   N Peru

○ *Hemitriccus griseipectus* White-bellied Tody-Tyrant[1]
  — *H.g.griseipectus* (E. Snethlage, 1907)   SE Peru, N Bolivia, C Brazil
  — *H.g.naumburgae* (J.T. Zimmer, 1945)   E Brazil

○ *Hemitriccus minimus* Todd, 1925 Zimmer's Tody-Tyrant[2]   C Brazil, E Bolivia

○ *Hemitriccus orbitatus* (Wied, 1831) Eye-ringed Tody-Tyrant   SE Brazil

○ *Hemitriccus iohannis* (E. Snethlage, 1907) Johannes's Tody-Tyrant   SW Amazonia

○ *Hemitriccus striaticollis* Stripe-necked Tody-Tyrant
  — *H.s.griseiceps* (Todd, 1925)   WC Brazil
  — *H.s.striaticollis* (Lafresnaye, 1853)   C Brazil to N Bolivia, E Peru, E Colombia

○ *Hemitriccus nidipendulus* Hangnest Tody-Tyrant
  — *H.n.nidipendulus* (Wied, 1831)   E Brazil
  — *H.n.paulistus* (Hellmayr, 1914)   SE Brazil

○ *Hemitriccus margaritaceiventer* Pearly-vented Tody-Tyrant
  — *H.m.impiger* (Sclater & Salvin, 1868)   NE Colombia, N Venezuela
  — *H.m.septentrionalis* (Chapman, 1914)   S Colombia
  — *H.m.chiribiquetensis* Stiles, 1996   S Colombia (Sierra Chiribiquete) (2342)
  — *H.m.duidae* (Chapman, 1929)   S Venezuela
  — *H.m.auyantepui* (Gilliard, 1941)   SE Venezuela
  — *H.m.breweri* (W.H. Phelps, Jr., 1977)   C Venezuela
  — *H.m.rufipes* (Tschudi, 1844)   C Peru, E Bolivia
  — *H.m.margaritaceiventer* (d'Orbigny & Lafresnaye, 1837)   WC Brazil to N Argentina
  — *H.m.wuchereri* (Sclater & Salvin, 1873)   E Brazil

○ *Hemitriccus inornatus* (Pelzeln, 1868) Pelzeln's Tody-Tyrant   NW Brazil

○ *Hemitriccus granadensis* Black-throated Tody-Tyrant
  — *H.g.lehmanni* (Meyer de Schauensee, 1945)   N Colombia
  — *H.g.intensus* (Phelps & Phelps, Jr., 1952)   NW Venezuela
  — *H.g.federalis* (Phelps & Phelps, Jr., 1950)   N Venezuela
  — *H.g.andinus* (Todd, 1952)   E Colombia, W Venezuela
  — *H.g.granadensis* (Hartlaub, 1843)   Colombia, NE Ecuador
  — *H.g.pyrrhops* (Cabanis, 1874)   S Ecuador, N Peru
  — *H.g.caesius* (Carriker, 1932)   C Peru, ?N Bolivia

○ *Hemitriccus mirandae* (E. Snethlage, 1925) Buff-breasted Tody-Tyrant
     NE Brazil

○ *Hemitriccus cinnamomeipectus* Fitzpatrick & O'Neill, 1979 Cinnamon-breasted Tody-Tyrant
     S Ecuador, N Peru (835)

○ *Hemitriccus kaempferi* (J.T. Zimmer, 1953) Kaempfer's Tody-Tyrant   SE Brazil

○ *Hemitriccus rufigularis* (Cabanis, 1873) Buff-throated Tody-Tyrant   Ecuador, Peru, N Bolivia

○ *Hemitriccus furcatus* (Lafresnaye, 1846) Fork-tailed Pygmy Tyrant   SE Brazil

MYIORNIS A. de W. Bertoni, 1901 M
○ *Myiornis auricularis* (Vieillot, 1818) Eared Pygmy Tyrant   SE Brazil, E Paraguay, N Argentina

○ *Myiornis albiventris* (Berlepsch & Stolzmann, 1894) White-bellied Pygmy Tyrant
     C Peru to C Bolivia

○ *Myiornis atricapillus* (Lawrence, 1875) Black-capped Pygmy Tyrant   E Costa Rica to NW Ecuador

---

[1] Separated from *H. zosterops* by Cohn-Haft *et al.* (1997) (543).
[2] Previously called *H. aenigma* but see Stotz (1992) (2356).

○ *Myiornis ecaudatus* Short-tailed Pygmy Tyrant
   — *M.e.miserabilis* (Chubb, 1919)   C Colombia, Trinidad, Guyana, Venezuela, Surinam, ?N Brazil

   — *M.e.ecaudatus* (d'Orbigny & Lafresnaye, 1837)   E Peru, E Bolivia, ?W Brazil

ONCOSTOMA P.L. Sclater, 1862 N
○ *Oncostoma cinereigulare* (P.L. Sclater, 1857) Northern Bentbill   S Mexico to W Panama

○ *Oncostoma olivaceum* (Lawrence, 1862) Southern Bentbill   E Panama, N Colombia

LOPHOTRICCUS Berlepsch, 1883 M
○ *Lophotriccus pileatus* Scale-crested Pygmy Tyrant
   — *L.p.luteiventris* Taczanowski, 1884   Costa Rica, W Panama
   — *L.p.sanctaeluciae* Todd, 1952   NE Colombia, N coast of Venezuela
   — *L.p.squamaecrista* (Lafresnaye, 1846)   S and C Colombia, W Ecuador, W Venezuela
   — *L.p.pileatus* (Tschudi, 1844)   E Ecuador, N and C Peru
   — *L.p.hypochlorus* Berlepsch & Stolzmann, 1906   SE Peru

○ *Lophotriccus eulophotes* Todd, 1925 Long-crested Pygmy Tyrant   W Brazil, SE Peru, N Bolivia

○ *Lophotriccus vitiosus* Double-banded Pygmy Tyrant
   — *L.v.affinis* J.T. Zimmer, 1940   SW Colombia to NE Peru, NW Brazil
   — *L.v.guianensis* J.T. Zimmer, 1940   SC Colombia, the Guianas, NE Brazil
   — *L.v.vitiosus* (Bangs & Penard, 1921)   E Peru
   — *L.v.congener* Todd, 1925   NW Brazil

○ *Lophotriccus galeatus* (Boddaert, 1783) Helmeted Pygmy Tyrant   The Guianas, N Brazil, Venezuela

ATALOTRICCUS Ridgway, 1905[1] M
○ *Atalotriccus pilaris* Pale-eyed Pygmy Tyrant
   — *A.p.wilcoxi* Griscom, 1924   W Panama
   — *A.p.pilaris* (Cabanis, 1847)   W Venezuela, N and E Colombia
   — *A.p.venezuelensis* Ridgway, 1906   N Venezuela
   — *A.p.griseiceps* (Hellmayr, 1911)   S Venezuela, W Guyana

POECILOTRICCUS Berlepsch, 1884[2] M
○ *Poecilotriccus ruficeps* Rufous-crowned Tody-Tyrant[3]
   — *P.r.melanomystax* Hellmayr, 1927   Andes of C and E Colombia and NW Venezuela
   — *P.r.ruficeps* (Kaup, 1852)   E Ecuador, S Colombia, SW Venezuela
   — *P.r.rufigenis* (Sclater & Salvin, 1877)   W Ecuador, W Colombia
   — *P.r.peruvianus* Chapman, 1924   NW Peru

○ *Poecilotriccus luluae* Johnson & Jones, 2001 Lulu's Tody-Tyrant #   NE Andes of Peru (Dept. Amazonas) (1214)

○ *Poecilotriccus albifacies* (Blake, 1959) White-cheeked Tody-Tyrant   SE Peru

○ *Poecilotriccus capitalis* (P.L. Sclater, 1857) Black-and-white Tody-Tyrant[4][5]
      SE Colombia, E Ecuador, NE Peru, SC Brazil

○ *Poecilotriccus senex* (Pelzeln, 1868) Buff-cheeked Tody-Flycatcher   C Brazil

○ *Poecilotriccus russatus* (Salvin & Godman, 1884) Ruddy Tody-Flycatcher
      SE Venezuela, NE Brazil

○ *Poecilotriccus plumbeiceps* Ochre-faced Tody-Flycatcher
   — *P.p.obscurus* (J.T. Zimmer, 1940)   SE Peru, N Bolivia
   — *P.p.viridiceps* (Salvadori, 1897)   C Bolivia, NW Argentina
   — *P.p.plumbeiceps* (Lafresnaye, 1846)   SE Brazil, Paraguay, NE Argentina
   — *P.p.cinereipectus* (Novaes, 1953)[6]   SC Brazil

---

[1] For recognition of this genus see Ridgely & Tudor (1994: 513) (2019).
[2] Includes several species from the genus *Todirostrum*, see Lanyon (1988) (1332).
[3] For a review see Johnson (2002) (1213).
[4] For correction of spelling see David & Gosselin (2002) (613).
[5] In last Edition we listed a species *P. tricolor*, but this is at best a weakly differentiated subspecies of *capitalis*, see Parker *et al.* (1997) (1751).
[6] For spelling see David & Gosselin (2002) (613).

○ *Poecilotriccus fumifrons* SMOKY-FRONTED TODY-FLYCATCHER
　　— *P.f.fumifrons* (Hartlaub, 1853)　　　　　　　　　NE Brazil
　　— *P.f.penardi* (Hellmayr, 1905)　　　　　　　　　　French Guiana, Surinam

○ *Poecilotriccus latirostris* RUSTY-FRONTED TODY-FLYCATCHER
　　— *P.l.mituensis* (Olivares, 1965)　　　　　　　　　Colombia
　　— *P.l.caniceps* (Chapman, 1924)　　　　　　　　　SE Colombia, E Ecuador, E Peru
　　— *P.l.latirostris* (Pelzeln, 1868)　　　　　　　　　C Brazil
　　— *P.l.mixtus* (J.T. Zimmer, 1940)　　　　　　　　SE Peru
　　— *P.l.ochropterus* (J.A. Allen, 1889)　　　　　　　S Brazil, SE Bolivia
　　— *P.l.austroriparius* (Todd, 1952)　　　　　　　　Santarem (Brazil)
　　— *P.l.senectus* (Griscom & Greenway, 1937)　　　Lower Amazon (Brazil)

○ *Poecilotriccus sylvia* SLATE-HEADED TODY-TYRANT
　　— *P.s.schistaceiceps* (P.L. Sclater, 1859)　　　　S Mexico to Panama
　　— *P.s.superciliaris* (Lawrence, 1871)　　　　　　N Colombia
　　— *P.s.griseolus* (Todd, 1913)　　　　　　　　　　N Venezuela
　　— *P.s.sylvia* (Desmarest, 1806)　　　　　　　　　Guyana, French Guiana, N Brazil
　　— *P.s.schulzi* (Berlepsch, 1907)　　　　　　　　　NE Brazil

○ *Poecilotriccus calopterus* GOLDEN-WINGED TODY-FLYCATCHER[1]
　　— *P.c.calopterus* (P.L. Sclater, 1857)　　　　　　S Colombia, E Ecuador
　　— *P.c.pulchellus* (P.L. Sclater, 1874)　　　　　　SE Peru

TAENIOTRICCUS Berlepsch & Hartert, 1902[2]　M
○ *Taeniotriccus andrei* BLACK-CHESTED TYRANT
　　— *T.a.andrei* Berlepsch & Hartert, 1902　　　　　SE Venezuela, NW Brazil
　　— *T.a.klagesi* Todd, 1925　　　　　　　　　　　　Rio Tapajos (C Brazil)

TODIROSTRUM Lesson, 1831　N
○ *Todirostrum maculatum* SPOTTED TODY-FLYCATCHER
　　— *T.m.amacurense* Eisenmann & Phelps, 1971　　NE Venezuela, Guyana, Trinidad
　　— *T.m.maculatum* (Desmarest, 1806)　　　　　　E Venezuela, the Guianas, N Brazil
　　— *T.m.signatum* Sclater & Salvin, 1881　　　　　E Ecuador, E Peru, W Brazil
　　— *T.m.diversum* J.T. Zimmer, 1940　　　　　　　WC Brazil
　　— *T.m.annectens* J.T. Zimmer, 1940　　　　　　　C Brazil

○ *Todirostrum poliocephalum* (Wied, 1831)　GREY-HEADED TODY-FLYCATCHER
　　　　　　　　　　　　　　　　　　　　　　　　　　　SE Brazil

● *Todirostrum cinereum* COMMON TODY-FLYCATCHER
　　— *T.c.virididorsale* Parkes, 1976　　　　　　　　SC Mexico
　　↙ *T.c.finitimum* Bangs, 1904　　　　　　　　　　S Mexico to Panama
　　— *T.c.wetmorei* Parkes, 1976　　　　　　　　　　C and E Costa Rica, Panama
　　— *T.c.sclateri* (Cabanis & Heine, 1859)　　　　　SW Colombia, W Ecuador, NW Peru
　　— *T.c.cinereum* (Linnaeus, 1766)　　　　　　　　The Guianas, N Brazil, S Venezuela, S Colombia
　　— *T.c.peruanum* J.T. Zimmer, 1930　　　　　　　E Ecuador, N and E Peru
　　— *T.c.coloreum* Ridgway, 1906　　　　　　　　　C Brazil, E Bolivia
　　— *T.c.cearae* Cory, 1916　　　　　　　　　　　　E Brazil

○ *Todirostrum viridanum* Hellmayr, 1927　MARACAIBO TODY-FLYCATCHER　NW Venezuela

○ *Todirostrum pictum* Salvin, 1897　PAINTED TODY-FLYCATCHER　The Guianas, N Brazil

○ *Todirostrum chrysocrotaphum* YELLOW-BROWED TODY-FLYCATCHER
　　— *T.c.guttatum* Pelzeln, 1868　　　　　　　　　　SW Venezuela, Colombia, NW Brazil, N Peru
　　— *T.c.neglectum* Carriker, 1932　　　　　　　　　NE Peru, N Bolivia, SW Brazil
　　— *T.c.chrysocrotaphum* Strickland, 1850　　　　　SE Peru, W Amazonia
　　— *T.c.simile* J.T. Zimmer, 1940[3]　　　　　　　　E Brazil (W Para)
　　— *T.c.illigeri* (Cabanis & Heine, 1859)　　　　　E Brazil (E Para, Maranhão)

---

[1] We prefer to await detailed publication before splitting this.
[2] Treated in the genus *Poecilotriccus* in last Edition, but see Ridgely & Tudor (1994: 539) (2019).
[3] For correction of spelling see David & Gosselin (2002) (613).

○ *Todirostrum nigriceps* P.L. Sclater, 1855 BLACK-HEADED TODY-FLYCATCHER

Costa Rica, Panama, N Colombia, W Ecuador

CNIPODECTES Sclater & Salvin, 1873 M
○ *Cnipodectes subbrunneus* BROWNISH FLYCATCHER
    __ *C.s.subbrunneus* (P.L. Sclater, 1860) — W Ecuador, W Colombia
    __ *C.s.panamensis* J.T. Zimmer, 1939 — E Panama, N Colombia
    __ *C.s.minor* P.L. Sclater, 1884 — SE Colombia, E Peru, W Brazil, NW Bolivia

RHYNCHOCYCLUS Cabanis & Heine, 1859 M
● *Rhynchocyclus brevirostris* EYE-RINGED FLATBILL
    √ *R.b.brevirostris* (Cabanis, 1847) — S Mexico to W Panama
    __ *R.b.pallidus* Binford, 1965 — S Mexico
    __ *R.b.hellmayri* Griscom, 1932 — E Panama, NW Colombia

○ *Rhynchocyclus olivaceus* OLIVACEOUS FLATBILL
    __ *R.o.bardus* (Bangs & Barbour, 1922) — E Panama, N Colombia
    __ *R.o.mirus* Meyer de Schauensee, 1950 — NW Colombia
    __ *R.o.tamborensis* Todd, 1952 — Colombia
    __ *R.o.jelambianus* Aveledo & Perez, 1994 — N Venezuela (69)
    __ *R.o.flavus* (Chapman, 1914) — NE Colombia, NW Venezuela
    __ *R.o.aequinoctialis* (P.L. Sclater, 1858) — S Colombia, E Ecuador, E Peru, NC Bolivia
    __ *R.o.guianensis* McConnell, 1911 — N and NW Amazonia
    __ *R.o.sordidus* Todd, 1952 — C Brazil
    __ *R.o.olivaceus* (Temminck, 1820) — SE Brazil

○ *Rhynchocyclus pacificus* (Chapman, 1914) PACIFIC FLATBILL[1] — NW Ecuador, W Colombia

○ *Rhynchocyclus fulvipectus* (P.L. Sclater, 1860) FULVOUS-BREASTED FLATBILL

Colombia, E Ecuador, SE Peru

TOLMOMYIAS Hellmayr, 1927 M
● *Tolmomyias sulphurescens* YELLOW-OLIVE FLYCATCHER
    √ *T.s.cinereiceps* (P.L. Sclater, 1859) — S Mexico to Costa Rica
    __ *T.s.flavoolivaceus* (Lawrence, 1863) — Panama
    __ *T.s.berlepschi* (Hartert & Goodson, 1917) — Trinidad
    __ *T.s.exortivus* (Bangs, 1908) — N Venezuela, N Colombia
    __ *T.s.asemus* (Bangs, 1910) — Colombia
    __ *T.s.confusus* J.T. Zimmer, 1939 — C Colombia, SW Venezuela, NE Ecuador
    __ *T.s.duidae* J.T. Zimmer, 1939 — SE Venezuela, NW Brazil
    __ *T.s.aequatorialis* (Berlepsch & Taczanowski, 1883) — W Ecuador, NW Peru
    __ *T.s.cherriei* (Hartert & Goodson, 1917) — E Colombia to the Guianas, N Brazil
    __ *T.s.peruvianus* (Taczanowski, 1874) — SE Ecuador, N Peru
    __ *T.s.insignis* J.T. Zimmer, 1939 — NE Peru, W Brazil
    __ *T.s.mixtus* J.T. Zimmer, 1939 — NE Brazil
    __ *T.s.inornatus* J.T. Zimmer, 1939 — SE Peru
    __ *T.s.pallescens* (Hartert & Goodson, 1917) — C Brazil to N Argentina
    __ *T.s.grisecens* (Chubb, 1910) — C Paraguay, N Argentina
    __ *T.s.sulphurescens* (Spix, 1825) — S Brazil, N Argentina

○ *Tolmomyias traylori* Schulenberg & Parker, 1997 ORANGE-EYED FLYCATCHER #

SE Colombia, E Ecuador, N Peru (2211)

○ *Tolmomyias assimilis* YELLOW-MARGINED FLYCATCHER
    __ *T.a.flavotectus* (E. Hartert, 1902) — Costa Rica, to NW Ecuador
    __ *T.a.neglectus* J.T. Zimmer, 1939 — E Colombia, S Venezuela, NE Brazil
    __ *T.a.obscuriceps* J.T. Zimmer, 1939 — S Colombia to NE Peru
    __ *T.a.examinatus* (Chubb, 1920) — SE Venezuela, the Guianas, N Brazil
    __ *T.a.assimilis* (Pelzeln, 1868) — C Brazil
    __ *T.a.clarus* J.T. Zimmer, 1939 — C Peru
    __ *T.a.paraensis* J.T. Zimmer, 1939 — N Brazil
    __ *T.a.calamae* J.T. Zimmer, 1939 — N Bolivia, SW Brazil

[1] Treated as a race of *brevirostris* in last Edition, but see A.O.U. (1998: 384) (3).

○ *Tolmomyias poliocephalus* Grey-crowned Flycatcher
    — *T.p.poliocephalus* (Taczanowski, 1884)   W Amazonia
    — *T.p.klagesi* (Ridgway, 1906)   S Venezuela
    — *T.p.sclateri* (Hellmayr, 1903)   The Guianas, N Brazil

○ *Tolmomyias flaviventris* Yellow-breasted Flycatcher
    — *T.f.aurulentus* (Todd, 1913)   N Colombia, N Venezuela
    — *T.f.collingwoodi* (Chubb, 1920)   C and E Colombia to Guyana, N Brazil, Trinidad
    — *T.f.dissors* J.T. Zimmer, 1939   NE Brazil, SE Venezuela
    — *T.f.viridiceps* (Sclater & Salvin, 1873)   SE Colombia, E Ecuador, E Peru, W Brazil
    — *T.f.zimmeri* Bond, 1947   NC Peru
    — *T.f.subsimilis* Carriker, 1935   SE Peru, NW Bolivia, SW Brazil
    — *T.f.flaviventris* (Wied, 1831)   E Bolivia to E Brazil

PLATYRINCHUS Desmarest, 1805   M
○ *Platyrinchus saturatus* Cinnamon-crested Spadebill
    — *P.s.saturatus* Salvin & Godman, 1882   S Venezuela, the Guianas, N Brazil
    — *P.s.pallidiventris* Novaes, 1968   C Brazil

○ *Platyrinchus cancrominus* Stub-tailed Spadebill
    — *P.c.cancrominus* Sclater & Salvin, 1860   S Mexico to E Nicaragua
    — *P.c.timothei* Paynter, 1954   SE Mexico to N Guatemala
    — *P.c.dilutus* (Miller & Griscom, 1925)   El Salvador to NW Costa Rica

○ *Platyrinchus mystaceus* White-throated Spadebill
    — *P.m.neglectus* (Todd, 1919)   E Costa Rica to E Colombia
    — *P.m.perijanus* Phelps & Phelps, Jr., 1954   NW Venezuela
    — *P.m.insularis* J.A. Allen, 1889   Trinidad, N Venezuela
    — *P.m.imatacae* Zimmer & Phelps, 1945   E Venezuela
    — *P.m.ventralis* Phelps & Phelps, Jr., 1955   S Venezuela
    — *P.m.duidae* J.T. Zimmer, 1939   SE Venezuela, N Brazil
    — *P.m.ptaritepui* Zimmer & Phelps, 1946   SE Venezuela
    — *P.m.albogularis* P.L. Sclater, 1860   W Ecuador, W Colombia
    — *P.m.zamorae* (Chapman, 1924)   E Ecuador, N Peru
    — *P.m.partridgei* Short, 1969   S Peru, N Bolivia
    — *P.m.mystaceus* Vieillot, 1818   E Brazil, Paraguay, N Argentina
    — *P.m.bifasciatus* J.A. Allen, 1889   SW Brazil
    — *P.m.cancromus* Temminck, 1820   E Brazil
    — *P.m.niveigularis* Pinto, 1954   NE Brazil

○ *Platyrinchus coronatus* Golden-crowned Spadebill
    — *P.c.superciliaris* Lawrence, 1863   Nicaragua to W Ecuador
    — *P.c.gumia* (Bangs & Penard, 1918)   The Guianas, N Brazil
    — *P.c.coronatus* P.L. Sclater, 1858   W Amazonia

○ *Platyrinchus flavigularis* Yellow-throated Spadebill
    — *P.f.flavigularis* P.L. Sclater, 1862   Colombia to Peru
    — *P.f.vividus* Phelps & Phelps, Jr., 1952   W Venezuela

○ *Platyrinchus platyrhynchos* White-crested Spadebill
    — *P.p.platyrhynchos* (J.F. Gmelin, 1788)   S Venezuela, the Guianas, N Brazil
    — *P.p.senex* Sclater & Salvin, 1880   E Ecuador, E Peru, N Bolivia
    — *P.p.nattereri* Hartert & Hellmayr, 1902   W Brazil
    — *P.p.amazonicus* Berlepsch, 1912   E Brazil

○ *Platyrinchus leucoryphus* Wied, 1831 Russet-winged Spadebill   SE Brazil, E Paraguay

ONYCHORHYNCHUS J.G. Fischer[1], 1810[2]   M
● *Onychorhynchus coronatus* Royal Flycatcher
    ↓ *O.c.mexicanus* (P.L. Sclater, 1857)   SE Mexico to E Panama

[1] Fischer is an abbreviation of Fischer von Waldheim.
[2] For reasons to treat the *mexicanus* group as conspecific with *coronatus* see Ridgely & Tudor (1994: 556-7) (2019).

— *O.c.fraterculus* Bangs, 1902	NE Colombia, E Venezuela
— *O.c.castelnaui* Deville, 1849	W Amazonia
— *O.c.coronatus* (Statius Müller, 1776)	E Venezuela, the Guianas, N Brazil
— *O.c.occidentalis* (P.L. Sclater, 1860)	W Ecuador
— *O.c.swainsoni* (Pelzeln, 1858)	SE Brazil

MYIOBIUS Darwin, 1839[1] M
○ *Myiobius villosus* Tawny-breasted Flycatcher

— *M.v.villosus* P.L. Sclater, 1860	E Panama, Colombia, W Ecuador
— *M.v.schaeferi* Aveledo & Pons, 1952	NW Venezuela, NE Colombia
— *M.v.clarus* J.T. Zimmer, 1939	E Ecuador
— *M.v.peruvianus* Todd, 1922	Peru, N Bolivia

○ *Myiobius barbatus* Sulphur-rumped Flycatcher[2]

— *M.b.sulphureipygius* (P.L. Sclater, 1857)	S Mexico to Honduras
— *M.b.aureatus* Bangs, 1908	S Honduras to W Ecuador
— *M.b.semiflavus* Todd, 1919	EC Colombia
— *M.b.barbatus* (J.F. Gmelin, 1789)	SE Colombia to the Guianas, N Brazil, N Peru
— *M.b.amazonicus* Todd, 1925	E Peru
— *M.b.insignis* J.T. Zimmer, 1939	NE Brazil
— *M.b.mastacalis* (Wied, 1821)	SE Brazil

○ *Myiobius atricaudus* Black-tailed Flycatcher

— *M.a.atricaudus* Lawrence, 1863	SW Costa Rica, Panama, W Colombia
— *M.a.portovelae* Chapman, 1924	W Ecuador, N Peru
— *M.a.modestus* Todd, 1912	E Venezuela
— *M.a.adjacens* J.T. Zimmer, 1939	S Colombia, E Ecuador, E Peru, W Brazil
— *M.a.connectens* J.T. Zimmer, 1939	Inland NE Brazil
— *M.a.snethlagei* Hellmayr, 1927	Coastal NE Brazil
— *M.a.ridgwayi* Berlepsch, 1888	SE Brazil

TERENOTRICCUS Ridgway, 1905 M
○ *Terenotriccus erythrurus* Ruddy-tailed Flycatcher

— *T.e.fulvigularis* (Salvin & Godman, 1889)	Guatemala to Panama, Colombia, Venezuela
— *T.e.signatus* J.T. Zimmer, 1939	C and S Colombia to NE Peru and W Brazil
— *T.e.hellmayri* (E. Snethlage, 1907)	Para and Maranhão (NE Brazil)
— *T.e.venezuelensis* J.T. Zimmer, 1939	E Colombia, W Venezuela, NW Brazil
— *T.e.erythrurus* (Cabanis, 1847)	The Guianas, S Venezuela, N Brazil
— *T.e.brunneifrons* Hellmayr, 1927	SW Brazil, E Peru, N Bolivia
— *T.e.purusianus* (Parkes & Panza, 1993)	Rio Purus (C Brazil) (1787)
— *T.e.amazonus* J.T. Zimmer, 1939	Rio Purus to R. Tapajos (C Brazil)

NEOPIPO Sclater & Salvin, 1869[3] F
○ *Neopipo cinnamomea* Cinnamon Tyrant

— *N.c.helenae* McConnell, 1911	Guyana, French Guiana, N Brazil
— *N.c.cinnamomea* (Lawrence, 1869)	E Ecuador, E Peru, W Brazil

PYRRHOMYIAS Cabanis & Heine, 1859 M[4]
○ *Pyrrhomyias cinnamomeus* Cinnamon Flycatcher

— *P.c.assimilis* (J.A. Allen, 1900)	NW Colombia
— *P.c.pyrrhopterus* (Hartlaub, 1843)	Colombia, Venezuela, Ecuador, N Peru
— *P.c.vieillotioides* (Lafresnaye, 1848)	Coastal ranges NW Venezuela
— *P.c.spadix* Wetmore, 1939	Coastal ranges NE Venezuela
— *P.c.pariae* Phelps & Phelps, Jr., 1949	Paria penin. (NE Venezuela)
— *P.c.cinnamomeus* (d'Orbigny & Lafresnaye, 1837)	Bolivia, Peru, NW Argentina

---

[1] See Lanyon (1988) (1331).
[2] For reasons to treat the *sulphureipygius* group as conspecific with *barbatus* see Ridgely & Tudor (1994: 558-9) (2019).
[3] Previously treated (e.g. in last Edition) in the family Pipridae, but see Mobley & Prum (1997) (1596).
[4] Gender addressed by David & Gosselin (2002) (614), multiple changes occur in specific names.

HIRUNDINEA d'Orbigny & Lafresnaye, 1837 F
○ *Hirundinea ferruginea* CLIFF FLYCATCHER
    — *H.f.ferruginea* (J.F. Gmelin, 1788)   Guyana, French Guiana, N Brazil
    — *H.f.sclateri* Reinhardt, 1870   E Colombia, Peru
    — *H.f.bellicosa* (Vieillot, 1819)   S Brazil, Paraguay, NE Argentina
    — *H.f.pallidior* Hartert & Goodson, 1917   Bolivia, NW Argentina

LATHROTRICCUS W.E. & S.M. Lanyon, 1986 (1334) M
○ *Lathrotriccus euleri* EULER'S FLYCATCHER[1]
    — †?*L.e.flaviventris* Lawrence, 1887[2]   Grenada
    — *L.e.lawrencei* (J.A. Allen, 1889)   Trinidad, N Venezuela, Surinam
    — *L.e.bolivianus* (J.A. Allen, 1889)   N and W Amazonia
    — *L.e.argentinus* (Cabanis, 1868)   SW to SE Amazonia >> E Peru, E Brazil
    — *L.e.euleri* (Cabanis, 1868)   SE Brazil, NE Argentina >> Peru, Bolivia

○ *Lathrotriccus griseipectus* (Lawrence, 1870) GREY-BREASTED FLYCATCHER[3]
                                     SW Ecuador, NW Peru

APHANOTRICCUS Ridgway, 1905 M
○ *Aphanotriccus capitalis* (Salvin, 1865) TAWNY-CHESTED FLYCATCHER   Nicaragua, Costa Rica

○ *Aphanotriccus audax* (Nelson, 1912) BLACK-BILLED FLYCATCHER   Panama, NW Colombia

CNEMOTRICCUS Hellmayr, 1927 M
○ *Cnemotriccus fuscatus* FUSCOUS FLYCATCHER
    — *C.f.cabanisi* (Léotaud, 1866)   Trinidad, Tobago I., Venezuela, Colombia
    — *C.f.fuscatior* (Chapman, 1926)   SW Venezuela, N Brazil, E Peru
    — *C.f.duidae* J.T. Zimmer, 1938   SE Venezuela
    — *C.f.fumosus* (Berlepsch, 1908)   The Guianas, N Brazil
    — *C.f.bimaculatus* (d'Orbigny & Lafresnaye, 1837)   W and C Brazil, E Bolivia, Paraguay, N Argentina
    — *C.f.beniensis* Gyldenstolpe, 1941   N Bolivia
    — *C.f.fuscatus* (Wied, 1831)   SE Brazil, N Argentina

XENOTRICCUS Dwight & Griscom, 1927 M
○ *Xenotriccus callizonus* Dwight & Griscom, 1927 BELTED FLYCATCHER   Mexico

○ *Xenotriccus mexicanus* (J.T. Zimmer, 1938) PILEATED FLYCATCHER   S Mexico

SAYORNIS Bonaparte, 1854 M[4]
● *Sayornis phoebe* (Latham, 1790) EASTERN PHOEBE   C and SE Canada, E USA >> E Mexico

● *Sayornis nigricans* BLACK PHOEBE
    ✓ *S.n.nigricans* (Swainson, 1827)   SW USA, Mexico
    ✓ *S.n.semiater* (Vigors, 1839)   W Mexico, W USA
    — *S.n.aquaticus* Sclater & Salvin, 1859   Guatemala, Nicaragua
    — *S.n.amnicola* Bangs, 1902   Costa Rica, W Panama
    — *S.n.angustirostris* Berlepsch & Stolzmann, 1896   C and N Colombia, E Panama, W Venezuela to Peru
    — *S.n.latirostris* (Cabanis & Heine, 1859)   Bolivia, NW Argentina

● *Sayornis saya* SAY'S PHOEBE
    — *S.s.pallidus* (Swainson, 1827)   S Mexico
    ✓ *S.s.saya* (Bonaparte, 1825)   WC Canada, W USA, N Mexico
    — *S.s.quiescens* Grinnell, 1926   NW Baja California

MITREPHANES Coues, 1882 M
● *Mitrephanes phaeocercus* TUFTED FLYCATCHER
    — *M.p.tenuirostris* Brewster, 1888   W Mexico
    — *M.p.phaeocercus* (P.L. Sclater, 1859)   S and E Mexico to Honduras
    — *M.p.burleighi* A.R. Phillips, 1966   Guerrero, SW Oaxaca

---

[1] See Lanyon & Lanyon (1986) (1334).
[2] See Banks (1997) (90) for use of this name in place of *johnstonei*.
[3] In last Edition we treated this in the genus *Empidonax* but we should have moved this with *euleri* into the newly erected genus *Lathrotriccus*.
[4] Gender addressed by David & Gosselin (2002) (614), multiple changes occur in specific names.

___ *M.p.nicaraguae* Miller & Griscom, 1925   S Mexico to Nicaragua

___ *M.p.aurantiiventris* (Lawrence, 1865)   Costa Rica, W Panama

___ *M.p.vividus* Griscom, 1927   C Panama

___ *M.p.eminulus* Nelson, 1912   E Panama, W Colombia

___ *M.p.berlepschi* E. Hartert, 1902   S Colombia, NE Ecuador

○ *Mitrephanes olivaceus* Berlepsch & Stolzmann, 1894   OLIVE FLYCATCHER

       Peru, N Bolivia

CONTOPUS Cabanis, 1855   M

● *Contopus cooperi* (Nuttall, 1831)   OLIVE-SIDED FLYCATCHER[1]   W Canada, N and W USA >> C America, N and W Sth. America

○ *Contopus pertinax*   GREATER PEWEE[2]

___ *C.p.pallidiventris* Chapman, 1897   S Arizona, N Mexico >> Belize

___ *C.p.pertinax* Cabanis & Heine, 1859   C and S Mexico, Guatemala

___ *C.p.minor* (Miller & Griscom, 1925)   Honduras, Belize, N Nicaragua

○ *Contopus lugubris* Lawrence, 1865   DARK PEWEE   Costa Rica, W Panama

○ *Contopus fumigatus*   SMOKE-COLOURED PEWEE

___ *C.f.cineraceus* (Lafresnaye, 1848)   N Venezuela

___ *C.f.duidae* (Chapman, 1929)   SE Venezuela

___ *C.f.ardosiacus* (Lafresnaye, 1844)   NE Peru, E Ecuador, Colombia, NW Venezuela

___ *C.f.zarumae* (Chapman, 1924)   W Ecuador, NE Peru

___ *C.f.fumigatus* (d'Orbigny & Lafresnaye, 1837)   SE Peru, Bolivia

___ *C.f.brachyrhynchus* Cabanis, 1883   NW Argentina

○ *Contopus ochraceus* Sclater & Salvin, 1869   OCHRACEOUS PEWEE   Costa Rica, W Panama

● *Contopus sordidulus*   WESTERN WOOD PEWEE

___ *C.s.sordidulus* P.L. Sclater, 1859   S Mexico >> Colombia, Ecuador

___ *C.s.veliei* Coues, 1866[3]   W and WC USA, W Mexico, >> Panama, N Sth. America

___ *C.s.saturatus* Bishop, 1900   W Nth. America >> N Sth. America

___ *C.s.peninsulae* Brewster, 1891   S Baja California, SE Mexico >> Sth. America

___ *C.s.griscomi* Webster, 1957   Mexico

● *Contopus virens* (Linnaeus, 1766)   EASTERN WOOD PEWEE   E Canada, E USA >> N Sth. America

● *Contopus cinereus*   TROPICAL PEWEE

___ *C.c.cinereus* (Spix, 1825)   SE Brazil, N Argentina, Paraguay

___ *C.c.pallescens* (Hellmayr, 1927)   S Brazil, N Paraguay

___ *C.c.surinamensis* Penard & Penard, 1910   The Guianas, S Venezuela, N Brazil

___ *C.c.bogotensis* (Bonaparte, 1850)   Trinidad, N Venezuela, N Colombia

___ *C.c.punensis* Lawrence, 1869   SW Ecuador, Peru, E Bolivia

___ *C.c.rhizophorus* (Dwight & Griscom, 1924)   W Costa Rica

___ *C.c.brachytarsus* (P.L. Sclater, 1859)   SE Mexico to Panama

___ *C.c.aithalodes* Wetmore, 1957   Coiba I. (Panama)

○ *Contopus albogularis* (Berlioz, 1962)   WHITE-THROATED PEWEE   French Guiana, NE Brazil

○ *Contopus nigrescens*   BLACKISH PEWEE[4]

___ *C.n.nigrescens* (Sclater & Salvin, 1880)   E Ecuador

___ *C.n.canescens* (Chapman, 1926)   NE Peru

○ *Contopus caribaeus*   CUBAN PEWEE[5]

___ *C.c.caribaeus* (d'Orbigny, 1839)   Cuba

___ *C.c.morenoi* Burleigh & Duvall, 1948   S Cuba

___ *C.c.nerlyi* Garrido, 1978   Islands off SC Cuba

___ *C.c.bahamensis* (H. Bryant, 1859)   Bahamas

---

[1] Previously *C. borealis* but see Banks & Browning (1995: 636) (91).
[2] In last Edition this species and *C. lugubris* were attached to the species *C. fumigatus*, but see A.O.U. (1998: 391) (3).
[3] Includes *amplus* Burleigh, 1960 (273), and *siccicola* Burleigh, 1960 (273), see Browning (1990) (249).
[4] Records from Guyana and E Brazil not identified to subspecies.
[5] Separated into three species by Reynard *et al.* (1993) (2013). The status and exact date of the name *florentinoi* Regalado Ruiz, 1977, is unclear and we give no reference (but see Olsen & Buden, 1989) (1707).

○ *Contopus hispaniolensis* Hispaniolan Pewee
    — *C.h.hispaniolensis* (H. Bryant, 1867)   Hispaniola
    — *C.h.tacitus* (Wetmore, 1928)   Gonave I.

○ *Contopus pallidus* (Gosse, 1847) Jamaican Pewee   Jamaica

○ *Contopus latirostris* Lesser Antillean Pewee
    — *C.l.latirostris* (J. Verreaux, 1866)   St. Lucia I.
    — *C.l.brunneicapillus* (Lawrence, 1878)   Dominica, Guadelupe, Martinique
    — *C.l.blancoi* (Cabanis, 1875)   Puerto Rico

EMPIDONAX Cabanis, 1855[1] M
○ *Empidonax flaviventris* (W.M. & S.F. Baird, 1843) Yellow-bellied Flycatcher
                                          S Canada, N USA >> S Mexico, C America

◐ *Empidonax virescens* (Vieillot, 1818) Acadian Flycatcher   E USA, E Mexico >> C America, Colombia, W Ecuador

● *Empidonax traillii* Willow Flycatcher
    — *E.t.brewstrei* Oberholser, 1918   S British Columbia to Sierra Nevada
    ✔ *E.t.adastus* Oberholser, 1932   C Rocky Mts. and Great Basin to Utah and Colorado
    — *E.t.extimus* A.R. Phillips, 1948   S California to New Mexico
    — *E.t.traillii* (Audubon, 1828)   SE Canada and NE USA to Great Plains >> C America
                                          to NW Sth. America

● *Empidonax alnorum* Brewster, 1895 Alder Flycatcher   E Canada, E USA >> C America and Sth. America

○ *Empidonax albigularis* White-throated Flycatcher
    — *E.a.timidus* Nelson, 1900   NW Mexico
    — *E.a.albigularis* Sclater & Salvin, 1859   C and S Mexico, Guatemala, Honduras
    — *E.a.australis* Miller & Griscom, 1925   Nicaragua to Panama

◐ *Empidonax minimus* (W.M. & S.F. Baird, 1843) Least Flycatcher   E Canada, E USA >> S Mexico, C America, NW Sth.
                                          America

● *Empidonax hammondii* (Xántus de Vesey, 1858) Hammond's Flycatcher
                                          W Canada, W USA >> Mexico, Guatemala

● *Empidonax oberholseri* A.R. Phillips, 1939 Dusky Flycatcher[2]   W USA, Mexico

● *Empidonax wrightii* S.F. Baird, 1858 Grey Flycatcher   W USA, California >> Mexico

○ *Empidonax affinis* Pine Flycatcher
    — *E.a.pulverius* Brewster, 1889   NW Mexico
    — *E.a.trepidus* Nelson, 1901   NE Mexico, Guatemala
    — *E.a.affinis* (Swainson, 1827)   C Mexico
    — *E.a.bairdi* P.L. Sclater, 1858   C and S Mexico
    — *E.a.vigensis* A.R. Phillips, 1942   E Mexico

● *Empidonax difficilis* Pacific-slope Flycatcher
    ✔ *E.d.difficilis* S.F. Baird, 1858   W Canada, W USA, NW Mexico >> Mexico
    — *E.d.cineritius* Brewster, 1888   Baja California >> ?
    — *E.d.insulicola* Oberholser, 1897   Channel Is. (off S California) >> ?

● *Empidonax occidentalis* Cordilleran Flycatcher[3]
    — *E.o.hellmayri* Brodkorb, 1935   N Mexico
    — *E.o.occidentalis* Nelson, 1897   S and C Mexico

● *Empidonax flavescens* Yellowish Flycatcher
    — *E.f.imperturbatus* Wetmore, 1942   S Mexico
    — *E.f.salvini* Ridgway, 1886   SE Mexico to Nicaragua
    ✔ *E.f.flavescens* Lawrence, 1865   Costa Rica, W Panama

○ *Empidonax fulvifrons* Buff-breasted Flycatcher
    — *E.f.pygmaeus* Coues, 1865   SW USA, W Mexico
    — *E.f.fulvifrons* (Giraud, 1841)   NE Mexico

---

[1] See Lanyon (1986) (1329) and Johnson & Marten (1988) (1215).
[2] Includes *spodius* Oberholser, 1974 (1673), see Browning (1978) (246).
[3] For reasons to treat this as distinct see A.O.U. (1998: 398) (3).

— *E.f.rubicundus* Cabanis & Heine, 1859 — C and S Mexico
— *E.f.brodkorbi* A.R. Phillips, 1966 — S Oaxaca (Mexico)
— *E.f.fusciceps* Nelson, 1904 — SE Mexico, Guatemala
— *E.f.inexpectatus* Griscom, 1932 — SC Honduras

○ *Empidonax atriceps* Salvin, 1870 BLACK-CAPPED FLYCATCHER — Costa Rica, W Panama

PYROCEPHALUS Gould, 1839 M
● *Pyrocephalus rubinus* VERMILION FLYCATCHER
✓ *P.r.mexicanus* P.L. Sclater, 1859 — SW USA, Mexico
— *P.r.flammeus* van Rossem, 1934 — N Mexico
— *P.r.blatteus* Bangs, 1911 — S Mexico to Honduras
— *P.r.pinicola* T.R. Howell, 1965 — Nicaragua
— *P.r.piurae* J.T. Zimmer, 1941 — W Colombia to NW Peru
— *P.r.saturatus* Berlepsch & Hartert, 1902 — N Colombia to Surinam, N Brazil
— *P.r.ardens* J.T. Zimmer, 1941 — NC Peru
— *P.r.cocachacrae* J.T. Zimmer, 1941 — SW Peru, N Chile
— *P.r.obscurus* Gould, 1839 — W Peru
— *P.r.rubinus* (Boddaert, 1783) — W and C Amazonia
— *P.r.major* Pelzeln, 1868 — SE Peru
— *P.r.nanus* Gould, 1839 — Galapagos Is.
— *P.r.dubius* Gould, 1839 — Chatham I. (Galapagos)

LESSONIA Swainson, 1832 F
● *Lessonia rufa* (J.F. Gmelin, 1789) AUSTRAL NEGRITO — Chile, Argentina >> Bolivia, S Brazil

○ *Lessonia oreas* (Sclater & Salvin, 1869) ANDEAN NEGRITO — S Peru, Chile, W Bolivia, NW Argentina

KNIPOLEGUS Boie, 1826 M
○ *Knipolegus striaticeps* (d'Orbigny & Lafresnaye, 1837) CINEREOUS TYRANT
   SE Bolivia, SC Brazil, W Paraguay, N Argentina

○ *Knipolegus hudsoni* (P.L. Sclater, 1872) HUDSON'S BLACK TYRANT — S Argentina >> Bolivia, S Brazil

○ *Knipolegus poecilocercus* (Pelzeln, 1868) AMAZONIAN BLACK TYRANT — S Venezuela and Guyana to NE Peru, C Brazil

○ *Knipolegus signatus* ANDEAN TYRANT
— *K.s.signatus* (Taczanowski, 1874) — N Peru
— *K.s.cabanisi* Schulz, 1882 — SE Peru, Bolivia, N Argentina

○ *Knipolegus cyanirostris* (Vieillot, 1818) BLUE-BILLED BLACK TYRANT — SE Brazil, Uruguay, Paraguay, E Argentina

○ *Knipolegus poecilurus* RUFOUS-TAILED TYRANT
— *K.p.poecilurus* (P.L. Sclater, 1862) — Colombia
— *K.p.peruanus* (Berlepsch & Stolzmann, 1896) — Peru, C Bolivia
— *K.p.venezuelanus* (Hellmayr, 1927) — Trinidad, Venezuela, NW Brazil
— *K.p.salvini* (P.L. Sclater, 1888) — Guyana
— *K.p.paraquensis* Phelps & Phelps, Jr., 1949 — S Venezuela

○ *Knipolegus orenocensis* RIVERSIDE TYRANT
— *K.o.orenocensis* Berlepsch, 1884 — Venezuela
— *K.o.xinguensis* Berlepsch, 1912 — N Brazil
— *K.o.sclateri* Hellmayr, 1906 — N Peru, C Brazil

○ *Knipolegus aterrimus* WHITE-WINGED BLACK TYRANT[1]
— *K.a.heterogyna* Berlepsch, 1907 — N Peru
— *K.a.anthracinus* Heine, 1859 — S Peru, W Bolivia
— *K.a.aterrimus* Kaup, 1853 — E Bolivia, W Argentina
— *K.a.franciscanus* E. Snethlage, 1928[2] — EC Brazil

○ *Knipolegus lophotes* Boie, 1828 CRESTED BLACK TYRANT — S Brazil, NE Paraguay, Uruguay

---

[1] The proposed "separation" of *franciscanus* seems to have been as a phylogenetic species.
[2] Probably 1927 see J.f. Orn. (1930: 134).

○ *Knipolegus nigerrimus* Velvety Black Tyrant
    — *K.n.hoflingi* Lenciono Neto, 1996   Bahia (Brazil) (1352)
    — *K.n.nigerrimus* (Vieillot, 1818)   SE Brazil

HYMENOPS Lesson, 1828 M[1]
○ *Hymenops perspicillatus* Spectacled Tyrant
    — *H.p.perspicillatus* (J.F. Gmelin, 1789)   Argentina, Paraguay, Uruguay
    — *H.p.andinus* (Ridgway, 1879)   C Chile, NW Argentina >> E Bolivia

OCHTHORNIS P.L. Sclater, 1888 M
○ *Ochthornis littoralis* (Pelzeln, 1868) Drab Water Tyrant[2]   E Colombia, S Venezuela, Guyana, W and C Brazil, E Ecuador, E Peru, N Bolivia

SATRAPA Strickland, 1844 M
○ *Satrapa icterophrys* (Vieillot, 1818) Yellow-browed Tyrant   E Bolivia, Brazil, Paraguay, Uruguay, N Argentina, N Venezuela

MUSCISAXICOLA d'Orbigny & Lafresnaye, 1837 M[3]
○ *Muscisaxicola fluviatilis* Sclater & Salvin, 1866 Little Ground Tyrant
     E Peru, N Bolivia, W Brazil

○ *Muscisaxicola maculirostris* Spot-billed Ground Tyrant
    — *M.m.niceforoi* J.T. Zimmer, 1947   Andes of C Colombia
    — *M.m.rufescens* Berlepsch & Stolzmann, 1896   Andes of Ecuador
    — *M.m.maculirostris* d'Orbigny & Lafresnaye, 1837   Andes of Peru, Bolivia, Chile, W Argentina

○ *Muscisaxicola griseus* Taczanowski, 1884 Taczanowski's Ground Tyrant[4]
     W Peru, W Bolivia

○ *Muscisaxicola juninensis* Taczanowski, 1884 Puna Ground Tyrant   S Peru, W Bolivia N Chile, N Argentina

○ *Muscisaxicola cinereus* Cinereous Ground Tyrant
    — *M.c.cinereus* Philippi & Landbeck, 1864   N and C Chile >> C Peru?
    — *M.c.argentina* Hellmayr, 1932   NW Argentina >> C Peru?

○ *Muscisaxicola albifrons* (Tschudi, 1844) White-fronted Ground Tyrant
     C Peru, W Bolivia, N Chile

● *Muscisaxicola flavinucha* Ochre-naped Ground Tyrant
    — *M.f.flavinucha* Lafresnaye, 1855   N Argentina, N Chile >> Bolivia, Peru
    — *M.f.brevirostris* Olrog, 1949   S Argentina, S Chile >> Bolivia, Peru

● *Muscisaxicola rufivertex* Rufous-naped Ground Tyrant
    — *M.r.occipitalis* Ridgway, 1887   Peru, NW Bolivia
    — *M.r.pallidiceps* Hellmayr, 1927[5]   SW Bolivia, NW Argentina, N Chile
    — *M.r.rufivertex* d'Orbigny & Lafresnaye, 1837   S Chile, SW Argentina

○ *Muscisaxicola maclovianus* Dark-faced Ground Tyrant
    — *M.m.mentalis* d'Orbigny & Lafresnaye, 1837   S Chile, S Argentina >> Peru, Bolivia, Uruguay, N and C and lowland Argentina, N and C Chile

    — *M.m.maclovianus* (Garnot, 1829)   Falkland Is. >> ?

● *Muscisaxicola albilora* Lafresnaye, 1855 White-browed Ground Tyrant
     Argentina, Chile >> Ecuador, Peru, Bolivia

○ *Muscisaxicola alpinus* Plain-capped Ground Tyrant
    — *M.a.columbianus* Chapman, 1912   WC Colombia
    — *M.a.quesadae* Meyer de Schauensee, 1942   C Colombia
    — *M.a.alpinus* (Jardine, 1849)   N Ecuador

○ *Muscisaxicola capistratus* (Burmeister, 1860) Cinnamon-bellied Ground Tyrant
     S Argentina, S Chile >> W Bolivia, S Peru, N and C Argentina, N and C Chile

---

[1] Gender addressed by David & Gosselin (2002) (614), multiple changes occur in specific names.
[2] Mistakenly placed in the genus *Ochthoeca* in last Edition.
[3] Gender addressed by David & Gosselin (2002) (614), multiple changes occur in specific names.
[4] For recognition of this at specific level, rather than treatment as a race of *alpinus*, see Chesser (2000) (346).
[5] The proposed race ¶ *achalensis* Nores & Yzurieta, 1983 (1660), awaits evaluation.

○ *Muscisaxicola frontalis* (Burmeister, 1860) Black-fronted Ground Tyrant

Argentina, Chile >> W Bolivia, S Peru

AGRIORNIS Gould, 1839 M[1]

○ *Agriornis montanus* Black-billed Shrike-Tyrant

— *A.m.solitarius* P.L. Sclater, 1858 Ecuador, Colombia

— *A.m.insolens* Sclater & Salvin, 1869 Peru

— *A.m.intermedius* Hellmayr, 1927 W Bolivia, N Chile

— *A.m.montanus* (d'Orbigny & Lafresnaye, 1837)[2] E Bolivia, NW Argentina

— *A.m.maritimus* (d'Orbigny & Lafresnaye, 1837) NC Chile

— *A.m.leucurus* Gould, 1839 C Chile, C Argentina

○ *Agriornis albicauda* White-tailed Shrike-Tyrant

— *A.a.pollens* P.L. Sclater, 1869[3] Ecuador

— *A.a.albicauda* (Philippi & Landbeck, 1863) Peru, W Bolivia, N Chile, NW Argentina

○ *Agriornis lividus* Great Shrike-Tyrant

— *A.l.lividus* (Kittlitz, 1835) SC Chile

— *A.l.fortis* Berlepsch, 1907 S Chile, S Argentina

○ *Agriornis micropterus* Grey-bellied Shrike-Tyrant

— *A.m.andecola* (d'Orbigny, 1840) Bolivia, S Peru, N Chile, NW Argentina >> Uruguay, Paraguay

— *A.m.micropterus* Gould, 1839 Argentina >> Uruguay, Paraguay

○ *Agriornis murinus* (d'Orbigny & Lafresnaye, 1837) Lesser Shrike-Tyrant

E Bolivia, Paraguay, Argentina (migrant from S to N of range)

XOLMIS Boie, 1826 M[4]

● *Xolmis pyrope* Fire-eyed Diucon

✓ *X.p.pyrope* (Kittlitz, 1830) S Chile, S Argentina >> N Chile

— *X.p.fortis* Philippi & Johnson, 1946 Chiloe I.

○ *Xolmis cinereus* Grey Monjita

— *X.c.cinereus* (Vieillot, 1816) S Brazil, Uruguay, N Argentina

— *X.c.pepoaza* (Vieillot, 1823) E Bolivia, Paraguay, N Argentina

○ *Xolmis coronatus* (Vieillot, 1823) Black-crowned Monjita Argentina >> E Bolivia, Uruguay, Paraguay

○ *Xolmis velatus* (M.H.K. Lichtenstein, 1823) White-rumped Monjita E Bolivia, S Brazil, Paraguay

○ *Xolmis irupero* White Monjita

— *X.i.niveus* (Spix, 1825) E Brazil

— *X.i.irupero* (Vieillot, 1823) E Bolivia, S Brazil, Uruguay, Paraguay, N Argentina

○ *Xolmis rubetra* Rusty-backed Monjita[5]

— *X.r.salinarum* Nores & Yzurieta, 1979 NC Argentina (1658)

— *X.r.rubetra* (Burmeister, 1860) C Argentina >> N Argentina

HETEROXOLMIS W.E. Lanyon, 1986 (1329) F

○ *Heteroxolmis dominicana* (Vieillot, 1823) Black-and-white Monjita[6]

S Brazil, Uruguay, Paraguay, E Argentina

MYIOTHERETES Reichenbach, 1850 M

○ *Myiotheretes striaticollis* Streak-throated Bush Tyrant

— *M.s.striaticollis* (P.L. Sclater, 1853) Venezuela, Colombia, Ecuador, Peru, W Bolivia

— *M.s.pallidus* Berlepsch, 1906 NW Argentina, E Peru, Bolivia

○ *Myiotheretes pernix* (Bangs, 1899) Santa Marta Bush Tyrant NE Colombia

---

[1] Gender addressed by David & Gosselin (2002) (614), multiple changes occur in specific names.
[2] The proposed race ¶ *fumosus* Nores & Yzurieta, 1983 (1660), awaits evaluation.
[3] Traylor (1979: 166) (2432) used the name *andicola* stating that it was not a homonym for *andecola*. That may have been so at the time as the 1961 version of the Code applied. Under the present Code (I.C.Z.N., 1999) Art. 58.12. does consider it a homonym and it may not be used.
[4] Gender addressed by David & Gosselin (2002) (614), multiple changes occur in specific names.
[5] Treated in *Neoxolmis* in last Edition, but see Lanyon (1986) (1329).
[6] Treated in the genus *Xolmis* in last Edition; but Lanyon (1986) (1329) now more often followed.

○ *Myiotheretes fumigatus* SMOKY BUSH TYRANT
   — *M.f olivaceus* (Phelps & Phelps, Jr., 1953)   N Colombia, W Venezuela
   — *M.f.fumigatus* (Boissonneau, 1840)   Colombia, Ecuador, Peru
   — *M.f.lugubris* (Berlepsch, 1883)   W Venezuela
   — *M.f.cajamarcae* (Chapman, 1927)   S Ecuador to S Peru

○ *Myiotheretes fuscorufus* (Sclater & Salvin, 1876) RUFOUS-BELLIED BUSH TYRANT
         C Peru to N Bolivia

CNEMARCHUS Ridgway, 1905[1] M
○ *Cnemarchus erythropygius* RED-RUMPED BUSH TYRANT
   — *C.e.orinomus* Wetmore, 1946   N Colombia
   — *C.e.erythropygius* (P.L. Sclater, 1853)   C Colombia to Ecuador, Peru, N Bolivia

POLIOXOLMIS W.E. Lanyon, 1986 (1329) F
○ *Polioxolmis rufipennis* RUFOUS-WEBBED TYRANT
   — *P.r.rufipennis* (Taczanowski, 1874)   SE Peru, N Bolivia
   — *P.r.bolivianus* Fjeldså, 1990   C Bolivia (846)

NEOXOLMIS Hellmayr, 1927 M
○ *Neoxolmis rufiventris* (Vieillot, 1823) CHOCOLATE-VENTED TYRANT   S Chile, S Argentina >> Uruguay, S Brazil

GUBERNETES Such, 1825 M
○ *Gubernetes yetapa* (Vieillot, 1818) STREAMER-TAILED TYRANT   S Brazil, Paraguay, E Bolivia, N Argentina

MUSCIPIPRA Lesson, 1831 F
○ *Muscipipra vetula* (M.H.K. Lichtenstein, 1823) SHEAR-TAILED GREY TYRANT
         SE Brazil, Paraguay, Argentina

FLUVICOLA Swainson, 1827 M
○ *Fluvicola pica* PIED WATER TYRANT
   — *F.p.pica* (Boddaert, 1783)   Trinidad, Venezuela, the Guianas, Colombia, N Brazil
   — *F.p.albiventer* (Spix, 1825)   E Brazil, SE Bolivia, Paraguay, N Argentina >> W Amazonia

○ *Fluvicola nengeta* MASKED WATER TYRANT
   — *F.n.atripennis* P.L. Sclater, 1860   SW Ecuador, NW Peru
   — *F.n.nengeta* (Linnaeus, 1766)   E Brazil

ARUNDINICOLA d'Orbigny, 1840[2] F
○ *Arundinicola leucocephala* (Linnaeus, 1764) WHITE-HEADED MARSH TYRANT
         NE Sth. America

ALECTRURUS Vieillot, 1816 M
○ *Alectrurus tricolor* (Vieillot, 1816) COCK-TAILED TYRANT   S Brazil, Paraguay, E Bolivia, N Argentina

○ *Alectrurus risora* (Vieillot, 1824) STRANGE-TAILED TYRANT   N Argentina, S Brazil, Paraguay, Uruguay

TUMBEZIA Chapman, 1925[3] F
○ *Tumbezia salvini* (Taczanowski, 1877) TUMBES TYRANT   NW Peru

OCHTHOECA Cabanis, 1847[4] F
○ *Ochthoeca frontalis* CROWNED CHAT-TYRANT[5]
   — *O.f.albidiadema* (Lafresnaye, 1848)   E Colombia
   — *O.f.frontalis* (Lafresnaye, 1847)   W Ecuador, W Colombia
   — *O.f.spodionota* Berlepsch & Stolzmann, 1896   C and SE Peru

○ *Ochthoeca pulchella* GOLDEN-BROWED CHAT-TYRANT
   — *O.p.similis* Carriker, 1933   C Peru
   — *O.p.pulchella* Sclater & Salvin, 1876   W Bolivia, SE Peru

---

[1] Treated in *Myiotheretes* in last Edition, but see Lanyon (1986) (1329).
[2] Treated in the genus *Fluvicola* in last Edition, but we now follow Lanyon (1986) (1329).
[3] For reasons to retain this genus see Garcia-Moreno *et al.* (1998) (912).
[4] Includes *Silvicultrix* see Garcia-Moreno *et al.* (1998) (912).
[5] We defer the split of *spodionota* proposed by Garcia-Moreno *et al.* (1998) (912) awaiting further evidence.

○ *Ochthoeca diadema* YELLOW-BELLIED CHAT-TYRANT
— *O.d.rubellula* Wetmore, 1946 — NE Colombia, NW Venezuela
— *O.d.jesupi* J.A. Allen, 1900 — N Colombia
— *O.d.tovarensis* Gilliard, 1940 — N Venezuela
— *O.d.diadema* (Hartlaub, 1843) — Colombia, W Venezuela
— *O.d.meridana* Phelps & Phelps, Jr., 1950 — NW Venezuela
— *O.d.gratiosa* (P.L. Sclater, 1862) — NW Peru, W Ecuador, W Colombia
— *O.d.cajamarcae* Carriker, 1934 — Cajamarca (Peru)

○ *Ochthoeca jelskii* Taczanowski, 1883 JELSKI'S CHAT-TYRANT — NW Peru, SW Ecuador

○ *Ochthoeca cinnamomeiventris* SLATY-BACKED CHAT-TYRANT[1]
— *O.c.nigrita* Sclater & Salvin, 1871 — W Venezuela
— *O.c.cinnamomeiventris* (Lafresnaye, 1843) — Colombia, E Ecuador
— *O.c.angustifasciata* Chapman, 1926 — N Peru
— *O.c.thoracica* Taczanowski, 1874 — Peru, Bolivia

○ *Ochthoeca rufipectoralis* RUFOUS-BREASTED CHAT-TYRANT
— *O.r.poliogastra* Salvin & Godman, 1880 — N Colombia
— *O.r.rubicundulus* Wetmore, 1946 — NE Colombia, NW Venezuela
— *O.r.obfuscata* J.T. Zimmer, 1942 — SW and W Colombia to Peru
— *O.r.rufopectus* (Lesson, 1844) — C Colombia, Ecuador, NW Peru
— *O.r.centralis* Hellmayr, 1927 — N Peru
— *O.r.tectricialis* Chapman, 1921 — S Peru
— *O.r.rufipectoralis* (d'Orbigny & Lafresnaye, 1837) — Bolivia, SE Peru

○ *Ochthoeca fumicolor* BROWN-BACKED CHAT-TYRANT
— *O.f.superciliosa* Sclater & Salvin, 1871 — W Venezuela
— *O.f.ferruginea* J.T. Zimmer, 1937 — C Colombia
— *O.f.fumicolor* P.L. Sclater, 1856 — E Colombia, W Venezuela
— *O.f.brunneifrons* Berlepsch & Stolzmann, 1896 — Peru, Ecuador, W Colombia
— *O.f.berlepschi* Hellmayr, 1914 — SE Peru, W Bolivia

○ *Ochthoeca oenanthoides* D'ORBIGNY'S CHAT-TYRANT
— *O.o.polionota* Sclater & Salvin, 1870 — Peru
— *O.o.oenanthoides* (d'Orbigny & Lafresnaye, 1837) — Bolivia, N Chile, NW Argentina

○ *Ochthoeca leucophrys* WHITE-BROWED CHAT-TYRANT
— *O.l.dissors* J.T. Zimmer, 1937 — N Peru
— *O.l.interior* J.T. Zimmer, 1930 — C Peru
— *O.l.urubambae* J.T. Zimmer, 1937 — C and S Peru
— *O.l.leucometopa* Sclater & Salvin, 1877 — W Peru, NW Chile
— *O.l.leucophrys* (d'Orbigny & Lafresnaye, 1837) — Bolivia
— *O.l.tucumana* Berlepsch, 1906 — W Argentina

○ *Ochthoeca piurae* Chapman, 1924 PIURA CHAT-TYRANT — NW Peru

COLORHAMPHUS Sundevall, 1872[2] M
○ *Colorhamphus parvirostris* (Darwin, 1839) PATAGONIAN TYRANT — S Chile, S Argentina

COLONIA J.E. Gray, 1827 F[3]
● *Colonia colonus* LONG-TAILED TYRANT
↙ *C.c.leuconota* (Lafresnaye, 1842) — S Honduras to Panama, W Colombia, Ecuador
— *C.c.fuscicapillus* (P.L. Sclater, 1861) — Colombia, E Ecuador, E Peru, W Brazil, Bolivia
— *C.c.poecilonota* (Cabanis, 1848) — The Guianas, SE Venezuela
— *C.c.niveiceps* J.T. Zimmer, 1930 — S Peru, N Bolivia
— *C.c.colonus* (Vieillot, 1818) — S Brazil, E Paraguay, N Argentina

MUSCIGRALLA d'Orbigny & Lafresnaye, 1837 F
○ *Muscigralla brevicauda* d'Orbigny & Lafresnaye, 1837 SHORT-TAILED FIELD TYRANT — SW Ecuador, W Peru, NW Chile

[1] We are aware that Garcia-Moreno *et al.* (1998) (912) recommended that *thoracica* be separated but we prefer to await further evidence.
[2] Treated in *Ochthoeca* in last Edition, but see Lanyon (1986) (1329).
[3] Gender addressed by David & Gosselin (2002) (614), multiple changes occur in specific names.

MACHETORNIS G.R. Gray, 1841 F[1]
○ *Machetornis rixosa* CATTLE TYRANT
    __ *M.r.flavigularis* Todd, 1912   N Colombia, N Venezuela
    __ *M.r.obscurodorsalis* Phelps & Phelps, Jr., 1948   SW Venezuela, E Colombia
    __ *M.r.rixosa* (Vieillot, 1819)   E Bolivia, S Brazil, Paraguay, Uruguay, N Argentina

LEGATUS P.L. Sclater, 1859 M
○ *Legatus leucophaius* PIRATIC FLYCATCHER
    __ *L.l.variegatus* (P.L. Sclater, 1857)   SE Mexico, Guatemala, Honduras >> ? Sth. America
    __ *L.l.leucophaius* (Vieillot, 1818)   Nicaragua to Panama, Trinidad, N and C Sth. America

PHELPSIA W.E. Lanyon, 1984 (1326) F
○ *Phelpsia inornata* (Lawrence, 1869) WHITE-BEARDED FLYCATCHER   N Venezuela

MYIOZETETES P.L. Sclater, 1859 M
○ *Myiozetetes cayanensis* RUSTY-MARGINED FLYCATCHER
    __ *M.c.rufipennis* Lawrence, 1869   N Venezuela, E Colombia, E Panama
    __ *M.c.hellmayri* Hartert & Goodson, 1917   Colombia, E Ecuador, NW Venezuela
    __ *M.c.cayanensis* (Linnaeus, 1766)   The Guianas, N Brazil, E Bolivia, S Venezuela
    __ *M.c.erythropterus* (Lafresnaye, 1853)   SE Brazil

● *Myiozetetes similis* SOCIAL FLYCATCHER
    __ *M.s.primulus* van Rossem, 1930   N Mexico
    __ *M.s.hesperis* A.R. Phillips, 1966   W Mexico
    √ *M.s.texensis* (Giraud, 1841)   S Mexico to N Costa Rica
    __ *M.s.columbianus* Cabanis & Heine, 1859   SW Costa Rica, Panama, N Colombia, N Venezuela
    __ *M.s.similis* (Spix, 1825)[2]   E Colombia, S Venezuela and W Amazonia to N Bolivia
    __ *M.s.grandis* Lawrence, 1871   W Ecuador to NW Peru
    __ *M.s.pallidiventris* Pinto, 1935   E Brazil south to E Paraguay, NE Argentina

◉ *Myiozetetes granadensis* GREY-CAPPED FLYCATCHER
    √ *M.g.granadensis* Lawrence, 1862   E Honduras to EC Panama
    __ *M.g.occidentalis* J.T. Zimmer, 1937   E Panama to NW Peru
    __ *M.g.obscurior* Todd, 1925   S Venezuela, E Colombia, E Ecuador, E Peru,
                                   N Bolivia, W Brazil

○ *Myiozetetes luteiventris* DUSKY-CHESTED FLYCATCHER
    __ *M.l.luteiventris* (P.L. Sclater, 1858)   E Colombia, E Ecuador, E Peru, NW Bolivia, C Brazil
    __ *M.l.septentrionalis* Blake, 1961   Surinam

PITANGUS Swainson, 1826 M
● *Pitangus sulphuratus* GREAT KISKADEE
    __ *P.s.texanus* van Rossem, 1940   S Texas to SE Mexico
    __ *P.s.derbianus* (Kaup, 1852)   W Mexico
    √ *P.s.guatimalensis* (Lafresnaye, 1852)   SE Mexico, Guatemala to Costa Rica
    __ *P.s.rufipennis* (Lafresnaye, 1851)   Coastal Colombia, N Venezuela
    __ *P.s.caucensis* Chapman, 1914   W Colombia
    __ *P.s.trinitatis* Hellmayr, 1906   Colombia, NW Brazil, Venezuela and Trinidad
    __ *P.s.sulphuratus* (Linnaeus, 1766)   S Colombia, the Guianas, N Brazil, NE Peru, E Ecuador
    __ *P.s.maximiliani* (Cabanis & Heine, 1859)   E and C Brazil, E Bolivia
    __ *P.s.bolivianus* (Lafresnaye, 1852)   E Bolivia
    __ *P.s.argentinus* Todd, 1952   Paraguay, Bolivia, N Argentina

○ *Pitangus lictor* LESSER KISKADEE[3]
    __ *P.l.panamensis* Bangs & Penard, 1918   E Panama to N Colombia
    __ *P.l.lictor* (M.H.K. Lichtenstein, 1823)   E Colombia, Venezuela, the Guianas, N Brazil to
                                   E Bolivia

---

[1] For correction of spelling see David & Gosselin (2002) (614).
[2] Includes *connivens* see Traylor (1979: 212) (2433).
[3] Treated in the genus *Philohydor* in last Edition; here submerged following A.O.U. (1998: 407) (3).

CONOPIAS Cabanis & Heine, 1859 M[1]
○ *Conopias albovittatus* WHITE-RINGED FLYCATCHER[2]
— *C.a.distinctus* (Ridgway, 1908)　　　　　　　E Costa Rica
— *C.a.albovittatus* (Lawrence, 1862)　　　　　　E Panama, W Colombia, NW Ecuador
— *C.a.parvus* (Pelzeln, 1868)　　　　　　　　　S Venezuela, the Guianas, NE Brazil

○ *Conopias trivirgatus* THREE-STRIPED FLYCATCHER
— *C.t.berlepschi* E. Snethlage, 1914　　　　　　Amazonia
— *C.t.trivirgatus* (Wied, 1831)　　　　　　　　SE Brazil, E Paraguay, NE Argentina

○ *Conopias cinchoneti* LEMON-BROWED FLYCATCHER
— *C.c.icterophrys* (Lafresnaye, 1845)　　　　　C Colombia, W Venezuela
— *C.c.cinchoneti* (Tschudi, 1844)　　　　　　　Ecuador to C Peru

MYIODYNASTES Bonaparte, 1857 M
○ *Myiodynastes hemichrysus* (Cabanis, 1861) GOLDEN-BELLIED FLYCATCHER

　　　　　　　　　　　　　　　　　　　　　　Costa Rica, W Panama

○ *Myiodynastes chrysocephalus* GOLDEN-CROWNED FLYCATCHER
— *M.c.cinerascens* Todd, 1912　　　　　　　　NW Venezuela, N Colombia
— *M.c.minor* Taczanowski & Berlepsch, 1885　　Colombia, Ecuador
— *M.c.chrysocephalus* (Tschudi, 1844)　　　　　Peru, Bolivia

○ *Myiodynastes bairdii* (Gambel, 1847) BAIRD'S FLYCATCHER[3]　　SW Ecuador, W Peru

◐ *Myiodynastes luteiventris* P.L. Sclater, 1859 SULPHUR-BELLIED FLYCATCHER

　　　　　　　　　　　　　　　　　　　　　　S USA to Costa Rica >> NW Sth. America

◐ *Myiodynastes maculatus* STREAKED FLYCATCHER
— *M.m.insolens* Ridgway, 1887　　　　　　　　SE Mexico to Honduras >> N Sth. America
— *M.m.nobilis* P.L. Sclater, 1859　　　　　　　NE Colombia
⌐ *M.m.difficilis* J.T. Zimmer, 1937　　　　　　W Costa Rica to Venezuela
— *M.m.chapmani* J.T. Zimmer, 1937　　　　　　SW Colombia to W Peru
— *M.m.maculatus* (Statius Müller, 1776)　　　　N Peru, Venezuela, the Guianas, N Brazil
— *M.m.tobagensis* J.T. Zimmer, 1937　　　　　　Guyana, Trinidad, Tobago I.
— *M.m.solitarius* (Vieillot, 1819)　　　　　　　S Peru to Uruguay, N Argentina >> ? Brazil

MEGARYNCHUS[4] Thunberg, 1824 M
◐ *Megarynchus pitangua* BOAT-BILLED FLYCATCHER
— *M.p.tardiusculus* R.T. Moore, 1941　　　　　NW Mexico
— *M.p.caniceps* Ridgway, 1906　　　　　　　　W Mexico
— *M.p.deserticola* Griscom, 1930　　　　　　　C Guatemala
— *M.p.mexicanus* (Lafresnaye, 1851)　　　　　SE Mexico to NW Colombia
— *M.p.chrysogaster* P.L. Sclater, 1860　　　　W Ecuador, NW Peru
— *M.p.pitangua* (Linnaeus, 1766)　　　　　　Trinidad, N and C Sth. America

TYRANNOPSIS Ridgway, 1905 F
○ *Tyrannopsis sulphurea* (Spix, 1825) SULPHURY FLYCATCHER　　Trinidad, the Guianas, Venezuela, Amazonian Brazil,
　　　　　　　　　　　　　　　　　　　　　　NE Peru, E Ecuador, N Bolivia

EMPIDONOMUS Cabanis & Heine, 1859 M
○ *Empidonomus varius* VARIEGATED FLYCATCHER
— *E.v.varius* (Vieillot, 1818)　　　　　　　　E Bolivia, SE Brazil, Paraguay, N Argentina >> W
　　　　　　　　　　　　　　　　　　　　　　Amazonia

— *E.v.rufinus* (Spix, 1825)　　　　　　　　　N and E Brazil, E Peru, E Venezuela, Guyana, French
　　　　　　　　　　　　　　　　　　　　　　Guiana, Colombia

○ *Empidonomus aurantioatrocristatus* CROWNED SLATY FLYCATCHER[5]
— *E.a.pallidiventris* (Hellmayr, 1929)　　　　NE Brazil
— *E.a.aurantioatrocristatus* (d'Orbigny & Lafresnaye, 1837)　E Bolivia, N Argentina, Uruguay, Paraguay and
　　　　　　　　　　　　　　　　　　　　　　S Brazil >> W Amazonia

---

[1] Gender addressed by David & Gosselin (2002) (614), multiple changes occur in specific names.
[2] This species was called *C. parva* in last Edition, but see Mees (1985) (1536) and A.O.U. (1998: 409) (3), where the taxon *parva* is retained in this species.
[3] Spelled *bairdi* in last Edition, but see Traylor (1979) (2433).
[4] Spelling not verified.
[5] Treated in the monotypic genus *Griseotyrannus* in last Edition, but a sister species of *E. varius* see Traylor (1979: 220) (2433) and better kept with that (J.V. Remsen, *in litt.*).

TYRANNUS Lacépède, 1799 M

○ *Tyrannus niveigularis* P.L. Sclater, 1860 Snowy-throated Kingbird — SW Colombia, W Ecuador, NW Peru

○ *Tyrannus albogularis* Burmeister, 1856 White-throated Kingbird — S Venezuela and the Guianas to E Bolivia and Amazonia (an intratropical migrant)

● *Tyrannus melancholicus* Tropical Kingbird[1]
    ⅃ *T.m.satrapa* (Cabanis & Heine, 1859) — S USA to N Colombia, N Venezuela
    — *T.m.despotes* (M.H.K. Lichtenstein, 1823) — NE Brazil
    — *T.m.melancholicus* Vieillot, 1819 — Tropical Sth. America >> C Argentina

● *Tyrannus couchii* S.F. Baird, 1858 Couch's Kingbird — S USA, E Mexico >> Guatemala

○ *Tyrannus vociferans* Cassin's Kingbird
    — *T.v.vociferans* Swainson, 1826 — SW USA and C Mexico >> Honduras
    — *T.v.xenopterus* Griscom, 1934 — SW Mexico

○ *Tyrannus crassirostris* Thick-billed Kingbird
    — *T.c.pompalis* Bangs & Peters, 1928 — SE Arizona, coastal W Mexico
    — *T.c.crassirostris* Swainson, 1826 — S Mexico, W Guatemala

● *Tyrannus verticalis* Say, 1823 Western Kingbird — SW Canada, W USA >> S Mexico, Guatemala

● *Tyrannus forficatus* (J.F. Gmelin, 1789) Scissor-tailed Flycatcher — SC USA >> C America

○ *Tyrannus savana* Fork-tailed Flycatcher
    — *T.s.monachus* Hartlaub, 1844 — S Mexico to NC Brazil
    — *T.s.sanctaemartae* (J.T. Zimmer, 1937) — N Colombia, NW Venezuela
    — *T.s.circumdatus* (J.T. Zimmer, 1937) — Amazon basin (EC Brazil)
    — *T.s.savana* Vieillot, 1808 — C and S Sth. America >> N Sth. America, W Indies

● *Tyrannus tyrannus* (Linnaeus, 1758) Eastern Kingbird — S Canada, C and E USA >> Sth. America

○ *Tyrannus dominicensis* Grey Kingbird
    — *T.d.dominicensis* (J.F. Gmelin, 1788) — SE USA to Colombia >> NC Sth. America
    — *T.d.vorax* Vieillot, 1819 — Lesser Antilles, Trinidad

○ *Tyrannus cubensis* Richmond, 1898 Giant Kingbird — Cuba, Gtr. Antilles

○ *Tyrannus caudifasciatus* Loggerhead Kingbird
    — *T.c.bahamensis* (H. Bryant, 1864) — Bahamas
    — *T.c.caudifasciatus* d'Orbigny, 1839 — Cuba
    — *T.c.flavescens* Parkes, 1963 — Isla de la Juventud, Gtr. Antilles
    — *T.c.caymanensis* (Nicoll, 1904) — Cayman Is.
    — *T.c.jamaicensis* (Chapman, 1892) — Jamaica
    — *T.c.taylori* (P.L. Sclater, 1864) — Puerto Rico
    — *T.c.gabbii* (Lawrence, 1876) — Haiti

RHYTIPTERNA Reichenbach, 1850[2] F

○ *Rhytipterna simplex* Greyish Mourner
    — *R.s.frederici* (Bangs & Penard, 1918) — Amazonia, S Venezuela and The Guianas
    — *R.s.simplex* (M.H.K. Lichtenstein, 1823) — SE Brazil

○ *Rhytipterna immunda* (Sclater & Salvin, 1873) Pale-bellied Mourner — Surinam, French Guiana, E Colombia, N Brazil

● *Rhytipterna holerythra* Rufous Mourner
    ⅃ *R.h.holerythra* (Sclater & Salvin, 1860) — Guatemala to Panama, N Colombia
    — *R.h.rosenbergi* (E. Hartert, 1905) — W Colombia, NW Ecuador

SIRYSTES Cabanis & Heine, 1859 M

○ *Sirystes sibilator* Sirystes[3]
    — *S.s.albogriseus* (Lawrence, 1863) — E Panama, NW Colombia
    — *S.s.albocinereus* Sclater & Salvin, 1880 — S Colombia, E Ecuador, E Peru, W Brazil

---

[1] Includes *chubbii* Davis, 1979 (618), see Vuilleumier & Mayr (1987) (2551).
[2] Our sequence from this genus to *Ramphotrigon* is based on Lanyon (1985) (1328).
[3] See Lanyon & Fitzpatrick (1983) (1333).

— *S.s.subcanescens* Todd, 1920   NE Brazil, S Surinam
— *S.s.sibilator* (Vieillot, 1818)   E Brazil, E Paraguay, NE Argentina
— *S.s.atimastus* Oberholser, 1902   SW Brazil

CASIORNIS Des Murs, 1856 M[1]

○ **Casiornis rufus** (Vieillot, 1816) RUFOUS CASIORNIS   SE Peru to E Bolivia, Paraguay, N Argentina and Amazonian and E Brazil

○ **Casiornis fuscus** Sclater & Salvin, 1873 ASH-THROATED CASIORNIS   NE Brazil

MYIARCHUS Cabanis, 1844 M

○ **Myiarchus semirufus** Sclater & Salvin, 1878 RUFOUS FLYCATCHER   Peru

○ **Myiarchus yucatanensis** YUCATÁN FLYCATCHER
— *M.y.yucatanensis* Lawrence, 1871   Yucatán Pen. (Mexico), Guatemala, Belize
— *M.y.navai* Parkes, 1982[2]   S Mexico (1774)
— *M.y.lanyoni* Parkes & Phillips, 1967   Cozumel I.

○ **Myiarchus barbirostris** (Swainson, 1827) SAD FLYCATCHER   Jamaica

● **Myiarchus tuberculifer** DUSKY-CAPPED FLYCATCHER
— *M.t.olivascens* Ridgway, 1884   SW USA, W Mexico
— *M.t.lawrenceii* (Giraud, 1841)   E and S Mexico
— *M.t.querulus* Nelson, 1904   SW Mexico
— *M.t.platyrhynchus* Ridgway, 1885   SE Mexico
— *M.t.manens* Parkes, 1982   S Yucatán (1774)
— *M.t.connectens* Miller & Griscom, 1925   Guatemala to N Nicaragua
— *M.t.littoralis* J.T. Zimmer, 1953   Pacific coast from SE Honduras to Costa Rica
— *M.t.nigricapillus* Cabanis, 1861   SE Nicaragua, Costa Rica, W Panama
— *M.t.brunneiceps* Lawrence, 1861   E Panama, W Colombia
— *M.t.pallidus* Zimmer & Phelps, 1946   N Colombia, W Venezuela
— *M.t.tuberculifer* (d'Orbigny & Lafresnaye, 1837)   Colombia and Surinam to Bolivia and S Brazil
— *M.t.nigriceps* P.L. Sclater, 1860   W Ecuador, W Colombia
— *M.t.atriceps* Cabanis, 1883   Andes of Peru, Bolivia, NW Argentina

○ **Myiarchus swainsoni** SWAINSON'S FLYCATCHER
— *M.s.swainsoni* Cabanis & Heine, 1859   E Paraguay, NE Argentina >> E Colombia, N Venezuela and Surinam
— *M.s.albimarginatus* Mees, 1985   N savanna of Surinam (1536)
— *M.s.phaeonotus* Salvin & Godman, 1883   S Guyana, S Venezuela
— *M.s.pelzelni* Berlepsch, 1883   E Peru, SE Brazil, N Bolivia
— *M.s.ferocior* Cabanis, 1883   SE Bolivia, N Argentina >> SE Colombia

○ **Myiarchus venezuelensis** Lawrence, 1865 VENEZUELAN FLYCATCHER   NE Colombia, NE Venezuela, Tobago I.

○ **Myiarchus panamensis** PANAMA FLYCATCHER
— *M.p.actiosus* Ridgway, 1906   Costa Rica
— *M.p.panamensis* Lawrence, 1860   Panama, N Colombia, NW Venezuela

○ **Myiarchus ferox** SHORT-CRESTED FLYCATCHER
— *M.f.brunnescens* Zimmer & Phelps, 1946   Venezuela
— *M.f.ferox* (J.F. Gmelin, 1789)   S Colombia to Peru, N Brazil, the Guianas
— *M.f.australis* Hellmayr, 1927   E Bolivia to S Brazil, N Argentina

○ **Myiarchus apicalis** Sclater & Salvin, 1881 APICAL FLYCATCHER   W Colombia

○ **Myiarchus cephalotes** PALE-EDGED FLYCATCHER
— *M.c.caribbaeus* Hellmayr, 1925   Venezuela
— *M.c.cephalotes* Taczanowski, 1879   Andes of Colombia, Ecuador, Peru, C Bolivia

○ **Myiarchus phaeocephalus** SOOTY-CROWNED FLYCATCHER
— *M.p.phaeocephalus* P.L. Sclater, 1860   W Ecuador, NW Peru
— *M.p.interior* J.T. Zimmer, 1938   NW Peru

---

[1] Gender addressed by David & Gosselin (2002) (614), multiple changes occur in specific names.
[2] Misspelled *navae* in last Edition.

● *Myiarchus cinerascens* Ash-throated Flycatcher
    ✓ *M.c.cinerascens* (Lawrence, 1851)   W USA, W Mexico >> Guatemala, NW Costa Rica
    — *M.c.pertinax* S.F. Baird, 1859   Baja California

○ *Myiarchus nuttingi* Nutting's Flycatcher
    — *M.n.inquietus* Salvin & Godman, 1889   W and C Mexico
    — *M.n.nuttingi* Ridgway, 1882   Mts. of C America
    — *M.n.flavidior* van Rossem, 1936   Pacific coast of C America

● *Myiarchus crinitus* (Linnaeus, 1758) Great Crested Flycatcher   E Canada, E USA >> Mexico, C America, NW Sth. America

● *Myiarchus tyrannulus* Brown-crested Flycatcher
    — *M.t.magister* Ridgway, 1884   SW USA, W Mexico, Tres Marias Is. >> S Mexico
    ✓ *M.t.cooperi* S.F. Baird, 1858   SC USA, E Mexico to Honduras
    — *M.t.cozumelae* Parkes, 1982   Cozumel I. (1774)
    — *M.t.tyrannulus* (Statius Müller, 1776)   Trinidad, Venezuela, the Guianas, N Brazil, N Colombia to N Argentina

    ✓ *M.t.brachyurus* Ridgway, 1887   El Salvador to NW Costa Rica
    — *M.t.insularum* Bond, 1936   Honduras coastal islands
    — *M.t.bahiae* Berlepsch & Leverkühn, 1890   N and E Brazil

○ *Myiarchus magnirostris* (Gould, 1838) Galapagos Flycatcher   Galapagos Is.

○ *Myiarchus nugator* Riley, 1904 Grenada Flycatcher   Lesser Antilles

○ *Myiarchus validus* Cabanis, 1847 Rufous-tailed Flycatcher   Jamaica

○ *Myiarchus sagrae* La Sagra's Flycatcher
    — *M.s.sagrae* (Gundlach, 1852)   Cuba, Grand Cayman I.
    — *M.s.lucaysiensis* (H. Bryant, 1867)   Bahamas

○ *Myiarchus stolidus* Stolid Flycatcher
    — *M.s.dominicensis* (H. Bryant, 1867)   Hispaniola
    — *M.s.stolidus* (Gosse, 1847)   Jamaica

○ *Myiarchus antillarum* (H. Bryant, 1866) Puerto Rican Flycatcher   Puerto Rico

○ *Myiarchus oberi* Lesser Antillean Flycatcher
    — *M.o.oberi* Lawrence, 1877   Dominica, Guadeloupe Is.
    — *M.o.sanctaeluciae* Hellmayr & Seilern, 1915   St Lucia I.
    — *M.o.berlepschii* Cory, 1888   Nevis, St Kitts Is.
    — *M.o.sclateri* Lawrence, 1879   Martinique

DELTARHYNCHUS Ridgway, 1893 M
○ *Deltarhynchus flammulatus* (Lawrence, 1875) Flammulated Flycatcher
       SW and S Mexico

RAMPHOTRIGON G.R. Gray, 1855[1] N[2]
○ *Ramphotrigon megacephalum* Large-headed Flatbill
    — *R.m.pectorale* Zimmer & Phelps, 1947   S Colombia, S Venezuela
    — *R.m.venezuelense* Phelps & Gilliard, 1941   W Venezuela
    — *R.m.bolivianum* J.T. Zimmer, 1939   SE Peru, W Brazil, N Bolivia
    — *R.m.megacephalum* (Swainson, 1836)   SE Brazil, E Paraguay, NE Argentina

○ *Ramphotrigon ruficauda* (Spix, 1825) Rufous-tailed Flatbill   Amazonia

○ *Ramphotrigon fuscicauda* Chapman, 1925 Dusky-tailed Flatbill   W Amazonia

ATTILA Lesson, 1830 M
○ *Attila phoenicurus* Pelzeln, 1868 Rufous-tailed Attila   Amazonian and S Brazil to NE Argentina

○ *Attila cinnamomeus* (J.F. Gmelin, 1789) Cinnamon Attila   Amazonia, S Venezuela, and The Guianas

○ *Attila torridus* P.L. Sclater, 1860 Ochraceous Attila   SW Colombia, W Ecuador, NW Peru

---

[1] See Lanyon (1988) (1332).
[2] Gender addressed by David & Gosselin (2002) (614), multiple changes occur in specific names.

○ *Attila citriniventris* P.L. Sclater, 1859 CITRON-BELLIED ATTILA   Venezuela, Ecuador, N Peru, NW Brazil

○ *Attila bolivianus* DULL-CAPPED ATTILA
— *A.b.nattereri* Hellmayr, 1902   N Brazil
— *A.b.bolivianus* Lafresnaye, 1848   SW Brazil, E Bolivia, E Peru

○ *Attila rufus* GREY-HOODED ATTILA
— *A.r.hellmayri* Pinto, 1935   E Brazil
— *A.r.rufus* (Vieillot, 1819)   SE Brazil

● *Attila spadiceus* BRIGHT-RUMPED ATTILA
— *A.s.pacificus* Hellmayr, 1929   NW Mexico
— *A.s.cozumelae* Ridgway, 1885   Cozumel I.
— *A.s.gaumeri* Salvin & Godman, 1891   SE Mexico
— *A.s.flammulatus* Lafresnaye, 1848   SE Mexico to El Salvador
— *A.s.salvadorensis* Dickey & van Rossem, 1929   El Salvador to NW Nicaragua
— *A.s.citreopyga* (Bonaparte, 1854)[1]   Nicaragua to W Panama
— *A.s.sclateri* Lawrence, 1862   E Panama, NW Colombia
— *A.s.caniceps* Todd, 1917   N Colombia
— *A.s.parvirostris* J.A. Allen, 1900   NE Colombia, NW Venezuela
— *A.s.parambae* E. Hartert, 1900   W Ecuador, W Colombia
— *A.s.spadiceus* (J.F. Gmelin, 1789)   The Guianas, Trinidad, N Venezuela, N Brazil, NE Peru, N Bolivia
— *A.s.uropygiatus* (Wied, 1831)   SE Brazil

THAMNOPHILIDAE[2] ANTBIRDS (46: 206)

CYMBILAIMUS G.R. Gray, 1840 M
○ *Cymbilaimus lineatus* FASCIATED[3] ANTSHRIKE
— *C.l.fasciatus* (Ridgway, 1884)   Nicaragua to NW Ecuador
— *C.l.intermedius* (Hartert & Goodson, 1917)   Upper Amazonia
— *C.l.lineatus* (Leach, 1814)[4]   SE Venezuela, the Guianas, NE Brazil

○ *Cymbilaimus sanctaemariae* Gyldenstolpe, 1941 BAMBOO ANTSHRIKE[5]
SE Peru, NE Bolivia, W Brazil

HYPOEDALEUS Cabanis & Heine, 1859 M
○ *Hypoedaleus guttatus* SPOT-BACKED ANTSHRIKE
— *H.g.leucogaster* Pinto, 1932   E Brazil
— *H.g.guttatus* (Vieillot, 1816)   S Brazil, Paraguay, NE Argentina

BATARA Lesson, 1831 F
○ *Batara cinerea* GIANT ANTSHRIKE
— *B.c.excubitor* Bond & Meyer de Schauensee, 1940   E Bolivia
— *B.c.argentina* Shipton, 1918   S Bolivia, NW Argentina
— *B.c.cinerea* (Vieillot, 1819)   SE Brazil, NE Argentina

MACKENZIAENA Chubb, 1918 F
○ *Mackenziaena leachii* (Such, 1825) LARGE-TAILED ANTSHRIKE   SE Brazil, NE Argentina

○ *Mackenziaena severa* (M.H.K. Lichtenstein, 1823) TUFTED ANTSHRIKE SE Brazil, NE Argentina

FREDERICKENA Chubb, 1918 F
○ *Frederickena viridis* (Vieillot, 1816) BLACK-THROATED ANTSHRIKE   E Venezuela, Guyana

○ *Frederickena unduligera* UNDULATED ANTSHRIKE
— *F.u.fulva* J.T. Zimmer, 1944   SE Colombia, E Ecuador
— *F.u.diversa* J.T. Zimmer, 1944   E and SE Peru

[1] For correction of spelling see David & Gosselin (2002) (613).
[2] For recognition of this family see A.O.U. (1998: 361) (3).
[3] As *fasciatus* just means striped perhaps this name could be simplified!
[4] Tentatively includes ¶ *Cymbilaimus l. brangeri* Aveledo & Perez, 1991 (68). See Rodner *et al.* (2000) (2096).
[5] For separation from *C. lineatus* see Pierpont & Fitzpatrick (1983) (1885).

__ *F.u.unduligera* (Pelzeln, 1869)	NW Brazil
__ *F.u.pallida* J.T. Zimmer, 1944	Brazil

TARABA Lesson, 1831 M
○ **Taraba major** GREAT ANTSHRIKE

__ *T.m.melanocrissus* (P.L. Sclater, 1860)	SE Mexico to W Panama
__ *T.m.obscurus* J.T. Zimmer, 1933	SW Costa Rica to NW Colombia
__ *T.m.transandeanus* (P.L. Sclater, 1855)	SW Colombia, W Ecuador, NW Peru
__ *T.m.granadensis* (Cabanis, 1872)	E Colombia, Venezuela
__ *T.m.semifasciatus* (Cabanis, 1872)	E Colombia to the Guianas, N and E Brazil
__ *T.m.duidae* Chapman, 1929	SE Venezuela
__ *T.m.melanurus* (P.L. Sclater, 1855)	Upper Amazonia
__ *T.m.borbae* (Pelzeln, 1869)	C Brazil
__ *T.m.stagurus* (M.H.K. Lichtenstein, 1823)	E and NE Brazil
__ *T.m.major* (Vieillot, 1816)	E Bolivia, S Brazil, Paraguay, N Argentina

SAKESPHORUS Chubb, 1918 M
○ **Sakesphorus canadensis** BLACK-CRESTED ANTSHRIKE

__ *S.c.pulchellus* (Cabanis & Heine, 1859)	N Colombia, W Venezuela
__ *S.c.paraguanae* Gilliard, 1940	NE Colombia, NW Venezuela
__ *S.c.intermedius* (Cherrie, 1916)	S Venezuela, N Brazil
__ *S.c.fumosus* J.T. Zimmer, 1933	S Venezuela
__ *S.c.trinitatis* (Ridgway, 1892)	NE Venezuela, Guyana, Trinidad
__ *S.c.canadensis* (Linnaeus, 1766)	Surinam, French Guiana
__ *S.c.loretoyacuensis* (E. Bartlett, 1882)	SE Colombia, NW Brazil

○ **Sakesphorus cristatus** (Wied, 1831) SILVERY-CHEEKED ANTSHRIKE   E Brazil

○ **Sakesphorus bernardi** COLLARED ANTSHRIKE

__ *S.b.bernardi* (Lesson, 1844)	W Ecuador
__ *S.b.piurae* (Chapman, 1923)	SW Ecuador, N Peru
__ *S.b.cajamarcae* (Hellmayr, 1917)	W Peru
__ *S.b.shumbae* Carriker, 1934	N Peru

○ **Sakesphorus melanonotus** (P.L. Sclater, 1855) BLACK-BACKED ANTSHRIKE
NE Colombia, W Venezuela
○ **Sakesphorus melanothorax** (P.L. Sclater, 1857) BAND-TAILED ANTSHRIKE
Surinam, French Guiana, NE Brazil
○ **Sakesphorus luctuosus** GLOSSY ANTSHRIKE

__ *S.l.luctuosus* (M.H.K. Lichtenstein, 1823)	NE Brazil
__ *S.l.araguayae* (Hellmayr, 1908)	C Brazil

BIATAS Cabanis & Heine, 1859 M
○ **Biatas nigropectus** (Lafresnaye, 1850) WHITE-BEARDED ANTSHRIKE   SE Brazil

THAMNOPHILUS Vieillot, 1816 M
● **Thamnophilus doliatus** BARRED ANTSHRIKE

__ *T.d.intermedius* Ridgway, 1888	E Mexico to E Costa Rica
__ *T.d.yucatanensis* Ridgway, 1908	S Mexico, N Guatemala
↓ *T.d.pacificus* Ridgway, 1908	Pacific slope of W Honduras to W Panama
__ *T.d.nesiotes* Wetmore, 1970	Pearl Islands (Panama) (2607)
__ *T.d.eremnus* Wetmore, 1957	Coibà I. (Panama) (2598)
__ *T.d.nigricristatus* Lawrence, 1865	E Panama, N Colombia
__ *T.d.albicans* Lafresnaye, 1844	N Colombia
__ *T.d.nigrescens* Lawrence, 1867	NE Colombia, NW Venezuela
__ *T.d.tobagensis* Hartert & Goodson, 1917	Tobago I.
__ *T.d.fraterculus* Berlepsch & Hartert, 1902	E Colombia, N Venezuela, Trinidad
__ *T.d.doliatus* (Linnaeus, 1764)	E Venezuela, the Guianas, N Brazil
__ *T.d.zarumae* Chapman, 1921	SW Ecuador, NW Peru
__ *T.d.palamblae* J.T. Zimmer, 1933	NW Peru
__ *T.d.subradiatus* Berlepsch, 1887	E Peru, W Brazil

— *T.d.signatus* J.T. Zimmer, 1933   NE Bolivia, SW Brazil
— *T.d.difficilis* Hellmayr, 1903   E Brazil
— *T.d.capistratus* Lesson, 1840   E Brazil
— *T.d.radiatus* Vieillot, 1816   E Bolivia, S Brazil, N Argentina
— *T.d.cadwaladeri* Bond & Meyer de Schauensee, 1940   S Bolivia

○ *Thamnophilus multistriatus* Bar-crested Antshrike
— *T.m.brachyurus* Todd, 1927   W Colombia
— *T.m.selvae* Meyer de Schauensee, 1950   W Colombia
— *T.m.multistriatus* Lafresnaye, 1844   C Colombia
— *T.m.oecotonophilus* Borrero & Hernández, 1958   NE Colombia, W Venezuela (199)

○ *Thamnophilus tenuepunctatus* Lined Antshrike[1]
— *T.t.tenuepunctatus* Lafresnaye, 1853   E Colombia
— *T.t.tenuifasciatus* Lawrence, 1867   SE Colombia, E Ecuador
— *T.t.berlepschi* Taczanowski, 1884   SE Ecuador, N Peru

○ *Thamnophilus palliatus* Chestnut-backed Antshrike
— *T.p.similis* J.T. Zimmer, 1933   C Peru
— *T.p.puncticeps* P.L. Sclater, 1890   SE Peru, Bolivia, W Brazil
— *T.p.palliatus* (M.H.K. Lichtenstein, 1823)   C and E Brazil

● *Thamnophilus bridgesi* P.L. Sclater, 1856 Black-hooded Antshrike   SW Costa Rica, W Panama

○ *Thamnophilus nigriceps* Black Antshrike
— *T.n.nigriceps* P.L. Sclater, 1869   E Panama, NW Colombia
— *T.n.magdalenae* Todd, 1927   N Colombia

○ *Thamnophilus praecox* J.T. Zimmer, 1937 Cocha Antshrike   S Ecuador

○ *Thamnophilus nigrocinereus* Blackish-grey Antshrike
— *T.n.cinereoniger* Pelzeln, 1869   Colombia, Venezuela, NW Brazil
— *T.n.kulczynskii* (Domaniewski & Stolzmann, 1922)[2]   French Guiana
— *T.n.nigrocinereus* P.L. Sclater, 1855   NE Brazil
— *T.n.tschudii* Pelzeln, 1869   W Brazil
— *T.n.huberi* E. Snethlage, 1907   N Brazil

○ *Thamnophilus cryptoleucus* (Ménégaux & Hellmayr, 1906) Castelnau's Antshrike
                                                                            NE Peru, W Brazil

○ *Thamnophilus aethiops* White-shouldered Antshrike
— *T.a.aethiops* P.L. Sclater, 1858   E Ecuador, NE Peru
— *T.a.wetmorei* Meyer de Schauensee, 1945   SE Colombia
— *T.a.polionotus* Pelzeln, 1869   S and E Venezuela, NW Brazil
— *T.a.kapouni* Seilern, 1913[3]   E and SE Peru, N Bolivia, W Brazil
— *T.a.juruanus* H. von Ihering, 1905   W Brazil
— *T.a.injunctus* J.T. Zimmer, 1933   NE Brazil
— *T.a.punctuliger* Pelzeln, 1869   C Brazil
— *T.a.atriceps* Todd, 1927   C Brazil
— *T.a.incertus* Pelzeln, 1869   NE Brazil

○ *Thamnophilus unicolor* Uniform Antshrike
— *T.u.unicolor* (P.L. Sclater, 1859)   E Ecuador
— *T.u.grandior* Hellmayr, 1924   Colombia, E Ecuador, N Peru
— *T.u.caudatus* Carriker, 1933   N Peru

○ *Thamnophilus schistaceus* Plain-winged Antshrike
— *T.s.capitalis* P.L. Sclater, 1858   Upper Amazonia
— *T.s.dubius* (Berlepsch & Stolzmann, 1894)   S Ecuador, N Peru
— *T.s.schistaceus* d'Orbigny, 1837   SE Peru, N Bolivia, W Brazil
— *T.s.heterogynus* (Hellmayr, 1907)   W Brazil
— *T.s.inornatus* Ridgway, 1888   C Brazil

[1] Split from *T. palliatus* based on Ridgely & Tudor (1994: 230) (2019) reverting to the treatment of Cory & Hellmayr (1924) (573).
[2] Spelled Sztolcman when he published in Polish journals.
[3] The usual abbreviated form of von Seilern-Aspang.

○ *Thamnophilus murinus* MOUSE-COLOURED ANTSHRIKE
— *T.m.murinus* Sclater & Salvin, 1867   E Ecuador, E Colombia to Surinam
— *T.m.cayennensis* Todd, 1927   French Guiana, N Brazil
— *T.m.canipennis* Todd, 1927   NE Peru, W Brazil

○ *Thamnophilus aroyae* (Hellmayr, 1904) UPLAND ANTSHRIKE   SE Peru, NW Bolivia

○ *Thamnophilus atrinucha* WESTERN SLATY ANTSHRIKE
— *T.a.atrinucha* Salvin & Godman, 1892[1]   SE Guatemala to NW Venezuela
— *T.a.gorgonae* Thayer & Bangs, 1905   Gorgona I.

○ *Thamnophilus punctatus* NORTHERN SLATY ANTSHRIKE[2]
— *T.p.interpositus* Hartert & Goodson, 1917   E Colombia, W Venezuela
— *T.p.punctatus* (Shaw, 1809)   E Venezuela, the Guianas, N Brazil
— *T.p.leucogaster* Hellmayr, 1924   Marañón Valley, Peru
— *T.p.huallagae* Carriker, 1934   Huallaga Valley, Peru

○ *Thamnophilus stictocephalus* NATTERER'S SLATY ANTSHRIKE
— *T.s.stictocephalus* Pelzeln, 1869[3]   C Brazil
— *T.s.parkeri* Isler, Isler & Whitney, 1997   E Bolivia (1197)

○ *Thamnophilus sticturus* Pelzeln, 1869 BOLIVIAN SLATY ANTSHRIKE   C Bolivia, SW Brazil

○ *Thamnophilus pelzelni* Hellmayr, 1924 PLANALTO SLATY ANTSHRIKE   C and E Brazil

○ *Thamnophilus ambiguus* Swainson, 1825 SOORETAMA SLATY ANTSHRIKE   Coastal SE Brazil

○ *Thamnophilus amazonicus* AMAZONIAN ANTSHRIKE
— *T.a.cinereiceps* Pelzeln, 1869   E Colombia, SC Venezuela, NC Brazil (west of the Rio Branco
— *T.a.divaricatus* Mees, 1974   SE Venezuela, the Guianas, NC Brazil (east of the Rio Branco) (1526)
— *T.a.amazonicus* P.L. Sclater, 1858   S Colombia to N Bolivia
— *T.a.obscurus* J.T. Zimmer, 1933   C Brazil (S of the Amazon)
— *T.a.paraensis* Todd, 1927   C Brazil (E of Rio Tocantins)

○ *Thamnophilus insignis* STREAK-BACKED ANTSHRIKE
— *T.i.insignis* Salvin & Godman, 1884   S Venezuela
— *T.i.nigrofrontalis* Phelps & Phelps, Jr., 1947   S Venezuela

○ *Thamnophilus caerulescens* VARIABLE ANTSHRIKE
— *T.c.subandinus* Taczanowski, 1882   N Peru
— *T.c.melanochrous* Sclater & Salvin, 1876   C and S Peru
— *T.c.aspersiventer* d'Orbigny & Lafresnaye, 1837   N Bolivia
— *T.c.connectens* Berlepsch, 1907   E Bolivia
— *T.c.dinellii* Berlepsch, 1906   C and S Bolivia, NW Argentina
— *T.c.paraguayensis* Hellmayr, 1904   S Brazil, N Paraguay
— *T.c.gilvigaster* Pelzeln, 1869   SE Brazil, NE Argentina
— *T.c.caerulescens* Vieillot, 1816   S Brazil, E Paraguay, NE Argentina
— *T.c.albonotatus* Spix, 1825   EC Brazil
— *T.c.ochraceiventer* E. Snethlage, 1928   SC Brazil
— *T.c.pernambucensis* Naumburg, 1937   E Brazil
— *T.c.cearensis* (Cory, 1919)   E Brazil

○ *Thamnophilus torquatus* Swainson, 1825 RUFOUS-WINGED ANTSHRIKE   Brazil, E Bolivia

○ *Thamnophilus ruficapillus* RUFOUS-CAPPED ANTSHRIKE
— *T.r.jaczewskii* Domaniewski, 1925   N Peru
— *T.r.marcapatae* Hellmayr, 1912   SE Peru
— *T.r.subfasciatus* Sclater & Salvin, 1876   W Bolivia
— *T.r.cochabambae* (Chapman, 1921)   W Bolivia, NW Argentina
— *T.r.ruficapillus* Vieillot, 1816   S and E Brazil, NE Argentina

---

[1] Includes *subcinereus* see Isler *et al.* (1997) (1197).
[2] The split of the broad species *T. punctatus* of last Edition, effected here, is based on Isler *et al.* (1997, 2001) (1197, 1199).
[3] Includes *saturatus* and *zimmeri* see Isler *et al.* (1997) (1197).

MEGASTICTUS Ridgway, 1909 M
○ *Megastictus margaritatus* (P.L. Sclater, 1855) PEARLY ANTSHRIKE   N Upper Amazonia

NEOCTANTES P.L. Sclater, 1869 M
○ *Neoctantes niger* (Pelzeln, 1859) BLACK BUSHBIRD   E Ecuador, N Peru, W Brazil

CLYTOCTANTES Elliot, 1870 M
○ *Clytoctantes alixii* Elliot, 1870 RECURVE-BILLED BUSHBIRD   Colombia

○ *Clytoctantes atrogularis* Lanyon, Stotz & Willard, 1991 RONDONIA BUSHBIRD

     SW Brazil [1325]

XENORNIS Chapman, 1924 M
○ *Xenornis setifrons* Chapman, 1924 SPINY-FACED ANTSHRIKE   E Panama, NW Colombia

THAMNISTES Sclater & Salvin, 1860 M
○ *Thamnistes anabatinus* RUSSET ANTSHRIKE

— *T.a.anabatinus* Sclater & Salvin, 1860	SE Mexico to Honduras
— *T.a.saturatus* Ridgway, 1908	Nicaragua to W Panama
— *T.a.coronatus* Nelson, 1912	C and E Panama
— *T.a.intermedius* Chapman, 1914	W Colombia, W Ecuador
— *T.a.gularis* Phelps & Phelps, Jr., 1956	C Colombia [1858]
— *T.a.aequatorialis* P.L. Sclater, 1862	SE Colombia, E Ecuador
— *T.a.rufescens* Cabanis, 1873	C and SE Peru, N Bolivia

DYSITHAMNUS Cabanis, 1847 M
○ *Dysithamnus stictothorax* (Temminck, 1823) SPOT-BREASTED ANTVIREO

     SE Brazil

○ *Dysithamnus mentalis* PLAIN ANTVIREO

— *D.m.septentrionalis* Ridgway, 1908	S Mexico to W Panama
— *D.m.suffusus* Nelson, 1912	E Panama, NW Colombia
— *D.m.extremus* Todd, 1916	C Colombia
— *D.m.semicinereus* P.L. Sclater, 1855	NE Colombia
— *D.m.viridis* Aveledo & Pons, 1952	NE Colombia, NW Venezuela [70]
— *D.m.cumbreanus* Hellmayr & Seilern, 1915	N Venezuela
— *D.m.andrei* Hellmayr, 1906	E Venezuela, Trinidad
— *D.m.oberi* Ridgway, 1908	Tobago I.
— *D.m.ptaritepui* Zimmer & Phelps, 1946	S Venezuela
— *D.m.spodionotus* Salvin & Godman, 1883	S and E Venezuela
— *D.m.aequatorialis* Todd, 1916	W Ecuador, NW Peru
— *D.m.napensis* Chapman, 1925	E Ecuador
— *D.m.tambillanus* Taczanowski, 1884	N Peru
— *D.m.olivaceus* (Tschudi, 1844)	C Peru
— *D.m.tavarae* J.T. Zimmer, 1932	SE Peru, Bolivia
— *D.m.emiliae* Hellmayr, 1912	E Brazil
— *D.m.affinis* Pelzeln, 1869	C Brazil
— *D.m.mentalis* (Temminck, 1823)	S and SE Brazil, E Paraguay, NE Argentina

○ *Dysithamnus striaticeps* Lawrence, 1865 STREAK-CROWNED ANTVIREO   E Nicaragua, E Costa Rica

○ *Dysithamnus puncticeps* SPOT-CROWNED ANTVIREO

— *D.p.puncticeps* Salvin, 1866	E Costa Rica, Panama, N Colombia
— *D.p.intensus* Griscom, 1932	S Panama, W Colombia
— *D.p.flemmingi* E. Hartert, 1900	SW Colombia, W Ecuador

○ *Dysithamnus xanthopterus* (Burmeister, 1856) RUFOUS-BACKED ANTVIREO

     SE Brazil

○ *Dysithamnus occidentalis* BICOLORED ANTVIREO

— *D.o.occidentalis* (Chapman, 1923)	S Colombia
— *D.o.punctitectus* Chapman, 1924	S Ecuador

○ *Dysithamnus plumbeus* PLUMBEOUS ANTVIREO[1]
— *D.p.tucuyensis* E. Hartert, 1894   Monagas to Lara (N Venezuela)
— *D.p.leucostictus* P.L. Sclater, 1858   E Colombia, E Ecuador
— *D.p.plumbeus* Wied, 1831)   SE Brazil

THAMNOMANES Cabanis, 1847 M
○ *Thamnomanes ardesiacus* DUSKY-THROATED ANTSHRIKE[2]
— *T.a.ardesiacus* (Sclater & Salvin, 1868)   SE Colombia, E Ecuador, N and C Peru
— *T.a.obidensis* (E. Snethlage, 1914)   S Venezuela, the Guianas, N Brazil

○ *Thamnomanes saturninus* SATURNINE ANTSHRIKE
— *T.s.huallagae* (Cory, 1916)   E Peru, W Brazil
— *T.s.saturninus* (Pelzeln, 1869)   WC Brazil

○ *Thamnomanes caesius* CINEREOUS ANTSHRIKE
— *T.c.glaucus* Cabanis, 1847   N and W Amazonia
— *T.c.intermedius* Carriker, 1935   C Peru
— *T.c.persimilis* Hellmayr, 1907[3]   C Brazil
— *T.c.hoffmannsi* Hellmayr, 1906   SC Brazil, E Bolivia
— *T.c.caesius* (Temminck, 1820)   E Brazil

○ *Thamnomanes schistogynus* Hellmayr, 1911 BLUISH-SLATE ANTSHRIKE   SE Peru, W Brazil

PYGIPTILA P.L. Sclater, 1858 F
○ *Pygiptila stellaris* SPOT-WINGED ANTSHRIKE
— *P.s.maculipennis* (P.L. Sclater, 1855)   SE Colombia to NE Peru
— *P.s.occipitalis* J.T. Zimmer, 1932   S Colombia to the Guianas, N Brazil
— *P.s.purusiana* Todd, 1927   W Brazil
— *P.s.stellaris* (Spix, 1825)   C Brazil

MYRMOTHERULA P.L. Sclater, 1858 F
○ *Myrmotherula brachyura* PYGMY ANTWREN
— *M.b.ignota* Griscom, 1929[4]   C and E Panama, NW Colombia
— *M.b.brachyura* (Hermann, 1783)   N and W Amazonia

○ *Myrmotherula obscura* J.T. Zimmer, 1932 SHORT-BILLED ANTWREN   SE Colombia, NE Peru, W Brazil

○ *Myrmotherula surinamensis* (J.F. Gmelin, 1788) GUIANAN STREAKED ANTWREN[5]
     S Venezuela, the Guianas, N Brazil

○ *Myrmotherula multostriata* P.L. Sclater, 1858 AMAZONIAN STREAKED ANTWREN
     W Amazonia

○ *Myrmotherula pacifica* Hellmayr, 1911 PACIFIC ANTWREN   E Panama to E Ecuador

○ *Myrmotherula cherriei* Belepsch & Hartert, 1902 CHERRIE'S ANTWREN   E Colombia, W Venezuela, N Brazil

○ *Myrmotherula klagesi* Todd, 1927 KLAGES'S ANTWREN   NE Brazil

○ *Myrmotherula longicauda* STRIPE-CHESTED ANTWREN
— *M.l.soderstromi* Gyldenstolpe, 1930   E Ecuador
— *M.l.pseudoaustralis* Gyldenstolpe, 1930   E Ecuador, N Peru
— *M.l.longicauda* Berlepsch & Stolzmann, 1894   E Peru
— *M.l.australis* Chapman, 1923   SE Peru, N Bolivia

○ *Myrmotherula ambigua* J.T. Zimmer, 1932 YELLOW-THROATED ANTWREN   E Colombia, S Venezuela, NW Brazil

○ *Myrmotherula sclateri* E. Snethlage, 1912 SCLATER'S ANTWREN   C Brazil

○ *Myrmotherula hauxwelli* PLAIN-THROATED ANTWREN
— *M.h.suffusa* J.T. Zimmer, 1932   SE Colombia, E Ecuador, NE Peru

---

[1] Split proposed by Ridgely & Tudor (1994) (2019) noted, but more evidence awaited.
[2] In moving this species to this genus we note that associated proposals to lump it with the next species are not usually followed.
[3] Thought to include ¶ *simillimus* Gyldenstolpe, 1951 (1031).
[4] Might be better treated conspecifically with *M. obscura*, see Meyer de Schauensee (1966) (1573), that then being the senior name, or, perhaps, as a separate species (M. Isler *in litt.*).
[5] The split here is based on Isler *et al.* (1999) (1198).

⎯ *M.h.hauxwelli* (P.L. Sclater, 1857)   E Peru, W Brazil

⎯ *M.h.clarior* J.T. Zimmer, 1932   C Brazil

⎯ *M.h.hellmayri* E. Snethlage, 1906   NE Brazil

◯ *Myrmotherula guttata* (Vieillot, 1825)   Rufous-bellied Antwren   N Amazonia

◯ *Myrmotherula gularis* (Spix, 1825)   Star-throated Antwren   SE Brazil

◯ *Myrmotherula fulviventris*   Chequer-throated Antwren

     ⎯ *M.f.costaricensis* Todd, 1927   S Honduras to W Panama

     ⎯ *M.f.fulviventris* Lawrence, 1862   E Panama to W Ecuador

     ⎯ *M.f.salmoni* (Chubb, 1918)   C Colombia

◯ *Myrmotherula gutturalis* Sclater & Salvin, 1881   Brown-bellied Antwren

      The Guianas, Venezuela, N Brazil

◯ *Myrmotherula leucophthalma*   White-eyed Antwren

     ⎯ *M.l.dissita* Bond, 1950   SE Peru, N Bolivia

     ⎯ *M.l.leucophthalma* (Pelzeln, 1869)   WC Brazil

     ⎯ *M.l.phaeonota* Todd, 1927   NC Brazil

     ⎯ *M.l.sordida* Todd, 1927   NE Brazil

◯ *Myrmotherula haematonota*   Stipple-throated Antwren

     ⎯ *M.h.pyrrhonota* Sclater & Salvin, 1873   SE Colombia, S Venezuela, NW Brazil

     ⎯ *M.h.haematonota* (P.L. Sclater, 1857)   NE Peru

     ⎯ *M.h.amazonica* H. von Ihering, 1905   W Brazil

◯ *Myrmotherula fjeldsaai* Krabbe *et al.*, 1999   Brown-backed Antwren[1] #

      E Ecuador, N Peru (1301)

◯ *Myrmotherula spodionota*   Foothill Antwren[2]

     ⎯ *M.s.spodionota* Sclater & Salvin, 1880   E Ecuador

     ⎯ *M.s.sororia* Berlepsch & Stolzmann, 1894   N and C Peru

◯ *Myrmotherula ornata*   Ornate Antwren

     ⎯ *M.o.ornata* (P.L. Sclater, 1853)   E Colombia

     ⎯ *M.o.saturata* (Chapman, 1923)   SE Colombia, E Ecuador, NE Peru

     ⎯ *M.o.atrogularis* Taczanowski, 1874   N and C Peru

     ⎯ *M.o.meridionalis* J.T. Zimmer, 1932   SE Peru, N Bolivia

     ⎯ *M.o.hoffmannsi* Hellmayr, 1906   C Brazil

◯ *Myrmotherula erythrura*   Rufous-tailed Antwren

     ⎯ *M.e.erythrura* P.L. Sclater, 1890   NW Amazonia

     ⎯ *M.e.septentrionalis* J.T. Zimmer, 1932   E Peru, W Brazil

◯ *Myrmotherula axillaris*   White-flanked Antwren

     ⎯ *M.a.albigula* Lawrence, 1865   C America, W Colombia, Ecuador

     ⎯ *M.a.melaena* (P.L. Sclater, 1857)   NW Amazonia

     ⎯ *M.a.heterozyga* J.T. Zimmer, 1932   E Peru, W Brazil

     ⎯ *M.a.axillaris* (Vieillot, 1817)   Trinidad, Venezuela, the Guianas, N Brazil

     ⎯ *M.a.fresnayana* (d'Orbigny, 1835)   SE Peru, Bolivia

     ⎯ *M.a.luctuosa* Pelzeln, 1869   E Brazil

◯ *Myrmotherula schisticolor*   Slaty Antwren

     ⎯ *M.s.schisticolor* (Lawrence, 1865)   Mexico to W Ecuador

     ⎯ *M.s.sanctaemartae* J.A. Allen, 1900   N Colombia, N Venezuela

     ⎯ *M.s.interior* (Chapman, 1914)   E Colombia, E Ecuador, N Peru

◯ *Myrmotherula sunensis*   Rio Suno Antwren

     ⎯ *M.s.sunensis* Chapman, 1925   E Ecuador, NE Peru

     ⎯ *M.s.yessupi* Bond, 1950   W Peru

◯ *Myrmotherula minor* Salvadori, 1864   Salvadori's Antwren   SE Brazil

◯ *Myrmotherula longipennis*   Long-winged Antwren

     ⎯ *M.l.longipennis* Pelzeln, 1869   N and W Amazonia

---

[1] We give all authors' names only when five or less are involved.
[2] The race *sororia* is associated here following Ridgely & Tudor (1994: 274) (2019).

_ *M.l.zimmeri* Chapman, 1925   E Ecuador, NE Peru
_ *M.l.garbei* H. von Ihering, 1905   NE Peru, W Brazil
_ *M.l.transitiva* Hellmayr, 1929   WC Brazil
_ *M.l.ochrogyna* Todd, 1927   C Brazil
_ *M.l.paraensis* (Todd, 1920)   EC Brazil

○ *Myrmotherula urosticta* (P.L. Sclater, 1857) BAND-TAILED ANTWREN   SE Brazil

○ *Myrmotherula iheringi* IHERING'S ANTWREN
_ *M.i.heteroptera* Todd, 1927   SC Brazil
_ *M.i.iheringi* E. Snethlage, 1914   C Brazil

○ *Myrmotherula fluminensis* Gonzaga, 1988 RIO DE JANEIRO ANTWREN #   Rio de Janeiro (Brazil) (949)

○ *Myrmotherula grisea* Carriker, 1935 ASHY ANTWREN   C Bolivia

○ *Myrmotherula unicolor* (Ménétriés, 1835) UNICOLOURED ANTWREN[1]   SE Brazil

○ *Myrmotherula snowi* Teixeira & Gonzaga, 1985 ALAGOAS ANTWREN   NE Brazil (2392)

○ *Myrmotherula behni* PLAIN-WINGED ANTWREN
_ *M.b.behni* Berlepsch & Leverkühn, 1890   E Colombia
_ *M.b.yavii* Zimmer & Phelps, 1948   S Venezuela
_ *M.b.inornata* P.L. Sclater, 1890   SE Venezuela, Guyana
_ *M.b.camanii* Phelps & Phelps, Jr., 1952   S Venezuela (1850)

○ *Myrmotherula menetriesii* GREY ANTWREN
_ *M.m.pallida* Berlepsch & Hartert, 1902   NW and W Amazonia
_ *M.m.cinereiventris* Sclater & Salvin, 1867   E Venezuela, the Guianas, N Brazil
_ *M.m.menetriesii* (d'Orbigny, 1837)   E Peru, N Bolivia, W Brazil
_ *M.m.berlepschi* Hellmayr, 1903   WC Brazil
_ *M.m.omissa* Todd, 1927   NE Brazil

○ *Myrmotherula assimilis* LEADEN ANTWREN
_ *M.a.assimilis* Pelzeln, 1868   NE Peru, W Brazil, N Bolivia (?)
_ *M.a.transamazonica* Gyldenstolpe, 1951   E Brazil (1031)

DICHROZONA Ridgway, 1888 F
○ *Dichrozona cincta* BANDED ANTBIRD
_ *D.c.cincta* (Pelzeln, 1868)   E Colombia, S Venezuela, NW Brazil
_ *D.c.stellata* (Sclater & Salvin, 1880)   E Ecuador, W Brazil
_ *D.c.zononota* Ridgway, 1888   C Brazil, N Bolivia

MYRMORCHILUS Ridgway, 1909 M
○ *Myrmorchilus strigilatus* STRIPE-BACKED ANTBIRD
_ *M.s.strigilatus* (Wied, 1831)   E Brazil
_ *M.s.suspicax* Wetmore, 1922   S Bolivia, S Brazil, W Paraguay, N Argentina

HERPSILOCHMUS Cabanis, 1847 M
○ *Herpsilochmus pileatus* (M.H.K. Lichtenstein, 1823) PILEATED ANTWREN[2]   E Brazil

○ *Herpsilochmus sellowi* Whitney, Pacheco, Buzzetti & Parrini, 2000 CAATINGA ANTWREN #   NE Brazil (2659)

○ *Herpsilochmus atricapillus* Pelzeln, 1869 BLACK-CAPPED ANTWREN   C and S Brazil, SE Bolivia, Paraguay, N Argentina

○ *Herpsilochmus motacilloides* Taczanowski, 1874 CREAMY-BELLIED ANTWREN   C Peru

○ *Herpsilochmus parkeri* Davis & O'Neill, 1986 ASH-THROATED ANTWREN[3]   N Peru (619)

○ *Herpsilochmus sticturus* Salvin, 1885 SPOT-TAILED ANTWREN[4]   E Venezuela, the Guianas, NE Brazil

[1] The elevation of *M. snowi* to specific status follows Whitney & Pacheco (1997) (2655).
[2] In last Edition treated as a species taking in *atricapillus* and *motacilloides* but all are here elevated to specific status following Davis & O'Neill (1986) (619).
[3] Of the suggestions offered by Vuilleumier *et al.* (1992) (2552) we prefer treatment as part of a superspecies *pileatus*.
[4] The separation of *dugandi*, based on Ridgely & Tudor (1994: 257) (2019), is tentatively accepted.

○ *Herpsilochmus dugandi* Meyer de Schauensee, 1945 DUGAND'S ANTWREN

SE Colombia, E Ecuador, NE Peru

○ *Herpsilochmus stictocephalus* Todd, 1927 TODD'S ANTWREN

E Venezuela, the Guianas

○ *Herpsilochmus dorsimaculatus* Pelzeln, 1869 SPOT-BACKED ANTWREN S Venezuela, E Colombia, NW Brazil

○ *Herpsilochmus roraimae* RORAIMAN ANTWREN
   — *H.r.kathleenae* Phelps & Dickerman, 1980 S Venezuela (1841)
   — *H.r.roraimae* Hellmayr, 1903 SE Venezuela, SW Guyana, N Brazil

○ *Herpsilochmus pectoralis* P.L. Sclater, 1857 PECTORAL ANTWREN E Brazil

○ *Herpsilochmus longirostris* Pelzeln, 1869 LARGE-BILLED ANTWREN SC Brazil, E Bolivia

○ *Herpsilochmus gentryi* Whitney & Alvarez-Alonso, 1998 ANCIENT ANTWREN #

SE Ecuador, N Peru (2654)

○ *Herpsilochmus axillaris* YELLOW-BREASTED ANTWREN
   — *H.a.senex* Bond & Meyer de Schauensee, 1940 SW Colombia
   — *H.a.aequatorialis* Taczanowski & Berlepsch, 1885 E Ecuador
   — *H.a.puncticeps* Taczanowski, 1882 N Peru
   — *H.a.axillaris* (Tschudi, 1844) S Peru

○ *Herpsilochmus rufimarginatus* RUFOUS-WINGED ANTWREN
   — *H.r.exiguus* Nelson, 1912 E Panama
   — *H.r.frater* Sclater & Salvin, 1880 N and W Amazonia
   — *H.r.scapularis* (Wied, 1831) E Brazil
   — *H.r.rufimarginatus* (Temminck, 1822) SE Brazil, E Paraguay, NE Argentina

MICRORHOPIAS P.L. Sclater, 1862 M[1]
● *Microrhopias quixensis* DOT-WINGED ANTWREN
   — *M.q.boucardi* (P.L. Sclater, 1858) S Mexico to SW Honduras
   √ *M.q.virgatus* (Lawrence, 1863) Nicaragua to W Panama
   — *M.q.consobrinus* (P.L. Sclater, 1860) E Panama to W Ecuador
   — *M.q.quixensis* (Cornalia, 1849) E Ecuador, NE Peru
   — *M.q.intercedens* J.T. Zimmer, 1932 N Peru
   — *M.q.nigriventris* Carriker, 1930 C Peru
   — *M.q.albicauda* Carriker, 1932 SE Peru
   — *M.q.microstictus* (Berlepsch, 1908) French Guiana
   — *M.q.bicolor* (Pelzeln, 1869) N Bolivia, C Brazil
   — *M.q.emiliae* Chapman, 1921 C Brazil

FORMICIVORA Swainson, 1824 F
○ *Formicivora iheringi* Hellmayr, 1909 NARROW-BILLED ANTWREN E Brazil

○ †?*Formicivora erythronotos* Hartlaub, 1852 BLACK-HOODED ANTWREN SE Brazil

○ *Formicivora grisea* WHITE-FRINGED ANTWREN
   — *F.g.alticincta* Bangs, 1902 Pearl I. (Panama)
   — *F.g.hondae* (Chapman, 1914) N Colombia
   — *F.g.fumosa* (Cory, 1913) E Colombia, W Venezuela
   — *F.g.intermedia* Cabanis, 1847 NE Colombia, NW Venezuela, Margarita I.
   — *F.g.tobagensis* Dalmas, 1900 Tobago I.
   — *F.g.orenocensis* Hellmayr, 1904 S Venezuela
   — *F.g.rufiventris* Carriker, 1936 E Colombia, S Venezuela
   — *F.g.grisea* (Boddaert, 1783) The Guianas, NE Brazil
   — *F.g.deluzae* Ménétriés, 1835 SE Brazil

○ *Formicivora serrana* SERRA ANTWREN[2]
   — *F.s.serrana* (Hellmayr, 1929) E Brazil
   — *F.s.interposita* Gonzaga & Pacheco, 1990 SE Brazil (951)

○ *Formicivora littoralis* Gonzaga & Pacheco, 1990 RESTINGA ANTWREN Rio de Janeiro (951)

[1] Gender addressed by David & Gosselin (2002) (614), multiple changes occur in specific names.
[2] Although Gonzaga & Pacheco placed *littoralis* in *serrana* we tentatively accept the separation proposed in Ridgely & Tudor (1994: 301) (2019).

○ *Formicivora melanogaster* BLACK-BELLIED ANTWREN
    __ *F.m.melanogaster* Pelzeln, 1869   E Bolivia, W and C Brazil
    __ *F.m.bahiae* Hellmayr, 1909   E Brazil

○ *Formicivora rufa* RUSTY-BACKED ANTWREN
    __ *F.r.urubambae* J.T. Zimmer, 1932   E Peru
    __ *F.r.chapmani* Cherrie, 1916   E Brazil
    __ *F.r.rufa* (Wied, 1831)   E Bolivia, E Paraguay, C and S Brazil

STYMPHALORNIS Bornschein, Reinert & Teixeira, 1995 (195) M
○ *Stymphalornis acutirostris* Bornschein, Reinert & Teixeira, 1995 PARANA ANTWREN #
                                        S Brazil (195)

DRYMOPHILA Swainson, 1824 F
○ *Drymophila ferruginea* (Temminck, 1822 FERRUGINOUS ANTBIRD   SE Brazil to NE Argentina

○ *Drymophila rubricollis* (A. Bertoni, 1901) BERTONI'S ANTWREN   SE Brazil (146)

○ *Drymophila genei* (de Filippi, 1847) RUFOUS-TAILED ANTBIRD   SE Brazil

○ *Drymophila ochropyga* (Hellmayr, 1906) OCHRE-RUMPED ANTBIRD   SE Brazil

○ *Drymophila malura* (Temminck, 1825) DUSKY-TAILED ANTBIRD   SE Brazil, N Argentina

○ *Drymophila squamata* (M.H.K. Lichtenstein, 1823) SCALED ANTBIRD   C and SE Brazil

○ *Drymophila devillei* STRIATED ANTBIRD
    __ *D.d.devillei* (Ménégaux & Hellmayr, 1906)   E Ecuador, E Peru, N Bolivia
    __ *D.d.subochracea* Chapman, 1921   C Brazil

○ *Drymophila caudata* LONG-TAILED ANTBIRD
    __ *D.c.hellmayri* Todd, 1915   NE Colombia
    __ *D.c.klagesi* Hellmayr & Seilern, 1912   N Venezuela
    __ *D.c.aristeguietana* Aveledo & Perez, 1994   Sierra de Perijá (Colombian/ Venezuelan border) (69)
    __ *D.c.caudata* (P.L. Sclater, 1855)   Colombia to N Bolivia

TERENURA Cabanis & Heine, 1859 F
○ *Terenura maculata* (Wied, 1831) STREAK-CAPPED ANTWREN   SE Brazil, NE Argentina, E Paraguay

○ *Terenura sicki* Teixeira & Gonzaga, 1983 ORANGE-BELLIED ANTWREN   NE Brazil (2390)

○ *Terenura callinota* RUFOUS-RUMPED ANTWREN
    __ *T.c.callinota* (P.L. Sclater, 1855)   Panama to N Peru
    __ *T.c.peruviana* Meyer de Schauensee, 1945   C Peru
    __ *T.c.venezuelana* Phelps & Phelps, Jr., 1954   NW Venezuela (1854)
    __ *T.c.guianensis* Blake, 1949[1]   Guyana

○ *Terenura humeralis* CHESTNUT-SHOULDERED ANTWREN
    __ *T.h.humeralis* Sclater & Salvin, 1880   E Ecuador, E Peru, W Brazil, N Bolivia
    __ *T.h.transfluvialis* Todd, 1927   C Brazil

○ *Terenura sharpei* Berlepsch, 1901 YELLOW-RUMPED ANTWREN   SE Peru, N Bolivia

○ *Terenura spodioptila* ASH-WINGED ANTWREN
    __ *T.s.signata* J.T. Zimmer, 1932   SE Colombia, NW Brazil
    __ *T.s.spodioptila* Sclater & Salvin, 1881   S Venezuela, Guyana, NW Brazil
    __ *T.s.elaopteryx* Leverkühn, 1889   French Guiana, NE Brazil
    __ *T.s.meridionalis* E. Snethlage, 1925   C Brazil

CERCOMACRA P.L. Sclater, 1858 F
○ *Cercomacra cinerascens* GREY ANTBIRD
    __ *C.c.cinerascens* (P.L. Sclater, 1857)   Upper Amazonia
    __ *C.c.immaculata* Chubb, 1918   E Venezuela, the Guianas, N Brazil
    __ *C.c.sclateri* Hellmayr, 1905   SW Amazonia
    __ *C.c.iterata* J.T. Zimmer, 1932   C Brazil

---

[1] Includes *Hylophilus puellus* Mees, 1974 (1526), see Mayr & Vuilleumier (1983) (1496).

○ *Cercomacra brasiliana* Hellmayr, 1905 RIO DE JANEIRO ANTBIRD   SE Brazil

○ *Cercomacra tyrannina* DUSKY ANTBIRD[1]
— *C.t.crepera* Bangs, 1901   S Mexico to W Panama
— *C.t.rufiventris* (Lawrence, 1865)   E Panama to W Ecuador
— *C.t.tyrannina* (P.L. Sclater, 1855)   E Colombia, S Venezuela, NW Brazil
— *C.t.vicina* Todd, 1927   NE Colombia, W Venezuela
— *C.t.saturatior* Chubb, 1918   E Venezuela, Guyana, Surinam

○ *Cercomacra laeta* WILLIS'S ANTBIRD
— *C.l.waimiri* Bierregaard, Cohn-Haft & Stotz, 1997   N Brazil (148)
— *C.l.laeta* Todd, 1920   C Brazil
— *C.l.sabinoi* Pinto, 1939   NE Brazil

○ *Cercomacra parkeri* Graves, 1997 PARKER'S ANTBIRD #   W Colombia (989)

○ *Cercomacra nigrescens* BLACKISH ANTBIRD
— *C.n.nigrescens* (Cabanis & Heine, 1859)   Surinam, French Guiana
— *C.n.aequatorialis* J.T. Zimmer, 1931   E Ecuador, N Peru
— *C.n.notata* J.T. Zimmer, 1931   C Peru
— *C.n.fuscicauda* J.T. Zimmer, 1931   E Peru, W Brazil, N Bolivia
— *C.n.approximans* Pelzeln, 1869   C Brazil
— *C.n.ochrogyna* E. Snethlage, 1928[2]   C Brazil

○ *Cercomacra serva* BLACK ANTBIRD
— *C.s.serva* (P.L. Sclater, 1858)   E Ecuador, NE Peru
— *C.s.hypomelaena* P.L. Sclater, 1890   E and SE Peru, N Bolivia, W Brazil

○ *Cercomacra nigricans* JET ANTBIRD
— *C.n.nigricans* P.L. Sclater, 1858   E Panama to N Brazil
— *C.n.atrata* Todd, 1927   NW Colombia

○ *Cercomacra carbonaria* Sclater & Salvin, 1873 RIO BRANCO ANTBIRD   Rio Branco (N Brazil)

○ *Cercomacra melanaria* (Ménétriés, 1835) MATO GROSSO ANTBIRD   E Bolivia, SW Brazil

○ *Cercomacra manu* Fitzpatrick & Willard, 1990 MANU ANTBIRD   SE Peru, NW Bolivia (838)

○ *Cercomacra ferdinandi* E. Snethlage, 1928[3] BANANAL ANTBIRD   C Brazil

PYRIGLENA Cabanis, 1847 F
○ *Pyriglena leuconota* WHITE-BACKED FIRE-EYE
— *P.l.pacifica* Chapman, 1923   W Ecuador
— *P.l.castanoptera* Chubb, 1916   E Ecuador, N Peru
— *P.l.picea* Cabanis, 1847   N and C Peru
— *P.l.similis* J.T. Zimmer, 1931   C Brazil
— *P.l.marcapatensis* Stolzmann[4] & Domaniewski, 1918   SE Peru
— *P.l.hellmayri* Stolzmann & Domaniewski, 1918   W Bolivia
— *P.l.maura* (Ménétriés, 1835)   SE Bolivia, SE Brazil
— *P.l.interposita* Pinto, 1947   EC Brazil
— *P.l.leuconota* (Spix, 1824)   S Brazil
— *P.l.pernambucensis* J.T. Zimmer, 1931   E Brazil

○ *Pyriglena atra* (Swainson, 1825) FRINGE-BACKED FIRE-EYE   E Brazil

○ *Pyriglena leucoptera* (Vieillot, 1818) WHITE-SHOULDERED FIRE-EYE   E Paraguay, NE Argentina, SE Brazil

RHOPORNIS Richmond, 1891 M
○ *Rhopornis ardesiacus* (Wied, 1831) SLENDER ANTBIRD[5]   SE Brazil

---

[1] The treatment of a species *laeta* follows Bierregaard *et al.* (1997) (148).
[2] Perhaps 1927 see J. f. Orn. (1930: 134).
[3] Perhaps 1927 see J. f. Orn. (1930: 134).
[4] He spelled his name Sztolcman when publishing in Poland.
[5] For correction of spelling see David & Gosselin (2002) (614).

MYRMOBORUS Cabanis & Heine, 1859 M
○ *Myrmoborus leucophrys* WHITE-BROWED ANTBIRD
   __ *M.l.erythrophrys* (P.L. Sclater, 1855)   E Colombia
   __ *M.l.leucophrys* (Tschudi, 1844)   N and W Amazonia
   __ *M.l.angustirostris* (Cabanis, 1848)   N Amazonia
   __ *M.l.koenigorum* O'Neill & Parker, 1997   C Peru (1717)
   __ *M.l.griseigula* J.T. Zimmer, 1932   C Brazil, N Bolivia

○ *Myrmoborus lugubris* ASH-BREASTED ANTBIRD
   __ *M.l.berlepschi* (Hellmayr, 1910)   NE Peru, W Brazil
   __ *M.l.stictopterus* Todd, 1927   C Brazil
   __ *M.l.femininus* (Hellmayr, 1910)   C Brazil
   __ *M.l.lugubris* (Cabanis, 1847)   C Brazil

○ *Myrmoborus myotherinus* BLACK-FACED ANTBIRD
   __ *M.m.elegans* (P.L. Sclater, 1857)   E Colombia, S Venezuela, NW Brazil
   __ *M.m.napensis* J.T. Zimmer, 1932   E Ecuador, NE Peru
   __ *M.m.myotherinus* (Spix, 1825)   S Peru, NE Bolivia, S and W Brazil
   __ *M.m.incanus* Hellmayr, 1929   WC Brazil
   __ *M.m.ardesiacus* Todd, 1927   W Brazil
   __ *M.m.proximus* Todd, 1927   W Brazil
   __ *M.m.ochrolaema* (Hellmayr, 1906)   C Brazil
   __ *M.m.sororius* (Hellmayr, 1910)   C and S Brazil

○ *Myrmoborus melanurus* (Sclater & Salvin, 1866) BLACK-TAILED ANTBIRD
                  NE Peru

HYPOCNEMIS Cabanis, 1847 F
○ *Hypocnemis cantator* WARBLING ANTBIRD
   __ *H.c.flavescens* P.L. Sclater, 1865   S Venezuela, E Colombia, NW Brazil
   __ *H.c.notaea* Hellmayr, 1920   SE Venezuela, Guyana, N Brazil
   __ *H.c.cantator* (Boddaert, 1783)   Surinam, French Guiana, N Brazil
   __ *H.c.saturata* Carriker, 1930   SE Colombia, E Ecuador, NE Peru
   __ *H.c.peruviana* Taczanowski, 1884   E Peru, W Brazil
   __ *H.c.implicata* J.T. Zimmer, 1932   C Brazil
   __ *H.c.striata* (Spix, 1825)   EC Brazil
   __ *H.c.affinis* J.T. Zimmer, 1932   EC Brazil
   __ *H.c.subflava* Cabanis, 1873   C Peru
   __ *H.c.collinsi* Cherrie, 1916   SE Peru, N Bolivia
   __ *H.c.ochrogyna* J.T. Zimmer, 1932   NE Bolivia, S Brazil

○ *Hypocnemis hypoxantha* YELLOW-BROWED ANTBIRD
   __ *H.h.hypoxantha* P.L. Sclater, 1869   SE Colombia, E Ecuador, NE Peru
   __ *H.h.ochraceiventris* Chapman, 1921   E Brazil

HYPOCNEMOIDES Bangs & Penard, 1918 M
○ *Hypocnemoides melanopogon* BLACK-CHINNED ANTBIRD
   __ *H.m.occidentalis* J.T. Zimmer, 1932   E and S Colombia, Venezuela, NW Brazil
   __ *H.m.melanopogon* (P.L. Sclater, 1857)   The Guianas, N Brazil
   __ *H.m.minor* Gyldenstolpe, 1941   C Brazil

○ *Hypocnemoides maculicauda* BAND-TAILED ANTBIRD
   __ *H.m.maculicauda* (Pelzeln, 1869)   Peru, N Bolivia, N Paraguay, W and S Brazil
   __ *H.m.orientalis* Gyldenstolpe, 1941   C and SE Brazil

MYRMOCHANES J.A. Allen, 1889 M
○ *Myrmochanes hemileucus* (Sclater & Salvin, 1866) BLACK-AND-WHITE ANTBIRD
                  Peru, N Bolivia, W Brazil

GYMNOCICHLA P.L. Sclater, 1858 F
○ *Gymnocichla nudiceps* BARE-CROWNED ANTBIRD
   __ *G.n.chiroleuca* Sclater & Salvin, 1869   E Guatemala to Costa Rica
   __ *G.n.erratilis* Bangs, 1907   SW Costa Rica, W Panama
   __ *G.n.nudiceps* (Cassin, 1850)   E Panama, NW Colombia
   __ *G.n.sanctamartae* Ridgway, 1908   N Colombia

SCLATERIA Oberholser, 1899 F
○ *Sclateria naevia* S<small>ILVERED</small> A<small>NTBIRD</small>
    — *S.n.naevia* (J.F. Gmelin, 1788)   NE Venezuela, the Guianas, N Brazil, Trinidad
    — *S.n.diaphora* Todd, 1913   Venezuela
    — *S.n.argentata* (Des Murs, 1856)   Upper Amazonia
    — *S.n.toddi* Hellmayr, 1924   C Brazil

PERCNOSTOLA Cabanis & Heine, 1859 F
○ *Percnostola rufifrons* B<small>LACK-HEADED</small> A<small>NTBIRD</small>
    — *P.r.rufifrons* (J.F. Gmelin, 1789)   The Guianas, NE Brazil
    — *P.r.subcristata* Hellmayr, 1908   N Brazil

○ *Percnostola minor* A<small>MAZONAS</small> A<small>NTBIRD</small>[1]
    — *P.m.minor* Pelzeln, 1869   NW Amazonia
    — *P.m.jensoni* Capparella, Rosenberg & Cardiff, 1997   NE Peru (290)

○ *Percnostola arenarum* Isler *et al.*, 2002[2] A<small>LLPAHUAYO</small> A<small>NTBIRD</small> #   Amazonian Peru (W of R Napo) (1199A)

○ *Percnostola lophotes* Hellmayr & Seilern, 1914 W<small>HITE-LINED</small> A<small>NTBIRD</small>[3]
                                   SE Peru

SCHISTOCICHLA Todd, 1927[4] F
○ *Schistocichla schistacea* (P.L. Sclater, 1858) S<small>LATE-COLOURED</small> A<small>NTBIRD</small>   SE Colombia, NE Peru, W Brazil

○ *Schistocichla leucostigma* S<small>POT-WINGED</small> A<small>NTBIRD</small>
    — *S.l.subplumbea* (Sclater & Salvin, 1880)   E Colombia, E Ecuador, NE Peru
    — *S.l.obscura* Zimmer & Phelps, 1946   C Venezuela
    — *S.l.saturata* (Salvin, 1885)   SE Venezuela
    — *S.l.leucostigma* (Pelzeln, 1869)   S Venezuela, the Guianas, N Brazil
    — *S.l.intensa* (J.T. Zimmer, 1927)   C Peru
    — *S.l.brunneiceps* J.T. Zimmer, 1931   SE Peru
    — *S.l.infuscata* Todd, 1927   S Venezuela, N and W Brazil
    — *S.l.humaythae* (Hellmayr, 1907)   C Brazil
    — *S.l.rufifacies* Hellmayr, 1929   C Brazil

○ *Schistocichla caurensis* (Hellmayr, 1906) C<small>AURA</small> A<small>NTBIRD</small>[5]   C and S Venezuela, N Brazil

MYRMECIZA G.R. Gray, 1841[6] F
○ *Myrmeciza disjuncta* Friedmann, 1945 Y<small>APACANA</small> A<small>NTBIRD</small>   S Venezuela, N Brazil

○ *Myrmeciza longipes* W<small>HITE-BELLIED</small> A<small>NTBIRD</small>
    — *M.l.panamensis* Ridgway, 1908   E Panama, N Colombia
    — *M.l.longipes* (Swainson, 1825)   E Colombia, N Venezuela, Trinidad
    — *M.l.boucardi* Berlepsch, 1888   NC Colombia
    — *M.l.griseipectus* Berlepsch & Hartert, 1902   SE Colombia, S Venezuela, Guyana, NE Brazil

○ *Myrmeciza exsul* C<small>HESTNUT-BACKED</small> A<small>NTBIRD</small>
    — *M.e.exsul* P.L. Sclater, 1859   E Nicaragua to W Panama
    — *M.e.occidentalis* Cherrie, 1891   W Costa Rica, S Panama
    — *M.e.cassini* (Ridgway, 1908)   E Panama, N Colombia
    — *M.e.niglarus* Wetmore, 1962   NW Colombia (2600)
    — *M.e.maculifer* (Hellmayr, 1906)   W Colombia, W Ecuador

○ *Myrmeciza ferruginea* F<small>ERRUGINOUS-BACKED</small> A<small>NTBIRD</small>
    — *M.f.ferruginea* (Statius Müller, 1776)   E Venezuela, The Guianas, N Brazil
    — *M.f.eluta* (Todd, 1927)   C Brazil

○ *Myrmeciza ruficauda* S<small>CALLOPED</small> A<small>NTBIRD</small>
    — *M.r.soror* Pinto, 1940   E Brazil
    — *M.r.ruficauda* (Wied, 1831)   SE Brazil

---

[1] Treated as a race of *P. rufifrons* in last Edition, but we tentatively follow Capparella *et al.* (1997) (290) who treated this as a species.
[2] We give all authors' names only when five or less are involved.
[3] Thought to include *Percnostola macrolopha* Berlioz, 1966 (141), see Mayr & Vuilleumier (1983) (1496).
[4] Submerged in *Percnostola* in last Edition. Recognition of this genus follows Ridgely & Tudor (1994: 327) (2019).
[5] Includes *australis* see Zimmer (1999) (2731).
[6] Includes *Sipia* see Robbins & Ridgely (1991) (2081).

○ *Myrmeciza loricata* (M.H.K. Lichtenstein, 1823) WHITE-BIBBED ANTBIRD

SE Brazil

○ *Myrmeciza squamosa* Pelzeln, 1869 SQUAMATE ANTBIRD

SE Brazil

○ *Myrmeciza laemosticta* DULL-MANTLED ANTBIRD[1]
— *M.l.laemosticta* Salvin, 1865   E Costa Rica, E Panama
— *M.l.palliata* Todd, 1917   NW and C Colombia, W Venezuela

○ *Myrmeciza nigricauda* Salvin & Godman, 1892 ESMERALDAS ANTBIRD[2]   W Colombia, W Ecuador

○ *Myrmeciza berlepschi* (E. Hartert, 1898) STUB-TAILED ANTWREN   W Colombia, NW Ecuador

○ *Myrmeciza pelzelni* P.L. Sclater, 1890 GREY-BELLIED ANTBIRD   E Colombia, Venezuela, N Brazil

○ *Myrmeciza castanea* NORTHERN CHESTNUT-TAILED ANTBIRD[3]
— *M.c.castanea* Zimmer, 1932   E Andean foothills of S Colombia to N Peru (2707A)
— *M.c.centunculorum* Isler *et al.*, 2002   Amazonian lowlands of E Ecuador and N Peru (1199B)

○ *Myrmeciza hemimelaena* SOUTHERN CHESTNUT-TAILED ANTBIRD
— *M.h.hemimelaena* P.L. Sclater, 1857   Amazonian lowlands of E Peru, NW Bolivia and SW
                                                                     Brazil (W of R Madeira)
— *M.h.pallens* Berlepsch & Hellmayr, 1905   Amazonian lowlands of NE Bolivia and SC Brazil
                                                                     (E of R Madeira)

○ *Myrmeciza atrothorax* BLACK-THROATED ANTBIRD
— *M.a.metae* Meyer de Schauensee, 1947   E Colombia
— *M.a.atrothorax* (Boddaert, 1783)[4]   S Venezuela, the Guianas, N Brazil
— *M.a.tenebrosa* J.T. Zimmer, 1932   NE Peru, W Brazil
— *M.a.maynana* Taczanowski, 1882   N Peru
— *M.a.obscurata* J.T. Zimmer, 1932   E Peru, W Brazil
— *M.a.griseiventris* Carriker, 1935   W Bolivia
— *M.a.melanura* (Ménétriés, 1835)   E Bolivia, SW Brazil

○ *Myrmeciza melanoceps* (Spix, 1825) WHITE-SHOULDERED ANTBIRD[5]   W Amazonia

○ *Myrmeciza goeldii* (E. Snethlage, 1908) GOELDI'S ANTBIRD   SW Brazil, SE Peru, NW Bolivia

○ *Myrmeciza hyperythra* (P.L. Sclater, 1855) PLUMBEOUS ANTBIRD   W Amazonia

○ *Myrmeciza fortis* SOOTY ANTBIRD
— *M.f.fortis* (Sclater & Salvin, 1868)   SE Colombia, E Ecuador, E Peru, W Brazil, NW Bolivia
— *M.f.incanescens* (Todd, 1927)   C Brazil

○ *Myrmeciza immaculata* IMMACULATE ANTBIRD
— *M.i.zeledoni* Ridgway, 1909   E Costa Rica, W Panama
— *M.i.macrorhyncha* Robbins & Ridgely, 1993[6]   E Panama to W Ecuador (2083)
— *M.i.immaculata* (Lafresnaye, 1845)   E Colombia, W Venezuela
— *M.i.brunnea* Phelps & Phelps, Jr., 1955   NW Venezuela (1855)

○ *Myrmeciza griseiceps* (Chapman, 1923) GREY-HEADED ANTBIRD   SW Ecuador, NW Peru

MYRMORNIS Hermann, 1783[7] F
○ *Myrmornis torquata* WING-BANDED ANTBIRD
— *M.t.stictoptera* (Salvin, 1893)   SE Nicaragua, Panama, NW Colombia
— *M.t.torquata* (Boddaert, 1783)   N and E Amazonia

PITHYS Vieillot, 1818 M[8]
○ *Pithys albifrons* WHITE-FACED ANTBIRD
— *P.a.albifrons* (Linnaeus, 1766)   S Venezuela, the Guianas, N Brazil
— *P.a.brevibarba* Chapman, 1928   NW Amazonia
— *P.a.peruvianus* Taczanowski, 1884   N and E Peru

[1] Revised following Robbins & Ridgely (1991) (2081).
[2] Called *Sipia rosenbergi* in last Edition, but see Robbins & Ridgely (1991) (2081).
[3] For recognition of this as a species wrongly placed in synonymy see Isler *et al.* (2002) (1199B).
[4] Including *stictothorax* following Schulenberg & Stotz (1991) (2212).
[5] Includes *Akletos peruvianus* Dunajewski, 1948 (769), a name given to the female of this species.
[6] A new name for *M.i.berlepschi* Ridgway, 1909, now preoccupied by *M. berlepschi* (Hartert, 1898).
[7] In last Edition treated between the Formicariid genera *Chamaeza* and *Pittasoma*; however this is a Thamnophilid see A.O.U. (1998: 368) (3).
[8] Gender addressed by David & Gosselin (2002) (614), multiple changes occur in specific names.

○ *Pithys castaneus* Berlioz, 1938 White-masked Antbird   N Peru

GYMNOPITHYS Bonaparte, 1857[1] M[2]
● *Gymnopithys leucaspis* Bicoloured Antbird
  ✔ *G.l.olivascens* (Ridgway, 1891)   Honduras to W Panama
  — *G.l.bicolor* (Lawrence, 1863)   E Panama, NW Colombia
  — *G.l.daguae* Hellmayr, 1906   W Colombia
  — *G.l.aequatorialis* (Hellmayr, 1902)   SW Colombia, W Ecuador
  — *G.l.ruficeps* Salvin & Godman, 1892   C Colombia
  — *G.l.leucaspis* (P.L. Sclater, 1855)   E Colombia
  — *G.l.castaneus* J.T. Zimmer, 1937   E Ecuador, NE Peru
  — *G.l.peruanus* J.T. Zimmer, 1937   N Peru
  — *G.l.lateralis* Todd, 1927   W Brazil

○ *Gymnopithys rufigula* Rufous-throated Antbird
  — *G.r.pallidus* (Cherrie, 1909)   S Venezuela
  — *G.r.pallidigula* Phelps & Phelps, Jr., 1947   S Venezuela
  — *G.r.rufigula* (Boddaert, 1783)   E Venezuela, the Guianas, N Brazil

○ *Gymnopithys salvini* White-throated Antbird
  — *G.s.maculatus* J.T. Zimmer, 1937   E Peru, W Brazil
  — *G.s.salvini* (Berlepsch, 1901)   Bolivia, SW Brazil

○ *Gymnopithys lunulatus* (Sclater & Salvin, 1873) Lunulated Antbird E Peru

RHEGMATORHINA Ridgway, 1888 F
○ *Rhegmatorhina gymnops* Ridgway, 1888 Bare-eyed Antbird   E Brazil

○ *Rhegmatorhina berlepschi* (E. Snethlage, 1907) Harlequin Antbird C Brazil

○ *Rhegmatorhina hoffmannsi* (Hellmayr, 1907) White-breasted Antbird SW Brazil

○ *Rhegmatorhina cristata* (Pelzeln, 1869) Chestnut-crested Antbird NW Brazil, SE Colombia

○ *Rhegmatorhina melanosticta* Hairy-crested Antbird
  — *R.m.melanosticta* (Sclater & Salvin, 1880)   E Ecuador
  — *R.m.brunneiceps* Chapman, 1928   N Peru
  — *R.m.purusiana* (E. Snethlage, 1908)   E Peru, W Brazil
  — *R.m.badia* J.T. Zimmer, 1932   SE Peru, N Bolivia, W Brazil

HYLOPHYLAX Ridgway, 1909 M[3]
○ *Hylophylax naevioides* Spotted Antbird
  — *H.n.capnitis* (Bangs, 1906)   E Nicaragua to W Panama
  — *H.n.naevioides* (Lafresnaye, 1847)   E Panama to W Ecuador
  — *H.n.subsimilis* Todd, 1917   WC Colombia

○ *Hylophylax naevius* Spot-backed Antbird
  — *H.n.theresae* (Des Murs, 1856)   W Amazonia
  — *H.n.peruvianus* Carriker, 1932   N Peru
  — *H.n.consobrinus* Todd, 1913   S Venezuela, NW Brazil
  — *H.n.naevius* (J.F. Gmelin, 1789)   S Venezuela, the Guianas
  — *H.n.obscurus* Todd, 1927   N Brazil
  — *H.n.ochraceus* (Berlepsch, 1912)   N Brazil

○ *Hylophylax punctulatus* Dot-backed Antbird
  — *H.p.punctulatus* (Des Murs, 1856)   S and E Venezuela, NE Peru, Brazil
  — *H.p.subochraceus* J.T. Zimmer, 1934   C Brazil

○ *Hylophylax poecilinotus* Scale-backed Antbird
  — *H.p.poecilinotus* (Cabanis, 1847)[4]   SE Venezuela, the Guianas, N Brazil
  — *H.p.duidae* Chapman, 1923   E Colombia, S Venezuela, NW Brazil

[1] Species sequence based on Hackett (1993) (1047).
[2] Gender addressed by David & Gosselin (2002) (614), multiple changes occur in specific names.
[3] Gender addressed by David & Gosselin (2002) (614), multiple changes occur in specific names.
[4] The emended name used here is judged to be in prevailing usage; see Art. 33.2 of the Code (I.C.Z.N., 1999) (1178) and p. 000 herein.

__ *H.p.lepidonota* (Sclater & Salvin, 1880)	SE Colombia, E Ecuador, NE Peru
__ *H.p.griseiventris* (Pelzeln, 1869)	SE Peru, N Bolivia, SW Brazil
__ *H.p.gutturalis* Todd, 1927	W Brazil
__ *H.p.nigrigula* (E. Snethlage, 1914)	C Brazil
__ *H.p.vidua* (Hellmayr, 1905)	EC Brazil

PHLEGOPSIS Reichenbach, 1850[1] F

○ *Phlegopsis nigromaculata* BLACK-SPOTTED BARE-EYE

__ *P.n.nigromaculata* (d'Orbigny & Lafresnaye, 1837)	E Peru, N Bolivia, W and SW Brazil
__ *P.n.bowmani* Ridgway, 1888	N Brazil
__ *P.n.confinis* J.T. Zimmer, 1932	N Brazil
__ *P.n.paraensis* Hellmayr, 1904	N Brazil

○ *Phlegopsis erythroptera* REDDISH-WINGED BARE-EYE

__ *P.e.erythroptera* (Gould, 1855)	N and W Amazonia
__ *P.e.ustulata* Todd, 1927	E Peru, W Brazil

SKUTCHIA Willis, 1968 F

○ *Skutchia borbae* (Hellmayr, 1907) PALE-FACED BARE-EYE   C Brazil

PHAENOSTICTUS Ridgway, 1909 M

○ *Phaenostictus mcleannani* OCELLATED ANTBIRD

__ *P.m.saturatus* (Richmond, 1896)	SE Nicaragua to W Panama
__ *P.m.mcleannani* (Lawrence, 1860)	EC Panama
__ *P.m.chocoanus* Bangs & Barbour, 1922	E Panama, NW Colombia
__ *P.m.pacificus* Hellmayr, 1924	SW Colombia, NW Ecuador

CONOPOPHAGIDAE GNATEATERS (1: 8)

CONOPOPHAGA Vieillot, 1816 F

○ *Conopophaga lineata* RUFOUS GNATEATER

__ *C.l.lineata* (Wied, 1831)	E and SE Brazil
__ *C.l.vulgaris* Ménétriés, 1835	SE Brazil, E Paraguay, NE Argentina
__ *C.l.cearae* Cory, 1916	E Brazil

○ *Conopophaga aurita* CHESTNUT-BELTED GNATEATER

__ *C.a.inexpectata* J.T. Zimmer, 1931	SE Colombia, NW Brazil
__ *C.a.aurita* (J.F. Gmelin, 1789)	The Guianas, N Brazil
__ *C.a.occidentalis* Chubb, 1917	E Ecuador, NE Peru
__ *C.a.australis* Todd, 1927	E Peru, W Brazil
__ *C.a.snethlageae* Berlepsch, 1912	C Brazil
__ *C.a.pallida* E. Snethlage, 1914	Brazil

○ *Conopophaga roberti* Hellmayr, 1905 HOODED GNATEATER   NE Brazil

○ *Conopophaga peruviana* Des Murs, 1856 ASH-THROATED GNATEATER   E Ecuador, E Peru, W Brazil

○ *Conopophaga ardesiaca* SLATY GNATEATER

__ *C.a.saturata* Berlepsch & Stolzmann, 1906	SE Peru
__ *C.a.ardesiaca* d'Orbigny & Lafresnaye, 1837	Bolivia

○ *Conopophaga castaneiceps* CHESTNUT-CROWNED GNATEATER

__ *C.c.chocoensis* Chapman, 1915	W Colombia
__ *C.c.castaneiceps* P.L. Sclater, 1857	E Colombia, NE Ecuador
__ *C.c.chapmani* Carriker, 1933	SE Ecuador, NE Peru
__ *C.c.brunneinucha* Berlepsch & Stolzmann, 1896	C Peru

○ *Conopophaga melanops* BLACK-CHEEKED GNATEATER

__ *C.m.perspicillata* (M.H.K. Lichtenstein, 1823)	E Brazil
__ *C.m.melanops* (Vieillot, 1818)	SE Brazil
__ ¶ *C.m.nigrifrons* Pinto, 1954	Alagaos, Brazil (1890)

○ *Conopophaga melanogaster* Ménétriés, 1835 BLACK-BELLIED GNATEATER N Bolivia, Brazil

---

[1] *Phlegopsis barringeri* Meyer de Schauensee, 1951 (1569), included in last Edition, was considered a hybrid by Graves (1992) (986); but see also Rodner *et al.* (2000) (2096).

RHINOCRYPTIDAE TAPACULOS (12: 55)

PTEROPTOCHOS Kittlitz, 1830 M
○ *Pteroptochos castaneus* Philippi & Landbeck, 1864 CHESTNUT-THROATED HUET-HUET
                                                         C Chile, W Argentina

○ *Pteroptochos tarnii* (P.P. King, 1831) BLACK-THROATED HUET-HUET   S Chile, W Argentina

● *Pteroptochos megapodius* MOUSTACHED TURCA
    — *P.m.atacamae* Philippi, 1946   N Chile
    ✓ *P.m.megapodius* Kittlitz, 1830   C Chile

SCELORCHILUS Oberholser, 1923 M
● *Scelorchilus albicollis* WHITE-THROATED TAPACULO
    — *S.a.atacamae* Hellmayr, 1924   N Chile
    ✓ *S.a.albicollis* (Kittlitz, 1830)   C Chile

○ *Scelorchilus rubecula* CHUCAO TAPACULO
    — *S.r.rubecula* (Kittlitz, 1830)   Chile, W Argentina
    — *S.r.mochae* Chapman, 1934   Mocha I.

RHINOCRYPTA G.R. Gray, 1840 F
○ *Rhinocrypta lanceolata* CRESTED GALLITO
    — *R.l.saturata* Brodkorb, 1939   Paraguay, S Bolivia
    — *R.l.lanceolata* (I. Geoffroy Saint-Hilaire, 1832)   Argentina

TELEDROMAS Wetmore & Peters, 1922 M
○ *Teledromas fuscus* (Sclater & Salvin, 1873) SANDY GALLITO   Argentina

LIOSCELES P.L. Sclater, 1865 M
○ *Liosceles thoracicus* RUSTY-BELTED TAPACULO
    — *L.t.dugandi* Meyer de Schauensee, 1950   SE Colombia, W Brazil
    — *L.t.erithacus* P.L. Sclater, 1890   E Ecuador, E Peru
    — *L.t.thoracicus* (P.L. Sclater, 1865)   SE Peru, SW Brazil

MELANOPAREIA Reichenbach, 1853 F
○ *Melanopareia torquata* COLLARED CRESCENT-CHEST
    — *M.t.torquata* (Wied, 1831)   E Brazil
    — *M.t.rufescens* Hellmayr, 1924   C Brazil

○ *Melanopareia maximiliani* OLIVE-CROWNED CRESCENT-CHEST
    — *M.m.maximiliani* (d'Orbigny, 1835)   W Bolivia
    — *M.m.bitorquata* (d'Orbigny & Lafresnaye, 1837)[1]   E Bolivia
    — *M.m.argentina* (Hellmayr, 1907)   Andes of S Bolivia, N Argentina
    — *M.m.pallida* Nores & Yzurieta, 1980   W Paraguay, N Argentina (1659)

○ *Melanopareia maranonica*[2] Chapman, 1924 MARAÑON CRESCENT-CHEST
                                                     N Peru

○ *Melanopareia elegans* ELEGANT CRESCENT-CHEST
    — *M.e.elegans* (Lesson, 1844)   W Ecuador
    — *M.e.paucalensis* (Taczanowski, 1884)   E Peru

PSILORHAMPHUS P.L. Sclater, 1855[3] M
○ *Psilorhamphus guttatus* (Ménétriés, 1835) SPOTTED BAMBOO-WREN   SE Brazil, NE Argentina

MERULAXIS Lesson, 1831 M
○ *Merulaxis ater* Lesson, 1831 SLATY BRISTLEFRONT   SE Brazil

○ †?*Merulaxis stresemanni* Sick, 1960 STRESEMANN'S BRISTLEFRONT   Coastal Bahia (E Brazil) (2268)

[1] This taxon accidentally omitted from last Edition.
[2] Corrected to agree in gender as advised by N. David (*in litt.*).
[3] This genus appears in "Peters" X on p. 456; this addendum is easily overlooked.

EUGRALLA Lesson, 1842 F
○ *Eugralla paradoxa* (Kittlitz, 1830) Ochre-flanked Tapaculo   Chile, S Argentina

MYORNIS Chapman, 1915 M
○ *Myornis senilis* (Lafresnaye, 1840) Ash-coloured Tapaculo   Colombia to Ecuador and C Peru

SCYTALOPUS Gould, 1837[1] M
○ *Scytalopus latrans* Blackish Tapaculo[2]
    — *S.l.latrans* Hellmayr, 1924   Colombia, W Venezuela, E Ecuador, N Peru
    — *S.l.subcinereus* J.T. Zimmer, 1939   SW Ecuador, NW Peru
    — *S.l.intermedius* J.T. Zimmer, 1939   NC Peru

○ *Scytalopus unicolor* Salvin, 1895 Unicoloured Tapaculo   NW Peru

○ *Scytalopus parvirostris* J.T. Zimmer, 1939 Trilling Tapaculo[3]   C and S Peru, N Bolivia

○ *Scytalopus speluncae* (Ménétriés, 1835) Mouse-coloured Tapaculo   SE Brazil, NE Argentina

○ *Scytalopus iraiensis* Bornschien, Reinert & Pichorim, 1998 Marsh Tapaculo #
    S Brazil (196)

○ *Scytalopus macropus* Berlepsch & Stolzmann, 1896 Large-footed Tapaculo
    Peru

○ *Scytalopus sanctaemartae* Chapman, 1915 Santa Marta Tapaculo   Santa Marta Mts. (NE Colombia)

○ *Scytalopus micropterus* (P.L. Sclater, 1858) Long-tailed Tapaculo   E Ecuador, N Peru

○ *Scytalopus femoralis* (Tschudi, 1844) Rufous-vented Tapaculo   C Peru

○ *Scytalopus atratus* White-crowned Tapaculo
    — *S.a.nigricans* Phelps & Phelps, Jr., 1953[4]   NW Venezuela (Perijá Mts.) (1853)
    — *S.a.atratus* Hellmayr, 1922   E Colombia, N Peru
    — *S.a.confusus* J.T. Zimmer, 1939   E Colombia

○ *Scytalopus bolivianus* J.A. Allen, 1889 Bolivian Tapaculo   SE Peru, N Bolivia

○ *Scytalopus argentifrons* Silvery-fronted Tapaculo
    — *S.a.argentifrons* Ridgway, 1891   Costa Rica, W Panama
    — *S.a.chiriquensis* Griscom, 1924   W Panama

○ *Scytalopus panamensis* Chapman, 1915 Pale-throated Tapaculo   NW Colombia (Cerro Tarcuna)

○ *Scytalopus chocoensis* Krabbe & Schulenberg, 1997 Chocó Tapaculo #
    E Panama, W Colombia, NW Ecuador (1302)

○ *Scytalopus robbinsi* Krabbe & Schulenberg, 1997 Ecuadorian Tapaculo #
    SW Ecuador (1302)

○ *Scytalopus vicinior* J.T. Zimmer, 1939 Nariño Tapaculo   W Colombia, NW Ecuador

○ *Scytalopus latebricola* Bangs, 1899 Brown-rumped Tapaculo[5]   Santa Marta Mts. (NE Colombia)

○ *Scytalopus fuscicauda* Hellmayr, 1922 Lara Tapaculo[6]   NW Venezuela

○ *Scytalopus meridanus* Hellmayr, 1922 Mérida Tapaculo   W Venezuela[7]

○ *Scytalopus caracae* Hellmayr, 1922 Caracas Tapaculo   N Venezuela

○ *Scytalopus spillmanni* Stresemann, 1937 Spillmann's Tapaculo   W Ecuador, W Colombia

○ *Scytalopus parkeri* Krabbe & Schulenberg, 1997 Chusquea Tapaculo #
    SE Ecuador (1302)

---

[1] Since the last Edition this genus has been reviewed by Krabbe & Schulenberg (1997) (1302) who we follow here.
[2] For this split see Coopmans, Krabbe & Schulenberg (2001) (571).
[3] For the separation of *parvirostris* from *S. unicolor* see Krabbe & Schulenberg (1997) (1302).
[4] Described as a race of a very broad species *femoralis* since split. Has also been listed as a race of the next species. No adequate evidence known to us for treatment as a species (but this may prove right).
[5] For the treatment of *meridanus*, *caracae* and *spillmanni* as species see Krabbe & Schulenberg (1997) (1302).
[6] "Distinction from *meridanus* far from clear" (N. Krabbe and T. Schulenberg, *in litt.*). In last Edition we treated this as a form of *S. magellanicus*. Colombian *infasciatus* Chapman, 1915 (315) may require review.
[7] The attribution of birds referred here but from the C Andes of Colombia is uncertain.

○ *Scytalopus novacapitalis* Sick, 1958 BRASILIA TAPACULO[1]   SE Brazil (2266)

○ *Scytalopus indigoticus* (Wied, 1831) WHITE-BREASTED TAPACULO   SE Brazil

○ *Scytalopus psychopompus* Teixeira & Carnevalli, 1989 BAHIA TAPACULO[2]

NE Brazil (2389)

○ *Scytalopus magellanicus* (J.F. Gmelin, 1789) MAGELLANIC TAPACULO   S Chile, S Argentina

○ *Scytalopus griseicollis* (Lafresnaye, 1840) MATTORAL TAPACULO   E Colombia

○ *Scytalopus altirostris* J.T. Zimmer, 1939 NEBLINA TAPACULO   N Peru

○ *Scytalopus affinis* J.T. Zimmer, 1939 ANCASH TAPACULO   W Peru

○ *Scytalopus acutirostris* (Tschudi, 1844) TSCHUDI'S TAPACULO   C and SE Peru

○ *Scytalopus urubambae* J.T. Zimmer, 1939 VILCABAMBA TAPACULO   S Peru

○ *Scytalopus simonsi* Chubb, 1917 PUNA TAPACULO   S Peru, N Bolivia

○ *Scytalopus zimmeri* Bond & Meyer de Schauensee, 1940 ZIMMER'S TAPACULO

S Bolivia

○ *Scytalopus superciliaris* WHITE-BROWED TAPACULO
　— *S.s.superciliaris* Cabanis, 1883   NW Argentina
　— *S.s.santabarbarae* Nores, 1986   Argentina (1657)

○ *Scytalopus fuscus* Gould, 1837 DUSKY TAPACULO   N Chile, W Argentina

○ *Scytalopus canus* PARAMO TAPACULO
　— *S.c.canus* Chapman, 1915   W Colombia
　— *S.c.opacus* J.T. Zimmer, 1939   E Ecuador

○ *Scytalopus schulenbergi* Whitney, 1994 DIADEMED TAPACULO #   SE Peru, EC Bolivia (2653)

ACROPTERNIS Cabanis & Heine, 1859 M
○ *Acropternis orthonyx* OCELLATED TAPACULO
　— *A.o.orthonyx* (Lafresnaye, 1843)   E Colombia, W Venezuela
　— *A.o.infuscatus* Salvadori & Festa, 1899[3]   Ecuador, N Peru

FORMICARIIDAE ANT-THRUSHES AND ANTPITTAS (7: 62)

FORMICARIUS Boddaert, 1783 M
○ *Formicarius colma* RUFOUS-CAPPED ANT-THRUSH
　— *F.c.colma* Boddaert, 1783   E Colombia, S Venezuela, the Guianas, N Brazil
　— *F.c.nigrifrons* Gould, 1855   E Ecuador, E Peru, NW Brazil
　— *F.c.amazonicus* Hellmayr, 1902   C Brazil
　— *F.c.ruficeps* (Spix, 1824)   E and SE Brazil

○ *Formicarius analis* BLACK-FACED ANT-THRUSH[4]
　— *F.a.moniliger* P.L. Sclater, 1857   S Mexico, E Guatamela
　— *F.a.pallidus* (Lawrence, 1882)   SE Mexico
　— *F.a.intermedius* Ridgway, 1908   Belize, Honduras
　— *F.a.umbrosus* Ridgway, 1893   E Honduras to W Panama
　— *F.a.hoffmanni* (Cabanis, 1861)   S Costa Rica, SW Panama
　— *F.a.panamensis* Ridgway, 1908   E Panama, NW Colombia
　— *F.a.virescens* Todd, 1915   N Colombia
　— *F.a.saturatus* Ridgway, 1893   NC Colombia, NW Venezuela, Trinidad
　— *F.a.griseoventris* Aveledo & Ginés, 1950   NE Colombia (64)
　— *F.a.connectens* Chapman, 1914   E Colombia
　— *F.a.zamorae* Chapman, 1923   E Ecuador, NE Peru, W Brazil
　— *F.a.olivaceus* J.T. Zimmer, 1931   N Peru

[1] Elevated to specific rank by Sick (1960) (2268).
[2] Accepted tentatively; see reservations expressed by Vuilleumier *et al.* (1992) (2552).
[3] For correction of spelling see David & Gosselin (2002) (614).
[4] The separation of the northern populations has been proposed, but A.O.U. (1998) (3) did not accept it.

__ *F.a.crissalis* (Cabanis, 1861)	SE Venezuela, the Guianas, N Brazil
__ *F.a.analis* (d'Orbigny & Lafresnaye, 1837)	Peru, W Brazil, E Bolivia
__ *F.a.paraensis* Novaes, 1957	E Amazonia, Brazil, south of the Amazon (1662)

○ *Formicarius rufifrons* Blake, 1957 RUFOUS-FRONTED ANT-THRUSH   SE Peru (158)

○ *Formicarius nigricapillus* BLACK-HEADED ANT-THRUSH
    __ *F.n.nigricapillus* Ridgway, 1893 — E Costa Rica, W Panama
    __ *F.n.destructus* E. Hartert, 1898 — W Colombia, W Ecuador

○ *Formicarius rufipectus* RUFOUS-BREASTED ANT-THRUSH
    __ *F.r.rufipectus* Salvin, 1866 — E Costa Rica, Panama
    __ *F.r.carrikeri* Chapman, 1912 — W Colombia, W Ecuador
    __ *F.r.lasallei* Aveledo & Ginés, 1952 — NW Venezuela (65)
    __ *F.r.thoracicus* Taczanowski & Berlepsch, 1885 — E Ecuador, E Peru

CHAMAEZA Vigors, 1825 F
○ *Chamaeza campanisona* SHORT-TAILED ANT-THRUSH
    __ *C.c.colombiana* Berlepsch & Stolzmann, 1896 — E Colombia
    __ *C.c.punctigula* Chapman, 1924 — E Ecuador, N Peru
    __ *C.c.olivacea* Tschudi, 1844 — C Peru
    __ *C.c.huachamacarii* Phelps & Phelps, Jr., 1951 — C Peru (1849)
    __ *C.c.berlepschi* Stolzmann, 1926 — SE Peru
    __ *C.c.venezuelana* Ménégaux & Hellmayr, 1906 — N Venezuela
    __ *C.c.yavii* Phelps & Phelps, Jr., 1947 — SC Venezuela
    __ *C.c.obscura* Zimmer & Phelps, 1944 — E Venezuela
    __ *C.c.fulvescens* Salvin & Godman, 1882 — SE Venezuela, Guyana
    __ *C.c.boliviana* Hellmayr & Seilern, 1912 — W Bolivia
    __ *C.c.campanisona* (M.H.K. Lichtenstein, 1823) — SE Brazil, E Paraguay

○ *Chamaeza nobilis* STRIATED ANT-THRUSH
    __ *C.n.rubida* J.T. Zimmer, 1932 — SE Colombia, E Ecuador, NE Peru
    __ *C.n.nobilis* Gould, 1855 — NE Peru, W Brazil
    __ *C.n.fulvipectus* Todd, 1927 — C Brazil

○ *Chamaeza meruloides* Vigors, 1825 CRYPTIC ANT-THRUSH[1]   SE Brazil

○ *Chamaeza ruficauda* (Cabanis & Heine, 1859) RUFOUS-TAILED ANT-THRUSH[2]
    SE Brazil

○ *Chamaeza turdina* SCALLOPED ANT-THRUSH
    __ *C.t.turdina* (Cabanis & Heine, 1859) — C Colombia
    __ *C.t.chionogaster* Hellmayr, 1906 — N Venezuela

○ *Chamaeza mollissima* BARRED ANT-THRUSH
    __ *C.m.mollissima* P.L. Sclater, 1855 — Colombia, Ecuador
    __ *C.m.yungae* Carriker, 1935 — S Peru, N Bolivia

PITTASOMA Cassin, 1860 N
○ *Pittasoma michleri* BLACK-CROWNED ANTPITTA
    __ *P.m.zeledoni* Ridgway, 1884 — Costa Rica, W Panama
    __ *P.m.michleri* Cassin, 1860 — E Panama, NW Colombia

○ *Pittasoma rufopileatum* RUFOUS-CROWNED ANTPITTA
    __ *P.r.rosenbergi* Hellmayr, 1911 — W Colombia
    __ *P.r.harterti* Chapman, 1917 — SW Colombia
    __ *P.r.rufopileatum* E. Hartert, 1901 — NW Ecuador

GRALLARIA Vieillot, 1816 F
○ *Grallaria squamigera* UNDULATED ANTPITTA
    __ *G.s.squamigera* Prévost & Des Murs, 1846 — Colombia, Venezuela, Ecuador
    __ *G.s.canicauda* Chapman, 1926 — Peru, NW Bolivia

---

[1] For elevation to species status see Willis (1992) (2670).
[2] Treated in last Edition as conspecific with the next species, but see Willis (1992) (2670) and Ridgely & Tudor (1994: 360) (2019).

○ *Grallaria gigantea* GIANT ANTPITTA
    — *G.g.lehmanni* Wetmore, 1945   C Colombia
    — *G.g.hylodroma* Wetmore, 1945   W Ecuador
    — *G.g.gigantea* Lawrence, 1866   E Ecuador

○ *Grallaria excelsa* GREAT ANTPITTA
    — *G.e.excelsa* Berlepsch, 1893   W Venezuela
    — *G.e.phelpsi* Gilliard, 1939   N Venezuela

○ *Grallaria varia* VARIEGATED ANTPITTA
    — *G.v.cinereiceps* Hellmayr, 1903   S Venezuela, NW Brazil
    — *G.v.varia* (Boddaert, 1783)   The Guianas, NW Brazil
    — *G.v.distincta* Todd, 1927   NC Brazil
    — *G.v.intercedens* Berlepsch & Leverkühn, 1890   E Brazil
    — *G.v.imperator* Lafresnaye, 1842   SE Brazil, Paraguay, NE Argentina

○ *Grallaria alleni* MOUSTACHED ANTPITTA
    — *G.a.alleni* Chapman, 1912   C Colombia
    — ¶ *G.a.andaquiensis* Hernández & Rodríguez, 1979[1]   S Colombia, N Ecuador (1106)

○ *Grallaria guatimalensis* SCALED ANTPITTA
    — *C.g.binfordi* Dickerman, 1990   C Mexico (722)
    — *G.g.ochraceiventris* Nelson, 1898   SW Mexico
    — *G.g.guatimalensis* Prévost & Des Murs, 1846   S Mexico to N Nicaragua
    — *G.g.princeps* Sclater & Salvin, 1869   Costa Rica, W Panama
    — *G.g.chocoensis* Chapman, 1917   Mts. of E Panama, NW Colombia
    — *G.g.regulus* P.L. Sclater, 1860   Ecuador, Peru
    — *G.g.carmelitae* Todd, 1915   NE Colombia, NW Venezuela
    — *G.g.aripoensis* Hellmayr & Seilern, 1912   N Trinidad
    — *G.g.roraimae* Chubb, 1921   S Venezuela, N Brazil

○ *Grallaria chthonia* Wetmore & Phelps, 1956 TACHIRA ANTPITTA   W Venezuela (2614)

○ *Grallaria haplonota* PLAIN-BACKED ANTPITTA
    — *G.h.haplonota* P.L. Sclater, 1877   N Venezuela
    — *G.h.pariae* Phelps & Phelps, Jr., 1949   NE Venezuela
    — *G.h.parambae* Rothschild, 1900   NW Ecuador
    — *G.h.chaplinae* Robbins & Ridgely, 1986   W Ecuador (2080)

○ *Grallaria dignissima* Sclater & Salvin, 1880 OCHRE-STRIPED ANTPITTA   E Ecuador, NE Peru

○ *Grallaria eludens* Lowery & O'Neill, 1969 ELUSIVE ANTPITTA   SE Peru (1395)

○ *Grallaria ruficapilla* CHESTNUT-CROWNED ANTPITTA
    — *G.r.ruficapilla* Lafresnaye, 1842   C Colombia, N Ecuador
    — *G.r.perijana* Phelps & Gilliard, 1940   NW Venezuela
    — *G.r.avilae* Hellmayr & Seilern, 1914   N Venezuela
    — *G.r.nigrolineata* P.L. Sclater, 1890   W Venezuela
    — *G.r.connectens* Chapman, 1923   SW Ecuador
    — *G.r.albiloris* Taczanowski, 1880   S Ecuador, NW Peru
    — *G.r.interior* J.T. Zimmer, 1934   N Peru

○ *Grallaria watkinsi* Chapman, 1919 WATKINS'S ANTPITTA   SW Ecuador, NW Peru

○ *Grallaria bangsi* J.A. Allen, 1900 SANTA MARTA ANTPITTA   NE Colombia

○ *Grallaria kaestneri* Stiles, 1992 CUNDINAMARCA ANTPITTA #   E Colombia (2339)

○ *Grallaria andicolus* STRIPE-HEADED ANTPITTA[2][3]
    — *G.a.andicolus* (Cabanis, 1873)   C Peru
    — *G.a.punensis* Chubb, 1918   SE Peru

○ *Grallaria griseonucha* GREY-NAPED ANTPITTA
    — *G.g.tachirae* Zimmer & Phelps, 1945   Venezuela
    — *G.g.griseonucha* Sclater & Salvin, 1871   W Venezuela

---

[1] This race seems to await review and to have been omitted in last Edition.
[2] The basis for treating this as two species in last edition cannot now be traced.
[3] For correction of spelling see David & Gosselin (2002) (613).

○ *Grallaria rufocinerea* Bicoloured Antpitta
— *G.r.rufocinerea* Sclater & Salvin, 1879   W slope of C Andes of Colombia
— ¶ *G.r.romeroana* Hernández & Rodriguez, 1979[1]   W slope of S Andes of Colombia (1106)

○ *Grallaria ridgelyi* Krabbe *et al.*, 1999[2] Jocotoco Antpitta #   S Ecuador (1300)

○ *Grallaria nuchalis* Chestnut-naped Antpitta
— *G.n.ruficeps* P.L. Sclater, 1874   C Colombia
— *G.n.obsoleta* Chubb, 1916   NW Ecuador
— *G.n.nuchalis* P.L. Sclater, 1859   E Ecuador

○ *Grallaria carrikeri* Schulenberg & Williams, 1982 Pale-billed Antpitta
     N Peru (2213)

○ *Grallaria albigula* White-throated Antpitta
— *G.a.albigula* Chapman, 1923   SE Peru, Bolivia
— ¶ *G.a.cinereiventris* Olrog & Contino, 1970   NW Argentina (1687)

○ *Grallaria flavotincta* P.L. Sclater, 1877 Yellow-breasted Antpitta[3]   W Colombia, NW Ecuador

○ *Grallaria hypoleuca* White-bellied Antpitta
— *G.h.castanea* Chapman, 1923   C Colombia, E Ecuador
— *G.h.hypoleuca* P.L. Sclater, 1855   W Colombia

○ *Grallaria przewalskii* Taczanowski, 1882 Rusty-tinged Antpitta   N Peru

○ *Grallaria capitalis* Chapman, 1926 Bay Antpitta   C Peru

○ *Grallaria erythroleuca* P.L. Sclater, 1874 Red-and-white Antpitta   S Peru

○ *Grallaria rufula* Rufous Antpitta
— *G.r.spatiator* Bangs, 1898   NE Colombia
— *G.r.saltuensis* Wetmore, 1946   NE Colombia
— *G.r.rufula* Lafresnaye, 1843   Colombia, W Venezuela, Ecuador
— *G.r.cajamarcae* (Chapman, 1927)   N Peru
— *G.r.obscura* Berlepsch & Stolzmann, 1896   C Peru
— *G.r.occabambae* (Chapman, 1923)   SE Peru
— *G.r.cochabambae* Bond & Meyer de Schauensee, 1940   N Bolivia

○ *Grallaria blakei* Graves, 1987 Chestnut Antpitta   Andes of Peru (983)

○ *Grallaria quitensis* Tawny Antpitta
— *G.q.quitensis* Lesson, 1844   C Colombia, Ecuador
— *G.q.alticola* Todd, 1919   E Colombia
— *G.q.atuensis* Carriker, 1933   N Peru

○ *Grallaria milleri* Chapman, 1912 Brown-banded Antpitta   C Colombia

○ *Grallaria erythrotis* Sclater & Salvin, 1876 Rufous-faced Antpitta   Yungas of N Bolivia

HYLOPEZUS Ridgway, 1909 M
○ *Hylopezus perspicillatus* Streak-chested Antpitta
— *H.p.intermedius* (Ridgway, 1884)   Caribbean slope from E Costa Rica to W Panama
— *H.p.lizanoi* (Cherrie, 1891)   W Costa Rica, W Panama
— *H.p.perspicillatus* (Lawrence, 1861)   E Panama, NW Colombia
— *H.p.periophthalmicus* (Salvadori & Festa, 1898)   Pacific coast of W Colombia, W Ecuador
— *H.p.pallidior* Todd, 1919   C Colombia

○ *Hylopezus macularius* Spotted Antpitta
— *H.m.diversus* J.T. Zimmer, 1934   SE Colombia, NE Peru, S Venezuela
— *H.m.macularius* (Temminck, 1830)   E Venezuela, the Guianas, N Brazil
— *H.m.paraensis* (E. Snethlage, 1910)   Brazil

○ *Hylopezus auricularis* (Gyldenstolpe, 1941) Masked Antpitta[4]   N Bolivia

---

[1] This race awaits review and seems to have been omitted in last Edition.
[2] We give all authors' names when five or less are involved.
[3] The arrangement of this species and the next four is tentative and follows Ridgely & Tudor (1994: 376) (2019).
[4] See Maijer (1998) (1413) for reasons for elevation to species level.

○ *Hylopezus dives* THICKET ANTPITTA[1]
 __ *H.d.dives* (Salvin, 1865)   Caribbean slope of Nicaragua, Costa Rica
 __ *H.d.flammulatus* Griscom, 1928   Caribbean slope of W Panama
 __ *H.d.barbacoae* Chapman, 1914   E Panama, W Colombia

○ *Hylopezus fulviventris* WHITE-LORED ANTPITTA
 __ *H.f.caquetae* Chapman, 1923   Colombia
 __ *H.f.fulviventris* (P.L. Sclater, 1858)   E Ecuador

○ *Hylopezus berlepschi* AMAZONIAN ANTPITTA
 __ *H.b.yessupi* (Carriker, 1930)   SE Peru
 __ *H.b.berlepschi* (Hellmayr, 1903)   N Bolivia, C and S Brazil

○ *Hylopezus ochroleucus* (Wied, 1831) WHITE-BROWED ANTPITTA   E Brazil

○ *Hylopezus nattereri* (Pinto, 1937) SPECKLE-BREASTED ANTPITTA[2]   SE Brazil, SE Paraguay, NE Argentina

MYRMOTHERA Vieillot, 1816 F
○ *Myrmothera campanisona* THRUSH-LIKE ANTPITTA
 __ *M.c.modesta* (P.L. Sclater, 1855)   SE Colombia
 __ *M.c.dissors* J.T. Zimmer, 1934   E Colombia, S Venezuela, NW Brazil
 __ *M.c.campanisona* (Hermann, 1783)   SE Venezuela, the Guianas, N Brazil
 __ *M.c.signata* J.T. Zimmer, 1934   E Ecuador, NE Peru
 __ *M.c.minor* (Taczanowski, 1882)   E Peru, W Brazil
 __ *M.c.subcanescens* Todd, 1927   N and NC Brazil

○ *Myrmothera simplex* TEPUI ANTPITTA
 __ ¶*M.s.pacaraimae* Phelps & Dickerman, 1980   Tepui of SW Bolivar (SE Venezuela) [1841]
 __ *M.s.guaiquinimae* Zimmer & Phelps, 1946   Tepuis of C and SC Bolivar (SE Venezuela)
 __ *M.s.simplex* (Salvin & Godman, 1884)   Tepuis of E Bolivar (SE Venezuela)
 __ *M.s.duidae* Chapman, 1929   Tepuis of Amazonas (S Venezuela)

GRALLARICULA P.L. Sclater, 1858 F
○ *Grallaricula flavirostris* OCHRE-BREASTED ANTPITTA
 __ *G.f.costaricensis* Lawrence, 1866   Costa Rica, Panama
 __ *G.f.brevis* Nelson, 1912   Mt. Pirri (E Panama)
 __ *G.f.ochraceiventris* Chapman, 1922   W Colombia
 __ *G.f.mindoensis* Chapman, 1925   N Ecuador
 __ *G.f.zarumae* Chapman, 1922   SW Ecuador
 __ *G.f.flavirostris* (P.L. Sclater, 1858)   E Colombia, E Ecuador
 __ *G.f.similis* Carriker, 1933   N Peru
 __ *G.f.boliviana* Chapman, 1919   N Bolivia

○ *Grallaricula loricata* (P.L. Sclater, 1857) SCALLOP-BREASTED ANTPITTA   N Venezuela

○ *Grallaricula cucullata* HOODED ANTPITTA
 __ *G.c.cucullata* (P.L. Sclater, 1856)   W Colombia
 __ *G.c.venezuelana* Phelps & Phelps, Jr., 1956   NW Venezuela [1857]

○ *Grallaricula peruviana* Chapman, 1923 PERUVIAN ANTPITTA   NW Peru

○ *Grallaricula ochraceifrons* Graves, O'Neill & Parker, 1983 OCHRE-FRONTED ANTPITTA
                   N Peru [995]

○ *Grallaricula ferrugineipectus* RUSTY-BREASTED ANTPITTA
 __ *G.f.rara* Hellmayr & Madarász, 1914   E Colombia, NW Venezuela
 __ *G.f.ferrugineipectus* (P.L. Sclater, 1857)   NE Colombia, N Venezuela
 __ *G.f.leymebambae* Carriker, 1933   Andes of N Peru and W Bolivia

○ *Grallaricula nana* SLATE-CROWNED ANTPITTA
 __ *G.n.occidentalis* Todd, 1927   W Colombia
 __ *G.n.nana* (Lafresnaye, 1842)   E Colombia, W Venezuela
 __ *G.n.olivascens* Hellmayr, 1917   N Venezuela

---

[1] For the separation of this see Ridgely & Gwynne (1989) (2017). See A.O.U. (1998: 371) (3).
[2] For elevation to species level see Whitney *et al.* (1995) (2657).

__ *G.n.cumanensis* E. Hartert, 1900	N Venezuela
__ *G.n.pariae* Phelps & Phelps, Jr., 1949	NE Venezuela
__ *G.n.kukenamensis* Chubb, 1918	SE Venezuela

○ *Grallaricula lineifrons* (Chapman, 1924) CRESCENT-FACED ANTPITTA S Colombia, Ecuador

FURNARIIDAE OVENBIRDS[1] (55: 236)

GEOSITTA Swainson, 1837 F
● *Geositta cunicularia* COMMON MINER

__ *G.c.juninensis* Taczanowski, 1884	Andes of C Peru
__ *G.c.titicacae* J.T. Zimmer, 1935	Andes of S Peru to N Chile, NW Argentina
__ *G.c.frobeni* (Philippi & Landbeck, 1864)	Inland S Peru and N Chile
__ *G.c.georgei* Koepcke, 1965	Coastal S Peru and N Chile (1291)
__ *G.c.deserticolor* Hellmayr, 1924	SW Peru, N Chile
☑ *G.c.fissirostris* (Kittlitz, 1835)	C Chile
__ ¶*G.c.contrerasi* Nores & Yzurieta, 1980[2]	C Argentina (Córdoba) (1659)
__ *G.c.hellmayri* J.L. Peters, 1925	W Argentina >> lower elevations
__ *G.c.cunicularia* (Vieillot, 1816)	S Argentina, S Chile >> S Brazil, Uruguay, Paraguay

○ *Geositta poeciloptera* (Wied, 1830) CAMPO MINER[3]   SC Brazil, E Bolivia

○ *Geositta antarctica* Landbeck, 1880 SHORT-BILLED MINER   S Chile, S Argentina >> C Chile, C Argentina

○ *Geositta tenuirostris* SLENDER-BILLED MINER

__ *G.t.kalimayae* Krabbe, 1992	Andes of N Ecuador (1299)
__ *G.t.tenuirostris* (Lafresnaye, 1836)	Andes of S Peru to NW Argentina

○ *Geositta maritima* (d'Orbigny & Lafresnaye, 1837) GREYISH MINER   W Peru, N Chile

○ *Geositta peruviana* COASTAL MINER

__ *G.p.paytae* Ménégaux & Hellmayr, 1906	N Peru
__ *G.p.peruviana* Lafresnaye, 1847	C Peru
__ *G.p.rostrata* Stolzmann, 1926	S Peru

○ *Geositta saxicolina* Taczanowski, 1875 DARK-WINGED MINER   C Peru

○ *Geositta punensis* Dabbene, 1917 PUNA MINER   W Bolivia, SW Peru, N Chile, NW Argentina

● *Geositta rufipennis* RUFOUS-BANDED MINER

☑ *G.r.fasciata* (Philippi & Landbeck, 1864)[4]	W Bolivia, N and C Chile >> lower elevations
__ *G.r.harrisoni* Marin, Kiff & Pena, 1989	N Chile (SW Antofagasta) (1433)
__ *G.r.rufipennis* (Burmeister, 1860)	NW Argentina
__ *G.r.giaii* Contreras, 1976	WC Argentina (San Juan, Mendoza) (565)
__ *G.r.ottowi* Hoy, 1968	C Argentina (Sierra de Córdoba) (1154)
__ *G.r.fragai* Nores, 1986	WC Argentina (La Rioja) (1657)
__ *G.r.hoyi* Contreras, 1980	SW Argentine and S Chile (568)

○ *Geositta isabellina* (Philippi & Landbeck, 1864) CREAMY-RUMPED MINER
C Chile, W Argentina

○ *Geositta crassirostris* THICK-BILLED MINER

__ *G.c.fortis* von Berlepsch & Stolzmann, 1901	SW Peru (2537)
__ *G.c.crassirostris* P.L. Sclater, 1866	Coastal W Peru

UPUCERTHIA I. Geoffroy Saint-Hilaire, 1832[5] F
○ *Upucerthia jelskii* PLAIN-BREASTED EARTHCREEPER

__ *U.j.saturata* Carriker, 1933	NW Peru
__ *U.j.jelskii* (Cabanis, 1874)	WC Peru
__ *U.j.pallida* Taczanowski, 1883	SW Peru, Bolivia, N Chile, NW Argentina

---

[1] Following Ridgely & Tudor (1994) (2019) and others we now, compared with last Edition, largely reverse the significant lumping of genera by Vaurie (1980) (2516).
[2] Tentatively accepted; requires evaluation.
[3] Historically treated in a monotypic genus *Geobates*, but see Vaurie (1980) (2516).
[4] Includes ¶*hellmayri* Esteban, 1949 (820A); this name is preoccupied by *G.c.hellmayri*.
[5] Includes the genus *Ochetorhynchus* see Meyer de Schauensee (1970) (1575) and Vaurie (1980) (2516).

○ *Upucerthia validirostris* Buff-breasted Earthcreeper
　　— *U.v.validirostris* (Burmeister, 1861)　　　　　　　　　S Bolivia, NW Argentina
　　— *U.v.rufescens* Nores, 1986　　　　　　　　　　　　　　C Argentina (1657)

○ *Upucerthia albigula* Hellmayr, 1932　White-throated Earthcreeper　SW Peru, N Chile

● *Upucerthia dumetaria* Scale-throated Earthcreeper
　　— *U.d.peruana* J.T. Zimmer, 1954　　　　　　　　　　　SE Peru, W Bolivia, N Chile, NW Argentina (2721)
　✓ *U.d.saturatior* W.E.D. Scott, 1900　　　　　　　　　　C Chile, SW Argentina
　　— *U.d.hypoleuca* Reichenbach, 1853[1]　　　　　　　　SW Bolivia, N Chile, W Argentina
　　— *U.d.dumetaria* I. Geoffroy Saint-Hilaire, 1832　　　　S Chile and C and S Argentina >> N Argentina

○ *Upucerthia serrana* Striated Earthcreeper
　　— *U.s.serrana* Taczanowski, 1875　　　　　　　　　　　WC Peru
　　— *U.s.huancavelicae* Morrison, 1938　　　　　　　　　　SW Peru

○ *Upucerthia ruficaudus* Straight-billed Earthcreeper[2]
　　— *U.r.montana* d'Orbigny & Lafresnaye, 1838　　　　　S Peru
　　— *U.r.ruficaudus* (Meyen, 1834)　　　　　　　　　　　W Bolivia, N Chile, NW Argentina
　　— *U.r.famatinae* Nores, 1986　　　　　　　　　　　　　N Argentina (1657)

○ *Upucerthia andaecola* d'Orbigny & Lafresnaye, 1838　Rock Earthcreeper
　　　　　　　　　　　　　　　　　　　　　　　　　　　　W Bolivia, NW Argentina, Chile

○ *Upucerthia harterti* Berlepsch, 1892　Bolivian Earthcreeper　　Andes of C Bolivia

○ *Upucerthia certhioides* Chaco Earthcreeper
　　— *U.c.luscinia* Burmeister, 1860　　　　　　　　　　　W Argentina
　　— *U.c.estebani* Wetmore & Peters, 1949　　　　　　　　S Paraguay, N Argentina
　　— *U.c.certhioides* (d'Orbigny & Lafresnaye, 1838)　　　N and C Argentina

CHILIA Salvadori, 1908 F
○ *Chilia melanura* Crag Chilia
　　— *C.m.atacamae* Hellmayr, 1925　　　　　　　　　　　N Chile
　　— *C.m.melanura* (G.R. Gray, 1846)　　　　　　　　　　C Chile

CINCLODES G.R. Gray, 1840 M
○ *Cinclodes excelsior* Stout-billed Cinclodes
　　— *C.e.colombianus* (Chapman, 1912)[3]　　　　　　　　Colombia (C Andes)
　　— *C.e.excelsior* P.L. Sclater, 1860　　　　　　　　　　SW Colombia, Ecuador

○ *Cinclodes aricomae* (Carriker, 1932)　Royal Cinclodes[4]　　S Peru, N Bolivia

● *Cinclodes fuscus* Bar-winged Cinclodes
　　— *C.f.heterurus* Madarász, 1903　　　　　　　　　　　W Venezuela
　　— *C.f.oreobates* W.E.D. Scott, 1900　　　　　　　　　　N Colombia
　　— *C.f.paramo* Meyer de Schauensee, 1945　　　　　　　S Colombia, N Ecuador
　　— *C.f.albidiventris* P.L. Sclater, 1860　　　　　　　　Ecuador
　　— *C.f.longipennis* (Swainson, 1838)　　　　　　　　　N Peru
　　— *C.f.rivularis* (Cabanis, 1873)　　　　　　　　　　　C and S Peru
　　— *C.f.albiventris* (Philippi & Landbeck, 1861)　　　　S Peru, Bolivia, N Chile, NW Argentina
　　— *C.f.riojanus* Nores, 1986　　　　　　　　　　　　　N Argentina (La Rioja) (1657)
　　— *C.f.rufus* Nores, 1986　　　　　　　　　　　　　　　C Argentina (Catamarca) (1657)
　　— *C.f.yzurietae* Nores, 1986　　　　　　　　　　　　　S Argentina (SE Catamarca) (1657)
　✓ *C.f.fuscus* (Vieillot, 1818)　　　　　　　　　　　　　C and S Chile, C and S Argentina >> S Brazil, Uruguay, NW Argentina

○ *Cinclodes comechingonus* Zotta & Gavio, 1944　Cordoba Cinclodes　C Argentina (Córdoba) >> N Argentina

○ *Cinclodes pabsti* Sick, 1969　Long-tailed Cinclodes　　SE Brazil (2269)

○ *Cinclodes olrogi* Nores & Yzurieta, 1979　Olrog's Cinclodes　　C Argentina (1658)

---

[1] Includes *hallinani* see Fjeldså & Krabbe (1990) (850).
[2] For correction of spelling see David & Gosselin (2002) (613).
[3] For correction of spelling see David & Gosselin (2002) (614).
[4] In separating this species from *C. excelsior* we follow Ridgely & Tudor (1994) (2019) and Fjeldså & Krabbe (1990) (850).

◑ *Cinclodes oustaleti* GREY-FLANKED CINCLODES
  ↙ *C.o.oustaleti* W.E.D. Scott, 1900 — Chile
  — *C.o.hornensis* Dabbene, 1917 — Cape Horn Is., Tierra del Fuego
  — *C.o.baeckstroemii* Lönnberg, 1921 — Juan Fernandez I.

● *Cinclodes patagonicus* DARK-BELLIED CINCLODES
  ↙ *C.p.chilensis* (Lesson, 1828) — C Chile, W Argentina
  — *C.p.patagonicus* (J.F. Gmelin, 1789) — S Chile, S Argentina

○ *Cinclodes antarcticus* BLACKISH CINCLODES
  — *C.a.maculirostris* Dabbene, 1917 — Cape Horn Is.
  — *C.a.antarcticus* (Garnot, 1826) — Falkland Is.

○ *Cinclodes taczanowskii* Berlepsch & Stolzmann, 1892 SURF CINCLODES
  Coast of C and S Peru

● *Cinclodes nigrofumosus* (d'Orbigny & Lafresnaye, 1838) SEASIDE CINCLODES
  Coast of N and C Chile

○ *Cinclodes atacamensis* WHITE-WINGED CINCLODES
  — *C.a.atacamensis* (Philippi, 1857) — Andes of C Peru, Bolivia, N Chile, NW Argentina
  — *C.a.schocolatinus* Reichenow, 1920 — C Argentina (Córdoba)

○ *Cinclodes palliatus* (Tschudi, 1844) WHITE-BELLIED CINCLODES — Andes of C Peru

FURNARIUS Vieillot, 1816 M
○ *Furnarius minor* Pelzeln, 1858 LESSER HORNERO — Amazonian islands in NE Peru, E Ecuador, S Colombia, W Brazil

○ *Furnarius figulus* BAND-TAILED HORNERO
  — *F.f.pileatus* Sclater & Salvin, 1878 — C Brazil
  — *F.f.figulus* (M.H.K. Lichtenstein, 1823) — E Brazil

○ *Furnarius leucopus* PALE-LEGGED HORNERO
  — *F.l.cinnamomeus* (Lesson, 1844) — SW Ecuador, NW Peru
  — *F.l.longirostris* Pelzeln, 1856 — N Colombia, NW Venezuela
  — *F.l.endoecus* Cory, 1919 — C Colombia, W Venezuela
  — *F.l.leucopus* Swainson, 1838 — Guyana, N Brazil
  — *F.l.tricolor* Giebel, 1868[1] — E Peru, W Brazil, N Bolivia
  — *F.l.assimilis* Cabanis & Heine, 1859 — E and S Brazil, SE Bolivia

○ *Furnarius torridus* Sclater & Salvin, 1866 PALE-BILLED HORNERO — Amazonian islands in NE Peru, E Ecuador, S Colombia, W Brazil

○ *Furnarius rufus* RUFOUS HORNERO
  — *F.r.commersoni* Pelzeln, 1868 — E Bolivia, W Brazil
  — *F.r.schuhmacheri* Laubmann, 1933 — S Bolivia
  — *F.r.paraguayae* Cherrie & Reichenberger, 1921 — Paraguay, NW Argentina
  — *F.r.rufus* (J.F. Gmelin, 1788) — S Brazil, Uruguay, C and E Argentina
  — *F.r.albogularis* (Spix, 1824) — SE Brazil

○ *Furnarius cristatus* Burmeister, 1888 CRESTED HORNERO — SC Bolivia, W Paraguay, N Argentina

SYLVIORTHORHYNCHUS Des Murs, 1847 M
○ *Sylviorthorhynchus desmursii* Des Murs, 1847 DES MURS'S WIRETAIL — S Chile, W Argentina

APHRASTURA Oberholser, 1899 F
◑ *Aphrastura spinicauda* THORN-TAILED RAYADITO
  ↙ *A.s.spinicauda* (J.F. Gmelin, 1789) — Tierra del Fuego, S Chile, W Argentina
  — *A.s.bullocki* Chapman, 1934 — Moncha I. (Chile)
  — *A.s.fulva* Angelini, 1905 — Chiloe I. (Chile)

○ *Aphrastura masafuerae* (Philippi & Landbeck, 1866) MASAFUERA RAYADITO
  Alejandro Selkirk I. (Juan Fernandez Is.).

---

[1] Treated as a separate species in last Edition, but evidence for this not now traced and we return it to *F. leucopus*. Presumably includes ¶*araguaiae* Pinto & Camargo, 1952 (1896).

LEPTASTHENURA Reichenbach, 1853 F
O ***Leptasthenura fuliginiceps*** Brown-capped Tit-Spinetail
    — *L.f.fuliginiceps* d'Orbigny & Lafresnaye, 1837   W Bolivia
    — *L.f.paranensis* P.L. Sclater, 1862   W Argentina >> lower elevations

O ***Leptasthenura yanacensis*** Carriker, 1933 Tawny Tit-Spinetail   Andes of W Peru, W Bolivia

O ***Leptasthenura platensis*** Reichenbach, 1853 Tufted Tit-Spinetail   SE Brazil, Uruguay, Argentina, W Parauguay

● ***Leptasthenura aegithaloides*** Plain-mantled Tit-Spinetail
    — *L.a.grisescens* Hellmayr, 1925   S Peru, N Chile
    — *L.a.berlepschi* E. Hartert, 1909   S Peru, W Bolivia, N Chile, W Argentina
    ✓ *L.a.aegithaloides* (Kittlitz, 1830)   C Chile
    — *L.a.pallida* Dabbene, 1920   S Chile, W and S Argentina

O ***Leptasthenura striolata*** (Pelzeln, 1856) Striolated Tit-Spinetail   SE Brazil

O ***Leptasthenura pileata*** Rusty-crowned Tit-Spinetail
    — *L.p.cajabambae* Chapman, 1921   NC Peru
    — *L.p.pileata* P.L. Sclater, 1881   WC Peru
    — *L.p.latistriata* Koepcke, 1965   SC Peru (1291)

O ***Leptasthenura xenothorax*** Chapman, 1921 White-browed Tit-Spinetail
                                                                 S Peru (Cuzco)

O ***Leptasthenura striata*** Streaked Tit-Spinetail
    — *L.s.superciliaris* Hellmayr, 1932   WC Peru
    — *L.s.albigularis* Morrison, 1938   C Peru
    — *L.s.striata* (Philippi & Landbeck, 1863)   SW Peru, N Chile

O ***Leptasthenura andicola*** Andean Tit-Spinetail
    — *L.a.certhia* (Madarász, 1903)   W Venezuela
    — *L.a.extima* Todd, 1916   N Colombia (Santa Marta Mts.)
    — *L.a.exterior* Todd, 1919   C Colombia (E Andes)
    — *L.a.andicola* P.L. Sclater, 1870   Colombia (C Andes), Ecuador
    — *L.a.peruviana* Chapman, 1919   C Peru, N Bolivia

O ***Leptasthenura setaria*** (Temminck, 1824) Araucaria Tit-Spinetail   SE Brazil, NE Argentina

SCHIZOEACA Cabanis, 1873 F
O ***Schizoeaca perijana*** W.H. Phelps, Jr., 1977 Perija Thistletail   NE Colombia, NW Venezuela (1836)

O ***Schizoeaca fuliginosa*** White-chinned Thistletail
    — *S.f.fuliginosa* (Lafresnaye, 1843)   W Venezuela, E Colombia
    — *S.f.fumigata* Borrero, 1960   C Andes of Colombia and NW Ecuador (197)
    — *S.f.peruviana* Cory, 1916   N Peru
    — *S.f.plengei* O'Neill & Parker, 1976   C Peru (1714)

O ***Schizoeaca vilcabambae*** Vilcabamba Thistletail[1]
    — *S.v.ayacuchensis* Vaurie, Weske & Terborgh, 1972   S Peru (N Ayacucho) (2518)
    — *S.v.vilcabambae* Vaurie, Weske & Terborgh, 1972   S Peru (Vilcabamba Mts.) (2518)

O ***Schizoeaca coryi*** (Berlepsch, 1888) Ochre-browed Thistletail   NW Venezuela

O ***Schizoeaca griseomurina*** (P.L. Sclater, 1882) Mouse-coloured Thistletail
                                                                 S Ecuador, N Peru

O ***Schizoeaca palpebralis*** Cabanis, 1873 Eye-ringed Thistletail   C Peru (Junin)

O ***Schizoeaca helleri*** Chapman, 1923 Puna Thistletail[2]   SE Peru

O ***Schizoeaca harterti*** Black-throated Thistletail[3]
    — *S.h.harterti* Berlepsch, 1901   N Bolivia
    — *S.h.bejaranoi* Remsen, 1981   C Bolivia (2004)

---

[1] For reasons to separate this from *S. fuliginosa* see Braun & Parker (1985) (210).
[2] For reasons to separate this from *S. palpebralis* see Braun & Parker (1985) (210).
[3] For reasons to separate this from *S. palpebralis* see Braun & Parker (1985) (210).

OREOPHYLAX Hellmayr, 1925[1] M
○ *Oreophylax moreirae* (Miranda-Ribeiro, 1906) I<small>TATIAIA</small> S<small>PINETAIL</small>   SE Brazil

ASTHENES Reichenbach, 1853[2] F
○ *Asthenes pyrrholeuca* S<small>HARP-BILLED</small> C<small>ANASTERO</small>
    — *A.p.pyrrholeuca* (Vieillot, 1817)[3]   C and S Argentina >> N Argentina, S Bolivia, Paraguay
                                and Uruguay

    — *A.p.sordida* (Lesson, 1839)   Chile, W Argentina

○ *Asthenes baeri* S<small>HORT-BILLED</small> C<small>ANASTERO</small>
    — *A.b.chacoensis* Brodkorb, 1938   Extreme SC Bolivia, NW Paraguay
    — *A.b.baeri* (Berlepsch, 1906)   S Bolivia, W Paraguay, extreme SE Brazil, Uruguay, N
                                and C Argentina

    — *A.b.neiffi* (Contreras, 1980)   W Argentina (567)

● *Asthenes humicola* D<small>USKY-TAILED</small> C<small>ANASTERO</small>
    — *A.h.goodalli* Marin, Kiff & Pena, 1989   N Chile (1433)
    ✓ *A.h.humicola* (Kittlitz, 1830)   N and C Chile, W Argentina
    — *A.h.polysticta* Hellmayr, 1925   S Chile

○ *Asthenes patagonica* (d'Orbigny, 1839) P<small>ATAGONIAN</small> C<small>ANASTERO</small>   S Argentina

○ *Asthenes pudibunda* C<small>ANYON</small> C<small>ANASTERO</small>
    — *A.p.neglecta* (Cory, 1916)   W Andes of NW Peru
    — *A.p.pudibunda* (P.L. Sclater, 1874)   W Andes of W Peru
    — *A.p.grisior* Koepcke, 1961   Andes of SW Peru (1288)

○ *Asthenes ottonis* (Berlepsch, 1901) R<small>USTY-FRONTED</small> C<small>ANASTERO</small>   Andes of SW Peru

○ *Asthenes heterura* (Berlepsch, 1901) M<small>AQUIS</small> C<small>ANASTERO</small>   Andes of N Bolivia

● *Asthenes modesta* C<small>ORDILLERAN</small> C<small>ANASTERO</small>
    — *A.m.proxima* (Chapman, 1921)   Andes of C and S Peru
    — *A.m.modesta* (Eyton, 1851)   Andes of SW Peru and W Bolivia to N Chile and NW
                                Argentina

    — *A.m.hilereti* (Dabenne, 1914)[4]   NW Argentina (Tucumán) (608A)
    — *A.m.rostrata* (Berlepsch, 1901)   Andes of N Bolivia
    — *A.m.serrana* Nores, 1986   C Argentina (La Rioja) (1657)
    — *A.m.cordobae* Nores & Yzurieta, 1980   C Argentina (Córdoba) (1659)
    ✓ *A.m.australis* Hellmayr, 1925   Andes of C Chile
    — *A.m.navasi* (Contreras, 1979)   S Chile, S Argentina (566)

○ *Asthenes cactorum* C<small>ACTUS</small> C<small>ANASTERO</small>
    — *A.c.cactorum* Koepcke, 1959   Coastal hills of W Peru (1287)
    — *A.c.monticola* Koepcke, 1965   W slope of W Andes of W Peru (1291)
    — *A.c.lachayensis* Koepcke, 1965   Coastal hills of SW Peru (1291)

○ *Asthenes humilis* S<small>TREAK-THROATED</small> C<small>ANASTERO</small>
    — *A.h.cajamarcae* J.T. Zimmer, 1936   Andes of NW Peru
    — *A.h.humilis* (Cabanis, 1873)   Andes of C Peru
    — *A.h.robusta* (Berlepsch, 1901)   Andes of S Peru and N Bolivia

○ *Asthenes dorbignyi* C<small>REAMY-BREASTED</small> C<small>ANASTERO</small>
    — *A.d.arequipae* (Sclater & Salvin, 1869)   Andes of SW Peru, N Chile, W Bolivia
    — *A.d.usheri* Morrison, 1947[5]   Andes of SC Peru (Apurímac)
    — *A.d.huancavelicae* Morrison, 1938   Andes of SC Peru (Huancavelico, Ayacucho)
    — *A.d.consobrina* Hellmayr, 1925   Andes of N Bolivia
    — *A.d.dorbignyi* (Reichenbach, 1853)   Andes of C Bolivia and NW Argentina

○ *Asthenes berlepschi* (Hellmayr, 1917) B<small>ERLEPSCH'S</small> C<small>ANASTERO</small>   N Bolivia

---

[1] Included in *Schizoeaca* in last Edition.
[2] Most species are here restored to this genus having been treated in *Thripophaga* in last Edition
[3] Provisionally includes *flavogularis* and *affinis* see Fjeldså & Krabbe (1990) (850); however *leptasthenuroides* Lillo, 1905, which is a senior synonym of *affinis*, requires revisiting (Olrog, 1958) (1678). See also Peters (1951: 110) (1828).
[4] For recognition see Fjeldså & Krabbe (1990) (850).
[5] In last Edition listed here, but also treated as a separate species! Separation has yet to be formally justified.

○ *Asthenes steinbachi* (E. Hartert, 1909) Steinbach's Canastero   Andes of W Argentina

○ *Asthenes luizae* Vielliard, 1990 Cipo Canastero #   Minas Gerais (Brazil) (2528)

○ *Asthenes wyatti* Streak-backed Canastero
   __ *A.w.wyatti* (Sclater & Salvin, 1871)   Eastern Andes of N Colombia
   __ *A.w.sanctaemartae* Todd, 1950   N Colombia (Santa Marta Mts.)
   __ *A.w.perijana* W.H. Phelps, Jr., 1977   NE Colombia, NW Venezuela (Perijá Mts.) (1836)
   __ *A.w.mucuchiesi* Phelps & Gilliard, 1941   Andes of W Venezuela (Mérida, Trujillo)
   __ *A.w.aequatorialis* (Chapman, 1921)   W Andes of C Ecuador
   __ *A.w.azuay* (Chapman, 1923)   Andes of S Ecuador
   __ *A.w.graminicola* (P.L. Sclater, 1874)   Andes of C and S Peru

○ *Asthenes sclateri* Puna Canastero[1]
   __ *A.s.punensis* (Berlepsch & Stolzmann, 1901)   Andes of extreme S Peru and N Bolivia
   __ *A.s.cuchacanchae* (Chapman, 1921)   Andes of C Bolivia
   __ *A.s.lilloi* (Oustalet, 1904)   Andes of NW Argentina
   __ *A.s.sclateri* (Cabanis, 1878)   C Argentina (Sierra de Córdoba)
   __ ¶*A.s.brunnescens* Nores & Yzurieta, 1983   C Argentina (Sierra de San Luis) (1660)

○ *Asthenes anthoides* (P.P. King, 1831) Austral Canastero   S Chile and SW Argentina S to Tierra del Fuego >> NC
                       Chile (Aconcagua)

○ *Asthenes hudsoni* (P.L. Sclater, 1874) Hudson's Canastero   Uruguay, E and S Argentina

○ *Asthenes urubambensis* Line-fronted Canastero
   __ *A.u.huallagae* (J.T. Zimmer, 1924)   Andes of C Peru
   __ *A.u.urubambensis* (Chapman, 1919)   Andes of S Peru and N Bolivia

○ *Asthenes flammulata* Many-striped Canastero
   __ *A.f.multostriata* (P.L. Sclater, 1858)   E Andes of Colombia
   __ *A.f.quindiana* (Chapman, 1915)   C Andes of Colombia
   __ *A.f.flammulata* (Jardine, 1850)   W Andes of S Colombia and Andes of Ecuador to
                       extreme N Peru
   __ *A.f.pallida* Carriker, 1933   Andes of NW Peru
   __ *A.f.taczanowskii* (Berlepsch & Stolzmann, 1894)   Andes of N and C Peru

○ *Asthenes virgata* (P.L. Sclater, 1874) Junin Canastero   Andes of C Peru

○ *Asthenes maculicauda* (Berlepsch, 1901) Scribble-tailed Canastero   Andes of S Peru, Bolivia and NW Argentina

SCHOENIOPHYLAX Ridgway, 1909 M
○ *Schoeniophylax phryganophilus* Chotoy Spinetail
   __ *S.p.phryganophilus* (Vieillot, 1817)[2]   E Bolivia, S Brazil, N Argentina
   __ *S.p.petersi* Pinto, 1949   E Brazil

SYNALLAXIS Vieillot, 1818 F
○ *Synallaxis candei* White-whiskered Spinetail
   __ *S.c.candei* d'Orbigny & Lafresnaye, 1838   N Colombia, W Venezuela
   __ *S.c.atrigularis* (Todd, 1917)   NC Colombia
   __ *S.c.venezuelensis* Cory, 1913   NE Colombia, NW Venezuela

○ *Synallaxis kollari* Pelzeln, 1856 Hoary-throated Spinetail   N Brazil

○ *Synallaxis scutatus* Ochre-cheeked Spinetail
   __ *S.s.scutatus* P.L. Sclater, 1859   E and C Brazil
   __ *S.s.whitii* P.L. Sclater, 1881   E Bolivia, S Brazil, NW Argentina
   __ *S.s.teretiala* (Oren, 1985)   Para (Brazil) (1720)

○ *Synallaxis unirufa* Rufous Spinetail
   __ *S.u.munoztebari* Phelps & Phelps, Jr., 1953   NE Colombia and NW Venezuela (Perija Mts.) (1853)
   __ *S.u.meridana* Hartert & Goodson, 1917   Extreme NW eastern Andes of Colombia and Andes of
                       W Venezuela (Tachira, Merida, and Trujillo)

---

[1] Treated as one species in last Edition, but with a note suggesting that a split should be expected following Vaurie (1980) (2516); however Navas & Bo (1982) (1639) argued against this and were followed by Ridgely & Tudor (1994) (2019).
[2] For correction of spelling see David & Gosselin (2002) (614).

___ *S.u.unirufa* Lafresnaye, 1843	Andes of N Colombia, Ecuador (on W slope only to Cotopaxi) and N Peru (Cajamarca)
___ *S.u.ochrogaster* J.T. Zimmer, 1935	Andes of Peru (Amazonas to N Cuzco in Cordillera Vilcabamba)
O *Synallaxis castanea* P.L. Sclater, 1856 BLACK-THROATED SPINETAIL	N Venezuela (Coastal Range)
O *Synallaxis fuscorufa* P.L. Sclater, 1882 RUSTY-HEADED SPINETAIL	N Colombia (Santa Marta Mts.)
O *Synallaxis ruficapilla* Vieillot, 1819 RUFOUS-CAPPED SPINETAIL	SE Brazil, E Paraguay to N Argentina
O *Synallaxis cinerea* Wied, 1831 BAHIA SPINETAIL[1]	NE Brazil
O *Synallaxis infuscata* Pinto, 1950 PINTO'S SPINETAIL[2]	Extreme E Brazil (1888)
O *Synallaxis cinnamomea* STRIPE-BREASTED SPINETAIL	
___ *S.c.carri* Chapman, 1895	Trinidad
___ *S.c.terrestris* Jardine, 1847	Tobago I.
___ *S.c.cinnamomea* Lafresnaye, 1843	Andes of E Colombia, NW Venezuela (Perija Mts.)
___ *S.c.aveledoi* Phelps & Phelps, Jr., 1946	Andes of extreme NE Colombia, W Venezuela
___ *S.c.bolivari* E. Hartert, 1917	Coastal range N Venezuela
___ *S.c.striatipectus* Chapman, 1899	Coastal range NE Venezuela
___ *S.c.pariae* Phelps & Phelps, Jr., 1949	Paria Pen. (Venezuela)
O *Synallaxis cinerascens* Temminck, 1823 GREY-BELLIED SPINETAIL	SE Brazil, Paraguay, NE Argentina
O *Synallaxis subpudica* P.L. Sclater, 1874 SILVERY-THROATED SPINETAIL	NE Colombia (E Andes)
O *Synallaxis frontalis* SOOTY-FRONTED SPINETAIL[3]	
___ *S.f.fuscipennis* Berlepsch, 1907	E Bolivia, NW Argentina
___ *S.f.frontalis* Pelzeln, 1859	S Brazil, Paraguay, Uruguay, N Argentina
O *Synallaxis azarae* AZARA'S SPINETAIL[4]	
___ *S.a.elegantior* P.L. Sclater, 1862	E Colombia, W Venezuela
___ *S.a.media* Chapman, 1914	W Colombia, N Ecuador
___ *S.a.ochracea* J.T. Zimmer, 1936	E Ecuador, NW Peru
___ *S.a.fruticicola* Taczanowski, 1880	N Peru
___ *S.a.infumata* J.T. Zimmer, 1925	NC Peru
___ *S.a.azarae* d'Orbigny, 1835[5]	C Peru to N Bolivia
___ *S.a.samaipatae* Bond & Meyer de Schauensee, 1941	S Bolivia
___ *S.a.superciliosa* Cabanis, 1883	NW Argentina
O *Synallaxis courseni* Blake, 1971 APURIMAC SPINETAIL	S Peru (Apurimac) (163)
O *Synallaxis albescens* PALE-BREASTED SPINETAIL	
___ *S.a.latitabunda* Bangs, 1907[6]	SW Costa Rica, SW Panama, NW Colombia
___ *S.a.insignis* J.T. Zimmer, 1935	C Colombia
___ *S.a.occipitalis* Madarász, 1903	E Colombia, NW Venezuela
___ *S.a.littoralis* Todd, 1948	Coast of N Colombia
___ *S.a.perpallida* Todd, 1916	NE Colombia, NW Venezuela,
___ *S.a.nesiotis* Clark, 1902	N Colombia, N Venezuela, Margarita I.
___ *S.a.trinitatis* J.T. Zimmer, 1935	C Venezuela, Trinidad
___ *S.a.josephinae* Chubb, 1919	S Venezuela, Guyana, Surinam, N Brazil
___ *S.a.inaequalis* J.T. Zimmer, 1935	French Guiana, NE Brazil
___ ¶ *S.a.pullata* Ripley, 1955	W Brazil (Rio Solimões) (2038)
___ *S.a.griseonota* Todd, 1948	C Brazil
___ *S.a.albescens* Temminck, 1823	E Brazil, Paraguay, NE Argentina
___ *S.a.australis* J.T. Zimmer, 1935	E Bolivia, W Paraguay, NW Argentina
O *Synallaxis albigularis* DARK-BREASTED SPINETAIL	
___ *S.a.rodolphei* Bond, 1956	S Colombia (Meta to Putumayo?) and E Ecuador (186)
___ *S.a.albigularis* P.L. Sclater, 1858	SE Colombia (extreme S Amazonas), E Peru and W Brazil

[1] *Synallaxis whitneyi* Pacheco & Gonzaga, 1995 (1736), is a synonym (Whitney *in litt.*).
[2] For recognition as a species see Vaurie (1980: 99) (2516).
[3] The name *poliophrys*, still in last Edition, is a synonym of *S. frontalis* (2515).
[4] Treated as three species (*S. superciliosa, S. azarae* and *S. elegantior*) in last Edition but see Ridgely & Tudor (1994) (2019).
[5] Includes *urubambae* and *carabayae* see Remsen *et al.* (1990) (2010).
[6] Includes *hypoleuca* see Wetmore (1972) (2608).

○ *Synallaxis hypospodia* P.L. Sclater, 1874 CINEREOUS-BREASTED SPINETAIL
SE Peru, N Bolivia, C Brazil

○ *Synallaxis spixi* P.L. Sclater, 1856 SPIX'S SPINETAIL S Brazil, E Paraguay to N Argentina

○ *Synallaxis rutilans* RUDDY SPINETAIL
— *S.r.caquetensis* Chapman, 1914 — SE Colombia, E Ecuador, NE Peru
— *S.r.confinis* J.T. Zimmer, 1935 — NW Brazil
— *S.r.dissors* J.T. Zimmer, 1935 — E Colombia to the Guianas, S Venezuela, N Brazil
— *S.r.amazonica* Hellmayr, 1907 — E Peru, N Bolivia, W and C Brazil
— *S.r.rutilans* Temminck, 1823 — C and S Brazil
— *S.r.omissa* E. Hartert, 1901 — EC Brazil
— *S.r.tertia* Hellmayr, 1907 — NE Bolivia, SW Brazil

○ *Synallaxis cherriei* CHESTNUT-THROATED SPINETAIL
— *S.c.napoensis* Gyldenstolpe, 1930 — SE Colombia, E Ecuador,
— *S.c.saturata* Carriker, 1934[1] — E Peru
— *S.c.cherriei* Gyldenstolpe, 1916 — C Amazonian Brazil

○ *Synallaxis erythrothorax* RUFOUS-BREASTED SPINETAIL
— *S.e.furtiva* Bangs & Peters, 1927 — SE Mexico
— *S.e.erythrothorax* P.L. Sclater, 1855 — SE Mexico to NW Honduras
— *S.e.pacifica* Griscom, 1930 — S Mexico to El Salvador

○ *Synallaxis brachyura* SLATY SPINETAIL[2]
— *S.b.nigrofumosa* Lawrence, 1865[3] — NC Honduras to Panama and NW Colombia south to NW Ecuador
— *S.b.griseonucha* Chapman, 1923[4] — SW Ecuador to extreme NW Peru
— *S.b.brachyura* Lafresnaye, 1843 — N Colombia (Magdalena Valley)
— *S.b.caucae* Chapman, 1914 — C Colombia (Cauca Valley)

○ *Synallaxis tithys* Taczanowski, 1877 BLACKISH-HEADED SPINETAIL SW Ecuador, NW Peru

○ *Synallaxis propinqua* Pelzeln, 1859 WHITE-BELLIED SPINETAIL Islands in Amazonia

○ *Synallaxis macconnelli* McCONNELL'S SPINETAIL
— *S.m.macconnelli* Chubb, 1919[5] — Venezuela (Tepui region of Bolivar State and Cerro Yaví)
— *S.m.obscurior* Todd, 1948[6] — Extreme SE Venezuela; French Guiana; Surinam; extreme NE Brazil

○ *Synallaxis moesta* DUSKY SPINETAIL
— *S.m.moesta* P.L. Sclater, 1856 — E Colombia
— *S.m.obscura* Chapman, 1914 — SE Colombia
— *S.m.brunneicaudalis* P.L. Sclater, 1858 — Extreme SE Colombia, E Ecuador, NE Peru

○ *Synallaxis cabanisi* CABANIS'S SPINETAIL[7]
— *S.c.cabanisi* Berlepsch & Leverkühn, 1890 — C and S Peru
— *S.c.fulviventris* Chapman, 1924 — N Bolivia

○ *Synallaxis gujanensis* PLAIN-CROWNED SPINETAIL
— *S.g.colombiana* Chapman, 1914 — E Colombia
— *S.g.gujanensis* (J.F. Gmelin, 1789) — E and S Venezuela, the Guianas, N and E Brazil
— *S.g.huallagae* Cory, 1919 — E Ecuador, NE Peru and N Bolivia
— *S.g.canipileus* Chapman, 1923 — SE Peru
— *S.g.inornata* Pelzeln, 1856 — NE Bolivia, W and C Brazil
— *S.g.certhiola* Todd, 1916 — E Bolivia

○ *Synallaxis maranonica* Taczanowski, 1879 MARAÑON SPINETAIL N Peru

[1] For recognition see Ridgely & Greenfield (2001) (2016A).
[2] In last Edition we listed a subspecies *jaraguana*; Sibley & Monroe (1990: 398) (2262) considered this a misidentification as to species, the specimens being *S. hypospodia*.
[3] Includes *chapmani* see Slud (1964) (2278A).
[4] For recognition see Ridgely & Greenfield (2001) (2016A).
[5] Includes *yavii* see Vaurie (1980: 102) (2516).
[6] This race was erroneously listed under *S. moesta* in last Edition (as was *yavii* – now in synonymy). Includes *griseipectus* see Vaurie (1980: 101) (2516) and Mees (1987) (1539A).
[7] For split from *moesta* see Vaurie (1980: 99) (2516).

○ *Synallaxis albilora* WHITE-LORED SPINETAIL
— *S.a.simoni* Hellmayr, 1907[1]   C Brazil
— *S.a.albilora* Pelzeln, 1856   SE Brazil, N Paraguay

○ *Synallaxis zimmeri* Koepcke, 1957 RUSSET-BELLIED SPINETAIL   C Peru (Ancash) (1286)

○ *Synallaxis stictothorax* NECKLACED SPINETAIL
— *S.s.stictothorax* P.L. Sclater, 1859   W Ecuador, Puna Is.
— *S.s.maculata* Lawrence, 1874   NW Peru
— *S.s.chinchipensis* Chapman, 1925   Upper Marañon Valley (NW Peru)

SIPTORNOPSIS Cory, 1919[2] F
○ *Siptornopsis hypochondriaca* (Salvin, 1895) GREAT SPINETAIL[3]   N Peru

GYALOPHYLAX J.L. Peters, 1950[4] M
○ *Gyalophylax hellmayri* (Reiser, 1905) RED-SHOULDERED SPINETAIL   NE Brazil

HELLMAYREA Stolzmann, 1926[5] F
○ *Hellmayrea gularis* WHITE-BROWED SPINETAIL
— *H.g.gularis* (Lafresnaye, 1843)   Andes of Colombia (all three ranges), Ecuador, and N Peru (Cajamarca)
— *H.g.brunneidorsalis* Phelps & Phelps, Jr., 1953   NE Colombia and NW Venezuela (1853)
— *H.g.cinereiventris* (Chapman, 1912)   Andes of W Venezuela (Tachira, Merida, and Trujillo)
— *H.g.rufiventris* (Berlepsch & Stolzmann, 1896)   N and C Peru

CRANIOLEUCA Reichenbach, 1853[6] [7] F
○ *Cranioleuca marcapatae* MARCAPATA SPINETAIL
— *C.m.weskei* Remsen, 1984   Vilcabamba Mts., Cuzco (SE Peru) (2005)
— *C.m.marcapatae* J.T. Zimmer, 1935   Marcapata, Cuzco (SE Peru)

○ *Cranioleuca albiceps* LIGHT-CROWNED SPINETAIL
— *C.a.albiceps* (d'Orbigny & Lafresnaye, 1837)   La Paz (Bolivia)
— *C.a.discolor* J.T. Zimmer, 1935   Cochabamba (Bolivia)

○ *Cranioleuca vulpina* RUSTY-BACKED SPINETAIL
— *C.v.dissita* (Wetmore, 1957)   Coiba I. (Panama) (2598)
— *C.v.apurensis* Zimmer & Phelps, 1948   SW Venezuela
— *C.v.alopecias* (Pelzeln, 1859)   E Colombia, W Venezuela, N Brazil
— *C.v.vulpina* (Pelzeln, 1856)   W and C Brazil
— *C.v.foxi* Bond & Meyer de Schauensee, 1940   W Bolivia
— *C.v.reiseri* (Reichenberger, 1922)   NE Brazil

○ *Cranioleuca vulpecula* (Sclater & Salvin, 1866) PARKER'S SPINETAIL[8]   NE Peru, W Brazil, NE Bolivia

○ *Cranioleuca sulphurifera* (Burmeister, 1869) SULPHUR-THROATED SPINETAIL
                                                              E Argentina, Uruguay, S Brazil

○ *Cranioleuca subcristata* CRESTED SPINETAIL
— *C.s.subcristata* (P.L. Sclater, 1874)   NE Colombia, N Venezuela
— *C.s.fuscivertex* Phelps & Phelps, Jr., 1955   W Venezuela (1855)

○ *Cranioleuca pyrrhophia* STRIPE-CROWNED SPINETAIL
— *C.p.rufipennis* (Sclater & Salvin, 1879)   Andes of N Bolivia
— *C.p.striaticeps* (d'Orbigny & Lafresnaye, 1837)   Andes of C Bolivia
— *C.p.pyrrhophia* (Vieillot, 1818)   S Bolivia to Uruguay, extreme S Brazil, N Argentina

○ *Cranioleuca henricae* Maijer & Fjeldså, 1997 BOLIVIAN SPINETAIL #   C Bolivia (1414)

---

[1] This subspecies is moved here from the species *S. gujanensis* see Ridgely & Tudor (1994) (2019).
[2] Included in the genus *Thripophaga* in last Edition.
[3] For correction of spelling see David & Gosselin (2002) (614).
[4] Included in the genus *Synallaxis* in the last Edition having been lumped by Vaurie (1980). We prefer the treatment of Peters (1951: 94) (1828).
[5] For recognition of this genus see Braun & Parker (1985) (210) and Garcia-Moreno *et al.* (1999) (913).
[6] Lumped in *Certhiaxis* in last Edition following Vaurie (1980) (2516).
[7] Sequence of species adapted from Garcia-Moreno *et al.* (1999) (913).
[8] For reasons to separate this from *C. vulpina* see Zimmer (1997) (2730).

○ *Cranioleuca obsoleta* (Reichenbach, 1853) OLIVE SPINETAIL   SE Brazil, E Paraguay, NE Argentina

○ *Cranioleuca pallida* (Wied, 1831) PALLID SPINETAIL   SE Brazil

○ *Cranioleuca semicinerea* GREY-HEADED SPINETAIL
    — *C.s.semicinerea* (Reichenbach, 1853)   Ceará, Bahia (E Brazil)
    — *C.s.goyana* Pinto, 1936   Goias (E Brazil)

○ *Cranioleuca albicapilla* CREAMY-CRESTED SPINETAIL
    — *C.a.albicapilla* (Cabanis, 1873)   C Peru
    — *C.a.albigula* J.T. Zimmer, 1924   S Peru

● *Cranioleuca erythrops* RED-FACED SPINETAIL
    ✓ *C.e.rufigenis* (Lawrence, 1868)   Mts. of Costa Rica, W Panama
    — *C.e.griseigularis* (Ridgway, 1909)   E Panama (Darien), W and C Andes of Colombia
    — *C.e.erythrops* (P.L. Sclater, 1860)   Andes of W Ecuador

○ *Cranioleuca demissa* TEPUI SPINETAIL[1]
    — ¶*C.d.cardonai* Phelps & Dickerman, 1980   Mts. of Amazonas and W Bolivar (S Venezuela) (1841)
    — *C.d.demissa* (Salvin & Godman, 1884)   Tepuis of SE Bolivar (SE Venezuela)

○ *Cranioleuca hellmayri* (Bangs, 1907) STREAK-CAPPED SPINETAIL   N Colombia (Santa Marta Mts.)

○ *Cranioleuca curtata* ASH-BROWED SPINETAIL
    — *C.c.curtata* (P.L. Sclater, 1870)   Eastern Andes of Colombia
    — *C.c.cisandina* (Taczanowski, 1882)   E Andes of S Colombia, E Ecuador and N Peru
    — *C.c.debilis* (Berlepsch & Stolzmann, 1906)   Andes of C Peru to C Bolivia

○ *Cranioleuca antisiensis* LINE-CHEEKED SPINETAIL[2]
    — *C.a.antisiensis* (P.L. Sclater, 1859)   S Ecuador
    — *C.a.palamblae* (Chapman, 1923)   N Peru

○ *Cranioleuca baroni* BARON'S SPINETAIL[3]
    — *C.b.baroni* (Salvin, 1895)   Amazonas and Cajamarca (Andes of Peru)
    — *C.b.capitalis* J.T. Zimmer, 1924   Huánuco (Andes of Peru)
    — *C.b.zaratensis* Koepcke, 1961   Pasco and Lima (Andes of S Peru) (1289)

○ *Cranioleuca gutturata* (d'Orbigny & Lafresnaye, 1838) SPECKLED SPINETAIL[4]
           S Venezuela, Surinam, French Guiana and NE Brazil south through Amazonian E Colombia, E Ecuador, and E Peru to E Bolivia and C Brazil

○ *Cranioleuca muelleri* (Hellmayr, 1911) SCALED SPINETAIL   N Brazil

CERTHIAXIS Lesson, 1844 M[5]
○ *Certhiaxis cinnamomeus* YELLOW-CHINNED SPINETAIL
    — *C.c.fuscifrons* (Madarász, 1913)   N Colombia
    — *C.c.marabinus* Phelps & Phelps, Jr., 1946   NW Venezuela
    — *C.c.valencianus* Zimmer & Phelps, 1944   WC Venezuela
    — *C.c.orenocensis* J.T. Zimmer, 1935   Orinoco valley (Venezuela)
    — *C.c.cinnamomeus* (J.F. Gmelin, 1788)   Trinidad, NE Venezuela, the Guianas, NE Brazil
    — *C.c.pallidus* J.T. Zimmer, 1935[6]   C Brazil
    — *C.c.cearensis* (Cory, 1916)   E Brazil
    — *C.c.russeolus* (Vieillot, 1817)   E Bolivia, S Brazil, Paraguay, NW Argentina

○ *Certhiaxis mustelinus* (P.L. Sclater, 1874) RED-AND-WHITE SPINETAIL   Amazon River of Peru, Colombia, Brazil

THRIPOPHAGA Cabanis, 1847 F
○ *Thripophaga cherriei* Berlepsch & Hartert, 1902 ORINOCO SOFT-TAIL   S Venezuela

○ *Thripophaga macroura* (Wied, 1821) STRIATED SOFT-TAIL   E Brazil

[1] For validation of this species see Vaurie (1971) (2515).
[2] In last Edition we listed a race *furcata*; this name was based on immature *C. curtata*.
[3] Species omitted in last Edition.
[4] In last Edition we "recognised" *peruviana* Cory, 1919 (572) and *hyposticta* Pelzeln, 1859 (2543) but we now prefer monotypic treatment as in Peters (1951) (1828).
[5] Gender addressed by David & Gosselin (2002) (614), multiple changes occur in specific names.
[6] If the genus *Cranioleuca* is lumped with *Certhiaxis* this name is preoccupied. In last Edition we missed this and used *pallida* as a species and again as a subspecies here.

○ *Thripophaga fusciceps* Plain Soft-tail[1]
— *T.f.dimorpha* Bond & Meyer de Schauensee, 1941   E Ecuador, S Peru
— *T.f.obidensis* Todd, 1925   C Brazil
— *T.f.fusciceps* P.L. Sclater, 1889   N Bolivia

○ *Thripophaga berlepschi* Hellmayr, 1905 Russet-mantled Soft-tail   Andes of N Peru

PHACELLODOMUS Reichenbach, 1853 M
○ *Phacellodomus rufifrons* Rufous-fronted Thornbird
— *P.r.inornatus* Ridgway, 1888   NC Venezuela
— *P.r.castilloi* Phelps & Aveledo, 1987   NE Colombia, W and C Venezuela (1839)
— *P.r.peruvianus* Hellmayr, 1925   Extreme S Ecuador, N Peru
— *P.r.specularis* Hellmayr, 1925   NE Brazil
— *P.r.rufifrons* (Wied, 1821)   E Brazil
— *P.r.sincipitalis* Cabanis, 1883[2]   E Bolivia, S Brazil, NC Paraguay and NW Argentina

○ *Phacellodomus sibilatrix* P.L. Sclater, 1879 Little Thornbird   S Bolivia, W Paraguay, N Argentina

○ *Phacellodomus striaticeps* Streak-fronted Thornbird
— *P.s.griseipectus* Chapman, 1919   Dry Andean valleys of S Peru
— *P.s.striaticeps* (d'Orbigny & Lafresnaye, 1838)   Andes of Bolivia, N Argentina

○ *Phacellodomus striaticollis* Freckle-breasted Thornbird
— *P.s.maculipectus* Cabanis, 1883   Andes from C Bolivia to NW Argentina
— *P.s.striaticollis* (d'Orbigny & Lafresnaye, 1838)   SE Brazil, Uruguay, NE Argentina

○ *Phacellodomus dorsalis* Salvin, 1895 Chestnut-backed Thornbird   Upper Marañon Valley in NW Peru

○ *Phacellodomus ruber* (Vieillot, 1817) Greater Thornbird   NE and E Bolivia, interior S Brazil. Paraguay, N Argentina

○ *Phacellodomus erythrophthalmus* Red-eyed Thornbird
— *P.e.erythrophthalmus* (Wied, 1821)   SE Brazil (S Bahia to NE São Paulo)
— *P.e.ferrugineigula* (Pelzeln, 1858)   SE Brazil (S São Paulo and S Rio Grande do Sul)

CLIBANORNIS Sclater & Salvin, 1873[3] M
○ *Clibanornis dendrocolaptoides* (Pelzeln, 1859) Canebrake Groundcreeper
    SE Brazil, E Paraguay, NE Argentina

SPARTONOICA J.L. Peters, 1950 F
○ *Spartonoica maluroides* (d'Orbigny & Lafresnaye, 1837) Bay-capped Wren-Spinetail
    S Brazil, Uruguay, Argentina

PHLEOCRYPTES Cabanis & Heine, 1859 M
○ *Phleocryptes melanops* Wren-like Rushbird
— *P.m.brunnescens* J.T. Zimmer, 1935   S Peru
— *P.m.juninensis* Carriker, 1932   C Peru
— *P.m.schoenobaenus* Cabanis & Heine, 1859   S Peru, W Bolivia, NW Argentina
— *P.m.loaensis* Philippi & Goodall, 1946   N Chile
— *P.m.melanops* (Vieillot, 1817)   S Brazil to C Chile, C Argentina

LIMNORNIS Gould, 1839 M
○ *Limnornis curvirostris* Gould, 1839 Curve-billed Reedhaunter   SE Brazil, S Uruguay, E Argentina

○ *Limnornis rectirostris* Gould, 1839 Straight-billed Reedhaunter[4]   SE Brazil, S Uruguay, E Argentina

ANUMBIUS d'Orbigny & Lafresnaye, 1838 M
○ *Anumbius annumbi* (Vieillot, 1817) Firewood-gatherer[5]   SE Brazil, C and E Paraguay, Uruguay, N and C Argentina

CORYPHISTERA Burmeister, 1860 F
○ *Coryphistera alaudina* Lark-like Brushrunner
— *C.a.campicola* Todd, 1915   SE Bolivia, W Paraguay
— *C.a.alaudina* Burmeister, 1860   S Bolivia, extreme SE Brazil and N Argentina

---

[1] This species and the next were treated in the genus *Phacellodromus* in last Edition.
[2] Includes *fargoi* see Gyldenstolpe (1945) (1030A).
[3] Submerged in *Phacellodomus* in last Edition.
[4] Once treated in a monotypic genus *Limnoctites*.
[5] Includes *machrisi* Stager, 1959 (2302) see Straube (1994) (2357A).

EREMOBIUS Gould, 1839 M
○ *Eremobius phoenicurus* Gould, 1839 Band-tailed Earthcreeper   S Argentina

SIPTORNIS Reichenbach, 1853 F
○ *Siptornis striaticollis* Spectacled Prickletail
    — *S.s.striaticollis* (Lafresnaye, 1843)   E Andes bordering Magdalena Valley of Colombia
    — *S.s.nortoni* Graves & Robbins, 1987   E slopes Andes of Ecuador and extreme N Peru (993)

METOPOTHRIX Sclater & Salvin, 1866 F[1]
○ *Metopothrix aurantiaca* Sclater & Salvin, 1866 Orange-fronted Plushcrown
                                    W Amazonia (SE Colombia, E Peru, E Ecuador, W
                                    Brazil and NE Bolivia)

XENERPESTES Berlepsch, 1886 M
○ *Xenerpestes minlosi* Double-banded Greytail
    — *X.m.minlosi* Berlepsch, 1886   Caribbean slope E Panama to Magdalena Valley (N
                                    Colombia)

    — *X.m.umbraticus* Wetmore, 1951   E Panama (Darién), NW Colombia (lowlands), NW
                                    Ecuador (2594)

○ *Xenerpestes singularis* (Taczanowski & Berlepsch, 1885) Equatorial Greytail
                                    Andes of Ecuador south to N Peru

PREMNORNIS Ridgway, 1909[2] M
○ *Premnornis guttuligera* Rusty-winged Barbtail
    — *P.g.venezuelana* Phelps & Phelps, Jr., 1956   Perijá Mts. (NW Venezuela) (1857)
    — *P.g.guttuligera* (P.L. Sclater, 1864)   Andes of W Venezuela, Colombia (not Pacific slope)
                                    and Peru

PREMNOPLEX Cherrie, 1891[3] M
● *Premnoplex brunnescens* Spotted Barbtail
    ↙ *P.b.brunneicauda* (Lawrence, 1865)[4]   Mts. of Costa Rica, W Panama
    — *P.b.albescens* Griscom, 1927   E Panama
    — *P.b.coloratus* Bangs, 1902   Santa Marta Mts. (N Colombia)
    — *P.b.rostratus* Hellmayr & Seilern, 1912   Coastal range N Venezuela
    — *P.b.brunnescens* (P.L. Sclater, 1856)   Cerros Tacarcuna, Pirre, and Mali (E Panama), Andes
                                    from Colombia and NW Venezuela to S Peru (Cuzco)
    — *P.b.stictonotus* (Berlepsch, 1901)   Andes from S Peru (Puno) to N Bolivia

○ *Premnoplex tatei* White-throated Barbtail
    — *P.t.tatei* Chapman, 1925   Mts. of NE Venezuela
    — *P.t.pariae* Phelps & Phelps, Jr., 1949   Paria Peninsula (NE Venezuela)

RORAIMIA Chapman, 1929[5] F
○ *Roraimia adusta* Roraiman Barbtail
    — *R.a.obscurodorsalis* Phelps & Phelps, Jr., 1948   SE Venezuela
    — *R.a.mayri* (W.H. Phelps, Jr., 1977)[6]   Tepuis of Jaua (Bolivar State, Venezuela) (1835)
    — *R.a.duidae* Chapman, 1939   Tepuis of C Amazonas (S Venezuela)
    — *R.a.adusta* (Salvin & Godman, 1884)   Tepuis of E Bolivar (E Venezuela) and W Guyana

ACROBATORNIS Pacheco, Whitney & Gonzaga, 1996 (1738) M
○ *Acrobatornis fonsecai* Pacheco, Whitney & Gonzaga, 1996 Pink-legged Graveteiro #
                                    SE Bahia (Brazil) (1738)

MARGARORNIS Reichenbach, 1853 M
● *Margarornis rubiginosus* Ruddy Treerunner
    ↙ *M.r.rubiginosus* Lawrence, 1865   Mts. of Costa Rica, W Panama
    — *M.r.boultoni* Griscom, 1924   Mts. of C Panama

---

[1] For correction of spelling see David & Gosselin (2002) (614).
[2] Submerged in *Margarornis* in last Edition.
[3] Submerged in *Margarornis* in last Edition.
[4] Includes *distinctus* and *mnionophilus* Wetmore, 1951 (2594) see Wetmore (1972) (2608).
[5] Submerged in *Margarornis* in last Edition, but see Rudge & Raikow (1992) (2126).
[6] Tentatively accepted; not known to have been evaluated.

○ *Margarornis stellatus* Sclater & Salvin, 1873 Fulvous-dotted Treerunner
W and C Andes of Colombia, NW Ecuador

○ *Margarornis bellulus* Nelson, 1912 Beautiful Treerunner[1]
Cerro Tacarcuna, Cerro Mali, Cerro Pirre, Altos de Quía, and Serranía de Majé (extreme E Panama)

○ *Margarornis squamiger* Pearled Treerunner
— *M.s.perlatus* (Lesson, 1844)
Perijá Mtns. and Andes of W Venezuela, Colombia, Ecuador, N Peru
— *M.s.peruvianus* Cory, 1913
Andes of N and C Peru
— *M.s.squamiger* (d'Orbigny & Lafresnaye, 1838)
Andes of S Peru, N Bolivia

PSEUDOSEISURA Reichenbach, 1853 F
○ *Pseudoseisura cristata* (Spix, 1824) Caatinga Cachalote
E Brazil (E Maranhão, Paraíba and Pernambuco south to C Minas Gerais)

○ *Pseudoseisura unirufa* (d'Orbigny & Lafresnaye, 1838) Rufous Cachalote[2]
E Bolivia, SW Brazil, N Paraguay

○ *Pseudoseisura lophotes* Brown Cachalote
— *P.l.lophotes* (Reichenbach, 1853)
S Bolivia, W Paraguay
— *P.l.argentina* Parkes, 1960
Uruguay, N and C Argentina, extreme SE Brazil (1753)

○ *Pseudoseisura gutturalis* White-throated Cachalote
— *P.g.ochroleuca* Olrog, 1959
Andean foothills N and C Argentina (1679)
— *P.g.gutturalis* (d'Orbigny & Lafresnaye, 1838)
Lowlands of C Argentina

PSEUDOCOLAPTES Reichenbach, 1853 M
● *Pseudocolaptes lawrencii* Buffy Tuftedcheek
↓ *P.l.lawrencii* Ridgway, 1878[3]
Mts. of Costa Rica to C Panama
— *P.l.johnsoni* Lönnberg & Rendahl, 1922
W Andes of Colombia, W Ecuador

○ *Pseudocolaptes boissonneautii* Streaked Tuftedcheek
— *P.b.striaticeps* Hellmayr & Seilern, 1912
Coastal range N Venezuela
— *P.b.meridae* Hartert & Goodson, 1917
Mts. of W Venezuela
— *P.b.boissonneautii* (Lafresnaye, 1840)[4]
Andes of Colombia, W Ecuador
— *P.b.orientalis* J.T. Zimmer, 1935
Andes of Ecuador
— *P.b.intermedianus* Chapman, 1923
Andes of NW Peru (temperate zone)
— *P.b.pallidus* J.T. Zimmer, 1935
Andes of NW Peru (subtropical zone)
— *P.b.medianus* Hellmayr, 1919
Andes of N Peru south of the Marañon River
— *P.b.auritus* (Tschudi, 1844)
Andes of C Peru
— *P.b.carabayae* J.T. Zimmer, 1936
Andes of S Peru to C Bolivia

BERLEPSCHIA Ridgway, 1887 F
○ *Berlepschia rikeri* (Ridgway, 1886) Point-tailed Palmcreeper
S Venezuela, the Guianas, Amazonia to E Brazil

ANABACERTHIA Lafresnaye, 1842[5] F
○ *Anabacerthia variegaticeps* Scaly-throated Foliage-gleaner
— *A.v.schaldachi* Winker, 1997
Mts. of Guerrero (SW Mexico) (2674)
— *A.v.variegaticeps* (P.L. Sclater, 1857)
Mts. of S Mexico, Guatemala, Honduras, Costa Rica and W Panama
— *A.v.temporalis* (P.L. Sclater, 1859)[6]
W Andes of Colombia and Ecuador

○ *Anabacerthia striaticollis* Montane Foliage-gleaner
— *A.s.anxia* (Bangs, 1902)
Santa Marta Mts. (N Colombia)
— *A.s.perijana* Phelps & Phelps, Jr., 1952
Perijá Mts. (NW Venezuela) (1851)
— *A.s.venezuelana* (Hellmayr, 1911)
Coastal range of N Venezuela
— *A.s.striaticollis* Lafresnaye, 1842
Andes of Colombia, W Venezuela
— *A.s.montana* (Tschudi, 1844)
Andes SE Colombia, E Ecuador to C Peru
— *A.s.yungae* (Chapman, 1923)
Andes of S Peru, N Bolivia

[1] "Reluctantly" maintained as a species by the A.O.U. (1998: 350) (3).
[2] For reasons to separate this from *P. cristata* see Zimmer & Whittaker (2000) (2732).
[3] Includes *panamensis* see Wetmore (1972) (2608).
[4] Includes *oberholseri* Cory, 1919 (572) see Deignan (1961) (662).
[5] Submerged in *Philydor* in last Edition.
[6] Winker (1997) (2674) considered that this is probably a separate species.

○ *Anabacerthia amaurotis* (Temminck, 1823) WHITE-BROWED FOLIAGE-GLEANER

SE Brazil, NE Argentina

SYNDACTYLA Reichenbach, 1853[1] F

○ *Syndactyla guttulata* GUTTULATED[2] FOLIAGE-GLEANER[3]

— *S.g.guttulata* (P.L. Sclater, 1858) Coastal range N Venezuela
— *S.g.pallida* Zimmer & Phelps, 1944 Mts. of NE Venezuela

○ *Syndactyla subalaris* LINEATED FOLIAGE-GLEANER

— *S.s.lineata* (Lawrence, 1865) Mts. Costa Rica, W Panama
— *S.s.tacarcunae* (Chapman, 1923) Mts. of E Panama, NW Colombia
— *S.s.subalaris* (P.L. Sclater, 1859) Andes of W Colombia, W Ecuador
— *S.s.striolata* (Todd, 1913) E Andes of Colombia, Andes W Venezuela (Lara and Barinas)
— *S.s.olivacea* Phelps & Phelps, Jr., 1956 Andes of W Venezuela (S Tachira) (1857)
— *S.s.mentalis* (Taczanowski & Berlepsch, 1885) E slope Andes of E Ecuador
— *S.s.colligata* J.T. Zimmer, 1935 E slope Andes of N Peru
— *S.s.ruficrissa* (Carriker, 1930) E slope Andes of C Peru

○ *Syndactyla rufosuperciliata* BUFF-BROWED FOLIAGE-GLEANER

— *S.r.similis* (Chapman, 1927) W Andes of NW Peru
— *S.r.cabanisi* (Taczanowski, 1875) Andes of S Ecuador, N Peru to C Bolivia
— *S.r.oleaginea* (P.L. Sclater, 1884) Andes of C Bolivia, NW Argentina
— *S.r.rufosuperciliata* (Lafresnaye, 1832) SE Brazil
— *S.r.acrita* (Oberholser, 1901) S Brazil, NC Paraguay, Uruguay, NE Argentina

○ *Syndactyla ruficollis* (Taczanowski, 1884) RUFOUS-NECKED FOLIAGE-GLEANER[4] [5]

Coastal hills and W slope Andes of SW Ecuador and NW Peru

SIMOXENOPS Chapman, 1928[6] M

○ *Simoxenops ucayalae* (Chapman, 1928) PERUVIAN RECURVEBILL[7] SE Peru, N Bolivia, S Amazonian Brazil

○ *Simoxenops striatus* (Carriker, 1935) BOLIVIAN RECURVEBILL Andean foothills N and C Bolivia

ANCISTROPS P.L. Sclater, 1862[8] M

○ *Ancistrops strigilatus* (Spix, 1825) CHESTNUT-WINGED HOOKBILL[9]

SE Colombia, E Ecuador, E Peru, N Bolivia and Amazonian Brazil (mostly south of the Amazon)

HYLOCTISTES Ridgway, 1909[10] M

○ *Hyloctistes subulatus* STRIPED WOODHAUNTER

— *H.s.nicaraguae* Miller & Griscom, 1925 E Nicaragua
— *H.s.virgatus* (Lawrence, 1867) Costa Rica, W Panama
— *H.s.assimilis* (Berlepsch & Taczanowski, 1884)[11] E Panama, W and N Colombia, W Ecuador
— *H.s.lemae* Phelps & Phelps, Jr., 1960 SE Venezuela (1862)
— *H.s.subulatus* (Spix, 1824) S Venezuela, W Amazonia

PHILYDOR Spix, 1824 N[12]

○ *Philydor ruficaudatum* RUFOUS-TAILED FOLIAGE-GLEANER

— *P.r.ruficaudatum* (d'Orbigny & Lafresnaye, 1838) Amazonia, the Guianas
— *P.r.flavipectus* Phelps & Gilliard, 1941 Tepui region of S Venezuela, N Brazil

○ *Philydor fuscipenne* SLATY-WINGED FOLIAGE-GLEANER[13]

— *P.f.fuscipenne* (Salvin, 1886) C Panama
— *P.f.erythronotum* Sclater & Salvin, 1873[14] E Panama, N Colombia and W Ecuador

[1] Submerged in *Philydor* in last Edition.
[2] The term *guttulata* means spotted and simplification of the English name seems possible.
[3] The name *mirandae*, included in last edition, was shown to be a synonym of *Philydor dimidiatus* see Novaes (1953) (1661A).
[4] For reasons to transfer this species from *Automolus* to *Syndactyla* see Parker *et al.* (1985) (1749).
[5] Includes *celicae* see Ridgely & Greenfield (2001) (2016A).
[6] Submerged in *Philydor* in last Edition.
[7] Includes *Megaxenops ferrugineus* Berlioz, 1966 (141); see Mayr & Vuilleumier, 1983 (1496).
[8] Submerged in *Philydor* in last Edition.
[9] Includes *cognitus* — JVR based on Gyldenstolpe, 1945 (1030) and 1951 (1031).
[10] Submerged in *Philydor* in last Edition.
[11] Includes *cordobae* Meyer de Schauensee, 1960 (1572), see Ridgely & Tudor (1994) (2019).
[12] Gender addressed by David & Gosselin (2002) (614), multiple changes occur in specific names.
[13] Treated as part of the species *P. erythrocercus* in last Edition but see A.O.U. (1998: 352) (3).
[14] Includes *Philydor fulvescens* Todd, 1950 (2413) see Mayr (1957) (1479). The full adult plumage.

○ *Philydor erythrocercum* RUFOUS-RUMPED FOLIAGE-GLEANER
    __ *P.e.subfulvum* P.L. Sclater, 1862   SE Colombia, E Ecuador, N Peru
    __ *P.e.ochrogaster* Hellmayr, 1917   Andes from C Peru to N Bolivia
    __ *P.e.lyra* Cherrie, 1916   E Peru, NE Bolivia, Amazonian Brazil
    __ *P.e.suboles* Todd, 1948   C Brazil
    __ *P.e.erythrocercum* (Pelzeln, 1859)   The Guianas, N Brazil

○ *Philydor erythropterum* CHESTNUT-WINGED FOLIAGE-GLEANER
    __ *P.e.erythropterum* (P.L. Sclater, 1856)   Extreme S Venezuela, W and S Amazonia
    __ *P.e.diluviale* Griscom & Greenway, 1937   E Brazil (N Pará, Maranhão)

○ *Philydor lichtensteini* Cabanis & Heine, 1859 OCHRE-BREASTED FOLIAGE-GLEANER
                                       SE Brazil, E Paraguay, NE Argentina

○ *Philydor novaesi* Teixeira & Gonzaga, 1983 ALAGOAS FOLIAGE-GLEANER #
                                       NE Brazil (Alagoas) [2391]

○ *Philydor atricapillus* (Wied, 1821) BLACK-CAPPED FOLIAGE-GLEANER   SE Brazil, E Paraguay, NE Argentina

○ *Philydor rufum* BUFF-FRONTED FOLIAGE-GLEANER
    __ *P.r.panerythrum* P.L. Sclater, 1862   Mts. Costa Rica to W Panama, W slope of E Andes and
                                       C Andes of Colombia
    __ *P.r.riveti* Ménégaux & Hellmayr, 1906   W Andes of Colombia and N Ecuador
    __ *P.r.colombianum* Cabanis & Heine, 1859   Coastal range of N Venezuela
    __ *P.r.cuchiverus* Phelps & Phelps, Jr., 1949   Tepui region of S Venezuela
    __ *P.r.bolivianum* Berlepsch, 1907   Andean foothills from E Ecuador to C Bolivia
    __ *P.r.chapadense* J.T. Zimmer, 1935   SW Brazil
    __ *P.r.rufum* (Vieillot, 1818)   SE Brazil, E Paraguay, NE Argentina

○ *Philydor pyrrhodes* (Cabanis, 1848) CINNAMON-RUMPED FOLIAGE-GLEANER
                                       N Amazonia, the Guianas

○ *Philydor dimidiatum* RUSSET-MANTLED FOLIAGE-GLEANER
    __ *P.d.dimidiatum* (Pelzeln, 1859)   SC Brazil
    __ *P.d.baeri* Hellmayr, 1911   SE Brazil, NC Paraguay

ANABAZENOPS Lafresnaye, 1842[1] M
○ *Anabazenops dorsalis* Sclater & Salvin, 1880 DUSKY-CHEEKED FOLIAGE-GLEANER[2]
                                 SE Colombia, E Ecuador, E Peru, N Bolivia, SW Brazil

○ *Anabazenops fuscus* (Vieillot, 1816) WHITE-COLLARED FOLIAGE-GLEANER
                                 SE Brazil

CICHLOCOLAPTES Reichenbach, 1853 M
○ *Cichlocolaptes leucophrus* (Jardine & Selby, 1830) PALE-BROWED TREEHUNTER
                                 SE Brazil

THRIPADECTES P.L. Sclater, 1862 M
○ *Thripadectes ignobilis* (Sclater & Salvin, 1879) UNIFORM TREEHUNTER W Andes of Colombia and Ecuador

● *Thripadectes rufobrunneus* (Lawrence, 1865) STREAK-BREASTED TREEHUNTER
                                 Mts. of Costa Rica and W Panama

○ *Thripadectes melanorhynchus* BLACK-BILLED TREEHUNTER
    __ *T.m.striaticeps* (Sclater & Salvin, 1875)   E Andes of Colombia
    __ *T.m.melanorhynchus* (Tschudi, 1844)   E Andes of Ecuador and Peru

○ *Thripadectes holostictus* STRIPED TREEHUNTER
    __ *T.h.striatidorsus* (Berlepsch & Taczanowski, 1884)   W Andes of SW Colombia and W Ecuador
    __ *T.h.holostictus* (Sclater & Salvin, 1876)   Andes from SW Venezuela and N Colombia to E
                                       Ecuador and N Peru
    __ *T.h.moderatus* J.T. Zimmer, 1935   Andes from C Peru to N Bolivia

○ *Thripadectes virgaticeps* STREAK-CAPPED TREEHUNTER
    __ *T.v.klagesi* (Hellmayr & Seilern, 1912)   Coastal range of N Venezuela
    __ *T.v.tachirensis* Phelps & Phelps, Jr., 1958   Andes of W Venezuela (1859)
    __ *T.v.magdalenae* Meyer de Schauensee, 1945   W Andes of NW Colombia and Andes of C Columbia

---

[1] Submerged in *Philydor* in last Edition.
[2] For reasons to move this from the genus *Automolus* see Kratter & Parker (1997) (1305).

— *T.v.sclateri* Berlepsch, 1907   W Andes of SW Colombia
— *T.v.virgaticeps* Lawrence, 1874   W Andes of N Ecuador
— *T.v.sumaco* Chapman, 1925   E Andes of N Ecuador

○ *Thripadectes flammulatus* FLAMMULATED TREEHUNTER
— *T.f.bricenoi* Berlepsch, 1907   Andes of Mérida, W Venezuela
— *T.f.flammulatus* (Eyton, 1849)   Santa Marta Mts. and Andes of SW Venezuela, Colombia, Ecuador, N Peru

○ *Thripadectes scrutator* Taczanowski, 1874 BUFF-THROATED TREEHUNTER
     Andes of C Peru and C Bolivia

AUTOMOLUS Reichenbach, 1853 M
○ *Automolus ochrolaemus* BUFF-THROATED FOLIAGE-GLEANER
— *A.o.cervinigularis* (P.L. Sclater, 1857)   S Mexico to Nicaragua and Guatemala (except SE)
— *A.o.hypophaeus* Ridgway, 1909[1]   Honduras to NW Panama
— *A.o.exsertus* Bangs, 1901   SW Costa Rica, W Panama
— *A.o.pallidigularis* Lawrence, 1862   E Panama, N and W Colombia, NW Ecuador
— *A.o.turdinus* (Pelzeln, 1859)   SE Colombia, S Venezuela, the Guianas, N and W Amazonia
— *A.o.ochrolaemus* (Tschudi, 1844)   E Peru, N Bolivia, W Brazil
— *A.o.auricularis* J.T. Zimmer, 1935   E Bolivia, SC Brazil

○ *Automolus infuscatus* OLIVE-BACKED FOLIAGE-GLEANER
— *A.i.infuscatus* (P.L. Sclater, 1856)   SE Colombia, E Ecuador, E Peru, N Bolivia
— *A.i.badius* J.T. Zimmer, 1935   S Venezuela, E Colombia, NW Brazil
— *A.i.cervicalis* (P.L. Sclater, 1889)   SE Venezuela, the Guianas, NE Brazil
— *A.i.purusianus* Todd, 1948   W Brazil
— *A.i.paraensis* E. Hartert, 1902   C Brazil

○ *Automolus leucophthalmus* WHITE-EYED FOLIAGE-GLEANER
— *A.l.lammi* J.T. Zimmer, 1947   NE Brazil
— *A.l.leucophthalmus* (Wied, 1821)   E Brazil
— *A.l.sulphurascens* (M.H.K. Lichtenstein, 1823)   S Brazil, E Paraguay, NE Argentina

○ *Automolus melanopezus* (P.L. Sclater, 1858) BROWN-RUMPED FOLIAGE-GLEANER
     SE Colombia, E Ecuador, E Peru, N Bolivia, W Brazil

○ *Automolus roraimae* WHITE-THROATED FOLIAGE-GLEANER
— ¶*A.r.paraquensis* Phelps & Phelps, Jr., 1947[2]   Mt. Paraque (SC Venezuela)
— *A.r.duidae* Chapman, 1939   Mt. Duida (SC Venezuela and extreme N Brazil)
— *A.r.roraimae* Hellmayr, 1917[3]   Mt. Roraima (SE Venezuela)
— *A.r.urutani* Phelps & Dickerman, 1980   SE Venezuela and extreme N Brazil (1841)

○ *Automolus rubiginosus* RUDDY FOLIAGE-GLEANER
— *A.r.guerrerensis* Salvin & Godman, 1891   Mts. of SW Mexico
— *A.r.rubiginosus* (P.L. Sclater, 1857)   Mts. of E Mexico
— *A.r.veraepacis* Salvin & Godman, 1891[4]   Mts. of S Mexico to N Guatemala
— *A.r.fumosus* Salvin & Godman, 1891   Mts. of SW Costa Rica and W Panama
— *A.r.saturatus* Chapman, 1915   Mts. of E Panama, NW Colombia
— *A.r.sasaimae* Meyer de Schauensee, 1947   W slope E Andes of C Colombia
— *A.r.nigricauda* E. Hartert, 1898   Andean foothills W Colombia, W Ecuador
— *A.r.rufipectus* Bangs, 1898   N Colombia (Sta. Marta Mts.)
— *A.r.venezuelanus* Zimmer & Phelps, 1947   Tepuis of S Venezuela
— *A.r.cinnamomeigula* Hellmayr, 1905   E Andean foothills in C Colombia
— *A.r.caquetae* Meyer de Schauensee, 1947   Andean foothills in SE Colombia, NE Ecuador
— *A.r.brunnescens* Berlioz, 1927   Andean foothills in E Ecuador
— *A.r.moderatus* J.T. Zimmer, 1935   Andean foothills in N Peru
— *A.r.watkinsi* Hellmayr, 1912   Andean foothills in SE Peru, N Bolivia
— *A.r.obscurus* (Pelzeln, 1859)   The Guianas, NE Brazil

[1] Includes *amusos* see Monroe (1968) (1604).
[2] Considered a synonym of *duidae* by Vaurie (1980) (2516), but he only examined the type.
[3] Includes *Philydor hylobius* Wetmore & Phelps, Jr., 1956 (2614); an erythristic specimen of this species (Vuilleumier et al., 1992) (2552).
[4] Includes *umbrinus* see Monroe (1968) (1604).

○ *Automolus rufipileatus* CHESTNUT-CROWNED FOLIAGE-GLEANER
    — *A.r.consobrinus* (P.L. Sclater, 1870)   N and W Amazonia
    — *A.r.rufipileatus* (Pelzeln, 1859)   Brazil S of the Amazon

HYLOCRYPTUS Chapman, 1919[1] M
○ *Hylocryptus erythrocephalus* HENNA-HOODED FOLIAGE-GLEANER
    — *H.e.erythrocephalus* Chapman, 1919   SW Ecuador, N Peru
    — *H.e.palamblae* J.T. Zimmer, 1935   NW Peru

○ *Hylocryptus rectirostris* (Wied, 1831) CHESTNUT-CAPPED FOLIAGE-GLEANER
                                      SC Brazil, NE Paraguay

SCLERURUS Swainson, 1827 M
○ *Sclerurus mexicanus* TAWNY-THROATED LEAFTOSSER
    — *S.m.mexicanus* P.L. Sclater, 1857   SE Mexico to Honduras
    — *S.m.pullus* Bangs, 1902   Mts. of Costa Rica to W Panama
    — *S.m.andinus* Chapman, 1914   E Panama, Colombia to Guyana
    — *S.m.obscurior* E. Hartert, 1901   W Colombia, W Ecuador
    — *S.m.peruvianus* Chubb, 1919   W Amazonia (mainly foothills)
    — *S.m.macconnelli* Chubb, 1919   E Venezuela, the Guianas, NE Brazil
    — *S.m.bahiae* Chubb, 1919   E Brazil

○ *Sclerurus rufigularis* SHORT-BILLED LEAFTOSSER
    — *S.r.fulvigularis* Todd, 1920   E and S Venezuela to French Guiana, NE Brazil
    — *S.r.furfurosus* Todd, 1948[2]   C Brazil (N of the Amazon)
    — *S.r.brunnescens* Todd, 1948   NW Brazil, SE Colombia, E Ecuador, NE Peru
    — *S.r.rufigularis* Pelzeln, 1869   E Peru, E Bolivia, W Brazil

○ *Sclerurus albigularis* GREY-THROATED LEAFTOSSER
    — *S.a.canigularis* Ridgway, 1889   Foothills of Costa Rica, W Panama
    — *S.a.propinquus* Bangs, 1899   N Colombia (Sta. Marta Mts.)
    — *S.a.albigularis* Sclater & Salvin, 1869[3]   E Colombia, N Venezuela, Trinidad, Tobago I.
    — *S.a.zamorae* Chapman, 1923   Andean foothills E Ecuador and C Peru
    — *S.a.albicollis* Carriker, 1935   Andean foothills N Bolivia and SE Peru
    — *S.a.kempffi* Kratter, 1997   E Bolivia and SW Brazil (1303)

○ *Sclerurus caudacutus* BLACK-TAILED LEAFTOSSER
    — *S.c.caudacutus* (Vieillot, 1816)   The Guianas
    — *S.c.insignis* J.T. Zimmer, 1934   S Venezuela, N Brazil
    — *S.c.brunneus* P.L. Sclater, 1857   W Amazonia
    — *S.c.olivascens* Cabanis, 1873   Peru, Bolivia
    — *S.c.pallidus* J.T. Zimmer, 1934   C Brazil (S of the Amazon)
    — *S.c.umbretta* (M.H.K. Lichtenstein, 1823) Coast of E Brazil

○ *Sclerurus scansor* RUFOUS-BREASTED LEAFTOSSER
    — *S.s.cearensis* E. Snethlage, 1924   NE Brazil
    — *S.s.scansor* (Ménétriés, 1835)   C and SE Brazil, E Paraguay, NE Argentina

○ *Sclerurus guatemalensis* SCALY-THROATED LEAFTOSSER
    — *S.g.guatemalensis* (Hartlaub, 1844)   S Mexico to C Panama
    — *S.g.salvini* Salvadori & Festa, 1899   E Panama, W Colombia, W Ecuador
    — *S.g.ennosiphyllus* Wetmore, 1951   C Colombia (2594)

LOCHMIAS Swainson, 1827 M
○ *Lochmias nematura* SHARP-TAILED STREAMCREEPER
    — *L.n.nelsoni* Aldrich, 1945   E Panama
    — *L.n.chimantae* Phelps & Phelps, Jr., 1947 Tepui region of SE Venezuela
    — *L.n.castanonotus* Chubb, 1918   Tepui region of E Venezuela
    — *L.n.sororius* Sclater & Salvin, 1873   Coastal range of N Venezuela, Andes of Colombia,
                                        Ecuador, N Peru

[1] Submerged in *Automolus* in last Edition.
[2] Requires confirmation.
[3] Includes *kunanensis* Aveledo & Ginés, 1950 (64), see Kratter (1997) (1303).

— *L.n.obscuratus* Cabanis, 1873   Andes from C Peru to NW Argentina

— *L.n.nematura* (M.H.K. Lichtenstein, 1823)[1]   SE Brazil, E Paraguay, Uruguay, NE Argentina

HELIOBLETUS Reichenbach, 1853[2] M

○ ***Heliobletus contaminatus*** Sharp-billed Treehunter

    — *H.c.contaminatus* Berlepsch, 1885   SE Brazil

    — *H.c.camargoi* Cardoso da Silva & Stotz, 1992 SE Brazil, E Paraguay, NE Argentina (298)

XENOPS Illiger, 1811 M

○ ***Xenops milleri*** (Chapman, 1914) Rufous-tailed Xenops   Amazonia, S Venezuela, the Guianas

○ ***Xenops tenuirostris*** Slender-billed Xenops

    — *X.t.acutirostris* Chapman, 1923   SE Colombia, S Venezuela, E Ecuador, NE Peru

    — *X.t.hellmayri* Todd, 1925   French Guiana, Surinam

    — *X.t.tenuirostris* Pelzeln, 1859   SE Peru, N Bolivia, W Brazil (south of the Amazon)

◑ ***Xenops minutus*** Plain Xenops

    — *X.m.mexicanus* P.L. Sclater, 1857   S Mexico to Honduras

    ⌐ *X.m.ridgwayi* Hartert & Goodson, 1917 Nicaragua to C Panama

    — *X.m.littoralis* P.L. Sclater, 1862   E Panama to NW Colombia, W Ecuador and NW Peru

    — *X.m.neglectus* Todd, 1913   NE Colombia, N Venezuela

    — *X.m.remoratus* J.T. Zimmer, 1935   E Colombia, S Venezuela, NW Brazil

    — *X.m.ruficaudus* (Vieillot, 1816)   E Colombia, Venezuela, the Guianas, N Brazil

    — *X.m.olivaceus* Aveledo & Pons, 1952   NE Colombia (70)

    — *X.m.obsoletus* J.T. Zimmer, 1924   E Ecuador, E Peru, N Bolivia, W Brazil

    — *X.m.genibarbis* Illiger, 1811   C Brazil (south of the Amazon)

    — *X.m.minutus* (Sparrman, 1788)   E and SE Brazil, E Paraguay, NE Argentina

○ ***Xenops rutilans*** Streaked Xenops

    — *X.r.septentrionalis* J.T. Zimmer, 1929 Costa Rica, W Panama

    — *X.r.incomptus* Wetmore, 1970   Cerro Pirre (E Panama) (2607)

    — *X.r.heterurus* Cabanis & Heine, 1859   Colombia, NE Ecuador, N Venezuela, Trinidad

    — *X.r.perijanus* Phelps & Phelps, Jr., 1954 Perijá Mtns (NE Colombia and NW Venezuela) (1854)

    — *X.r.phelpsi* Meyer de Schauensee, 1959 N Colombia (1571)

    — *X.r.guayae* Hellmayr, 1920   Lowland W Ecuador, NW Peru

    — *X.r.peruvianus* J.T. Zimmer, 1935   Andean foothills E Ecuador to S Peru

    — *X.r.connectens* Chapman, 1919   Andean foothills N Bolivia to NW Argentina

    — *X.r.purusianus* Todd, 1925   S Amazonian Brazil

    — *X.r.chapadensis* J.T. Zimmer, 1935   E Bolivia, SW Brazil

    — *X.r.rutilans* Temminck, 1821   SE Brazil, Paraguay, NE Argentina

MEGAXENOPS Reiser, 1905 M

○ ***Megaxenops parnaguae*** Reiser, 1905 Great Xenops   E Brazil

PYGARRHICHAS Burmeister, 1837 M

◕ ***Pygarrhichas albogularis*** (P.P. King, 1831) White-throated Treerunner

                                      C and S Chile, SW Argentina

DENDROCOLAPTIDAE WOODCREEPERS[3] (13: 50)

DENDROCINCLA G.R. Gray, 1840 F

○ ***Dendrocincla tyrannina*** Tyrannine Woodcreeper

    — *D.t.tyrannina* (Lafresnaye, 1851)   Colombia

    — *D.t.hellmayri* Cory, 1913   E Colombia, W Venezuela

○ ***Dendrocincla fuliginosa*** Plain-brown Woodcreeper[4]

    — *D.f.ridgwayi* Oberholser, 1904   E Central America, W Colombia, W Ecuador

---

[1] The name *nematura* is invariable as it is a noun phrase (N. David *in litt.*).

[2] Submerged in *Xenops* in last Edition.

[3] Species and subspecies limits not well understood in some genera (C. Marantz *in litt.*).

[4] We follow A.O.U. (1998: 355) (3) in keeping *turdina* as a subspecies.

__ *D.f.lafresnayei* Ridgway, 1888   N and E Colombia, NW Venezuela

__ *D.f.meruloides* (Lafresnaye, 1851)   Coast of N Venezuela, Trinidad

__ *D.f.deltana* Phelps & Phelps, Jr., 1950   Orinoco delta (1848)

__ *D.f.barinensis* Phelps & Phelps, Jr., 1949   C Venezuela

__ *D.f.phaeochroa* Berlepsch & Hartert, 1902   S Colombia, E Ecuador, E Peru

__ *D.f.neglecta* Todd, 1948   W Brazil

__ *D.f.atrirostris* (d'Orbigny & Lafresnaye, 1838)   NE Bolivia, SW Brazil

__ *D.f.fuliginosa* (Vieillot, 1818)   E Venezuela, the Guianas, N Brazil

__ *D.f.rufoolivacea* Ridgway, 1888   EC Brazil

__ *D.f.taunayi* Pinto, 1939   NE Brazil

__ *D.f.trumaii* Sick, 1950   C Brazil (2265)

__ *D.f.turdina* (M.H.K. Lichtenstein, 1820)   E and S Brazil, E Paraguay, NE Argentina

○ **Dendrocincla anabatina** TAWNY-WINGED WOODCREEPER

__ *D.a.anabatina* P.L. Sclater, 1859   S Mexico to Costa Rica

__ *D.a.typhla* Oberholser, 1904   SE Mexico

○ **Dendrocincla merula** WHITE-CHINNED WOODCREEPER

__ *D.m.bartletti* Chubb, 1919   S Venezuela, W Brazil, NE Paraguay

__ *D.m.merula* (M.H.K. Lichtenstein, 1820)   The Guianas, N Brazil

__ *D.m.obidensis* Todd, 1948   Brazil

__ *D.m.remota* Todd, 1925   EC Bolivia

__ *D.m.olivascens* J.T. Zimmer, 1934   Brazil

__ *D.m.castanoptera* Ridgway, 1888   C Brazil

__ *D.m.badia* J.T. Zimmer, 1934   C Brazil

○ **Dendrocincla homochroa** RUDDY WOODCREEPER

__ *D.h.homochroa* (P.L. Sclater, 1859)   S Mexico to Honduras

__ *D.h.acedesta* Oberholser, 1904   SW Nicaragua, W Costa Rica

__ *D.h.ruficeps* Sclater & Salvin, 1868   Panama, W Venezuela

__ *D.h.meridionalis* Phelps & Phelps, Jr., 1953   NW Venezuela, NE Colombia (1852)

DECONYCHURA Cherrie, 1891 F

○ **Deconychura longicauda** LONG-TAILED WOODCREEPER

__ *D.l.typica* Cherrie, 1891   SW Costa Rica, W Panama

__ *D.l.darienensis* Griscom, 1929   E Panama

__ *D.l.minor* Todd, 1917   N Colombia

__ *D.l.longicauda* (Pelzeln, 1868)   The Guianas, N Brazil

__ *D.l.connectens* J.T. Zimmer, 1929   NW Amazonia

__ *D.l.pallida* J.T. Zimmer, 1929   SE Peru, N Bolivia, W Brazil

__ *D.l.zimmeri* Pinto, 1974   E Para, Brazil (1894)

○ **Deconychura stictolaema** SPOT-THROATED WOODCREEPER

__ *D.s.clarior* J.T. Zimmer, 1929   French Guiana, NE Brazil

__ *D.s.secunda* Hellmayr, 1904   E Ecuador, NE Peru, W Brazil, S Venezuela

__ *D.s.stictolaema* (Pelzeln, 1868)   C Brazil

SITTASOMUS Swainson, 1827 M

● **Sittasomus griseicapillus** OLIVACEOUS WOODCREEPER

__ *S.g.harrisoni* Sutton, 1955   SW Tamaulipas and E San Luis Potosí to Puebla (NE Mexico) (2377)

__ *S.g.jaliscensis* Nelson, 1900   Jalisco (SW Mexico)

⊥ *S.g.sylvioides* Lafresnaye, 1850   Veracruz and Chiapas (SE Mexico) to Costa Rica

__ *S.g.gracileus* Bangs & Peters, 1928   Yucatán Pen. (E Mexico) and Belize

__ *S.g.levis* Bangs, 1902   W Panama, N Colombia

__ *S.g.veraguensis* Aldrich, 1937   E Panama

__ *S.g.aequatorialis* Ridgway, 1891   W Ecuador, NW Peru

__ *S.g.perijanus* Phelps & Gilliard, 1940   NW Venezuela

__ *S.g.tachirensis* Phelps & Phelps, Jr., 1956   W Venezuela (1858)

__ *S.g.griseus* Jardine, 1847   Coast of N Venezuela, Tobago I.

__ *S.g.enochrus* Wetmore, 1970   Colombia (2607)

__ *S.g.amazonus* Lafresnaye, 1850	SE Colombia, S Venezuela, E Ecuador, W Brazil
__ *S.g.axillaris* J.T. Zimmer, 1934	E Venezuela, N Brazil
__ *S.g.viridis* Carriker, 1935	N and E Bolivia
__ *S.g.viridior* Todd, 1948	E Bolivia
__ *S.g.transitivus* Pinto & Camargo, 1948	C Brazil
__ *S.g.griseicapillus* (Vieillot, 1818)	S Bolivia, W Brazil, W Paraguay, NW Argentina
__ *S.g.reiseri* Hellmayr, 1917	NE Brazil
__ *S.g.olivaceus* Wied, 1831	E Brazil
__ *S.g.sylviellus* (Temminck, 1821)	SE Brazil, NE Argentina

GLYPHORYNCHUS [1] Wied, 1831 M
⬤ *Glyphorynchus spirurus* WEDGE-BILLED WOODCREEPER

__ *G.s.pectoralis* Sclater & Salvin, 1860	S Mexico to Nicaragua
✔ *G.s.sublestus* J.L. Peters, 1929	Costa Rica to W Ecuador, W Venezuela
__ *G.s.subrufescens* Todd, 1948	W Colombia
__ *G.s.integratus* J.T. Zimmer, 1946	NE Colombia
__ *G.s.rufigularis* J.T. Zimmer, 1934	E Colombia, S Venezuela, N Ecuador
__ *G.s.amacurensis* Phelps & Phelps, Jr., 1952	NE Venezuela (1850)
__ *G.s.coronobscurus* Phelps & Phelps, Jr., 1955	S Venezuela (1856)
__ *G.s.spirurus* (Vieillot, 1819)	E Venezuela, the Guianas, N Brazil
__ *G.s.pallidulus* Wetmore, 1970	Panama (2607)
__ *G.s.castelnaudii* Des Murs, 1856	E Ecuador, N Peru
__ *G.s.albigularis* Chapman, 1923	SE Peru, N Bolivia
__ *G.s.inornatus* J.T. Zimmer, 1934	C Brazil
__ *G.s.cuneatus* (M.H.K. Lichtenstein, 1820)	EC Brazil
__ *G.s.paraensis* Pinto, 1974	Belem (Brazil) (1894)

DRYMORNIS Eyton, 1853 M
◯ *Drymornis bridgesii* (Eyton, 1849) SCIMITAR-BILLED WOODHEWER | Paraguay, Uruguay, Argentina

NASICA Lesson, 1830 M
◯ *Nasica longirostris* (Vieillot, 1818) LONG-BILLED WOODCREEPER | W Amazonia to French Guiana

DENDREXETASTES Eyton, 1851 M
◯ *Dendrexetastes rufigula* CINNAMON-THROATED WOODCREEPER

__ *D.r.devillei* (Lafresnaye, 1850)	W Amazonia
__ *D.r.rufigula* (Lesson, 1844)	The Guianas, N Brazil
__ *D.r.moniliger* J.T. Zimmer, 1934	W Brazil
__ *D.r.paraensis* Lorenz, 1895	NE Brazil

HYLEXETASTES P.L. Sclater, 1889 M
◯ *Hylexetastes perrotii* RED-BILLED WOODCREEPER[2]

__ *H.p.perrotii* (Lafresnaye, 1844)	E Venezuela, Guyana, French Guiana, N Brazil
__ *H.p.uniformis* Hellmayr, 1909	N Brazil
__ *H.p.brigidai* (Cardoso da Silva, Novaes & Oren, 1995)	N Brazil (295)

◯ *Hylexetastes stresemanni* BAR-BELLIED WOODCREEPER

__ *H.s.insignis* J.T. Zimmer, 1934	NW Brazil
__ *H.s.stresemanni* E. Snethlage, 1925	NW Brazil
__ *H.s.undulatus* Todd, 1925	E Peru, W Brazil

XIPHOCOLAPTES Lesson, 1840 M
◯ *Xiphocolaptes promeropirhynchus* STRONG-BILLED WOODCREEPER

__ *X.p.omiltemensis* Nelson, 1903	SW Mexico
__ *X.p.sclateri* Ridgway, 1890	S Mexico
__ *X.p.emigrans* Sclater & Salvin, 1859	S Mexico to N Nicaragua
__ *X.p.costaricensis* Ridgway, 1889	Costa Rica

[1] Original spelling verified.
[2] Ridgely & Tudor (1994) (2019) split this, but we are unaware of detailed evidence yet published.

   — *X.p.panamensis* Griscom, 1927   S Panama
   — *X.p.rostratus* Todd, 1917   N Colombia
   — *X.p.sanctaemartae* Hellmayr, 1925   N Colombia, W Venezuela
   — *X.p.virgatus* Ridgway, 1890   C Colombia
   — *X.p.macarenae* Blake, 1959   WC Colombia (160)
   — *X.p.promeropirhynchus* (Lesson, 1840)   EC Colombia, W Venezuela
   — *X.p.procerus* Cabanis & Heine, 1859   N Venezuela
   — *X.p.tenebrosus* Zimmer & Phelps, 1948   E Venezuela
   — *X.p.neblinae* Phelps & Phelps, Jr., 1955   S Venezuela (1856)
   — *X.p.ignotus* Ridgway, 1890   W Ecuador
   — *X.p.crassirostris* Taczanowski & Berlepsch, 1885   SW Ecuador, NW Peru
   — *X.p.compressirostris* Taczanowski, 1882   N Peru
   — *X.p.phaeopygus* Belepsch & Stolzmann, 1896   E Peru
   — *X.p.solivagus* Bond, 1950   E Peru
   — *X.p.lineatocephalus* (G.R. Gray, 1847)   SE Peru, N and W Bolivia
   — *X.p.orenocensis* Berlepsch & Hartert, 1902   Venezuela, E Ecuador, E Peru, W Brazil
   — *X.p.berlepschi* E. Snethlage, 1908   E Peru, W Brazil
   — *X.p.paraensis* Pinto, 1945   C Brazil
   — *X.p.obsoletus* Todd, 1917   N and E Bolivia

○ *Xiphocolaptes albicollis* WHITE-THROATED WOODCREEPER[1]
   — *X.a.bahiae* Cory, 1919   NE Brazil
   — *X.a.villanovae* Lima, 1920[2]   E Brazil (1362)
   — *X.a.albicollis* (Vieillot, 1818)   SE Brazil, E Paraguay, NE Argentina

○ *Xiphocolaptes falcirostris* MOUSTACHED WOODCREEPER[3]
   — *X.f.falcirostris* (Spix, 1824)   NE Brazil
   — *X.f.franciscanus* E. Snethlage, 1927   C Brazil

○ *Xiphocolaptes major* GREAT RUFOUS WOODCREEPER
   — *X.m.remoratus* Pinto, 1945   C Brazil
   — *X.m.castaneus* Ridgway, 1890   S Brazil, E and S Bolivia, NW Argentina
   — *X.m.estebani* Cardoso da Silva & Oren, 1991   NW Argentina (296)
   — *X.m.major* (Vieillot, 1818)   Paraguay, N Argentina

DENDROCOLAPTES Hermann, 1804 M
○ *Dendrocolaptes sanctithomae* NORTHERN BARRED WOODCREEPER[4]
   — *D.s.sheffleri* Binford, 1965   SW Mexico (149)
   — *D.s.sanctithomae* (Lafresnaye, 1852)[5]   Pacific Oaxaca and Guerrero (Mexico) along Pacific
                                     coast to NW Colombia
   — *D.s.hesperius* Bangs, 1907   SW Costa Rica and NW Panama
   — *D.s.punctipectus* Phelps & Gilliard, 1940[6] NE Colombia to NW Venezuela

○ *Dendrocolaptes certhia* AMAZONIAN BARRED WOODCREEPER
   — *D.c.certhia* (Boddaert, 1783)   S Venezuela, the Guianas, N Brazil
   — *D.c.radiolatus* Sclater & Salvin, 1868   SE Colombia, E Ecuador, NE Peru, NW Brazil
   — *D.c.juruanus* H. von Ihering, 1905   E Peru, E Bolivia, W Brazil
   — *D.c.polyzonus* Todd, 1913   N Bolivia, SE Peru
   — *D.c.medius* Todd, 1920   NE Brazil
   — *D.c.concolor* Pelzeln, 1868[7]   NE Bolivia and SC Brazil[8]

○ *Dendrocolaptes hoffmannsi* Hellmayr, 1909 HOFFMANNS'S WOODCREEPER
                                     SC Brazil (S of the Amazon between Rio Madeira and
                                     Rio Xingu)

[1] Although Ridgely & Tudor (1994) (2019) separated *villanovae* (and we followed this in last Edition), we believe the case to elevate it from a subspecies is not proven.
[2] This name appeared on the plate but *villadenovae* appeared with the descriptive text; an erratum slip stated that the name *villanovae* was correct. For placement in this species see Cory & Hellmayr (1925: 278) (574) and Pinto (1978: 285) (1895).
[3] In last Edition we treated *franciscanus* as a separate species; Ridgely & Tudor (1994) (2019) associated it with this species after comments by Texiera *et al.* (1989) (2394); see also Cardoso da Silva & Oren (1997) (297).
[4] Separation from *certhia* follows Willis (1992) (2669); see also Marantz (1997) (1423).
[5] Includes *nigrirostris* and *colombianus* see Marantz (1997) (1423) and *legtersi* Paynter, 1954 (1806); see Parkes (1999) (1784).
[6] Includes *hyleorus* see Marantz (1997) (1423).
[7] Includes *ridgwayi* see Marantz (1997) (1423).
[8] Sibley & Monroe (1990) (2262) reduced this to a subspecies. Marantz (1997) (1423) confirmed this.

⭘ *Dendrocolaptes picumnus* BLACK-BANDED WOODCREEPER
   — *D.p.puncticollis* Sclater & Salvin, 1868   Mts. of S Mexico to Honduras
   — *D.p.costaricensis* Ridgway, 1909[1]   Mts. of Costa Rica, W Panama
   — *D.p.multistrigatus* Eyton, 1851   Andes of E Colombia, W Venezuela
   — *D.p.seilerni* Hartert & Goodson, 1917   N Colombia (Santa Marta Mts.), N Venezuela (coastal ranges)
   — *D.p.picumnus* M.H.K. Lichtenstein, 1820   E Venezuela, the Guianas, N Brazil
   — *D.p.validus* Tschudi, 1844   W Amazonia
   — *D.p.transfasciatus* Todd, 1925   SC Brazil
   — *D.p.olivaceus* J.T. Zimmer, 1934   Andean foothills of C Bolivia
   — *D.p.pallescens* Pelzeln, 1868[2]   SW Brazil, S Bolivia, Paraguay
   — *D.p.casaresi* Steullet & Deautier, 1950   NW Argentina

⭘ *Dendrocolaptes platyrostris* PLANALTO WOODCREEPER
   — *D.p.intermedius* Berlepsch, 1883   C and NE Brazil, NE Paraguay
   — *D.p.platyrostris* Spix, 1824   SE Brazil, Paraguay, NE Argentina

XIPHORHYNCHUS Swainson, 1827 M
⭘ *Xiphorhynchus picus* STRAIGHT-BILLED WOODCREEPER
   — *X.p.extimus* (Griscom, 1927)   S Panama
   — *X.p.dugandi* (Wetmore & Phelps, 1946)   N Colombia
   — *X.p.picirostris* (Lafresnaye, 1847)   NW Colombia, W Venezuela
   — *X.p.saturatior* (Hellmayr, 1925)   E Colombia, W Venezuela
   — *X.p.borreroi* Meyer de Schauensee, 1959   SW Colombia (1571)
   — *X.p.choicus* (Wetmore & Phelps, 1946)   N Venezuela
   — *X.p.paraguanae* Phelps & Phelps, Jr., 1962   NW Venezuela (1864)
   — *X.p.longirostris* (Richmond, 1896)   Margarita I.
   — *X.p.altirostris* (Léotaud, 1866)   Trinidad
   — *X.p.phalara* (Wetmore, 1939)   S Venezuela
   — *X.p.deltanus* Phelps & Phelps, Jr., 1952   NE Venezuela (1850)
   — *X.p.picus* (J.F. Gmelin, 1788)   E Colombia to the Guianas, N Brazil
   — *X.p.duidae* (J.T. Zimmer, 1934)   S Venezuela, NW Brazil
   — *X.p.peruvianus* (J.T. Zimmer, 1934)   E Peru, W Brazil, N Bolivia
   — *X.p.kienerii* (Des Murs, 1856)   W Brazil
   — *X.p.rufescens* (Todd, 1948)   C Brazil
   — *X.p.bahiae* (Bangs & Penard, 1921)   NE Brazil

⭘ *Xiphorhynchus necopinus* (J.T. Zimmer, 1934) ZIMMER'S WOODCREEPER
                                                            C and NE Brazil

⭘ *Xiphorhynchus obsoletus* STRIPED WOODCREEPER
   — *X.o.palliatus* (Des Murs, 1856)   W Amazonia, Bolivia
   — *X.o.notatus* (Eyton, 1853)   E Colombia, SW Venezuela, NW Brazil
   — *X.o.obsoletus* (M.H.K. Lichtenstein, 1820)   E Venezuela to French Guiana, N Brazil
   — *X.o.caicarae* Zimmer & Phelps, 1955   NE Venezuela (2729)

⭘ *Xiphorhynchus fuscus* LESSER WOODCREEPER[3]
   — *X.f.atlanticus* (Cory, 1916)   E Brazil
   — *X.f.brevirostris* (Pinto, 1938)   NE Brazil
   — *X.f.tenuirostris* (M.H.K. Lichtenstein, 1820) E Brazil
   — *X.f.fuscus* (Vieillot, 1818)   SE Brazil, E Paraguay, NE Argentina

⭘ *Xiphorhynchus ocellatus* OCELLATED WOODCREEPER
   — *X.o.napensis* Chapman, 1924   SE Colombia, E Ecuador, NE Peru
   — *X.o.lineatocapilla* (Berlepsch & Leverkühn, 1890)[4] E Venezuela
   — *X.o.ocellatus* (Spix, 1824)   E Colombia, S Venezuela, NE Peru, NW Brazil
   — *X.o.perplexus* J.T. Zimmer, 1934   NE Peru, W Brazil
   — *X.o.chunchotambo* (Tschudi, 1844)   E Peru
   — *X.o.brevirostris* J.T. Zimmer, 1934   SE Peru, NE Bolivia

---

[1] Includes *veraguensis* see Marantz (1997) (1423).
[2] Includes *extimus* see Marantz (1997) (1423) (the type locality of which is open to question).
[3] For reasons to transfer this species here from the genus *Lepidocolaptes* see Garcia-Moreno & Cardoso da Silva (1997) (910).
[4] For correction of spelling see David & Gosselin (2002) (613).

○ *Xiphorhynchus elegans* Elegant Woodcreeper[1]

  — *X.e.buenavistae* J.T. Zimmer, 1948   E Colombia
  — *X.e.ornatus* J.T. Zimmer, 1934   SE Colombia, E Ecuador, NE Peru, W Brazil
  — *X.e.insignis* (Hellmayr, 1905)   EC Peru
  — *X.e.juruanus* (H. von Ihering, 1905)   SE Peru, NE Bolivia, W Brazil
  — *X.e.elegans* (Pelzeln, 1868)   SW Brazil

○ *Xiphorhynchus spixii* (Lesson, 1830) Spix's Woodcreeper   SC Brazil

○ *Xiphorhynchus pardalotus* Chestnut-rumped Woodcreeper

  — *X.p.caurensis* Todd, 1948   SE Venezuela, W Guyana
  — *X.p.pardalotus* (Vieillot, 1818)   The Guianas, N Brazil

● *Xiphorhynchus guttatus* Buff-throated Woodcreeper[2]

  — *X.g.polystictus* (Salvin & Godman, 1883)   E Columbia to NE Brazil (outside the Amazon drainage)
  — *X.g.connectens* Todd, 1948   NE Brazil (Amazon drainage)
  — *X.g.guttatoides* (Lafresnaye, 1850)   W Amazonia
  — *X.g.vicinalis* Todd, 1948   Western SC Amazonian Brazil
  — *X.g.eytoni* (P.L. Sclater, 1854)   Eastern SC Amazonian Brazil
  — *X.g.gracilirostris* Pinto & Camargo, 1957   E Brazil (Ceará) (1898)
  — *X.g.dorbignyanus* (Lafresnaye, 1850)   NE Bolivia, SW Brazil
  — *X.g.guttatus* (M.H.K. Lichtenstein, 1820)   Atlantic forest of EC Brazil

○ *Xiphorhynchus susurrans* Cocoa Woodcreeper

  — *X.s.confinis* (Bangs, 1903)   E Guatemala, N Honduras
  — *X.s.costaricensis* (Ridgway, 1888)   SW Honduras to Panama
  — *X.s.marginatus* Griscom, 1927   E Panama
  — *X.s.nanus* (Lawrence, 1863)[3]   E Panama, N and E Colombia, W Venezuela
  — *X.s.rosenbergi* Bangs, 1910   W Colombia
  — *X.s.susurrans* (Jardine, 1847)   NE Venezuela, Trinidad, Tobago I.
  — *X.s.jardinei* (Dalmas, 1900)   NE Venezuela
  — *X.s.margaritae* Phelps & Phelps, Jr., 1949   Margarita I.

○ *Xiphorhynchus flavigaster* Ivory-billed Woodcreeper

  — *X.f.tardus* Bangs & Peters, 1928   NW Mexico
  — *X.f.mentalis* (Lawrence, 1867)   W Mexico
  — *X.f.flavigaster* Swainson, 1827   SW Mexico
  — *X.f.saltuarius* Wetmore, 1942[4]   NE Mexico
  — *X.f.yucatanensis* Ridgway, 1909   Yucatán Pen., Meco I.
  — *X.f.ascensor* Wetmore & Parkes, 1962   S Mexico (2610)
  — *X.f.eburneirostris* (Des Murs, 1847)   SE Mexico to NW Costa Rica
  — *X.f.ultimus* Bangs & Griscom, 1932   NW Costa Rica

○ *Xiphorhynchus lachrymosus* Black-striped Woodcreeper

  — *X.l.lachrymosus* (Lawrence, 1862)   E Nicaragua to W Ecuador
  — *X.l.alarum* Chapman, 1915   N Colombia

● *Xiphorhynchus erythropygius* Spotted Woodcreeper

  — *X.e.erythropygius* (P.L. Sclater, 1859)   S Mexico
  — *X.e.parvus* Griscom, 1937   S Mexico to N Nicaragua
  ⊥ *X.e.puncticula* (Ridgway, 1889)   S Nicaragua to W Panama
  — *X.e.insolitus* Ridgway, 1909   E Panama, NW Colombia
  — *X.e.aequatorialis* (Berlepsch & Taczanowski, 1884)   W Colombia, E Ecuador

○ *Xiphorhynchus triangularis* Olive-backed Woodcreeper

  — *X.t.triangularis* (Lafresnaye, 1842)   Colombia, W Venezuela, E Ecuador, N Peru
  — *X.t.hylodromus* Wetmore, 1939   N Venezuela
  — *X.t.intermedius* Carriker, 1935   C and S Peru
  — *X.t.bangsi* Chapman, 1919   Bolivia

---

[1] For treatment of *X. spixii* as monotypic (with transfer of associated forms to *X. elegans*) see Haffer (1997) (1054).
[2] For separation of *X. susurrans* see Willis (1983) (2667).
[3] Includes *demonstratus* see Phelps & Phelps (1963: 35) (1865) (Rodner & Restall, *in litt.*).
[4] Includes *striatigularis* see Winker (1995) (2673).

LEPIDOCOLAPTES Reichenbach, 1853 M

○ *Lepidocolaptes leucogaster* W<small>HITE-STRIPED</small> W<small>OODCREEPER</small>

    — *L.l.umbrosus* R.T. Moore, 1934   NW Mexico
    — *L.l.leucogaster* (Swainson, 1827)   C and S Mexico

● *Lepidocolaptes souleyetii* S<small>TREAK-HEADED</small> W<small>OODCREEPER</small>

    — *L.s.guerrerensis* van Rossem, 1939   W Mexico
    — *L.s.insignis* (Nelson, 1897)   SE Mexico to N Honduras
    ⤵ *L.s.compressus* (Cabanis, 1861)   S Mexico to W Panama
    — *L.s.lineaticeps* (Lafresnaye, 1850)   E Panama, N Colombia, W Venezuela
    — *L.s.littoralis* (Hartert & Goodson, 1917)   N Colombia to Guyana, Trinidad, N Brazil
    — *L.s.uaireni* Phelps & Phelps, Jr., 1950   SE Venezuela
    — *L.s.esmeraldae* Chapman, 1923   SW Colombia, W Ecuador
    — *L.s.souleyetii* (Des Murs, 1849)   SW Ecuador, NW Peru

○ *Lepidocolaptes angustirostris* N<small>ARROW-BILLED</small> W<small>OODCREEPER</small>

    — *L.a.griseiceps* Mees, 1974   Sipaliwini (Surinam) (1526)
    — *L.a.coronatus* (Lesson, 1830)   N Brazil
    — *L.a.bahiae* (Hellmayr, 1903)   NE Brazil
    — *L.a.bivittatus* (M.H.K. Lichtenstein, 1822)   E Bolivia, C and E Brazil
    — *L.a.hellmayri* Naumburg, 1925   WC Bolivia
    — *L.a.certhiolus* (Todd, 1913)   C Bolivia to Paraguay, NW Argentina
    — *L.a.dabbenei* Esteban, 1948   SW Paraguay, N Argentina
    — *L.a.angustirostris* (Vieillot, 1818)   E Paraguay, SW Brazil, N Argentina
    — *L.a.praedatus* (Cherrie, 1916)   W Uruguay, E and S Argentina

○ *Lepidocolaptes affinis* S<small>POT-CROWNED</small> W<small>OODCREEPER</small> [1]

    — *L.a.lignicida* (Bangs & Penard, 1919)   NE Mexico
    — *L.a.affinis* (Lafresnaye, 1839)   S Mexico to N Nicaragua
    — *L.a.neglectus* (Ridgway, 1909)   Costa Rica, W Panama

○ *Lepidocolaptes lacrymiger* M<small>ONTANE</small> W<small>OODCREEPER</small>

    — *L.l.sanctaemartae* (Chapman, 1912)   N Colombia
    — *L.l.sneiderni* Meyer de Schauensee, 1945   W Colombia
    — *L.l.lacrymiger* (Des Murs, 1849)   E Colombia, W Venezuela
    — *L.l.lafresnayi* (Cabanis & Heine, 1859)   N Venezuela
    — *L.l.aequatorialis* (Ménégaux, 1912)   SW Colombia, Ecuador
    — *L.l.frigidus* Meyer de Schauensee, 1951   SW Colombia (1568)
    — *L.l.warscewiczi* (Cabanis & Heine, 1859)   N and C Peru
    — *L.l.carabayae* Hellmayr, 1920   SE Peru
    — *L.l.bolivianus* (Chapman, 1919)   Bolivia

○ *Lepidocolaptes squamatus* S<small>CALED</small> W<small>OODCREEPER</small>

    — *L.s.wagleri* (Spix, 1824)   NE Brazil
    — *L.s.squamatus* (M.H.K. Lichtenstein, 1822)   E Brazil

○ *Lepidocolaptes falcinellus* (Cabanis & Heine, 1859) S<small>CALLOPED</small> W<small>OODCREEPER</small>[2]

                                               SE Brazil, Paraguay

○ *Lepidocolaptes albolineatus* L<small>INEATED</small> W<small>OODCREEPER</small>

    — *L.a.albolineatus* (Lafresnaye, 1846)   E Venezuela, the Guianas, N Brazil
    — *L.a.duidae* J.T. Zimmer, 1934   S Venezuela, NW Brazil
    — *L.a.fuscicapillus* (Pelzeln, 1868)   SW Amazonia
    — *L.a.madeirae* (Chapman, 1919)   C Brazil
    — *L.a.layardi* (P.L. Sclater, 1873)   C Brazil

CAMPYLORHAMPHUS Bertoni, 1901 M

○ *Campylorhamphus pucherani* (Des Murs, 1849) G<small>REATER</small> S<small>CYTHEBILL</small>   S Colombia, E Ecuador

○ *Campylorhamphus trochilirostris* R<small>ED-BILLED</small> S<small>CYTHEBILL</small>

    — *C.t.brevipennis* Griscom, 1932   E Panama, W Colombia

---

[1] Reasons for separation appear in A.O.U. (1998: 360) (3).
[2] For reasons to separate this from *L. squamatus* see Cardoso da Silva & Straube (1996) (299).

—— *C.t.venezuelensis* (Chapman, 1889)	N Colombia, N Venezuela
—— *C.t.thoracicus* (P.L. Sclater, 1860)	SW Colombia, W Ecuador
—— *C.t.zarumillanus* Stolzmann, 1926	NW Peru
—— *C.t.napensis* Chapman, 1925	E Ecuador, E Peru
—— *C.t.notabilis* J.T. Zimmer, 1934	W Brazil
—— *C.t.snethlageae* J.T. Zimmer, 1934	C Brazil
—— *C.t.major* Ridgway, 1911	E Brazil (Piauí and Ceará)
—— *C.t.omissus* Pinto, 1933	E Brazil (Goias, N Minas Gerais and most of Bahia)
—— *C.t.trochilirostris* (M.H.K. Lichtenstein, 1820)	E Brazil (SE Bahia)
¶ *C.t.guttistriatus* Pinto & Camargo, 1955	SE Brazil (1897)
—— *C.t.devius* J.T. Zimmer, 1934	N Bolivia
—— *C.t.lafresnayanus* (d'Orbigny, 1846)	E Bolivia, N Paraguay
—— *C.t.hellmayri* Laubmann, 1930	N Argentina

O ***Campylorhamphus falcularius*** (Vieillot, 1822) Black-billed Scythebill
SE Brazil, Paraguay, NE Argentina

● ***Campylorhamphus pusillus*** Brown-billed Scythebill

⊥ *C.p.borealis* Carriker, 1910	Costa Rica, W Panama
—— *C.p.olivaceus* Griscom, 1927	C Panama
—— *C.p.tachirensis* Phelps & Phelps, Jr., 1956	NE Colombia (1857)
¶ *C.p.guapiensis* Romero-Zambrano, 1980[1]	WC Colombia (2099)
—— *C.p.pusillus* (P.L. Sclater, 1860)	Colombia, W Ecuador

O ***Campylorhamphus procurvoides*** Curve-billed Scythebill[2]

—— *C.p.sanus* J.T. Zimmer, 1934	E Colombia, Venezuela, W Guyana, N Brazil
—— *C.p.procurvoides* (Lafresnaye, 1850)	French Guiana, N Brazil
—— *C.p.probatus* J.T. Zimmer, 1934	C Brazil (Amazonas)
—— *C.p.multostriatus* (E. Snethlage, 1907)	C Brazil (Para)

MENURIDAE LYREBIRDS (1: 2)

MENURA Latham, 1802[3] F
O ***Menura alberti*** Bonaparte, 1850 Prince Albert's Lyrebird
SE Queensland, NE New South Wales

● ***Menura novaehollandiae*** Superb Lyrebird

—— *M.n.edwardi* Chisholm, 1921	SE Queensland, NE New South Wales
—— *M.n.novaehollandiae* Latham, 1802[4]	EC and SE New South Wales
—— *M.n.victoriae* Gould, 1865[5]	SE New South Wales, E Victoria

ATRICHORNITHIDAE SCRUBBIRDS (1: 2)

ATRICHORNIS Stejneger, 1885 M
O ***Atrichornis rufescens*** Rufous Scrubbird

—— *A.r.rufescens* (E.P. Ramsay, 1867)	SE Queensland, NE New South Wales
—— *A.r.ferrieri* Schodde & Mason, 1999	EC New South Wales (2189)

O ***Atrichornis clamosus*** (Gould, 1844) Noisy Scrubbird
SW Western Australia (Two Peoples Bay)

PTILONORHYNCHIDAE BOWERBIRDS (8: 18)

AILUROEDUS Cabanis, 1851 M
O ***Ailuroedus buccoides*** White-eared Catbird

—— *A.b.buccoides* (Temminck, 1836[6])	W Papuan islands, SW and NW New Guinea
—— *A.b.geislerorum* A.B. Meyer, 1891	Yapen I., N New Guinea
—— *A.b.stonii* Sharpe, 1876	SE New Guinea
—— *A.b.cinnamomeus* Mees, 1964	S New Guinea (1513)

---

[1] Recognised tentatively; not yet independently evaluated.
[2] The proposed race ¶ *successor* Todd, 1948 (2412) requires evaluation.
[3] For reasons to use 1802 not 1801 see Browning & Monroe (1991) (258).
[4] For reasons to use 1802 not 1801 see Browning & Monroe (1991) (258).
[5] For recognition of this see Schodde & Mason (1999) (2189).
[6] Mees (1994) (1541) has shown that Livr. 97 appeared in Apr. 1836 or later.

○ *Ailuroedus crassirostris* GREEN CATBIRD

   — *A.c.misoliensis* Mayr & Meyer de Schauensee, 1939   Misool I. (W Papuan Is.)

   — *A.c.arfakianus* A.B. Meyer, 1874   Mts. of NW New Guinea (Vogelkop)

   — *A.c.jobiensis* Rothschild, 1895   Mts. of W and N New Guinea

   — *A.c.facialis* Mayr, 1936   Mts. of WC New Guinea

   — *A.c.guttaticollis* Stresemann, 1922   Mts. of EC New Guinea

   — *A.c.astigmaticus* Mayr, 1931   Mts. of NE New Guinea (Huon Pen.)

   — *A.c.melanocephalus* E.P. Ramsay, 1882   Mts. of SE New Guinea

   — *A.c.melanotis* (G.R. Gray, 1858)   Aru Is., SC New Guinea (Trans-Fly)

   — *A.c.joanae* Mathews, 1941[1]   E Cape York Pen.

   — *A.c.maculosus* E.P. Ramsay, 1874   NE Queensland

   — *A.c.crassirostris* (Paykull, 1815)[2]   SE Queensland, E New South Wales

SCENOPOEETES Coues, 1891 M

○ *Scenopoeetes dentirostris* (E.P. Ramsay, 1876) TOOTH-BILLED BOWERBIRD

     Mts. of NE Queensland

ARCHBOLDIA Rand, 1940 F

○ *Archboldia papuensis* ARCHBOLD'S BOWERBIRD

   — *A.p.papuensis* Rand, 1940   Mts. of WC New Guinea

   — *A.p.sanfordi* Mayr & Gilliard, 1950   Mts. of EC New Guinea

AMBLYORNIS Elliot, 1872 F[3]

○ *Amblyornis inornata* (Schlegel, 1871) VOGELKOP BOWERBIRD   Mts. of NW New Guinea (Vogelkop)

○ *Amblyornis macgregoriae* MACGREGOR'S BOWERBIRD

   — *A.m.mayri* E. Hartert, 1930   Mts. of WC New Guinea

   — *A.m.kombok* Schodde & McKean, 1973   Mts. of EC New Guinea (2195)

   — *A.m.amati* T.K. Pratt, 1982   Adelbert Mts. (NE New Guinea) (1921)

   — *A.m.germana* Rothschild, 1910   Mts. of Huon Pen. (NE New Guinea)

   — *A.m.macgregoriae* De Vis, 1890   Mts. of EC New Guinea

   — *A.m.nubicola* Schodde & McKean, 1973   Mts. of extreme SE New Guinea (2195)

   — *A.m.lecroyae* C.B. & D.W. Frith, 1997   Mt. Bosavi (SC New Guinea) (886)

○ *Amblyornis subalaris* Sharpe, 1884 STREAKED BOWERBIRD   Mts. of SE New Guinea

○ *Amblyornis flavifrons* Rothschild, 1895 GOLDEN-FRONTED BOWERBIRD   Foia Mts. (western NC New Guinea)

PRIONODURA De Vis, 1883 F

○ *Prionodura newtoniana* De Vis, 1883 GOLDEN BOWERBIRD   Mts. of NE Queensland

SERICULUS Swainson, 1825 M

○ *Sericulus aureus* FLAME BOWERBIRD

   — *S.a.aureus* (Linnaeus, 1758)   N and W New Guinea

   — *S.a.ardens* (D'Albertis & Salvadori, 1879)   SW to SC New Guinea

○ *Sericulus bakeri* (Chapin, 1929) FIRE-MANED BOWERBIRD   Adelbert Mts. (NE New Guinea)

○ *Sericulus chrysocephlaus* (Lewin, 1808) REGENT BOWERBIRD[4]   NE New South Wales, E Queensland

PTILONORHYNCHUS Kuhl, 1820 M

● *Ptilonorhynchus violaceus* SATIN BOWERBIRD

   ✓ *P.v.violaceus* (Vieillot, 1816)   SE Queensland to E and S Victoria

   — *P.v.minor* A.J. Campbell, 1912   NE Queensland

[1] For recognition see Schodde & Mason (1999: 622) (2189).
[2] Should not have been separated as a monotypic species in last Edition; if there were to be separation four species would probably emerge, see Schodde & Mason (1999: 623) (2189).
[3] Gender addressed by David & Gosselin (2002) (614), multiple name changes occur.
[4] For monotypic treatment see Schodde & Mason (1999: 629) (2189); *rothschildi* is a synonym.

CHLAMYDERA Gould, 1837 F

○ *Chlamydera maculata* (Gould, 1837) SPOTTED BOWERBIRD   EC Australia

○ *Chlamydera guttata* WESTERN BOWERBIRD[1]
   — *C.g.guttata* Gould, 1862   WC and C Australia
   — *C.g.carteri* Mathews, 1920[2]   NW Cape Pen., (WC Western Australia)

○ *Chlamydera nuchalis* GREAT BOWERBIRD
   — *C.n.nuchalis* (Jardine & Selby, 1830)[3]   NW to NC Australia, Melville I., Groote Eylandt
   — *C.n.orientalis* (Gould, 1879)[4]   Cape York Pen. to NE Queensland

○ *Chlamydera lauterbachi* YELLOW-BREASTED BOWERBIRD
   — *C.l.lauterbachi* (Reichenow, 1897)   NC New Guinea
   — *C.l.uniformis* Rothschild, 1931   C and SW New Guinea

○ *Chlamydera cerviniventris* Gould, 1850 FAWN-BREASTED BOWERBIRD   E to NC and SC New Guinea, NE Cape York Pen.

CLIMACTERIDAE AUSTRALIAN TREECREEPERS (2: 7)

CORMOBATES Mathews, 1922 F[5] [6]

● *Cormobates leucophaea* WHITE-THROATED TREECREEPER
   — *C.l.minor* (E.P. Ramsay, 1891)   NE Queensland
   — *C.l.intermedia* (Boles & Longmore, 1983)   EC Queensland (176)
   — *C.l.metastasis* Schodde, 1989   SE Queensland, NE New South Wales (2177)
   — *C.l.leucophaea* (Latham, 1802)[7]   SE Australia
   — *C.l.grisescens* (Mathews, 1912)   EC South Australia (Mt. Lofty Range)

○ *Cormobates placens* PAPUAN TREECREEPER
   — *C.p.placens* (P.L. Sclater, 1874)   NW New Guinea (Mts. of Vogelkop)
   — *C.p.steini* (Mayr, 1936)   NW New Guinea (Weyland Mts.)
   — *C.p.inexpectata* (Rand, 1940)   WC New Guinea
   — *C.p.meridionalis* (E. Hartert, 1907)   SE New Guinea

CLIMACTERIS Temminck, 1820 M[8]

○ *Climacteris erythrops* Gould, 1841 RED-BROWED TREECREEPER[9]   SE Queensland, E New South Wales, E Victoria

○ *Climacteris affinis* WHITE-BROWED TREECREEPER
   — *C.a.superciliosus* North, 1895   WC Australia
   — *C.a.affinis* Blyth, 1864   EC Australia

○ *Climacteris rufus* Gould, 1841 RUFOUS TREECREEPER   SW to SC Australia (to Eyre Pen.)

○ *Climacteris picumnus* BROWN TREECREEPER
   — *C.p.melanotus* Gould, 1847   Cape York Pen.
   — *C.p.picumnus* Temminck, 1824[10]   NE to inland E Australia
   — *C.p.victoriae* Mathews, 1912[11]   SE Queensland, E New South Wales to E and S
                           Victoria

○ *Climacteris melanurus* BLACK-TAILED TREECREEPER
   — *C.m.melanurus* Gould, 1843   NW to NC Australia
   — *C.m.wellsi* Ogilvie-Grant, 1909   NC Western Australia (Pilbara)

---

[1] For reasons to separate this from *C. maculata* see Schodde & Mason (1999: 634-635) (2189).
[2] For recognition see Frith & Frith (1997) (885).
[3] Includes *melvillensis* and *oweni*; the illustration of the type of *nuchalis* shows the western form, see Schodde & Mason (1999: 637) (2189).
[4] Includes *yorki* see Schodde & Mason (1999: 636) (2189).
[4] We follow Christidis & Boles (1994) (351) in the use of this genus.
[6] Gender addressed by David & Gosselin (2002) (614), multiple changes occur in specific names.
[7] For reasons to use 1802 not 1801 see Browing & Monroe (1991) (258).
[8] Gender addressed by David & Gosselin (2002) (614), multiple changes occur in specific names.
[9] Includes *olinda*; for monotypic treatment see Schodde & Mason (1999: 73) (2189).
[10] Attributable to Temminck alone not Temminck & Laugier as given by Greenway (1967: 163) (1008). See Dickinson (2001) (739).
[11] For recognition see Schodde & Mason (1999: 78) (2189).

MALURIDAE AUSTRALASIAN WRENS[1] (5: 28)

SIPODOTUS Mathews, 1928[2] M

○ *Sipodotus wallacii* WALLACE'S FAIRY-WREN

— *S.w.wallacii* (G.R. Gray, 1862) — Misool I., Yapen I., NW and N New Guinea, Goodenough Is.

— *S.w.coronatus* (Gould, 1878)[3] — C, SC and SE New Guinea, Aru Is.

MALURUS Vieillot, 1816[4] M

○ *Malurus grayi* (Wallace, 1862) BROAD-BILLED FAIRY-WREN — NW and N New Guinea

○ *Malurus campbelli* Schodde, 1982 CAMPBELL'S FAIRY-WREN[5] — SE New Guinea (2198)

○ *Malurus cyanocephalus* EMPEROR FAIRY-WREN

— *M.c.cyanocephalus* (Quoy & Gaimard, 1830) — NW and N New Guinea

— *M.c.mysorensis* (A.B. Meyer, 1874) — Biak I. (Geelvink Bay)

— *M.c.bonapartii* (G.R. Gray, 1859) — S New Guinea

○ *Malurus amabilis* Gould, 1852 LOVELY FAIRY-WREN[6] — N and NE Queensland

○ *Malurus lamberti* VARIEGATED FAIRY-WREN

— *M.l.dulcis* Mathews, 1908 — N Northern Territory (Arnhem Land)

— *M.l.rogersi* Mathews, 1912 — N Western Australia (Kimberley)

— *M.l.assimilis* North, 1901 — W, C, SC and inland E Australia

— *M.l.bernieri* Ogilvie-Grant, 1909[7] — Bernier Is. (Western Australia)

— *M.l.lamberti* Vigors & Horsfield, 1827 — SE Queensland, E New South Wales

○ *Malurus pulcherrimus* Gould, 1844 BLUE-BREASTED FAIRY-WREN — S Western Australia and SW South Australia (Eyre Pen.)

○ *Malurus elegans* Gould, 1837 RED-WINGED FAIRY-WREN — SW Western Australia

● *Malurus cyaneus* SUPERB FAIRY-WREN[8]

— *M.c.cyaneus* (Ellis, 1782) — Tasmania

— *M.c.samueli* Mathews, 1912 — Flinders I. (Bass Strait)

— *M.c.elizabethae* A.J. Campbell, 1901 — King I. (Bass Strait)

— *M.c.cyanochlamys* Sharpe, 1881 — SE Queensland to C Victoria

√ *M.c.leggei* Mathews, 1912 — SC and SE South Australia, SW Victoria

— *M.c.ashbyi* Mathews, 1912 — Kangaroo I.

○ *Malurus splendens* SPLENDID FAIRY-WREN[9]

— *M.s.splendens* (Quoy & Gaimard, 1830) — SW to C Western Australia

— *M.s.musgravi* Mathews, 1922[10] — C and SC Australia

— *M.s.melanotus* Gould, 1841 — SC Queensland, C New South Wales, NW Victoria, EC South Australia

— *M.s.emmottorum* Schodde & Mason, 1999 — C Queensland (2189)

○ *Malurus coronatus* PURPLE-CROWNED FAIRY-WREN

— *M.c.coronatus* Gould, 1858 — NW Northern Territory, N Western Australia (Kimberley)

— *M.c.macgillivrayi* Mathews, 1913 — NE Northern Territory, NW Queensland

○ *Malurus alboscapulatus* WHITE-SHOULDERED FAIRY-WREN[11]

— *M.a.alboscapulatus* A.B. Meyer, 1874 — NW New Guinea (Vogelkop)

— *M.a.aida* E. Hartert, 1930 — NC New Guinea

— *M.a.lorentzi* van Oort, 1909 — SC New Guinea

— *M.a.kutubu* Schodde & Hitchcock, 1968 — EC New Guinea

[1] Based on Rowley & Russell (1997) (2115) and Schodde & Mason (1999) (2189).
[2] Treated as a species in the genus *Malurus* in last Edition, but see Schodde (1982) (2198).
[3] The name *coronatus* is preoccupied in *Malurus*. Mayr (1986) (1486) provided the substitute name *capillatus* used in last Edition.
[4] Sequence arranged to follow Schodde (1982) (2198) and Schodde & Mason (1999) (2189).
[5] We prefer to recognise this, but see also LeCroy & Diamond (1995) (1344).
[6] For recognition as a species see Schodde (1982: 68) (2198).
[7] For recognition of this see Schodde & Mason (1999: 85-86) (2189).
[8] For recognition of *samueli*, *elizabethae*, *leggei* and *ashbyi* see Schodde & Mason (1999: 90-92) (2189).
[9] Comments on the names *callainus* and *whitei* are given by Schodde & Mason (1999: 94) (2189); *aridus* is treated as a synonym of *splendens*.
[10] For recognition see Schodde & Mason (1999: 93) (2189).
[11] Treatment follows Schodde (1982) (2198), which see for placement of synonyms *tappenbecki*, *randi*, *balim*, *dogwa* and *mafalu*.

__ *M.a.moretoni* De Vis, 1892	SE New Guinea
__ *M.a.naimii* Salvadori & D'Albertis, 1876	NE New Guinea

○ *Malurus melanocephalus* RED-BACKED FAIRY-WREN

__ *M.m.cruentatus* Gould, 1840	N Australia
__ *M.m.melanocephalus* (Latham, 1802)[1]	EC Queensland to NE New South Wales

○ *Malurus leucopterus* WHITE-WINGED FAIRY-WREN

__ *M.l.leucopterus* Dumont, 1824	Dirk Hartog I. (Western Australia)
__ *M.l.edouardi* A.J. Campbell, 1901	Barrow I. (Western Australia)
__ *M.l.leuconotus* Gould, 1865	W, C and inland S and E Australia

CLYTOMYIAS Sharpe, 1879 M

○ *Clytomyias insignis* ORANGE-CROWNED WREN

__ *C.i.insignis* Sharpe, 1879	Mts. of NW New Guinea (Vogelkop)
__ *C.i.oorti* Rothschild & Hartert, 1907	Mts. of WC to SE New Guinea

STIPITURUS Lesson, 1831 M

○ *Stipiturus malachurus* SOUTHERN EMU-WREN

__ *S.m.malachurus* (Shaw, 1798)	S Victoria, E New South Wales, SE Queensland
__ *S.m.littleri* Mathews, 1912	Tasmania
__ *S.m.polionotum* Schodde & Mason, 1999	SE South Australia, SW Victoria (2189)
__ *S.m.intermedius* Ashby, 1920	SC South Australia (Mt. Lofty Range)
__ *S.m.halmaturinus* Parsons, 1920	Kangaroo I.
__ *S.m.parimeda* Schodde & Weatherly, 1981	S Eyre Pen. (South Australia) (2197)
__ *S.m.westernensis* A.J. Campbell, 1912	SW Western Australia
__ *S.m.hartogi* Carter, 1916	Dirk Hartog I. (Western Australia)

○ *Stipiturus mallee* A.J. Campbell, 1908 MALLEE EMU-WREN   NW Victoria, SE South Australia

○ *Stipiturus ruficeps* A.J. Campbell, 1899 RUFOUS-CROWNED EMU-WREN   WC and C Australia

AMYTORNIS Stejneger, 1885 M

○ *Amytornis barbatus* GREY GRASSWREN

__ *A.b.barbatus* Favaloro & McEvey, 1968	NW New South Wales, SW Queensland (Bulloo River system)
__ *A.b.diamantina* Schodde & Christidis, 1987	EC Australia (Diamantina River system) (2181)

○ *Amytornis housei* (Milligan, 1902) BLACK GRASSWREN   N Western Australia (Kimberley)

○ *Amytornis woodwardi* E. Hartert, 1905 WHITE-THROATED GRASSWREN   N Northern Territory (Arnhem Land)

○ *Amytornis dorotheae* (Mathews, 1914) CARPENTARIAN GRASSWREN   NE Northern Territory, NW Queensland

○ *Amytornis merrotsyi* Mellor, 1913 SHORT-TAILED GRASSWREN[2]   EC and SC South Australia

○ *Amytornis striatus* STRIATED GRASSWREN

__ *A.s.rowleyi* Schodde & Mason, 1999	C Queensland (2189)
__ *A.s.whitei* Mathews, 1910	WC Western Australia
__ *A.s.striatus* (Gould, 1840)[3]	EC Western Australia, SW Northern Territory, inland South Australia, W New South Wales, NW Victoria

○ *Amytornis goyderi* (Gould, 1875) EYREAN GRASSWREN   EC Australia (Simpson-Strzelecki Desert)

○ *Amytornis textilis* THICK-BILLED GRASSWREN

__ *A.t.textilis* (Dumont, 1824)	WC Western Australia
__ *A.t.myall* (Mathews, 1916)	SW South Australia
__ *A.t.modestus* (North, 1902)	C and EC Australia

○ *Amytornis purnelli* (Mathews, 1914) DUSKY GRASSWREN   C Australia

○ *Amytornis ballarae* Condon, 1969 KALKADOON GRASSWREN[4]   NW Queensland

---

[1] For reasons to use 1802 not 1801 see Browing & Monroe (1991) (258).
[2] For reasons to elevate this from a race of *striatus* to a species see Schodde & Mason (1999: 111) (2189).
[3] Includes *oweni* see Schodde & Mason (1999: 113) (2189).
[4] For reasons to elevate this from a race of *purnelli* to a species see Schodde & Mason (1999: 119) (2189).

MELIPHAGIDAE HONEYEATERS[1] (44: 174)

MOHO Lesson, 1831 M

○ †?*Moho braccatus* (Cassin, 1855) KAUAI O-O — Kauai I. (Hawaiian Is.)

○ †*Moho apicalis* Gould, 1860 OAHU O-O — Oahu I. (Hawaiian Is.)

○ †?*Moho bishopi* (Rothschild, 1893) BISHOP'S O-O — Molokai and Maui Is. (Hawaiian Is.)

○ †*Moho nobilis* (Merrem, 1786) HAWAII O-O — Hawaii I. (Hawaiian Is.)

CHAETOPTILA P.L. Sclater, 1871 F

○ †*Chaetoptila angustipluma* (Peale, 1848) KIOEA — Hawaii I. (Hawaiian Is.)

NOTIOMYSTIS Richmond, 1908 F

○ *Notiomystis cincta* STITCH-BIRD
    — *N.c.hautura* Mathews, 1935 — Little Barrier I. (New Zealand)
    — †?*N.c.cinta* (Du Bus, 1839)[2] — North Island (New Zealand)

XANTHOTIS Reichenbach, 1852[3] M[4]

○ *Xanthotis polygrammus* SPOTTED HONEYEATER
    — *X.p.polygrammus* (G.R. Gray, 1861) — Waigeo I. (W Papuan Is.)
    — *X.p.keuhni* E. Hartert, 1930 — Misool I. (W Papuan Is.)
    — *X.p.poikilosternos* A.B. Meyer, 1875 — Salawati I., NW to SW New Guinea
    — *X.p.septentrionalis* Mayr, 1931 — N New Guinea
    — *X.p.lophotis* Mayr, 1931 — NE to SE New Guinea
    — *X.p.candidior* Mayr & Rand, 1935 — SC New Guinea

○ *Xanthotis macleayanus* (E.P. Ramsay, 1875) MACLEAY'S HONEYEATER — NE Queensland

○ *Xanthotis flaviventer* TAWNY-BREASTED HONEYEATER
    — *X.f.fusciventris* Salvadori, 1876 — Waigeo, Batanta Is. (W Papuan Is.)
    — *X.f.flaviventer* (Lesson, 1828)[5] — W Papuan islands, NW New Guinea
    — *X.f.saturatior* (Rothschild & Hartert, 1903)[6] — Aru Is., SW to eastern SC New Guinea
    — *X.f.visi* (E. Hartert, 1896)[7] — SE New Guinea (N and S slopes)
    — *X.f.madaraszi* (Rothschild & Hartert, 1903) — NE New Guinea (Huon Pen. region)
    — *X.f.philemon* Stresemann, 1921 — N New Guinea
    — *X.f.meyeri* Salvadori, 1876 — Yapen I. (Geelvink Bay)
    — *X.f.spilogaster* (Ogilvie-Grant, 1896) — Trobriand Is.
    — *X.f.filiger* (Gould, 1851) — Cape York Pen.

○ *Xanthotis provocator* (E.L. Layard, 1875) KANDAVU HONEYEATER — Kandavu I. (Fiji Is.)

LICHENOSTOMUS Cabanis, 1851[8] M

○ *Lichenostomus reticulatus* (Temminck, 1820)[9] STREAKY-BREASTED HONEYEATER — Timor

○ *Lichenostomus subfrenatus* BLACK-THROATED HONEYEATER
    — *L.s.subfrenatus* (Salvadori, 1876) — Mts. of NW New Guinea (Vogelkop)
    — *L.s.melanolaemus* (Reichenow, 1915) — Mts. of NC New Guinea
    — *L.s.utakwensis* (Ogilvie-Grant, 1915) — Mts. of SC New Guinea
    — *L.s.salvadorii* (E. Hartert, 1896) — Mts. of EC to SE New Guinea

○ *Lichenostomus obscura* OBSCURE HONEYEATER
    — *L.o.viridifrons* (Salomonsen, 1966) — Mts. of NW New Guinea (Vogelkop)
    — *L.o.obscurus* (De Vis, 1897) — Mts. of WC to SE New Guinea

---

[1] Sequence, by R. Schodde, based on that of Wolters (1979: 255-263) (2685).
[2] Du Bus is an abbreviation of Du Bus de Gisignies.
[3] Treated in the genus *Meliphaga* in last Edition. For the use of this genus see Christidis & Schodde (1993) (352).
[4] Gender addressed by David & Gosselin (2002) (614), multiple changes occur in specific names.
[5] The name *rubiensis* is treated as unidentifiable by Schodde & Mason (1999: 228) (2189).
[6] Apparently includes *tararae* see Schodde & Mason (1999: 228) (2189).
[7] The names *giulianetti* and *kumusi* are treated as unidentifiable in Schodde & Mason (1999: 228) (2189).
[8] The species herein were treated in the genus *Meliphaga* in last Edition. For reasons to use this genus see Schodde (1975) (2175) and Christidis & Schodde (1993) (352).
[9] Livraison 5 appeared in 1820 with a wrapper having this name see Froriep (1821) (888) and Dickinson (2001) (739).

○ *Lichenostomus frenatus* (E.P. Ramsay, 1875) BRIDLED HONEYEATER   NE Queensland

○ *Lichenostomus hindwoodi* (Longmore & Boles, 1983) EUNGELLA HONEYEATER
EC Queensland (1370)

● *Lichenostomus chrysops* YELLOW-FACED HONEYEATER
__ *L.c.barroni* (Mathews, 1912)[1]   NE Queensland
✓ *L.c.chrysops* (Latham, 1802)[2]   E and SE Australia
__ *L.c.samueli* (Mathews, 1912)   SC South Australia (Mt. Lofty, Flinders Range)

○ *Lichenostomus virescens* SINGING HONEYEATER
__ *L.v.cooperi* (Mathews, 1912)[3]   N Northern Territory, Melville I.
__ *L.v.sonorus* (Gould, 1841)[4]   Inland E to SC Australia
__ *L.v.virescens* (Vieillot, 1817)[5]   WC and SW Western Australia
__ *L.v.forresti* (Ingram, 1906)   WC and C Australia

○ *Lichenostomus versicolor* VARIED HONEYEATER
__ *L.v.sonoroides* (G.R. Gray, 1861)   W Papuan islands, coastal NW New Guinea
                                                          (Vogelkop)
                                                          Yapen I., coastal N New Guinea, Fergusson I.
__ *L.v.vulgaris* (Salomonsen, 1966)   W Louisiade Arch. (E New Guinea)
__ *L.v.intermedius* (Mayr & Rand, 1935)   Coastal S New Guinea, Torres Strait, coastal NE
__ *L.v.versicolor* (Gould, 1843)   Queensland

○ *Lichenostomus fasciogularis* (Gould, 1854) MANGROVE HONEYEATER   Coastal EC Queensland to NE New South Wales

○ *Lichenostomus unicolor* (Gould, 1843) WHITE-GAPED HONEYEATER   N Australia

○ *Lichenostomus flavus* YELLOW HONEYEATER
__ *L.f.flavus* (Gould, 1843)   NE Queensland
__ *L.f.addendus* (Mathews, 1912)[6]   Coastal EC Queensland

○ *Lichenostomus leucotis* WHITE-EARED HONEYEATER
__ *L.l.leucotis* (Latham, 1802)[7]   Coastal SE Australia
__ *L.l.novaenorciae* (Milligan, 1904)   SW to SC and inland E Australia
__ *L.l.thomasi* (Mathews, 1912)[8]   Kangaroo I. (South Australia)

○ *Lichenostomus flavicollis* (Vieillot, 1817) YELLOW-THROATED HONEYEATER
Tasmania, Bass Strait Is.

○ *Lichenostomus melanops* YELLOW-TUFTED HONEYEATER
__ *L.m.meltoni* (Mathews, 1912)[9]   SE Queensland, E New South Wales to C Victoria
__ *L.m.melanops* (Latham, 1802)[10]   Coastal New South Wales to SE Victoria
__ *L.m.cassidix* (Gould, 1867)   Central SE Victoria

○ *Lichenostomus cratitius* PURPLE-GAPED HONEYEATER
__ *L.c.occidentalis* Cabanis, 1851[11]   Central SW and SC Australia
__ *L.c.cratitius* (Gould, 1841)[12]   Kangaroo I.

○ *Lichenostomus keartlandi* (North, 1895) GREY-HEADED HONEYEATER   WC, C and central NE Australia

○ *Lichenostomus ornatus* (Gould, 1838) YELLOW-PLUMED HONEYEATER   Inland SW and SC to inland SE Australia

○ *Lichenostomus plumulus* GREY-FRONTED HONEYEATER
__ *L.p.plumulus* (Gould, 1841)   Inland WC to C Australia
__ *L.p.planasi* (A.J. Campbell, 1910)   Inland N Australia
__ *L.p.graingeri* (Mathews, 1912)[13]   Inland SC to EC Australia

[1] For recognition see Schodde & Mason (1999: 232) (2189).
[2] For reasons to use 1802 not 1801 see Browing & Monroe (1991) (258).
[3] Includes *westwoodia* see Schodde & Mason (1999: 234) (2189).
[4] For recognition see Schodde & Mason (1999: 234) (2189).
[5] Includes *insularis* see Schodde & Mason (1999: 234) (2189).
[6] For recognition see Schodde & Mason (1999: 239) (2189).
[7] For reasons to use 1802 not 1801 see Browing & Monroe (1991) (258).
[8] For recognition see Schodde & Mason (1999: 240) (2189).
[9] For recognition see Schodde & Mason (1999: 243) (2189).
[10] For reasons to use 1802 not 1801 see Browing & Monroe (1991) (258).
[11] For recognition see Schodde & Mason (1999: 245) (2189).
[12] Includes *halmaturina*, as the lectotype of *cratitius* is from Kangaroo I., see Schodde & Mason (1999: 246) (2189).
[13] For recognition see Schodde & Mason (1999: 250) (2189) who gave reasons for using this name in place of *ethelae*.

○ *Lichenostomus fuscus* Fuscous Honeyeater
— *L.f.subgermana* (Mathews, 1912)   NE to EC Queensland
— *L.f.fuscus* (Gould, 1837)[1]   SE Queensland to E New South Wales, Victoria and SE
           South Australia

○ *Lichenostomus flavescens* Yellow-tinted Honeyeater[2]
— *L.f.germana* (E.P. Ramsay, 1879)   SE New Guinea
— *L.f.flavescens* (Gould, 1840)[3]   N Australia
— *L.f.melvillensis* (Mathews, 1912)   Melville I.

● *Lichenostomus penicillatus* White-plumed Honeyeater[4]
∠ *L.p.penicillatus* (Gould, 1837)[5]   SE and EC Australia
— *L.p.leilavalensis* (North, 1899)   C to inland NC and NE Australia
— *L.p.carteri* (A.J. Campbell, 1899)[6]   W and WC Australia
— *L.p.calconi* (Mathews, 1912)[7]   NW Australia

GUADALCANARIA E. Hartert, 1929[8] F
○ *Guadalcanaria inexpectata* E. Hartert, 1929 Guadalcanal Honeyeater
           Mts. of Guadalcanal I. (SE Solomons)

OREORNIS van Oort, 1910 M
○ *Oreornis chrysogenys* van Oort, 1910 Orange-cheeked Honeyeater   Mts. of WC New Guinea

MELIPHAGA Lewin, 1808 F
○ *Meliphaga mimikae* Mottle-breasted Honeyeater
— *M.m.rara* Salomonsen, 1966   Western NC New Guinea
— *M.m.mimikae* (Ogilvie-Grant, 1911)   Mts. of WC New Guinea
— *M.m.bastille* Diamond, 1967   Mts. of EC New Guinea (697)
— *M.m.granti* Rand, 1936   Mts. of SE New Guinea

○ *Meliphaga montana* White-marked Forest Honeyeater[9]
— *M.m.margaretae* Greenway, 1966   Batanta I. (W Papuan Is.)
— *M.m.montana* (Salvadori, 1880)   NW New Guinea (Vogelkop)
— *M.m.steini* Stresemann & Paludan, 1932   Yapen I. (Geelvink Bay)
— *M.m.sepik* Rand, 1936   Northern mts. of C New Guinea
— *M.m.germanorum* E. Hartert, 1930   Mts. of N New Guinea
— *M.m.huonensis* Rand, 1936   Mts. of NE New Guinea (Huon Pen.)
— *M.m.aicora* Rand, 1936   Mts. of north and east SE New Guinea

○ *Meliphaga orientalis* Hill-forest Honeyeater
— *M.o.facialis* Rand, 1936   Mts. of W New Guinea
— *M.o.citreola* Rand, 1941[10]   Northern mts. of C New Guinea
— *M.o.becki* Rand, 1936   Mts. of NE New Guinea (Huon Pen.)
— *M.o.orientalis* (A.B. Meyer, 1894)   Mts. of SE New Guinea

○ *Meliphaga albonotata* (Salvadori, 1876) White-marked Scrub Honeyeater
           New Guinea

○ *Meliphaga analoga* Mimic Honeyeater
— *M.a.longirostris* (Ogilvie-Grant, 1911)   Aru Is.
— *M.a.analoga* (Reichenbach, 1852)[11]   W Papuan islands, W to SE New Guinea
— *M.a.flavida* Stresemann & Paludan, 1932[12]   Western NC to NE New Guinea, Yapen I. and Meos
           Num Is. (Geelvink Bay)

○ *Meliphaga vicina* (Rothschild & Hartert, 1912) Tagula Honeyeater   Tagula I.

---

[1] Includes *dawsoni*, named for an intergradient population, see Schodde & Mason (1999: 251) (2189).
[2] For reasons to separate this from *fuscus* see Ford (1986) (862).
[3] Includes *zanda* and *deserticola* see Schodde & Mason (1999: 253) (2189).
[4] The names *centralia* and *interioris* belong to intergrading populations see Schodde & Mason (1999: 255) (2189).
[5] Includes *mellori* see Schodde & Mason (1999: 255) (2189).
[6] Includes *geraldtonensis* and *ladasi* see Schodde & Mason (1999: 255) (2189).
[7] For recognition see Schodde & Mason (1999: 256) (2189).
[8] Treated in the genus *Meliphaga* in last Edition. This genus was reintroduced by Sibley & Monroe (1990) (2262).
[9] Compared to Salomonsen (1967: 367) this species lacks the races *auga*, *gretae* and *setekwa*. This follows the revelation by Diamond (1972) (699) that *auga* is the same as *albonotata* to which monotypic species these three names are now transferred; in last Edition we recognised *setekwa*.
[10] Transferred from the species *analoga* by Diamond (1969) (698).
[11] Includes *papuae* see Diamond (1969) (698).
[12] Includes *connectens* see Diamond (1969) (698).

○ *Meliphaga gracilis* GRACEFUL HONEYEATER
　　__ *M.g.stevensi* Rand, 1936　　　　　　　　N coast of SE New Guinea
　　__ *M.g.cinereifrons* Rand, 1936　　　　　　S coast of SE New Guinea
　　__ *M.g.gracilis* (Gould, 1866)　　　　　　S New Guinea, Aru Is., Torres Strait Is., Cape York Pen.
　　__ *M.g.imitatrix* (Mathews, 1912)　　　　　NE Queensland

○ *Meliphaga flavirictus* YELLOW-GAPED HONEYEATER
　　__ *M.f.flavirictus* (Salvadori, 1880)　　　SC to SE New Guinea
　　__ *M.f.crockettorum* Mayr & Meyer de Schauensee, 1939　　W, N and NE New Guinea

○ *Meliphaga albilineata* WHITE-LINED HONEYEATER
　　__ *M.a.albilineata* (H.L. White, 1917)　　　N Northern Australia (Arnhem Land)
　　__ *M.a.fordiana* Schodde, 1989　　　　　N Western Australia (Kimberley) (2177)

○ *Meliphaga aruensis* PUFF-BACKED HONEYEATER[1]
　　__ *M.a.sharpei* (Rothschild & Hartert, 1903)　　Waigeo I., north coast of W, N and E New Guinea,
　　　　　　　　　　　　　　　　　　　　　　　　　　　D'Entrecasteaux Arch.
　　__ *M.a.aruensis* (Sharpe, 1884)　　　　　South coast of S and SE New Guinea, Aru Is.

○ *Meliphaga notata* YELLOW-SPOTTED HONEYEATER
　　__ *M.n.notata* (Gould, 1867)　　　　　　Torres Strait Is., Cape York Pen.
　　__ *M.n.mixta* (Mathews, 1912)　　　　　NE Queensland

○ *Meliphaga lewinii* LEWIN'S HONEYEATER
　　__ *M.l.amphochlora* Schodde, 1989　　　EC Cape York Pen. (2177)
　　__ *M.l.mab* (Mathews, 1912)[2]　　　　　NE to EC Queensland
　　__ *M.l.lewinii* (Swainson, 1837)[3]　　　EC to SE Australia

FOULEHAIO Reichenbach, 1852[4] M
◉ *Foulehaio carunculatus* WATTLED HONEYEATER[5]
　　__ *F.c.procerior* (Finsch & Hartlaub, 1867)　　W Fiji Is.
　　__ *F.c.taviunensis* (Wiglesworth, 1891)　　Taveuni, Vanua Levu Is. (NC Fiji)
　　✓ *F.c.carunculatus* (J.F. Gmelin, 1788)　　E Fiji Is., Tonga, Horne Is., Samoa

GYMNOMYZA Reichenow, 1914 F
○ *Gymnomyza viridis* GIANT GREEN HONEYEATER
　　__ *G.v.viridis* (E.L. Layard, 1875)　　　Taveuni, Vanua Levu Is. (Fiji)
　　__ *G.v.brunneirostris* (Mayr, 1932)　　　Viti Levu I. (Fiji)

○ *Gymnomyza samoensis* (Hombron & Jacquinot, 1841) MAO/BLACK-BREASTED HONEYEATER
　　　　　　　　　　　　　　　　　　　　　　　　Mts. of Samoa

○ *Gymnomyza aubryana* (Verreaux & Des Murs, 1860) CROW-HONEYEATER
　　　　　　　　　　　　　　　　　　　　　　　　Mts. of New Caledonia

MANORINA Vieillot, 1818 F
○ *Manorina melanophrys* (Latham, 1802)[6] BELL MINER　　SE Australia

● *Manorina melanocephala* NOISY MINER[7]
　　__ *M.m.titaniota* Schodde & Mason, 1999　　NE Queensland (2189)
　　__ *M.m.lepidota* Schodde & Mason, 1999　　EC to inland SE Australia (2189)
　　__ *M.m.melanocephala* (Latham, 1802)[8]　　Coastal SE Australia
　　__ *M.m.leachi* (Mathews, 1912)[9]　　　　E Tasmania

○ *Manorina flavigula* YELLOW-THROATED MINER
　　__ *M.f.melvillensis* (Mathews, 1912)　　　Melville I.
　　__ *M.f.lutea* (Gould, 1840)[10]　　　　　NW Australia (Kimberley and Arnhem Land)

---

[1] This species belongs with the *lewini* group, see Christidis & Schodde (1993) (352).
[2] For recognition see Schodde & Mason (1999: 257) (2189).
[3] Includes *nea* see Schodde & Mason (1999: 257) (2189).
[4] The species *provocator* is now treated in the genus *Xanthotis*.
[5] For correction of spelling see David & Gosselin (2002) (614).
[6] For reasons to use 1802 not 1801 see Browing & Monroe (1991) (258).
[7] The name *crassirostris* relates to a type locality in the zone where intergradation takes places between *titaniota* and *lepidota* see Schodde & Mason (1999: 268) (2189).
[8] For reasons to use 1802 not 1801 see Browing & Monroe (1991) (258).
[9] For recognition see Schodde & Mason (1999: 267) (2189).
[10] Includes *casuarina* and *alligator* see Schodde & Mason (1999: 269) (2189).

— *M.f.flavigula* (Gould, 1840) Inland E Australia
— *M.f.wayensis* (Mathews, 1912)[1] W and C Australia
— *M.f.obscura* (Gould, 1841)[2] SW Western Australia
— *M.f.melanotis* (F.E. Wilson, 1911)[3] NW Victoria, central SE South Australia

ENTOMYZON Swainson, 1825 M
○ ***Entomyzon cyanotis*** BLUE-FACED HONEYEATER
— *E.c.albipennis* (Gould, 1841)[4] NW Australia (Kimberley and Arnhem Land)
— *E.c.griseigularis* van Oort, 1909[5] Cape York Pen., SC New Guinea
— *E.c.cyanotis* (Latham, 1802)[6] E Australia south to inland SE Australia

MELITHREPTUS Vieillot, 1816 M
○ ***Melithreptus gularis*** BLACK-CHINNED HONEYEATER[7]
— *M.g.laetior* Gould, 1875[8] Inland NW to NC Australia and Cape York Pen.
— *M.g.gularis* (Gould, 1837) E to inland SE and SC Australia

○ ***Melithreptus validirostris*** (Gould, 1837) STRONG-BILLED HONEYEATER[9]
Tasmania, Bass Strait Is.

○ ***Melithreptus brevirostris*** BROWN-HEADED HONEYEATER[10]
— *M.b.brevirostris* (Vigors & Horsfield, 1827) Coastal SE Australia
— *M.b.wombeyi* Schodde & Mason, 1999 S Victoria (2189)
— *M.b.pallidiceps* Mathews, 1912[11] Inland E, SE and SC Australia
— *M.b.magnirostris* North, 1905 Kangaroo I.
— *M.b.leucogenys* Milligan, 1903[12] Western SC and SW Australia

○ ***Melithreptus albogularis*** WHITE-THROATED HONEYEATER
— *M.a.albogularis* Gould, 1848[13] N Australia, SC and SE New Guinea
— *M.a.inopinatus* Schodde 1989[14] NE Queensland to NE New South Wales (2177)

● ***Melithreptus lunatus*** WHITE-NAPED HONEYEATER
✔ *M.l.lunatus* (Vieillot, 1802) NE to SE and eastern SC Australia
— *M.l.chloropsis* Gould, 1848 SW Western Australia

○ ***Melithreptus affinis*** (Lesson, 1839) BLACK-HEADED HONEYEATER[15] Tasmania, Bass Strait is.

ANTHORNIS G.R. Gray, 1840 F
● ***Anthornis melanura*** NEW ZEALAND BELLBIRD
✔ *A.m.melanura* (Sparrman, 1786)[16] North Island, South Island, Stewart I., Auckland Is. (New Zealand)
— *A.m.obscura* Falla, 1948 Three Kings I. (New Zealand)
— ¶*A.m.oneho* Bartle & Sagar, 1987 Poor Knights Is. (102)
— †*A.m.melanocephala* G.R. Gray, 1843 Chatham Is.

PROSTHEMADERA G.R. Gray, 1840 F
● ***Prosthemadera novaeseelandiae*** TUI/PARSON BIRD
✔ *P.n.novaeseelandiae* (J.F. Gmelin, 1788) New Zealand, Auckland, Stewart Is.
— *P.n.kermadecensis* Mathews & Iredale, 1914 Kermadec Is.
— *P.n.chathamensis* E. Hartert, 1928 Chatham I.

---

[1] For recognition see Schodde & Mason (1999: 269) (2189); *pallida* is a junior synonym.
[2] Includes *clelandi* see map in Schodde & Mason (1999: 269) (2189).
[3] For a discussion on whether or not this, with *obscura* as a race, should be treated as a distinct species see Schodde & Mason (1999: 271-273) (2189).
[4] Includes *apsleyi* see Schodde & Mason (1999: 274) (2189).
[5] For recognition see Schodde & Mason (1999: 274) (2189).
[6] For reasons to use 1802 not 1801 see Browing & Monroe (1991) (258). The name *harterti* applies to intergrades between this form and *griseigularis* see Schodde & Mason (1999: 275) (2189).
[7] For a discussion of the treatment of the two taxa united here see Schodde & Mason (1999: 277) (2189). The names *carpentarius* and *normantoniensis* relate to intergrading populations.
[8] Includes *parus* see Schodde & Mason (1999: 276) (2189).
[9] Includes *kingi* see Schodde & Mason (1999: 278) (2189).
[10] The name *augustus* relates to an intergrading population see Schodde & Mason (1999: 280) (2189).
[11] For recognition see Schodde & Mason (1999: 280) (2189).
[12] This form was recognised by Salomonsen (1967: 395), but was in the synonymy of *augustus* in our last Edition.
[13] Includes *subalbogularis* see Schodde & Mason (1999: 282) (2189).
[14] Includes *schoddei* McAllan & Bruce, 1989 (1499).
[15] Includes *alisteri* see Schodde & Mason (1999: 286) (2189).
[16] Includes *dumerilii* and *incoronata* see O.S.N.Z. (1990) (1668).

PYCNOPYGIUS Salvadori, 1880 M

○ *Pycnopygius ixoides* PLAIN HONEYEATER
    __ *P.i.ixoides* (Salvadori, 1878)   NW New Guinea (Vogelkop)
    __ *P.i.cinereifrons* Salomonsen, 1966   SW New Guinea
    __ *P.i.simplex* (Reichenow, 1915)   Western NC New Guinea
    __ *P.i.proximus* (Madarász, 1900)   Eastern NC New Guinea
    __ *P.i.unicus* Mayr, 1931   NE New Guinea
    __ *P.i.finschi* (Rothschild & Hartert, 1903) SE New Guinea

○ *Pycnopygius cinereus* MARBLED HONEYEATER
    __ *P.c.cinereus* (P.L. Sclater, 1873)   Mts. of NW New Guinea (Vogelkop)
    __ *P.c.dorsalis* Stresemann & Paludan, 1934 Mts. of WC New Guinea
    __ *P.c.marmoratus* (Sharpe, 1882)   Mts. of C to E New Guinea

○ *Pycnopygius stictocephalus* (Salvadori, 1876) STREAK-HEADED HONEYEATER
                        Aru Is., New Guinea, Salawati I.

MELITOGRAIS Sundevall, 1872[1] F
○ *Melitograis gilolensis* (Bonaparte, 1850) WHITE-STREAKED FRIARBIRD   N Moluccas

PHILEMON Vieillot, 1816 M
○ *Philemon meyeri* Salvadori, 1878 MEYER'S FRIARBIRD   E to N and to SC New Guinea

○ *Philemon brassi* Rand, 1940 BRASS'S FRIARBIRD   Inland western NC New Guinea

○ *Philemon citreogularis* LITTLE FRIARBIRD
    __ *P.c.kisserensis* A.B. Meyer, 1885[2]   Kisar, Leti Is. (EC Lesser Sundas)
    __ *P.c.papuanus* Mayr & Rand, 1935   SC New Guinea
    __ *P.c.sordidus* (Gould, 1848)   N Australia
    __ *P.c.citreogularis* (Gould, 1837)   E to inland SE Australia

○ *Philemon inornatus* (G.R. Gray, 1846) TIMOR FRIARBIRD   Timor (Lesser Sundas)

○ *Philemon fuscicapillus* (Wallace, 1862) DUSKY FRIARBIRD   N Moluccas

○ *Philemon subcorniculatus* (Hombron & Jacquinot, 1841) GREY-NECKED FRIARBIRD
                      Seram (S Moluccas)

○ *Philemon moluccensis* BLACK-FACED FRIARBIRD
    __ *P.m.moluccensis* (J.F. Gmelin, 1788)   Buru (S Moluccas)
    __ *P.m.plumigenis* (G.R. Gray, 1858)   Kai Is., Tanimbar Is. (E Banda Sea)

○ *Philemon buceroides* HELMETED FRIARBIRD[3]
    __ *P.b.neglectus* (Büttikofer, 1891)   Lombok to Sumba and Alor Is. (W to C Lesser Sundas)
    __ *P.b.buceroides* (Swainson, 1838)   Roti, Timor, Savu, Wetar Is. (EC Lesser Sundas)
    __ *P.b.novaeguineae* (S. Müller, 1843)   W Papuan Is., NW New Guinea (Vogelkop)
    __ *P.b.aruensis* (A.B. Meyer, 1884)   Aru Is.
    __ *P.b.jobiensis* (A.B. Meyer, 1875)   Yapen I., N New Guinea
    __ *P.b.brevipennis* Rothschild & Hartert, 1913   SW to south coast of SE New Guinea
    __ *P.b.trivialis* Salomonsen, 1966   N coast of SE New Guinea
    __ *P.b.subtuberosus* E. Hartert, 1896   Trobriand Is., D'Entrecasteaux Arch.
    __ *P.b.tagulanus* Rothschild & Hartert, 1918 Tagula I.
    __ *P.b.gordoni* Mathews, 1912   Melville I., coastal N Northern Territory
    __ *P.b.ammitophilus* Schodde, Mason & McKean, 1979 Inland NC Northern Territory (Arnhem Land) (2190)
    __ *P.b.yorki* Mathews, 1912[4]   Torres Strait, NE Queensland

○ *Philemon cockerelli* NEW BRITAIN FRIARBIRD
    __ *P.c.umboi* E. Hartert, 1926   Umboi I. (Bismarck Arch.)
    __ *P.c.cockerelli* P.L. Sclater, 1877   New Britain (Bismarck Arch.)

○ *Philemon eichhorni* Rothschild & Hartert, 1924 NEW IRELAND FRIARBIRD
                      New Ireland (Bismarck Arch.)

---

[1] Treated in the genus *Philemon* in last Edition. We follow White & Bruce (1986: 396) (2652) in using this genus.
[2] This disjunct form might be a separate species (Grey Friarbird) see White & Bruce (1986: 397) (2652), but a review is awaited.
[3] We tentatively follow the decision in Schodde & Mason (1999: 290) (2189) to retain a broad species, requiring that we here unite *gordoni* and *novaeguineae* with *bucerotides*.
[4] Includes *confusus* see Schodde & Mason (1999: 289) (2189).

○ *Philemon albitorques* P.L. Sclater, 1877 WHITE-NAPED FRIARBIRD   Manus I. (Admiralty Is.)

○ *Philemon argenticeps* SILVER-CROWNED FRIARBIRD
  — *P.a.argenticeps* (Gould, 1840)[1]   NC Australia (Kimberley to S Gulf of Carpentaria)
  — *P.a.kempi* Mathews, 1912   Cape York Pen.

● *Philemon corniculatus* NOISY FRIARBIRD
  — *P.c.corniculatus* (Latham, 1790)   NE Queensland
  ✓ *P.c.monachus* (Latham, 1802)[2]   EC and SE Australia

○ *Philemon diemenensis* (Lesson, 1831) NEW CALEDONIAN FRIARBIRD   Loyalty Is., New Caledonia

PLECTORHYNCHA Gould, 1838 F
○ *Plectorhyncha lanceolata* Gould, 1838 STRIPED HONEYEATER   EC to inland E and SC Australia

ACANTHAGENYS Gould, 1838 F
○ *Acanthagenys rufogularis* Gould, 1838 SPINY-CHEEKED HONEYEATER[3]   W, C and SC to inland E Australia, Kangaroo I.

ANTHOCHAERA Vigors & Horsfield, 1827 F
● *Anthochaera chrysoptera* LITTLE WATTLEBIRD[4]
  ✓ *A.c.chrysoptera* (Latham, 1802)[5]   SE to SC Australia
  — *A.c.halmaturina* (Mathews, 1912)   Kangaroo I.
  — *A.c.tasmanica* (Mathews, 1912)   N and E Tasmania

○ *Anthochaera lunulata* Gould, 1838 WESTERN WATTLEBIRD   SW Western Australia

● *Anthochaera carunculata* RED WATTLEBIRD
  ✓ *A.c.carunculata* (Shaw, 1790)[6]   SE Australia
  — *A.c.clelandi* (Mathews, 1923)[7]   Kangaroo I.
  — *A.c.woodwardi* Mathews, 1912   SW and SC Australia

○ *Anthochaera paradoxa* YELLOW WATTLEBIRD
  — *A.p.paradoxa* (Daudin, 1800)   Tasmania, Flinders I.
  — *A.p.kingi* (Mathews, 1925)[8]   King I. (W Bass Strait)

XANTHOMYZA Swainson, 1837[9] F
○ *Xanthomyza phrygia* (Shaw, 1794) REGENT HONEYEATER   SE Australia to SC South Australia, Kangaroo I.

MELIPOTES P.L. Sclater, 1873 M
○ *Melipotes gymnops* P.L. Sclater, 1873 ARFAK HONEYEATER   Mts. of NW New Guinea (Vogelkop)

○ *Melipotes fumigatus* SMOKY HONEYEATER
  — *M.f.kumawa* Diamond, 1985   Mts. of SW New Guinea (Bombarai Pen.) (701)
  — *M.f.goliathi* Rothschild & Hartert, 1911   Mts. of WC to EC and N New Guinea
  — *M.f.fumigatus* A.B. Meyer, 1886   Mts. of SE New Guinea

○ *Melipotes ater* Rothschild & Hartert, 1911 SPANGLED HONEYEATER   Mts. of NE New Guinea (Huon Pen.)

MACGREGORIA De Vis, 1897[10] F
○ *Macgregoria pulchra* MACGREGOR'S BIRD-OF-PARADISE
  — *M.p.carolinae* Junge, 1939   Mts. of WC and C New Guinea
  — *M.p.pulchra* De Vis, 1897   Mts. of SE New Guinea

[1] Includes *melvillensis* and *alexis* see Schodde & Mason (1999: 291) (2189).
[2] For reasons to use 1802 not 1801 see Browing & Monroe (1991) (258). Includes *clamans* see Schodde & Mason (1999: 292) (2189); the name *ellioti* relates to intergrades between the two recognised forms.
[3] Includes *parkeri* Parkes, 1980 (1772) see Schodde & Mason (1999: 296) (2189).
[4] For the separation of *lunulata* see Schodde & Mason (1999: 299) (2189).
[5] For reasons to use 1802 not 1801 see Browing & Monroe (1991) (258).
[6] Ascribed to White by Salomonsen (1967: 446) (2159) but names in White's Journal were given by Shaw. See Schodde & Mason (1997: 127) (2188).
[7] For recognition see Schodde & Mason (1999: 301) (2189).
[8] For recognition see Schodde & Mason (1999: 303) (2189).
[9] For the use of this spelling rather than *Zanthomyza* see Salomonsen (1967: 436 fn) (2159) and Christidis & Boles (1994) (351).
[10] For the affinities of this species see Cracraft & Feinstein (2000) (584).

MELIDECTES P.L. Sclater, 1873 M
○ *Melidectes fuscus* SOOTY HONEYEATER
    — *M.f.occidentalis* Junge, 1939   Mts. of WC to EC New Guinea
    — *M.f.fuscus* (De Vis, 1897)[1]   Mts. of EC and SE New Guinea

○ *Melidectes whitemanensis* (Gilliard, 1960) GILLIARD'S HONEYEATER   Mts. of New Britain

○ *Melidectes nouhuysi* (van Oort, 1910) SHORT-BEARDED HONEYEATER   Mts. of WC New Guinea

○ *Melidectes princeps* Mayr & Gilliard, 1951 LONG-BEARDED HONEYEATER   Mts. of EC New Guinea

○ *Melidectes ochromelas* CINNAMON-BROWED HONEYEATER
    — *M.o.ochromelas* (A.B. Meyer, 1875)   Mts. of NW New Guinea (Vogelkop)
    — *M.o.batesi* (Sharpe, 1886)   Mts. of WC and SE New Guinea
    — *M.o.lucifer* Mayr, 1931   Mts. of NE New Guinea (Huon Pen.)

○ *Melidectes leucostephes* (A.B. Meyer, 1875) VOGELKOP HONEYEATER   Mts. of NW New Guinea (Vogelkop)

○ *Melidectes rufocrissalis* YELLOW-BROWED HONEYEATER
    — *M.r.rufocrissalis* (Reichenow, 1915)   S slopes of WC New Guinea Mts. to N slopes of EC
                                New Guinea Mts.
    — *M.r.thomasi* Diamond, 1969[2]   S slopes of EC New Guinea Mts. (698)

○ *Melidectes foersteri* (Rothschild & Hartert, 1911) HUON HONEYEATER   Mts. of NE New Guinea (Huon Pen.)

○ *Melidectes belfordi* BELFORD'S HONEYEATER
    — *M.b.joiceyi* (Rothschild, 1921)   Mts. of central W New Guinea
    — *M.b.kinneari* Mayr, 1936   Mts. of western C New Guinea
    — *M.b.schraderensis* Gilliard & LeCroy, 1968   Mts. of EC New Guinea (942)
    — *M.b.belfordi* (De Vis, 1890)[3]   Mts. of SE New Guinea

○ *Melidectes torquatus* CINNAMON-BREASTED HONEYEATER
    — *M.t.torquatus* P.L. Sclater, 1873   Mts. of NW New Guinea (Vogelkop)
    — *M.t.nuchalis* Mayr, 1936   Mts. of WC New Guinea (S slopes)
    — *M.t.mixtus* Rand, 1941   Mts. of WC to C New Guinea (N slopes)
    — *M.t.polyphonus* Mayr, 1931   Mts. of EC New Guinea
    — *M.t.cahni* Mertens, 1923   Mts. of NE New Guinea (Huon Pen.)
    — *M.t.emilii* A.B. Meyer, 1886   Mts. of SE New Guinea

MELIARCHUS Salvadori, 1880 M
○ *Meliarchus sclateri* (G.R. Gray, 1870) SAN CRISTOBAL HONEYEATER   San Cristobal I. (SE Solomons)

PTILOPRORA De Vis, 1894 F
○ *Ptiloprora plumbea* LEADEN HONEYEATER
    — *P.p.granti* Mayr, 1931   Mts. of WC New Guinea
    — *P.p.plumbea* (Salvadori, 1894)   Mts. of SE New Guinea

○ *Ptiloprora meekiana* OLIVE-STREAKED HONEYEATER
    — *P.m.occidentalis* Rand, 1940   Mts. of WC New Guinea
    — *P.m.meekiana* (Rothschild & Hartert, 1907)   Mts. of NE and SE New Guinea

○ *Ptiloprora erythropleura* RUFOUS-SIDED HONEYEATER
    — *P.e.erythropleura* (Salvadori, 1876)   Mts. of NW New Guinea (Vogelkop)
    — *P.e.dammermani* Stresemann & Paludan, 1934   Mts. of SW and WC New Guinea

○ *Ptiloprora guisei* RUFOUS-BACKED HONEYEATER
    — *P.g.umbrosa* Mayr, 1931   Mts. of E, NC and NE New Guinea
    — *P.g.guisei* (De Vis, 1894)   Mts. of eastern SC to SE New Guinea

○ *Ptiloprora mayri* MAYR'S HONEYEATER
    — *P.m.mayri* E. Hartert, 1930   Cyclops Mts. of west central N New Guinea
    — *P.m.acrophila* Diamond, 1969[4]   Bewani Mts. of east central N New Guinea (698)

[1] Includes *gilliardi* Salomonsen see Diamond (1969) (698).
[2] Originally named *M.r.gilliardi* Diamond, 1969, see Salomonsen (1967: 418 fn) (2159) but this was preoccupied.
[3] Tentatively includes *brassi* see Salomonsen (1967: 419) (2159). But LeCroy (*in litt.*) suspects this lumping is erroneous.
[4] This was incorrectly placed in *P. guisei* in last Edition.

○ *Ptiloprora perstriata* Black-backed Honeyeater
  — *P.p.praedicta* E. Hartert, 1930   Mts. of central W New Guinea
  — *P.p.incerta* Junge, 1952[1]   Mts. of WC New Guinea
  — *P.p.perstriata* (De Vis, 1898)   Mts. of C and E New Guinea

MYZA Meyer & Wiglesworth, 1895   F
○ *Myza celebensis* Lesser Streaked Honeyeater
  — *M.c.celebensis* (Meyer & Wiglesworth, 1895)   Mts. of N, C and SE Sulawesi
  — *M.c.meridionalis* (Meyer & Wiglesworth, 1896)   Mts. of SW Sulawesi

○ *Myza sarasinorum* Greater Streaked Honeyeater
  — *M.s.sarasinorum* Meyer & Wiglesworth, 1895   Mts. of N Sulawesi
  — *M.s.chionogenys* Stresemann, 1931   Mts. of SC Sulawesi
  — *M.s.pholidota* Stresemann, 1932   Mts. of SE Sulawesi

MELILESTES Salvadori, 1876   M
○ *Melilestes megarhynchus* Long-billed Honeyeater
  — *M.m.vagans* (Bernstein, 1864)   Batanta, Waigeo Is.
  — *M.m.megarhynchus* (G.R. Gray, 1858)[2]   Aru Is., W Papuan Is., W (Vogelkop) to SE New Guinea and NE New Guinea (Huon Pen.)
  — *M.m.stresemanni* E. Hartert, 1930   NC New Guinea, Yapen I.

STRESEMANNIA Meise, 1950[3]   F
○ *Stresemannia bougainvillei* (Mayr, 1932) Bougainville Honeyeater   Bougainville I. (N Solomons)

GLYCIFOHIA Mathews, 1929[4]   F
○ *Glycifohia undulata* (Sparrman, 1787) Barred Honeyeater   New Caledonia

○ *Glycifohia notabilis* White-bellied Honeyeater
  — *G.n.notabilis* (Sharpe, 1899)   Banks Is., NW Vanuatu
  — *G.n.superciliaris* (Mayr, 1932)   N Vanuatu

LICHMERA Cabanis, 1851   F
○ *Lichmera lombokia* (Mathews, 1926) Scaly-crowned Honeyeater   Lombok, Sumbawa, Flores (Lesser Sundas)

○ *Lichmera argentauris* (Finsch, 1870) Olive Honeyeater   W New Guinea islands, Halmahera, Gebe and small islands near Seram (Moluccas)

○ *Lichmera indistincta* Brown Honeyeater
  — *L.i.limbata* (S. Müller, 1843)   Bali, Lombok, Timor (west to C Lesser Sundas)
  — *L.i.nupta* (Stresemann, 1912)   Aru Is.
  — *L.i.melvillensis* (Mathews, 1912)   Melville I. (Arnhem Land)
  — *L.i.indistincta* (Vigors & Horsfield, 1827)   N, C and W Australia
  — *L.i.ocularis* (Gould, 1838)   NE to EC Australia, SC New Guinea

○ *Lichmera incana* Dark-brown Honeyeater
  — *L.i.incana* (Latham, 1790)   New Caledonia
  — *L.i.poliotis* (G.R. Gray, 1859)   Loyalty Is.
  — *L.i.mareensis* Salomonsen, 1966   Mare I. (Loyalty Is.)
  — *L.i.griseoviridis* Salomonsen, 1966   C Vanuata
  — *L.i.flavotincta* (G.R. Gray, 1870)   Erromanga I. (Vanuatu)

○ *Lichmera alboauricularis* Silver-eared Honeyeater
  — *L.a.alboauricularis* (E.P. Ramsay, 1879)   SE New Guinea
  — *L.a.olivacea* Mayr, 1938   NC New Guinea

○ *Lichmera squamata* (Salvadori, 1878) White-tufted Honeyeater   E Lesser Sundas to Kai and Tanimbar Is. (S Banda Sea)

○ *Lichmera deningeri* (Stresemann, 1912) Buru Honeyeater   Buru (S Moluccas)

○ *Lichmera monticola* (Stresemann, 1912) Spectacled Honeyeater   Seram (S Moluccas)

---

[1] For recognition see Rand & Gilliard (1967) (1975). However, Diamond (1969) (698) thought it might be a hybrid.
[2] Includes *brunneus* see Gilliard & LeCroy (1970) (943).
[3] Treated in the genus *Melilestes* in last Edition. This genus was reintroduced by Sibley & Monroe (1990) (2262).
[4] Treated in the genus *Phylidonyris* in last Edition. The generic name *Glycifohia* was used by Wolters (1979: 260) (2685); see also Boles & Longmore (1985) (177).

○ *Lichmera flavicans* (Vieillot, 1817) YELLOW-EARED HONEYEATER   Timor (Lesser Sundas)

○ *Lichmera notabilis* (Finsch, 1898) BLACK-NECKLACED HONEYEATER   Wetar I. (Lesser Sundas)

TRICHODERE North, 1912[1] F
○ *Trichodere cockerelli* (Gould, 1869) WHITE-STREAKED HONEYEATER   N and E Cape York Pen.

GRANTIELLA Mathews, 1911[2] F
○ *Grantiella picta* (Gould, 1838) PAINTED HONEYEATER   Inland NC to E and SE Australia

PHYLIDONYRIS Lesson, 1831 M[3]
● *Phylidonyris pyrrhopterus* CRESCENT HONEYEATER
 ⅃ *P.p.pyrrhopterus* (Latham, 1802)[4]   SE Australia, Tasmania, Bass Strait Is.
 — *P.p.halmaturinus* (A.G. Campbell, 1906)[5]   SC South Australia (Mt. Lofty range), Kangaroo I.

● *Phylidonyris novaehollandiae* NEW HOLLAND HONEYEATER
 ⅃ *P.n.novaehollandiae* (Latham, 1790)   SE Australia to SC South Australia
 — *P.n.caudatus* Salomonsen, 1966   Bass Strait Is.
 — *P.n.canescens* (Latham, 1790)   Tasmania
 — *P.n.campbelli* (Mathews, 1923)   Kangaroo I. (SC Australia)
 — *P.n.longirostris* (Gould, 1846)   SW and S Western Australia

○ *Phylidonyris niger* WHITE-CHEEKED HONEYEATER
 — *P.n.niger* (Bechstein, 1811)   NE Queensland to E New South Wales
 — *P.n.gouldii* (Schlegel, 1872)   SW Western Australia

○ *Phylidonyris albifrons* (Gould, 1841) WHITE-FRONTED HONEYEATER   W and C Australia to inland SE Australia

GLYCIPHILA[6] Swainson, 1837 F
○ *Glyciphila melanops* TAWNY-CROWNED HONEYEATER
 — *G.m.melanops* (Latham, 1802)[7]   SW, SC and SE Australia, Bass Strait Is., E Tasmania
 — *G.m.chelidonia* Schodde & Mason, 1999   W Tasmania (2189)

RAMSAYORNIS Mathews, 1912 M
○ *Ramsayornis modestus* (G.R. Gray, 1858) BROWN-BACKED HONEYEATER   NW to S New Guinea, Torres Strait Is., NE Queensland

○ *Ramsayornis fasciatus* (Gould, 1843) BAR-BREASTED HONEYEATER[8]   N Australia and EC Queensland

CONOPOPHILA Reichenbach, 1852 F
○ *Conopophila albogularis* (Gould, 1843) RUFOUS-BANDED HONEYEATER[9]   NW and S New Guinea, Aru Is., Cape York Pen., N Northern Territory (Arnhem Land), Melville I.

○ *Conopophila rufogularis* (Gould, 1843) RUFOUS-THROATED HONEYEATER[10]   N and NE Australia

○ *Conopophila whitei* (North, 1910) GREY HONEYEATER   C Western Australia to SW Northern Territory and NW South Australia

ACANTHORHYNCHUS Gould, 1837 M
● *Acanthorhynchus tenuirostris* EASTERN SPINEBILL
 — *A.t.cairnsensis* Mathews, 1912   NE Queensland
 ⅃ *A.t.tenuirostris* (Latham, 1802)[11]   E and SE Australia
 — *A.t.dubius* Gould, 1837[12]   Tasmania, Bass Strait Is.
 — *A.t.halmaturinus* A.G. Campbell, 1906   SE South Australia, Kangaroo I.

○ *Acanthorhynchus superciliosus* Gould, 1837 WESTERN SPINEBILL   SW Western Australia

[1] Treated in the genus *Lichmera* in last Edition. For the use of this genus see Christidis & Boles (1994) (351).
[2] Treated in the genus *Conopophila* in last Edition. For the use of this genus see Christidis & Boles (1994) (351).
[3] Gender addressed by David & Gosselin (2002) (614), multiple changes occur in specific names.
[4] For reasons to use 1802 not 1801 see Browing & Monroe (1991) (258). Includes *rex* and *inornatus* see Schodde & Mason (1999: 309) (2189).
[5] Includes *indistinctus* see Schodde & Mason (1999: 309) (2189).
[6] Not *Glyciphilus* as given in last Edition. See Christidis & Boles (1994) (351).
[7] For reasons to use 1802 not 1801 see Browing & Monroe (1991) (258). Includes *braba* and *crassirostris* see Schodde & Mason (1999: 316) (2189).
[8] For monotypic treatment see Ford (1986) (862); *apsleyi* and *broomei* are synonyms.
[9] For monotypic treatment see Ford (1983) (860); *mimikae* is a synonym.
[10] For monotypic treatment see Schodde & Mason (1999: 323) (2189); *queenslandica* is a synonym.
[11] For reasons to use 1802 not 1801 see Browing & Monroe (1991) (258). Includes *trochiloides* and *loftyi* see Schodde & Mason (1999: 325) (2189).
[12] Includes *regius* see Schodde & Mason (1999: 325) (2189).

CERTHIONYX Lesson, 1830 M

○ *Certhionyx variegatus* Lesson, 1830 Pied Honeyeater   WC to inland EC Australia

○ *Certhionyx pectoralis* (Gould, 1841) Banded Honeyeater[1]   N Australia

○ *Certhionyx niger* (Gould, 1838) Black Honeyeater   WC to inland EC Australia

MYZOMELA Vigors & Horsfield, 1827 F

○ *Myzomela blasii* (Salvadori, 1882) Drab Honeyeater   Seram, Ambon Is. (S Moluccas)

○ *Myzomela albigula* White-chinned Honeyeater
  — *M.a.pallidior* E. Hartert, 1898   W Louisiade Arch.
  — *M.a.albigula* E. Hartert, 1898   Rossel I. (E Louisiade Arch.)

○ *Myzomela cineracea* Bismarck Honeyeater
  — *M.c.cineracea* P.L. Sclater, 1879   New Britain
  — *M.c.rooki* E. Hartert, 1926   Umboi I. (Bismarck Arch.)

○ *Myzomela eques* Red-throated Honeyeater
  — *M.e.eques* (Lesson & Garnot, 1827)   W Papuan Is., NW New Guinea (Vogelkop)
  — *M.e.primitiva* Stresemann & Paludan, 1932   N New Guinea
  — *M.e.nymani* Rothschild & Hartert, 1903   S and E New Guinea
  — *M.e.karimuiensis* Diamond, 1967   EC New Guinea (697)

○ *Myzomela obscura* Dusky Honeyeater
  — *M.o.mortyana* E. Hartert, 1903   Morotai I. (N Moluccas)
  — *M.o.simplex* G.R. Gray, 1860   Damar, Ternate, Bacan, Halmahera (N Moluccas)
  — *M.o.rubrotincta* Salvadori, 1878   Obi (C Moluccas)
  — *M.o.rubrobrunnea* A.B. Meyer, 1875   Biak I. (Geelvink Bay)
  — *M.o.aruensis* Kinnear, 1924   Aru Is.
  — *M.o.fumata* (Bonaparte, 1850)   S New Guinea and Boigu and Saibai (N Torres Strait Is.)
  — *M.o.harterti* Mathews, 1911[2]   Torres Strait Is., NE and E Queensland
  — *M.o.obscura* Gould, 1843   N Northern Territory, Melville I.

○ *Myzomela cruentata* Red Honeyeater
  — *M.c.cruentata* A.B. Meyer, 1875   Yapen I., New Guinea
  — *M.c.coccinea* E.P. Ramsay, 1878   New Britain, Duke of York Is. (Bismarck Arch.)
  — *M.c.erythrina* E.P. Ramsay, 1878[3]   New Ireland (Bismarck Arch.)
  — *M.c.lavongai* Salomonsen, 1966   New Hanover (Bismarck Arch.)
  — *M.c.cantans* Mayr, 1955   Tabar I. (Bismarck Arch.)
  — *M.c.vinacea* Salomonsen, 1966   Dyaul I. (Bismarck Arch.)

○ *Myzomela nigrita* Blackened Honeyeater
  — *M.n.steini* Stresemann & Paludan, 1932   Waigeo I. (W Papuan Is.)
  — *M.n.nigrita* G.R. Gray, 1858   Aru Is., W, S and E New Guinea
  — *M.n.meyeri* Salvadori, 1881   Yapen I., N New Guinea
  — *M.n.pluto* Forbes, 1879   Meos Num I. (Geelvink Bay)
  — *M.n.forbesi* E.P. Ramsay, 1880   D'Entrecasteaux Arch.
  — *M.n.louisiadensis* E. Hartert, 1898   Louisiade Arch.

○ *Myzomela pulchella* Salvadori, 1891 New Ireland Honeyeater   New Ireland

○ *Myzomela kuehni* Rothschild, 1903 Crimson-hooded Honeyeater   Wetar I. (Lesser Sundas)

○ *Myzomela erythrocephala* Red-headed Honeyeater
  — *M.e.dammermani* Siebers, 1928[4]   Sumba I. (Lesser Sundas)
  — *M.e.erythrocephala* Gould, 1840   N Australia, Melville I.
  — *M.e.infuscata* Forbes, 1879   N Cape York Pen., Torres Strait Is., Aru Is., S New Guinea

○ *Myzomela adolphinae* Salvadori, 1876 Red-headed Mountain Honeyeater
                     Mts. of New Guinea

---

[1] The need to use the generic name *Cissomela* (used in last Edition) has not been substantiated; see Schodde & Mason (1999: 329) (2189).
[2] Includes *munna* see Schodde & Mason (1999: 337) (2189).
[3] The emended name used here is judged to be in prevailing usage; see Art. 33.2 of the Code (I.C.Z.N., 1999) (1178) and p. 000 herein.
[4] Treated as conspecific with *erythrocephala* by White & Bruce (1986: 401) (2652).

○ *Myzomela boiei* WALLACEAN HONEYEATER[1]
    __ *M.b.chloroptera* Walden, 1872   N, C and SE Sulawesi
    __ *M.b.juga* Riley, 1921   SW Sulawesi
    __ *M.b.eva* Meise, 1929   Tanahjampea, Saleyer Is. (Flores Sea)
    __ *M.b.batjanensis* E. Hartert, 1903   Bacan (N Moluccas)
    __ *M.b.elisabethae* van Oort, 1911   Seram (S Moluccas)
    __ *M.b.wakoloensis* Forbes, 1883   Buru (S Moluccas)
    __ *M.b.boiei* (S. Müller, 1843)   Banda I. (Banda Sea)
    __ *M.b.annabellae* P.L. Sclater, 1883   Babar, Tanimbar Is. (E Lesser Sundas)

○ *Myzomela sanguinolenta* SCARLET HONEYEATER[2]
    __ *M.s.caledonica* Forbes, 1879   New Caledonia
    __ *M.s.sanguinolenta* (Latham, 1802)[3]   E Australia

● *Myzomela cardinalis* CARDINAL HONEYEATER
    __ *M.c.pulcherrima* E.P. Ramsay, 1881   San Cristobal, Ugi Is. (SE Solomons)
    __ *M.c.sanfordi* Mayr, 1931   Rennell I. (SE Solomons)
    __ *M.c.sanctaecrucis* F. Sarasin, 1913   Torres, Santa Cruz Is.
    __ *M.c.tucopiae* Mayr, 1937   Tikopia I. (NE Banks Is.)
    __ *M.c.tenuis* Mayr, 1937   N Vanuatu, Banks Is.
    __ *M.c.cardinalis* (J.F. Gmelin, 1788)   S Vanuatu
    __ *M.c.lifuensis* E.L. Layard, 1878   Loyalty Is.
    ✔ *M.c.nigriventris* Peale, 1848   Samoa Is.
    __ *M.c.chermesina* G.R. Gray, 1846   Rotuma Is. (N of Fiji)

○ *Myzomela rubratra* MICRONESIAN HONEYEATER[4]
    __ *M.r.asuncionis* Salomonsen, 1966   N Marianas
    __ *M.r.saffordi* Wetmore, 1917   S Marianas
    __ *M.r.kobayashii* Momiyama, 1922   Palau Is.
    __ *M.r.kurodai* Momiyama, 1922   Yap I. (Caroline Is.)
    __ *M.r.major* Bonaparte, 1854   Truk I. (Caroline Is.)
    __ *M.r.dichromata* Wetmore, 1919   Pohnpei I. (Caroline Is.)
    __ *M.r.rubratra* (Lesson, 1827)   Kusaie I. (Caroline Is.)

○ *Myzomela sclateri* Forbes, 1879 RED-BIBBED HONEYEATER   N New Guinea islands and New Britain islands
                                                         (Bismarck Arch.)

○ *Myzomela pammelaena* EBONY HONEYEATER[5]
    __ *M.p.ernstmayri* Meise, 1929   Islets W of Admiralty Is.
    __ *M.p.pammelaena* P.L. Sclater, 1877   Admiralty Is.
    __ *M.p.hades* Meise, 1929   St. Matthias Is. (Bismarck Arch.)
    __ *M.p.ramsayi* Finsch, 1886   Tingwon I., islets off NW New Ireland (Bismarck Arch.)
    __ *M.p.nigerrima* Salomonsen, 1966   Long I. (NE New Guinea)

○ *Myzomela lafargei* Pucheran, 1853 SCARLET-NAPED HONEYEATER   Bougainville, Choiseul, Santa Ysabel Is. (N Solomons)

○ *Myzomela eichhorni* YELLOW-VENTED HONEYEATER
    __ *M.e.eichhorni* Rothschild & Hartert, 1901   New Georgia group (C Solomons)
    __ *M.e.ganongae* Mayr, 1932   Ganonga I. (C Solomons)
    __ *M.e.atrata* E. Hartert, 1908   Vellalavella, Baga Is. (C Solomons)

○ *Myzomela malaitae* Mayr, 1931 MALAITA HONEYEATER   Malaita I. (NE Solomons)

○ *Myzomela melanocephala* E.P. Ramsay, 1879 BLACK-COWLED HONEYEATER
                                               Guadalcanal, Florida group (EC and SE Solomons)

○ *Myzomela tristrami* E.P. Ramsay, 1881 TRISTRAM'S HONEYEATER   San Cristobal, Santa Ana Is. (SE Solomons)

○ *Myzomela jugularis* Peale, 1848 ORANGE-BREASTED HONEYEATER   Fiji Is.

---

[1] We follow Wolters (1979: 262) (2685) in separating the Wallacean forms from *M. sanguinolenta* and like him see three groups of subspecies; three species may be correct, but we have seen no published detailed proposal. White (1986) (2652) included the forms of Australia and New Caledonia so that the name *sanguinolenta* that he applied to a broader species is not here accepted.
[2] Schodde & Mason (1999: 334) (2189) treated these two forms as one species.
[3] For reasons to use 1802 not 1801 see Browing & Monroe (1991) (258). The name *dibapha* was used by some authors, such as Mathews, who believed *sanguinolenta* to be indeterminate. See McAllan (1990) (1498). This is based on *Certhia dibapha* Latham, 1802.
[4] Tentatively split from *M. cardinalis* following Pratt (1987: 277) (1920).
[5] For separation of this from *M. nigrita* see Diamond (1976) (700).

○ *Myzomela erythromelas* Salvadori, 1881 BLACK-BELLIED HONEYEATER   New Britain (Bismarck Arch.)

○ *Myzomela vulnerata* (S. Müller, 1843) RED-RUMPED HONEYEATER   Timor (Lesser Sundas)

○ *Myzomela rosenbergii* RED-COLLARED HONEYEATER
   — *M.r.rosenbergii* Schlegel, 1871   Mts. of NW New Guinea (Vogelkop)
   — *M.r.wahgiensis* Gyldenstolpe, 1955   Mts. of WC to SE New Guinea
   — *M.r.longirostris* Mayr & Rand, 1935   Goodenough I. (D'Entrecasteaux Arch.)

TIMELIOPSIS Salvadori, 1876 F

○ *Timeliopsis fulvigula* OLIVE STRAIGHT-BILL
   — *T.f.fulvigula* (Schlegel, 1871)   Mts. of NW New Guinea (Vogelkop)
   — *T.f.meyeri* (Salvadori, 1896)   Mts. of C to SE New Guinea
   — *T.f.fuscicapilla* Mayr, 1931   Mts. of NE New Guinea (Huon Pen.)

○ *Timeliopsis griseigula* TAWNY STRAIGHT-BILL
   — *T.g.griseigula* (Schlegel, 1871)   W New Guinea
   — *T.g.fulviventris* (E.P. Ramsay, 1882)   SC to SE New Guinea

○ *Timeliopsis fallax* GREEN-BACKED HONEYEATER[1]
   — *T.f.pallida* (Stresemann & Paludan, 1932)   Batanta, Waigeo Is.
   — *T.f.poliocephala* (Salvadori, 1878)   NW to NE New Guinea, Yapen I. (Geelvink Bay)
   — *T.f.fallax* (Salvadori, 1878)   Misool and Aru Is., SW to SE New Guinea
   — *T.f.claudi* (Mathews, 1914)   NE Cape York Pen. (Queensland)

EPTHIANURA Gould, 1838[2] F

○ *Epthianura tricolor* Gould, 1841 CRIMSON CHAT   W, C and inland E Australia

○ *Epthianura aurifrons* Gould, 1838 ORANGE CHAT   W, C and inland E Australia

○ *Epthianura crocea* YELLOW CHAT
   — *E.c.tunneyi* Mathews, 1912[3]   N Northern Territory (W Arnhem Land)
   — *E.c.crocea* Castelnau & Ramsay, 1877   N Western Australia (S Kimberley) to inland W Queensland and NE South Australia
   — †?*E.c.macgregori* Keast, 1958   Fitzroy River (EC Queensland)

○ *Epthianura albifrons* (Jardine & Selby, 1828) WHITE-FRONTED CHAT[4]   S Western Australia to S Queensland, New South Wales, Victoria and Tasmania

ASHBYIA North, 1911 F
○ *Ashbyia lovensis* (Ashby, 1911) GIBBERBIRD   SC Northern Territory, NE South Australia, SW Queensland, NW New South Wales

DASYORNITHIDAE BRISTLEBIRDS (1: 3)

DASYORNIS Vigors & Horsfield, 1827 M

○ *Dasyornis brachypterus* EASTERN BRISTLEBIRD
   — *D.b.monoides* Schodde & Mason, 1999   SE Queensland, NE New South Wales (2189)
   — *D.b.brachypterus* (Latham, 1802)[5]   SE New South Wales, extreme E Victoria

○ *Dasyornis longirostris* Gould, 1841 WESTERN BRISTLEBIRD[6]   SW Western Australia

○ *Dasyornis broadbenti* RUFOUS BRISTLEBIRD
   — †?*D.b.litoralis* (Milligan, 1902)   Cape Mentelle (SW Western Australia)
   — *D.b.broadbenti* (McCoy, 1867)[7]   SE South Australia, SW Victoria
   — *D.b.caryochrous* Schodde & Mason, 1999   Otway Pen., SC Victoria (2189)

---

[1] For reasons to place this in this genus, instead of in a monotypic genus *Glychichaera*, see Schodde & Mason (1999: 339) (2189).
[2] For a note on this spelling see McAllan & Bruce (1989) (1499). The genus was attached to the Acanthizidae with the next one as *incertae sedis* by Mayr (1986) (1487) and in our last Edition, but see Christidis & Boles (1994) (351).
[3] Includes *boweri* see Schodde & Mason (1999: 344) (2189).
[4] For treatment as monotypic see Schodde & Mason (1999: 346) (2189); *tasmanica* is a synonym.
[5] For reasons to use 1802 not 1801 see Browing & Monroe (1991) (258).
[6] Treated as separate from *D. brachypterus* by Schodde (1975) (2175) and by Christidis & Boles (1994) (351).
[7] Includes *whitei* see Schodde & Mason (1999: 139) (2189).

PARDALOTIDAE[1] PARDALOTES (1: 4)

PARDALOTUS Vieillot, 1816 M
- ● ***Pardalotus punctatus*** SPOTTED PARDALOTE[2]
    - __ *P.p.millitaris* Mathews, 1912 — NE Queensland
    - ✓ *P.p.punctatus* (Shaw, 1792) — E, SE and SW Australia, Tasmania
    - __ *P.p.xanthopyge* McCoy, 1866 — SC Australia (inland SW Australia to NW Victoria)

- ○ ***Pardalotus quadragintus*** Gould, 1838 FORTY-SPOTTED PARDALOTE — Tasmania, Furneaux Group (Bass Strait)

- ○ ***Pardalotus rubricatus*** RED-BROWED PARDALOTE
    - __ *P.r.rubricatus* Gould, 1838[3] — NW and C Australia to EC Queensland
    - __ *P.r.yorki* Mathews, 1913[4] — Cape York Pen. (Queensland)

- ○ ***Pardalotus striatus*** STRIATED PARDALOTE[5]
    - __ *P.s.uropygialis* Gould, 1840 — N Australia
    - __ *P.s.melvillensis* Mathews, 1912 — Melville I.
    - __ *P.s.melanocephalus* Gould, 1838[6] — EC to SE Queensland
    - __ *P.s.ornatus* Temminck, 1826 — Coastal SE Australia
    - __ *P.s.substriatus* Mathews, 1912 — W, C and S to inland SE Australia
    - __ *P.s.striatus* (J.F. Gmelin, 1789) — Tasmania, Flinders and King Is.

ACANTHIZIDAE THORNBILLS, GERYGONES (14: 60)

PYCNOPTILUS Gould, 1851 M
- ○ ***Pycnoptilus floccosus*** PILOTBIRD
    - __ *P.f.sandlandi* Mathews, 1912[7] — Montane SE New South Wales, E Victoria
    - __ *P.f.floccosus* Gould, 1851 — Lowlands and foothills in SE New South Wales, NE Victoria

ACANTHORNIS Legge, 1887[8] F[9]
- ○ ***Acanthornis magna*** SCRUBTIT
    - __ *A.m.magna* (Gould, 1855) — Tasmania
    - __ *A.m.greeniana* Schodde & Mason, 1999 — King I. (2189)

ORIGMA Gould, 1838 F
- ● ***Origma solitaria*** (Lewin, 1808) ROCK WARBLER — EC New South Wales

CALAMANTHUS Gould, 1838[10] M
- ○ ***Calamanthus pyrrhopygius*** CHESTNUT-RUMPED HEATHWREN
    - __ *C.p.pyrrhopygius* (Vigors & Horsfield, 1827) — SE Australia
    - __ *C.p.parkeri* Schodde & Mason, 1999 — Mt. Lofty Range (South Australia) (2189)
    - __ *C.p.pedleri* Schodde & Mason, 1999 — S Flinders Range (South Australia) (2189)

- ○ ***Calamanthus cautus*** SHY HEATHWREN
    - __ *C.c.macrorhynchus* Schodde & Mason, 1999 — C New South Wales (2189)
    - __ *C.c.cautus* (Gould, 1843) — SC Australia
    - __ *C.c.halmaturinus* (Mathews, 1912)[11] — Kangaroo I.
    - __ *C.c.whitlocki* (Mathews, 1912)[12] — S Western Australia

- ○ ***Calamanthus fuliginosus*** STRIATED FIELDWREN
    - __ *C.f.albiloris* North, 1902[13] — SE New South Wales, SE Victoria
    - __ *C.f.bourneorum* Schodde & Mason, 1999 — SW Victoria, SE South Australia (2189)

[1] For reasons to treat this group as a family, close to the Acanthizidae, see Schodde & Mason (1999: 121) (2189).
[2] For polytypic treatment see Schodde & Mason (1999: 122) (2189).
[3] Implicitly includes *parryi* see Schodde & Mason (1999: 126) (2189).
[4] Implicitly includes *carpentariae* see Schodde & Mason (1999: 126) (2189).
[5] The component parts of this broad species, treated separately in last Edition, were united by Schodde (1975) (2175).
[6] Implicitly includes *restrictus*, *barroni* and *bowensis* see map in Mason & Schodde (1999: 128) (2189).
[7] For recognition see Schodde & Mason (1999: 140) (2189).
[8] Treated in the genus *Sericornis* in last Edition. For a diagnosis supporting the recognition of this genus see Schodde & Mason (1999: 142) (2189).
[9] For correct gender and spelling see David & Gosselin (2002) (614).
[10] Species sequence follows Schodde & Mason (1999), in whose work the genus *Hylacola* is submerged as it is herein.
[11] For recognition see Schodde & Mason (1999: 149) (2189).
[12] For recognition see Schodde & Mason (1999: 149) (2189).
[13] For recognition see Schodde & Mason (1999: 150) (2189).

— *C.f.fuliginosus* (Vigors & Horsfield, 1827) E Tasmania
— *C.f.diemenensis* North, 1904[1] W Tasmania

○ *Calamanthus montanellus* Milligan, 1903 WESTERN FIELDWREN[2] SW Australia

○ *Calamanthus campestris* RUFOUS FIELDWREN
— *C.c.winiam* A.J. & A.G. Campbell, 1927 W Victoria, SE South Australia
— *C.c.campestris* (Gould, 1841)[3] SC Australia
— *C.c.rubiginosus* A.J. Campbell, 1899 WC Western Australia
— *C.c.hartogi* Carter, 1916[4] Dirk Hartog I.
— *C.c.dorrie* Mathews, 1912 Shark Bay Is. (W Australia)
— *C.c.wayensis* Mathews, 1912[5] C Western Australia
— *C.c.isabellinus* North, 1896 Extreme E New South Wales, C South Australia

PYRRHOLAEMUS Gould, 1841 M
○ *Pyrrholaemus brunneus* Gould, 1841 REDTHROAT Inland W, C and S Australia

○ *Pyrrholaemus sagittatus* (Latham, 1802)[6] SPECKLED WARBLER[7] EC to SE Australia

OREOSCOPUS North, 1905[8] M
○ *Oreoscopus gutturalis* (De Vis, 1889) FERNWREN NE Queensland

CRATEROSCELIS Sharpe, 1883 F
○ *Crateroscelis murina* LOWLAND MOUSE WARBLER
— *C.m.murina* (P.L. Sclater, 1858) Salwati I., Yapen I., New Guinea
— *C.m.monacha* (G.R. Gray, 1858) Aru Is.
— *C.m.pallida* Rand, 1938 SC New Guinea
— *C.m.capitalis* Stresemann & Paludan, 1932 Waigeo I.
— *C.m.fumosa* Ripley, 1957 Misool I. (Geelvink Bay)

○ *Crateroscelis nigrorufa* BICOLOURED MOUSE WARBLER
— *C.n.blissi* Stresemann & Paludan, 1934 Midmontane WC New Guinea
— *C.n.nigrorufa* (Salvadori, 1894) Midmontane E New Guinea

○ *Crateroscelis robusta* MOUNTAIN MOUSE WARBLER
— *C.r.peninsularis* E. Hartert, 1930 NW New Guinea (Arfak Mts.)
— *C.r.ripleyi* Mayr & Meyer de Schauensee, 1939 NW New Guinea (Tamrau Mts.)
— *C.r.bastille* Diamond, 1969 Mts. of coastal N New Guinea
— *C.r.deficiens* E. Hartert, 1930 N New Guinea (Cyclops Mts.)
— *C.r.sanfordi* E. Hartert, 1930 Montane WC and C New Guinea
— *C.r.robusta* (De Vis, 1898)[9] Montane E New Guinea

SERICORNIS Gould, 1838[10] M
○ *Sericornis spilodera* PALE-BILLED SERICORNIS
— *S.s.ferrugineus* Stresemann & Paludan, 1932 Waigeo I.
— *S.s.batantae* Mayr, 1986 Batanta I. (W Papuan Is.)
— *S.s.spilodera* (G.R. Gray, 1859) Yapen I., NW and N New Guinea
— *S.s.granti* (E. Hartert, 1930) WC New Guinea
— *S.s.wuroi* Mayr, 1937 S New Guinea
— *S.s.guttatus* (Sharpe, 1882) SE New Guinea
— *S.s.aruensis* Ogilvie-Grant, 1911 Aru Is.

○ *Sericornis papuensis* PAPUAN SERICORNIS
— *S.p.meeki* Rothschild & Hartert, 1913 Montane W New Guinea
— *S.p.buergersi* Stresemann, 1921 Montane C New Guinea
— *S.p.papuensis* (De Vis, 1894) Montane SE New Guinea

---

[1] For recognition see Schodde & Mason (1999: 151) (2189).
[2] For a discussion of the separation of this see Schodde & Mason (1999: 152-153) (2189).
[3] Includes *ethelae* see Schodde & Mason (1999: 154) (2189).
[4] For recognition see Schodde & Mason (1999: 155) (2189).
[5] For recognition see Schodde & Mason (1999: 156) (2189).
[6] For reasons to use 1802 not 1801 see Browing & Monroe (1991) (258).
[7] For the submergence of the genus *Chthonicola* see Schodde & Mason (1999: 158) (2189).
[8] Treated in *Crateroscelis* in last Edition; recognition of this monotypic genus follows Christidis & Boles (1994) (351).
[9] Includes *pratti* Engilis & Cole, 1991 (801) (R. Schodde).
[10] Species sequence follows Christidis *et al.* (1988) (353).

○ *Sericornis keri* Mathews, 1920 ATHERTON SCRUBWREN — NE Queensland

● *Sericornis frontalis* WHITE-BROWED SCRUBWREN[1]

    __ *S.f.laevigaster* Gould, 1847[2] — E Queensland

    __ *S.f.tweedi* Mathews, 1922[3] — NE New South Wales

    __ *S.f.frontalis* (Vigors & Horsfield, 1827)[4] — SE Australia

    __ *S.f.harterti* Mathews, 1912[5] — SC Victoria, Kent group (Bass Strait)

    __ *S.f.rosinae* Mathews, 1912[6] — Mt. Lofty Range (SC South Australia)

    __ *S.f.ashbyi* Mathews, 1912[7] — Kangaroo I.

    __ *S.f.mellori* Mathews, 1912[8] — SC and SW South Australia, SE and SC Western Australia

    __ *S.f.maculatus* Gould, 1847[9] — SW Western Australia

    __ *S.f.balstoni* Ogilvie-Grant, 1909 — C coast of W Australia

    __ *S.f.flindersi* White & Mellor, 1913[10] — Flinders I.

    __ *S.f.humilis* Gould, 1838 — Tasmania

    __ *S.f.tregellasi* (Mathews, (1914) — King I.

○ *Sericornis citreogularis* YELLOW-THROATED SCRUBWREN

    __ *S.c.cairnsi* Mathews, 1912 — N Queensland

    __ *S.c.intermedius* Mathews, 1912[11] — SE Queensland, NE New South Wales

    __ *S.c.citreogularis* Gould, 1838 — EC and SE New South Wales

○ *Sericornis magnirostra* LARGE-BILLED SCRUBWREN[12] [13]

    __ *S.m.imitator* Mayr, 1937 — NW New Guinea (Arfak Mts.)

    __ *S.m.wondiwoi* Mayr, 1937 — NW New Guinea (Wandammen Pen.)

    __ *S.m.weylandi* Mayr, 1937 — WC New Guinea (Weyland Mts.)

    __ *S.m.idenburgi* Rand, 1941 — NC New Guinea

    __ *S.m.jobiensis* Stresemann & Paludan, 1932 — Yapen I. (Geelvink Bay)

    __ *S.m.cyclopum* E. Hartert, 1930 — NC New Guinea (Cyclops Mts.)

    __ *S.m.boreonesioticus* Diamond, 1969 — NC New Guinea (N Sepik Mts.)

    __ *S.m.virgatus* (Reichenow, 1915) — C New Guinea

    __ *S.m.pontifex* Stresemann, 1921 — C New Guinea

    __ *S.m.randi* Mayr, 1937 — SC New Guinea

    __ *S.m.beccarii* Salvadori, 1874 — Aru Is.

    __ *S.m.minimus* Gould, 1875 — N Cape York Pen.

    __ *S.m.dubius* Mayr, 1937 — EC to SE Cape York Pen.

    __ *S.m.viridior* Mathews, 1912 — NE Queensland

    __ *S.m.magnirostra* (Gould, 1838) — E Australia

    __ *S.m.howei* Mathews, 1912[14] — SE Victoria

○ *Sericornis nouhuysi* LARGE SCRUBWREN

    __ *S.n.cantans* Mayr, 1930 — Mts. of NW New Guinea (Vogelkop)

    __ *S.n.nouhuysi* van Oort, 1909 — Mts. of WC New Guinea

    __ *S.n.stresemanni* Mayr, 1930 — Mts. of C and EC New Guinea

    __ *S.n.adelberti* T.K. Pratt, 1983 — Adelbert Mts. (NE New Guinea) (1921)

    __ *S.n.oorti* Rothschild & Hartert, 1913 — Mts. of SE New Guinea and Huon Pen.

    __ *S.n.monticola* Mayr & Rand, 1936 — Subalpine SE New Guinea

○ *Sericornis perspicillatus* Salvadori, 1896 BUFF-FACED SCRUBWREN — Mts. of WC to SE New Guinea

○ *Sericornis rufescens* (Salvadori, 1876) VOGELKOP SCRUBWREN — Mts. of NW New Guinea

○ *Sericornis arfakianus* (Salvadori, 1876) GREY-GREEN SCRUBWREN — Mts. throughout New Guinea

[1] In last Edition we recognised three species *S. maculatus*, *humilis* and *frontalis* as in Mayr (1986) (1487). Here we follow Schodde & Mason (1999: 169) (2189). The name *osculans* relates to intergrading birds (op. cit. p. 171).
[2] Includes *herbertoni* see Schodde & Mason (1999: 165) (2189).
[3] For recognition see Schodde & Mason (1999: 166) (2189).
[4] Includes *longirostris*, *gularis* and *insularis* see Schodde & Mason (1999: 165, 169) (2189).
[5] For recognition see Schodde & Mason (1999: 166) (2189).
[6] For recognition see Schodde & Mason (1999: 166) (2189).
[7] For recognition see Schodde & Mason (1999: 167) (2189).
[8] For recognition see Schodde & Mason (1999: 167) (2189).
[9] Includes *mondraini* see Schodde & Mason (1999: 165) (2189).
[10] For recognition see Schodde & Mason (1999: 168) (2189).
[11] For recognition see Schodde & Mason (1999: 163) (2189).
[12] For correction of spelling see David & Gosselin (2002) (613).
[13] In last Edition this was treated as two species, *S. beccarii* and *S. magnirostris* as in Mayr (1986) (1487); but the two have been found to intergrade, for a discussion see Ford & Schodde in Schodde & Mason (1999: 174) (2189).
[14] For recognition see Schodde & Mason (1999: 174) (2189).

SMICRORNIS Gould, 1843 M
○ *Smicrornis brevirostris* WEEBILL[1]
    — *S.b.flavescens* Gould, 1843   N and NC Australia
    — *S.b.brevirostris* (Gould, 1838)   E Australia
    — *S.b.occidentalis* Bonaparte, 1850[2]   W and C southern Australia
    — *S.b.ochrogaster* Schodde & Mason, 1999   WC Australia (2189)

GERYGONE Gould, 1841[3] F
○ *Gerygone mouki* BROWN GERYGONE
    — *G.m.mouki* Mathews, 1912   NE Queensland
    — *G.m.amalia* Meise, 1931   EC Queensland
    — *G.m.richmondi* (Mathews, 1915)   SE Queensland, E New South Wales, SE Victoria

● *Gerygone igata* GREY WARBLER
    — *G.i.modesta* Pelzeln, 1860   Norfolk I.
    — †*G.i.insularis* E.P. Ramsay, 1878   Lord Howe Is.
    ✓ *G.i.igata* (Quoy & Gaimard, 1830)   New Zealand

○ *Gerygone albofrontata* G.R. Gray, 1844 CHATHAM ISLAND GERYGONE Chatham Is.

○ *Gerygone flavolateralis* FAN-TAILED GERYGONE
    — *G.f.flavolateralis* (G.R. Gray, 1859)   New Caledonia and Mare I. (Loyalty Is.)
    — *G.f.lifuensis* (F. Sarasin, 1913)   Lifu I. (Loyalty Is.)
    — *G.f.rouxi* (F. Sarasin, 1913)   Uvea I. (Loyalty Is.)
    — *G.f.correiae* Mayr, 1931   N Vanuatu, Banks Is.
    — *G.f.citrina* Mayr, 1931   Rennell I. (Solomons)

○ *Gerygone ruficollis* TREEFERN GERYGONE
    — *G.r.ruficollis* Salvadori, 1876   Mts. of NW New Guinea
    — *G.r.insperata* De Vis, 1892   Mts. of WC to SE New Guinea

○ *Gerygone sulphurea* GOLDEN-BELLIED GERYGONE
    — *G.s.flaveola* Cabanis, 1873   Sulawesi, Peleng and Salayer Is.
    — *G.s.sulphurea* Wallace, 1864   Vietnamese to eastern Thai coastline, Malay Pen., Gtr.
                                      Sundas and Lesser Sunda to Alor I.
    — *G.s.muscicapa* Oberholser, 1912[4]   Enggano I. (off Sumatra)
    — *G.s.simplex* Cabanis, 1872[5]   N and C Philippines
    — *G.s.rhizophorae* Mearns, 1905   Mindanao, Basilan, Sulu Arch.

○ *Gerygone dorsalis* RUFOUS-SIDED GERYGONE
    — *G.d.senex* Meise, 1929   Kalaotua, Madu Is. (Flores Sea)
    — *G.d.kuehni* E. Hartert, 1900   Damar I. (Lesser Sundas)
    — *G.d.fulvescens* A.B. Meyer, 1885   Roma, Moa, Kisar, Babar Is. (Lesser Sundas)
    — *G.d.keyensis* Büttikofer, 1893   Kai Is. (Banda Sea)
    — *G.d.dorsalis* P.L. Sclater, 1883   Tamimbar Is. (Banda Sea)

○ *Gerygone levigaster* MANGROVE GERYGONE
    — *G.l.levigaster* Gould, 1843   Coastal N Australian
    — *G.l.cantator* (Weatherill, 1908)[6]   Coastal E Australian
    — *G.l.pallida* Finsch, 1898   S New Guinea

○ *Gerygone inornata* Wallace, 1864 PLAIN GERYGONE   Timor, Wetar, Roti Is. (Lesser Sundas)

○ *Gerygone fusca* WHITE-TAILED GERYGONE
    — *G.f.fusca* (Gould, 1838)   SW and SC Western Australia
    — *G.f.exsul* Mathews, 1912[7]   E Australia
    — *G.f.mungi* Mathews, 1912   WC to EC Australia

---

[1] Several areas of intergradation have led to the use of the names *cairnsi, pallescens, stirlingi* and *mallee*; see map in Schodde & Mason (1999: 176) (2189); all now in synonymy.
[2] Schodde & Mason (1999: 178) (2189) explained the availability of this name.
[3] Species sequence arranged by R. Schodde.
[4] For recognition see Mees (1986) (1539).
[5] Recognition of *simplex* and *rhizophorae* by Dickinson *et al.* (1991) (745) did not review whether *simplex* is distinct from Bornean birds. Hence arrangement tentative.
[6] For correct spelling see David & Gosselin (2002) (613).
[7] For recognition see Schodde & Mason (1999: 185) (2189).

○ *Gerygone tenebrosa* DUSKY GERYGONE
 __ *G.t.tenebrosa* (R. Hall, 1901) — NW Western Australia
 __ *G.t.christophori* Mathews, 1912[1] — Coastal WC Western Australia

○ *Gerygone magnirostris* LARGE-BILLED GERYGONE
 __ *G.m.cobana* (Mathews, 1926) — Waigeo, Batanta, Salwati Is. (off NW New Guinea)
 __ *G.m.conspicillata* (G.R. Gray, 1859) — NW New Guinea
 __ *G.m.hypoxantha* Salvadori, 1878[2] — Biak I. (Geelvink Bay)
 __ *G.m.affinis* A.B. Meyer, 1874 — Yapen, Dampier Is., N New Guinea
 __ *G.m.occasa* Ripley, 1957 — Kafiau I. (off W New Guinea)
 __ *G.m.proxima* Rothschild & Hartert, 1918 — D'Entrecasteaux Arch.
 __ *G.m.onerosa* E. Hartert, 1899 — Misima I. (Louisiade Arch.)
 __ *G.m.tagulana* Rothschild & Hartert, 1918 — Tagula I. (Louisiade Arch.)
 __ *G.m.rosseliana* E. Hartert, 1899 — Sudest I. (Louisiade Arch.)
 __ *G.m.brunneipectus* (Sharpe, 1879)[3] — Aru Is., SW to SE New Guinea
 __ *G.m.magnirostris* Gould, 1843 — N Northern Territory (Arnhem Land), N Western Australia (W Kimberly)
 __ *G.m.cairnsensis* Mathews, 1912 — NE Queensland

○ *Gerygone chrysogaster* YELLOW-BELLIED GERYGONE
 __ *G.c.neglecta* Wallace, 1865 — Waigeo I.
 __ *G.c.notata* Salvadori, 1878 — NW New Guinea (Vogelkop), Misool, Batanta Is.
 __ *G.c.leucothorax* Mayr, 1940 — NW New Guinea (head of Geelvink Bay)
 __ *G.c.dohertyi* Rothschild & Hartert, 1903 — SW New Guinea
 __ *G.c.chrysogaster* G.R. Gray, 1858[4] — Aru Is., S, E and NC New Guinea, Yapen I.

○ *Gerygone cinerea* Salvadori, 1876 GREY GERYGONE — New Guinea

○ *Gerygone chloronota* GREEN-BACKED GERYGONE[5]
 __ *G.c.cinereiceps* (Sharpe, 1886) — Waigeo I. (?), New Guinea
 __ *G.c.aruensis* Büttikofer, 1893 — Aru Is.
 __ *G.c.chloronota* Gould, 1843 — N Northern Territory (Arnhem Land)
 __ *G.c.darwini* Mathews, 1912[6] — N Western Australia (Kimberley)

○ *Gerygone olivacea* WHITE-THROATED GERYGONE
 __ *G.o.cinerascens* Sharpe, 1878 — SE New Guinea, Cape York Pen.
 __ *G.o.rogersi* Mathews, 1911 — NW to NC Australia
 __ *G.o.olivacea* (Gould, 1838) — E Australia

○ *Gerygone palpebrosa* FAIRY GERYGONE
 __ *G.p.palpebrosa* Wallace, 1865 — Aru Is., W Papuan Is., NW New Guinea (Vogelkop)
 __ *G.p.wahnesi* (A.B. Meyer, 1899) — Yapen I., N New Guinea
 __ *G.p.inconspicua* E.P. Ramsay, 1879 — SE New Guinea
 __ *G.p.tarara* Rand, 1941 — SC New Guinea
 __ *G.p.personata* Gould, 1866 — Cape York Pen.
 __ *G.p.flavida* E.P. Ramsay, 1877 — E Queensland

ACANTHIZA Vigors & Horsfield, 1827[7] F
○ *Acanthiza katherina* De Vis, 1905 MOUNTAIN THORNBILL — NE Queensland

● *Acanthiza pusilla* BROWN THORNBILL
 __ *A.p.dawsonensis* A.G. Campbell, 1922[8] — EC Queensland
 ✔ *A.p.pusilla* (Shaw, 1790)[9] — Coastal and subcoastal SE Australia
 __ *A.p.diemenensis* Gould, 1838 — Tasmania, Kent group (Bass Strait)
 __ †?*A.p.archibaldi* Mathews, 1910 — King I.
 __ *A.p.zietzi* North, 1904 — Kangaroo I.

---

[1] Includes *whitlocki* see Ford (1983) (860).
[2] Sometimes separated but see Ford (1986) (861).
[3] Includes *mimikae* see Ford (1983) (860).
[4] The name *ruficauda* Ford & Johnstone, 1983 (865) was based on foxed specimens of this taxon.
[5] For correct spelling see David & Gosselin (2002) (613).
[6] For recognition see Schodde & Mason (1999: 190) (2189).
[7] Species sequence arranged by R. Schodde.
[8] The name *mcgilli* is a synonym see Schodde & Mason (1999: 198) (2189).
[9] The name *bunya* relates to a population intergrading between this and *dawsonensis* see Schodde & Mason (1999: 198) (2189).

○ *Acanthiza apicalis* INLAND THORNBILL
 — *A.a.cinerascens* Schodde & Mason, 1999   WC Queensland (2189)
 — *A.a.whitlocki* North, 1909[1]   WC and C Australia
 — *A.a.apicalis* Gould, 1847[2]   SW and SC Australia
 — *A.a.albiventris* North, 1904   Inland SE Australia

○ *Acanthiza ewingii* TASMANIAN THORNBILL
 — *A.e.ewingii* Gould, 1844   Tasmania, Flinders I.
 — *A.e.rufifrons* A.J. Campbell, 1903[3]   King I.

○ *Acanthiza murina* (De Vis, 1897) PAPUAN THORNBILL   High montane C and SE New Guinea

○ *Acanthiza uropygialis* Gould, 1838 CHESTNUT-RUMPED THORNBILL[4]   Inland W, SC and E Australia

○ *Acanthiza reguloides* BUFF-RUMPED THORNBILL
 — *A.r.squamata* De Vis, 1890   Lower NE to EC Queensland
 — *A.r.nesa* (Mathews, 1920)[5]   SE Queensland
 — *A.r.reguloides* Vigors & Horsfield, 1827   SE Australia
 — *A.r.australis* (North, 1904)[6]   SW Victoria, SE South Australia

○ *Acanthiza inornata* Gould, 1841 WESTERN THORNBILL[7]   SW Western Australia

○ *Acanthiza iredalei* SLENDER-BILLED THORNBILL
 — *A.i.iredalei* Mathews, 1911   WC and SC Australia
 — *A.i.hedleyi* Mathews, 1912   W Victoria, SE South Australia
 — *A.i.rosinae* Mathews, 1913   Upper Gulf St. Vincent, South Australia

○ *Acanthiza chrysorrhoa* YELLOW-RUMPED THORNBILL[8]
 — *A.c.normantoni* (Mathews, 1913)   Inland C and NE Australia
 — *A.c.leighi* Ogilvie-Grant, 1909[9]   SE Australia
 — *A.c.leachi* Mathews, 1912[10]   C and E Tasmania
 — *A.c.chrysorrhoa* (Quoy & Gaimard, 1830)[11]   SW and WC Australia

○ *Acanthiza nana* YELLOW THORNBILL
 — *A.n.flava* H.L. White, 1922   NE Queensland
 — *A.n.modesta* De Vis, 1905   Inland E and SE Australia
 — *A.n.nana* Vigors & Horsfield, 1827   Coastal SE Queensland, E New South Wales

○ *Acanthiza lineata* STRIATED THORNBILL
 — *A.l.alberti* Mathews, 1920   SE Queensland
 — *A.l.lineata* Gould, 1838   SE Australia
 — *A.l.clelandi* Mathews, 1912[12]   SE South Australia, W Victoria
 — *A.l.whitei* Mathews, 1912[13]   Kangaroo I.

○ *Acanthiza robustirostris* Milligan, 1903 SLATY-BACKED THORNBILL   WC, C and inland EC Australia

APHELOCEPHALA Oberholser, 1899 F
○ *Aphelocephala leucopsis* SOUTHERN WHITEFACE
 — *A.l.castaneiventris* (Milligan, 1903)   WC and C Western Australia
 — *A.l.leucopsis* (Gould, 1841)[14]   C, S and inland SE Australia

○ *Aphelocephala pectoralis* (Gould, 1871) CHESTNUT-BREASTED WHITEFACE
     C South Australia

---

[1] Includes *tanami* see Schodde & Mason (1999: 201) (2189).
[2] The name *leeuwinensis* relates to a population intergrading between this and *whitlocki* see Schodde & Mason (1999: 201) (2189).
[3] For recognition see Schodde & Mason (1999: 204) (2189).
[4] Treated as monotypic see Schodde & Mason (1999: 206) (2189); *augusta* is a synonym.
[5] For recognition see Schodde & Mason (1999: 207) (2189).
[6] For recognition see Schodde & Mason (1999: 208) (2189).
[7] Treated as monotypic see Schodde & Mason (1999: 206) (2189); *mastersi* is a synonym.
[8] The name *ferdinandi* appears to relate to intergrading populations.
[9] For recognition see Schodde & Mason (1999: 212) (2189). This is the prior name once *chrysorrhoa* is moved to Western Australia; *sandlandi* and *addenda* are synonyms.
[10] For recognition see Schodde & Mason (1999: 213) (2189).
[11] Schodde & Mason (1999: 213) (2189) restricted the type locality to King George Sound so this name cannot apply to the New South Wales population; *multi* and *pallida* are synonyms.
[12] Includes *chandleri* see Schodde & Mason (1999: 216) (2189).
[13] For recognition see Schodde & Mason (1999: 217) (2189).
[14] Includes *whitei* see Schodde & Mason (1999: 219) (2189).

○ *Aphelocephala nigricincta* (North, 1895) BANDED WHITEFACE   WC and C Australia

MOHOUA Lesson, 1835 F
○ *Mohoua ochrocephala* YELLOWHEAD[1]
     — *M.o.albicilla* (Lesson, 1830)   North Island (New Zealand)
     — *M.o.ochrocephala* (J.F. Gmelin, 1789)   South Island (New Zealand)

FINSCHIA Hutton, 1903 F
● *Finschia novaeseelandiae* (J.F. Gmelin, 1789) PIPIPI/NEW ZEALAND CREEPER
     South Island (New Zealand)

POMATOSTOMIDAE AUSTRALASIAN BABBLERS (2: 5)

GARRITORNIS Iredale, 1956 M
○ *Garritornis isidorei* ISIDORE'S RUFOUS BABBLER
     — *G.i.isidorei* (Lesson, 1827)   C and S New Guinea, Misool I.
     — *G.i.calidus* (Rothschild, 1931)   N New Guinea

POMATOSTOMUS Cabanis, 1851 M
○ *Pomatostomus temporalis* GREY-CROWNED BABBLER
     — *P.t.strepitans* (Mayr & Rand, 1935)   S New Guinea
     — *P.t.temporalis* (Vigors & Horsfield, 1827)[2]   E and SE Australia
     — *P.t.rubeculus* (Gould, 1840)[3]   NW, N and C Australia

○ *Pomatostomus halli* Cowles, 1964 HALL'S BABBLER   C and SC Queensland, NW New South Wales (576A)

○ *Pomatostomus superciliosus* WHITE-BROWED BABBLER
     — *P.s.gilgandra* (Mathews, 1912)   SE Australia
     — *P.s.superciliosus* (Vigors & Horsfield, 1827)[4]   W and S Australia
     — *P.s.ashbyi* Mathews, 1911   SW Western Australia
     — *P.s.centralis* Schodde & Mason, 1999   C Australia (2189)

○ *Pomatostomus ruficeps* (Hartlaub, 1852) CHESTNUT-CROWNED BABBLER
     SW Queensland, W New South Wales, NW Victoria, E South Australia

ORTHONYCHIDAE LOGRUNNERS (1: 3)

ORTHONYX Temminck, 1820 M
○ *Orthonyx novaeguineae* PAPUAN LOGRUNNER
     — *O.n.novaeguineae* A.B. Meyer, 1874[5]   Mts. of W New Guinea (Vogelkop)
     — *O.n.victorianus* van Oort, 1909[6]   Mts. of C and E New Guinea

○ *Orthonyx temminckii* Ranzani, 1822 AUSTRALIAN LOGRUNNER[7]   SE Queensland, E New South Wales

○ *Orthonyx spaldingii* CHOWCHILLA
     — *O.s.melasmenus* Schodde & Mason, 1999   NE Queensland (north of *spaldingii*) (2189)
     — *O.s.spaldingii* E.P. Ramsay, 1868   Cairns-Atherton region to Paluma Range (NE Queensland)

CNEMOPHILIDAE SATIN BIRDS (2: 3)

CNEMOPHILUS De Vis, 1890 M
○ *Cnemophilus loriae* LORIA'S CNEMOPHILUS[8]
     — *C.l.inexpectata* (Junge, 1939)   Mts. of WC New Guinea

[1] Sometimes treated as two species.
[2] Includes *tregallasi*, *trivirgatus* and *cornwalli* see Schodde & Mason (1999: 400-402) (2189).
[3] Includes *intermedius*, *mountfordae*, *browni*, *bamba* and *nigrescens* see Schodde & Mason (1999: 400-402) (2189).
[4] Includes *gwendolenae* see Schodde & Mason (1999: 404-405) (2189).
[5] Includes *dorsalis* see Joseph *et al.* (2001) (1227). Note however that there is a significant genetic gap between *novaeguineae* and *victorianus* and that if specific distinctness is accorded then *victorianus* has a subspecies *dorsalis*.
[6] For correction of spelling see David & Gosselin (2002) (614).
[7] For treatment at specific level see Joseph *et al.* (2001) (1227).
[8] Treated in a monotypic genus *Loria* in last Edition.

— *C.l.amethystina* (Stresemann, 1934)	Mts. of EC New Guinea
— *C.l.loriae* (Salvadori, 1894)	Mts. of SE New Guinea

○ *Cnemophilus macgregorii* CRESTED CNEMOPHILUS

— *C.m.sanguineus* Iredale, 1948	Mts. of EC New Guinea
— *C.m.macgregorii* De Vis, 1890	Mts. of SE New Guinea

LOBOPARADISEA Rothschild, 1896 F
○ *Loboparadisea sericea* YELLOW-BREASTED CNEMOPHILUS

— *L.s.sericea* Rothschild, 1896	Mts. of C New Guinea
— *L.s.aurora* Mayr, 1930	Mts. of E New Guinea

MELANOCHARITIDAE BERRYPECKERS (6: 12)

MELANOCHARIS P.L. Sclater, 1858 F
○ †?*Melanocharis arfakiana* (Finsch, 1900) OBSCURE BERRYPECKER — Midmontane New Guinea

○ *Melanocharis nigra* BLACK BERRYPECKER

— *M.n.pallida* Stresemann & Paludan, 1932	Waigeo I.
— *M.n.nigra* (Lesson, 1830)	Misool and Salawatti Is., W New Guinea
— *M.n.unicolor* Salvadori, 1878	Yapen I., N and E New Guinea
— *M.n.chloroptera* Salvadori, 1876	Aru Is., S New Guinea

○ *Melanocharis longicauda* LEMON-BREASTED BERRYPECKER

— *M.l.longicauda* Salvadori, 1876	NW New Guinea (incl. Vogelkop)
— *M.l.chloris* Stresemann & Paludan, 1934	WC and SW New Guinea
— *M.l.umbrosa* Rand, 1941	NC New Guinea
— *M.l.captata* Mayr, 1931	C and NE New Guinea
— *M.l.orientalis* Mayr, 1931	SE New Guinea

○ *Melanocharis versteri* FAN-TAILED BERRYPECKER

— *M.v.versteri* (Finsch, 1876)	Mts. of NW New Guinea
— *M.v.meeki* (Rothschild & Hartert, 1911)	Mts. of WC and C New Guinea
— *M.v.virago* (Stresemann, 1923)	Mts. of N, EC and NE New Guinea
— *M.v.maculiceps* (De Vis, 1898)	Mts. of SE New Guinea

○ *Melanocharis striativentris* STREAKED BERRYPECKER

— *M.s.axillaris* (Mayr, 1931)	Mts. of NW and WC New Guinea
— *M.s.striativentris* Salvadori, 1894[1]	Mts. of EC and SE New Guinea
— *M.s.chrysocome* (Mayr, 1931)	NE New Guinea (Mts. of Huon Pen.)

RHAMPHOCHARIS Salvadori, 1876[2] F
○ *Rhamphocharis crassirostris* SPOTTED BERRYPECKER

— *R.c.crassirostris* Salvadori, 1876	NW to C New Guinea
— *R.c.viridescens* Mayr, 1931	EC New Guinea
— *R.c.piperata* (De Vis, 1898)	SE New Guinea

OEDISTOMA Salvadori, 1876[3] N
○ *Oedistoma iliolophum* PLUMED LONGBILL

— *O.i.cinerascens* (Stresemann & Paludan, 1932)	Waigeo I. (W Papuan Is.)
— *O.i.affine* (Salvadori, 1876)	Mts of NW New Guinea (Vogelkop)
— *O.i.iliolophum* (Salvadori, 1876)	Yapen I., N New Guinea
— *O.i.flavum* (Mayr & Rand, 1935)	S and SE New Guinea to Huon Pen.
— *O.i.fergussone* (E. Hartert, 1896)	D'Entrecasteaux Arch.

○ *Oedistoma pygmaeum* PYGMY LONGBILL

— *O.p.waigeuense* Salomonsen, 1966	Waigeo I. (W Papuan Is. except Misool)
— *O.p.pygmaeum* Salvadori, 1876[4]	Misool I., New Guinea
— *O.p.meeki* (E. Hartert, 1896)	D'Entrecasteaux Arch.

---

[1] Includes *prasina* see Schodde (1978) (2176).
[2] This genus was submerged in *Melanocharis* in Sibley & Monroe (1990) (2262).
[3] We place this genus here following Sibley & Ahlquist (1990) (2261).
[4] Includes *flavipectus* and *olivascens* see Diamond (1969) (698).

TOXORHAMPHUS Stresemann, 1914[1] M
○ **Toxorhamphus novaeguineae** YELLOW-BELLIED LONGBILL
    __ *T.n.novaeguineae* (Lesson, 1827)   NW to SW and NE New Guinea, W Papuan Is., Yapen I.
    __ *T.n.flaviventris* (Rothschild & Hartert, 1911)   Aru Is., S New Guinea

○ **Toxorhamphus poliopterus** SLATY-CHINNED LONGBILL
    __ *T.p.maximus* Rand, 1941   Mts. of NC New Guinea
    __ *T.p.poliopterus* (Sharpe, 1882)[2]   Mts. of C to SE New Guinea

OREOCHARIS Salvadori, 1876 F
○ **Oreocharis arfaki** (A.B. Meyer, 1875) TIT-BERRYPECKER   Mts. of New Guinea

PARAMYTHIA De Vis, 1892 F
○ **Paramythia montium** CRESTED BERRYPECKER
    __ *P.m.olivacea* Van Oort, 1910[3]   Mts. of WC New Guinea
    __ *P.m.montium* De Vis, 1892   Mts. of EC and SE New Guinea
    __ *P.m.brevicauda* Mayr & Gilliard, 1954   Mts of NE New Guinea (Huon Pen.)

CALLAEATIDAE WATTLED CROWS (3: 3)

CALLAEAS J.R. Forster, 1788 M
○ **Callaeas cinereus** KOKAKO
    __ *C.c.wilsoni* (Bonaparte, 1850)   North Island (New Zealand)
    __ †?*C.c.cinereus* (J.F. Gmelin, 1788)[4]   South Island and Stewart I. (New Zealand)

PHILESTURNUS I. Geoffroy Saint-Hilaire, 1832[5]
○ **Philesturnus carunculatus** SADDLEBACK
    __ *P.c.rufusater* (Lesson, 1828)   North Island (New Zealand)
    __ *P.c.carunculatus* (J.F. Gmelin, 1789)   South Island (New Zealand), Stewart I.

HETERALOCHA Cabanis, 1851 F
○ †*Heteralocha acutirostris* (Gould, 1837) HUIA   North Island (New Zealand)

EUPETIDAE WHIPBIRDS, WEDGEBILLS AND JEWELBABBLERS[6] (4: 10)

ANDROPHOBUS Hartert & Paludan, 1934 M
○ **Androphobus viridis** (Rothschild & Hartert, 1911) PAPUAN WHIPBIRD
                                     WC New Guinea

PSOPHODES Vigors & Horsfield, 1827[7] M
◑ **Psophodes olivaceus** EASTERN WHIPBIRD
    __ *P.o.lateralis* North, 1897   NE Queensland
    ↙ *P.o.olivaceus* (Latham, 1802)[8]   EC to SE Australia

○ **Psophodes nigrogularis** Gould, 1844 WESTERN WHIPBIRD   SW Western Australia

○ **Psophodes leucogaster** MALLEE WHIPBIRD[9]
    __ *P.l.leucogaster* Howe & Ross, 1933[10]   NW Victoria, SE and SC South Australia
    __ *P.l.oberon* Schodde & Mason, 1991   SW Western Australia (2187)
    __ *P.l.lashmari* Schodde & Mason, 1991   Kangaroo I. (2187)

○ **Psophodes cristatus** (Gould, 1838) CHIRRUPING WEDGEBILL   SE Northern Territory, SW Queensland, NW New
                                        South Wales, NE South Australia

○ **Psophodes occidentalis** (Mathews, 1912) CHIMING WEDGEBILL[11]   WC and C Australia, NW South Australia

---

[1] We place this genus here following Sibley & Ahlquist (1990) (2261).
[2] Dated from 1883 by Salomonsen (1967: 342) (2159). This is the volume date, but see Howes (1896: vii) (1153) for the part dates.
[3] For correction of spelling see David & Gosselin (2002) (613).
[4] For correction of spelling see David & Gosselin (2002) (614).
[5] We use this generic name in preference to *Creadion* following the O.S.N.Z. (1990) (1668). However *Creadion* is the older name and requires I.C.Z.N. action to determine its validity.
[6] The genus *Eupetes* may deserve a monotypic family; the remaining species would then belong to the Psophodidae.
[7] We now follow Ford (1971) (856) in submerging *Sphenostoma* in *Psophodes*.
[8] For reasons to use 1802 not 1801 see Browing & Monroe (1991) (258).
[9] For reasons to treat this as a species separate from *P. nigrogularis* see Schodde & Mason (1999: 411) (2189).
[10] Includes *pondalowiensis* see Schodde & Mason (1991) (2187).
[11] For recognition of this species as distinct from *cristatus* see Ford (1987) (863).

PTILORRHOA J.L. Peters, 1940 F
○ ***Ptilorrhoa leucosticta*** SPOTTED JEWEL-BABBLER
     — *P.l.leucosticta* (P.L. Sclater, 1874)   Mts. of W New Guinea (Vogelkop)
     — *P.l.mayri* (E. Hartert, 1930)   Wandammen Mts. of W New Guinea
     — *P.l.centralis* (Mayr, 1936)   Mts. of WC New Guinea
     — *P.l.sibilans* (Mayr, 1931)   Cyclops Mts. of N New Guinea
     — *P.l.menawa* (Diamond, 1969)   Sepik Mts. of N New Guinea (698)
     — *P.l.amabilis* (Mayr, 1931)   Mts. of NE New Guinea (Huon Pen.)
     — *P.l.loriae* (Salvadori, 1896)   Mts. of SE New Guinea

○ ***Ptilorrhoa caerulescens*** BLUE JEWEL-BABBLER
     — *P.c.caerulescens* (Temminck, 1836[1])   W New Guinea
     — *P.c.neumanni* (Mayr & Meyer de Schauensee, 1939)   NC New Guinea
     — *P.c.nigricrissus* (Salvadori, 1876)[2]   S New Guinea
     — *P.c.geislerorum* (A.B. Meyer, 1892)   NE New Guinea

○ ***Ptilorrhoa castanonota*** CHESTNUT-BACKED JEWEL-BABBLER
     — *P.c.gilliardi* (Greenway, 1966)   Batanta I., W New Guinea (1007)
     — *P.c.castanonota* (Salvadori, 1876)   Mts. of W New Guinea (Vogelkop)
     — *P.c.saturata* (Rothschild & Hartert, 1911)   Mts. of SW New Guinea
     — *P.c.uropygialis* (Rand, 1940)   Mts. of NW New Guinea
     — *P.c.buergersi* (Mayr, 1931)   Mts. of C New Guinea
     — *P.c.par* (Meise, 1930)   Mts. of NE New Guinea (Huon Pen.)
     — *P.c.pulcher* (Sharpe, 1883)   Mts. of SE New Guinea

EUPETES Temminck, 1831 M
○ ***Eupetes macrocerus*** MALAY RAIL-BABBLER
     — *E.m.macrocerus* Temminck, 1831   Malay Pen., Sumatra
     — *E.m.borneensis* Robinson & Kloss, 1921   N Borneo

CINCLOSOMATIDAE QUAIL-THRUSHES (1: 5)

CINCLOSOMA Vigors & Horsfield, 1827 N
○ ***Cinclosoma punctatum*** SPOTTED QUAIL-THRUSH
     — *C.p.punctatum* (Shaw, 1794)   EC to SE Australia
     — *C.p.dovei* Mathews, 1912   E Tasmania
     — *C.p.anachoreta* Schodde & Mason, 1999   SE South Australia (Mt. Lofty Range) (2189)

○ ***Cinclosoma castanotum*** CHESTNUT-BACKED QUAIL-THRUSH[3]
     — *C.c.fordianum* Schodde & Mason, 1999   SW and SC Australia (2189)
     — *C.c.clarum* Morgan, 1926   WC and C Western Australia, SW Northern Territory,
                                      NW and C South Australia
     — *C.c.castanotum* Gould, 1840[4]   SE South Australia, NW Victoria, SW and C New South
                                      Wales

○ ***Cinclosoma cinnamomeum*** CINNAMON QUAIL-THRUSH
     — *C.c.tirariensis* Schodde & Mason, 1999   SE Northern Territory, SW Queensland, NE South
                                      Australia (2189)
     — *C.c.cinnamomeum* Gould, 1846[5]   SC Northern Territory, C South Australia, NW New
                                      South Wales, SE Queensland
     — *C.c.alisteri* Mathews, 1910[6]   SE Western Australia, SW South Australia

○ ***Cinclosoma castaneothorax*** CHESTNUT-BREASTED QUAIL-THRUSH
     — *C.c.castaneothorax* Gould, 1849   C and SC Queensland, NC New South Wales
     — *C.c.marginatum* Sharpe, 1883[7]   C Western Australia, SW Northern Territory

[1] Mees (1994) (1541) has shown that this dates from 1836; Sherborn (1898) (2242) and Deignan (1964: 234) (664) gave 1835.
[2] For correction of spelling see David & Gosselin (2002) (613).
[3] For correction of spelling see David & Gosselin (2002) (613).
[4] Includes *mayri* and *morgani* see Schodde & Mason (1999: 421) (2189); *dundasi* was considered to apply to birds intergrading between the unnamed southwestern population and *clarum* (its position in synonymy is not clear. ECD).
[5] Includes *samueli* see Schodde & Mason (1999: 424) (2189).
[6] Attached to this species by Ford (1983) (859).
[7] Attached to this species by Ford (1983) (859).

○ *Cinclosoma ajax* PAINTED QUAIL-THRUSH
    — *C.a.ajax* (Temminck, 1836[1])   W New Guinea
    — *C.a.muscale* Rand, 1940   S of C New Guinea
    — *C.a.alare* Mayr & Rand, 1935   C of S New Guinea
    — *C.a.goldiei* (E.P. Ramsay, 1879)   SE New Guinea

PLATYSTEIRIDAE SHRIKE-FLYCATCHERS, WATTLE-EYES AND BATISES (6: 28)

MEGABYAS J. & E. Verreaux, 1855[2] M
○ *Megabyas flammulatus* RED-EYED SHRIKE-FLYCATCHER
    — *M.f.flammulatus* J. & E. Verreaux, 1855   Sierra Leone to W Zaire
    — *M.f.aequatorialis* Jackson, 1904[3]   C and E Zaire to S Sudan, Uganda, W Kenya, NW
                                         Angola

BIAS Lesson, 1830 M
○ *Bias musicus* BLACK-AND-WHITE SHRIKE-FLYCATCHER
    — *B.m.musicus* (Vieillot, 1818)   Gambia and Sierra Leone to Zaire, Uganda and N Angola
    — *B.m.changamwensis* van Someren, 1919   C Kenya, E Tanzania
    — *B.m.clarens* Clancey, 1966   Mozambique, S Malawi, E Zimbabwe

DYAPHOROPHYIA Bonaparte, 1854[4] F
○ *Dyaphorophyia castanea* CHESTNUT WATTLE-EYE
    — *D.c.hormophora* Reichenow, 1901   Sierra Leone to Togo
    — *D.c.castanea* (Fraser, 1843)   Benin and Nigeria to W Kenya, NW Tanzania and N
                                         Angola

○ *Dyaphorophyia tonsa* Bates, 1911 WHITE-SPOTTED WATTLE-EYE   S Nigeria to NE Zaire

○ *Dyaphorophyia blissetti* Sharpe, 1872 RED-CHEEKED WATTLE-EYE[5]   Guinea to W Cameroon

○ *Dyaphorophyia chalybea* Reichenow, 1897 REICHENOW'S WATTLE-EYE   W and S Cameroon to Gabon and W Angola, Bioko I.

○ *Dyaphorophyia jamesoni* Sharpe, 1809 JAMESON'S WATTLE-EYE   NE Zaire to Sudan, W Kenya, NW Tanzania

○ *Dyaphorophyia concreta* YELLOW-BELLIED WATTLE-EYE
    — *D.c.concreta* (Hartlaub, 1855)   Sierra Leone to Ghana
    — *D.c.ansorgei* E. Hartert, 1905   W Angola
    — *D.c.graueri* E. Hartert, 1908[6]   Nigeria to Gabon, Zaire and W Kenya
    — *D.c.kungwensis* Moreau, 1941   W Tanzania

BATIS Boie, 1833 F
○ *Batis diops* Jackson, 1905 RUWENZORI BATIS   E Zaire, Rwanda, Burundi, SW Uganda

○ *Batis margaritae* MARGARET'S BATIS
    — *B.m.margaritae* Boulton, 1934   W Angola
    — *B.m.kathleenae* C.M.N. White, 1941   NW Zambia, SE Zaire

○ *Batis mixta* FOREST BATIS
    — *B.m.ultima* Lawson, 1962   SE Kenya
    — *B.m.mixta* (Shelley, 1889)   S Kenya, N Tanzania to N Malawi

○ *Batis capensis* CAPE BATIS[7]
    — *B.c.reichenowi* Grote, 1911   SE Tanzania
    — *B.c.sola* Lawson, 1964   N Malawi
    — *B.c.dimorpha* (Shelley, 1893)   S Malawi, N Mozambique
    — *B.c.kennedyi* Smithers & Paterson, 1956   SW and C Zimbabwe
    — *B.c.erythrophthalma* Swynnerton, 1907   E Zimbabwe, W Mozambique
    — *B.c.hollidayi* Clancey, 1952   S Mozambique, Zululand, Swaziland
    — *B.c.capensis* (Linnaeus, 1766)   S Cape Province to Orange Free State and Natal

---

[1] Mees (1994) (1541) has shown that this dates from 1836; Sherborn (1898) (2242) and Deignan (1964: 234) (664) gave 1835.
[2] A member of the genus *Bias* in last Edition; this monotypic genus recognised by Érard & Fry (1997: 548) (811).
[3] Includes *carolathi* see Érard & Fry (1997: 549) (811).
[4] Submerged in *Platysteira* in last Edition, but see Érard & Fry (1997: 555) (811).
[5] We follow Dowsett & Dowsett-Lemaire (1993) (761) and Harris & Franklin (2000) (1073) in splitting this and the next two; but see also Érard & Fry (1997: 563) (811).
[6] Includes *kumbaensis*, *harterti* and *silvae* see Érard & Fry (1997: 565) (811).
[7] In last Edition we treated a species *dimorpha* with a race *sola*. We now unite these and *reichenowi* (treated in *B. mixta* in last Edition) with *capensis* following Érard & Fry (1997: 580) (811).

○ *Batis fratrum* (Shelley, 1900) Woodward's Batis[1]   Mozambique to S Malawi and N Natal

○ *Batis molitor* Chinspot Batis
    — *B.m.puella* Reichenow, 1893   E Zaire to W and C Kenya, Tanzania
    — *B.m.pintoi* Lawson, 1966   N and C Angola, NW Zambia
    — *B.m.palliditergum* Clancey, 1955   S Angola, S Zambia to N Namibia, Transvaal
    — *B.m.molitor* (Küster, 1850)   S Mozambique to E Cape Province

○ *Batis senegalensis* (Linnaeus, 1766) Senegal Batis   Senegal and Gambia to Nigeria and Cameroon

○ *Batis orientalis* Grey-headed Batis
    — *B.o.chadensis* Alexander, 1908[2]   S Chad and Central African Republic to Sudan and W
        Ethiopia
    — *B.o.lynesi* Grant & Mackworth-Praed, 1940   NE Sudan
    — *B.o.orientalis* (Heuglin, 1871)[3]   Ethiopia and Eritrea to Somalia and N Kenya

○ *Batis soror* Reichenow, 1903 East Coast Batis   SE Kenya to S Mozambique

○ *Batis pririt* Pririt Batis
    — *B.p.affinis* (Wahlberg, 1856)   SW Angola, Namibia, W Botswana, NW Cape Province
    — *B.p.pririt* (Vieillot, 1818)   Transvaal to W Cape Province

○ *Batis minor* Black-headed Batis
    — *B.m.erlangeri* Neumann, 1907   Cameroon and Angola, Somalia and NW Tanzania
    — *B.m.minor* Erlanger, 1901[4]   S Somalia, SE Kenya, E Tanzania

○ *Batis perkeo* Neumann, 1907 Pygmy Batis   S Ethiopia, N and E Kenya, S Somalia, NE Tanzania

○ *Batis minulla* (Bocage, 1874) Angola Batis   W Angola, Congo, W Zaire

○ *Batis minima* (J. & E. Verreaux, 1855) Verreaux's Batis   Gabon, S Cameroon

○ *Batis ituriensis* Chapin, 1921 Ituri Batis   NE and E Zaire, W Uganda

○ *Batis poensis* Bioko Batis
    — *B.p.occulta* Lawson, 1984[5]   Liberia to Cameroon
    — *B.p.poensis* Alexander, 1903   Bioko I.

LANIOTURDUS Waterhouse, 1838[6] M
○ *Lanioturdus torquatus* Waterhouse, 1838 White-tailed Shrike[7]   SW Angola, NW and C Namibia

PLATYSTEIRA Jardine & Selby, 1830 F
○ *Platysteira cyanea* Brown-throated Wattle-eye
    — *P.c.cyanea* (Statius Müller, 1776)   Senegal and Gambia to SW Zaire and NW Angola
    — *P.c.nyansae* Neumann, 1905[8]   Central African Republic and Zaire to W SW Sudan
        and W Kenya
    — *P.c.aethiopica* Neumann, 1905   SE Sudan, E Ethiopia

○ *Platysteira albifrons* Sharpe, 1873 White-fronted Wattle-eye   W Angola

○ *Platysteira peltata* Black-throated Wattle-eye
    — *P.p.laticincta* Bates, 1926[9]   Highlands of W Cameroon
    — *P.p.cryptoleuca* Oberholser, 1905   Somalia to N Mozambique and E Zimbabwe
    — *P.p.mentalis* Bocage, 1878   Uganda and W Kenya to Angola and Zambia
    — *P.p.peltata* Sundevall, 1850   SE Zambia to C and S Mozambique and Natal

---

[1] Includes *sheppardi* see Érard & Fry (1997: 583) (811).
[2] For recognition see Érard & Fry (1997: 594) (811).
[3] Includes *bella* see White (1963) (2648) and Zimmerman *et al.* (1996) (2735). Recognised by Érard & Fry (1997) (811).
[4] Includes *suahelicus* see White (1963: 29) (2648).
[5] Érard & Fry (1997: 602) (811) placed *occulta* in *B. poensis* following Érard & Colston (1988) (809).
[6] Placed in the Malaconotinae in our last Edition, but see Harris & Arnott (1988) (1072) and Fry (1997: 604) (891).
[7] Treated as monotypic, with *mesicus* placed in synonymy by Érard & Fry (1997: 605) (811).
[8] Not *nyanzae* (as in last Edition) nor *nyansea* as sometimes given.
[9] In last Edition treated as a species (Bamenda Wattle-eye), and so treated by Harris & Franklin (2000) (1073); for subspecific treatment see Érard & Fry (1997: 572) (811).

GENERA INCERTAE SEDIS (2: 4)

TEPHRODORNIS Swainson, 1832 M
○ ***Tephrodornis virgatus*** LARGE WOODSHRIKE
  — *T.v.sylvicola* Jerdon, 1839   W Ghats (India)
  — *T.v.pelvicus* (Hodgson, 1837)   E Himalayas, E Ghats, N Burma
  — *T.v.jugans* Deignan, 1948   E Burma, N Thailand, SW Yunnan
  — *T.v.vernayi* Kinnear, 1924   SE Burma, SW Thailand, N Malay Pen.
  — *T.v.annectens* Robinson & Kloss, 1918   C Malay Pen.
  — *T.v.fretensis* Robinson & Kloss, 1920   S Malay Pen., N Sumatra
  — *T.v.virgatus* (Temminck, 1824)[1]   S Sumatra, Java
  — *T.v.frenatus* Büttikofer, 1887   Borneo
  — *T.v.mekongensis* Meyer de Schauensee, 1946   E and SE Thailand, S Laos, S Vietnam, Cambodia
  — *T.v.hainanus* Ogilvie-Grant, 1910   N Laos, N Vietnam, Hainan (China)
  — *T.v.latouchei* Kinnear, 1925   E Yunnan to Fujian (S China)

○ ***Tephrodornis pondicerianus*** COMMON WOODSHRIKE
  — *T.p.affinis* Blyth, 1847   Sri Lanka
  — *T.p.pallidus* Ticehurst, 1920   Pakistan, NW India
  — *T.p.pondicerianus* (J.F. Gmelin, 1789)   E India, Burma, N Thailand S Laos
  — *T.p.orientis* Deignan, 1948   Cambodia, S Vietnam

PHILENTOMA Eyton, 1845 F[2]
○ ***Philentoma pyrhoptera*** RUFOUS-WINGED PHILENTOMA
  — *P.p.pyrhoptera* (Temminck, 1836)   Malay Pen., S Vietnam, Sumatra, Borneo
  — *P.p.dubia* E. Hartert, 1894   Natuna Is.

○ ***Philentoma velata*** MAROON-BREASTED PHILENTOMA
  — *P.v.caesia* (Lesson, 1839)   Malay Pen., Sumatra, Borneo
  — *P.v.velata* (Temminck, 1825)   Java

MALACONOTIDAE HELMET-SHRIKES, BUSH SHRIKES AND PUFFBACKS (10: 52)

PRIONOPS Vieillot, 1816 M[3] [4]
○ ***Prionops plumatus*** WHITE HELMET-SHRIKE
  — *P.p.plumatus* (Shaw, 1809)[5]   Senegal to N Cameroon
  — *P.p.concinnatus* Sundevall, 1850   C Cameroon to Sudan, NW Ethiopia, NE Zaire and N Uganda
  — *P.p.cristatus* Rüppell, 1836   Eritrea, N, W and C Ethiopia, SE Sudan to NW Kenya
  — *P.p.vinaceigularis* Richmond, 1897[6]   Somalia, E and S Ethiopia, N and E Kenya, NE Tanzania
  — *P.p.poliocephalus* (Stanley, 1814)[7]   S Uganda, C and S Kenya to Angola, N Namibia, Transvaal and Natal

○ ***Prionops poliolophus*** Fischer & Reichenow, 1884 GREY-CRESTED HELMET-SHRIKE
       SW Kenya, N Tanzania

○ ***Prionops alberti*** Schouteden, 1933 YELLOW-CRESTED HELMET-SHRIKE   E Zaire

○ ***Prionops caniceps*** RED-BILLED HELMET-SHRIKE
  — *P.c.caniceps* (Bonaparte, 1850)[8]   Guinea and Mali to Togo
  — *P.c.harterti* (Neumann, 1908)   Benin to W Cameroon

○ ***Prionops rufiventris*** RUFOUS-BELLIED HELMET-SHRIKE[9]
  — *P.r.rufiventris* (Bonaparte, 1853)   Cameroon, Central African Republic to Congo and NW Zaire
  — *P.r.mentalis* (Sharpe, 1884)   NE and C Zaire, W Uganda

[1] The older name *Lanius gularis* Raffles, 1822, is preoccupied by *Lanius gularis* Bechstein, 1811.
[2] Gender addressed by David & Gosselin (2002) (614), multiple name changes occur.
[3] Harris & Franklin (2000) (1073) appeared to place the helmet-shrikes as part of the Malaconotidae.
[4] Gender addressed by David & Gosselin (2002) (614), multiple changes occur in specific names.
[5] Includes *adamauae* see White (1962: 9) (2646).
[6] Includes *melanopterus* see White (1962: 9) (2646).
[7] Includes *talacoma* see Rand (1960: 312) (1971) and *angolicus* see White (1962: 9) (2646).
[8] Given as 1851 by Rand (1960: 312) (1971) but see Zimmer (1926) (2707).
[9] For split from *P. caniceps* see Urban (2000: 496) (2453). Not split by Harris & Franklin (2000) (1073).

○ *Prionops retzii* RETZ'S HELMET-SHRIKE
　— *P.r.graculinus* Cabanis, 1868[1]
　— *P.r.tricolor* G.R. Gray, 1864[2]

　— *P.r.nigricans* (Neumann, 1899)
　— *P.r.retzii* Wahlberg, 1856

S Somalia to NE Tanzania
W and S Tanzania to E Zambia, Mozambique, E Transvaal and N Natal
SE Zaire and NW Zambia to Angola
Zimbabwe, N and W Transvaal and N Botswana to S Angola

○ *Prionops gabela* Rand, 1957 GABELA HELMET-SHRIKE

Gabela (W Angola)

○ *Prionops scopifrons* CHESTNUT-FRONTED HELMET-SHRIKE
　— *P.s.kirki* (W.L. Sclater, 1924)
　— *P.s.keniensis* (van Someren, 1923)
　— *P.s.scopifrons* (W.K.H. Peters, 1854)

Coastal S Somalia, Kenya and NE Tanzania
C Kenya
SE Tanzania, E Mozambique, E Zimbabwe

MALACONOTUS Swainson, 1824　M
○ *Malaconotus cruentus* (Lesson, 1830) FIERY-BREASTED BUSH SHRIKE[3]

Sierra Leone, Gabon and E Zaire

○ *Malaconotus monteiri* MONTEIRO'S BUSH SHRIKE
　— *M.m.perspicillatus* (Reichenow, 1894)
　— *M.m.monteiri* (Sharpe, 1870)

Mt. Cameroon
NW Angola

○ *Malaconotus blanchoti* GREY-HEADED BUSH SHRIKE
　— *M.b.approximans* (Cabanis, 1869)
　— *M.b.blanchoti* Stephens, 1826
　— *M.b.catharoxanthus* Neumann, 1899

　— *M.b.hypopyrrhus* Hartlaub, 1844
　— *M.b.interpositus* E. Hartert, 1911
　— *M.b.citrinipectus* Meise, 1968[4]
　— *M.b.extremus* Clancey, 1957

E and S Ethiopia, Somalia, C and E Kenya to N Tanzania
Senegal to N Cameroon
Central African Republic to N and W Ethiopia, Uganda, W Kenya
Tanzania to Natal, west to Rwanda and C and S Zambia
Angola, SE Zaire, W Zambia
SW Angola (1554)
E Cape Province

○ *Malaconotus lagdeni* LAGDEN'S BUSH SHRIKE
　— *M.l.lagdeni* (Sharpe, 1884)
　— *M.l.centralis* Neumann, 1920

Liberia to Ghana
E Zaire, Rwanda, W Uganda

○ *Malaconotus gladiator* (Reichenow, 1892) GREEN-BREASTED BUSH SHRIKE

SE Nigeria, W Cameroon

○ *Malaconotus alius* Friedmann, 1927 ULUGURU BUSH SHRIKE

Uluguru Mts. (E Tanzania)

CHLOROPHONEUS Cabanis, 1850[5]　M
○ *Chlorophoneus kupeensis* Serle, 1951 MOUNT KUPÉ BUSH SHRIKE

Mt. Kupé (Cameroon)

○ *Chlorophoneus multicolor* MANY-COLOURED BUSH SHRIKE
　— *C.m.multicolor* (G.R. Gray, 1849)
　— *C.m.batesi* Sharpe, 1908

　— *C.m.graueri* (E. Hartert, 1908)

Sierra Leone to Mt. Cameroon
S Cameroon and Central African Republic to NW Angola, NE Zaire to W Uganda
E Zaire

○ *Chlorophoneus nigrifrons* BLACK-FRONTED BUSH SHRIKE
　— *C.n.nigrifrons* (Reichenow, 1896)
　— *C.n.manningi* (Shelley, 1899)
　— *C.n.sandgroundi* Bangs, 1931

W and C Kenya, Tanzania, N Malawi
SE Zaire, N Zambia
SE Malawi, Mozambique, E Zimbabwe, NE Transvaal

○ *Chlorophoneus olivaceus* OLIVE BUSH SHRIKE
　— *C.o.makawa* Benson, 1945
　— *C.o.bertrandi* (Shelley, 1894)
　— *C.o.vitorum* (Clancey, 1967)
　— *C.o.interfluvius* (Clancey, 1969)
　— *C.o.olivaceus* (Shaw, 1809)[6]

SW Malawi, E Zimbabwe, N Transvaal
SE Malawi
SE Mozambique (411)
E Zimbabwe, W Mozambique (418)
S and E Cape Province, Natal, Swaziland, E and N Transvaal

[1] Includes *neumanni* see White (1962: 12) (2646).
[2] Includes *intermedius* see Urban (2000: 497) (2453).
[3] Includes *gabonensis* and *adolfifriederici* see White (1962: 36) (2646). Harris & Kimball (2000) (1073) accepted a race *gabonensis* and emphasised the existence of two or more colour morphs.
[4] Accepted tentatively.
[5] For reasons to use this genus see Harris & Franklin (2000) (1073). In our last Edition all these species were treated in the genus *Telophorus*.
[6] Includes *taylori* see Rand (1960: 335) (1971).

○ *Chlorophoneus bocagei* GREY-GREEN BUSH SHRIKE
  __ *C.b.bocagei* (Reichenow, 1894)[1] S Cameroon to NW Angola
  __ *C.b.jacksoni* (Sharpe, 1901) C and NE Zaire, Uganda, W Kenya

○ *Chlorophoneus sulfureopectus* ORANGE-BREASTED BUSH SHRIKE
  __ *C.s.sulfureopectus* (Lesson, 1831) Senegal and Guinea to NE Zaire and W Uganda
  __ *C.s.similis* (A. Smith, 1836)[2] Ethiopia to E Cape Province, W to SE Zaire and Angola

○ *Chlorophoneus viridis* GORGEOUS BUSH SHRIKE[3]
  __ *C.v.viridis* (Vieillot, 1817) S Zaire, N Angola, NW Zambia
  __ *C.v.nigricauda* (Stephenson Clarke, 1913) SE Kenya, E Tanzania
  __ *C.v.quartus* (Clancey, 1960) Mozambique, E Zimbabwe, S Malawi (386)
  __ *C.v.quadricolor* (Cassin, 1851) Natal, Swaziland, E and N Transvaal

○ *Chlorophoneus dohertyi* (Rothschild, 1901) DOHERTY'S BUSH SHRIKE W and C Kenya, W Uganda, W Rwanda, W Burundi, E Zaire

TELOPHORUS Swainson, 1831[4] M
○ *Telophorus zeylonus* BOKMAKIERIE
  __ *T.z.restrictus* Irwin, 1968 Zimbabwe (1188)
  __ *T.z.phanus* (E. Hartert, 1920) S Angola, Namibia
  __ *T.z.thermophilus* Clancey, 1960 Namibia, W Cape Province (388)
  __ *T.z.zeylonus* (Linnaeus, 1766) Transvaal, Cape Province

RHODOPHONEUS Heuglin, 1871[5] M
○ *Rhodophoneus cruentus* ROSY-PATCHED SHRIKE
  __ *R.c.cruentus* (Hemprich & Ehrenberg, 1828) SE Egypt, NE Sudan, Eritrea, N Ethiopia
  __ *R.c.hilgerti* (Neumann, 1903) Somalia, E and S Ethiopia, N and E Kenya
  __ *R.c.cathemagmenus* (Reichenow, 1887)[6] S Kenya, NE Tanzania

BOCAGIA Shelley, 1894[7] F
○ *Bocagia minuta* BLACKCAP BUSH SHRIKE
  __ *B.m.minuta* (Hartlaub, 1858)[8] Sierra Leone to Lower Congo R, Ethiopia and W Kenya
  __ *B.m.reichenowi* (Neumann, 1900)[9] E Tanzania to E Zimbabwe and Mozambique (formerly coastal Kenya)
  __ *B.m.anchietae* (Bocage, 1870) Angola to SW Tanzania, N Malawi

TCHAGRA Lesson, 1831 M[10]
○ *Tchagra australis* BROWN-CROWNED TCHAGRA
  __ *T.a.ussheri* (Sharpe, 1882) Sierra Leone to SW Nigeria
  __ *T.a.emini* (Reichenow, 1893)[11] SE Nigeria to Lower Congo R, E to S Sudan and C Kenya
  __ *T.a.minor* (Reichenow, 1887)[12] SE Kenya, E and S Tanzania to N Zimbabwe and C Mozambique
  __ *T.a.ansorgei* (Neumann, 1909) W Angola
  __ *T.a.bocagei* da Rosa Pinto, 1968 C Cuando (SE Angola) (607)
  __ *T.a.souzae* (Bocage, 1892) C Angola, S Zaire, NW Zambia
  __ *T.a.rhodesiensis* (Roberts, 1932) NE Namibia, NW Botswana, SE Angola (S Cuando), SW Zambia
  __ *T.a.australis* (A. Smith, 1836) SE Zimbabwe, NE Sth Africa, S Mozambique
  __ *T.a.damarensis* (Reichenow, 1915) NW and C Namibia, Botswana, SW Zimbabwe, N Sth. Africa

[1] Includes *ansorgei* see Rand (1960: 334) (1971).
[2] Includes *terminus* Clancey, 1959 (381) see Fry (2000: 406) (892).
[3] DJP would prefer to recognise two species, monotypic *viridis* and polytypic *T. quadricolor*. This was our treatment in last Edition; but we defer to Fry (2000: 410) (892) and Harris & Kimball (2000) (1073).
[4] We follow Harris & Kimball (2000) (1073) in treating a monotypic genus.
[5] Treated in the genus *Tchagra* in last Edition. We prefer to follow Zimmerman (1996) (2735) and employ this monotypic genus.
[6] Harris & Kimball (2000) (1073) emphasised the distinctness of this form.
[7] Treated in the genus *Tchagra* in last Edition. The generic name *Bocagia* was previously considered preoccupied by *Bocageia*. Now the one letter difference is enough.
[8] Should the use of *Antichromus* be shown to be necessary this name becomes *minutus*.
[9] Includes *remota* Clancey, 1959 (383) see White (1962: 19) (2646).
[10] Gender addressed by David & Gosselin (2002) (614), multiple changes occur in specific names.
[11] Includes *frater* see Rand (1960: 324) (1971).
[12] Includes *littoralis* and *congener* see White (1962: 21) (2646).

○ *Tchagra jamesi* Three-streaked Tchagra
— *T.j.jamesi* (Shelley, 1885)
— *T.j.mandanus* (Neumann, 1903)

Somalia, S Ethiopia, N and E Kenya, NE Tanzania
Coastal Kenya, Manda and Lamu Is.

○ *Tchagra tchagra* Southern Tchagra
— *T.t.tchagra* (Vieillot, 1816)
— *T.t.natalensis* (Reichenow, 1903)
— *T.t.caffrariae* Quickelberge, 1967

S Cape Province
Natal to W Swaziland, SE Transvaal
E Cape Province (1959)

○ *Tchagra senegalus* Black-crowned Tchagra
— *T.s.cucullatus* (Temminck, 1840)
— *T.s.percivali* (Ogilvie-Grant, 1900)
— *T.s.remigialis* (Hartlaub & Finsch, 1870)
— *T.s.nothus* (Reichenow, 1920)[1]
— *T.s.senegalus* (Linnaeus, 1766)[2]

— *T.s.habessinicus* (Hemprich & Ehrenberg, 1833)
— *T.s.warsangliensis* Stephenson Clarke, 1919
— *T.s.armenus* (Oberholser, 1906)[3]

— *T.s.orientalis* (Cabanis, 1869)[4]

— *T.s.kalahari* (Roberts, 1932)

N Africa
S Arabia
C Chad to C Sudan
Mali to W Chad
Senegal and Sierra Leone to S Chad, N Central
    African Republic
Ethiopia, Eritrea, N Somalia, SE Sudan
NE Somalia
S Cameroon to SW Sudan, south to C Angola and Tete
    (W Mozambique), S Uganda to SE Zaire, Zambia
S Somalia to E Tanzania, Mozambique, E Transvaal
    and E Cape Province
S Angola, N Namibia and SW Zambia to W Transvaal

DRYOSCOPUS Boie, 1826 M
○ *Dryoscopus sabini* Sabine's Puffback
— *D.s.sabini* (J.E. Gray, 1831)
— *D.s.melanoleucus* (J. & E. Verreaux, 1851)

Sierra Leone to S Nigeria
Cameroon to NE Zaire and N Angola

○ *Dryoscopus angolensis* Pink-footed Puffback
— *D.a.boydi* Bannerman, 1915
— *D.a.angolensis* Hartlaub, 1860
— *D.a.nandensis* Sharpe, 1900
— *D.a.kungwensis* Moreau, 1941

Cameroon
SW Zaire, N Angola
E Zaire to Sudan and W Kenya
W Tanzania

○ *Dryoscopus senegalensis* (Hartlaub, 1857) Red-eyed Puffback

S Nigeria to W Uganda and N Angola

○ *Dryoscopus cubla* Black-backed Puffback
— *D.c.affinis* (G.R. Gray, 1837)
— *D.c.nairobiensis* Rand, 1958
— *D.c.hamatus* Hartlaub, 1863[5]

— *D.c.chapini* Clancey, 1954
— *D.c.okavangensis* Roberts, 1932
— *D.c.cubla* (Shaw, 1809)

E Kenya, Zanzibar
S Kenya, N Tanzania
SW Kenya, W and S Tanzania to N Angola, N Zambia,
    E Zimbabwe and Mozambique
S Mozambique to N Transvaal
N Botswana, S Zambia to S Angola, Namibia
Transvaal, Natal, E and S Cape Province

○ *Dryoscopus gambensis* Northern Puffback
— *D.g.gambensis* (M.H.K. Lichtenstein, 1823)
— *D.g.congicus* Sharpe, 1901
— *D.g.malzacii* (Heuglin, 1871)[6]

— *D.g.erythreae* Neumann, 1899

Senegal to Cameroon and Gabon
SW Congo, W Zaire
Chad and Central African Republic to W and S Sudan
    and Kenya
E Sudan, Ethiopia, E Eritrea

○ *Dryoscopus pringlii* Jackson, 1893 Pringle's Puffback

S Somalia, S Ethiopia to NE Tanzania

LANIARIUS Vieillot, 1816 M
○ *Laniarius leucorhynchus* (Hartlaub, 1848) Sooty Boubou

Sierra Leone to Ghana, SE Nigeria to Congo, east to S
    Uganda; formerly W Kenya

[1] Includes *timbuktana* see Rand (1960: 321) (1971).
[2] Includes *pallidus* see White (1962: 22) (2646).
[3] Includes *camerunensis*, *sudanensis* and *rufofuscus* see Pearson (2000: 426) (1814).
[4] Includes *mozambicus* and *confusus* see Pearson (2000: 427) (1814).
[5] Includes *chapini* see White (1962: 16) (2646).
[6] Includes *erwini* see Fry (2000: 437) (892).

○ *Laniarius poensis* MOUNTAIN SOOTY BOUBOU[1]
    __ *L.p.camerunensis* Eisentraut, 1968   SE Nigeria, W Cameroon (797)
    __ *L.p.poensis* (Alexander, 1903)   Bioko I.
    __ *L.p.holomelas* (Jackson, 1906)   E Zaire, W Rwanda, W Uganda

○ *Laniarius fuelleborni* FÜLLEBORN'S BOUBOU
    __ *L.f.usambaricus* Rand, 1957[2]   E Tanzania
    __ *L.f.fuelleborni* (Reichenow, 1900)   S Tanzania, N Malawi, NE Zambia

○ *Laniarius funebris* (Hartlaub, 1863) SLATE-COLOURED BOUBOU[3]   S and E Ethiopia, SE Sudan and Somalia to N and W
                                                 Tanzania

○ *Laniarius luehderi* LÜHDER'S BUSH SHRIKE[4]
    __ *L.l.luehderi* Reichenow, 1874[5]   Cameroon to Congo east to S Uganda and W Kenya
    __ *L.l.brauni* Bannerman, 1939   NW Angola
    __ *L.l.amboimensis* Moltoni, 1932   W Angola

○ *Laniarius ruficeps* RED-NAPED BUSH SHRIKE
    __ *L.r.ruficeps* (Shelley, 1885)   NW Somalia
    __ *L.r.rufinuchalis* (Sharpe, 1895)   E and S Ethiopia, S Somalia, E and SE Kenya
    __ *L.r.kismayensis* (Erlanger, 1901)   Coastal S Somalia, E Kenya

○ *Laniarius liberatus* Smith, Arctander, Fjeldså & Amir, 1991 BULO BURTI BUSH SHRIKE[6] #
                         C Somalia (2281)

○ *Laniarius aethiopicus* TROPICAL BOUBOU
    __ *L.a.major* (Hartlaub, 1848)   Sierra Leone to Sudan and W Kenya, south to N
                                Zambia, N Malawi
    __ *L.a.aethiopicus* (J.F. Gmelin, 1789[7])   Eritrea, Ethiopia, NW Somalia
    __ *L.a.erlangeri* Reichenow, 1905   S Somalia
    __ *L.a.sublacteus* (Cassin, 1851)   Coastal S Somalia, Kenya, NE Tanzania, Zanzibar
    __ *L.a.ambiguus* Madarász, 1904   C and E Kenya, NE Tanzania
    __ *L.a.limpopoensis* Roberts, 1922   SE Zimbabwe, N Transvaal
    __ *L.a.mossambicus* (Fischer & Reichenow, 1880)   S Zambia and E Botswana to N and C Mozambique

○ *Laniarius ferrugineus* SOUTHERN BOUBOU
    __ *L.f.transvaalensis* Roberts, 1922   S and E Transvaal
    __ *L.f.tongensis* Roberts, 1931   S Mozambique, NE Natal
    __ *L.f.natalensis* Roberts, 1922   E Cape Province
    __ *L.f.pondoensis* Roberts, 1922   Coastal E Cape Province
    __ *L.f.savensis* da Rosa Pinto, 1963   SE Mozambique (605)
    __ *L.f.ferrugineus* (J.F. Gmelin, 1788)   SW and S Cape Province

○ *Laniarius bicolor* SWAMP BOUBOU
    __ *L.b.bicolor* (Hartlaub, 1857)[8]   Coastal Cameroon, Gabon
    __ *L.b.guttatus* (Hartlaub, 1865)   Coastal Congo to W Angola
    __ *L.b.sticturus* Finsch & Hartlaub, 1879   S Angola, N Botswana, W Zambia

○ *Laniarius turatii* (J. Verreaux, 1858) TURATI'S BOUBOU   Guinea Bissau to Sierra Leone

○ *Laniarius barbarus* YELLOW-CROWNED GONOLEK
    __ *L.b.helenae* Kelsall, 1913   Coastal Sierra Leone
    __ *L.b.barbarus* (Linnaeus, 1766)   Senegal to S Chad

○ *Laniarius mufumbiri* Ogilvie-Grant, 1911 PAPYRUS GONOLEK   Uganda, Rwanda, W Kenya

○ *Laniarius erythrogaster* (Cretzschmar, 1829) BLACK-HEADED GONOLEK   N Cameroon to W Ethiopia and N Tanzania

○ *Laniarius atrococcineus* (Burchell, 1822) CRIMSON-BREASTED GONOLEK
                               S Angola to W Zimbabwe and N Sth. Africa

[1] For separation from *fuelleborni* see Dowsett-Lemaire & Dowsett (1990) (764).
[2] Includes *ulugurensis* see Fry (2000: 444) (892).
[3] Includes *degener* see White (1962: 31) (2646).
[4] We prefer to follow White (1962: 25) (2646) and Harris & Franklin (2000) (1073) and retain a broad species here (*contra* Fry, 2000: 449) (892).
[5] Includes *castaneiceps* see White (1962: 25) (2646).
[6] This is accepted with reservations; if a sufficient viable population is found confirmatory evidence is desirable, as the DNA sample is supported by a very limited "voucher specimen" of just some moulted feathers.
[7] Rand (1960: 328) (1971) gave 1788, but see Zimmer (1926) (2707).
[8] Rand (1960: 329) (1971) cited the right source, but attributed this name to Verreaux, however Hartlaub implied it was a MS name.

○ *Laniarius atroflavus* Shelley, 1887 YELLOW-BREASTED BOUBOU[1] Highlands of Cameroon

NILAUS Swainson, 1827 M
○ *Nilaus afer* BRUBRU/BRUBRU SHRIKE
— *N.a.afer* (Latham, 1802)[2] Senegal to Ethiopia
— *N.a.camerunensis* Neumann, 1907 Cameroon, Central African Republic
— *N.a.hilgerti* Neumann, 1907 C Ethiopia
— *N.a.minor* Sharpe, 1895 E Eritrea to E Kenya
— *N.a.massaicus* Neumann, 1907 SW Kenya to Rwanda, E Zaire
— *N.a.nigritemporalis* Reichenow, 1892 SC Zaire to Tanzania and Natal
— *N.a.brubru* (Latham, 1802)[3] S Angola to N Cape Province
— *N.a.solivagus* Clancey, 1958 Natal
— *N.a.affinis* Bocage, 1878 N Angola
— *N.a.miombensis* Clancey, 1971 Sul do Save (Mozambique) (434)

MACHAERIRHYNCHIDAE BOATBILLS (1: 2)

MACHAERIRHYNCHUS Gould, 1851 M
○ *Machaerirhynchus flaviventer* YELLOW-BREASTED BOATBILL
— *M.f.albifrons* G.R. Gray, 1862 Waigeo I.
— *M.f.albigula* Mayr & Meyer de Schauensee, 1939 W Papuan Is., SW, and NW to NC New Guinea
— *M.f.novus* Rothschild & Hartert, 1912 NE New Guinea
— *M.f.xanthogenys* G.R. Gray, 1858 S to SE New Guinea, Aru Is.
— *M.f.flaviventer* Gould, 1851 N and E Cape York Pen.
— *M.f.secundus* Mathews, 1912 NE Queensland

○ *Machaerirhynchus nigripectus* BLACK-BREASTED BOATBILL
— *M.n.nigripectus* Schlegel, 1871 Mts. of NW New Guinea (Vogelkop)
— *M.n.saturatus* Rothschild & Hartert, 1913 Mts. of C New Guinea
— *M.n.harterti* van Oort, 1909 Mts. of SE New Guinea and Huon Pen.

VANGIDAE VANGAS (15: 22)

CALICALICUS Bonaparte, 1854 M
○ *Calicalicus madagascariensis* (Linnaeus, 1766) RED-TAILED VANGA   W, N and E Madagascar

○ *Calicalicus rufocarpalis* Goodman, Hawkins & Domergue, 1997 RED-SHOULDERED VANGA #
SW Madagascar (956)

VANGA Vieillot, 1816 F
○ *Vanga curvirostris* HOOK-BILLED VANGA
— *V.c.curvirostris* (Linnaeus, 1766) W, N and E Madagascar
— *V.c.cetera* Bangs, 1928 Extreme S Madagascar

ORIOLIA I. Geoffroy Saint-Hilaire, 1838 F
○ *Oriolia bernieri* I. Geoffroy Saint-Hilaire, 1838 BERNIER'S VANGA   NE Madagascar

XENOPIROSTRIS Bonaparte, 1850 M
○ *Xenopirostris xenopirostris* (Lafresnaye, 1850) LAFRESNAYE'S VANGA   SW Madagascar

○ *Xenopirostris damii* Schlegel, 1866 VAN DAM'S VANGA   NW Madagascar

○ *Xenopirostris polleni* (Schlegel, 1868) POLLEN'S VANGA   NW, NE and E Madagascar

FALCULEA I. Geoffroy Saint-Hilaire, 1836 F
○ *Falculea palliata* I. Geoffroy Saint-Hilaire, 1836 SICKLE-BILLED VANGA
N, W and S Madagascar

---

[1] Includes *craterum* see Rand (1960: 332) (1971).
[2] For reasons to use 1802 not 1801 see Browing & Monroe (1991) (258).
[3] For reasons to use 1802 not 1801 see Browing & Monroe (1991) (258).

ARTAMELLA W.L. Sclater, 1924 F
○ *Artamella viridis* White-headed Vanga   N and E Madagascar
    — *A.v.viridis* (Statius Müller, 1776)   W and S Madagascar
    — *A.v.annae* (Stejneger, 1878)

LEPTOPTERUS Bonaparte, 1854 M
○ *Leptopterus chabert* Chabert's Vanga
    — *L.c.chabert* (Statius Müller, 1776)   W, N and E Madagascar
    — *L.c.schistocercus* (Neumann, 1908)   SW Madagascar

CYANOLANIUS Bonaparte, 1854[1] M
○ *Cyanolanius madagascarinus* Blue Vanga
    — *C.m.madagascarinus* (Linnaeus, 1766)   Madagascar (except S)
    — *C.m.comorensis* (Shelley, 1894)   Mwali (Moheli) I. (Comoros)
    — *C.m.bensoni* Louette & Herremans, 1982   Grande Comore I. (1381)

SCHETBA Lesson, 1830 F
○ *Schetba rufa* Rufous Vanga
    — *S.r.rufa* (Linnaeus, 1766)   N and E Madagascar
    — *S.r.occidentalis* Delacour, 1931   W Madagascar

EURYCEROS Lesson, 1831 M   NE Madagascar
○ *Euryceros prevostii* Lesson, 1831 Helmet Vanga

TYLAS Hartlaub, 1862[2] F
○ *Tylas eduardi* Tylas Vanga
    — *T.e.eduardi* Hartlaub, 1862   NW, NE and E Madagascar
    — *T.e.albigularis* Hartlaub, 1877   WC Madagascar

HYPOSITTA A. Newton, 1881 F
○ *Hypositta corallirostris* (A. Newton, 1863) Nuthatch-Vanga   NW, NE and E Madagascar

○ †?*Hypositta perdita* D.S. Peters, 1996 Bluntschli's Vanga[3] #   Madagascar (1820)

NEWTONIA Schlegel & Pollen, 1868[4] F
○ *Newtonia amphichroa* Reichenow, 1891 Dark Newtonia   NE and E Madagascar

○ *Newtonia brunneicauda* Common Newtonia
    — *N.b.brunneicauda* (A. Newton, 1863)   Madagascar
    — *N.b.monticola* Salomonsen, 1934   Ankaratra Mts. (Madagascar)

○ *Newtonia archboldi* Delacour & Berlioz, 1931 Archbold's Newtonia
                                   SW Madagascar

○ *Newtonia fanovanae* Gyldenstolpe, 1933 Red-tailed Newtonia   E Madagascar[5]

PSEUDOBIAS Sharpe, 1870[6] M
○ *Pseudobias wardi* Sharpe, 1870 Ward's Flycatcher   E Madagascar

MYSTACORNIS Sharpe, 1870[7] M
○ *Mystacornis crossleyi* (A. Grandidier, 1870) Crossley's Babbler   NW, NE and E Madagascar

---

[1] For recognition of this genus see Louette & Herremans (1982) (1381). T. Schulenberg (*in litt.*) agrees.
[2] For reasons to place this genus with the vangas see Yamagishi *et al.* (2001) (2692).
[3] This may prove to be an immature plumage of *H. corallirostris* see Goodman *et al.* (1997) (957).
[4] For treatment of this group as vangas not warblers see Yamagishi *et al.* (2001) (2692).
[5] Recently rediscovered see Goodman *et al.* (1997) (957); there are a few other recent sight records.
[6] DNA sampling sugests this belongs in the Vangidae (Schulenberg, In Press) (2208).
[7] Placed near the babblers by Deignan (1964) (666). Cibois *et al.* (1999) (357) considered this a member of a corvoid clade, but had insufficient data to determine final placement.Cibois *et al.* (1999); it seems best treated here pending further evidence.

CRACTICIDAE BUTCHER BIRDS (4: 13)

CRACTICUS Vieillot, 1816 M

○ *Cracticus quoyi* BLACK BUTCHERBIRD
- — *C.q.quoyi* (Lesson & Garnot, 1827)
- — *C.q.spaldingi* E.P. Ramsay, 1878
- — *C.q.alecto* Schodde & Mason, 1999
- — *C.q.jardini* Mathews, 1912[1]
- — *C.q.rufescens* De Vis, 1883

New Guinea (except Trans-Fly)
Coastal W and C Northern Territory, Melville I.
Aru Is., SC New Guinea (Trans-Fly), Torres Straits Is. (2189)
Coastal Cape York Pen.
NE to EC Queensland

○ *Cracticus torquatus* GREY BUTCHERBIRD
- — *C.t.leucopterus* Gould, 1848[2]
- — *C.t.torquatus* (Latham, 1802)[3]
- — *C.t.cinereus* (Gould, 1837)

W, C, S and inland E Australia
Coastal SE Australia
Tasmania

○ *Cracticus argenteus* SILVER-BACKED BUTCHERBIRD[4]
- — *C.a.argenteus* Gould, 1841
- — *C.a.colletti* Mathews, 1912[5]

N Western Australia (Kimberley)
N Northern Territory (Arnhem Land)

○ *Cracticus mentalis* BLACK-BACKED BUTCHERBIRD
- — *C.m.mentalis* Salvadori & D'Albertis, 1876
- — *C.m.kempi* Mathews, 1912

SE New Guinea
Cape York Pen.

○ *Cracticus nigrogularis* PIED BUTCHERBIRD
- — *C.n.picatus* Gould, 1848[6]
- — *C.n.nigrogularis* (Gould, 1837)

W, C, NW and NC Australia
E Australia

○ *Cracticus cassicus* HOODED BUTCHERBIRD
- — *C.c.cassicus* (Boddaert, 1783)

- — *C.c.hercules* Mayr, 1940

W Papuan Is., New Guinea, islands in Geelvink Bay, Aru Is.
Trobriand Is., D'Entrecasteaux Arch.

○ *Cracticus louisiadensis* Tristram, 1889 TAGULA BUTCHERBIRD

Tagula I. (Louisiade Arch.)

GYMNORHINA G.R. Gray, 1840 F

◑ *Gymnorhina tibicen* AUSTRALIAN MAGPIE
- — *G.t.papuana* Bangs & Peters, 1926
- — *G.t.eylandtensis* H.L. White, 1922
- — *G.t.terraereginae* (Mathews, 1912)
- ✔ *G.t.tibicen* (Latham, 1802)[8]
- — *G.t.tyrannica* Schodde & Mason, 1999
- ✔ *G.t.hypoleuca* (Gould, 1837)
- — *G.t.telonocua* Schodde & Mason, 1999[9]
- — *G.t.dorsalis* A.J. Campbell, 1895
- — *G.t.longirostris* Milligan, 1903

SC New Guinea (Trans-Fly)
Inland NC Australia[7]
E Australia (inland in SE)
Coastal SE Australia
S Victoria, SE South Australia, King I. (2189)
E Tasmania, Flinders I.
SC South Australia (2189)
SW Australia
WC Western Australia

STREPERA Lesson, 1830 F

● *Strepera graculina* PIED CURRAWONG
- — *S.g.magnirostris* H.L. White, 1923[10]
- — *S.g.robinsoni* Mathews, 1912
- — *S.g.graculina* (Shaw, 1790)
- — *S.g.crissalis* Sharpe, 1877
- ✔ *S.g.nebulosa* Schodde & Mason, 1999
- — *S.g.ashbyi* Mathews, 1913

E Cape York Pen.
NE Queensland
EC Australia >> NE Australia
Lord Howe I.
SE Australia (2189)
SW Victoria

[1] For recognition see Schodde & Mason (1999: 534) (2189).
[2] Includes *latens* Ford, 1979 (857) see Schodde & Mason (1999: 537-538) (2189).
[3] For reasons to use 1802 not 1801 see Browing & Monroe (1991) (258).
[4] For reason to separate this from *C. torquatus* see Schodde & Mason (1999: 540) (2189).
[5] For recognition see Schodde & Mason (1999: 540) (2189).
[6] Includes *kalgoorli* see Schodde & Mason (1999: 543) (2189).
[7] The prior name *finki* relates to an intergrading population further south see Schodde & Mason (1999: 545) (2189).
[8] For reasons to use 1802 not 1801 see Browing & Monroe (1991) (258).
[9] This name applies to the birds of the Yorke and Eyre peninsulas; those of "South Australia", for which the name *leuconota* was employed, belong to an intergrading population see Schodde & Mason (1999: 549) (2189).
[10] For recognition see Schodde & Mason (1999: 551) (2189).

○ *Strepera fuliginosa* BLACK CURRAWONG
      — *S.f.fuliginosa* (Gould, 1837)   Tasmania
      — *S.f.parvior* Schodde & Mason, 1999   Flinders I. (2189)
      — *S.f.colei* Mathews, 1916[1]   King I.

● *Strepera versicolor* GREY CURRAWONG
      — *S.v.versicolor* (Latham, 1802)[2]   SE Australia
      — *S.v.arguta* Gould, 1846   E Tasmania
      — *S.v.melanoptera* Gould, 1846[3]   SE South Australia, SW New South Wales, W Victoria
      — *S.v.halmaturina* Mathews, 1912[4]   Kangaroo I. (SC Australia)
      — *S.v.intermedia* Sharpe, 1877   SC South Australia (SA Gulfs)
      — *S.v.plumbea* Gould, 1846[5]   SW Australia

PELTOPS Wagler, 1829[6] M
○ *Peltops blainvillii* (Lesson & Garnot, 1827) LOWLAND PELTOPS   Lowland New Guinea

○ *Peltops montanus* Stresemann, 1921 MOUNTAIN PELTOPS   Mts. of New Guinea

ARTAMIDAE WOODSWALLOWS (1: 10)

ARTAMUS Vieillot, 1816 M
○ *Artamus fuscus* Vieillot, 1817 ASHY WOODSWALLOW   Sri Lanka, India, S China, mainland SE Asia

○ *Artamus leucorynchus* WHITE-BREASTED WOODSWALLOW
        — *A.l.pelewensis* Finsch, 1876   Palau Is.
        — *A.l.leucorynchus* (Linnaeus, 1771)   Philippines to Borneo
        — *A.l.amydrus* Oberholser, 1917   Sumatra, Bangka I., Java, Bali, Kangean Is.
        — *A.l.humei* Stresemann, 1913   Andaman, Cocos Is.
        — *A.l.albiventer* (Lesson, 1830)   Sulawesi and islands in Flores Sea, Lesser Sundas
        — *A.l.musschenbroeki* A.B. Meyer, 1884   Tanimbar Is. (E Lesser Sundas)
        — *A.l.leucopygialis* Gould, 1842   Moluccas, Aru Is., New Guinea, N Australia
        — *A.l.melaleucus* (Wagler, 1827)   New Caledonia, Loyalty Is.
        — *A.l.tenuis* Mayr, 1943   Vanuatu, Banks Is.
        — *A.l.mentalis* Jardine, 1845   W and N Fiji Is.

○ *Artamus monachus* Bonaparte, 1850 IVORY-BACKED WOODSWALLOW   Sulawesi, Banggai, Sula Is.

○ *Artamus maximus* A.B. Meyer, 1874 GREAT WOODSWALLOW   Mts. of New Guinea

○ *Artamus insignis* P.L. Sclater, 1877 BISMARCK WOODSWALLOW   New Britain, New Ireland (Bismarck Arch.)

○ *Artamus personatus* (Gould, 1841) MASKED WOODSWALLOW   Inland Australia

○ *Artamus superciliosus* (Gould, 1837) WHITE-BROWED WOODSWALLOW   Inland E to SC Australia

○ *Artamus cinereus* BLACK-FACED WOODSWALLOW
        — *A.c.perspicillatus* Bonaparte, 1850   Timor, Letti, Sermatta Is. (Lesser Sundas)
        — *A.c.normani* (Mathews, 1923)   SC New Guinea, C and S Cape York Pen., inland NE
                                         Queensland
        — *A.c.dealbatus* Schodde & Mason, 1999[7]   EC Queensland (2189)
        — *A.c.melanops* Gould, 1865[8]   WC and NW to inland SE Australia
        — *A.c.cinereus* Vieillot, 1817   SW Australia

○ *Artamus cyanopterus* DUSKY WOODSWALLOW
        — *A.c.cyanopterus* (Latham, 1802)[9]   SC to SE Australia, Tasmania, King I., Flinders I. >> EC
                                         Australia
        — *A.c.perthi* (Mathews, 1915)   SW to SC Australia

---

[1] For recognition see Schodde & Mason (1999: 555) (2189).
[2] For reasons to use 1802 not 1801 see Browing & Monroe (1991) (258).
[3] The name *howei* attaches to intergrades on the east of the range of this form, see Schodde & Mason (1999: 557) (2189).
[4] For recognition see Schodde & Mason (1999: 558) (2189).
[5] Includes *plumbea* see Schodde & Mason (1999: 557) (2189).
[6] For reasons to place *Peltops* close to the genus *Artamus* see Sibley & Ahlquist (1984) (2260).
[7] The name *hypoleucos* (a new name for *albiventris*) is based on a lectotype that is from an intergrading population see Schodde & Mason (1999: 567)
   (2189); *inkermani* Keast, 1958 (1236) applies to an intergrading population in W Australia.
[8] For recognition see Schodde & Mason (1999: 566) (2189).
[9] For reasons to use 1802 not 1801 see Browing & Monroe (1991) (258).

○ *Artamus minor* Little Woodswallow
— *A.m.derbyi* Mathews, 1912[1]   N to EC Australia
— *A.m.minor* Vieillot, 1817   WC and C Australia

AEGITHINIDAE IORAS (1: 4)

AEGITHINA Vieillot, 1816 F
○ *Aegithina tiphia* Common Iora
— *A.t.multicolor* (J.F. Gmelin, 1789)   Kerala (S India), Sri Lanka
— *A.t.deignani* B.P. Hall, 1957   S India (not Kerala), N and C Burma
— *A.t.humei* E.C.S. Baker, 1922   C India
— *A.t.tiphia* (Linnaeus, 1758)   C and E Himalayas, NE India, SW Burma NE India,
                                          SW Burma
— *A.t.septentrionalis* Koelz, 1939   NW Himalayas
— *A.t.philipi* Oustalet, 1885   S Yunnan, E Burma, N Thailand, N and C Indochina
— *A.t.cambodiana* B.P. Hall, 1957   Cambodia, SE Thailand, S Vietnam
— *A.t.horizoptera* Oberholser, 1912[2]   SE Burma, SW Thailand, Malay Pen., Sumatra and
                                          islands, Bangka
— *A.t.scapularis* (Horsfield, 1821)   Java, Bali
— *A.t.viridis* (Bonaparte, 1850)[3]   S Borneo
— *A.t.aequanimis* Bangs, 1922   N Borneo[4], Palawan
○ *Aegithina nigrolutea* (G.F.L. Marshall, 1876) Marshall's Iora   Pakistan, NW India
○ *Aegithina viridissima* Green Iora
— *A.v.viridissima* (Bonaparte, 1850)[5] Malay Pen., Sumatra, Borneo, Bangka
— *A.v.thapsina* Oberholser, 1917   Anamba and Natuna Is.
○ *Aegithina lafresnayei* Great Iora
— *A.l.innotata* (Blyth, 1847)   SW Burma, Thailand south to N Malay Pen., N and C
                                          Indochina
— *A.l.xanthotis* (Sharpe, 1881)   SE Thailand, S Indochina
— *A.l.lafresnayei* (Hartlaub, 1844)   C and S Malay Pen.

PITYRIASIDAE[6] BRISTLEHEAD (1: 1)

PITYRIASIS Lesson, 1839 F
○ *Pityriasis gymnocephala* (Temminck, 1836[7]) Bristlehead   Borneo

CAMPEPHAGIDAE CUCKOO-SHRIKES (7: 81)

CORACINA Vieillot, 1816 F
○ *Coracina maxima* (Rüppell, 1839) Ground Cuckoo-shrike[8]   Australia

○ *Coracina macei* Large Cuckoo-shrike[9]
— *C.m.nipalensis* (Hodgson, 1836)[10]   Himalayas, Assam
— *C.m.macei* (Lesson, 1831)   C and S India
— *C.m.layardi* (Blyth, 1866)   Sri Lanka
— *C.m.andamana* (Neumann, 1915)   Andamans
— *C.m.rexpineti* (Swinhoe, 1863)   SE China, Taiwan, N Vietnam, N Laos
— *C.m.larvivora* (E. Hartert, 1910)[11] Hainan (China)

---

[1] For recognition see Schodde & Mason (1999: 570) (2189).
[2] Includes *micromelaena*, *singapurensis* and *djungkulanensis* Hoogerwerf, 1962 (1140), see Mees (1986) (1539).
[3] Given as 1851 by Delacour (1960: 302) (681) but see Zimmer (1926) (2707).
[4] Davison in Smythies (2000: 457) (2289) considered *trudiae* Prescott, 1970 (1924), of Brunei Bay an intergradient population.
[5] Given as 1851 by Delacour (1960: 302) (681) but see Zimmer (1926) (2707).
[6] Hachisuka (1953) (1044) suggested an affinity with the *Cracticidae*. Ahlquist *et al.* (1984) (18) offered molecular evidence in support.
[7] Mees (1994) (1541) has shown that Livr. 97 appeared in Apr. 1836 or later.
[8] In last Edition treated in a monotypic genus *Pteropodocys*, here tentatively placed in *Coracina* following Schodde & Mason, (1999: 572) (2189).
[9] The taxa comprising this species were part of a broad species *C. novaehollandiae* in last Edition.
[10] Given as "1836 (1837)" by Peters *et al.* (1960) (1832), but see their dating of *Coracina melaschistos*.
[11] For correction of spelling see David & Gosselin (2002) (613).

___ *C.m.siamensis* (E.C.S. Baker, 1918)   Burma, Yunnan, Thailand (south to Isthmus of Kra), S Indochina

___ *C.m.larutensis* (Sharpe, 1887)   S Malay Pen.

○ *Coracina javensis* (Horsfield, 1821) Javan Cuckoo-shrike[1]   Java, Bali

● *Coracina novaehollandiae* Black-faced Cuckoo-shrike

___ *C.n.subpallida* Mathews, 1912   WC Australia (Pilbara)

___ *C.n.melanops* (Latham, 1802)[2]   Australia >> New Guinea, S Moluccas, Lesser Sundas, W Solomons

___ *C.n.novaehollandiae* (J.F. Gmelin, 1789)   Tasmania, Flinders I. >> E Australia

○ *Coracina personata* Wallacean Cuckoo-shrike

___ *C.p.floris* (Sharpe, 1878)[3]   Sumbawa, Flores

___ *C.p.alfrediana* (E. Hartert, 1898)   Lomblen, Alor

___ *C.p.sumbensis* (A.B. Meyer, 1882)   Sumba

___ *C.p.personata* (S. Müller, 1843)[4]   Roti, Timor, Wetar, Leti to Sermata

___ *C.p.unimoda* (P.L. Sclater, 1883)   Tanimbar

___ *C.p.pollens* (Salvadori, 1874)   Kai Is.

○ *Coracina fortis* (Salvadori, 1878) Buru Cuckoo-shrike   S Moluccas

○ *Coracina atriceps* Moluccan Cuckoo-shrike

___ *C.a.atriceps* (S. Müller, 1843)   Seram (S Moluccas)

___ *C.a.magnirostris* (Bonaparte, 1850)[5]   Bacan, Ternate, Halmahera (N Moluccas)

○ *Coracina schistacea* (Sharpe, 1878[6]) Slaty Cuckoo-shrike   Peleng, Banggai, Sula Is.

○ *Coracina caledonica* Melanesian Cuckoo-shrike

___ *C.c.bougainvillei* (Mathews, 1928)   Bougainville I. (Solomons)

___ *C.c.kulambangrae* Rothschild & Hartert, 1916   Kulambangara I., New Georgia group (C Solomons)

___ *C.c.welchmani* (Tristram, 1892)   Santa Ysabel I. (C Solomons)

___ *C.c.amadonis* Cain & Galbraith, 1955   Guadalcanal I. (S Solomons)

___ *C.c.thilenii* (Neumann, 1915)   Espiritu Santo, Malekula Is. (Vanuatu)

___ *C.c.seiuncta* Mayr & Ripley, 1941   Erromanga (Vanuatu)

___ *C.c.lifuensis* (Tristram, 1879)   Lifu, Loyalty Is.

___ *C.c.caledonica* (J.F. Gmelin, 1788)   New Caledonia

○ *Coracina caeruleogrisea* Stout-billed Cuckoo-shrike

___ *C.c.strenua* (Schlegel, 1871)   W, N and C New Guinea, Yapen

___ *C.c.caeruleogrisea* (G.R. Gray, 1858)   Aru Is., SC New Guinea

___ *C.c.adamsoni* Mayr & Rand, 1936   SE New Guinea

○ *Coracina temminckii* Cerulean Cuckoo-shrike

___ *C.t.temminckii* (S. Müller, 1843)   Mts. of N Sulawesi

___ *C.t.rileyi* Meise, 1931   Mts. of C and SE Sulawesi

___ *C.t.tonkeana* (A.B. Meyer, 1903)   Mts. of E Sulawesi

○ *Coracina larvata* Sunda Cuckoo-shrike

___ *C.l.melanocephala* (Salvadori, 1879)   Sumatra

___ *C.l.larvata* (S. Müller, 1843)   Java

___ *C.l.normani* (Sharpe, 1887)   Borneo

○ *Coracina striata* Bar-bellied Cuckoo-shrike

___ *C.s.dobsoni* (Ball, 1872)   Andamans

___ *C.s.sumatrensis* (S. Müller, 1843)   Malay Pen., Sumatra, Siberut, Sipora and Pagi Is., Borneo and islands, Rhio Arch.

___ *C.s.simalurensis* (Richmond, 1903)   Simeuluë (off W Sumatra)

___ *C.s.babiensis* (Richmond, 1903)   Babi (off W Sumatra)

---

[1] Treated as part of *C. novaehollandiae* in last Edition, and should perhaps not be separated from *macei*; the position of *larutensis* requires evaluation, it may belong with *javensis*.

[2] For reasons to use 1802 not 1801 see Browing & Monroe (1991) (258).

[3] Peters (1960) (1832) cited 1879, but the name appeared in 1878 in Mitt. K. Zool. Mus. Dresden, 1, pt. 3, p. 363.

[4] Includes *lettiensis* see White & Bruce (1986: 302) (2652); treated as close to *javensis* in last Edition.

[5] Given as 1851 by Peters *et al.* (1960: 173) (1832) but see Zimmer (1926) (2707).

[6] Peters (1960) (1832) cited 1879, but the name appeared in 1878 in Mitt. K. Zool. Mus. Dresden, 1, pt. 3, p. 363.

___ *C.s.kannegieteri* (Büttikofer, 1897) Nias (off W Sumatra)
___ *C.s.enganensis* (Salvadori, 1892) Enggano (off W Sumatra)
___ *C.s.bungurensis* (E. Hartert, 1894) Anamba and Natuna Is.
___ *C.s.vordermani* (E. Hartert, 1901) Kangean (Java)
___ *C.s.difficilis* (E. Hartert, 1895) Palawan, Balabac
___ *C.s.striata* (Boddaert, 1783) Luzon, Lubang (N Philippines)
___ *C.s.mindorensis* (Steere, 1890) Mindoro (WC Philippines)
___ *C.s.panayensis* (Steere, 1890) Masbate, Panay, Negros
___ *C.s.boholensis* Rand & Rabor, 1959 Bohol, Leyte, Samar (EC Philippines) (1976)
___ †?*C.s.cebuensis* (Ogilvie-Grant, 1896)[1] Cebu (C Philippines)
___ *C.s.kochii* (Kutter, 1882) Mindanao, Basilan (S Philippines)
___ *C.s.guillemardi* (Salvadori, 1886) Sulu Arch. (S Philippines)

○ *Coracina bicolor* (Temminck, 1824) Pied Cuckoo-shrike N Sulawesi and islands nearby

○ *Coracina lineata* Barred Cuckoo-shrike
___ *C.l.axillaris* (Salvadori, 1876) Waigeo, Mts. of New Guinea
___ *C.l.maforensis* (A.B. Meyer, 1874) Numfor I. (Geelvink Bay)
___ *C.l.sublineata* (P.L. Sclater, 1879) New Ireland, New Britain (Bismarck Arch.)
___ *C.l.nigrifrons* (Tristram, 1892) Bougainville, Santa Ysabel Is. (N Solomons)
___ *C.l.ombriosa* (Rothschild & Hartert, 1905) Kolombangara, New Georgia, Rendova Is. (C Solomons)
___ *C.l.pusilla* (E.P. Ramsay, 1879) Guadalcanal I. (S Solomons)
___ *C.l.malaitae* Mayr, 1931 Malaita I. (NE Solomons)
___ *C.l.makirae* Mayr, 1935 San Cristobal I. (SE Solomons)
___ *C.l.gracilis* Mayr, 1931 Rennell I. (S Solomons)
___ *C.l.lineata* (Swainson, 1825) E Queensland, NE New South Wales

○ *Coracina boyeri* Boyer's Cuckoo-shrike
___ *C.b.boyeri* (G.R. Gray, 1846) Yapen, NW and N New Guinea
___ *C.b.subalaris* (Sharpe, 1878) S and SE New Guinea

○ *Coracina leucopygia* (Bonaparte, 1850)[2] White-rumped Cuckoo-shrike
  Sulawesi and nearby islands

○ *Coracina papuensis* White-bellied Cuckoo-shrike
___ *C.p.papuensis* (J.F. Gmelin, 1788)[3] N Moluccas, Misool, Yapen, W and N New Guinea
___ *C.p.intermedia* Rothschild, 1931 Western SC New Guinea
___ *C.p.angustifrons* (Sharpe, 1878) Northern SC to SE New Guinea
___ *C.p.louisiadensis* (E. Hartert, 1898) Louisiade Arch.
___ *C.p.oriomo* Mayr & Rand, 1936[4] Sthn. SC New Guinea, Cape York Pen.
___ *C.p.timorlaoensis* (A.B. Meyer, 1884) Tanimbar Is.
___ *C.p.hypoleuca* (Gould, 1848) N Australia (Kimberley to Gulf of Carpentaria), ? Kai and Aru Is.
___ *C.p.apsleyi* Mathews, 1912 Melville I. (Australia)
___ *C.p.artamoides* Schodde & Mason, 1999 NE to SE Queensland, NE New South Wales (2189)
___ *C.p.robusta* (Latham, 1802)[5] SE Australia
___ *C.p.sclaterii* (Salvadori, 1878) Bismarck Arch.
___ *C.p.perpallida* Rothschild & Hartert, 1916 Bougainville, Choiseul, Santa Ysabel Is. (N Solomons)
___ *C.p.elegans* (E.P. Ramsay, 1881) New Georgia, Rendova, Guadalcanal Is. (C, S Solomons)
___ *C.p.eyerdami* Mayr, 1931 Malaita I. (NE Solomons)

○ *Coracina ingens* (Rothschild & Hartert, 1914) Manus Cuckoo-shrike[6]
  Admiralty Is.

○ *Coracina longicauda* Hooded Cuckoo-shrike
___ *C.l.grisea* Junge, 1939 Mts. of WC New Guinea
___ *C.l.longicauda* (De Vis, 1890) Mts. of EC and S New Guinea

○ *Coracina parvula* (Salvadori, 1878) Halmahera Cuckoo-shrike Halmahera

---

[1] Previously omitted as extinct. See Dickinson *et al.* (1991: 276) (745).
[2] Given as 1851 by Peters *et al.* (1960: 179) (1832) but see Zimmer (1926) (2707).
[3] Includes *melanolora* see White & Bruce (1986: 304) (2652).
[4] The name *stalkeri* antedates both *oriomo* and *artamoides*; it attaches to an intergradient population see Schodde & Mason (1999: 584) (2189).
[5] For reasons to use 1802 not 1801 see Browning & Monroe (1991) (258). Treated as a separate species in last Edition but see Galbraith (1969) (906).
[6] Treated as an allospecies of *C. papuensis* by Schodde & Mason (1999: 583) (2189).

○ *Coracina abbotti* (Riley, 1918) PYGMY CUCKOO-SHRIKE — Mts. of C Sulawesi

○ *Coracina analis* (Verreaux & Des Murs, 1860) NEW CALEDONIAN CUCKOO-SHRIKE
New Caledonia

○ *Coracina caesia* GREY CUCKOO-SHRIKE
   — *C.c.pura* (Sharpe, 1891)[1] — Cameroon, Bioko, Ethiopia to E Zaire and Malawi
   — *C.c.caesia* (M.H.K. Lichtenstein, 1823) — Zimbabwe and C Mozambique to E and S Sth. Africa

○ *Coracina pectoralis* (Jardine & Selby, 1828) WHITE-BREASTED CUCKOO-SHRIKE
W Africa to Ethiopia and Uganda, W and S Tanzania to Angola, Zimbabwe and NE Sth. Africa

○ *Coracina graueri* Neumann, 1908 GRAUER'S CUCKOO-SHRIKE — E Zaire

○ *Coracina cinerea* ASHY CUCKOO-SHRIKE
   — *C.c.cucullata* (Milne-Edwards & Oustalet, 1885) — Grande Comore (Comoros)
   — *C.c.moheliensis* Benson, 1960 — Mwali (Moheli) I. (Comoros) (122)
   — *C.c.cinerea* (Statius Müller, 1776) — N and E Madagascar
   — *C.c.pallida* Delacour, 1931 — W and S Madagascar

○ *Coracina azurea* (Cassin, 1852) BLUE CUCKOO-SHRIKE — Sierra Leone to Zaire

○ *Coracina typica* (Hartlaub, 1865) MAURITIUS CUCKOO-SHRIKE — Mauritius

○ *Coracina newtoni* (Pollen, 1866) REUNION CUCKOO-SHRIKE — Reunion I. (Indian Ocean)

○ *Coracina coerulescens* BLACKISH CUCKOO-SHRIKE
   — *C.c.coerulescens* (Blyth, 1842) — Luzon (N Philippines)
   — *C.c.deschauenseei* duPont, 1972 — Marinduque (NC Philippines) (774)
   — †?*C.c.altera* (Wardlaw Ramsay, 1881) — Cebu (C Philippines)

○ *Coracina dohertyi* (E. Hartert, 1896) SUMBA CUCKOO-SHRIKE — Sumba, Flores (Lesser Sundas)

○ *Coracina dispar* (Salvadori, 1878) KAI CUCKOO-SHRIKE[2] — Seram Laut, Banda, Kai, Tanimbar and associated islands

○ *Coracina tenuirostris* CICADABIRD[3]
   — *C.t.monacha* (Hartlaub & Finsch, 1872) — Palau Is.
   — *C.t.nesiotis* (Hartlaub & Finsch, 1872) — Yap I. (Caroline Is.)
   — *C.t.insperata* (Finsch, 1875) — Pohnpei (Caroline Is.)
   — *C.t.edithae* (Stresemann, 1932) — S Sulawesi
   — *C.t.pererrata* (E. Hartert, 1918)[4] — Tukangbesi (off Sulawesi)
   — *C.t.kalaotuae* (Meise, 1929) — Kalaotua (off Sulawesi)
   — *C.t.emancipata* (E. Hartert, 1896) — Djampea I. (Java Sea)
   — *C.t.timoriensis* (Sharpe, 1878) — Timor, Lomblen (Lesser Sundas)
   — *C.t.pelingi* (E. Hartert, 1918)[5] — Peleng (NW Moluccas)
   — *C.t.grayi* (Salvadori, 1879) — N Moluccas
   — *C.t.obiensis* (Salvadori, 1878) — Obi, Bisa (C Moluccas)
   — *C.t.amboinensis* (Hartlaub, 1865) — Ambon, Seram (S Moluccas)
   — *C.t.admiralitatis* (Rothschild & Hartert, 1914) — Admiralty Is.
   — *C.t.matthiae* (Sibley, 1946) — Storm, St. Matthias Is. (Bismarck Arch.)
   — *C.t.remota* (Sharpe, 1878) — New Ireland, New Hanover (Bismarck Arch.)
   — *C.t.ultima* Mayr, 1955 — Lihir, Tanga Is. (Bismarck Arch.) (1477)
   — *C.t.heinrothi* (Stresemann, 1922) — New Britain (Bismarck Arch.)
   — *C.t.rooki* (Rothschild & Hartert, 1914) — Umboi I. (Bismarck Arch.)
   — *C.t.nehrkorni* (Salvadori, 1890) — Waigeo I. (off W New Guinea)
   — *C.t.numforana* Peters & Mayr, 1960 — Numfor I. (Geelvink Bay)
   — *C.t.meyerii* (Salvadori, 1878) — Biak I. (Geelvink Bay)
   — *C.t.aruensis* (Sharpe, 1878)[6] — Aru Is., SW New Guinea
   — *C.t.muellerii* (Salvadori, 1876) — Kafiau, Misool Is., New Guinea, D'Entrecasteaux Arch.
   — *C.t.tagulana* (E. Hartert, 1898) — Tagula and Misima Is., Louisiade Arch.)

[1] Includes *preussi* see White (1962: 70) (2646).
[2] White & Bruce (1986: 307) (2652) treated this as a species pending investigation of the reported sympatry of *C. tenuirostris amboinensis* with *C.* [*t.*] *dispar* in Maar.
[3] No doubt better treated as several allospecies, but limits unclear, see Schodde & Mason (1999: 575) (2189).
[4] Given as 1917 by Peters *et al.* (1960: 185) (1832), but see Hartert (1922) (1084).
[5] Given as 1917 by Peters *et al.* (1960: 186) (1832), but see Hartert (1922) (1084).
[6] Possibly a synonym of *muelleri*: see identity of SE New Guinea specimens in Schodde & Mason (1999: 576) (2189).

— *C.t.rostrata* (E. Hartert, 1898) — Rossel I. (Louisiade Arch.)

— *C.t.saturatior* (Rothschild & Hartert, 1902) — N and C Solomon Is.

— *C.t.nisoria* (Mayr, 1950) — Russell Is. (SC Solomons)

— *C.t.erythropygia* (Sharpe, 1888) — Guadalcanal, Malaita Is. (S Solomons)

— *C.t.melvillensis* Mathews, 1912 — N Australia, incl. Melville I.

— *C.t.tenuirostris* (Jardine, 1831) — E Australia >> New Guinea

○ *Coracina salomonis* (Tristram, 1879) MAKIRA CUCKOO-SHRIKE[1] San Cristobal (S Solomons)

○ *Coracina mindanensis* BLACK-BIBBED CUCKOO-SHRIKE

— *C.m.lecroyae* Parkes, 1971 — Luzon (N Philippines) [1763]

— *C.m.elusa* (McGregor, 1905) — Mindoro (NC Philippines)

— *C.m.ripleyi* Parkes, 1971 — Bohol, Samar, Leyte (EC Philippines) [1763]

— *C.m.mindanensis* (Tweeddale, 1879[2]) — Mindanao, Basilan (S Philippines)

— *C.m.everetti* (Sharpe, 1893) — Sulu Arch. (SW Philippines)

○ *Coracina morio* SULAWESI CUCKOO-SHRIKE

— *C.m.morio* (S. Müller, 1843) — Sulawesi, Muna I.

— *C.m.salvadorii* (Sharpe, 1878) — Sangihe Is.

— *C.m.talautensis* (Meyer & Wiglesworth, 1895) — Talaut Is.

○ *Coracina sula* (E. Hartert, 1918)[3] SULA CUCKOO-SHRIKE Sula Is.

○ *Coracina ceramensis* PALE CUCKOO-SHRIKE

— *C.c.ceramensis* (Bonaparte, 1850) — Seram, Buru, Boano

— *C.c.hoogerwerfi* (Jany, 1955) — Obi

○ *Coracina incerta* (A.B.Meyer, 1874) BLACK-SHOULDERED CUCKOO-SHRIKE Waigeo, Yapen, New Guinea

○ *Coracina schisticeps* GREY-HEADED CUCKOO-SHRIKE

— *C.s.schisticeps* (G.R. Gray, 1846) — Misool I., NW New Guinea

— *C.s.reichenowi* (Neumann, 1917) — N New Guinea

— *C.s.poliopsa* (Sharpe, 1882) — S New Guinea

— *C.s.vittata* (Rothschild & Hartert, 1914) — D'Entrecasteaux Arch.

○ *Coracina melas* NEW GUINEA CUCKOO-SHRIKE

— *C.m.waigeuensis* (Stresemann & Paludan, 1932) — Waigeo I. (off W New Guinea)

— *C.m.tommasonis* (Rothschild & Hartert, 1903) — Yapen I. (Geelvink Bay)

— *C.m.melas* (Lesson, 1828)[4] — W and N New Guinea, W Papuan Is.

— *C.m.meeki* (Rothschild & Hartert, 1903) — E and S New Guinea

— *C.m.goodsoni* (Mathews, 1928) — Aru Is.

— *C.m.batantae* (Gyldenstolpe & Mayr, 1955) — Batanta (W Papuan Is.) [1033]

○ *Coracina montana* BLACK-BELLIED CUCKOO-SHRIKE

— *C.m.montana* (A.B. Meyer, 1874) — Mts. of Vogelkop and central cordillera of New Guinea

— *C.m.bicinia* Diamond, 1969 — North Sepik Mts. (New Guinea) [698]

○ *Coracina holopolia* SOLOMON ISLANDS CUCKOO-SHRIKE

— *C.h.holopolia* (Sharpe, 1888) — Bougainville, Choiseul, Guadalcanal (N Solomons)

— *C.h.tricolor* (Mayr, 1931) — Malaita (NE Solomons)

— *C.h.pygmaea* (Mayr, 1931) — Kolombangara, Vangunu (C Solomons)

○ *Coracina ostenta* Ripley, 1952 WHITE-WINGED CUCKOO-SHRIKE Guimaras, Panay and Negros (EC Philippines)

○ *Coracina mcgregori* (Mearns, 1907) MCGREGOR'S CUCKOO-SHRIKE Mts. W Mindanao (S Philippines)

○ *Coracina polioptera* INDOCHINESE CUCKOO-SHRIKE

— *C.p.jabouillei* Delacour, 1951 — NC Vietnam

— *C.p.indochinensis* (Kloss, 1925) — W, C and E Burma, NW and E Thailand, C and S Laos, SC Vietnam

— *C.p.polioptera* (Sharpe, 1878)[5] — SE Burma, SW Thailand, extreme S Vietnam, Cambodia

[1] For reasons to treat this as a separate species see Galbraith & Galbraith (1962) (907). Makira is an alternate name for San Cristobal.
[2] Given as 1878 by Peters *et al.* (1960: 190) (1832), but see Duncan (1937) (770).
[3] Given as 1917 by Peters *et al.* (1960: 189) (1832), but see Hartert (1922) (1084).
[4] For correction of spelling see David & Gosselin (2002) (613).
[5] Peters (1960) (1832) cited 1879, but the name appeared in 1878 in Mitt. K. Zool. Mus. Dresden, 1, pt. 3, p. 370.

○ *Coracina melaschistos* BLACK-WINGED CUCKOO-SHRIKE
    __ *C.m.melaschistos* (Hodgson, 1836)

    __ *C.m.avensis* (Blyth, 1852)

    __ *C.m.intermedia* (Hume, 1877)

    __ *C.m.saturata* (Swinhoe, 1870)[1]

Himalayas to W Yunnan, N Burma >> N and NE India
W China, S and E Burma, N Thailand, N Laos, N
    Vietnam; further south in winter
C and S China, Taiwan >> S Thailand, S Vietnam,
    S Burma
NE and C Vietnam, Hainan (China) >> Thailand

○ *Coracina fimbriata* LESSER CUCKOO-SHRIKE
    __ *C.f.neglecta* (Hume, 1877)
    __ *C.f.culminata* (Hay, 1845)
    __ *C.f.schierbrandii* (Pelzeln, 1865)
    __ *C.f.compta* (Richmond, 1903)
    __ *C.f.fimbriata* (Temminck, 1824)

S Burma, S Thailand
Malay Pen.
Sumatra, Borneo
Simeuluë and Siberut (off W Sumatra)
Java, Bali

○ *Coracina melanoptera* BLACK-HEADED CUCKOO-SHRIKE
    __ *C.m.melanoptera* (Rüppell, 1839)
    __ *C.m.sykesi* (Strickland, 1844)

NW India
NC and E India south to Sri Lanka

CAMPOCHAERA Sharpe, 1878 F
○ *Campochaera sloetii* GOLDEN CUCKOO-SHRIKE
    __ *C.s.sloetii* (Schlegel, 1866)
    __ *C.s.flaviceps* Salvadori, 1880

NW New Guinea
SW to SE New Guinea

LALAGE Boie, 1826 F
○ *Lalage melanoleuca* BLACK-AND-WHITE TRILLER
    __ *L.m.melanoleuca* (Blyth, 1861)
    __ *L.m.minor* (Steere, 1890)

Luzon, Mindoro (N Philippines)
Samar, Leyte, Mindanao (EC and S Philippines)

○ *Lalage nigra* PIED TRILLER
    __ *L.n.davisoni* Kloss, 1926
    __ *L.n.striga* (Horsfield, 1821)
    __ *L.n.nigra* J.R. Forster, 1781[2]
    __ *L.n.leucopygialis* Walden, 1872[3]

Nicobars
Malaysia, Sumatra, Java
Borneo, Philippines
Sulawesi, Sula Is.

○ *Lalage sueurii* (Vieillot, 1818) WHITE-SHOULDERED TRILLER

E Java, Lesser Sundas, S Sulawesi

○ *Lalage tricolor* (Swainson, 1825) WHITE-WINGED TRILLER[4]

SE New Guinea, Australia incl. Melville I.

○ *Lalage aurea* (Temminck, 1825[5]) RUFOUS-BELLIED TRILLER

N Moluccas

○ *Lalage atrovirens* BLACK-BROWED TRILLER
    __ *L.a.moesta* P.L. Sclater, 1883[6]
    __ *L.a.atrovirens* (G.R. Gray, 1862)
    __ *L.a.leucoptera* (Schlegel, 1871)

Tanimbar Is.
Misool and Waigeo Is., N New Guinea
Biak I. (Geelvink Bay)

○ *Lalage leucomela* VARIED TRILLER
    __ *L.l.keyensis* Rothschild & Hartert, 1917
    __ *L.l.polygrammica* (G.R. Gray, 1858)
    __ *L.l.obscurior* Rothschild & Hartert, 1917
    __ *L.l.trobriandi* Mayr, 1936
    __ *L.l.pallescens* Rothschild & Hartert, 1917
    __ *L.l.falsa* E. Hartert, 1925
    __ *L.l.karu* (Lesson & Garnot, 1827)
    __ *L.l.albidior* E. Hartert, 1924
    __ *L.l.ottomeyeri* Stresemann, 1933
    __ *L.l.tabarensis* Mayr, 1955

Kai Is.
Aru Is., E and S New Guinea
D'Entrecasteaux Arch. (E New Guinea)
Trobriand Is. (E New Guinea)
Louisiade Arch.
New Britain, Umboi I. (Bismarck Arch.)
New Ireland (Bismarck Arch.)
New Hanover (Bismarck Arch.)
Lihir Is. (Bismarck Arch.)
Tabar I. (Bismarck Arch.)

---

[1] We tentatively consider *quyi* Dao Van Tien, 1961 (2404), to be a synonym of this. See Dickinson & Dekker (2002) (743A).
[2] Last Edition took no account of Stresemann (1952) (2366), who shifted the type locality of *nigra* to Manila.
[3] Treated as a species by Andrew (1992) (50). Discussed by Inskipp *et al.* (1996; 126) (1182), but the reason they give for separation is flawed, the sympatry is only between *sueurii* and *nigra* we have not found it. See Dickinson & Dekker (2002) (743A).
[4] Separated from *L. sueurii* following White & Bruce (1986: 310) (2652) and Schodde & Mason (1999: 590) (2189).
[5] Peters *et al.* (1960) (1832) cited 1827, but pl. 382 is in livr. 64, which appeared in Dec. 1825 (Sherborn, 1898; Dickinson, 2001) (739, 2242).
[6] For reasons to retain this within *L. atrovirens* see Inskipp *et al.* (1996: 126) (1182).

— *L.l.conjuncta* Rothschild & Hartert, 1924	St. Matthias I. (Bismarck Arch.)
— *L.l.sumunae* Salomonsen, 1964	Dyaul I. (Bismarck Arch.) (2157)
— *L.l.macrura* Schodde, 1989	N Western Australia (Kimberley) (2177)
— *L.l.rufiventris* (G.R. Gray, 1846)	N Northern Territory (Arnhem Land) incl. Melville I.
— *L.l.yorki* Mathews, 1912	Cape York Pen. (Queensland)
— *L.l.leucomela* (Vigors & Horsfield, 1827)	E Queensland, NE New South Wales

○ *Lalage maculosa* POLYNESIAN TRILLER

— *L.m.ultima* Mayr & Ripley, 1941	Efate I. (C Vanuatu)
— *L.m.modesta* Mayr & Ripley, 1941	N and C Vanuatu
— *L.m.melanopygia* Mayr & Ripley, 1941	Utupa, Ndeni Is. (Santa Cruz Is.)
— *L.m.vanikorensis* Mayr & Ripley, 1941	Vanikoro (Santa Cruz Is.)
— *L.m.soror* Mayr & Ripley, 1941	Kandavu I. (Fiji Is.)
— *L.m.pumila* Neumann, 1927	Viti Levu (Fiji Is.)
— *L.m.mixta* Mayr & Ripley, 1941	C and NW Fiji Is.
— *L.m.woodi* Wetmore, 1925	Vanua Levu (Fiji Is.)
— *L.m.rotumae* Neumann, 1927	Rotuma I.
— *L.m.nesophila* Mayr & Ripley, 1941	Lau Arch.
— *L.m.tabuensis* Mayr & Ripley, 1941	Tonga
— *L.m.vauana* Mayr & Ripley, 1941	Vavau group, Tonga Is.
— *L.m.keppeli* Mayr & Ripley, 1941	Keppel, Boscawen Is.
— *L.m.futunae* Mayr & Ripley, 1941	Futuna, Horne Is.
— *L.m.whitmeei* Sharpe, 1878	Niue, Savage Is.
— *L.m.maculosa* (Peale, 1848)	Upolu, Savaii Is. (Samoa)

○ *Lalage sharpei* SAMOAN TRILLER

— *L.s.sharpei* Rothschild, 1900	Upolu I. (Samoa)
— *L.s.tenebrosa* Mayr & Ripley, 1941	Savaii I. (Samoa)

○ *Lalage leucopyga* LONG-TAILED TRILLER

— *L.l.affinis* (Tristram, 1879)	San Cristobal I. (SE Solomons)
— *L.l.deficiens* Mayr & Ripley, 1941	Torres, Banks Is.
— *L.l.albiloris* Mayr & Ripley, 1941	C and N Vanuatu
— *L.l.simillima* (F. Sarasin, 1913)	S Vanuatu, Loyalty Is.
— *L.l.montrosieri* Verreaux & Des Murs, 1860	New Caledonia
— †*L.l.leucopyga* (Gould, 1838)	Norfolk I.

CAMPEPHAGA Vieillot, 1816 F

○ *Campephaga flava* Vieillot, 1817 BLACK CUCKOO-SHRIKE

S Sudan and S Ethiopia to Angola, N and E Botswana and E and S Sth. Africa

○ *Campephaga phoenicea* (Latham, 1790) RED-SHOULDERED CUCKOO-SHRIKE

Senegal, Gambia and Guinea to Ethiopia, N Zaire and Uganda

○ *Campephaga petiti* Oustalet, 1884 PETIT'S CUCKOO-SHRIKE

SE Nigeria to NW Angola, NE Zaire, SW Uganda, W Kenya

○ *Campephaga quiscalina* PURPLE-THROATED CUCKOO-SHRIKE

— *C.q.quiscalina* Finsch, 1869	Guinea and Sierra Leone to NW Angola
— *C.q.martini* (Jackson, 1912)	Zambia to SE Zaire, NE Zaire to C Kenya
— *C.q.muenzneri* Reichenow, 1915	E Tanzania

LOBOTOS Reichenbach, 1850[1] M

○ *Lobotos lobatus* (Temminck, 1824) WESTERN WATTLED CUCKOO-SHRIKE E Sierra Leone, Liberia, W Ivory Coast, Ghana

○ *Lobotos oriolinus* Bates, 1909 EASTERN WATTLED CUCKOO-SHRIKE

S Cameroon, SW Central African Republic, Gabon, E Zaire

PERICROCOTUS Boie, 1826 M

○ *Pericrocotus roseus* (Vieillot, 1818) ROSY MINIVET[2]

E Afghanistan, Himalayas, Assam, Burma, S China, NE Vietnam >> Thailand, Laos, NW and S Vietnam

---

[1] Treated as part of *Campephaga* in last Edition. We recognise this genus following Pearson & Keith (1992: 271) (1816).
[2] Cheng (1987) (333) retained *cantonensis* in *roseus*; where the two meet a small population occurs with mixed characters and this has been called *stanfordi*; this requires further study to determine whether the characteristics suggest limited hybridisation between two species as our treatment implies.

471

○ *Pericrocotus cantonensis* Swinhoe, 1861 SWINHOE'S MINIVET — S and E China >> S Burma, Thailand, Laos, S Vietnam, Cambodia

○ *Pericrocotus divaricatus* ASHY MINIVET
— *P.d.divaricatus* (Raffles, 1822) — NE China, Korea, Amurland, Japan (except as below), E China, >> SE Asia, Philippines

— *P.d.tegimae* Stejneger, 1887[1] — Tokara and Ryukyu Is. (Japan)

○ *Pericrocotus cinnamomeus* SMALL MINIVET
— *P.c.pallidus* E.C.S. Baker, 1920 — Pakistan, NW India
— *P.c.peregrinus* (Linnaeus, 1766) — NW Himalayas, N India
— *P.c.malabaricus* (J.F. Gmelin, 1789) — W India
— *P.c.cinnamomeus* (Linnaeus, 1766) — C and S India, Sri Lanka
— *P.c.vividus* E.C.S. Baker, 1920 — C Himalayas, E India, Bangladesh, W and C Burma, SW and C Thailand, Andamans[2]
— *P.c.thai* Deignan, 1947[3] — NE Burma, N and NE Thailand, Laos
— *P.c.sacerdos* Riley, 1940 — Cambodia, S Vietnam
— *P.c.separatus* Deignan, 1947 — SE Burma, NW Pen. Thailand
— *P.c.saturatus* E.C.S. Baker, 1920 — Java, Bali

○ *Pericrocotus igneus* FIERY MINIVET[4]
— *P.i.igneus* Blyth, 1846 — S Malay Pen., Sumatra, Borneo, Palawan
— *P.i.trophis* Oberholser, 1912 — Simeuluĕ (off W Sumatra)

○ *Pericrocotus lansbergei* Büttikofer, 1886 FLORES MINIVET — Sumbawa, Flores Is.

○ *Pericrocotus erythropygius* WHITE-BELLIED MINIVET
— *P.e.erythropygius* (Jerdon, 1840) — W and C India
— *P.e.albifrons* Jerdon, 1862 — C Burma

○ *Pericrocotus solaris* GREY-CHINNED MINIVET
— *P.s.solaris* Blyth, 1846 — C and E Himalayas, Assam, W and N Burma
— *P.s.rubrolimbatus* Salvadori, 1887 — E and SE Burma, N Thailand
— *P.s.montpellieri* La Touche, 1922 — N Yunnan (SW China)
— *P.s.griseogularis* Gould, 1863 — SE China, Taiwan, Hainan, NE Laos, N Vietnam
— *P.s.deignani* Riley, 1940 — S Laos, SC and S Vietnam
— *P.s.nassovicus* Deignan, 1938 — SE Thailand, S Cambodia
— *P.s.montanus* Salvadori, 1879 — W Malaysia, W Sumatra
— *P.s.cinereigula* Sharpe, 1889 — N Borneo

○ *Pericrocotus ethologus* LONG-TAILED MINIVET
— *P.e.favillaceus* Bangs & Phillips, 1914 — NE Afghanistan, NW Himalayas to W Nepal >> W India
— *P.e.laetus* Mayr, 1940 — E Nepal, N and W Assam, SE Xizang
— *P.e.ethologus* Bangs & Phillips, 1914 — E Assam, N Burma, C and NE China to Hebei >> N Thailand, N Indochina
— *P.e.yvettae* Bangs, 1921 — NE Burma, W and C Yunnan
— *P.e.mariae* Ripley, 1952 — SE Assam, W Burma
— *P.e.ripponi* E.C.S. Baker, 1924 — E Burma, NW Thailand
— *P.e.annamensis* Robinson & Kloss, 1923 — SC Vietnam

○ *Pericrocotus brevirostris* SHORT-BILLED MINIVET
— *P.b.brevirostris* (Vigors, 1831) — C and E Himalayas, W Assam
— *P.b.affinis* (McClelland, 1840) — E Assam, N Burma, W and N Yunnan, SW Sichuan
— *P.b.neglectus* Hume, 1877 — SE Burma, NW Thailand
— *P.b.anthoides* Stresemann, 1923 — S China, N Vietnam, Laos

○ *Pericrocotus miniatus* (Temminck, 1822) SUNDA MINIVET — W Sumatra, Java

---

[1] Sometimes treated as a separate species, a view favoured by CSR based on evidence of moult see Stresemann & Stresemann (1972) (2370); we tentatively retain the views in Orn. Soc. Japan (2000: 297) (1727).
[2] The name *osmastoni* Roselaar & Prins, 2000 (2108), has recently been put forward as a "new name" for Andamans birds, but the review of their affinities has not yet been provided.
[3] Given as 1948 in Peters *et al.* (1960) (1832) but correctly 1947; in same paper as race *separatus*.
[4] Whether this species is safely separated from *P. cinnamomeus* is not satisfactorily resolved. Zoogeography suggests that the present arrangement is not ideal (perhaps *saturatus* is a third species).

○ *Pericrocotus flammeus* SCARLET MINIVET

— *P.f.flammeus* (J.R. Forster, 1781)	W and S India, Sri Lanka
— *P.f.siebersi* Rensch, 1928	Java, Bali
— *P.f.exul* Wallace, 1864	Lombok
— *P.f.andamanensis* Beavan, 1867	Andamans
— *P.f.minythomelas* Oberholser, 1912	Simeuluĕ (Sumatra)
— *P.f.modiglianii* Salvadori, 1892	Enggano (Sumatra)
— *P.f.speciosus* (Latham, 1790)[1]	NW, C and E Himalayas >> N India
— *P.f.fraterculus* Swinhoe, 1870	Yunnan, Assam (hills south of the Brahmaputra), N Burma, N Thailand, N Indochina, ? Hainan (2382A)
— *P.f.fohkiensis* Buturlin, 1910	SE China
— *P.f.semiruber* Whistler & Kinnear, 1933	EC India, S Burma, Thailand
— *P.f.flammifer* Hume, 1875	SE Burma, SW Thailand, N and C Malay Pen.
— *P.f.xanthogaster* (Raffles, 1822)	S Malay Pen., Sumatra
— *P.f.insulanus* Deignan, 1946	Borneo
— *P.f.novus* McGregor, 1904	Luzon, Negros
— *P.f.leytensis* Steere, 1890	Samar, Leyte
— *P.f.johnstoniae* Ogilvie-Grant, 1905	Mt. Apo (SE Mindanao)
— *P.f.gonzalesi* Ripley & Rabor, 1961	N and E Mindanao (2073)
— *P.f.nigroluteus* Parkes, 1981[2]	SC Mindanao (1773)
— *P.f.marchesae* Guillemard, 1885	Jolo (Sulu Arch.)

HEMIPUS Hodgson, 1845 M

○ *Hemipus picatus* BAR-WINGED FLYCATCHER-SHRIKE

— *H.p.capitalis* (Horsfield, 1840)[3]	Himalayas, Assam, SW Yunnan, N and E Burma, NW Thailand, N Indochina
— *H.p.picatus* (Sykes, 1832)	C and S India, S Burma, S, C, and W Thailand, C and S Indochina
— *H.p.intermedius* Salvadori, 1879	Malay Pen., Sumatra, N Borneo
— *H.p.leggei* Whistler, 1939	Sri Lanka

○ *Hemipus hirundinaceus* (Temminck, 1822) BLACK-WINGED FLYCATCHER-SHRIKE

Malay Pen., Sumatra, Java, Bali, Borneo

NEOSITTIDAE SITELLAS (1: 3)

DAPHOENOSITTA De Vis, 1897 F

○ *Daphoenositta chrysoptera* VARIED SITTELLA[4]

— *D.c.leucoptera* (Gould, 1840)	NW to NC Australia
— *D.c.striata* (Gould, 1869)	Cape York Pen., NE Queensland
— *D.c.leucocephala* (Gould, 1838)[5]	EC and SE Queensland
— *D.c.chrysoptera* (Latham, 1802)[6]	SE Australia
— *D.c.pileata* (Gould, 1838)	WC, SW, C and SC Australia

○ *Daphoenositta papuensis* PAPUAN SITTELLA[7]

— *D.p.papuensis* (Schlegel, 1871)	NW New Guinea (Mts. of Vogelkop)
— *D.p.intermedia* (Junge, 1952)	Mts. of WC New Guinea (Nassau Range)
— *D.p.toxopeusi* (Rand, 1940)	Mts. of WC New Guinea (Oranje Range)
— *D.p.alba* (Rand, 1940)	Mts. of NC New Guinea
— *D.p.wahgiensis* (Gyldenstolpe, 1955)	Mts. of EC New Guinea
— *D.p.albifrons* (E.P. Ramsay, 1883)	Mts. of SE New Guinea

[1] In last Edition we listed both *elegans* and *fraterculus* however Deignan (1946: 524) (639) placed *fraterculus* in the synonymy of *elegans*. Ripley (1961: 325) (2042) modified this, placing Sadiya birds (and thus the name *elegans*) within the range and synonymy of *speciosus*. This is reviewed by Dickinson & Dekker (2002) (743A) and Ripley's view is accepted; *elegans* is thus a synonym.
[2] Originally named *neglectus* Parkes, 1974 (1766), but this name is preoccupied.
[3] Peters *et al.* (1960: 218) (1832) credited this name to McClelland, however there is only a Latin diagnosis by Horsfield, the description in English usually supplied by McClelland is absent.
[4] This species and the next treated within the genus *Neositta* in last Edition. But see Schodde & Mason (1999) (2189).
[5] Includes *lumholtzi* see Schodde & Mason (1999) (2189).
[6] For reasons to use 1802 not 1801 see Browing & Monroe (1991) (258). Includes *lathami, albata, magnirostris* and *rothschildi* see Schodde & Mason (1999) (2189).
[7] Sometimes united with *chrysoptera*.

○ *Daphoenositta miranda* BLACK SITTELLA
    — *D.m.frontalis* van Oort, 1910   Mts. of WC New Guinea
    — *D.m.kuboriensis* Mayr & Gilliard, 1952   Mts. of EC New Guinea
    — *D.m.miranda* De Vis, 1897   Mts. of SE New Guinea

FALCUNCULIDAE SHRIKE-TITS (2: 4)

EULACESTOMA De Vis, 1894 N
○ *Eulacestoma nigropectus* WATTLED PLOUGHBILL
    — *E.n.clara* Stresemann & Paludan, 1934   Mts. of WC to EC New Guinea
    — *E.n.nigropectus* De Vis, 1894   Mts. of SE New Guinea

FALCUNCULUS Vieillot, 1816 M
○ *Falcunculus whitei* A.J. Campbell, 1910 NORTHERN SHRIKE-TIT   N Northern Territory, NW Western Australia

○ *Falcunculus leucogaster* Gould, 1838 WESTERN SHRIKE-TIT   S Western Australia

○ *Falcunculus frontatus* (Latham, 1802)[1] EASTERN SHRIKE-TIT[2]   E and SE Australia

GENUS INCERTAE SEDIS (1: 1)

TURNAGRA Lesson, 1837[3] F
○ *Turnagra capensis* PIOPIO/NEW ZEALAND THRUSH
    — †*T.c.turnagra* (Schlegel, 1865)   North Island (New Zealand)
    — †*T.c.capensis* (Sparrman, 1787)   South Island (New Zealand)

PACHYCEPHALIDAE WHISTLERS (6: 41)

PACHYCARE Gould, 1876 N[4]
○ *Pachycare flavogriseum* GOLDENFACE
    — *P.f.flavogriseum* (A.B. Meyer, 1874)   Montane NW New Guinea
    — *P.f.randi* Gilliard, 1961   Montane NC New Guinea
    — *P.f.subaurantium* Rothschild & Hartert, 1911   Montane C New Guinea
    — *P.f.subpallidum* E. Hartert, 1930   Montane E New Guinea

RHAGOLOGUS Stresemann & Paludan, 1934 M
○ *Rhagologus leucostigma* MOTTLED WHISTLER
    — *R.l.leucostigma* (Salvadori, 1876)   Mts. of NW New Guinea
    — *R.l.novus* Rand, 1940   Mts. of N New Guinea
    — *R.l.obscurus* Rand, 1940   Mts. of C and E New Guinea

HYLOCITREA Mathews, 1925 F
○ *Hylocitrea bonensis* OLIVE-FLANKED WHISTLER
    — *H.b.bonensis* (Meyer & Wiglesworth, 1894)   Mts. of N, C and SE Sulawesi
    — *H.b.bonthaina* (Meyer & Wiglesworth, 1896)   Mts. of SC and SW Sulawesi

CORACORNIS Riley, 1918 M
○ *Coracornis raveni* Riley, 1918 MAROON-BACKED WHISTLER   Mts. of C and SE Sulawesi

ALEADRYAS Iredale, 1956 F
○ *Aleadryas rufinucha* RUFOUS-NAPED WHISTLER
    — *A.r.rufinucha* (P.L. Sclater, 1874)   Mts. of NW New Guinea (Vogelkop)
    — *A.r.niveifrons* (E. Hartert, 1930)   Mts. of WC to C New Guinea
    — *A.r.lochmia* (Mayr, 1931)   Mts. of NE New Guinea (Huon Pen.)
    — *A.r.gamblei* (Rothschild, 1897)[5]   Mts. of SE New Guinea

---

[1] For reasons to use 1802 not 1801 see Browing & Monroe (1991) (258).
[2] For separation into three species see Schodde & Mason (1999: 468) (2189).
[3] According to further comparative osteological and feather studies, *Turnagra* is not a bowerbird, and is probably a pachycephalid (R. Schodde & I.J. Mason *in litt.*).
[4] Gender addressed by David & Gosselin (2002) (614), multiple changes occur in names.
[5] Includes *prasinonota* see Mayr (1967: 10) (1481).

PACHYCEPHALA Vigors, 1825 F
○ *Pachycephala olivacea* Olive Whistler
    — *P.o.macphersoniana* H.L. White, 1920
    — *P.o.olivacea* Vigors & Horsfield, 1827
    — *P.o.bathychroa* Schodde & Mason, 1999
    — *P.o.apatetes* Schodde & Mason, 1999
    — *P.o.hesperus* Schodde & Mason, 1999

	NE New South Wales, SE Queensland
	SE Australia
	SC Victoria (2189)
	Tasmania, King I., Flinders I. (2189)
	SW Victoria, SE South Australia (2189)

○ *Pachycephala rufogularis* Gould, 1841 Red-lored Whistler   S Australia

○ *Pachycephala inornata* Gould, 1841 Gilbert's Whistler[1]   S Western Australia, SE Australia

○ *Pachycephala grisola* Mangrove Whistler[2]
    — *P.g.grisola* (Blyth, 1843)   Bangladesh to Indochina, Gtr. Sundas
    — *P.g.plateni* (W.H. Blasius, 1888)   Palawan

○ *Pachycephala albiventris* Green-backed Whistler
    — *P.a.albiventris* (Ogilvie-Grant, 1894)   N Luzon
    — *P.a.crissalis* (J.T. Zimmer, 1918)   C and S Luzon
    — *P.a.mindorensis* (Bourns & Worcester, 1894)   Mindoro

○ *Pachycephala homeyeri* White-vented Whistler
    — *P.h.homeyeri* (W.H. Blasius, 1890)   Mindanao, Sulu Arch., and Siamil I. (off E Borneo)
    — *P.h.major* (Bourns & Worcester, 1894)   Cebu
    — *P.h.winchelli* (Bourns & Worcester, 1894)   W Visayas (C Philippines)

○ *Pachycephala phaionota* (Bonaparte, 1850) Island Whistler[3]   Aru Is. and islands in Banda Sea to N Moluccas, islands off NW New Guinea and Geelvink Bay

○ *Pachycephala hyperythra* Rusty-breasted Whistler
    — *P.h.hyperythra* Salvadori, 1875   W New Guinea
    — *P.h.sepikiana* Stresemann, 1921   NC New Guinea
    — *P.h.reichenowi* Rothschild & Hartert, 1911   NE New Guinea
    — *P.h.salvadorii* Rothschild, 1897   SC to SE New Guinea

○ *Pachycephala modesta* Brown-backed Whistler
    — *P.m.modesta* (De Vis, 1894)   Mts. of SE New Guinea
    — *P.m.hypoleuca* Reichenow, 1915   Mts. of NC to NE New Guinea
    — *P.m.telefolminensis* Gilliard & Lecroy, 1961   Mts. of C New Guinea

○ *Pachycephala philippinensis* Yellow-bellied Whistler
    — *P.p.fallax* (McGregor, 1904)   Calayan I. (off N Luzon)
    — *P.p.illex* (McGregor, 1907)   Camiguin Norte I. (off N Luzon)
    — *P.p.philippinensis* (Walden, 1872)   Luzon, Catanduanes
    — *P.p.siquijorensis* Rand & Rabor, 1957   Siquijor
    — *P.p.apoensis* (Mearns, 1905)   Samar, Leyte, Mindanao
    — *P.p.basilanica* (Mearns, 1909)   Basilan
    — *P.p.boholensis* Parkes, 1966   Bohol (1758A)

○ *Pachycephala sulfuriventer* (Walden, 1872) Sulphur-bellied Whistler[4]   Mts. of Sulawesi

○ *Pachycephala hypoxantha* Bornean Whistler
    — *P.h.hypoxantha* (Sharpe, 1887)   Mts. of Sabah to N Sarawak
    — *P.h.sarawacensis* Chasen, 1935   Poi Mts. (W Sarawak)

○ *Pachycephala meyeri* Salvadori, 1890 Vogelkop Whistler   NW New Guinea

○ *Pachycephala simplex* Grey Whistler[5]
    — *P.s.rufipennis* G.R. Gray, 1858   Kai Is.
    — *P.s.gagiensis* Mayr, 1940   Gagi I. (W of Waigeo)
    — *P.s.waigeuensis* Stresemann & Paludan, 1932   Waigeo I., Gebe I.?

---

[1] Includes *gilberti* see Schodde & Mason (1999: 437) (2189).
[2] For the treatment as given here see Parkes (1989) (1777).
[3] For correct spelling see David & Gosselin (2002) (613).
[4] Including *meridionalis* see White & Bruce (1986) (2652).
[5] For a discussion of the potential split of *simplex* and *griseiceps*, neither of which would be monotypic see Schodde & Mason (1999: 439-440) (2189).

___ *P.s.griseiceps* G.R. Gray, 1858   Aru Is., W and NW New Guinea

___ *P.s.miosnomensis* Salvadori, 1879   Meos Num I. (Geelvink Bay)

___ *P.s.jobiensis* A.B. Meyer, 1874   Yapen I. and NC New Guinea

___ *P.s.perneglecta* E. Hartert, 1930   S New Guinea

___ *P.s.dubia* E.P. Ramsay, 1879   SE New Guinea, D'Entrecasteaux Arch.

___ *P.s.sudestensis* (De Vis, 1892)   Tagula I., Louisiade Arch.

___ *P.s.peninsulae* E. Hartert, 1899   E Cape York, NE Queensland

___ *P.s.simplex* Gould, 1843   N Northern Territory, Melville I.

○ *Pachycephala orpheus* Jardine, 1849   Fawn-breasted Whistler   Timor, Wetar Is. (Lesser Sundas)

○ *Pachycephala soror*   Sclater's Whistler

___ *P.s.soror* P.L. Sclater, 1874   NW New Guinea

___ *P.s.klossi* Ogilvie-Grant, 1915   C and E New Guinea

___ *P.s.octogenarii* Diamond, 1985   Kumawa Mts., New Guinea (701)

___ *P.s.bartoni* Ogilvie-Grant, 1915   SE New Guinea and Goodenough I.

○ *Pachycephala fulvotincta*   Fulvous-tinted Whistler[1]

___ *P.f.teysmanni* Büttikofer, 1893   Saleyer I. (Flores Sea)

___ *P.f.everetti* E. Hartert, 1896   Djampea, Kalao Tua, Madu Is. (Flores Sea)

___ *P.f.javana* E. Hartert, 1928   E Java, Bali

___ *P.f.fulvotincta* Wallace, 1864   Sumbawa I. to Alor I. (W Lesser Sundas)

___ *P.f.fulviventris* E. Hartert, 1896   Sumba I. (SC Lesser Sundas)

○ *Pachycephala macrorhyncha*   Banda Sea Whistler[2]

___ *P.m.calliope* Bonaparte, 1850[3]   Timor, Semau, Wetar Is.

___ *P.m.sharpei* A.B. Meyer, 1885   Babar I. (E Lesser Sundas)

___ *P.m.dammeriana* E. Hartert, 1900   Damar I. (E Lesser Sundas)

___ *P.m.par* E. Hartert, 1904   Roma I. (E Lesser Sundas)

___ *P.m.compar* E. Hartert, 1904   Leti, Moa Is. (E Lesser Sundas)

___ *P.m.fuscoflava* P.L. Sclater, 1883   Tanimbar Is.

___ *P.m.macrorhyncha* Strickland, 1849[4]   Ambon, Seram (S Moluccas)

___ *P.m.buruensis* E. Hartert, 1899   Buru (S Moluccas)

___ *P.m.clio* Wallace, 1863   Sula Is.

___ *P.m.pelengensis* Neumann, 1941   Banggai Is.

○ *Pachycephala mentalis*   Black-chinned Whistler[5]

___ *P.m.tidorensis* van Bemmel, 1939   Tidore, Ternate Is. (N Moluccas)

___ *P.m.mentalis* Wallace, 1863   Bacan, Halmahera, Morotai Is. (N Moluccas)

___ *P.m.obiensis* Salvadori, 1878   Obi (C Moluccas)

○ *Pachycephala pectoralis*   Golden Whistler[6]

___ *P.p.balim* Rand, 1940[7]   Mts. of WC New Guinea

___ *P.p.pectoralis* (Latham, 1802)[8]   E Queensland, NE New South Wales

___ *P.p.xanthoprocta* Gould, 1838   Norfolk I.

___ *P.p.contempta* E. Hartert, 1898   Lord Howe I.

___ *P.p.youngi* Mathews, 1912   SE Australia

___ *P.p.glaucura* Gould, 1845   Tasmania, King I., Flinders I.

___ *P.p.fuliginosa* Vigors & Horsfield, 1827[9]   SW and SC Australia

○ *Pachycephala citreogaster*   Bismarck Whistler[10]

___ *P.c.collaris* E.P. Ramsay, 1878[11]   Islands off SE New Guinea, Louisiade Arch. (excl. Rossel I.)

---

[1] The groups making up *Pachycephala pectoralis* sensu Galbraith (1956) (904) are here arranged, as proposed by R. Schodde (*in litt.*), based on the premise that they were over-lumped. See also Schodde & Mason (1999: 443) (2189). The form Galbraith called *melanops* is already treated as a species (now called *jacquinoti*). This tentative arrangement logically expands the treatment discussed by Schodde & Mason (1999: 444) (2189); it requires further validation. This is Galbraith's group A. *fulvotincta*.

[2] This is the Banda Sea element of Galbraith's group H. *calliope* for which *macrorhynchus* is an older name.

[3] Given as "1851?" by Mayr (1967: 20) (1481) but see Zimmer (1926) (2707).

[4] For correction of spelling see David & Gosselin (2002) (613).

[5] This is Galbraith's group B. *mentalis*.

[6] This is essentially Galbraith's group F. *pectoralis* (with *balim* added).

[7] Treated as a race of *melanura* in last Edition.

[8] For reasons to use 1802 not 1801 see Browing & Monroe (1991) (258). Includes *queenslandica* and *ashbyi* see Schodde & Mason (1999: 443) (2189).

[9] Includes *occidentalis* see Schodde & Mason (1999: 444) (2189).

[10] This is the Bismarck Arch. element of Galbraith's group H. May be conspecific with *pectoralis* sensu stricto.

[11] Includes *misimae* see Mayr (1967: 25) (1481).

— *P.c.rosseliana* E. Hartert, 1898   Rossel I. (Louisiade Arch.)
— *P.c.citreogaster* E.P. Ramsay, 1876   New Hanover, New Britain, New Ireland
— *P.c.sexuvaria* Rothschild & Hartert, 1924   St. Matthias Is. (N Bismarck Arch.)
— *P.c.goodsoni* Rothschild & Hartert, 1914   Admiralty Is.
— *P.c.tabarensis* Mayr, 1955   Tabar I. (E Bismarck Arch.)
— *P.c.ottomeyeri* Stresemann, 1933   Lihir I. (E Bismarck Arch.)

○ *Pachycephala orioloides* Yellow-throated Whistler[1]
— *P.o.bougainvillei* Mayr, 1932   Buka, Bougainville Is. (N Solomons)
— *P.o.orioloides* Pucheran, 1853   Choiseul, Ysabel, Florida Is. (NC Solomons)
— *P.o.centralis* Mayr, 1932   E New Georgia Is. (SC Solomons)
— *P.o.melanoptera* Mayr, 1932   S New Georgia Is. (SC Solomons)
— *P.o.melanonota* E. Hartert, 1908   Ganonga, Vellalavella Is. (SC Solomons)
— *P.o.pavuvu* Mayr, 1932   Pavuvu Is. (EC Solomons)
— *P.o.sanfordi* Mayr, 1931   Malaita I. (NE Solomons)
— *P.o.cinnamomea* (E.P. Ramsay, 1879)   Beagle, Guadalcanal Is. (SE Solomons)
— *P.o.christophori* Tristram, 1879   Santa Ana, San Cristobal Is. (E Solomons)
— *P.o.feminina* Mayr, 1931   Rennell I. (far SE Solomons)

○ *Pachycephala caledonica* New Caledonian Whistler[2]
— *P.c.vanikorensis* Oustalet, 1877   Vanikoro (Santa Cruz Is.)
— *P.c.intacta* Sharpe, 1900   Banks Is., N and C Vanuatu
— *P.c.cucullata* (G.R. Gray, 1859)   Aneiteum I. (S Vanuatu)
— *P.c.chlorura* G.R. Gray, 1859   Erromango I. (S Vanuatu)
— *P.c.littayei* E.L. Layard, 1878   Loyalty Is.
— *P.c.caledonica* (J.F. Gmelin, 1789)   New Caledonia

○ *Pachycephala vitiensis* Fiji Whistler[3]
— *P.v.utupuae* Mayr, 1932   Utupua I. (Santa Cruz Is.)
— *P.v.ornata* Mayr, 1932   N Santa Cruz Is.
— *P.v.kandavensis* E.P. Ramsay, 1876   Kandavu Is. (Fiji)
— *P.v.lauana* Mayr, 1932   S Lau Arch. (Fiji)
— *P.v.vitiensis* G.R. Gray, 1859   Ngau I. (Fiji)
— *P.v.bella* Mayr, 1932   Vatu Vara I. (Fiji)
— *P.v.koroana* Mayr, 1932   Karo I. (Fiji)
— *P.v.torquata* E.L. Layard, 1875   Taviuni I. (Fiji)
— *P.v.aurantiiventris* Seebohm, 1891   Yanganga (Vanua Levu Is., Fiji)
— *P.v.ambigua* Mayr, 1932   Rambi, Kio Is. (Vanua Levu Is., Fiji)
— *P.v.optata* Hartlaub, 1866   Ovalau (SE Viti Levu Is., Fiji)
— *P.v.graeffii* Hartlaub, 1866   Waia (Viti Levu Is., Fiji)

○ *Pachycephala jacquinoti* Bonaparte, 1850[4] Tongan Whistler   Vavau, Tonga Is. (178)

○ *Pachycephala melanura* Mangrove Golden Whistler[5]
— *P.m.dahli* Reichenow, 1897[6]   SE New Guinea (E of Hall Sound), islands off SE New Guinea to E Bismarck Arch.
— *P.m.spinicaudus* (Pucheran, 1853)[7]   SC to SE New Guinea (Hall Sound), Torres Straits Is.
— *P.m.melanura* Gould, 1843[8]   NW to WC Western Australia
— *P.m.robusta* Masters, 1876   NW to NE Australia
— *P.m.whitneyi* E. Hartert, 1929[9]   Whitney I. (NW Solomons)

○ *Pachycephala flavifrons* (Peale, 1848) Samoan Whistler   W Samoa Is.

○ *Pachycephala implicata* Hooded Whistler
— *P.i.richardsi* Mayr, 1932   Bougainville I. (N Solomons)
— *P.i.implicata* E. Hartert, 1929   Guadalcanal I. (SE Solomons)

[1] This is Galbraith's group C. *orioloides*.
[2] This is Galbraith's group G. *caledonica* (treated as a monotypic species by Mayr, 1967 (1481)).
[3] This is Galbraith's group D. *graeffi*, to which we attach *vitiensis* which is an older name.
[4] A prior name of *Pachycephala melanops* (Pucheran, 1853) see duPont (1976: 155) (776).
[5] This is Galbraith's group E. *melanura* (without *balim*).
[6] Includes *fergussonis* see Mayr (1967: 25) (1481).
[7] For correction of spelling see David & Gosselin (2002) (613).
[8] Includes *bynoei*, *hilli* and *violatae* see Schodde & Mason (1999: 446) (Schodde, 1999: 446).
[9] Tentatively placed in this group; may belong in *orioloides*.

○ *Pachycephala nudigula* BARE-THROATED WHISTLER
  — *P.n.ilsa* Rensch, 1928   Sumbawa (Lesser Sundas)
  — *P.n.nudigula* E. Hartert, 1897   Flores (Lesser Sundas)

○ *Pachycephala lorentzi* Mayr, 1931 LORENTZ'S WHISTLER   Mts. of WC New Guinea

○ *Pachycephala schelgelii* REGENT WHISTLER
  — *P.s.schlegelii* Schlegel, 1871   Mts. of NW New Guinea
  — *P.s.cyclopum* E. Hartert, 1930   NC New Guinea (Cyclops Mts.)
  — *P.s.obscurior* E. Hartert, 1896   Mts. of WC to E New Guinea

○ *Pachycephala aurea* Reichenow, 1899 GOLDEN-BACKED WHISTLER   Mts. of SW to SE New Guinea

● *Pachycephala rufiventris* RUFOUS WHISTLER[1]
  — *P.r.minor* Zeitz, 1914   Melville I.
  — *P.r.falcata* Gould, 1843[2]   N Northern Territory, N Western Australia
  — *P.r.pallida* E.P. Ramsay, 1878[3]   N Queensland
  ✔ *P.r.rufiventris* (Latham, 1802)[4]   W, C, S and E Australia
  — *P.r.xanthetraea* (J.R. Forster, 1844)[5]   New Caledonia

○ *Pachycephala monacha* WHITE-BELLIED WHISTLER
  — *P.m.kebirensis* A.B. Meyer, 1885   Moa, Roma, Damar, Wetar Is. (Lesser Sundas)
  — *P.m.arctitorquis* P.L. Sclater, 1883   Tanimbar Is.
  — *P.m.tianduana* E. Hartert, 1901   Tiandou I., (W Kai Is.)
  — *P.m.monacha* G.R. Gray, 1858   Aru Is.
  — *P.m.dorsalis* Ogilvie-Grant, 1911   Mts. of C and E New Guinea
  — *P.m.leucogastra* Salvadori & D'Albertis, 1875[6]   Lowland SE New Guinea
  — *P.m.meeki* E. Hartert, 1898   Rossel I. (Louisiade Arch.)

○ *Pachycephala griseonota* DRAB WHISTLER
  — *P.g.lineolata* Wallace, 1863   Sula Is.
  — *P.g.cinerascens* Salvadori, 1878   N Moluccas, Ternate I.
  — *P.g.johni* E. Hartert, 1903   Obi Major I. (Moluccas)
  — *P.g.examinata* E. Hartert, 1898   Buru I. (Moluccas)
  — *P.g.griseonota* G.R. Gray, 1862   Seram I. (Moluccas)
  — *P.g.kuehni* E. Hartert, 1898   Kai Is.

○ *Pachycephala lanioides* WHITE-BREASTED WHISTLER
  — *P.l.carnaroni* (Mathews, 1913)[7]   Coastal WC Western Australia
  — *P.l.lanioides* Gould, 1840   Coastal N Western Australia
  — *P.l.fretorum* De Vis, 1889   Coastal NE Western Australia, N Northern Territory,
                     NW Queensland

LANIIDAE SHRIKES (4: 30)

CORVINELLA Lesson, 1831 F
○ *Corvinella corvina* YELLOW-BILLED SHRIKE
  — *C.c.corvina* (Shaw, 1809)   Senegal and Gambia to SW Niger
  — *C.c.togoensis* Neumann, 1900   Guinea to Nigeria, Central African Republic, S Chad,
                     W and C Sudan

  — *C.c.caliginosa* Friedmann & Bowen, 1933   Bahr al Ghazal (SW Sudan)
  — *C.c.affinis* Hartlaub, 1857[8]   S Sudan, W Kenya, N Uganda, NE Zaire

UROLESTES Cabanis, 1850[9] M
○ *Urolestes melanoleucus* MAGPIE-SHRIKE
  — *U.m.aequatorialis* Reichenow, 1887   SW Kenya, Tanzania

---

[1] Our arrangement of this species follows Schodde & Mason (1999) (2189).
[2] Implicitly includes *colletti* see Schodde & Mason (1999: 449) (2189).
[3] Implicitly includes *dulcior* see Schodde & Mason (1999: 449) (2189).
[4] For reasons to use 1802 not 1801 see Browing & Monroe (1991) (258). Implicitly includes *maudeae* see Schodde & Mason (1999: 449) (2189).
[5] Drawings with Forster's proposed names were published many years after his death (d. 1798).
[6] For correct spelling see David & Gosselin (2002) (613).
[7] Implicitly includes *bulleri* see Schodde & Mason (1999: 451) (2189).
[8] Includes *chapini* see White (1962: 39) (2646).
[9] Treated in the genus *Corvinella* in last Edition. We follow Fry *et al.* (2000: 376) (899) in using this monotypic genus. Harris & Franklin (2000) (1073)
 placed it in *Corvinella*.

— *U.m.expressus* (Clancey, 1961)	SE Zimbabwe, E Transvaal, S Mozambique to N Natal (390)
— *U.m.melanoleucus* (Jardine, 1831)[1]	S Angola to N Mozambique, Zimbabwe, Botswana, Transvaal, Orange Free State

EUROCEPHALUS A. Smith, 1836[2] M

○ ***Eurocephalus rueppelli*** Bonaparte, 1853 WHITE-RUMPED SHRIKE | E and S Ethiopia to SE Sudan and Tanzania

○ ***Eurocephalus anguitimens*** WHITE-CROWNED SHRIKE

— *E.a.anguitimens* A. Smith, 1836	SW Angola, N Namibia to W Zimbabwe, W Transvaal
— *E.a.niveus* Clancey, 1965	E Zimbabwe, E Transvaal, S Mozambique (407)

LANIUS Linnaeus, 1758 M

○ ***Lanius tigrinus*** Drapiez, 1828 TIGER SHRIKE | EC and E China to NE China, Korea, S Ussuriland and Japan >> S Thailand and S Indochina to Gtr. Sundas

○ ***Lanius souzae*** SOUZA'S SHRIKE

— *L.s.souzae* Bocage, 1878	Congo, N and C Angola to N Zambia, SE Zaire
— *L.s.tacitus* Clancey, 1970	SE Angola, S Zambia, Malawi, W Mozambique (419)
— *L.s.burigi* Chapin, 1950	Rwanda, W and S Tanzania, E Zambia

○ ***Lanius bucephalus*** BULL-HEADED SHRIKE

— *L.b.bucephalus* Temminck & Schlegel, 1845[3]	E and NE China, Korea, Ussuriland, Japan, S Sakhalin, S Kurils, Daito Is.
— *L.b.sicarius* Bangs & Peters, 1928	Gansu (NC China)

○ ***Lanius cristatus*** BROWN SHRIKE

— *L.c.cristatus* Linnaeus, 1758	C and E Siberia to N Mongolia, Baikal area and Kamchatka >> S and SE Asia
— *L.c.confusus* Stegmann, 1929	NE China, Russian Far East, Sakhalin >> Thailand, Malay Pen., Sumatra
— *L.c.lucioniensis* Linnaeus, 1766	C, S and E China to Liaoning, Korea and Kyushu >> Philippines, N Borneo, N Sulawesi, N Moluccas
— *L.c.superciliosus* Latham, 1802[4]	N and C Honshu, Hokkaido, S Sakhalin, coastal Russia Far East >> Sundas

○ ***Lanius collurio*** RED-BACKED SHRIKE

— *L.c.collurio* Linnaeus, 1758[5]	Europe and W Siberia to W Asia Minor, N Kazakhstan and Altai Mts. >> E and S Africa
— *L.c.kobylini* (Buturlin, 1906)[6]	E Turkey, Caucasus area, Levant, N and NW Iran >> E Africa

○ ***Lanius isabellinus*** ISABELLINE SHRIKE[7]

— *L.i.phoenicuroides* (Schalow, 1875)	S, E and NE Iran and Afghanistan to S Kazakhstan, Tien Shan, NW Xinjiang >> Arabia, NE and E Africa
— *L.i.isabellinus* Hemprich & Ehrenberg, 1833	NE Xinjiang and Mongolia to S and E Transbaikalia and C and N Inner Mongolia >> Arabia, NE Africa
— *L.i.arenarius* Blyth, 1846	S and C Xinjiang to SW Inner Mongolia (W China) >> Pakistan, N India
— *L.i.tsaidamensis* Stegmann, 1930	Qaidam Basin (N Qinghai, China) >> Pakistan, N India

○ ***Lanius collurioides*** BURMESE SHRIKE

— *L.c.collurioides* Lesson, 1831[8]	S Assam and W and N Burma to S China, N Thailand and N and C Indochina
— *L.c.nigricapillus* Delacour, 1926	S Vietnam

○ ***Lanius gubernator*** Hartlaub, 1882 EMIN'S SHRIKE | Mali and N Ivory Coast to S Sudan, N Uganda

○ ***Lanius vittatus*** BAY-BACKED SHRIKE

— *L.v.nargianus* Vaurie, 1955	SE Iran, S Turkmenistan, Afghanistan, W Pakistan
— *L.v.vittatus* Valenciennes, 1826	C and S Pakistan to India (except NE) and SW Nepal

---

[1] Includes *angolensis* see Fry (2000: 376) (892).
[2] We follow Lefranc & Worfolk (1997) (1347) and Fry *et al.* (2000) (899) in placing this genus in this family.
[3] Page 39 appeared in 1845, not 1847 as given by Rand (1960: 343) (1971), see Holthuis & Sadai (1970) (1128).
[4] For reasons to use 1802 not 1801 see Browing & Monroe (1991) (258). There are some reports of breeding overlap with *confusus* in the Russian Far East.
[5] Includes *juxtus* which, whether a valid form or not, is now extirpated. See Cramp *et al.* (1993: 477) (594). Also includes *pallidifrons* see Stepanyan (1990: 377) (2326).
[6] For recognition see Roselaar (1995) (2107).
[7] See Pearson (2000) (1813). We suggest the name Red-tailed Shrike be reserved for the *phoenicuroides* component of this species if it should be split.
[8] Given as 1834 by Rand (1960) (1971), but see Browning & Monroe (1991) (258).

○ *Lanius schach* Long-tailed Shrike

  — *L.s.erythronotus* (Vigors, 1831)[1] [2]   WC Asia from SC Kazakhstan and Uzbekistan to Afghanistan, Pakistan and NW India

  — *L.s.caniceps* Blyth, 1846   C, S and W India, Sri Lanka

  — *L.s.tricolor* Hodgson, 1837   E India, Himalayas, Burma, N Thailand, N Indochina, W and S Yunnan

  — *L.s.schach* Linnaeus, 1758   E and C China, Taiwan, Hainan, N Vietnam

  — *L.s.longicaudatus* Ogilvie-Grant, 1902   C and SE Thailand

  — *L.s.bentet* Horsfield, 1822   S Malay Pen., Sumatra to Timor

  — *L.s.nasutus* Scopoli, 1786   Philippines, N Borneo

  — *L.s.suluensis* (Mearns, 1905)   Sulu Arch.

  — *L.s.stresemanni* Mertens, 1923   E New Guinea

○ *Lanius tephronotus* Grey-backed Shrike

  — *L.t.lahulensis* Koelz, 1950   NW India

  — *L.t.tephronotus* (Vigors, 1831)   S and E Tibetan Plateau from Kumaun to N Yunnan, Guizhou, S Shaanxi and Ningxia >> mainland SE Asia

○ *Lanius validirostris* Mountain Shrike

  — *L.v.validirostris* Ogilvie-Grant, 1894   Mts. of N Luzon

  — *L.v.tertius* Salomonsen, 1953   Mts. of Mindoro

  — *L.v.hachisuka* Ripley, 1949[3]   Mts. of Mindanao

○ *Lanius mackinnoni* Sharpe, 1891 Mackinnon's Shrike   Cameroon to Angola and W Kenya

○ *Lanius minor* (J.F. Gmelin, 1788) Lesser Grey Shrike[4]   C and S Europe, SW and WC Asia >> E and S Africa

● *Lanius ludovicianus* Loggerhead Shrike

  ✓ *L.l.excubitorides* Swainson, 1832[5]   C Canada, W and C USA >> E Mexico

  ✓ *L.l.migrans* Palmer, 1898   E Nth. America >> NE Mexico

  — *L.l.ludovicianus* Linnaeus, 1766   SE USA

  — *L.l.miamensis* Bishop, 1933   S Florida

  — *L.l.anthonyi* Mearns, 1898   Santa Barbara I.

  — *L.l.mearnsi* Ridgway, 1903   San Clemente I.

  — *L.l.grinnelli* Oberholser, 1919   N Baja California

  — *L.l.mexicanus* C.L. Brehm, 1854[6]   C Mexico; S Baja California

● *Lanius excubitor* Great Grey Shrike/Northern Shrike

  ✓ *L.e.borealis* Vieillot, 1808[7]   Alaska, Canada >> to C USA

  — *L.e.excubitor* Linnaeus, 1758   N, C and E Europe, NW Siberia

  — *L.e.sibiricus* Bogdanov, 1881   C and E Siberia

  — *L.e.bianchii* E. Hartert, 1907   S Kuril Is., Sakhalin >> N Japan

  — *L.e.mollis* Eversmann, 1853   Altai and Sayan Mts., N Mongolia

  — *L.e.funereus* Menzbier, 1894   S and E Tien Shan and Dzhungarian Alatau Mts.

  — *L.e.homeyeri* Cabanis, 1873[8]   E Balkans, S European Russia, SW Siberia >> SW and C Asia

○ *Lanius meridionalis* Southern Grey Shrike[9]

  — *L.m.meridionalis* Temminck, 1820   S France, Iberia

  — *L.m.koenigi* E. Hartert, 1901   Canary Is.

  — *L.m.algeriensis* Lesson, 1839   NW Africa

  — *L.m.elegans* Swainson, 1832   W Sahara E to N Chad, Libya, C and N Egypt, SW Israel

  — *L.m.leucopygos* Hemprich & Ehrenberg, 1833   S Sahara, Mali to W and C Sudan

  — *L.m.aucheri* Bonaparte, 1853   Coastal Sudan to NW Somalia, SE Israel, Jordan, Arabia (except SW), SW Iran

  — *L.m.theresae* R. Meinertzhagen, 1953[10]   N Israel, S Lebanon

---

[1] CSR believes *jaxartensis* should be recognised, but it is almost invariably in synonymy, see Stepanyan (1990) (2326) and Cramp *et al.* (1993) (594).
[2] Ripley (1981: 266) (2055) considered *nigriceps* had been applied to a "hybrid" between this and *tricolor*.
[3] Including *quartus* see Dickinson *et al.* (1991: 371) (745).
[4] Treated as monotypic following Cramp *et al.* (1993: 482) (594).
[5] Includes *gambeli* and *sonoriensis* see Phillips (1986) (1875).
[6] Includes *nelsoni* see Phillips (1986) (1875).
[7] Includes *invictus* see Phillips (1986) (1875).
[8] Includes *leucopterus* see Stepanyan (1990: 383) (2326).
[9] Separated from *L. excubitor* following Isenmann & Bouchet (1993) (1195). See also Sangster *et al.* (2002) (2167).
[10] For recognition see Cramp *et al.* (1993: 500) (594).

— *L.m.buryi* Lorenz & Hellmayr, 1901   SW Arabia

— *L.m.uncinatus* Sclater & Hartlaub, 1881   Socotra

— *L.m.jebelmarrae* Lynes, 1923[1]   Darfur (W Sudan).

— *L.m.lahtora* (Sykes, 1832)   Pakistan, NW and C India, S Nepal, W Bangladesh

— *L.m.pallidirostris* Cassin, 1852   NE Iran and N Afghanistan to lower Volga R., S Kazakhstan, W Tien Shan, N and E Xinjiang, Mongolia and Inner Mongolia >> SW Asia and NE Africa

○ *Lanius sphenocercus* CHINESE GREY SHRIKE[2]

— *L.s.sphenocercus* Cabanis, 1873   Inner Mongolia, Ningxia and Shaanxi to NE China, N Mongolia, and Russian Far East >> NE and E China

— *L.s.giganteus* Przevalski, 1887   E Tibetan Plateau to W Sichuan and SW Gansu >> E China

○ *Lanius excubitoroides* GREY-BACKED FISCAL

— *L.e.excubitoroides* Prévost & Des Murs, 1847[3]   Mali, L. Chad, W and C Sudan to N and E Uganda, WC Kenya

— *L.e.intercedens* Neumann, 1905   C Ethiopia, W Kenya

— *L.e.boehmi* Reichenow, 1902   W Tanzania to E Zaire, SW Uganda, SW Kenya

○ *Lanius cabanisi* E. Hartert, 1906 LONG-TAILED FISCAL   S Somalia, Kenya, E and S Tanzania

○ *Lanius dorsalis* Cabanis, 1878 TAITA FISCAL   S Ethiopia, S Somalia, Kenya, NE Tanzania

○ *Lanius somalicus* Hartlaub, 1859 SOMALI FISCAL   Somalia, E and S Ethiopia, N Kenya

○ *Lanius collaris* FISCAL SHRIKE

— *L.c.smithii* (Fraser, 1843)   Sierra Leone to W Uganda, NW Tanzania

— *L.c.marwitzi* Reichenow, 1901[4]   EC to SC Tanzania

— *L.c.humeralis* Stanley, 1814   Ethiopia, L Victoria basin to S Tanzania to Zambia, N Mozambique

— *L.c.capelli* (Bocage, 1879)   S Congo and Angola to SW Uganda, SE Zaire, Zambia, Malawi

— *L.c.aridicolus* Clancey, 1955   Coastal NW Namibia, SW Angola

— *L.c.pyrrhostictus* Holub & Pelzeln, 1882[5]   E Botswana and Zimbabwe to S Mozambique and Natal

— *L.c.subcoronatus* A. Smith, 1841   C and N Namibia, S Botswana

— *L.c.collaris* Linnaeus, 1766   S Namibia, Cape Province to S Transvaal and W Swaziland

○ *Lanius newtoni* Bocage, 1891 NEWTON'S FISCAL   São Tomé I.

○ *Lanius senator* WOODCHAT SHRIKE

— *L.s.senator* Linnaeus, 1758   C and SE Europe to Italy, Greece, NW Asia Minor, N Tunisia and N Libya >> W and NC Africa

— *L.s.rutilans* Temminck, 1840   Iberia, NW Africa >> W Africa

— *L.s.badius* Hartlaub, 1854   Balearic Is., Corsica, Sardinia >> W and NC Africa

— *L.s.niloticus* (Bonaparte, 1853)   SC Turkey and Levant to Transcaucasia and Iran >> NE Africa

○ *Lanius nubicus* M.H.K. Lichtenstein, 1823 MASKED SHRIKE   SE Europe and Turkey to Iran and Jordan, ? S Turkmenistan >> NE Africa

VIREONIDAE VIREOS AND ALLIES[6] (4: 52)

CYCLARHIS Swainson, 1824 F[7]

○ *Cyclarhis gujanensis* RUFOUS-BROWED PEPPERSHRIKE

— *C.g.septentrionalis* A.R. Phillips, 1991[8]   E Mexico (1876)

— *C.g.flaviventris* Lafresnaye, 1842   SE Mexico, E Guatemala, N Honduras

— *C.g.yucatanensis* Ridgway, 1886   SE Mexico (Yucatán Pen.)

---

[1] For recognition see Pearson (2000: 361) (1814).
[2] There are reports that these two forms overlap, but proof of sympatric breeding is still awaited.
[3] A later replacement name *excubitorius* is considered to date from 1851; see Sherborn & Woodward (1902) (2242A).
[4] We follow Fry (2000: 342) (892) in retaining this as a race of *L. collaris*.
[5] For recognition see White (1962: 45) (2646).
[6] In this Edition we follow A.O.U. (1998: 429) (3) in not recognising subfamilies.
[7] Gender addressed by David & Gosselin (2002) (614), multiple changes occur in specific names.
[8] Tentatively accepted; see Dickerman & Parkes (1997) (729).

___ *C.g.insularis* Ridgway, 1885 — Cozumel I. (off Yucatán Pen.)
___ *C.g.nicaraguae* Miller & Griscom, 1925 — S Mexico, Guatemala, El Salvador, Honduras, Nicaragua
___ *C.g.subflavescens* Cabanis, 1861 — Costa Rica, W Panama
___ *C.g.perrygoi* Wetmore, 1950 — WC Panama
___ *C.g.flavens* Wetmore, 1950 — E Panama
___ *C.g.coibae* E. Hartert, 1901 — Coiba I. (Panama)
___ *C.g.cantica* Bangs, 1898 — N and E Colombia
___ *C.g.flavipectus* P.L. Sclater, 1858 — NE Venezuela, Trinidad
___ *C.g.parva* Chapman, 1917 — E Colombia, N Venezuela
___ *C.g.gujanensis* (J.F. Gmelin, 1789) — E Colombia, S Venezuela, the Guianas, Brazil, E Peru, NE Bolivia

___ *C.g.cearensis* S.F. Baird, 1866 — E Brazil
___ *C.g.ochrocephala* Tschudi, 1845 — SE Brazil, Paraguay, Uruguay, NE Argentina
___ *C.g.viridis* (Vieillot, 1822) — Paraguay, N Argentina
___ *C.g.virenticeps* P.L. Sclater, 1860 — Ecuador, NW Peru
___ *C.g.contrerasi* Taczanowski, 1879 — N Peru
___ *C.g.saturata* J.T. Zimmer, 1925 — C Peru
___ *C.g.pax* Bond & Meyer de Schauensee, 1942 — EC Bolivia
___ *C.g.dorsalis* J.T. Zimmer, 1942 — C Bolivia
___ *C.g.tarijae* Bond & Meyer de Schauensee, 1942 — SE Bolivia, NW Argentina

○ *Cyclarhis nigrirostris* BLACK-BILLED PEPPERSHRIKE
___ *C.n.nigrirostris* Lafresnaye, 1842 — C Colombia, E Ecuador
___ *C.n.atrirostris* P.L. Sclater, 1887 — SW Colombia, W Ecuador

VIREOLANIUS Bonaparte, 1850 M
○ *Vireolanius melitophrys* CHESTNUT-SIDED SHRIKE-VIREO
___ *V.m.melitophrys* Bonaparte, 1850 — S Mexico
___ *V.m.crossini* A.R. Phillips, 1991[1] — SW Mexico (1876)
___ *V.m.quercinus* Griscom, 1935 — S Mexico, Guatemala

○ *Vireolanius pulchellus* GREEN SHRIKE-VIREO[2]
___ *V.p.ramosi* A.R. Phillips, 1991[3] — SE Mexico (1876)
___ *V.p.pulchellus* Sclater & Salvin, 1859 — Guatemala to Honduras
___ *V.p.verticalis* Ridgway, 1885 — Nicaragua, Costa Rica, W Panama
___ *V.p.viridiceps* Ridgway, 1903 — C Panama

○ *Vireolanius eximius* YELLOW-BROWED SHRIKE-VIREO
___ *V.e.mutabilis* Nelson, 1912 — E Panama
___ *V.e.eximius* S.F. Baird, 1866 — N Colombia, NW Venezuela

○ *Vireolanius leucotis* SLATY-CAPPED SHRIKE-VIREO
___ *V.l.mikettae* E. Hartert, 1900 — W Colombia, NW Ecuador
___ *V.l.leucotis* (Swainson, 1838) — N and W Amazonia
___ *V.l.simplex* Berlepsch, 1912 — N Brazil, S Peru
___ *V.l.bolivianus* Berlepsch, 1901 — SE Peru, N Bolivia

VIREO Vieillot, 1808 M
○ *Vireo brevipennis* SLATY VIREO
___ *V.b.browni* (Miller & Ray, 1944) — SW Mexico
___ *V.b.brevipennis* (P.L. Sclater, 1858) — SE Mexico

● *Vireo griseus* WHITE-EYED VIREO
✓ *V.g.griseus* (Boddaert, 1783)[4] — SE USA >> E Mexico to N Honduras, Cuba
___ *V.g.maynardi* Brewster, 1887 — S Florida
___ *V.g.bermudianus* Bangs & Bradlee, 1901 — Bermuda I.
✓ *V.g.micrus* Nelson, 1899 — S Texas, NE Mexico
___ *V.g.perquisitor* Nelson, 1900 — E Mexico
___ *V.g.marshalli* A.R. Phillips, 1991 — EC Mexico (1876)

[1] Tentatively accepted; see Dickerman & Parkes (1997) (729).
[2] The proposed race ¶*dearborni* A.R. Phillips, 1991 (1876), requires substantiation.
[3] For recognition see Parkes in Dickerman & Parkes (1997) (729).
[4] Includes *noveboracensis* J.F. Gmelin, 1789 (J.V. Remsen *in litt.*).

○ *Vireo crassirostris* THICK-BILLED VIREO
— *V.c.crassirostris* (H. Bryant, 1859) — Bahamas
— *V.c.stalagmium* Buden, 1985 — Caicos Is. (266)
— *V.c.tortugae* Richmond, 1917 — Tortue I., Haiti
— *V.c.cubensis* Kirkconnell & Garrido, 2000 — Cayo Paredón Grande, Cuba (1270)
— *V.c.alleni* Cory, 1886[1] — Cayman Is. (571A)

○ *Vireo pallens* MANGROVE VIREO[2]
— *V.p.paluster* R.T. Moore, 1938 — NW Mexico
— *V.p.ochraceus* Salvin, 1863 — SW Mexico, W Guatemala, W El Salvador
— *V.p.pallens* Salvin, 1863 — S Honduras, W Nicaragua, W Costa Rica
— *V.p.nicoyensis* Parkes, 1990 — NW Costa Rica (1778)
— *V.p.salvini* van Rossem, 1934 — SE Mexico, Belize
— *V.p.semiflavus* Salvin, 1863 — N Guatemala to SE Nicaragua
— *V.p.wetmorei* A.R. Phillips, 1991[3] — El Cayo (off Guatemala) (1876)
— *V.p.angulensis* Parkes, 1990 — Bahia I. (off N Honduras) (1778)
— *V.p.browningi* A.R. Phillips, 1991[4] — SE Nicaragua (1876)
— *V.p.approximans* Ridgway, 1884 — Isla Providencia, Sta. Catalina I. (W Caribbean)

○ *Vireo bairdi* Ridgway, 1885 COZUMEL VIREO — Cozumel I. (off Yucatán Pen.)

○ *Vireo caribaeus* Bond & Meyer de Schauensee, 1942 ST. ANDREW VIREO
Isla Andrès (W Caribbean)

○ *Vireo modestus* P.L. Sclater, 1860 JAMAICAN VIREO — Jamaica

○ *Vireo gundlachii* CUBAN VIREO
— *V.g.magnus* Garrido, 1971 — Cuba (Cayo Cantiles) (917)
— *V.g.sanfelipensis* Garrido, 1973 — Cuba (Cayo Real) (920)
— *V.g.gundlachii* Lembeye, 1850 — C and E Cuba
— *V.g.orientalis* Todd, 1910 — SE Cuba

○ *Vireo latimeri* S.F. Baird, 1866 PUERTO RICAN VIREO — W Puerto Rico

○ *Vireo nanus* (Lawrence, 1875) FLAT-BILLED VIREO — Hispaniola

◑ *Vireo bellii* BELL'S VIREO
— *V.b.pusillus* Coues, 1866 — C California to N Baja California>> S Baja California
— *V.b.arizonae* Ridgway, 1903 — SW USA, NW Mexico >> W Mexico
✓ *V.b.medius* Oberholser, 1903 — SC USA >> Mexico
— *V.b.bellii* Audubon, 1844 — C and S USA >> Mexico, Guatemala, El Salvador, Honduras, N Nicaragua

◑ *Vireo atricapilla* Woodhouse, 1852 BLACK-CAPPED VIREO[5] — C and S USA >> N and W Mexico

○ *Vireo nelsoni* Bond, 1936 DWARF VIREO — SW Mexico

◑ *Vireo vicinior* Coues, 1866 GREY VIREO — SW USA, NW Mexico >> NW Mexico, SW Texas

○ *Vireo osburni* (P.L. Sclater, 1861) BLUE MOUNTAIN VIREO — Jamaica

◑ *Vireo flavifrons* Vieillot, 1808 YELLOW-THROATED VIREO — S Canada, E and C USA >> Colombia, Venezuela, Central America

◑ *Vireo plumbeus* PLUMBEOUS VIREO[6]
✓ *V.p.plumbeus* Coues, 1866[7] — WC USA to SW Mexico >> NW Mexico
— *V.p.gravis* A.R. Phillips, 1991[8] — EC Mexico (1876)
— *V.p.notius* van Tyne, 1933 — Belize
— *V.p.montanus* van Rossem, 1933 — S Mexico, to Honduras

---

[1] For recognition of this see Buden (1985) (266).
[2] The proposed race *olsoni* A.R. Phillips, 1991 (1876), is not accepted here, but see Dickerman & Parkes (1997) (729).
[3] Tentatively accepted; see Dickerman & Parkes (1997) (729).
[4] Tentatively accepted; see Dickerman & Parkes (1997) (729).
[5] For correction of spelling see David & Gosselin (2002) (613).
[6] This species and the next separated here following A.O.U. (1998: 434) (3), but note reservations therein as to subspecific attributions.
[7] Includes *jacksoni* Oberholser, 1974 (1673), see Browning (1978) (246).
[8] Tentatively accepted; see Dickerman & Parkes (1997) (729).

● *Vireo cassinii* CASSIN'S VIREO
    ☑ *V.c.cassinii* Xantús de Vesey, 1858   SW Canada, W USA >> Mexico, Guatemala
    __ *V.c.lucasanus* Brewster, 1891   S Baja California

● *Vireo solitarius* BLUE-HEADED VIREO
    ☑ *V.s.solitarius* (A. Wilson, 1810)   Canada, NC and E USA >> E Mexico, C America, W Cuba
    __ *V.s.alticola* Brewster, 1886   E USA (Appalachian Mts.) >> SE USA

○ *Vireo carmioli* S.F. Baird, 1866 YELLOW-WINGED VIREO   Costa Rica, W Panama

○ *Vireo masteri* Salaman & Stiles, 1996 CHOCO VIREO #   W Colombia (2148)

● *Vireo huttoni* HUTTON'S VIREO
    __ *V.h.obscurus* Anthony, 1890   SW Canada to NW California
    __ *V.h.parkesi* Rea, 1991   N California (1876)
    __ *V.h.sierrae* Rea, 1991   Sierra Nevada Mts. (California) (1876)
    __ *V.h.huttoni* Cassin, 1851   C California
    __ *V.h.unitti* Rea, 1991   Santa Catalina I. (California) (1876)
    __ *V.h.cognatus* Ridgway, 1903   S Baja California
    __ *V.h.stephensi* Brewster, 1882   SW USA, NW Mexico
    __ *V.h.carolinae* H.W. Brandt, 1938   SW Texas, NE Mexico
    __ *V.h.pacificus* A.R. Phillips, 1966   W and SW Mexico
    __ *V.h.mexicanus* Ridgway, 1903   C and S Mexico
    __ *V.h.vulcani* Griscom, 1930   W Guatemala

○ *Vireo hypochryseus* GOLDEN VIREO
    __ *V.h.nitidus* van Rossem, 1934   S Sonora (NW Mexico)
    __ *V.h.hypochryseus* P.L. Sclater, 1862   W Mexico
    __ *V.h.sordidus* Nelson, 1898   Tres Marias Is.

● *Vireo gilvus* WARBLING VIREO[1]
    ☑ *V.g.swainsonii* S.F. Baird, 1858[2]   W Canada, W USA >> W Mexico, Guatemala, Honduras, Nicaragua
    __ *V.g.brewsteri* Ridgway, 1903[3]   W USA (mainly Rocky Mts.) to W Mexico >> Mexico
    __ *V.g.victoriae* Sibley, 1940   S Baja California
    __ *V.g.gilvus* (Vieillot, 1808)   SW Canada, C, S and NE USA >> Middle America
    __ *V.g.sympatricus* A.R. Phillips, 1991   C Mexico (1876)

○ *Vireo leucophrys* BROWN-CAPPED VIREO[4]
    __ *V.l.eleanorae* Sutton & Burleigh, 1940   NE Mexico
    __ *V.l.dubius* A.R. Phillips, 1991   EC Mexico (1876)
    __ *V.l.amauronotus* Salvin & Godman, 1881   SE Mexico
    __ *V.l.strenuus* Nelson, 1900   S Mexico (Chiapas)
    __ *V.l.bulli* J.S. Rowley, 1968   SE Oaxaca (2116)
    __ *V.l.palmeri* A.R. Phillips, 1991[5]   Honduras (1876)
    __ *V.l.costaricensis* Ridgway, 1903   C Costa Rica
    __ *V.l.dissors* J.T. Zimmer, 1941[6]   S Costa Rica to NW Colombia
    __ *V.l.mirandae* E. Hartert, 1917   N Colombia, NW Venezuela
    __ *V.l.josephae* P.L. Sclater, 1859   SW Colombia, W Ecuador
    __ *V.l.leucophrys* (Lafresnaye, 1844)   E Andes from Colombia to C Peru
    __ *V.l.maranonicus* J.T. Zimmer, 1941   N Peru
    __ *V.l.laetissimus* (Todd, 1924)   S Peru, N Bolivia, C Colombia, Ecuador, N Peru

● *Vireo philadelphicus* (Cassin, 1851) PHILADELPHIA VIREO   W Canada, N USA >> Mexico, C America, Colombia

● *Vireo olivaceus* RED-EYED VIREO
    ☑ *V.o.olivaceus* (Linnaeus, 1766)   Canada, WC and E USA >> Sth. America
    __ *V.o.caucae* (Chapman, 1912)   W Colombia

[1] Our treatment of this species and the next follows Phillips (1991) (1876).
[2] Includes *leucopolius* see Phillips (1991) (1876).
[3] Includes *petrorus* Oberholser, 1974 (1673), see Browning (1990) (249) and *connectens* see Phillips (1991) (1876).
[4] We follow A.O.U. (1998: 436) (3) in treating *V. leucophrys* as an allospecies of *V. gilvus* and in retaining *swainsonii* within the latter; in last Edition *leucophrys* was understood to be monotypic.
[5] Tentatively accepted; see Dickerman & Parkes (1997) (729).
[6] Includes *disjunctus* and *chiriquensis* see Olson (1981) (1697).

— *V.o.griseobarbatus* (Berlepsch & Taczanowski, 1883)   W Ecuador, NW Peru

— *V.o.pectoralis* J.T. Zimmer, 1941   N Peru

— *V.o.solimoensis* Todd, 1931   E Ecuador, NE Peru

— *V.o.vividior* Hellmayr & Seilern, 1913   Colombia, Venezuela, the Guianas, N Brazil, Trinidad

— *V.o.tobagensis* Hellmayr, 1935   Tobago I.

— *V.o.agilis* (M.H.K. Lichtenstein, 1823)   NE Brazil

— *V.o.diversus* J.T. Zimmer, 1941   SE Brazil, E Paraguay

— *V.o.chivi* (Vieillot, 1817)   W and SW Amazonia

○ *Vireo gracilirostris* Sharpe, 1890   NORONHA VIREO[1]   Fernando de Noronha I. (Brazil)

○ *Vireo flavoviridis* YELLOW-GREEN VIREO

— *V.f.flavoviridis* (Cassin, 1851)   Mexico to Panama >> W Amazonia

— *V.f.forreri* Madarász, 1885   Tres Marias Is., N Mexico >> Upper Amazonia

— *V.f.perplexa* A.R. Phillips, 1991[2]   N Guatemala (1876)

○ *Vireo altiloquus* BLACK-WHISKERED VIREO

— *V.a.barbatulus* (Cabanis, 1855)   S Florida, Cuba, Haiti >> Sth. America

— *V.a.altiloquus* (Vieillot, 1808)   Gtr. Antilles >> N Sth. America

— *V.a.barbadensis* (Ridgway, 1874)   St. Croix, Barbados Is.

— *V.a.bonairensis* Phelps & Phelps, Jr., 1948   Aruba, Curaçao, Bonaire Is.

— *V.a.grandior* (Ridgway, 1884)   Isla Providencia, Sta. Catalina I. (W Caribbean)

— *V.a.canescens* (Cory, 1887)   Isla Andrès

○ *Vireo magister* YUCATÁN VIREO

— *V.m.magister* (Lawrence, 1871)   SE Mexico, Belize

— *V.m.decoloratus* (A.R. Phillips, 1991)[3]   Islands off Belize (1876)

— *V.m.stilesi* (A.R. Phillips, 1991)[4]   Islands off S Belize and N Honduras (1876)

— *V.m.caymanensis* Cory, 1887   Grand Cayman I.

HYLOPHILUS Temminck, 1822   M

○ *Hylophilus poicilotis* Temminck, 1822   RUFOUS-CROWNED GREENLET   SE Brazil, E Paraguay, NE Argentina

○ *Hylophilus amaurocephalus* (Nordmann, 1835)   GREY-EYED GREENLET[5]

E Brazil

○ *Hylophilus thoracicus* LEMON-CHESTED GREENLET

— *H.t.aemulus* (Hellmayr, 1920)   SE Colombia, E Ecuador, E Peru, N Bolivia

— *H.t.griseiventris* Berlepsch & Hartert, 1902   E Venezuela, the Guianas, N Brazil

— *H.t.thoracicus* Temminck, 1822   SE Brazil

○ *Hylophilus semicinereus* GREY-CHESTED GREENLET

— *H.s.viridiceps* (Todd, 1929)   S Venezuela, the Guianas, N Brazil

— *H.s.semicinereus* Sclater & Salvin, 1867   N Brazil

— *H.s.juruanus* Gyldenstolpe, 1941   NW Brazil

○ *Hylophilus pectoralis* P.L. Sclater, 1866   ASHY-HEADED GREENLET   NE Venezuela, the Guianas, E Brazil

○ *Hylophilus sclateri* Salvin & Godman, 1883   TEPUI GREENLET   S Venezuela, W Guyana, NC Brazil

○ *Hylophilus brunneiceps* P.L. Sclater, 1866   BROWN-HEADED GREENLET[6]   E Colombia, S Venezuela, NW Brazil

○ *Hylophilus semibrunneus* Lafresnaye, 1845   RUFOUS-NAPED GREENLET   N Colombia, NW Venezuela, E Ecuador

○ *Hylophilus aurantiifrons* GOLDEN-FRONTED GREENLET

— *H.a.aurantiifrons* Lawrence, 1862   E Panama, N Colombia

— *H.a.helvinus* Wetmore & Phelps, 1956   NW Venezuela

— *H.a.saturatus* (Hellmayr, 1906)   E Colombia, N Venezuela, Trinidad

○ *Hylophilus hypoxanthus* DUSKY-CAPPED GREENLET

— *H.h.hypoxanthus* Pelzeln, 1868   SE Colombia

---

[1] Separated from the above following Olson (1994) Olson (1994) and A.O.U. (1998: 438) (3).
[2] Tentatively accepted, see Dickerman & Parkes (1997) (729); but we have not accepted *vanrossemi* A.R. Phillips, 1991 (1876).
[3] Tentatively accepted, see Dickerman & Parkes (1997) (729).
[4] Tentatively accepted, see Dickerman & Parkes (1997) (729).
[5] For reasons to separate this from *H. poicilotis* see Willis (1991) (2668).
[6] The race *inornatus* has been moved to *H. hypoxanthus*.

__ *H.h.fuscicapillus* Sclater & Salvin, 1880	E Ecuador, N Peru
__ *H.h.flaviventris* Cabanis, 1873	C Peru
__ *H.h.ictericus* Bond, 1953	W Brazil, NE Peru, N Bolivia
__ *H.h.albigula* (Chapman, 1921)	N Brazil
__ *H.h.inornatus* (E. Snethlage, 1914)[1]	N Brazil

○ *Hylophilus muscicapinus* BUFF-CHEEKED GREENLET
__ *H.m.muscicapinus* Sclater & Salvin, 1873	S Venezuela, the Guianas, E Brazil, E Bolivia
__ *H.m.griseifrons* (E. Snethlage, 1907)	N Brazil

○ *Hylophilus flavipes* SCRUB GREENLET
__ *H.f.viridiflavus* Lawrence, 1862	S Costa Rica to C Panama
__ *H.f.xuthus* Wetmore, 1957	Coiba I. (Panama)
__ *H.f.flavipes* Lafresnaye, 1845	C and N Colombia
__ *H.f.melleus* Wetmore, 1941	N Colombia
__ *H.f.galbanus* Wetmore & Phelps, 1956	NE Colombia, NW Venezuela
__ *H.f.acuticauda* Lawrence, 1865	N Venezuela
__ *H.f.insularis* P.L. Sclater, 1861	Tobago I.

○ *Hylophilus olivaceus* Tschudi, 1844 OLIVACEOUS GREENLET E Ecuador, N Peru

○ *Hylophilus ochraceiceps* TAWNY-CROWNED GREENLET
__ *H.o.ochraceiceps* P.L. Sclater, 1859	S Mexico, Guatemala
__ *H.o.pallidipectus* (Ridgway, 1903)	Guatemala to Costa Rica
__ *H.o.pacificus* Parkes, 1991	SE Costa Rica, NW Panama (1876)
__ *H.o.nelsoni* (Todd, 1929)	W to E Panama
__ *H.o.bulunensis* E. Hartert, 1902	E Panama, W Colombia, W Ecuador
__ *H.o.ferrugineifrons* P.L. Sclater, 1862	SE Colombia, S Venezuela, Guyana, Ecuador, Peru, NW Brazil
__ *H.o.viridior* (Todd, 1929)	S Peru, N Bolivia
__ *H.o.luteifrons* P.L. Sclater, 1881	E Venezuela, the Guianas, N Brazil
__ *H.o.lutescens* (E. Snethlage, 1914)	N Brazil
__ *H.o.rubrifrons* Sclater & Salvin, 1867	NE Brazil

○ *Hylophilus decurtatus* LESSER GREENLET
__ *H.d.brevipennis* (Giraud, 1852)	E Mexico
__ *H.d.dickermani* Parkes, 1991	SE Mexico (1876)
__ *H.d.phillipsi* Parkes, 1991	SE Mexico (Yucatán Pen.) (1876)
__ *H.d.decurtatus* (Bonaparte, 1838)	C America
__ *H.d.darienensis* (Griscom, 1927)	C Panama to N Colombia
__ *H.d.minor* Berlepsch & Taczanowski, 1883	SW Colombia, W Ecuador, NW Peru

ORIOLIDAE (ORIOLES AND FIGBIRDS) (2: 29)

SPHECOTHERES Vieillot, 1816[2] M
○ *Sphecotheres viridis* Vieillot, 1816 TIMOR FIGBIRD	Timor, Roti Is. (Lesser Sundas)
○ *Sphecotheres hypoleucus* Finsch, 1898 WETAR FIGBIRD	Wetar I. (Lesser Sundas)

○ *Sphecotheres vieilloti* AUSTRALASIAN FIGBIRD
__ *S.v.salvadorii* Sharpe, 1877	SE New Guinea
__ *S.v.cucullatus* (Rosenberg, 1867)	Kai Is.
__ *S.v.ashbyi* Mathews, 1912[3]	N Western Australia (Kimberley), N Northern Territory (Arnhem Land), Melville I.
__ *S.v.flaviventris* Gould, 1850	N and NE Queensland, Torres Strait Is.
__ *S.v.vieilloti* Vigors & Horsfield, 1827	Coastal EC Australia

ORIOLUS Linnaeus, 1766 M
○ *Oriolus szalayi* (Madarász, 1900) BROWN ORIOLE	W Papuan Is., New Guinea

---

[1] Placed here following Ridgely & Tudor (1989: 157) (2018).
[2] We follow Greenway (1962: 137) (1006) in treating *viridis* and *hypoleucos* as separate species, but we unite *vieilloti* and *flaviventris* following Schodde & Mason (1999: 595-596) (2189).
[3] For recognition see Schodde & Mason (1999: 595) (2189).

○ *Oriolus phaeochromus* G.R. Gray, 1861 DUSKY-BROWN ORIOLE   Halmahera (N Moluccas)

○ *Oriolus forsteni* (Bonaparte, 1850) GREY-COLLARED ORIOLE   Seram (S Moluccas)

○ *Oriolus bouroensis* BLACK-EARED ORIOLE
　　— *O.b.bouroensis* (Quoy & Gaimard, 1830)   Buru (S Moluccas)
　　— *O.b.decipiens* (P.L. Sclater, 1883)   Tanimbar Is. (Lesser Sundas)

○ *Oriolus melanotis* OLIVE-BROWN ORIOLE
　　— *O.m.finschi* E. Hartert, 1904   Wetar I. (Lesser Sundas)
　　— *O.m.melanotis* Bonaparte, 1850   Timor, Roti Is. (Lesser Sundas)

○ *Oriolus sagittatus* OLIVE-BACKED ORIOLE
　　— *O.s.magnirostris* van Oort, 1910   SC New Guinea (Trans-Fly)
　　— *O.s.affinis* Gould, 1848   NW to NC Australia (to S Gulf of Carpentaria)
　　— *O.s.grisescens* Schodde & Mason, 1999   Cape York Pen. [2189]
　　— *O.s.sagittatus* (Latham, 1802)[1]   NE to SE Australia

○ *Oriolus flavocinctus* GREEN ORIOLE[2]
　　— *O.f.migrator* E. Hartert, 1904   Leti, Moa, Roma (E Lesser Sundas)
　　— *O.f.muelleri* Bonaparte, 1850   SC New Guinea (Trans-Fly)
　　— *O.f.flavocinctus* (P.P. King, 1826)   N Western Australia (Kimberley), N Northern
　　　　　　　Territory (Arnhem Land)
　　— *O.f.tiwi* Schodde & Mason, 1999   Melville I. [2189]
　　— *O.f.flavotinctus* Schodde & Mason, 1999   Cape York Pen. [2189]
　　— *O.f.kingi* Mathews, 1912   NE Queensland

○ *Oriolus xanthonotus* DARK-THROATED ORIOLE
　　— *O.x.xanthonotus* Horsfield, 1821   Malay Pen., Sumatra, Java, SW Borneo
　　— *O.x.mentawi* Chasen & Kloss, 1926   W Sumatran islands
　　— *O.x.consobrinus* Wardlaw Ramsay, 1880   N, C and E Borneo
　　— *O.x.persuasus* Bangs, 1922   Palawan I.

○ *Oriolus steerii* PHILIPPINE ORIOLE[3]
　　— *O.s.albiloris* Ogilvie-Grant, 1894   WC and NE Luzon
　　— *O.s.samarensis* Steere, 1890   Samar, Leyte, E Mindanao
　　— †?*O.s.assimilis* Tweeddale, 1878   Cebu
　　— *O.s.steerii* Sharpe, 1877   Masbate, Negros
　　— *O.s.basilanicus* Ogilvie-Grant, 1896   Basilan, W Mindanao
　　— *O.s.cinereogenys* Bourns & Worcester, 1894   Sulu Is.

○ *Oriolus isabellae* Ogilvie-Grant, 1894 ISABELA ORIOLE   WC and NE Luzon

○ *Oriolus oriolus* EURASIAN GOLDEN ORIOLE
　　— *O.o.oriolus* (Linnaeus, 1758)   Europe, NW Africa, W and WC Asia >> C, E and S
　　　　　　　Africa, NW India
　　— *O.o.kundoo* Sykes, 1832   C Asia, N India

○ *Oriolus auratus* AFRICAN GOLDEN ORIOLE
　　— *O.a.auratus* Vieillot, 1817   S Senegal, Guinea to W Ethiopia >> NE Zaire, N Uganda
　　— *O.a.notatus* W.K.H. Peters, 1868   Angola to Malawi and W Mozambique >> Uganda and
　　　　　　　S Kenya

○ *Oriolus tenuirostris* SLENDER-BILLED ORIOLE[4]
　　— *O.t.tenuirostris* Blyth, 1846   E Nepal, C Burma, SW Yunnan, N Laos >> S Burma, W
　　　　　　　and SW Thailand
　　— *O.t.invisus* Riley, 1940   S Vietnam

○ *Oriolus chinensis* BLACK-NAPED ORIOLE
　　— *O.c.diffusus* Sharpe, 1877   E Asia south to N Laos >> India, Sri Lanka, mainland
　　　　　　　SE Asia

[1]　For reasons to use 1802 not 1801 see Browing & Monroe (1991) (258).
[2]　Treated as monotypic in last Edition; *migrator* and *muelleri* were recognised by Greenway (1962: 124) (1006), Australian races follow Schodde & Mason (1999: 598-599) (2189).
[3]　Although split from *O. xanthonotus* by Dickinson *et al.* (1991) a detailed comparative study of the two has not yet been done. Subspecific names "corrected" in Dickinson *et al.* (1991) (745) were restored by Dickinson & Kennedy (2000) (744).
[4]　Both *O.t.tenuriostris* and *O.c.diffusus* have been reported summering in N Laos, but proof of sympatric breeding there or in SW Yunnan is lacking. Treatment as two species remains hypothetical.

    — *O.c.andamanensis* Beavan, 1867   Andamans

    — *O.c.macrourus* Blyth, 1846   Nicobars

    — *O.c.maculatus* Vieillot, 1817   S Malay Pen., Sumatra, Nias I. (off W Sumatra), Borneo, Java, Bali

    — *O.c.mundus* Richmond, 1903   Simeuluë I. (off Sumatra)

    — *O.c.sipora* Chasen & Kloss, 1926   Sipora I. (off Sumatra)

    — *O.c.richmondi* Oberholser, 1912   Siberut, Pagi Is. (off Sumatra)

    — *O.c.lamprochryseus* Oberholser, 1917   Solombo Besar I. (off Sumatra)

    — *O.c.insularis* Vorderman, 1893   Kangean I. (Java Sea)

    — *O.c.melanisticus* Meyer & Wiglesworth, 1894   Talaut Is. (Indonesia)

    — *O.c.sanghirensis* Meyer & Wiglesworth, 1898   Sangihe and Tabukan (Sangihe Is., N of Sulawesi)

    — *O.c.formosus* Cabanis, 1872   Siau, Ruang, Mayu Is. (Sangihe Is., N of Sulawesi)

    — *O.c.celebensis* (Walden, 1872)   Sulawesi

    — *O.c.frontalis* Wallace, 1863   Peling, Banggai, Sula Is.

    — *O.c.boneratensis* Meyer & Wiglesworth, 1896   Flores Sea Is.

    — *O.c.broderipii* Bonaparte, 1852   Lombok to Alor, Sumba (Lesser Sundas)

    — *O.c.chinensis* Linnaeus, 1766   Philippines

    — *O.c.suluensis* Sharpe, 1877   Sulu Is.

○ ***Oriolus chlorocephalus*** GREEN-HEADED ORIOLE

    — *O.c.amani* Benson, 1947   SE Kenya, E Tanzania

    — *O.c.chlorocephalus* Shelley, 1896   Malawi, C Mozambique

    — *O.c.speculifer* Clancey, 1969[1]   Mt. Gorongoza (S Mozambique) (418)

○ ***Oriolus crassirostris*** Hartlaub, 1857 SÃO TOMÉ ORIOLE   São Tomé I.

○ ***Oriolus brachyrhynchus*** WESTERN BLACK-HEADED ORIOLE

    — *O.b.brachyrhynchus* Swainson, 1837   Sierra Leone to Togo

    — *O.b.laetior* Sharpe, 1897   Nigeria to Uganda, W Kenya

○ ***Oriolus monacha*** ABYSSINIAN BLACK-HEADED ORIOLE

    — *O.m.monacha* (J.F. Gmelin, 1789)   N Ethiopia

    — *O.m.meneliki* Blundell & Lovat, 1899   S Ethiopia

○ ***Oriolus percivali*** Ogilvie-Grant, 1903 MOUNTAIN ORIOLE   E Zaire to Uganda, W and C Kenya, W Tanzania

○ ***Oriolus larvatus*** EASTERN BLACK-HEADED ORIOLE

    — *O.l.rolleti* Salvadori, 1864   S Sudan, Ethiopia to E Zaire, N Tanzania

    — *O.l.reichenowi* Zedlitz, 1916   S Somalia, E Kenya, E Tanzania

    — *O.l.angolensis* Neumann, 1905   W Tanzania to Angola and Zimbabwe

    — *O.l.additus* Lawson, 1969[2]   Coastal S Mozambique to S Tanzania (1340)

    — *O.l.larvatus* M.H.K. Lichtenstein, 1823   Inland S Mozambique, E South Africa

○ ***Oriolus nigripennis*** J. & E. Verreaux, 1855 BLACK-WINGED ORIOLE[3]   Sierra Leone to S Sudan and NW Angola

○ ***Oriolus xanthornus*** BLACK-HOODED ORIOLE

    — *O.x.xanthornus* (Linnaeus, 1758)[4]   N India, Thailand, Indochina, N of W Malaysia

    — *O.x.maderaspatanus* Franklin, 1831   S India

    — *O.x.ceylonensis* Bonaparte, 1850   Sri Lanka

    — *O.x.reubeni* Abdulali, 1977[5]   Andamans (12)

    — *O.x.tanakae* N. Kuroda, Sr., 1925   E Sabah (Borneo)

○ ***Oriolus hosii*** Sharpe, 1892 BLACK ORIOLE   Borneo (mainly Sarawak)

○ ***Oriolus cruentus*** BLACK-AND-CRIMSON ORIOLE

    — *O.c.malayanus* Robinson & Kloss, 1923   W Malaysia

    — *O.c.consanguineus* (Wardlaw Ramsay, 1881)   Sumatra

    — *O.c.cruentus* (Wagler, 1827)   Java

    — *O.c.vulneratus* Sharpe, 1887   N Borneo

---

[1] Clancey (1969) appeared in April; it was described with type details by Wolters & Clancey (1969) (2687), but not till August (R.J. Dowsett *in litt.*).
[2] Proposed to replace *tibicen* Lawson, 1962 (1337), preoccupied.
[3] For monotypic treatment see White (1962: 48) (2646).
[4] Includes *thaiocous* see Deignan (1963) (663).
[5] Originally described as *andamanensis* Abdulali, 1967 (8).

○ *Oriolus traillii* MAROON ORIOLE
   — *O.t.traillii* (Vigors, 1832)   Himalayas, Burma, Thailand, SW China
   — *O.t.robinsoni* Delacour, 1927   S Indochina
   — *O.t.nigellicauda* (Swinhoe, 1870)   N Vietnam, SE Yunnan, Hainan
   — *O.t.ardens* (Swinhoe, 1862)   Taiwan

○ *Oriolus mellianus* Stresemann, 1922 SILVER ORIOLE[1]   SC China

COLLURICINCLIDAE SHRIKE-THRUSHES AND ALLIES (3: 14)

COLLURICINCLA Vigors & Horsfield, 1827 F
○ *Colluricincla boweri* (E.P. Ramsay, 1885) BOWER'S SHRIKE-THRUSH   NE Queensland

○ *Colluricincla umbrina* SOOTY SHRIKE-THRUSH[2]
   — *C.u.atra* (Rothschild, 1931)   N face mts. of C New Guinea
   — *C.u.umbrina* (Reichenow, 1915)   S face mts. of C New Guinea

○ *Colluricincla megarhyncha* LITTLE SHRIKE-THRUSH
   — *C.m.affinis* (G.R. Gray, 1862)   Waigeo I.
   — *C.m.batantae* (Meise, 1929)   Batanta I.
   — *C.m.misoliensis* (Meise, 1929)   Misool I.
   — *C.m.megarhyncha* (Quoy & Gaimard, 1830)   Salawati I., W New Guinea
   — *C.m.ferruginea* (Hartert & Paludan, 1936)   NW New Guinea (head of Geelvink Bay)
   — *C.m.aruensis* (G.R. Gray, 1858)   Aru Is.
   — *C.m.goodsoni* (E. Hartert, 1930)[3]   S New Guinea coastal SC New Guinea (lower Trans-Fly)
   — *C.m.palmeri* (Rand, 1938)   SC New Guinea (upper Trans-Fly)
   — *C.m.despecta* (Rothschild & Hartert, 1903)   S coast of SE New Guinea
   — *C.m.superflua* (Rothschild & Hartert, 1912)   N coast of SE New Guinea
   — *C.m.neos* (Mayr, 1931)[4]   Mts. of EC New Guinea
   — *C.m.madaraszi* (Rothschild & Hartert, 1903)   NE New Guinea (Huon Pen.)
   — *C.m.tappenbecki* Reichenow, 1899   NE central New Guinea
   — *C.m.maeandrina* (Stresemann, 1921)   N central New Guinea
   — *C.m.idenburgi* (Rand, 1940)   Inland west NC New Guinea
   — *C.m.hybridus* (Meise, 1929)[5]   Coastal west NC New Guinea
   — *C.m.obscura* (A.B. Meyer, 1874)   Yapen I. (Geelvink Bay)
   — *C.m.melanorhyncha* (A.B. Meyer, 1874)   Biak I. (Geelvink Bay)
   — *C.m.fortis* (Gadow, 1883)   D'Entrecasteaux Arch.
   — *C.m.trobriandi* (E. Hartert, 1896)   Trobriand Is.
   — *C.m.discolor* De Vis, 1890   Tagula I.
   — *C.m.parvula* Gould, 1845[6]   N Western Australia, N Northern Territory (Arnhem Land), Melville I.
   — *C.m.aelptes* Schodde & Mason, 1976   Coastal NE Northern Territory (2184)
   — *C.m.normani* (Mathews, 1914)[7]   Cape York Pen., Torres Straits Is.
   — *C.m.griseata* (G.R. Gray, 1858)   NE Queensland
   — *C.m.synaptica* Schodde & Mason, 1999   Northern EC Queensland (2189)
   — *C.m.gouldii* (G.R. Gray, 1858)   Southern EC Queensland
   — *C.m.rufogaster* Gould, 1845   SE Queensland, NE New South Wales

○ *Colluricincla sanghirensis* (Oustalet, 1881) SANGIHE SHRIKE-THRUSH[8]   Sangihe Is.

○ *Colluricincla tenebrosa* (Hartlaub & Finsch, 1868) MORNING BIRD[9]   Palau Is.

● *Colluricincla harmonica* GREY SHRIKE-THRUSH[10]
   — *C.h.brunnea* Gould, 1841   NW to NC Australia, Melville I.

[1] This may be better treated as a race of *O. traillii* as an albinistic form also in *O. steerii*.
[2] Called *Pachycephala tenebrosa* in last Edition; however, the transfer of that species to *Colluricincla* and of the species *Pitohui tenebrosa* also to *Colluricincla* required the former, as the younger name to take the next available name (Diamond, 1972) (699).
[3] Mees (1982) (1534) placed this and *palmeri* and *wuroi* (the latter here treated as a synonym) in the synonymy of the nominate form.
[4] For correction of spelling see David & Gosselin (2002) (613).
[5] For correction of spelling see David & Gosselin (2002) (613).
[6] Includes *conigravi* see Schodde & Mason (1999: 456) (2189).
[7] Includes *parvissima* see Schodde & Mason (1999: 456) (2189).
[8] For separation from the preceding species see Rozendaal & Lambert (1999) (2124).
[9] From the genus *Pitohui* see Diamond (1972) (699).
[10] Revision follows Schodde & Mason (1999: 458-462) (2189); for names listed last Edition see their treatment, in which the recently proposed race *kolichisi* Ford, 1987 (864), is a synonym of *rufiventris*.

○ *C.h.superciliosa* Masters, 1876   Cape York Pen., Torres Straits Is., coastal and settled E New Guinea

— *C.h.harmonica* (Latham, 1802)[1]   NE to SE Australia

— *C.h.strigata* Swainson, 1837   Tasmania, King I., Flinders I.

— *C.h.rufiventris* Gould, 1841   W, S and C Australia

○ **Colluricincla woodwardi** E. Hartert, 1905 SANDSTONE SHRIKE-THRUSH[2]   N Western Australia (Kimberley), N Northern Territory (Arnhem Land to SW Gulf of Carpentaria)

***PITOHUI* Lesson, 1830 M**

○ **Pitohui kirhocephalus** VARIABLE PITOHUI[3]

— *P.k.uropygialis* (G.R. Gray, 1862)   Salawati, Misool Is. (W Papuan Is.)

— *P.k.pallidus* van Oort, 1907   Sagewin, Batanta Is.(W Papuan Is.)

— *P.k.cerviniventris* (G.R. Gray, 1862)   Waigeo, Gemien Is. (W Papuan Is.)

— *P.k.tibialis* (Sharpe, 1877)   NW New Guinea (W Vogelkop)

— *P.k.kirhocephalus* (Lesson & Garnot, 1827)   NW New Guinea (E Vogelkop)

— *P.k.dohertyi* Rothschild & Hartert, 1903   NW New Guinea (Wandammen district)

— *P.k.rubiensis* (A.B. Meyer, 1884)   NW New Guinea (head of Geelvink Bay)

— *P.k.brunneivertex* Rothschild, 1931   NW New Guinea (E coast Geelvink Bay)

— *P.k.decipiens* (Salvadori, 1878)   SW New Guinea (S Vogelkop)

— *P.k.stramineipectus* van Oort, 1907   SW New Guinea (Triton Bay district)

— *P.k.adiensis* Mees, 1964   Adi I. (SW New Guinea)

— *P.k.carolinae* Junge, 1952   SW New Guinea (Etna Bay district)

— *P.k.jobiensis* (A.B. Meyer, 1874)   Kurudu and Yapen Is. (Geelvink Bay)

— *P.k.meyeri* Rothschild & Hartert, 1903   WC northern New Guinea (Mamberamo coast)

— *P.k.senex* Stresemann, 1922   EC northern New Guinea (upper Sepik basin)

— *P.k.brunneicaudus* (A.B. Meyer, 1891)   EC northern New Guinea (lower Sepik to Astrolabe Bay)

— *P.k.meridionalis* (Sharpe, 1888)   Coastal SE New Guinea

— *P.k.brunneiceps* (D'Albertis & Salvadori, 1879)   S New Guinea (Gulf of Papua to Fly River)

— *P.k.nigripectus* van Oort, 1909   Western SC New Guinea

— *P.k.aruensis* (Sharpe, 1877)   Aru Is.

○ **Pitohui dichrous** (Bonaparte, 1850) HOODED PITOHUI[4]   Hill forests of New Guinea

○ **Pitohui incertus** van Oort, 1909 WHITE-BELLIED PITOHUI   SC New Guinea

○ **Pitohui ferrugineus** RUSTY PITOHUI

— *P.f.leucorhynchus* (G.R. Gray, 1862)   Waigeo I. (W Papuan Is.)

— *P.f.fuscus* Greenway, 1966   Batanta I. (W Papuan Is.)

— *P.f.brevipennis* (E. Hartert, 1896)   Aru Is.

— *P.f.ferrugineus* (Bonaparte, 1850)   Misool I., NW New Guinea (Vogelkop)

— *P.f.holerythrus* (Salvadori, 1878)   Yapen I., NC New Guinea

— *P.f.clarus* (A.B. Meyer, 1894)   E and S New Guinea

○ **Pitohui cristatus** CRESTED PITOHUI

— *P.c.cristatus* (Salvadori, 1876)   W New Guinea

— *P.c.arthuri* E. Hartert, 1930   N and S New Guinea

— *P.c.kodonophonos* Mayr, 1931   SE New Guinea

○ **Pitohui nigrescens** BLACK PITOHUI

— *P.n.nigrescens* (Schlegel, 1871)   NW New Guinea (Vogelkop Mts.)

— *P.n.wandamensis* E. Hartert, 1930   Mts. of W New Guinea (head of Geelvink Bay)

— *P.n.meeki* Rothschild & Hartert, 1913   Mts. of WC New Guinea

— *P.n.buergersi* Stresemann, 1922   Mts. of N and EC New Guinea

— *P.n.harterti* (Reichenow, 1911)   Mts. of NE New Guinea (Huon Pen.)

— *P.n.schistaceus* (Reichenow, 1900)   Mts. of SE New Guinea

***OREOICA* Gould, 1838 F**

○ **Oreoica gutturalis** CRESTED BELLBIRD

— *O.g.pallescens* Mathews, 1912   W, C and inland N Australia

— *O.g.gutturalis* (Vigors & Horsfield, 1827)   SW, SC and inland E Australia

---

[1] For reasons to use 1802 not 1801 see Browing & Monroe (1991) (258).
[2] Includes *assimilis*, for monotypic treatment see Schodde & Mason (1999: 463) (2189).
[3] The name *salvadori* relates to an intergrading population.
[4] Includes *monticola* see Mayr (1967: 49) (1481).

DICRURIDAE DRONGOS (2: 22)

CHAETORHYNCHUS A.B. Meyer, 1874 M
○ *Chaetorhynchus papuensis* A.B. Meyer, 1874 PYGMY DRONGO — Mts. of New Guinea

DICRURUS Vieillot, 1816 M
○ *Dicrurus ludwigii* SQUARE-TAILED DRONGO
— *D.l.sharpei* Oustalet, 1879 — Senegal to N Zaire, NW Uganda, S Sudan, W Kenya south to Lower Congo R., NW Angola

— *D.l.saturnus* Clancey, 1976 — C Angola, N Zambia, SE Zaire (455)
— *D.l.tephrogaster* Clancey, 1975 — Mozambique, E Zimbabwe, S Malawi (450)
— *D.l.ludwigii* (A. Smith, 1834) — S Mozambique, E Transvaal to Natal, E Cape Province

○ *Dicrurus atripennis* Swainson, 1837 SHINING DRONGO — Sierra Leone to NE Zaire

○ *Dicrurus adsimilis* FORK-TAILED DRONGO
— *D.a.divaricatus* (M.H.K. Lichtenstein, 1823) — Senegal, N Guinea to C and S Sudan, Ethiopia, Eritrea, Somalia, N Uganda, N Zaire

— *D.a.fugax* W.K.H. Peters, 1868 — S Uganda, Kenya to Zimbabwe, E Transvaal, Mozambique
— *D.a.apivorus* Clancey, 1976 — Angola, S Zaire to Namibia, N Cape Province, W Transvaal (454)

— *D.a.adsimilis* (Bechstein, 1794) — C Transvaal and Natal to E and S Cape Province

○ *Dicrurus modestus* VELVET-MANTLED DRONGO[1]
— *D.a.atactus* Oberholser, 1899 — Sierra Leone to S Ghana, SW Nigeria
— *D.a.coracinus* J. & E. Verreaux, 1851 — Bioko, S Nigeria to W Kenya south to C Zaire, NW Angola

— *D.a.modestus* Hartlaub, 1849 — Principe I.

○ *Dicrurus fuscipennis* (Milne-Edwards & Oustalet, 1887) COMORO DRONGO — Grande Comore I.

○ *Dicrurus aldabranus* (Ridgway, 1893) ALDABRA DRONGO — Aldabra I.

○ *Dicrurus forficatus* CRESTED DRONGO
— *D.f.forficatus* (Linnaeus, 1766) — Madagascar
— *D.f.potior* (Bangs & Penard, 1922) — Anjouan I.

○ *Dicrurus waldenii* Schlegel, 1866 MAYOTTE DRONGO — Mayotte I.

○ *Dicrurus macrocercus* BLACK DRONGO
— *D.m.albirictus* (Hodgson, 1836) — SE Iran, E Afghanistan, Pakistan, N India
— *D.m.macrocercus* Vieillot, 1817 — S India
— *D.m.minor* Blyth, 1850 — Sri Lanka
— *D.m.cathoecus* Swinhoe, 1871 — Burma (except SE), S, E and NE China (? S Russian Far East), N Thailand, N and C Indochina, >> sthn. SE Asia

— *D.m.thai* Kloss, 1921 — SE Burma, SW, C and SE Thailand, Cambodia, S Vietnam

— *D.m.harterti* E.C.S. Baker, 1918 — Taiwan
— *D.m.javanus* Kloss, 1921 — Java, Bali

○ *Dicrurus leucophaeus* ASHY DRONGO
— *D.l.leucogenis* (Walden, 1870) — C, N and E China >> mainland SE Asia
— *D.l.salangensis* Reichenow, 1890 — SE China >> Hainan I., mainland SE Asia
— *D.l.longicaudatus* Jerdon, 1862 — NE Afghanistan, W and C Himalayas >> S India, Sri Lanka

— *D.l.hopwoodi* E.C.S. Baker, 1918 — E Himalayas, Assam, W, N and NE Burma, S China to Sichuan and Guizhou, NE Indochina >> mainland SE Asia

— *D.l.innexus* (Swinhoe, 1870) — Hainan I.
— *D.l.mouhoti* (Walden, 1870) — SW, C and E Burma, N Thailand, NW and C Indochina >> sthn. mainland SE Asia

[1] For reasons for separation see Pearson (2000: 530) (1814).

___ *D.l.bondi* Meyer de Schauensee, 1937 — C Thailand, S Indochina

___ *D.l.nigrescens* Oates, 1889 — S and SE Burma, Malay Pen.

___ *D.l.batakensis* (Robinson & Kloss, 1919) — N Sumatra

___ *D.l.phaedrus* (Reichenow, 1904) — S Sumatra

___ *D.l.periophthalmicus* (Salvadori, 1894) — Sipora I. (W Sumatra)

___ *D.l.celaenus* Oberholser, 1912 — Simeuluĕ I (W Sumatra)

___ *D.l.siberu* Chasen & Kloss, 1926 — S Mentawi Is. (W Sumatra)

___ *D.l.stigmatops* (Sharpe, 1879) — Mts. of N Borneo

___ *D.l.leucophaeus* Vieillot, 1817 — Java, Bali, Lombok, Palawan

○ **Dicrurus caerulescens** WHITE-BELLIED DRONGO

___ *D.c.caerulescens* (Linnaeus, 1758) — W, C and S India (not Himalayas or NE), S Nepal

___ *D.c.insularis* (Sharpe, 1877) — N Sri Lanka

___ *D.c.leucopygialis* Blyth, 1846 — S Sri Lanka

○ **Dicrurus annectans** (Hodgson, 1836) CROW-BILLED DRONGO — C and E Himalayas, S China, Burma, N Thailand, N and C Indochina, Hainan >> S Thailand and S Indochina to Gtr. Sundas

○ **Dicrurus aeneus** BRONZED DRONGO

___ *D.a.aeneus* Vieillot, 1817 — India, C and E Himalayas, Burma and S and SE China south to N Malay Pen.

___ *D.a.malayensis* (Blyth, 1846) — S Malay Pen., Sumatra, Borneo

___ *D.a.braunianus* (Swinhoe, 1863) — Taiwan

○ **Dicrurus remifer** LESSER RACKET-TAILED DRONGO

___ *D.r.tectirostris* (Hodgson, 1836) — Lower Himalayas, Assam, Burma, S China, N mainland SE Asia

___ *D.r.peracensis* (E.C.S. Baker, 1918) — Most of S mainland SE Asia incl. W Malaysia

___ *D.r.lefoli* (Delacour & Jabouille, 1928) — S Cambodia

___ *D.r.remifer* (Temminck, 1823) — Sumatra, Java

○ **Dicrurus balicassius** BALICASSIAO

___ *D.b.abraensis* Vaurie, 1947 — N Luzon

___ *D.b.balicassius* (Linnaeus, 1766) — Lubang, C and S Luzon, Mindoro

___ *D.b.mirabilis* Walden & Layard, 1872 — Panay, Cebu, Negros, Guimaras, Masbate

○ **Dicrurus hottentottus** HAIR-CRESTED DRONGO[1]

___ *D.h.brevirostris* (Cabanis, 1851) — China (north to Hebei), N Burma, N Laos, N Vietnam >> C Malay Pen.

___ *D.h.hottentottus* (Linnaeus, 1766) — India, Burma, W Yunnan, Thailand (south to N Malay Pen.), S Indochina

___ *D.h.sumatranus* Wardlaw Ramsay, 1880 — Sumatra

___ *D.h.viridinitens* (Salvadori, 1894) — Mentawei Is.

___ *D.h.jentincki* (Vorderman, 1893) — E Java, Bali, Nusa Penida and islands in Java Sea (Solombo Besar, Kangean)

___ *D.h.borneensis* (Sharpe, 1879) — N Borneo

___ *D.h.palawanensis* Tweeddale, 1878 — Cagayan Sulu I., Palawan, Balabac

___ *D.h.samarensis* Vaurie, 1947 — Samar, Leyte, Bohol Is.

___ *D.h.striatus* Tweeddale, 1877 — Mindanao, Basilan Is.

___ *D.h.cuyensis* (McGregor, 1903) — Cuyo, Semirara (Philippines)

___ *D.h.menagei* (Bourns & Worcester, 1894) — Tablas (C Philippines

___ *D.h.suluensis* E. Hartert, 1902 — Sibutu I., Sulu Arch.

___ *D.h.guillemardi* (Salvadori, 1890) — Obi Is. (C Moluccas)

___ *D.h.pectoralis* Wallace, 1863[2] — Banggai, Sula Is.

___ *D.h.leucops* Wallace, 1865 — Sulawesi

___ *D.h.bimaensis* Wallace, 1864[3] — Lesser Sundas (except Timor and Sumba)

___ *D.h.densus* (Bonaparte, 1850) — Timor I. (Lesser Sundas)

---

[1] The proposal by Bruce in White & Bruce (1986) (2652) to split this was not accepted by Dickinson *et al.* (1991) (745) nor by Inskipp *et al.* (1996) (1182). The species concept used here is zoogeographically more logical.

[2] Vaurie (1962) (2508) listed 1862, which is the volume year, but see Duncan (1937) (770).

[3] Vaurie (1962) (2508) recognised *renschi*, but Mees (1965) (1515) has shown that the type locality of *bimaensis* must be Sumbawa not Flores. Birds from Lombok and other islands may need recognition as *vicinus* (here, tentatively, a synonym).

— *D.h.sumbae* Rensch, 1931	Sumba (Lesser Sundas)
— *D.h.kuehni* E. Hartert, 1901	Tanimbar Is. (E Lesser Sundas)
— *D.h.megalornis* G.R. Gray, 1858	Kai Is.

○ ***Dicrurus montanus*** (Riley, 1919) Sᴜʟᴀᴡᴇsɪ Dʀᴏɴɢᴏ

Mts. of Sulawesi

○ ***Dicrurus bracteatus*** Sᴘᴀɴɢʟᴇᴅ Dʀᴏɴɢᴏ[1]

— *D.b.amboinensis* G.R. Gray, 1860	Ambon, Seram (S Moluccas)
— *D.b.buruensis* E. Hartert, 1919	Buru I. (S Moluccas)
— *D.b.morotensis* Vaurie, 1946	Morotai I. (N Moluccas)
— *D.b.atrocaeruleus* G.R. Gray, 1860	Halmahera Is. (N Moluccas), Kafiau (W Papuan Is.)
— *D.b.carbonarius* Bonaparte, 1850	W Papuan Is., New Guinea, Aru Is., Geelvink Bay islands, Trobriand Is., D'Entrecasteaux Arch., Louisiade Arch., N Torres Straits Is.
— *D.b.baileyi* Mathews, 1912	N Western Australia (N Kimberley), N Northern Territory (Arnhem Land)
— *D.b.atrabectus* Schodde & Mason, 1999	Cape York Pen., NE Queensland >> S New Guinea (2189)
— *D.b.bracteatus* Gould, 1843	EC Australia >> NE Australia
— *D.b.laemostictus* P.L. Sclater, 1877	New Britain (Bismarck Arch.)
— *D.b.meeki* Rothschild & E. Hartert, 1901	Guadalcanal I. (SE Solomons)
— *D.b.longirostris* E.P. Ramsay, 1882[2]	San Cristobal I. (SE Solomons)

○ ***Dicrurus megarhynchus*** (Quoy & Gaimard, 1830) Rɪʙʙᴏɴ-ᴛᴀɪʟᴇᴅ Dʀᴏɴɢᴏ

New Ireland (Bismarck Arch.)

○ ***Dicrurus andamanensis*** Aɴᴅᴀᴍᴀɴ Dʀᴏɴɢᴏ

— *D.a.dicruriformis* (Hume, 1872)	Gt. Cocos, Table Is., N Andamans
— *D.a.andamanensis* Beavan, 1867	S Andamans

○ ***Dicrurus paradiseus*** Gʀᴇᴀᴛᴇʀ Rᴀᴄᴋᴇᴛ-ᴛᴀɪʟᴇᴅ Dʀᴏɴɢᴏ

— *D.p.grandis* (Gould, 1836)	N India, Assam, W and N Burma, Yunnan, N Indochina
— *D.p.rangoonensis* (Gould, 1836)	C India, Bangladesh, SW, C and E Burma, N Thailand, C Indochina
— *D.p.paradiseus* (Linnaeus, 1766)	S India, Tenasserim, SW and SE Thailand, S Indochina, N Malay Pen.
— *D.p.johni* (E. Hartert, 1902)	Hainan I.
— *D.p.ceylonicus* Vaurie, 1949	Dry zone Sri Lanka
— *D.p.lophorinus* Vieillot, 1817	Wet zone Sri Lanka
— *D.p.otiosus* (Richmond, 1903)	Andamans
— *D.p.nicobariensis* (E.C.S. Baker, 1918)	Nicobars
— *D.p.hypoballus* (Oberholser, 1926)[3]	C Malay Pen. (Trang to Perak)
— *D.p.platurus* Vieillot, 1817	S Malay Pen., Sumatra, NW Sumatra Is.
— *D.p.microlophus* (Oberholser, 1917)	Anambas, Natunas and Pulau Tioman
— *D.p.brachyphorus* (Bonaparte, 1850)	Borneo
— *D.p.banguey* (Chasen & Kloss, 1929)	N Borneo islands
— *D.p.formosus* (Cabanis, 1851)	Java

RHIPIDURIDAE FANTAILS (1: 43)

RHIPIDURA Vigors & Horsfield, 1827 F

○ ***Rhipidura hypoxantha*** Blyth, 1843 Yᴇʟʟᴏᴡ-ʙᴇʟʟɪᴇᴅ Fᴀɴᴛᴀɪʟ[4]

Himalayas, Assam, SC China, W, N and E Burma, N Thailand, NW Indochina

○ ***Rhipidura superciliaris*** Bʟᴜᴇ Fᴀɴᴛᴀɪʟ

— *R.s.samarensis* (Steere, 1890)	Bohol, Samar, Leyte (E Visayas)
— *R.s.apo* Hachisuka, 1930	E and C Mindanao
— *R.s.superciliaris* (Sharpe, 1877)	Basilan, W Mindanao

[1] Apart from the inclusion of the Solomons forms this species is that used by Schodde & Mason (1999: 491) (2189).
[2] The name *solomenensis* Doughty *in* Doughty *et al.*, 1999, is based on the type of this name and thus a synonym; if *meeki* and *longirostris* represent a distinct species as may well be the case, the name *longirostris* must be used, however fuller reasoning should be published.
[3] The name *malayensis* used in last Edition is preoccupied.
[4] This species used to be treated in the monotypic genus *Chelidorynx*.

○ *Rhipidura cyaniceps* Blue-headed Fantail
    __ *R.c.pinicola* Parkes, 1958   NW Luzon
    __ *R.c.cyaniceps* (Cassin, 1855)   NE, C, S Luzon
    __ *R.c.sauli* Bourns & Worcester, 1894   Tablas I. (C Philippines)
    __ *R.c.albiventris* (Sharpe, 1877)   Guimaras, Masbate, Negros, Panay (W Visayas)

○ *Rhipidura albicollis* White-throated Fantail
    __ *R.a.canescens* (Koelz, 1939)   NW Himalayan foothills
    __ *R.a.albicollis* (Vieillot, 1818)   C Himalayan foothills (Nepal-Sikkim), Bangladesh
    __ *R.a.stanleyi* E.C.S. Baker, 1913   E Himalayas, Assam, W Burma
    __ *R.a.orissae* Ripley, 1955   NE central India
    __ *R.a.vernayi* (Whistler, 1931)   SE India
    __ *R.a.albogularis* (Lesson, 1831)[1][2]   SW India
    __ *R.a.celsa* Riley, 1929   E Xizang, NE and E Burma, S China, N, NE and SW Thailand, N Indochina

    __ *R.a.cinerascens* Delacour, 1927   S Indochina, SE Thailand
    __ *R.a.atrata* Salvadori, 1879   Malay Pen., Sumatra
    __ *R.a.sarawacensis* Chasen, 1941   Sarawak (NW Borneo)
    __ *R.a.kinabalu* Chasen, 1941   Sabah, Brunei (N Borneo)

○ *Rhipidura euryura* S. Müller, 1843 White-bellied Fantail   Java

○ *Rhipidura aureola* White-browed Fantail
    __ *R.a.aureola* Lesson, 1830   Pakistan, N and C India
    __ *R.a.compressirostris* (Blyth, 1849)   S India, Sri Lanka
    __ *R.a.burmanica* (Hume, 1880)   S Assam, Burma, W Thailand, S Indochina

○ *Rhipidura javanica* Pied Fantail
    __ *R.j.longicauda* Wallace, 1865   SE Burma, S Thailand, Cambodia and S Vietnam to Sumatra, Bangka, Belitung, Borneo
    __ *R.j.javanica* (Sparrman, 1788)   Java, Bali
    __ *R.j.nigritorquis* Vigors, 1831   Philippines (incl. Sulu Arch.)

○ *Rhipidura perlata* S. Müller, 1843 Spotted Fantail   C and S Malay Pen., Gtr. Sundas

● *Rhipidura leucophrys* Willie-wagtail
    __ *R.l.melaleuca* (Quoy & Gaimard, 1830)   Moluccas, New Guinea, N Torres Strait Is., Bismarck Arch., Solomon Is.
    __ *R.l.picata* Gould, 1848   N Australia
    ✓ *R.l.leucophrys* (Latham, 1802)[3]   Australia (except N)

○ *Rhipidura diluta* Brown-capped Fantail
    __ *R.d.sumbawensis* Büttikofer, 1892   Sumbawa I. (Lesser Sundas)
    __ *R.d.diluta* Wallace, 1864   Flores, Lomblen Is. (Lesser Sundas)

○ *Rhipidura fuscorufa* P.L. Sclater, 1883 Cinnamon-tailed Fantail   Tanimbar, Babar Is. (E Lesser Sundas)

○ *Rhipidura rufiventris* Northern Fantail
    __ *R.r.tenkatei* Büttikofer, 1892   Roti I. (Lesser Sundas)
    __ *R.r.rufiventris* (Vieillot, 1818)   Timor (Lesser Sundas)
    __ *R.r.pallidiceps* E. Hartert, 1904   Wetar I. (Lesser Sundas)
    __ *R.r.hoedti* Büttikofer, 1892   Romang, , Leti, Sermata Is. (Lesser Sundas)
    __ *R.r.assimilis* G.R. Gray, 1858[4]   Kai, Tayandu Is. (E Banda Sea)
    __ *R.r.finitima* E. Hartert, 1918   Watubela Is. (E Banda Sea)
    __ *R.r.bouruensis* Wallace, 1863   Buru I. (S Moluccas)
    __ *R.r.cinerea* Wallace, 1865   Seram, Ambon Is. (S Moluccas)
    __ *R.r.obiensis* Salvadori, 1876   Obi I. (C Moluccas)
    __ *R.r.vidua* Salvadori & Turati, 1874   Kafiau I. (W Papuan Is.)
    __ *R.r.gularis* S. Müller, 1843   W Papuan Is., New Guinea, islands in Geelvink Bay, D'Entrecasteaux Arch., Torres Strait Is.

[1] Given as 1832 by Watson (1979) (2566), but see Browning & Monroe (1991) (258).
[2] The distinctness of this requires evaluation.
[3] For reasons to use 1802 not 1801 see Browing & Monroe (1991) (258).
[4] Includes *perneglecta* see Schodde & Mathews (1977) (2193).

— *R.r.kordensis* A.B. Meyer, 1874   Biak I. (Geelvink Bay)
— *R.r.nigromentalis* E. Hartert, 1898   Louisiade Arch.
— *R.r.finschii* Salvadori, 1882   New Britain, Duke of York I. (Bismarck Arch.)
— *R.r.setosa* (Quoy & Gaimard, 1830)   New Ireland, New Hanover, Dyaul Is. (Bismarck Arch.)
— *R.r.gigantea* Stresemann, 1933   Lihir, Tabar Is. (Bismarck Arch.)
— *R.r.tangensis* Mayr, 1955   Boang, Tanga Is. (Bismarck Arch.)
— *R.r.niveiventris* Rothschild & Hartert, 1914   Admiralty Is. (Bismarck Arch.)
— *R.r.mussai* Rothschild & Hartert, 1924   St. Matthias Is. (Bismarck Arch.)
— *R.r.isura* Gould, 1841[1]   N Australia

○ **Rhipidura cockerelli** WHITE-WINGED FANTAIL
— *R.c.septentrionalis* Rothschild & Hartert, 1916   Bougainville Is. (N Solomons)
— *R.c.interposita* Rothschild & Hartert, 1916   Ysabel I. (NC Solomons)
— *R.c.lavellae* Rothschild & Hartert, 1916   N New Georgia group (SC Solomons)
— *R.c.albina* Rothschild & Hartert, 1901   C and S New Georgia group (SC Solomons)
— *R.c.floridana* Mayr, 1931   Florida Is. (EC Solomons)
— *R.c.coultasi* Mayr, 1931   Malaita I. (E Solomons)
— *R.c.cockerelli* (E.P. Ramsay, 1879)   Guadalcanal I. (SE Solomons)

○ **Rhipidura threnothorax** SOOTY THICKET FANTAIL
— *R.t.threnothorax* S. Müller, 1843   W Papuan Is., New Guinea, Aru Is.
— *R.t.fumosa* Schlegel, 1871   Yapen I. (Geelvink Bay)

○ **Rhipidura maculipectus** G.R. Gray, 1858 BLACK THICKET FANTAIL   W and all S New Guinea, Aru Is.

○ **Rhipidura leucothorax** WHITE-BELLIED THICKET FANTAIL
— *R.l.leucothorax* Salvadori, 1874   W to NC and SE New Guinea
— *R.l.episcopalis* E.P. Ramsay, 1878   E New Guinea
— *R.l.clamosa* Diamond, 1967   EC New Guinea

○ **Rhipidura atra** BLACK FANTAIL
— *R.a.atra* Salvadori, 1876   Mts. of NW and WC to SE New Guinea
— *R.a.vulpes* Mayr, 1931   Mts. of N New Guinea

○ **Rhipidura hyperythra** CHESTNUT-BELLIED FANTAIL
— *R.h.hyperythra* G.R. Gray, 1858   Aru Is.
— *R.h.mulleri* A.B. Meyer, 1874   W to NC and SC New Guinea, Yapen I. (Geelvink Bay)
— *R.h.castaneothorax* E.P. Ramsay, 1879   E New Guinea

○ **Rhipidura albolimbata** FRIENDLY FANTAIL
— *R.a.albolimbata* Salvadori, 1874   Mid mts. NW to SE New Guinea
— *R.a.lorentzi* van Oort, 1909   High mts. WC to EC New Guinea

● **Rhipidura albiscapa** GREY FANTAIL
— *R.a.brenchleyi* Sharpe, 1879[2]   San Cristobal I. (SE Solomons), Banks Is., Vanuatu
— *R.a.bulgeri* E.L. Layard, 1877   Lifu I. (Loyalty Is.), New Caledonia
— *R.a.keasti* Ford, 1981[3]   NE to EC Queensland
— *R.a.pelzelni* G.R. Gray, 1862   Norfolk I.
— *R.a.alisteri* Mathews, 1911   EC to SE Australia >> N Australia
— *R.a.albiscapa* Gould, 1840   Tasmania, Bass Strait Is. >> SE Australia
— *R.a.preissi* Cabanis, 1850   SW Western Australia
— *R.a.albicauda* North, 1895   WC to C Australia

● **Rhipidura fuliginosa** NEW ZEALAND FANTAIL[4]
— *R.f.fuliginosa* (Sparrman, 1787)   South Island (New Zealand), Stewart I.
— *R.f.placabilis* Bangs, 1921   North Island (New Zealand)
— *R.f.penitus* Bangs, 1911   Chatham Is.
— †*R.f.cervina* E.P. Ramsay, 1879   Lord Howe I.

---

[1] Includes *superciliosa* see Mayr in Watson *et al.* (1986: 538) (2566).
[2] This race and the next were attached to *R. fuliginosa* is last Edition, as they were treated by Mayr in Watson *et al.* (1986: 526) (2566), but see Schodde & Mason (1999: 479) (2189).
[3] This race was accidentally omitted in last Edition.
[4] For reasons to separate this species from *R. albiscapa* see Schodde & Mason (1999: 479-480) (2189).

○ *Rhipidura phasiana* De Vis, 1884 MANGROVE FANTAIL[1]   Coastal S New Guinea, coastal WC to NC Australia (to SE Gulf of Carpentaria)

○ *Rhipidura drownei* MOUNTAIN FANTAIL
— *R.d.drownei* Mayr, 1931   Bougainville I. (N Solomons)
— *R.d.ocularis* Mayr, 1931   Guadalcanal I. (SE Solomons)

○ *Rhipidura tenebrosa* E.P. Ramsay, 1882 DUSKY FANTAIL   San Cristobal I. (SE Solomons)

○ *Rhipidura rennelliana* Mayr, 1931 RENNELL FANTAIL   Rennell I. (SE Solomons)

○ *Rhipidura spilodera* STREAKED FANTAIL
— *R.s.spilodera* G.R. Gray, 1870   Banks Is., N and C Vanuatu
— *R.s.verreauxi* Marié, 1870   Loyalty Is., New Caledonia
— *R.s.rufilateralis* Sharpe, 1879   Taviuni I. (W Fiji)
— *R.s.layardi* Salvadori, 1877   Ovalau, Viti Levu Is. (W Fiji)
— *R.s.erythronota* Sharpe, 1879   Yanganga, Vanua Levu Is. (NC Fiji)

○ *Rhipidura personata* E.P. Ramsay, 1876 KANDAVU FANTAIL   Kandavu I. (S Fiji Is.)

○ *Rhipidura nebulosa* SAMOAN FANTAIL
— *R.n.nebulosa* Peale, 1848   Upolu I. (W Samoa)
— *R.n.altera* Mayr, 1931   Savaii I. (W Samoa)

○ *Rhipidura phoenicura* S. Müller, 1843 RUFOUS-TAILED FANTAIL   Java

○ *Rhipidura nigrocinnamomea* BLACK-AND-CINNAMON FANTAIL
— *R.n.hutchinsoni* Mearns, 1907   W and S Mindanao
— *R.n.nigrocinnamomea* E. Hartert, 1903   SE Mindanao

○ *Rhipidura brachyrhyncha* DIMORPHIC FANTAIL
— *R.b.brachyrhyncha* Schlegel, 1871   Mts. of NW New Guinea (Vogelkop)
— *R.b.devisi* North, 1897   Mts. of WC to E New Guinea

○ *Rhipidura lepida* Hartlaub & Finsch, 1868 PALAU FANTAIL   Palau Is.

○ *Rhipidura dedemi* van Oort, 1911 STREAKY-BREASTED FANTAIL   Mts. of Seram (S Moluccas)

○ *Rhipidura superflua* E. Hartert, 1899 TAWNY-BACKED FANTAIL   Mts. of Buru (S Moluccas)

○ *Rhipidura teysmanni* RUSTY-BELLIED FANTAIL
— *R.t.teysmanni* Büttikofer, 1892   SW Sulawesi
— *R.t.toradja* Stresemann, 1931   N, C and SE Sulawesi
— *R.t.sulaensis* Neumann, 1939   Sula Is.

○ *Rhipidura opistherythra* P.L. Sclater, 1883 LONG-TAILED FANTAIL   Tanimbar Is. (E Lesser Sundas)

○ *Rhipidura rufidorsa* RUFOUS-BACKED FANTAIL
— *R.r.rufidorsa* A.B. Meyer, 1874   Misool and Yapen Is., W to NC and SC New Guinea
— *R.r.kumusi* Mathews, 1928   N coast of E New Guinea
— *R.r.kubuna* Rand, 1938   S coast of E New Guinea

○ *Rhipidura dahli* BISMARCK FANTAIL
— *R.d.dahli* Reichenow, 1897   New Britain (Bismarck Arch.)
— *R.d.antonii* E. Hartert, 1926   New Ireland (Bismarck Arch.)

○ *Rhipidura matthiae* Heinroth, 1902 MATTHIAS FANTAIL   St. Matthias Is. (Bismarck Arch.)

○ *Rhipidura malaitae* Mayr, 1931 MALAITA FANTAIL   Malaita I. (NC Solomons)

○ *Rufifrons semirubra* P.L. Sclater, 1877 MANUS FANTAIL   Admiralty Is.

○ *Rhipidura rufifrons* RUFOUS FANTAIL[2]
— *R.r.torrida* Wallace, 1865   N and C Moluccas
— *R.r.uraniae* Oustalet, 1881   Guam (Marianas)
— *R.r.saipanensis* E. Hartert, 1898   Saipan, Tinian Is. (Marianas)
— *R.r.mariae* R.H. Baker, 1946   Rota I. (Marianas)

---

[1]   Treatment of this as a species is now generally accepted see Schodde & Mason (1999: 479-480) (2189).
[2]   For reasons to separate *dryas* see Schodde & Mason (1999: 477) (2189).

— *R.r.versicolor* Hartlaub & Finsch, 1872   Yap I. (Caroline Is.)

— *R.r.kubaryi* Finsch, 1876[1]   Pohnpei I. (E Caroline Is.)

— *R.r.agilis* Mayr, 1931   Santa Cruz I. (Santa Cruz Is.)

— *R.r.melaenolaema* Sharpe, 1879   Vanikoro I. (Santa Cruz Is.)

— *R.r.utupuae* Mayr, 1931   Utupua I. (Santa Cruz Is.)

— *R.r.commoda* E. Hartert, 1918   Bougainville (N Solomons)

— *R.r.granti* E. Hartert, 1918   Rendova I. (C Solomons)

— *R.r.brunnea* Mayr, 1931   Malaita I. (NE Solomons)

— *R.r.rufofronta* E.P. Ramsay, 1879   Guadalcanal (SE Solomons)

— *R.r.ugiensis* Mayr, 1931   Ugi I. (E Solomons)

— *R.r.russata* Tristram, 1879   San Cristobal I. (SE Solomons)

— *R.r.kuperi* Mayr, 1931   Santa Ana I. (SE Solomons)

— *R.r.louisiadensis* E. Hartert, 1899   Louisiade Arch. (E New Guinea)

— *R.r.intermedia* North, 1902   E Queensland

— *R.r.rufifrons* (Latham, 1802)[2]   SE Australia >> E Queensland

○ **Rhipidura dryas** Gould, 1843 ARAFURA FANTAIL[3]

— *R.d.celebensis* Büttikofer, 1892   Tanahjampea, Kalao Is. (Flores Sea)

— *R.d.mimosae* Meise, 1929   Kalaotoa I. (Flores Sea)

— *R.d.sumbensis* E. Hartert, 1896   Sumba I.

— *R.d.semicollaris* S. Müller, 1843   Flores to Timor and Wetar (C Lesser Sundas)

— *R.d.elegantula* Sharpe, 1879   Roma, Leti, Moa, Damar (EC Lesser Sundas)

— *R.d.reichenowi* Finsch, 1901   Babar (E Lesser Sundas)

— *R.d.hamadryas* P.L. Sclater, 1883   Tanimbar (E Lesser Sundas)

— *R.d.squamata* S. Müller, 1843[4]   W Papuan Is., Banda, Seram Laut, Tayandu, Kai Is., Aru Is. (Babi) (E Banda Sea)

— *R.d.streptophora* Ogilvie-Grant, 1911   Coastal SC New Guinea

— *R.d.dryas* Gould, 1843   Coastal N Australia (Kimberley to W Cape York Pen.)

MONARCHIDAE MONARCHS[5] (15: 87)

HYPOTHYMIS Boie, 1826 F

○ **Hypothymis azurea** BLACK-NAPED MONARCH

— *H.a.styani* (Hartlaub, 1898)   India to S China and Vietnam

— *H.a.oberholseri* Stresemann, 1913   Taiwan

— *H.a.ceylonensis* Sharpe, 1879   Sri Lanka

— *H.a.tytleri* (Beavan, 1867)   Andamans

— *H.a.idiochroa* Oberholser, 1911   Car Nicobar I.

— *H.a.nicobarica* Bianchi, 1907   S Nicobar Is.

— *H.a.montana* Riley, 1929   N and C Thailand

— *H.a.galerita* (Deignan, 1956)   SW and SE Thailand

— *H.a.forrestia* Oberholser, 1911   Mergui Arch., Burma

— *H.a.prophata* Oberholser, 1911   Malay Pen., Sumatra, Borneo

— *H.a.javana* Chasen & Kloss, 1929   Java

— *H.a.penidae* Meise, 1941   Penida I. (near Bali)

— *H.a.karimatensis* Chasen & Kloss, 1932   Karimata I.

— *H.a.opisthocyanea* Oberholser, 1911   Anamba Is.

— *H.a.gigantoptera* Oberholser, 1911   Bunguran, Natuna Is.

— *H.a.aeria* Bangs & Peters, 1927   Maratua Is. (off Borneo)

— *H.a.consobrina* Richmond, 1902   Simeuluě I. (off Sumatra)

— *H.a.leucophila* Oberholser, 1911   Siberut I. (off Sumatra)

— *H.a.richmondi* Oberholser, 1911   Enggano I. (off Sumatra)

— *H.a.abbotti* Richmond, 1902   Babi, Reusam (off Sumatra)

---

[1] This is treated as a distinct species by Pratt *et al.* (1987) (1920); in the light of the separation of *R. dryas* we await a more detailed study.

[2] For reasons to use 1802 not 1801 see Browing & Monroe (1991) (258).

[3] Schodde & Mason (1999: 476-477) (2189) explained both the separation of this species from *rufifrons*, and the priority accorded this name over *semicollaris* and *squamata*.

[4] Includes *henrici* see White & Bruce (1986: 373) (2652).

[5] Some prefer the prior name Myiagridae, see Boles (1981) (174), but Bock (1994) (173) argued for the preservation of Monarchidae based on prevailing usage.

— *H.a.symmixta* Stresemann, 1913   W and C Lesser Sundas.
— *H.a.azurea* (Boddaert, 1783)   Philippines (except Camiguin Sur I.)
— *H.a.catamanensis* Rand & Rabor, 1969   Camiguin Sur I. (off Mindanao)
— *H.a.puella* (Wallace, 1863)   Sulawesi and nearby islands
— *H.a.blasii* E. Hartert, 1898   Sula Is.

○ *Hypothymis helenae* SHORT-CRESTED MONARCH
— *H.h.personata* (McGregor, 1907)   Camiguin Norte I. (off N Luzon)
— *H.h.helenae* (Steere, 1890)   Luzon, Polillo, Samar
— *H.h.agusanae* Rand, 1970   NE Mindanao, Dinagat

○ *Hypothymis coelestis* CELESTIAL MONARCH
— *H.c.coelestis* Tweeddale, 1877   Luzon, Samar, Mindanao, Basilan
— *H.c.rabori* Rand, 1970   Negros, Sibuyan

EUTRICHOMYIAS Meise, 1939   M
○ *Eutrichomyias rowleyi* (A.B. Meyer, 1878)   CERULEAN PARADISE-FLYCATCHER
    Sangihe I. (N of Sulawesi)

TROCHOCERCUS Cabanis, 1850[1]   M
○ *Trochocercus cyanomelas* BLUE-MANTLED PARADISE-FLYCATCHER
— *T.c.vivax* Neave, 1909   Uganda to S Zaire and Zambia
— *T.c.bivittatus* Reichenow, 1879   Somalia to E Tanzania and Zanzibar
— *T.c.megalolophus* Swynnerton, 1907   Malawi and N Mozambique to Zimbabwe and E Zululand
— *T.c.segregus* Clancey, 1975   E Transvaal to Natal and W Zululand
— *T.c.cyanomelas* (Vieillot, 1818)   E and S Cape Province

○ *Trochocercus nitens* BLUE-HEADED PARADISE-FLYCATCHER
— *T.n.reichenowi* Sharpe, 1904   Sierra Leone to Togo
— *T.n.nitens* Cassin, 1859   Nigeria and Cameroon to Uganda and Angola

TERPSIPHONE Gloger, 1827   F
○ *Terpsiphone rufiventer* RED-BELLIED PARADISE-FLYCATCHER
— *T.r.rufiventer* (Swainson, 1837)   Senegal and Gambia to W Guinea
— *T.r.nigriceps* (Hartlaub, 1855)   Sierra Leone and Guinea to Togo and SW Benin
— *T.r.fagani* (Bannerman, 1921)   Benin and SW Nigeria
— *T.r.tricolor* (Fraser, 1843)   Bioko I.
— *T.r.smithii* (Fraser, 1843)   Pagalu I. (Gulf of Guinea)
— *T.r.neumanni* Stresemann, 1924   SE Nigeria, Cameroon, Gabon, S Congo, Cabinda (Angola)
— *T.r.schubotzi* (Reichenow, 1911)   SE Cameroon, SW Central African Republic
— *T.r.mayombe* (Chapin, 1932)   Congo and W Zaire
— *T.r.somereni* Chapin, 1948   W and S Uganda
— *T.r.emini* Reichenow, 1893   SE Uganda, W Kenya, NW Tanzania
— *T.r.ignea* (Reichenow, 1901)   E Central African Republic, Zaire to NE Angola, NW Zambia

○ *Terpsiphone bedfordi* (Ogilvie-Grant, 1907)   BEDFORD'S PARADISE-FLYCATCHER
    NE and E Zaire

○ *Terpsiphone rufocinerea* Cabanis, 1875   RUFOUS-VENTED PARADISE-FLYCATCHER
    S Cameroon to W Zaire and N Angola

○ *Terpsiphone batesi* BATES'S PARADISE-FLYCATCHER[2]
— *T.b.batesi* Chapin, 1921   S Cameroon and Gabon to E Zaire
— *T.b.bannermani* Chapin, 1948   N Angola, Congo, SW Zaire

○ *Terpsiphone viridis* AFRICAN PARADISE-FLYCATCHER
— *T.v.harterti* (R. Meinertzhagen, 1923)   SW Arabia to S Oman
— *T.v.viridis* (Statius Müller, 1776)   Senegal and Gambia to Sierra Leone
— *T.v.speciosa* (Cassin, 1859)   S Cameroon to Sudan, Zaire and NE Angola

---

[1] We follow Érard (1997) (808) in recognising this genus as distinct from *Terpsiphone*.
[2] For treatment as a species see Érard (1997: 540) (808).

— *T.v.ferreti* (Guérin-Méneville, 1843)	Mali and Ivory Coast to Somalia, Kenya and NE Tanzania
— *T.v.restricta* (Salomonsen, 1933)	S Uganda
— *T.v.kivuensis* Salomonsen, 1949	SW Uganda, Rwanda, Burundi and Kivu (Zaire) to NW Tanzania
— *T.v.suahelica* Reichenow, 1898	W Kenya, N Tanzania
— *T.v.ungujaensis* (Grant & Mackworth-Praed, 1947)	E Tanzania, Zanzibar, Mafia and Pemba Is.
— *T.v.plumbeiceps* Reichenow, 1898	N Namibia and Angola to SW Tanzania, Mozambique and N Sth. Africa >> Cameroon, S Zaire, SE Kenya
— *T.v.granti* (Roberts, 1948)	Natal to SW Cape Province

○ ***Terpsiphone paradisi*** Asian Paradise-flycatcher

— *T.p.leucogaster* (Swainson, 1838)	W Tien Shan to Afghanistan, N Pakistan, NW and NC India, W and C Nepal >> S India
— *T.p.saturatior* (Salomonsen, 1933)	E Nepal, NE India, E Bangladesh, N Burma >> Malay Pen.
— *T.p.paradisi* (Linnaeus, 1758)	C and S India, C Bangladesh and SW Burma >> Sri Lanka
— *T.p.ceylonensis* (Zarudny & Härms, 1912)	Sri Lanka
— *T.p.nicobarica* Oates, 1890	Nicobar Is. >> Andaman Is.
— *T.p.incei* (Gould, 1852)	C, E and NE China, C Russian Far East, N Korea >> SE Asia
— *T.p.burmae* (Salomonsen, 1933)	C Burma
— *T.p.indochinensis* (Salomonsen, 1933)	E Burma and Thailand to S China and Indochina
— *T.p.affinis* (Blyth, 1846)[1]	Malay Pen., Sumatra
— *T.p.procera* (Richmond, 1903)	Simeuluë I. (off Sumatra)
— *T.p.insularis* Salvadori, 1887	Nias I. (off Sumatra)
— *T.p.borneensis* (E. Hartert, 1916)	Borneo
— *T.p.sumbaensis* A.B. Meyer, 1894	Sumba I. (Lesser Sundas)
— *T.p.floris* Büttikofer, 1894	Sumbawa, Alor, Flores Is. (Lesser Sundas)

○ ***Terpsiphone atrocaudata*** Japanese Paradise-flycatcher

— *T.a.atrocaudata* (Eyton, 1839)	Japan, C and S Korea >> SE Asia
— *T.a.illex* Bangs, 1901	Ryukyu Is.
— *T.a.periophthalmica* (Ogilvie-Grant, 1895)	Lanyu I. (off S Taiwan), Batan Is. (off N Luzon) >> C Luzon, Mindoro

○ ***Terpsiphone cyanescens*** (Sharpe, 1877) Blue Paradise-flycatcher Palawan

○ ***Terpsiphone cinnamomea*** Rufous Paradise-flycatcher

— *T.c.unirufa* Salomonsen, 1937	N Philippines
— *T.c.cinnamomea* (Sharpe, 1877)	Mindanao, Basilan, Sulu Arch.
— *T.c.talautensis* (Meyer & Wiglesworth, 1894)	Talaud Is. (Indonesia)

○ ***Terpsiphone atrochalybea*** (Thomson, 1842) São Thomé Paradise-flycatcher

São Thomé I. (Gulf of Guinea)

○ ***Terpsiphone mutata*** Madagascan Paradise-flycatcher

— *T.m.mutata* (Linnaeus, 1766)	E Madagascar
— *T.m.singetra* Salomonsen, 1933	W Madagascar
— *T.m.pretiosa* (Lesson, 1847)	Mayotte I.
— *T.m.vulpina* (E. Newton, 1877)	Anjouan I.
— *T.m.voeltzkowiana* Stresemann, 1924	Mwali (Moheli) I.
— *T.m.comoroensis* Milne-Edwards & Oustalet, 1885	Grand Comoro I.

○ ***Terpsiphone corvina*** (E. Newton, 1867) Seychelles Paradise-flycatcher

La Digue (Seychelles)

○ ***Terpsiphone bourbonnensis*** Mascarene Paradise-flycatcher

— *T.b.bourbonnensis* (Statius Müller, 1776)	Réunion
— *T.b.desolata* (Salomonsen, 1933)	Mauritius

CHASIEMPIS Cabanis, 1847 F

◗ ***Chasiempis sandwichensis*** Elepaio

∠ *C.s.sandwichensis* (J.F. Gmelin, 1789)	Hawaii (drier areas)

---

[1] Includes *madzoedi* and *australis* see van Marle & Voous (1988) (2468).

___ *C.s.ridgwayi* Stejneger, 1888   Hawaii (wet slopes Hilo Dist.)
___ *C.s.bryani* H.D. Pratt, 1979   Mauna Kea I. (Hawaiian Is.)
✓ *C.s.sclateri* Ridgway, 1882   Kauai I. (Hawaiian Is.)
___ *C.s.gayi* S.B. Wilson, 1891   Oahu I. (Hawaiian Is.)

POMAREA Bonaparte, 1854 F
○ *Pomarea dimidiata* (Hartlaub & Finsch, 1871) RAROTONGAN MONARCH
          Rarotonga I. (Cook Is.)

○ *Pomarea nigra* TAHITIAN MONARCH
___ *P.n.nigra* (Sparrman, 1786)   Tahiti (Society Is.)
___ †*P.n.pomarea* (Garnot, 1828)[1]   Maupiti I. (Society Is.)

○ *Pomarea mendozae* MARQUESAN MONARCH
___ †?*P.m.mendozae* (Hartlaub, 1854)   Tahuata, Hivaoa Is. (Marquesas)
___ *P.m.motanensis* Murphy & Mathews, 1928   Motane I. (Marquesas)
___ †?*P.m.mira* Murphy & Mathews, 1928   Huapu I. (Marquesas)
___ †?*P.m.nukuhivae* Murphy & Mathews, 1928   Nukuhiva I. (Marquesas)

○ *Pomarea iphis* IPHIS MONARCH
___ *P.i.iphis* Murphy & Mathews, 1928   Huahuna I. (Marquesas)
___ †?*P.i.fluxa* Murphy & Mathews, 1928   Eioa I. (Marquesas)

○ *Pomarea whitneyi* Murphy & Mathews, 1928 LARGE MONARCH   Fatuhiva I. (Marquesas)

MAYRORNIS Wetmore, 1932 M
○ *Mayrornis schistaceus* Mayr, 1933 VANIKORO MONARCH   Vanikoro I. (Santa Cruz Is.)

○ *Mayrornis versicolor* Mayr, 1933 VERSICOLOUR MONARCH   Ogea Levu I. (E Fiji)

○ *Mayrornis lessoni* SLATY MONARCH
___ *M.l.lessoni* (G.R. Gray, 1846)   W and C Fiji Is.
___ *M.l.orientalis* Mayr, 1933   E Fiji Is.

NEOLALAGE Mathews, 1928 F
○ *Neolalage banksiana* (G.R. Gray, 1870) BUFF-BELLIED MONARCH   Banks Is. (Vanuatu)

CLYTORHYNCHUS Elliot, 1870 M
○ *Clytorhynchus pachycephaloides* SOUTHERN SHRIKEBILL
___ *C.p.pachycephaloides* Elliot, 1870   New Caledonia
___ *C.p.grisescens* Sharpe, 1899   Banks Is. (Vanuatu)

○ *Clytorhynchus vitiensis* FIJI SHRIKEBILL
___ *C.v.wiglesworthi* Mayr, 1933   Rotuma I. (NW of Fiji)
___ *C.v.compressirostris* (E.L. Layard, 1876)   Kandavu, Ono, Vanuakula Is. (S Fiji)
___ *C.v.vitiensis* (Hartlaub, 1866)   Viti Levu, Ovalau Is. (Fiji)
___ *C.v.buensis* (E.L. Layard, 1876)   Vanua Levu I. (Fiji)
___ *C.v.layardi* Mayr, 1933   Taviuni I. (W Fiji)
___ *C.v.pontifex* Mayr, 1933   Ngambia, Rambi Is. (W Fiji)
___ *C.v.vatuanus* Mayr, 1933[2]   N Lau Arch. (E Fiji)
___ *C.v.nesiotes* (Wetmore, 1919)   S Lau Arch. (E Fiji)
___ *C.v.fortunae* (E.L. Layard, 1876)   Fotuna, Alofa Is. (NE of Fiji)
___ *C.v.heinei* (Finsch & Hartlaub, 1870)   C Tonga Is.
___ *C.v.keppeli* Mayr, 1933   Keppel, Boscawen Is. (E of Tonga)
___ *C.v.powelli* (Salvin, 1879)   Manua I. (Samoa)

○ *Clytorhynchus nigrogularis* BLACK-THROATED SHRIKEBILL
___ *C.n.nigrogularis* (E.L. Layard, 1875)   Fiji Is.
___ *C.n.sanctaecrucis* Mayr, 1933   Santa Cruz I.

○ *Clytorhynchus hamlini* (Mayr, 1931) RENNELL SHRIKEBILL   Rennell I. (SE Solomons)

[1] Considered a separate species by Holyoak & Thibault (1984) (1133).
[2] For correction of spelling see David & Gosselin (2002) (613).

METABOLUS Bonaparte, 1854 M

○ *Metabolus rugensis* (Hombron & Jacquinot, 1841) TRUK MONARCH

Truk Is. (Caroline Is.)

MONARCHA Vigors & Horsfield, 1827 M

○ *Monarcha axillaris* BLACK MONARCH

— *M.a.axillaris* Salvadori, 1876 — Mts. of NW New Guinea

— *M.a.fallax* (E.P. Ramsay, 1885) — Mts. of WC to SE New Guinea

○ *Monarcha rubiensis* (A.B. Meyer, 1874) RUFOUS MONARCH — N, WC and SW New Guinea

○ *Monarcha cinerascens* ISLAND MONARCH

— *M.c.commutatus* Brüggemann, 1876 — Sangihe, Siau, Maju and Tifore Is. (N of Sulawesi)

— *M.c.cinerascens* (Temminck, 1827) — Sulawesi, Moluccas, Lesser Sundas

— *M.c.inornatus* (Garnot, 1829) — W Papuan Is., NW New Guinea (NW coast of Vogelkop), Aru Is.

— *M.c.steini* Stresemann & Paludan, 1932 — Numfor I. (Geelvink Bay)

— *M.c.geelvinkianus* A.B. Meyer, 1885 — Yapen, Biak Is. (Geelvink Bay)

— *M.c.fuscescens* A.B. Meyer, 1885 — N New Guinea islands

— *M.c.nigrirostris* Neumann, 1929 — NE New Guinea and islands

— *M.c.fulviventris* Hartlaub, 1868 — Anchorite and Admiralty Is. (W Bismarck Arch.)

— *M.c.perpallidus* Neumann, 1924 — St. Matthias, Emirau Is., New Hanover, New Ireland

— *M.c.impediens* E. Hartert, 1926 — Islets off E Bismarck Arch. to Solomon Is.

— *M.c.rosselianus* Rothschild & Hartert, 1916 — Trobriand Is., D'Entrecasteaux and Louisiade Arch.

○ *Monarcha melanopsis* (Vieillot, 1818) BLACK-FACED MONARCH — E Australia >> E Cape York Pen. to SE New Guinea

○ *Monarcha frater* BLACK-WINGED MONARCH

— *M.f.frater* P.L. Sclater, 1874 — NW New Guinea (Vogelkop)

— *M.f.kunupi* Hartert & Paludan, 1934 — WC New Guinea

— *M.f.periophthalmicus* Sharpe, 1882 — C to SE New Guinea

— *M.f.canescens* Salvadori, 1876 — NE Cape York Pen., Torres Straits Is. >> SC New Guinea

○ *Monarcha erythrostictus* (Sharpe, 1888) BOUGAINVILLE MONARCH — Bougainville I. (N Solomons)

○ *Monarcha castaneiventris* CHESTNUT-BELLIED MONARCH

— *M.c.castaneiventris* J. Verreaux, 1858 — Choiseul, Ysabel, Guadalcanal, Malaita Is. (C and E Solomons)

— *M.c.obscurior* Mayr, 1935 — Russell Is. (EC Solomons)

— *M.c.megarhynchus* Rothschild & Hartert, 1908 — San Cristobal I. (SE Solomons)

— *M.c.ugiensis* (E.P. Ramsay, 1882) — Ugi Is. (SE Solomons)

○ *Monarcha richardsii* (E.P. Ramsay, 1881) WHITE-CAPPED MONARCH — SC Solomon Is.

○ *Monarcha leucotis* Gould, 1850 WHITE-EARED MONARCH — E Queensland, NE New South Wales

○ *Monarcha pileatus* WHITE-NAPED MONARCH[1]

— *M.p.pileatus* Salvadori, 1878 — Halmahera (N Moluccas)

— *M.p.buruensis* A.B. Meyer, 1885 — Buru (S Moluccas)

— *M.p.castus* P.L. Sclater, 1883 — Tanimbar Is.

○ *Monarcha guttula* (Garnot, 1829) SPOT-WINGED MONARCH — New Guinea and satellite islands

○ *Monarcha mundus* P.L. Sclater, 1883 BLACK-BIBBED MONARCH — Tanimbar, Babar, Damar Is. (E Lesser Sundas)

○ *Monarcha sacerdotum* Mees, 1973 FLORES MONARCH — Flores I. (Lesser Sundas)

○ *Monarcha boanensis* van Bemmel, 1939 BLACK-CHINNED MONARCH — Boano I. (S Moluccas)

○ *Monarcha trivirgatus* SPECTACLED MONARCH

— *M.t.trivirgatus* (Temminck, 1826) — C and E Lesser Sundas

— *M.t.nigrimentum* G.R. Gray, 1860 — Watubela Is. to Seram, Ambon Is. (S Moluccas)

— *M.t.diadematus* Salvadori, 1878 — Obi I. (C Moluccas)

— *M.t.bimaculatus* G.R. Gray, 1860 — Halmahara, Bacan, Morotai Is. (N Moluccas)

— *M.t.bernsteini* Salvadori, 1878 — Salawati I. (NW New Guinea)

— *M.t.melanopterus* G.R. Gray, 1858 — Louisiade Arch. and satellite islands

---

[1] We tentatively follow White & Bruce (1986: 368) (2652) in treating this as separate from *M. leucotis*. We would prefer to see a more detailed analysis.

    __ *M.t.albiventris* Gould, 1866

    __ *M.t.melanorrhous* Schodde & Mason, 1999

    __ *M.t.gouldii* G.R. Gray, 1860

Torres Strait Is., NE Cape York Pen., S New Guinea

NE Queensland (2189)

EC and SE Queensland, NE New South Wales >> Cape York Pen.

○ *Monarcha leucurus* G.R. Gray, 1858 WHITE-TAILED MONARCH — Kai Is.

○ *Monarcha everetti* E. Hartert, 1896 WHITE-TIPPED MONARCH — Tanahjampea I. (Flores Sea)

○ *Monarcha loricatus* Wallace, 1863 BLACK-TIPPED MONARCH — Buru I. (S Moluccas)

○ *Monarcha julianae* Ripley, 1959 KAFIAU MONARCH — Kafiau I. (W Papuan Is.)

○ *Monarcha brehmii* Schlegel, 1871 BIAK MONARCH — Biak I. (Geelvink Bay)

○ *Monarcha manadensis* (Quoy & Gaimard, 1830) HOODED MONARCH — New Guinea

○ *Monarcha infelix* MANUS MONARCH

    __ *M.i.infelix* P.L. Sclater, 1877 — Manus I. (Admiralty Is.)

    __ *M.i.coultasi* Mayr, 1955 — Rambutyo I.

○ *Monarcha menckei* Heinroth, 1902 WHITE-BREASTED MONARCH — St. Matthias Is. (Bismarck Arch.)

○ *Monarcha verticalis* BLACK-TAILED MONARCH

    __ *M.v.ateralbus* Salomonsen, 1964 — Dyaul I. (Bismarck Arch.)

    __ *M.v.verticalis* P.L. Sclater, 1877 — Main islands of Bismarck Arch.

○ *Monarcha barbatus* BLACK-AND-WHITE MONARCH

    __ *M.b.barbatus* E.P. Ramsay, 1879 — Bougainville, Guadalcanal, Choiseul, Ysabel Is. (main Solomon Is.)

    __ *M.b.malaitae* Mayr, 1931 — Malaita I. (E Solomons)

○ *Monarcha browni* KOLOMBANGARA MONARCH

    __ *M.b.browni* E.P. Ramsay, 1883 — Kolombangara, New Georgia, Vangunu Is. (SC Solomons)

    __ *M.b.ganongae* Mayr, 1935 — Ganonga I. (SC Solomons)

    __ *M.b.nigrotectus* E. Hartert, 1908 — Vellalavella I. (SC Solomons)

    __ *M.b.meeki* Rothschild & Hartert, 1905 — Rendova I. (SC Solomons)

○ *Monarcha viduus* WHITE-COLLARED MONARCH

    __ *M.v.viduus* (Tristram, 1879) — San Cristobal I. (SE Solomons)

    __ *M.v.squamulatus* (Tristram, 1882) — Ugi I. (SE Solomons)

○ *Monarcha godeffroyi* Hartlaub, 1868 YAP MONARCH — Yap I. (Caroline Is.)

○ *Monarcha takatsukasae* (Yamashina, 1931) TINIAN MONARCH — Tinian I. (Marianas)

○ *Monarcha chrysomela* GOLDEN MONARCH

    __ *M.c.aruensis* Salvadori, 1874 — Aru Is., SW New Guinea

    __ *M.c.melanonotus* P.L. Sclater, 1877 — W Papuan Is., NW New Guinea (Vogelkop)

    __ *M.c.kordensis* A.B. Meyer, 1874 — Biak I. (Geelvink Bay)

    __ *M.c.aurantiacus* A.B. Meyer, 1891 — N New Guinea

    __ *M.c.nitidus* (De Vis, 1897)[1] — E and S New Guinea, D'Entrecasteaux Arch.

    __ *M.c.pulcherrimus* Salomonsen, 1964[2] — Dyaul I. (Bismarck Arch.)

    __ *M.c.chrysomela* (Garnot, 1827) — New Hanover, New Ireland (Bismarck Arch.)

    __ *M.c.whitneyorum* Mayr, 1955 — Lihir Is. (E Bismarck Arch.)

    __ *M.c.tabarensis* Mayr, 1955 — Tabar I. (E Bismarck Arch.)

ARSES Lesson, 1830 M

○ *Arses insularis* (A.B. Meyer, 1874) RUFOUS-COLLARED MONARCH[3] — N New Guinea, Yapen I. (Geelvink Bay)

○ *Arses telescopthalmus* FRILLED MONARCH

    __ *A.t.batantae* Sharpe, 1879 — Batanta, Waigeo (W Papuan Is.)

    __ *A.t.telescopthalmus* (Lesson & Garnot, 1827)[4] — Misool I., NW New Guinea

---

[1] For correction of spelling see David & Gosselin (2002) (613).
[2] For correction of spelling see David & Gosselin (2002) (613).
[3] For treatment of this and *lorealis* at specific level see Schodde & Mason (1999: 503) (2189).
[4] For correct spelling and for attribution to Lesson & Garnot see Schodde & Mason (1999: 502) (2189); in this specific case publication in the Atlas (attributed to the two authors) preceded publication of a description usually attributed to Garnot alone.

— *A.t.aruensis* Sharpe, 1879   Aru Is.

— *A.t.harterti* van Oort, 1909   SW to W and SE New Guinea, Torres Strait Is.

— *A.t.lauterbachi* Reichenow, 1897[1]   Eastern SE to NE New Guinea

— *A.t.henkei* A.B. Meyer, 1886   SE New Guinea

○ *Arses lorealis* De Vis, 1895   Frill-necked Monarch[2]   N and E Cape York Pen.

○ *Arses kaupi*   Pied Monarch

— *A.k.terraereginae* A.J. Campbell, 1895[3]   NE Queensland (Endeavour to Daintree Rivers)

— *A.k.kaupi* Gould, 1851   NE Queensland (Daintree River to Paluma Range)

GRALLINA Vieillot, 1816[4]   F

● *Grallina cyanoleuca*   Magpie-lark

— *G.c.neglecta* Mathews, 1912[5]   N Australia

⤸ *G.c.cyanoleuca* (Latham, 1802)[6]   W, C, S and E Australia

○ *Grallina bruijni* Salvadori, 1875   Torrent-lark   Mts. of New Guinea

MYIAGRA Vigors & Horsfield, 1827[7]   F

○ *Myiagra oceanica*   Oceanic Flycatcher[8]

— *M.o.erythrops* Hartlaub & Finsch, 1868   Palau Is.

— †?*M.o.freycineti* Oustalet, 1881   Guam I. (Marianas Is.)

— *M.o.oceanica* Pucheran, 1853   Truk I. (W Caroline Is.)

— *M.o.pluto* Finsch, 1876   Pohnpei I.

○ *Myiagra galeata*   Moluccan Flycatcher

— *M.g.galeata* G.R. Gray, 1860   N Moluccas

— *M.g.goramensis* Sharpe, 1879   Ambon, Seram, Goram, Little Kai Is. (S and SE Moluccas)

— *M.g.buruensis* E. Hartert, 1903   Buru I. (S Moluccas)

○ *Myiagra atra* A.B. Meyer, 1874   Biak Flycatcher   Numfor, Biak Is. (Geelvink Bay)

● *Myiagra rubecula*   Leaden Flycatcher

— *M.r.sciurorum* Rothschild & Hartert, 1918   D'Entrecasteaux and Louisiade Arch.

— *M.r.papuana* Rothschild & Hartert, 1918   S and SE New Guinea, Torres Strait Is.

— *M.r.concinna* Gould, 1848   NW to NC Australia

— *M.r.okyri* Schodde & Mason, 1999   Cape York Pen. (2189)

— *M.r.yorki* Mathews, 1912   NE to EC Australia >> Cape York Pen.

⤸ *M.r.rubecula* (Latham, 1802)[9]   SE Australia >> E Queensland to Cape York Pen.

○ *Myiagra ferrocyanea*   Steel-blue Flycatcher

— *M.f.cinerea* (Mathews, 1928)   Bougainville I. (N Solomons)

— *M.f.ferrocyanea* E.P. Ramsay, 1879   Ysabel, Choiseul, Guadalcanal Is. (NC Solomons)

— *M.f.feminina* Rothschild & Hartert, 1901   New Georgia group (SC Solomons)

— *M.f.malaitae* Mayr, 1931   Malaita I. (NE Solomons)

○ *Myiagra cervinicauda* Tristram, 1879   Makira Flycatcher   San Cristobal I. (SE Solomons)

○ *Myiagra caledonica*   Melanesian Flycatcher

— *M.c.caledonica* Bonaparte, 1857   New Caledonia

— *M.c.viridinitens* G.R. Gray, 1859   Loyalty Is.

— *M.c.melanura* G.R. Gray, 1860   S Vanuatu

— *M.c.marinae* Salomonsen, 1934   N and C Vanuatu

— *M.c.occidentalis* Mayr, 1931   Rennell I. (SE Solomons)

---

[1]   The name *henkei* was applied to an intergrading population, see Schodde & Mason (1999: 502) (2189).

[2]   For reasons to separate this species see Schodde & Mason (1999: 503) (2189).

[3]   For recognition see Schodde & Mason (1999: 505) (2189).

[4]   Placement between *Arses* and *Myiagra* follows Schodde & Mason (1999: 507) (2189), it having been shown that a family Grallidae is unfounded (Olson, 1989; Baverstock *et al.,* 1992) (108, 1700).

[5]   For recognition see Schodde & Mason (1999: 507) (2189).

[6]   For reasons to use 1802 not 1801 see Browing & Monroe (1991) (258).

[7]   We follow Australian practice in using Flycatcher not Monarch as the generic vernacular.

[8]   Although split into several species in Pratt *et al.* (1987) (1920) we are unaware of a detailed review.

[9]   For reasons to use 1802 not 1801 see Browing & Monroe (1991) (258).

○ *Myiagra vanikorensis* VANIKORO FLYCATCHER
    —— *M.v.vanikorensis* (Quoy & Gaimard, 1830)   Vanikoro I. (Santa Cruz Is.)
    —— *M.v.rufiventris* Elliot, 1859   Navigator I. (NW Fiji)
    —— *M.v.kandavensis* Mayr, 1933   Kandavu I. (SW Fiji)
    —— *M.v.dorsalis* Mayr, 1933   N Lau Is. (SC Fiji)
    —— *M.v.townsendi* Wetmore, 1919   S Lau Is. (E Fiji)

○ *Myiagra albiventris* (Peale, 1848) SAMOAN FLYCATCHER   W Samoa

○ *Myiagra azureocapilla* BLUE-CRESTED FLYCATCHER
    —— *M.a.azureocapilla* E.L. Layard, 1875   Taviuni I. (W Fiji)
    —— *M.a.castaneigularis* E.L. Layard, 1876   Vanua Levu, Kambara Is. (NC Fiji)
    —— *M.a.whitneyi* Mayr, 1933   Viti Levu I. (W Fiji)

○ *Myiagra ruficollis* BROAD-BILLED FLYCATCHER
    —— *M.r.ruficollis* (Vieillot, 1818)   C to E Lesser Sundas, islands in Flores Sea
    —— *M.r.fulviventris* P.L. Sclater, 1883   Tanimbar Is.
    —— *M.r.mimikae* Ogilvie-Grant, 1911   Coastal S New Guinea, Torres Strait Is., coastal NW to EC Australia

○ *Myiagra cyanoleuca* (Vieillot, 1818) SATIN FLYCATCHER   SE Australia, Bass Strait Is., Tasmania >> NE Australia, Torres Strait Is., S New Guinea

○ *Myiagra alecto* SHINING FLYCATCHER
    —— *M.a.alecto* (Temminck, 1827)   N and C Moluccas, islands in Banda Sea
    —— *M.a.longirostris* (Mathews, 1928)   Tanimbar Is.
    —— *M.a.rufolateralis* (G.R. Gray, 1858)   Aru Is.
    —— *M.a.chalybeocephala* (Lesson & Garnot, 1828)[1]   W Papuan Is., all New Guinea to Bismarck Arch. (excl. St. Matthias and Lihir groups)
    —— *M.a.lucida* G.R. Gray, 1858)   Trobriand Is. to D'Entrecasteaux Arch. and Louisiade Arch.
    —— *M.a.manumudari* (Rothschild & Hartert, 1915)   Vulcan I. (NE New Guinea)
    —— *M.a.*subsp?[2]   N Australia, Melville I.
    —— *M.a.wardelli* (Mathews, 1911)   NE and E Australia

○ *Myiagra hebetior* DULL FLYCATCHER
    —— *M.h.hebetior* (E. Hartert, 1924)   St. Matthais Is. (Bismarck Arch.)
    —— *M.h.eichhorni* (E. Hartert, 1924)   New Hanover, New Ireland, New Britain (Bismarck Arch.)
    —— *M.h.cervinicolor* (Salomonsen, 1964)   Dyaul I. (Bismarck Arch.)

○ *Myiagra nana* (Gould, 1870) PAPERBARK FLYCATCHER[3]   SC New Guinea (Trans-Fly) to N Australia

○ *Myiagra inquieta* (Latham, 1802) RESTLESS FLYCATCHER[4]   E, S and SW Australia

LAMPROLIA Finsch, 1874 F
○ *Lamprolia victoriae* SILKTAIL
    —— *L.v.victoriae* Finsch, 1874   Taviuni I. (W Fiji)
    —— *L.v.kleinschmidti* E.P. Ramsay, 1876   Vanua Levu I. (NC Fiji)

CORVIDAE CROWS, JAYS[5] (24: 117)

PLATYLOPHUS Swainson, 1832 M
○ *Platylophus galericulatus* CRESTED JAY
    —— *P.g.ardesiacus* (Bonaparte, 1850)   Malay Pen.
    —— *P.g.coronatus* (Raffles, 1822)   Sumatra, Borneo (except N)
    —— *P.g.lemprieri* Nicholson, 1883[6]   N Borneo
    —— *P.g.galericulatus* (Cuvier, 1816)[7]   Java

---

[1] The reason for attribution to Lesson & Garnot is similar to that of Schodde & Mason (1999: 502) (2189) for *Arses telescopthalmus*; this relates to the Atlas in 1827/28 rather than the later description by Garnot alone.
[2]
[3] For treatment at specific level see Schodde & Mason (1999: 518) (2189).
[4] For reasons to use 1802 not 1801 see Browing & Monroe (1991) (258). Mayr in Watson *et al.* (1986: 526) (2566) treated *westralensis* as a synonym.
[5] The sequence of New World jays follows Espinosa de los Monteros & Cracraft (1997) (820). We believe the interesting findings of Cibois & Pasquet (1999) (356) need extending to genera not therein represented before we act upon them.
[6] Included with reservations, still listed by Smythies (2000) (2289).
[7] Dated 1817 by Blake & Vaurie (1962: 205) (169), but see Browning & Monroe (1991) (258).

PLATYSMURUS Reichenbach, 1850 M
○ *Platysmurus leucopterus* BLACK MAGPIE
    — *P.l.leucopterus* (Temminck, 1824)   Malay Pen., Sumatra
    — *P.l.aterrimus* (Temminck, 1829[1])   Borneo

PERISOREUS Bonaparte, 1831 M
○ *Perisoreus infaustus* SIBERIAN JAY
    — *P.i.infaustus* (Linnaeus, 1758)   Scandinavia and NW European Russia
    — *P.i.ostjakorum* Sushkin & Stegmann, 1929   NE European Russia and N Siberia east to the W Lena basin
    — *P.i.yakutensis* Buturlin, 1916   NE Siberia (lower Lena basin to W Anadyrland and N Sea of Okhotsk)
    — *P.i.ruthenus* Buturlin, 1916   WC to SE European Russia, SW Siberia
    — *P.i.opicus* Bangs, 1913   Altai, Sayan and Tannu Ola Mts. (SC Siberia, Tuva Rep.)
    — *P.i.sibericus* (Boddaert, 1783)[2]   Mid-Yenisey basin to Baikal area and N Mongolia
    — *P.i.tkatchenkoi* Sushkin & Stegmann, 1929[3]   Middle Lena and Aldan basins to NW Manchuria and W Amurland
    — *P.i.maritimus* Buturlin, 1915[4]   E Amurland and NE Manchuria to Russian Far East and Sakhalin
○ *Perisoreus internigrans* (Thayer & Bangs, 1912) SICHUAN JAY   E Xizang, SE Qinghai, W Gansu, N Sichuan

● *Perisoreus canadensis* GREY JAY
    ✓ *P.c.pacificus* (J.F. Gmelin, 1788)   NC and S Alaska
    — *P.c.canadensis* (Linnaeus, 1766)[5]   NE Alaska, C Canada, NC and NE USA
    — *P.c.nigricapillus* Ridgway, 1882   NE Canada
    — *P.c.albescens* J.L. Peters, 1920   W Canada
    — *P.c.bicolor* A.H. Miller, 1933   SW Canada, NW USA
    — *P.c.capitalis* Ridgway, 1874   WC and SC USA
    — *P.c.griseus* Ridgway, 1899   SW Canada, NW USA
    — *P.c.obscurus* Ridgway, 1874   NW USA

CYANOLYCA Cabanis, 1851 F
○ *Cyanolyca armillata* BLACK-COLLARED JAY
    — *C.a.meridana* (Sclater & Salvin, 1876)   NW Venezuela
    — *C.a.armillata* (G.R. Gray, 1845)   E Colombia, W Venezuela
    — *C.a.quindiuna* (Sclater & Salvin, 1876)   S Colombia, N Ecuador
○ *Cyanolyca viridicyanus* WHITE-COLLARED JAY[6]
    — *C.v.joylaea* (Bonaparte, 1852)   N and C Peru
    — *C.v.cyanolaema* Hellmayr, 1917   S Peru
    — *C.v.viridicyanus* (d'Orbigny & Lafresnaye, 1838) N Bolivia
○ *Cyanolyca turcosa* (Bonaparte, 1853) TURQUOISE JAY   S Colombia, N Peru
○ *Cyanolyca pulchra* (Lawrence, 1876) BEAUTIFUL JAY   W Colombia, W Ecuador, NW Peru
○ *Cyanolyca cucullata* AZURE-HOODED JAY
    — *C.c.mitrata* Ridgway, 1899   E and S Mexico
    — *C.c.guatemalae* Pitelka, 1951   S Mexico (Chiapas), Guatemala
    — *C.c.hondurensis* Pitelka, 1951   W Honduras
    — *C.c.cucullata* (Ridgway, 1885)   N Costa Rica to WC Panama
○ *Cyanolyca pumilo* (Strickland, 1849) BLACK-THROATED JAY   S Mexico to Honduras
○ *Cyanolyca nana* (Du Bus, 1847)[7] DWARF JAY   S Mexico
○ *Cyanolyca mirabilis* Nelson, 1903 WHITE-THROATED JAY[8]   SW Mexico

[1] The section of text containing this name seems best dated from Livr. 80 in 1829 (see Dickinson, 2001) (739).
[2] Includes *rogosowi* see Cramp *et al.* (1994: 42) (595).
[3] Includes *varnak* see Cramp *et al.* (1994: 42) (595).
[4] Includes *sakhalinensis* see Cramp *et al.* (1994: 42) (595).
[5] Includes *arcus* see Phillips (1986) (1875).
[6] For correction of spelling see David & Gosselin (2002) (613).
[7] Du Bus is an abbreviation of Du Bus de Gisignies.
[8] Includes *hardyi* A.R. Phillips, 1966 (1870) see Phillips (1986) (1875).

○ *Cyanolyca argentigula* SILVERY-THROATED JAY
    — *C.a.albior* Pitelka, 1951   C Costa Rica
    — *C.a.argentigula* (Lawrence, 1875)   S Costa Rica, W Panama

CYANOCORAX Boie, 1826[1] M
○ *Cyanocorax melanocyanea* BUSHY-CRESTED JAY
    — *C.m.melanocyanea* (Hartlaub, 1844)   C Guatemala to SC El Salvador
    — *C.m.chavezi* (Miller & Griscom, 1925)   NE Guatemala, Honduras, NW El Salvador, N Nicaragua

○ *Cyanocorax sanblasiana* SAN BLAS JAY
    — *C.s.nelsoni* (Bangs & Penard, 1919)   SW Mexico (Nayarit to W Guerrero)
    — *C.s.sanblasiana* (Lafresnaye, 1842)   SW Mexico (C Guerrero)

● *Cyanocorax yucatanica* YUCATÁN JAY
    — *C.y.yucatanica* (A.J.C. Dubois, 1875)   SE Mexico, Guatemala
    — *C.y.rivularis* (Brodkorb, 1940)   SE Mexico

○ *Cyanocorax beecheii* (Vigors, 1829) PURPLISH-BACKED JAY   NW Mexico

○ *Cyanocorax violaceus* VIOLACEOUS JAY
    — *C.v.pallidus* Zimmer & Phelps, 1944   N Venezuela
    — *C.v.violaceus* Du Bus, 1847[2]   W Amazonia

○ *Cyanocorax caeruleus* (Vieillot, 1818) AZURE JAY   SE Brazil, E Paraguay, NE Argentina

○ *Cyanocorax cyanomelas* (Vieillot, 1818) PURPLISH JAY   SE Peru to N Argentina, SW Brazil

○ *Cyanocorax cristatellus* (Temminck, 1823) CURL-CRESTED JAY   C and E Brazil, NE Paraguay, E Bolivia

○ *Cyanocorax dickeyi* R.T. Moore, 1935 TUFTED JAY   W Mexico

○ *Cyanocorax affinis* BLACK-CHESTED JAY
    — *C.a.zeledoni* Ridgway, 1899   S Costa Rica, Panama
    — *C.a.affinis* Pelzeln, 1856   N Colombia, NW Venezuela

○ *Cyanocorax mystacalis* (I. Geoffroy Saint-Hilaire, 1835) WHITE-TAILED JAY
                                   SW Ecuador, NW Peru

○ *Cyanocorax cayanus* (Linnaeus, 1766) CAYENNE JAY   SE Venezuela, the Guianas, N Brazil

○ *Cyanocorax heilprini* Gentry, 1885 AZURE-NAPED JAY   E Colombia, S Venezuela, NW Brazil

○ *Cyanocorax chrysops* PLUSH-CRESTED JAY
    — *C.c.diesingii* Pelzeln, 1856   N Brazil
    — *C.c.insperatus* Pinto & Camargo, 1961[3]   NE Brazil (1899)
    — *C.c.chrysops* (Vieillot, 1818)   E Bolivia to SE Brazil
    — *C.c.tucumanus* Cabanis, 1883   NW Argentina

○ *Cyanocorax cyanopogon* (Wied, 1821) WHITE-NAPED JAY   E Brazil

● *Cyanocorax yncas* GREEN JAY
    ⦟ *C.y.glaucescens* Ridgway, 1900   S Texas, NE Mexico
    — *C.y.luxuosus* (Lesson, 1839)   NE Mexico
    — *C.y.speciosus* (Nelson, 1900)   W Mexico
    — *C.y.vividus* (Ridgway, 1900)   SW Mexico
    — *C.y.maya* (van Rossem, 1934)   SE Mexico (Yucatán Pen.)
    — *C.y.confusus* A.R. Phillips, 1966[4]   Extreme S Mexico, W Guatemala (1870)
    — *C.y.centralis* (van Rossem, 1934)   SE Mexico (Tabasco), E Guatemala, Honduras
    — *C.y.cozumelae* (van Rossem, 1934)   Cozumel I.
    — *C.y.galeatus* (Ridgway, 1900)   Colombia
    — *C.y.cyanodorsalis* (A.J.C. Dubois, 1874)   E Colombia
    — *C.y.andicolus* Hellmayr & von Seilern, 1912   NE Columbia, NW Venezuela (1105)
    — *C.y.guatimalensis* (Bonaparte, 1850)   N Venezuela

[1] We follow A.O.U. (1998: 443) (3) in submerging *Cissilopha* and *Psilorhinus* herein.
[2] Du Bus is an abbreviation of Du Bus de Gisignies.
[3] Requires evaluation.
[4] Includes *persimilis* A.R. Phillips, 1966 (1870) see Phillips (1986) (1875).

_ *C.y.yncas* (Boddaert, 1783)   SW Colombia, Ecuador, Peru, N Bolivia
_ *C.y.longirostris* (Carriker, 1933)   N Peru (Marañon valley)

● *Cyanocorax morio* BROWN JAY
_ *C.m.palliatus* (van Rossem, 1934)   S Texas, NE and C Mexico
_ *C.m.morio* (Wagler, 1829)[1]   C Mexico to W Panama
↙ *C.m.vociferus* (Cabot, 1843)   SE Mexico (Yucatán Pen.)

CALOCITTA G.R. Gray, 1841 F
● *Calocitta colliei* (Vigors, 1829) BLACK-THROATED MAGPIE-JAY   W Mexico

● *Calocitta formosa* WHITE-THROATED MAGPIE-JAY
_ *C.f.formosa* (Swainson, 1827)   SW Mexico
_ *C.f.azurea* Nelson, 1897   SE Mexico, Guatemala
↙ *C.f.pompata* Bangs, 1914   S Mexico to NW Costa Rica

CYANOCITTA Strickland, 1845 F
● *Cyanocitta cristata* BLUE JAY
_ *C.c.bromia* Oberholser, 1921   S Canada, C USA >> SE USA
✓ *C.c.cristata* (Linnaeus, 1758)[2]   EC and SE USA
_ *C.c.semplei* Todd, 1928   S Florida
_ *C.c.cyanotephra* Sutton, 1935   SC USA

● *Cyanocitta stelleri* STELLER'S JAY
_ *C.s.stelleri* (J.F. Gmelin, 1788)   S Alaska to NW Canada (Oregon)
_ *C.s.carlottae* Osgood, 1901   Queen Charlotte Is.
↯ *C.s.frontalis* (Ridgway, 1873)   W USA (S Oregon to S California)
↯ *C.s.carbonacea* Grinnell, 1900   C California
_ *C.s.annectens* (S.F. Baird, 1874)   W Canada, interior NW USA
_ *C.s.macrolopha* S.F. Baird, 1854   W USA (Rocky Mts., Great Basin region)
✓ *C.s.diademata* (Bonaparte, 1850)   SW USA to C Mexico
_ *C.s.phillipsi* Browning, 1993   C Mexico (255)
_ *C.s.azteca* Ridgway, 1899   C Mexico
_ *C.s.coronata* (Swainson, 1827)[3]   SC Mexico to El Salvador
_ *C.s.purpurea* Aldrich, 1944   SW Mexico (Michoacan)
_ *C.s.restricta* A.R. Phillips, 1966[4]   S Mexico (Oaxaca) (1870)
_ *C.s.suavis* Miller & Griscom, 1925   C Honduras to N Nicaragua

APHELOCOMA Cabanis, 1851[5] F
● *Aphelocoma ultramarina* MEXICAN JAY
∠ *A.u.arizonae* (Baird & Ridgway, 1874)   SW USA, N Mexico
_ *A.u.wollweberi* Kaup, 1854   NW and W Mexico
_ *A.u.gracilis* G.S. Miller, 1896   WC Mexico
_ *A.u.colimae* Nelson, 1899   SW Mexico
_ *A.u.ultramarina* (Bonaparte, 1825)   SC Mexico
_ *A.u.potosina* Nelson, 1899   EC Mexico
_ *A.u.couchii* (S.F. Baird, 1858)   S Texas, NE Mexico

○ *Aphelocoma unicolor* UNICOLOURED JAY
_ *A.u.guerrerensis* Nelson, 1903   C Mexico
_ *A.u.concolor* (Cassin, 1848)   SE Mexico
_ *A.u.oaxacae* Pitelka, 1946   S Mexico
_ *A.u.unicolor* (Du Bus, 1847)[6]   SE Mexico, Guatemala
_ *A.u.griscomi* van Rossem, 1928   El Salvador, W Honduras

---

[1] Includes *cyanogenys* and *mexicanus* see Phillips (1986) (1875).
[2] Includes *burleighi* Bond, 1962 (187), see Browning (1990) (249).
[3] Includes *teotepecensis* and *ridgwayi* see Phillips (1986) (1875).
[4] For recognition Browning (1993) (255), but not by Binford (1989) (150).
[5] The sequence of species follows Peterson (1992) (1833).
[6] Du Bus is an abbreviation of Du Bus de Gisignies.

● *Aphelocoma californica* Western Scrub Jay[1]
  ☑ *A.c.californica* (Vigors, 1839)[2]   W USA (S Washington to C California)
  ☑ *A.c.obscura* Anthony, 1889   SW California, N Baja California
  — *A.c.cactophila* Huey, 1942   C Baja California
  — *A.c.hypoleuca* Ridgway, 1887   C and S Baja California
  — *A.c.suttoni* (A.R. Phillips, 1966)[3][4]   SC USA, NC Mexico (1872)
  ☑ *A.c.woodhouseii* (S.F. Baird, 1858)[5]   WC and SC USA to NW Mexico
  — *A.c.texana* Ridgway, 1902   WC Texas
  — *A.c.grisea* Nelson, 1899   NW Mexico
  — *A.c.cyanotis* Ridgway, 1887   EC Mexico
  — *A.c.sumichrasti* (Baird & Ridgway, 1874)   SC Mexico
  — *A.c.remota* Griscom, 1934   SW Mexico
  ☑ *A.c.cana* Pitelka, 1951   California
  — *A.c.nevadae* Pitelka, 1945   WC USA, N Mexico

○ *Aphelocoma insularis* Henshaw, 1886 Island Scrub Jay   Santa Cruz I. (off S California)

○ *Aphelocoma coerulescens* (Bosc, 1795) Florida Scrub Jay   S Florida

GYMNORHINUS Wied, 1841 M
● *Gymnorhinus cyanocephalus* Wied, 1841 Pinyon Jay   W USA, NW Mexico

GARRULUS Brisson, 1760 M
● *Garrulus glandarius* Eurasian Jay[6]
  ☑ *G.g.rufitergum* E. Hartert, 1903   Scotland, England, W France
  — *G.g.hibernicus* Witherby & Hartert, 1911   Ireland
  — *G.g.glandarius* (Linnaeus, 1758)   N and C Europe east to the Urals and south to the
                     Pyrenees, the Alps, N Croatia, N Serbia, Romania
  — *G.g.fasciatus* (A.E. Brehm, 1857)[7]   Iberia
  — *G.g.ichnusae* O. Kleinschimdt, 1903   Sardinia
  ☑ *G.g.corsicanus* Laubmann, 1912   Corsica
  — *G.g.albipectus* O. Kleinschmidt, 1920[8]   Mainland Italy, Sicily, W Croatia, W Albania, Ionian Is.
  — *G.g.graecus* Kleiner, 1939[9]   E Albania, S Serbia, W and C Bulgaria, mainland Greece
  — *G.g.ferdinandi* Keve, 1943[10]   E Bulgaria, N European Turkey
  — *G.g.cretorum* R. Meinertzhagen, 1920   Crete
  — *G.g.glaszneri* Madarász, 1902   Cyprus
  — *G.g.cervicalis* Bonaparte, 1853   N and E Algeria, N Tunisia
  — *G.g.whitakeri* E. Hartert, 1903   N Morocco, NW Algeria
  — *G.g.minor* J. Verreaux, 1857   C Morocco, Saharan Atlas Mts. of Algeria
  — *G.g.atricapillus* I. Geoffroy Saint-Hilaire, 1832   W Syria to Israel and W Jordan
  — *G.g.anatoliae* Seebohm, 1883[11]   Asia Minor (except NE), NW Syria, NE Iraq, W and
                     SW Iran
  — *G.g.samios* Kleiner, 1940   Samos
  — *G.g.krynicki* Kaleniczenko, 1839   NE Turkey, Caucasus area
  — *G.g.iphigenia* Sushkin & Ptuschenko, 1914   Crimea
  — *G.g.hyrcanus* Blanford, 1873   SE Azerbaijan, N Iran
  — *G.g.brandtii* Eversmann, 1842[12]   S Siberia from Ural Mts. to Sakhalin, S Kuril Is., and
                     Hokkaido south to Mongolia, N China and Korea

[1] For the separation of the Florida and Santa Cruz taxa as species see A.O.U. (1998: 446) (3).
[2] Includes *immanis*, *caurina* and *oocleptica* see Phillips (1986) (1875).
[3] Includes *mesolega* Oberholser, 1974 (1673), see Browning (1978) (246).
[4] For correction of date of publication see Dickerman & Parkes (1997: 226) (729).
[5] Includes *cana* and *nevadae* see Phillips (1986) (1875).
[6] Treatments differ. We combine those of Cramp *et al.* (1994: 7) (595) and Roselaar (1995: 160-162) (2107), for their areas, with that of Vaurie (1959)
  (2502) further east, with complementary footnotes here. Populations in the Aegean require the publication of revisionary work not reflected here.
[7] Includes *lusitanicus* as here we follow Vaurie (1959: 137) (2502).
[8] Implicitly includes *jordansi* Keve, 1966 (1263) see Vaurie (1959: 138) (2502). This was recognised in Cramp *et al.* (1994: 7) (595).
[9] For recognition see Cramp *et al.* (1994: 7) (595).
[10] For recognition see Cramp *et al.* (1994: 7) (595).
[11] For recognition see Cramp *et al.* (1994: 7) (595) and Roselaar (1995) (2107). Here we provisionally include *rhodius*, *zervasi* and *susianae* Keve, 1973
  (1265) see Cramp *et al.* (1994: 30) (595); the name *hansguentheri* Keve, 1967 (1264), was given to intergrading birds. Revision needed, but hampered by
  lost types.
[12] The name *severzowi* was applied to a small intergrading population just west of the Ural Mts. In addition *bambergi* is treated as a synonym, see Vaurie
  (1959: 142) (2502).

— *G.g.kansuensis* Stresemann, 1928 — Qinghai, Gansu, NW Sichuan

— *G.g.pekingensis* Reichenow, 1905 — Shanxi, Hebei, Beijing, S Liaoning

— *G.g.sinensis* Swinhoe, 1871 — C, S and E China (except S Yunnan), N Burma

— *G.g.taivanus* Gould, 1862 — Taiwan

— *G.g.bispecularis* Vigors, 1831 — W Himalayas to W Nepal

— *G.g.interstinctus* E. Hartert, 1918[1] — E Himalayas, SE Xizang

— *G.g.persaturatus* E. Hartert, 1918 — S Assam

— *G.g.oatesi* Sharpe, 1896[2] — NW Burma

— *G.g.haringtoni* Rippon, 1905 — W Burma

— *G.g.leucotis* Hume, 1874 — C and E Burma, S Yunnan, W, N and E Thailand, C and S Indochina

— *G.g.japonicus* Temminck & Schlegel, 1847[3] — Honshu, Shikoku, Tsushima, Oshima Is. (W and C Japan)

— *G.g.tokugawae* Taka-Tsukasa, 1931 — Sado I. (W Japan)

— *G.g.hiugaensis* Momiyama, 1927 — Kyushu (S Japan)

— *G.g.orii* N. Kuroda, Sr., 1923 — Yakushima Is. (S Japan)

○ *Garrulus lanceolatus* Vigors, 1831 BLACK-HEADED JAY — NE Afghanistan, W and N Pakistan, NW Himalayas to C Nepal

○ *Garrulus lidthi* Bonaparte, 1850 LIDTH'S JAY — Amami O-shima (Ryukyu Is.)

CYANOPICA Bonaparte, 1850 F
○ *Cyanopica cyanus* AZURE-WINGED MAGPIE[4]

— *C.c.cooki* Bonaparte, 1850 — Spain, Portugal

— *C.c.cyanus* (Pallas, 1776)[5] — Mongolia, Baikal area and Yakutia to N Manchuria and Russian Far East >> Nei Mongol

— *C.c.japonica* Parrot, 1905 — N and C Honshu (Japan)

— *C.c.stegmanni* Meise, 1932 — S Manchuria

— *C.s.koreensis* Yamashina, 1939 — Korea

— *C.c.interposita* E. Hartert, 1917 — Ningxia, E Gansu and Shaanxi to Shandong and Hebei (N China)

— *C.c.swinhoei* E. Hartert, 1903 — C and N Sichuan to Zhejiang and Jiangxi (EC and E China)

— *C.c.kansuensis* Meise, 1937 — NE Qinghai, W Gansu, NW Sichuan (NC China)

UROCISSA Cabanis, 1850 F
○ *Urocissa ornata* (Wagler, 1829) SRI LANKA BLUE MAGPIE — Sri Lanka

○ *Urocissa caerulea* Gould, 1863 TAIWAN BLUE MAGPIE — Taiwan

○ *Urocissa flavirostris* YELLOW-BILLED BLUE MAGPIE

— *U.f.cucullata* Gould, 1861 — W Himalayas to W Nepal

— *U.f.flavirostris* (Blyth, 1846) — E Himalayas, SE Xizang, N Burma, W Yunnan

— *U.f.schaeferi* Sick, 1939 — W Burma

— *U.f.robini* Delacour & Jabouille, 1930 — NW Vietnam

○ *Urocissa erythrorhyncha* RED-BILLED BLUE MAGPIE

— *U.e.occipitalis* (Blyth, 1846) — W and C Himalayas

— *U.e.magnirostris* (Blyth, 1846) — S Assam, Burma, Thailand, S Indochina

— *U.e.alticola* Birckhead, 1938 — W and N Yunnan, NE Burma

— *U.e.brevivexilla* Swinhoe, 1873 — NE China (S Gansu to S Liaoning)

— *U.e.erythrorhyncha* (Boddaert, 1783) — C, E and S China, N Indochina, Hainan

○ *Urocissa whiteheadi* WHITE-WINGED MAGPIE

— *U.w.whiteheadi* Ogilvie-Grant, 1899 — Hainan I.

— *U.w.xanthomelana* (Delacour, 1927) — S Sichuan, Guangxi, C Laos, N and C Vietnam

---

[1] Includes *azureitinctus* see Ripley (1982: 285) (2055).
[2] For recognition see Smythies (1953) (2285).
[3] Plate 43 appeared in 1847 see Holthuis & Sakai (1970) (1128); Blake & Vaurie (1962: 234) (169) gave 1848; *namiyei* is a synonym see Vaurie (1959: 144) (2502).
[4] For correction of spelling see David & Gosselin (2002) (613).
[5] Includes *pallescens* see Stepanyan (1990: 400) (2326).

CISSA Boie, 1826 F
○ ***Cissa chinensis*** COMMON GREEN MAGPIE
  ___ *C.c.chinensis* (Boddaert, 1783)   Himalayas, Assam, Burma, S China, W, N and E
                       Thailand, N Indochina

  ___ *C.c.klossi* Delacour & Jabouille, 1924   C Indochina
  ___ *C.c.margaritae* Robinson & Kloss, 1919   SC and S Vietnam
  ___ *C.c.robinsoni* Ogilvie-Grant, 1906   W Malaysia
  ___ *C.c.minor* Cabanis, 1850   Sumatra, NW Borneo

○ ***Cissa hypoleuca*** INDOCHINESE GREEN MAGPIE
  ___ *C.h.jini* Delacour, 1930   S Sichuan, Guangxi
  ___ *C.h.concolor* Delacour & Jabouille, 1928 N and NC Vietnam
  ___ *C.h.chauleti* Delacour, 1926   C Vietnam
  ___ *C.h.hypoleuca* Salvadori & Giglioli, 1885 E, SE Thailand, S Indochina
  ___ *C.h.katsumatae* Rothschild, 1903   Hainan I.

○ ***Cissa thalassina*** SHORT-TAILED GREEN MAGPIE
  ___ *C.t.thalassina* (Temminck, 1826)   Java
  ___ *C.t.jeffreyi* Sharpe, 1888   NW Borneo

DENDROCITTA Gould, 1833 F
○ ***Dendrocitta vagabunda*** RUFOUS TREEPIE
  ___ *D.v.bristoli* Paynter, 1961[1]   N and C Pakistan, NW India
  ___ *D.v.pallida* (Blyth, 1846)   W and WC India
  ___ *D.v.vagabunda* (Latham, 1790)   EC and NE India, Nepal, Bangladesh
  ___ *D.v.parvula* Whistler & Kinnear, 1932 SW India
  ___ *D.v.vernayi* Kinnear & Whistler, 1930 SE India
  ___ *D.v.sclateri* E.C.S. Baker, 1922   W and N Burma
  ___ *D.v.kinneari* E.C.S. Baker, 1922   C and E Burma, NW Thailand, SW Yunnan
  ___ *D.v.saturatior* Ticehurst, 1922   SE Burma, SW Thailand
  ___ *D.v.sakeratensis* Gyldenstolpe, 1920   E, C and SE Thailand, S Indochina

○ ***Dendrocitta occipitalis*** (S. Müller, 1835) SUMATRAN TREEPIE Sumatra

○ ***Dendrocitta cinerascens*** Sharpe, 1879 BORNEAN TREEPIE[2] Borneo

○ ***Dendrocitta formosae*** GRAY TREEPIE
  ___ *D.f.occidentalis* Ticehurst, 1925   NW Himalayas to W Nepal
  ___ *D.f.himalayensis* Blyth, 1865[3]   E Himalayas, Assam, NW, N and NE Burma, SW and S
                     Yunnan, N and C Laos, NW Vietnam

  ___ *D.f.sarkari* Kinnear & Whistler, 1930 S India
  ___ *D.f.assimilis* Hume, 1877   SW and E Burma, Thailand
  ___ *D.f.sinica* Stresemann, 1913   EC Sichuan to E and SE China and NE Vietnam
  ___ *D.f.sapiens* (Deignan, 1955)   WC Sichuan
  ___ *D.f.formosae* Swinhoe, 1863   Taiwan
  ___ *D.f.insulae* E. Hartert, 1910   Hainan I.

○ ***Dendrocitta leucogastra*** Gould, 1833 WHITE-BILLED TREEPIE S India

○ ***Dendrocitta frontalis*** Horsfield, 1840 COLLARED TREEPIE E Himalayas, N and NE Burma, W Yunnan, NW Vietnam

○ ***Dendrocitta baylyi*** Blyth, 1863 ANDAMAN TREEPIE Andaman Is.

CRYPSIRINA Vieillot, 1816 F
○ ***Crypsirina temia*** (Daudin, 1800) RACQUET-TAILED TREEPIE S and SE Burma, Thailand, Indochina, Java

○ ***Crypsirina cucullata*** Jerdon, 1862 HOODED TREEPIE N and C Burma

TEMNURUS Lesson, 1830 M
○ ***Temnurus temnurus*** (Temminck, 1825) RATCHET-TAILED TREEPIE SE Burma, W Thailand, C Laos, Vietnam, Hainan I.

[1] For recognition see Ripley (1982: 288) (2055).
[2] For treatment as a separate species from *D. occipitalis* see Smythies (2000) (2289) and references therein.
[3] Deignan (1963) (663) used *himalayana* Jerdon, 1864. The case for this is being investigated separately.

PICA Brisson, 1760 F

● *Pica pica* COMMON MAGPIE

    — *P.p.fennorum* Lönnberg, 1927   N Scandinavia and E Baltic to W and N European Russia

    ✓ *P.p.pica* (Linnaeus, 1758)[1]   S Scandinavia and British Isles to S and SE Europe, W and N Asia Minor and Caucasus area

    — *P.p.melanotos* A.E. Brehm, 1857   Spain, Portugal

    — *P.p.mauretanica* Malherbe, 1845   NW Africa

    — *P.p.asirensis* Bates, 1936   SW Saudi Arabia

    — *P.p.bactriana* Bonaparte, 1850[2]   SE Turkey, Iran, and Ural R. and Mts., E through Transcaspia and S Siberia to W Pakistan, Kashmir, Xinjiang, Mongolia and Transbaikalia

    — *P.p.hemileucoptera* Stegmann, 1928   W and C Siberia, WC Asia incl. Xinjiang

    — *P.p.leucoptera* Gould, 1862   Mongolia and Nei Mongol

    — *P.p.camtschatica* Stejneger, 1884   Anadyrland to Kamchatka (NE Siberia)

    — *P.p.sericea* Gould, 1845   Amurland, Korea, and Kyushu through NE and E China to Taiwan, C Indochina, and N and E Burma

    — *P.p.bottanensis* Delessert, 1840   E Tibetan Plateau (S and E Xizang and Qinghai to N Bhutan and W Sichuan)

● *Pica hudsonia* (Sabine, 1823) BLACK-BILLED MAGPIE[3]   S Alaska, W Canada, W USA

● *Pica nuttalli* (Audubon, 1837) YELLOW-BILLED MAGPIE   W California

ZAVATTARIORNIS Moltoni, 1938[4] M

○ *Zavattariornis stresemanni* Moltoni, 1938 STRESEMANN'S BUSH CROW   S Ethiopia

PODOCES J.G. Fischer, 1821[5] M

○ *Podoces hendersoni* Hume, 1871 MONGOLIAN GROUND JAY   Xinjiang, S Tuva Rep. and W and S Mongolia to N Qinghai and Ningxia, ? E Kazakhstan

○ *Podoces biddulphi* Hume, 1874 XINJIANG GROUND JAY   Tarim Basin (SW Xinjiang, China), ? W Gansu

○ *Podoces panderi* TURKESTAN GROUND JAY

    — *P.p.panderi* J.G. Fischer, 1821[6]   Turkmenistan, Uzbekistan, SW and SC Kazakhstan

    — *P.p.ilensis* Menzbier & Schnitnikov, 1915[7]   SE Balkhash area (E Kazakhstan)

○ *Podoces pleskei* Zarudny, 1896 PLESKE'S GROUND JAY   C Iran, ? S Afghanistan, ? W Pakistan

PSEUDOPODOCES Zarudny & Loudon, 1902[8] M

○ *Pseudopodoces humilis* (Hume, 1871) HUME'S GROUNDPECKER   Tibetan Plateau (S Xinjiang and NE Kashmir to SW Gansu and W Sichuan)

NUCIFRAGA Brisson, 1760 F

● *Nucifraga columbiana* (A. Wilson, 1811) CLARK'S NUTCRACKER   SW Canada, W USA >> N Mexico

○ *Nucifraga caryocatactes* EURASIAN NUTCRACKER

    — *N.c.caryocatactes* (Linnaeus, 1758)   N, C and SE Europe to C European Russia >> S Russia

    — *N.c.macrorhynchos* C.L. Brehm, 1823   E European Russia to Kamchatka and Sakhalin south to N Mongolia, NE China and Korea >> W Europe to C China

    — *N.c.rothschildi* E. Hartert, 1903   C Kyrgyzstan to WC Xinjiang and SE Kazakhstan

    — *N.c.japonica* E. Hartert, 1897   S Kurils, Japan

    — *N.c.owstoni* Ingram, 1910   Taiwan

    — *N.c.interdicta* Kleinschmidt & Weigold, 1922   Henan and Shanxi to Hebei and SW Liaoning (N China)

    — *N.c.multipunctata* Gould, 1849   E Afghanistan, W and N Pakistan, Kashmir E to Lahul

    — *N.c.hemispila* Vigors, 1831   NW and C Himalayas

    — *N.c.macella* Thayer & Bangs, 1909[9]   E Himalayas, N Burma, C China

---

[1] Includes *galliae* see Cramp *et al.* (1994: 75) (595).
[2] Includes *hemileucoptera* and *leucoptera* see Stepanyan (1990: 401) (2326).
[3] For separation of this from *P. pica* see A.O.U. (2000) (4). However, we note that molecular studies seem not to have included Asian forms of *P. pica* that are probably the closest relatives.
[4] We leave this genus here pending molecular studies, but note that Fry (2000: 646) (892) treated it with the Sturnidae.
[5] Fischer is an abbreviation of Fischer von Waldheim.
[6] Fischer is an abbreviation of Fischer von Waldheim.
[7] For recognition see Stepanyan (1990: 403) (2326).
[8] Proposals to treat this within the Paridae based on molecular studies have been submitted for publication (H. James *in litt.*). We await publication for assessment, but field experience suggests the species is not a crow.
[9] Includes *yunnanensis* see Cheng (1987: 546) (333).

511

PYRRHOCORAX Tunstall, 1771 M
○ ***Pyrrhocorax pyrrhocorax*** RED-BILLED CHOUGH
    __ *P.p.pyrrhocorax* (Linnaeus, 1758)   Parts of W Britain, Ireland
    __ *P.p.erythrorhamphos* (Vieillot, 1817)[1]   W France; Iberia to Alps and Italy
    __ *P.p.barbarus* Vaurie, 1954   Palma (Canary Is.), NW Africa
    __ *P.p.docilis* (S.G. Gmelin, 1774)   SE Europe to Caucasus area, Afghanistan and W Pakistan
    __ *P.p.centralis* Stresemann, 1928   NW Himalayas and Pamir-Alai Mts to Tien Shan, Altai, Mongolia, NW Manchuria
    __ *P.p.himalayanus* (Gould, 1862)   Fringe of S and E Tibetan Plateau (Garhwal to N Yunnan, north to Gansu)
    __ *P.p.brachypus* (Swinhoe, 1871)   Ningxia and Shanxi to Shandong and Hebei (N China)
    __ *P.p.baileyi* Rand & Vaurie, 1955   Highlands of N and C Ethiopia

● ***Pyrrhocorax graculus*** ALPINE CHOUGH/YELLOW-BILLED CHOUGH
    ⅃ *P.g.graculus* (Linnaeus, 1766)   Europe, N Africa, N Turkey, Caucasus area, N Iran
    __ *P.g.digitatus* Hemprich & Ehrenberg, 1833   S Turkey, Lebanon, SW Iran
    __ *P.g.forsythi* Stolickza, 1874[2]   Mts. of C Asia (Afghanistan and Pakistan to Altai, Tuva Rep., and S and E Tibetan Plateau

PTILOSTOMUS Swainson, 1837 M
○ ***Ptilostomus afer*** (Linnaeus, 1766) PIAPIAC   Senegal to Ethiopia, Uganda

CORVUS Linnaeus, 1758 M
◑ ***Corvus monedula*** EURASIAN JACKDAW
    __ *C.m.monedula* Linnaeus, 1758   N and C Europe (Scandinavia to N Alps and Carpathian Alps)
    ⅃ *C.m.spermologus* Vieillot, 1817   W and S Europe east to Italy; Morocco, NW Algeria
    __ *C.m.soemmerringii* J.G. Fischer, 1811[3]   SE and E Europe and C Asia E to Kashmir, NW Xinjiang, W Mongolia, and SC Siberia >> NE Africa, SW Asia E Europe, N and C Asia >> Iran, W India
    __ *C.m.cirtensis* (Rothschild & Hartert, 1912)   NE Algeria, Tunisia

○ ***Corvus dauuricus*** Pallas, 1776 DAURIAN JACKDAW   EC and E Asia (Altai and E Tibetan Plateau to N China, Korea, middle Amur Valley) >> S and E China and Japan

○ ***Corvus splendens*** HOUSE CROW
    __ *C.s.zugmayeri* Laubmann, 1913   Pakistan, NW India
    __ *C.s.splendens* Vieillot, 1817   India (except NW), Nepal, Bangladesh, Bhutan
    __ *C.s.protegatus* Madarász, 1904   Sri Lanka
    __ *C.s.maledivicus* Reichenow, 1904   Maldive Is.
    __ *C.s.insolens* Hume, 1874   Burma, S Yunnan, SW Thailand[4]

○ ***Corvus moneduloides*** Lesson, 1830 NEW CALEDONIAN CROW   New Caledonia, Loyalty Is.

○ †?***Corvus unicolor*** (Rothschild & Hartert, 1900) BANGGAI CROW[5]   Banggai I.

○ ***Corvus enca*** SLENDER-BILLED CROW
    __ *C.e.compilator* Richmond, 1903   W Malaysia, Sumatra, Nias, Simeuluë, Borneo
    __ *C.e.enca* (Horsfield, 1822)   Java, Bali, Mentawi Is.[6]
    __ *C.e.celebensis* Stresemann, 1936   Sulawesi and nearer satellite islands
    __ *C.e.mangoli* Vaurie, 1958[7]   Sula Is.
    __ *C.e.violaceus* Bonaparte, 1850   Ambon, Seram, Buru (S Moluccas)
    __ *C.e.pusillus* Tweeddale, 1878   Balabac, Palawan, Mindoro
    __ *C.e.sierramadrensis* Rand & Rabor, 1961   Luzon
    __ *C.e.samarensis* Steere, 1890   Samar, Mindanao

○ ***Corvus typicus*** (Bonaparte, 1853) PIPING CROW   C and S Sulawesi, Muna and Butung Is.

---

[1] For correction of spelling see David & Gosselin (2002) (613).
[2] For recognition see Stepanyan (1990: 407) (2326).
[3] Fischer is an abbreviation of Fischer von Waldheim.
[4] Not this race in W Malaysia but *protegatus* (introduced).
[5] Perhaps only a race of *C. enca* see Vaurie (1958) (2501).
[6] It is improbable that the Mentawi Is. to the west of Sumatra are home to the same form as is found in Java, as *compilator* is interposed between them. Van Marle & Voous (1988) (2468) suggested the Simeuluë birds may also need naming.
[7] For continued recognition see White & Bruce (1986: 321) (2652).

○ *Corvus florensis* Büttikofer, 1894 FLORES CROW | Flores I. (Lesser Sundas)

○ *Corvus kubaryi* Reichenow, 1885 MARIANA CROW | Guam, Rota Is. (Marianas)

○ *Corvus validus* Bonaparte, 1850 LONG-BILLED CROW | N and C Moluccas

○ *Corvus woodfordi* WHITE-BILLED CROW
— *C.w.meeki* Rothschild, 1904 | Bougainville I. (N Solomons)
— *C.w.woodfordi* (Ogilvie-Grant, 1887)[1] | Choiseul, Ysabel and Guadalcanal I. (NC and SE Solomons)

○ *Corvus fuscicapillus* BROWN-HEADED CROW
— *C.f.fuscicapillus* G.R. Gray, 1859 | Aru Is., W New Guinea
— *C.f.megarhynchus* Bernstein, 1864 | Waigeo, Geimen Is. (W Papuan Is.)

○ *Corvus tristis* Lesson & Garnot, 1827 GREY CROW | W Papuan Is., New Guinea, D'Entrecasteaux Arch.

○ *Corvus capensis* CAPE CROW
— *C.c.kordofanensis* Laubmann, 1919[2] | Sudan, Eritrea and Somalia to Uganda, Kenya, Tanzania
— *C.c.capensis* M.H.K. Lichtenstein, 1823 | Angola, Zambia and Zimbabwe to South Africa

● *Corvus frugilegus* ROOK
✓ *C.f.frugilegus* Linnaeus, 1758 | Europe, W and C Asia east to Tien Shan and N Altai >> N Africa, SW Asia
— *C.f.pastinator* Gould, 1845 | Mongolia and C and E China to S Yakutia, Russian Far East and Korea >> E Asia

● *Corvus brachyrhynchos* AMERICAN CROW
✓ *C.b.hesperis* Ridgway, 1887[3] | W Canada, W and SE USA
— *C.b.hargravei* A.R. Phillips, 1942[4] | Montane W USA
✓ *C.b.brachyrhynchos* C.L. Brehm, 1822 | C and E Canada, C and NE USA >> SE USA
— *C.b.pascuus* Coues, 1899 | S Florida

● *Corvus caurinus* S.F. Baird, 1858 NORTH-WESTERN CROW | W Canada, NW USA

○ *Corvus imparatus* J.L. Peters, 1929 TAMAULIPAS CROW | NE Mexico >> to S Texas

○ *Corvus sinaloae* J. Davis, 1958 SINALOA CROW | NW Mexico

● *Corvus ossifragus* A. Wilson, 1812 FISH CROW | E USA

○ *Corvus palmarum* PALM CROW[5]
— *C.p.palmarum* Württemberg, 1835 | Hispaniola
— *C.p.minutus* Gundlach, 1852 | Cuba

○ *Corvus jamaicensis* J.F. Gmelin, 1788 JAMAICAN CROW | Jamaica

○ *Corvus nasicus* Temminck, 1826 CUBAN CROW | Cuba, Grand Caicos I.

○ *Corvus leucognaphalus* Daudin, 1800 WHITE-NECKED CROW | Hispaniola, Puerto Rico

○ *Corvus hawaiiensis* Peale, 1848 HAWAIIAN CROW | Hawaii Is.

● *Corvus corone* CARRION CROW[6]
✓ *C.c.corone* Linnaeus, 1758 | SW and W Europe >> N Africa
✓ *C.c.cornix* Linnaeus, 1758 | N and E Europe and W Siberia south to N Balkans, N Ukraine and Ob' basin >> W and S Europe, SW Asia
— *C.c.sharpii* Oates, 1889[7] | Italy and S and E Balkans to Turkey and Iran and through S Ukraine and N Kazakhstan to NW Altai >> NE Africa to SW Asia
— *C.c.pallescens* (Madarász, 1904) | Egypt, Cyprus and S Turkey to Jordan and N Iraq
— *C.c.capellanus* P.L. Sclater, 1876 | C and S Iraq, SW Iran
— *C.c.orientalis* Eversmann, 1841 | WC and E Asia (Afghanistan, E Turkmenistan, Uzbekistan, SC Kazakhstan and Yenisey Valley east to C China, Korea, Japan and Kamchatka >> NW India, S China

---

[1] Includes *vegetus* see Blake & Vaurie (1962: 266) (169).
[2] For recognition see Urban (2000: 533) (2453).
[3] Includes *paulus* see Rea in Phillips, 1986 (1875).
[4] Recognised by Rea in Phillips, 1986 (1875).
[5] For reasons to retain *minutus* in this species see A.O.U. (1998) (3).
[6] Taxonomic treatment controversial; evolving concepts suggest several species could be recognised, but some evidence is contradictory. Under review by the B.O.U. Records Committee (B.O.U., 2002) (74). For now we stay with one polytypic species as in Cramp *et al.* (1994: 172) (595).
[7] Includes *sardonius* see Cramp *et al.* (1994: 172) (595); presumably also includes *italicus* Trischitta, 1939 (2439).

○ *Corvus pectoralis* Gould, 1836 Collared Crow[1]   E and C China, Hainan, N Vietnam (965)

○ *Corvus macrorhynchos* Large-billed Crow/Jungle Crow[2]
     __ *C.m.japonensis* Bonaparte, 1850   C and S Sakhalin I., S Kuril Is., Japan
     __ *C.m.connectens* Stresemann, 1916   Amami Oshima and N Ryukyu Is.
     __ *C.m.osai* Ogawa, 1905   S Ryukyu Is.
     __ *C.m.mandschuricus* Buturlin, 1913   N Sakhalin, Russian Far East and NE China to Hebei and Korea
     __ *C.m.colonorum* Swinhoe, 1864[3]   C and S China, Taiwan, Hainan, N Indochina
     __ *C.m.tibetosinensis* Kleinschmidt & Weigold, 1922   SE and E Tibetan Plateau, N Bhutan, N and NE Burma, W and N Yunnan to Qinghai
     __ *C.m.intermedius* Adams, 1859   Afghanistan, W and N Pakistan and through NW Himalayas to S Xizang and N Nepal
     __ *C.m.culminatus* Sykes, 1832   N India (except Himalayas) and SW Nepal to Sri Lanka
     __ *C.m.levaillantii* Lesson, 1830   SE Nepal, Bangladesh, NE India, Andamans, Burma and Thailand east to C and S Indochina and south to N Malay Pen.
     __ *C.m.macrorhynchos* Wagler, 1827   C and S Malay Pen., Gtr. and Lesser Sundas
     __ *C.m.philippinus* (Bonaparte, 1853)   Philippines

○ *Corvus orru* Torresian Crow
     __ *C.o.latirostris* A.B. Meyer, 1884   Tanimbar, Babar Is. (E Lesser Sundas)
     __ *C.o.orru* Bonaparte, 1850[4]   N and C Moluccas, New Guinea and satellite islands
     __ *C.o.insularis* Heinroth, 1903   New Britain, New Ireland, New Hanover and satellite islands (Bismarck Arch.)
     __ *C.o.ceciliae* Mathews, 1912   N Australia (from WC to EC)

○ *Corvus bennetti* North, 1901 Little Crow   Inland Australia

○ *Corvus tasmanicus* Forest Raven
     __ *C.t.boreus* I. Rowley, 1970[5]   NE New South Wales (2114)
     __ *C.t.tasmanicus* Mathews, 1912   SE South Australia, S Victoria, Tasmania

○ *Corvus mellori* Mathews, 1912 Little Raven   SE Australia, King I. (Bass Strait)

● *Corvus coronoides* Australian Raven
     ✓ *C.c.coronoides* Vigors & Horsfield, 1827   EC, E and SE Australia
     __ *C.c.perplexus* Mathews, 1912[6]   SW and SC Australia

○ *Corvus albus* Statius Müller, 1776 Pied Crow   W, C, E and Sthn. Africa, Madagascar

○ *Corvus ruficollis* Brown-necked Raven
     __ *C.r.ruficollis* Lesson, 1830   Cape Verde Is. and Sahara to Socotra and Sinai, deserts of SW and WC Asia from Arabia, Iran and W Pakistan to W Xinjiang and Kazakhstan
     __ *C.r.edithae* Lort Phillips, 1895   N Kenya, Somalia, E Ethiopia, Djibouti, Eritrea

● *Corvus corax* Common Raven
     ✓ *C.c.principalis* Ridgway, 1887   Alaska, Canada, N USA, Greenland
     ✓ *C.c.sinuatus* Wagler, 1829   SW Canada, WC USA, Middle America (Mexico to Nicaragua)
     ✓ *C.c.clarionensis* Rothschild & Hartert, 1902[7]   SW USA, Baja California, Revillagigedo Is. (Mexico)
     __ *C.c.varius* Brünnich, 1764   Iceland, Faroe Is.
     ✓ *C.c.corax* Linnaeus, 1758   N, C and SE Europe, W Siberia to Yenisey basin, N Kazakhstan, NE Turkey and Caucasus area to N Iran
     __ *C.c.hispanus* Hartert & Kleinschmidt, 1901[8]   Iberia and Balearic Is.

---

[1] The name *Corvus torquatus* Lesson, 1830, is preoccupied by *C. monedula torquata* Bechstein, 1791 (S. Eck *in litt.*).
[2] There is little doubt that there are two species or more here (see e.g. Martens & Eck, 1995 (1443)) but the attribution of many races is unclear. We defer a split until more is known.
[3] Includes *hainanus* and *mengtszensis* see Cheng (1987: 555) (333).
[4] Given as 1851 by Blake & Vaurie (1962: 275) (169), but see Zimmer (1926) (2707) whom we follow.
[5] Includes *novaanglica* Goodwin, 1976 (960); Goodwin believed that Rowley had published this name, having apparently seen a draft of Rowley's paper.
[6] For recognition see Schodde & Mason (1999: 612) (2189).
[7] For recognition see A.M. Rea in Phillips (1986) (1875).
[8] For recognition see Cramp *et al.* (1994: 206) (595).

— *C.c.laurenci* Hume, 1873[1]
S and E Greece, Turkey (except NE) and Levant to NW India, Xinjiang and S and E Kazakhstan

— *C.c.tingitanus* Irby, 1874
N Africa

— *C.c.canariensis* Hartert & Kleinschmidt, 1901[2]
Canary is.

— *C.c.tibetanus* Hodgson, 1849
Mts. of C Asia

— *C.c.kamtschaticus* Dybowski, 1883
NE Siberia south to W and N Mongolia, NE China, Russian Far East, S Kuril Is. >> N Japan, Aleutian Is., SW Alaska

● *Corvus cryptoleucus* Couch, 1854 CHIHUAHUAN RAVEN[3]
SW USA, N Mexico

○ *Corvus rhipidurus* FAN-TAILED RAVEN

— *C.r.stanleyi* Roselaar, 1993
Sinai Peninsula, Israel, and Jordan through Arabia to Yemen and S Oman (2105)

— *C.r.rhipidurus* E. Hartert, 1918
N Afrotropics (Niger to Sudan and Somalia)

○ *Corvus albicollis* Latham, 1790 WHITE-NECKED RAVEN
E and S Africa

○ *Corvus crassirostris* Rüppell, 1836 THICK-BILLED RAVEN
Ethiopia, E Sudan

CORCORACIDAE AUSTRALIAN MUDNESTERS (2: 2)

CORCORAX Lesson, 1830 M
○ *Corcorax melanorhamphos* WHITE-WINGED CHOUGH

— *C.m.melanorhamphos* (Vieillot, 1817)
E and SE Australia

— *C.m.whiteae* Mathews, 1912[4]
SC and SE South Australia

STRUTHIDEA Gould, 1837 F
○ *Struthidea cinerea* APOSTLEBIRD

— *S.c.cinerea* Gould, 1837
Inland E Australia

— *S.c.dalyi* Mathews, 1923
Inland NE to NC Australia

GENERA INCERTAE SEDIS (2: 3)

MELAMPITTA Schlegel, 1871 F
○ *Melampitta lugubris* LESSER MELAMPITTA

— *M.l.lugubris* Schlegel, 1871
Mts. of NW New Guinea (Vogelkop)

— *M.l.rostrata* (Ogilvie-Grant, 1913)
Mts. of WC New Guinea

— *M.l.longicauda* Mayr & Gilliard, 1952
Mts. of C and E New Guinea

○ *Melampitta gigantea* (Rothschild, 1899) GREATER MELAMPITTA
Mts. of New Guinea

IFRITA Rothschild, 1898 F
○ *Ifrita kowaldi* BLUE-CAPPED IFRIT

— *I.k.brunnea* Rand, 1940
Mts. of WC New Guinea

— *I.k.kowaldi* (De Vis, 1890)
Mts. of C and E New Guinea

PARADISAEIDAE MANUCODES AND BIRDS OF PARADISE (16: 40)

LYCOCORAX Bonaparte, 1853 M
○ *Lycocorax pyrrhopterus* PARADISE-CROW

— *L.p.morotensis* Schlegel, 1863
Morotai, Rau (N Moluccas)

— *L.p.pyrrhopterus* (Bonaparte, 1850)
Bacan, Halmahera (N Moluccas)

— *L.p.obiensis* Bernstein, 1864
Obi, Bisa (C Moluccas)

MANUCODIA Boddaert, 1783 M[5]
○ *Manucodia ater* (Lesson, 1830) GLOSSY-MANTLED MANUCODE[6]
W Papuan Is., lowland N, W and S to SE New Guinea, Louisuade Arch. (Tagula I.)

---

[1] For recognition see Cramp *et al.* (1994: 206) (595). The name *subcorax* has been used for this population, but wrongly so as its type is identifiable with *C. ruficollis*, see Cramp *et al.* (1994: 206) (595).
[2] For recognition see Cramp *et al.* (1994: 206) (595).
[3] Includes *reai* A.R. Phillips, 1986 (1875), see Browning (1990) (249).
[4] For recognition see Schodde & Mason (1999: 616) (2189).
[5] Gender determined by I.C.Z.N. Direction 26 see David & Gosselin (2002) (614).
[6] Varies clinally and treated as monotypic by Cracraft (1992) (581); *subalter* and *alter* are synonyms.

○ *Manucodia jobiensis* Salvadori, 1876 Jobi Manucode[1]   W and N New Guinea, Yapen I.

○ *Manucodia chalybatus* (Pennant, 1781) Crinkle-collared Manucode   New Guinea, Misool I.

○ *Manucodia comrii* Curl-crested Manucode
    — *M.c.comrii* P.L. Sclater, 1876   D'Entrecasteaux Arch.
    — *M.c.trobriandi* Mayr, 1936   Trobriand Is.

PHONYGAMMUS Lesson & Garnot, 1826 M
○ *Phonygammus keraudrenii* Trumpet Manucode
    — *P.k.keraudrenii* (Lesson & Garnot, 1826)   W and NW New Guinea
    — *P.k.neumanni* (Reichenow, 1918)   EC New Guinea (mts. of Sepik drainage)
    — *P.k.adelberti* Gilliard & LeCroy, 1967   NE New Guinea (Adelbert Mts.) (941)
    — *P.k.diamondi* Cracraft, 1992   Eastern Highlands EC New Guinea (581) #
    — *P.k.purpureoviolaceus* (A.B. Meyer, 1885)[2]   SE New Guinea and NE New Guinea (Huon Pen.)
    — *P.k.hunsteini* (Sharpe, 1882)   D'Entrecasteaux Arch.
    — *P.k.jamesii* (Sharpe, 1877)   S New Guinea, Torres Straits Is.
    — *P.k.aruensis* Cracraft, 1992   Aru Is. (581) #
    — *P.k.gouldii* (G.R. Gray, 1859)   N Cape York Pen., Queensland.

PARADIGALLA Lesson, 1835 F
○ *Paradigalla carunculata* Long-tailed Paradigalla
    — *P.c.carunculata* Lesson, 1835   NW New Guinea (Vogelkop)
    — *P.c.intermedia* Ogilvie-Grant, 1913   Mts. of WC New Guinea

○ *Paradigalla brevicauda* Rothschild & Hartert, 1911 Short-tailed Paradigalla
                                    Mts. of WC to EC New Guinea

ASTRAPIA Vieillot, 1816 F
○ *Astrapia nigra* (J.F. Gmelin, 1788) Arfak Astrapia   Mts. of Vogelkop (NW New Guinea)

○ *Astrapia splendidissima* Rothschild, 1895 Splendid Astrapia[3]   W and WC New Guinea

○ *Astrapia mayeri* Stonor, 1939 Ribbon-tailed Astrapia   Mts. of EC New Guinea

○ *Astrapia stephaniae* Princess Stephanie's Astrapia
    — *A.s.feminina* Neumann, 1922   Owen Stanley Mts. of NC New Guinea
    — *A.s.stephaniae* (Finsch, 1885)[4]   Mts. of EC to SE New Guinea

○ *Astrapia rothschildi* Foerster, 1906 Huon Astrapia   Mts. of Huon Pen. (NE New Guinea)

PAROTIA Vieillot, 1816 F
○ *Parotia sefilata* (Pennant, 1781) Western Parotia   Arfak Mts. (New Guinea)

○ *Parotia carolae* Carola's Parotia
    — *P.c.carolae* A.B. Meyer, 1894   Weyland Mts. to Paniai (N scarps, W New Guinea)
    — *P.c.chalcothorax* Stresemann, 1934   Doorman Mts. (N scarps, WC New Guinea)
    — *P.c.clelandiae* Gilliard, 1961   Victor Emanuel Mts. (N scarps, C New Guinea)
    — *P.c.meeki* Rothschild, 1910   Snow Mts. (S scarps, WC New Guinea)
    — *P.c.berlepschi* O. Kleinschmidt, 1897   Foya Mts. (NC New Guinea)
    — *P.c.chrysenia* Stresemann, 1934   N scarps, EC New Guinea Mts.

○ *Parotia lawesii* E.P. Ramsay, 1885 Lawes's Parotia   Mts. of EC to SE New Guinea (S scarps)

○ *Parotia helenae* De Vis, 1897 Eastern Parotia[5]   Mts. of SE New Guinea (N scarps)

○ *Parotia wahnesi* Rothschild, 1906 Wahnes's Parotia   Mts. of NE coastal ranges (NE New Guinea)

PTERIDOPHORA A.B. Meyer, 1894 F
○ *Pteridophora alberti* King of Saxony Bird-of-Paradise
    — *P.a.alberti* A.B. Meyer, 1894   Mts. of WC New Guinea
    — *P.a.buergersi* Rothschild, 1931[6]   Mts. of EC New Guinea

---

[1] Includes *rubiensis* see Cracraft (1992) (581).
[2] Includes *mayri* see Cracraft (1992) (581), but like many other Huon Peninsula forms this may prove distinct.
[3] Treated as monotypic by Cracraft (1992) (581); *helios* and *elliottsmithi* are synonyms.
[4] Includes *ducalis* see Cracraft (1992) (581).
[5] For reasons to separate this from *P. lawesii* see Schodde & McKean (1972, 1973) (2194, 2195).
[6] Includes *hallstromi* see Cracraft (1992) (581); he recognises that western *alberti* is more distinct than were the two eastern forms.

LOPHORINA Vieillot, 1816 F
○ *Lophorina superba* SUPERB BIRD-OF-PARADISE[1]
— *L.s.superba* (Pennant, 1781) — Mts. of Vogelkop (NW New Guinea)
— *L.s.niedda* Mayr, 1930 — Mt. Wondiwoi (W New Guinea)
— *L.s.feminina* Ogilvie-Grant, 1915[2] — Mts. of WC to EC New Guinea
— *L.s.latipennis* Rothschild, 1907[3] — Mts. of Huon Pen. (NE New Guinea)
— *L.s.minor* E.P. Ramsay, 1885 — Mts. of SE New Guinea

PTILORIS Swainson, 1825 M[4]
○ *Ptiloris paradiseus* Swainson, 1825 PARADISE RIFLEBIRD — SE Queensland, NE New South Wales

○ *Ptiloris victoriae* Gould, 1850 VICTORIA'S RIFLEBIRD — NE Queensland

○ *Ptiloris magnificus* MAGNIFICENT RIFLEBIRD
— *P.m.magnificus* (Vieillot, 1819) — W to NC and in the south to SC New Guinea
— *P.m.alberti* Elliot, 1871 — N Cape York Pen.

○ *Ptiloris intercedens* Sharpe, 1882 GROWLING RIFLEBIRD[5] — NE and SE New Guinea

EPIMACHUS Cuvier, 1816[6] M
○ *Epimachus fastuosus* BLACK SICKLEBILL
— *E.f.fastuosus* (Hermann, 1783) — Mts. of Vogelkop (NW New Guinea)
— *E.f.ultimus* Diamond, 1969 — Bewani-Torricelli Mts. of NC New Guinea (698)
— *E.f.atratus* (Rothschild & Hartert, 1911)[7] — N and C New Guinea

○ *Epimachus meyeri* BROWN SICKLEBILL
— *E.m.albicans* (van Oort, 1915)[8] — WC New Guinea
— *E.m.bloodi* Mayr & Gilliard, 1951 — EC New Guinea
— *E.m.meyeri* Finsch, 1885 — SE New Guinea

DREPANORNIS P.L. Sclater, 1873 M
○ *Drepanornis albertisi* BLACK-BILLED SICKLEBILL
— *D.a.albertisi* (P.L. Sclater, 1873) — Mts. of Vogelkop (NW New Guinea)
— *D.a.cervinicauda* P.L. Sclater, 1883 — Mts. of WC to SE New Guinea
— *D.a.geisleri* A.B. Meyer, 1893[9] — Mts. of Huon Pen. (NE New Guinea)

○ *Drepanornis bruijnii* Oustalet, 1880 PALE-BILLED SICKLEBILL — NC New Guinea

DIPHYLLODES Lesson, 1834[10] M
○ *Diphyllodes magnificus* MAGNIFICENT BIRD-OF-PARADISE
— *D.m.magnificus* (Pennant, 1781) — Salawati I., NW New Guinea (Vogelkop)
— *D.m.chrysopterus* Elliot, 1873[11] — Yapen I., W and C New Guinea
— *D.m.hunsteini* A.B. Meyer, 1885 — E New Guinea

○ *Diphyllodes respublica* (Bonaparte, 1850) WILSON'S BIRD-OF-PARADISE
Waigeo and Batanta (W Papuan Is.)

CICINNURUS Vieillot, 1816 M
○ *Cicinnurus regius* KING BIRD-OF-PARADISE
— *C.r.regius* (Linnaeus, 1758)[12] — Aru Is., Misool and Salawati Is., NW and S New Guinea
— *C.r.coccineifrons* Rothschild, 1896[13] — N, C and E New Guinea, Yapen I.

[1] The name *sphinx* was given to a single specimen from an unknown locality; it requires validation.
[2] Includes *pseudoparotia*; based on a unique specimen, requires validation.
[3] Includes *connectens* see Cracraft (1992) (581).
[4] Gender determined by I.C.Z.N. Direction 26 see David & Gosselin (2002) (614).
[5] Separation of this from *P. magnificens*, see Beehler & Swaby (1991) (110), is tentatively accepted.
[6] Given as 1817 by Mayr (1962: 190) (1480) but see Browning & Monroe (1991) (258).
[7] Includes *stresemanni* see Cracraft (1992) (581).
[8] Includes *megarhynchus* see Cracraft (1992) (581).
[9] We believe a sufficient sample would show this to be distinct from *albertisi* and have retained it.
[10] There is some support in Nunn & Cracraft (1996) (1665) for retaining this genus.
[11] Includes *intermedius* see Cracraft (1992) (581).
[12] Includes *rex* and *gymnorhynchus* see Mees (1982) (1534).
[13] Includes *similis* and *cryptorhynchus* see Mees (1982) (1534).

SEMIOPTERA G.R. Gray, 1859[1] F
○ ***Semioptera wallacii*** STANDARDWING
　　— *S.w.halmaherae* Salvadori, 1881　　　　　　　Halmahera (N Moluccas)
　　— *S.w.wallacii* (G.R. Gray, 1859)　　　　　　　Bacan (N Moluccas)

SELEUCIDIS Lesson, 1835 M
○ ***Seleucidis melanoleucus*** TWELVE-WIRED BIRD-OF-PARADISE
　　— *S.m.melanoleucus* (Daudin, 1800)[2]　　　　　Salawati I., W, S and E New Guinea
　　— *S.m.auripennis* Schlüter, 1911　　　　　　　N New Guinea

PARADISAEA Linnaeus, 1758 F
○ ***Paradisaea apoda*** Linnaeus, 1758 GREATER BIRD-OF-PARADISE[3]　　Aru Is., SW to SC New Guinea

○ ***Paradisaea raggiana*** RAGGIANA BIRD-OF-PARADISE
　　— *P.r.raggiana* P.L. Sclater, 1873[4]　　　　　Eastern SC to SE New Guinea (south coast)
　　— *P.r.intermedia* De Vis, 1894　　　　　　　E New Guinea (north coast)
　　— *P.r.granti* North, 1906[5]　　　　　　　　SE New Guinea (north coast)
　　— *P.r.augustaevictoriae* Cabanis, 1888　　　　NE New Guinea

○ ***Paradisaea minor*** LESSER BIRD-OF-PARADISE
　　— *P.m.finschi* A.B. Meyer, 1885　　　　　　Eastern NC New Guinea
　　— *P.m.minor* Shaw, 1809[6]　　　　　　　　Misool I., NW to SW and western NC and W New Guinea
　　— *P.m.jobiensis* Rothschild, 1897　　　　　Yapen I. (Geelvink Bay)

○ ***Paradisaea decora*** Salvin & Godman, 1883 GOLDIE'S BIRD-OF-PARADISE
　　　　　　　　　　　　　　　　　　　　　　D'Entrecasteaux Arch.

○ ***Paradisaea rubra*** Daudin, 1800 RED BIRD-OF-PARADISE　　Batanta, Waigeo and Saonek (W Papuan Is.)

○ ***Paradisaea guilielmi*** Cabanis, 1888 EMPEROR BIRD-OF-PARADISE　　Mts. of Huon Pen. (NE New Guinea)

○ ***Paradisaea rudolphi*** BLUE BIRD-OF-PARADISE
　　— *P.r.margaritae* Mayr & Gilliard, 1951　　　　EC New Guinea
　　— *P.r.rudolphi* (Finch, 1885)[7]　　　　　　　SE New Guinea

PETROICIDAE AUSTRALASIAN ROBINS[8] (13: 45)

POECILODRYAS Gould, 1865 F
○ ***Poecilodryas albispecularis*** ASHY ROBIN[9]
　　— *P.a.albispecularis* (Salvadori, 1876)　　　　NW New Guinea (Vogelkop)
　　— *P.a.rothschildi* (E. Hartert, 1930)　　　　　SC New Guinea
　　— *P.a.centralis* (Rand, 1940)　　　　　　　　NC New Guinea
　　— *P.a.atricapilla* (Mayr, 1931)[10]　　　　　　NE New Guinea (Huon Pen.)
　　— *P.a.armiti* De Vis, 1894　　　　　　　　　SE New Guinea
　　— *P.a.cinereifrons* E.P. Ramsay, 1876[11]　　　NE Queensland

○ ***Poecilodryas brachyura*** BLACK-CHINNED ROBIN
　　— *P.b.brachyura* (P.L. Sclater, 1874)　　　　NW New Guinea (Vogelkop)
　　— *P.b.albotaeniata* (A.B. Meyer, 1874)　　　Yapen I., western NC New Guinea
　　— *P.b.dumasi* Ogilvie-Grant, 1915　　　　　Eastern NC New Guinea

○ ***Poecilodryas hypoleuca*** BLACK-SIDED ROBIN
　　— *P.h.steini* Stresemann & Paludan, 1932　　　Waigeo I. (off W New Guinea)
　　— *P.h.hypoleuca* (G.R. Gray, 1859)　　　　　Misool I., W to SE New Guinea
　　— *P.h.hermani* Madarász, 1894　　　　　　　N New Guinea

○ ***Poecilodryas superciliosa*** (Gould, 1847) WHITE-BROWED ROBIN　　Cape York Pen. to NE Queensland

[1] For spelling see Opinion 1606, I.C.Z.N. (1990) (1176).
[2] For correction of spelling see David & Gosselin (2002) (614).
[3] Includes *novaeguineae* see Cracraft (1992) (581).
[4] Includes *salvadorii* see Cracraft (1992) (581).
[5] Accepted by Cracraft (1992) (581) but possibly a name given to intergrading birds.
[6] Includes *pulchra* see Mees (1965) (1515), but see also Cracraft (1992) (581).
[7] Includes *ampla* see Cracraft (1992) (581).
[8] Previously Eopsaltriidae, but see Bock (1994) (173).
[9] Treated in the genus *Heteromyias* in last Edition, but that is submerged, see Schodde & Mason (1999: 351) (2189). The name Grey-headed Robin is better reserved for the Australian form if treated as a species.
[10] For correction of spelling see David & Gosselin (2002) (614).
[11] For the replacement of this taxon within the species *albispecularis* see Schodde & Mason (1999: 352) (2189).

○ *Poecilodryas cerviniventris* (Gould, 1858) Buff-sided Robin[1] N Western Australia, mid N Northern Territory
(S Arnhem Land), NW Queensland

○ *Poecilodryas placens* (E.P. Ramsay, 1879) Olive-yellow Robin C New Guinea cordillera from Vogelkop to SE New Guinea

○ *Poecilodryas albonotata* Black-bibbed Robin
— *P.a.albonotata* (Salvadori, 1875) NW New Guinea (Vogelkop)
— *P.a.griseiventris* Rothschild & Hartert, 1913 SW to C New Guinea
— *P.a.correcta* E. Hartert, 1930 SE New Guinea (Huon Pen.)

PENEOTHELLO Mathews, 1920 F[2]
○ *Peneothello sigillata* White-winged Robin
— *P.s.saruwagedi* (Mayr, 1931) NE New Guinea (Huon Pen.)
— *P.s.quadrimaculata* (van Oort, 1910) WC New Guinea
— *P.s.hagenensis* Mayr & Gilliard, 1952 EC New Guinea
— *P.s.sigillata* (De Vis, 1890) SE New Guinea

○ *Peneothello cryptoleuca* Smoky Robin
— *P.c.cryptoleuca* (E. Hartert, 1930) NW New Guinea (Vogelkop)
— *P.c.albidior* (Rothschild, 1931) WC New Guinea
— *P.c.maxima* Diamond, 1985 Kumawa Mts. (W New Guinea) (701)

○ *Peneothello cyanus* Blue-grey Robin
— *P.c.cyanus* (Salvadori, 1874) NW New Guinea (Vogelkop)
— *P.c.atricapilla* (Hartert & Paludan, 1934) NC and WC New Guinea
— *P.c.subcyanea* (De Vis, 1897) SE to EC New Guinea

○ *Peneothello bimaculata* White-rumped Robin
— *P.b.bimaculata* (Salvadori, 1874) Yapen I., NW to southern SE New Guinea
— *P.b.vicaria* (De Vis, 1892) Northern SE New Guinea

PENEOENANTHE Mathews, 1920 F
○ *Peneoenanthe pulverulenta* Mangrove Robin
— *P.p.pulverulenta* (Bonaparte, 1850) NC and S New Guinea
— *P.p.leucura* (Gould, 1869) Aru Is., N Queensland
— *P.p.alligator* (Mathews, 1912) N Northern Territory (Arnhem Land)
— *P.p.cinereiceps* (E. Hartert, 1905) WC to N Western Australia

TREGELLASIA Mathews, 1912 F
○ *Tregellasia leucops* White-faced Robin
— *T.l.leucops* (Salvadori, 1876) NW New Guinea (Vogelkop)
— *T.l.mayri* (E. Hartert, 1930) W New Guinea (neck of the Vogelkop)
— *T.l.heurni* (E. Hartert, 1932) North of WC New Guinea
— *T.l.nigroorbitalis* (Rothschild & Hartert, 1913) South of WC New Guinea
— *T.l.nigriceps* (Neumann, 1922) C New Guinea
— *T.l.melanogenys* (A.B. Meyer, 1894) Western NC to NE New Guinea
— *T.l.wahgiensis* Mayr & Gilliard, 1952 EC New Guinea
— *T.l.albifacies* (Sharpe, 1882) SE New Guinea
— *T.l.auricularis* (Mayr & Rand, 1935) SC New Guinea
— *T.l.albigularis* (Rothschild & Hartert, 1907) NE Cape York Pen.

○ *Tregellasia capito* Pale-yellow Robin
— *T.c.nana* (E.P. Ramsay, 1878) NE Queensland
— *T.c.capito* (Gould, 1854) SE Queensland, NE New South Wales

EOPSALTRIA Swainson, 1832 F
● *Eopsaltria australis* Eastern Yellow Robin
— *E.a.chrysorrhos* Gould, 1869[3][4] EC to NE Australia
✓ *E.a.australis* (Shaw, 1790)[5] SE Australia

---

[1] For reasons to separate this from the preceding species see Schodde & Mason (1999: 354) (2189).
[2] Gender addressed by David & Gosselin (2002) (614), multiple changes occur in specific names.
[3] For correction of spelling see David & Gosselin (2002) (613).
[4] Includes *magnirostris* see Schodde & Mason (1999: 362) (2189); the names *austina* and *coomooboolaroo* relate to intergrading populations.
[5] Includes *viridior* see Schodde & Mason (1999: 362) (2189). Credited to White by Mayr (1986) (1488), but Schodde & Mason (1997) (2188) have shown that scientific names in this travel journal were given by Shaw.

○ *Eopsaltria griseogularis* WESTERN YELLOW ROBIN[1]
   — *E.g.griseogularis* Gould, 1838   SW Western Australia
   — *E.g.rosinae* (Mathews, 1912)   S Western Australia, SC South Australia (Eyre Pen.)

○ *Eopsaltria georgiana* (Quoy & Gaimard, 1830) WHITE-BREASTED ROBIN
                                      SW Western Australia

○ *Eopsaltria flaviventris* Sharpe, 1903 YELLOW-BELLIED ROBIN   New Caledonia

MELANODRYAS Gould, 1865[2] F
○ *Melanodryas cucullata* HOODED ROBIN
   — *M.c.melvillensis* (Zietz, 1914)[3]   Melville I.
   — *M.c.picata* Gould, 1865   N to EC Australia
   — *M.c.cucullata* (Latham, 1802)[4]   SE Australia
   — *M.c.westralensis* (Mathews, 1912)[5]   SW and WC to SC Australia

○ *Melanodryas vittata* DUSKY ROBIN
   — *M.v.vittata* (Quoy & Gaimard, 1830)   Tasmania, Flinders I.
   — *M.v.kingi* (Mathews, 1914)[6]   King I.

PACHYCEPHALOPSIS Salvadori, 1879 F
○ *Pachycephalopsis hattamensis* GREEN-BACKED ROBIN
   — *P.h.hattamensis* (A.B. Meyer, 1874)   NW New Guinea (Vogelkop)
   — *P.h.ernesti* E. Hartert, 1930   NW New Guinea (neck of Vogelkop)
   — *P.h.axillaris* Mayr, 1931   WC New Guinea
   — *P.h.insularis* Diamond, 1985   Yapen I. (Geelvink Bay) (2045)
   — *P.h.lecroyae* Boles, 1989   SC New Guinea (175)

○ *Pachycephalopsis poliosoma* WHITE-EYED ROBIN
   — *P.p.albigularis* (Rothschild, 1931)   WC New Guinea (Weyland Mts.)
   — *P.p.approximans* (Ogilvie-Grant, 1911)   S slope WC New Guinea
   — *P.p.idenburgi* Rand, 1940   N slope WC New Guinea
   — *P.p.balim* Rand, 1940   WC New Guinea
   — *P.p.hunsteini* (Neumann, 1922)   EC New Guinea
   — *P.p.hypopolia* Salvadori, 1899   NE New Guinea (Huon Pen.)
   — *P.p.poliosoma* Sharpe, 1882   SE New Guinea

MONACHELLA Salvadori, 1874 F
○ *Monachella muelleriana* TORRENT FLYCATCHER
   — *M.m.muelleriana* (Schlegel, 1871)   New Guinea
   — *M.m.coultasi* Mayr, 1934   New Britain

MICROECA Gould, 1841[7] F
○ *Microeca papuana* A.B. Meyer, 1875 MONTANE FLYCATCHER   Montane New Guinea

○ *Microeca griseoceps* YELLOW-LEGGED FLYCATCHER
   — *M.g.occidentalis* Rothschild & Hartert, 1903   NC to NW New Guinea
   — *M.g.griseoceps* De Vis, 1894   SE New Guinea
   — *M.g.kempi* (Mathews, 1913)[8]   SC New Guinea (?), NE Cape York Pen.

○ *Microeca flavovirescens* OLIVE-YELLOW FLYCATCHER
   — *M.f.flavovirescens* G.R. Gray, 1858   Aru Is., extreme SC New Guinea
   — *M.f.cuicui* (De Vis, 1897)   W, N and E New Guinea

---

[1] For a discussion on the merits of separating this from *E. australis* see Schodde & Mason (1999: 365-366) (2189). As regards the distribution, the map captions in Schodde & Mason (1999: 365) (2189) were transposed.
[2] Treated in the genus *Petroica* in last Edition. For recognition of *Melanodryas* see Schodde & Mason (1999: 369) (2189).
[3] For recognition see Schodde & Mason (1999: 368) (2189).
[4] For reasons to use 1802 not 1801 see Browing & Monroe (1991) (258).
[5] For recognition see Schodde & Mason (1999: 369) (2189).
[6] For recognition see Schodde & Mason (1999: 371) (2189).
[7] Some may prefer the generic vernacular Flyrobin, an abbreviation of Flycatcher-Robin; for a brief name, as opposed to a traditional and less accurate one, I would prefer Robin! (Ed.)
[8] For recognition see Schodde & Mason (1999: 373) (2189).

○ *Microeca flavigaster* LEMON-BELLIED FLYCATCHER
    — *M.f.laeta* Salvadori, 1878   N New Guinea
    — *M.f.tarara* Rand, 1940   SC New Guinea
    — *M.f.flavissima* Schodde & Mason, 1999[1]   Cape York Pen. (2189)
    — *M.f.laetissima* Rothschild, 1916   EC Queensland
    — *M.f.flavigaster* Gould, 1843   N Northern Territory (Arnhem Land)
    — *M.f.tormenti* Mathews, 1916   N Western Australia (Kimberley)

○ *Microeca hemixantha* P.L. Sclater, 1883   GOLD-BELLIED FLYCATCHER   Tanimbar Is.

○ *Microeca fascinans*[2] JACKY WINTER
    — *M.f.zimmeri* Mayr & Rand, 1935   New Guinea
    — *M.f.pallida* De Vis, 1884   N Western Australia to C Queensland
    — *M.f.fascinans* (Latham, 1802)[3]   E and SE Australia
    — *M.f.assimilis* Gould, 1841   SW to SC Australia

EUGERYGONE Finsch, 1901   F
○ *Eugerygone rubra* GARNET ROBIN
    — *E.r.rubra* (Sharpe, 1879)   NW New Guinea (Vogelkop Mts.)
    — *E.r.saturatior* Mayr, 1931   Montane WC to SE New Guinea

PETROICA Swainson, 1830   F
○ *Petroica rosea* Gould, 1840   ROSE ROBIN   SE Australia

○ *Petroica rodinogaster* PINK ROBIN
    — *P.r.inexpectata* Mathews, 1912[4]   SE New South Wales to S Victoria
    — *P.r.rodinogaster* (Drapiez, 1820)   Tasmania, Bass Strait Is.

○ *Petroica archboldi* Rand, 1940   SNOW MOUNTAIN ROBIN   WC New Guinea

○ *Petroica bivittata* CLOUD-FOREST ROBIN
    — *P.b.caudata* Rand, 1940   WC New Guinea
    — *P.b.bivittata* De Vis, 1897   SE and EC New Guinea

○ *Petroica phoenicea* Gould, 1837   FLAME ROBIN   SE Australia, Tasmania, Bass Strait Is.

● *Petroica multicolor* PACIFIC ROBIN
    — *P.m.multicolor* (J.F. Gmelin, 1789)   Norfolk I.
    — *P.m.septentrionalis* Mayr, 1934   Bougainville I. (N Solomons)
    — *P.m.kulambangrae* Mayr, 1934   Kolombangara I. (C Solomons)
    — *P.m.dennisi* Cain & Galbraith, 1955   Guadalcanal I. (S Solomons)
    — *P.m.polymorpha* Mayr, 1934   San Cristobal I. (SE Solomons)
    — *P.m.soror* Mayr, 1934   Vanua Leva (Banks Is.)
    — *P.m.ambrynensis* Sharpe, 1900   N and C Vanuatu
    — *P.m.feminina* Mayr, 1934   Efate, Mai Is. (Vanuatu)
    — *P.m.cognata* Mayr, 1938   Erromanga I. (Vanuatu)
    — *P.m.similis* G.R. Gray, 1860   Tanna I. (Vanuatu)
    — *P.m.kleinschmidti* Finsch, 1876   Fiji Is.
    — *P.m.taveunensis* Holyoak, 1979   Taveuni I. (Fiji)
    — *P.m.becki* Mayr, 1934   Kandavu I. (Fiji)
    — *P.m.pusilla* Peale, 1848   W Samoa

○ *Petroica boodang* SCARLET ROBIN[5]
    — *P.b.leggii* Sharpe, 1879[6]   Flinders I., E Tasmania
    — *P.b.campbelli* Sharpe, 1898   SW Western Australia
    — *P.b.boodang* (Lesson, 1838)   SE Australia, Kangaroo I.

○ *Petroica goodenovii* (Vigors & Horsfield, 1827)   RED-CAPPED ROBIN   Inland W, C, SC and E Australia

---

[1] The name *terraereginae* attaches to a type that is intermediate between the Cape York population and that of EC Queensland and is considered indeterminate. R. Schodde (*in litt.*).
[2] For reasons to replace the name *M. leucophaea* see Schodde (1992) (2178).
[3] For reasons to use 1802 not 1801 see Browing & Monroe (1991) (258). The name *barcoo* relates to the intergrading population in SE Australia.
[4] For recognition see Schodde & Mason (1999: 382) (2189).
[5] For reasons to separate this from *P. multicolor* see Schodde & Mason (1999: 388) (2189).
[6] For recognition see Schodde & Mason (1999: 385) (2189).

● *Petroica macrocephala* NEW ZEALAND TIT/TOMTIT
    — *P.m.toitoi* (Lesson, 1828)   North Island (New Zealand)
    — *P.m.macrocephala* (J.F. Gmelin, 1789)   South Island (New Zealand)
    — *P.m.marrineri* (Mathews & Iredale, 1913)   Auckland Is.
    — *P.m.chathamensis* C.A. Fleming, 1950   Chatham I.
    — *P.m.dannefaerdi* (Rothschild, 1894)   Snares I.

● *Petroica australis* NEW ZEALAND ROBIN
    — *P.a.longipes* (Garnot, 1827)   North Island (New Zealand)
    ⊿ *P.a.australis* (Sparrman, 1788)   South Island (New Zealand)
    — *P.a.rakiura* C.A. Fleming, 1950   Stewart I.

○ *Petroica traversi* (Buller, 1872) CHATHAM ISLAND ROBIN   Chatham I.

DRYMODES Gould, 1840[1] F[2]
○ *Drymodes superciliaris* NORTHERN SCRUB ROBIN
    — *D.s.beccarii* (Salvadori, 1876)   NW New Guinea
    — *D.s.nigriceps* Rand, 1940   WC and NC New Guinea
    — *D.s.brevirostris* (De Vis, 1897)   Aru Is., S to SE New Guinea
    — *D.s.superciliaris* Gould, 1850[3]   N Cape York Pen.

○ *Drymodes brunneopygia* Gould, 1841 SOUTHERN SCRUB ROBIN[4]   SC and SW New South Wales, NW Victoria, S South
                                       Australia, SW Western Australia

AMALOCICHLA De Vis, 1892[5] F
○ *Amalocichla sclateriana* GREATER GROUND ROBIN
    — *A.s.occidentalis* Rand, 1940   WC New Guinea
    — *A.s.sclateriana* De Vis, 1892   SE New Guinea

○ *Amalocichla incerta* LESSER GROUND ROBIN
    — *A.i.incerta* (Salvadori, 1876)   Arfak Mts. (W New Guinea)
    — *A.i.olivascentior* E. Hartert, 1930   WC New Guinea
    — *A.i.brevicauda* (De Vis, 1894)   E and SE New Guinea

PICATHARTIDAE BALD CROWS (1: 2)

PICATHARTES Lesson, 1828 M
○ *Picathartes gymnocephalus* (Temminck, 1825) WHITE-NECKED PICATHARTES
                                       Guinea to Ghana

○ *Picathartes oreas* Reichenow, 1899 GREY-NECKED PICATHARTES   SE Nigeria, Cameroon, NE Gabon

BOMBYCILLIDAE WAXWINGS (5: 8)

BOMBYCILLINAE

BOMBYCILLA Vieillot, 1808 F
● *Bombycilla garrulus* BOHEMIAN WAXWING
    — *B.g.garrulus* (Linnaeus, 1758)[6]   N Europe, N Asia >> W Europe to C and E Asia
    ⊿ *B.g.pallidiceps* Reichenow, 1908   W Canada, N and W USA

○ *Bombycilla japonica* (P.F. Siebold, 1824) JAPANESE WAXWING   SE Siberia to Russian Far East >> E China, Korea, Japan

● *Bombycilla cedrorum* Vieillot, 1808 CEDAR WAXWING[7]   Canada, USA, Mexico, Central America, Colombia,
                                       Venezuela

---

[1] For the transfer of this genus from the thrushes, in last Edition and in Ripley (1964) (2045) see Sibley & Ahlquist (1982) (2259).
[2] For correct gender and spellings see David & Gosselin (2002) (614).
[3] The purported race *colcloughi* is of doubtful provenance see Schodde & Mason (1999: 391) (2189) and not accepted.
[4] For monotypic treatment see Schodde & Mason (1999: 392) (2189).
[5] We place *Amalocichla* in the Petroicidae following Sibley & Monroe (1990) (2262).
[6] Includes *centralasiae* see Stepanyan (1990: 418) (2326).
[7] Includes *aquilonia* Burleigh, 1963 (276), placed in synonymy by Browning (1990) (251) and *larifuga* Burleigh, 1963 (276), doubtfully recognised by Browning (1990) (251), so the species is here treated as monotypic.

PTILOGONATINAE

PHAINOPTILA Salvin, 1877 F
○ *Phainoptila melanoxantha* Black-and-yellow Silky Flycatcher
    — *P.m.melanoxantha* Salvin, 1877   SE Costa Rica, NW Panama
    — *P.m.parkeri* Barrantes & Sánchez, 2000   NW Costa Rica (99)

PTILOGONYS Swainson, 1827 M
○ *Ptilogonys cinereus* Grey Silky Flycatcher
    — *P.c.otofuscus* R.T. Moore, 1935   NW Mexico
    — *P.c.cinereus* Swainson, 1827[1]   C and E Mexico
    — *P.c.pallescens* Griscom, 1934   SW Mexico
    — *P.c.molybdophanes* Ridgway, 1887   S Mexico, W Guatemala

○ *Ptilogonys caudatus* Cabanis, 1861 Long-tailed Silky Flycatcher   Costa Rica, W Panama

PHAINOPEPLA S.F. Baird, 1858 F
◉ *Phainopepla nitens* Phainopepla
    ✓ *P.n.lepida* Van Tyne, 1925   SW USA, NW Mexico
    — *P.n.nitens* (Swainson, 1838)   S Texas, NC Mexico

HYPOCOLIINAE

HYPOCOLIUS Bonaparte, 1850 M
○ *Hypocolius ampelinus* Bonaparte, 1850[2] Grey Hypocolius   Iraq, S Iran, Afghanistan, S Turkmenistan, ? W
                                                Pakistan >> SW Asia, Arabia, NE Africa

DULIDAE PALMCHAT (1: 1)

DULUS Vieillot, 1816 M
○ *Dulus dominicus* (Linnaeus, 1766) Palmchat   Gonave I, Hispaniola

PARIDAE TITS, CHICKADEES (3: 54)

PARUS Linnaeus, 1758 M
○ *Parus funereus* Dusky Tit
    — *P.f.funereus* (J. & E. Verreaux, 1885)   Guinea to Zaire and W Kenya
    — *P.f.gabela* Traylor, 1961   W Angola

○ *Parus niger* Southern Black Tit
    — *P.n.ravidus* Clancey, 1964   E Zambia and Zimbabwe to Mozambique
    — *P.n.xanthostomus* Shelley, 1892   S Angola, S and C Zambia, N Namibia to Transvaal
    — *P.n.niger* Vieillot, 1818   S Mozambique to E Cape Province

○ *Parus leucomelas* Northern Black Tit[3]
    — *P.l.guineensis* Shelley, 1900   Senegal to S Sudan, Uganda, W Ethiopia
    — *P.l.leucomelas* Rüppell, 1840   C and SE Ethiopia
    — *P.l.insignis* Cabanis, 1880   S Congo to Zambia and Tanzania

○ *Parus carpi* Macdonald & Hall, 1957 Carp's Black Tit   SW Angola, N Namibia

○ *Parus albiventris* Shelley, 1881 White-bellied Black Tit   E Nigeria, Cameroon, S Sudan to Tanzania

○ *Parus leuconotus* Guérin-Méneville, 1843 White-backed Black Tit   Ethiopia

○ *Parus fasciiventer* Stripe-breasted Tit
    — *P.f.fasciiventer* Reichenow, 1893   E Zaire, Rwanda, Burundi, SW Uganda
    — *P.f.tanganjicae* Reichenow, 1909   Itombwe Mts. (E Zaire)
    — *P.f.kaboboensis* Prigogine, 1956   Mt. Kabobo (E Zaire)

---

[1] Includes *schistaceus* A.R. Phillips, 1966 (1871) see Binford (1989) (150).
[2] Although the generic name was dated 1850 the specific name was dated 1851 by Greenway (1960: 373) (1005), but see Zimmer (1926) (2707).
[3] We follow Fry (2000: 98) (892) in retaining *guineensis* as part of *leucomelas*.

○ *Parus fringillinus* Fischer & Reichenow, 1884 RED-THROATED TIT   S Kenya, N and C Tanzania

○ *Parus rufiventris* RUFOUS-BELLIED TIT[1]
     — *P.r.rufiventris* Bocage, 1877   W and S Zaire to C Angola and C Zambia
     — *P.r.masukuensis* Shelley, 1900   SE Zaire, E Zambia, Malawi
     — *P.r.diligens* Clancey, 1979   NE Namibia, S Angola, SW Zambia (473)
     — *P.r.pallidiventris* Reichenow, 1885   Tanzania, SE Malawi, N Mozambique
     — *P.r.stenotopicus* Clancey, 1989   E Zimbabwe, W Mozambique (515)

○ *Parus thruppi* NORTHERN GREY TIT[2]
     — *P.t.thruppi* Shelley, 1885   Ethiopia, N Somalia
     — *P.t.barakae* Jackson, 1899   E Uganda, Kenya, NE Tanzania

○ *Parus afer* SOUTHERN GREY TIT
     — *P.a.arens* Clancey, 1963   Lesotho, E Cape Province
     — *P.a.afer* J.F. Gmelin, 1789   SW Namibia, W and S Cape Province

○ *Parus cinerascens* ASHY TIT
     — *P.c.benguelae* Hall & Taylor, 1959[3]   SW Angola, NW Namibia
     — *P.c.cinerascens* Vieillot, 1818   Namibia, Botswana, Zimbabwe, N Cape Province
     — *P.c.orphnus* Clancey, 1958   Transvaal, Orange Free State

○ *Parus griseiventris* Reichenow, 1882 MIOMBO GREY TIT[4]   W Tanzania to Zimbabwe and Angola

○ *Parus nuchalis* Jerdon, 1845 WHITE-NAPED TIT   W and SC India

● *Parus major* GREAT TIT[5]
     — *P.m.newtoni* Prazák, 1894   British Isles
     ✓ *P.m.major* Linnaeus, 1758[6]   Europe, Asia Minor, N and E Kazakhstan, S Siberia and N Mongolia to mid-Amur valley

     — *P.m.excelsus* Buvry, 1857   NW Africa
     ✓ *P.m.corsus* O. Kleinschmidt, 1903   Portugal, S Spain, Corsica
     — *P.m.mallorcae* von Jordans, 1913[7]   Balearic Is.
     — *P.m.ecki* von Jordans, 1970[8]   Sardinia (2540)
     — *P.m.niethammeri* von Jordans, 1970[9]   Crete (2540)
     — *P.m.aphrodite* Madarász, 1901   S Italy, S Greece, Aegean Is., Cyprus
     — *P.m.terraesanctae* E. Hartert, 1910   Lebanon, Israel, Jordan, Syria
     — *P.m.karelini* Zarudny, 1910   SE Azerbaijan, NW Iran
     — *P.m.blanfordi* Prazák, 1894   NC and SW Iran
     — *P.m.intermedius* Zarudny, 1890   NE Iran
     — *P.m.caschmirensis* E. Hartert, 1905   N Pakistan, NW Himalayas
     — *P.m.decolorans* Koelz, 1939   NE Afghanistan, NW Pakistan
     — *P.m.ziaratensis* Whistler, 1929   C and S Afghanistan, W Pakistan
     — *P.m.mahrattarum* E. Hartert, 1905   SW India, Sri Lanka
     — *P.m.stupae* Koelz, 1939   W, C and SE India
     — *P.m.nipalensis* Hodgson, 1838   NW and N India, Nepal, Assam (except E), W and C Burma

     — *P.m.vauriei* Ripley, 1950   E Assam
     — *P.m.templorum* Meyer de Schauensee, 1946   W and C Thailand, S Indochina
     — *P.m.commixtus* Swinhoe, 1868   S and E China, N Vietnam, Taiwan
     — *P.m.nigriloris* Hellmayr, 1900   Ishigaki, Iriomote (S Ryukyu Is.)
     — *P.m.hainanus* E. Hartert, 1905   Hainan
     — *P.m.ambiguus* (Raffles, 1822)   Malay Pen., Sumatra
     — *P.m.sarawacensis* Slater, 1885   W Sarawak

[1] We follow Wiggins (2000: 96) (2663) in treating *pallidiventris* within a broad species *rufiventris*.
[2] Now generally treated as a species see Wiggins (2000: 86) (2663).
[3] Erroneously attached to *afer* in last Edition.
[4] Includes *parvirostris* see White (1963: 45) (2648).
[5] Stepanyan (1990) (2326) treated this as two species *P. major* and *P. minor*. We await a definitive review of the entire complex. We are not sure whether the *cinereus* group should be treated at specific level too, as we do not understand where some taxa should be attached. Our range statements show some overlap. If we recognised three species we should attach the taxa *newtoni* through *blanfordi* to *major*, *caschmirensis* to *sarawakensis* to *cinereus*, and the rest to *minor*.
[6] Includes *kapustini* see Stepanyan (1990: 579) (2326) and *bargaensis* see Vaurie (1959: 510) (2502).
[7] For recognition see Cramp *et al.* (1993: 255) (594).
[8] For recognition see Cramp *et al.* (1993: 255) (594).
[9] For recognition see Cramp *et al.* (1993: 255) (594).

— *P.m.cinereus* Vieillot, 1818	Java, Lesser Sundas
— *P.m.tibetanus* E. Hartert, 1905	SE and E Xizang, N Burma, E Qinghai, W Sichuan
— *P.m.nubicolus* Meyer de Schauensee, 1946	E Burma, N Thailand, NW Indochina
— *P.m.minor* Temminck & Schlegel, 1848	NC and NE China, Russian Far East, S Sakhalin, Korea, S Kuril Is., Japan south to N Kyushu
— *P.m.kagoshimae* Taka-Tsukasa, 1919	S Kyushu, Goto Is.
— *P.m.dageletensis* Kuroda & Mori, 1920	Ullung-do I. (= Dagelet I.) (Korea)
— *P.m.amamiensis* O. Kleinschmidt, 1922	Amami O-shima, Tokunoshima (N Ryukyu Is.)
— *P.m.okinawae* E. Hartert, 1905	Okinawa (C Ryukyu Is.)

○ **Parus bokharensis** TURKESTAN TIT

— *P.b.bokharensis* M.H.K. Lichtenstein, 1823	Plains of Turkmenistan and N Afghanistan to Uzbekistan and SC Kazakhstan
— *P.b.turkestanicus* Zarudny & Loudon, 1905	SE Kazakhstan, N Xinjiang, extreme SW Mongolia
— *P.b.ferghanensis* Buturlin, 1912[1]	Pamir-Alai Mts. to W Tien Shan

○ **Parus monticolus** GREEN-BACKED TIT

— *P.m.monticolus* Vigors, 1831	W and C Himalayas (N Pakistan to W Nepal)
— *P.m.yunnanensis* La Touche, 1922	E Himalayas, W, N and NE Burma, C China north to Gansu, NW Vietnam
— *P.m.legendrei* Delacour, 1927	Langbian (SC Vietnam)
— *P.m.insperatus* Swinhoe, 1866	Taiwan

○ **Parus venustulus** Swinhoe, 1870 YELLOW-BELLIED TIT — W Sichuan, Gansu and SE Nei Mongol to Hebei, Zhejiang and N Guangdong

○ **Parus elegans** ELEGANT TIT

— *P.e.edithae* (McGregor, 1907)	Calayan, Camiguin Norte Is.
— *P.e.montigenus* (Hachisuka, 1930)	NW Luzon
— *P.e.gilliardi* Parkes, 1958	Bataan Pen. (WC Luzon)
— *P.e.elegans* Lesson, 1831	EC and S Luzon, Panay, Mindoro, Catanduanes
— *P.e.visayanus* (Hachisuka, 1930)	Cebu
— *P.e.albescens* (McGregor, 1907)	Guimaras, Masbate, Negros
— *P.e.mindanensis* (Mearns, 1905)	Mindanao, Samar, Leyte
— *P.e.suluensis* (Mearns, 1916)	Sulu Arch. (not Bongao)
— *P.e.bongaoensis* Parkes, 1958	Bongao I.

○ **Parus amabilis** Sharpe, 1877 PALAWAN TIT — Balabac, Palawan

○ **Parus xanthogenys** BLACK-LORED TIT

— *P.x.xanthogenys* Vigors, 1831	NW and C Himalayas to E Nepal
— *P.x.aplonotus* Blyth, 1847	C India
— *P.x.travancoreensis* (Whistler & Kinnear, 1932)	W Ghats (S India)

○ **Parus spilonotus** YELLOW-CHEEKED TIT

— *P.s.spilonotus* Bonaparte, 1850	E Himalayas (E Nepal to E Assam and SE Xizang)
— *P.s.subviridis* Blyth, 1855	S Assam, Burma, W Yunnan, N Thailand
— *P.s.rex* David, 1874	S and E Yunnan and S Sichuan to SE China and N and C Indochina
— *P.s.basileus* (Delacour, 1932)	S Laos and SC Vietnam

○ **Parus holsti** Seebohm, 1894 YELLOW TIT — Taiwan

◑ **Parus caeruleus** BLUE TIT

— *P.c.obscurus* Prazák, 1894	British Isles
✓ *P.c.caeruleus* Linnaeus, 1758	N, C, and E Europe south to N Spain, Italy, N Greece, W and N Asia Minor
✓ *P.c.ogliastrae* E. Hartert, 1905	S Iberia, Corsica, Sardinia
— *P.c.balearicus* von Jordans, 1913	Mallorca I.
— *P.c.calamensis* Parrot, 1908[2]	S Greece, Cyclades, Crete, Rhodes
— *P.c.orientalis* (Zarudny & Loudon, 1905)	SE European Russia
— *P.c.satunini* (Zarudny, 1908)	Crimea, Caucasus area, E Turkey, NW Iran

---

[1] For recognition see Stepanyan (1990: 582) (2326).
[2] For recognition see Cramp *et al.* (1993: 225) (594).

__ *P.c.raddei* Zarudny, 1908	N Iran
__ *P.c.persicus* Blanford, 1873	SW Iran
__ *P.c.ultramarinus* Bonaparte, 1841	NW Africa
__ *P.c.cyrenaicae* E. Hartert, 1922	NE Libya
__ *P.c.ombriosus* Meade-Waldo, 1890	Hierro I. (W Canary Is.)
__ *P.c.palmensis* Meade-Waldo, 1889	Palma I. (W Canary Is.)
__ *P.c.teneriffae* Lesson, 1831	Gran Canaria, Tenerife Is. (C Canary Is.)
__ *P.c.degener* E. Hartert, 1901	Fuerteventura, Lanzarote Is. (E Canary Is.)

○ ***Parus cyanus*** AZURE TIT

__ *P.c.cyanus* Pallas, 1770	Belarus, W and C European Russia to C Ural Mts.
__ *P.c.hyperriphaeus* Dementiev & Heptner, 1932[1]	SW Ural Mts., SW Siberia, N Kazakhstan
__ *P.c.yenisseensis* Buturlin, 1911	Altai Mts. and Mongolia to NE China and Russian Far East
__ *P.c.tianschanicus* (Menzbier, 1884)	C and E Tien Shan, W and N Tarim Basin (Xinjiang)
__ *P.c.koktalensis* (Portenko, 1954)[2]	SE Kazakhstan

○ ***Parus flavipectus*** YELLOW-BREASTED TIT[3]

__ *P.f.flavipectus* Severtsov, 1873	W Tien Shan to Pamir-Alai Mts. and C and E Afghanistan >> NE Iran, NW Pakistan
__ *P.f.berezowskii* (Pleske, 1893)[4]	NE Qinghai (NC China)

○ ***Parus rubidiventris*** RUFOUS-VENTED TIT

__ *P.r.rubidiventris* Blyth, 1847	C Himalayas (Kumaun to Nepal)
__ *P.r.beavani* (Jerdon, 1863)	E Himalayas, NE Burma and WC China N to S Shaanxi
__ *P.r.saramatii* Ripley, 1961	Mt. Saramati (NW Burma)

○ ***Parus rufonuchalis*** Blyth, 1849 RUFOUS-NAPED TIT — SW Xinjiang and Tien Shan to C Afghanistan, W and N Pakistan and NW Himalayas

○ ***Parus melanolophus*** Vigors, 1831 SPOT-WINGED TIT[5] — E Afghanistan and NW Pakistan to NW Himalayas

● ***Parus ater*** COAL TIT

__ *P.a.britannicus* Sharpe & Dresser, 1871	Britain
__ *P.a.hibernicus* Ogilvie-Grant, 1910	Ireland
⊥ *P.a.ater* Linnaeus, 1758	N, C and SE Europe, W and S Asia Minor, Siberia east to Kamchatka and south to Altai, NE Mongolia, NE China, Korea and Hokkaido
__ *P.a.vieirae* Nicholson, 1906	Spain, Portugal
__ *P.a.sardus* O. Kleinschmidt, 1903	Corsica, Sardinia
__ *P.a.atlas* Meade-Waldo, 1901	Morocco
__ *P.a.ledouci* Malherbe, 1845	N Algeria, NW Tunisia
__ *P.a.cypriotes* Dresser, 1888	Cyprus
__ *P.a.moltchanovi* Menzbier, 1903	S Crimea
__ *P.a.michalowskii* Bogdanov, 1879	Caucasus (excl. SW), C and E Transcaucasia
__ *P.a.derjugini* (Zarudny & Loudon, 1903)	NE Turkey to SW Caucasus
__ *P.a.gaddi* (Zarudny, 1911)	SE Azerbaijan, N Iran
__ *P.a.chorassanicus* Zarudny & Bilkevitsch, 1911	NE Iran, SW Turkmenistan
__ *P.a.phaeonotus* Blanford, 1873	SW Iran
__ *P.a.rufipectus* Severtsov, 1873	C and E Tien Shan (C Asia)
__ *P.a.martensi* Eck, 1998[6]	Kali Gandaki area (Nepal) (791)
__ *P.a.aemodius* Blyth, 1844	E Himalayas, C China, N and NE Burma
__ *P.a.pekinensis* David, 1870	N China (Shanxi to Hebei and Shandong)
__ *P.a.insularis* Hellmayr, 1902	S Kurils and Japan
__ *P.a.kuatunensis* La Touche, 1923	SE China
__ *P.a.ptilosus* Ogilvie-Grant, 1912	Taiwan

---

[1] For recognition see Stepanyan (1990: 577) (2326).
[2] Not *kotkalensis* as in Snow (1967: 117) (2292).
[3] Treated as a separate species following Stepanyan (1990: 575) (2326). Birds called *carruthersi* are included in *flavipectus*.
[4] CSR believes this is better treated as a separate species, but sufficient detailed support for this has yet to be published.
[5] For arguments in favour of treating this as a form of *P. ater* see Martens & Eck (1995) (1443).
[6] We list this very tentatively. We do not yet treat *melanolophus* (the adjoining population to the west) as part of this species.

○ *Parus varius* VARIED TIT
   __ *P.v.varius* Temminck & Schlegel, 1848   S Kurils, Japan, Korea, Liaoning (NE China)
   __ *P.v.sunsunpi* N. Kuroda, Sr., 1919   Tanegashima, Yakushima (off S Kyushu, S Japan)
   __ *P.v.namiyei* N. Kuroda, Sr., 1918   N Izu Is.
   __ *P.v.owstoni* Ijima, 1893   S Izu Is.
   __ *P.v.amamii* (N. Kuroda, Sr., 1922)   Amami to Okinawa Is. (N Ryukyu Is.)
   __ †?*P.v.orii* (N. Kuroda, Sr., 1923)   Daito-jima Is. (Borodino Is.)
   __ *P.v.olivaceus* (N. Kuroda, Sr., 1923)   Iriomote I. (S Ryukyu Is.)
   __ *P.v.castaneoventris* Gould, 1862   Taiwan

○ *Parus semilarvatus* WHITE-FRONTED TIT
   __ *P.s.snowi* Parkes, 1971   NE Luzon (1763)
   __ *P.s.semilarvatus* (Salvadori, 1865)   C and S Luzon
   __ *P.s.nehrkorni* (W.H. Blasius, 1890)   Mindanao

● *Parus palustris* MARSH TIT
   __ *P.p.dresseri* Stejneger, 1886[1]   C and S England, Wales, W France
   __ *P.p.palustris* Linnaeus, 1758   Most of Europe from Spain to Scandinavia east to C
               Poland, W Balkans and Greece
   __ *P.p.italicus* Tschusi & Hellmayr, 1900[2]   French Alps, mainland Italy, Sicily
   __ *P.p.stagnatilis* C.L. Brehm, 1855[3]   E Europe to S Urals and NW Turkey
   __ *P.p.kabardensis* Buturlin, 1929[4]   Caucasus area, NE Turkey?
   __ *P.p.brevirostris* (Taczanowski, 1872)   SC and E Siberia south to N Mongolia, Transbaikalia,
               W and N Manchuria and extreme N Korea
   __ *P.p.ernsti* Yamashina, 1933[5]   Sakhalin
   __ *P.p.hensoni* Stejneger, 1892   S Kuril Is., N Japan
   __ *P.p.jeholicus* O. Kleinschmidt, 1922[6]   N Hebei, Liaoning, N Korea
   __ *P.p.hellmayri* Bianchi, 1902   N of E China, S Korea

○ *Parus hypermelaenus* (Berezowski & Bianchi, 1891) BLACK-BIBBED TIT[7]
               C China south to N and W Burma

● *Parus lugubris* SOMBRE TIT[8]
   __ *P.l.lugubris* Temminck, 1820   Hungary, N Greece
   __ *P.l.lugens* C.L. Brehm, 1855   C and S Greece
   __ *P.l.anatoliae* E. Hartert, 1905   Asia Minor
   __ *P.l.hyrcanus* (Zarudny & Loudon, 1905)[9]   SE Azerbaijan, N Iran
   __ *P.l.dubius* Hellmayr, 1901   SW Iran
   __ *P.l.kirmanensis* Koelz, 1950   S Iran

○ *Parus montanus* WILLOW TIT
   __ *P.m.kleinschmidti* Hellmayr, 1900   Britain
   __ *P.m.rhenanus* O. Kleinschmidt, 1900   W Europe south to France and N Switzerland
   __ *P.m.montanus* Conrad von Baldenstein, 1827   Alps and Carpathian Alps to Greece
   __ *P.m.salicarius* C.L. Brehm, 1831   C Europe
   __ *P.m.borealis* Selys-Longchamps, 1843   N and EC Europe, Siberia
   __ *P.m.uralensis* Grote, 1927   SE European Russia, W Siberia, N Kazakhstan
   __ *P.m.baicalensis* (Swinhoe, 1871)   Yenisey, Altai and N Mongolia to Kolyma basin,
               Russian Far East, NE China and N Korea
   __ *P.m.anadyrensis* Belopolski, 1932   Koryakland and Anadyrland (NE Siberia)
   __ *P.m.kamtschatkensis* (Bonaparte, 1850)   Kamchatka, Kuril Is.
   __ *P.m.sachalinensis* Lönnberg, 1908   Sakhalin
   __ *P.m.restrictus* Hellmayr, 1900[10]   Japan

---

[1] For recognition see Cramp *et al.* (1993: 146) (594).
[2] For recognition see Cramp *et al.* (1993: 146) (594).
[3] For recognition see Cramp *et al.* (1993: 146) (594).
[4] This name was used by Cramp *et al.* (1993:146) (594) although Vaurie (1957, 1959) (2499, 2502) and Snow (1967) (2292) had used the name *brandtii*. The type of *brandtii*, once in ZISP, is now lost. Bogdanov's description, however, is incompatible with the species *P. palustris* and the name is best considered indeterminate (V. Loskot *in litt.*).
[5] For recognition see Stepanyan (1990: 564) (2326).
[6] For recognition see Eck (1980) (789). Close to *hellmayri* but has longer wing and relatively longer tail (S. Eck *in litt.*).
[7] We recognise this based on the findings of Eck (1980) (789); three apparently disjunct populations. We understand the spelling *hypermelaena* is required if the genus *Poecile* is used.
[8] We treat *hyrcanus* as part of this species following Eck (1980) (789), but note that acoustic evidence may in due course upset this view.
[9] Includes *talischensis* Stepanyan, 1974 (2319); see Eck (1980) (789).
[10] Includes *abei* Mishima, 1961 (1594) see Orn. Soc. Japan (1974: 362) (1726).

○ *Parus songarus* Songar Tit[1]
    __ *P.s.songarus* Severtsov, 1873   C and E Tien Shan, NW Xinjiang (C Asia)
    __ *P.s.affinis* (Przevalski, 1876)   NC China (NE Qinghai, Gansu, SW Shaanxi to Ningxia)
    __ *P.s.stoetzneri* O. Kleinschmidt, 1921   NE China (Shanxi and SE Nei Mongol to Hebei)
    __ *P.s.weigoldicus* O. Kleinschmidt, 1921   SC China (E Xizang, W Sichuan, N Yunnan)

● *Parus carolinensis* Carolina Chickadee[2]
    __ *P.c.atricapilloides* Lunk, 1952   SC USA (Kansas, Texas, Oklahoma)
    __ *P.c.agilis* Sennett, 1888   SC USA (SC Texas, W Louisiana, S Arkansas)
    ✓ *P.c.carolinensis* Audubon, 1834[3]   SE USA
    __ *P.c.extimus* (Todd & Sutton, 1936)   EC USA

● *Parus atricapillus* Black-capped Chickadee
    ✓ *P.a.turneri* Ridgway, 1884   C and S Alaska
    __ *P.a.occidentalis* S.F. Baird, 1858   Coastal SW Canada, NW USA
    __ *P.a.fortuitus* (Dawson & Bowles, 1909)   Inland SW Canada, NW USA
    __ *P.a.septentrionalis* Harris, 1846   NC Canada to C USA
    __ *P.a.bartletti* (Aldrich & Nutt, 1939)   Newfoundland
    __ *P.a.atricapillus* Linnaeus, 1766   E Canada to C USA
    __ *P.a.nevadensis* (Linsdale, 1938)   W USA
    __ *P.a.garrinus* Behle, 1951   WC USA
    __ *P.a.practicus* (Oberholser, 1937)   E USA (Appalachian Mts.)

● *Parus gambeli* Mountain Chickadee
    ✓ *P.g.baileyae* Grinnell, 1908[4]   Pacific slope of the Rocky Mts. (W Canada, W USA)
    __ *P.g.inyoensis* (Grinnell, 1918)   Rocky Mts. WC USA
    ✓ *P.g.gambeli* (Ridgway, 1886)   Great Basin and SW USA
    __ *P.g.atratus* (Grinnell & Swarth, 1926)   N Baja California

○ *Parus sclateri* Mexican Chickadee
    __ *P.s.eidos* (J.L. Peters, 1927)   SW USA, N Mexico
    __ *P.s.garzai* A.R. Phillips, 1986[5]   NE Mexico (1875)
    __ *P.s.sclateri* O. Kleinschmidt, 1897   C Mexico
    __ *P.s.rayi* Miller & Storer, 1950   S Mexico

○ *Parus superciliosus* (Przevalski, 1876) White-browed Tit   SE, E and NE Tibetan Plateau (S and SE Xizang to N Qinghai and Gansu)

○ *Parus davidi* (Berezowski & Bianchi, 1891) Père David's Tit   NC China (NE Qinghai and Gansu to W Sichuan and Hubei)

○ *Parus cinctus* Siberian Tit
    __ *P.c.lapponicus* Lundahl, 1848   Lapland, N European Russia
    __ *P.c.cinctus* Boddaert, 1783   NE European Russia, Siberia east to W Anadyrland and south to Baikal area
    __ *P.c.sayanus* (Sushkin, 1903)   Altai to S Baikal area and NE Mongolia
    __ *P.c.lathami* Stephens, 1817   W and N Alaska

● *Parus hudsonicus* Boreal Chickadee
    __ *P.h.stoneyi* Ridgway, 1887   N Alaska, NW Canada
    __ *P.h.columbianus* Rhoads, 1893[6]   S Alaska, W Canada, NW USA
    __ *P.h.hudsonicus* J.R. Forster, 1772   Alaska to E Canada
    __ *P.h.littoralis* H. Bryant, 1865   SE Canada, NE USA
    __ *P.h.farleyi* Godfrey, 1951   SC Canada

● *Parus rufescens* Chestnut-backed Chickadee
    ✓ *P.r.rufescens* J.K. Townsend, 1837   Alaska, W Canada, NW USA
    ✓ *P.r.neglectus* Ridgway, 1879   C California
    __ *P.r.barlowi* Grinnell, 1900   S California

---

[1] We recognise this based on Eck (1980) (789), but see also Kvist *et al.* (2001) (1311).
[2] Treatment based tentatively on Phillips (1986) (1875).
[3] Includes *impiger* see Phillips (1986) (1875).
[4] Includes *abbreviatus* see Phillips (1986) (1875).
[5] For recognition see Parkes in Dickerman & Parkes (1997) (729).
[6] Includes *cascadensis* see Phillips (1986) (1875).

○ *Parus dichrous* Grey-crested Tit
    — *P.d.kangrae* (Whistler, 1932)   NW Himalayas (Kashmir to Kumaun)
    — *P.d.dichrous* Blyth, 1844   C and E Himalayas (incl. SE Xizang)
    — *P.d.dichroides* (Przevalski, 1877)   NE Xizang to Gansu and S Shaanxi (NC China)
    — *P.d.wellsi* E.C.S. Baker, 1917   E Xizang and SC Sichuan to W and N Yunnan,
                                            NE Burma

● *Parus cristatus* Crested Tit
    — *P.c.scoticus* (Prazák, 1897)   NC Scotland
    — *P.c.abadiei* Jouard, 1929   W France
    — *P.c.weigoldi* Tratz, 1914   S and W Iberia
    — *P.c.cristatus* Linnaeus, 1758   N and E Europe south to E Carpathian Alps
    — *P.c.baschkirikus* (Snigirewski, 1931)   SW and C Ural Mts.
    — *P.c.mitratus* C.L. Brehm, 1831   C Europe south to NE Spain, Alps, Croatia, N Serbia
    — *P.c.bureschi* von Jordans, 1940[1]   Albania to Bulgaria and Greece

○ *Parus wollweberi* Bridled Titmouse
    — *P.w.vandevenderi* Rea, 1986[2]   Arizona (SW USA) (1875)
    — *P.w.phillipsi* van Rossem, 1947   SW USA, NW Mexico
    — *P.w.wollweberi* (Bonaparte, 1850)   C and S Mexico
    — *P.w.caliginosus* van Rossem, 1947   SW Mexico

● *Parus inornatus* Oak Titmouse[3]
    ✓ *P.i.inornatus* Gambel, 1845[4]   SW Oregon to C California
    — *P.i.affabilis* (Grinnell & Swarth, 1926)[5]   SW California, N Baja California
    — *P.i.mohavensis* A.H. Miller, 1946   Little San Bernardino Mts. (SE California)
    — *P.i.cineraceus* (Ridgway, 1883)   S Baja California

● *Parus ridgwayi* Juniper Titmouse
    ✓ *P.r.ridgwayi* Richmond, 1902[6] [7]   WC USA, N Mexico
    — *P.r.zaleptus* (Oberholser, 1932)   SE Oregon, E California, W Nevada

● *Parus bicolor* Tufted Titmouse[8]
    ✓ *P.b.bicolor* Linnaeus, 1766   E USA
    — *P.b.castaneifrons* Sennett, 1887[9]   SW Oklahoma and C Texas (SC USA)
    — *P.b.paloduro* (Stevenson, 1940)[10]   N Texas (panhandle and trans-Pecos)
    ✓ *P.b.atricristatus* Cassin, 1850   S Texas, NE Mexico

MELANOCHLORA Lesson, 1839 F
○ *Melanochlora sultanea* Sultan Tit
    — *M.s.sultanea* (Hodgson, 1837)   E Himalayas, Assam, Burma, N Thailand, W and S
                                              Yunnan
    — *M.s.flavocristata* (Lafresnaye, 1837)   Malay Pen., Sumatra
    — *M.s.seorsa* Bangs, 1924   SE China, Hainan, N Indochina to C Laos
    — *M.s.gayeti* Delacour & Jabouille, 1925   SC Vietnam

SYLVIPARUS E. Burton, 1836 M
○ *Sylviparus modestus* Yellow-browed Tit
    — *S.m.simlaensis* E.C.S. Baker, 1917   NW Himalayas (Kashmir to Kumaun)
    — *S.m.modestus* E. Burton, 1836   C and E Himalayas, Burma, SC and SE China, N
                                              Indochina, N Thailand
    — *S.m.klossi* Delacour & Jabouille, 1930   SE Laos and SC Vietnam

---

[1] For recognition see Cramp *et al.* (1993: 195) (594).
[2] For recognition see Browning (1990) (251).
[3] We follow Cicero (1996) (360) and A.O.U. (1998) in splitting *ridgwayi*
[4] Includes *sequestratus* and *kernensis* see Cicero (1996) (360).
[5] Includes *transpositus* see Cicero (1996) (360).
[6] Includes *plumbescens* see Cicero (1996) (360).
[7] If treated in the genus *Baeolophus* the name *griseus* Ridgway, 1882, is probably not preoccupied, but see A.O.U. (2000: 852) (4).
[8] For now we follow A.O.U. (1998: 467) (3) in retaining *atricristatus* within the species *bicolor*.
[9] Includes *sennetti* see Phillips (1986) (1875).
[10] Includes *dysleptus* see Phillips (1986) (1875).

REMIZIDAE PENDULINE TITS (5: 10)

REMIZ Jarocki, 1819 M
○ *Remiz pendulinus* PENDULINE TIT[1]

— *R.p.pendulinus* (Linnaeus, 1758)	W and S Europe east to C European Russia and W Asia Minor >> S Europe, SW Asia
— *R.p.menzbieri* (Zarudny, 1913)[2]	C Asia Minor and Levant to Transcaucasia and SW Iran
— *R.p.caspius* (Pelzam, 1870)	SE European Russia, N and W Caspian Sea >> SW Asia
— *R.p.jaxarticus* (Severtsov, 1872)[3]	SW Siberia and N Kazakhstan east to foot of Altai Mts. >> WC and SW Asia
— *R.p.macronyx* (Severtsov, 1873)	Lower Ural R. and SW and SC Kazakhstan to foot of Tien Shan Mts. >> SW Asia
— *R.p.altaicus* (Radde, 1899)[4]	SE Azerbaijan and NW Iran
— *R.p.neglectus* (Zarudny, 1908)[5]	NE Iran, SW and S Turkmenistan
— †?*R.p.nigricans* (Zarudny, 1908)	SE Iran, SW Afghanistan
— *R.p.ssaposhnikowi* (H.E. Johansen, 1907)[6]	Balkhash area (SE Kazakhstan)
— *R.p.coronatus* (Severtsov, 1873)	E Turkmenistan and N Afghanistan to SC and SE Kazakhstan and Kyrgyzstan >> SW Asia
— *R.p.stoliczkae* (Hume, 1874)	Xinjiang and NE Kazaskstan to upper Yenisey, Mongolia, Transbaikalia and NW Manchuria
— *R.p.consobrinus* (Swinhoe, 1870)	NE Manchuria, E China.

ANTHOSCOPUS Cabanis, 1850 M
○ *Anthoscopus punctifrons* (Sundevall, 1850) SENNAR PENDULINE TIT   Mauritania to C Sudan and Eritrea

○ *Anthoscopus parvulus* (Heuglin, 1864) YELLOW PENDULINE TIT   Senegal to S Chad, SW Sudan

○ *Anthoscopus musculus* (Hartlaub, 1882) MOUSE-COLOURED PENDULINE TIT
Somalia to S Sudan and NE Tanzania

○ *Anthoscopus flavifrons* FOREST PENDULINE TIT

— *A.f.waldroni* Bannerman, 1935	Liberia to Ghana
— *A.f.flavifrons* (Cassin, 1855)	Gabon, Cameroon, N Zaire
— *A.f.ruthae* Chapin, 1958	E Zaire

○ *Anthoscopus caroli* AFRICAN PENDULINE TIT

— *A.c.roccattii* Salvadori, 1906	S Uganda to W Kenya, E Burundi, NW Tanzania
— *A.c.pallescens* Ulfstrand, 1960	W Tanzania
— *A.c.ansorgei* E. Hartert, 1905	W and C Angola, W Zaire
— *A.c.rhodesiae* W.L. Sclater, 1932	SE Zaire, NE Zambia, SW Tanzania
— *A.c.robertsi* Haagner, 1909[7]	SE Kenya to NE Zambia, N and C Mozambique
— *A.c.caroli* (Sharpe, 1871)	N Namibia, S Angola, SW Zambia, N Botswana
— *A.c.winterbottomi* C.M.N. White, 1946	NW Zambia
— *A.c.rankinei* Irwin, 1963	NE Zimbabwe
— *A.c.hellmayri* Roberts, 1914	E and S Zimbabwe, S Mozambique, NE Sth. Africa
— *A.c.sylviella* Reichenow, 1904	SC Kenya to NE and C Tanzania
— *A.c.sharpei* E. Hartert[8]	SW Kenya to N Tanzania

○ *Anthoscopus minutus* CAPE PENDULINE TIT

— *A.m.damarensis* Reichenow, 1905	N Namibia, N and E Botswana, W Zimbabwe, W Transvaal
— *A.m.gigi* Winterbottom, 1959	S Cape Province to W Orange Free State
— *A.m.minutus* (Shaw & Nodder, 1812)	C and S Namibia, SW Botswana, W and N Cape Province

[1] Russian authors favour treating this complex as two or three species; Harrap (1996) (1071) treated four. If three are accepted, which CSR favours, the first four races are *R. pendulinus*, the next five are *R. macronyx* and the last three are *R. coronatus*.
[2] For recognition see Stepanyan (1990: 560) (2326).
[3] For recognition see Stepanyan (1990: 560) (2326).
[4] Listed *pro memoria*; very variable and in a multi-species treatment said to be a "hybrid" form. For recognition see Stepanyan (1990: 562) (2326).
[5] For recognition see Stepanyan (1990: 562) (2326).
[6] For recognition see Cramp *et al.* (1993: 377) (594).
[7] Includes *taruensis* see White (1963: 49) (2648).
[8] For treatment within the species *caroli* see White (1963: 50) (2648).

AURIPARUS S.F. Baird, 1864 M
● *Auriparus flaviceps* Verdin
    — *A.f.acaciarum* Grinnell, 1931   SW USA, NW Mexico
    — *A.f.ornatus* (Lawrence, 1852)   SC USA, NE Mexico
    — *A.f.flaviceps* (Sundevall, 1850)[1]   N Baja California, NW Mexico
    — *A.f.lamprocephalus* Oberholser, 1897   S Baja California
    — *A.f.sinaloae* A.R. Phillips, 1986[2]   NW Mexico (1875)
    — *A.f.hidalgensis* A.R. Phillips, 1986[3]   NC Mexico (1875)

CEPHALOPYRUS Bonaparte, 1854 M
○ *Cephalopyrus flammiceps* Fire-capped Tit
    — *C.f.flammiceps* (E. Burton, 1836)   N and E Pakistan, NW Himalayas to W Nepal >> N and
                                               C India
    — *C.f.olivaceus* Rothschild, 1923   E Himalayas, E Assam, SC China >> N Thailand, N Laos

PHOLIDORNIS Hartlaub, 1857[4] M
○ *Pholidornis rushiae* Tit-Hylia
    — *P.r.ussheri* Reichenow, 1905   Sierra Leone to Ghana
    — *P.r.rushiae* (Cassin, 1855)   S Nigeria to N Angola
    — *P.r.bedfordi* Ogilvie-Grant, 1904   Bioko I.
    — *P.r.denti* Ogilvie-Grant, 1907   C and E Zaire, Uganda

HIRUNDINIDAE SWALLOWS, MARTINS[5] (20: 84)

PSEUDOCHELIDONINAE

PSEUDOCHELIDON Hartlaub, 1861 F
○ *Pseudochelidon eurystomina* Hartlaub, 1861 African River Martin   Gabon, Congo, W and NC Zaire

○ *Pseudochelidon sirintarae* Thonglongya, 1968. White-eyed River Martin
                                            Breeding range unknown >> C Thailand (2401)

HIRUNDININAE

PSALIDOPROCNE Cabanis, 1850 F
○ *Psalidoprocne nitens* Square-tailed Saw-wing
    — *P.n.nitens* (Cassin, 1857)   Guinea to C Zaire
    — *P.n.centralis* Neumann, 1904   NE Zaire

○ *Psalidoprocne fuliginosa* Shelley, 1887 Mountain Saw-wing   Cameroon, Bioko I.

○ *Psalidoprocne albiceps* White-headed Saw-wing
    — *P.a.albiceps* P.L. Sclater, 1864   Uganda and W Kenya to Zambia and N Malawi
    — *P.a.suffusa* Ripley, 1960   N Angola (2041)

○ *Psalidoprocne pristoptera* Black Saw-wing[6]
    — *P.p.pristoptera* (Rüppell, 1836)   Eritrea, N and NW Ethiopia
    — *P.p.blanfordi* Blundell & Lovat, 1899   Highlands of WC Ethiopia
    — *P.p.antinorii* Salvadori, 1884   C and S Ethiopia
    — *P.p.oleaginea* Neumann, 1904   SW Ethiopia, SE Sudan
    — *P.p.mangbettorum* Chapin, 1923   S Sudan to NE Zaire (Uelle, Lendu)
    — *P.p.chalybea* Reichenow, 1892   N and C Cameroon to N and C Zaire, W Sudan
    — *P.p.petiti* Sharpe & Bouvier, 1876   E Nigeria to Gabon
    — *P.p.ruwenzori* Chapin, 1932   E Zaire
    — *P.p.orientalis* Reichenow, 1889   S Tanzania and E Zambia to E Zimbabwe and C
                                               Mozambique
    — *P.p.reichenowi* Neumann, 1904   Angola, SW Zaire, Zambia

---

[1] Includes *fraterculus* see Phillips (1986) (1875).
[2] Tentatively accepted see Dickerman & Parkes (1997) (729).
[3] Tentatively accepted see Dickerman & Parkes (1997) (729).
[4] This genus has been placed in this family by Harrap (1996) (1071) and by Fry (2000: 121) (892).
[5] Generic sequence based on Sheldon *et al.* (1999) (2238) and related work.
[6] This is essentially the treatment of White (1961) (2645), followed by Urban & Keith (1992: 130-131) (2454); however, some authors treat as many as
   seven or eight species. See, for example, Zimmerman *et al.* (1996) (2735). A full review is needed.

— *P.p.massaica* Neumann, 1904   Kenya, N and C Tanzania

— *P.p.holomelas* (Sundevall, 1850)[1]   S Mozambique and S Malawi to E and S Sth. Africa

○ *Psalidoprocne obscura* (Hartlaub, 1855) FANTI SAW-WING   S Senegal and Gambia to C and S Nigeria and W Cameroon

PSEUDHIRUNDO Roberts, 1922 F

○ *Pseudhirundo griseopyga* GREY-RUMPED SWALLOW[2]

— *P.g.melbina* (J. & E. Verreaux, 1851)[3] Gambia River to Liberia and Gabon

— *P.g.griseopyga* (Sundevall, 1850)[4] Nigeria to Ethiopia, south to Natal

CHERAMOECA Cabanis, 1850 F

○ *Cheramoeca leucosterna* (Gould, 1841) WHITE-BACKED SWALLOW[5] W, C, and inland S and E Australia

PHEDINA Bonaparte, 1857 F

○ *Phedina borbonica* MASCARENE MARTIN

— *P.b.borbonica* (J.F. Gmelin, 1789) Mauritius, Reunion Is.

— *P.b.madagascariensis* Hartlaub, 1860 Madagascar >> E Africa

○ *Phedina brazzae* Oustalet, 1886 BRAZZA'S MARTIN Congo, SW Zaire, NE Angola

RIPARIA T. Forster, 1817 F

○ *Riparia paludicola* PLAIN MARTIN

— *R.p.maurItanica* (Meade-Waldo, 1901) W Morocco

— *R.p.minor* (Cabanis, 1850)[6] Senegal and Gambia to Sudan and NW Ethiopia

— *R.p.schoensis* Reichenow, 1920[7] Ethiopian Highlands

— *R.p.newtoni* Bannerman, 1937 SE Nigeria, Cameroon

— *R.p.ducis* Reichenow, 1908 Uganda, E Zaire, Kenya, N and C Tanzania

— *R.p.paludicola* (Vieillot, 1817) S Tanzania to Angola and Sth. Africa

— *R.p.cowani* (Sharpe, 1882) Madagascar

— *R.p.chinensis* (J.E. Gray, 1830) SW Tajikistan and E Afghanistan through S Asia to Indochina and Taiwan

— *R.p.tantilla* Riley, 1935 Luzon (Philippines)

○ *Riparia congica* (Reichenow, 1887) CONGO SAND MARTIN C Zaire

● *Riparia riparia* COLLARED SAND MARTIN/BANK SWALLOW

 √ *R.r.riparia* (Linnaeus, 1758)[8] a) Nth. America >> N Sth. America

                 b) Europe to C Asia >> Africa

— *R.r.innominata* Zarudny, 1916[9] SE Kazakhstan >> Africa/SW or S Asia ?

— *R.r.taczanowskii* Stegmann, 1925[10] Baikal area and C Mongolia to Russian Far East >> SE Asia

— *R.r.ijimae* (Lönnberg, 1908) Sakhalin, Kuril Is. and Japan >> SE Asia incl. Philippines

— *R.r.shelleyi* (Sharpe, 1885) Egypt >> south to Sudan and NE Ethiopia

— *R.r.eilata* Shirihai & Colston, 1992[11] SW Asia ? >> NE Africa? (2246)

○ *Riparia diluta* PALE SAND MARTIN[12]

— *R.d.gavrilovi* Loskot, 2001 C Siberia E to the Lena River and Cisbaikalia S to the Altai and Tuva Republic >> E Kazakhstan and ?? (1372)

— *R.d.transbaykalica* Goroshko, 1993 Transbaikalia (964)

— *R.d.diluta* (Sharpe & Wyatt, 1893) S and SE Kazakhstan >> NW India, Nepal

— *R.d.indica* Ticehurst, 1916[13] River systems N Pakistan and NW India

[1] For correction of spelling see David & Gosselin (2002) (613).
[2] A single individual has been named *andrewi* Williams, 1966 (2665); breeding quarters unknown.
[3] Implicitly includes *liberiae* see Urban & Keith (1992: 146) (2454).
[4] Includes *gertrudis* see White (1961: 60) (2645).
[5] For correction of spelling see David & Gosselin (2002) (613).
[6] It was suggested that *palidibula* (Rüppell, 1835) has priority, see Brooke (1975) (235) but that author admitted the name had not been in use since 1894 or earlier; we prefer to retain *minor* as did Cramp *et al.* (1988) (592).
[7] For recognition see White (1961: 53) (2645).
[8] Includes *kolymensis* see Stepanyan (1990) (2326), but more recent work suggests this may be separable.
[9] Includes *dolgushini* Gavrilov & Savtchenko, 1991 (931) see Loskot & Dickinson (2001) (1375).
[10] Recognised by Goroshko (1993) (964) and tentatively accepted by Dickinson *et al.* (2001) (749).
[11] Obtained on passage at Eilat (Israel).
[12] For reasons to separate this from *R. riparia* see Gavrilov & Savtchenko (1991) (931) and Goroshko (1993) (964).
[13] This and the next two forms require review following separation of *diluta* from *riparia*.

— *R.d.tibetana* Stegmann, 1925 — C Asia

— *R.d.fohkienensis* (La Touche, 1908)[1] — C and S China

○ *Riparia cincta* BANDED MARTIN

    — *R.c.erlangeri* Reichenow, 1905 — Ethiopia

    — *R.c.suahelica* van Someren, 1922 — S Sudan and Kenya to W Mozambique, Malawi and Zimbabwe

    — *R.c.parvula* Amadon, 1954 — N Angola, NW Zambia, SW Zaire

    — *R.c.xerica* Clancey & Irwin, 1966 — N Namibia, N Botswana, S and W Angola (528)

    — *R.c.cincta* (Boddaert, 1783) — SW Zimbabwe, Sth. Africa

TACHYCINETA Cabanis, 1850 F

◕ *Tachycineta bicolor* (Vieillot, 1808) TREE SWALLOW — North America >> Middle America, West Indies, N Sth. America

● *Tachycineta albilinea* (Lawrence, 1863) MANGROVE SWALLOW[2] — NW, E and S Mexico, Central America

○ *Tachycineta stolzmanni* (Philippi, 1902) TUMBES SWALLOW[3] — W Peru

○ *Tachycineta albiventer* (Boddaert, 1783) WHITE-WINGED SWALLOW — Trinidad, Sth. America, mainly E of Andes from Colombia to N Argentina

○ *Tachycineta leucorrhoa* (Vieillot, 1817) WHITE-RUMPED SWALLOW — E Bolivia, S Brazil south to C Argentina >> northwards

◕ *Tachycineta meyeni* (Cabanis, 1850) CHILEAN SWALLOW[4] — C and S Chile, SW Argentina >> E Bolivia, Paraguay, S Brazil

○ *Tachycineta euchrysea* GOLDEN SWALLOW

    — *T.e.euchrysea* (Gosse, 1847) — Jamaica

    — *T.e.sclateri* (Cory, 1884) — Hispaniola

◕ *Tachycineta thalassina* VIOLET-GREEN SWALLOW

    ↙ *T.t.thalassina* (Swainson, 1827)[5] — NW Canada, W and SW USA, NW Mexico >> N Mexico to Costa Rica

    — *T.t.brachyptera* Brewster, 1902 — C and S Baja California, NW Mexico >> to SW Mexico

○ *Tachycineta cyaneoviridis* (H. Bryant, 1859) BAHAMA SWALLOW — N Bahama Is. >> S Bahamas, Cuba

PROGNE Boie, 1826 F

● *Progne subis* PURPLE MARTIN

    ↙ *P.s.subis* (Linnaeus, 1758) — S Canada, E USA, E Mexico >> Brazil

    — *P.s.hesperia* Brewster, 1889 — Arizona, Baja California, NW Mexico >> Sth America

    ↙ *P.s.arboricola* Behle, 1968 — W USA, N Mexico >> Sth America (112)

○ *Progne cryptoleuca* S.F. Baird, 1865 CUBAN MARTIN — Cuba, Isla de la Juventud >> presumably Sth. America

○ *Progne dominicensis* (J.F. Gmelin, 1789) CARIBBEAN MARTIN — Jamaica, Hispaniola, Lesser Antilles >> presumably Sth. America

○ *Progne sinaloae* Nelson, 1898 SINALOA MARTIN — NW Mexico >> presumably Sth. America

○ *Progne chalybea* GREY-BREASTED MARTIN

    — *P.c.chalybea* (J.F. Gmelin, 1789) — Middle America, N Sth. America >> southwards

    — *P.c.warneri* A.R. Phillips, 1986 — W Mexico (1875)

    — *P.c.macrorhamphus* Brooke, 1974[6] — C Sth. America >> northwards

○ *Progne modesta* Gould, 1838 GALAPAGOS MARTIN[7] — C and S Galapagos Is.

○ *Progne murphyi* Chapman, 1925 PERUVIAN MARTIN — W Peru, N Chile

○ *Progne elegans* S.F. Baird, 1865 SOUTHERN MARTIN — E Bolivia, Argentina >> winters northward

---

[1] Original spelling verified, see Dickinson *et al.* (2001) (749).
[2] United with *bicolor* in last Edition, but see A.O.U. (1983) (2).
[3] No longer included in *albilinea* see range in A.O.U. (1998) (3).
[4] For reasons to use *meyeni* in place of *leucopyga* Meyen see Brooke (1974) (234).
[5] Includes *lepida* see Phillips (1986) (1875).
[6] For reasons to use this name in place of *domestica* Vieillot see Brooke (1974) (234).
[7] In last Edition we treated this and the next two species as one species *P. modesta*; A.O.U. (1998: 456) (3) have preferred to treat the three as members of a broader superspecies, we see no reason to differ.

○ *Progne tapera* BROWN-CHESTED MARTIN[1]
    — *P.t.tapera* (Linnaeus, 1766)   Colombia to the Guianas, Brazil and Peru
    — *P.t.fusca* (Vieillot, 1817)   C Sth. America

PYGOCHELIDON S.F. Baird, 1865 F
● *Pygochelidon cyanoleuca* BLUE-AND-WHITE SWALLOW
    ⤴ *P.c.cyanoleuca* (Vieillot, 1817)   Costa Rica, Panama, N and C Sth. America
    — *P.c.peruviana* Chapman, 1922   W Peru
    — *P.c.patagonica* (d'Orbigny & Lafresnaye, 1837)   S Sth. America >> N Sth America, Panama

NOTIOCHELIDON S.F. Baird, 1865 F
○ *Notiochelidon murina* BROWN-BELLIED SWALLOW
    — *N.m.meridensis* (Zimmer & Phelps, 1947)   W Venezuela
    — *N.m.murina* (Cassin, 1853)   Andes from Colombia to C Peru
    — *N.m.cyanodorsalis* (Carriker, 1935)   S Peru, N Bolivia

○ *Notiochelidon flavipes* (Chapman, 1922) PALE-FOOTED SWALLOW   Andes from Venezuela and Colombia to N Bolivia

○ *Notiochelidon pileata* (Gould, 1858) BLACK-CAPPED SWALLOW   S Mexico to W Honduras

HAPLOCHELIDON Todd, 1929[2] F
○ *Haplochelidon andecola* ANDEAN SWALLOW
    — *H.a.oroyae* (Chapman, 1924)   C Peru
    — *H.a.andecola* (d'Orbigny & Lafresnaye, 1837)   S Peru, W Bolivia, N Chile

ATTICORA Boie, 1844 F
○ *Atticora fasciata* (J.F. Gmelin, 1789) WHITE-BANDED SWALLOW   Amazonia, the Guianas

○ *Atticora melanoleuca* (Wied, 1820) BLACK-COLLARED SWALLOW   E Amazonia

NEOCHELIDON P.L. Sclater, 1862 F
○ *Neochelidon tibialis* WHITE-THIGHED SWALLOW
    — *N.t.minima* Chapman, 1924[3]   E Panama, W Colombia, W Ecuador
    — *N.t.griseiventris* Chapman, 1924   S Colombia, SE Venezuela, E Ecuador, E Peru,
                                                      N Bolivia, W Brazil
    — *N.t.tibialis* (Cassin, 1853)   SE Brazil

STELGIDOPTERYX S.F. Baird, 1858 F
◉ *Stelgidopteryx serripennis* NORTHERN ROUGH-WINGED SWALLOW
    ⤴ *S.s.serripennis* (Audubon, 1838)   S Canada, USA, >> S USA to Panama
    — *S.s.psammochrous* Griscom, 1929   SW USA, Baja California to SW Mexico >> C Mexico to
                                                       Panama
    — *S.s.fulvipennis* (P.L. Sclater, 1859)   C and S Mexico to Costa Rica
    ⤴ *S.s.ridgwayi* Nelson, 1901   Yucatán
    — *S.s.stuarti* Brodkorb, 1942   SE Mexico, E Guatemala
    — *S.s.burleighi* A.R. Phillips, 1986   Belize, Guatemala (1875)

● *Stelgidopteryx ruficollis* SOUTHERN ROUGH-WINGED SWALLOW
    ⤴ *S.r.decolor* Griscom, 1929   W Costa Rica, W Panama
    — *S.r.uropygialis* (Lawrence, 1863)   W Colombia, Ecuador, NW Peru
    — *S.r.aequalis* Bangs, 1901   N Colombia, W Venezuela, Trinidad
    — *S.r.ruficollis* (Vieillot, 1817)[4]   E Venezuela, the Guianas, SE Colombia to N Argentina

ALOPOCHELIDON Ridgway, 1903 F
○ *Alopochelidon fucata* (Temminck, 1822) TAWNY-HEADED SWALLOW   SE Bolivia, S Brazil to C Argentina >> northward to
                                                           Colombia, Venezuela

---

[1] In last Edition treated in the monotypic genus *Phaeoprogne* but see A.O.U. (1998: 456) (3).
[2] This species was treated in *Hirundo* in last Edition, but it does not belong in the narrow genus of that name. The use of this generic name was suggested by F. Sheldon (*in litt.*) as the relationships of this species remain unclear.
[3] For correction of spelling see David & Gosselin (2002) (613).
[4] Includes *cacabata* see Haverschmidt (1982) (1090).

HIRUNDO Linnaeus, 1758 F

⬤ *Hirundo rustica* BARN SWALLOW[1]

⤺ *H.r.rustica* Linnaeus, 1758 — Europe and NW and NC Africa E to the Yenisey river, W Mongolia, Xinjiang and C Himalayas >> Africa, S Asia

— *H.r.transitiva* (E. Hartert, 1910) — Israel, Lebanon, S Syria, W Jordan >> Egypt

— *H.r.savignii* Stephens, 1817 — Egypt

⤺ *H.r.gutturalis* Scopoli, 1786 — E Himalayas, S, C, E China, and Taiwan to Japan, Kuril Is., Korea and lower Amur R. >> SE Asia to N Australia

— *H.r.tytleri* Jerdon, 1864 — SC Siberia and C Mongolia to N Inner Mongolia >> E India to SE Asia

— *H.r.saturata* Ridgway, 1883[2] — Kamtchatka (E Siberia), Manchuria and mid-Amur basin E to Sea of Okhotsk and S to Hebei (NE China) >> SE Asia[3]

⤺ *H.r.erythrogaster* Boddaert, 1783 — Nth. America to C Mexico >> to S Sth. America

◯ *Hirundo lucida* RED-CHESTED SWALLOW

— *H.l.lucida* Hartlaub, 1858[4] — W Africa

— *H.l.subalaris* Reichenow, 1905 — Congo, Zaire

— *H.l.rothschildi* Neumann, 1904 — Ethiopia

◯ *Hirundo angolensis* Bocage, 1868 ANGOLAN SWALLOW[5] — W Gabon to W Angola, NE Namibia, Zambia to Uganda and W Kenya

◯ *Hirundo tahitica* PACIFIC SWALLOW

— *H.t.domicola* Jerdon, 1844[6] — S India, Sri Lanka

— *H.t.javanica* Sparrman, 1789[7] — Andamans, SE Burma, SE Thailand, Cambodia, southernmost Vietnam, Malay Pen., Borneo, Sumatra, Belitung, Java, Lesser Sundas, Sulawesi, Moluccas, Philippines and small islands across this range

— *H.t.namiyei* (Stejneger, 1887) — Ryukyu Is., Taiwan and nearby Lanyu I.

— *H.t.frontalis* Quoy & Gaimard, 1830 — W and N New Guinea,

— *H.t.albescens* Schodde & Mason, 1999 — S and E New Guinea (2189)

— *H.t.ambiens* Mayr, 1934 — New Britain

— *H.t.subfusca* Gould, 1856 — Polynesia, Melanesia, Fiji, Tonga Is.

— *H.t.tahitica* J.F. Gmelin, 1789 — Society Is.

⬤ *Hirundo neoxena* WELCOME SWALLOW[8]

⤺ *H.n.carteri* (Mathews, 1912) — SW Australia

⤺ *H.n.neoxena* Gould, 1842[9] — NE to SE and SC Australia, Tasmania, Norfolk and Lord Howe Is., New Zealand >> Cape York Pen., Torres Strait

◯ *Hirundo albigularis* Strickland, 1849 WHITE-THROATED SWALLOW[10] — Botswana to C and S Mozambique and Sth. Africa >> north to Angola and S Zaire

◯ *Hirundo aethiopica* ETHIOPIAN SWALLOW

— *H.a.amadoni* C.M.N. White, 1956 — E Ethiopia, Somalia, NE Kenya (2635)

— *H.a.aethiopica* Blanford, 1869[11] — Senegal and Gambia to Ethiopia and NE Tanzania

◯ *Hirundo smithii* WIRE-TAILED SWALLOW

— *H.s.smithii* Leach, 1818 — Gambia to Ethiopia south to Angola and Natal.

— *H.s.filifera* Stephens, 1826[12] — S Uzbekistan, S Tajikistan, Afghanistan, Pakistan, India east to Burma, NW and NE Thailand, Laos, C Vietnam and Cambodia

[1] For Siberian populations we tentatively follow Vaurie (1959) (2502) as does Orn. Soc. Japan (2000) (1727). Roselaar (*in litt.*) would prefer to sink *saturata* but apparently not, as Dementiev (1936) (685) did, in *erythrogaster*.
[2] Tentatively includes *mandschurica* (but likely to be separable).
[3] The treatment of Vaurie (1951, 1954) (2492, 2495) is in conflict with those of Russian authors, and there is a view that Kamtchatka has no permanent breeding form and that opportunistic breeding occurs by sometimes one form sometimes another.
[4] Includes *clara* see White (1961: 54) (2645).
[5] Includes *arcticincta* see White (1961: 54) (2645).
[6] This may well be a separate species, but no thorough study has yet been published.
[7] Includes *mallopega* see Parkes (1971) (1763) and *abbotti* see Ripley (1944) (2027).
[8] For reasons to separate this from *tahitica* see Schodde & Mason (1999: 670) (2189).
[9] Includes *parsonsi* of Tasmania see Schodde & Mason (1999: 669) (2189).
[10] Includes *ambigua* see White (1961: 56) (2645).
[11] Includes *fulvipectus* see White (1961: 55) (2645).
[12] For reasons to date this from 1826 see Zimmer (1926) (2707) and Dickinson *et al.* (2001) (749).

○ *Hirundo atrocaerulea* Sundevall, 1850 MONTANE BLUE SWALLOW | SE Zaire and SW Tanzania to Natal and E Transvaal >> north to NE Zaire, Uganda and W Kenya

○ *Hirundo nigrita* G.R. Gray, 1845 WHITE-THROATED BLUE SWALLOW | Sierra Leone to Zaire and N Angola

○ *Hirundo leucosoma* Swainson, 1837 PIED-WINGED SWALLOW | Senegal and Gambia to Nigeria

○ *Hirundo megaensis* Benson, 1942 WHITE-TAILED SWALLOW | S Ethiopia

○ *Hirundo nigrorufa* Bocage, 1877 BLACK-AND-RUFOUS SWALLOW | C Angola, SE Zaire, N Zambia

○ *Hirundo dimidiata* PEARL-BREASTED SWALLOW
— *H.d.marwitzi* Reichenow, 1903 | Angola to Malawi, Zimbabwe and N Transvaal
— *H.d.dimidiata* Sundevall, 1850 | Cape Province to S Transvaal, Botswana and NE Namibia

PTYONOPROCNE Reichenbach, 1850[1] F
◉ *Ptyonoprogne rupestris* (Scopoli, 1769) EURASIAN CRAG MARTIN[2] | NW Africa and S Europe to C Asia and NE China >> Mediterranean basin to S Asia

○ *Ptyonoprogne obsoleta* PALE CRAG MARTIN
— *P.o.spatzi* (Geyr von Schweppenburg, 1916) | S Algeria, SW Libya, N Chad
— *P.o.presaharica* (Vaurie, 1953) | NC Algeria to S Morocco and N Mauritania
— *P.o.buchanani* (E. Hartert, 1921) | Aïr Massif (Niger)
— *P.o.obsoleta* (Cabanis, 1850) | Egypt, Sinai, N, C and E Arabia, Turkey and Iran
— *P.o.arabica* (Reichenow, 1905) | Ennedi (Chad) to N Sudan, coastal Eritrea, SW Arabia, N Somalia, Socotra
— *P.o.perpallida* Vaurie, 1951 | Al Hufuf area (NE Arabia)
— *P.o.pallida* (Hume, 1872)[3] | E Iran to Afghanistan and NW India

○ *Ptyonoprogne fuligula* ROCK MARTIN
— *P.f.pusilla* (Zedlitz, 1908)[4] | S Mali to C Sudan, C Ethiopia and Eritrea
— *P.f.bansoensis* (Bannerman, 1923) | Sierra Leone to Nigeria and Cameroon
— *P.f.fusciventris* Vincent, 1933[5] | S Chad to W Kenya, S to Zimbabwe and Mozambique
— *P.f.pretoriae* Roberts, 1922 | S Mozambique, SW Zimbabwe, E Sth. Africa
— *P.f.anderssoni* (Sharpe & Wyatt, 1887) | SW Angola, N and C Namibia
— *P.f.fuligula* (M.H.K. Lichtenstein, 1842) | S Namibia, W Sth. Africa

○ *Ptyonoprogne concolor* DUSKY CRAG MARTIN
— *P.c.concolor* (Sykes, 1832) | India
— *P.c.sintaungensis* (E.C.S. Baker, 1933) | Burma, N and E Thailand, S Yunnan, N Laos, N Vietnam

DELICHON Moore, 1854[6] N[7]
◉ *Delichon urbicum* NORTHERN HOUSE MARTIN
— *D.u.urbicum* (Linnaeus, 1758) | W, C, and N Europe, W Siberia >> W, C and S Africa
⤴ *D.u.meridionale* (E. Hartert, 1910) | N Africa, S Europe and WC Asia to Tien Shan and Kashmir >> Africa and SW Asia
— *D.u.lagopodum* (Pallas, 1811) | E Asia from Yenisey valley and C Mongolia to NE Siberia >> SE Asia

○ *Delichon dasypus* ASIAN HOUSE MARTIN
— *D.d.cashmeriense* (Gould, 1858) | Himalayas, C China >> India, SE Asia
— *D.d.nigrimentale* (E. Hartert, 1910) | S and E China, Taiwan >> SE Asia
— *D.d.dasypus* (Bonaparte, 1850)[8] | SC Siberia to Japan, Korea and NE China > SE Asia to Gtr. Sundas and Philippines

○ *Delichon nipalense* NEPAL HOUSE MARTIN
— *D.n.nipalense* Moore, 1854[9] | Himalayas, Assam, NW Burma
— *D.n.cuttingi* Mayr, 1941 | NE Burma, NE Thailand, Laos, NW Vietnam

[1] Seen as a subgenus of *Hirundo* in last Edition; for recognition see Cramp *et al.* (1988: 248)(592).
[2] Includes *theresae* as implied by monotypic treatment in Cramp *et al.* (1988) (592).
[3] If treated in a broad genus *Hirundo* the name *pallida* is preoccupied; *peloplasta* Brooke, 1974 (234) must then be used. C.S. Roselaar (*in litt.*) considers this a synonym of nominate *obsoleta*.
[4] Presumably this range, used by Fry (1992: 169) (890), includes *birwae* although this was placed in *bansoensis* by White (1961: 62) (2645).
[5] Includes *rufigula* see White (1961: 62) (2645).
[6] This name has usually been attributed to Horsfield & Moore but see Dickinson *et al.* (2001) (749), where the name *nipalensis* is discussed; the reasoning is identical for the generic name.
[7] Gender addressed by David & Gosselin (2002) (614), multiple changes occur in specific names.
[8] Dated 1851 by Peters (1960: 125) (1830) but see Zimmer (1926) (2707).
[9] This name has usually been attributed to Horsfield & Moore but see Dickinson *et al.* (2001) (749).

CECROPIS Boie, 1826 F

○ *Cecropis cucullata* (Boddaert, 1783) GREATER STRIPED SWALLOW   Sthn. Africa >> Angola and S Zaire

○ *Cecropis abyssinica* LESSER STRIPED SWALLOW

    — *C.a.puella* (Temminck & Schlegel, 1845)[1]   Senegal and Gambia to NE Nigeria and N Cameroon

    — *C.a.maxima* (Bannerman, 1923)   SE Nigeria, S Cameroon, SW Central African Republic

    — *C.a.bannermani* (Grant & Mackworth-Praed, 1942)   W and SW Sudan, NE Central African Republic

    — *C.a.abyssinica* (Guérin-Méneville, 1843)   Ethiopia, E Sudan

    — *C.a.unitatis* (Sclater & Mackworth-Praed, 1918)   S Sudan, Zaire and E Africa to Gabon, N Angola, Botswana and E Sth. Africa

    — *C.a.ampliformis* (Clancey, 1969)   S Angola and N Namibia to W Zambia and NW Zimbabwe (416)

○ *Cecropis semirufa* RUFOUS-CHESTED SWALLOW

    — *C.s.gordoni* (Jardine, 1851)   Senegal and Gambia to S Sudan, W Kenya N Angola and N Zaire

    — *C.s.semirufa* (Sundevall, 1850)   S and E Angola to W Malawi and NE Sth. Africa

○ *Cecropis senegalensis* MOSQUE SWALLOW

    — *C.s.senegalensis* (Linnaeus, 1766)   Senegal, Gambia and Mauritania to N Cameroon and W Sudan

    — *C.s.saturatior* (Bannerman, 1923)   S Ghana to S Cameroon, S Sudan, Ethiopia, Uganda and N Kenya

    — *C.s.monteiri* (Hartlaub, 1862)   Tanzania to Angola, N Namibia and Mozambique

● *Cecropis daurica* RED-RUMPED SWALLOW

    — *C.d.daurica* (Laxmann, 1769)[2]   NE Kazakhstan and N Mongolia to W Amurland and N Inner Mongolia; C China south to N Yunnan >> S and SE Asia (1341)

    — *C.d.japonica* (Temminck & Schlegel, 1845)[3]   S, E, and NE China to E Amurland, Korea and Japan >> SE Asia to N Australia

    — *C.d.nipalensis* (Hodgson, 1837)   Himalayas, SE Xizang, N Burma >> India

    — *C.d.erythropygia* (Sykes, 1832)   Plains of India >> S India, Sri Lanka

    — *C.d.hyperythra* (Blyth, 1849)   Sri Lanka

    — *C.d.rufula* (Temminck, 1835)   S Europe and N Africa to Arabia, N Somalia, Pakistan, N Kashmir, and Tien Shan >> Africa, SW Asia

    — *C.d.domicella* (Hartlaub & Finsch, 1870)   Senegal and Gambia to Sudan and SW Ethiopia

    — *C.d.melanocrissus* Rüppell, 1845[4]   Highlands of Ethiopia and Eritrea

    — *C.d.kumboensis* (Bannerman, 1923)[5]   Birwa Plateau (Sierra Leone), Bamenda highland (Cameroon)

    — *C.d.emini* (Reichenow, 1892)   SE Sudan and Kenya to Malawi

○ *Cecropis striolata* STRIATED SWALLOW

    — *C.s.striolata* (Schlegel, 1844)[6]   Taiwan, Philippines, Borneo, Sumatra, Java, Lesser Sundas

    — *C.s.mayri* (B.P. Hall, 1953)   Assam, N Burma, SW Yunnan, NW Thailand

    — *C.s.stanfordi* (Mayr, 1941)   Burma, N Thailand, N Laos

    — *C.s.vernayi* (Kinnear, 1924)   SW Thailand

○ *Cecropis badia* Cassin, 1853 RUFOUS-BELLIED SWALLOW[7]   Malay Pen.

PETROCHELIDON Cabanis, 1850[8] F

○ *Petrochelidon rufigula* (Bocage, 1878) RED-THROATED SWALLOW   Gabon and N and C Angola to SE Zaire and NW Zambia

○ *Petrochelidon preussi* (Reichenow, 1898) PREUSS'S SWALLOW   Mali to L Chad and Cameroon, NW and NE Zaire

○ *Petrochelidon perdita* (Fry & Smith, 1985) RED SEA SWALLOW #   Red Sea (897)

○ *Petrochelidon spilodera* (Sundevall, 1850) SOUTH AFRICAN SWALLOW   Sthn. Africa >> W Zaire

---

[1] The page containing this name was issued in 1845 not 1847 as given by Peters (1960: 113) (1830) see Holthuis & Sakai (1970) (1128).
[2] Includes *gephyra* see Stepanyan (1990: 333) (2326).
[3] Both plate and text appeared in 1845 see Holthuis & Sakai (1970) (1128) *contra* Peters (1960: 115) (1830).
[4] For correction of spelling see David & Gosselin (2002) (613).
[5] Includes *disjuncta* see White (1961: 59) (2645).
[6] Mees (1971) (1523) has shown that Schlegel published this name before the "Fauna Japonica" appeared. Status in Borneo unclear.
[7] For reasons to treat this at specific level see Dickinson & Dekker (2001) (743).
[8] For reasons to recognise this genus see Sheldon & Winkler (1993) (2236).

○ *Petrochelidon fuliginosa* (Chapin, 1925) Forest Swallow | E Nigeria, W Cameroon, Gabon

○ *Petrochelidon fluvicola* (Blyth, 1855) Streak-throated Swallow | Afghanistan, Himalayas, N India

○ *Petrochelidon ariel* (Gould, 1843) Fairy Martin | Australia, Melville I., Groote Eylandt

○ *Petrochelidon nigricans* Tree Martin
— *P.n.timoriensis* Sharpe, 1885 | Timor, Lesser Sundas
— *P.n.neglecta* Mathews, 1912 | Australia
— *P.n.nigricans* (Vieillot, 1817) | Tasmania >> E and C Australia, New Guinea to Solomon Is.

● *Petrochelidon pyrrhonota* Cliff Swallow
⤶ *P.p.pyrrhonota* (Vieillot, 1817)[1] | Nth. America >> to S Sth. America
— *P.p.ganieri* (A.R. Phillips, 1986)[2] | SC USA (Tennessee to Texas) >> Mexico, C and S America (1875)
— *P.p.hypopolia* Oberholser, 1920 | W USA >> Sth. America
— *P.p.tachina* Oberholser, 1903 | SW USA, WC America >> ??
— *P.p.minima* van Rossem & Hachisuka, 1938 | SW USA, NW Mexico >> Sth America

● *Petrochelidon fulva* Cave Swallow[3]
⤶ *P.f.pallida* Nelson, 1902[4] | SC USA, N Mexico >> Middle America (234)
— *P.f.citata* Van Tyne, 1938 | SE Mexico
— *P.f.fulva* (Vieillot, 1808)[5] | Hispaniola
— *P.f.cavicola* Barbour & Brooks, 1917[6] | Cuba, Isla de la Juventud
— *P.f.poeciloma* (Gosse, 1847) | Jamaica
— *P.f.puertoricensis* Garrido, Peterson & Komar, 1999 | Puerto Rico (927)

○ *Petrochelidon rufocollaris* Chestnut-collared Swallow[7]
— *P.r.aequatorialis* Chapman, 1924[8] | SW Ecuador
— *P.r.rufocollaris* (Peale, 1848) | W Peru

AEGITHALIDAE LONG-TAILED TITS (4:11)

AEGITHALOS Hermann, 1804 M
● *Aegithalos caudatus* Long-tailed Tit
— *A.c.caudatus* (Linnaeus, 1758) | N and E Europe, N Asia east to Kamchatka and south to N Mongolia, NE China, N Korea, Hokkaido
— *A.c.rosaceus* Mathews, 1937 | British Isles
— *A.c.europaeus* (Hermann, 1804) | Denmark and E France to NE Croatia and W and S Romania
— *A.c.aremoricus* Whistler, 1929 | W France
— *A.c.taiti* Ingram, 1913 | N Portugal, NW Spain, Pyrenees
⤶ *A.c.irbii* (Sharpe & Dresser, 1871) | S Iberia, Corsica
— *A.c.italiae* (Jourdain, 1910) | C and S mainland Italy
— *A.c.siculus* (Whitaker, 1901) | Sicily
— *A.c.macedonicus* (Dresser, 1892) | SW Croatia, Albania, Macedonia and N Greece
— *A.c.tephronotus* (Gunther, 1865) | S and E Greece, W and C Turkey
— *A.c.tauricus* (Menzbier, 1903) | Crimea
— *A.c.major* (Radde, 1884) | NE Turkey, Caucasus area
— *A.c.alpinus* (Hablizl, 1783) | SE Azerbaijan, N Iran
— *A.c.passekii* (Zarudny, 1904) | SE Turkey, SW Iran
— *A.c.vinaceus* (J. Verreaux, 1871) | C and NE China (E Qinghai to Gansu and Hebei)
— *A.c.glaucogularis* (F. Moore, 1854)[9] | N of E China (Shaanxi and Hubei to Zhejiang)

---

[1] Includes *melanogaster* see Phillips (1986) (1875).
[2] For recognition see Browning (1990) (251).
[3] We follow A.O.U. (1998: 461) (3) in treating C. American taxa as conspecific with *pallida* (which A.O.U. called *pelodoma*). We note the views of Garrido et al. (1999) (927), but defer the separation of *pallida* pending an A.O.U. decision.
[4] Sometimes called *pelodoma* Brooke, 1974 (234), however Brooke proposed this epithet to use in the genus *Hirundo*; it is only when this name is placed in *Hirundo* that Naumann's name takes priority and Nelson's must be supplanted. See A.O.U. (2000: 852) (4).
[5] For the date correction from 1807 see Browning & Monroe (1991) (258).
[6] For recognition of this race and the next see Garrido *et al.* (1999) (927)
[7] For recognition of this species as distinct from *fulva* see Ridgely & Tudor (1989: 64) (2018).
[8] Parkes (1993) (1780) has explained why Chapman's name should stand, rather than give way to *chapmani* Brooke, 1974 (234).
[9] See footnote to *Aegithalos niveogularis*.

— *A.c.trivirgatus* (Temminck & Schlegel, 1848)[1]   Honshu (Japan)
— *A.c.kiusiuensis* N. Kuroda, Sr., 1923   Kyushu and Shikoku (Japan)
— *A.c.magnus* (Clark, 1907)   C and S Korea, Tsushima I. (Japan)

○ *Aegithalos leucogenys* (Moore, 1854)[2] WHITE-CHEEKED TIT   E Afghanistan to W and N Pakistan

○ *Aegithalos concinnus* BLACK-THROATED TIT
— *A.c.iredalei* (E.C.S. Baker, 1920)[3]   Himalayas (NE Pakistan to N Assam and SE Xizang)
— *A.c.manipurensis* (Hume, 1888)   S and SE Assam, W Burma
— *A.c.talifuensis* (Rippon, 1903)   NE Burma, Yunnan, SW Sichuan, N Indochina to C Laos
— *A.c.pulchellus* (Rippon, 1900)   E Burma
— *A.c.concinnus* (Gould, 1855)   NC and E China, Taiwan
— *A.c.annamensis* (Robinson & Kloss, 1919)   S Laos, S Vietnam

○ *Aegithalos niveogularis* (F. Moore, 1854)[4] WHITE-THROATED TIT[5]   N Pakistan to C Nepal (NW Himalayas)

○ *Aegithalos iouschistos* (Blyth, 1844) BLACK-BROWED TIT   C and E Himalayas (C Nepal to SE Xizang)

○ *Aegithalos bonvaloti* PÈRE BONVALOT'S TIT[6]
— *A.b.bonvaloti* (Oustalet, 1891)   N and NE Burma, N Yunnan, NW Guizhou, SW Sichuan
— *A.b.obscuratus* (Mayr, 1940)   WC Sichuan
— *A.b.sharpei* (Rippon, 1904)   Chin Hills (W Burma)

○ *Aegithalos fuliginosus* (J. Verreaux, 1870) SOOTY TIT   NC China (Gansu and Sichuan to Hubei)

LEPTOPOECILE Severtsov, 1873 F
○ *Leptopoecile sophiae* WHITE-BROWED TIT-WARBLER
— *L.s.sophiae* Severtsov, 1873   Pakistan, NW India, C Asia
— *L.s.stoliczkae* (Hume, 1874)   SC Asia, W Gobi Desert
— *L.s.major* Menzbier, 1885   E Tien Shan
— *L.s.obscura* Przevalski, 1887   NW China, Tibet, E Himalayas

○ *Leptopoecile elegans* CRESTED TIT-WARBLER
— *L.e.meissneri* Schäfer, 1937   SE Tibet
— *L.e.elegans* Przevalski, 1887   S China

PSALTRIA Temminck, 1836 F
○ *Psaltria exilis* Temminck, 1836 PYGMY TIT   W and C Java

PSALTRIPARUS Bonaparte, 1850 M
● *Psaltriparus minimus* BUSHTIT[7] [8]
— *P.m.saturatus* Ridgway, 1903   SW British Columbia, NW Washington
✓ *P.m.minimus* J.K. Townsend, 1837   Pacific Coast of W USA
— *P.m.californicus* Ridgway, 1884   SC Oregon, NC California
— *P.m.plumbeus* (S.F. Baird, 1854)[9]   WC and S USA, N Mexico
— *P.m.melanurus* Grinnell & Swarth, 1926   NW Baja California
— *P.m.grindae* Ridgway, 1883   S Baja California
— *P.m.dimorphicus* van Rossem & Hachisuka, 1938   SW Texas, NE Mexico
— *P.m.iulus* Jouy, 1893   W and C Mexico
— *P.m.personatus* Bonaparte, 1850   C Mexico
— *P.m.melanotis* (Hartlaub, 1844)   S Mexico, Guatemala

---

[1] Includes *pallidolumbo* Mishima, 1961 (1594) see Orn. Soc. Japan (1974: 361) (1726).
[2] Ascribed to Horsfield & Moore by Snow (1967: 56) (2291) but see footnote on *Aegithalos niveogularis*.
[3] Includes *rubricapillus* see Snow (1967: 57) (2291).
[4] It has been found that Gould and Moore both named this and both named the related taxon *glaucogularis* in April. 1855. In fact Moore's paper in 1855 (1608) included a third taxon (*leucogenys*) in the same genus. Snow (1967: 56) (2291) credited this third name to Horsfield & Moore, 1854; this citation is correct for all three names and has priority, however the name is attributable to Moore alone, see Art. 50.1. of the Code (I.C.Z.N., 1999) (1178).
[5] Martens & Eck (1995) (1443) considered western *niveogularis* to be conspecific with eastern *bonvaloti*; however, we do not feel we can unite the taxa, as they did, with *iouschistos* clearly interposed although it may be morphologically rather distinct.
[6] Treated as distinct from *niveogularis* but conspecific with *iouschistos* by Harrap (1996) (1071), who discussed reports of overlap between the latter and *bonvaloti*. In the absence of better information the choice seems to be between uniting the three species, as in Snow (1967) (2291), or recognising three, as uniting any two seems to suggest a knowledge we lack. We prefer to recognise three.
[7] For the combination of *minimus* and *melanotis* see Raitt (1967) (1960).
[8] Subspecies as arranged by Rea & Phillips *in* Phillips (1986) (1875).
[9] Includes *sociabilis, providentialis, lloydi* and *cecaumenorum* see Rea & Phillips *in* Phillips (1986) (1875).

ALAUDIDAE LARKS[1] (19: 92)

MIRAFRA Horsfield, 1821 F

○ *Mirafra cantillans* SINGING BUSHLARK[2]

__ *M.c.marginata* Hawker, 1898	S Sudan to Somalia, Kenya and NE Tanzania
__ *M.c.chadensis* Alexander, 1908	Senegal to C Sudan and W Ethiopia
__ *M.c.simplex* (Heuglin, 1868)	W and S Arabia
__ *M.c.cantillans* Blyth, 1845[3]	Pakistan, N and C India

○ *Mirafra javanica* HORSFIELD'S BUSHLARK

__ *M.j.williamsoni* E.C.S. Baker, 1915	C Burma, C and E Thailand, Cambodia, Laos, C Vietnam, S China
__ *M.j.beaulieui* Delacour, 1932	Southernmost Vietnam
__ *M.j.philippinensis* Wardlaw Ramsay, 1886	Luzon, Mindoro
__ *M.j.mindanensis* Hachisuka, 1931	Mindanao
__ *M.j.javanica* Horsfield, 1821	S Borneo, Java, Bali
__ *M.j.parva* Swinhoe, 1871	Lombok, Sumbawa, Sumba, Flores
__ *M.j.timorensis* Mayr, 1944	Savu, Timor
__ *M.j.aliena* Greenway, 1935[4]	N and NE New Guinea
__ *M.j.woodwardi* Milligan, 1901	WC Western Australia
__ *M.j.halli* Bianchi, 1906[5]	NW Western Australia
__ *M.j.forresti* Mayr & McEvey, 1960	N Western Australia (1490)
__ *M.j.melvillensis* Mathews, 1912	Melville and Bathurst Is. (Northern Territory)
__ *M.j.soderbergi* Mathews, 1921	N Northern Territory (Arnhem Land)
__ *M.j.rufescens* Ingram, 1906[6]	E Northern Territory, W Queensland, NE South Australia
__ *M.j.athertonensis* Schodde & Mason, 1999	NE Queensland (2189)
__ *M.j.horsfieldii* Gould, 1847	C and S Queensland, New South Wales, Victoria, SE South Australia
__ *M.j.secunda* Sharpe, 1890	SC South Australia

○ *Mirafra hova* Hartlaub, 1860 MADAGASCAN BUSHLARK — Madagascar

○ *Mirafra passerina* Gyldenstolpe, 1926 MONOTONOUS BUSHLARK — SW Angola, N Namibia, E and C Botswana, Zimbabwe, N Cape Province, Transvaal

○ *Mirafra albicauda* Reichenow, 1891 WHITE-TAILED BUSHLARK[7] — W Chad, E Sudan, N and W Uganda, W and S Kenya to C and SW Tanzania

○ *Mirafra cheniana* A. Smith, 1843 MELODIOUS BUSHLARK — Zimbabwe, Botswana, Transvaal, E Orange Free State, N and E Cape Province, C Natal

○ *Mirafra cordofanica* Strickland, 1852 KORDOFAN BUSHLARK — Mauritania to Niger, WC Sudan

○ *Mirafra williamsi* Macdonald, 1956 WILLIAMS'S BUSHLARK — N Kenya

○ *Mirafra pulpa* Friedmann, 1930 FRIEDMANN'S BUSHLARK — S Ethiopia, S Kenya

○ *Mirafra africana* RUFOUS-NAPED BUSHLARK

__ *M.a.henrici* Bates, 1930	Guinea, Liberia
__ *M.a.batesi* Bannerman, 1923	Niger, Nigeria
__ *M.a.stresemanni* Bannerman, 1923	N Cameroon
__ *M.a.bamendae* Serle, 1959	W Cameroon (2225)
__ *M.a.kurrae* Lynes, 1923	W Sudan
__ *M.a.tropicalis* E. Hartert, 1900	SC Uganda, W Kenya, N and NW Tanzania
__ *M.a.sharpii* Elliot, 1897	NW Somalia
__ *M.a.ruwenzoria* Kinnear, 1921	SW Uganda, E Zaire
__ *M.a.athi* E. Hartert, 1900	NE Tanzania, C Kenya
__ *M.a.harterti* Neumann, 1908	E Kenya

---

[1] For Asian species see Dickinson & Dekker (2001) (742) and Dickinson *et al.* (2001) (748).
[2] In Dickinson *et al.* (2001) (748) this is lumped in *M. javanica* on the grounds that no evaluation has yet appeared of the suggested conspecificity of these four forms. As it was recognised in last Edition we prefer not to confuse readers and here we retain it as separate.
[3] Given as 1844 by Peters (1960: 5) (1829), but see Dickinson *et al.* (2001) (748).
[4] Includes *sepikiana* see Mees (1982: 109) (1534).
[5] Includes *subrufescens* see Schodde & Mason (1999: 717) (2189).
[6] Includes *normantoni* Mayr & McEvey, 1960 (1490), see Schodde & Mason (1999: 717) (2189).
[7] Includes *rukwensis* see White (1961) (2645).

__ *M.a.malbranti* Chapin, 1946	S and C Zaire
__ *M.a.chapini* Grant & Mackworth-Praed, 1939	SE Zaire, NW Zambia
__ *M.a.occidentalis* (Hartlaub, 1857)[1]	W Angola
__ *M.a.kabalii* C.M.N. White, 1943	NW Zambia (Balovale), NE Angola
__ *M.a.gomesi* C.M.N. White, 1944[2]	W Zambia, E Angola
__ *M.a.grisescens* Sharpe, 1902	SW Zambia, NW Zimbabwe, N Botswana
__ *M.a.pallida* Sharpe, 1902	SW Angola, NW Namibia
__ *M.a.ghansiensis* (Roberts, 1932)	W Botswana, E Namibia
__ *M.a.nigrescens* Reichenow, 1900	NE Zambia, S Tanzania
__ *M.a.isolata* Clancey, 1956[3]	Malawi
__ *M.a.nyikae* Benson, 1939	E Zambia, N Malawi
__ *M.a.transvaalensis* E. Hartert, 1900[4]	SE Botswana, N Sth Africa and Zimbabwe to SE Zambia, S Malawi, Mozambique and E Tanzania
__ *M.a.africana* A. Smith, 1836[5]	Natal, E Cape Province

○ *Mirafra hypermetra* RED-WINGED BUSHLARK

__ *M.h.kathangorensis* Cave, 1940	SE Sudan
__ *M.h.kidepoensis* Macdonald, 1940	SE Sudan border, NE Uganda
__ *M.h.gallarum* E. Hartert, 1907	C and S Ethiopia
__ *M.h.hypermetra* (Reichenow, 1879)	S Somalia, Kenya, NE Tanzania

○ *Mirafra somalica* SOMALI BUSHLARK

__ *M.s.somalica* (Witherby, 1903)	N Somalia
__ *M.s.rochei* Colston, 1982	C and S Somalia (551)

○ *Mirafra ashi* Colston, 1982 ASH'S BUSHLARK | S Somalia (551)

○ *Mirafra angolensis* ANGOLAN BUSHLARK

__ *M.a.marungensis* B.P. Hall, 1958	SE Zaire (1061)
__ *M.a.angolensis* Bocage, 1880	WC and N Angola
__ *M.a.antonii* B.P. Hall, 1958[6]	E Angola, S Zaire, NW Zambia (1061)

○ *Mirafra rufocinnamomea* FLAPPET LARK

__ *M.r.buckleyi* (Shelley, 1873)	Senegal and Gambia to Ghana, N Nigeria and N Cameroon
__ *M.r.serlei* C.M.N. White, 1960	SE Nigeria (2641)
__ *M.r.tigrina* Oustalet, 1892	W Cameroon to N Zaire
__ *M.r.furensis* Lynes, 1923	W Sudan
__ *M.r.sobatensis* Lynes, 1914	E Sudan
__ *M.r.rufocinnamomea* (Salvadori, 1865)	NW and C Ethiopia
__ *M.r.omoensis* Neumann, 1928	SW Ethiopia
__ *M.r.torrida* Shelley, 1882	S Ethiopia to N and C Uganda, C and S Kenya, NE and C Tanzania
__ *M.r.kawirondensis* van Someren, 1921	E Zaire, W and S Uganda, W Kenya
__ *M.r.fischeri* (Reichenow, 1878)[7]	S Somalia to N Mozambique, Malawi, N Zambia and C Angola
__ *M.r.schoutedeni* C.M.N. White, 1956[8]	Gabon and W Zaire to Central African Republic, NE Angola
__ *M.r.twenarum* C.M.N. White, 1945[9]	NW Zambia
__ *M.r.smithersi* C.M.N. White, 1956	S Zambia, Zimbabwe, NE Botswana, N Transvaal
__ *M.r.pintoi* C.M.N. White, 1956	E Transvaal, NE Natal
__ *M.r.mababiensis* (Roberts, 1932)	W Zambia to C Botswana

[1] Includes *anchietae* da Rosa Pinto, 1967 (606) (R.J. Dowsett *in litt.*).
[2] Includes *irwini* da Rosa Pinto, 1968 (607) (W.R.J. Dean *in litt.*).
[3] For recognition see Dean *et al.* (1992: 23) (630).
[4] Includes *zuluensis* see White (1961) (2645).
[5] Includes *rostratus* see White (1961) (2645).
[6] Includes *minyanyae* White, 1958 (2637), see Dean *et al.* (1992: 28) (630). It also includes *niethammeri* da Rosa Pinto, 1968 (607) but a specimen needs re-examination to be sure (W.R.J. Dean *in litt.*).
[7] Includes *zombae* see White (1961) (2645).
[8] For recognition see White (1961) (2645).
[9] For recognition see White (1961) (2645).

○ *Mirafra apiata* CLAPPER LARK
  — *M.a.reynoldsi* Benson & Irwin, 1965   N Botswana, SW Zambia (132)
  — *M.a.jappi* Traylor, 1962   W Zambia (2424)
  — *M.a.nata* Smithers, 1955   NE Botswana
  — *M.a damarensis* Sharpe, 1875[1]   N Namibia, W and C Botswana
  — *M.a.fasciolata* (Sundevall, 1850)[2]   SW Botswana to N and C South Africa
  — *M.a.apiata* (Vieillot, 1816)   SW Cape Province to Namaqualand and Port Elizabeth
  — *M.a.marjoriae* Winterbottom, 1956   Cape Peninsula to Knysna (2677)

○ *Mirafra africanoides* FAWN-COLOURED BUSHLARK
  — *M.a.intercedens* Reichenow, 1895[3]   SE and S Ethiopia to SW Somalia, Kenya and NE Tanzania
  — *M.a.alopex* Sharpe, 1890   NW Somalia, E Ethiopia
  — *M.a.macdonaldi* C.M.N. White, 1953   S Ethiopia
  — *M.a.trapnelli* C.M.N. White, 1943   W Zambia, E Angola
  — *M.a.harei* Roberts, 1917[4]   NW Namibia
  — *M.a.makarikari* (Roberts, 1932)   SE Angola, N Namibia, SW Zambia, N and C Botswana, NW Zimbabwe
  — *M.a.sarwensis* (Roberts, 1932)   NE Namibia, W Botswana
  — *M.a.vincenti* (Roberts, 1938)   C Zimbabwe, S Mozambique
  — *M.a.austinrobertsi* C.M.N. White, 1947   SW Zimbabwe, E Botswana, W Transvaal, NE Cape Province
  — *M.a.africanoides* A. Smith, 1836   S and E Namibia to S Botswana and N Cape Province

○ *Mirafra collaris* Sharpe, 1896 COLLARED BUSHLARK   SE Ethiopia, Somalia, NE and E Kenya

○ *Mirafra assamica* Horsfield, 1840 RUFOUS-WINGED BUSHLARK   N India, Nepal, Assam, Bangladesh, W Burma

○ *Mirafra affinis* Blyth, 1845 JERDON'S BUSHLARK[5]   S and SE India, Sri Lanka

○ *Mirafra microptera* Hume, 1873 BURMESE BUSHLARK   C Burma

○ *Mirafra erythrocephala* Salvadori & Giglioli, 1885 INDOCHINESE BUSHLARK
                                                    S Burma, Thailand, S Indochina

○ *Mirafra rufa* RUSTY BUSHLARK
  — *M.r.nigriticola* Bates, 1932   Mali, Niger
  — *M.r.rufa* Lynes, 1920   Chad, W Sudan
  — *M.r.lynesi* Grant & Mackworth-Paed, 1933   C Sudan

○ *Mirafra gilletti* Sharpe, 1895 GILLETT'S BUSHLARK
  — *M.g.gilletti* Sharpe, 1895   E and SE Ethiopia, NW Somalia
  — *M.g.arorihensis* Érard, 1975   C Somalia to NE Kenya (804)

○ *Mirafra degodiensis* Érard, 1975 DEGODI BUSHLARK   S Ethiopia (804)

○ *Mirafra poecilosterna* (Reichenow, 1879) PINK-BREASTED BUSHLARK[6]   SW Ethiopia and S Somalia to E Uganda, Kenya and N Tanzania

○ *Mirafra sabota* SABOTA BUSHLARK
  — *M.s.plebeja* (Cabanis, 1875)   Cabinda coast (W Angola)
  — *M.s.ansorgei* W.L. Sclater, 1926   W Angola
  — *M.s.naevia* (Strickland, 1853)   NW Namibia
  — *M.s.waibeli* Grote, 1922   N Namibia, N Botswana
  — *M.s.herero* (Roberts, 1936)   S and E Namibia, NW Cape Province
  — *M.s.sabota* A. Smith, 1836   C Zimbabwe, E Botswana, N Cape Province to Natal
  — *M.s.sabotoides* (Roberts, 1932)   W Zimbabwe, C and S Botswana, NW Transvaal
  — *M.s.suffusca* Clancey, 1958[7]   SE Zimbabwe, S Mozambique, NE Natal, E Transvaal, Swaziland (378)
  — *M.s.bradfieldi* (Roberts, 1928)   C, E and N Cape Province

[1] The name *deserti* is here treated as a synonym, but it may be distinct (W.R.J. Dean *in litt.*).
[2] White (1961) (2645) considered the name *rufipilea*, of which *hewitti* was thought by Peters (1960) (1829) to be a synonym, indeterminate. The name *hewitti* is a junior synonym.
[3] Includes *longonotensis* see White (1961) (2645).
[4] Includes *omaruru* see White (1961) (2645) and *quaesita* Clancey, 1958 (378) (R.J. Dowsett, *in litt.*).
[5] This and the next two forms were *tentatively* united by Dickinson & Dekker (2001) (742) pending publication of molecular studies. Here they are *tentatively* split!
[6] Treated as polytypic in last Edition, but generally treated as monotypic since White (1961) (2645).
[7] Presumably includes *fradei* da Rosa Pinto, 1963 (605).

○ *Mirafra erythroptera* INDIAN BUSHLARK
   — *M.e.sindiana* Ticehurst, 1920[1]   Pakistan, NW India
   — *M.e.erythroptera* Blyth, 1845   S and C India

HETEROMIRAFRA Ogilvie-Grant, 1913 F
○ *Heteromirafra ruddi* (Ogilvie-Grant, 1908) RUDD'S LARK   SE Transvaal, E Orange Free State

○ *Heteromirafra archeri* Stephenson Clarke, 1920 ARCHER'S LARK   NW Somalia

○ *Heteromirafra sidamoensis* (Érard, 1975) SIDAMO LARK[2]   S Ethiopia (803)

CERTHILAUDA Swainson, 1827 F
○ *Certhilauda benguelensis* BENGUELA LONG-BILLED LARK[3]
   — *C.b.benguelensis* (Sharpe, 1904)   Coastal SW Angola
   — *C.b.kaokensis* Bradfield, 1944   NW Namibia

○ *Certhilauda subcoronata* KAROO LONG-BILLED LARK
   — *C.s.damarensis* (Sharpe, 1904)[4]   N Namibia
   — *C.s.bradshawi* (Sharpe, 1904)   S Namibia, NW Cape Province
   — *C.s.subcoronata* A. Smith, 1843   E Cape Province
   — *C.s.gilli* Roberts, 1936   SC Cape Province

○ *Certhilauda semitorquata* EASTERN LONG-BILLED LARK
   — *C.s.semitorquata* A. Smith, 1836   E Cape Province, W Natal, Orange Free State
   — *C.s.transvaalensis* Roberts, 1936[5]   N Natal, Transvaal, N Cape Province
   — *C.s.algida* Quickelberge, 1967   E Cape Province (1958)

○ *Certhilauda curvirostris* CAPE LONG-BILLED LARK
   — *C.c.curvirostris* (Hermann, 1783)   SW Cape Province
   — *C.c.falcirostris* Reichenow, 1916   W Cape Province

○ *Certhilauda brevirostris* Roberts, 1941 AGULHAS LONG-BILLED LARK Coastal W Cape Province

○ *Certhilauda chuana* (A. Smith, 1836) SHORT-CLAWED LARK   N Cape Province, SE Botswana, W Transvaal

○ *Certhilauda albescens* KAROO LARK
   — *C.a.codea* A. Smith, 1843   Coastal W and NW Cape Province, SW Namibia
   — *C.a.albescens* (Lafresnaye, 1839)   SW Cape Province
   — *C.a.guttata* (Lafresnaye, 1839)   C and W Cape Province

○ *Certhilauda erythrochlamys* DUNE LARK[6]
   — *C.e.erythrochlamys* (Strickland, 1853)   WC Namibia
   — *C.e.barlowi* (Roberts, 1937)   SW Namibia

○ *Certhilauda burra* (Bangs, 1930) RED LARK   NW Cape Province

PINAROCORYS Shelley, 1902 F
○ *Pinarocorys nigricans* DUSKY LARK
   — *P.n.nigricans* (Sundevall, 1850)   SE Zaire, N Zambia, W Tanzania >> S to Sth. Africa
   — *P.n.occidentis* Clancey, 1968[7]   Angola, SW Zaire >> to Namibia (414)

○ *Pinarocorys erythropygia* (Strickland, 1852) RUFOUS-RUMPED LARK   Ivory Coast to S Sudan >> Senegal and Gambia to
                                                      N Sudan

CHERSOMANES Cabanis, 1851 F
○ *Chersomanes albofasciata* SPIKE-HEELED LARK[8]
   — *C.a.beesleyi* Benson, 1966   N Tanzania (125)

---

[1] Includes *furva* see Ripley (1982: 247) (2055), but tentatively recognised by Dickinson & Dekker (2001) (742).
[2] Described in the genus *Mirafra*. Placed in *Heteromirafra* by Dean *et al.* (1992: 45) (630).
[3] For recognition of this as a coastal Angolan race see White (1961) (2645). Now a product of the split of *C. curvirostris* into five species (*benguelensis,
subcoronata, semitorquata, curvirostris* and *brevirostris*) by Ryan & Bloomer (1999) (2144).
[4] If treated in the genus *Mirafra* the name *damarensis* is preoccupied. White (1961) (2645) used the name *kaokensis* Bradfield, 1944, in its place; in the light
of the separation of species that may need to be re-examined.
[5] The alternate name *infelix* C.M.N. White, 1960 (2642) was proposed for use if this species was placed in *Mirafra*. It is not required in *Certhilauda*.
[6] Includes *cavei*. For monotypic treatment see Dean *et al.* (1992: 52) (630). Since then Ryan *et al.* (1998 (2145) proposed that this be split. This will
probably require recognition, but intergrades occur and field studies are needed to determine the dynamics.
[7] For recognition of this see Dean *et al.* (1992: 58) (630).
[8] We follow Dean *et al.* (1992: 59-60) (630) in reducing the number of forms recognised.

__ *C.a.obscurata* (E. Hartert, 1907)[1]	SW, C and E Angola
__ *C.a.erikssoni* (E. Hartert, 1907)	N Namibia
__ *C.a.kalahariae* (Ogilvie-Grant, 1912)[2]	S and W Botswana, NW Cape Province
__ *C.a.boweni* (Meyer de Schauensee, 1931)	NW Namibia
__ *C.a.arenaria* (Reichenow, 1904)	S Namibia, N Cape Province
__ *C.a.barlowi* C.M.N. White, 1961[3]	E Botswana (2644)
__ *C.a.alticola* Roberts, 1932[4]	C and S Transvaal
__ *C.a.albofasciata* (Lafresnaye, 1836)[5]	S Botswana to Orange Free State and NE, E and C Cape Province
__ *C.a.garrula* (A. Smith, 1846)[6]	NW and W Cape Province
__ *C.a.macdonaldi* Winterbottom, 1958	S and E Karoo (Cape Province) (2678)

ALAEMON Keyserling & Blasius, 1840 M
○ *Alaemon alaudipes* GREATER HOOPOE-LARK

__ *A.a.boavistae* E. Hartert, 1917	E Cape Verde Is.
__ *A.a.alaudipes* (Desfontaines, 1789)	N Africa, Sahara to Sinai
__ *A.a.desertorum* (Stanley, 1814)	NE Sudan to N Somalia, W Arabia
__ *A.a.doriae* (Salvadori, 1868)	E Arabia, Iraq to NW India

○ *Alaemon hamertoni* LESSER HOOPOE-LARK

__ *A.h.altera* Witherby, 1905	NE Somalia
__ *A.h.tertia* Stephenson Clarke, 1919	N Somalia
__ *A.h.hamertoni* Witherby, 1905	E Somalia

RAMPHOCORIS Bonaparte, 1850 M
○ *Ramphocoris clotbey* (Bonaparte, 1850) THICK-BILLED LARK — N Africa, N and C Arabia, E Jordan, S Syria

MELANOCORYPHA Boie, 1828 F
○ *Melanocorypha calandra* CALANDRA LARK

__ *M.c.calandra* (Linnaeus, 1766)	S Europe and NW Africa to Turkey (except SC and SE), Caucasus area, NW Kazakhstan
__ *M.c.psammochroa* E. Hartert, 1904	NE Iran and C Kazakhstan to SW Mongolia and SC Siberia >> Arabia and SW Asia
__ *M.c.gaza* R. Meinertzhagen, 1919[7]	E Syria, SE Turkey, Iraq, SW Iran
__ *M.c.hebraica* R. Meinertzhagen, 1920[8]	Israel and W Jordan to SC Turkey and W Syria

○ *Melanocorypha bimaculata* BIMACULATED LARK

__ *M.b.bimaculata* (Ménétriés, 1832)	E Turkey, Transcaucasia, N Iran >> NE Africa and W Arabia
__ *M.b.rufescens* C.L. Brehm, 1855	C and S Turkey, W Syria, Lebanon, Iraq >> Eritrea and W Arabia
__ *M.b.torquata* Blyth, 1847	NE Iran and Kazakhstan to N Xinjiang and Afghanistan >> Arabia to NW India

○ *Melanocorypha mongolica* (Pallas, 1776) MONGOLIAN LARK — Mongolia and E Qinghai to Transbaikalia, Nei Mongol and Hebei >> C and E Asia

○ *Melanocorypha maxima* TIBETAN LARK

__ *M.m.flavescens* Stegmann, 1937[9]	Qaidam Basin to W Qinghai (N Tibetan Plateau)
__ *M.m.maxima* Blyth, 1867	Sikkim and SE Tibet to W Sichuan and S Shaanxi
__ *M.m.holdereri* Reichenow, 1911	SW, C, and NE Tibetan Plateau to E Qinghai and SW Gansu

○ *Melanocorypha leucoptera* (Pallas, 1811) WHITE-WINGED LARK — SE European Russia and N Kazakhstan to SC Siberia and SW Mongolia >> S Russia, Transcaspia, Iran

○ *Melanocorypha yeltoniensis* (J.R. Forster, 1767) BLACK LARK — SE European Russia and N Kazakhstan to SW Siberia and E Kazakhstan >> SE Europe and SW Asia

---

[1] Implicitly includes *longispina* Rudebeck, 1970 (2125) see Dean *et al.* (1992: 60) (630).
[2] Includes *bathoeni* Winterbottom, 1958 (2678), and *bathoeni* Paterson, 1958 (1793), see White (1961) (2645).
[3] This name is preoccupied in *Certhilauda* when *salinicola* Clancey, 1962 (392A), must be used.
[4] Includes *subpallida* see Peters (1960: 28) (1829).
[5] Includes *latimerae* Winterbottom, 1958 (2678) see White (1960) (2645).
[6] Includes *bushmanensis* see White (1961) (2645).
[7] Includes *dathei* Kumerloeve, 1970 (1308), see Roselaar (1995) (2107).
[8] For recognition see Roselaar (1995) (2107).
[9] Recognition advised by C.S. Roselaar (*in litt.*).

AMMOMANES Cabanis, 1851 F[1]

○ ***Ammomanes cinctura*** BAR-TAILED LARK

  — *A.c.cinctura* (Gould, 1839)[2]   Cape Verde Is.

  — *A.c.arenicolor* Sundevall, 1851[3]   Sahara, Middle East, Arabia

  — *A.c.zarudnyi* E. Hartert, 1902   C Iran, S Afghanistan, W Pakistan

○ ***Ammomanes phoenicura*** RUFOUS-TAILED LARK

  — *A.p.phoenicura* (Franklin, 1831)   C India

  — *A.p.testacea* Koelz, 1951   S India

○ ***Ammomanes deserti*** DESERT LARK

  — *A.d.payni* E. Hartert, 1924   S Morocco, W Algeria

  — *A.d.algeriensis* Sharpe, 1890[4]   C Algeria, Tunisia and W Libya to Tibisti (N Chad)

  — *A.d.whitakeri* E. Hartert, 1911   SW Libya

  — *A.d.mya* E. Hartert, 1912   W and C Sahara

  — *A.d.geyri* E. Hartert, 1924[5]   Mauritania to N Nigeria and SE Algeria

  — *A.d.kollmanspergeri* Niethammer, 1955   NE Chad, W Sudan

  — *A.d.deserti* (M.H.K. Lichtenstein, 1823)[6]   S Egypt to N Sudan

  — *A.d.erythrochroa* Reichenow, 1904   C Sudan

  — *A.d.isabellina* (Temminck, 1823)   NC and E Egypt, N and C Arabia, SW Iraq

  — *A.d.samharensis* Shelley, 1902   NE Sudan, N Eritrea, S Arabia

  — *A.d.taimuri* Meyer de Schauensee & Ripley, 1953   N Oman and E UAE

  — *A.d.assabensis* Salvadori, 1902   S Eritrea, Djibouti, NW Somalia

  — *A.d.akeleyi* Elliot, 1897   N Somalia

  — *A.d.azizi* Ticehurst & Cheesman, 1924   EC Arabia

  — *A.d.saturata* Ogilvie-Grant, 1900   S Arabia

  — *A.d.insularis* Ripley, 1951   Bahrain

  — *A.d.annae* R. Meinertzhagen, 1923   Transjordan

  — *A.d.coxi* R. Meinertzhagen, 1923[7]   SC Turkey, Syria, N Iraq

  — *A.d.cheesmani* R. Meinertzhagen, 1923   E Iraq, SW Iran

  — *A.d.darica* Koelz, 1951[8]   S Zagros Mts. (S Iran)

  — *A.d.parvirostris* E. Hartert, 1890   NE Iran and W Turkmenistan

  — *A.d.orientalis* Zarudny & Loudon, 1904   N Afghanistan, E Turkmenistan, E Uzbekistan, S
                      Tajikistan

  — *A.d.iranica* Zarudny, 1911   C and E Iran, S Afghanistan

  — *A.d.phoenicuroides* (Blyth, 1853)   E Afghanistan, Pakistan, NW India

○ ***Ammomanes grayi*** GRAY'S LARK

  — *A.g.hoeschi* Niethammer, 1955   NW Namibia, SW Angola

  — *A.g.grayi* (Wahlberg, 1855)[9]   W Namibia

CALANDRELLA Kaup, 1829 F

○ ***Calandrella brachydactyla*** GREATER SHORT-TOED LARK

  — *C.b.brachydactyla* (Leisler, 1814)   S Europe, W Asia Minor, Cyprus >> N Africa and
                       Mauritania to Somalia

  — *C.b. hungarica* Horváth, 1956[10]   Hungary, S Slovakia, N Serbia, W Romania

  — *C.b.rubiginosa* Fromholz, 1913   N Africa

  — *C.b.hermonensis* Tristram, 1864   NE Sinai to W Syria, W Iraq and N Arabia

  — *C.b.woltersi* Kumerloeve, 1969   SC Turkey (1307)

  — *C.b.artemisiana* Banjkowski, 1913   C and E Turkey, Transcaucasia, NE Iraq, SW and N
                       Iran >> SW Asia

  — *C.b.longipennis* (Eversmann, 1848)   Ukraine, N Caucasus area and Iran to SC Siberia, W
                       Mongolia, and Tien Shan >> Arabia and S Asia

[1] Gender addressed by David & Gosselin (2002) (614), multiple changes occur in specific names.
[2] Peters (1960: 32) (1829) dated this 1841, but see Zimmer (1926: 159) (2707) and www.zoonomen.net (A.P. Peterson *in litt.*).
[3] Includes *pallens* see White (1961) (2645) and most recent authors.
[4] Range includes that of *mirei* see Dean *et al.* (1992: 73) (630), although this is often placed in *whitakeri* (Roselaar, *in litt.*).
[5] Includes *janeti* see Dean *et al.* (1992: 73) (630).
[6] Includes *borosi* Horváth, 1958 (1149) see Dean *et al.* (1992: 73) (630).
[7] Recognition advised by C.S. Roselaar (*in litt.*).
[8] Recognition advised by C.S. Roselaar (*in litt.*). Dickinson *et al.* (2001) (748) recognised *iranica* and listed *darica* as a synonym of that.
[9] Peters (1960: 38) (1829) dated this 1856, but see Gyldenstolpe (1926: 26) (1027) and www.zoonomen.net (A.P. Peterson *in litt.*).
[10] For recognition see Roselaar in Cramp *et al.* (1988: 134) (592).

___ *C.b.orientalis* Sushkin, 1925	C Asia, Mongolia, Manchuria
___ *C.b.dukhunensis* (Sykes, 1832)	Tibetan Plateau, Inner Mongolia, C and E Mongolia, Transbaikalia >> SW S and E Asia

○ *Calandrella blanfordi* BLANFORD'S SHORT-TOED LARK[1]
___ *C.b.eremica* (Reichenow & Peters, 1935)[2]	SW Arabia
___ *C.b.blanfordi* (Shelley, 1902)	Eritrea
___ *C.b.erlangeri* (Neumann, 1906)	C Ethiopia
___ *C.b.daroodensis* C.M.N. White, 1960[3]	N Somalia (2640)

○ *Calandrella cinerea* RED-CAPPED LARK[4]
___ *C.c.saturatior* Reichenow, 1904[5]	Uganda and W Kenya to Angola, Zambia and Zimbabwe
___ *C.c.williamsi* Clancey, 1952	C Kenya
___ *C.c.spleniata* (Strickland, 1853)	W Namibia
___ *C.c.millardi* Paterson, 1958[6]	S Botswana (1793)
___ *C.c.cinerea* (J.F. Gmelin, 1789)[7]	S Namibia, Cape Province, W Orange Free State
___ *C.c.anderssoni* (Tristram, 1869)[8]	NE Namibia, N Botswana, S Mozambique, Transvaal, Natal and E Cape Province

○ *Calandrella acutirostris* HUME'S SHORT-TOED LARK
___ *C.a.acutirostris* Hume, 1873	NE Iran, Afghanistan and Pakistan to Tien Shan and W and N Xinjiang >> SW Asia, India
___ *C.a.tibetana* W.E. Brooks, 1880	Tibetan Plateau >> S Asia

○ *Calandrella raytal* SAND LARK
___ *C.r.adamsi* (Hume, 1871)	S Iran, E Afghanistan, Pakistan, NW India
___ *C.r.raytal* (Blyth, 1845)	NC and NE India, N Burma
___ *C.r.krishnarkumarsinhji* Vaurie & Dharmakumarsinhji, 1954	Kathiawar Pen. (WC India)

○ *Calandrella rufescens* LESSER SHORT-TOED LARK
___ *C.r.rufescens* (Vieillot, 1820)	Tenerife
___ *C.r.polatzeki* E. Hartert, 1904	Gran Canaria, Fuerteventura, Lanzarote
___ *C.r.apetzii* (A.E. Brehm, 1857)	S Portugal, S Spain
___ *C.r.minor* (Cabanis, 1851)	N Africa to NW Egypt, Sinai to SW Turkey, SW Iran, N Arabia >> Sahara and SW Asia
___ *C.r.nicolli* E. Hartert, 1909	Nile Delta (Egypt)
___ *C.r.pseudobaetica* Stegmann, 1932	E Turkey, Transcaucasia, NW and N Iran
___ *C.r.heinei* (Homeyer, 1873)	E Bulgaria, S Ukraine, and N Kazakhstan to W Altai

○ *Calandrella cheleensis* ASIAN SHORT-TOED LARK[9]
___ *C.c.niethammeri* Kumerloeve, 1963	C Asia Minor (1306)
___ *C.c.persica* (Sharpe, 1890)[10]	C and E Iran to W and S Afghanistan
___ *C.c.leucophaea* Severtsov, 1872	W and S Kazakhstan to NW Xinjiang
___ *C.c.seebohmi* (Sharpe, 1890)	Tarim Basin (Xinjiang, W China)
___ *C.c.tuvinica* Stepanyan, 1975	Tuva Republic and NW Mongolia (2321)
___ *C.c.cheleensis* (Swinhoe, 1871)	C and E Mongolia, Transbaikalia, NE China south to Shandong and S Shaanxi
___ *C.c.kukunoorensis* (Przevalski, 1876)[11]	E Tibetan Plateau and Qaidam Basin
___ *C.c.beicki* Meise, 1933	NE Qinghai, Gansu, W Inner Mongolia, S Mongolia

○ *Calandrella somalica* SOMALI SHORT-TOED LARK
___ *C.s.perconfusa* C.M.N. White, 1960[12]	NW Somalia (2643)
___ *C.s.somalica* (Sharpe, 1895)[13]	N Somalia to E Ethiopia

[1] Treated as a monotypic species by Peters (1960: 47) (1829), but Hall & Moreau (1970) (1066) implied the scope recognised here.
[2] This taxon is tentatively attached to *C. blanfordi* and may belong with *brachydactyla*.
[3] This taxon may belong to *C. cinerea*.
[4] We follow White (1961) (2645) except that we segregate *blanfordi* and allies.
[5] Thought to include *fulvida* Clancey, 1978 (464).
[6] For recognition see White (1961) (2645).
[7] Implicitly includes *vagilans* Clancey, 1978 (464).
[8] For recognition see White (1961) (2645). Implicitly includes *alluvia* Clancey, 1971 (433).
[9] Probably, but not certainly, separate from *C. rufescens*; the races listed here may not all belong together and may represent two or more species. Dickinson *et al.* (2001) (748) placed *persica* and by inference *niethammeri* in *C. rufescens*.
[10] Thought to include *aharonii* based on winter-taken specimens.
[11] Includes *tangutica* see Vaurie (1959: 33) (2502).
[12] Omitted in last Edition.
[13] Includes *vulpecula* see White (1961) (2645).

___ *C.s.megaensis* Benson, 1946	S Ethiopia, N Kenya
___ *C.s.athensis* (Sharpe, 1900)	S Kenya, NE Tanzania

SPIZOCORYS Sundevall, 1872 F
○ *Spizocorys conirostris* PINK-BILLED LARK

___ *S.c.damarensis* Roberts, 1922	NW Namibia
___ *S.c.crypta* (Irwin, 1957)	NE Botswana (1186)
___ *S.c.makawai* (Traylor, 1962)	W Zambia (2424)
___ *S.c.harti* (Benson, 1964)	SW Zambia (124)
___ *S.c.barlowi* Roberts, 1942	S Botswana, S Namibia, NW Cape Province
___ *S.c.transiens* (Clancey, 1959)[1]	N and E Cape Province to W Orange Free State, SW Transvaal and SE Botswana (383)
___ *S.c.conirostris* (Sundevall, 1850)	C Natal, S Transvaal, E Orange Free State

○ *Spizocorys fringillaris* (Sundevall, 1850) BOTHA'S LARK[2] — SE Transvaal, Orange Free State

○ *Spizocorys sclateri* (Shelley, 1902) SCLATER'S LARK[3] — S Namibia, NW and C Cape Province

○ *Spizocorys obbiensis* Witherby, 1905 OBBIA LARK — Obbia to Hal Hambo (coastal Somalia)

○ *Spizocorys personata* MASKED LARK

___ *S.p.personata* Sharpe, 1895	E Ethiopia
___ *S.p.yavelloensis* (Benson, 1947)	S Ethiopia to N Kenya (Dida Galgalu Desert)
___ *S.p.mcchesneyi* (Williams, 1957)	N Kenya (Marsabit)
___ *S.p.intensa* (Rothschild, 1931)	NC Kenya

EREMALAUDA W.L. Sclater, 1926 F
○ *Eremalauda dunni* DUNN'S LARK

___ *E.d.dunni* (Shelley, 1904)	S Sahara (Mauritania to C Sudan
___ *E.d.eremodites* R. Meinertzhagen, 1923	Syria and Israel to Arabia

○ *Eremalauda starki* (Shelley, 1902) STARK'S LARK[4] — SW Angola, Namibia, S and SW Botswana, W Cape Province

CHERSOPHILUS Sharpe, 1890 M
○ *Chersophilus duponti* DUPONT'S LARK

___ *C.d.duponti* (Vieillot, 1820)	Spain, N Morocco, N Algeria, NW Tunisia
___ *C.d.margaritae* (Koenig, 1888)	NC Algeria, S Tunisia, Libya, NW Egypt

PSEUDALAEMON Lort Philips, 1898 M
○ *Pseudalaemon fremantlii* SHORT-TAILED LARK

___ *P.f.fremantlii* (Lort Phillips, 1897)	Somalia, SE Ethiopia
___ *P.f.megaensis* Benson, 1946	S Ethiopia, N Kenya
___ *P.f.delamerei* Sharpe, 1900	S Kenya, N Tanzania

GALERIDA Boie, 1828 F
● *Galerida cristata* CRESTED LARK[5]

___ *G.c.pallida* (C.L. Brehm, 1858)	Portugal, Spain
___ *G.c.cristata* (Linnaeus, 1758)	C Europe to Baltic countries and N Ukraine
___ *G.c.neumanni* Hilgert, 1907	Tuscany to Rome (C Italy)
___ *G.c.apuliae* von Jordans, 1935	S and SE Italy, Sicily
___ *G.c.meridionalis* C.L. Brehm, 1841	W Croatia to S Bulgaria and C Greece
___ *G.c.cypriaca* Bianchi, 1907	Cyprus, Rhodes, Kárpathos
___ *G.c.tenuirostris* (C.L. Brehm, 1850)	E Croatia and E Hungary to Crimea, N Caucasus and SW Urals
___ *G.c.caucasica* Taczanowski, 1888	Caucasus, N Asia Minor, E Aegean Is.
___ *G.c.kleinschmidti* (Erlanger, 1899)	NW Morocco
___ *G.c.riggenbachi* E. Hartert, 1902	W Morocco

---

[1] Implicitly includes *griseovinacea* Clancey, 1972 (438), see Dean *et al.* (1992: 84) (630).
[2] Listed in a monotypic genus *Botha* in last Edition. For treatment in *Spizocorys* see Dean *et al.* (1992: 85) (630).
[3] Includes *capensis* see Dean *et al.* (1992: 86) (630).
[4] Includes *gregaria* Clancey, 1959 (383), see White (1961) (2645).
[5] European and Mediterranean races revised based on Cramp *et al.* (1988) (592) and similarly grouped. Eight more are recognised than in last Edition: *kleinschmidti, neumanni, apuliae, cypriaca, tenuirostris, helenae, brachyura* and *halfae*.

— *G.c.carthaginis* Kleinschmidt & Hilgert, 1905	Coastal NE Morocco to Tunisia
— *G.c.randoni* Loche, 1860	High plateaux of E Morocco to N Algeria
— *G.c.macrorhyncha* Tristram, 1859	S Morocco, N Algerian Sahara
— *G.c.arenicola* Tristram, 1859	S Tunisia, NW Libya
— *G.c.festae* E. Hartert, 1922	Coastal NE Libya
— *G.c.brachyura* Tristram, 1864	NE Egypt, Sinai, Israel, N Arabia and Iraq
— *G.c.helenae* Lavauden, 1926	SE Algeria, SW Libya
— *G.c.jordansi* Niethammer, 1955	Air Mts. (Niger)
— *G.c.nigricans* (C.L. Brehm, 1855)	Nile Delta (N Egypt)
— *G.c.maculata* C.L. Brehm, 1858	C Egypt
— *G.c.halfae* Nicoll, 1921	S Egypt, N Sudan
— *G.c.altirostris* (C.L. Brehm, 1855)	NC and NE Sudan, Eritrea
— *G.c.somaliensis* Reichenow, 1907	N Somalia, SE Ethiopia, N Kenya
— *G.c.balsaci* De Keyser & Villiers, 1950	Coastal Mauritania
— *G.c.senegallensis* (Statius Müller, 1776)	Senegal and Gambia to Sierra Leone and Niger
— *G.c.alexanderi* Neumann, 1908[1]	Nigeria to W Sudan
— *G.c.isabellina* Bonaparte, 1850	C Sudan
— *G.c.cinnamomina* E. Hartert, 1904	W Lebanon, NW Israel
— *G.c.zion* R. Meinertzhagen, 1920	S Turkey, Syria, E Lebanon, NE Israel
— *G.c.subtaurica* (Kollibay, 1912)	C Turkey to SW and N Iran and SW Turkmenistan
— *G.c.magna* Hume, 1871	NE Arabia to Kazakhstan, Nei Mongol and NW China
— *G.c.leautungensis* (Swinhoe, 1861)	S Manchuria, E China
— *G.c.coreensis* (Taczanowski, 1888)	Korea
— *G.c.lynesi* Whistler, 1928	N Pakistan (Gilgit)
— *G.c.chendoola* (Franklin, 1831)	Pakistan and N India

○ *Galerida theklae* THEKLA LARK

— *G.t.theklae* (A.E. Brehm, 1857)[2]	Portugal, Spain, Balearic Is
— *G.t.erlangeri* E. Hartert, 1904	N Morocco
— *G.t.ruficolor* Whitaker, 1898[3]	C and coastal NE Morocco, N Algeria and N Tunisia
— *G.t.theresae* R. Meinertzhagen, 1939	SW Morocco and Mauritania
— *G.t.superflua* E. Hartert, 1897	N Algeria (Atlas range) east to N Libya and NW Egypt
— *G.t.carolinae* Erlanger, 1897[4]	Algeria south of the Atlas range (N Sahara) to N Libya, NW Egypt
— *G.t.harrarensis* Érard & Jarry, 1973	Harar and Jijiga (E Ethiopia), NW Somalia (812)
— *G.t.huei* Érard & de Naurois, 1973	Bale Mts. (EC Ethiopia) (810)
— *G.t.praetermissa* (Blanford, 1869)	Highlands of Eritrea to C Ethiopia
— *G.t.ellioti* E. Hartert, 1897	N and C Somalia
— *G.t.mallablensis* Colston, 1982	Coastal S Somalia (551)
— *G.t.huriensis* Benson, 1947	S Ethiopia, N Kenya

○ *Galerida malabarica* (Scopoli, 1786) MALABAR LARK[5] — W India

○ *Galerida deva* (Sykes, 1832) SYKES'S LARK — WC and SC India

○ *Galerida modesta* SUN LARK

— *G.m.modesta* Heuglin, 1864[6]	Burkina Faso to S Sudan and NW Uganda
— *G.m.nigrita* (Grote, 1920)	Senegal, Gambia, Sierra Leone, Mali
— *G.m.struempelli* (Reichenow, 1901)	Cameroon
— *G.m.bucolica* (Hartlaub, 1887)	N Zaire

○ *Galerida magnirostris* LARGE-BILLED LARK[7]

— *G.m.magnirostris* (Stephens, 1826)	W and SW Cape Province
— *G.m.sedentaria* Clancey, 1993	Namibia to West Griqualand (524)
— *G.m.harei* (Roberts, 1924)	SW Transvaal, Orange Free State, Cape Province
— *G.m.montivaga* (Vincent, 1948)	Lesotho, SW Natal

---

[1] Includes *zalingei* see White (1961) (2645).
[2] Not C.L. Brehm as usually cited (S. Eck, *in litt.*).
[3] Includes *aguirrei contra* Peters (1960) (1829) and Cramp *et al.* (1988) (592) as its type locality is not within the range of *erlangeri* nor in Spanish Sahara (Roselaar, *in litt.*).
[4] Includes *deichleri* see Roselaar in Cramp *et al.* (1988: 172) (592).
[5] In last Edition united with *G. theklae* however we follow Cramp *et al.* (1988) (592) who did not do this.
[6] Includes *giffardi* see White (1961) (2645). Not *giffordi* as in Peters (1960:63) (1829).
[7] Treated in a monotypic genus *Calendula* in last Edition. For treatment in *Galerida* see Dean *et al.* (1992: 98) (630).

LULLULA Kaup, 1829 F
● ***Lullula arborea*** WOODLARK
    — *L.a.arborea* (Linnaeus, 1758)   N, C and NW Europe
    ☑ *L.a.pallida* Zarudny, 1902   S Europe, NW Africa, Middle East, Caucasus area, Iran, SW Turkmenistan

ALAUDA Linnaeus, 1758 F
● ***Alauda arvensis*** EURASIAN SKYLARK[1]
    ☑ *A.a.arvensis* Linnaeus, 1758   N, W and C Europe
    — *A.a.sierrae* Weigold, 1913   Portugal, NW, C and S Spain
    — *A.a.harterti* Whitaker, 1904   NW Africa
    — *A.a.cantarella* Bonaparte, 1850   S Europe from NE Spain and S France to C and N Turkey and Caucasus Mts.
    — *A.a.armenica* Bogdanov, 1879[2]   Mts. of SE Turkey, Transcaucasia, SW and N Iran
    — *A.a.dulcivox* Hume, 1872[3]   SE European Russia to W Siberia, Kazakhstan, N Xinjiang, SW Mongolia >> SE Europe and SW Asia
    — *A.a.kiborti* Zaliesski, 1917   Altai and NW Mongolia to Baikal area, Transbaikalia, NE and E Mongolia, NW Manchuria >> China
    — *A.a.intermedia* Swinhoe, 1863[4]   N Siberia in basins of mid-Lena, Vilyuy, Yana, Indigirka, and Kolyma rivers south to N China and Korea and the Amur valley >> E and SE Asia
    — *A.a.pekinensis* Swinhoe, 1863   NE Siberia from Magadan area and Koryakland to Kamchatka and N Kuril Is. >> Japan and E China
    — *A.a.lonnbergi* Hachisuka, 1926   N Sakhalin I., Korea, NE China, Japan
    — *A.a.japonica* Temminck & Schlegel, 1848[5]   S Sakhalin, S Kuril Is., Japan, Ryukyu Is.

○ ***Alauda gulgula*** ORIENTAL SKYLARK
    — *A.g.lhamarum* R. & A. Meinertzhagen, 1926   Pamir Mts. and W Tibetan Plateau from Kashmir to SW Xizang and N Nepal
    — *A.g.inopinata* Bianchi, 1904   N, E and SE Tibetan Plateau
    — *A.g.vernayi* Mayr, 1941   N Bhutan, SE Xizang, N Burma, W Yunnan
    — *A.g.inconspicua* Severtsov, 1873[6]   W Turkmenistan and E Iran to S Kazakhstan, SW Tajikistan and Aghanistan, Pakistan and NW India >> SW Asia
    — *A.g.gulgula* Franklin, 1831   NC and NE India, Sri Lanka, Burma, NW Indochina
    — *A.g.dharmakumarsinhjii* Abdulali, 1976   WC India (11)
    — *A.g.australis* W.E. Brooks, 1873   SW India
    — *A.g.weigoldi* E. Hartert, 1922   C and E China (S and E Sichuan to S Shaanxi and Yangtze valley)
    — *A.g.coelivox* Swinhoe, 1859   SE China, N Vietnam
    — *A.g.sala* Swinhoe, 1870   Hainan I.
    — *A.g.herberti* E. Hartert, 1923   C and SE Thailand, Cambodia, S Vietnam
    — *A.g.wattersi* Swinhoe, 1871   Taiwan
    — *A.g.wolfei* Hachisuka, 1930   Luzon (Philippines)

○ ***Alauda razae*** (Alexander, 1898) RASO ISLAND LARK[7]   Raso I., Cape Verde Is.

EREMOPTERIX Kaup, 1836 M[8]
○ ***Eremopterix australis*** (A. Smith, 1836) BLACK-EARED SPARROW-LARK   S Namibia, S Botswana, Cape Province

○ ***Eremopterix leucotis*** CHESTNUT-BACKED SPARROW-LARK
    — *E.l.melanocephalus* (M.H.K. Lichtenstein, 1823)   Senegal and Gambia to Nile Valley
    — *E.l.leucotis* (Stanley, 1814)   E and S Sudan to Ethiopia

---

[1] The limits of Eastern populations are disputed and various authors place synonyms differently.
[2] For recognition see Cramp *et al.* (1988) (592).
[3] For a discussion of the various authors of this name see Dickinson *et al.* (2001) (748), Hume was the first to use the name with a description. Includes *dementievi* Korelov, 1953 (1298) see Stepanyan (1990: 351) (2326).
[4] Includes *nigrescens* see Dickinson *et al.* (2001) (748), although C.S. Roselaar (*in litt.*) would recognise this.
[5] Some authors, including CSR, believe that *lonnbergi* and *japonica* breed sympatrically in Sakhalin and that *japonica* must either be placed with *gulgula* or treated as a third species. Orn. Soc. Japan (2000) (1727) did not accept this and considered both forms part of *arvensis*.
[6] We include *punjaubi* here, but based on his studies CSR believes a breeding population of Pakistan and NW India should be recognised under this name. This awaits publication.
[7] For notes supporting treatment in *Alauda* see Hazevoet (1989) (1091).
[8] Gender addressed by David & Gosselin (2002) (614), multiple changes occur in specific names.

___ *E.l.madaraszi* (Reichenow, 1902) S Somalia, Kenya, NE and C Tanzania, N Mozambique, N Malawi

___ *E.l.hoeschi* C.M.N. White, 1959 N Namibia, S Angola, N Botswana to W Zimbabwe (2638)

___ *E.l.smithi* (Bonaparte, 1850)[1] S Malawi and S Zambia to NE Sth. Africa

◯ *Eremopterix signatus* CHESTNUT-HEADED SPARROW-LARK
___ *E.s.harrisoni* (Ogilvie-Grant, 1901) SE Sudan, NW Kenya
___ *E.s.signatus* (Oustalet, 1886) S and E Ethiopia, Somalia, N and E Kenya

◯ *Eremopterix verticalis* GREY-BACKED SPARROW-LARK[2]
___ *E.v.khama* Irwin, 1957 NE Botswana, Zimbabwe, S Zambia
___ *E.v.harti* Benson & Irwin, 1965 SW Zambia (131)
___ *E.v.damarensis* Roberts, 1931 W Angola, Namibia, NW Botswana, SW Zambia, NW Cape Province
___ *E.v.verticalis* (A. Smith, 1836) Cape Province to W Transvaal, S Botswana and SW Zimbabwe

◯ *Eremopterix nigriceps* BLACK-CROWNED SPARROW-LARK
___ *E.n.nigriceps* (Gould, 1839)[3] Cape Verde Is.
___ *E.n.albifrons* Sundevall, 1850 Mauritania to Nile valley
___ *E.n.melanauchen* (Cabanis, 1851) E Sudan, Eritrea, Ethiopia, Somalia, Arabia, S Iraq, S Iran
___ *E.n.forbeswatsoni* Ripley & Bond, 1966 Socotra I.[4] (2063)
___ *E.n.affinis* (Blyth, 1867) Pakistan, NW India

◯ *Eremopterix griseus* (Scopoli, 1786) ASHY-CROWNED SPARROW-LARK Pakistan, India, S Nepal, Bangladesh, Sri Lanka

◯ *Eremopterix leucopareia* (Fischer & Reichenow, 1884) FISCHER'S SPARROW-LARK NE Uganda, Kenya, Tanzania, NE Zambia, N Malawi

EREMOPHILA Boie, 1828 F
● *Eremophila alpestris* HORNED LARK/SHORELARK
___ *E.a.arcticola* (Oberholser, 1902) N Alaska, W Canada, NW USA
___ *E.a.alpina* (Jewett, 1943) High Mts. of W Washington State
___ *E.a.hoyti* (Bishop, 1896) Arctic coast N Canada >> N USA
___ *E.a.alpestris* (Linnaeus, 1758) NE Canada >> SE USA
___ *E.a.leucolaema* Coues, 1874 S Canada, C and S USA >> N Mexico
___ *E.a.enthymia* (Oberholser, 1902) C Canada, C USA >> N Mexico
___ *E.a.praticola* (Henshaw, 1884) SE Canada, C and EC USA
___ *E.a.strigata* (Henshaw, 1884) Coastal Brit. Columbia and NW USA
___ *E.a.merrilli* (Dwight, 1890) Pacific slope W Canada, W USA
___ *E.a.lamprochroma* (Oberholser, 1932) Inland montane W and SW USA
___ *E.a.utahensis* (Behle, 1938) Montane WC USA
___ *E.a.sierrae* (Oberholser, 1920) Montane NE California
___ *E.a.rubea* (Henshaw, 1884) C California
___ *E.a.actia* (Oberholser, 1902) Coastal range S California, N Baja California
___ *E.a.insularis* (Dwight, 1890) Islands off S California
___ *E.a.ammophila* (Oberholser, 1902) Deserts of SE California and nearby Nevada >> NW Mexico
___ *E.a.leucansiptila* (Oberholser, 1902) Colorado desert SW Nevada, W Arizona, NE Baja California, and NW Sonora (Mexico)
___ *E.a.occidentalis* (McCall, 1851) N Arizona, N New Mexico >> N Mexico
___ *E.a.adusta* (Dwight, 1890) S Arizona, SW New Mexico, N Sonora (Mexico)
___ *E.a.giraudi* (Henshaw, 1884) Coastal regions of S Texas, NC Mexico
___ *E.a.enertera* (Oberholser, 1907) WC Baja California
___ *E.a.aphrasta* (Oberholser, 1902) NW Mexico
___ *E.a.lactea* A.R. Phillips, 1970 Coahuila (NE Mexico) (1874)
___ *E.a.diaphora* (Oberholser, 1902) NE Mexico (south of *lactea*)
___ *E.a.chrysolaema* (Wagler, 1831) SC Mexico

---

[1] Given as 1851 by Peters (1960: 30) (1829). Zimmer (1926) (2707) accepted this but see Hartlaub (1951) (1087).
[2] Three races are said to overlap in SW Zambia, but species nomadic and breeding ranges may be discrete.
[3] Peters (1960: 31) (1829) dated this 1841 but see Zimmer (1926: 159) (2707) and www.zoonomen.net (A.P. Peterson *in litt.*).
[4] For recognition see Dean *et al.* (1992: 115) (630).

— *E.a.oaxacae* (Nelson, 1897)   S Mexico

— *E.a.peregrina* (P.L. Sclater, 1855)   Colombia

— *E.a.atlas* (Whitaker, 1898)   Mts. of Morocco

— *E.a.flava* (J.F. Gmelin, 1789)   N Europe, N Asia S to Lake Baikal and NW Amurland

— *E.a.brandti* (Dresser, 1874)   SE European Russia to N Turkmenistan, Mongolia, Inner Mongolia, Manchuria, and the Baikal area

— *E.a.przewalskii* (Bianchi, 1904)   Qaidam Basin (N Qinghai, China)

— *E.a.balcanica* (Reichenow, 1895)   Balkans to Greece

— *E.a.bicornis* (C.L. Brehm, 1842)   Lebanon

— *E.a.kumerloevei* Roselaar, 1995   W and C Asia Minor [2107]

— *E.a.penicillata* (Gould, 1838)   E Turkey, Caucasus area, SW and N Iran

— *E.a.albigula* (Bonaparte, 1850)   NE Iran and SW Turkmenistan to NW Pakistan and Tien Shan Mts.

— *E.a.longirostris* (F. Moore, 1856)   NE Pakistan, Kashmir, NW Himalayas

— *E.a.teleschowi* (Przevalski, 1887)   C and E Kun Lun Mts. (S Xinjiang, N Xizang and NW Qinghai, China)

— *E.a.argalea* (Oberholser, 1902)   W Kun Lun Mts., N Ladakh, and W Tibetan Plateau

— *E.a.elwesi* (Blanford, 1872)   S and E Tibetan Plateau

— *E.a.khamensis* (Bianchi, 1904)   SE Xizang and W Sichuan

— *E.a.nigrifrons* (Przevalski, 1876)   NE Qinghai (China)

○ ***Eremophila bilopha*** (Temminck, 1823)   TEMMINCK'S LARK   N Africa, N and NE Arabia, W Iraq, S Syria, Jordan

CISTICOLIDAE CISTICOLAS AND ALLIES (21: 110)

CISTICOLA Kaup, 1829   M[1]

○ ***Cisticola erythrops***   RED-FACED CISTICOLA

— *C.e.erythrops* (Hartlaub, 1857)   Gambia to Central African Republic

— *C.e.pyrrhomitra* Reichenow, 1916[2]   Ethiopia, SE Sudan

— *C.e.niloticus* Madarász, 1914   SW Sudan

— *C.e.sylvia* Reichenow, 1904   NE Zaire and S Sudan to Tanzania and S Zaire

— *C.e.nyasa* Lynes, 1930[3]   S Tanzania to N Botswana and Natal

— *C.e.lepe* Lynes, 1930[4]   Angola

○ ***Cisticola cantans***   SINGING CISTICOLA

— *C.c.swanzii* (Sharpe, 1870)   Gambia to S Nigeria

— *C.c.concolor* (Heuglin, 1896)   N Nigeria to Sudan

— *C.c.adamauae* Reichenow, 1910[5]   Cameroon, Congo, NW Zaire

— *C.c.cantans* (Heuglin, 1869)   W and C Ethiopia, Eritrea

— *C.c.belli* Ogilvie-Grant, 1908   Central African Republic to E Zaire, Uganda and NW Tanzania

— *C.c.pictipennis* Madarász, 1904   Kenya, N Tanzania

— *C.c.muenzneri* Reichenow, 1916   S Tanzania to Zimbabwe

○ ***Cisticola lateralis***   WHISTLING CISTICOLA

— *C.l.lateralis* (Fraser, 1843)   Gambia to Cameroon

— *C.l.antinorii* (Heuglin, 1869)   Central African Republic to W Kenya

— *C.l.modestus* (Bocage, 1880)   Gabon to N Angola and S Zaire

○ ***Cisticola woosnami***   TRILLING CISTICOLA

— *C.w.woosnami* Ogilvie-Grant, 1908   E Zaire, Uganda to N Tanzania

— *C.w.lufira* Lynes, 1930   SW Tanzania, N Malawi, N and C Zambia, SE Zaire

○ ***Cisticola anonymus*** (J.W. von Müller, 1855)   CHATTERING CISTICOLA   SE Sierra Leone, S Nigeria to E Zaire and N Angola border

○ ***Cisticola bulliens***   BUBBLING CISTICOLA

— *C.b.bulliens* Lynes, 1930   W Angola

— *C.b.septentrionalis* Tye, 1992   Lower Congo R., NW Angola [2446]

[1] See examples under Art. 30.1.4.2. in I.C.Z.N. (1999: 35) [1178] for *Petricola*; a parallel case.
[2] For correct spelling see David & Gosselin (2002) [614].
[3] Includes *arcanus* and *elusus* see Tye (1997: 141) [2448].
[4] For comments on suggestions that this be treated as a separate species see Tye (1997: 141) [2448].
[5] For recognition see Tye (1997: 143) [2448].

○ **Cisticola chubbi** CHUBB'S CISTICOLA
 — *C.c.chubbi* Sharpe, 1892   E Zaire to W Kenya
 — *C.c.marungensis* Chapin, 1932   Marungu Plateau (SE Zaire)
 — *C.c.adametzi* Reichenow, 1910   Nigeria and Cameroon
 — *C.c.discolor* Sjöstedt, 1893   Mt. Cameroon

○ **Cisticola hunteri** Shelley, 1889 HUNTER'S CISTICOLA   W and C Kenya, N Tanzania

○ **Cisticola nigriloris** Shelley, 1897 BLACK-LORED CISTICOLA   S Tanzania, N Malawi, NE Zambia

○ **Cisticola aberrans** ROCK-LOVING CISTICOLA
 — *C.a.petrophilus* Alexander, 1907   N Nigeria to SW Sudan and NE Zaire
 — *C.a.admiralis* Bates, 1930   Guinea, Mali, Sierra Leone, S Ghana
 — *C.a.emini* Reichenow, 1892[1]   S Kenya, N Tanzania
 — *C.a.bailunduensis* Neumann, 1931   C Angola
 — *C.a.lurio* Vincent, 1933   NW Mozambique, E Malawi
 — *C.a.nyika* Lynes, 1930   SW Tanzania, E Zambia to Zimbabwe and
                                                 W Mozambique

 — *C.a.aberrans* (A. Smith, 1843)   SE Botswana, Transvaal, W Swaziland, highland Natal,
                                                 E Orange Free State

 — *C.a.minor* Roberts, 1913   E Cape Province, lowland Natal, E Swaziland,
                                                 S Mozambique

○ **Cisticola chiniana** RATTLING CISTICOLA[2]
 — *C.c.simplex* (Heuglin, 1869)   S Sudan to N Uganda
 — *C.c.fricki* Mearns, 1913   S Ethiopia, N Kenya
 — *C.c.fortis* Lynes, 1930   Gabon, S Congo, S Zaire, N and C Angola, Zambia
 — *C.c.humilis* Madarász, 1904   W Kenya, NE Uganda
 — *C.c.fischeri* Reichenow, 1891   NC Tanzania
 — *C.c.ukamba* Lynes, 1930   E Kenya, NE Tanzania
 — *C.c.victoria* Lynes, 1930   SW Kenya, N Tanzania
 — *C.c.heterophrys* Oberholser, 1906   Coastal Kenya and Tanzania
 — *C.c.keithi* Parkes, 1987   SC Tanzania (1775)
 — *C.c.mbeya* Parkes, 1987   S Tanzania (1775)
 — *C.c.emendatus* Vincent, 1944[3]   SE Tanzania, N Mozambique, Malawi, E Zambia
 — *C.c.procerus* W.K.H. Peters, 1868   Tete (Mozambique), S Malawi
 — *C.c.frater* Reichenow, 1916   C Namibia
 — *C.c.bensoni* Traylor, 1964   S Zambia
 — *C.c.smithersi* B.P. Hall, 1956[4]   W Zimbabwe, NW Botswana, SW Zambia, N Namibia,
                                                 S Angola

 — *C.c.chiniana* (A. Smith, 1843)[5]   Zimbabwe and S Mozambique to Transvaal and
                                                 SE Botswana

 — *C.c.campestris* Gould, 1845   Natal, Swaziland, coastal S Mozambique

○ **Cisticola bodessa** BORAN CISTICOLA
 — *C.b.kaffensis* Érard, 1974   Kaffa (Ethiopia)
 — *C.b.bodessa* Mearns, 1913   SE Sudan, C and S Ethiopia, N and C Kenya

○ **Cisticola njombe** CHURRING CISTICOLA
 — *C.n.njombe* Lynes, 1933   S Tanzania
 — *C.n.mariae* Benson, 1945[6]   N Malawi, NE Zambia

○ **Cisticola cinereolus** ASHY CISTICOLA
 — *C.c.cinereolus* Salvadori, 1888   NE Ethiopia, N Somalia
 — *C.c.schillingsi* Reichenow, 1905[7]   S Ethiopia and SE Sudan to N Tanzania

○ **Cisticola restrictus** Traylor, 1967 TANA RIVER CISTICOLA   Lower Tana River basin (E Kenya)

---

[1] Includes *teitensis* see Tye (1997: 157) (2448).
[2] For correct spelling see David & Gosselin (2002) (613).
[3] For recognition see Tye (1997: 162) (2448).
[4] Includes *huilensis* see Tye (1997: 162) (2448).
[5] Includes *vulpiniceps* Clancey, 1992 (523), see Tye (1997: 162) (2448).
[6] For recognition see Tye (1997: 165) (2448).
[7] For recognition see Tye (1997: 166) (2448).

○ *Cisticola rufilatus* TINKLING CISTICOLA
  — *C.r.ansorgei* Neumann, 1906[1]   Angola to S Zaire, N Zambia and Malawi
  — *C.r.vicinior* Clancey, 1973   C Zimbabwe
  — *C.r.rufilatus* (Hartlaub, 1870)   N Namibia to N Cape Province and S Zambia

○ *Cisticola subruficapilla* GREY-BACKED CISTICOLA[2]
  — *C.s.newtoni* da Rosa Pinto, 1967   NW Namibia, SW Angola
  — *C.s.windhoekensis* (Roberts, 1937)   C Namibia
  — *C.s.karasensis* (Roberts, 1937)   S Namibia, NW Cape Province
  — *C.s.namaqua* Lynes, 1930   NW Cape Province
  — *C.s.subruficapilla* (A. Smith, 1843)   SW Cape Province
  — *C.s.jamesi* Lynes, 1930[3]   SE Cape Province

○ *Cisticola lais* WAILING CISTICOLA
  — *C.l.distinctus* Lynes, 1930   E Uganda, W and C Kenya
  — *C.l.namba* Lynes, 1931   Angola
  — *C.l.semifasciatus* Reichenow, 1905   S Tanzania, E Zambia, Malawi, NW Mozambique
  — *C.l.mashona* Lynes, 1930   Zimbabwe, S Mozambique, N Transvaal, Swaziland
  — *C.l.oreobates* Irwin, 1966   Mt. Gorongoza (Mozambique)
  — *C.l.monticola* Roberts, 1913   SW Transvaal
  — *C.l.lais* (Hartlaub & Finsch, 1870)[4]   SE Transvaal, E Orange Free State to Natal, E Cape
                                       Province
  — *C.l.maculatus* Lynes, 1930   S Cape Province

○ *Cisticola galactotes* WINDING CISTICOLA
  — *C.g.amphilectus* Reichenow, 1875[5]   Senegal to W Congo basin
  — *C.g.zalingei* Lynes, 1930   N Nigeria to W Sudan
  — *C.g.lugubris* (Rüppell, 1840)   Ethiopia
  — *C.g.marginatus* (Heuglin, 1869)   S Sudan, N Uganda
  — *C.g.haematocephala* Cabanis, 1868   Coastal S Somalia, Kenya, N Tanzania
  — *C.g.nyansae* Neumann, 1905   C and E Zaire, Uganda, W Kenya
  — *C.g.suahelicus* Neumann, 1905   C Tanzania, SE Zaire, N Zambia, Malawi,
                                       ?N Mozambique
  — *C.g.schoutedeni* C.M.N. White, 1954   W and NW Zambia
  — *C.g.luapula* Lynes, 1933   NE Zambia
  — *C.g.stagnans* Clancey, 1969   N Botswana, S Zambia, W Zimbabwe
  — *C.g.isodactylus* W.K.H. Peters, 1868[6]   S Malawi, SE Zimbabwe, W Mozambique
  — *C.g.galactotes* (Temminck, 1821)   Coastal Natal and Zululand

○ *Cisticola pipiens* CHIRPING CISTICOLA
  — *C.p.congo* Lynes, 1936   E Angola, Zaire, Zambia, Burundi, Tanzania
  — *C.p.pipiens* Lynes, 1930   W Angola
  — *C.p.arundicola* Clancey, 1969   Zimbabwe, Botswana, SE Angola, Namibia

○ *Cisticola carruthersi* Ogilvie-Grant, 1909 CARRUTHERS'S CISTICOLA   E and NE Zaire, Rwanda, Burundi, Uganda, W Kenya,
                                       NW Tanzania

○ *Cisticola tinniens* LEVAILLANT'S CISTICOLA
  — *C.t.dyleffi* Prigogine, 1952   E Zaire
  — *C.t.oreophilus* van Someren, 1922   C and W Kenya
  — *C.t.shiwae* C.M.N. White, 1947   NE Zambia, SE Zaire, SW Tanzania
  — *C.t.perpullus* E. Hartert, 1920   Angola, NW Zambia, S Zaire
  — *C.t.tinniens* (M.H.K. Lichtenstein, 1842)   Zimbabwe to Sth. Africa
  — *C.t.brookei* Herremans *et al.*[7], 1999   W Cape Province (1109)

○ *Cisticola robustus* STOUT CISTICOLA
  — *C.r.robustus* (Rüppell, 1845)   N Ethiopian plateau

---

[1] Includes *venustulus* see Irwin (1997: 168) (1192).
[2] For correct spelling see David & Gosselin (2002) (613).
[3] Includes *euroa* see Irwin (1997: 169) (1192).
[4] Implicitly includes *oreodytes* Clancey, 1982 (490) — not in Watson *et al.* (1986) (2564).
[5] Includes *griseus* see Tye (1997: 173) (2448).
[6] Tye (1997: 174) (2448) placed this in *suahelicus*, but this union should require use of the older name. We retain both pending review.
[7] When the number of authors exceeds five we have listed only the first author.

__ *C.r.schraderi* Neumann, 1906[1]	C Eritrea, N Ethiopia
__ *C.r.omo* Neumann & Lynes, 1928	SW Ethiopia
__ *C.r.santae* Bates, 1926	W Cameroon
__ *C.r.nuchalis* Reichenow, 1893[2]	E Zaire, S Sudan to W and S Kenya, N Tanzania
__ *C.r.awemba* Lynes, 1933	SW Tanzania, SE Zaire, NE Zambia
__ *C.r.angolensis* (Bocage, 1877)[3]	S Zaire, NE and C Angola, NW Zambia

O *Cisticola aberdare* Lynes, 1930 ABERDARE CISTICOLA — C Kenya

O *Cisticola natalensis* CROAKING CISTICOLA

__ *C.n.strangei* (Fraser, 1843)[4]	Senegal and Gambia to W Zaire east to S Sudan, Uganda and C and S Kenya
__ *C.n.inexpectatus* Neumann, 1906	Ethiopia
__ *C.n.argenteus* Reichenow, 1905	SE Somalia, S Ethiopia, N Kenya
__ *C.n.tonga* Lynes, 1930	C Sudan
__ *C.n.katanga* Lynes, 1930	NE Angola, SE Zaire, NE Zambia, N Malawi, SW Tanzania
__ *C.n.huambo* Lynes, 1930	NW and C Angola
__ *C.n.natalensis* (A. Smith, 1843)[5]	S Tanzania to E Zimbabwe, Mozambique and Natal
__ *C.n.holubii* (Pelzeln, 1882)	N Botswana, W Zimbabwe, S Zambia

O *Cisticola ruficeps* RED-PATE CISTICOLA

__ *C.r.guinea* Lynes, 1930	Senegal and Gambia to N Nigeria and Cameroon
__ *C.r.ruficeps* (Cretzschmar, 1827)	Chad to W and SW Sudan
__ *C.r.scotoptera* (Sundevall, 1850)	C and E Sudan to Eritrea
__ *C.r.mongalla* Lynes, 1930[6]	S Sudan, N Uganda

O *Cisticola dorsti* Chappuis & Érard, 1991 DORST'S CISTICOLA # — NW Nigeria, N Cameroon, S Chad (317)

O *Cisticola nanus* Fischer & Reichenow, 1884 TINY CISTICOLA — C and S Ethiopia to C and S Kenya, NE and E Tanzania

O *Cisticola brachypterus* SHORT-WINGED CISTICOLA

__ *C.b.brachypterus* (Sharpe, 1870)	W Africa to SW Sudan south to NW Angola
__ *C.b.hypoxanthus* Hartlaub, 1881[7]	N Uganda, NE Zaire, SE Sudan
__ *C.b.zedlitzi* Reichenow, 1909	Eritrea, Ethiopia
__ *C.b.reichenowi* Mearns, 1911	S Somalia, Coastal Kenya and N Tanzania
__ *C.b.ankole* Lynes, 1930	S Uganda, Rwanda, Burundi, NW Tanzania, E Zaire
__ *C.b.kericho* Lynes, 1930	SW Kenya
__ *C.b.katonae* Madarász, 1904	C and S Kenya, N Tanzania
__ *C.b.loanda* Lynes, 1930	C Angola, SE Zaire, W Zambia
__ *C.b.isabellinus* Reichenow, 1907[8]	C Tanzania to Mozambique and Zimbabwe west to E Zambia

O *Cisticola rufus* (Fraser, 1843) RUFOUS CISTICOLA — Senegal and Gambia to S Chad

O *Cisticola troglodytes* FOXY CISTICOLA

__ *C.t.troglodytes* (Antinori, 1864)	Central African Republic to S Sudan and NW Kenya
__ *C.t.ferrugineus* Heuglin, 1864	E Sudan, W Ethiopia

O *Cisticola fulvicapilla* PIPING CISTICOLA/NEDDICKY[9]

__ *C.f.dispar* Sousa, 1887	W Zaire, C Angola, NW Zambia
__ *C.f.muelleri* Alexander, 1899	NW Zambia to Mozambique
__ *C.f.hallae* Benson, 1955	S Angola, SW Zambia, NW Zimbabwe, N Namibia, N Botswana
__ *C.f.dexter* Clancey, 1971	E Botswana, S Zimbabwe, N Transvaal
__ *C.f.ruficapilla* (A. Smith, 1842)	S Transvaal to N and NE Cape Province
__ *C.f.lebombo* (Roberts, 1936)	S Mozambique to E Transvaal, W Natal

---

[1] For recognition see Tye (1997: 183) (2448).
[2] Includes *ambiguus* see Tye (1997: 183) (2448).
[3] For comments on the proposal to treat *angolensis* as a species (mentioned in Note 211 at the end of our previous Edition), and the subsequent retraction of that proposal, see Tye (1997: 183) (2448).
[4] Includes *kapitensis* and *littoralis* see Tye (1997: 186) (2448).
[5] Includes *matengorum* see White (1960: 671) (2639) and *vigilax* Clancey, 1994 (526).
[6] Tye (1997: 188) (2448) explained that calls ascribed to this (leading to Note 209 at the end of our previous Edition) belong instead to *C. dorsti*.
[7] For recognition see Tye (1997: 191) (2448).
[8] Includes *tenebricosus* see Type (1997: 192) (2448).
[9] For correct spelling see David & Gosselin (2002) (613).

— *C.f.fulvicapilla* (Vieillot, 1817) — Interior E Cape Province to W Lesotho
— *C.f.dumicola* Clancey, 1983 — W Zululand and E Natal to S and E Cape Province
— *C.f.silberbaueri* (Roberts, 1919) — SW Cape Province

○ *Cisticola angusticauda* Reichenow, 1891 LONG-TAILED CISTICOLA[1] — SW Kenya to N Zambia

○ *Cisticola melanurus* (Cabanis, 1882) BLACK-TAILED CISTICOLA — N Angola, S Zaire

● *Cisticola juncidis* FAN-TAILED CISTICOLA/ZITTING CISTICOLA
— *C.j.cisticola* (Temminck, 1820)[2] — W France, Iberia, NW Africa
— *C.j.juncidis* (Rafinesque, 1810) — S France and Italy to Balkans and W Asia Minor; Egypt
— *C.j.uropygialis* (Fraser, 1843)[3] — Senegal and Gambia to Ethiopia south to Nigeria, Rwanda and N Tanzania
— *C.j.terrestris* (A. Smith, 1842) — Sthn. Africa
— *C.j.neuroticus* R. Meinertzhagen, 1920 — SC Turkey, Cyprus and Levant to SW Iran
— *C.j.cursitans* (Franklin, 1831) — Pakistan, India (except SW) to dry zone Sri Lanka, east to N and C Burma, Yunnan
— *C.j.salimalii* Whistler, 1936 — SW India
— *C.j.omalurus* Blyth, 1851 — Wet zone Sri Lanka
— *C.j.brunniceps* (Temminck & Schlegel, 1850) — S Korea, Japan >> Philippines
— *C.j.tinnabulans* (Swinhoe, 1859)[4] — C and NE China to Thailand, Indochina, Philippines (except Palawan group) and Taiwan
— *C.j.nigrostriatus* Parkes, 1971 — Palawan and Culion I.
— *C.j.malaya* Lynes, 1930[5] — Nicobars, SE Burma and SW Thailand to Malay Pen., Sumatra and islands, Java[6]
— *C.j.fuscicapilla* Wallace, 1864 — Kangean I. (Java Sea), Lesser Sundas, Moluccas
— *C.j.constans* Lynes, 1938 — Sulawesi, Buton and Peleng Is.
— *C.j.leanyeri* Givens & Hitchcock, 1953 — NE Western Australia, N Northern Territory (Arnhem Land), Melville I.
— *C.j.normani* Mathews, 1914 — NW Queensland (S and SE Gulf of Carpentaria)
— *C.j.laveryi* Schodde & Mason, 1979 — N and E Cape York Pen., NE Queensland coast

○ *Cisticola haesitatus* (Sclater & Hartlaub, 1881) SOCOTRA CISTICOLA — Socotra I.

○ *Cisticola cherina* (A. Smith, 1842) MADAGASCAN CISTICOLA[7] — Madagascar

○ *Cisticola aridulus* DESERT CISTICOLA
— *C.a.aridulus* Witherby, 1900 — Mali, Niger, N Nigeria to C Sudan
— *C.a.lavendulae* Ogilvie-Grant & Reid, 1901 — Ethiopia to coastal Eritrea, NW Somalia
— *C.a.tanganyika* Lynes, 1930 — Kenya, N Tanzania
— *C.a.lobito* Lynes, 1930 — Coastal Angola
— *C.a.perplexus* C.M.N. White, 1947 — N Zambia
— *C.a.kalahari* Ogilvie-Grant, 1910 — C Namibia to Transvaal
— *C.a.traylori* Benson & Irwin, 1966 — E Angola, W Zambia
— *C.a.caliginus* Clancey, 1955 — S Mozambique, NE Transvaal, Swaziland and Natal
— *C.a.eremicus* Clancey, 1984 — S Angola and N Namibia to Zimbabwe

○ *Cisticola textrix* CLOUD CISTICOLA
— *C.t.bulubulu* Lynes, 1931 — W Angola
— *C.t.anselli* C.M.N. White, 1960 — E Angola, NW Zambia
— *C.t.major* (Roberts, 1913) — Transvaal to E Cape Province
— *C.t.marleyi* (Roberts, 1932) — S Mozambique, NE Zululand, coastal Natal
— *C.t.textrix* (Vieillot, 1817) — S Cape Province

○ *Cisticola eximius* BLACK-BACKED CISTICOLA
— *C.e.occidens* Lynes, 1930 — S Senegal to Nigeria
— *C.e.winneba* Lynes, 1931 — S Ghana
— *C.e.eximius* (Heuglin, 1869) — Ethiopia to Uganda and W to S Central African Republic

[1] For correct spelling see David & Gosselin (2002) (613).
[2] For recognition see Cramp *et al.* (1992: 20) (593); not recognised by Fry (1997: 200) (891).
[3] Includes *perennius* see Fry (1997: 200) (891).
[4] Original spelling of Swinhoe (1859) (2381) verified. Has later been misspelled *tintinnabulans*.
[5] Lynes's MS name appeared a year earlier in Bartels & Stresemann (1929) (101) see Mees (1996) (1543). No description was attached to it.
[6] All of Java, see Mees (1996) (1543).
[7] For correct spelling see David & Gosselin (2002) (613).

○ *Cisticola dambo* DAMBO CISTICOLA
    — *C.d.kasai* Lynes, 1936   NW Kasai (Zaire)
    — *C.d.dambo* Lynes, 1931   E Angola, S Zaire, NW Zambia

○ *Cisticola brunnescens* PECTORAL-PATCH CISTICOLA
    — *C.b.mbangensis* Chappuis & Érard, 1973   N Cameroon
    — *C.b.lynesi* Bates, 1926   W Cameroon
    — *C.b.wambera* Lynes, 1930   NW Ethiopia
    — *C.b.brunnescens* Heuglin, 1862   Ethiopia, NW Somalia
    — *C.b.nakuruensis* van Someren, 1922   WC Kenya to N Tanzania
    — *C.b.hindii* Sharpe, 1896   EC Kenya to NE Tanzania

○ *Cisticola cinnamomeus* PALE-CROWNED CISTICOLA[1]
    — *C.c.midcongo* Lynes, 1938   SE Gabon to middle Congo R.
    — *C.c.cinnamomeus* Reichenow, 1904   SE Zaire, S Tanzania, Zambia, Zimbabwe, N Botswana,
                                        Angola
    — *C.c.egregius* (Roberts, 1913)[2]   S Mozambique to E Sth. Africa

○ *Cisticola ayresii* WING-SNAPPING CISTICOLA
    — *C.a.gabun* Lynes, 1931   Gabon, Congo, NW Zaire
    — *C.a.imatong* Cave, 1938   S Sudan
    — *C.a.entebbe* Lynes, 1930   E Zaire, Rwanda, Uganda, NW Tanzania
    — *C.a.itombwensis* Prigogine, 1957   E Zaire
    — *C.a.mauensis* van Someren, 1922   W and C Kenya
    — *C.a.ayresii* Hartlaub, 1863   Angola, S Zaire, S Tanzania to Zimbabwe and E Sth.
                                        Africa

○ *Cisticola exilis* GOLDEN-HEADED CISTICOLA
    — *C.e.tytleri* Jerdon, 1863   N and NE India, Nepal, Bangladesh, W, N and
                                  C Burma, W Yunnan
    — *C.e.erythrocephalus* Blyth, 1851   SW India
    — *C.e.equicaudatus* E.C.S. Baker, 1924   E Burma, Thailand, Laos, Cambodia, S Vietnam
    — *C.e.courtoisi* La Touche, 1926   S and E China, ?NW Vietnam
    — *C.e.volitans* (Swinhoe, 1859)   Taiwan
    — *C.e.semirufus* Cabanis, 1872   Philippines, Sulu Arch.
    — *C.e.rusticus* Wallace, 1863   Sulawesi, Moluccas
    — *C.e.lineocapilla* (Gould, 1847)   Java, Bali, Lesser Sundas, NW Australia (N Kimberley,
                                        Arnhem Land)
    — *C.e.diminutus* Mathews, 1922   NE Queensland, Cape York Pen., New Guinea[3]
    — *C.e.alexandrae* Mathews, 1912   Inland N Australia
    — *C.e.exilis* (Vigors & Horsfield, 1827)[4]   E and SE Australia
    — *C.e.polionotus* Mayr, 1934   Bismarck Arch.

INCANA Lynes, 1930[5] F
○ *Incana incana* (Sclater & Hartlaub, 1881) SOCOTRA WARBLER   Socotra I.

SCOTOCERCA Sundevall, 1872 F
○ *Scotocerca inquieta* STREAKED SCRUB WARBLER
    — *S.i.theresae* R. Meinertzhagen, 1939   SW and C Morocco
    — *S.i.saharae* (Loche, 1858)   SE Morocco to NW Libya
    — *S.i.inquieta* (Cretzschmar, 1827)   NE Libya to Syria and NW Arabia
    — *S.i.grisea* Bates, 1936   WC Arabia to E Yemen and NE Arabia
    — *S.i.buryi* Ogilvie-Grant, 1902   SW Saudi Arabia, W Yemen
    — *S.i.montana* Stepanyan, 1970   Mts. of Turkmenistan and NE Iran to N Afghanistan
                                        and W Tien Shan
    — *S.i.platyura* (Severtsov, 1873)   NW Turkmenistan and S Kazakhstan to foot of Tien Shan
    — *S.i.striata* (W.E. Brooks, 1872)   S Afghanistan, Pakistan >> NW India

---

[1] For separation of this species from *brunnescens* see Urban *et al.* (1997: 138) (2455).
[2] Includes *taciturnus* Clancey, 1992 (523) see Tye (1997: 212) (2448).
[3] Mention of the Solomons in the range in last Edition seems to have been an error.
[4] Includes *mixta* see Lynes (1930) (1404).
[5] Treated as a member of the genus *Cisticola* in last Edition, but see Urban *et al.* (1997: 215) (2455).

RHOPOPHILUS Giglioli & Salvadori, 1870 M
○ *Rhopophilus pekinensis* WHITE-BROWED CHINESE WARBLER
    — *R.p.albosuperciliaris* (Hume, 1873)   Tarim Basin to Lop Nor
    — *R.p.leptorhynchus* Meise, 1933   Gansu, NE Tsinghai
    — *R.p.pekinensis* (Swinhoe, 1868)   S Manchuria, Korea, Shanxi, Ningxia

MALCORUS A. Smith, 1829[1] M
○ *Malcorus pectoralis* RUFOUS-EARED WARBLER
    — *M.p.etoshae* (Winterbottom, 1965)   C and N Namibia
    — *M.p.ocularius* (A. Smith, 1843)   S Namibia, SW Botswana, N Cape Province to Transvaal
    — *M.p.pectoralis* A. Smith, 1829   W Cape Province to E Cape and Orange Free State

PRINIA Horsfield, 1821 F
○ *Prinia burnesii* RUFOUS-VENTED PRINIA
    — *P.b.burnesii* (Blyth, 1844)   Pakistan, NW India
    — *P.b.cinerascens* (Walden, 1874)   NE India, N Bangladesh

○ *Prinia crinigera* STRIATED PRINIA[2]
    — *P.c.striatula* (Hume, 1873)   Afghanistan, Pakistan (except NE)
    — *P.c.crinigera* Hodgson, 1836   Lower Himalayas from NE Pakistan to Bhutan
    — *P.c.catharia* Reichenow, 1908   S and E Assam, W and N Burma, N Yunnan, WC China
                                  to S Shaanxi
    — *P.c.parvirostris* (La Touche, 1922)   SE Yunnan
    — *P.c.parumstriata* (David & Oustalet, 1877) EC and E China
    — *P.c.striata* Swinhoe, 1859   Taiwan

○ *Prinia polychroa* BROWN PRINIA
    — *P.p.cooki* (Harington, 1913)   C and E Burma, S Yunnan, Thailand, Laos, Cambodia
    — *P.p.bangsi* (La Touche, 1922)   SE Yunnan
    — *P.p.rocki* Deignan, 1957   SC Vietnam
    — *P.p.polychroa* (Temminck, 1828)   Java

○ *Prinia atrogularis* HILL PRINIA
    — *P.a.atrogularis* (F. Moore, 1854)   E Himalayas, SE Xizang
    — *P.a.khasiana* (Godwin-Austen, 1876)   Assam, W Burma
    — *P.a.erythropleura* (Walden, 1875)   E Burma, NW Thailand
    — *P.a.superciliaris* (Anderson, 1871)   NE Burma, S and SE China to W Sichuan, NE Thailand,
                                    N Indochina
    — *P.a.klossi* (Hachisuka, 1926)   C Vietnam
    — *P.a.waterstradti* (E. Hartert, 1902)   Gunong Tahan (W Malaysia)
    — *P.a.dysancrita* (Oberholser, 1912)   W Sumatra

○ *Prinia cinereocapilla* F. Moore, 1854 GREY-CROWNED PRINIA   N Pakistan, N and E India, S Nepal, Bhutan

○ *Prinia buchanani* Blyth, 1844 RUFOUS-FRONTED PRINIA   Pakistan, W and C India

○ *Prinia rufescens* RUFESCENT PRINIA
    — *P.r.rufescens* Blyth, 1847   EC and NE India to W, C and N Burma and S China
    — *P.r.beavani* Walden, 1867   SE Burma, Thailand (except SE and penin.) east to N
                                    and C Indochina
    — *P.r.dalatensis* (Riley, 1940)   SC and S Vietnam
    — *P.r.objurgans* Deignan, 1942   SE Thailand, S Cambodia
    — *P.r.peninsularis* Deignan, 1942   N and C Malay Pen.
    — *P.r.extrema* Deignan, 1942   S Malay Pen.

○ *Prinia hodgsonii* GREY-BREASTED PRINIA
    — *P.h.rufula* Godwin-Austen, 1874   N Pakistan along Himalayas to W Yunnan; Assam hills
    — *P.h.confusa* Deignan, 1942   SC China, N Indochina
    — *P.h.hodgsonii* Blyth, 1844   W and C India, E Bangladesh, S Assam, W Burma
    — *P.h.erro* Deignan, 1942   C, E and SE Burma through Thailand to SW and east
                                    to C and S Indochina

---

[1] For recognition of this genus see Maclean (1974) (1412).
[2] For correction of spelling see David & Gosselin (2002) (613).

— *P.h.albogularis* Walden, 1870	S and SE India
— *P.h.pectoralis* Legge, 1874	Sri Lanka

○ *Prinia gracilis* GRACEFUL WARBLER

— *P.g.natronensis* Nicoll, 1917	Wadi Natrun (Egypt)
— *P.g.deltae* Reichenow, 1904	Nile delta to W Israel
— *P.g.gracilis* (M.H.K. Lichtenstein, 1823)	Cairo (Egypt) to N Sudan
— *P.g.carlo* Zedlitz, 1911	C and E Sudan, Eritrea to N and E Somalia
— *P.g.yemenensis* E. Hartert, 1909	SW Saudi Arabia, Yemen, S Oman
— *P.g.hufufae* Ticehurst & Cheesman, 1924	E Saudi Arabia, Bahrein
— *P.g.carpenteri* Meyer de Schauensee & Ripley, 1953	United Arab Emirates and N Oman
— *P.g.palaestinae* Zedlitz, 1911	S Syria, E Israel, Lebanon
— *P.g.irakensis* R. Meinertzhagen, 1923	E Syria, Iraq, SW Iran
— *P.g.akyildizi* Watson, 1961	SC Turkey, N Syria
— *P.g.lepida* Blyth, 1844	SE Iran to NW India and SC Nepal
— *P.g.stevensi* E. Hartert, 1923	SE Nepal to NE India and Bangladesh

○ *Prinia sylvatica* JUNGLE PRINIA

— *P.s.insignis* (Hume, 1872)	W India
— *P.s.gangetica* (Blyth, 1867)	N India to NW Nepal, S Bhutan, W Bangladesh
— *P.s.mahendrae* Koelz, 1939	E India (Orissa)
— *P.s.sylvatica* Jerdon, 1840	C and S India
— *P.s.valida* (Blyth, 1851)	Sri Lanka

○ *Prinia familiaris* Horsfield, 1821 BAR-WINGED PRINIA[1] | S Sumatra, Java, Bali

○ *Prinia flaviventris* YELLOW-BELLIED PRINIA

— *P.f.sindiana* Ticehurst, 1920	NW India, Pakistan
— *P.f.flaviventris* (Delessert, 1840)	Lower Himalayas to NE India, W and N Burma
— *P.f.sonitans* Swinhoe, 1860	S China, Taiwan, Hainan, NE Vietnam
— *P.f.delacouri* Deignan, 1942	S and E Burma, Thailand, S Yunnan, NW, C and S Indochina
— *P.f.rafflesi* Tweeddale, 1877	Malay Pen., Sumatra, Java
— *P.f.halistona* (Oberholser, 1912)	Nias I. (Sumatra)
— *P.f.latrunculus* (Finsch, 1905)	Borneo

○ *Prinia socialis* ASHY PRINIA

— *P.s.stewarti* Blyth, 1847	NE Pakistan, W and N India, Nepal
— *P.s.inglisi* Whistler & Kinnear, 1933	NE India, Bhutan, Assam, Bangladesh
— *P.s.socialis* Sykes, 1832	S India
— *P.s.brevicauda* Legge, 1879	Sri Lanka

○ *Prinia subflava* TAWNY-FLANKED PRINIA

— *P.s.subflava* (J.F. Gmelin, 1789)	Senegal to S Sudan, SC Ethiopia and S Eritrea
— *P.s.pallescens* Madarász, 1914	Sahel from Mali to N Ethiopia and NW Eritrea
— *P.s.tenella* (Cabanis, 1869)	Coastal S Somalia and Kenya, E Tanzania
— *P.s.melanorhyncha* (Jardine & Fraser, 1852)	Sierra Leone to Cameroon, N Zaire, C Kenya and NW Tanzania
— *P.s.graueri* E. Hartert, 1920	C Angola, E Zaire, Rwanda
— *P.s.affinis* (A. Smith, 1843)	SE Zaire to S Mozambique south to E Botswana and Transvaal
— *P.s.kasokae* C.M.N. White, 1946	E Angola, W Zambia
— *P.s.mutatrix* Meise, 1936[2]	S Tanzania, Malawi, E Zambia, N and C Mozambique, E Zimbabwe
— *P.s.bechuanae* Macdonald, 1941	N Botswana, SW Angola, N Namibia, SW Zambia, W Zimbabwe
— *P.s.pondoensis* Roberts, 1922	S Mozambique to E Cape Province

○ *Prinia inornata* PLAIN PRINIA[3]

— *P.i.terricolor* (Hume, 1874)	Pakistan, NW India

[1] For reasons to treat this as monotypic see Mees (1996) (1543).
[2] For recognition see Irwin (1997: 222) (1192).
[3] We split this very tentatively from *subflava*; for a brief discussion see Inskipp *et al.* (1996: 163) (1182).

— *P.i.fusca* Hodgson, 1845	C and E Himalayan hills to Assam, Bangladesh, SW Burma
— *P.i.blanfordi* (Walden, 1875)	N, C, and E Burma, NW Thailand, W Yunnan
— *P.i.extensicauda* (Swinhoe, 1860)	N and E Yunnan and W Sichuan to E China, N Indochina, Hainan
— *P.i.herberti* E.C.S. Baker, 1918	C and E Thailand, C and S Indochina
— *P.i.flavirostris* (Swinhoe, 1863)	Taiwan
— *P.i.inornata* Sykes, 1832	SE and C India
— *P.i.franklinii* Blyth, 1844	SW India
— *P.i.insularis* (Legge, 1879)	Sri Lanka
— *P.i.blythi* (Bonaparte, 1850)[1]	Java (178)

○ *Prinia somalica* PALE PRINIA

— *P.s.somalica* (Elliot, 1897)	N Somalia
— *P.s.erlangeri* Reichenow, 1905	SE Sudan, S Ethiopia, S Somalia, NE Uganda, N and E Kenya

○ *Prinia fluviatilis* Chappuis, 1974 RIVER PRINIA — NW Senegal to Mali and Lake Chad area

○ *Prinia flavicans* BLACK-CHESTED PRINIA

— *P.f.bihe* Boulton & Vincent, 1936	SW and C Angola, W Zambia
— *P.f.ansorgei* W.L. Sclater, 1927	Coastal SW Angola, NW Namibia
— *P.f.flavicans* (Vieillot, 1820)	S and E Namibia, W and C Botswana, N and NW Cape Province
— *P.f.nubilosa* Clancey, 1957	SW Zambia, E Botswana, SW Zimbabwe, Transvaal
— *P.f.ortleppi* (Tristram, 1869)	NE Cape Province, SW Transvaal, W Orange Free State

○ *Prinia maculosa* KAROO PRINIA

— *P.m.exultans* Clancey, 1982	Lesotho to adjacent NE Cape Province and W Natal (490)
— *P.m.psammophila* Clancey, 1963	SW Namibia, W Cape Province
— *P.m.maculosa* (Boddaert, 1783)	S Namibia to S Orange Free State and Cape Province
— *P.m.hypoxantha* Sharpe, 1877	Transvaal, W Swaziland and E Orange Free State to Natal and E Cape Province

○ *Prinia molleri* Bocage, 1887 SÃO TOMÉ PRINIA — São Tomé I.

○ *Prinia bairdii* BANDED PRINIA

— *P.b.bairdii* (Cassin, 1855)	SE Nigeria to Lower Congo R. and NE Zaire
— *P.b.melanops* (Reichenow & Neumann, 1895)	W Kenya
— *P.b.obscura* (Neumann, 1908)	E Zaire, W Uganda, Rwanda, Burundi
— *P.b.heinrichi* Meise, 1958	NW Angola

SCHISTOLAIS Wolters, 1980[2] F
○ *Schistolais leucopogon* WHITE-CHINNED PRINIA

— *S.l.leucopogon* (Cabanis, 1875)	SE Nigeria to E Zaire and Zambia
— *S.l.reichenowi* (Hartlaub, 1890)	NE Zaire to W Kenya and W Tanzania

○ *Schistolais leontica* (Bates, 1930) SIERRA LEONE PRINIA — E Sierra Leone to Ivory Coast

PHRAGMACIA Brooke & Dean, 1990[3] F
○ *Phragmacia substriata* NAMAQUA WARBLER[4]

— *P.s.confinis* Clancey, 1991	NW Cape Province (521)
— *P.s.substriata* (A. Smith, 1842)	W and C Cape Province, SW Orange Free State

OREOPHILAIS Clancey, 1991[5] M
○ *Oreophilais robertsi* (Benson, 1946) BRIAR WARBLER — E Zimbabwe, W Mozambique

HELIOLAIS Sharpe, 1903[6] M[7]
○ *Heliolais erythropterus* RED-WINGED WARBLER

— *H.e.erythropterus* (Jardine, 1849)	Senegal to N Cameroon

[1] Omitted by Watson *et al.* (1986) (2564).
[2] Treated within the genus *Prinia* in last Edition, but see Irwin (1997: 245) (1192).
[3] See Brooke & Dean (1990) (240). Genus recognised by Urban *et al.* (1997: 219) (2455).
[4] Called White-breasted Prinia in last Edition.
[5] See Clancey (1991) (522). Genus recognised by Urban *et al.* (1997: 236) (2455).
[6] See Urban *et al.* (1997: 238) (2455) for acceptance of this genus.
[7] For correct gender and spellings see David & Gosselin (2002) (614).

— *H.e.jodopterus* (Heuglin, 1864)	C Cameroon to S Sudan
— *H.e.major* (Blundell & Lovat, 1899)	Ethiopia
— *H.e.rhodopterus* (Shelley, 1880)	Kenya to Mozambique

UROLAIS Alexander, 1903 M[1]
○ **Urolais epichlorus** GREEN LONGTAIL

— *U.e.epichlorus* (Reichenow, 1892)	SE Nigeria, W Cameroon
— *U.e.cinderella* Bates, 1928	Mt. Manenguba (W Cameroon)
— *U.e.mariae* Alexander, 1903	Bioko I.

DRYMOCICHLA Hartlaub, 1881 F
○ **Drymocichla incana** Hartlaub, 1881 RED-WINGED GREY WARBLER   E Nigeria to SW Sudan and NW Uganda

SPILOPTILA Sundevall, 1872 F
○ **Spiloptila clamans** (Temminck, 1828) CRICKET WARBLER   S Mauritania and N Senegal to C Sudan and W Eritrea

PHYLLOLAIS Hartlaub, 1881 F
○ **Phyllolais pulchella** (Cretzschmar, 1827) BUFF-BELLIED WARBLER   NE Nigeria, to S Chad, C Sudan and Eritrea south to NE Zaire, Uganda and N Tanzania

APALIS Swainson, 1833 F
○ **Apalis thoracica** BAR-THROATED APALIS

— *A.t.fuscigularis* Moreau, 1938	Taita Hills (SE Kenya)
— *A.t.griseiceps* Reichenow & Neumann, 1895[2]	Chyulu Hills (SE Kenya), N and SC Tanzania
— *A.t.uluguru* Neumann, 1914	Uluguru Mts. (NE Tanzania)
— *A.t.lynesi* Vincent, 1933	N Mozambique
— *A.t.quarta* Irwin, 1966	NE Zimbabwe, Mozambique (Mt. Gorongoza)
— *A.t.pareensis* Ripley & Heinrich, 1966	S Pare Mts. (NE Tanzania)
— *A.t.youngi* Kinnear, 1936	SW Tanzania, N Malawi
— *A.t.murina* Reichenow, 1904	NE, E and S Tanzania, NE Zambia, N Malawi
— *A.t.bensoni* Grant & Mackworth-Praed, 1937	E Zambia, S Malawi
— *A.t.flavigularis* Shelley, 1893	SE Malawi
— *A.t.rhodesiae* Gunning & Roberts, 1911	W Zimbabwe, NE Botswana
— *A.t.arnoldi* Roberts, 1936	E Zimbabwe, W Mozambique
— *A.t.drakensbergensis* Roberts, 1937	SE Transvaal, N Natal, E Orange Free State, W Swaziland
— *A.t.flaviventris* Gunning & Roberts, 1911	SE Botswana, N and W Transvaal
— *A.t.spelonkensis* Gunning & Roberts, 1911	NE and E Transvaal
— *A.t.lebomboensis* Roberts, 1931	NE Zululand, E Swaziland, S Mozambique
— *A.t.capensis* Roberts, 1929	S and SW Cape Province
— *A.t.venusta* Gunning & Roberts, 1911[3]	Zululand, Durban, Natal, E Griqualand
— *A.t.thoracica* (Shaw, 1811)	SE Cape Province
— *A.t.claudei* W.L. Sclater, 1910	S Cape Province
— *A.t.griseopyga* Lawson, 1965	Coastal W Cape Province

○ **Apalis pulchra** BLACK-COLLARED APALIS

— *A.p.pulchra* Sharpe, 1891	SE Nigeria, Cameroon, S Sudan, Kenya, NE Zaire
— *A.p.murpheyi* Chapin, 1932	SE Zaire

○ **Apalis ruwenzorii** Jackson, 1904 RUWENZORI APALIS[4]   E Zaire, SW Uganda, Rwanda, Burundi

○ **Apalis ruddi** RUDD'S APALIS

— *A.r.ruddi* C.H.B. Grant, 1908	SE Mozambique
— *A.r.caniviridis* Hanmer, 1979	S Malawi
— *A.r.fumosa* Clancey, 1966	S Mozambique (Maputo) to N Natal

○ **Apalis flavida** YELLOW-BREASTED APALIS

— *A.f.caniceps* (Cassin, 1859)	Gambia to W Kenya and N Angola
— *A.f.viridiceps* Hawker, 1898	NE Ethiopia, N Somalia
— *A.f.abyssinica* Érard, 1974	SW Ethiopia

[1] For correct gender and spellings see David & Gosselin (2002) (614).
[2] Includes *iringae* see Irwin (1997: 254) (1192).
[3] Includes *darglensis* see White (1962: 697) (2647).
[4] Includes *catiodes* see White (1962: 700) (2647).

— *A.f.flavocincta* (Sharpe, 1882) | SE Sudan, N Uganda, S Ethiopia, S Somalia, NE and E Kenya

— *A.f.pugnax* Lawson, 1968 | W and C Kenya
— *A.f.golzi* (Fischer & Reichenow, 1884) | SE Kenya, C Tanzania, Rwanda
— *A.f.neglecta* (Alexander, 1899)[1] | E Angola to SE Kenya and N Natal
— *A.f.flavida* (Strickland, 1852) | W Angola, N Namibia, N Botswana, SW Zambia
— *A.f.florisuga* (Reichenow, 1898) | C Natal to E Cape Province

○ *Apalis binotata* Reichenow, 1895 MASKED APALIS | S Cameroon, NE Gabon, NW Angola, E Zaire, Uganda, NW Tanzania

○ *Apalis personata* MOUNTAIN APALIS
— *A.p.personata* Sharpe, 1902 | SW Uganda, E Zaire, Rwanda, Burundi
— *A.p.marungensis* Chapin, 1932 | SE Zaire

○ *Apalis jacksoni* BLACK-THROATED APALIS
— *A.j.bambuluensis* Serle, 1949 | Nigeria, W Cameroon
— *A.j.minor* Ogilvie-Grant, 1917 | S Cameroon, NE Zaire
— *A.j.jacksoni* Sharpe, 1891 | S Sudan, Uganda, Kenya, E Zaire, N Angola

○ *Apalis chariessa* WHITE-WINGED APALIS
— *A.c.chariessa* Reichenow, 1879 | Tana River (SE Kenya)
— *A.c.macphersoni* Vincent, 1934 | C Tanzania, NC Mozambique

○ *Apalis nigriceps* BLACK-CAPPED APALIS
— *A.n.nigriceps* (Shelley, 1873) | Sierra Leone to Central African Republic and Gabon
— *A.n.collaris* van Someren, 1915 | E Zaire, SW Uganda

○ *Apalis melanocephala* BLACK-HEADED APALIS
— *A.m.nigrodorsalis* Granvik, 1923 | C Kenya
— *A.m.moschi* van Someren, 1931 | SE Kenya, E Tanzania
— *A.m.muhuluensis* Grant & Mackworth-Praed, 1947 | SE Tanzania
— *A.m.melanocephala* (Fischer & Reichenow, 1884) | Coastal S Somalia to NE Tanzania
— *A.m.lightoni* Roberts, 1938 | C Mozambique to SE Zimbabwe
— *A.m.fuliginosa* Vincent, 1933 | Mulanje and Thyolo hills (SE Malawi)
— *A.m.tenebricosa* Vincent, 1933 | N Mozambique
— *A.m.adjacens* Clancey, 1969 | SE Malawi
— *A.m.addenda* Clancey, 1968 | SE Mozambique

○ *Apalis chirindensis* CHIRINDA APALIS
— *A.c.vumbae* Roberts, 1936 | E Zimbabwe
— *A.c.chirindensis* Shelley, 1906 | SE Zimbabwe, Mt. Gorongoza (Mozambique)

○ *Apalis porphyrolaema* CHESTNUT-THROATED APALIS
— *A.p.porphyrolaema* Reichenow & Neumann, 1895 | E Zaire to Uganda, W and C Kenya and N Tanzania
— *A.p.kaboboensis* Prigogine, 1955 | Mt. Kobobo (E Zaire)

○ *Apalis chapini* CHAPIN'S APALIS[2]
— *A.c.chapini* Friedmann, 1928 | C Tanzania
— *A.c.strausae* Boulton, 1931 | SW Tanzania, Malawi, NE Zambia

○ *Apalis sharpii* Shelley, 1884 SHARPE'S APALIS | Sierra Leone to Ghana

○ *Apalis rufogularis* BUFF-THROATED APALIS
— *A.r.rufogularis* (Fraser, 1843) | Bioko I., S Nigeria to Gabon
— *A.r.sanderi* Serle, 1951 | SW Nigeria
— *A.r.nigrescens* (Jackson, 1906) | SW Sudan to W Kenya south to NW Zambia and NE Angola
— *A.r.angolensis* (Bannerman, 1922) | NW Angola
— *A.r.brauni* Stresemann, 1934 | W Angola
— *A.r.kigezi* Keith, Twomey & Friedmann, 1967 | SW Uganda
— *A.r.argentea* Moreau, 1941[3] | E Zaire, Rwanda, Burundi, W Tanzania

[1] Includes *renata*, *niassae* and *tenerrima* see Irwin (1997: 263) (1192).
[2] For separation from *A. porphyrolaemus* see Dowsett & Dowsett-Lemaire (1980) (760).
[3] Includes *eidos* see Irwin (1997: 280) (1192).

561

◯ *Apalis karamojae* Karamoja Apalis
— *A.k.karamojae* (van Someren, 1921)   NE Uganda
— *A.k.stronachi* Stuart & Collar, 1985   NC Tanzania (2372)

◯ *Apalis bamendae* Bannerman, 1922   Bamenda Apalis   Bamenda Mts. (Cameroon)

◯ *Apalis goslingi* Alexander, 1908   Gosling's Apalis   S Cameroon to N Angola and NE Zaire

◯ *Apalis cinerea* Grey Apalis
— *A.c.funebris* Bannerman, 1937[1]   Nigeria, Cameroon
— *A.c.cinerea* (Sharpe, 1891)   E Zaire to W and C Kenya and N Tanzania
— *A.c.sclateri* (Alexander, 1903)   Mt. Cameroon, Bioko I.
— *A.c.grandis* Boulton, 1931   W Angola

◯ *Apalis alticola* Brown-headed Apalis
— *A.a.dowsetti* Prigogine, 1973   E Zaire
— *A.a.alticola* (Shelley, 1899)   Angola, Zambia and N Malawi to SW Kenya

URORHIPIS Heuglin, 1871[2] F
◯ *Urorhipis rufifrons* Red-faced Apalis
— *U.r.rufifrons* (Rüppell, 1840)   Chad, N Sudan, Eritrea, to Djibouti and NW Somalia
— *U.r.smithi* (Sharpe, 1895)   Somalia and S Ethiopia to Kenya and N Tanzania
— *U.r.rufidorsalis* (Sharpe, 1897)   Tsavo (SE Kenya)

HYPERGERUS Reichenbach, 1850[3] M
◯ *Hypergerus atriceps* (Lesson, 1831)   Oriole-Warbler   Senegal, Gambia and Guinea to Central African Republic

EMINIA Hartlaub, 1881[4] F
◯ *Eminia lepida* Hartlaub, 1881   Grey-capped Warbler   NE Zaire, S Sudan, Uganda, Rwanda, Burundi, Kenya, N Tanzania

CAMAROPTERA Sundevall, 1850 F
◯ *Camaroptera brachyura* Bleating Warbler[5]
— *C.b.brevicaudata* (Cretzschmar, 1827)   Senegal to C Sudan and NW Ethiopia
— *C.b.tincta* (Cassin, 1855)   Liberia to W Kenya south to N Angola, NW Zambia and W Tanzania

— *C.b.harterti* Zedlitz, 1911   NW Angola escarpment
— *C.b.aschani* Granvik, 1934   Highlands of W Kenya, SW Uganda and E Zaire
— *C.b.abessinica* Zedlitz, 1911   S Sudan to NW Somalia, Kenya and E NE Zaire
— *C.b.insulata* Desfayes, 1975   W Ethiopia
— *C.b.erlangeri* Reichenow, 1905   Coastal S Somalia, E Kenya, NE Tanzania
— *C.b.griseigula* Sharpe, 1892   W Kenya, E Uganda, N Tanzania
— *C.b.intercalata* C.M.N. White, 1960   SE Zaire, W Tanzania, W, C and E Angola, N, NW and E Zambia

— *C.b.bororensis* Gunning & Roberts, 1911   SE Tanzania to N Mozambique
— *C.b.pileata* Reichenow, 1891   Coastal S Kenya to SE Tanzania, Zanzibar and Mafia Is.
— *C.b.fugglescouchmani* Moreau, 1939   E Tanzania, N Malawi
— *C.b.beirensis* Roberts, 1932   C Mozambique to NE Zululand
— *C.b.constans* Clancey, 1952   S Mozambique to E Zululand
— *C.b.transitiva* Clancey, 1974   Zimbabwe, Transvaal, SE Botswana
— *C.b.sharpei* Zedlitz, 1911   S Angola, Namibia and NE Botswana to Malawi and W Transvaal

— *C.b.brachyura* (Vieillot, 1820)   SE Cape Province to Natal and W Zululand

◯ *Camaroptera superciliaris* (Fraser, 1843)   Yellow-browed Camaroptera
     Sierra Leone to Uganda and NW Angola

◯ *Camaroptera chloronota* Olive-green Camaroptera
— *C.c.kelsalli* W.L. Sclater, 1927   Senegal to Ghana
— *C.c.chloronota* Reichenow, 1895   Togo to S Cameroon

---

[1] For recognition see Irwin (1997: 284) (1192).
[2] Treated in the genus *Apalis* in last Edition. For the use of this genus see Irwin (1989) (1191).
[3] For a discussion on the relationships of *Hypergerus* and *Eminia* see Cibois *et al.* (1999) (357).
[4] Treated in *Hypergerus* in last Edition, but this genus is recognised by Urban *et al.* (1997: 372) (2455).
[5] Treatment of this, as one species not three, tentatively follows Grimes, Fry & Keith (1997: 296) (1013); their account suggests their own treatment is tentative.

— *C.c.granti* Alexander, 1903   Bioko I.
— *C.c.kamitugaensis* Prigogine, 1961   Itombwe (E Zaire)
— *C.c.toroensis* (Jackson, 1905)   Central African Republic, N and C Zaire to Uganda
                                                        and W Kenya

CALAMONASTES Sharpe, 1883 M
○ ***Calamonastes simplex*** (Cabanis, 1878) GREY WREN-WARBLER[1]   Somalia to SE Sudan and NE Tanzania

○ ***Calamonastes undosus*** MIOMBO WREN-WARBLER
— *C.u.undosus* (Reichenow, 1882)   SW Kenya, W and C Tanzania, Rwanda
— *C.u.cinereus* Reichenow, 1887   Congo, W and SW Zaire, NE Angola, NW Zambia
— *C.u.katangae* Neave, 1909   SE Zaire, N Zambia
— *C.u.huilae* (Meise, 1958)   WC Angola
— *C.u.stierlingi* Reichenow, 1901[2]   SE Angola, NE Namibia, S Zambia, S Malawi,
                                                        E Tanzania, N Mozambique
— *C.u.irwini* (Smithers & Paterson, 1956)   Botswana, Zimbabwe, E Zambia, Malawi, C Mozambique
— *C.u.olivascens* (Clancey, 1969)   Coastal Mozambique
— *C.u.pintoi* (Irwin, 1960)   NE Sth. Africa, Swaziland, S Mozambique

○ ***Calamonastes fasciolatus*** BARRED WREN-WARBLER
— *C.f.pallidior* E. Hartert, 1907   SW Angola
— *C.f.fasciolatus* (A. Smith, 1847)   C Namibia, Botswana, SW Zimbabwe, N Cape Province
— *C.f.europhilus* (Clancey, 1970)   W Transvaal, S Zimbabwe, SE Botswana

EURYPTILA Sharpe, 1883 F
○ ***Euryptila subcinnamomea*** CINNAMON-BREASTED WARBLER
— *E.s.subcinnamomea* (A. Smith, 1847)   W and C Cape Province
— *E.s.petrophila* Clancey, 1990   NW Cape Province, S Namibia (519)

GENERA INCERTAE SEDIS (4: 17)

NEOMIXIS Sharpe, 1881[3] F
○ ***Neomixis tenella*** COMMON JERY
— *N.t.tenella* (Hartlaub, 1866)   NW, N and NE Madagascar
— *N.t.decaryi* Delacour, 1931   WC and C Madagascar
— *N.t.orientalis* Delacour, 1931   EC and SE Madagascar
— *N.t.debilis* Delacour, 1931   S Madagascar

○ ***Neomixis viridis*** GREEN JERY
— *N.v.delacouri* Salomonsen, 1934   NE Madagascar
— *N.v.viridis* (Sharpe, 1883)   SE Madagascar

○ ***Neomixis striatigula*** STRIPE-THROATED JERY
— *N.s.sclateri* Delacour, 1931   NE Madagascar
— *N.s.striatigula* Sharpe, 1881   E Madagascar
— *N.s.pallidior* Salomonsen, 1934   SW, S and SE Madagascar

ORTHOTOMUS Horsfield, 1821 M
○ ***Orthotomus cucullatus*** MOUNTAIN TAILORBIRD
— *O.c.coronatus* Blyth, 1861   E Himalayas, Assam, E Bangladesh, Burma, S China,
                                                        W and N Thailand, Indochina
— *O.c.thais* (Robinson & Kloss, 1923)   Pen. Thailand
— *O.c.malayanus* (Chasen, 1938)   Mts. of W Malaysia
— *O.c.cucullatus*[4] Temminck, 1836   Sumatra, Java, Bali
— *O.c.cinereicollis* (Sharpe, 1888)   NE Borneo
— *O.c.viridicollis* Salomonsen, 1962   Mts. of Palawan
— *O.c.philippinus* (E. Hartert, 1897)   Highlands N Luzon
— *O.c.heterolaemus* (Mearns, 1905)   Mts. of Mindanao

---

[1] For treatment as a monotypic species, stripped of its 'races' see Grimes, Fry & Keith (1997: 304) (1013).
[2] Includes *buttoni* see Grimes, Fry & Keith (1997: 305) (1013). For reasons to lump this 'species' see Dowsett & Dowsett-Lemaire (1980) (760).
[3] Cibois *et al.* (1999) (357) considered this genus to belong to the "*Cisticola* group".
[4] The emended name used here is judged to be in prevailing usage; see Art. 33.2 of the Code (I.C.Z.N., 1999) (1178) and p. 000 herein.

__ *O.c.everetti* (E. Hartert, 1897)	Flores I.
__ *O.c.riedeli* (Meyer & Wiglesworth, 1895)	Mts. of N Sulawesi
__ *O.c.stentor* (Stresemann, 1938)	Mts. of C and SE Sulawesi
__ *O.c.meisei* (Stresemann, 1931)	Mts. of SC Sulawesi
__ *O.c.hedymeles* (Stresemann, 1932)	Lompobatang, S Sulawesi
__ *O.c.dumasi* (E. Hartert, 1899)	Buru, Seram
__ *O.c.batjanensis* (E. Hartert, 1912)	Bacan

○ *Orthotomus sutorius* COMMON TAILORBIRD

__ *O.s.guzuratus* (Latham, 1790)	Pakistan, W, C and S India
__ *O.s.patia* Hodgson, 1845	C and E Himalayas and Bangladesh to SW and C Burma
__ *O.s.luteus* Ripley, 1948	E Assam
__ *O.s.inexpectatus* La Touche, 1922	E Burma, Laos, Yunnan, Thailand (except penin.)
__ *O.s.longicauda* (J.F. Gmelin, 1789)	SE China, Hainan, NE Indochina
__ *O.s.maculicollis* F. Moore, 1855	SE Burma, C and S Malay Pen., S Indochina
__ *O.s.edela* Temminck, 1836	Java
__ *O.s.fernandonis* Whistler, 1939	Montane Sri Lanka
__ *O.s.sutorius* (Pennant, 1769)	Lowland Sri Lanka

○ *Orthotomus atrogularis* DARK-NECKED TAILORBIRD[1]

__ *O.a.nitidus* Hume, 1874	SE India, Bangladesh, Burma and Thailand (except Malay Pen.), S Yunnan, Indochina
__ *O.a.atrogularis* Temminck, 1836	Malay Pen., Sumatra, most Borneo
__ *O.a.anambensis* Watson, 1986	Tioman I., Anambas, Natunas
__ *O.a.humphreysi* Chasen & Kloss, 1929	N Borneo

○ *Orthotomus castaneiceps* PHILIPPINE TAILORBIRD

__ *O.c.chloronotos* Ogilvie-Grant, 1895	N and C Luzon
__ *O.c.castaneiceps* Walden, 1872	Guimaras, Masbate, Panay, Ticao
__ *O.c.rabori* Parkes, 1961	Negros, Cebu?
__ *O.c.frontalis* Sharpe, 1877[2]	Samar, Leyte, Bohol, Mindanao
__ *O.c.mearnsi* McGregor, 1907	Basilan

○ *Orthotomus derbianus* GREY-BACKED TAILORBIRD

__ *O.d.derbianus* F. Moore, 1855	C Luzon
__ *O.d.nilesi* Parkes, 1988	Catanduanes I. (1776)

○ *Orthotomus sericeus* RUFOUS-TAILED TAILORBIRD

__ *O.s.hesperius* Oberholser, 1932	Malay Pen., Sumatra, Belitung
__ *O.s.sericeus* Temminck, 1836[3]	Borneo, Balabac, Palawan, Cagayan Sulu I.
__ *O.s.rubicundulus* Chasen & Kloss, 1931	Sirhassen, Natunas

○ *Orthotomus ruficeps* ASHY TAILORBIRD[4]

__ *O.r.cineraceus* Blyth, 1845	SE Burma, Malay Pen., Sumatra, Bangka, Belitung
__ *O.r.baeus* Oberholser, 1912	Nias, N and S Pagi Is.
__ *O.r.concinnus* Riley, 1927	Siberut, Sipura Is.
__ *O.r.ruficeps* (Lesson, 1830)	Java (mainly mangroves)
__ *O.r.palliolatus* Chasen & Kloss, 1932	Kangean, Karimon Java Is.
__ *O.r.baweanus* Hoogerwerf, 1962	Bawean I.
__ *O.r.borneoensis* Salvadori, 1874	Borneo
__ *O.r.cagayanensis* Riley, 1935	Cagayan Sulu I.

○ *Orthotomus sepium* OLIVE-BACKED TAILORBIRD

__ *O.s.sundaicus* Hoogerwerf, 1962	Panaitan I. (Java)
__ *O.s.sepium* Horsfield, 1821	Java, Bali, Lombok

○ *Orthotomus cinereiceps* WHITE-EARED TAILORBIRD

__ *O.c.obscurior* Mayr, 1947	Mindanao
__ *O.c.cinereiceps* Sharpe, 1877	Basilan

---

[1] Although no thorough review has been published to support the split accepted by Dickinson *et al.* (1991: 332) (745) we tentatively follow in accepting *O. castaneiceps*.

[2] Dickinson *et al.* (1991: 332) (745) retained this within the species *castaneiceps* awaiting good acoustic evidence of all the races before considering a split.

[3] Includes *nuntius* see Dickinson *et al.* (1991: 335) (745).

[4] Much pre-1986 literature used this name or '*sepium*' for taxa now treated as *O. sericeus* see Dickinson *et al.* (1991: 334-5) (745). Hence, so far, records from Palawan relate to *sericeus* not to *ruficeps*. Situation as regards 'Cochinchine' unclear.

○ *Orthotomus nigriceps* Tweeddale, 1878 Black-headed Tailorbird   Mindanao

○ *Orthotomus samarensis* Steere, 1890 Yellow-breasted Tailorbird   Samar, Bohol, Leyte

ARTISORNIS Friedmann, 1928[1] M
○ *Artisornis moreaui* Moreau's Tailorbird
  — *A.m.moreaui* (W.L. Sclater, 1931)   NE Tanzania
  — *A.m.sousae* (Benson, 1945)   NW Mozambique

○ *Artisornis metopias* African Tailorbird
  — *A.m.metopias* (Reichenow, 1907)[2]   NE, E and SE Tanzania, NW Mozambique
  — *A.m.altus* (Friedmann, 1927)   Uluguru Mts (E Tanzania)

POLIOLAIS Alexander, 1903 M
○ *Poliolais lopesi* White-tailed Warbler[3]
  — *P.l.lopesi* (Alexander, 1903)   Bioko I.
  — *P.l.alexanderi* Bannerman, 1915   Mt. Cameroon
  — *P.l.manengubae* Serle, 1949   SE Nigeria, W Cameroon

PYCNONOTIDAE BULBULS (22: 118)[4]

SPIZIXOS Blyth, 1845 M
○ *Spizixos canifrons* Crested Finchbill
  — *S.c.canifrons* Blyth, 1845   S Assam, W Burma
  — *S.c.ingrami* Bangs & Phillips, 1914   E Burma to SW China, N Indochina

○ *Spizixos semitorques* Collared Finchbill
  — *S.s.semitorques* Swinhoe, 1861   C and S China, N Vietnam
  — *S.s.cinereicapillus* Swinhoe, 1871   Taiwan

PYCNONOTUS Boie, 1826 M
○ *Pycnonotus zeylanicus* (J.F. Gmelin, 1789) Straw-headed Bulbul   Malay Pen., Sumatra, Nias, Java, Borneo

○ *Pycnonotus striatus* Striated Bulbul
  — *P.s.striatus* (Blyth, 1842)   C and E Himalayas, S Assam, W Burma
  — *P.s.arctus* Ripley, 1948   NE Assam
  — *P.s.paulus* (Bangs & Phillips, 1914)   NE and E Burma to N Indochina, SW China

○ *Pycnonotus leucogrammicus* (S. Müller, 1836)[5] Cream-striped Bulbul   W Sumatra

○ *Pycnonotus tympanistrigus* (S. Müller, 1836) Spot-necked Bulbul   W Sumatra

○ *Pycnonotus melanoleucos* (Eyton, 1839) Black-and-white Bulbul   Malay Pen., Sumatra, Borneo

○ *Pycnonotus priocephalus* (Jerdon, 1839) Grey-headed Bulbul   SW India

○ *Pycnonotus atriceps* Black-headed Bulbul
  — *P.a.fuscoflavescens* (Hume, 1873)   Andaman Is.
  — *P.a.atriceps* (Temminck, 1822)   NE India east to mainland SE Asia and south to Gtr. Sundas, Bali and Palawan
  — *P.a.hyperemnus* (Oberholser, 1912)   W Sumatran islands
  — *P.a.baweanus* (Finsch, 1901)   Bawean I.
  — *P.a.hodiernus* (Bangs & Peters, 1927)   Maratua I.

○ *Pycnonotus melanicterus* Black-crested Bulbul
  — *P.m.melanicterus* (J.F.Gmelin, 1789)   Sri Lanka
  — *P.m.gularis* (Gould, 1836)   SW India
  — *P.m.flaviventris* (Tickell, 1833)   Himalayas, C and NE India, W and N Burma

[1] Treated within the genus *Orthotomus* in last Edition, but see Urban *et al.* (1997: 290) (2455).
[2] Includes *pallidus* see Irwin (1997: 291) (1192).
[3] The emended name used here is judged to be in prevailing usage; see Art. 33.2 of the Code (I.C.Z.N., 1999) (1178) and p. 000 herein.
[4] In breaking up the broad genus *Hypsipetes* we have not followed Sibley & Monroe (1990) (2262) due to comments in Bock (1994) (173). Instead we follow Dickinson & Gregory (2002) (743C).
[5] The date of this paper was established by Richmond (1926) (2015A).

__ *P.m.vantynei* Deignan, 1948	E Burma to N Indochina and S China
__ *P.m.xanthops* Deignan, 1948	C and SE Burma, NW Thailand
__ *P.m.auratus* Deignan, 1948	NE Thailand, NW Laos
__ *P.m.johnsoni* (Gyldenstolpe, 1913)	SE Thailand, S Indochina
__ *P.m.elbeli* Deignan, 1954	Islands off SE Thailand
__ *P.m.negatus* Deignan, 1954	SE Burma, SW Thailand
__ *P.m.caecilii* Deignan, 1948	Malay Pen.
__ *P.m.dispar* (Horsfield, 1821)	Sumatra, Java
__ *P.m.montis* (Sharpe, 1879)	N Borneo

○ *Pycnonotus squamatus* SCALY-BREASTED BULBUL

__ *P.s.weberi* (Hume, 1879)	Malay Pen., Sumatra
__ *P.s.squamatus* (Temminck, 1828)	W and C Java
__ *P.s.borneensis* Chasen, 1941	Borneo

○ *Pycnonotus cyaniventris* GREY-BELLIED BULBUL

__ *P.c.cyaniventris* Blyth, 1842	Malay Pen., Sumatra
__ *P.c.paroticalis* (Sharpe, 1878)	Borneo

● *Pycnonotus jocosus* RED-WHISKERED BULBUL

__ *P.j.fuscicaudatus* (Gould, 1866)	W and C India
__ *P.j.abuensis* (Whistler, 1931)	NW India
__ *P.j.pyrrhotis* (Bonaparte, 1850	Nepal, N India
__ *P.j.emeria* (Linnaeus, 1758)	E India, Bangladesh, S Burma, SW Thailand
__ *P.j.whistleri* Deignan, 1948	Andamans, Nicobars
__ *P.j.monticola* (McClelland, 1840)	E Himalayas, Assam, N Burma, Yunnan
__ *P.j.jocosus* (Linnaeus, 1758)[1]	S China and N Vietnam
__ *P.j.pattani* Deignan, 1948	S Indochina, Thailand south to SC Malay Pen.

○ *Pycnonotus xanthorrhous* BROWN-BREASTED BULBUL

__ *P.x.xanthorrhous* Anderson, 1869	NE and E Burma, NW Thailand, SC China to W Sichuan, N Laos, NW Vietnam
__ *P.x.andersoni* Swinhoe, 1870	S and C China

○ *Pycnonotus sinensis* LIGHT-VENTED BULBUL

__ *P.s.sinensis* (J.F. Gmelin, 1789)[2]	C and E China
__ *P.s.hainanus* (Swinhoe, 1870)	S China (incl. Hainan), N Vietnam
__ *P.s.formosae* E. Hartert, 1910	N and W Taiwan
__ *P.s.orii* N. Kuroda, Sr., 1923	S Ryukyu Is.

○ *Pycnonotus taivanus* Styan, 1893 STYAN'S BULBUL | S and E Taiwan

○ *Pycnonotus leucogenys* WHITE-CHEEKED BULBUL[3]

__ *P.l.leucogenys* (J.E. Gray, 1835)	NE Afghanistan, Himalayas, Assam
__ *P.l.mesopotamia* Ticehurst, 1918[4] [5]	Iraq, NE Arabia, SW and S Iran
__ *P.l.humii* (Oates, 1889)[6]	NW Pakistan
__ *P.l.leucotis* (Gould, 1836)	S Iran through S Afghanistan and S Pakistan to NW India

● *Pycnonotus cafer* RED-VENTED BULBUL[7]

__ *P.c.humayuni* Deignan, 1951	Pakistan, NW India
__ *P.c.intermedius* Blyth, 1846	NW Himalayas
__ *P.c.bengalensis* Blyth, 1845[8]	C and E Himalayas, NE India, S Assam, Bangladesh
__ *P.c.stanfordi* Deignan, 1949	N Burma, W Yunnan
__ *P.c.melanchimus* Deignan, 1949	SC Burma
__ *P.c.wetmorei* Deignan, 1960	NE of C India

[1] Includes *hainanensis* see Cheng (1987) (333).
[2] Includes *hoyi* see Mauersberger & Fischer (1992) (1461).
[3] We follow Paludan (1959) (1741) and Roberts (1992) (2090A) in treating *leucotis* and *leucogenys* as conspecific as reported sympatry in Afghanistan was implicitly rejected by Paludan. Although Vaurie (1958) (2500) considered the two distinct, *humii* was considered an intergradient form by Sibley & Short (1959) (2263) and Paludan. The case for separate treatment needs to be argued convincingly.
[4] Includes *dactylus* see Cramp *et al.* (1988: 479) (592).
[5] The trinomial given by Ticehurst was a place name and thus a noun and should not have been emended (Bruce *in litt.*).
[6] Recognised by Ripley (1982: 312) (2055).
[7] Stresemann (1952) (2366) corrected the type locality of the nominate form from Ceylon to Pondicherry. Subspecific names thus change compared to Rand & Deignan (1960: 235) (1972), who did not follow Stresemann.
[8] Includes *primrosei* see Ripley (1982: 313) (2055).

— *P.c.cafer* (Linnaeus, 1766)[1]   S and C India

— *P.c.haemorrhousus* (J.F. Gmelin, 1789)   Sri Lanka

○ *Pycnonotus aurigaster* SOOTY-HEADED BULBUL

— *P.a.chrysorrhoides* (Lafresnaye, 1845)   SE China

— *P.a.resurrectus* Deignan, 1952   S China, NE Vietnam

— *P.a.dolichurus* Deignan, 1949   C Vietnam

— *P.a.latouchei* Deignan, 1949   E Burma to Yunnan, Sichuan, N Indochina

— *P.a.klossi* (Gyldenstolpe, 1920)   SE Burma, N Thailand

— *P.a.schauenseei* Delacour, 1943   S Burma, SW Thailand

— *P.a.thais* (Kloss, 1924)   SC and SE Thailand

— *P.a.germani* (Oustalet, 1878)   SE of E Thailand, S Indochina

— *P.a.aurigaster* (Vieillot, 1818)   Java, Bali; introd. Singapore, Sumatra, S Sulawesi

○ *Pycnonotus xanthopygos* (Hemprich & Ehrenberg, 1833) WHITE-SPECTACLED BULBUL

                         S Turkey, Levant, Sinai, Arabia

○ *Pycnonotus nigricans* BLACK-FRONTED BULBUL

— *P.n.nigricans* (Vieillot, 1818)[2]   SW Angola, Namibia, Botswana, N Cape Province and NW Transvaal

— *P.n.superior* Clancey, 1959   E Cape Province to Orange Free State and S Transvaal (382)

○ *Pycnonotus capensis* (Linnaeus, 1766) CAPE BULBUL   S and W Cape Province

○ *Pycnonotus barbatus* COMMON BULBUL

— *P.b.barbatus* (Desfontaines, 1789)   Morocco to Tunisia

— *P.b.inornatus* (Fraser, 1843)[3]   Senegal to W Chad and N Cameroon

— *P.b.arsinoe* (M.H.K. Lichtenstein, 1823) Nile valley (Egypt, N and C Sudan); E Chad

— *P.b.gabonensis* Sharpe, 1871[4]   C Nigeria and C Cameroon to Gabon and S Congo

— *P.b.tricolor* (Hartlaub, 1862)[5]   E Cameroon to S Sudan, W and C Kenya, south to Angola, N Namibia, NW Botswana and N Zambia

— *P.b.schoanus* Neumann, 1905   Eritrea to W, C and E Ethiopia, SE Sudan

— *P.b.somaliensis* Reichenow, 1905[6]   Djibouti, SE Ethiopia, NW Somalia

— *P.b.spurius* Reichenow, 1905   S Ethiopia

— *P.b.dodsoni* Sharpe, 1895[7]   Somalia, SE Ethiopia, N and E Kenya

— *P.b.layardi* J.H. Gurney, 1879[8]   SE Kenya, E Tanzania and E and S Zambia to E Sth. Africa

○ *Pycnonotus eutilotus* (Jardine & Selby, 1836)[9] PUFF-BACKED BULBUL   Malay Pen., Sumatra, Borneo

○ *Pycnonotus nieuwenhuisii* BLUE-WATTLED BULBUL[10]

— *P.n.inexpectatus* (Chasen, 1939)   Lesten (Sumatra)

— *P.n.nieuwenhuisii* (Finsch, 1901)   Kayan river (Borneo)

○ *Pycnonotus urostictus* YELLOW-WATTLED BULBUL

— *P.u.ilokensis* Rand & Rabor, 1967   N Luzon (1978)

— *P.u.urostictus* (Salvadori, 1870)   C and S Luzon, Polillo, Catanduanes Is.

— *P.u.atricaudatus* Parkes, 1967   Bohol, Samar, Leyte (1759)

— *P.u.philippensis* (Hachisuka, 1934)   Mindanao (excl. SW tip),

— *P.u.basilanicus* (Steere, 1890)   S Zamboanga Pen. (Mindanao), Basilan

○ *Pycnonotus bimaculatus* ORANGE-SPOTTED BULBUL

— *P.b.snouckaerti* Siebers, 1928   NW Sumatra

— *P.b.bimaculatus* Horsfield, 1821[11]   C and S Sumatra, W and C Java

— *P.b.tenggerensis* (van Oort, 1911)   E Java, Bali

---

[1] Includes *pusillus* see Ripley (1982: 313) (2055).
[2] Includes *grisescentior* Clancey, 1975 (451) see Keith (1992: 372) (1239).
[3] Includes *goodi* see White (1962: 73) (2646).
[4] Includes *nigeriae* see White (1962: 74) (2646).
[5] Includes *minor, fayi* and *ngamii* see White (1962: 74) (2646).
[6] If the genus *Andropadus* is submerged in *Pycnonotus* it is necessary to use the substitute name *zeilae* Deignan, 1960. See addendum to Rand & Deignan (1960: 458) (1972).
[7] Includes *peasei* see White (1962: 75) (2646).
[8] Includes *naumanni* and *pallidus* see Rand & Deignan (1960: 242) (1972) and *tenebrior* and *micrus* see White (1962: 75) (2646).
[9] Dated 1837 by Rand & Deignan (1960: 243) (1972) but see Sherborn (1894) (2241).
[10] Both forms may well be natural wild hybrids.
[11] Includes *barat* see Mees (1996: 42) (1543).

○ *Pycnonotus finlaysoni* STRIPE-THROATED BULBUL
    — *P.f.davisoni* (Hume, 1875)   SW and C Burma
    — *P.f.eous* Riley, 1940   Thailand (except Malay Pen.), C and S Indochina
    — *P.f.finlaysoni* Strickland, 1844   Malay Pen.

○ *Pycnonotus xantholaemus* (Jerdon, 1845) YELLOW-THROATED BULBUL   S and E India

○ *Pycnonotus penicillatus* Blyth, 1851 YELLOW-EARED BULBUL   Sri Lanka

○ *Pycnonotus flavescens* FLAVESCENT BULBUL
    — *P.f.flavescens* Blyth, 1845   S Assam, W Burma
    — *P.f.vividus* (E.C.S. Baker, 1917)   NE Burma, W Yunnan, Thailand, N Indochina
    — *P.f.sordidus* (Robinson & Kloss, 1919)   S Indochina
    — *P.f.leucops* (Sharpe, 1888)   N Borneo

○ *Pycnonotus goiavier* YELLOW-VENTED BULBUL
    — *P.g.jambu* Deignan, 1955   C and SE Thailand to S Indochina
    — *P.g.analis* (Horsfield, 1821)   Malay Pen., Sumatra, Riau Arch., Bangka and
                                              Belitung, Java, Bali, Lombok, S Sulawesi
    — *P.g.gourdini* G.R. Gray, 1847[1]   Borneo, Maratua I., Karimundjawa Is.
    — *P.g.goiavier* (Scopoli, 1786)   N and NC Philippines
    — *P.g.samarensis* Rand & Rabor, 1960   Bohol, Cebu, Leyte, Samar (EC Philippines) (1977)
    — *P.g.suluensis* Mearns, 1909   Mindanao, Basilan, Camiguin Sur and Sulu Arch.

○ *Pycnonotus luteolus* WHITE-BROWED BULBUL
    — *P.l.luteolus* (Lesson, 1841)   C and S India
    — *P.l.insulae* Whistler & Kinnear, 1933   Sri Lanka

○ *Pycnonotus plumosus* OLIVE-WINGED BULBUL
    — *P.p.porphyreus* Oberholser, 1912   W Sumatra and islands
    — *P.p.plumosus* Blyth, 1845[2]   Malay Pen., E Sumatra, Java, Bawean, Belitung, W and
                                              S Borneo and Anamba Is.
    — *P.p.hutzi* Stresemann, 1938   N and E Borneo
    — *P.p.hachisukae* Deignan, 1952   Islands off N Borneo
    — *P.p.cinereifrons* (Tweeddale, 1878)   Palawan

○ *Pycnonotus blanfordi* STREAK-EARED BULBUL
    — *P.b.blanfordi* Jerdon, 1862   C and SW Burma
    — *P.b.conradi* (Finsch, 1873)[3]   SE Burma, Thailand, M of W Malaysia, S Indochina

○ *Pycnonotus simplex* CREAM-VENTED BULBUL
    — *P.s.simplex* Lesson, 1839[4]   Malay Pen., Sumatra, Bangka, Belitung, Borneo,
                                              S Natuna Is.
    — *P.s.prillwitzi* E. Hartert, 1902   Java
    — *P.s.halizonus* Oberholser, 1917   Anamba, N Natuna Is.

○ *Pycnonotus brunneus* RED-EYED BULBUL
    — *P.b.brunneus* Blyth, 1845   Malay Pen., Sumatra (incl. islands), Borneo
    — *P.b.zapolius* Oberholser, 1917   Tioman and Anamba Is.

○ *Pycnonotus erythrophthalmus* (Hume, 1878) SPECTACLED BULBUL[5]   Malay Pen., Belitung I., Sumatra and Borneo

ANDROPADUS Swainson, 1832[6] M
○ *Andropadus masukuensis* SHELLEY'S GREENBUL
    — *A.m.kakamegae* (Sharpe, 1900)   E Zaire, Uganda, W Kenya, NW Tanzania
    — *A.m.roehli* Reichenow, 1905   E Tanzania
    — *A.m.masukuensis* Shelley, 1897   S Tanzania, N Malawi
    — *A.m.kungwensis* (Moreau, 1941)[7]   W Tanzania (Kungwe-Mahari Mt.)

---

[1] Includes *karimuniensis* Hoogerwerf, 1963 (1143), see Mees (1986) (1539).
[2] Includes *billitonis*, *chiropethis* and *sibergi* Hoogerwerf, 1965 (1146), see Mees (1986) (1539).
[3] Includes *robinsoni* see Rand & Deignan (1960: 249) (1972) and Medway & Wells (1976) (1507).
[4] Includes *oblitus* and *perplexus* see Mees (1986) (1539).
[5] For submergence of *salvadorii* see Dickinson & Dekker 9220) (743B).
[6] Submerged in *Pycnonotus* in previous Editions following Rand & Deignan (1960) (1972); however African ornithology generally has not followed this. See Keith (1992: 279) (1239).
[7] This taxon was reassigned to this species by Hall & Moreau (1964) (1065).

○ *Andropadus montanus* Reichenow, 1892 CAMEROON MONTANE GREENBUL
    SE Nigeria, W Cameroon

○ *Andropadus tephrolaemus* GREY-THROATED GREENBUL[1]
    __ *A.t.bamendae* Bannerman, 1923   SE Nigeria, Cameroon
    __ *A.t.tephrolaemus* (G.R. Gray, 1862)   Mt. Cameroon, Bioko I.
    __ *A.t.kikuyuensis* (Sharpe, 1891)[2]   Zaire to C Kenya

○ *Andropadus nigriceps* MOUNTAIN GREENBUL
    __ *A.n.nigriceps* (Shelley, 1890)   S Kenya, N Tanzania
    __ *A.n.usambarae* (Grote, 1919)   SE Kenya, NE Tanzania
    __ *A.n.neumanni* (E. Hartert, 1922)   Uluguru Mts. (NE Tanzania)
    __ *A.n.fusciceps* (Shelley, 1893)   S Tanzania, Malawi, NW Mozambique

○ *Andropadus chlorigula* (Reichenow, 1899) GREEN-THROATED GREENBUL
    E and C Tanzania

○ *Andropadus milanjensis* STRIPE-CHEEKED GREENBUL
    __ *A.m.striifacies* (Reichenow & Neumann, 1895)   SE Kenya, NE and C Tanzania
    __ *A.m.olivaceiceps* (Shelley, 1896)   S Tanzania, Malawi, NW Mozambique
    __ *A.m.milanjensis* (Shelley, 1894)[3]   S Malawi, W Mozambique, E Zimbabwe

○ *Andropadus virens* LITTLE GREENBUL[4]
    __ *A.v.erythropterus* Hartlaub, 1858   Gambia to S Nigeria
    __ *A.v.amadoni* Dickerman, 1997[5]   Bioko I. (725)
    __ *A.v.virens* Cassin, 1858[6]   Cameroon to SW Sudan, W Kenya and N Angola
    __ *A.v.zombensis* Shelley, 1894[7]   SE Zaire, N Zambia, Malawi and N Mozambique to SE Kenya, Mafia I.
    __ *A.v.zanzibaricus* (Pakenham, 1935)   Zanzibar

○ *Andropadus gracilis* GREY GREENBUL
    __ *A.g.extremus* E. Hartert, 1922   Sierra Leone to SW Nigeria
    __ *A.g.gracilis* Cabanis, 1880   SE Nigeria to C and S Zaire and NE Angola
    __ *A.g.ugandae* van Someren, 1915   E Zaire to W Kenya

○ *Andropadus ansorgei* ANSORGE'S GREENBUL
    __ *A.a.ansorgei* E. Hartert, 1907[8]   Sierra Leone to Zaire
    __ *A.a.kavirondensis* (van Someren, 1920)   W Kenya

○ *Andropadus curvirostris* PLAIN GREENBUL
    __ *A.c.leoninus* Bates, 1930   Sierra Leone to C Ghana
    __ *A.c.curvirostris* Cassin, 1860   C and S Ghana to W Kenya and N Angola

○ *Andropadus gracilirostris* SLENDER-BILLED GREENBUL
    __ *A.g.gracilirostris* Strickland, 1844[9]   Guinea to Zaire, W Kenya and N Angola
    __ *A.g.percivali* (Neumann, 1903)   C Kenya

○ *Andropadus latirostris* YELLOW-WHISKERED GREENBUL
    __ *A.l.congener* Reichenow, 1897   Senegal to SW Nigeria
    __ *A.l.latirostris* Strickland, 1844[10]   SE Nigeria to Kenya and N Angola
    __ *A.l.australis* Moreau, 1941[11]   Ufipa Plateau (Tanzania)

○ *Andropadus importunus* SOMBRE GREENBUL
    __ *A.i.insularis* Hartlaub, 1861[12]   S Somalia to E Tanzania, Zanzibar I.
    __ *A.i.importunus* (Vieillot, 1818)[13]   Cape Province to Natal and E Transvaal, W Swaziland

---

[1] Although Keith (1992) (1239) retained a broad species *tephrolaemus* we follow the split proposed by Short *et al.* (1990) (2257) adding the species *nigriceps* and *chlorigula*.
[2] This may be a separate monotypic species see Keith (1992: 283) (1239) or it may belong to *A. nigriceps* see Zimmerman *et al.* (1996) (2735).
[3] Includes *disjunctus* Clancey, 1969 (418), see Keith (1992: 286) (1239); perhaps a valid race (R.J. Dowsett *in litt.*).
[4] *Andropadus hallae* Prigogine, 1972 (1931), has since been considered by its author to be a melanistic specimen of this species, see Keith (1992: 288) (1239).
[5] Originally named *Pycnonotus virens poensis* Dickerman, 1994 (724).
[6] Includes *holochlorus*, a name given to an intermediate population (Keith, 1992: 287) (1239).
[7] Includes *marwitzi* see White (1962: 80) (2646).
[8] In last Edition we listed *muniensis*, this is a synonym see Rand & Deignan (1960: 253) (1972).
[9] Includes *chagwensis* see White (1962: 78) (2646) and *congensis* see Keith (1992: 288) (1239).
[10] Includes *eugenius* and *saturatus* see Keith (1992: 296) (1239).
[11] This may relate to a sibling species see Keith (1992: 296) (1239).
[12] Includes *somaliensis* and *subalaris* see White (1962: 79) (2646) and *fricki* see Pearson in Keith (1992: 299) (1239).
[13] Includes *noomei* see White (1962: 79) (2646) and *errolius* Clancey, 1975 (447), see Keith (1992: 299) (1239).

—— *A.i.oleaginus* W.K.H. Peters, 1868[1] S Mozambique and S Zimbabwe to N Zululand and E Swaziland

—— *A.i.hypoxanthus* Sharpe, 1876[2] SE Tanzania and Mafia I., N and C Mozambique, Malawi and E Zimbabwe

CALYPTOCICHLA Oberholser, 1905 F
○ *Calyptocichla serina* (J. & E. Verreaux, 1855) GOLDEN GREENBUL Sierra Leone to N Zaire and N Angola

BAEOPOGON Heine, 1860 M
○ *Baeopogon indicator* HONEYGUIDE GREENBUL
—— *B.i.leucurus* (Cassin, 1856)[3] Sierra Leone to Togo
—— *B.i.indicator* (J. & E. Verreaux, 1855)[4] Nigeria to W Kenya and N Angola

○ *Baeopogon clamans* (Sjöstedt, 1893) SJÖSTEDT'S GREENBUL SE Nigeria to S Congo, NE and E Zaire

IXONOTUS J. & E. Verreaux, 1851 M
○ *Ixonotus guttatus* J. & E. Verreaux, 1851 SPOTTED GREENBUL[5] Liberia to Zaire and S Uganda

CHLOROCICHLA Sharpe, 1881[6] F
○ *Chlorocichla laetissima* JOYFUL GREENBUL
—— *C.l.laetissima* (Sharpe, 1899) NE Zaire to S Sudan and W Kenya
—— *C.l.schoutedeni* Prigogine, 1954 E Zaire, NE Zambia

○ *Chlorocichla prigoginei* DeRoo, 1967 PRIGOGINE'S GREENBUL Lendu Plateau and Butembo (E Zaire) (628)

○ *Chlorocichla flaviventris* YELLOW-BELLIED GREENBUL
—— *C.f.centralis* Reichenow, 1887 S Somalia, S Kenya, E Tanzania, N Mozambique
—— *C.f.occidentalis* Sharpe, 1881[7] Angola, SE Zaire and W Tanzania to Zimbabwe, C Mozambique and Transvaal
—— *C.f.flaviventris* (A. Smith, 1834) Natal, S Mozambique

○ *Chlorocichla falkensteini* (Reichenow, 1874) YELLOW-NECKED GREENBUL[8]
  S Cameroon, N Gabon and SW Cent. Afr. Rep., S Congo to W Angola

○ *Chlorocichla simplex* (Hartlaub, 1855) SIMPLE GREENBUL Guinea-Bissau to Zaire and N Angola

○ *Chlorocichla flavicollis* YELLOW-THROATED GREENBUL
—— *C.f.flavicollis* (Swainson, 1837) Senegal to N Cameroon
—— *C.f.soror* (Neumann, 1914)[9] C Cameroon to W and N Zaire, SW Sudan and W Ethiopia
—— *C.f.flavigula* (Cabanis, 1880)[10] W Kenya and W Tanzania to E and SE Zaire, N Zambia and Angola

THESCELOCICHLA Oberholser, 1905 F
○ *Thescelocichla leucopleura* (Cassin, 1856) SWAMP PALM BULBUL Senegal and Gambia to Gabon and Zaire

PYRRHURUS Cassin, 1859[11] M
○ *Pyrrhurus scandens* LEAF-LOVE
—— *P.s.scandens* (Swainson, 1837) Senegal to N Cameroon
—— *P.s.orientalis* (Hartlaub, 1883)[12] Cameroon to S Sudan, NW Tanzania and S Zaire

PHYLLASTREPHUS Swainson, 1831[13] M
○ *Phyllastrephus terrestris* TERRESTRIAL BROWNBUL
—— *P.t.suahelicus* Reichenow, 1904[14] S Somalia, E Kenya, E Tanzania, N Mozambique

---

[1] The substitution of Clancey's name *mentor* in last Edition is reversed, see Keith (1992: 299) (1239).
[2] Includes *loquax* Clancey, 1975 (447), see Keith (1992: 299) (1239).
[3] Includes *togoensis* see White (1962: 85) (2646).
[4] Includes *chlorosaturatus* see Rand & Deignan (1960: 259) (1972).
[5] Includes *bugoma* see Keith (1992: 307) (1239).
[6] Our species sequence follows Keith (1992) (1239).
[7] Includes *zambesiae* see Rand & Deignan (1960: 262) (1972) and *ortiva* Clancey, 1979 (475) see Keith (1992: 311) (1239).
[8] Includes *viridescentior* see Rand & Deignan (1960: 261) (1972).
[9] Includes *simplicicolor* see White (1962: 87) (2646).
[10] Includes *pallidigula* see Keith (1992: 316) (1239).
[11] Treated within the genus *Phyllastrephus* in last Edition, but see Keith (1992: 322) (1239).
[12] Includes *acedis* and *upembae* see Keith (1992: 322) (1239).
[13] Our species sequence follows Keith (1992) (1239). Madagascan species, other than *xanthophrys* which was earlier transferred to *Oxylabes*, have been transferred to the warbler genera *Bernieria* and *Xanthomixis* see Cibois *et al.* (2001) (358).
[14] Includes *bensoni* see Keith (1992: 325) (1239).

— *P.t.intermedius* Gunning & Roberts, 1911[1]   SW Angola, Zambia, SE Zaire and SW Tanzania to
S Mozambique and E Zululand

— *P.t.terrestris* Swainson, 1837[2]   N Transvaal to S and E Cape Province

○ *Phyllastrephus strepitans* (Reichenow, 1879) NORTHERN BROWNBUL   W Sudan (Darfur), SE Sudan, S Ethiopia and
S Somalia to NE Tanzania

○ *Phyllastrephus cerviniventris* GREY-OLIVE GREENBUL
   — *P.c.cerviniventris* (Shelley, 1894)   C Kenya to Zambia and C Mozambique
   — *P.c.schoutedeni* Prigogine, 1969   Katanga (Zaire) (1930)

○ *Phyllastrephus fulviventris* (Cabanis, 1876) PALE-OLIVE GREENBUL   W Zaire (Lower Congo river), W Angola

○ *Phyllastrephus baumanni* (Reichenow, 1895) BAUMANN'S GREENBUL   Sierra Leone to S Nigeria

○ *Phyllastrephus hypochloris* (Jackson, 1906) TORO OLIVE GREENBUL   S Sudan, NE and E Zaire, S Uganda, W Kenya

○ *Phyllastrephus lorenzi* Sassi, 1914 SASSI'S GREENBUL   E Zaire

○ *Phyllastrephus fischeri* (Reichenow, 1879) FISCHER'S GREENBUL[3]   S Somalia to NE Mozambique

○ *Phyllastrephus cabanisi* CABANIS'S GREENBUL[4]
   — *P.c.sucosus* Reichenow, 1903[5]   S Sudan, E Zaire to NW Tanzania and W Kenya
   — *P.c.placidus* (Shelley, 1889)   C Kenya to Malawi and NW Mozambique
   — *P.c.cabanisi* (Sharpe, 1881)   C Angola to N Zambia and W Tanzania

○ *Phyllastrephus poensis* (Alexander, 1903) CAMEROON OLIVE GREENBUL
   SE Nigeria, W Cameroon, Bioko I.

○ *Phyllastrephus icterinus* (Bonaparte, 1850) ICTERINE GREENBUL[6]   Sierra Leone to Zaire

○ *Phyllastrephus xavieri* XAVIER'S GREENBUL
   — *P.x.serlei* Chapin, 1949   Cameroon (N and W of Mt. Cameroon)
   — *P.x.xavieri* (Oustalet, 1892)[7]   Cameroon to Uganda and S Congo

○ *Phyllastrephus leucolepis* Gatter, 1985 LIBERIAN GREENBUL   SE Liberia (Grand Gedeh County) (929)

○ *Phyllastrephus albigularis* WHITE-THROATED GREENBUL
   — *P.a.albigularis* (Sharpe, 1881)   SW Senegal to Zaire, S Sudan and Uganda
   — *P.a.viridiceps* Rand, 1955   N Angola

○ *Phyllastrephus flavostriatus* YELLOW-STREAKED GREENBUL
   — *P.f.graueri* Neumann, 1908   NE and E Zaire
   — *P.f.olivaceogriseus* Reichenow, 1908[8]   Ruwenzori Mts. to SW Uganda, W Rwanda, N Burundi,
Itombwe and Mt Kobobo (E Zaire)
   — *P.f.kungwensis* Moreau, 1941   Kungwe-Mahari Mts. (W Tanzania)
   — *P.f.uzungwensis* Jensen & Stuart, 1982   Morogoro (E Tanzania) (1207)
   — *P.f.tenuirostris* (Fischer & Reichenow, 1884)   SE Kenya to NE Mozambique
   — *P.f.alfredi* (Shelley, 1903)   SW Tanzania, E Zambia, N Malawi
   — *P.f.vincenti* Grant & Mackworth-Praed, 1940   SE Malawi, W Mozambique
   — *P.f.flavostriatus* (Sharpe, 1876)[9]   S Mozambique, E Zimbabwe, E Sth. Africa

○ *Phyllastrephus poliocephalus* (Reichenow, 1892) GREY-HEADED GREENBUL
   SE Nigeria, W Cameroon

○ *Phyllastrephus debilis* TINY GREENBUL
   — *P.d.rabai* Hartert & van Someren, 1921   SE Kenya, NE Tanzania
   — *P.d.albigula* (Grote, 1919)   Usambara and Nguru Mts. (Tanzania)
   — *P.d.debilis* (W.L. Sclater, 1899)   SE Tanzania, E Zimbabwe, Mozambique

---

[1] Includes *rhodesiae* see Rand & Deignan (1960: 265) (1972) and *katangae* Prigogine, 1969 (1930), see Keith (1992: 325) (1239).
[2] Includes *montanus* see Rand & Deignan (1960: 265) (1972).
[3] The separation of *fischeri* (from the later recombination of *placidus* and *cabanisi*) was proposed by Ripley & Heinrich (1966) (2068).
[4] In last Edition *sucosus* (and the two names in synonymy herein) was erroneously attached to *fischeri*; the recombination of *placidus* and *cabanisi* was proposed by Dowsett & Dowsett-Lemaire (1980) (760).
[5] Includes *nandensis* Cunningham-van Someren & Schifter, 1981 (602), and *ngurumanensis* Cunningham-van Someren & Schifter, 1981 (602) (DRP).
[6] Includes *tricolor* see Keith (1992: 338) (1239).
[7] Includes *sethsmithi* see Rand & Deignan (1960: 271) (1972).
[8] Includes *ruwenzorii* Prigogine, 1975 (1932), and *itombwensis* Prigogine, 1975 (1932), see Keith (1992: 344) (1239).
[9] Includes *distans* Clancey, 1962 (393), *dendrophilus* Clancey, 1962 (393), and *dryobates* Clancey, 1975 (447), see Keith (1992: 344) (1239).

BLEDA Bonaparte, 1857 M[1]
○ **Bleda syndactylus** COMMON BRISTLEBILL
    — *B.s.syndactylus* (Swainson, 1837)[2]   Sierra Leone to W Zaire
    — *B.s.woosnami* Ogilvie-Grant, 1907   C and E Zaire to S Sudan, Uganda, W Kenya and
                                             NW Zambia
    — *B.s.nandensis* Cunningham-van Someren & Schifter, 1981[3]   N Nandi Forest (W Kenya) (602)

○ **Bleda eximius** GREEN-TAILED BRISTLEBILL[4]
    — *B.e.eximius* (Hartlaub, 1855)   Sierra Leone to Ghana
    — *B.e.notatus* (Cassin, 1857)   S Nigeria to lower Congo, Bioko I.
    — *B.e.ugandae* van Someren, 1915   C and N Zaire to SW Sudan, W and S Uganda

○ **Bleda canicapillus** GREY-HEADED BRISTLEBILL[5]
    — *B.c.canicapillus* (Hartlaub, 1854)   Guinea-Bissau to S Nigeria
    — *B.c.morelorum* Érard, 1992[6]   Senegal, Gambia (807)

CRINIGER Temminck, 1820[7] M
○ **Criniger barbatus** WESTERN BEARDED GREENBUL[8]
    — *C.b.barbatus* (Temminck, 1821)   Sierra Leone to Togo
    — *C.b.ansorgeanus* E. Hartert, 1907   S Nigeria

○ **Criniger chloronotus** (Cassin, 1860) EASTERN BEARDED GREENBUL[9]   Cameroon to Cabinda, N and E Zaire

○ **Criniger calurus** RED-TAILED GREENBUL
    — *C.c.verreauxi* Sharpe, 1871   Senegal to SW Nigeria
    — *C.c.calurus* (Cassin, 1857)   S Nigeria to W Zaire, Bioko I.
    — *C.c.emini* Chapin, 1948   WC Zaire and NE Angola to Uganda

○ **Criniger ndussumensis** Reichenow, 1904 WHITE-BEARDED GREENBUL   SE Nigeria to N and E Zaire

○ **Criniger olivaceus** (Swainson, 1837) YELLOW-BEARDED GREENBUL   S Senegal and SW Mali to Liberia and Ghana

○ **Criniger finschii** Salvadori, 1871 FINSCH'S BULBUL   Malay Pen., Sumatra, Borneo

○ **Criniger flaveolus** WHITE-THROATED BULBUL
    — *C.f.flaveolus* (Gould, 1836)   C and E Himalayas to E, NE and C Burma
    — *C.f.burmanicus* Oates, 1889   SE Burma, NW and W Thailand, SW Yunnan

○ **Criniger pallidus** PUFF-THROATED BULBUL
    — *C.p.griseiceps* Hume, 1873   Pegu Yoma (C Burma)
    — *C.p.robinsoni* Ticehurst, 1932   N Tenasserim (SE Burma)
    — *C.p.henrici* Oustalet, 1896   E Burma, N Thailand, SE Yunnan, NW Laos, N Vietnam
    — *C.p.pallidus* Swinhoe, 1870   Hainan (China)
    — *C.p.isani* Deignan, 1956   E Thailand
    — *C.p.annamensis* Delacour & Jabouille, 1924   C Laos and C Vietnam
    — *C.p.khmerensis* Deignan, 1956   S Laos and S Vietnam

○ **Criniger ochraceus** OCHRACEOUS BULBUL
    — *C.o.hallae* Deignan, 1956   S Vietnam
    — *C.o.cambodianus* Delacour & Jabouille, 1928   SE Thailand, SW Cambodia
    — *C.o.ochraceus* F. Moore, 1854   S Burma, SW Thailand
    — *C.o.sordidus* Richmond, 1900   C Malay Pen.
    — *C.o.sacculatus* Robinson, 1915   Mts. of W Malaysia
    — *C.o.sumatranus* Wardlaw Ramsay, 1882   W Sumatra
    — *C.o.fowleri* Amadon & Harrisson, 1957   Sarawak (Borneo)
    — *C.o.ruficrissus* Sharpe, 1879   NE Borneo

---

[1] Gender addressed by David & Gosselin (2002) (614), multiple changes occur in specific names.
[2] Includes *multicolor* see Keith (1992: 349) (1239).
[3] Validity requires confirmation.
[4] These three forms may belong to two species not one see Keith (1992: 351) (1239).
[5] For correction of spelling see David & Gosselin (2002) (613).
[6] Described as *moreli* Érard, 1991 (806).
[7] Following Hall & Moreau, 1970 (1066) Asian species are often treated under the generic name *Alophoixus*. However, although it is clear that *Criniger* as here employed is polyphyletic (Pasquet *et al.* 2001) (1791), the generic limits within the Asian species are not yet adequately studied and we, like Wolters, prefer to avoid confusing the nomenclature by extending *Alophoixus* beyond its type species and it is thus kept submerged waiting for more conclusive molecular studies.
[8] Keith (1992: 358) (1239) restored *chloronotus* to specific rank.
[9] Includes *weileri* see White (1962: 97) (2646).

○ *Criniger bres* Grey-cheeked Bulbul
    — *C.b.tephrogenys* (Jardine & Selby, 1833)   Malay Pen., E Sumatra
    — *C.b.bres* (Lesson, 1831)[1] [2]   Java and Bali
    — *C.b.gutturalis* (Bonaparte, 1850)   Borneo
    — *C.b.frater* Sharpe, 1877   Palawan

○ *Criniger phaeocephalus* Yellow-bellied Bulbul[3]
    — *C.p.phaeocephalus* (Hartlaub, 1844)   Malay Pen., Sumatra, Bangka, Belitung
    — *C.p.connectens* (Chasen & Kloss, 1929)   NE Borneo
    — *C.p.diardi* Finsch, 1867   W Borneo
    — *C.p.sulphuratus* (Bonaparte, 1850)   C and S Borneo[4]

ACRITILLAS Oberholser, 1905 F
○ *Acritillas indica* Yellow-browed Bulbul
    — *A.i.icterica* (Strickland, 1844)   W India
    — *A.i.indica* (Jerdon, 1839)   SW India, Sri Lanka (not SW)
    — *A.i.guglielmi* (Ripley, 1946)   SW Sri Lanka

SETORNIS Lesson, 1839 M
○ *Setornis criniger* Lesson, 1839 Hook-billed Bulbul   E Sumatra, Bangka I., Borneo

TRICHOLESTES Salvadori, 1874 M
○ *Tricholestes criniger* Hairy-backed Bulbul
    — *T.c.criniger* (Blyth, 1845)   Malay Pen., E Sumatra, N Natuna Is.
    — *T.c.sericeus* Robinson & Kloss, 1924   W Sumatra
    — *T.c.viridis* (Bonaparte, 1854)   Borneo

IOLE Blyth, 1844 F
○ *Iole virescens* Olive Bulbul[5]
    — *I.v.cacharensis* (Deignan, 1948)   S Assam
    — *I.v.myitkyinensis* (Deignan, 1948)   NE and E Burma
    — *I.v.virescens* Blyth, 1845   S Burma, SW Thailand

○ *Iole propinqua* Grey-eyed Bulbul
    — *I.p.aquilonis* (Deignan, 1948)   NE Vietnam, S China
    — *I.p.propinqua* (Oustalet, 1903)   E Burma, N Thailand, N Laos, NW Vietnam
    — *I.p.simulator* (Deignan, 1948)   SE Thailand, S Indochina
    — *I.p.innectens* (Deignan, 1948)   Extreme S Vietnam
    — *I.p.lekhakuni* (Deignan, 1954)   S Burma, SW Thailand
    — *I.p.cinnamomeoventris* E.C.S. Baker, 1917   Malay Pen. south to Trang

○ *Iole olivacea* Buff-vented Bulbul[6]
    — *I.o.olivacea* Blyth, 1844   Malay Pen., Sumatra (incl. islands), Bangka, Belitung, Anamba Is. and N Natuna Is.
    — *I.o.charlottae* (Finsch, 1867)   S and W Borneo
    — *I.o.perplexa* Riley, 1939   N and E Borneo (and islands)

○ *Iole palawanensis* (Tweeddale, 1878) Sulphur-bellied Bulbul   Palawan

IXOS Temminck, 1825[7] M
○ *Ixos nicobariensis* (F. Moore, 1854)[8] Nicobar Bulbul   Nicobars

---

[1] Given as 1832 by Rand & Deignan (1960) (1972), but see Browning & Monroe (1991) (258).
[2] Includes *balicus* see Mees (1986) (1539).
[3] Type species of the genus *Alophoixus* here retained in *Cringer*.
[4] Variable. Bornean birds require further review.
[5] The name *virescens* Blyth (introduced in the combination *Iole virescens*) is preoccupied in a broad genus *Hypsipetes*. Blyth later proposed the name *viridescens*, which was available and used in the context of that broad genus. The same would apply with a broad genus *Ixos* (which is now the oldest accepted generic name, see below); although it is not preoccupied in *Iole*, Art. 59.3 of the Code (I.C.Z.N., 1999) (1178) describes circumstances in which it would remain unavailable. Were that the case the name *viridescens* Blyth, 1867 would replace it. An explanation of the treatment herein of the old broad genus *Hypsipetes* is given by Dickinson & Gregory, (2002) (743C).
[6] The name *olivaceus* Blyth was preoccupied in a broad genus *Hypsipetes*. The next available name for the population of the Asian mainland was *crypta* of the Anamba Islands, which goes into synonymy when the generic name *Iole* is used. The oldest name for the species, when treated within the broad genus *Hypsipetes* (or *Ixos*), is then that of the form from S Borneo. Again Art. 59.3 of the Code (I.C.Z.N., 1999) (1178) might require the retention of *cryptus* and of the specific name *charlottae* in place of *olivacea*.
[7] This name, for the validity of which see Bock (1994: 202) (173), is the oldest name for the broad genus for which Deignan (1960) (1972) used the name *Hypsipetes*. See also Mees (1969: 302-303) (1520) and Dickinson & Gregory (2002) (743C).
[8] Originally called *Ixocincla virescens* Blyth, 1845, but, in the broad genus *Hypsipetes* and here in *Ixos*, this name is preoccupied by *virescens* Temminck, 1825. The latter is the type species of the genus *Ixos* and if placed in a different group of species would impose that generic name on such a group, as it is the oldest bulbul name. Sibley & Monroe (1990) (2262), and others since, have overlooked this point. See Bock (1994) (173) and Dickinson & Gregory (2002) (743C).

○ *Ixos mcclellandii* Mountain Bulbul
    — *I.m.mcclellandii* (Horsfield, 1840)   Himalayas, Assam
    — *I.m.ventralis* (Stresemann, 1940)   SW Burma
    — *I.m.tickelli* (Blyth, 1855)   E Burma, NW Thailand
    — *I.m.similis* (Rothschild, 1921)   NE Burma to N Indochina and Yunnan
    — *I.m.holtii* (Swinhoe, 1861)   C and S China
    — *I.m.loquax* (Deignan, 1940)   NC and EC Thailand, S Laos
    — *I.m.griseiventer* (Robinson & Kloss, 1919)   SC Vietnam (Langbian plateau)
    — *I.m.canescens* (Riley, 1933)   SE Thailand
    — *I.m.peracensis* (Hartert & Butler, 1898)   Malay Pen.

○ *Ixos malaccensis* (Blyth, 1845) Streaked Bulbul   Malay Pen., Sumatra, Borneo

○ *Ixos virescens* Sunda Bulbul
    — *I.v.sumatranus* (Wardlaw Ramsay, 1882)   W Sumatra
    — *I.v.virescens* Temminck, 1825   Java

○ *Ixos philippinus* Philippine Bulbul
    — *I.p.philippinus* (J.R. Forster, 1795)   Luzon, Samar, Leyte, Bohol
    — *I.p.parkesi* (duPont, 1980)   Burias (777)
    — *I.p.guimarasensis* (Steere, 1890)   Masbate, Panay, Negros
    — *I.p.saturatior* (E. Hartert, 1916)   E Mindanao
    — *I.p.mindorensis* (Steere, 1890)   Mindoro

○ *Ixos rufigularis* (Sharpe, 1877) Zamboanga Bulbul[1]   W Mindanao, Basilan

○ *Ixos siquijorensis* Streak-breasted Bulbul
    — *I.s.cinereiceps* (Bourns & Worcester, 1894)   Tablas, Romblon
    — †?*I.s.monticola* (Bourns & Worcester, 1894)   Cebu
    — *I.s.siquijorensis* (Steere, 1890)   Siquijor

○ *Ixos everetti* Yellowish Bulbul
    — *I.e.everetti* (Tweeddale, 1877)[2]   Dinagat, Leyte, E and C Mindanao and Samar
    — *I.e.catarmanensis* (Rand & Rabor, 1969)   Camiguin Sur I. (off N Mindanao) (1979)
    — *I.e.haynaldi* (W.H. Blasius, 1890)   Sulu Arch.

THAPSINILLAS Oberholser, 1905 F
○ *Thapsinillas affinis* Golden Bulbul[3]
    — *T.a.platenae* (W.H. Blasius, 1888)   Sangihe Is.
    — *T.a.aurea* (Walden, 1872)   Togian I.
    — *T.a.harterti* (Stresemann, 1912)   Peleng and Banggai
    — *T.a.longirostris* (Wallace, 1863)   Sula Is.
    — *T.a.chloris* (Finsch, 1867)   Morotai, Halmahera, Bacan (Moluccas)
    — *T.a.lucasi* E. Hartert, 1903   Obi (Moluccas)
    — *T.a.mystacalis* (Wallace, 1863)   Buru (Moluccas)
    — *T.a.affinis* (Hombron & Jacquinot, 1841)   Seram (Moluccas)
    — *T.a.flavicaudus* (Bonaparte, 1850)   Ambon (Moluccas)

MICROSCELIS G.R. Gray, 1840 M
○ *Microscelis amaurotis* Brown-eared Bulbul
    — *M.a.amaurotis* (Temminck, 1830)[4]   S Sakhalin; Hokkaido, Honshu, Kyushu and N Izu Is.
                                 (Japan) >> S Korea, NE China, Ryukyu Is.
    — *M.a.matchiae* Momiyama, 1923   Hachijo-jima (S Izu Is.), Tanegashima and Yaku-shima
                                   (S of Kyushu).
    — *M.a.ogawae* (E. Hartert, 1907)   N Ryukyu Is.
    — *M.a.pryeri* (Stejneger, 1887)[5]   C Ryukyu Is.

---

[1] Treated as a distinct species by Dickinson *et al.* (1991: 288) (745).
[2] Includes *samarensis* see Dickinson *et al.* (1991: 289) (745).
[3] We do not believe this assembly of forms, which may be more than one species (Delacour, 1943) (675), belongs in *Alophoixus* (which we have, for the moment, left in *Criniger*). Deignan (1960: 288) (1972) placed this between *everetti* and *indicus* but these two have not been posited to be close relatives so no relationship for *affinis* can be construed from the sequence and we prefer to return *affinis* to a monotypic genus pending molecular studies. See Dickinson & Gregory (2002) (743C).
[4] Includes *hensoni* see Orn. Soc. Japan (2000: 298) (1727).
[5] Includes *insignis* see Orn. Soc. Japan (1974: 358) (1726).

___ *M.a.stejnegeri* (E. Hartert, 1907)   S Ryukyu Is.

___ *M.a.squamiceps* (Kittlitz, 1831)   Ogasawara-shoto (Bonin) Is.

___ *M.a.magnirostris* (E. Hartert, 1905)   Iwo (Volcano) Is.

___ *M.a.borodinonis* N. Kuroda, Sr., 1923   Borodino Is.

___ *M.a.harterti* N. Kuroda, Sr., 1922[1]   S Taiwan and nearby Lanyu I.

___ *M.a.batanensis* (Mearns, 1907)   Batan I. (N Philippines)

___ *M.a.fugensis* (Ogilvie-Grant, 1895)   Calayan, Fuga Is. (N of Luzon)

___ *M.a.camiguinensis* (McGregor, 1907)   Camiguin Norte I. (N of Luzon)

HEMIXOS Blyth, 1845 M
○ *Hemixos flavala* ASHY BULBUL[2]

     ___ *H.f.flavala* Blyth, 1845[3]   Himalayas, W and N Burma, W Yunnan

     ___ *H.f.hildebrandi* (Hume, 1874)   E Burma, NW Thailand

     ___ *H.f.davisoni* (Hume, 1877)   SE Burma, SW Thailand

     ___ *H.f.bourdellei* (Delacour, 1926)   E Thailand, N and C Laos

     ___ *H.f.remotus* (Deignan, 1957)   S Laos and S Vietnam

     ___ *H.f.canipennis* (Seebohm, 1890)   S China, NE Vietnam

     ___ *H.f.castanonotus* (Swinhoe, 1870)   S Guangdong and Hainan

     ___ *H.f.cinereus* (Blyth, 1845)   Malay Pen., Sumatra

     ___ *H.f.connectens* (Sharpe, 1887)   N Borneo

HYPSIPETES Vigors, 1831 M
○ *Hypsipetes crassirostris* E. Newton, 1867 THICK-BILLED BULBUL   Seychelles

○ *Hypsipetes borbonicus* OLIVACEOUS BULBUL

     ___ *H.b.borbonicus* (J.R. Forster, 1781)   Reunion

     ___ *H.b.olivaceus* Jardine & Selby, 1836   Mauritius

○ *Hypsipetes madagascariensis* MADAGASCAN BULBUL[4]

     ___ *H.m.madagascariensis* (Statius Müller, 1776)   Madagascar; Comoro Is. (lowlands)

     ___ *H.m.grotei* (Friedmann, 1929)   Glorioso I.

     ___ *H.m.rostratus* (Ridgway, 1893)   Aldabra I.

○ *Hypsipetes parvirostris* COMORO BULBUL[5]

     ___ *H.p.parvirostris* Milne-Edwards & Oustalet, 1885   Grande Comore, Anjouan

     ___ *H.p.moheliensis* Benson, 1960   Mwali (Moheli) I. (122)

○ *Hypsipetes leucocephalus* BLACK BULBUL

     ___ *H.l.humii* (Whistler & Kinnear, 1933)   Sri Lanka

     ___ *H.l.ganeesa* Sykes, 1832   SW India

     ___ *H.l.psaroides* Vigors, 1831   NW and C Himalayas

     ___ *H.l.nigrescens* E.C.S. Baker, 1917   S and E Assam, W Burma

     ___ *H.l.concolor* Blyth, 1849   E Burma and S Yunnan south to SE Burma, N and EC Thailand, Laos and C and S Vietnam

     ___ *H.l.ambiens* (Mayr, 1942)   NE Burma, NW Yunnan

     ___ *H.l.sinensis* (La Touche, 1922)   N Yunnan and SE Xizang (China) >> Thailand, Laos

     ___ *H.l.stresemanni* (Mayr, 1942)   C Yunnan >> Thailand, Laos

     ___ *H.l.leucothorax* (Mayr, 1942)   Sichuan and S Shaanxi to middle Yangtze valley >> Thailand, Laos, Vietnam

     ___ *H.l.leucocephalus* (J.F. Gmelin, 1789)   SE China

     ___ *H.l.nigerrimus* Gould, 1863   Taiwan

     ___ *H.l.perniger* Swinhoe, 1870   Hainan, S Shandong (S China)

CERASOPHILA Bingham, 1900 F
○ *Cerasophila thompsoni* Bingham, 1900 WHITE-HEADED BULBUL   E Burma, NW Thailand

---

[1] The names *nagamichii* and *kurodae* Mishima, 1960 (1593) are synonyms. See also Orn. Soc. Japan (2000: 298) (1727).
[2] This probably comprises two or more species, but detailed evidence for a split awaits presentation and should deal with the whole "species".
[3] For correction of spelling see David & Gosselin (2002) (613).
[4] This name has been applied to a broader species including *leucocephalus*, but this is now generally rejected.
[5] We follow Cheke & Diamond (1986) (325) in transferring *moheliensis* from the species *H. crassirostris*.

GENUS SEDIS INCERTAE (4: 12)

NEOLESTES Cabanis, 1875[1] M
○ *Neolestes torquatus* Cabanis, 1875 Black-collared Bulbul   Gabon, Zaire, Angola

NICATOR Hartlaub & Finsch, 1870 M
○ *Nicator chloris* (Valenciennes, 1826) Western Nicator   Senegal to Sudan, Zaire, Uganda

○ *Nicator gularis* Hartlaub & Finsch, 1870 Eastern Nicator[2]   Kenya and Zambia to Natal

○ *Nicator vireo* Cabanis, 1876 Yellow-throated Nicator   Cameroon to N Angola, Uganda

ELMINIA Bonaparte, 1854 F
○ *Elminia longicauda* Blue Crested Flycatcher
    — *E.l.longicauda* (Swainson, 1838)   Senegal and Gambia to Nigeria
    — *E.l.teresita* Antinori, 1864   Cameroon to S Sudan and W Kenya, Angola

○ *Elminia albicauda* Bocage, 1877 Blue-and-white Crested Flycatcher[3]
                                                              Angola to E Zaire, Tanzania and Malawi

○ *Elminia nigromitrata* Dusky Crested Flycatcher
    — *E.n.colstoni* (Dickerman, 1994)   Liberia to Nigeria (731)
    — *E.n.nigromitrata* (Reichenow, 1874)   S Cameroon, Gabon to W Kenya

○ *Elminia albiventris* White-bellied Crested Flycatcher
    — *E.a.albiventris* (Sjöstedt, 1893)   Nigeria, Cameroon, Bioko I.
    — *E.a.toroensis* (Jackson, 1906)   NE Zaire, Rwanda, W Uganda

○ *Elminia albonotata* White-tipped Crested Flycatcher
    — *E.a.albonotata* (Sharpe, 1891)   E Zaire to Kenya, SW Tanzania, NE Zambia and N Malawi
    — *E.a.swynnertoni* (Neumann, 1908)   E Zimbabwe, C Mozambique
    — *E.a.subcaerulea* (Grote, 1923)   E and S Tanzania to S Malawi and N Mozambique

ERYTHROCERCUS Hartlaub, 1857[4] M
○ *Erythrocercus holochlorus* Erlanger, 1901 Little Yellow Flycatcher   S Somalia, coastal Kenya, Tanzania

○ *Erythrocercus mccallii* Chestnut-capped Flycatcher
    — *E.m.nigeriae* Bannerman, 1920   Sierra Leone to SW Nigeria
    — *E.m.mccallii* (Cassin, 1855)   SE Nigeria and Cameroon to Zaire
    — *E.m.congicus* Ogilvie-Grant, 1907   E and S Zaire, Uganda

○ *Erythrocercus livingstonei* Livingstone's Flycatcher
    — *E.l.thomsoni* Shelley, 1882   S Tanzania, N Malawi, N Mozambique
    — *E.l.livingstonei* G.R. Gray, 1870   Zambia, Zimbabwe, NW Mozambique
    — *E.l.francisi* W.L. Sclater, 1898   S Malawi, C and S Mozambique

SYLVIIDAE OLD WORLD WARBLERS (48: 265)

MEGALURINAE

AMPHILAIS S. Parker, 1984 M
○ *Amphilais seebohmi* (Sharpe, 1879) Madagascan Grassbird   NE and EC Madagascar

MEGALURUS Horsfield, 1821 M
○ *Megalurus pryeri* Japanese Swamp Warbler[5]
    — *M.p.sinensis* (Witherby, 1912)   NE China to SE Transbaikalia and Lake Khanka
                                                              (Russian Far East) >> NE Mongolia, EC China, Korea

    — *M.p.pryeri* Seebohm, 1884   N and C Honshu (Japan)

---

[1] For comments on this see Dowsett *et al.* (1999) (763).
[2] Implicitly includes *phyllophilus* Clancey, 1980 (478) see Keith (2000: 483) (1240).
[3] We prefer to keep the name Blue Flycatcher for members of the Asian genus *Cyornis*; we extend the name Crested Flycatcher to all the genus *Elminia* (removing it from those of the genus *Trochocercus* not transferred into *Elminia*), and we propose a new English name for this species (a course we have generally sought to avoid).
[4] For reasons to remove this genus from the Monarchidae see Pasquet *et al.* (2002) (1792).
[5] Morioka & Shigeta (1993) (1626) suggested this belongs in *Locustella*. We believe that just moves the problem and we suggest a new genus should be erected.

○ *Megalurus timoriensis* Tawny Grassbird[1]
   — *M.t.tweeddalei* McGregor, 1908   Luzon, Samar, Negros and Panay and small islands
                                                            between them

   — *M.t.alopex* Parkes, 1970   Bohol, Cebu, Leyte
   — *M.t.crex* Salomonsen, 1953   Mindanao, Basilan
   — *M.t.mindorensis* Salomonsen, 1953   Mindoro
   — *M.t.amboinensis* (Salvadori, 1876)   Ambon I. (N Moluccas)
   — *M.t.celebensis* Riley, 1919   Sulawesi
   — *M.t.inquirendus* Siebers, 1928   Sumba I. (Lesser Sundas)
   — *M.t.timoriensis* Wallace, 1864   Timor I. (Lesser Sundas)
   — *M.t.muscalis* Rand, 1938   S New Guinea
   — *M.t.alisteri* Mathews, 1912[2]   N and E Australia

○ *Megalurus macrurus* Papuan Grassbird[3]
   — *M.m.stresemanni* E. Hartert, 1930   NW New Guinea
   — *M.m.mayri* E. Hartert, 1930   N New Guinea
   — *M.m.wahgiensis* Mayr & Gilliard, 1951   Lower montane C New Guinea
   — *M.m.macrurus* (Salvadori, 1876)   Lower montane SC and SE New Guinea
   — *M.m.harterti* Mayr, 1931   Montane Huon Pen. (NE New Guinea)
   — *M.m.alpinus* Mayr & Rand, 1935[4]   Alpine SE and WC New Guinea
   — *M.m.interscapularis* P.L. Sclater, 1880   Bismarck Arch.

○ *Megalurus palustris* Striated Grassbird
   — *M.p.toklao* (Blyth, 1843)   NE Pakistan, C and N India to S China and mainland
                                                            SE Asia south to C Thailand

   — *M.p.palustris* Horsfield, 1821   Java
   — *M.p.forbesi* Bangs, 1919   Philippines (except Palawan and Sulu groups)

○ *Megalurus albolimbatus* (D'Albertis & Salvadori, 1879) Fly River Grassbird
                                                            SE New Guinea

○ *Megalurus gramineus* Little Grassbird
   — *M.g.papuensis* Junge, 1952   New Guinea
   — *M.g.goulburni* Mathews, 1912[5]   NE Western Australia, C, E and SE Australia
   — *M.g.thomasi* Mathews, 1912[6]   SW Western Australia
   — *M.g.gramineus* (Gould, 1845)   Tasmania, King I., Flinders I.

○ *Megalurus punctatus* Fernbird
   — *M.p.vealeae* (Kemp, 1912)   North Island (New Zealand)
   — *M.p.punctatus* (Quoy & Gaimard, 1830)   South Island (New Zealand)
   — *M.p.stewartianus* (Oliver, 1930)   Stewart I.
   — *M.p.wilsoni* (Stead, 1936)   Codfish I.
   — †*M.p.rufescens* (Buller, 1869)   Chatham Is.
   — *M.p.caudatus* (Buller, 1894)   Snares I.

CINCLORAMPHUS Gould, 1838 M
○ *Cincloramphus mathewsi* Iredale, 1911 Rufous Songlark   Australia

○ *Cinclorhamphus cruralis* (Vigors & Horsfield, 1827) Brown Songlark
                                                            Australia

EREMIORNIS North, 1900 M
○ *Eremiornis carteri* North, 1900 Spinifex-bird   WC Western Australia (Pilbara), C and S Northern
                                                            Territory, W Queensland, NE South Australia

BUETTIKOFERELLA Stresemann, 1928[7] F
○ *Buettikoferella bivittata* (Bonaparte, 1850) Buff-banded Grassbird   Timor I.

MEGALURULUS J. Verreaux, 1869[8] M
○ *Megalurulus mariei* J. Verreaux, 1869 New Caledonian Grassbird   New Caledonia

[1] In last Edition we treated a broad species. Here we treat two species (*timoriensis* and *macrurus*) following Schodde & Mason (1999: 701-702) (2189).
[2] Includes *oweni* see Schodde & Mason (1999: 702) (2189).
[3] We suspect that large alpine birds may prove to be specifically distinct.
[4] Tentatively includes *montanus*. The range ascribed to *montanus* in Watson *et al.* (1986: 40) (2564) falls within that assigned to *alpinus* in the same work.
[5] For recognition see Schodde & Mason (1999: 704) (2189).
[6] For recognition see Schodde & Mason (1999: 704) (2189).
[7] Treated in the genus *Megalurulus* in last Edition, but see Bruce in White & Bruce (1986: 342) (2652).
[8] Includes *Cichlornis* and *Ortygocichla* see Ripley (1985) (2058). Ripley also suggested *Buettikoferella* be submerged here, but we prefer to retain it.

○ *Megalurulus whitneyi* MELANESIAN THICKETBIRD
　　　— *M.w.grosvenori* (Gilliard, 1960)　　　　　　New Britain (Bismarck Arch.)
　　　— *M.w.llaneae* (Hadden, 1983)　　　　　　　Bougainville I. (N Solomons)
　　　— *M.w.turipavae* (Cain & Galbraith, 1955)　　Guadalcanal I. (Solomons)
　　　— *M.w.whitneyi* (Mayr, 1933)　　　　　　　Espiritu Santo I. (N Vanuatu)

○ *Megalurulus rubiginosus* (P.L. Sclater, 1881)　RUSTY THICKETBIRD　　New Britain

○ *Megalurulus rufus* LONG-LEGGED THICKETBIRD[1]
　　　— *M.r.rufus* (Reichenow, 1890)　　　　　　Viti Levu (Fiji Is.)
　　　— *M.r.cluniei* (Kinsky, 1975)　　　　　　　Vanua Levu I. (Fiji Is.)

CHAETORNIS G.R. Gray, 1848　F
○ *Chaetornis striata* (Jerdon, 1841)　BRISTLED GRASS WARBLER[2]　　NE Pakistan, India, S Nepal, Bangladesh

GRAMINICOLA Jerdon, 1863　M[3]
○ *Graminicola bengalensis* LARGE GRASS WARBLER
　　　— *G.b.bengalensis* Jerdon, 1863　　　　　　N and NE India, S Nepal, S Bhutan, N Bangladesh
　　　— *G.b.sinicus* Stresemann, 1923　　　　　　SE China
　　　— *G.b.striatus* Styan, 1892　　　　　　　E Burma, N Vietnam, Hainan (formerly C Thailand)

SCHOENICOLA Blyth, 1844　M
○ *Schoenicola platyurus* (Jerdon, 1844)　BROAD-TAILED WARBLER[4]　　SW India, Sri Lanka

○ *Schoenicola brevirostris* FAN-TAILED GRASSBIRD[5]
　　　— *S.b.alexinae* (Heuglin, 1863)　　　　　　Guinea to Ethiopia, Angola, Tanzania
　　　— *S.b.brevirostris* (Sundevall, 1850)　　　　Malawi to E Sth. Africa.

ACROCEPHALINAE

HEMITESIA Chapin, 1948　F
○ *Hemitesia neumanni* (Rothschild, 1908)　NEUMANN'S WARBLER　　E Zaire, SW Uganda

OLIGURA Hodgson, 1844　F
○ *Oligura castaneocoronata* CHESTNUT-HEADED TESIA
　　　— *O.c.castaneocoronata* (E. Burton, 1836)　　Himalayas, Assam, Burma, NW Thailand, S China
　　　— *O.c.ripleyi* Deignan, 1951[6]　　　　　　NW Yunnan
　　　— *O.c.abadiei* (Delacour & Jabouille, 1930)　N Vietnam

TESIA Hodgson, 1837　F
○ *Tesia olivea* (McClelland, 1840)　SLATY-BELLIED TESIA　　E Himalayas, Burma, N Thailand, Yunnan, N and
　　　　　　　　　　　　　　　　　　　　　　　　　　　C Indochina

○ *Tesia cyaniventer* Hodgson, 1837　GREY-BELLIED TESIA　　Himalayas, Burma, S China, NW Thailand, Laos,
　　　　　　　　　　　　　　　　　　　　　　　　　　　NW and SC Vietnam

○ *Tesia superciliaris* (Bonaparte, 1850)　JAVAN TESIA　　Java

○ *Tesia everetti* RUSSET-CAPPED STUBTAIL[7]
　　　— *T.e.everetti* (E. Hartert, 1897)　　　　　Flores I.
　　　— *T.e.sumbawana* (Rensch, 1928)　　　　　Sumbawa I.

UROSPHENA Swinhoe, 1877[8]　F
○ *Urosphena squameiceps* ASIAN STUBTAIL
　　　— *U.s.squameiceps* (Swinhoe, 1863)　　　　S Sakhalin, S Kuril Is., Japan >> E China to mainland
　　　　　　　　　　　　　　　　　　　　　　　　　　　SE Asia

　　　— *U.s.ussurianus* (Seebohm, 1881)[9]　　　Russian Far East, Korea, NE China south to Hebei >>
　　　　　　　　　　　　　　　　　　　　　　　　　　　E China to mainland SE Asia

---

[1] Type species of the genus *Trichocichla* here as elsewhere submerged.
[2] For correction of spelling see David & Gosselin (2002) (614).
[3] Gender addressed by David & Gosselin (2002) (614), multiple changes occur in specific names.
[4] For correction of spelling see David & Gosselin (2002) (614).
[5] For separation from *platyura* see Prigogine (1985) (1943).
[6] Perhaps not distinct from *abadiei*. CSR.
[7] For reasons to place this in *Tesia* see King (1989) (1267).
[8] For changes to this genus since Watson *et al.* (1986) (2564) see King (1989) (1267).
[9] For recognition see Stepanyan (1990: 435) (2326).

○ *Urosphena whiteheadi* (Sharpe, 1888) BORNEAN STUBTAIL   NE Borneo

○ *Urosphena subulata* TIMOR STUBTAIL
     __ *U.s.subulata* (Sharpe, 1884)   Timor I.
     __ *U.s.advena* (E. Hartert, 1906)   Babar I.

CETTIA Bonaparte, 1834[1] F
○ *Cettia pallidipes* PALE-FOOTED BUSH WARBLER
     __ *C.p.pallidipes* (Blanford, 1872)   W and C Himalayas, Assam, N Burma >> S India
     __ *C.p.laurentei* La Touche, 1921   E Burma, S China, N and C Indochina, NW Thailand
                                         >> mainland SE Asia
     __ *C.p.osmastoni* (E. Hartert, 1908)   S Andaman

○ *Cettia diphone* JAPANESE BUSH WARBLER[2]
     __ *C.d.borealis* C.W. Campbell, 1892   Shandong to NE China, Korea, S Russian Far East >>
                                         S China to mainland SE Asia
     __ *C.d.canturians* (Swinhoe, 1860)   C and E China >> S China to mainland SE Asia
     __ *C.d.sakhalinensis* (Yamashina, 1927)[3]   C and S Sakhalin, S Kuril Is. >> SE China
     __ *C.d.cantans* (Temminck & Schlegel, 1847)   C and S Japan
     __ *C.d.riukiuensis* (N. Kuroda, Sr., 1925)   Ryukyu Is.
     __ †?*C.d.restricta* (N. Kuroda, Sr., 1923)   Daito (Borodino) Is.
     __ *C.d.diphone* (Kittlitz, 1830)   Ogasawara-shoto (Bonin) Is.

○ *Cettia seebohmi* Ogilvie-Grant, 1894 LUZON BUSH WARBLER[4]   N Luzon

○ *Cettia annae* (Hartlaub & Finsch, 1868) PALAU BUSH WARBLER   Palau Is.

○ *Cettia parens* (Mayr, 1935) SHADE WARBLER   San Cristobal I. (Solomon Is.)

○ *Cettia ruficapilla* FIJI BUSH WARBLER
     __ *C.r.ruficapilla* (E.P. Ramsay, 1876)   Kandavu I.
     __ *C.r.badiceps* (Finsch, 1876)   Viti Levu I.
     __ *C.r.castaneoptera* (Mayr, 1935)   Vanua Levu I.
     __ *C.r.funebris* (Mayr, 1935)   Taveuni I.

○ *Cettia fortipes* BROWNISH-FLANKED BUSH WARBLER
     __ *C.f.pallida* (W.E. Brooks, 1871)   NW Himalayas to W Nepal
     __ *C.f.fortipes* (Hodgson, 1845)   E Himalayas, Assam, Burma, W Yunnan
     __ *C.f.davidiana* (J. Verreaux, 1871)   C and E China, N Indochina
     __ *C.f.robustipes* (Swinhoe, 1866)[5]   Taiwan

○ *Cettia vulcania* SUNDA BUSH WARBLER
     __ *C.v.sepiaria* Kloss, 1931   N Sumatra
     __ *C.v.flaviventris* (Salvadori, 1879)   C and S Sumatra
     __ *C.v.vulcania* (Blyth, 1870)   E Sumatra, Java, Bali, Lombok
     __ *C.v.kolichisi* Johnstone & Darnell, 1997   Alor (1220)
     __ *C.v.everetti* E. Hartert, 1898   Timor I.
     __ *C.v.banksi* Chasen, 1935   NW Borneo
     __ *C.v.oreophila* Sharpe, 1888   NE Borneo
     __ *C.v.palawana* Ripley & Rabor, 1962   Palawan

○ *Cettia major* CHESTNUT-CROWNED BUSH WARBLER
     __ *C.m.major* (Moore, 1854)[6]   C and E Himalayas, SE Xizang, NE Burma, C China
     __ *C.m.vafer* (Koelz, 1954)   Hills of S Assam

○ *Cettia carolinae* Rozendaal, 1987 TANIMBAR BUSH WARBLER   Tanimbar Is. (2120)

○ *Cettia flavolivacea* ABERRANT BUSH WARBLER
     __ *C.f.flavolivacea* (Blyth, 1845)   C and E Himalayas

---

[1] Much work remains to be done to elucidate this genus.
[2] Reviewed by Vaurie (1954) (2494). Despite his views, it is very probable that *diphone* comprises two or even three species, and King & Dickinson (1975) (1268) separated *canturians* from *diphone*, but no detailed explanation has been published. Dickinson *et al.* (1991: 339) (745) reverted to one broad species, which we follow here, but we recommend further study.
[3] Treated as a synonym of *cantans* by Watson *et al.* (1986: 10) (2564); we place *viridis* in synonymy here and the name *sakhalinensis* has priority.
[4] For reasons for separation see Orenstein & Pratt (1983) (1723).
[5] For reasons to attach *robustipes* here see Dickinson *et al.* (1991: 340) (745).
[6] Ascribed to Horsfield & Moore, 1854, by Watson (1986: 13) (2564), however Moore made clear that this was his new name by indicating that he had it in press with the Proc. Zool. Soc.

— *C.f.stresemanni* (Koelz, 1954)	SW Assam
— *C.f.alexanderi* (Ripley, 1951)	SE Assam
— *C.f.intricata* (E. Hartert, 1909)[1]	N and E Burma, NW and NC China (Yunnan and Sichuan to Shanxi)
— *C.f.weberi* (Mayr, 1941)	Chin hills (W Burma)
— *C.f.oblita* (Mayr, 1941)	N Indochina

○ **Cettia acanthizoides** YELLOWISH-BELLIED BUSH WARBLER

— *C.a.brunnescens* (Hume, 1872)	C and E Himalayas to N Burma
— *C.a.acanthizoides* (J. Verreaux, 1871)	E Burma, C and E China
— *C.a.concolor* Ogilvie-Grant, 1912[2]	Mountains of Taiwan

○ **Cettia brunnifrons** (Hodgson, 1845) GREY-SIDED BUSH WARBLER[3] — Himalayas to N Burma and C China >> NE India, W Burma

● **Cettia cetti** CETTI'S WARBLER

— *C.c.cetti* (Temminck, 1820)	NW Africa, SW and W Europe east to S and W Italy and Sicily
— *C.c.sericea* (Temminck, 1820)[4]	NE Italy and Balkans to W Asia Minor
— *C.c.orientalis* Tristram, 1867	C Asia Minor, Cyprus and Levant to N and E Iran and NW Kazakhstan
— *C.c.albiventris* Severtsov, 1873	SW Iran and C Turkmenistan and Uzbekistan to Aghanistan and NW Xinjiang >> Pakistan

BRADYPTERUS Swainson, 1837 M

○ **Bradypterus baboecala** LITTLE RUSH WARBLER

— *B.b.centralis* Neumann, 1908	Nigeria, S Cameroon to NE Zaire to SW Uganda
— *B.b.chadensis* Bannerman, 1936	W Chad
— *B.b.sudanensis* Grant & Mackworth-Praed, 1941	S Sudan
— *B.b.abyssinicus* (Blundell & Lovat, 1899)	Ethiopia
— *B.b.elgonensis* Madarász, 1912	W and C Kenya, SE Uganda
— *B.b.tongensis* Roberts, 1931	SE Kenya to E South Africa
— *B.b.msiri* Neave, 1909	E Angola, SE Zaire, NE Botswana, W and N Zambia
— *B.b.benguellensis* Bannerman, 1927	W Angola
— *B.b.transvaalensis* Roberts, 1919[5]	C Zimbabwe, W Swaziland to W Natal and Zululand
— *B.b.baboecala* (Vieillot, 1817)	S South Africa

○ **Bradypterus graueri** Neumann, 1908 GRAUER'S SWAMP WARBLER — E Zaire, SW Uganda, Rwanda, N Burundi

○ **Bradypterus grandis** Ogilvie-Grant, 1917 JA RIVER SCRUB WARBLER — S Cameroon, Gabon

○ **Bradypterus carpalis** Chapin, 1916 WHITE-WINGED SWAMP WARBLER — NE and E Zaire, Uganda, Rwanda, Burundi, W Kenya, NW Tanzania

○ **Bradypterus alfredi** BAMBOO WARBLER

— *B.a.kungwensis* Moreau, 1942	NW Zambia, W Tanzania
— *B.a.alfredi* Hartlaub, 1890	W Ethiopia to E Zaire

○ **Bradypterus sylvaticus** KNYSNA SCRUB WARBLER

— *B.s.sylvaticus* Sundevall, 1860	S Cape Province coast
— *B.s.pondoensis* Haagner, 1910	E Cape Province and Natal coast

○ **Bradypterus lopesi** EVERGREEN-FOREST WARBLER[6]

— *B.l.lopesi* (Alexander, 1903)[7]	Bioko I.
— *B.l.camerunensis* Alexander, 1909	Mt. Cameroon
— *B.l.barakae* Sharpe, 1906	Zaire, Rwanda, SW Uganda
— *B.l.mariae* Madarász, 1905	Kenya, NE Tanzania
— *B.l.usambarae* Reichenow, 1917	SE Kenya (Taita Hills), E and SW Tanzania, N Malawi, N Mozambique

[1] Includes *dulcivox* which was treated as a synonym of *C. fortipes davidiana* by Watson *et al.* (1986: 12) (2564). Cheng (1987: 769) (333) followed Vaurie (1969: 224) (2502) in considering that it belonged here; however, as Vaurie noted, the unique type has been destroyed so that this treatment is speculative (even if topotypes were available to Cheng). May best be considered unidentifiable.
[2] See Dickinson *et al.* (1991: 340) (745).
[3] Includes *whistleri* and *umbratica*. We follow Vaurie (1959: 226) (2502) in treating this as monotypic.
[4] For recognition see Roselaar (1995) (2107).
[5] For recognition see Grimes, Fry & Keith (1997: 64) (1013).
[6] For treatment of *mariae* not as a species, but within this broad species, see Grimes, Fry & Keith (1997: 72) (1013).
[7] The emended name used here is judged to be in prevailing usage; see Art. 33.2 of the Code (I.C.Z.N., 1999) (1178) and p. 000 herein.

○ _B.l.ufipae_ (Grant & Mackworth-Praed, 1941)	SW Tanzania, SE Zaire, N Zambia
○ _B.l.granti_ Benson, 1939[1]	Malawi, N Mozambique
○ _B.l.boultoni_ Chapin, 1948	Angola

○ **_Bradypterus barratti_** BARRATT'S SCRUB WARBLER

○ _B.b.barratti_ Sharpe, 1876	SW Mozambique to E Transvaal, N Natal
○ _B.b.godfreyi_ (Roberts, 1922)	Cape Province, Natal
○ _B.b.cathkinensis_ Vincent, 1948	Drakensberg Mts. (Sth. Africa)
○ _B.b.priesti_ Benson, 1946[2]	E Zimbabwe and W Mozambique

○ **_Bradypterus victorini_** Sundevall, 1860 VICTORIN'S SCRUB WARBLER — S Cape Province

○ **_Bradypterus cinnamomeus_** CINNAMON BRACKEN WARBLER

○ _B.c.bangwaensis_ Delacour, 1943[3]	E Nigeria, Cameroon
○ _B.c.cavei_ Macdonald, 1939	SE Sudan
○ _B.c.cinnamomeus_ (Rüppell, 1840)	Ethiopia to E Zaire and N Tanzania
○ _B.c.mildbreadi_ Reichenow, 1908	Ruwenzori Mts.
○ _B.c.nyassae_ Shelley, 1893	NE and SW Tanzania, SE Zaire, NE Zambia, Malawi

○ **_Bradypterus thoracicus_** SPOTTED BUSH WARBLER[4]

○ _B.t.davidi_ (La Touche, 1923)	SE Transbaikalia, Russian Far East, NE China south to N Hubei >> mainland SE Asia
○ _B.t.suschkini_ (Stegmann, 1929)[5]	SC Siberia (Altai to SW Transbaikalia >> mainland SE Asia
○ _B.t.przewalskii_ (Sushkin, 1925)	E Tibetan Plateau (NE Xizang to SW Gansu and S Shaanxi)
○ _B.t.kashmirensis_ (Sushkin, 1925)	NW Himalayas east to Kumaun
○ _B.t.thoracicus_ (Blyth, 1845)	C and E Himalayas (Nepal to N Burma and SE Zixang), Yunnan to NC Sichuan, Guizhou and Dayao Shan (Juangxi)

○ **_Bradypterus major_** LONG-BILLED BUSH WARBLER

○ _B.m.major_ (W.E. Brooks, 1872)	E Pamir Mts. and Kashmir to Garhwal (NW Himalayas)
○ _B.m.innae_ (Portenko, 1955)	W and S Xinjiang (W China)

○ **_Bradypterus tacsanowskius_** (Swinhoe, 1871) CHINESE BUSH WARBLER — SC Siberia, Transbaikalia, S Russian Far East, NE China, and NC China south to Sichuan >> NE India and Nepal to Indochina

○ **_Bradypterus luteoventris_** (Hodgson, 1845) BROWN BUSH WARBLER[6] — E Himalayas, W and N Burma, S and SE China, N Vietnam

○ **_Bradypterus alishanensis_** Rasmussen, Round, Dickinson & Rozendaal, 2000 TAIWAN BUSH WARBLER # — N and C Taiwan (1993)

○ **_Bradypterus mandelli_** RUSSET BUSH WARBLER[7]

○ _B.m.mandelli_ (W.E. Brooks, 1875)	E Himalayas, Assam, W, N and NE Burma, SC and S China, N Thailand, N Indochina (242)
○ _B.m.idoneus_ (Riley, 1940)	Langbian Plateau (SC Vietnam)
○ _B.m.melanorhynchus_ (Rickett, 1898)	E China (Hubei to Fujian) >> SE China, N Indochina

○ **_Bradypterus seebohmi_** (Ogilvie-Grant, 1895) BENGUET BUSH WARBLER[8] — N Luzon

○ **_Bradypterus montis_** (E. Hartert, 1896) JAVAN BUSH WARBLER[9] — E and C Java (Bali ?)

○ **_Bradypterus timorensis_** Mayr, 1944 TIMOR BUSH WARBLER[10] — Timor

---

[1] For recognition see Grimes, Fry & Keith (1997: 71) (1013).
[2] For treatment in this species see Grimes, Fry & Keith (1997: 76) (1013).
[3] We prefer to leave this taxon attached to this species, as in White (1960: 407) (2639), rather than attach it to _lopesi_ as in Urban _et al._ (1997) (2455).
[4] Round & Loskot (1995) (2110) said that this might constitute two species; we understand a paper formally proposing the split is going to press and that _davidi_ (to be called Siberian Bush Warbler and with subspecies _suschkini_) will be separated from _thoracicus_.
[5] Includes _shanensis_ see Round & Loskot (1995) (2110).
[6] Includes _ticehursti_ see Dickinson _et al._ (2000) (746).
[7] For the use of the name _mandelli_ see Dickinson _et al._ (2000) (746).
[8] For restriction to a monotypic Philippine endemic see Dickinson _et al._ (2000) (746).
[9] For treatment as a monotypic species endemic to Java (but with sight records and recordings from Bali) see Dickinson _et al._ (2000) (746).
[10] For reasons not to lump with _montis_ see Dickinson _et al._ (2000) (746).

○ *Bradypterus caudatus* Long-tailed Bush Warbler
    — *B.c.caudatus* (Ogilvie-Grant, 1895)   N Luzon
    — *B.c.unicolor* (E. Hartert, 1904)   Mindanao (except Zamboanga Pen.)
    — *B.c.malindangensis* (Mearns, 1909)   Mindanao (N Zamboanga Pen.)

○ *Bradypterus accentor* (Sharpe, 1888)   Friendly Bush Warbler   N Borneo

○ *Bradypterus castaneus* Chestnut-backed Bush Warbler
    — *B.c.disturbans* (E. Hartert, 1900)   Buru
    — *B.c.musculus* (Stresemann, 1914)   Seram
    — *B.c.castaneus* (Büttikofer, 1893)   C Moluccas

DROMAEOCERCUS Sharpe, 1877[1] M
○ *Dromaeocercus brunneus* Sharpe, 1877 Brown Emu-tail[2]   NE and EC Madagascar

ELAPHRORNIS Legge, 1879[3] M
○ *Elaphrornis palliseri* (Blyth, 1851) Sri Lankan Bush Warbler   Sri Lanka

BATHMOCERCUS Reichenow, 1895 M
○ *Bathmocercus cerviniventris* (Sharpe, 1877) Black-headed Rufous Warbler
                                       Sierra Leone to Ghana

○ *Bathmocercus rufus* Black-faced Rufous Warbler
    — *B.r.rufus* Reichenow, 1895   Cameroon, Gabon, Central African Republic
    — *B.r.vulpinus* Reichenow, 1895   E Zaire, Sudan, Uganda, Kenya, Tanzania

○ *Bathmocercus winifredae* (Moreau, 1938) Mrs. Moreau's Warbler   E Tanzania

NESILLAS Oberholser, 1899 F
○ *Nesillas typica* Madagascan Brush Warbler
    — *N.t.moheliensis* Benson, 1960   Mwali (Moheli) I. (Comoro Is.)
    — *N.t.obscura* Delacour, 1931   W and N Madagascar (limestone hills)
    — *N.t.typica* (Hartlaub, 1860)   WC, S, E and EC Madagascar

○ *Nesillas lantzii* (A. Grandidier, 1867) Subdesert Brush Warbler[4]   S and W Madagascar

○ *Nesillas longicaudata* (E. Newton, 1877) Anjouan Brush Warbler[5]   Anjouan I. (Comoro Is.)

○ *Nesillas brevicaudata* (Milne-Edwards & Oustalet, 1888) Grande Comore Brush Warbler[6]
                                        Grande Comore I. (Comoro Is.)

○ *Nesillas mariae* Benson, 1960 Moheli Brush Warbler   Mwali (Moheli) I.

○ †?*Nesillas aldabrana* Benson & Penny, 1968 Aldabra Brush Warbler[7]
                                        Aldabra I.

MELOCICHLA Hartlaub, 1857 F
○ *Melocichla mentalis* Moustached Grass Warbler
    — *M.m.mentalis* (Fraser, 1843)   S Senegal to Central African Republic,
                                        N Zambia and Angola
    — *M.m.amauroura* (Pelzeln, 1883)[8]   S Sudan, SW Ethiopia and EC Africa to C Zambia
    — *M.m.orientalis* (Sharpe, 1883)   E Kenya to Zimbabwe and C Mozambique
    — *M.m.incana* Diesselhorst, 1959[9]   NE Tanzania (Mt. Meru)
    — *M.m.luangwae* Benson, 1958   E Zambia, W Malawi

SPHENOEACUS Strickland, 1841 M
○ *Sphenoeacus afer* Cape Grassbird
    — *S.a.excisus* Clancey, 1973   E Zimbabwe, W Mozambique

---

[1] Treated within the genus *Bradypterus* in last Edition. We prefer to revert to the separation employed by Watson *et al.* (1986) (2564), but we do not lump *Amphilais*.
[2] The use of the name Emu-tail for this species is misleading as its previous congener *Amphilais seebohmi* is now treated within a different subfamily. See Parker (1984) (1745).
[3] This is entirely distinct from the genus *Bradypterus*, but perhaps related; further subdivision of *Bradypterus* is probably warranted. ECD.
[4] Separated from *N. typica* by Schulenberg *et al.* (1993) (2215).
[5] Separated from *N. typica* by Louette (1988) (1385).
[6] Separated from *N. typica* by Louette (1988) (1385).
[7] For correct spelling see David & Gosselin (2002) (613).
[8] For correction of spelling see David & Gosselin (2002) (613).
[9] For correction of spelling see David & Gosselin (2002) (613).

___ *S.a.natalensis* Shelley, 1882 — Transvaal, Natal, Zululand, W Swaziland, Orange Free State, N Lesotho

___ *S.a.intermedius* Shelley, 1882 — E Cape Province

___ *S.a.afer* (J.F. Gmelin, 1789) — SW Cape Province

LOCUSTELLA Kaup, 1829 F

○ *Locustella lanceolata* LANCEOLATED WARBLER

___ *L.l.hendersonii* (Cassin, 1858)[1] — Sakhalin and S Kuril Is. to Hokkaido and Honshu (Japan) >> Philippines to E Indonesia

___ *L.l.lanceolata* (Temminck, 1840) — N Europe and N Asia to Kamchatka, south to N Mongolia and Russian Far East. >> Philippines, SE Asia south to Gtr. Sundas

○ *Locustella naevia* GRASSHOPPER WARBLER

___ *L.n.naevia* (Boddaert, 1783) — Europe E to Ukraine and W European Russia >> W Africa

___ *L.n.straminea* Seebohm, 1881 — E European Russia, SW Siberia, N and E Kazakhstan to Tien Shan and NE Afghanistan >> NE Africa, SW and S Asia

___ *L.n.obscurior* Buturlin, 1929 — E Turkey, Caucasus area >> NE or E Africa

___ *L.n.mongolica* Sushkin, 1925 — E Kazakhstan and W and NW Mongolia to SC Siberia >> S Asia

○ *Locustella certhiola* RUSTY-RUMPED WARBLER

___ *L.c.certhiola* (Pallas, 1811)[2] — E Transbaikalia, Russian Far East, NE China >> SE China to Gtr. Sundas

___ *L.c.sparsimstriata* Meise, 1934[3] — SW Siberia to Transbaikalia >> India to Gt Sundas

___ *L.c.centralasiae* Sushkin, 1925 — NW China to W and C Mongolia and C Nei Mongol >> NE India to Indochina

___ *L.c.rubescens* Blyth, 1845 — N Siberia (lower Ob' to Kolyma basin) >> S and SE Asia

○ *Locustella ochotensis* (Middendorff, 1853) MIDDENDORFF'S WARBLER[4] — Commander Is., Kamchatka and Magadan area to N Russian Far East and Hokkaido >> E China and Philippines to Indonesia

○ *Locustella pleskei* Taczanowski, 1889 PLESKE'S WARBLER — Small islands off S Russian Far East and Korea and off C and S Japan >> SE China and Sundas

○ *Locustella fluviatilis* (Wolf, 1810) RIVER WARBLER — C Europe to SW Siberia and Finland, Bulgaria, NW Kazakhstan >> SE Africa

● *Locustella luscinioides* SAVI'S WARBLER

⊥ *L.l.luscinioides* (Savi, 1824) — Europe and NW Africa to C Ukraine and Belarus >> N Afrotropics

___ *L.l.sarmatica* Kazakov, 1973 — S European Russia >> NE Africa

___ *L.l.fusca* (Severtsov, 1873) — Asia Minor and NE Arabia to W Mongolia and Kazakhstan >> NE and E Africa

○ *Locustella fasciolata* GRAY'S WARBLER

___ *L.f.fasciolata* (G.R. Gray, 1860) — SC Siberia, Transbaikalia, NE China, Russian Far East >> SE Asia to New Guinea

___ *L.f.amnicola* Stepanyan, 1972[5] — Sakhalin, S Kuril Is., N Japan >> Indonesia (? Philippines)

PHRAGMATICOLA[6] Jerdon, 1845 F

○ *Phragmaticola aedon* THICK-BILLED WARBLER

___ *P.a.aedon* (Pallas, 1776) — NC Asia >> Burma, Thailand, Indonesia

___ *P.a.rufescens* Stegmann, 1929 — E Siberia, Mongolia >> SE China, Thailand, Indochina

ACROCEPHALUS J.A. & J.F. Naumann, 1811[7] M

○ *Acrocephalus griseldis* (Hartlaub, 1981) BASRA REED WARBLER — Iraq, SW Iraq, Kuwait >> E and SE Africa

---

[1] For recognition see Loskot & Sokolov (1993) (1376).
[2] Includes *minor* see Watson *et al.* (1986: 53) (2564).
[3] For recognition see Stepanyan (1990: 443) (2326).
[4] Includes *subcerthiola*. For monotypic treatment see Kalyakin *et al.* (1993) (1233). These authors suggested that the name *ochotensis* cannot be used. However see Loskot (2002) (1374).
[5] For treatment within *L. fasciolata* see Cramp *et al.* (1992: 102) (593).
[6] For the use of this spelling see Bond (1975) (180); we use this name following the molecular findings of Helbig & Seibold (1999) (1098). Treated in *Acrocephalus* in last Edition.
[7] Arranged after considering the findings of Leisler *et al.* (1997) (1350) and of Helbig & Seibold (1999) (1098).

○ *Acrocephalus brevipennis* (Keulemans, 1866) CAPE VERDE SWAMP WARBLER

Cape Verde Is.

○ *Acrocephalus rufescens* GREATER SWAMP WARBLER
— *A.r.senegalensis* Colston & Morel, 1985
— *A.r.rufescens* (Sharpe & Bouvier, 1876)
— *A.r.chadensis* (Alexander, 1907)
— *A.r.ansorgei* (E. Hartert, 1906)

Senegal (554)
Ghana to NW Zaire
Lake Chad
S Sudan to NW Angola, N Botswana and Zambia

○ *Acrocephalus gracilirostris* LESSER SWAMP WARBLER
— *A.g.neglectus* (Alexander, 1908)
— *A.g.tsanae* (Bannerman, 1937)
— *A.g.jacksoni* (Neumann, 1901)
— *A.g.parvus* (Fischer & Reichenow, 1884)
— *A.g.leptorhynchus* (Reichenow, 1879)
— *A.g.winterbottomi* (C.M.N. White, 1947)
— *A.g.cunenensis* (E. Hartert, 1903)

— *A.g.gracilirostris* (Hartlaub, 1864)[1]

W Chad
NW Ethiopia
NE Zaire, Uganda, W Kenya, S Sudan
SW Ethiopia, S Kenya, N Tanzania, Rwanda, Burundi
E and C Tanzania to coastal Natal and E Transvaal
N and NW Zambia to E Angola
SW Angola, N Namibia, N Botswana, SW Zambia, W Zimbabwe
S Mozambique, SE Zimbabwe, Sth. Africa

○ *Acrocephalus newtoni* (Hartlaub, 1863) MADAGASCAN SWAMP WARBLER   Madagascar

○ *Acrocephalus sechellensis* (Oustalet, 1877) SEYCHELLES BRUSH WARBLER[2]

Cousin I.

○ *Acrocephalus rodericanus* (A. Newton, 1865) RODRIGUEZ BRUSH WARBLER

Rodriguez I.

● *Acrocephalus arundinaceus* GREAT REED WARBLER
⤓ *A.a.arundinaceus* (Linnaeus, 1758)

— *A.a.zarudnyi* E. Hartert, 1907

Europe, Asia Minor, N Iran, NW Arabia >> W, C, E and S Africa
SE European Russia and SW Siberia to Xinjiang and W Mongolia >> E and S Africa

○ *Acrocephalus orientalis* (Temminck & Schlegel, 1847) ORIENTAL REED WARBLER

S Transbaikalia and C Mongolia to Amurland, Japan and E China >> SE Asia to N and NE Australia

○ *Acrocephalus stentoreus* CLAMOROUS REED WARBLER
— *A.s.stentoreus* (Hemprich & Ehrenberg, 1833)
— *A.s.levantina* Roselaar, 1994
— *A.s.brunnescens* (Jerdon, 1839)

— *A.s.amyae* E.C.S. Baker, 1922
— *A.s.meridionalis* (Legge, 1875)
— *A.s.siebersi* Salomonsen, 1928
— *A.s.harterti* Salomonsen, 1928
— *A.s.celebensis* Heinroth, 1903
— *A.s.lentecaptus* E. Hartert, 1924
— *A.s.sumbae* E. Hartert, 1924[3]

Egypt, NC Sudan
Israel, Jordan, Syria, NW Arabia (2105A)
C and S Red Sea, Oman and Iran to SC Kazakhstan and NW India >> SW Asia to Sri Lanka
NE India, Thailand, Burma, S China >> Indochina
S India, Sri Lanka
Java
Philippines
Sulawesi
Lombok, Sumbawa Is.
Moluccas, Sumba I., New Guinea, New Britain (Bismarck Arch.) and the Solomon Is.

○ *Acrocephalus australis* AUSTRALIAN REED WARBLER
— *A.a.gouldi* A.J.C. Dubois, 1901
— *A.a.australis* (Gould, 1838)

Western Australia >> NW Australia
C, E and SE Australia, E Tasmania, Flinders I. >> N Australia

○ *Acrocephalus familiaris* MILLERBIRD/HAWAIIAN REED WARBLER
— †*A.f.familiaris* (Rothschild, 1892)
— *A.f.kingi* (Wetmore, 1924)

Laysan (Hawaii)
Nihoa (Hawaii)

○ *Acrocephalus luscinius* NIGHTINGALE REED WARBLER[4]
— †?*A.l.yamashinae* (Taka-Tsukasa, 1931)

Pagan I.

[1] Includes *zuluensis* see Pearson (1997: 121) (1812).
[2] The use of a genus *Bebrornis* is not supported by molecular studies (Helbig & Seibold, 1999) (1098).
[3] It is unlikely that this is a single taxon. Schodde & Mason (1999: 699) (2189) have shown that there is no evidence of breeding from NE Australia except in the case of *A. australis* which they separate on the evidence of Leisler *et al.* (1997) (1350).
[4] For correction of spelling see David & Gosselin (2002) (613).

— *A.l.luscinius* (Quoy & Gaimard, 1830)[1] Almagan, Saipan, Aguijan, Guam Is.

— †?*A.l.astrolabii* Holyoak & Thibault, 1978 ? Yap I.

— *A.l.syrinx* (Kittlitz, 1835)[2] Chuuk (= Truk) and Pohnpei Is. (Caroline Is.)

— †?*A.l.rehsei* (Finsch, 1883)[3] Nauru I.

○ *Acrocephalus aequinoctialis* KIRITIMATI REED WARBLER/BOKIKOKIKO

— *A.a.aequinoctialis* (Latham, 1790) Kiritimati I. (Kiribati)

— *A.a.pistor* Tristram, 1883 Washington I., (formerly Fanning I.)

○ *Acrocephalus caffer* LONG-BILLED REED WARBLER[4]

— †*A.c.garretti* Holyoak & Thibault, 1978 Huahine I. (Society Is.)

— †?*A.c.longirostris* (J.F. Gmelin, 1789) Moorea I. (Society Is.)

— *A.c.caffer* (Sparrman, 1786) Tahiti (Society Is.)

— *A.c.flavidus* (Murphy & Mathews, 1929) Pukapuka I. (Tuamotu Is.)

— *A.c.atyphus* (Wetmore, 1919) NW Tuamotu Is.

— *A.c.eremus* (Wetmore, 1919) Makatea I. (Tuamotu Is.)

— *A.c.niauensis* (Murphy & Mathews, 1929) Niau I. (Tuamotu Is.)

— *A.c.palmarum* (Murphy & Mathews, 1929) Anaa I. (Tuamotu Is.)

— *A.c.ravus* (Wetmore, 1919) SE Tuamotu Is.

— *A.c.consobrinus* (Murphy & Mathews, 1928) Motu One I. (Marquesas Is.)

— *A.c.aquilonis* (Murphy & Mathews, 1928) Eiao I. (Marquesas Is.)

— *A.c.percernis* (Wetmore, 1919) Nuku Hiva I. (Marquesas Is.)

— *A.c.idae* (Murphy & Mathews, 1928) Ua Huka I. (Marquesas Is.)

— *A.c.dido* (Murphy & Mathews, 1928) Ua Pou I. (Marquesas Is.)

— *A.c.mendanae* Tristram, 1883 Hiva Oa, Tahuata Is. (Marquesas Is.)

— *A.c.fatuhivae* (Murphy & Mathews, 1928) Fatu Hiva I. (Marquesas Is.)

— *A.c.postremus* (Murphy & Mathews, 1928) Hatutu I.[5] (Marquesas Is.)

○ *Acrocephalus kerearako* COOK ISLANDS REED WARBLER[6]

— *A.k.kerearako* Holyoak, 1974 Mangaia I.

— *A.k.kaoko* Holyoak, 1974 Mitiaro I.

○ *Acrocephalus rimitarae* (Murphy & Mathews, 1929) RIMATARA REED WARBLER[7]

Rimatara I. (Iles Australes)

○ *Acrocephalus taiti* Ogilvie-Grant, 1913 HENDERSON ISLAND REED WARBLER

Henderson I.

○ *Acrocephalus vaughanii* (Sharpe, 1900) PITCAIRN REED WARBLER Pitcairn I.

○ *Acrocephalus bistrigiceps* Swinhoe, 1860 BLACK-BROWED REED WARBLER

E Transbaikalia to Sakhalin, Japan, Korea and
NE China >> SE China to Gtr. Sundas

● *Acrocephalus melanopogon* MOUSTACHED WARBLER

↲ *A.m.melanopogon* (Temminck, 1823) S Europe to SW Ukraine, NW Africa, W Asia Minor

— *A.m.mimicus* (Madarász, 1903)[8] S European Russia, C Turkey, and Levant to E
Kazakhstan and Afghanistan >> Arabia to NW India

○ *Acrocephalus paludicola* (Vieillot, 1817) AQUATIC WARBLER C and E Europe, SW Siberia >> W and C Africa

● *Acrocephalus schoenobaenus* (Linnaeus, 1758) SEDGE WARBLER Europe to NW Iran, N Kazakhstan and W Siberia >>
sub-Saharan Africa

○ *Acrocephalus sorghophilus* (Swinhoe, 1863) STREAKED REED WARBLER NE China >> SE China, Philippines

○ *Acrocephalus concinens* BLUNT-WINGED WARBLER

— *A.c.haringtoni* Witherby, 1920 Afghanistan, N Pakistan, Kashmir >> ?

---

[1] Includes †?*nijoi* from Aguijan, which lies within the spread of islands credited to the nominate form.
[2] For reasons to separate this see Pratt *et al.* (1987: 254) (1920); we have not seen a detailed proposal.
[3] For reasons to separate this see Pratt *et al.* (1987: 254) (1920); we have not seen a detailed proposal.
[4] Watson *et al.* (1986) treated this as two species (*caffra* and *atyphus*), but Mayr added an editorial footnote which we follow. We do so because the Tuamotos lie between the Society Islands and the Marquesas. Pratt *et al.* (1987: 256) (1920), citing an unpublished thesis, split *caffra* (without *atyphus*) into two species (the second from the Marquesas would take the name *mendanae*). This resulted in three species and may be the best treatment but we have not seen a detailed proposal including acoustic evidence.
[5] Not located on maps and possibly not in geographic sequence here.
[6] Described as a new species. An open question whether conspecific with *vaughani* as treated by Watson *et al.* (1964) (2564), but that treatment has been modified by Graves (1992) (987). In the circumstances we tentatively separate this too.
[7] This species and the next separated from *vaughani* by Graves (1992) (987).
[8] Includes *albiventris* see Stepanyan (1990: 447) (2326).

— *A.c.stevensi* E.C.S. Baker, 1922   SE Nepal, Bangladesh, Assam, Burma

— *A.c.concinens* (Swinhoe, 1870)   C and E China north to Hebei >> E Burma to Indochina (may breed N Vietnam)

○ *Acrocephalus tangorum* La Touche, 1912 MANCHURIAN REED WARBLER[1]   SE Transbaikalia and E Mongolia through NE China to Lake Khanka (Russian Far East) >> Thailand, Indochina

○ †?*Acrocephalus orinus* Oberholser, 1905 HUME'S REED WARBLER[2]   Sutlej Valley, Himachal Pradesh (N India)

○ *Acrocephalus agricola* PADDYFIELD WARBLER

— *A.a.septimus* Gavrilenko, 1954   NE Bulgaria to Ukraine, east to lower Volga and E Turkey >> SW Asia

— *A.a.agricola* (Jerdon, 1845)   Kazakhstan to NE Iran to W Mongolia and NC China >> SW and S Asia east to Burma

○ *Acrocephalus dumetorum* Blyth, 1849 BLYTH'S REED WARBLER   Poland and Finland to W and C Siberia, NE Iran, Tien Shan, W Mongolia >> SW and S Asia

◉ *Acrocephalus scirpaceus* EURASIAN REED WARBLER

✓ *A.s.scirpaceus* (Hermann, 1804)   Europe and NW Africa to W European Russia, Ukraine and W Turkey >> W and NC Africa

— *A.s.fuscus* (Hemprich & Ehrenberg, 1833)   Egypt, C Asia Minor and SE European Russia to Iran and E Kazakhstan, C Asia >> SW Asia, E and S Africa

— *A.s.avicenniae* Ash, Pearson, Nikolaus & Colston, 1989[3]   Coastal mangroves of Saudi Arabia, Yemen, Sudan, Eritrea and NW Somalia (61)

○ *Acrocephalus baeticatus* AFRICAN REED WARBLER[4]

— *A.b.guiersi* Colston & Morel, 1984   N Senegal

— *A.b.cinnamomeus* Reichenow, 1908[5]   S Senegal to C Ethiopia and S Somalia, south to Mozambique and Zambia

— *A.b.suahelicus* Grote, 1926   E Tanzania and offshore islands

— *A.b.hallae* C.M.N. White, 1960   SW Angola, N Namibia, SW Botswana, SW Zambia, N and W Cape Province

— *A.b.baeticatus* (Vieillot, 1817)   N and NE Botswana, Zimbabwe, Sth. Africa

○ *Acrocephalus palustris* (Bechstein, 1798) MARSH WARBLER[6]   Europe to SC Siberia, Kazakhstan and Caucasus area >> E and SE Africa,

IDUNA Keyserling & Blasius, 1840[7] F

○ *Iduna caligata* (M.H.K. Lichtenstein, 1823) BOOTED WARBLER   European Russia, N Kazakhstan >> SW and SC India

○ *Iduna rama* SYKES'S WARBLER[8]

— *I.r.annectens* (Sushkin, 1925)   E Kazakhstan and N Xinjiang to NW Mongolia and Tuva Rep. >> N India

— *I.r.rama* (Sykes, 1832)   Lower Volga, WC Iran and NE Arabia to S Kazakhstan, Tien Shan, SW Xinjiang and Pakistan >> SW and S Asia

○ *Iduna pallida* OLIVACEOUS WARBLER

— *I.p.opaca* (Cabanis, 1850)[9]   S Spain, SE Portugal, NW Africa to N Libya >> W Africa

— *I.p.elaeica* (Lindermayer, 1843)   Balkans to W Asia Minor >> NE and E Africa

— *I.p.tamariceti* (Severtsov, 1873)   C Turkey and Levant to Afghanistan and NW Xinjiang >> NE and E Africa

— *I.p.reiseri* (Hilgert, 1908)   NW Sahara (S Morocco and Mauritania to S Tunisia) >> SW Sahara

— *I.p.pallida* (Hemprich & Ehrenberg, 1833)   Egypt, ? S Arabia, ? N Somalia >> NE Africa

— *I.p.laeneni* Niethammer, 1955   Niger, Chad, NE Nigeria, C Sudan

---

[1] Historically attached to *A. agricola* (see Vaurie, 1959) (2502), but placed in *A. bistrigiceps* by Watson *et al.* (1986: 60) (2564). Alstrom *et al.* (1991) (35) recommended reattachment to *agricola*. Molecular studies show it deserves treatment at specific level (Helbig & Seibold, 1999) (1098).

[2] Originally named *Acrocephalus macrorhynchus* Hume, 1869 (1162A), but this name is preoccupied (Oberholser, 1905) (1670). Known from one specimen; for findings of molecular study see Bensch & Pearson (2002) (114A).

[3] Treated as a race of *baeticatus* by Pearson (1997: 108) (1812); however, molecular findings cluster it with *scirpaceus* (Helbig & Seibold, 1999) (1098).

[4] For reasons to reunite *cinnamomeus* (split by Clancey) see Pearson (1997: 108) (1812).

[5] Implicitly includes *fraterculus.*

[6] Here treated as including *laricus*, but there appear to be two forms; *laricus* or the much older name *turcomana* may apply to the second. Until types can be compared, and related to differing African migrants, the species is best treated as monotypic.

[7] Helbig & Seibold (1999) (1098) showed that the species treated here do not match true *Hippolais* and show a closer relationship to the genus *Acrocephalus.* They used *Iduna* as a subgenus. We do not use subgenera in this work and feel it is helpful to use it as a full genus.

[8] For separation from *I. caligata* see Stepanyan (1990: 457) (2326). We believe *annectens*, if valid, is closer to *rama.*

[9] For suggestions that this should be separated from *pallida* see Helbig & Seibold (1999) (1098). We await studies including more of these races before splitting.

HIPPOLAIS Conrad von Baldenstein[1], 1827 F
○ *Hippolais languida* (Hemprich & Ehrenberg, 1833) UPCHER'S WARBLER[2]

C Turkey and Levant to SC Kazakhstan, Afghanistan and NW Pakistan

○ *Hippolais olivetorum* (Strickland, 1837) OLIVE-TREE WARBLER

SE Europe, Asia Minor, Levant >> E and S Africa

○ *Hippolais polyglotta* (Vieillot, 1817) MELODIOUS WARBLER

SW Europe, NW Africa >> W Africa

○ *Hippolais icterina* (Vieillot, 1817) ICTERINE WARBLER[3]

C, N and E Europe to SW Siberia, Caucasus area and N Iran >> Sth. Africa

CHLOROPETA A. Smith, 1847 F
○ *Chloropeta natalensis* AFRICAN YELLOW WARBLER
   — *C.n.batesi* Sharpe, 1905

Nigeria to N Zaire and SW Sudan

   — *C.n.massaica* Fischer & Reichenow, 1884

Ethiopia and SW Sudan to S Tanzania

   — *C.n.major* E. Hartert, 1904

Gabon and Angola to S Zaire and N Zambia

   — *C.n.natalensis* A. Smith, 1847

S Tanzania to Sth. Africa

○ *Chloropeta similis* Richmond, 1897 MOUNTAIN YELLOW WARBLER

SE Sudan, W Kenya and E Zaire to N Malawi and NE Zambia

○ *Chloropeta gracilirostris* PAPYRUS YELLOW WARBLER
   — *C.g.gracilirostris* Ogilvie-Grant, 1906

E Zaire to Burundi, W Kenya

   — *C.g.bensoni* Amadon, 1954

Luapula River (Zambia)

GENERA INCERTAE SEDIS (included in the family; subfamily unclear)

MACROSPHENUS Cassin, 1859 M
○ *Macrosphenus flavicans* YELLOW LONGBILL
   — *M.f.flavicans* Cassin, 1859

Cameroon to W Zaire and NW Angola

   — *M.f.hypochondriacus* (Reichenow, 1893)

E Zaire, Uganda to SW Sudan

○ *Macrosphenus kempi* KEMP'S LONGBILL
   — *M.k.kempi* (Sharpe, 1905)

Sierra Leone to SW Nigeria

   — *M.k.flammeus* Marchant, 1950

SE Nigeria and W Cameroon

○ *Macrosphenus concolor* (Hartlaub, 1857) GREY LONGBILL[4]

S Guinea to Zaire and Uganda

○ *Macrosphenus pulitzeri* Boulton, 1931 PULITZER'S LONGBILL

W Angola

○ *Macrosphenus kretschmeri* KRETSCHMER'S LONGBILL
   — *M.k.kretschmeri* (Reichenow & Neumann, 1895)

NE and EC Tanzania

   — *M.k.griseiceps* Grote, 1911

SE Tanzania, NE Mozambique

HYLIOTA Swainson, 1837 F
○ *Hyliota flavigaster* YELLOW-BELLIED HYLIOTA
   — *H.f.flavigaster* Swainson, 1837

Senegal to Ethiopia, Kenya

   — *H.f.barbozae* Hartlaub, 1883[5]

Angola to Tanzania, N Mozambique

○ *Hyliota australis* SOUTHERN HYLIOTA
   — *H.a.slatini* Sassi, 1914

NE Zaire, W Uganda, W Kenya

   — *H.a.usambara* W.L. Sclater, 1932[6]

NE Tanzania

   — *H.a.inornata* Vincent, 1933[7]

Angola, S Zaire, Zambia, Malawi

   — *H.a.australis* Shelley, 1882

Zimbabwe, C and S Mozambique

○ *Hyliota violacea* VIOLET-BACKED HYLIOTA
   — *H.v.nehrkorni* Hartlaub, 1892

Liberia to Ghana and Togo

   — *H.v.violacea* J. & E. Verreaux, 1851

Nigeria and Cameroon to Zaire

---

[1] All this was the family name; sometimes, wrongly, shortened.
[2] Includes *magnirostris*. We treat this as monotypic following Stepanyan (1990: 458) (2326).
[3] Includes *alaris*. We treat this as monotypic following Stepanyan (1990: 456) (2326).
[4] Implicitly includes *grisescens* not *griscens* as in our last Edition.
[5] Includes *marginalis* see Fry (1997: 423) (891).
[6] We follow Zimmerman *et al.* (1996) (2735) in retaining this within the species *H. australis*.
[7] Includes *pallidipectus* see Fry (1997: 425) (891).

HYLIA Cassin, 1859 F

○ *Hylia prasina* Green Hylia
— *H.p.poensis* Alexander, 1903   Bioko I.
— *H.p.prasina* (Cassin, 1855)   W Gambia to C Angola and W Kenya

AMAUROCICHLA Sharpe, 1892 F

○ *Amaurocichla bocagii* Sharpe, 1892 Bocage's Longbill   São Tomé I.

OXYLABES Sharpe, 1870[1] M

○ *Oxylabes madagascariensis* (J.F. Gmelin, 1789) White-throated Oxylabes
                                                                  NW, NE and E Madagascar

BERNIERIA Bonaparte, 1854[2] F

○ *Bernieria madagascariensis* Long-billed Bernieria
— *B.m.madagascariensis* (J.F. Gmelin, 1789)   E Madagascar
— *B.m.inceleber* Bangs & Peters, 1926   N and W Madagascar

CRYPTOSYLVICOLA Goodman, Langrand & Whitney, 1996 (955) M

○ *Cryptosylvicola randrianasoloi* Goodman, Langrand & Whitney, 1996 Cryptic Warbler #
                                                                  E Madagascar (955)

HARTERTULA Stresemann, 1925[3] F

○ *Hartertula flavoviridis* (E. Hartert, 1924) Wedge-tailed Jery   E Madagascar

THAMNORNIS Milne-Edwards & Grandidier, 1882[4] M

○ *Thamnornis chloropetoides* (A. Grandidier, 1867) Kiritika Warbler   SW and S Madagascar

XANTHOMIXIS Sharpe, 1881[5] (2228)[6] F

○ *Xanthomixis zosterops* Short-billed Tetraka
— *X.z.fulvescens* (Delacour, 1931)   Montagne d'Ambre (N Madagascar)
— *X.z.andapae* (Salomonsen, 1934)   NE Madagascar
— *X.z.zosterops* (Sharpe, 1875)   Madagascar (except N)
— *X.z.ankafanae* (Salomonsen, 1934)   SE Madagascar

○ *Xanthomixis apperti* (Colston, 1972) Appert's Tetraka   SW Madagascar (550)

○ *Xanthomixis tenebrosus* (Stresemann, 1925) Dusky Tetraka   E Madagascar

○ *Xanthomixis cinereiceps* (Sharpe, 1881) Grey-crowned Tetraka E Madagascar

CROSSLEYIA Hartlaub, 1877[7] F

○ *Crossleyia xanthophrys* (Sharpe, 1875) Madagascan Yellowbrow EC Madagascar (2226)

RANDIA Delacour & Berlioz, 1931[8] F

○ *Randia pseudozosterops* Delacour & Berlioz, 1931 Marvantsetra Warbler
                                                                  E Madagascar (rain forests)

PHYLLOSCOPINAE

PHYLLOSCOPUS Boie, 1826 M

○ *Phylloscopus ruficapilla* Yellow-throated Woodland Warbler[9]
— *P.r.ochrogularis* (Moreau, 1941)   W Tanzania
— *P.r.minullus* (Reichenow, 1905)   SE Kenya, E Tanzania
— *P.r.johnstoni* (W.L. Sclater, 1927)   S Tanzania, Malawi, NW Mozambique, NE Zambia
— *P.r.quelimanensis* (Vincent, 1933)   Mt. Namule (N Mozambique)

[1] This genus was placed near the babblers by Deignan (1964: 429) (666), but Cibois *et al.* (1999) (357) found evidence that it is a warbler.
[2] For the move from the bulbuls (genus *Phyllastrephus*) to the warblers and for use of this generic name see Cibois *et al.* (2001) (358).
[3] Previously treated within the genus *Neomixis* but belongs in this monotypic genus see Cibois *et al.* (1999) (357).
[4] Treated as a warbler by Watson *et al.* (1986) (2564).
[5] For the move from the bulbuls (genus *Phyllastrephus*) to the warblers and for use of this generic name see Cibois *et al.* (2001) (358).
[6] The paper cited is believed to have appeared before planned publication elsewhere as the title implies that only the genus *Neomixis* was a new genus. No earlier publication has been traced.
[7] Previously treated within the genus *Phyllastrephus* by Rand & Deignan (1960) (1972), but placed in the genus *Oxylabes* in our last Edition, but see Cibois et al. (2001) (358).
[8] Not examined by Cibois *et al.* (2001) (358), but seems best kept with *Cryptosylvicola* pending availability of further evidence.
[9] For correct spelling see David & Gosselin (2002) (613). The suffix of this name is "invariable" i.e. the masculinity of the generic name does not affect it.

___ *P.r.alacris* (Clancey, 1969)   E Zimbabwe, W Mozambique

___ *P.r.ruficapilla* (Sundevall, 1850)[1]   Natal, E Transvaal

___ *P.r.voelckeri* (Roberts, 1941)   Coastal Cape Province

○ **Phylloscopus laurae** Laura's Woodland Warbler

___ *P.l.eustacei* (Benson, 1954)   SE Zaire, N and NW Zambia, SW Tanzania

___ *P.l.laurae* (Boulton, 1931)   Mt. Moco (W Angola)

○ **Phylloscopus laetus** Red-faced Woodland Warbler

___ *P.l.schoutedeni* (Prigogine, 1955)   Mt. Kabobo (E Zaire)

___ *P.l.laetus* (Sharpe, 1902)   NE and E Zaire, Rwanda, Burundi, W Uganda

○ **Phylloscopus herberti** Black-capped Woodland Warbler

___ *P.h.herberti* (Alexander, 1903)   Bioko I.

___ *P.h.camerunensus* (Ogilvie-Grant, 1909) SE Nigeria, W Cameroon

○ **Phylloscopus budongoensis** (Seth-Smith, 1907) Uganda Woodland Warbler

                                    E Zaire, Uganda, W Kenya

○ **Phylloscopus umbrovirens** Brown Woodland Warbler

___ *P.u.yemenensis* (Ogilvie-Grant, 1913)   SW Saudi Arabia, W Yemen

___ *P.u.williamsi* Clancey, 1956   NC Somalia

___ *P.u.umbrovirens* (Rüppell, 1840)   Eritrea, N and C Ethiopian highlands, NW Somalia

___ *P.u.omoensis* (Neumann, 1905)[2]   W and S Ethiopia

___ *P.u.mackensianus* (Sharpe, 1892)   SE Sudan, E Uganda, Kenya

___ *P.u.wilhelmi* (Gyldenstolpe, 1922)   E Zaire, Rwanda, SW Uganda

___ *P.u.alpinus* (Ogilvie-Grant, 1906)   Ruwenzori Mts.

___ *P.u.dorcadichrous* (Reichenow & Neumann, 1895)   SE Kenya, N and NE Tanzania

___ *P.u.fugglescouchmani* (Moreau, 1941)   Uluguru Mts. (E Tanzania)

● **Phylloscopus trochilus** Willow Warbler

___ *P.t.trochilus* (Linnaeus, 1758)   S Britain and S Sweden to N Spain, Alps and
                                   Carpathian Alps >> W Africa

___ *P.t.acredula* (Linnaeus, 1758)   Scandinavia (except S Sweden), C and E Europe,
                                   W Siberia >> C, E and S Africa

___ *P.t.yakutensis* Ticehurst, 1935   N and NE Siberia >> E and S Africa

● **Phylloscopus collybita** Common Chiffchaff[3]

___ *P.c.abietinus* (Nilsson, 1819)   N and E Europe >> Asia Minor, Arabia, Africa south to
                                   N Nigeria, Ethiopia and Kenya

✓ *P.c.collybita* (Vieillot, 1817)[4]   NW, C and SE Europe >> S Europe, N and W Africa

___ *P.c.brevirostris* (Strickland, 1837)   N and W Asia Minor

___ *P.c.caucasicus* Loskot, 1991   Caucasus, Transcaucasus, NW and N Iran (1371)

___ *P.c.menzbieri* (Shestoperov, 1937)[5]   Turkmenistan, NE Iran

___ *P.c.tristis* Blyth, 1843[6]   E European Russia, Siberia, N Kazakhstan, NW
                                   Mongolia >> SW and C Asia

○ **Phylloscopus ibericus** Ticehurst, 1937 Iberian Chiffchaff[7]   SW France, Iberia (except NE Spain), NW Africa >> W
                                    Africa

○ **Phylloscopus canariensis** Canary Islands Chiffchaff

___ *P.c.canariensis* (Hartwig, 1886)   W and C Canary Is.

___ †?*P.c.exsul* E. Hartert, 1907   Lanzarote I.

○ **Phylloscopus sindianus** Mountain Chiffchaff

___ *P.s.lorenzii* (Lorenz, 1887)[8]   Caucasus area and NE Turkey

___ *P.s.sindianus* W.E. Brooks, 1879   Pamir-Alai Mts. to NW Himalayas, Kun Lun Mts. and S
                                   Tien Shan >> E Afghanistan, Pakistan and N India

---

[1] Includes *ochraceiceps* see Pearson (1997: 364) (1812).
[2] For recognition see Pearson (1997: 363) (1812).
[3] Our arrangement of this and associated species is tentative. See also Sangster *et al.* (2002) (2167) and references given there.
[4] Implicitly includes *sardus* Trischitta, 1939 (2440).
[5] For recognition see Stepanyan (1990: 474) (2326), who also, perhaps correctly, recognised *fulvescens* (here a synonym of *tristis*). See also Marova-Kleinbub
& Leonovitch (1998) (1437).
[6] This may be a separate species see Martens & Eck (1995: 256) (1443).
[7] For reasons to separate this species and *P. canariensis* from *P. collybita* see Helbig *et al.* (1996) (1099); for the use of the name *ibericus* in place of *brehmii* see
Svensson (2001) (2379).
[8] Stepanyan (1990: 474) (2326) treated this as a separate species. We follow the B.O.U. (2000: 28) (73) in waiting for further evidence.

○ *Phylloscopus neglectus* Hume, 1870 PLAIN LEAF WARBLER — Mts. of Iran, Turkmenistan, E Uzbekistan, W Tajikistan and W Pakistan >> NE Arabia to NW India

○ *Phylloscopus bonelli* (Vieillot, 1819) WESTERN BONELLI'S WARBLER — SW and C Europe to W Ukraine, Slovenia and Italy, NW Africa >> W Africa

○ *Phylloscopus orientalis* (C.L. Brehm, 1855) EASTERN BONELLI'S WARBLER[1] — Balkans, Asia Minor, Levant >> NE Africa

● *Phylloscopus sibilatrix* (Bechstein, 1793) WOOD WARBLER — Europe, SW Siberia to Altai, W Caucasus >> W and C Africa (sometimes E Africa)

○ *Phylloscopus fuscatus* DUSKY WARBLER
  — *P.f.fuscatus* (Blyth, 1842) — C and E Siberia to N Mongolia, NE China and Russian Far East >> S Asia and mainland SE Asia

  — *P.f.weigoldi* Stresemann, 1924 — C and E Tibetan Plateau >> SE Xizang, Yunnan, Himalayan foothills

  — *P.f.robustus* Stresemann, 1924 — S Gansu to N Sichuan >> S China, Indochina

○ *Phylloscopus fuligiventer* SMOKY WARBLER[2]
  — *P.f.fuligiventer* (Hodgson, 1845) — C Himalayas, S Xizang >> N and NE India
  — *P.f.tibetanus* Ticehurst, 1937 — E Himalayas, SE Xizang >> N Assam

○ *Phylloscopus affinis* (Tickell, 1833) TICKELL'S LEAF WARBLER — N Pakistan, Himalayas to C China (N Yunnan to E Qinghai and C Gansu) >> India to Burma

○ *Phylloscopus subaffinis* Ogilvie-Grant, 1900 BUFF-THROATED WARBLER[3] — N and E Yunnan and W Sichuan to E Guandong, Fujian and Shaanxi >> S Assam (once)[4], S China and SE Asia

○ *Phylloscopus griseolus* Blyth, 1847 SULPHUR-BELLIED WARBLER — Mts. of WC Asia from C Afghanistan SE to N Pakistan and Kashmir and NE to Tien Shan, Altai, C Mongolia and Tuva area >> N and C India

○ *Phylloscopus armandii* YELLOW-STREAKED WARBLER
  — *P.a.armandii* (A. Milne-Edwards, 1865) — NC and NE China (NE Xizang and E Qinghai to S Liaoning) >> S China, Burma, N Thailand, N Indochina

  — *P.a.perplexus* Ticehurst, 1934 — SC China (SE Xizang, Yunnan, SW Sichuan) >> Yunnan, Burma

○ *Phylloscopus schwarzi* (Radde, 1863) RADDE'S WARBLER — SC and SE Siberia, NE China, N Korea, Sakhalin, ? N Mongolia >> C and S China to mainland SE Asia

○ *Phylloscopus pulcher* BUFF-BARRED WARBLER
  — *P.p.kangrae* Ticehurst, 1923 — NW Himalayas
  — *P.p.pulcher* Blyth, 1845 — C and E Himalayas, Assam, W, N and E Burma, Yunnan to W Gansu, S Shaanxi and NW Vietnam >> N India to mainland SE Asia

○ *Phylloscopus maculipennis* ASHY-THROATED WARBLER
  — *P.m.virens* Ticehurst, 1926 — Breeding range unknown >> NW Himalayas
  — *P.m.maculipennis* (Blyth, 1867) — C and E Himalayas, W and N Burma, Yunnan, SW Sichuan, S Laos, SC Vietnam, ? N Indochina

○ *Phylloscopus kansuensis* Meise, 1933 GANSU LEAF WARBLER — NE Qinghai and SW Gansu (W China)

○ *Phylloscopus yunnanensis* La Touche, 1922 CHINESE LEAF WARBLER[5] — C and NE China (NE Qinghai, N Sichuan, Gansu and Shaanxi to Hebei and S Liaoning), ? Nepal

○ *Phylloscopus proregulus* (Pallas, 1811) PALLAS'S LEAF WARBLER — SW to SE Siberia, Sakhalin I., N China >> S China, Indochina

○ *Phylloscopus chloronotus* LEMON-RUMPED WARBLER[6]
  — *P.c.simlaensis* Ticehurst, 1920 — N Pakistan to W Himalayas >> N India
  — *P.c.chloronotus* (J.E. & G.R. Gray, 1846) — E Himalayas and C China (N Yunnan to E Qinghai and S Gansu) >> E India and mainland SE Asia

○ *Phylloscopus subviridis* (W.E. Brooks, 1872) BROOKS'S LEAF WARBLER — NE Afghanistan, NW and N Pakistan >> C Pakistan and NW India

---

[1] For separation see Helbig *et al.* (1995) (1100) and Sangster *et al.* (2002) (2167).
[2] Sometimes treated as conspecific with *P. fuscatus* see Cheng (1987: 801) (333).
[3] For treatment as a species separate from *P. affinis* see Alström & Olsson (1992) (32).
[4] We do not routinely include vagrants, but the occurrence of this species in Indian limits has been a matter of disagreement, see Rasmussen (1996) (1984).
[5] *Phylloscopus sichuanensis* Alström, Olsson & Colston, 1992 (36), is apparently a junior synonym, see Martens (2000) (1441); we understand the holotype in the MCZ has been re-examined, and that this is now confirmed (Martens *in litt.*).
[6] For separation of this from *P. proregulus* see Alström & Olsson (1990) (31).

● *Phylloscopus inornatus* (Blyth, 1842) YELLOW-BROWED WARBLER

NE European Russia to E Siberia, NE Altai, Baikal Mts., NE Mongolia, NE China, N Korea >> S and SE Asia

○ *Phylloscopus humei* HUME'S LEAF WARBLER[1]

— *P.h.humei* (W.E. Brooks, 1878)

Mts. of WC Asia from Afghanistan SE to C Nepal, NE to Tien Shan, Altai, Sayans, S Baikal Mts., W Mongolia >> Indian sub-continent

— *P.h.mandellii* (W.E. Brooks, 1879)

E Himalayas, C China (N Yunnan to Ningxia and Shanxi) >> NE India to N Indochina

● *Phylloscopus borealis* ARCTIC WARBLER

— *P.b.borealis* (J.H. Blasius, 1858)[2]

N Europe and N Siberia to Kolyma R., Russian Far East, N Mongolia and NE China >> SE Asia to Indonesia

— *P.b.xanthodryas* (Swinhoe, 1863)

Chukotka and Kamchatka to Sakhalin and Japan >> Indochina, Philippines, Indonesia

√ *P.b.kennicotti* (S.F. Baird, 1869)

W Alaska >> Philippines, Indonesia

○ *Phylloscopus nitidus* Blyth, 1843 GREEN LEAF WARBLER[3]

Caucasus to N Turkey, N Iran and NW Afghanistan >> S India

○ *Phylloscopus trochiloides* GREENISH WARBLER

— *P.t.viridanus* Blyth, 1843

EC and NE Europe to W Siberia and C Mongolia; Mts. of WC Asia south to Pamir >> Pakistan, N India

— *P.t.ludlowi* Whistler, 1931

C and E Afghanistan, N Pakistan, NW Himalayas >> S India

— *P.t.trochiloides* (Sundevall, 1837)

C and E Himalayas, SE Xizang >> N India to Indochina

— *P.t.obscuratus* Stresemann, 1929

E Tibetan Plateau to Yunnan, W Sichuan, S Shaanxi >> C China to nthn. SE Asia

○ *Phylloscopus plumbeitarsus* Swinhoe, 1861 TWO-BARRED GREENISH WARBLER

EC and E Siberia to NE Mongolia, NE China, N Korea and N Sakhalin >> SE Asia

○ *Phylloscopus tenellipes* Swinhoe, 1860 PALE-LEGGED LEAF WARBLER

NE Mongolia, Transbaikalia, NE China, Russian Far East >> SE Asia to W Malaysia

○ *Phylloscopus borealoides* Portenko, 1950 SAKHALIN LEAF WARBLER

S Kurils, Sakhalin, N Japan >> Indochina

○ *Phylloscopus magnirostris* Blyth, 1843 LARGE-BILLED LEAF WARBLER

NE Afghanistan, Himalayas to N Burma and C China >> S Indian subcontinent and mainland SE Asia

○ *Phylloscopus tytleri* W.E. Brooks, 1872 TYTLER'S LEAF WARBLER

NE Afghanistan, N Pakistan, Kashmir >> SW India

○ *Phylloscopus occipitalis* (Blyth, 1845) WESTERN CROWNED WARBLER

E Uzbekistan, W Tajikistan, and E Afghanistan to NW Himalayas >> C and S India

○ *Phylloscopus coronatus* (Temminck & Schlegel, 1847) EASTERN CROWNED WARBLER

C and NE China, Korea and Russian Far East to Japan and S Sakhalin >> E India to SE Asia and Gtr. Sundas

○ *Phylloscopus ijimae* (Stejneger, 1892) IJIMA'S LEAF WARBLER[4]

Izu Is. and isles off S Kyushu (Japan) >> N Philippines

○ *Phylloscopus reguloides* BLYTH'S LEAF WARBLER

— *P.r.kashmirensis* Ticehurst, 1933

NW Himalayas >> foothills

— *P.r.reguloides* (Blyth, 1842)

C Himalayas >> India, Bangladesh, Burma

— *P.r.assamensis* E. Hartert, 1921

E Himalayas, SE Xizang, Assam, W, N and E Burma, NW Yunnan, W and C Indochina >> E India to mainland SE Asia

— *P.r.claudiae* (La Touche, 1922)

C China (Sichuan, Gansu, S Shaanxi) >> S China, SE Asia

— *P.r.fokiensis* E. Hartert, 1917

E China (Hubei, Guizhou, Guangxi, Fujian)

— *P.r.goodsoni* E. Hartert, 1910[5]

Hainan I.

— *P.r.ticehursti* Delacour & Greenway, 1939

Langbian Plateau (SC Vietnam)

○ *Phylloscopus emeiensis* Alström & Olsson, 1995 EMEI LEAF WARBLER #

Mt. Emei Shan (Sichuan), NE Yunnan (SC China) (33)

---

[1] For basis for separation of this from *P. inornatus* see B.O.U. (1997) (72).
[2] Includes *hylebata*, *transbaicalica* and *talovka* see Stepanyan (1990: 476) (2326).
[3] Separation based on Helbig *et al.* (1995) (1100).
[4] Dated 1882 by Watson *et al.* (1986: 246) (2564) but see www.zoonomen.net (A.P. Peterson *in litt.*).
[5] For the allocation of this form to this species see Olsson *et al.* (1993) (1711).

○ *Phylloscopus davisoni* WHITE-TAILED WILLOW WARBLER

   — *P.d.davisoni* (Oates, 1889)   N and E Burma, NW Thailand, N, W and C Yunnan,
                                                N and C Laos, C Vietnam

   — *P.d.disturbans* (La Touche, 1922)   S Sichuan to SE Yunnan and NW Vietnam

   — *P.d.ogilviegranti* (La Touche, 1922)   Fujian

   — *P.d.intensior* Deignan, 1956   SE Thailand, SW Cambodia (?)

   — *P.d.klossi* (Riley, 1922)   S Laos, SC and S Vietnam

○ *Phylloscopus hainanus* Olsson, Alström & Colston, 1993 HAINAN LEAF WARBLER #
                                                  Hainan (1711)

○ *Phylloscopus cantator* YELLOW-VENTED LEAF WARBLER

   — *P.c.cantator* (Tickell, 1833)   E Nepal, Sikkim, N Assam, Burma, W Yunnan >>
                                                  NE India to NW Thailand

   — *P.c.pernotus* Bangs & Van Tyne, 1930   N and C Laos

○ *Phylloscopus ricketti* (Slater, 1897) SULPHUR-BREASTED LEAF WARBLER   EC and S China, C Laos >> N and E Thailand, N and C
                                                  Indochina

○ *Phylloscopus olivaceus* (Moseley, 1891) PHILIPPINE LEAF WARBLER   Philippines (excl. Luzon, Mindoro, Palawan group)

○ *Phylloscopus cebuensis* LEMON-THROATED LEAF WARBLER

   — *P.c.luzonensis* Rand & Rabor, 1952   N Luzon

   — *P.c.sorsogonensis* Rand & Rabor, 1967   S Luzon

   — *P.c.cebuensis* (A.J.C. Dubois, 1900)   Cebu, Negros

○ *Phylloscopus trivirgatus* MOUNTAIN LEAF WARBLER

   — *P.t.parvirostris* Stresemann, 1912   Gunong Tahan (W Malaysia)

   — *P.t.trivirgatus* Strickland, 1849   Sumatra, Java, Bali, NW Borneo

   — *P.t.kinabaluensis* (Sharpe, 1901)   Mt. Kinabalu (N Borneo)

   — *P.t.sarawacensis* (Chasen, 1938)   Poi Mts. (W Borneo)

   — *P.t.peterseni* Salomonsen, 1962   Palawan

   — *P.t.benguetensis* Ripley & Rabor, 1958[1] N Luzon

   — *P.t.nigrorum* (Moseley, 1891)   S Luzon, Negros, Mindoro

   — *P.t.diutae* Salomonsen, 1953   Diuata Mts. (NE Mindanao)

   — *P.t.mindanensis* (E. Hartert, 1903)   Montane S Mindanao

   — *P.t.malindangensis* (Mearns, 1909)   Mt. Malindang (NW Mindanao)

   — *P.t.flavostriatus* Salomonsen, 1953   Mt. Katanglad (NW Mindanao)

○ *Phylloscopus sarasinorum* SULAWESI LEAF WARBLER

   — *P.s.sarasinorum* (Meyer & Wiglesworth, 1896)   S Sulawesi

   — *P.s.nesophilus* (Riley, 1918)   N and C Sulawesi

○ *Phylloscopus presbytes* TIMOR LEAF WARBLER

   — *P.p.presbytes* (Blyth, 1870)   Timor

   — *P.p.floris* (E. Hartert, 1898)   Flores I.

○ *Phylloscopus poliocephalus* ISLAND LEAF WARBLER

   — *P.p.henrietta* Stresemann, 1931   N Halmahera

   — *P.p.waterstradti* (E. Hartert, 1903)   Moluccas

   — *P.p.everetti* (E. Hartert, 1899)   Buru (Moluccas)

   — *P.p.ceramensis* (Ogilvie-Grant, 1910) Seram (Moluccas)

   — *P.p.avicola* E. Hartert, 1924   Great Kai Is.

   — *P.p.matthiae* Rothschild & Hartert, 1924 St. Matthias Is. (Bismarck Arch.)

   — *P.p.moorhousei* Gilliard & LeCroy, 1967 New Britain (Bismarck Arch.)

   — *P.p.leletensis* Salomonsen, 1965   New Ireland (Bismarck Arch.)

   — *P.p.maforensis* (A.B. Meyer, 1874)   Numfor I. (Geelvink Bay)

   — *P.p.misoriensis* Meise, 1931   Biak I. (Geelvink Bay)

   — *P.p.poliocephalus* (Salvadori, 1876)   NW New Guinea

   — *P.p.albigularis* Hartert & Paludan, 1936 WC New Guinea

   — *P.p.paniaiae* Junge, 1952   W New Guinea

   — *P.p.cyclopum* E. Hartert, 1930   N New Guinea

   — *P.p.giulianettii* (Salvadori, 1896)   C and SE New Guinea

   — *P.p.hamlini* Mayr & Rand, 1935   Goodenough I. (D'Entrecasteaux Arch.)

[1] For recognition see Dickinson *et al.* (1991: 328) (745).

— *P.p.becki* E. Hartert, 1929   Guadalcanal, Malaita Is. (Solomon Is.)

— *P.p.bougainvillei* Mayr, 1935   Bougainville I. (Solomon Is.)

— *P.p.pallescens* Mayr, 1935   Kolombangara I. (Solomon Is.)

○ *Phylloscopus makirensis* Mayr, 1935   SAN CRISTOBAL LEAF WARBLER   San Cristobal I.

○ *Phylloscopus amoenus* (E. Hartert, 1929)   KOLOMBANGARA LEAF WARBLER

Kolombangara I. (Solomon Is.)

SEICERCUS Swainson, 1837   M

○ *Seicercus affinis*   WHITE-SPECTACLED WARBLER

— *S.a.affinis* (Hodgson, 1854)   E Himalayas, Assam >> Indochina

— *S.a.intermedius* (La Touche, 1898)   C and E China >> S China, Indochina

○ *Seicercus burkii* (E. Burton, 1836)   GOLDEN-SPECTACLED WARBLER[1]   W and C Himalayas to Bhutan and SE Xizang >> C India

○ *Seicercus tephrocephalus* (Anderson, 1871)   GREY-CROWNED WARBLER   S and E Assam, W and N Burma, C China (Yunnan, W Sichuna, W Hubei, S Shaanxi), NW Vietnam >> most of mainland SE Asia

○ *Seicercus whistleri*   WHISTLER'S WARBLER

— *S.w.whistleri* Ticehurst, 1925   W and C Himalayas (may winter lower)

— *S.w.nemoralis* Koelz, 1954   E Himalayas, S Assam, W Burma (may winter lower)

○ *Seicercus valentini*   BIANCHI'S WARBLER

— *S.v.valentini* (E. Hartert, 1907)   N Yunnan, Sichuan, S Gansu, S Shaanxi, NW Vietnam

— *S.v.latouchei* Bangs, 1929   N Hubei and Fujian, ? Guizhou, ? Guangdong, ? Guangxi >> S China

○ *Seicercus omeiensis* Martens, Eck, Päckert & Sun, 1999   MARTENS'S WARBLER #[2]

W Sichuan, S Gansu and S Shaanxi, ? Hubei (1445)

○ *Seicercus soror* Alström & Olsson, 1999   PLAIN-TAILED WARBLER #   N Yunnan, Sichuan, S Shaanxi, Guizhou, Jiangxi and Fujian (34)

○ *Seicercus xanthoschistos*   GREY-HOODED WARBLER

— *S.x.xanthoschistos* (J.E. & G.R. Gray, 1846)   NW Himalayas

— *S.x.jerdoni* (W.E. Brooks, 1871)   C Himalayas

— *S.x.flavogularis* (Godwin-Austen, 1877)   NE Assam, N Burma, SE Xizang

— *S.x.tephrodiras* Sick, 1939   S Assam, W and SW Burma

○ *Seicercus poliogenys* (Blyth, 1847)   GREY-CHEEKED WARBLER   N India, N Burma >> N Vietnam

○ *Seicercus castaniceps*   CHESTNUT-CROWNED WARBLER

— *S.c.castaniceps* (Hodgson, 1845)   C and E Himalayas, Assam, N and W Burma, SE Xizang, W Yunnan

— *S.c.laurentei* (La Touche, 1922)   SE Yunnan

— *S.c.sinensis* (Rickett, 1898)   C and E China, N and C Indochina

— *S.c.collinsi* Deignan, 1943   E Burma, NW Thailand

— *S.c.stresemanni* Delacour, 1932   Bolovens (S Laos)

— *S.c.annamensis* (Robinson & Kloss, 1919)   Langbian (S Vietnam)

— *S.c.youngi* (Robinson, 1915)   Pen. Thailand

— *S.c.butleri* (E. Hartert, 1898)   W Malaysia

— *S.c.muelleri* (Robinson & Kloss, 1916)   Sumatra

○ *Seicercus montis*   YELLOW-BREASTED WARBLER

— *S.m.davisoni* (Sharpe, 1888)   W Malaysia

— *S.m.inornatus* (Robinson & Kloss, 1920)   Sumatra

— *S.m.montis* (Sharpe, 1887)   Borneo

— *S.m.xanthopygius* (J. Whitehead, 1893)   Palawan

— *S.m.floris* (E. Hartert, 1897)   S Flores I.

— *S.m.paulinae* Mayr, 1944   Timor

○ *Seicercus grammiceps*   SUNDA WARBLER

— *S.g.sumatrensis* (Robinson & Kloss, 1916)   Sumatra

— *S.g.grammiceps* (Strickland, 1849)   Java, Bali

---

[1] For the separation of this into several species see Alström & Olsson (1999) (34). Some parts of their arrangement, as they stated, require confirmation.
[2] This taxon proves to be part of a broader species, with representation in Burma, and a revision will appear shortly (J. Martens *in litt.*).

TICKELLIA Blyth, 1861 F
○ *Tickellia hodgsoni* BROAD-BILLED WARBLER
    __ *T.h.hodgsoni* (F. Moore, 1854) — E Himalayas, Assam, W Burma
    __ *T.h.tonkinensis* (Delacour & Jabouille, 1930) — W and SE Yunnan, NE Laos, NW Vietnam

ABROSCOPUS E.C.S. Baker, 1930 M
○ *Abroscopus albogularis* RUFOUS-FACED WARBLER
    __ *A.a.albogularis* (F. Moore, 1854) — E Himalayas, Assam, E Bangladesh, W and N Burma
    __ *A.a.hugonis* Deignan, 1938 — E Burma, NW Thailand
    __ *A.a.fulvifacies* (Swinhoe, 1870) — S China, N and C Indochina

○ *Abroscopus schisticeps* BLACK-FACED WARBLER
    __ *A.s.schisticeps* (J.E. & G.R. Gray, 1846) — C Himalayas
    __ *A.s.flavimentalis* (E.C.S. Baker, 1924) — E Himalayas, Assam, W Burma
    __ *A.s.ripponi* (Sharpe, 1902) — N, E Burma, W and S Yunnan, NW Vietnam

○ *Abroscopus superciliaris* YELLOW-BELLIED WARBLER
    __ *A.s.flaviventris* (Jerdon, 1863) — C Himalayas, S Assam, E Bangladesh
    __ *A.s.drasticus* Deignan, 1947 — N and E Assam, N and E Burma >> SW Thailand
    __ *A.s.smythiesi* Deignan, 1947 — Irrawady basin (C Burma)
    __ *A.s.superciliaris* (Blyth, 1859) — Most of Burma, N and W Thailand, N and C Laos and S Yunnan
    __ *A.s.euthymus* Deignan, 1947 — Vietnam
    __ *A.s.bambusarum* Deignan, 1947 — Pen. Thailand (south to Trang)
    __ *A.s.sakaiorum* (Stresemann, 1912) — S Malay Pen.
    __ *A.s.papilio* Deignan, 1947 — Sumatra
    __ *A.s.schwaneri* (Blyth, 1870) — Borneo
    __ *A.s.vordermani* (Büttikofer, 1893) — Java

GRAUERIA E. Hartert, 1908 F
○ *Graueria vittata* E. Hartert, 1908 GRAUER'S WARBLER — E Zaire, W Rwanda, SW Uganda

EREMOMELA Sundevall, 1850 F
○ *Eremomela icteropygialis* YELLOW-BELLIED EREMOMELA
    __ *E.i.alexanderi* Sclater & Mackworth-Praed, 1918 — Senegal and Gambia to W Sudan
    __ *E.i.salvadorii* Reichenow, 1891[1] — SE Gabon, W and S Zaire, N and C Angola, W Zambia
    __ *E.i.griseoflava* Heuglin, 1862[2] — C Sudan, Eritrea and NW Somalia to Uganda, Rwanda and W Kenya
    __ *E.i.abdominalis* Richenow, 1905 — N, C and SE Kenya, N Tanzania
    __ *E.i.polioxantha* Sharpe, 1883 — S Zaire, S Tanzania to SW Zimbabwe, E Transvaal and Mozambique
    __ *E.i.helenorae* Alexander, 1899 — Caprivi Strip (Namibia), SW Zambia and Zimbabwe
    __ *E.i.puellula* Grote, 1929 — SW Angola
    __ *E.i.icteropygialis* (Lafresnaye, 1839) — Namibia, W Botswana
    __ *E.i.perimacha* Oberholser, 1920[3] — S Botswana, W Transvaal, NW Cape Province
    __ *E.i.saturatior* Ogilvie-Grant, 1910[4] — Cape Province to Orange Free State

○ *Eremomela flavicrissalis* Sharpe, 1895 YELLOW-VENTED EREMOMELA — Somalia, Kenya, S Ethiopia

○ *Eremomela pusilla* Hartlaub, 1857 SENEGAL EREMOMELA — Senegal, Gambia and Guinea to NW Central African Republic

○ *Eremomela canescens* GREEN-BACKED EREMOMELA
    __ *E.c.elegans* Heuglin, 1864 — W, C and NE Sudan
    __ *E.c.abyssinica* Bannerman, 1911 — Eritrea, Ethiopia, SE Sudan
    __ *E.c.canescens* Antinori, 1864[5] — Central African Republic, S Chad, S Sudan, Uganda and W Kenya

[1] Said not to intergrade where it meets *icteropygialis* in Angola and separated Lack (1997: 318) (1313). On the advice of R.J. Dowsett (*in litt.*) we prefer to retain the previous treatment.
[2] Includes *karamojensis* and *crawfurdi* see Lack (1997: 319) (1313).
[3] For recognition see Lack (1997: 322) (1313).
[4] Includes *sharpei* see Lack (1997: 322) (1313).
[5] We prefer to place *elgonensis* in here. DJP.

○ *Eremomela scotops* Green-capped Eremomela
   — *E.s.kikuyuensis* van Someren, 1931
   — *E.s.citriniceps* (Reichenow, 1882)
   — *E.s.congensis* Reichenow, 1905[1]
   — *E.s.pulchra* (Bocage, 1878)
   — *E.s.scotops* Sundevall, 1850[2]

                    C Kenya
                    W Kenya, Uganda, W Tanzania
                    Congo, W Zaire, N Angola
                    C and S Angola, S Zaire, Zambia to W Malawi
                    E Kenya to Transvaal and Natal

○ *Eremomela gregalis* Karoo Eremomela
   — *E.g.damarensis* Wahlberg, 1855[3]
   — *E.g.gregalis* (A. Smith, 1829)[4]

                    W Namibia
                    S Namibia, W and C Cape Province

○ *Eremomela usticollis* Burnt-necked Eremomela
   — *E.u.rensi* Benson, 1943
   — *E.u.usticollis* Sundevall, 1850[5]

                    S Zambia, S Malawi, N and E Zimbabwe, C Mozambique
                    S Angola and Namibia to S Zimbabwe, N Transvaal
                       and S Mozambique

○ *Eremomela badiceps* Rufous-crowned Eremomela
   — *E.b.fantiensis* Macdonald, 1940
   — *E.b.latukae* B.P. Hall, 1949[6]
   — *E.b.badiceps* (Fraser, 1843)

                    Sierra Leone to Benin
                    S Sudan
                    Bioko I., S Nigeria to N Angola, Zaire and W Uganda

○ *Eremomela turneri* Turner's Eremomela
   — *E.t.kalindei* Prigogine, 1958
   — *E.t.turneri* van Someren, 1920

                    EC Zaire and SW Uganda
                    W Kenya

○ *Eremomela atricollis* Bocage, 1894 Black-necked Eremomela[7]
                    Angola, S Zaire, Zambia

SYLVIETTA Lafresnaye, 1839 F

○ *Sylvietta brachyura* Northern Crombec
   — *S.b.brachyura* Lafresnaye, 1839

   — *S.b.carnapi* (Reichenow, 1900)
   — *S.b.leucopsis* (Reichenow, 1879)

                    Senegal, Gambia and Sierra Leone to Sudan and
                     N Eritrea
                    Cameroon to Uganda and W Kenya
                    S Eritrea, Ethiopia, SE Sudan and Somalia to
                     NE Tanzania

○ *Sylvietta whytii* Red-faced Crombec
   — *S.w.loringi* Mearns, 1911
   — *S.w.jacksoni* (Sharpe, 1897)
   — *S.w.minima* (Ogilvie-Grant, 1900)
   — *S.w.whytii* (Shelley, 1894)[8]

                    SE Sudan and Ethiopia to NW Kenya
                    S and E Uganda, W, C and S Kenya, W Tanzania, N Malawi
                    Coastal Kenya, E Tanzania
                    S Tanzania, S Malawi, Mozambique, Zimbabwe

○ *Sylvietta philippae* Williams, 1955 Philippa's Crombec
                    NW and W Somalia, S and E Ethiopia

○ *Sylvietta rufescens* Long-billed Crombec
   — *S.r.adelphe* Grote, 1927
   — *S.r.flecki* (Reichenow, 1900)[9]

   — *S.r.ansorgei* E. Hartert, 1907
   — *S.r.pallida* (Alexander, 1899)

   — *S.r.resurga* Clancey, 1953
   — *S.r.rufescens* (Vieillot, 1817)
   — *S.r.diverga* Clancey, 1954

                    SE Zaire, Zambia, N Malawi
                    S Angola, E Namibia, N and E Botswana, SW Zambia,
                     Zimbabwe, Transvaal
                    Coastal Angola
                    S Malawi and C and S Mozambique to E Zimbabwe
                     and N Zululand
                    Natal
                    S Botswana, W and NW Cape Province, SW Transvaal
                    SW to E Cape Province, Orange Free State, S Transvaal

○ *Sylvietta isabellina* (Elliot, 1897) Somali Crombec
                    E and S Ethiopia, NW and S Somalia, N and E Kenya

○ *Sylvietta ruficapilla* Red-capped Crombec
   — *S.r.schoutedeni* C.M.N. White, 1953

                    E Zaire

---

[1] Includes *angolensis* see Lack (1997: 328) (1313).
[2] Includes *chlorochlamys* and *occipitalis* see Lack (1997: 328) (1313).
[3] For recognition see Lack (1997: 330) (1313).
[4] Includes *albigularis* see Lack (1997: 330) (1313).
[5] Includes *baumgarti* see Lack (1997: 331) (1313).
[6] For recognition see Lack (1997: 332) (1313).
[7] Monotypic treatment by Lack (1997: 334) (1313) implies inclusion of *venustula*.
[8] Includes *nemorivaga* see Lack (1997: 340) (1313).
[9] Includes *ochrocara* see Lack (1997: 344) (1313).

— *S.r.rufigenis* (Reichenow, 1887)	W and S Zaire
— *S.r.chubbi* (Ogilvie-Grant, 1910)	SE Zaire to Malawi and N Mozambique
— *S.r.mackayii* C.M.N. White, 1953	N Angola
— *S.r.ruficapilla* Bocage, 1877	C and E Angola
— *S.r.gephyra* C.M.N. White, 1953	S Zaire, NW and W Zambia

○ **Sylvietta virens** GREEN CROMBEC

— *S.v.flaviventris* (Sharpe, 1877)	Sierra Leone to SW Nigeria
— *S.v.virens* Cassin, 1859	SE Nigeria to NW Zaire
— *S.v.baraka* (Sharpe, 1897)	C and E Zaire to S Sudan, Uganda and W Kenya
— *S.v.tando* W.L. Sclater, 1927[1]	NW Angola, SW Zaire, Congo

○ **Sylvietta denti** LEMON-BELLIED CROMBEC

— *S.d.hardyi* (Bannerman, 1911)	Sierra Leone to Ghana
— *S.d.denti* (Ogilvie-Grant, 1906)	Cameroon to Zaire and W Uganda

○ **Sylvietta leucophrys** WHITE-BROWED CROMBEC

— *S.l.leucophrys* (Sharpe, 1891)	W Uganda, W and C Kenya
— *S.l.chloronata* E. Hartert, 1920[2]	E Zaire to SW Uganda, W Tanzania
— *S.l.chapini* Schouteden, 1947	Lendu Plateau (NE Zaire)

SYLVIINAE[3]

SYLVIA Scopoli, 1769[4] F

● **Sylvia atricapilla** BLACKCAP

— *S.a.gularis* Alexander, 1898[5]	Cape Verde Is., Azores
— *S.a.heineken* (Jardine, 1830)	Madeira, Canary Is., Morocco, NW Algeria, Portugal, W Spain
✓ *S.a.pauluccii* Arrigoni, 1902[6]	NE Algeria, Tunisia, E Spain, Balearic Is., Corsica, Sardinia, S and C Italy
✓ *S.a.atricapilla* (Linnaeus, 1758)	Europe (except SW) to SW Siberia and W Asia Minor, W Europe, W Russia >> W and S Europe, >> N, W and E Africa
— *S.a.dammholzi* Stresemann, 1928	Caucasus area, E Asia Minor, N Iran >> E Africa

○ **Sylvia borin** GARDEN WARBLER

— *S.b.borin* (Boddaert, 1783)	W, C and N Europe >> W, C and S Africa
— *S.b.woodwardi* (Sharpe, 1877)	E Balkans, S Ukraine and S European Russia to SW Siberia and Caucasus area >> E and S Africa

○ **Sylvia nisoria** BARRED WARBLER

— *S.n.nisoria* (Bechstein, 1795)	C and E Europe, Turkey, Caucasus area >> E Africa
— *S.n.merzbacheri* Schalow, 1907	NE Iran and N Kazakhstan to Tien Shan, Mongolia and SW Siberia >> E Africa

○ **Sylvia curruca** LESSER WHITETHROAT[7]

— *S.c.curruca* (Linnaeus, 1758)	NW, C and E Europe, W Siberia >> NC and NE Africa, SW Asia and N India.
— *S.c.halimodendri* Sushkin, 1904	N Kazakhstan to C Siberia, N Mongolia, Transbaikalia >> SW Asia, N India
— *S.c.jaxartica* Snigirewski, 1927	S Kazakstan and Turkmenistan to E Uzbekistan and W Tajikistan >> SW Asia
— *S.c.telengitica* Sushkin, 1925	NE Xinjiang, SE Russian Altai, S Mongolia >> SW Asia to NW India
— *S.c.minula* Hume, 1873	W Nei Mongol, S and C Xinjiang >> SW Asia to NW India

---

[1] Includes *meridionalis* see Lack (1997: 248) (1313).
[2] Includes *arileuca* Parkes, 1987 (1775).
[3] The genera placed here are probably babblers not warblers see Sibley & Ahlquist (1990) (2261), Fjeldså *et al.* (1999) (851) and Barker *et al.* (2002) (98). Steps need to be taken in relation to the family name, in conjunction with the ICZN, to minimise the nomenclatural consequences of such a shift.
[4] We follow the species sequence adopted by Shirihai *et al.* (2001) (2247).
[5] Includes *atlantis* see Roselaar in Cramp *et al.* (1992: 515) (593).
[6] Includes *koenigi* see Roselaar in Cramp *et al.* (1992: 515) (593).
[7] This complex *may* represent two or more species; see Martens & Steil (1997) (1444) and Shirihai *et al.* (2001) (2247). Races listed here largely follow Cramp *et al.* (1992: 515) (593). The proposed arrangement of Shirihai *et al.* (2001), as they stated, requires further study.

— *S.c.margelanica* Stolzmann, 1898[1]

Kun Lun Mts., east to NE Qinghai, SW Gansu, Ningxia
>> SW Asia and NW India

— *S.c.caucasica* Ognev & Bankovski, 1910[2]

C and E Turkey, Caucasus area, SW and N Iran >>
Arabia, NE Africa

○ *Sylvia althaea* HUME'S LESSER WHITETHROAT
— *S.a.althaea* Hume, 1878[3]

NE and SE Iran and SW Turkmenistan to W and
N Pakistan and Kashmir

— *S.a.monticola* Portenko, 1955[4]

Pamir-Alai area to W and C Tien Shan

○ *Sylvia hortensis* ORPHEAN WARBLER
— *S.h.hortensis* (J.F. Gmelin, 1789)

SW Europe to Italy, NW Africa to NW Libya >> S Sahara

— *S.h.crassirostris* Cretzschmar, 1827

NE Libya, SE Europe to C Turkey and Levant >>
NE Africa

— *S.h.balchanica* Zarudny & Bilkevich, 1918[5]

E Turkey, Caucasus area, Iran, SW Turkmenistan >>
NW India

— *S.h.jerdoni* (Blyth, 1847)

SE Iran and Afghanistan to W Tien Shan and Pakistan
>> W and N India

○ *Sylvia leucomelaena* RED SEA WARBLER
— *S.l.leucomelaena* (Hemprich & Ehrenberg, 1833)

W Arabian coast to S Oman

— *S.l.negevensis* Shirihai, 1988

SE Israel, SW Jordan (2244)

— *S.l.blanfordi* Seebohm, 1879

SE Egypt, E Sudan, Eritrea

— *S.l.somaliensis* (Sclater & Mackworth-Praed, 1918)

Djibouti, N Somalia

○ *Sylvia nana* DESERT WARBLER
— *S.n.nana* (Hemprich & Ehrenberg, 1833)

Iran and lower Volga R. to W and S Mongolia, W Inner
Mongolia, Xinjiang and N Afghanistan >>
NE Africa, SW Asia

— *S.n.deserti* (Loche, 1858)

Sahara (S Morocco and Mauritania to S Tunisia)

○ *Sylvia communis* GREATER WHITETHROAT
— *S.c.communis* Latham, 1787

Europe to Ukraine and C and N European Russia,
N Africa, N Asia Minor >> N tropical Africa

— *S.c.volgensis* Domaniewski, 1915[6]

SE European Russia, W Siberia, N Kazakhstan >> NC,
E and S Africa

— *S.c.icterops* Ménétriés, 1832[7]

S and SE Turkey and Levant to Caucasus, Iran and
SW Turkmenistan >> E and S Africa

— *S.c.rubicola* Stresemann, 1928

Mts. of WC Asia from Pamir-Alai and Tien Shan to
Altai, Mongolia and S Transbaikalia >> E and S Africa

○ *Sylvia undata* DARTFORD WARBLER
— *S.u.dartfordiensis* Latham, 1787

S England, W France, NW Spain

— *S.u.undata* (Boddaert, 1783)

NE Spain, S France, Menorca, Italy

— *S.u.toni* E. Hartert, 1909

S Portugal, C and S Spain, NW Africa

○ *Sylvia sarda* MARMORA'S WARBLER
— *S.s.balearica* von Jordans, 1913

Balearic, Elba, Pantelleria, Zembra, and Kuriate Is.
(W Mediterranean)

— *S.s.sarda* Temminck, 1820

Corsica, Sardinia, S France >> NW Africa

○ *Sylvia deserticola* TRISTRAM'S WARBLER
— *S.d.deserticola* Tristram, 1859

N Algeria, N Tunisia >> NW Sahara

— *S.d.maroccana* E. Hartert, 1917

Morocco to NW Algeria >> NW Sahara

— †?*S.d.ticehursti* R. Meinertzhagen, 1939[8]

E Morocco (status?)

[1] Includes *chuancheica* see Cramp *et al.* (1992: 468) (593).
[2] Stepanyan (1990: 467) (2326) considered this a race of *althaea*, but see Loskot (2001) (1373).
[3] Includes *monticola* see Stepanyan (1990: 467) (2326), but see also Loskot (2001) (1373).
[4] For recognition see Loskot (2001) (1373).
[5] For recognition see Roselaar (1995) (2107).
[6] For recognition see Cramp *et al.* (1992: 459) (593).
[7] Includes *traudeli* Kumerloeve, 1969 (1307A).
[8] Attachment to this species is still tentative. This may not be a valid form.

○ *Sylvia conspicillata* Spectacled Warbler
— *S.c.conspicillata* Temminck, 1820[1]   SW Europe, NW Africa, Crete, Cyprus, Levant >> W Sahara, Egypt

— *S.c.orbitalis* (Wahlberg, 1854)[2]   Madeira, Canary, Cape Verde Is.

● *Sylvia cantillans* Subalpine Warbler
— *S.c.cantillans* (Pallas, 1764)   Iberia, S France, N Italy >> NW and W African Sahel
— *S.c.moltonii* Orlando, 1937[3]   Corsica, Sardinia, Balearic Is., ? S Italy, ? Sicily >> N and W Africa

— *S.c.albistriata* (C.L. Brehm, 1855)   NE Italy and SE Europe to W Asia Minor >> NE Africa
— *S.c.inornata* Tschudi, 1906   NW Africa >> Senegal to W Niger

● *Sylvia melanocephala* Sardinian Warbler
⤳ *S.m.melanocephala* (J.F. Gmelin, 1789)   E Canary Is., S Europe, NW Africa to Libya, NW and N Asia Minor >> S Sahara

— *S.m.leucogastra* (Ledru, 1810)   W Canary Is.
— *S.m.pasiphae* Stresemann & Schiebel, 1925   Rhodes, Crete, S Aegean Is., SW Asia Minor
— †?*S.m.norrisae* Nicoll, 1917   Faiyum (Egypt)
— *S.m.momus* (Hemprich & Ehrenberg, 1833)   SC Turkey and NE Sinai to Israel and W Jordan

○ *Sylvia mystacea* Ménétriés's Warbler
— *S.m.mystacea* Ménétriés, 1832   NE Turkey, E Transcaucasia, N Iran, W Caspian shore to lower Volga R. >> Arabia and NE Africa

— *S.m.rubescens* Blanford, 1874[4]   SE Turkey, Syria and E Jordan to Iraq and SW Iran >> Arabia and NE Africa

— *S.m.turcmenica* Zarudny & Bilkevich, 1918[5]   NE Iran and W Pakistan to SC Kazakhstan, E Uzbekistan and W Tajikistan >> Arabia and NE Africa

● *Sylvia rueppelli* Temminck, 1823 Rüppell's Warbler   Greece, W and S Turkey, NW Syria >> Chad, Sudan

○ *Sylvia melanothorax* Tristram, 1872 Cyprus Warbler   Cyprus >> E Egypt, NE Sudan

PARISOMA Swainson, 1832 N
○ *Parisoma buryi* Ogilvie-Grant, 1913 Yemen Parisoma   SW Saudi Arabia, W Yemen

○ *Parisoma lugens* Brown Parisoma
— *P.l.lugens* (Rüppell, 1840)[6]   W and C Ethiopia
— *P.l.griseiventris* Érard, 1978   Bale Mts. (SC Ethiopia)
— *P.l.jacksoni* Sharpe, 1899   SE Sudan, Kenya, N Tanzania, SE Zaire, Malawi
— *P.l.clara* Meise, 1934[7]   Matengo Highlands (Tanzania)
— *P.l.prigoginei* Schouteden, 1952   Itombwe (E Zaire)

○ *Parisoma subcaeruleum* Chestnut-vented Warbler
— *P.s.ansorgei* Zedlitz, 1921   S Angola
— *P.s.cinerascens* Reichenow, 1902   Namibia, Botswana
— *P.s.orpheanum* Clancey, 1954   Zimbabwe to Orange Free State and Natal
— *P.s.subcaeruleum* (Vieillot, 1817)   Cape Province, SW Orange Free State

○ *Parisoma layardi* Layard's Warbler
— *P.l.aridicola* Winterbottom, 1958   NW Cape Province
— *P.l.barnesi* Vincent, 1948   Lesotho
— *P.l.subsolanum* Clancey, 1963[8]   Interior Cape Province
— *P.l.layardi* Hartlaub, 1862   SW Cape Province

○ *Parisoma boehmi* Banded Parisoma
— *P.b.somalicum* Friedmann, 1928   Ethiopia, N Somalia
— *P.b.marsabit* van Someren, 1931   NC Kenya
— *P.b.boehmi* Reichenow, 1882   S Kenya, Tanzania

---

[1] Implicitly includes *extratipica* Trischitta, 1939 (2440).
[2] Includes *bella* see Vaurie (1959: 268) (2502).
[3] For recognition see Gargallo (1994) (914).
[4] For recognition see Cramp et al (1992: 358) (593).
[5] For recognition see Cramp et al (1992: 358) (593).
[6] Watson *et al.* (1986: 267) (2564) had 1804, a *lapsus* as Rüppell was then 10!
[7] For recognition see Fry (1997: 420) (891).
[8] For correction of spelling see David & Gosselin (2002) (613).

TIMALIIDAE BABBLERS AND PARROTBILLS (50: 273)

LEONARDINA Mearns, 1906[1] F
- ○ **Leonardina woodi** (Mearns, 1905) BAGOBO BABBLER   Mindanao

PELLORNEUM Swainson, 1832 N
- ○ **Pellorneum albiventre** SPOT-THROATED BABBLER
  - — *P.a.ignotum* Hume, 1877   Mishmi hills (NE Assam)
  - — *P.a.albiventre* (Godwin-Austen, 1877)[2]   Bhutan, NW and S Assam, W Burma
  - — *P.a.cinnamomeum* (Rippon, 1900)   C and E Burma, NW Thailand, S Yunnan, S and C Indochina
  - — *P.a.pusillum* (Delacour, 1927)   NW Vietnam, N Laos

- ○ **Pellorneum palustre** Gould, 1872 MARSH BABBLER   C and E Assam, SE Bangladesh

- ○ **Pellorneum ruficeps** PUFF-THROATED BABBLER
  - — *P.r.olivaceum* Jerdon, 1839   Extreme SW India
  - — *P.r.ruficeps* Swainson, 1832   S and C India
  - — *P.r.pallidum* Abdulali, 1982   SE India (15)
  - — *P.r.punctatum* (Gould, 1838)   W Himalayas
  - — *P.r.mandellii* Blanford, 1871   E Himalayas (C Nepal to NC Assam)
  - — *P.r.chamelum* Deignan, 1947[3]   S Assam
  - — *P.r.pectorale* Godwin-Austen, 1877   Mishmi hills (NE Assam)
  - — *P.r.ripleyi* Deignan, 1947[4]   Lakhimpur (SE Assam)
  - — *P.r.vocale* Deignan, 1951   Manipur (E India)
  - — *P.r.stageri* Deignan, 1947   NE Burma
  - — *P.r.shanense* Deignan, 1947   E Burma, SW Yunnan
  - — *P.r.hilarum* Deignan, 1947   C Burma
  - — *P.r.victoriae* Deignan, 1947   Chin Hills (W Burma)
  - — *P.r.minus* Hume, 1873   C Burma
  - — *P.r.subochraceum* Swinhoe, 1871   Upper Tenasserim (Burma)
  - — *P.r.insularum* Deignan, 1947   Lower Tenasserim (Burma)
  - — *P.r.acrum* Deignan, 1947   Pen. and WC Thailand
  - — *P.r.chthonium* Deignan, 1947   NW Thailand
  - — *P.r.indistinctum* Deignan, 1947   E part of N Thailand
  - — *P.r.elbeli* Deignan, 1956   NW part of E Thailand
  - — *P.r.dusiti* Dickinson & Chaiyaphun, 1970   W slope of Dong Phaya Fai range EC Thailand (740)
  - — *P.r.oreum* Deignan, 1947   S China, N Indochina
  - — *P.r.vividum* La Touche, 1921   N Vietnam
  - — *P.r.ubonense* Deignan, 1947   E part of E Thailand, S Laos
  - — *P.r.deignani* Delacour, 1951   S Vietnam
  - — *P.r.dilloni* Delacour, 1951   Extreme S Vietnam
  - — *P.r.euroum* Deignan, 1947   W Cambodia, SC and SE Thailand
  - — *P.r.smithi* Riley, 1924   Coastal islands off SE Thailand and Cambodia

- ○ **Pellorneum fuscocapillus** BROWN-CAPPED BABBLER[5]
  - — *P.f.babaulti* (T. Wells, 1919)   N and E Sri Lanka
  - — *P.f.fuscocapillus* (Blyth, 1849)   SW Sri Lanka
  - — *P.f.scortillum* Ripley, 1946   SW Sri Lanka

- ○ **Pellorneum tickelli** BUFF-BREASTED BABBLER
  - — *P.t.assamense* (Sharpe, 1883)[6]   Assam, NW Burma
  - — *P.t.grisescens* (Ticehurst, 1932)   Arakan Yomas (SW Burma)
  - — *P.t.fulvum* (Walden, 1875)[7]   NE Burma, W and S Yunnan, N Thailand, N and C Indochina

---

[1] Genus retained *contra* Ripley *et al.* (1985) (2059), see Dickinson *et al.* (1991) (745).
[2] Includes *nagaense* see Deignan (1964: 246) (665).
[3] Perhaps better united with *mandelli* see Ripley (1982: 320) (2055).
[4] Perhaps better united with *mandelli* see Ripley (1982: 320) (2055).
[5] For correction of spelling see David & Gosselin (2002) (613).
[6] This name is used with reservations; a senior synonym seems to exist. Also subsumes *Brachypterx cryptica* Ripley, 1980 (2054), see Ripley (1984) (2057).
[7] Provisionally this includes N Vietnamese birds listed as *ochracea* in last Edition (but more correctly *ochraceum*).

__ *P.t.annamense* (Delacour, 1926)	S and C Vietnam, S Laos
__ *P.t.tickelli* Blyth, 1859[1]	SE Burma, W Thailand and Malay Pen.

○ *Pellorneum buettikoferi* (Vorderman, 1892) SUMATRAN BABBLER[2]   Lowland and submontane Sumatra

○ *Pellorneum pyrrogenys* TEMMINCK'S BABBLER

__ *P.p.pyrrogenys* (Temminck, 1827)[3]	Java
__ *P.p.erythrote* (Sharpe, 1883)	Mts. Poi and Penrissen (Sarawak)
__ *P.p.longstaffi* (Harrisson & Hartley, 1934)	Mts. Dulit and Taman (Sarawak)
__ *P.p.canicapillus* (Sharpe, 1887)[4]	Mt. Kinabalu (Sabah)

○ *Pellorneum capistratum* BLACK-CAPPED BABBLER

__ *P.c.nigrocapitatum* (Eyton, 1839)[5]	Malay Pen., Sumatra, Bangka, Belitung and N Natuna Is.
__ *P.c.capistratoides* (Strickland, 1849)	W and S Borneo
__ *P.c.morrelli* Chasen & Kloss, 1929	N and E Borneo and Banggi I.
__ *P.c.capistratum* (Temminck, 1823)	Java

TRICHASTOMA Blyth, 1842[6] N

○ *Trichastoma rostratum* WHITE-CHESTED BABBLER

__ *T.r.rostratum* Blyth, 1842	Malay Pen., Sumatra, Belitung
__ *T.r.macropterum* (Salvadori, 1868)	Borneo and Banggi I.

○ *Trichastoma celebense* SULAWESI BABBLER

__ *T.c.celebense* (Strickland, 1849)	N Sulawesi
__ *T.c.connectens* (Mayr, 1938)	NC Sulawesi
__ *T.c.rufofuscum* (Stresemann, 1931)	C Sulawesi
__ *T.c.finschi* (Walden, 1876)	SW Sulawesi
__ *T.c.sordidum* Stresemann, 1938	E and S Sulawesi, Pula I.
__ *T.c.togianense* (Voous, 1952)	Togian I.

○ *Trichastoma bicolor* (Lesson, 1839) FERRUGINOUS BABBLER   Malay Pen., E Sumatra, Bangka, Borneo

MALACOCINCLA Blyth, 1845 F

○ *Malacocincla abbotti* ABBOTT'S BABBLER

__ *M.a.abbotti* Blyth, 1845[7]	C and E Himalayas, Assam, SW, SC and SE Burma, W and SW Thailand, NW of W Malaysia
__ *M.a.krishnarajui* Ripley & Beehler, 1985	E Ghats (India) (2060)
__ *M.a.williamsoni* Deignan, 1948	E Thailand, NW Cambodia
__ *M.a.obscurior* Deignan, 1948	SE Thailand, SW Cambodia
__ *M.a.altera* (Sims, 1957)	N Laos, C Vietnam
__ *M.a.olivacea* (Strickland, 1847)	Extreme S Thailand, W Malaysia, Sumatra
__ *M.a.concreta* Büttikofer, 1895[8]	Borneo, Belitung, Matu Siri I.
__ *M.a.baweana* Oberholser, 1917	Bawean I.

○ *Malacocincla sepiaria* HORSFIELD'S BABBLER[9]

__ *M.s.tardinata* E. Hartert, 1915	Malay Pen.
__ *M.s.barussana* Robinson & Kloss, 1921[10]	Sumatra
__ *M.s.sepiaria* (Horsfield, 1821)[11]	Java and Bali
__ *M.s.rufiventris* Salvadori, 1874	W and S Borneo
__ *M.s.harterti* Chasen & Kloss, 1929	N and E Borneo

○ †?*Malacocincla perspicillata* (Bonaparte, 1850) BLACK-BROWED BABBLER[12]

Martapoera or Bandjermasin District of Kalimantan (Borneo)

---

[1] Includes *australis* see Deignan (1964: 249) (665); if recognised the spelling *australe* must be used.
[2] For reasons to treat this as a species see Wells *et al.* (2001) (2582).
[3] Includes *besuki* see Mees (1996) (1543).
[4] For correction of spelling see David & Gosselin (2002) (613).
[5] Includes *nyctilampis* see Mees (1986) (1539).
[6] The review by Ripley & Beehler (1985) (2059) has been considered. Molecular studies are needed.
[7] Includes *rufescentior* see Deignan (1964: 256) (665).
[8] For reasons to use this name in place of *finschi* (used in last Edition) or *sirensis* (used by Deignan, 1964) see Mees (1971) (1523).
[9] For correction of spelling see David & Gosselin (2002) (613).
[10] Includes *liberalis* and *vanderbilti* see Mees (1995) (1542).
[11] Includes *minor* see Mees (1996) (1543); records from Bangka (*bangkae*) not accepted by Mees (1986: 15) (1539).
[12] In last Edition we listed a subspecies *vanderbilti*. Mees (1995) (1542) has shown this to be a synonym of nominate *Malacocincla sepiaria*.

○ *Malacocincla malaccensis* Short-tailed Babbler
    — *M.m.malaccensis* (Hartlaub, 1844)   Malay Pen., Sumatra and islands, Anamba and Natuna Is.
    — *M.m.poliogenys* (Strickland, 1849)[1]   Bangka, Belitung Is., W and S Borneo
    — *M.m.sordida* (Chasen & Kloss, 1929)   NE Borneo
    — *M.m.feriata* (Chasen & Kloss, 1931)   Gunong Mulu (Sarawak)[2]

○ *Malacocincla cinereiceps* (Tweeddale, 1878) Ashy-headed Babbler   Balabac, Palawan

MALACOPTERON Eyton, 1839 N
○ *Malacopteron magnirostre* Moustached Babbler
    — *M.m.magnirostre* (F. Moore, 1854)[3]   Malay Pen., Sumatra, Anamba Is.
    — *M.m.cinereocapilla* (Salvadori, 1868)[4]   Borneo

○ *Malacopteron affine* (Blyth, 1842) Sooty-capped Babbler[5]   Malay Pen., Sumatra incl. Banyak Is., Borneo

○ *Malacopteron cinereum* Scaly-crowned Babbler
    — *M.c.indochinense* (Robinson & Kloss, 1921)   SE Thailand, NE, C and S Indochina
    — *M.c.rufifrons* Cabanis, 1850   Java
    — *M.c.cinereum* Eyton, 1839[6]   Malay Pen., Sumatra, Bangka, Borneo, N Natunas
    — *M.c.niasense* (Riley, 1937)   Nias I.

○ *Malacopteron magnum* Rufous-crowned Babbler
    — *M.m.magnum* Eyton, 1839   Malay Pen., Sumatra, Borneo (except NE), Natunas
    — *M.m.saba* Chasen & Kloss, 1930   NE Borneo

○ *Malacopteron palawanense* Büttikofer, 1895 Melodious Babbler   Balabac, Palawan

○ *Malacopteron albogulare* Grey-breasted Babbler
    — *M.a.albogulare* (Blyth, 1844)   Malay Pen., NE Sumatra and Batu Is.
    — *M.a.moultoni* (Robinson & Kloss, 1919)   NW Borneo

ILLADOPSIS Heine, 1860 F[7]
○ *Illadopsis cleaveri* Black-capped Thrush-Babbler
    — *I.c.johnsoni* (Büttikofer, 1889)   Sierra Leone, Liberia
    — *I.c.cleaveri* (Shelley, 1874)   Ghana
    — *I.c.marchanti* Serle, 1956   S Nigeria
    — *I.c.batesi* (Sharpe, 1901)   SE Nigeria, Gabon, Cameroon
    — *I.c.poensis* Bannerman, 1934   Bioko I.

○ *Illadopsis albipectus* Scaly-breasted Thrush-Babbler
    — *I.a.barakae* (Jackson, 1906)   N Zaire, Uganda, S Sudan, W Kenya
    — *I.a.albipectus* (Reichenow, 1887)   W Zaire
    — ¶*I.a.trensei* Meise, 1978   N Angola (1555)

○ *Illadopsis rufescens* (Reichenow, 1878) Rufous-winged Thrush-Babbler
                                          S Senegal to Ghana

○ *Illadopsis puveli* Puvel's Thrush-Babbler
    — *I.p.puveli* (Salvadori, 1901)   S Senegal to Togo
    — *I.p.strenuipes* (Bannerman, 1920)   S Nigeria, Cameroon, NE Zaire and W Uganda

○ *Illadopsis rufipennis* Pale-breasted Thrush-Babbler
    — *I.r.bocagei* (Salvadori, 1903)   Bioko I.
    — *I.r.extrema* (Bates, 1930)   Sierra Leone to Ghana
    — *I.r.rufipennis* (Sharpe, 1872)   S Nigeria to Congo, Uganda, W Kenya
    — *I.r.distans* (Friedmann, 1928)   S Kenya, E Tanzania, Zanzibar

○ *Illadopsis fulvescens* Brown Thrush-Babbler
    — *I.f.gularis* Sharpe, 1870   Sierra Leone to Ghana
    — *I.f.moloneyana* (Sharpe, 1892)   E Ghana, Togo

[1] Includes *saturata* see Mees (1986) (1539).
[2] Known only from the type and not certainly this species.
[3] Includes *flavum* see Deignan (1964: 263) (665).
[4] For correction of spelling see David & Gosselin (2002) (613).
[5] Treated as monotypic following Mees (1986) (1539).
[6] Includes *bungurense* see Deignan (1964: 264) (665).
[7] Gender addressed by David & Gosselin (2002) (614), multiple changes occur in specific names.

— *I.f.iboensis* (E. Hartert, 1907)   S Nigeria

— *I.f.fulvescens* (Cassin, 1859)   Cameroon to Congo, W Zaire

— *I.f.ugandae* (van Someren, 1915)   C and E Zaire, Uganda, S Sudan, W Kenya

— *I.f.dilutior* (C.M.N. White, 1953)[1]   NW Angola

○ *Illadopsis pyrrhoptera* MOUNTAIN THRUSH-BABBLER

    — *I.p.pyrrhoptera* (Reichenow & Neumann, 1895)   E Zaire to W Uganda, Rwanda, Burundi, W Tanzania, W Kenya

    — *I.p.nyasae* (Benson, 1939)   N Malawi

KAKAMEGA Mann, Burton & Lennerstedt, 1978 (1416)[2] F

○ ***Kakamega poliothorax*** (Reichenow, 1900) GREY-CHESTED THRUSH-BABBLER

     S Cameroon, Bioko I., E Zaire, SW Kenya

PSEUDOALCIPPE Bannerman, 1923[3] F

○ ***Pseudoalcippe abyssinica*** AFRICAN HILL BABBLER

    — *P.a.monachus* (Reichenow, 1892)   Mt. Cameroon

    — *P.a.claudei* (Alexander, 1903)   Bioko I.

    — *P.a.atriceps* (Sharpe, 1902)[4]   Nigeria, Cameroon, E Zaire to Rwanda and W Uganda

    — *P.a.stierlingi* (Reichenow, 1898)   E and S Tanzania

    — *P.a.stictigula* (Shelley, 1903)   Malawi, NW Mozambique

    — *P.a.abyssinica* (Rüppell, 1840)[5]   Ethiopia, S Sudan, Kenya, N and W Tanzania, SE Zaire, W Angola

PTYRTICUS Hartlaub, 1883 M

○ ***Ptyrticus turdinus*** SPOTTED THRUSH-BABBLER

    — *P.t.harterti* Stresemann, 1921   C Cameroon

    — *P.t.turdinus* Hartlaub, 1883   SW Sudan, NE Zaire

    — *P.t.upembae* Verheyen, 1951   SE Zaire, NE Angola, N Zambia

POMATORHINUS Horsfield, 1821 M

○ ***Pomatorhinus hypoleucos*** LARGE SCIMITAR BABBLER

    — *P.h.hypoleucos* (Blyth, 1844)   Assam, Bangladesh, W and N Burma

    — *P.h.tickelli* Hume, 1877   Thailand, N Indochina, SW China

    — *P.h.brevirostris* Robinson & Kloss, 1919   S Indochina

    — *P.b.wrayi* Sharpe, 1887   Mts. of W Malaysia (2231)

    — *P.b.hainanus* Rothschild, 1903   Hainan I.

○ ***Pomatorhinus erythrogenys*** RUSTY-CHEEKED SCIMITAR BABBLER

    — *P.e.erythrogenys* Vigors, 1832   NW Himalayas

    — *P.e.ferrugilatus* Hodgson, 1836[6]   C Himalayas (Nepal to Bhutan)

    — *P.e.mcclellandi* Godwin-Austin, 1870   S Assam, W Burma

    — *P.e.imberbis* Salvadori, 1889   Karenni (E Burma)

    — *P.e.celatus* Deignan, 1941[7]   Shan States (E Burma), NW Thailand

    — *P.e.odicus* Bangs & Phillips, 1914   NE and E Burma, N Indochina, S Yunnan to Guizhou (S China)

    — *P.e.decarlei* Deignan, 1952   SE Xizang, SW Sichuan, NE Yunnan

    — *P.e.dedekeni* Oustalet, 1892   E Xizang, W Sichuan, NW Yunnan

    — *P.e.gravivox* David, 1873[8]   N Sichuan, S Gansu, Shaanxi

    — *P.e.cowensae* Deignan, 1952   E Sichuan, SW Hubei

    — *P.e.swinhoei* David, 1874   S and SE China

    — *P.e.erythrocnemis* Gould, 1863   Taiwan

○ ***Pomatorhinus horsfieldii*** INDIAN SCIMITAR BABBLER

    — *P.h.melanurus* Blyth, 1847   Sri Lanka (wet lowlands and western hills)

---

[1] For correction of spelling see David & Gosselin (2002) (613).

[2] Evidence now suggests this may be a thrush (Cibois, 2000) (355).

[3] We use this genus following Keith (2000: 29) (1240).

[4] Prigogine (1985) (1943) proposed reversion to a treatment in which this is a separate species, but see Dowsett & Dowsett-Lemaire (1993) (762). Thought to include *kivuensis*.

[5] Includes *ansorgei* see White (1962: 161) (2646) and *hildegardae* Ripley & Heinrich, 1966 (2068), *poliothorax* Cunningham-van Someren & Schifter, 1981 (602), and *loima* Cunningham-van Someren & Schifter, 1981 (602) (these last three perhaps await review).

[6] Includes *haringtoni* see Abdulali (1982) (14).

[7] There is insufficient evidence of sympatry between *celatus* and *odicus* to require treatment as two species.

[8] Tentatively includes *sowerbyi* see Cheng (1987) (333).

— *P.h.holdsworthi* Whistler, 1942[1]   Sri Lanka (dry lowlands and eastern hills)
— *P.h.travancoreensis* Harington, 1914   SW India
— *P.h.horsfieldii* Sykes, 1832   W India
— *P.h.obscurus* Hume, 1872   NW India
— *P.h.maderaspatensis* Whistler, 1936   EC India

○ *Pomatorhinus schisticeps* WHITE-BROWED SCIMITAR BABBLER
— *P.s.leucogaster* Gould, 1838   NW Himalayas
— *P.s.schisticeps* Hodgson, 1836   E Himalayas, S Assam, NW Burma
— *P.s.salimalii* Ripley, 1948   Mishmi hills (NE Assam)
— *P.s.cryptanthus* E. Hartert, 1915   Lakhimpur (SE Assam)
— *P.s.mearsi* Ogilvie-Grant, 1905   W Burma
— *P.s.ripponi* Harington, 1910   Extreme eastern Burma, N Thailand, N Laos
— *P.s.nuchalis* Wardlaw Ramsay, 1877   W of Shan States (E Burma)
— *P.s.difficilis* Deignan, 1956   SE Burma, W Thailand
— *P.s.olivaceus* Blyth, 1847   S Burma, SW Thailand
— *P.s.fastidiosus* E. Hartert, 1916   Malay Pen. (south to Trang)
— *P.s.humilis* Delacour, 1932   NE and E Thailand, S Laos, C Vietnam
— *P.s.annamensis* Robinson & Kloss, 1919   S Vietnam
— *P.s.klossi* E.C.S. Baker, 1917   SE Thailand, SW Cambodia

○ *Pomatorhinus montanus* CHESTNUT-BACKED SCIMITAR BABBLER
— *P.m.occidentalis* Robinson & Kloss, 1923   W Malaysia, Sumatra
— *P.m.montanus* Horsfield, 1821   W and C Java
— *P.m.ottolanderi* Robinson, 1918   E Java, Bali
— *P.m.bornensis* Cabanis, 1851   Borneo, Bangka

○ *Pomatorhinus ruficollis* STREAK-BREASTED SCIMITAR BABBLER
— *P.r.ruficollis* Hodgson, 1836   C Himalayas
— *P.r.godwini* Kinnear, 1944   E Himalayas, N Assam, SE Xizang
— *P.r.bakeri* Harington, 1914   S and SE Assam, W Burma
— *P.r.similis* Rothschild, 1926   NE Burma, NW Yunnan
— *P.r.bhamoensis* Mayr, 1941   Bhamo area (NE Burma)
— *P.r.albipectus* La Touche, 1923   SC Yunnan, N Laos
— *P.r.beaulieui* Delacour & Greenway, 1940   NE and C Laos
— *P.r.laurentei* La Touche, 1921   NE Yunnan (Kunming)
— *P.r.reconditus* Bangs & Phillips, 1914   SE Yunnan, N Vietnam
— *P.r.stridulus* Swinhoe, 1861   NC and E China
— *P.r.hunanensis* Cheng, Tso-hsin, 1974[2]   SC China (331)
— *P.r.eidos* Bangs, 1930   S Sichuan
— *P.r.musicus* Swinhoe, 1859   Taiwan
— *P.r.nigrostellatus* Swinhoe, 1870   Hainan I.

○ *Pomatorhinus ochraceiceps* RED-BILLED SCIMITAR BABBLER
— *P.o.stenorhynchus* Godwin-Austen, 1877   NE Assam, N Burma, SW Yunnan
— *P.o.austeni* Hume, 1881   E and S Assam
— *P.o.ochraceiceps* Walden, 1873   E Burma, N Thailand, N and C Indochina
— *P.o.alius* Riley, 1940   NW of E Thailand, S Laos (Bolovens), SC Vietnam (Langbian)

○ *Pomatorhinus ferruginosus* CORAL-BILLED SCIMITAR BABBLER
— *P.f.ferruginosus* Blyth, 1845   E Himalayas
— *P.f.namdapha* Ripley, 1980   Easternmost Assam (2054)
— *P.f.formosus* Koelz, 1952   S Assam and Manipur
— *P.f.phayrei* Blyth, 1847   SW Burma
— *P.f.stanfordi* Ticehurst, 1935   NE Burma
— *P.f.albogularis* Blyth, 1855[3]   E Burma, NW Thailand
— *P.f.orientalis* Delacour, 1927   N Indochina, S Yunnan
— *P.f.dickinsoni* Eames, 2002   C Vietnam (780A)

---

[1] For recognition of two subspecies in Sri Lanka see Phillips (1978) (1881) and Ripley (1982: 324) (2055).
[2] Misspelled in last Edition; a new name for *intermedius* which was preoccupied.
[3] Including *mariae* see Deignan (1964: 278) (665).

XIPHIRHYNCHUS Blyth, 1842 M
○ *Xiphirhynchus superciliaris* SLENDER-BILLED SCIMITAR BABBLER
— *X.s.superciliaris* Blyth, 1842   C Himalayas
— *X.s.intextus* Ripley, 1948   S and E Assam, W Burma
— *X.s.forresti* Rothschild, 1926   N and NE Burma, W Yunnan
— *X.s.rothschildi* Delacour & Jabouille, 1930   NW Vietnam, SE Yunnan

JABOUILLEIA Delacour, 1927 F
○ *Jabouilleia danjoui* SHORT-TAILED SCIMITAR BABBLER
— *J.d.parvirostris* Delacour, 1927   C Laos and NC Vietnam
— *J.d.danjoui* (Robinson & Kloss, 1919)   SC Vietnam

RIMATOR Blyth, 1847 M
○ *Rimator malacoptilus* LONG-BILLED WREN-BABBLER
— *R.m.malacoptilus* Blyth, 1847   E Himalayas, S and E Assam, N and NE Burma,
         NW Yunnan
— *R.m.pasquieri* Delacour & Jabouille, 1930   NW Vietnam
— *R.m.albostriatus* Salvadori, 1879   W Sumatra

PTILOCICHLA Sharpe, 1877 F
○ *Ptilocichla leucogrammica* (Bonaparte, 1850) BORNEAN WREN-BABBLER
         Borneo
○ *Ptilocichla mindanensis* STRIATED WREN-BABBLER[1]
— *P.m.minuta* Bourns & Worcester, 1894   Leyte, Samar
— *P.m.fortichi* Rand & Rabor, 1957   Bohol
— *P.m.mindanensis* (W.H. Blasius, 1890)   Mindanao
— *P.m.basilanica* Steere, 1890   Basilan
○ *Ptilocichla falcata* Sharpe, 1877 FALCATED WREN-BABBLER   Balabac, Palawan Is.

KENOPIA G.R. Gray, 1869 F
○ *Kenopia striata* (Blyth, 1842) STRIPED WREN-BABBLER   Malay Pen., lowland Sumatra, Borneo

NAPOTHERA G.R. Gray, 1842 F
○ *Napothera rufipectus* (Salvadori, 1879) RUSTY-BREASTED WREN-BABBLER
         Highlands W Sumatra
○ *Napothera atrigularis* (Bonaparte, 1850) BLACK-THROATED WREN-BABBLER
         Borneo
○ *Napothera macrodactyla* LARGE WREN-BABBLER
— *N.m.macrodactyla* (Strickland, 1844)   Malay Pen.
— *N.m.beauforti* (Voous, 1949)   NE Sumatra
— *N.m.lepidopleura* (Bonaparte, 1850)   Java
○ *Napothera marmorata* MARBLED WREN-BABBLER
— *N.m.grandior* (Voous, 1949)   Highlands W Malaysia
— *N.m.marmorata* (Wardlaw Ramsay, 1880)   Highlands W Sumatra
○ *Napothera crispifrons* LIMESTONE WREN-BABBLER
— *N.c.annamensis* (Delacour & Jabouille, 1928)   SE Yunnan, N Indochina
— *N.c.calcicola* Deignan, 1939   SW part of NE Thailand
— *N.c.crispifrons* (Blyth, 1855)   N and W Thailand, SW Burma
○ *Napothera brevicaudata* STREAKED WREN-BABBLER
— *N.b.striata* (Walden, 1871)   S Assam, Manipur, SW Burma
— *N.b.venningi* (Harington, 1913)   W Yunnan, NE and E Burma
— *N.b.brevicaudata* (Blyth, 1855)   N and W Thailand, SW Burma
— *N.b.stevensi* (Kinnear, 1925)   N Vietnam, N and C Laos, SE Yunnan, Guangxi
— *N.b.proxima* Delacour, 1930   C Vietnam, S Laos
— *N.b.rufiventer* (Delacour, 1927)   Langbian (SC Vietnam)

[1] This English name is necessary to avoid duplication if the group name Ground Babbler, preferred in Kennedy *et al.*, 2000 (1259), is dropped.

   — *N.b.griseigularis* (Delacour & Jabouille, 1928)   SE Thailand, SW Cambodia

   — *N.b.leucosticta* (Sharpe, 1887)   Malay Pen.

○ *Napothera crassa* (Sharpe, 1888)   MOUNTAIN WREN-BABBLER   Highlands of N Borneo

○ *Napothera rabori*   RABOR'S WREN-BABBLER

   — *N.r.rabori* Rand, 1960   N Luzon

   — *N.r.mesoluzonica* duPont, 1971   C Luzon (771)

   — *N.r.sorsogonensis* Rand & Rabor, 1967   S Luzon (1978)

○ *Napothera epilepidota*   EYEBROWED WREN-BABBLER

   — *N.e.guttaticollis* (Ogilvie-Grant, 1895)   E Bhutan, N Assam

   — *N.e.roberti* (Godwin-Austen & Walden, 1875)   S and E Assam, NW Burma

   — *N.e.bakeri* (Harington, 1913)   E Burma

   — *N.e.davisoni* (Ogilvie-Grant, 1910)   N and W Thailand, SW Burma

   — *N.e.amyae* (Kinnear, 1925)   Extreme S China, N Indochina

   — *N.e.delacouri* Yen, 1934   Da Yao Shan (Guangxi)

   — *N.e.hainana* (E. Hartert, 1910)   Hainan

   — *N.e.clara* (Robinson & Kloss, 1919)   C and SC Vietnam

   — *N.e.granti* (Richmond, 1900)   Malay Pen.

   — *N.e.diluta* (Robinson & Kloss, 1916)[1]   Highlands of N Sumatra

   — *N.e.mendeni* Neumann, 1937   Highlands of S Sumatra

   — *N.e.epilepidota* (Temminck, 1828[2])   W and C Java

   — *N.e.exsul* (Sharpe, 1888)   Highlands of Borneo

PNOEPYGA Hodgson, 1844   F

○ *Pnoepyga albiventer*   SCALY-BREASTED WREN-BABBLER

   — *P.a.pallidior* Kinnear, 1924   NW and WC Himalayas

   — *P.a.albiventer* (Hodgson, 1837)   EC and E Himalayas, Assam, W and N Burma,
                                           SC China, NW Vietnam

   — *P.a.formosana* Ingram, 1909[3]   Taiwan

○ *Pnoepyga immaculata* Martens & Eck, 1991   NEPAL WREN-BABBLER #   WC and E Nepal (1442)

○ *Pnoepyga pusilla*   PYGMY WREN-BABBLER

   — *P.p.pusilla* Hodgson, 1845   C and E Himalayas, Assam, Burma, NW Thailand,
                                           S China, N Indochina (south to C Laos)

   — *P.p.annamensis* Robinson & Kloss, 1919   S Indochina

   — *P.p.harterti* Robinson & Kloss, 1918   Malay Pen.

   — *P.p.lepida* Salvadori, 1879   W Sumatra

   — *P.p.rufa* Sharpe, 1882   Java

   — *P.p.everetti* Rothschild, 1897   Flores I.

   — *P.p.timorensis* Mayr, 1944   Timor I.

SPELAEORNIS David & Oustalet, 1877   M

○ *Spelaeornis caudatus*   RUFOUS-THROATED WREN-BABBLER

   — *S.c.caudatus* (Blyth, 1845)   E Nepal, Sikkim, Bhutan and NW Assam

   — *S.c.badeigularis* Ripley, 1948[4]   Mishmi hills (NE Assam)

○ *Spelaeornis troglodytoides*   BAR-WINGED WREN-BABBLER

   — *S.t.sherriffi* Kinnear, 1934   E Bhutan, N Assam

   — *S.t.indiraji* Ripley, Saha & Beehler, 1991   Easternmost Assam (2076)

   — *S.t.souliei* Oustalet, 1898   NE Burma, W Yunnan

   — *S.t.rocki* Riley, 1929   N Yunnan (Mekong valley)

   — *S.t.troglodytoides* (J. Verreaux, 1870)   W Sichuan

   — *S.t.nanchuanensis* Li, Yang & Yu, 1992   C Sichuan, Hunan, Hubei (1357)

   — *S.t.halsueti* (David, 1875)   Shaanxi

○ *Spelaeornis formosus* (Walden, 1874)   SPOTTED WREN-BABBLER   E Himalayas, Assam, W Burma, SE Yunnan and
                                           NW Fujian, NW and C Vietnam, C Laos

---

[1] Including *lucilleae* following van Marle & Voous (1988) (2468).

[2] Pl. 448 appeared in Livr. 75 in 1828 (Dickinson, 2001) (739); Deignan (1964: 293) (665) cited Livr. 74 with date 1827.

[3] Attached to this species by Harrap (1989) (1070), not as previously to *pusilla*.

[4] As this is still essentially unknown (based on one specimen) we prefer not to treat this as a species.

○ *Spelaeornis chocolatinus* Long-tailed Wren-Babbler[1]

— *S.c.chocolatinus* (Godwin-Austen & Walden, 1875)   S and E Assam, Manipur

— *S.c.oatesi* (Rippon, 1904)   SE Assam, W Burma

— *S.c.reptatus* (Bingham, 1903)   NE and E Burma, N and SW Yunnan, SW Sichuan

— *S.c.kinneari* Delacour & Jabouille, 1930   NW Vietnam

○ *Spelaeornis longicaudatus* (F. Moore, 1854) Tawny-breasted Wren-Babbler

    S Assam, Manipur

SPHENOCICHLA Godwin-Austen & Walden, 1875 F

○ *Sphenocichla humei* Wedge-billed Wren-Babbler

— *S.h.humei* (Mandelli, 1873)   E Himalayas

— *S.h.roberti* Godwin-Austen & Walden, 1875   S and E Assam, N and NE Burma and NW Yunnan

STACHYRIS Hodgson, 1844[2] F

○ *Stachyris rodolphei* Deignan, 1939 Deignan's Babbler[3]   Doi Chiang Dao (N Thailand)

○ *Stachyris rufifrons* Rufous-fronted Babbler

— *S.r.pallescens* (Ticehurst, 1932)   W and SW Burma

— *S.r.rufifrons* Hume, 1873   E Burma, W and NW Thailand

— *S.r.obscura* (E.C.S. Baker, 1917)   SE Burma, SW Thailand

— *S.r.poliogaster* Hume, 1880   W Malaysia, Sumatra

— *S.r.sarawacensis* (Chasen, 1939)   Borneo

○ *Stachyris ambigua* Buff-chested Babbler[4]

— *S.a.ambigua* (Harington, 1915)   E Himalayas, Assam

— *S.a.planicola* Mayr, 1941   NE Burma, NW Yunnan

— *S.a.adjuncta* Deignan, 1939   N and E Thailand, N Indochina (south to C Laos)

— *S.a.insuspecta* Deignan, 1939   Bolovens (S Laos)

○ *Stachyris ruficeps* Rufous-capped Babbler

— *S.r.ruficeps* Blyth, 1847[5]   SE Xizang, Assam, W and NW Burma, E Himalayas, SE Tibet

— *S.r.bhamoensis* (Harington, 1908)   NE Burma, W Yunnan

— *S.r.davidi* (Oustalet, 1899)   C, E and S China, N Indochina to S Laos

— *S.r.pagana* (Riley, 1940)   Langbian (SC Vietnam)

— *S.r.praecognita* Swinhoe, 1866   Taiwan

— *S.r.goodsoni* (Rothschild, 1903)   Hainan I.

○ *Stachyris pyrrhops* Blyth, 1844 Black-chinned Babbler   NW and C Himalayas to E Nepal

○ *Stachyris chrysaea* Golden Babbler

— *S.c.chrysaea* Blyth, 1844   C and E Himalayas, Assam, N Burma, W Yunnan

— *S.c.binghami* Rippon, 1904   SE Assam, W Burma

— *S.c.aurata* Meyer de Schauensee, 1938   S Burma, extreme N Thailand, N and C Indochina

— *S.c.assimilis* (Walden, 1875)   E and SE Burma, W Thailand

— *S.c.chrysops* Richmond, 1902   Malay Pen.

— *S.c.frigida* (Hartlaub, 1865)   W Sumatra

○ *Stachyris plateni* Pygmy Babbler

— *S.p.pygmaea* (Ogilvie-Grant, 1896)   Samar, Leyte

— *S.p.plateni* (W.H. Blasius, 1890)   Mindanao

○ *Stachyris dennistouni* (Ogilvie-Grant, 1895) Golden-crowned Babbler[6]

    N Luzon

○ *Stachyris nigrocapitata* Black-crowned Babbler

— *S.n.affinis* (McGregor, 1907)   S Luzon

— *S.n.nigrocapitata* (Steere, 1890)   Samar, Leyte

— *S.n.boholensis* Rand & Rabor, 1957   Bohol

[1] This is the name preferred by Inskipp *et al.* (1996) (1182). Perhaps it should be reconsidered, as, in Latin, it is a homonym for the next species.
[2] Molecular studies confirm that this genus requires revision and splitting (Cibois, 2000) (355).
[3] This species is of doubtful validity, but a detailed review is needed; it is perhaps aberrant *rufifrons*.
[4] May be conspecific with *rufifrons* see Robson (2000) (2095).
[5] Includes *rufipectus* see Ripley (1982: 334) (2055).
[6] This and the next species treated as conspecific with *capitalis* in last Edition, but see Dickinson *et al.* (1991) (745).

○ *Stachyris capitalis* Rusty-crowned Babbler
    — *S.c.capitalis* (Tweeddale, 1877)   Dinagat
    — *S.c.euroaustralis* Parkes, 1988   Mindanao (1776)
    — *S.c.isabelae* Parkes, 1963   Basilan (1756)

○ *Stachyris speciosa* (Tweeddale, 1878) Flame-templed Babbler   Negros

○ *Stachyris whiteheadi* Chestnut-faced Babbler
    — *S.w.whiteheadi* (Ogilvie-Grant, 1894)   N Luzon
    — *S.w.sorsogonensis* Rand & Rabor, 1967   S Luzon (1978)

○ *Stachyris striata* (Ogilvie-Grant, 1894) Luzon Striped Babbler   N Luzon

○ *Stachyris latistriata* Gonzales & Kennedy, 1990 Panay Striped Babbler
                                           Panay (954)

○ *Stachyris nigrorum* Rand & Rabor, 1952 Negros Striped Babbler   Negros

○ *Stachyris hypogrammica* Salomonsen, 1961 Palawan Striped Babbler   Palawan

○ *Stachyris grammiceps* (Temminck, 1828[1]) White-breasted Babbler   W Java

○ *Stachyris herberti* (E.C.S. Baker, 1920) Sooty Babbler   C Laos, C Vietnam

○ *Stachyris nigriceps* Grey-throated Babbler
    — *S.n.nigriceps* Blyth, 1844[2]   C and E Himalayas, E Assam, SE Xizang
    — *S.n.coltarti* Harington, 1913[3]   S and SE Assam, W, N and NE Burma SW Yunnan,
                                         NW Thailand
    — *S.n.yunnanensis* La Touche, 1921   N Thailand (not NW), extreme E Burma, N and
                                           C Indochina, SE Yunnan
    — *S.n.rileyi* Chasen, 1936   Langbian (S Vietnam) and ? Bolovens (S Laos)
    — *S.n.dipora* Oberholser, 1922   Malay Pen. (south to Trang)
    — *S.n.davisoni* Sharpe, 1892   Sthn. Malay Pen.
    — *S.n.larvata* (Bonaparte, 1850)   Lingga Arch., Sumatra
    — *S.n.natunensis* E. Hartert, 1894   N Natunas
    — *S.n.tionis* Robinson & Kloss, 1927   Tioman I
    — *S.n.hartleyi* Chasen, 1935   Mts. of W Sarawak
    — *S.n.borneensis* Sharpe, 1887   Borneo (excl. W Sarawak)

○ *Stachyris poliocephala* (Temminck, 1836) Grey-headed Babbler[4]   Malay Pen., Sumatra, Borneo

○ *Stachyris striolata* Spot-necked Babbler
    — *S.s.helenae* Delacour & Greenway, 1939   N Thailand, N Laos, S Yunnan
    — *S.s.guttata* (Blyth, 1859)   W Thailand, SE Burma
    — *S.s.tonkinensis* Kinnear, 1924   NE Burma, SE Yunnan, Guangxi, N Vietnam, C Laos
    — *S.s.swinhoei* Rothschild, 1903   Hainan I.
    — *S.s.nigrescentior* Deignan, 1947   Pen. Thailand
    — *S.s.umbrosa* (Kloss, 1921)   NE Sumatra
    — *S.s.striolata* (S. Müller, 1835)   W Sumatra

○ *Stachyris oglei* (Godwin-Austen, 1877) Snowy-throated Babbler   E Assam, NE Burma

○ *Stachyris maculata* Chestnut-rumped Babbler
    — *S.m.maculata* (Temminck, 1836)[5]   Malay Pen., Sumatra, Borneo
    — *S.m.banjakensis* Richmond, 1902   Banyak Is.
    — *S.m.hypopyrrha* Oberholser, 1912   Batu Is.

○ *Stachyris leucotis* White-necked Babbler
    — *S.l.leucotis* (Strickland, 1848)   Malay Pen.
    — *S.l.sumatrensis* Chasen, 1939   Sumatra
    — *S.l.obscurata* Mayr, 1942   Borneo

○ *Stachyris nigricollis* (Temminck, 1836) Black-throated Babbler[6]   Malay Pen., E Sumatra, Borneo

---

[1] Pl. 448 appeared in Livr. 75 in 1828 see Dickinson (2001) (739); Deignan (1964: 309) (665) cited Livr. 74 with date 1827.
[2] No date given by Deignan (1964: 309) (665). Including *coei* see Ripley (1982: 335) (2055).
[3] Including *spadix* see Ripley (1982: 335) (2055).
[4] Includes *pulla* see van Marle & Voous (1988) (2468).
[5] Implicitly includes *pectoralis* (although not listed as a synonym by Deignan (1964: 313) (665).
[6] The name *erythronotus* is restored to synonymy. Treated as monotypic by Deignan (1964: 314) (665) and not seriously challenged since.

○ *Stachyris thoracica* WHITE-BIBBED BABBLER
— *S.t.thoracica* (Temminck, 1821)   S Sumatra, W and C Java
— *S.t.orientalis* Robinson, 1918   E Java

○ *Stachyris erythroptera* CHESTNUT-WINGED BABBLER
— *S.e.erythroptera* (Blyth, 1842)   Malay Pen., N Natunas
— *S.e.pyrrhophaea* (Hartlaub, 1844)[1]   Sumatra and Batu Is., Bangka, Belitung
— *S.e.fulviventris* (Richmond, 1903)   Banyak I.
— *S.e.bicolor* (Blyth, 1865)   N and E Borneo and Banggi I.
— *S.e.rufa* (Chasen & Kloss, 1927)   SW Borneo

○ *Stachyris melanothorax* CRESCENT-CHESTED BABBLER
— *S.m.melanothorax* (Temminck, 1823)[2] W and C Java
— *S.m.intermedia* (Robinson, 1918)   E Java
— *S.m.baliensis* (E. Hartert, 1915)   Bali

DUMETIA Blyth, 1849 F
○ *Dumetia hyperythra* TAWNY-BELLIED BABBLER
— *D.h.hyperythra* (Franklin, 1831)   Himalayan foothills and NC India
— *D.h.abuensis* Harington, 1915[3]   W India
— *D.h.albogularis* (Blyth, 1847)   S India
— *D.h.phillipsi* Whistler, 1941   Sri Lanka

RHOPOCICHLA Oates, 1889 F
○ *Rhopocichla atriceps* DARK-FRONTED BABBLER
— *R.a.atriceps* (Jerdon, 1839)   W India south to Nilgiris
— *R.a.bourdilloni* (Hume, 1876)   Kerala (SW India)
— *R.a.siccata* Whistler, 1941   N and E Sri Lanka
— *R.a.nigrifrons* (Blyth, 1849)   SW Sri Lanka

MACRONOUS Jardine & Selby, 1835 M
○ *Macronous gularis* STRIPED TIT-BABBLER
— *M.g.rubicapilla* (Tickell, 1833)[4][5]   Nepal, NE and EC India, Assam and Bangladesh
— *M.g.ticehursti* (Stresemann, 1940)   W Burma
— *M.g.sulphureus* (Rippon, 1900)   E and NE Burma, N and W Thailand, SW Yunnan
— *M.g.lutescens* (Delacour, 1926)   SE Yunnan, N and E Thailand, Laos, N Vietnam
— *M.g.kinneari* (Delacour & Jabouille, 1924) C Vietnam
— *M.g.versuricola* (Oberholser, 1922)   S Vietnam, E Cambodia
— *M.g.saraburiensis* Deignan, 1956   SW part of E Thailand, W Cambodia
— *M.g.connectens* (Kloss, 1918)   Coastal Cambodia and Thailand (east side) and SE Burma

— *M.g.inveteratus* (Oberholser, 1922)   Coastal islands off SE Thailand and Cambodia
— *M.g.condorensis* (Robinson, 1920)   Con Son I. (S Vietnam)
— *M.g.archipelagicus* (Oberholser, 1922) Mergui Arch. (SE Burma)
— *M.g.chersonesophilus* (Oberholser, 1922) Malay Pen. (Trang to Trengganu)
— *M.g.gularis* (Horsfield, 1822)   S Malay Pen., Sumatra, Banyak and Batu Is.
— *M.g.zopherus* (Oberholser, 1917)   Anambas
— *M.g.zaperissus* (Oberholser, 1932)   N Natunas (except Bunguran)
— *M.g.everetti* (E. Hartert, 1894)   Bungaran (N Natuna Is.)
— *M.g.javanicus* (Cabanis, 1851)   Java (not E ?)
— *M.g.bornensis* (Bonaparte, 1850)   Borneo (except Sabah area)
— *M.g.montanus* (Sharpe, 1887)   Sabah area
— *M.g.cagayanensis* (Guillemard, 1885) Cagayan Sulu I.
— *M.g.argenteus* (Chasen & Kloss, 1930) Islands off N Borneo
— *M.g.woodi* (Sharpe, 1877)   Palawan I.

---

[1] Including *apega* see Mees (1986) (1539).
[2] Including *albigula* and *mendeni* see Mees (1986) (1539).
[3] Tentatively includes *navarroi* Abdulali, 1959 (5), see Ripley (1982: 336) (2055). See also Abdulali (1982) (14).
[4] For correction of spelling see David & Gosselin (2002) (613).
[5] Birds from S Karnataka require review.

○ *Macronus flavicollis* Grey-cheeked Tit-Babbler[1]
 — *M.f.flavicollis* (Bonaparte, 1850)   Java
 — *M.f.prillwitzi* (E. Hartert, 1901)   Kangean I.

○ *Macronous kelleyi* (Delacour, 1932) Grey-faced Tit-Babbler   S Laos, C and SC Vietnam

○ *Macronous striaticeps* Brown Tit-Babbler
 — *M.s.mindanensis* Steere, 1890   Samar, Leyte, Bohol, Mindanao
 — *M.s.alcasidi* duPont & Rabor, 1973   Dinagat and Siargao Is. (778)
 — *M.s.striaticeps* Sharpe, 1877   Basilan, Malamaui I.
 — *M.s.kettlewelli* Guillemard, 1885   Sulu Arch.

○ *Macronous ptilosus* Fluffy-backed Tit-Babbler
 — *M.p.ptilosus* Jardine & Selby, 1835   Malay Pen.
 — *M.p.trichorrhos* (Temminck, 1836)   Sumatra, Batu Is
 — *M.p.reclusus* E. Hartert, 1915[2]   Borneo, Bangka, Belitung

MICROMACRONUS Amadon, 1962 M
○ *Micromacronus leytensis* Miniature Tit-Babbler
 — *M.l.leytensis* Amadon, 1962   Leyte (and probably Samar)
 — *M.l.sordidus* Ripley & Rabor, 1968   Mts. of S Mindanao (2075)

TIMALIA Horsfield, 1821 F
○ *Timalia pileata* Chestnut-capped Babbler
 — *T.p.bengalensis* Godwin-Austen, 1872   Nepal, Assam, Bangladesh, W Burma
 — *T.p.smithi* Deignan, 1955   NE and E Burma, S China, N Thailand, N and
                 C Indochina
 — *T.p.intermedia* Kinnear, 1924   C and S Burma, SW Thailand
 — *T.p.patriciae* Deignan, 1955   WC and N Pen. Thailand
 — *T.p.dictator* Kinnear, 1930   E and SE Thailand, S Indochina
 — *T.p.pileata* Horsfield, 1821   Java

CHRYSOMMA Blyth, 1843 N
○ *Chrysomma sinense* Yellow-eyed Babbler
 — *C.s.nasale* (Legge, 1879)   Sri Lanka
 — *C.s.hypoleucum* (Franklin, 1831)   Pakistan, NW India
 — *C.s.sinense* (J.F. Gmelin, 1789)   C and NE India and Nepal to S China

MOUPINIA David & Oustalet, 1877 F
○ *Moupinia altirostris* Jerdon's Babbler
 — *M.a.scindica* (Harington, 1916)   Pakistan
 — *M.a.griseigularis* (Hume, 1877)   S Nepal, NE India, S Assam, NE Burma
 — †?*M.a.altirostris* (Jerdon, 1862)   SC Burma

○ *Moupinia poecilotis* (J. Verreaux, 1870) Rufous-tailed Babbler   C China (N Yunnan to N and E Sichuan)

CHAMAEA Gambel, 1847 F
● *Chamaea fasciata* Wren-Tit
 — *C.f.phaea* Osgood, 1899   Coast of Oregon
 ∠ *C.f.margra* Browning, 1992   S Oregon (254)
 — *C.f.rufula* Ridgway, 1903   Coast of N California
 — *C.f.fasciata* (Gambel, 1845)[3]   Coast of C California
 — *C.f.henshawi* Ridgway, 1882[4]   Inland C California and NW Baja California

TURDOIDES Cretzschmar, 1827 F[5]
○ *Turdoides nipalensis* (Hodgson, 1836) Spiny Babbler   Nepal

---

[1] Treated as two forms of *M. gularis* in last Edition, but apparent sympatry on Java precludes this see van Balen (1993) (2458). The form named *prillwitzi* may require reassignment, but situation complex and review must include Philippine forms.
[2] Including *sordidus* see Mees (1986) (1539).
[3] Includes *intermedia* see Phillips (1986) (1875).
[4] Includes *canicauda* see Phillips (1986) (1875).
[5] Gender addressed by David & Gosselin (2002) (614), multiple changes occur in specific names.

○ *Turdoides altirostris* (E. Hartert, 1909) Iraq Babbler   C and SE Iraq, SW Iran

○ *Turdoides caudata* Common Babbler
    — *T.c.salvadorii* (De Filippi, 1865)   C Iraq to SW Iran
    — *T.c.huttoni* (Blyth, 1847)   S Afghanistan, SE Iran, W Pakistan
    — *T.c.eclipes* (Hume, 1877)   N Pakistan
    — *T.c.caudata* (Dumont, 1823)   SE Pakistan and India (except NE)

○ *Turdoides earlei* Striated Babbler
    — *T.e.sonivia* (Koelz, 1954)   Pakistan, extreme NW India
    — *T.e.earlei* (Blyth, 1844)   N and NE India, Assam, SW, C and S Burma

○ *Turdoides gularis* (Blyth, 1855) White-throated Babbler   C and S Burma

○ *Turdoides longirostris* (F. Moore, 1854) Slender-billed Babbler   C Nepal, Assam, SW Burma

○ *Turdoides malcolmi* (Sykes, 1832) Large Grey Babbler   W, C and S India

○ *Turdoides squamiceps* Arabian Babbler
    — *T.s.squamiceps* (Cretzschmar, 1827)   Sinai, Israel, W Jordan, NW and inland C and S Arabia
    — *T.s.yemensis* (Neumann, 1904)   SW Saudi Arabia, W Yemen
    — *T.s.muscatensis* Meyer de Schauensee & Ripley, 1953   Eastern United Arab Emirates

○ *Turdoides fulva* Fulvous Babbler
    — *T.f.maroccana* Lynes, 1925   S Morocco
    — *T.f.fulva* (Desfontaines, 1789)   N Algeria, Tunisia, NW Libya
    — *T.f.buchanani* (E. Hartert, 1921)   C Sahara (S Mauritania to C Chad)
    — *T.f.acaciae* (M.H.K. Lichtenstein, 1823)   N Chad, C and N Sudan to S Egypt and N Eritrea

○ *Turdoides aylmeri* Scaly Chatterer
    — *T.a.aylmeri* (Shelley, 1885)   Somalia, E Ethiopia
    — *T.a.boranensis* (Benson, 1947)   SE Ethiopia, N Kenya
    — *T.a.keniana* (Jackson, 1910)[1]   C and SE Kenya, NE Tanzania
    — *T.a.mentalis* (Reichenow, 1887)   NC Tanzania

○ *Turdoides rubiginosa* Rufous Chatterer
    — *T.r.bowdleri* Deignan, 1964   SE Ethiopia
    — *T.r.rubiginosa* (Rüppell, 1845)   N Uganda, N, C and S Kenya, C and S Ethiopia, S Sudan
    — *T.r.heuglini* (Sharpe, 1883)   S Somalia, SE Kenya, NE Tanzania
    — *T.r.schnitzeri* Deignan, 1964   N Tanzania

○ *Turdoides subrufa* Rufous Babbler
    — *T.s.subrufa* (Jerdon, 1839)   W India south to the Nilgiris
    — *T.s.hyperytha* (Sharpe, 1883)   Kerala and SW Madras

○ *Turdoides striata* Jungle Babbler
    — *T.s.sindiana* (Ticehurst, 1920)   Pakistan, NW India
    — *T.s.striata* (Dumont, 1823)   N India to C Assam and Bangladesh
    — *T.s.somervillei* (Sykes, 1832)   W coast of India
    — *T.s.malabarica* (Jerdon, 1845)   SW India
    — *T.s.orientalis* (Jerdon, 1845)   C and S India
    — *T.s.orissae* (Jerdon, 1847)[2]   E India

○ *Turdoides rufescens* (Blyth, 1847) Orange-billed Babbler   Sri Lanka

○ *Turdoides affinis* Yellow-billed Babbler
    — *T.a.affinis* (Jerdon, 1845)   S India
    — *T.a.taprobanus* Ripley, 1958   Sri Lanka

○ *Turdoides melanops* Black-faced Babbler
    — *T.m.angolensis* da Rosa Pinto, 1967[3]   Huila (SW Angola) (606)
    — *T.m.melanops* (Hartlaub, 1867)   SW Angola, N Namibia
    — *T.m.querula* Clancey, 1979   NW Botswana, Caprivi (NE Namibia) (474)

[1] Including *loveridgei* see Fry (2000: 63) (892).
[2] See Ripley (1969) (2048) for reasons to use the name *orissae*.
[3] May be a synonym of nominate *melanops*. Not mentioned by Fry (2000: 50) (892).

○ *Turdoides sharpei* Black-lored Babbler[1]
  — *T.s.vepres* R. Meinertzhagen, 1936   Slopes of Mt. Kenya (EC Kenya)
  — *T.s.sharpei* (Reichenow, 1891)[2]   S Uganda, SW Kenya, W Tanzania

○ *Turdoides tenebrosa* (Hartlaub, 1883) Dusky Babbler   NE Zaire, S Sudan, NW Uganda, SW Ethiopia

○ *Turdoides reinwardtii* Blackcap-Babbler
  — *T.r.reinwardtii* (Swainson, 1831)   Senegal to Sierra Leone
  — *T.r.stictilaema* (Alexander, 1901)[3]   Ghana to N Cameroon, Central African Republic and
         N Zaire

○ *Turdoides plebejus* Brown Babbler
  — *T.p.platycirca* (Swainson, 1837)   Senegal and Guinea to W Nigeria
  — *T.p.plebejus* (Cretzschmar, 1828)   NE Nigeria to W and C Sudan
  — *T.p.cinerea* (Heuglin 1856)[4]   SE Nigeria to S Sudan, SW Ethiopia, W Kenya

○ *Turdoides leucocephala* Cretzschmar, 1827 White-headed Babbler[5]   E Sudan, NW Ethiopia, N Eritrea

○ *Turdoides jardineii* Arrow-marked Babbler
  — *T.j.hyposticta* (Cabanis & Reichenow, 1877)   W and S Zaire to W Angola
  — *T.j.tanganjicae* (Reichenow, 1886)   SE Zaire, N Zambia
  — *T.j.emini* (Neumann, 1904)[6]   E Zaire to SW Kenya, NW and N Tanzania
  — *T.j.kirkii* (Sharpe, 1876)   SE Kenya to C Mozambique, Malawi, E Zambia
  — *T.j.tamalakanei* Meyer de Schauensee, 1932   SW Zambia, N Botswana, S Angola
  — *T.j.jardineii* (A. Smith, 1836)[7]   NW and C Zambia to Zimbabwe, S Mozambique,
         Transvaal, Natal

○ *Turdoides squamulata* Scaly Babbler[8]
  — *T.s.jubaensis* van Someren, 1931   Jubba R. (S Somalia), Doua R. (NE Kenya/S Ethiopia)
  — *T.s.carolinae* Ash, 1981   R. Shebeelei (S Somalia), SE Ethiopia (58)
  — *T.s.squamulata* (Shelley, 1884)   Coast of Kenya

○ *Turdoides leucopygia* White-rumped Babbler
  — *T.l.leucopygia* (Rüppell, 1840)   C and E Eritrea to N Ethiopia
  — *T.l.limbata* (Rüppell, 1845)   W Eritrea and NW Ethiopia
  — *T.l.smithii* (Sharpe, 1895)   NW Somalia to E and SE Ethiopia
  — *T.l.lacuum* (Neumann, 1903)   C Ethiopia
  — *T.l.omoensis* (Neumann, 1903)   S and SW Ethiopia, SE Sudan

○ *Turdoides hartlaubii* Hartlaub's Babbler[9]
  — *T.h.hartlaubi* (Bocage, 1868)[10]   Angola to SE and E Zaire, C and NE Zambia,
         SW Tanzania
  — *T.h.griseosquamata* Clancey, 1974   N Botswana to SW Zambia (445)

○ *Turdoides hindei* (Sharpe, 1900) Hinde's Babbler   C Kenya

○ *Turdoides hypoleuca* Northern Pied Babbler
  — *T.h.hypoleuca* (Cabanis, 1878)   C and S Kenya to Kilimanjaro (N Tanzania)
  — *T.h.rufuensis* (Neumann, 1906)   N and NE Tanzania

○ *Turdoides bicolor* (Jardine, 1831) Southern Pied Babbler   Namibia to W and S Zimbabwe, N Sth. Africa,
         SW Mozambique

○ *Turdoides gymnogenys* Bare-cheeked Babbler
  — *T.g.gymnogenys* (Hartlaub, 1865)   SW Angola
  — *T.g.kaokensis* (Roberts, 1937)   N Namibia

---

[1] For reasons to split this from *melanops* see Hall & Moreau (1970) (1066) and Fry (2000: 48) (892).
[2] Including *clamosa* see White (1962: 172) (2646).
[3] Including *houyi* see Deignan (1964: 340) (665).
[4] Including *gularis* Reichenow, 1910, Orn. Monatsber. 18: 7, see Fry (2000: 38) (892) a preoccupied name mistakenly used in last Edition; also includes *uamensis* and *elberti* its available junior synonyms – see Deignan (1964: 341) (665).
[5] In last Edition, and in Peters Check-list, treated as a race of *plebejus* (although *plebejus* is a younger name than *leucocephala*). For split from *T. plebejus* see Hall & Moreau (1970) (1066) and Fry (2000: 43) (892).
[6] Implicitly includes *kikuyuensis* see Fry (2000: 40) (892).
[7] Includes *convergens* see Fry (2000: 40) (892).
[8] Two additional unnamed subspecies were listed by Fry (2000: 57) (892), but hybrid origin may not yet be excluded.
[9] This species separated from *leucopygius* by Clancey (1984) (497).
[10] Includes *atra* see Fry (2000: 49) (892).

BABAX David, 1875 M
○ ***Babax lanceolatus*** CHINESE BABAX
    — *B.l.bonvaloti* Oustalet, 1892[1]   E Xizang to SW Sichuan
    — *B.l.lanceolatus* (J. Verreaux, 1870)   NE and E Burma, Yunnan to S Gansu, S Shaanxi,
                                     Hubei and Guizhou

    — *B.l.woodi* Finn, 1902   SE Assam, W Burma
    — *B.l.latouchei* Stresemann, 1929   SE China

○ ***Babax waddelli*** GIANT BABAX
    — *B.w.waddelli* Dresser, 1905[2]   SE Xizang
    — *B.w.jomo* Vaurie, 1955   Jomo area (SC Xizang)

○ ***Babax koslowi*** TIBETAN BABAX
    — *B.k.koslowi* (Bianchi, 1906)   S Qinghai and NE Xizang
    — *B.k.yuquensis* Li & Wang, 1979   SE Xizang (1354)

GARRULAX Lesson, 1831 M
○ ***Garrulax cinereifrons*** Blyth, 1851 ASHY-HEADED LAUGHING-THRUSH   SW Sri Lanka

○ ***Garrulax palliatus*** SUNDA LAUGHING-THRUSH
    — *G.p.palliatus* (Bonaparte, 1850)   W Sumatra
    — *G.p.schistochlamys* Sharpe, 1888   N Borneo

○ ***Garrulax rufifrons*** RUFOUS-FRONTED LAUGHING-THRUSH
    — *G.r.rufifrons* Lesson, 1831   W Java
    — *G.r.slamatensis* Siebers, 1929   C and E Java

○ ***Garrulax perspicillatus*** (J.F. Gmelin, 1789) MASKED LAUGHING-THRUSH
                                     C and E China to N and C Vietnam

○ ***Garrulax albogularis*** WHITE-THROATED LAUGHING-THRUSH
    — *G.a.whistleri* E.C.S. Baker, 1921   NW Himalayas
    — *G.a.albogularis* (Gould, 1836)   C and E Himalayas
    — *G.a.eous* Riley, 1930   SE Xizang to E Qinghai, N Sichuan, S Shaanxi and
                                     Yunnan, NW Vietnam

    — *G.a.ruficeps* Gould, 1863   Taiwan

○ ***Garrulax leucolophus*** WHITE-CRESTED LAUGHING-THRUSH
    — *G.l.leucolophus* (Hardwicke, 1815)   Himalayas, SE Xizang
    — *G.l.patkaicus* Reichenow, 1913   S Assam, W Burma, W and SW Yunnan
    — *G.l.belangeri* Lesson, 1831[3]   C and SE Burma, SW Thailand
    — *G.l.diardi* (Lesson, 1831)   E Burma, Thailand (except SW), SE Yunnan, Indochina
    — *G.l.bicolor* Hartlaub, 1844   W Sumatra

○ ***Garrulax monileger*** LESSER NECKLACED LAUGHING-THRUSH
    — *G.m.monileger* (Hodgson, 1836)   E Himalayas, W and SW Yunnan, W and N Burma
    — *G.m.badius* Ripley, 1948   Mishmi hills (NE Assam)
    — *G.m.stuarti* Meyer de Schauensee, 1955   SC and E Burma, NW Thailand
    — *G.m.fuscatus* E.C.S. Baker, 1918   SE Burma, SW Thailand
    — *G.m.mouhoti* Sharpe, 1883   SE Thailand, S Indochina
    — *G.m.pasquieri* Delacour & Jabouille, 1924   C Vietnam
    — *G.m.schauenseei* Delacour & Greenway, 1939   N of E Burma, NE Thailand, N Laos, SE Yunnan
    — *G.m.tonkinensis* Delacour, 1927   Guangxi, N Vietnam
    — *G.m.melli* Stresemann, 1923   Guangdong to Anhui (SE China)
    — *G.m.schmackeri* Hartlaub, 1898   Hainan I.

○ ***Garrulax pectoralis*** GREATER NECKLACED LAUGHING-THRUSH
    — *G.p.pectoralis* (Gould, 1836)   C Nepal
    — *G.p.melanotis* Blyth, 1843   E Himalayas, Assam, N Burma
    — *G.p.pingi* Cheng, Tso-hsin, 1963   C Yunnan (330)
    — *G.p.subfusus* Kinnear, 1924   SE Burma, N and W Thailand, NW Laos

---

[1] For reasons to accept this see Traylor (1967) (2427).
[2] Includes *lumsdeni* see Vaurie (1955) (2497) and Ripley (1982: 349) (2055).
[3] Given as 1832 in Deignan (1964) (665), but see Browning & Monroe (1991) (285).

— *G.p.robini* Delacour, 1927   NE and C Laos, N and C Vietnam
— *G.p.picticollis* Swinhoe, 1872   Guangdong to Shaanxi and Anhui
— *G.p.semitorquatus* Ogilvie-Grant, 1900   Hainan I.

○ *Garrulax lugubris* BLACK LAUGHING-THRUSH
— *G.l.lugubris* (S. Müller, 1835)   W Malaysia, W Sumatra
— *G.l.calvus* (Sharpe, 1888)   NE Borneo

○ *Garrulax striatus* STRIATED LAUGHING-THRUSH
— *G.s.striatus* (Vigors, 1831)   NW Himalayas
— *G.s.vibex* Ripley, 1950   W and C Nepal
— *G.s.sikkimensis* (Ticehurst, 1924)   E Nepal to Bhutan
— *G.s.cranbrooki* (Kinnear, 1932)   Assam, SE Xizang, W, N and NE Burma

○ *Garrulax strepitans* WHITE-NECKED LAUGHING-THRUSH
— *G.s.strepitans* Blyth, 1855   E and SE Burma, W and N Thailand, S Yunnan, NW Laos
— *G.s.ferrarius* Riley, 1930[1]   SW Cambodia (SE Thailand?)

○ *Garrulax milleti* BLACK-HOODED LAUGHING-THRUSH
— *G.m.sweeti* Eames, 2002   C Vietnam (780A)
— *G.m.milleti* Robinson & Kloss, 1919   SC Vietnam, SE Laos

○ *Garrulax maesi* GREY LAUGHING-THRUSH
— *G.m.grahami* (Riley, 1922)   SE Xizang, SW Sichuan, NE Yunnan
— *G.m.maesi* (Oustalet, 1890)   Guangxi, N Vietnam
— *G.m.varennei* (Delacour, 1926)   NE and C Laos
— *G.m.castanotis* (Ogilvie-Grant, 1899)   Hainan I.

○ *Garrulax nuchalis* Godwin-Austen, 1876 CHESTNUT-BACKED LAUGHING-THRUSH
                                                                NE Assam, NW and NE Burma

○ *Garrulax chinensis* BLACK-THROATED LAUGHING-THRUSH
— *G.c.lochmius* Deignan, 1941   SW Yunnan, E Burma, N and E Thailand, NW Laos
— *G.c.propinquus* (Salvadori, 1915)   C and SE Burma, SW Thailand
— *G.c.germaini* (Oustalet, 1890)   S Vietnam, SE Cambodia
— *G.c.chinensis* (Scopoli, 1786)   S China, N Indochina (except NW Laos)
— *G.c.monachus* Swinhoe, 1870   Hainan I.

○ *Garrulax vassali* (Ogilvie-Grant, 1906) WHITE-CHEEKED LAUGHING-THRUSH
                                                                S Laos, C and S Vietnam, E Cambodia

○ *Garrulax galbanus* YELLOW-THROATED LAUGHING-THRUSH
— *G.g.galbanus* Godwin-Austen, 1874   S and E Assam, W Burma, E Bangladesh
— *G.g.courtoisi* Ménégaux, 1923   NE Jiangxi
— *G.g.simaoensis* Cheng & Tang, 1982   Simao (S Yunnan) (336)

○ *Garrulax delesserti* RUFOUS-VENTED LAUGHING-THRUSH[2]
— *G.d.gularis* (McClelland, 1840)   Bhutan, Assam, N Burma, N and C Laos
— *G.d.delesserti* (Jerdon, 1839)   SW India

○ *Garrulax variegatus* VARIEGATED LAUGHING-THRUSH
— *G.v.nuristani* Paludan, 1959   Afghanistan, NW Pakistan (1741)
— *G.v.similis* (Hume, 1871)   NC Pakistan
— *G.v.variegatus* (Vigors, 1831)   NW Himalayas (NE Pakistan to C Nepal)

○ *Garrulax davidi* PLAIN LAUGHING-THRUSH
— *G.d.chinganicus* (Meise, 1934)   N Hebei to NW Manchuria
— *G.d.davidi* (Swinhoe, 1868)   NE Qinghai and E Gansu to Shanxi and S Hebei
— *G.d.experrectus* (Bangs & Peters, 1928)   NW Gansu
— *G.d.concolor* (Stresemann, 1923)   NW Sichuan

○ *Garrulax sukatschewi* (Berezowski & Bianchi, 1891) SNOWY-CHEEKED LAUGHING-THRUSH
                                                                S Gansu, N Sichuan

---

[1] Treated as a separate species named Cambodian Laughing-thrush by Robson (2000) (2095). See also Round & Robson (2001) (2111), but evidence presented against conspecificity limited.
[2] If treated as two species use this name for *gularis* and call *delesserti* the Wynaad Laughing-thrush. This split is probably valid, but it has not been sufficiently defended since the two were lumped.

○ *Garrulax cineraceus* MOUSTACHED LAUGHING-THRUSH
　— *G.c.cineraceus* (Godwin-Austen, 1874)　　　S Assam, W Burma
　— *G.c.strenuus* Deignan, 1957　　　　　　　　E Burma, Yunnan, S Sichuan, W Guangxi
　— *G.c.cinereiceps* (Styan, 1887)　　　　　　　W Sichuan and Shanxi to E China

○ *Garrulax rufogularis* RUFOUS-CHINNED LAUGHING-THRUSH
　— *G.r.occidentalis* (E. Hartert, 1909)[1]　　　NW Himalayas (to W Nepal)
　— *G.r.rufogularis* (Gould, 1835)　　　　　　　C Nepal to N Assam
　— *G.r.assamensis* (E. Hartert, 1909)　　　　　SE Assam, E Bangladesh, W Burma
　— *G.r.rufitinctus* (Koelz, 1952)　　　　　　　Garo and Khasi hills (SW Assam)
　— *G.r.rufiberbis* (Koelz, 1954)　　　　　　　Patkai Range (SE Assam), N Burma
　— *G.r.intensior* Delacour & Jabouille, 1930　NW Vietnam

○ *Garrulax konkakinhensis* J.C. & C. Eames, 2001 CHESTNUT-EARED LAUGHING-THRUSH #
　　　　　　　　　　　　　　　　　　　　　　　Central Highlands, Vietnam (781)

○ *Garrulax lunulatus* BARRED LAUGHING-THRUSH
　— *G.l.lunulatus* (J. Verreaux, 1870)　　　　　S Gansu, S Shaanxi, W Sichuan, W Hubei
　— *G.l.liangshanensis* Li, Zhang & Zhang, 1979　SW Sichuan (1359)

○ *Garrulax bieti* (Oustalet, 1897) WHITE-SPECKLED LAUGHING-THRUSH　NW Yunnan, WSW Sichuan

○ *Garrulax maximus* (J. Verreaux, 1870) GIANT LAUGHING-THRUSH　SE and E Tibetan Plateau (SE and E Xizang and
　　　　　　　　　　　　　　　　　　　　　　　　　　　　E Qinghai to S Gansu and W Sichuan)

○ *Garrulax ocellatus* SPOTTED LAUGHING-THRUSH
　— *G.o.griseicauda* Koelz, 1950　　　　　　　W Himalayas (Kumaum to W Nepal)
　— *G.o.ocellatus* (Vigors, 1831)　　　　　　　E Himalayas (C Nepal to SE Xizang)
　— *G.o.maculipectus* Hachisuka, 1953　　　　W Yunnan, NE Burma
　— *G.o.artemisiae* (David, 1871)　　　　　　　Sichuan, W and N Guizhou, W Hubei

○ *Garrulax caerulatus* GREY-SIDED LAUGHING-THRUSH
　— *G.c.caerulatus* (Hodgson, 1836)　　　　　E Himalayas (C Nepal and SE Xizang to N and E Assam)
　— *G.c.subcaerulatus* Hume, 1878　　　　　　SW Assam
　— *G.c.livingstoni* Ripley, 1952　　　　　　　SE Assam, NW Burma
　— *G.c.kaurensis* (Rippon, 1901)　　　　　　　NC Burma
　— *G.c.latifrons* (Rothschild, 1926)　　　　　W Yunnan, NE Burma

○ *Garrulax poecilorhynchus* RUSTY LAUGHING-THRUSH[2]
　— *G.p.ricinus* (Riley, 1930)　　　　　　　　　NW Yunnan
　— *G.p.berthemyi* (Oustalet, 1876)　　　　　　NW Fujian, N Guangdong, Zhejiang, Hunan, SC Sichuan
　— *G.p.poecilorhynchus* Gould, 1863　　　　　Taiwan

○ *Garrulax mitratus* CHESTNUT-CAPPED LAUGHING-THRUSH
　— *G.m.major* (Robinson & Kloss, 1919)　　　W Malaysia
　— *G.m.mitratus* (S. Müller, 1835)　　　　　　W Sumatra
　— *G.m.damnatus* (Harrisson & Hartley, 1934)　E Sarawak
　— *G.m.griswoldi* (J.L. Peters, 1940)　　　　　C Borneo
　— *G.m.treacheri* (Sharpe, 1879)　　　　　　　Kinabalu (N Borneo)

○ *Garrulax ruficollis* (Jardine & Selby, 1838) RUFOUS-NECKED LAUGHING-THRUSH
　　　　　　　　　　　　　　　　　　　　　　　C and E Himalayas, Assam, W, N and NE Burma,
　　　　　　　　　　　　　　　　　　　　　　　　SW Yunnan

○ *Garrulax merulinus* SPOT-BREASTED LAUGHING-THRUSH
　— *G.m.merulinus* Blyth, 1851　　　　　　　　S and SE Assam, W Yunnan, W and N Burma, N Thailand
　— *G.m.obscurus* Delacour & Jabouille, 1930　SE Yunnan, NE Laos, NW and NC Vietnam
　— *G.m.annamensis* (Robinson & Kloss, 1919)[3]　Langbian (SC Vietnam)

● *Garrulax canorus* MELODIOUS LAUGHING-THRUSH/HWAMEI
　— *G.c.canorus* (Linnaeus, 1758)[4]　　　　　S and EC China, NE and C Laos, N and C Vietnam
　— *G.c.owstoni* (Rothschild, 1903)　　　　　　Hainan
　— *G.c.taewanus* Swinhoe, 1859　　　　　　　Taiwan

[1] Includes *grosvenori* see Ripley (1982: 354) (2055).
[2] For reasons to separate from *G. caerulatus* see Vaurie (1965) (2511).
[3] May be a separate species as suggested by Robson (2000) (2095). Full review needed.
[4] Includes *mengliensis* Cheng & Yang, 1980 (327), see Cheng (1987) (333).

○ *Garrulax sannio* White-browed Laughing-thrush
— *G.s.albosuperciliaris* Godwin-Austen, 1874 | SE Assam
— *G.s.comis* Deignan, 1952 | Yunnan, SE Xizang, NE and E Burma, N Indochina, N Thailand
— *G.s.sannio* Swinhoe, 1867 | NE Vietnam, SE China
— *G.s.oblectans* Deignan, 1952 | WC China (W Sichuan to Gansu, east to Guizhou and Hubei)

○ *Garrulax cachinnans* Jerdon, 1839 Nilgiri Laughing-thrush | Nilgiri Hills (SW India)

○ *Garrulax jerdoni* Grey-breasted Laughing-thrush
— *G.j.jerdoni* Blyth, 1851 | Coorg (WC India)
— *G.j.fairbanki* (Blanford, 1869) | Palni hills (S India)
— *G.j.meridionalis* (Blanford, 1880) | S Kerala (SW India)

○ *Garrulax lineatus* Streaked Laughing-thrush
— *G.l.bilkevitchi* (Zarudny, 1910) | S Tadzhikistan, Afghanistan, NW Pakistan
— *G.l.schachdarensis* Stepanyan, 1998 | Extreme SE Tajikistan (2330)
— *G.l.gilgit* (E. Hartert, 1909) | NW and N Pakistan (Chitral and Gilgit areas)
— *G.l.lineatus* (Vigors, 1831) | NW Himalayas (NE Pakistan to W Kumaun)
— *G.l.setafer* (Hodgson, 1836) | C Himalayas (Nepal to Sikkim and Darjeeling)
— *G.l.imbricatus* Blyth, 1843 | E Himalayas (Bhutan to NW Assam)

○ *Garrulax virgatus* (Godwin-Austen, 1874) Striped Laughing-thrush | SE Assam, W Burma

○ *Garrulax austeni* Brown-capped Laughing-thrush
— *G.a.austeni* (Godwin-Austen, 1870) | S Assam
— *G.a.victoriae* (Rippon, 1906) | Chin Hills (W Burma)

○ *Garrulax squamatus* (Gould, 1835) Blue-winged Laughing-thrush | C and E Himalayas, Assam, W, N and NE Burma, W and SE Yunnan, NW Vietnam

○ *Garrulax subunicolor* Scaly Laughing-thrush
— *G.s.subunicolor* (Blyth, 1843) | C and E Himalayas, NE Assam
— *G.s.griseatus* (Rothschild, 1921) | NE Burma, W Yunnan
— *G.s.fooksi* Delacour & Jabouille, 1930 | NW Vietnam, SE Yunnan

○ *Garrulax elliotii* Elliot's Laughing-thrush
— *G.e.prjevalskii* (Menzbier, 1887) | E Qinghai and Gansu to N Yunnan, Hubei and Guizhou
— *G.e.elliotii* (J. Verreaux, 1870) | EC Tibetan Plateau (E Xizang, S Qinghai)

○ *Garrulax henrici* Brown-cheeked Laughing-thrush
— *G.h.henrici* (Oustalet, 1892) | S Tibetan Plateau in EC and S Xizang
— *G.h.gucenensis* Li, Wang & Jiang, 1978[1] | Qamdo area, SE Xizang (1355)

○ *Garrulax affinis* Black-faced Laughing-thrush
— *G.a.affinis* Blyth, 1843 | W and C Nepal, S Xizang
— *G.a.bethelae* Rand & Fleming, 1956 | E Nepal to NW Assam
— *G.a.oustaleti* (E. Hartert, 1909) | NE Assam, SE Xizang, N Burma, W Yunnan
— *G.a.muliensis* Rand, 1953 | NC Yunnan, SW Sichuan
— *G.a.blythii* (J. Verreaux, 1870) | SE Gansu, NW and SC Sichuan
— *G.a.saturatus* Delacour & Jabouille, 1930 | NW Vietnam, SE Yunnan

○ *Garrulax morrisonianus* (Ogilvie-Grant, 1906) White-whiskered Laughing-thrush[2] | Taiwan

○ *Garrulax erythrocephalus* Chestnut-crowned Laughing-thrush
— *G.e.erythrocephalus* (Vigors, 1832) | NW Himalayas
— *G.e.kali* Vaurie, 1953 | C Himalayas (W and C Nepal)
— *G.e.nigrimentum* (Oates, 1889)[3] | E Himalayas (E Nepal to N and E Assam)
— *G.e.chrysopterus* (Gould, 1835) | SW Assam
— *G.e.godwini* (Harington, 1914) | N Cachar Hills (SE Assam)
— *G.e.erythrolaemus* (Hume, 1881)[4] | E Manipur, W and SW Burma

[1] Cheng (1987) (333) suggested this might be a hybrid. Review required.
[2] Cheng (1987) (333) followed Vaurie (1954) (2493) in separating this from *affinis*.
[3] Includes *imprudens* see Ripley (1982: 362) (2055).
[4] For correction of spelling see David & Gosselin (2002) (614).

__ *G.e.woodi* (E.C.S. Baker, 1914)	N and NE Burma, W Yunnan
__ *G.e.connectens* (Delacour, 1929)	NE and C Laos, SE Yunnan, NW Vietnam
__ *G.e.ngoclinhensis* Eames, Trai & Cu, 1999	Mt. Ngoc Linh (C Vietnam) (783) #
__ *G.e.subconnectens* Deignan, 1938	Doi Phu Kha (N Thailand)
__ *G.e.schistaceus* Deignan, 1938	E Burma, northernmost Thailand
__ *G.e.melanostigma* Blyth, 1855	E and SE Burma, S part of NW Thailand
__ *G.e.ramsayi* (Ogilvie-Grant, 1904)	Sthn. SE Burma
__ *G.e.peninsulae* (Sharpe, 1887)	Malay Pen.

○ *Garrulax yersini* (Robinson & Kloss, 1919) COLLARED LAUGHING-THRUSH

SC and S Vietnam

○ *Garrulax formosus* RED-WINGED LAUGHING-THRUSH

__ *G.f.formosus* (J. Verreaux, 1869)	SW Sichuan, NE Yunnan
__ *G.f.greenwayi* Delacour & Jabouille, 1930	Fansipan Mts. (NW Vietnam)

○ *Garrulax milnei* RED-TAILED LAUGHING-THRUSH

__ *G.m.sharpei* (Rippon, 1901)	N and E Burma, W and S Yunnan, NW Thailand, NW Vietnam, NE and C Laos
__ *G.m.vitryi* Delacour, 1932	S Laos and C Vietnam
__ *G.m.sinianus* (Stresemann, 1930)	Guizhou to Guangxi and N Guangdong
__ *G.m.milnei* (David, 1874)	NW Fujian

LIOCICHLA Swinhoe, 1877 F
○ *Liocichla phoenicea* RED-FACED LIOCICHLA

__ *L.p.phoenicea* (Gould, 1837)	E Himalayas (Nepal to NE Assam)
__ *L.p.bakeri* (E. Hartert, 1908)	S Assam, W, N and NE Burma, NW Yunnan
__ *L.p.ripponi* (Oates, 1900)	E Burma, NW Thailand, SW Yunnan
__ *L.p.wellsi* (La Touche, 1921)	SE Yunnan, N Indochina

○ *Liocichla omeiensis* Riley, 1926 EMEI SHAN LIOCICHLA

SC Sichuan and NE Yunnan

○ *Liocichla steerii* Swinhoe, 1877 STEERE'S LIOCICHLA

Mountains of S Taiwan

LEIOTHRIX Swainson, 1832 F
○ *Leiothrix argentauris* SILVER-EARED MESIA

__ *L.a.argentauris* (Hodgson, 1837)	C and E Himalayas, N Assam
__ *L.a.aureigularis* (Koelz, 1953)[1]	S Assam, W Burma
__ *L.a.vernayi* (Mayr & Greenway, 1938)	NE Assam, N Burma, W Yunnan
__ *L.a.galbana* (Mayr & Greenway, 1938)	E and SE Burma, NW Thailand, SW Yunnan
__ *L.a.ricketti* (La Touche, 1923)	SE Yunnan, N Indochina south to C Laos
__ *L.a.cunhaci* (Robinson & Kloss, 1919)	S Laos, C and S Vietnam and SE Cambodia (this race?)
__ *L.a.tahanensis* (Yen, 1934)	Malay Pen.
__ *L.a.rookmakeri* (Junge, 1948)	NW Sumatra
__ *L.a.laurinae* Salvadori, 1879	Sumatra (except NW)

● *Leiothrix lutea* RED-BILLED LEIOTHRIX[2]

__ *L.l.kumaiensis* Whistler, 1943	NW Himalayas
__ *L.l.calipyga* (Hodgson, 1837)[3]	C and E Himalayas, S Assam, Manipur, W and SW Burma, SE Xizang
__ *L.l.yunnanensis* Rothschild, 1921	NE Burma, W Yunnan
__ *L.l.kwangtungensis* Stresemann, 1923	SE China (SE Yunnan, Guanxi, Guangdong), NE Vietnam
__ *L.l.lutea* (Scopoli, 1786)	C and E China

CUTIA Hodgson, 1837 F
○ *Cutia nipalensis* CUTIA

__ *C.n.nipalensis* Hodgson, 1837	C and E Himalayas, Assam, SE Xizang, W and N Burma, W Yunnan, SW Sichuan
__ *C.n.melanchima* Deignan, 1947	E Burma, NW Thailand, N Indochina to C Laos, S Yunnan (S Laos and C Vietnam: this race?)

---

[1] Recognised by Ripley (1982: 363) (2055).
[2] The name *astleyi*, given to aviary birds of unknown origin, probably relates to hybrids.
[3] Includes *luteola* see Ripley (1982: 364) (2055).

— *C.n.cervinicrissa* Sharpe, 1888   Mts. of W Malaysia

— *C.n.hoae* Eames, 2002   C Vietnam (780A)

— *C.n.legalleni* Robinson & Kloss, 1919   SC Vietnam

PTERUTHIUS Swainson, 1832[1] M

○ *Pteruthius rufiventer* BLACK-HEADED SHRIKE-BABBLER

— *P.r.rufiventer* Blyth, 1842   C and E Himalayas, Assam, W and NE Burma, W Yunnan

— *P.r.delacouri* Mayr, 1941   Fansipan Mts. (NW Vietnam)

○ *Pteruthius flaviscapis* WHITE-BROWED SHRIKE-BABBLER

— *P.f.validirostris* Koelz, 1951   N Pakistan, Himalayas, Assam, W Burma

— *P.f.ricketti* Ogilvie-Grant, 1904   NE Burma, N Yunnan and W Sichuan to SE China,
                                 NE Thailand, N and C Indochina

— *P.f.aeralatus* Blyth, 1855[2]   E Burma, NW and W Thailand (and E and SE Thailand
                                 and SW Cambodia: this race?)

— *P.f.lingshuiensis* Cheng, Tso-hsin, 1963 Hainan (329)

— *P.f.annamensis* Robinson & Kloss, 1919 Langbian (SC Vietnam)

— *P.f.schauenseei* Deignan, 1946   N Pen. Thailand

— *P.f.cameranoi* Salvadori, 1879   W Malaysia, W Sumatra

— *P.f.flaviscapis* (Temminck, 1836[3])   Java

— *P.f.robinsoni* Chasen & Kloss, 1931   N Borneo

○ *Pteruthius xanthochlorus* GREEN SHRIKE-BABBLER

— *P.x.occidentalis* Harington, 1913   NW Himalayas to W Nepal

— *P.x.xanthochlorus* J.E. & G.R. Gray, 1846 E Himalayas (E Nepal to NW Assam)

— *P.x.hybrida* Harington, 1913[4]   SE Assam, W Burma

— *P.x.pallidus* (David, 1871)   NE Burma, C China (Yunnan to Gansu and S Shaanxi)

○ *Pteruthius melanotis* BLACK-EARED SHRIKE-BABBLER

— *P.m.melanotis* Hodgson, 1847   C and E Himalayas, SE Xizang, Burma, W and
                                 S Yunnan, N Thailand, N and C Indochina

— *P.m.tahanensis* E. Hartert, 1902   Mts. W Malaysia

○ *Pteruthius aenobarbus* CHESTNUT-FRONTED SHRIKE-BABBLER

— *P.a.aenobarbulus* Koelz, 1954   Garo hills (W Assam), W Burma?

— *P.a.intermedius* (Hume, 1877)   NE and E Burma, NW and W Thailand, W and
                                 S Yunnan, N Indochina to S Laos

— *P.a.yaoshanensis* Stresemann, 1929 Da Yao Shan (Guangxi)

— *P.a.indochinensis* Delacour, 1927   Langbian (SC Vietnam)

— *P.a.aenobarbus* (Temminck, 1836[5]) W Java

GAMPSORHYNCHUS Blyth, 1844 M

○ *Gampsorhynchus rufulus* WHITE-HOODED BABBLER

— *G.r.rufulus* Blyth, 1844   Himalayas to SW Burma and W Yunnan

— *G.r.torquatus* Hume, 1874   C and SE Burma, SE Yunnan, Thailand, Laos, Vietnam

— *G.r.saturatior* Sharpe, 1888   Mts. of W Malaysia

ACTINODURA Gould, 1836 F

○ *Actinodura egertoni* RUSTY-FRONTED BARWING

— *A.e.egertoni* Gould, 1836   C and E Himalayas (C Nepal to N Assam)

— *A.e.lewisi* Ripley, 1948   Mishmi hills (NE Assam)

— *A.e.khasiana* (Godwin-Austen, 1876) S Assam, NW Burma

— *A.e.ripponi* Ogilvie-Grant & La Touche, 1907 SW and NE Burma, W Yunnan

○ *Actinodura ramsayi* SPECTACLED BARWING

— *A.r.yunnanensis* Bangs & Phillips, 1914 SE Yunnan, S Guizhou, Guangxi, N Vietnam

— *A.r.radcliffei* Harington, 1910   N part E Burma, NE Laos (and NW and C Laos ?)

— *A.r.ramsayi* (Walden, 1875)   S part E Burma, NW Thailand

---

[1] Evidence now suggests this genus is not a timaliid one and will need to move (Cibois, 2000) (355).
[2] Omitted from last Edition, a *lapsus*.
[3] Livr. 99 must be dated from 1836 see Dickinson (2001) (739). Deignan (1964: 387) (665) gave 1835 following Sherborn (1898) (2242).
[4] For correction of spelling see David & Gosselin (2002) (613).
[5] Livr. 99 must be dated from 1836 see Dickinson (2001) (739). Deignan (1964: 389) (665) used 1835 following Sherborn (1898) (2242).

○ *Actinodura sodangorum* Eames, Trai, Cu & Eve, 1999 BLACK-CROWNED BARWING #
Mt. Ngoc Linh (C Vietnam) (784)

○ *Actinodura nipalensis* (Hodgson, 1836) HOARY-THROATED BARWING[1]   C Himalayas (Nepal to Bhutan)

○ *Actinodura waldeni* STREAK-THROATED BARWING
__ *A.w.daflaensis* (Godwin-Austen, 1875)   N and E Assam, SE Xizang
__ *A.w.waldeni* Godwin-Austen, 1874   SE Assam, NW Burma
__ *A.w.poliotis* (Rippon, 1905)   W Burma
__ *A.w.saturatior* (Rothschild, 1921)   Nthn. E Burma, W Yunnan

○ *Actinodura souliei* STREAKED BARWING
__ *A.s.souliei* Oustalet, 1897   NW Yunnan to SW Sichuan
__ *A.s.griseinucha* Delacour & Jabouille, 1930   SE Yunnan, NW Vietnam

○ *Actinodura morrisoniana* Ogilvie-Grant, 1906 TAIWAN BARWING   Taiwan

MINLA Hodgson, 1837 F
○ *Minla cyanouroptera* BLUE-WINGED MINLA
__ *M.c.cyanouroptera* (Hodgson, 1838)   C and E Himalayas, E Assam
__ *M.c.aglae* (Deignan, 1942)   SE Assam, W Burma
__ *M.c.wingatei* (Ogilvie-Grant, 1900)[2]   N part of E Burma, eastern N Thailand, S China to S Sichuan, Laos, N Vietnam, Hainan
__ *M.c.sordida* (Hume, 1877)   E and SE Burma, NW Thailand
__ *M.c.orientalis* (Robinson & Kloss, 1919)   Langbian (SC Indochina)
__ *M.c.rufodorsalis* (Engelbach, 1946)   SE Thailand, SW Cambodia
__ *M.c.sordidior* (Sharpe, 1888)   Malay Pen.

○ *Minla strigula* CHESTNUT-TAILED MINLA
__ *M.s.simlaensis* (R. Meinertzhagen, 1926)   NW Himalayas to W Nepal
__ *M.s.strigula* (Hodgson, 1837)   C Himalayas (C Nepal to Bhutan)
__ *M.s.cinereigenae* (Ripley, 1952)   Mt. Japvo (Naga Hills, SW Assam)
__ *M.s.yunnanensis* (Rothschild, 1921)   N and E Assam, SE Xizang, W and N Burma, Yunnan, SW Sichuan, N Indochina to C Laos
__ *M.s.traii* Eames, 2002   C Vietnam (780A)
__ *M.s.castanicauda* (Hume, 1877)   SE Burma, NW Thailand
__ *M.s.malayana* (E. Hartert, 1902)   Mts. W Malaysia

○ *Minla ignotincta* RED-TAILED MINLA
__ *M.i.ignotincta* Hodgson, 1837   C and E Himalayas, Assam, N and W Burma, N and W Yunnan
__ *M.i.mariae* La Touche, 1921   SE Yunnan, NW and NC Vietnam, C and S Laos
__ *M.i.sini* Stresemann, 1929   Da Yao Shan (C Guangxi)
__ *M.i.jerdoni* J. Verreaux, 1870   SW Sichuan, NE Yunnan, Guizhou, Hunan, N Guangxi

ALCIPPE Blyth, 1844 F
○ *Alcippe chrysotis* GOLDEN-BREASTED FULVETTA
__ *A.c.chrysotis* (Blyth, 1845)   C and E Himalayas, E Assam
__ *A.c.albilineata* (Koelz, 1954)   SE Assam
__ *A.c.forresti* (Rothschild, 1926)   NE Burma, NW Yunnan
__ *A.c.swinhoii* (J. Verreaux, 1870)   C and E China (W Sichuan and S Gansu to Guangxi and Guangdong)
__ *A.c.amoena* (Mayr, 1941)   SE Yunnan, NW Vietnam
__ *A.c.robsoni* Eames, 2002   C Vieynam (780A)

○ *Alcippe variegaticeps* Yen, 1932 GOLD-FRONTED FULVETTA   Mt. Omei and Dafending (Sichuan), Da Yao Shan (Guangxi)

○ *Alcippe cinerea* (Blyth, 1847) YELLOW-THROATED FULVETTA   E Himalayas, SE Xizang, Assam, E Bangladesh, NW Yunnan, N Burma, N Laos, NC Vietnam

○ *Alcippe castaneceps* RUFOUS-WINGED FULVETTA
__ *A.c.castaneceps* (Hodgson, 1837)   E Himalayas, SE Xizang, N, E and SE Assam, W, N and E Burma, NW and W Thailand, W Yunnan

---

[1] Includes *vinctura* see Ripley (1979) (2053) and Abdulali (1983) (16).
[2] Includes *croizati* see Cheng (1987) (333).

— *A.c.exul* Delacour, 1932	Eastern part N Thailand, S and E Yunnan, Laos, NW Vietnam
— *A.c.stepanyani* Eames, 2002	C Vietnam (780A)
— *A.c.klossi* Delacour & Jabouille, 1919	Langbian (S Vietnam)
— *A.c.soror* (Sharpe, 1887)	Mts. of W Malaysia

○ *Alcippe vinipectus* WHITE-BROWED FULVETTA
— *A.v.kangrae* (Ticehurst & Whistler, 1924)	NW Himalayas
— *A.v.vinipectus* (Hodgson, 1838)	C Himalayas (W and C Nepal)
— *A.v.chumbiensis* (Kinnear, 1939)	E Himalayas (E Nepal to Bhutan)
— *A.v.austeni* (Ogilvie-Grant, 1895)	SE and E Assam, NW and N Burma
— *A.v.ripponi* (Harington, 1913)	Chin hills (W Burma)
— *A.v.perstriata* (Mayr, 1941)	NE Burma, W Yunnan
— *A.v.valentinae* Delacour & Jabouille, 1930	Fansipan Mts. (N Vietnam)
— *A.v.bieti* Oustalet, 1891	NW Yunnan, S Sichuan

○ *Alcippe striaticollis* (J. Verreaux, 1870) CHINESE FULVETTA
	SE and E Tibetan Plateau (SE and E Xizang and SE Qinghai to Gansu, W Sichuan, N Yunnan)

○ *Alcippe ruficapilla* SPECTACLED FULVETTA[1]
— *A.r.ruficapilla* (J. Verreaux, 1870)	S Gansu and S Shaanxi to C Sichuan
— *A.r.sordidior* (Rippon, 1903)	SW, C and NE Yunnan to S Sichuan
— *A.r.danisi* Delacour & Greenway, 1941	SE Yunnan, NE and C Laos, NW Vietnam, W Guangxi
— *A.r.bidoupensis* Eames, Robson & Cu, 1994	SE Laos, SC and S Vietnam (782)

○ *Alcippe ludlowi* (Kinnear, 1935) BROWN-THROATED FULVETTA[2]
	SE Xizang, E Bhutan, N and E Assam, NW Burma

○ *Alcippe cinereiceps* STREAK-THROATED FULVETTA
— *A.c.manipurensis* (Ogilvie-Grant, 1906)	S Assam, W, N and NE Burma, W Yunnan
— *A.c.tonkinensis* Delacour & Jabouille, 1930	SE Yunnan, NW Vietnam
— *A.c.guttaticollis* (La Touche, 1897)	Fujian, N Guangdong
— *A.c.formosana* (Ogilvie-Grant, 1906)	Taiwan
— *A.c.fucata* (Styan, 1899)	C Hubei to Hunan (C China)
— *A.c.cinereiceps* (J. Verreaux, 1870)	W Hubei to SW Sichuan and Guizhou (EC China)
— *A.c.fessa* (Bangs & Peters, 1928)	Gansu, S Shaanxi, NE Qinghai

○ *Alcippe rufogularis* RUFOUS-THROATED FULVETTA
— *A.r.rufogularis* (Mandelli, 1873)	E Himalayas, N Assam
— *A.r.collaris* Walden, 1874	E and S Assam, E Bangladesh
— *A.r.major* (E.C.S. Baker, 1920)	N and E Burma, N and E Thailand, NW and C Laos
— *A.r.stevensi* (Kinnear, 1924)	NE Laos, NW and NC Vietnam, SE Yunnan
— *A.r.kelleyi* (Bangs & Van Tyne, 1930)	C Vietnam
— *A.r.khmerensis* (Meyer de Schauensee, 1938)	SE Thailand, SW Cambodia

○ *Alcippe dubia* RUSTY-CAPPED FULVETTA[3]
— *A.d.mandellii* (Godwin-Austen, 1876)	E Himalayas, Assam, W Burma
— *A.d.intermedia* (Rippon, 1900)	NE and E Burma, W Yunnan
— *A.d.genestieri* Oustalet, 1897	C and SE Yunnan to Hunan and W Guangdong, N and C Indochina
— *A.d.cui* Eames, 2002	C Vietnam (780A)
— *A.d.dubia* (Hume, 1874)	SE Burma

○ *Alcippe brunnea* DUSKY FULVETTA
— *A.b.olivacea* Styan, 1896	SE Sichuan and S Shaanxi to Hubei and Guizhou
— *A.b.weigoldi* Stresemann, 1923[4]	Red Basin of Sichuan
— *A.b.superciliaris* (David, 1874)	E and SE China
— *A.b.brunnea* Gould, 1863	Taiwan
— *A.b.arguta* (E. Hartert, 1910)	Hainan I.

---

[1] Robson (2000) (2095) suggested the two southern populations constitute a separate species. Deatialed argument remains to be published.
[2] For treatment as a species see Ripley *et al.* (1991) (2076).
[3] The separation of *dubia* from *brunnea*, not accepted in last Edition, deserves better explanation, but has been consistently used by Cheng in his various Check-lists.
[4] For recognition see Cheng (1987) (333).

○ *Alcippe brunneicauda* BROWN FULVETTA
  __ *A.b.brunneicauda* (Salvadori, 1879)   Malay Pen., Sumatra, NW Borneo, N Natunas
  __ *A.b.eriphaea* (Oberholser, 1922)   Borneo (excl. NW)

○ *Alcippe poioicephala* BROWN-CHEEKED FULVETTA
  __ *A.p.poioicephala* (Jerdon, 1844)   W Ghats (India)
  __ *A.p.brucei* Hume, 1870   C and S India
  __ *A.p.fusca* Godwin-Austen, 1876   S Assam, NW Burma
  __ *A.p.phayrei* Blyth, 1845   W and SW Burma
  __ *A.p.haringtoniae* E. Hartert, 1909   NE Burma to NW Thailand
  __ *A.p.alearis* (Bangs & Van Tyne, 1930)   N and E Thailand, SE Yunnan, N and C Laos, N Vietnam.
  __ *A.p.grotei* Delacour, 1926[1]   C Vietnam, S Laos
  __ *A.p.karenni* Robinson & Kloss, 1923   SE Burma, SW Thailand
  __ *A.p.davisoni* Harington, 1915   S Thailand, Mergui Arch.

○ *Alcippe pyrrhoptera* (Bonaparte, 1850) JAVAN FULVETTA   W and C Java

○ *Alcippe peracensis* MOUNTAIN FULVETTA
  __ *A.p.annamensis* Robinson & Kloss, 1919   Bolovens (S Laos), S Vietnam
  __ *A.p.eremita* Riley, 1936[2]   SE Thailand
  __ *A.p.peracensis* Sharpe, 1887   Mts. W Malaysia

○ *Alcippe morrisonia* GREY-CHEEKED FULVETTA
  __ *A.m.yunnanensis* Harington, 1913   NW Yunnan, SW Sichuan, NE Burma
  __ *A.m.fratercula* Rippon, 1900[3]   SW Yunnan, E and SE Burma, N Thailand, N and C Laos
  __ *A.m.schaefferi* La Touche, 1923   SE Yunnan, NW Vietnam
  __ *A.m.davidi* Styan, 1871   W Sichuan to S Shaanxi, Hubei, Hunan and N Guangxi
  __ *A.m.hueti* David, 1874   NE Guangdong to Anhui and Zhejiang (SE China)
  __ *A.m.rufescentior* (E. Hartert, 1910)   Hainan I.
  __ *A.m.morrisonia* Swinhoe, 1863   Taiwan

○ *Alcippe nipalensis* NEPAL FULVETTA
  __ *A.n.nipalensis* (Hodgson, 1838)[4]   C and E Himalayas, Assam, NE Burma
  __ *A.n.stanfordi* Ticehurst, 1930   E Bangladesh, SW Burma

LIOPTILUS Bonaparte, 1850[5] M
○ *Lioptilus nigricapillus* (Vieillot, 1818) BUSH BLACKCAP   E Cape Province, Natal, N Transvaal

KUPEORNIS Serle, 1949 M
○ *Kupeornis gilberti* Serle, 1949 WHITE-THROATED MOUNTAIN BABBLER   Mt. Kupe (Cameroon)

○ *Kupeornis rufocinctus* (Rothschild, 1908) RED-COLLARED BABBLER   E Zaire

○ *Kupeornis chapini* CHAPIN'S BABBLER
  __ *K.c.chapini* Schouteden, 1949   Ituri river (E Zaire)
  __ *K.c.nyombensis* (Prigogine, 1960)   Mt. Nyombe (E Zaire)
  __ *K.c.kalindei* (Prigogine, 1964)   E Zaire (1929)

PAROPHASMA Reichenow, 1905 N
○ *Parophasma galinieri* (Guérin-Méneville, 1843) ABYSSINIAN CATBIRD   C and S Ethiopia

PHYLLANTHUS Lesson, 1844 M
○ *Phyllanthus atripennis* CAPUCHIN BABBLER
  __ *P.a.atripennis* (Swainson, 1837)   Senegal to Liberia
  __ *P.a.rubiginosus* (Blyth, 1865)   Ivory Coast to S Nigeria
  __ *P.a.bohndorffi* (Sharpe, 1884)   NE Zaire to W Uganda

---

[1] Included in *peracensis* in Deignan (1964) (665) and in our last Edition. Provisionally treated here as in Inskipp *et al.* (1996) (1182). Robson (2000) (2095) treated it as a species and included *eremita* from *A. peracensis*. His detailed findings, which are important, require publication and will then deserve following.
[2] Robson (2000) (2095) treated this as a form of *grotei* (see above).
[3] For correction of spelling see David & Gosselin (2002) (613).
[4] Includes *commoda* see Ripley (1982: 379) (2055).
[5] Deignan (1964: 413) (665) gave the date as 1851, but see Zimmer (1926) (2707).

CROCIAS Temminck, 1836 M

○ *Crocias langbianis* Gyldenstolpe, 1939 Grey-crowned Crocias — Langbian (SC Vietnam)

○ *Crocias albonotatus* (Lesson, 1831)[1] Spotted Crocias — Mts. of W and C Java

HETEROPHASIA Blyth, 1842 F

○ *Heterophasia annectans* Rufous-backed Sibia[2]

— *H.a.annectans* (Blyth, 1847) — E Himalayas, Assam, W, N and NE Burma, SW Yunnan

— *H.a.mixta* Deignan, 1948 — Extreme E Burma, SE Yunnan, E part of NW Thailand, N and C Indochina

— *H.a.saturata* (Walden, 1875) — E and SE Burma, NW and W Thailand

— *H.a.roundi* Eames, 2002 — C Vietnam (780A)

— *H.a.eximia* (Riley, 1940) — SC Vietnam

○ *Heterophasia capistrata* Rufous Sibia

— *H.c.capistrata* (Vigors, 1831) — NW Himalayas

— *H.c.nigriceps* (Hodgson, 1839) — C Himalayas (W and C Nepal)

— *H.c.bayleyi* (Kinnear, 1939) — E Himalayas (E Nepal to NW Assam)

○ *Heterophasia gracilis* (McClelland, 1840) Grey Sibia — S and SE Assam, W, NW and NE Burma, SW Yunnan

○ *Heterophasia melanoleuca* Black-headed Sibia[3]

— *H.m.desgodinsi* (Oustalet, 1877) — NE Burma, SC China (Yunnan to SW Sichuan and W Guangxi)

— *H.m.castanoptera* (Salvadori, 1889) — West part E Burma

— *H.m.tonkinensis* (Yen, 1934) — NW and NC Vietnam

— *H.m.melanoleuca* (Blyth, 1859) — East part E Burma, NW and W Thailand, NW Laos

— *H.m.engelbachi* (Delacour, 1930) — Bolovens (SE Laos)

— *H.m.kingi* Eames, 2002 — Kon Tum (C Vietnam) (780A)

— *H.m.robinsoni* (Rothschild, 1921) — Langbian (SC Vietnam)

○ *Heterophasia auricularis* (Swinhoe, 1864) White-eared Sibia — Taiwan

○ *Heterophasia pulchella* (Godwin-Austen, 1874) Beautiful Sibia — SE Xizang, Assam, NE Burma, W Yunnan

○ *Heterophasia picaoides* Long-tailed Sibia

— *H.p.picaoides* (Hodgson, 1839) — C and E Himalayas, NE Burma

— *H.p.cana* (Riley, 1929) — E and SE Burma, N and W Thailand, N Indochina (south to C Laos)

— *H.p.wrayi* (Ogilvie-Grant, 1910) — Mts. of W Malaysia

— *H.p.simillima* Salvadori, 1879 — W Sumatra

YUHINA Hodgson, 1836 F

○ *Yuhina castaniceps* Striated Yuhina

— *Y.c.rufigenis* (Hume, 1877) — E Himalayas

— *Y.c.plumbeiceps* (Godwin-Austen, 1877) — E Assam, NW and NE Burma, W Yunnan

— *Y.c.castaniceps* (F. Moore, 1854) — S Assam, W and SW Burma

— *Y.c.striata* (Blyth, 1859) — E and SE Burma, NW Thailand

— *Y.c.torqueola* (Swinhoe, 1870) — S Sichuan, C Yunnan and E part of N Thailand to E China and north and C Indochina, N to C Yunnan and S Sichuan

— *Y.c.everetti* (Sharpe, 1887)[4] — N Borneo

○ *Yuhina bakeri* Rothschild, 1926 White-naped Yuhina — E Himalayas, NW Yunnan, Assam, N Burma

○ *Yuhina flavicollis* Whiskered Yuhina

— *Y.f.albicollis* (Ticehurst & Whistler, 1924) — NW Himalayas to W Nepal

— *Y.f.flavicollis* Hodgson, 1836 — E Himalayas, SE Xizang

[1] Given as 1832 in Deignan (1964) (665), but see Browning & Monroe (1991) (285).
[2] In last Edition treated in the genus *Minla* following Harrison (1986) (1074); not accepted here, molecular evidence awaited. The spelling *annectens* used in Peters Check-list and in our last Edition is now considered an unjustified emendation.
[3] The *desgodinsi* group of forms may well be a separate species as Robson (2000) (2095) treated them. If so call it Black-headed and *melanoleuca* Dark-backed Sibia. However detailed substantiation is needed.
[4] Very distinct. May well be a separate species as treated by Cranbrook in Smythies (1981) (2288). If so call it Chestnut-crested Yuhina.

— *Y.f.rouxi* (Oustalet, 1896)   NE and S Assam, W, N and NE Burma, Yunnan, N Laos, NW Vietnam

— *Y.f.clarki* (Oates, 1894)[1]   E Burma
— *Y.f.humilis* (Hume, 1877)   SE Burma, W Thailand
— *Y.f.rogersi* Deignan, 1937   Northernmost Thailand
— *Y.f.constantiae* Ripley, 1953   NW and C Laos

○ *Yuhina gularis* STRIPE-THROATED YUHINA
— *Y.g.vivax* Koelz, 1954   WC Himalayas (Kumaon area)
— *Y.g.gularis* Hodgson, 1836   C and E Himalayas, Assam, SE Xizang, W, C and SE Yunnan, W and NE Burma
— *Y.g.omeiensis* Riley, 1930   NE Yunnan, SW and C Sichuan
— *Y.g.uthaii* Eames, 2002   C Vietnam (780A)

○ *Yuhina diademata* WHITE-COLLARED YUHINA
— *Y.d.ampelina* Rippon, 1900[2]   NE Burma, Yunnan, NW Vietnam
— *Y.d.diademata* J. Verreaux, 1869   Sichuan to S Gansu, Hubei, Guizhou

○ *Yuhina occipitalis* RUFOUS-VENTED YUHINA
— *Y.o.occipitalis* Hodgson, 1836   C and E Himalayas, SE Xizang
— *Y.o.obscurior* Rothschild, 1921   NE Burma, Yunnan (and ? SW Sichuan)

○ *Yuhina brunneiceps* Ogilvie-Grant, 1906 TAIWAN YUHINA   Taiwan

○ *Yuhina nigrimenta* BLACK-CHINNED YUHINA
— *Y.n.nigrimenta* Blyth, 1845   C and E Himalayas, Assam, E Bangladesh
— *Y.n.intermedia* Rothschild, 1922   NE Burma, SE Xizang and SW Sichuan to Hubei, Hunan and Guizhou, N Indochina to C Laos and E Cambodia
— *Y.n.pallida* La Touche, 1897   Guangdong, Fujian (SE China)

ERPORNIS Hodgson, 1844[3] F
○ *Erpornis zantholeuca* WHITE-BELLIED YUHINA
— *E.z.zantholeuca* Blyth, 1844   C and E Himalayas, Assam, W Yunnan, Burma, W and SW Thailand
— *E.z.tyrannulus* (Swinhoe, 1870)   East part of N Thailand and along the Dong Phraya Fai range, N and C Indochina, S Yunnan, Hainan I.
— *E.z.griseiloris* (Stresemann, 1923)   SE Yunnan and Guizhou to SE China, Taiwan
— *E.z.sordida* (Robinson & Kloss, 1919)   E Thailand, S Indochina
— *E.z.canescens* Delacour & Jabouille, 1928   SE Thailand, W Cambodia
— *E.z.interposita* (E. Hartert, 1917)   Mts. of W Malaysia
— *E.z.saani* Chasen, 1939   NW Sumatra
— *E.z.brunnescens* (Sharpe, 1876)   Borneo

PANURUS Koch, 1816 M
○ *Panurus biarmicus* BEARDED TIT
— *P.b.biarmicus* (Linnaeus, 1758)[4]   W, WC and S Europe to W Poland, Italy, Greece and W Asia Minor
— †?*P.b.kosswigi* Kumerloeve, 1958   SC Turkey (1305B)
— *P.b.russicus* (C.L. Brehm, 1831)   EC and E Europe to E Asia, south to C Turkey, NE Iran, Xinjiang and Qaidam Basin

CONOSTOMA Hodgson, 1842 N
○ *Conostoma oemodium* Hodgson, 1842 GREAT PARROTBILL   C Himalayas, NE Burma, C China (N Yunnan to SW Gansu and S Shaanxi)

PARADOXORNIS Gould, 1836 M
○ *Paradoxornis paradoxus* THREE-TOED PARROTBILL
— *P.p.paradoxus* (J. Verreaux, 1870)   S Gansu to SW and E Sichuan
— *P.p.taipaiensis* Cheng, Lo & Chao, 1973   Qin Ling Mts. (S Shaanxi) (341)

---

[1] This was treated with the next form as a species *Y. humilis* in King & Dickinson (1975) (1268). This may well be correct, but a thorough study remains to be done. If accepted usually called Burmese (but not endemic).
[2] Recognised following Traylor (1967) (2427).
[3] Treated in the genus *Yuhina* by Deignan (1964: 426) (665), but in *Stachyris* in our last Edition following Harrison (1986) (1074). We do not now accept that and prefer to reintroduce this generic name. Molecular evidence supports its separation, but it may not be a babbler (Cibois *et al.*, 2002) (359).
[4] Includes *occidentalis* see Cramp *et al.* (1993: 100) (594).

◯ *Paradoxornis unicolor* (Hodgson, 1843) Brown Parrotbill   E Himalayas, E Assam, NE Burma, W and N Yunnan to
SW Sichuan

◯ *Paradoxornis flavirostris* Gould, 1836 Black-breasted Parrotbill[1] E Himalayas, N part of S Assam

◯ *Paradoxornis guttaticollis* David, 1871 Spot-breasted Parrotbill[2] S Assam, N and E Burma, W and NW Yunnan, W Sichuan
to S Shaanxi, N Guangdong and Fujian, N Indochina

◯ *Paradoxornis conspicillatus* Spectacled Parrotbill
— *P.c.conspicillatus* (David, 1871)   NE Qinghai, SW Gansu, S Shaanxi, Sichuan
— *P.c.rocki* (Bangs & Peters, 1928)   W and SW Hubei

◯ *Paradoxornis webbianus* Vinous-throated Parrotbill[3]
— *P.w.mantschuricus* (Taczanowski, 1885)   NE China, S Ussuriland
— *P.w.fulvicauda* (C.W. Campbell, 1892)   NE Hebei, N Henan, Korea
— *P.w.webbianus* (Gould, 1852)   Coast of S Jiangsu, N Zhejiang
— *P.w.suffusus* (Swinhoe, 1871)   C China (S Gansu and S Shanxi to Guangdong and
Fujian)
— *P.w.ganluoensis* Li & Zhang, 1980   Ganluao (SW Sichuan) (1358)
— *P.w.bulomachus* (Swinhoe, 1866)   Taiwan
— *P.w.alphonsianus* (J. Verreaux, 1870)   W Sichuan
— *P.w stresemanni* Yen, 1934   Guizhou
— *P.w.yunnanensis* (La Touche, 1921)   S and SE Yunnan, NW Vietnam

◯ *Paradoxornis brunneus* Brown-winged Parrotbill[4]
— *P.b.brunneus* (Anderson, 1871)   NE Burma, W Yunnan
— *P.b.ricketti* Rothschild, 1922   SW Sichuan and N Yunnan (east of Yunling Mts.)
— *P.b.styani* Rippon, 1903   N Yunnan (west of Yunling Mts.)

◯ *Paradoxornis zappeyi* Grey-hooded Parrotbill
— *P.z.zappeyi* (Thayer & Bangs, 1912)   S Sichuan (Ebian, Ganluo, Washan and Mt. Emei),
NE Yunnan (Weining)
— *P.z.erlangshanicus* Cheng, Li & Zhang, 1983   SW Sichuan (Erlang Shan) (344)

◯ *Paradoxornis przewalskii* (Berezowski & Bianchi, 1891) Rusty-throated Parrotbill
SW and S Gansu

◯ *Paradoxornis fulvifrons* Fulvous-fronted Parrotbill
— *P.f.fulvifrons* (Hodgson, 1845)   C Himalayas
— *P.f.chayulensis* (Kinnear, 1940)   N Assam, SE Xizang
— *P.f.albifacies* (Mayr & Birckhead, 1937)   NE Burma, W Yunnan, SW Sichuan
— *P.f.cyanophrys* (David, 1874)   W Sichuan, SW Shaanxi

◯ *Paradoxornis nipalensis* Black-throated Parrotbill
— *P.n.garhwalensis* Fleming & Traylor, 1964   WC Himlayas (854)
— *P.n.nipalensis* (Hodgson, 1837)   W and C Nepal
— *P.n.humii* (Sharpe, 1883)   E Nepal, Sikkim, W Bhutan
— *P.n.crocotius* Kinnear, 1954   SE Xizang, E Bhutan
— *P.n.poliotis* (Blyth, 1851)   Assam, NE Burma, NW Yunnan
— *P.n.patriciae* (Koelz, 1954)   SE Assam
— *P.n.ripponi* (Sharpe, 1905)   Mt. Victoria (W Burma)
— *P.n.feae* (Salvadori, 1889)   E Burma, NW Thailand
— *P.n.craddocki* (Bingham, 1903)   Extreme NE Burma, NW Vietnam, Davao Shan
(Guangxi), Guizhou?
— *P.n.beaulieu* Ripley, 1953   N and C Laos (S Laos ?)
— *P.n.kamoli* Eames, 2002   C Vietnam (780A)

◯ *Paradoxornis verreauxi* Golden Parrotbill[5]
— *P.v.verreauxi* (Sharpe, 1883)   NE Yunnan, Sichuan, S Shaanxi, Hubei

---

[1] Birds from W Burma assigned here by Deignan (1964) (667) have been reassigned to *guttaticollis* by Robson (2000) (2095), but no explanatory note on this has been traced; specimens examined suggest Robson is right and we follow.

[2] Treated as conspecific with *flavirostris* by Cheng (1987) (333). Old reports of altitudinal replacement in the hills of S Assam need validation (taking potential winter wandering into account; no specimens of *flavirostris* in Tring date from May to Nov.). Races require review: *gongshanensis* Cheng, 1984 (332) needs comparison with birds from Assam, Burma, N Thailand and N Indochina. Meanwhile we prefer to recognise no races. Cheng's new race appeared again, as if it was a first description, in Cheng *et al.* (1987) (345).

[3] This arrangement, validating the separation of *brunneus*, but making *alphonsianus* a form of this species, follows the large, careful study of Han (1991) (1067), and maps in earlier publications were corrected by Cheng (1993) (334), who endorsed Han's findings.

[4] Despite Han's study, breeding sympatry with *P. webbianus* requires more evidence.

[5] We tentatively follow Vaurie (1959) (2502) in separating *verreauxi* from *nipalensis*.

__ *P.v.pallidus* (La Touche, 1922)	NW Fujian
__ *P.v.morrisonianus* (Ogilvie-Grant, 1906)	Taiwan

○ *Paradoxornis davidianus* SHORT-TAILED PARROTBILL

__ *P.d.davidianus* (Slater, 1897)	Fujian, S Hunan, ? N Guangdong
__ *P.d.tonkinensis* (Delacour, 1927)	NW Vietnam
__ *P.d.thompsoni* (Bingham, 1903)	E Burma, NW Laos, east part of N Thailand

○ *Paradoxornis atrosuperciliaris* LESSER RUFOUS-HEADED PARROTBILL

__ *P.a.oatesi* (Sharpe, 1903)	E Himalayas
__ *P.a.atrosuperciliaris* (Godwin-Austen, 1877)	S and E Assam, N Burma, N Laos, W Yunnan, extreme N Thailand

○ *Paradoxornis ruficeps* GREATER RUFOUS-HEADED PARROTBILL

__ *P.r.ruficeps* Blyth, 1842	E Himalayas
__ *P.r.bakeri* (E. Hartert, 1900)	S and E Assam, SE Xizang, W Yunnan, N and E Burma, N Laos
__ *P.r.magnirostris* (Delacour, 1927)	N Vietnam

○ *Paradoxornis gularis* GREY-HEADED PARROTBILL

__ *P.g.gularis* G.R. Gray, 1845	E Himalayas, N and NE Assam
__ *P.g.transfluvialis* (E. Hartert, 1900)	S and SE Assam, E Bangladesh, N and E Burma, SW Yunnan, NW Thailand
__ *P.g.rasus* (Stresemann, 1940)	Chin hills (W Burma)
__ *P.g.laotianus* (Delacour, 1926)	Extreme E Burma, east part of N Thailand, N and C Indochina to S Laos
__ *P.g.fokiensis* (David, 1874)	SW Sichuan to SE China
__ *P.g.hainanus* (Rothschild, 1903)	Hainan I.
__ *P.g.margaritae* (Delacour, 1927)	S Vietnam

○ *Paradoxornis polivanovi* NORTHERN PARROTBILL[1]

__ *P.p.mongolicus* Stepanyan, 1979	NE Mongolia and NW Manchuria (2322)
__ *P.p.polivanovi* Stepanyan, 1974	Lake Khanka (NE China and S Russian Far East) (2320)

○ *Paradoxornis heudei* David, 1872 REED PARROTBILL   E and NE China from lower Yangtze R. to N Hebei and S Liaoning

GENERA INCERTAE SEDIS (5: 7)

MYZORNIS Blyth, 1843 M
○ *Myzornis pyrrhoura* Blyth, 1843 FIRE-TAILED MYZORNIS   C and E Himalayas, E Assam, W Yunnan, N and NE Burma

MALIA Schlegel, 1880[2] F
○ *Malia grata* MALIA

__ *M.g.recondita* Meyer & Wiglesworth, 1894	N Sulawesi
__ *M.g.stresemanni* Meise, 1931	C and SE Sulawesi
__ *M.g.grata* Schlegel, 1880	SW Sulawesi

HORIZORHINUS Oberholser, 1899 M
○ *Horizorhinus dohrni* (Hartlaub, 1866) DOHRN'S THRUSH-BABBLER   Principe I.

CHAETOPS Swainson, 1832[3] M
○ *Chaetops pycnopygius* ROCKRUNNER

__ *C.p.pycnopygius* (P.L. Sclater, 1852)	S Angola, N Namibia
__ *C.p.spadix* (Clancey, 1972)	Huila (Angola)

○ *Chaetops frenatus* ROCKJUMPER[4]

__ *C.f.frenatus* (Temminck, 1826)	W Cape Province
__ *C.f.aurantius* E.L. Layard, 1867	Natal, E Cape Province

---

[1] Substantially larger than *P. heudei*; for split see Stepanyan (1998) (2331).
[2] The affinities of this genus remain unclear; we prefer to retain it near the babblers.
[3] For reasons to submerge the monotypic genus *Achaetops* in *Chaetops* see Olson (1998) (1705).
[4] In last Edition we had this split based on Clancey (1980) (482); we now re-unite the forms following Fry (2000: 10) (892).

MODULATRIX Ripley, 1952[1] F
○ **Modulatrix stictigula** SPOT-THROAT
    — M.s.stictigula (Reichenow, 1906)   Usambara and Nguru Mts. (NE Tanzania)
    — M.s.pressa (Bangs & Loveridge, 1931)   E and S Tanzania, N Malawi

○ **Modulatrix orostruthus** DAPPLED MOUNTAIN ROBIN[2]
    — M.o.amani (Sclater & Moreau, 1935)   E Usambara Mts. (NE Tanzania)
    — M.o.orostruthus (Vincent, 1933)   Mt. Namuli (N Mozambique)
    — M.o.sanjei Jensen & Stuart, 1982   Uzungwa Mts. (Tanzania) (1207)

ZOSTEROPIDAE WHITE EYES (14: 95)

ZOSTEROPS Vigors & Horsfield, 1826 M[3]
○ **Zosterops erythropleurus** Swinhoe, 1863 CHESTNUT-FLANKED WHITE-EYE   Russian Far East, NE China, N Korea, ?C China >>
                                                    S China and mainland SE Asia

● **Zosterops japonicus** JAPANESE WHITE-EYE
    — Z.j.yesoensis N. Kuroda, Jr., 1951   S Sakhalin, Hokkaido
    — Z.j.japonicus Temminck & Schlegel, 1845[4] Honshu, Kyushu, Shikoku, Tsushima
    — Z.j.stejnegeri Seebohm, 1891   Izu Is., Ogasawara-shoto (Bonin) Is. (introd.)
    — Z.j.alani E. Hartert, 1905   Ogasawara-shoto (Bonin) Is., Iwo (Volcano) Is. ?
    — Z.j.insularis Ogawa, 1905   Tanegashima, Yakushima Is. (off S Kyushu)
    — Z.j.loochooensis Tristram, 1889   Ryukyu Is.
    — Z.j.daitoensis N. Kuroda, Sr., 1923   Borodino Is.
    — Z.j.simplex Swinhoe, 1861   China, N Vietnam, Taiwan >> northern mainland
                                    SE Asia
    — Z.j.hainanus E. Hartert, 1923   Hainan

○ **Zosterops meyeni** LOWLAND WHITE-EYE
    — Z.m.batanis McGregor, 1907   Lanyu and Kashoto Is. (off S Taiwan), Batan Is. (N of
                                    Babuyan Is., Philippines)
    — Z.m.meyeni Bonaparte, 1850   Luzon and Babuyan Is. and small islands towards
                                    Mindoro

○ **Zosterops palpebrosus** ORIENTAL WHITE-EYE
    — Z.p.palpebrosus (Temminck, 1824)[5]   NE Afghanistan, Pakistan, N and C India, Bangladesh,
                                    W Assam, Nepal, Sikkim, Bhutan to SC China and
                                    N, C and E Burma
    — Z.p.nilgiriensis Ticehurst, 1927   Nilgiri and Palni Hills (SW India)
    — Z.p.salimalii Whistler, 1933   SE India
    — Z.p.egregius Madarász, 1911[6]   Sri Lanka
    — Z.p.siamensis Blyth, 1867   E Burma, N, E and SE Thailand, Indochina
    — Z.p.nicobaricus Blyth, 1845   Andamans, Nicobars
    — Z.p.williamsoni Robinson & Kloss, 1919 East coast of Thailand from the Gulf down the E of
                                      the Malay Pen.
    — Z.p.auriventer Hume, 1878   SE Burma, W coast of Malay Pen., coastal Sumatra,
                                    Bangka, coastal Borneo (Kuching area only?)
    — Z.p.buxtoni Nicholson, 1879   Submontane Sumatra, N tip of W Java
    — Z.p.melanurus Hartlaub, 1865   Java (except NW tip), Bali
    — Z.p.unicus E. Hartert, 1897   Sumbawa, Flores (Lesser Sundas)

○ **Zosterops ceylonensis** Holdsworth, 1872 SRI LANKA WHITE-EYE   Sri Lanka

○ **Zosterops rotensis** Taka-Tsukasa & Yamashina, 1931 ROTA BRIDLED WHITE-EYE
                                      Rota I.

---

[1] Fry *et al.* (2000: 8) (899) noted that this genus was now considered to belong to the babblers rather than the thrushes. However, fresh evidence appears to suggest otherwise (A. Cibois *in litt.*).
[2] Treated as a species of the bulbul genus *Phyllastrephus* in Rand & Deignan (1960: 270) (1972) and in our first Edition. But as a thrush in our last Edition following Benson & Irwin (1975) (133) and Irwin & Clancey (1986) (1194), however the genus *Arcanator* proposed by the latter is here treated as unnecessary following Dowsett & Dowsett-Lemaire (1993) (761).
[3] Gender addressed by David & Gosselin (2002) (614), multiple changes occur in specific names.
[4] The plate appeared in 1845 before the text, see Holthuis & Sadai (1970) (1128).
[5] Includes *occidentis* (although on characters this would be seen as a synonym of *egregius*, see below) and *joannae*.
[6] The recognition of *egregius* and its restriction here to Sri Lanka is artificial and done for convenience. Ripley (1982) (2055) lumped this population in *palpebrosa* in spite of the *apparently* interposed populations of southern India. Mees (1957) (1509) recognised *egregius* with a range statement suggesting possible continuity, but wrote that ranges needed "working out". Listing this as we do may stimulate more research. Possibly *egregius, aureiventer* and *williamsoni* represent a separate lowland species which has failed to invade the ranges of *salimalii* and *nilgiriensis*.

○ *Zosterops conspicillatus* Marianas Bridled White-eye[1]
— *Z.c.saypani* A.J.C. Dubois, 1902   Tinian, Saipan Is.
— †*Z.c.conspicillatus* (Kittlitz, 1832)   Guam I.

○ *Zosterops semperi* Carolines White-eye
— *Z.s.semperi* Hartlaub, 1868   Palau Is.
— *Z.s.owstoni* E. Hartert, 1900   Truk I.
— *Z.s.takatsukasai* Momiyama, 1922   Pohnpei I.

○ *Zosterops hypolais* Hartlaub & Finsch, 1872 Yap White-eye   Yap I.

○ *Zosterops salvadorii* Meyer & Wiglesworth, 1894 Enggano White-eye
         Enggano I.

○ *Zosterops atricapilla* Black-capped White-eye[2]
— *Z.a.viridicatus* Chasen, 1941   N Sumatra
— *Z.a.atricapilla* Salvadori, 1879   C and S Sumatra, N Borneo

○ *Zosterops everetti* Everett's White-eye
— *Z.e.boholensis* McGregor, 1908   Bohol, Leyte, Samar
— *Z.e.everetti* Tweeddale, 1878   Cebu
— *Z.e.basilanicus* Steere, 1890   Mindanao, Basilan
— *Z.e.siquijorensis* Bourns & Worcester, 1894   Siquijor
— *Z.e.mandibularis* Stresemann, 1931   Sulu Arch.
— *Z.e.babelo* Meyer & Wiglesworth, 1895   Talaud Is.
— *Z.e.tahanensis* Ogilvie-Grant, 1906   SE Thailand, S Malay Pen., Borneo
— *Z.e.wetmorei* Deignan, 1943   N Malay Pen.

○ *Zosterops nigrorum* Yellowish White-eye
— *Z.n.richmondi* McGregor, 1904   Cagayancillo I. (Sulu Sea)
— *Z.n.meyleri* McGregor, 1907   Camiguin Norte I. (off N Luzon)
— *Z.n.aureiloris* Ogilvie-Grant, 1895   NW Luzon
— *Z.n.innominatus* Finsch, 1901[3]   NE and C Luzon (833)
— *Z.n.luzonicus* Ogilvie-Grant, 1895   S Luzon, Catanduanes
— *Z.n.catarmanensis* Rand & Rabor, 1969   Camiguin Sur I. (off N Mindanao) (1979)
— *Z.n.nigrorum* Tweeddale, 1878   Masbate, Negros, Panay
— *Z.n.mindorensis* Parkes, 1971   Mindoro (1763)

○ *Zosterops montanus* Mountain White-eye[4]
— *Z.m.parkesi* duPont, 1971   Palawan (771)
— *Z.m.whiteheadi* E. Hartert, 1903   N Luzon
— *Z.m.gilli* duPont, 1971   Marinduque (771)
— *Z.m.diuatae* Salomonsen, 1953   NC and NE Mindanao
— *Z.m.vulcani* E. Hartert, 1903   Mt. Apo and Mt. Katanglad (W and S Mindanao)
— *Z.m.pectoralis* Mayr, 1945   Negros
— *Z.m.halconensis* Mearns, 1907   Mindoro
— *Z.m.obstinatus* E. Hartert, 1900   Bacan, Ternate, Seram
— *Z.m.montanus* Bonaparte, 1850[5]   Gtr. and Lesser Sundas, Timor, Sulawesi
— *Z.m.difficilis* Robinson & Kloss, 1918   S Sumatra

○ *Zosterops wallacei* Finsch, 1901 Yellow-spectacled White-eye   Sumbawa, Sumba, Flores Is. (W Lesser Sundas)

○ *Zosterops flavus* (Horsfield, 1821) Javan White-eye   NW Java, S Borneo

○ *Zosterops chloris* Lemon-bellied White-eye
— *Z.c.maxi* Finsch, 1907   Lombok I.
— *Z.c.intermedius* Wallace, 1864   S and E Sulawesi (and islands offshore), Sumbawa, Flores
— *Z.c.mentoris* Meise, 1952   NC Sulawesi
— *Z.c.flavissimus* E. Hartert, 1903   Tukangbesi I. (off SE Sulawesi)
— *Z.c.chloris* Bonaparte, 1850[6]   Banda, Seram Laut, Tayandu, Aru, Kai and Misool Is.

---

[1] Pratt *et al.* (1987) (1920) postulated several species were involved in the one treated by Mees (1969) (1520). Recent molecular studies by Slikas *et al.* (2000) (2278) largely confirmed this. We follow them in treating four species (*conspicillatus*, *semperi*, *hypolais* and *rotensis*).
[2] For correct spelling see David & Gosselin (2002) (613).
[3] Includes *sierramadrensis* Rand & Rabor, 1969 (1979), see Dickinson *et al.* (1991: 400) (745).
[4] Based on White & Bruce (1986) (2652) for Wallacea and Dickinson *et al.* (1991) (745) for the Philippines; the two books took different views on the value of recognition of the less well-marked subspecies.
[5] Dated '1851?' by Mayr (1967: 301) (1482) (1623), but see Zimmer (1926) (2707).
[6] Dated '1851?' by Mayr (1967: 304) (1482), but see Zimmer (1926) (2707).

○ *Zosterops citrinella* Ashy-bellied White-eye[1]
   — *Z.c.citrinella* Bonaparte, 1850[2]   Sumba to Timor (Lesser Sundas)
   — *Z.c.harterti* Stresemann, 1912   Alor I. (C Lesser Sundas)
   — *Z.c.albiventris* Reichenbach, 1852   Southwest and Southeast Is., Tanimbar Is., Torres Straits Is.

○ *Zosterops consobrinorum* A.B. Meyer, 1904 Pale-bellied White-eye   SE Sulawesi

○ *Zosterops grayi* Wallace, 1864 Pearl-bellied White-eye   Gt. Kai I. (E Banda Sea)

○ *Zosterops uropygialis* Salvadori, 1874 Golden-bellied White-eye   Little Kai I. (E Banda Sea)

○ *Zosterops anomalus* Meyer & Wiglesworth, 1896 Lemon-throated White-eye
    S Sulawesi

○ *Zosterops atriceps* Cream-throated White-eye
   — *Z.a.dehaani* van Bemmel, 1939   Morotai (N Moluccas)
   — *Z.a.fuscifrons* Salvadori, 1878   Halmahera (N Moluccas)
   — *Z.a.atriceps* G.R. Gray, 1860   Bacan (N Moluccas)

○ *Zosterops nehrkorni* W.H. Blasius, 1888 Sangihe White-eye[3]   Sangihe Is.

○ *Zosterops atrifrons* Black-crowned White-eye[4]
   — *Z.a.atrifrons* Wallace, 1864   N Sulawesi
   — *Z.a.surdus* Riley, 1919   C Sulawesi
   — *Z.a.*subsp?   SC Sulawesi[5]
   — *Z.a.subatrifrons* Meyer & Wiglesworth, 1896   Peling and Banggai Is.
   — *Z.a.sulaensis* Neumann, 1939   Sula Is.

○ *Zosterops stalkeri* Ogilvie-Grant, 1910 Seram White-eye[6]   Seram (S Moluccas)

○ *Zosterops minor* Black-fronted White-eye
   — *Z.m.minor* A.B. Meyer, 1874   Yapen I., Mts. of N New Guinea
   — *Z.m.chrysolaemus* Salvadori, 1876   NW New Guinea (Mts. of Vogelkop)
   — *Z.m.rothschildi* Stresemann & Paludan, 1934   Mts. of WC New Guinea
   — *Z.m.gregarius* Mayr, 1933   NE New Guinea (Mts. of Huon Pen.)
   — *Z.m.tenuifrons* Greenway, 1934   Mts. of C and EC New Guinea
   — *Z.m.delicatulus* Sharpe, 1882   Mts. of SE New Guinea
   — *Z.m.pallidogularis* Stresemann, 1931   Fergusson, Goodenough Is. (D'Entrecasteaux Arch.)

○ *Zosterops meeki* E. Hartert, 1898 White-throated White-eye[7]   Tagula I. (Louisiade Arch.)

○ *Zosterops hypoxanthus* Bismarck White-eye
   — *Z.h.hypoxanthus* Salvadori, 1881   New Britain
   — *Z.h.ultimus* Mayr, 1955   New Hanover, New Ireland
   — *Z.h.admiralitatis* Rothschild & Hartert, 1914   Manus I. (Admiralty Is.)

○ *Zosterops mysorensis* A.B. Meyer, 1874 Biak White-eye   Biak I., New Guinea

○ *Zosterops fuscicapilla* Capped White-eye[8]
   — *Z.f.fuscicapilla* Salvadori, 1876   Mts. of W and C New Guinea
   — *Z.f.crookshanki* Mayr & Rand, 1935   Goodenough I. (D'Entrecasteaux Arch.)

○ *Zosterops buruensis* Salvadori, 1878 Buru White-eye   Buru

○ *Zosterops kuehni* E. Hartert, 1906 Ambon White-eye   Ambon, Seram Is. (S Moluccas)

○ *Zosterops novaeguineae* New Guinea White-eye
   — *Z.n.novaeguineae* Salvadori, 1878   NW New Guinea (Mts. of Vogelkop)
   — *Z.n.aruensis* Mees, 1953   Aru Is.
   — *Z.n.wuroi* Mayr & Rand, 1935   Coastal S New Guinea
   — *Z.n.wahgiensis* Mayr & Gilliard, 1951   Mts. of EC New Guinea

---

[1] For correct spelling see David & Gosselin (2002) (613).
[2] Dated '1851?' by Mayr (1967: 304) (1482), but see Zimmer (1926) (2707).
[3] For recognition as a species see Rasmussen *et al.* (2000) (1994).
[4] Rasmussen *et al.* (2000) (1994), stated that specific status was probably warranted for the group formed by subspecies 3 to 5 based on acoustic evidence, but decided further study was needed.
[5] A name proposed conditionally in 1985 was not made current by that proposal, see Art. 15 of the Code, I.C.Z.N. (1999: 19) (1178).
[6] For treatment as a species see Rasmussen *et al.* (2000) (1994).
[7] For a discussion of this species see Mees (1961: 62-64, 91-92) (1511) and Mees (1969: 284-285) (1520).
[8] For correct spelling see David & Gosselin (2002) (613).

    — *Z.n.crissalis* Sharpe, 1884   Mts. of SE New Guinea

    — *Z.n.oreophilus* Mayr, 1931   NE New Guinea (Mts. of Huon Pen.)

    — *Z.n.magnirostris* Mees, 1955   Coastal N New Guinea

○ **Zosterops metcalfii** Yellow-throated White-eye

    — *Z.m.exiguus* Murphy, 1929   Buka, Bougainville, Choiseul Is. (N Solomons)

    — *Z.m.metcalfii* Tristram, 1894   Ysabel, St George Is. (NC Solomons)

    — *Z.m.floridanus* Rothschild & Hartert, 1901   Florida I. (EC Solomons)

○ **Zosterops natalis** Lister, 1889 Christmas Island White-eye   Christmas I. (Indian Ocean)

○ **Zosterops luteus** Australian Yellow White-eye

    — *Z.l.balstoni* Ogilvie-Grant, 1909   NW Western Australia

    — *Z.l.luteus* Gould, 1843   N Australia, NE Queensland

○ **Zosterops griseotinctus** Louisiade White-eye

    — *Z.g.pallidipes* De Vis, 1890   Rossel I. (Louisiade Arch.)

    — *Z.g.aignani* E. Hartert, 1899   Misima I. (Louisiade Arch.)

    — *Z.g.griseotinctus* G.R. Gray, 1858   Satellite islands (Louisiade Arch.)

    — *Z.g.longirostris* E.P. Ramsay, 1879   Bonvouloir, Alchester Is. (off SE New Guinea)

    — *Z.g.eichhorni* E. Hartert, 1926   Bismarck Arch. (satellite islands)

○ **Zosterops rennellianus** Murphy, 1929 Rennell Island White-eye   Rennell I. (SE Solomons)

○ **Zosterops vellalavella** E. Hartert, 1908 Banded White-eye   Bagga, Vellalavalla Is. (C Solomons)

○ **Zosterops luteirostris** Splendid White-eye

    — *Z.l.luteirostris* E. Hartert, 1904   Gizo I. (C Solomons)

    — *Z.l.splendidus* E. Hartert, 1929   Ganonga I. (C Solomons)

○ **Zosterops kulambangrae** Solomon Islands White-eye

    — *Z.k.kulambangrae* Rothschild & Hartert, 1901   Kolombangara, Vangunu, New Georgia Is. (C Solomons)

    — *Z.k.paradoxus* Mees, 1955[1]   Rendova I. (C Solomons)

    — *Z.k.tetiparius* Murphy, 1929   Tetipari I. (C Solomons)

○ **Zosterops murphyi** E. Hartert, 1929 Hermit White-eye   Kolombangara I. (WC Solomons)

○ **Zosterops ugiensis** Grey-throated White-eye[2]

    — *Z.u.ugiensis* (E.P. Ramsay, 1881)   San Cristobal I. (SE Solomons)

    — *Z.u.oblitus* E. Hartert, 1929   Guadalcanal I. (E Solomons)

    — *Z.u.hamlini* Murphy, 1929   Bougainville I. (N Solomons)

○ **Zosterops stresemanni** Mayr, 1931 Malaita White-eye   Malaita I. (E Solomons)

○ **Zosterops sanctaecrucis** Tristram, 1894 Santa Cruz White-eye   Santa Cruz I. (SW Pacific)

○ **Zosterops samoensis** Murphy & Mathews, 1929 Samoan White-eye   Savaii (Samoa Is.)

○ **Zosterops explorator** E.L. Layard, 1875 Layard's White-eye   Fiji Is.

○ **Zosterops flavifrons** Yellow-fronted White-eye

    — *Z.f.gauensis* Murphy & Mathews, 1929   Gaua I. (Banks Is.)

    — *Z.f.perplexus* Murphy & Mathews, 1929   Vanua Levu and Meralav Is. (Banks Is.), NE Vanuatu

    — *Z.f.brevicauda* Murphy & Mathews, 1929   Malo, Espiritu Santo Is. (NW Vanuatu)

    — *Z.f.macgillivrayi* Sharpe, 1900   Malekula I. (NC Vanuatu)

    — *Z.f.efatensis* Mayr, 1937   Nguna, Efate, Erromanga Is. (SC Vanuatu)

    — *Z.f.flavifrons* (J.F. Gmelin, 1789)   Tanna I. (S Vanuatu)

    — *Z.f.majusculus* Murphy & Mathews, 1929   Aneitum I. (S Vanuatu)

○ **Zosterops minutus** E.L. Layard, 1878 Small Lifu White-eye   Lifu, Loyalty Is.

○ **Zosterops xanthochroa** G.R. Gray, 1859 Green-backed White-eye[3]   New Caledonia

---

[1] Mayr (1967: 312) (1482) used the name *rendovae* Tristram for this. We have chosen not to, and nor do we use it for what Mayr called, and we call, *Z. ugiensis* (see below). We believe the name is now a source of confusion and we anticipate an Application will be made for its suppression.

[2] Mees (1961, 1969) (1511, 1520) used the name *rendovae* for this species (transferring the name to a bird from Ugi). This was discussed by Galbraith (1957) (905) who disagreed, and Galbraith's views are reflected in the treatment of Mayr (1967) (1482). We believe that whichever author is followed this trio of forms makes up one species. As used by Mayr (1967) (1482) the name *rendovae* was attached to one of six forms, which he treated as three species, from which six forms Mees made four species.

[3] Spelling per (N. David *in litt.*) see also David & Gosselin (2002) (613).

● **Zosterops lateralis** Silver-eye[1]

__ _Z.l.vegetus_ E. Hartert, 1899	E Cape York Pen., NE Queensland
__ _Z.l.cornwalli_ Mathews, 1912	EC and SE Queensland, NE New South Wales
__ _Z.l.chlorocephalus_ Campbell & White, 1910	Capricorn, Bunker Is.
__ _Z.l.westernensis_ (Quoy & Gaimard, 1930)	SE New South Wales, E Victoria
__ _Z.l.tephropleurus_ Gould, 1855[2]	Lord Howe I.
⊿ _Z.l.lateralis_ (Latham, 1802)[3]	Tasmania, Flinders I. (Furneaux Group), Norfolk I., New Zealand
__ _Z.l.ochrochrous_ Schodde & Mason, 1999	King I. (2189)
__ _Z.l.pinarochrous_ Schodde & Mason, 1999	S South Australia, SW New South Wales, W Victoria (2189)
__ _Z.l.chloronotus_ Gould, 1841[4]	SW and S Western Australia, SW South Australia
__ _Z.l.griseonota_ G.R. Gray, 1859	New Caledonia
__ _Z.l.nigrescens_ F. Sarasin, 1913	Mare, Uvea Is. (Loyalty Is.)
__ _Z.l.melanops_ G.R. Gray, 1860	Lifu I. (Loyalty Is.)
__ _Z.l.macmillani_ Mayr, 1937	Tanna, Aniwa Is. (S Vanuatu)
__ _Z.l.tropicus_ Mees, 1969	Espiritu Santo I. (1520)
__ _Z.l.vatensis_ Tristram, 1879	N Vanuatu, Banks Is., Torres Is.
__ _Z.l.valuensis_ Murphy & Mathews, 1929	Valua I. (Banks Is.)
__ _Z.l.flaviceps_ Peale, 1848	Fiji Is.

○ **Zosterops tenuirostris** Slender-billed White-eye

__ _Z.t.tenuirostris_ Gould, 1837	Norfolk I.
__ †_Z.t.strenuus_ Gould, 1855[5]	Lord Howe I.

○ **Zosterops albogularis** Gould, 1837 White-chested White-eye   Norfolk I.

○ **Zosterops inornatus** E.L. Layard, 1878 Large Lifu White-eye   Lifu I. (Loyalty Is.)

○ **Zosterops cinereus** Grey White-eye

__ _Z.c.finschii_ (Hartlaub, 1868)	Palau Is. (Caroline Is.)
__ _Z.c.ponapensis_ Finsch, 1876	Pohnpei I. (Caroline Is.)
__ _Z.c.cinereus_ (Kittlitz, 1832)	Kusaie I. (Caroline Is.)

○ **Zosterops abyssinicus** Abyssinian White-eye

__ _Z.a.abyssinicus_ Guérin-Méneville, 1843	N and C Ethiopia, Eritrea, NE Sudan
__ _Z.a.socotranus_ Neumann, 1908	Socotra I., N Somalia
__ _Z.a.arabs_ Lorenz & Hellmayr, 1901	SW Saudi Arabia, Yemen, S Oman
__ _Z.a.omoensis_ Neumann, 1904	W Ethiopia
__ _Z.a.jubaensis_ Erlanger, 1901	S Ethiopia, S Somalia, N Kenya
__ _Z.a.flavilateralis_ Reichenow, 1892	E and S Kenya, N and NE Tanzania

○ **Zosterops pallidus** Cape White-eye[6]

__ _Z.p.pallidus_ Swainson, 1838	Namibia, NW Cape Province
__ _Z.p.sundevalli_ Hartlaub, 1865	NE Cape Province, SW Transvaal, Orange Free State
__ _Z.p.caniviridis_ Clancey, 1962	Transvaal, SE Botswana
__ _Z.p.capensis_ Sundevall, 1850	W Cape Province
__ _Z.p.atmorii_ Sharpe, 1877	Inland S Cape Province
__ _Z.p.virens_ Sundevall, 1850	E Cape Province to SE Transvaal

○ **Zosterops senegalensis** African Yellow White-eye

__ _Z.s.senegalensis_ Bonaparte, 1850[7]	Senegal to NW Ethiopia
__ _Z.s.demeryi_ Büttikofer, 1890	Sierra Leone, Liberia, Ivory Coast
__ _Z.s.stenocricotus_ Reichenow, 1892	Bioko I., SE Nigeria to Gabon
__ _Z.s.gerhardi_ van den Elzen & König, 1983	S Sudan and NE Uganda (2463A)
__ _Z.s.stuhlmanni_ Reichenow, 1892	E Zaire, S Uganda, NW Tanzania
__ _Z.s.reichenowi_ A.J.C. Dubois, 1911	E Zaire

---

[1] Treatment based on Schodde & Mason (1999: 687-691) (2189) several names become synonyms and other names are newly recognised. The detailed synonymy is apparently to follow.
[2] Suggestions that this might be a distinct species were discussed by Mees (1969: 90) (1520).
[3] For reasons to use 1802 not 1801 see Browing & Monroe (1991) (258).
[4] Includes _gouldi_ a nomen novum now considered unnecessary for this name, and perhaps _halmaturinus_ given to birds of Kangaroo I. which are intermediate.
[5] Omitted from last Edition as extinct.
[6] These taxa are united in one species by Fry (2000: 318) (892); _atmorii_ is also recognised therein.
[7] Dated '1851?' by Moreau (1967: 328) (1623), but see Zimmer (1926) (2707).

__ *Z.s.toroensis* Reichenow, 1904	NE Zaire to W Uganda
__ *Z.s.jacksoni* Neumann, 1899	W Kenya, N Tanzania
__ *Z.s.kasaicus* Chapin, 1932	SW Zaire, NE Angola
__ *Z.s.heinrichi* Meise, 1958	NW Angola
__ *Z.s.quanzae* Meyer de Schauensee, 1932	C Angola
__ *Z.s.anderssoni* Shelley, 1892	S Angola, SE Zaire and W Mozambique (Tete) to N Namibia, Zimbabwe and E Transvaal
__ *Z.s.tongensis* Roberts, 1931	SE Zimbabwe, NE Natal, S Mozambique
__ *Z.s.stierlingi* Reichenow, 1899	E and S Tanzania, E Zambia, Malawi, NW Mozambique

○ **Zosterops poliogastrus** BROAD-RINGED WHITE-EYE

__ *Z.p.poliogastrus* Heuglin, 1861[1]	Eritrea, N, C and E Ethiopia
__ *Z.p.kaffensis* Neumann, 1902	W and SW Ethiopia
__ *Z.p.kulalensis* Williams, 1947	Mt. Kulal (N Kenya)
__ *Z.p.kikuyuensis* Sharpe, 1891	C Kenya
__ *Z.p.silvanus* Peters & Loveridge, 1935	Taita Hills (SE Kenya)
__ *Z.p.eurycricotus* Fischer & Reichenow, 1884	NE Tanzania
__ *Z.p.mbuluensis* Sclater & Moreau, 1936	S Kenya, N and NE Tanzania
__ *Z.p.winifredae* W.L. Sclater, 1935	S Pare Mts (NE Tanzania)

○ **Zosterops borbonicus** MASCARENE WHITE-EYE

__ *Z.b.mauritianus* (J.F. Gmelin, 1789)	Mauritius
__ *Z.b.borbonicus* (J.F. Gmelin, 1789)	Réunion
__ *Z.b.alopekion* Storer & Gill, 1966	Cilaos (Réunion)
__ *Z.b.xerophilus* Storer & Gill, 1966	Etang les Bains (Réunion)

○ **Zosterops ficedulinus** PRINCIPE WHITE-EYE

__ *Z.f.ficedulinus* Hartlaub, 1866	Principe I.
__ *Z.f.feae* Salvadori, 1901	São Tomé I.

○ **Zosterops griseovirescens** Bocage, 1893 ANNOBON WHITE-EYE — Annobon I.

○ **Zosterops maderaspatanus** MADAGASCAN WHITE-EYE

__ *Z.m.aldabransis* Ridgway, 1894	Aldabra I.
__ *Z.m.maderaspatanus* (Linnaeus, 1766)	Madagascar, Glorioso I.
__ *Z.m.kirki* Shelley, 1880	Grande Comore I.
__ *Z.m.anjouanensis* E. Newton, 1877	Anjouan I.
__ *Z.m.comorensis* Shelley, 1900	Mwali (Moheli) I.
__ *Z.m.voeltzkowi* Reichenow, 1905	Europa I.
__ *Z.m.menaiensis* Benson, 1969	Cosmoledo Atoll (127)

○ **Zosterops mayottensis** CHESTNUT-SIDED WHITE-EYE

__ †*Z.m.semiflavus* E. Newton, 1867	Marianne I. (Seychelles)
__ *Z.m.mayottensis* Schlegel, 1866	Mayotte I. (Comoros)

○ **Zosterops modestus** E. Newton, 1867 SEYCHELLES WHITE-EYE — Mahe I.

○ **Zosterops mouroniensis** Milne-Edwards & Oustalet, 1885 GRAND COMORE WHITE-EYE — Grande Comore I.

○ **Zosterops olivaceus** (Linnaeus, 1766) RÉUNION OLIVE WHITE-EYE — Réunion I.

○ **Zosterops chloronothos** (Vieillot, 1817) MAURITIUS OLIVE WHITE-EYE — Mauritius I.

○ **Zosterops vaughani** Bannerman, 1924 PEMBA WHITE-EYE — Pemba I.

WOODFORDIA North, 1906 F

○ ***Woodfordia superciliosa*** North, 1906 BARE-EYED WHITE-EYE — Rennell I. (SE Solomons)

○ ***Woodfordia lacertosa*** (Murphy & Mathews, 1929) SANFORD'S WHITE-EYE — Santa Cruz I. (SW Pacific)

RUKIA Momiyama, 1922 F

○ ***Rukia oleaginea*** (Hartlaub & Finsch, 1872) YAP RUKIA[2] — Yap I.

---

[1] For correction of spelling see David & Gosselin (2002) (613).
[2] Sibley & Monroe (1990: 607) (2262) and Slikas *et al.* (2000) (2278) suggested that this was should be placed in *Zosterops*. However, Mees (1969: 227) (1520) noted that the most distinctive of these three species is *longirostris* and we feel it is best to defer change until the affinities of all three members of this genus are understood.

○ *Rukia ruki* (E. Hartert, 1897) Truk White-eye — Truk I. (Caroline Is.)

○ *Rukia longirostra* (Taka-Tsukasa & Yamashina, 1931) Long-billed White-eye
Pohnpei I. (Caroline Is.)

CLEPTORNIS Oustalet, 1889[1] M
○ *Cleptornis marchei* (Oustalet, 1889) Golden White-eye — Mariana Is.

APALOPTERON Bonaparte, 1854[2] N
○ *Apalopteron familiare* Bonin Honeyeater
— *A.f.familiare* (Kittlitz, 1831) — Muko-jima I. (N Ogasawara-shoto or Bonin Is.)
— *A.f.hahasima* Yamashina, 1930 — Hahajima and Chichijima Is. (S Ogasawara-shoto or Bonin Is.)

TEPHROZOSTEROPS Stresemann, 1931 M
○ *Tephrozosterops stalkeri* (Ogilvie-Grant, 1910) Bicolored White-eye
Seram (S Moluccas)

MADANGA Rothschild & Hartert, 1923 F
○ *Madanga ruficollis* Rothschild & Hartert, 1923 Rufous-throated White-eye
Mts. of Buru (S Moluccas)

LOPHOZOSTEROPS E. Hartert, 1896 M[3]
○ *Lophozosterops pinaiae* (Stresemann, 1912) Grey-hooded White-eye — Mts. of Seram (S Moluccas)

○ *Lophozosterops goodfellowi* Black-masked White-eye
— *L.g.gracilis* Mees, 1969 — Diuata Mts. (NE Mindanao) (1520)
— *L.g.goodfellowi* (E. Hartert, 1903) — Mt. Apo (SE Mindanao)
— *L.g.malindangensis* (Mearns, 1909) — Mt. Malindang (NW Mindanao)

○ *Lophozosterops squamiceps* Streak-headed White-eye
— *L.s.heinrichi* (Stresemann, 1931) — Matinan Mts. (NW Sulawesi)
— *L.s.striaticeps* Riley, 1918 — Mts. of NC Sulawesi
— *L.s.stresemanni* (van Marle, 1940) — Mt. Saputan (NE Sulawesi)
— *L.s.stachyrinus* (Stresemann, 1932) — Latimojong Mts. (SC Sulawesi)
— *L.s.squamiceps* (E. Hartert, 1896) — Lompobattang Mts. (S Sulawesi)
— *L.s.analogus* (Stresemann, 1932) — Mengkoka Mts. (SE Sulawesi)

○ *Lophozosterops javanicus* Javan Grey-throated White-eye
— *L.j.frontalis* (Reichenbach, 1852) — Mts. of W Java
— *L.j.javanicus* (Horsfield, 1821) — Mts. of C and E Java
— *L.j.elongatus* (Stresemann, 1913) — Mts. of E Java, Bali

○ *Lophozosterops superciliaris* Yellow-browed White-eye
— *L.s.hartertianus* (Rensch, 1928) — Mts. of W Sumbawa (Lesser Sundas)
— *L.s.superciliaris* (E. Hartert, 1897) — Mts. of Flores (Lesser Sundas)

○ *Lophozosterops dohertyi* Crested White-eye
— *L.d.dohertyi* E. Hartert, 1896 — Sumbawa (Lesser Sundas)
— *L.d.subcristatus* E. Hartert, 1897 — Flores (Lesser Sundas)

OCULOCINCTA Mees, 1953 F
○ *Oculocincta squamifrons* (Sharpe, 1892) Pygmy White-eye — N and W Borneo

HELEIA Hartlaub, 1865 F
○ *Heleia muelleri* Hartlaub, 1865 Spot-breasted White-eye — W Timor (Lesser Sundas)

○ *Heleia crassirostris* (E. Hartert, 1897) Thick-billed White-eye — Flores, Sumbawa (Lesser Sundas)

CHLOROCHARIS Sharpe, 1888 F
○ *Chlorocharis emiliae* Mountain Black-eye
— *C.e.emiliae* Sharpe, 1888 — Mt. Kinabalu (N Borneo)
— *C.e.trinitae* Harrisson, 1957 — Mt. Trus Madi (N Borneo)
— *C.e.fusciceps* Mees, 1954 — Mts. of NE Sarawak
— *C.e.moultoni* Chasen & Kloss, 1927 — Mts. of Sarawak

[1] In last Edition we treated this genus as meliphagid; here we follow Sibley & Ahlquist (1990) (2261) and Slikas *et al.* (2000) (2278).
[2] We place this in the Zosteropidae following Springer *et al.* (1995) (2301).
[3] Gender addressed by David & Gosselin (2002) (614), multiple changes occur in specific names.

MEGAZOSTEROPS Stresemann, 1930[1] M
○ ***Megazosterops palauensis*** (Reichenow, 1915) GIANT WHITE-EYE   Palau Is.

HYPOCRYPTADIUS E. Hartert, 1903 M
○ ***Hypocryptadius cinnamomeus*** E. Hartert, 1903 CINNAMON IBON   Mts. of Mindanao

SPEIROPS Reichenbach, 1852 M[2]
○ ***Speirops brunneus*** Salvadori, 1903 FERNANDO PO SPEIROPS   Bioko I.

○ ***Speirops leucophoeus*** (Hartlaub, 1857) PRINCIPE SPEIROPS   Principe I.

○ ***Speirops lugubris*** (Hartlaub, 1848) BLACK-CAPPED SPEIROPS   São Tomé I.

○ ***Speirops melanocephalus*** (G.R. Gray, 1862) MT. CAMEROON SPEIROPS Mt. Cameroon

IRENIDAE FAIRY-BLUEBIRDS (1: 2)

IRENA Horsfield, 1821 F
○ ***Irena puella*** ASIAN FAIRY-BLUEBIRD[3]
  — *I.p.puella* (Latham, 1790)   S India
  — *I.p.sikkimensis* Whistler & Kinnear, 1933[4]   E Nepal and NE India to Assam and Burma, Thailand, Indochina, Andamans, Nicobars

  — *I.p.malayensis* F. Moore, 1854   Malay Pen.
  — *I.p.crinigera* Sharpe, 1877[5]   Sumatra and nearby islands, Borneo and nearby islands
  — *I.p.turcosa* Walden, 1870   Java
  — *I.p.tweeddalei* Sharpe, 1877   Palawan

○ ***Irena cyanogastra*** PHILIPPINE FAIRY-BLUEBIRD[6]
  — *I.c.cyanogastra* Vigors, 1831   Luzon, Polillo, Catanduanes
  — *I.c.ellae* Steere, 1890   Bohol, Leyte, Samar
  — *I.c.hoogstraali* Rand, 1948   Mindanao, Dinagat
  — *I.c.melanochlamys* Sharpe, 1877   Basilan

REGULIIDAE GOLDCRESTS/KINGLETS[7] (1: 5)

REGULUS Cuvier, 1800 M
○ ***Regulus ignicapilla*** FIRECREST[8] [9]
  — *R.i.ignicapilla* (Temminck, 1820)   C and S Europe, Asia Minor
  — ¶*R.i.caucasicus* Stepanyan, 1998   W Caucasus (2332)
  — ¶*R.r.tauricus* Redkin, 2001   Crimea (1997A)
  — *R.i.madeirensis* Harcourt, 1851   Madeira
  — *R.i.balearicus* von Jordans, 1924   Balearic Is., NW Africa

○ ***Regulus goodfellowi*** Ogilvie-Grant, 1906 FLAMECREST   Taiwan

● ***Regulus regulus*** GOLDCREST
  ∠ *R.r.regulus* (Linnaeus, 1758)[10]   Europe and W Siberia >> Mediterranean islands
  — *R.r.teneriffae* Seebohm, 1883[11]   W Canary Is.
  — *R.r.azoricus* Seebohm, 1883   San Miguel (E Azores)
  — *R.r.sanctaemariae* Vaurie, 1954   Santa Maria (E Azores)
  — *R.r.inermis* Murphy & Chapin, 1929   W and C Azores
  — *R.r.buturlini* Loudon, 1911   Asia Minor, Crimea, Caucasus area >> N Iran
  — *R.r.hyrcanus* Zarudny, 1910   N Iran
  — *R.r.coatsi* Sushkin, 1904   SC Siberia to Altai and Baikal area

[1] Treated in the genus *Rukia* in last Edition. For reasons to use this generic name see Mees (1969: 233) (1520).
[2] Gender addressed by David & Gosselin (2002) (614), multiple changes occur in specific names.
[3] Authors differ greatly over subspecific ranges and those given here are tentative. A fresh review is required.
[4] Includes *andamanica* Abdulali, 1964 (6), placed in synonymy (of nominate form) by Ripley (1982: 308) (2055).
[5] For correction of spelling see David & Gosselin (2002) (613).
[6] For correction of spelling see David & Gosselin (2002) (613).
[7] For reasons to treat this at family level see references in A.O.U. (1998: 487) (3).
[8] For correction of spelling see David & Gosselin (2002) (613).
[9] Treatment follows Cramp *et al.* (1992) (593), but *madeirensis* may belong with *R. regulus* or may deserve specific status see Päckert *et al.* (2001) (1739).
[10] Includes *anglorum* and *interni* see Vaurie (1959: 299) (2502).
[11] Placement here is tentative.

— *R.r.tristis* Pleske, 1892	Tien Shan Mts. >> SW Asia
— *R.r.himalayensis* Bonaparte, 1856	E Afghanistan to NW and WC Himalayas
— *R.r.sikkimensis* R. & A. Meinertzhagen, 1926	E Nepal, E Himalayas, NC China (E and NE Qinghai to SW Gansu)
— *R.r.japonensis* Blakiston, 1862	E Transbaikalia, Amurland and NE China to Sakhalin, S Kuril Is., Japan and N Korea >> E China
— *R.r.yunnanensis* Rippon, 1906	SE Xizang and NE Burma to N Yunnan, Sichuan, S Shaanxi and Guizhou

● *Regulus satrapa* GOLDEN-CROWNED KINGLET

⊥ *R.s.satrapa* M.H.K. Lichtenstein, 1823	C and E Canada, NE USA >> S Canada to S and SE USA
⊻ *R.s.olivaceus* S.F. Baird, 1864	SE Alaska to Oregon >> S California
— *R.s.apache* Jenks, 1936[1]	Interior Alaska south through Cascades and Rocky Mts. to N Mexico
— *R.s.aztecus* Lawrence, 1887	C Mexico
— *R.s.clarus* Dearborn, 1907	S Mexico, Central America

● *Regulus calendula* RUBY-CROWNED KINGLET

⊻ *R.c.calendula* (Linnaeus, 1766)[2]	N and E Canada, E USA >> Central America, S USA
⊻ *R.c.grinnelli* Palmer, 1897	Coast of Alaska, British Columbia >> W USA
— †?*R.c.obscurus* Ridgway, 1876	Guadalupe I. (Mexico)

TROGLODYTIDAE WRENS (16: 76)

CAMPYLORHYNCHUS Spix, 1824 M
○ *Campylorhynchus albobrunneus* WHITE-HEADED WREN

— *C.a.albobrunneus* (Lawrence, 1862)	C Panama
— *C.a.harterti* (Berlepsch, 1907)	E Panama, W Colombia

● *Campylorhynchus zonatus* BAND-BACKED WREN[3]

— *C.z.zonatus* (Lesson, 1832)[4]	EC Mexico
— *C.z.restrictus* (Nelson, 1901)	S Mexico, Guatemala
— *C.z.vulcanius* (Brodkorb, 1940)	S Mexico to Nicaragua
⊥ *C.z.costaricensis* Berlepsch, 1888	Costa Rica, W Panama
— *C.z.panamensis* (Griscom, 1927)	WC Panama
— *C.z.brevirostris* Lafresnaye, 1845	NW Colombia, NW Ecuador
— *C.z.curvirostris* Ridgway, 1888	NE Colombia (Santa Marta Mts.)
— ¶*C.z.imparilis* Borrero & Hernández, 1958	NE Colombia (lowlands) (199)

○ *Campylorhynchus megalopterus* GREY-BARRED WREN

— *C.m.megalopterus* Lasfresnaye, 1845	C Mexico
— *C.m.nelsoni* (Ridgway, 1903)	SC Mexico

○ *Campylorhynchus nuchalis* STRIPE-BACKED WREN

— *C.n.pardus* P.L. Sclater, 1858	N Colombia
— *C.n.brevipennis* Lawrence, 1866	N Venezuela
— *C.n.nuchalis* Cabanis, 1847	C Venezuela

○ *Campylorhynchus fasciatus* FASCIATED WREN

— *C.f.pallescens* Lafresnaye, 1846	SW Ecuador, NW Peru
— *C.f.fasciatus* (Swainson, 1838)	W Peru

○ *Campylorhynchus chiapensis* Salvin & Godman, 1891 GIANT WREN   Chiapas (Mexico)

○ *Campylorhynchus griseus* BICOLOURED WREN

— *C.g.albicilius* (Bonaparte, 1854)	N Colombia, NW Venezuela
— *C.g.bicolor* (Pelzeln, 1875)	Colombia
— *C.g.zimmeri* Borrero & Hernández, 1958	C Columbia (199)
— *C.g.minor* (Cabanis, 1851)	E Colombia, N Venezuela
— *C.g.pallidus* Phelps & Phelps, Jr., 1947	S Venezuela
— *C.g.griseus* (Swainson, 1838)	E Venezuela, Guyana, N Brazil

[1] Tentatively includes *amoenus* A.R. Phillips, 1991 (1876).
[2] Watson *et al.* (1986: 292) (2564) included *arizonensis* as a synonym and dated it from 1964. The name in fact dates from 1966 see Dickerman & Parkes (1997) (729).
[3] We are unaware of any re-evalutaion of this.
[4] Includes *vonbloekeri* J.S. Rowley, 1968 (2116) see Phillips (1986) (1875).

● *Campylorhynchus rufinucha* RUFOUS-NAPED WREN
— *C.r.rufinucha* (Lesson, 1838)   E Mexico (Veracruz)
— *C.r.rufum* Nelson, 1897   WC Mexico
— *C.r.humilis* P.L. Sclater, 1856   SW Mexico
— *C.r.nigricaudatus* (Nelson, 1897)   S Mexico, W Guatemala
— *C.r.capistratus* (Lesson, 1842)   El Salvador to N Costa Rica
— *C.r.xerophilus* Griscom, 1930   Guatemala (Motagua Valley)
— *C.r.castaneus* Ridgway, 1888   Guatemala to Honduras
— *C.r.nicoyae* A.R. Phillips, 1986[1]   NW Costa Rica (Nicoya Pen.) (1875)
— *C.r.nicaraguae* (Miller & Griscom, 1925)   Nicaragua (1584)

○ *Campylorhynchus gularis* P.L. Sclater, 1860 SPOTTED WREN   W and C Mexico

○ *Campylorhynchus jocosus* P.L. Sclater, 1859 BOUCARD'S WREN   SC Mexico

○ *Campylorhynchus yucatanicus* (Hellmayr, 1934) YUCATÁN WREN   SE Mexico

● *Campylorhynchus brunneicapillus* CACTUS WREN
— *C.b.sandiegensis* Rea, 1986   SW California, NW Baja California (1875)
— *C.b.bryanti* (Anthony, 1894)   N Baja California
— *C.b.affinis* Xantús de Vesey, 1859[2]   S Baja California
— *C.b.seri* (van Rossem, 1932)   Tiburon I.
✓ *C.b.anthonyi* Mearns, 1902   SW USA, NW Mexico
— *C.b.brunneicapillus* (Lafresnaye, 1835)   NW Mexico
— *C.b.guttatus* (Gould, 1837)[3]   SC Texas to C Mexico

○ *Campylorhynchus turdinus* THRUSH-LIKE WREN
— *C.t.aenigmaticus* Meyer de Schauensee, 1948   SW Colombia
— *C.t.hypostictus* Gould, 1855   NW and W Amazonia
— *C.t.turdinus* (Wied, 1821)   EC Brazil
— *C.t.unicolor* Lafresnaye, 1846   E Bolivia, W Brazil

ODONTORCHILUS Richmond, 1915 M
○ *Odontorchilus branickii* GREY-MANTLED WREN
— *O.b.branickii* (Taczanowski & Berlepsch, 1885)   Colombia, Ecuador, Peru, N Bolivia
— *O.b.minor* (Hartert, 1900)   N Ecuador

○ *Odontorchilus cinereus* (Pelzeln, 1868) TOOTH-BILLED WREN   C Brazil, E Bolivia

SALPINCTES Cabanis, 1847 M
● *Salpinctes obsoletus* ROCK WREN
✓ *S.o.obsoletus* (Say, 1823)   SW Canada, W USA, N and C Mexico
— *S.o.pulverius* Grinnell, 1898   San Clemente and San Nicolas Is., California
— *S.o.proximus* Swarth, 1914   San Martín I. (Mexico)
— *S.o.tenuirostris* van Rossem, 1943   San Benito I. (Mexico)
— *S.o.guadeloupensis* Ridgway, 1876   Guadeloupe I.
†? *S.o.exsul* Ridgway, 1903   San Benedicto I.
— *S.o.neglectus* Nelson, 1897   S Mexico to Honduras
— *S.o.guttatus* Salvin & Godman, 1891   El Salvador
— *S.o.fasciatus* Salvin & Godman, 1891   NW Nicaragua
— *S.o.costaricensis* van Rossem, 1941   NW Costa Rica

CATHERPES S.F. Baird, 1858 M
● *Catherpes mexicanus* CANYON WREN
— *C.m.griseus* Aldrich, 1946   SW Canada to Oregon, Utah and Idaho
— *C.m.pallidior* A.R. Phillips, 1986[4]   Montana to NW Colorado, NE Utah (1875)
✓ *C.m.conspersus* Ridgway, 1873   W USA to N Mexico
— *C.m.punctulatus* Ridgway, 1882   California

[1] For recognition see Parkes in Dickerman & Parkes (1997) (729).
[2] Includes *purus* see Rea (1986) (1875).
[3] Includes *couesi* Sharpe, 1882 (2228A) see Rea (1986) (1875).
[4] For recognition see Browning (1990) (251).

— *C.m.croizati* A.R. Phillips, 1986[1]   S Baja California (1875)

— *C.m.mexicanus* (Swainson, 1829)   C and S Mexico

— *C.m.meliphonus* Oberholser, 1930   NW Mexico

— *C.m.cantator* A.R. Phillips, 1966[2]   SW Mexico (1871)

HYLORCHILUS Nelson, 1897 M

○ **Hylorchilus sumichrasti** (Lawrence, 1871) SUMICHRAST'S WREN   WC Veracruz and NC Oaxaca (Mexico)

○ **Hylorchilus navai** Crossin & Ely, 1973 NAVA'S WREN[3]   SE Veracruz and W Chiapas (Mexico) (598)

CINNYCERTHIA Lesson, 1844 F

○ **Cinnycerthia unirufa** RUFOUS WREN

— *C.u.unirufa* (Lafresnaye, 1840)   NE Colombia

— *C.u.chakei* Aveledo & Ginés, 1952   NW Venezuela

— *C.u.unibrunnea* (Lafresnaye, 1853)   S and C Colombia, Ecuador, N Peru

○ **Cinnycerthia olivascens** SHARPE'S WREN[4]

— *C.o.bogotensis* (Matschie, 1885)   E Andes of Colombia

— *C.o.olivascens* Sharpe, 1881   W and C Andes of Colombia, to Ecuador and N Peru

○ **Cinnycerthia peruana** (Cabanis, 1873) PERUVIAN WREN   C Peru

○ **Cinnycerthia fulva** FULVOUS WREN

— *C.f.fitzpatricki* Remsen & Brumfield, 1998   SE Peru (Vilcabamba Mtns.) (2008)

— *C.f.fulva* (P.L. Sclater, 1874)   S Peru

— *C.f.gravesi* Remsen & Brumfield, 1998   S Peru, N Bolivia (2008)

CISTOTHORUS Cabanis, 1850 M

○ **Cistothorus platensis** SEDGE WREN[5]

— *C.p.stellaris* (J.F. Naumann, 1823)   E Canada, E USA, NE Mexico

— *C.p.tinnulus* R.T. Moore, 1941   W Mexico

— *C.p.potosinus* Dickerman, 1975   San Luis Potosí (Mexico) (710)

— *C.p.jalapensis* Dickerman, 1975   Veracruz (Mexico) (710)

— *C.p.warneri* Dickerman, 1975   S Veracruz, NE Chiapas (Mexico) (710)

— *C.p.russelli* Dickerman, 1975   Belize (710)

— *C.p.elegans* Sclater & Salvin, 1859   S Mexico, Guatemala

— *C.p.graberi* Dickerman, 1975   SE Honduras, NE Nicaragua (710)

— *C.p.lucidus* Ridgway, 1903   Costa Rica, W Panama

— *C.p.aequatorialis* Lawrence, 1871   Andes of Colombia, Venezuela and C Ecuador

— *C.p.graminicola* Taczanowski, 1874   Andes of S Ecuador to Peru and N Bolivia

— *C.p.tucumanus* E. Hartert, 1909[6]   Andes of S Bolivia, NW Argentina

— *C.p.platensis* (Latham, 1790)   C and E Argentina

— *C.p.hornensis* (Lesson, 1834)   S Chile, S Argentina

— *C.p.falklandicus* Chapman, 1934   Falkland Is.

— *C.p.polyglottus* (Vieillot, 1819)   SE Brazil to NE Argentina

— *C.p.alticola* Salvin & Godman, 1883   E Colombia, N Venezuela, N Guyana

— *C.p.minimus* Carriker, 1935   S Peru

○ **Cistothorus meridae** Hellmayr, 1907 PARAMO WREN   NW Venezuela

○ **Cistothorus apolinari** Chapman, 1914 APOLINAR'S WREN   C Colombia

● **Cistothorus palustris** MARSH WREN

— *C.p.browningi* Rea, 1986   SW British Columbia, WC Washington (1875)

— *C.p.dissaeptus* Bangs, 1902[7]   E USA

↙ *C.p.iliacus* Ridgway, 1903[8]   C Canada, NC to SE USA

— *C.p.laingi* (Harper, 1926)   WC Canada >> N Mexico

---

[1] For recognition see Browning (1990) (251).
[2] Tentatively accepted see Parkes in Dickerman & Parkes (1997) (729).
[3] For elevation to species level see Atkinson *et al.* (1993) (62) and A.O.U. (1998) (3).
[4] In last Edition this and the next two species were lumped as *C. peruana*, but see Brumfield & Remsen (1996) (262).
[5] Races recognised follow the revision by Traylor (1988) (2435).
[6] In Hartert & Venturi, see p. 160 (M. LeCroy *in litt.*).
[7] Includes *canniphonus* Oberholser, 1974 (1673) see Browning (1978, 1990) (246, 251).
[8] Includes *cryphius* Oberholser, 1974 (1673) see Browning (1978, 1990) (246, 251).

— *C.p.palustris* (A. Wilson, 1810)[1]	Coastal E USA
— *C.p.waynei* (Dingle & Sprunt, 1932)	SE USA (Virginia, N. Carolina)
— *C.p.griseus* Brewster, 1893	SE USA (S. Carolina to Florida)
— *C.p.marianae* W.E.D. Scott, 1888	SE USA (Gulf Coast)
— *C.p.plesius* Oberholser, 1897	W USA >> NW Mexico
— *C.p.pulverius* Aldrich, 1946	W USA
— *C.p.tolucensis* (Nelson, 1904)	C Mexico
— *C.p.paludicola* S.F. Baird, 1864	S Washington, N Oregon
— *C.p.aestuarinus* (Swarth, 1917)	Coastal C California
— *C.p.deserticola* Rea, 1986	SECalifornia, SW Arizona (1875)
— *C.p.clarkae* Unitt, Messer & Thery, 1996	SW California (2450)

THRYOMANES P.L. Sclater, 1862 M
◐ **Thryomanes bewickii** BEWICK'S WREN

— *T.b.calophonus* Oberholser, 1898	SW Canada, NW USA
— *T.b.marinensis* Grinnell, 1910	N California
— *T.b.spilurus* (Vigors, 1839)	C California
— *T.b.drymoecus* Oberholser, 1898	S Oregon, California
— *T.b.charienturus* Oberholser, 1898	SW California, NW Baja California
— *T.b.cerroensis* (Anthony, 1897)	WC Baja California
— *T.b.magdalenensis* Huey, 1942	SW Baja California
— † *T.b.leucophrys* (Anthony, 1895)[2]	San Clemente I. (California)
— † *T.b.brevicaudus* Ridgway, 1876	Guadeloupe I. (Mexico)
— *T.b.bewickii* (Audubon, 1827)	EC USA
— *T.b.pulichi* A.R. Phillips, 1986[3]	Kansas, Oklahoma >>Texas (1875)
— *T.b.cryptus* Oberholser, 1898	SE Colorado, W Oklahoma
— *T.b.eremophilus* Oberholser, 1898	W USA to C Mexico
— *T.b.sadai* A.R. Phillips, 1986[4]	S Texas to NE Mexico (1875)
— *T.b.mexicanus* (Deppe, 1830)	SW and WC Mexico

○ **Thryomanes sissonii** (Grayson, 1868) SOCORRO WREN — Socorro I., Revillagigedo Group

FERMINIA Barbour, 1926 F
○ **Ferminia cerverai** Barbour, 1926 ZAPATA WREN — Cuba

THRYOTHORUS Vieillot, 1816 M
○ **Thryothorus atrogularis** Salvin, 1864 BLACK-THROATED WREN — Nicaragua, Costa Rica, W Panama

○ **Thryothorus spadix** (Bangs, 1910) SOOTY-HEADED WREN — E Panama, W Colombia

○ **Thryothorus fasciatoventris** BLACK-BELLIED WREN

— *T.f.melanogaster* Sharpe, 1881	SW Costa Rica, W Panama
— *T.f.albigularis* (P.L. Sclater, 1855)	C and E Panama, W Colombia
— *T.f.fasciatoventris* Lafresnaye, 1845	N Colombia

○ **Thryothorus euophrys** PLAIN-TAILED WREN

— *T.e.euophrys* P.L. Sclater, 1860	SW Colombia, W Ecuador
— *T.e.longipes* J.A. Allen, 1889	E Ecuador
— *T.e.atriceps* (Chapman, 1924)	NW Peru
— *T.e.schulenbergi* Parker & O'Neill, 1985	Amazonas (Peru) (1747)

○ **Thryothorus eisenmanni** Parker & O'Neill, 1985 INCA WREN — Cuzco (Peru) (1747)

○ **Thryothorus genibarbis** MOUSTACHED WREN[5]

— *T.g.macrurus* J.A. Allen, 1889	Colombia
— *T.g.amaurogaster* (Chapman, 1914)	E Colombia
— *T.g.saltuensis* (Bangs, 1910)	W Colombia
— *T.g.yananchae* Meyer de Schauensee, 1951	SW Colombia

---

[1] Cited as 1807 by Paynter & Vaurie (1960: 394) (1810), but see Burtt & Peterson (1995) (280).
[2] This name is preoccupied in *Troglodytes* and the substitute name *anthonyi* Rea, 1986 (1875), is available should the genus be submerged.
[3] For recognition see Browning (1990) (251).
[4] Tentatively accepted see Parkes in Dickerman & Parkes (1997) (729).
[5] We await published details to justify separating *T. mystacalis* from this species.

— *T.g.consobrinus* Madarász, 1904 — NW Venezuela

— *T.g.ruficaudatus* Berlepsch, 1883 — N Venezuela

— *T.g.tachirensis* Phelps & Gilliard, 1941 — NW Venezuela

— *T.g.mystacalis* P.L. Sclater, 1860 — Ecuador

— *T.g.genibarbis* Swainson, 1838 — NC Brazil

— *T.g.juruanus* H. von Ihering, 1905 — W Brazil, NE Bolivia

— *T.g.intercedens* Hellmayr, 1908 — C Brazil

— *T.g.bolivianus* (Todd, 1913) — C Bolivia

○ **Thryothorus coraya** CORAYA WREN

— *T.c.obscurus* Zimmer & Phelps, 1947 — E Venezuela

— *T.c.caurensis* Berlepsch & Hartert, 1902 — E Colombia, S Venezuela, N Brazil

— *T.c.barrowcloughiana* Aveledo & Perez, 1994 — Mt. Roraima and Mt. Cuquenan (S Venezuela) (69)

— *T.c.ridgwayi* Berlepsch, 1889 — E Venezuela, W Guyana

— *T.c.coraya* (J.F. Gmelin, 1789) — The Guianas, N Brazil

— *T.c.herberti* Ridgway, 1888 — N Brazil

— *T.c.griseipectus* Sharpe, 1881 — Colombia, Ecuador, W Brazil, N Peru

— *T.c.amazonicus* Sharpe, 1881 — E Peru

— *T.c.albiventris* Taczanowski, 1882 — N Peru

— *T.c.cantator* Taczanowski, 1874 — Junín (Peru)

○ **Thryothorus felix** HAPPY WREN

— *T.f.sonorae* (van Rossem, 1930) — NW Mexico

— *T.f.pallidus* Nelson, 1899 — W Mexico

— *T.f.lawrencii* Ridgway, 1878 — Maria Madre I. (Mexico)

— *T.f.magdalenae* Nelson, 1898 — Maria Magdalena I. (Mexico)

— *T.f.grandis* Nelson, 1900 — C Mexico

— *T.f.felix* P.L. Sclater, 1859 — SW Mexico

○ **Thryothorus maculipectus** SPOT-BREASTED WREN

— *T.m.microstictus* (Griscom, 1930) — NE Mexico

— *T.m.maculipectus* Lafresnaye, 1845 — E Mexico

— *T.m.canobrunneus* Ridgway, 1887 — SE Mexico, Guatemala

— *T.m.umbrinus* Ridgway, 1887 — S Mexico, Belize, Guatemala, Honduras, El Salvador, E Nicaragua

— *T.m.petersi* (Griscom, 1930) — N Honduras

○ **Thryothorus rutilus** RUFOUS-BREASTED WREN[1]

— *T.r.hyperythrus* Salvin & Godman, 1880 — W Costa Rica, W Panama

— *T.r.tobagensis* (Hellmayr, 1921) — Tobago I.

— *T.r.rutilus* Vieillot, 1819 — Trinidad, N Venezuela

— *T.r.intensus* (Todd, 1932) — NW Venezuela

— *T.r.laetus* Bangs, 1898 — NW Venezuela, N Colombia

— *T.r.interior* (Todd, 1932) — C Colombia

— *T.r.hypospodius* Salvin & Godman, 1880 — NC Colombia

○ **Thryothorus sclateri** SPECKLE-BREASTED WREN

— *T.s.columbianus* (Chapman, 1924) — W Colombia

— *T.s.paucimaculatus* Sharpe, 1881 — W Ecuador, NW Peru

— *T.s.sclateri* Taczanowski, 1879 — N Peru

○ **Thryothorus semibadius** Salvin, 1870 RIVERSIDE WREN — SW Costa Rica, SW Panama

● **Thryothorus nigricapillus** BAY WREN

└ *T.n.costaricensis* (Sharpe, 1881) — SE Nicaragua to NW Panama

— *T.n.castaneus* Lawrence, 1861 — C Panama

— *T.n.odicus* Wetmore, 1959 — Isla Escudo (Panama) (2599A)

— *T.n.reditus* (Griscom, 1932) — C and E Panama

— *T.n.schottii* (S.F. Baird, 1864) — E Panama, NW Colombia

— *T.n.nigricapillus* P.L. Sclater, 1860 — W Ecuador

● **Thryothorus thoracicus** Salvin, 1864 STRIPE-BREASTED WREN — E Nicaragua, Costa Rica, W Panama

[1] The next species is separated as in A.O.U. (1998) (3).

○ **_Thryothorus leucopogon_** STRIPE-THROATED WREN
— _T.l.grisescens_ (Griscom, 1932)　　　　　　　　　E Panama
— _T.l.leucopogon_ (Salvadori & Festa, 1899)　　　　W Colombia, NW Ecuador

○ **_Thryothorus pleurostictus_** BANDED WREN
— _T.p.nisorius_ P.L. Sclater, 1869　　　　　　　　　SW Mexico (Michoacan to N Guerrero)
— _T.p.oaxacae_ Brodkorb, 1942　　　　　　　　　　SW Mexico (Guerrero to Oaxaca)
— _T.p.acaciarum_ Brodkorb, 1942　　　　　　　　　S Mexico
— _T.p.oblitus_ (van Rossem, 1934)　　　　　　　　　S Mexico, Guatemala, W El Salvador
— _T.p.pleurostictus_ P.L. Sclater, 1860　　　　　　Guatemala
— _T.p.lateralis_ (Dickey & van Rossem, 1927)　　　El Salvador, S Honduras
— _T.p.ravus_ (Ridgway, 1903)　　　　　　　　　　　Nicaragua, NW Costa Rica

● **_Thryothorus ludovicianus_** CAROLINA WREN
↙ _T.l.ludovicianus_ (Latham, 1790)　　　　　　　　C, S, SE USA
— _T.l.burleighi_ Lowery, 1940　　　　　　　　　　　Cat I. (Mississippi)
— _T.l.nesophilus_ Stevenson, 1973　　　　　　　　　Dog I. (NW Florida) (2336)
— _T.l.miamensis_ Ridgway, 1875　　　　　　　　　　SE Georgia, Florida
✓ _T.l.lomitensis_ Sennett, 1890　　　　　　　　　　S Texas, NE Mexico
— _T.l.oberholseri_ Lowery, 1940　　　　　　　　　　SW Texas, N Mexico
— _T.l.berlandieri_ S.F. Baird, 1858　　　　　　　　NC and NE Mexico
— _T.l.tropicalis_ Lowery & Newman, 1949　　　　　NC Mexico
— _T.l.albinucha_ (Cabot, 1847)　　　　　　　　　　SE Mexico, N Guatemala
— _T.l.subfulvus_ Miller & Griscom, 1925　　　　　　Guatemala, Nicaragua

○ **_Thryothorus rufalbus_** RUFOUS-AND-WHITE WREN
— _T.r.rufalbus_ Lafresnaye, 1845　　　　　　　　　　S Chiapas, Guatemala, El Salvador
— _T.r.sylvus_ A.R. Phillips, 1986[1]　　　　　　　　N Guatemala, Honduras (1875)
— _T.r.castanonotus_ (Ridgway, 1888)[2]　　　　　　Nicaragua and Pacific slope of Costa Rica to W Panama
— _T.r.cumanensis_ (Cabanis, 1860)　　　　　　　　N Colombia, N Venezuela
— _T.r.minlosi_ (Berlepsch, 1884)　　　　　　　　　EC Colombia, NW Venezuela

○ **_Thryothorus nicefori_** Meyer de Schauensee, 1946　NICEFORO'S WREN　N Colombia

○ **_Thryothorus sinaloa_** SINALOA WREN
— _T.s.cinereus_ (Brewster, 1889)　　　　　　　　　NW Mexico
— _T.s.sinaloa_ (S.F. Baird, 1864)　　　　　　　　　WC Mexico
— _T.s.russeus_ (Nelson, 1903)　　　　　　　　　　SW Mexico

● **_Thryothorus modestus_** PLAIN WREN
↙ _T.m.modestus_ Cabanis, 1860[3]　　　　　　　　S Mexico to Costa Rica
— _T.m.vanrossemi_ A.R. Phillips, 1986[4]　　　　　El Salvador (1875)
— _T.m.zeledoni_ (Ridgway, 1878)　　　　　　　　　E Nicaragua, E Costa Rica, NW Panama
— _T.m.elutus_ (Bangs, 1902)　　　　　　　　　　　SW Costa Rica, W Panama

○ **_Thryothorus leucotis_** BUFF-BREASTED WREN
— _T.l.galbraithii_ Lawrence, 1861　　　　　　　　　E Panama, NW Colombia
— _T.l.conditus_ (Bangs, 1903)　　　　　　　　　　　Pearl Islands (Panama)
— _T.l.leucotis_ Lafresnaye, 1845　　　　　　　　　N Colombia
— _T.l.collinus_ (Wetmore, 1946)　　　　　　　　　　NE Colombia
— _T.l.venezuelanus_ Cabanis, 1851　　　　　　　　NE Colombia, NW Venezuela
— _T.l.zuliensis_ Hellmayr, 1934　　　　　　　　　　Colombia, Venezuela
— _T.l.hypoleucus_ (Berlepsch & Hartert, 1901)　　NC Venezuela
— _T.l.bogotensis_ (Hellmayr, 1901)　　　　　　　　E Colombia, C Venezuela
— _T.l.albipectus_ Cabanis, 1849　　　　　　　　　　NE Venezuela, the Guianas, NE Brazil
— _T.l.peruanus_ (Hellmayr, 1921)　　　　　　　　　W Brazil, SE Colombia, E Ecuador, E Peru, N Bolivia
— _T.l.rufiventris_ P.L. Sclater, 1870　　　　　　　C Brazil

---

[1] For recognition see Parkes in Dickerman & Parkes (1997) (729).
[2] A substitute name _skutchi_ A.R. Phillips, 1986 (1875) has been proposed for _castanonotus_ of others not of Ridgway, but appears to be unnecessary see Parkes in Dickerman & Parkes (1997) (729).
[3] Includes _roberti_ A.R. Phillips, 1986 (1875) see Parkes in Dickerman & Parkes (1997) (729).
[4] Tentatively accepted see Parkes in Dickerman & Parkes (1997) (729).

○ *Thryothorus superciliaris* SUPERCILIATED WREN
    __ *T.s.superciliaris* Lawrence, 1869   Ecuador
    __ *T.s.baroni* (Hellmayr, 1902)   S Ecuador, N Peru

○ *Thryothorus guarayanus* (d'Orbigny & Lafresnaye, 1837) FAWN-BREASTED WREN
                                         Bolivia, SW Brazil

○ *Thryothorus longirostris* LONG-BILLED WREN
    __ *T.l.bahiae* (Hellmayr, 1903)   NE Brazil
    __ *T.l.longirostris* Vieillot, 1818   E Brazil

○ *Thryothorus griseus* (Todd, 1925) GREY WREN   W Brazil

TROGLODYTES Vieillot, 1809 M
● *Troglodytes troglodytes* WINTER WREN[1]

__ *T.t.hiemalis* Vieillot, 1819	Canada, N USA >> S USA
__ *T.t.pullus* (Burleigh, 1935)	E USA (Appalachian Mtns.) >> S USA
__ *T.t.ochroleucus* Rea, 1986[2]	Islands off SE Alaska >> British Columbia (1875)
__ *T.t.pacificus* S.F. Baird, 1864	Prince of Wales I, Queen Charlotte Islands >> California
↓ *T.t.salebrosus* Burleigh, 1959	British Columbia to Oregon, NE California >> S California, SW Arizona (271)
__ *T.t.obscurior* Rea, 1986[3]	California (Sierra Nevada Mtns. and C coast) (1875)
↓ *T.t.muiri* Rea, 1986[4]	SW Oregon, N California coast (1875)
__ *T.t.helleri* (Osgood, 1901)	Kodiak I.
__ *T.t.semidiensis* (W.P. Brooks, 1915)	Semidi I.
__ *T.t.kiskensis* (Oberholser, 1919)	Aleutian Is. (Kiska eastward)
__ *T.t.meligerus* (Oberholser, 1900)	Aleutian Is. (Attu, Agatta)
__ *T.t.islandicus* E. Hartert, 1907	Iceland
__ *T.t.borealis* Fischer, 1861	Faroe Is.
__ *T.t.zetlandicus* E. Hartert, 1910	Shetland Is. (Scotland)
__ *T.t.fridariensis* Williamson, 1951	Fair Isle (Scotland)
__ *T.t.hebridensis* R. Meinertzhagen, 1924	Outer Hebrides (Scotland)
__ *T.t.hirtensis* Seebohm, 1884	St Kilda I. (Scotland)
__ *T.t.indigenus* Clancey, 1937	Ireland, Scotland, England
↓ *T.t.troglodytes* (Linnaeus, 1758)	Europe south to C Spain, Sicily and C Greece
__ *T.t.koenigi* Schiebel, 1910	Corsica, Sardinia
__ *T.t.kabylorum* E. Hartert, 1910	NW Africa, Balearic Is., S Spain
__ *T.t.juniperi* E. Hartert, 1922	NE Libya
__ *T.t.cypriotes* (Bate, 1903)	Aegean Is., W and S Turkey, Cyprus, Levant
__ *T.t.hyrcanus* Zarudny & Loudon, 1905	Caucasus area, N Iran, N Turkey, Crimea
__ *T.t.subpallidus* Zarudny & Loudon, 1905	NE Iran, SW Turkmenistan, S Uzbekistan, NW Afghanistan
__ *T.t.zagrossiensis* Zarudny & Loudon, 1908	SW Iran
__ *T.t.tianschanicus* Sharpe, 1881	NE Afghanistan and Pamir to Tien Shan, ? Altai
__ *T.t.neglectus* W.E. Brooks, 1872	E Afghanistan, W Himalayas
__ *T.t.magrathi* (C.H.T. Whitehead, 1907)	SE Afghanistan, W Pakistan
__ *T.t.nipalensis* Blyth, 1845	C and E Himalayas
__ *T.t.talifuensis* (Sharpe, 1902)	Yunnan, NE Burma
__ *T.t.szetschuanus* E. Hartert, 1910	E Xizang, SE Qinghai, W Sichuan (WC China)
__ *T.t.idius* (Richmond, 1907)	NE Qinghai and Gansu to Hebei (NC China)
__ *T.t.dauricus* Dybowski & Taczanowski, 1884	SE Siberia, NE China, Korea, Tsushima I. (SW Japan)
__ *T.t.alascensis* S.F. Baird, 1869	Pribilof Is.
__ *T.t.pallescens* (Ridgway, 1883)	Commander Is., ? Kamchatka
__ *T.t.kurilensis* Stejneger, 1889	N Kuril Is.
__ *T.t.fumigatus* Temminck, 1835	S Kuril Is., Sakhalin, Japan, N Izu Is.
__ *T.t.mosukei* Momiyama, 1923	S Izu and Borodino Is.
__ *T.t.ogawae* E. Hartert, 1910	Tanegashima, Yakushima Is.
__ *T.t.taivanus* E. Hartert, 1910	Taiwan

---

[1] For the North American taxa we tentatively follow Phillips (1986) (1875), pending review by the A.O.U.
[2] Recognition tentative; provisionally treated as a synonym of *pacificus* by Browning (1990) (251).
[3] Recognition tentative; provisionally treated as a synonym of *pacificus* by Browning (1990) (251).
[4] Recognition tentative; provisionally treated as a synonym of *pacificus* by Browning (1990) (251).

○ *Troglodytes tanneri* C.H. Townsend, 1890 Clarion Wren | Isla Clarion (off W Mexico)

● *Troglodytes aedon* House Wren[1] [2]
    __ *T.a.aedon* Vieillot, 1809[3] | SE Canada, E USA >> S USA, E Mexico
    ✓ *T.a.parkmanii* Audubon, 1839 | SW Canada, C and W USA, N Mexico >> SW and SE USA
    __ *T.a.cahooni* Brewster, 1888 | N Mexico
    __ *T.a.brunneicollis* P.L. Sclater, 1858 | C and S Mexico
    __ *T.a.peninsularis* Nelson, 1901 | Outer Yucatán Pen. (Mexico)
    __ *T.a.intermedius* Cabanis, 1860 | S Mexico to Costa Rica
    __ *T.a.inquietus* S.F. Baird, 1864 | SW Costa Rica, Panama, N Colombia
    __ *T.a.carychrous* Wetmore, 1957 | Coiba I., Panama
    __ *T.a.pallidipes* A.R. Phillips, 1986 | Pearl Islands (Panama) (1875)
    __ *T.a.beani* Ridgway, 1885 | Cozumel I., SE Mexico
    __ †? *T.a.guadeloupensis* (Cory, 1886) | Guadeloupe I. (Lesser Antilles)
    __ *T.a.rufescens* (Lawrence, 1877) | Dominica I.
    __ † *T.a.martinicensis* (P.L. Sclater, 1866) | Martinique I.
    __ *T.a.mesoleucus* (P.L. Sclater, 1876) | St Lucia I.
    __ *T.a.musicus* (Lawrence, 1878) | St Vincent I.
    __ *T.a.grenadensis* (Lawrence, 1878) | Grenada I.
    __ *T.a.atopus* Oberholser, 1904 | N Colombia
    __ *T.a.striatulus* (Lafresnaye, 1845) | Colombia, Venezuela
    __ *T.a.columbae* Stone, 1899 | E Colombia
    __ *T.a.effutitus* Wetmore, 1958 | NW and W Venezuela (2599)
    __ *T.a.albicans* Berlepsch & Taczanowski, 1884 | Trinidad, Colombia, Venezuela, W Ecuador, N Peru, the Guianas, Brazil

    __ *T.a.tobagensis* Lawrence, 1888 | Tobago I.
    __ *T.a.audax* Tschudi, 1844 | W Peru
    __ *T.a.puna* Berlepsch & Stolzmann, 1896 | W Bolivia, Peru
    __ *T.a.carabayae* Chapman & Griscom, 1924 | Peru
    __ *T.a.tecellatus* d'Orbigny & Lafresnaye, 1837 | Peru, N Chile
    __ *T.a.rex* Berlepsch & Leverkühn, 1890 | Bolivia, Paraguay, Argentina
    ✓ *T.a.atacamensis* Hellmayr, 1924 | N and C Chile
    __ *T.a.musculus* J.F. Naumann, 1823 | C and S Brazil, E Paraguay, N Argentina
    __ *T.a.bonariae* Hellmayr, 1919 | S Brazil, Uruguay, NE Argentina
    __ *T.a.chilensis* Lesson, 1830 | S Chile, S Argentina
    __ *T.a.cobbi* Chubb, 1909[4] | Falkland Is.

○ *Troglodytes rufociliatus* Rufous-browed Wren[5]
    __ *T.r.rehni* Stone, 1932 | S Mexico to Honduras
    __ *T.r.rufociliatus* Sharpe, 1881 | S Guatemala
    __ *T.r.nannoides* Dickey & van Rossem, 1929 | SW El Salvador

● *Troglodytes ochraceus* Ochraceous Wren
    ✓ *T.o.ochraceus* Ridgway, 1882 | Costa Rica, W Panama
    __ *T.o.festinus* Nelson, 1912 | E Panama (Cerro Pirre)

○ *Troglodytes solstitialis* Mountain Wren
    __ *T.s.solitarius* Todd, 1912 | Colombia, Venezuela
    __ *T.s.solstitialis* P.L. Sclater, 1859 | Colombia, Ecuador, N Peru
    __ *T.s.macrourus* Berlepsch & Stolzmann, 1902 | EC Peru
    __ *T.s.frater* Sharpe, 1881 | SE Peru, Bolivia
    __ *T.s.auricularis* Cabanis, 1883 | N Argentina

○ *Troglodytes monticola* Bangs, 1899 Santa Marta Wren[6] | Santa Marta Mts. (N Colombia)

○ *Troglodytes rufulus* Tepui Wren
    __ *T.r.rufulus* Cabanis, 1849 | SE Venezuela

---

[1] For comments on the prior name *domesticus* see Parkes in Dickerman & Parkes (1997) (729).
[2] We follow A.O.U. (1998: 481) (3) in maintaining a broad species, but recognise various Caribbean forms may be shown to be species.
[3] Dated '1808?' by Paynter & Vaurie (1960: 422) (1810), but see Browning & Monroe (1991) (258).
[4] See Woods (1993) (2690) for reasons to consider this at or near to the status of a separate species. Acoustic evidence is needed.
[5] Three species (*rufociliatus*, *ochraceus* and *monticola*) derive from *T. solstitialis* (A.O.U., 1998) (3).
[6] Treated as a subspecies of *solstitialis* in last Edition, but see Ridgely & Tudor (1989: 86) (2018).

_ *T.r.fulvigularis* Zimmer & Phelps, 1945   SE Venezuela
_ *T.r.yavii* Phelps & Phelps, Jr., 1949   S Venezuela
_ *T.r.duidae* Chapman, 1929   S Venezuela
_ *T.r.wetmorei* Phelps & Phelps, Jr., 1955   S Venezuela
_ *T.r.marahuacae* Phelps & Aveledo, 1984   Amazonas (Venezuela) (1838)

THRYORCHILUS Oberholser, 1904 M
○ **Thryorchilus browni** Bangs, 1902 TIMBERLINE WREN[1]   Costa Rica, W Panama

UROPSILA Sclater & Salvin, 1873 F
○ **Uropsila leucogastra** WHITE-BELLIED WREN
_ *U.l.pacifica* (Nelson, 1897)   SW Mexico
_ *U.l.grisescens* (Griscom, 1928)   C Mexico
_ *U.l.leucogastra* (Gould, 1837)   EC and E Mexico
_ *U.l.centralis* A.R. Phillips, 1986[2]   EC Mexico (1875)
_ *U.l.restricta* A.R. Phillips, 1986[3]   Yucatán Penin (Mérida), Mexico (1875)
_ *U.l.brachyura* (Lawrence, 1887)   Yucatán Penin of SE Mexico, Guatemala
_ *U.l.australis* (van Rossem, 1938)[4]   S Quintana Roo (Mexico), Belize, Honduras

HENICORHINA Sclater & Salvin, 1868 F
○ **Henicorhina leucosticta** WHITE-BREASTED WOOD WREN
_ *H.l.decolorata* A.R. Phillips, 1986[5]   C Mexico (1875)
_ *H.l.prostheleuca* (P.L. Sclater, 1857)   S Mexico to NW Costa Rica
_ *H.l.smithei* Dickerman, 1973   Yucatán Pen. (Mexico), Peten (Guatemala) (707)
_ *H.l.costaricensis* Dickerman, 1973   C Costa Rica (707)
_ *H.l.pittieri* Cherrie, 1893   SW Costa Rica to C Panama
_ *H.l.alexandri* A.R. Phillips, 1986[6]   E Panama, NW Colombia (1875)
_ *H.l.darienensis* Hellmayr, 1921   E Panama (Darién, NW Colombia
_ *H.l.albilateralis* Chapman, 1917   C Colombia
_ *H.l.leucosticta* (Cabanis, 1847)   S Venezuela, Guyana, Surinam, N Brazil
_ *H.l.eucharis* Bangs, 1910   Colombia (Dagua Valley)
_ *H.l.inornata* Hellmayr, 1903   W Colombia to NW Ecuador
_ *H.l.hauxwelli* Chubb, 1920   S Colombia, E Ecuador, C Peru

● **Henicorhina leucophrys** GREY-BREASTED WOOD WREN
_ *H.l.minuscula* A.R. Phillips, 1966[7]   SC Mexico (1871)
_ *H.l.festiva* Nelson, 1903   SW Mexico
_ *H.l.mexicana* Nelson, 1897   E Mexico
_ *H.l.castanea* Ridgway, 1903   S Mexico (N Chiapas, N Guatemala
_ *H.l.capitalis* Nelson, 1897   S Mexico, W Guatemala
_ *H.l.composita* Griscom, 1932   Honduras, El Salvador
_ *H.l.collina* Bangs, 1902   Costa Rica, W Panama
_ *H.l.anachoreta* Bangs, 1899   N Colombia
_ *H.l.bangsi* Ridgway, 1903   N Colombia
_ *H.l.manastarae* Aveledo & Ginés, 1952   NW Venezuela
_ *H.l.sanluisensis* Phelps & Phelps, 1959   NW Venezuela (1861)
_ *H.l.venezuelensis* Hellmayr, 1903   N Venezuela
_ *H.l.meridana* Todd, 1932   W Venezuela
_ *H.l.tamae* Zimmer & Phelps, 1944   E Colombia, NW Venezuela
_ *H.l.brunneiceps* Chapman, 1914   SW Colombia, N Ecuador
_ *H.l.hilaris* Berlepsch & Taczanowski, 1884   SW Ecuador
_ *H.l.leucophrys* (Tschudi, 1844)   W Colombia, Ecuador, Peru
_ *H.l.boliviana* Todd, 1932   N Bolivia

[1] For monotypic treatment see Phillips (1986) (1875).
[2] For comments see Parkes in Dickerman & Parkes (1997) (729).
[3] For comments see Parkes in Dickerman & Parkes (1997) (729).
[4] For recognition of this population, if not its name, see Phillips (1986) (1875); *hawkinsi* Monroe, 1963 (1603) is a synonym. However see also Parkes in Dickerman & Parkes (1997) (729).
[5] Tentatively accepted see Parkes in Dickerman & Parkes (1997) (729).
[6] Tentatively accepted see Parkes in Dickerman & Parkes (1997) (729).
[7] Tentatively accepted see Parkes in Dickerman & Parkes (1997) (729).

○ *Henicorhina leucoptera* Fitzpatrick, Terborgh & Willard, 1977 BAR-WINGED WOOD WREN
                                         NC Peru (839)

MICROCERCULUS P.L. Sclater, 1862 M
○ *Microcerculus philomela* (Salvin, 1861) NIGHTINGALE WREN   S Mexico to N Costa Rica

○ *Microcerculus marginatus* SCALY-BREASTED WREN
    __ *M.m.luscinia* Salvin, 1866   Costa Rica to E Panama (2218)
    __ *M.m.corrasus* Bangs, 1902   N Colombia
    __ *M.m.squamulatus* Sclater & Salvin, 1875   NE Colombia, N Venezuela
    __ *M.m.taeniatus* Salvin, 1881   C and W Colombia, W Ecuador
    __ *M.m.marginatus* (P.L. Sclater, 1855)   W Amazonia

○ *Microcerculus ustulatus* FLUTIST WREN
    __ *M.u.duidae* Chapman, 1929   S Venezuela
    __ *M.u.lunatipectus* Zimmer & Phelps, 1946   S Venezuela
    __ *M.u.obscurus* Zimmer & Phelps, 1946   SE Venezuela
    __ *M.u.ustulatus* Salvin & Godman, 1883   SE Venezuela, W Guyana

○ *Microcerculus bambla* WING-BANDED WREN
    __ *M.b.albigularis* (P.L. Sclater, 1858)   E Ecuador, W Brazil
    __ *M.b.caurensis* Berlepsch & Hartert, 1902 S Venezuela
    __ *M.b.bambla* (Boddaert, 1783)   SE Venezuela, the Guianas, NE Brazil

CYPHORHINUS Cabanis, 1844 M
○ *Cyphorhinus thoracicus* CHESTNUT-BREASTED WREN
    __ *C.t.thoracicus* Tschudi, 1844   C? Peru to N Bolivia
    __ *C.t.dichrous* Sclater & Salvin, 1879   S Colombia, Ecuador, N? Peru

○ *Cyphorhinus arada* MUSICIAN WREN[1]
    __ *C.a.urbanoi* (Zimmer & Phelps, 1946)   S Venezuela
    __ *C.a.arada* (Hermann, 1783)   S Venezuela, the Guianas, NE Brazil
    __ *C.a.faroensis* (Zimmer & Phelps, 1946) N Brazil
    __ *C.a.griseolateralis* Ridgway, 1888   N Brazil
    __ *C.a.interpositus* (Todd, 1932)   S Brazil
    __ *C.a.transfluvialis* (Todd, 1932)   N Brazil, SE Colombia
    __ *C.a.salvini* Sharpe, 1881   SE Colombia, E Ecuador, NE Peru
    __ *C.a.modulator* (d'Orbigny, 1838)   E Peru, W Brazil, N Bolivia

○ *Cyphorinus phaeocephalus* SONG WREN
    __ *C.p.richardsoni* Salvin, 1893   Honduras to W Panama
    __ *C.p.lawrencii* Lawrence, 1863   C Panama to NW Colombia
    __ *C.p.propinquus* (Todd, 1919)   N Colombia
    __ *C.p.chocoanus* (Meyer de Schauensee, 1946) W Colombia
    __ *C.p.phaeocephalus* P.L. Sclater, 1860 SW Colombia, Ecuador

GENUS INCERTAE SEDIS (1: 1)

DONACOBIUS Swainson, 1832[2] M
○ *Donacobius atricapilla* BLACK-CAPPED DONACOBIUS
    __ *D.a.brachypterus* Madarász, 1913   E Panama, N Colombia
    __ *D.a.nigrodorsalis* Traylor, 1948   SE Colombia, E Ecuador, E Peru
    __ *D.a.atricapilla* (Linnaeus, 1766)[3]   Venezuela, the Guianas to NE Argentina
    __ *D.a.albovittatus* d'Orbigny & Lafresnaye, 1837 E Bolivia

---

[1] We follow A.O.U. (1998) (3) in treating *phaeocephalus* as distinct from *arada*; the name *arada* is based on a Cayenne Indian name and may not be changed to *aradus* see David & Gosselin (2002) (613).
[2] Unpublished molecular data indicate that this species does not belong in the Troglodytidae or the Mimidae (Barker 1999) (96).
[3] For correction of spelling see David & Gosselin (2002) (613).

POLIOPTILIDAE GNATCATCHERS (3: 14)

MICROBATES Sclater & Salvin, 1873 M
○ *Microbates collaris* COLLARED GNATWREN

— *M.c.paraguensis* W.H. Phelps, Sr., 1946	S Venezuela
— *M.c.collaris* (Pelzeln, 1868)	SE Colombia to the Guianas
— *M.c.perlatus* Todd, 1927	N Brazil, NE Peru

○ *Microbates cinereiventris* TAWNY-FACED GNATWREN

— *M.c.semitorquatus* (Lawrence, 1862)	Caribbean slope of S Nicaragua to NW Colombia (Gulf of Urubá)
— *M.c.albapiculus* Olson, 1980	N Colombia (Cauca Valley) (1692)
— *M.c.magdalenae* Chapman, 1915	N Colombia (Magdalena Valley)
— *M.c.cinereiventris* (P.L. Sclater, 1855)	Pacific slope from SW Panama to W Colombia and SW Ecuador
— *M.c.unicus* Olson, 1980	E Colombia (Cundinamarca) (1692)
— *M.c.hormotus* Olson, 1980	S Colombia to E Ecuador and NE Peru (1692)
— *M.c.peruvianus* Chapman, 1923	E Peru from Amazonas to Puno.

RAMPHOCAENUS Vieillot, 1819 M
● *Ramphocaenus melanurus* LONG-BILLED GNATWREN

— *R.m.rufiventris* (Bonaparte, 1838)[1]	SE Mexico
— *R.m.ardeleo* Van Tyne & Trautman, 1941	Yucatán (Mexico), N Guatemala
— *R.m.panamensis* A.R. Phillips, 1991[2]	C and E Panama (incl. Azuero Pen.) (1876)
— *R.m.sanctaemarthae* P.L. Sclater, 1861	N Colombia, NW Venezuela
— *R.m.griseodorsalis* Chapman, 1912	WC Colombia
— *R.m.trinitatis* Lesson, 1839	E Colombia to N Venezuela and Trinidad
— *R.m.pallidus* Todd, 1913	EC Colombia to NC Venezuela
— *R.m.duidae* J.T. Zimmer, 1937	S Venezuela to NE Ecuador
— *R.m.badius* J.T. Zimmer, 1937	NE Peru, SE Ecuador
— *R.m.amazonum* Hellmayr, 1907	EC Peru, NW Brazil
— *R.m.obscurus* J.T. Zimmer, 1931	E Peru to N Bolivia
— *R.m.albiventris* P.L. Sclater, 1883	E Venezuela, the Guianas, NE Brazil
— *R.m.austerus* J.T. Zimmer, 1937	E Brazil (S of the Amazon)
— *R.m.melanurus* Vieillot, 1819	Coastal EC Brazil
— *R.m.sticturus* Hellmayr, 1902	SW Brazil

POLIOPTILA P.L. Sclater, 1855 F
● *Polioptila caerulea* BLUE-GREY GNATCATCHER[3]

✔ *P.c.caerulea* (Linnaeus, 1766)[4]	SE Canada and C and E USA >> SE USA south to Central America
— *P.c.obscura* Ridgway, 1883[5]	W USA to S Baja California >> S Mexico
— *P.c.perplexa* A.R. Phillips, 1991[6]	NC Mexico (1876)
— *P.c.deppei* van Rossem, 1934	E Mexico to Belize >> Guatemala
— *P.c.nelsoni* Ridgway, 1903	Oaxaca (SW Mexico)
— *P.c.cozumelae* Griscom, 1926	Cozumel I. (E Mexico)
— *P.c.caesiogaster* Ridgway, 1887[7]	Bahama Is.

● *Polioptila melanura* BLACK-TAILED GNATCATCHER

— *P.m.lucida* van Rossem, 1931[8]	SE California, S Nevada, C Arizona to NE Baja California and NW Sonora (Mexico)
— *P.m.melanura* Lawrence, 1857	C and S New Mexico and S Texas S to Tamaulipas (Mexico)
— *P.m.curtata* van Rossem, 1932	Tiburon I.

---

[1] Phillips (1991) (1876) treated *rufiventris* as a separate species with *panamensis* as a race.
[2] For recognition and comments see Dickerman & Parkes (1997) (729).
[3] Revision based on Phillips (1991) (1876) but *comiteca* A.R. Phillips, 1991 (1876), with a range apparently intersected by that of *nelsoni*, awaits evaluation see Dickerman & Parkes (1997) (729).
[4] Includes *mexicana* see Phillips (1991) (1876).
[5] Includes *amoenissima* and *gracilis* see Phillips (1991) (1876).
[6] Tentatively accepted see Dickerman & Parkes (1997) (729).
[7] Omitted in last Edition.
[8] Includes *pontilis* see Mellink & Rea (1994) (1556).

○ *Polioptila californica* CALIFORNIA GNATCATCHER[1]
  __ *P.c.californica* Brewster, 1881   SW California
  __ *P.c.atwoodi* Mellink & Rea, 1994   N Baja California (1556)
  __ *P.c.margaritae* Ridgway, 1904   S Baja California, Santa Margarita I.

○ *Polioptila lembeyei* (Gundlach, 1858) CUBAN GNATCATCHER   E Cuba

○ *Polioptila albiloris* WHITE-LORED GNATCATCHER
  __ *P.a.vanrossemi* Brodkorb, 1944   S and W Mexico
  __ *P.a.albiventris* Lawrence, 1885   N Yucatán Pen. (Mexico)
  __ *P.a.albiloris* Sclater & Salvin, 1860   Guatemala to Costa Rica

○ *Polioptila nigriceps* BLACK-CAPPED GNATCATCHER
  __ *P.n.restricta* Brewster, 1889   Sonora to N Sinaloa (Mexico)
  __ *P.n.nigriceps* S.F. Baird, 1864   C Sinaloa to Colima (Mexico)

● *Polioptila plumbea* TROPICAL GNATCATCHER
  √ *P.p.brodkorbi* Parkes, 1979   SE Mexico to N Costa Rica (1771)
  __ *P.p.superciliaris* Lawrence, 1861   Costa Rica to N Colombia
  __ *P.p.cinericia* Wetmore, 1957   Coiba I. (Panama)
  __ *P.p.bilineata* (Bonaparte, 1850)[2]   NW Colombia to W Peru
  __ *P.p.daguae* Chapman, 1915   WC Colombia
  __ *P.p.anteocularis* Hellmayr, 1900   C Colombia
  __ *P.p.plumbiceps* Lawrence, 1865   NE Colombia, N Venezuela
  __ *P.p.innotata* Hellmayr, 1901   E Colombia, S Venezuela, N Brazil
  __ *P.p.plumbea* (J.F. Gmelin, 1788)   The Guianas, NE Brazil
  __ *P.p.maior* Hellmayr, 1900[3]   NE Peru
  __ *P.p.parvirostris* Sharpe, 1885   EC Peru
  __ *P.p.atricapilla* (Swainson, 1832)   NE Brazil

○ *Polioptila lactea* Sharpe, 1885 CREAMY-BELLIED GNATCATCHER   SE Brazil to E Paraguay and NE Argentina

○ *Polioptila guianensis* GUIANAN GNATCATCHER
  __ *P.g.facilis* J.T. Zimmer, 1942   Venezuela, NE Brazil
  __ *P.g.guianensis* Todd, 1920   The Guianas
  __ *P.g.paraensis* Todd, 1937   E Brazil

○ *Polioptila schistaceigula* E. Hartert, 1898 SLATE-THROATED GNATCATCHER
                     E Panama to NW Ecuador

○ *Polioptila dumicola* MASKED GNATCATCHER
  __ *P.d.saturata* Todd, 1946   Highlands of Bolivia
  __ *P.d.berlepschi* Hellmayr, 1901   E Brazil, E Bolivia
  __ *P.d.dumicola* (Vieillot, 1817)   SE Bolivia to Uruguay, N Argentina

SITTIDAE NUTHATCHES AND WALLCREEPER (2: 25)

SITTINAE

SITTA Linnaeus, 1758 F
● *Sitta europaea* EURASIAN NUTHATCH
  __ *S.e.europaea* Linnaeus, 1758   N and E Europe
  __ *S.e.arctica* Buturlin, 1907   NE Siberia (Lena basin to Anadyr and N Koryakland)
  __ *S.e.albifrons* Taczanowski, 1882   Kamchatka, SE Koryakland
  √ *S.e.caesia* Wolf, 1810   W, C and SE Europe (Britain to Greece)
  __ *S.e.hispaniensis* Witherby, 1913   Spain, Portugal, NW Africa
  __ *S.e.cisalpina* Sachtleben, 1919   Italy, Sicily, W Slovenia, W Croatia
  __ *S.e.levantina* E. Hartert, 1905   Israel, Lebanon, S Turkey
  __ *S.e.persica* Witherby, 1903   SE Turkey, N Iraq, SW Iran

---

[1] For separation from *P. melanura* see Atwood (1991) (63); however since reviewed by Mellink & Rea (1994) (1556).
[2] Given as 1851 by Paynter (1964: 452) (1807) but see Zimmer (1926) (2707).
[3] Recent authors seem to suggest that birds of the upper Marañon valley may be a distinct species. The origin of this idea appears to be Hellmayr (1934: 493-4 fn) (1103). Although the name *andina* may have been given to a female of this taxon it is not the oldest name and there should be no cause to employ a name other than *maior*.

___ *S.e.caucasica* Reichenow, 1901   NE Turkey, Caucasus area

___ *S.e.rubiginosa* Tschusi & Zarudny, 1905   N Iran, Azerbaijan

___ *S.e.asiatica* Gould, 1837   SW Urals Mts. and S Siberia east to Sea of Okhotsk, south to N Mongolia, NW Manchuria, Hokkaido

___ *S.e.seorsa* Portenko, 1955   E Xinjiang (NW China)

___ *S.e.sinensis* J. Verreaux, 1871   NC and E China to Hebei, Taiwan and N Guangdong

___ *S.e.amurensis* Swinhoe, 1871   N Hebei to NE Manchuria, S Russian Far East, Korea, Honshu, Shikoku, N Kyushu

___ *S.e.roseilia* Bonaparte, 1850   S Kyushu

___ *S.e.bedfordi* Ogilvie-Grant, 1909   Cheiu Do (Quelpart) I. (Korea)

○ *Sitta nagaensis* CHESTNUT-VENTED NUTHATCH[1]

___ *S.n.montium* La Touche, 1899[2]   E Xizang, SW Sichuan, N and E Yunnan, Guizhou, NW Fujian, NW Vietnam

___ *S.n.nagaensis* Godwin-Austen, 1874   SE Xizang, Assam, NW, N and E Burma, W and S Yunnan, NW Thailand

___ *S.n.grisiventris* Kinnear, 1920   W Burma; SE Laos, Langbian (SC Vietnam) [3]

○ *Sitta cashmirensis* W.E. Brooks, 1871 KASHMIR NUTHATCH[4]   Afghanistan, SW Pakistan, NW and central Himalayas

○ *Sitta castanea* CHESTNUT-BELLIED NUTHATCH

___ *S.c.almorae* Kinnear & Whistler, 1930   WC Himalayan foothills

___ *S.c.cinnamoventris* Blyth, 1842   EC and E Himalayan foothills, N, E and SW Assam, E Bangladesh

___ *S.c.castanea* Lesson, 1830   N and C India (plains and terai)

___ *S.c.prateri* Whistler & Kinnear, 1932[5]   E Ghats (EC India)

___ *S.c.koelzi* Vaurie, 1950   SE Assam, W and N Burma

___ *S.c.neglecta* Walden, 1870   C, S and E Burma, W, N and E Thailand, SW Yunnan and S Indochina

___ *S.c.tonkinensis* Kinnear, 1936   Eastern N Thailand, SE Yunnan, N Laos, N Vietnam

○ *Sitta himalayensis* Jardine & Selby, 1835 WHITE-TAILED NUTHATCH[6]   Himalayas, Assam, W, N and E Burma, N Laos, NW Vietnam

○ *Sitta victoriae* Rippon, 1904 WHITE-BROWED NUTHATCH   Chin Hills (W Burma)

◐ *Sitta pygmaea* PYGMY NUTHATCH

√ *S.p.melanotis* van Rossem, 1929[7]   SW Canada, W USA, NW Mexico

___ *S.p.pygmaea* Vigors, 1839   C California

___ *S.p.leuconucha* Anthony, 1889   SW California, Baja California

___ *S.p.elii* A.R. Phillips, 1986[8]   N Mexico (1875)

___ *S.p.flavinucha* van Rossem, 1939   E Mexico

___ *S.p.brunnescens* Norris, 1958   W Mexico

● *Sitta pusilla* BROWN-HEADED NUTHATCH

√ *S.p.pusilla* Latham, 1790[9]   S USA

___ *S.p.insularis* Bond, 1931   Grand Bahama I.

● *Sitta whiteheadi* Sharpe, 1884 CORSICAN NUTHATCH   Corsica

○ *Sitta ledanti* Vielliard, 1976 ALGERIAN NUTHATCH   N Algeria (2526)

○ *Sitta krueperi* Pelzeln, 1863 KRÜPER'S NUTHATCH   Samos, Lesbos (E Greece), Asia Minor, W Caucasus area

○ *Sitta yunnanensis* Ogilvie-Grant, 1900 YUNNAN NUTHATCH   south to C China (SE Xizang and S Sichuan to Yunnan and Guizhou)

● *Sitta canadensis* Linnaeus, 1766 RED-BREASTED NUTHATCH   Canada, USA

---

[1] Separation of *nagaensis* is open to question; breeding sympatry with *europaea* not yet demonstrated. Altitudinal replacement seems to occur and the two may be one species (Harrap, 1996) (1071).
[2] Includes *nebulosa* see Harrap (1996) (1071).
[3] Two apparently unconnected populations, no doubt subtly different.
[4] Harrap (1996) (1071) treated this as a separate species based mainly on voice.
[5] Doubtfully distinct. Erroneously dated 1936 by Greenway (1967: 133) (1007A).
[6] Includes *australis* (and *whistleri* of NW Vietnam – erroneously placed in *S. europaea* in last Edition), see Harrap (1996) (1071).
[7] Includes *canescens* and *chihuahuae* see Phillips (1986) (1875).
[8] Tentatively accepted see Dickerman & Parkes (1997) (729).
[9] Includes *caniceps* see Phillips (1986) (1875).

○ *Sitta villosa* CHINESE NUTHATCH
— *S.v.bangsi* Stresemann, 1929 — NW Sichuan, SW Gansu, E Qinghai, Ningxia, S Shaanxi
— *S.v.villosa* J. Verreaux, 1865 — Shanxi to E Manchuria, Korea and S Russian Far East

○ *Sitta leucopsis* WHITE-CHEEKED NUTHATCH
— *S.l.leucopsis* Gould, 1850 — E Afghanistan, N Pakistan, NW Himalayas to W Nepal
— *S.l.przewalskii* Berezowski & Bianchi, 1891 — SE and E Tibetan plateau to W Sichuan and Gansu

◐ *Sitta carolinensis* WHITE-BREASTED NUTHATCH
— *S.c.aculeata* Cassin, 1857 — W USA
— *S.c.alexandrae* Grinnell, 1926 — N Baja California (Mexico)
— *S.c.lagunae* Brewster, 1891 — S Baja California
— *S.c.tenuissima* Grinnell, 1918 — SW Canada, NW and W USA
— *S.c.atkinsi* W.E.D. Scott, 1890 — Florida
— *S.c.nelsoni* Mearns, 1902 — C and S USA (mainly Rocky Mts.), N Mexico
— *S.c.umbrosa* van Rossem, 1939 — NW Mexico
— *S.c.oberholseri* H.W. Brandt, 1938 — NE Mexico
— *S.c.mexicana* Nelson & Palmer, 1894 — N and C Mexico
— *S.c.kinneari* van Rossem, 1939 — SW Mexico
⌁ *S.c.carolinensis* Latham, 1790 — E Canada, E USA

◐ *Sitta neumayer* WESTERN ROCK NUTHATCH
⌁ *S.n.neumayer* Michahelles, 1830 — SE Europe
— *S.n.syriaca* Temminck, 1835 — W and S Asia Minor and Levant to N Israel
— *S.n.rupicola* Blanford, 1873 — NE Turkey, Transcaucasia, N Iran
— *S.n.tschitscherini* Zarudny, 1904 — Iraq, SW Iran
— *S.n.plumbea* Koelz, 1950 — SE Iran

○ *Sitta tephronota* EASTERN ROCK NUTHATCH
— *S.t.dresseri* Zarudny & Buturlin, 1906 — SE Turkey to SW Iran and Iraq
— *S.t.obscura* Zarudny & Loudon, 1905 — S Transcaucasia to N and C Iran and NE Turkey
— *S.t.tephronota* Sharpe, 1872 — E Turkmenistan to SE Kazakhstan, Afghanistan and W Pakistan
— *S.t.iranica* (Buturlin, 1916)[1] — SW Turkmenistan and NE Iran to Kyzylkum Desert (Uzbekistan)

○ *Sitta frontalis* VELVET-FRONTED NUTHATCH[2]
— *S.f.frontalis* Swainson, 1820 — C and E Himalayas, India, Burma, N Thailand, Indochina
— *S.f.saturatior* E. Hartert, 1902 — Malay Pen., N Sumatra
— *S.f.corallipes* (Sharpe, 1888) — Borneo
— *S.f.palawana* E. Hartert, 1905 — Palawan
— *S.f.velata* Temminck, 1821 — S Sumatra[3], Java (2395)

○ *Sitta solangiae* YELLOW-BILLED NUTHATCH[4]
— *S.s.solangiae* (Delacour & Jabouille, 1930) — NW and NC Vietnam
— *S.s.fortior* Delacour & Greenway, 1939 — SE Laos, C and S Vietnam
— *S.s.chienfengensis* Cheng, Ting & Wang, 1964 — Hainan (339)

○ *Sitta oenochlamys* SULPHUR-BILLED NUTHATCH
— *S.o.mesoleuca* (Ogilvie-Grant, 1894) — N Luzon
— *S.o.isarog* Rand & Rabor, 1967 — NE, C and S Luzon (1978)
— *S.o.lilacea* (J. Whitehead, 1897) — Samar, Leyte
— *S.o.apo* (Hachisuka, 1930) — Mindanao (except Zamboanga Pen.)
— *S.o.zamboanga* Rand & Rabor, 1957 — W Mindanao, Basilan
— *S.o.oenochlamys* (Sharpe, 1877) — Cebu, Guimaras, Panay, Negros

○ *Sitta azurea* BLUE NUTHATCH
— *S.a.expectata* (E. Hartert, 1914) — W Malaysia, Sumatra
— *S.a.nigriventer* (Robinson & Kloss, 1919) — W Java
— *S.a.azurea* Lesson, 1830 — C and E Java

[1] Recognized in Cramp *et al.* (1993: 314) (594).
[2] The relationships of this species, *solangiae* and *oenochlamys* are uncertain. Harrap (1996) (1071) preferred to treat them as three species, pending further research. We adopt this hypothesis, not that of Dickinson *et al.* (1991) (745).
[3] See Mees (1986) (1539); but van Marle & Voous (1988) (2468) listed nominate *frontalis* from there.
[4] Quite possibly conspecific with the next species, but note on *frontalis*.

○ **Sitta magna** GIANT NUTHATCH

   — *S.m.ligea* Deignan, 1938   Yunnan to S Sichuan, Guizhou and Guangxi (S and C China)

   — *S.m.magna* Wardlaw Ramsay, 1876   C and E Burma, NW Thailand

○ **Sitta formosa** Blyth, 1843 BEAUTIFUL NUTHATCH   E Himalayas, Assam, W, N and E Burma, extreme N Thailand, SE Yunnan, N and C Laos, NW Vietnam

TICHODROMADINAE

TICHODROMA Illiger, 1811 F

○ **Tichodroma muraria** WALLCREEPER

   — *T.m.muraria* (Linnaeus, 1766)   S and E Europe, Turkey, Caucasus area and N Iran

   — *T.m.nepalensis* Bonaparte, 1850   Afghanistan to Himalayas, Tien Shan, N and C China, and S Mongolia

CERTHIIDAE   TREECREEPERS (2: 8)

CERTHIINAE

CERTHIA Linnaeus, 1758 F

● **Certhia familiaris** EURASIAN TREECREEPER

   — *C.f.britannica* Ridgway, 1882   Britain, Ireland

   — *C.f.macrodactyla* C.L. Brehm, 1831[1]   W and C Europe to N Spain, Italy and W Croatia

   — *C.f.familiaris* Linnaeus, 1758   N and E Europe to Balkan countries

   ✓ *C.f.corsa* E. Hartert, 1905   Corsica

   — *C.f.caucasica* Buturlin, 1907   N Turkey, Caucasus area

   — *C.f.persica* Zarudny & Loudon, 1905   SE Azerbaijan, N Iran

   — *C.f.tianschanica* E. Hartert, 1905   S and E Tien Shan to Hami area (E Xinjiang)

   — *C.f.hodgsoni* W.E. Brooks, 1873   N Pakistan, Kashmir

   — *C.f.mandellii* W.E. Brooks, 1874   Himalayas (NW India to NW Assam)

   — *C.f.bianchii* E. Hartert, 1905   E and NE Qinghai to C Gansu and S Shaanxi (N, C China)

   — *C.f.khamensis* Bianchi, 1903   SE and E Xizang, N Burma, N Yunnan, W Sichuan, S Gansu

   — *C.f.daurica* Domaniewski, 1922[2]   S Siberia (Urals to Sakhalin), NE China, S Kurils Is., Hokkaido (N Japan)

   — *C.f.japonica* E. Hartert, 1897   Honshu and Shikoku (Japan)

● **Certhia americana** BROWN CREEPER/AMERICAN TREECREEPER

   — *C.a.montana* Ridgway, 1882[3]   W Canada, W USA

   — *C.a.occidentalis* Ridgway, 1882   NW Canada, W USA

   ✓ *C.a.alascensis* Webster, 1986[4]   Alaska (1875)

   — *C.a.stewarti* Webster, 1986[5]   British Columbia (1875)

   — *C.a.zelotes* Osgood, 1901   N California

   — *C.a.phillipsi* Unitt & Rea, 1997   C California (2449)

   — *C.a.leucosticta* van Rossem, 1931   S Nevada, Utah

   — *C.a.albescens* Berlepsch, 1888   SW USA, NW Mexico

   — *C.a.jaliscensis* Miller & Griscom, 1925   W Mexico

   — *C.a.guerrerensis* van Rossem, 1939   SW Mexico

   — *C.a.alticola* G.S. Miller, 1895[6]   C and SE Mexico

   — *C.a.pernigra* Griscom, 1935   S Mexico, Guatemala

   — *C.a.extima* Miller & Griscom, 1925   Nicaragua

   ✓ *C.a.americana* Bonaparte, 1838   C and E Canada, EC and SE USA

   — *C.a.nigrescens* Burleigh, 1935   EC USA

---

[1] Implicitly includes *pyrenaica* see Vaurie (1959: 540) (2502).

[2] Includes *orientalis* (considered very poorly differentiated by Vaurie, 1959 (2502); since then reported ranges of this and of *daurica* conflict).

[3] Includes *iletica* Oberholser, 1974 (1673), see Browning (1978) (246), and *idahoensis* Webster, 1986 (1875), see Browning (1990) (251).

[4] For recognition see Browning (1990) (251).

[5] For recognition see Browning (1990) (251).

[6] Includes *molinensis* see Dickerman & Parkes (1997) (729).

● *Certhia brachydactyla* SHORT-TOED TREECREEPER
    __ *C.b.mauritanica* Witherby, 1905   NW Africa
    ✓ *C.b.megarhyncha* C.L. Brehm, 1831   Portugal, W Spain and W and N France to W Germany
    __ *C.b.brachydactyla* C.L. Brehm, 1820[1]   E Spain, Italy, Sicily to C and SE Europe and NW Turkey
    __ *C.b.stresemanni* Kummerlöwe & Niethammer, 1934   Asia Minor
    __ *C.b.rossocaucasica* Stepanyan, 2000   SW Caucasus Mts. (2333)
    __ *C.b.dorotheae* E. Hartert, 1904   Cyprus, Crete

○ *Certhia himalayana* BAR-TAILED TREECREEPER
    __ *C.h.taeniura* Severtsov, 1873   W Tien Shan, E Uzbekistan, W Tajikistan, N Afghanistan
    __ *C.h.himalayana* Vigors, 1832[2]   SE Afghanistan, N and W Pakistan east to W Nepal
    __ *C.h.yunnanensis* Sharpe, 1902   E Assam, N and NE Burma, SE Xizang and N Yunnan
                                         to W Sichuan, S Shaanxi and SW Guizhou (C China)
    __ *C.h.ripponi* Kinnear, 1929   W Burma (? S Assam)

○ *Certhia nipalensis* Blyth, 1845   RUSTY-FLANKED TREECREEPER   C and E Himalayas to SE Xizang, N, E and SE Assam,
                                         N and NE Burma

○ *Certhia discolor* BROWN-THROATED TREECREEPER
    __ *C.d.discolor* Blyth, 1845   C and E Himalayas to N Assam and SE Xizang
    __ *C.d.manipurensis* Hume, 1881   S and SE Assam, W and SW Burma
    __ *C.d.shanensis* E.C.S. Baker, 1930   N Burma, W Yunnan, W and N Thailand, NW Vietnam
    __ *C.d.laotiana* Delacour, 1951   N and C Laos
    __ *C.d.meridionalis* Robinson & Kloss, 1919   Langbian (SC Vietnam)

○ *Certhia tianquanensis*[3] Li, 1995   SICHUAN TREECREEPER[4]   Tianquan (Sichuan) (1356)

SALPORNITHINAE

SALPORNIS G.R. Gray, 1847   M
○ *Salpornis spilonotus* SPOTTED CREEPER
    __ *S.s.emini* Hartlaub, 1884   Gambia to NE Zaire, NW Uganda
    __ *S.s.erlangeri* Neumann, 1907   W and S Ethiopia
    __ *S.s.salvadori* (Bocage, 1878)   W Kenya, Angola to S Tanzania and N Mozambique
    __ *S.s.xylodromus* Clancey, 1975   E Zimbabwe, W Mozambique (448)
    __ *S.s.rajputanae* R. Meinertzhagen, 1926   Rajasthan (W India)
    __ *S.s.spilonotus* (Franklin, 1831)   C India

MIMIDAE   MOCKINGBIRDS, THRASHERS (12: 34)

DUMETELLA "S.D.W." = C. T. Wood, 1837   F
◕ *Dumetella carolinensis* (Linnaeus, 1766)   GRAY CATBIRD[5]   S Canada, USA, Bermuda >>C America, West Indies

MELANOPTILA P.L. Sclater, 1858   F
○ *Melanoptila glabrirostris* P.L. Sclater, 1858   BLACK CATBIRD   SE Mexico, N Guatemala, N Honduras

MIMUS Boie, 1826   M
● *Mimus polyglottos* NORTHERN MOCKINGBIRD
    ✓ *M.p.polyglottos* (Linnaeus, 1758)[6]   S Canada, USA to S Mexico
    __ *M.p.orpheus* (Linnaeus, 1758)   Cuba, Hispaniola, Puerto Rico, Jamaica

◕ *Mimus gilvus* TROPICAL MOCKINGBIRD
    ✓ *M.g.gracilis* Cabanis, 1852   S Mexico (incl. Yucatán Pen.), Guatemala, Honduras,
                                         El Salvador
    __ *M.g.leucophaeus* Ridgway, 1888   Cozumel I. (Mexico)
    __ *M.g.antillarum* Hellmayr & Seilern, 1915   Martinique, Windward Is.
    __ *M.g.tobagensis* Dalmas, 1900   Trinidad, Tobago I.

---

[1] Includes *harterti* see Roselaar (1995) (2107).
[2] Includes *limes* and *infima* following Vaurie (1959: 544) (2502), but see Ripley (1982: 511) (2055).
[3] This spelling is from the main paper in Chinese, the English summary once spells the name *tianguanensis* – a *lapsus*.
[4] Occurs in sympatry with *C. familiaris*. For treatment at specific level see Martens *et al.* In press. (1445a).
[5] Includes *meridianus* Burleigh, 1960 (274), see Browning (1990) (251).
[6] Includes *leucopterus* see Phillips (1986) (1875).

— *M.g.rostratus* Ridgway, 1884   N Venezuela islands
— *M.g.tolimensis* Ridgway, 1904   W and C Colombia
— *M.g.melanopterus* Lawrence, 1849   N and E Colombia, Venezuela, Guyana, N Brazil
— *M.g.gilvus* (Vieillot, 1808)   French Guiana, Surinam
— *M.g.antelius* Oberholser, 1919   N and E Brazil
— *M.g.magnirostris* Cory, 1887   San Andres I. (Colombia)

○ *Mimus gundlachii* BAHAMA MOCKINGBIRD
— *M.g.gundlachii* Cabanis, 1855   Bahamas, cays off N Cuba
— *M.g.hillii* March, 1863   Jamaica

● *Mimus thenca* (Molina, 1782) CHILEAN MOCKINGBIRD   C Chile

○ *Mimus longicaudatus* LONG-TAILED MOCKINGBIRD
— *M.l.platensis* Chapman, 1924   La Plata I., W Ecuador
— *M.l.albogriseus* Lesson, 1844   SW Ecuador, N Peru
— *M.l.longicaudatus* Tschudi, 1844   W Peru
— *M.l.maranonicus* Carriker, 1933   NC Peru (Marañon Valley)

○ *Mimus saturninus* CHALK-BROWED MOCKINGBIRD
— *M.s.saturninus* (M.H.K. Lichtenstein, 1823)   N Brazil
— *M.s.arenaceus* Chapman, 1890   NE Brazil
— *M.s.frater* Hellmayr, 1903   N Bolivia, SW Brazil
— *M.s.modulator* (Gould, 1836)   E Bolivia, S Brazil, Uruguay, N Argentina

○ *Mimus patagonicus* (d'Orbigny & Lafresnaye, 1837) PATAGONIAN MOCKINGBIRD
      W and S Argentina, S Chile >> N Argentina

○ *Mimus triurus* (Vieillot, 1818) WHITE-BANDED MOCKINGBIRD   Argentina >> E Bolivia, S Brazil, Paraguay, Uruguay

○ *Mimus dorsalis* (d'Orbigny & Lafresnaye, 1837) BROWN-BACKED MOCKINGBIRD
      Andes of Bolivia, NW Argentina

NESOMIMUS Ridgway, 1890 M
○ *Nesomimus parvulus* GALAPAGOS MOCKINGBIRD
— *N.p.parvulus* (Gould, 1837)   Narborough, Albemarle, Daphne, Seymour, Indefatigable Is.
— *N.p.barringtoni* Rothschild, 1898   Barrington I.
— *N.p.personatus* Ridgway, 1890   Abingdon, Bindloe, James, Jervis Is.
— *N.p.wenmani* Swarth, 1931   Wenman I.
— *N.p.hulli* Rothschild, 1898   Culpepper I.
— *N.p.bauri* Ridgway, 1894   Tower I.

○ *Nesomimus trifasciatus* (Gould, 1837) CHARLES MOCKINGBIRD   Gardner, Champion (Galapagos Is.)

○ *Nesomimus macdonaldi* Ridgway, 1890 HOOD MOCKINGBIRD   Hood, Gardner Is.

○ *Nesomimus melanotis* (Gould, 1837) CHATHAM MOCKINGBIRD   Chatham/San Cristobal I.

OREOSCOPTES S.F. Baird, 1858 M
● *Oreoscoptes montanus* (J.K. Townsend, 1837) SAGE THRASHER   W and SW USA, Baja California >> C Mexico

MIMODES Ridgway, 1882 M
○ *Mimodes graysoni* (Lawrence, 1871) SOCORRO MOCKINGBIRD   Socorro I., Revillagigedo Group

TOXOSTOMA Wagler, 1831 N
● *Toxostoma rufum* BROWN THRASHER
↙ *T.r.rufum* (Linnaeus, 1758)   SE Canada, NC, E and SE USA
— *T.r.longicauda* (S.F. Baird, 1858)   SC Canada, C and S USA

● *Toxostoma longirostre* LONG-BILLED THRASHER
↙ *T.l.sennetti* (Ridgway, 1888)   S Texas, NE Mexico
— *T.l.longirostre* (Lafresnaye, 1838)   E Mexico

○ *Toxostoma guttatum* (Ridgway, 1885) COZUMEL THRASHER   Cozumel I. (Mexico)

○ *Toxostoma cinereum* Grey Thrasher
　　— *T.c.mearnsi* (Anthony, 1895)　　　　　　　　　N Baja California
　　— *T.c.cinereum* (Xantús de Vesey, 1860)　　　　　S Baja California

○ *Toxostoma bendirei* Coues, 1873 Bendire's Thrasher[1]　SW USA, NW Mexico

○ *Toxostoma ocellatum* Ocellated Thrasher
　　— *T.o.ocellatum* (P.L. Sclater, 1862)　　　　　　SC Mexico
　　— *T.o.villai* A.R. Phillips, 1986[2]　　　　　　　S Mexico (1875)

● *Toxostoma curvirostre* Curve-billed Thrasher
　　— *T.c.palmeri* (Coues, 1872)　　　　　　　　　S Arizona, N Sonora (Mexico)
　　— *T.c.insularum* van Rossem, 1930　　　　　　　San Esteban, Tiburon Is.
　　— *T.c.maculatum* (Nelson, 1900)　　　　　　　　NW Mexico
　　— *T.c.occidentale* (Ridgway, 1882)　　　　　　　WC Mexico
　　— *T.c.celsum* R.T. Moore, 1941　　　　　　　　　S USA, NC Mexico
　　— *T.c.curvirostre* (Swainson, 1827)　　　　　　C and SC Mexico
　　— *T.c.oberholseri* Law, 1928　　　　　　　　　　S Texas, NE Mexico

● *Toxostoma redivivum* California Thrasher
　　— *T.r.sonomae* Grinnell, 1915　　　　　　　　　N California
　　— *T.r.redivivum* (Gambel, 1845)　　　　　　　　S California, NW Baja California

○ *Toxostoma crissale* Crissal Thrasher[3]
　　— *T.c.coloradense* van Rossem, 1946　　　　　　Colorado Desert of SE California and SE Arizona
　　— *T.c.crissale* Henry, 1858[4]　　　　　　　　　C Arizona and NW Mexico (N Baja California,
　　　　　　　　　　　　　　　　　　　　　　　　　　　N Sonora, N Coahuila)
　　— *T.c.dumosum* R.T. Moore, 1941　　　　　　　NC Mexico

○ *Toxostoma lecontei* Le Conte's Thrasher
　　— *T.l.lecontei* Lawrence, 1851　　　　　　　　　SW USA, NW Mexico, N Baja California
　　— *T.l.macmillanorum* A.R. Phillips, 1966[5]　　　S California (1872)
　　— *T.l.arenicola* (Anthony, 1897)[6]　　　　　　W Baja California

RAMPHOCINCLUS Lafresnaye, 1843 M
○ *Ramphocinclus brachyurus* White-breasted Thrasher
　　— *R.b.brachyurus* (Vieillot, 1818)　　　　　　　Martinique I.
　　— *R.b.sanctaeluciae* Cory, 1887　　　　　　　　St. Lucia I.

MELANOTIS Bonaparte, 1850 M
○ *Melanotis caerulescens* Blue Mockingbird
　　— *M.c.caerulescens* (Swainson, 1827)　　　　　C and S Mexico
　　— *M.c.longirostris* Nelson, 1898　　　　　　　　Tres Marias Is.

○ *Melanotis hypoleucus* Hartlaub, 1852 Blue-and-white Mockingbird　SE Mexico, Guatemala, Honduras, El Salvador

ALLENIA Cory, 1891 F
○ *Allenia fusca* Scaly-breasted Thrasher[7]
　　— *A.f.hypenema* (Buden, 1993)　　　　　　　　N Lesser Antilles (268)
　　— *A.f.vincenti* (Kratter & Garrido, 1996)　　　St. Vincent I. (1304)
　　— †?*A.f.atlanticus* (Buden, 1993)　　　　　　　Barbados (268)
　　— *A.f.schwartzi* (Buden, 1993)　　　　　　　　St. Lucia I. (268)
　　— *A.f.fusca* (Statius Müller, 1776)　　　　　　S Lesser Antilles

MARGAROPS P.L. Sclater, 1859 M
○ *Margarops fuscatus* Pearly-eyed Thrasher
　　— *M.f.fuscatus* (Vieillot, 1808)　　　　　　　　S Bahamas, Hispaniola, Puerto Rico, N Leeward Is.

---

[1] Includes *candidum* and *rubricatum* see Phillips (1986) (1875).
[2] Tentatively accepted see Parkes in Dickerman & Parkes (1997) (729).
[3] The name *dorsale* Henry, 1858, has been suppressed in favour of *crissale*; see Opinion 1249, I.C.Z.N. (1983) (1175).
[4] The original citation for *dorsale* in Davis & Miller (1960) (617) applies also to *crissale*. Includes *trinitatis* see Phillips (1986) (1875).
[5] For recognition see Browning (1990) (251). For correct date see Dickerman & Parkes (1997: 226) (729).
[6] A.O.U. (1998) (3) considered *arenicola* best treated within the species *T. lecontei*.
[7] See Buden (1993) (268) for a revision; however, Hunt *et al.* (2001) (1165) provided evidence for maintaining *Allenia* not submerging it in *Margarops*.

— *M.f.klinikowskii* Garrido & Remsen, 1996   St. Lucia I. (925)
— *M.f.densirostris* (Vieillot, 1818)   S Leeward Is.
— *M.f.bonairensis* Phelps & Phelps, Jr., 1948   Bonaire, Los Hermanos Is., N Venezuela

CINCLOCERTHIA G.R. Gray, 1840 F
○ *Cinclocerthia ruficauda* BROWN TREMBLER
    — *C.r.pavida* Ridgway, 1904   NW Leeward Is.
    — *C.r.tremula* (Lafresnaye, 1843)   Guadeloupe I.
    — *C.r.ruficauda* (Gould, 1836)   Dominica I.
    — *C.r.tenebrosa* Ridgway, 1904   St. Vincent I.

○ *Cinclocerthia gutturalis* GREY TREMBLER
    — *C.g.gutturalis* (Lafresnaye, 1843)   Martinique I.
    — *C.g.macrorhyncha* P.L. Sclater, 1866   St. Lucia I.

RHABDORNITHIDAE PHILIPPINE CREEPERS (1: 2)

RHABDORNIS Reichenbach, 1853 M
○ *Rhabdornis mystacalis* STRIPE-HEADED CREEPER
    — *R.m.mystacalis* (Temminck, 1825)   Luzon, Masbate, Negros, Panay
    — *R.m.minor* Ogilvie-Grant, 1896   Basilan, Bohol, Samar, Leyte, Mindanao

○ *Rhabdornis inornatus* STRIPE-BREASTED CREEPER
    — *R.i.grandis* Salomonsen, 1952[1]   N and C Luzon
    — *R.i.inornatus* Ogilvie-Grant, 1896   Samar
    — *R.i.leytensis* Parkes, 1973   Leyte, Biliran (1765)
    — *R.i.rabori* Rand, 1950   Negros
    — *R.i.alaris* Rand, 1948[2]   Mindanao

STURNIDAE STARLINGS[3] [4] (25: 115)

STURNINAE

APLONIS Gould, 1836[5] F
○ *Aplonis metallica* SHINING STARLING
    — *A.m.circumscripta* (A.B. Meyer, 1884)   Tanimbar, Damar Is.
    — *A.m.metallica* (Temminck, 1824)   Moluccas, Kai Is., Aru Is., New Guinea and satellite
                                                 islands, NE Queensland
    — *A.m.nitida* (G.R. Gray, 1858)   New Britain and New Ireland (Bismarck Arch.),
                                                 Solomon Is.
    — *A.m.purpureiceps* (Salvadori, 1878)   Admiralty Is. (W Bismarck Arch.)
    — *A.m.inornata* (Salvadori, 1880)   Biak, Numfor Is. (Geelvink Bay)

○ *Aplonis mystacea* (Ogilvie-Grant, 1911) YELLOW-EYED STARLING   WC and SC New Guinea

○ *Aplonis cantoroides* (G.R. Gray, 1862) SINGING STARLING   New Guinea and satellite islands, Bismarck Arch.,
                                                 Solomon Is.

○ *Aplonis crassa* (P.L. Sclater, 1883) TANIMBAR STARLING   Tanimbar Is. (E Lesser Sundas)

○ *Aplonis feadensis* ATOLL STARLING
    — *A.f.heureka* Meise, 1929   Islets in E Bismarck Arch.
    — *A.f.feadensis* (E.P. Ramsay, 1882)   Islets in NW Solomon Is.

○ *Aplonis insularis* Mayr, 1931 RENNELL STARLING   Rennell I. (SE Solomons)

○ *Aplonis magna* LONG-TAILED STARLING
    — *A.m.magna* (Schlegel, 1871)   Biak I. (Geelvink Bay)
    — *A.m.brevicauda* (van Oort, 1908)   Numfor I. (Geelvink Bay)

---

[1] Treated as a separate species in Kennedy *et al.* (2000) (1259), but with a rider attached. ECD believes this to be simply a large subspecies.
[2] Includes *zamboanga* see Dickinson *et al.* (1991: 303) (745).
[3] As far as possible (there are some differences in generic assignments) we follow the sequence of genera and sequences of species used by Feare & Craig (1998) (830).
[4] Amadon (1962: 103) (45) listed *Necropsar leguati*. No specimens exist. The associated *rodericanus* is apparently not a starling, but its corrected identity awaits publication.
[5] The species sequence follows Feare & Craig (1998) (830).

○ *Aplonis brunneicapillus* (Danis, 1938) White-eyed Starling[1]   Bougainville, Rendova, Guadalcanal Is. (Solomons)

○ *Aplonis grandis* Brown-winged Starling
— *A.g.grandis* (Salvadori, 1881)   N and C Solomons
— *A.g.malaitae* Mayr, 1931   Malaita I. (E Solomons)
— *A.g.macrura* Mayr, 1931   Guadalcanal I. (SE Solomons)

○ *Aplonis dichroa* (Tristram, 1895) Makira Starling   San Cristobal I. (SE Solomons)

○ *Aplonis zelandica* Rufous-winged Starling
— *A.z.maxwellii* Forbes, 1900   Santa Cruz (Santa Cruz Is.)
— *A.z.zelandica* (Quoy & Gaimard, 1830)   Vanikoro I. (Santa Cruz Is.)
— *A.z.rufipennis* E.L. Layard, 1881   Banks Is., C and N Vanuatu

○ *Aplonis striata* Striated Starling
— *A.s.striata* (J.F. Gmelin, 1788)   New Caledonia
— *A.s.atronitens* G.R. Gray, 1859   Loyalty Is.

○ *Aplonis santovestris* Harrisson & Marshall, 1937 Mountain Starling   Mts. of Espiritu Santo I. (Vanuatu)

○ *Aplonis panayensis* Asian Glossy Starling
— *A.p.affinis* (Blyth, 1846)   S Assam, SW and SE Burma >> E Bangladesh
— *A.p.tytleri* (Hume, 1873)   Andamans, Car Nicobar
— *A.p.albiris* Abdulali, 1967[2]   C and S Nicobars (9)
— *A.p.strigata* (Horsfield, 1821)   Malay Pen., Sumatra, Java, N Borneo
— *A.p.altirostris* (Salvadori, 1887)   Simeuluë, Nias, Banyak Is. (off W Sumatra)
— *A.p.pachistorhina* (Oberholser, 1912)[3]   Batu, Mentawi Is., (W Sumatran islands)
— *A.p.enganensis* (Salvadori, 1892)   Enggano I.
— *A.p.heterochlora* (Oberholser, 1917)   Anamba, Natuna Is.
— *A.p.eustathis* (Oberholser, 1926)   Borneo (except N)
— *A.p.alipodis* (Oberholser, 1926)   Maratua Is. (off E Borneo)
— *A.p.gusti* Stresemann, 1913   Bali
— *A.p.sanghirensis* (Salvadori, 1876)   Sangihe, Talaud Is. (Indonesia)
— *A.p.panayensis* (Scopoli, 1783)   Sulawesi, Philippines

○ *Aplonis mysolensis* Moluccan Starling
— *A.m.mysolensis* (G.R. Gray, 1862)   W Papuan islands, Moluccas
— *A.m.sulaensis* (Sharpe, 1890)   Banggai, Sula Is.

○ *Aplonis minor* Short-tailed Starling
— *A.m.minor* (Bonaparte, 1850)[4]   Small islands in C Lesser Sundas and Flores Sea, Sulawesi
— *A.m.todayensis* (Mearns, 1905)[5]   Mindanao

○ *Aplonis opaca* Micronesian Starling
— *A.o.aenea* (Taka-Tsukasa & Yamashina, 1931)   N Marianas
— *A.o.guami* Momiyama, 1922   S Marianas
— *A.o.orii* (Taka-Tsukasa & Yamashina, 1931)   Palau Is.
— *A.o.kurodai* Momiyama, 1920   Yap I. (W Caroline Is.)
— *A.o.ponapensis* Taka-Tsukasa & Yamashina, 1931   Pohnpei I. (EC Caroline Is.)
— *A.o.opaca* (Kittlitz, 1833)   Kusaie I. (E Caroline Is.)
— *A.o.anga* Momiyama, 1922   Truk Is. (C Caroline Is.)

○ *Aplonis pelzelni* Finsch, 1876 Pohnpei Starling   Pohnpei I. (Caroline Is.)

◉ *Aplonis tabuensis* Polynesian Starling
— *A.t.pachyrampha* Mayr, 1942[6]   Santa Cruz Is.
— *A.t.tucopiae* Mayr, 1942   Tucopia I. (Santa Cruz Is.)
— *A.t.rotumae* Mayr, 1942   Rotuma I. (C Polynesia)
— *A.t.vitiensis* E.L. Layard, 1876   Fiji Is.

[1] For correction of spelling see David & Gosselin (2002) (613).
[2] For recognition see Ripley (1982: 277) (2055).
[3] Includes *leptorrhyncha* see van Marle & Voous (1988: 195) (2468).
[4] Given as 1851 by Amadon (1962: 82) (45), but we follow Hartlaub (1851) (1087) in accepting 1850 as the date of all 543 pp. of Vol. 1.
[5] Recognised by Dickinson *et al.* (1991: 372) (745), but requires comparative review.
[6] For correction of spelling see David & Gosselin (2002) (614).

— *A.t.manuae* Mayr, 1942   Manua Is.

— *A.t.tabuensis* (J.F. Gmelin, 1788)   Tonga Is., Lau Arch. (E Fiji)

— *A.t.fortunae* E.L. Layard, 1876   Fotuna, Alofa, Uea Is.

— *A.t.tenebrosa* Mayr, 1942   Keppel, Boscawen Is.

— *A.t.nesiotes* Mayr, 1942   Niuafou I. (C Polynesia)

— *A.t.brunnescens* Sharpe, 1890   Niue I. (C Polynesia)

↙ *A.t.tutuilae* Mayr, 1942   Tutuila I. (Samoa)

— *A.t.brevirostris* (Peale, 1848)   Upolu, Savaii Is. (Samoa)

◉ *Aplonis atrifusca* (Peale, 1848) SAMOAN STARLING   Samoa

○ †*Aplonis corvina* (Kittlitz, 1833) KOSRAE STARLING   Kosrae I. (Caroline Is.)

○ *Aplonis cinerascens* Hartlaub & Finsch, 1871 RAROTONGAN STARLING   Cook Is.

○ †*Aplonis mavornata* Buller, 1887 MYSTERIOUS STARLING   Cook I. (Society Is.?)

○ *Aplonis fusca* TASMAN STARLING

— †*A.f.fusca* Gould, 1836   Norfolk I.

— †?*A.f.hulliana* Mathews, 1912[1]   Lord Howe Is.

MINO Lesson, 1827 M

○ *Mino dumontii* Lesson, 1827 YELLOW-FACED MYNA   W Papuan Is., New Guinea, Aru Is.

○ *Mino kreffti* LONG-TAILED MYNA[2]

— *M.k.giliau* Stresemann, 1922[3]   New Britain, Umboi I. (SW Bismarck Arch.)

— *M.k.kreffti* P.L. Sclater, 1869   New Ireland, New Hanover, main N and C Solomons

— *M.k.sanfordi* E. Hartert, 1929   Guadalcanal, Malaita Is. (E Solomons)

○ *Mino anais* GOLDEN MYNA

— *M.a.anais* (Lesson, 1839)   NW New Guinea

— *M.a.orientalis* (Schlegel, 1871)   N New Guinea

— *M.a.robertsoni* D'Albertis, 1877   S New Guinea

BASILORNIS Bonaparte, 1850[4] M

○ *Basilornis celebensis* G.R. Gray, 1861 SULAWESI MYNA   Sulawesi, Lembeh (to NE), and Muna and Butung (SE)

○ *Basilornis galeatus* A.B. Meyer, 1894 HELMETED MYNA   Banggai, Sula Is.

○ *Basilornis corythaix* (Wagler, 1827) LONG-CRESTED MYNA   Seram I. (S Moluccas)

○ *Basilornis mirandus* (E. Hartert, 1903) APO MYNA[5]   Mts. of C and S Mindanao

SARCOPS Walden, 1877 M

○ *Sarcops calvus* COLETO

— *S.c.calvus* (Linnaeus, 1766)   N Philippines

— *S.c.melanotus* Ogilvie-Grant, 1906   C and SE Philippines (not Palawan group)

— *S.c.lowii* Sharpe, 1877   Sulu Is.

STREPTOCITTA Bonaparte, 1850[6] F

○ *Streptocitta albicollis* WHITE-NECKED MYNA

— *S.a.torquata* (Temminck, 1828)   N and E Sulawesi, Lembeh, Togian

— *S.a.albicollis* (Vieillot, 1818)   S and SE Sulawesi, Muna, Butung

○ *Streptocitta albertinae* (Schlegel, 1866) BARE-EYED MYNA   Sula Is.

ENODES Temminck, 1839 M

○ *Enodes erythrophris* (Temminck, 1824) FIERY-BROWED MYNA   Mts. of Sulawesi

---

[1] Recognised by Schodde & Mason (1999: 654) (2189).
[2] Treated as a species separate from *M. dumontii* by Feare & Craig (1998) (830).
[3] Significantly shorter-tailed than nominate *kreffti*, see Amadon (1956) (43). Eck (2001) (792) suggested that it is closer to *M. dumontii* and that uniting these two species would be better.
[4] Dated 1851 by Amadon (1962: 116) (45), but see Zimmer (1926) (2707).
[5] For correction of spelling see David & Gosselin (2002) (614).
[6] Dated 1851 by Amadon (1962: 117) (45), but see Zimmer (1926) (2707).

SCISSIROSTRUM Lafresnaye, 1845 N
○ *Scissirostrum dubium* (Latham, 1802)[1] FINCH-BILLED MYNA
Sulawesi and nearby islands (Bangka, Lembeh, Togian, Butung, Banggai Is.)

SAROGLOSSA Hodgson, 1844 F
○ *Saroglossa spiloptera* (Vigors, 1831) SPOT-WINGED STARLING
Himalayan foothills >> Assam, Bangladesh, NE, E, SE Burma, SW Thailand

○ *Saroglossa aurata* (Statius Müller, 1776) MADAGASCAN STARLING
Madagascar

AMPELICEPS Blyth, 1842 M
○ *Ampeliceps coronatus* Blyth, 1842 GOLDEN-CRESTED MYNA
NE India, C, S and SE Burma, Thailand (south to Trang), C and S Indochina

GRACULA Linnaeus, 1758 F
○ *Gracula ptilogenys* Blyth, 1846 SRI LANKA MYNA
Sri Lanka

○ *Gracula religiosa* HILL MYNA[2]
    — *G.r.indica* (Cuvier, 1829)
SW India, Sri Lanka
    — *G.r.peninsularis* Whistler, 1933
EC India (Orissa)
    — *G.r.intermedia* Hay, 1845[3]
N and NE India, Burma, Thailand, Indochina, S Yunnan, Hainan
    — *G.r.andamanensis* (Beavan, 1867)
Andamans, Nicobars
    — *G.r.religiosa* Linnaeus, 1758
Malay Pen., Sumatra, Simeuluĕ, Bangka, Borneo, Java, Bali, and Kangean and Bawean (Java Sea)
    — *G.r.enganensis* Salvadori, 1892
Enggano (off W Sumatra)
    — *G.r.batuensis* Finsch, 1899
Batu and Mentawi Is. (off W Sumatra)
    — *G.r.robusta* Salvadori, 1887
Banyak, Nias Is. (off W Sumatra)
    — *G.r.palawanensis* (Sharpe, 1890)
Palawan
    — *G.r.venerata* Bonaparte, 1850[4]
Sumbawa to Alor (Lesser Sundas)

ACRIDOTHERES Vieillot, 1816 M
○ *Acridotheres grandis* F. Moore, 1858 GREAT MYNA[5]
E and S Assam, W, N and E Burma, S Yunnan, Guangxi, Thailand (except Pen.), Indochina

○ *Acridotheres cristatellus* CRESTED MYNA
    — *A.c.cristatellus* (Linnaeus, 1766)[6]
SC, S and E China
    — *A.c.brevipennis* E. Hartert, 1910
Hainan I., Indochina
    — *A.c.formosanus* (E. Hartert, 1912)
Taiwan

○ *Acridotheres javanicus* Cabanis, 1851 WHITE-VENTED MYNA[7]
Java

○ *Acridotheres cinereus* Bonaparte, 1850 PALE-BELLIED MYNA[8]
S Sulawesi

◒ *Acridotheres fuscus* JUNGLE MYNA
    — *A.f.mahrattensis* (Sykes, 1832)
W and S India
    — *A.f.fuscus* (Wagler, 1827)
N Pakistan, N India, W, N and S Assam, SW, C and SE Burma
    — *A.f.fumidus* Ripley, 1950
E and NE Assam
    — *A.f.torquatus* W.R. Davison, 1892
Malay Pen.

○ *Acridotheres albocinctus* Godwin-Austen & Walden, 1875 COLLARED MYNA
SE Assam, W and N Burma, W and N Yunnan

○ *Acridotheres ginginianus* (Latham, 1790) BANK MYNA
Pakistan, N and C India, Nepal, Bangladesh >> S India

---

[1] For reasons to use 1802 not 1801 see Browing & Monroe (1991) (258).
[2] Feare & Craig (1998) (830) favoured splitting this species into four. More research is required to show the extent of sympatry, if any, between a postulated *G. indica* (including *penisularis*) and a restricted *G. religiosa* in northern India. The status of all taxa from the islands off West Sumatra is confused; at least here *enganensis* is accorded subspecific rank, based on Hoogerwerf (1963) (1145) and Feare & Craig, *contra* Ripley (1944) (2027), Amadon (1962: 119) (45) and van Marle & Voous (1988: 197) (2468). All these populations should be re-examined, but few specimens may be available, see Hoogerwerf (1963) (1145).
[3] Given as 1844 by Amadon (1962: 118) (45) but given as "1844-45" in Tweeddale's Collected Ornithological Works (1881).
[4] Dated 1851 by Amadon (1962: 120) (45), but see Zimmer (1926) (2707).
[5] Includes *infuscatus* see Amadon (1962: 114) (45). Ripley (1982) (2055) used a different species concept and confused matters with his choice of vernacular name.
[6] A Burmese record (Smythies, 1953; Amadon, 1962) (2285, 45) needs re-examination; it may be a valid record of a vagrant, but it is more likely to have been *A. grandis*.
[7] This and the next two species are treated as allospecies by Feare & Craig (1998) (830); we tentatively accept them at species level, but detailed studies are needed. If the three are united the prior name is thought to be *cinereus* see Browning & Monroe (1991) (258).
[8] Dated 1851 by Amadon (1962: 120) (45) but see Zimmer (1926) (2707).

● *Acridotheres tristis* COMMON MYNA
    ✔ *A.t.tristis* (Linnaeus, 1766)[1] [2]

SC Kazakhstan, Turkmenistan and E Iran through mainland S Asia to W Malaya and Indochina; introduced
Arabia, Caucasus area, E China, Australia, S Africa,
islands in Indian and Pacific Oceans

    — *A.t.melanosternus* Legge, 1879
Sri Lanka

○ *Acridotheres melanopterus* BLACK-WINGED MYNA[3]
    — *A.m.melanopterus* (Daudin, 1800)
    — *A.m.tricolor* (Horsfield, 1821)
    — *A.m.tertius* (E. Hartert, 1896)

W and C Java
E Java
Bali, ?Lombok

LEUCOPSAR Stresemann, 1912 M
○ *Leucopsar rothschildi* Stresemann, 1912 BALI MYNA

Bali

STURNUS Linnaeus, 1758[4] M
○ *Sturnus burmannicus* VINOUS-BREASTED STARLING
    — *S.b.burmannicus* (Jerdon, 1862)
    — *S.b.leucocephalus* (Giglioli & Salvadori, 1870)

C Burma, SW Yunnan
Thailand (not S Pen.), Cambodia, S Indochina

○ *Sturnus nigricollis* (Paykull, 1807) BLACK-COLLARED STARLING

S and SE China, Burma, Thailand (not S penin.),
Indochina

○ *Sturnus contra* ASIAN PIED STARLING
    — *S.c.contra* Linnaeus, 1758
    — *S.c.sordidus* Ripley, 1950
    — *S.c.superciliaris* (Blyth, 1863)
    — *S.c.floweri* (Sharpe, 1897)
    — *S.c.jalla* (Horsfield, 1821)

N and C India, Bangladesh, W Assam
E and SE Assam
Manipur, N, C and SE Burma, SW Yunnan
E Burma, Thailand, SE Yunnan, NW Laos, Cambodia
Sumatra, Java, Bali

○ *Sturnus sturninus* (Pallas, 1776) PURPLE-BACKED STARLING/DAURIAN STARLING

NE Mongolia, S Transbaikalia and Amur Valley to NC
and NE China and N Korea >> SE Asia to Gtr. Sundas

○ *Sturnus philippensis* (J.R. Forster, 1781) CHESTNUT-CHEEKED STARLING

SE Russian Far East, S Sakhalin, S Kuril Is., Japan >>
Borneo, Philippines, N Sulawesi

○ *Sturnus sinensis* (J.F. Gmelin, 1788) WHITE-SHOULDERED STARLING[5]

S and SE China, N and C Vietnam >> mainland
SE Asia, Taiwan

○ *Sturnus malabaricus* CHESTNUT-TAILED STARLING
    — *S.m.blythii* (Jerdon, 1844)
    — *S.m.malabaricus* (J.F. Gmelin, 1789)
    — *S.m.nemoricola* (Jerdon, 1862)

SW India
India (except SW) to S Nepal, Bangladesh, N Assam
S Assam, Burma, S China, NW Thailand, N and
C Indochina >> mainland SE Asia

○ *Sturnus erythropygius* WHITE-HEADED STARLING
    — *S.e.andamanensis* (Tytler, 1867)
    — *S.e.erythropygius* (Blyth, 1846)
    — *S.e.katchalensis* (Richmond, 1902)

Andamans
Car Nicobar I.
Katchal I., Nicobars

○ *Sturnus albofrontatus* (E.L. Layard, 1854) WHITE-FACED STARLING[6]

Hills of SW Sri Lanka

○ *Sturnus pagodarum* (J.F. Gmelin, 1789) BRAHMINY STARLING[7]

E Afghanistan, N Pakistan, India, Sri Lanka, Bangladesh

○ *Sturnus roseus* (Linnaeus, 1758) ROSY STARLING[8]

SE Europe to SW and WC Asia >> India, Sri Lanka

○ *Sturnus sericeus* J.F. Gmelin, 1789 RED-BILLED STARLING

EC, E and S China

○ *Sturnus cineraceus* Temminck, 1835 WHITE-CHEEKED STARLING

NC China, NE Mongolia and Transbaikalia to Shandong,
Korea, Japan, S and C Kuril and S and C Sakhalin
>> S China

---

[1] Includes *tristoides* "Hodgson, 1836"; perhaps a *nomen nudum* until described by Brooke (1976) (236). Under review.
[2] Includes *naumanni* Dementiev, 1957 (691).
[3] For transfer to this genus from *Sturnus* see Feare & Nee (1992) (831).
[4] Although this genus has been split by Feare & Craig (1998) (830) we prefer to wait for molecular studies as boundaries are not very clear.
[5] Type species of the genus *Sturnia*. Not here recognised.
[6] Treated under the name *senex* in last Edition, but see Mees (1997) (1544).
[7] Type species of the genus *Temenuchus*. Not here recognised.
[8] Type species of the genus *Pastor*. Not here recognised.

● *Sturnus vulgaris* EUROPEAN STARLING[1]
    __ *S.v.faroensis* Feilden, 1872   Faroe Is.
    __ *S.v.zetlandicus* E. Hartert, 1918   Outer Hebrides, Shetland Is.
    __ *S.v.granti* E. Hartert, 1903   Azores
    ✓ *S.v.vulgaris* Linnaeus, 1758   Europe (except SW and shore Black Sea) >> N Africa.
                                               Introd. Nth. America
    __ *S.v.poltaratskyi* Finsch, 1878   SE European Russia, W and C Siberia and N Kazakhstan
                                             to W and N Mongolia, Baikal area and mid-Lena
                                           >> SW and S Asia
    __ *S.v.tauricus* Buturlin, 1904   W and N shore of Black Sea and Sea of Azov, Asia
                                           Minor (except E) >> Middle East
    __ *S.v.purpurascens* Gould, 1868   E Turkey, W Transcaucasia >> Egypt, Middle East
    __ *S.v.caucasicus* Lorenz, 1887   N Caucasus area to lower Volga, E Transcaucasia to N,
                                           W, and SW Iran >> SW Asia
    __ *S.v.oppenheimi* Neumann, 1915[2]   SE Turkey, N Iraq >> Middle East
    __ *S.v.nobilior* Hume, 1879   NE Iran, S Turkmenistan, Afghanistan >> NW India
    __ *S.v.porphyronotus* Sharpe, 1888   E Uzbekistan and Tajikistan to E Kazakhstan and W
                                           Xinjiang >> Nepal, N India
    __ *S.v.humii* W.E. Brooks, 1876[3]   Kashmir >> N India
    __ *S.v.minor* Hume, 1873   C and S Pakistan

○ *Sturnus unicolor* Temminck, 1820 SPOTLESS STARLING   NW Africa, Iberia, S France, Corsica, Sardinia, Sicily

CREATOPHORA Lesson, 1847 F
○ *Creatophora cinerea* (Meuschen, 1787) WATTLED STARLING   Ethiopia to Angola and Cape Province

FREGILUPUS Lesson, 1831 M
○ †*Fregilupus varius* (Boddaert, 1783) RÉUNION STARLING   Réunion I.

LAMPROTORNIS Temminck, 1820[4] M
○ *Lamprotornis nitens* CAPE GLOSSY STARLING
    __ *L.n.nitens* (Linnaeus, 1766)   Gabon to W Angola
    __ *L.n.phoenicopterus* Swainson, 1838   Namibia to Zimbabwe, N Cape Province, N Natal
    __ *L.n.culminator* (Clancey & Holliday, 1951)   E Cape Province, S Natal

○ *Lamprotornis chalybaeus* GREATER BLUE-EARED STARLING
    __ *L.c.chalybaeus* Hemprich & Ehrenberg, 1828   Senegal to C and E Sudan
    __ *L.c.cyaniventris* Blyth, 1856   Eritrea, Ethiopia, NW Somalia to C and W Kenya,
                                           E Uganda, E Zaire
    __ *L.c.sycobius* (Hartlaub, 1859)   SW Uganda, SE Kenya, Tanzania to N and E Zambia,
                                           Malawi, W Mozambique
    __ *L.c.nordmanni* (Hartert & Neumann, 1914)   S Angola, S Zambia to Botswana, N Transvaal,
                                           S Mozambique

○ *Lamprotornis chloropterus* LESSER BLUE-EARED STARLING
    __ *L.c.chloropterus* Swainson, 1838[5]   Senegal to Eritrea, Uganda, NW Kenya
    __ *L.c.elisabeth* (Stresemann, 1924)   SE Kenya, Tanzania to Zambia, Zimbabwe, Mozambique

○ *Lamprotornis chalcurus* BRONZE-TAILED GLOSSY STARLING
    __ *L.c.chalcurus* Nordmann, 1835   Senegal to Nigeria
    __ *L.c.emini* (Neumann, 1920)   N Cameroon, Central African Republic to Uganda,
                                         W Kenya

○ *Lamprotornis splendidus* SPLENDID GLOSSY STARLING
    __ *L.s.chrysonotis* Swainson, 1837   Senegal to Togo
    __ *L.s.splendidus* (Vieillot, 1822)   Nigeria and Ethiopia to Angola, W Tanzania
    __ *L.s.lessoni* (Pucheran, 1859)   Bioko I.
    __ *L.s.bailundensis* (Neumann, 1920)   S Angola, SE Zaire, N Zambia

---

[1] Essentially follows Cramp *et al.* (1994: 238) (595).
[2] For recognition see Roselaar (1995) (2107).
[3] For use of this name in place of *indicus* see Biswas (1963) (157).
[4] Species sequence follows Fry *et al.* (2000) (899); their genus is also broader than that of Feare & Craig (1998) (830) and differences are discussed.
   Includes the genera *Cosmopsarus*, used in last Edition and by Amadon (1962: 103) (45) and *Hylopsar*, used by Feare & Craig (1998) (830).
[5] Includes *cyanogenys* see White (1962: 57) (2646).

○ *Lamprotornis ornatus* (Daudin, 1800) Principe Glossy Starling   Principe I.

○ *Lamprotornis iris* (Oustalet, 1879) Emerald Starling[1]   Guinea to Ivory Coast

○ *Lamprotornis purpureus* Purple Glossy Starling
     — *L.p.purpureus* (Statius Müller, 1776)   Senegal, S Mali to Nigeria
     — *L.p.amethystinus* (Heuglin, 1863)   N Cameroon to S Sudan, Uganda, W Kenya

○ *Lamprotornis purpuroptera* Ruppell's Glossy Starling[2]
     — *L.p.aeneocephalus* Heuglin, 1863   E Sudan, Eritrea, N Ethiopia
     — *L.p.purpuroptera* Rüppell, 1845   S Ethiopia, S Sudan, Uganda, Kenya, W Tanzania

○ *Lamprotornis caudatus* (Statius Müller, 1776) Long-tailed Glossy Starling
     Senegal to C Sudan

○ *Lamprotornis regius* Golden-breasted Starling
     — *L.r.regius* (Reichenow, 1879)   S Ethiopia, S Somalia, Kenya
     — *L.r.magnificus* (van Someren, 1924)   E Kenya

○ *Lamprotornis mevesii* Meves's Long-tailed Starling
     — *L.m.benguelensis* Shelley, 1906[3]   SW and W Angola (2240)
     — *L.m.violacior* Clancey, 1973   SW Angola, NW Namibia (440)
     — *L.m.mevesii* (Wahlberg, 1856)   N and E Botswana, S and E Zambia, Zimbabwe, S Malawi, N Transvaal

○ *Lamprotornis australis* (A. Smith, 1836) Burchell's Starling[4]   SE Angola, SW Zambia to N Cape Province, Transvaal, SW Mozambique

○ *Lamprotornis acuticaudus* Sharp-tailed Starling
     — *L.a.acuticaudus* (Bocage, 1870)   Angola, Zambia, SE Zaire
     — *L.a.ecki* Clancey, 1980   S Angola, N Namibia (477)

○ *Lamprotornis corruscus* Black-bellied Glossy Starling
     — *L.c.corruscus* Nordmann, 1835[5]   Coasts of S Somalia to Cape Province
     — *L.c.vaughani* (Bannerman, 1926)   Pemba I.

○ *Lamprotornis superbus* Rüppell, 1845 Superb Starling[6]   S Sudan, Ethiopia and Somalia to Tanzania

○ *Lamprotornis hildebrandti* (Cabanis, 1878) Hildebrandt's Starling   S Kenya, N Tanzania

○ *Lamprotornis shelleyi* (Sharpe, 1890) Shelley's Starling   SE Sudan, S Ethiopia, N Somalia >> E Kenya, S Somalia

○ *Lamprotornis pulcher* (Statius Müller, 1776) Chestnut-bellied Starling[7]
     S Mauritania, Senegal to C and NE Sudan, Eritrea, NW Ethiopia

○ *Lamprotornis purpureiceps* (J. & E. Verreaux, 1851) Purple-headed Glossy Starling
     S Nigeria to Congo and Uganda

○ *Lamprotornis cupreocauda* (Hartlaub, 1857) Copper-tailed Glossy Starling
     Sierra Leone to Ghana

○ *Lamprotornis unicolor* (Shelley, 1881) Ashy Starling[8]   S Kenya, Tanzania

○ *Lamprotornis fischeri* (Reichenow, 1884) Fischer's Starling   S Somalia, Kenya, N Tanzania

CINNYRICINCLUS Lesson, 1840 M
○ *Cinnyricinclus femoralis* (Richmond, 1897) Abbott's Starling[9]   S Kenya, N Tanzania

○ *Cinnyricinclus leucogaster* Amethyst Starling
     — *C.l.leucogaster* (Boddaert, 1783)[10]   Senegal to S Ethiopia, N Zaire, NW Uganda
     — *C.l.arabicus* Grant & Mackworth-Praed, 1942   SW Arabia, NE Sudan, Eritrea to NW Somalia, N Ethiopia
     — *C.l.verreauxi* (Bocage, 1870)   Angola to Kenya and E Cape Province

[1] Type species of the genus *Coccycolius*. Not here recognised.
[2] For correction of spelling see David & Gosselin (2002) (613).
[3] Includes *chalceus* see White (1962: 60) (2646), a new name for *purpureus* for which the available name *benguelensis* was overlooked.
[4] Includes *degener* see White (1962: 61) (2646).
[5] Includes *mandanus* see Craig (2000: 602) (587).
[6] Implictly includes *excelsior* Clancey, 1987 (507) see Wilkinson (2000: 626) (2664).
[7] Treated as monotypic by Fry *et al.* (2000: 628) (899); *rufiventris* is a synonym see White (1962: 63) (2646).
[8] In last Edition we used the genus *Cosmopsarus* for this species and the next. Here we follow Wilkinson (2000) (2664).
[9] We follow White (1962) (2646) in placing this species in this genus.
[10] Includes *friedmanni* see White (1962: 61) (2646).

SPREO Lesson, 1831 M

○ *Spreo bicolor* (J.F. Gmelin, 1789) African Pied Starling   Ethiopia, E and Sth. Africa

○ *Spreo albicapillus* White-crowned Starling
     — *S.a.albicapillus* Blyth, 1856   S Ethiopia, N and C Somalia
     — *S.a.horeensis* Keith, 1964   N Kenya (1237)

ONYCHOGNATHUS Hartlaub, 1849[1] M

○ *Onychognathus morio* Red-winged Starling
     — *O.m.rueppellii* (J. Verreaux, 1865)   Ethiopia, SE Sudan
     — *O.m.morio* (Linnaeus, 1766)[2]   Kenya to Zimbabwe, W Mozambique, E and S Sth. Africa

○ *Onychognathus tenuirostris* Slender-billed Starling
     — *O.t.tenuirostris* (Rüppell, 1836)   Eritrea, Ethiopia,
     — *O.t.theresae* R. Meinertzhagen, 1937   E Zaire to Uganda, Kenya, Tanzania, Malawi

○ *Onychognathus fulgidus* Chestnut-winged Starling
     — *O.f.fulgidus* Hartlaub, 1849   São Tomé I.
     — *O.f.hartlaubii* G.R. Gray, 1858   Guinea to Uganda
     — *O.f.intermedius* E. Hartert, 1895   Gabon to NW Angola (1079)

○ *Onychognathus walleri* Waller's Starling
     — *O.w.preussi* Reichenow, 1892   SE Nigeria, Cameroon, Bioko
     — *O.w.elgonensis* (Sharpe, 1891)   SE Sudan, Uganda to E Zaire, W Kenya
     — *O.w.walleri* (Shelley, 1880)   C Kenya, Tanzania, N Malawi

○ *Onychognathus blythii* (Hartlaub, 1859) Somali Starling   Ethiopia, Eritrea to N Somalia, Socotra I.

○ *Onychognathus frater* (Sclater & Hartlaub, 1881) Socotra Chestnut-winged Starling
     Socotra I.

○ *Onychognathus tristramii* (P.L. Sclater, 1858) Tristram's Starling   Sinai, Israel and Jordan, W and SW Arabia to S Oman

○ *Onychognathus nabouroup* (Daudin, 1800) Pale-winged Starling   SW Angola, Namibia, Cape Province

○ *Onychognathus salvadorii* (Sharpe, 1891) Bristle-crowned Starling   Somalia, E and S Ethiopia, N Kenya

○ *Onychognathus albirostris* (Rüppell, 1836) White-billed Starling   Eritrea, Ethiopia

○ *Onychognathus neumanni* Neumann's Starling[3]
     — *O.n.modicus* Bates, 1932   Senegal to W Mali
     — *O.n.neumanni* (Alexander, 1908)   E Mali, Nigeria, Cameroon to Central African Republic and W Sudan

POEOPTERA Bonaparte, 1854 F

○ *Poeoptera stuhlmanni* (Reichenow, 1893) Stuhlmann's Starling   SW Ethiopia, SE Sudan, W Kenya, W Uganda to E Zaire

○ *Poeoptera kenricki* Kenrick's Starling
     — *P.k.bensoni* (van Someren, 1945)   Kenya
     — *P.k.kenricki* Shelley, 1894   N and E Tanzania

○ *Poeoptera lugubris* Bonaparte, 1854 Narrow-tailed Starling[4]   Sierra Leone to W Uganda, N Angola

PHOLIA Reichenow, 1900[5] F

○ *Pholia sharpii* (Jackson, 1898) Sharpe's Starling   E Zaire to S Ethiopia and Tanzania

GRAFISIA Bates, 1926 F

○ *Grafisia torquata* (Reichenow, 1909) White-collared Starling   Cameroon, N Zaire, Central African Republic

SPECULIPASTOR Reichenow, 1879 M

○ *Speculipastor bicolor* Reichenow, 1879 Magpie-Starling   N and C Somalia, S Ethiopia, N Kenya >> SE Kenya

---

[1] Species sequence follows Fry *et al.* (2000) (899).
[2] Includes *shelleyi* see Craig (2000: 582) (587).
[3] For a note regarding separation from *O. morio* see Craig (2000: 586) (587).
[4] Includes *webbi* Keith, 1968 (1238) see Fry (2000: 574) (892).
[5] This genus has been used by Pearson (2000: 639) (1814) for *sharpii* and for *femoralis*. We prefer, however, to place *femoralis* with *Cinnyricinclus leucogaster* as the species is sexually dimorphic.

NEOCICHLA Sharpe, 1876 F
○ *Neocichla gutturalis* White-winged Babbling Starling
    — *N.g.gutturalis* (Bocage, 1871)   S Angola
    — *N.g.angusta* Friedmann, 1930   E Zambia, Tanzania, Malawi

BUPHAGINAE

BUPHAGUS Brisson, 1760 M
○ *Buphagus erythrorhynchus* (Stanley, 1814) Red-billed Oxpecker[1]   Ethiopia, Somalia and E Africa to Botswana, Transvaal
                                          and Natal

○ *Buphagus africanus* Yellow-billed Oxpecker
    — *B.a.africanus* Linnaeus, 1766[2]   Senegal to Ethiopia, E Africa, Angola and Zambia to
                                          NE Sth. Africa

    — *B.a.langi* Chapin, 1921   Gabon, Congo

TURDIDAE THRUSHES (24: 165)

NEOCOSSYPHUS Fischer & Reichenow, 1884 M
○ *Neocossyphus rufus* Red-tailed Ant Thrush
    — *N.r.rufus* (Fischer & Reichenow, 1884)   Coastal Kenya and Tanzania, Zanzibar
    — *N.r.gabunensis* Neumann, 1908   S Cameroon and Gabon to N Zaire and W Uganda

○ *Neocossyphus poensis* White-tailed Ant Thrush
    — *N.p.poensis* (Strickland, 1844)   Sierra Leone to Cameroon, Gabon and S Congo,
                                          Bioko I.

    — *N.p.praepectoralis* Jackson, 1906[3]   W Zaire to NW Angola, Uganda and W Kenya

STIZORHINA Oberholser, 1899 F
○ *Stizorhina fraseri* Rufous Flycatcher-Thrush
    — *S.f.fraseri* (Strickland, 1844)   Bioko I.
    — *S.f.rubicunda* (Hartlaub, 1860)   Nigeria to Angola and Zambia
    — *S.f.vulpina* Reichenow, 1902   S Sudan, Uganda, NE Zaire

○ *Stizorhina finschii* (Sharpe, 1870) Finsch's Flycatcher-Thrush   Sierra Leone to Nigeria

MYOPHONUS Temminck, 1822[4] M
○ *Myophonus blighi* (Holdsworth, 1872) Sri Lanka Whistling Thrush   Sri Lanka

○ *Myophonus melanurus* (Salvadori, 1879) Shiny Whistling Thrush   Sumatra

○ *Myophonus glaucinus* Sunda Whistling Thrush
    — *M.g.glaucinus* (Temminck, 1823)   Java, Bali
    — *M.g.castaneus* (Wardlaw Ramsay, 1880)   Sumatra
    — *M.g.borneensis* P.L. Sclater, 1885   Borneo

○ *Myophonus robinsoni* (Ogilvie-Grant, 1905) Malayan Whistling Thrush
                                          Mts. of W Malaysia

○ *Myophonus horsfieldii* Vigors, 1831 Malabar Whistling Thrush   S India

○ *Myophonus insularis* (Gould, 1862) Taiwan Whistling Thrush   Taiwan

○ *Myophonus caeruleus* Blue Whistling Thrush[5]
    — *M.c.temminckii* Vigors, 1832[6]   Afghanistan to W Tien Shan and through Himalyas and
                                          E Assam to N and NE Burma, W and N Sichuan
                                          and SW Sichuan >> north of S and SE Asia (2529)

[1] Treated as monotypic by Feare & Craig (1998) (830) and by Fry (2000: 666) (892). Synonyms include *caffer, scotinus, invictus* Clancey, 1962 (391), *angolensis* da Rosa Pinto, 1968 (607), *bestiarum* Brooke, 1970 (134), and *archeri* Cunningham-van Someren, 1984 (601).
[2] Implicitly includes *haematophagus* Clancey, 1980 (479).
[3] Includes *pallidigularis, kakamegoes* Cunningham-van Someren & Schifter, 1981 (602), and *nigridorsalis* Cunningham-van Someren & Schifter, 1981 (602), see Zimmerman *et al.* (1996) (2735).
[4] For reasons to use this spelling see Deignan (1965) (669) and Dickinson (2001) (739).
[5] Stepanyan (1996) (2329) has proposed splitting this species, but adequate evidence of sympatry is still required, however the southern heavy-billed *flavirostris* group may prove to overlap in Indochina.
[6] Wrongly credited to "Gray in Temminck, 1822" by Ripley (1964: 142) (2045), but see Deignan (1965) (669).

— *M.c.caeruleus* (Scopoli, 1786)	C and E China >> S China to N Indochina
— *M.c.eugenei* (Hume, 1873)	C, E and SE Burma, W, N and E Thailand, C and S Yunnan, N and C Indochina
— *M.c.crassirostris* (Robinson, 1910)	SE Thailand, Cambodia, the Malay Pen. south to Kedah on W coast and Haadyai on E
— *M.c.dichrorhynchus* Salvadori, 1879	Malay Pen. (S of Kedah and Pattani), Sumatra
— *M.c.flavirostris* (Horsfield, 1821)[1]	Java

GEOMALIA Stresemann, 1931 F
○ ***Geomalia heinrichi*** Stresemann, 1931 GEOMALIA — Sulawesi

ZOOTHERA Vigors, 1832 F
○ ***Zoothera schistacea*** (A.B. Meyer, 1884) SLATY-BACKED THRUSH — Tanimbar Is.

○ ***Zoothera dumasi*** MOLUCCAN THRUSH
— *Z.d.dumasi* (Rothschild, 1898)	Buru
— *Z.d.joiceyi* (Rothschild & Hartert, 1921)	Seram

○ ***Zoothera interpres*** CHESTNUT-CAPPED THRUSH
— *Z.i.interpres* (Temminck, 1828[2])	S Thailand, Sumatra to Flores I., Borneo, Basilan
— *Z.i.leucolaema* (Salvadori, 1892)	Enggano I.

○ ***Zoothera erythronota*** RED-BACKED THRUSH
— *Z.e.erythronota* (P.L. Sclater, 1859)	Sulawesi
— *Z.e.mendeni* (Neumann, 1939)	Peleng I.

○ ***Zoothera dohertyi*** (E. Hartert, 1896) CHESTNUT-BACKED THRUSH — Lombok to Timor I.

○ ***Zoothera wardii*** (Blyth, 1842) PIED THRUSH — W and C Himalayas, S Assam >> S India, Sri Lanka

○ ***Zoothera cinerea*** (Bourns & Worcester, 1894) ASHY THRUSH — Luzon, Mindoro

○ ***Zoothera peronii*** ORANGE-SIDED THRUSH
— *Z.p.peronii* (Vieillot, 1818)	W Timor I.
— *Z.p.audacis* (E. Hartert, 1899)	E Timor I., Wetar I. to Babar I.

○ ***Zoothera citrina*** ORANGE-HEADED THRUSH
— *Z.c.citrina* (Latham, 1790)	Himalayas to NE India and W and N Burma >> S India, Sri Lanka
— *Z.c.cyanota* (Jardine & Selby, 1828)[3]	C and S India
— *Z.c.innotata* (Blyth, 1846)	S Yunnan, E Burma, N and SE Thailand, NW and S Indochina >> Malay Pen.
— *Z.c.melli* (Stresemann, 1923)	S China
— *Z.c.courtoisi* (E. Hartert, 1919)	Anhui (E China)
— *Z.c.aurimacula* (E. Hartert, 1910)	NE and C Indochina, Hainan
— *Z.c.andamanensis* (Walden, 1874)	Andamans
— *Z.c.albogularis* (Blyth, 1847)	Nicobars
— *Z.c.gibsonhilli* (Deignan, 1950)	C and SE Burma, SW Thailand
— *Z.c.aurata* (Sharpe, 1888)	N Borneo
— *Z.c.rubecula* (Gould, 1836)[4]	Java, Bali

○ ***Zoothera everetti*** (Sharpe, 1892) EVERETT'S THRUSH — N Borneo

○ ***Zoothera sibirica*** SIBERIAN THRUSH
— *Z.s.sibirica* (Pallas, 1776)	C and E Siberia to N Mongolia, N Manchuria and Russian Far East >> SE Asia to Gtr. Sundas
— *Z.s.davisoni* (Hume, 1877)	Sakhalin, S Kuril Is., Japan >> SE China to Gtr. Sundas

○ ***Zoothera piaggiae*** ABYSSINIAN THRUSH[5]
— *Z.p.piaggiae* (Bouvier, 1877)	Ethiopia to E Zaire and N and W Kenya

[1] Apparently *dichrorhynchus* is inseparable from *flavirostris* and could be submerged (C.S. Roselaar *in litt.*). CSR would also separate a monotypic species *flavirostris*. Publication awaited.
[2] Pl. 458 appeared in Livr. 77 in 1828 (Dickinson, 2001) (739); Ripley (1964: 145) (2045) wrongly cited Livr. 78 and dated it 1826.
[3] For correction of spelling see David & Gosselin (2002) (613).
[4] Includes *orientis* see Mees (1996) (1543).
[5] For reasons to unite *piaggae* and *tanganjicae* see Keith & Prigogine (1997: 24) (1241).

— *Z.p.hadii* (Macdonald, 1940)   SE Sudan

— *Z.p.ruwenzorii* Prigogine, 1985   Ruwenzori Mts. (1941)

— *Z.p.kilimensis* (Neumann, 1900)   C and S Kenya, N Tanzania

— *Z.p.tanganjicae* (Sassi, 1914)   E Zaire, Rwanda, N Burundi, SW Uganda

— *Z.p.rowei* (Grant & Mackworth-Praed, 1937)   N Tanzania

○ **Zoothera crossleyi** CROSSLEY'S THRUSH

— *Z.c.crossleyi* (Sharpe, 1871)   SE Nigeria, Cameroon, S Congo

— *Z.c.pilettei* (Schouteden, 1918)   E Zaire

○ **Zoothera gurneyi** ORANGE THRUSH

— *Z.g.chuka* (van Someren, 1930)   Mt. Kenya and Kikuyu escarpment

— *Z.g.raineyi* (Mearns, 1913)[1]   SE Kenya

— *Z.g.otomitra* (Reichenow, 1904)   Tanzania and N Malawi to SE Zaire, W Angola

— *Z.g.gurneyi* (Hartlaub, 1864)   Natal to E Cape Province

— *Z.g.disruptans* (Clancey, 1955)   C Malawi to Mozambique, E Zimbabwe, Transvaal

○ **Zoothera oberlaenderi** (Sassi, 1914) OBERLÄNDER'S THRUSH   NE Zaire, W Uganda

○ **Zoothera cameronensis** BLACK-EARED THRUSH[2]

— *Z.c.cameronensis* (Sharpe, 1905)   Cameroon, Gabon

— *Z.c.graueri* Sassi, 1914[3]   NE Zaire, Uganda

— *Z.c.kibalensis* Prigogine, 1978   Kibale Forest (W Uganda) (1936)

○ **Zoothera princei** GREY THRUSH

— *Z.p.princei* (Sharpe, 1873)   Liberia to Ghana

— *Z.p.batesi* (Sharpe, 1905)   S Cameroon to Uganda

○ **Zoothera guttata** SPOTTED THRUSH

— *Z.g.maxis* (Nikolaus, 1982)   S Sudan (1655)

— *Z.g.fischeri* (Hellmayr, 1901)   Coastal Kenya and Tanzania

— *Z.g.belcheri* (Benson, 1950)   Malawi[4]

— *Z.g.lippensi* Prigogine & Louette, 1984   SE Zaire (1947)

— *Z.g.guttata* (Vigors, 1831)[5]   Natal to E Cape Rovince

○ **Zoothera spiloptera** (Blyth, 1847) SPOT-WINGED THRUSH   Sri Lanka

○ **Zoothera andromedae** (Temminck, 1826) SUNDA THRUSH   Sumatra, Enggano, Java, the Lesser Sundas and the Philippines (Luzon, Mindoro, Panay, Negros and Mindanao)

○ **Zoothera mollissima** PLAIN-BACKED THRUSH

— *Z.m.whiteheadi* (E.C.S. Baker, 1913)   NW Himalayas

— *Z.m.mollissima* (Blyth, 1842)   C and E Himalayas, N Yunnan, SW Sichuan >> Burma to Indochina

— *Z.m.griseiceps* (Delacour, 1930)   N Sichuan to C and E Yunnan and NW Vietnam >> mainland SE Asia

○ **Zoothera dixoni** (Seebohm, 1881) LONG-TAILED THRUSH   SE Xizang and Assam hills to NE Burma, Yunnan and W Sichuan >> Burma, Thailand, N Vietnam

○ **Zoothera aurea** WHITE'S THRUSH[6]

— *Z.a.aurea* (Holandre, 1825)[7]   E European Russia and S Siberia to N Mongolia and the Sea of Okhotsk >> S China, Taiwan, mainland SE Asia

— *Z.a.toratugumi* (Momiyama, 1940)   Amurland and Korea to Sakhalin, S Kuril Is and Japan >> E China and Taiwan

○ **Zoothera major** (Ogawa, 1905) AMAMI THRUSH   Amami O-shima (N Ryukyu Is.)

---

[1] For recognition of this see Keith & Prigogine (1997: 21) (1241).

[2] For reasons to include *kibalensis* see Keith & Prigogine (1997: 17) (1241).

[3] A substitute name *prigoginei* was proposed for use in a broad genus *Turdus* see Hall (1966) (1063).

[4] Harebottle *et al.* (1997) (1068) rejected the view that the population of N Zululand is identifiable with *belcheri* and perhaps Malawi birds are not truly distinct.

[5] The name *natalicus* (not *nateliens* as in last Edition) was used for *guttata* when that was thought preoccupied.

[6] We treat *aurea* as distinct from *dauma* on size and on voice, see Martens & Eck (1995: 202) (1443).

[7] Includes the name *hancii* see Mees (1977) (1528) (type examined), but not the population for which Ripley (1964: 157) (2045) used it.

● *Zoothera dauma* SCALY THRUSH[1]
  — *Z.d.dauma* (Latham, 1790)[2]   Himalayas, Assam, Burma, N and W Thailand, Yunnan to Sichuan and Guangxi (SC China) >> Taiwan, mainland S and SE Asia

  — *Z.d.*subsp. ?   Iriomote Jima (S Ryukyu Is.)[3]
  — *Z.d.horsfieldi* (Bonaparte, 1857)   Sumatra, Java, Lombok I.
  — *Z.d.neilgherriensis* (Blyth, 1847)   SW India
  — *Z.d.imbricata* E.L. Layard, 1854   Sri Lanka

○ *Zoothera machiki* (Forbes, 1883) FAWN-BREASTED THRUSH   Tanimbar Is.

○ *Zoothera heinei* RUSSET-TAILED THRUSH
  — *Z.h.papuensis* (Seebohm, 1881)   WC to SE New Guinea
  — *Z.h.eichhorni* (Rothschild & Hartert, 1924)   St. Matthias Is. (Bismarck Arch.)
  — *Z.h.choiseuli* (E. Hartert, 1924)   Choiseul I. (N Solomons)
  — *Z.h.heinei* (Cabanis, 1850)[4]   E Queensland, NE New South Wales

○ *Zoothera lunulata* BASSIAN THRUSH
  — *Z.l.cuneata* (De Vis, 1890)   NE Queensland
  — *Z.l.lunulata* (Latham, 1802)[5]   SE Australia, Tasmania, Bass Strait Is.
  — *Z.l.halmaturina* (A.G. Campbell, 1906)   SC South Australia, Kangaroo I.

○ *Zoothera talaseae* SOUTH SEAS THRUSH
  — *Z.t.talaseae* (Rothschild & Hartert, 1926) New Britain
  — *Z.t.atrigena* Ripley & Hadden, 1982   Bougainville I (N Solomons) (2065)
  — *Z.t.turipavae* Cain & Galbraith, 1955   Guadalcanal I. (S Solomons)
  — *Z.t.margaretae* (Mayr, 1935)   San Cristobal I. (S Solomons)

○ *Zoothera monticola* LONG-BILLED THRUSH
  — *Z.m.monticola* Vigors, 1832   Himalayas, Assam, E Bangladesh, NE Burma
  — *Z.m.atrata* Delacour & Greenway, 1939   NW Vietnam

○ *Zoothera marginata* Blyth, 1847 DARK-SIDED THRUSH   Himalayas, Assam and Burma to S Yunnan and Indochina

○ †*Zoothera terrestris* (Kittlitz, 1831) BONIN THRUSH   Ogasawara-shoto (Bonin) Is. (Japan)

IXOREUS Bonaparte, 1854[6] M
● *Ixoreus naevius* VARIED THRUSH
  — *I.n.meruloides* (Swainson, 1832)   S Alaska, NW Canada >> WC USA
  ✓ *I.n.naevius* (J.F. Gmelin, 1789)   SE Alaska south to NW California >> SW California
  ✓ *I.n.carlottae* (A.R. Phillips, 1991)[7]   Queen Charlotte Is. (Brit. Columbia) >> C California (1876)
  — *I.n.godfreii* (A.R. Phillips, 1991)[8]   Interior Brit. Columbia S to Idaho and Montana (1876)

RIDGWAYIA Stejneger, 1883[9] F
○ *Ridgwayia pinicola* AZTEC THRUSH
  — *R.p.maternalis* A.R. Phillips, 1991[10]   SW Chihuahua to S Jalisco (NW Mexico) (1876)
  — *R.p.pinicola* (P.L. Sclater, 1859)   Veracruz and Michoacán south to S Guerrero and S Oaxaca (SW Mexico)

CATAPONERA E. Hartert, 1896 F
○ *Cataponera turdoides* SULAWESI THRUSH
  — *C.t.abditiva* Riley, 1918   NC Sulawesi
  — *C.t.tenebrosa* Stresemann, 1938   S Sulawesi
  — *C.t.turdoides* E. Hartert, 1896   SW Sulawesi
  — *C.t.heinrichi* Stresemann, 1938   SE Sulawesi

---

[1] This has usually been seen as a broad species including not just *aurea*, but Australasian taxa; we believe some recent authors, as noted by Schodde & Mason (1999: 641) (2189), erred in treating *horsfieldi* as a separate species. However the forms of southern India and Sri Lanka will probably prove to constitute one or two separate species.
[2] Provisionally, but after examining the types, this includes *socia* from Yunnan based on a first winter bird, as well as Taiwanese breeding birds discussed by Mees (1977) (1528), and *affinis* applied to wintering Malay Pen. birds. The name *hancii* does not belong here (see above).
[3] Breeding birds await description.
[4] Includes *paraheinei* Schodde, 1989 (2177), see Schodde & Mason (1999: 642) (2189).
[5] For reasons to use 1802 not 1801 see Browing & Monroe (1991) (258). Implicitly includes *macrorhyncha* see Schodde & Mason (1999: 645) (2189).
[6] We follow A.O.U. (1998: 513) (3) in retaining this genus.
[7] Tentatively accepted see Dickerman & Parkes (1997) (729).
[8] Tentatively accepted see Dickerman & Parkes (1997) (729).
[9] We follow A.O.U. (1998: 513) (3) in retaining this genus.
[10] Tentatively accepted see Dickerman & Parkes (1997) (729).

SIALIA Swainson, 1827 F

● **Sialia sialis** EASTERN BLUEBIRD

    ✓ *S.s.sialis* (Linnaeus, 1758) | E USA >> N Mexico

    — *S.s.grata* Bangs, 1898 | S Florida

    — *S.s.nidificans* A.R. Phillips, 1991[1] | Caribbean slope of central Mexico (1876)

    — *S.s.fulva* Brewster, 1885 | Arizona, N Mexico

    — *S.s.guatemalae* Ridgway, 1882 | SE Mexico, Guatemala

    — *S.s.meridionalis* Dickey & van Rossem, 1930 | Honduras, El Salvador

    — *S.s.caribaea* T.R. Howell, 1965 | E Honduras, N Nicaragua (1151)

    — *S.s.bermudensis* Verrill, 1901 | Bermuda

● **Sialia mexicana** WESTERN BLUEBIRD

    ✓ *S.m.occidentalis* J.K. Townsend, 1837[2] | W Canada, W USA

    — *S.m.bairdi* Ridgway, 1894 | SW USA, NW Mexico

    — *S.m.jacoti* A.R. Phillips, 1991[3] | S Texas, NE Mexico (1876)

    — *S.m.amabilis* R.T. Moore, 1939 | NW Mexico

    — *S.m.nelsoni* A.R. Phillips, 1991[4] | N Mexico (1876)

    — *S.m.mexicana* Swainson, 1832[5] | C Mexico

● **Sialia currucoides** (Bechstein, 1798) MOUNTAIN BLUEBIRD | W Canada, W USA >> SW USA, W Mexico

MYADESTES Swainson, 1838 M

○ **Myadestes obscurus** (J.F. Gmelin, 1789) OMAO/HAWAIIAN THRUSH | Hawaii

○ †?**Myadestes myadestinus** (Stejneger, 1887) KAMAO/LARGE KAUAI THRUSH
                                     Kauai I.

○ †**Myadestes woahensis** (Bloxam, 1899) AMAUI[6] | Oahu I. (2672)

○ **Myadestes palmeri** (Rothschild, 1893) PUAIOHI/SMALL KAUAI THRUSH | Kauai I.

○ **Myadestes lanaiensis** OLOMAO/LANAI THRUSH

    — †*M.l.lanaiensis* (S.B. Wilson, 1891) | Lanai I.

    — †?*M.l.rutha* (Bryan, 1908) | Molokai I.

● **Myadestes townsendi** TOWNSEND'S SOLITAIRE

    ✓ *M.t.townsendi* (Audubon, 1838) | W Canada, W USA >> N Mexico

    — *M.t.calophonus* R.T. Moore, 1937 | N and C Mexico

○ **Myadestes occidentalis** BROWN-BACKED SOLITAIRE

    — *M.o.occidentalis* Stejneger, 1882[7] | NW Mexico

    — *M.o.insularis* Stejneger, 1882 | Tres Marias Is.

    — *M.o.oberholseri* Dickey & van Rossem, 1925[8] | C and S Mexico to Guatemala, El Salvador, Honduras

○ **Myadestes elisabeth** CUBAN SOLITAIRE

    — *M.e.elisabeth* (Lembeye, 1850) | E and W Cuba

    — †?*M.e.retrusus* Bangs & Zappey, 1905 | Isla de la Juventud

○ **Myadestes genibarbis** RUFOUS-THROATED SOLITAIRE

    — *M.g.solitarius* S.F. Baird, 1866 | Jamaica

    — *M.g.montanus* Cory, 1881 | Hispaniola

    — *M.g.dominicanus* Stejneger, 1882 | Dominica I.

    — *M.g.genibarbis* Swainson, 1838 | Martinique I.

    — *M.g.sanctaeluciae* Stejneger, 1882 | St. Lucia I.

    — *M.g.sibilans* Lawrence, 1878 | St. Vincent I.

● **Myadestes melanops** Salvin, 1865 BLACK-FACED SOLITAIRE | Costa Rica, W Panama

---

[1] Tentatively recognised see Dickerman & Parkes (1997) (729).
[2] Tentatively includes *annabelae* see Phillips (1991) (1876).
[3] Tentatively recognised see Dickerman & Parkes (1997) (729).
[4] For recognition see Parkes in Dickerman & Parkes, 1997 (729).
[5] Tentatively includes *australis* see Phillips (1991) (1876).
[6] Our numbered reference is to Wilson & Evans, in this connection see Olson (1996) (1703) for explanation of Bloxham.
[7] Pratt (1988) (1916) noting that the name *obscurus* Lafresnaye, 1839, was preoccupied in *Myadestes* proposed a new name for the "eastern" and formerly nominate subspecies. Includes *cinereus* see Phillips (1991) (1876).
[8] Includes *deignani* A.R. Phillips, 1966 (1871) see Phillips (1991) (1876) and *orientalis* H.D. Pratt, 1988 (1916) — the nom. nov. for *obscurus* (see above); however this new name is antedated by *oberholseri* and not needed unless both are to be recognised.

○ *Myadestes coloratus* Nelson, 1912 VARIED SOLITAIRE E Panama

○ *Myadestes ralloides* ANDEAN SOLITAIRE
— *M.r.plumbeiceps* Hellmayr, 1921   W Colombia, W Ecuador
— *M.r.candelae* Meyer de Schauensee, 1947   NC Colombia
— *M.r.venezuelensis* P.L. Sclater, 1856   N Venezuela, E Colombia to N Peru
— *M.r.ralloides* (d'Orbigny, 1840)   C and S Peru, W Bolivia

○ *Myadestes unicolor* SLATE-COLOURED SOLITAIRE
— *M.u.unicolor* P.L. Sclater, 1857   S Mexico, Guatemala, N Honduras
— *M.u.pallens* Miller & Griscom, 1925   Nicaragua, E Honduras

CICHLOPSIS Cabanis, 1851[1] F
○ *Cichlopsis leucogenys* RUFOUS-BROWN SOLITAIRE
— *C.l.gularis* Salvin & Godman, 1882   Guyana
— *C.l.chubbi* Chapman, 1924   W Ecuador
— *C.l.peruvianus* Hellmayr, 1930   C Peru
— *C.l.leucogenys* Cabanis, 1851   SE Brazil

CATHARUS Bonaparte, 1850[2] M
○ *Catharus gracilirostris* BLACK-BILLED NIGHTINGALE-THRUSH
— *C.g.gracilirostris* Salvin, 1865   Costa Rica
— *C.g.accentor* Bangs, 1902   W Panama

○ *Catharus aurantiirostris* ORANGE-BILLED NIGHTINGALE-THRUSH
— *C.a.aenopennis* R.T. Moore, 1938   NW Mexico
— *C.a.clarus* Jouy, 1894   NC Mexico
— *C.a.melpomene* (Cabanis, 1850)   C and S Mexico
— *C.a.bangsi* Dickey & van Rossem, 1925   El Salvador, Honduras, Guatemala
— *C.a.costaricensis* Hellmayr, 1902   Costa Rica
— *C.a.russatus* Griscom, 1924   SW Costa Rica, W Panama
— *C.a.griseiceps* Salvin, 1866   WC and C Panama
— *C.a.phaeoplurus* Sclater & Salvin, 1875   NC Colombia
— *C.a.aurantiirostris* (Hartlaub, 1850)   NE Colombia, NW Venezuela
— *C.a.birchalli* Seebohm, 1881   NE Venezuela, Trinidad
— *C.a.barbaritoi* Aveledo & Ginés, 1952   C Venezuela (Perijá Mtns.)
— *C.a.sierrae* Hellmayr, 1919   Santa Marta Mts. (Colombia) (1102)
— *C.a.inornatus* J.T. Zimmer, 1944   EC Colombia
— *C.a.insignis* J.T. Zimmer, 1944   N Colombia

● *Catharus fuscater* SLATY-BACKED NIGHTINGALE-THRUSH
— *C.f.hellmayri* Berlepsch, 1902   Costa Rica, W Panama
— *C.f.mirabilis* Nelson, 1912   E Panama
— *C.f.sanctaemartae* Ridgway, 1904   N Colombia
— *C.f.fuscater* (Lafresnaye, 1845)   Ecuador, Colombia, W Venezuela
— *C.f.opertaneus* Wetmore, 1955   W Colombia
— *C.f.caniceps* Chapman, 1924   N and C Peru
— *C.f.mentalis* Sclater & Salvin, 1876   SE Peru, N Bolivia

○ *Catharus occidentalis* RUSSET NIGHTINGALE-THRUSH
— *C.o.olivascens* Nelson, 1899   NW Mexico
— *C.o.lambi* A.R. Phillips, 1969[3]   W and C Mexico (1873)
— *C.o.fulvescens* Nelson, 1897   SW Mexico
— *C.o.occidentalis* P.L. Sclater, 1859   Oaxaca (S Mexico)

● *Catharus frantzii* RUDDY-CAPPED NIGHTINGALE-THRUSH
— *C.f.omiltemensis* Ridgway, 1905   SW Mexico (2021)
— *C.f.nelsoni* A.R. Phillips, 1969[4]   SC Mexico (1873)

---

[1] Treated in *Myadestes* in last Edition, but for recognition of the genus *Cichlopsis* see Ridgely & Tudor (1989: 106) (2018).
[2] Given as 1851 by Ripley (1964: 164) (2045) but see Zimmer (1926) (2707).
[3] Tentatively accepted see Dickerman & Parkes (1997) (729). Includes *durangensis* A.R. Phillips, 1969 (1873) see Phillips (1991) (1876).
[4] Includes *confusus* A.R. Phillips, 1969 (1873) see Dickerman & Parkes (1997) (729).

— *C.f.chiapensis* A.R. Phillips, 1969[1] S Mexico to W Guatemala (1873)
— *C.f.alticola* Salvin & Godman, 1879 SE Chiapas to S Guatemala
— *C.f.juancitonis* Stone, 1931 Honduras (2345)
— *C.f.waldroni* A.R. Phillips, 1969[2] N Nicaragua (1873)
√ *C.f.frantzii* Cabanis, 1861 Costa Rica
— *C.f.wetmorei* A.R. Phillips, 1969[3] W Panama (1873)

O **Catharus mexicanus** BLACK-HEADED NIGHTINGALE-THRUSH
— *C.m.mexicanus* (Bonaparte, 1856) EC Mexico
— *C.m.cantator* Griscom, 1930 S Mexico, E Guatemala
— ¶ *C.m.yaegeri* A.R. Phillips, 1991[4] Honduras (1876)
— *C.m.carrikeri* A.R. Phillips, 1991[5] Nicaragua, Costa Rica, W Panama (1876)

O **Catharus dryas** SPOTTED NIGHTINGALE-THRUSH
— *C.d.harrisoni* Phillips & Rook, 1965[6] Oaxaca (Mexico) (1879)
— *C.d.ovandensis* Brodkorb, 1938 Chiapas (Mexico)
— *C.d.dryas* (Gould, 1855) W Guatemala, Honduras, W Ecuador
— *C.d.maculatus* (P.L. Sclater, 1858) E Colombia, E Ecuador, Peru, Bolivia
— *C.d.ecuadoreanus* Carriker, 1935 W Ecuador
— *C.d.blakei* Olrog, 1973[7] Jujuy (Argentina) (1683)

O **Catharus fuscescens** VEERY[8]
— *C.f.fuscescens* (Stephens, 1817) E Canada, E USA >> SC Mexico
— *C.f.fuliginosus* (Howe, 1900)[9] SE Canada >> SC Mexico
— *C.f.salicicola* (Ridgway, 1882) W Canada, W USA >> SC Mexico
— *C.f.subpallidus* (Burleigh & Duvall, 1959) NW USA >> SC Mexico

● **Catharus minimus** GREY-CHEEKED THRUSH
√ *C.m.minimus* (Lafresnaye, 1848) Extreme NE Siberia, Alaska, Canada >> N Sth. America
— *C.m.aliciae* (S.F. Baird, 1858) SE Canada >> N Sth. America

O **Catharus bicknelli** (Ridgway, 1882) BICKNELL'S THRUSH[10] Newfoundland and adjacent Canada to NE USA >> Hispaniola

● **Catharus ustulatus** SWAINSON'S THRUSH[11]
— *C.u.swainsoni* (Tschudi, 1845) C and E Canada, N USA >> E Sth America
— *C.u.appalachiensis* Ramos, 1991 E USA >> Andes from Colombia to Peru (1876)
— *C.u.incanus* Godfrey, 1952 Alaska, W Canada >> Colombia to Peru
√ *C.u.ustulatus* (Nuttall, 1840)[12] SE Alaska along Pacific Coast to California >> Mexico to Panama
— *C.u.phillipsi* Ramos, 1991 Queen Charlotte Islands, Canada >> S. Mexico to Nicaragua (1876)
— *C.u.oedicus* (Oberholser, 1899) California>> Mexico to Nicaragua

● **Catharus guttatus** HERMIT THRUSH[13]
— *C.g.guttatus* (Pallas, 1811)[14] Alaska, W Canada >> W USA, N and C Mexico
— *C.g.nanus* (Audubon, 1839)[15] SE Alaska, W Canada >> W USA, Baja California
— *C.g.munroi* A.R. Phillips, 1962[16] WC Canada and N Montana >> WC USA and NC Mexico

---

[1] Tentatively accepted see Dickerman & Parkes (1997) (729).
[2] Tentatively accepted see Dickerman & Parkes (1997) (729).
[3] For recognition see Dickerman & Parkes (1997) (729).
[4] Tentatively accepted see Dickerman & Parkes (1997) (729).
[5] Tentatively accepted see Dickerman & Parkes (1997) (729). The name *fumosus* is apparently based on hybrid *C. fuscater* and *C. mexicanus* and is thus unavailable, see Phillips (1991) (1876).
[6] See Dickerman & Parkes (1997) (729).
[7] The name proposed was spelled *blakey* – a *lapsus*.
[8] We defer decision on *levyi* A.R. Phillips, 1991 (1876), and *pulichorum* A.R. Phillips, 1991 (1876), pending their further review. See also Parkes in Dickerman & Parkes (1997) (729).
[9] For correction of spelling see David & Gosselin (2002) (613).
[10] Submerged , in last Edition, within *C. minimus*, but see Ouellet (1993) (1734).
[11] Revision based on Ramos *in* Phillips (1991) (1876).
[12] Includes *almae* see Ramos *in* Phillips (1991) (1876).
[13] We follow Browning (1990) (251) in accepting the two races proposed by Phillips (1962) (1868); however the views of Phillips (1991) (1876) on the correct application of the names *guttatus* and *nanus* have been rejected see Parkes in Dickerman & Parkes (1997) (729). His revision is thus largely footnoted here and not followed.
[14] Phillips (1991) (1876) considered this name applicable to birds breeding in S Alaska that migrate to WC and SW USA, and treated this population as *euborius*.
[15] Phillips (1991) (1876) considered this name applicable to birds breeding in E Canada that migrate to SC and SE USA. In his arrangement *faxoni* and *crymophilus* are synonyms of that. He treated birds from the islands off British Columbia as *osgoodi* A.R. Phillips, 1991 (1876) in the northern islands (and SC Alaska) and *verecundus* in the southern ones. Both these names are treated here as synonyms, see Parkes in Dickerman & Parkes (1997) (729).
[16] For recognition see Browning (1990) (251).

√ *C.g.faxoni* (Bangs & Penard, 1921)[1]   EC and E Canada, E USA >> SE USA and NE and
     EC Mexico

— *C.g.jewetti* A.R. Phillips, 1962[2]   N of W coast USA >> SW USA
— *C.g.auduboni* (S.F. Baird, 1864)   Rocky Mts. of NW USA >> Mexico south to Guatemala
— *C.g.polionotus* (Grinnell, 1918)   Rocky Mts. of NC USA >> W Mexico
— *C.g.slevini* (Grinnell, 1901)   S of W coast USA >> NW Mexico
— *C.g.sequoiensis* (Belding, 1889)   Mountains of SW USA >> N Mexico, SC USA

HYLOCICHLA S.F. Baird, 1864[3] F
◉ **Hylocichla mustelina** (J.F. Gmelin, 1789) Wood Thrush   SE Canada, E USA >> E Mexico, Cuba, Central America

ENTOMODESTES Stejneger, 1883[4] M
○ **Entomodestes coracinus** (Berlepsch, 1897) Black Solitaire   W Colombia, W Ecuador

○ **Entomodestes leucotis** (Tschudi, 1844) White-eared Solitaire   C Peru to C Bolivia

PLATYCICHLA S.F. Baird, 1864 F
○ **Platycichla flavipes** Yellow-legged Thrush
— *P.f.venezuelensis* (Sharpe, 1902)   Colombia, N and W Venezuela
— *P.f.melanopleura* (Sharpe, 1902)   NE Venezuela, Trinidad
— *P.f.xanthoscela* (Jardine, 1847)[5]   Tobago I.
— *P.f.polionota* (Sharpe, 1902)   S Venezuela, Guyana
— *P.f.flavipes* (Vieillot, 1818)   SE Brazil, Argentina, NE Paraguay

○ **Platycichla leucops** (Taczanowski, 1877) Pale-eyed Thrush   N Sth. America

PSOPHOCICHLA Cabanis, 1860 F
○ **Psophocichla litsipsirupa** Groundscraper Thrush
— *P.l.simensis* (Rüppell, 1840)   Eritrea and Ethiopia
— *P.l.litsipsirupa* (A. Smith, 1836)   C Namibia and S Zambia to N Sth. Africa, Swaziland
— *P.l.pauciguttatus* Clancey, 1956   NW Botswana, N Namibia, S Angola
— *P.l.stierlingi* (Reichenow, 1900)   C and E Angola to SW Tanzania

TURDUS Linnaeus, 1758 M
○ **Turdus pelios** African Thrush
— *T.p.chiguancoides* Seebohm, 1881   Senegal to N Ghana
— *T.p.saturatus* (Cabanis, 1882)   W Ghana to C Cameroon, W Congo and Gabon
— *T.p.nigrilorum* Reichenow, 1892   Mt. Cameroon
— *T.p.poensis* Alexander, 1903   Bioko I.
— *T.p.pelios* Bonaparte, 1850[6]   E Cameroon to Sudan, W Eritrea and W, C and E Ethiopia
— *T.p.bocagei* (Cabanis, 1882)   W Zaire, W and NW Angola
— *T.p.centralis* Reichenow, 1905   E Congo and S Central African Republic to S Ethiopia,
     W Kenya and NW Tanzania
— *T.p.graueri* Neumann, 1908   W Tanzania, Burundi, Rwanda, E Zaire
— *T.p.stormsi* Hartlaub, 1886[7]   SE Zaire, E Angola, N Zambia

○ **Turdus tephronotus** Cabanis, 1878 Bare-eyed Thrush   S Ethiopia, S Somalia, C and E Kenya, NE Tanzania

○ **Turdus libonyanus** Kurrichane Thrush
— *T.l.tropicalis* W.K.H. Peters, 1881   Tanzania to C and E Zambia, Malawi, E Zimbabwe and
     Mozambique
— *T.l.verreauxi* Bocage, 1869[8]   SC Zaire, Angola, Namibia, W Zambia, W Zimbabwe,
     N Botswana
— *T.l.libonyanus* (A. Smith, 1836)   Botswana, N Cape Province, Transvaal, Swaziland
— *T.l.peripheris* Clancey, 1952   Natal, SE Swaziland, S Mozambique

---

[1] Includes *crymophilus* — based on foxed specimens, see Phillips (1991) (1876).
[2] For recognition see Browning (1990) (251).
[3] For notes on generic recognition see A.O.U. (1998: 506) (3).
[4] Pasquet at al. (1999) (1790) found this to be a true thrush, and not close to *Myadestes* (a "proto-thrush").
[5] For correction of spelling see David & Gosselin (2002) (613).
[6] Given as 1851 by Ripley (1964: 180) (2045) but see Zimmer (1926) (2707). Includes *adamauae* see White (1952: 153) (2646).
[7] Includes *williami* see White (1962: 152) (2646).
[8] Includes *chobiensis* see Urban (1997: 42) (2452).

○ *Turdus olivaceofuscus* GULF OF GUINEA THRUSH
   — *T.o.olivaceofuscus* Hartlaub, 1852   São Tomé I.
   — *T.o.xanthorhynchus* Salvadori, 1901   Principe I.

○ *Turdus olivaceus* OLIVE THRUSH[1]
   — *T.o.ludoviciae* (Lort Phillips, 1895)   N Somalia
   — *T.o.abyssinicus* J.F. Gmelin, 1789[2]   Ethiopia to N Uganda, N, W and C Kenya and
                                               Loliondo (N Tanzania)

   — *T.o.helleri* (Mearns, 1913)   SE Kenya
   — *T.o.deckeni* Cabanis, 1868   Kilimanjaro to Monduli (N Tanzania)
   — *T.o.oldeani* Sclater & Moreau, 1935   Mbulu and Crater Highlands (N Tanzania)
   — *T.o.bambusicola* Neumann, 1908   E Zaire, Burundi, Rwanda, SW Uganda, NW Tanzania
   — *T.o.baraka* (Sharpe, 1903)   E Zaire (Virunga Park), Ruwenzori Mts.
   — *T.o.roehli* Reichenow, 1905   NE Tanzania
   — *T.o.nyikae* Reichenow, 1904   E and S Tanzania, N Malawi, NE Zambia
   — *T.o.milanjensis* Shelley, 1893   S Malawi, NW Mozambique
   — *T.o.swynnertoni* Bannerman, 1913   E Zimbabwe and W Mozambique
   — *T.o.transvaalensis* (Roberts, 1936)   N and E Transvaal, W Swaziland
   — *T.o.smithi* Bonaparte, 1850[3]   S Namibia, N Cape Province to Orange Free State and
                                               SW Transvaal, SE Botswana

   — *T.o.culminans* Clancey, 1982   W Natal (489)
   — *T.o.olivaceus* Linnaeus, 1766   SW Cape Province
   — *T.o.pondoensis* Reichenow, 1917   E Cape Province and E Natal

○ *Turdus menachensis* Ogilvie-Grant, 1913 YEMEN THRUSH   SW Saudi Arabia, W Yemen

○ *Turdus bewsheri* COMORO THRUSH
   — *T.b.comorensis* Milne-Edwards & Oustalet, 1885   Grande Comore I.
   — *T.b.moheliensis* Benson, 1960   Mwali (Moheli) I.
   — *T.b.bewsheri* E. Newton, 1877   Anjouan I.

○ *Turdus hortulorum* P.L. Sclater, 1863 GREY-BACKED THRUSH   SE Siberia, Manchuria, Korea >> SE China to Indochina

○ *Turdus unicolor* Tickell, 1833 TICKELL'S THRUSH   W and C Himalayas (N Pakistan to Bhutan) >>
                                               C Pakistan, N India

○ *Turdus dissimilis* Blyth, 1847 BLACK-BREASTED THRUSH   NE India, Burma, Yunnan, N Indochina

○ *Turdus cardis* Temminck, 1831 JAPANESE THRUSH[4]   E China, Korea, Japan >> S and E China to Indochina

○ *Turdus albocinctus* Royle, 1839 WHITE-COLLARED BLACKBIRD   Himalayas, S and SE Xizang, SW Sichuan >> N India
                                               to N Burma

○ *Turdus torquatus* RING OUSEL
   — *T.t.torquatus* Linnaeus, 1758   N Europe, British Isles, W France >> S Europe,
                                               NW Africa

   — *T.t.alpestris* (C.L. Brehm, 1831)   N Spain and C Europe to Balkans, Greece and W Asia
                                               Minor >> N Africa

   — *T.t.amicorum* Hartert, 1923   C and E Turkey, Caucasus area, N Iran, SW Turkmenistan
                                               >> SW Asia

○ *Turdus boulboul* (Latham, 1790) GREY-WINGED BLACKBIRD   Himalayas, Assam, W Sichuan, S China, N Indochina
                                               >> N Burma

● *Turdus merula* EURASIAN BLACKBIRD[5]
   ✓ *T.m.merula* Linnaeus, 1758   N, W and C Europe>> W and S Europe
   — *T.m.azorensis* E. Hartert, 1905   Azores
   — *T.m.cabrerae* E. Hartert, 1901   Madeira, W Canary Is.
   — *T.m.mauritanicus* E. Hartert, 1902[6]   NW Africa

---

[1] For a discussion of the taxonomy of this species see Urban (1997: 36) (2452).
[2] Includes *polius* see Ripley (1964: 182) (2045) and *mwaki* see Urban (1997: 35) (2452), who implicitly also included *porini* Cunningham-van Someren & Schifter, 1981 (602); a replacement for the name *fuscatus* given it earlier in the same paper.
[3] Given as 1851 by Ripley (1964: 181) (2045) but see Zimmer (1926) (2707).
[4] Includes *yessoensis* Fushihara, 1959 (901), see Orn. Soc. Japan (1974) (1726).
[5] May merit treatment as three species. Group "A" (*merula*) contains the first 7 forms, group "B" (*mandarinus*) the next 4 and group "C" (*simillimus*) the last 5. Detailed studies not yet presented. CSR.
[6] The range of this form could be confined to SW and C Morocco to SW Tunisia. Birds from N Morocco, coastal N Algeria and N Tunisia and of Iberia could be recognised as *algirus*. CSR.

—— *T.m.aterrimus* (Madarász, 1903)	SE Europe, W and N Turkey, Caucasus area, N Iran >> Middle East
—— *T.m.syriacus* Hemprich & Ehrenberg, 1833[1]	Crete, S Turkey, Levant and Egypt to Iran (except N)
—— *T.m.intermedius* (Richmond, 1896)	Tien Shan to N Afghanistan >> Iraq to NW India
—— *T.m.maximus* (Seebohm, 1881)	NW Himalayas (N Pakistan to Kumaun)
—— *T.m.buddae* R. & A. Meinertzhagen, 1926	SE Xizang, Sikkim, N Bhutan >> Nepal, N Assam
—— *T.m.sowerbyi* Deignan, 1951	W Sichuan to S Gansu (C China)
—— *T.m.mandarinus* Bonaparte, 1850[2]	Guizhou and E Sichuan to E China
—— *T.m.nigropileus* (Lafresnaye, 1840)	W and C India
—— *T.m.spencei* Whistler & Kinnear, 1932	SE India
—— *T.m.simillimus* Jerdon, 1839	SW India
—— *T.m.bourdilloni* (Seebohm, 1881)	S India
—— *T.m.kinnisii* (Kelaart, 1851)	Sri Lanka

○ ***Turdus poliocephalus*** ISLAND THRUSH

—— *T.p.erythropleurus* Sharpe, 1887	Christmas I. (Indian Ocean)
—— *T.p.loeseri* Meyer de Schauensee, 1939	N Sumatra
—— *T.p.indrapurae* Robinson & Kloss, 1916	SC Sumatra
—— *T.p.fumidus* S. Müller, 1843[3]	Mts. Papandajan, Pangrango and Gedeh (W Java)
—— *T.p.javanicus* Horsfield, 1821	Mts. of C Java
—— *T.p.stresemanni* M. Bartels, Jr., 1938	Mt. Lawoe (EC Java)
—— *T.p.whiteheadi* (Seebohm, 1893)	Mts. of E Java
—— *T.p.seebohmi* (Sharpe, 1888)	N Borneo
—— *T.p.niveiceps* (Hellmayr, 1919)	Taiwan
—— *T.p.thomassoni* (Seebohm, 1894)	N Luzon
—— *T.p.mayonensis* (Mearns, 1907)	S Luzon
—— *T.p.mindorensis* Ogilvie-Grant, 1896	Mindoro
—— *T.p.nigrorum* Ogilvie-Grant, 1896	Negros
—— *T.p.malindangensis* (Mearns, 1907)	Mt. Malindang (NW Mindanao)
—— *T.p.katanglad* Salomonsen, 1953	Mt. Katanglad (C Mindanao)
—— *T.p.kelleri* (Mearns, 1905)	Mt. Apo (SE Mindanao)
—— *T.p.hygroscopus* Stresemann, 1931	S Sulawesi
—— *T.p.celebensis* (Büttikofer, 1893)	SW Sulawesi
—— *T.p.schlegelii* P.L. Sclater, 1861	W Timor I.
—— *T.p.sterlingi* Mayr, 1944	E Timor I.
—— *T.p.deningeri* Stresemann, 1912	Seram
—— *T.p.versteegi* Junge, 1939	Mts. of W New Guinea
—— *T.p.carbonarius* Mayr & Gilliard, 1951	Mts. of EC New Guinea
—— *T.p.keysseri* Mayr, 1931	Mts. of Huon Pen. (NE New Guinea)
—— *T.p.papuensis* (De Vis, 1890)	Mts. of SE New Guinea
—— *T.p.tolokiwae* Diamond, 1989	Tolokiwa I. (Bismarck Arch.) (702)
—— *T.p.beehleri* Ripley, 1977	New Ireland (Bismarck Arch.) (2052)
—— *T.p.heinrothi* Rothschild & Hartert, 1924	St.Matthias Is. (Bismarck Arch.)
—— *T.p.canescens* (De Vis, 1894)	Goodenough I. (D'Entrecasteaux Is.)
—— *T.p.bougainvillei* Mayr, 1941	Bougainville I. (N Solomons)
—— *T.p.kulambangrae* Mayr, 1941	Kolombangara I. (C Solomons)
—— *T.p.sladeni* Cain & Galbraith, 1955	Guadalcanal I. (S Solomons)
—— *T.p.rennellianus* Mayr, 1931	Rennell I. (S Solomons)
—— *T.p.vanikorensis* Quoy & Gaimard, 1830	Vanikoro, Santa Cruz, Espiritu Santo Is. (N Vanuatu)
—— *T.p.placens* Mayr, 1941	Ureparapara, Vanue Lava Is. (Banks Group)
—— *T.p.whitneyi* Mayr, 1941	Gaua I. (Banks Group)
—— *T.p.malekulae* Mayr, 1941	Pentecost, Malekula, Ambrim Is. (Vanuatu)
—— *T.p.becki* Mayr, 1941	Paema, Lopevi, Epi, Mai Is. (Vanuatu)
—— *T.p.efatensis* Mayr, 1941	Efate, Nguna Is. (Vanuatu)
—— *T.p.albifrons* (E.P. Ramsay, 1879)	Erromanga I. (Vanuatu)
—— †*T.p.pritzbueri* E.L. Layard, 1878	Tana (Vanuatu), Lifu Is. (Loyalty Is.)
—— †?*T.p.mareensis* Layard & Tristram, 1879	Mare I. (Loyalty Is.)

---

[1] Includes *insularum*, treated as a synonym of *aterrimus* by Cramp *et al.* (1988: 949) (592), but better placed here. CSR.
[2] Given as 1851 by Ripley (1964: 191) (2045) but see Zimmer (1926) (2707).
[3] Includes *biesenbachi* see Mees (1996) (1543).

— *T.p.xanthopus* J.R. Forster, 1844[1]   New Caledonia I.
— †*T.p.poliocephalus* Latham, 1802[2]   Norfolk I.
— †*T.p.vinitinctus* (Gould, 1855)   Lord Howe I.
— *T.p.layardi* (Seebohm, 1890)   Viti Levu, Ovalau, Yasawa, Koro Is. (W Fiji Is.)
— *T.p.ruficeps* (E.P. Ramsay, 1876)   Kandavu I. (S Fiji Is.)
— *T.p.vitiensis* E.L. Layard, 1876   Vanua Levu I. (E Fiji Is.)
— *T.p.hades* Mayr, 1941   Ngau I. (E Fiji Is.)
— *T.p.tempesti* E.L. Layard, 1876   Taveuni I. (E Fiji Is.)
— *T.p.samoensis* Tristram, 1879   Savaii, Upolu Is. (Samoa)

○ *Turdus rubrocanus* CHESTNUT THRUSH
— *T.r.rubrocanus* J.E. & G.R. Gray, 1846   E Afghanistan, NW Himalayas, S and SE Xizang >> N and NE India
— *T.r.gouldi* (J. Verreaux, 1871)   E and NE Tibetan Plateau to C China and N and W Yunnan >> Burma

○ *Turdus kessleri* (Przevalski, 1876) KESSLER'S THRUSH   E and NE Tibetan Plateau (E Qinghai and SW Gansu to N Yunnan) >> S Himalayas, SE Xizang

○ *Turdus feae* (Salvadori, 1887) GREY-SIDED THRUSH   W and N Hebei (N China) >> Burma, Assam, NW Thailand and N Laos

○ *Turdus obscurus* J.F. Gmelin, 1789 EYEBROWED THRUSH   C and E Siberia to Kamchatka, S to NE Mongolia and Amurland >> E India and S China to W Indonesia

○ *Turdus pallidus* J.F. Gmelin, 1789 PALE THRUSH   SE Siberia, NE China, Korea, SW Honshu >> China, Japan, Taiwan

○ *Turdus chrysolaus* BROWN-HEADED THRUSH
— *T.c.orii* Yamashina, 1929   Sakhalin, N and C Kurils >> Japan, Ryukyu Is.
— *T.c.chrysolaus* Temminck, 1832[3]   N Japan, S Korea >> SE China, N Philippines

○ *Turdus celaenops* Stejneger, 1887 IZU ISLANDS THRUSH   Izu Is. and islands off S Kyushu

○ *Turdus atrogularis* Jarocki, 1819 BLACK-THROATED THRUSH[4][5]   E European Russia and W and NC Siberia, S to Tarbagatay Mts. and NW Mongolia >> SW, S, and E Asia

○ *Turdus ruficollis* Pallas, 1776 RED-THROATED THRUSH   SC Siberia (E Altai and Baikal area to N Mongolia and Transbaikalia) >> N India to N Indochina and China

○ *Turdus naumanni* Temminck, 1820 NAUMANN'S THRUSH   SC Siberia (mid-Yenisey to middle and upper Lena basin) >> E Asia

○ *Turdus eunomus* Temminck, 1831 DUSKY THRUSH[6]   NC and NE Siberia (lower Yenisey to Chukotka and Kamchatka) >> E and mainland SE Asia

● *Turdus pilaris* Linnaeus, 1758 FIELDFARE   N and C Europe, W and C Siberia to Aldan basin and Transbaikalia, N Asia >> W and S Europe, SW Asia

● *Turdus iliacus* REDWING
— *T.i.coburni* Sharpe, 1901   Iceland, Faroe Is. >> W Europe
↙ *T.i.iliacus* Linnaeus, 1766   N and E Europe, N Siberia to lower Kolyma R. and Altai >> W and S Europe, N Africa, SW Asia

● *Turdus philomelos* SONG THRUSH
— *T.p.hebridensis* W.E. Clarke, 1913   Outer Hebrides, Isle of Skye >> British Is.
— *T.p.clarkei* E. Hartert, 1909   British Isles, W Europe >> SW Europe, NW Africa
— *T.p.philomelos* C.L. Brehm, 1831   Europe (except W), N Turkey, Caucasus area, N Iran >> W and S Europe, N Africa, SW Asia
— *T.p.nataliae* Buturlin, 1929   W and C Siberia >> NE Africa, SW Asia

○ *Turdus mupinensis* Laubmann, 1920 CHINESE THRUSH   Yunnan to SW Gansu and Hebei (C China)

○ *Turdus viscivorus* MISTLE THRUSH
— *T.v.deichleri* Erlanger, 1897[7]   NW Africa, Corsica, Sardinia
— *T.v.viscivorus* Linnaeus, 1758   Europe to W Siberia and N Iran >> N Africa to SW Asia
— *T.v.bonapartei* Cabanis, 1860   Turkmenistan and S-C Siberia to W Nepal and Altai >> C and S Asia

[1] Drawings with Forster's proposed names were published many years after his death (d. 1798).
[2] For reasons to use 1802 not 1801 see Browing & Monroe (1991) (258).
[3] Pl. 537 appeared in Livr. 91 in 1832 (Dickinson, 2001); Ripley (1964: 199) (2045) wrongly cited Livr. 87 and dated it 1831.
[4] Separated from *ruficollis* based on Russian work; see Stepanyan (1983: 203; 1990: 541) (2323, 2326).
[5] Includes *vogulorum* Portenko, 1981 (1911). CSR.
[6] Separated from *naumanni* based on Russian work; see Stepanyan (1983: 205; 1990: 542) (2323, 2326).
[7] For recognition see Cramp *et al.* (1988: 1011) (592).

○ *Turdus fuscater* GREAT THRUSH[1]
— *T.f.cacozelus* (Bangs, 1898)   N Colombia (Santa Marta Mts.)
— *T.f.clarus* Phelps & Phelps, Jr., 1953   E Colombia, W Venezuela (Perijá Mtns)
— *T.f.quindio* Chapman, 1925   S and W Colombia, N Ecuador
— *T.f.gigas* Fraser, 1841   E Colombia, W Venezuela
— *T.f.gigantodes* Cabanis, 1873   S Ecuador to C Peru
— *T.f.ockendeni* Hellmayr, 1906   SE Peru
— *T.f.fuscater* d'Orbigny & Lafresnaye, 1837   W Bolivia

○ *Turdus chiguanco* CHIGUANCO THRUSH
— *T.c.chiguanco* d'Orbigny & Lafresnaye, 1837   Coastal Peru, NW Bolivia
— *T.c.conradi* Salvadori & Festa, 1899   S Ecuador, C Peru
— *T.c.anthracinus* Burmeister, 1858   S Bolivia, NE Chile, W Argentina

○ *Turdus nigrescens* Cabanis, 1860 SOOTY ROBIN   Costa Rica, W Panama

○ *Turdus infuscatus* (Lafresnaye, 1844) BLACK ROBIN   C Mexico to Honduras

○ *Turdus serranus* GLOSSY-BLACK THRUSH
— *T.s.cumanensis* (Hellmayr, 1919)   NE Venezuela
— *T.s.atrosericeus* (Lafresnaye, 1848)   NE Colombia, N Venezuela
— *T.s.fuscobrunneus* (Chapman, 1912)   C and S Colombia, Ecuador
— *T.s.serranus* Tschudi, 1844[2]   Peru, W Bolivia, NW Argentina

○ *Turdus nigriceps* Cabanis, 1874 ANDEAN SLATY THRUSH[3]
— *T.n.nigriceps* Cabanis, 1874   Andes from S Ecuador to N Argentina
— *T.n.subalaris* (Seebohm, 1887)   S Brazil, Paraguay, NE Argentina

○ *Turdus reevei* Lawrence, 1870 PLUMBEOUS-BACKED THRUSH   W Ecuador, NW Peru

○ *Turdus olivater* BLACK-HOODED THRUSH
— *T.o.sanctaemartae* (Todd, 1913)   N Colombia
— *T.o.olivater* (Lafresnaye, 1848)   E Colombia, Venezuela
— *T.o.paraquensis* Phelps & Phelps, Jr., 1946   S Venezuela
— *T.o.kemptoni* Phelps & Phelps, Jr., 1955   C Venezuela
— *T.o.duidae* Chapman, 1929   Mt. Duida (S Venezuela)
— *T.o.roraimae* Salvin & Godman, 1884   S Venezuela, S Guyana
— *T.o.caucae* (Chapman, 1914)   C Colombia
— *T.o.ptaritepui* Phelps & Phelps, Jr., 1946   SE Venezuela

○ *Turdus maranonicus* Taczanowski, 1880 MARAÑON THRUSH   N Peru

○ *Turdus fulviventris* P.L. Sclater, 1857 CHESTNUT-BELLIED THRUSH   N Colombia, NW Venezuela, Ecuador, Peru

○ *Turdus rufiventris* RUFOUS-BELLIED THRUSH
— *T.r.juensis* (Cory, 1916)   NE Brazil
— *T.r.rufiventris* Vieillot, 1818   S Brazil, Uruguay, Paraguay, N Argentina

● *Turdus falcklandii* AUSTRAL THRUSH[4]
— *T.f.falcklandii* Quoy & Gaimard, 1824   Falkland Is.
↙ *T.f.magellanicus* P.P. King, 1831   S Chile, S Argentina

○ *Turdus leucomelas* PALE-BREASTED THRUSH
— *T.l.leucomelas* Vieillot, 1818   S Brazil, SE Bolivia, E Peru, Paraguay, NE Argentina
— *T.l.albiventer* Spix, 1824   N Colombia, Venezuela, NE Brazil, the Guianas
— *T.l.cautor* Wetmore, 1946   N Colombia

○ *Turdus amaurochalinus* Cabanis, 1850 CREAMY-BELLIED THRUSH   C Sth. America

● *Turdus plebejus* MOUNTAIN ROBIN
— *T.p.differens* (Nelson, 1901)   SE Mexico, Guatemala
↘ *T.p.rafaelensis* Miller & Griscom, 1925   Honduras, Nicaragua, El Salvador, NW Costa Rica
— *T.p.plebejus* Cabanis, 1861   Costa Rica, W Panama

---

[1] In previous Editions we wrongly listed *opertaneus* in this species as well as in *Catharus fuscater* (where it belongs).
[2] Presumably includes *unicolor* Olrog & Contino, 1970 (1688), from Prov. de Jujuy, Argentina, but the name is preoccupied and no replacement name seems to have been proposed.
[3] Although a split was proposed by Ridgely & Tudor (1989) (2018), reported in last Ed., we have elected to await further evidence.
[4] For *pembertoni*, listed in last edition see Fjeldså & Krabbe (1990) (850).

○ *Turdus ignobilis* Black-billed Thrush
— *T.i.goodfellowi* Hartert & Hellmayr, 1901 — Andes of W Colombia
— *T.i.ignobilis* P.L. Sclater, 1857 — Andes of E and C Colombia
— *T.i.debilis* Hellmayr, 1902 — E Colombia, W Venezuela, W Amazonia
— *T.i.murinus* Salvin, 1885 — S Venezuela, Guyana
— *T.i.arthuri* (Chubb, 1914) — Tepuis of SE Venezuela, Guyana, French Guiana

○ *Turdus lawrencii* Coues, 1880 Lawrence's Thrush — Upper Amazonia

○ *Turdus fumigatus* Cocoa Thrush [1]
— *T.f.aquilonalis* (Cherrie, 1909) — NE Colombia, N Venezuela, Trinidad
— *T.f.orinocensis* Zimmer & Phelps, 1955 — E Colombia, W Venezuela
— *T.f.fumigatus* M.H.K. Lichtenstein, 1823 — N and E Brazil, the Guianas
— *T.f.bondi* Deignan, 1951 — St. Vincent I.
— *T.f.personus* (Barbour, 1911) — Grenada I.

○ *Turdus obsoletus* Pale-vented Thrush
— *T.o.obsoletus* Lawrence, 1862 — Costa Rica, Panama, NW Colombia
— *T.o.parambanus* E. Hartert, 1920 — W Colombia, W Ecuador
— *T.o.colombianus* Hartert & Hellmayr, 1901 — C Colombia

○ *Turdus hauxwelli* Lawrence, 1869 Hauxwell's Thrush[2] — Upper Amazonia

○ *Turdus haplochrous* Todd, 1931 Unicoloured Thrush — E Bolivia

● *Turdus grayi* Clay-coloured Robin
— *T.g.tamaulipensis* (Nelson, 1897) — Nuevo Leon, Tamaulipas and N Vera Cruz (E Mexico)
— *T.g.microrhynchus* Lowery & Newman, 1949 — SC San Luis Potosí (C Mexico)
— *T.g.lanyoni* Dickerman, 1981 — Caribbean drainage of S Mexico, N Guatemala and S Belize (712)
— *T.g.yucatanensis* A.R. Phillips, 1991[3] — N Yucatán Peninsula, Mexico and N Belize (1876)
— *T.g.linnaei* A.R. Phillips, 1966[4] — Pacific Oaxaca and lowland S Chiapas to the Guatemala border (1876)
— *T.g.grayi* Bonaparte, 1838 — S Mexico (Pacific slope S Chiapas) to W Guatemala
— *T.g.megas* Miller & Griscom, 1925 — Pacific slope of W Guatemala to Honduras Nicaragua
↙ *T.g.casius* (Bonaparte, 1855)[5] — Costa Rica to NW Colombia

○ *Turdus nudigenis* Bare-eyed Robin[6]
— *T.n.nudigenis* Lafresnaye, 1848 — Lesser Antilles, Trinidad, NE Sth. America
— *T.n.extimus* Todd, 1931 — N Brazil
— *T.n.maculirostris* Berlepsch & Taczanowski, 1883 — W Ecuador, NW Peru

○ *Turdus jamaicensis* J.F. Gmelin, 1789 White-eyed Thrush — Jamaica

○ *Turdus assimilis* White-throated Thrush[7]
— *T.a.calliphthongus* R.T. Moore, 1937 — NW Mexico
— *T.a.lygrus* Oberholser, 1921[8] — C and S Mexico
— *T.a.suttoni* A.R. Phillips, 1991[9] — SW Tamaulipas and NE San Luis Potosí (Mexico) (1876)
— *T.a.assimilis* Cabanis, 1850 — C Mexico
— *T.a.leucauchen* P.L. Sclater, 1858[10] — S Mexico to Costa Rica
— *T.a.hondurensis* A.R. Phillips, 1991[11] — Lake Yojoa area, C Honduras (1876)
— *T.a.benti* A.R. Phillips, 1991[12] — SW El Salvador (1876)
— *T.a.rubicundus* (Dearborn, 1907) — S Mexico, W Guatemala, El Salvador
— *T.a.atrotinctus* Miller & Griscom, 1925 — E Honduras, N Nicaragua
— *T.a.cnephosus* (Bangs, 1902) — SW Costa Rica, W Panama

[1] We tentatively reunite *fumigatus* and *personus*. Our previous separation seems to have been in advance of sufficient published evidence.
[2] This split from *obsoletus*, from Ridgely & Tudor (1989) (2018), is recognised by A.O.U. (1998) (3).
[3] Tentatively recognised see Dickerman & Parkes (1997) (729).
[4] For recognition see Parkes in Dickerman & Parkes (1997) (729).
[5] Implicitly includes *incomptus* see review by Dickerman (1981) (712).
[6] Although a split was proposed by Ridgely & Tudor (1989) (2018), reported in last Edition, we await further evidence.
[7] Split following Ridgeley & Tudor (1989) (2018) and A.O.U. (1998) (3). For *assimilis* we tentatively accept the new taxa recently proposed.
[8] Includes *renominatus* see Ripley (1964: 222) (2045) and *oaxacae* Orr & Webster, 1968 (1728) see Phillips (1991) (1876).
[9] Tentatively recognised see Dickerman & Parkes (1997) (729).
[10] Tentatively includes *oblitus* see Phillips (1991) (1876).
[11] Tentatively recognised see Dickerman & Parkes (1997) (729).
[12] Tentatively recognised see Dickerman & Parkes (1997) (729).

— *T.a.campanicola* A.R. Phillips, 1991[1]   Cerro Campana (C Panama) (1876)

— *T.a.croizati* A.R. Phillips, 1991[2]   Azuero Pen. (SC Panama) (1876)

— *T.a.coibensis* Eisenmann, 1950   Coiba I.

— *T.a.daguae* Berlepsch, 1897   E Panama to NW Ecuador

○ **Turdus albicollis** WHITE-NECKED THRUSH

— *T.a.phaeopygoides* Seebohm, 1881[3]   NE Colombia, N Venezuela, Trinidad

— *T.a.phaeopygus* Cabanis, 1849   E Colombia to the Guianas, N Brazil

— *T.a.spodiolaemus* Berlepsch & Stolzmann, 1896[4] E Ecuador to N Bolivia, W Brazil

— *T.a.contemptus* Hellmayr, 1902   S Bolivia

— *T.a.crotopezus* M.H.K. Lichtenstein, 1823   E Brazil

— *T.a.albicollis* Vieillot, 1818   SE Brazil

— *T.a.paraguayensis* (Chubb, 1910)   SW Brazil, Paraguay, N Argentina

○ **Turdus rufopalliatus** RUFOUS-BACKED ROBIN

— *T.r.rufopalliatus* Lafresnaye, 1840[5]   W Mexico

— *T.r.interior* A.R. Phillips, 1991[6]   SC Mexico (1876)

— *T.r.graysoni* (Ridgway, 1882)   Tres Marias Is.

○ **Turdus rufitorques** Hartlaub, 1844 RUFOUS-COLLARED ROBIN   SE Mexico, Guatemala, El Salvador

● **Turdus migratorius** AMERICAN ROBIN[7]

✓ *T.m.migratorius* Linnaeus, 1766   Alaska, Canada, C USA >> E USA, E Mexico

✓ *T.m.nigrideus* Aldrich & Nutt, 1939   E Canada (Newfoundland, Labrador >> E USA

— *T.m.achrusterus* (Batchelder, 1900)   S USA >> SE Mexico

✓ *T.m.caurinus* (Grinnell, 1909)   SE Alaska, W Canada >> SW USA

✓ *T.m.propinquus* Ridgway, 1877[8]   W Canada, W USA, C Mexico >> Guatemala

— *T.m.phillipsi* Bangs, 1915[9]   SW Mexico

— *T.m.confinis* S.F. Baird, 1864   S Baja California

○ **Turdus swalesi** LA SELLE THRUSH

— *T.s.dodae* Graves & Olson, 1986   C Dominica (990)

— *T.s.swalesi* (Wetmore, 1927)   Haiti

○ **Turdus aurantius** J.F. Gmelin, 1789 WHITE-CHINNED THRUSH   Jamaica

○ †?**Turdus ravidus** (Cory, 1886) GRAND CAYMAN THRUSH   Grand Cayman

○ **Turdus plumbeus** RED-LEGGED THRUSH

— *T.p.plumbeus* Linnaeus, 1758   N Bahama Is.

— *T.p.schistaceus* (S.F. Baird, 1864)   E Cuba

— *T.p.rubripes* Temminck, 1826   C and W Cuba, Isla de la Juventud

— *T.p.coryi* (Sharpe, 1902)   Cayman Brac I.

— *T.p.ardosiaceus* Vieillot, 1823   Hispaniola, Puerto Rico, Gonave I.

— *T.p.albiventris* (P.L. Sclater, 1889)   Dominica I.

NESOCICHLA Gould, 1855 F

○ **Nesocichla eremita** TRISTAN DA CUNHA THRUSH

— *N.e.eremita* Gould, 1855   Tristan da Cunha I.

— *N.e.gordoni* Stenhouse, 1924   Inaccessible I.

— *N.e.procax* Elliott, 1954   Nightingale I.

CICHLHERMINIA Bonaparte, 1854 F

○ **Cichlherminia lherminieri** FOREST THRUSH

— *C.l.lherminieri* (Lafresnaye, 1844)   Guadeloupe I.

— *C.l.lawrencii* Cory, 1891   Montserrat I.

— *C.l.dominicensis* (Lawrence, 1880)   Dominica I.

— *C.l.sanctaeluciae* (P.L. Sclater, 1880) St.Lucia I.

[1] Tentatively recognised see Dickerman & Parkes (1997) (729).
[2] Tentatively recognised see Dickerman & Parkes (1997) (729).
[3] Includes *minuscula* see Ripley (1964: 223) (2045).
[4] Includes *berlepschi* see Ripley (1964: 224) (2045).
[5] Includes *grisior* see Ripley (1964: 225) (2045).
[6] Tentatively recognised see Dickerman & Parkes (1997) (729).
[7] The slight rearrangement follows Phillips (1991) (1876).
[8] Includes *aleucus* Oberholser, 1974 (1673), see Browning (1978) (246).
[9] Includes *permixtus* see Phillips (1991) (1876).

COCHOA Hodgson, 1836 F
○ *Cochoa purpurea* Hodgson, 1836 PURPLE COCHOA | Himalayas, Assam, N and E Burma, SC China, NW Thailand, N Indochina

○ *Cochoa viridis* Hodgson, 1836 GREEN COCHOA | C and E Himalayas, Assam, E Burma, S Yunnan, Thailand, Indochina, ? Fujian

○ *Cochoa azurea* (Temminck, 1824) JAVAN COCHOA | W and C Java

○ *Cochoa beccarii* Salvadori, 1879 SUMATRAN COCHOA | W Sumatra

CHLAMYDOCHAERA Sharpe, 1887[1] F
○ *Chlamydochaera jefferyi* Sharpe, 1887 FRUITHUNTER | Borneo

BRACHYPTERYX Horsfield, 1821 F
○ *Brachypteryx stellata* GOULD'S SHORTWING
— *B.s.stellata* Gould, 1868 | C Himalayas, SE Xizang
— *B.s.fusca* Delacour & Jabouille, 1930 | NE Burma to NW Vietnam

○ *Brachypteryx hyperythra* Jerdon & Blyth, 1861 RUSTY-BELLIED SHORTWING
E Himalayas, N Assam

○ *Brachypteryx major* WHITE-BELLIED SHORTWING
— *B.m.major* (Jerdon, 1844) | Mysore, W Madras
— *B.m.albiventris* (Blanford, 1867) | Kerala, SW Madras

○ *Brachypteryx leucophrys* LESSER SHORTWING
— *B.l.nipalensis* F. Moore, 1854 | Himalayas, W and N Burma, W Yunnan, SW Sichuan
— *B.l.carolinae* La Touche, 1898 | S and SE China, E Burma, NW Thailand, N Indochina
— *B.l.langbianensis* Delacour & Greenway, 1939 | S Indochina
— *B.l.wrayi* Ogilvie-Grant, 1906 | C and S Malay Pen.
— *B.l.leucophrys* (Temminck, 1828[2]) | Sumatra, Java, Lesser Sundas, Timor I.

○ *Brachypteryx montana* WHITE-BROWED SHORTWING
— *B.m.cruralis* (Blyth, 1843) | Himalayas, Burma, W China, NW Thailand, N Indochina
— *B.m.sinensis* Rickett & La Touche, 1897 | SE China
— *B.m.goodfellowi* Ogilvie-Grant, 1912 | Taiwan
— *B.m.sillimani* Ripley & Rabor, 1962 | S Palawan
— *B.m.poliogyna* Ogilvie-Grant, 1895 | N Luzon
— *B.m.andersoni* Rand & Rabor, 1967 | S Luzon (1978)
— *B.m.mindorensis* E. Hartert, 1916[3] | Mindoro (1082)
— *B.m.brunneiceps* Ogilvie-Grant, 1896 | Negros, Panay
— *B.m.malindangensis* Mearns, 1909 | Mt. Malindang (Mindanao)
— *B.m.mindanensis* Mearns, 1905 | Mt. Apo (Mindanao)[4]
— *B.m.erythrogyna* Sharpe, 1888 | N Borneo
— *B.m.saturata* Salvadori, 1879 | Sumatra
— *B.m.montana* Horsfield, 1822 | Java
— *B.m.floris* E. Hartert, 1897 | Flores I.

HEINRICHIA Stresemann, 1931 F
○ *Heinrichia calligyna* GREAT SHORTWING
— *H.c.simplex* Stresemann, 1931 | N Sulawesi
— *H.c.calligyna* Stresemann, 1931 | SC Sulawesi
— *H.c.picta* Stresemann, 1932 | SE Sulawesi

ALETHE Cassin, 1859 F
○ *Alethe diademata* FIRE-CRESTED ALETHE
— *A.d.diademata* (Bonaparte, 1850)[5] | Senegal to Togo
— *A.d.castanea* (Cassin, 1856) | Nigeria to W Zaire, Bioko I.
— *A.d.woosnami* Ogilvie-Grant, 1906 | C and N Zaire to SW Sudan and Uganda

[1] For reasons to move this species from the Campephagidae to here see Ahlquist *et al.* (1984) (18) and Olson (1987) (1699A).
[2] Ripley (1964: 16) (2045) cited 1827, Livr. 76. But plate 448 appeared in Livr. 75 in 1828 see Dickinson (2001) (739). We judge *leucophrys* to be the spelling in prevailing usage.
[3] For recognition see Dickinson *et al.* (1991: 314) (745).
[4] Reports that a sibling species occurs in Mindanao have yet to be validated.
[5] Given as 1851 by Ripley (1964: 61) (2045), but see Zimmer (1926) (2707).

○ *Alethe poliophrys* RED-THROATED ALETHE
    ___ *A.p.poliophrys* Sharpe, 1902   NE and E Zaire, W Uganda, W Rwanda, W Burundi
    ___ *A.p.kaboboensis* Prigogine, 1957   Mt. Kabobo (E Zaire)

○ *Alethe poliocephala* BROWN-CHESTED ALETHE
    ___ *A.p.castanonota* Sharpe, 1871[1]   Sierra Leone to Ghana
    ___ *A.p.giloensis* Cunningham-van Someren & Schifter, 1981   S Sudan (602)
    ___ *A.p.poliocephala* (Bonaparte, 1850)   Bioko I., Nigeria to NW Angola (305)
    ___ *A.p.vandeweghei* Prigogine, 1984[2]   Rwanda, Burundi (1940)
    ___ *A.p.carruthersi* Ogilvie-Grant, 1906[3]   E Zaire, Uganda, W Kenya
    ___ *A.p.akeleyae* Dearborn, 1909[4]   C Kenya
    ___ *A.p.kungwensis* Moreau, 1941   Mt. Kungwe (W Tanzania)
    ___ *A.p.ufipae* Moreau, 1942   SE Zaire, SW Tanzania
    ___ *A.p.hallae* Traylor, 1961   Gabela (Angola)

○ *Alethe fuelleborni* Reichenow, 1900 WHITE-CHESTED ALETHE[5]   NE Tanzania to N Malawi, C Mozambique

○ *Alethe choloensis* CHOLO ALETHE
    ___ *A.c.choloensis* W.L. Sclater, 1927   S Malawi, Mt. Chiperone (Mozambique)
    ___ *A.c.namuli* Vincent, 1933   Mt. Namuli (Mozambique)

MUSCICAPIDAE CHATS AND OLD WORLD FLYCATCHERS[6] (48: 275)

SAXICOLINAE (CHATS)

POGONOCICHLA Cabanis, 1847 F
○ *Pogonocichla stellata* WHITE-STARRED ROBIN[7]
    ___ *P.s.intensa* Sharpe, 1901   N and C Kenya, N Tanzania
    ___ *P.s.ruwenzorii* (Ogilvie-Grant, 1906)[8]   SW Uganda, W Rwanda, NE Zaire
    ___ *P.s.elgonensis* (Ogilvie-Grant, 1911)   Mt. Elgon (Uganda/Kenya)
    ___ *P.s.guttifer* (Reichenow & Neumann, 1895)   Mt. Kilimanjaro (Tanzania) Imatong Mts. (S Sudan) (602)
    ___ *P.s.macarthuri* van Someren, 1939   Chyulu Hills (SE Kenya)
    ___ *P.s.helleri* Mearns, 1913   Taita Hills (SE Kenya), NE Tanzania
    ___ *P.s.orientalis* (Fischer & Reichenow, 1884)   W, E and S Tanzania to Malawi and N Mozambique
    ___ *P.s.transvaalensis* (Roberts, 1912)[9]   E Zimbabwe and W Mozambique to N Transvaal
    ___ *P.s.stellata* (Vieillot, 1818)[10]   Natal, E and S Cape Province

SWYNNERTONIA Roberts, 1922 F
○ *Swynnertonia swynnertoni* SWYNNERTON'S ROBIN
    ___ *S.s.swynnertoni* (Shelley, 1906)[11]   E Zimbabwe, W Mozambique
    ___ *S.s.rodgersi* Jensen & Stuart, 1982   Mwanihana forest and Chita (C Tanzania) (1207)

STIPHRORNIS Hartlaub, 1855 M
○ *Stiphrornis erythrothorax* FOREST ROBIN
    ___ *S.e.erythrothorax* Hartlaub, 1855   Sierra Leone to S Nigeria
    ___ *S.e.gabonensis* Sharpe, 1883   Gabon, Bioko I.
    ___ *S.e.xanthogaster* Sharpe, 1903[12]   S Cameroon to N and C Zaire, S Uganda

○ *Stiphrornis sanghensis* Beresford & Cracraft, 1999 SANGHA ROBIN #
                              Dzanga-Sangha Dense Forest Reserve (Central African Republic) (135)

[1] Some errors in subspecific nomenclature and range transposition crept into Keith (1992: 446) (1240).
[2] Should perhaps be a synonym of *carruthersi*.
[3] Includes *nandensis* Cunningham-van Someren & Schifter, 1981 (602), see Zimmerman *et al.* (1996) (2735).
[4] Not *akeleyi* as given by Keith (1992: 446); named for Mrs. Akeley.
[5] See Keith (1992: 447) (1240) for treatment as a monotypic species. Includes *xuthera* Clancey & Lawson, 1969 (530).
[6] We follow Sibley & Monroe (1990) (2262) in placing the chats with the flycatchers.
[7] We follow Oatley *et al.* (1992) (1669).
[8] Includes *friedmanni* Clancey, 1972 (436) see Oatley *et al.* (1992: 388) (1669).
[9] Includes *hygrica* Clancey, 1969 (418) Oatley *et al.* (1992: 389) (1669).
[10] Includes *margaritata* see Oatley *et al.* (1992: 388) (1669).
[11] Includes *umbratica* Clancey, 1974 (446), see Oatley *et al.* (1992: 391) (1669).
[12] Includes *mabirae* see Oatley *et al.* (1992: 393) (1669).

SHEPPARDIA Haagner, 1909 F
○ ***Sheppardia bocagei*** BOCAGE'S AKALAT
    — *S.b.granti* (Serle, 1949)   W Cameroon
    — *S.b.poensis* (Alexander, 1903)[1]   Bioko I.
    — *S.b.kungwensis* (Moreau, 1941)   Mt. Kungwe (W Tanzania)
    — *S.b.ilyae* Prigogine, 1987   E of Mt. Kungwe (W Tanzania) (1944)
    — *S.b.kaboboensis* (Prigogine, 1955) Mt. Kabobo (E Zaire)
    — *S.b.schoutedeni* (Prigogine, 1952) E Zaire
    — *S.b.chapini* (Benson, 1955)[2]   N Zambia, SE Zaire
    — *S.b.bocagei* (Finsch & Hartlaub, 1870) W Angola

○ ***Sheppardia cyornithopsis*** LOWLAND AKALAT
    — *S.c.houghtoni* Bannerman, 1931   Sierra Leone, Liberia
    — *S.c.cyornithopsis* (Sharpe, 1901)   S Cameroon, Gabon
    — *S.c.lopezi* (Alexander, 1907)   N and E Zaire, W and S Uganda

○ ***Sheppardia aequatorialis*** EQUATORIAL AKALAT
    — *S.a.acholiensis* Macdonald, 1940[3]   Imatong Mts. (S Sudan)
    — *S.a.aequatorialis* (Jackson, 1906)[4] E Zaire, S Uganda, Rwanda, Burundi, W Kenya

○ ***Sheppardia sharpei*** SHARPE'S AKALAT
    — *S.s.usambarae* Macdonald, 1940   Usambara and Nguru Mts. (Tanzania)
    — *S.s.sharpei* (Shelley, 1903)   S Tanzania, N Malawi

○ ***Sheppardia gunningi*** EAST COAST AKALAT
    — *S.g.sokokensis* (van Someren, 1921) Coastal SE Kenya, NE Tanzania
    — *S.g.alticola* Fjeldså, Roy & Kiure, 2000 Nguru Mts. (E Tanzania) (852)
    — *S.g.bensoni* Kinnear, 1938   NC Malawi
    — *S.g.gunningi* Haagner, 1909   C Mozambique

○ ***Sheppardia gabela*** (Rand, 1957) GABELA AKALAT   WC Angola

○ ***Sheppardia montana*** (Reichenow, 1907) USAMBARA AKALAT   W Usambara Mts. (NE Tanzania)

○ ***Sheppardia lowei*** (Grant & Mackworth-Praed, 1941) IRINGA AKALAT   Uzungwa Mts. to Livingstone Mts. (S Tanzania)

ERITHACUS Cuvier, 1800 M
● ***Erithacus rubecula*** EUROPEAN ROBIN
    — *E.r.melophilus* E. Hartert, 1901   British Isles >> SW Europe
    ↙ *E.r.rubecula* (Linnaeus, 1758)[5]   Continental Europe to Ural Mts. and W Asia Minor,
                                    NW Morocco, W Canary Is., Madeira, Azores >>
                                    Europe and NE Africa
    — *E.r.superbus* Koenig, 1889   Tenerife, Gran Canaria I.
    — *E.r.witherbyi* E. Hartert, 1910   N Algeria, N Tunisia
    — *E.r.sardus* O. Kleinschmidt, 1906 S Spain, Corsica, Sardinia
    — *E.r.valens* Portenko, 1954[6]   S Crimea
    — *E.r.caucasicus* Buturlin, 1907[7]   E Turkey, Caucasus area >> Middle East
    — *E.r.hyrcanus* Blanford, 1874   SE Azerbaijan, N Iran >> Middle East
    — *E.r.tataricus* Grote, 1928   Urals and SW Siberia >> SW Asia

LUSCINIA T. Forster, 1817 F
○ ***Luscinia akahige*** JAPANESE ROBIN
    — *L.a.akahige* (Temminck, 1835)[8]   Sakhalin, S Kuril Is., Japan >> Ryukyu Is and E China
    — *L.a.tanensis* (N. Kuroda, Sr., 1923) Izu Is. and islands off S Kyushu

[1] If treated in the genus *Cossypha* the name *poensis* is preoccupied and *insulana* Grote, 1935, must be used.
[2] Implicitly includes *hallae* Prigogine, 1969 (1930) see Oatley *et al.* (1992: 395) (1669).
[3] This taxon was previously listed in *S. cyornithopsis* but see White (1962: 135) (2646).
[4] Includes *pallidigularis* Cunningham-van Someren & Schifter, 1981 (602), see Oatley *et al.* (1992: 397) (1669) (previously listed in *S. cyornithopsis*).
[5] Includes *balcanicus* see Roselaar (1995) (2107).
[6] For recognition see Cramp *et al.* (1988: 596) (592).
[7] For recognition see Cramp *et al.* (1988: 596) (592).
[8] Includes *.rishirensis* N. Kuroda, Jr., 1965 (1310).

○ *Luscinia komadori* RYUKYU ROBIN

— *L.k.komadori* (Temminck, 1835)[1] | Tanegashima to Tokunoshima (N Ryukyu Is.) >> S Ryukyu Is

— *L.k.namiyei* (Stejneger, 1886) | Okinawa I.

○ *Luscinia svecica* BLUETHROAT

— *L.s.svecica* (Linnaeus, 1758)[2] | Scandinavia and Finland to N Asia, Alaska >> SW Europe, N and N tropical Africa, Arabia, S and E Asia

— *L.s.namnetum* Mayaud, 1934 | SW and C France >> ??

— *L.s.cyanecula* (Meisner, 1804) | N and E France and Netherlands to NW Ukraine and Belarus >> S Europe, N and N tropical Africa

— *L.s.volgae* (O. Kleinschmidt, 1907) | NE Ukraine and C and E European Russia >> NE Africa, SW Asia

— *L.s.magna* (Zarudny & Loudon, 1904)[3] | E Turkey, Caucasus area, NW Iran >> Arabia, NE Africa

— *L.s.pallidogularis* (Zarudny, 1897) | Plains of Kazakhstan and Transcaspia >> SW and S Asia

— *L.s.abbotti* (Richmond, 1896) | SE Afghanistan, Pakistan, NW Himalayas >> NW India

— *L.s.saturatior* (Sushkin, 1925)[4] | Mts. of C Asia (Pamir and Tien Shan to Altai and Sayans) >> S Asia

— *L.s.kobdensis* (Tugarinov, 1929)[5] | Xinjiang and Mongolia >> S and SE Asia

— *L.s.przevalskii* (Tugarinov, 1929)[6] | C and E Qinghai to Nan Shan and east to Ala Shan and Ordos >> E China

○ *Luscinia calliope* SIBERIAN RUBYTHROAT

— *L.c.calliope* (Pallas, 1776) | C Urals and Siberia to Anadyr, N Korea, NE China, N Mongolia >> S and SE Asia

— *L.c.camtschatkensis* (J.F. Gmelin, 1789)[7] | Kamchatka, Commander and Kuril Is., N Japan >> SE Asia, Philippines

— *L.c.beicki* Meise, 1937[8] | NE Qinghai, SW Gansu, N Sichuan (NC China)

○ *Luscinia pectoralis* WHITE-TAILED RUBYTHROAT

— *L.p.ballioni* Severtsov, 1873 | NE Afghanistan and Pamir to E Tien Shan >> S Asia

— *L.p.pectoralis* (Gould, 1837) | SE Afghanistan and NW Himalayas to C Nepal >> S Asia

— *L.p.confusa* E. Hartert, 1910 | E Nepal and Sikkim >> NE India

— *L.p.tschebaiewi* (Przevalski, 1876) | S, E and NE Tibetan Plateau, N Kashmir, N Burma >> S Asia to Burma

○ *Luscinia ruficeps* (E. Hartert, 1907) RUFOUS-HEADED ROBIN | S Shaanxi to N Sichuan >> SE Asia

○ *Luscinia obscura* (Berezowsky & Bianchi, 1891) BLACK-THROATED ROBIN[9] | S Shaanxi, SW Gansu, N Sichuan >> NE Burma, N Thailand

○ *Luscinia pectardens* (David, 1871) FIRE-THROATED ROBIN | SE Xizang, N Yunnan, W Sichuan, SW Gansu, S Shaanxi >> NE India, SE Asia

○ *Luscinia brunnea* INDIAN BLUE ROBIN

— *L.b.brunnea* (Hodgson, 1837) | E Afghanistan, Himalayas, Assam, C China >> S India, Sri Lanka

— *L.b.wickhami* (E.C.S. Baker, 1916) | Chin hills (W Burma)

○ *Luscinia cyane* SIBERIAN BLUE ROBIN

— *L.c.cyane* (Pallas, 1776) | SC Siberia, N Mongolia >> SE Asia

— *L.c.bochaiensis* (Shulpin, 1928) | E Siberia, NE China, N Korea, Japan >> Malay Pen., Gtr. Sundas

○ *Luscinia indica* WHITE-BROWED BUSH ROBIN

— *L.i.indica* (Vieillot, 1817) | C and E Himalayas, Assam hills

— *L.i.yunnanensis* (Rothschild, 1922) | N Yunnan to W Sichuan >> N Burma to N Indochina

— *L.i.formosana* (E. Hartert, 1909) | Taiwan

---

[1] Includes *subrufus* see Orn. Soc. Japan (2000: 301) (1727).
[2] The subspecies *gaetkei*, included here, could be recognised (CSR). Unlike the nominate form it migrates through W and SW Europe to NW Africa. No recent authority now recognises this, but publication is foreseen.
[3] Includes *luristanica* see Vaurie (1959: 384) (2502).
[4] Includes *tianshanica* see Vaurie (1959: 383) (2502).
[5] For recognition see Stepanyan (1990: 534) (2326).
[6] Omitted in last Edition following Ripley (1964: 44) (2045), but see Vaurie (1959: 384) (2502) and Cheng (1987) (333).
[7] For recognition see Kistchinski (1988) (1272).
[8] For recognition see Eck (1996) (790).
[9] The name *Heteroxenicus joannae* has been said to relate to this, see Cheng (1987: 583 fn) (333), but this is nowhere reviewed in detail and is under investigation.

○ *Luscinia hyperythra* (Blyth, 1847) RUFOUS-BREASTED BUSH ROBIN — C and E Himalayas, SE Xizang, N Yunnan >> NE India, N Burma

○ *Luscinia johnstoniae* (Ogilvie-Grant, 1906) COLLARED BUSH ROBIN — Taiwan

○ *Luscinia cyanura* ORANGE-FLANKED BUSH ROBIN
— *L.c.cyanura* (Pallas, 1773)[1] — Finland and N Russia to N Mongolia, NE China, Japan >> S and SE Asia
— *L.c.pallidior* (E.C.S. Baker, 1924) — NW Himalayas >> India
— *L.c.rufilata* (Hodgson, 1845) — C and E Himalayas, C China >> Burma to Indochina

○ *Luscinia chrysaea* GOLDEN BUSH ROBIN[2]
— *L.c.whistleri* (Ticehurst, 1922) — NW Himalayas
— *L.c.chrysaea* (Hodgson, 1845)[3] — C and E Himalayas, Assam, N and W Yunnan, W Sichuan, S Shaanxi >> N India to N Indochina

○ *Luscinia sibilans* (Swinhoe, 1863) RUFOUS-TAILED ROBIN — C and E Siberia, NE China, N Korea >> E China, SE Asia

○ *Luscinia luscinia* (Linnaeus, 1758) THRUSH NIGHTINGALE — N, C and E Europe, SW Siberia, N Kazakhstan >> SE Africa

● *Luscinia megarhynchos* COMMON NIGHTINGALE
√ *L.m.megarhynchos* C.L. Brehm, 1831 — NW Africa, W and C Europe to Poland, Czech Rep., W Croatia >> and C W Africa
— *L.m.baehrmanni* Eck, 1975[4] — SE Poland, Slovakia, E Austria, E Croatia to Levant, C Turkey, Crimea >> NE and E Africa (787)
— *L.m.africana* (Fischer & Reichenow, 1884) — Caucasus area and E Turkey to SW and N Iran >> NE and E Africa
— *L.m.hafizi* Severtsov, 1873 — E Iran to Kazakhstan, Xinjiang, SW Mongolia, C Asia >> E Africa

IRANIA De Filippi, 1863 F
○ *Irania gutturalis* (Guérin-Méneville, 1843) WHITE-THROATED IRANIA — Asia Minor and Levant to S Kazakhstan, Tajikistan, Afghanistan >> NE and E Africa

COSSYPHICULA Grote, 1934[5] F
○ *Cossyphicula roberti* WHITE-BELLIED ROBIN-CHAT
— *C.r.roberti* (Alexander, 1903) — W Cameroon, Bioko I.
— *C.r.rufescentior* (E. Hartert, 1908) — E Zaire

COSSYPHA Vigors, 1825 F
○ *Cossypha isabellae* MOUNTAIN ROBIN-CHAT
— *C.i.batesi* (Bannerman, 1922) — E Nigeria, W Cameroon
— *C.i.isabellae* G.R. Gray, 1862 — Mt. Cameroon

○ *Cossypha archeri* ARCHER'S ROBIN-CHAT
— *C.a.archeri* Sharpe, 1902[6] — E Zaire, SW Uganda, W Rwanda, W Burundi
— *C.a.kimbutui* Prigogine, 1955 — Mt. Kabobo (E Zaire)

○ *Cossypha anomala* OLIVE-FLANKED ROBIN-CHAT
— *C.a.grotei* (Reichenow, 1932) — E and S Tanzania
— *C.a.anomala* (Shelley, 1893)[7] — Mt. Milanje (S Malawi), Mt. Namule (N Mozambique)
— *C.a.macclounii* (Shelley, 1903) — SW Tanzania, N Malawi
— *C.a.mbuluensis* (Grant & Mackworth-Praed, 1936) — Mbulu Highlands (NC Tanzania)

○ *Cossypha caffra* CAPE ROBIN-CHAT
— *C.c.iolaema* Reichenow, 1900[8] — S Sudan to Malawi and N Mozambique
— *C.c.kivuensis* Schouteden, 1937 — SW Uganda, E Zaire
— *C.c.caffra* (Linnaeus, 1771)[9] — Zimbabwe, C and E Transvaal and Natal to SW Cape Province
— *C.c.namaquensis* W.L. Sclater, 1911 — S Namibia and NW Cape to Orange Free State and W Transvaal

[1] Includes *pacificus* Portenko, 1954 (1907), see Vaurie (1959: 390).
[2] Type species of the genus *Tarsiger*. Not recognised here.
[3] The proposed race *vitellinus* from Sichuan may prove recognisable. CSR.
[4] The same name was published slightly later in Zool. Abh. Staatl. Mus. Tierk. Dresden, 33.
[5] We employ this genus following Oatley *et al.* (1992: 415) (1669).
[6] Includes *albimentalis* see Stuart (1992: 420) (2371).
[7] Includes *albigularis* see Ripley (1964: 57) (2045) and *gurue* see White (1962: 134) (2646).
[8] Implicitly includes *vespera* Clancey, 1972 (439) see Oatley *et al.* (1992: 423) (1669).
[9] Includes *drakensbergi* see White (1962: 143) (2646) and we believe *ardens* Clancey, 1981 (486).

○ *Cossypha humeralis* (A. Smith, 1836) WHITE-THROATED ROBIN-CHAT[1]   Zimbabwe to S Mozambique and NE Sth. Africa

○ *Cossypha polioptera* GREY-WINGED ROBIN-CHAT[2]
— *C.p.polioptera* Reichenow, 1892[3]   S Sudan to Burundi and W Kenya, N Angola to
                                         NW Zambia
— *C.p.nigriceps* Reichenow, 1910   Sierra Leone to N Cameroon
— *C.p.tessmanni* Reichenow, 1921   E Cameroon

○ *Cossypha cyanocampter* BLUE-SHOULDERED ROBIN-CHAT
— *C.c.cyanocampter* (Bonaparte, 1850)   Sierra Leone to Gabon
— *C.c.bartteloti* Shelley, 1890[4]   NE Zaire to S Sudan and W Kenya

○ *Cossypha semirufa* RÜPPELL'S ROBIN-CHAT
— *C.s.semirufa* (Rüppell, 1840)   S and W Ethiopia to SE Sudan and N Kenya
— *C.s.donaldsoni* Sharpe, 1895   E Ethiopia
— *C.s.intercedens* (Cabanis, 1878)   C and SE Kenya, N Tanzania

○ *Cossypha heuglini* WHITE-BROWED ROBIN-CHAT
— *C.h.subrufescens* Bocage, 1869   Gabon to W Angola
— *C.h.heuglini* Hartlaub, 1866[5]   S Chad and S Sudan to E Angola, Zimbabwe and
                                      E Transvaal
— *C.h.intermedia* (Cabanis, 1868)[6]   S Somalia, coastal E Africa to Zululand

○ *Cossypha natalensis* RED-CAPPED ROBIN-CHAT
— *C.n.larischi* Meise, 1957   Nigeria to N Angola
— *C.n.intensa* Mearns, 1913[7] [8]   S Somalia, S Sudan and E Africa to E Angola and
                                       E Transvaal
— *C.n.natalensis* A. Smith, 1840   Natal to E Cape Province

○ *Cossypha dichroa* CHORISTER ROBIN-CHAT
— *C.d.dichroa* (J.F. Gmelin, 1789)   E Transvaal to S Cape Province
— *C.d.mimica* Clancey, 1981   NE Transvaal (485)

○ *Cossypha heinrichii* Rand, 1955 WHITE-HEADED ROBIN-CHAT   N Angola, W Zaire

○ *Cossypha niveicapilla* SNOWY-CROWNED ROBIN-CHAT
— *C.n.niveicapilla* (Lafresnaye, 1838)   Senegal, Gambia and Sierra Leone to Nigeria, Sudan
                                           and W Ethiopia
— *C.n.melanonota* (Cabanis, 1875)   Lake Victoria basin

○ *Cossypha albicapilla* WHITE-CROWNED ROBIN-CHAT
— *C.a.albicapilla* (Vieillot, 1818)   Senegal and Gambia to NE Guinea
— *C.a.giffardi* E. Hartert, 1899   N Ghana to N Cameroon
— *C.a.omoensis* Sharpe, 1900   SE Sudan, SW Ethiopia

XENOCOPSYCHUS E. Hartert, 1907 M
○ *Xenocopsychus ansorgei* E. Hartert, 1907 ANGOLA CAVE CHAT   W Angola

CICHLADUSA W.K.H. Peters, 1863 F
○ *Cichladusa arquata* W.K.H. Peters, 1863 COLLARED PALM THRUSH   SE Zaire, S Uganda and coastal Kenya to S Mozambique

○ *Cichladusa ruficauda* (Hartlaub, 1857) RED-TAILED PALM THRUSH   S Gabon and W Zaire to W Angola

○ *Cichladusa guttata* SPOTTED PALM THRUSH
— *C.g.guttata* (Heuglin, 1862)   S Sudan, NE Zaire, Uganda, NW Kenya
— *C.g.intercalans* Clancey, 1986   SW Ethiopia to C Kenya and C Tanzania (503)
— *C.g.rufipennis* Sharpe, 1901   S Somalia, E Kenya, NE Tanzania

---

[1] Includes *crepuscula* see Oatley *et al.* (1992: 425) (1669).
[2] Treated within the genus *Sheppardia* in last Edition, but see Oatley *et al.* (1992: 418) (1669).
[3] Includes *grimwoodi* see Oatley *et al.* (1992: 427) (1669).
[4] Includes *pallidiventris* Cunningham-van Someren & Schifter, 1981 (602), see Oatley *et al.* (1992: 428) (1669).
[5] Includes *pallidior* see White (1962: 145) (2646) and implicitly *orphea* Clancey, 1979 (470) see Oatley *et al.* (1992: 431) (1669).
[6] Includes *euronota* see White (1962: 145) (2646).
[7] Includes *hylophona* and *egregior* see White (1962: 142) (2646) and *garguensis*, *tennenti* and *seclusa* Clancey, 1981 (487), see Oatley *et al.* (1992: 433) (1669).
[8] Thought to include ¶*clanceyi* Prigogine, 1988 (1945).

CERCOTRICHAS Boie, 1831 F
○ *Cercotrichas leucosticta* FOREST SCRUB ROBIN
   __ *C.l.leucosticta* (Sharpe, 1883)   Ghana
   __ *C.l.colstoni* Tye, 1991   Kambui Hills (Sierra Leone), Liberia (2445)
   __ *C.l.collsi* (Alexander, 1907)   W Uganda, NE Zaire, SW Central African Republic
   __ *C.l.reichenowi* (E. Hartert, 1907)   Huila escarpment (NW Angola)

○ *Cercotrichas barbata* (Hartlaub & Finsch, 1870) MIOMBO BEARDED SCRUB ROBIN[1]
              W Tanzania to Angola

○ *Cercotrichas quadrivirgata* EASTERN BEARDED SCRUB ROBIN
   __ *C.q.quadrivirgata* (Reichenow, 1879)[2]   Somalia to NE Namibia and E Sth. Africa
   __ *C.q.greenwayi* (Moreau, 1938)   Mafia and Zanzibar Is.

○ *Cercotrichas signata* BROWN SCRUB ROBIN
   __ *C.s.tongensis* (Roberts, 1931)   S Mozambique and coastal Zululand
   __ *C.s.signata* (Sundevall, 1850)[3]   E Transvaal to E Cape Province

○ *Cercotrichas hartlaubi* (Reichenow, 1891) BROWN-BACKED SCRUB ROBIN[4]
              S Cameroon to SW Central African Republic, NE Zaire
               to C Kenya, N Angola

○ *Cercotrichas leucophrys* WHITE-BROWED SCRUB ROBIN[5]
   __ *C.l.leucoptera* (Rüppell, 1845)   SE Sudan, S Ethiopia, N Somalia, N Kenya
   __ *C.l.eluta* (Bowen, 1934)   S Somalia, NE Kenya
   __ *C.l.vulpina* (Reichenow, 1891)   E Kenya, E Tanzania
   __ *C.l.brunneiceps* (Reichenow, 1891)   C and S Kenya, N Tanzania
   __ *C.l.sclateri* (Grote, 1930)   C Tanzania
   __ *C.l.zambesiana* (Sharpe, 1882)   S Sudan, N and E Zaire, Uganda, and W Kenya to
               E and S Zambia, E Zimbabwe and N Mozambique
   __ *C.l.munda* (Cabanis, 1880)   S Gabon to C Angola and W Zaire
   __ *C.l.ovamboensis* (Neumann, 1920)   S Angola and N Namibia to SW Zambia and
               W Zimbabwe
   __ *C.l.simulator* (Clancey, 1964)   S Mozambique (402)
   __ *C.l.leucophrys* (Vieillot, 1817)[6]   S Zimbabwe and Transvaal to E Cape Province

○ *Cercotrichas galactotes* RUFOUS SCRUB ROBIN
   __ *C.g.galactotes* (Temminck, 1820)   S and E Iberia, N Africa to S Algeria, Israel, and
               SW Syria >> Sahel zone (W Africa)
   __ *C.g.syriaca* (Hemprich & Ehrenberg, 1833)   Balkans to W and S Turkey, W Syria and Lebanon >>
               NE and E Africa
   __ *C.g.familiaris* (Ménétriés, 1832)   SE Turkey, Transcaucasia, NE Arabia and Iraq to
               S Kazakhstan and W Pakistan >> NE Africa and
               NW India
   __ *C.g.minor* (Cabanis, 1850)   Senegal and Gambia to N Somalia
   __ *C.g.hamertoni* (Ogilvie-Grant, 1906)   E Somalia

○ *Cercotrichas paena* KALAHARI SCRUB ROBIN
   __ *C.p.benguellensis* (E. Hartert, 1907)   SW Angola
   __ *C.p.paena* (A. Smith, 1836)   Botswana, N Cape Province
   __ *C.p.damarensis* (E. Hartert, 1907)   Namibia
   __ *C.p.oriens* (Clancey, 1957)   S Zimbabwe, Orange Free State, Transvaal

○ *Cercotrichas coryphaeus* KARROO SCRUB ROBIN
   __ *C.c.coryphaeus* (Lesson, 1831)[7]   S Namibia, NW Cape Province,to W Orange Free State
               and E Cape
   __ *C.c.cinerea* (Macdonald, 1952)   SW Cape Province

---

[1] Treated as monotypic by Oatley *et al.* (1992: 472) (1669). Includes *thamnodytes* Clancey, 1974 (443).
[2] Includes *erlangeri*, *rovumae* and *wilsoni* see White (1962: 128) (2646) and implicitly *interna* and *brunnea* Ripley & Heinrich, 1966 (2068) see Oatley *et al.*
   (1992: 473) (1669).
[3] Includes *oatleyi* and *reclusa* Clancey, 1966 (408), see Oatley *et al.* (1992: 475) (1669).
[4] Includes *kenia* see White (1962: 127) (2646).
[5] Type species of the genus *Erythropygia* (here lumped with otherwise monotypic *Cercotrichas* — type species *podobe*).
[6] Oatley *et al.* (1992: 478) (1669) included *pectoralis*, and implicitly *makalaka* and *strepitans* Clancey, 1975 (453).
[7] Probably includes *eurina* Clancey, 1969 (417), see Oatley *et al.* (1992: 484) (1669).

○ *Cercotrichas podobe* Black Scrub Robin

    — *C.p.podobe* (Statius Müller, 1776)   N Senegal, Gambia and S Mauritania to N NE Sudan and Somalia

    — *C.p.melanoptera* (Hemprich & Ehrenberg, 1833)   S Israel, W, C and S Saudi Arabia, Yemen, Aden

NAMIBORNIS Bradfield, 1935 M

○ *Namibornis herero* (Meyer de Schauensee, 1931) Herero Chat   W Namibia, SW Angola

COPSYCHUS Wagler, 1827 M

○ *Copsychus albospecularis* Madagascan Magpie-Robin

    — *C.a.pica* Pelzeln, 1858   W and N Madagascar
    — *C.a.albospecularis* (Eydoux & Gervais, 1836)   NE Madagascar
    — *C.a.inexpectatus* Richmond, 1897[1]   EC to SE Madagascar

○ *Copsychus sechellarum* A. Newton, 1865 Seychelles Magpie-Robin   Seychelles

○ *Copsychus saularis* Oriental Magpie-Robin

    — *C.s.saularis* (Linnaeus, 1758)   NE Pakistan, N, C and W India
    — *C.s.ceylonensis* P.L. Sclater, 1861   S India, Sri Lanka
    — *C.s.erimelas* Oberholser, 1923   NE India to Thailand and Indochina
    — *C.s.andamanensis* Hume, 1874   Andaman Is.
    — *C.s.prosthopellus* Oberholser, 1923   S and E China, Hainan I.
    — *C.s.musicus* (Raffles, 1822)[2]   S Thailand, Malaysia, Sumatra, Belitung and Bangka Is., W Java, S and W Borneo
    — *C.s.zacnecus* Oberholser, 1912   Simeuluë I.
    — *C.s.nesiarchus* Oberholser, 1923   Nias I.
    — *C.s.masculus* Ripley, 1943   Batu Is.
    — *C.s.pagiensis* Richmond, 1912   Mentawei Is.
    — *C.s.amoenus* (Horsfield, 1821)   E Java, Bali
    — *C.s.adamsi* Elliot, 1890   N Borneo, Banggi I.
    — *C.s.pluto* Bonaparte, 1850[3]   Maratua I., E and SE Borneo
    — *C.s.deuteronymus* Parkes, 1963   Luzon
    — *C.s.mindanensis* (Boddaert, 1783)   S Philippines

◉ *Copsychus malabaricus* White-rumped Shama[4]

    — *C.m.malabaricus* (Scopoli, 1786[5])   W and S India
    — *C.m.leggei* (Whistler, 1941)   Sri Lanka
    — *C.m.albiventris* (Blyth, 1859)   Andaman Is.
    — *C.m.indicus* (E.C.S. Baker, 1924)   Nepal, N, C-E and NE India, Burma, S Yunnan, NW Thailand, N Indochina
    — *C.m.interpositus* (Robinson & Kloss, 1922)   Thailand (except NW and Pen., S Indochina (except Con Son I.
    — *C.m.macrourus* (J.F. Gmelin, 1789)[6]   Con Son I. (S Vietnam)
    — *C.m.minor* (Swinhoe, 1870)   Hainan I.
    — *C.m.pellogynus* (Oberholser, 1923)[7]   SE Burma, Pen. Thailand
    — *C.m.mallopercnus* (Oberholser, 1923)   W Malaysia
    — *C.m.tricolor* (Vieillot, 1818)[8]   Sumatra, W Java, Bangka, Belitung Is.
    — *C.m.mirabilis* Hoogerwerf, 1962   Prinsen I.
    — *C.m.melanurus* (Salvadori, 1887)   West Sumatra Is. except Banyak and Batu Is., and Enggano
    — *C.m.opisthisus* (Oberholser, 1912)[9]   Banyak Is.
    — *C.m.opisthopelus* (Oberholser, 1912)   Batu Is.

---

[1] Thought to include *winterbottomi* Farkas, 1972 (826).
[2] Includes *javensis*, *nesiotes* and *problematicus* see Mees (1986) (1539).
[3] Given as 1851 by Ripley (1964: 67) (2045) but see Zimmer (1926) (2707).
[4] Mees (1996) (1543) discussed the interbreeding between *C.m.suavis* and *C.s.stricklandi* that caused Smythies to treat these two as members of one species. True sympatry seems unproven.
[5] The use of 1788 by Ripley (1964) (2045) appears to be a *lapsus* see Ripley (1982: 455) (2055).
[6] For mention of this see Mees (1996) (1543); its validity needs confirmation, it may apply to mainland birds and have priority over *interpositus*.
[7] For recognition see Medway & Wells (1976: 317) (1507).
[8] Includes *abbotti* see Ripley (1964: 70) (2045) and Mees (1986) (1539).
[9] For recognition see Van Marle & Voous (1988) (2468), but although they cite Ripley (1944) (2027) he treated this as a synonym of *opisthopelus* and this may be correct.

— *C.m.omissus* (E. Hartert, 1902)[1]   WC, C and E Java

— *C.m.nigricauda* (Vorderman, 1893)   Kangean I.

— *C.m.ochroptilus* (Oberholser, 1917)   Anamba Is.

— *C.m.eumesus* (Oberholser, 1932)   Natuna Is.

— *C.m.suavis* P.L. Sclater, 1861   Sarawak and Kalimantan (Borneo)

— *C.m.stricklandii* Motley & Dillwyn, 1855[2]   Sabah (N Borneo) incl. Banggi I.

— *C.m.barbouri* (Bangs & Peters, 1927)   Maratua Is. (NE Borneo)

○ *Copsychus luzoniensis* WHITE-BROWED SHAMA

— *C.l.luzonensis* (Kittlitz, 1832)   Luzon and Catanduanes Is.

— *C.l.parvimaculatus* (McGregor, 1910)   Polillo Is.

— *C.l.shemleyi* duPont, 1976[3]   Marinduque (775)

— *C.l.superciliaris* (Bourns & Worcester, 1894)   Masbate, Negros, Panay and Ticao

○ *Copsychus niger* (Sharpe, 1877) WHITE-VENTED SHAMA   Balabac, Palawan and Calamianes

○ *Copsychus cebuensis* (Steere, 1890) BLACK SHAMA[4]   Cebu

TRICHIXOS Lesson, 1839[5] M

○ *Trichixos pyrrhopygus* Lesson, 1839 RUFOUS-TAILED SHAMA[6]   W Malaysia, Sumatra, Borneo

SAXICOLOIDES Lesson, 1832 M[7]

○ *Saxicoloides fulicatus* INDIAN ROBIN

— *S.f.cambaiensis* (Latham, 1790)   Pakistan, N and W India

— *S.f.erythrurus* (Lesson, 1831)[8]   NE India

— *S.f.intermedius* Whistler & Kinnear, 1932   C India

— *S.f.fulicatus* (Linnaeus, 1766)   S India

— *S.f.leucopterus* (Lesson, 1840)   Sri Lanka

PHOENICURUS T. Forster, 1817 M

○ *Phoenicurus alaschanicus* (Przevalski, 1876) PRZEVALSKI'S REDSTART   NC China >> S Shaanxi, Hebei and Shanxi

○ *Phoenicurus erythronotus* (Eversmann, 1841) RUFOUS-BACKED REDSTART

N Tajikistan and W and N Xinjiang to Baikal area and Mongolia >> SW and C Asia to N India

○ *Phoenicurus caeruleocephala* Vigors, 1831 BLUE-CAPPED REDSTART[9]   E Afghanistan to Tien Shan and W and C Himalayas

● *Phoenicurus ochruros* BLACK REDSTART

↙ *P.o.gibraltariensis* (J.F. Gmelin, 1789)   Europe (except SW), NW Africa >> W and S Europe to N Africa and Middle East

— *P.o.aterrimus* von Jordans, 1923[10]   Portugal, S and C Spain

— *P.o.ochruros* (S.G. Gmelin, 1774)   Asia Minor, Caucasus area, NW Iran >> Middle East

— *P.o.semirufus* (Hemprich & Ehrenberg, 1833)   W Syria, Lebanon, NE Israel

— *P.o.phoenicuroides* (F. Moore, 1854)   C Kazakhstan and Tien Shan to Tuva Rep. and Mongolia >> NE Africa, SW, S and C Asia

— *P.o.rufiventris* (Vieillot, 1818)   Turkmenistan and Pamir-Alai Mts. to Himalayas and C China >> SW, WC and S Asia

● *Phoenicurus phoenicurus* COMMON REDSTART

— *P.p.phoenicurus* (Linnaeus, 1758)[11]   NW Africa and Europe to C Siberia and N Mongolia >> E Africa

— *P.p.samamisicus* (Hablizl, 1783)   S Balkans and Greece to Turkmenistan, S Uzbekiastan and Iran >> NE Africa to SW Asia

○ *Phoenicurus hodgsoni* (F. Moore, 1854) HODGSON'S REDSTART   SE, E and NE Tibetan Plateau >> Burma, NE India

[1] Includes *javanus* see Mees (1996) (1543).
[2] This race and *barbouri* have striking white crowns and are seen by some authors to comprise a distinct species.
[3] For recognition see Dickinson *et al.* (1991: 316) (745).
[4] For recognition at specific level see Dickinson *et al.* (1991: 317) (745).
[5] Treated in *Copsychus* in last Edition; but see Smythies (2000) (2289).
[6] For correction of spelling see David & Gosselin (2002) (613).
[7] Gender addressed by David & Gosselin (2002) (614), multiple changes occur in specific names.
[8] Given as 1832 by Ripley (1964) (2045), but see Browning & Monroe (1991) (258).
[9] For correction of spelling see David & Gosselin (2002) (613).
[10] For recognition see Cramp *et al.* (1988: 683) (592).
[11] Includes *algeriensis* see Vaurie (1959: 366) (2502).

O *Phoenicurus schisticeps* (J.E. & G.R. Gray, 1846) WHITE-THROATED REDSTART

    C and E Himalayas, N Yunnan to Qinghai (C China)
        >> Assam, N Burma

O *Phoenicurus auroreus* DAURIAN REDSTART

   — *P.a.leucopterus* Blyth, 1843   E Himalayas to C and E China >> NE India, N Indochina
   — *P.a.auroreus* (Pallas, 1776)   SC Siberia and Mongolia to Amurland, Korea and
        Hebei >> Japan to S China

O *Phoenicurus moussieri* (Olph-Galliard, 1852) MOUSSIER'S REDSTART   NW Africa

O *Phoenicurus erythrogastrus* GÜLDENSTADT'S REDSTART[1]

   — *P.e.erythrogastrus* (Güldenstädt, 1775)   Greater Caucasus Mts. >> Transcaucasia, N Iran
   — *P.e.grandis* (Gould, 1850)   E Afghanistan to Baikal area, C Asia, C China, Tibet,
        Himalayas, Pakistan, N India >> N India to E and
        NE China

O *Phoenicurus frontalis* Vigors, 1832 BLUE-FRONTED REDSTART   E Afghanistan, Himalayas, N Burma, S China >>
        N India to N Indochina

HODGSONIUS Bonaparte, 1850[2] M
O *Hodgsonius phaenicuroides* WHITE-BELLIED REDSTART

   — *H.p.phaenicuroides* (J.E. & G.R. Gray, 1846)   Himalayas, W Burma>> to N and NE India, N Burma
   — *H.p.ichangensis* E.C.S. Baker, 1922   NE and E Burma, C and NE China >> Thailand to
        N Indochina

RHYACORNIS Blanford, 1872 F
O *Rhyacornis fuliginosa* PLUMBEOUS WATER REDSTART

   — *R.f.fuliginosa* Vigors, 1831   E Afghanistan and Himalayas to NE and E China,
        Hainan, Indochina

   — *R.f.affinis* (Ogilvie-Grant, 1906)   Taiwan

O *Rhyacornis bicolor* (Ogilvie-Grant, 1894) LUZON WATER REDSTART   N Luzon

CHAIMARRORNIS Hodgson, 1844[3] M
O *Chaimarrornis leucocephalus* (Vigors, 1831) WHITE-CAPPED WATER REDSTART

        E Uzbekistan, Tajikistan and Afghanistan to C and NE
        China, C Laos, N Burma >> N India to Indochina

MYIOMELA G.R. Gray, 1846[4] F
O *Myiomela leucura* WHITE-TAILED ROBIN

   — *M.l.leucura* (Hodgson, 1845)   C and E Himalayas, Assam, Burma, C China, N and
        C Indochina, Hainan

   — *M.l.montium* Swinhoe, 1864[5]   Taiwan (2382)
   — *M.l.cambodiana* (Delacour & Jabouille, 1928)   S Indochina and SE Thailand

O *Myiomela diana* SUNDA ROBIN

   — *M.d.sumatrana* (Robinson & Kloss, 1918)   N and WC Sumatra
   — *M.d.diana* (Lesson, 1831)[6]   Java

CINCLIDIUM Blyth, 1842 N
O *Cinclidium frontale* BLUE-FRONTED ROBIN

   — *C.f.frontale* Blyth, 1842   Nepal, Sikkim, Bhutan
   — *C.f.orientale* Delacour & Jabouille, 1930   E Assam, SW Sichuan, N Laos, NW Vietnam >>
        Thailand

GRANDALA Hodgson, 1843 F
O *Grandala coelicolor* Hodgson, 1843 GRANDALA   Himalayas and C China (N Yunnan to E Qinghai and
        SW Gansu)

ENICURUS Temminck, 1822 M
O *Enicurus scouleri* Vigors, 1832 LITTLE FORKTAIL[7]   W Tien Shan and Pamir-Alai through Himalayas to C
        and E China, Assam, W Burma, NW Vietnam, Taiwan

---

[1] For correction of spelling see David & Gosselin (2002) (613).
[2] Given as 1851 by Ripley (1964) (2045), but for the dates of Bonaparte's "Conspectus" see Zimmer (1926) (2707).
[3] Treated in the genus *Phoenicurus* in last Edition; but see Ripley (1982: 471) (2055).
[4] We believe *Muscisylvia* Hodgson to be preoccupied by *Muscisylvia* (sic) Agassiz, this spelling not differing as suggested by Ripley (1964: 81) (2045); see
   Agassiz, 1842-46, Nomenclator Zool., fasc 2., pp. 49, 88 (per M. Walters *in litt.*).
[5] For recognition see Cheng (1987: 604) (333).
[6] Given as 1834 in Ripley (1964) (2045), but see Browning & Monroe (1991) (258).
[7] Includes *fortis* see Cheng (1987: 606) (333).

○ *Enicurus velatus* SUNDA FORKTAIL
— *E.v.sumatranus* (Robinson & Kloss, 1923) — Sumatra
— *E.v.velatus* Temminck, 1822 — Java

○ *Enicurus ruficapillus* Temminck, 1832[1] CHESTNUT-NAPED FORKTAIL — Malay Pen., Borneo, Sumatra

○ *Enicurus immaculatus* (Hodgson, 1836) BLACK-BACKED FORKTAIL — C and E Himalayas, Burma, NW Thailand, W Yunnan

○ *Enicurus schistaceus* (Hodgson, 1836) SLATY-BACKED FORKTAIL — C and E Himalayas, Burma, C and E China, Thailand, Indochina

○ *Enicurus leschenaulti* WHITE-CROWNED FORKTAIL
— *E.l.indicus* E. Hartert, 1909 — NE India, Burma, Yunnan, W, N and E Thailand, Indochina
— *E.l.sinensis* Gould, 1865 — W Sichuan and E Qinghai to S and E China incl. Hainan
— *E.l.frontalis* Blyth, 1847 — C and S Malay Pen., Sumatra, Nias I., Borneo (lowlands?)[2]
— *E.l.borneensis* (Sharpe, 1889) — Borneo (higher elevations?)
— *E.l.chaseni* Meyer de Schauensee, 1940 — Batu Is. (W Sumatra)
— *E.l.leschenaulti* (Vieillot, 1818) — Java, Bali

○ *Enicurus maculatus* SPOTTED FORKTAIL
— *E.m.maculatus* Vigors, 1831 — E Afghanistan, W and C Himalayas
— *E.m.guttatus* Gould, 1865 — E Himalayas, Burma, W and C Yunnan, SW Sichuan
— *E.m.bacatus* Bangs & Phillips, 1914 — SE Yunnan, Fujian, N Indochina
— *E.m.robinsoni* E.C.S. Baker, 1922 — SC Vietnam (Langbian Plateau)

SAXICOLA Bechstein, 1803 M[3]

◓ *Saxicola rubetra* (Linnaeus, 1758) WHINCHAT[4] — Europe, Asia Minor, W Siberia >> W, C and NE Africa

○ *Saxicola macrorhynchus* (Stoliczka, 1872) STOLICZKA'S BUSHCHAT — S Afghanistan, Pakistan, NW India

○ *Saxicola insignis* J.E. & G.R. Gray, 1846 HODGSON'S BUSHCHAT — Russian Altai, W and N Mongolia >> foot of C and E Himalayas

○ *Saxicola dacotiae* CANARY ISLANDS CHAT
— *S.d.dacotiae* (Meade-Waldo, 1889) — Fuerteventura I. (Canary Is.)
— †*S.d.murielae* Bannerman, 1913 — Allegranza I. (Canary Is.)

◓ *Saxicola torquatus* COMMON STONECHAT[5]
— *S.t.hibernans* (E. Hartert, 1910) — British Is., W France, Portugal
— *S.t.rubicola* (Linnaeus, 1766) — W, C and S Europe, Turkey to W Caucasus area, NW Africa >> N Africa to Middle East
— *S.t.variegatus* (S.G. Gmelin, 1774) — E Caucasus area to lower Ural R. and NW Iran >> NE Africa
— *S.t.armenicus* Stegmann, 1935 — SE Turkey, Transcaucasia, SW Iran >> SW Asia, NE Africa
— *S.t.maurus* (Pallas, 1773) — E Finland and N and E European Russia to Pakistan, E Tien Shan and Mongolia >> SW and S Asia
— *S.t.indicus* (Blyth, 1847) — NW and C Himalayas >> C India
— *S.t.przewalskii* (Pleske, 1889) — Tibetan Plateau to C China >> N and NE India to E China and SE Asia
— *S.t.stejnegeri* (Parrot, 1908) — E Siberia and E Mongolia to NE China, Korea, Japan, and Anadyrland >> E and SE Asia S to the Malay Pen.
— *S.t.felix* Bates, 1936 — SW Saudi Arabia, W Yemen
— *S.t.albofasciatus* Rüppell, 1845 — SE Sudan, Ethiopian Highlands, NE Uganda
— *S.t.jebelmarrae* Lynes, 1920 — Darfur (W Sudan), E Chad
— *S.t.moptanus* Bates, 1932 — Inner Niger Delta (S Mali), Senegal Delta
— *S.t.nebularum* Bates, 1930 — Sierra Leone to W Ivory Coast
— *S.t.axillaris* (Shelley, 1884) — E Zaire to Kenya and N and W Tanzania

[1] 1823 as given by Ripley (1964: 86) (2045) was a *lapsus* for 1832.
[2] The two Bornean "races" attributed to *Enicurus leschenaulti* seem not to be separated altitudinally but to overlap: a problem to be resolved here!
[3] Gender addressed by David & Gosselin (2002) (614), multiple changes occur in specific names.
[4] Implicitly includes *incertus* Trischitta, 1939 (2440).
[5] Proposed splits of this complex species seem unsatisfactory if only in how the forms are ascribed to species.

__ *S.t.promiscuus* E. Hartert, 1922[1]	S Tanzania to W Mozambique and E Zimbabwe
__ *S.t.salax* (J. & E. Verreaux, 1851)[2]	E Nigeria to NW Angola, Bioko I
__ *S.t.stonei* Bowen, 1932	SW Tanzania to S and E Angola and Sth. Africa
__ *S.t.clanceyi* Latimer, 1961	Coastal W Cape Province
__ *S.t.torquatus* (Linnaeus, 1776)	SW Cape Province to Transvaal and W Swaziland
__ *S.t.oreobates* Clancey, 1956	Lesotho Highlands
__ *S.t.sibilla* (Linnaeus, 1766)	Madagascar[3]
__ *S.t.voeltzkowi* Grote, 1926	Grande Comore I.
__ *S.t.tectes* (J.F. Gmelin, 1789)	Réunion I.

○ **Saxicola leucurus** (Blyth, 1847) WHITE-TAILED STONECHAT   Pakistan, N India

○ **Saxicola caprata** PIED BUSHCHAT

__ *S.c.rossorum* (E. Hartert, 1910)	NE Iran and Afghanistan to SC Kazakhstan >> SW Asia
__ *S.c.bicolor* Sykes, 1832	SE Iran, Pakistan, N India >> C India
__ *S.c.burmanicus* E.C.S. Baker, 1923	C and SE India, Burma, Yunnan, S Sichuan, N Thailand, Indochina
__ *S.c.nilgiriensis* Whistler, 1940	SW India
__ *S.c.atratus* (Kelaart, 1851)	Sri Lanka
__ *S.c.caprata* (Linnaeus, 1766)	Luzon, Mindoro,
__ *S.c.randi* Parkes, 1960	Bohol, Cebu, Negros, Panay, Siquijor
__ *S.c.anderseni* Salomonsen, 1953	Leyte, Mindanao
__ *S.c.fruticola* Horsfield, 1821	Java to Flores and Alor Is.
__ *S.c.pyrrhonotus* (Vieillot, 1818)	Kisser, Wetar, Savu, Timor Is.
__ *S.c.francki* Rensch, 1931	Sumba I.
__ *S.c.albonotatus* (Stresemann, 1912)	Saleyer I., Sulawesi
__ *S.c.cognatus* Mayr, 1944	Babar I.
__ *S.c.aethiops* (P.L. Sclater, 1880)	New Britain, N New Guinea
__ *S.c.belensis* Rand, 1940	WC New Guinea
__ *S.c.wahgiensis* Mayr & Gilliard, 1951	EC and E New Guinea

○ **Saxicola jerdoni** (Blyth, 1867) JERDON'S BUSHCHAT   N and NE India, Bangladesh, Burma, N Thailand, N Indochina

○ **Saxicola ferreus** J.E. & G. R. Gray, 1846 GREY BUSHCHAT   Himalayas to C and E China, N Thailand, NW Vietnam >> N India, S Burma, S Indochina

○ **Saxicola gutturalis** WHITE-BELLIED BUSHCHAT

__ *S.g.gutturalis* (Vieillot, 1818)	Timor I., (Roti I.?)
__ *S.g.luctuosus* Bonaparte, 1850[4]	Semau I., near Timor

CAMPICOLOIDES Roberts, 1922[5] M

○ **Campicoloides bifasciatus** (Temminck, 1829) BUFF-STREAKED CHAT   S Transvaal, Natal, Cape Province, Lesotho, Swaziland

OENANTHE Vieillot, 1816[6] F

○ **Oenanthe moesta** BUFF-RUMPED WHEATEAR

__ *O.m.moesta* (M.H.K. Lichtenstein, 1823)	N Africa
__ *O.m.brooksbanki* R. Meinertzhagen, 1923	Sinai and NW Saudi Arabia to S Syria and W Iraq >> C Arabia

○ **Oenanthe pileata** CAPPED WHEATEAR

__ *O.p.neseri* Macdonald, 1952	S Angola, N Namibia
__ *O.p.livingstonii* (Tristram, 1867)	C Kenya to Botswana and NE Cape Province
__ *O.p.pileata* (J.F. Gmelin, 1789)	S Namibia, Sth. Africa

○ **Oenanthe bottae** RED-BREASTED WHEATEAR

__ *O.b.bottae* (Bonaparte, 1854)	Highlands of Yemen and SW Saudi Arabia
__ *O.b.frenata* (Heuglin, 1869)	Highlands of Eritrea and N and C Ethiopia

[1] Includes *altivagus* Clancey, 1988 (511), see Pearson (1992: 495) (1811).
[2] Includes *adamauae* and *pallidigula* see White (1962: 102) (2646).
[3] Montane birds of Madagascar require detailed evaluation and may be valid relict forms deserving specific status see Morris & Hawkins (1998: 227) (1627). So may *S.t.tectes*.
[4] Given as 1851 by Ripley (1964: 115) (2045) but see Zimmer (1926) (2707).
[5] Treated in *Saxicola* in last Edition when we followed Tye (1989) (2444). However, we are not now convinced that it belongs there.
[6] Species sequence as advised by C.S. Roselaar (*in litt.*).

○ *Oenanthe heuglini* (Finsch & Hartlaub, 1870) Heuglin's Wheatear[1]

Mali to Sudan, W Ethiopia and NE Uganda

○ *Oenanthe isabellina* (Temminck, 1829) Isabelline Wheatear

Greece, E Balkans and S and E Ukraine to Lake Baikal, Inner Mongolia and N-C China >> NC and NE Africa to SW and S Asia

● *Oenanthe oenanthe* Northern Wheatear
— *O.o.leucorhoa* (J.F. Gmelin, 1789)

NE Canada, Greenland, Iceland >> W Europe, W Africa

☑ *O.o.oenanthe* (Linnaeus, 1758)

N and C Europe, Siberia and N Kazakhstan to Alaska >> N and C Africa

— *O.o.libanotica* (Hemprich & Ehrenberg, 1833)[2]

S Europe, Asia Minor, Levant, Transcaucasia, Iran, Turkmenistan, N Afghanistan, Tien Shan, Xinjiang, Mongolia, S Transbaikalia >> N Afrotropics and SW Asia

— *O.o.seebohmi* (Dixon, 1882)

NW Africa >> W Africa

○ *Oenanthe phillipsi* (Shelley, 1885) Somali Wheatear

N and E Somalia, SE Ethiopia

○ *Oenanthe xanthoprymna* (Hemprich & Ehrenberg, 1833) Red-rumped Wheatear

SE Turkey, SW Iran >> NE Africa

○ *Oenanthe chrysopygia* (De Filippi, 1863) Red-tailed Wheatear[3]

NE Turkey, Armenia and Iran to Pamir Mts and Afghanistan >> Arabia to NW India

○ *Oenanthe pleschanka* (Lepechin, 1770) Pied Wheatear

SE Europe to Transbaikalia, N China and NW Himalayas >> NE Africa, SW Asia

○ *Oenanthe cypriaca* (Homeyer, 1884) Cyprus Wheatear

Cyprus >> N and NE Africa

● *Oenanthe hispanica* Black-eared Wheatear
— *O.h.hispanica* (Linnaeus, 1758)

NW Africa and SW Europe to C Italy and Croatia >> W African Sahel

— *O.h.melanoleuca* (Güldenstädt, 1775)

SE Europe, Asia Minor andf Levat to S Iran and E shore of Caspian Sea >> C and E Sahel and NE Africa

○ *Oenanthe deserti* Desert Wheatear
— *O.d.homochroa* (Tristram, 1859)

N Africa east to NW Egypt

— *O.d.deserti* (Temminck, 1825)

NE Egypt and Levant to S Kazakstan, Afghanistan and Mongolia >> NE Africa and SW Asia

— *O.d.oreophila* (Oberholser, 1900)

Pamir, Tibetan Plateau and S Xinjiang >> Arabia, SW Asia, N India

○ *Oenanthe lugens* Mourning Wheatear
— *O.l.halophila* (Tristram, 1859)

NW Africa to N Libya

— *O.l.lugens* (M.H.K. Lichtenstein, 1823)

E Egypt and NE Sudan to Syria and NW Arabia

— *O.l.persica* (Seebohm, 1881)

NC, WC and SW Iran >> Arabia, NE Africa

○ *Oenanthe lugubris* Abyssinian Black Wheatear[4]
— *O.l.lugentoides* (Seebohm, 1881)

SW Saudi Arabia, W Yemen

— *O.l.boscaweni* Bates, 1937

NE Yemen and S Oman

— *O.l.vauriei* R. Meinertzhagen, 1949

NE Somalia

— *O.l.lugubris* (Rüppell, 1837)

Highlands of Eritrea and N and C Ethiopia

— *O.l.schalowi* (Fischer & Reichenow, 1884)

S Kenya, NE Tanzania

○ *Oenanthe finschii* Finsch's Wheatear
— *O.f.finschii* (Heuglin, 1869)

Asia Minor and Levant to Transcaucasia and N and SW Iran

— *O.f.barnesi* (Oates, 1890)

NE and E Iran and W Turkmenistan to W Pakistan, E Afghanistan and SC Kazakhstan >> SW Asia

○ *Oenanthe picata* (Blyth, 1847) Variable Wheatear[5]

Iran and Turkmenistan to W Tien Shan, Pamirs, and W and N Pakistan >> SW Asia

---

[1] For reasons to separate *O. heuglini* from *O. bottae* see Zimmermann *et al.* (1996) (2735).
[2] Includes *virago* and *nivea* see Roselaar in Cramp *et al.* (1988: 792) (592).
[3] For reasons to treat as a species separate from *O. xanthoprymna* see Bates (1935) (103) and Nikolaus (1987) (1656). Reported intergrades have been re-identified as 1st. winter *xanthoprymna*. Includes *kingi* see Vaurie (1959: 350) (2502).
[4] For separation of *O. lugubris* from *O. lugens* see Zimmerman *et al.* (1996) (2735).
[5] Three apparent colour morphs, perhaps races (*capistrata*, *opistholeuca* and *picata*), present problems not yet resolved. See Panov (1992) (1742).

○ *Oenanthe monticola* MOUNTAIN WHEATEAR
— *O.m.albipileata* (Bocage, 1867)   Coastal Benguela (Angola)
— *O.m.nigricauda* Traylor, 1961   Huambo (Angola)
— *O.m.atmorii* Tristram, 1869   W Namibia
— *O.m.monticola* Vieillot, 1818[1]   S Namibia and Cape Province to Swaziland and C Transvaal

○ *Oenanthe albonigra* (Hume, 1872) HUME'S WHEATEAR[2]   Iraq and Iran to S Afghanistan, SW Pakistan and NE Arabia

○ *Oenanthe leucopyga* WHITE-CROWNED BLACK WHEATEAR
— *O.l.aegra* E. Hartert, 1913   W and C Sahara (S Morocco and Mauritania to W Egypt and W Sudan)
— *O.l.leucopyga* (C.L. Brehm, 1855)   C and E Egypt and Sudan, Eritrea, Djibouti
— *O.l.ernesti* R. Meinertzhagen, 1930   Sinai and Israel to C Arabia and Yemen

○ *Oenanthe leucura* BLACK WHEATEAR
— *O.l.leucura* (J.F. Gmelin, 1789)   Spain, Portugal, S France, ?Italy
— *O.l.syenitica* (Heuglin, 1869)   NW Africa

○ *Oenanthe monacha* (Temminck, 1825) HOODED WHEATEAR   E Egypt, S Israel, S Jordan, Arabia, Iran, SW Pakistan

CERCOMELA Bonaparte, 1856 F
○ *Cercomela sinuata* SICKLE-WINGED CHAT
— *C.s.hypernephela* Clancey, 1956   Lesotho
— *C.s.ensifera* Clancey, 1958   S Namibia to SW Transvaal
— *C.s.sinuata* (Sundevall, 1858)   W and S Cape Province

○ *Cercomela schlegelii* KAROO CHAT
— *C.s.benguellensis* (W.L. Sclater, 1928)   SW Angola, NW Namibia
— *C.s.schlegelii* (Wahlberg, 1855)   Coastal Namibia
— *C.s.namaquensis* (W.L. Sclater, 1928)   S Namibia, NW Cape Province
— *C.s.pollux* (Hartlaub, 1866)   W Orange Free State to S and W Cape Province

○ *Cercomela fusca* (Blyth, 1851) BROWN ROCKCHAT   NE Pakistan, N and C India

○ *Cercomela tractrac* TRACTRAC CHAT
— *C.t.hoeschi* (Niethammer, 1955)   SW Angola, NW Namibia
— *C.t.albicans* (Wahlberg, 1855)   NW Namibia
— *C.t.barlowi* (Roberts, 1937)   C and S Great Namaqualand (Namibia)
— *C.t.nebulosa* Clancey, 1962   SW Namibia
— *C.t.tractrac* (Wilkes, 1817)   Karoo to NE Cape and SW Orange Free State (S Africa)

○ *Cercomela familiaris* FAMILIAR CHAT
— *C.f.falkensteini* (Cabanis, 1875)[3]   W Africa to NW Ethiopia, south to Zambezi valley
— *C.f.omoensis* (Neumann, 1904)   SE Sudan, SW Ethiopia, NW Kenya, NE Uganda
— *C.f.angolensis* Lynes, 1926   W Angola, N Namibia
— *C.f.galtoni* (Strickland, 1852)   E Namibia and W Botswana to N and NW Cape Province
— *C.f.hellmayri* (Reichenbach, 1902)   SE Botswana to S Mozambique, NE Cape Province and Transvaal
— *C.f.actuosa* Clancey, 1966   Drakensbergs to W Natal and Lesotho (408)
— *C.f.familiaris* (Stephens, 1826)   S Cape Province to Great Kei River

○ *Cercomela scotocerca* BROWN-TAILED ROCKCHAT
— *C.s.furensis* Lynes, 1926   Darfur (Sudan)
— *C.s.scotocerca* (Heuglin, 1869)   NE Sudan, N Ethiopia
— *C.s.turkana* van Someren, 1920   S Ethiopia, N and NC Kenya, NE Uganda
— *C.s.spectatrix* Stephenson Clarke, 1919   Awash valley (E Ethiopia), N Somalia
— *C.s.validior* Berlioz & Roche, 1970   NE Somalia (145)

○ *Cercomela dubia* (Blundell & Lovat, 1899) SOMBRE ROCKCHAT   N Somalia, EC Ethiopia

---

[1] Includes *griseiceps* see Tye (1992: 506) (2447).
[2] For correction of spelling see David & Gosselin (2002) (613).
[3] Includes *modesta* see White (1962: 112) (2646).

○ *Cercomela melanura* Blackstart
— *C.m.melanura* (Temminck, 1824)   Israel, Jordan and Sinai to NW and interior C and S Arabia

— *C.m.neumanni* Ripley, 1952   SW Saudi Arabia, W and S Yemen, SW Oman
— *C.m.lypura* (Hemprich & Ehrenberg, 1833)   NC and NE Sudan to Eritrea
— *C.m.aussae* Thesiger & Meynell, 1934   NE Ethiopia, Djibouti, N Somalia
— *C.m.airensis* E. Hartert, 1921   Aïr (N Niger) to C Sudan
— *C.m.ultima* Bates, 1933[1]   E Mali, W Niger

○ *Cercomela sordida* Moorland Chat
— *C.s.sordida* (Rüppell, 1837)   Ethiopia
— *C.s.ernesti* (Sharpe, 1900)[2]   Kenya, Uganda
— *C.s.olimotiensis* (Elliott, 1945)   Crater Highlands (N Tanzania)
— *C.s.hypospodia* (Shelley, 1885)   Mt. Kilimanjaro (Tanzania)

MYRMECOCICHLA Cabanis, 1850   F
○ *Myrmecocichla tholloni* (Oustalet, 1886) Congo Moorchat   Central African Republic, Congo, W Zaire, C Angola

○ *Myrmecocichla aethiops* Northern Anteater Chat
— *M.a.aethiops* Cabanis, 1850   Senegal and Gambia to Chad
— *M.a.sudanensis* Lynes, 1920   WC and C Sudan
— *M.a.cryptoleuca* Sharpe, 1891   W and C Kenya, N Tanzania

○ *Myrmecocichla formicivora* (Vieillot, 1818) Southern Anteater Chat[3]
    Namibia to SW Zimbawe and Sth. Africa

○ *Myrmecocichla nigra* (Vieillot, 1818) Sooty Chat   Nigeria to W Kenya and E Tanzania, south to Angola and Zambia

○ *Myrmecocichla melaena* (Rüppell, 1837) Rüppell's Black Chat   Eritrea and N Ethiopia

○ *Myrmecocichla albifrons* White-fronted Black Chat
— *M.a.frontalis* (Swainson, 1837)   Senegal and Gambia to N Cameroon
— *M.a.limbata* (Reichenow, 1921)   E Cameroon to Central African Republic
— *M.a.albifrons* (Rüppell, 1837)   Eritrea and N Ethiopia
— *M.a.pachyrhyncha* (Neumann, 1906)   SW Ethiopia
— *M.a.clericalis* (Hartlaub, 1882)   S Sudan, Uganda

○ *Myrmecocichla arnotii* White-headed Black Chat
— *M.a.arnotii* (Tristram, 1869)[4]   Rwanda and Burundi to Zambia, C Mozambique and NE Sth. Africa

— *M.a.harterti* Neunzig, 1926   Angola

THAMNOLAEA Cabanis, 1850[5]   F
○ *Thamnolaea cinnamomeiventris* Mocking Cliffchat
— *T.c.cavernicola* Bates, 1933   Mopti (Mali)
— *T.c.bambarae* Bates, 1928   E Senegal, SW Mali
— *T.c.coronata* Reichenow, 1902   N Ghana and E Burkina Faso to W Sudan
— *T.c.kordofanensis* Wettstein, 1916   C Sudan
— *T.c.albiscapulata* (Rüppell, 1837)   N Eritrea, N and C Ethiopia
— *T.c.subrufipennis* Reichenow, 1887   SE Sudan and SW Ethiopia to Zambia and Malawi
— *T.c.odica* Clancey, 1962   E Zimbabwe
— *T.c.cinnamomeiventris* (Lafresnaye, 1836)   E Sth. Africa, Lesotho, W Swaziland
— *T.c.autochthones* Clancey, 1952   S Mozambique to E Swaziland, Transkei, Natal and E Transvaal

○ *Thamnolaea semirufa* (Rüppell, 1837) White-winged Cliffchat   Ethiopia and Eritrea

PINARORNIS Sharpe, 1876[6]   M
○ *Pinarornis plumosus* Sharpe, 1876 Boulder Chat   E Zambia, W Malawi, Zimbabwe, E Botswana

---

[1] May be better treated as a synonym of *airensis* (C.S. Roselaar *in litt.*).
[2] Implicitly includes *rudolfi* from Mt. Elgon, see Ripley (1964: 103) (2045); not mentioned by White (1962) (2646) or Fry (1992: 543) (890).
[3] Treated as monotypic by White (1962: 115) (2646) including *orestes* and *minor*.
[4] Includes *leucolaema* see White (1962: 116) (2646).
[5] Treated in *Myrmecocichla* in last Edition; but we now follow Ripley (1964) (2045) in retaining the generic name *Thamnolaea*.
[6] This may belong with the "proto-thrushes" not with the chats, see Olson (1998) (1705).

MONTICOLA Boie, 1822[1] M

○ *Monticola rupestris* (Vieillot, 1818) Cape Rock Thrush   South Africa

○ *Monticola explorator* Sentinel Rock Thrush
    — *M.e.explorator* (Vieillot, 1818)   SW and S Cape Province to Transvaal and Natal
    — *M.e.tenebriformis* Clancey, 1952   Lesotho >> N Natal, S Mozambique

○ *Monticola brevipes* Short-toed Rock Thrush
    — *M.b.brevipes* (Waterhouse, 1838)[2]   W Angola, Namibia, N Cape Province
    — *M.b.pretoriae* Gunning & Roberts, 1911[3]   SE Botswana and NE Cape Province to Transvaal and Swaziland

○ *Monticola angolensis* Miombo Rock Thrush
    — *M.a.angolensis* Sousa, 1888[4]   Angola, N Zambia and S Zaire to SW Tanzania and Rwanda
    — *M.a.hylophilus* Clancey, 1965   Zimbabwe, S Zambia, W Malawi, Mozambique (405)

○ *Monticola saxatilis* (Linnaeus, 1766) Common Rock Thrush[5]   NW Africa, S and C Europe and SW and C Asia to Inner Mongolia and SC China

○ *Monticola rufocinereus* Little Rock Thrush
    — *M.r.rufocinereus* (Rüppell, 1837)   Ethiopia and N Somalia to Kenya and NE Tanzania
    — *M.r.sclateri* E. Hartert, 1917   W Saudi Arabia

● *Monticola solitarius* Blue Rock Thrush
    — *M.s.solitarius* (Linnaeus, 1758)   NW Africa, S Europe to Italy and N Balkans, N Turkey, Transcaucasia >> N Africa and Arabia
    — *M.s.longirostris* (Blyth, 1847)   Greece, W and S Turkey and Levant to Tien Shan and NW Himalayas >> NE Africa to N India
    — *M.s.pandoo* (Sykes, 1832)   C Himalayas, C and E China, NW Vietnam, Hainan >> S and SE Asia to Gtr. Sundas
    — *M.s.philippensis* (Statius Müller, 1776)[6]   Russian Far East, S Sakhalin, S Kuril Is., Japan to Ryukyu Is., NE China and Korea >> SE China to Philippines and Indonesia
    — *M.s.madoci* Chasen, 1940   Malay Pen., N Sumatra

○ *Monticola rufiventris* (Jardine & Selby, 1833) Chestnut-bellied Rock Thrush   Himalayas, Assam, Burma, S and C, N Indochina >> Thailand and Indochina

○ *Monticola cinclorhynchus* (Vigors, 1832) Blue-capped Rock Thrush   Afghanistan, W and C Himalayas, S Assam hills >> W Burma, S India

○ *Monticola gularis* (Swinhoe, 1863) White-throated Rock Thrush   Lake Baikal to Russian Far East, N Korea, NE China >> S China to mainland SE Asia

PSEUDOCOSSYPHUS Sharpe, 1883 M

○ *Pseudocossyphus imerinus* (Hartlaub, 1860) Littoral Rock Thrush   Coastal SW and S Madagascar

○ *Pseudocossyphus sharpei* Madagascan Robin-Chat
    — *P.s.erythronotus* (Lavauden, 1929)[7]   Montagne d'Ambre (Madagascar)
    — *P.s.sharpei* (G.R. Gray, 1871)[8]   E Madagascar
    — *P.s.salomonseni* (Farkas, 1973)   EC Madagascar (827)
    — *P.s.bensoni* (Farkas, 1971)[9]   SW Madagascar (Isalo Massif) (825)

MUSCICAPINAE (OLD WORLD FLYCATCHERS)

FRASERIA Bonaparte, 1854 F

○ *Fraseria ocreata* Fraser's Forest Flycatcher
    — *F.o.kelsalli* Bannerman, 1922   Sierra Leone

---

[1] This genus may require splitting, see for example Stepanyan (1990) (2326).
[2] Includes *leucocapilla* see Fry (1997: 4) (891) and, implictly, *niveiceps* Clancey, 1972 (435). Regarding its type material see Farkas (1979) (828).
[3] Suggestions by Farkas (1979) (828) that this is a separate species based on seasonal dimorphism were not accepted by Fry (1997: 5) (891).
[4] Includes *niassae* see Fry (1997: 6) (891).
[5] Includes *coloratus* Stepanyan, 1965 (2317), see Roselaar (1995) (2107).
[6] The population of NE Asia named *magnus* is separable (CSR), but no authority is yet citable.
[7] Treated as a species by Morris & Hawkins (1998: 258) (1627) and Sinclair & Langrand (1998) (2277); field and genetic studies apparently 'on-going'.
[8] Thought to include *tsaratananae* Milon, Petter & Randrianasolo, 1973 (1586).
[9] Treated as part of *P. sharpei* by Sinclair & Langrand (1998) (2277); the molecular studies they refer to are believed to be appearing in Ostrich.

__ *F.o.prosphora* Oberholser, 1899   Liberia to Ghana

__ *F.o.ocreata* (Strickland, 1844)   Bioko I. and Nigeria to Uganda and N Angola

○ *Fraseria cinerascens* WHITE-BROWED FOREST FLYCATCHER

__ *F.c.cinerascens* Hartlaub, 1857   Senegal and Gambia to Ghana

__ *F.c.ruthae* Dickerman, 1994   S Nigeria and Cameroon to Zaire (724)

MELAENORNIS G.R. Gray, 1840 M

○ *Melaenornis brunneus* ANGOLAN SLATY FLYCATCHER

__ *M.b.brunneus* (Cabanis, 1886)   W Angola lowlands

__ *M.b.bailunduensis* (Neumann, 1929)   W Angola highlands

○ *Melaenornis fischeri* WHITE-EYED SLATY FLYCATCHER[1]

__ *M.f.toruensis* (E. Hartert, 1900)   E Zaire, SW Uganda, Rwanda, Burundi

__ *M.f.semicinctus* (E. Hartert, 1916)   NE Zaire

__ *M.f.nyikensis* (Shelley, 1899)[2]   Marungu Highlands (E Zaire), E and S Tanzania, N Malawi

__ *M.f.fischeri* (Reichenow, 1884)   SE Sudan to Kenya and N Tanzania

○ *Melaenornis chocolatinus* ABYSSINIAN SLATY FLYCATCHER

__ *M.c.chocolatinus* (Rüppell, 1840   N and C Ethiopia, S Eritrea

__ *M.c.reichenowi* (Neumann, 1902)   W Ethiopia

○ *Melaenornis annamarulae* Forbes-Watson, 1970 NIMBA FLYCATCHER   Guinea, Liberia, SW Ivory Coast

○ *Melaenornis ardesiacus* Berlioz, 1936 YELLOW-EYED BLACK FLYCATCHER   E Zaire

○ *Melaenornis edolioides* NORTHERN BLACK FLYCATCHER

__ *M.e.edolioides* (Swainson, 1837)   Senegal and Gambia to W Cameroon

__ *M.e.lugubris* (J.W. von Müller, 1851)   E Cameroon to Eritrea, W Ethiopia, W Kenya and N Tanzania

__ *M.e.schistaceus* Sharpe, 1895   N and E Ethiopia, N Kenya

○ *Melaenornis pammelaina* SOUTHERN BLACK FLYCATCHER

__ *M.p.pammelaina* (Stanley, 1814)   Kenya to W and S Zaire, C Angola, E Transvaal and E Cape Province

__ *M.p.diabolicus* (Sharpe, 1877)   S Angola, SW Zambia, N Namibia, N and E Botswana, W Zimbabwe, W Transvaal

○ *Melaenornis pallidus* PALE FLYCATCHER

__ *M.p.pallidus* (J.W. von Müller, 1851)   Senegal and Gambia to C and S Sudan and W Ethiopia

__ *M.p.parvus* (Reichenow, 1907)   SW Ethiopia to NW Uganda

__ *M.p.bowdleri* (Collin & Hartert, 1927)   Eritrea, C Ethiopia

__ *M.p.bafirawari* (Bannerman, 1924)   S Ethiopia, NE Kenya

__ *M.p.duyerali* (Traylor, 1970)   NE Ethiopia and C Somalia

__ *M.p.subalaris* (Sharpe, 1874)   E Kenya, NE Tanzania

__ *M.p.erlangeri* Reichenow, 1905   S Somalia

__ *M.p.modestus* (Shelley, 1873)   Guinea to Central African Republic

__ *M.p.murinus* (Hartlaub & Finsch, 1870)   W and S Kenya, E and S Uganda to Angola, N Namibia, N Botswana and W and S Zambia

__ *M.p.aquaemontis* (Stresemann, 1937)   C Namibia

__ *M.p.griseus* (Reichenow, 1882)   SE Kenya and C Tanzania to N and E Zambia and N and C Malawi

__ *M.p.divisus* (Lawson, 1961)   SE Zambia, S Malawi, SE Zimbabwe, Mozambique, Swaziland

__ *M.p.sibilans* (Clancey, 1966)   S Mozambique, N Natal

○ *Melaenornis infuscatus* CHAT-FLYCATCHER

__ *M.i.benguellensis* (Sousa, 1886)   S Angola, NW Namibia

__ *M.i.namaquensis* (Macdonald, 1957)   Namibia

__ *M.i.placidus* (Clancey, 1958)   Botswana, W Transvaal, N Cape Province, NW Orange Free State

---

[1] Type species of the genus *Dioptrornis*. Not here recognised.
[2] Includes *upifae* see Fry (1997: 440) (891).

— *M.i.seimundi* (Ogilvie-Grant, 1913)   SC and E Cape Province to SW Orange Free State

— *M.i.infuscatus* (A. Smith, 1839)   SW Namibia to SW Cape Province

○ *Melaenornis microrhynchus* African Grey Flycatcher

— *M.m.microrhynchus* (Reichenow, 1887)   SW Kenya, N, C and W Tanzania

— *M.m.pumilus* (Sharpe, 1895)   C Ethiopia, N Somalia

— *M.m.neumanni* (Hilgert, 1908)   S Sudan, S Ethiopia to N Kenya, C Somalia

— *M.m.burae* (Traylor, 1970)   SE Somalia, E Kenya

— *M.m.taruensis* (van Someren, 1921)   SE Kenya

○ *Melaenornis mariquensis* Mariqua Flycatcher[1]

— *M.m.acaciae* (Irwin, 1957)   S Angola, Namibia, SW Botswana, NW Cape Province

— *M.m.mariquensis* (A. Smith, 1847)   SW Zambia and C Zimbabwe to NE Cape Province and W Transvaal

— *M.m.territinctus* Clancey, 1979   NE Namibia, NW Botswana

○ *Melaenornis silens* Fiscal-Flycatcher[2]

— *M.s.lawsoni* Clancey, 1966   N Cape Province to W Transvaal and SE Botswana

— *M.s.silens* (Shaw, 1809)   SW, S and E Cape to Orange Free State, E Transvaal

EMPIDORNIS Reichenow, 1901[3] M

○ *Empidornis semipartitus* (Rüppell, 1840) Silverbird[4]   W Sudan to N Tanzania

RHINOMYIAS Sharpe, 1879 M[5]

○ *Rhinomyias additus* (E. Hartert, 1900) Streaky-breasted Jungle Flycatcher

     Buru (Moluccas)

○ *Rhinomyias oscillans* Russet-backed Jungle Flycatcher

— *R.o.oscillans* (E. Hartert, 1897)   Flores (Lesser Sundas)

— *R.o.stresemanni* (Siebers, 1928)   Sumba (Lesser Sundas)

○ *Rhinomyias brunneatus* Brown-chested Jungle Flycatcher

— *R.b.brunneatus* (Slater, 1897)   SE China >> Thailand, W Malaysia

— *R.b.nicobaricus* Richmond, 1902   Breeding range ? >> Nicobars

○ *Rhinomyias olivaceus* Fulvous-chested Jungle Flycatcher

— *R.o.olivaceus* (Hume, 1877)   Malay Pen., Sumatra, Java, Bali, Belitung, N Borneo

— *R.o.perolivaceus* Chasen & Kloss, 1929   Islands off N Borneo

○ *Rhinomyias umbratilis* (Strickland, 1849) Grey-chested Jungle Flycatcher

     Malay Pen., Sumatra, Belitung, Borneo

○ *Rhinomyias ruficauda*[6] Rufous-tailed Jungle Flycatcher

— *R.r.ruficrissa* Sharpe, 1887   Mt. Kinabalu

— *R.r.isola* Hachisuka, 1932   Mts. of Borneo (excl. Kinabalu)

— *R.r.boholensis* Rand & Rabor, 1957   Bohol

— *R.r.samarensis* (Steere, 1890)   Leyte, Samar, E Mindanao

— *R.r.zamboanga* Rand & Rabor, 1957   W Mindanao

— *R.r.ruficauda* (Sharpe, 1877)   Basilan

— *R.r.occularis*[7] Bourns & Worcester, 1894   Sulu Arch.

○ *Rhinomyias colonus* Henna-tailed Jungle Flycatcher

— *R.c.colonus* E. Hartert, 1898   Sula Is.

— *R.c.pelingensis* Vaurie, 1952   Peleng I.

— *R.c.subsolanus* Meise, 1932   E Sulawesi ?

○ *Rhinomyias gularis* Sharpe, 1888 Eyebrowed Jungle Flycatcher   Mts. of Borneo

○ *Rhinomyias albigularis* Bourns & Worcester, 1894 White-throated Jungle Flycatcher[8]

     Guimaras, Negros, Panay

---

[1] Type species of the genus *Bradornis*. Not here recognised.
[2] Type species of the genus *Sigelus*. Not here recognised.
[3] Treated as a species within the genus *Melaenornis* as in Watson *et al.* (1986: 297) (2565), but this has not been generally accepted.
[4] Includes *kavirondensis* see White (1963: 17) (2648).
[5] Gender addressed by David & Gosselin (2002) (614), multiple changes occur in specific names.
[6] The suffix of this name is "invariable" i.e. the masculinity of the generic name does not affect it. Similarly the two subspecific epithets one might expect to change are each in their own way invariable too.
[7] The original spelling; perhaps justifiably emended to *ocularis*.
[8] Treated as a separate species from *R. gularis* by Dickinson *et al.* (1991: 343) (745).

○ *Rhinomyias insignis* Ogilvie-Grant, 1895 WHITE-BROWED JUNGLE FLYCATCHER
                                                                N Luzon

○ *Rhinomyias goodfellowi* Ogilvie-Grant, 1905 SLATY-BACKED JUNGLE FLYCATCHER
                                                                Mindanao

MUSCICAPA Brisson, 1760 F
● *Muscicapa striata* SPOTTED FLYCATCHER
    ✓ *M.s.striata* (Pallas, 1764)   Europe, W Siberia, NW Africa >> W, E and S Africa
    — *M.s.balearica* von Jordans, 1913   Balearic Is. >> W and SW Africa
    ✓ *M.s.tyrrhenica* Schiebel, 1910   Corsica, Sardinia >> Africa ?
    — *M.s.inexpectata* Dementiev, 1932   Crimea
    — *M.s.neumanni* Poche, 1904   Aegean Is., Cyprus and Levant to Caucasus area and N
                                                                and SW Iran; also C Siberia east to Lake Baikal >>
                                                                E and S Africa

    — *M.s.sarudnyi* Snigirewski, 1928   E Iran and Turkmenistan to Tien Shan, Pamir and N
                                                                and W Pakistan >> E and S Africa ?

    — *M.s.mongola* Portenko, 1955   SE Transbaikalia, Mongolia

○ *Muscicapa gambagae* (Alexander, 1901) GAMBAGA FLYCATCHER   SW Arabia; N tropical Africa, S Sudan to Kenya

○ *Muscicapa griseisticta* (Swinhoe, 1861) GREY-STREAKED FLYCATCHER   NE China, N Korea and Russian Far East to SE Yakutia
                                                                and Kamchatka >> Philippines, Sulawesi, Moluccas
                                                                and New Guinea

○ *Muscicapa sibirica* DARK-SIDED FLYCATCHER
    — *M.s.sibirica* J.F. Gmelin 1789   C and SE Siberia to Kamchatka, Japan, N Korea, NE
                                                                China and N Mongolia Japan >> S China, mainland
                                                                and archipelagic SE Asia

    — *M.s.gulmergi* (E.C.S. Baker, 1923)   E Afghanistan, NW Himalayas
    — *M.s.cacabata* Penard, 1919   C and E Himalayas, SE Xizang, Assam >> S Burma,
                                                                S Thailand
    — *M.s.rothschildi* (E.C.S. Baker, 1923)   N Burma and C China to SW Gansu and NW Vietnam
                                                                >> S China, Indochina, Malay Pen.

○ *Muscicapa dauurica* ASIAN BROWN FLYCATCHER
    — *M.d.dauurica* Pallas, 1811   SC Siberia and N Mongolia to NE China, N Korea,
                                                                Japan S. Kuril Is., Sakhalin; Himalayan foothills to
                                                                W Sichuan >> India, S China to the Gtr. Sundas

    — *M.d.williamsoni* Deignan, 1957   SE Burma, SW Thailand >> Malay Pen. Sumatra and
                                                                Borneo
    — *M.d.siamensis* (Gyldenstolpe, 1916)   Yunnan, E Burma, NW Thailand, SC Vietnam,
                                                                ?N Vietnam
    — *M.d.umbrosa* D.R. Wells, 1982   NE Borneo

○ *Muscicapa randi* Amadon & duPont, 1970 ASHY-BREASTED FLYCATCHER[1]
                                                                Philippines

○ *Muscicapa segregata* (Siebers, 1928) SUMBA BROWN FLYCATCHER   Sumba I.

○ *Muscicapa muttui* (E.L. Layard, 1854) BROWN-BREASTED FLYCATCHER   NE India, Burma, C China (Yunnan to E Gansu and
                                                                Guangxi), NW Thailand, NW Vietnam >> SW India,
                                                                Sri Lanka

○ *Muscicapa ruficauda* Swainson, 1838 RUSTY-TAILED FLYCATCHER   W Tien Shan to E Afghanistan, W and C Himalayas >>
                                                                SW India

○ *Muscicapa ferruginea* (Hodgson, 1845) FERRUGINOUS FLYCATCHER   C and E Himalayas to C China and Taiwan >>
                                                                mainland and archipelagic SE Asia

○ *Muscicapa caerulescens* ASHY FLYCATCHER
    — *M.c.nigrorum* (Collin & Hartert, 1927)   Guinea to Togo
    — *M.c.brevicauda* Ogilvie-Grant, 1907   S Nigeria to S Sudan, W Kenya, NW Angola and S Zaire
    — *M.c.cinereola* Hartlaub & Finsch, 1870   S Somalia, E Kenya, E Tanzania
    — *M.c.impavida* Clancey, 1957   SW and C Angola to W Tanzania, N Mozambique,
                                                                W Transvaal

[1] For provisional recognition of this and the next as separate species from *M. dauurica* see Dickinson *et al.* (1991: 346) (745). Still provisional!

— *M.c.vulturna* Clancey, 1957   C Mozambique and S Malawi to E and S Zimbabwe, E Transvaal and N Swaziland

— *M.c.caerulescens* (Hartlaub, 1865)   S Mozambique to E Cape Province

○ *Muscicapa aquatica* SWAMP FLYCATCHER
— *M.a.aquatica* Heuglin, 1864   Gambia to SW Sudan
— *M.a.infulata* Hartlaub, 1881   S Sudan to E Zaire and NE Zambia
— *M.a.lualabae* (Chapin, 1932)   Shaba (Zaire)
— *M.a.grimwoodi* Chapin, 1952   S Zambia

○ *Muscicapa cassini* Heine, 1859 CASSIN'S FLYCATCHER   Sierra Leone to W Uganda, N Angola, N Zambia

○ *Muscicapa olivascens* OLIVACEOUS FLYCATCHER
— *M.o.olivascens* (Cassin, 1859)   Ghana to E Zaire
— *M.o.nimbae* Colston & Curry-Lindahl, 1986   Liberia and Ivory Coast (553)

○ *Muscicapa lendu* CHAPIN'S FLYCATCHER
— *M.l.lendu* (Chapin, 1932)   NE Zaire, SW Uganda, W Kenya
— *M.l.itombwensis* Prigogine, 1957   Itombwe Mts. (E Zaire)

○ *Muscicapa adusta* AFRICAN DUSKY FLYCATCHER
— *M.a.poensis* (Alexander, 1903)[1]   Bioko I., Cameroon
— *M.a.pumila* (Reichenow, 1892)[2]   E Zaire to C Kenya, S Sudan
— *M.a.minima* Heuglin, 1862   Eritrea, Ethiopia
— *M.a.subadusta* (Shelley, 1897)[3]   Angola to NW Mozambique, E Zimbabwe
— *M.a.marsabit* (van Someren, 1931)   N Kenya
— *M.a.murina* (Fischer & Reichenow, 1884)[4]   NE Tanzania, SE Kenya
— *M.a.fuelleborni* Reichenow, 1900   S and W Tanzania
— *M.a.mesica* Clancey, 1974   Zimbabwe
— *M.a.fuscula* Sundevall, 1850   Natal, Swaziland to N Transvaal
— *M.a.adusta* (Boie, 1828)   S and E Cape Province

○ *Muscicapa epulata* (Cassin, 1855) LITTLE GREY FLYCATCHER   SE Guinea and Liberia to Gabon, Congo and NE Zaire

○ *Muscicapa sethsmithi* (van Someren, 1922) YELLOW-FOOTED FLYCATCHER
SE Nigeria to SW Congo, E Zaire and W Uganda

○ *Muscicapa comitata* DUSKY-BLUE FLYCATCHER
— *M.c.aximensis* (W.L. Sclater, 1924)   Sierra Leone to S Nigeria
— *M.c.camerunensis* (Reichenow, 1892)   Mt. Cameroon
— *M.c.comitata* (Cassin, 1857)   Cameroon to NW Angola, Zaire, SW Sudan and Uganda

○ *Muscicapa tessmanni* (Reichenow, 1907) TESSMANN'S FLYCATCHER   Ivory Coast to S Congo and NE Zaire

○ *Muscicapa infuscata* SOOTY FLYCATCHER
— *M.i.infuscata* (Cassin, 1855)   Nigeria to C Zaire and Angola
— *M.i.minuscula* (Grote, 1922)[5]   E Zaire to Uganda

○ *Muscicapa ussheri* (Sharpe, 1871) USSHER'S FLYCATCHER   Guinea to Nigeria

○ *Muscicapa boehmi* (Reichenow, 1884) BÖHM'S FLYCATCHER   Angola to W Tanzania and W Malawi

MYIOPARUS Roberts, 1922[6] M
○ *Myioparus griseigularis* GREY-THROATED TIT-FLYCATCHER
— *M.g.parelii* (Traylor, 1970)   Liberia to S Ghana
— *M.g.griseigularis* (Jackson, 1906)   S Nigeria to Uganda and NW Angola

○ *Myioparus plumbeus* GREY TIT-FLYCATCHER
— *M.p.plumbeus* (Hartlaub, 1858)   Senegal and Gambia to Ethiopia, NW Tanzania and W Kenya

[1] The name *obscura* used by Fry (1997: 477) (891) is preoccupied. In the context treated *poensis* is valid. The name *albiventris*, *kumboensis* and *okuensis* are also synonyms, see Fry (1997: 477) (891).
[2] Includes *grotei* see Fry (1997: 477) (891).
[3] Includes *angolensis* see Fry (1997: 477) (891).
[4] Includes *murina* see Fry (1997: 477) (891).
[5] For recognition see Érard (1997: 488) (808).
[6] For comments on the use of this name see Érard (1997: 492) (808).

— *M.p.orientalis* (Reichenow & Neumann, 1895) E Kenya to E Transvaal and N Natal
— *M.p.catoleucum* (Reichenow, 1900) Angola to SE Tanzania, N Botswana, Transvaal and Natal

STENOSTIRA Cabanis & Bonaparte, 1850[1] F
◯ ***Stenostira scita*** Fairy Flycatcher
 — *S.s.scita* (Vieillot, 1818) W and NW Cape Province >> S Namibia
 — *S.s.saturatior* Lawson, 1962 C, E and S Cape Province >> to W Orange Free State
 — *S.s.rudebecki* Clancey, 1955 Lesotho >> Transvaal

HUMBLOTIA Milne-Edwards & Oustalet, 1885 F
◯ ***Humblotia flavirostris*** Milne-Edwards & Oustalet, 1885 Humblot's Flycatcher
  Comoro Is.

FICEDULA Brisson, 1760 F
◉ ***Ficedula hypoleuca*** Pied Flycatcher
 ✓ *F.h.hypoleuca* (Pallas, 1764) W, N and C Europe to Ural Mts. >> W and WC Africa
 — *F.h.iberiae* (Witherby, 1928)[2] Iberia >> W Africa
 — *F.h.sibirica* Khakhlov, 1915[3] W and SC Siberia >> W? and WC Africa
 — *F.h.speculigera* (Bonaparte, 1850)[4] NW Africa >> W Africa

◯ ***Ficedula albicollis*** (Temminck, 1815) Collared Flycatcher SE Sweden; E France and Italy to N Macedonia, Ukraine and C European Russia >> SC Africa

◯ ***Ficedula semitorquata*** (Homeyer, 1885) Semi-collared Flycatcher[5] Albania and Greece to Bulgaria, Turkey, Caucasus area and Iran >> EC Africa

◯ ***Ficedula zanthopygia*** (Hay, 1845) Yellow-rumped Flycatcher E and NE China to E Transbaikalia, Russian Far East and Korea >> SE Asia to Gtr. Sundas

◯ ***Ficedula narcissina*** Narcissus Flycatcher
 — *F.n.narcissina* (Temminck, 1836[6])[7] Sakhalin, S Kuril Is., Japan (except S Ryukyu Is.) >> Hainan, S Indochina, Philippines, Borneo
 — *F.n.owstoni* (Bangs, 1901) S Ryukyu Is.
 — *F.n.elisae* (Weigold, 1922)[8] NE Hebei and Shanxi (E China) >> Thailand, Malay Pen.

◯ ***Ficedula mugimaki*** (Temminck, 1836[9]) Mugimaki Flycatcher SC and SE Siberia, NE China, N Korea, Russian Far East, Sakhalin >> S China, mainland SE Asia, Gtr. Sundas, Philippines

◯ ***Ficedula hodgsonii*** (J. Verreaux, 1871) Slaty-backed Flycatcher C and E Himalayas, Assam, N Burma, E Tibetan Plateau from Yunnan to S Gansu >> E India to north mainland SE Asia

◯ ***Ficedula dumetoria*** Rufous-chested Flycatcher
 — *F.d.muelleri* (Sharpe, 1879) Malay Pen., Sumatra, Borneo
 — *F.d.dumetoria* (Wallace, 1864) Java, Lombok, Sumbawa, Flores
 — *F.d.riedeli* (Büttikofer, 1886) Tanimbar Is.

◯ ***Ficedula strophiata*** Rufous-gorgetted Flycatcher
 — *F.s.strophiata* (Hodgson, 1837) Himalayas to W and N Burma, C China (W and N Yunnan to SW Gansu, Shaanxi, Hubei, Guizhou) >> NE India to north mainland SE Asia
 — *F.s.fuscogularis* (E.C.S. Baker, 1923) S Laos, S Vietnam

◯ ***Ficedula parva*** Red-breasted Flycatcher
 — *F.p.parva* (Bechstein, 1792) C, SE, E Europe, SW Siberia, N Asia Minor and Caucasus area to N Iran >> Pakistan, N India
 — *F.p.albicilla* (Pallas, 1811)[10] E European Russia, Siberia to Kamchatka, N Mongolia, Amurland >> NE India and SE Asia to Gtr. Sundas

---

[1] Disputed affinities discussed by Fry (1997: 498) (891).
[2] For recognition see Cramp *et al.* (1993: 64) (594).
[3] This name is preoccupied in *Muscicapa*. If treated in that genus the substitute name *tomensis* Johansen, 1916, is to be used.
[4] Saetre *et al.* (2001) (2146) have suggested that this deserves specific rank.
[5] For reasons to separate this from the two preceding species see Cramp *et al.* (1993: 48) (594).
[6] Mees (1994) (1541) has shown that Livr. 97 appeared in or after Apr, 1836. Watson (1986: 338) (2565) gave 1835.
[7] Although not recognised by Orn. Soc. Japan (2000: 230) (1727) *jakuschima* and *shonis* may deserve recognition.
[8] We have felt it best to include *beijingnica* Zheng, Song, Zhang, Zhang & Guo, 2000 (2706) here, as a species *inquirenda*. It was proposed as a species but critical acoustic information is missing, and the description might apply to a subadult form of *elisae*. Listing it as a subspecies would be misleading.
[9] Mees (1994) (1727) has shown that Livr. 97 appeared in or after Apr, 1836. Watson (1986: 339) (2565) gave 1835.
[10] For the present we continue to treat this as conspecific.

○ *Ficedula subrubra* (Hartert & Steinbacher, 1934) KASHMIR FLYCATCHER

Kashmir >> Sri Lanka

○ *Ficedula monileger* WHITE-GORGETTED FLYCATCHER
     — *F.m.monileger* (Hodgson, 1845)   C and E Himalayas
     — *F.m.leucops* (Sharpe, 1888)   S and E Assam, E Burma, N Thailand, N and C Indochina
     — *F.m.gularis* (Blyth, 1847)   Arakan (SW Burma)

○ *Ficedula solitaris* RUFOUS-BROWED FLYCATCHER
     — *F.s.submoniliger* (Hume, 1877)   SE Burma E to SE Laos and S Vietnam S to N Malay Pen.
     — *F.s.malayana* (Sharpe, 1888)   C and S Malay Pen.
     — *F.s.solitaris* (S. Müller, 1835)   Sumatra

○ *Ficedula hyperythra* SNOWY-BROWED FLYCATCHER
     — *F.h.hyperythra* (Blyth, 1843)   C and E Himalayas, SE Xixang, Assam, W, N and E
                                                Burma, SC China, N Thailand, N and C Indochina

     — *F.h.annamensis* (Robinson & Kloss, 1919)   Langbian (SC Vietnam)
     — *F.h.innexa* (Swinhoe, 1866)[1]   Taiwan
     — *F.h.sumatrana* (Hachisuka, 1926)   Malay Pen., Sumatra, Borneo
     — *F.h.mjobergi* (E. Hartert, 1925)   Poi Mts. (Sarawak, Borneo)
     — *F.h.vulcani* (Robinson, 1918)   Java and west Lesser Sundas
     — *F.h.clarae* (Mayr, 1944)   Timor
     — *F.h.audacis* (E. Hartert, 1906)   Babar I.
     — *F.h.rara* (Salomonsen, 1977)   Palawan
     — *F.h.calayensis* (McGregor, 1921)   Calayan (off N Luzon)
     — *F.h.luzoniensis* (Ogilvie-Grant, 1894)   Luzon
     — *F.h.mindorensis* (Hachisuka, 1935)   Mindoro
     — *F.h.nigrorum* (J. Whitehead, 1897)   Negros
     — *F.h.montigena* (Mearns, 1905)   Mt. Apo, Mt. Katanglad (Mindanao)
     — *F.h.matutumensis* Kennedy, 1987   Mt. Busa, Mt. Matutum (Mindanao) [1256]
     — *F.h.daggayana* Meyer de Schauensee & duPont, 1962   NC Mindanao
     — *F.h.malindangensis* Rand & Rabor, 1957   Mt. Malindang (NW Mindanao)
     — *F.h.jugosae* (Riley, 1921)   C, SE and S Sulawesi
     — *F.h.annalisa* (Stresemann, 1931)   N Sulawesi
     — *F.h.negroides* (Stresemann, 1914)   Seram (Moluccas)
     — *F.h.pallidipectus* (E. Hartert, 1903)   Bacan (Moluccas)
     — *F.h.alifura* (Stresemann, 1912)   Buru (Moluccas)

○ *Ficedula basilanica* LITTLE SLATY FLYCATCHER
     — *F.b.samarensis* (Bourns & Worcester, 1894)   Leyte, Samar
     — *F.b.basilanica* (Sharpe, 1877)   Basilan, Mindanao

○ *Ficedula rufigula* (Wallace, 1865) RUFOUS-THROATED FLYCATCHER   Sulawesi

○ *Ficedula buruensis* CINNAMON-CHESTED FLYCATCHER
     — *F.b.buruensis* (E. Hartert, 1899)   Buru (Moluccas)
     — *F.b.ceramensis* (Ogilvie-Grant, 1910)   Seram (Moluccas)
     — *F.b.siebersi* (E. Hartert, 1924)   Kai Is.

○ *Ficedula henrici* (E. Hartert, 1899) DAMAR FLYCATCHER   Damar I. (Lesser Sundas)

○ *Ficedula harterti* (Siebers, 1928) SUMBA FLYCATCHER   Sumba I. (Lesser Sundas)

○ *Ficedula platenae* (W.H. Blasius, 1888) PALAWAN FLYCATCHER   Palawan

○ *Ficedula crypta* (Vaurie, 1951) CRYPTIC FLYCATCHER   Mindanao

○ *Ficedula disposita* (Ripley & Marshall, 1967) FURTIVE FLYCATCHER[2]   Luzon

○ *Ficedula bonthaina* (E. Hartert, 1896) LOMPOBATTANG FLYCATCHER   S Sulawesi

○ *Ficedula westermanni* LITTLE PIED FLYCATCHER
     — *F.w.collini* (Rothschild, 1925)   W and C Himalayas >> N India
     — *F.w.australorientis* (Ripley, 1952)   E Himalayas, Assam, SC China, Burma, N Thailand, N and
                                                C Laos, NW Vietnam >> N India, mainland SE Asia

---

[1] Omitted from last Edition.
[2] Dutson (1993) (779) reported the male plumage of this, thus proving it distinct from *F. crypta*.

— *F.w.langbianis* (Kloss, 1927)   Bolovens (S Laos), Langbian (SC Vietnam)
— *F.w.westermanni* (Sharpe, 1888)   Malay Pen., N Sumatra, Borneo, Mindanao, Sulawesi (except S), Seram, Bacan

— *F.w.rabori* (Ripley, 1952)[1]   Luzon, Negros, Panay
— *F.w.palawanensis* (Ripley & Rabor, 1962)   Palawan
— *F.w.hasselti* (Finsch, 1898)   S Sumatra, Java, western Lesser Sundas, S Sulawesi
— *F.w.mayri* (Ripley, 1952)   Timor, Wetar (Lesser Sundas)

○ *Ficedula superciliaris* ULTRAMARINE FLYCATCHER
— *F.s.superciliaris* (Jerdon, 1840)   E Afghanistan, NW Himalayas >> C India
— *F.s.aestigma* (J.E. & G.R. Gray, 1846)   E Himalayas, Assam, SC China (SE Xizang to SW Sichuan and N Yunnan) >> E India, nthn. mainland SE Asia

○ *Ficedula tricolor* SLATY-BLUE FLYCATCHER
— *F.t.tricolor* (Hodgson, 1845)   W and C Himalayas
— *F.t.minuta* (Hume, 1872)   E Himalayas, W Assam, SE Xizang
— *F.t.cerviniventris* (Sharpe, 1879)   SE Assam, W Burma
— *F.t.diversa* Vaurie, 1953   C China (N Yunnan to Gansu and Guizhou) >> S China, N Thailand, N Indochina

○ *Ficedula sapphira* SAPPHIRE FLYCATCHER
— *F.s.sapphira* (Blyth, 1843)   E Himalayas, N Burma, N Yunnan, SW Sichuan
— *F.s.laotiana* (Delacour & Greenway, 1939)   NW Thailand, N and C Laos, NW Vietnam
— *F.s.tienchuanensis* Cheng, Tso-hsin, 1964   NC China (WC Sichuan to S Shaanxi)

○ *Ficedula nigrorufa* (Jerdon, 1839) BLACK-AND-ORANGE FLYCATCHER   SW India

○ *Ficedula timorensis* (Hellmayr, 1919) BLACK-BANDED FLYCATCHER   Timor I. (Lesser Sundas)

CYANOPTILA Blyth, 1847 F
○ *Cyanoptila cyanomelaena* BLUE-AND-WHITE FLYCATCHER
— *C.c.cumatilis* Thayer & Bangs, 1909   NE China and Russian Far East to N Korea >> SE China, mainland SE Asia and Gtr. Sundas Is.
— *C.c.cyanomelaena* (Temminck, 1829)   S Kuril Is., Japan, S Korea >> SE China, Indochina, Philippines, Borneo

EUMYIAS Cabanis, 1850[2] M[3]
○ *Eumyias sordidus* (Walden, 1870) DULL VERDITER FLYCATCHER   Sri Lanka

○ *Eumyias thalassinus* ASIAN VERDITER FLYCATCHER
— *E.t.thalassinus* (Swainson, 1838)   Himalayas to C and S China, Thailand (except C and S Pen.) and Indochina >> SW India and SE Asia
— *E.t.thalassoides* (Cabanis, 1850)   C and S Malay Pen., Sumatra

○ *Eumyias panayensis* ISLAND VERDITER FLYCATCHER
— *E.p.nigrimentalis* (Ogilvie-Grant, 1894)   Luzon, Mindoro
— *E.p.panayensis* Sharpe, 1877   Negros, Panay
— *E.p.nigriloris* (E. Hartert, 1904)   Mts. of Mindanao
— *E.p.septentrionalis* (Büttikofer, 1893)   N, C Sulawesi
— *E.p.meridionalis* (Büttikofer, 1893)   S Sulawesi
— *E.p.obiensis* (E. Hartert, 1912)   Obi (Moluccas)
— *E.p.harterti* (van Oort, 1911)   Seram (Moluccas)

○ *Eumyias albicaudatus* (Jerdon, 1840) NILGIRI VERDITER FLYCATCHER   SW India

○ *Eumyias indigo* INDIGO FLYCATCHER
— *E.i.ruficrissa* (Salvadori, 1879)   Sumatra
— *E.i.indigo* (Horsfield, 1821)   Java
— *E.i.cerviniventris* (Sharpe, 1887)   Borneo

CYORNIS Blyth, 1843[4] M[5]
○ *Cyornis hainanus* (Ogilvie-Grant, 1900) HAINAN BLUE FLYCATCHER   E and SE Burma to S China, Hainan, N and C Indochina, Thailand (except C and S peninsular)

---

[1] For recognition of this and *palawanensis* see Dickinson *et al.* (1991: 351) (745).
[2] Treated within a broader genus *Muscicapa* in last Edition.
[3] For correct gender and spellings see David & Gosselin (2002) (614).
[4] Submerged in *Niltava* in last Edition. We have placed this genus before *Niltava* (and have resequenced both) in an attempt to better exhibit relationships. Molecular studies will eventually suggest corrections. ECD.
[5] For correct gender and spellings see David & Gosselin (2002) (614).

○ *Cyornis unicolor* PALE BLUE FLYCATCHER
    — *C.u.unicolor* Blyth, 1843   C and E Himalayas, Assam, Burma, Thailand, N and
                        C Laos, S China

    — *C.u.diaoluoensis* (Cheng, Yang & Lu, 1981)[1]   Hainan
    — *C.u.harterti* Robinson & Kinnear, 1928   Malay Pen., Sumatra, Java, Borneo

○ *Cyornis ruecki* (Oustalet, 1881) RÜCK'S BLUE FLYCATCHER   NE Sumatra[2]

○ *Cyornis herioti* BLUE-BREASTED BLUE FLYCATCHER
    — *C.h.herioti* Wardlaw Ramsay, 1886   N and C Luzon
    — *C.h.camarinensis* (Rand & Rabor, 1967)   S Luzon

○ *Cyornis pallipes* (Jerdon, 1840) WHITE-BELLIED BLUE FLYCATCHER   W Ghats (SW India)

○ *Cyornis poliogenys* PALE-CHINNED BLUE FLYCATCHER
    — *C.p.poliogenys* W.E. Brooks, 1879   C Himalayas, W and S Assam, E Bangladesh, SW Burma
    — *C.p.cachariensis* (Madarász, 1884)   E Himalayas, E and SE Assam, NE, NW Burma, N Yunnan
    — *C.p.laurentei* (La Touche, 1921)   S Yunnan
    — *C.p.vernayi* Whistler, 1931   E Ghats (EC India)

○ *Cyornis banyumas* HILL BLUE FLYCATCHER
    — *C.b.magnirostris* Blyth, 1849   C and E Himalayas, Assam >> S Burma, and Malay Pen.
    — *C.b.whitei* Harington, 1908   N and E Burma, SC China, N Thailand, N and
                        C Indochina
    — *C.b.lekhakuni* (Deignan, 1956)   Sthn. hills of E Thailand
    — *C.b.deignani* Meyer de Schauensee, 1939   SE Thailand
    — *C.b.coerulifrons* E.C.S. Baker, 1918[3]   C and S Malay Pen.
    — *C.b.ligus* (Deignan, 1947)   W Java
    — *C.b.banyumas* (Horsfield, 1821)   C and E Java
    — *C.b.coeruleatus* (Büttikofer, 1900)   Borneo
    — *C.b.lemprieri* (Sharpe, 1884)[4]   Palawan, Balabac

○ *Cyornis tickelliae* TICKELL'S BLUE FLYCATCHER
    — *C.t.tickelliae* Blyth, 1843   NE, C and S India, Bangladesh, N and W Burma
    — *C.t.jerdoni* Holdsworth, 1872   Sri Lanka
    — *C.t.sumatrensis* (Sharpe, 1879)   Malay Pen., Sumatra
    — *C.t.indochina* Chasen & Kloss, 1928   S Burma, Thailand, C and S Indochina
    — *C.t.lamprus* Oberholser, 1917   Anamba Is.

○ *Cyornis caerulatus* LARGE-BILLED BLUE FLYCATCHER
    — *C.c.albiventer* Junge, 1933   Sumatra
    — *C.c.rufifrons* Wallace, 1865   W Borneo
    — *C.c.caerulatus* (Bonaparte, 1857)   N, E and S Borneo

○ *Cyornis superbus* Stresemann, 1925 BORNEAN BLUE FLYCATCHER   Borneo

○ *Cyornis rubeculoides* BLUE-THROATED BLUE FLYCATCHER
    — *C.r.rubeculoides* (Vigors, 1831)   Himalayas, Assam, W, N and NE Burma >> S India,
                        Sri Lanka, S Burma
    — *C.r.dialilaemus* Salvadori, 1889   E and SE Burma, N and W Thailand
    — *C.r.rogersi* Robinson & Kinnear, 1928   C and SW Burma
    — *C.r.glaucicomans* Thayer & Bangs, 1909   N and E Yunnan to S Shaanxi, Hubei and Guangxi >>
                        Malay Pen.
    — *C.r.klossi* Robinson, 1921   E Thailand, C and S Indochina

○ *Cyornis turcosus* MALAYSIAN BLUE FLYCATCHER
    — *C.t.rupatensis* Oberholser, 1920   Malay Pen., Sumatra, W Borneo
    — *C.t.turcosus* Brüggeman, 1877   E Borneo

○ *Cyornis rufigastra* MANGROVE BLUE FLYCATCHER[5]
    — *C.r.rufigastra* (Raffles, 1822)   Malay Pen., Sumatra, Borneo

---

[1] At this period Cheng Tso-hsin used the transliterated form Zheng Zuoxin for his name. ECD.
[2] Types from Malacca are thought to be "trade skins" and not from Malaysia.
[3] Dated 1919 by Watson *et al.* (1986: 366) (2565), presumably a *lapsus*, "No. CCXXXVI" bears the date "October 29th. 1918".
[4] Treated as a separate species by Dickinson *et al.* (1991: 353) (745), but with reservations.
[5] For correct spelling see David & Gosselin (2002) (613).

— *C.r.longipennis* Chasen & Kloss, 1930[1]	Karimunjawa I.
— *C.r.rhizophorae* Stresemann, 1925	W Java
— *C.r.karimatensis* Oberholser, 1924	Karimata Is. (off W Borneo)
— *C.r.blythi* (Giebel, 1875)	Luzon, Polillo, Catanduanes
— *C.r.marinduquensis* duPont, 1972	Marinduque
— *C.r.philippinensis* Sharpe, 1877[2]	C and S Philippines, Palawan, Sulu Arch.
— *C.r.mindorensis* Mearns, 1907	Mindoro
— *C.r.omissus* (E. Hartert, 1896)[3]	Sulawesi
— *C.r.peromissus* E. Hartert, 1920	Salayer I.
— *C.r.djampeanus* (E. Hartert, 1896)	Djampea I.
— *C.r.kalaoensis* (E. Hartert, 1896)	Kalao I.

○ *Cyornis hyacinthinus* TIMOR BLUE FLYCATCHER

— *C.h.hyacinthinus* (Temminck, 1820)	Timor, Semau Is. (Lesser Sundas)
— *C.h.kuhni* E. Hartert, 1904	Wetar I. (Lesser Sundas)

○ *Cyornis hoevelli* (A.B. Meyer, 1903) BLUE-FRONTED BLUE FLYCATCHER | C and SE Sulawesi

○ *Cyornis sanfordi* Stresemann, 1931 MATINAN BLUE FLYCATCHER | N Sulawesi

○ *Cyornis concretus* WHITE-TAILED BLUE FLYCATCHER[4]

— *C.c.cyaneus* (Hume, 1877)	NE India, NW and SE Burma, W Thailand, W and S Yunnan, N and C Indochina
— *C.c.concretus* (S. Müller, 1835)	Malay Pen., Sumatra
— *C.c.everetti* (Sharpe, 1890)	Borneo

NILTAVA Hodgson, 1837 F

○ *Niltava davidi* La Touche, 1907 FUJIAN NILTAVA | C and E China, ? NW Vietnam >> Indochina, SE Thailand

○ *Niltava sundara* RUFOUS-BELLIED NILTAVA

— *N.s.whistleri* Ticehurst, 1926	NW Himalayas
— *N.s.sundara* Hodgson, 1837	C and E Himalayas, Assam, Burma (except E and SE), W Yunnan >> N India to Burma
— *N.s.denotata* Bangs & Phillips, 1914	E Burma, E Yunnan to Sichuan, Shaanxi and Guizhou >> N Thailand, N Indochina

○ *Niltava sumatrana* Salvadori, 1879 RUFOUS-VENTED NILTAVA | Mts. of W Malaysia, Sumatra

○ *Niltava vivida* VIVID NILTAVA

— *N.v.oatesi* Salvadori, 1887	E Himalayas and SE Xizang to SW Sichuan, Yunnan, Burma and NW Thailand
— *N.v.vivida* (Swinhoe, 1864)	Taiwan

○ *Niltava grandis* LARGE NILTAVA

— *N.g.grandis* (Blyth, 1842)	E Himalayas, Assam, Burma, N and W Thailand
— *N.g.griseiventris* La Touche, 1921	SE Yunnan, NE and C Laos, NW Vietnam
— *N.g.decorata* Robinson & Kloss, 1919	Langbian (SC Vietnam)
— *N.g.decipiens* Salvadori, 1892	Malay Pen., Sumatra

○ *Niltava macgregoriae* SMALL NILTAVA

— *N.m.macgregoriae* (E. Burton, 1836)	W and C Himalayas
— *N.m.signata* (Horsfield, 1840)	E Himalayas, Assam, N and E Burma, N and W Thailand, N and C Indochina, S China

MUSCICAPELLA Bianchi, 1907 F

○ *Muscicapella hodgsoni* PYGMY BLUE FLYCATCHER

— *M.h.hodgsoni* (F. Moore, 1854)	C and E Himalayas, Assam, W and N Burma, NW Thailand, W Yunnan, C Indochina
— *M.h.sondaica* (Robinson & Kloss, 1923)	Malay Pen., Sumatra, Borneo

---

[1] The name *lepidula* Deignan, 1947 (642) (= *lepidulus*), was proposed against the possibility that the whole genus would be submerged in *Muscicapa*; within *Cyornis* the original name, used above, is not preoccupied.
[2] Includes *litoralis* see Dickinson *et al.* (1991: 353) (745).
[3] This and the next three races form a group sometimes treated as a separate species; but this is a matter of taste rather than proved.
[4] Historically sometimes treated within the genus *Niltava* even when most *Cyornis* species were not. It fits well in neither, and probably deserves its own genus.

CULICICAPA Swinhoe, 1871 F
○ ***Culicicapa ceylonensis*** Grey-headed Canary-Flycatcher
    __ *C.c.calochrysea* Oberholser, 1923   Himalayas, ? C and SE India, Bangladesh, Assam to
                                                N and C Indochina and C China

    __ *C.c.antioxantha* Oberholser, 1923   SE Burma, SW and peninsular Thailand, S Indochina
    __ *C.c.ceylonensis* (Swainson, 1820)   SW India, Sri Lanka, S Malay Pen., to Sumatra, Java,
                                                  Bali, Borneo

    __ *C.c.sejuncta* E. Hartert, 1897   Flores, Lombok (Lesser Sundas)
    __ *C.c.connectens* Rensch, 1931   Sumba (Lesser Sundas)

○ ***Culicicapa helianthea*** Citrine Canary-Flycatcher
    __ *C.h.septentrionalis* Parkes, 1960   NW Luzon
    __ *C.h.zimmeri* Parkes, 1960   C and S Luzon
    __ *C.h.panayensis* (Sharpe, 1877)   Cebu, Leyte, Mindanao, Negros, Palawan, Panay
    __ *C.h.mayri* Deignan, 1947   Sulu Arch.
    __ *C.h.helianthea* Wallace, 1865   Sulawesi, Banggai I., Salayer I.

CINCLIDAE DIPPERS (1: 5)

CINCLUS Borkhausen, 1797 M
● ***Cinclus cinclus*** White-throated Dipper
    __ *C.c.hibernicus* E. Hartert, 1910   Ireland, W Scotland
    __ *C.c.gularis* (Latham, 1802)[1]   Scotland (excl. W), W and C England
    ✓ *C.c.cinclus* (Linnaeus, 1758)   N Europe, W and C France, N Spain, Corsica
    ✓ *C.c.aquaticus* Bechstein, 1803   C and SE Europe, S and E Spain, Sicily
    __ *C.c.olympicus* Madarász, 1903[2]   Crete, Cyprus, W and S Asia Minor
    __ *C.c.minor* Tristram, 1870   NW Africa
    __ *C.c.rufiventris* Tristram, 1884   W Syria, Lebanon
    __ *C.c.uralensis* Serebrovski, 1927[3]   Ural Mts. (Russia)
    __ *C.c.caucasicus* Madarász, 1903   N Asia Minor, Caucasus area, N Iran
    __ *C.c.persicus* Witherby, 1906   SE Turkey, NW and SW Iran, Iraq
    __ *C.c.leucogaster* Bonaparte, 1850   Afghanistan, Tien Shan and Kun Lun Mts., N Pakistan
    __ *C.c.baicalensis* Dresser, 1892[4]   Altai Mts. and Mongolia to Transbaikalia and SE Siberia
    __ *C.c.cashmeriensis* Gould, 1860   W and C Himalayas
    __ *C.c.przewalskii* Bianchi, 1900   E Himalayas, S Tibet, W China

○ ***Cinclus pallasii*** Brown Dipper
    __ *C.p.tenuirostris* Bonaparte, 1850   Afghanistan to W Tien Shan and W and C Himalayas
    __ *C.p.dorjei* Kinnear, 1937[5]   E Himalayas, Assam, W, N and E Burma, W Yunnan,
                                                  N Thailand

    __ *C.p.pallasii* Temminck, 1820[6]   NE, E and C China, Korea, Sakhalin, E and NE Siberia,
                                                  S Kuril Is., Japan, Taiwan, N Indochina

● ***Cinclus mexicanus*** American Dipper
    ✓ *C.m.unicolor* Bonaparte, 1827   W Canada, W USA
    __ *C.m.mexicanus* Swainson, 1827   N and C Mexico
    __ *C.m.anthonyi* Griscom, 1930   S Mexico, Guatemala
    __ *C.m.dickermani* A.R. Phillips, 1966   S Mexico (1871)
    __ *C.m.ardesiacus* Salvin, 1867   Costa Rica, W Panama

○ ***Cinclus leucocephalus*** White-capped Dipper
    __ *C.l.rivularis* Bangs, 1899   N Colombia
    __ *C.l.leuconotus* P.L. Sclater, 1858   W Venezuela, S Colombia, Ecuador
    __ *C.l.leucocephalus* Tschudi, 1844   Peru, Bolivia

○ ***Cinclus schulzi*** Cabanis, 1882 Rufous-throated Dipper   NW Argentina

---

[1] For reasons to use 1802 not 1801 see Browing & Monroe (1991) (258).
[2] For recognition see Cramp *et al.* (1988: 510) (592).
[3] For recognition see Cramp *et al.* (1988: 510) (592).
[4] For recognition see Stepanyan (1990: 422) (2326).
[5] In last Edition the race *marila* was listed erroneously from the Khasi Hills. This was a *lapsus* for *undina*, in synonymy here *undina* is not recognised by Ripley (1982: 489) (2055). A review of the species is in hand. CSR.
[6] Includes *marila* from Taiwan and *yesoensis* Mishima, 1958 (1590), see Orn. Soc. Japan (1974: 361) (1726).

CHLOROPSEIDAE LEAFBIRDS (1: 8)

CHLOROPSIS Jardine & Selby, 1827 F

○ *Chloropsis flavipennis* (Tweeddale, 1878) PHILIPPINE LEAFBIRD   Cebu, Mindanao

○ *Chloropsis palawanensis* (Sharpe, 1876)[1] YELLOW-THROATED LEAFBIRD

Palawan (2227)

○ *Chloropsis sonnerati* GREATER GREEN LEAFBIRD
— *C.s.zosterops* Vigors & Horsfield, 1830[2]   S Thailand, Malaysia, Sumatra, Nias, Borneo, Bangka
— *C.s.sonnerati* Jardine & Selby, 1827   Java

○ *Chloropsis cyanopogon* LESSER GREEN LEAFBIRD
— *C.c.septentrionalis* Robinson & Kloss, 1918   S Burma, S Thailand
— *C.c.cyanopogon* (Temminck, 1830[3])   Malay Pen., Sumatra, Borneo

○ *Chloropsis cochinchinensis* BLUE-WINGED LEAFBIRD
— *C.c.jerdoni* (Blyth, 1844)   C and S India, Sri Lanka
— *C.c.chlorocephala* Walden, 1871[4]   S Assam, Bangladesh, Burma, SW Yunnan, W Thailand
— *C.c.cochinchinensis* (J.F. Gmelin, 1789)   SE Thailand, S Indochina
— *C.c.kinneari* Hall & Deignan, 1956   N of E Thailand, N Indochina, SE Yunnan
— *C.c.serithai* Deignan, 1946   Peninsular Thailand
— *C.c.moluccensis* (J.E. Gray, 1832)   Extreme S Thailand, W Malaysia
— *C.c.icterocephala* (Lesson, 1840)[5]   Sumatra, Bangka, Belitung.
— *C.c.natunensis* Chasen, 1938   Natuna Is.
— *C.c.viridinucha* (Sharpe, 1877)   Lowland Borneo
— *C.c.flavocincta* (Sharpe, 1887)[6]   Mts. of Borneo (2230)
— *C.c.nigricollis* (Vieillot, 1818)   Java

○ *Chloropsis aurifrons* GOLDEN-FRONTED LEAFBIRD
— *C.a.aurifrons* (Temminck, 1829)   Himalayas, NE India, N and W Burma
— *C.a.frontalis* (Pelzeln, 1856)   W and S India
— *C.a.insularis* Whistler & Kinnear, 1933   Kerala (S India), Sri Lanka
— *C.a.pridii* Deignan, 1946   E and S Burma, N (and ? W)[7] Thailand, N and C Laos, SC Yunnan
— *C.a.inornata* Kloss, 1918   C, E and SE Thailand, Cambodia, S Vietnam
— *C.a.incompta* Deignan, 1948   S Laos, SC and S Vietnam
— *C.a.media* (Bonaparte, 1850)   Sumatra

○ *Chloropsis hardwickei* ORANGE-BELLIED LEAFBIRD
— *C.h.hardwickei* Jardine & Selby, 1830   Himalayas, Burma, N Thailand, N and C Laos, NW Vietnam
— *C.h.melliana* Stresemann, 1923   SE China, NE and C Vietnam, C Laos
— *C.h.lazulina* (Swinhoe, 1870)   Hainan (China)
— *C.h.malayana* Robinson & Kloss, 1923   Mts. of W Malaysia

○ *Chloropsis venusta* (Bonaparte, 1850) BLUE-MASKED LEAFBIRD   Sumatra

DICAEIDAE FLOWERPECKERS (2: 44)

PRIONOCHILUS Strickland, 1841 M
○ *Prionochilus olivaceus* OLIVE-BACKED FLOWERPECKER
— *P.o.parsonsi* McGregor, 1927   Luzon
— *P.o.samarensis* Steere, 1890   Bohol, Samar, Leyte
— *P.o.olivaceus* Tweeddale, 1877   Basilan, Mindanao

---

[1] Delacour, 1960 (681), erred in citing the original description, which we date correctly here.
[2] Includes *parvirostris* see Mees (1986) (1539).
[3] Pl. 512 appeared in Livr. 86 and must be dated from 1830 (Dickinson, 2001) (739).
[4] We accept this following Hall & Deignan (1956) (1064); Ripley (1982: 308) (2055) placed it in the synonymy of the nominate form, but we have seen no review rebutting Hall & Deignan.
[5] Includes *billitonis* see Mees (1986) (1539).
[6] Omitted in last Edition and by Delacour (1960) (681). We agree with Mees (1986) (1539) that this should be treated as a species (Bornean Montane Leafbird) but, under our own rules, cannot do so until evidence is formally presented!
[7] The population of SW Thailand requires review.

○ *Prionochilus maculatus* Yellow-breasted Flowerpecker
    — *P.m.septentrionalis* Robinson & Kloss, 1921   N and C Malay Pen.
    — *P.m.oblitus* (Mayr, 1938)   W Malaysia
    — *P.m.maculatus* (Temminck, 1836)   Sumatra, Belitung, Nias I., Borneo
    — *P.m.natunensis* (Chasen, 1935)   Great Natuna I.

○ *Prionochilus percussus* Crimson-breasted Flowerpecker
    — *P.p.ignicapilla* (Eyton, 1839)[1]   Malay Pen., Sumatra and Borneo
    — *P.p.regulus* (Meyer de Schauensee, 1940)   Tanahmasa, Batu Is.
    — *P.p.percussus* (Temminck, 1826)   Java

○ *Prionochilus plateni* Palawan Flowerpecker
    — *P.p.plateni* W.H. Blasius, 1888   Palawan
    — *P.p.culionensis* (Rand, 1948)[2]   Calamianes Group (Philippines)

○ *Prionochilus xanthopygius* Salvadori, 1868 Yellow-rumped Flowerpecker
    Borneo

○ *Prionochilus thoracicus* (Temminck, 1836) Scarlet-breasted Flowerpecker
    S Malay Pen., Belitung, Borneo

DICAEUM Cuvier, 1817 N
○ *Dicaeum annae* (Büttikofer, 1894) Golden-rumped Flowerpecker   Flores, Sumbawa

○ *Dicaeum agile* Thick-billed Flowerpecker[3]
    — *D.a.agile* (Tickell, 1833)   S India to S Nepal, NW Pakistan and W Bangladesh
    — *D.a.zeylonicum* (Whistler, 1944)   Sri Lanka
    — *D.a.modestum* (Hume, 1875)[4]   E Bangladesh, Bhutan and NE India east to C and
        S Indochina and south to N Malay Pen.
    — *D.a.remotum* (Robinson & Kloss, 1915)   S Malay Pen. and Borneo
    — *D.a.atjehense* Delacour, 1946[5]   Sumatra
    — *D.a.finschi* M. Bartels, Sr., 1914   W Java
    — *D.a.tinctum* (Mayr, 1944)   Sumba, Flores, Alor Is.
    — *D.a.obsoletum* (S. Müller, 1843)   Timor I.
    — *D.a.striatissimum* Parkes, 1962   Luzon, Lubang, Romblon, Sibuyan, Catanduanes
    — *D.a.aeruginosum* (Bourns & Worcester, 1894)   Cebu, Negros, Mindoro, Mindanao
    — *D.a.affine* (J.T. Zimmer, 1918)   Palawan

○ *Dicaeum everetti* Brown-backed Flowerpecker[6]
    — *D.e.sordidum* (Robinson & Kloss, 1918)[7]   W Malaysia, Riau Arch.
    — *D.e.everetti* (Sharpe, 1877)   Borneo
    — *D.e.bungurense* (Chasen, 1934)[8]   Gt. Natuna I.

○ *Dicaeum proprium* Ripley & Rabor, 1966 Whiskered Flowerpecker   Mts. of Mindanao

○ *Dicaeum chrysorrheum* Yellow-vented Flowerpecker
    — *D.c.chrysochlore* Blyth, 1843   C Himalayas; S and SE Assam, S Yunnan and Burma to
        Indochina south to N Malay Pen.
    — *D.c.chrysorrheum* Temminck, 1829   C and S Malay Pen. to Gtr. Sundas

○ *Dicaeum melanoxanthum* (Blyth, 1843) Yellow-bellied Flowerpecker
    C and E Himalayas to S China and W and E Burma >>
    N Thailand, N Indochina

○ *Dicaeum vincens* (P.L. Sclater, 1872) Legge's Flowerpecker   Sri Lanka

○ *Dicaeum aureolimbatum* Yellow-sided Flowerpecker
    — *D.a.aureolimbatum* (Wallace, 1865)   Sulawesi, Muna, Buton Is.
    — *D.a.laterale* Salomonsen, 1960   Great Sangihe I.

---

[1] For correct spelling see David & Gosselin (2002) (613).
[2] For recognition see Dickinson *et al.* (1991: 388) (745).
[3] Sheldon (1985) (2234) considered that the separation of the Philippine forms in a species *aeruginosus* was not justified. Future molecular studies may eventually disagree, and Mann (2002) (1415A) does disagree, but for we now follow Sheldon.
[4] Includes *pallescens* see Salomonsen (1967: 176) (2158); however Cheke & Mann (2001) (326) recognised this as a broad ranging form engulfing *remotum* and may be correct to do so, but details need publication.
[5] The name *sumatranum* Chasen, 1939, is preoccupied in *Dicaeum* by *Dicaeum sumatranum* Cabanis, 1878. In rejecting the use of *atjehense*, as opposed to its distinctness, Cheke & Mann (2001) (326) overlooked this.
[6] For a discussion of the validity of this species see Sheldon (1985) (2234).
[7] For recognition see Medway & Wells (1976: 383) (1507); not recognised by Cheke & Mann (2001) (326).
[8] This taxon shows some features of *D. agile* and may be more closely related to that species, see Cheke & Mann (2001) (326).

O *Dicaeum nigrilore* Olive-capped Flowerpecker
 — *D.n.nigrilore* E. Hartert, 1904   Mts. of Mindanao (except NE)
 — *D.n.diuatae* Salomonsen, 1953[1]   Diuata Mts. (NE Mindanao)

O *Dicaeum anthonyi* Flame-crowned Flowerpecker
 — *D.a.anthonyi* (McGregor, 1914)   NE Luzon
 — *D.a.masawan* Rand & Rabor, 1957   Mt. Malindang (NW Mindanao)
 — *D.a.kampalili* Manuel & Gilliard, 1953   Mts. of NC and S Mindanao

O *Dicaeum bicolor* Bicoloured Flowerpecker
 — *D.b.inexpectatum* (E. Hartert, 1895)   Luzon, Mindoro
 — *D.b.bicolor* (Bourns & Worcester, 1894)   Bohol, Leyte, Samar, Mindanao
 — *D.b.viridissimum* Parkes, 1971   Guimaras, Negros (1763)

O *Dicaeum australe* (Hermann, 1783) Red-keeled Flowerpecker   Philippines (except Guimaras, Panay and Negros)

O *Diaceum haematostictum* Sharpe, 1876 Black-belted Flowerpecker[2]   Guimaras, Panay, Negros

O *Dicaeum retrocinctum* Gould, 1872 Scarlet-collared Flowerpecker   Mindoro

O *Dicaeum quadricolor* (Tweeddale, 1877) Cebu Flowerpecker[3]   Cebu

O *Dicaeum trigonostigma* Orange-bellied Flowerpecker
 — *D.t.rubropygium* E.C.S. Baker, 1921   S Bangladesh, S Burma, SW Thailand and extreme
                 N Malay Pen. (? Assam)

 — *D.t.trigonostigma* (Scopoli, 1786)[4]   NC to S Malay Pen., Sumatra incl. islands off W coast,
                 Belitung, Bangka

 — *D.t.megastoma* E. Hartert, 1918   Great Natuna I.
 — *D.t.flaviclunis* E. Hartert, 1918   Java, Bali
 — *D.t.dayakanum* Chasen & Kloss, 1929   Borneo and islands offshore
 — *D.t.xanthopygium* Tweeddale, 1877   Marinduque, Mindoro, Luzon
 — *D.t.intermedium* Bourns & Worcester, 1894   Romblon
 — *D.t.cnecolaemum* Parkes, 1989   Tablas I. (1777)
 — *D.t.sibuyanicum* Bourns & Worcester, 1894   Sibuyan I.
 — *D.t.dorsale* Sharpe, 1876   Masbate, Panay, Negros
 — *D.t.besti* Steere, 1890   Siquijor I.
 — †?*D.t.pallidius* Bourns & Worcester, 1894   Cebu
 — *D.t.cinereigulare* Tweeddale, 1877   Mindanao, Samar, Leyte, Bohol
 — *D.t.isidroi* Rand & Rabor, 1969   Camiguin South I. (1979)
 — *D.t.assimile* Bourns & Worcester, 1894   Tawitawi, Jolo, Siasi (Sulu Arch.)
 — *D.t.sibutuense* Sharpe, 1893   Sibutu I.

O *Dicaeum hypoleucum* Buzzing Flowerpecker
 — *D.h.cagayanense* Rand & Rabor, 1967[5]   NE Luzon (1978)
 — *D.h.obscurum* Ogilvie-Grant, 1894[6]   N and C Luzon, S Luzon?
 — *D.h.pontifex* Mayr, 1946   Bohol, Samar, Leyte, Mindanao
 — *D.h.mindanense* Tweeddale, 1877[7]   Zamboanga Pen. (Mindanao)
 — *D.h.hypoleucum* Sharpe, 1876   Basilan, Sulu Arch.

O *Dicaeum erythrorhynchos* Pale-billed Flowerpecker
 — *D.e.erythrorhynchos* (Latham, 1790)   Foothills of W and C Himalayas, S and C India to
                 Assam and W and SW Burma

 — *D.e.ceylonense* Babault, 1920   Sri Lanka

O *Dicaeum concolor* Plain Flowerpecker
 — *D.c.olivaceum* Walden, 1875   E Himalayas, Assam and Burma to SC and S China and
                 N and C Indochina

 — *D.c.concolor* Jerdon, 1840   SW India

---

[1] Recognised by Salomonsen (1967: 180) (2158).
[2] This population is very distinct and now been separated see Brooks *et al.* (1992) (241), Cheke & Mann (2001) (326) and Mann (2002) (1415A).
[3] Omitted from last Edition; then believed to be extinct.
[4] Includes *antioproctum* see Salomonsen (1967: 183) (2158). The name *melanostigma* applied in last Edition to part of this population was a *lapsus*, no such name has been traced.
[5] The original spelling *cagayanensis* has been widely used (Dickinson et al., 1991; Cheke & Mann, 2001) (745, 326) but requires correction to accord with the gender of the generic name.
[6] Includes *lagunae* see Dickinson *et al.* (1991: 395) (745).
[7] For recognition see Salomonsen (1967: 186) (2158).

___ *D.c.virescens* Hume, 1873	S Andamans
___ *D.c.minullum* Swinhoe, 1870	Hainan
___ *D.c.uchidai* N. Kuroda, Sr., 1920	Taiwan
___ *D.c.borneanum* Lönnberg, 1925	W Malaysia, Sumatra, Borneo, Natunas
___ *D.c.sollicitans* E. Hartert, 1901	Java, Bali

○ *Dicaeum pygmaeum* PYGMY FLOWERPECKER
___ *D.p.palawanorum* Hachisuka, 1926	Balabac, Palawan
___ *D.p.fugaense* Parkes, 1988	Fuga and Calayan (off N Luzon) (1776)
___ *D.p.salomonseni* Parkes, 1962	NW Luzon
___ *D.p.pygmaeum* (Kittlitz, 1833)	C and S Luzon, Mindoro, Negros, Leyte, Samar, Cebu
___ *D.p.davao* Mearns, 1905	Mindanao, Camiguin South I.

○ *Dicaeum nehrkorni* W.H. Blasius, 1886 CRIMSON-CROWNED FLOWERPECKER[1]

Mts. of Sulawesi

○ *Dicaeum erythrothorax* FLAME-BREASTED FLOWERPECKER
___ *D.e.schistaceiceps* G.R. Gray, 1860	N and C Moluccas
___ *D.e.erythrothorax* Lesson, 1828	Buru

○ *Dicaeum vulneratum* Wallace, 1863 ASHY FLOWERPECKER   S Moluccas

○ *Dicaeum pectorale* OLIVE-CROWNED FLOWERPECKER
___ *D.p.ignotum* Mees, 1964	Gebe I. (W Papuan Is.)
___ *D.p.pectorale* S. Müller, 1843	W Papuan Is. (except Gebe), NW New Guinea (Vogelkop)

○ *Dicaeum geelvinkianum* RED-CAPPED FLOWERPECKER
___ *D.g.maforense* Salvadori, 1876	Numfor I. (Geelvink Bay)
___ *D.g.misoriense* Salvadori, 1876	Biak I. (Geelvink Bay)
___ *D.g.geelvinkianum* A.B. Meyer, 1874	Yapen and Kurudu Is. (Geelvink Bay)
___ *D.g.obscurifrons* Junge, 1952	WC New Guinea
___ *D.g.setekwa* Rand, 1941	SW New Guinea
___ *D.g.diversum* Rothschild & Hartert, 1903	N New Guinea
___ *D.g.centrale* Rand, 1941	WC New Guinea
___ *D.g.albopunctatum* D'Albertis & Salvadori, 1879	Far SC New Guinea
___ *D.g.rubrigulare* D'Albertis & Salvadori, 1879	S New Guinea
___ *D.g.rubrocoronatum* Sharpe, 1876	SE and NE to EC New Guinea
___ *D.g.violaceum* Mayr, 1936	D'Entrecasteaux Arch.

○ *Dicaeum nitidum* LOUISIADE FLOWERPECKER
___ *D.n.nitidum* Tristram, 1889	Tagula, Misima Is. (Louisiade Arch.)
___ *D.n.rosseli* Rothschild & Hartert, 1914	Rossel I. (Louisiade Arch.)

○ *Dicaeum eximium* RED-BANDED FLOWERPECKER
___ *D.e.layardorum* Salvadori, 1880	New Britain
___ *D.e.eximium* P.L. Sclater, 1877	New Ireland, New Hanover
___ *D.e.phaeopygium* Salomonsen, 1964	Dyaul I.

○ *Dicaeum aeneum* MIDGET FLOWERPECKER
___ *D.a.aeneum* Pucheran, 1853	N Solomon Is.
___ *D.a.malaitae* Salomonsen, 1960	Malaita I. (NE Solomons)
___ *D.a.becki* E. Hartert, 1929	Guadalcanal I. (SE Solomons)

○ *Dicaeum tristrami* Sharpe, 1884 MOTTLED FLOWERPECKER   San Cristobal I. (E Solomons)

○ *Dicaeum igniferum* Wallace, 1863 BLACK-FRONTED FLOWERPECKER   Sumbawa, Flores to Alor (Lesser Sundas)

○ *Dicaeum maugei* RED-CHESTED FLOWERPECKER
___ *D.m.splendidum* Büttikofer, 1893	Saleyer, Djampea Is. (Flores Sea)
___ *D.m.maugei* Lesson, 1830	Samau, Timor, Sawu, Roma, Damar (Lesser Sundas)
___ *D.m.salvadorii* A.B. Meyer, 1884	Babar, Moa (Lesser Sundas)
___ *D.m.neglectum* E. Hartert, 1897	Lombok (Lesser Sundas)

---

[1] The citation in Salomonsen (1967: 189) (2158) should be corrected to read Braunschweigische Anzeigen, 52: 457.

○ *Dicaeum hirundinaceum* MISTLETOEBIRD

— *D.h.hirundinaceum* (Shaw, 1792)	Australia
— *D.h.ignicolle* G.R. Gray, 1858	Aru Is.
— *D.h.keiense* Salvadori, 1875	Kai Is.
— *D.h.fulgidum* P.L. Sclater, 1883	Tanimbar Is.

○ *Dicaeum celebicum* GREY-SIDED FLOWERPECKER

— *D.c.talautense* Meyer & Wiglesworth, 1895	Talaud Is.
— *D.c.sanghirense* Salvadori, 1876	Sangihe Is.
— *D.c.celebicum* S. Müller, 1843	Sulawesi; Muna, Buton Is.,
— *D.c.sulaense* Sharpe, 1884	Banggai, Sula Is.
— *D.c.kuehni* E. Hartert, 1903	Tukangbesi I.

○ *Dicaeum monticolum* Sharpe, 1887 BLACK-SIDED FLOWERPECKER   Borneo

○ *Dicaeum ignipectum* FIRE-BREASTED FLOWERPECKER

— *D.i.ignipectum* (Blyth, 1843)	Himalayas and Assam to C and E China, N and C Vietnam, Laos, N Thailand
— *D.i.cambodianum* Delacour & Jabouille, 1928	Cambodia, E and SE Thailand
— *D.i.dolichorhynchum* Deignan, 1938	C and S Malay Pen.
— *D.i.beccarii* Robinson & Kloss, 1916	N Sumatra
— *D.i.formosum* Ogilvie-Grant, 1912	Taiwan
— *D.i.luzoniense* Ogilvie-Grant, 1894	N and C Luzon
— *D.i.bonga* E. Hartert, 1904	Samar
— *D.i.apo* E. Hartert, 1904	Leyte, Negros, Mindanao

○ *Dicaeum sanguinolentum* BLOOD-BREASTED FLOWERPECKER

— *D.s.sanguinolentum* Temminck, 1829	Java, Bali
— *D.s.rhodopygiale* Rensch, 1928	Flores I.
— *D.s.wilhelminae* Büttikofer, 1892	Sumba
— *D.s.hanieli* Hellmayr, 1912	Timor

○ *Dicaeum cruentatum* SCARLET-BACKED FLOWERPECKER

— *D.c.cruentatum* (Linnaeus, 1758)[1]	EC Himalayas and E Bangladesh to SE China (incl. Hainan), Indochina and Thailand S to Malay Pen.
— *D.c.sumatranum* Cabanis, 1878	Sumatra
— *D.c.niasense* Meyer de Schauensee & Ripley, 1939	Nias I.
— *D.c.batuense* Richmond, 1912	Mentawai Is.
— *D.c.simalurense* Salomonsen, 1961	Simeuluë I.
— *D.c.nigrimentum* Salvadori, 1874[2]	Borneo

○ *Dicaeum trochileum* SCARLET-HEADED FLOWERPECKER

— *D.t.trochileum* (Sparrman, 1789)	Java, Bali, SE Borneo, Kangean Is.
— *D.t.stresemanni* Rensch, 1928	Lombok I.

NECTARINIIDAE SUNBIRDS (16: 127)

CHALCOPARIA Cabanis, 1851[3] F

○ *Chalcoparia singalensis* RUBYCHEEK/RUBY-CHEEKED SUNBIRD

— *C.s.assamensis* Kloss, 1930	C and E Nepal, Bangladesh, Assam, Burma (except S), SW Yunnan, N Thailand
— *C.s.koratensis* Kloss, 1918	E and SE Thailand, SE Yunnan, Indochina
— *C.s.internota* (Deignan, 1955)	N Malay Pen.
— *C.s.interposita* Robinson & Kloss, 1921	C Malay Pen.
— *C.s.singalensis* (J.F. Gmelin, 1788)	S Malay Pen.
— *C.s.sumatrana* Kloss, 1921	Sumatra, Belitung
— *C.s.panopsia* Oberholser, 1912	Islands W of Sumatra
— *C.s.pallida* Chasen, 1934	N Natunas

---

[1] Includes *siamense* and *ignitum* see Salomonsen (1967: 198) (2158).
[2] The name *hosei* (sic) was used by Cheke & Mann (2001) (326) as the label for a Bornean colour morph. However the holotype of *hosii* (original spelling, verified) has been shown to match *nigrimentum*, see White (1974) (2651), and the name is unavailable for use in a different context. Cheke & Mann rightly used quotation marks.
[3] Treated in a broad genus *Anthreptes* in last Edition but see Cheke & Mann (2001) (326).

— *C.s.borneana* Kloss, 1921	Borneo and Banggi I.
— *C.s.bantenensis* (Hoogerwerf, 1967)	W Java
— *C.s.phoenicotis* (Temminck, 1822)	E and C Java

DELEORNIS Wolters, 1977[1] M
○ *Deleornis fraseri* FRASER'S SUNBIRD

— *D.f.idius* (Oberholser, 1899)	Sierra Leone to Togo
— *D.f.cameroonensis* (Bannerman, 1921)	S Nigeria to NW Angola and W Zaire
— *D.f.fraseri* (Jardine & Selby, 1843)	Bioko I.
— *D.f.axillaris* (Reichenow, 1893)[2]	C Zaire to Uganda

ANTHREPTES Swainson, 1832 M
○ *Anthreptes reichenowi* PLAIN-BACKED SUNBIRD

— *A.r.yokanae* E. Hartert, 1921	Coastal Kenya and Tanzania
— *A.r.reichenowi* Gunning, 1909	E Zimbabwe, Mozambique

○ *Anthreptes anchietae* (Bocage, 1878) ANCHIETA'S SUNBIRD — Angola to N Zambia, SW Tanzania, W Malawi, NW Mozambique

○ *Anthreptes simplex* (S. Müller, 1843) PLAIN SUNBIRD — Malay Pen., Sumatra, Borneo

○ *Anthreptes malacensis* BROWN-THROATED SUNBIRD

— *A.m.malacensis* (Scopoli, 1786)[3]	SW and SE Burma east to Indochina and south to Sumatra, islands off W Sumatra, Java, S Borneo, Tioman and Anamba Is.
— *A.m.bornensis* Riley, 1920	N Borneo
— *A.m.mjobergi* Bangs & Peters, 1927	Maratua Is.
— *A.m.paraguae* Riley, 1920	Palawan
— *A.m.birgitae* Salomonsen, 1953	Luzon
— *A.m.griseigularis* (Tweeddale, 1877)	Samar, Leyte, NE Mindanao
— *A.m.heliolusius* Oberholser, 1923	W and C Mindanao, Basilan
— *A.m.wiglesworthi* E. Hartert, 1902	Sulu Arch. (except Sibutu)
— *A.m.iris* Parkes, 1971	Sibutu I. (1763)
— *A.m.chlorigaster* Sharpe, 1877	WC Philippines, SW Mindanao
— *A.m.cagayanensis* Mearns, 1905	Cagayan Sulu I.
— *A.m.heliocalus* Oberholser, 1923	Sangi Is.
— *A.m.celebensis* Shelley, 1877	Sulawesi
— *A.m.nesophilus* Eck, 1976	Banggai I. (788)
— *A.m.extremus* Mees, 1966	Sula Is.
— *A.m.convergens* Rensch, 1929	Lesser Sundas (except Sumba)
— *A.m.rubrigena* Rensch, 1931	Sumba I.

○ *Anthreptes rhodolaemus* Shelley, 1878 RED-THROATED SUNBIRD[4] — Malay Pen., Sumatra, Borneo

○ *Anthreptes gabonicus* (Hartlaub, 1861) BROWN SUNBIRD — Gambia to Cabinda (Angola)

○ *Anthreptes longuemarei* WESTERN VIOLET-BACKED SUNBIRD

— *A.l.longuemarei* (Lesson, 1833)[5][6]	Senegal to N Zaire, S Sudan and W Kenya
— *A.l.angolensis* Neumann, 1906	S Zaire, Angola, Zambia, Malawi, Tanzania
— *A.l.nyassae* Neumann, 1906	SE Tanzania, N Mozambique, E Zimbabwe

○ *Anthreptes orientalis* Hartlaub, 1880 EASTERN VIOLET-BACKED SUNBIRD[7] — Somalia to SE Sudan and NE and C Tanzania

○ *Anthreptes neglectus* Neumann, 1922 ULUGURU VIOLET-BACKED SUNBIRD — SE Kenya, E Tanzania, N Mozambique

○ *Anthreptes aurantium* J. & E. Verreaux, 1851 VIOLET-TAILED SUNBIRD — Cameroon to Congo, Zaire, S Central African Republic, NE Angola

---

[1] Treated in a broad genus *Anthreptes* in last Edition but see Cheke & Mann (2001) (326).
[2] For comments on the inclusion of this taxon here see Fry *et al.* (2000: 154) (899).
[3] Birds of the Anamba Is. named *anambae* were placed here by Chasen (1935: 279). Cheke & Mann (2001) (326) accepted this form.
[4] For correction of spelling see David & Gosselin (2002) (614).
[5] Dated 1831 by Rand (1967: 216) (1961) but see Zimmer (1926) (2707) and www.zoonomen.net (A.P. Peterson *in litt.*).
[6] Includes *haussarum* see White (1963: 55) (2648).
[7] Treated as monotypic by White (1963: 56) (2648) and by Irwin (2000: 140) (1193) (thus *neumanni* implicitly included).

○ *Anthreptes seimundi* Little Green Sunbird
— *A.s.kruensis* (Bannerman, 1911)   Sierra Leone to Togo
— *A.s.seimundi* (Ogilvie-Grant, 1908)   Bioko I.
— *A.s.minor* Bates, 1926[1]   Nigeria to S Sudan, Uganda, N Angola

○ *Anthreptes rectirostris* Green Sunbird
— *A.r.rectirostris* (Shaw, 1812)[2]   Sierra Leone to Ghana
— *A.r.tephrolaemus* (Jardine & Selby, 1851)[3]   Bioko I., S Nigeria to NW Angola and W Kenya

○ *Anthreptes rubritorques* Reichenow, 1905 Banded Green Sunbird   E Tanzania

HEDYDIPNA Cabanis, 1851[4] F
○ *Hedydipna collaris* Collared Sunbird
— *H.c.subcollaris* (Hartlaub, 1857)   Senegal and Gambia to SW Nigeria
— *H.c.hypodila* (Jardine & Selby, 1851)   Bioko I.
— *H.c.somereni* (Chapin, 1949)   SE Nigeria, N and W Zaire, N Angola, SW Sudan
— *H.c.djamdjamensis* (Benson, 1942)   SW Ethiopia
— *H.c.garguensis* (Mearns, 1915)[5]   SE Sudan, Uganda and W Kenya to N Zambia, W Tanzania
— *H.c.elachior* (Mearns, 1910)[6]   E Kenya, NE Tanzania, S Somalia, S Ethiopia
— *H.c.zambesiana* (Shelley, 1876)[7]   SE Angola, N Botswana, S Zambia, SE Tanzania, Mozambique, E Transvaal NE Natal
— *H.c.patersonae* (Irwin, 1960)[8]   E Zimbabwe, W Mozambique
— *H.c.collaris* (Vieillot, 1819)   S and E Cape Province, Natal, Swaziland

○ *Hedydipna platura* (Vieillot, 1819) Pygmy Sunbird   Senegal to W and S Sudan and NW Kenya

○ *Hedydipna metallica* (M.H.K. Lichtenstein, 1823) Nile Valley Sunbird[9]
E Chad, N and E Sudan and Egypt to SW Saudi Arabia, S Oman >> N Somalia

○ *Hedydipna pallidigaster* (Sclater & Moreau, 1935) Amani Sunbird   E Kenya, NE and EC Tanzania

HYPOGRAMMA Reichenbach, 1853 N
○ *Hypogramma hypogrammicum* Purple-naped Sunbird
— *H.h.lisettae* (Delacour, 1926)   NW and NE Burma, W and N Thailand, N and C Indochina
— *H.h.mariae* (Deignan, 1943)   Cambodia, S Laos, S Vietnam
— *H.h.nuchale* (Blyth, 1843)   Malay Pen.
— *H.h.hypogrammicum* (S. Müller, 1843)   Sumatra, Borneo
— *H.h.natunense* (Chasen, 1935)   N Natunas

ANABATHMIS Reichenow, 1905[10] F
○ *Anabathmis reichenbachii* (Hartlaub, 1857) Reichenbach's Sunbird   Liberia to Cabinda (Angola) and N Zaire

○ *Anabathmis hartlaubii* (Hartlaub, 1857) Principe Sunbird   Principe I.

○ *Anabathmis newtonii* (Bocage, 1887) Newton's Sunbird   São Tomé I.

DREPTES Reichenow, 1914[11] M
○ *Dreptes thomensis* Bocage, 1889 Giant Sunbird   São Tomé I

ANTHOBAPHES Cabanis, 1850[12] F
○ *Anthobaphes violacea* (Linnaeus, 1766) Orange-breasted Sunbird   Cape Province

[1] When placed in a broad genus *Nectarinia* this name is preoccupied and the name *traylori* must be used see Rand (1967: 225) (1961).
[2] Includes *pujoli* see Érard (1979) (805).
[3] For correction of spelling see David & Gosselin (2002) (614). Includes *amadoni* Eisentraut, 1973 (798) (R.A. Cheke *in litt.*).
[4] Treated in a broad genus *Anthreptes* in last Edition but see Cheke & Mann (2001) (326).
[5] Includes *phillipsi* see White (1963: 58) (2648).
[6] Includes *jubaensis* see White (1963: 57) (2648).
[7] Includes *chobiensis* see Fry (2000: 220) (892).
[8] Emended to *patersonae* by Rand (1967: 220) (1961), but the Code (I.C.Z.N., 1999) (1178) although making clear that this would have been correct (Art. 31.1.2) does not require that it be emended. However, the original name was emended after only 7 years, and the emendation has had over 40 years to achieve prevalence, and we believe it best to use that.
[9] For a discussion of separation from *platura* see Fry (2000: 225-228) (892).
[10] Treated in a broad genus *Nectarinia* in last Edition but see Cheke & Mann (2001) (326).
[11] Treated in a broad genus *Nectarinia* in last Edition but see Cheke & Mann (2001) (326).
[12] Treated in a broad genus *Nectarinia* in last Edition but see Cheke & Mann (2001) (326).

CYANOMITRA Reichenbach, 1854[1] F

○ *Cyanomitra verticalis* GREEN-HEADED SUNBIRD

    — *C.v.verticalis* (Latham, 1790)   Senegal to Nigeria

    — *C.v.bohndorffi* (Reichenow, 1887)   Cameroon to Central African Republic, N Angola, S Zaire

    — *C.v.cyanocephala* (Shaw, 1812)   W Gabon to the River Congo

    — *C.v.viridisplendens* (Reichenow, 1892)   S Sudan, E Zaire to W Kenya, NE Zambia, N Malawi.

○ *Cyanomitra bannermani* Grant & Mackworth-Praed, 1943 BANNERMAN'S SUNBIRD

                                             Angola, S Zaire, NW Zambia

○ *Cyanomitra cyanolaema* BLUE-THROATED BROWN SUNBIRD

    — *C.c.magnirostrata* (Bates, 1930)   Sierra Leone to Ghana

    — *C.c.cyanolaema* (Jardine & Fraser, 1851)   Bioko I.

    — *C.c.octaviae* Amadon, 1953   Nigeria to Uganda and N Angola

○ *Cyanomitra oritis* CAMEROON SUNBIRD

    — *C.o.poensis* Alexander, 1903   Mts. of Bioko I.

    — *C.o.oritis* (Reichenow, 1892)   Mt. Cameroon

    — *C.o.bansoensis* Bannerman, 1922   W Cameroon

○ *Cyanomitra alinae* BLUE-HEADED SUNBIRD

    — *C.a.derooi* (Prigogine, 1975)   NE Zaire (1933)

    — *C.a.kaboboensis* (Prigogine, 1975)   Mt. Kobobo (E Zaire) (1933)

    — *C.a.alinae* Jackson, 1904   SW Uganda, NW Rwanda

    — *C.a.tanganjicae* (Reichenow, 1915)   E Zaire to SW Rwanda, W Burundi

    — *C.a.marunguensis* (Prigogine, 1975)   Marungu Mts. (SE Zaire) (1933)

○ *Cyanomitra olivacea* OLIVE SUNBIRD[2]

    — *C.o.guineensis* Bannerman, 1921   Senegal to Togo

    — *C.o.cephaelis* (Bates, 1930)   Benin to Zaire, N Angola

    — *C.o.obscura* (Jardine, 1843)   Principe, Bioko Is.

    — *C.o.vincenti* Grant & Mackworth-Praed, 1943   S Sudan, W Kenya, Uganda, W Tanzania, E Zaire

    — *C.o.ragazzii* (Salvadori, 1888)   SW Ethiopia

    — *C.o.changamwensis* Mearns, 1910   S Somalia to NE Tanzania

    — *C.o.neglecta* Neumann, 1900   C Kenya, N Tanzania

    — *C.o.granti* Vincent, 1934   Pemba, Zanzibar

    — *C.o.lowei* Vincent, 1934   SE Zaire, N Zambia

    — *C.o.alfredi* Vincent, 1934[3]   S Tanzania, Malawi, SE Zambia, N Mozambique

    — *C.o.sclateri* Vincent, 1934   E Zimbabwe, W Mozambique

    — *C.o.olivacina* (W.K.H. Peters, 1881)   NE Natal, coastal S Mozambique

    — *C.o.olivacea* (A. Smith, 1840)   E Natal to coastal E Cape Province

○ *Cyanomitra veroxii* MOUSE-COLOURED SUNBIRD

    — *C.v.fischeri* (Reichenow, 1880)   Coastal S Somalia to Mozambique and N Zululand

    — *C.v.zanzibarica* Grote, 1932   Zanzibar

    — *C.v.veroxii* (A. Smith, 1831)   S Zululand to E Cape Province

CHALCOMITRA Reichenbach, 1854[4] F

○ *Chalcomitra adelberti* BUFF-THROATED SUNBIRD

    — *C.a.adelberti* (Gervais, 1834)   Sierra Leone to Ghana

    — *C.a.eboensis* (Jardine, 1843)   Togo to SE Nigeria

○ *Chalcomitra fuliginosa* CARMELITE SUNBIRD

    — *C.f.aurea* (Lesson, 1847)   Coastal Liberia to Gabon

    — *C.f.fuliginosa* (Shaw, 1812)   Lower Congo R., NW Angola

○ *Chalcomitra rubescens* GREEN-THROATED SUNBIRD

    — *C.r.crossensis* Serle, 1963   SE Nigeria, W Cameroon

    — *C.r.stangerii* (Jardine, 1842)   Bioko I.

    — *C.r.rubescens* (Vieillot, 1819)   Cameroon to Zaire, SW Sudan, W Kenya, N Angola and NW Zambia

---

[1] Treated in a broad genus *Nectarinia* in last Edition but see Cheke & Mann (2001) (326).
[2] This broad species follows the treatment in Zimmerman *et al*. Treated as two species by Fry (2000: 175-180) (892) and by Cheke & Mann (2001) (326).
[3] Includes *intercalans* Clancey, 1978 (465) see Fry (2000: 179) (892)
[4] Treated in a broad genus *Nectarinia* in last Edition, but see Cheke & Mann (2001) (326).

○ *Chalcomitra amethystina* Amethyst Sunbird

— *C.a.kalckreuthi* (Cabanis, 1878)   S Somalia, E Kenya, NE Tanzania

— *C.a.doggetti* (Sharpe, 1902)   W Kenya, Uganda, NW Tanzania

— *C.a.kirkii* (Shelley, 1876)   SE Sudan, W and C Kenya, W and S Tanzania to E
    Zambia, Zimbabwe and C Mozambique

— *C.a.deminuta* Cabanis, 1880   S Zaire, W Zambia, Angola, N Namibia, N Botswana

— *C.a.adjuncta* (Clancey, 1975)   Transvaal, S Mozambique, N Natal (452)

— *C.a.amethystina* (Shaw, 1812)   W Natal, E and S Cape Province

○ *Chalcomitra senegalensis* Scarlet-chested Sunbird

— *C.s.senegalensis* (Linnaeus, 1766)   Senegal to N Nigeria

— *C.s.acik* (Hartmann, 1866)[1]   Cameroon to W and S Sudan, NE Zaire, NW Uganda

— *C.s.cruentata* (Rüppell, 1845)   SE Sudan, Ethiopia, Eritrea

— *C.s.lamperti* (Reichenow, 1897)   E Zaire, S and E Uganda, W and C Kenya, W and
    C Tanzania

— *C.s.saturatior* (Reichenow, 1891)   S Zaire, Angola, W Zambia, to Namibia, N Botswana,

— *C.s.gutturalis* (Linnaeus, 1766)[2]   S Somalia to E Tanzania, E Zambia, E Zimbabwe,
    Mozambique and Natal

○ *Chalcomitra hunteri* (Shelley, 1889) Hunter's Sunbird[3]   Somalia, E Ethiopia, Kenya, NE Tanzania

○ *Chalcomitra balfouri* (Sclater & Hartlaub, 1881) Socotra Sunbird   Socotra I.

LEPTOCOMA Cabanis, 1850[4] F

○ *Leptocoma zeylonica* Purple-rumped Sunbird

— *L.z.flaviventris* (Hermann, 1804)   India to Bangladesh and SW Burma

— *L.z.zeylonica* (Linnaeus, 1766)   Sri Lanka

○ *Leptocoma minima* (Sykes, 1832) Crimson-backed Sunbird   W and S India

○ *Leptocoma sperata* Purple-throated Sunbird[5]

— *L.s.emmae* Delacour & Jabouille, 1928   Cambodia, S Laos, S Vietnam

— *L.s.brasiliana* (J.F. Gmelin, 1788)[6]   S and SE Assam, E Bangladesh, southern Burma and
    southern Thailand S to Malay Pen., Gtr. Sundas
    and islands off west Sumatra (except Simeuluë)

— *L.s.mecynorhyncha* (Oberholser, 1912)   Simeuluë I.

— *L.s.eumecis* (Oberholser, 1917)   Anambas

— *L.s.axantha* (Oberholser, 1932)   Natunas

— *L.s.henkei* (A.B. Meyer, 1884)   N Luzon and islands of it

— *L.s.sperata* (Linnaeus, 1766)   S and C Luzon, Catanduanes

— *L.s.trochilus* (Salomonsen, 1953)   Palawan group, Mindoro, E and W Visayas, E Mindanao;
    Maratua Is. (off Borneo)

— *L.s.juliae* (Tweeddale, 1877)[7]   W and S Mindanao, Basilan Is

○ *Leptocoma sericea* Black Sunbird[8]

— *L.s.talautensis* (Meyer & Wiglesworth, 1894)   Talaud Is.

— *L.s.sangirensis* (A.B. Meyer, 1874)   Sangihe Is.

— *L.s.grayi* (Wallace, 1865)   N Sulawesi

— *L.s.porphyrolaema* (Wallace, 1865)   C and S Sulawesi

— *L.s.auriceps* (G.R. Gray, 1860)   Sula Is. to N Moluccas

— *L.s.auricapilla* (Mees, 1965)   Kajoa I. (NW Moluccas)

— *L.s.aspasioides* (G.R. Gray, 1860)[9]   Seram, Ambon to Watubela Is. (S Moluccas) and Aru Is.

— *L.s.proserpina* (Wallace, 1863)   Buru I. (S Moluccas)

— *L.s.chlorolaema* (Salvadori, 1874)   Kai Is.

— *L.s.mariae* (Ripley, 1959)   Kafiau I. (off W. New Guinea)

---

[1] Includes *adamauae* see Rand (1967: 234) (1961).
[2] Includes *inaestimata* see White (1963: 68) (2648).
[3] Treated as monotypic by Fry (2000: 197) (892), including *siccata* Clancey, 1986 (504).
[4] Treated in a broad genus *Nectarinia* in last Edition, but see Cheke & Mann (2001) (326).
[5] *N. s. marinduquensis* duPont, 1971 (771) is now thought to have been based on a female specimen of another species (*Aethopyga siparaja*) see Cheke & Mann (2001: 243) (326); it is not in synonymy in this species.
[6] Includes *phayrei* see Baker (1930: 287) (79), may need review as used by Smythies, 1953 (2285), but was not in Rand (1967) (1961).
[7] An unstable hybrid population named *davaoensis* can be found in synonymy here see Dickinson (1991: 379) (745).
[8] For details of the change of name from *sericea* to *aspasia* see Mees (1966) (1517). Thus when placed in a broad genus *Nectarinia* the name *sericea* is preoccupied and the name *aspasia* must be used (Mees, 1965) (1516). It is not needed in the context used here.
[9] Includes *chlorocephala* see Mees (1965) (1516).

__ *L.s.cochrani* (Stresemann & Paludan, 1932)	Misool, Waigeo Is. (off W. New Guinea)
__ *L.s.sericea* (Lesson, 1827)[1]	New Guinea
__ *L.s.maforensis* (A.B. Meyer, 1874)	Numfor I. (Geelvink Bay)
__ *L.s.salvadorii* (Shelley, 1877)	Islands off W Yapen I. (Geelvink Bay)
__ *L.s.nigriscapularis* (Salvadori, 1876)	Meos Num, Rani Is. (Geelvink Bay)
__ *L.s.mysorensis* (A.B. Meyer, 1874)	Biak I. (Geelvink Bay)
__ *L.s.veronica* (Mees, 1965)	Liki I. (off NW New Guinea)
__ *L.s.cornelia* (Salvadori, 1878)	Tarawai I. (off NC New Guinea)
__ *L.s.christianae* (Tristram, 1889)	D'Entrecasteaux and Louisiade Arch.
__ *L.s.caeruleogula* (Mees, 1965)	New Britain, Umboi I.
__ *L.s.corinna* (Salvadori, 1878)	Bismarck Arch.
__ *L.s.eichhorni* (Rothschild & Hartert, 1926)	Feni I (Bismarck Arch.)

○ ***Leptocoma calcostetha*** (Jardine, 1843) Copper-throated Sunbird[2]   Coastal SE Burma, SE and SW Thailand and S Indochina to Gtr. Sundas and Palawan

NECTARINIA Illiger, 1811 F
○ ***Nectarinia bocagii*** Shelley, 1879[3] Bocage's Sunbird   Angola, S Zaire

○ ***Nectarinia purpureiventris*** (Reichenow, 1893) Purple-breasted Sunbird
SW Uganda, E Zaire, Rwanda, Burundi

○ ***Nectarinia tacazze*** Tacazze Sunbird
| __ *N.t.tacazze* (Stanley, 1814) | Ethiopia, Eritrea |
| __ *N.t.jacksoni* Neumann, 1899 | SE Sudan, N and E Uganda, W Kenya, N Tanzania |

○ ***Nectarinia kilimensis*** Bronzy Sunbird
__ *N.k.kilimensis* Shelley, 1884	E Zaire to S Uganda, W and C Kenya, N Tanzania
__ *N.k.arturi* P.L. Sclater, 1906	S Tanzania, Malawi, NE Zambia, E Zimbabwe
__ *N.k.gadowi* Bocage, 1892	WC Angola

○ ***Nectarinia famosa*** Yellow-tufted Malachite Sunbird
| __ *N.f.cupreonitens* Shelley, 1876[4] | Ethiopia to E Zaire, Tanzania and N Malawi |
| __ *N.f.famosa* (Linnaeus, 1766)[5] | South Africa, E Zimbabwe, W Mozambique |

○ ***Nectarinia johnstoni*** Scarlet-tufted Malachite Sunbird
__ *N.j.johnstoni* Shelley, 1885	W and C Kenya, N Tanzania
__ *N.j.dartmouthi* Ogilvie-Grant, 1906	E Zaire to W Rwanda, SW Uganda
__ *N.j.salvadorii* Shelley, 1903[6]	S Tanzania, NE Zambia, N Malawi
__ *N.j.itombwensis* Prigogine, 1977	Itombwe Mts. (E Zaire) (1935)

DREPANORHYNCHUS Fischer & Reichenow, 1884 M
○ ***Drepanorhynchus reichenowi*** Golden-winged Sunbird
__ *D.r.shellyae* (Prigogine, 1952)	E Zaire
__ *D.r.lathburyi* (Williams, 1956)	N Kenya
__ *D.r.reichenowi* Fischer, 1884	S Uganda to C and S Kenya, NE Tanzania

CINNYRIS Cuvier, 1817 M[7]
○ ***Cinnyris chloropygius*** Olive-bellied Sunbird
__ *C.c.kempi* Ogilvie-Grant, 1910	Senegal to SW Nigeria
__ *C.c.chloropygius* (Jardine, 1842)[8]	Bioko, SE Nigeria to NW Angola and W Zaire
__ *C.c.orphogaster* Reichenow, 1899[9]	NE Angola, C and E Zaire, SW Sudan, SW Ethiopia, Uganda, NW Tanzania, W Kenya

○ ***Cinnyris minullus*** Tiny Sunbird
| __ *C.m.amadoni* (Eisentraut, 1965) | Bioko I. |
| __ *C.m.minullus* Reichenow, 1899 | Liberia to Zaire and W Uganda |

[1] Includes *vicina* see Mees (1965) (1516).
[2] Type species of the genus *Chalcostetha* Cabanis, 1850 (which is sometimes recognised as a monotypic genus), a name thought not to have priority over *Leptocoma*.
[3] The original name was restored by Rand (1967) (1961) and we believe has regained prevailing usage and must be used; see Art. 33.2 of the Code (I.C.Z.N., 1999) (1178) and p. 000 herein.
[4] Includes *aeneigularis* see White (1963: 84) (2648).
[5] Includes *major* see White (1963: 83) (2648).
[6] If used in a broad genus *Nectarinia* the name *salvadorii* is preoccupied and must be replaced by *nyikensis* Delacour, 1944. Note the subspecies in *Leptocoma sericea*.
[7] Gender addressed by David & Gosselin (in press) (614), multiple name changes occur.
[8] Includes *insularis* and *luhderi* see White (1963: 75) (2648).
[9] Includes *bineschensis* see Fry (2000: 234) (892).

○ *Cinnyris manoensis* MIOMBO DOUBLE-COLLARED SUNBIRD

   — *C.m.manoensis* Reichenow, 1907[1]   SE Tanzania, SE Zambia, Zimbabwe, N Mozambique

   — *C.m.pintoi* (Wolters, 1965)   Angola, S Zaire, Zambia, N Malawi

   — *C.m.amicorum* (Clancey, 1970)   Mt. Gorongoza (S Mozambique) (426)

○ *Cinnyris chalybeus* SOUTHERN DOUBLE-COLLARED SUNBIRD

   — *C.c.subalaris* Reichenow, 1899[2]   Transvaal, Natal, E Cape Province

   — *C.c.chalybeus* (Linnaeus, 1766)   W and S Cape Province

   — *C.c.albilateralis* (Winterbottom, 1963)   Coastal W Cape Province

○ *Cinnyris neergaardi* C.H.B. Grant, 1908 NEERGAARD'S SUNBIRD   S Mozambique, N Natal

○ *Cinnyris stuhlmanni* RUWENZORI DOUBLE-COLLARED SUNBIRD[3]

   — *C.s.stuhlmanni* Reichenow, 1893   Ruwenzori Mts. (NE Zaire/W Uganda)

   — *C.s.graueri* Neumann, 1908   E Zaire, Rwanda, SW Uganda

   — *C.s.chapini* Prigogine, 1952   E Zaire

   — *C.s.prigoginei* Macdonald, 1958   Murungu Mts. (SE Zaire)

   — *C.s.schubotzi* Reichenow, 1908   E Zaire, Rwanda, W Burundi

○ *Cinnyris ludovicensis* LUDWIG'S DOUBLE-COLLARED SUNBIRD

   — *C.l.ludovicensis* (Bocage, 1870)[4]   Angola

   — *C.l.whytei* Benson, 1948   Zambia, Malawi

○ *Cinnyris reichenowi* NORTHERN DOUBLE-COLLARED SUNBIRD

   — *C.r.preussi* Reichenow, 1892   Bioko I., SE Nigeria, Cameroon

   — *C.r.reichenowi* Sharpe, 1891[5]   NE Zaire to Rwanda, W Uganda, S Sudan, W and C Kenya

○ *Cinnyris afer* GREATER DOUBLE-COLLARED SUNBIRD[6]

   — *C.a.saliens* (Clancey, 1962)   E Cape Province to W Natal and Transvaal

   — *C.a.afer* (Linnaeus, 1766)   S and SW Cape Province

○ *Cinnyris regius* REGAL SUNBIRD

   — *C.r.regius* Reichenow, 1893   Rwenzori Mts. (NE Zaire/W Uganda)

   — *C.r.kivuensis* Schouteden, 1937   E Zaire to SW Uganda

   — *C.r.anderseni* Williams, 1950   W Tanzania

○ *Cinnyris rockefelleri* Chapin, 1932 ROCKEFELLER'S SUNBIRD   E Zaire

○ *Cinnyris mediocris* EASTERN DOUBLE-COLLARED SUNBIRD

   — *C.m.mediocris* Shelley, 1885   W and C Kenya, N Tanzania

   — *C.m.usambaricus* Grote, 1922   SE Kenya, NE Tanzania

   — *C.m.fuelleborni* Reichenow, 1899   C and S Tanzania, N Malawi, NE Zambia

   — *C.m.bensoni* Williams, 1953   Malawi, N Mozambique

○ *Cinnyris moreaui* W.L. Sclater, 1933 MOREAU'S SUNBIRD   NE and E Tanzania

○ *Cinnyris loveridgei* E. Hartert, 1922 LOVERIDGE'S SUNBIRD   Uluguru Mts. (E Tanzania)

○ *Cinnyris pulchellus* BEAUTIFUL SUNBIRD

   — *C.p.pulchellus* (Linnaeus, 1766)[7]   Senegal to W Ethiopia, Uganda and NW Kenya

   — *C.p.melanogastrus* (Fischer & Reichenow, 1884)   W and S Kenya, Tanzania

○ *Cinnyris mariquensis* MARIQUA SUNBIRD

   — *C.m.osiris* (Finsch, 1872)   Ethiopia and Eritrea to N Kenya, N Uganda

   — *C.m.suahelicus* Reichenow, 1891   S Uganda, W and C Kenya, Tanzania, Rwanda, NE Zambia

   — *C.m.mariquensis* A. Smith, 1836   Namibia, Botswana, W Transvaal, Zimbabwe

   — *C.m.lucens* (Clancey, 1973)   SE Zimbabwe, S Mozambique to Natal (441)

   — *C.m.ovamboensis* Reichenow, 1904[8]   NE Namibia, S Angola, SW Zambia, NW Botswana

---

[1] Includes *gertrudis* see Cheke & Mann (2001) (326).

[2] Includes *capricornensis* see White (1963: 73) (2648).

[3] For a discussion of the species concept we follow here see Fry (2000: 244-246) (892).

[4] Brooke (1993) (237) placed *erikssoni* Trimen, 1882 (2438A) in synonymy. We mentioned this form in note 237 in the Appendix to our last Edition.

[5] In a broad genus *Nectarinia* the name *reichenowi* is preoccupied. The next oldest name for the southern form is *kikuyuensis* Mearns, 1915, and the specific name will be that of the northern form.

[6] Our treatment of the previous broad species follows the division in Fry (2000: 245-246) (892).

[7] Includes *aegra* [= *aeger*] and *lucidipectus* see White (1963: 83) (2648).

[8] For recognition see Fry (2000: 263) (892).

○ *Cinnyris shelleyi* SHELLEY'S SUNBIRD
— *C.s.hofmanni* Reichenow, 1915   E Tanzania
— *C.s.shelleyi* Alexander, 1899   SC Tanzania to C Zambia and N Mozambique

○ *Cinnyris congensis* (van Oort, 1910) CONGO SUNBIRD   Zaire

○ *Cinnyris erythrocercus* (Hartlaub, 1857) RED-CHESTED SUNBIRD   S Sudan, Uganda to E Zaire, W Kenya and NW Tanzania

○ *Cinnyris nectarinioides* BLACK-BELLIED SUNBIRD
— *C.n.erlangeri* (Reichenow, 1905)   S Somalia, NE Kenya
— *C.n.nectarinioides* Richmond, 1897   E Kenya to NE Tanzania

○ *Cinnyris bifasciatus* PURPLE-BANDED SUNBIRD
— *C.b.bifasciatus* (Shaw, 1812)   Gabon to W Angola
— *C.b.microrhynchus* Shelley, 1876   S Somalia, coastal Kenya, E Tanzania
— *C.b.strophium* Clancey & Williams, 1957   Uganda and S Tanzania to Zambia, N and E Zimbabwe and Natal

○ *Cinnyris tsavoensis* van Someren, 1922 TSAVO PURPLE-BANDED SUNBIRD[1]
     SE Kenya, NE Tanzania

○ *Cinnyris chalcomelas* Reichenow, 1905 VIOLET-BREASTED SUNBIRD   S Somalia, SE Kenya

○ *Cinnyris pembae* Reichenow, 1905 PEMBA SUNBIRD[2]   Pemba I.

○ *Cinnyris bouvieri* Shelley, 1877 ORANGE-TUFTED SUNBIRD   Cameroon to W Kenya, N Angola and NW Zambia

○ *Cinnyris osea* PALESTINE SUNBIRD
— *C.o.osea* Bonaparte, 1856   Lebanon and SW Syria to Sinai, and W and S Arabia to S Oman
— *C.o.decorsei* Oustalet, 1904   Lake Chad, E Cameroon to W and S Sudan, NE Zaire

○ *Cinnyris habessinicus* SHINING SUNBIRD
— *C.h.kinneari* Bates, 1935   W Saudi Arabia
— *C.h.hellmayri* Neumann, 1904   Extreme SW Saudi Arabia, Yemen, S Oman
— *C.h.habessinicus* (Hemprich & Ehrenberg, 1828)   NE Sudan to N and C Ethiopia
— *C.h.alter* Neumann, 1906[3]   E Ethiopia, N Somalia
— *C.h.turkanae* van Someren, 1920   S Ethiopia, SE Sudan, S Somalia, N Kenya, NE Uganda

○ *Cinnyris coccinigastrus* (Latham, 1802)[4] SPLENDID SUNBIRD   Senegal to NE Zaire

○ *Cinnyris johannae* JOHANNA'S SUNBIRD
— *C.j.fasciatus* (Jardine & Fraser, 1852)   Sierra Leone to Benin
— *C.j.johannae* J. & E. Verreaux, 1851   S Cameroon to Zaire

○ *Cinnyris superbus* SUPERB SUNBIRD
— *C.s.ashantiensis* Bannerman, 1922   Sierra Leone to Togo
— *C.s.nigeriae* (Rand & Traylor, 1959)   S Nigeria
— *C.s.superbus* (Shaw, 1812)   S Cameroon to Zaire, Angola
— *C.s.buvuma* van Someren, 1932   NE Zaire, Uganda

○ *Cinnyris rufipennis* (Jensen, 1983) RUFOUS-WINGED SUNBIRD   Udzungwa Mts. (EC Tanzania) (1206)

○ *Cinnyris oustaleti* OUSTALET'S SUNBIRD
— *C.o.oustaleti* (Bocage, 1878)   C Angola
— *C.o.rhodesiae* Benson, 1955   NE Zambia

○ *Cinnyris talatala* WHITE-BELLIED SUNBIRD
— *C.t.talatala* A. Smith, 1836[5]   SE Tanzania and Mozambique to S Angola, N and E Botswana, N Transvaal
— *C.t.arestus* (Clancey, 1962)   Natal and S Transvaal

○ *Cinnyris venustus* VARIABLE SUNBIRD
— *C.v.venustus* (Shaw & Nodder, 1799)   Senegal to Cameroon
— *C.v.fazoqlensis* (Heuglin, 1871)[6]   SE Sudan, N and W Ethiopia, Eritrea

---

[1] Separation follows Lack (1985) (1312).
[2] This species and the next were treated as distinct by Rand (1967: 264-265) (1961) and this treatment has been broadly adopted, see Fry (2000: 275-276) (892).
[3] For correct spelling see David & Gosselin (2002) (613).
[4] For reasons to use 1802 not 1801 see Browing & Monroe (1991) (258).
[5] Original spelling verified. Includes *anderssoni* see Fry (2000: 282) (892).
[6] Presumably includes *blicki* recognised by Rand (1967: 251) (1961).

— *C.v.albiventris* (Strickland, 1852)   Somalia, E Ethiopia, NE Kenya

— *C.v.falkensteini* Fischer & Reichenow, 1884[1]   W and S Kenya, Tanzania and Zambia to Angola and N Zimbabwe

— *C.v.igneiventris* Reichenow, 1899   Uganda, Rwanda, E Zaire

○ *Cinnyris fuscus* DUSKY SUNBIRD

— *C.f.fuscus* (Vieillot, 1819)   S Angola to S Cape Province

— *C.f.inclusus* (Clancey, 1970)   Mossamedes (Angola) (424)

○ *Cinnyris ursulae* (Alexander, 1903) URSULA'S SUNBIRD   Bioko I., Cameroon Mt.

○ *Cinnyris batesi* Ogilvie-Grant, 1908 BATES'S SUNBIRD   Bioko I., Ivory Coast Zaire, NW Zambia

○ *Cinnyris cupreus* COPPER SUNBIRD

— *C.c.cupreus* (Shaw, 1812)   Senegal to N and E Zaire, W Ethiopia, W Kenya, NW Tanzania

— *C.c.chalceus* (Hartlaub, 1862)   S Zaire, S Tanzania to Angola, Zambia, N Zimbabwe

○ *Cinnyris asiaticus* PURPLE SUNBIRD

— *C.a.brevirostris* (Blanford, 1873)   E United Arab Emirates, N Oman, SE Iran, NE Afghanistan, Pakistan, NW India

— *C.a.asiaticus* (Latham, 1790)   Nepal, NC, C and S India, Sri Lanka

— *C.a.intermedius* (Hume, 1870)   Bangladesh, Assam, Burma, N, E, and SW Thailand, S Indochina

○ *Cinnyris jugularis* OLIVE-BACKED SUNBIRD

— *C.j.andamanicus* (Hume, 1873)   Andamans and Coco Is.

— *C.j.proselius* Oberholser, 1923   Car Nicobar I.

— *C.j.klossi* (Richmond, 1902)   S Nicobars

— *C.j.rhizophorae* (Swinhoe, 1869)[2]   S China (SE Yunnan to Guangdong), Hainan, N Indochina

— *C.j.flammaxillaris* (Blyth, 1845)   Burma, S Indochina, Thailand and N Malay Pen.

— *C.j.ornatus* Lesson, 1827   S Malay Pen., Sumatra, Java, Borneo, Lesser Sundas

— *C.j.polyclystus* Oberholser, 1912   Enggano

— *C.j.aurora* (Tweeddale, 1878)   Palawan and nearby islands

— *C.j.obscurior* Ogilvie-Grant, 1894   N Luzon

— *C.j.jugularis* (Linnaeus, 1766)   S Luzon and rest of Philippines

— *C.j.woodi* (Mearns, 1909)   Sulu Arch.

— *C.j.plateni* (W.H. Blasius, 1885)[3]   Sulawesi, Salayer I.

— *C.j.infrenatus* E. Hartert, 1903   Tukangbesi I.

— *C.j.robustirostris* (Mees, 1964)   Banggai, Sula Is.

— *C.j.teysmanni* Büttikofer, 1893   Djampea, Kalao Is. (off SE Sulawesi)

— *C.j.frenatus* (S. Müller, 1843)[4]   N Moluccas, Aru Is., New Guinea (except WC north) and N and E Queensland

— *C.j.buruensis* E. Hartert, 1910   Buru I.

— *C.j.clementiae* Lesson, 1827   S Moluccas to Watubela Is. (except Buru)

— *C.j.keiensis* Stresemann, 1913   Kai Is.

— *C.j.idenburgi* Rand, 1940   WC northern New Guinea

— *C.j.flavigastra* (Gould, 1843)[5]   Bismarck Arch., Solomon Is.

○ *Cinnyris buettikoferi* E. Hartert, 1896 APRICOT-BREASTED SUNBIRD   Sumba I.

○ *Cinnyris solaris* FLAME-BREASTED SUNBIRD

— *C.s.solaris* (Temminck, 1825)   Sumbawa to Alor and Timor (Lesser Sundas)

— *C.s.exquisita* E. Hartert, 1904[6]   Wetar (Lesser Sundas)

○ *Cinnyris souimanga* MADAGASCAN SUNBIRD[7]

— *C.s.souimanga* (J.F. Gmelin, 1788)   Madagascar (except S), Glorioso I.

---

[1] Includes *niassae* see White (1963: 70) (2648).
[2] Probably includes ¶ *tamdaoensis* Vo Quy, 1981 (2532), described in Vietnamese and not yet evaluated.
[3] Includes *meyeri* see White & Bruce (1986: 405) (2652).
[4] This may well comprise two or more recognisable populations.
[5] For correct spelling see David & Gosselin (2002) (613).
[6] White in White & Bruce (1986: 406) (2652) considered *exquisitus* better treated as a synonym (Bruce disagreed). It is recognised by Cheke & Mann (2001) (326).
[7] Sometimes called Souimanga Sunbird, but Souimanga is the French generic term for sunbird.

__ *C.s.apolis* E. Hartert, 1920	S Madagascar
__ *C.s.aldabrensis* Ridgway, 1895	Aldabra I.
__ *C.s.abbotti* Ridgway, 1895	Assumption I. (Aldabra)
__ *C.s.buchenorum* Williams, 1953	Cosmoledo I. (Aldabra)

○ *Cinnyris notatus* LONG-BILLED GREEN SUNBIRD

__ *C.n.notatus* (Statius Müller, 1776)	Madagascar
__ *C.n.moebii* Reichenow, 1887	Grande Comore I.
__ *C.n.voeltzkowi* Reichenow, 1905	Mwali (Moheli) I. (Comoro Is.)

○ *Cinnyris dussumieri* (Hartlaub, 1860) SEYCHELLES SUNBIRD   Seychelles

○ *Cinnyris humbloti* HUMBLOT'S SUNBIRD

__ *C.h.humbloti* Milne-Edwards & Oustalet, 1885	Grande Comore I.
__ *C.h.mohelicus* Stresemann & Grote, 1926	Mwali (Moheli) I. (Comoro Is.)

○ *Cinnyris comorensis* W.K.H. Peters, 1864 ANJOUAN SUNBIRD   Anjouan I. (Comoro Is.)

○ *Cinnyris coquerellii* (Hartlaub, 1860) MAYOTTE SUNBIRD   Mayotte I. (Comoro Is.)

○ *Cinnyris lotenius* LOTEN'S SUNBIRD

__ *C.l.hindustanicus* Whistler, 1944	S India
__ *C.l.lotenius* (Linnaeus, 1766)	Sri Lanka

AETHOPYGA Cabanis, 1851 F

○ *Aethopyga primigenia* GREY-HOODED SUNBIRD[1]

__ *A.p.diuatae* Salomonsen, 1953	NE Mindanao
__ *A.p.primigenia* (Hachisuka, 1941)	WC and E Mindanao

○ *Aethopyga boltoni* APO SUNBIRD

__ *A.b.malindangensis* Rand & Rabor, 1957	Mts. Katanglad and Malindang (W Mindanao)
__ *A.b.boltoni* Mearns, 1905	Mt. Apo (E Mindanao)
__ *A.b.tibolii* Kennedy, Gonzales & Miranda, 1997	Mts. Busa and Matutum (S Mindanao) (1258)

○ *Aethopyga linaraborae* Kennedy, Gonzales & Miranda, 1997 LINA'S SUNBIRD #

     Mts. Pasian, Putting Bato and Mayo (E Mindanao) (1258)

○ *Aethopyga flagrans* FLAMING SUNBIRD

__ *A.f.decolor* Parkes, 1963	NE Luzon
__ *A.f.flagrans* Oustalet, 1876	C and S Luzon, Catanduanes
__ *A.f.guimarasensis* (Steere, 1890)	Panay, Guimaras
__ *A.f.daphoenonota* Parkes, 1963	Negros

○ *Aethopyga pulcherrima* METALLIC-WINGED SUNBIRD

__ *A.p.jeffreyi* (Ogilvie-Grant, 1894)	N and C Luzon
__ *A.p.pulcherrima* Sharpe, 1876	Samar, Leyte, Mindanao, Basilan
__ *A.p.decorosa* (McGregor, 1907)	Bohol

○ *Aethopyga duyvenbodei* (Schlegel, 1871) ELEGANT SUNBIRD   Sangihe Is

○ *Aethopyga shelleyi* Sharpe, 1876 LOVELY SUNBIRD   Balabac, Palawan

○ *Aethopyga bella* HANDSOME SUNBIRD[2]

__ *A.b.flavipectus* Ogilvie-Grant, 1894	N Luzon
__ *A.b.minuta* Bourns & Worcester, 1894	C and S Luzon, Mindoro, Polillo
__ *A.b.rubrinota* McGregor, 1905	Lubang I.
__ *A.b.bella* Tweeddale, 1877	Samar, Leyte, Mindanao
__ *A.b.bonita* Bourns & Worcester, 1894	Ticao, Masbate, Panay, Negros, Cebu
__ *A.b.arolasi* Bourns & Worcester, 1894	Sulu Arch.

○ *Aethopyga gouldiae* MRS. GOULD'S SUNBIRD

__ *A.g.gouldiae* (Vigors, 1831)	Himalayas to N Assam and SE Xizang
__ *A.g.isolata* E.C.S. Baker, 1925	S and SE Assam, E Bangladesh, W Burma
__ *A.g.dabryii* (J. Verreaux, 1867)	EC and S China, N and C Indochina >> C Burma, N Thailand
__ *A.g.annamensis* Robinson & Kloss, 1919	Bolovens (S Laos), Langbian (S Vietnam)

---

[1] For correction of spelling see David & Gosselin (2002) (613). Type species of the genus *Philippinia*. Not recognised here.
[2] Separation from *A. shelleyi* follows Mann (2002) (1415A).

○ *Aethopyga nipalensis* Green-tailed Sunbird
— *A.n.horsfieldi* (Blyth, 1843)[1] — W Himalayas to W Nepal
— *A.n.nipalensis* (Hodgson, 1836)[2] — C Himalayas (C and E Nepal and Sikkim)
— *A.n.koelzi* Ripley, 1948 — E Bangladesh, Bhutan, and SE Xizang to SW Sichuan, NW Vietnam, and N Burma

— *A.n.victoriae* Rippon, 1904 — W Burma
— *A.n.karenensis* Ticehurst, 1939 — Karen hills (E Burma)
— *A.n.angkanensis* Riley, 1929 — NW Thailand
— *A.n.australis* Robinson & Kloss, 1923 — Pen. Thailand
— *A.n.blanci* Delacour & Greenway, 1939 — NC and C Laos
— *A.n.ezrai* Delacour, 1926 — SE Cambodia, S Vietnam

○ *Aethopyga eximia* (Horsfield, 1821) White-flanked Sunbird — Java

○ *Aethopyga christinae* Fork-tailed Sunbird
— *A.c.latouchii* Slater, 1891 — C, S and SE China, N and C Vietnam, C Laos
— *A.c.sokolovi* Stepanyan, 1985 — S Vietnam (2324)
— *A.c.christinae* Swinhoe, 1869 — Hainan I.

○ *Aethopyga saturata* Black-throated Sunbird
— *A.s.saturata* (Hodgson, 1836) — C Himalayas (Kumaun to Bhutan)
— *A.s.assamensis* (McClelland, 1839) — Assam, Bangladesh, W and N Burma, W Yunnan
— *A.s.galenae* Deignan, 1948 — NW Thailand
— *A.s.petersi* Deignan, 1948 — E Burma, N and C Laos, N Vietnam, S and SE Yunnan to W Guangdong

— *A.s.sanguinipectus* Walden, 1875 — SE Burma
— *A.s.anomala* Richmond, 1900 — S Pen. Thailand
— *A.s.wrayi* Sharpe, 1887 — W Malaysia
— *A.s.ochra* Deignan, 1948 — Bolovens (S Laos), C Vietnam
— *A.s.johnsi* Robinson & Kloss, 1919 — Langbian (S Vietnam)
— *A.s.cambodiana* Delacour, 1948 — SW Cambodia, SE Thailand

○ *Aethopyga siparaja* Crimson Sunbird
— *A.s.vigorsii* (Sykes, 1832)[3] — WC India
— *A.s.seheriae* (Tickell, 1833) — W and C Himalayan foothills to Sikkim and N India
— *A.s.labecula* (Horsfield, 1840)[4][5] — E Himalayas, Bangladesh, NE India, SW, W, N and NE Burma, S Yunnan, NW Laos

— *A.s.owstoni* Rothschild, 1910 — Naozhou I. (S Guangdong)
— *A.s.tonkinensis* E. Hartert, 1917 — Guangxi (S China), NE Vietnam
— *A.s.mangini* Delacour & Jabouille, 1924 — SE Thailand, C and S Indochina
— *A.s.insularis* Delacour & Jabouille, 1928 — Dao Phu Quoc I. (Extreme SW Vietnam)
— *A.s.cara* Hume, 1874 — C, S, E and SE Burma, NW, N and NE Thailand
— *A.s.trangensis* Meyer de Schauensee, 1946 — Pen. Thailand
— *A.s.siparaja* (Raffles, 1822) — S Malay Pen., Sumatra, Borneo
— *A.s.nicobarica* Hume, 1873 — Nicobars
— *A.s.heliogona* Oberholser, 1923 — Java
— *A.s.natunae* Chasen, 1934 — N Natunas
— *A.s.magnifica* Sharpe, 1876 — WC Philippines
— *A.s.flavostriata* (Wallace, 1865) — N Sulawesi
— *A.s.beccarii* Salvadori, 1875 — S Sulawesi

○ *Aethopyga mystacalis* (Temminck, 1822) Javan Sunbird — Java

○ *Aethopyga temminckii* (S. Müller, 1843) Temminck's Sunbird[6] — S Malay Pen., Sumatra, Borneo

---

[1] Dated 1844-45 by Rand (1967: 275) (1961) presumably taken from Baker (1930) (79), but Mayr (1979) (1485) gave 1843 for taxa described in the same part of this article (however, 1844 is possible, and is being researched). Rand also erred in spelling this *horsfieldii*.
[2] Dated 1837 by Rand (1967: 275) (1961), but see www.zoonomen.net (A.P. Peterson *in litt.*).
[3] Treated as a monotypic species in Cheke & Mann (2001) (326); we await the promised explanatory paper before following this.
[4] For recognition see Ripley (1961: 590; 1982: 533) (2042, 2055) who conceded intergradation with *seheriae*, for which reason Deignan (1963) (663) and others lumped the two taxa.
[5] Includes *terglanei* er-Rashid, 1966 (817A) see Ripley (1982: 533) (2055).
[6] For reasons to separate this from *A. mystacalis* see Mees (1986) (1539).

○ *Aethopyga ignicauda* FIRE-TAILED SUNBIRD
 — *A.i.ignicauda* (Hodgson, 1836)[1]   Himalayas, E Bangladesh, Assam, N Burma, W Yunnan
 — *A.i.flavescens* E.C.S. Baker, 1921   Chin Hills (W Burma)

ARACHNOTHERA Temminck, 1826 F
○ *Arachnothera longirostra* LITTLE SPIDERHUNTER
 — *A.l.longirostra* (Latham, 1790)   SW and SE India, C Himalayas and Bangladesh to
                                        Assam, Burma, N and W Thailand, SW Yunnan

 — *A.l.sordida* La Touche, 1921   S Yunnan, NE Thailand, N Indochina
 — *A.l.pallida* Delacour, 1932   SE Thailand, S Indochina
 — *A.l.cinereicollis* (Vieillot, 1819)   Malay Pen., Sumatra and islands off W coast (except
                                            as shown)
 — *A.l.zharina* Oberholser, 1912[2]   Banjak Is. (off W Sumatra)
 — *A.l.niasensis* van Oort, 1910   Nias I. (off W Sumatra)
 — *A.l.prillwitzi* E. Hartert, 1901   Java
 — *A.l.rothschildi* van Oort, 1910   N Natunas
 — *A.l.atita* Oberholser, 1932   S Natunas
 — *A.l.buettikoferi* van Oort, 1910   Borneo
 — *A.l.dilutior* Sharpe, 1876   Palawan
 — *A.l.flammifera* Tweeddale, 1878   Samar, Leyte, Bohol, Mindanao
 — *A.l.randi* Salomonsen, 1953   Basilan

○ *Arachnothera crassirostris* (Reichenbach, 1854) THICK-BILLED SPIDERHUNTER
                                        Malay Pen., Sumatra, Borneo

○ *Arachnothera robusta* LONG-BILLED SPIDERHUNTER
 — *A.r.robusta* Müller & Schlegel, 1845   Malay Pen., Sumatra, Borneo
 — *A.r.armata* Müller & Schlegel, 1845   Java

○ *Arachnothera flavigaster* (Eyton, 1839) SPECTACLED SPIDERHUNTER   Malay Pen., Sumatra, Borneo

○ *Arachnothera chrysogenys* YELLOW-EARED SPIDERHUNTER
 — *A.c.chrysogenys* (Temminck, 1826)   Malay Pen., Sumatra, Java, W Borneo
 — *A.c.harrissoni* Deignan, 1957   E Borneo

○ *Arachnothera clarae* NAKED-FACED SPIDERHUNTER
 — *A.c.luzonensis* Alcasid & Gonzales, 1968   EC and S Luzon (21)
 — *A.c.philippinensis* (Steere, 1890)   Samar, Leyte
 — *A.c.clarae* W.H. Blasius, 1890   Davao (E Mindanao)
 — *A.c.malindangensis* Rand & Rabor, 1957   W Mindanao

○ *Arachnothera modesta* GREY-BREASTED SPIDERHUNTER[3]
 — *A.m.caena* Deignan, 1956   S Burma, SW Thailand, S Indochina
 — *A.m.modesta* (Eyton, 1839)   Malay Pen., W Borneo
 — *A.m.concolor* Snelleman, 1887   Sumatra

○ *Arachnothera affinis* (Sharpe, 1893) STREAKY-BREASTED SPIDERHUNTER
 — *A.a.affinis* (Horsfield, 1821)[4]   Java and Bali
 — *A.a.everetti* (Sharpe, 1893)[5]   N and C Borneo

○ *Arachnothera magna* STREAKED SPIDERHUNTER
 — *A.m.magna* Hodgson, 1836[6]   C and E Himalayas, Assam, E Bangladesh, W and
                                    N Burma, W Yunnan
 — *A.m.aurata* Blyth, 1855   C Burma
 — *A.m.musarum* Deignan, 1956   E Burma, N Thailand, S Yunnan, Guangxi, N and
                                   C Indochina
 — *A.m.remota* Riley, 1940   SC and S Vietnam
 — *A.m.pagodarum* Deignan, 1956   SE Burma, SW Thailand

○ *Arachnothera juliae* Sharpe, 1887 WHITEHEAD'S SPIDERHUNTER   N Borneo

---

[1] Dated 1837 by Rand (1967: 275) (1961) but see www.zoonomen.net (A.P. Peterson *in litt.*).
[2] For recognition see van Marle & Voous (1988) (2468).
[3] For separation of this species and the next see Davison (2000) (622).
[4] Dated 1822 by Rand (1967: 287) (1961), but see Raphael (1970) (1981).
[5] Probably includes *pars* see Davison (2000) (622).
[6] Dated 1837 by Rand (1967: 275) (1961) but see www.zoonomen.net (A.P. Peterson *in litt.*).

PROMEROPIDAE SUGARBIRDS[1] (1: 2)

PROMEROPS Brisson, 1760 M

○ *Promerops cafer* (Linnaeus, 1758) Cape Sugarbird   W and S Cape Province

○ *Promerops gurneyi* Gurney's Sugarbird
    — *P.g.ardens* Friedmann, 1952   E Zimbabwe, W Mozambique
    — *P.g.gurneyi* J. Verreaux, 1871   E Cape Province, Natal, E Transvaal

PASSERIDAE (SPARROWS, SNOWFINCHES AND ALLIES) (11: 40)

PLOCEPASSER A. Smith, 1836 M

○ *Plocepasser mahali* White-browed Sparrow Weaver
    — *P.m.melanorhynchus* Bonaparte, 1850   S Sudan, S Ethiopia, Uganda, Kenya
    — *P.m.propinquatus* Shelley, 1887   S Somalia
    — *P.m.pectoralis* (W.K.H. Peters, 1868)[2]   E Zambia to Malawi, SE Tanzania, Mozambique
    — *P.m.ansorgei* E. Hartert, 1907   S Angola
    — *P.m.terricolor* Clancey, 1979   NE Namibia to W Zimbabwe, NW Transvaal (471)
    — *P.m.stentor* Clancey, 1957   Namibia, NW and N Cape Province, S and SW Botswana
    — *P.m.mahali* A. Smith, 1836   SE Botswana, Transvaal to Orange Free State, NE Cape Province

○ *Plocepasser superciliosus* (Cretzschmar, 1827) Chestnut-crowned Sparrow Weaver[3]
                       Senegal to Ethiopia, Uganda and NW Kenya

○ *Plocepasser donaldsoni* Sharpe, 1895 Donaldson-Smith's Sparrow Weaver
                       SW Ethiopia, N Kenya

○ *Plocepasser rufoscapulatus* Büttikofer, 1888 Chestnut-mantled Sparrow Weaver
                       S and C Angola, SE Zaire, Zambia, Malawi

HISTURGOPS Reichenow, 1887 M

○ *Histurgops ruficauda* Reichenow, 1887 Rufous-tailed Weaver   N Tanzania

PSEUDONIGRITA Reichenow, 1903 F

○ *Pseudonigrita arnaudi* Grey-headed Social Weaver
    — *P.a.australoabyssinicus* Benson, 1942   S Ethiopia
    — *P.a.arnaudi* (Bonaparte, 1850)[4]   SW Sudan, Kenya, Uganda, N Tanzania
    — *P.a.dorsalis* (Reichenow, 1887)[5]   NC Tanzania

○ *Pseudonigrita cabanisi* (Fischer & Reichenow, 1884) Black-capped Social Weaver
                       S Ethiopia, E Kenya, NE Tanzania

PHILETAIRUS A. Smith, 1837 M

○ *Philetairus socius* Sociable Weaver
    — *P.s.geminus* Grote, 1922   N Namibia
    — *P.s.xericus* Clancey, 1989   W Namibia (516)
    — *P.s.socius* (Latham, 1790)[6]   S Namibia, S Botswana, W Transvaal to W Orange Free State, N and C Cape Province

PASSER Brisson, 1760 M

○ *Passer ammodendri* Saxaul Sparrow
    — *P.a.ammodendri* Gould, 1872[7]   Iran and N Afghanistan to SC Kazakhstan
    — *P.a.nigricans* Stepanyan, 1961   SE and E Kazakhstan, N Xinjiang and extreme SW Mongolia (2316)
    — *P.a.stoliczkae* Hume, 1874[8]   Tarim Basin (S Xinjiang) to S Mongolia, W Nei Mongol, Ningxia and N Gansu

---

[1] The affinities of these birds await resolution (R. Bowie *in litt.*). Sibley & Ahlquist (1990) (2261) treated them as related to the nectarinids, but see also Farquhar *et al.* (1996) (829). We follow Fry (2000) (892) and for now accord them family status.
[2] Includes *stridens* Clancey, 1968 (413).
[3] Includes *brunnescens* see White (1963: 162) (2648) for monotypic treatment.
[4] Dated 1851 by Moreau & Greenway (1962: 7) (1625), but see Zimmer (1926) (2707).
[5] Probably includes ¶ *iringae* Ripley & Heinrich, 1966 (2068).
[6] Includes *lepidus* see Clancey (1989) (516). Here we also include *eremnus* as Clancey (*op. cit.*) showed that it is polytopic and that across part of the intervening range, which he ascribed to *socius*, there are intermediate birds. It might be better yet to treat the species as monotypic with mosaic variation according to substrate colour.
[7] Includes *korejewi* see Stepanyan (1990: 600) (2326).
[8] Includes *timidus* see Stepanyan (1961) (2316).

● *Passer domesticus* HOUSE SPARROW[1]
    ∠ *P.d.domesticus* (Linnaeus, 1758)[2]   W and N Europe and N Kazakhstan through N Asia to Kamchatka, N Sakhalin, N Hokkaido and NW Manchuria; introduced Americas, Sth. Africa, Australia

    ∠ *P.d.balearoibericus* von Jordans, 1923   S Europe east to C Turkey (except Italy)
    — *P.d.biblicus* E. Hartert, 1904   SE Turkey and Levant to NW Iran
    — *P.d.hyrcanus* Zarudny & Kudashev, 1916   SE Azerbaijan, N Iran
    — *P.d.persicus* Zarduny & Kudashev, 1916   C Iran east to SW Afghanistan
    — *P.d.indicus* Jardine & Selby, 1835   S Israel and Arabia through S Asia to Laos
    — *P.d.bactrianus* Zarudny & Kudashev, 1916   NE Iran, W Turkmenistan and S Kazakhstan to NW Pakistan, Tien Shan and NW Xinjiang
    — *P.d.parkini* Whistler, 1920   NW and C Himalayas
    — *P.d.hufufae* Ticehurst & Cheesman, 1924   NE Arabia
    — *P.d.tingitanus* Loche, 1867[3]   NW Africa to NE Libya
    — *P.d.niloticus* Nicoll & Bonhote, 1909   Egypt
    — *P.d.rufidorsalis* C.L. Brehm, 1855   Sudan

○ *Passer hispaniolensis* SPANISH SPARROW
    — *P.h.hispaniolensis* (Temminck, 1820)   SW Europe, Nth. Africa, W Asia Minor and the Balkans >> NE Africa, SW Asia
    — *P.h.transcaspicus* Tschusi, 1902   Levant, Cyprus and E Turkey to S Kazakhstan, Xinjiang and Afghanistan >> SW Asia

○ *Passer pyrrhonotus* Blyth, 1845 SIND SPARROW[4]   SE Iran, Pakistan, NW India

○ *Passer castanopterus* SOMALI SPARROW
    — *P.c.fulgens* Friedmann, 1931   Ethiopia, N Kenya
    — *P.c.castanopterus* Blyth, 1855   Somalia

○ *Passer rutilans* RUSSET SPARROW
    — *P.r.cinnamomeus* (Gould, 1836)   NE Afghanistan, Himalayas, S and SE Xizang
    — *P.r.intensior* Rothschild, 1922[5]   S Assam, W, N and E Burma, SC and S China, NW Indochina
    — *P.r.rutilans* (Temminck, 1836)[6]   NC and E China, Taiwan, Korea, N Japan, S Sakhalin, S Kuril Is.

○ *Passer flaveolus* Blyth, 1845[7] PLAIN-BACKED SPARROW   C and E Burma, Thailand (except penin.), S Indochina

○ *Passer moabiticus* DEAD SEA SPARROW
    — *P.m.moabiticus* Tristram, 1864   Israel, Jordan
    — *P.m.mesopotamicus* Zarudny, 1904[8]   S Turkey, N Syria, Iraq, SW Iran
    — *P.m.yatii* Sharpe, 1888   SE Iran, S Afghanistan >> W Pakistan

○ *Passer iagoensis* (Gould, 1838) CAPE VERDE SPARROW[9]   Cape Verde Is.

○ *Passer motitensis* RUFOUS SPARROW[10]
    — *P.m.insularis* Sclater & Hartlaub, 1881   Socotra I. (Yemen)
    — *P.m.hemileucus* Ogilvie-Grant & Forbes, 1900   Abd el Kuri I. (W of Socotra)
    — *P.m.cordofanicus* Heuglin, 1871   W Sudan
    — *P.m.shelleyi* Sharpe, 1891   S Sudan, S Ethiopia, N Uganda, NW Kenya
    — *P.m.rufocinctus* Finsch & Reichenow, 1884   C Kenya to N Tanzania
    — *P.m.subsolanus* Clancey, 1964   SW Zimbabwe, Transvaal, W Orange Free State, Swaziland (400)
    — *P.m.benguellensis* Lynes, 1926   SW Angola
    — *P.m.motitensis* A. Smith, 1848   Namibia, Botswana, N Cape Province

[1] It has been recommended that the name *italiae* should be used for stabilised hybrids found where *domesticus* and *hispaniolensis* meet; the names *maltae* and *brutius* have been used similarly. So, recently, have *payni* Hachisuka, 1926 (1034), and *africanus* Stephan, 1999 (2334). However, Stephan (2000) (2335) considered that *italiae* might prove to be a valid species.
[2] Includes *plecticus* Oberholser, 1974 (1673), see Browning (1978) (246).
[3] Thought to include *maroccanus* Stephan, 1999 (2334).
[4] For reasons to correct the date see earlier footnote about *Mirafra cantillans*.
[5] Probably includes *batangensis* Cheng & Tan, 1963 (338), but this needs confirmation.
[6] Livr. 99 must be dated from 1836 see Dickinson (2001) (739).
[7] For reasons to correct the date see earlier footnote about *Mirafra cantillans*.
[8] For recognition see Cramp *et al.* (1994: 320) (595).
[9] Moreau & Greenway (1962: 15) (1625) and White (1963: 165) (2648) treated a broad species, including the next five listed here. Cramp *et al.* (1994: 331) (595) separated this and treated the others as allospecies. We agree that this should be separated and discuss the others below.
[10] As early as 1980 we split *P. insularis* and *P. rufocinctus*. Since then Summers-Smith (1984) (2374) separated *P. iagoensis* from *P. motitensis*, but treated *insularis* and *rufocinctus* as forms of *motitensis*. Dowsett & Dowsett-Lemaire (1993) (761) preferred not to recognise *insularis* as a species either. We follow them.

○ *Passer melanurus* CAPE SPARROW/MOSSIE
    — *P.m.damarensis* Reichenow, 1902   SW Angola, Namibia, Botswana, SW Zimbabwe, N and
                                              W Transvaal

    — *P.m.melanurus* (Statius Müller, 1776)[1]   Transvaal and Natal south to Cape Province

○ *Passer griseus* NORTHERN GREY-HEADED SPARROW[2]
    — *P.g.swainsonii* (Rüppell, 1840)   NE Sudan, Eritrea, Ethiopia, N Somalia
    — *P.g.gongonensis* (Oustalet, 1890)   SW Ethiopia, N and E Kenya, NE Tanzania
    — *P.g.suahelicus* Reichenow, 1904   SW and S Kenya to C Tanzania
    — *P.g.laeneni* Niethammer, 1955[3]   E Chad
    — *P.g.griseus* (Vieillot, 1817)   Senegal to C Sudan, W Eritrea S to C Ghana, C Nigeria
    — *P.g.ugandae* Reichenow, 1904   S Ghana, S Nigeria to S Sudan S to Angola, Zambia,
                                              Malawi and E to W Kenya, W and S Tanzania

○ *Passer diffusus* SOUTHERN GREY-HEADED SPARROW[4]
    — *P.d.luangwae* Benson, 1956   SE Zambia
    — *P.d.mosambicus* van Someren, 1921   E Tanzania, N Malawi, NE Mozambique
    — *P.d.diffusus* (A. Smith, 1836)   W and S Angola, Namibia, Botswana, W Zimbabwe to
                                              N Cape Province

    — *P.d.stygiceps* Clancey, 1953   N and E Zimbabwe, S Malawi, C and S Mozambique,
                                              E Sth. Africa

○ *Passer simplex* DESERT SPARROW
    — *P.s.zarudnyi* Pleske, 1896   E Turkmenistan, C Uzbekistan (?E Iran)
    — *P.s.saharae* Erlanger, 1899   NW Sahara (Mauritania, Algeria, S Tunisia, W Libya)
    — *P.s.simplex* (M.H.K. Lichtenstein, 1823)   S Sahara (C Mali, Niger, N Chad, NW and C Sudan)

● *Passer montanus* EURASIAN TREE SPARROW
    ✓ *P.m.montanus* (Linnaeus, 1758)   W Europe to Kamchatka and Commander Is. south to
                                              W Asia Minor, N Kazakhstan, Altai, NE Mongolia,
                                              NW Manchuria

    — *P.m.dybowskii* Domaniewski, 1915[5]   C and E Manchuria, Amurland, N Korea
    — *P.m.transcaucasicus* Buturlin, 1906   N and E Asia Minor and Caucasus area to N Iran and
                                              lower Volga R.

    — *P.m.kansuensis* Stresemann, 1932   N and E Qinghai, W Gansu
    — *P.m.dilutus* Richmond, 1895[6]   E Iran and Pakistan N to S Kazakhstan E through
                                              Xinjiang to NW Gansu and W and S Mongolia

    — *P.m.tibetanus* E.C.S. Baker, 1925   S and E Tibetan Plateau to W Sichuan and SE Qinghai
    — *P.m.saturatus* Stejneger, 1885[7]   EC and E China to Hebei; Taiwan, S Korea, N Philip-
                                              pines, Ryukyu Is., Japan, Sakhalin, S Kuril Is.

    — *P.m.hepaticus* Ripley, 1948   NE Assam, NW Burma, SE Xizang
    — *P.m.malaccensis* A.J.C. Dubois, 1885   Himalayan foothills, S Assam, mainland SE Asia,
                                              S Yunnan, Hainan, Sumatra, Java, S Philippines,
                                              Sulawesi and Lesser Sundas

○ *Passer luteus* (M.H.K. Lichtenstein, 1823) SUDAN GOLDEN SPARROW[8]
                                              Mali, N Nigeria to N Sudan, Eritrea

○ *Passer euchlorus* (Bonaparte, 1850)[9] ARABIAN GOLDEN SPARROW   SW Arabia, N Somalia

○ *Passer eminibey* (Hartlaub, 1880) CHESTNUT SPARROW[10]   W and C Sudan to W and S Ethiopia, Uganda, Kenya,
                                              N Tanzania

CARPOSPIZA J.W. von Müller, 1854[11]   F
○ *Carpospiza brachydactyla* (Bonaparte, 1850)[12] PALE ROCK SPARROW   C and N Arabia, Lebanon, and SE Turkey to Iran and
                                              SW and S Turkmenistan >> NE Africa

---

[1] Includes *vicinus* see White (1963: 166) (2648).
[2] We agree with Dowsett & Dowsett-Lemaire (1993) (761) that firm evidence seems to be lacking that would support treating this group as several species. In our previous Edition we had retained the *diffusus* group within *griseus*, but had accepted the split of *swainsonii*, *gongonensis* and *suahelicus* such that we had four species.
[3] May be better treated as a synonym of the nominate form.
[4] For recognition see Dowsett & Dowsett-Lemaire (1993) (761).
[5] For recognition see Cramp *et al.* (1994: 336) (595).
[6] Includes *zaissanensis* see Cramp *et al.* (1994: 351) (595).
[7] Includes *iubilaeus* see Cheng (1987: 938) (333).
[8] Treated in *Auripasser* in last Edition, like *P. euchlorus*, but see Moreau & Greenway (1962: 22) (1625) and Cramp *et al.* (1994: 351) (595).
[9] Dated 1851 by Moreau & Greenway (1962: 22) (1625), but see Zimmer (1926) (2707).
[10] Treated in the genus *Sorelia* in last Edition, but see Moreau & Greenway (1962: 22) (1625) and White (1963: 170) (2648).
[11] Treated in the genus *Petronia* in last Edition, but see Cramp *et al.* (1994: 357) (595).
[12] Dated 1851 by Moreau & Greenway (1962: 23) (1625), but see Zimmer (1926) (2707).

PETRONIA Kaup, 1829 F

○ *Petronia petronia* Rock Sparrow

— *P.p.petronia* (Linnaeus, 1766)   S Europe, Canary Is. and Madeira to W Asia Minor

— *P.p.barbara* Erlanger, 1899   NW Africa to NW Libya

— *P.p.puteicola* Festa, 1894   S Turkey to Syria, N Israel and Jordan

— *P.p.exigua* (Hellmayr, 1902)   C Turkey to N Caucasus, N Iraq and N Iran

— *P.p.kirhizica* Sushkin, 1925[1]   Lower Volga River to Aral Sea

— *P.p.intermedia* E. Hartert, 1901   SW, E and NE Iran and W Turkmenistan to
                     E Kazakhstan and Xinjiang >> Pakistan

— *P.p.brevirostris* Taczanowski, 1874   Altai and Mongolia to Transbaikalia, Nei Mongol and
                     W China

GYMNORIS J.E. Gray, 1831 F

○ *Gymnoris superciliaris* Yellow-throated Petronia

— *G.s.rufitergum* (Clancey, 1964)   S Zaire to SW Tanzania, Angola, NE Namibia,
                     NW Botswana (399)

— *G.s.flavigula* (Sundevall, 1850)   SE Zambia, Zimbabwe to E Botswana, N Sth Africa

— *G.s.bororensis* (Roberts, 1912)   E Tanzania to Mozambique, E Transvaal, NE Natal

— *G.s.superciliaris* Blyth, 1845   Natal to E Cape Province

○ *Gymnoris dentata* (Sundevall, 1850) Bush Petronia[2]   Senegal to Ethiopia, SW Arabia

○ *Gymnoris pyrgita* Yellow-spotted Petronia

— *G.p.pallida* Neumann, 1908   Senegal, Mauritania to SW Sudan

— *G.p.pyrgita* (Heuglin, 1862)   E Sudan to Somalia, Kenya, NE Tanzania

○ *Gymnoris xanthocollis* Yellow-throated Sparrow[3]

— *G.x.transfuga* E. Hartert, 1904   SE Turkey, Kuwait and NE Arabia to S Afghanistan and
                     C and S Pakistan >> NW India

— *G.x.xanthocollis* (E. Burton, 1838)   NE Afghanistan, N Pakistan, S Nepal, India, ? Sri Lanka

MONTIFRINGILLA C.L. Brehm, 1828 F

○ *Montifringilla nivalis* White-winged Snowfinch[4]

— *M.n.nivalis* (Linnaeus, 1766)   S Europe (Spain and Alps to Greece)

— *M.n.leucura* Bonaparte, 1855   S and E Asia Minor

— *M.n.alpicola* (Pallas, 1811)   Caucasus area, NW and N Iran, Afghanistan,
                     W Pamir Mts.

— *M.n.gaddi* Zarudny & Loudon, 1904   SW and S Iran

— *M.n.tianschanica* Keve, 1943   N Tajikistan to W and C Tien Shan

— *M.n.groumgrzimaili* Zarudny & Loudon, 1904[5]   E Tien Shan, Bogdo Shan, Altai, Tuva Rep., Mongolia

— *M.n.kwenlunensis* Bianchi, 1908   S Xinjiang, N Tibet

○ *Montifringilla henrici* (Oustalet, 1891) Tibetan Snowfinch   E Tibetan Plateau (NE Xizang, E Qinghai)

○ *Montifringilla adamsi* Black-winged Snowfinch[6]

— *M.a.xerophila* Stegmann, 1932   SE Xinjiang, W and N Qinghai

— *M.a.adamsi* Adams, 1858   S and E fringe of Tibetan Plateau (Ladakh and
                     N Nepal to W Sichuan, E Qinghai and SW Gansu)

ONYCHOSTRUTHUS Richmond, 1917[7] M

○ *Onychostruthus taczanowskii* (Przevalski, 1876) White-rumped Snowfinch
                     Tibet, Qinghai, SW Gansu, NW Sichuan

PYRGILAUDA J. Verreaux, 1871[8] F

○ *Pyrgilauda davidiana* Père David's Snowfinch

— *P.d.potanini* Sushkin, 1925   SE Russian Altai, Mongolia, SE Transbaikalia

— *P.d.davidiana* J. Verreaux, 1871   E Qinghai to extreme S Mongolia and W and
                     C Nei Mongolia

---

[1] For recognition see Stepanyan (1990: 604) (2326).
[2] Includes *buchanani* see Moreau & Greenway (1962: 26) (1625).
[3] For reasons to separate this from *P. pyrgita* see Cramp *et al.* (1994: 370-371) (595).
[4] Situation controversial. Moreau & Greenway (1962: 28) (1625) accepted two species, *nivalis* and *adamsi* and placed *henrici* within the former. Cramp *et al.* (1994: 400) (595), whom we follow in our recognition of subspecies, lumped *nivalis* and *adamsi* but recognised *henrici*. Support for *henrici* as a species is offered by Martens & Eck (1995) (1443). We tentatively accept three species. See also Gebauer *et al.* (1999) (932).
[5] The spelling *groum-grzimaili* is not permissible with a hyphen see Art. 27 (I.C.Z.N., 1999) (1178).
[6] The distinctness of this species has been discussed by Gebauer *et al.* (1999) (932).
[7] For reasons to use this generic name see Eck (1996) (790).
[8] We follow Stepanyan (1990) (2326) in accepting this genus.

○ *Pyrgilauda ruficollis* Rufous-necked Snowfinch

    — *P.r.isabellina* Stegmann, 1932   S Xinjiang

    — *P.r.ruficollis* (Blanford, 1871)   S and E Tibetan Plateau to N Nepal and W Sichuan

○ *Pyrgilauda blanfordi* Blanford's Snowfinch

    — *P.b.barbata* Przevalski, 1877   E Qinghai

    — *P.b.ventorum* Stegmann, 1932   SE Xinjiang, NW Qinghai

    — *P.b.blanfordi* (Hume, 1876)   Tibetan Plateau (Ladakh and N Nepal to SW Xinjiang and S Qinghai)

○ *Pyrgilauda theresae* (R. Meinertzhagen, 1937) Afghan Snowfinch   C Afghanistan, S Tajikistan >> S Turkmenistan

PLOCEIDAE WEAVERS, SPARROWS (11: 108)

BUBALORNITHINAE

BUBALORNIS A. Smith, 1836 M

○ *Bubalornis albirostris* (Vieillot, 1817) White-billed Buffalo Weaver   Senegal to Eritrea, S Sudan, N Uganda, NW Kenya

○ *Bubalornis niger* Red-billed Buffalo Weaver

    — *B.n.intermedius* (Cabanis, 1868)[1]   SE Sudan, Ethiopia, Somalia, Kenya, Tanzania

    — *B.n.niger* A. Smith, 1836   S Angola, N Namibia to W Zimbabwe, W Transvaal

    — *B.n.militaris* Clancey, 1977   S Zambia to S Mozambique, E Transvaal, N Natal (462)

DINEMELLIA Reichenbach, 1863 F

○ *Dinemellia dinemelli* White-headed Buffalo Weaver

    — *D.d.dinemelli* (Rüppell, 1845)   S Sudan, S Ethiopia, Somalia, NE Uganda, Kenya

    — *D.d.boehmi* (Reichenow, 1885)   SE Zaire, Tanzania

PLOCEINAE

SPOROPIPES Cabanis, 1847 M

○ *Sporopipes squamifrons* Scaly-fronted Weaver

    — *S.s.fuligescens* Clancey, 1957   NE Cape Province, to Transvaal, Zimbabwe, E and N Botswana

    — *S.s.squamifrons* (A. Smith, 1836)[2]   SW Angola to NW and N Cape Province

○ *Sporopipes frontalis* Speckle-fronted Weaver

    — *S.f.frontalis* (Daudin, 1800)[3]   Senegal to Eritrea and N Ethiopia

    — *S.f.pallidior* E. Hartert, 1921   S Sahara

    — *S.f.emini* Neumann, 1900   S Sudan, NE Uganda, Kenya, N Tanzania

AMBLYOSPIZA Sundevall, 1850 F

○ *Amblyospiza albifrons* Grosbeak-Weaver

    — *A.a.capitalba* (Bonaparte, 1850)   Sierra Leone to SW Nigeria

    — *A.a.saturata* Sharpe, 1908   S Nigeria to W and N Zaire

    — *A.a.melanota* (Heuglin, 1863)   NE Zaire to Ethiopia, Uganda, NW Kenya

    — *A.a.montana* van Someren, 1921   E Zaire to Tanzania, SW and C Kenya

    — *A.a.unicolor* (Fischer & Reichenow, 1878)   E Kenya, E Tanzania

    — *A.a.tandae* Bannerman, 1921   NW Angola

    — *A.a.kasaica* Schouteden, 1953   S Zaire

    — *A.a.maxima* Roberts, 1932   N Botswana, SW Zambia

    — *A.a.woltersi* Clancey, 1956   W Mozambique, E Zimbabwe

    — *A.a.albifrons* (Vigors, 1831)   SE Zaire to Malawi, Zimbabwe, S Mozambique, E Sth. Africa

PLOCEUS Cuvier, 1816[4][5] M

○ *Ploceus baglafecht* Baglafecht Weaver

    — *P.b.baglafecht* (Daudin, 1802)   Eritrea, W and C Ethiopia, S Sudan

---

[1] Misplaced in the preceding species in our last Edition.
[2] Presumably includes *pallidus* da Rosa Pinto, 1967 (606), but may need evaluation.
[3] Dated 1802 by Moreau & Greenway (1962: 30) (1625), but this is believed to be wrong as other authors in "Peters" used 1800, see www.zoonomen.net (A.P. Peterson *in litt.*).
[4] Dated 1817 by Moreau & Greenway (1962: 32) (1625), but see Browning & Monroe (1991) (258) and www.zoonomen.net (A.P. Peterson *in litt.*).
[5] We have not listed *P. victoriae* Ash, 1986 (59). The validity of this has been debated, see Louette (1987) (1378A) and Ash (1987) (59A). It may be better considered a *species inquirenda*, see Vuilleumier *et al.* (1992) (2552), but we provisionally treat it as of probable hybrid origin.

__ *P.b.neumanni* (Bannerman, 1923)	Cameroon
__ *P.b.eremobius* (Hartlaub, 1887)	NE Zaire, SW Sudan
__ *P.b.emini* (Hartlaub, 1882)	S Sudan, N Uganda
__ *P.b.reichenowi* (Fischer, 1884)	Kenya, N Tanzania
__ *P.b.stuhlmanni* (Reichenow, 1893)	E Zaire, C and S Uganda, NW Tanzania
__ *P.b.sharpii* (Shelley, 1890)	S and SW Tanzania
__ *P.b.nyikae* Benson, 1938)	NE Zambia, N Malawi

◯ *Ploceus bannermani* Chapin, 1932 BANNERMAN'S WEAVER — W Cameroon

◯ *Ploceus batesi* (Sharpe, 1908) BATES'S WEAVER — S Cameroon

◯ *Ploceus nigrimentus* Reichenow, 1904 BLACK-CHINNED WEAVER[1] — W Angola, W Zaire

◯ *Ploceus bertrandi* (Shelley, 1893) BERTRAM'S WEAVER — Tanzania, Malawi, NW Mozambique

◯ *Ploceus pelzelni* SLENDER-BILLED WEAVER
__ *P.p.pelzelni* (Hartlaub, 1887)[2]	Uganda, W Kenya to NW Tanzania, E Zaire
__ *P.p.monacha* (Sharpe, 1890)[3]	Ghana to Gabon, NW Angola, C and S Zaire

◯ *Ploceus subpersonatus* (Cabanis, 1876) LOANGA WEAVER — W Gabon to Congo mouth

◯ *Ploceus luteolus* LITTLE MASKED WEAVER
__ *P.l.luteolus* (M.H.K. Lichtenstein, 1823)	Senegal to Eritrea south to NE Zaire, N Kenya
__ *P.l.kavirondensis* (van Someren, 1921)	Uganda, W Kenya, NW Tanzania

◯ *Ploceus ocularis* SPECTACLED WEAVER
__ *P.o.crocatus* (Hartlaub, 1881)	Cameroon to W Ethiopia south to Angola, W Zambia, NW Tanzania, W Kenya
__ *P.o.suahelicus* Neumann, 1905	E Kenya and E Tanzania to Zambia, Zimbabwe to Mozambique
__ *P.o.brevior* Lawson, 1962	N Transkei, Natal to E Transvaal (1338)
__ *P.o.tenuirostris* Traylor, 1964	N Botswana, N Namibia (2426)
__ *P.o.ocularis* A. Smith, 1839	E Cape Province

◯ *Ploceus nigricollis* BLACK-NECKED WEAVER
__ *P.n.brachypterus* Swainson, 1837	Senegal to Nigeria
__ *P.n.nigricollis* (Vieillot, 1805)	S Cameroon to S Sudan, Zaire, NW Angola, W Kenya
__ *P.n.po* E. Hartert, 1907	Bioko I.
__ *P.n.melanoxanthus* (Cabanis, 1878)	S Ethiopia, S Somalia, C and E Kenya, NE Tanzania

◯ *Ploceus alienus* (Sharpe, 1902) STRANGE WEAVER — E Zaire, W Uganda

◯ *Ploceus melanogaster* BLACK-BILLED WEAVER
__ *P.m.melanogaster* Shelley, 1887	E Nigeria, W Cameroon, Bioko I.
__ *P.m.stephanophorus* (Sharpe, 1891)	S Sudan, E Zaire, Uganda, W Kenya

◯ *Ploceus capensis* CAPE WEAVER
__ *P.c.rubricomus* Clancey, 1966	E Transvaal, N Natal (409)
__ *P.c.olivaceus* (Hahn, 1822)	E Cape Province, Orange Free State W Natal
__ *P.c.capensis* (Linnaeus, 1766)	W and S Cape Province

◯ *Ploceus temporalis* (Bocage, 1880) BOCAGE'S WEAVER — C Angola to NW Zambia

◯ *Ploceus subaureus* AFRICAN GOLDEN WEAVER
__ *P.s.aureoflavus* A. Smith, 1839	E Kenya, E Tanzania, Malawi, Mozambique
__ *P.s.tongensis* (Roberts, 1931)	S Mozambique
__ *P.s.subaureus* A. Smith, 1839	Natal, E Cape Province

◯ *Ploceus xanthops* (Hartlaub, 1862) HOLUB'S GOLDEN WEAVER — Gabon, S and E Zaire, Uganda and S Kenya south to Angola, N Botswana, Zimbabwe, E Sth. Africa

◯ *Ploceus aurantius* ORANGE WEAVER
__ *P.a.aurantius* (Vieillot, 1805)	Liberia to Cameroon, Gabon, Zaire
__ *P.a.rex* Neumann, 1908	Uganda, NW Tanzania

---

[1] For correction of spelling see David & Gosselin (2002) (613).
[2] Includes *tuta* see White (1963: 129) (2648).
[3] For correction of spelling see David & Gosselin (2002) (613).

○ *Ploceus heuglini* Reichenow, 1886 HEUGLIN'S MASKED WEAVER | Senegal to W and S Sudan, Uganda, NW Kenya

○ *Ploceus bojeri* (Cabanis, 1869) GOLDEN PALM WEAVER | S Somalia, C and SE Kenya, NE Tanzania

○ *Ploceus castaneiceps* (Sharpe, 1890) TAVETA GOLDEN WEAVER | SE Kenya, NE Tanzania

○ *Ploceus princeps* (Bonaparte, 1850)[1] PRINCIPE GOLDEN WEAVER | Principe I.

○ *Ploceus castanops* Shelley, 1888 NORTHERN BROWN-THROATED WEAVER | Uganda to E Zaire, NW Tanzania, W Kenya

○ *Ploceus xanthopterus* SOUTHERN BROWN-THROATED WEAVER
— *P.x.castaneigula* (Cabanis, 1884) | NE Namibia, SE Angola, N Botswana, SW Zambia, NW Zimbabwe
— *P.x.marleyi* (Roberts, 1929) | Natal to S Mozambique
— *P.x.xanthopterus* (Finsch & Hartlaub, 1870) | NE Zimbabwe, Malawi, Mozambique

○ *Ploceus burnieri* N.E. & E.M. Baker, 1990 KILOMBERO WEAVER | EC Tanzania (81)

○ *Ploceus galbula* Rüppell, 1840 RÜPPELL'S WEAVER | NE Sudan to N Ethiopia, N Somalia, SW Arabia to S Oman

○ *Ploceus taeniopterus* NORTHERN MASKED WEAVER
— *P.t.furensis* Lynes, 1923 | W Sudan
— *P.t.taeniopterus* Reichenbach, 1863 | C and S Sudan to N Uganda, SW Ethiopia, NW Kenya

○ *Ploceus intermedius* LESSER MASKED WEAVER
— *P.i.intermedius* Rüppell, 1845 | S Sudan, Ethiopia, Somalia to E Zaire, W and C Tanzania
— *P.i.cabanisii* (W.K.H. Peters, 1868)[2] | SE Zaire, Malawi, E Tanzania to Zimbabwe, Mozambique, Transvaal, Natal, Namibia, S Angola, N Botswana
— *P.i.beattyi* Traylor, 1959 | W Angola

○ *Ploceus velatus* AFRICAN MASKED WEAVER
— *P.v.velatus* Vieillot, 1819[3] | S Angola, S and E Zambia, Malawi and Mozambique to Botswana, Zimbabwe, N, W and S Cape Province, SW and E Transvaal, N Natal
— *P.v.nigrifrons* (Cabanis, 1850) | E Cape Province to Orange Free State, Natal, C Transvaal

○ *Ploceus katangae* KATANGA MASKED WEAVER[4]
— *P.k.upembae* (Verheyen, 1953) | SE Zaire
— *P.k.katangae* (Verheyen, 1947) | NW Zambia, SE Zaire

○ *Ploceus ruweti* Louette & Benson, 1982 LUFIRA MASKED WEAVER | S Zaire (1380)

○ *Ploceus reichardi* Reichenow, 1886 TANZANIAN MASKED WEAVER | SW Tanzania

○ *Ploceus vitellinus* VITELLINE MASKED WEAVER
— *P.v.vitellinus* (M.H.K. Lichtenstein, 1823) | Senegal to Chad, SW Sudan
— *P.v.peixotoi* Frade & de Naurois, 1964 | São Tomé I. (870)
— *P.v.uluensis* (Neumann, 1900) | SE Sudan and Somalia to Tanzania

○ *Ploceus spekei* (Heuglin, 1861) SPEKE'S WEAVER | S and E Ethiopia, Somalia, Kenya, NE Tanzania

○ *Ploceus spekeoides* Grant & Mackworth-Praed, 1947 FOX'S WEAVER | Uganda

○ *Ploceus cucullatus* VILLAGE WEAVER
— *P.c.cucullatus* (Statius Müller, 1776) | Senegal to Cameroon, S Chad, Bioko I.
— *P.c.abyssinicus* (J.F. Gmelin, 1789) | N Sudan to Eritrea, W and C Ethiopia
— *P.c.bohndorffi* Reichenow, 1887 | N Zaire, Uganda, S Sudan, W Kenya, NW Tanzania
— *P.c.frobenii* Reichenow, 1923 | S and SE Zaire
— *P.c.collaris* Vieillot, 1819 | Gabon, W Zaire, N Angola
— *P.c.graueri* E. Hartert, 1911 | E Zaire, Rwanda, W Tanzania
— *P.c.paroptus* Clancey, 1959 | C Mozambique to E Zambia, SE Zaire, S and E Tanzania, E Kenya and S Somalia
— *P.c.nigriceps* (E.L. Layard, 1867)[5] | S Angola, W Zambia, NE Namibia, N Botswana, N Zimbabwe

[1] Dated 1851 by Moreau & Greenway (1962: 42) (1625), but see Zimmer (1926) (2707).
[2] Includes *luebberti* see White (1963: 43) (2648).
[3] Includes *tahatali, caurinus, shelleyi* and *finschi* see Moreau & Greenway (1962: 45) (1625).
[4] For reasons to separate this from *P. velatus* see Louette & Benson (1982) (1380) and Vuilleumier *et al.* (1992) (2552). Both this species and the next require re-evaluation.
[5] Mistakenly treated as a separate species in last Edition.

— *P.c.dilutescens* Clancey, 1956	NE Natal to SE Botswana, S Mozambique
— *P.c.spilonotus* Vigors, 1831	Natal to E Cape Province

○ *Ploceus grandis* (G.R. Gray, 1844) GIANT WEAVER | São Tomé I.

○ *Ploceus nigerrimus* VIEILLOT'S WEAVER

— *P.n.castaneofuscus* Lesson, 1840	Sierra Leone to W Nigeria
— *P.n.nigerrimus* Vieillot, 1819	E Nigeria, Cameroon to W Kenya

○ *Ploceus weynsi* (A.J.C. Dubois, 1900) WEYNS'S WEAVER | N Zaire, S Uganda, NW Tanzania

○ *Ploceus golandi* (Stephenson Clarke, 1913) CLARKE'S WEAVER | E Kenya

○ *Ploceus dicrocephalus* (Salvadori, 1896) JUBA WEAVER | S Ethiopia, S Somalia, NE Kenya

○ *Ploceus melanocephalus* BLACK-HEADED WEAVER

— *P.m.melanocephalus* (Linnaeus, 1758)	Gambia, Guinea-Bissau
— *P.m.capitalis* (Latham, 1790)	N Senegal to S Chad, Nigeria, Central African Republic
— *P.m.duboisi* Hartlaub, 1886	Zaire, N Zambia
— *P.m.dimidiatus* (Antinori & Salvadori, 1873)	NE Sudan, W Eritrea
— *P.m.fischeri* Reichenow, 1887	Uganda to E Zaire, W Kenya, NW Tanzania

○ *Ploceus jacksoni* Shelley, 1888 GOLDEN-BACKED WEAVER | SE Sudan, W and S Kenya, Uganda, N and C Tanzania

○ *Ploceus badius* CINNAMON WEAVER

— *P.b.badius* (Cassin, 1850)	E Sudan
— *P.b.axillaris* (Heuglin, 1867)	S Sudan

○ *Ploceus rubiginosus* CHESTNUT WEAVER

— *P.r.rubiginosus* Rüppell, 1840	Ethiopia, Somalia, SE Sudan, NE Uganda, Kenya, NE Tanzania
— *P.r.trothae* Reichenow, 1905	SW Angola, N Namibia

○ *Ploceus aureonucha* Sassi, 1920 GOLDEN-NAPED WEAVER | NE Zaire

○ *Ploceus tricolor* YELLOW-MANTLED WEAVER

— *P.t.tricolor* (Hartlaub, 1854)	Guinea to Cameroon, Gabon, Angola
— *P.t.interscapularis* Reichenow, 1893	Zaire, S Uganda, NW Angola

○ *Ploceus albinucha* MAXWELL'S WEAVER

— *P.a.albinucha* (Bocage, 1876)	Sierra Leone to Ghana
— *P.a.maxwelli* (Alexander, 1903)	Bioko I.
— *P.a.holomelas* Sassi, 1920	E Nigeria to Gabon, Central African Republic, N Zaire, W Uganda

○ *Ploceus nelicourvi* (Scopoli, 1786) NELICOURVI WEAVER | N and E Madagascar

○ *Ploceus sakalava* SAKALAVA WEAVER[1]

— *P.s.sakalava* Hartlaub, 1861	W and N Madagascar
— *P.s.minor* (Delacour & Berlioz, 1931)	SW Madagascar

○ *Ploceus hypoxanthus* ASIAN GOLDEN WEAVER

— *P.h.hymenaicus* Deignan, 1947	C, S and SE Burma, C Thailand, S Vietnam, Cambodia, S Laos?
— *P.h.hypoxanthus* (Sparrman, 1788)	Sumatra, Java

○ *Ploceus superciliosus* (Shelley, 1873) COMPACT WEAVER[2] | Sierra Leone to SW Ethiopia, Uganda, W Kenya, NW Tanzania, NW Zambia and N Angola

○ *Ploceus benghalensis* (Linnaeus, 1758) BLACK-BREASTED WEAVER | Pakistan, N, NE and C India, Bangladesh, S Yunnan

○ *Ploceus manyar* STREAKED WEAVER

— *P.m.flaviceps* (Lesson, 1831)	Pakistan, India (except NE), Sri Lanka
— *P.m.peguensis* E.C.S. Baker, 1925	NE India, E Bangladesh, Burma
— *P.m.williamsoni* B.P. Hall, 1957	Thailand, S Vietnam
— *P.m.manyar* (Horsfield, 1821)	Java, Bali, Bawean

[1] Treated in the genus *Foudia* in last Edition, but see Moreau & Greenway (1962: 51) (1625).
[2] Sometimes treated in a monotypic genus *Pachyphantes* but see Moreau & Greenway (1962: 52) (1625).

○ *Ploceus philippinus* Baya Weaver
    — *P.p.philippinus* (Linnaeus, 1766)   Pakistan, S Nepal, NW, C and SE India, Sri Lanka
    — *P.p.travencoreensis* Ali & Whistler, 1936   SW India
    — *P.p.burmanicus* Ticehurst, 1932   NE India, Bangladesh, Burma, NW Thailand,
                                          W Yunnan
    — *P.p.angelorum* Deignan, 1956   C Thailand
    — *P.p.infortunatus* E. Hartert, 1902   S Vietnam, Malay Pen., Sumatra, Nias

○ *Ploceus megarhynchus* Finn's Weaver
    — *P.m.megarhynchus* Hume, 1869   WC Himalayas (N Uttar Pradesh and SW Nepal)
    — *P.m.salimalii* Abdulali, 1961[1]   NE India, S Bhutan

○ *Ploceus bicolor* Dark-backed Weaver
    — *P.b.tephronotus* (Reichenow, 1892)[2]   E Nigeria, S Cameroon, Gabon, Bioko I.
    — *P.b.amaurocephalus* (Cabanis, 1880)[3]   N and C Angola
    — *P.b.mentalis* (Hartlaub, 1891)   S Sudan, E Zaire, Uganda, W Kenya
    — *P.b.kigomaensis* (Grant & Mackworth-Praed, 1956)   SE Zaire, E Angola, N Zambia, W Tanzania
    — *P.b.kersteni* (Finsch & Hartlaub, 1870)   S Somalia, E Kenya, E Tanzania
    — *P.b.stictifrons* (Fischer & Reichenow, 1885)   SE Zimbabwe, SE Tanzania, Mozambique, S Malawi
    — *P.b.sylvanus* Clancey, 1977   E Zimbabwe, W Mozambique (457)
    — *P.b.lebomboensis* (Roberts, 1936)   S Mozambique, Swaziland, N Natal
    — *P.b.sclateri* (Roberts, 1931)   E Natal, SE Mozambique
    — *P.b.bicolor* Vieillot, 1819   S Mozambique to E Cape Province

○ *Ploceus preussi* (Reichenow, 1892) Preuss's Weaver   Sierra Leone, S Cameroon to Zaire

○ *Ploceus dorsomaculatus* (Reichenow, 1893) Yellow-capped Weaver   S Cameroon, Congo, E Zaire

○ *Ploceus olivaceiceps* Olive-headed Weaver
    — *P.o.olivaceiceps* (Reichenow, 1899)   SE Tanzania, Malawi, NW Mozambique
    — *P.o.vicarius* Clancey & Lawson, 1966   S Mozambique (529)

○ *Ploceus nicolli* Usambara Weaver
    — *P.n.nicolli* W.L. Sclater, 1931   Usambara (Tanzania)
    — *P.n.anderseni* Franzmann, 1983   Morogoro (Tanzania) (871)

○ *Ploceus insignis* (Sharpe, 1891) Brown-capped Weaver[4]   SE Nigeria to S Sudan, Bioko, Angola, Zaire, W Kenya,
                                            W Tanzania

○ *Ploceus angolensis* (Bocage, 1878) Bar-winged Weaver   Angola, N Namibia, S Zaire, N Zambia

○ *Ploceus sanctaethomae* (Hartlaub, 1848) São Tomé Weaver[5]   São Tomé I. (1086)

○ *Ploceus flavipes* (Chapin, 1916) Yellow-legged Weaver[6]   NE Zaire

MALIMBUS Vieillot, 1805 M
○ *Malimbus coronatus* Sharpe, 1906 Red-crowned Malimbe   S Cameroon to N and NE Zaire

○ *Malimbus cassini* (Elliot, 1859) Black-throated Malimbe   S Cameroon, Gabon, Congo to W and N Zaire

○ *Malimbus racheliae* (Cassin, 1857) Rachel's Malimbe   SE Nigeria to Gabon

○ *Malimbus ballmanni* Wolters, 1974 Tai Malimbe   SW Ivory Coast (2684)

○ *Malimbus scutatus* Red-vented Malimbe
    — *M.s.scutatus* (Cassin, 1849)   Sierra Leone to Ghana
    — *M.s.scutopartitus* Reichenow, 1894   S Nigeria, W Cameroon

○ *Malimbus ibadanensis* Elgood, 1958 Ibadan Malimbe   SW Nigeria

○ *Malimbus nitens* (J.E. Gray, 1831) Blue-billed Malimbe[7]   Guinea Bissau to S Nigeria, Zaire, W Uganda

---

[1] Date 1960 in Moreau & Greenway (1962) (1625); this issue of the JBNHS was published in January, 1961 (A. Pittie *in litt.*).
[2] Includes *analogus* see Moreau & Greenway (1962: 54) (1625).
[3] Thought to include *albidigularis* Ripley & Heinrich, 1966 (2067), but needs evaluation.
[4] Includes *unicus* see White (1963: 145) (2648).
[5] May deserve a monotypic genus *Thomasophantes*.
[6] Treated in the genus *Malimbe* in last Edition, but both Moreau & Greenway (1962: 55) (1625) and White (1963: 146) (2648) treated this as a member of the genus *Ploceus* and we now follow.
[7] We follow Moreau & Greenway (1962: 58) (1625) in treating this as monotypic; *microrhynchus* and *moreaui* are synonyms.

○ *Malimbus rubricollis* RED-HEADED MALIMBE
— *M.r.bartletti* Sharpe, 1890   Sierra Leone to Ghana
— *M.r.nigeriae* Bannerman, 1921   Benin, SW Nigeria
— *M.r.rubricollis* (Swainson, 1838)   SE Nigeria to Zaire, S Sudan, Uganda, W Kenya
— *M.r.rufovelatus* (Fraser, 1842)   Bioko I.
— *M.r.praedi* Bannerman, 1921   NW Angola

○ *Malimbus erythrogaster* Reichenow, 1893 RED-BELLIED MALIMBE   SE Nigeria to Gabon, N, C and E Zaire, W Uganda

○ *Malimbus malimbicus* CRESTED MALIMBE
— *M.m.nigrifrons* (Hartlaub, 1855)   Sierra Leone to S Nigeria
— *M.m.malimbicus* (Daudin, 1800)[1]   S Cameroon to Zaire, W Uganda, NW Angola

ANAPLECTES Reichenbach, 1863 M
○ *Anaplectes melanotis* RED-HEADED WEAVER[2]
— *A.m.melanotis* Lafresnaye, 1840   Senegal to Ethiopia, N Somalia, Uganda, Kenya,
                                        N Tanzania; also W Angola
— *A.m.jubaensis* van Someren, 1920   S Somalia, NE Kenya
— *A.m.rubriceps* (Sundevall, 1850)[3]   N Namibia, S Angola to Transvaal, Mozambique,
                                        SE Zaire, S Tanzania

QUELEA Reichenbach, 1850 F
○ *Quelea cardinalis* CARDINAL QUELEA
— *Q.c.cardinalis* (Hartlaub, 1880)   S Sudan, S Ethiopia, Uganda, NW Kenya, Rwanda,
                                        NW Tanzania
— *Q.c.rhodesiae* Grant & Mackworth-Praed, 1944   SE Kenya, Tanzania, Zambia

○ *Quelea erythrops* (Hartlaub, 1848) RED-HEADED QUELEA[4]   Senegal to Ethiopia, Angola, Natal and E Cape Province

○ *Quelea quelea* RED-BILLED QUELEA
— *Q.q.quelea* (Linnaeus, 1758)   Senegal to Chad, Central African Republic
— *Q.q.aethiopica* (Sundevall, 1850)   Sudan to Somalia, E Zaire to Kenya, Tanzania,
                                        NE Zambia
— *Q.q.lathamii* (A. Smith, 1836)[5]   Angola, S Zaire, Malawi, Mozambique to Sth. Africa

FOUDIA Reichenbach, 1850 F
○ *Foudia madagascariensis* (Linnaeus, 1766) RED FODY   Madagascar, Mauritius, Réunion Is., Chagos Arch.

○ *Foudia eminentissima* RED-HEADED FODY[6]
— *F.e.aldabrana* Ridgway, 1893   Aldabra I.
— *F.e.consobrina* Milne-Edwards & Oustalet, 1885   Grande Comore I.
— *F.e.anjuanensis* (Milne-Edwards & Oustalet, 1888)   Anjouan I.
— *F.e.eminentissima* Bonaparte, 1850[7]   Mwali (Moheli) I.
— *F.e.algondae* (Schlegel, 1866)   Mayotte I.

○ *Foudia omissa* Rothschild, 1912 FOREST FODY   NW, NE and E Madagascar

○ *Foudia rubra* (J.F. Gmelin, 1789) MAURITIUS FODY   Mauritius

○ *Foudia sechellarum* E. Newton, 1867 SEYCHELLES FODY   Seychelles

○ *Foudia flavicans* E. Newton, 1865 RODRIGUEZ FODY   Rodriguez I.

BRACHYCOPE Reichenow, 1900 F
○ *Brachycope anomala* (Reichenow, 1887) BOB-TAILED WEAVER   SE Cameroon, Congo

EUPLECTES Swainson, 1829 M
○ *Euplectes afer* YELLOW-CROWNED BISHOP
— *E.a.strictus* Hartlaub, 1857   Highlands of Ethiopia

---

[1] Dated 1802 by Moreau & Greenway (1962: 60) (1625), but this is believed to be wrong as other authors in "Peters" used 1800, see www.zoonomen.net (A.P. Peterson *in litt.*).
[2] Although the specific name *rubriceps* was used by Moreau & Greenway this was due to treatment in a broadened genus *Malimbus*; we do not believe the name *melanotis* is preoccupied in *Anaplectes*. Use of *Malimbus* requires that the western form be named *leuconotus* see Moreau & Greenway (1962: 60) (1625).
[3] Includes *gurneyi* see Moreau & Greenway (1962: 60) (1625).
[4] When treated as monotypic includes *viniceps* Clancey, 1986 (502).
[5] Includes *spoliator* see Moreau & Greenway (1962: 62) (1625).
[6] Proposals to split this require more detailed argument.
[7] Dated 1851 by Moreau & Greenway (1962: 63) (1625), but see Zimmer (1926) (2707).

— *E.a.afer* (J.F. Gmelin, 1789)	Senegal to W Sudan, south to Zaire and NW Angola
— *E.a.ladoensis* Reichenow, 1885[1]	S Sudan, SW Ethiopia, Uganda, W and C Kenya, N Tanzania
— *E.a.taha* A. Smith, 1836	SW Angola, Zambia and SW Tanzania S to Sth. Africa

○ *Euplectes diadematus* Fischer & Reichenow, 1878 Fire-fronted Bishop
  S Somalia, E Kenya, NE Tanzania

○ *Euplectes gierowii* Black Bishop
  — *E.g.ansorgei* (E. Hartert, 1899) S Sudan, C Ethiopia to E Zaire, Uganda, W Kenya
  — *E.g.friederichseni* Fischer & Reichenow, 1884 S Kenya, N Tanzania
  — *E.g.gierowii* Cabanis, 1880 Angola

○ *Euplectes nigroventris* Cassin, 1848 Zanzibar Red Bishop SE Kenya, E Tanzania, N Mozambique, Zanzibar

○ *Euplectes hordeaceus* Black-winged Bishop
  — *E.h.hordeaceus* (Linnaeus, 1758) Senegal to W Sudan, Zaire, Tanzania, SE Kenya, south to Angola, Zimbabwe, N Mozambique
  — *E.h.craspedopterus* (Bonaparte, 1850) S Sudan, SW Ethiopia, Uganda, W Kenya

○ *Euplectes orix* Southern Red Bishop
  — *E.o.nigrifrons* (Böhm, 1884) E Zaire to SW Kenya, Tanzania, E Zambia, N and C Mozambique
  — *E.o.sundevalli* Bonaparte, 1850 N Transvaal, Natal, Zimbabwe, S Mozambique
  — *E.o.orix* (Linnaeus, 1758) N Namibia, N Botswana to Angola, W Zambia, S Zaire
  — *E.o.turgidus* Clancey, 1958 S Namibia, Cape Province to W Natal, S Transvaal

○ *Euplectes franciscanus* Northern Red Bishop[2]
  — *E.f.franciscanus* (Isert, 1789) Senegal to Ethiopia, Uganda, NW Kenya
  — *E.f.pusillus* (E. Hartert, 1901) SE Ethiopia, Somalia

○ *Euplectes aureus* (J.F. Gmelin, 1789) Golden-backed Bishop São Tomé I., W Angola

○ *Euplectes capensis* Yellow Bishop
  — *E.c.phoenicomerus* G.R. Gray, 1862 E Nigeria, Cameroon
  — *E.c.xanthomelas* Rüppell, 1840 Ethiopia
  — *E.c.angolensis* Neunzig, 1928[3] Angola
  — *E.c.crassirostris* (Ogilvie-Grant, 1907) E Zaire to Kenya, south to NE Botswana, Zimbabwe, Transvaal, Mozambique
  — *E.c.approximans* (Cabanis, 1851) E Cape Province to Natal, W Swaziland, SE Transvaal
  — *E.c.capensis* (Linnaeus, 1766) SW and S Cape Province
  — *E.c.macrorhynchus* Roberts, 1919 NW Cape Province

○ *Euplectes axillaris* Fan-tailed Widowbird
  — *E.a.bocagei* (Sharpe, 1871)[4] Niger, Cameroon, W and S Zaire, Angola, W Zambia, N Botswana
  — *E.a.traversii* (Salvadori, 1888) N and C Ethiopia
  — *E.a.phoeniceus* (Heuglin, 1862) Sudan, Uganda, W Kenya, W Tanzania, E Zaire, NE Zambia
  — *E.a.zanzibaricus* (Shelley, 1881) S Somalia, E Kenya, E Tanzania and offshore islands
  — *E.a.quanzae* (E. Hartert, 1903) W Angola
  — *E.a.axillaris* (A. Smith, 1838) E Zambia, Malawi, Mozambique and E South Africa

○ *Euplectes macroura* Yellow-mantled Widowbird[5]
  — *E.m.macrocercus* (M.H.K. Lichtenstein, 1823) Ethiopian Highlands, W Kenya
  — *E.m.macroura* (J.F. Gmelin, 1789)[6] Senegal to S Sudan, Uganda, SW Kenya south to Angola, Zimbabwe, Mozambique
  — *E.m.conradsi* (Berger, 1908) Ukerewe I. (Lake Victoria)

---

[1] Submerged in *taha* by Craig (1993) (585). We defer this until 'Birds of Africa' appears.
[2] For reasons to separate this from *E. orix* see Craig (1993) (585) and Zimmerman *et al.* (1996) (2735).
[3] For recognition see Moreau & Greenway (1962: 69) (1625).
[4] Includes *batesi* see Moreau & Greenway (1962: 70) (1625).
[5] For correction of spelling see David & Gosselin (2002) (613).
[6] Includes *intermedius* see White (1963: 157) (2648)

○ *Euplectes hartlaubi* HARTLAUB'S MARSH WIDOWBIRD
— *E.h.humeralis* (Sharpe, 1901)   Cameroon to W Zaire, Uganda, W Kenya
— *E.h.hartlaubi* (Bocage, 1878)   Angola, S Zaire, N Zambia

○ *Euplectes psammocromius* (Reichenow, 1900) MONTANE MARSH WIDOWBIRD
                                                     E Zaire, NE Zambia, N Malawi, SW Tanzania

○ *Euplectes albonotatus* WHITE-WINGED WIDOWBIRD
— *E.a.eques* (Hartlaub, 1863)[1]   W Sudan, C Ethiopia to Uganda, E Zaire, Kenya,
                                                       N and C Tanzania
— *E.a.asymmetrurus* (Reichenow, 1892)   Gabon to Angola and N Namibia
— *E.a.albonotatus* (Cassin, 1848)   Zambia, SE Zaire, S Tanzania to E Botswana, Transvaal
                                                       and Natal

○ *Euplectes ardens* RED-COLLARED WIDOWBIRD[2]
— *E.a.concolor* (Cassin, 1848)   Gambia, Sierra Leone to S Sudan, NW Uganda, N and
                                                       W Zaire
— *E.a.laticauda* (M.H.K. Lichtenstein, 1823)   SE Sudan, Ethiopia
— *E.a.suahelicus* (van Someren, 1921)   W and C Kenya, N Tanzania
— *E.a.tropicus* (Reichenow, 1904)[3]   Uganda, Tanzania, SE Kenya to Angola, S Zaire,
                                                       Zimbabwe, C Mozambique
— *E.a.ardens* (Boddaert, 1783)   Transvaal, S Mozambique to E Cape Province

○ *Euplectes progne* LONG-TAILED WIDOWBIRD
— *E.p.delamerei* (Shelley, 1903)   C Kenya
— *E.p.delacouri* Wolters, 1953[4]   Angola, S Zaire, Zambia
— *E.p.progne* (Boddaert, 1783)   E South Africa

○ *Euplectes jacksoni* (Sharpe, 1891) JACKSON'S WIDOWBIRD   W and C Kenya, N Tanzania

ESTRILDIDAE WAXBILLS, GRASS FINCHES, MUNIAS AND ALLIES (26: 130)

ESTRILDINAE

PARMOPTILA Cassin, 1859 F
○ *Parmoptila woodhousei* WOODHOUSE'S ANTPECKER
— *P.w.woodhousei* Cassin, 1859   SE Nigeria, Cameroon, W Zaire
— *P.w.ansorgei* E. Hartert, 1904   NW Angola

○ *Parmoptila rubrifrons* RED-FRONTED ANTPECKER[5]
— *P.r.rubrifrons* (Sharpe & Ussher, 1872)   Liberia and Ghana
— *P.r.jamesoni* (Shelley, 1890)   N and E Zaire, W Uganda

NIGRITA Strickland, 1843 M[6]
○ *Nigrita fusconotus* WHITE-BREASTED NEGROFINCH
— *N.f.uropygialis* Sharpe, 1869   Guinea to SW Nigeria
— *N.f.fusconotus* Fraser, 1843   Bioko I., SE Nigeria to Uganda, W Kenya, N Angola

○ *Nigrita bicolor* CHESTNUT-BREASTED NEGROFINCH
— *N.b.bicolor* (Hartlaub, 1844)   Guinea to Ghana
— *N.b.brunnescens* Reichenow, 1902   S Nigeria to W Uganda and N Angola

○ *Nigrita luteifrons* PALE-FRONTED NEGROFINCH
— *N.l.luteifrons* J. & E. Verreaux, 1851   S Nigeria to N Zaire and NW Angola
— *N.l.alexanderi* Ogilvie-Grant, 1907   Bioko I.

○ *Nigrita canicapillus* GREY-HEADED NEGROFINCH[7]
— *N.c.emilae* Sharpe, 1869   Guinea to Togo
— *N.c.canicapillus* (Strickland, 1841)   S Nigeria to Gabon, N and C Zaire

[1] Includes *sassii* see White (1963: 152) (2648).
[2] Colour morphs occur and the nomenclature may need to change to better reflect real racial variation.
[3] For recognition see Clancey (1968) (415), but for a contrary view see Craig (1993) (586).
[4] Thought to include *definita* Clancey, 1970 (420), but needs re-evaluation.
[5] In last Edition we erroneously used the specific name *jamesoni*; *rubrifrons* has priority.
[6] Gender addressed by David & Gosselin (2002) (614), multiple name changes occur.
[7] For correction of spelling see David & Gosselin (2002) (613).

___ *N.c.schistaceus* Sharpe, 1891[1]   S Sudan, E Zaire, Uganda, W Kenya, NW Tanzania
___ *N.c.angolensis* Bannerman, 1921   S Zaire, NW Angola
___ *N.c.diabolicus* (Reichenow & Neumann, 1895)   C Kenya, N Tanzania
___ *N.c.candidus* Moreau, 1942   W Tanzania

NESOCHARIS Alexander, 1903   F
○ *Nesocharis shelleyi* SHELLEY'S OLIVEBACK
___ *N.s.shelleyi* Alexander, 1903   Bioko I., Mt. Cameroon
___ *N.s.bansoensis* Bannerman, 1923   SE Nigeria, Cameroon

○ *Nesocharis ansorgei* (E. Hartert, 1899) WHITE-COLLARED OLIVEBACK   E Zaire, Rwanda, SW Uganda

○ *Nesocharis capistrata* (Hartlaub, 1861) GREY-HEADED OLIVEBACK   Gambia to SW Sudan, NW Uganda

PYTILIA Swainson, 1837   F
○ *Pytilia lineata* Heuglin, 1863 ETHIOPIAN AURORA FINCH[2]   Highlands of Ethiopia

○ *Pytilia phoenicoptera* RED-WINGED PYTILIA
___ *P.p.phoenicoptera* Swainson, 1837   Senegal to N Nigeria
___ *P.p.emini* E. Hartert, 1899   N Cameroon to N Uganda and S Sudan

○ *Pytilia hypogrammica* Sharpe, 1870 YELLOW-WINGED/RED-FACED PYTILIA
     Sierra Leone to Central African Republic

○ *Pytilia afra* (J.F. Gmelin, 1789) ORANGE-WINGED PYTILIA   Ethiopia, S Sudan to Tanzania, S Zaire, Angola, N Transvaal and Mozambique

○ *Pytilia melba* GREEN-WINGED PYTILIA[3]
___ *P.m.citerior* Strickland, 1852   Senegal to C Sudan
___ *P.m.jessei* Shelley, 1903   Coastal Sudan, Eritrea, NW Somalia
___ *P.m.soudanensis* (Sharpe, 1890)   SE Sudan, Ethiopia, C and S Somalia to N Kenya
___ *P.m.percivali* van Someren, 1919   C and SW Kenya, NC Tanzania
___ *P.m.belli* Ogilvie-Grant, 1907   E Zaire to W Uganda, W and N Tanzania, SW Kenya
___ *P.m.grotei* Reichenow, 1919   E Tanzania, N Mozambique, E Malawi
___ *P.m.hygrophila* Irwin & Benson, 1967   N Zambia, N Malawi
___ *P.m.melba* (Linnaeus, 1758)[4]   W and S Zaire, SW Tanzania to S Mozambique, N Sth. Africa

AMADINA Swainson, 1827[5]   F
○ *Amadina erythrocephala* (Linnaeus, 1758) RED-HEADED FINCH[6]   S Angola, Namibia, Botswana, Sth. Africa

○ *Amadina fasciata* CUT-THROAT
___ *A.f.fasciata* (J.F. Gmelin, 1789)   Senegal and N Nigeria to Sudan, Uganda
___ *A.f.alexanderi* Neumann, 1908   Ethiopia, Somalia, Kenya, NE Tanzania
___ *A.f.meridionalis* Neunzig, 1910   S Angola, N Namibia to W Zimbabwe, N Mozambique
___ *A.f.contigua* Clancey, 1970   Transvaal, N Natal to S Mozambique (427)

MANDINGOA E. Hartert, 1919   F
○ *Mandingoa nitidula* GREEN-BACKED TWINSPOT
___ *M.n.schlegeli* (Sharpe, 1870)   Sierra Leone to Zaire, Uganda, NW Angola
___ *M.n.virginiae* (Amadon, 1953)   Bioko I.
___ *M.n.chubbi* (Ogilvie-Grant, 1912)   S Ethiopia, Kenya, Tanzania to NE Zambia, N Malawi
___ *M.n.nitidula* (Hartlaub, 1865)   SE Zaire to S Malawi, Mozambique and E Cape Province

CRYPTOSPIZA Salvadori, 1884   F
○ *Cryptospiza reichenovii* RED-FACED CRIMSON-WING
___ *C.r.reichenovii* (Hartlaub, 1874)   Bioko, Cameroon, W Angola
___ *C.r.australis* Shelley, 1896[7]   E Zaire, SW Uganda and Tanzania to Malawi, E Zimbabwe, N Mozambique

[1] Implicitly includes *sparsimguttatus*; although recognised by Mayr *et al.* (1968: 310) (1497) later authors such as Goodwin (1982) (963) have withheld recognition.
[2] For reasons to separate see Nicolai (1968) (1648).
[3] The name *flavicaudata* Welch & Welch, 1988, (2571), has been considered an invalid name by Payne (1979) (1796). This view, which we follow here, may be open to challenge.
[4] Includes *thamnophila* see White (1963: 188) (2648).
[5] Placement here follows Goodwin (1982) (963) and Baptista *et al.* (1999) (95).
[6] Includes *dissita* see White (1963: 186) (2648).
[7] Includes *ocularis* and *homogenes* Clancey, 1971 (430) (DJP).

○ *Cryptospiza salvadorii* ETHIOPIAN CRIMSON-WING
    — *C.s.salvadorii* Reichenow, 1892[1]   S Ethiopia, N Kenya
    — *C.s.ruwenzori* W.L. Sclater, 1925   E Zaire, SW Uganda
    — *C.s.kilimensis* Moreau & Sclater, 1934   SE Sudan, Kenya, N Tanzania

○ *Cryptospiza jacksoni* Sharpe, 1902 DUSKY CRIMSON-WING   E Zaire, Rwanda, SW Uganda

○ *Cryptospiza shelleyi* Sharpe 1902 SHELLEY'S CRIMSON-WING   E Zaire, Rwanda, SW Uganda

PYRENESTES Swainson, 1837 M
○ *Pyrenestes ostrinus* BLACK-BELLIED SEEDCRACKER[2]
    — *P.o.sanguineus* Swainson, 1837   Senegal to Guinea-Bissau
    — *P.o.coccineus* Cassin, 1848   Sierra Leone, Liberia
    — *P.o.ostrinus* (Vieillot, 1805)[3]   Ghana to Uganda, N Angola, N Zambia
    — *P.o.minor* Shelley, 1894   E Tanzania, Malawi, E Zimbabwe, N Mozambique

SPERMOPHAGA Swainson, 1837 F
○ *Spermophaga poliogenys* (Ogilvie-Grant, 1906) GRANT'S BLUEBILL   E Zaire, W Uganda

○ *Spermophaga haematina* WESTERN BLUEBILL
    — *S.h.haematina* (Vieillot, 1805)   Gambia to Ghana
    — *S.h.togoensis* (Neumann, 1910)   Togo to SW Nigeria
    — *S.h.pustulata* (Voigt, 1831)   SE Nigeria, Cameroon to C and S Zaire, NW Angola

○ *Spermophaga ruficapilla* RED-HEADED BLUEBILL
    — *S.r.ruficapilla* (Shelley, 1888)[4]   NW Angola, S and E Zaire east to S Sudan, Uganda,
                                        W Kenya
    — *S.r.cana* (Friedmann, 1927)   NE Tanzania

CLYTOSPIZA Shelley, 1896 F
○ *Clytospiza monteiri* (Hartlaub, 1860) BROWN TWINSPOT   E Nigeria to S Sudan, Uganda

HYPARGOS Reichenbach, 1862-63 M
○ *Hypargos margaritatus* (Strickland, 1844) PINK-THROATED TWINSPOT   N Natal, S Mozambique

○ *Hypargos niveoguttatus* PETERS'S TWINSPOT
    — *H.n.macrospilotus* Mearns, 1913[5]   Kenya, Tanzania to Zambia, Malawi, N Mozambique
    — *H.n.niveoguttatus* (W.K.H. Peters, 1868)   S Mozambique, E Zimbabwe, S Malawi

EUSCHISTOSPIZA Wolters, 1943 F
○ *Euschistospiza dybowskii* (Oustalet, 1892) DYBOWSKI'S TWINSPOT   Sierra Leone to SW Sudan and extreme W Uganda

○ *Euschistospiza cinereovinacea* DUSKY TWINSPOT
    — *E.c.cinereovinacea* (Sousa, 1889)   W Angola
    — *E.c.graueri* (Rothschild, 1909)   E Zaire to SW Uganda

LAGONOSTICTA Cabanis, 1851 F
○ *Lagonosticta rara* BLACK-BELLIED FIREFINCH
    — *L.r.forbesi* Neumann, 1908   Sierra Leone to Nigeria
    — *L.r.rara* (Antinori, 1864)   N Cameroon to S Sudan, Uganda, W Kenya

○ *Lagonosticta rufopicta* BAR-BREASTED FIREFINCH
    — *L.r.rufopicta* (Fraser, 1843)   Senegal to N Cameroon, Central African Republic
    — *L.r.lateritia* Heuglin, 1864   Sudan, NE Zaire, N Uganda, W Kenya

○ *Lagonosticta nitidula* Hartlaub, 1886 BROWN FIREFINCH[6]   E and S Angola, SE Zaire, N and SW Zambia to
                                          N Botswana, NW Zimbabwe

---

[1] Presumably includes *crystallochresta* Desfayes, 1975 (695), but may need evaluation.
[2] For reasons to treat this as a broad species see Smith (1990) (2282).
[3] Includes *rothschildi* and *frommi* see White (1963: 180) (2648).
[4] Includes *kilgoris* Cunningham-van Someren & Schifter, 1981 (602) see Zimmerman *et al.* (1996) (2735).
[5] Includes *idius* and *interior* see White (1963: 177) (2648) and probably ¶*baddeleyi* Wolters, 1972 (2683). Indeed the species may be better treated as monotypic.
[6] Includes *plumbaria* see White (1963: 202) (2648).

○ *Lagonosticta senegala* RED-BILLED FIREFINCH
   — *L.s.senegala* (Linnaeus, 1766)[1]   Senegal to Cameroon
   — *L.s.rhodopsis* (Heuglin, 1863)   S Mauritania, N Senegal to Sudan, W Ethiopia and Eritrea
   — *L.s.brunneiceps* Sharpe, 1890   Highlands of Ethiopia
   — *L.s.somaliensis* Salvadori, 1894   SE Ethiopia to Somalia, E Kenya, E Tanzania
   — *L.s.ruberrima* Reichenow, 1903[2]   E and S Zaire, Uganda to C Kenya, Tanzania, NE Zambia, N Malawi
   — *L.s.rendalli* E. Hartert, 1898[3]   Angola, SE Zaire, Zambia to S Tanzania, Mozambique, Namibia, Transvaal and Natal

○ *Lagonosticta sanguinodorsalis* Payne, 1998 ROCK FIREFINCH #   N Nigeria (1801)

○ *Lagonosticta umbrinodorsalis* Reichenow, 1910 CHAD FIREFINCH[4]   E Nigeria, Cameroon, Chad

○ *Lagonosticta virata* Bates, 1932 MALI FIREFINCH   Mali, Senegal

○ *Lagonosticta rubricata* AFRICAN FIREFINCH
   — *L.r.polionota* Shelley, 1873   Guinea to Nigeria
   — *L.r.ugandae* Salvadori, 1906   Cameroon to C Kenya, N Tanzania, Ethiopia
   — *L.r.congica* Sharpe, 1890   Gabon to S Zaire, NW Zambia
   — *L.r.haematocephala* Neumann, 1907   S Tanzania to Malawi, E Zimbabwe, C Mozambique
   — *L.r.rubricata* (M.H.K. Lichtenstein, 1823)   S Mozambique to Cape Province

○ *Lagonosticta landanae* Sharpe, 1890 PALE-BILLED FIREFINCH[5]   Lower Congo, NW Angola

○ *Lagonosticta rhodopareia* JAMESON'S FIREFINCH[6]
   — *L.r.rhodopareia* (Heuglin, 1868)   S Ethiopia, NW Kenya
   — *L.r.jamesoni* Shelley, 1882[7]   S Kenya to Zambia, Transvaal, N Natal
   — *L.r.ansorgei* Neumann, 1908   Lower Congo to W Angola

○ *Lagonosticta larvata* BLACK-FACED FIREFINCH[8]
   — *L.l.larvata* (Rüppell, 1840)   W and C Ethiopia, SE Sudan
   — *L.l.vinacea* (Hartlaub, 1857)   Senegal to Guinea
   — *L.l.togoensis* (Neumann, 1907)   Ghana, Togo to N Cameroon, W Sudan
   — *L.l.nigricollis* Heuglin, 1863   Central African Republic to S Sudan and N Uganda

URAEGINTHUS Cabanis, 1851 M
○ *Uraeginthus angolensis* BLUE-BREASTED CORDON-BLEU/BLUE WAXBILL
   — *U.a.angolensis* (Linnaeus, 1758)   SW Zaire, N Angola, NW Zambia
   — *U.a.cyanopleurus* Wolters, 1963   S Angola, SW Zambia, W Zimbabwe, N Botswana, W Transvaal
   — *U.a.niassensis* Reichenow, 1911   E and S Tanzania, to C and E Zambia, Malawi, Mozambique
   — *U.a.natalensis* Zedlitz, 1911   S Zimbabwe, Transvaal, Natal

○ *Uraeginthus bengalus* RED-CHEEKED CORDON-BLEU
   — *U.b.bengalus* (Linnaeus, 1766)   Senegal to Ethiopia, Uganda, W Kenya
   — *U.b.brunneigularis* Mearns, 1911   C Kenya
   — *U.b.littoralis* van Someren, 1922   SE Kenya, NE Tanzania
   — *U.b.ugogoensis* Reichenow, 1911   N, W and C Tanzania
   — *U.b.katangae* Vincent, 1934   S Zaire, N Zambia

○ *Uraeginthus cyanocephalus* (Richmond, 1897) BLUE-CAPPED CORDON-BLEU[9]   S Somalia, S Ethiopia to NE Tanzania

○ *Uraeginthus granatinus* VIOLET-EARED WAXBILL[10] [11]
   — *U.g.granatinus* (Linnaeus, 1766)   S Angola, W Zambia, N and E Botswana to Transvaal, NE Cape Province

[1] Includes *guineensis* Hald-Mortensen, 1970 (1056) (R.B. Payne *in litt.*).
[2] Includes *kikuyuensis* see White (1963: 203) (2648).
[3] Includes *pallidicrissa* and *confidens* Clancey, 1979 (476).
[4] In last Edition listed as *L. rhodopareia bruneli* Érard & Roche, 1977 (813), for the applicability of Reichenow's prior name see Payne & Louette (1983) (1802).
[5] May be better treated as a race of *L. rubricata* (R.B. Payne *in litt.*).
[6] For *bruneli* see *umbrinodorsalis*.
[7] Includes *taruensis* see White (1963: 203) (2648).
[8] We revert to a broad species including *vinacea* (prematurely split in last Edition).
[9] For correction of spelling see David & Gosselin (2002) (613).
[10] For correction of spellings of all three taxa see David & Gosselin (2002) (613).
[11] The generic name *Granatina* may deserve reinstatement for this species and the next.

___ *U.g.siccatus* (Clancey, 1959)   W Angola to W Botswana, NW Cape Province

___ *U.g.retusus* (Clancey, 1961)   S Mozambique

○ *Uraeginthus ianthinogaster* Reichenow, 1879 PURPLE GRENADIER[1]   Somalia, E and S Ethiopia to N and C Tanzania

ESTRILDA Swainson, 1827 F

○ *Estrilda caerulescens* (Vieillot, 1817) RED-TAILED LAVENDER WAXBILL   Senegal to Central African Republic

○ *Estrilda perreini* BLACK-TAILED LAVENDER WAXBILL

___ *E.p.perreini* (Vieillot, 1817)   Gabon, S Zaire to N Angola, Zambia and S Tanzania

___ *E.p.incana* Sundevall, 1850[2]   S Malawi, Mozambique, E Zimbabwe to Natal

○ *Estrilda thomensis* Sousa, 1888 CINDERELLA WAXBILL   São Tomé I.

○ *Estrilda melanotis* SWEE WAXBILL

___ *E.m.bocagei* (Shelley, 1903)   W Angola

___ *E.m.melanotis* (Temminck, 1823)   Zimbabwe, Transvaal to E and S Cape Province

○ *Estrilda quartinia* YELLOW-BELLIED WAXBILL

___ *E.q.quartinia* Bonaparte, 1850[3]   Ethiopia, Eritrea, SE Sudan

___ *E.q.kilimensis* (Sharpe, 1890)[4]   E Zaire to C Kenya, N Tanzania

___ *E.q.stuartirwini* Clancey, 1969   E and S Tanzania, E Zambia to E Zimbabwe (418)

○ *Estrilda poliopareia* Reichenow, 1902 ANAMBRA WAXBILL   S Nigeria

○ *Estrilda paludicola* FAWN-BREASTED WAXBILL

___ *E.p.paludicola* Heuglin, 1863   N Zaire, N Uganda, S Sudan, W Kenya

___ *E.p.ochrogaster* Salvadori, 1897   Ethiopia, SE Sudan

___ *E.p.roseicrissa* Reichenow, 1892   E Zaire to SW Uganda, NW Tanzania

___ *E.p.marwitzi* Reichenow, 1900   C Tanzania

___ *E.p.benguellensis* Neumann, 1908   S Zaire, Angola, N Zambia

___ *E.p.ruthae* Chapin, 1950   C Zaire

○ *Estrilda melpoda* ORANGE-CHEEKED WAXBILL

___ *E.m.melpoda* (Vieillot, 1817)   Gambia to Zaire, N Angola, NE Zambia

___ *E.m.tschadensis* Grote, 1922   N Cameroon, Chad

○ *Estrilda rhodopyga* CRIMSON-RUMPED WAXBILL

___ *E.r.rhodopyga* Sundevall, 1850   N Sudan, Eritrea, N Ethiopia, NW Somalia

___ *E.r.centralis* Kothe, 1911   S Ethiopia and SE Sudan to Malawi and E Zaire

○ *Estrilda rufibarba* (Cabanis, 1851) ARABIAN WAXBILL   SW Arabia

○ *Estrilda troglodytes* (M.H.K. Lichtenstein, 1823) BLACK-RUMPED WAXBILL
                                Senegal to W Ethiopia, Uganda and W Kenya

○ *Estrilda astrild* COMMON WAXBILL

___ *E.a.kempi* Bates, 1930   Sierra Leone, Guinea, Liberia

___ *E.a.occidentalis* Jardine & Fraser, 1851   Bioko I., Ivory Coast to N Zaire

___ *E.a.sousae* Reichenow, 1904   São Tomé I.

___ *E.a.peasei* Shelley, 1903   Ethiopia

___ *E.a.macmillani* Ogilvie-Grant, 1907   Sudan

___ *E.a.adesma* Reichenow, 1916[5]   E Zaire to Uganda, NW Tanzania, W Kenya

___ *E.a.massaica* Neumann, 1907   C Kenya to N Tanzania

___ *E.a.minor* (Cabanis, 1878)   S Somalia, E Kenya, NE Tanzania, Zanzibar

___ *E.a.cavendishi* Sharpe, 1900   SE Zaire, S Tanzania to Zambia, Zimbabwe, Mozambique

___ *E.a.schoutedeni* Wolters, 1962   S Zaire

___ *E.a.neidiecki* Reichenow, 1916[6]   C Angola to SW Zambia, N Botswana, W Zimbabwe

___ *E.a.angolensis* Reichenow, 1902   W Angola

___ *E.a.jagoensis* Alexander, 1898   W Angola, Cape Verde Is.

---

[1] Treated as monotypic by White (1963: 198) (2648) and by Zimmerman *et al.* (1986) (2735); *hawkeri*, *roosevelti* and *rothschildi* are synonyms.

[2] Includes *poliogastra* see Goodwin (1982) (963) and implicitly *torrida* Clancey, 1974 (444) which appears to have been a name for the same population.

[3] Dated '1851?' in Mayr *et al.* (1968: 338) (1497), but see Zimmer (1926) (2707).

[4] Listed under *E. melanotis* in last Edition; but see Goodwin (1982) (963) for an indication of its affinity.

[5] A prior name *nyansae* Neumann, 1907, is preoccupied; if the broad genus *Estrilda* used here is narrowed the preoccupation preventing its use may no longer apply.

[6] Includes *ngamiensis* see White (1963: 194) (2648).

— *E.a.rubriventris* (Vieillot, 1823)   Coastal Gabon to NW Angola
— *E.a.damarensis* Reichenow, 1902   Namibia
— *E.a.astrild* (Linnaeus, 1758)   S Botswana, W Transvaal, Orange Free State, W Cape Province
— *E.a.tenebridorsa* Clancey, 1957   E Cape Province, SE Transvaal, Natal

○ *Estrilda nigriloris* Chapin, 1928   BLACK-LORED WAXBILL   SE Zaire

○ *Estrilda nonnula*   BLACK-CROWNED WAXBILL
— *E.n.elizae* Alexander, 1903   Bioko I.
— *E.n.eisentrauti* Wolters, 1964   Mt. Cameroon
— *E.n.nonnula* Hartlaub, 1883   E Cameroon to S Sudan, E Zaire, NW Tanzania and W Kenya

○ *Estrilda atricapilla*   BLACK-HEADED WAXBILL
— *E.a.atricapilla* J. & E. Verreaux, 1851   SE Nigeria to Gabon, NW Zaire
— *E.a.marungensis* Prigogine, 1975   Marungu Mts. (SE Zaire) (1934)
— *E.a.avakubi* Traylor, 1964   S and C Zaire, NE Angola
— *E.a.kandti* Reichenow, 1902   E Zaire to Uganda, Rwanda and Kenya

○ *Estrilda erythronotos*   BLACK-FACED WAXBILL
— *E.e.delamerei* Sharpe, 1900   Uganda, Kenya, Tanzania
— *E.e.soligena* Clancey, 1964[1]   Angola, Namibia to N Transvaal, N Zimbabwe
— *E.e.erythronotos* (Vieillot, 1817)   S Zimbabwe, Transvaal, N Cape Province

○ *Estrilda charmosyna*   BLACK-CHEEKED WAXBILL/RED-RUMPED WAXBILL
— *E.c.charmosyna* (Reichenow, 1881)[2]   Somalia, S Ethiopia, S Sudan, Uganda, N Kenya
— *E.c.kiwanukae* van Someren, 1919   S Kenya, N Tanzania

AMANDAVA Blyth, 1836   F
○ *Amandava amandava*   RED AVADAVAT
— *A.a.amandava* (Linnaeus, 1758)   Pakistan, India, S Nepal
— *A.a.flavidiventris* (Wallace, 1863)   S Yunnan, Burma, NW, C Thailand, W and C Lesser Sundas
— *A.a.punicea* (Horsfield, 1821)[3]   Cambodia, S Vietnam, Java, Bali

○ *Amandava formosa* (Latham, 1790)   GREEN AVADAVAT   C India

○ *Amandava subflava*   ZEBRA WAXBILL
— *A.s.subflava* (Vieillot, 1819)   Senegal to Ethiopia, Uganda
— *A.s.niethammeri* Wolters, 1969   S Zaire to Zambia, N Malawi, NW Botswana, NE Namibia (2682)
— *A.s.clarkei* (Shelley, 1903)   Kenya and Tanzania to SE Zambia, S Malawi, Mozambique, Zimbabwe, SE Botswana and E South Africa

ORTYGOSPIZA Sundevall, 1850   F
○ *Ortygospiza atricollis*   AFRICAN QUAILFINCH
— *O.a.atricollis* (Vieillot, 1817)   Senegal to Chad, N Zaire
— *O.a.ansorgei* Ogilvie-Grant, 1910   Guinea to Ivory Coast
— *O.a.ugandae* van Someren, 1921   S Sudan, Uganda, W Kenya
— *O.a.fuscocrissa* Heuglin, 1863   Eritrea, Ethiopia
— *O.a.muelleri* Zedlitz, 1911[4]   S Kenya, Tanzania, Malawi to C Namibia, S Botswana, W Transvaal
— *O.a.smithersi* Benson, 1955   NE Zambia
— *O.a.pallida* Roberts, 1932   N Botswana
— *O.a.digressa* Clancey, 1958   S, E and N Cape Province to S Mozambique, E Zimbabwe

○ *Ortygospiza gabonensis*   BLACK-CHINNED QUAILFINCH/RED-BILLED QUAILFINCH
— *O.g.gabonensis* Lynes, 1914   Gabon to C Zaire
— *O.g.fuscata* W.L. Sclater, 1932   N Angola, S Zaire, W Zambia
— *O.g.dorsostriata* van Someren, 1921   E Zaire to S and W Uganda

[1] Doubtfully distinct. (DJP)
[2] Includes *pallidior* see Zimmermann *et al.* (1996) (2735).
[3] Dated 1831 by Mayr *et al.* (1968: 348) (1497), a typographical error.
[4] Includes *bradfieldi* see White (1963: 209) (2648).

731

○ *Ortygospiza locustella* Locust Finch[1]
  — *O.l.uelensis* (Chapin, 1916)   N Zaire
  — *O.l.locustella* (Neave, 1909)   S Zaire, Zambia, Malawi, NW Mozambique, Zimbabwe

POEPHILINAE[2]

EMBLEMA Gould, 1842 N
○ *Emblema pictum* Gould, 1842 Painted Finch[3]   WC to EC Australia

STAGONOPLEURA Reichenbach, 1850[4] F
○ *Stagonopleura bella* Beautiful Firetail
  — *S.b.bella* (Latham, 1802)[5]   SE New South Wales, SE Victoria, Tasmania, Flinders I.
  — *S.b.interposita* Schodde & Mason, 1999 SE South Australia, SW Victoria (2189)
  — *S.b.samueli* (Mathews, 1912)[6]   Mt. Lofty Range to Kangaroo I. (SC South Australia)

○ *Stagonopleura oculata* (Quoy & Gaimard, 1830) Red-eared Firetail
                    SW Western Australia

○ *Stagonopleura guttata* (Shaw, 1796) Diamond Firetail E to SC Australia

OREOSTRUTHUS De Vis, 1898 M
○ *Oreostruthus fuliginosus* Mountain Firetail
  — *O.f.pallidus* Rand, 1940   Mts. of WC New Guinea
  — *O.f.hagenensis* Mayr & Gilliard, 1954 Mts. of EC New Guinea
  — *O.f.fuliginosus* (De Vis, 1897)   Mts. of SE New Guinea

NEOCHMIA G.R. Gray, 1849 F
◉ *Neochmia temporalis* Red-browed Finch[7]
  — *N.t.minor* (A.J. Campbell, 1901)   Cape York Pen.
  ✓ *N.t.temporalis* (Latham, 1802)[8]   E to SC Australia, Kangaroo I.

○ *Neochmia phaeton* Crimson Finch
  — *N.p.evangelinae* D'Albertis & Salvadori, 1879 S New Guinea, Cape York Pen.
  — *N.p.phaeton* (Hombron & Jacquinot, 1841)[9] N Western Australia (Kimberley), N Northern Territory
                    (Arnhem Land), NW and NE Queensland

○ *Neochmia ruficauda* Star Finch
  — *N.r.subclarescens* (Mathews, 1912)[10] WC and NW Western Australia, N Northern Territory
                    (Arnhem Land), S Gulf of Carpentaria
  — *N.r.clarescens* (E. Hartert, 1899)   S Cape York Pen. to NE Queensland
  — †?*N.r.ruficauda* (Gould, 1837)   EC Queensland

○ *Neochmia modesta* (Gould, 1837) Plum-headed Finch[11] EC Queensland, E New South Wales

POEPHILA Gould, 1842 F
○ *Poephila personata* Masked Finch
  — *P.p.personata* Gould, 1842   N Western Australia (Kimberley), N Northern Territory
                    (Arnhem Land)
  — *P.p.leucotis* Gould, 1847   Cape York Pen.

○ *Poephila acuticauda* Long-tailed Finch
  — *P.a.acuticauda* (Gould, 1840)   N Western Australia (Kimberley)
  — *P.a.hecki* Heinroth, 1900   N Northern Territory (Arnhem Land), NW Queensland

[1] In previous Editions we included a 'race' *rendalli*; this was a misinterpretation of a *lapsus* by Mackworth-Praed & Grant. It has no relevance to this species or indeed this family.
[2] For elevation of the tribes recognised by Mayr *et al.* (1968) (1497) to subfamilies, and for the resurrection of several of the generic names used herewith and some related changes of association and sequence, see Christidis (1987a, b) (349, 350).
[3] For correction of spelling see David & Gosselin (2002) (613).
[4] Placed within the genus *Emblema* in last Edition, but see Schodde & Mason (1999: 744) (2189).
[5] For reasons to use 1802 not 1801 see Browing & Monroe (1991) (258).
[6] For recognition see Schodde & Mason (1999: 745) (2189).
[7] Treated in the monotypic genus *Aegintha* in last Edition, but see Schodde & Mason (1999: 748) (2189).
[8] For reasons to use 1802 not 1801 see Browing & Monroe (1991) (258). Includes *loftyi* see Schodde & Mason (1999: 749) (2189).
[9] Includes *albiventer* Schodde & Mason (1999: 751) (2189).
[10] For recognition see Schodde & Mason (1999: 752) (2189).
[11] Treated in the monotypic genus *Aidemosyne* in last Edition, but see Schodde & Mason (1999: 754) (2189).

○ *Poephila cincta* BLACK-THROATED FINCH
    — *P.c.atropygialis* Diggles, 1876[1]   Cape York Pen., NE Queensland
    — *P.c.cincta* (Gould, 1837)   EC Queensland, NE New South Wales

TAENIOPYGIA Reichenbach, 1862-63[2] F
○ *Taeniopygia guttata* ZEBRA FINCH
    — *T.g.guttata* (Vieillot, 1817)   Lesser Sundas
    — *T.g.castanotis* (Gould, 1837)[3]   Inland Australia

○ *Taeniopygia bichenovii* DOUBLE-BARRED FINCH[4]
    — *T.b.annulosa* (Gould, 1840)   N Western Australia (Kimberley), N Northern Territory
                                       (Arnhem Land)

    — *T.b.bichenovii* (Vigors & Horsfield, 1827)   E Australia

LONCHURINAE[5]

ERYTHRURA Swainson, 1837 F
○ *Erythrura hyperythra* TAWNY-BREASTED PARROTFINCH
    — *E.h.brunneiventris* (Ogilvie-Grant, 1894)   N and C Luzon, Mindoro
    — *E.h.borneensis* (Sharpe, 1889)   Borneo
    — *E.h.malayana* (Robinson, 1928)   Mts. W Malaysia
    — *E.h.hyperythra* (Reichenbach, 1863)   W Java
    — *E.h.microrhyncha* (Stresemann, 1931)   Sulawesi
    — *E.h.intermedia* (E. Hartert, 1896)   W Lesser Sundas

○ *Erythrura prasina* PIN-TAILED PARROTFINCH
    — *E.p.prasina* (Sparrman, 1788)   NE Thailand, N Laos, C Vietnam, Malay Pen.,
                                   Sumatra, Java[6]

    — *E.p.coelica* E.C.S. Baker, 1925   Borneo

○ *Erythrura viridifacies* Hachisuka & Delacour, 1937 GREEN-FACED PARROTFINCH
                                    Luzon, Negros

○ *Erythrura tricolor* (Vieillot, 1817) TRICOLOR PARROTFINCH   Timor to Tanimbar (E Lesser Sundas)

○ *Erythrura coloria* Ripley & Rabor, 1961 RED-EARED PARROTFINCH   Mts. Apo and Katanglad (Mindanao)

○ *Erythrura trichroa* BLUE-FACED PARROTFINCH
    — *E.t.sanfordi* Stresemann, 1931   NC and SC Sulawesi
    — *E.t.modesta* Wallace, 1862   N Moluccas
    — *E.t.pinaiae* Stresemann, 1914   S Moluccas
    — *E.t.sigillifera* (De Vis, 1897)   Mts. of New Guinea, D'Entrecasteaux and Louisiade
                                     Arch., New Britain, New Ireland
    — *E.t.macgillivrayi* Mathews, 1914[7]   NE Queensland
    — *E.t.eichhorni* E. Hartert, 1924   St. Matthias Is. (Bismarck Arch.)
    — *E.t.pelewensis* N. Kuroda, Sr., 1922   Palau Is.
    — *E.t.clara* Taka-Tsukasa & Yamashina, 1931   Truk, Pohnpei (Caroline Is.)
    — *E.t.trichroa* (Kittlitz, 1835)   Kusaie (Caroline Is.)
    — *E.t.woodfordi* E. Hartert, 1900   Guadalcanal I. (SE Solomons)
    — *E.t.cyanofrons* E.L. Layard, 1878   Banks Is., Vanuatu, Loyalty Is.

○ *Erythrura papuana* E. Hartert, 1900 PAPUAN PARROTFINCH   Mts. of New Guinea

○ *Erythrura psittacea* (J.F. Gmelin, 1789) RED-THROATED PARROTFINCH New Caledonia

○ *Erythrura cyaneovirens* RED-HEADED PARROTFINCH[8]
    — *E.c.regia* (P.L. Sclater, 1881)   Banks Is., N Vanuatu
    — *E.c.efatensis* Mayr, 1931   Efate I. (S Vanuatu)

---

[1] Includes *nigrotecta* see Schodde & Mason (1999: 760) (2189).
[2] Treated in the genus *Poephila* in last Edition, but see Schodde & Mason (1999: 761) (2189).
[3] We tentatively follow Schodde & Mason (1999: 760) (2189) who view this as "approaching speciation".
[4] This species may be better placed in the genus *Poephila* (R.B. Payne *in litt.*), or perhaps *Stizoptera* should be retained as a monotypic genus.
[5] We tentatively include the parrotfinches here following Christidis (1987a, b) (349, 350).
[6] Movements or migrations not yet understood.
[7] For recognition see Schodde & Mason (1999: 765) (2189).
[8] Pending further evidence we prefer to retain one species for the birds of Vanuatu, Fiji and Samoa.

— *E.c.serena* (P.L. Sclater, 1881) — Aneiteum I. (S Vanuatu)
— *E.c.pealii* Hartlaub, 1852 — Fiji Is.
— *E.c.gaughrani* duPont, 1972 — Savaii I. (W Samoa) (772)
— *E.c.cyaneovirens* (Peale, 1848) — Upolu I. (W. Samoa)

○ *Erythrura kleinschmidti* (Finsch, 1879) PINK-BILLED PARROTFINCH — Viti Levu I. (W Fiji)

○ *Erythrura gouldiae* (Gould, 1844) GOULDIAN FINCH[1] — N Western Australia (Kimberley), N Northern Territory (Arnhem Land), N Queensland (W Cape York Pen.)

LONCHURINAE (MUNIAS OR MANNIKINS)

LONCHURA Sykes, 1832[2] F
○ *Lonchura nana* (Pucheran, 1845) MADAGASCAN MUNIA[3] — Madagascar

○ *Lonchura cantans* AFRICAN SILVERBILL
— *L.c.cantans* (J.F. Gmelin, 1789) — Senegal to W Sudan
— *L.c.orientalis* (Lorenz & Hellmayr, 1901) — SW Arabia to S Oman, Somalia, Eritrea, Ethiopia to NE Tanzania

○ *Lonchura malabarica* (Linnaeus, 1758) INDIAN SILVERBILL/WHITE-THROATED MUNIA — C and NE Sauda Arabia, N Oman, and SE Iran to Indian subcontinent and Sri Lanka

○ *Lonchura griseicapilla* Delacour, 1943 GREY-HEADED SILVERBILL[4] — S Ethiopia to Tanzania

○ *Lonchura cucullata* BRONZE MANNIKIN
— *L.c.cucullata* (Swainson, 1837) — Senegal to S Sudan, Uganda, W Kenya
— *L.c.scutata* (Heuglin, 1863)[5] — Ethiopia to E Angola and E Sth. Africa

○ *Lonchura bicolor* BLACK-AND-WHITE MANNIKIN
— *L.b.bicolor* (Fraser, 1843) — Guinea to Cameroon
— *L.b.poensis* (Fraser, 1843)[6] — Bioko to N Angola, S Ethiopia, W Kenya, NW Tanzania
— *L.b.nigriceps* (Cassin, 1852)[7] — S Somalia, C and E Kenya, Tanzania, E Zaire to NE Sth. Africa
— *L.b.woltersi* Schouteden, 1956 — SE Zaire, NW Zambia

○ *Lonchura fringilloides* (Lafresnaye, 1835) MAGPIE-MANNIKIN[8] — Senegal to Uganda, Angola, Mozambique and Natal

○ *Lonchura striata* WHITE-RUMPED MUNIA
— *L.s.acuticauda* (Hodgson, 1836) — N India, Bangladesh, Nepal, Burma, N Thailand, SW Yunnan
— *L.s.striata* (Linnaeus, 1766) — C and S India, Sri Lanka
— *L.s.fumigata* (Walden, 1873) — Andamans
— *L.s.semistriata* (Hume, 1874) — Nicobars
— *L.s.subsquamicollis* (E.C.S. Baker, 1925)[9] — SE Yunnan, Hainan, SE Thailand, Malay Pen., Sumatra, Indochina
— *L.s.swinhoei* (Cabanis, 1882) — EC and E China, Taiwan

○ *Lonchura leucogastroides* (Moore, 1856)[10] JAVAN MUNIA — S Sumatra, Java, Bali, Lombok

○ *Lonchura fuscans* (Cassin, 1852) DUSKY MUNIA — Borneo; Cagayan Sulu[11] (Philippines)

○ *Lonchura molucca* (Linnaeus, 1766) BLACK-FACED MUNIA — Talaud, Sangihe Is., Sulawesi, islands in Flores Sea, Moluccas to Kai Is., Lesser Sundas (Sumbawa to Tanimbar)

[1] Treated in a monotypic genus *Chloebia* in last Edition, but see Christidis (1987a, b) (349, 350).
[2] Various African species may be better treated as a genus *Spermestes* see Baptista *et al.* (1999) (95).
[3] With *Padda* considered to belong within *Lochura* (see below) it seems better to treat this within a broad genus *Lonchura*, see also Dowsett & Dowsett-Lemaire (1993) (761); the generic name, *Lepidopygia*, used in last Edition was thought to be preoccupied and *Lemuresthes* Wolters, 1949, was proposed as a substitute. But apparently *Lepidopygia* is not preoccupied, see Wolters (1979: 288) (2685).
[4] If this is treated in the genus *Spermestes* the specific name *caniceps* (Reichenow, 1879) is no longer preoccupied and should be used.
[5] Includes *tesselatus* treated with a query by Mayr *et al.* (1968: 370) (1497).
[6] Includes *stigmatophorus* see White (1963: 211) (2648).
[7] Includes *minor* see White (1963: 211) (2648).
[8] Includes *pica* Clancey, 1987 (506) (DJP).
[9] Includes *sumatrensis* Restall, 1995 (2011), a name which is anyway preoccupied.
[10] Ascribed to Horsfield & Moore (1856) by Mayr *et al.* (1968: 374) (1497), but reference to the original shows that Moore clearly claimed sole authorship of this name.
[11] But not the main Sulu archipelago.

● *Lonchura punctulata* Scaly-breasted Munia
    — *L.p.punctulata* (Linnaeus, 1758)   Sri Lanka, N Pakistan, India (except NE), Nepal terai
    — *L.p.subundulata* (Godwin-Austen, 1872)   Bhutan, Assam, Bangladesh, W Burma
    — *L.p.yunnanensis* Parkes, 1958   SE Xizang, S Sichuan, Yunnan, N and NE Burma
    — *L.p.topela* (Swinhoe, 1863)   SE Burma, Thailand, SE and S China, Taiwan, Hainan, Indochina
    — *L.p.cabanisi* (Sharpe, 1890)   Palawan, Luzon, Mindoro, Negros, Panay, Cebu (W, N and WC Philippines)
    — *L.p.fretensis* (Kloss, 1931)   S Malay Pen., Sumatra, Nias I.
    — *L.p.nisoria* (Temminck, 1830)[1]   Java, Bali, Lombok, Sumbawa
    — *L.p.sumbae* Mayr, 1944   Sumba (Lesser Sundas)
    — *L.p.blasii* (Stresemann, 1912)   Flores to Tanimbar (Lesser Sundas)
    — *L.p.baweana* Hoogerwerf, 1963[2]   Bawean I.
    — *L.p.holmesi* Restall, 1992[3]   W Kalimantan (2011)
    — *L.p.particeps* (Riley, 1920)   Sulawesi

○ *Lonchura kelaarti* Black-throated Munia
    — *L.k.vernayi* (Whistler & Kinnear, 1933)   E India
    — *L.k.jerdoni* (Hume, 1873)   SW India
    — *L.k.kelaarti* (Jerdon, 1863)   Sri Lanka

○ *Lonchura leucogastra* White-bellied Munia
    — *L.l.leucogastra* (Blyth, 1846)   Malay Pen., Sumatra
    — *L.l.palawana* Ripley & Rabor, 1962   Palawan, N and E Borneo
    — *L.l.smythiesi* Parkes, 1958   SW Sarawak
    — *L.l.castanonota* Mayr, 1938   S Borneo (Kalimantan)
    — *L.l.everetti* (Tweeddale, 1877)   Luzon, Mindoro, Catandaunes
    — *L.l.manueli* Parkes, 1958   C and S Philippines

○ *Lonchura tristissima* Streak-headed Mannikin
    — *L.t.tristissima* (Wallace, 1865)   NW New Guinea (Vogelkop)
    — *L.t.hypomelaena* Stresemann & Paludan, 1934   W New Guinea
    — *L.t.calaminoros* (Reichenow, 1916)   SW to N and N slopes of SE New Guinea
    — *L.t.moresbyae* Restall, 1995[4]   S slopes from SE to eastern SC New Guinea (2011)
    — *L.t.leucosticta* (D'Albertis & Salvadori, 1879)[5]   SC New Guinea

○ *Lonchura quinticolor* (Vieillot, 1807) Five-coloured Munia[6]   Lesser Sundas

● *Lonchura malacca* Black-headed Munia[7][8]
    — *L.m.rubronigra* (Hodgson, 1836)[9]   N India, Nepal terai
    — *L.m.malacca* (Linnaeus, 1766)   S India, Sri Lanka
    — *L.m.atricapilla* (Vieillot, 1807)   NE India, Bangladesh, Assam, Burma, SW Yunnan
    — *L.m.deignani* Parkes, 1958   N, C Thailand, Indochina
    — *L.m.sinensis* (Blyth, 1852)   Malay Pen., Sumatran lowlands
    — *L.m.batakana* (Chasen & Kloss, 1929)   N Sumatran highlands
    — *L.m.formosana* (Swinhoe, 1865)   Taiwan, N Luzon,
    — *L.m.jagori* (C.E. Martens, 1866)[10]   Philippines (excl. N Luzon, but incl. Palawan and the Sulu Arch.), Borneo, Sulawesi, Muna, Butung, Halmahera
    — *L.m.ferruginosa* (Sparrman, 1789)   Java

[1] Includes *fortior* see White & Bruce (1986: 423) (2652).
[2] Treated with a query by Mayr *et al.* (1968) (1497); needs re-evaluation.
[3] This population like those in Brunei may be introduced or be composed of recent colonists like those in Sabah which match *cabanisi* from the Philippines (Smythies, 2000) (2289). It is listed with reservations.
[4] Although it has been suggested by LeCroy (1999) (1343) that this is an erroneous spelling Art. 31.1.2 of the Code (I.C.Z.N., 1999) (1178) does not seem to mandate its correction. LeCroy also suggested that its wild origins were still uncertain; however material in the ANWC allows the above range statement to be given (R. Schodde). The name *bigilalae* Restall, 1995 (2011) (and similarly uncorrectable) is not included pending further evidence of wild provenance.
[5] For reasons to include this taxon here see Coates (1990) (539).
[6] We follow White & Bruce (1986: 423) (2652) in treating this as monotypic (variation being either individual or perhaps clinal). Includes *sumbae* Restall, 1995 (2011), this name however is preoccupied (LeCroy, 1999) (1343).
[7] The reasons given by Restall (1995) (2011) for recognising *malacca* and *atricapilla* as distinct species require further evaluation. We believe molecular studies should soon clarify the situation; for the moment we prefer to retain a broad species including *ferruginosa* too.
[8] The "species" *Lonchura pallidiventer* Restall, 1996 (2012) was thought by van Balen (1998) (2459) to be a hybrid. Not since found in the wild (Smythies, 2000) (2289). R. Payne (*in litt.*) confirms the hybrid origin, with *L. atricapilla* providing the maternal parent, and reports that Restall agrees with this conclusion.
[9] For correction of spelling see David & Gosselin (2002) (613).
[10] Includes *obscura* Restall, 1995 (2011), and *selimbaue* Restall, 1995 (2011). If a name, other than or in addition to *jagori*, is to be used for Bornean birds it will be necessary to consider for the reasons explained by LeCroy (1999) (1343) – a point not discussed in Smythies (2000) (2289).

○ *Lonchura maja* (Linnaeus, 1766) WHITE-HEADED MUNIA[1]   Malay Pen., Sumatra, Simeuluë, Nias, Java, Bali and
     ? S Vietnam

○ *Lonchura pallida* (Wallace, 1863) PALE-HEADED MUNIA   Sulawesi, islands in Flores Sea, W to EC Lesser Sundas

○ *Lonchura grandis* GRAND MANNIKIN

   — *L.g.heurni* E. Hartert, 1932   Western NC New Guinea

   — *L.g.destructa* (E. Hartert, 1930)   NC New Guinea

   — *L.g.ernesti* (Stresemann, 1921)   Eastern NC New Guinea

   — *L.g.grandis* (Sharpe, 1882)   E New Guinea

○ *Lonchura vana* (E. Hartert, 1930) GREY-BANDED MANNIKIN   Mts. of NW New Guinea (Vogelkop)

○ *Lonchura caniceps* GREY-HEADED MANNIKIN

   — *L.c.caniceps* (Salvadori, 1876)   S coast of SE New Guinea

   — *L.c.scratchleyana* (Sharpe, 1898)   Mts. of SE New Guinea

   — *L.c.kumusii* (E. Hartert, 1911)   N coast of SE New Guinea

○ *Lonchura nevermanni* Stresemann, 1934 GREY-CROWNED MANNIKIN   SC New Guinea

○ *Lonchura spectabilis* HOODED MANNIKIN

   — *L.s.wahgiensis* Mayr & Gilliard, 1952   Mts. of EC New Guinea

   — *L.s.gajduseki* Diamond, 1967   EC New Guinea (Karimui basin)

   — *L.s.sepikensis* Jonkers & Roersma, 1990   Eastern NC New Guinea (Sepik basin) (1224)

   — *L.s.mayri* (E. Hartert, 1930)   NC New Guinea

   — *L.s.spectabilis* (P.L. Sclater, 1879)   New Britain (Bismarck Arch.)

○ *Lonchura forbesi* (P.L. Sclater, 1879) NEW IRELAND MANNIKIN   New Ireland

○ *Lonchura hunsteini* WHITE-COWLED MANNIKIN

   — *L.h.hunsteini* (Finsch, 1886)   N New Ireland (Bismarck Arch.)

   — *L.h.nigerrima* (Rothschild & Hartert, 1899)   New Hanover (Bismarck Arch.)

   — *L.h.minor* (Yamashina, 1931)[2]   Pohnpei I. (Caroline Is.)

○ *Lonchura flaviprymna* (Gould, 1845) YELLOW-RUMPED MANNIKIN   NE Western Australia, NW Northern Territory
     (W Arnhem Land)

○ *Lonchura castaneothorax* CHESTNUT-BREASTED MANNIKIN

   — *L.c.uropygialis* Stresemann & Paludan, 1934   NW New Guinea

   — *L.c.boschmai* Junge, 1952   WC New Guinea

   — *L.c.sharpii* (Madarász, 1894)   N and NE New Guinea

   — *L.c.ramsayi* Delacour, 1943   E New Guinea

   — *L.c.castaneothorax* (Gould, 1837)[3]   NW to coastal E Australia

○ *Lonchura stygia* Stresemann, 1934 BLACK MANNIKIN   SC New Guinea

○ *Lonchura teerinki* Rand, 1940 BLACK-BREASTED MANNIKIN[4]   NC New Guinea

○ *Lonchura monticola* (De Vis, 1897) ALPINE MANNIKIN[5]   SE New Guinea

○ *Lonchura montana* Junge, 1939 SNOW MOUNTAIN MANNIKIN   High mts. of WC New Guinea

○ *Lonchura melaena* BISMARCK MANNIKIN

   — *L.m.melaena* (P.L. Sclater, 1880)   New Britain

   — *L.m.bukaensis* Restall, 1995   Buka I. (N of Bougainville, Solomons) (2011)

○ *Lonchura fuscata* (Vieillot, 1807) TIMOR SPARROW[6]   Timor, Semau, Roti (Lesser Sundas)

● *Lonchura oryzivora* (Linnaeus, 1758) JAVA SPARROW   Java and Bali; Lombok, Sumbawa ? introduced

HETEROMUNIA Mathews, 1913[7] F

○ *Heteromunia pectoralis* (Gould, 1841) PICTORELLA FINCH   N Western Australia to NW Queensland

---

[1] Includes *vietnamensis* Restall, 1995 (2011). The types have been destroyed (Restall pers. comm.). The taxon can be evaluated when fresh material is collected.
[2] Validity questionable as reported introduced as late as the 1920s, but accepted by Baker (1951) (82).
[3] Includes *assimilis* see Schodde & Mason (1999: 771) (2189).
[4] Includes *mariae*; Mayr *et al.* (1968: 386) (1497) noted that confirmation was required.
[5] The name *myolae* Restall, 1995, has been discussed by LeCroy (1999); at least one type is a topotype of *monticola*. Birds from Myola, if different, require a new name.
[6] We tentatively submerge the genus *Padda* see Christidis (1987a, b) (349, 350).
[7] Treated within the genus *Lonchura* in last Edition, but see Christidis (1987a, b) (349, 350).

VIDUIDAE[1] INDIGOBIRDS AND ALLIES (2: 20)

VIDUA Cuvier, 1816[2] F

○ *Vidua chalybeata* VILLAGE INDIGOBIRD
— *V.c.chalybeata* (Statius Müller, 1776)   Senegal to Sierra Leone
— *V.c.neumanni* (Alexander, 1908)   Mali to Sudan
— *V.c.ultramarina* (J.F. Gmelin, 1789)   Ethiopia
— *V.c.centralis* (Neunzig, 1928)   Kenya, Uganda, W Tanzania
— *V.c.amauropteryx* (Sharpe, 1890)   Somalia to Zambia, Mozambique
— *V.c.okavangoensis* Payne, 1973   W Zambia, Botswana, Angola (1795)

○ *Vidua purpurascens* (Reichenow, 1883) PURPLE INDIGOBIRD   Kenya to Angola, Transvaal

○ *Vidua raricola* Payne, 1982 JAMBANDU INDIGOBIRD   Sierra Leone to S Sudan (1798)

○ *Vidua larvaticola* Payne, 1982 BARKA INDIGOBIRD   Guinea Bissau to W Ethiopia (1798)

○ *Vidua funerea* DUSKY INDIGOBIRD
— *V.f.nigerrima* (Sharpe, 1871)   W Kenya and S Congo south to Zimbabwe
— *V.f.funerea* (Tarragon, 1847)   Swaziland and Sth. Africa

○ *Vidua codringtoni* (Neave, 1907) PETERS'S TWINSPOT INDIGOBIRD[3]   S Zambia, Malawi, W Zimbabwe

○ *Vidua wilsoni* (E. Hartert, 1901) WILSON'S INDIGOBIRD[4]   N Nigeria to W Sudan

○ *Vidua nigeriae* (Alexander, 1908) QUAILFINCH INDIGOBIRD[5]   S Nigeria, Cameroon, S Sudan

○ *Vidua maryae* Payne, 1982 JOS PLATEAU INDIGOBIRD[6]   Jos Plateau (Nigeria) (1798)

○ *Vidua camerunensis* (Grote, 1922) FONIO INDIGOBIRD[7]   Gambia to Ethiopia

○ *Vidua macroura* (Pallas, 1764) PIN-TAILED WHYDAH[8]   Senegal to Ethiopia and Cape Province

○ *Vidua hypocherina* J. & E. Verreaux, 1856 STEEL-BLUE WHYDAH   Ethiopia and Somalia to Uganda, Kenya and NE Tanzania

○ *Vidua fischeri* (Reichenow, 1882) STRAW-TAILED WHYDAH   Somalia to Uganda and N Tanzania

○ *Vidua regia* (Linnaeus, 1766) SHAFT-TAILED WHYDAH[9]   S Angola to S Mozambique

○ *Vidua paradisaea* (Linnaeus, 1758) EASTERN PARADISE WHYDAH   E Sudan to S Angola and Natal

○ *Vidua orientalis* SAHEL PARADISE WHYDAH[10]
— *V.o.aucupum* (Neumann, 1908)   Senegal to N Nigeria
— *V.o.orientalis* Heuglin, 1870[11]   Lake Chad to Eritrea

○ *Vidua interjecta* (Grote, 1922) EXCLAMATORY WHYDAH   N Cameroon to S Sudan

○ *Vidua togoensis* (Grote, 1923) TOGO PARADISE WHYDAH   Sierra Leone to Togo

○ *Vidua obtusa* (Chapin, 1922) BROAD-TAILED PARADISE WHYDAH   Angola and S Zaire to Kenya and Mozambique

ANOMALOSPIZA Shelley, 1901[12] F

○ *Anomalospiza imberbis* (Cabanis, 1868) PARASITIC WEAVER   Sierra Leone to S Sudan, Ethiopia south to Zimbabwe and Transvaal

PRUNELLIDAE ACCENTORS (1: 13)

PRUNELLA Vieillot, 1816 F

○ *Prunella collaris* ALPINE ACCENTOR
— *P.c.collaris* (Scopoli, 1769)   SW Europe and NW Africa to Carpathian Mts. and Slovenia

[1] We use this rank on the urging of R.B. Payne, who has papers in press circumscribing this.
[2] Dated 1817 by Traylor (1968: 394) (2428) but see Browning & Monroe (1991) (258) and www.zoonomen.net (A.P. Peterson *in litt.*).
[3] For recognition as a species see Payne *et al.* (1992) (1804); *lusituensis* Payne 1973 (1795) showing variation, but now known to intergrade, is a synonym (R.B. Payne *in litt.*).
[4] Includes *lorenzi* Nicolai, 1972 (1649) as a synonymy see Payne (1982) (1798); Payne (*op. cit.*) also discussed the name *incognita* Nicolai, 1972 (1649) which had been applied to a captive bird of unknown origin. See also Payne *et al.* (2002) (1804A).
[5] For treatment as a species see Payne & Payne (1994) (1803).
[6] For treatment as a species see Payne (1998) (1801).
[7] Treated by Payne at species level since 1994; includes *sorora* Payne, 1982 (1798) (R.B. Payne *in litt.*).
[8] Includes *arenosa* Clancey, 1977 (460) (R.B. Payne *in litt.*).
[9] Includes *woltersi* (R.B. Payne *in litt.*).
[10] For the separation of the next three species from this see Payne (1985) (1799).
[11] Dated 1871 by Traylor (1968: 396) (2428), but see Zimmer (1926) (2707) and www.zoonomen.net (A.P. Peterson *in litt.*).
[12] For reasons to place this genus next to *Vidua* see Sorenson & Payne (2001) (2299A).

— *P.c.subalpina* (C.L. Brehm, 1831)   SE Europe, Crete, W and S Turkey

— *P.c.montana* (Hablizl, 1783)   E Turkey, Caucasus area, Iran

— *P.c.rufilata* (Severtsov, 1879)   Afghanistan and N Pakistan to E Tien Shan, W and S Xinjiang (W China)

— *P.c.whymperi* (E.C.S. Baker, 1915)   NW Himalayas

— *P.c.nipalensis* (Blyth, 1843)   C and E Himalayas to SW Sichuan, Yunnan, N Burma

— *P.c.tibetana* (Bianchi, 1904)   E Tibetan Plateau

— *P.c.erythropygia* (Swinhoe, 1870)   Altai to NE Siberia and Japan, south to S Shaanxi (China) and Korea

— *P.c.fennelli* Deignan, 1964   Taiwan (668)

○ *Prunella himalayana* (Blyth, 1842) ALTAI ACCENTOR   Pamir and Tien Shan to Altai and Baikal area >> Himalayas

○ *Prunella rubeculoides* ROBIN ACCENTOR

— *P.r.muraria* R. & A. Meinertzhagen, 1926[1]   NW Himalayas

— *P.r.rubeculoides* (F. Moore, 1854)[2]   S, E, and NE fringe of Tibetan Plateau, Himalayas

○ *Prunella strophiata* RUFOUS-BREASTED ACCENTOR

— *P.s.jerdoni* (W.E. Brooks, 1872)   E Afghanistan, NW Himalayas

— *P.s.strophiata* (Blyth, 1843)   C and E Himalayas, N Burma, Yunnan to Gansu (C China)

○ *Prunella montanella* SIBERIAN ACCENTOR

— *P.m.montanella* (Pallas, 1776)   NE Europe, NW Siberai to Khatanga R.; C Asia from Altai to SE Siberia >> C and E Asia

— *P.m.badia* Portenko, 1929[3]   NE Siberia (Lower Lena to Chukotka and Anadyrland) >> E Asia

○ *Prunella fulvescens* BROWN ACCENTOR

— *P.f.fulvescens* (Severtsov, 1873)   Tien Shan, Afghanistan, Pakistan, NW India

— *P.f.dahurica* (Taczanowski, 1874)   E Kazakhstan, Altai, Mongolia to Transbaikalia, Nei Mongol and Ningxia (China)

— *P.f.dresseri* E. Hartert, 1910   S Xinjiang and Qaidam Basin (W China)

— *P.f.nanschanica* Sushkin, 1925   NE Qinghai, W Sichuan, S Gansu (NC China)

— *P.f.khamensis* Sushkin, 1925[4]   C and E Himalayas, C and E Xizang, W Gansu

○ *Prunella ocularis* (Radde, 1884) RADDE'S ACCENTOR   S and E Turkey, Transcaucasia, Iran

○ *Prunella fagani* (Ogilvie-Grant, 1913) YEMEN ACCENTOR[5]   W Yemen

○ *Prunella atrogularis* BLACK-THROATED ACCENTOR

— *P.a.atrogularis* (J.F. Brandt, 1844)   NE Europe, N and C Ural Mts. >> SW Asia

— *P.a.huttoni* (F. Moore, 1854)   Pamir and Tien Shan to Altai and Karlik Tagh Mts. (WC Asia) >> SW Asia

○ *Prunella koslowi* (Przevalski, 1887) MONGOLIAN ACCENTOR   W and C Mongolia >> NC China

● *Prunella modularis* DUNNOCK/HEDGE SPARROW

— *P.m.hebridium* R. Meinertzhagen, 1934   Ireland, Hebrides (Scotland)

— *P.m.occidentalis* (E. Hartert, 1910)   Scotland (except Hebrides), England, W France

— *P.m.modularis* (Linnaeus, 1758)   N and C Europe >> S Europe, N Africa

— *P.m.mabbotti* Harper, 1919   Iberia, SW France, Italy

— *P.m.meinertzhageni* Harrison & Pateff, 1937   S Balkans and Greece

— *P.m.fuscata* Mauersberger, 1971   Mts. of S Crimea (1459)

— *P.m.euxina* Watson, 1961   NW Asia Minor >> W and S Turkey

— *P.m.obscura* (Hablizl, 1783)   E Turkey, Caucasus area, N Iran >> Middle East

○ *Prunella rubida* (Temminck & Schlegel, 1848) JAPANESE ACCENTOR[6]   Sakhalin, S Kuril Is., Hokkaido, Honshu, Kyushu

○ *Prunella immaculata* (Hodgson, 1845) MAROON-BACKED ACCENTOR   C Himalayas, SE and E Xizang, N Burma, W and N Yunnan to S Gansu (C China)

---

[1] For recognition see Vaurie (1959: 211) (2502).
[2] Includes *fusca* see Vaurie (1959: 211) (2502).
[3] For recognition see Vaurie (1959: 214) (2502).
[4] Includes *sushkini*. Cheng (op. cit.), included *sushkini* in *nanshanica*, perhaps misled by Vaurie (1959) (2502) who did not mention *khamensis*, but it belongs here. Studies show *nanshanica* is clearly distinct from *khamensis contra* Cheng (1987) (333). CSR.
[5] Included in *ocularis* in last Edition, but see Vaurie (1955) (2496).
[6] Stepanyan (1990) (2326) recognised *fervida* but this is not accepted by Orn. Soc. Japan (2000) (1727). Perhaps Sakhalin birds *are* distinct.

PEUCEDRAMIDAE OLIVE WARBLER[1] (1: 1)

PEUCEDRAMUS Henshaw, 1875 M
○ *Peucedramus taeniatus* OLIVE WARBLER

__ *P.t.arizonae* Miller & Griscom, 1925	SW USA, N Mexico
__ *P.t.jaliscensis* Miller & Griscom, 1925	NW Mexico
__ *P.t.giraudi* J.T. Zimmer, 1948	C Mexico
__ *P.t.aurantiacus* Ridgway, 1896	Chiapas (S Mexico)
__ *P.t.taeniatus* (Du Bus, 1847)[2]	S Mexico, W Guatemala
__ *P.t.micrus* Miller & Griscom, 1925	El Salvador, Honduras, N Nicaragua

MOTACILLIDAE WAGTAILS, PIPITS[3] (5: 64)

DENDRONANTHUS Blyth, 1844 M
○ *Dendronanthus indicus* (J.F. Gmelin, 1789) FOREST WAGTAIL   C and E China to S Japan, Korea and Amurland >> S and SE Asia to Sri Lanka and Gtr. Sundas

MOTACILLA Linnaeus, 1758 F
◉ *Motacilla flava* YELLOW WAGTAIL

__ *M.f.flavissima* Blyth, 1834	British Is. and coastal W Europe >> Senegal and Gambia to Ivory Coast
__ *M.f.flava* Linnaeus, 1758	W and C Europe to C European Russia and N Ukraine >> sub-Saharan Africa
__ *M.f.iberiae* E. Hartert, 1921	SW France, Iberia, NW Africa >> W Africa
✓ *M.f.cinereocapilla* Savi, 1831	SE France, Italy, Sicily, Sardinia, W Croatia >> Mali to Lake Chad
__ *M.f.pygmaea* (A.E. Brehm, 1854)	Egypt
__ *M.f.beema* (Sykes, 1832)	SE European Russia, N Kazakhstan and SW Siberia to N Altai and SC Siberia >> E and SE Africa, India
__ *M.f.leucocephala* (Przevalski, 1887)	NW Mongolia and N Xinjiang >> E Africa and SW Asia
__ *M.f.lutea* (S.G. Gmelin, 1774)	Lower Volga R. to Kazakhstan and SW Siberia >> E and S Africa, India
__ *M.f.zaissanensis* (Poliakov, 1911)	Zaisan basin and upper Irtysh R (E Kazakhstan/NW Xinjiang) >> India
__ *M.f.thunbergi* Billberg, 1828	N Europe, NW Siberia to lower Yenisey basin >> sub-Saharan Africa
__ *M.f.plexa* (Thayer & Bangs, 1914)	NE Siberia from Khatanga to Kolyma R. >> India to SE Asia
__ *M.f.macronyx* (Stresemann, 1920)[4]	Baikal area and NE Mongolia to NE China and Russian Far East >> SE Asia to N Australia
__ *M.f.simillima* E. Hartert, 1905	C and S Kamchatka and N Kuril Is. >> Philippines to N Australia
✓ *M.f.tschutschensis* J.F. Gmelin, 1789	N Kamchatka to Chukotka, Alaska >> E Asia to Gtr. Sundas
__ *M.f.taivana* (Swinhoe, 1863)	Vilyuy and upper Lena basins to Sea of Okhotsk, Sakhalin, Amurland and N Hokkaido >> Philippines to N Australia
__ *M.f.feldegg* Michahelles, 1830	Balkans, Levant and S Ukraine to Caucasus area and N and SW Iran >> Sudan to Nigeria and Zaire
__ *M.f.melanogrisea* (Homeyer, 1878)	Volga Delta to S Kazakhstan, NE Iran and Afghanistan >> NE Africa and SW Asia

○ *Motacilla citreola* CITRINE WAGTAIL

__ *M.c.citreola* Pallas, 1776[5]	Finland and N European Russia to C Siberia, Mongolia and NW Manchuria >> to India

---

[1]   Given familial status in A.O.U. (1998: 532) (3). But see also Groth (1998) (1021).
[2]   Du Bus is an abbreviation for Du Bus de Gisignies.
[3]   For reports on phylogeny see Groth (1998) (1021) and Ericson *et al.* (2000) (816).
[4]   Includes *angarensis* (CSR).
[5]   Includes *quassitrix* Portenko, 1960 (1908), see Cramp *et al.* (1988: 441) (592).

—— *M.c.werae* (Buturlin, 1907)[1]	N Ukraine, Poland and Baltic Republics E through S Russia and N Kazakhstan to Altai and NW Xinjiang >> SW and S Asia
—— *M.c.calcarata* Hodgson, 1836	E Iran and mountains of C Asia to Tien Shan, Himalayas and C China >> SW and S Asia, Burma, N and C Thailand, Laos

○ *Motacilla capensis* CAPE WAGTAIL

—— *M.c.simplicissima* Neumann, 1929	Angola, and NE Namibia to Zambia and W Zimbabwe
—— *M.c.capensis* Linnaeus, 1766[2]	Sth. Africa to S and W Namibia, C Zimbabwe and Mozambique
—— *M.c.wellsi* Ogilvie-Grant, 1911	E Zaire to Kenya

○ *Motacilla flaviventris* Hartlaub, 1860 MADAGASCAN WAGTAIL — Madagascar

○ *Motacilla cinerea* GREY WAGTAIL

—— *M.c.patriciae* Vaurie, 1957	Azores
—— *M.c.schmitzi* Tschusi, 1900	Madeira
—— *M.c.canariensis* E. Hartert, 1901	Canary Is.
—— *M.c.cinerea* Tunstall, 1771[3]	Europe and NW Africa to Iran >> W Europe, W and E Africa
—— *M.c.melanope* Pallas, 1776	Ural Mts. and Afghanistan to Koryak region (Russia), Transbaikalia, C China, Taiwan and Himalayas >> NE Africa, S and E Asia
—— *M.c.robusta* (C.L. Brehm, 1857)	S and C Kamchatka, NE China and Amurland to Korea and Japan >> E and SE Asia to Indonesia and Philippines

○ *Motacilla clara* MOUNTAIN WAGTAIL

—— *M.c.chapini* Amadon, 1954	Sierra Leone to W Uganda
—— *M.c.clara* Sharpe, 1908	Ethiopia
—— *M.c.torrentium* Ticehurst, 1940	E Uganda and Kenya to Angola and E Sth. Africa

◉ *Motacilla alba* WHITE WAGTAIL

—— *M.a.yarrelli* Gould, 1837	British Isles >> south to NW Africa
✓ *M.a.alba* Linnaeus, 1758	Iceland and Europe to Ural Mts., W Asia Minor and Levant >> Europe, Arabia and N and tropical Africa
—— *M.a.dukhunensis* Sykes, 1832	C Asia Minor and Caucasus area to SW and N Iran; N Kazakhstan and W Siberia >> NE Africa and SW Asia
—— *M.a.persica* Blanford, 1876	WC and S Iran
—— *M.a.personata* Gould, 1861	Siberia, W Asia, Afghanistan, Iran, N India
—— *M.a.baicalensis* Swinhoe, 1871	Baikal area and C Mongolia to N Inner Mongolia and W Amurland >> E China and SE Asia
—— *M.a.ocularis* Swinhoe, 1860	Lower and middle Yenisey to W Alaska, N Kamchatka and Sea of Okhotsk >> E and SE Asia to Philippines
—— *M.a.lugens* Gloger, 1829	S Koryak Region, Kamchatka, Kuril Is., Japan and Russian Far East >> E Asia to Taiwan
—— *M.a.leucopsis* Gould, 1838	NC, E and NE China, Korea and middle Amur valley to SW Japan and Taiwan >> E and SE Asia
—— *M.a.alboides* Hodgson, 1836	SW China, Himalayas, Burma
—— *M.a.subpersonata* Meade-Waldo, 1901	Morocco

○ *Motacilla aguimp* AFRICAN PIED WAGTAIL

—— *M.a.vidua* Sundevall, 1850	W, C, E and SE Africa
—— *M.a.aguimp* Dumont, 1821	Orange and lower Vaal Rivers to Lesotho and SW Transvaal

○ *Motacilla samveasnae* Duckworth *et al.*, 2001 MEKONG WAGTAIL[4] # — Lower Mekong River (Cambodia, S Laos, extreme E Thailand) (765)

○ *Motacilla grandis* Sharpe, 1885 JAPANESE PIED WAGTAIL — Japan, Korea >> E China, Taiwan

[1] Includes *sindzianicus* Portenko, 1960 (1908), see Stepanyan (1990: 367) (2326).
[2] Includes *bradfieldi* see White (1961: 69) (2645).
[3] For validation of this name see Opinion 882, I.C.Z.N. (1969) (1172).
[4] We include all authors' names when five or fewer are involved.

○ *Motacilla maderaspatensis* J.F. Gmelin, 1789 WHITE-BROWED WAGTAIL   Pakistan, India, Nepal, Bhutan and Bangladesh

TMETOTHYLACUS Cabanis, 1879 M
○ *Tmetothylacus tenellus* (Cabanis, 1878) GOLDEN PIPIT   SE Sudan, NE Uganda, Somalia, N and E Kenya,
NE and C Tanzania

MACRONYX Swainson, 1827 M
○ *Macronyx sharpei* Jackson, 1904 SHARPE'S LONGCLAW[1]   W and C Kenya
○ *Macronyx flavicollis* Rüppell, 1840 ABYSSINIAN LONGCLAW   Ethiopia

○ *Macronyx fuellebornii* FÜLLEBORN'S LONGCLAW
— *M.f.fuellebornii* Reichenow, 1900   S Tanzania
— *M.f.ascensi* Salvadori, 1907   Angola to SW Tanzania

○ *Macronyx capensis* CAPE LONGCLAW
— *M.c.capensis* (Linnaeus, 1766)   S and SW Cape Province
— *M.c.colletti* Schouteden, 1908[2]   E Zimbabwe to Orange Free State and E Cape Province

○ *Macronyx croceus* (Vieillot, 1816) YELLOW-THROATED LONGCLAW[3]   W and NC Africa, Uganda and Kenya to Mozambique
and E Cape Province

○ *Macronyx aurantiigula* Reichenow, 1891 PANGANI LONGCLAW   SE Somalia, S and SE Kenya, NE Tanzania

○ *Macronyx ameliae* de Tarragon, 1845 ROSY-BREASTED LONGCLAW[4]   SW Kenya to N Tanzania, SW Tanzania to C Angola,
NE Namibia, Malawi and S Mozambique

○ *Macronyx grimwoodi* Benson, 1955 GRIMWOOD'S LONGCLAW[5]   C and NE Angola to S Zaire and NW Zambia

ANTHUS Bechstein, 1805[6] M
○ *Anthus richardi* RICHARD'S PIPIT[7]
— *A.r.richardi* Vieillot, 1818   W Siberia to Yenisey basin >> VN Africa, S Europe,
SW Asia

— *A.r.dauricus* H. Johansen, 1952   N Mongolia to middle Lena R and upper Amur R. >>
S and SE Asia

— *A.r.centralasiae* (Kistiakovsky, 1928)   Xinjiang, W and S Mongolia, W Iner Mongolia >>
S and SE Asia

— *A.r.ussuriensis* H. Johansen, 1952[8]   E Qinghai to Shandong to lower Amur, NE China and
Korea >> C and E China, SE Asia

— *A.r.sinensis* (Bonaparte, 1850)   C and E China >> mainland SE Asia and Sumatra
— *A.r.waitei* Whistler, 1936   SE Afghanistan, Pakistan, NW India
— *A.r.rufulus* Vieillot, 1818   C and SE India and C Nepal to S China and mainland
SE Asia (except Malay Pen.)

— *A.r.malayensis* Eyton, 1839[9]   S India, Sri Lanka, Malay Pen., Sumatra, Java, Borneo
— *A.r.lugubris* (Walden, 1875)   Philippines
— *A.r.albidus* Stresemann, 1912   Lesser Sundas, S Sulawesi
— *A.r.medius* Wallace, 1863   Savu, Timor Is.

○ *Anthus australis* AUSTRALIAN PIPIT
— *A.a.exiguus* Greenway, 1935   EC New Guinea
— *A.a.rogersi* Mathews, 1913   Coastal NW Australia to Cape York Pen.
— *A.a.bilbali* Mathews, 1912   SW Western Australia, SC South Australia
— *A.a.australis* Vieillot, 1818[10]   WC, C, E and SE Australia
— *A.a.bistriatus* (Swainson, 1837)   Tasmania, Flinders I., King I.

◉ *Anthus novaeseelandiae* NEW ZEALAND PIPIT
— *A.n.reischeki* Lorenz-Liburnau, 1902   North Island (New Zealand)
— *A.n.novaeseelandiae* (J.F. Gmelin, 1789)   South Island (New Zealand)
— *A.n.chathamensis* Lorenz-Liburnau, 1902   Chatham I.

[1] Treated in the genus *Hemimacronyx* in last Edition, but see Fry *et al.* (1992: 246) (900).
[2] Includes *stabilior* see White (1961: 83) (2645) and *latimerae* Clancey, 1963 (395), see White (1963: 215) (2648).
[3] Includes *vulturinus* and *tertius* Clancey, 1958 (379), see White (1961: 82) (2645), and *hygricus* Clancey, 1984 (496), Taylor (1992: 254) (2386).
[4] Includes *altanus* Clancey, 1966 (409), see Taylor (1992: 260) (2386).
[5] Implicitly includes *cuandocubangensis* da Rosa Pinto, 1968 (607), see Taylor (1992: 262) (2386).
[6] For a molecular study see Voelker (1999a, b) (2533, 2534).
[7] Although often split between *A. richardi* and *A. rufulus* we believe these are conspecific. See Roselaar in Cramp *et al.* (1988: 309) (592).
[8] For recognition see Hall (1961) (1062) and Cramp *et al.* (1988: 301) (592).
[9] Provisionally includes *idjenensis* Hoogerwerf, 1962 (1142), see Mees (1996) (1543).
[10] Includes *subaustralis* see Schodde & Mason (1999) (2189).

741

— *A.n.aucklandicus* G.R. Gray, 1862   Auckland Is.

— *A.n.steindachneri* Reischek, 1889   Antipodes Is.

○ **Anthus cinnamomeus** AFRICAN PIPIT[1]

— *A.c.eximius* Clancey, 1986   Yemen (505)

— *A.c.annae* R. Meinertzhagen, 1921   Eritrea and Djibouti to NE Tanzania

— *A.c.lynesi* Bannerman & Bates, 1926   SE Nigeria and E Cameroon to W Sudan

— *A.c.camaroonensis* Shelley, 1900   Mt. Cameroon and Mt. Manenguba (Cameroon)

— *A.c.stabilis* Clancey, 1986   C and SE Sudan (505)

— *A.c.cinnamomeus* Rüppell, 1840   W and SE Ethiopian Highlands

— *A.c.itombwensis* Prigogine, 1981   E Zaire highlands (1938)

— *A.c.latistriatus* Jackson, 1899   SW Uganda, W Kenya, E Zaire, NW Tanzania (breeding range unknown)

— *A.c.lacuum* R. Meinertzhagen, 1920   SE Uganda, W and S Kenya, N and C Tanzania

— *A.c.spurium* Clancey, 1951   SE Tanzania to Zambezi valley and NE Namibia

— *A.c.bocagei* Nicholson, 1884   W and S Angola and Namibia to N Cape Province

— *A.c.grotei* Niethammer, 1957   N Botswana, N Namibia

— *A.c.lichenya* Vincent, 1933   W Uganda to C Malawi, NW Mozambique, N and E Zambia, S Zaire and NE Angola

— *A.c.rufuloides* Roberts, 1936   Sth. Africa

○ **Anthus hoeschi** Stresemann, 1938 MOUNTAIN PIPIT[2]   Lesotho, NE Cape Province >> E Angola, S Zaire, NW Zambia

○ **Anthus godlewskii** (Taczanowski, 1876) BLYTH'S PIPIT   Mongolia and Tuva Rep. to Transbaikalia and Inner Mongolia >> S Asia

◑ **Anthus campestris** TAWNY PIPIT

✓ *A.c.campestris* (Linnaeus, 1758)[3]   Europe and NW Africa to Iran and SW Turkmenistan >> N tropical Africa to India

— *A.c.griseus* Nicoll, 1920   Kazakhstan and N Turkmenistan to Tien Shan and SW Mongolia >> SW Asia

— *A.c.kastschenkoi* H. Johansen, 1952   NE Kazakhstan and S-C Siberia to Baikal area >> S Asia

○ **Anthus similis** LONG-BILLED PIPIT

— *A.s.nicholsoni* Sharpe, 1884[4]   SE Botswana to S Transvaal and NE Cape Province

— *A.s.petricolus* Clancey, 1956   Lesotho, S and W Cape Province to E Transvaal

— *A.s.leucocraspedon* Reichenow, 1915   NW Cape Province to S and W Namibia

— *A.s.palliditinctus* Clancey, 1956   NW Namibia, SW Angola

— *A.s.moco* Traylor, 1962   Mt. Moco (C Angola) (2425)

— *A.s.dewittei* Chapin, 1937[5]   SE and E Zaire, SW Uganda, Rwanda, Burundi

— *A.s.hararensis* Neumann, 1906   Ethiopian Highlands

— *A.s.chyuluensis* van Someren, 1939   N, W and S Kenya, N Tanzania

— *A.s.nivescens* Reichenow, 1905   NE Sudan and Eritrea to N Somalia

— *A.s.sokotrae* E. Hartert, 1917   Socotra I.

— *A.s.jebelmarrae* Lynes, 1920   C and W Sudan

— *A.s.asbenaicus* Rothschild, 1920   C Niger, E and C Mali

— *A.s.bannermani* Bates, 1930[6]   Sierra Leone to Nigeria and Cameroon

— *A.s.captus* E. Hartert, 1905   Lebanon, Syria, Jordan, Israel

— *A.s.arabicus* E. Hartert, 1917   SW Arabia

— *A.s.decaptus* R. Meinertzhagen, 1920   NE Arabia, Iran, SE Afghanistan, Pakistan >> NW India

— *A.s.jerdoni* Finsch, 1870   NE Afghanistan, NW Himalayas to W Nepal >> Pakistan, N India

— *A.s.yamethini* B.P. Hall, 1957   C Burma

— *A.s.similis* Jerdon, 1840   W India

— *A.s.travancoriensis* Ripley, 1953   SW India

---

[1] In last Edition treated as part of *Anthus novaeseelandiae*. Here, with reservations, we follow Prigogine (1981, 1985) (1938, 1943), and list five allospecies including African *cinnamomeus*, but see Pearson (1992: 218) (1811) for an alternative view. We follow the subspecies given by Pearson (op. cit.).

[2] Includes *lwenarum* and *editus* (listed under *Anthus novaeseelandiae* in last Edition and so treated by White (1961) (2645), for whom *hoeschi* was a synonym of *A. n. bocagei*). For reasons to elevate *hoeschi* to specific status see Clancey (1990) (518).

[3] Includes *boehmii* Portenko, 1960 (1908), see Cramp *et al.* (1988: 326) (592).

[4] Tentatively includes *primarius* Clancey, 1990 (518).

[5] Includes *hallae* see Pearson (1992: 224) (1811).

[6] Includes *josensis* see Pearson (1992: 224) (1811).

○ *Anthus nyassae* WOODLAND PIPIT
    — *A.n.frondicolus* Clancey, 1964   Zimbabwe to NE Botswana and W Mozambique (401)
    — *A.n.schoutedeni* Chapin, 1937[1]   S Zaire to S Congo, Angola, and W Zambia
    — *A.n.nyassae* Neumann, 1906[2]   E Zambia, SW Tanzania, Malawi, NW Mozambique

○ *Anthus vaalensis* BUFFY PIPIT
    — *A.v.chobiensis* (Roberts, 1932)   S Zaire, SW Tanzania and Malawi to Zimbabwe and
                                            N Botswana
    — *A.v.neumanni* R. Meinertzhagen, 1920   C Angola
    — *A.v.namibicus* Clancey, 1990   NE and C Namibia (520)
    — *A.v.clanceyi* Winterbottom, 1963   N Cape Province, SW Transvaal, SE Botswana (2679)
    — *A.v.exasperatus* Winterbottom, 1963   NE Botswana (2679)
    — *A.v.vaalensis* Shelley, 1900[3]   S Botswana to S Africa

○ *Anthus longicaudatus* Liversidge, 1996 LONG-TAILED PIPIT #   South Africa (1363)

○ *Anthus leucophrys* PLAIN-BACKED PIPIT[4]
    — *A.l.ansorgei* C.M.N. White, 1948   Senegal, Gambia, S Mauritania, Guinea Bissau
    — *A.l.gouldii* Fraser, 1843   Sierra Leone, Liberia, Ivory Coast
    — *A.l.zenkeri* Neumann, 1906   S Mali, Ghana and Nigeria to S Sudan, N Zaire,
                                            Uganda and W Kenya
    — *A.l.saphiroi* Neumann, 1906   SE Ethiopia, NW Somalia
    — *A.l.goodsoni* R. Meinertzhagen, 1920   C and SW Kenya, N Tanzania
    — *A.l.omoensis* Neumann, 1906   N, C and W Ethiopia, SE Sudan
    — *A.l.bohndorffi* Neumann, 1906   N and C Angola to Tanzania and N Malawi
    — *A.l.leucophrys* Vieillot, 1818[5]   Sth. Africa, Lesotho, Swaziland
    — *A.l.tephridorsus* Clancey, 1967   S Angola, NE Namibia, SW Zambia, NW Botswana (411)

○ *Anthus pallidiventris* LONG-LEGGED PIPIT
    — *A.p.pallidiventris* Sharpe, 1885   Equatorial Guinea to NW Angola
    — *A.p.esobe* Chapin, 1937   Middle basin of Congo River

◉ *Anthus pratensis* MEADOW PIPIT
    — *A.p.whistleri* Clancey, 1942[6]   Ireland, Scotland (N of the Great Glen) >> W Europe
    ✓ *A.p.pratensis* (Linnaeus, 1758)   Greenland, Iceland, Britain (except N Scotland),
                                            continental Europe, NW Siberia >> W and S Europe,
                                            N Africa, SW Asia

○ *Anthus trivialis* TREE PIPIT
    — *A.t.trivialis* (Linnaeus, 1758)[7]   Europe, Turkey and the Caucasus to W Siberia >>
                                            sub-Saharan Africa
    — *A.t.sibiricus* Sushkin, 1925[8]   S Siberia to N Mongolia >> NE Africa and SW Asia
    — *A.t.haringtoni* Witherby, 1917   Mts. of WC Asia from Afghanistan to Tien Shan area
                                            and NW Himalayas >> S Asia

○ *Anthus hodgsoni* OLIVE-BACKED PIPIT
    — *A.h.yunnanensis* Uchida & Kuroda, 1916   N European Russia and N Siberia south to N Mongolia,
                                            N Manchuria and Hokkaido >> S and SE Asia
    — *A.h.hodgsoni* Richmond, 1907   Himalayas and WC China to S Mongolia, Korea and
                                            Honshu to NW China, Korea and Japan >> to SE Asia

○ *Anthus gustavi* PECHORA PIPIT
    — *A.g.gustavi* Swinhoe, 1863   NE European Russia and N Asia to Chukotka and Kamt-
                                            chatka >> E and SE Asia to Philippines and Indonesia
    — *A.g.stejnegeri* Ridgway, 1883[9]   Commander Is. >> ? E Asia (2020)
    — *A.g.menzbieri* Shulpin, 1928[10]   Middle Amur valley, S Ussuriland (Russia) and
                                            NE China, perhaps formerly Korea >> ??

---

[1] Includes *chersophilus* Clancey, 1989 (517) see Pearson (1992: 224) (1811).
[2] Includes *winterbottomi* Clancey, 1985 (498) see Pearson (1992: 224) (1811).
[3] Includes *daviesi* see Vaurie *et al.* (1960: 151) (2517).
[4] The races *saphiroi* and *goodsoni* attached here were placed there by Hall (1961) (1062), but not listed in our last Edition. Clancey (1990) placed these in *Anthus vaalensis*, but see Pearson (1992: 226) (1811).
[5] Includes *enunciator* see Vaurie (1960: 150) (2517).
[6] Includes *theresae* see Cramp *et al.* (1988: 365); their range included "western Scotland". This is misleading and suggests it does not include the type locality of *whistleri* which is Dornoch on the east coast.
[7] Includes *differens* Clancey, 1987 (510), see Roselaar (1995) (2107).
[8] For recognition see Hall (1961) (1062).
[9] This name, which has priority over *commandorensis*, appeared in Hartert (1903) (1080), but not in Sharpe (1885) (2229), Vaurie (1959) (2502) nor Vaurie *et al.* (1960) (2517).
[10] This may be a separate species see Leonovich *et al.* (1997) (1353).

○ *Anthus roseatus* Blyth, 1847 Rosy Pipit

Mts. in E Pamir and NE Pakistan over S Xinjiang and Himalayas to SW and C China >> lower elevations

○ *Anthus cervinus* Red-throated Pipit
— *A.c.rufogularis* C.L.Brehm, 1824[1]

N Europe and NW Siberia to C Taymyr >> Africa and SW Asia

— *A.c.cervinus* (Pallas, 1811)

NE Siberia from E Taymyr to N Kuril and Commander Is. and NW Alaska >> S and SE Asia

◉ *Anthus rubescens* Buff-bellied Pipit
— *A.r.rubescens* (Tunstall, 1771)
— *A.r.pacificus* Todd, 1935
— *A.r.alticola* Todd, 1935
— *A.r.japonicus* Temminck & Schlegel, 1847

N and NE Asia, Nth. America, Mexico
W Canada, W USA, W Mexico
SW USA, NW Mexico
C and E Siberia south to Baikal area and Sakhalin >> Middle East, S and E Asia

◉ *Anthus spinoletta* Water Pipit
— *A.s.spinoletta* Linnaeus, 1758
— *A.s.coutellii* Audouin, 1828

Mts. of C and SW Europe >> W and S Europe, N Africa
Asia Minor and Caucasus area to N Iran >> NE Africa and SW Asia

— *A.s.blakistoni* Swinhoe, 1863[2]

NE Afghanistan to Transbaicalia and C China >> N India, N Indochina

○ *Anthus petrosus* Rock Pipit[3]
— *A.p.kleinschmidti* E. Hartert, 1905
— *A.p.petrosus* (Montagu, 1798)

Faroe Is., Shetlands and St. Kilda
W France, British Isles (except Shetland Is. and St. Kilda) >> W Europe

— *A.p.littoralis* C.L. Brehm, 1823

Scandinavia, NW Russia, Estonia, Denmark >> W Europe

○ *Anthus nilghiriensis* Sharpe, 1885 Nilgiri Pipit

S India

○ *Anthus sylvanus* (Hodgson, 1845)[4] Upland Pipit

NE Afghanistan to Nepal; NE Burma to SE China

○ *Anthus berthelotii* Berthelot's Pipit
— *A.b.berthelotii* Bolle, 1862
— *A.b.madeirensis* E. Hartert, 1905

Canary Is.
Madeira I.

○ *Anthus lineiventris*[5] Sundevall, 1850 Striped Pipit[6]

S Kenya to NE and E Sth. Africa, W Angola

○ *Anthus brachyurus* Short-tailed Pipit
— *A.b.leggei* Ogilvie-Grant, 1906[7]

SW Uganda to Rwanda, S Gabon and S Zaire to NW Zambia, S Tanzania

— *A.b.brachyurus* Sundevall, 1850[8]

E Sth. Africa

○ *Anthus caffer* Bush Pipit
— *A.c.australoabyssinicus* Benson, 1942
— *A.c.blayneyi* van Someren, 1919
— *A.c.mzimbaensis* Benson, 1955
— *A.c.traylori* Clancey, 1964
— *A.c.caffer* Sundevall, 1850[9]
— *A.c.*subsp?

S Ethiopia
S Kenya, N Tanzania
NE Botswana to Zambia and W Malawi
NE Natal, S Mozambique (398)
SE Botswana to Transvaal, N Natal and W Swaziland
W Angola

○ *Anthus sokokensis* van Someren, 1921 Sokoke Pipit

Coastal Kenya and NE Tanzania

○ *Anthus melindae* Malindi Pipit
— *A.m.mallablensis* Colston, 1987[10]
— *A.m.melindae* Shelley, 1900

SC Somalia (552)
S Somalia, coastal Kenya

---

[1] Not recognised by Cramp *et al.* (1988: 392), but accepted by Stepanyan (1990: 360) (2326).
[2] For recognition see Hall (1961) (1062).
[3] We previously split this from *spinoletta* citing Knox (1988) (1277). We now correct the allocation of subspecies between the three species recognised based on Cramp *et al.* (1988: 393) (592). See also Sangster *et al.* (2002) (2167).
[4] Hall (1961) (1062) noted that Hodgson's name appeared before the same name appeared from Blyth. Baker (1930) (79) also had evidence that this paper by Blyth appeared late.
[5] Vaurie *et al.* (1960: 167) (2517) dated this 1851, but see Gyldenstolpe (1926: 29) (1027) and www.zoonomen.net (A.P. Peterson *in litt.*).
[6] Includes *stygium* see White (1961: 81) (2645); the listing of *angolensis* in last Edition related to Bocage, 1870, which was in synonymy as early as 1885. Also implicitly includes *sylvivagus* Clancey, 1984 (495), see Pearson (1992: 242) (1811).
[7] Includes *eludens* Clancey, 1985 (499), see Pearson (1992: 230) (1811).
[8] Vaurie *et al.* (1960: 156) (2517) dated this 1851, but see Gyldenstolpe (1926: 29) (1027) and www.zoonomen.net (A.P. Peterson *in litt.*).
[9] Vaurie *et al.* (1960: 156) (2517) dated this 1851, but see Gyldenstolpe (1926: 29) (1027) and www.zoonomen.net (A.P. Peterson *in litt.*).
[10] This is a new name for *pallidus* Colston, 1982 (551).

○ *Anthus pseudosimilis* Liversidge & Voelker, 2002 KIMBERLEY PIPIT #   SW Namibia, S Africa (1363A)

○ *Anthus crenatus* Finsch & Hartlaub, 1870 YELLOW-TUFTED PIPIT   S and E Cape Province to SE Transvaal, Lesotho, Swaziland

○ *Anthus chloris* M.H.K. Lichtenstein, 1842 YELLOW-BREASTED PIPIT[1]   E Sth. Africa, Lesotho

○ *Anthus gutturalis* ALPINE PIPIT
  — *A.g.gutturalis* De Vis, 1894   SE New Guinea
  — *A.g.rhododendri* Mayr, 1931   EC New Guinea
  — *A.g.wollastoni* Ogilvie-Grant, 1913   WC New Guinea

○ *Anthus spragueii* (Audubon, 1844) SPRAGUE'S PIPIT   NC and S USA, S Mexico

○ *Anthus lutescens* YELLOWISH PIPIT
  — *A.l.parvus* Lawrence, 1865   W Panama
  — *A.l.peruvianus* Nicholson, 1878   Peru, N Chile
  — *A.l.lutescens* Pucheran, 1855   Colombia, Venezuela, the Guianas, Brazil, Argentina

○ *Anthus furcatus* SHORT-BILLED PIPIT
  — *A.f.brevirostris* Taczanowski, 1874   Peru, Bolivia
  — *A.f.furcatus* d'Orbigny & Lafresnaye, 1837   Brazil, Paraguay, Uruguay, Argentina

○ *Anthus chacoensis* J.T. Zimmer, 1952 CHACO PIPIT   Paraguay, Argentina

○ *Anthus correndera* CORRENDERA PIPIT
  — *A.c.calcaratus* Taczanowski, 1874   Peru
  — *A.c.correndera* Vieillot, 1818   S Brazil, Uruguay, Paraguay, N Argentina
  — *A.c.catamarcae* Hellmayr, 1921   Bolivia, N Chile, NW Argentina
  — *A.c.chilensis* (Lesson, 1839)   S Chile, S Argentina
  — *A.c.grayi* Bonaparte, 1850   Falkland Is.

○ *Anthus antarcticus* Cabanis, 1884 SOUTH GEORGIA PIPIT   S Georgia I.

○ *Anthus nattereri* P.L. Sclater, 1878 OCHRE-BREASTED PIPIT   SE Brazil, Paraguay

○ *Anthus hellmayri* HELLMAYR'S PIPIT
  — *A.h.hellmayri* E. Hartert, 1909   Peru, Bolivia, NW Argentina
  — *A.h.dabbenei* Hellmayr, 1921   Chile, W Argentina
  — *A.h.brasilianus* Hellmayr, 1921   SE Brazil, Uruguay, N Argentina

○ *Anthus bogotensis* PARAMO PIPIT
  — *A.b.meridae* J.T. Zimmer, 1953   NW Venezuela
  — *A.b.bogotensis* P.L. Sclater, 1855   Colombia, Ecuador
  — *A.b.immaculatus* Cory, 1916   Bolivia, Peru
  — *A.b.shiptoni* (Chubb, 1923)   NW Argentina >> Bolivia

FRINGILLIDAE FINCHES AND HAWAIIAN HONEYCREEPERS (42: 168)

FRINGILLINAE

FRINGILLA Linnaeus, 1758 F
● *Fringilla coelebs* CHAFFINCH
  — *F.c.gengleri* O. Kleinschmidt, 1909   British Isles
  ✓ *F.c.coelebs* Linnaeus, 1758   Europe (except SE), W and C Siberia to Lake Baikal >> N Africa, SW Asia

  — *F.c.solomkoi* Menzbier & Sushkin, 1913   Crimea, SW Caucasus
  — *F.c.balearica* von Jordans, 1923[2]   Iberia and the Balearic Is.
  ✓ *F.c.tyrrhenica* Schiebel, 1910[3]   Corsica
  — *F.c.sarda* Rapine, 1925   Sardinia
  — *F.c.schiebeli* Stresemann, 1925   S Greece, Crete, W Asia Minor
  — *F.c.syriaca* Harrison, 1945[4]   Cyprus, Levant, SE Turkey, N Iraq

---

[1] Treated in the genus *Hemimacronyx* in last Edition, but see Fry *et al.* (1992: 246) (900).
[2] For recognition see Cramp *et al.* (1994: 448) (595).
[3] For recognition see Cramp *et al.* (1994: 448) (595).
[4] For recognition see Cramp *et al.* (1994: 448) (595).

— *F.c.caucasica* Serebrowski, 1925[1]   N Turkey, NW Iran, Caucasus area
— *F.c.alesandrovi* Zarudny, 1916   N Iran >> Middle East
— *F.c.transcaspia* Zarudny, 1916   NE Iran, SW Turkmenistan
— *F.c.africana* Levaillant, 1850   NW Africa east to extreme NW Tunisia
— *F.c.spodiogenys* Bonaparte, 1841   Tunisia, NW Libya
— *F.c.moreletti* Pucheran, 1859   Azores Is.
— *F.c.maderensis* Sharpe, 1888   Madeira I.
— *F.c.canariensis* Vieillot, 1817   Gran Canaria, Tenerife, Gomera Is. (Canary Is.)
— *F.c.ombriosa* E. Hartert, 1913   Hierro I. (Canary Is.)
— *F.c.palmae* Tristram, 1889   La Palma I. (Canary Is.)

○ *Fringilla teydea* BLUE CHAFFINCH
— *F.t.teydea* Webb, Berthelot & Moquin-Tandon, 1841   Tenerife I.
— *F.t.polatzeki* E. Hartert, 1905   Gran Canaria I.

● *Fringilla montifringilla* Linnaeus, 1758   BRAMBLING   N and E Europe, N Siberia to Kamchatka and Sakhalin >> N Africa, SW, C and E Asia to Japan

CARDUELINAE

SERINUS Koch, 1816[2]   M
○ *Serinus pusillus* (Pallas, 1811)   RED-FRONTED SERIN   Asia Minor and Caucasus area to Tien Shan and NW Himalayas >> Israel, Aegean Is.

● *Serinus serinus* (Linnaeus, 1766)   EUROPEAN SERIN[3]   W and C Europe, Asia Minor, N Africa

○ *Serinus syriacus* Bonaparte, 1850[4]   SYRIAN SERIN   Lebanon, Syria >> Iraq, Egypt

○ *Serinus canaria* (Linnaeus, 1758)   ISLAND CANARY   Canary, Azores, Madeira Is.

○ *Serinus thibetanus* (Hume, 1872)   TIBETAN SISKIN   Nepal, SE Tibet, NE Burma and SW China

○ *Serinus canicollis* YELLOW-CROWNED CANARY / CAPE CANARY
— *S.c.flavivertex* (Blanford, 1869)   Ethiopia and Eritrea to N Tanzania
— *S.c.sassii* Neumann, 1922   E Zaire, SW Uganda to N Malawi, SW Tanzania
— *S.c.huillensis* Sousa, 1889   C Angola
— *S.c.griseitergum* Clancey, 1967   E Zimbabwe
— *S.c.canicollis* (Swainson, 1838)[5]   Transvaal, Natal to Cape Province

○ *Serinus nigriceps* Rüppell, 1840   ETHIOPIAN SISKIN   N and C Ethiopia

○ *Serinus citrinelloides* AFRICAN CITRIL[6]
— *S.c.citrinelloides* Rüppell, 1840   Eritrea, Ethiopia, SE Sudan
— *S.c.kikuyensis* (Neumann, 1905)   W Kenya
— *S.c.brittoni* Traylor, 1970   S Sudan to W Kenya (2430)
— *S.c.frontalis* Reichenow, 1904   E Zaire to W Uganda, W Tanzania, NE Zambia
— *S.c.hypostictus* (Reichenow, 1904)   W Ethiopia, S Kenya, Tanzania, E Zambia to Mozambique
— *S.c.martinsi* da Rosa Pinto, 1962   Angola

○ *Serinus capistratus* BLACK-FACED CANARY
— *S.c.capistratus* (Finsch & Hartlaub, 1870)   Gabon to N Angola, S and E Zaire, N Zambia
— *S.c.hildegardae* Rand & Traylor, 1959   C Angola

○ *Serinus koliensis* Grant & Mackworth-Praed, 1952   PAPYRUS CANARY   Uganda, W Kenya, Rwanda

○ *Serinus scotops* FOREST CANARY
— *S.s.transvaalensis* Roberts, 1940   N and E Transvaal
— *S.s.umbrosus* Clancey, 1964   SE Transvaal, W Natal to S Cape Province
— *S.s.scotops* (Sundevall, 1850)   Lowland S Natal, E Cape Province

○ *Serinus leucopygius* WHITE-RUMPED SEEDEATER
— *S.l.riggenbachi* Neumann, 1908   Senegal to Chad, Central African Republic, W Sudan

[1] For recognition see Cramp *et al.* (1994: 448) (595).
[2] For preliminary views on sequencing this genus see Arnaiz-Villena *et al.* (1999) (53); not followed here as too many species not yet determined.
[3] Includes *flaviserinus* Trischitta, 1939 (2440).
[4] Dated '1851?' by Howell *et al.* (1968: 210) (1152), but see Zimmer (1926) (2707).
[5] Includes *thompsonae* see White (1963: 112) (2648).
[6] Treated as two species by Howell *et al.* (1968) (1152), but as one in our last Edition. Revised by van den Elzen (1985) (2463); since when some authors treat two species and some three. We will modify this *next* Edition after *Birds of Africa* takes a position on this.

— *S.l.pallens* Vaurie, 1956	Niger
— *S.l.leucopygius* (Sundevall, 1850)	E Sudan, N Ethiopia, Eritrea, NW Uganda

○ *Serinus atrogularis* YELLOW-RUMPED SEEDEATER/BLACK-THROATED CANARY

— *S.a.xanthopygius* Rüppell, 1840	Eritrea, N and W Ethiopia
— *S.a.reichenowi* Salvadori, 1888	SE Sudan, S and C Ethiopia, Somalia to Kenya, NE and C Tanzania
— *S.a.somereni* E. Hartert, 1912	E Zaire, W Uganda, W Kenya
— *S.a.lwenarum* C.M.N. White, 1944	S Zaire, C Angola, N and C Zambia
— *S.a.atrogularis* (A. Smith, 1836)	Zimbabwe, SE Botswana, W Transvaal
— *S.a.impiger* Clancey, 1959	SE Transvaal, W Natal, N Cape Province
— *S.a.semideserti* Roberts, 1932	S Angola, N Namibia, S Zambia, Botswana
— *S.a.deserti* (Reichenow, 1918)	SW Angola, NW Namibia

○ *Serinus rothschildi* Ogilvie-Grant, 1902 OLIVE-RUMPED SERIN[1] W Saudi Arabia, W Yemen

○ *Serinus flavigula* Salvadori, 1888 YELLOW-THROATED SEEDEATER[2] C Ethiopia

○ *Serinus xantholaemus* Salvadori, 1896 SALVADORI'S SEEDEATER[3] [4] S Ethiopia (2162)

○ *Serinus citrinipectus* Clancey & Lawson, 1960 LEMON-BREASTED CANARY

  S Malawi, SE Zimbabwe, S Mozambique, N Natal

○ *Serinus mozambicus* YELLOW-FRONTED CANARY

— *S.m.caniceps* (d'Orbigny, 1839)	Senegal to Nigeria, N Cameroon
— *S.m.punctigula* Reichenow, 1898	C and S Cameroon
— *S.m.grotei* Sclater & Mackworth-Praed, 1918	E Sudan to Eritrea, W Ethiopia
— *S.m.gommaensis* Grant & Mackworth-Praed, 1945	W Ethiopia
— *S.m.barbatus* (Heuglin, 1864)	S Chad, W and S Sudan to N Zaire, W Kenya, NW and C Tanzania
— *S.m.santhome* Bannerman, 1921	São Tomé I.
— *S.m.tando* Sclater & Mackworth-Praed, 1918	Gabon, SW Zaire, N Angola
— *S.m.samaliyae* C.M.N. White, 1947	SE Zaire, SW Tanzania, NE Zambia
— *S.m.vansoni* Roberts, 1932	SE Angola, NE Namibia, SW Zambia
— *S.m.mozambicus* (Statius Müller, 1776)	E Kenya to C Mozambique, Zimbabwe and N Sth. Africa
— *S.m.granti* Clancey, 1957	S Mozambique

○ *Serinus donaldsoni* Sharpe, 1895 NORTHERN GROSBEAK-CANARY

  Somalia, E and S Ethiopia, N Kenya

○ *Serinus buchanani* E. Hartert, 1919 SOUTHERN GROSBEAK-CANARY[5]

  S Kenya, NE Tanzania

○ *Serinus flaviventris* YELLOW CANARY

— *S.f.damarensis* (Roberts, 1922)	SW Angola, Namibia, W and S Botswana to extreme N Cape Province
— *S.f.flaviventris* (Swainson, 1828)[6] [7]	W, S and C Cape Province
— *S.f.guillarmodi* (Roberts, 1936)	Lesotho
— *S.f.marshalli* Shelley, 1902[8]	NW, NE Cape Province, Orange Free State, S and W Transvaal, SE Botswana

○ *Serinus dorsostriatus* WHITE-BELLIED CANARY

— *S.d.maculicollis* Sharpe, 1859	S Ethiopia, Somalia, N, C and SE Kenya, NE Tanzania
— *S.d.dorsostriatus* (Reichenow, 1887)	SE Uganda, SW Kenya, NW Tanzania

○ *Serinus sulphuratus* BRIMSTONE CANARY

— *S.s.sharpii* Neumann, 1900[9]	Angola to Kenya, Tanzania and N Mozambique
— *S.s.wilsoni* (Roberts, 1936)[10]	Zimbabwe, S Mozambique, E Natal and Sth. Africa (except S Cape Province)
— *S.s.sulphuratus* (Linnaeus, 1766)	S Cape Province

---

[1] For treatment as a separate species see Hall & Moreau (1970: 273) (1066); further substantiation would be desirable.
[2] For the extraction of this from the synonymy of *S. atrogularis xanthopygius* see Érard (1974) (802).
[3] For correction of spelling see David & Gosselin (2002) (613).
[4] For reasons to separate this from *S. flavigula* see Érard (1974) (802) and Ash & Gullick (1990) (60). Not listed in last Edition.
[5] For treatment as separate from *S. donaldsoni* see Zimmerman *et al.* (1996) (2735).
[6] Includes *hesperus*, this name being given to an intergrading form see White (1965: 298) (2649).
[7] Includes *quintoni* see White (1963: 110) (2648), but see also White (1965: 298) (2649).
[8] Includes *aurescens* see White (1963: 110) (2648).
[9] Includes *loveridgei*, not mentioned by White (1963: 111) (2648).
[10] Includes *languens* see White (1963: 111) (2648).

○ *Serinus albogularis* White-throated Canary
    — *S.a.crocopygius* (Sharpe, 1871)   SW Angola, N Namibia
    — *S.a.sordahlae* (Friedmann, 1932)   S Namibia, NW Cape Province
    — *S.a.orangensis* (Roberts, 1937)   W Orange Free State, SW Transvaal
    — *S.a.albogularis* (A. Smith, 1833)   Coastal W Cape Province
    — *S.a.hewitti* (Roberts, 1937)   C to E Cape Province

○ *Serinus reichardi* Reichard's Seedeater / Stripe-breasted Seedeater
    — *S.r.striatipectus* (Sharpe, 1891)   SE Sudan, S Ethiopia, N Kenya
    — *S.r.reichardi* (Reichenow, 1882)   S Zaire, Zambia to C Tanzania, N Mozambique

○ *Serinus gularis* Streaky-headed Seedeater
    — *S.g.canicapilla*[1] (Du Bus, 1855)[2]   Ghana to N Cameroon
    — *S.g.montanorum* (Bannerman, 1923) SE Nigeria and W Cameroon
    — *S.g.elgonensis* (Ogilvie-Grant, 1912)[3] S Chad to N Zaire, S Sudan, Uganda, W Kenya
    — *S.g.benguellensis* (Reichenow, 1904) C Angola, W Zambia
    — *S.g.mendosus* Clancey, 1966   NE Botswana, Zimbabwe, NW Transvaal
    — *S.g.gularis* (A. Smith, 1836)   N and W Transvaal, SE Botswana, N Cape Province
    — *S.g.endemion* (Clancey, 1952)   S Mozambique, E Sth. Africa
    — *S.g.humilis* (Bonaparte, 1850)   SW and S Cape Province

○ *Serinus mennelli* (Chubb, 1908) Black-eared Seedeater   E Angola to SE Tanzania, Mozambique

○ *Serinus tristriatus* Rüppell, 1840 Brown-rumped Seedeater   Eritrea, Ethiopia, NW Somalia

○ *Serinus ankoberensis* Ash, 1979 Ankober Serin   C Ethiopia (57)

○ *Serinus menachensis* (Ogilvie-Grant, 1913) Yemen Serin   SW Saudi Arabia, W Yemen

○ *Serinus striolatus* Streaky Seedeater
    — *S.s.striolatus* Rüppell, 1840[4]   Eritrea, Ethiopiato Kenya, N Tanzania
    — *S.s.graueri* E. Hartert, 1907   E Zaire, SW Uganda, Rwanda
    — *S.s.whytii* Shelley, 1897   S Tanzania, N Malawi

○ *Serinus burtoni* Thick-billed Seedeater
    — *S.b.burtoni* (G.R. Gray, 1862)   Cameroon
    — *S.b.tanganjicae* Granvik, 1923   E Zaire to W Uganda and C Angola
    — *S.b.kilimensis* (Richmond, 1897)   W Kenya, N Tanzania
    — *S.b.albifrons* (Sharpe, 1891)   E Kenya
    — *S.b.melanochrous* Reichenow, 1900 S Tanzania

○ *Serinus rufobrunneus* Principe Seedeater
    — *S.r.rufobrunneus* (G.R. Gray, 1862) Principe I. (Gulf of Guinea)
    — ¶*S.r.fradei* (de Naurois, 1975)   Ilot Bone de Joquei, near Principe (626)
    — *S.r.thomensis* (Bocage, 1888)   São Tomé I. (Gulf of Guinea)

○ *Serinus leucopterus* (Sharpe, 1871) Protea Canary   SW Cape Province

○ *Serinus totta* (Sparrman, 1786) Cape Siskin   SW and S Cape Province

○ *Serinus symonsi* (Roberts, 1916) Drakensburg Siskin   NE Cape Province, Lesotho, W Natal

○ *Serinus alario* Black-headed Canary
    — *S.a.leucolaemus* (Sharpe, 1903)[5]   NW and N Cape Province to W Orange Free State,
                                      S Botswana and S Namibia
    — *S.a.alario* (Linnaeus, 1758)   W Cape Province to S Orange Free State

○ *Serinus estherae* Mountain Serin
    — *S.e.vanderbilti* Meyer de Schauensee, 1939 N Sumatra
    — *S.e.estherae* (Finsch, 1902)   W Java
    — *S.e.orientalis* Chasen, 1940   E Java
    — *S.e.renatae* Schuchmann & Wolters, 1982 Mt. Rantekombola (Sulawesi) (2204)
    — *S.e.mindanensis* Ripley & Rabor, 1961   Mts. Apo and Katanglad (Mindanao)

[1] For correct spelling see David & Gosselin (2002) (613).
[2] Du Bus is an abbreviation of Du Bus de Gisignies.
[3] Includes *uamensis* see White (1963: 120) (2648).
[4] White (1963) (2648) inferentially included *affinis*.
[5] For correction of spelling see David & Gosselin (2002) (613).

NEOSPIZA Salvadori, 1903 F
○ *Neospiza concolor* (Bocage, 1888) São Tomé Grosbeak   São Tomé I. (Gulf of Guinea)

LINURGUS Reichenbach, 1850 M
○ **Linurgus olivaceus** Oriole-Finch
    — *L.o.olivaceus* (Fraser, 1842)   SE Nigeria, Cameroon, Bioko I.
    — *L.o.prigoginei* Schouteden, 1950   E Zaire
    — *L.o.elgonensis* van Someren, 1918   SE Sudan, W and C Kenya
    — *L.o.kilimensis* (Reichenow & Neumann, 1895)   S Kenya, Tanzania, N Malawi

RHYNCHOSTRUTHUS Sclater & Hartlaub, 1881 M
○ **Rhynchostruthus socotranus** Golden-winged Grosbeak
    — *R.s.louisae* Lort Phillips, 1897   N Somalia
    — *R.s.percivali* Ogilvie-Grant, 1900   SW Arabia
    — *R.s.socotranus* Sclater & Hartlaub, 1881   Socotra I

CARDUELIS Brisson, 1760[1] F
● **Carduelis chloris** European Greenfinch
    — *C.c.harrisoni* (Clancey, 1940)[2]   Britain (excl. N Scotland), Ireland
    — *C.c.chloris* (Linnaeus, 1758)   N France, N Scotland and Norway to W Siberia >>
            S Europe
    — *C.c.muehlei* Parrot, 1905[3]   Serbia, Macedonia and Romania to E Greece, Cyprus
            and W Asia Minor
    — *C.c.aurantiiventris* (Cabanis, 1851)   S Europe (C Spain and S France to Croatia and
            W Greece)
    — *C.c.madaraszi* (Tschusi, 1911)[4]   Corsica, Sardinia
    — *C.c.vanmarli* Voous, 1951[5]   NW Morocco, Portugal, NW Spain
    — *C.c.voousi* Roselaar, 1993   Algeria, C Morocco (2105)
    — *C.c.chlorotica* (Bonaparte, 1850)   SC Turkey, Levant, W Jordan, NE Egypt
    — *C.c.bilkevitschi* (Zarudny, 1911)[6]   NE Turkey, Caucasus area, N Iran, SW Turkmenistan
    — *C.c.turkestanica* (Zarudny, 1907)   W and N Tien Shan >> Afghanistan

○ **Carduelis sinica** Oriental Greenfinch
    — *C.s.ussuriensis* (E. Hartert, 1903)[7]   NE China and Korea to E Transbaikalia, Russian Far
            East and S Sea of Okhotsk
    — *C.s.kawarahiba* (Temminck, 1836)[8]   Kamchatka, Koryakland, Sakhalin, Kuril Is., NE
            Hokkaido >> Japan, SE China and Taiwan
    — *C.s.minor* (Temminck & Schlegel, 1848)[9]   S Hokkaido to Kyushu, Izu Is. and Tsushima (Japan)
    — *C.s.kittlitzi* (Seebohm, 1890)   Ogasawara-shoto (Bonin) Is. and Iwo (Volcano) Is.
    — *C.s.sinica* (Linnaeus, 1766)   E and C China, Vietnam

○ **Carduelis spinoides** Yellow-breasted Greenfinch
    — *C.s.spinoides* Vigors, 1831   N Pakistan, Himalayas to NW Assam and SE Xizang
    — *C.s.heinrichi* Stresemann, 1940   SE Assam, W Burma

○ **Carduelis monguilloti** (Delacour, 1926) Vietnamese Greenfinch[10]   Langbian (SC Vietnam)

○ **Carduelis ambigua** Black-headed Greenfinch
    — *C.a.taylori* (Kinnear, 1939)   SE Xizang >> N Assam
    — *C.a.ambigua* (Oustalet, 1896)   E and NE Burma, N Laos, S China north to S Sichuan
            >> N Thailand, N Indochina

● **Carduelis spinus** (Linnaeus, 1758) Eurasian Siskin   Europe and N Asia E to Kamchatka and S to N Turkey,
            N Iran, Altai, N Japan >> N Africa, S Japan and China

---

[1] In last Edition we separated some of these species in a genus *Acanthis*, but see Cramp *et al.* (1994) (595).
[2] For recognition see Cramp *et al.* (1994: 548) (595).
[3] For recognition see Cramp *et al.* (1994: 548) (595).
[4] For recognition see Cramp *et al.* (1994: 548) (595).
[5] For recognition see Cramp *et al.* (1994: 548) (595).
[6] For recognition see Cramp *et al.* (1994: 548) (595); see also Roselaar (1995) (2107).
[7] Includes *chabarovi* see Stepanyan (1990: 613) (2326).
[8] Given as 1835 by Howell *et al.* (1968: 237) (1152), but see Dickinson (2001) (739). Tentatively includes *sichitoensis* as in Howell *et al.* (1968: 237) (1152) and Orn. Soc. Japan (2000) (1727), but CSR has reservations.
[9] Includes *tokumii* Mishima, 1961 (1594) see Orn. Soc. Japan (1974: 361) (1726).
[10] This was separated from *C. spinoides* by Clement (1993) (538). No separate detailed review noted.

● *Carduelis pinus* PINE SISKIN
    ⊥ *C.p.pinus* (A. Wilson, 1810)   Canada, USA >> C Mexico
    — *C.p.macroptera* (Bonaparte, 1850)   N Baja California, NW and C Mexico
    — *C.p.perplexa* (van Rossem, 1938)   S Mexico, W Guatemala

○ *Carduelis atriceps* (Salvin, 1863) BLACK-CAPPED SISKIN   S Mexico, W Guatemala

○ *Carduelis spinescens* ANDEAN SISKIN
    — *C.s.spinescens* (Bonaparte, 1850)[1]   Colombia, W Venezuela
    — *C.s.capitanea* (Bangs, 1898)   N Colombia
    — *C.s.nigricauda* (Chapman, 1912)   N Colombia

○ *Carduelis yarrellii* Audubon, 1839 YELLOW-FACED SISKIN   N Venezuela, NE Brazil

○ *Carduelis cucullata* Swainson, 1820 RED SISKIN   NE Colombia, N Venezuela, Guyana

○ *Carduelis crassirostris* THICK-BILLED SISKIN
    — *C.c.amadoni* (George, 1964)   SE Peru
    — *C.c.crassirostris* (Landbeck, 1877)   S Bolivia, C Chile, W Argentina

○ *Carduelis magellanica* HOODED SISKIN
    — *C.m.capitalis* (Cabanis, 1866)   S Colombia, Ecuador, NW Peru
    — *C.m.paula* (Todd, 1926)   S Ecuador, W Peru
    — *C.m.peruana* (Berlepsch & Stolzmann, 1896)   C Peru
    — *C.m.urubambensis* (Todd, 1926)   S Peru, N Chile
    — *C.m.santaecrucis* (Todd, 1926)   EC Bolivia
    — *C.m.boliviana* (Sharpe, 1888)   S Bolivia
    — ¶ *C.m.hoyi* Koenig, 1981   Central Andes of NW Argentina (Jujuy Prov.) (1293)
    — *C.m.tucumana* (Todd, 1926)   W Andes and foothills of NW Argentina
    — *C.m.alleni* (Ridgway, 1899)   SE Bolivia, Paraguay, NE Argentina
    — *C.m.icterica* (M.H.K. Lichtenstein, 1823)   SE Brazil, E and S Paraguay
    — *C.m.magellanica* (Vieillot, 1805)   Uruguay, E Argentina
    — *C.m.longirostris* (Sharpe, 1888)   SE Venezuela, Guyana, N Brazil

○ *Carduelis dominicensis* (H. Bryant, 1868) ANTILLEAN SISKIN   Hispaniola

○ *Carduelis siemiradzkii* (Berlepsch & Taczanowski, 1883) SAFFRON SISKIN
                                           SW Ecuador, NW Peru

○ *Carduelis olivacea* (Berlepsch & Stolzmann, 1894) OLIVACEOUS SISKIN
                                           E Ecuador, Peru, N Bolivia

○ *Carduelis notata* BLACK-HEADED SISKIN
    — *C.n.notata* Du Bus, 1847[2]   E and C Mexico, N Guatemala
    — *C.n.forreri* (Salvin & Godman, 1886)   W Mexico
    — *C.n.oleacea* (Griscom, 1932)   Belize to N Nicaragua

○ *Carduelis xanthogastra* YELLOW-BELLIED SISKIN
    — *C.x.xanthogastra* (Du Bus, 1855)[3]   Costa Rica to Colombia, Venezuela
    — *C.x.stejnegeri* (Sharpe, 1888)   S Peru, C Bolivia

○ *Carduelis atrata* d'Orbigny & Lafresnaye, 1837 BLACK SISKIN   S Peru to N Chile, W Argentina

● *Carduelis uropygialis* (P.L. Sclater, 1862) YELLOW-RUMPED SISKIN   S Peru to Chile, W Argentina

● *Carduelis barbata* (Molina, 1782) BLACK-CHINNED SISKIN   S Chile, W Argentina

● *Carduelis tristis* AMERICAN GOLDFINCH
    ⊯ *C.t.tristis* (Linnaeus, 1758)   C USA >> SE USA, E Mexico
    — *C.t.pallida* (Mearns, 1890)   W Canada, WC USA >> N Mexico
    ↧ *C.t.jewetti* (van Rossem, 1943)   SW Canada, NW USA
    — *C.t.salicamans* (Grinnell, 1897)   SW USA, N Baja California

● *Carduelis psaltria* LESSER GOLDFINCH
    ↓ *C.p.hesperophila* (Oberholser, 1903)   W USA, NW Mexico

---

[1] Dated '1851?' by Howell *et al.* (1968: 240) (1152), but see Zimmer (1926) (2707).
[2] Du Bus is an abbreviation of Du Bus de Gisignies.
[3] Du Bus is an abbreviation of Du Bus de Gisignies.

__ *C.p.witti* (P.R. Grant, 1964)   Tres Marias Is. (Mexico)

☑ *C.p.psaltria* (Say, 1823)   SC USA, N Mexico

__ *C.p.jouyi* (Ridgway, 1898)   SE Mexico

__ *C.p.colombiana* Lafresnaye, 1843   S Mexico to Peru, Venezuela

● ***Carduelis lawrencei*** Cassin, 1852 LAWRENCE'S GOLDFINCH   SW USA >> NW Mexico

● ***Carduelis carduelis*** EUROPEAN GOLDFINCH

☑ *C.c.britannica* (E. Hartert, 1903)   British Isles, W Netherlands to NW France >> SW Europe

__ *C.c.carduelis* (Linnaeus, 1758)   S Scandinavia and C European Russia to C France and Italy >> S Europe

__ *C.c.parva* Tschusi, 1901   W Mediterranean, Azores, Canary Is.

☑ *C.c.tschusii* Arrigoni, 1902   Corsica, Sardinia, Sicily

__ *C.c.balcanica* Sachtleben, 1919   Balkan countries, Greece

__ *C.c.niediecki* Reichenow, 1907   Cyprus, W and SC Asia Minor, Levant >> SW Iran, Egypt

__ *C.c.loudoni* Zarudny, 1906[1]   E Turkey, Transcaucasia, N Iran

__ *C.c.colchicus* Koudashev, 1915   Crimea, N Caucasus area

__ *C.c.volgensis* Buturlin, 1906   S Ukraine, SE European Russia, NW Kazakhstan

__ *C.c.frigoris* Wolters, 1953[2]   SW and SC Siberia >> E Europe, SW Asia

__ *C.c.paropanisi* Kollibay, 1910   W Turkmenistan and E Iran to N Afghanistan, W Pamir, Tien Shan and NW Xinjiang >> SW Asia

__ *C.c.subulata* (Gloger, 1833)   Altai and W Mongolia to Lake Baikal >> SW and C Asia

__ *C.c.caniceps* Vigors, 1831   W and N Pakistan, NW Himalayas to C Nepal and SW Xizang

__ *C.c.ultima* Koelz, 1949[3]   SE Fars, Kirman (S Iran)

○ ***Carduelis citrinella*** CITRIL FINCH[4]

__ *C.c.citrinella* (Pallas, 1764)   Spain, France, Alps

__ *C.c.corsicana* (Koenig, 1899)[5]   Corsica, Sardinia

● ***Carduelis flammea*** COMMON REDPOLL[6]

__ *C.f.cabaret* (Statius Müller, 1776)   Alps and N Carpathians to Britain, Denmark, SE Norway, SW Sweden, Poland >> W and S Europe

☑ *C.f.flammea* (Linnaeus, 1758)   N Europe, Siberia, Nth. America >> S Europe, C and E Asia

__ *C.f.rostrata* (Coues, 1862)[7]   NE Canada, Greenland, Iceland >> NE USA, NW Europe

● ***Carduelis hornemanni*** HOARY REDPOLL

☑ *C.f.exilipes* (Coues, 1862)   N Eurasia, N Nth. America >> C Europe, N China

__ *C.f.hornemanni* (Holboell, 1843)   Greenland, N Canada >> British Isles, S Canada

● ***Carduelis flavirostris*** TWITE

__ *C.f.bensonorum* (R. Meinertzhagen, 1934)[8]   Outer Hebrides (Scotland)

__ *C.f.pipilans* (Latham, 1787)   N British Isles, Ireland

☑ *C.f.flavirostris* (Linnaeus, 1758)   Norway, Kola Pen. (Russia) >> W, C and SE Europe

__ *C.f.brevirostris* (F. Moore, 1856)   S and E Turkey, Caucasus area, NW and N Iran

__ *C.f.kirghizorum* Sushkin, 1925[9]   N Kazakhstan

__ *C.f.korejevi* (Zarudny & Härms, 1914)   C and E Tien Shan, E Kazakhstan, N Xinjiang

__ *C.f.altaica* (Sushkin, 1925)   Russian Altai, Mongolia, Tuva Rep.

__ *C.f.montanella* (Hume, 1873)   W and S fringe of Tarim Basin S (Xinjiang)

__ *C.f.pamirensis* Zarudny & Härms, 1914[10]   Pamir-Alai Mts. to NW Pakistan

__ *C.f.miniakensis* (Jacobi, 1923)   E Tibetan Plateau (E Xizang to Qinghai and Gansu)

__ *C.f.rufostrigata* (Walton, 1905)   W and S Tibetan Plateau (Ladakh to W and S Xizang and N Nepal)

---

[1] Serves to replace the name *brevirostris* Zarudny, 1889, which is preoccupied when the genus *Acanthis* is submerged in *Carduelis*.
[2] A new name to replace *major* Taczanowski, 1879, which is preoccupied when the genus *Acanthis* is submerged in *Carduelis*.
[3] For recognition see Cramp *et al.* (1994: 569) (595).
[4] Moved here from the genus *Serinus* following Arnaiz-Villena *et al.* (1998) (52). Sangster *et al.* (2002) (2167) preferred to retain this in *Serinus*.
[5] We are not convinced that this should be split, but see Sangster *et al.* (2002) (2167).
[6] Due to post-irruptive breeding the relationships within this group are still unclear. But see Knox *et al.* (2001) (1278) for reasons to split.
[7] Includes *islandica* see Herremans (1990) (1108)
[8] For recognition see Cramp *et al.* (1994: 625) (595).
[9] For recognition see Cramp *et al.* (1994: 625) (595).
[10] For recognition see Cramp *et al.* (1994: 625) (595).

● *Carduelis cannabina* COMMON LINNET
  — *C.c.autochthona* (Clancey, 1946)   Scotland
  ↙ *C.c.cannabina* (Linnaeus, 1758)   W, C and N Europe, W and SC Siberia to N Altai >>
                          N Africa, SW Asia
  — *C.c.bella* (C.L. Brehm, 1845)   C Turkey, Cyprus and Levant to Caucasus area and
                          Iran; Turkmenistan and N Afghanistan through
                          Tien Shan to SW Altai
  — *C.c.mediterranea* (Tschusi, 1903)[1]   Iberia and NW Africa to W Asia Minor, Crete and Libya
  — *C.c.guentheri* Wolters, 1953[2]   Madeira I.
  — *C.c.meadewaldoi* (E. Hartert, 1901)   W and C Canary Is.
  — *C.c.harterti* (Bannerman, 1913)   E Canary Is.

○ *Carduelis yemensis* (Ogilvie-Grant, 1913) YEMENI LINNET   SW Saudi Arabia, W Yemen

○ *Carduelis johannis* (Stephenson Clarke, 1919) WARSANGLI LINNET   NE Somalia

LEUCOSTICTE Swainson, 1832 F
○ *Leucosticte nemoricola* PLAIN MOUNTAIN FINCH
  — *L.n.altaica* (Eversmann, 1848)   E Afghanistan through Tien Shan and Altai to S Baikal
                          area, to Kun Lun Mts., and through NW Himalayas
                          to Kumaun
  — *L.n.nemoricola* (Hodgson, 1836)   C and E Himalayas, E Tibetan Plateau to N Qinghai
                          and S Shaanxi >> NE Burma

○ *Leucosticte brandti* BRANDT'S MOUNTAIN FINCH[3]
  — *L.b.margaritacea* (Madarász, 1904)   Tarbagatay Mts., SE Russian Altai, W Mongolia
  — *L.b.brandti* Bonaparte, 1850[4]   Tien Shan to NW and C Xinjiang >> N Pakistan
  — *L.b.pamirensis* Severtsov, 1883   C and NE Afghanistan, Pamir-Alai Mts., SW Tien Shan,
                          NE Afghanistan, SW Xinjiang >> N Pakistan
  — *L.b.pallidior* Bianchi, 1909   S Xinjiang, N Qinghai, Ningxia
  — *L.b.haematopygia* (Gould, 1853)   N Pakistan, NW Himalayas, W Xizang
  — *L.b.audreyana* Stresemann, 1939   SC Tibetan Plateau (S Xizang, NE Nepal, N Bhutan)
  — *L.b.walteri* (E. Hartert, 1904)   SE Tibetan Plateau (E Xizang, N Yunnan, W Sichuan)
  — *L.b.intermedia* Stegmann, 1932   NE Tibetan Plateau (Qinghai, SW Gansu)

○ *Leucosticte sillemi* Roselaar, 1992 SILLEM'S MOUNTAIN FINCH #   W Tibetan Plateau (N India/SW China) (2103)

○ *Leucosticte arctoa* ASIAN ROSY FINCH[5]
  — *L.a.arctoa* (Pallas, 1811)   Altai, W Sayan, Tannu Ola Mts., NW Mongolia >>
                          NW Xinjiang
  — *L.a.cognata* (Madarász, 1909)   E Sayan to S Baikal Mts. and Hövsgöl area (N Mongolia)
  — *L.a.sushkini* Stegmann, 1932   Hangayn Nuruu (N Mongolia)
  — *L.a.gigliolii* Salvadori, 1868   N Baikal Mts. to W Stanovoy Mts. (SW Yakutia)
  — *L.a.brunneonucha* (J.F. Brandt, 1842)   NE Siberia from C Stanovoy Mts. north to the Lena
                          delta, east to Chukotka, Kuril Is. and EC Sakhalin
                          >> Russian Far East, Japan

○ *Leucosticte tephrocotis* GREY-CROWNED ROSY FINCH
  — *L.t.maxima* W.S. Brooks, 1915[6]   Commander Is. (NE Siberia)
  — *L.t.griseonucha* (J.F. Brandt, 1842)   Aleutian, Kodiak Is., Alaska
  — *L.t.umbrina* Murie, 1944   St. Matthew, Pribilof Is.
  — *L.t.irvingi* Feinstein, 1958   N Alaska
  — *L.t.littoralis* S.F. Baird, 1869   E Alaska, W Canada >> SW USA
  — *L.t.tephrocotis* (Swainson, 1832)   WC Canada >> WC USA
  — *L.t.dawsoni* Grinnell, 1913   E California
  — *L.t.wallowa* A.H. Miller, 1939   NE Oregon >> W Nevada

○ *Leucosticte atrata* Ridgway, 1874 BLACK ROSY FINCH   WC USA >> SC USA

○ *Leucosticte australis* Ridgway, 1873 BROWN-CAPPED ROSY FINCH   W USA (mainly Rocky Mtns. of Colorado)

---

[1] For recognition see Cramp *et al.* (1994: 604) (595).
[2] A new name for *nana* which is preoccupied in *Carduelis* see Howell *et al.* (1968: 255) (1152).
[3] Races recognised follow Howell *et al.* (1968) (1152).
[4] Dated '1851?' by Howell *et al.* (1968: 258) (1152), but see Zimmer (1926) (2707).
[5] In last Edition a broad species concept treated this and the three species after it as one. We now follow A.O.U. (1998: 659) (3) in accepting four species.
[6] For recognition see Stepanyan (1990: 627) (2326).

CALLACANTHIS Bonaparte, 1850[1] F
○ *Callacanthis burtoni* (Gould, 1838) RED-SPECTACLED FINCH — NW Pakistan, W and C Himalayas

RHODOPECHYS Cabanis, 1851[2] M[3]
○ *Rhodopechys sanguineus* CRIMSON-WINGED FINCH
— *R.s.alienus* Whitaker, 1897 — Morocco, NE Algeria
— *R.s.sanguineus* (Gould, 1838) — Asia Minor, Levant, Caucasus area, Iran, mts. of WC Asia

BUCANETES Cabanis, 1851 M
○ *Bucanetes githagineus* TRUMPETER FINCH
— *B.g.amantum* (E. Hartert, 1903) — C and E Canary Is.
— *B.g.zedlitzi* (Neumann, 1907) — S Spain, NW Africa to Libya and NW Sudan
— *B.g.githagineus* (M.H.K. Lichtenstein, 1823) — Egypt, NC and NE Sudan
— *B.g.crassirostris* (Blyth, 1847) — Sinai, Arabia, S and E Turkey, Iran to Uzbekistan and W Pakistan >> NW India

○ *Bucanetes mongolicus* (Swinhoe, 1870) MONGOLIAN FINCH — E Turkey and N and E Iran through WC Asia to Qinghai, W Hebei, Mongolia and S and E Kazakhstan

RHODOSPIZA Sharpe, 1888 F
○ *Rhodospiza obsoleta* (M.H.K. Lichtenstein, 1823) DESERT FINCH — N Arabia, Levant and SE Turkey east to S Kazakhstan, Xinjiang and W Pakistan >> SW Asia

URAGUS Keyserling & Blasius, 1840 M
○ *Uragus sibiricus* LONG-TAILED ROSEFINCH
— *U.s.sibiricus* (Pallas, 1773) — SW and SC Siberia, N Mongolia, NW Manchuria, upper Amur valley >> SE European Russia, N and E Kazakhstan, NW and NC China
— *U.s.ussuriensis* Buturlin, 1915 — NE Manchuria, middle Amur valley, Ussuriland, Korea >> E China
— *U.s.sanguinolentus* (Temminck & Schlegel, 1848) — Sakhalin, S Kurils, Hokkaido, N Honshu >> S Japan
— *U.s.lepidus* David & Oustalet, 1877 — NE Xizang and E Qinghai to Gansu, Shaanxi and Shanxi (NC China)
— *U.s.henrici* Oustalet, 1891 — E Xizang, NW Yunnan, W and N Sichuan

UROCYNCHRAMUS Przevalski, 1876[4] M
○ *Urocynchramus pylzowi* Przevalski, 1876 PRZEVALSKI'S ROSEFINCH — E Xizang and W Sichuan to E Qinghai and SW Gansu (C China)

CARPODACUS Kaup, 1829 M
○ *Carpodacus rubescens* (Blanford, 1872) BLANFORD'S ROSEFINCH — C and E Himalayas, SE Xizang, WC China

○ *Carpodacus nipalensis* DARK-BREASTED ROSEFINCH
— *C.n.kangrae* (Whistler, 1939) — NW Himalayas
— *C.n.nipalensis* (Hodgson, 1836)[5] — C and E Himalayas, N and NE Burma, WC China (N Yunnan to E Qinghai and Henan)

○ *Carpodacus erythrinus* COMMON ROSEFINCH
— *C.e.erythrinus* (Pallas, 1770) — N and C Europe, W and C Siberia south to Mongolia and east to Lena and Kolyma basins >> India
— *C.e.grebnitskii* Stejneger, 1885 — NE China and N Korea to Kamchatka and NE Siberia >> SE China
— *C.e.kubanensis* Laubmann, 1915 — Balkans, Turkey, Caucasus area, N and NE Iran >> India
— *C.e.ferghanensis* (Kozlova, 1939) — NW Himalayas, W and N Pakistan, Afghanistan, Tien Shan, Xinjiang >> NW India
— *C.e.roseatus* (Blyth, 1842) — C and E Himalayas and E Tibetan Plateau to W and N Yunnan, Guizhou, Shanxi and S Nei Mongol >> S India, nthn. mainland SE Asia

---

[1] Dated '1851?' by Howell *et al.* (1968: 262) (1152), but see Zimmer (1926) (2707).
[2] In last Edition we followed Howell *et al.* (1968) (1152) in using the broad genus circumscribed by Vaurie (1956, 1959) (2498, 2502). However this has generally not been adopted in Europe, see for example Voous (1977) (2548) and Cramp *et al.* (1994) (595). We therefore follow Voous (1977) (2548).
[3] Gender addressed by David & Gosselin (2002) (614), multiple name changes occur.
[4] For a recent molecular study see Groth (2000) (1022).
[5] Includes *intensicolor* see Cheng (1987: 975) (333).

● *Carpodacus purpureus* PURPLE FINCH
    ✓ *C.p.purpureus* (J.F. Gmelin, 1789)   Canada, NE USA >> SE USA
    ✓ *C.p.californicus* S.F. Baird, 1858   SW Canada, W USA >> to SW USA, Baja California

● *Carpodacus cassinii* S.F. Baird, 1854 CASSIN'S FINCH   SW Canada, W USA >> to N Mexico

● *Carpodacus mexicanus* HOUSE FINCH
    ⊥ *C.m.frontalis* (Say, 1823)[1]   SW Canada, W USA, NW Mexico
    ___ *C.m.clementis* Mearns, 1898   San Clemente, Los Coronados Is. (off California and
                                        NW Mexico)

    ___ †?*C.m.mcgregori* Anthony, 1897   San Benito I. (Baja California)
    ___ *C.m.amplus* Ridgway, 1876   Guadeloupe I. (Baja California)
    ___ *C.m.ruberrimus* Ridgway, 1887   S Baja California, NW Mexico
    ___ *C.m.rhodopnus* R.T. Moore, 1936   C Sinaloa (Mexico)
    ___ *C.m.coccineus* R.T. Moore, 1939   SW Mexico
    ___ *C.m.potosinus* Griscom, 1928   NC Mexico
    ___ *C.m.centralis* R.T. Moore, 1937   C Mexico
    ___ *C.m.mexicanus* (Statius Müller, 1776)   SC Mexico
    ___ *C.m.roseipectus* Sharpe, 1888[2]   Oaxaca valley (Mexico)
    ___ *C.m.griscomi* R.T. Moore, 1939   Guerrero (Mexico)

○ *Carpodacus pulcherrimus* BEAUTIFUL ROSEFINCH
    ___ *C.p.pulcherrimus* (F. Moore, 1856)   C Himalayas (Kumaun to Bhutan)
    ___ *C.p.waltoni* (Sharpe, 1905)   S and E Xizang
    ___ *C.p.argyrophrys* Berlioz, 1929   WC China (Qinghai and W Sichuan to SW Gansu)
    ___ *C.p.davidianus* A. Milne-Edwards, 1866   C Mongolia, EC China (Shaanxi and Ningxia to
                                          W Hebei and S and SE Nei Mongol)

○ *Carpodacus eos* (Stresemann, 1930) PINK-RUMPED ROSEFINCH   E Qinghai, W Sichuan, N Yunnan >> E Xizang
                                                NW Thailand

○ *Carpodacus rodochroa* (Vigors, 1831) PINK-BROWED ROSEFINCH[3]   NW and C Himalayas (N Pakistan to W Bhutan

○ *Carpodacus vinaceus* VINACEOUS ROSEFINCH
    ___ *C.v.vinaceus* J. Verreaux, 1871   Nepal, SE Xizang and Yunnan north to Gansu, Henan
                                          and Guizhou >> N India, N and E Burma
    ___ *C.v.formosanus* Ogilvie-Grant, 1911   Taiwan

○ *Carpodacus edwardsii* DARK-RUMPED ROSEFINCH
    ___ *C.e.edwardsii* J. Verreaux, 1871   W and N Yunnan, W Sichuan >> NE Burma
    ___ *C.e.rubicundus* (Greenway, 1933)[4]   E Himalayas, SE Xizang >> N Burma, Nepal

○ *Carpodacus synoicus* PALE ROSEFINCH
    ___ *C.s.synoicus* (Temminck, 1825)   Sinai, S Israel, SW Jordan, NW Saudi Arabia
    ___ *C.s.salimalii* (R. Meinertzhagen, 1938) C Afghanistan
    ___ *C.s.stoliczkae* (Hume, 1874)   S Xinjiang to NW Qinghai
    ___ *C.s.beicki* (Stresemann, 1930)   NE Qinghai, SW Gansu

○ *Carpodacus roseus* PALLAS'S ROSEFINCH
    ___ *C.r.roseus* (Pallas, 1776)   C and E Siberia (Altai and middle Yenisey to Stanovoy
                                        and Kolyma Mts.) >> NE, EC China, Korea
    ___ *C.r.portenkoi* Browning, 1988   Sakhalin I. >> N Japan (247)

○ *Carpodacus trifasciatus* J. Verreaux, 1871 THREE-BANDED ROSEFINCH   E Tibetan Plateau E to N Yunnan and S Shaanxi >> Nepal

○ *Carpodacus rodopeplus* SPOT-WINGED ROSEFINCH
    ___ *C.r.rodopeplus* (Vigors, 1831)   Himalayas
    ___ *C.r.verreauxii* (David & Oustalet, 1877) NE Burma and N Yunnan to SW Sichuan

○ *Carpodacus thura* WHITE-BROWED ROSEFINCH
    ___ *C.t.blythi* (Biddulph, 1882)   NE Afghanistan, NW Himalayas
    ___ *C.t.thura* Bonaparte & Schlegel, 1850 C Himalayas
    ___ *C.t.femininus* Rippon, 1906   SE Xizang, N Yunnan, W Sichuan

---

[1] Includes *anconophila* Oberholser, 1974 (1673) see Browning (1978) (246).
[2] For recognition of *roseipectus* see Binford (2000) (151).
[3] For correction of spelling see David & Gosselin (2002) (613).
[4] For correction of spelling see David & Gosselin (2002) (613).

___ *C.t.dubius* Przevalski, 1876 | NE Xizang, E Qinghai, N Sichuan, Gansu, Ningxia

___ *C.t.deserticolor* (Stegmann, 1931) | NC Qinghai

○ *Carpodacus rhodochlamys* (J.F. Brandt, 1843) RED-MANTLED ROSEFINCH[1]

Tien Shan and Xinjiang to SE Altai and N Mongolia

○ *Carpodacus grandis* Blyth, 1849 BLYTH'S ROSEFINCH[2]

NW Himalayas, W and N Pakistan, Afghanistan, Pamir-Alai and SW Tien Shan Mts.

○ *Carpodacus rubicilloides* STREAKED ROSEFINCH

___ *C.r.lucifer* R. & A. Meinertzhagen, 1926 | S Tibetan Plateau (Ladakh to E Xizang and N Nepal)

___ *C.r.rubicilloides* Przevalski, 1876 | E Tibetan Plateau (NE Xizang, Qinghai, W Sichuan, Gansu)

○ *Carpodacus rubicilla* GREAT ROSEFINCH

___ *C.r.rubicilla* (Güldenstädt, 1775) | Caucasus

___ *C.r.diabolicus* (Koelz, 1939)[3] | NE Afghanistan, W Pamir and Alai Mts.

___ *C.r.kobdensis* (Sushkin, 1925) | C and N Xinjiang, Altai, Sayans, W and S Mongolia

___ *C.r.severtzovi* Sharpe, 1886 | Tien Shan, E Pamir, N Pakistan, S Xinjiang and Tibetan Plateau

○ *Carpodacus puniceus* RED-FRONTED ROSEFINCH

___ *C.p.kilianensis* Vaurie, 1956 | Mts. at W and S fringe of Tarim Basin (SW Xinjiang)

___ *C.p.humii* (Sharpe, 1888) | Tien Shan and Pamir-Alai Mts. to Ladakh, N Pakistan and NW Himalayas

___ *C.p.puniceus* (Blyth, 1845) | C Himalayas, S Xizang

___ *C.p.sikangensis* Vaurie, 1956 | SW Sichuan, NW Yunnan

___ *C.p.longirostris* (Przevalski, 1876)[4] | Qinghai, Gansu, N Sichuan

KOZLOWIA Bianchi, 1907[5] F

○ *Kozlowia roborowskii* (Przevalski, 1887) ROBOROVSKI'S ROSEFINCH | S and C Qinghai

CHAUNOPROCTUS Bonaparte, 1850[6] M

○ †*Chaunoproctus ferreorostris* (Vigors, 1829)[7] BONIN GROSBEAK | Ogasawara-shoto (Bonin) Is. (Japan)

PINICOLA Vieillot, 1808 F[8]

● *Pinicola enucleator* PINE GROSBEAK

___ *P.e.enucleator* (Linnaeus, 1758) | Scandinaviato NW Siberia >> S Sweden to SW Siberia

___ *P.e.kamtschatkensis* (Dybowski, 1883)[9] | C Siberia (Yenisey and Altai to Kamchatka) >> N China

___ *P.e.sakhalinensis* Buturlin, 1915 | Sakhalin, Kurils, Hokkaido

___ *P.e.alascensis* Ridgway, 1898 | Alaska, W Canada, NW USA

___ *P.e.flammula* Homeyer, 1880 | W Canada >> NW USA

___ *P.e.carlottae* A. Brooks, 1922 | Queen Charlotte, Vancouver Is.

___ *P.e.montana* Ridgway, 1898 | SW Canada, WC USA

___ *P.e.californica* Price, 1897 | E California

___ *P.e.leucura* (Statius Müller, 1776) | C and E Canada >> NE USA

___ *P.e.eschatosa* Oberholser, 1914 | SE Canada >> NE USA

○ *Pinicola subhimachala* (Hodgson, 1836) CRIMSON-BROWED FINCH | E Himalayas, S and SE Xizang, S Sichuan, N Yunnan >> Nepal, Assam, N Burma

HAEMATOSPIZA Blyth, 1845 F

○ *Haematospiza sipahi* (Hodgson, 1836) SCARLET FINCH | C Himalayas, S Assam, W and NE Burma, W Yunnan >> E Himalayas, N Burma, NW Thailand, NW Indochina

LOXIA Linnaeus, 1758 F

○ *Loxia pytyopsittacus* Borkhausen, 1793 PARROT CROSSBILL | N Europe, W Siberia

○ *Loxia scotica* E. Hartert, 1904 SCOTTISH CROSSBILL | Scotland

---

[1] Includes *kotschubeii* seen as doubtfully distinct in Howell *et al.* (1968) (1152).
[2] Treated as a species following Stepanyan (1983; 1990: 632) (2523, 2526); there is considerable breeding sympatry between this and *rhodochlamys*.
[3] For correction of spelling see David & Gosselin (2002) (613).
[4] Tentatively includes *szetschuanus* which was recognised by Cheng (1987) (333).
[5] We agree with Vaurie (1959) (2502) and Cheng (1987) (333) that this genus should be retained.
[6] Dated '1851?' by Howell *et al.* (1968: 283) (1152), but see Zimmer (1926) (2707).
[7] Dated "1828 (= 1829?)" by Howell *et al.* (1968: 283) (1152). The issue in which this article appears is that for Oct. 28, 1828 to Jan. 29, 1829. The name can safely be dated 1829.
[8] Gender addressed by David & Gosselin (2002) (614), multiple changes occur in names.
[9] Includes *pacatus* see Stepanyan (1990: 637) (2326).

● *Loxia curvirostra* RED CROSSBILL
— *L.c.curvirostra* Linnaeus, 1758   W, C and N Europe south to N Spain and C Italy, N and NE Siberia to N Amurland

— *L.c.corsicana* Tschusi, 1912   Corsica
— *L.c.balearica* (Homeyer, 1862)   Balearic Is.
— *L.c.poliogyna* Whitaker, 1898   NW Africa
— *L.c.guillemardi* Madarász, 1903[1]   E Balkans, Cyprus, Crimea, Turkey, Caucasus area
— *L.c.altaiensis* Sushkin, 1925   Altai, Sayans, W and N Mongolia
— *L.c.tianschanica* Laubmann, 1927   Tien Shan, NW Xinjiang >> N China
— *L.c.himalayensis* Blyth, 1845   Himalayas, WC China >> N Burma
— *L.c.meridionalis* Robinson & Kloss, 1919   S Vietnam
— *L.c.japonica* Ridgway, 1885   NE China, Russian Far East, Japan, ? Korea, ? Sakhalin, ? S Kuril Is. >> EC China, S Japan

— *L.c.luzoniensis* Ogilvie-Grant, 1894   NW Luzon
— *L.c.pusilla* Gloger, 1834   Newfoundland >> NE USA
— *L.c.minor* (C.L. Brehm, 1846)   S Canada, N USA >> E USA
✓ *L.c.bendirei* Ridgway, 1884   SW Canada, W USA >> S USA
— *L.c.vividior* A.R. Phillips, 1981[2]   Colorado (1607)
— *L.c.reai* A.R. Phillips, 1981   Idaho (1607)
✓ *L.c.grinnelli* Griscom, 1937   SW USA
— *L.c.stricklandi* Ridgway, 1885   S USA, Mexico
— *L.c.mesamericana* Griscom, 1937   Guatemala to N Nicaragua

● *Loxia leucoptera* WHITE-WINGED CROSSBILL
— *L.l.bifasciata* (C.L. Brehm, 1827)   N Europe, N Asia south to Altai, Baikal area and Amurland >> C Europe to C Asia and Japan

— *L.l.leucoptera* J.F. Gmelin, 1789   Canada, N USA
— *L.l.megaplaga* Riley, 1916   Hispaniola

PYRRHULA Brisson, 1760 F
○ *Pyrrhula nipalensis* BROWN BULLFINCH
— *P.n.nipalensis* Hodgson, 1836   Himalayas east to NW Assam
— *P.n.ricketti* La Touche, 1905   SE Assam, W and N Burma, SE Xizang, W and N Yunnan, SE China, NW Vietnam

— *P.n.*subsp.?   SC Vietnam
— *P.n.victoriae* Rippon, 1906   Chin Hills (W Burma)
— *P.n.waterstradti* E. Hartert, 1902   Mts. of W Malaysia
— *P.n.uchidai* N. Kuroda, Sr., 1916   Taiwan

○ *Pyrrhula leucogenis* WHITE-CHEEKED BULLFINCH
— *P.l.leucogenis* Ogilvie-Grant, 1895[3]   N Luzon
— *P.l.steerei* Mearns, 1909[4]   Mindanao

○ *Pyrrhula aurantiaca* Gould, 1858 ORANGE BULLFINCH   N Pakistan, NW Himalayas to Garhwal

○ *Pyrrhula erythrocephala* Vigors, 1832 RED-HEADED BULLFINCH   Himalayas (Kashmir to Bhutan)

○ *Pyrrhula erythaca* GREY-HEADED BULLFINCH
— *P.e.erythaca* Blyth, 1862[5]   E Himalayas and N Burma to C China
— *P.e.owstoni* Hartert & Rothschild, 1907   Taiwan

○ *Pyrrhula pyrrhula* EURASIAN BULLFINCH[6]
— *P.p.pileata* W. MacGillivray, 1837   British Isles
— *P.p.pyrrhula* (Linnaeus, 1758)   N, EC and SE Europe, W and C Siberia east to Yakutia, south to N Mongolia and Baikal area >> S Europe, SW and C Asia

— *P.p.europoea* Vieillot, 1816   W Europe (W and C France and Italy to W Denmark)

---

[1] Includes *mariae* see Cramp *et al.* (1994: 707) (595).
[2] This subspecies was reviewed by Browning (1990) (251) and tentatively accepted, as was the next.
[3] For spelling see Dickinson *et al.* (1991: 410) (745).
[4] Includes *apo* and *coriaria* see Dickinson *et al.* (1991) (745).
[5] Includes *wilderi* see Cheng (1987: 997) (333).
[6] Subspecies sequence geographical; not based on discernible groups, but this is a "ring species" with overlap. Treated as three species (*pyrrhula*, *cineracea* and *griseiventris*) by Stepanyan (1990) (2326).

___ *P.p.iberiae* Voous, 1951 — N Iberia, S France

___ *P.p.paphlagoniae* Roselaar, 1995 — NW Asia Minor (2107)

___ *P.p.rossikowi* Derjugin & Bianchi, 1900 — NE Asia Minor, Caucasus area >> NW Iran

___ *P.p.caspica* Witherby, 1908 — N Iran

___ *P.p.cineracea* Cabanis, 1872 — SC Siberia to N Mongolia and Transbaikalia

___ *P.p.cassinii* S.F. Baird, 1869 — S Koryakland, Kamchatka, N and W shore of Sea of Okhotsk >> Japan, NE China

___ *P.p.griseiventris* Lafresnaye, 1841 — C and S Kurils Is., Japan

___ *P.p.rosacea* Seebohm, 1882[1] — Sakhalin, Russian Far East, NE China, Korea >> E China, Japan

___ *P.p.murina* Godman, 1866[2] — San Miguel (Azores)

COCCOTHRAUSTES Brisson, 1760[3] M
● *Coccothraustes coccothraustes* HAWFINCH
 ↙ *C.c.coccothraustes* (Linnaeus, 1758) — Europe, S Siberia E to Transbaikalia and N Mongolia >> N Africa, C and E Asia

___ *C.c.buvryi* Cabanis, 1862[4] — NW Africa

___ *C.c.nigricans* Buturlin, 1908 — Crimea, Caucasus area, N and NE Iran >> Middle East

___ *C.c.humii* Sharpe, 1886 — E Uzbekistan, W Tajikistan, W Kyrgyzstan >> Pakistan, NW India

___ *C.c.schulpini* H. Johansen, 1944[5] — NE China, Russian Far East, N Korea >> E China

___ *C.c.japonicus* Temminck & Schlegel, 1848 — Sakhalin I., S Kamchatka, N and C Japan >> Ogasawara-shoto (Bonin) Is., Taiwan

EOPHONA Gould, 1851 F
○ *Eophona migratoria* YELLOW-BILLED GROSBEAK
___ *E.m.migratoria* E. Hartert, 1903 — NE China, E Transbaikalia, middle Amur and Ussuri valleys, Korea >> E China, nthn. SE Asia

___ *E.m.sowerbyi* Riley, 1915 — C and EC China >> Yunnan

○ *Eophona personata* JAPANESE GROSBEAK
___ *E.p.personata* (Temminck & Schlegel, 1848) — N and C Japan >> S Japan

___ *E.p.magnirostris* E. Hartert, 1896 — NE China, Korea, Russian Far East >> S China

MYCEROBAS Cabanis, 1847 M
○ *Mycerobas icterioides* (Vigors, 1831) BLACK-AND-YELLOW GROSBEAK — NE Afghanistan to NW Himalayas

○ *Mycerobas affinis* (Blyth, 1855) COLLARED GROSBEAK — Himalayas, NE Burma and N Yunnan to W Sichuan

○ *Mycerobas melanozanthos* (Hodgson, 1836) SPOT-WINGED GROSBEAK — NE Pakistan, Himalayas, E and S Assam, W, N and NE Burma, SC China, nthn. SE Asia

○ *Mycerobas carnipes* WHITE-WINGED GROSBEAK
___ *M.c.speculigerus* (J.F. Brandt, 1841) — NE Iran and SW Turkmenistan to EC Afghanistan and W Pakistan

___ *M.c.merzbacheri* Schalow, 1908[6] — Pamir-Alai and Tien Shan Mts. to SW, C and NW Xinjiang

___ *M.c.carnipes* (Hodgson, 1836) — NE Afghanistan, N Pakistan, Himalayas and C China from S and E Xizang, N Burma and N Yunnan north to Gansu and Ningxia

HESPERIPHONA Bonaparte, 1850[7] F
● *Hesperiphona vespertina* EVENING GROSBEAK
___ *H.v.vespertina* (W. Cooper, 1825) — C and E Canada >> NE USA
___ *H.v.brooksi* Grinnell, 1917 — W Canada >> SW USA
___ *H.v.montana* Ridgway, 1874 — W and SW Mexico

○ *Hesperiphona abeillei* HOODED GROSBEAK
___ *H.a.pallida* Nelson, 1928 — NW Mexico

[1] For recognition see Cramp *et al.* (1994: 815) (595) and Orn. Soc. Japan (2000) (1727).
[2] May be a distinct species, but requires confirmatory publication.
[3] We prefer to recognise distinct genera, as for *Hesperiphona* in Cramp *et al.* (1994: 847) (595) *contra* Paynter in Howell *et al.* (1968: 299) (1152); the "presumed" relationships are now open to discernment using molecular techniques.
[4] The spelling *burryi* in Howell *et al.* (1968) (1152) was erroneous.
[5] For recognition see Cramp *et al.* (1994: 832) (595). Howell *et al.* (1968) (1152) used an emended spelling.
[6] For recognition see Stepanyan (1990: 649) (2326).
[7] Treated in *Coccothraustes* in last Edition, but see Cramp *et al.* (1994: 847) (595).

    — *H.a.saturata* Sutton & Burleigh, 1939   W Mexico
    — *H.a.abeillei* (Lesson, 1839)   C and S Mexico
    — *H.a.cobanensis* Nelson, 1928   S Mexico, Guatemala

PYRRHOPLECTES Hodgson, 1844 M
○ *Pyrrhoplectes epauletta* (Hodgson, 1836) Gold-naped Finch   Himalayas, SE Xizang, NW Yunnan, NE Burma >> lower Burma

DREPANIDINAE

TELESPYZA S.B. Wilson, 1890 F
○ *Telespyza cantans* S.B. Wilson, 1890 Laysan Finch   Laysan I. (Hawaiian Is.)

○ *Telespyza ultima* Bryan, 1917 Nihoa Finch   Nihoa I. (Hawaiian Is.)

PSITTIROSTRA Temminck, 1820 F
○ †?*Psittirostra psittacea* (J.F. Gmelin, 1789) Ou   Kauai, Maui, Hawaii, Oahu, Molokai and Lanai Is. (Hawaiian Is.)

DYSMORODREPANIS Perkins, 1919 F
○ †*Dyrmorodrepanis munroi* Perkins, 1919 Lanai Hookbill   Lanai I. (Hawaiian Is.)

LOXIOIDES Oustalet, 1877 M
○ *Loxioides bailleui* (Oustalet, 1877) Palila   Hawaii I. (Hawaiian Is.)

RHODACANTHIS Rothschild, 1892 F
○ †*Rhodacanthis flaviceps* Rothschild, 1892 Lesser Koa Finch   Hawaii I. (Hawaiian Is.)

○ †*Rhodacanthis palmeri* Rothschild, 1892 Greater Koa Finch   Hawaii I. (Hawaiian Is.)

CHLORIDOPS S.B. Wilson, 1888 M
○ †*Chloridops kona* S.B. Wilson, 1888 Kona Grosbeak   Hawaii I. (Hawaiian Is.)

PSEUDONESTOR Rothschild, 1893 M
○ *Pseudonestor xanthophrys* Rothschild, 1893 Maui Parrotbill   Maui I. (Hawaiian Is.)

HEMIGNATHUS M.H.K. Lichtenstein, 1839 M
○ *Hemignathus virens* Hawaii Amakihi
    — *H.v.wilsoni* (Rothschild, 1893)   Molokai, Lanai and Maui Is. (Hawaiian Is.)
    — *H.v.virens* (J.F. Gmelin, 1788)   Hawaii I. (Hawaiian Is.)

○ †*Hemignathus flavus* (Bloxham, 1827) Oahu Amakihi[1]   Oahu I. (Hawaiian Is.)

◉ †*Hemignathus kauaiensis* H.D. Pratt, 1989 Kauai Amakihi[2]   Kauai I. (Hawaiian Is.) (1917)

◉ *Hemignathus parvus* (Stejneger, 1887) Anianiau   Kauai I. (Hawaiian Is.)

○ †*Hemignathus sagittirostris* (Rothschild, 1892) Greater Amakihi   Hawaii I. (Hawaiian Is.)

○ †*Hemignathus obscurus* (J.F. Gmelin, 1788) Lesser Akialoa   Hawaii I. (Hawaiian Is.)

○ *Hemignathus ellisianus* Greater Akialoa[3]
    — †*H.e.ellisianus* (G.R. Gray, 1860)   Oahu I. (Hawaiian Is.)
    — †*H.e.lanaiensis* (Rothschild, 1893)   Lanai I. (Hawaiian Is.)
    — †?*H.e.stejnegeri* (S.B. Wilson, 1889)[4]   Kauai I. (Hawaiian Is.)

○ *Hemignathus lucidus* Nukupuu
    — †*H.l.lucidus* M.H.K. Lichtenstein, 1839   Oahu I. (Hawaiian Is.)
    — †?*H.l.hanapepe* S.B. Wilson, 1889   Kauai I. (Hawaiian Is.)
    — †?*H.l.affinis* Rothschild, 1893   Maui I. (Hawaiian Is.)

○ *Hemignathus munroi* H.D. Pratt, 1979 Akiapolaau[5]   Hawaii I. (Hawaiian Is.) (1915)

[1] Greenway (1968: 96) (1009) treated this in the genus *Viridonia* in which the earlier name *flavus* is unavailable and *chloris* must be used.
[2] Greenway (1968: 96) (1009) treated this in the genus *Viridonia* in which the earlier name *stejnegeri* is available.
[3] For the separation of *ellisianus* from *obscurus* see A.O.U. (1998: 674) (3).
[4] In last Edition we listed *Hemignathus procerus*, but Olson & James (1988) (1708) noted that *stejnegeri* is a prior name and when united, as in A.O.U. (1998: 675) (3), *stejnegeri* yields priority to *ellisianus* as the specific epithet.
[5] Originally named *Heterorhynchus wilsoni* Rothschild, 1893, but this is preoccupied in a broad genus *Hemignathus*.

OREOMYSTIS Stejneger, 1903 F
○ *Oreomystis bairdi* Stejneger, 1887 Akikiki   Kauai I. (Hawaiian Is.)

● *Oreomystis mana* (S.B. Wilson, 1891) Hawaii Creeper   Hawaii I. (Hawaiian Is.)

PAROREOMYZA Perkins, 1901 F
○ †?*Paroreomyza maculata* (Cabanis, 1850)[1] Oahu Alauahio   Oahu I. (Hawaiian Is.)

○ †?*Paroreomyza flammea* (S.B. Wilson, 1890) Kakawahie   Molokai I. (Hawaiian Is.)

○ *Paroreomyza montana* Maui Alauahio
    — †?*P.m.montana* (S.B. Wilson, 1890)   Lanai I. (Hawaiian Is.)
    — *P.m.newtoni* (Rothschild, 1893)   Maui I. (Hawaiian Is.)

LOXOPS Cabanis, 1847 M
○ *Loxops caeruleirostris* (S.B. Wilson, 1890) Akekee[2]   Kauai I. (Hawaiian Is.)

○ *Loxops coccineus* Akepa
    — *L.c.coccineus* (J.F. Gmelin, 1789)   Hawaii I. (Hawaiian Is.)
    — †?*L.c.ochraceus* Rothschild, 1893[3]   Maui I. (Hawaiian Is.)
    — †*L.c.wolstenholmei* Rothschild, 1893[4]   Oahu I. (Hawaiian Is.)

CIRIDOPS A. Newton, 1892 M
○ †*Ciridops anna* (Dole, 1879) Ula-ai-hawane   Hawaii I. (Hawaiian Is.)

VESTIARIA Jarocki, 1821 F
◑ *Vestiaria coccinea* (J.R. Forster, 1780) Iiwi   Kauai, Oahu, Molokai, Maui, Lanai, Hawaii Is.
                                                                      (Hawaiian Is.)

DREPANIS Temminck, 1820 F
○ †*Drepanis pacifica* (J.F. Gmelin, 1788) Hawaii Mamo   Hawaii I. (Hawaiian Is.)

○ †*Drepanis funerea* A. Newton, 1894[5] Black Mamo   Molokai I. (Hawaiian Is.)

PALMERIA Rothschild, 1893 F
○ *Palmeria dolei* (S.B. Wilson, 1891) Akohekohe/Crested Honeycreeper
                                                                      Maui I. (Hawaiian Is.)

HIMATIONE Cabanis, 1851 F
● *Himatione sanguinea* Apapane
    — †*H.s.freethii* Rothschild, 1892   Laysan I. (Hawaiian Is.)
    ✓ *H.s.sanguinea* (J.F. Gmelin, 1788)   Kauai, Oahu, Molokai, Maui, Hawaii Is. (Hawaiian Is.)

MELAMPROSOPS Casey & Jacobi, 1974 M
○ *Melamprosops phaeosoma* Casey & Jacobi, 1974 Poo-uli[6]   Maui I. (Hawaiian Is.) (304)

PARULIDAE NEW WORLD WARBLERS (24: 112)

VERMIVORA Swainson, 1827 F
○ †?*Vermivora bachmanii* (Audubon, 1833) Bachman's Warbler   C and SE USA >> Cuba

● *Vermivora chrysoptera* (Linnaeus, 1766) Golden-winged Warbler E USA >> C America, Nth. S America

○ *Vermivora pinus* (Linnaeus, 1766) Blue-winged Warbler   E USA >> E Mexico, C America, Nth. S America

● *Vermivora peregrina* (A. Wilson, 1811) Tennessee Warbler   NW, C and SE Canada, E USA >> S Mexico to N Sth.
                                                                      America

● *Vermivora celata* Orange-crowned Warbler
    — *V.c.celata* (Say, 1823)   N and NW Canada, S USA >>Mexico, Guatemala
    ✓ *V.c.lutescens* (Ridgway, 1872)   W Canada, W USA >> W Mexico
    — *V.c.orestera* Oberholser, 1905   WC Canada, WC USA >> C Mexico
    — *V.c.sordida* (C.H. Townsend, 1891)   S California, N Baja California and islands

---

[1] Dated 1851 by Greenway (1968: 100) (1009), but see Browning & Monroe (1991) (258) and www.zoonomen.net (A.P. Peterson, *in litt.*).
[2] For reasons to assign this specific rank see Pratt (1989) (1918).
[3] For correction of spelling see David & Gosselin (2002) (614).
[4] The earlier name *rufus* based on *Fringilla rufa* Bloxham, 1827, is preoccupied in *Fringilla* and unavailable see A.O.U. (1998: 677) (3).
[5] Dated 1893 by Greenway (1968: 95) (1009), but see Duncan (1937) (770).
[6] This may not belong in this subfamily, see Pratt (1992) (1919).

● *Vermivora ruficapilla* Nashville Warbler
    ☑ *V.r.ridgwayi* van Rossem, 1929   W USA >> W Mexico, Guatemala
    — *V.r.ruficapilla* (A. Wilson, 1811)   S Canada, C and E USA >> Mexico, Guatemala

○ *Vermivora virginiae* (S.F. Baird, 1860) Virginia's Warbler   SW USA >> W Mexico

○ *Vermivora crissalis* (Salvin & Godman, 1889) Colima Warbler   S USA >> EC Mexico

○ *Vermivora luciae* (J.G. Cooper, 1861) Lucy's Warbler   SW USA >> W Mexico

PARULA Bonaparte, 1838 F
○ *Parula gutturalis* (Cabanis, 1860) Flame-throated Warbler[1]   Costa Rica, W Panama

○ *Parula superciliosa* Crescent-chested Warbler
    — *P.s.sodalis* (R.T. Moore, 1941)   NC Mexico
    — *P.s.mexicana* Bonaparte, 1850   E Mexico
    — *P.s.palliata* (van Rossem, 1939)   SW Mexico
    — *P.s.superciliosa* (Hartlaub, 1844)   S Mexico, Guatemala, W Honduras
    — *P.s.parva* (Miller & Griscom, 1925)   E Honduras, Nicaragua

● *Parula americana* (Linnaeus, 1758) Northern Parula   SE Canada, E USA, E Mexico >> C America, West Indies

● *Parula pitiayumi* Tropical Parula
    — *P.p.graysoni* (Ridgway, 1887)   Socorro I., Revillagigedo Is. (off Mexico)
    — *P.p.insularis* Lawrence, 1873   Tres Marias Is. (off Mexico)
    — *P.p.pulchra* (Brewster, 1889)   NW Mexico
    ☑ *P.p.nigrilora* Coues, 1878   S Texas, NE Mexico
    — *P.p.inornata* S.F. Baird, 1864   S Mexico, E Guatemala, N Honduras
    — *P.p.speciosa* (Ridgway, 1902)   S Honduras, Nicaragua, Costa Rica, W Panama
    — *P.p.cirrha* Wetmore, 1957   Coiba I. (Panama)
    — *P.p.nana* (Griscom, 1927)   E Panama, NW Colombia
    — *P.p.elegans* (Todd, 1912)   Colombia, N Venezuela, N Brazil, Trinidad
    — *P.p.roraimae* (Chapman, 1929)   S Venezuela, N Brazil
    — *P.p.alarum* (Chapman, 1924)   E Ecuador, N Peru
    — *P.p.pacifica* Berlepsch & Taczanowski, 1885   SW Colombia, W Ecuador, NW Peru
    — *P.p.melanogenys* (Todd, 1924)   S Peru, N? Bolivia
    — *P.p.pitiayumi* (Vieillot, 1817)   S? Bolivia, C and S Brazil, Uruguay, Paraguay,
                                                    N Argentina

DENDROICA G.R. Gray, 1842 F
● *Dendroica pensylvanica* (Linnaeus, 1766) Chestnut-sided Warbler

                                               S Canada, E USA >> Central America, Nth. S America

● *Dendroica petechia* Yellow Warbler[2]
    — *D.p.rubiginosa* (Pallas, 1811)   W Canada >> Mexico, Central America
    ☑ *D.p.banksi* Browning, 1994   Alaska, NW Canada >> Central America, N Sth.
                                            America (256)

    — *D.p.parkesi* Browning, 1994   N Canada >> C America, N Sth. America (256)
    — *D.p.amnicola* Batchelder, 1918   Canada >> Mexico, Central America, N Sth. America
    — *D.p.aestiva* (J.F. Gmelin, 1789)[3]   S Canada, C USA >> Cental America, N Sth. America
    — *D.p.morcomi* Coale, 1887   W Canada, W USA >> Central America, N Sth. America
    ☑ *D.p.brewsteri* Grinnell, 1903   W USA, NW Baja California >> C America
    — *D.p.sonorana* Brewster, 1888[4]   SW USA >> C America, Colombia, Ecuador
    — *D.p.dugesi* Coale, 1887   C Mexico
    — *D.p.rufivertex* Ridgway, 1885   Cozumel I. (off Yucatán Pen.)
    — *D.p.flavida* Cory, 1887   Isla Andrès (W Caribbean)
    — *D.p.armouri* Greenway, 1933   Isla Providencia (W Caribbean)
    — *D.p.flaviceps* Chapman, 1892[5]   Bahama Is.
    — *D.p.gundlachi* S.F. Baird, 1865   Cuba, Isla de la Juventud, S Florida

---

[1] The A.O.U. (1983, 1998) (2, 3) treated this and the next species in this genus following Eisenmann (1955) (794). We now follow.
[2] The treatment of this species has been brought into line with the review by Browning (1994) (256). We regret that we previously overlooked the review by, and new races of, Olson (1980) (1693).
[3] Includes *ineditus* see Browning (1994) (256).
[4] Includes *hypochlora* Oberholser, 1974 (1673), see Browning (1974) (244).
[5] For recognition see Browning (1994) (256).

__ *D.p.chlora* Browning, 1994	Siete Hermanos Is., off Hispaniola (256)
__ *D.p.solaris* Wetmore, 1929[1]	Gonave Is. (Haiti)
__ *D.p.albicollis* (J.F. Gmelin, 1789)	Hispaniola
__ *D.p.eoa* (Gosse, 1847)	Jamaica, Cayman Is.
__ *D.p.bartholemica* Sundevall, 1870[2]	Puerto Rico, Virgin Is., N Lesser Antilles
__ *D.p.melanoptera* Lawrence, 1879	C Lesser Antilles
__ *D.p.ruficapilla* (J.F. Gmelin, 1789)	Martinique I.
__ *D.p.babad* Bond, 1927	St. Lucia I.
__ *D.p.petechia* (Linnaeus, 1766)	Barbados I.
__ *D.p.alsiosa* J.L. Peters, 1926	Grenadine Is. (Lesser Antilles)
__ *D.p.rufopileata* Ridgway, 1885	Aruba, Curaçao, Bonaire Is.
__ *D.p.obscura* Cory, 1909	Los Roques Is. (off Venezuela)
__ *D.p.aurifrons* Phelps & Phelps, Jr., 1950	NC Venezuela and islands
__ *D.p.oraria* Parkes & Dickerman, 1967	E Mexico
__ *D.p.bryanti* Ridgway, 1873	Caribbean coast, SE Mexico to Costa Rica
?__ *D.p.erithachorides* S.F. Baird, 1858	E Costa Rica, E Panama, N Colombia
__ *D.p.chrysendeta* Wetmore, 1946	NE Colombia, NW Venezuela
__ *D.p.paraguanae* Phelps & Gilliard, 1941	NW Venezuela
__ *D.p.cienagae* Zimmer & Phelps, 1944	NC Venezuela
__ *D.p.castaneiceps* Ridgway, 1886[3]	S Baja California
__ *D.p.rhizophorae* van Rossem, 1935	NW Mexico
__ *D.p.phillipsi* Browning, 1994	W Mexico to Honduras (256)
__ *D.p.xanthotera* Todd, 1924	Pacific coast, Guatemala to Costa Rica
__ *D.p.aithocorys* Olson, 1980	NW Panama and Coiba I. (1693)
__ *D.p.iguanae* Olson, 1980	Isla Iguana (off W Panama) (1693)
__ *D.p.aequatorialis* Sundevall, 1870	SW Panama and Pearl Is.
__ *D.p.jubaris* Olson, 1980	SW Panama, NW Colombia (1693)
__ *D.p.peruviana* Sundevall, 1870	SW Colombia, W Ecuador, NW Peru
__ *D.p.aureola* (Gould, 1839)	Isla de Coco (EC Pacific), Galapagos Is.

● *Dendroica striata* (J.R. Forster, 1772) BLACKPOLL WARBLER
Canada, C and E USA >> N and C Sth. America

○ *Dendroica castanea* (A. Wilson, 1810) BAY-BREASTED WARBLER
C and SE Canada, E USA >> C America, Colombia, Venezuela

○ *Dendroica fusca* (Statius Müller, 1776) BLACKBURNIAN WARBLER
SE Canada, E USA >> C America, Venezuela, Colombia, Ecuador, Peru

● *Dendroica magnolia* (A. Wilson, 1811) MAGNOLIA WARBLER
S Canada, E USA >> Middle America, Gtr. Antilles, Nth. S America

○ *Dendroica cerulea* (A. Wilson, 1810) CERULEAN WARBLER
E USA >> Colombia to S Peru

○ *Dendroica tigrina* (J.F. Gmelin, 1789) CAPE MAY WARBLER
C and SE Canada, NC and E USA >> West Indies

○ *Dendroica caerulescens* BLACK-THROATED BLUE WARBLER

__ *D.c.caerulescens* (J.F. Gmelin, 1789)	SE Canada, NE USA >> Bahamas, Gtr. Antilles, Nth. S America
__ *D.c.cairnsi* Coues, 1897	EC USA >> Gtr. Antilles

● *Dendroica coronata* YELLOW-RUMPED WARBLER

✔ *D.c.coronata* (Linnaeus, 1766)	Canada, C and E USA >> C America, West Indies, Nth. S America
✔ *D.c.auduboni* (J.K. Townsend, 1837)	SW Canada, W USA >> Mexico, Guatemala, W Honduras
__ *D.c.nigrifrons* Brewster, 1889	NC Mexico
__ *D.c.goldmani* Nelson, 1897	W Guatemala

● *Dendroica nigrescens* BLACK-THROATED GREY WARBLER

__ *D.n.nigrescens* (J.K. Townsend, 1837)	SW Canada, W USA >> N Mexico, Guatemala
__ *D.n.halseii* (Giraud, 1841)	SW USA, NW Mexico, N Baja California >> ? C Mexico

● *Dendroica chrysoparia* (Sclater & Salvin, 1860)[4] GOLDEN-CHEEKED WARBLER
S USA >> Mexico, Guatemala, Honduras, Nicaragua

---

[1] For recognition see Browning (1994) (256).
[2] Includes *cruciana* see Browning (1994) (256).
[3] Includes *hueyi* see Lowery & Monroe (1968: 18) (1394).
[4] Dated 1861 by Lowery & Monroe (1968: 26) (1394), but see Duncan (1937) (770).

● *Dendroica virens* (J.F. Gmelin, 1789) Black-throated Green Warbler
C and SE Canada, E USA >> Mexico, C America, West Indies

● *Dendroica townsendi* (J.K. Townsend, 1837) Townsend's Warbler
W Canada, W USA >> Mexico, Guatemala, Honduras, Nicaragua

● *Dendroica occidentalis* (J.K. Townsend, 1837) Hermit Warbler
W USA >> W Mexico, Guatemala, Honduras, Nicaragua

● *Dendroica dominica* Yellow-throated Warbler
  ⌄ *D.d.albilora* Ridgway, 1873[1] EC and SE USA >> E Mexico, C America, Cuba, Jamaica
  — *D.d.dominica* (Linnaeus, 1766) E and SE USA >> Gtr. Antilles
  — *D.d.stoddardi* Sutton, 1951 SE USA
  — *D.d.flavescens* Todd, 1909 Bahamas

○ *Dendroica graciae* Grace's Warbler
  — *D.g.graciae* S.F. Baird, 1865 SW USA >> W Mexico
  — *D.g.yaegeri* Phillips & Webster, 1961 W Mexico
  — *D.g.remota* Griscom, 1935 S Mexico, Guatemala, El Salvador, W Honduras
  — *D.g.decora* Ridgway, 1873 Belize, E Honduras, Nicaragua

● *Dendroica discolor* Prairie Warbler
  ⌄ *D.d.discolor* (Vieillot, 1809)[2] E USA, West Indies
  — *D.d.paludicola* A.H. Howell, 1930 SE USA >> Gtr. Antilles

○ *Dendroica vitellina* Vitelline Warbler
  — *D.v.crawfordi* Nicoll, 1904 Little Cayman I.
  — *D.v.vitellina* Cory, 1886 Grand Cayman I.
  — *D.v.nelsoni* Bangs, 1919 Swan I. (W Caribbean)

○ *Dendroica adelaidae* S.F. Baird, 1865 Adelaide's Warbler[3] Puerto Rico

○ *Dendroica subita* Riley, 1904 Barbuda Warbler Barbuda (Lesser Antilles)

○ *Dendroica delicata* Ridgway, 1883 St. Lucia Warbler St. Lucia (Lesser Antilles)

○ *Dendroica pityophila* (Gundlach, 1858) Olive-capped Warbler Cuba, Bahamas

● *Dendroica pinus* Pine Warbler
  ⌄ *D.p.pinus* (A. Wilson, 1811) SE Canada >> SE USA
  — *D.p.florida* (Maynard, 1906) S Florida
  — *D.p.achrustera* Bangs, 1900 Bahamas
  — *D.p.chrysoleuca* Griscom, 1923 Hispaniola

○ *Dendroica kirtlandii* (S.F. Baird, 1852) Kirtland's Warbler Michigan >> Bahamas

● *Dendroica palmarum* Palm Warbler
  — *D.p.palmarum* (J.F. Gmelin, 1789) C and E Canada, E USA >> Gtr. Antilles, C America, Nth. S America
  ⌄ *D.p.hypochrysea* Ridgway, 1876 SE Canada, NE USA >> SE USA

○ *Dendroica plumbea* Lawrence, 1878 Plumbeous Warbler Dominica, Guadeloupe Is.

○ *Dendroica pharetra* (Gosse, 1847) Arrow-headed Warbler Jamaica

○ *Dendroica angelae* Kepler & Parkes, 1972 Elfin Woods Warbler Puerto Rico (1261)

CATHAROPEZA P.L. Sclater, 1880 F
○ *Catharopeza bishopi* (Lawrence, 1878) Whistling Warbler St. Vincent I. (Lesser Antilles)

MNIOTILTA Vieillot, 1816 F
◕ *Mniotilta varia* (Linnaeus, 1766) Black-and-white Warbler NW, C and SE Canada, C and E USA >> C America, West Indies, Venezuela, Colombia

SETOPHAGA Swainson, 1827 F
● *Setophaga ruticilla* (Linnaeus, 1758) American Redstart S Canada, C and E USA >> Middle America, West Indies, N Sth. America

[1] Includes *axantha* Oberholser, 1974 (1673), see Browning (1978) (246).
[2] Dated 1808 or 1809 by Lowery & Monroe (1968: 27) (1394), but see Browning & Monroe (1991) (258).
[3] The next two species have been split from this, see A.O.U. (2000: 852) (4) and sources given there.

762

PROTONOTARIA S.F. Baird, 1858 F
● ***Protonotaria citrea*** (Boddaert, 1783) PROTHONOTARY WARBLER   E USA >> C America, West Indies, N Sth. America

HELMITHEROS Rafinesque, 1819 N
○ ***Helmitheros vermivorum*** (J.F. Gmelin, 1789) WORM-EATING WARBLER[1]
    E USA >> Middle America, West Indies

LIMNOTHLYPIS Stone, 1914 F
○ ***Limnothlypis swainsonii*** (Audubon, 1834) SWAINSON'S WARBLER   SE USA >> E Mexico, West Indies

SEIURUS Swainson, 1827 M
● ***Seiurus aurocapilla*** OVENBIRD[2]
    — *S.a.aurocapilla* (Linnaeus, 1766)   C and SE Canada, E USA >> West Indies, Mexico to Colombia and Venezuela

    — *S.a.cinereus* A.H. Miller, 1942   WC USA >> S Mexico, El Salvador, Honduras, Costa Rica

    — *S.a.furvior* Batchelder, 1918   Newfoundland >> Bahamas, Cuba, E C America

● ***Seiurus noveboracensis*** (J.F. Gmelin, 1789) NORTHERN WATERTHRUSH[3]   Canada, E USA >> West Indies, Middle America, N Sth. America

● ***Seiurus motacilla*** (Vieillot, 1809)[4] LOUISIANA WATERTHRUSH   E USA >> Mexico, West Indies, C America, Colombia, Venezuela

OPORORNIS S.F. Baird, 1858[5] M
● ***Oporornis formosus*** (A. Wilson, 1811) KENTUCKY WARBLER[6]   SE USA >> E Mexico, Central America, Colombia, Venezuela

○ ***Oporornis agilis*** (A. Wilson, 1812) CONNECTICUT WARBLER   EC Canada, NC USA >> Brazil

○ ***Oporornis philadelphia*** (A. Wilson, 1810) MOURNING WARBLER   C and E Canada, NE USA >> Nicaragua, Costa Rica, Colombia, Venezuela

● ***Oporornis tolmiei*** MACGILLIVRAY'S WARBLER
    ⊻ *O.t.tolmiei* (J.K. Townsend, 1839)   SW Canada, W USA >> Central America
    — *O.t.monticola* A.R. Phillips, 1947[7]   WC USA, NE Mexico >> S Mexico to Guatemala and Honduras

GEOTHLYPIS Cabanis, 1847 F
● ***Geothlypis trichas*** COMMON YELLOWTHROAT
    — *G.t.trichas* (Linnaeus, 1766)[8]   SE Canada, EC USA >> Mexico, West Indies, N Sth. America

    — *G.t.typhicola* Burleigh, 1934   SC USA, NE Mexico
    — *G.t.ignota* Chapman, 1890   SE USA
    ⊻ *G.t.insperata* Van Tyne, 1933   S Texas
    ⊻ *G.t.campicola* Behle & Aldrich, 1947   W Canada, NW and SW USA >> N Mexico
    — *G.t.arizela* Oberholser, 1899   W Canada, W USA and NW Mexico
    — *G.t.occidentalis* Brewster, 1883   WC USA >> Mexico to Honduras
    — *G.t.sinuosa* Grinnell, 1901   N California
    — *G.t.scirpicola* Grinnell, 1901   S California, N Baja California
    — *G.t.chryseola* van Rossem, 1930   W Texas, NW Mexico
    — *G.t.modesta* Nelson, 1900   W Sonora, Mexico
    — *G.t.melanops* S.F. Baird, 1865   C Mexico
    — *G.t.chapalensis* Nelson, 1903   Jalisco (Mexico)
    — *G.t.riparia* van Rossem, 1941   S Sonora (Mexico)

○ ***Geothlypis beldingi*** BELDING'S YELLOWTHROAT
    — *G.b.goldmani* Oberholser, 1917   C Baja California
    — *G.b.beldingi* Ridgway, 1883   S Baja California

○ ***Geothlypis flavovelata*** Ridgway, 1896 ALTAMIRA YELLOWTHROAT   E Mexico

---

[1] For correct gender and spelling see David & Gosselin (2002) (614).
[2] For correction of spelling see David & Gosselin (2002) (613).
[3] For treatment as a monotypic species see Molina *et al.* (2000) (1597).
[4] Dated 1808 by Lowery & Monroe (1968: 36) (1394), but see Browning & Monroe (1991) (258).
[5] The species herein were treated in the genus *Geothlypis* in last Edition; we recognize *Oporornis* following A.O.U. (1998: 556) (3).
[6] Includes *umbraticus* Oberholser, 1974 (1673), see Browning (1978) (246).
[7] For recognition see Lowery & Monroe (1968: 48) (1394) and Dickerman & Parkes (1997) (729).
[8] Includes *brachidactylus* see Lowery & Monroe (1968: 39) (1394).

○ *Geothlypis rostrata* BAHAMA YELLOWTHROAT
— *G.r.tanneri* Ridgway, 1886 — N Bahamas
— *G.r.rostrata* H. Bryant, 1867 — W Bahamas
— *G.r.coryi* Ridgway, 1886 — Eleuthera, Cat Is. (Bahamas)

○ *Geothlypis semiflava* OLIVE-CROWNED YELLOWTHROAT
— *G.s.bairdi* Ridgway, 1884 — S Honduras, Nicaragua, Costa Rica, NW Panama
— *G.s.semiflava* P.L. Sclater, 1861 — W Colombia, W Ecuador

○ *Geothlypis speciosa* BLACK-POLLED YELLOWTHROAT
— *G.s.speciosa* P.L. Sclater, 1859 — C Mexico
— *G.s.limnatis* Dickerman, 1970 — Guanajuata (Mexico) (706)

○ *Geothlypis nelsoni* HOODED YELLOWTHROAT
— *G.n.nelsoni* Richmond, 1900 — E Mexico
— *G.n.karlenae* R.T. Moore, 1946 — SW Mexico

○ *Geothlypis aequinoctialis* MASKED YELLOWTHROAT[1]
— *G.a.chiriquensis* Salvin, 1872 — W Panama
— *G.a.aequinoctialis* (J.F. Gmelin, 1789) — NE Colombia, Venezuela, the Guianas, Surinam, N Brazil
— *G.a.auricularis* Salvin, 1884 — W Ecuador, W Peru
— *G.a.peruviana* Taczanowski, 1884 — N Peru
— *G.a.velata* (Vieillot, 1809)[2] — S Peru, Bolivia, Brazil, Paraguay, Uruguay, N Argentina

● *Geothlypis poliocephala* GREY-CROWNED YELLOWTHROAT
— *G.p.poliocephala* S.F. Baird, 1865 — N and W Mexico
— *G.p.ralphi* Ridgway, 1894 — NE Mexico
↘ *G.p.palpebralis* Ridgway, 1887 — E and S Mexico, Guatemala, Honduras, Nicaragua, Costa Rica
— *G.p.caninucha* Ridgway, 1872 — SW Mexico, W Guatemala, S Honduras, El Salvador
— *G.p.icterotis* Ridgway, 1889 — W Nicaragua, W Costa Rica
— *G.p.pontilis* (Brodkorb, 1943) — W Mexico
— *G.p.ridgwayi* (Griscom, 1930) — SW Costa Rica, W Panama

MICROLIGEA Cory, 1884 F
○ *Microligea palustris* GREEN-TAILED WARBLER
— *M.p.palustris* (Cory, 1884) — Hispaniola
— *M.p.vasta* Wetmore & Lincoln, 1931 — SW Dominica I.

TERETISTRIS Cabanis, 1855 M
○ *Teretistris fernandinae* (Lembeye, 1850) YELLOW-HEADED WARBLER W Cuba

○ *Teretistris fornsi* ORIENTE WARBLER
— *T.f.fornsi* Gundlach, 1858 — N Cuba (Matanzas to Oriente)
— ¶ *T.f.turquinensis* Garrido, 2000 — SE Cuba (Pico Turquin) (923)

LEUCOPEZA P.L. Sclater, 1877 F
○ †?*Leucopeza semperi* P.L. Sclater, 1877 SEMPER'S WARBLER St. Lucia I. (Lesser Antilles)

WILSONIA Bonaparte, 1838 F
● *Wilsonia citrina* (Boddaert, 1783) HOODED WARBLER E USA >> E Mexico, Central America

● *Wilsonia pusilla* WILSON'S WARBLER
↘ *W.p.pileolata* (Pallas, 1811) — W Canada, WC USA >> C Mexico, C America
↙ *W.p.chryseola* Ridgway, 1902 — SW USA >> W and S Mexico, Guatemala
— *W.p.pusilla* (A. Wilson, 1811) — S and E Canada, NE USA >> E Mexico, C America

○ *Wilsonia canadensis* (Linnaeus, 1766) CANADA WARBLER SE Canada, NE USA >> Central America, N Sth. America

CARDELLINA Bonaparte, 1850 F
○ *Cardellina rubrifrons* (Giraud, 1841) RED-FACED WARBLER SW USA >> W and S Mexico, Guatemala, W Honduras

[1] We now follow the more conservative position of the A.O.U. (1998: 560) (3) in keeping *chiriquensis* within this species.
[2] Dated 1808 by Lowery & Monroe (1968: 45) (1394), but see Browning & Monroe (1991) (258).

ERGATICUS S.F. Baird, 1865 M
○ *Ergaticus ruber* RED WARBLER
    — *E.r.melanauris* R.T. Moore, 1937   NW Mexico
    — *E.r.ruber* (Swainson, 1827)   W and SW Mexico
    — *E.r.rowleyi* Orr & Webster, 1968   Oaxaca (S Mexico) (1728)

○ *Ergaticus versicolor* (Salvin, 1864) PINK-HEADED WARBLER   S Mexico, W Guatemala

MYIOBORUS S.F. Baird, 1865 M
○ *Myioborus pictus* PAINTED REDSTART
    — *M.p.pictus* (Swainson, 1829)   SW USA, N Mexico
    — *M.p.guatemalae* (Sharpe, 1885)   S Mexico to N Nicaragua

● *Myioborus miniatus* SLATE-THROATED REDSTART
    — *M.m.miniatus* (Swainson, 1827)   W and SW Mexico
    — *M.m.molochinus* Wetmore, 1942   E Mexico
    — *M.m.intermedius* (Hartlaub, 1852)   S Mexico, E Guatemala
    — *M.m.hellmayri* van Rossem, 1936   W Guatemala, El Salvador
    — *M.m.connectens* Dickey & van Rossem, 1928   El Salvador, Honduras
    ✓ *M.m.comptus* Wetmore, 1944   W Costa Rica
    — *M.m.aurantiacus* (S.F. Baird, 1865)   E Costa Rica, W Panama
    — *M.m.ballux* Wetmore & Phelps, 1944   E Panama, Colombia, W Venezuela, NW Ecuador
    — *M.m.sanctaemartae* J.T. Zimmer, 1949   N Colombia
    — *M.m.pallidiventris* (Chapman, 1899)   N Venezuela
    — *M.m.subsimilis* J.T. Zimmer, 1949   SW Ecuador, NW Peru
    — *M.m.verticalis* (d'Orbigny & Lafresnaye, 1837)   SE Ecuador, Peru, Bolivia, SE Venezuela, Guyana, NW Brazil

○ *Myioborus brunniceps* BROWN-CAPPED REDSTART[1]
    — *M.b.castaneocapilla* (Cabanis, 1849)[2]   SE Venezuela, W Guyana, N Brazil
    — *M.b.duidae* Chapman, 1929   SE Venezuela
    — *M.b.maguirei* Phelps & Phelps, Jr., 1961   SE Venezuela
    — *M.b.brunniceps* (d'Orbigny & Lafresnaye, 1837)   Bolivia, N Argentina

○ *Myioborus pariae* Phelps & Phelps, Jr., 1949 YELLOW-FACED REDSTART   NE Venezuela (Paria Pen.)

○ *Myioborus cardonai* Zimmer & Phelps, 1945 SAFFRON-BREASTED REDSTART
                                            SE Venezuela

● *Myioborus torquatus* (S.F. Baird, 1865) COLLARED REDSTART   Costa Rica, W Panama

○ *Myioborus ornatus* GOLDEN-FRONTED REDSTART
    — *M.o.ornatus* (Boissonneau, 1840)   E Colombia, SW Venezuela
    — *M.o.chrysops* (Salvin, 1878)   W Colombia

○ *Myioborus melanocephalus* SPECTACLED REDSTART
    — *M.m.ruficoronatus* (Kaup, 1852)   SW Colombia, S Ecuador
    — *M.m.griseonuchus* Chapman, 1927   NW Peru
    — *M.m.malaris* J.T. Zimmer, 1949   N Peru
    — *M.m.melanocephalus* (Tschudi, 1844)   C Peru
    — *M.m.bolivianus* Chapman, 1919   S Peru, N Bolivia

○ *Myioborus albimfrons* (Sclater & Salvin, 1871) WHITE-FRONTED REDSTART
                                            Andes of W Venezuela

○ *Myioborus flavivertex* (Salvin, 1887) YELLOW-CROWNED REDSTART   N Colombia (Santa Marta Mts.)

○ *Myioborus albifacies* Phelps & Phelps, Jr., 1946 WHITE-FACED REDSTART
                                           S Venezuela

EUTHLYPIS Cabanis, 1850[3] F
○ *Euthlypis lachrymosa* FAN-TAILED WARBLER
    — *E.l.tephra* Ridgway, 1902   W Mexico
    — *E.l.schistacea* Dickey & van Rossem, 1926   W Chiapas
    — *E.l.lachrymosa* (Bonaparte, 1850)[4]   S Mexico to N Nicaragua

---

[1] We defer separation of the "Tepui Whitestart" (the first three subspecies here) awaiting more evidence.
[2] For correction of spelling see David & Gosselin (2002) (613).
[3] Dated 1851 by Lowery & Monroe (1968: 59) (1394), but see Browning & Monroe (1991) (258) and www.zoonomen.net (A.P. Peterson *in litt.*).
[4] Dated '1851?' by Lowery & Monroe (1968: 59) (1394), but see Zimmer (1926) (2707).

BASILEUTERUS Cabanis, 1849 M
○ *Basileuterus fraseri* Grey-and-gold Warbler
— *B.f.ochraceicrista* Chapman, 1921   W Ecuador
— *B.f.fraseri* P.L. Sclater, 1884   SW Ecuador, NW Peru

○ *Basileuterus bivittatus* Two-banded Warbler
— *B.b.roraimae* Sharpe, 1885   Guyana, SE Venezuela, N Brazil
— *B.b.bivittatus* (d'Orbigny & Lafresnaye, 1837)   S Peru, N Bolivia
— *B.b.argentinae* J.T. Zimmer, 1949   S Bolivia, NW Argentina

○ *Basileuterus chrysogaster* Golden-bellied Warbler[1]
— *B.c.chlorophrys* Berlepsch, 1907   SW Colombia, NW Ecuador
— *B.c.chrysogaster* (Tschudi, 1844)   C and S Peru

○ *Basileuterus flaveolus* Flavescent Warbler
— *B.f.pallidirostris* Oren, 1985   NE Colombia, N Venezuela (1719)
— *B.f.flaveolus* (S.F. Baird, 1865)   Colombia, Venezuela; C Brazil, E Bolivia

○ *Basileuterus luteoviridis* Citrine Warbler
— *B.l.luteoviridis* (Bonaparte, 1845)   SW Venezuela, E Colombia, E Ecuador
— *B.l.quindianus* Meyer de Schauensee, 1946   C Colombia
— *B.l.richardsoni* Chapman, 1912   W Colombia
— *B.l.striaticeps* (Cabanis, 1873)   N Peru
— *B.l.euophrys* Sclater & Salvin, 1877   S Peru, N Bolivia

○ *Basileuterus signatus* Pale-legged Warbler
— *B.s.signatus* Berlepsch & Stolzmann, 1906   C Peru
— *B.s.flavovirens* Todd, 1929   S Peru, Bolivia, NW Argentina

○ *Basileuterus nigrocristatus* (Lafresnaye, 1840) Black-crested Warbler
                                                        Venezuela, Colombia, Ecuador, N Peru

○ *Basileuterus griseiceps* Sclater & Salvin, 1869 Grey-headed Warbler   NE Venezuela (coastal range)

○ *Basileuterus basilicus* (Todd, 1913) Santa Marta Warbler   NE Colombia (Santa Marta Mts.)

○ *Basileuterus cinereicollis* Grey-throated Warbler
— *B.c.pallidulus* Wetmore, 1941   W Venezuela, NE Colombia
— *B.c.zuliensis* Aveledo & Perez, 1994   Sierra de Perijá (Colombia/Venezuela border) (69)
— *B.c.cinereicollis* P.L. Sclater, 1865   E Colombia

○ *Basileuterus conspicillatus* Salvin & Godman, 1880 White-lored Warbler[2]
                                                        N Colombia

○ *Basileuterus coronatus* Russet-crowned Warbler
— *B.c.regulus* Todd, 1929   Venezuela, Colombia
— *B.c.elatus* Todd, 1929   SW Colombia, W Ecuador
— *B.c.orientalis* Chapman, 1924   E Ecuador
— *B.c.castaneiceps* Sclater & Salvin, 1878   SW Ecuador, NW Peru
— *B.c.chapmani* Todd, 1929   NW Peru
— *B.c.inaequalis* J.T. Zimmer, 1949   N Peru
— *B.c.coronatus* (Tschudi, 1844)   S Peru, N Bolivia
— *B.c.notius* Todd, 1929   C Bolivia

○ *Basileuterus culicivorus* Golden-crowned Warbler[3]
— *B.c.flavescens* Ridgway, 1902   WC Mexico
— *B.c.brasherii* (Giraud, 1841)   EC Mexico
— *B.c.culicivorus* (Deppe, 1830)   SC Mexico to Costa Rica
— *B.c.godmani* Berlepsch, 1888   S Costa Rica, W Panama
— *B.c.occultus* J.T. Zimmer, 1949   W Colombia
— *B.c.austerus* J.T. Zimmer, 1949   C Colombia
— *B.c.indignus* Todd, 1916   N Colombia
— *B.c.cabanisi* Berlepsch, 1879   NW Venezuela, NE Colombia

[1] Sometimes treated as two species.
[2] The separation of *conspicillatus* from the next species follows the summary of treatment and the views of Ridgely & Tudor (1989: 189) (2018).
[3] Local hybridization occurs between this species and the next.

__ *B.c.olivascens* Chapman, 1893   Venezuela, Colombia, Trinidad

__ *B.c.segrex* Zimmer & Phelps, 1949   SE Venezuela, W Guyana, N Brazil

__ *B.c.auricapillus* (Swainson, 1838)   C Brazil

__ *B.c.azarae* J.T. Zimmer, 1949   S Brazil, Paraguay, Uruguay, NE Argentina

__ *B.c.viridescens* Todd, 1913   E Bolivia

○ *Basileuterus hypoleucus* Bonaparte, 1850[1] White-bellied Warbler   C and SE Brazil, E Paraguay

● *Basileuterus rufifrons* Rufous-capped Warbler

__ *B.r.caudatus* Nelson, 1899   NW Mexico

__ *B.r.dugesi* Ridgway, 1893   W and C Mexico

__ *B.r.jouyi* Ridgway, 1893   E Mexico

__ *B.r.rufifrons* (Swainson, 1838)   S Mexico, N Guatemala

__ *B.r.salvini* Cherrie, 1892   SW Mexico, N Guatemala

✓ *B.r.delattrii* Bonaparte, 1854   W Guatemala to N Costa Rica

__ *B.r.mesochrysus* P.L. Sclater, 1861   S Costa Rica, Panama, N Colombia, W Venezuela

__ *B.r.actuosus* Wetmore, 1957   Coiba I. (Panama)

○ *Basileuterus belli* Golden-browed Warbler

__ *B.b.bateli* R.T. Moore, 1946   W Mexico

__ *B.b.belli* (Giraud, 1841)   C and E Mexico

__ *B.b.clarus* Ridgway, 1902   SW Mexico

__ *B.b.scitulus* Nelson, 1900   SE Mexico, Guatemala, W Honduras

__ *B.b.subobscurus* Wetmore, 1940   C Honduras

○ *Basileuterus melanogenys* Black-cheeked Warbler

__ *B.m.melanogenys* S.F. Baird, 1865   Costa Rica

__ *B.m.eximus* Nelson, 1912   Panama

__ *B.m.bensoni* Griscom, 1927   Panama

○ *Basileuterus ignotus* Nelson, 1912 Pirre Warbler   E Panama, NW Colombia

● *Basileuterus tristriatus* Three-striped Warbler

__ *B.t.chitrensis* Griscom, 1927   W Panama

__ *B.t.tacarcunae* Chapman, 1924   E Panama, NW Colombia

__ *B.t.daedalus* Bangs, 1908   W Colombia, W Ecuador

__ *B.t.auricularis* Sharpe, 1885   E Colombia, SW Venezuela

__ *B.t.meridanus* Sharpe, 1885   W Venezuela

__ *B.t.bessereri* Hellmayr, 1922   N Venezuela

__ *B.t.pariae* Phelps & Phelps, Jr., 1949   NE Venezuela

__ *B.t.baezae* Chapman, 1924   E Ecuador

__ *B.t.tristriatus* (Tschudi, 1844)   SE Ecuador, C Peru

__ *B.t.inconspicuus* J.T. Zimmer, 1949   S Peru, N Bolivia

__ *B.t.punctipectus* Chapman, 1924   C Bolivia

__ *B.t.canens* J.T. Zimmer, 1949   E Bolivia

○ *Basileuterus trifasciatus* Three-banded Warbler

__ *B.t.nitidior* Chapman, 1924   SW Ecuador, NW Peru

__ *B.t.trifasciatus* Taczanowski, 1881   NW Peru

○ *Basileuterus leucoblepharus* White-browed Warbler

__ *B.l.leucoblepharus* (Vieillot, 1817)   SE Brazil to NE Argentina

__ *B.l.lemurum* Olson, 1975   Uruguay (1691)

○ *Basileuterus leucophrys* Pelzeln, 1868 White-striped Warbler   SC Brazil

PHAEOTHLYPIS Todd, 1929[2] F

● *Phaeothlypis fulvicauda* Buff-rumped Warbler[3]

✓ *P.f.leucopygia* (Sclater & Salvin, 1873)   Honduras to W Panama

__ *P.f.veraguensis* (Sharpe, 1885)   SW Costa Rica, C Panama

---

[1] Dated '1851?' by Lowery & Monroe (1968: 74) (1394), but see Zimmer (1926) (2707).

[2] These two species, then treated as one, placed in *Basileuterus* in last Edition; but this genus has been brought back into use by A.O.U. (1998: 567) (3) apparently persuaded by evidence of the relationships of *Catharopeza*.

[3] This species and the next now thought distinct and forming a superspecies (Ridgely & Tudor, 1989: 197) (2018).

— *P.f.semicervina* (P.L. Sclater, 1861) E Panama to NW Peru
— *P.f.motacilla* (A.H. Miller, 1952) N Colombia
— *P.f.fulvicauda* (Spix, 1825) E Colombia, E Ecuador, NE Peru, W Brazil
— *P.f.significans* (J.T. Zimmer, 1949) SE Peru, NW Bolivia

○ *Phaeothlypis rivularis* RIVERBANK WARBLER
— *P.r.mesoleuca* (P.L. Sclater, 1866) E Venezuela, the Guianas, N Brazil
— *P.r.rivularis* (Wied, 1821) SE Brazil, E Paraguay, NE Argentina
— *P.r.boliviana* (Sharpe, 1885) E Bolivia

ZELEDONIA Ridgway, 1889 F
○ *Zeledonia coronata* Ridgway, 1889 WREN-THRUSH Costa Rica, W Panama

ICTERIA Vieillot, 1807[1] F
◉ *Icteria virens* YELLOW-BREASTED CHAT
⤶ *I.v.auricollis* (Deppe, 1830) SW Canada, W USA >> W Mexico, Guatemala
— *I.v.virens* (Linnaeus, 1758)[2] E USA >> E Mexico, Central America
— *I.v.tropicalis* van Rossem, 1939 S Sonora (Mexico)

GENERA INCERTAE SEDIS (2: 4)

GRANATELLUS Bonaparte, 1850 M
○ *Granatellus venustus* RED-BREASTED CHAT
— *G.v.francescae* S.F. Baird, 1865 Tres Marias Is. (Mexico)
— *G.v.venustus* Bonaparte, 1850[3] W and SW Mexico
— *G.v.melanotis* van Rossem, 1940 W coast of Mexico

○ *Granatellus sallaei* GREY-THROATED CHAT
— *G.s.sallaei* (Bonaparte, 1856) E Mexico
— *G.s.boucardi* Ridgway, 1886 SE Mexico, E Guatemala, Belize

○ *Granatellus pelzelni* ROSE-BREASTED CHAT
— *G.p.pelzelni* P.L. Sclater, 1865 SE Venezuela, Guyana, Surinam, NW Brazil
— *G.p.paraensis* Rothschild, 1906 C Brazil, E Bolivia

XENOLIGEA Bond, 1967 F
○ *Xenoligea montana* (Chapman, 1917) WHITE-WINGED WARBLER Mts. of Hispaniola

ICTERIDAE NEW WORLD BLACKBIRDS[4] (26: 98)

PSAROCOLIUS Wagler, 1827 M
○ *Psarocolius oseryi* (Deville, 1849) CASQUED OROPENDOLA E Ecuador, E Peru, W Brazil

○ *Psarocolius decumanus* CRESTED OROPENDOLA
— *P.d.melanterus* (Todd, 1917) Panama, N Colombia
— *P.d.insularis* (Dalmas, 1900) Trinidad, Tobago I.
— *P.d.decumanus* (Pallas, 1769) N Sth. America
— *P.d.maculosus* (Chapman, 1920) E Peru to Paraguay, N Argentina

○ *Psarocolius viridis* (Statius Müller, 1776) GREEN OROPENDOLA N Amazonia, Venezuela, the Guianas

○ *Psarocolius atrovirens* (d'Orbigny & Lafresnaye, 1838) DUSKY-GREEN OROPENDOLA
C Peru to N Bolivia

○ *Psarocolius angustifrons* RUSSET-BACKED OROPENDOLA
— *P.a.salmoni* (P.L. Sclater, 1883) C Colombia
— *P.a.atrocastaneus* (Cabanis, 1873) W Ecuador
— *P.a.sincipitalis* (Cabanis, 1873) NC Colombia
— *P.a.neglectus* (Chapman, 1914) E Colombia, NW Venezuela

---

[1] Dated 1808 by Lowery & Monroe (1968: 81) (1394), but see Browning & Monroe (1991) (258).
[2] Includes *danotia* Oberholser, 1974 (1673), see Browning (1978) (246).
[3] Dated '1851?' by Lowery & Monroe (1968: 79) (1394), but see Zimmer (1926) (2707).
[4] Our species sequence is based on Lanyon & Omland (1999) (1323).

— *P.a.oleagineus* (P.L. Sclater, 1883)   NC Venezuela

— *P.a.angustifrons* (Spix, 1824)   W Amazonia

— *P.a.alfredi* (Des Murs, 1856)   SE Ecuador, E Peru, E Bolivia

● *Psarocolius wagleri* CHESTNUT-HEADED OROPENDOLA

   — *P.w.wagleri* (G.R. Gray, 1845)   SE Mexico to NE Nicaragua

   ✓ *P.w.ridgwayi* (van Rossem, 1934)   S Nicaragua to Panama, W Ecuador

● *Psarocolius montezuma* (Lesson, 1830) MONTEZUMA OROPENDOLA   S Mexico to C Panama

○ *Psarocolius cassini* (Richmond, 1898) BAUDO OROPENDOLA   NW Colombia

○ *Psarocolius bifasciatus* OLIVE OROPENDOLA

   — *P.b.yuracares* (d'Orbigny & Lafresnaye, 1838)   W Amazonia

   — *P.b.bifasciatus* (Spix, 1824)   Amazonas (C Brazil)

   — *P.b.neivae* (E. Snethlage, 1925)   Para (C Brazil)

○ *Psarocolius guatimozinus* (Bonaparte, 1853) BLACK OROPENDOLA   E Panama, NW Colombia

OCYALUS Waterhouse, 1841 M

○ *Ocyalus latirostris* (Swainson, 1838) BAND-TAILED OROPENDOLA[1]   E Ecuador, NE Peru, W Brazil

CACICUS Lacépède, 1799 M

○ *Cacicus cela* YELLOW-RUMPED CACIQUE

   — *C.c.vitellinus* (Lawrence, 1864)   Panama, N Colombia

   — *C.c.flavicrissus* (P.L. Sclater, 1860)   W Ecuador, NW Peru

   — *C.c.cela* (Linnaeus, 1758)   Venezuela, E Colombia, the Guianas, Trinidad, Amazonia, E Brazil

○ *Cacicus haemorrhous* RED-RUMPED CACIQUE

   — *C.h.haemorrhous* (Linnaeus, 1766)   E Colombia, SE Venezuela, the Guianas, N Brazil

   — *C.h.pachyrhynchus* von Berlepsch, 1889   Amazon Basin and southern tributaries (2536)

   — *C.h.affinis* (Swainson, 1834)   E Brazil, E Paraguay, NE Argentina

● *Cacicus uropygialis* SCARLET-RUMPED CACIQUE

   ✓ *C.u.microrhynchus* (Sclater & Salvin, 1864)   S Honduras to Panama

   — *C.u.pacificus* Chapman, 1915   E Panama, Colombia, E Ecuador

   — *C.u.uropygialis* (Lafresnaye, 1843)   S Venezuela to C Peru

○ *Cacicus chrysopterus* (Vigors, 1825) GOLDEN-WINGED CACIQUE   E Bolivia, S Brazil, Paraguay, NE Argentina

○ *Cacicus chrysonotus* MOUNTAIN CACIQUE

   — *C.c.leucoramphus* (Bonaparte, 1844)   NW Venezuela to E Ecuador

   — *C.c.peruvianus* (J.T. Zimmer, 1924)   N Peru

   — *C.c.chrysonotus* (d'Orbigny & Lafresnaye, 1838)   S Peru, N Bolivia

○ *Cacicus sclateri* (A.J.C. Dubois, 1887) ECUADORIAN CACIQUE   E Ecuador, N Peru

○ *Cacicus koepckeae* Lowery & O'Neill, 1965 SELVA CACIQUE   SE Peru

○ *Cacicus solitarius* (Vieillot, 1816) SOLITARY CACIQUE   C Amazonia to N Argentina, S Brazil

○ *Cacicus melanicterus* (Bonaparte, 1825) YELLOW-WINGED CACIQUE   W and SW Mexico

AMBLYCERCUS Cabanis, 1851 M

● *Amblycercus holosericeus* YELLOW-BILLED CACIQUE[2]

   ✓ *A.h.holosericeus* (Deppe, 1830)   SE Mexico to W Colombia

   — *A.h.flavirostris* (Chapman, 1915)   N Colombia to N Peru

   — *A.h.australis* (Chapman, 1919)   C Peru to N Bolivia

ICTERUS Brisson, 1760[3] M

○ *Icterus icterus* TROUPIAL

   — *I.i.ridgwayi* (E. Hartert, 1902)   N Colombia, NW Venezuela, Aruba, Curaçao Is.

   — *I.i.icterus* (Linnaeus, 1766)   E Colombia, NW Venezuela

---

[1] Treated in the genus *Psarocolius* in last Edition, but see Ridgely & Tudor (1989: 373) (2018).
[2] Treated in the genus *Cacicus* in last Edition, but see Ridgely & Tudor (1989: 366) (2018).
[3] *I. xantholemus* see Blake (1968: 154) (162) seems to still require determination.

— *I.i.metae* Phelps & Aveledo, 1966   SW Venezuela
— *I.i.croconotus* (Wagler, 1829)   SW Guyana to SE Peru
— *I.i.strictifrons* Todd, 1924   N and E Bolivia, SW Brazil, N Argentina
— *I.i.jamaicaii* (J.F. Gmelin, 1788)   E Brazil

● ***Icterus pectoralis*** SPOTTED-BREASTED ORIOLE
    — *I.p.carolynae* Dickerman, 1981   Guerrero and Oaxaca (Mexico) (713)
    — *I.p.pectoralis* (Wagler, 1829)   E Oaxaca to C Chiapas (Mexico)
    — *I.p.guttulatus* Lafresnaye, 1844   Chiapas (Mexico) to NW Nicaragua (1315)
    — *I.p.espinachi* Ridgway, 1882   S Nicaragua to NW Costa Rica

○ ***Icterus graceannae*** Cassin, 1867   WHITE-EDGED ORIOLE   SW Ecuador, NW Peru

○ ***Icterus mesomelas*** YELLOW-TAILED ORIOLE
    — *I.m.mesomelas* (Wagler, 1829)   SE Mexico to Honduras
    — *I.m.salvinii* Cassin, 1867   Nicaragua to W Panama (Caribbean slope)
    — *I.m.carrikeri* Todd, 1917   Panama, N and W Colombia, NW Venezuela
    — *I.m.taczanowskii* Ridgway, 1901   W Ecuador, NW Peru

○ ***Icterus cayanensis*** EPAULET ORIOLE[1]
    — *I.c.cayanensis* (Linnaeus, 1766)   Surinam, French Guiana
    — *I.c.chrysocephalus* (Linnaeus, 1766)   NE Peru, SE Colombia, Venezuela, the Guianas, N Brazil
    — *I.c.tibialis* Swainson, 1838   E Brazil
    — *I.c.valenciobuenoi* H. von Ihering, 1902   SE Brazil
    — *I.c.periporphyrus* (Bonaparte, 1850)   NE Bolivia, W Brazil
    — *I.c.pyrrhopterus* (Vieillot, 1819)   SE Bolivia to Uruguay

○ ***Icterus bonana*** (Linnaeus, 1766)   MARTINIQUE ORIOLE   Martinique I.

○ ***Icterus laudabilis*** P.L. Sclater, 1871   ST. LUCIA ORIOLE   St. Lucia I.

○ ***Icterus oberi*** Lawrence, 1880   MONTSERRAT ORIOLE   Montserrat I.

Icterus dominicensis GREATER ANTILLEAN ORIOLE[2]
    — *I.d.northropi* J.A. Allen, 1890   Andros I. (Bahamas)
    — *I.d.melanopsis* (Wagler, 1829)   Cuba
    — *I.d.dominicensis* (Linnaeus, 1766)   Hispaniola
    — *I.d.portoricensis* H. Bryant, 1866   Puerto Rico

● ***Icterus prosthemelas*** BLACK-COWLED ORIOLE
    — *I.p.prosthemelas* (Strickland, 1850)   SE Mexico to Nicaragua
    — *I.p.praecox* Phillips & Dickerman, 1965   E Costa Rica, W Panama

● ***Icterus spurius*** ORCHARD ORIOLE
    ✓ *I.s.spurius* (Linnaeus, 1766)   S Canada, E USA >> Middle America, Colombia
    — *I.s.phillipsi* Dickerman & Warner, 1962   C Mexico
    — *I.s.fuertesi* Chapman, 1911   NE Mexico

● ***Icterus cucullatus*** HOODED ORIOLE
    ✓ *I.c.nelsoni* Ridgway, 1885[3]   SW USA, NW Mexico >> W Mexico
    — *I.c.trochiloides* Grinnell, 1927   S Baja California
    — *I.c.sennetti* Ridgway, 1901   S Texas, E Mexico >> S Mexico
    — *I.c.cucullatus* Swainson, 1827   SW Texas, NC and C Mexico
    — *I.c.igneus* Ridgway, 1885[4]   SE Mexico, Belize

○ ***Icterus wagleri*** BLACK-VENTED ORIOLE
    — *I.w.castaneopectus* Brewster, 1888   NW Mexico
    — *I.w.wagleri* P.L. Sclater, 1857   W and S Mexico to Nicaragua

○ ***Icterus maculialatus*** Cassin, 1848   BAR-WINGED ORIOLE   S Mexico to El Salvador

○ ***Icterus parisorum*** Bonaparte, 1838   SCOTT'S ORIOLE   SC USA, C and W Mexico

---

[1] In last Edition we separated *chrysocephalus* from *cayanensis*; here we revert to following Blake (1968) (162) in a broad species. See also Omland *et al.* (1999) (1712).
[2] This and the next species split following Lanyon & Omland (1999) (1323).
[3] Includes *restrictus* see Blake (1968: 157) (162).
[4] Includes *masoni*, *duplexus* and *cozumelae* see Blake (1968: 158) (162).

○ *Icterus auricapillus* Cassin, 1848 ORANGE-CROWNED ORIOLE   E Panama to N Venezuela

○ *Icterus chrysater* YELLOW-BACKED ORIOLE
— *I.c.chrysater* (Lesson, 1844)   S Mexico to Nicaragua
— *I.c.mayensis* van Rossem, 1938   SE Mexico
— *I.c.giraudii* Cassin, 1848   Panama, N and C Colombia, N Venezuela
— *I.c.hondae* Chapman, 1914[1]   Magdalena valley (NC Colombia)

○ *Icterus graduacauda* AUDUBON'S ORIOLE
— *I.g.audubonii* Giraud, 1841   N Mexico
— *I.g.nayaritensis* van Rossem, 1938   WC Mexico
— *I.g.dickeyae* van Rossem, 1938   Guerrero
— *I.g.graduacauda* Lesson, 1839[2]   C and S Mexico

● *Icterus galbula* (Linnaeus, 1758) BALTIMORE ORIOLE[3]   Canada, E USA >> Colombia

○ *Icterus abeillei* (Lesson, 1839) BLACK-BACKED ORIOLE   SC Mexico

● *Icterus bullockii* (Swainson, 1827) BULLOCK'S ORIOLE[4]   SW Canada, W USA >> W Mexico to Nicaragua

○ *Icterus pustulatus* STREAK-BACKED ORIOLE
— *I.p.microstictus* Griscom, 1934   NC Sonora to Jalisco (Mexico)
— *I.p.graysonii* Cassin, 1867   Tres Marias Is. (Mexico)
— *I.p.yaegeri* A.R. Phillips, 1995[5]   S Sinaloa to S Nayarit (Mexico) (1878)
— *I.p.dickermani* A.R. Phillips, 1995[6]   SW Jalisco to S Guerrero (Mexico) (1878)
— *I.p.interior* A.R. Phillips, 1995[7]   C and S Mexico (1878)
— *I.p.pustulatus* (Wagler, 1829)[8]   SE Oaxaca (Mexico) to W Guatemala
— *I.p.alticola* Miller & Griscom, 1925   Guatemala, E Honduras
— *I.p.pustuloides* (Wagler, 1829)   Volcán San Miguel (El Salvador)
— *I.p.sclateri* Cassin, 1867   El Salvador to NW Costa Rica
— *I.p.maximus* Griscom, 1930   NC Guatemala (1015)

○ *Icterus leucopteryx* JAMAICAN ORIOLE
— *I.l.lawrencii* Cory, 1887   Isla Andrés (W Caribbean)
— †?*I.l.bairdi* Cory, 1886   Grand Cayman I.
— *I.l.leucopteryx* (Wagler, 1827)   Jamaica

● *Icterus auratus* Bonaparte, 1850[9] ORANGE ORIOLE   SE Mexico

○ *Icterus nigrogularis* YELLOW ORIOLE
— *I.n.nigrogularis* (Hahn, 1819)   NE Sth. America
— *I.n.curasoensis* Ridgway, 1884   Aruba, Curaçao, Bonaire Is.
— *I.n.helioeides* Clark, 1902   Margarita I. (off Venezuela)
— *I.n.trinitatis* E. Hartert, 1914   NE Venezuela, Trinidad

● *Icterus gularis* ALTAMIRA ORIOLE
— *I.g.tamaulipensis* Ridgway, 1901   S Texas, E Mexico
— *I.g.yucatanensis* Berlepsch, 1888   SE Mexico, N Belize
— *I.g.flavescens* A.R. Phillips, 1966   SW Mexico (Guerrero)
— *I.g.gularis* (Wagler, 1829)   SW Mexico to El Salvador
— *I.g.troglodytes* Griscom, 1930   S Mexico, W Guatemala
— *I.g.gigas* Griscom, 1930   S Guatemala, Honduras

NESOPSAR P.L. Sclater, 1859 M
○ *Nesopsar nigerrimus* (Osburn, 1859) JAMAICAN BLACKBIRD   Jamaica

GYMNOMYSTAX Reichenbach, 1850 M
○ *Gymnomystax mexicanus* (Linnaeus, 1766) ORIOLE-BLACKBIRD   N Sth. America

---

[1] For restriction of range see Olson (1981) (1697).
[2] Includes *richardsoni* see Blake (1968: 165) (162).
[3] A.O.U. (1998: 654-5) (3) treated this and the next two species as distinct. In last Edition we listed *parvus*; but this was placed in synonymy by Rising (1970) (2077).
[4] Includes *eleutherus* Oberholser, 1974 (1673), see Browning (1978) (246).
[5] Tentatively accepted see Dickerman & Parkes (1997) (729).
[6] For recognition see Parkes in Dickerman & Parkes (1997) (729).
[7] Tentatively accepted see Dickerman & Parkes (1997) (729).
[8] Thought to include *formosus* see Phillips (1995) (1878).
[9] Dated '1951?' by Blake (1968: 153) (162), but see Zimmer (1926) (2707).

MACROAGELAIUS Cassin, 1866 M
○ *Macroagelaius subalaris* (Boissonneau, 1840) MOUNTAIN GRACKLE   C Colombia

○ *Macroagelaius imthurni* (P.L. Sclater, 1881) GOLDEN-TUFTED GRACKLE
                                                      S Venezuela, N Brazil, W Guyana

HYPOPYRRHUS Bonaparte, 1850[1] M
○ *Hypopyrrhus pyrohypogaster* (de Tarragon, 1847) RED-BELLIED GRACKLE
                                                      Colombian Andes

LAMPROPSAR Cabanis, 1847 M
○ *Lampropsar tanagrinus* VELVET-FRONTED GRACKLE
    — *L.t.guianensis* Cabanis, 1849   NE Venezuela, NW Guyana
    — *L.t.tanagrinus* (Spix, 1824)   E Ecuador, NE Peru, W Brazil
    — *L.t.macropterus* Gyldenstolpe, 1945   W Brazil
    — *L.t.boliviensis* Gyldenstolpe, 1942   E Bolivia
    — *L.t.violaceus* Hellmayr, 1906   W Brazil

GNORIMOPSAR Richmond, 1908 M
○ *Gnorimopsar chopi* CHOPI BLACKBIRD
    — *G.c.sulcirostris* (Spix, 1824)   E Bolivia, NW Argentina, E Brazil
    — *G.c.chopi* (Vieillot, 1819)   SE Bolivia to Uruguay, NE Argentina

CURAEUS P.L. Sclater, 1862 M
● *Curaeus curaeus* AUSTRAL BLACKBIRD
    ↙ *C.c.curaeus* (Molina, 1782)   S Argentina, Chile
    — *C.c.recurvirostris* Markham, 1971   Magellanes (S Chile) (1435)
    — *C.c.reynoldsi* W.L. Sclater, 1939   Tierra del Fuego

○ *Curaeus forbesi* (P.L. Sclater, 1886) FORBES'S BLACKBIRD   E Brazil

AMBLYRAMPHUS Leach, 1814 M
○ *Amblyramphus holosericeus* (Scopoli, 1786) SCARLET-HEADED BLACKBIRD
                                                      E Bolivia, S Brazil, Paraguay, Uruguay, N Argentina

CHRYSOMUS Swainson, 1837[2] M
○ *Chrysomus xanthophthalmus* (Short, 1969) PALE-EYED BLACKBIRD   E Ecuador, E Peru (2248)

○ *Chrysomus cyanopus* UNICOLOURED BLACKBIRD
    — *C.c.xenicus* (Parkes, 1966)   NE Brazil
    — *C.c.atroolivaceus* (Wied, 1831)   E Brazil
    — *C.c.beniensis* (Parkes, 1966)   NE Bolivia
    — *C.c.cyanopus* (Vieillot, 1819)   SE Bolivia, Paraguay, SE Brazil, N Argentina

● *Chrysomus thilius* YELLOW-WINGED BLACKBIRD
    — *C.t.alticola* (Todd, 1932)   S Peru, Bolivia
    ↙ *C.t.thilius* (Molina, 1782)   S Chile, SW Argentina
    — *C.t.petersii* (Laubmann, 1934)   SE Brazil, Uruguay, N Argentina

○ *Chrysomus ruficapillus* CHESTNUT-CAPPED BLACKBIRD
    — *C.r.frontalis* (Vieillot, 1819)   French Guiana, E Brazil
    — *C.r.ruficapillus* (Vieillot, 1819)   SE Bolivia to Uruguay, N Argentina

○ *Chrysomus icterocephalus* YELLOW-HOODED BLACKBIRD
    — *C.i.bogotensis* (Chapman, 1914)   E Colombia (Andes)
    — *C.i.icterocephalus* (Linnaeus, 1766)   Surinam to NE Peru

XANTHOPSAR Ridgway, 1901 M
○ *Xanthopsar flavus* (J.F. Gmelin, 1788) SAFFRON-COWLED BLACKBIRD   SE Brazil, SE Paraguay, NE Argentina

PSEUDOLEISTES P.L. Sclater, 1862 M
○ *Pseudoleistes guirahuro* (Vieillot, 1819) YELLOW-RUMPED MARSHBIRD   SE Brazil, E Paraguay, NE Argentina

---

[1] Dated '1851?' by Blake (1968: 182) (162), but see Zimmer (1926) (2707).
[2] The South American species previously in the genus *Agelaius* are here separated following the suggestion of Lanyon & Omland (1999) (1323).

○ *Pseudoleistes virescens* (Vieillot, 1819) Brown-and-yellow Marshbird

SE Brazil, Uruguay, NE Argentina

OREOPSAR W.L. Sclater, 1939 M

○ *Oreopsar bolivianus* W.L. Sclater, 1939 Bolivian Blackbird[1]   C Bolivia

AGELAIOIDES Cassin, 1866[2] M

○ *Agelaioides badius* Bay-winged Cowbird

— *A.b.fringillarius* (Spix, 1824)   NE Brazil
— *A.b.badius* (Vieillot, 1819)   Bolivia to S Brazil, N Argentina
— *A.b.bolivianus* Hellmayr, 1917   S Bolivia

MOLOTHRUS Swainson, 1832 M

○ *Molothrus rufoaxillaris* Cassin, 1866 Screaming Cowbird   S Bolivia, S Brazil to C Argentina

○ *Molothrus oryzivorus* Giant Cowbird[3]

— *M.o.impacifa* (J.L. Peters, 1929)   S Mexico to W Panama
— *M.o.oryzivorus* (J.F. Gmelin, 1788)   E Panama, Trinidad, N Sth. America

● *Molothrus aeneus* Bronzed Cowbird

— *M.a.loyei* Parkes & Blake, 1965   SW USA, NW Mexico
— *M.a.assimilis* (Nelson, 1900)   S and SW Mexico
↙ *M.a.aeneus* (Wagler, 1829)   S Texas, E Mexico to Panama
— *M.a.armenti* Cabanis, 1851   N Colombia

● *Molothrus bonariensis* Shiny Cowbird

— *M.b.cabanisii* Cassin, 1866   E Panama, Colombia
— *M.b.aequatorialis* Chapman, 1915   SW Colombia, W Ecuador
— *M.b.occidentalis* Berlepsch & Stolzmann, 1892   SW Ecuador, W Peru
— *M.b.venezuelensis* Stone, 1891   E Colombia, N Venezuela
— *M.b.minimus* Dalmas, 1900   West Indies, S Florida, the Guianas, N Brazil
— *M.b.riparius* Griscom & Greenway, 1937   E Peru
— *M.b.bonariensis* (J.F. Gmelin, 1789)   C Sth. America

● *Molothrus ater* Brown-headed Cowbird

✓ *M.a.artemisiae* Grinnell, 1909   W Canada, W USA >> C Mexico
— *M.a.obscurus* (J.F. Gmelin, 1789)[4]   SW USA >> S Mexico
✓ *M.a.ater* (Boddaert, 1783)   C and S USA >> SE USA

DIVES Deppe, 1830 M

○ *Dives atroviolaceus* (d'Orbigny, 1839) Cuban Blackbird[5]   Cuba

○ *Dives dives* (Deppe, 1830) Melodious Blackbird[6]   E Mexico to Nicaragua

○ *Dives warszewiczi* Scrub Blackbird

— *D.w.warszewiczi* (Cabanis, 1861)   SW Ecuador, NW Peru
— *D.w.kalinowskii* Berlepsch & Stolzmann, 1892   W Peru

AGELAIUS Vieillot, 1816 M

● *Agelaius phoeniceus* Red-winged Blackbird[7]

— *A.p.arctolegus* Oberholser, 1907   NW and C Canada to N Great Plains >> N Colorado,
                                        Texas and Louisiana

— *A.p.nevadensis* Grinnell, 1914   SW Canada to SW USA
— *A.p.fortis* Ridgway, 1901[8]   WC USA (E of the Rocky Mts.) south to New Mexico
— *A.p.caurinus* Ridgway, 1901   SW British Columbia to N California
↙ *A.p.phoeniceus* (Linnaeus, 1766)   NC USA to SE Canada south to Texas and EC Georgia
                                        >> coast of Gulf of Mexico and Florida

[1] It has been suggested by Lowther (2001) (1401) that this species belongs in the same genus as the next. If so it must be called *Agelaioides oreopsar* Lowther.
[2] Treated in the genus *Molothrus* in last Edition, but this species is believed not to belong there; see Lanyon (1992) (1322) and Jaramillo & Burke (1999) (1202).
[3] For the submergence of *Scaphidura* in *Molothrus* see Lanyon (1992) (1322) and A.O.U. (2000: 853) (4).
[4] Including *californicus* see Blake (1968: 200) (162).
[5] For correction of spelling see David & Gosselin (2002) (614).
[6] The race *kalinowskii* has been moved to the following species see Ridgely & Tudor (1989: 352) (2018).
[7] Requires revision; here based on Blake (1968: 168-172) (162).
[8] Tentatively includes *stereus* Oberholser, 1974 (1673), *zastereus* Oberholser, 1974 (1673), and *heterurus* Oberholser, 1974 (1673), although see Browning (1978) (246).

__ *A.p.mearnsi* Howell & van Rossem, 1928	SE Georgia to NE Florida
__ *A.p.floridanus* Maynard, 1895	S Florida
__ *A.p.littoralis* Howell & van Rossem, 1928	C coastal Texas to NW Florida
__ *A.p.mailliardorum* van Rossem, 1926	WC California
✓ *A.p.californicus* Nelson, 1897	C California
__ *A.p.aciculatus* Mailliard, 1915	SC California
__ *A.p.neutralis* Ridgway, 1901	S California, NW Baja California
__ *A.p.sonoriensis* Ridgway, 1887	SW USA, NW Mexico
✓ *A.p.megapotamus* Oberholser, 1919	S Texas, NE Mexico
__ *A.p.gubernator* (Wagler, 1832)	NC Mexico
__ *A.p.nelsoni* Dickerman, 1965	SC Mexico
__ *A.p.nyaritensis* Dickey & van Rossem, 1925	SW Mexico to W El Salvador
__ *A.p.richmondi* Nelson, 1897[1]	S and SE Mexico to N Costa Rica
__ *A.p.arthuralleni* Dickerman, 1974	N Guatemala (708)
__ *A.p.brevirostris* R.T. Moore, 1963	N Honduras to SE Nicaragua
__ *A.p.grinnelli* A.H. Howell, 1917	El Salvador, W Nicaragua, NW Costa Rica
__ *A.p.bryanti* Ridgway, 1887	NW Bahamas

○ *Agelaius assimilis* RED-SHOULDERED BLACKBIRD[2]

__ *A.a.assimilis* Lembeye, 1850	W Cuba
__ *A.a.subniger* Bangs, 1913	Isla de la Juventud (Cuba)

○ *Agelaius tricolor* (Audubon, 1837) TRICOLOURED BLACKBIRD — W USA (mainly California)

○ *Agelaius humeralis* TAWNY-SHOULDERED BLACKBIRD

__ *A.h.humeralis* (Vigors, 1827)	Hispaniola
__ *A.h.scopulus* Garrido, 1970	Cuba (916)

○ *Agelaius xanthomus* YELLOW-SHOULDERED BLACKBIRD

__ *A.x.xanthomus* (P.L. Sclater, 1862)	Puerto Rico
__ *A.x.monensis* Barnés, 1945	Mona I. (Puerto Rico)

EUPHAGUS Cassin, 1867 M
● *Euphagus carolinus* RUSTY BLACKBIRD

__ *E.c.carolinus* (Statius Müller, 1776)	Canada, NE and C USA >> SE USA
__ *E.c.nigrans* Burleigh & Peters, 1948	Newfoundland >> SE USA

● *Euphagus cyanocephalus* (Wagler, 1829) BREWER'S BLACKBIRD — SW Canada, W USA >> Mexico

QUISCALUS Vieillot, 1816 M
● *Quiscalus quiscula* COMMON GRACKLE

__ *Q.q.versicolor* Vieillot, 1819	C and SE Canada, NE and C USA >> S USA
__ *Q.q.stonei* Chapman, 1935	NC USA >> SE USA
✓ *Q.q.quiscula* (Linnaeus, 1758)	SE USA

○ *Qusicalus lugubris* CARIB GRACKLE

__ *Q.l.guadeloupensis* Lawrence, 1879	Monserrat, Guadeloupe, Martinique Is.
__ *Q.l.inflexirostris* Swainson, 1838	St. Lucia I.
__ *Q.l.contrusus* (J.L. Peters, 1925)	St. Vincent I.
__ *Q.l.luminosus* Lawrence, 1879	Grenada I.
__ *Q.l.fortirostris* Lawrence, 1868	Barbados, Antigua Is.
__ *Q.l.orquillensis* (Cory, 1909)	Los Hermanos Is. (off Venezuela)
__ *Q.l.insularis* Richmond, 1896	Margarita I. (off Venezuela)
__ *Q.l.lugubris* Swainson, 1838	Trinidad, N Venezuela, the Guianas, NE Brazil

● *Quiscalus mexicanus* GREAT-TAILED GRACKLE

__ *Q.m.nelsoni* (Ridgway, 1901)	SW USA, NW Mexico
__ *Q.m.graysoni* P.L. Sclater, 1884	NW Mexico
__ *Q.m.obscurus* Nelson, 1900	W Mexico
✓ *Q.m.monsoni* (A.R. Phillips, 1950)	S USA, C Mexico
__ *Q.m.prosopidicola* (Lowery, 1938)	S USA, NE Mexico

[1] Includes *matudae* and *pallidulus* see Blake (1968: 172) (162).
[2] We follow A.O.U. (1998: 641) (3) in treating this separately from *A. phoeniceus*.

___ *Q.m.mexicanus* (J.F. Gmelin, 1788)	C and S Mexico to N Nicaragua
___ *Q.m.loweryi* (Dickerman & Phillips, 1966)	Belize
___ *Q.m.peruvianus* Swainson, 1838	Costa Rica to Peru, Venezuela

● *Quiscalus major* BOAT-TAILED GRACKLE

___ *Q.m.torreyi* (Harper, 1935)	New Jersey to extreme NE Florida
✓ *Q.m.westoni* Sprunt, 1934	Florida
___ *Q.m.alabamensis* Stevenson, 1978[1]	Alabama, SE Mississippi (2337)
✓ *Q.m.major* Vieillot, 1819	SE Texas and Louisiana

○ †*Quiscalus palustris* (Swainson, 1827) SLENDER-BILLED GRACKLE C Mexico

○ *Quiscalus nicaraguensis* Salvin & Godman, 1891 NICARAGUAN GRACKLE

Nicaragua

○ *Quiscalus niger* GREATER ANTILLEAN GRACKLE

___ *Q.n.caribaeus* (Todd, 1916)	W Cuba
___ *Q.n.gundlachii* Cassin, 1866	C and E Cuba
___ *Q.n.caymanensis* Cory, 1886	Grand Cayman I.
___ *Q.n.bangsi* (J.L. Peters, 1921)	Little Cayman I.
___ *Q.n.crassirostris* Swainson, 1838	Jamaica
___ *Q.n.niger* (Boddaert, 1783)	Hispaniola
___ *Q.n.brachypterus* Cassin, 1866	Puerto Rico

STURNELLA Vieillot, 1816 F

○ *Sturnella militaris* (Linnaeus, 1758) RED-BREASTED BLACKBIRD[2] N Sth. America

○ *Sturnella superciliaris* (Bonaparte, 1850) WHITE-BROWED BLACKBIRD S Peru, S Brazil to C Argentina

○ *Sturnella bellicosa* PERUVIAN MEADOWLARK

___ *S.b.bellicosa* Filippi, 1847	W Ecuador, NW Peru
___ *S.b.albipes* (Philippi & Landbeck, 1861)	SW Peru, N Chile

○ *Sturnella defilippii* (Bonaparte, 1850) PAMPAS MEADOWLARK SE Brazil, Uruguay, NE Argentina

● *Sturnella loyca* LONG-TAILED MEADOWLARK

✓ *S.l.loyca* (Molina, 1782)[3]	S Chile, S Argentina
___ *S.l.catamarcana* (Zotta, 1937)	NW Argentina
___ *S.l.obscura* Nores & Yzurieta, 1979	NC Argentina (1658)
___ *S.l.falklandica* (Leverkühn, 1889)	Falkland Is.

● *Sturnella magna* EASTERN MEADOWLARK

___ *S.m.magna* (Linnaeus, 1758)	SE Canada, C and E USA
✓ *S.m.argutula* Bangs, 1899	SC and SE USA
___ *S.m.hippocrepis* (Wagler, 1832)	Cuba
___ *S.m.lilianae* Oberholser, 1930	SW USA, NW Mexico
✓ *S.m.hoopesi* Stone, 1897	S Texas, NE Mexico
___ *S.m.auropectoralis* Saunders, 1934	C and SW Mexico
___ *S.m.saundersi* Dickerman & Phillips, 1970	Oaxaca, Mexico (733)
___ *S.m.alticola* Nelson, 1900	S Mexico to Nicaragua
___ *S.m.mexicana* P.L. Sclater, 1861	SE Mexico
___ *S.m.griscomi* van Tyne & Trautman, 1941	SE Mexico
___ *S.m.inexpectata* Ridgway, 1888	E Guatemala, Honduras
___ *S.m.subulata* Griscom, 1934	W Panama
___ *S.m.meridionalis* P.L. Sclater, 1861	N Colombia, NW Venezuela
___ *S.m.paralios* Bangs, 1901	N Colombia, W Venezuela
___ *S.m.praticola* Chubb, 1921[4]	S Venezuela, N Guyana
___ *S.m.quinta* Dickerman, 1989	NE Brazil, Surinam (721)

● *Sturnella neglecta* Audubon, 1844 WESTERN MEADOWLARK SW Canada, W USA >> NW Mexico

[1] For recognition see Browning (1990) (251).
[2] Type species of the genus *Leistes*.
[3] The name *Sturnus militaris* Linnaeus, 1771, is available in the genus *Pezites* as in Blake (1968: 176) (162), but when this genus and *Leistes* are submerged in *Sturnella* is preoccupied by *Emberiza militaris* Linnaeus, 1758 (applicable to another taxon), and the name *loyca* must be used.
[4] Includes *monticola* see Blake (1968: 180) (162).

XANTHOCEPHALUS Bonaparte, 1850 M

● *Xanthocephalus xanthocephalus* (Bonaparte, 1826) YELLOW-HEADED BLACKBIRD

SW Canada, W USA, W Mexico

DOLICHONYX Swainson, 1827 M

● *Dolichonyx oryzivorus* (Linnaeus, 1758) BOBOLINK   S Canada, N USA >> C Sth. America

COEREBIDAE BANANAQUIT (1: 1)

COEREBA Vieillot, 1809[1] F

● *Coereba flaveola* BANANAQUIT

⚓ *C.f.mexicana* (P.L. Sclater, 1857)	SE Mexico, C America
— *C.f.cerinoclunis* Bangs, 1901	Pearl Is. (Panama)
— *C.f.columbiana* (Cabanis, 1865)	E Panama, C Colombia, SC Venezuela
— *C.f.gorgonae* Thayer & Bangs, 1905	Gorgona I. (W Colombia)
— *C.f.caucae* Chapman, 1914	W Colombia
— *C.f.intermedia* (Salvadori & Festa, 1899)	SW Venezuela to Ecuador, W Brazil
— *C.f.magnirostris* (Taczanowski, 1881)	N Peru
— *C.f.pacifica* P.R. Lowe, 1912	NW Peru
— *C.f.dispar* J.T. Zimmer, 1942	SE Peru, NW Bolivia
— *C.f.caboti* (Ridgway, 1873)	Cozumel, Holbox Is. (off Yucatán Pen.)
— *C.f.tricolor* (Ridgway, 1885)	Isla Providencia (W Caribbean)
— *C.f.oblita* Griscom, 1923	Isla Andrès (W Caribbean)
— *C.f.sharpei* (Cory, 1886)	Cayman Is.
— *C.f.bahamensis* (Reichenbach, 1853)	Bahamas
— *C.f.flaveola* (Linnaeus, 1758)	Jamaica
— *C.f.bananivora* (J.F. Gmelin, 1789)	Hispaniola
— *C.f.nectarea* Wetmore, 1929	Tortue I., Haiti
— *C.f.portoricensis* (H. Bryant, 1866)	Puerto Rico
— *C.f.sanctithomae* (Sundevall, 1870)	Virgin Is.
— *C.f.newtoni* (Ridgway, 1873)	St. Croix I. (Virgin Is.)
— *C.f.bartholemica* (Sparrman, 1788)	N Lesser Antilles
— *C.f.martinicana* (Reichenbach, 1853)	Martinique, St. Lucia Is.
— *C.f.barbadensis* (Ridgway, 1873)	Barbados I.
— *C.f.atrata* (Lawrence, 1878)	St. Vincent I.
— *C.f.aterrima* (Lesson, 1830)	Grenada I.
— *C.f.uropygialis* Berlepsch, 1892	Aruba, Curaçao Is.
— *C.f.bonairensis* Voous, 1955	Bonaire I.
— *C.f.melanornis* Phelps & Phelps, Jr., 1954	Cayo Sal I. (off Venezuela)
— *C.f.lowii* Cory, 1909	Los Roques I. (off Venezuela)
— *C.f.ferryi* Cory, 1909	La Tortuga I. (off Venezuela)
— *C.f.frailensis* Phelps & Phelps, Jr., 1946	Los Frailes, Los Hermanos Is. (off Venezuela)
— *C.f.laurae* P.R. Lowe, 1908	Los Testigos I. (off Venezuela)
— *C.f.luteola* (Cabanis, 1851)	N Colombia to Trinidad and Tobago I.
— *C.f.obscura* Cory, 1913	NE Colombia, W Venezuela
— *C.f.montana* P.R. Lowe, 1912	W Venezuela
— *C.f.bolivari* Zimmer & Phelps, 1946	E Venezuela
— *C.f.guianensis* (Cabanis, 1850)	E Venezuela, Guyana
— *C.f.roraimae* Chapman, 1929	SE Venezuela, NW Brazil, SW Guyana
— *C.f.minima* (Bonaparte, 1854)	N Brazil, French Guiana, Surinam
— *C.f.chloropyga* (Cabanis, 1850)	S Peru, Bolivia to NE Argentina
— *C.f.alleni* P.R. Lowe, 1912	C Brazil, E Bolivia

EMBERIZIDAE BUNTINGS, AMERICAN SPARROWS AND ALLIES (73: 308)

EMBERIZINAE

MELOPHUS Swainson, 1837 M

○ *Melophus lathami* (J.E. Gray, 1831) CRESTED BUNTING   Himalayan foothills from N Pakistan east to SE China,
south to C India, Burma and C Laos >> Thailand

[1] Dated 1808 by Lowery & Monroe (1968: 87) (1394), but see Browning & Monroe (1991) (258).

LATOUCHEORNIS Bangs, 1931 M

○ *Latoucheornis siemsseni* (G.H. Martens, 1906) SLATY BUNTING
SE Gansu, Ningxia, S Shaanxi, Sichuan >> Guizhou, W Hubei, Anhui, NW Fujian and N Guangdong

EMBERIZA Linnaeus, 1758 F

◑ *Emberiza calandra* CORN BUNTING[1]

  ↙ *E.c.calandra* Linnaeus, 1758
Canary Is., N Africa, Europe, Asia Minor, Caucasus area, N Iran

  — *E.c.buturlini* H.E. Johansen, 1907[2]
Levant, SE Turkey, Syria, SW and NE Iran to WC Xinjiang and SE Kazakhstan >> SW Asia

◑ *Emberiza citrinella* YELLOWHAMMER
  — *E.c.caliginosa* Clancey, 1940
N and W British Isles

  — *E.c.citrinella* Linnaeus, 1758
W Europe to W European Russia, W Balkans and Greece >> N Africa

  — *E.c.erythrogenys* C.L. Brehm, 1855
E Balkans, C Ukraine and E European Russia to Lake Baikal >> SW and WC Asia, Mongolia

○ *Emberiza leucocephalos* PINE BUNTING[3]
  — *E.l.leucocephalos* S.G. Gmelin, 1771
E European Russia, Siberia east to Kolyma R. and Sakhalin, south to C and E Tien Shan, N Mongolia, NE China and Russian Far East, Xinjiang, Nei Mongol >> SE Europe, SW, S and SE Asia

  — *E.l.fronto* Stresemann, 1930
N Qinghai, SW Gansu (NC China)

○ *Emberiza cia* WESTERN ROCK BUNTING[4]
  — *E.c.cia* Linnaeus, 1766[5]
C and S Europe to Balkans and N Asia Minor, N Africa

  — *E.c.hordei* C.L. Brehm, 1831[6]
Greece, SW and SC Asia Minor, Levant

  — *E.c.prageri* Laubmann, 1915[7]
Crimea, Caucasus area, E Turkey, SW and N Iran

  — *E.c.par* E. Hartert, 1904
NE Iran through WC Asia to SW Altai, N Xinjiang and W and N Pakistan

  — *E.c.stracheyi* F. Moore, 1856
NW Himalayas and SW Xizang to W Nepal

  — *E.c.flemingorum* J. Martens, 1972
C Nepal (1440)

○ *Emberiza godlewskii* EASTERN ROCK BUNTING
  — *E.g.decolorata* Sushkin, 1925
S and E Tien Shan, W Xinjiang

  — *E.g.godlewskii* Taczanowski, 1874
N and E Qinghai and Gansu to Mongolia, Transbaikalia, S Baikal Mts. and Altai

  — *E.g.khamensis* Sushkin, 1925
S and E Xizang and SE Qinghai to N Assam and SW Sichuan

  — *E.g.yunnanensis* Sharpe, 1902
SE Xizang, Yunnan, S Sichuan, Guizhou, Guangxi (SC China)

  — *E.g.omissa* Rothschild, 1921
N Sichuan, Henan, and Shaanxi to Hebei (EC China)

○ *Emberiza cioides* MEADOW BUNTING
  — *E.c.tarbagataica* Sushkin, 1925
C and E Tien Shan, N Xinjiang, SW Altai

  — *E.c.cioides* J.F. Brandt, 1843
SC Siberia, Transbaikalia Mongolia; E Qinghai

  — *E.c.weigoldi* Jacobi, 1923
NE China, Russian Far East, N Korea >> C China and C Korea

  — *E.c.castaneiceps* F. Moore, 1856
EC and E China, S Korea >> S China, Taiwan

  — *E.c.ciopsis* Bonaparte, 1850[8]
Japan >> S Japan

○ *Emberiza stewarti* (Blyth, 1854) WHITE-CAPPED BUNTING
W Tien Shan, Pamir-Alai Mts., Afghanistan, W and N Pakistan, NW Himalayas >> Pakistan, NW India

○ *Emberiza jankowskii* Taczanowski, 1888 JANKOWSKI'S BUNTING
NE China, S Russian Far East >> Korea

○ *Emberiza buchanani* GREY-NECKED BUNTING
  — *E.b.cerrutii* de Filippi, 1863
E Turkey, Transcaucasia, Iran, SW Turkmenistan >> India

---

[1] For reasons to retain this in *Emberiza* rather than use a monotypic genus *Miliaria*, pending broader review, see Lee *et al.* (2001) (1345).
[2] For recognition see Cramp *et al.* (1994: 324) (596).
[3] For correction of spelling see David & Gosselin (2002) (613).
[4] See Mauersberger (1972) (1460) for reasons to separate *godlewskii* which we accept tentatively. More evidence is desirable.
[5] Inferentially includes *africana* see Cramp *et al.* (1994: 182) (596).
[6] For recognition see Roselaar (1995) (2107).
[7] For recognition see Cramp *et al.* (1994: 182) (596).
[8] Includes *shiretokoensis* Mishima, 1959 (1591) see Orn. Soc. Japan (2000) (1727).

— *E.b.buchanani* Blyth, 1845[1]   Afghanistan and NW Pakistan to Pamir-Alai Mts. >> W and C India

— *E.b.neobscura* Paynter, 1970   Kazakhstan, Tien Shan, W and N Xinjiang, W Mongolia >> India

○ *Emberiza cineracea* CINEREOUS BUNTING

— *E.c.cineracea* C.L. Brehm, 1855   Lesbos, Khios (E Greece), W, SW and SC Asia Minor >> NE Africa

— *E.c.semenowi* Zarudny, 1904   SE Turkey, SW Iran >> NE Africa, SW Arabia

○ *Emberiza hortulana* Linnaeus, 1758 ORTOLAN BUNTING   Europe to SC Siberia and Mongolia S to NW Africa, Levant and Iran >> N Afrotropics, SW Asia

○ *Emberiza caesia* Cretzschmar, 1828 CRETZSCHMAR'S BUNTING   Greece, W and S Turkey, Cyprus, Levant >> Eritrea, W, C and NE Sudan

● *Emberiza cirlus* Linnaeus, 1766 CIRL BUNTING[2]   NW Africa, W and S Europe to W and N Asia Minor

○ *Emberiza striolata* HOUSE BUNTING

— *E.s.sanghae* Traylor, 1960   N Senegal, Mauritania, S Mali

— *E.s.sahari* Levaillant, 1850   NW Africa and C Sahara east to N and NE Chad

— *E.s.saturatior* (Sharpe, 1901)   C and S Sudan, Ethiopia, NW Kenya, N Somalia

— *E.s.jebelmarrae* (Lynes, 1920)   Darfur, Kordofan (W Sudan)

— *E.s.striolata* (M.H.K. Lichtenstein, 1823)   W Red Sea coast, Arabia, Sinai, SE Israel, SW Jordan, and S Iran to E Afghanistan and W India >> NW and C India

○ *Emberiza impetuani* LARK-LIKE BUNTING

— *E.i.eremica* Clancey, 1989   N Namibia, S Angola, NW Cape Province (514)

— *E.i.impetuani* A. Smith, 1836   Botswana, W Cape Province

— *E.i.sloggetti* (Macdonald, 1957)   C Cape Province

○ *Emberiza tahapisi* CINNAMON-BREASTED BUNTING

— *E.t.arabica* (Lorenz & Hellmayr, 1902)   SW Arabia to S Oman

— *E.t.insularis* (Ogilvie-Grant & Forbes, 1899)   Socotra I.

— *E.t.goslingi* (Alexander, 1906)   Senegal and S Mauritania to W Sudan and N Zaire

— *E.t.septemstriata* Rüppell, 1840   E Sudan, N Ethiopia, Eritrea

— *E.t.tahapisi* A. Smith, 1836   S Ethiopia, Kenya, Uganda, Gabon and S Zaire to Angola, NE Namibia and E Sth. Africa

— *E.t.nivenorum* Winterbottom, 1965   NW Namibia

○ *Emberiza socotrana* (Ogilvie-Grant & Forbes, 1899) SOCOTRA BUNTING

    Socotra I.

○ *Emberiza capensis* CAPE BUNTING

— *E.c.vincenti* (P.R. Lowe, 1932)   N Mozambique, S Malawi, E Zambia

— *E.c.nebularum* (Rudebeck, 1958)   SW Angola

— *E.c.bradfieldi* (Roberts, 1928)   N Namibia

— *E.c.smithersii* (Plowes, 1951)   E Zimbabwe, W Mozambique

— *E.c.plowesi* (Vincent, 1950)   NE Botswana, S Zimbabwe

— *E.c.limpopoensis* (Roberts, 1924)   C and SW Transvaal, SE Botswana

— *E.c.reidi* (Shelley, 1902)   S Transvaal, Natal, Orange Free State, W Swaziland, N Lesotho

— *E.c.basutoensis* (Vincent, 1950)   Lesotho highlands

— *E.c.cinnamomea* (M.H.K. Lichtenstein, 1842)[3]   Karoo to W Orange Free State

— *E.c.vinacea* Clancey, 1963   N Cape Province

— *E.c.capensis* Linnaeus, 1766   S Namibia, W and SW Cape Province

○ *Emberiza tristrami* Swinhoe, 1870 TRISTRAM'S BUNTING   Russian Far East, NE China >> S and SE China, N of SE Asia

○ *Emberiza fucata* CHESTNUT-EARED BUNTING

— *E.f.arcuata* Sharpe, 1888   N Pakistan, NW Himalayas to W Nepal; C China >> N India, N of SE Asia

---

[1] Dated 1844 by Paynter (1970: 13) (1808), but see footnote herein for *Mirafra cantillans*.
[2] Treated as monotypic by Cramp *et al.* (1994: 170) (596).
[3] Includes *media* see Clancey (1980) (482).

— *E.f.fucata* Pallas, 1776   Baikal Mts. and NE Mongolia to S Sakhalin, Japan, Korea and NE China >> SE China, N of SE Asia

— *E.f.kuatunensis* La Touche, 1925   Fujian and Zhejiang >> SE China

○ *Emberiza pusilla* Pallas, 1776 LITTLE BUNTING   N Eurasia >> N India, S China, N of SE Asia

○ *Emberiza chrysophrys* Pallas, 1776 YELLOW-BROWED BUNTING   C and EC Siberia (mid-Yenisey valley and Baikal Mts. to Vilyuy and Aldan basins) >> E and S China

○ *Emberiza rustica* Pallas, 1776 RUSTIC BUNTING[1]   N Europe, Siberia east to Kamchatka and Sakhalin, south to Altai and N Baikal Mts. >> E China, Japan

○ *Emberiza elegans* YELLOW-THROATED BUNTING
— *E.e.elegans* Temminck, 1836[2]   Russian Far East and NE China to Hebei, Korea, Tsushima (W Japan) >> S Japan, S and E China
— *E.e.elegantula* Swinhoe, 1870   C China >> NE Burma

○ *Emberiza aureola* YELLOW-BREASTED BUNTING
— *E.a.aureola* Pallas, 1773   E Europe and N Siberia (N Ukraine and E Finland to Kolyma basin), south to W and C Mongolia and Baikal area >> India, mainland SE Asia
— *E.a.ornata* Shulpin, 1928   NE Mongolia, E Transbaikalia and NE China to N Japan, Kuril Is., Kamchtaka, Anadyrland >> S China

○ *Emberiza poliopleura* (Salvadori, 1888) SOMALI BUNTING   EC Africa

○ *Emberiza flaviventris* GOLDEN-BREASTED BUNTING
— *E.f.flavigaster* Cretzschmar, 1828   Mali to Eritrea
— *E.f.kalaharica* Roberts, 1932[3]   SE Sudan, Uganda, W Kenya, E and S Zaire to C Namibia, Botswana, Mozambique, N Sth. Africa
— *E.f.flaviventris* Stephens, 1815   S and E Cape Province to Natal, E Swaziland and Maputo (S Mozambique)

○ *Emberiza affinis* BROWN-RUMPED BUNTING
— *E.a.nigeriae* Bannerman & Bates, 1926   Gambia to Nigeria, N Cameroon
— *E.a.vulpecula* Grote, 1921   Cameroon, S Chad, Central African Republic
— *E.a.affinis* Heuglin, 1867   S Sudan, SW Ethiopia, N Uganda, NE Zaire
— *E.a.omoensis* Neumann, 1905[4]   S Ethiopia

○ *Emberiza cabanisi* CABANIS'S BUNTING
— *E.c.cabanisi* (Reichenow, 1875)   Sierra Leone to S Sudan, N Uganda
— *E.c.orientalis* (Shelley, 1882)[5]   S Zaire, Tanzania to Angola, Zambia, Zimbabwe, N Mozambique

○ *Emberiza rutila* Pallas, 1776 CHESTNUT BUNTING   Mid-Yenisey R. and Baikal Mts. to S Yakutia and Russian Far East (C and SE Siberia), and NE China >> NE India, S China, mainland SE Asia

○ *Emberiza koslowi* Bianchi, 1904 TIBETAN BUNTING   SC Qinghai >> E Xizang, SE Qinghai

○ *Emberiza melanocephala* Scopoli, 1769 BLACK-HEADED BUNTING   SE Europe, Asia Minor, Caucasus area, Levant, Iran >> N and C India

○ *Emberiza bruniceps* J.F. Brandt, 1841 RED-HEADED BUNTING   NE Iran, W Turkmenistan and lower Volga R. to Afghanistan, Xinjiang, SW Mongolia and E Kazakhstan >> S India

○ *Emberiza sulphurata* Temminck & Schlegel, 1848 JAPANESE YELLOW BUNTING   Honshu (C Japan), S Korea >> S Japan, E China, N Philippines

○ *Emberiza spodocephala* BLACK-FACED BUNTING
— *E.s.spodocephala* Pallas, 1776   C and E Siberia east to N Sakhalin and N Korea, south to N Altai, NE Mongolia and NE China >> E and S China, Taiwan
— *E.s.personata* Temminck, 1836[6]   S Sakhalin, S Kuril Is., N and C Japan >> S Japan
— *E.s.sordida* Blyth, 1845   EC China >> N and E India, nthn. SE Asia

[1] For treatment as monotypic see Stepanyan (1990: 669) (2326).
[2] Livr. 98 must be dated from 1836 see Dickinson (2001) (739). Includes *ticehursti* see Stepanyan (1990: 668) (2326).
[3] Includes *princeps* see White (1963: 96) (2648) and *carychroa* see Zimmerman *et al.* (1996) (2735).
[4] For recognition see White (1963: 97) (2648) s. n. *E. forbesi omoensis.*
[5] Includes *cognominata* see White (1963: 96) (2648).
[6] Livr. 98 must be dated from 1836 see Dickinson (2001) (739).

○ *Emberiza variabilis* Grey Bunting
  _ *E.v.variabilis* Temminck, 1836[1]   Sakhalin, Kuril Is., N and C Japan >> S Japan
  _ *E.v.musica* (Kittlitz, 1858)[2]   C Kamchatka >> Japan

○ *Emberiza pallasi* Pallas's Bunting
  _ *E.p.pallasi* (Cabanis, 1851)   E Tien Shan, Altai, Sayan Mts., NW Mongolia,
                                                 SW Transbaikalia
  _ *E.p.minor* Middendorff, 1853[3]   N and E Transbaikalia to Russian Far East >> C and
                                                 E China
  _ *E.p.polaris* Middendorff, 1851   NE European Russia, N Siberia to Anadyrland and
                                                 N Kamchatka >> E Asia
  _ *E.p.lydiae* Portenko, 1929[4]   NW, C and NE Mongolia, Tuva Rep., SE Transbaikalia,
                                                 ? NW Manchuria

○ *Emberiza yessoensis* Ochre-rumped Bunting
  _ *E.y.yessoensis* (Swinhoe, 1874)   N and C Japan
  _ *E.y.continentalis* Witherby, 1913   SE Siberia, NE China >> S Korea, E China

◉ *Emberiza schoeniclus* Reed Bunting[5]
  _ *E.s.schoeniclus* (Linnaeus, 1758)   W, N and NC Europe east to N European Russia >>
                                                 S Europe, N Africa, SW Asia
  _ *E.s.passerina* Pallas, 1771   NW Siberia >> SW and C Asia to N India and China
  _ *E.s.parvirostris* Buturlin, 1910   C Siberia (mid-Yenisey to C Yakutia) >> Xinjiang,
                                                 Qinghai, Mongolia
  _ *E.s.pyrrhulina* Swinhoe, 1876   Transbaikalia, NE Mongolia and NE China to N Japan,
                                                 Kuril Is., Kamchatka >> C Japan, Korea, E China
  _ *E.s.pallidior* E. Hartert, 1904   SW and SC Siberia and NC Kazakhstan to Lake Baikal
                                                 >> SW and C Asia to NW India
  _ *E.s.stresemanni* F. Steinbacher, 1930   E Austria, Hungary, N Serbia, NW Romania
  _ *E.s.ukrainae* (Zarudny, 1917)   Moldavia and Ukraine to C European Russia >>
                                                 basin of Black Sea
  _ *E.s.incognita* (Zarudny, 1917)   NW, C and NE Kazakhstan >> WC Asia, Xinjiang
  _ *E.s.pyrrhuloides* Pallas, 1811   N and E Caspian Sea to C Tien Shan and SE Kazakhstan
                                                 >> SW and C Asia
  _ *E.s.harterti* Sushkin, 1906   E Kazakhstan, W and S Mongolia, NW Xinjiang
  _ *E.s.centralasiae* E. Hartert, 1904   Tarim Basin (SW Xinjiang)
  _ *E.s.zaidamensis* Portenko, 1929   Qaidam Basin (SE Xinjiang/N Qinghai)
  _ *E.s.witherbyi* von Jordans, 1923   Iberia, S France, Sardinia, Balearic Is.
  _ *E.s.intermedia* Degland, 1849   Italy to coastal Croatia and NW Albania
  _ *E.s.tschusii* Reiser & Almásy, 1898   N Bulgaria, S and E Romania, shores of N Black Sea
                                                 and Sea of Azov
  _ *E.s.reiseri* E. Hartert, 1904   SE Albania, Macedonia, Greece, W and C Turkey
  _ *E.s.caspia* Ménétriés, 1832   E Turkey, E Transcaucasia, NW and N Iran
  _ *E.s.korejewi* (Zarudny, 1907)   SE Turkey, Syria, SW and E Iran, S Turkmenistan

CALCARIUS Bechstein, 1802 M
○ *Calcarius mccownii* (Lawrence, 1851) McCown's Longspur   SC Canada, WC USA >> S USA, N Mexico

◉ *Calcarius lapponicus* Lapland Longspur
  _ *C.l.lapponicus* (Linnaeus, 1758)   N Europe, N Asia E to C Chukotka and N Anadyrland
                                                 >> W and S Europe, C Asia
  _ *C.l.subcalcaratus* C.L. Brehm, 1826   Greenland, N Canada >> N America, NW Europe (211)
  _ *C.l.kamtschaticus* Portenko, 1937   S Anadyrland, Koryakland, N Sea of Okhotsk,
                                                 Kamchatka >> E Asia
  _ *C.l.coloratus* Ridgway, 1898   Commander Is. >> EC China
  √ *C.l.alascensis* Ridgway, 1898   E Chukotka (Siberia), Alaska, W Canada >> W USA

○ *Calcarius pictus* (Swainson, 1832) Smith's Longspur   C Alaska, N Canada >> SC USA

---

[1] Livr. 98 must be dated from 1836 see Dickinson (2001) (739).
[2] For recognition see Stepanyan (1990: 655) (2326); not accepted by Orn. Soc. Japan (2000) (1727).
[3] For attachment to this species see Stepanyan (1990: 667) (2326). Treated as a form of *schoeniclus* in Paynter (1970: 33) (1808) and in our last Edition.
[4] Requires further research to determine whether conspecific; atypical in habitat.
[5] Revision based on Cramp *et al.* (1994: 276) (596), whom see for recognition of *stresemanni*, *tschusii*, *intermedia*, *harterti* and *centralasiae*.

○ *Calcarius ornatus* (J.K. Townsend, 1837) Chestnut-collared Longspur
S Canada, NC USA >> S USA, N Mexico

PLECTROPHENAX Stejneger, 1882 M
○ *Plectrophenax nivalis* Snow Bunting
— *P.n.nivalis* (Linnaeus, 1758)
N North America, N Europe >> C USA, W, C and
SE Europe

— *P.n.insulae* Salomonsen, 1931
Iceland, ? N Scotland >> W and C Europe
— *P.n.vlasowae* Portenko, 1937
NE European Russia to NE Siberia >> C and E Asia
— *P.n.townsendi* Ridgway, 1887
W Aleutians, Commander Is.

○ *Plectrophenax hyperboreus* Ridgway, 1884 McKay's Bunting
Bering Sea islands (Hall, St. Matthew and the
Pribilofs) >> W Alaska

CALAMOSPIZA Bonaparte, 1838 F
◓ *Calamospiza melanocorys* Stejneger, 1885 Lark-Bunting
S Canada, WC USA >> SC USA, N Mexico

PASSERELLA Swainson, 1837 F
◑ *Passerella iliaca* Fox Sparrow
✓ *P.i.iliaca* (Merrem, 1786)
C and E Canada >> E USA
— *P.i.chilcatensis* Webster, 1983[1]
SE Alaska to NW British Columbia >> Oregon to
C California (2568)

✓ *P.i.zaboria* Oberholser, 1946
N and W Alaska, W Canada >> C and S USA
— *P.i.altivagans* Riley, 1911
SW Canada >> California
— *P.i.unalaschensis* (J.F. Gmelin, 1789)
Aleutian Is., Alaska Pen. >> British Columbia to
California

— *P.i.insularis* Ridgway, 1900[2]
Kodiak Is., Alaska >> British Columbia to California
— *P.i.sinuosa* Grinnell, 1910
S Alaska >> >> British Columbia to California
— *P.i.annectens* Ridgway, 1900
SC Alaska >> >> British Columbia to California
— *P.i.townsendi* (Audubon, 1838)
SE Alaska, Queen Charlotte Is. >> California
— *P.i.fuliginosa* Ridgway, 1899
SE Alaska to NW USA >> >> British Columbia to
California

— *P.i.olivacea* Aldrich, 1943
SW Canada, NW USA >> S California, NW Mexico
— *P.i.schistacea* S.F. Baird, 1858
SW Canada, NC USA >> SW USA, NW Mexico
— *P.i.swarthi* Behle & Selander, 1951
NW Utah, SE Idaho >> unknown
✓ *P.i.megarhyncha* S.F. Baird, 1858[3]
C and SW Oregon to California >> C California to
NW Mexico

— *P.i.canescens* Swarth, 1918
E California, C Nevada >> S California, S Arizona,
NW Mexico

MELOSPIZA S.F. Baird, 1858 F
● *Melospiza melodia* Song Sparrow
— *M.m.melodia* (A. Wilson, 1810)[4]
SE Canada, NE USA >> SE USA
✓ *M.m.atlantica* Todd, 1924
Coastal EC USA >> E USA
— *M.m.euphonia* Wetmore, 1936[5]
NC USA >> SC and SE USA
— *M.m.juddi* Bishop, 1896
WC and C Canada, WC USA >> WC, SC and SE USA
— *M.m.montana* Henshaw, 1884
Montane WC USA >> SW USA
— *M.m.fallax* (S.F. Baird, 1854)
SE Nevada, SW Utah, Arizona, NC Mexico
— *M.m.saltonis* Grinnell, 1909
SE California, S Nevada, W Arizona, NW Baja California
— *M.m.inexspectata* Riley, 1911
SE Alaska and W Canada >> SW Canada, Oregon
— *M.m.rufina* (Bonaparte, 1850)
Islands off SE Alaska and British Columbia >>
W Washington

— *M.m.morphna* Oberholser, 1899
SW Canada and NW USA >> N California
— *M.m.merrilli* Brewster, 1896
Montane WC Canada and WC USA >> S California,
Arizona, New Mexico
✓ *M.m.fisherella* Oberholser, 1911
NE Oregon, SW Idaho, EC California, W Nevada >>
S California

— *M.m.maxima* Gabrielson & Lincoln, 1951
Attu to Atka Is. (Aleutian Is.)

---

[1] Not mentioned by Zink (1986) (2736), but recognised by Browning (1990) (251).
[2] Paynter (1970: 43) (1808) provided the substitute name *ridgwayi* because *insularis* is preoccupied in *Zonotrichia*.
[3] Includes *fulva, monoensis, brevicauda* and *stephensi* see Zink (1986) (2736).
[4] Includes *melanchra* Oberholser, 1974 (1673) see Browning (1978) (246).
[5] Includes *callima* Oberholser, 1974 (1673) see Browning (1978) (246).

—— *M.m.sanaka* McGregor, 1901 — Seguam, Sanak, Unimak Is. (Aleutian Is.), W Alaska Pen. and islands

—— *M.m.amaka* Gabrielson & Lincoln, 1951 — Amak I. (Alaska Pen.)

—— *M.m.insignis* S.F. Baird, 1869 — Kodiak I., E Alaska Pen.

—— *M.m.kenaiensis* Ridgway, 1900 — Coastal S Alaska >> SE Alaska

—— *M.m.caurina* Ridgway, 1899 — Coastal SE Alaska >> N California

—— *M.m.cleonensis* McGregor, 1899 — SW Oregon, NW California

—— *M.m.gouldii* S.F. Baird, 1858 — Coastal C California

—— *M.m.mailliardi* Grinnell, 1911 — Central valley of California

—— *M.m.samuelis* (S.F. Baird, 1858) — Coastal C California

—— *M.m.maxillaris* Grinnell, 1909 — Coastal C California

—— *M.m.pusillula* Ridgway, 1899 — Coastal C California

—— *M.m.heermani* S.F. Baird, 1858 — Inland C California

—— *M.m.cooperi* Ridgway, 1899 — Coastal SW California (and deserts inland), N Baja California

—— *M.m.micronyx* Grinnell, 1928 — San Miguel I. (California)

—— *M.m.clementae* C.H. Townsend, 1890 — Santa Roza, Santa Cruz Is. (California)

—— *M.m.graminea* C.H. Townsend, 1890 — Santa Barbara I. (California)

—— *M.m.coronatorum* Grinnell & Daggett, 1903 — Coronados I. (N Baja California)

—— *M.m.rivularis* W.E. Bryant, 1888 — SC Baja California

—— *M.m.goldmani* Nelson, 1899 — Sierra Madre Occidental (Mexico)

—— *M.m.niceae* Dickerman, 1963 — NE Mexico

—— *M.m.mexicana* Ridgway, 1874 — Tlaxcala and Puebla, Mexico

—— *M.m.azteca* Dickerman, 1963 — Central valley and Mexico City area

—— *M.m.villai* Phillips & Dickerman, 1957 — WC Mexico (Upper Lerma valley)

—— *M.m.yuriria* Phillips & Dickerman, 1957 — C Mexico (Lower Lerma valley)

—— *M.m.adusta* Nelson, 1899 — N Michoacán (WC Mexico)

—— *M.m.zacapu* Dickerman, 1963 — N Michoacán (WC Mexico)

● *Melospiza lincolnii* LINCOLN'S SPARROW

—— *M.l.lincolnii* (Audubon, 1834) — Canada >> SW USA, S Mexico, Guatemala

—— *M.l.gracilis* (Kittlitz, 1858) — S Alaska, W Canada >> California

—— *M.l.alticola* (Miller & McCabe, 1935) — NW USA >> Mexico, Guatemala

● *Melospiza georgiana* SWAMP SPARROW

—— *M.g.ericrypta* Oberholser, 1938 — W and C Canada >> SW USA, NW Mexico

—— *M.g.georgiana* (Latham, 1790) — NE USA >>SE USA

ZONOTRICHIA Swainson, 1832 F

● *Zonotrichia capensis* RUFOUS-COLLARED SPARROW

—— *Z.c.septentrionalis* Griscom, 1930 — S Mexico, Guatemala, El Salvador, Honduras

—— *Z.c.antillarum* (Riley, 1916) — Dominican Republic

—— *Z.c.orestera* Wetmore, 1951 — Cerra Campana (W Panama)

—— *Z.c.costaricensis* J.A. Allen, 1891[1] — Costa Rica, Panama, Venezuela, Colombia

—— *Z.c.insularis* (Ridgway, 1898) — Curaçao, Aruba (Nether. Antilles)

—— *Z.c.venezuelae* Chapman, 1939 — N Venezuela

—— *Z.c.inaccessibilis* Phelps & Phelps, Jr., 1955 — Cerro de la Neblina (Venezuela)

—— *Z.c.perezchinchillorum* Phelps & Aveledo, 1984[2] — Cerro Marahuaca (Venezuela) (1838)

—— *Z.c.roraimae* (Chapman, 1929) — S Colombia, E Venezuela (including slopes of Mt. Roraima), Guyana, N Brazil

—— *Z.c.macconelli* Sharpe, 1900 — Summit of Mt. Roraima (Venezuela)

—— *Z.c.capensis* (Statius Müller, 1776) — French Guiana

—— *Z.c.bonnetiana* Stiles, 1995 — Sierra Chiribiquete (S Colombia) (2340)

—— *Z.c.tocantinsi* Chapman, 1940 — EC Brazil (south of the Amazon)

—— *Z.c.novaesi* Oren, 1985 — Serra dos Carajás, Pará (E Brazil) (1719)

—— *Z.c.matutina* (M.H.K. Lichtenstein, 1823) — NE Brazil to WC and SW Brazil and E Bolivia

—— *Z.c.subtorquata* Swainson, 1837 — E and C Brazil, Paraguay, Uruguay

—— *Z.c.illescasensis* Koepcke, 1963 — NW Peru

[1] Includes *orestera* see Olson (1981) (1697).
[2] Published as *perezchinchillae* but named for two men; see Art. 31.1.2. of the Code (I.C.Z.N., 1999) (1178).

— _Z.c.markli_ Koepcke, 1971   Lowland NW Peru (1292)

— _Z.c.huancabambae_ Chapman, 1940   Andes of N and C Peru

— _Z.c.peruviensis_ (Lesson, 1834)   Coast and Pacific Andean slope of Peru

— _Z.c.pulacayensis_ (Ménégaux, 1909)   Andes of C Peru to NW and C Bolivia and extreme
        NW Argentina

— _Z.c.carabayae_ Chapman, 1940   E slopes of Peruvian Andes and NW and C Bolivia

— _Z.c.antofagastae_ Chapman, 1940   NW Chile

— _Z.c.arenalensis_ Nores, 1986   NW Argentina (Catarmarca) (1657)

— _Z.c.sanborni_ Hellmayr, 1932   High Andes of Chile, NW Argentina (San Juan)

— _Z.c.hypoleuca_ (Todd, 1915)[1]   E and S Bolivia, NC and NE Argentina, C Paraguay

— _Z.c.choraules_ (Wetmore & Peters, 1922)   WC Argentina (Mendoza, Neuquen)

☑ _Z.c.chilensis_ (Meyen, 1834)   C Chile, SC Argentina

— _Z.c.australis_ (Latham, 1790)   S Chile, S Argentina >> N Bolivia

○ **_Zonotrichia querula_** (Nuttall, 1840) Harris's Sparrow   WC Canada >> W USA

◐ **_Zonotrichia leucophrys_** White-crowned Sparrow

  — _Z.l.leucophrys_ (J.R. Forster, 1772)[2]   C and E Canada >> SE USA, Cuba

  ☑ _Z.l.gambelii_ (Nuttall, 1840)   NW and W Canada >> W USA, N Mexico

  ☑ _Z.l.oriantha_ Oberholser, 1932   SW Canada >> WC USA, N Mexico

  — _Z.l.pugetensis_ Grinnell, 1928   SW Canada >> SW California

  — _Z.l.nuttalli_ Ridgway, 1899   WC California

◐ **_Zonotrichia albicollis_** (J.F. Gmelin, 1789) White-throated Sparrow   Canada, N and E USA >> S USA, E Mexico

◐ **_Zonotrichia atricapilla_** (J.F. Gmelin, 1789) Golden-crowned Sparrow

    Alaska, W Canada >> W USA, NW Mexico

JUNCO Wagler, 1831 M

○ **_Junco vulcani_** (Boucard, 1878) Volcano Junco   Costa Rica, W Panama

◐ **_Junco hyemalis_** Dark-eyed Junco

  ☑ _J.h.hyemalis_ (Linnaeus, 1758)   N Canada, NC USA >> S USA, N Mexico

  — _J.h.carolinensis_ Brewster, 1886   EC USA

  — _J.h.aikeni_ Ridgway, 1873   WC USA >> S USA

  ☑ _J.h.oreganus_ (J.K. Townsend, 1837)   S Alaska, W Canada >> C California

  — _J.h.cismontanus_ Dwight, 1918   W Canada >> W USA

  — _J.h.shufeldti_ Coale, 1887[3]   NW USA >> S California

  ☑ _J.h.montanus_ Ridgway, 1898   W Canada, W USA >> NW Mexico

  — _J.h.mearnsi_ Ridgway, 1897   SW Canada >> WC USA, NW Mexico

  — _J.h.thurberi_ Anthony, 1890   S Oregon >> California, N Baja California

  — _J.h.pinosus_ Loomis, 1893   C and S California

  — _J.h.pontilus_ Oberholser, 1919   N Baja California

  — _J.h.townsendi_ Anthony, 1889   N Baja California

  — _J.h.insularis_ Ridgway, 1876[4]   Guadalupe I. (Baja California)

  — _J.h.caniceps_ (Woodhouse, 1853)   C and SC USA >> S USA and N Mexico

  ☑ _J.h.dorsalis_ Henry, 1858   New Mexico, N Arizona

  — _J.h.mutabilis_ van Rossem, 1931   S Nevada, SE California

○ **_Junco phaeonotus_** Yellow-eyed Junco

  — _J.p.palliatus_ Ridgway, 1885   SW USA, N Mexico

  — _J.p.phaeonotus_ Wagler, 1831   C and S Mexico

  — _J.p.bairdi_ Ridgway, 1883   S Baja California

  — _J.p.fulvescens_ Nelson, 1897   S Mexico

  — _J.p.alticola_ Salvin, 1863   S Mexico, W Guatemala

PASSERCULUS Bonaparte, 1838 M

◐ **_Passerculus sandwichensis_** Savannah Sparrow

  — _P.s.labradorius_ Howe, 1901   E Canada >> SE USA

---

[1] Includes _mellea_ considered doubtfully distinct by Paynter (1970) (1808).

[2] Includes _aphaea_ Oberholser, 1974 (1673) see Browning (1978) (246).

[3] Includes _eumesus_ Oberholser, 1974 (1673) see Browning (1978) (246). There are differences of opinion on the availability of the name _shufeldti_, see Paynter (1970: 65) (1808) and Parkes in Dickerman & Parkes (1997) (729).

[4] This and the _caniceps_ group are retained in the species _hyemalis_ following A.O.U. (1998: 625-626) (3).

___ *P.s.savanna* (A. Wilson, 1811)	SE Canada >> SE USA, SE Mexico
⏚ *P.s.princeps* Maynard, 1872	Sable I. >> SE USA
___ *P.s.mediogriseus* Aldrich, 1940	SE Canada, NE USA >> SE USA
___ *P.s.oblitus* Peters & Griscom, 1938	C Canada, C USA, NE Mexico
⏚ *P.s.nevadensis* Grinnell, 1910	SW Canada, WC USA >> SC USA, N Mexico
___ *P.s.brooksi* Bishop, 1915	SW Canada >> California, Baja California
___ *P.s.athinus* Bonaparte, 1853	NE Siberia, Alaska, W Canada, W USA >> SW USA, W Mexico
___ *P.s.sandwichensis* (J.F. Gmelin, 1789)	Alaska >> W USA
___ *P.s.crassus* Peters & Griscom, 1938	Aleutian Is., W Alaska >> C California
___ *P.s.alaudinus* Bonaparte, 1853	N and C California
___ *P.s.beldingi* Ridgway, 1885	S California, N Baja California
___ *P.s.anulus* Huey, 1930	WC Baja California
___ *P.s.sanctorum* Coues, 1884	San Benito I., Baja California
___ *P.s.guttatus* Lawrence, 1867	W and S Baja California
___ *P.s.magdalenae* van Rossem, 1947	S Baja California
___ *P.s.rostratus* (Cassin, 1852)	S California, Baja California, W Mexico
___ *P.s.rufofuscus* Camras, 1940	Arizona, New Mexico, N Mexico
___ *P.s.atratus* van Rossem, 1930	NW Mexico
___ *P.s.brunnescens* (Butler, 1888)	NC Mexico
___ *P.s.wetmorei* van Rossem, 1938	SW Guatemala

AMMODRAMUS Swainson, 1827 M
○ *Ammodramus maritimus* SEASIDE SPARROW

___ *A.m.maritimus* (A. Wilson, 1811)	Massachusetts to N North Carolina >> to NE Florida
___ *A.m.macgillivrayi* (Audubon, 1834)	North Carolina to S Georgia
___ *A.m.pelonotus* (Oberholser, 1931)[1]	NE Florida
___ *A.m.mirabilis* (A.H. Howell, 1919)	SW Florida
___ *A.m.peninsulae* J.A. Allen, 1888	W Florida
___ *A.m.junicola* (Griscom & Nichols, 1920)	N Florida
___ †?*A.m.nigrescens* Ridgway, 1873	EC Florida
___ *A.m.fisheri* Chapman, 1899	Gulf Coast (SW and S USA)
___ *A.m.sennetti* J.A. Allen, 1888	Gulf Coast (S Texas)

○ *Ammodramus nelsoni* NELSON'S SHARP-TAILED SPARROW[2]

___ *A.n.nelsoni* J.A. Allen, 1875	W and C Canada, NC USA >> SE USA
___ *A.n.alter* (Todd, 1938)[3]	NC and NE Canada >> SE USA
___ *A.n.subvirgatus* Dwight, 1887	NE Canda, NE USA >> SE USA

○ *Ammodramus caudacutus* SALTMARSH SHARP-TAILED SPARROW

___ *A.c.caudacutus* (J.F. Gmelin, 1788)	NE USA >> SE USA
___ *A.c.diversus* Bishop, 1901	E USA >> SE USA

● *Ammodramus leconteii* (Audubon, 1844)[4] LE CONTE'S SPARROW

WC Canada >> C and SE USA

○ *Ammodramus bairdii* (Audubon, 1844) BAIRD'S SPARROW

WC Canada, NC USA >> SC USA, N Mexico

● *Ammodramus henslowii* HENSLOW'S SPARROW

___ *A.h.susurrans* (Brewster, 1918)	NE USA >> SE USA
___ *A.h.henslowii* (Audubon, 1829)[5]	SC and C USA >> SE USA

● *Ammodramus savannarum* GRASSHOPPER SPARROW

___ *A.s.pratensis* (Vieillot, 1817)	S Canada, E USA >> SE Mexico, West Indies
___ *A.s.floridanus* (Mearns, 1902)	Florida
⏚ *A.s.perpallidus* (Coues, 1872)	SW Canada, C and SW USA >> Mexico, Guatemala
___ *A.s.ammolegus* Oberholser, 1942	S Arizona, NW Mexico >> Guatemala
___ *A.s.bimaculatus* Swainson, 1827	S Mexico, Honduras, Nicaragua, NW Costa Rica, W Panama

[1] For correction of spelling see David & Gosselin (2002) (613).
[2] For reasons to separate this see Greenlaw (1993) (1002) and A.O.U. (1998: 619) (3).
[3] For correction of spelling see David & Gosselin (2002) (613).
[4] Dated 1843 by Paynter (1970: 77) (1808) but 1844 by A.O.U. (1983, 1998) (2, 3) and www.zoonomen.net (A.P. Peterson *in litt.*).
[5] Includes *houstonensis* Arnold, 1983 (56) see Browning (1990) (251).

___ *A.s.beatriceae* Olson, 1980   C Panama (1694)

___ *A.s.cracens* (Bangs & Peck, 1908)   Guatemala, E Honduras, NE Nicaragua

___ *A.s.caucae* Chapman, 1912   Colombia

___ *A.s.savannarum* (J.F. Gmelin, 1789)   Jamaica

___ *A.s.intricatus* E. Hartert, 1907   Hispaniola

___ *A.s.borinquensis* J.L. Peters, 1917   Puerto Rico

___ *A.s.caribaeus* (E. Hartert, 1902)   Bonaire, Curacao Is.

XENOSPIZA Bangs, 1931 F

○ *Xenospiza baileyi* Bangs, 1931 SIERRA MADRE SPARROW   N and C Mexico

MYOSPIZA Ridgway, 1898 F

○ *Myospiza humeralis* GRASSLAND SPARROW

___ *M.h.humeralis* (Bosc, 1792)   Colombia, Venezuela, Guyana, French Guiana, Brazil

___ *M.h.pallidulus* Wetmore, 1949   Guajira Pen. of NE Colombia, NW Venezuela

___ *M.h.xanthornus* (Darwin, 1839)   NE Bolivia, S Brazil, Paraguay, Uruguay, Argentina

___ *M.h.tarijensis* Bond & Meyer de Shauensee, 1939   SE Bolivia

○ *Myospiza aurifrons* YELLOW-BROWED SPARROW

___ *M.a.apurensis* Phelps & Gilliard, 1941   NE Colombia, W Venezuela

___ *M.a.cherriei* Chapman, 1914   E Colombia

___ *M.a.tenebrosus* Zimmer & Phelps, 1949   Venezuela, Colombia

___ *M.a.aurifrons* (Spix, 1825)   SE Colombia, E Ecuador, E Peru, E Bolivia, Amazonian Brazil

SPIZELLA Bonaparte, 1831 F

● *Spizella arborea* AMERICAN TREE SPARROW

___ *S.a.ochracae* Brewster, 1882   NW and W Canada >> W USA

___ *S.a.arborea* (A. Wilson, 1810)   NC and NE Canada >> C USA

● *Spizella passerina* CHIPPING SPARROW

✓ *S.p.passerina* (Bechstein, 1798)   SE Canada, C USA >> S USA

___ *S.p.arizonae* Coues, 1872   W Canada >> SW USA, W Mexico

___ *S.p.atremaeus* R.T. Moore, 1937[1]   NC and WC to EC Mexico

___ *S.p.mexicana* Nelson, 1899[2]   C and S Mexico, Guatemala

___ *S.p.pinetorum* Salvin, 1863   S Guatemala to NE Nicaragua

● *Spizella pusilla* FIELD SPARROW

✓ *S.p.pusilla* (A. Wilson, 1810)   SE Canada, E USA >> C and SE USA

___ *S.p.arenacea* Chadbourne, 1886[3]   C USA >> SE USA, NE Mexico

○ *Spizella wortheni* Ridgway, 1884 WORTHEN'S SPARROW   NE and E Mexico

○ *Spizella atrogularis* BLACK-CHINNED SPARROW

___ *S.a.evura* Coues, 1866   SW USA >> NW Mexico

___ *S.a.caurina* A.H. Miller, 1929   C California

___ *S.a.cana* Coues, 1866   SW California >> S Baja California

___ *S.a.atrogularis* (Cabanis, 1851)   NC Mexico

● *Spizella pallida* (Swainson, 1832) CLAY-COLOURED SPARROW   SC Canada, C and SC USA >> W Mexico

● *Spizella breweri* BREWER'S SPARROW

___ *S.b.taverneri* Swarth & Brooks, 1925   N Canada >> SW USA

✓ *S.b.breweri* Cassin, 1856   SW Canada, W and SW USA >> NW Mexico

POOECETES S.F. Baird, 1858 M

● *Pooecetes gramineus* VESPER SPARROW

___ *P.g.gramineus* (J.F. Gmelin, 1789)   SE Canada, E USA >> S Mexico

___ *P.g.confinis* S.F. Baird, 1858   SW Canada, WC USA >> SW USA, W Mexico

___ *P.g.affinis* G.S. Miller, 1888   W USA >> NW Baja California

---

[1] Thought to include *comparanda* see Paynter (1970: 83) (1808), but see also Dickerman & Parkes (1997) (729).
[2] Includes *repetens* see Paynter (1970: 83) (1808).
[3] Includes *perissuria* Oberholser, 1974 (1673) and *vernonia* Oberholser, 1974 (1673) see Browning (1978) (246).

CHONDESTES Swainson, 1827 M
● *Chondestes grammacus* Lark-Sparrow
    — *C.g.grammacus* (Say, 1823)   N and C USA >> SE USA, E and S Mexico
    — *C.g.strigatus* Swainson, 1827[1]   SW Canada, W USA >> Mexico, Guatemala

AMPHISPIZA Coues, 1874 F
● *Amphispiza bilineata* Black-throated Sparrow
    — *A.b.bilineata* (Cassin, 1850)   NC Texas, NE Mexico
    — *A.b.opuntia* Burleigh & Lowery, 1939[2]   SC USA, N Mexico
    — *A.b.deserticola* Ridgway, 1898   WC USA >> NW Mexico, Baja California
    — *A.b.bangsi* Grinnell, 1927   S Baja California
    — *A.b.tortugae* van Rossem, 1930   Tortuga I.
    — *A.b.belvederei* Banks, 1963   Cerralvo I.
    — *A.b.pacifica* Nelson, 1900   NW Mexico
    — *A.b.cana* van Rossem, 1930   San Esteban I.
    — *A.b.grisea* Nelson, 1898   C Mexico

○ *Amphispiza belli* Sage Sparrow
    — *A.b.nevadensis* (Ridgway, 1873)   W USA >> SW USA, N Baja California, NW Mexico
    — *A.b.canescens* Grinnell, 1905   SW California, W Nevada, NE Baja California
    — *A.b.belli* (Cassin, 1850)   S California, NW Baja California
    — *A.b.clementeae* Ridgway, 1898   San Clemente I.
    — *A.b.cinerea* C.H. Townsend, 1890   C Baja California

AIMOPHILA Swainson, 1837 F
● *Aimophila ruficauda* Stripe-headed Sparrow
    — *A.r.acuminata* (Salvin & Godman, 1886) WC Mexico
    — *A.r.lawrencii* (Salvin & Godman, 1886) S Mexico
    — *A.r.connectens* Griscom, 1930   E Guatemala
    — ¶*A.r.ibarrorum* Dickerman, 1987   SE Guatemala, El Salvador (717)
    — *A.r.ruficauda* (Bonaparte, 1853)   Honduras, Nicaragua

○ *Aimophila humeralis* (Cabanis, 1851) Black-chested Sparrow   WC Mexico

○ *Aimophila mystacalis* (Hartlaub, 1852) Bridled Sparrow   SC Mexico

○ *Aimophila sumichrasti* (Lawrence, 1871) Cinnamon-tailed Sparrow   S Mexico

○ *Aimophila stolzmanni* (Taczanowski, 1877) Tumbes Sparrow[3]   SW Ecuador, N Peru

○ *Aimophila strigiceps* Stripe-capped Sparrow
    — *A.s.dabbenei* (Hellmayr, 1912)   NW Argentina, SE Bolivia
    — *A.s.strigiceps* (Gould, 1841)[4]   E Argentina

○ *Aimophila carpalis* Rufous-winged Sparrow
    — *A.c.carpalis* (Coues, 1873)   SC Arizona, NW Mexico
    — *A.c.cohaerens* R.T. Moore, 1946[5]   NW Mexico

● *Aimophila cassinii* (Woodhouse, 1852) Cassin's Sparrow   SC USA >> C Mexico

○ *Aimophila aestivalis* Bachman's Sparrow
    — *A.a.bachmani* (Audubon, 1833)   SC USA >> SE USA
    — *A.a.illinoensis* (Ridgway, 1879)   NC USA >> S USA
    — *A.a.aestivalis* (M.H.K. Lichtenstein, 1823) S California, Georgia, Florida

○ *Aimophila botterii* Botteri's Sparrow
    — *A.b.arizonae* (Ridgway, 1873)   SE Arizona, NW Mexico
    — *A.b.texana* A.R. Phillips, 1943   S Texas, NE Mexico
    — *A.b.mexicana* (Lawrence, 1867)   NC Mexico
    — *A.b.goldmani* A.R. Phillips, 1943   W Mexico
    — *A.b.botterii* (P.L. Sclater, 1858)   SC and S Mexico

---

[1] Includes *quillini* Oberholser, 1974 (1673), see Browning (1978) (246).
[2] Includes *dapolia* Oberholser, 1974 (1673), see Browning (1978) (246).
[3] Treated in a monotypic genus *Rhynchospiza* in last Edition, but see Paynter (1970: 94) (1808).
[4] Paynter (1970: 94) (1808) dated this 1841, but see Zimmer (1926: 159) (2707) and www.zoonomen.net (A.P. Peterson *in litt.*).
[5] The name *distinguenda* (used in last Edition) is a synonym of *bangsi* (which is preoccupied) see Paynter (1970: 98) (1808); but the population is probably
    inseparable from this form see Dickerman & Parkes (1997) (729).

— *A.b.petenica* (Salvin, 1863)[1]	E and SE Mexico, Guatemala, Honduras
— *A.b.spadiconigrescens* T.R. Howell, 1965	N Honduras, NE Nicaragua
— *A.b.vantynei* Webster, 1959	C Guatemala
— *A.b.vulcanica* Miller & Griscom, 1925	Nicaragua, N Costa Rica

● *Aimophila ruficeps* RUFOUS-CROWNED SPARROW

— *A.r.ruficeps* (Cassin, 1852)	C California
— *A.r.canescens* Todd, 1922	SW California, NE Baja California
— *A.r.obscura* Dickey & van Rossem, 1923	Santa Catalina I.
— *A.r.sanctorum* van Rossem, 1947	Todos Santos I.
— *A.r.soraria* Ridgway, 1898	S Baja California
— *A.r.rupicola* van Rossem, 1946	SW Arizona
— *A.r.scottii* (Sennett, 1888)	W Oklahoma, extreme W Texas, N Arizona, C New Mexico to NC Sonora and NW Chihuahua
⅃ *A.r.eremoeca* (N.C. Brown, 1882)	C Oklahoma to WC and C Texas >> E Mexico
— *A.r.simulans* van Rossem, 1934	EC Sonora south to NE Sinaloa, SW Chihuahua and W Durango (NW Mexico)
— *A.r.pallidissima* A.R. Phillips, 1966[2]	S Coahuila south to N San Luis Potosí
— *A.r.phillipsi* Hubbard & Crossin, 1974	SE Sinaloa (1158)
— *A.r.boucardi* (P.L. Sclater, 1867)	S San Luis Potosí, SW to N Michoacán and south to E Puebla (Mexico)
— *A.r.suttoni* Hubbard, 1975	E Nayarit, W and N Jalisco (1157)
— *A.r.laybournae* Hubbard, 1975	WC Veracruz to SE Puebla (1157)
— *A.r.duponti* Hubbard, 1975	Mexico City area (1157)
— *A.r.fusca* (Nelson, 1897)	E Jalisco and N Michoacán (W Mexico)
— *A.r.australis* (Nelson, 1897)	C Oaxaca (S Mexico)
— *A.r.extima* A.R. Phillips, 1966[3]	Guerrero and Puebla south to S Oaxaca

● *Aimophila rufescens* RUSTY SPARROW

— *A.r.antonensis* van Rossem, 1942	NW Mexico
— *A.r.mcleodii* Brewster, 1888	NW Mexico
— *A.r.rufescens* (Swainson, 1827)[4]	W and SW Mexico to SW Chiapas
— *A.r.pyrgitoides* (Lafresnaye, 1839)[5]	Guatemala, Honduras, El Salvador, E and SE Mexico
— *A.r.discolor* Ridgway, 1887	N Honduras, NE Nicaragua
— *A.r.pectoralis* Dickey & van Rossem, 1927	SE Mexico, Guatemala, El Salvador
— *A.r.hypaethrus* Bangs, 1909	NW Costa Rica

○ *Aimophila notosticta* (Sclater & Salvin, 1868) OAXACA SPARROW — S Mexico

○ *Aimophila quinquestriata* FIVE-STRIPED SPARROW

— *A.q.septentrionalis* van Rossem, 1934	NW Mexico
— *A.q.quinquestriata* (Sclater & Salvin, 1868)	W Mexico

TORREORNIS Barbour & Peters, 1927 F
○ *Torreornis inexpectata* ZAPATA SPARROW

— *T.i.inexpectata* Barbour & Peters, 1927	SW Cuba
— *T.i.sigmani* Spence & Smith, 1961	S Cuba
— ¶ *T.i.varonai* Regalado Ruiz, 1981[6]	Cayo Coco (1999)

ORITURUS[7] Bonaparte, 1850 M
○ *Oriturus superciliosus* STRIPED SPARROW

— *O.s.palliatus* (van Rossem, 1938)	NW and W Mexico
— *O.s.superciliosus* (Swainson, 1838)	C and SW Mexico

PORPHYROSPIZA Sclater & Salvin, 1873[8] F
○ *Porphyrospiza caerulescens* (Wied, 1830) BLUE FINCH — Brazil, extreme E Bolivia

[1] Includes *tabascensis* see Paynter (1970: 96) (1808); but also see Dickerman & Parkes (1997) (729).
[2] For recognition see Hubbard (1975) (1157).
[3] For recognition see Hubbard (1975) (1157).
[4] Includes *brodkorbi* see Paynter (1970: 101) (1808) and *disjuncta* see Binford (1989) (150). See also Dickerman & Parkes (1997) (729).
[5] Includes *newmani* see Paynter (1970: 101) (1808), but see also Dickerman & Parkes (1997) (729).
[6] There seems to be an element of doubt whether the paper referenced here, which is reputed to contain this name, was ever published. We could not trace it.
[7] Dated '1851?' by Paynter (1970: 103) (1808), but see Zimmer (1926) (2707).
[8] For a comment agreeing this is not related closely to *Passerina* see Bates *et al.* (1992) (106).

PHRYGILUS Cabanis, 1844 M

○ *Phrygilus atriceps* (d'Orbigny & Lafresnaye, 1837) BLACK-HOODED SIERRA FINCH

SW Peru (Arequipa, Tacna), W Bolivia, N Chile, NW Argentina

○ *Phrygilus punensis* PERUVIAN SIERRA FINCH[1]

— *P.p.chloronotus* Berlepsch & Stolzmann, 1896   N and C Peru
— *P.p.punensis* Ridgway, 1887   S Peru (Puno), N Bolivia (La Paz)

● *Phrygilus gayi* GREY-HOODED SIERRA FINCH

— *P.g.gayi* (Gervais, 1834)   N Chile
✔ *P.g.minor* Philippi & Goodall, 1957   C Chile
— *P.g.caniceps* Burmeister, 1860   S Chile, Argentina

● *Phrygilus patagonicus* P.R. Lowe, 1923 PATAGONIAN SIERRA FINCH   S Chile, Argentina >> C Chile

○ *Phrygilus fruticeti* MOURNING SIERRA FINCH

— *P.f.peruvianus* J.T. Zimmer, 1924   Peru to N Bolivia
— *P.f.coracinus* P.L. Sclater, 1891[2]   SW Bolivia, N Chile (2219)
— *P.f.fruticeti* (Kittlitz, 1833)   S Chile and W Argentina

● *Phrygilus unicolor* PLUMBEOUS SIERRA FINCH

— *P.u.nivarius* (Bangs, 1899)   N Colombia, NW Venezuela
— *P.u.geospizopsis* (Bonaparte, 1853)   S Colombia, Ecuador
— *P.u.inca* J.T. Zimmer, 1929   Peru, N Bolivia
✔ *P.u.unicolor* (d'Orbigny & Lafresnaye, 1837)   SW Peru, Chile, WC and SW Argentina
— *P.u.tucumanus* Chapman, 1925   C Bolivia to NW Argentina
— ¶*P.u.cyaneus* Nores & Yzurieta, 1983   NC Argentina (1660)
— *P.u.ultimus* Ripley, 1950   S Argentina (Tierra del Fuego)

○ *Phrygilus dorsalis* Cabanis, 1883 RED-BACKED SIERRA FINCH   SW Bolivia, Chile, NW Argentina

○ *Phrygilus erythronotus* (Philippi & Landbeck, 1861) WHITE-THROATED SIERRA FINCH

Peru, Bolivia, Chile

○ *Phrygilus plebejus* ASH-BREASTED SIERRA FINCH

— *P.p.ocularis* P.L. Sclater, 1859   Andes of Ecuador, N Peru
— *P.p.plebejus* Tschudi, 1844   Andes of Peru, Chile, Bolivia and NW and WC Argentina
— ¶*P.p.naroskyi* Nores & Yzurieta, 1983   NC Argentina (Sierras de Córdoba) (1660)

○ *Phrygilus carbonarius* (d'Orbigny & Lafresnaye, 1837) CARBONATED SIERRA FINCH[3]

C Argentina

○ *Phrygilus alaudinus* BAND-TAILED SIERRA FINCH

— *P.a.bipartitus* J.T. Zimmer, 1924   W Ecuador, Peru
— *P.a.humboldti* Koepcke, 1963   S Ecuador, N Peru
— *P.a.bracki* O'Neill & Parker, 1997   C Peru (1717)
— *P.a.excelsus* Berlepsch, 1906   S Peru, Bolivia
— *P.a.alaudinus* (Kittlitz, 1833)   C Chile
— *P.a.venturii* E. Hartert, 1909   NW Argentina

MELANODERA Bonaparte, 1850[4] F

○ *Melanodera melanodera* CANARY-WINGED FINCH

— *M.m.princetoniana* (W.E.D. Scott, 1900)   S Chile, S Argentina
— *M.m.melanodera* (Quoy & Gaimard, 1824)   Falkland Is.

○ *Melanodera xanthogramma* YELLOW-BRIDLED FINCH

— *M.x.barrosi* Chapman, 1923   Chile, W Argentina
— *M.x.xanthogramma* (G.R. Gray, 1839)   S Argentina, Falkland Is.

HAPLOSPIZA Cabanis, 1851 F

○ *Haplospiza rustica* SLATY FINCH

— *H.r.uniformis* Sclater & Salvin, 1873   Highlands of S Mexico
— *H.r.barrilesensis* (Davidson, 1932)   Highlands of Honduras, Costa Rica, W Panama

---

[1] For reasons to separate this from *P. atriceps* see Ridgely & Tudor (1989: 444) (2018).
[2] For recognition see Fjeldså (1993) (847).
[3] Presumably a sparkling bird!
[4] Dated '1851?' by Paynter (1970: 108) (1808), but see Zimmer (1926) (2707).

— *H.r.arcana* (Wetmore & Phelps, 1949)   Tepuis of S Venezuela

— *H.r.rustica* (Tschudi, 1844)   Andes from N Venezuela, Colombia, Ecuador, Peru to C Bolivia

○ *Haplospiza unicolor* Cabanis, 1851   UNIFORM FINCH   SE Brazil, E Paraguay, NE Argentina

ACANTHIDOPS Ridgway, 1882 M

○ *Acanthidops bairdii* Ridgway, 1882   PEG-BILLED SPARROW   Highlands of Costa Rica, W Panama

LOPHOSPINGUS Cabanis, 1878 M

○ *Lophospingus pusillus* (Burmeister, 1860)   BLACK-CRESTED FINCH   S Bolivia, W Paraguay, W Argentina

○ *Lophospingus griseocristatus* (d'Orbigny & Lafresnaye, 1837)   GREY-CRESTED FINCH
     S Bolivia, NW Argentina

DONACOSPIZA Cabanis, 1851 F

○ *Donacospiza albifrons* (Vieillot, 1817)   LONG-TAILED REED FINCH   E Bolivia, S Brazil, Paraguay, Uruguay, NE Argentina

ROWETTIA P.R. Lowe, 1923 F

○ *Rowettia goughensis* (W.E. Clarke, 1904)   GOUGH ISLAND FINCH   Gough I.

NESOSPIZA Cabanis, 1873 F

○ *Nesospiza acunhae*   NIGHTINGALE ISLAND FINCH

     — *N.a.acunhae* Cabanis, 1873   Inaccessible I. (South Atlantic)

     — *N.a.questi* P.R. Lowe, 1923   Nightingale I., Tristan da Cunha I.

○ *Nesospiza wilkinsi*   WILKINS'S FINCH

     — *N.w.wilkinsi* P.R. Lowe, 1923   Nightingale I., Tristan da Cunha I.

     — *N.w.dunnei* Hagen, 1952   Inaccessible I. (South Atlantic)

DIUCA Reichenbach, 1850 F

○ *Diuca speculifera*   WHITE-WINGED DIUCA FINCH

     — *D.s.magnirostris* Carriker, 1935   Andes of C Peru

     — *D.s.speculifera* (d'Orbigny & Lafresnaye, 1837)   Andes of S Peru, N Chile, N Bolivia

● *Diuca diuca*   COMMON DIUCA FINCH

     — *D.d.crassirostris* Hellmayr, 1932   N Chile, Argentina

     ✓ *D.d.diuca* (Molina, 1782)   NC Chile, Argentina

     — *D.d.chiloensis* Philippi & Peña, 1964   SC Chile

     — *D.d.minor* Bonaparte, 1850   C Argentina > N Argentina, S Brazil

IDIOPSAR Cassin, 1866 M

○ *Idiopsar brachyurus* Cassin, 1866   SHORT-TAILED FINCH   Andes of SE Peru, Bolivia, NW Argentina

PIEZORHINA Lafresnaye, 1843 F

○ *Piezorhina cinerea* (Lafresnaye, 1843)   CINEREOUS FINCH   NW Peru

XENOSPINGUS Cabanis, 1867 M

○ *Xenospingus concolor* (d'Orbigny & Lafresnaye, 1837)   SLENDER-BILLED FINCH
     SW Peru, NW Chile

INCASPIZA Ridgway, 1898 F

○ *Incaspiza pulchra* (P.L. Sclater, 1886)   GREAT INCA FINCH   Andes of WC Peru (Ancash, Lima)

○ *Incaspiza personata* (Salvin, 1895)   RUFOUS-BACKED INCA FINCH   Upper Marañon Valley in N Peru

○ *Incaspiza ortizi* J.T. Zimmer, 1952   GREY-WINGED INCA FINCH   N Peru (Cajamarca, Piura)

○ *Incaspiza laeta* (Salvin, 1895)   BUFF-BRIDLED INCA FINCH   Upper Marañon Valley in N Peru

○ *Incaspiza watkinsi* Chapman, 1925   LITTLE INCA FINCH   Middle Marañon Valley in N Peru

POOSPIZA Cabanis, 1847 F

○ *Poospiza thoracica* (Nordmann[1], 1835)   BAY-CHESTED WARBLING FINCH
     SE Brazil

---

[1] Erroneously spelled Nordman by Paynter (1970: 117) (1808).

○ *Poospiza boliviana* Sharpe, 1888 Bolivian Warbling Finch   Andes of Bolivia

○ *Poospiza alticola* Salvin, 1895 Plain-tailed Warbling Finch   Andes of W Peru (Cajamarca to La Libertad and Ancash)

○ *Poospiza hypochondria* Rufous-sided Warbling Finch
   — *P.h.hypochondria* (d'Orbigny & Lafresnaye, 1837)   Andes of Bolivia
   — *P.h.affinis* Berlepsch, 1906   Andes of NW Argentina

○ *Poospiza erythrophrys* Rusty-browed Warbling Finch
   — *P.e.cochabambae* Gyldenstolpe, 1942   Andes of C Bolivia
   — *P.e.erythrophrys* P.L. Sclater, 1881   Andes of S Bolivia, NW Argentina

○ *Poospiza ornata* (Leybold, 1865) Cinnamon Warbling Finch   WC Argentina >> NW Argentina

○ *Poospiza nigrorufa* Black-and-rufous Warbling Finch[1]
   — *P.n.nigrorufa* (d'Orbigny & Lafresnaye, 1837)   S Brazil, Uruguay, Paraguay
   — *P.n.whitii* P.L. Sclater, 1883   Bolivia, NW Argentina
   — *P.n.wagneri* Stolzmann, 1926   Bolivia

○ *Poospiza lateralis* Red-rumped Warbling Finch
   — *P.l.lateralis* (Nordmann, 1835)   E Brazil
   — *P.l.cabanisi* Bonaparte, 1850   SE Brazil, Uruguay, SE Paraguay, NE Argentina

○ *Poospiza rubecula* Salvin, 1895 Rufous-breasted Warbling Finch   W Andes of N and C Peru

○ *Poospiza caesar* Sclater & Salvin, 1869 Chestnut-breasted Mountain Finch
     Andes of S Peru (Cuzco, Puno)

○ *Poospiza hispaniolensis* Bonaparte, 1850[2] Collared Warbling Finch
     SW Ecuador, W Peru

○ *Poospiza torquata* Ringed Warbling Finch
   — *P.t.torquata* (d'Orbigny & Lafresnaye, 1837)   E Bolivia
   — *P.t.pectoralis* Todd, 1922   N and C Argentina >> S Bolivia, W Paraguay

○ *Poospiza melanoleuca* (d'Orbigny & Lafresnaye, 1837) Black-capped Warbling Finch
     S Bolivia, Paraguay, W Uruguay, S Brazil, N Argentina

○ *Poospiza cinerea* Bonaparte, 1850[3] Cinereous Warbling Finch[4]   SC and SW Brazil

○ *Poospiza garleppi* (Berlepsch, 1893) Cochabamba Mountain Finch[5]   Andes of C Bolivia

○ *Poospiza baeri* (Oustalet, 1904) Tucuman Mountain Finch   Andes of NW Argentina

SICALIS Boie, 1828 F
○ *Sicalis citrina* Stripe-tailed Yellow Finch
   — *S.c.browni* Bangs, 1898   Colombia, Venezuela, Guyana, NE Brazil
   — *S.c.citrina* Pelzeln, 1870   E Brazil
   — *S.c.occidentalis* Carriker, 1932   S Peru

○ *Sicalis lutea* (d'Orbigny & Lafresnaye, 1837) Puna Yellow Finch   S Peru, W Bolivia, NW Argentina

○ *Sicalis uropygialis* Bright-rumped Yellow Finch
   — *S.u.sharpei* (Berlepsch & Stolzmann, 1894)   N and C Peru
   — *S.u.connectens* (Chapman, 1919)   S Peru (Cuzco)
   — *S.u.uropygialis* (d'Orbigny & Lafresnaye, 1837)   S Peru (Puno), W Bolivia, N Chile, NW Argentina

○ *Sicalis luteocephala* (d'Orbigny & Lafresnaye, 1837) Citron-headed Yellow Finch
     Andes of C Bolivia

● *Sicalis auriventris* Philippi & Landbeck, 1864 Greater Yellow Finch
     Andes of N Chile, WC Argentina

○ *Sicalis olivascens* Greenish Yellow Finch
   — *S.o.salvini* (Chubb, 1919)   Andes of N and C Peru
   — *S.o.chloris* Tschudi, 1846   W Andes from C Peru to C Chile
   — *S.o.olivascens* (d'Orbigny & Lafresnaye, 1837)   Andes in S Peru, W Bolivia, NW Argentina
   — *S.o.mendozae* (Sharpe, 1888)   Andes of W Argentina

[1] For an opinion that this represents two species see Ridgely & Tudor (1989: 457) (2018).
[2] Dated '1851?' by Paynter (1970: 121) (1808), but see Zimmer (1926) (2707).
[3] Dated '1851?' by Paynter (1970: 122) (1808), but see Zimmer (1926) (2707).
[4] For separation of this species from *P. melanoleuca* see Ridgely & Tudor (1989: 455) (2018).
[5] This species and *P. baeri* were retained in the genus *Compsospiza* in last Edition; we now follow Paynter (1970: 120) (1808).

○ *Sicalis lebruni* (Oustalet, 1891) PATAGONIAN YELLOW FINCH   S Argentina, S Chile

○ *Sicalis columbiana* ORANGE-FRONTED YELLOW FINCH
    — *S.c.columbiana* Cabanis, 1851   N Venezuela, E Colombia
    — *S.c.leopoldinae* Hellmayr, 1906   E Brazil
    — *S.c.goeldii* Berlepsch, 1906   E Peru, E Brazil

● *Sicalis flaveola* SAFFRON FINCH
    — *S.f.flaveola* (Linnaeus, 1766)   Colombia, Venezuela, the Guianas, Trinidad
    — *S.f.valida* Bangs & Penard, 1921   Ecuador, NW Peru
    — *S.f.brasiliensis* (J.F. Gmelin, 1789)   NE Brazil
    — *S.f.pelzelni* P.L. Sclater, 1872   SE Brazil, E Bolivia, Paraguay, Uruguay, N Argentina
    — ¶*S.f.koenigi* Hoy, 1978   E Andes of NW Argentina (1155)

● *Sicalis luteola* GRASSLAND YELLOW FINCH
    — *S.l.chrysops* P.L. Sclater, 1862   S Mexico, Guatemala, E Honduras, Nicaragua
    — *S.l.mexicana* Brodkorb, 1943   S Mexico
    — *S.l.eisenmanni* Wetmore, 1953   Panama
    — *S.l.bogotensis* Chapman, 1924   Colombia, Ecuador, Peru, Venezuela
    — *S.l.luteola* (Sparrman, 1789)   Colombia, Venezuela, Guyana, Brazil
    — *S.l.flavissima* Todd, 1922   N Brazilian islands
    — *S.l.chapmani* Ridgway, 1899   NE Brazil
    ☑ *S.l.luteiventris* (Meyen, 1834)   C and SC South America

○ *Sicalis raimondii* Taczanowski, 1874 RAIMONDI'S YELLOW FINCH   Andes of W Peru

○ *Sicalis taczanowskii* Sharpe, 1888 SULPHUR-THROATED FINCH   SW Ecuador, NW Peru

EMBERIZOIDES Temminck, 1822 M
○ *Emberizoides herbicola* WEDGE-TAILED GRASS FINCH
    — *E.h.lucaris* Bangs, 1908   SW Costa Rica
    — *E.h.hypochondriacus* Hellmayr, 1906   W and C Panama
    — *E.h.floresae* Griscom, 1924   Panama
    — *E.h.apurensis* Gilliard, 1940   E Colombia, W Venezuela
    — *E.h.sphenurus* (Vieillot, 1818)   N Colombia to the Guianas, N Brazil
    — *E.h.herbicola* (Vieillot, 1817)   E and S Brazil, E Bolivia, NE Argentina

○ *Emberizoides ypiranganus* H. von Ihering, 1908 LESSER GRASS FINCH   SE Brazil, E Paraguay, NE Argentina (2538)

○ *Emberizoides duidae* Chapman, 1929 MT. DUIDA GRASS FINCH   Tepuis of SE Venezuela

EMBERNAGRA Lesson, 1831 F
○ *Embernagra platensis* GREAT PAMPA FINCH
    — *E.p.platensis* (J.F. Gmelin, 1789)   SE Brazil, Paraguay, Uruguay, E Argentina
    — *E.p.olivascens* d'Orbigny, 1839   SE Bolivia, W Paraguay, NW Argentina
    — *E.p.catamarcanus* Nores, 1986   W Argentina (1657)

○ *Embernagra longicauda* Strickland, 1844 PALE-THROATED PAMPA FINCH
    E Brazil

VOLATINIA Reichenbach, 1850 F
● *Volatinia jacarina* BLUE-BLACK GRASSQUIT
    ☑ *V.j.splendens* (Vieillot, 1817)   Middle America, N South America, Trinidad
    — *V.j.jacarina* (Linnaeus, 1766)   Amazonian Brazil to C Argentina
    — *V.j.peruviensis* (Peale, 1848)   W Ecuador, W Peru, NW Chile

SPOROPHILA Cabanis, 1844 F[1]
○ *Sporophila frontalis* (J. Verreaux, 1869) BUFFY-FRONTED SEEDEATER   SE Brazil, E Paraguay, NE Argentina

○ *Sporophila falcirostris* (Temminck, 1820) TEMMINCK'S SEEDEATER   SE Brazil

○ *Sporophila schistacea* SLATE-COLOURED SEEDEATER
    — *S.s.subconcolor* Berlioz, 1959   S Mexico
    — *S.s.schistacea* (Lawrence, 1862)   Costa Rica, Panama, N Colombia

[1] See Ouellet (1992) (1733) for comments on taxonomic problems.

— *S.s.incerta* Riley, 1914   W Colombia, Ecuador

— *S.s.longipennis* Chubb, 1921   Venezuela, E Colombia, N Brazil

○ *Sporophila plumbea* PLUMBEOUS SEEDEATER

— *S.p.colombiana* (Sharpe, 1888)   N Colombia

— *S.p.whiteleyana* (Sharpe, 1888)   E Colombia, S Venezuela, the Guianas, N Brazil

— *S.p.plumbea* (Wied, 1831)   C and S Brazil, Paraguay, NW Bolivia, N Argentina

● *Sporophila corvina* VARIABLE SEEDEATER[1]

∠ *S.c.corvina* (P.L. Sclater, 1859)   Central America from E Mexico to Panama

— *S.c.hoffmannii* Cabanis, 1861   SW Panama

— *S.c.hicksii* Lawrence, 1865[2]   Pacific slope of E Panama and NW Colombia

— *S.c.ophthalmica* (P.L. Sclater, 1860)   SW Colombia, W Ecuador, Peru

○ *Sporophila intermedia* GREY SEEDEATER[3]

— *S.i.intermedia* Cabanis, 1851   N Colombia, N Venezuela, Guyana, Trinidad

— *S.i.bogotensis* (Gilliard, 1946)[4]   W Andes and upper Magdalena valley (Colombia)

● *Sporophila americana* WING-BARRED SEEDEATER

— *S.a.americana* (J.F. Gmelin, 1789)   NE Venezuela, the Guianas, Brazil, Tobago I.

— *S.a.dispar* Todd, 1922   Central Amazonian Brazil

○ *Sporophila murallae* Chapman, 1915 CAQUETÁ SEEDEATER[5]   SE Colombia

● *Sporophila torqueola* WHITE-COLLARED SEEDEATER

— *S.t.sharpei* Lawrence, 1889   S Texas, NE Mexico

— *S.t.torqueola* (Bonaparte, 1850)[6]   C and SW Mexico

∠ *S.t.morelleti* (Bonaparte, 1850)[7]   Atlantic slopes of Central America, S Mexico to Panama

— *S.t.mutanda* Griscom, 1930   Pacific slopes from SW Mexico to El Salvador

○ *Sporophila collaris* RUSTY-COLLARED SEEDEATER

— *S.c.ochrascens* Hellmayr, 1904   E Bolivia, W Brazil

— *S.c.collaris* (Boddaert, 1783)   S Brazil

— *S.c.melanocephala* (Vieillot, 1817)   SW Brazil, Paraguay, N Argentina

○ *Sporophila bouvronides* LESSON'S SEEDEATER[8]

— *S.b.restricta* Todd, 1917   NC Colombia

— *S.b.bouvronides* (Lesson, 1831)   NE Colombia, N Venezuela, Trinidad, Tobago I. >> N Amazonia

○ *Sporophila lineola* (Linnaeus, 1758) LINED SEEDEATER   SE Bolivia, N Argentina, Paraguay, S Brazil >> Amazonia, N South America

○ *Sporophila luctuosa* (Lafresnaye, 1843) BLACK-AND-WHITE SEEDEATER   Andes from Venezuela and Colombia to C Bolivia

○ *Sporophila nigricollis* YELLOW-BELLIED SEEDEATER

— *S.n.nigricollis* (Vieillot, 1823)   Costa Rica, Panama and N South America to Bolivia

— *S.n.vivida* (Hellmayr, 1938)   SW Colombia, W Ecuador

— *S.n.inconspicua* Berlepsch & Stolzmann, 1906   W Peru

○ *Sporophila ardesiaca* (A.J.C. Dubois, 1894) DUBOIS'S SEEDEATER   SE Brazil

○ †?*Sporophila melanops* (Pelzeln, 1870) HOODED SEEDEATER   SC Brazil

○ *Sporophila caerulescens* DOUBLE-COLLARED SEEDEATER

— *S.c.caerulescens* (Vieillot, 1823)   C Brazil, E Bolivia, Paraguay, Uruguay, C Argentina >> S Amazonia

— *S.c.hellmayri* Wolters, 1939   E Brazil (Bahia)

— *S.c.yungae* Gyldenstolpe, 1941   N Bolivia

○ *Sporophila albogularis* (Spix, 1825) WHITE-THROATED SEEDEATER   E Brazil

---

[1] Stiles (1996) (2343) agreed with Olson (1981) (1695) that the name *aurita* applied to a hybrid between this and the allospecies *americana*. His arrangement of an *americana* superspecies with four allospecies (*corvina, intermedia, americana* and *murallae*) is treated tentatively here by listing the allospecies.
[2] Includes *chocoana* see Stiles (1996) (2343).
[3] Stiles (1996) (2343) considered *anchicayae* to be a hybrid parented by *S. intermedia* and *S. c. hicksi*.
[4] Includes *agustini* see Stiles (1996) (2343).
[5] For treatment as a species see Stiles (1996) (2343).
[6] Dated '1851?' by Paynter (1970: 138) (1808), but see Zimmer (1926) (2707).
[7] Dated '1851?' by Paynter (1970: 138) (1808), but see Zimmer (1926) (2707).
[8] For reasons to separate this from *S. lineola* see Ridgely & Tudor (1989: 409) (2018).

○ *Sporophila leucoptera* WHITE-BELLIED SEEDEATER
    — *S.l.mexianae* Hellmayr, 1912   Mexiana I. (Brazil)
    — *S.l.cinereola* (Temminck, 1820)   E Brazil
    — *S.l.leucoptera* (Vieillot, 1817)   C and SW Brazil, Paraguay, N Argentina
    — *S.l.bicolor* (d'Orbigny & Lafresnaye, 1837)   NE Bolivia

○ *Sporophila peruviana* PARROT-BILLED SEEDEATER
    — *S.p.devronis* (J. Verreaux, 1852)   W Ecuador, NW Peru
    — *S.p.peruviana* (Lesson, 1842)   W Peru

○ *Sporophila simplex* (Taczanowski, 1874) DRAB SEEDEATER   S Ecuador, W Peru

○ *Sporophila nigrorufa* (d'Orbigny & Lafresnaye, 1837) BLACK-AND-TAWNY SEEDEATER
                                           SW Brazil, E Bolivia

○ *Sporophila bouvreuil* CAPPED SEEDEATER
    — *S.b.bouvreuil* (Statius Müller, 1776)   E Brazil
    — *S.b.saturata* Hellmayr, 1904   SE Brazil (Sao Paulo)
    — *S.b.crypta* Sick, 1968   SE Brazil (Rio de Janiero)
    — *S.b.pileata* (P.L. Sclater, 1864)   S Brazil, E Paraguay, N Argentina

○ *Sporophila insulata* Chapman, 1921 TUMACO SEEDEATER   Tumaco I. off SW Colombia

○ *Sporophila minuta* RUDDY-BREASTED SEEDEATER
    — *S.m.parva* (Lawrence, 1883)   Central America from SW Mexico to Nicaragua
    — *S.m.centralis* Bangs & Penard, 1918   SW Costa Rica, W Panama
    — *S.m.minuta* (Linnaeus, 1758)   Trinidad and N South America

○ *Sporophila hypoxantha* Cabanis, 1851 TAWNY-BELLIED SEEDEATER   E Bolivia, Paraguay, E and C Brazil, Uruguay,
                                           N Argentina

○ *Sporophila hypochroma* Todd, 1915 RUFOUS-RUMPED SEEDEATER   E Bolivia, SW Brazil, NE Argentina

○ *Sporophila ruficollis* Cabanis, 1851 DARK-THROATED SEEDEATER   S Brazil, E Bolivia to N Argentina

○ *Sporophila palustris* (Barrows, 1883) MARSH SEEDEATER   SE Brazil, Paraguay, Uruguay, N Argentina

○ *Sporophila castaneiventris* Cabanis, 1849 CHESTNUT-BELLIED SEEDEATER
                                           Amazonia, the Guianas

○ *Sporophila cinnamomea* (Lafresnaye, 1839) CHESTNUT SEEDEATER   S Brazil, E Paraguay, NE Argentina

○ *Sporophila melanogaster* (Pelzeln, 1870) BLACK-BELLIED SEEDEATER   SE Brazil

○ *Sporophila telasco* (Lesson, 1828) CHESTNUT-THROATED SEEDEATER   W Ecuador, W Peru, extreme NW Chile

○ *Sporophila zelichi* Narosky, 1977 NAROSKY'S SEEDEATER[1]   NE Argentina (Entre Rios) >> ?? (1636)

ORYZOBORUS Cabanis, 1851 M
○ *Oryzoborus funereus* THICK-BILLED SEED FINCH
    — *O.f.funereus* P.L. Sclater, 1859   SE Mexico, Guatemala, Belize and Honduras
    — *O.f.salvini* (Ridgway, 1884)   S Nicaragua, Costa Rica (excl. SW), NW Panama (1667)
    — *O.f.ochrogyne* (Olson, 1981)   SW Costa Rica and Pacific slope S to NW Colombia (1695)
    — *O.f.aethiops* (P.L. Sclater, 1860)   W Ecuador, SW Colombia (2216)

○ *Oryzoborus angolensis* CHESTNUT-BELLIED SEED FINCH
    — *O.a.theobromae* (Olson, 1981)   NC Colombia (Upper Magdalena Valley) (1695)
    — *O.a.angolensis* (Linnaeus, 1766)[2]   E Colombia, Venezuela, the Guianas, Trinidad and E
                                        of the Andes S to E Bolivia, Paraguay, N Argentina

○ *Oryzoborus nuttingi* Ridgway, 1884 NICARAGUAN SEED FINCH   Nicaragua to Panama

○ *Oryzoborus crassirostris* (J.F. Gmelin, 1789) LARGE-BILLED SEED FINCH
                                           Colombia, Venezuela, the Guianas, N Brazil

○ *Oryzoborus maximiliani* GREAT-BILLED SEED FINCH[3]
    — *O.m.magnirostris* Phelps & Phelps, Jr., 1950[4]   Trinidad, E Venezuela

---

[1] For comments on the status of this see Ridgely & Tudor (1989: 425) (2018).

[2] Includes *torridus* see Olson (1981) (1695).

[3] Meyer de Schauensee (1970) (1576) showed that monotypic *crassirostris* and polytypic *maximiliani* occur in sympatry (in this context *atrirostris* was kept within *maximiliani*).

[4] Olson (1981) (1695) considered *Oryzoborus* should be submerged in *Sporophila*; this name, *magnirostris*, is preoccupied in *Sporophila* and for it he proposed the substitute name *parkesi*.

— *O.m.maximiliani* (Cabanis, 1851)   NE Brazil, the Guianas

— *O.m.occidentalis* P.L. Sclater, 1860   W Colombia, NW Ecuador

○ *Oryzoborus atrirostris* BLACK-BILLED SEED FINCH

— *O.a.atrirostris* Sclater & Salvin, 1878   N Peru

— *O.a.gigantirostris* Bond & Meyer de Schauensee, 1939   N Bolivia

AMAUROSPIZA Cabanis, 1861   F

○ *Amaurospiza concolor* BLUE SEEDEATER

— *A.c.relicta* (Griscom, 1934)   S Mexico

— *A.c.concolor* Cabanis, 1861   Honduras, Nicaragua, Costa Rica, Panama

— *A.c.aequatorialis* Sharpe, 1888   SW Colombia, Ecuador

○ *Amaurospiza moesta* (Hartlaub, 1853)   BLACKISH-BLUE SEEDEATER   E and SE Brazil, E Paraguay, NE Argentina

MELOPYRRHA Bonaparte, 1853   F

○ *Melopyrrha nigra* CUBAN BULLFINCH

— *M.n.nigra* (Linnaeus, 1758)   Cuba, Isla de la Juventud

— *M.n.taylori* E. Hartert, 1896   Grand Cayman I.

DOLOSPINGUS Elliot, 1871   M

○ *Dolospingus fringilloides* (Pelzeln, 1870)   WHITE-NAPED SEEDEATER   E Colombia, S Venezuela, N Brazil

CATAMENIA Bonaparte, 1850[1]   F

○ *Catamenia analis* BAND-TAILED SEEDEATER

— *C.a.alpica* Bangs, 1902   N Colombia (Santa Marta Mts.)

— *C.a.schistaceifrons* Chapman, 1915   Eastern Andes of Colombia

— *C.a.soderstromi* Chapman, 1924   N Ecuador

— *C.a.insignis* J.T. Zimmer, 1930   N Peru

— *C.a.analoides* (Lafresnaye, 1847)   W Andes of Peru

— *C.a.griseiventris* Chapman, 1919   S Peru

— *C.a.analis* (d'Orbigny & Lafresnaye, 1837)   Bolivia, NW Argentina, N Chile

○ *Catamenia inornata* PLAIN-COLOURED SEEDEATER

— *C.i.mucuchiesi* Phelps & Gilliard, 1941   Andes of W Venezuela (Mérida)

— *C.i.minor* Berlepsch, 1885   Andes of W Venezuela (Táchira) and Colombia to C Peru

— *C.i.inornata* (Lafresnaye, 1847)   Andes from S Peru to NW Argentina

— ¶*C.i.cordobensis* Nores & Yzurieta, 1983   NC Argentina (1660)

○ *Catamenia homochroa* PARAMO SEEDEATER

— *C.h.oreophila* Todd, 1913[2]   N Colombia (Santa Marta Mts.)

— *C.h.homochroa* P.L. Sclater, 1858   Andes from W Venezuela, Colombia to N Bolivia

— *C.h.duncani* (Chubb, 1921)   Tepui region of Venezuela, N Brazil

TIARIS Swainson, 1827   M[3]

○ *Tiaris canorus* (J.F. Gmelin, 1789)   CUBAN GRASSQUIT   Cuba

● *Tiaris olivaceus* YELLOW-FACED GRASSQUIT

⊥ *T.o.pusillus* Swainson, 1827   E Mexico, Central America, Colombia, Venezuela

— *T.o.intermedius* (Ridgway, 1885)   Cozumel, Holbox Is., E Mexico

— *T.o.ravidus* Wetmore, 1957   Panama

— *T.o.olivaceus* (Linnaeus, 1766)   Cuba, Jamaica, Cayman Is.

— *T.o.bryanti* (Ridgway, 1898)   Puerto Rico

○ *Tiaris obscurus* DULL-COLOURED GRASSQUIT[4]

— *T.o.haplochroma* (Todd, 1912)   N Colombia, NW Venezuela

— *T.o.pauper* (Berlepsch & Taczanowski, 1884)   S Colombia, Ecuador, NW Peru

— *T.o.obscurus* (d'Orbigny & Lafresnaye, 1837)   C Peru, Bolivia, Argentina

— *T.o.pacificus* (Koepcke, 1963)   W Peru

---

[1] Dated '1851?' by Paynter (1970: 152) (1808), but see Zimmer (1926) (2707).
[2] For reasons to treat *oreophila* here see Romero-Zambrano (1977) (2098A) and Ridgely & Tudor (1989: 438) (2018).
[3] Gender addressed by David & Gosselin (in press) (614), multiple changes occur in specific names.
[4] Often previously placed in *Sporophila* see Ouellet (1992) (1733).

○ *Tiaris bicolor* BLACK-FACED GRASSQUIT
 — *T.b.bicolor* (Linnaeus, 1766)   Bahamas, Cuba
 — *T.b.marchii* (S.F. Baird, 1863)   Jamaica, Hispaniola
 — *T.b.omissus* Jardine, 1847   Puerto Rico, Tobago I., Colombia, Venezuela
 — *T.b.huilae* A.H. Miller, 1952   Colombia
 — *T.b.grandior* (Cory, 1887)   Isla Andrès and Isla Providencia (W Caribbean)
 — *T.b.johnstonei* (P.R. Lowe, 1906)   La Blanquilla I.
 — *T.b.sharpei* (E. Hartert, 1893)   Aruba, Curacao I.
 — *T.b.tortugensis* Cory, 1909   La Tortuga I.

○ *Tiaris fuliginosus* SOOTY GRASSQUIT
 — *T.f.zuliae* Aveledo & Ginés, 1948[1]   NW Venezuela
 — *T.f.fumosus* (Lawrence, 1874)   NC, NE and SE Venezuela, Trinidad
 — *T.f.fuliginosus* (Wied, 1830)   NE and C Brazil, E Bolivia

LOXIPASSER H. Bryant, 1866 M
○ *Loxipasser anoxanthus* (Gosse, 1847) YELLOW-SHOULDERED GRASSQUIT   Jamaica

LOXIGILLA Lesson, 1831 F
○ *Loxigilla portoricensis* PUERTO RICAN BULLFINCH
 — *L.p.portoricensis* (Daudin, 1800)   Puerto Rico
 — †*L.p.grandis* Lawrence, 1881   St. Kitts (Lesser Antilles)

○ *Loxigilla violacea* GREATER ANTILLEAN BULLFINCH
 — *L.v.violacea* (Linnaeus, 1758)   Bahama Is.
 — *L.v.ofella* Buden, 1986   Middle and East Caicos Is. (267)
 — *L.v.maurella* Wetmore, 1919   Tortue I. (off north Hispaniola)
 — *L.v.affinis* (Ridgway, 1898)[2]   Hispaniola and Gonâve, Saona and Catalina,
                                                Ile-à-Vache, Beata Is. (off S Hispaniola)
 — *L.v.ruficollis* (J.F. Gmelin, 1789)   Jamaica

○ *Loxigilla noctis* LESSER ANTILLEAN BULLFINCH
 — *L.n.coryi* (Ridgway, 1898)   St. Kitts, Monserrat Is.
 — *L.n.ridgwayi* (Cory, 1892)   Anguilla, Antigua, Barbuda Is.
 — *L.n.desiradensis* Danforth, 1937   Desirade I.
 — *L.n.dominicana* (Ridgway, 1898)   Guadeloupe, Dominica Is.
 — *L.n.noctis* (Linnaeus, 1766)   Martinique I.
 — *L.n.sclateri* J.A. Allen, 1880   St. Lucia I.
 — *L.n.crissalis* (Ridgway, 1898)   St. Vincent I.
 — *L.n.grenadensis* (Cory, 1892)   Grenada I.
 — *L.n.barbadensis* Cory, 1886   Barbados I.

EUNEORNIS Fitzinger, 1856[3] M
○ *Euneornis campestris* (Linnaeus, 1758) ORANGEQUIT   Jamaica

MELANOSPIZA Ridgway, 1897 F
○ *Melanospiza richardsoni* (Cory, 1886) ST. LUCIA BLACK FINCH   St. Lucia I.

GEOSPIZA Gould, 1837 F
○ *Geospiza magnirostris* Gould, 1837 LARGE GROUND FINCH   Galapagos

○ *Geospiza fortis* Gould, 1837 MEDIUM GROUND FINCH   Galapagos

○ *Geospiza fuliginosa* Gould, 1837 SMALL GROUND FINCH   Galapagos

○ *Geospiza difficilis* SHARP-BEAKED GROUND FINCH
 — *G.d.difficilis* Sharpe, 1888   Tower, Abingdon Is.
 — *G.d.debilirostris* Ridgway, 1894   James, Albemarle, Narborough Is.
 — *G.d.septentrionalis* Rothschild & Hartert, 1899   Culpepper, Wenman Is.

[1] Cited as Hostos & Ginés by Paynter (1970: 157) (1808).
[2] Includes *parishi* see Buden (1986) (267).
[3] In last Edition treated in the Thraupinae. We place this here following A.O.U. (1998: 596) (3).

○ *Geospiza scandens* COMMON CACTUS FINCH
   — *G.s.scandens* (Gould, 1837)   James, Jervis Is.
   — *G.s.intermedia* Ridgway, 1894   Barrington, Charles, Duncan, Indefatigable, Albemarle Is.

   — *G.s.abingdoni* (Sclater & Salvin, 1870)   Abingdon I.
   — *G.s.rothschildi* Heller & Snodgrass, 1901   Bindloe I.

○ *Geospiza conirostris* LARGE CACTUS FINCH
   — *G.c.conirostris* Ridgway, 1890   Hood I.
   — *G.c.propinqua* Ridgway, 1894   Tower I.
   — *G.c.darwini* Rothschild & Hartert, 1899   Culpepper I.

CAMARHYNCHUS Gould, 1837   M
○ *Camarhynchus crassirostris* Gould, 1837   VEGETARIAN FINCH   Galapagos

○ *Camarhynchus psittacula* LARGE TREE FINCH
   — *C.p.habeli* Sclater & Salvin, 1870   Abingdon, Bindloe Is.
   — *C.p.affinis* Ridgway, 1894   Albemarle, Narborough Is.
   — *C.p.psittacula* Gould, 1837   Seymour, Barrington, Indefatigable, Charles, Duncan, Jervis, James Is.

○ *Camarhynchus pauper* Ridgway, 1890   MEDIUM TREE FINCH   Charles I.

○ *Camarhynchus parvulus* SMALL TREE FINCH
   — *C.p.parvulus* (Gould, 1837)   James, Jervis, Indefatigable, Seymour, Barrington, Albemarle, Duncan, Charles, Narborough Is.

   — *C.p.salvini* Ridgway, 1894   Chatham I.

○ *Camarhynchus pallidus* WOODPECKER FINCH
   — *C.p.pallidus* (Sclater & Salvin, 1870)   James, Jervis, Seymour, Duncan, Indefatigable, Charles Is.

   — *C.p.productus* Ridgway, 1894   Albemarle, Narborough Is.
   — *C.p.striatipectus* (Swarth, 1931)   Chatham I.

○ *Camarhynchus heliobates* (Snodgrass & Heller, 1901)   MANGROVE FINCH
     Albemarle, Narborough Is.

CERTHIDEA Gould, 1837   F
○ *Certhidea olivacea* WARBLER-FINCH
   — *C.o.becki* Rothschild, 1898   Culpepper, Wenman Is.
   — *C.o.mentalis* Ridgway, 1894   Tower I.
   — *C.o.fusca* Sclater & Salvin, 1870   Abingdon, Bindloe Is.
   — *C.o.olivacea* Gould, 1837   James, Jervis, Seymour, Duncan, Albemarle, Narborough Is.

   — *C.o.bifasciata* Ridgway, 1894   Barrington I.
   — *C.o.luteola* Ridgway, 1894   Chatham I.
   — *C.o.cinerascens* Ridgway, 1890   Hood I.
   — *C.o.ridgwayi* Rothschild & Hartert, 1899   Charles I.

PINAROLOXIAS Sharpe, 1885   F
○ *Pinaroloxias inornata* (Gould, 1843)   COCOS ISLAND FINCH   Isla de Coco (off Costa Rica)

PIPILO Vieillot, 1816   M
● *Pipilo chlorurus* (Audubon, 1839)   GREEN-TAILED TOWHEE[1]   W USA >> C Mexico

○ *Pipilo ocai* COLLARED TOWHEE
   — *P.o.alticola* (Salvin & Godman, 1889)   W Mexico
   — *P.o.nigrescens* (Salvin & Godwin, 1889)   W Mexico
   — *P.o.guerrerensis* van Rossem, 1938   SW Mexico
   — *P.o.brunnescens* van Rossem, 1938   S Mexico
   — *P.o.ocai* (Lawrence, 1865)[2]   EC Mexico

---

[1] For reasons to place this in *Pipilo* rather than retain the monotypic genus *Chlorura* (not *Chlorurus* as given in last Edition) see Sibley (1955) (2258).
[2] Dated 1867 by Paynter (1970: 170) (1808), but see A.O.U. (1998: 605) (3) and www.zoonomen.net (A.P. Peterson *in litt.*).

● *Pipilo maculatus* SPOTTED TOWHEE[1]

  — *P.m.montanus* Swarth, 1905   SW USA >> N Mexico

  — *P.m.gaigei* van Tyne & Sutton, 1937   Texas

  — *P.m.curtatus* Grinnell, 1911   SW Canada, W USA >> SE California

  — *P.m.oregonus* Bell, 1849   W USA >> S California

  — *P.m.falcinellus* Swarth, 1913   W USA

  — *P.m.falcifer* McGregor, 1900   SW USA

  — *P.m.megalonyx* S.F. Baird, 1858   SW USA, NW Baja California

  — *P.m.clementae* Grinnell, 1897   San Clemente I.

  — *P.m.umbraticola* Grinnell & Swarth, 1926   NW Baja California

  — †*P.m.consobrinus* Ridgway, 1876   Isla Guadalupe, Baja California

  — *P.m.magnirostris* Brewster, 1891   S Baja California

  — *P.m.griseipygius* van Rossem, 1934   W Mexico

  — *P.m.orientalis* Sibley, 1950   EC Mexico

  — *P.m.maculatus* Swainson, 1827   E Mexico

  — *P.m.macronyx* Swainson, 1827   SC Mexico

  — *P.m.vulcanorum* Sibley, 1951   SC Mexico

  — *P.m.oaxacae* Sibley, 1950   S Mexico

  — *P.m.repetens* Griscom, 1930   S Mexico, W Guatemala

  — *P.m.chiapensis* van Rossem, 1938   S Mexico

  — *P.m.socorroensis* Grayson, 1867   Socorro I., Revillagigedo Group

● *Pipilo erythrophthalmus* EASTERN TOWHEE[2]

  ✓ *P.e.erythrophthalmus* (Linnaeus, 1758)   S Canada, E USA >> S USA

  ✓ *P.e.rileyi* Koelz, 1939   SE USA

  — *P.e.alleni* Coues, 1871   Florida

  ✓ *P.e.canaster* A.H. Howell, 1913   SC USA >> SE USA

  — *P.e.arcticus* (Swainson, 1832)   S Canada, NC, C and SC USA >> N Mexico

○ *Pipilo albicollis* WHITE-THROATED TOWHEE

  — *P.a.marshalli* Parkes, 1974   S Puebla and N Oaxaca (Mexico) (1767)

  — *P.a.albicollis* P.L. Sclater, 1858[3]   S Oaxaca (Mexico)

○ *Pipilo fuscus* CANYON TOWHEE

  — *P.f.mesoleucus* S.F. Baird, 1854   SW USA, N Mexico

  — *P.f.intermedius* Nelson, 1899   N Mexico

  — *P.f.jamesi* C.T. Townsend, 1923   Tiburon I.

  — *P.f.mesatus* Oberholser, 1937   SC USA

  — *P.f.texanus* van Rossem, 1934[4]   W and C Texas, N Mexico

  — *P.f.perpallidus* van Rossem, 1934   N and C Mexico

  — *P.f.fuscus* Swainson, 1827   E and C Mexico

  — *P.f.potosinus* Ridgway, 1899   N and C Mexico

  — *P.f.campoi* R.T. Moore, 1949   EC Mexico

  — *P.f.toroi* R.T. Moore, 1942   SC Mexico

● *Pipilo crissalis* CALIFORNIA TOWHEE[5]

  — *P.c.bullatus* Grinnell & Swarth, 1926   SW Oregon, NC California

  — *P.c.carolae* McGregor, 1899   Interior California

  ✓ *P.c.petulans* Grinnell & Swarth, 1926   Coastal N and C California

  — *P.c.crissalis* (Vigors, 1839)   WC California

  — *P.c.eremophilus* van Rossem, 1935   C and SC California

  ✓ *P.c.senicula* Anthony, 1895   Coastal S California, NW Baja California

  — *P.c.aripolius* Oberholser, 1919   C Baja California

  — *P.c.albigula* S.F. Baird, 1860   S Baja California

○ *Pipilo aberti* ABERT'S TOWHEE

  — *P.a.aberti* S.F. Baird, 1852   SW USA

  — *P.a.dumeticolus* van Rossem, 1946[6]   NE Baja California, NW Mexico

---

[1] For references to this split from *P. erythrophthalmus* see A.O.U. (1998: 605) (3).
[2] In last Edition we listed a subspecies *sympatricus*, this is a synonym of *P. ocai ocai* see Paynter (1970: 170) (1808).
[3] Includes *parvirostris* see Parkes (1974) (1767).
[4] Includes *aimophilus* Oberholser, 1974 (1673) see Browning (1978) (246).
[5] For references to this split from *P. fuscus* see A.O.U. (1998: 606) (3).
[6] Includes *vorhiesi* placed in synonymy by Browning (1990) (251).

MELOZONE Reichenbach, 1850 F[1]

○ *Melozone kieneri* RUSTY-CROWNED GROUND SPARROW
 — *M.k.grisior* van Rossem, 1933   NW Mexico
 — *M.k.kieneri* (Bonaparte, 1850)[2]   W Mexico
 — *M.k.rubricata* (Cabanis, 1851)[3]   C and SW Mexico

○ *Melozone biarcuata* PREVOST'S GROUND SPARROW
 — *M.b.hartwegi* Brodkorb, 1938[4]   S Mexico
 — *M.b.biarcuata* (Prévost & Des Murs, 1846)   S Mexico to W Honduras
 — *M.b.cabanisi* (Sclater & Salvin, 1868)   C Costa Rica

○ *Melozone leucotis* WHITE-EARED GROUND SPARROW
 — *M.l.occipitalis* (Salvin, 1878)   S Mexico, Guatemala, El Salvador
 — *M.l.nigrior* Miller & Griscom, 1925   Nicaragua
 — *M.l.leucotis* Cabanis, 1860   Costa Rica

ARREMONOPS Ridgway, 1896 M

● *Arremonops rufivirgatus* OLIVE SPARROW
 ✓ *A.r.rufivirgatus* (Lawrence, 1851)   S Texas, NE Mexico
 — *A.r.ridgwayi* (Sutton & Burleigh, 1941)   E Mexico
 — *A.r.crassirostris* (Ridgway, 1878)   SE Mexico
 — *A.r.rhyptothorax* Parkes, 1974   Coastal N Yucatán Pen. (1768)
 — *A.r.verticalis* (Ridgway, 1878)   SE Mexico (Incl. S Yucatán Pen.), Guatemala, Belize
 — *A.r.sinaloae* Nelson, 1899   W Mexico
 — *A.r.sumichrasti* (Sharpe, 1888)   SW Mexico
 — *A.r.chiapensis* Nelson, 1904   S Mexico
 — *A.r.superciliosus* (Salvin, 1864)   W Costa Rica

○ *Arremonops tocuyensis* Todd, 1912 TOCUYO SPARROW   NE Colombia, NW Venezuela

○ *Arremonops chloronotus* GREEN-BACKED SPARROW
 — *A.c.chloronotus* (Salvin, 1861)   SE Mexico, Guatemala, NW Honduras
 — *A.c.twomeyi* Monroe, 1963   NC Honduras

● *Arremonops conirostris* BLACK-STRIPED SPARROW
 ✓ *A.c.richmondi* Ridgway, 1898   E Honduras, Nicaragua, Costa Rica, Panama
 — *A.c.striaticeps* (Lafresnaye, 1853)   C and E Panama, Colombia, W Ecuador
 — *A.c.viridicatus* Wetmore, 1957[5]   Isla Coiba
 — *A.c.inexpectatus* Chapman, 1914   Colombia
 — *A.c.conirostris* (Bonaparte, 1850)[6]   E Colombia, N Venezuela, N Brazil
 — *A.c.umbrinus* Todd, 1923   E Colombia, W Venezuela

ARREMON Vieillot, 1816 M

○ *Arremon taciturnus* PECTORAL SPARROW
 — *A.t.axillaris* P.L. Sclater, 1855   NE Colombia, W Venezuela
 — *A.t.taciturnus* (Hermann, 1783)   Extreme E Colombia, SE Venezuela, the Guianas,
                  Amazonian Brazil, NE Bolivia
 — *A.t.nigrirostris* P.L. Sclater, 1886   SE Peru, N Bolivia
 — *A.t.semitorquatus* Swainson, 1838[7]   EC Brazil

○ *Arremon franciscanus* Raposo, 1997 SÃO FRANCISCO SPARROW #   EC Brazil (1982)

○ *Arremon flavirostris* SAFFRON-BILLED SPARROW
 — *A.f.flavirostris* Swainson, 1838   EC Brazil
 — *A.f.dorbignii* P.L. Sclater, 1856   Andes of C Bolivia to NW Argentina
 — *A.f.devillii* Des Murs, 1856   SW Brazil, E Bolivia
 — *A.f.polionotus* Bonaparte, 1850   SE Brazil, Paraguay, NE Argentina

---

[1] Gender addressed by David & Gosselin (2002) (614), multiple changes occur in specific names.
[2] Dated '1851?' by Paynter (1970: 180) (1808), but see Zimmer (1926) (2707).
[3] Includes *obscurior* see Paynter (1970: 181) (1808).
[4] For recognition see Paynter (1970: 181) (1808).
[5] Omitted in last Edition perhaps because Paynter (1970: 189) (1808) attached a query to it (doubting its distinctness).
[6] Dated '1851?' by Paynter (1970: 189) (1808), but see Zimmer (1926) (2707).
[7] Raposo & Parrini (1997) (1983) demonstrated that there is no intergradation with *taciturnus* and argued for recognition of *semitorquatus* as a species. They confirm, however, that there is no sympatry.

● *Arremon aurantiirostris* ORANGE-BILLED SPARROW
    — *A.a.saturatus* Cherrie, 1891   SE Mexico, Guatemala, Belize
    ⊿ *A.a.rufidorsalis* Cassin, 1865   Honduras, Nicaragua, Costa Rica
    — *A.a.aurantiirostris* Lafresnaye, 1847   W Costa Rica to C Panama
    — *A.a.strictocollaris* Todd, 1922   E Panama to NW Colombia
    — *A.a.occidentalis* Hellmayr, 1911   W Colombia, NW Ecuador
    — *A.a.santarosae* Chapman, 1925   SW Ecuador
    — *A.a.erythrorhynchus* P.L. Sclater, 1855   N Colombia
    — *A.a.spectabilis* P.L. Sclater, 1855   SE Colombia, E Ecuador, NE Peru

○ *Arremon schlegeli* GOLDEN-WINGED SPARROW
    — *A.s.fratruelis* Wetmore, 1946   N Colombia
    — *A.s.canidorsum* J.T. Zimmer, 1941   NE Colombia
    — *A.s.schegeli* Bonaparte, 1850[1]   NE Colombia, Venezuela

○ *Arremon abeillei* BLACK-CAPPED SPARROW
    — *A.a.abeillei* Lesson, 1844   W Ecuador, NW Peru
    — *A.a.nigriceps* Taczanowski, 1880   NC Peru in Marañon Valley

BUARREMON Bonaparte, 1850[2] M
● *Buarremon brunneinuchus* CHESTNUT-CAPPED BRUSH FINCH[3]
    — *B.b.apertus* (Wetmore, 1942)   E Mexico
    — *B.b.brunneinuchus* (Lafresnaye, 1839)[4]   E Mexico
    — *B.b.suttoni* (Parkes, 1954)   S Mexico
    — *B.b.nigrilatera* (J.S. Rowley, 1968)   Oaxaca (Mexico)
    — *B.b.macrourus* (Parkes, 1954)   S Mexico, SW Guatemala
    — *B.b.alleni* (Parkes, 1954)   El Salvador, Honduras, W Nicaragua
    ⊿ *B.b.elsae* (Parkes, 1954)   Costa Rica, W and C Panama
    — *B.b.frontalis* (Tschudi, 1844)   E Panama, Colombia, Venezuela, Ecuador, Peru
    — *B.b.allinornatus* (Phelps & Phelps, Jr., 1949)   NW Venezuela
    — *B.b.inornatus* Sclater & Salvin, 1879   WC Ecuador

○ *Buarremon virenticeps* Bonaparte, 1855 GREEN-STRIPED BRUSH FINCH   W and C Mexico

○ *Buarremon torquatus* STRIPE-HEADED BRUSH FINCH[5]
    — *B.t.costaricensis* Bangs, 1907   Highlands of SW Costa Rica, W Panama
    — *B.t.tacarcunae* Chapman, 1923   Highlands of E Panama
    — *B.t.atricapillus* Lawrence, 1874   Andes of N Colombia
    — *B.t.basilicus* Bangs, 1898   N Colombia (Santa Marta Mts.)
    — *B.t.perijanus* (Phelps & Gilliard, 1940)   E Colombia, W Venezuela (Perijá Mts.)
    — *B.t.larensis* (Phelps & Phelps, Jr., 1949)   W Venezuela (Lara, Táchira)
    — *B.t.phaeopleurus* P.L. Sclater, 1856   N Venezuela (C Coastal Range)
    — *B.t.phygas* Berlepsch, 1912   NE Venezuela (E Coastal Range)
    — *B.t.assimilis* (Boissonneau, 1840)   Andes of W Venezuela and Colombia to N Peru
    — *B.t.nigrifrons* Chapman, 1923   Andes of SW Ecuador, NW Peru
    — *B.t.poliophrys* Berlepsch & Stolzmann, 1896 Andes of C and SE Peru
    — *B.t.torquatus* (d'Orbigny & Lafresnaye, 1837) Andes of N Bolivia
    — *B.t.fimbriatus* Chapman, 1923   Andes of Bolivia
    — *B.t.borelli* Salvadori, 1897   Andes of S Bolivia, NW Argentina

OREOTHRAUPIS P.L. Sclater, 1856 F
○ *Oreothraupis arremonops* (P.L. Sclater, 1855) TANAGER-FINCH   W Andes from N Colombia to NW Ecuador

PEZOPETES Cabanis, 1860 M
○ *Pezopetes capitalis* Cabanis, 1860 LARGE-FOOTED FINCH   Highlands of Costa Rica, W Panama

LYSURUS Ridgway, 1898 M
○ *Lysurus crassirostris* (Cassin, 1865) SOOTY-FACED FINCH   Costa Rica, Panama

---

[1] Dated '1851?' by Paynter (1970: 185) (1808), but see Zimmer (1926) (2707).
[2] For recognition of this genus see A.O.U. (1998: 601) (3); these species were treated in *Atlapetes* in last Edition.
[3] In last Edition we treated *apertus* at specific level; for treatment as a race of *brunneinucha* see A.O.U. (1998: 601) (3).
[4] Includes *parkesi* see Binford (1989) (150).
[5] For reasons to unite *torquatus* and *atricapillus* see Remsen & Graves (1995) (2009).

○ *Lysurus castaneiceps* (P.L. Sclater, 1859) OLIVE FINCH   E Andes from Colombia to S Peru

ATLAPETES Wagler, 1831[1] M
○ *Atlapetes pileatus* RUFOUS-CAPPED BRUSH FINCH
    — *A.p.dilutus* Ridgway, 1898   N and C Mexico
    — *A.p.pileatus* Wagler, 1831   S Mexico

○ *Atlapetes albofrenatus* MOUSTACHED BRUSH FINCH
    — *A.a.meridae* (Sclater & Salvin, 1870)   Andes of Venezuela
    — *A.a.albofrenatus* (Boissonneau, 1840)   Andes of NE Colombia

○ *Atlapetes semirufus* OCHRE-BREASTED BRUSH FINCH
    — *A.s.denisei* (Hellmayr, 1911)   N Venezuela (Sucre to Aragua)
    — *A.s.benedettii* Phelps & Gilliard, 1941   N Venezuela (Falcón, Lara, Trujillo)
    — *A.s.albigula* Zimmer & Phelps, 1946   W Venezuela (Táchira)
    — *A.s.zimmeri* Meyer de Schauensee, 1947   Andes of Venezuela, NE Colombia
    — *A.s.majusculus* Todd, 1919   E Andes of Colombia (Boyacá)
    — *A.s.semirufus* (Boissonneau, 1840)   E Andes of Colombia (Cundinamarca)

○ *Atlapetes personatus* TEPUI BRUSH FINCH
    — *A.p.personatus* (Cabanis, 1848)   Tepui region of S Venezuela (Mt. Roraima)
    — *A.p.collaris* Chapman, 1939   Tepui region of S Venezuela (Auyan-tepui)
    — *A.p.duidae* Chapman, 1929   Tepui region of S Venezuela (Duida, Guaiquinima)
    — *A.p.parui* Phelps & Phelps, Jr., 1950   Tepui region of S Venezuela (Cerro Parú)
    — *A.p.paraquensis* Phelps & Phelps, Jr., 1946   Tepui region of S Venezuela (Paraque, Yavi)
    — *A.p.jugularis* Phelps & Phelps, Jr., 1955   Tepui region of S Venezuela, N Brazil (Cerro Neblina)

● *Atlapetes albinucha* WHITE-NAPED BRUSH FINCH[2]
    — *A.a.albinucha* (d'Orbigny & Lafresnaye, 1838)   E Mexico
    — *A.a.griseipectus* Dwight & Griscom, 1921   S Mexico, W Guatemala, El Salvador
    — *A.a.fuscipygius* Dwight & Griscom, 1921   Honduras, El Salvador, NW Nicaragua
    ⟂ *A.a.parvirostris* Dwight & Griscom, 1921   Costa Rica
    — *A.a.brunnescens* Chapman, 1915   W Chiriquí (Panama)
    — *A.a.coloratus* Griscom, 1924   E Chiriquí, Veraguas (Panama)
    — *A.a.azuerensis* Aldrich, 1937   Azuero Pen. (S Panama)
    — *A.a.gutturalis* (Lafresnaye, 1843)   N Colombia

○ *Atlapetes melanocephalus* (Salvin & Godman, 1880) SANTA MARTA BRUSH FINCH
N Colombia (Santa Marta Mts.)

○ *Atlapetes pallidinucha* PALE-NAPED BRUSH FINCH
    — *A.p.pallidinucha* (Boissonneau, 1840)   Andes of E Colombia, SW Venezuela
    — *A.p.papallactae* Hellmayr, 1913   Andes of Colombia (Central Andes), Ecuador, N Peru

○ *Atlapetes flaviceps* Chapman, 1912 OLIVE-HEADED BRUSH FINCH   Central Andes of Colombia (Tolima, Huila)

○ *Atlapetes fuscoolivaceus* Chapman, 1914 DUSKY-HEADED BRUSH FINCH   Andes of S Colombia (upper Magdalena Valley)

○ *Atlapetes crassus* Bangs, 1908 DUSKY BRUSH FINCH[3]   W Andes of Colombia and NW Ecuador

○ *Atlapetes tricolor* (Taczanowski, 1874) TRICOLOURED BRUSH FINCH   Andes of C Peru (La Libertad, Junín)

○ *Atlapetes leucopis* (Sclater & Salvin, 1878) WHITE-RIMMED BRUSH FINCH
Andes of S Colombia, N Ecuador

○ *Atlapetes latinuchus* YELLOW-BREASTED BRUSH FINCH
    — *A.l.phelpsi* Paynter, 1970   NE Colombia, W Venezuela (Perijá Mts.)
    — *A.l.elaeoprorus* (Sclater & Salvin, 1879)   Colombia (N Central Andes)
    — *A.l.simplex* (Berlepsch, 1888)   Colombia (Bogota area)
    — *A.l.caucae* Chapman, 1927   Colombia (W and S central Andes)
    — *A.l.spodionotus* (Sclater & Salvin, 1879)   S Colombia, N Ecuador
    — *A.l.comptus* (Sclater & Salvin, 1879)   SW Ecuador, NW Peru
    — *A.l.latinuchus* (Du Bus, 1855)[4]   E Andes of SE Ecuador, NE Peru

[1] The arrangement of this genus is based on Garcia-Moreno & Fjeldså (1999) (911).
[2] For reasons to accept the lumping of *albinucha* and *gutturalis* see A.O.U. (1998: 601) (3).
[3] For reasons to treat this as separate from *A. tricolor* see Garcia-Moreno & Fjeldså (1999) (911).
[4] Du Bus is an abbreviation for Du Bus de Gisignies.

○ — *A.l.chugurensis* Chapman, 1927   NW Peru (Cajamarca)

    — *A.l.baroni* (Salvin, 1895)   N Peru (upper Marañon Valley)

○ *Atlapetes rufigenis* (Salvin, 1895)   RUFOUS-EARED BRUSH FINCH   Andes of NW Peru (Cajamarca to Huánuco)

○ *Atlapetes forbesi* Morrison, 1947   APURÍMAC BRUSH FINCH   Andes of SW Peru (Apurímac, Cuzco)

○ *Atlapetes melanopsis*[1] Valqui & Fjeldså, 2002   BLACK-SPECTACLED BRUSH FINCH #

     C Peru (2457A)

○ *Atlapetes schistaceus* SLATY BRUSH FINCH
    — *A.s.fumidus* Wetmore & Phelps, 1953   W Venezuela, NE Colombia (Perijá Mts.)
    — *A.s.castaneifrons* (Sclater & Salvin, 1875)   Andes of Venezuela
    — *A.s.tamae* Cory, 1913   Andes of Venezuela (Táchira), NE Colombia
    — *A.s.schistaceus* (Boissonneau, 1840)   Andes of Colombia, Ecuador
    — *A.s.taczanowskii* (Sclater & Salvin, 1875)   Andes of C Peru (Huánuco, Pasco, Junín)

○ *Atlapetes leucopterus* WHITE-WINGED BRUSH FINCH
    — *A.l.leucopterus* (Jardine, 1856)   Andes of Ecuador
    — *A.l.dresseri* (Taczanowski, 1883)   Andes of SW Ecuador, NW Peru
    — *A.l.paynteri* Fitzpatrick, 1980   Andes of NC Peru (834)

○ *Atlapetes albiceps* (Taczanowski, 1884)   WHITE-HEADED BRUSH FINCH   Andes of SW Ecuador, NW Peru

○ *Atlapetes pallidiceps* (Sharpe, 1900)   PALE-HEADED BRUSH FINCH   Andes of S Ecuador (Azuay)

○ *Atlapetes seebohmi* BAY-CROWNED BRUSH FINCH
    — *A.s.celicae* Chapman, 1925   Andes of SW Ecuador
    — *A.s.simonsi* (Sharpe, 1900)   Andes of SW Ecuador
    — *A.s.seebohmi* (Taczanowski, 1883)   Andes of W Peru

○ *Atlapetes nationi* RUSTY-BELLIED BRUSH FINCH
    — *A.n.nationi* (P.L. Sclater, 1881)   Andes of W Peru
    — *A.n.brunneiceps* (Berlepsch & Stolzmann, 1906)   Andes of SW Peru

○ *Atlapetes canigenis* Chapman, 1919   SOOTY BRUSH FINCH   Andes of S Peru (Cuzco)

○ *Atlapetes terborghi* Remsen, 1993   VILCABAMBA BRUSH FINCH   S Peru (Vilcabamba Mts.) (2006)

○ *Atlapetes melanolaemus* (Sclater & Salvin, 1879)[2]   BLACK-FACED BRUSH FINCH

     S Peru (Cuzco and Puno)

○ *Atlapetes rufinucha* RUFOUS-NAPED BRUSH FINCH
    — *A.r.rufinucha* (d'Orbigny & Lafresnaye, 1837)   N Bolivia
    — *A.r.carrikeri* Bond & Meyer de Schauensee, 1939   C Bolivia

○ *Atlapetes fulviceps* (d'Orbigny & Lafresnaye, 1837)   FULVOUS-HEADED BRUSH FINCH

     Andes from C Bolivia to NW Argentina

○ *Atlapetes citrinellus* (Cabanis, 1883)   YELLOW-STRIPED BRUSH FINCH   Argentina

PSELLIOPHORUS Ridgway, 1898   M

○ *Pselliophorus tibialis* (Lawrence, 1864)   YELLOW-THIGHED FINCH   Costa Rica, W Panama

○ *Pselliophorus luteoviridis* Griscom, 1924   YELLOW-GREEN FINCH   E Panama

UROTHRAUPIS Taczanowski & Berlepsch, 1885   F

○ *Urothraupis stolzmanni* Taczanowski & Berlepsch, 1885   BLACK-BACKED BUSH TANAGER

     Central Andes of Colombia to C Ecuador

CHARITOSPIZA Oberholser, 1905   F

○ *Charitospiza eucosma* Oberholser, 1905   COAL-CRESTED FINCH   C and E Brazil, extreme E Bolivia, NE Argentina

SALTATRICULA Burmeister, 1861   F

○ *Saltatricula multicolor* (Burmeister, 1860)   MANY-COLOURED CHACO FINCH

     SE Bolivia, W Paraguay, W Uruguay, N Argentina

[1]   The name *melanops* Valqui & Fjeldså, 1999 (2457) was preoccupied.
[2]   The earlier name *Buarremon melanops* Sclater & Salvin, 1876 (2219A), may be applicable instead of this name.

CORYPHASPIZA G.R. Gray, 1840 F
○ ***Coryphaspiza melanotis*** Black-masked Finch
    — *C.m.marajoara* Sick, 1967   Marajó I. (Brazil)
    — *C.m.melanotis* (Temminck, 1822)   Extreme SE Peru, NE Bolivia, S Brazil, E Paraguay,
                                        NE Argentina

CORYPHOSPINGUS Cabanis, 1851 M
○ ***Coryphospingus pileatus*** Pileated Finch
    — *C.p.rostratus* A.H. Miller, 1947   C Colombia
    — *C.p.brevicaudus* Cory, 1916   N Colombia, N Venezuela
    — *C.p.pileatus* (Wied, 1821)   EC Brazil

○ ***Coryphospingus cucullatus*** Red-crested Finch
    — *C.c.cucullatus* (Statius Müller, 1776)   The Guianas, Brazil
    — *C.c.rubescens* (Swainson, 1825)   S Brazil, E Paraguay, Uruguay, Argentina
    — *C.c.fargoi* Brodkorb, 1938   E Peru, E Bolivia, N Argentina, W Paraguay

RHODOSPINGUS Sharpe, 1888 M
○ ***Rhodospingus cruentus*** (Lesson, 1844) Crimson-breasted Finch   W Ecuador, NW Peru

GUBERNATRIX Lesson, 1837 F
○ ***Gubernatrix cristata*** (Vieillot, 1817) Yellow Cardinal   Uruguay, N and E Argentina

PAROARIA Bonaparte, 1832 F
● ***Paroaria coronata*** (J.F. Miller, 1776) Red-crested Cardinal   E Bolivia, Paraguay, Uruguay, Argentina

○ ***Paroaria dominicana*** (Linnaeus, 1758) Red-cowled Cardinal   NE Brazil

○ ***Paroaria gularis*** Red-capped Cardinal
    — *P.g.nigrogenis* (Lafresnaye, 1846)   Trinidad, E Colombia, Venezuela
    — *P.g.gularis* (Linnaeus, 1766)   Colombia, Venezuela, the Guianas, Ecuador, Peru,
                                        W Brazil
    — *P.g.cervicalis* P.L. Sclater, 1862   E Bolivia, Brazil

○ ***Paroaria baeri*** Crimson-fronted Cardinal
    — *P.b.baeri* Hellmayr, 1907   SC Brazil (NE Mato Grosso, N Goiás)
    — *P.b.xinguensis* Sick, 1950   SC Brazil (N Mato Grosso)

○ ***Paroaria capitata*** Yellow-billed Cardinal
    — *P.c.fuscipes* Bond & Meyer de Schauensee, 1939   SE Bolivia
    — *P.c.capitata* (d'Orbigny & Lafresnaye, 1837)   S Brazil, Paraguay, N Argentina

THRAUPIDAE TANAGERS[1] (50: 202)

ORCHESTICUS Cabanis, 1851 M
○ ***Orchesticus abeillei*** (Lesson, 1839) Brown Tanager   SE Brazil

SCHISTOCHLAMYS Reichenbach, 1850[2] F
○ ***Schistochlamys ruficapillus*** Cinnamon Tanager
    — *S.r.capistrata* (Wied, 1821)   NE Brazil
    — *S.r.sicki* Pinto & Camargo, 1952   E Mato Grosso (Brazil)
    — *S.r.ruficapillus* (Vieillot, 1817)   SE Brazil

○ ***Schistochlamys melanopis*** Black-faced Tanager
    — *S.m.aterrima* Todd, 1912   NE Colombia, Venezuela, W Guyana
    — *S.m.melanopis* (Latham, 1790)   The Guianas, NE Brazil
    — *S.m.grisea* Cory, 1916   EC Peru
    — *S.m.olivina* (P.L. Sclater, 1864)   E Bolivia, SC Brazil
    — *S.m.amazonica* J.T. Zimmer, 1947   SE Brazil

---

[1] Most changes in the sequence of genera in this family come from Burns (1997) (278).
[2] For the conservation of this name see Opinion 2004 (I.C.Z.N., 2002) (1178A).

CISSOPIS Vieillot, 1816 M
○ *Cissopis leverianus* Magpie-Tanager
    — *C.l.leverianus* (J.F. Gmelin, 1788)[1]   E Colombia, S Venezuela, the Guianas, Upper Amazonia
                                                to N Bolivia

    — *C.l.major* Cabanis, 1851   Paraguay, SE Brazil, N Argentina

CONOTHRAUPIS P.L. Sclater, 1880 F
○ *Conothraupis speculigera* (Gould, 1855) Black-and-white Tanager   W Ecuador, N and E Peru

○ †?*Conothraupis mesoleuca* (Berlioz, 1939) Cone-billed Tanager[2]   Mato Grosso

LAMPROSPIZA Cabanis, 1847 F
○ *Lamprospiza melanoleuca* (Vieillot, 1817) Red-billed Pied Tanager   The Guianas, N and E Brazil, E Peru, N Bolivia

COMPSOTHRAUPIS Richmond, 1915 F
○ *Compsothraupis loricata* (M.H.K. Lichtenstein, 1819) Scarlet-throated Tanager
                                                E Brazil

SERICOSSYPHA Lesson, 1844 F
○ *Sericossypha albocristata* (Lafresnaye, 1843) White-capped Tanager   SW Venezuela, Colombia, Ecuador, N and C Peru

NEMOSIA Vieillot, 1816 F
○ *Nemosia pileata* Hooded Tanager
    — *N.p.hypoleuca* Todd, 1916   N Colombia, N Venezuela
    — *N.p.surinamensis* J.T. Zimmer, 1947   Guyana, Surinam
    — *N.p.pileata* (Boddaert, 1783)   French Guiana, Brazil, N Bolivia
    — *N.p.interna* J.T. Zimmer, 1947   N Brazil
    — *N.p.nana* Berlepsch, 1912   NE Peru, W Brazil
    — *N.p.caerulea* (Wied, 1831)   S and E Brazil, E Bolivia, Paraguay, N Argentina

○ †?*Nemosia rourei* Cabanis, 1870 Cherry-throated Tanager   SE Brazil

CREURGOPS P.L. Sclater, 1858 M
○ *Creurgops verticalis* P.L. Sclater, 1858 Rufous-crested Tanager   SW Venezuela to C Peru

○ *Creurgops dentatus* (Sclater & Salvin, 1876) Slaty Tanager[3]   C Peru to C Bolivia

MITROSPINGUS Ridgway, 1898 M
○ *Mitrospingus cassinii* Dusky-faced Tanager
    — *M.c.costaricensis* Todd, 1922   E Costa Rica, W Panama
    — *M.c.cassinii* (Lawrence, 1861)   E Panama, W Colombia, W Ecuador

○ *Mitrospingus oleagineus* Olive-backed Tanager
    — *M.o.obscuripectus* Zimmer & Phelps, 1945)   SE Venezuela, N Brazil
    — *M.o.oleagineus* (Salvin, 1886)   SE Venezuela, Guyana

ORTHOGONYS Strickland, 1844 M
○ *Orthogonys chloricterus* (Vieillot, 1819) Olive-green Tanager   SE Brazil

HEMISPINGUS Cabanis, 1851 M
○ *Hemispingus atropileus* Black-capped Hemispingus[4]
    — *H.a.atropileus* (Lafresnaye, 1842)   SW Venezuela, Colombia, Ecuador
    — *H.a.auricularis* (Cabanis, 1873)   N and C Peru

○ *Hemispingus calophrys* (Sclater & Salvin, 1876) Orange-browed Hemispingus
                                                S Peru, N Bolivia

○ *Hemispingus parodii* Weske & Terborgh, 1974 Parodi's Hemispingus
                                                Peru (Cuzco area) (2586)

---

[1] For correction of spelling see David & Gosselin (2002) (614).
[2] For some comments on this see Bond (1951) (183).
[3] For correction of spelling see David & Gosselin (2002) (614).
[4] The taxon *calophrys* is separated following Weske & Terborgh (1974) (2586).

○ *Hemispingus superciliaris* SUPERCILIARIED HEMISPINGUS
___ *H.s.chrysophrys* (Sclater & Salvin, 1875) SW Venezuela
___ *H.s.superciliaris* (Lafresnaye, 1840) C Colombia
___ *H.s.nigrifrons* (Lawrence, 1875) Colombia, Ecuador
___ *H.s.maculifrons* J.T. Zimmer, 1947 SW Ecuador, NW Peru
___ *H.s.insignis* J.T. Zimmer, 1947 N Peru
___ *H.s.leucogastrus* (Taczanowski, 1874)[1] C Peru
___ *H.s.urubambae* J.T. Zimmer, 1947 S Peru, N Bolivia

○ *Hemispingus reyi* (Berlepsch, 1885) GREY-CAPPED HEMISPINGUS SW Venezuela

○ *Hemispingus frontalis* OLEAGINOUS HEMISPINGUS
___ *H.f.ignobilis* (P.L. Sclater, 1861) W Venezuela
___ *H.f.flavidorsalis* Phelps & Phelps, Jr., 1953 W Venezuela (Perijá Mts.)
___ *H.f.hanieli* Hellmayr & Seilern, 1914 N Venezuela
___ *H.f.iteratus* Chapman, 1925 NE Venezuela
___ *H.f.frontalis* (Tschudi, 1844) Colombia, Ecuador, N and C Peru

○ *Hemispingus melanotis* BLACK-EARED HEMISPINGUS
___ *H.m.melanotis* (P.L. Sclater, 1855) SW Venezuela, C and E Colombia, E Ecuador
___ *H.m.ochraceus* (Berlepsch & Taczanowski, 1884) SW Colombia, W Ecuador
___ *H.m.piurae* Chapman, 1923 NW Peru
___ *H.m.macrophrys* Koepcke, 1961 W Peru
___ *H.m.berlepschi* (Taczanowski, 1880) C Peru
___ *H.m.castaneicollis* (P.L. Sclater, 1858) S Peru to C Bolivia

○ *Hemispingus goeringi* (Sclater & Salvin, 1870) SLATY-BACKED HEMISPINGUS
SW Venezuela

○ *Hemispingus rufosuperciliaris* Blake & Hocking, 1974 RUFOUS-BROWED HEMISPINGUS
N and C Peru (166)

○ *Hemispingus verticalis* (Lafresnaye, 1840) BLACK-HEADED HEMISPINGUS
S and C Colombia, Ecuador, N Peru

○ *Hemispingus xanthophthalmus* (Taczanowski, 1874) DRAB HEMISPINGUS
N Peru to N Bolivia

○ *Hemispingus trifasciatus* (Taczanowski, 1874) THREE-STRIPED HEMISPINGUS
C Peru to N Bolivia

CNEMOSCOPUS Bangs & Penard, 1919 M
○ *Cnemoscopus rubrirostris* GREY-HOODED BUSH TANAGER
___ *C.r.rubrirostris* (Lafresnaye, 1840) SW Venezuela, Colombia, E Ecuador
___ *C.r.chrysogaster* (Taczanowski, 1874) N and C Peru

THLYPOPSIS Cabanis, 1851 F
○ *Thlypopsis fulviceps* FULVOUS-HEADED TANAGER
___ *T.f.fulviceps* Cabanis, 1851 NE Colombia, NW Venezuela
___ *T.f.obscuriceps* Phelps & Phelps, Jr., 1953 W Venezuela
___ *T.f.meridensis* Phelps & Phelps, Jr., 1962 W Venezuela
___ *T.f.intensa* Todd, 1917 NE Colombia

○ *Thlypopsis ornata* RUFOUS-CHESTED TANAGER
___ *T.o.ornata* (P.L. Sclater, 1859) SW Colombia, W Ecuador
___ *T.o.media* J.T. Zimmer, 1930 S Ecuador, N and C Peru
___ *T.o.macropteryx* Berlepsch & Stolzmann, 1896 C and S Peru

○ *Thlypopsis pectoralis* (Taczanowski, 1884) BROWN-FLANKED TANAGER C Peru

○ *Thlypopsis sordida* ORANGE-HEADED TANAGER
___ *T.s.orinocensis* Friedmann, 1942 EC Venezuela
___ *T.s.chrysopis* (Sclater & Salvin, 1880) E Ecuador, E Peru, W Brazil
___ *T.s.sordida* (d'Orbigny & Lafresnaye, 1837) E and S Brazil, E Bolivia, Paraguay, N Argentina

---

[1] For correction of spelling see David & Gosselin (2002) (613).

○ *Thlypopsis inornata* (Taczanowski, 1879) Buff-bellied Tanager   N Peru

○ *Thlypopsis ruficeps* (d'Orbigny & Lafresnaye, 1837) Rust-and-yellow Tanager
  C Peru, Bolivia, NW Argentina

PYRRHOCOMA Cabanis, 1851 F
○ *Pyrrhocoma ruficeps* (Strickland, 1844) Chestnut-headed Tanager   SE Brazil, E Paraguay, NE Argentina

CYPSNAGRA Lesson, 1831 F
○ *Cypsnagra hirundinacea* White-rumped Tanager
    — *C.h.pallidigula* Hellmayr, 1907   C Brazil, NE Bolivia
    — *C.h.hirundinacea* (Lesson, 1831)   S Brazil, SE Bolivia, NE Paraguay

NEPHELORNIS Lowery & Tallman, 1976[1] M
○ *Nephelornis oneilli* Lowery & Tallman, 1976 Pardusco   C Peru (1400)

TRICHOTHRAUPIS Cabanis, 1850 F
○ *Trichothraupis melanops* (Vieillot, 1818) Black-goggled Tanager   N Peru, Bolivia, N Argentina, E Paraguay, SE Brazil

EUCOMETIS P.L. Sclater, 1856 F
○ *Eucometis penicillata* Grey-headed Tanager
    — *E.p.pallida* Berlepsch, 1888   SE Mexico to E Guatemala
    — *E.p.spodocephala* (Bonaparte, 1854)   Nicaragua, W Costa Rica
    — *E.p.stictothorax* Berlepsch, 1888   SW Costa Rica, W Panama
    — *E.p.cristata* (Du Bus, 1855)[2]   E Panama to W Venezuela
    — *E.p.affinis* Berlepsch, 1888   N Venezuela
    — *E.p.penicillata* (Spix, 1825)   SE Colombia, E Ecuador, E Peru, the Guianas, N Brazil
    — *E.p.albicollis* (d'Orbigny & Lafresnaye, 1837)   E Bolivia, SC Brazil, N Paraguay

TACHYPHONUS Vieillot, 1816 M
○ *Tachyphonus cristatus* Flame-crested Tanager
    — *T.c.cristatus* (Linnaeus, 1766)   French Guiana, NE Brazil
    — *T.c.intercedens* Berlepsch, 1880   E Venezuela, Guyana, Surinam
    — *T.c.orinocensis* Zimmer & Phelps, 1945   E Colombia, S Venezuela
    — *T.c.cristatellus* P.L. Sclater, 1862   S Venezuela to N Peru
    — *T.c.fallax* J.T. Zimmer, 1945   S Colombia to NE Peru
    — *T.c.huarandosae* Chapman, 1925   N Peru
    — *T.c.madeirae* Hellmayr, 1910   C Brazil
    — *T.c.pallidigula* J.T. Zimmer, 1945   NE Brazil
    — *T.c.brunneus* (Spix, 1825)   E Brazil
    — *T.c.nattereri* Pelzeln, 1870[3]   SW Brazil

○ *Tachyphonus rufiventer* (Spix, 1825) Yellow-crested Tanager   E Peru, N Bolivia, W Brazil

○ *Tachyphonus surinamus* Fulvous-crested Tanager
    — *T.s.surinamus* (Linnaeus, 1766)   E and S Venezuela, the Guianas, N Brazil
    — *T.s.brevipes* Lafresnaye, 1846   S Venezuela to NE Peru
    — *T.s.napensis* Lawrence, 1864   E Peru, NW Brazil
    — *T.s.insignis* Hellmayr, 1906   N Brazil

○ *Tachyphonus luctuosus* White-shouldered Tanager
    — *T.l.axillaris* (Lawrence, 1874)   E Honduras to W Panama
    — *T.l.nitidissimus* Salvin, 1870   SW Costa Rica, W Panama
    — *T.l.panamensis* Todd, 1917   E Panama to W Ecuador, W Venezuela
    — *T.l.luctuosus* d'Orbigny & Lafresnaye, 1837   Tropical South America
    — *T.l.flaviventris* (P.L. Sclater, 1856)   NE Venezuela, Trinidad

○ *Tachyphonus delatrii* Lafresnaye, 1847 Tawny-crested Tanager   Nicaragua to W Ecuador

○ *Tachyphonus coronatus* (Vieillot, 1822) Ruby-crowned Tanager   SE Brazil, E Paraguay, NE Argentina

---

[1] Treated as a Parulid in last Edition, but see Vuilleumier & Mayr (1987: 145) (2551).
[2] Du Bus is an abbreviation for Du Bus de Gisignies.
[3] We revert to placing this here following Ridgely & Tudor (1989) (2018).

○ *Tachyphonus rufus* (Boddaert, 1783) WHITE-LINED TANAGER — Costa Rica, Panama, N Sth. America south to E Brazil, Paraguay, NE Argentina

○ *Tachyphonus phoenicius* Swainson, 1838 RED-SHOULDERED TANAGER — E Colombia, S Venezuela, the Guianas, N and C Brazil, E Bolivia; NE Peru

LANIO Vieillot, 1816 M

○ *Lanio aurantius* Lafresnaye, 1846 BLACK-THROATED SHRIKE-TANAGER — SE Mexico to Honduras

○ *Lanio leucothorax* WHITE-THROATED SHRIKE-TANAGER
— *L.l.leucothorax* Salvin, 1864 — E Honduras to E Costa Rica
— *L.l.reversus* Bangs & Griscom, 1932 — NW Costa Rica
— *L.l.melanopygius* Salvin & Godman, 1883 — SW Costa Rica, W Panama
— *L.l.ictus* Kennard & Peters, 1927 — NW Panama

○ *Lanio fulvus* FULVOUS SHRIKE-TANAGER
— *L.f.peruvianus* Carriker, 1934 — S Colombia to NE Peru
— *L.f.fulvus* (Boddaert, 1783) — S Venezuela, the Guianas, N Brazil

○ *Lanio versicolor* WHITE-WINGED SHRIKE-TANAGER
— *L.v.versicolor* (d'Orbigny & Lafresnaye, 1837) — E Peru, N Bolivia, W Brazil
— *L.v.parvus* Berlepsch, 1912 — S Brazil

RAMPHOCELUS Desmarest, 1805 M

● *Ramphocelus sanguinolentus* CRIMSON-COLLARED TANAGER
— *R.s.sanguinolentus* (Lesson, 1831) — SE Mexico to Honduras
↧ *R.s.apricus* (Bangs, 1908) — E Honduras to NW Panama

○ *Ramphocelus nigrogularis* (Spix, 1825) MASKED CRIMSON TANAGER — SE Colombia to E Peru, NE Bolivia, C Brazil

○ *Ramphocelus dimidiatus* CRIMSON-BACKED TANAGER
— *R.d.isthmicus* Ridgway, 1901 — W and C Panama
— *R.d.arestus* Wetmore, 1957 — Coiba I. (Panama)
— *R.d.limatus* Bangs, 1901 — Pearl Arch. (Panama)
— *R.d.dimidiatus* Lafresnaye, 1837 — E Panama, N Colombia, W Venezuela
— *R.d.molochinus* Meyer de Schauensee, 1950 — N Colombia

○ *Ramphocelus melanogaster* BLACK-BELLIED TANAGER
— *R.m.melanogaster* (Swainson, 1838) — N Peru
— *R.m.transitus* J.T. Zimmer, 1929 — EC Peru

○ *Ramphocelus carbo* SILVER-BEAKED TANAGER
— *R.c.unicolor* P.L. Sclater, 1856 — E Colombia
— *R.c.capitalis* J.A. Allen, 1892 — NE Venezuela
— *R.c.magnirostris* Lafresnaye, 1853 — Trinidad
— *R.c.carbo* (Pallas, 1764) — E Peru to Surinam
— *R.c.venezuelensis* Lafresnaye, 1853 — E Colombia, W Venezuela
— *R.c.connectens* Berlepsch & Stolzmann, 1896 — SE Peru, NW Bolivia
— *R.c.atrosericeus* d'Orbigny & Lafresnaye, 1837 — N and E Bolivia
— *R.c.centralis* Hellmayr, 1920 — EC Brazil, N Paraguay

○ *Ramphocelus bresilius* BRAZILIAN TANAGER
— *R.b.bresilius* (Linnaeus, 1766) — NE Brazil
— *R.b.dorsalis* P.L. Sclater, 1855 — SE Brazil

● *Ramphocelus passerinii* SCARLET-RUMPED TANAGER[1]
↧ *R.p.passerinii* Bonaparte, 1831 — SE Mexico to W Panama
— *R.p.costaricensis* Cherrie, 1891 — W Costa Rica

○ *Ramphocelus flammigerus* FLAME-RUMPED TANAGER[2]
— *R.f.icteronotus* Bonaparte, 1838 — Panama, W Colombia, W Ecuador
— *R.f.flammigerus* (Jardine & Selby, 1833) — WC Colombia

[1] Although A.O.U. (1998) (3) split this, we defer judgement due to a pending study.
[2] Like Ridgely & Tudor (1989) (2018) we tentatively lump these two which seem to hybridize over a broad area.

THRAUPIS Boie, 1826 F
● ***Thraupis episcopus*** Blue-grey Tanager
    ↙ *T.e.cana* (Swainson, 1836)   SE Mexico to N Venezuela
    — *T.e.caesitia* Wetmore, 1959   W Panama
    — *T.e.cumatilis* Wetmore, 1957   Coiba I. (Panama)
    — *T.e.nesophilus* Riley, 1912   E Colombia to Trinidad
    — *T.e.berlepschi* (Dalmas, 1900)   Tobago I.
    — *T.e.mediana* J.T. Zimmer, 1944   SE Colombia, NW Brazil, N Bolivia
    — *T.e.episcopus* (Linnaeus, 1766)   The Guianas, N Brazil
    — *T.e.ehrenreichi* Reichenow, 1915[1]   Rio Purús (N Brazil)
    — *T.e.leucoptera* (P.L. Sclater, 1886)   C Colombia
    — *T.e.quaesita* Bangs & Noble, 1918   SW Colombia, W Ecuador, NW Peru
    — *T.e.caerulea* J.T. Zimmer, 1929   SE Ecuador, N Peru
    — *T.e.major* (Berlepsch & Stolzmann, 1896)   C Peru
    — *T.e.urubambae* J.T. Zimmer, 1944   SE Peru
    — *T.e.coelestis* (Spix, 1825)   SE Colombia to C Peru, W Brazil

○ ***Thraupis sayaca*** Sayaca Tanager
    — *T.s.boliviana* Bond & Meyer de Schauensee, 1941   NW Bolivia
    — *T.s.obscura* Naumburg, 1924   C and S Bolivia, W Argentina
    — *T.s.sayaca* (Linnaeus, 1766)   E and S Brazil, Paraguay to Uruguay

○ ***Thraupis glaucocolpa*** Cabanis, 1850 Glaucous Tanager   N Colombia, NW Venezuela

○ ***Thraupis cyanoptera*** (Vieillot, 1817) Azure-shouldered Tanager   E Paraguay to SE Brazil

○ ***Thraupis ornata*** (Sparrman, 1789) Golden-chevroned Tanager   SE Brazil

● ***Thraupis abbas*** (Deppe, 1830) Yellow-winged Tanager   E Mexico to Nicaragua

● ***Thraupis palmarum*** Palm Tanager
    ↙ *T.p.atripennis* Todd, 1922   E Nicaragua to NW Venezuela
    — *T.p.violilavata* (Berlepsch & Taczanowski, 1883)   SW Colombia, W Ecuador
    — *T.p.melanoptera* (P.L. Sclater, 1857)   Amazonia, Trinidad
    — *T.p.palmarum* (Wied, 1821)   E Bolivia, E Paraguay, E and S Brazil

○ ***Thraupis cyanocephala*** Blue-capped Tanager
    — *T.c.cyanocephala* (d'Orbigny & Lafresnaye, 1837)   W Ecuador to N Bolivia
    — *T.c.annectens* J.T. Zimmer, 1944   C Colombia
    — *T.c.auricrissa* (P.L. Sclater, 1856)   NC Colombia, W Venezuela
    — *T.c.margaritae* (Chapman, 1912)   N Colombia
    — *T.c.hypophaea* (Todd, 1917)   NW Venezuela
    — *T.c.olivicyanea* (Lafresnaye, 1843)   N Venezuela
    — *T.c.subcinerea* (P.L. Sclater, 1861)   NE Venezuela
    — *T.c.buesingi* (Hellmayr & Seilern, 1913)   NE Venezuela, Trinidad

○ ***Thraupis bonariensis*** Blue-and-yellow Tanager
    — *T.b.darwinii* (Bonaparte, 1838)   Ecuador to N Chile
    — *T.b.composita* J.T. Zimmer, 1944   E and C Bolivia
    — *T.b.schulzei* Brodkorb, 1938   Paraguay, NW Argentina
    — *T.b.bonariensis* (J.F. Gmelin, 1789)   S Brazil to EC Argentina

CALOCHAETES P.L. Sclater, 1879 M
○ ***Calochaetes coccineus*** (P.L. Sclater, 1858) Vermilion Tanager   S Colombia to E Peru

CYANICTERUS Bonaparte, 1850[2] M
○ ***Cyanicterus cyanicterus*** (Vieillot, 1819) Blue-backed Tanager   E Venezuela, the Guianas

BANGSIA Penard, 1919[3] F
○ ***Bangsia arcaei*** Blue-and-gold Tanager
    — *B.a.caeruleigularis* (Ridgway, 1893)   E Costa Rica
    — *B.a.arcaei* (Sclater & Salvin, 1869)   W Panama

[1] For recognition see Parkes (1993) (1781).
[2] We follow Hartlaub (1851) (1087) in accepting 1850 as the date of all 543 pp. of Vol. 1.
[3] Ridgely & Tudor (1989: 293) (2018) favoured keeping *Bangsia* and we follow. These species were listed in the genus *Buthraupis* in last Edition.

○ *Bangsia melanochlamys* (Hellmayr, 1910) Black-and-gold Tanager   W Colombia

○ *Bangsia rothschildi* (Berlepsch, 1897) Golden-chested Tanager   SW Colombia, NW Ecuador

○ *Bangsia edwardsi* (Elliot, 1865) Moss-backed Tanager   SW Colombia, NW Ecuador

○ *Bangsia aureocincta* (Hellmayr, 1910) Gold-ringed Tanager   W Colombia

WETMORETHRAUPIS Lowery & O'Neill, 1964 F
○ *Wetmorethraupis sterrhopteron* Lowery & O'Neill, 1964 Orange-throated Tanager
                                                                                  N Peru

BUTHRAUPIS Cabanis, 1850 F
○ **Buthraupis montana** Hooded Mountain Tanager
  —— ¶*B.m.venezuelana* Aveledo & Perez, 1989   NW Venezuela (Sierra de Perija) and NE Colombia (67)
  —— *B.m.gigas* (Bonaparte, 1851)   NC Colombia
  —— *B.m.cucullata* (Jardine & Selby, 1842)   W Colombia, Ecuador
  —— *B.m.cyanonota* Berlepsch & Stolzmann, 1896   N and C Peru
  —— *B.m.saturata* Berlepsch & Stolzmann, 1906   SE Peru
  —— *B.m.montana* (d'Orbigny & Lafresnaye, 1837)   N Bolivia

○ **Buthraupis eximia** Black-chested Mountain Tanager
  —— *B.e.eximia* (Boissonneau, 1840)   NC Colombia, SW Venezuela
  —— *B.e.zimmeri* R.T. Moore, 1934   WC Colombia
  —— *B.e.chloronota* P.L. Sclater, 1855   SE Colombia, NW Ecuador
  —— *B.e.cyanocalyptra* R.T. Moore, 1934   SC Ecuador, NW Peru

○ **Buthraupis aureodorsalis** Blake & Hocking, 1974 Golden-backed Mountain Tanager
                                                                                  C Peru (166)

○ **Buthraupis wetmorei** (R.T. Moore, 1934) Masked Mountain Tanager
                                                                                  SW Colombia, SC Ecuador, NW Peru

ANISOGNATHUS Reichenbach, 1850 M
○ *Anisognathus melanogenys* (Salvin & Godman, 1880) Black-cheeked Mountain Tanager
                                                                                  N Colombia

○ **Anisognathus lacrymosus** Lacrimose Mountain Tanager
  —— *A.l.pallididorsalis* Phelps & Phelps, Jr., 1952   E Colombia, Venezuela
  —— *A.l.melanops* (Berlepsch, 1893)   W Venezuela
  —— *A.l.tamae* (Phelps & Gilliard, 1941)   NC Colombia, SW Venezuela
  —— *A.l.intensus* Meyer de Schauensee, 1951   SW Colombia
  —— *A.l.olivaceiceps* (Berlepsch, 1912)   W Colombia
  —— *A.l.palpebrosus* (Lafresnaye, 1847)   SW Colombia, E Ecuador
  —— *A.l.caerulescens* (Taczanowski & Berlepsch, 1885)   S Ecuador, N Peru
  —— *A.l.lacrymosus* (Du Bus, 1846)[1]   C Peru

○ **Anisognathus igniventris** Scarlet-bellied Mountain Tanager
  —— *A.i.lunulatus* (Du Bus, 1839)[2]   NC Colombia, W Venezuela
  —— *A.i.erythrotus* (Jardine & Selby, 1840)   S Colombia, Ecuador
  —— *A.i.ignicrissa* (Cabanis, 1873)[3]   NC Peru
  —— *A.i.igniventris* (d'Orbigny & Lafresnaye, 1837)   SE Peru, N Bolivia

○ **Anisognathus somptuosus** Blue-winged Mountain Tanager[4]
  —— *A.s.venezuelanus* (Hellmayr, 1913)   N Venezuela
  —— *A.s.virididorsalis* (Phelps & Phelps, Jr., 1949)   Venezuela
  —— *A.s.antioquiae* (Berlepsch, 1912)   Colombia
  —— *A.s.victorini* (Lafresnaye, 1842)   C Colombia, SW Venezuela
  —— *A.s.cyanopterus* (Cabanis, 1866)   SW Colombia, W Ecuador
  —— *A.s.baezae* (Chapman, 1925)   S Colombia, E Ecuador
  —— *A.s.alamoris* (Chapman, 1925)   SW Ecuador
  —— *A.s.somptuosus* (Lesson, 1831)   SE Ecuador, E Peru
  —— *A.s.flavinuchus* (d'Orbigny & Lafresnaye, 1837)   SE Peru, Bolivia

---

[1] Du Bus is an abbreviation for Du Bus de Gisignies.
[2] Du Bus is an abbreviation for Du Bus de Gisignies.
[3] For correction of spelling see David & Gosselin (2002) (613).
[4] Called *A. flavinuchus* in last Edition and by Storer (1970: 334-5) (2348), but erroneously as the oldest subspecific name listed then was *somptuosus*.

○ *Anisognathus notabilis* (P.L. Sclater, 1855) Black-chinned Mountain Tanager
SW Colombia, NW Ecuador

CHLORORNIS Reichenbach, 1850 M[1]
○ *Chlorornis riefferii* Grass-green Tanager
    — *C.r.riefferii* (Boissonneau, 1840) — Colombia, Ecuador
    — *C.r.dilutus* J.T. Zimmer, 1947 — N Peru
    — *C.r.elegans* (Tschudi, 1844) — C Peru
    — *C.r.celatus* J.T. Zimmer, 1947 — S Peru
    — *C.r.bolivianus* (Berlepsch, 1912) — N Bolivia

DUBUSIA Bonaparte, 1850 F
○ *Dubusia taeniata* Buff-breasted Mountain Tanager
    — *D.t.carrikeri* Wetmore, 1946 — N Colombia
    — *D.t.taeniata* (Boissonneau, 1840) — W Venezuela to NW Peru
    — *D.t.stictocephala* Berlepsch & Stolzmann, 1894 — N and C Peru

DELOTHRAUPIS P.L. Sclater, 1886 F
○ *Delothraupis castaneoventris* (P.L. Sclater, 1851) Chestnut-bellied Mountain Tanager[2]
C Peru to N Bolivia

STEPHANOPHORUS Strickland, 1841 M
○ *Stephanophorus diadematus* (Temminck, 1823) Diademed Tanager — SE Brazil, SE Paraguay, NE Argentina

IRIDOSORNIS Lesson, 1844 M[3]
○ *Iridosornis porphyrocephalus* (P.L. Sclater, 1856) Purplish-mantled Tanager
W Colombia, W Ecuador

○ *Iridosornis analis* (Tschudi, 1844) Yellow-throated Tanager — S Colombia to S Peru

○ *Iridosornis jelskii* Golden-collared Tanager
    — *I.j.jelskii* (Cabanis, 1873) — N and C Peru
    — *I.j.bolivianus* Berlepsch, 1912 — S Peru, N Bolivia

○ *Iridosornis rufivertex* Golden-crowned Tanager
    — *I.r.rufivertex* (Lafresnaye, 1842) — W Venezuela to NW Peru
    — *I.r.caeruleoventris* Chapman, 1915 — NW Colombia
    — *I.r.ignicapillus* Chapman, 1915 — SW Colombia
    — *I.r.subsimilis* J.T. Zimmer, 1944 — W Ecuador

○ *Iridosornis reinhardti* (P.L. Sclater, 1865) Yellow-scarfed Tanager — N and C Peru

PIPRAEIDEA Swainson, 1827 F
○ *Pipraeidea melanonota* Fawn-breasted Tanager
    — *P.m.venezuelensis* (P.L. Sclater, 1857) — Venezuela to W Bolivia, NW Argentina
    — *P.m.melanonota* (Vieillot, 1819) — Paraguay, SE Brazil to NE Argentina

NEOTHRAUPIS Hellmayr, 1936[4] F
○ *Neothraupis fasciata* (M.H.K. Lichtenstein, 1823) White-banded Tanager
E and S Brazil, SE Bolivia, NE Paraguay

CHLOROCHRYSA Bonaparte, 1851 F
○ *Chlorochrysa phoenicotis* (Bonaparte, 1851) Glistening-green Tanager
W Colombia, W Ecuador

○ *Chlorochrysa calliparaea* Orange-eared Tanager
    — *C.c.bourcierci* (Bonaparte, 1851) — Colombia to NE Peru
    — *C.c.calliparaea* (Tschudi, 1844) — C Peru
    — *C.c.fulgentissima* Chapman, 1901 — S Peru, N Bolivia

○ *Chlorochrysa nitidissima* P.L. Sclater, 1873 Multicoloured Tanager — W Colombia

---

[1] Gender addressed by David & Gosselin (2002) (614), multiple changes occur in specific names.
[2] Includes *peruviana* see Isler & Isler (1987) (1196).
[3] Gender addressed by David & Gosselin (2002) (614), multiple changes occur in specific names.
[4] For the conservation of this name see Opinion 2004 (I.C.Z.N., 2002) (1178A).

TANGARA Brisson, 1760 F

○ ***Tangara inornata*** Plain-coloured Tanager
 — *T.i.rava* Wetmore, 1963   Costa Rica, W Panama
 — *T.i.languens* Bangs & Barbour, 1922   Panama, NW Colombia
 — *T.i.inornata* (Gould, 1855)   N Colombia

○ ***Tangara cabanisi*** (P.L. Sclater, 1868) Azure-rumped Tanager   S Mexico, SW Guatemala

○ ***Tangara palmeri*** (Hellmayr, 1909) Grey-and-gold Tanager   E Panama to W Ecuador

○ ***Tangara mexicana*** Turquoise Tanager
 — *T.m.vieilloti* (P.L. Sclater, 1857)   Trinidad
 — *T.m.media* (Berlepsch & Hartert, 1902)   S and E Venezuela, NW Brazil
 — *T.m.mexicana* (Linnaeus, 1766)   The Guianas
 — *T.m.boliviana* (Bonaparte, 1851)   Western Amazonia
 — *T.m.brasiliensis* (Linnaeus, 1766)   SE Brazil

○ ***Tangara chilensis*** Paradise Tanager
 — *T.c.paradisea* (Swainson, 1837)   The Guianas, N Brazil
 — *T.c.coelicolor* (P.L. Sclater, 1851)   E Colombia, S Venezuela, NW Brazil
 — *T.c.chlorocorys* J.T. Zimmer, 1929   NC Peru
 — *T.c.chilensis* (Vigors, 1832)   Western Amazonia

○ ***Tangara fastuosa*** (Lesson, 1831) Seven-coloured Tanager   E Brazil

○ ***Tangara seledon*** (Statius Müller, 1776) Green-headed Tanager   SE Brazil, E Paraguay, NE Argentina

○ ***Tangara cyanocephala*** Red-necked Tanager
 — *T.c.cearensis* Cory, 1916   NE Brazil
 — *T.c.corallina* (Berlepsch, 1903)   E Brazil
 — *T.c.cyanocephala* (Statius Müller, 1776)   SE Brazil, E Paraguay, N Argentina

○ ***Tangara desmaresti*** (Vieillot, 1819) Brassy-breasted Tanager   SE Brazil

○ ***Tangara cyanoventris*** (Vieillot, 1819) Gilt-edged Tanager   SE Brazil

○ ***Tangara johannae*** (Dalmas, 1900) Blue-whiskered Tanager   W Colombia, NW Ecuador

○ ***Tangara schrankii*** Green-and-gold Tanager
 — *T.s.venezuelana* Phelps & Phelps, Jr., 1957   S Venezuela
 — *T.s.schrankii* (Spix, 1825)   W Amazonia

○ ***Tangara florida*** Emerald Tanager
 — *T.f.florida* (Sclater & Salvin, 1869)   Costa Rica, W Panama
 — *T.f.auriceps* Chapman, 1914   E Panama, W Colombia

○ ***Tangara arthus*** Golden Tanager
 — *T.a.arthus* Lesson, 1832   N and E Venezuela
 — *T.a.palmitae* Meyer de Schauensee, 1947   Magdalena (NE Colombia)
 — *T.a.sclateri* (Lafresnaye, 1854)   E Colombia
 — *T.a.aurulenta* (Lafresnaye, 1843)   C Colombia, NW Venezuela
 — *T.a.occidentalis* Chapman, 1914   W Colombia
 — *T.a.goodsoni* E. Hartert, 1913   W Ecuador
 — *T.a.aequatorialis* (Taczanowski & Berlepsch, 1885)   E Ecuador, N Peru
 — *T.a.pulchra* (Tschudi, 1844)   C Peru
 — *T.a.sophiae* (Berlepsch, 1901)   S Peru, W Bolivia

○ ***Tangara icterocephala*** Silver-throated Tanager
 — *T.i.frantzii* (Cabanis, 1861)   Costa Rica, W Panama
 — *T.i.oresbia* Wetmore, 1962   WC Panama
 — *T.i.icterocephala* (Bonaparte, 1851)   E Panama, W Colombia, W Ecuador

○ ***Tangara xanthocephala*** Saffron-crowned Tanager
 — *T.x.venusta* (P.L. Sclater, 1855)   W Venezuela to N and C Peru
 — *T.x.xanthocephala* (Tschudi, 1844)   C Peru
 — *T.x.lamprotis* (P.L. Sclater, 1851)   S Peru, N Bolivia

○ *Tangara chrysotis* (Du Bus, 1846)[1] GOLDEN-EARED TANAGER S Colombia to N Bolivia

○ *Tangara parzudakii* FLAME-FACED TANAGER
 — *T.p.parzudakii* (Lafresnaye, 1843) SW Venezuela to Peru
 — *T.p.urubambae* J.T. Zimmer, 1943 S Peru
 — *T.p.lunigera* (P.L. Sclater, 1851) W Colombia, W Ecuador

○ *Tangara xanthogastra* YELLOW-BELLIED TANAGER
 — *T.x.xanthogastra* (P.L. Sclater, 1851) S Venezuela to N Bolivia
 — *T.x.phelpsi* J.T. Zimmer, 1943 Tepui region of S Venezuela, N Brazil

○ *Tangara punctata* SPOTTED TANAGER
 — *T.p.punctata* (Linnaeus, 1766) S Venezuela, the Guianas, N Brazil
 — *T.p.zamorae* Chapman, 1925 E Ecuador, N Peru
 — *T.p.perenensis* Chapman, 1925 Chanchamayo area (C Peru)
 — *T.p.annectens* J.T. Zimmer, 1943 S Peru
 — *T.p.punctulata* (Sclater & Salvin, 1876) N Bolivia

● *Tangara guttata* SPECKLED TANAGER
 ✓ *T.g.eusticta* Todd, 1912 Costa Rica, W Panama
 — *T.g.tolimae* Chapman, 1914 Tolima (Colombia)
 — *T.g.bogotensis* Hellmayr & Seilern, 1912 E Colombia, W Venezuela
 — *T.g.chrysophrys* (P.L. Sclater, 1851) Venezuela, NW Brazil
 — *T.g.guttata* (Cabanis, 1850) SE Venezuela, N Brazil
 — *T.g.trinitatis* Todd, 1912 N Trinidad

○ *Tangara varia* (Statius Müller, 1776) DOTTED TANAGER S Venezuela, the Guianas, N Brazil

○ *Tangara rufigula* (Bonaparte, 1851) RUFOUS-THROATED TANAGER W Colombia, NW Ecuador

● *Tangara gyrola* BAY-HEADED TANAGER
 ✓ *T.g.bangsi* (Hellmayr, 1911) Costa Rica, W Panama
 — *T.g.deleticia* (Bangs, 1908) E Panama, W Colombia
 — *T.g.nupera* Bangs, 1917 SW Colombia, W Ecuador
 — *T.g.toddi* Bangs & Penard, 1921 N Colombia, NW Venezuela
 — *T.g.viridissima* (Lafresnaye, 1847) Trinidad, NE Venezuela
 — *T.g.catharinae* (Hellmayr, 1911) E Colombia to C Bolivia
 — *T.g.parva* J.T. Zimmer, 1943 SW Venezuela to NE Peru, NW Brazil
 — *T.g.gyrola* (Linnaeus, 1758) SE Venezuela, the Guianas, N Brazil
 — *T.g.albertinae* (Pelzeln, 1877) C Brazil

○ *Tangara lavinia* RUFOUS-WINGED TANAGER
 — *T.l.cara* (Bangs, 1905) E Guatemala to Costa Rica
 — *T.l.dalmasi* (Hellmayr, 1910) W Panama
 — *T.l.lavinia* (Cassin, 1858) E Panama to NW Ecuador

○ *Tangara cayana* BURNISHED-BUFF TANAGER
 — *T.c.fulvescens* Todd, 1922 C Colombia
 — *T.c.cayana* (Linnaeus, 1766) The Guianas to E Peru, N Brazil
 — *T.c.huberi* (Hellmayr, 1910) NE Brazil
 — *T.c.flava* (J.F. Gmelin, 1789) NE Brazil
 — *T.c.sincipitalis* (Berlepsch, 1907) C Brazil
 — *T.c.chloroptera* (Vieillot, 1819) SE Brazil, E Paraguay, NE Argentina
 — *T.c.margaritae* (J.A. Allen, 1891) C Brazil

○ *Tangara cucullata* LESSER ANTILLEAN TANAGER
 — *T.c.versicolor* (Lawrence, 1878) St. Vincent I.
 — *T.c.cucullata* (Swainson, 1834) Grenada I.

○ *Tangara peruviana* (Desmarest, 1806) BLACK-BACKED TANAGER SE Brazil

○ *Tangara preciosa* (Cabanis, 1850) CHESTNUT-BACKED TANAGER SE Brazil, E Paraguay, Uruguay, NE Argentina

○ *Tangara vitriolina* (Cabanis, 1850) SCRUB TANAGER W Colombia, NW Ecuador

[1] Du Bus is an abbreviation for Du Bus de Gisignies.

○ *Tangara meyerdeschauenseei* Schulenberg & Binford, 1985 Green-capped Tanager
                                                                     S Peru (2209)

○ *Tangara rufigenis* (P.L. Sclater, 1857) Rufous-cheeked Tanager   N Venezuela

○ *Tangara ruficervix* Golden-naped Tanager
   — *T.r.ruficervix* (Prévost & Des Murs, 1846)   Colombia
   — *T.r.leucotis* (P.L. Sclater, 1851)   W Ecuador
   — *T.r.taylori* (Taczanowski & Berlepsch, 1885)   SE Colombia, E Ecuador
   — *T.r.amabilis* J.T. Zimmer, 1943   N Peru
   — *T.r.inca* Parkes, 1969   S Peru
   — *T.r.fulvicervix* (Sclater & Salvin, 1876)   N Bolivia

○ *Tangara labradorides* Metallic-green Tanager
   — *T.l.labradorides* (Boisonneau, 1840)   W Colombia, W Ecuador
   — *T.l.chaupensis* Chapman, 1925   NW Peru

○ *Tangara cyanotis* Blue-browed Tanager
   — *T.c.lutleyi* Hellmayr, 1917   S Colombia, Ecuador, Peru
   — *T.c.cyanotis* (P.L. Sclater, 1858)   N Bolivia

○ *Tangara cyanicollis* Blue-necked Tanager
   — *T.c.granadensis* (Berlepsch, 1884)   W Colombia
   — *T.c.caeruleocephala* (Swainson, 1838)   C Colombia to N Peru
   — *T.c.cyanicollis* (d'Orbigny & Lafresnaye, 1837)   E Peru, E Bolivia
   — *T.c.cyanopygia* (Berlepsch & Taczanowski, 1883) W Ecuador
   — *T.c.hannahiae* (Cassin, 1864)   E Colombia, W Venezuela
   — *T.c.melanogaster* Cherrie & Reichenberger, 1923 C Brazil, E Bolivia
   — *T.c.albotibialis* Traylor, 1950   Goias (Brazil)

● *Tangara larvata* Golden-hooded Tanager
   ⊥ *T.l.larvata* (Du Bus, 1846)[1]   S Mexico to N Costa Rica
   — *T.l.centralis* (Berlepsch, 1912)   E Costa Rica, W Panama
   — *T.l.franciscae* (P.L. Sclater, 1856)   W Costa Rica, W Panama
   — *T.l.fanny* (Lafresnaye, 1847)   E Panama to NW Ecuador

○ *Tangara nigrocincta* (Bonaparte, 1838) Masked Tanager   N and W Amazonia

● *Tangara dowii* (Salvin, 1863) Spangle-cheeked Tanager   Costa Rica, W Panama

○ *Tangara fucosa* Nelson, 1912 Green-naped Tanager   E Panama

○ *Tangara nigroviridis* Beryl-spangled Tanager
   — *T.n.cyanescens* (P.L. Sclater, 1857)   NW Venezuela to W Ecuador
   — *T.n.consobrina* Hellmayr, 1921   C Colombia
   — *T.n.nigroviridis* (Lafresnaye, 1843)   E Colombia, E Ecuador
   — *T.n.lozanoana* Aveledo & Perez, 1994   NW Venezuela (69)
   — *T.n.berlepschi* (Taczanowski, 1884)   Peru, N Bolivia

○ *Tangara vassorii* Blue-and-black Tanager
   — *T.v.vassorii* (Boissoneau, 1840)   NW Venezuela to NW Peru
   — *T.v.branickii* (Taczanowski, 1882)   N Peru
   — *T.v.atrocoerulea* (Tschudi, 1844)   S Peru to C Bolivia

○ *Tangara heinei* (Cabanis, 1850) Black-capped Tanager   NW Venezuela to E Ecuador

○ *Tangara phillipsi* Graves & Weske, 1987 Sira Tanager   E Peru (994)

○ *Tangara viridicollis* Silver-backed Tanager
   — *T.v.fulvigula* (Berlepsch & Stolzmann, 1906)   S Ecuador, N Peru
   — *T.v.viridicollis* (Taczanowski, 1884)   C and S Peru

○ *Tangara argyrofenges* Green-throated Tanager
   — *T.a.caeruleigularis* Carriker, 1935   N and C Peru
   — *T.a.argyrofenges* (Sclater & Salvin, 1876)   N and C Bolivia

---

[1] Du Bus is an abbreviation for Du Bus de Gisignies.

○ *Tangara cyanoptera* Black-headed Tanager
　　— *T.c.whitelyi* (Salvin & Godman, 1884)　　　　S Venezuela, Guyana
　　— *T.c.cyanoptera* (Swainson, 1834)　　　　N Colombia, N and W Venezuela

○ *Tangara velia* Opal-rumped Tanager
　　— *T.v.velia* (Linnaeus, 1758)　　　　The Guianas, N Brazil
　　— *T.v.iridina* (Hartlaub, 1841)　　　　W Amazonia
　　— *T.v.signata* (Hellmayr, 1905)　　　　NE Brazil
　　— *T.v.cyanomelas* (Wied, 1830)[1]　　　　SE Brazil

○ *Tangara callophrys* (Cabanis, 1849) Opal-crowned Tanager　　SE Colombia to E Peru, W Brazil

TERSINA Vieillot, 1819[2] F
○ *Tersina viridis* Swallow-Tanager
　　— *T.v.grisescens* Griscom, 1929　　　　N Colombia
　　— *T.v.occidentalis* (P.L. Sclater, 1855)　　　　E Panama, Colombia, Venezuela, the Guianas, Ecuador, NE Peru, N Bolivia, N Brazil
　　— *T.v.viridis* (Illiger, 1811)　　　　E and S Brazil, E Bolivia, Paraguay, NE Argentina

DACNIS Cuvier, 1816 F
○ *Dacnis albiventris* (P.L. Sclater, 1852) White-bellied Dacnis　　S Venezuela to NE Peru and C Brazil

○ *Dacnis lineata* Black-faced Dacnis
　　— *D.l.egregia* P.L. Sclater, 1855　　　　C Colombia
　　— *D.l.aequatorialis* Berlepsch & Taczanowski, 1883　　W Ecuador
　　— *D.l.lineata* (J.F. Gmelin, 1789)　　　　N and W Amazonia

○ *Dacnis flaviventer* d'Orbigny & Lafresnaye, 1837 Yellow-bellied Dacnis
　　　　W and C Amazonia

○ *Dacnis hartlaubi* P.L. Sclater, 1855 Turquoise Dacnis[3]　　W Colombia

○ *Dacnis nigripes* Pelzeln, 1856 Black-legged Dacnis　　SE Brazil

○ *Dacnis venusta* Scarlet-thighed Dacnis
　　— *D.v.venusta* Lawrence, 1862　　　　Costa Rica, W Panama
　　— *D.v.fuliginata* Bangs, 1908　　　　E Panama to NW Ecuador

○ *Dacnis cayana* Blue Dacnis
　　— *D.c.callaina* Bangs, 1905　　　　W Costa Rica, W Panama
　　— *D.c.ultramarina* Lawrence, 1864　　　　E Nicaragua to NW Colombia
　　— *D.c.napaea* Bangs, 1898　　　　N Colombia
　　— *D.c.baudoana* Meyer de Schauensee, 1946　　SW Colombia, W Ecuador
　　— *D.c.coerebicolor* P.L. Sclater, 1851　　　　C Colombia
　　— *D.c.cayana* (Linnaeus, 1766)　　　　E Colombia to French Guiana, N and C Brazil
　　— *D.c.glaucogularis* Berlepsch & Stolzmann, 1896　　S Colombia to N and E Bolivia
　　— *D.c.paraguayensis* Chubb, 1910　　　　S and E Brazil, Paraguay, NE Argentina

○ *Dacnis viguieri* Salvin & Godman, 1883 Viridian Dacnis　　E Panama, NE Colombia

○ *Dacnis berlepschi* E. Hartert, 1900 Scarlet-breasted Dacnis　　SW Colombia, NW Ecuador

CYANERPES Oberholser, 1899 M
○ *Cyanerpes nitidus* (Hartlaub, 1847) Short-billed Honeycreeper　　NW Amazonia

○ *Cyanerpes lucidus* Shining Honeycreeper
　　— *C.l.lucidus* (Sclater & Salvin, 1859)　　　　S Mexico to N Nicaragua
　　— *C.l.isthmicus* Bangs, 1917　　　　Costa Rica to NW Colombia

○ *Cyanerpes caeruleus* Purple Honeycreeper
　　— *C.c.chocoanus* Hellmayr, 1920　　　　W Colombia, W Ecuador
　　— *C.c.caeruleus* (Linaeus, 1758)　　　　Colombia to the Guianas, NE Brazil

[1] For correction of spelling see David & Gosselin (2002) (613).
[2] Treated in a subfamily of its own in last Edition, but reduced to a tribe see Ridgely & Tudor (1989) (2018).
[3] Type species of the genus *Pseudodacnis*. Not here recognised.

— *C.c.hellmayri* Gyldenstolpe, 1945 — Guyana
— *C.c.longirostris* (Cabanis, 1850) — Trinidad
— *C.c.microrhynchus* (Berlepsch, 1884) — W and C Amazonia

○ *Cyanerpes cyaneus* RED-LEGGED HONEYCREEPER
— *C.c.carneipes* (P.L. Sclater, 1859) — E and S Mexico
— *C.c.pacificus* Chapman, 1915 — W Colombia, W Ecuador
— *C.c.gigas* Thayer & Bangs, 1905 — Gorgona I. (W Colombia)
— *C.c.gemmeus* Wetmore, 1941 — N Colombia (Guajira Pen.)
— *C.c.eximius* (Cabanis, 1850) — N Colombia, N Venezuela
— *C.c.tobagensis* Hellmayr & Seilern, 1914 — Tobago I.
— *C.c.cyaneus* (Linnaeus, 1766) — SE Venezuela, Trinidad, the Guianas, N Brazil
— *C.c.dispar* J.T. Zimmer, 1942 — SW Venezuela to NE Peru, W Brazil
— *C.c.violaceus* J.T. Zimmer, 1942 — C Bolivia, W Brazil
— *C.c.brevipes* (Cabanis, 1850) — C Brazil
— *C.c.holti* Parkes, 1977 — E Brazil (1770)

CHLOROPHANES Reichenbach, 1853 M
○ *Chlorophanes spiza* GREEN HONEYCREEPER
— *C.s.guatemalensis* P.L. Sclater, 1861 — S Mexico to Honduras
— *C.s.arguta* Bangs & Barbour, 1922 — E Honduras to NW Colombia
— *C.s.exsul* Berlepsch & Taczanowski, 1883 — SW Colombia, W Ecuador
— *C.s.subtropicalis* Todd, 1924 — C Andes of Colombia to W Venezuela
— *C.s.spiza* (Linnaeus, 1758) — Venezuela, Trinidad, the Guianas, N Brazil
— *C.s.caerulescens* Cassin, 1864 — Andes of SE Colombia to Bolivia
— *C.s.axillaris* J.T. Zimmer, 1929 — E Brazil

IRIDOPHANES Ridgway, 1901 M
○ *Iridophanes pulcherrimus* GOLDEN-COLLARED HONEYCREEPER[1]
— *I.p.pulcherrimus* (P.L. Sclater, 1853)[2] — Colombia to S Peru
— *I.p.aureinucha* (Ridgway, 1879) — W Ecuador

HETEROSPINGUS Ridgway, 1898 M
○ *Heterospingus rubrifrons* (Lawrence, 1865) SULPHUR-RUMPED TANAGER[3]
— E Costa Rica, Panama

○ *Heterospingus xanthopygius* SCARLET-BROWED TANAGER
— *H.x.xanthopygius* (P.L. Sclater, 1855) — E Panama, N Colombia
— *H.x.berliozi* Wetmore, 1965 — W Colombia, NW Ecuador

HEMITHRAUPIS Cabanis, 1850 F
○ *Hemithraupis guira* GUIRA TANAGER
— *H.g.nigrigula* (Boddaert, 1783) — NC Colombia to NE Brazil
— *H.g.roraimae* (Hellmayr, 1910) — SE Venezuela, Guyana
— *H.g.guirina* (P.L. Sclater, 1856) — W Colombia to NW Peru
— *H.g.huambina* Stolzmann, 1926 — SE Colombia to NE Peru, W Brazil
— *H.g.boliviana* J.T. Zimmer, 1947 — E Bolivia, NW Argentina
— *H.g.amazonica* J.T. Zimmer, 1947 — C Brazil
— *H.g.guira* (Linnaeus, 1766) — E Brazil
— *H.g.fosteri* (Sharpe, 1905) — SE Brazil, E Paraguay, NE Argentina

○ *Hemithraupis ruficapilla* RUFOUS-HEADED TANAGER
— *H.r.bahiae* J.T. Zimmer, 1947 — E Brazil
— *H.r.ruficapilla* (Vieillot, 1818) — SE Brazil

○ *Hemithraupis flavicollis* YELLOW-BACKED TANAGER
— *H.f.ornata* Nelson, 1912 — E Panama, NW Colombia
— *H.f.albigularis* (P.L. Sclater, 1855) — Colombia
— *H.f.peruana* Bonaparte, 1851 — SC Colombia to NE Peru

[1] Treated as a species within the genus *Tangara* in last Edition, but see Ridgely & Tudor (1989: 242) (2018).
[2] For correction of spelling see David & Gosselin (2002) (614).
[3] We follow A.O.U. (1998: 575) (3) in separating this from *H. xanthopygius*.

__ *H.f.sororia* J.T. Zimmer, 1947	N Peru
__ *H.f.centralis* (Hellmayr, 1907)	SE Peru, N Bolivia, C Brazil
__ *H.f.aurigularis* Cherrie, 1916	SE Colombia, SW Venezuela, N Brazil
__ *H.f.hellmayri* Berlepsch, 1912	SE Venezuela, W Guyana
__ *H.f.flavicollis* (Vieillot, 1818)	Surinam, French Guiana, NE Brazil
__ *H.f.obidensis* Parkes & Humphrey, 1963	N Brazil
__ *H.f.melanoxantha* (M.H.K. Lichtenstein, 1823)	E Brazil
__ *H.f.insignis* (P.L. Sclater, 1856)	SE Brazil

CHRYSOTHLYPIS Berlepsch, 1912 F
○ *Chrysothlypis chrysomelas* BLACK-AND-YELLOW TANAGER

__ *C.c.titanota* Olson, 1981	E Costa Rica, W Panama (1697)
__ *C.c.chrysomelas* (Sclater & Salvin, 1869)[1]	WC Panama
__ *C.c.ocularis* Nelson, 1912	E Panama

○ *Chrysothlypis salmoni* (P.L. Sclater, 1886) SCARLET-AND-WHITE TANAGER
W Colombia, NW Ecuador

XENODACNIS Cabanis, 1873 F
○ *Xenodacnis parina* TIT-LIKE DACNIS

__ *X.p.bella* Bond & Meyer de Schauensee, 1939	N Peru
__ *X.p.petersi* Bond & Meyer de Schauensee, 1939	C Peru
__ *X.p.parina* Cabanis, 1873	S Peru

CONIROSTRUM d'Orbigny & Lafresnaye[2], 1838[3] N
○ *Conirostrum speciosum* CHESTNUT-VENTED CONEBILL

__ *C.s.guaricola* Phelps & Phelps, Jr., 1949	C Venezuela
__ *C.s.amazonum* (Hellmayr, 1917)	The Guianas to Ecuador >> N Peru
__ *C.s.speciosum* (Temminck, 1824)	SE Peru and E Bolivia to N Argentina

○ *Conirostrum leucogenys* WHITE-EARED CONEBILL

__ *C.l.panamense* (Griscom, 1927)	E Panama, NW Colombia
__ *C.l.leucogenys* (Lafresnaye, 1852)	N Colombia, NE Venezuela
__ *C.l.cyanochroum* (Todd, 1924)[4]	W Venezuela

○ *Conirostrum bicolor* BICOLOURED CONEBILL

__ *C.b.bicolor* (Vieillot, 1809)[5]	N Colombia to the Guianas, N Brazil
__ *C.b.minus* (Hellmayr, 1935)[6]	W Brazil, E Ecuador, E Peru

○ *Conirostrum margaritae* (Holt, 1931) PEARLY-BREASTED CONEBILL   N Brazil, NE Peru

○ *Conirostrum cinereum* CINEREOUS CONEBILL

__ *C.c.fraseri* P.L. Sclater, 1859	SW Colombia, E Ecuador
__ *C.c.littorale* Berlepsch & Stolzmann, 1897	W Peru, N Chile
__ *C.c.cinereum* d'Orbigny & Lafresnaye, 1838	S Peru, N Bolivia

○ *Conirostrum tamarugense* Johnson & Millie, 1972 TAMARUGO CONEBILL
SC Peru, N Chile (1212)

○ *Conirostrum ferrugineiventre* P.L. Sclater, 1856 WHITE-BROWED CONEBILL
S Peru, W Bolivia

○ *Conirostrum rufum* Lafresnaye, 1843 RUFOUS-BROWED CONEBILL   N Colombia

○ *Conirostrum sitticolor* BLUE-BACKED CONEBILL

__ ¶*C.s.pallidus* Aveledo & Perez, 1989[7]	NE Colombia, NW Venezuela (Perijá Mts.) (67)
__ *C.s.intermedium* Berlepsch, 1893	W Venezuela (Andes of Merida, N Tachira)
__ *C.s.sitticolor* Lafresnaye, 1840	S Colombia, Ecuador, NW Peru
__ *C.s.cyaneum* Taczanowski, 1875	Peru, W Bolivia

---

[1] For correct spelling see David & Gosselin (2002) (613).
[2] We are indebted to A.P. Peterson (*in litt.*) for noting that Lowery & Monroe (1968: 82) (1394) gave the authors of this generic name (and of the specific name *cinereum*) as Lafresnaye & d'Orbigny. Throughout Peters Check-list all other names cited from this paper treated the authors as d'Orbigny & Lafresnaye. We adjust the names here to provide consistency. Which is right is not resolved, as the nature of the authorship of each author is not clear.
[3] For reasons to place this genus in the Thraupidae see A.O.U. (1998: 569) (3).
[4] For correction of spelling see David & Gosselin (2002) (613).
[5] Dated 1808 by Lowery & Monroe (1968: 83) (1394), but see Browning & Monroe (1991) (258).
[6] For correction of spelling see David & Gosselin (2002) (613).
[7] Included tentatively as not known to have been re-evaluated.

○ *Conirostrum albifrons* CAPPED CONEBILL
   — *C.a.cyanonotum* Todd, 1932   N Venezuela
   — *C.a.albifrons* Lafresnaye, 1842   W Venezuela, E Colombia
   — *C.a.centralandium* Meyer de Schauensee, 1946   C Colombia
   — *C.a.atrocyaneum* Lafresnaye, 1848   SW Colombia, Ecuador, N Peru
   — *C.a.sordidum* Berlepsch, 1901   S Peru, N Bolivia
   — *C.a.lugens* Berlepsch, 1901   E Bolivia (Cochabamba, Santa Cruz)

OREOMANES P.L. Sclater, 1860 M
○ *Oreomanes fraseri* P.L. Sclater, 1860 GIANT CONEBILL   SW Colombia to C Bolivia

DIGLOSSA Wagler, 1832 F
○ *Diglossa baritula* CINNAMON-BELLIED FLOWERPIERCER[1]
   — *D.b.baritula* Wagler, 1832   C Mexico
   — *D.b.montana* Dearborn, 1907   S Mexico to El Salvador
   — *D.b.parva* Griscom, 1932   E Guatemala, Honduras

○ *Diglossa plumbea* SLATY FLOWERPIERCER
   — *D.p.plumbea* Cabanis, 1860   Costa Rica, W Panama
   — *D.p.veraguensis* Griscom, 1927   W Panama

○ *Diglossa sittoides* RUSTY FLOWERPIERCER
   — *D.s.hyperythra* Cabanis, 1850   NE Colombia, N Venezuela
   — *D.s.mandeli* Blake, 1940   NE Venezuela
   — *D.s.coelestis* Phelps & Phelps, Jr., 1953   W Venezuela
   — *D.s.dorbignyi* (Boissonneau, 1840)   E Colombia, W Venezuela
   — *D.s.decorata* J.T. Zimmer, 1930   Ecuador, Peru
   — *D.s.sittoides* (d'Orbigny & Lafresnaye, 1838)   Bolivia, NW Argentina

○ *Diglossa gloriosissima* CHESTNUT-BELLIED FLOWERPIERCER
   — *D.g.gloriosissima* Chapman, 1912   W Colombia (Cauca)
   — *D.g.boylei* Graves, 1990   W Colombia (Antioquía) (985)

○ *Diglossa lafresnayii* (Boissonneau, 1840) GLOSSY FLOWERPIERCER[2]   W Venezuela to Ecuador, N Peru

○ *Diglossa mystacalis* MOUSTACHED FLOWERPIERCER
   — *D.m.unicincta* Hellmayr, 1905   N Peru
   — *D.m.pectoralis* Cabanis, 1873   C Peru
   — *D.m.albilinea* Chapman, 1919   SE Peru
   — *D.m.mystacalis* Lafresnaye, 1846   N Bolivia

○ *Diglossa gloriosa* Sclater & Salvin, 1870 MERIDA FLOWERPIERCER[3]   W Venezuela

○ *Diglossa humeralis* BLACK FLOWERPIERCER
   — *D.h.nocticolor* Bangs, 1898   N Colombia, W Venezuela
   — *D.h.humeralis* (Fraser, 1840)   C Colombia, SW Venezuela
   — *D.h.aterrima* Lafresnaye, 1846   W Colombia, Ecuador, NW Peru

○ *Diglossa brunneiventris* BLACK-THROATED FLOWERPIERCER
   — *D.b.vuilleumieri* Graves, 1980   NW Colombia, Ecuador (978)
   — *D.b.brunneiventris* Lafresnaye, 1846   NW Colombia to N Chile

○ *Diglossa carbonaria* (d'Orbigny & Lafresnaye, 1838) GREY-BELLIED FLOWERPIERCER
     Bolivia

○ *Diglossa venezuelensis* Chapman, 1925 VENEZUELAN FLOWERPIERCER   NE Venezuela (Coastal Range)

○ *Diglossa albilatera* WHITE-SIDED FLOWERPIERCER
   — *D.a.federalis* Hellmayr, 1922   N Venezuela
   — *D.a.albilatera* Lafresnaye, 1843   W Venezuela to Ecuador
   — *D.a.schistacea* Chapman, 1925   SW Ecuador to NW Peru
   — *D.a.affinis* J.T. Zimmer, 1942   N and C Peru

[1] For the treatment of this species and the next two see Hackett (1995) (1048).
[2] Like Ridgely & Tudor (1989) (2018) we follow Vuilleumier (1969) (2550) in treating *gloriosissima* and *mystacalis* as separate species from *lafresnayii*.
[3] Again we follow Ridgely & Tudor (1989) (2018), here accepting the views of Graves (1982) (980) in making four species, this and the next three, of what in last Edition was *D. carbonaria*.

○ *Diglossa duidae* SCALED FLOWERPIERCER
 — *D.d.hitchcocki* Phelps & Phelps, Jr., 1948   S Venezuela
 — *D.d.duidae* Chapman, 1929[1]   S Venezuela, N Brazil
 — *D.d.georgebarrowcloughi* Dickerman, 1987   Cerra Jime (S Venezuela) (716)

○ *Diglossa major* GREATER FLOWERPIERCER
 — *D.m.gilliardi* Chapman, 1939   SE Venezuela
 — *D.m.disjuncta* Zimmer & Phelps, 1944   SE Venezuela
 — *D.m.chimantae* Phelps & Phelps, Jr., 1947   SE Venezuela
 — *D.m.major* Cabanis, 1849   SE Venezuela, N Brazil

○ *Diglossa indigotica* P.L. Sclater, 1856 INDIGO FLOWERPIERCER   SW Colombia, W Ecuador

○ *Diglossa glauca* DEEP-BLUE FLOWERPIERCER[2]
 — *D.g.tyrianthina* Hellmayr, 1930   S Colombia, E Ecuador
 — *D.g.glauca* Sclater & Salvin, 1876   Peru, N Bolivia

○ *Diglossa caerulescens* BLUISH FLOWERPIERCER
 — *D.c.caerulescens* (P.L. Sclater, 1856)   N Venezuela
 — *D.c.ginesi* Phelps & Phelps, Jr., 1952   NW Venezuela
 — *D.c.saturata* (Todd, 1917)   SW Venezuela, Colombia
 — *D.c.media* Bond, 1955   S Ecuador, NW Peru
 — *D.c.pallida* (Berlepsch & Stolzmann, 1896)   C Peru
 — *D.c.mentalis* J.T. Zimmer, 1942   SE Peru, NW Bolivia

○ *Diglossa cyanea* MASKED FLOWERPIERCER
 — *D.c.tovarensis* Zimmer & Phelps, 1952   N Venezuela
 — *D.c.obscura* Phelps & Phelps, Jr., 1952   NW Venezuela
 — *D.c.cyanea* (Lafresnaye, 1840)   W Venezuela to Ecuador
 — *D.c.dispar* J.T. Zimmer, 1942   SW Ecuador, NW Peru
 — *D.c.melanopis* Tschudi, 1844   Peru to C Bolivia

GENERA INCERTAE SEDIS (12: 69)

CHLOROSPINGUS Cabanis, 1851 M
● *Chlorospingus ophthalmicus* COMMON BUSH TANAGER
 — *C.o.albifrons* Salvin & Godman, 1889[3]   SW Mexico
 — *C.o.wetmorei* Lowery & Newman, 1949   E Mexico
 — *C.o.persimilis* A.R. Phillips, 1966   S Oaxaca
 — *C.o.ophthalmicus* (Du Bus, 1847)[4]   SE Mexico
 — *C.o.dwighti* Underdown, 1931   S Mexico, E Guatemala
 — *C.o.postocularis* Cabanis, 1866   S Mexico, W Guatemala
 — *C.o.honduratius* Berlepsch, 1912   El Salvador, Honduras
 — *C.o.regionalis* Bangs, 1906   Nicaragua, E Costa Rica, W Panama
 — *C.o.novicius* Bangs, 1902   SW Costa Rica, Panama (Boquete region and E slope
                   of Volcan de Chiriqui)

 — *C.o.punctulatus* Sclater & Salvin, 1869   W Panama
 — *C.o.jaqueti* Hellmayr, 1921   NE Colombia, N Venezuela
 — *C.o.falconensis* Phelps & Gilliard, 1941   NW Venezuela
 — *C.o.venezuelanus* Berlepsch, 1893   SW Venezuela
 — *C.o.ponsi* Phelps & Phelps, Jr., 1952   W Venezuela
 — *C.o.eminens* J.T. Zimmer, 1946   NE Colombia
 — *C.o.trudis* Olson, 1983   Santander (Colombia) (1698)
 — *C.o.exitelus* Olson, 1983   North end of C Andes (Antiquoia, Colombia) (1698)
 — *C.o.flavopectus* (Lafresnaye, 1840)   C Colombia
 — *C.o.macarenae* J.T. Zimmer, 1947   E Colombia
 — *C.o.nigriceps* Chapman, 1912   C Colombia
 — *C.o.phaeocephalus* Sclater & Salvin, 1877   Ecuador

---

[1] Includes *parui* Phelps & Phelps, 1950 (1847) see Phelps & Phelps Jr. (1963) (1865).
[2] We prefer to follow the consensus against recognition of a genus *Diglossopsis* P.L. Sclater, 1856. Both generic names are feminine.
[3] Includes *persimilis* see Storer (1970: 254) (2348) and Binford (1989) (150).
[4] Du Bus is an abbreviation for Du Bus de Gisignies.

— *C.o.cinereocephalus* Taczanowski, 1874   C Peru
— *C.o.hiaticolus* O'Neill & Parker, 1981   C Peru (1716)
— *C.o.peruvianus* Carriker, 1933   S Peru
— *C.o.bolivianus* Hellmayr, 1921   N Bolivia
— *C.o.fulvigularis* Berlepsch, 1901   C Bolivia
— *C.o.argentinus* Hellmayr, 1921   C Bolivia, N Argentina

○ *Chlorospingus tacarcunae* Griscom, 1924 TACARCUNA BUSH TANAGER   E Panama, NW Colombia

○ *Chlorospingus inornatus* (Nelson, 1912) PIRRE BUSH TANAGER   E Panama

○ *Chlorospingus semifuscus* DUSKY BUSH TANAGER
— *C.s.livingstoni* Bond & Meyer de Schauensee, 1940   W Colombia
— *C.s.semifuscus* Sclater & Salvin, 1873   SW Colombia, W Ecuador

● *Chlorospingus pileatus* SOOTY-CAPPED BUSH TANAGER
⊥ *C.p.pileatus* Salvin, 1864   Costa Rica, W Panama
— *C.p.diversus* Griscom, 1924   W Panama

○ *Chlorospingus parvirostris* SHORT-BILLED BUSH TANAGER
— *C.p.huallagae* Carriker, 1933   S Colombia, Peru
— *C.p.medianus* J.T. Zimmer, 1947   C Peru
— *C.p.parvirostris* Chapman, 1901   S Peru, N Bolivia

○ *Chlorospingus flavigularis* YELLOW-THROATED BUSH TANAGER
— *C.f.hypophaeus* Sclater & Salvin, 1868   W Panama
— *C.f.marginatus* Chapman, 1914   SW Colombia, W Ecuador
— *C.f.flavigularis* (P.L. Sclater, 1852)   S Colombia, E Ecuador, E Peru

○ *Chlorospingus flavovirens* (Lawrence, 1867) YELLOW-GREEN BUSH TANAGER
    W Colombia, W Ecuador

○ *Chlorospingus canigularis* ASHY-THROATED BUSH TANAGER
— *C.c.olivaceiceps* Underwood, 1898   W Costa Rica
— *C.c.canigularis* (Lafresnaye, 1848)   C Colombia, SW Venezuela
— *C.c.conspicillatus* Todd, 1922   W Colombia
— *C.c.paulus* J.T. Zimmer, 1947   SW Ecuador
— *C.c.signatus* Taczanowski & Berlepsch, 1885   E Ecuador, N and C Peru

PIRANGA Vieillot, 1807[1][2] F
○ *Piranga bidentata* FLAME-COLOURED TANAGER
— *P.b.bidentata* (Swainson, 1827)   W Mexico
— *P.b.flammea* Ridgway, 1887   Tres Marias Is.
— *P.b.sanguinolenta* (Lafresnaye, 1839)   E Mexico to El Salvador
— *P.b.citrea* van Rossem, 1934   Costa Rica, W Panama

○ *Piranga flava* HEPATIC TANAGER
— *P.f.hepatica* (Swainson, 1827)   SW USA, W Mexico
— *P.f.dextra* Bangs, 1907   SW USA, E Mexico >> W Guatemala
— *P.f.figlina* (Salvin & Godman, 1883)   E Guatemala, Belize
— *P.f.savannarum* T.R. Howell, 1965   Honduras, NE Nicaragua
— *P.f.albifacies* J.T. Zimmer, 1929   W Guatemala to N Nicaragua
— *P.f.testacea* (Sclater & Salvin, 1868)   Costa Rica, Panama
— *P.f.faceta* Bangs, 1898   N Colombia, N Venezuela, Trinidad
— *P.f.haemalea* (Salvin & Godman, 1883)   S Venezuela, W Guyana, N Brazil
— *P.f.macconnelli* Chubb, 1921   S Guyana, N Brazil
— *P.f.toddi* Parkes, 1969   W slope of E Andes (Colombia)
— *P.f.desidiosa* Bangs & Noble, 1918   SW Colombia
— *P.f.lutea* (Lesson, 1834)   W Ecuador to NW Bolivia
— *P.f.rosacea* Todd, 1922   E Bolivia
— *P.f.saira* (Spix, 1825)   E Brazil
— *P.f.flava* (Vieillot, 1822)   S Bolivia to Uruguay, N Argentina

[1] Dated 1808 by Storer (1970: 301) (2348), but see Browning & Monroe (1991) (258).
[2] Species sequence based on Burns (1998) (279).

● *Piranga rubra* SUMMER TANAGER
    — *P.r.cooperi* (Ridgway, 1869)[1]   SW USA >> C Mexico
    ✓ *P.r.rubra* (Linnaeus, 1758)   SE USA >> Central and Sth. America

○ *Piranga roseogularis* ROSE-THROATED TANAGER
    — *P.r.roseogularis* (Cabot, 1846)   SE Mexico
    — *P.r.tincta* Paynter, 1950   SE Mexico, N Guatemala
    — *P.r.cozumelae* Ridgway, 1901   Cozumel I. (Mexico)

● *Piranga olivacea* (J.F. Gmelin, 1789) SCARLET TANAGER   SE Canada, E USA >> NW Sth. America

● *Piranga ludoviciana* (A. Wilson, 1811) WESTERN TANAGER[2]   W Nth. America >> W Mexico, W Central America

○ *Piranga leucoptera* WHITE-WINGED TANAGER
    — *P.l.leucoptera* (Trudeau, 1839)   E Mexico to Nicaragua
    — *P.l.latifasciata* Ridgway, 1887   Costa Rica, W Panama
    — *P.l.venezuelae* J.T. Zimmer, 1947   Colombia, Venezuela, N Brazil
    — *P.l.ardens* (Tschudi, 1844)   SW Colombia to Bolivia

○ *Piranga erythrocephala* RED-HEADED TANAGER
    — *P.e.candida* Griscom, 1934   NW Mexico
    — *P.e.erythrocephala* (Swainson, 1827)   SC and S Mexico

○ *Piranga rubriceps* (G.R. Gray, 1844) RED-HOODED TANAGER   W Colombia to N Peru

HABIA Blyth, 1840 F
○ *Habia rubica* RED-CROWNED ANT TANAGER
    — *H.r.holobrunnea* Griscom, 1930   E Mexico
    — *H.r.rosea* (Nelson, 1898)   SW Mexico
    — *H.r.affinis* (Nelson, 1897)   S Mexico
    — *H.r.nelsoni* (Ridgway, 1902)   SE Mexico
    — *H.r.rubicoides* (Lafresnaye, 1844)   S Mexico to El Salvador
    — *H.r.vinacea* (Lawrence, 1867)   W Costa Rica, W Panama
    — *H.r.alfaroana* (Ridgway, 1905)   NW Costa Rica
    — *H.r.perijana* Phelps & Phelps, Jr., 1957 NE Colombia, NW Venezuela
    — *H.r.coccinea* (Todd, 1919)   NC Colombia, W Venezuela
    — *H.r.crissalis* Parkes, 1969   NE Venezuela
    — *H.r.rubra* (Vieillot, 1819)   Trinidad
    — *H.r.mesopotamia* Parkes, 1969   E Bolivar (SE Venezuela)
    — *H.r.rhodinolaema* (Salvin & Godman, 1883) SE Colombia to NE Peru, NW Brazil
    — *H.r.peruviana* (Taczanowski, 1884)   E Peru, NC Bolivia
    — *H.r.hesterna* Griscom & Greenway, 1937 C Brazil
    — *H.r.bahiae* Hellmayr, 1936   E Brazil
    — *H.r.rubica* (Vieillot, 1817)   SE Brazil, Paraguay, N Argentina

○ *Habia fuscicauda* RED-THROATED ANT TANAGER
    — *H.f.salvini* (Berlepsch, 1883)   SE Mexico to El Salvador
    — *H.f.insularis* (Salvin, 1888)   SE Mexico, N Guatemala
    — *H.f.discolor* (Ridgway, 1901)   Nicaragua
    — *H.f.fuscicauda* (Cabanis, 1861)   S Nicaragua to W Panama
    — *H.f.willisi* Parkes, 1969   C Panama
    — *H.f.erythrolaema* (P.L. Sclater, 1862)   N Colombia

○ *Habia atrimaxillaris* (Dwight & Griscom, 1924) BLACK-CHEEKED ANT TANAGER
                                SW Costa Rica

○ *Habia gutturalis* (P.L. Sclater, 1854) SOOTY ANT TANAGER   NW Colombia

○ *Habia cristata* (Lawrence, 1875) CRESTED ANT TANAGER   W Colombia

CHLOROTHRAUPIS Salvin & Godman, 1883 F
○ *Chlorothraupis carmioli* CARMIOL'S TANAGER[3]
    — *C.c.carmioli* (Lawrence, 1868)   Nicaragua to NW Panama

---

[1] Tentatively includes *ochracea* which was in synonymy in Storer (1970: 306) (2348), but recognized by Browning (1990) (251).
[2] *P.l.zephyrica* Oberholser, 1974 (1673), was placed in synonymy by Browning (1978) (246).
[3] We find the vernacular name Olive suggestive of the scientific name of the next species and prefer Carmiol's.

      __ *C.c.magnirostris* Griscom, 1927   W Panama

      __ *C.c.lutescens* Griscom, 1927   E Panama, NW Colombia

      __ *C.c.frenata* Berlepsch, 1907   S Colombia, Peru, N Bolivia

○ *Chlorothraupis olivacea* (Cassin, 1860) Lemon-spectacled Tanager   E Panama to NW Ecuador

○ *Chlorothraupis stolzmanni* Ochre-breasted Tanager

      __ *C.s.dugandi* Meyer de Schauensee, 1948   SW Colombia

      __ *C.s.stolzmanni* (Berlepsch & Taczanowski, 1883)   W Ecuador

NESOSPINGUS P.L. Sclater, 1885 M

○ *Nesospingus speculiferus* (Lawrence, 1875) Puerto Rican Tanager   Puerto Rico

PHAENICOPHILUS Strickland, 1851 M

○ *Phaenicophilus palmarum* (Linnaeus, 1766) Black-crowned Palm Tanager

                                        Hispaniola, Soana I.

○ *Phaenicophilus poliocephalus* Grey-crowned Palm Tanager

      __ *P.p.poliocephalus* (Bonaparte, 1851)   S Haiti and Hispaniola

      __ *P.p.coryi* Richmond & Swales, 1924   Gonave I. (Hispaniola)

CALYPTOPHILUS Cory, 1884 M

○ *Calyptophilus tertius* Western Chat-Tanager[1]

      __ *C.t.tertius* Wetmore, 1929   Mts. of S Haiti

      __ *C.t.neibei* Bond & Dod, 1977   C Dominican Republic (188)

○ *Calyptophilus frugivorus* Eastern Chat-Tanager

      __ *C.f.frugivorus* (Cory, 1883)   Dominica I.

      __ *C.f.abbotti* Richmond & Swales, 1924   Gonave I. (Hispaniola)

SPINDALIS Jardine & Selby, 1837 M

○ *Spindalis zena* Western Spindalis[2]

      __ *S.z.townsendi* Ridgway, 1887   N Bahama Is.

      __ *S.z.zena* (Linnaeus, 1766)   C Bahama Is.

      __ *S.z.pretrei* (Lesson, 1831)   Cuba, Isla de la Juventud

      __ *S.z.salvini* Cory, 1886   Grand Cayman I.

      __ *S.z.benedicti* Ridgway, 1885   Cozumel I.

○ *Spindalis dominicensis* (H. Bryant, 1867)[3] Hispaniolan Spindalis   Hispaniola

○ *Spindalis portoricensis* (H. Bryant, 1866) Puerto Rican Spindalis   Puerto Rico

○ *Spindalis nigricephala* (Jameson, 1835) Jamaican Spindalis   Jamaica

RHODINOCICHLA Hartlaub, 1853 F

○ *Rhodinocichla rosea* Rosy Thrush-Tanager

      __ *R.r.schistacea* Ridgway, 1878   W Mexico

      __ *R.r.eximia* Ridgway, 1902   SW Costa Rica, W Panama

      __ *R.r.harterti* Hellmayr, 1918   C Colombia

      __ *R.r.beebei* Phelps & Phelps, Jr., 1949   NE Colombia, NW Venezuela

      __ *R.r.rosea* (Lesson, 1832)   NW Venezuela

EUPHONIA Desmarest, 1806 F

○ *Euphonia jamaica* (Linnaeus, 1766) Jamaican Euphonia   Jamaica

○ *Euphonia plumbea* Du Bus, 1855[4] Plumbeous Euphonia   S Venezuela to Surinam, N Brazil

○ *Euphonia affinis* Scrub Euphonia

      __ *E.a.godmani* Brewster, 1889   W Mexico

      __ *E.a.olmecorum* Dickerman, 1981   S Mexico (711)

      __ *E.a.affinis* (Lesson, 1842)   E Mexico to Costa Rica

---

[1] For references to the divided opinions over whether this and the next species are two or one see A.O.U. (1998: 572) (3).

[2] See Garrido *et al.* (1997) (926) for the treatment of the forms of Hispaniola and Puerto Rico as separate species from *S. zena*.

[3] Erroneously dated 1866 by Storer (1970: 317) (2348) but see Bangs (1930: 405) (83).

[4] Du Bus is an abbreviation for Du Bus de Gisignies.

○ *Euphonia luteicapilla* (Cabanis, 1860) Yellow-crowned Euphonia   E Nicaragua to Panama

○ *Euphonia chlorotica* Purple-throated Euphonia
   — *E.c.cynophora* (Oberholser, 1918)   E Colombia, S Venezuela, N Brazil
   — *E.c.chlorotica* (Linnaeus, 1766)   The Guianas, N and NE Brazil
   — *E.c.amazonica* Parkes, 1969   Amazonian Brazil
   — *E.c.taczanowskii* P.L. Sclater, 1886   E Peru, N Bolivia
   — *E.c.serrirostris* d'Orbigny & Lafresnaye, 1837   SE Bolivia to Uruguay, S Brazil

○ *Euphonia trinitatis* Strickland, 1851 Trinidad Euphonia   N Colombia to Trinidad

○ *Euphonia concinna* P.L. Sclater, 1855 Velvet-fronted Euphonia   C Colombia

○ *Euphonia saturata* (Cabanis, 1860) Orange-crowned Euphonia   W Colombia to NW Peru

○ *Euphonia finschi* Sclater & Salvin, 1877 Finsch's Euphonia   E Venezuela, the Guianas

○ *Euphonia violacea* Violaceous Euphonia
   — *E.v.rodwayi* (Penard, 1919)   E Venezuela, Trinidad
   — *E.v.violacea* (Linnaeus, 1758)   The Guianas, N Brazil
   — *E.v.aurantiicollis* A. Bertoni, 1901   SE Brazil, Paraguay

○ *Euphonia laniirostris* Thick-billed Euphonia
   — *E.l.crassirostris* P.L. Sclater, 1857   Costa Rica to N Venezuela
   — *E.l.melanura* P.L. Sclater, 1851   Colombia to N Peru, W Brazil
   — *E.l.hypoxantha* Berlepsch & Taczanowski, 1883   E Ecuador, NW Peru
   — *E.l.zopholega* (Oberholser, 1918)   EC Peru
   — *E.l.laniirostris* d'Orbigny & Lafresnaye, 1837   E Bolivia, SW Brazil

● *Euphonia hirundinacea* Yellow-throated Euphonia
   — *E.h.suttoni* A.R. Phillips, 1966[1]   E Mexico
   — *E.h.caribbaea* A.R. Phillips, 1966[2]   SE Mexico
   — *E.h.hirundinacea* Bonaparte, 1838[3]   E Mexico to E Nicaragua
   ↙ *E.h.gnatho* (Cabanis, 1860)   NW Nicaragua to W Panama

○ *Euphonia chalybea* (Mikan, 1825) Green-throated Euphonia   SE Brazil, E Paraguay, NE Argentina

○ *Euphonia elegantissima* Elegant Euphonia
   — *E.e.rileyi* (van Rossem, 1941)   NW Mexico
   — *E.e.elegantissima* (Bonaparte, 1838)   C and S Mexico to Honduras

○ *Euphonia cyanocephala* Golden-rumped Euphonia
   — *E.c.pelzelni* P.L. Sclater, 1886   S Colombia, W Ecuador
   — *E.c.insignis* Sclater & Salvin, 1877   S Ecuador
   — *E.c.cyanocephala* (Vieillot, 1819)[4]   SE Venezuela to Surinam (2525)

○ *Euphonia musica* Antillean Euphonia
   — *E.m.musica* (J.F. Gmelin, 1789)   Hispaniola
   — *E.m.sclateri* (P.L. Sclater, 1854)   Puerto Rico
   — *E.m.flavifrons* (Sparrman, 1789)   Lesser Antilles

○ *Euphonia fulvicrissa* Fulvous-vented Euphonia
   — *E.f.fulvicrissa* P.L. Sclater, 1857   Panama, NW Colombia
   — *E.f.omissa* E. Hartert, 1901   C Colombia
   — *E.f.purpurascens* E. Hartert, 1901   SW Colombia, NW Ecuador

○ *Euphonia imitans* (Hellmayr, 1936) Spot-crowned Euphonia   W Costa Rica, W Panama

● *Euphonia gouldi* Olive-backed Euphonia
   — *E.g.gouldi* P.L. Sclater, 1857[5]   SE Mexico to Honduras
   ↙ *E.g.praetermissa* (J.L. Peters, 1929)   E Honduras to Panama

---

[1] For recognition see Dickermann & Parkes (1997) (729).
[2] Includes *russelli* see Parkes in Dickermann & Parkes (1997) (729).
[3] See comments on nomenclature in A.O.U. (1998: 584) (3).
[4] For use of the name *cyanocephala* in place of *aureata* see A.O.U. (1998: 584) (3). When *Euphonia* was subsumed within *Tanagra* (now suppressed), the name *T. cyanocephala* Statius Müller, 1776, had priority so that *cyanocephala* was unavailable. This is no longer the case due to the suppression of *Tanagra* Linnaeus by Opinion 852, I.C.Z.N. (1968) (1171). See Art. 59.3 of the Code (I.C.Z.N., 1999) (1178).
[5] Includes *loetscheri* see Storer (1970: 350) (2348) but see also Dickermann & Parkes (1997) (729).

○ *Euphonia chrysopasta* GOLDEN-BELLIED EUPHONIA
    — *E.c.chrysopasta* Sclater & Salvin, 1869   Western Amazonia
    — *E.c.nitida* (Penard, 1923)   E Colombia to French Guiana, N Brazil

○ *Euphonia mesochrysa* BRONZE-GREEN EUPHONIA
    — *E.m.mesochrysa* Salvadori, 1873   C Colombia, E Ecuador
    — *E.m.media* (J.T. Zimmer, 1943)   N Peru
    — *E.m.tavarae* (Chapman, 1925)   S Peru to C Bolivia

○ *Euphonia minuta* WHITE-VENTED EUPHONIA
    — *E.m.humilis* (Cabanis, 1860)   S Mexico to W Ecuador
    — *E.m.minuta* Cabanis, 1849   The Guianas to C Bolivia, W Brazil

○ *Euphonia anneae* TAWNY-CAPPED EUPHONIA
    — *E.a.anneae* Cassin, 1865   W Costa Rica, W Panama
    — *E.a.rufivertex* Salvin, 1866   W Panama, NW Colombia

○ *Euphonia xanthogaster* ORANGE-BELLIED EUPHONIA
    — *E.x.oressinoma* Olson, 1981   W Panama to NW Colombia (1696)
    — *E.x.chocoensis* Hellmayr, 1911   E Panama to NW Ecuador
    — *E.x.badissima* Olson, 1981[1]   N Venezuela (1696)
    — *E.x.quitensis* (Nelson, 1912)   W Ecuador
    — *E.x.dilutior* (J.T. Zimmer, 1943)   S Colombia, NE Peru
    — *E.x.cyanonota* Parkes, 1969   W Brazil
    — *E.x.brunneifrons* Chapman, 1901   SE Peru
    — *E.x.ruficeps* d'Orbigny & Lafresnaye, 1837   W Bolivia
    — *E.x.brevirostris* Bonaparte, 1851   N and W Amazonia
    — *E.x.exsul* Berlepsch, 1912   NE Colombia, N Venezuela
    — *E.x.xanthogaster* (Sundevall, 1834)   S and E Brazil

○ *Euphonia rufiventris* RUFOUS-BELLIED EUPHONIA
    — *E.r.rufiventris* (Vieillot, 1819)   Western Amazonia
    — *E.r.carnegiei* Dickerman, 1988   S Venezuela (718)

○ *Euphonia pectoralis* (Latham, 1802)[2] CHESTNUT-BELLIED EUPHONIA   SE Brazil, E Paraguay, NE Argentina

○ *Euphonia cayennensis* (J.F. Gmelin, 1789) GOLDEN-SIDED EUPHONIA   SE Venezuela, the Guianas, N Brazil

CHLOROPHONIA Bonaparte, 1851 F
○ *Chlorophonia flavirostris* P.L. Sclater, 1861 YELLOW-COLLARED CHLOROPHONIA
          SW Colombia, NW Ecuador

○ *Chlorophonia cyanea* BLUE-NAPED CHLOROPHONIA
    — *C.c.psittacina* Bangs, 1902   N Colombia
    — *C.c.frontalis* (P.L. Sclater, 1851)   N Venezuela
    — *C.c.minuscula* Hellmayr, 1922   NE Venezuela
    — *C.c.roraimae* Salvin & Godman, 1884   S Venezuela, Guyana
    — *C.c.intensa* J.T. Zimmer, 1943   W Colombia
    — *C.c.longipennis* (Du Bus, 1855)[3]   W Venezuela to W Bolivia
    — *C.c.cyanea* (Thunberg, 1822)   SE Brazil, E Paraguay, NE Argentina

○ *Chlorophonia pyrrhophrys* (P.L. Sclater, 1851) CHESTNUT-BREASTED CHLOROPHONIA
          W Venezuela, N Colombia to C Peru

○ *Chlorophonia occipitalis* (Du Bus, 1847)[4] BLUE-CROWNED CHLOROPHONIA
          SE Mexico to Nicaragua

● *Chlorophonia callophrys* (Cabanis, 1860) GOLDEN-BROWED CHLOROPHONIA
          Costa Rica, W Panama

CATAMBLYRHYNCHUS Lafresnaye, 1842 M
○ *Catamblyrhynchus diadema* PLUSH-CAPPED FINCH
    — *C.d.federalis* Phelps & Phelps, Jr., 1953   N Venezuela
    — *C.d.diadema* Lafresnaye, 1842   NW Venezuela, Colombia, Ecuador
    — *C.d.citrinifrons* Berlepsch & Stolzmann, 1896   Peru, Bolivia, NW Argentina

---

[1] Includes *lecroyana* Aveledo & Perez, 1994 (69) see Rodner *et al.* (2000) (2096).
[2] For reasons to use 1802 not 1801 see Browing & Monroe (1991) (258).
[3] Du Bus is an abbreviation for Du Bus de Gisignies.
[4] Du Bus is an abbreviation for Du Bus de Gisignies.

CARDINALIDAE CARDINAL, GROSBEAKS, SALTATORS AND ALLIES (11: 42)

SPIZA Bonaparte, 1824 F

● *Spiza americana* (J.F. Gmelin, 1789) DICKCISSEL E North America >> Central America, Trinidad, Colombia, Venezuela

PHEUCTICUS Reichenbach, 1850 M

○ *Pheucticus chrysopeplus* YELLOW GROSBEAK
— *P.c.dilutus* van Rossem, 1934 Highlands of NW Mexico
— *P.c.chrysopeplus* (Vigors, 1832)[1] Highlands of W and C Mexico
— *P.c.aurantiacus* Salvin & Godman, 1891 Highlands of S Mexico, Guatemala

○ *Pheucticus tibialis* Lawrence, 1867 BLACK-THIGHED GROSBEAK C Costa Rica, W Panama

○ *Pheucticus chrysogaster* GOLDEN-BELLIED GROSBEAK
— *P.c.laubmanni* Hellmayr & Seilern, 1915 N Colombia, N Venezuela
— *P.c.chrysogaster* (Lesson, 1832) SW Colombia, Ecuador, Peru

○ *Pheucticus aureoventris* BLACK-BACKED GROSBEAK
— *P.a.meridensis* Riley, 1905 Andes of W Venezuela
— *P.a.uropygialis* Sclater & Salvin, 1871 E and C Andes of Colombia
— *P.a.crissalis* Sclater & Salvin, 1877 Andes of SW Colombia, Ecuador
— *P.a.terminalis* Chapman, 1919 Andes of E Peru
— *P.a.aureoventris* d'Orbigny & Lafresnaye, 1837 S Peru, Bolivia, Brazil, Paraguay, N Argentina

● *Pheucticus ludovicianus* (Linnaeus, 1766) ROSE-BREASTED GROSBEAK[2] S Canada, E USA >> Mexico, Central America, N Sth America

● *Pheuticus melanocephalus* BLACK-HEADED GROSBEAK
⤶ *P.m.maculatus* (Audubon, 1837) SW Canada, W USA, Baja California
— *P.m.melanocephalus* (Swainson, 1827)[3] C and E Rocky Mts. from W Canada and C USA to the Mexican plateau

CARDINALIS Bonaparte, 1838 M

● *Cardinalis cardinalis* NORTHERN CARDINAL
⤶ *C.c.cardinalis* (Linnaeus, 1758) E USA
— *C.c.floridanus* Ridgway, 1896 SE Georgia, Florida
✓ *C.c.magnirostris* Bangs, 1903 SE Texas, Louisiana
⤶ *C.c.canicaudus* Chapman, 1891 SC USA, C and E Mexico
— *C.c.coccineus* Ridgway, 1873 E Mexico (San Luis Potosí and N and C Veracruz)
— *C.c.littoralis* Nelson, 1897 E Mexico (S Veracruz and Tabasco)
— *C.c.yucatanicus* Ridgway, 1887 SE Mexico (base of Yucatán Pen.)
— *C.c.phillipsi* Parkes, 1997 Yucatán Pen. (Mexico) (1783)
— *C.c.flammiger* J.L. Peters, 1913[4] SE Mexico, Guatemala, Belize
— *C.c.sinaloensis* Nelson, 1899 W Mexico
— *C.c.saturatus* Ridgway, 1885 Cozumel I.
— *C.c.superbus* Ridgway, 1885 SW USA, NW Mexico
— *C.c.townsendi* (van Rossem, 1932) Tiburon I., NW Mexico
— *C.c.affinis* Nelson, 1899 WC Mexico
— *C.c.mariae* Nelson, 1898 Tres Marias Is
— *C.c.carneus* (Lesson, 1842) W and S Mexico
— *C.c.seftoni* (Huey, 1940) C Baja California
— *C.c.igneus* S.F. Baird, 1860 S Baja California
— *C.c.clintoni* (Banks, 1963) Cerralvo I.

○ *Cardinalis phoeniceus* Bonaparte, 1838 VERMILION CARDINAL Colombia, Venezuela

○ *Cardinalis sinuatus* PYRRHULOXIA[5]
— *C.s.sinuatus* Bonaparte, 1838 S USA, N and C Mexico
— *C.s.fulvescens* (van Rossem, 1934) S Arizona, NW Mexico
— *C.s.peninsulae* (Ridgway, 1887) Baja California

---

[1] Includes *rarissimus* see Paynter (1970: 217) (1809).
[2] We do not lump this species and the next, but follow the A.O.U. (1998: 635) (3) in treating them as a superspecies.
[3] Here presumed to include *rostratus* A.R. Phillips, 1994 (1877), but see Dickerman & Parkes (1997) (729).
[4] For correction of spelling see David & Gosselin (2002) (613).
[5] This species still appeared in the genus *Pyrrhuloxia* in last Edition, as oddly did the preceding species. Paynter (1970: 224) (1809) submerged the genus and A.O.U. (1983, 1998) (2, 3) followed.

CARYOTHRAUSTES Reichenbach, 1850 M

● *Caryothraustes poliogaster* BLACK-FACED GROSBEAK

  — *C.p.poliogaster* (Du Bus, 1847)[1]   SE Mexico, Guatemala, Honduras

  ⊥ *C.p.scapularis* (Ridgway, 1886)   Nicaragua, Costa Rica, W Panama

  — *C.p.simulans* Nelson, 1912   E Panama

○ *Caryothraustes canadensis* YELLOW-GREEN GROSBEAK

  — *C.c.canadensis* (Linnaeus, 1766)   Colombia, Venezuela, the Guianas, Brazil

  — *C.c.frontalis* (Hellmayr, 1905)   NE Brazil

  — *C.c.brasiliensis* Cabanis, 1851   EC Brazil

PARKERTHRAUSTES Remsen, 1997 (2007)[2] M

○ *Parkerthraustes humeralis* (Lawrence, 1867) YELLOW-SHOULDERED GROSBEAK[3]

                     W Amazonia

RHODOTHRAUPIS Ridgway, 1898 F

○ *Rhodothraupis celaeno* (Deppe, 1830) CRIMSON-COLLARED GROSBEAK   NE Mexico

PERIPORPHYRUS Reichenbach, 1850 M

○ *Periporphyrus erythromelas* (J.F. Gmelin, 1789) RED-AND-BLACK GROSBEAK

                     Venezuela, Guyana, French Guiana, Brazil

SALTATOR Vieillot, 1816[4] M

○ *Saltator grossus* SLATE-COLOURED GROSBEAK

  — *S.g.saturatus* (Todd, 1922)   Nicaragua to Ecuador

  — *S.g.grossus* (Linnaeus, 1766)   Venezuela, Guyana, Brazil, W Colombia, W Ecuador,
                     Peru, Bolivia

○ *Saltator fuliginosus* (Daudin, 1800) BLACK-THROATED GROSBEAK   Brazil, Paraguay, N Argentina

○ *Saltator atriceps* BLACK-HEADED SALTATOR

  — *S.a.atriceps* (Lesson, 1832)   E Mexico, Guatemala, Honduras, Costa Rica

  — *S.a.suffuscus* Wetmore, 1942   SE Vera Cruz (Mexico)

  — *S.a.flavicrissus* Griscom, 1937   Guerrero (Mexico)

  — *S.a.peeti* Brodkorb, 1940   S Mexico

  — *S.a.raptor* (Cabot, 1845)   SE Mexico

  — *S.a.lacertosus* Bangs, 1900   W Costa Rica, Panama

● *Saltator maximus* BUFF-THROATED SALTATOR

  — *S.m.gigantodes* Cabanis, 1851   E and S Mexico

  ↓ *S.m.magnoides* Lafresnaye, 1844   S Mexico to Panama

  — *S.m.intermedius* Lawrence, 1864   SW Costa Rica, NW Panama

  — *S.m.iungens* Griscom, 1929   E Panama, NW Colombia

  — *S.m.maximus* (Statius Müller, 1776)   Colombia, Venezuela, the Guianas, Ecuador, Peru,
                     Bolivia, Paraguay

○ *Saltator atripennis* BLACK-WINGED SALTATOR

  — *S.a.atripennis* P.L. Sclater, 1856   W and C Andes of Colombia and NW Ecuador

  — *S.a.caniceps* Chapman, 1914   W slope of E Andes of Colombia and W Ecuador

○ *Saltator similis* GREEN-WINGED SALTATOR

  — *S.s.similis* d'Orbigny & Lafresnaye, 1837   S Brazil, SE Bolivia, Paraguay, Uruguay, NE Argentina

  — *S.s.ochraceiventris* Berlepsch, 1912   SE Brazil

○ *Saltator coerulescens* GREYISH SALTATOR

  — *S.c.vigorsii* G.R. Gray, 1844   NW Mexico

  — *S.c.plumbiceps* Lawrence, 1867[5]   WC Mexico

  — *S.c.grandis* (Deppe, 1830)   E Mexico, Guatemala, Honduras, Nicaragua, Costa Rica

  — *S.c.yucatanensis* Berlepsch, 1912   SE Mexico

  — *S.c.hesperis* Griscom, 1930   Guatemala, El Salvador, Honduras, Nicaragua

  — *S.c.brevicaudus* van Rossem, 1931   W Costa Rica

---

[1] Du Bus is an abbreviation for Du Bus de Gisignies.
[2] This species was previously treated in the genus *Caryothraustes*.
[3] Relationships were discussed by Demastes & Remsen (1994) (684).
[4] For placement of *Pitylus* in *Saltator* see Hellack & Schnell (1977) (1101), Tamplin *et al.* (1993) (2385), Demastes & Remsen (1994) (684) and A.O.U. (1998: 631) (3).
[5] Includes *richardsoni*; the type locality of *plumbiceps* has now been elucidated and the name is not a synonym of *vigorsii* see Ludwig (1998) (1403).

    — *S.c.plumbeus* Bonaparte, 1853   N Colombia
    — *S.c.brewsteri* Bangs & Penard, 1918   NE Colombia, Venezuela, Trinidad
    — *S.c.olivascens* Cabanis, 1849   Venezuela, the Guianas, N Brazil
    — *S.c.azarae* d'Orbigny, 1839   E Colombia, Ecuador, E Peru, Bolivia, Brazil
    — *S.c.mutus* P.L. Sclater, 1856   N Brazil
    — *S.c.superciliaris* (Spix, 1825)   NE Brazil
    — *S.c.coerulescens* Vieillot, 1817   E Bolivia, SW Brazil, Paraguay, N Argentina

○ **Saltator orenocensis** ORINOCAN SALTATOR
    — *S.o.rufescens* Todd, 1912   NE Colombia, NW Venezuela
    — *S.o.orenocensis* Lafresnaye, 1846   NE Venezuela

○ **Saltator maxillosus** Cabanis, 1851 THICK-BILLED SALTATOR   SE Brazil, Paraguay, N Argentina

○ **Saltator nigriceps** (Chapman, 1914) BLACK-COWLED SALTATOR[1]   SW Ecuador, NW Peru

○ **Saltator aurantiirostris** GOLDEN-BILLED SALTATOR
    — *S.a.iteratus* Chapman, 1927   S Ecuador, NW Peru
    — *S.a.albociliaris* (Philippi & Landbeck, 1861)   N Peru south to N Chile
    — *S.a.hellmayri* Bond & Meyer de Schauensee, 1939   Bolivia
    — *S.a.aurantiirostris* Vieillot, 1817   NE Bolivia, NW Argentina, Paraguay
    — *S.a.parkesi* Cardoso da Silva, 1990   SE Brazil, Uruguay, NE Argentina (293)
    — *S.a.nasica* Wetmore & Peters, 1922   W Argentina

○ **Saltator cinctus** J.T. Zimmer, 1943 MASKED SALTATOR   E Ecuador, N Peru

○ **Saltator atricollis** Vieillot, 1817 BLACK-THROATED SALTATOR   SE Bolivia, Paraguay, S Brazil

○ **Saltator rufiventris** d'Orbigny & Lafresnaye, 1837 RUFOUS-BELLIED SALTATOR
         C and S Bolivia, NW Argentina

○ **Saltator albicollis** LESSER ANTILLEAN SALTATOR
    — *S.a.albicollis* Vieillot, 1817   Martinique, St Lucia Is.
    — *S.a.guadelupensis* Lafresnaye, 1844   Guadeloupe, Dominica Is.

○ **Saltator striatipectus** STREAKED SALTATOR[2]
    — *S.s.furax* Bangs & Penard, 1919   SW Costa Rica, Panama
    — *S.s.isthmicus* P.L. Sclater, 1861   W Panama
    — *S.s.scotinus* Wetmore, 1957   Islands off Veraguas (Panama)
    — *S.s.melicus* Wetmore, 1952   Taboga I. (Bay of Panama)
    — *S.s.speratus* Bangs & Penard, 1919   Pearl Is. (Gulf of Panama)
    — *S.s.striatipectus* Lafresnaye, 1847   E Panama, W Colombia
    — *S.s.perstriatus* Parkes, 1959   NE Colombia, Venezuela, Trinidad
    — *S.s.flavidicollis* P.L. Sclater, 1860   SW Colombia, Ecuador, NW Peru
    — *S.s.immaculatus* Berlepsch & Stolzmann, 1892   W Peru
    — *S.s.peruvianus* Cory, 1916   N Peru

CYANOLOXIA Bonaparte, 1850[3] F
○ **Cyanoloxia glaucocaerulea** (d'Orbigny & Lafresnaye, 1837) INDIGO GROSBEAK
         SE Brazil, E Paraguay, Uruguay, NE Argentina

CYANOCOMPSA Cabanis, 1861 F
○ **Cyanocompsa cyanoides** BLUE-BLACK GROSBEAK
    — *C.c.concreta* (Du Bus, 1855)[4]   SE Mexico, Guatemala, Honduras
    — *C.c.caerulescens* Todd, 1923[5]   Nicaragua, Costa Rica, W Panama
    — *C.c.cyanoides* (Lafresnaye, 1847)   E Panama, Colombia, W Venezuela, Ecuador
    — *C.c.rothschildii* (E. Bartlett, 1890)   Western Amazonia

○ **Cyanocompsa brissonii** ULTRAMARINE GROSBEAK
    — *C.b.caucae* Chapman, 1912   W Colombia
    — *C.b.minor* Cabanis, 1861   N Venezuela
    — *C.b.brissonii* (M.H.K. Lichtenstein, 1823)   NE Brazil

[1] Tentative separation of *nigriceps* from *S. aurantiirostris* follows Ridgely & Tudor (1989: 388) (2018).
[2] See A.O.U. (1998: 631) (3) for the split of *striatipectus* from *S. albicollis*.
[3] Dated '1851?' by Paynter (1970: 238) (1809), but see Zimmer (1926) (2707).
[4] Du Bus is an abbreviation for Du Bus de Gisignies.
[5] Paynter (1970: 239) (1809) submerged this genus in *Passerina* and, noting that this name was then preoccupied, provided the substitute name *toddi*.

__ *C.b.sterea* Oberholser, 1901	E and S Brazil, NE Argentina, W Paraguay
__ *C.b.argentina* (Sharpe, 1888)	W Brazil, E Bolivia, E Paraguay, N Argentina

○ *Cyanocompsa parellina* BLUE BUNTING

__ *C.p.beneplacita* Bangs, 1915	NE Mexico
__ *C.p.indigotica* (Ridgway, 1887)	W and SW Mexico
__ *C.p.lucida* Sutton & Burleigh, 1939	NE Mexico
__ *C.p.parellina* (Bonaparte, 1850)[1]	E and S Mexico to Nicaragua

PASSERINA Vieillot, 1816 F

● *Passerina caerulea* BLUE GROSBEAK[2]

☑ *P.c.caerulea* (Linnaeus, 1758)[3]	SE USA >> E Mexico and Central America
__ *P.c.interfusa* (Dwight & Griscom, 1927)	SW USA >> W Mexico, Guatemala, Honduras
__ *P.c.salicaria* (Grinnell, 1911)	SW USA >> W Mexico, Baja California
__ *P.c.eurhyncha* (Coues, 1874)	C and S Mexico
__ *P.c.chiapensis* (Nelson, 1898)	S Mexico
__ *P.c.deltarhyncha* (van Rossem, 1938)	W coast of Mexico
__ *P.c.lazula* (Lesson, 1842)	Honduras, Nicaragua, Costa Rica

● *Passerina cyanea* (Linnaeus, 1766) INDIGO BUNTING — S Canada, E USA >> Central America, Cuba, Jamaica, Colombia, Venezuela

● *Passerina amoena* (Say, 1823) LAZULI BUNTING — W USA >> W Mexico, Baja California

○ *Passerina versicolor* VARIED BUNTING

__ *P.v.versicolor* (Bonaparte, 1838)	S USA, C and S Mexico
__ *P.v.dickeyae* van Rossem, 1934	S Arizona, W Mexico
__ *P.v.pulchra* Ridgway, 1887	S Baja California, NW Mexico
__ *P.v.purpurascens* Griscom, 1930	S Mexico, Guatemala

● *Passerina ciris* PAINTED BUNTING

__ *P.c.ciris* (Linnaeus, 1758)	SE USA >> SE Mexico, Bahamas
☑ *P.c.pallidior* Mearns, 1911	S USA >> Mexico, Central America

○ *Passerina rositae* (Lawrence, 1874) ROSE-BELLIED BUNTING — SE Oaxaca, W Chiapas (S Mexico)

○ *Passerina leclancherii* ORANGE-BREASTED BUNTING

__ *P.l.grandior* Griscom, 1934	Colima to W Chiapas (S Mexico)
__ *P.l.leclancherii* Lafresnaye, 1840	Coastal Guerrero (SW Mexico)

---

[1] Dated '1851?' by Paynter (1970: 241) (1809), but see Zimmer (1926) (2707).
[2] For reasons to submerge the genus *Guiraca* see Klicka *et al.* (2001) (1276).
[3] Includes *mesophila* Oberholser, 1974 (1673), see Browning (1978) (246).

APPENDIX I
Matters of Nomenclature

Spelling issues

Whether taxon names are spelled correctly or not depends first on the original spelling and use. If the name is still used in combination with the same generic name then the spelling we use today should be unchanged. This is true, unless the original author used a binomen (i.e. specific epithet) that is by nature variable (i.e. varies in accord with the gender of the genus) and got the name wrong in the original context. Apart from that, the original spelling may not be in use today due to emendations. These may have been deliberate or may have been mere mistakes, in works which have been influential.

Most of the rules relating to spelling and gender are discussed in two papers by David & Gosselin (2002) (2803, 2804). They seem to be simple. For example, if the generic name terminates in –us, and thus seems to be masculine, then the specific epithet should change to agree. In practice it is much less simple, as nouns are generally invariable. So making a change requires a knowledge of whether the binomen is adjectival in character or not. This is just one example of the complexities, but it is perhaps the most fundamental.

Nonetheless the starting point is knowing the gender of the generic name that is now in use, or is the one you want to use. This is why we have included M, F and N designations after each generic name throughout the list.

Emendations

The International Code for Zoological Nomenclature (I.C.Z.N., 1999) (1178) seeks to distinguish between emendations that are justified and those that are not. Only in defined circumstances is an emendation justified, in principle names once given should be maintained. However, the continued repetition of unjustified emendations eventually creates an environment in which their uniform acceptance is considered beneficial to stability.

In considering any example it is first essential to determine whether the changed name is just an emended name or whether it is a different name. If it is a different name then the newer (younger) name is a synonym. It must then be considered under the rules that govern synonyms, where priority is of greater importance than it is in the case of emendations.

The key to survival of an emendation is not whether it was initially justified or not. Instead it is whether it is in 'prevailing usage'. To be in prevailing usage it must be adopted by "at least a substantial majority of the most recent authors concerned with the relevant taxon, irrespective of how long ago their work was published" (I.C.Z.N., 1999: Glossary: 121) (1178). It must be remembered that Code applies to the whole field of zoology; to narrow fields and to broad. In the field of ornithology, bursting with literature ranging from broad works of popular interest to narrowly focussed specialist reviews, some in obscure journals, a detailed assessment would demand many hours careful searching of the literature. It is hard to see who now would consider the time invested worthwhile. Nonetheless a number of troublesome names are contested due to emendations. In the table below we examine 17 names that we reviewed when deciding what to use.

We considered that the two most influential works, upon which others would have drawn, since the mid 19th century were the *Catalogue of the birds in the British Museum* and 'Peters' Checklist. We made the presumption that following each work we could assume there were long periods in which they would have been broadly followed, even if, in exceptional cases, spellings were soon disputed. To assess modern usage we then included, where we could, spellings employed in the published volumes of the *Handbook of the Birds of the World*. Taken together with the original spelling, which we usually verified, this gave us three or four points of reference as a basis for consideration.

Because writers accustomed to certain names are likely to continue to use them, we presumed that for a proposed spelling to reach a level of use that might be assumed to be 'prevailing' would take about 20 years. We chose this number for reasons of pragmatism; we felt that the reason that the Code favours stability must be for the benefit of the living not the dead. Relatively recent usage is thus considered to be more significant. The Code does *not* say this. It is nevertheless open to interpretation. Here we simply offer a view on how to apply a pragmatic, but consistent, judgement that does not require an exhaustive search.

Table I. EMENDATIONS: Recommended usage is given in bold type.

Original name	Emended name	Cat. Bds. Brit. Mus. Name O or E?	Peters Checklist Name O or E?	HBW Name O or E?	COMMENTS [In our column 3 "V" implies that we have verified the original spelling]
1 *Diomedea melanophris* Temminck, 1828	V ***melanophrys*** Temminck, 1839	E.	1931 O; 1979 Ed. E.	O.	HBW usage represents a correction in 1992 that runs counter to prevailing usage.
2 *Anser Gambeli* Hartlaub, 1852	V ***gambeli*** Hartlaub, 1852	E.	1931 O; 1979 Ed. E.	E.	HBW usage consistent with prevailing usage.
3 *Sparvius cirrhocephalus* Vieillot, 1817	V ***cirrhocephalus*** Vieillot, 1817	E.	1931 E; 1979 Ed. E.	O.	HBW usage represents a correction in 1994 that runs counter to prevailing usage.
4 *Columba reinwardtsi* Temminck, 1824	V ***Reinwardtii*** Temminck, 1839	Neither; one - i only	O.	E.	See Mees (1964) [1514A]. Prevailing usage established since then.
5 *Columba Maugeus* Temminck, 1811	V ***Maugei*** Temminck, 1811 (in the Index)	E.	O.	E.	See Mees (1975) [1527]. Prevailing usage reestablished since then.
6 *Columba naina* Temminck, 1835	V ***nana*** Temminck, 1839	*nanus* (in Ptilinopus)	O.	E.	HBW usage represents a correction in 1997 that seems to run counter to prevailing usage since 1937. We thus use *nainus.*
7 *Palaeornis Calthrapae* Blyth, 1849	V ***calthropae***	E.	Neither, ***calthorpae***	E.	The *lapsus* by Blyth was emended in or before Salvadori (1891) [2161B]; this taxon was named for a Miss. Calthrop and not as Peters (1937) [1823] wrote a Miss Calthorp. Use of this accurate emendation, from 1891 to 1937, has been succeeded by over 50 years of use of *calthorpae.* Sibley & Monroe (1990) [2262] resurrected *calthropae,* but we judge that *calthorpae* must benefit from the rules on prevailing usage.
8 *Carpococcyx radiceus* Temminck, 1832	V ***radiatus*** Temminck, 1832	E.	O.	E.	Peters (1940) [1825] wrote that *radiatus* was "often used", but the 1839 name is a synonym.
9 *Ornismya mulsant* Bourcier, 1842 *fide* Wolters (1976) [2684A]	***mulsanti*** (source of emendation, apparently the original author)	E.	E.	O.	Wolters (1976) [2684A] reintroduced the spelling *mulsant,* Schuchmann (1999) [2203] explained that this spelling was used about one month earlier than the use of *mulsanti* by the same author. Nonetheless *mulsanti* was in prevailing use from 1892 until at least 1979 and must be followed.
10 *Hypocnemis poecilinota* Cabanis, 1847	V ***poecilonota***	E.	E.	n/a	Original spelling used by Pinto (1978) [1895]. Otherwise the emendation has been in prevailing use.
11 *Pitta arquata* Gould, 1871	V ***arcuata.*** Emended by Salvadori (1874) [2161A]	E.	E.	n/a	See Cranbrook (1982) [596A] and Dickinson & Dekker (2000) [741]. We accept that *arquata* has regained prevalence.
12 *Phlexis lopezi* Alexander, 1903	V ***lopesi*** Alexander, 1903.	n/a	E.	n/a	Emendation by the author in Ibis (6 months later). The emendation is in unchallenged use.
13 *Orthotomus euculatus* Temminck, 1836	V ***cuculatus*** Temminck, 1839, but to ***cucullatus*** by Bonaparte, 1850	E.	E.	n/a	Bonaparte's emendation is in prevailing use.
14 *Apalis lopezi* Alexander, 1903	V ***lopesi*** Alexander, 1903.	n/a	E.	n/a	Emendation by the author in Ibis (6 months later). The emendation is in unchallenged use.
15 *Anthreptes collaris patersoni* Irwin, 1960	V ***patersonae*** see Rand, (1967: 220) [1971A]	n/a	E.	n/a	The emendation is now in prevailing use.
16 *Nectarinia bocagii* Shelley, 1879	V ***bocagei***	E.	O.	n/a	The restoration of the original name by Rand (1967) [1971A] is now in prevailing usage.
17 *Myzomela erythrina* E.P. Ramsay, 1878	V ***erythrina***	E.	E.	n/a	The emendation is in unchallenged use.

We have included one name (no.8) in this table that cannot be decided by the rules applying to emendations. This is because the case involves not an emendation but two different names.

Names and their priority

The first scientific name given to a bird is the name it should have, or at least the basis for that name. Younger names are said to be junior synonyms, but only when two names are both dated can one be shown to be junior. Some early names passed almost unnoticed. They then sank into obscurity and were shown, years later, to really exist and to be earlier i.e. senior names.

During the 19th and much of the 20th century the discovery of such names was not infrequent, and with each such discovery came the need for change, the acceptance of the older name. And while newer and older names overlapped in usage the resultant confusion harmed common understanding. The consequence was that the Principle of Priority, upon which the Code rested, was seen to work against stable and easy communication to such a point that some re-balancing was needed.

The current Code, the 4th Edition (I.C.Z.N., 1999) (1178), places higher value on stability than its predecessors. The replacement of a name that is in current use by a prior name is no longer automatic. Indeed, if the prior name has been totally out of use since 1900, it is almost certain that replacement will not be allowed. On the other hand if it has been used since 1899 then the reassertion of its priority can hardly be opposed (Art. 23.9).

By way of illustration two cases treated in preparing this edition will assist us. We can see in Table 1, that the name *radiatus* has to be considered a junior synonym of *radiceus* (and not an emendation). As *radiceus* was still used after 1899 it is protected against rejection. As a mere emendation the name was open to displacement based solely on prevailing usage. Second, within the list the name *Polyplectron napoleonis* has been used in place of *Polyplectron emphanum*. The details are given by Dickinson (2001) (738), but, in a nutshell, had the name *napoleonis* not still been in use within the 20th century it would have been too late to resurrect it, despite proof of its priority.

Other nomenclatural legalities

In certain circumstances names can be set aside. Usually such suppressions are restricted in application. Junior names are admitted in their place and so the older names are suppressed for the Principle of Priority. On the other hand the 'suppressed' name most often remains in force in so far as the Principle of Homonymy concerns it.

In its simplest usage a homonym is either of two names having identical spellings.

So the continued validity of a name suppressed, in respect of the Principle of Priority, and not in relation to the Principle of Homonymy, implies that its existence in synonymy means that the proposal to use a name spelled the same way — whether a new proposal or one made some years ago — is blocked. This, of course, also serves stability. Should it not remain the obstacle that it was before there could be a domino effect, and other names be required to move into synonymy. So track needs to be kept of these names and the I.C.Z.N. maintains lists. For a period after any such suppression we need to remember these 'suppressed' names; a literature search that seeks a complete bibliography of a species could lose much of the early history of the species simply because the search did not seek the suppressed name. It follows that lists like Peters Checklist, and this one, try to help by using footnotes to flag these situations. There are therefore a handful of such footnotes in this Edition.

APPENDIX II
Notes on Bibliographic Details

THE LIST OF REFERENCES

In so far as we have been able every reference is cited with full pagination. Those references that relate to taxa that have been described since the different volumes of Peters Check-list have all been verified. We have not found it possible to attach to such references a list of the new taxa included and the page numbers concerned. We expect to make this information available in conjunction with the Museum nationale d'Histoire naturelle, Paris.

AUTHOR AND DATE ENTRIES

Space within the checklist itself precludes listing all authors when more than five have authored a name.

We wish this book to be as accurate as possible. We have corrected many errors found in the pages of Peters Checklist. As this book is revised in future Editions we plan to add to such corrections. We shall be working in collaboration with Alan Peterson (web site www.zoonomen.net). All amendments of this kind in this Checklist are footnoted as to source, or to this Appendix. The following have received particular attention in this Edition.

Names

● **Rendering of names** P.L. Statius Müller has in the past been given as P.L.S. Müller; however it is now clear that his family name was Statius Muller (Kooiman, 1950) (1297A). He used the umlaut at least when publishing in Germany. Du Bus de Gisignies is here always abbreviated to Du Bus (but with the full name footnoted). In Peters the abbreviated name was used without comment by some authors. We have consistently used Conrad von Baldenstein as all this was his family name (Meyer de Schauensee is a similar case); in Peters it appeared in different formats.

● **Attribution of scientific names to authors** There were varied attributions, in the different volumes of Peters Checklist, attached to the names given in Hemprich & Ehrenberg's *Symbolae Physiciae*. We have examined this work and find that Ehrenberg, the surviving author at the time of publication, was precise; in citing each new name he added the initials "H. et E." except in the three cases where he named birds for his deceased fellow traveller. Except for these three names we attribute every name in this work to both authors, in accord with Art. 50.1. of the "Code" (I.C.Z.N., 1999) (1178). The Code, however, also stipulates when it is incorrect (Art. 50.1.1.) to attribute a name within a work to the authors of the work as a whole. Some such cases, where we have corrected the attribution, are footnoted.

● Many early names were based on manuscript (MS.) names. It seems clear that authors who deliberately reused these generally intended the original authors to derive credit from the name. Only subsequently has the scientific community devised rules that attribute the credit to the publisher of the name *and* the related description. We believe it is erroneous to suppose that authors, like Blyth, Temminck and many others, all deliberately sought to take credit themselves for the names of others. The recognition of a MS name by the author of a published name was elegantly treated in Peters Checklist. Here we lack the space to do this, and such information, while historically interesting, does not contribute to the formally recognised name.

Initials

Initials are given in every case where we have found that two or more authors with the same surname have named birds listed here[1]. When they wrote together in pairs we give them as, for example, I.C.J. & E.H. Galbraith. When both have the same initials we use Sr. and Jr. as for W.H. Phelps, father and son, (in the case of the Kurodas, Nagamichi is given as N. Kuroda, Sr. and Nagahisa as N. Kuroda, Jr. Any of these authors writing alone is given his or her initials (this was not done consistently in Peters Checklist). When these individuals were or are co-authors with others with differing surnames the initials are dropped, as little or no confusion is likely.

---

[1] We shall be pleased to hear of further such cases that we may have missed.

Dates

Many dates given in Peters Checklist need revision, but we have so far made few changes and only limited use of www.zoonomen.net. This is because all changes of this kind need to be discussed.

We have made some, but not all, of the changes that relate to the *Proceedings of the Zoological Society of London*, see Duncan (1937) (770), the rest will be changed after a focussed search. A number of changes have been made based on Browning & Monroe (1991) (258). One of the works about which these authors wrote was the 1801 or 1802 Supplement to Latham's *Index ornithologicus*. As the 1802 date has been accepted by the A.O.U. (1998) (3) and by Schodde & Mason (1999) (2189) we have standardised on it here. We are aware that an argument can be made for retaining 1801, but this has not yet been published as a rebuttal of Browning & Monroe (1991). In the case of Bonaparte's *Conspectus Generum Avium*, volume 1, we have used 1850 for all names from pages 1-543 based on the information contained in the report by Hartlaub (1851) (1087). Note however that later authors, e.g. Zimmer (1926) (2707) have not found proof of this view.

The dates of all names derived from Temminck & Laugier's *Planches Coloriées* have been brought into line with Dickinson (2001) (739); the authorship has been corrected to Temminck alone (fully stated this would be Temminck *in* Temminck & Laugier). This is an example of Art. 50.1.1. of the Code at work. In this instance Cuvier's introduction to this work makes clear that all responsibility for nomenclature lay with Temminck.

List of References[1][2]

1   A.O.U., 1957. *Check-list of North American Birds*: i-xiii, 1-691. — American Ornithologists' Union, Baltimore, Md.

2   A.O.U., 1983. *Check-list of North American Birds*: i-xxix, 1-877. — American Ornithologists' Union, Washington D.C.

3   A.O.U., 1998. *Check-list of North American Birds*: i-liv, 1-829. — American Ornithologists' Union, Washington D.C.

4   A.O.U., 2000. 42nd. Supplement to the American Ornithologists' Union Check-list of North American Birds. — *Auk*, 117 (3): 847-858.

5   Abdulali, H., 1959. A new white-throated race of the babbler *Dumetia hyperythra*. — *J. Bombay Nat. Hist. Soc.*, 56 (2): 333-335 (1958).

6   Abdulali, H., 1964. Four new races of birds from the Andaman and Nicobar islands. — *J. Bombay Nat. Hist. Soc.*, 61: 410-417.

7   Abdulali, H., 1965. Notes on Indian birds. 3 - the Alpine swift, *Apus melba* (Linnaeus), with a description of one new race. — *J. Bombay Nat. Hist. Soc.*, 62 (1): 153-160.

8   Abdulali, H., 1967. More new races of birds from the Andaman and Nicobar Islands. — *J. Bombay Nat. Hist. Soc.*, 63 (2): 420-422 (1966).

9   Abdulali, H., 1967. A new race of *Aplonis panayensis* (Scopoli) based on eye colour. — *Bull. Brit. Orn. Cl.*, 87: 33-34.

10   Abdulali, H., 1975. A Catalogue of the Birds in the Collection of the Bombay Nat. Hist. Soc. - 17. — *J. Bombay Nat. Hist. Soc.*, 72 (1): 113-131.

11   Abdulali, H., 1976. On a new subspecies of the skylark *Alauda gulgula* Franklin in Gujarat. — *J. Bombay Nat. Hist. Soc.*, 72: 448-449 (1975).

12   Abdulali, H., 1977. A new name for the Andaman Blackheaded Oriole. — *J. Bombay Nat. Hist. Soc.*, 73: 395 (1976).

13   Abdulali, H., 1979. The birds of Great and Car Nicobars with some notes on wildlife conservation in the islands. — *J. Bombay Nat. Hist. Soc.*, 75 (3): 744-772 (1978).

14   Abdulali, H., 1982. A Catalogue of the Birds in the collection of the Bombay Natural History Society — 25. — *J. Bombay Nat. Hist. Soc.*, 79 (2): 336-360.

15   Abdulali, H., 1982. On a new subspecies of *Pellorneum ruficeps* (Swainson) in peninsular. — *J. Bombay Nat. Hist. Soc.*, 79 (1): 152-154.

16   Abdulali, H., 1983. A Catalogue of the Birds in the collection of the Bombay Natural History Society - 28. — *J. Bombay Nat. Hist. Soc.*, 80 (2): 349-369.

17   Abdulali, H. & R. Reuben, 1964. The Jungle Bush-Quail *Perdicula asiatica* (Latham): a new record from south India. — *J. Bombay Nat. Hist. Soc.*, 61 (3): 698-691.

18   Ahlquist, J.E. *et al.*, 1984. The relationships of the Bornean Bristlehead (*Pityriasis gymnocephala*) and the Black-collared Thrush (*Chlamydochaera jefferyi*). — *J. f. Orn.*, 125: 129-140.

19   Ainley, D.G., 1980. Geographic variation in Leach's Storm-Petrel. — *Auk*, 97 (4): 837-853.

20   Ainley, D.G., 1983. Further notes on variation in Leach's Storm-Petrel. — *Auk*, 100: 200-233.

21   Alcasid, G.L. & P. Gonzales, 1968. A new race of the Naked-faced Spider-hunter (*Arachnothera clarae*) from Luzon. — *Bull. Brit. Orn. Cl.*, 88: 129-130.

22   Aldrich, J.W., 1942. New Bobwhite from northeastern Mexico. — *Proc. Biol. Soc. Wash.*, 55: 67-70.

23   Aldrich, J.W., 1946. New subspecies of birds from western North America. — *Proc. Biol. Soc. Wash.*, 59: 129-136.

24   Aldrich, J.W., 1972. A new subspecies of Sandhill Crane from Mississippi. — *Proc. Biol. Soc. Wash.*, 85: 63-70.

25   Aldrich, J.W. & B.P. Bole, Jr., 1937. Birds and mammals of the western slope of the Azuero Peninsula (Republic of Panama). — *Sci. Publ. Cleveland Mus. Nat. Hist.*, 7: 1-196.

26   Aldrich, J.W. & H. Friedmann, 1943. A revision of the Ruffed Grouse. — *Condor*, 45 (3): 85-103.

27   Ali, S., 1943. The birds of Mysore. — *J. Bombay Nat. Hist. Soc.*, 44 (2): 206-220.

28   Ali, S. & S.D. Ripley, 1983. *Handbook of the Birds of India and Pakistan*. 4: i-xvi, 1-267. 2nd. edition. — Oxford University Press, Bombay.

29   Allen, G.A., Jr. *et al.*, 1977. New species of curassow discovered. — *Game bird breeders, aviculturists, zoologists and conservationists' gazette*, 26 (6): 6.

30   Allison, A., 1946. Note d'Ornithologie No. 2. — *Notes d'Orn., Université L'Aurore, Shanghai*,, 1 (2): 1-7.

31   Alström, P. & U. Olsson, 1990. Taxonomy of the *Phylloscopus proregulus* group. — *Bull. Brit. Orn. Cl.*, 110 (1): 38-47.

32   Alström, P. & U. Olsson, 1992. On the taxonomic status of *Phylloscopus affinis* and *Phylloscopus subaffinis*. — *Bull. Brit. Orn. Cl.*, 112 (2): 111-125.

33   Alström, P. & U. Olsson, 1995. A new species of *Phylloscopus* warbler from Sichuan Province, China. — *Ibis*, 137 (4): 459-468.

34   Alström, P. & U. Olsson, 1999. The Golden-spectacled Warbler: a complex of sibling species, including a previously undescribed species. — *Ibis*, 141 (4): 545-568.

35   Alström, P. *et al.*, 1991. The taxonomic status of *Acrocephalus agricola tangorum*. — *Forktail*, 6: 3-13.

36   Alström, P. *et al.*, 1992. A new species of *Phylloscopus* warbler from central China. — *Ibis*, 134 (4): 329-334.

37   Altshuler, D.L., 1999. Species accounts within the Family Trochilidae (Hummingbirds). Pages 549-680. In *Handbook of the Birds of the World* (J. del Hoyo, A. Elliott & J. Sargatal, eds.). — Lynx Edicions, Barcelona.

38   Alvarez Alonso, J. & B.M. Whitney, 2001. A new *Zimmerius* tyrannulet (Aves: Tyrannidae) from white sand forests of northern Amazonian Peru. — *Wilson Bull.*, 113: 1-9.

39   Alviola, P.L., 1997. A new species of frogmouth (Podargidae - Caprimulgiformes) from Busuanga Island, Palawan, Philippines. — *Asia Life Sciences*, 6 (1/2): 51-55.

40   Amadon, D., 1942. Birds collected during the Whitney South Sea Expedition. 50. Notes on some non-passerine genera. 2. — *Amer. Mus. Novit.*, 1176: 1-21.

41   Amadon, D., 1943. Birds collected during the Whitney South Sea Expedition. 52. Notes on some non-passerine genera. 3. — *Amer. Mus. Novit.*, 1237: 1-22.

42   Amadon, D., 1953. Avian systematics and evolution in the Gulf of Guinea. — *Bull. Amer. Mus. Nat. Hist.*, 100 (3): 393-452.

43   Amadon, D., 1956. Remarks on the starlings, family Sturnidae. — *Amer. Mus. Novit.*, 1803: 1-41.

44   Amadon, D., 1959. Remarks on the subspecies of the Grass Owl *Tyto capensis*. — *J. Bombay Nat. Hist. Soc.*, 56: 344-346.

45   Amadon, D., 1962. Family Sturnidae. Pages 75-121. In *Check-list of Birds of the World*, Vol. 15 (E. Mayr & J.C. Greenway, Jr., eds.). — Mus. Comp. Zool., Cambridge, Mass.

46   Amadon, D., 1974. Taxonomic notes on the Serpent-eagles of the genus *Spilornis*. — *Bull. Brit. Orn. Cl.*, 94: 159-163.

47   Amadon, D. & J. Bull, 1988. Hawks and owls of the world; a distributional and taxonomic list; with the genus *Otus* by Joe Marshall and Ben King. — *Proc. Western Found. Vert. Zool.*, 3 (4): 295-357.

48   Amadon, D. & S.G. Jewett, Jr., 1946. Notes on Philippine Birds. — *Auk*, 63: 542-559.

49   Amadon, D. & G. Woolfenden, 1952. Notes on the Mathews collection of Australian birds. The order Ciconiiformes. — *Amer. Mus. Novit.*, 1564: 1-16.

50   Andrew, P., 1992. *The birds of Indonesia*: i-xii, 1-83. — Indonesian Orn. Soc., Jakarta.

51   Archibald, G.W. & C.D. Meine, 1996. Family Gruidae (Cranes). Pages 60-89. In *Handbook of the Birds of the World*. Vol. 3. (J. del Hoyo, A. Elliott & J. Sargatal, eds.). — Lynx Edicions, Barcelona.

---

[1] The chosen bibliographic software found the Check-list too big to handle. A partial manual solution was used which led to a few errors of duplication or omission. The numbering is thus not quite consecutive and some references have been inserted with a number using an "A" suffix.

[2] Where there are more than two authors we name the first and use '*et al.*'.

52   Arnaiz-Villena, A. *et al.*, 1998. Phylogeny and rapid Northern and Southern Hemisphere speciation of goldfinches during Miocene and Pleistocene Epochs. — *Cell. Mol. Life Sci.*, 54: 1031-1041.

53   Arnaiz-Villena, A. *et al.*, 1999. Rapid radiation of canaries (genus *Serinus*). — *Mol. Biol. Evol.*, 16: 2-11.

54   Arndt, K., 1996. *Lexicon of Parrots.* Vol. 7. — Arndt-Verlag, Bretten.

55   Arndt, K., 1996. *Lexikon der Papageien.* Vol. 3. — Arndt-Verlag, Bretten.

56   Arnold, K.A., 1983. A new subspecies of Henslow's Sparrow (*Ammodramus henslowi*). — *Auk*, 100: 504-505.

57   Ash, J.S., 1979. A new species of serin from Ethiopia. — *Ibis*, 121 (1): 1-7.

58   Ash, J.S., 1981. A new race of Scaly Babbler *Turdoides squamulatus* from Somalia. — *Bull. Brit. Orn. Cl.*, 101: 399-403.

59   Ash, J.S., 1986. A *Ploceus* sp. nov. from Uganda. — *Ibis*, 128: 330-336.

59A   Ash, J.S., 1987. *Ploceus victoriae.* — *Ibis*, 129: 406-407.

60   Ash, J.S. & T. Gullick, 1990. *Serinus flavigula* rediscovered. — *Bull. Brit. Orn. Cl.*, 110 (2): 81-83.

61   Ash, J.S. *et al.*, 1989. The mangrove reed warblers of the Red Sea and Gulf of Aden coasts, with description of a new subspecies of the African Reed Warbler *Acrocephalus baeticatus*. — *Bull. Brit. Orn. Cl.*, 109: 36-43.

62   Atkinson, P.W. *et al.*, 1983. Notes on the ecology, conservation and taxonomic status of *Hylorchilus* wrens. — *Bird Conserv. Int.*, 3: 75-85.

63   Atwood, J.L., 1991. Subspecies limits and geographic patterns of morphological variation in California Gnatcatchers (*Polioptila californica*). — *Bull. Southern California Acad. Sci.*, 90 (3): 118-133.

63A   Aveledo, H.R., 1998. Nueva subspecie de la familia Picidae del Estado Lara. - *Bol. Soc. Venez. Cienc. Nat.*, 46: 3-7.

64   Aveledo, R. & Ginés, Rev. Hmno., 1950. Descripción de cuatro aves nuevas de Venezuela. — *Mem. Soc. Cienc. Nat. La Salle*, 10 (26): 59-71.

65   Aveledo, R. & Ginés, Rev. Hmno., 1952. Cuatro aves nuevas y dos extensiones de distribución para Venezuela, de Perijá. — *Nov. Cien. Mus. Hist. Nat. La Salle, Zool.*, 6: 1-15.

66   Aveledo, R. & Ginés, Rev. Hmno., 1953. Ave nueva para la Ciencia y comentarios sobre *Tiaris fuliginosa zuliae*. — *Nov. Cien. Mus. Hist. Nat., Caracas*, 10: 1-8.

67   Aveledo, R. & C.L. Pérez, 1989. Tres nuevas subspecies de aves (Picidae, Parulidae, Thraupidae) de la Sierra de Perijá, Venezuela y lista hipotetica para la avifauna Colombiana de Perijá. — *Bol. Soc. Venez. Cienc. Nat.*, 43: 7-28.

68   Aveledo, R. & L.A. Pérez, 1991. Dos nuevas subspecies de aves (Trochilidae y Formicariidae) de la region oriental y occidental de Venezuela. — *Bol. Soc. Venez. Cienc. Nat.*, 44: 15-25.

69   Aveledo, R. & L.A. Pérez, 1994. Descripción de nueve subspecies nuevas y comentarios sobre dos especies de aves de Venezuela. — *Bol. Soc. Venez. Cienc. Nat.*, 44: 229-257.

70   Aveledo, H.R. & A.R. Pons, 1952. Aves nuevas y extensiones de distribucion a Venezuela. — *Nov. Cien. Mus. Hist. Nat. La Salle, Zool.*, 7: 3-21.

71   Avise, J.C. *et al.*, 1994. DNA sequence support for a close phylogenetic relationship between some storks and New World vultures. — *Proc. Acad. Nat. Sci. Philad.*, 91: 5173-5177.

72   B.O.U., 1997. Records Committee: Twenty-third report. — *Ibis*, 139 (1): 197-201.

73   B.O.U., 2000. *The British List 2000*: 1-34. — B.O.U., Tring.

74   B.O.U., 2002. Records Committee: Twenty-eighth report. — *Ibis*, 144 (1): 181-1844.

75   Bailey, H.H., 1935. A revision of the genus *Coturnicops* Gray. — *Bull. Bailey Mus. Miami*, 10: 1-2.

76   Bailey, H.H., 1941. An undescribed race of Eastern Ruffled [sic] Grouse. — *Bull. Bailey Mus. Miami*, 14: 1.

77   Baillon, F., 1992. *Streptopelia cf. hypopyrrha*, nouvelle espèce de tourterelle pour le Sénégal. — *L'Oiseau*, 62: 320-334.

78   Baker, A.J. *et al.*, 1995. Flightless Brown Kiwis of New Zealand possess extremely subdivided population structure and cryptic species like small mammals. — *Proc. Acad. Nat. Sci. Philad.*, 92: 8254-8258.

79   Baker, E.C.S., 1930. *Fauna of British India Birds.* 7: i-viii, 1-484. 2nd. edition. — Taylor and Francis, London.

80   Baker, E.C.S., 1934. *Nidification of Birds of the Indian Empire.* 3: i-vii 1-568. — Taylor and Francis, London.

81   Baker, N.E. & E.M. Baker, 1990. A new species of weaver from Tanzania. — *Bull. Brit. Orn. Cl.*, 110: 51-58.

82   Baker, R.H., 1951. The Avifauna of Micronesia, its origin, Evolution and distribution. — *Univ. Kansas Publ. Mus. Nat. Hist.*, 3 (1): 1-359.

83   Bangs, O.C., 1930. Types of birds now in the Museum of Comparative Zoology. — *Bull. Mus. Comp. Zool.*, 70 (4): 145-426.

84   Banks, R.C., 1963. New birds from Cerralvo Island, Baja California, Mexico. — *Occ. Pap. Calif. Acad. Sci.*, 37: 1-5.

85   Banks, R.C., 1986. Subspecies of the Greater Scaup and their names. — *Wilson Bull.*, 98 (3): 433-444.

86   Banks, R.C., 1987. Taxonomic notes on Singing Quail (Genus *Dactylortyx*) from western and southern Mexico. — *Occ. Pap. Western Found. Vert. Zool.*, 4: 1-6.

87   Banks, R.C., 1990. Taxonomic status of the coquette hummingbird of Guerrero, Mexico. — *Auk*, 107: 191-192.

88   Banks, R.C., 1990. Taxonomic status of the Rufous-bellied Chachalaca (*Ortalis wagleri*). — *Condor*, 92: 749-753.

89   Banks, R.C., 1993. The generic name of the Crested Argus *Rheinardia ocellata*. — *Forktail*, 8: 3-6.

90   Banks, R.C., 1997. The name of Lawrence's Flycatcher. Pages 21-24. In *The Era of Allan R. Phillips: A Festschrift* (R.W. Dickerman, ed.) — Albuquerque.

91   Banks, R.C. & M.R. Browning, 1995. Comments on the status of revived old names for some North American birds. — *Auk*, 112: 633-648.

92   Bannerman, D.A., 1916. A revision of the genus *Haplopelia*. — *Ibis*, (10) 4 (1): 1-16.

93   Bannerman, D.A., 1934. [A new race of the Ahanta Francolin from Gambia and Portuguese Guinea.]. — *Bull. Brit. Orn. Cl.*, 55: 5-6.

94   Baptista, L.F. *et al.*, 1997. Family Columbidae (Pigeons and Doves). Pages 60-243. In *Handbook of the Birds of the World.* Vol. 4. (J. del Hoyo, A. Elliott & J. Sargatal, eds.). — Lynx Edicions, Barcelona.

95   Baptista, L.L.F. *et al.*, 1999. Relationships of some mannikins and waxbills in the Estrildidae. — *J. f. Orn.*, 140 (2): 179-192.

96   Barker, F.K., 1999. The evolution of co-operative breeding in *Campylorhynchus* wrens: a comparative approach., University of Chicago, Evolutionary Biology, Ph. D. thesis, Chicago, 1-286.

97   Barker, F.K. & S.M. Lanyon, 2000. The impact of parsimony weighting schemes on inferred relationships among toucans and Neotropical barbets (Aves: Piciformes). — *Molec. Phylogen. Evol.*, 15: 215-234.

98   Barker, F.K. *et al.*, 2002. A phylogenetic hypothesis for passerine birds: taxonomic and biogeographic implications of an analysis of 15 nuclear DNA sequence data. — *Proc. Roy. Soc., Lond.*, 269 B: 295-308.

99   Barrantes, G. & J.E. Sanchez, 2000. A new subspecies of Black and Yellow Silky Flycatcher, *Phainoptila melanoxantha*, from Costa Rica. — *Bull. Brit. Orn. Cl.*, 120 (1): 40-46.

100   Bartels, M., Jr., 1938. Eine neue Rasse von *Arborophila brunneopectus* aus Java. — *Treubia*, 16 (3): 321-322.

101   Bartels, M., Jr. & E. Stresemann, 1929. Systematische Übersicht der bisher von Java nachgewiesenen Vögel. — *Treubia*, 11 (1): 89-146.

102   Bartle, J.A. & P.M. Sagar, 1987. Intraspecific variation in the New Zealand Bellbird *Anthornis melanura*. — *Notornis*, 34: 253-306.

103   Bates, G.L., 1935. On *Oenanthe xanthoprymna* and *O. chrysopygia*. — *Ibis*, (13) 5: 198-201.

104   Bates, G.L. & N.B. Kinnear, 1937. [A new race of Sandgrouse.]. — *Bull. Brit. Orn. Cl.*, 57: 142.

105   Bates, J.M. & R.M. Zink, 1994. Evolution in the Andes: molecular evidence for species relationships in the genus *Leptopogon*. — *Auk*, 111 (3): 507-515.

106   Bates, J.M. *et al.*, 1992. Observations of the campo, cerrado, and forest avifaunas of eastern Dpto. Santa Cruz, Bolivia, including 21 species new to the country. — *Bull. Brit. Orn. Cl.*, 112 (2): 86-98.

107   Bauer, K., 1960. Variabilität und Rassengliederung des Haselhuhns (*Tetrastes bonasia*) in Mitteleuropa. — *Bonn. Zool. Beitr.*, 11: 1-18.

108   Baverstock, P.R. *et al.*, 1992. Evolutionary relationships of the Australasian mud-nesters (Grallinidae, Corcoracidae): immunological evidence. — *Austral. J. Zool.*, 40: 173-179.

109   Becking, J.H., 1975. New evidence of the specific affinity of *Cuculus lepidus* Müller. — *Ibis*, 117 (3): 275-284.
110   Beehler, B.M. & R.J. Swaby, 1991. Phylogeny and biogeography of the *Ptiloris* riflebirds (Aves: Paradisaeidae). — *Condor*, 93: 738-745.
111   Begbie, P.F., 1834. *The Malayan Peninsula*: i-xvii, 1-523. — The author/Vespery Mission Press, Madras.
112   Behle, W.H., 1968. A new race of the Purple Martin from Utah. — *Condor*, 70: 166-169.
113   Behle, W.H. & R.K. Selander, 1951. A new race of Dusky Grouse (*Dendragapus obscurus*) from the Great Basin. — *Proc. Biol. Soc. Wash.*, 64: 125-127.
114   Beintema, A.J., 1972. The history of the Island Hen (*Gallinula nesiotis*) the extinct flightless gallinule of Tristan da Cunha. — *Bull. Brit. Orn. Cl.*, 92: 106-113.
114A   Bensch, S. & D. Pearson, 2002. The Large-billed Warbler *Acrocephalus orinus* revisited. — *Ibis*, 144 (2): 259-267.
115   Benson, C.W., 1939. A new Francolin from Nyasaland. — *Bull. Brit. Orn. Cl.*, 59: 42-43.
116   Benson, C.W., 1942. A new species and ten new races from Southern Abyssinia. — *Bull. Brit. Orn. Cl.*, 63: 8-19.
117   Benson, C.W., 1947. A new race of Double-banded Sand-Grouse from Portuguese East Africa. — *Bull. Brit. Orn. Cl.*, 67: 44-45.
118   Benson, C.W., 1947. A new race of the Double-banded Sandgrouse from Angola. — *Bull. Brit. Orn. Cl.*, 67: 79-80.
119   Benson, C.W., 1948. A new race of Coucal from Nyasaland. — *Bull. Brit. Orn. Cl.*, 68: 127-128.
120   Benson, C.W., 1948. A new race of barbet from south-western Tanganyika Territory and Northern Nysasland. — *Bull. Brit. Orn. Cl.*, 68: 144-145.
121   Benson, C.W., 1956. The races of Whyte's Barbet. — *Bull. Brit. Orn. Cl.*, 76: 14-16.
122   Benson, C.W., 1960. The birds of the Comoro Islands: results of the British Ornithologists' Union Centenary Expedition 1958. — *Ibis*, 103b: 5-106.
123   Benson, C.W., 1964. A further revision of the races of Whyte's Barbet *Buccanodon whytii* Shelley. — *Arnoldia* (Rhod.), 1 (6): 1-4.
124   Benson, C.W., 1964. A new subspecies of Pink-billed Lark, *Calandrella conirostris* from Barotseland, Northern Rhodesia. — *Bull. Brit. Orn. Cl.*, 84: 106-107.
125   Benson, C.W., 1966. The Spike-heeled Lark *Chersomanes albofasciata* in East Africa. — *Bull. Brit. Orn. Cl.*, 86: 76-77.
126   Benson, C.W., 1967. A new subspecies of Black Swift *Apus barbatus* from Sierra Leone. — *Bull. Brit. Orn. Cl.*, 87: 125-126.
127   Benson, C.W., 1969. The white-eye *Zosterops maderaspatana* (Linn.) of Menai Island, Cosmoledo Atoll. — *Bull. Brit. Orn. Cl.*, 89: 24-27.
128   Benson, C.W., 1974. Une race nouvelle du Martin-chasseur Malgache. — *L'Oiseau*, 44: 186-187.
129   Benson, C.W. & M.P.S. Irwin, 1960. A new form of *Apus barbatus* from the Victoria Falls. — *Bull. Brit. Orn. Cl.*, 80: 98-99.
130   Benson, C.W. & M.P.S. Irwin, 1965. A new species of tinker-barbet from northern Rhodesia. — *Bull. Brit. Orn. Cl.*, 85: 5-9.
131   Benson, C.W. & M.P.S. Irwin, 1965. The grey-backed sparrow-lark, *Eremopterix verticalis* (Smith). — *Arnoldia* (Rhod.), 1 (36): 1-9.
132   Benson, C.W. & M.P.S. Irwin, 1965. A new subspecies of clapper lark, *Mirafra apiata* from Barotseland. — *Arnoldia*, 1 (37): 1-3.
133   Benson, C.W. & M.P.S. Irwin, 1975. The systematic position of *Phyllastrephus orostruthus* and *Phyllastrephus xanthophrys*, two species incorrectly placed in the family Pycnonotidae (Aves). — *Arnoldia* (Rhod.), 7 (10): 1-10.
134   Benson, C.W. *et al.*, 1970. Notes on the birds of Zambia. Part V. — *Arnoldia*, 4 (40): 1-59.
135   Beresford, P. & J. Cracraft, 1999. Speciation in African Forest Robins (*Stiphrornis*): Species Limits, Phylogenic Relationships and Molecular Biogeography. — *Amer. Mus. Novit.*, 3270: 1-22.
136   Berla, H.F., 1954. Um novo "Psittacidae" do Nordeste Brasiliero. — *Rev. Bras. Biol.*, 14: 59-60.
137   Berlioz, J., 1941. Note critique sur une sous-espèce de Trochilidé *Schistes a. bolivianus* Simon. — *L'Oiseau*, NS 11: 233-236.
138   Berlioz, J., 1949. Note sur une espèce de Trochilidé *Chlorostilbon aureoventris* Lafr. et d'Orb. — *Bull. Mus. Hist. Nat. Paris*, (2) 21: 51-55.
139   Berlioz, J., 1953. Notes systématique sur le torcol à gorge rousse. — *L'Oiseau*, 23 (1): 63-67.
140   Berlioz, J., 1965. Note critique sur les Trochilidés des genres *Timolia* et *Augasma*. — *L'Oiseau*, 35: 1-8.
141   Berlioz, J., 1966. Descriptions de deux espèces nouvelles d'oiseaux de Pérou. — *L'Oiseau*, 36 (1): 1-3.
142   Berlioz, J., 1974. Notes critiques sur quelques trochilidés. — *L'Oiseau*, 44 (4): 281-290.
143   Berlioz, J. & J. Dorst, 1956. Quelle est l'identité du *Bolborhynchus orbignesius* (Souancé)? — *L'Oiseau*, 26 (2): 81-86.
144   Berlioz, J. & W.H. Phelps, 1953. Description d'une sous-espèce nouvelle de Trochilidé du Venezuela. — *L'Oiseau*, 23 (1): 1-4.
145   Berlioz, J. & J. Roche, 1970. Note sur le *Cercomela spectatrix* Clarke et description d'une sous-espèce nouvelle. — *Monitore Zool.*, 3 (12, Italiana Suppl.): 267-271.
146   Bertoni, A. de W., 1901. *Aves nuevas del Paraguay*: 1-206.
147   Bierregaard, R.O., 1994. Various species accounts (Falconiformes). Pages 1-638. In *Handbook of the Birds of the World*. Vol. 2. (J. del Hoyo, A. Elliott & J. Sargatal, eds.). — Lynx Edicions, Barcelona.
148   Bierregaard, R.O. *et al.*, 1997. Cryptic biodiversity : An overlooked species and new subspecies of antbird (Aves : Formicariidae) with a revision of *Cercomacra tyrannina* in northeastern South America. — *Orn. Monogr.*, 48: 111-128.
149   Binford, L.C., 1965. Two new subspecies of birds from Oaxaca, Mexico. — *Occ. Pap. Mus. Zool., Louisiana St. Univ.*, 30: 1-6.
150   Binford, L.C., 1989. *A distributional survey of the birds of the Mexican State of Oaxaca*: 43. i-viii, 1-418. — A.O.U.
151   Binford, L.C., 2000. Re-evaluation of the House Finch subspecies *Carpodacus mexicanus roseipectus* from Oaxaca, Mexico. — *Bull. Brit. Orn. Cl.*, 120 (2): 120-128.
152   Birckhead, H., 1937. The birds of the Sage West China Expedition. — *Amer. Mus. Novit.*, 966: 1-17.
153   Biswas, B., 1950. The generic limits of *Treron* Vieillot. — *Bull. Brit. Orn. Cl.*, 70 (5): 34.
154   Biswas, B., 1951. On some larger Spine-tailed Swifts, with the description of a new subspecies from Nepal. — *Ardea*, 39 (4): 318-321.
155   Biswas, B., 1951. Revisions of Indian birds. — *Amer. Mus. Novit.*, 1500: 1-12.
156   Biswas, B., 1952. Geographical variation in the Woodpecker *Picus flavinucha* Gould. — *Ibis*, 94: 210-219.
157   Biswas, B., 1963. The birds of Nepal. Part 10. — *J. Bombay Nat. Hist. Soc.*, 60 (2): 388-399.
158   Blake, E.R., 1957. A new species of ant-thrush from Peru. — *Fieldiana, Zool.*, 39 (7): 51-53.
159   Blake, E.R., 1959. Two new game birds from Peru. — *Fieldiana, Zool.*, 39 (32): 373-376.
160   Blake, E.R., 1959. New and rare Colombian birds. — *Lozania* (Acta. Zool. Colomb.), 11: 1-10.
161   Blake, E.R., 1963. A new race of *Penelope montagnii* from southeastern Peru. — *Fieldiana, Zool.*, 44: 121-122.
162   Blake, E.R., 1968. Family Icteridae. Pages 138-202. In *Check-list of Birds of the World*. Vol. 14. (R.A. Paynter, Jr., ed.) — Mus. Comp. Zool., Cambridge, Mass.
163   Blake, E.R., 1971. A new species of spinetail (*Synallaxis*) from Peru. — *Auk*, 88 (1): 179.
164   Blake, E.R., 1977. *Manual of Neotropical Birds:* 1. i-l, 1-674. — Univ. Chicago Press, Chicago.
165   Blake, E.R., 1979. Order Tinamiformes. Pages 12-47. In *Check-list of Birds of the World*. Vol. 1 (2nd. Edit.) (E. Mayr & G.W. Cottrell, eds.). — Mus. Comp. Zool., Cambridge, Mass.
166   Blake, E.R. & P. Hocking, 1974. Two new species of tanager from Peru. — *Wilson Bull.*, 86 (4): 321-324.
167   Blake, E.R. *et al.*, 1961. Variation in the Quail Dove *Geotrygon frenata*. — *Fieldiana, Zool.*, 39 (50): 567-572.
168   Blake, E.R. & M.A. Traylor, 1947. The subspecies of *Aratinga acuticaudata*. — *Fieldiana, Zool.*, 31 (21): 163-169.
169   Blake, E.R. & C. Vaurie, 1962. Family Corvidae. Pages 204-282. In *Check-list of Birds of the World*. Vol. 15. (E. Mayr & J.C. Greenway, Jr., eds.). — Mus. Comp. Zool., Cambridge, Mass.
170   Bleiweiss, R. *et al.*, 1994. DNA-DNA hybridization-based phylogeny for "higher" nonpasserines: re-evaluating a key portion of the avian family tree. — *Molec. Phylogen. Evol.*, 3: 248-255.
171   Blyth, E., 1844. Further observations on the ornithology of the neighbourhood of Calcutta. — *Ann. Mag. Nat. Hist.*, 14: 114-125.
172   Bochenski, Z.M., 1994. The comparative osteology of grebes (Aves: Podicipediformes) and its systematic implications. — *Acta Zool. Cracov.*, 37: 191-346.

173 Bock, W.J., 1994. History and nomenclature of Avian Family-group Names. — *Bull. Amer. Mus. Nat. Hist.*, 222: 1-281.
174 Boles, W.E., 1981. The sub-family name of the monarch flycatchers. — *Emu*, 81: 50.
175 Boles, W.E., 1989. A new subspecies of the Green-backed Robin *Pachycephalopsis hattamensis*, comprising the first record from Papua New Guinea. — *Bull. Brit. Orn. Cl.*, 109: 119-121.
176 Boles, W.E. & N.W. Longmore, 1983. A new subspecies of treecreeper in the *Climacteris leucophaea* superspecies. — *Emu*, 83: 272-275.
177 Boles, W.E. & N.W. Longmore, 1985. Generic allocation of the Tawny-crowned Honeyeater. — *South Australian Orn.*, 29: 221-223.
178 Bonaparte, C.L.J.L., 1850. *Conspectus generum avium*. Vol. 1: 1-543. — Lugduni Batavorum.
179 Bonaparte, C.L.J.L., 1856. Tableaux paralléliques de l'ordre des Gallinacés. — *Comptes Rend. Acad. Sci. Paris*, 42: 874-884.
180 Bond, G.M., 1975. The correct spelling of Jerdon's generic name for the Thick-billed Warbler. — *Bull. Brit. Orn. Cl.*, 95 (2): 50-51.
181 Bond, J., 1936. Resident birds of the Bay Islands of Spanish Honduras. — *Proc. Acad. Nat. Sci. Philad.*, 88: 353-364.
182 Bond, J., 1945. Additional notes on West Indian birds. — *Notulae Naturae*, 148: 1-4.
183 Bond, J., 1951. Taxonomic notes on South American birds. — *Auk*, 68 (4): 527-529.
184 Bond, J., 1954. A new race of *Gallinula chloropus* from Barbados. — *Notulae Naturae*, 264: 1.
185 Bond, J., 1954. Notes on Peruvian Piciformes. — *Proc. Acad. Nat. Sci. Philad.*, 106: 45-61.
186 Bond, J., 1956. Additional notes on Peruvian birds. II. — *Proc. Acad. Nat. Sci. Philad.*, 108: 227-247.
187 Bond, J., 1962. A new Blue Jay (*Cyanocitta cristata*) from Newfoundland. — *Proc. Biol. Soc. Wash.*, 75: 205-206.
188 Bond, J. & A. Dod, 1977. A new race of Chat Tanager (*Calyptophilus frugivorus*) from the Dominican Republic. — *Notulae Naturae*, 451: 1-4.
189 Bond, J. & R. Meyer de Schauensee, 1939. Descriptions of new birds from Bolivia. Part II - A new species of the genus *Pauxi*. — *Notulae Naturae*, 29: 1-3.
190 Bond, J. & R. Meyer de Schauensee, 1941. On some birds from Southern Colombia. — *Proc. Acad. Nat. Sci. Philad.*, 92: 153-169 (1940).
191 Bond, J. & R. Meyer de Schauensee, 1941. Descriptions of new birds from Bolivia. Part IV. — *Notulae Naturae*, 93: 1-7.
192 Bond, J. & R. Meyer de Schauensee, 1943. A new species of dove of the genus *Leptotila* from Columbia. — *Notulae Naturae*, 122: 1-2.
193 Bond, J. & R. Meyer de Schauensee, 1944. A new race of *Pyrrhura rupicola* from Peru. — *Notulae Naturae*, 138: 1-2.
194 Bornschein, M.R. & B.L. Reinert, 1996. On the diagnosis of *Aramides cajanea avicenniae* Stotz, 1992. — *Bull. Brit. Orn. Cl.*, 116 (4): 272.
195 Bornschein, M.R. *et al.*, 1995. Um nova Formicariidae do sul do Brasil (Aves, Passeriformes). — *Publ. Técnico-Cientifica Inst. Iguaçu de Pesquisa e Preservação Ambiental*, 1: 1-18.
196 Bornschein, M.R. *et al.*, 1998. Descriçao, ecologia e conservaçao de um nova *Scytalopus* (Rhinocryptidae) do sul do Brasil, com comentarios sobre a morfologia da familia. — *Ararajuba*, 6 (1): 3-36.
197 Borrero, J.I., 1960. Notas sobre *Schizoeaca fuliginosa* y descripcion de una nueva subspecie. — *Noved. Colomb.*, 1 (5): 238-242.
198 Borrero, J.I. & J.H. Camacho, 1961. Notas sobre aves de Colombia y descripcion de una nueva subspecie de *Forpus conspicillatus*. — *Noved. Colomb.*, 1: 430-445.
199 Borrero, J.I. & J. Hernández, 1958. Apuntes sobre aves colombianas. — *Caldasia*, 8 (37): 253-294.
200 Boulton, R., 1934. New birds from Angola. — *Proc. Biol. Soc. Wash.*, 47: 45-48.
201 Bourne, W.R.P., 1955. The birds of Cape Verde Islands. — *Ibis*, 97 (3): 508-556.
202 Bourne, W.R.P., 1955. On the status and appearance of the races of Cory's Shearwater *Procellaria diomedea*. — *Ibis*, 97 (1): 145-148.
203 Bourne, W.R.P., 1987. The classification and nomenclature of the petrels. — *Ibis*, 129: 404.
204 Bourne, W.R.P., 2001. The status of the genus *Lugensa* Mathews and the birds collected by Carmichael on Tristan da Cunha in 1816-1817. — *Bull. Brit. Orn. Cl.*, 121 (3): 215-216.
205 Bourne, W.R.P. & J.R. Jehl, 1982. Variation and nomenclature of Leach's Storm-Petrels. — *Auk*, 99 (4): 793-797.
206 Bourne, W.R.P. *et al.*, 1988. The Yelkouan Shearwater *Puffinus (puffinus?) yelkouan*. — *Brit. Birds*, 81: 306-319.
207 Bourret, R., 1944. Liste des Oiseaux de la Collection du Laboratoire de Zoologie. — *Notes et travaux de l'École Supérieure des Sciences, Hanoi*, 3: 20-36.
208 Bradbury, J.W. & S.L. Vehrencamp, 1998. *Principles of Animal Communication*: i-xvi, 1-882. — Sinauer Assoc. Inc., Sunderland, Mass.
209 Bradley, D., 1962. Additional records of birds from Rennell and Bellona Islands. Pages 1-120. In *The Natural History of Rennell Island, British Solomon Islands*. Vol. 4. (T. Wolff, ed.) — Danish Science Press Ltd., Copenhagen.
210 Braun, M.T. & T.A. Parker, III., 1985. Molecular, morphological and behavioral evidence concerning the taxonomic relationships of "*Synallaxis*" *gularis* and other Synallaxines. — *Orn. Monogr.*, 36: 333-346.
211 Brehm, C.L., 1826. Eine Bergleichung. — *Isis von Oken*, 19: 927-935.
212 Bretagnolle, V. *et al.*, 1998. Relationships of *Pseudobulweria* and *Bulweria* (Procellaridae). — *Auk*, 115: 188-195.
213 Bretagnolle, V. *et al.*, 2000. Audubon's Shearwaters *Puffinus lherminieri* on Réunion Island, Indian Ocean: behaviour, census, distribution, biometrics and breeding biology. — *Ibis*, 142 (3): 399-412.
214 Briggs, M.A., 1954. Apparent neoteny in the Saw-Whet Owls of Mexico and Central America. — *Proc. Biol. Soc. Wash.*, 67: 179-182.
215 Britton, P.L., 1986. Various species accounts. In *The Birds of Africa*. Vol. 2: i-xvi, 1-552. (E.K. Urban, C.H. Fry & S. Keith, eds.). — Academic Press, London.
216 Brodkorb, P., 1933. Remarks on the genus *Limnodromus*. — *Proc. Biol. Soc. Wash.*, 46: 123-128.
217 Brodkorb, P., 1934. Geographical variation in *Belonopterus chilensis* (Molina). — *Occ. Pap. Mus. Zool., Univ. Michigan*, 293: 1-13.
218 Brodkorb, P., 1936. *Larus canus stegmanni* nom. nov. for *Larus canus major* Middendorff. — *Orn. Monatsber.*, 44 (4): 122.
219 Brodkorb, P., 1939. New subspecies of birds from the district of Soconusco, Chiapas. — *Occ. Pap. Mus. Zool., Univ. Michigan*, 401: 1-7.
220 Brodkorb, P., 1939. A southern race of the Jacana. — *Proc. Biol. Soc. Wash.*, 52: 185-186.
221 Brodkorb, P., 1940. New birds from Southern Mexico. — *Auk*, 57 (4): 542-549.
222 Brodkorb, P., 1941. The Pygmy Owl of the district of Soconusco, Chiapas. — *Occ. Pap. Mus. Zool., Univ. Michigan*, 450: 1-2.
223 Brodkorb, P., 1942. The Chachalaca of Interior Chiapas. — *Proc. Biol. Soc. Wash.*, 55: 181-182.
224 Brodkorb, P., 1942. A new race of Bob-white from Interior Chiapas. — *Occ. Pap. Mus. Zool., Univ. Michigan*, 467: 1-4.
225 Brooke, M.d.L. & G. Rowe, 1996. Behavioural and molecular evidence for specific status of light and dark morphs of the Herald Petrel *Pterodroma heraldica*. — *Ibis*, 138 (3): 420-432.
226 Brooke, R.K., 1967. *Apus aequatorialis* (von Müller) (Aves) in Rhodesia and adjacent areas with description of a new race. — *Arnoldia* (Rhod.), 3 (7): 1-8.
227 Brooke, R.K., 1969. *Apus berliozi* Ripley, its races and siblings. — *Bull. Brit. Orn. Cl.*, 89: 11-16.
228 Brooke, R.K., 1969. The tropical African population of *Apus affinis*. — *Bull. Brit. Orn. Cl.*, 89: 166-167.
229 Brooke, R.K., 1969. *Hemiprocne coronata* is a good species. — *Bull. Brit. Orn. Cl.*, 89: 168-169.
230 Brooke, R.K., 1970. Geographical variation and distribution in *Apus barbatus, A.bradfieldi* and *A. niansae* (Aves: Apodidae). — *Durban Mus. Novit.*, 8: 363-374.
231 Brooke, R.K., 1971. Geographical variation in the swifts *Apus horus* and *A. caffer* (Aves: Apodidae). — *Durban Mus. Novit.*, 9 (4): 29-38.
232 Brooke, R.K., 1971. Geographical variation in the Little Swift *Apus affinis* (Aves: Apodidae). — *Durban Mus. Novit.*, 9 (7): 93-103.
233 Brooke, R.K., 1972. Geographical variation in Palm Swifts *Cypsiurus* spp. (Aves: Apodidae). — *Durban Mus. Novit.*, 9 (15): 217-231.
234 Brooke, R.K., 1974. Nomenclatural notes on and the type-localities of some taxa in the Apodidae and Hirundinidae (Aves). — *Durban Mus. Novit.*, 10 (9): 127-137.
235 Brooke, R.K., 1975. *Cotyle palidibula* Rüppell, 1835. — *Bull. Brit. Orn. Cl.*, 95 (3): 90-91.
236 Brooke, R.K., 1976. Morphological notes on *Acridotheres tristis* in Natal. — *Bull. Brit. Orn. Cl.*, 96 (1): 8-13.

237 Brooke, R.K., 1993. Annotated catalogue of the Aves type specimens in the South African Museum. — *Ann. Mus. South Afr. Mus.*, 102: 327-349.

238 Brooke, R.K. & P.A. Clancey, 1981. The authorship of the generic and specific names of the Bat Hawk. — *Bull. Brit. Orn. Cl.*, 101 (4): 371-372.

239 Brooke, R.K. & J. Cooper, 1979. The distinctiveness of southern African *Larus dominicanus* (Aves; Laridae). — *Durban Mus. Novit.*, 12 (3): 27-35.

240 Brooke, R.K. & W.R.J. Dean, 1990. On the biology and taxonomic position of *Drymoica substriata* Smith, the so-called Namaqua Prinia. — *Ostrich*, 61 (1/2): 50-55.

241 Brooks, T.M. *et al.*, 1992. The conservation status of the birds of Negros, Philippines. — *Bird Conserv. Int.*, 2: 273-302.

242 Brooks, W.E., 1875. Notes on a new *Dumeticola* and on *Tribura luteoventris*, Hodgson, and *Dumeticola affinis*, Hodgson. — *Stray Feathers*, 3: 284-287.

243 Brosset, A. & C.H. Fry, 1988. Various species accounts. In *The Birds of Africa*. Vol. 3: i-xvi, 1-611 (C.H. Fry, S. Keith & E.K. Urban, eds.). — Academic Press, London.

244 Browning, M.R., 1974. Taxonomic remarks on recently described subspecies of birds that occur in the northwestern United States. — *Murrelet*, 55 (3): 32-38.

245 Browning, M.R., 1977. The types and type localities of *Oreortyx pictus* (Douglas) and *Oreortyx plumiferus* (Gould). — *Proc. Biol. Soc. Wash.*, 90: 808-812.

246 Browning, M.R., 1978. An evaluation of the new species and subspecies proposed in Oberholser's *Bird Life of Texas*. — *Proc. Biol. Soc. Wash.*, 91 (1): 85-122.

247 Browning, M.R., 1988. A new subspecies of *Carpodacus roseus*. — *Bull. Brit. Orn. Cl.*, 108: 177-179.

248 Browning, M.R., 1989. The type specimens of Hekstra's owls. — *Proc. Biol. Soc. Wash.*, 102 (2): 515-519.

249 Browning, M.R., 1990. Taxa of North American birds described from 1957-1987. — *Proc. Biol. Soc. Wash.*, 103 (2): 432-451.

250 Browning, M.R. & R.C. Banks, 1990. The identity of Pennant's "Wapacuthu Owl" and the subspecific name of the population of *Bubo virginianus* from west of Hudson Bay. — *J. Raptor Res.*, 24: 80-83.

251 Browning, M.R., 1990. Taxa of North American birds described from 1957-1987. — *Proc. Biol. Soc. Wash.*, 103 (2): 432-451.

252 Browning, M.R., 1991. Taxonomic comments on the Dunlin *Calidris alpina* from northern Alaska and eastern Siberia. — *Bull. Brit. Orn. Cl.*, 111 (3): 140-145.

253 Browning, M.R., 1992. Comments on the nomenclature and dates of publication of some taxa in Bucerotidae. — *Bull. Brit. Orn. Cl.*, 112 (1): 22-27.

254 Browning, M.R., 1992. A new subspecies of *Chamaea fasciata* (Wrentit) from Oregon (Aves: Timaliinae). — *Proc. Biol. Soc. Wash.*, 105 (3): 414-419.

255 Browning, M.R., 1993. Taxonomy of the blue-crested group of *Cyanocitta stelleri* (Steller's Jay) with a description of a new subspecies. — *Bull. Brit. Orn. Cl.*, 113 (1): 34-41.

256 Browning, M.R., 1994. A taxonomic review of *Dendroica petechia* (Yellow Warbler) (Aves: Parulinae). — *Proc. Biol. Soc. Wash.*, 107 (1): 27-51.

257 Browning, M.R., 1997. Taxonomy of *Picoides pubescens* (Downy Woodpecker) from the Pacific northwest. Pages 25-34. In *The Era of Allan R. Phillips: A Festschrift* (R.W. Dickerman, ed.) — Albuquerque.

258 Browning, M.R. & B.L. Monroe, Jr., 1991. Clarifications and corrections of the dates of issue of some publications containing descriptions of North American birds. — *Arch. Nat. Hist.*, 18 (3): 381-405.

259 Bruce, M.D., 1989. A reappraisal of species limits in the Pied Imperial Pigeon *Ducula bicolor* (Scopoli, 1786) superspecies. — *Riv. Ital. Orn.*, 59: 218-222.

260 Bruce, M.D. & I.A.W. McAllan, 1989 (1988). *The birds of New South Wales. A working list.* — Biocon Research Group, Turramurra.

261 Brumfield, R.T. & M.J. Braun, 2001. Phylogenetic relationships in Bearded Manakins (Pipridae: *Manacus*) indicate that male plumage color is a misleading taxonomic marker. — *Condor*, 103 (2): 248-258.

262 Brumfield, R.T. & J.V. Remsen, Jr., 1996. Geographic variation and species limits in *Cinnycerthia* wrens of the Andes. — *Wilson Bull.*, 108 (2): 205-227.

263 Brumfield, R.T. *et al.*, 2001. Evolutionary implications of divergent clines in an avian (*Manacus*: Aves) hybrid zone. — *Evolution*, 18: 2070-2087.

264 Bruun, B. *et al.*, 1981. A new subspecies of Lappet-faced Vulture *Torgos tracheliotus* from the Negev Desert, Israel. — *Bull. Brit. Orn. Cl.*, 101 (1): 244-247.

265 Buden, D.W., 1985. A new subspecies of Common Ground-Dove from Île de la Tortue, Haiti, with taxonomic reappraisal of Bahaman populations (Aves: Columbidae). — *Proc. Biol. Soc. Wash.*, 98: 790-798.

266 Buden, D.W., 1985. New subspecies of Thick-billed Vireo (Aves: Vireonidae) from the Caicos Islands, with remarks on taxonomic status of other populations. — *Proc. Biol. Soc. Wash.*, 98 (3): 591-597.

267 Buden, D.W., 1986. A new subspecies of Greater Antillean Bullfinch *Loxigilla violacea* from the Caicos Islands with notes on other populations. — *Bull. Brit. Orn. Cl.*, 106: 156-161.

268 Buden, D.W., 1993. Geographic variation in the Scaly-breasted Thrasher *Margarops fuscus* with descriptions of three new subspecies. — *Bull. Brit. Orn. Cl.*, 113 (2): 75-84.

269 Bündgen, R., 1999. Species accounts within the Family Trochilidae (Hummingbirds). Pages 549-680. In *Handbook of the Birds of the World*. Vol. 5. (J. del Hoyo, A. Elliott & J. Sargatal, eds.). — Lynx Edicions, Barcelona.

270 Burger, J. & M. Gochfeld, 1996. Family Laridae (Gulls). Pages 572-623. In *Handbook of the Birds of the World*. Vol. 3. (J. del Hoyo, A. Elliott & J. Sargatal, eds.). — Lynx Edicions, Barcelona.

271 Burleigh, T.D., 1959. Two new subspecies of birds from western North America. — *Proc. Biol. Soc. Wash.*, 72: 15-18.

272 Burleigh, T.D., 1960. Three new subspecies of birds from western North America. — *Auk*, 77 (2): 210-215.

273 Burleigh, T.D., 1960. Geographic variation in the Western Wood Peewee. — *Proc. Biol. Soc. Wash.*, 73: 141-146.

274 Burleigh, T.D., 1960. Geographical variation in the Catbird. — *Oriole*, 24 (3/4): 29-32 (1959).

275 Burleigh, T.D., 1961. A new subspecies of Downy Woodpecker from the Northwest. — *Murrelet*, 4 (3): 42-44 (1960).

276 Burleigh, T.D., 1963. Geographic variation in the Cedar Waxwing (*Bombycilla cedrorum*). — *Proc. Biol. Soc. Wash.*, 76: 177-180.

277 Burleigh, T.D. & G. Lowery, 1942. An inland race of *Sterna albifrons*. — *Occ. Pap. Mus. Zool.*, *Louisiana St. Univ.*, 10: 173-177.

278 Burns, K.J., 1997. Molecular systematics of tanagers (Thraupinae): Evolution and biogeography of a diverse radiation of neotropical birds. — *Molec. Phylogen. Evol.*, 8 (3): 334-348.

279 Burns, K.J., 1998. Molecular phylogenetics of the genus *Piranga*: Implications for biogeography and the evolution of morphology and behavior. — *Auk*, 115 (3): 621-634.

280 Burtt, E.H. & A.P. Peterson, 1995. Alexander Wilson and the founding of North American Ornithology. Pages 359-386. In *Contributions to the history of North American Ornithology* (W.E.J. Davis & J.A. Jackson, eds.). — Nuttall Ornithological Club, Cambridge, Mass.

281 Buturlin, S.A., 1933. *Opreditel promyslovkh ptits* (Key to the identification of gamebirds): 92 pp.

282 Buturlin, S.A., 1934. *Larus taimyrensis armenicus* subsp. nov. — *Ibis*, 13 (4): 171-172.

283 Buturlin, S.A., 1934. Polnyi Opredelitel Ptits S.S.S.R., 1: 1-255.

284 Buturlin, S.A., 1935. Ptitsy Tazovskoi ekspeditsii Vserossiiskogo geograficheskogo obshchestva 1926-1927. [Birds of the Tazovskoi Expedition of the All-Russian Geographical Society]. — *Sbornik Trudy Zool. Muz. MGU* [Archives Zool. Mus. Moscow Univ.], 1: 61-100 (1934).

285 Buturlin, S.A., 1938. Polnyi Opredelitel Ptitsy S.S.S.R., 2: 1-280.

286 Cabot, J., 1992. Family Tinamidae (Tinamous). Pages 112-138. In *Handbook of the Birds of the World* (J. del Hoyo, A. Elliott & J. Sargatal, eds.). — Lynx Edicions, Barcelona.

287 Cabot, J., 1997. Dos nuevas subespecies de *Nothoprocta ornata* y sobre la distribución de *N. ornata rostrata* (Aves, Tinamidae). — *Iheringia, Ser. Zool.*, 82: 119-125.

288 Cain, A.J., 1955. A revision of *Trichoglossus haematodus* and of the Australian Platycercine parrots. — *Ibis*, 97: 432-479.

289 Calder, W.A., 1999. Species accounts within the Family Trochilidae (Hummingbirds). Pages 549-680. In *Handbook of the Birds of the World*. Vol. 5. (J. del Hoyo, A. Elliott & J. Sargatal, eds.). — Lynx Edicions, Barcelona.

290 Capparella, A.P. *et al.*, 1997. A new subspecies of *Percnostola rufifrons* (Formicariidae) from northeastern Amazonian Peru, with a revision of the *rufifrons* complex. — *Orn. Monogr.*, 48: 165-170.

291 Carboneras, C., 1992. Family Diomedeidae (Albatrosses). Pages 198-215. In *Handbook of the Birds of the World*. Vol. 1. (J. del Hoyo, A. Elliott & J. Sargatal, eds.). — Lynx Edicions, Barcelona.

292 Carboneras, C., 1992. Family Procellariidae (Petrels and Shearwaters). Pages 216-257. In *Handbook of the Birds of the World*. Vol. 1. (J. del Hoyo, A. Elliott & J. Sargatal, eds.). — Lynx Edicions, Barcelona.

293 Cardoso da Silva, J.M., 1990. Description of a new subspecies of *Saltator aurantiirostris*, with comments on *S. maxillosus*. — *Bull. Brit. Orn. Cl.*, 110: 171-175.

294 Cardoso da Silva, J.M., 1996. New data support the specific status of Reiser's Tyrannulet, a central Brazilian endemic. — *Bull. Brit. Orn. Cl.*, 116 (2): 109-113.

295 Cardoso da Silva, J.M. *et al.*, 1995. A new species of the genus *Hylexetastes* (Dendrocolaptidae) from eastern Amazonia. — *Bull. Brit. Orn. Cl.*, 115 (4): 200-206.

296 Cardoso da Silva, J.M. & D.C. Oren, 1991. A new subspecies of *Xiphocolaptes major* (Vieillot) from Argentina. — *Bull. Brit. Orn. Cl.*, 111 (3): 147-149.

297 Cardoso da Silva, J.M. & D.C. Oren, 1997. Geographic variation and conservation of the Moustached Woodcreeper *Xiphocolaptes falcirostris*, an endemic and threatened species of north-eastern Brazil. — *Bird Conserv. Int.*, 7: 263-274.

298 Cardoso da Silva, J.M. & D.F. Stotz, 1992. Geographic variation in the Sharp-billed Treehunter *Heliobletus contaminatus*. — *Bull. Brit. Orn. Cl.*, 112 (2): 98-101.

299 Cardoso da Silva, J.M. & F.C. Straube, 1996. Systematics and biogeography of Scaled Woodcreepers (Aves : Dendrocolaptidae). — *Studies Neotrop. Fauna Environ.*, 31 (1): 3-10.

300 Carnaby, I.C., 1948. Variation in the White-tailed Black Cockatoo. — *West. Austral. Nat.*, 1: 136-138.

301 Carriker, M.R., 1934. Description of new birds from Peru, with notes on the nomenclature and status of other little-known species. — *Proc. Acad. Nat. Sci. Philad.*, 86: 317-334.

302 Carriker, M.R., 1935. Descriptions of new birds from Peru and Ecuador, with critical notes on other little-known species. — *Proc. Acad. Nat. Sci. Philad.*, 87: 343-359.

303 Carroll, J.P., 1994. Family Odontophoridae (New World Quails). Pages 412-433. In *Handbook of the Birds of the World*. Vol. 2. (J. del Hoyo, A. Elliott & J. Sargatal, eds.). — Lynx Edicions, Barcelona.

304 Casey, T.L.C. & J.D. Jacobi, 1974. A new genus and species of bird from the island of Maui, Hawaii (Passeriformes: Drepanididae). — *Occ. Pap. Bernice Bishop Mus.*, 24 (2): 215-226.

305 Cassin, J., 1859. Catalogue of birds collected on the Rivers Camma and Ogobai, western Africa by Mr. P.B. Duchaillu in 1858 with notes and descriptions of new species. — *Proc. Acad. Nat. Sci. Philad.*, 11: 30-55.

306 Castroviejo, J., 1967. Eine neue Auerhuhnrasse von der Iberischen Halbinsel. — *J. f. Orn.*, 108: 220-221.

307 Cave, F.M., 1940. New races of a Francolin and a Lark from the Southern Sudan. — *Bull. Brit. Orn. Cl.*, 60: 96-97.

308 Chantler, P., 1999. Family Apodidae (Swifts). Pages 388-457. In *Handbook of the Birds of the World*. Vol. 5. (J. del Hoyo, A. Elliott & J. Sargatal, eds.). — Lynx Edicions, Barcelona.

309 Chantler, P. & G. Driessens, 1995. Swifts. *A guide to the swifts and treeswifts of the world*: i-x, 1-237. — Pica Press, Robertsbridge.

310 Chapin, J.P., 1936. A new peacock-like bird from the Belgian Congo. — *Rev. Zool. Bot. Afr.*, 29 (1): 1-6.

311 Chapin, J.P., 1937. A new race of *Francolinus albogularis* from Marungu, Belgian Congo. — *Rev. Zool. Bot. Afr.*, 29: 395-396.

312 Chapin, J.P., 1954. A Juba River race of Klaas's Cuckoo. — *Auk*, 71: 89.

313 Chapin, J.P., 1954. Races of the African Finfoot (Aves, Heliornithidae). — *Amer. Mus. Novit.*, 1659: 1-10.

314 Chapin, J.P., 1958. A new honey-guide from the Kivu District, Belgian Congo. — *Bull. Brit. Orn. Cl.*, 78: 46-48.

315 Chapman, F.M., 1915. The more northern species of the genus *Scytalopus*. — *Auk*, 32: 406-423.

316 Chapman, F.M., 1939. The upper zonal birds of Mt. Auyantepui, Venezuela. — *Amer. Mus. Novit.*, 1051: 1-15.

317 Chappuis, C. & C. Érard, 1991. A new cisticola from west-central Africa. — *Bull. Brit. Orn. Cl.*, 111 (2): 59-70.

318 Charlemagne, M., 1934. Ornithologische Notizen. — *Journ. Cycle Bio-zoologique Akad. Sci. Ukraine, Kiev.*, 4 (8): 21-29.

319 Chasen, F.N., 1935. A Handlist of Malaysian Birds. — *Bull. Raffles Mus.*, 11: i-xx, 1-389.

320 Chasen, F.N., 1937. The birds of Billiton Island. — *Treubia*, 16: 205-238.

321 Chasen, F.N., 1938. Vier neue Vogelrassen aus Malaysia. — *Orn. Monatsber.*, 46 (1): 5-8.

322 Chasen, F.N., 1939. Preliminary diagnoses of new birds from north Sumatra II. — *Treubia*, 17 (2): 183-184.

323 Chasen, F.N., 1940. Notes on some Javan birds. — *Treubia*, 17 (4): 263-266.

324 Chasen, F.N., 1941. The birds of the Netherlands Indian Mt. Leuser Expedition 1937 to North Sumatra; with a general survey, an itinerary and field notes by A. Hoogerwerf. — *Treubia*, 18 (Suppl): 1-125.

325 Cheke, A.S. & A.W. Diamond, 1986. Birds on Moheli and Grande Comore (Comoro Islands) in February, 1975. — *Bull. Brit. Orn. Cl.*, 106 (4): 138-148.

326 Cheke, R.A. & C.F. Mann, 2001. *Sunbirds: a guide to the sunbirds, flowerpeckers, spiderhunters and sugarbirds of the world*: 1-384. — A. & C. Black, London.

327 Cheng, Pao-lai & Yang, Lan, 1980. A new subspecies of *Garrulax canorus - Garrulax canorus mengliensis*. — *Zool. Res.* [China], 1 (3): 391-395.

328 Cheng, Tso-hsin, 1956. A new form of the White-backed Woodpecker (*Dendrocopus leucotos tangi* subsp. nov.) from Szechwan, China. — *Acta Zool. Sinica*, 8 (2): 133-142.

329 Cheng, Tso-hsin, 1963. Subspecific differentiation of the Red-winged Shrike-Babbler (*Pteruthius flaviscapis*) in China including a new subspecies *P.f.lingshuiensis*. — *Acta Zool. Sinica*, 15 (4): 639-647.

330 Cheng, Tso-hsin, 1963. Subspecific differentiation of the two sibling species of the Necklaced Laughing Thrushes (*Garrulax pectoralis* and *Garrulax monilegerus*) in China, including a new subspecies *Garrulax pectoralis pingi*. — *Acta Zool. Sinica*, 15 (3): 471-478.

331 Cheng, Tso-hsin, 1974. A new name for *Pomatorhinus ruficollis intermedius* Cheng, 1962. — *Acta Zool. Sinica*, 20 (1): 108.

332 Cheng, Tso-hsin, 1984. A systematic review of crow-tits (*Paradoxornis*) hitherto recorded from China. — *Acta Zool. Sinica*, 30: 278-285.

333 Cheng, Tso-hsin, 1987. *A synopsis of the Avifauna of China*: i-xvi, 1-1233. — Science Press, Beijing.

334 Cheng, Tso-hsin, 1993. A discussion of the taxonomy of the Rufous-headed Crowtit. — *Acta Zootax. Sinica*, 18: 108-113.

335 Cheng, Tso-hsin, 1997. On the systematic status *Crossoptilon harmani*. — *Ann. Rev. World Pheasant Assoc.*, 1996/1997: 31-32.

336 Cheng, Tso-hsin & Tang, Rui-chang, 1982. A new subspecies of *Garrulax galbanus* from Yunnan, China — *Garrulax galbanus simaoensis*. — *Sinozoologia*, 2 (2): 1-2.

337 Cheng, Tso-hsin & Wu, Minchuan, 1979. A new subspecies of *Tragopan caboti - Tragopan caboti guangxiensis*. — *Acta Zool. Sinica*, 25 (3): 292-294.

338 Cheng, Tso-hsin *et al.*, 1963. Taxonomic studies on birds from southwestern Szechwan and northwestern Yunnan. Pt. III. Passeriformes (cont.). — *Acta Zool. Sinica*, 15 (2): 295-316.

339 Cheng, Tso-hsin *et al.*, 1964. A new subspecies of the Velvet-fronted Nuthatch from Hainan — *Sitta frontalis chienfengensis*. — *Acta Zootax. Sinica*, 1 (1): 1-5.

340 Cheng, Tso-hsin *et al.*, 1964. A new subspecies of the Silver Pheasant from Szechwan — *Lophura nycthemera omeiensis*. — *Acta Zootax. Sinica*, 1 (2): 221-228.

341   Cheng, Tso-hsin *et al.*, 1973. A new Three-toed Parrotbill from Tsinling Range, Shensi, China. — *Acta Zool. Sinica*, 19 (1): 48-50.

342   Cheng, Tso-hsin *et al.*, 1973. A new subspecies of *Treron sieboldii* from the Tsingling Range, Shensi, China. — *Acta Zool. Sinica*, 19 (1): 51-53.

343   Cheng, Tso-hsin *et al.*, 1975. A new subspecies of *Dendrocopos major* from the Inner Mongolian Autonomous Region - *Dendrocopos major wulashanicus*. — *Acta Zool. Sinica*, 21 (4): 385-388.

344   Cheng, Tso-hsin *et al.*, 1983. A new subspecies of *Paradoxornis zappeyi - P. z. erlangshanicus*. — *Acta Zootax. Sinica*, 8 (3): 327-329.

345   Cheng, Tso-hsin *et al.*, 1987. *Fauna Sinica Aves XI: Passeriformes; Muscicapidae II. Timaliinae.* i-v, 1-307. — Science Press, Beijing.

346   Chesser, R.T., 2000. Evolution in the high Andes: the phylogenetics of *Muscisaxicola* ground-tyrants. — *Molec. Phylogen. Evol.*, 15 (3): 369-380.

347   Chou, Yu-yuah, 1955. On a new Green Imperial Pigeon from the district of Lo-fao-shan, Kwangtung, China. — *J. Zhongshan Univ.* (Zhongshandaxue Xuebao), 3: 128-130.

348   Christian, P.D. *et al.*, 1992. Biochemical systematics of the Australian dotterels and plovers (Charadriiforms, Charadriidae). — *Austral. J. Zool.*, 40 (2): 225-233.

349   Christidis, L., 1987. Biochemical systematics within palaeotropic finches (Aves: Estrildidae). — *Auk*, 104: 380-392.

350   Christidis, L., 1987. Phylogeny and systematics of estrildine finches and their relationships to other seed-eating passerines. — *Emu*, 87: 117-123.

351   Christidis, L. & W.E. Boles, 1994. *The taxonomy and species of birds of Australia and its territories.* R.A.O.U. Monograph 2. i-iv, 1-112. — R.A.O.U., Melbourne.

352   Christidis, L. & R. Schodde, 1993. Relationships and radiations in the meliphagine honeyeaters, *Meliphaga, Lichenostomus* and *Xanthotis* (Aves: Meliphagidae); protein evidence and its integration with morphology and ecogeography. — *Austral. J. Zool.*, 41: 293-316.

353   Christidis, L. *et al.*, 1988. Genetic and morphological differentiation and phylogeny in the Australo-Papuan scrubwrens (*Sericornis*, Acanthizidae). — *Auk*, 105: 616-629.

354   Chu, P.C., 1995. Phylogenetic reanalysis of Strauch's osteological data set for the Charadriiformes. — *Condor*, 97: 174-196.

355   Cibois, A., 2000. Phylogénie et biogéographie des Timaliidae. Museum national d'Histoire naturelle; Zoology: Mammals and birds, PhD. thesis, Paris, 1-243.

356   Cibois, A. & E. Pasquet, 1999. Molecular analysis of the phylogeny of 11 genera of the Corvidae. — *Ibis*, 141 (2): 297-306.

357   Cibois, A. *et al.*, 1999. Molecular systematics of the Malagasy Babblers (Passeriformes: Timaliidae) and Warblers (Passeriformes: Sylviidae), based on Cytochrome *b* and 16S rRNA sequences. — *Molec. Phylogen. Evol.*, 13 (3): 581-595.

358   Cibois, A. *et al.*, 2001. An endemic radiation of Malagasy songbirds is revealed by mitochondrial DNA sequence data. — *Evolution*, 55 (6): 1198-1206.

359   Cibois, A. *et al.*, 2002. Molecular phylogenetics of babblers (Timaliidae): re-evaluation of the genera *Yuhina* and *Stachyris*. — *J. Avian Biol.*, 33: 380-390.

360   Cicero, C., 1996. Sibling species of Titmice in the *Parus inornatus* complex (Aves : Paridae). — *Univ. Calif. Publ. Zool.*, 128: 1-217.

361   Clancey, P.A., 1939. A new race of Moorhen from Scotland. — *Bull. Brit. Orn. Cl.*, 59: 69-70.

362   Clancey, P.A., 1949. Some remarks on *Charadrius hiaticula* Linnaeus in the Western Palaearctic Region with special reference to the western populations formerly known as *Ch. h. major* Seebohm, 1885. — *Limosa*, 22 (3): 318-319.

363   Clancey, P.A., 1950. A new race of *Columba palumbus* Linné from the western Palaearctic. Pages 89-92 In Syllegomena biologica (Kleinschmidt-Festschrift) (A. von Jordans & F. Peus, eds.). — Geest & Portig Akad. Verlagsgesellsch., Leipzig and A. Ziemsen Verlag, Wittenburg.

364   Clancey, P.A., 1951. The characters of a new race of *Alcedo semitorquata* Swainson from the low country of Portuguese East Africa. — *Ostrich*, 22: 176-178.

365   Clancey, P.A., 1951. Notes on birds of the South African subcontinent. — *Ann. Natal Mus.*, 12: 137-152.

366   Clancey, P.A., 1952. Miscellaneous taxonomic notes on African birds. I. 2. The South African races of Wahlberg's Honeyguide *Prodotiscus regulus* Sundevall. — *Durban Mus. Novit.*, 4 (1): 8-11.

367   Clancey, P.A., 1952. Miscellaneous taxonomic notes on African birds. I. 3. The South African races of the Red-throated Wryneck *Jynx ruficollis* Wagler, 1830: Kaffirland, South Africa. — *Durban Mus. Novit.*, 4 (1): 11-13.

368   Clancey, P.A., 1952. A systematic account of the birds collected on the Natal Museum Expedition to the Lebombo Mountains and Tongaland, July, 1951. — *Ann. Natal Mus.*, 12: 227-274.

369   Clancey, P.A., 1952. Taxonomic studies of South African birds. 1. Geographical variation in the Ground Woodpecker *Geocolaptes olivaceus* (Gmelin), a unique South African avian endemism. — *J. Sci. Soc. Univ. Natal*, 8: 3-8.

370   Clancey, P.A., 1953. Miscellaneous taxonomic notes on African Birds. III. Four new races of Birds from eastern and south-eastern Africa. — *Durban Mus. Novit.*, 4 (4): 57-64.

371   Clancey, P.A., 1953. A new geographical race of the Golden-tailed Woodpecker from south-eastern Africa; with notes on the other South African forms. — *Ostrich*, 24: 167-170.

372   Clancey, P.A., 1956. Miscellaneous taxonomic notes on African Birds. VI. 1. On the races of *Buccanodon whytii* (Shelley). — *Durban Mus. Novit.*, 4 (15): 245-251.

373   Clancey, P.A., 1956. Miscellaneous taxonomic notes on African birds. VII. 1. The South African races of the Black-collared Barbet *Lybius torquatus* (Dumont). — *Durban Mus. Novit.*, 4 (17): 273-280.

374   Clancey, P.A., 1957. A new race of *Francolinus africanus* Stephens from the Drakensberg Mountains of South Africa. — *Bull. Brit. Orn. Cl.*, 77: 58-59.

375   Clancey, P.A., 1957. The South African races of the Speckled Mousebird. — *Bull. Brit. Orn. Cl.*, 77: 26-29.

376   Clancey, P.A., 1958. Geographical variation in the Knysa Woodpecker *Campethera notata* (Lichtenstein). — *Bull. Brit. Orn. Cl.*, 78: 31-35.

377   Clancey, P.A., 1958. The South African races of the Bearded Woodpecker *Thripias namaquus* (Lichtenstein). — *Bull. Brit. Orn. Cl.*, 78: 35-43.

378   Clancey, P.A., 1958. Miscellaneous taxonomic notes on African birds. X. 1. Four new races of passerine birds from South Africa. — *Durban Mus. Novit.*, 5 (8): 99-105.

379   Clancey, P.A., 1958. The South African races of the Yellow-throated Longclaw *Macronyx croceus* (Vieillot). — *Ostrich*, 29: 75-78.

380   Clancey, P.A., 1959. Miscellaneous taxonomic notes on African birds. XII. 1. Geographical variation in the Narina Trogon *Apaloderma narina* (Stephens) of Africa. — *Durban Mus. Novit.*, 5 (12): 151-165.

381   Clancey, P.A., 1959. Miscellaneous taxonomic notes on African birds. XII. 2. The South African races of the Orange-breasted Bush-Shrike *Malaconotus sulfureopectus* (Lesson). — *Durban Mus. Novit.*, 5 (12): 166-172.

382   Clancey, P.A., 1959. Geographical variation in the South African populations of the Red-eyed Bulbul *Pycnonotus nigricans* (Vieillot). — *Bull. Brit. Orn. Cl.*, 79: 166-170.

383   Clancey, P.A., 1959. Miscellaneous taxonomic notes on African birds. XIII. 2. Five new races of southern African birds. — *Durban Mus. Novit.*, 5 (16): 208-218.

384   Clancey, P.A., 1960. Miscellaneous taxonomic notes on African birds. XIV. 2. Notes on variation in the South African populations of the Yellow-billed Hornbill *Tockus flavirostris* (Ruppell), with the characters of a new race. — *Durban Mus. Novit.*, 5 (18): 238-242.

385   Clancey, P.A., 1960. Miscellaneous taxonomic notes on African Birds. XV. 1. A new race of Shelley's Francolin *Francolinus shelleyi* Ogilvie-Grant from Natal and Zululand. — *Durban Mus. Novit.*, 6 (2): 13-14.

386   Clancey, P.A., 1960. Miscellaneous taxonomic notes on African Birds. XV. 8. The South African races of the Gorgeous Bush Shrike *Telophorus quadricolor* (Cassin). — *Durban Mus. Novit.*, 6 (2): 38-41.

387   Clancey, P.A., 1960. A new race of Crowned Plover *Vanellus* (*Stephanibyx*) *coronatus* (Boddaert) from South-West Africa. — *Bull. Brit. Orn. Cl.*, 80: 13-14.

388 Clancey, P.A., 1960. The races of the Bokmakierie *Telphorus zeylonus* (Linnaeus), with the characters of a new form from South-West Africa. — *Bull. Brit. Orn. Cl.*, 80: 121-124.

389 Clancey, P.A., 1961. Miscellaneous taxonomic notes on African birds. XVI. 2. Polytypic variation in the South African sub-continental populations of the Yellow-fronted Tinker Barbet *Pogoniulus chrysoconus* (Temminck). — *Durban Mus. Novit.*, 6 (6): 80-85.

390 Clancey, P.A., 1961. Geographical variation in the South African populations of the Magpie-Shrike *Lanius melanoleucos* Jardine. — *Bull. Brit. Orn. Cl.*, 81: 52-54.

391 Clancey, P.A., 1962. An additional race of *Buphagus erythrorhynchus* (Stanley) from the Somali Arid District. — *Bull. Brit. Orn. Cl.*, 82: 19-20.

392 Clancey, P.A., 1962. Miscellaneous taxonomic notes on African birds. XIX. 1. Two new geographical races of the Yellow-bill *Ceuthmochares aereus* (Vieillot). — *Durban Mus. Novit.*, 6 (15): 181-185.

392A Clancey, P.A., 1962. Miscellaneous taxonomic notes on African birds. XIX. 2. On the name of the Lake Dow, Bechuanaland, form of *Certhilauda albofasciata* Lafresnaye. — *Durban Mus. Novit.*, 6 (15): 185-186.

393 Clancey, P.A., 1962. Miscellaneous taxonomic notes on African Birds. XVIII. Six new races of birds from southern Africa. — *Durban Mus. Novit.*, 6 (13): 149-160.

394 Clancey, P.A., 1963. Miscellaneous taxonomic notes on African birds. XX. 1. The South African races of the broadbill *Smithornis capensis* (Smith). — *Durban Mus. Novit.*, 6 (19): 231-241.

395 Clancey, P.A., 1963. Miscellaneous taxonomic notes on African birds. XX. 3. Notes, mainly systematic, on some birds from the Cape Province. — *Durban Mus. Novit.*, 6 (19): 244-264.

396 Clancey, P.A., 1964. Miscellaneous taxonomic notes on African birds. XXI. [1.] Geographical variation in the Go-away Bird *Corythaixoides concolor* (Smith). — *Durban Mus. Novit.*, 7 (5): 125-130.

397 Clancey, P.A., 1964. Miscellaneous taxonomic notes on African birds. XXI. [2.] A new subspecies of Red-billed Hornbill *Tockus erythrorhynchus* (Temminck) from the south-eastern lowlands of Africa. — *Durban Mus. Novit.*, 7 (5): 130-132.

398 Clancey, P.A., 1964. *The Birds of Natal, Zululand:* i-xxxiv, 1-511. — Oliver & Boyd, Edinburgh.

399 Clancey, P.A., 1964. Miscellaneous taxonomic notes on African birds. XXI. [3.] The geographical races of the Yellow-throated Sparrow *Petronia superciliaris* (Blyth). — *Durban Mus. Novit.*, 7 (5): 132-137.

400 Clancey, P.A., 1964. Miscellaneous taxonomic notes on African birds. XXI. [4.] New subspecies of the Greater Sparrow *Passer iagoensis* (Gould) and Black-cheeked Waxbill *Estrilda erythronotos* (Vieillot) from South Africa. — *Durban Mus. Novit.*, 7 (5): 137-140.

401 Clancey, P.A., 1964. Miscellaneous taxonomic notes on African birds. XXII. 3. On the South African races of the Long-billed Pipit *Anthus similis* Jerdon. — *Durban Mus. Novit.*, 7 (7): 177-182.

402 Clancey, P.A., 1964. Subspeciation in the South African populations of the Scrub Robin *Erythropygia leucophrys* (Vieillot). — *Arnoldia* (Rhod.), 1 (11): 1-12.

403 Clancey, P.A., 1965. A catalogue of birds of the South African Sub-Region. Pt. II. Families Glareolidae - Pittidae. — *Durban Mus. Novit.*, 7: 305-388.

404 Clancey, P.A., 1965. Miscellaneous taxonomic notes on African birds. XXIII. 1. The austral races of the nightjar *Caprimulgus fossii* Hartlaub. — *Durban Mus. Novit.*, 8 (1): 1-9.

405 Clancey, P.A., 1965. Miscellaneous taxonomic notes on African birds. XXIII. 2. Racial variation in the rock-thrush *Monticola angolensis* Sousa. — *Durban Mus. Novit.*, 8 (1): 9-17.

406 Clancey, P.A., 1965. On the type locality of *Campethera abingoni abingoni* (Smith), 1836. — *Bull. Brit. Orn. Cl.*, 85: 64-65.

407 Clancey, P.A., 1965. Variation in the white-crowned shrike *Eurocephalus anguitimens* Smith, 1836. — *Arnoldia* (Rhod.), 1 (23): 1-3.

408 Clancey, P.A., 1966. A catalogue of birds of the South African sub-region (Part III: families Alaudidae - Turdidae). — *Durban Mus. Novit.*, 7 (11): 389-464.

409 Clancey, P.A., 1966. A catalogue of birds of the South African sub-region (Part IV. Families Sylviidae - Prionopidae). — *Durban Mus. Novit.*, 7 (13): 465-544.

410 Clancey, P.A., 1967. Miscellaneous taxonomic notes on African birds. XXIV. Subspeciation in the green pigeon *Treron australis* (Linnaeus) in southern Africa. — *Durban Mus. Novit.*, 8 (7): 53-67.

411 Clancey, P.A., 1967. Miscellaneous taxonomic notes on African birds. XXV. Formal descriptions of four new races of African birds. — *Durban Mus. Novit.*, 8 (10): 109-114.

412 Clancey, P.A., 1967. A new race of the Little Bee-eater from the South West Arid District of Africa. — *Bull. Brit. Orn. Cl.*, 87: 166-167.

413 Clancey, P.A., 1968. On geographical variation in the Whitebrowed Sparrow-Weaver *Plocepasser mahali* Smith of Africa. — *Bonn. Zool. Beitr.*, 19: 257-268.

414 Clancey, P.A., 1968. Seasonal movement and variation in the southern populations of the Dusky Lark *Pinarocorys nigricans* (Sundevall). — *Bull. Brit. Orn. Cl.*, 88: 166-171.

415 Clancey, P.A., 1968. Subspeciation in some birds from Rhodesia. Part 1. — *Durban Mus. Novit.*, 8: 115-152.

416 Clancey, P.A., 1969. Miscellaneous taxonomic notes on African birds. XXVII. 3. The southern forms of *Hirundo abyssinica* Guerin-Méneville. — *Durban Mus. Novit.*, 8 (15): 231-239.

417 Clancey, P.A., 1969. Miscellaneous taxonomic notes on African birds. XXVII. 4. An adjustment to the names of *Erythropygia coryphaeus* (Lesson) forms. — *Durban Mus. Novit.*, 8 (15): 240-241.

418 Clancey, P.A., 1969. Miscellaneous taxonomic notes on African birds. XXVII. 6. Systematic and distributional notes on Moçambique birds. — *Durban Mus. Novit.*, 8 (15): 243-274.

419 Clancey, P.A., 1970. Miscellaneous taxonomic notes on African birds. XXVIII. [5.] A further subspecies of *Lanius souzae* Bocage. — *Durban Mus. Novit.*, 8 (17): 340-344.

420 Clancey, P.A., 1970. Miscellaneous taxonomic notes on African birds. XXVIII. [6.] On the status and range of *Euplectes progne delacouri* Wolters, 1953. — *Durban Mus. Novit.*, 8 (17): 344-348.

421 Clancey, P.A., 1970. On *Smithornis capensis suahelicus* Grote, 1926. — *Bull. Brit. Orn. Cl.*, 90: 164-166.

422 Clancey, P.A., 1970. Miscellaneous taxonomic notes on African birds. XXIX. [2.] A name for a form of Greater Honeyguide. — *Durban Mus. Novit.*, 8 (20): 377-378.

423 Clancey, P.A., 1970. Miscellaneous taxonomic notes on African birds. XXX. [1.] Variation in the southern African populations of the Laughing Dove *Streptopelia senegalensis* (Linnaeus). — *Durban Mus. Novit.*, 9 (1): 1-8.

424 Clancey, P.A., 1970. Miscellaneous taxonomic notes on African birds. XXX. [2.] Two new subspecies of passerine birds from western Angola. — *Durban Mus. Novit.*, 9 (1): 8-11.

425 Clancey, P.A., 1970. On the South African race of the Little Spotted Woodpecker. — *Ostrich*, 42: 119-122.

426 Clancey, P.A., 1970. Miscellaneous taxonomic notes on African birds. XXXI. A new isolate subspecies of *Nectarinia afra* (Linnaeus) from Moçambique. — *Durban Mus. Novit.*, 9 (3): 25-28.

427 Clancey, P.A., 1970. The southern populations of *Amadina fasciata* (Gmelin) (Aves, Ploceidae). — *Zool. Abh. Staatl. Mus. Tierk. Dresden*, 31: 51-54.

428 Clancey, P.A., 1971. Miscellaneous taxonomic notes on African birds. XXXII. 1. On the present nominate subspecies of *Merops superciliosus* Linnaeus. — *Durban Mus. Novit.*, 9 (5): 39-44.

429 Clancey, P.A., 1971. Miscellaneous taxonomic notes on African birds. XXXII. 2. A name for an undescribed race of *Pogoniulus bilineatus* (Sundevall) from Malawi. — *Durban Mus. Novit.*, 9 (5): 44-46.

430 Clancey, P.A., 1971. Miscellaneous taxonomic notes on African birds. XXXII. 7. A name for the Rhodesian isolates of *Cryptospiza reichenovii* (Hartlaub). — *Durban Mus. Novit.*, 9 (5): 57.

431 Clancey, P.A., 1971. Miscellaneous taxonomic notes on African birds. XXXIII. [1.] Variation in Kittlitz's Sandplover *Charadrius pecuarius* Temminck. — *Durban Mus. Novit.*, 9 (9): 109-112.

432 Clancey, P.A., 1971. Miscellaneous taxonomic notes on African birds. XXXIII. [2.] The southern African races of the Whitefronted Sandplover *Charadrius marginatus* Vieillot. — *Durban Mus. Novit.*, 9 (9): 113-118.

433 Clancey, P.A., 1971. Miscellaneous taxonomic notes on African birds. XXXIII. [4.] A new flood-plain race of the Red-capped Lark *Calandrella cinerea* (Gmelin) from southern Moçambique. — *Durban Mus. Novit.*, 9 (9): 120-122.

434 Clancey, P.A., 1971. Miscellaneous taxonomic notes on African birds. XXXIII. [5.] On the southern range limits of *Nilaus afer nigritemporalis* Reichenow, 1892. — *Durban Mus. Novit.*, 9 (9): 122-129.

435 Clancey, P.A., 1972. Miscellaneous taxonomic notes on African birds. XXXIV. [2.] New races of a rockthrush and a warbler from Angola. — *Durban Mus. Novit.*, 9 (11): 146-152.

436 Clancey, P.A., 1972. Miscellaneous taxonomic notes on African birds. XXXIV. [3.] On the western Equatorial peripheral isolates of *Pogonocichla stellata* (Vieillot). — *Durban Mus. Novit.*, 9 (11): 152-154.

437 Clancey, P.A., 1972. A catalogue of birds of the South African sub-region, Supplement. No. II. — *Durban Mus. Novit.*, 9: 165-200.

438 Clancey, P.A., 1972. Miscellaneous taxonomic notes on African birds. XXXV. [1.] An undescribed race of the Pinkbilled Lark *Calandrella conirostris* (Sundevall) from the Transvaal. — *Durban Mus. Novit.*, 9 (16): 233-236.

439 Clancey, P.A., 1972. Miscellaneous taxonomic notes on African birds. XXXV. [3.] A name for an undescribed race of *Cossypha caffra* (Linnaeus). — *Durban Mus. Novit.*, 9 (16): 244-245.

440 Clancey, P.A., 1973. Miscellaneous taxonomic notes on African birds. XXXVI. A new race of *Lamprotornis mevesii* (Wahlberg) from north-western South West Africa and adjacent Angola. — *Durban Mus. Novit.*, 9 (18): 279-284.

441 Clancey, P.A., 1973. Miscellaneous taxonomic notes on African birds. XXXVII. [3.] An undescribed race of *Nectarinia mariquensis* (Smith) from the south-east African lowlands. — *Durban Mus. Novit.*, 10 (1): 12-13.

442 Clancey, P.A., 1974. Subspeciation studies in some Rhodesian birds. — *Arnoldia*, 6 (28): 1-43.

443 Clancey, P.A., 1974. Miscellaneous taxonomic notes on African birds. XXXIX. [3.] Variation in the Bearded Scrub Robin *Erythropygia barbata* (Hartlaub and Finsch), 1870. — *Durban Mus. Novit.*, 10 (7): 95-98.

444 Clancey, P.A., 1974. Miscellaneous taxonomic notes on African birds. XXXIX. [7.] On the validity of *Habropyga poliogastra* Reichenow, 1886. — *Durban Mus. Novit.*, 10 (7): 107-108.

445 Clancey, P.A., 1974. Miscellaneous taxonomic notes on African birds. XL. The *hartlaubii* subspecies-group of *Turdoides leucopygius* (Rüppell), with the characters of a new race from Botswana. — *Durban Mus. Novit.*, 10 (11): 147-150.

446 Clancey, P.A., 1974. Subspeciation studies in some Rhodesian birds. — *Arnoldia*, 6 (28): 1-43.

447 Clancey, P.A., 1975. On the endemic birds of the montane evergreen forest biome of the Transvaal. — *Durban Mus. Novit.*, 10 (12): 151-180.

448 Clancey, P.A., 1975. Miscellaneous taxonomic notes on African birds. XLI. [3.] The austral African races of *Salpornis spilonotus* Franklin. — *Durban Mus. Novit.*, 10 (14): 206-208.

449 Clancey, P.A., 1975. Miscellaneous taxonomic notes on African birds. XLIII. [2.] Subspeciation in the Sharpbilled or Wahlberg's Honeyguide *Prodotiscus regulus* Sundevall. — *Durban Mus. Novit.*, 11 (1): 9-15.

450 Clancey, P.A., 1975. Miscellaneous taxonomic notes on African birds. XLIII. [3.] An undescribed race of the forest-dwelling drongo *Dicrurus ludwigii* (Smith). — *Durban Mus. Novit.*, 11 (1): 15-17.

451 Clancey, P.A., 1975. Miscellaneous taxonomic notes on African birds. XLIII. [4.] An additional subspecies of the Redeyed Bulbul *Pycnonotus nigricans* (Vieillot). — *Durban Mus. Novit.*, 11 (1): 17-20.

452 Clancey, P.A., 1975. Miscellaneous taxonomic notes on African birds. XLIII. [5.] On the present nominate subspecies of *Nectarinia amethystina* (Shaw). — *Durban Mus. Novit.*, 11 (1): 20-24.

453 Clancey, P.A., 1975. The southern African races of the Whitebrowed Scrub Robin *Erythropygia leucophrys* (Vieillot) - some further considerations. — *Bonn. Zool. Beitr.*, 26: 175-182.

454 Clancey, P.A., 1976. Miscellaneous taxonomic notes on African birds. XLIV. [2.] Further on the nominate subspecies of *Dicrurus adsimilis* (Bechstein), 1794: Duiwenhok R., Swellendam, south-western Cape. — *Durban Mus. Novit.*, 11 (4): 88-92.

455 Clancey, P.A., 1976. Miscellaneous taxonomic notes on African birds. XLIV. [3.] Subspeciation in the Squaretailed Drongo *Dicrurus ludwigii* (A. Smith), 1834. — *Durban Mus. Novit.*, 11 (4): 92-101.

456 Clancey, P.A., 1977. Miscellaneous taxonomic notes on African birds. XLVII. [1.] The characters and range limits of the nominate subspecies of *Indicator minor* Stephens. — *Durban Mus. Novit.*, 11 (10): 181-187.

457 Clancey, P.A., 1977. Miscellaneous taxonomic notes on African birds. XLVII. [5.] The south-eastern brown-backed subspecies of the Forest Weaver *Ploceus bicolor* Vieillot. — *Durban Mus. Novit.*, 11 (10): 201-208.

458 Clancey, P.A., 1977. Miscellaneous taxonomic notes on African birds. XLVIII. [1.] On southern African *Indicator variegatus* Lesson. — *Durban Mus. Novit.*, 11 (11): 213-215.

459 Clancey, P.A., 1977. Miscellaneous taxonomic notes on African birds. XLVIII. [2.] A further subspecies of *Indicator exilis* (Cassin). — *Durban Mus. Novit.*, 11 (11): 215-217.

460 Clancey, P.A., 1977. Miscellaneous taxonomic notes on African birds. XLVIII. [3.] Preliminary steps in the study of the polytypic status of the Pintailed Whydah *Vidua macroura* (Pallas). — *Durban Mus. Novit.*, 11 (11): 217-222.

461 Clancey, P.A., 1977. Miscellaneous taxonomic notes on African birds. XLIX. [1.] Subspecific variation in the Redcrested Korhaan *Eupodotis (Lophotis) ruficrista* (Smith), 1836. — *Durban Mus. Novit.*, 11 (12): 223-227.

462 Clancey, P.A., 1977. Miscellaneous taxonomic notes on African birds. XLIX. [3.] Variation in the southern populations of *Bubalornis niger* A. Smith, 1836. — *Durban Mus. Novit.*, 11 (12): 229-238.

463 Clancey, P.A., 1977. Miscellaneous taxonomic notes on African birds. L. [3.] A name for *Lybius zombae* auct., nec. Shelley, 1893. — *Durban Mus. Novit.*, 11 (14): 251-252.

464 Clancey, P.A., 1978. Miscellaneous taxonomic notes on African birds. LI. [4.] The tropical African races of the Redcapped Lark *Calandrella cinerea* (Gmelin). — *Durban Mus. Novit.*, 11 (16): 281-286.

465 Clancey, P.A., 1978. Miscellaneous taxonomic notes on African birds. LII. [4.] On the southern and eastern races of *Nectarinia olivacea* (Smith), 1840. — *Durban Mus. Novit.*, 11 (19): 317-327.

466 Clancey, P.A., 1979. Miscellaneous taxonomic notes on African birds. LIII. [2.] An overlooked race of Wattled Plover. — *Durban Mus. Novit.*, 12 (1): 5-6.

467 Clancey, P.A., 1979. Miscellaneous taxonomic notes on African birds. LIII. [3.] A name for an undescribed subspecies of *Sterna bergii* Lichtenstein. — *Durban Mus. Novit.*, 12 (1): 6.

468 Clancey, P.A., 1979. Miscellaneous taxonomic notes on African birds. LIII. [5.] An additional subspecies of *Colius striatus* Gmelin from southern Malawi. — *Durban Mus. Novit.*, 12 (1): 8-11.

469 Clancey, P.A., 1979. Miscellaneous taxonomic notes on African birds. LIII. [6.] The subspecies of the Scaly-throated Honeyguide *Indicator variegatus* Lesson. — *Durban Mus. Novit.*, 12 (1): 11-15.

470 Clancey, P.A., 1979. Miscellaneous taxonomic notes on African birds. LIV. [2.] A further race of *Cossypha heuglini* Hartlaub from the South West Africa/Angola border country. — *Durban Mus. Novit.*, 12 (4): 41-42.

471 Clancey, P.A., 1979. Miscellaneous taxonomic notes on African birds. LIV. [5.] An overlooked subspecies of the Whitebrowed Sparrow-Weaver *Plocepasser mahali* Smith. — *Durban Mus. Novit.*, 12 (4): 45-46.

472   Clancey, P.A., 1979. Miscellaneous taxonomic notes on African birds. LV. [3.] A new isolate subspecies of Woodward's Barbet *Cryptolybia woodwardi* (Shelley) from south-eastern Tanzania. — *Durban Mus. Novit.*, 12 (5): 50-52.

473   Clancey, P.A., 1979. Miscellaneous taxonomic notes on African birds. LV. [4.] A further race of *Parus rufiventris* Bocage from the middle and lower Okavango R. drainage. — *Durban Mus. Novit.*, 12 (5): 52-54.

474   Clancey, P.A., 1979. Miscellaneous taxonomic notes on African birds. LV. [5.] A second southern race of *Turdoides melanops* (Hartlaub) of the Afrotropical region. — *Durban Mus. Novit.*, 12 (5): 54-55.

475   Clancey, P.A., 1979. Miscellaneous taxonomic notes on African birds. LV. [6.] An adjustment to the southern races of *Chlorocichla flaviventris* (Smith). — *Durban Mus. Novit.*, 12 (5): 55-57.

476   Clancey, P.A., 1979. Miscellaneous taxonomic notes on African birds. LV. [9.] An undescribed race of *Lagonosticta senegala* (Linnaeus). — *Durban Mus. Novit.*, 12 (5): 60-61.

477   Clancey, P.A., 1980. On birds of the mid-Okavango Valley on the South West Africa/Angola border. — *Durban Mus. Novit.*, 12: 87-127.

478   Clancey, P.A., 1980. Miscellaneous taxonomic notes on African birds. LVI. [1.] Variation in *Nicator gularis* Hartlaub and Finsch. — *Durban Mus. Novit.*, 12 (10): 129-134.

479   Clancey, P.A., 1980. Miscellaneous taxonomic notes on African birds. LVII. [2.] The subspecific status of the southern African Yellowbilled Oxpecker *Buphagus africanus* Linnaeus population. — *Durban Mus. Novit.*, 12 (12): 145-149.

480   Clancey, P.A., 1980. Miscellaneous taxonomic notes on African birds. LVIII. [1.] The mainland Afrotropical subspecies of the Little Swift *Apus affinis* (Gray). — *Durban Mus. Novit.*, 12 (13): 151-156.

481   Clancey, P.A., 1980. Miscellaneous taxonomic notes on African birds. LVIII. [3.] On the availability of the name *Dendropicos hartlaubi noomei* Roberts, 1924, for the Mozambique race of the Cardinal Woodpecker. — *Durban Mus. Novit.*, 12 (13): 158-161.

482   Clancey, P.A. (ed) 1980. S.A.O.S. *Checklist of southern African birds.* i-xiii, 1-325. — Southern African Ornithological Society.

483   Clancey, P.A., 1981. Miscellaneous taxonomic notes on African birds. LIX. [2.] Variation in the present nominate race of *Glareola nuchalis* Gray. — *Durban Mus. Novit.*, 12 (20): 224-227.

484   Clancey, P.A., 1981. Miscellaneous taxonomic notes on African birds. LX. [1.] Variation in the Afrotropical populations of *Charadrius tricollaris* Vieillot. — *Durban Mus. Novit.*, 13 (1): 1-6.

485   Clancey, P.A., 1981. Miscellaneous taxonomic notes on African birds. LX. [2.] Variation in the Chorister Robin *Cossypha dichroa* (Gmelin), 1789. — *Durban Mus. Novit.*, 13 (1): 6-10.

486   Clancey, P.A., 1981. Miscellaneous taxonomic notes on African birds. LXI. [1.] Further on the South African sub-region races of *Cossypha caffra* (Linnaeus). — *Durban Mus. Novit.*, 13 (4): 41-47.

487   Clancey, P.A., 1981. Miscellaneous taxonomic notes on African birds. LXI. [2.] An additional subspecies of *Cossypha natalensis* Smith. — *Durban Mus. Novit.*, 13 (4): 47-48.

488   Clancey, P.A., 1982. Namibian ornithological miscellanea. — *Durban Mus. Novit.*, 13: 55-62.

489   Clancey, P.A., 1982. Miscellaneous taxonomic notes on African birds. LXII. [1.] The Olive Thrush *Turdus olivaceus* Linnaeus in southern Africa. — *Durban Mus. Novit.*, 13 (7): 65-70.

490   Clancey, P.A., 1982. Miscellaneous taxonomic notes on African birds. LXII. [2.] Two new montane subspecies of warblers from the high Drakensberg. — *Durban Mus. Novit.*, 13 (7): 70-73.

491   Clancey, P.A., 1983. An overlooked subspecies of the African Palm Swift *Cypsiurus parvus*. — *Bull. Brit. Orn. Cl.*, 103: 80-81.

492   Clancey, P.A., 1984. Miscellaneous taxonomic notes on African birds. LXIV. [1.] An undescribed Equatorial rainforest race of the Pygmy Kingfisher. — *Durban Mus. Novit.*, 13 (14): 169-172.

493   Clancey, P.A., 1984. Miscellaneous taxonomic notes on African birds. LXIV. [5.] Variation in *Pogoniulus atroflavus* (Sparrman), 1798, of forested Upper and Lower Guinea. — *Durban Mus. Novit.*, 13 (14): 177-178.

494   Clancey, P.A., 1984. Miscellaneous taxonomic notes on African birds. LXIV. [6.] Subspeciation in *Pogoniulus simplex* (Fischer & Reichenow), 1888, of tropical eastern Africa. — *Durban Mus. Novit.*, 13 (14): 179-180.

495   Clancey, P.A., 1984. Miscellaneous taxonomic notes on African birds. 66. [3.] Subspeciation in the Striped Pipit *Anthus lineiventris* Sundevall. — *Durban Mus. Novit.*, 13 (18): 227-232.

496   Clancey, P.A., 1984. Miscellaneous taxonomic notes on African birds. 66. [4.] Further on subspeciation in the Yellowthroated Longclaw *Macronyx croceus* (Vieillot). — *Durban Mus. Novit.*, 13 (18): 232-238.

497   Clancey, P.A., 1984. The relationship of the white-rumped babblers *Turdoides leucopygius* (Rüppell) and *T. hartlaubii* (Bocage). — *Ostrich*, 55: 28-30.

498   Clancey, P.A., 1985. Species limits in the long-billed pipits of the southern Afrotropics. — *Ostrich*, 56: 157-169.

499   Clancey, P.A., 1985. Subspeciation in *Anthus brachyurus* Sundevall, 1850. — *Bull. Brit. Orn. Cl.*, 105: 133-135.

500   Clancey, P.A., 1986. Miscellaneous taxonomic notes on African birds. 67. [3.] Further comment on variation in the Afrotropical Redeyed Dove *Streptopelia semitorquata* (Rüppell), 1837. — *Durban Mus. Novit.*, 14 (2): 9-13.

501   Clancey, P.A., 1986. On the Equatorial populations of *Halcyon albiventris* (Scopoli). — *Bull. Brit. Orn. Cl.*, 106: 78-79.

502   Clancey, P.A., 1986. Breeding season and subspecific variation in the Red-headed Quelea. — *Ostrich*, 57: 207-210.

503   Clancey, P.A., 1986. Miscellaneous taxonomic notes on African birds. 67. [6.] Subspeciation in *Cichladusa guttata* (Heuglin). — *Durban Mus. Novit.*, 14 (2): 17-19.

504   Clancey, P.A., 1986. Miscellaneous taxonomic notes on African birds. 67. [8.] Variation in Hunter's Sunbird *Nectarinia hunteri* (Shelley), 1889. — *Durban Mus. Novit.*, 14 (2): 23-25.

505   Clancey, P.A., 1986. Subspeciation in the pipit *Anthus cinnamomeus* Rüppell of the Afrotropics. — *Gerfaut*, 76: 187-211.

506   Clancey, P.A., 1987. Subspeciation in the Pied Mannikin *Spermestes fringilloides* (Lafresnaye), of the Afrotropics. — *Gerfaut*, 76: 301-305 (1986).

507   Clancey, P.A., 1987. Subspeciation in the Afrotropical Superb Starling *Lamprotornis superbus*. — *Bull. Brit. Orn. Cl.*, 107: 25-27.

508   Clancey, P.A., 1987. On the Red-throated Wryneck *Jynx ruficollis* Wagler, 1830, in East Africa. — *Bull. Brit. Orn. Cl.*, 107 (3): 107-112.

509   Clancey, P.A., 1987. Subspeciation in the Afrotropical Gabar Goshawk *Micronisus gabar*. — *Bull. Brit. Orn. Cl.*, 107: 173-177.

510   Clancey, P.A., 1987. The Tree Pipit *Anthus trivialis* (Linnaeus) in southern Africa. — *Durban Mus. Novit.*, 14 (3): 29-42.

511   Clancey, P.A., 1988. A previously undescribed race of the Stonechat *Saxicola torquata*. — *Bull. Brit. Orn. Cl.*, 108: 62-64.

512   Clancey, P.A., 1988. Relationships in the *Campethera notata*, *C. abingoni* and *C. (a). mombassica* complex of the Afrotropics. — *Bull. Brit. Orn. Cl.*, 108 (4): 169-172.

513   Clancey, P.A., 1989. The status of (*Cursorius temminckii*) *damarensis* Reichenow, 1901. — *Bull. Brit. Orn. Cl.*, 109: 51-53.

514   Clancey, P.A., 1989. Subspeciation in the Larklike Bunting of the southwestern Afrotropics. — *Bull. Brit. Orn. Cl.*, 109: 130-134.

515   Clancey, P.A., 1989. The southern isolate of *Parus rufiventris pallidiventris* Reichenow, 1885. — *Bull. Brit. Orn. Cl.*, 109: 134-137.

516   Clancey, P.A., 1989. Subspeciation in the Sociable Weaver *Philetairus socius* of the South West Arid Zone of Africa. — *Bull. Brit. Orn. Cl.*, 109 (4): 228-232.

517   Clancey, P.A., 1989. The Wood Pipit - a species new to the South West African avifauna. — *Cimbebasia*, 10: 47-50 (1988).

518   Clancey, P.A., 1990. A review of the indigenous pipits (Genus *Anthus* Bechstein: Motacillidae) of the Afrotropics. — *Durban Mus. Novit.*, 15: 42-72.

519   Clancey, P.A., 1990. Variation in the Cinnamon-breasted Warbler of the South West Arid Zone of the Afrotropics. — *Bonn. Zool. Beitr.*, 41: 109-112.

520   Clancey, P.A., 1990. Taxonomic and distributional findings on some birds from Namibia. — *Cimbebasia*, 11: 111-133 (1989).

521   Clancey, P.A., 1991. On the generic status and geographical variation of the Namaqua Prinia. — *Bull. Brit. Orn. Cl.*, 111 (2): 101-104.
522   Clancey, P.A., 1991. The generic status of Roberts' Prinia of the south-eastern Afrotropics. — *Bull. Brit. Orn. Cl.*, 111 (4): 217-222.
523   Clancey, P.A., 1992. Taxonomic comment on southeastern representatives of two wide-ranging African cisticolas. — *Bull. Brit. Orn. Cl.*, 112 (4): 218-225.
524   Clancey, P.A., 1993. Subspeciation in the austral African Thick-billed Lark. — *Bull. Brit. Orn. Cl.*, 113 (3): 173-178.
525   Clancey, P.A., 1994. The austral races of the afrotropical Fiery-necked Nightjar *Caprimulgus pectoralis* Cuvier 1816. — *Bull. Brit. Orn. Cl.*, 114 (1): 48-55.
526   Clancey, P.A., 1994. An additional subspecies of the Croaking Cisticola from the temperate uplands of southern Africa. — *Bull. Brit. Orn. Cl.*, 114 (2): 86-88.
527   Clancey, P.A., 1996. Further on subspeciation in the Red-billed Francolin *Pternistis adspersus* (Waterhouse), 1838. — *Bull. Brit. Orn. Cl.*, 116 (2): 104-108.
528   Clancey, P.A. & M.P.S. Irwin, 1966. The South African races of the Banded Sand Martin *Riparia cincta* (Boddaert). — *Durban Mus. Novit.*, 8 (3): 25-33.
529   Clancey, P.A. & W. Lawson, 1966. A new subspecies of the olive-headed golden weaver from southern Moçambique. — *Durban Mus. Novit.*, 8 (4): 35-37.
530   Clancey, P.A. & W.J. Lawson, 1969. A new race of White-breasted Alethe from Moçambique. — *Bull. Brit. Orn. Cl.*, 89: 4-6.
531   Clancey, P.A. *et al.*, 1981. Variation in the current nominate subspecies of *Pterodroma mollis* (Gould) (Aves: Procellariidae). — *Durban Mus. Novit.*, 12 (18): 203-213.
532   Clark, W.S., 1992. The taxonomy of Steppe and Tawny Eagles, with criteria for separation of museum specimens and live eagles. — *Bull. Brit. Orn. Cl.*, 112 (3): 150-157.
533   Clark, W.S., 1994. Various species accounts (Falconiformes). In *Handbook of the Birds of the World.* Vol. 2: 1-638 (J. del Hoyo, A. Elliott & J. Sargatal, eds.). — Lynx Edicions, Barcelona.
534   Clark, W.S., 1999. Plumage differences and taxonomic status of three similar *Circaetus* snake-eagles. — *Bull. Brit. Orn. Cl.*, 119 (1): 56-59.
535   Cleere, N., 1998. *Nightjars. A guide to nightjars and related birds*: 1-317. — Pica Press, Mountfield, Sussex, U.K.
536   Cleere, N., 1999. Family Caprimulgidae (Nightjars). Pages 302-386. In *Handbook of the Birds of the World.* Vol. 5 (J. del Hoyo, A. Elliott & J. Sargatal, eds.). — Lynx Edicions, Barcelona.
537   Cleere, N., 2001. The validity of the genus *Veles* Bangs, 1918 (Caprimulgidae). — *Bull. Brit. Orn. Cl.*, 121 (4): 278-279.
538   Clement, P., 1993. *Finches and sparrows; an identification guide*: i-ix, 1-500. — A. & C. Black, London.
539   Coates, B.J., 1990. *The birds of New Guinea including the Bismarck Archipelago and Bougainville.* 2. Passerines: 1-576. — Dove Publications, Alderley, Queensland.
540   Coelho, G. & W. Silva, 1998. A new species of *Antilophia* (Passeriformes: Pipridae) from Chapada do Araripe, Ceara, Brazil. — *Ararajuba*, 6 (2): 81-84.
541   Cohn-Haft, M., 1996. Why the Yungas Tody-Tyrant (*Hemitriccus spodiops*) is a *Snethlagea*, and why it matters. — *Auk*, 113 (3): 709-714.
542   Cohn-Haft, M., 1999. Family Nyctibiidae (Potoos). Pages 288-301. In *Handbook of the Birds of the World.* Vol. 5 (J. del Hoyo, A. Elliott & J. Sargatal, eds.). — Lynx Edicions, Barcelona.
543   Cohn-Haft, M. *et al.*, 1997. A new look at the "species poor" Central Amazon: the avifauna north of Manaus, Brazil. — *Orn. Monogr.*, 48: 205-235.
544   Collar, N.J., 1996. Family Otididae (Bustards). Pages 240-273. In *Handbook of the Birds of the World.* Vol. 3 (J. del Hoyo, A. Elliott & J. Sargatal, eds.). — Lynx Edicions, Barcelona.
545   Collar, N.J., 1997. Family Psittacidae (Parrots). Pages 280-477. In *Handbook of the Birds of the World.* Vol. 4 (J. del Hoyo, A. Elliott & J. Sargatal, eds.). — Lynx Edicions, Barcelona.
546   Collar, N.J., 2001. Family Trogonidae (Trogons). Pages 80-127 In *Handbook of the Birds of the World.* Vol. 6 (J. del Hoyo, A. Elliott & J. Sargatal, eds.). — Lynx Edicions, Barcelona.
547   Collar, N.J. & A.J. Long, 1996. Taxonomy and names of *Carpococcyx* cuckoos from the Greater Sundas. — *Forktail*, 11: 135-150 (1995).
548   Collar, N.J. & A.J. Pittman, 1996. *Amazona kawalli* is a valid name for a valid species. — *Bull. Brit. Orn. Cl.*, 116 (4): 256-265.
549   Collins, C.T., 1972. A new species of swift of the genus *Cypseloides* from north-eastern South America (Aves: Apodidae). — *Los Angeles Cty. Mus., Contrib. Sci.*, (229): 1-9.
549A Collinson, M., 2001. Shifting sands: taxonomic changes in the world of the field ornithologist. — *Brit. Birds*, 94: 2-27.
550   Colston, P.R., 1972. A new bulbul from southwestern Madagascar. — *Ibis*, 114 (1): 89-92.
551   Colston, P.R., 1982. A new species of *Mirafra* (Alaudidae) and new races of the Somali Long-billed Lark *Mirafra somalica*, Thekla Lark *Galerida malabarica* and Malindi Pipit *Anthus melindae* from southern coastal Somalia. — *Bull. Brit. Orn. Cl.*, 102: 106-114.
552   Colston, P.R., 1987. A new name for *Anthus melindae pallidus* Colston (1982). — *Bull. Brit. Orn. Cl.*, 107: 92.
553   Colston, P.R. & K. Curry-Lindahl, 1986. The Birds of Mount Nimba, Liberia. — *Publ. Brit. Mus. (Nat. Hist.)*, 982: 1-129.
554   Colston, P.R. & G.J. Morel, 1985. A new subspecies of rufous swamp warbler *Acrocephalus rufescens* from Senegal. — *Malimbus*, 7 (2): 61-62.
555   Condon, H., 1941. The Australian Broad-tailed parrots (subfamily Platycercinae). — *Rec. S. Austral. Mus.*, 7: 117-144.
556   Condon, H., 1975. *Checklist of the birds of Australia. Pt. 1. Non-Passerines*: i-xx, 1-311. — Royal Australian Ornithologists' Union, Melbourne.
557   Conover, H.B., 1934. A new species of Rail from Paraguay. — *Auk*, 51: 365-366.
558   Conover, H.B., 1934. A new Trumpeter from Brazil. — *Proc. Biol. Soc. Wash.*, 47: 119-120.
559   Conover, H.B., 1935. A new race of Ruffed Grouse from Vancouver Island. — *Condor*, 37 (4): 204-206.
560   Conover, H.B., 1937. A new race of *Dactylortyx* from Honduras. — *Proc. Biol. Soc. Wash.*, 50: 73-74.
561   Conover, H.B., 1938. A new Pigeon from Colombia. — *Field Mus. Nat. Hist., Zool.*, 20 (36): 477-478.
562   Conover, H.B., 1938. A new race of Bob White from the Cauca valley, Colombia. — *Proc. Biol. Soc. Wash.*, 51: 53-54.
563   Conover, H.B., 1945. A new race of *Penelope argyrotis* from Colombia. — *Proc. Biol. Soc. Wash.*, 58: 125-126.
564   Conover, H.B., 1949. A new race of *Rallus nigricans* from Columbia. — *Proc. Biol. Soc. Wash.*, 62: 173-174.
565   Contreras, J.R., 1976. Una nueva subespecie de *Geositta rufipennis* procedente de las cercanias de San Carlos de Bariloche, Provincia de Rio Negro, Argentina (Aves, Furnariidae). — *Physis, Sec. C, Buenos Aires*, 35: 213-220.
566   Contreras, J.R., 1979. Una nueva subespecie de Furnariidae *Tripophaga modesta navasi* (Aves, Passeriformes). — *Hist. Nat., Mendoza*, 1: 13-16.
567   Contreras, J.R., 1980. Furnariidae argentinos. IV. Aportes al conocimiento de *Tripophaga steinbachi* (Hartert y Venturi), con la descripcion de *Tripophaga steinbachi neiffi*, nueva subespecie. — *Hist. Nat., Mendoza*, 1: 29-32.
568   Contreras, J.R., 1980. *Geositta rufipennis hoyi*, nueva subespecie y consideraciones sobre *G. rufipennis* en el centro y el oeste Argentinos (Aves: Furnariidae). — *Hist. Nat., Mendoza*, 1 (19): 137-148.
569   Cooper, A. *et al.*, 2001. Complete mitochondrial genome sequences of two extinct moas clarify ratite Evolution. — *Nature*, 409: 704-707.
570   Coopmans, P. & N. Krabbe, 2000. A new species of flycatcher (Tyrannidae: *Myiopagis*) from eastern Ecuador and eastern Peru. — *Wilson Bull.*, 112 (3): 305-312.
571   Coopmans, P. *et al.*, 2001. Vocal evidence of species rank for nominate Unicolored Tapaculo *Scytalopus unicolor*. — *Bull. Brit. Orn. Cl.*, 121 (3): 208-213.
571A Cory, C.B., 1886. Descriptions of thirteen new species of birds from the island of Grand Cayman, West Indies. — *Auk*, 3: 497-501.
572   Cory, C.B., 1919. New forms of South American birds and proposed new subgenera. — *Auk*, 36 (2): 273-276.
573   Cory, C.B. & C.E. Hellmayr, 1924. Catalogue of the Birds of the Americas. — *Field Mus. Nat. Hist., Zool.*, 13 (3): i-vii, 1-369.

574 Cory, C.B. & C.E. Hellmayr, 1925. Catalogue of the Birds of the Americas. — *Field Mus. Nat. Hist., Zool.*, 13 (4): 1-390.

575 Cottam, P.A., 1957. The pelecaniform characters of the skeleton of the Shoebill Stork, *Balaeniceps rex.* — *Bull. Brit. Mus. (Nat. Hist.), Zool.*, 5: 49-72.

576 Cowan, I.M., 1939. The white-tailed Ptarmigan of Vancouver Island. — *Condor*, 41: 82-83.

576A Cowles, G.S., 1964. A new Australian babbler. *Emu*, 64: 1-5.

577 Cowles, G.S., 1980. A new subspecies of *Halcyon chloris* from an isolated population in eastern Arabia. — *Bull. Brit. Orn. Cl.*, 100: 226-230.

578 Cracraft, J., 1981. Towards a phylogenetic classification of the Recent birds of the world (Class Aves). — *Auk*, 98: 681-714.

579 Cracraft, J., 1985. Monophyly and phylogenetic relationships of the Pelecaniformes: a numerical cladistic analysis. — *Auk*, 102: 834-853.

580 Cracraft, J., 1988. The major clades of birds. Pages 339-361. In *The Phylogeny and Classification of the Tetrapods* (M.J. Benton, ed.) — Clarendon Press, Oxford.

581 Cracraft, J., 1992. The species of the birds-of-paradise (Paradisaeidae): applying the phylogenetic species concept to a complex pattern of diversification. — *Cladistics*, 8: 1-43.

582 Cracraft, J., 2001. Avian Evolution, Gondwana biogeography, and the Cretaceous-Tertiary mass extinction event. — *Proc. Roy. Soc., Lond.*, 268B: 459-469.

583 Cracraft, J. & J. Clarke, The basal clades of modern birds. In Proc. Internat. Symp. in honor of John H. Ostrom. *Yale Univerity Press, Peabody Mus. Nat. Hist., New Haven.* 2001: 143-156.

584 Cracraft, J. & J. Feinstein, 2000. What is not a bird-of-paradise? Molecular and morphological evidence places *Macgregoria* in the Meliphagidae and the Cnemophilinae near the base of the corvoid tree. — *Proc. Roy. Soc., Lond.*, B 267: 233-241.

585 Craig, A.J.F.K., 1993. Geographical variation and taxonomy of the genus *Euplectes* (Aves, Ploceidae) Part 1: the short-tailed bishop birds. — *J. African Zool.*, 107 (1): 83-96.

586 Craig, A.J.F.K., 1993. Geographical variation and taxonomy of the genus *Euplectes* (Aves, Ploceidae) Part 2: the long-tailed widow birds. — *J. African Zool.*, 107 (2): 139-151.

587 Craig, A.J.F.K., 2000. Various species accounts. In *The Birds of Africa*. Vol. 6: i-xvii, 1-724 (C.H. Fry, S. Keith & E.K. Urban, eds.). — Academic Press, London.

588 Cramp, S. *et al.*, 1977. *Handbook of the Birds of Europe, the Middle East and North Africa; The Birds of the Western Palearctic*: 1. *Ostrich* to Ducks. i-viii, 1-722. — Oxford University Press, Oxford.

589 Cramp, S. *et al.*, 1980. *Handbook of the Birds of Europe, the Middle East and North Africa; The Birds of the Western Palearctic*: 2. Hawks to Bustards. i-viii, 1-695. — Oxford University Press, Oxford.

590 Cramp, S. *et al.*, 1983. *Handbook of the Birds of Europe, the Middle East and North Africa; The Birds of the Western Palearctic*: 3. Waders to Gulls. i-viii, 1-913. — Oxford University Press, Oxford.

591 Cramp, S. *et al.*, 1985. *Handbook of the Birds of Europe, the Middle East and North Africa; The Birds of the Western Palearctic*: 4. Terns to Woodpeckers. i-ix, 1-960. — Oxford University Press, Oxford.

592 Cramp, S. *et al.*, 1988. *Handbook of the Birds of Europe, the Middle East and North Africa; The Birds of the Western Palearctic*: 5. Tyrant Flycatchers to Thrushes. i-viii, 1-1063. — Oxford University Press, Oxford.

593 Cramp, S. *et al.*, 1992. H*andbook of the Birds of Europe, the Middle East and North Africa; The Birds of the Western Palearctic*: 6. Warblers. [i-vi], 1-728. — Oxford University Press, Oxford.

594 Cramp, S. *et al.*, 1993. *Handbook of the Birds of Europe, the Middle East and North Africa; The Birds of the Western Palearctic*: 7. Flycatchers to Shrikes. [i-vi], 1-577. — Oxford University Press, Oxford.

595 Cramp, S. *et al.*, 1994. *Handbook of the Birds of Europe, the Middle East and North Africa; The Birds of the Western Palearctic*: 8. Crows to Finches. [i-vii], 1-899. — Oxford University Press, Oxford.

596 Cramp, S. *et al.*, 1994. *Handbook of the Birds of Europe, the Middle East and North Africa; The Birds of the Western Palearctic*: 9. Buntings and New World Warblers. [i-viii], 1-488. — Oxford University Press, Oxford.

596A Cranbrook, Lord, 1982. Birds of Borneo by B.E. Smythies. Editorial notes on the 3rd. Edition. — *Sabah Soc. J.*, 7: 148-150.

597 Crome, F.H.J. *et al.*, 1980. An analysis of phenotypic variation in the Spinifex Piegons (*Geophaps plumifera*). — *Austral. J. Zool.*, 28: 135-150.

598 Crossin, R.S. & C.A. Ely, 1973. A new race of Sumichrast's Wren from Chiapas, México. — *Condor*, 75: 137-139.

599 Crowe, T.M. *et al.*, 1986. Various species accounts. In *The Birds of Africa*. Vol. 2: i-xvi, 1-552 (E.K. Urban, C.H. Fry & S. Keith, eds.). — Academic Press, London.

600 Cuervo, A.M. *et al.*, 2001. A new species of piha (Cotingidae: *Lipaugus*) from the Cordillera central of Colombia. — *Ibis*, 143 (3): 353-368.

601 Cunningham-van Someren, G.R., 1984. A new race of Red-billed Oxpecker *Buphagus erythrorhynchus* from Kenya. — *Bull. Brit. Orn. Cl.*, 104: 120-121.

602 Cunningham-van Someren, G.R. & H. Schifter, 1981. New races of montane birds from Kenya and southern Sudan. — *Bull. Brit. Orn. Cl.*, 101 (3): 347-354, (4) 355-363.

603 da Rosa Pinto, A., 1959. Algunas novos do aves para o Sul do Save e Moçambique. — *Bol. Soc. Estud. Moçamb.*, 28 (118): 15-25.

604 da Rosa Pinto, A., 1962. As observaçoes de maior destaque das expediçoes do Instituto de Investigação de Angola. — *Bol. Inst. Invest. Cient. Angola*, 1: 21-38.

605 da Rosa Pinto, A., 1963. Notas sobre uma recente colecção de aves de Moçambique do Museu Dr. Álvaro de Castro, com a descrição de duas novas subespécies. — *Mem. Inst. Invest. Cient. Moçamb.*, 5: 31-49.

606 da Rosa Pinto, A., 1967. Descrição de quatro novas subespécies de aves de Angola. — *Bol. Inst. Invest. Cient. Angola*, 4 (2): 29-31.

607 da Rosa Pinto, A., 1968. Algumas formas novas para Angola e outras para a Ciencia descobertas no distrito do Cuando-Cubango. (Angola). — *Bonn. Zool. Beitr.*, 19: 280-288.

608 da Rosa Pinto, A., 1972. Contribuição para o estudo da avifauna do distrito de Cabinda (Angola). — *Mem. Trab. Inst. Invest. Cient. Ang.*, 10: 1-103.

608A Dabbene, R., 1914. Distribution des oiseaux en Argentine d'après ouvrage de Lord Brabourne et Chubb "The Birds of South America". – *Buenos Aires Bol. Soc. Physis*, 1: 241-261, 293-366 (1913).

609 Danforth, C.H., 1934. A new Clapper Rail from Antigua, British West Indies. — *Proc. Biol. Soc. Wash.*, 47: 19-20.

610 Danforth, C.H., 1938. The races of *Oreopeleia mystacea* (Temminck). — *Proc. Biol. Soc. Wash.*, 51: 73-74.

611 Daugherty, C.H. *et al.*, 1999. Genetic differentiation, taxonomy and conservation of Australasian teals *Anas* spp. — *Bird Conserv. Int.*, 9: 29-42.

612 David, N. & M. Gosselin, 2000. The supposed significance of originally capitalized species-group names. — *Bull. Brit. Orn. Cl.*, 120 (4): 261-266.

613 David, N. & M. Gosselin, 2002. Gender agreement of avian species names. — *Bull. Brit. Orn. Cl.*, 122: 14-49.

614 David, N. & M. Gosselin, 2002. The grammatical gender of avian genera. – *Bull. Brit. Orn. Cl.*, 122: 257-282.

615 David-Beaulieu, A., 1944. *Les oiseaux de Trannink*: 1-225. — Université Indochinoise, Hanoi.

616 Davis, J., 1959. A new race of the Mexican Pootoo from Western Mexico. — *Condor*, 61 (4): 300-301.

617 Davis, J. & A.H. Miller, 1960. Family Mimidae. Pages 440-458. In *Check-list of Birds of the World*. Vol. 9 (E. Mayr & J.C. Greenway, Jr., eds.). — Mus. Comp. Zool., Cambridge, Mass.

618 Davis, L.I., 1979. Acoustic evidence of relationship in some American kingbirds. — *Pan American Studies*, 2 (1): 36-63.

619 Davis, T.J. & J.P. O'Neill, 1986. A new species of antwren (Formicariidae: *Herpsilochmus*) from Peru, with comments on the systematics of other members of the genus. — *Wilson Bull.*, 98 (3): 337-352.

620   Davison, G.W.H., 1974. Geographical variation in *Lophophorus sclateri*. — *Bull. Brit. Orn. Cl.*, 94: 163-164.
621   Davison, G.W.H., 1982. Systematics within the genus *Arborophila* Hodgson. — *Fed. Mus. J.*, 27: 125-234.
622   Davison, G.W.H., 2000. Notes on the taxonomy of some Bornean birds. — *Sarawak Mus. J.*, (75): 289-299 (1999).
623   de Juana, E., 1994. Family Tetraonidae (Grouse). Pages 376-410. In *Handbook of the Birds of the World.* Vol. 2 (J. del Hoyo, A. Elliott & J. Sargatal, eds.). — Lynx Edicions, Barcelona.
624   de Juana, E., 1997. Family Pteroclidae (Sandgrouse). Pages 30-57. In *Handbook of the Birds of the World.* Vol. 4 (J. del Hoyo, A. Elliott & J. Sargatal, eds.). — Lynx Edicions, Barcelona.
625   de la Paz, R.M., 1976. *Phapitreron amethystina imeldae*, a new subspecies of the Amethyst Brown Fruit Dove from Marinduque Island and notes on related subspecies (Aves: Columbidae). — *Nat. & Applied. Sci. Bull.*, 28 (1): 1-6.
626   de Naurois, R., 1975. Les Carduelinae des iles de São Tome et Principe (Golfe de Guinée). — *Ardeola*, 21: 903-931.
627   de Naurois, R., 1978. Procellariidae reproducteurs en Nouvelle-Calédonie pendant l'été austral. — *Comptes Rend. Acad. Sci. Paris, Ser. D.*, 287 (4): 269-271.
628   De Roo, A., 1967. A new species of *Chlorocichla* from north-eastern Congo (Aves: Pycnonotidae). — *Rev. Zool. Bot. Afr.*, 75 (3/4): 392-395.
629   De Roo, A.E.M., 1970. A new race of the African Black Swift *Apus barbatus* (Sclater) from the Republic of Cameroon (Aves: Apodidae). — *Rev. Zool. Bot. Afr.*, 81: 156-162.
630   Dean, W.R.J. *et al.*, 1992. Various species accounts. In *The Birds of Africa.* Vol. 4: i-xv, 1-609 (S. Keith, E.K. Urban & C.H. Fry, eds.). — Academic Press Inc., San Diego.
631   Debus, S.J.S., 1994. Various species accounts (Falconiformes). In *Handbook of the Birds of the World.* Vol. 2: 1-638 (J. del Hoyo, A. Elliott & J. Sargatal, eds.). — Lynx Edicions, Barcelona.
632   Debus, S.J.S., 1996. Family Turnicidae (Buttonquails). Pages 44-59. In *Handbook of the Birds of the World.* Vol. 3 (J. del Hoyo, A. Elliott & J. Sargatal, eds.). — Lynx Edicions, Barcelona.
633   Decoux, J.P., 1988. Various species accounts. In *The Birds of Africa.* Vol. 3: i-xvi, 1-611 (C.H. Fry, S. Keith & E.K. Urban, eds.). — Academic Press, London.
634   Deignan, H.G., 1941. New birds from the Indo-chinese sub-region. — *Auk*, 58: 396-398.
635   Deignan, H.G., 1941. Remarks on the Kentish Plovers of the Extreme Orient, with a separation of a new subspecies. — *J. Wash. Acad. Sci.*, 31 (3): 105-107.
636   Deignan, H.G., 1943. Remarks on *Phasianus Crawfurdii* and other birds named by J. E. Gray. — *Auk*, 60: 88-89.
637   Deignan, H.G., 1945. The birds of northern Thailand. — *U. S. Nat. Mus. Bull.*, 186: i-v, 1-616.
638   Deignan, H.G., 1946. Corrections and additions to the published record of Siamese birds. — *Auk*, 63: 243-245.
639   Deignan, H.G., 1946. The races of the Scarlet Minivet (*Pericrocotus flammeus*) (Forster). — *Auk*, 63 (4): 511-533.
640   Deignan, H.G., 1946. Two new hemipodes from southeastern Asia. — *J. Wash. Acad. Sci.*, 36 (11): 390-391.
641   Deignan, H.G., 1946. New subspecies of birds from peninsular Siam. — *J. Wash. Acad. Sci.*, 36 (12): 428.
642   Deignan, H.G., 1947. Some untenable names in the Old World flycatchers. — *Proc. Biol. Soc. Wash.*, 60: 165-168.
643   Deignan, H.G., 1948. The races of the Silver-breasted Broadbill *Serilophus lunatus* (Gould). — *J. Wash. Acad. Sci.*, 38 (3): 108-111.
644   Deignan, H.G., 1950. Five new races of birds from south-eastern Asia. — *Zoologica*, 35 (2): 127-128.
645   Deignan, H.G., 1950. The races of the Collared Scops Owl, *Otus bakkamoena* Pennant. — *Auk*, 67 (2): 189-200.
646   Deignan, H.G., 1950. Remarks on some recently named Chinese birds. — *Ibis*, 92: 318.
647   Deignan, H.G., 1950. Two new races of the Spotted Nightjar *Eurostopodus guttatus* (Vigors and Horsfield). — *Emu*, 50: 21-23.
648   Deignan, H.G., 1951. A new frogmouth from Groote Eylandt, Gulf of Carpentaria. — *Emu*, 51: 71.
649   Deignan, H.G., 1951. A new race of the Hawk Owl (*Ninox scutulata*) from the Philippines. — *Proc. Biol. Soc. Wash.*, 64: 41-42.
650   Deignan, H.G., 1952. The correct name for the Malayo-Sumatran race of the Chestnut-breasted Malkoha (Cuculidae). — *Bull. Raffles Mus.*, 24: 219.
651   Deignan, H.G., 1955. Remarks on *Picus vittatus* Vieillot and some of its allies. — *Ibis*, 97: 18-24.
652   Deignan, H.G., 1955. Eastern Asiatic races of the bee-eater, *Merops philippinus* Linnaeus. — *Bull. Brit. Orn. Cl.*, 75: 57-59.
653   Deignan, H.G., 1955. The identity of *Collocalia maxima* Hume. — *Bull. Brit. Orn. Cl.*, 75: 82.
654   Deignan, H.G., 1955. The races of the swiftlet, *Collocalia brevirostris* (McClelland). — *Bull. Brit. Orn. Cl.*, 75: 116-118.
655   Deignan, H.G., 1955. Four new races of birds from East Asia. — *Proc. Biol. Soc. Wash.*, 68: 145-148.
656   Deignan, H.G., 1955. The Long-tailed Nightjars of North Borneo and Palawan. — *Sarawak Mus. J.*, 6: 314-315.
657   Deignan, H.G., 1956. New races of birds from Laem Thong, the Golden Chersonese. — *Proc. Biol. Soc. Wash.*, 69: 207-214.
658   Deignan, H.G., 1957. A trio of new birds from Tropical Asia. — *Proc. Biol. Soc. Wash.*, 70: 43-44.
659   Deignan, H.G., 1958. Two new birds from Eastern Asia. — *Proc. Biol. Soc. Wash.*, 71: 161-162.
660   Deignan, H.G., 1960. A new race of the Brown Barbet from Thailand. — *Bull. Brit. Orn. Cl.*, 80: 121.
661   Deignan, H.G., 1960. The oldest name for the Bat-eating Pern. — *Bull. Brit. Orn. Cl.*, 80: 121.
662   Deignan, H.G., 1961. Type specimens of birds in the United States National Museum. — *U. S. Nat. Mus. Bull.*, 221: i-x, 1-718.
663   Deignan, H.G., 1963. Checklist of the Birds of Thailand. — *U. S. Nat. Mus. Bull.*, 226: i-x, 1-263.
664   Deignan, H.G., 1964. Family Muscicapidae, Subfamily Orthonychinae. Pages 228-240. In *Check-list of Birds of the World.* Vol. 10 (E. Mayr & R.A. Paynter, Jr., eds.). — Mus. Comp. Zool., Harvard, Cambridge, Mass.
665   Deignan, H.G., 1964. Family Muscicapidae, Subfamily Timaliinae. Pages 240-427. In *Check-list of Birds of the World.* Vol. 10 (E. Mayr & R.A. Paynter, Jr., eds.). — Mus. Comp. Zool., Harvard, Cambridge, Mass.
666   Deignan, H.G., 1964. Family Muscicapidae, Genera sedis incertae. Pages 427-429. In *Check-list of birds of the world.* Vol. 10 (E. Mayr & R.A. Paynter, Jr., eds.). — Mus. Comp. Zool., Harvard, Cambridge, Mass.
667   Deignan, H.G., 1964. Family Muscicapidae, Subfamily Panurinae. Pages 430-442. In *Check-list of birds of the world.* Vol. 10 (E. Mayr & R.A. Paynter, Jr., eds.). — Mus. Comp. Zool., Harvard, Cambridge, Mass.
668   Deignan, H.G., 1964. A new race of the Alpine Accentor *Prunella collaris* from Formosa. — *Bull. Brit. Orn. Cl.*, 84: 39-40.
669   Deignan, H.G., 1965. Notes on the nomenclature of the whistling-thrushes. — *Bull. Brit. Orn. Cl.*, 85 (1): 3-4.
670   Dekker, R.W.R.J. & E.C. Dickinson, 2000. Systematic notes on Asian birds. 2. A preliminary review of the Eurylaimidae. — *Zool. Verhand.*, Leiden, 331: 65-76.
671   Dekker, R.W.R.J. *et al.*, 2000. Systematic notes on Asian birds. 3. Types of the Eurylaimidae. — *Zool. Verhand.*, Leiden, 331: 77-88.
672   Dekker, R.W.R.J. *et al.*, 2001. Systematic notes on Asian Birds. 18. Some nomenclatural issues relating to Japanese taxa described in the Planches Coloriées (1820-1839) and Fauna Japonica, Aves (1844-1850). — *Zool. Verhand., Leiden*, 335: 199-214.
673   del Hoyo, J., 1994. Family Cracidae (Chachalacas, Guans and Curassows). Pages 310-363. In *Handbook of the Birds of the World.* Vol. 2 (J. del Hoyo, A. Elliott & J. Sargatal, eds.). — Lynx Edicions, Barcelona.
674   del Hoyo, J. *et al.* (eds), 1994. *Handbook of the Birds of the World.* New World vultures to guineafowl. 2: 1-638. — Lynx Edicions, Barcelona.
675   Delacour, J., 1943. A revision of the genera and species of the family Pyconotidae (Bulbuls). — *Zoologica*, 28: 17-28.
676   Delacour, J., 1945. Note on the Eared Pheasants (*Crossoptilon*) with the description of a new subspecies. — *Zoologica*, 31 (1): 43-45.
677   Delacour, J., 1947. *Birds of Malaysia*: i-xvi, 1-382. — Macmillan, New York.
678   Delacour, J., 1948. The subspecies of *Lophura nycthemera*. — *Amer. Mus. Novit.*, 1377: 1-12.
679   Delacour, J., 1949. A new subspecies of *Pavo muticus*. — *Ibis*, 91: 348-349.
680   Delacour, J., 1951. Commentaires, modifications et additions à la liste des oiseaux de l'Indochine Française. — *L'Oiseau*, 21: 1-32, 81-119.

681 Delacour, J., 1960. Family Irenidae. Pages 300-308. In *Check-list of Birds of the World*. Vol. 9 (E. Mayr & J.C. Greenway, Jr., eds.). — Mus. Comp. Zool., Cambridge, Mass.

682 Delacour, J. & E. Mayr, 1945. Notes on the taxonomy of the birds of the Philippines. — *Zoologica*, 30 (3): 105-117.

683 Delgado, B.F.S., 1985. A new subspecies of the Painted Parakeet (*Pyrrhura picta*) from Panama. — *Orn. Monogr.*, 36: 16-20.

684 Demastes, J.W. & J.V. Remsen, Jr., 1994. The genus *Caryothraustes* (Cardinalinae) is not monophyletic. — *Wilson Bull.*, 106: 733-738.

685 Dementiev, G.P., 1933. Les variations géographiques et individuelles des hirondelles de cheminée *Hirundo rustica* L. dans l'Asie Orientale. — *Alauda*, 8: 49-53.

686 Dementiev, G.P., 1940. Materiali k avifaune Koryatskoi zemli. [Information on the avifauna of Koryak Land.]. — *Materiali k poznaniyu fauni I flori SSSR, NS* 17 (2): 1-83.

687 Dementiev, G.P., 1941. Dopolneniya k toman Pervomy, Vtoromu, Tret'emu I Chetvertomu 'Polnogo opredelitel'ya ptits S.S.S.R. [Additions to the first, second, third and fourth volumes of the Complete Guide to the Birds of the USSR.]. Pages 13-94. In *Polnyi Opredelitel Ptitsy SSSR* (S.A. Buturlin & G.P. Dementiev, eds.).

688 Dementiev, G.P., 1945. Novaya forma kamennoi kuropatki *Alectoris graeca* iz Turkmenii. [A new form of the Rock Partridge *Alectoris graeca* from Turkmenistan.]. — *Izvest. AN Turkmenskoi S.S.R.*, (5-6): 178-179.

689 Dementiev, G.P., 1951. Otryad Sovi. Pages 329-429 In *Ptitsi Sovetskogo Soyuza*. 1. (G.P. Dementiev & N.A. Gladkov, eds.). —, Moscow.

690 Dementiev, G.P., 1952. O balkanskoi forme filina. [On the Balkan form of the Eagle owl.]. — *Byull. Mosk. O-va Ispyt. Prir. (Otd. Biol.)*, 57 (2): 91-92.

691 Dementiev, G.P., 1957. Die Ausbreitung einiger Vogelarten in Mittelasien. — *Der Falke*, 4 (3): 13-16.

692 Deraniyagala, P.E.P., 1951. A new race of Broad-billed Roller from Ceylon. — *Spolia Zeylanica*, 26 (2): 155-157.

693 Deraniyagala, P.E.P., 1955. Admin. Rept. of the Dir. Nat. Museums, Ceylon, for 1954, Pt. IV, Educ. Sci., & Art (E). Page E5, Colombo.

694 Deraniyagala, P.E.P., 1955. A new race of Trogon from Ceylon. — *Spolia Zeylanica*, 27 (2): 281-283.

695 Desfayes, M., 1975. Birds from Ethiopia. — *Rev. Zool. Afr.*, 89: 505-535.

696 Devillers, P., 1977. Observations at a breeding colony of *Larus* (*belcheri*) *atlanticus*. — *Gerfaut*, 67: 22-43.

697 Diamond, J.M., 1967. New subspecies and records of birds from the Karimui Basin, New Guinea. — *Amer. Mus. Novit.*, 2284: 1-17.

698 Diamond, J.M., 1969. Preliminary results of an ornithological exploration of the North Coastal range, New Guinea. — *Amer. Mus. Novit.*, 2362: 1-57.

699 Diamond, J.M., 1972. *Avifauna of the Eastern Highlands of New Guinea*: i-viii, 1-438. — Nuttall Orn. Cl., Cambridge, Mass., USA.

700 Diamond, J.M., 1976. Preliminary results of an ornithological exploration of the islands of Vitiaz and Dampier Straits, Papua New Guinea. — *Emu*, 76: 1-7.

701 Diamond, J.M., 1985. New distributional records and taxa from the outlying mountain ranges of New Guinea. — *Emu*, 85 (2): 65-91.

702 Diamond, J.M., 1989. A new subspecies of the Island Thrush *Turdus poliocephalus* from Tolokiwa Island in the Bismarck Archipelago. — *Emu*, 89: 58-60.

703 Diamond, J.M., 1991. A new species of rail from the Solomon islands and convergent evolution of insular flightlessness. — *Auk*, 108 (3): 461-470.

703A Diamond, J.M., 2002. Dispersal, mimicry, and geographic variation in northern Melanesian birds. — *Pacific Science*, 56 (1): 1-22.

704 Diamond, J.M. & M. LeCroy, 1979. Birds of Karkar and Bagabag islands, New Guinea. — *Bull. Amer. Mus. Nat. Hist.*, 164 (4): 467-531.

705 Dickerman, R.W., 1966. A new subspecies of the Virginia Rail from Mexico. — *Condor*, 68: 215-216.

706 Dickerman, R.W., 1970. A systematic revision of *Geothlypis speciosa*, the Black-polled Yellowthroat. — *Condor*, 72 (1): 95-98.

707 Dickerman, R.W., 1973. A review of the White-breasted Woodwrens of México and Central America. — *Condor*, 75 (3): 361-363.

708 Dickerman, R.W., 1974. Review of Red-winged Blackbirds (*Agelaius phoeniceus*) of eastern, west-central, and southern Mexico and Central America. — *Amer. Mus. Novit.*, 2538: 1-18.

709 Dickerman, R.W., 1975. New subspecies of *Caprimulgus sericocaudatus* from the Amazon River Basin. — *Bull. Brit. Orn. Cl.*, 95 (1): 18-19.

710 Dickerman, R.W., 1975. Revision of the Short-billed Marsh Wren (*Cistothorus platensis*) of Mexico and Central America. — *Amer. Mus. Novit.*, 2569: 1-8.

711 Dickerman, R.W., 1981. Geographic variation in the Scrub Euphonia. — *Occ. Pap. Mus. Zool., Louisiana St. Univ.*, 59: 1-6.

712 Dickerman, R.W., 1981. Preliminary review of the Clay-coloured Robin *Turdus grayi* with redesignation of the type locality of the nominate form and description of a new subspecies. — *Bull. Brit. Orn. Cl.*, 101 (2): 285-289.

713 Dickerman, R.W., 1981. A taxonomic review of the Spotted-breasted Oriole. — *Nemouria*, 26: 1-10.

714 Dickerman, R.W., 1985. A new subspecies of *Mecocerculus leucophrys* from Venezuela. — *Bull. Brit. Orn. Cl.*, 105 (2): 73-75.

715 Dickerman, R.W., 1986. Two hitherto unnamed populations of *Aechmophorus* (Aves: Podicipitidae). — *Proc. Biol. Soc. Wash.*, 99: 435-436.

716 Dickerman, R.W., 1987. Notes on the plumages of *Diglossa duidae* with the description of a new subspecies. — *Bull. Brit. Orn. Cl.*, 107: 42-44.

717 Dickerman, R.W., 1987. Two new subspecies of birds from Guatemala. — *Occ. Pap. Western Found. Vert. Zool.*, 3: 1-6.

718 Dickerman, R.W., 1988. An unnamed subspecies of *Euphonia rufiventris* from Venezuela and northern Brazil. — *Bull. Brit. Orn. Cl.*, 108 (1): 20-22.

719 Dickerman, R.W., 1988. A review of the Least Nighthawk *Chordeiles pusillus*. — *Bull. Brit. Orn. Cl.*, 108: 120-125.

720 Dickerman, R.W., 1989. Notes on the Malachite Kingfisher *Corythornis* (*Alcedo*) *cristata*. — *Bull. Brit. Orn. Cl.*, 109: 158-159.

721 Dickerman, R.W., 1989. Notes on *Sturnella magna* in South America with a description of a new subspecies. — *Bull. Brit. Orn. Cl.*, 109: 160-162.

722 Dickerman, R.W., 1990. The Scaled Antpitta, *Grallaria guatimalensis*, in Mexico. — *Southwestern Naturalist*, 35 (4): 460-463.

723 Dickerman, R.W., 1993. In C.R. Preston & R.D. Beane, No. 52. Red-tailed Hawk (*Buteo jamaicensis*). Pages 1-24. In *The Birds of North America* (A. Poole & F. Gill, eds.). — AOU, Philadelphia.

724 Dickerman, R.W., 1994. Notes on birds from Africa with descriptions of three new subspecies. — *Bull. Brit. Orn. Cl.*, 114 (4): 274-278.

725 Dickerman, R.W., 1997. A substitute name for the Bioko race of *Pycnonotus virens*. — *Bull. Brit. Orn. Cl.*, 117 (1): 75.

726 Dickerman, R.W., 1997. Geographic variation in the southwestern and Mexican populations of the Spotted Owl, with the description of a new subspecies. Pages 45-48. In *The Era of Allan R. Phillips: A Festschrift* (R.W. Dickerman, ed.) — Albuquerque.

727 Dickerman, R.W. & J. Gustafson, 1996. The Prince of Wales Spruce Grouse : A new subspecies from southeastern Alaska. — *Western Birds*, 27 (1): 41-47.

728 Dickerman, R.W. & J.P. Hubbard, 1994. An extinct subspecies of Sharp-tailed Grouse from New Mexico. — *Western Birds*, 25: 128-136.

729 Dickerman, R.W. & K.C. Parkes, 1997. Taxa described by Allan R. Phillips, 1939-1994: a critical list. Pages 210-233. In *The Era of Allan R. Phillips: A Festschrift* (R.W. Dickerman, ed.) — Albuquerque.

730 Dickerman, R.W. & W.H. Phelps, Jr., 1987. Tres nuevos atrapamoscas (Tyrannidae) del Cerro de la Neblina Territorio Amazonas, Venezuela. — *Bol. Soc. Venez. Cienc. Nat.*, 41 (144): 27-32.

731 Dickerman, R.W. *et al.*, 1994. Report on three collections of birds from Liberia. — *Bull. Brit. Orn. Cl.*, 114 (4): 267-274.

733[2] Dickerman, R.W. & A.R. Phillips, 1970. Taxonomy of the Common Meadowlark (*Sturnella magna*) in central and southern México and Caribbean Central America. — *Condor*, 72 (3): 305-309.

734 Dickinson, E.C., 1970. Notes upon a collection of birds from Indochina. — *Ibis*, 112 (4): 481-487.

[2] No. 732 has been deleted.

735   Dickinson, E.C., 1975. The identity of *Ninox scutulata* Raffles. — *Bull. Brit. Orn. Cl.*, 95 (3): 104-105.

736   Dickinson, E.C., 1989. A review of larger Philippine swiftlets of the genus *Collocalia*. — *Forktail*, 4: 19-53.

737   Dickinson, E.C., 1990. A review of smaller Philippine swiftlets of the genus *Collocalia*. — *Forktail*, 5: 23-34 (1989).

738   Dickinson, E.C., 2001. The correct scientific name of the Palawan Peacock-pheasant is *Polyplectron napoleonis* Lesson, 1831. — *Bull. Brit. Orn. Cl.*, 121 (4): 266-272.

739   Dickinson, E.C., 2001. Systematic notes on Asian birds. 9. The "Nouveau recueil de planches coloriées" of Temminck & Laugier (1820-1839). — *Zool. Verhand., Leiden*, 335: 7-54.

740   Dickinson, E.C. & S. Chaiyaphun, 1970. Notes on Thai birds. 2. A first contribution to our knowledge on the birds of Thung Salaeng Luang National Park, Phisanulok Province. — *Nat. Hist. Bull., Siam Soc.*, 23: 515-523.

741   Dickinson, E.C. & R.W.R.J. Dekker, 2000. Systematic notes on Asian Birds. 4. A preliminary review of the Pittidae. — *Zool. Verhand., Leiden*, 331: 89-99.

742   Dickinson, E.C. & R.W.R.J. Dekker, 2001. Systematic notes on Asian Birds. 11. A preliminary review of the Alaudidae. — *Zool. Verhand., Leiden*, 335: 61-84.

743   Dickinson, E.C. & R.W.R.J. Dekker, 2001. Systematic notes on Asian Birds. 13. A preliminary review of the Hirundinidae. — *Zool. Verhand., Leiden*, 335: 127-144.

743A   Dickinson, E.C. & R.W.R.J. Dekker, 2002. Systematic notes on Asian Birds. 22. A preliminary review of the Campephagidae. — *Zool. Verhand., Leiden*, 340: 7-30.

743B   Dickinson, E.C. & R.W.R.J. Dekker, 2002. Systematic notes on Asian Birds. 25. A preliminary review of the Pycnonotidae. – *Zool. Verhand., Leiden*, 340: 93-114.

743C   Dickinson, E.C. & S.M.S. Gregory, in press. Systematic notes on Asian birds. 24. On the priority of the name *Hypsipetes* Vigors, 1831, and the division of the broad genus of that name. — *Zool. Verhand., Leiden*, 340: 75-91.

744   Dickinson, E.C. & R.S. Kennedy, 2000. Systematic notes on Asian Birds. 6. A re-examination of the application of the name *Oriolus steerii* Sharpe, 1877. — *Zool. Verhand., Leiden*, 331: 127-130.

745   Dickinson, E.C. et al., 1991. *The Birds of the Philippines. An annotated Check-list*: 1-507. BOU Check-list Ser. No. 12 — British Ornithologists' Union, Tring, Herts.

746   Dickinson, E.C. et al., 2000. Systematic notes on Asian birds. 1. A review of the russet bush-warbler *Bradypterus seebohmi* (Ogilvie-Grant, 1895). — *Zool. Verhand., Leiden*, 331: 11-64.

747   Dickinson, E.C. et al., 2000. Systematic notes on Asian Birds. 5. Types of the Pittidae. — *Zool. Verhand., Leiden*, 331: 101-119.

748   Dickinson, E.C. et al., 2001. Systematic notes on Asian Birds. 12. Types of the Alaudidae. — *Zool. Verhand., Leiden*, 335: 85-126.

749   Dickinson, E.C. et al., 2001. Systematic notes on Asian Birds. 14. Types of the Hirundinidae. — *Zool. Verhand., Leiden*, 335: 145-166.

750   Dickinson, E.C. et al., 2001. Systematic notes on Asian birds. 19. Type material from Japan in the Natural History Museum, Tring, U.K. — *Zool. Verhand., Leiden*, 335: 215-228.

751   Dimcheff, D.E. et al., 2000. Cospeciation and horizontal transmission of avian sarcoma and leukosis virus *gag* genes in galliform birds. — *J. Virology*, 74: 3984-3995.

752   Dinesen, L. et al., 1994. A new genus and species of perdicine bird (Phasianidae, Perdicini) from Tanzania; a relict form with Indo-Malayan affinities. — *Ibis*, 136 (1): 3-11.

753   Domaniewski, J., 1933. Contribution à la connaissance des oiseaux de la Transbaïkalie du sud-ouest et de Mongolie du nord. — *Acta Orn. Mus. Zool. Pol.*, 1 (6): 147-179.

753A   Donázar, J.A. et al., 2002. Description of a new subspecies of the Egyptian Vulture (Accipitridae: *Neophron percnopterus*) from the Canary Islands. — *J. Raptor Res.*, 36: 17-23.

754   Dorst, J. & C. Jouanin, 1952. Description d'une espèce nouvelle de francolin d'Afrique orientale. — *L'Oiseau*, 22: 71-74.

755   Dorst, J. & J.-L. Mougin, 1979. Order Pelecaniformes. Pages 155-193. In *Check-list of Birds of the World*. Vol. 1 (2nd. Ed.) (E. Mayr & G.W. Cottrell, eds.). — Mus. Comp. Zool., Cambridge, Mass.

756   Doughty, C. et al., 1999. *Birds of the Solomons, Vanuatu & New Caledonia*: 1-206. — Christopher Helm, London.

757   Dove, C.J. & R.C. Banks, 1999. A taxonomic study of Crested Caracaras (Falconidae). — *Wilson Bull.*, 111 (3): 330-339.

758   Dowding, J.E., 1994. Morphometrics and ecology of the New Zealand Dotterel (*Charadrius obscurus*), with a description of a new subspecies. — *Notornis*, 41: 221-233.

759   Dowsett, R.J., 1989. The nomenclature of some African barbets of the genus *Tricholaema*. — *Bull. Brit. Orn. Cl.*, 109 (3): 180-181.

760   Dowsett, R.J. & F. Dowsett-Lemaire, 1980. The systematic status of some Zambian birds. — *Gerfaut*, 70: 151-199.

761   Dowsett, R.J. & F. Dowsett-Lemaire, 1993. Comments on the taxonomy of some Afrotropical bird species. — *Tauraco Research Report*, 5: 323-389.

762   Dowsett, R.J. & F. Dowsett-Lemaire, 1993. *A contribution to the distribution and taxonomy of Afrotropical and Malagasy birds*: 5. 1-389. — Tauraco Press, Liège.

763   Dowsett, R.J. et al., 1999. Systematic status of the Black-collared Bulbul *Neolestes torquatus*. — *Ibis*, 141: 22-28.

764   Dowsett-Lemaire, F. & R.J. Dowsett, 1990. Zoogeography and taxonomic relationships of the forest birds of the Albertine Rift Afromontane region. Pages 87-109. In *Tauraco Research Report 3*, Tauraco Press, Liège.

765   Duckworth, J.W. et al., 2001. A new species of wagtail from the lower Mekong basin. — *Bull. Brit. Orn. Cl.*, 121 (3): 152-182.

766   Dugand, A., 1943. Dos nuevas aves de Colombia. — *Caldasia*, 2: 191-198.

767   Dunajewski, A., 1937. Bemerkungen über einige mittelasiatische Vögel. — *Acta Orn. Mus. Zool. Pol.*, 1 (7): 181-251.

768   Dunajewski, A., 1938. Zwei neue Vogelformen. — *Acta Orn. Mus. Zool. Pol.*, 2: 157-160.

768A   Dunajewski, A., 1940. Beiträge zur systematische Stellung der karpatischen Habichtseule. — *Ann. Hist. Nat. Mus. Nat. Hung.*, 33: 98-100.

769   Dunajewski, A., 1948. New races of the Brown owl, Hedge-Sparrow and a species of *Attila*; also a new genus of Cotingidae. — *Bull. Brit. Orn. Cl.*, 68: 130-132.

770   Duncan, F.M., 1937. On the dates of publication of the Society's "Proceedings", 1859 -1926. — *Proc. Zool. Soc., Lond.*: 71-83.

771   duPont, J.E., 1971. Notes on Philippine Birds (No. 1). — *Nemouria*, 3: 1-6.

772   duPont, J.E., 1972. Notes from Western Samoa, including the description of a new parrot-finch (*Erythrura*). — *Wilson Bull.*, 84: 375-376.

773   duPont, J.E., 1972. Notes on Philippine birds (No. 2). Birds of Ticao. — *Nemouria*, 6: 1-13.

774   duPont, J.E., 1972. Notes on Philippine birds (No. 3). Birds of Marinduque. — *Nemouria*, 7: 1-14.

775   duPont, J.E., 1976. Notes on Philippine birds (No. 4). Additions and corrections to Philippine Birds. — *Nemouria*, 17: 1-13.

776   duPont, J.E., 1976. *South Pacific Birds*: i-xii, 1-218. — Delaware Mus. Nat. Hist., Greenville, Del.

777   duPont, J.E., 1980. Notes on Philippine birds (No. 5). Birds of Burias. — *Nemouria*, 24: 1-6.

778   duPont, J.E. & D.S. Rabor, 1973. Birds of Dinagat and Siargao, Philippines. An expedition report. — *Nemouria*, 10: 1-111.

779   Dutson, G.C.L., 1993. A sighting of *Ficedula* (*crypta*) *disposita* in Luzon, Philippines. — *Forktail*, 8: 144-147.

780   Dziadosz, V.M. & K.C. Parkes, 1984. Two new Philippine subspecies of the Crimson-breasted Barbet (Aves: Capitonidae). — *Proc. Biol. Soc. Wash.*, 97: 788-791.

780A   Eames, J.C., 2002. Eleven new sub-species of babbler (Passeriformes: Timaliinae) from Kon Tum Province, Vietnam. — *Bull. Brit. Orn. Cl.*, 122 (2): 109-141.

781   Eames, J.C. & C. Eames, 2001. A new species of Laughingthrush (Passeriformes: Garrulacinae) from the Central Highlands of Vietnam. — *Bull. Brit. Orn. Cl.*, 121 (1): 10-23.

782   Eames, J.C. et al., 1994. A new subspecies of Spectacled Fulvetta *Alcippe ruficapilla* from Vietnam. — *Forktail*, 10: 141-157.

783 Eames, J.C. *et al.*, 1999. A new species of Laughingthrush (Passeriformes: Garrulacinae) from the Western Highlands of Vietnam. — *Bull. Brit. Orn. Cl.*, 119 (1): 4-15.

784 Eames, J.C. *et al.*, 1999. New species of Barwing *Actinodura* (Passeriformes: Sylvinae: Timaliini) from the western highlands of Vietnam. — *Ibis*, 141 (1): 1-10.

784A Earlé, R.A., 1986. Reappraisal of variation in the Ground Woodpecker *Geocolaptes olivaceus* (Gmelin) (Aves: Picidae) with notes on its moult. — *Navors. Nas. Mus. Bloemfontein*, 5: 79-92.

785 Eck, S., 1971. Katalog der Eulen des Staatlichen Museums für Tierkunde Dresden (Aves, Strigidae). — *Zool. Abh. Staatl. Mus. Tierk. Dresden*, 30 (15): 173-218.

786 Eck, S., 1973. Katalog der ornithologischen Sammlung des Zoologischen Institutes der Karl-Marx-Universität Leipzig, übernommen vom Staatlichen Museum für Tierkunde Dresden. I. — *Strigidae. Zool. Abh. Staatl. Mus. Tierk. Dresden*, 32: 155-169.

787 Eck, S., 1975. Über die Nachtigallen (*Luscinia megarhynchos*) Mittel- und Südosteuropas. — *Beitr. Vogelk.*, 21: 21-30.

788 Eck, S., 1976. Die Vögel der Banggai-Inseln, insbesondere Pelengs (Aves). — *Zool. Abh. Staatl. Mus. Tierk. Dresden*, 34 (5): 53-100.

789 Eck, S., 1980. Intraspezifische Evolution bei Graumeisen (Aves, Paridae: *Parus*, Subgenus *Poecile*). — *Zool. Abh. Staatl. Mus. Tierk. Dresden*, 36: 135-219.

790 Eck, S., 1996. Die palaearktischen Vögel — Geospezies und Biospezies. — *Zool. Abh. Staatl. Mus. Tierk. Dresden*, 49 (Suppl.): 1-103.

791 Eck, S., 1998. *Parus ater martensi* subspec. nov., die Tannenmeise der Thakkhola, Nepal (Aves: Passeriformes: Paridae). — *Zool. Abh. Staatl. Mus. Tierk. Dresden*, 50: 129-132.

792 Eck, S., 2001. *Mino dumontii giliau* Stresemann, 1922, der "Papuaatzel" von New Britain, ist eine valide Subspezies (Aves: Passeriformes: Sturnidae). — *Zool. Abh. Staatl. Mus. Tierk. Dresden*, 51 (11): 133-138.

793 Eck, S. & H. Busse, 1973. Eulen. Die rezenten und fossilen Formen. Aves, Strigidae. — *Die Neue Brehm-Büch*, 1: 1-196.

794 Eisenmann, E., 1955. The species of Middle American birds. — *Trans. Linn. Soc. New York*, 7: 1-128.

795 Eisenmann, E., 1962. Notes on nighthawks of the genus *Chordeiles* in Southern and Middle American with a description of a new race of *Chordeiles minor* breeding in Panama. — *Amer. Mus. Novit.*, 2094: 1-21.

796 Eisenmann, E. & V.F.C. Lehmann, 1962. A new species of swift of the genus *Cypseloides* from Colombia. — *Amer. Mus. Novit.*, 2117: 1-16.

797 Eisentraut, M., 1968. Beitrag zur Vogelfauna von Fernando Poo und Westkamerun. — *Bonn. Zool. Beitr.*, 19: 49-68.

798 Eisentraut, M., 1973. Die Wirbeltierfauna von Fernando Poo und Westkamerun. — *Bonn. Zool. Monogr.*, 3: 1-428.

799 Ellis, D.H. & G.C. Peres, 1983. The Pallid Falcon *Falco kreyenborgi* is a color phase of the Austral Peregrine Falcon (*Falco peregrinus cassini*). — *Auk*, 100: 269-271.

800 Engelmoer, M. & C.S. Roselaar, 1998. *Geographical variation in waders*: 1-331. — Kluwer Acad. Publ., Dordrecht.

801 Engilis, A., Jr. & R.E. Cole, 1997. Avifaunal observations from the Bishop Museum Expedition to Mt. Dayman, Milne Bay Province, Papua New Guinea. — *Bishop Mus. Occas. Pap.*, 52: 1-19.

802 Érard, C., 1974. Taxonomie des serins à gorge jaune d'Ethiopie. — *L'Oiseau*, 44 (4): 308-323.

803 Érard, C., 1975. Une nouvelle alouette du Sud de l'Ethiopie. — *Alauda*, 43 (2): 115-124.

804 Érard, C., 1975. Variation geographique de *Mirafra gilletti* Sharpe. Description d'une espèce jumelle. — *L'Oiseau*, 45 (4): 293-312.

805 Érard, C., 1979. What in reality is *Anthreptes pujoli* Berlioz? — *Bull. Brit. Orn. Cl.*, 99 (4): 142-143.

806 Érard, C., 1991. Variation géographique de *Bleda canicapilla* (Hartlaub) 1854 (Aves, Pycnonotidae). Description d'une sous-espèce nouvelle en Séné-gambie. — *L'Oiseau*, 61 (1): 66-67.

807 Érard, C., 1992. *Bleda canicapilla morelorum*, émendation du nom subspécifique d'une bulbul récemment décrit du Sénégal. — *L'Oiseau*, 62 (3): 288.

808 Érard, C., 1997. Various species accounts. In *The Birds of Africa*. Vol. 5: i-xix, 1-669 (E.K. Urban, C.H. Fry & S. Keith, eds.). — Academic Press, London.

809 Érard, C. & P.R. Colston, 1988. *Batis minima* (Verreaux) new for Cameroon. — *Bull. Brit. Orn. Cl.*, 108: 182-184.

810 Érard, C. & R. de Naurois, 1973. A new race of Thekla Lark in Bale, Ethiopia. — *Bull. Brit. Orn. Cl.*, 93: 141-142.

811 Érard, C. & C.H. Fry, 1997. Various species accounts. In *The Birds of Africa*. Vol. 5: i-xix, 1-669 (E.K. Urban, C.H. Fry & S. Keith, eds.). — Academic Press, London.

812 Érard, C. & G. Jarry, 1973. A new race of Thekla Lark in Harrar, Ethiopia. — *Bull. Brit. Orn. Cl.*, 93: 139-140.

813 Érard, C. & J. Roche, 1977. Un nouveau *Lagonosticta* du Tchad méridional. — *L'Oiseau*, 47 (4): 335-343.

814 Érard, C. & F. Roux, 1983. La Chevechette du Cap *Glaucidium capense* dans l'ouest africain. Description d'une race géographique nouvelle. — *L'Oiseau*, 53 (2): 97-104.

815 Érard, C. *et al.*, 1993. Variation géographique de *Cursorius cinctus* (Heuglin, 1863). — *Bonn. Zool. Beitr.*, 44: 165-192.

816 Ericson, P.G.P. *et al.*, 2000. Major divisions in oscines revealed by insertions in the nuclear gene c-*myc*: a novel gene in avian phylogenetics. — *Auk*, 117: 1077-1086.

817 Ericson, P.G.P. *et al.*, 2002. A Gondwanan origin of passerine birds supported by DNA sequences of the endemic New Zealand wrens. — *Proc. Roy. Soc., Lond.*, Ser. B. 269: 235-241.

817A er-Rashid, H., 1966. Observations on some birds of East Pakistan (including the description of a new subspecies). — *J. Asiatic Soc. Pakistan*, 11 (3): 113-117.

818 Espinosa de los Monteros, A., 1998. Phylogenetic relationships among the trogons. — *Auk*, 115 (4): 937-954.

819 Espinosa de los Monteros, A., 2000. Higher-level phylogeny of Trogoniformes. — *Molec. Phylogen. Evol.*, 14 (1): 20-34.

820 Espinosa de los Monteros, A. & J. Cracraft, 1997. Intergeneric relationships of the New World jays inferred from cytochrome *b* gene sequences. — *Condor*, 99: 490-502.

820A Esteban, J.G., 1949. Una nueva subspecie chilena de *Geositta rufipennis*. — *Acta Zool. Lilloana*, 8: 203-207.

821 Etchecopar, R.D. & F. Hue, 1964. *Les Oiseaux du Nord de l'Afrique*: 1-606. — Editions N. Boubee et Cie., Paris.

822 Falla, R.A., 1936. Rep. B.A.N.Z. Antarctic Res. Exp. 1929-31: 2B. 1-288.

823 Falla, R.A., 1978. Banded Dotterel at the Auckland Islands: description of a new sub-species. — *Notornis*, 25 (2): 101-108.

824 Falla, R.A. & J.-L. Mougin, 1979. Order Sphenisciformes. Pages 121-134. In *Check-list of Birds of the World*. Vol. 1. (2nd. Ed.) (E. Mayr & G.W. Cottrell, eds.). — Mus. Comp. Zool., Cambridge, Mass.

825 Farkas, T., 1971. *Monticola bensoni*, a new species from south-western Madagascar. — *Ostrich*, (Suppl. 9: 10th. Anniv. Commem. Percy Fitzpatrick Inst. of African Ornith.): 83-90.

826 Farkas, T., 1972. *Copsychus albospecularis winterbottomi*, a new subspecies from South-east of Madagascar. — *Ostrich*, 43 (4): 228-230.

827 Farkas, T., 1973. The biology and a new subspecies of *Monticola sharpei*. — *Bull. Brit. Orn. Cl.*, 93: 145-155.

828 Farkas, T., 1979. A further note on the status of *Monticola pretoriae* Gunning & Roberts, 1911. — *Bull. Brit. Orn. Cl.*, 99 (1): 20-21.

829 Farquhar, M.R. *et al.*, 1996. Feather ultrastructure and skeleton morphology as taxonomic characters in African sunbirds (Nectariniidae) and sugarbirds (Promeropidae). — *J. African Zool.*, 110: 321-331.

830 Feare, C.J. & A.J.F.K. Craig, 1998. Starlings and Mynas: 1-285. — A. & C. Black, London.

831 Feare, C.J. & Kang Nee, 1992. Allocation of *Sturnus melanopterus* to *Acridotheres*. — *Bull. Brit. Orn. Cl.*, 112 (2): 126-129.

832 Fernandez-Yepez, A.F., 1945. Lista parcial de las aves de la Isla Tortuga. — *Mem. Soc. Cienc. Nat. La Salle*, 4 (12): 47-48.

833 Finsch, O., 1901. Zosteropidae. — *Das Tierreich*, 15: i-xiv, 1-55.

834 Fitzpatrick, J.W., 1980. A new race of *Atlapetes leucopterus*, with comments on widespread albinism in *A. l. dresseri* (Taczanowski). — *Auk*, 97 (4): 883-887.

835   Fitzpatrick, J.W. & J.P. O'Neill, 1979. A new tody-tyrant from northern Peru. — *Auk*, 96 (3): 443-447.
836   Fitzpatrick, J.W. & J.P. O'Neill, 1986. *Otus petersoni*, a new screech-owl from the Eastern Andes, with systematic notes on *O. colombianus* and *O. ingens*. — *Wilson Bull.*, 98 (1): 1-14.
837   Fitzpatrick, J.W. & D.F. Stotz, 1997. A new species of tyrannulet (*Phylloscartes*) from the Andean foothills of Peru and Bolivia. — *Orn. Monogr.*, 48: 37-44.
838   Fitzpatrick, J.W. & D.E. Willard, 1990. *Cercomacra manu*, a new species of antbird from southwestern Amazonia. — *Auk*, 107 (2): 239-245.
839   Fitzpatrick, J.W. *et al.*, 1977. A new species of wood-wren from Peru. — *Auk*, 94 (2): 195-201.
840   Fitzpatrick, J.W. *et al.*, 1979. A new species of hummingbird from Peru. — *Wilson Bull.*, 91 (2): 177-186.
841   Fjeldså, J., 1982. Some behavior patterns of four closely related grebes *Podiceps nigricollis*, *P. gallardoi*, *P. occidentalis* and *P. taczanowskii*, with reflections on phylogeny and adaptive aspects of the Evolution of displays. — *Dansk Orn. Foren. Tidsskr.*, 76: 37-68.
842   Fjeldså, J., 1983. A Black Rail from Junin, Central Peru: *Laterallus jamaicensis tuerosi* ssp. n. (Aves, Rallidae). — *Steenstrupia*, 8 (13): 277-282.
843   Fjeldså, J., 1983. Geographic variation in the Andean Coot *Fulica ardesiaca*. — *Bull. Brit. Orn. Cl.*, 103: 18-22.
844   Fjeldså, J., 1986. Color variation in the Ruddy Duck *Oxyura jamaicensis andina*. — *Wilson Bull.*, 98: 592-594.
845   Fjeldså, J., 1990. Systematic relations of an assembly of allopatric rails from western South America (Aves: Rallidae). — *Steenstrupia*, 16: 109-116.
846   Fjeldså, J., 1990. Geographic variation in the Rufous-webbed Tyrant *Polioxolmis rufipennis*, with description of a new subspecies. — *Bull. Brit. Orn. Cl.*, 110 (1): 26-31.
847   Fjeldså, J., 1993. *Phrygilus coracinus* Sclater, 1891, is a valid taxon. — *Bull. Brit. Orn. Cl.*, 113 (2): 121-126.
848   Fjeldså, J., 1999. Species accounts within the Family Trochilidae (Hummingbirds). Pages 549-680. In *Handbook of the Birds of the World*. Vol. 5 (J. del Hoyo, A. Elliott & J. Sargatal, eds.). — Lynx Edicions, Barcelona.
849   Fjeldså, J., 2001. Family Thinocoridae (Seedsnipes). Pages 538-545. In *Handbook of the Birds of the World*. Vol. 6 (J. del Hoyo, A. Elliott & J. Sargatal, eds.). — Lynx Edicions, Barcelona.
850   Fjeldså, J. & N. Krabbe, 1990. Birds of the High Andes. A manual to the birds of the temperate zone of the Andes and Patagonia, South America: 1-880. — Zool. Mus., Univ. Copenhagen, Copenhagen.
851   Fjeldså, J. *et al.*, 1999. Molecular evidence for relationships of Malagasy birds. In *Proc. 22nd. Internat. Ornith. Congr. BirdLife South Africa, Durban.* 1998: 3084-3094.
852   Fjeldså, J. *et al.*, 2000. A new montane subspecies of *Sheppardia gunningi* (East-coast Akalat) from Tanzania. — *Bull. Brit. Orn. Cl.*, 120 (1): 27-33.
853   Fleming, R.L., 1947. A new race of Koklas pheasant. — *Fieldiana, Zool.*, 31 (11): 93-96.
854   Fleming, R.L. & M.A. Traylor, 1964. Further notes on Nepal birds. — *Fieldiana, Zool.*, 35 (9): 489-558.
855   Flint, V.E. & A.A. Kistchinski, 1982. Taxonomische Wechselbeziehungen innerhalb der Gruppe der Prachttaucher (Gaviidae, Aves). — *Beitr. Vogelk.*, 28: 193-206.
856   Ford, J., 1971. Distribution, ecology and taxonomy of some Western Australian passerine birds. — *Emu*, 71: 103-120.
857   Ford, J., 1979. A new subspecies of Grey Butcherbird from the Kimberley, Western Australia. — *Emu*, 79: 191-194.
858   Ford, J., 1981. Hybridization and migration in Australian populations of the Little and Rufous-breasted Bronze-Cuckoos. — *Emu*, 81 (4): 209-222.
859   Ford, J., 1983. Evolutionary and ecological relationships between Quail-thrushes. — *Emu*, 83: 152-172.
860   Ford, J., 1983. Taxonomic notes on some mangrove-inhabiting birds in Australia. — *Rec. West. Austr. Mus.*, 10: 381-415.
861   Ford, J., 1986. Phylogeny of the acanthizid warbler genus *Gerygone* based on numerical analyses of morphological characters. — *Emu*, 86: 12-22.
862   Ford, J., 1986. Avian hybridization and allopatry in the region of the Einasleigh Uplands and Burdekin-Lynd Divide, north-eastern Queensland. — *Emu*, 86: 87-110.
863   Ford, J., 1987. Hybrid zones in Australian birds. — *Emu*, 87: 158-178.
864   Ford, J., 1987. New subspecies of Grey Shrike-thrush and Long-billed Corella from Western Australia. — *West. Austral. Nat.*, 16: 172-176.
865   Ford, J. & R.E. Johnstone, 1983. The Rusty-tailed Flyeater, a new species from Queensland. — *West. Austral. Nat.*, 15: 133-135.
866   Forshaw, J.M., 1971. A new parrot from the Sula Islands, Indonesia. — *Bull. Brit. Orn. Cl.*, 91: 163-164.
867   Forshaw, J.M. & W.T. Cooper, 1978. *Parrots of the World*: 1-616. — David & Charles, Newton Abbot, Devon.
868   Forshaw, J.M. & W.T. Cooper, 1989. *Parrots of the World*: 1-672. 3rd. edition. — Lansdowne Editions, Sydney.
869   Frade, F., 1976. Aves do arquipélago de Cabo Verde (Colecção do Centro de Zoologia da J. I. C. U.). — *Garcia de Orta (Zool.)*, 5: 47-57.
870   Frade, F. & R. de Naurois, 1964. Une nouvelle sous-espèce de Tisserin, *Ploceus velatus peixotoi* (Ile de S. Tomé, Golfe de Guinée). — *Garcia de Orta (Zool.)*, 12: 621-626.
871   Franzmann, N.-E., 1983. A new subspecies of the Usambara Weaver *Ploceus nicolli*. — *Bull. Brit. Orn. Cl.*, 103 (2): 49-51.
872   Fraser, L., 1840. Characters of new species of humming-birds. — *Proc. Zool. Soc., Lond.*, 14-19.
873   Freitag, S. & T.J. Robinson, 1993. Phylogeographic patterns in mitochondrial DNA of the Ostrich (*Struthio camelus*). — *Auk*, 110 (3): 614-622.
874   Friedmann, H., 1943. A new race of Sharp-tailed Grouse. — *J. Wash. Acad. Sci.*, 33 (6): 189-191.
875   Friedmann, H., 1943. A new wood quail of the genus *Dendrortyx*. — *J. Wash. Acad. Sci.*, 33 (9): 272-273.
876   Friedmann, H., 1943. Critical notes on the avian genus *Lophortyx*. — *J. Wash. Acad. Sci.*, 33 (12): 369-371.
877   Friedmann, H., 1944. A review of the forms of *Colinus leucopogon*. — *Proc. Biol. Soc. Wash.*, 57: 15-16.
878   Friedmann, H., 1945. Two new birds from the upper Rio Negro, Brazil. — *Proc. Biol. Soc. Wash.*, 58: 113-116.
879   Friedmann, H., 1945. The genus *Nyctiprogne*. — *Proc. Biol. Soc. Wash.*, 58: 117-120.
880   Friedmann, H., 1949. A new heron and a new owl from Venezuela. — *Smithsonian Misc. Coll.*, 111 (9): 1-3.
881   Friedmann, H., 1964. A new swift from Mt. Moroto, Uganda. — *Contrib. Sci. Los Angeles Co. Mus.*, 83: 1-4.
882   Friedmann, H., 1968. The evolutionary history of the avian genus *Chrysococcyx*. — *U. S. Nat. Mus. Bull.*, 265: i-viii, 1-137.
883   Friedmann, H. & H.G. Deignan, 1939. Notes on some Asiatic owls of the genus *Otus*, with description of a new form. — *J. Wash. Acad. Sci.*, 29 (7): 287-291.
884   Frith, C.B. & D.W. Frith, 1983. A systematic review of the hornbill genus *Anthracoceros* (Aves, Bucerotidae). — *Zool. J. Linn. Soc. Lond.*, 78: 29-71.
885   Frith, C.B. & D.W. Frith, 1997. *Chlamydera guttata carteri* Mathews, 1920 — an overlooked subspecies of Western Bowerbird (Ptilonorhynchidae) from North West Cape, Western Australia. — *Rec. West. Austr. Mus.*, 18: 225-231.
886   Frith, C.B. & D.W. Frith, 1997. A distinctive new subspecies of Macgregor's Bowerbird (Ptilonorhynchidae) of New Guinea. — *Bull. Brit. Orn. Cl.*, 117 (3): 199-205.
887   Frith, H.J., 1982. *Pigeons and Doves of Australia*: 1-304. — Rigby, Adelaide.
888   Froriep, L.F., 1821. Notizen aus dem Gebiete der Natur- und Heilkunde. Cols. 20-22, Erfurt.
889   Fry, C.H., 1988. Various species accounts. In *The Birds of Africa*. Vol. 3: i-xvi, 1-611 (C.H. Fry, S. Keith & E.K. Urban, eds.). — Academic Press, London.
890   Fry, C.H., 1992. Various species accounts. In *The Birds of Africa*. Vol. 4: i-xv, 1-609 (S. Keith, E.K. Urban & C.H. Fry, eds.). — Academic Press Inc., San Diego.
891   Fry, C.H., 1997. Various species accounts. In *The Birds of Africa*. Vol. 5: i-xix, 1-669 (E.K. Urban, C.H. Fry & S. Keith, eds.). — Academic Press, London.
892   Fry, C.H., 2000. Various species accounts. In *The Birds of Africa*. Vol. 6: i-xvii, 1-724 (C.H. Fry, S. Keith & E.K. Urban, eds.). — Academic Press, London.
893   Fry, C.H., 2001. Family Meropidae (Bee-eaters). Pages 286-341. In *Handbook of the Birds of the World*. Vol. 6 (J. del Hoyo, A. Elliott & J. Sargatal, eds.). — Lynx Edicions, Barcelona.

894   Fry, C.H., 2001. Family Coraciidae (Rollers). Pages 342-376. In *Handbook of the Birds of the World*. Vol. 6 (J. del Hoyo, A. Elliott & J. Sargatal, eds.). — Lynx Edicions, Barcelona.

895   Fry, C.H. & K. Fry, 1992. *Kingfishers, bee-eaters and rollers. A handbook*: i-xii, 1-324. – A. & C. Black, London.

896   Fry, C.H. & R.M. Harwin, 1988. Various species accounts. In *The Birds of Africa*. Vol. 3: i-xvi, 1-611 (C.H. Fry, S. Keith & E.K. Urban, eds.). — Academic Press, London.

897   Fry, C.H. & D.A. Smith, 1985. A new swallow from the Red Sea. — *Ibis*, 127: 1-6.

898   Fry, C.H. *et al.*, 1985. Evolutionary expositions from "The Birds of Africa": *Halcyon* song phylogeny; cuckoo host partitioning; systematics of *Aplopelia* and *Bostrychia*. In *Proc. Int. Symp. Afr. Vertebr. Mus. A. Koenig, Bonn* (ed. K.-L. Schuchmann), 1984: 163-180.

899   Fry, C.H. *et al.* (eds), 2000. *The Birds of Africa*. Vol. 6: i-xvii, 1-724. — Academic Press, London.

900   Fry, C.H. *et al.*, 1992. Various species accounts. In *The Birds of Africa*. Vol. 4: i-xv, 1-609 (S. Keith, E.K. Urban & C.H. Fry, eds.). — Academic Press Inc., San Diego.

901   Fushihara, H., 1959. Anting by the Japanese Grey Thrush, *Turdus cardis cardis* Temminck. — *Tori*, 15 (72): 61-70.

902   Gabrielson, I.N. & F.C. Lincoln, 1949. A new race of Ptarmigan in Alaska. — *Proc. Biol. Soc. Wash.*, 62: 175-176.

903   Gabrielson, I.N. & F.C. Lincoln, 1951. A new race of Ptarmigan from Alaska. — *Proc. Biol. Soc. Wash.*, 64: 63-64.

904   Galbraith, I.C.J., 1956. Variation, relationships and evolution in the *Pachycephala pectoralis* superspecies. — *Bull. Brit. Mus. (Nat. Hist.), Zool.*, 4 (4): 133-222.

905   Galbraith, I.C.J., 1957. On the application of the name *Zostrops rendovae* Tristram, 1882. — *Bull. Brit. Orn. Cl.*, 77: 10-16.

906   Galbraith, I.C.J., 1969. The Papuan and Little Cuckoo-shrikes, *Coracina papuensis* and *robusta*, as races of a single species. — *Emu*, 69: 9-29.

907   Galbraith, I.C.J. & E.H. Galbraith, 1962. Land birds of Guadalcanal and the San Cristoval group, Eastern Solomon Islands. — *Bull. Brit. Mus. (Nat. Hist.), Zool.*, 9 (1): 1-80.

908   Gamauf, A. & M. Preleuthner, 1998. A new taxon of the Barred Honeybuzzard *Pernis celebensis* from the Philippines. — *Bull. Brit. Orn. Cl.*, 118 (2): 90-101.

909   Ganier, A.F., 1954. A new race of the Yellow-bellied Sapsucker. — *Migrant*, 25 (3): 37-41.

910   Garcia-Moreno, J. & J.M. Cardoso da Silva, 1997. An interplay between forest and non-forest South American avifaunas suggested by a phylogeny of *Lepidocolaptes* woodcreepers (Dendrocolaptinae). — *Stud. Neotrop. Fauna Envir.*, 32: 164-173.

911   Garcia-Moreno, J. & J. Fjeldså, 1999. Re-evaluation of species limits in the genus *Atlapetes* based on mtDNA sequence data. — *Ibis*, 141 (2): 199-207.

912   Garcia-Moreno, J. *et al.*, 1998. Pre-Pleistocene differentiation among chat-tyrants. — *Condor*, 100: 629-640.

913   Garcia-Moreno, J. *et al.*, 1999. A case of rapid diversification in the Neotropics: phylogenetic relationships among *Cranioleuca* Spinetails (Aves, Furnariidae). — *Molec. Phylogen. Evol.*, 12 (3): 273-281.

914   Gargallo, G., 1994. On the taxonomy of the western Mediterranean islands populations of Subalpine Warbler *Sylvia cantillans*. — *Bull. Brit. Orn. Cl.*, 114 (1): 31-36.

915   Garrido, O.H., 1966. Nueva subespecie del Carpintero Jabad, *Centurus superciliaris* (Aves: Picidae). — *Poeyana* (Ser. A), 29: 1-4.

916   Garrido, O.H., 1970. Variación del género *Agelaius* (Aves: Icteridae) en Cuba. — *Poeyana* (Ser. A), 68: 1-18.

917   Garrido, O.H., 1971. Una nueva subespecie del *Vireo gundlachii* (Aves: Vireonidae), para Cuba. — *Poeyana*, 81: 1-8.

918   Garrido, O.H., 1971. Variacion del genero monotipico *Xiphidiopicus* (Aves: Picidae) en Cuba. — *Poeyana*, 83: 1-12.

919   Garrido, O.H., 1971. Nueva raza del arriero, *Saurothera merlini* (Aves: Cuculidae) para Cuba. — *Poeyana*, 87: 1-12.

920   Garrido, O.H., 1973. Anfibios, reptiles y aves de Cayo Real (Cayos de San Felipe), Cuba. — *Poeyana* (Ser. A), 119: 1-50.

921   Garrido, O.H., 1978. Nueva subespecie de Carpintero Verde (Aves: Picidae) para Cayo Coco, Cuba. — *Acad. Cienc. Cuba, Inst. Zool. Informe Cien. Tec.*, 67: 1-6.

922   Garrido, O.H., 1983. A new subspecies of *Caprimulgus cubanensis* (Aves: Caprimulgidae) from the Isle of Pines, Cuba. — *Auk*, 100: 988-991.

923   Garrido, O.H., 2000. A new subspecies of Oriente Warbler *Teretistris fornsi* from Pico Turquino, Cuba, with ecological comments on the genus. — *Cotinga*, 14: 88-93.

924   Garrido, O.H., 2001. Una nueva subespecie del Sijú de Sabana *Speotyto cunicularia* para Cuba. — *Cotinga*, 15: 75-78.

925   Garrido, O.H. & J.V. Remsen, Jr., 1996. A new subspecies of the Pearly-eyed Thrasher *Margarops fuscatus* (Mimidae) from the island of St. Lucia, Lesser Antilles. — *Bull. Brit. Orn. Cl.*, 116 (2): 75-80.

926   Garrido, O.H. *et al.*, 1997. Taxonomy of the Stripe-headed Tanager, genus *Spindalis* (Aves: Thraupidae) of the West Indies. — *Wilson Bull.*, 109 (4): 561-594.

927   Garrido, O.H. *et al.*, 1999. Geographic variation and taxonomy of the Cave Swallow (*Petrochelidon fulva*) complex, with the description of a new subspecies from Puerto Rico. — *Bull. Brit. Orn. Cl.*, 119 (2): 80-91.

928   Garson, P., 2000. Pheasant Specialist Group [Report on the]. — *Ann. Rev. World Pheasant Assoc.*, 1999/2000: 31-34.

929   Gatter, W., 1985. Ein neuer Bülbül aus Westafrika (Aves, Pycnonotidae). — *J. f. Orn.*, 126 (2): 155-161.

930   Gaucher, P. *et al.*, 1996. Taxonomy of the Houbara Bustard *Chlamydotis undulata* subspecies considered on the basis of sexual display and genetic divergence. — *Ibis*, 138 (2): 273-282.

931   Gavrilov, E.I. & A.P. Savtchenko, 1991. On species validity of the Pale Sand Martin (*Riparia diluta* Sharpe et Wyatt, 1893). — *Byull. Mosk. O-va Ispyt. Prir. (Otd. Biol.)*, 96 (4): 34-44.

932   Gebauer, A. *et al.*, 1999. New aspects of biology and systematics of snowfinches (genus *Montifringilla*) and Mountain-steppe Sparrows (genus *Pyrgilauda*) in China. — *The Ring*, 21 (1): 207-208.

933   Gerwin, J.A. & R.M. Zink, 1989. Phylogenetic patterns in the genus *Heliodoxa* (Aves, Trochilidae): an allozymic perspective. — *Wilson Bull.*, 101: 525-544.

934   Gerwin, J.A. & R.M. Zink, 1998. Phylogenetic patterns in the Trochilidae. — *Auk*, 115 (1): 105-118.

935   Gibbs, D. *et al.*, 2001. *Pigeons and doves. A guide to the pigeons and doves of the world*: 1-615. — Pica Press, Mountsfield, Sussex.

936   Gibson, D.D. & B. Kessel, 1989. Geographic variation in the Marbled Godwit and description of an Alaska subspecies. — *Condor*, 91: 436-443.

937   Gilliard, E.T., 1940. Descriptions of seven new birds from Venezuela. — *Amer. Mus. Novit.*, 1071: 1-13.

938   Gilliard, E.T., 1949. Five new birds from the Philippines. — *Auk*, 66: 275-280.

939   Gilliard, E.T., 1949. A new puff-bird from Colombia. — *Amer. Mus. Novit.*, 1438: 1-3.

940   Gilliard, E.T., 1961. Four new birds from the mountains of central New Guinea. — *Amer. Mus. Novit.*, 2031: 1-7.

941   Gilliard, E.T. & M. LeCroy, 1967. Annotated list of birds of the Adelbert Mountains, New Guinea. Results of the 1959 Gilliard Expedition. — *Bull. Amer. Mus. Nat. Hist.*, 138 (2): 51-81.

942   Gilliard, E.T. & M. LeCroy, 1968. Birds of the Schrader Mountain region, New Guinea. Results of the American Museum of Natural History Expedition to New Guinea in 1964. — *Amer. Mus. Novit.*, 2343: 1-41.

943   Gilliard, E.T. & M. LeCroy, 1970. Notes on bird from the Tamrau Mountains, New Guinea. — *Amer. Mus. Novit.*, 2420: 1-28.

944   Gladkov, N.A., 1951. Otrad Dyatli. Pages 548-617. In *Ptitsi Sovetskogo Soyuza*. Vol. 1. (G.P. Dementiev & N.A. Gladkov, eds.). — Sovetskaya Nauka., Moscow.

945   Glauert, L.A., 1945. A Western Australian Grass Owl. — *Emu*, 44: 229-230.

946   Gochfeld, M. & J. Burger, 1996. Family Sternidae (Terns). Pages 624-667. In *Handbook of the Birds of the World*. Vol. 3. (J. del Hoyo, A. Elliott & J. Sargatal, eds.). — Lynx Edicions, Barcelona.

947   Godfrey, W.E., 1947. A new Long-eared Owl. — *Can. Field-Nat.*, 61 (6): 196-197.

948   Godfrey, W.E., 1986. *The Birds of Canada*: 1-595. Revd. edition. — National Museums of Canada, Ottawa.

949 Gonzaga, L.P., 1988. A new antwren (*Myrmotherula*) from southeastern Brazil. — *Bull. Brit. Orn. Cl.*, 108: 132-135.
950 Gonzaga, L.P., 1996. Family Cariamidae (Seriemas). Pages 234-239. In *Handbook of the Birds of the World*. Vol. 3 (J. del Hoyo, A. Elliott & J. Sargatal, eds.). — Lynx Edicions, Barcelona.
951 Gonzaga, L.P. & J.F. Pacheco, 1990. Two new subspecies of *Formicivora serrana* (Hellmayr) from southeastern Brazil, and notes on the type locality of *Formicivora deluzae* Ménétriés. — *Bull. Brit. Orn. Cl.*, 110: 187-193.
952 Gonzaga, L.P. & J.F. Pacheco, 1995. A new species of *Phylloscartes* (Tyrannidae) from the mountains of southern Bahia, Brazil. — *Bull. Brit. Orn. Cl.*, 115 (2): 88-97.
953 Gonzales, P.C., 1983. Birds of Catanduanes (Revised Edition). — *Zool. Pap. Natl. Museum Manila*, 2: 1-125.
954 Gonzales, P.C. & R.S. Kennedy, 1990. A new species of *Stachyris* babbler (Aves: Timaliidae) from the island of Panay, Philippines. — *Wilson Bull.*, 102 (3): 367-379.
955 Goodman, S.M. *et al.*, 1996. A new genus and species of passerine from the eastern rain forest of Madagascar. — *Ibis*, 138 (2): 153-159.
956 Goodman, S.M. *et al.*, 1997. A new species of vanga (Vangidae, *Calicalicus*) from southwestern Madagascar. — *Bull. Brit. Orn. Cl.*, 117 (1): 5-10.
957 Goodman, S.M. *et al.*, 1997. The birds of southeastern Madagascar. — *Fieldiana, Zool.*, 87 (1487): 1-132.
957A Goodwin, D., 1963. On a new race of *Streptopelia lugens*. — *Bull. Brit. Orn. Cl.*, 83: 125-126.
958 Goodwin, D., 1963. A new name for *Streptopelia lugens arabica*. — *Bull. Brit. Orn. Cl.*, 83: 153.
959 Goodwin, D., 1967. *Pigeons and Doves of the World*: i-vi, 1-446. — Trustees of the British Museum (Nat. Hist.), London.
960 Goodwin, D., 1968. Notes on woodpeckers (Picidae). — *Bull. Brit. Mus. (Nat. Hist.), Zool.*, 17 (1): 1-44.
961 Goodwin, D., 1969. A new subspecies of the White-quilled Rock Pigeon. — *Bull. Brit. Orn. Cl.*, 89: 131-133.
962 Goodwin, D., 1976. *Crows of the World*: i-vi, 1-354. — Comstock Publishing Associates, Ithaca, New York.
963 Goodwin, D., 1982. *Estrildid finches of the world*: 1-328. — British Museum (Nat. Hist.), London.
964 Goroshko, O.A., 1993. Taxonomic status of the pale (sand?) martin *Riparia* (*riparia*?) *diluta* (Sharpe et Wyatt, 1893). — *Russ. J. Orn.*, 2 (3): 303-323.
965 Gould, J., 1836. [Species also characterised by Mr. Gould]. — *Proc. Zool. Soc., Lond.*, 4: 18.
966 Gould, J., 1856. On a new turkey *Meleagris mexicana*. — *Proc. Zool. Soc., Lond.*, 61-63.
967 Grant, C.H.B. & C.W. Mackworth-Praed, 1934. [Eastern African races of *Francolinus sephaena* (Smith)]. — *Bull. Brit. Orn. Cl.*, 54: 170-173.
968 Grant, C.H.B. & C.W. Mackworth-Praed, 1934. [Descriptions of two new races of Francolin.]. — *Bull. Brit. Orn. Cl.*, 54: 173-174.
969 Grant, C.H.B. & C.W. Mackworth-Praed, 1934. [Two new races of Francolin, a new race of Sarothrura Rail, and a note on the type locality of the Purple-Heron.]. — *Bull. Brit. Orn. Cl.*, 55: 15-19.
970 Grant, C.H.B. & C.W. Mackworth-Praed, 1934. On a new subspecies of Francolin from eastern Africa. — *Bull. Brit. Orn. Cl.*, 55: 62-63.
971 Grant, C.H.B. & C.W. Mackworth-Praed, 1953. A new race of Woodpecker from Portuguese East Africa. — *Bull. Brit. Orn. Cl.*, 73: 55-56.
972 Grantsau, R., 1967. Sobre o genero *Augastes*, com a descrição de una subespécie nova (Aves, Trochiliidae). — *Pap. Dep. Zool. Sec. Agric., S. Paulo*, 21 (3): 21-31.
973 Grantsau, R., 1968. Uma nova espécie de *Phaethornis* (Aves, Trochilidae). — *Pap. Avul. Dep. Zool., Sao Paulo*, 22 (7): 57-59.
974 Grantsau, R., 1969. Uma nova espécie de *Threnetes* (Aves, Trochilidae). — *Pap. Avul. Dep. Zool., Sao Paulo*, 22 (23): 245-247.
975 Grantsau, R., 1988. *Os Beija-flores do Brasil*. — Rio de Janeiro.
976 Grantsau, R. & H.F. de Almeida Camargo, 1989. Nova espécie brasileira de *Amazona* (Aves, Psittacidae). — *Rev. Bras. Biol.*, 49 (4): 1017-1020.
977 Graves, G.R., 1980. A new species of metaltail hummingbird from northern Peru. — *Wilson Bull.*, 92 (1): 1-7.
978 Graves, G.R., 1980. A new subspecies of *Diglossa* (*Carbonaria*) *brunneiventris*. — *Bull. Brit. Orn. Cl.*, 100: 230-232.
979 Graves, G.R., 1981. A new subspecies of Coppery Metaltail (*Metallura theresiae*) from northern Peru. — *Auk*, 98: 382.
980 Graves, G.R., 1982. Speciation in the Carbonated Flower-piercer (*Diglossa carbonaria*) complex of the Andes. — *Condor*, 84: 1-14.
981 Graves, G.R., 1986. Geographic variation in the White-mantled Barbet (*Capito hypoleucus*) of Colombia (Aves: Capitonidae). — *Proc. Biol. Soc. Wash.*, 99: 61-64.
982 Graves, G.R., 1986. Systematics of the Gorgetted Woodstars (Aves: Trochilidae: *Acestrura*). — *Proc. Biol. Soc. Wash.*, 99: 218-224.
983 Graves, G.R., 1987. A cryptic new species of antpitta (Formicariidae: *Grallaria*) from the Peruvian Andes. — *Wilson Bull.*, 99 (3): 313-321.
984 Graves, G.R., 1988. *Phylloscartes lanyoni*, a new species of bristle-tyrant (Tyrannidae) from the lower Cauca Valley of Colombia. — *Wilson Bull.*, 100 (4): 529-534.
985 Graves, G.R., 1990. A new subspecies of *Diglossa gloriosissima* (Aves: Thraupinae) from the western Andes of Colombia. — *Proc. Biol. Soc. Wash.*, 103 (4): 962-965.
986 Graves, G.R., 1992. Diagnosis of a hybrid antbird (*Phlegopsis nigromaculata x Phlegopsis erythroptera*) and the rarity of hybridization among suboscines. — *Proc. Biol. Soc. Wash.*, 105 (4): 834-840.
987 Graves, G.R., 1992. The endemic land birds of Henderson Island, southeastern Polynesia: notes on natural history and conservation. — *Wilson Bull.*, 104: 32-43.
988 Graves, G.R., 1993. Relic of a lost world: a new species of Sunangel (Trochilidae: *Heliangelus*) from 'Bogota'. — *Auk*, 110 (1): 1-8.
989 Graves, G.R., 1997. Colorimetric and morphometric gradients in Colombian populations of Dusky Antbirds (*Cercomacra tyrannina*), with a description of a new species, *Cercomacra parkeri*. — *Orn. Monogr.*, 48: 21-35.
990 Graves, G.R. & S.L. Olson, 1986. A new subspecies of *Turdus swalesi* (Aves:Passeriformes: Muscicapidae) from the Dominican Republic. — *Proc. Biol. Soc. Wash.*, 99: 580-583.
991 Graves, G.R. & S.L. Olson, 1987. *Chlorostilbon bracei* Lawrence, an extinct species of hummingbird from New Providence Island, Bahamas. — *Auk*, 104: 296-302.
992 Graves, G.R. & D.U. Restrepo, 1989. A new allopatric taxon in the *Hapalopsittaca amazonina* (Psittacidae) superspecies from Colombia. — *Wilson Bull.*, 101 (3): 369-376.
993 Graves, G.R. & M.B. Robbins, 1987. A new subspecies of *Siptornis striaticollis* (Aves: Furnariidae) from the eastern slope of the Andes. — *Proc. Biol. Soc. Wash.*, 100: 121-124.
994 Graves, G.R. & J.S. Weske, 1987. *Tangara phillipsi*, a new species of tanager from the Cerros del Sira, eastern Peru. — *Wilson Bull.*, 99 (1): 1-6.
995 Graves, G.R. *et al.*, 1983. *Grallaricula ochraceifrons*, a new species of antpitta from northern Peru. — *Wilson Bull.*, 95 (1): 1-6.
996 Gray, G.R., 1860. List of birds collected by Mr. Wallace at the Molucca Islands, with descriptions of new species etc. — *Proc. Zool. Soc., Lond.*, 341-366.
997 Gray, G.R., 1861. List of species comprising the family Megapodiidae with descriptions of new species and some account of the habits of the species. — *Proc. Zool. Soc., Lond.*, 288-296.
998 Gray, J.E., 1829. [On new birds]. In *The Animal Kingdom arranged in conformity with its organization*: 1-690 (E. Griffiths & E. Pidgeon, eds.). — Whitaker, Treacher & Co., London.
999 Green, A.J., 1992. The status and conservation of the White-winged Wood Duck *Cairina scutulata*. IWRB Special Publication, 17. 1-115. — I.W.R.B., Slimbridge, Glos.
1000 Green, A.J., 1993. The biology of the White-winged Duck *Cairina scutulata*. — *Forktail*, 8: 65-82.
1001 Green, J.E. & L.W. Arnold, 1939. An Unrecognised Race of Murrelet on the Pacific Coast of North America. — *Condor*, 41 (1): 25-29.
1002 Greenlaw, J.S., 1993. Behavioral and morphological diversification in Sharp-tailed Sparrows (*Ammodramus caudacutus*) of the Atlantic coast. — *Auk*, 110 (2): 286-303.
1003 Greenway, J.C., Jr., 1933. Birds from northwest Yunnan. — *Bull. Mus. Comp. Zool.*, 74: 109-168.

1004 Greenway, J.C., Jr., 1935. Birds from the coastal range between the Markham and the Waria rivers, northeastern New Guinea. — *Proc. New England Zool. Cl.*, 14: 15-106.

1005 Greenway, J.C., Jr., 1960. Family Bombycillidae. Pages 369-373. In *Check-list of Birds of the World*. Vol. 9. (E. Mayr & J.C. Greenway, Jr., eds.). — Mus. Comp. Zool., Cambridge, Mass.

1006 Greenway, J.C., Jr., 1962. Family Oriolidae. Pages 122-137. In *Check-list of Birds of the World*. Vol. 15 (E. Mayr & J.C. Greenway, Jr., eds.). — Mus. Comp. Zool., Cambridge, Mass.

1007 Greenway, J.C., Jr., 1966. Birds collected on Batanta, off Western New Guinea, by E. Thomas Gilliard, in 1961. — *Amer. Mus. Novit.*, 2258: 1-27.

1007A Greenway, J.C., Jr., 1967. Family Sittidae. Pages 125-149 In *Check-list of Birds of the World* (R.A. Paynter, Jr., ed.) – Mus. Comp. Zool., Cambridge, Mass.

1008 Greenway, J.C., Jr., 1967. Family Rhabdornithidae. Pages 161-162. In *Check-list of Birds of the World*. Vol. 12. (R.A. Paynter, Jr., ed.) — Mus. Comp. Zool., Cambridge, Mass.

1009 Greenway, J.C., Jr., 1968. Family Drepanididae. Pages 93-103. In *Check-list of Birds of the World*. Vol. 14 (R.A. Paynter, Jr., ed.) — Mus. Comp. Zool., Cambridge, Mass.

1010 Greenway, J.C., Jr., 1978. Type specimens of birds in the American Museum of Natural History. 2. — *Bull. Amer. Mus. Nat. Hist.*, 161 (1): 1-306.

1011 Gregory, S.M.S., 1998. The correct citation of *Coragyps* (Cathartinae) and *Ardeotis* (Otididae). — *Bull. Brit. Orn. Cl.*, 118 (2): 126-127.

1012 Griffiths, C.S., 1994. Monophyly of the Falconiformes based on syringeal morphology. — *Auk*, 117: 787-805.

1013 Grimes, L.G. *et al.*, 1997. Various species accounts. In *The Birds of Africa*. Vol. 5: i-xix, 1-669 (E.K. Urban, C.H. Fry & S. Keith, eds.). — Academic Press, London.

1014 Grimmett, R. *et al.*, 1998. *Birds of the Indian Subcontinent*: 1-888. — Christopher Helm: A. & C. Black, London.

1015 Griscom, L., 1930. Studies from the Dwight collection of Guatemala birds. III. — *Amer. Mus. Novit.*, 438: 1-18.

1016 Griscom, L., 1935. Critical Notes on Central American Birds in the British Museum. — *Ibis*, 13 (5): 541-554.

1017 Griscom, L. & J.C. Greenway, Jr., 1937. Critical notes on new Neotropical Birds. — *Bull. Mus. Comp. Zool.*, 81 (2): 417-437.

1018 Griscom, L. & J.C. Greenway, Jr., 1941. Birds of Lower Amazonia. — *Bull. Mus. Comp. Zool.*, 88 (3): 83-344.

1019 Grote, H., 1936. Das Perlhuhn von Nordostkamerun. — *Orn. Monatsber.*, 44 (5): 155-158.

1020 Grote, H., 1948. Über eine neue Frankolinrasse aus Südkamerun: *Francolinus bicalcaratus molunduensis*. — *Orn. Ber., Heidelberg*, 1: 145.

1021 Groth, J.G., 1998. Molecular phylogenetics of finches and sparrows: Consequences of character state removal in cytochrome *b* sequences. — *Molec. Phylogen. Evol.*, 10 (3): 377-390.

1022 Groth, J.G., 2000. Molecular evidence for the systematic position of *Urocynchramus pylzowi*. — *Auk*, 117: 787-791.

1023 Groth, J.G. & G.F. Barrowclough, 1999. Basal divergences in birds and the phylogenetic utility of the nuclear RAG-1 gene. — *Molec. Phylogen. Evol.*, 12 (2): 115-123.

1024 Guan, Guanxun, 1989. New records to South China avifauna with description of a new subspecies. — *J. Guandong Ecol. Soc.,*: 68-80.

1025 Gundlach, J., 1874. Beitrag zur Ornithologie der Insel Portorico. — *J. f. Orn.*, 22: 304-315.

1026 Gutiérrez, R.J. *et al.*, 2000. A classification of the grouse (Aves: Tetraoninae) based on mitochondrial DNA sequences. — *Wildlife Biology*, 6 (4): 205-211.

1027 Gyldenstolpe, N., 1926. Types of birds in the Royal Natural History Museum in Stockholm. — *Arkiv for Zool.*, 19A (1): 1-116.

1028 Gyldenstolpe, N., 1941. Preliminary descriptions of some new birds from the Brazilian Amazonas. — *Arkiv for Zool.*, 33B (12): 1-10.

1029 Gyldenstolpe, N., 1941. Preliminary diagnoses of some new birds from Bolivia. — *Arkiv for Zool.*, 33B (13): 1-10.

1030 Gyldenstolpe, N., 1945. The bird fauna of Rio Juruá in western Brazil. — *K. Svenska Vetensk. Akad. Handl.*, (3) 22 (3): 1-338.

1030A Gyldenstolpe, N., 1945. A contribution to the ornithology of northern Bolivia. — *K. Svenska Vetensk. Akad. Handl.*, 23 (1): 1-300.

1031 Gyldenstolpe, N., 1951. The ornithology of the Rio Purús region in western Brazil. — *Arkiv for Zool.*, (2) 2 (1): 1-320.

1032 Gyldenstolpe, N., 1955. Notes on a collection of birds made in the Western Highlands, Central New Guinea, 1951. — *Arkiv for Zool.*, 8 (1): 1-181.

1033 Gyldenstolpe, N., 1955. Birds collected by Dr. Sten Bergman during his expedition to Dutch New Guinea 1948-1949. — *Arkiv for Zool.*, 8 (2): 183-397.

1034 Hachisuka, M., 1926. [A new race of the Italian Sparrow from Corsica.]. — *Bull. Brit. Orn. Cl.*, 47: 76.

1035 Hachisuka, M., 1937. Description d'une nouvelle race de faisan des Balkans. — *L'Oiseau*, 7 (1): 3-6.

1036 Hachisuka, M., 1938. Description of a new Kaleege Pheasant. — *Bull. Brit. Orn. Cl.*, 58: 91-93.

1037 Hachisuka, M., 1939. A new race of Bronze-winged Dove. — *Bull. Brit. Orn. Cl.*, 59: 45-47.

1038 Hachisuka, M., 1939. New races of a rail and a fruit pigeon from Micronesia and the Philippines. — *Bull. Brit. Orn. Cl.*, 59: 151-153.

1039 Hachisuka, M., 1939. The red junglefowl from the Pacific Islands. — *Tori*, 10 (49): 599-601.

1040 Hachisuka, M., 1941. Further contributions to the ornithology of the Philippine Islands. — *Tori*, 11 (51/52): 61-89.

1041 Hachisuka, M., 1941. New race of swift from the Philippine Islands. — *Proc. Biol. Soc. Wash.*, 54: 169-170.

1042 Hachisuka, M., 1952. Changes of names among sunbirds and a woodpecker. — *Bull. Brit. Orn. Cl.*, 72: 22-23.

1043 Hachisuka, M., 1952. A change of name for a Green Pigeon from the Philippine Islands. — *Bull. Brit. Orn. Cl.*, 72: 95.

1044 Hachisuka, M., 1953. The affinities of *Pityriasis* of Borneo In *Proc. 7th. Pacific Sci. Congr.* Whitcombe & Tombs, Auckland and Christchurch, New Zealand. 1949: 67-69.

1045 Hachisuka, M. & N. Taka-Tsukasa, 1939. A new race of partridge from Manchuria. — *Bull. Brit. Orn. Cl.*, 59: 88.

1046 Hachler, E., 1950. A new local race of the Hazel-grouse (*Tetrastes bonasia* [L.]) from the East-Carpathian mountains. — *Aquila*, 51-54: 81-84.

1047 Hackett, S.J., 1993. Phylogenetic and biogeographic relationships in the Neotropical genus *Gymnopithys* (Formicariidae). — *Wilson Bull.*, 105: 301-315.

1048 Hackett, S.J., 1995. Molecular systematics and zoogeography of flowerpiercers in the *Diglossa baritula* complex. — *Auk*, 112: 156-170.

1049 Hackett, S.J. & C.A. Lehn, 1997. Lack of generic divergence in a genus (*Pteroglossus*) of neotropical birds: the connection between life-history characteristics and levels of genetic divergence. — *Orn. Monogr.*, 48: 267-279.

1050 Haddrath, O. & A.J. Baker, 2001. Complete mitochondrial DNA genome sequences of extinct birds: ratite phylogenetics and the vicariance biogeography hypothesis. — *Proc. Roy. Soc., Lond.*, 268: 939-945.

1051 Haffer, J., 1961. A new subspecies of woodpecker from Northern Colombia: *Picumnus cinnamomeus persaturatus* subsp. nova. — *Noved. Colomb.*, 1 (6): 397-400.

1052 Haffer, J., 1962. Zum Vorkommen von *Brachygalba salmoni* Sclater & Salvin mit Beschreibung einer neuen Form. — *J. f. Orn.*, 103: 38-46.

1053 Haffer, J., 1974. Avian speciation in tropical South America. — *Publ. Nuttall Orn. Cl.*, 14: i-viii, 1-390.

1053A Haffer, J., 1992. The history of species concepts and species limits in ornithology. — *Bull. Brit. Orn. Cl.*, 112A Centenary Supplement: 107-158.

1054 Haffer, J., 1997. Contact zones between birds of southern Amazonia. — *Orn. Monogr.*, 48: 281-305.

1054A Haffer, J., 1997. Species concepts and species limits in ornithology. Pages 11-24. In *Handbook of the Birds of the World*. Vol. 4 (J. del Hoyo, A. Elliott, and J. Sargatal, eds.). — Lynx Edicions, Barcelona.

1055 Hagen, Y., 1952. No. 20. Birds of Tristan da Cunha. Pages 1-248. In *Results of the Norwegian Scientific Expedition to Tristan da Cunha 1937-1938*. Vol. 3 (E. Christophersen, ed.) — I kommisjon hos Jacob Dybwad, Oslo.

1056 Hald-Mortensen, P., 1970. A new subspecies of the Senegal Firefinch (*Lagonosticta senegala* (L.)) from West Africa. — *Dansk Orn. Foren. Tidsskr.*, 64: 113-117.

1057   Hale, W.G., 1971. A revision of the taxonomy of the Redshank *Tringa totanus*. — *Zool. J. Linn. Soc. Lond.*, 50 (3): 199-268.

1058   Hale, W.G., 1973. The distribution of the Redshank *Tringa totanus* in the winter range. — *Zool. J. Linn. Soc. Lond.*, 53 (3): 177-236.

1059   Hall, B.P., 1952. A new race of Agapornis from Angola. — *Bull. Brit. Orn. Cl.*, 72: 25.

1060   Hall, B.P., 1958. A new race of Honeyguide from Mt. Moco. — *Bull. Brit. Orn. Cl.*, 78: 151-152.

1061   Hall, B.P., 1958. Variation in the Angola Lark *Mirafra angolensis* Bocage with a description of two new races. — *Bull. Brit. Orn. Cl.*, 78: 152-154.

1062   Hall, B.P., 1961. The taxonomy and identification of pipits (genus *Anthus*). — *Bull. Brit. Mus. (Nat. Hist.), Zool.*, 7 (5): 245-289.

1063   Hall, B.P., 1966. A new name for *Geocichla princei graueri* Sassi. — *Bull. Brit. Orn. Cl.*, 86: 123-124.

1064   Hall, B.P. & H.G. Deignan, 1956. A new race of leafbird from Indochina. — *Bull. Brit. Orn. Cl.*, 76: 96.

1065   Hall, B.P. & R.E. Moreau, 1964. Notes on *Andropus masukuensis* Shelley and the status of *Andropadus tephrolaema kungwensis* (Moreau). — *Bull. Brit. Orn. Cl.*, 84: 133-134.

1066   Hall, B.P. & R.E. Moreau, 1970. *An atlas of speciation in African Passerine Birds*: i-xv, 1-423. — Trustees of the British Museum (Nat. Hist.), London.

1067   Han, Lian-xian, 1991. A taxonomic study on Rufous-headed Crowtit in China. — *Zool. Res.* [China], 12 (2): 117-124.

1068   Harebottle, D.M. *et al.*, 1997. The subspecies status of the spotted ground thrush *Zoothera guttata guttata* (Aves: Turdidae) in South Africa — a multivariate analysis. — *Durban Mus. Novit.*, 22: 32-36.

1069   Harper, P.C., 1980. The field identification and distribution of the prions (genus *Pachyptila*), with particular reference to the identification of storm-cast material. — *Notornis*, 27: 235-286.

1070   Harrap, S., 1989. Identification, vocalizations and taxonomy of *Pnoepyga* wren-babblers. — *Forktail*, 5: 61-70.

1071   Harrap, S., 1996. *Tits, Nuthatches and Treecreepers*: 1-464. – A. & C. Black, London.

1072   Harris, T. & G. Arnott, 1988. *Shrikes of southern Africa. True shrikes, helmet-shrikes and bush-shrikes, including the batises and Black-throated Wattle-Eye*: 1-224. — Struik Winchester, Cape Town.

1073   Harris, T. & K. Franklin, 2000. *Shrikes and Bush-Shrikes*: 1-392. — A. & C. Black, London.

1074   Harrison, C.J.O., 1986. A re-assessment of the affinities of some small Oriental babblers Timaliidae. — *Forktail*, 1: 81-83.

1075   Harrison, J.M., 1952. On the history of the partridge in the German Friesian islands with the description of a new race from the island of Borkum. — *Bull. Brit. Orn. Cl.*, 72: 18-21.

1076   Harrison, J.M., 1952. The name of the Borkum partridge. — *Bull. Brit. Orn. Cl.*, 72: 47.

1077   Harrison, J.M., 1957. Exhibition of a new race of the Little Owl from the Iberian Peninsula. — *Bull. Brit. Orn. Cl.*, 77: 2-3.

1078   Harshman, J., 1994. Reweaving the tapestry - what can we learn from Sibley & Ahlquist (1990)? — *Auk*, 111 (2): 377-388.

1079   Hartert, E., 1895. On some birds from the Congo region. — *Novit. Zool.*, 2: 55-56.

1080   Hartert, E., 1903. *Die Vögel der paläarktischen Fauna*. Heft 1. Pages i-xi, 1-112 (Vol. 1.) R. Friedlander & Sohn, Berlin.

1081   Hartert, E., 1904. The birds of the South West Islands: Wetter, Roma, Kisser, Letti and Moa. — *Novit. Zool.*, 11: 174-221.

1082   Hartert, E., 1916. [*Brachypteryx poliogyna mindorensis* described from Mindoro.]. — *Bull. Brit. Orn. Cl.*, 36: 48-50.

1083   Hartert, E., 1921. *Die Vögel der paläarktischen Fauna*. Heft 16. Pages 1893-2020 (Vol. 3). R. Friedlander & Sohn, Berlin.

1084   Hartert, E., 1922. Types of birds in the Tring Museum. B. Types in the General Collection (cont.) — *Novit. Zool.*, 29: 365-412.

1085   Hartert, E. *et al.*, 1936. Die Vögel des Weyland-Gebirges und seines Vorlandes. — *Mitt. Zool. Mus. Berlin*, 21 (2): 165-240.

1086   Hartlaub, G., 1848. Description de cinq nouvelles espèces d'oiseau de l'Afrique occidentale. — *Rev. Zool. Soc. Cuvier, Paris*, 11: 108-110.

1087   Hartlaub, G., 1851. Bericht über die Leistung in der Naturgeschichte der Vögel während des Jahres 1850. — *Archiv für Naturg.*, 17 (2): 33-67.

1088   Harwin, R.M., 1983. Reappraisal of variation in the nightjar *Caprimulgus natalensis* Smith. — *Bull. Brit. Orn. Cl.*, 103 (4): 140-144.

1089   Hawkins, R.W., 1948. A new western race of the Nighthawk. — *Condor*, 50 (3): 131-132.

1090   Haverschmidt, F., 1982. The status of the Rough-winged Swallow *Stelgidopteryx ruficollis* in Suriname. — *Bull. Brit. Orn. Cl.*, 102 (2): 75-77.

1090A   Hayes, F.E., 2001. Geographic variation, hybridization, and the leapfrog pattern of Evolution in the Suiriri Flycatcher complex (*Suiriri suiriri*) complex. — *Auk*, 118 (2): 457-471.

1091   Hazevoet, C.J., 1989. Notes on behaviour and breeding of the Razo Lark *Alauda razae*. — *Bull. Brit. Orn. Cl.*, 109: 82-86.

1092   Hazevoet, C.J., 1995. *The birds of the Cape Verde Islands*: 1-192. BOU Check-list Ser. No. 13 — British Ornithologists' Union, Tring, Herts.

1093   Hedges, S.B. & C.G. Sibley, 1994. Molecules vs. morphology in avian systematics: the case of the "pelecaniform" birds. — *Proc. Nat. Acad. Sci.*, 91: 9861-9865.

1094   Heidrich, P. *et al.*, 1995. Bioakustik, Taxonomie and molekulare Systematik amerikanischer Sperlingskäuze (Strigidae: *Glaucidium* spp.). — *Stuttgarter Beitr. Naturk.*, Ser. A, 534: 1-47.

1095   Heindl, M., 1999. Species accounts within the Family Trochilidae (Hummingbirds). Pages 549-680. In *Handbook of the Birds of the World*. Vol. 5 (J. del Hoyo, A. Elliott & J. Sargatal, eds.). — Lynx Edicions, Barcelona.

1096   Heindl, M. & K.-L. Schuchmann, 1998. Biogeography, geographical variation and taxonomy of the Andean hummingbird genus *Metallura* Gould 1847. — *J. f. Orn.*, 139 (4): 425-473.

1097   Hekstra, G.P., 1982. Description of twenty four new subspecies of American *Otus* (Aves, Strigidae). — *Bull. Zool. Mus. Univ. Amsterdam*, 9 (7): 49-63.

1098   Helbig, A. & I. Seibold, 1999. Molecular phylogeny of Palearctic-African *Acrocephalus* and *Hippolais* warblers (Aves: Sylviidae). — *Molec. Phylogen. Evol.*, 11: 246-260.

1099   Helbig, A.J. *et al.*, 1996. Phylogeny and species limits in the Palaearctic chiffchaff *Phylloscopus collybita* complex: mitochondrial genetic differentiation and bioacoustic evidence. — *Ibis*, 138 (4): 650-666.

1100   Helbig, A.J. *et al.*, 1995. Genetic differentiation and phylogenetic relationships of Bonelli's Warbler *Phylloscopus bonelli* and Green Warbler *P. nitidus*. — *J. Avian Biol.*, 26 (2): 139-153.

1101   Hellack, J.J. & G.D. Schnell, 1977. Phenetic analysis of the subfamily Cardinalinae using external and skeletal characteristics. — *Wilson Bull.*, 89: 130-148.

1102   Hellmayr, C.E., 1919. Miscellanea ornithologica. IV. xii. Vier neue Formen aus dem tropischen Amerika. — *Verh. Orn. Ges. Bayern*, 14: 126-133.

1103   Hellmayr, C.E., 1934. Catalogue of the Birds of the Americas. — *Field Mus. Nat. Hist., Zool.*, 13 (7): i-vi, 1-530.

1104   Hellmayr, C.E. & H.B. Conover, 1942. Catalogue of the birds of the Americas and adjacent islands. 1. — *Field Mus. Nat. Hist., Zool.*, 13 (1): i-vi, 1-636.

1105   Hellmayr, C.E. & J. von Seilern, 1912. Beiträge zur Ornithologie von Venezuela. — *Arch. Naturg.*, 78A (5): 34-166.

1106   Hernández-Camacho, J.I. & J.V. Rodriguez-M, 1979. Dos nuevos taxa del genero *Grallaria* (Aves: Formicariidae) del alto Valle del Magdalena (Colombia). — *Caldasia*, 12: 573-580.

1107   Hernández-Camacho, J.I. & H. Romero-Zambrano, 1978. Descripcion de una nueva subespecie de *Momotus momota* para Colombia. — *Caldasia*, 12 (58): 353-358.

1108   Herremans, M., 1990. Taxonomy and evolution in Redpolls *Carduelis flammea - hornemanni*; a multivariate study of their biometry. — *Ardea*, 78: 441-458.

1109   Herremans, M. *et al.*, 1999. Description of a new taxon *brookei* of Levaillant's Cisticola *Cisticola tinniens* from the Western Cape, South Africa. — *Ostrich*, 70: 164-172.

1110   Herzog, S.K., 2001. A re-evaluation of Straneck's (1993) data on the taxonomic status of *Serpophaga subcristata* and *S. munda* (Passeriformes: Tyrannidae): conspecifics or semispecies? — *Bull. Brit. Orn. Cl.*, 121 (4): 273-277.

1111   Heynen, I., 1999. Species accounts within the Family Trochilidae (Hummingbirds). Pages 549-680. In *Handbook of the Birds of the World*. Vol. 5 (J. del Hoyo, A. Elliott & J. Sargatal, eds.). — Lynx Edicions, Barcelona.

1112   Higgins, P.J. (ed) 1999. *Handbook of Australian, New Zealand and Antarctic birds*. Vol. 4. Parrots to Dollarbird. 1-1248. — Oxford Univ. Press, Melbourne.

1113   Higgins, P.J. & S.J.J.F. Davies (eds), 1996. *Handbook of Australian, New Zealand and Antarctic birds*. Vol. 3. Snipe to Pigeons. 1-1028. — Oxford Univ. Press, Melbourne.

1114   Higgins, P.J. *et al.* (eds), 2001. *Handbook of Australian, New Zealand and Antarctic birds*. Vol. 5. Tyrant-flycatchers to Chats. 1-1269. — Oxford Univ. Press, Melbourne.

1115   Hillcoat, B. *et al.*, 1997. *Diomedea cauta* Shy Albatross. — BWP Update, 1 (1): 57-59.

1116   Hilty, S.L. & W.L. Brown, 1986. *Birds of Colombia*: i-xii, 1-836. — Princeton Univ. Press, Princeton, N.J.

1117   Hinkelmann, C., 1988. Comments on recently described new species of hermit hummingbirds. — *Bull. Brit. Orn. Cl.*, 108 (4): 159-169.

1118   Hinkelmann, C., 1988. On the identity of *Phaethornis maranhaoensis* Grantsau, 1968 (Trochilidae). — *Bull. Brit. Orn. Cl.*, 108 (1): 14-18.

1119   Hinkelmann, C., 1989. Notes on the taxonomy and geographic variation of *Phaethornis bourcieri* (Aves: Trochilidae) with the description of a new subspecies. — *Bonn. Zool. Beitr.*, 40 (2): 99-107.

1120   Hinkelmann, C., 1996. Systematics and geographic variation in Long-tailed Hermit Hummingbirds, the *Phaethornis superciliosus - malaris - longirostris* species group (Trochilidae) with notes on their biogeography. — *Orn. Neotrop.*, 7: 119-148.

1121   Hinkelmann, C., 1999. Species accounts within the Family Trochilidae (Hummingbirds). Pages 537-547. In *Handbook of the Birds of the World*. Vol. 5 (J. del Hoyo, A. Elliott & J. Sargatal, eds.). — Lynx Edicions, Barcelona.

1122   Hinkelmann, C. & K.-L. Schuchmann, 1997. Phylogeny of the hermit hummingbirds (Trochilidae: Phaeornithinae). — *Stud. Neotrop. Fauna Envir.*, 32: 142-163.

1123   Hiraldo, F. *et al.*, 1976. Sobre el status taxonómico del Aquila *Imperial iberica*. — *Doñana Acta Vertebrata*, 3: 171-182.

1124   Hirschfeld, E. *et al.*, 2000. Identification, taxonomy and distribution of Greater and Lesser Sand Plovers. — *Brit. Birds*, 93: 162-189.

1125   Hockey, P.A.R., 1996. Family Haematopodidae (Oystercatchers). Pages 308-325. In *Handbook of the Birds of the World*. Vol. 3 (J. del Hoyo, A. Elliott & J. Sargatal, eds.). — Lynx Edicions, Barcelona.

1126   Hoesch, W. & G. Niethammer, 1940. Die Vogelwelt Deutsch-Südwestafrikas namentlich des Damara- und Namalandes. — *J. f. Orn.*, 88 (Sonderh.): 1-404.

1127   Holt, D.W. *et al.*, 1999. Species accounts within the Family Strigidae (Typical Owls). Pages 153-242. In *Handbook of the Birds of the World*. Vol. 5 (J. del Hoyo, A. Elliott & J. Sargatal, eds.). — Lynx Edicions, Barcelona.

1128   Holthuis, L.B. & T. Sakai, 1970. Ph. F. von Siebold and Fauna Japonica. A history of early Japanese Zoology: 323 pp. — Academic Press of Japan, Tokyo.

1129   Holyoak, D.T., 1974. Undescribed land birds from the Cook Islands, Pacific Ocean. — *Bull. Brit. Orn. Cl.*, 94: 145-150.

1130   Holyoak, D.T., 1976. *Halcyon* "*ruficollaris*". — *Bull. Brit. Orn. Cl.*, 96: 40.

1131   Holyoak, D.T., 1999. Family Podargidae (Frogmouths). Pages 266-287. In *Handbook of the Birds of the World*. Vol. 5 (J. del Hoyo, A. Elliott & J. Sargatal, eds.). — Lynx Edicions, Barcelona.

1131A   Holyoak, D.T., 2001. *Nightjars and their Allies. The Caprimulgiformes*: i-xxii, 1-773. – O.U.P., Oxford.

1132   Holyoak, D.T. & J.-C. Thibault, 1976. La variation géographique de *Gygis alba*. — *Alauda*, 44 (4): 457-473.

1133   Holyoak, D.T. & J.-C. Thibault, 1984. Contribution à l'étude des oiseaux de Polynésie orientale. — *Mém. Mus. Nat. d'Hist. Nat., A. Zool.*, 127: 1-209.

1134   Homberger, D.G., 1980. Funktionell-morphologische Untersuchungen zur Radiation der Ernährungs- und Trinkmethoden der Papageien (Psittaci). [Functional morphological studies on the radiation of the feeding and drinking methods of the parrots]. — *Bonn. Zool. Monogr.*, 19: 1-192.

1135   Homberger, D.G., 1980. Functional morphology and evolution of the feeding apparatus in parrots, with special reference to the Pesquet's Parrot, *Psittrichas fulgidus* (Lesson). Pages 471-485 In *Conservation of New World Parrots. Technical Paper No. 1.* (R.F. Pasquier, ed.) — International Council for Bird Preservation, Washington, D.C.

1136   Homberger, D.G., 1985. Parrot. Pages 437-439. In *A New Dictionary of Birds* (B. Campbell & E. Lack, eds.). — British Ornithologists' Union/Poyser, Calton.

1137   Homberger, D.G., The evolutionary history of parrots and cockatoos: A model for evolution in the Australasian avifauna. In *Acta XX Congr. Int. Ornithol., Christchurch, New Zealand.* 1991: 398-403.

1138   Homberger, D.G., 2002. The comparative biomechanics of a prey-predator relationship: The adaptive morphologies of the feeding apparatus of Australian Black-Cockatoos and their foods as a basis for the reconstruction of the Evolutionary history of Psittaciformes pp.203-228. In *Vertebrate Biomechanics and* Evolution (V.L. Bels, J.-P. Gasc & A. Casinos, eds.). — BIOS Scientific Publishers, Oxford.

1139   Hoogerwerf, A., 1962. Ornithological notes on the Sunda Strait area and the Karimundjawa, Bawean and Kangean islands. — *Bull. Brit. Orn. Cl.*, 82: 142-147.

1140   Hoogerwerf, A., 1962. On *Aegithina tiphia* (Linn.), the Common Iora from Udjong Kulon, western Java. — *Bull. Brit. Orn. Cl.*, 82: 160-165.

1141   Hoogerwerf, A., 1962. Some ornithological notes on the smaller islands around Java (with the descriptions of seven new subspecies). — *Ardea*, 50 (3/4): 180-206.

1142   Hoogerwerf, A., 1962. Notes on Indonesian birds with special reference to the avifauna of Java and surrounding small islands (1). — *Treubia*, 26 (1): 11-38.

1143   Hoogerwerf, A., 1963. On the Yellow-vented Bulbul, *Pycnonotus goiavier* (Scop). — *Bull. Brit. Orn. Cl.*, 83: 56-60.

1144   Hoogerwerf, A., 1963. The Golden-backed Woodpecker, *Chrysocolaptes lucidus* (Scopoli) in the Kangean Archipelago. — *Bull. Brit. Orn. Cl.*, 83: 112-114.

1145   Hoogerwerf, A., 1963. Some subspecies of *Gracula religiosa* (Linn.) living in Indonesia. — *Bull. Brit. Orn. Cl.*, 83: 155-158.

1146   Hoogerwerf, A., 1965. *Pycnonotus plumosus* subspp. with the description of a new subspecies from Bawean Island. — *Bull. Brit. Orn. Cl.*, 85: 47-53.

1147   Hoogerwerf, A. & L. de Boer, 1947. Eine nieuwe uil van Billiton *Strix leptogrammica chaseni* subsp. nov. — *Chronica Naturae*, 103 (7): 140-142.

1148   Hørring, R., 1937. 6. Birds collected on the Fifth Thule Expedition. Pages 1-134. In *Report of the 5th. Thule Expedition 1921-24; the Danish Expedition to Arctic North America in charge of Knud Rasmussen.* Vol. 2. Gyldenalske Boghandel Nordisk Verlag, Copenhagen.

1149   Horváth, L., 1958. A new race of the Desert Lark from Egypt. — *Bull. Brit. Orn. Cl.*, 78: 124-125.

1150   Howell, S.N.G., 1993. A taxonomic review of the Green-fronted Hummingbird. — *Bull. Brit. Orn. Cl.*, 113 (3): 179-187.

1151   Howell, T.R., 1965. New subspecies of birds from the lowland pine savanna of north-eastern Nicaragua. — *Auk*, 82: 438-464.

1152   Howell, T.R. *et al.*, 1968. Family Fringillidae, Subfamily Carduelinae. Pages 207-306. In *Check-list of Birds of the World.* Vol. 14 (R.A. Paynter, Jr., ed.) — Mus. Comp. Zool., Cambridge, Mass.

1153   Howes, G.B., 1896. *General index of the first twenty volumes of the Journal (Zoology) and the zoological portion of the Proceedings, November 1838 to 1890, of the Linnean Society*: i-viii, 1-437. — The Linnean Society, London.

1154   Hoy, G., 1968. *Geositta rufipennis ottowi* eine neue Subspecies aus der Sierra de Cordoba. — *J. f. Orn.*, 109: 228-229.

1155   Hoy, G., 1978. *Sicalis flaveola königi* (Aves, Fringillidae), eine neue Subspecies des Safranfinken aus den Anden. — *Stuttgarter Beitr. Naturk. Ser. A (Biol.)*, 305: 1-4.

1156   Hu, Da-Shih *et al.*, 2000. Distribution, variation and taxonomy of *Topaza* Hummingbirds (Aves: Trochilidae). — *Orn. Neotrop.*, 11 (2): 123-142.

1157   Hubbard, J.P., 1975. Geographic variation in non-California populations of the Rufous-crowned Sparrow. — *Nemouria*, 15: 1-28.

1158   Hubbard, J.P. & R.S. Crossin, 1974. Notes on northern Mexican birds. An expedition report. — *Nemouria*, 14: 1-41.

1159   Hubbard, J.P. & J.E. duPont, 1974. A revision of the Ruddy Kingfisher, *Halcyon coromanda* (Latham). — *Nemouria*, 13: 1-29.

1160   Hübner, S.M. *et al.*, 2001. Molekulargenetische Untersuchungen zur Phylogenie der Hornvögel (Bucerotiformes). — *J. f. Orn.*, 142 (Sonderheft 1): 196.

1161   Hughes, J.M., 1996. No. 244. Greater Roadrunner (*Geococcyx californianus*). Pages 1-24. In *The Birds of North America* (A. Poole & F. Gill, eds.). — The Academy of Natural Sciences, Philadelphia, Philadelphia.

1162   Hughes, J.M. & A.J. Baker, 1999. Phylogenetic relationships of the enigmatic Hoatzin (*Opisthocomus hoazin*) resolved using mitochondrial and nuclear gene sequences. — *Mol. Biol. Evol.*, 16 (9): 1300-1307.

1162A   Hume, A.O., 1869. [Untitled]. — *Ibis*,: 355-357.

1163   Hume, R., 1996. Family Burhininidae (Thick-knees). Pages 348-363. In *Handbook of the Birds of the World*. Vol. 3 (J. del Hoyo, A. Elliott & J. Sargatal, eds.). — Lynx Edicions, Barcelona.

1164   Humphrey, P.S. & M.C. Thompson, 1981. A new species of steamer-duck (*Tachyeres*) from Argentina. — *Occ. Pap. Mus. Nat. Hist., Univ. Kansas*, 95: 1-12.

1165   Hunt, J.S. *et al.*, 2001. Molecular systematics and biogeography of Antillean thrashers, tremblers, and mockingbirds (Aves : Mimidae). — *Auk*, 118 (1): 35-55.

1166   Husain, K.Z., 1958. Subdivisions and zoogeography of the genus *Treron* (green Fruit Pigeons). — *Ibis*, 100 (3): 334-348.

1167   Husain, K.Z., 1959. Taxonomic status of the Burmese Slaty-headed Parakeet. — *Ibis*, 101 (2): 249-250.

1168   Hussain, S.A. & M.A. Reza Khan, 1978. A new subspecies of Bay Owl (*Phodilus badius* (Horsfield)) from peninsular India. — *J. Bombay Nat. Hist. Soc.*, 74 (2): 334-336 (1977).

1169   Husson, A.M. & L.B. Holthuis, 1955. The dates of publication of "Verhandelingen over de natuurlijke geschiedens der Nederlandsche Overzeesche Bezittingen" edited by C. J. Temminck. — *Zool. Meded., Leiden*, 34 (2): 17-24.

1170   I.C.Z.N., 1957. Opinion 497. Suppression under plenary powers of the specific name *munda* by Kuhl, 1820, as published in the combination *Procellaria munda* and on the same occasion in the combination *Nectris munda* (Class: Aves). — *Opinions and Declarations I.C.Z.N.*, 17: 349-360.

1171   I.C.Z.N., 1968. Opinion 852. *Tanagra* Linnaeus, 1764 (Aves): suppressed under the plenary powers. — *Bull. zool. Nomencl.*, 25 (2/3): 74-79.

1172   I.C.Z.N., 1969. Opinion 882; *Ornithologica Britannica, 1771:* validation of four specific names of birds. — *Bull. zool. Nomencl.*, 26 (1): 26-27.

1173   I.C.Z.N., 1970. Opinion 895. *Strix capensis* Daudin, 1800 (Aves): suppressed under the plenary powers. — *Bull. zool. Nomencl.*, 26 (5/6): 194-195.

1174   I.C.Z.N., 1976. Opinion 1056. *Eudyptes atratus* Finsch, 1875 ex Hutton MS (Aves): suppressed under the plenary powers. — *Bull. zool. Nomencl.*, 33: 16-18.

1175   I.C.Z.N., 1983. Opinion 1249. *Toxostoma crissale* ruled to be the correct original spelling of the name first published as *Toxostoma dorsalis* Baird, 1858 (Aves). — *Bull. zool. Nomencl.*, 40: 83-84.

1176   I.C.Z.N., 1990. Opinion 1606. *Semioptera wallacii* Gray, 1859 (Aves, Paradisaeidae): conserved as the correct spelling of the generic and specific names. — *Bull. zool. Nomencl.*, 47 (2): 169-170.

1177   I.C.Z.N., 1991. Opinion 1648. *Micropterus patachonicus* King, 1831 and *Anas pteneres* Forster, 1844 (both currently in *Tachyeres* Owen, 1875; Aves, Anseriformes): specific names conserved. — *Bull. zool. Nomencl.*, 48: 187-188.

1178   I.C.Z.N., 1999. *International Code of Zoological Nomenclature*: i-xxix, 1-306. (4th. Ed.) — The International Trust for Zoological Nomenclature.

1178A   I.C.Z.N., 2002. Opinion 2004 (Case 3167). *Schistochlamys* Reichenbach, 1850, and *Neothraupis* Hellmayr, 1936 (Aves, Passeriformes): conserved. — *Bull. zool. Nomencl.*, 59 (2): 151-152.

1179   Imber, M.J., 1985. Origins, phylogeny and taxonomy of the gadfly petrels *Pterodroma* spp. — *Ibis*, 127: 197-229.

1180   Imber, M.J. & J.D. Tennyson, 2001. A new petrel species (Procellariidae) from the south-west Pacific. — *Emu*, 101: 123-127.

1181   Inskipp, T. & P.D. Round, 1989. A review of the Black-tailed Crake *Porzana bicolor*. — *Forktail*, 5: 3-15.

1182   Inskipp, T.P. *et al.*, 1996. *An Annotated Checklist of the Birds of the Oriental Region*: [i-x], 1-294. — The Oriental Bird Club, Sandy, Beds., UK.

1183   Iredale, T., 1946. A new Australian parrot. — *Emu*, 46 (1): 1-2.

1184   Irestedt, M. *et al.*, 2001. Phylogeny of major lineages of suboscines (Passeriformes) analysed by nuclear DNA sequence data. — *J. Avian Biol.*, 32: 15-25.

1185   Irwin, M.P.S., 1957. Description of a new race of *Buccanodon leucotis* from southern Portuguese East Africa and eastern Southern Rhodesia. — *Bull. Brit. Orn. Cl.*, 77: 57-58.

1186   Irwin, M.P.S., 1957. A new race of Pink-billed Lark *Calandrella conirostris* from Northern Bechuanaland Protectorate. — *Bull. Brit. Orn. Cl.*, 77: 117.

1187   Irwin, M.P.S., 1963. Systematic and distributional notes on Southern African birds. (1) A revision of the south-west arid races of the courser *Rhinoptilus africanus* (Temminck). — *Durban Mus. Novit.*, 7 (1): 1-17.

1188   Irwin, M.P.S., 1968. A new race of the bokmakierie *Telophorus zeylonus* (Linnaeus) (Aves) from the Chimanimani Mountains, Rhodesia. — *Arnoldia*, 3 (26): 1-5.

1189   Irwin, M.P.S., 1981. *The birds of Zimbabwe*: i-xvi, 1-464. — Quest Publishing, Bulawayo.

1190   Irwin, M.P.S., 1988. Various species accounts. In *The Birds of Africa*. Vol. 3: i-xvi, 1-611 (C.H. Fry, S. Keith & E.K. Urban, eds.). — Academic Press, London.

1191   Irwin, M.P.S., 1989. The number of rectrices in some Afrotropical warblers (Sylviidae). — *Honeyguide*, 35: 179-180.

1192   Irwin, M.P.S., 1997. Various species accounts. In *The Birds of Africa*. Vol. 5: i-xix, 1-669 (E.K. Urban, C.H. Fry & S. Keith, eds.). — Academic Press, London.

1193   Irwin, M.P.S., 2000. Various species accounts. In *The Birds of Africa*. Vol. 6: i-xvii, 1-724 (C.H. Fry, S. Keith & E.K. Urban, eds.). — Academic Press, London.

1194   Irwin, M.P.S. & P.A. Clancey, 1986. A new generic status for the Dappled Mountain Robin. — *Bull. Brit. Orn. Cl.*, 106 (3): 111-115.

1195   Isenmann, P. & M.A. Bouchet, 1993. French distribution area and taxonomic status of the Southern Great Grey Shrike *Lanius elegans meridionalis*. — *Alauda*, 61: 223-227.

1196   Isler, M.L. & P.R. Isler, 1987. *The Tanagers. Natural history, distribution and identification*: 1-404. — Smithsonian Institution Press, Washington.

1197   Isler, M.L. *et al.*, 1997. Biogeography and systematics of the *Thamnophilus punctatus* (Thamnophilidae) complex. — *Orn. Monogr.*, 48: 355-381.

1198   Isler, M.L. *et al.*, 1999. Species limits in antbirds (Passeriformes : Thamnophilidae): the *Myrmotherula surinamensis* complex. — *Auk*, 116 (1): 83-96.

1199   Isler, M.L. *et al.*, 2001. Species limits in antbirds: the *Thamnophilus punctatus* complex continued. — *Condor*, 103: 278-286.

1199A   Isler, M.L. *et al.*, 2002. A new species of *Percnostola* antbird (Passeriformes: Thamnophilidae) from Amazonian Peru, and an analysis of species limits within *Percnostola rufifrons*. — *Wilson Bull.*, 113: 164-176.

1199B   Isler, M.L. *et al.*, 2002. Rediscovery of a cryptic species and description of a new subspecies in the *Myrmeciza hemimelaena* complex (Thamnophilidae) of the Neotropics. — *Auk*, 119: 362-378.

1200   Jany, E., 1955. Neue Vogel-Formen von den Nord-Molukken. — *J. f. Orn.*, 96: 102-106.

1201   Jany, E., 1955. *Loriculus stigmatus croconotus* subsp. nova. — *J. f. Orn.*, 96: 220.

1202   Jaramillo, A. & P. Burke, 1999. *New World blackbirds; the Icterids*: 1-431. — A. & C. Black, London.

1203   Jehl, J.R., 1968. The systematic position of the Surfbird *Aphriza virgata*. — *Condor*, 70: 206-210.

1204   Jehl, J.R., Jr., 1987. Geographic variation and evolution in the California Gull (*Larus californicus*). — *Auk*, 104: 421-428.

1205   Jenni, D.A., 1996. Family Jacanidae (Jacanas). Pages 276-291. In *Handbook of the Birds of the World*. Vol. 3 (J. del Hoyo, A. Elliott & J. Sargatal, eds.). — Lynx Edicions, Barcelona.

1206   Jensen, F.P., 1983. A new species of sunbird from Tanzania. — *Ibis*, 125 (4): 447-449.

1207   Jensen, F.P. & S.N. Stuart, 1982. New subspecies of forest birds from Tanzania. — *Bull. Brit. Orn. Cl.*, 102: 95-99.

1208   Johansen, H., 1957. Rassen und Populationen des Auerhuhns (*Tetrao urogallus*). — *Viltrevy*, 1 (3): 233-366.

1209   Johansen, H., 1960. Die Vogelfauna Westsibiriens. III. Teil (Non-Passeres). 9. Fortsetzung: Alcidae, Laridae. — *J. f. Orn.*, 101 (3): 316-339.

1210   Johansson, U.S. *et al.*, 2001. Clades within the 'higher land birds', evaluated by nuclear DNA sequences. — *J. Zool. Syst. Evol. Res.*, 39: 37-51.

1211   Johnsgard, P.A., 1979. Order Anseriformes. Pages 425-506. In *Check-list of Birds of the World*. Vol. 1 (2nd. Ed.) (E. Mayr & G.W. Cottrell, eds.). — Mus. Comp. Zool., Cambridge, Mass.

1212   Johnson, A.W. & W.R. Millie, 1972. A new species of conebill (*Conirostrum*) from northern Chile. — Birds of Chile, (Suppl.): 4-8.

1213   Johnson, N.K., 2002. Leapfrogging revisited in Andean birds: geographical variation in the Tody-Tyrant superspecies *Poecilotriccus ruficeps* and *P. luluae*. — *Ibis*, 144 (1): 69-84.

1214   Johnson, N.K. & R.E. Jones, 2001. A new species of tody-tyrant (Tyrannidae: *Poecilotriccus*) from northern Peru. — *Auk*, 118: 334-341.

1215   Johnson, N.K. & J.A. Marten, 1988. Evolutionary genetics of flycatchers. 2. Differentiation in the *Empidonax difficilis* complex. — *Auk*, 105 (1): 177-191.

1216   Johnson, N.K. *et al.*, 1998. Refined colorimetry validates endangered subspecies of the Least Tern. — *Condor*, 100: 18-26.

1216A   Johnson, N.K. *et al.*, 1999. Resolution of the debate over species concepts in ornithology: a new comprehensive biologic species concept. In *Proc. 22nd. Internat. Ornith. Congr. BirdLife SouthAfrica, Durban*. 1998: 1470-1482.

1217   Johnstone, R.E., 1981. Notes on the distribution, ecology and taxonomy of the Red-crowned Pigeon (*Ptilinopus regina*) and Torres Strait Pigeon (*Ducula bicolor*) in Western Australia. — *Rec. West. Austr. Mus.*, 9: 7-22.

1218   Johnstone, R.E., 1981. Notes on the distribution, ecology and taxonomy of the the Partridge Pigeon (*Geophaps smithii*) and Spinifex Pigeon (*Geophaps plumifera*) in Western Australia. — *Rec. West. Austr. Mus.*, 9: 49-64.

1219   Johnstone, R.E., 1983. A review of the Mangrove Kingfisher, *Halcyon chloris* (Boddaert), in Australia, with a description of a new subspecies from Western Australia. — *Rec. West. Austr. Mus.*, 11 (1): 25-31.

1220   Johnstone, R.E. & J.C. Darnell, 1997. Description of a new subspecies of bush-warbler of the genus *Cettia* from Alor Island, Indonesia. — *West. Austral. Nat.*, 21: 145-151.

1221   Johnstone, R.E. & J.C. Darnell, 1997. Description of a new subspecies of Boobook Owl *Ninox novaeseelandiae* (Gmelin) from Roti Island, Indonesia. — *West. Austral. Nat.*, 21: 161-173.

1222   Joiris, C., 1978. Le Goéland argenté portugais (*Larus argentatus lusitanius*), nouvelle forme de Goéland argenté à pattes jaunes. — *Aves*, 15: 17-18.

1223   Jones, D.N. *et al.*, 1995. *The Megapodes. Megapodiidae*: i-xx, 1-262. — Oxford Univ. Press, Oxford.

1124   Jonkers, B. & H. Roersma, 1990. New subspecies of *Lonchura spectabilis* from East Sepik Province, Papua New Guinea. — *Dutch Birding*, 12: 22-25.

1225   Joseph, L., 1999. A curious quandary concerning questionable curassows: can the DNA toolbox solve a century-old problem? — *WPA News*, 60: 9-11.

1226   Joseph, L., 2000. Beginning an end to 63 years of uncertainty: the Neotropical parakeets known as *Pyrrhura picta* and *P. leucotis* comprise more than two species. — *Proc. Acad. Nat. Sci. Philad.*, 150: 279-292.

1227   Joseph, L. *et al.*, 2001. Molecular systematics and phylogeography of New Guinean logrunners (Orthonychidae). — *Emu*, 101: 273-280.

1228   Jouanin, C., 1959. Les emeus de l'Expedition Baudin. — *L'Oiseau*, 29 (3): 169-203.

1229   Jouanin, C. & J.-L. Mougin, 1979. Order Procellariiformes. Pages 48-121. In *Check-list of Birds of the World*. Vol. 1. (2nd. Ed.) (E. Mayr & G.W. Cottrell, eds.). — Mus. Comp. Zool., Cambridge, Mass.

1230   Junge, G.C.A., 1939. Description of a new bird from Simalur. — *Zool. Meded., Leiden*, 22: 120.

1231   Junge, G.C.A., 1948. Notes on some Sumatran birds. — *Zool. Meded., Leiden*, 29: 311-326.

1232   Junge, G.C.A., 1952. New subspecies of birds from New Guinea. — *Zool. Meded., Leiden*, 31 (22): 247-249.

1233   Kalyakin, M.V. *et al.*, 1993. On the problem of systematic relations between Pallas' Grasshopper (*Locustella certhiola*) and Middendorff' Grasshopper (*Locustella ochotensis*). Pages 164-182. In *Hibridizaciya i problema vida u pozvonochnikh* (O.D. Rossolino, ed.): 1-223 — Moscow State University Press, Moscow.

1234   Kasparek, M., 1996. On the identity of *Ceryle rudis syriaca*. — *J. f. Orn.*, 137: 357-358.

1235   Kear, J. (ed) In Press. *The ducks, geese and swans*. — Oxford Univ. Press, Oxford.

1236   Keast, A., 1958. Seasonal movements and geographic variations in the Australian Wood-Swallows. — *Emu*, 58: 207-218.

1237   Keith, S., 1964. A new subspecies of *Spreo albicapillus* (Blyth) from Kenya. — *Bull. Brit. Orn. Cl.*, 84: 162-163.

1238   Keith, S., 1968. A new subspecies of *Poeoptera lugubris* Bonaparte from Uganda. — *Bull. Brit. Orn. Cl.*, 88: 119-120.

1239   Keith, S., 1992. Pycnonotidae, bulbuls. Pages 279-377. In *The Birds of Africa*. Vol. 4 (S. Keith, E.K. Urban & C.H. Fry, eds.). — Academic Press, San Diego.

1240   Keith, S., 2000. Various species accounts. In *The Birds of Africa*. Vol. 6: i-xvii, 1-724 (C.H. Fry, S. Keith & E.K. Urban, eds.). — Academic Press, London.

1241   Keith, S. & A. Prigogine, 1997. Various species accounts. In *The Birds of Africa*. Vol. 5: i-xix, 1-669 (E.K. Urban, C.H. Fry & S. Keith, eds.). — Academic Press, London.

1242   Keith, S. & P.B. Taylor, 1986. Various species accounts. In *The Birds of Africa*. Vol. 2: i-xvi, 1-552 (E.K. Urban, C.H. Fry & S. Keith, eds.). — Academic Press, London.

1243   Keith, S. & A. Twomey, 1968. New distributional records of some East African birds. — *Ibis*, 110: 537-548.

1243A   Keith, S. *et al.*, 1970. The genus *Sarothrura* (Aves, Rallidae). — *Bull. Amer. Mus. Nat. Hist.*, 143: 1-84.

1244   Kelso, L., 1939. Additional races of American owls. — *Biol. Leaflet*, 11: [1-2].

1245   Kelso, L., 1940. Additional races of American owls. — *Biol. Leaflet*, 12: [1].

1246   Kelso, L., 1941. Additional races of American owls. — *Biol. Leaflet*, 13: [1-2].

1247   Kelso, L., 1942. The ear of *Otus asio* and the Trinidad Screech Owl. — *Biol. Leaflet*, 14: [1-2].

1248   Kemp, A.C., 1988. The systematics and zoogeography of Oriental and Australasian hornbills (Aves: Bucerotidae). — *Bonn. Zool. Beitr.*, 39: 315-345.

1249   Kemp, A.C., 1988. Various species accounts. In *The Birds of Africa*. Vol. 3: i-xvi, 1-611 (C.H. Fry, S. Keith & E.K. Urban, eds.). — Academic Press, London.

1250   Kemp, A.C., 1994. Various species accounts. In *Handbook of the Birds of the World*. Vol. 2: 1-638 (J. del Hoyo, A. Elliott & J. Sargatal, eds.). — Lynx Edicions, Barcelona.

1251   Kemp, A.C., 1995. *The Hornbills: Bucerotiformes*: i-xvi, 1-302. — Oxford Univ. Press, Oxford.

1252   Kemp, A.C., 2001. Family Bucerotidae (Hornbills). Pages 436-523. In *Handbook of the Birds of the World*. Vol. 6 (J. del Hoyo, A. Elliott & J. Sargatal, eds.). — Lynx Edicions, Barcelona.

1253   Kemp, A.C. & T.M. Crowe, 1985. The systematics and zoogeography of Afrotropical hornbills (Aves: Bucerotidae). In *Proc. Int. Symp. Afr. Vertebr. Mus. A. Koenig, Bonn* (ed. K.-L. Schuchmann). 1984: 279-324.

1254   Kennedy, M. & H.G. Spencer, 2000. Phylogeny, biogeography and taxonomy of Australasian teals. — *Auk*, 117 (1): 154-163.

1255   Kennedy, M. *et al.*, 1996. Hop, step and gape: do the social displays of the Pelecaniformes reflect phylogeny? — *Anim. Behav.*, 51: 273-291.

1256   Kennedy, R.S., 1987. New subspecies of *Dryocopus javensis* (Aves: Picidae) and *Ficedula hyperythra* (Aves: Muscicapidae) from the Philippines. — *Proc. Biol. Soc. Wash.*, 100: 40-43.

1257   Kennedy, R.S. & C.A. Ross, 1987. A new subspecies of *Rallina eurizonoides* (Aves: Rallidae) from the Batan Islands, Philippine. — *Proc. Biol. Soc. Wash.*, 100 (3): 459-461.

1258   Kennedy, R.S. *et al.*, 1997. New *Aethopyga* sunbirds (Aves : Nectariniidae) from the island of Mindanao, Philippines. — *Auk*, 114 (1): 1-10.

1259   Kennedy, R.S. *et al.*, 2000. *A guide to the birds of the Philippines*: i-xx, 1-369, 72 col. pls with texts opp. — Oxford Univ. Press, Oxford.

1260   Kennedy, R.S. *et al.*, 2001. A new species of woodcock (Aves: Scolopacidae) from the Philippines and a re-evaluation of other Asian/Papuasian woodcock. — *Forktail*, 17: 1-12.

1261   Kepler, C.B. & K.C. Parkes, 1972. A new species of warbler (Parulidae) from Puerto Rico. — *Auk*, 89 (1): 1-18.

1261A   Keve, A., 1948. Preliminary Note on the Geographical Variation of the Hazel-Grouse (*Tetrastes bonasia* [*L.*]). – *Dansk Orn. Foren. Tidsskr.*, 42 (3): 162-164.

1262   Keve, A., 1959. Some notes on the bird-life of the valley of the Lurio River, Moçambique. — Proc. 1st. Pan-African Orn. Cong. In: *Ostrich.*, (Suppl. 3): 84-85.

1263   Keve, A., 1966. Studi sulle variazioni nella Ghiandaia (*Garrulus glandarius* L.) d'Italia. — *Riv. Ital. Orn.*, 36: 315-323.

1264   Keve, A., 1967. A new form of *Garrulus glandarius* (Linn.). — *Bull. Brit. Orn. Cl.*, 87: 39-40.

1265   Keve, A., 1973. Über einige taxonomische Fragen der Eichelhäher des Nahen Ostens (Aves, Corvidae). — *Zool. Abh. Staatl. Mus. Tierk. Dresden*, 32: 175-198.

1266   Keve, A. & S. Kohl, 1961. A new race of the Little Owl from Transylvania. — *Bull. Brit. Orn. Cl.*, 81: 51-52.

1267   King, B.F., 1989. The avian genera *Tesia* and *Urosphena*. — *Bull. Brit. Orn. Cl.*, 109: 162-166.

1268   King, B.F. & E.C. Dickinson, 1975. *A field guide to the birds of south-east Asia*: 1-480. — Collins, London.

1269   Kinghorn, J.R., 1937. Notes on some Pacific island birds. I. — *Proc. Zool. Soc., Lond.*, 107B: 177-184.

1269A   Kirkconnell, A., 2000. Varación morfológica del Carpintero Verde *Xiphidiopicus percussus* en Cuba. — *Cotinga*, 14: 94-98.

1270   Kirkconnell, A. & O.H. Garrido, 2000. Nueva subespecie del Vireo de Bahamas *Vireo crassirostris* de Cayo Paredón Grande, archipiélago de Sabana-Camagüey, Cuba. — *Cotinga*, 14: 79-87.

1271   Kirchman, J.J. *et al.*, 2001. Phylogeny and systematics of ground rollers (Brachypteraciidae) of Madagascar. — *Auk*, 118: 849-863.

1272   Kistchinski, A.A., 1988. *Ornitofauna severo-vostoka Azii* [The bird fauna of north-east Asia]: 1-288. — Nauka, Moskva.

1273   Kleinschmidt, A., 1970. Wesen und Bedeutung von Variations-Studien, im besonderen in den Arbeiten von Otto Kleinschmidt, für die Klärung genealogischer Zusammenhänge. — *Zool. Abh. Staatl. Mus. Tierk. Dresden*, 31: 231-262.

1274   Kleinschmidt, O., 1943. *Katalog meiner ornithologischen Sammlung*: i-xii, 1-236.

1275   Kleinschmidt, O., 1943. Zu K 8279: Haselhühner. — *Falco*, 37: 18.

1276   Klicka, J. *et al.*, 2001. A cytochrome-*b* perspective on *Passerina* bunting relationships. — *Auk*, 118: 611-623.

1277   Knox, A.G., 1988. The taxonomy of the Rock/Water Pipit superspecies *Anthus petrosus*, *spinoletta* and *rubescens*. — *Brit. Birds*, 81: 206-211.

1278   Knox, A.G. *et al.*, 2001. The taxonomic status of Lesser Redpoll. — *Brit. Birds*, 94: 260-267.

1279   Koelz, W.N., 1939. New birds from Asia, chiefly from India. — *Proc. Biol. Soc. Wash.*, 52: 61-82.

1280   Koelz, W.N., 1950. New subspecies of birds from southwestern Asia. — *Amer. Mus. Novit.*, 1452: 1-10.

1281   Koelz, W.N., 1951. Four new subspecies of birds from southwestern Asia. — *Amer. Mus. Novit.*, 1510: 1-3.

1282   Koelz, W.N., 1952. New races of Indian birds. — *J. Zool. Soc. India*, 4 (1): 37-46.

1283   Koelz, W.N., 1953. New races of Assam birds. — *J. Zool. Soc. India*, 4 (1952) (2): 155.

1284   Koelz, W.N., 1954. Ornithological studies. I. New birds from Iran, Afghanistan, and India. — *Contrib. Inst. Reg. Expl.*, 1: 1-32.

1285   Koelz, W.N., 1954. Ornithological studies. II. A new subspecies of Red-bellied Woodpecker from Texas. — *Contrib. Inst. Reg. Expl.*, 1: 32.

1286   Koepcke, M., 1957. Una nueva especie de *Synallaxis* (Furnariidae, Aves) de las vertientes occidentales Andinas del Peru central. — *Publ. Mus. Hist. Nat. Javier Prado, Ser. A. Zool.*, 18: 1-8.

1287   Koepcke, M., 1959. Ein neuer *Asthenes* (Aves, Furnariidae) von der Küste und dem westlichen Andenabhang Südperus. — *Beitr. Neotrop. Fauna*, 1 (3): 243-248.

1288   Koepcke, M., 1961. Birds of the western slope of the Andes of Peru. — *Amer. Mus. Novit.*, 2028: 1-31.

1289   Koepcke, M., 1961. Las razas geograficas de *Cranioleuca antisiensis* (Furnariidae, Aves) con la descripcion de una nueva subespecie. — *Publ. Mus. Hist. Nat. Javier Prado, Ser. A. Zool.*, 20: 1-17.

1290   Koepcke, M., 1962. Zur Kenntnis der in Peru lebenden Tauben der Gattung *Columbigallina* Boie. — *Beitr. Neotrop. Fauna*, 2 (4): 295-301.

1291   Koepcke, M., 1965. Zur Kenntnis einiger Furnariiden (Aves) der Küste und des westlichen Andenabhanges Perus (mit Beschreibung neuer Subspezies). — *Beitr. Neotrop. Fauna*, 4 (3): 150-173.

1292   Koepcke, M., 1971. *Zonotrichia capensis markli* nov. subspec. (Fringillidae, Aves), una raza geografica nueva del Gorrion Americano de la Costa Norte del Peru. — *Publ. Mus. Hist. Nat. Javier Prado, Ser. A. Zool.*, 23: 1-11.

1293   König, C., 1981. Formenaufspaltung des Magellanzeisigs (*Carduelis magellanica*) im zentralen Andenraum. — *Stuttgarter Beitr. Naturk. Ser. A (Biol.)*, 350: 1-10.

1294   König, C., 1991. Zur Taxonomie und Ökologie der Sperlingskäuze (*Glaucidium* spp) des Andenraumes. — *Ökol. Vögel*, 13 (1): 15-75.

1295   König, C. & R. Straneck, 1989. Eine neue Eule (Aves: Strigidae) aus Nordargentinien. — *Stuttgarter Beitr. Naturk. Ser. A (Biol.)*, (428): 1-20.

1296   König, C. & M. Wink, 1995. Eine neue Unterart des Brasil - Sperlingskauzes aus Zentralargentinien: *Glaucidium brasilianum stranecki* n.ssp. — *J. f. Orn.*, 136 (4): 461-465.

1297   König, C. *et al.*, 1999. *Owls. A guide to the owls of the world*: 1-462. — Pica Press, Sussex.

1297A   Kooiman, W.J., 1950. Philippus Ludovicus Statius Muller. Pages 74-130 In *Earebondel ta de tachtichste jierdei fan Dr. G.A. Wumkes op 4 septimber 1949*. Oanbean Troch de Fryske Akademy. Utjowerij Fa. A.J. Osinga, Boalsert.

1298   Korelov, M.N., 1953. O formakh kazakhstanskikh polevykh zhavoronkov. [On the forms of the Kazakhstan field larks]. — *Vestnik Akademii Nauk Kazakhskoi SSR*, 5 (98): 113-115.

1299   Krabbe, N., 1992. A new subspecies of the Slender-billed Minor *Geositta tenuirostris* (Furnariidae) from Ecuador. — *Bull. Brit. Orn. Cl.*, 112 (3): 166-169.

1300   Krabbe, N. *et al.*, 1999. A new species of Antpitta (Formicariidae : *Grallaria*) from the southern Ecuadorian Andes. — *Auk*, 116 (4): 882-890.

1301   Krabbe, N. *et al.*, 1999. A new species in the *Myrmotherula haematonota* superspecies (Aves: Thamnophilidae) from the western Amazonian lowlands of Ecuador and Peru. — *Wilson Bull.*, 111 (2): 157-165.

1302   Krabbe, N. & T.S. Schulenberg, 1997. Species limits and natural history of *Scytalopus* Tapaculos (Rhinocryptidae), with descriptions of the Ecuadorian taxa, including three new species. — *Orn. Monogr.*, 48: 47-88.

1303   Kratter, A.W., 1997. A new subspecies of *Sclerurus albigularis* (Gray-throated Leaftosser) from northeastern Bolivia, with notes on geographic variation. — *Orn. Neotrop.*, 8: 23-30.

1304   Kratter, A.W. & O. Garrido, 1996. A new subspecies of *Margarops fuscus* (Scaly-breasted Thrasher) from St. Vincent, Lesser Antilles. — *Bull. Brit. Orn. Cl.*, 116 (3): 189-193.

1305   Kratter, A.W. & T.A. Parker, III., 1997. Relationship of two bamboo-specialized foliage-gleaners: *Automolus dorsalis* and *Anabazenops fuscus* (Furnariidae). — *Orn. Monogr.*, 48: 383-397.

1305A Kratter, A.W. *et al.*, 2001. Avifauna of a lowland forest site on Isabel, Solomon Islands. — *Auk*, 118: 472-283.

1305B Kumerloeve, H., 1958. Eine neue Bartmeisenform vom Amik Gölü (See von Antiochia). — *Bonn. Zool. Beitr.*, 9: 194-199.

1306 Kumerloeve, H., 1963. *Calandrella rufescens niethammeri*, eine neue Stummellerchenform aus Inneranatolien (Türkei). — *Die Vogelwelt*, 84: 146-148.

1307 Kumerloeve, H., 1969. Zur Rassenbildung der Kurzzehenlerche, *Calandrella brachydactyla*, im vorderasiatischen Raum. — *J. f. Orn.*, 110 (3): 324-325.

1307A Kumerloeve, H., 1969. Zur Avifauna des Van Gölü- und Hakkari-Gebietes (E/SE-Kleinasien). — *Istanbul Univ. Fen. Fak. Mecmuasi*, B 34: 245-312.

1308 Kumerloeve, H., 1970. Zur Vogelwelt im Raume Ceylanpinar (türkisch-syrisches Grenzgebiet). — *Beitr. Vogelk.*, 16: 239-249.

1309 Kuroda, N., Sr., 1939. *Geese & Ducks of the World*. — Privately published, Tokyo. (See text p. 52).

1310 Kuroda, N., Jr., 1965. A new race of the Japanese Robin, *Erithacus akahige*, from Rishiri I., Hokkaido. — *Misc. Rep. Yamashina Inst. Orn.*, 4: 221-223.

1311 Kvist, L. *et al.*, 2001. Phylogeography of a Palaearctic sedentary passerine, the willow tit (*Parus montanus*). — *J. Evol. Biol.*, 14: 930-941.

1311A Lack, D., 1958. A new race of the White-rumped Swift. – *J. Bombay Nat. Hist. Soc.*, 55 (1): 160-161.

1312 Lack, P.C., 1985. The ecology of the land-birds of Tsavo East National Park, Kenya. — *Scopus*, 9: 2-23, 57-96.

1313 Lack, P.C., 1997. Various species accounts. In *The Birds of Africa*. Vol. 5: i-xix, 1-669 (E.K. Urban, C.H. Fry & S. Keith, eds.). — Academic Press, London.

1314 Lafontaine, R.M. & N. Moulaert, 1998. Une nouvelle espèce de petit-duc (*Otus*, Aves) aux Comores : taxonomie et statut de conservation. — *J. African Zool.*, 112 (2): 163-169.

1315 Lafresnaye, N.F.A.A., 1844. *Icterus guttulatus*. — *Mag. de Zool.*, (2) 6: 1-3 and pl. 52.

1316 Lambert, F.R., 1998. A new species of *Amaurornis* from Talaud Islands, Indonesia, and a review of taxonomy of bush hens occurring from the Philippines to Australasia. — *Bull. Brit. Orn. Cl.*, 118 (2): 67-82.

1317 Lambert, F.R., 1998. A new species of *Gymnocrex* from the Talaud Islands, Indonesia. — *Forktail*, 13: 1-6.

1318 Lambert, F.R. & P.C. Rasmussen, 1998. A new Scops Owl from Sangihe Island, Indonesia. — *Bull. Brit. Orn. Cl.*, 118 (4): 204-217.

1319 Lambert, F. & M.W. Woodcock, 1996. *Pittas, Broadbills and Asities*: i-viii, 1-271. — Pica Press, Robertsbridge.

1320 Land, H.C. & W.L. Schultz, 1963. A proposed subspecies of the Great Potoo, *Nyctibius grandis* (Gmelin). — *Auk*, 80: 195-196.

1321 Lanyon, S.M., 1985. A molecular perspective on higher-level relationships in the Tyrannoidea (Aves). — *Syst. Zool.*, 34: 404-418.

1322 Lanyon, S.M., 1992. Interspecific brood parasitism in blackbirds (Icterinae): a phylogenetic perspective. — *Science*, 255: 77-79.

1323 Lanyon, S.M. & K.E. Omland, 1999. A molecular phylogeny of the blackbirds (Icteridae): five lineages revealed by cytochrome-*b* sequence data. — *Auk*, 116 (3): 629-639.

1324 Lanyon, S.M. & W.E. Lanyon, 1989. The systematic position of the Plantcutters, *Phytotoma*. — *Auk*, 106: 422-432.

1325 Lanyon, S.M. *et al.*, 1991. *Clytoctantes atrogularis* — a new species of antbird from western Brazil. — *Wilson Bull.*, 102 (4): 571-580 (1990).

1326 Lanyon, W.E., 1984. A phylogeny of the Kingbirds and their allies. — *Amer. Mus. Novit.*, 2797: 1-28.

1327 Lanyon, W.E., 1984. The systematic position of Cocos flycatcher. — *Condor*, 86 (1): 42-47.

1328 Lanyon, W.E., 1985. A phylogeny of the Myiarchine flycatchers. — *Orn. Monogr.*, 36: 361-380.

1329 Lanyon, W.E., 1986. A phylogeny of the thirty-three genera in the *Empidonax* assemblage of Tyrant Flycatchers. — *Amer. Mus. Novit.*, 2846: 1-64.

1330 Lanyon, W.E., 1988. A phylogeny of the thirty-two genera in the *Elaenia* assemblage of Tyrant Flycatchers. — *Amer. Mus. Novit.*, 2914: 1-57.

1331 Lanyon, W.E., 1988. The phylogenetic affinities of the flycatcher genera *Myiobius* Darwin and *Terenotriccus* Ridgway. — *Amer. Mus. Novit.*, 2915: 1-11.

1332 Lanyon, W.E., 1988. A phylogeny of the flatbill and tody-tyrant assemblage of tyrant flycatchers. — *Amer. Mus. Novit.*, 2923: 1-41.

1333 Lanyon, W.E. & J.W. Fitzpatrick, 1983. Behavior, morphology and systematics of *Sirystes sibilator* (Tyrannidae). — *Auk*, 100 (1): 98-104.

1334 Lanyon, W.E. & S.M. Lanyon, 1986. Generic status of Euler's flycatcher: A morphological and biochemical study. — *Auk*, 103 (2): 341-350.

1334A Latham, J., 1790. Index ornithologicus, sive systema ornithologiæ; complectens avium divisionem in classes, ordines, genera, species, ipsarumque varietates: adjectis synonymis, locis, descriptionibus, & c.:2. 467-920. –, London.

1335 Laubmann, A., 1950. Bemerkungen zur Nomenklatur des Formenkreises *"Halcyon princeps"*. Pages pp. 229-230. In *Syllegomena biologica* (Kleinschmidt-Festschrift) (A. von Jordans & F. Peus, eds.): i-viii, 221-472 — Geest & Portig Akad. Verlagsgesellsch., Leipzig and A. Ziemsen Verlag, Wittenburg.

1336 Lavauden, L., 1930. Une nouvelle perdrix des montagnes du Sahara central. — *Alauda*, 2 (3/4): 241-245.

1337 Lawson, W.J., 1962. On the geographical variation in the blackheaded oriole *Oriolus larvatus* Lichtenstein of Africa. — *Durban Mus. Novit.*, 6: 195-201.

1338 Lawson, W.J., 1962. Systematic notes on African birds. 1. A new race of *Ploceus ocularis* Smith from southern Moçambique. — *Durban Mus. Novit.*, 6 (18): 224-225.

1339 Lawson, W.J., 1963. A contribution to the ornithology of Sul do Save, southern Moçambique. — *Durban Mus. Novit.*, 7: 73-124.

1340 Lawson, W.J., 1969. A new name for a race of the Black-headed Oriole. — *Bull. Brit. Orn. Cl.*, 89: 16.

1341 Laxmann, E., 1769. Hirundo daurica, area temporali rubra, uropygio luteo rufescente. — *Kungl. Svenska Veteskapad. Handl.*, 30: 209-213.

1342 Le Corre, M. & P. Jouventin, 1999. Geographical variation in the White-tailed Tropicbird *Phaethon lepturus*, with the description of a new subspecies endemic to Europa Island, southern Mozambique Channel. — *Ibis*, 141 (2): 233-239.

1343 LeCroy, M., 1999. Type specimens of new forms of *Lonchura*. — *Bull. Brit. Orn. Cl.*, 119 (4): 214-220.

1344 LeCroy, M. & J.M. Diamond, 1995. Plumage variation in the Broad-billed Fairy-wren *Malurus grayi*. — *Emu*, 95: 185-193.

1345 Lee, P.L.M. *et al.*, 2001. The phylogenetic status of the Corn Bunting *Miliaria calandra* based on mitochondrial DNA sequences. — *Ibis*, 143 (2): 299-303.

1346 Lee, K. *et al.*, 1997. Phylogenetic relationships of the ratite birds: resolving conflicts between molecular and morphological data sets. Pages 173-211. In *Avian Molecular Evolution and Systematics* (D.P. Mindell, ed.) — Academic Press, New York.

1347 Lefranc, N. & T. Worfolk, 1997. *Shrikes. A Guide to the shrikes of the world*: 1-192. — Pica Press, Mountfield, Sussex.

1348 Legge, W.V., 1880. *Birds of Ceylon*. — Pt. 3: 731-1237.

1349 Lehmann, V.F.C., 1946. Two new birds from the Andes of Columbia. — *Auk*, 63: 218-223.

1350 Leisler, B. *et al.*, 1997. Taxonomy and phylogeny of reed warblers (genus *Acrocephalus*) based on mtDNA sequences and morphology. — *J. f. Orn.*, 138: 469-496.

1351 Lencioni-Neto, F., 1994. Une nouvelle especie de *Chordeiles* (Aves, Caprimulgidae) de Bahia (Brazil). — *Alauda*, 62: 241-245.

1352 Lencioni-Neto, F., 1996. Uma nova subspecie de *Knipolegus* (Aves, Tyrannidae) do estado da Bahia, Brasil. — *Rev. Bras. Biol.*, 56 (2): 197-201.

1353 Leonovich, V.V. *et al.*, 1997. On the taxonomy and phylogeny of pipits (genus *Anthus*, Motacillidae, Aves) in Eurasia. — *Byull. Mosk. O-va Ispyt. Prir.* (Otd. Biol.), 102 (2): 14-22.

1354 Li, De-hao & Wang, Zu-xiang, 1979. A new subspecies of Babaoc (*Babax koslowi yuquensis*) from Xizang, China. — *Acta Zootax. Sinica*, 4 (3): 304-305.

1355 Li, De-hao *et al.*, 1978. Studies on the birds of southeastern Xizang, with notes on their vertical distribution. — *Acta Zool. Sinica*, 24 (3): 231-250.

1356 Li, Gui-yuan, 1995. A new subspecies of *Certhia familiaris* (Passeriformes: Certhiidae). — *Acta Zootax. Sinica*, 20 (3): 373-376.

1357   Li, Gui-yuan *et al.*, 1992. A new subspecies of *Spelaeornis troglodytoides*. — *Zool. Res.* [China], 13 (1): 31-35.

1358   Li, Gui-yuan & Zhang, Qing-mao, 1980. A new subspecies of *Paradoxornis webbianus* from Sichuan - *Paradoxornis webbianus ganluoensis*. — *Acta Zootax. Sinica*, 5 (3): 312-314.

1359   Li, Gui-yuan *et al.*, 1979. A new subspecies of *Garrulax lunulatus* from Sichuan - *G. l. liangshanensis*. — *Acta Zootax. Sinica*, 4 (1): 93-94.

1360   Ligon, J.D., 1967. Relationships of the cathartid vultures. — *Occ. Pap. Mus. Zool., Univ. Michigan*, 651: 1-26.

1361   Ligon, J.D., 2001. Family Phoeniculidae (Woodhoopoes). Pages 412-434. In *Handbook of the Birds of the World*. Vol. 6 (J. del Hoyo, A. Elliott & J. Sargatal, eds.). — Lynx Edicions, Barcelona.

1362   Lima, J.L., 1920. Aves colligidas nos Estados de Sao Paulo, Mattao Grosso e Bahia com algunas formas novas. — *Rev. Mus. Paul.*, 12 (2): 91-106.

1363   Liversidge, R., 1996. A new species of pipit in southern Africa. — *Bull. Brit. Orn. Cl.*, 116 (4): 211-215.

1363A   Liversidge, R. & G. Voelker, 2002. The Kimberley Pipit: a new African species. — *Bull. Brit. Orn. Cl.*, 122 (2): 93-109.

1364   Livezey, B.C., 1986. A phylogenetic analysis of Recent anseriform genera using morphological characters. — *Auk*, 105: 681-698.

1365   Livezey, B.C., 1991. A phylogenetic analysis and classification of Recent dabbling ducks (Tribe Anatini) based on comparative morphology. — *Auk*, 108: 471-507.

1366   Livezey, B.C., 1995. Phylogeny and comparative ecology of stiff-tailed ducks (Anatidae: Oxyurini). — *Wilson Bull.*, 107: 214-234.

1367   Livezey, B.C., 1997. A phylogenetic analysis of basal Anseriforms, the fossil *Presbyornis*, and the interordinal relationships of waterfowl. — *Zool. J. Linn. Soc. Lond.*, 121: 361-428.

1368   Livezey, B.C., 1998. A phylogenetic analysis of the Gruiformes (Aves) based on morphological characters, with an emphasis on the rails (Rallidae). — *Phil. Trans. Roy. Soc. London*, 353B: 2077-2151.

1369   Livezey, B.C. & R.L. Zusi, 2001. Higher-order phylogenetics of modern Aves based on comparative anatomy. — *Netherlands J. Zool.*, 51: 179-205.

1370   Longmore, N.W. & W.E. Boles, 1983. Description and systematics of the Eungella Honeyeater *Meliphaga hindwoodi*, a new species of honeyeater from central eastern Queensland, Australia. — *Emu*, 83 (2): 59-65.

1371   Loskot, V.M., 1991. A new Chiffchaff subspecies (Aves, Sylviidae) from Caucasus. — *Vestnik Zoologii,* (3): 76-77.

1372   Loskot, V.M., 2001. A new subspecies of the Pale Sand Martin, *Riparia diluta* (Sharpe & Wyatt), from the Altai and Middle Siberia. — *Zoosystematica Rossica*, 9: 461-462.

1373   Loskot, V.M., 2001. Taxonomic revision of the Hume's Whitethroat *Sylvia althaea* Hume, 1878. — *Proc. Biol. Station "Rybachy"*, 6: 41-42.

1374   Loskot, V.M., 2002. On the type specimens of *Locustella ochotensis* (Middendorff, 1853) in the collection of the Zoological Institute, St. Petersurg (Aves: Sylviidae). — *Zoosystematica Rossica*. 11(1): 239-242.

1375   Loskot, V.M. & E.C. Dickinson, 2001. Systematic notes on Asian birds. 15. Nomenclatural issues concerning the common sand martin *Riparia riparia* (Linnaeus, 1758) and the pale sand martin *R. diluta* (Sharpe & Wyatt, 1893), with a new synonymy. — *Zool. Verhand., Leiden*, 335: 167-174.

1376   Loskot, V.M. & E.P. Sokolov, 1993. Taxonomy of the mainland and insular Lanceolated Warblers, *Locustella lanceolata* (Temminck) (Aves: Sylviidae). — *Zoosystematica Rossica*, 2 (1): 189-200.

1377   Louette, M., 1981. A new species of honeyguide from West Africa (Aves, Indicatoridae). — *Rev. Zool. Afr.*, 95 (1): 131-135.

1378   Louette, M., 1986. Geographical contacts between taxa of *Centropus* in Zaire, with the description of a new race. — *Bull. Brit. Orn. Cl.*, 106: 126-133.

1378A   Louette, M., 1987. A new weaver from Uganda? – *Ibis*, 129: 405-406.

1379   Louette, M., 1990. A new species of nightjar from Zaire. — *Ibis*, 132: 349-353.

1380   Louette, M. & C.W. Benson, 1982. Swamp-dwelling weavers of the *Ploceus velatus/vitellinus* complex, with the description of a new species. — *Bull. Brit. Orn. Cl.*, 102: 24-31.

1381   Louette, M. & M. Herremans, 1982. The Blue Vanga *Cyanolanius madagascarinus* on Grand Comoro. — *Bull. Brit. Orn. Cl.*, 102: 132-135.

1382   Louette, M. & M. Herremans, 1985. A new race of Audubon's Shearwater *Puffinus lherminieri* breeding at Moheli, Comoro Islands. — *Bull. Brit. Orn. Cl.*, 105: 42-49.

1383   Louette, M. & A. Prigogine, 1982. An appreciation of the distribution of *Dendropicos goertae* and the description of a new race (Aves: Picidae). — *Rev. Zool. Afr.*, 96 (3): 461-492.

1384   Louette, M. *et al.*, 2000. A reassessment of the subspecies in the Ruwenzori Turaco *Ruwenzorornis johnstoni*. — *Bull. Brit. Orn. Cl.*, 120 (1): 34-39.

1385   Louette, M. *et al.*, 1988. Taxonomy and Evolution in the brush warblers *Nesillas* on the Comoro Islands. — *Tauraco*, 1 (1): 110-129.

1386   Lousada, S., 1989. *Amazona auropalliata caribaea*: a new subspecies of parrot from the Bay Islands, northern Honduras. — *Bull. Brit. Orn. Cl.*, 109: 232-235.

1387   Lousada, S.A. & S.N.G. Howell, 1997. *Amazona oratrix hondurensis*: a new subspecies of parrot from the Sula Valley of northern Honduras. — *Bull. Brit. Orn. Cl.*, 117 (3): 205-209.

1388   Løvenskiold, H.L., 1950. Den geografiske variasjon hos Fjæreplytten (*Calidris maritima* (Brünn.)). — *Dansk Orn. Foren. Tidsskr.*, 44 (3): 161-167.

1389   Lovette, I.J. & E. Bermingham, 2000. *c-mos* variation in songbirds: molecular evolution, phylogenetic implications, and comparisons with mitochondrial differentiation. — *Mol. Biol. Evol.*, 17 (10): 1569-1577.

1390   Lowe, P.R., 1934. [A new form of the Red-legged Partridge.]. — *Bull. Brit. Orn. Cl.*, 55: 8-9.

1391   Lowe, W.P., 1944. A new Banded Rail from the Philippines. — *Bull. Brit. Orn. Cl.*, 65: 5.

1392   Lowery, G.H., Jr. & W.W. Dalquest, 1951. Birds from the State of Veracruz, Mexico. — *Univ. Kansas Publ. Mus. Nat. Hist.*, 3 (4): 531-649.

1393   Lowery, G.H., Jr. & R.J. Newman, 1949. New birds from the state of San Luis Potosi and the Tuxtla mountains of Veracruz, Mexico. — *Occ. Pap. Mus. Zool., Louisiana St. Univ.*, 22: 1-10.

1394   Lowery, G.H., Jr. & B.L. Monroe, Jr., 1968. Family Parulidae. Pages 3-93. In *Check-list of Birds of the World*. Vol. 14 (R.A. Paynter, Jr., ed.) — Mus. Comp. Zool., Cambridge, Mass.

1395   Lowery, G.H., Jr. & J.P. O'Neill, 1969. A new species of antpitta from Peru and a revision of the subfamily Grallariinae. — *Auk*, 86 (1): 1-12.

1400   Lowery, G.H., Jr. & D.A. Tallman, 1976. A new genus and species of nine-primaried oscine of uncertain affinities from Peru. — *Auk*, 93 (3): 415-428.

1401   Lowther, P.E., 2001. New name for the Bolivian Blackbird. — *Bull. Brit. Orn. Cl.*, 121 (4): 280-281.

1402   Ludlow, F., 1951. The birds of Kongbo and Pome, south-east Tibet. — *Ibis*, 93: 547-578.

1403   Ludwig, C.A., 1998. Type locality and taxonomic status of *Saltator plumbiceps* 'Baird, MS' Lawrence, 1867 (Aves: Passeriformes: Cardinalidae). — *Proc. Biol. Soc. Wash.*, 111 (2): 418-419.

1404   Lynes, H., 1930. Review of the genus *Cisticola*. — *Ibis*, (12) 6 (Suppl.): i-ii, 1-673, i-vii.

1405   Lysaght, A.M., 1953. A rail from Tonga, *Rallus philippensis ecaudata* Miller, 1783. — *Bull. Brit. Orn. Cl.*, 73: 74-75.

1406   Macdonald, J.D., 1940. A new race of Francolin and a new race of Lark from the Sudan. — *Bull. Brit. Orn. Cl.*, 60: 57-59.

1407   Macdonald, J.D., 1946. A new species of Wood Hoopoe from the Sudan. — *Bull. Brit. Orn. Cl.*, 67: 5.

1408   Macdonald, J.D., 1953. The races in South West Africa of the Orange River Francolin. — *Bull. Brit. Orn. Cl.*, 73: 34-36.

1409   Macdonald, J.D., 1954. Note on the Double-Banded Sandgrouse, *Pterocles bicinctus*. — *Bull. Brit. Orn. Cl.*, 74 (1): 6-8.

1410   Macdonald, J.D., 1957. *Contributions to the Ornithology of Western South Africa. Results of the British Museum (Natural History) South West African Expedition 1949-50*: i-xi, 1-174. — Trustees of the British Museum, London.

1411 Mackworth-Praed, C.W. & C.H.B. Grant, 1957. *Birds of Eastern and North-eastern Africa*: 1. i-xxvi, 1-846. 2nd. edition. — Longmans, Green and Co.

1412 Maclean, G.L., 1974. The breeding biology of the Rufous-eared Warbler and its bearing on the genus *Prinia*. — *Ostrich*, 45: 9-14.

1413 Maijer, S., 1998. Rediscovery of *Hylopezus* (*macularius*) *auricularis*: distinctive song and habitat indicate species rank. — *Auk*, 115 (4): 1072-1073.

1414 Maijer, S. & J. Fjeldså, 1997. Description of a new *Cranioleuca* spinetail from Bolivia and a "leapfrog pattern" of geographic variation in the genus. — *Ibis*, 139 (4): 606-616.

1415 Maijer, S. *et al.*, 1998. A distinctive new subspecies of the Green-cheeked Parakeet (*Pyrrhura molinae*, Psittacidae) from Bolivia. – *Orn. Neotrop.*, 9: 185-191.

1415A Mann, C.F., 2002. Systematic notes on Asian birds. 28. Taxonomic comments on some south-east Asian members of the family Nectariniidae. — *Zool. Verhand., Leiden*, 340: 179-189.

1416 Mann, C.F. *et al.*, 1978. A re-appraisal of the systematic position of *Trichastoma poliothorax* (Timaliinae, Muscicapidae). — *Bull. Brit. Orn. Cl.*, 98 (4): 131-140.

1417 Manghi, M.S., 1984. Una nueva subespecie de *Podiceps major* Boddaert (Aves, Podicipedidae). — *Com. Mus. Argentino Cienc. Nat. "Bernardino Rivadavia", Zool.*, 4: 115-119.

1418 Manuel, C.G., 1936. A review of Philippine Pigeons I. — *Phil. J. Sci.*, 59: 289-305.

1419 Manuel, C.G., 1936. New Philippine Fruit Pigeons. — *Phil. J. Sci.*, 59: 307-310.

1420 Manuel, C.G., 1936. A review of Philippine Pigeons IV. — *Phil. J. Sci.*, 60: 407-419.

1421 Manuel, C.G., 1957. Resident birds of Polillo Island. — *Phil. J. Sci.*, 86: 1-11.

1422 Manuel, C.G. & E.T. Gilliard, 1952. Undescribed and newly recorded Philippine birds. — *Amer. Mus. Novit.*, 1545: 1-9.

1423 Marantz, C.A., 1997. Geographic variation of plumage patterns in the woodcreeper genus *Dendrocolaptes* (Dendrocolaptidae). — *Orn. Monogr.*, 48: 399-430.

1424 Marchant, S., 1950. A new race of Honey Guide, a new race of *Macrosphenus* from Nigeria and a note on *Illadopsis*. — *Bull. Brit. Orn. Cl.*, 70: 25-28.

1425 Marchant, S., 1972. Evolution of the genus *Chrysococcyx*. — *Ibis*, 114 (2): 219-233.

1426 Marchant, S. & P.J. Higgins (eds), 1990. *Handbook of Australian, New Zealand and Antarctic birds*. Vol. 1A. Ratites to Petrels. 1-734. — Oxford Univ. Press, Melbourne.

1427 Marchant, S. & P.J. Higgins (eds), 1990. *Handbook of Australian, New Zealand and Antarctic birds*. Vol. 1B. Australian Pelican to Ducks. 735-1400. — Oxford Univ. Press, Melbourne.

1428 Marchant, S. & P.J. Higgins (eds), 1993. *Handbook of Australian, New Zealand and Antarctic birds*. Vol. 2. Raptors to Lapwings. 1-984. — Oxford Univ. Press, Melbourne.

1429 Marien, D., 1950. Notes on some Asiatic Meropidae (Birds). — *J. Bombay Nat. Hist. Soc.*, 49 (2): 151-164.

1430 Marin, M., 1997. Species limits and distribution of some new world Spine-tailed Swifts (*Chaetura* spp.). — *Orn. Monogr.*, 48: 431-443.

1431 Marin, M., 2000. Species limits, distribution and biogeography of some new world Gray-rumped Spine-tailed Swifts (*Chaetura*, Apodidae). — *Orn. Neotrop.*, 11 (2): 93-107.

1432 Marin, M. & F.G. Stiles, 1992. On the biology of five species of swifts (Apodidae, Cypseloidinae) in Costa Rica. — *Proc. Western Found. Vert. Zool.*, 4: 287-351.

1433 Marin, M. *et al.*, 1989. Notes on Chilean birds, with descriptions of two new subspecies. — *Bull. Brit. Orn. Cl.*, 109: 66-82.

1434 Marion, L. *et al.*, 1985. Coexistence progressive de la reproduction de *Larus argentatus* et de *Larus cachinnans* sur les côtes atlantiques françaises. — *Alauda*, 53: 81-87.

1435 Markham, B.J., 1971. Descripcion de una nueva subespecie de "tordo", *Curaeus curaeus recurvirostris* subsp. nov. — *Anales Inst. Patagonia Ser. Cienc. Nat.*, 2: 158-159.

1436 Marks, J.S. *et al.*, 1999. Family Strigidae (Typical Owls). Pages 76-151. In *Handbook of the Birds of the World*. Vol. 5 (J. del Hoyo, A. Elliott & J. Sargatal, eds.). — Lynx Edicions, Barcelona.

1437 Marova-Kleinbub, I.M. & V.L. Leonovitch, 1998. The mysterious Chiffchaff of Kopetdagh: ecology, vocalization, and relations of *Phylloscopus collybita menzbieri* (Shest.). — *Biol. Cons. Fauna*, 102: 231.

1438 Marshall, J.T., Jr., 1967. Parallel variation in North and Middle American Screech-Owls. — *Monog. Western Found. Verteb. Zool.*, 1: 1-72.

1439 Marshall, J.T., Jr., 1978. Systematics of smaller Asian night birds based on voice. — *Orn. Monogr.*, 25: i-v, 1-58 + record.

1440 Martens, J., 1972. Brutverbreitung paläarktischer Vögel im Nepal-Himalaya. — *Bonn. Zool. Beitr.*, 23: 95-121.

1441 Martens, J., 2000. *Phylloscopus yunnanensis* la Touche, 1922 Alströmlaubsänger. Pages 1-3. In *Atlas der Verbreitung Palaearktischer Vögel* (J. Martens, ed.) — Erwin-Stresemann-Gesellschaft für paläarktische Avifaunistik e.V, Berlin.

1442 Martens, J. & S. Eck, 1991. *Pnoepyga immaculata* n. sp., eine neue bodenbewohnende Timalie aus dem Nepal-Himalaya. — *J. f. Orn.*, 132 (2): 179-198.

1443 Martens, J. & S. Eck, 1995. Towards an ornithology of the Himalayas: Systematics, Ecology and Vocalizations of Nepal Birds. — *Bonn. Zool. Monogr.*, 38: 1-445.

1444 Martens, J. & B. Steil, 1997. Reviergesänge und Speziesdifferenzierung in der Zaungrasmücken-Gruppe *Sylvia*. — *J. f. Orn.*, 138: 1-23.

1445 Martens, J. *et al.*, 1999. The Golden-spectacled Warbler *Seicercus burkii* - a species swarm (Aves: Passeriformes : Sylviidae) Part 1. — *Zool. Abh. Staatl. Mus. Tierk. Dresden*, 50 (2 (18)): 281-327.

1445A Martens, J. *et al.* 2002. *Certhia tianquanensis* Li, a treecreeper with relict distribution in Sichuan, China. — *J. f. Orn.*, 143: 440-456.

1446 Martínez, I., 1994. Family Numididae (Guineafowl). Pages 554-567. In *Handbook of the Birds of the World*. Vol. 2. (J. del Hoyo, A. Elliott & J. Sargatal, eds.). — Lynx Edicions, Barcelona.

1447 Martínez-Vilalta, A. & A. Motis, 1992. Family Ardeidae (Herons). Pages 376-429. In *Handbook of the Birds of the World*. Vol. 1 (J. del Hoyo, A. Elliott & J. Sargatal, eds.). — Lynx Edicions, Barcelona.

1448 Mason, I.J., 1982. The identity of certain early Australian types referred to the Cuculidae. — *Bull. Brit. Orn. Cl.*, 102 (3): 99-106.

1449 Mason, I.J., 1983. A new subspecies of Masked Owl *Tyto novaehollandiae* (Stephens) from southern New Guinea. — *Bull. Brit. Orn. Cl.*, 103: 122-128.

1450 Mason, I.J. & R.I. Forrester, 1996. Geographical differentiation in the Channel-billed Cuckoo *Scythrops novaehollandiae* Latham, with description of two new subspecies from Sulawesi and the Bismarck Archipelago. — *Emu*, 96: 217-233.

1451 Mason, I.J. & R. Schodde, 1980. Subspeciation in the Rufous Owl *Ninox rufa* (Gould). — *Emu*, 80: 141-144.

1452 Mason, I.J. *et al.*, 1984. Geographical variation in the Pheasant Coucal *Centropus phasianinus* (Latham) and a description of a new subspecies from Timor. — *Emu*, 84 (1): 1-15.

1453 Mathews, G.M., 1911. On some necessary alterations in the nomenclature of birds. — *Novit. Zool.*, 18: 1-22.

1454 Mathews, G.M., 1927. *Systema Avium Australasianarum: A systematic list of the birds of the Australasian Region*: 1. i-x, 1-426. — British Ornithologists' Union, London.

1455 Mathews, G.M., 1933. *Chionarchus minor* and its subspecies. — *Emu*, 33 (2): 138.

1456 Mathews, G.M., 1935. Remarks on *Apteryx australis* and *Stictapteryx owenii*, with necessary changes of nomenclature. — *Bull. Brit. Orn. Cl.*, 60: 180.

1456A Mathews, G.M., 1935. A new name for the British Redshank. — *Brit. Birds*, 29: 152.

1457 Mathews, G.M., 1937. [Some overlooked generic names.]. — *Bull. Brit. Orn. Cl.*, 58: 13.

1458   Mathews, G.M. & O. Neumann, 1939. Six new races of Australian birds from North Queensland. — *Bull. Brit. Orn. Cl.*, 59: 153-155.

1459   Mauersberger, G., 1971. Über die östlichen Formen von *Prunella modularis* (L.). — *J. f. Orn.*, 112: 438-450.

1460   Mauersberger, G., 1972. Über den taxonomischen Rang von *Emberiza godlewskii* Tackzanowski. — *J. f. Orn.*, 113 (1): 53-59.

1461   Mauersberger, G. & S. Fischer, 1992. Intraspecific variability of the Light-vented Bulbul, *Pycnonotus sinensis* (Gmelin, 1789). — *Mitt. Zool. Mus. Berlin*, 68 (Suppl. Ann. Orn. 16): 167-177.

1462   Maynard, C.J., 1889. Description of two supposed new subspecies of birds from Vancouver's Island. — *Ornithol. and Oologist*, 14: 58-59.

1463   Mayr, E., 1933. Birds collected during the Whitney South Sea Expedition. XXII. Three new genera from Polynesia and Melanesia. — *Amer. Mus. Novit.*, 590: 1-7.

1464   Mayr, E., 1933. Birds collected during the Whitney South Sea Expedition. XXIII. Two new birds from Melanesia. — *Amer. Mus. Novit.*, 609: 1-4.

1465   Mayr, E., 1937. Birds collected during the Whitney South Sea Expedition. XXXVI. Notes on New Guinea Birds 3. — *Amer. Mus. Novit.*, 947: 1-11.

1466   Mayr, E., 1938. Birds collected during the Whitney South Sea Expedition XXXIX. Notes on New Guinea birds IV. — *Amer. Mus. Novit.*, 1006: 1-16.

1467   Mayr, E., 1938. Birds collected during the Whitney South Sea Expedition. XL. Notes on New Guinea Birds V. — *Amer. Mus. Novit.*, 1007: 1-16

1468   Mayr, E., 1938. Notes on a collection of birds from south Borneo. — *Bull. Raffles Mus.*, 14: 5-46.

1469   Mayr, E., 1940. Birds collected during the Whitney South Sea Expedition. XLI. Notes on New Guinea Birds 6. — *Amer. Mus. Novit.*, 1056: 1-12.

1470   Mayr, E., 1940. Birds collected during the Whitney South Sea Expedition. XLIII. Notes on New Guinea Birds, VII. — *Amer. Mus. Novit.*, 1091: 1-3.

1471   Mayr, E., 1941. Birds collected during the Whitney South Sea Expedition. XLVII. Notes on the genera *Halcyon*, *Turdus* and *Eurostopodus*. — *Amer. Mus. Novit.*, 1152: 1-7.

1472   Mayr, E., 1941. *List of New Guinea Birds*: i-xi, 1-260. — American Museum of Natural History, New York.

1473   Mayr, E., 1943. Notes on Australian Birds (II). — *Emu*, 43: 3-17.

1474   Mayr, E., 1944. The birds of Timor and Sumba. — *Bull. Amer. Mus. Nat. Hist.*, 83 (2): 123-194.

1475   Mayr, E., 1949. Birds collected during the Whitney South Sea Expedition. LVII. Notes on the Birds of Northern Melanesia II. — *Amer. Mus. Novit.*, 1417: 1-38.

1476   Mayr, E., 1951. Notes on some pigeons and parrots from Western Australia. — *Emu*, 51: 137-145.

1477   Mayr, E., 1955. Notes on the Birds of Northern Melanesia 3. Passeres. — *Amer. Mus. Novit.*, 1707: 1-46.

1478   Mayr, E., 1956. The names *Treron griseicauda* and *Treron pulverulenta*. — *Bull. Brit. Orn. Cl.*, 76: 62.

1479   Mayr, E., 1957. New species of birds described from 1941 to 1955. — *J. f. Orn.*, 98 (1): 22-35.

1480   Mayr, E., 1962. Family Paradisaeidae. Pages 181-204. In *Check-list of Birds of the World*. Vol. 15 (E. Mayr & J.C. Greenway, Jr., eds.). — Mus. Comp. Zool., Cambridge, Mass.

1481   Mayr, E., 1967. Family Muscicapidae, Subfamily Pachycephalinae. Pages 3-51. In *Check-list of Birds of the World*. Vol. 12 (R.A. Paynter, Jr., ed.). — Mus. Comp. Zool., Cambridge, Mass.

1482   Mayr, E., 1967. Family Zosteropidae, Indo-Australian taxa. Pages 289-325. In *Check-list of Birds of the World*. Vol. 12 (R.A. Paynter, Jr., ed.). — Mus. Comp. Zool., Cambridge, Mass.

1483   Mayr, E., 1971. New species of birds described from 1956 to 1965. — *J. f. Orn.*, 112: 302-316.

1484   Mayr, E., 1979. Order Struthioniformes. Pages 3-11. In *Check-list of Birds of the World*. Vol. 1. (2nd. Edit.) (E. Mayr & G.W. Cottrell, eds.). — Mus. Comp. Zool., Cambridge, Mass.

1485   Mayr, E., 1979. Family Pittidae. Pages 310-329. In *Check-list of Birds of the World*. Vol. 8 (M.A. Traylor, Jr., ed.). — Mus. Comp. Zool., Cambridge, Mass.

1486   Mayr, E., 1986. Family Maluridae. Pages 390-409. In *Check-list of Birds of the World*. Vol. 11 (E. Mayr & G.W. Cottrell, eds.). — Mus. Comp. Zoology, Cambridge, Mass.

1487   Mayr, E., 1986. Family Acanthizidae. Pages 409-464. In *Check-list of Birds of the World*. Vol. 11 (E. Mayr & G.W. Cottrell, eds.). — Mus. Comp. Zoology, Cambridge, Mass.

1488   Mayr, E., 1986. Family Eopsaltridae. Pages 556-583. In *Check-list of Birds of the World*. Vol. 11 (E. Mayr & G.W. Cottrell, eds.). — Mus. Comp. Zoology, Cambridge, Mass.

1489   Mayr, E. & E.T. Gilliard, 1951. New species and subspecies of birds from the highlands of New Guinea. — *Amer. Mus. Novit.*, 1524: 1-15.

1490   Mayr, E. & A. McEvey, 1960. The distribution and variation of *Mirafra javanica* in Australia. — *Emu*, 60 (3): 155-192.

1491   Mayr, E. & R. Meyer de Schauensee, 1939. Zoological Results of the Denison-Crockett Expedition to the South Pacific for the Academy of Natural Sciences of Philadelphia. 1937-1938. I. The birds of the island of Biak. — *Proc. Acad. Nat. Sci. Philad.*, 91: 1-37.

1492   Mayr, E. & A.L. Rand, 1935. Results of the Archbold Expeditions. 6. Twenty-four apparently undescribed birds from New Guinea. — *Amer. Mus. Novit.*, 814: 1-17.

1493   Mayr, E. & A.L. Rand, 1936. Neue Unterarten von Vögeln aus Neuguinea. — *Mitt. Zool. Mus. Berlin*, 21: 241-248.

1494   Mayr, E. & A.L. Rand, 1936. Results of the Archbold Expeditions. No. 10. Two new subspecies of birds from New Guinea. — *Amer. Mus. Novit.*, 868: 1-3.

1495   Mayr, E. & L.L. Short, 1970. *Species taxa of North American birds*: i-vi, 1-127. — Nuttall Orn. Cl., Cambridge, Mass. Publ. No. 9.

1496   Mayr, E. & F. Vuilleumier, 1983. New species of birds described from 1966 to 1975. — *J. f. Orn.*, 124 (3): 217-232.

1497   Mayr, E. *et al.*, 1968. Family Estrilididae. Pages 306-390. In *Check-list of Birds of the World*. Vol. 14 (R.A. Paynter, Jr., ed.). — Mus. Comp. Zool., Cambridge, Mass.

1498   McAllan, I.A.W., 1990. The Cochineal Creeper and the Fascinating Grosbeak: a re-examination of some names of John Latham. — *Bull. Brit. Orn. Cl.*, 110 (3): 153-159.

1499   McAllan, I.A.W. & M.D. Bruce, 1989. *The birds of New South Wales. A working list*: i-viii, 1-103 (1988) + errata sheet. — Biocon Research Group, Turramurra, N.S.W.

1500   McCracken, K.G. & F.H. Sheldon, 1998. Molecular and osteological heron phylogenies: sources of incongruence. — *Auk*, 115: 127-141.

1501   McCracken, K.G. *et al.*, 1999. Data set incongruence and correlated character Evolution: an example of functional convergence in the hind-limbs of stifftail diving ducks. — *Syst. Biol.*, 48 (4): 683-714.

1502   McGowan, P.J.K., 1994. Family Phasianidae (Pheasants and Partridges). Pages 434-553. In *Handbook of the Birds of the World*. Vol. 2 (J. del Hoyo, A. Elliott & J. Sargatal, eds.). — Lynx Edicions, Barcelona.

1503   McGowan, P.J.K. & A.L. Panchen, 1994. Plumage variation and geographical distribution in the Kalij and Silver Pheasants. — *Bull. Brit. Orn. Cl.*, 114 (2): 113-123.

1504   Medway, Lord, 1962. The swiftlets (*Collocalia*) of Java and their relationships. — *J. Bombay Nat. Hist. Soc.*, 59: 146-153.

1505   Medway, Lord, 1965. Nomenclature of the Asian Palm Swift. — *J. Bombay Nat. Hist. Soc.*, 72 (2): 539-543.

1506   Medway, Lord, 1966. Field characters as a guide to the specific relationships of swiftlets. — *Proc. Linn. Soc. London*, 177 (2): 151-172.

1507   Medway, Lord, & D.R. Wells, 1976. *The Birds of the Malay Peninsula*: 5. i-xxxi, 1-448. — H.F. & G. Witherby Ltd, London.

1508   Mees, G.F., 1957. Over het belang van Temminck's "Discours preliminaire" voor de Zoologiche Nomenclatuur. — *Zool. Meded., Leiden*, 35 (15): 205-227.

1509   Mees, G.F., 1957. A systematic review of the Indo-Australian Zosteropidae. Part I. — *Zool. Verhand., Leiden*, (35): 1-204.

1510   Mees, G.F., 1961. An annotated catalogue of a collection of bird skins from West Pilbarra, Western Australia. — *J. Roy. Soc. W. Aust.*, 44: 97-143.
1511   Mees, G.F., 1961. A systematic review of the Indo-Australian Zosteropidae. Part II. — *Zool. Verhand., Leiden*, (50): 1-168.
1512   Mees, G.F., 1964. Geographical variation in *Bubo sumatranus* (Raffles) (Aves, Strigidae). — *Zool. Meded., Leiden*, 40: 115-117.
1513   Mees, G.F., 1964. Four new subspecies of birds from the Moluccas and New Guinea. — *Zool. Meded., Leiden*, 40 (15): 125-130.
1514   Mees, G.F., 1964. A revision of the Australian owls (Strigidae and Tytonidae). — *Zool. Meded., Leiden*, 65: 1-62.
1514A   Mees, G.F., 1964. Notes on two small collections of birds from New Guinea. — *Zool. Meded., Leiden*, 66: 1-37.
1515   Mees, G.F., 1965. The avifauna of Misool. — *Nova Guinea, Zoology*, 31: 139-203.
1516   Mees, G.F., 1965. Revision of *Nectarinia sericea* (Lesson). — *Ardea*, 53 (1/2): 38-56.
1517   Mees, G.F., 1966. A new subspecies of *Anthreptes malacensis* (Scopoli) from the Soela Islands (Aves: Nectarinidae). — *Zool. Meded., Leiden*, 41 (18): 255-257.
1518   Mees, G.F., 1967. Zur Nomenklatur einiger Raubvögel und Eulen. — *Zool. Meded., Leiden*, 42 (14): 143-146.
1519   Mees, G.F., 1968. Enige voor de avifauna van Suriname nieuwe vogelsoorten. — *Gerfaut*, 58: 101-107.
1520   Mees, G.F., 1969. A systematic review of the Indo-Australian Zosteropidae (Part III). — *Zool. Meded., Leiden*, 102: 1-390.
1521   Mees, G.F., 1970. Notes on some birds from the Island of Formosa (Taiwan). — *Zool. Meded., Leiden*, 44 (20): 285-304.
1522   Mees, G.F., 1971. The Philippine subspecies of *Centropus bengalensis* (Gmelin) (Aves: Cuculidae). — *Zool. Meded., Leiden*, 45 (18): 189-191.
1523   Mees, G.F., 1971. Systematic and faunistic remarks on birds from Borneo and Java, with new records. — *Zool. Meded., Leiden*, 45 (21): 225-244.
1524   Mees, G.F., 1973. Once more: the identity and authorship of *Treron griseicauda*. — *Bull. Brit. Orn. Cl.*, 93: 119-120.
1525   Mees, G.F., 1973. The status of two species of migrant swifts in Java and Sumatra. — *Zool. Meded., Leiden*, 46 (15): 196-207.
1526   Mees, G.F., 1974. Additions to the avifauna of Suriname. — *Zool. Meded., Leiden*, 48: 55-67.
1527   Mees, G.F., 1975. A list of the birds known from Roti and adjacent islands (Lesser Sunda islands). — *Zool. Meded., Leiden*, 49 (12): 115-140.
1528   Mees, G.F., 1977. Additional records of birds from Formosa (Taiwan). — *Zool. Meded., Leiden*, 51 (15): 243-264.
1529   Mees, G.F., 1977. Geographical variation of *Caprimulgus macrurus* Horsfield. — *Zool. Meded., Leiden*, (155): 1-47.
1530   Mees, G.F., 1977. The subspecies of *Chlidonias hybridus* (Pallas), their breeding, distribution and migrations. — *Zool. Meded., Leiden*, (157): 1-64.
1531   Mees, G.F., 1980. Supplementary notes on the avifauna of Misool. — *Zool. Meded., Leiden*, 55 (1): 1-10.
1532   Mees, G.F., 1981. The sparrow-hawks (*Accipiter*) of the Andaman Islands. — *J. Bombay Nat. Hist. Soc.*, 77 (1980) (3): 371-412.
1533   Mees, G.F., 1982. Bird records from the Moluccas. — *Zool. Meded., Leiden*, 56 (7): 91-111.
1534   Mees, G.F., 1982. Birds from the lowlands of southern New Guinea (Merauke and Koembe). — *Zool. Meded., Leiden*, 191: 1-188.
1535   Mees, G.F., 1984. A new subspecies of *Accipiter virgatus* (Temminck) from Flores, Lesser Sunda Islands, Indonesia (Aves: Accipitridae). — *Zool. Meded., Leiden*, 58 (18): 313-321.
1536   Mees, G.F., 1985. Nomenclature and systematics of birds from Suriname. — *Proc. Kon. Ned. Akad. v. Wetensch., Ser. C.*, 88: 75-91.
1537   Mees, G.F., 1985. *Caprimulgus macrurus* Horsfield and related forms, a re-evaluation (Aves, Caprimulgidae). — *Proc. Kon. Ned. Akad. v. Wetensch.*, 88 (4): 419-428.
1538   Mees, G.F., 1985. Comments on species of the genus *Hirundapus* (Apodidae). — *Proc. Kon. Ned. Akad. v. Wetensch.*, 88 (1): 63-73.
1539   Mees, G.F., 1986. A list of the birds recorded from Bangka Island, Indonesia. — *Zool. Meded., Leiden*, (232): 1-176.
1539A   Mees, G.F., 1987. The juvenile plumage, systematic position and range of *Synallaxis macconnelli* Chubb (Aves: Furnariidae). — *Proc. Kon. Akad. v. Wetensch.*, 90 (3): 303-309.
1540   Mees, G.F., 1989. Remarks on the ornithological parts of Horsfield's "Zoological Researches in Java". — *Proc. Kon. Ned. Akad. v. Wetensch.*, 92 (3): 367-378.
1541   Mees, G.F., 1994. Vogelkundig onderzoek op Nieuw Guinea in 1828. — *Zool. Bijd.*, 40: 1-64.
1542   Mees, G.F., 1995. On *Malacocincla vanderbilti* de Schauensee & Ripley and *Malacocincla perspicillata* Bonaparte. — *Proc. Kon. Ned. Akad. v. Wetensch.*, 98 (1): 63-68.
1543   Mees, G.F., 1996. *Geographical variation in birds of Java.*: i-viii, 1-119. — Nuttall Ornithological Club, Cambridge, Mass., USA. Publ. No. 26.
1544   Mees, G.F., 1997. On the identity of *Heterornis senex* Bonaparte. — *Bull. Brit. Orn. Cl.*, 117 (1): 67-68.
1545   Meinertzhagen, R.C., 1933. [Descriptions of new subspecies of wren and francolin.]. — *Bull. Brit. Orn. Cl.*, 54: 20-21.
1546   Meinertzhagen, R.C., 1937. [Six new races from Mt. Kenya, Kenya Colony.]. — *Bull. Brit. Orn. Cl.*, 57: 67-70.
1547   Meinertzhagen, R.C., 1939. New species and races from Morocco. — *Bull. Brit. Orn. Cl.*, 59: 63-69.
1548   Meinertzhagen, R.C., 1948. A new race of Stone Curlew from Morocco. — *Bull. Brit. Orn. Cl.*, 68: 52.
1549   Meinertzhagen, R.C., 1949. New races of a Courser, Woodpecker, Swift, Lark, Wheatear and Serin from Africa. — *Bull. Brit. Orn. Cl.*, 69: 104-108.
1550   Meinertzhagen, R.C., 1951. A new race of *Alectoris melanocephala* Rüppell. — *Bull. Brit. Orn. Cl.*, 71: 29.
1551   Meise, W., 1933. Vorläufiges über die Expedition des Dresdner Museums für Tierkunde nach dem Matengo-Hochland am Njassasee. — *Orn. Monatsber.*, 41: 141-145.
1552   Meise, W., 1941. Ueber die Vogelwelt von Noesa Penida bei Bali nach einer Sammlung von Baron Viktor von Plessen. — *J. f. Orn.*, 89 (4): 345-376.
1553   Meise, W., 1958. Über neue Hühner-, Specht- und Singvogelrassen von Angola. — *Abhandl. Verh. Naturw. Ver. Hamburg*, 2: 63-83 (1957).
1554   Meise, W., 1968. Zur Speciation afrikanischer, besonders angolanischer Singvögel der Gattungen *Terpsiphone, Dicrurus* und *Malaconotus*. — *Zool. Beitr. (N. F.)*, 14: 1-60.
1555   Meise, W., 1978. Afrikanische Arten der Gattung *Trichastoma* (Aves, Timaliidae). — *Rev. Zool. Afr.*, 92: 789-804.
1556   Mellink, E. & A.M. Rea, 1994. Taxonomic status of the California Gnatcatchers of northwestern Baja California, Mexico. — *Western Birds*, 25 (1): 50-62.
1557   Meyer, B., 1822. *Zusätze und Berichtigungen zu Meyers und Wolfs Taschenbuch der deutschen Vögelkunde, nebst kurzer Beschreibung derjenigen Vögel, welche ausser Deutschland, in den übrigen Theilen von Europa vorkommen, als dritter Theil jenes Taschenbuchs*: 1-264. — M. Brönner, Frankfurt-am-Main.
1558   Meyer de Schauensee, R., 1937. First preliminary report on the results of the Second Dolan Expedition to West China and Tibet: two new birds from Tibet. — *Proc. Acad. Nat. Sci. Philad.*, 89: 339-340.
1559   Meyer de Schauensee, R., 1941. A new subspecies of *Arborophila*. — *Notulae Naturae*, 82: 1.
1560   Meyer de Schauensee, R., 1944. Notes on Colombian parrots. — *Notulae Naturae*, 140: 1-5.
1561   Meyer de Schauensee, R., 1944. Notes on Colombian birds, with a description of a new form of *Zenaida*. — *Notulae Naturae*, 144: 1-4.
1562   Meyer de Schauensee, R., 1945. Notes on Columbian birds. — *Proc. Acad. Nat. Sci. Philad.*, 97: 1-39.
1563   Meyer de Schauensee, R., 1946. Columbian Zoological Survey. Part III. - Notes on Columbian birds. — *Notulae Naturae*, 163: 1-9.
1564   Meyer de Schauensee, R., 1946. On Siamese birds. — *Proc. Acad. Nat. Sci. Philad.*, 98: 1-82.
1565   Meyer de Schauensee, R., 1949. The Birds of the Republic of Colombia (Contd.). — *Caldasia*, 5 (23): 381-644.
1566   Meyer de Schauensee, R., 1950. Columbian Zoological Survey. Part V. - New birds from Columbia. — *Notulae Naturae*, 221: 1-13.
1567   Meyer de Schauensee, R., 1951. Columbian Zoological Survey. Part VIII. On birds from Nariño, Colombia, with descriptions of four new subspecies. — *Notulae Naturae*, 232: 1-6.
1569[3]   Meyer de Schauensee, R., 1951. Colombian zoological survey. Part IX. A new species of ant-bird (*Phlegopsis*) from Colombia. — *Notulae Naturae*, 241: 1-3.

---

[3] No. 1568 has been deleted.

1570   Meyer de Schauensee, R., 1958. The birds of the island of Bangka, Indonesia. — *Proc. Acad. Nat. Sci. Philad.*, 110: 279-299.

1571   Meyer de Schauensee, R., 1959. Additions to the "Birds of the Republic of Colombia". — *Proc. Acad. Nat. Sci. Philad.*, 111: 53-75.

1572   Meyer de Schauensee, R., 1960. A new race of *Hyloctistes subulatus* (Furnariidae) from Colombia. — *Notulae Naturae*, 332: 1.

1573   Meyer de Schauensee, R., 1966. *The species of birds of South America and their distribution*: i-xvii, 1-577. — Livingstone Publ. Co., Narbeth, Penn.

1574   Meyer de Schauensee, R., 1967. *Eriocnemis mirabilis*, a new species of hummingbird from Colombia. — *Notulae Naturae*, 402: 1-2.

1575   Meyer de Schauensee, R., 1970. *A guide to the Birds of South America*: 1-470. — Livingston Press, Narbeth, Pa.

1576   Meyer de Schauensee, R., 1970. A review of the South American finch *Oryzoborus crassirostris*. — *Notulae Naturae*, 418: 1-6.

1577   Meyer de Schauensee, R. & W.H. Phelps, 1978. *A guide to the Birds of Venezuela*: i-xxii, 1-424. — Princeton Univ. Press, Princeton.

1578   Meyer de Schauensee, R. & S.D. Ripley, 1940. Zoological results of the George Vanderbilt Sumatran Expedition, 1936-1939. Part I - Birds from Atjeh. — *Proc. Acad. Nat. Sci. Philad.*, 91: 311-368.

1579   Meyer de Schauensee, R. & S.D. Ripley, 1940. Zoological results from the George Vanderbilt Sumatran Expedition, 1936-1939. Pt. III. Birds from Nias island. — *Proc. Acad. Nat. Sci. Philad.*, 91: 399-413.

1580   Miller, A.H., 1946. Endemic birds of the Little San Bernadino Mountains, California. — *Condor*, 48: 75-79.

1581   Miller, A.H., 1948. A new subspecies of eared poor-will from Guerrero, Mexico. — *Condor*, 50: 224-225.

1582   Miller, A.H., 1959. A new race of nighthawk from the Upper Magdalena Valley of Columbia. — *Proc. Biol. Soc. Wash.*, 72: 155-157.

1583   Miller, A.H. & L. Miller, 1951. Geographic variation of the Screech Owls of the deserts of western North America. — *Condor*, 53 (4): 161-177.

1584   Miller, W. deW. & L. Griscom, 1925. Descriptions of new birds from Nicaragua. — *Amer. Mus. Novit.*, 159: 1-9.

1585   Milon, P., 1950. Description d'une sous-espèce nouvelle d'oiseau de Madagascar. — *Bull. Mus. Hist. Nat. Paris*, (2) 22: 65-66.

1586   Milon, P. *et al.*, 1973. Faune de Madagascar. No. 35. Oiseaux. 1-263. — ORSTOM, Tananarive C.N.R.S., Paris.

1587   Miranda, J., Hector C. *et al.*, 1997. Phylogenetic placement of *Mimizuku gurneyi* (Aves: Strigidae) inferred from mitochondrial DNA. — *Auk*, 114 (3): 315-323.

1588   Miranda-Ribeiro, A. de, 1938. Notas ornithologicas. XII. — *Rev. Mus. Paul.*, 23: 35-90.

1589   Mishima, T., 1956. A new race of *Charadrius placidus* from Japan. — *Tori*, 14 (67): 15-16.

1590   Mishima, T., 1958. Diagnosis of the Dipper from Hokkaido. — *Japan Wildlife Bull.*, 16 (2): 123-124.

1591   Mishima, T., 1959. A small collection of birds from Hokkaido, gathered in summer of 1958. — J*Japan Wildlife Bull.*, 17 (1): 1-6.

1592   Mishima, T., 1960. A new race of *Apus pacificus* breeding in the southern parts of Izu Islands, Japan. — *Japan Wildlife Bull.*, 17 (2): 201-202.

1593   Mishima, T., 1960. On the Brown-eared Bulbul *Hypsipetes amaurotis*. — *Japan Wildlife Bull.*, 17 (2): 165-172.

1594   Mishima, T., 1961. Notes on some birds of Shikoku. — *Japan Wildlife Bull.*, 18 (1): 157-172.

1595   Mlikovsky, J., 1989. Note on the osteology and taxonomic position of Salvadori's Duck *Salvadorina waigiuensis* (Aves: Anseridae). — *Bull. Brit. Orn. Cl.*, 109 (1): 22-25.

1596   Mobley, J.A. & R.O. Prum, 1997. Phylogenetic relationships of the cinnamon tyrant, *Neopipo cinnamomea*, to the tyrant flycatchers (Tyrannidae). — *Condor*, 97 (3): 650-662.

1597   Molina, P. *et al.*, 2000. Geographic variation and taxonomy of the northern waterthrush. — *Wilson Bull.*, 112 (3): 337-346.

1598   Moltoni, E., 1935. Contributo alla conoscenza degli uccelli della Somalia italiana. — *Atti. Soc. Ital. Sci. Nat., Milano.*, 74: 333-371.

1599   Momiyama, T.T., 1926. On the specimens of birds collected in Quelpart island south of Korea (Part 1). — *Tori*, 5 (22): 101-126.

1600   Momiyama, T.T., 1939. New addition to the birds of Miyagi Prefecture, north-eastern Hondo, with description of a new Jack Snipe, *Limnocryptes minimus nipponensis*. — *Shikubutsu oyobi Dobutsu* (Botany & Zoology), 7 (7): 1219-1227.

1601   Monard, A., 1934. Ornithologie de l'Angola. — *Arch. Mus. Bocage*, 5: 1-110.

1602   Monard, A., 1949. Vertébrés nouveaux du Cameroun. — *Rev. Suisse Zool.* (Geneve), 56: 731-745.

1603   Monroe, B.L., Jr., 1963. Three new subspecies of birds from Honduras. — *Occ. Pap. Mus. Zool., Louisiana St. Univ.*, 26: 1-7.

1604   Monroe, B.L., Jr., 1968. *A distributional survey of the birds of Honduras*: Orn. Monog. 7: 1-458. — A.O.U.

1605   Monroe, B.L., Jr. & M.R. Browning, 1992. A re-analysis of *Butorides*. — *Bull. Brit. Orn. Cl.*, 112 (2): 81-85.

1606   Monroe, B.L., Jr. & T.R. Howell, 1966. Geographic variation in Middle American parrots of the *Amazona ochrocephala* complex. — *Occ. Pap. Mus. Zool., Louisiana St. Univ.*, 34: 1-18.

1607   Monson, G. & A.R. Phillips, 1981. The races of Red Crossbill, *Loxia curvirostris*, in Arizona. Appendix pp. 223-230. In *Annotated Checklist of the Birds of Arizona*. — The University of Arizona Press, Tucson, Arizona.

1608   Moore, F., 1855. Descriptions of three new species of titmice. — *Proc. Zool. Soc., Lond.*, 139-140 (1854).

1609   Moore, R.T., 1937. Four new birds from north-western Mexico. — *Proc. Biol. Soc. Wash.*, 50: 95-102.

1610   Moore, R.T., 1937. New race of *Chubbia jamesoni* from Colombia. — *Proc. Biol. Soc. Wash.*, 50: 151-152.

1611   Moore, R.T., 1937. A new race of Finsch's Parrot. — *Auk*, 54: 528-529.

1612   Moore, R.T., 1938. A new race of Wild Turkey. — *Auk*, 55: 112-115.

1613   Moore, R.T., 1941. Three new races in the genus *Otus* from central Mexico. — *Proc. Biol. Soc. Wash.*, 54: 151-160.

1614   Moore, R.T., 1946. A new woodpecker from Mexico. — *Proc. Biol. Soc. Wash.*, 59: 103-105.

1615   Moore, R.T., 1947. New species of parrot and race of quail from Mexico. — *Proc. Biol. Soc. Wash.*, 60: 27-28.

1616   Moore, R.T., 1947. New owls of the genera *Otus* and *Glaucidium*. — *Proc. Biol. Soc. Wash.*, 60: 31-35.

1617   Moore, R.T., 1947. Two new owls, a swift and a Poorwill from Mexico. — *Proc. Biol. Soc. Wash.*, 60: 141-146.

1618   Moore, R.T., 1949. A new hummingbird of the genus *Lophornis* from southern Mexico. — *Proc. Biol. Soc. Wash.*, 62: 103-104.

1619   Moore, R.T., 1950. A new race of the species *Amazilia beryllina* from southern Mexico. — *Proc. Biol. Soc. Wash.*, 63: 59-60.

1620   Moore, R.T., 1951. A new race of *Melanerpes chrysogenys* from central Mexico. — *Proc. Biol. Soc. Wash.*, 63: 109-110.

1621   Moore, R.T. & J.T. Marshall, Jr., 1959. A new race of screech owl from Oaxaca. — *Condor*, 61 (3): 224-225.

1622   Moore, R.T. & D.R. Medina, 1957. The status of the chachalacas in western Mexico. — *Condor*, 59: 230-234.

1623   Moreau, R.E., 1967. Family Zosteropidae, African and Indian Ocean taxa. Pages 326-337. In *Check-list of Birds of the World*. Vol. 12 (R.A. Paynter, Jr., ed.). — Mus. Comp. Zool., Cambridge, Mass.

1624   Moreau, R.E. & J.P. Chapin, 1951. The African Emerald Cuckoo *Chrysococcyx cupreus*. — *Auk*, 68: 174-189.

1625   Moreau, R.E. & J.C. Greenway, Jr., 1962. Family Ploceidae, Weaverbirds. Pages 3-75. In *Check-list of Birds of the World*. Vol. 15 (E. Mayr & J.C. Greenway, Jr., eds.). — Mus. Comp. Zool., Cambridge, Mass.

1626   Morioka, H. & Y. Shigeta, 1993. Generic allocation of the Japanese Marsh Warbler *Megalurus pryeri* (Aves: Sylviidae). — *Bull. Nat. Sci. Mus., Tokyo, 'A',* 19 (1): 37-43.

1627   Morris, P. & F. Hawkins, 1998. *Birds of Madagascar. A photographic guide*: i-xi, 1-316. — Pica Press, Mountfield, Sussex.

1628   Morrison, A., 1939. A new Coot from Peru. — *Bull. Brit. Orn. Cl.*, 59: 56-57.

1629   Mougin, J.-L. & R. de Naurois, 1981. Le Noddi bleu des iles Gambier *Procelsterna caerulea murphyi* ssp. nov. — *L'Oiseau*, 51 (3): 201-204.

1630   Mukherjee, A.K., 1952. Taxonomic notes on the Lineated Barbet and the Green Barbets of India, with a description of a new race. — *Bull. Brit. Orn. Cl.*, 72: 34-36.

1631   Mukherjee, A.K., 1958. A new race of the Striated Scops Owl. — *Rec. Indian Mus.*, 53 (1/2): 301-302.

1632   Mukherjee, A.K., 1960. A new race of the Emerald Dove *Chalcophaps indica* (Linnaeus) from India. — *Bull. Brit. Orn. Cl.*, 80: 6-7.

1633   Murie, O.J., 1944. Two new subspecies of birds from Alaska. — *Condor*, 46: 121-123.

1634   Murphy, R.C., 1938. Birds collected during the Whitney South Sea Expedition. XXXVII. On pan-Antarctic terns. — *Amer. Mus. Novit.*, 977: 1-17.

1635   Nardelli, P.M., 1993. *A Preservação do Mutum-de-Alagoas* Mitu mitu. The Preservation of the Alagoas Curassow *Mitu mitu*: i-xxviii, 1-251. — Zôo-botânica Mário Nardelli, Nilópolis - Rio de Janeiro, Brazil.
1636   Narosky, S., 1977. Una nueva especie del genero *Sporophila* (Emberizidae). — *Hornero*, 11: 345-348.
1637   Navarro, A.G. *et al.*, 1992. *Cypseloides storeri* a new species of swift from Mexico. — *Wilson Bull.*, 104 (1): 55-64.
1638   Navarro-Siguenza, A. & A.T. Peterson, 1999. Comments on the taxonomy of the genus *Cynanthus* (Swainson), with a restricted type locality for *C. doubledayi*. — *Bull. Brit. Orn. Cl.*, 119 (2): 109-112.
1639   Navas, J.R. & N.A. Bo, 1982. La posicion taxionomica de *Thripophaga sclateri* y *T. punensis* (Aves, Furnariidae). — *Com. Mus. Argentino Cienc. Nat. "Bernardino Rivadavia"*, *Zool.*, 4 (11): 85-93.
1640   Nechaev, V.A. & P.S. Tomkovich, 1987. A new subspecies of the Dunlin, *Calidris alpina litoralis* ssp. n. (Charadriidae, Aves), from the Sakhalin Island. — *Zool. Zhurn.*, 66: 1110-1113.
1641   Nechaev, V.A. & P.S. Tomkovich, 1988. A new name for Sakhalin Dunlin (Aves: Charadriidae). — *Zool. Zhurn.*, 67: 1596.
1642   Neumann, O., 1933. [Three new geographical races of the White-eyebrowed Guan.] — *Bull. Brit. Orn. Cl.*, 53: 93-95.
1643   Neumann, O., 1933. Unbenannte geographische Rassen aus Kamerun und Angola. — *Verh. Orn. Ges. Bayern*, 20: 225-228.
1644   Neumann, O., 1934. Drei neue geographische Rassen aus dem paläarktischen Gebiet. — *Anz. Ornith. Ges. Bayern*, 2 (8): 331-334.
1645   Neumann, O., 1939. A new species and eight new races from Peleng and Taliaboe. — *Bull. Brit. Orn. Cl.*, 59: 89-94.
1646   Neumann, O., 1939. [Six new races from Peling.]. — *Bull. Brit. Orn. Cl.*, 59: 104-108.
1647   Neumann, O., 1941. Neue Subspecies von Vögeln aus Niederländisch-Indien. — *Zool. Meded., Leiden*, 23: 109-113.
1648   Nicolai, J., 1968. Die Schnabelfärbung als potentieller Isolationsfaktor zwischen *Pytilia phoenicoptera* Swainson und *Pytilia lineata* Heuglin (Familie: Estrildidae). — *J. f. Orn.*, 109 (4): 450-461.
1649   Nicolai, J., 1972. Zwei neue *Hypochera*-Arten aus West-Afrika (Ploceidae, Viduinae). — *J. f. Orn.*, 113 (3): 229-240.
1650   Niethammer, G., 1953. Zur Vogelwelt Boliviens. — *Bonn. Zool. Beitr.*, 4 (3/4): 195-303.
1651   Niethammer, G., 1954. *Choriotis arabs geyri* subspec. nova. — *Bonn. Zool. Beitr.*, 5 (3/4): 193-194.
1652   Niethammer, G., 1955. Zur Vogelwelt des Ennedi-Gebirges (Französich Äquatorial-Afrika). — *Bonn. Zool. Beitr.*, 6 (1/2): 29-80.
1653   Niethammer, G., 1957. Ein weiterer Beitrag zur Vogelwelt des Ennedi-Gebirges. — *Bonn. Zool. Beitr.*, 8: 275-284.
1654   Niethammer, G. & H.E. Wolters, 1966. Kritische Bemerkungen über einige südafrikanische Vögel im Museum A. Koenig, Bonn. — *Bonn. Zool. Beitr.*, 17 (3/4): 157-185.
1655   Nikolaus, G., 1982. A new race of the Spotted Ground Thrush *Turdus fischeri* from South Sudan. — *Bull. Brit. Orn. Cl.*, 102 (2): 45-47.
1656   Nikolaus, G., 1987. Distribution atlas of Sudan's birds with notes on habitat and status. — *Bonn. Zool. Monogr.*, 29: 1-322.
1656A   Nitsch, C.L., 1840. *System der Pterylographie*. — Eduard Anton, Halle.
1657   Nores, M., 1986. Diez nuevas subspecies de aves provenientes de islas ecologicas Argentinas. — *Hornero*, 12 (4): 262-273.
1658   Nores, M. & D. Yzurieta, 1979. Una nueva especie y dos nuevas subspecies de aves (Passeriformes). — *Acad. Nac. Cienc. (Cordoba), Misc.* No. 61: 4-8.
1659   Nores, M. & D. Yzurieta, 1980. Nuevas aves de la Argentina (1). — *Hist. Nat., Mendoza*, 1: 169-172.
1660   Nores, M. & D. Yzurieta, 1983. Especiacion en las sierras Pampeanas de Cordoba y San Luis (Argentina), con descripcion de siete nuevas subspecies de aves. — *Hornero*, (No. Extraordinario): 88-102.
1661   Norman, J.A. *et al.*, 1998. Molecular data confirms the species status of the Christmas Island hawk-owl *Ninox natalis*. — *Emu*, 98 (3): 197-208.
1661A   Novaes, F.C., 1953. Sobre a validade de *Syndactyla mirandae* (Snethlage, 1928) (Furnariidae, Aves). — *Rev. Bras. Biol.*, 14 (1): 75-76.
1662   Novaes, F.C., 1957. Notas de Ornithologia Amazônica. 1. Gêneros *Formicarius* e *Phlegopsis*. — *Bol. Mus. Para. Emílio Goeldi, Nov. Ser., Zool.*, 8: 1-9.
1663   Novaes, F.C., 1991. A new subspecies of Grey-cheeked Nunlet *Nonnula ruficapilla* from Brazilian Amazonia. — *Bull. Brit. Orn. Cl.*, 111 (4): 187-188.
1664   Novaes, F.C. & M.F.C. Lima, 1991. Variação geográfica e anotações sobre morfologia de *Selenidera gouldii* (Piciformes : Ramphastidae). — *Ararajuba*, 2: 59-63.
1665   Nunn, G.B. & J. Cracraft, 1996. Phylogenetic relationships among the major lineages of the Birds-of-Paradise (Paradisaeidae) using mitochondrial DNA gene sequences. — *Molec. Phylogen. Evol.*, 5 (3): 445-449.
1666   Nunn, G.B. *et al.*, 1996. Evolutionary relationships among extant albatrosses (Procellariiformes: Diomedeidae) established from complete cytochrome-*b* sequences. — *Auk*, 113: 784-801.
1667   Nutting, C.C. & R. Ridgway, 1884. On a collection of birds from Nicaragua. — *Proc. U. S. Nat. Mus.*, 6: 372-410.
1668   O.S.N.Z., 1990. *Checklist of the birds of New Zealand and the Ross Dependency, Antarctica*: i-xv, 1-247. 3rd. edition. — Random Century, Auckland, New Zealand.
1669   Oatley, T.B. *et al.*, 1992. Various species accounts. In *The Birds of Africa*. Vol. 4: i-xv, 1-609 (S. Keith, E.K. Urban & C.H. Fry, eds.). — Academic Press Inc., San Diego.
1670   Oberholser, H.C., 1905. Birds collected by Dr. W.L. Abbott in the Kilimanjaro Region, East Africa. — *Proc. U. S. Nat. Mus.*, 28: 823-936.
1671   Oberholser, H.C., 1937. A revision of the clapper rails (*Rallus longirostris* Boddaert). — *Proc. U. S. Nat. Mus.*, 84: 313-354.
1672   Oberholser, H.C., 1974. *The Bird Life of Texas*: 1. 1-530. — Univ. Texas Press, Austin & London.
1673   Oberholser, H.C., 1974. *The Bird Life of Texas*: 2. 531-1069. — Univ. of Texas Press, Austin & London.
1674   O'Brien, R.M. & J. Davies, 1990. A new subspecies of Masked Booby *Sula dactylatra* from Lord Howe, Norfolk and Kermadec Islands. — *Marine Orn.*, 18 (1): 1-7.
1675   Ogilvie-Grant, W.R., 1893. *A Catalogue of the Birds in the British Museum*. Vol. XXII: i-xvi, 1-585. — Trustees of the British Museum (Natural History), London.
1676   Oliver, W.R.B., 1955. *New Zealand Birds*: 1-661. 2nd. edition. — A.H. & A.W. Reed, Wellington.
1677   Olmos, F. *et al.*, 1997. Distribution and dry-season ecology of Pfrimer's Conure *Pyrrhura pfrimeri*, with a reappraisal of Brazilian *Pyrrhura 'leucotis'*. — *Orn. Neotrop.*, 8 (2): 121-132.
1678   Olrog, C.C., 1958. Notas ornitologicas sobre la coleccion del Instituto Miguel Lillo Tucuman. III. — *Acta Zool. Lilloana*, 15: 5-18.
1679   Olrog, C.C., 1959. Tres nuevas subspecies de aves Argentinas. — Neotropica, 5: 39-44.
1680   Olrog, C.C., 1962. Notas ornitologicas sobre la coleccion del Instituto Miguel Lillo (Tucuman). VI. — *Acta Zool. Lilloana*, 18: 111-120.
1681   Olrog, C.C., 1963. Una nueva subspecie de *Celeus lugubris* de Bolivia (Aves, Piciformes, Picidae). — *Neotropica*, 9 (29): 87-88.
1682   Olrog, C.C., 1972. Adiciones a la avifauna Argentina (1° Suplemento de "La lista y distribución de las aves argentinas", Opera Lilloana, IX, 1963). — *Acta Zool. Lilloana*, 26: 257-264.
1683   Olrog, C.C., 1973. Notas ornitologicas. IX. Sobre la coleccion del Instituto Miguel Lillo de Tucuman. — *Acta Zool. Lilloana*, 30: 7-11.
1684   Olrog, C.C., 1975. *Uropsalis lyra* nueva para la fauna Argentina (Aves, Caprimulgidae). — *Neotropica*, 21 (66): 147-148.
1685   Olrog, C.C., 1976. Sobre una subspecie de *Athene cunicularia* de Argentina (Aves, Strigidae). — *Neotropica*, 68: 107-108.
1686   Olrog, C.C., 1979. Notas ornitologicas. 11. Sobre la coleccion del Instituto Miguel Lillo. — *Acta Zool. Lilloana*, 33 (2): 5-7.
1687   Olrog, C.C. & F. Contino, 1970. Una nueva subspecie de *Grallaria albigula* Chapman (Aves, Formicariidae). — *Neotropica*, 16: 51-52.
1688   Olrog, C.C. & F. Contino, 1970. Dos especies nuevas para la avifauna Argentina. — *Neotropica*, 16: 94-95.
1688A   Olsen, J. *et al.*, 2002. A new *Ninox* owl from Sumba, Indonesia. — *Emu*, 102: 223-231.
1689   Olsen, K.M. & H. Larsson, 1995. *Terns of Europe and North America*: 1-175. — A. & C. Black, London.
1690   Olson, S.L., 1973. A study of the Neotropical rail *Anurolimnas castaneiceps* (Aves: Rallidae) with a description of a new subspecies. — *Proc. Biol. Soc. Wash.*, 86 (34): 403-412.

1691   Olson, S.L., 1975. Geographic variation and other notes on *Basileuterus leucoblepharus* (Parulidae). — *Bull. Brit. Orn. Cl.*, 95 (3): 101-104.
1692   Olson, S.L., 1980. Revision of the Tawny-faced Antwren, *Microbates cinereiventris* (Aves, Passeriformes). — *Proc. Biol. Soc. Wash.*, 93: 68-74.
1693   Olson, S.L., 1980. Geographic variation in the Yellow Warbler (*Dendroica petechia:* Parulidae) of the Pacific coast of Middle and South America. — *Proc. Biol. Soc. Wash.*, 93: 473-480.
1694   Olson, S.L., 1980. The subspecies of Grasshopper Sparrow (*Ammodramus savannarum*) in Panama (Aves: Emberizinae). — *Proc. Biol. Soc. Wash.*, 93: 757-759.
1695   Olson, S.L., 1981. A revision of the subspecies of *Sporophila* ("*Oryzoborus*") *angolensis* (Aves: Emberizinae). — *Proc. Biol. Soc. Wash.*, 94 (1): 43-51.
1696   Olson, S.L., 1981. A revision of the northern forms of *Euphonia xanthogaster* (Aves: Thraupidae). — *Proc. Biol. Soc. Wash.*, 94 (1): 101-106.
1697   Olson, S.L., 1981. Systematic notes on certain oscines from Panama and adjacent areas (Aves: Passeriformes). — *Proc. Biol. Soc. Wash.*, 94 (2): 363-373.
1698   Olson, S.L., 1983. Geographic variation in *Chlorospingus ophthalmicus* in Colombia and Venezuela (Aves: Thraupidae). — *Proc. Biol. Soc. Wash.*, 96 (1): 103-109.
1699   Olson, S.L., 1987. *Gallirallus sharpei* (Büttikofer) nov. comb. A valid species of rail (Rallidae) of unknown origin. — *Gerfaut*, 76: 263-269 (1986).
1699A   Olson, S.L., 1987. More on the affinities of the Black-breasted Thrush (*Chlamydochaera jefferyi*). — *J. f. Orn.*, 128: 246-248.
1700   Olson, S.L., 1989. Preliminary systematic notes on some Old World Passerines. — *Riv. Ital. Orn.*, 59: 183-195.
1701   Olson, S.L., 1995. The genera of owls in the Asioninae. — *Bull. Brit. Orn. Cl.*, 115 (1): 35-39.
1702   Olson, S.L., 1995. Types and nomenclature of two Chilean parrots from the voyage of HMS Blonde (1825). — *Bull. Brit. Orn. Cl.*, 115 (4): 235-239.
1703   Olson, S.L., 1996. The contribution of the voyage of H.M.S. Blonde (1825) to Hawaiian ornithology. — *Arch. Nat. Hist.*, 23: 1-42.
1704   Olson, S.L., 1997. Towards a less imperfect understanding of the systematics and biogeography of the Clapper and King Rail complex (*Rallus longirostris* and *R. elegans*). Pages 93-111. In *The Era of Allan R. Phillips: A Festschrift* (R.W. Dickerman, ed.) — Albuquerque.
1705   Olson, S.L., 1998. Notes on the systematics of the Rockrunner *Achaetops* (Passeriformes, Timaliidae) and its presumed relatives. — *Bull. Brit. Orn. Cl.*, 118 (1): 47-52.
1706   Olson, S.L., 2000. A new genus for the Kerguelen Petrel. — *Bull. Brit. Orn. Cl.*, 120 (1): 59-62.
1707   Olson, S.L. & D. Buden, 1989. The avifaunas of the cayerias of southern Cuba, with the ornithological results of the Paul Bartsch Expedition of 1930. — *Smithsonian Contrib. Zool.*, 477: i-iii, 1-34.
1708   Olson, S.L. & H.F. James, 1988. Nomenclature of the Kauai Amakihi and Kauai Akialoa (Drepanidini). — *Elepaio*, 48: 13-14.
1709   Olson, S.L. & D.W. Steadman, 1981. The relationships of the Pedionomidae (Aves, Charadriiformes). — *Smithsonian Contrib. Zool.*, 337: 1-25.
1710   Olson, S.L. & K.I. Warheit, 1988. A new genus for *Sula abbotti*. — *Bull. Brit. Orn. Cl.*, 108: 9-12.
1711   Olsson, U. *et al.*, 1993. A new species of *Phylloscopus* warbler from Hainan Island, China. — *Ibis*, 135 (1): 3-7.
1712   Omland, K.E. *et al.*, 1999. A molecular phylogeny of the New World Orioles (Icterus): The importance of dense taxon sampling. — *Molec. Phylogen. Evol.*, 12 (2): 224-239.
1713   O'Neill, J.P. & G.R. Graves, 1977. A new genus and species of owl (Aves: Strigidae) from Peru. — *Auk*, 94 (3): 409-416.
1714   O'Neill, J.P. & T.A. Parker, III., 1976. New subspecies of *Schizoeaca fuliginosa* and *Uromyias agraphia* from Peru. — *Bull. Brit. Orn. Cl.*, 96: 136-141.
1715   O'Neill, J.P. & T.A. Parker, III., 1977. Taxonomy and range of *Pionus seniloides* in Peru. — *Condor*, 79 (2): 274.
1716   O'Neill, J.P. & T.A. Parker, III., 1981. New subspecies of *Pipreola riefferii* and *Chlorospingus ophthalmicus* from Peru. — *Bull. Brit. Orn. Cl.*, 101: 294-299.
1717   O'Neill, J.P. & T.A. Parker, III., 1997. New subspecies of *Myrmoborus leucophrys* (Formicariidae) and *Phrygilus alaudinus* (Emberizidae) from the Upper Huallaga Valley, Peru. — *Orn. Monogr.*, 48: 485-491.
1718   O'Neill, J.P. *et al.*, 1991. *Nannopsittaca dachilleae*, a new species of Parrotlet from eastern Peru. — *Auk*, 108 (2): 225-229.
1719   O'Neill, J.P. *et al.*, 2000. A striking new species of barbet (Capitoninae: *Capito*) from the eastern Andes of Peru. — *Auk*, 117: 569-577.
1720   Oren, D.C., 1985. Two new subspecies of birds from the canga vegetation, Serra dos Carajás, Pará, Brasil, and one from Venezuela. — *Publ. Avulsas Mus. Paraense Emilio Goeldi*, 40: 93-100.
1721   Oren, D.C., 1993. *Celeus torquatus pieteroyensi*, a new subspecies of Ringed Woodpecker (Aves, Picidae) from eastern Para and western Maranhao, Brazil. — *Bol. Mus. Para. Emilio Goeldi, Zool.*, 8 (1992) (2): 385-389.
1722   Oren, D.C. & F.C. Novaes, 1985. A new subspecies of the White Bellbird *Procnias alba* (Hermann) from southeastern Amazonia. — *Bull. Brit. Orn. Cl.*, 105: 23-25.
1723   Orenstein, R.I. & H.D. Pratt, 1983. The relationships and evolution of the Southwest Pacific Warbler genera *Vitia* and *Psamathia* (Sylviinae). — *Wilson Bull.*, 95 (2): 184-198.
1724   Orlando, C., 1957. Contributo allo studio delle forme europee del *Bubo bubo* (L.). — *Riv. Ital. Orn.*, 27: 42-54.
1725   Orn. Soc. Japan., 1958. *A hand-list of the Japanese birds*: i-xii, 1-264. 4th. edition. — Orn. Soc. Japan.
1726   Orn. Soc. Japan., 1974. *Check-list of the Japanese birds*: i-viii, 1-364. 5th. edition. — Gakken Co. Ltd., Tokyo.
1727   Orn. Soc. Japan., 2000. *Check-list of Japanese birds*: i-xii, 1-345. 6th. edition. — Orn. Soc. Japan.
1728   Orr, R.T. & J.D. Webster, 1968. New subspecies of birds from Oaxaca (Aves: Phasianidae, Turdidae, Parulidae). — *Proc. Biol. Soc. Wash.*, 81: 37-40.
1729   Orta, J., 1992. Family Phalacrocoracidae (Cormorants). Pages 326-353. In *Handbook of the Birds of the World*. Vol. 1 (J. del Hoyo, A. Elliott & J. Sargatal, eds.). — Lynx Edicions, Barcelona.
1730   Orta, J., 1992. Family Anhingidae (Darters). Pages 354-361. In *Handbook of the Birds of the World*. Vol. 1 (J. del Hoyo, A. Elliott & J. Sargatal, eds.). — Lynx Edicions, Barcelona.
1731   Orta, J., 1994. Various species accounts. In *Handbook of the Birds of the World*. Vol. 2: 1-638 (J. del Hoyo, A. Elliott & J. Sargatal, eds.). Lynx Edicions, Barcelona.
1732   Ouellet, H., 1990. A new Ruffed Grouse, Aves: Phasianidae: *Bonasa umbellus*, from Labrador, Canada. — *Can. Field-Nat.*, 104 (3): 445-449.
1733   Ouellet, H., 1992. Speciation, zoogeography and taxonomic problems in the Neotropical genus *Sporophila* (Aves: Emberizinae). — *Bull. Brit. Orn. Cl.*, 112A (Centenary Suppl.): 225-235.
1734   Ouellet, H., 1993. Bicknell's Thrush : Taxonomic status and distribution. — *Wilson Bull.*, 105 (4): 545-572.
1735   Oustalet, E., 1890. Description d'un nouveau Martin Pêcheur des Iles Philippines. — *Le Naturaliste*, 12: 62-63.
1736   Pacheco, J.F. & L.P. Gonzaga, 1995. A new species of *Synallaxis* of the *ruficapilla/infuscata* complex from eastern Brazil (Passeriformes: Furnariidae). — *Ararajuba*, 3: 3-11.
1737   Pacheco, J.F. & B.M. Whitney, 1998. Correction of the specific name of Long-trained Nightjar. — *Bull. Brit. Orn. Cl.*, 118 (4): 259-261.
1738   Pacheco, J.F. *et al.*, 1996. A new genus and species of Furnariid (Aves: Furnariidae) from the cocoa-growing region of southeastern Bahia, Brazil. — *Wilson Bull.*, 108 (3): 397-433.
1739   Päckert, M. *et al.*, 2001. Lautäusserungen der Sommergoldhähnchen von den Inseln Madeira und Mallorca (*Regulus ignicapillus madeirensis, R. i. balearicus*). — *J. f. Orn.*, 142 (1): 16-29.
1740   Pakenham, R.H.W., 1940. A new Green Pigeon from Pemba Island. — *Bull. Brit. Orn. Cl.*, 60: 94-95.
1741   Paludan, K., 1959. On the birds of Afghanistan. — *Vidensk. Medd. fra Dansk naturh. Foren.*, 122: 1-332.
1742   Panov, E.N., 1992. Emergence of hybridogenous polymorphism in the *Oenanthe picata* complex. — *Bull. Brit. Orn. Cl.*, 112A (Centenary Suppl.): 237-249.

1743   Parker, R.H. & C.W. Benson, 1971. Variation in *Caprimulgus tristigma* Rüppell especially in West Africa. — *Bull. Brit. Orn. Cl.*, 91: 113-119.
1744   Parker, S.A., 1981. Prolegomenon to further studies in the *Chrysococcyx "malayanus"* group (Aves: Cuculidae). — *Zool. Verhand., Leiden,* 187: 3-56.
1745   Parker, S.A., 1984. The relationships of the Madagascan genus *Dromaeocercus* (Sylviidae). — *Bull. Brit. Orn. Cl.*, 104 (1): 11-18.
1746   Parker, S.A., 1984. The extinct Kangaroo Island *Emu,* a hitherto-unrecognized species. — *Bull. Brit. Orn. Cl.*, 104: 19-22.
1747   Parker, T.A., III. & J.P. O'Neill, 1985. A new species and a new subspecies of *Thryothorus* wren from Peru. Acta Zool. Lilloana. — *AOU Monogr.*, 36: 9-15.
1748   Parker, T.A., III. & J.V. Remsen, Jr., 1987. Fifty-two Amazonian bird species new to Bolivia. — *Bull. Brit. Orn. Cl.*, 107 (3): 94-107.
1749   Parker, T.A., III. *et al.*, 1985. The avifauna of Huancabamba Region, northern Peru. — *Orn. Monogr.*, 36: 169-197.
1750   Parker, T.A., III. *et al.*, 1991. Records of new and unusual birds from northern Bolivia. — *Bull. Brit. Orn. Cl.*, 111 (3): 120-138.
1751   Parker, T.A., III. *et al.*, 1997. Notes on avian bamboo specialists in southwestern Amazonian Brazil. — *Orn. Monogr.*, 48: 543-547.
1752   Parkes, K.C., 1949. A new button-quail from New Guinea. — *Auk,* 66: 84-86.
1753   Parkes, K.C., 1960. The Brown Cachalote *Pseudoseisura lophotes* in Bolivia. — *Auk,* 77: 226-227.
1754   Parkes, K.C., 1960. Notes on Philippine races of *Dryocopus javensis.* — *Bull. Brit. Orn. Cl.*, 80: 59-61.
1755   Parkes, K.C., 1962. New subspecies of birds from Luzon, Philippines. — *Postilla,* 67: 1-8.
1756   Parkes, K.C., 1963. A new subspecies of tree-babbler from the Philippines. — *Auk,* 80: 543-544.
1757   Parkes, K.C., 1965. The races of the Pompadour Green Pigeon, *Treron pompadora,* in the Philippine Islands. — *Bull. Brit. Orn. Cl.*, 85: 137-139.
1758   Parkes, K.C., 1966. Geographic variation in Winchell's Kingfisher *Halcyon winchelli,* of the Philippines. — *Bull. Brit. Orn. Cl.*, 86: 82-86.
1758A   Parkes, K.C., 1966. A new subspecies of the Yellow-bellied Whistler *Pachycephala philippensis.* – *Bull. Brit. Orn. Cl.*, 86: 170-171.
1759   Parkes, K.C., 1967. A new subspecies of the Wattled Bulbul *Pycnonotus urostictus* of the Philippines. — *Bull. Brit. Orn. Cl.*, 87 (2): 23-25.
1760   Parkes, K.C., 1968. An undescribed subspecies of button-quail from the Philippines. — *Bull. Brit. Orn. Cl.*, 88: 24-25.
1761   Parkes, K.C., 1970. The races of the Rusty-breasted Nunlet (*Nonnula rubecula*). — *Bull. Brit. Orn. Cl.*, 90: 154-157.
1762   Parkes, K.C., 1970. A revision of the Philippine Trogon (*Harpactes ardens*). — *Nat. Hist. Bull., Siam Soc.*, 23: 345-352.
1763   Parkes, K.C., 1971. Taxonomic and distributional notes on Philippine birds. — *Nemouria,* 4: 1-67.
1764   Parkes, K.C., 1971. Two new parrots from the Philippines. — *Bull. Brit. Orn. Cl.*, 91: 96-98.
1765   Parkes, K.C., 1973. Annotated List of the birds of Leyte Island, Philippines. — *Nemouria,* 11: 1-73.
1766   Parkes, K.C., 1974. Geographic variation in the Flame Minivet (*Pericrocotus flammeus*) on the island of Mindanao, Philippines (Aves: Campephagidae). — Ann. Carnegie Mus., 45: 35-41.
1767   Parkes, K.C., 1974. Systematics of the White-throated Towhee (*Pipilo albicollis*). — *Condor,* 76: 457-459.
1768   Parkes, K.C., 1974. Variation in the Olive Sparrow in the Yucatan Peninsula. — *Wilson Bull.*, 86: 293-295.
1769   Parkes, K.C., 1976. The status of *Aratinga astec melloni* Twomey. — *Bull. Brit. Orn. Cl.*, 96 (1): 13-15.
1770   Parkes, K.C., 1977. An undescribed subspecies of the Red-legged Honeycreeper *Cyanerpes cyaneus.* — *Bull. Brit. Orn. Cl.*, 97 (2): 65-68.
1771   Parkes, K.C., 1979. A new northern subspecies of the Tropical Gnatcatcher *Polioptila plumbea.* — *Bull. Brit. Orn. Cl.*, 99: 72-75.
1772   Parkes, K.C., 1980. A new subspecies of the Spiny-cheeked Honeyeater *Acanthagenys rufogularis,* with note on generic relationships. — *Bull. Brit. Orn. Cl.*, 100: 143-147.
1773   Parkes, K.C., 1981. A substitute name for a Philippine minivet (*Pericrocotus*). — *Bull. Brit. Orn. Cl.*, 101: 370.
1774   Parkes, K.C., 1982. Parallel geographic variation in three *Myiarchus* flycatchers in the Yucatan Peninsula and adjacent areas (Aves: Tyrannidae). — *Ann. Carnegie Mus.*, 51 (1): 1-16.
1775   Parkes, K.C., 1987. Taxonomic notes on some African warblers (Aves: Sylviinae). — *Ann. Carnegie Mus.*, 56 (13): 231-243.
1776   Parkes, K.C., 1988. Three new subspecies of Philippine birds. — *Nemouria,* 30: 1-8.
1777   Parkes, K.C., 1989. Notes on the Menage collection of Philippine birds 1. Revision of *Pachycephala cinerea* (Pachycephalidae) and an overlooked subspecies of *Dicaeum trigonostigma* (Dicaeidae). — *Nemouria,* 33: 1-9.
1778   Parkes, K.C., 1990. A revision of the Mangrove Vireo (*Vireo pallens*) (Aves: Vireonidae). — *Ann. Carnegie Mus.*, 59: 49-60.
1779   Parkes, K.C., 1990. A critique of the description of *Amazona auropalliata caribaea* Lousada, 1989. — *Bull. Brit. Orn. Cl.*, 110 (4): 175-179.
1780   Parkes, K.C., 1993. The name of the Ecuadorean subspecies of the Chestnut-collared Swallow *Hirundo rufocollaris.* — *Bull. Brit. Orn. Cl.*, 113 (2): 119-120.
1781   Parkes, K.C., 1993. *Thraupis episcopus ehrenreichi* (Reichenow) is a valid subspecies. — *Bol. Mus. Para. Emilio Goeldi, Zool.,* 9 (2): 313-316.
1782   Parkes, K.C., 1994. Taxonomic notes on the White-collared Swift. — *Avocetta,* 17: 95-100.
1783   Parkes, K.C., 1997. The Northern Cardinals of the Caribbean slope of Mexico, with the description of an additional subspecies from Yucatan. Pages 129-138. In *The Era of Allan R. Phillips: A Festschrift* (R.W. Dickerman, ed.) — Albuquerque.
1784   Parkes, K.C., 1999. On the status of the Barred Woodcreeper *Dendrocolaptes certhia* in the Yucatan Peninsula. — *Bull. Brit. Orn. Cl.*, 119 (1): 65-68.
1785   Parkes, K.C. & D. Amadon, 1959. A new species of rail from the Philippine Islands. — *Wilson Bull.*, 71 (4): 303-306.
1786   Parkes, K.C. & D.M. Niles, 1988. Notes on Philippine birds, 12. An undescribed subspecies of *Centropus viridis.* — *Bull. Brit. Orn. Cl.*, 108 (4): 193-194.
1787   Parkes, K.C. & R.K. Panza, 1993. A new Amazonian subspecies of the Ruddy-tailed Flycatcher Myiobus (Terenotriccus) erythurus. — *Bull. Brit. Orn. Cl.*, 113 (1): 21-23.
1788   Parkes, K.C. & A.R. Phillips, 1978. Two new Caribbean subspecies of Barn Owl (*Tyto alba*), with remarks on variation in other populations. — *Ann. Carnegie Mus.*, 47: 479-492.
1789   Parkes, K.C. & A.R. Phillips, 1999. A new subspecies of the Northern Beardless-Tyrannulet *Camptostoma imberbe.* — *Bull. Brit. Orn. Cl.*, 119 (1): 59-62.
1790   Pasquet, E. *et al.*, 1999. Relationships between the ant-thrushes *Neocossyphus* and the flycatcher-thrushes *Stizorhina,* and their position relative to *Myadestes, Entomodestes* and some other Turdidae (Passeriformes). — *J. Zool. Syst. Evol. Res.,* 37 (4): 177-183.
1791   Pasquet, E. *et al.*, 2001. Towards a molecular systematics of the genus *Criniger* and a preliminary phylogeny of the bulbuls (Aves, Passeriformes, Pycnonotidae). — *Zoosystema,* 24 (4): 857-863.
1792   Pasquet, E. *et al.*, 2002. What are African monarchs? A phylogenetic analysis of mitochondrial genes. — *Comptes Rend. Acad. Sci. Paris,* 325: 1-12.
1793   Paterson, M.L., 1958. Two new races of larks from Bechuanaland Protectorate. — *Bull. Brit. Orn. Cl.*, 78: 125-162.
1794   Paton, T. *et al.*, 2002. Complete mitochondrial DNA genome sequences show that modern birds are not descended from transitional shorebirds. — *Proc. Roy. Soc., Lond.,* Ser. B. 269: 839-846.
1795   Payne, R.B., 1973. Behaviour, mimetic songs and song dialects, and relationships of the parasitic indigobirds (*Vidua*) of Africa. — *Orn. Monogr.,* 11: 1-333 (1972).
1796   Payne, R.B., 1979. Commentary on the Melba Finches *Pytilia melba* of Djibouti and the requirement of a specimen for a taxonomic description. — *Bull. Brit. Orn. Cl.*, 109 (2): 117-119.
1797   Payne, R.B., 1979. Order Ciconiiformes, family Ardeidae. Pages 193-244. In *Check-list of Birds of the World.* Vol. 1. (2nd. Edit.) (E. Mayr & G.W. Cottrell, eds.). — Mus. Comp. Zool., Cambridge, Mass.
1798   Payne, R.B., 1982. Species limits in the indigobirds (Ploceidae, *Vidua*) of West Africa: Mouth mimicry, song mimicry, and description of new species. — *Misc. Publ. Mus. Zool. Univ. Michigan,* 162: 1-96.

1799   Payne, R.B., 1985. The species of parasitic finches in West Africa. — *Malimbus*, 7: 103-113.
1800   Payne, R.B., 1997. Family Cuculidae (Cuckoos). Pages 508-607. In *Handbook of the Birds of the World*. Vol. 4. (J. del Hoyo, A. Elliott & J. Sargatal, eds.). — Lynx Edicions, Barcelona.
1801   Payne, R.B., 1998. A new species of firefinch *Lagnosticta* from northern Nigeria and its association with the Jos Plateau Indigobird *Vidua maryae*. — *Ibis*, 140 (3): 368-381.
1802   Payne, R.B. & M. Louette, 1983. What is *Lagonosticta umbrinodorsalis* Reichenow, 1910? — *Mitt. Zool. Mus. Berlin*, 59 (Ann. Orn. 7): 157-161.
1803   Payne, R.B. & L.L. Payne, 1994. Song mimicry and species associations of West African indigobirds *Vidua* with Quail-finch *Ortygospiza atricollis*, Goldbreast *Amadana subflava* and Brown Twinspot *Clytospiza montieri*. — *Ibis*, 136 (3): 291-304.
1804   Payne, R.B. *et al.*, 1992. Song mimicry and species status of the green widowfinch *Vidua codringtoni*. — *Ostrich*, 63: 86-97.
1804A   Payne, R.B. *et al.*, 2002. Behaviour and genetic evidence of a recent population switch to a novel host species in brood-parasitic indigobirds *Vidua chalybeata*. — *Ibis*, 144: 373-383.
1805   Paynter, R.A., Jr., 1950. A new Clapper Rail from the territory of Quintana Roo, Mexico. — *Condor*, 52 (3): 139-140.
1806   Paynter, R.A., Jr., 1954. Three new birds from the Yucatan Peninsula. — *Postilla*, 18: 1-4.
1807   Paynter, R.A., Jr., 1964. Family Muscicapidae, Subfamily Polioptilinae. Pages 443-455. In *Check-list of Birds of the World*. Vol. 10 (E. Mayr & R.A. Paynter, Jr., eds.). — Mus. Comp. Zool., Harvard, Cambridge, Mass.
1808   Paynter, R.A., Jr., 1970. Family Emberizidae, Subfamily Emberizinae. Pages 3-214. In *Check-list of Birds of the World*. Vol. 13 (R.A. Paynter, Jr., ed.). — Mus. Comp. Zool., Cambridge, Mass.
1809   Paynter, R.A., Jr., 1970. Family Emberizidae, Subfamily Cardinalinae. Pages 216-245. In *Check-list of Birds of the World*. Vol. 13 (R.A. Paynter, Jr., ed.). — Mus. Comp. Zool., Cambridge, Mass.
1810   Paynter, R.A., Jr. & C. Vaurie, 1960. Family Troglodytidae. Pages 379-440. In *Check-list of Birds of the World*. Vol. 9 (E. Mayr & J.C. Greenway, Jr., eds.). — Mus. Comp. Zool., Cambridge, Mass.
1811   Pearson, D.J., 1992. Various species accounts. In *The Birds of Africa*. Vol. 4: i-xv, 1-609 (S. Keith, E.K. Urban & C.H. Fry, eds.). — Academic Press Inc., San Diego.
1812   Pearson, D.J., 1997. Various species accounts. In *The Birds of Africa*. Vol. 5: i-xix, 1-669 (E.K. Urban, C.H. Fry & S. Keith, eds.). — Academic Press, London.
1813   Pearson, D.J., 2000. The races of the Isabelline Shrike *Lanius isabellinus* and their nomenclature. — *Bull. Brit. Orn. Cl.*, 120 (1): 22-27.
1814   Pearson, D.J., 2000. Various species accounts. In *The Birds of Africa*. Vol. 6: i-xvii, 1-724 (C.H. Fry, S. Keith & E.K. Urban, eds.). — Academic Press, London.
1815   Pearson, D.J. & J.S. Ash, 1996. The taxonomic position of the Somali courser *Cursorius* (*cursor*) *somalensis*. — *Bull. Brit. Orn. Cl.*, 116 (4): 225-229.
1816   Pearson, D.J. & S. Keith, 1992. Campephagidae, cuckoo-shrikes. Pages 263-278. In *The Birds of Africa*. Vol. 4. (S. Keith, E.K. Urban & C.H. Fry, eds.). — Academic Press, San Diego.
1817   Pecotich, L., 1982. Speciation of the Grey Swiftlet *Aerodramus spodiopygius* in Australia. — *Tower Karst*, 4: 53-57.
1818   Penhallurick, J., 2001. *Primolius* Bonaparte, 1857, has priority over *Propyrrhura* Ribiero, 1920. — *Bull. Brit. Orn. Cl.*, 121 (1): 38-39.
1819   Pereyra, J.A., 1944. Descripcion de un nuevo ejemplar de ralido de la isla Georgica del Sud. — *Hornero*, 8 (3): 484-489.
1820   Peters, D.S., 1996. *Hypositta perdita*, n.sp., eine neue Vogelart aus Madagaskar (Aves : Passeriformes : Vangidae). — *Senckenbergiana biologica*, 76 (1/2): 7-14.
1821   Peters, J.L., 1931. *Check-list of Birds of the World*: 1. i-xviii, 1-345. 1st edition. — Harvard University Press, Cambridge, Mass.
1822   Peters, J.L., 1934. *Check-list of Birds of the World*: 2. i-xvii, 1-401. — Harvard University Press, Cambridge, Mass.
1823   Peters, J.L., 1937. *Check-list of Birds of the World*: 3. i-xiii, 1-311. — Harvard University Press, Cambridge, Mass.
1824   Peters, J.L., 1939. Collection from the Philippine Islands: Birds. — *Bull. Mus. Comp. Zool.*, 86 (2): 74-122.
1825   Peters, J.L., 1940. *Check-list of Birds of the World*: 4. i-xii, 1-291. — Harvard University Press, Cambridge, Mass.
1826   Peters, J.L., 1945. *Check-list of Birds of the World*: 5. i-xi, 1-306. — Harvard University Press, Cambridge, Mass.
1827   Peters, J.L., 1948. *Check-list of Birds of the World*: 6. i-xi, 1-259. — Harvard University Press, Cambridge, Mass.
1828   Peters, J.L., 1951. *Check-list of Birds of the World*: 7. i-x, 1-318. — Mus. Comp. Zool., Cambridge, Mass.
1829   Peters, J.L., 1960. Family Alaudidae. Pages 3-80. In *Check-list of Birds of the World*. Vol. 9 (E. Mayr & J.C. Greenway, Jr., eds.). — Mus. Comp. Zool., Cambridge, Mass.
1830   Peters, J.L., 1960. Family Hirundinidae. Pages 80-129. In *Check-list of Birds of the World*. Vol. 9. (E. Mayr & J.C. Greenway, Jr., eds.). — Mus. Comp. Zool., Cambridge, Mass.
1831   Peters, J.L. & J.A. Griswold, Jr., 1943. Birds of the Harvard Peruvian Expedition. — *Bull. Mus. Comp. Zool.*, 92: 279-327.
1832   Peters, J.L. *et al.*, 1960. Family Campephagidae. Pages 167-221. In *Check-list of Birds of the World*. Vol. 9 (E. Mayr & J.C. Greenway, Jr., eds.). — Mus. Comp. Zool., Cambridge, Mass.
1833   Peterson, A.T., 1992. Phylogeny and rates of molecular evolution in the *Aphelocoma* jays (Corvidae). — *Auk*, 109: 133-147.
1834   Peterson, A.T., 1993. Species status of *Geotrygon carrikeri*. — *Bull. Brit. Orn. Cl.*, 113 (3): 166-168.
1834A   Peterson, A.T., 1998. New species and new species limits in birds. — *Auk*, 115 (3): 555-558.
1834B   Peterson, A.T. & A.G. Navarro-Siguenza, 2000. A new taxon in the "*Amazilia viridifrons*" complex of southern Mexico. — *Proc. Biol. Soc. Wash.*, 113 (864-870).
1835   Phelps, W.H., Jr., 1977. Aves colectadas en las Mesetas de Sarisariñama y Jaua durante tres expediciones al Macizo de Jaua, Estado Bolivar. Descripciones de dos nuevas subespecies. — *Bol. Soc. Venez. Cienc. Nat.*, 33 (134): 15-42.
1836   Phelps, W.H., Jr., 1977. Una nueva especie y dos nuevas subespecies de aves (Psittacidae, Furnariidae) de la Sierra de Perija cerca de la divisoria Colombo-Venezolana. — *Bol. Soc. Venez. Cienc. Nat.*, 33 (134): 43-53.
1837   Phelps, W.H., Jr. & H.R. Aveledo, 1966. A new subspecies of *Icterus icterus* and other notes on the birds of northern South America. — *Amer. Mus. Novit.*, 2270: 1-14.
1838   Phelps, W.H., Jr. & R. Aveledo, 1984. Dos nuevas subespecies de aves (Troglodytidae, Fringillidae) del Cerro Marahuaca, Territorio Amazonas. — *Bol. Soc. Venez. Cienc. Nat.*, 39 (142): 5-10.
1839   Phelps, W.H., Jr. & R. Aveledo, 1987. Cinco nuevas subespecies de aves (Rallidae, Trochilidae, Picidae, Furnariidae) y tres extensiones de distribucion para Venezuela. — *Bol. Soc. Venez. Cienc. Nat.*, 41 (144): 7-26.
1840   Phelps, W.H., Jr. & H.R. Aveledo, 1988. Una nueva subespecie de aves de la familia (Trochilidae) de la Serrania Tapirapeco, Territorio Amazonas, Venezuela. — *Bol. Soc. Venez. Cienc. Nat.*, 42 (145): 7-10.
1841   Phelps, W.H., Jr. & R. Dickerman, 1980. Cuatro subespecies nuevas de aves (Furnariidae, Formicariidae) de la region de Pantepui, Estado Bolivar y Territorio Amazonas, Venezuela. — *Bol. Soc. Venez. Cienc. Nat.*, 33 (138): 139-147.
1842   Phelps, W.H. & E.T. Gilliard, 1940. Six new birds from the Perija Mountains of Venezuela. — *Amer. Mus. Novit.*, 1100: 1-8.
1843   Phelps, W.H. & W.H. Phelps, Jr., 1947. Ten new subspecies of birds from Venezuela. — *Proc. Biol. Soc. Wash.*, 60: 149-163.
1844   Phelps, W.H. & W.H. Phelps, Jr., 1948. Descripción de seis aves nuevas de Venezuela y notas sobre vienticuatro adiciones a la avifauna del Brasil. — Bol. Soc. Venez. Cienc. Nat., 11 (71): 53-74 (1947).
1845   Phelps, W.H. & W.H. Phelps, Jr., 1949. Eleven new subspecies of birds from Venezuela. — *Proc. Biol. Soc. Wash.*, 62: 109-124.
1846   Phelps, W.H. & W.H. Phelps, Jr., 1949. Seven new subspecies of birds from Venezuela. — *Proc. Biol. Soc. Wash.*, 62: 185-196.
1847   Phelps, W.H. & W.H. Phelps, Jr., 1950. Three new subspecies of birds from Venezuela. — *Proc. Biol. Soc. Wash.*, 63: 43-49.
1848   Phelps, W.H. & W.H. Phelps, Jr., 1950. Seven new subspecies of Venezuelan birds. — *Proc. Biol. Soc. Wash.*, 63: 115-126.

1849   Phelps, W.H. & W.H. Phelps, Jr., 1951. Four new Venezuelan birds. — *Proc. Biol. Soc. Wash.*, 64: 65-72.

1850   Phelps, W.H. & W.H. Phelps, Jr., 1952. Nine new subspecies of birds from Venezuela. — *Proc. Biol. Soc. Wash.*, 65: 39-54.

1851   Phelps, W.H. & W.H. Phelps, Jr., 1952. Nine new birds from the Perija Mountains and eleven extensions of ranges to Venezuela. — *Proc. Biol. Soc. Wash.*, 65: 89-108.

1852   Phelps, W.H. & W.H. Phelps, Jr., 1953. Eight new birds and thirty-three extensions of ranges to Venezuela. — *Proc. Biol. Soc. Wash.*, 66: 125-144.

1853   Phelps, W.H. & W.H. Phelps, Jr., 1953. Eight new subspecies of birds from the Perija Mountains, Venezuela. — *Proc. Biol. Soc. Wash.*, 66: 1-12.

1854   Phelps, W.H. & W.H. Phelps, Jr., 1954. Notes on Venezuelan birds and descriptions of six new subspecies. — *Proc. Biol. Soc. Wash.*, 67: 103-114.

1855   Phelps, W.H. & W.H. Phelps, Jr., 1955. Five new Venezuelan birds and nine extensions of ranges to Colombia. — *Proc. Biol. Soc. Wash.*, 68: 47-58.

1856   Phelps, W.H. & W.H. Phelps, Jr., 1955. Seven new birds from Cerro de la Neblina, Territorio Amazonas, Venezuela. — *Proc. Biol. Soc. Wash.*, 68: 113-123.

1857   Phelps, W.H. & W.H. Phelps, Jr., 1956. Five new birds from Rio Chiquito, Tachira, Venezuela and two extensions of ranges from Colombia. — *Proc. Biol. Soc. Wash.*, 69: 157-166.

1858   Phelps, W.H. & W.H. Phelps, Jr., 1956. Three new birds from Cerro El Teteo, Venezuela, and extensions of ranges to Venezuela and Colombia. — *Proc. Biol. Soc. Wash.*, 69: 127-134.

1859   Phelps, W.H. & W.H. Phelps, Jr., 1958. Descriptions of two new Venezuelan birds and distributional notes. — *Proc. Biol. Soc. Wash.*, 71: 119-124.

1860   Phelps, W.H. & W.H. Phelps, Jr., 1958. Lista de las Aves de Venezuela con su Distribución. Tomo 2 (Parte 1) No Passeriformes. — *Bol. Soc. Venez. Cienc. Nat.*, 19: 1-317.

1861   Phelps, W.H. & W.H. Phelps, Jr., 1959. Two new subspecies of birds from the San Luis mountains of Venezuela and distributional notes. — *Proc. Biol. Soc. Wash.*, 72: 121-126.

1862   Phelps, W.H. & W.H. Phelps, Jr., 1960. A new subspecies of Furnariidae from Venezuela and extension of ranges. — *Proc. Biol. Soc. Wash.*, 73: 1-3.

1863   Phelps, W.H. & W.H. Phelps, Jr., 1961. Notes on Venezuelan birds and description of a new subspecies of Trochilidae. — *Proc. Biol. Soc. Wash.*, 74: 3-6.

1864   Phelps, W.H. & W.H. Phelps, Jr., 1962. Two new subspecies of birds from Venezuela, the rufous phase of *Pauxi pauxi*, and other notes. — *Proc. Biol. Soc. Wash.*, 75: 199-204.

1865   Phelps, W.H. & W.H. Phelps, Jr., 1963. Lista de las Aves de Venezuela con su distribucion. Tomo I. Parte II. Passeriformes. (2nd. Ed.). — *Bol. Soc. Venez. Cienc. Nat.*, 24 (104/105): 1-479.

1866   Phillips, A.R., 1947. The Button Quails and Tree Sparrows of the Riu Kiu Islands. — *Auk*, 64 (1): 126-127.

1867   Phillips, A.R., 1959. Las subespecies de la Cordoniz de Gambel y el problema de los cambios climaticos en Sonora. — *Anal. Inst. Biol. Univ. Mex.*, 29 (1/2): 361-374.

1868   Phillips, A.R., 1962. Notas sistematicas sobre aves Mexicanas. I. — *Anales del Instituto de Biología Mexicana*, 32 (1/2): 333-381 (1961).

1869   Phillips, A.R., 1966. Further systematic notes on Mexican birds. — *Bull. Brit. Orn. Cl.*, 86: 86-94.

1870   Phillips, A.R., 1966. Further systematic notes on Mexican birds. — *Bull. Brit. Orn. Cl.*, 86: 103-112.

1871   Phillips, A.R., 1966. Further systematic notes on Mexican birds. — *Bull. Brit. Orn. Cl.*, 86: 125-131.

1872   Phillips, A.R., 1966. Notas sistematicas sobre aves Mexicanas. III. — *Rev. Soc. mex. Hist. Nat.*, 25: 217-242 (1964).

1873   Phillips, A.R., 1969. An ornithological comedy of errors: *Catharus occidentalis* and *C. frantzii*. — *Auk*, 86: 605-623.

1874   Phillips, A.R., 1970. A northern race of lark supposedly breeding in Mexico. — *Bull. Brit. Orn. Cl.*, 90: 115-116.

1875   Phillips, A.R., 1986. *The Known Birds of North and Middle America. (Distributions and variation, migrations, changes, hybrids, etc. Part 1. Hirundinidae to Mimidae; Certhiidae)*: 1. i-lxi, 1-259. – Privately published, Denver, Colorado.

1876   Phillips, A.R., 1991. *The Known Birds of North and Middle America. (Distributions and variation, migrations, changes, hybrids, etc. Part II Bombycillidae; Sylviidae to Sturnidae; Vireonidae)*: 2. i-liv, 1-249. – Privately published, Denver, Colorado.

1877   Phillips, A.R., 1994. A review of the northern *Pheucticus* grosbeaks. — *Bull. Brit. Orn. Cl.*, 114 (3): 162-170.

1878   Phillips, A.R., 1995. The northern races of *Icterus pustulatus* (Icteridae), Scarlet-headed or Streaked-backed Oriole. — *Bull. Brit. Orn. Cl.*, 115 (2): 98-105.

1879   Phillips, A.R. & W. Rook, 1965. A new race of the Spotted Nightingale-Thrush from Oaxaca, Mexico. — *Condor*, 67: 3-5.

1880   Phillips, W.W.A., 1949. A new race of the Common Hawk-Cuckoo from Ceylon. — *Bull. Brit. Orn. Cl.*, 69: 56-57.

1881   Phillips, W.W.A., 1978. *Annotated Checklist of the birds of Ceylon*: i-xix, 1-93. — The Wildlife and Nature Protection Society of Sri Lanka, Colombo.

1882   Phillips, W.W.A. & R.W. Sims, 1958. Two new races of birds from the Maldive Archipelago. — *Bull. Brit. Orn. Cl.*, 78: 51-53.

1883   Pierce, R.J., 1996. Family Recurvirostridae (Stilts and Avocets). Pages 332-347. In *Handbook of the Birds of the World*. Vol. 3 (J. del Hoyo, A. Elliott & J. Sargatal, eds.). — Lynx Edicions, Barcelona.

1884   Piersma, T., 1996. Family Charadriidae (Plovers). Pages 384-409. In *Handbook of the Birds of the World*. Vol. 3 (J. del Hoyo, A. Elliott & J. Sargatal, eds.). — Lynx Edicions, Barcelona.

1885   Pierpont, N. & J.W. Fitzpatrick, 1983. Specific status and behaviour of *Cymbilaimus sanctaemariae*, the Bamboo Antshrike, from southwestern Amazonia. — *Auk*, 100 (3): 645-652.

1886   Pinchon, R., 1963. Une sous-espèce nouvelle d'Engoulevent à la Martinique: *Stenopsis cayennensis manati* subsp. nov. — *L'Oiseau*, 33 (2): 107-110.

1887   Pinto, O., 1938. Nova contribuição a ornithologia Amazonica. Estudo critico de uma collecção de Aves do baixo Solimões e do alto Rio Negro. — *Rev. Mus. Paul.*, 23 (1937): 495-604.

1888   Pinto, O., 1950. Miscelanea ornitologica. V. Descriçao de uma subespecie nordestina em *Synallaxis ruficapilla* Vieillot (Fam. Furnariidae). — *Pap. Avul. Dep. Zool., Sao Paulo*, 9 (24): 361-365.

1889   Pinto, O., 1950. Da classifição e nomenclatura dos surucuás Brasileiros (Trogonidae). — *Pap. Avul. Dep. Zool., Sao Paulo*, 9 (9): 89-136.

1890   Pinto, O., 1954. Resultados ornitologicas de duas viagens cientificas ao estado de Alagoes. — *Pap. Dep. Zool. Sec. Agric., S. Paulo*, 12: 1-98.

1891   Pinto, O., 1960. Algunas adendas à avifauna Brasiliera. — *Pap. Dep. Zool. Sec. Agric., S. Paulo*, 14 (2): 11-15.

1892   Pinto, O., 1962. Miscelânea ornitológica. VII. Notas sobre a variação geográfica nas populaçoes Brasilieras de *Neomorphus geoffroyi*, com a descrição de uma subespecie nova. — *Pap. Avul. Dep. Zool., Sao Paulo*, 15 (22): 299-301.

1893   Pinto, O., 1962. Miscelânea ornitológica. VIII. Nome novo para as populaçoes este-brasilieras de *Pionus menstruus* Linn. (Aves, Psittacidae). — *Pap. Avul. Dep. Zool., Sao Paulo*, 15 (22): 301.

1894   Pinto, O., 1974. Miscelânea ornitológica. IX. Duas subespécies novas de Furnariidae brasileiras. — *Pap. Avul. Dep. Zool., Sao Paulo*, 27 (14): 177-178.

1895   Pinto, O., 1978. *Novo catálogo das aves do Brasil. Primeira parte. Aves não Passeriformes e Passeriformes não Oscines, com exclusão da familia Tyrannidae*: i-xvi, 1-446. — Empresa Gráfica da Revista dos Tribunais, São Paulo.

1896   Pinto, O. & E.A. de Camargo, 1952. Nova contribuição à ornitologia do Rio das Mortes. Restulados do Expedição conjunta de Instituta Bantantan e Departamentida Zoologica. — *Pap. Avul. Dep. Zool., Sao Paulo*, 10 (11): 213-234.

1897   Pinto, O. & E.A. de Camargo, 1955. Lista anotada de aves colecionadas nos limites occidentais do Estado do Parana. — *Pap. Dep. Zool. Sec. Agric., S. Paulo*, 12: 215-234.

1898   Pinto, O. & E.A. de Camargo, 1957. Sobre uma colecao de aces da regiao e cachimbo (Sul do Estado do Para). — *Pap. Avul. Dep. Zool., Sao Paulo,* 13 (4): 51-69.

1899   Pinto, O. & E.A. de Camargo, 1961. Resultados ornitologicos de quatro recentes e pediçães do Departemento de Zoologia ao Nordeste do Brasil, com a descrição de seis novas subespecies. — *Arch. Zool. S. Paulo,* 11: 193-284.

1900   Pitelka, F.A., 1950. Geographic variation and the species problem in the shorebird genus *Limnodromus.* — *Univ. Calif. Publ. Zool.,* 50 (1): 1-108.

1901   Pitman, R.L. & J.R. Jehl, 1998. Geographic variation and reassessment of species limits in the 'Masked' Boobies of the eastern Pacific Ocean. — *Wilson Bull.,* 110 (2): 155-170.

1901A   Pittie, A. 2001. *A bibliographic index to the birds of the Indian Subcontinent.* CD-ROM. Privately published. Enquiries to aasheesh@vsnl.in

1902   Portenko, L.A., 1936. The bar-tailed godwit and its races. — *Auk,* 53: 194-197.

1903   Portenko, L.A., 1937. Einige neue Unterarten palaearktischer Vögel. — *Mitt. Zool. Mus. Berlin,* 22: 219-229.

1904   Portenko, L.A., 1939. *Fauna Anadyrskogo Kraya. Ptitsy (Fauna of the Anadyr Territory. Birds) Part 1:* i-v, 1-211. — Glavsevmorput' Publ, Leningrad.

1905   Portenko, L.A., 1939. On some new forms of Arctic Gulls. — *Ibis,* 14 (3): 264-269.

1906   Portenko, L.A., 1944. Novye formy ptits s ostrova Vrangelya. [New forms of birds from Wrangel Island]. — *Dokl. Akad. Nauk. SSSR,* 43 (5): 237-240.

1907   Portenko, L.A., 1954. *Birds of the USSR:* 3. 1-254. — Moscow. [In Russian.]

1908   Portenko, L.A., 1960. *Birds of the USSR:* 4. 1-414. — Moscow. [In Russian.]

1909   Portenko, L.A., 1963. The taxonomic value and systematic status of the slaty-backed gull (*Larus argentatus schistisagus* Stejn.). Pages 61-64. In *Trudy Kamchatskoi komplex. Exsped. Akad. Nauk S.S.S.R.*

1910   Portenko, L.A., 1972. *Birds Tschukotsk Peninsula and Wrangel Island:* 1. 1-424. Nauka Leningrad.

1911   Portenko, L.A., 1981. Geographical variation in Dark-throated Thrushes (*Turdus ruficollis* Pallas) and its taxonomical value. — *Trudy Zool. Inst. Leningrad,* 102: 72-109.

1912   Potapov, R.L., 1985. Otryad kuroobraznie (Galliformes). Chast' 2. Semeistvo teterevinie (Tetraonidae) (Order Galliformes, Pt. 2. The grouse family, Tetraonidae). — *Fauna SSSR, No. 133,* Vol. 3 (1): 637.

1913   Potapov, R.L., 1993. New subspecies of the Himalayan Snowcock, *Tetraogallus himalayensis sauricus* subsp. nova. — *Russ. J. Orn.,* 2 (1): 3-5.

1914   Powers, D.R., 1999. Species accounts within the Family Trochilidae (Hummingbirds). Pages 549-680. In *Handbook of the Birds of the World.* Vol. 5 (J. del Hoyo, A. Elliott & J. Sargatal, eds.). — Lynx Edicions, Barcelona.

1915   Pratt, H.D., 1979. A systematic analysis of the endemic avifauna of the Hawaiian Islands. The Louisiana State Univ. and Agricult. and Mech. Coll., Zoology, Ph. D. thesis (Dissert. Abstracts, 40 B: 1581) 1-245.

1916   Pratt, H.D., 1988. A new name for the eastern subspecies of the Brown-backed Solitaire *Myadestes occidentalis.* — *Bull. Brit. Orn. Cl.,* 108: 135-136.

1917   Pratt, H.D., 1989. A new name for the Kauai Amakihi (Drepanidinae: *Hemignathus*). — *Elepaio,* 49 (3): 13-14.

1918   Pratt, H.D., 1989. Species limits in the akepas (Drepanidinae: *Loxops*). — *Condor,* 91: 933-940.

1919   Pratt, H.D., 1992. Is the Poo-uli a Hawaiian honeycreeper (Drepanidinae)? — *Condor,* 94: 172-180.

1920   Pratt, H.D. *et al.,* 1987. *A Field Guide to the birds of Hawaii and the Tropical Pacific:* i-xx, 1-409. — Princeton Univ. Press, Princeton, N. J.

1921   Pratt, T.K., 1983. Additions to the avifauna of the Adelbert Range, Papua New Guinea. — *Emu,* 82 (3): 117-125 (1982).

1922   Pratt, T.K., 2000. Evidence for a previously unrecognised species of owlet-nightjar. — *Auk,* 117 (1): 1-11.

1923   Preleuthner, M. & A. Gamauf, 1998. A possible new subspecies of the Philippine Hawk-Eagle (*Spizaëtus philippensis*) and its future prospects. — *J. Raptor Res.,* 32 (2): 126-135.

1924   Prescott, K.W., 1970. A new subspecies of the Common Iora from north Borneo. — *Bull. Brit. Orn. Cl.,* 90: 39-40.

1925   Preston, C.R. & R.D. Beane, 1993. Red-tailed Hawk (*Buteo jamaicensis*) No. 52. Pages 1-24. In T*he Birds of North America* (A. Poole & F. Gill, eds.). — Acad. Nat. Sci., Philadelphia.

1926   Prevost, Y.A., 1983. Osprey distribution and subspecies taxonomy. In *1st. Internat. Symp. on Bald Eagles and Ospreys.* Macdonald Raptor Res. Center, McGill Univ., Montreal. 1981: i-x, 1-325.

1927   Prigogine, A., 1957. Trois nouveaux oiseaux de l'est du Congo belge. — *Rev. Zool. Bot. Afr.,* 55 (1/2): 39-46.

1928   Prigogine, A., 1960. Un nouveau Martinet du Congo. — *Rev. Zool. Bot. Afr.,* 62 (1/2): 103-105.

1929   Prigogine, A., 1964. Un nouvel oiseau de la république du Congo. — *Rev. Zool. Bot. Afr.,* 70: 401-404.

1930   Prigogine, A, 1969. Trois nouveaux Oiseaux du Katanga. Republique democratique du Congo. — *Rev. Zool. Bot. Afr.,* 79: 110-116.

1931   Prigogine, A., 1972. Description of a new green bulbul from the Republic of Zaire. — *Bull. Brit. Orn. Cl.,* 92: 138-141.

1932   Prigogine, A., 1975. Etude taxonomique des populations de *Phyllastrephus flavostriatus* de l'Afrique Centrale et description de deux nouvelles races de la République du Zaïre. — *Gerfaut,* 63: 219-234 (1973).

1933   Prigogine, A., 1975. Etude taxonomique de *Nectarinia alinae* et description de trois nouvelles formes de la République du Zaire. — *Rev. Zool. Afr.,* 89: 455-480.

1934   Prigogine, A., 1975. Les populations de *Estrilda atricapilla* (Verreaux) de l'Afrique centrale et description d'une nouvelle race (Aves). — *Rev. Zool. Afr.,* 89: 600-617.

1935   Prigogine, A., 1977. Populations of the Scarlet-tufted Malachite Sunbird, *Nectarinia johnstoni* Shelley, in Central Africa and description of a new subspecies from the Republic of Zaire. — *Mitt. Zool. Mus. Berlin,* 53 (Suppl. Ann. Orn., 1): 117-125.

1936   Prigogine, A., 1978. A new ground-thrush from Africa. — *Gerfaut,* 68: 482-492.

1937   Prigogine, A., 1980. Etude de quelques contacts secondaires au Zaïre oriental. — *Gerfaut,* 70: 305-384.

1938   Prigogine, A., 1982. The status of *Anthus latistriatus* Jackson, and the description of a new subspecies of *Anthus cinnamomeus* from Itombwe. — *Gerfaut,* 71: 537-573 (1981).

1939   Prigogine, A., 1983. Un nouveau *Glaucidium* de l'Afrique centrale (Aves, Strigidae). — *Rev. Zool. Afr.,* 97 (4): 886-895.

1940   Prigogine, A., 1984. L'Alethe à poitrine brune, *Alethe poliocephala,* au Rwanda et au Burundi. — *Gerfaut,* 74: 181-184.

1941   Prigogine, A., 1985. Les populations occidentales de la Grive terrestre d'Abyssinie, *Zoothera piaggiae,* et description d'une nouvelle sous-espèce du Ruwenzori. — *Gerfaut,* 74: 383-389 (1984).

1942   Prigogine, A., 1985. Statut de quelques chevêchettes africaines et description d'une nouvelle race de *Glaucidium scheffleri* du Zaïre. — *Gerfaut,* 75: 131-139.

1943   Prigogine, A., 1985. Recently recognised bird species in the Afrotropical region — a critical review. In *Proc. Int. Symp. Afr. Vertebr. Mus. A. Koenig, Bonn* (ed. K.-L. Schuchmann). 1984: 91-114.

1944   Prigogine, A., 1987. Non-conspecificity of *Cossypha insulana* Grote and *Cossypha bocagei* Finsch & Hartlaub, with the description of a new subspecies of *Cossypha bocagei* from western Tanzania. — *Bull. Brit. Orn. Cl.,* 107: 49-55.

1945   Prigogine, A., 1988. Une nouvelle sous-espèce du Cossyphe du Natal, *Cossypha natalensis,* du nord-est de la région Afrotropicale. — *Gerfaut,* 77: 477-484 (1987).

1946   Prigogine, A. & M. Louette, 1983. Contacts secondaires entres les taxons appartenant à la super-espèce *Dendropicos goertae.* — *Gerfaut,* 73: 9-83.

1947   Prigogine, A. & M. Louette, 1984. A new race of Spotted Ground Thrush, *Zoothera guttata,* from Upemba, Zaire. — *Gerfaut,* 74: 185-186.

1948   Priolo, A., 1984. Variabilita' in *Alectoris graeca* e descrizione di *A. graeca orlandoi* subsp. nova degli Appennini. — *Riv. Ital. Orn.,* 54: 45-76.

1949   Prum, R.O., 1988. Phylogenetic interrelationships of the barbets (Aves: Capitonidae) and toucans (Aves: Ramphastidae) based on morphology with comparisons to DNA-DNA hybridization. — *Zool. J. Linn. Soc. Lond.,* 92 (4): 313-343.

1950   Prum, R.O., 1990. Phylogenetic analysis of the evolution of display behaviour in the Neotropical manakins (Aves: Pipridae). — *Ethology*, 84 (3): 202-231.

1951   Prum, R.O., 1992. Syringeal morphology, phylogeny and evolution of the Neotropical manakins (Aves: Pipridae). — *Amer. Mus. Novit.*, 3043: 1-65.

1952   Prum, R.O., 1993. Phylogeny, biogeography and evolution of the broadbills (Eurylaimidae) and asities (Philepittidae) based on morphology. — *Auk*, 110 (2): 304-324.

1953   Prum, R.O., 1994. Phylogenetic analysis of the evolution of alternative social behavior in the manakins (Aves: Pipridae). — *Evolution*, 48 (5): 1657-1675.

1954   Prum, R.O., 1994. Species status of the White-fronted Manakin, *Lepidothrix serena* (Pipridae), with comments on conservation biology. — *Condor*, 96 (3): 692-702.

1955   Prum, R.O., 2001. A new genus for the Andean Green Pihas (Cotingidae). — *Ibis*, 143 (2): 307-309.

1956   Prum, R.O. & W.E. Lanyon, 1989. Monophyly and phylogeny of the *Schiffornis* group (Tyrannoidea). — *Condor*, 91: 444-461.

1957   Prum, R.O. *et al.*, 2000. A preliminary phylogenetic hypothesis for the *Cotinga*s (Cotingidae) based on mitochondrial DNA. — *Auk*, 117 (1): 236-241.

1958   Quickelberge, C.D., 1967. The racial taxonomy of South-east African populations of the Long-billed Lark. — *Ann. Cape Prov. Mus.*, 6 (3): 39-46.

1959   Quickelberge, C.D., 1967. A systematic revision of the Tchagra Shrike. — *Ann. Cape Prov. Mus.*, 6: 47-54.

1959A   Raikow, R.J., 1982. Monophyly of the Passeriformes: test of a phylogenetic hypothesis. – *Auk*, 99: 431-445.

1960   Raitt, R.J., 1967. Relationships between black-eared and plain-eared forms of bushtits (*Psaltriparus*). — *Auk*, 84: 503-528.

1961   Rand, A.L., 1967. Family Nectariniidae. Pages 208-289. In *Check-list of Birds of the World*. Vol. 12 (R.A. Paynter, Jr., ed.). — Mus. Comp. Zool., Cambridge, Mass.

1962   Rand, A.L., 1938. Results of the Archbold Expeditions No. 19. On some non-passerine New Guinea birds. — *Amer. Mus. Novit.*, 990: 1-15.

1963   Rand, A.L., 1940. Results of the Archbold Expeditions No. 25. New birds from the 1938-1939 Expedition. — *Amer. Mus. Novit.*, 1072: 1-14.

1964   Rand, A.L., 1941. Results of the Archbold Expeditions No. 32. New and interesting birds from New Guinea. — *Amer. Mus. Novit.*, 1102: 1-15.

1965   Rand, A.L., 1941. Results of the Archbold Expeditions No. 33. A new race of quail from New Guinea: with notes on the origin of the grassland avifauna. — *Amer. Mus. Novit.*, 1122: 1-2.

1966   Rand, A.L., 1942. Results of the Archbold Expeditions No. 43. Birds of the 1938-1939 New Guinea Expedition. — *Bull. Amer. Mus. Nat. Hist.*, 79 (7): 425-515.

1967   Rand, A.L., 1948. Five new birds from the Philippines. — *Fieldiana, Zool.*, 31 (25): 201-205.

1968   Rand, A.L., 1949. The races of the African Wood-Dove *Turtur afer*. — *Fieldiana, Zool.*, 31: 307-312.

1969   Rand, A.L., 1950. A new race of owl, *Otus bakkamoena*, from Negros, Philippine Islands. — *Nat. Hist. Misc., Chicago Acad. Sci.*, 72: 1-5.

1970   Rand, A.L., 1953. A new barbet from French Indochina. — *Fieldiana, Zool.*, 34 (21): 211-212.

1971   Rand, A.L., 1960. Family Laniidae. Pages 309-365. In *Check-list of Birds of the World*. Vol. 9. (E. Mayr & J.C. Greenway, Jr., eds.). — Mus. Comp. Zool., Cambridge, Mass.

1971A   Rand, A.L., 1967. Family Nectariniidae. Pages 208-289. In *Check-list of Birds of the World*. Vol. 12. (R.A. Paynter, Jr., ed.). — Mus. Comp. Zool., Cambridge, Mass.

1972   Rand, A.L. & H.G. Deignan, 1960. Family Pycnonotidae. Pages 221-300. In *Check-list of Birds of the World*. Vol. 9 (E. Mayr & J.C. Greenway, Jr., eds.). — Mus. Comp. Zool., Cambridge, Mass.

1973   Rand, A.L. & R.L. Fleming, 1953. A new fruit pigeon from Nepal. — *Fieldiana, Zool.*, 34 (19): 201-202.

1974   Rand, A.L. & R.L. Fleming, 1956. Two new birds from Nepal. — *Fieldiana, Zool.*, 39 (1): 1-3.

1975   Rand, A.L. & E.T. Gilliard, 1967. *Handbook of New Guinea birds*: i-x, 1-612. — Weidenfeld & Nicolson, London.

1976   Rand, A.L. & D.S. Rabor, 1959. Three new birds from the Philippine Islands. — *Fieldiana, Zool.*, 39 (26): 275-277.

1977   Rand, A.L. & D.S. Rabor, 1960. Birds of the Philippine Islands: Siquijor, Mt. Malindang, Bohol and Samar. — *Fieldiana, Zool.*, 35 (7): 221-441.

1978   Rand, A.L. & D.S. Rabor, 1967. New birds from Luzon, Philippine Islands. — *Fieldiana, Zool.*, 51 (6): 85-89.

1979   Rand, A.L. & D.S. Rabor, 1969. New birds from Camiguin South, Philippines. — *Fieldiana, Zool.*, 51 (13): 157-168.

1980   Rand, A.L. & M.A. Traylor, 1959. Three new birds from West Africa. — *Fieldiana, Zool.*, 39: 269-273.

1981   Raphael, S., 1970. The publication dates of the Transactions of the Linnean Society, Series I., 1791-1875. — *Biol. J. Linn. Soc., Lond.*, 2: 61-76.

1982   Raposo, M.A., 1997. A new species of *Arremon* (Passeriformes : Emberizidae) from Brazil. — *Ararajuba*, 5 (1): 3-9.

1983   Raposo, M.A. & R. Parrini, 1997. On the validity of the Half-collared Sparrow *Arremon semitorquatus* Swainson, 1837. — *Bull. Brit. Orn. Cl.*, 117 (4): 294-298.

1984   Rasmussen, P.C., 1996. Buff-throated Warbler *Phylloscopus affinis* restored to the avifauna of the Indian subcontinent. — *Forktail*, 11: 173-175.

1985   Rasmussen, P.C., 1998. A new Scops-owl from Great Nicobar Island. — *Bull. Brit. Orn. Cl.*, 118 (3): 141-153.

1986   Rasmussen, P.C., 1999. A new species of Hawk-Owl *Ninox* from north Sulawesi, Indonesia. — *Wilson Bull.*, 111 (4): 457-464.

1987   Rasmussen, P.C., 2000. On the status of the Nicobar Sparrowhawk *Accipiter butleri* on Great Nicobar Island, India. — *Forktail*, 16: 185-186.

1988   Rasmussen, P.C., 2000. A review of the taxonomy and status of the Plain-pouched Hornbill *Aceros subruficollis*. — *Forktail*, 16: 83-91.

1989   Rasmussen, P.C., 2000. Streak-breasted Woodpecker *Picus viridanus* in Bangladesh: re-identification of the region's sole specimen recorded as Laced Woodpecker *P. vittatus*. — *Forktail*, 16: 183-184.

1990   Rasmussen, P.C. & N.J. Collar, 1999. Major specimen fraud in the Forest Owlet *Heteroglaux* (*Athene* auct.) *blewitti*. — *Ibis*, 141 (1): 11-21.

1991   Rasmussen, P.C. & N.J. Collar, 1999. On the hybrid status of Rothschild's Parakeet *Psittacula intermedia* (Aves, Psittacidae). — *Bull. Nat. Hist. Mus. Lond.* (Zool.), 65 (1): 31-50.

1992   Rasmussen, P.C. & S.J. Parry, 2001. The taxonomic status of the "Long-billed" Vulture *Gyps indicus*. — *Vulture News*, 44: 18-21.

1993   Rasmussen, P.C. *et al.*, 2000. A new Bush Warbler (Sylviidae, *Bradypterus*) from Taiwan. — *Auk*, 117 (2): 279-289.

1994   Rasmussen, P.C. *et al.*, 2000. On the specific status of the Sangihe White-eye *Zosterops nehrkorni*, and the taxonomy of the Black-crowned *Z. atrifrons* complex. — *Forktail*, 16: 69-81.

1995   Rasmussen, P.C. *et al.*, 2000. Geographical variation in the Malagasy Scops-Owl (*Otus rutilus* auct.): the existence of an unrecognized species on Madagascar and the taxonomy of other Indian Ocean taxa. — *Bull. Brit. Orn. Cl.*, 120 (2): 75-102.

1996   Rasmussen, P.C. *et al.*, 2001. Field identification of 'Long-billed' Vultures (Indian and Slender-billed Vultures). — *OBC Bull.*, 34: 24-29.

1997   Rea, A.M., 1973. The Scaled Quail (*Callipepla squamata*) of the Southwest: Systematic and historical consideration. — *Condor*, 75: 322-329.

1997A   Redkin, Y.A., 2001. A new subspecies of Firecrest *Regulus ignicapillus* (Temminck, 1820) (Regulidae, Passeriformes) from the mountains of the Crimea. — *Ornitologia*, 29: 98-104.

1998   Regalado Ruiz, P., 1977. *Xiphidiopicus percussus marthae*, nueva subespecie. — *Rev. Forestal*, 3: 34-37.

1999   Regalado Ruiz, P., 1981. El género *Torreornis* (aves, Fringillidae), descripción de una nueva subespecie en Cayo Coco, Cuba. — *Centro Agricola*, 2: 87-112.

2000   Reich, G.C., 1795. Kurze Beschreibung verschiedener neuen, oder doch wenig bekannten Thiere, welche Herr LeBlond der naturforschenden Gesellschaft zu Paris aus Cayenne als Geschenk ueberschickt hat. — *Mag. des Thierreichs (Erlangen)*, 1 (3): 128-134.

2001   Reichenbach, L., 1850. Vorläufer einer Iconographie der Arten der Vögel aller Welttheile. Pages i-xxxi. In *Avium Systema Naturale. Das Natürlich System der Vögel. Vorläufer einer Iconographie der Arten der Vögel aller Welttheile Friedrich Hofmeister, Leipzig*.

2002   Reichenow, A., 1893. Neue africanische Arten. — *Orn. Monatsber.*, 1: 177-178.

2003   Reichenow, A., 1900. *Die Vogel Afrikas*: 1. i-civ, 1-706. — J. Neumann, Neudamm.

2004   Remsen, J.V., Jr., 1981. A new subspecies of *Schizoeaca harterti* with notes on taxonomy and natural history of *Schizoeaca* (Aves: Furnariidae). — *Proc. Biol. Soc. Wash.*, 94: 1068-1075.

2005   Remsen, J.V., Jr., 1984. Geographic variation, zoogeography, and possible rapid evolution in some *Cranioleuca* spinetails (Furnariidae) of the Andes. — *Wilson Bull.*, 96: 515-523.

2006   Remsen, J.V., Jr., 1993. Zoogeography and geographic variation of *Atlapetes rufinucha* (Aves : Emberizinae), including a distinctive new subspecies, in southern Peru and Bolivia. — *Proc. Biol. Soc. Wash.*, 106 (3): 429-435.

2007   Remsen, J.V., Jr., 1997. A new genus for the Yellow-shouldered Grosbeak. — *Orn. Monogr.*, 48: 89-90.

2008   Remsen, J.V., Jr. & R.T. Brumfield, 1998. Two new subspecies of *Cinnycerthia fulva* (Aves : Troglodytidae) from the southern Andes. — *Proc. Biol. Soc. Wash.*, 111 (4): 1008-1015.

2009   Remsen, J.V., Jr & W.S. Graves, IV., 1995. Distribution patterns of *Buarremon* brush-finches (Emberizidae) and interspecific competition in Andean birds. — *Auk*, 112: 225-236.

2010   Remsen, J.V., Jr. *et al.*, 1990. Natural history notes on some poorly known Bolivian birds. Part 3. — *Gerfaut*, 78: 363-381 (1988).

2011   Restall, R.L., 1995. Proposed additions to the genus *Lonchura* (Estrildinae). — *Bull. Brit. Orn. Cl.*, 115 (3): 140-157.

2012   Restall, R.L., 1996. A proposed new species of munia, genus *Lonchura* (Estrildinae). — *Bull. Brit. Orn. Cl.*, 116 (3): 137-142.

2013   Reynard, G.B. *et al.*, 1993. Taxonomic revision of the Greater Antillean Pewee. — *Wilson Bull.*, 105 (2): 217-227.

2014   Ribon, R., 1995. Nova subspecie de *Caprimulgus* (Linnaeus) (Aves, Caprimulgidae) do Espirito Santo, Brasil. — *Rev. Bras. Zool.*, 12: 333-337.

2015   Richmond, C.W., 1896. Catalogue of a collection of birds made by Dr. W. L. Abbott in eastern Turkestan, the Thian Shan Mountains and Tagdumbash Pamirs, central Asia with notes on some of the species. — *Proc. U. S. Nat. Mus.*, 28 (1083): 569-591.

2015A   Richmond, C.W., 1926. Note on *Myiothera loricata* S. Müller. — *Proc. Biol. Soc. Wash.*, 39: 141.

2016   Ridgely, R.S. & M.B. Robbins, 1988. *Pyrrhura orcesi*, a new parakeet from south-western Ecuador, with systematic notes on the *P. melanura* complex. — *Wilson Bull.*, 100 (2): 173-182.

2016A   Ridgely, R.S. & P.J. Greenfield., 2001. *Birds of Ecuador. I. Status, distribution, and taxonomy*. 1-848. — Cornell Univ. Press, Ithaca, N.Y.

2017   Ridgely, R.S. & J.A. Gwynne, 1989. *A guide to the birds of Panama, with Costa Rica, Nicaragua, and Honduras*: i-xvi, 1-534. 2nd. edition. — Princeton Univ. Press, Princeton, N.J.

2018   Ridgely, R.S. & G. Tudor, 1989. *The Birds of South America*: Vol. 1, The Oscine Passerines. i-xvi, 1-516. — Oxford University Press, Oxford.

2019   Ridgely, R.S. & G. Tudor, 1994. *The Birds of South America*: Vol. 2. The Suboscine Passerines. i-xii, 1-814. — Oxford University Press, Oxford.

2020   Ridgway, R., 1883. Descriptions of some birds supposed to be undescribed, from the Commander Islands and Petropaulovski, collected by Dr. Leonhard Stejneger. — *Proc. U. S. Nat. Mus.*, 6: 90-96.

2021   Ridgway, R., 1905. New genera of Tyrannidae and Turdidae, and new forms of Tanagridae and Turdidae. — *Proc. Biol. Soc. Wash.*, 18: 211-214.

2022   Ridgway, R., 1914. The birds of North and Middle America. — *Bull. U.S. Nat. Mus.*, 50 (6): i-xx, 1-882.

2023   Riley, J.H., 1938. Three new birds from Bangka and Borneo. — *Proc. Biol. Soc. Wash.*, 51: 95-96.

2024   Ripley, S.D., 1941. Notes on a collection of birds from Northern Celebes. — *Occ. Pap. Boston Soc. Nat. Hist.*, 8: 343-358.

2025   Ripley, S.D., 1942. Notes on Malaysian Cuckoos. — *Auk*, 59: 575-576.

2026   Ripley, S.D., 1942. The species *Eurystomus orientalis*. — *Proc. Biol. Soc. Wash.*, 55: 169-176.

2027   Ripley, S.D., 1944. The bird fauna of the West Sumatran islands. — *Bull. Mus. Comp. Zool.*, 94 (8): 307-430.

2028   Ripley, S.D., 1945. A new race of nightjar from Ceylon. — *Bull. Brit. Orn. Cl.*, 65: 40-41.

2029   Ripley, S.D., 1946. The koels of the Bay of Bengal. — *Auk*, 63: 240-241.

2030   Ripley, S.D., 1947. A report on the birds collected by Logan J. Bennett on Nissan Island and the Admiralty Islands. — *J. Wash. Acad. Sci.*, 37 (3): 95-102.

2031   Ripley, S.D., 1948. New birds from the Mishmi Hills. — *Proc. Biol. Soc. Wash.*, 61: 99-107.

2032   Ripley, S.D., 1948. Notes on Indian Birds. II. The species *Glaucidium cuculoides*. — *Zoologica*, 33 (4): 199-202.

2033   Ripley, S.D., 1949. A new race of southern Indian Green Pigeon. — *Proc. Biol. Soc. Wash.*, 62: 9-10.

2034   Ripley, S.D., 1950. New birds from Nepal and the Indian region. — *Proc. Biol. Soc. Wash.*, 63: 101-106.

2035   Ripley, S.D., 1951. Notes on Indian Birds IV. — *Postilla*, 6: 1-6.

2036   Ripley, S.D., 1953. A new race and new records for the Riu Kiu Islands. — *Tori*, 13 (63): 49-50.

2037   Ripley, S.D., 1953. Notes on Indian Birds V. — *Postilla*, 17: 1-4.

2038   Ripley, S.D., 1955. A new White-throated Spinetail from western Brazil. — *Postilla*, 23: 1-2.

2039   Ripley, S.D., 1957. New birds from the western Papuan islands. — *Postilla*, 31: 1-4.

2040   Ripley, S.D., 1959. Birds from Djailolo, Halmahera. — *Postilla*, 41: 1-8.

2041   Ripley, S.D., 1960. Two new birds from Angola. — *Postilla*, 43: 1-3.

2042   Ripley, S.D., 1961. *A synopsis of the birds of India and Pakistan*: i-xxxvi, 1-703. 1st. edition. — Bombay Nat. Hist. Soc., Bombay.

2043   Ripley, S.D., 1962. A new subspecies of the Black-chinned Fruit Pigeon. — *Proc. Biol. Soc. Wash.*, 75: 315-316.

2044   Ripley, S.D., 1964. A systematic and ecological study of birds of New Guinea. — *Bull. Peabody Mus.*, 19: (i-vi) 1-85.

2045   Ripley, S.D., 1964. Family Muscicapidae, Subfamily Turdidae. Pages 13-227. In *Check-list of Birds of the World*. Vol. 10 (E. Mayr & R.A. Paynter, Jr., eds.). — Mus. Comp. Zool., Harvard, Cambridge, Mass.

2046   Ripley, S.D., 1965. Le Martinet Pale de Socotra (*Apus pallidus Berliozi*). — *L'Oiseau*, 35 (spécial): 101-102.

2047   Ripley, S.D., 1966. A notable owlet from Kenya. — *Ibis*, 108: 136-137.

2048   Ripley, S.D., 1969. The name of the Jungle Babbler *Turdoides striatus* (Aves) from Orissa. — *J. Bombay Nat. Hist. Soc.*, 66 (1): 167-168.

2049   Ripley, S.D., 1970. A new form of rail from the Celebes. — *Nat. Hist. Bull.*, Siam Soc., 23 (3): 367-368.

2050   Ripley, S.D., 1977. *Rails of the world*: i-xx, 1-406. — David R. Godine, Boston.

2051   Ripley, S.D., 1977. A revision of the subspecies of *Strix leptogrammica* Temminck, 1831. — *Proc. Biol. Soc. Wash.*, 90 (4): 993-1001.

2052   Ripley, S.D., 1977. A new subspecies of Island Thrush, *Turdus poliocephalus,* from New Ireland. — *Auk*, 94: 772-773.

2053   Ripley, S.D., 1979. A comment on *Actinodura nipalensis* (and *waldeni*). — *J. Bombay Nat. Hist. Soc.*, 76 (1): 21-23.

2054   Ripley, S.D., 1980. A new species, and a new subspecies of bird from Tirap District, Arunachal Pradesh, and comments on the subspecies of *Stachyris nigriceps* Blyth. — *J. Bombay Nat. Hist. Soc.*, 77 (1): 1-5.

2055   Ripley, S.D., 1982. *A synopsis of the birds of India and Pakistan*: i-xxvi, 1-653. 2nd edition. — Bombay Nat. Hist. Soc., Bombay.

2056   Ripley, S.D., 1983. The subspecific name of the Common Paradise Kingfisher *Tanysiptera galatea* from Halmahera Island, North Moluccas (Maluku Utara), Indonesia. — *Bull. Brit. Orn. Cl.*, 103: 145-146.

2057   Ripley, S.D., 1984. A note on the status of *Brachypteryx cryptica*. — *J. Bombay Nat. Hist. Soc.*, 81 (3): 700-701.

2058   Ripley, S.D., 1985. Relationships of the Pacific warbler *Cichlornis* and its allies. — *Bull. Brit. Orn. Cl.*, 105 (3): 109-112.

2059   Ripley, S.D. & B. Beehler, 1985. A revision of the babbler genus *Trichastoma* and its allies (Aves: Timaliinae). — *Ibis*, 127 (4): 495-509.

2060   Ripley, S.D. & B.M. Beehler, 1985. A new subspecies of the babbler *Malacocincla abbotti* from the Eastern Ghats. — *Bull. Brit. Orn. Cl.*, 105: 66-67.

2061   Ripley, S.D. & B.M. Beehler, 1987. New evidence for sympatry in the sibling species *Caprimulgus atripennis* Jerdon and *C. macrurus* Horsfield. — *Bull. Brit. Orn. Cl.*, 107 (2): 47-49.

2061A Ripley, S.D. & B.M. Beehler, 1989. Orthithogeographic affinities of the Andaman and Nicobar Islands. — *J. Biogeography*, 16: 323-332.
2062 Ripley, S.D. & H. Birckhead, 1942. Birds collected during the Whitney South Sea Expedition. 51. On the fruit pigeons of the *Ptilinopus purpuratus* group. — *Amer. Mus. Novit.*, 1192: 1-14.
2063 Ripley, S.D. & G.M. Bond, 1966. The birds of Socotra and Abd-el-Kuri. — *Smithsonian Misc. Coll.*, 151 (7): 1-37.
2064 Ripley, S.D. & G.M. Bond, 1971. Systematic notes on a collection of birds from Kenya. — *Smithsonian Contrib. Zool.*, 111: 1-21.
2065 Ripley, S.D. & D. Hadden, 1982. A new subspecies of *Zoothera* (Aves: Muscicapidae: Turdinae) from the northern Solomon Islands. — *J. Yamashina Inst. Orn.*, 14: 103-107.
2066 Ripley, S.D. & G. Heinrich, 1960. Additions to the avifauna of northern Angola I. — *Postilla*, 47: 7.
2067 Ripley, S.D. & G. Heinrich, 1965. Additions to the avifauna of northern Angola. II. — *Postilla*, 95: 1-29.
2068 Ripley, S.D. & G. Heinrich, 1966. Comments on the avifauna of Tanzania I. — *Postilla*, 96: 1-45.
2069 Ripley, S.D. & G. Heinrich, 1969. Comments on the avifauna of Tanzania II. — *Postilla*, 134: 1-21.
2070 Ripley, S.D. & S.L. Olson, 1973. Re-identification of *Rallus pectoralis deignani*. — *Bull. Brit. Orn. Cl.*, 93: 115.
2071 Ripley, S.D. & D.S. Rabor, 1955. A new fruit pigeon from the Philippines. — *Postilla*, 21: 1-2.
2072 Ripley, S.D. & D.S. Rabor, 1958. Notes on a collection of birds from Mindoro island. — *Peabody Mus. Nat. Hist. Bull.*, 13: 1-83.
2073 Ripley, S.D. & D.S. Rabor, 1961. The avifauna of Mount Katanglad. — *Postilla*, 50: 1-20.
2074 Ripley, S.D. & D.S. Rabor, 1962. New birds from Palawan and Culion Islands, Philippines. — *Postilla*, 73: 1-16.
2075 Ripley, S.D. & D.S. Rabor, 1968. Two new subspecies of birds from the Philippines and comments on the validity of two others. — *Proc. Biol. Soc. Wash.*, 81: 31-36.
2076 Ripley, S.D. *et al.*, 1991. Notes on birds from the Upper Nao Dihing, Arunachal Pradesh, Northeastern India. — *Bull. Brit. Orn. Cl.*, 111: 19-28.
2077 Rising, J.D., 1970. Morphological variation and evolution in some North American orioles. — *Syst. Zool.*, 19: 315-351.
2079[4] Robbins, M.B. & S.N.G. Howell, 1995. A new species of Pygmy-Owl (Strigidae : *Glaucidium*) from the eastern Andes. — *Wilson Bull.*, 107 (1): 1-6.
2080 Robbins, M.B. & R.S. Ridgely, 1986. A new race of *Grallaria haplonota* (Formicariidae) from Ecuador. — *Bull. Brit. Orn. Cl.*, 106: 101-104.
2081 Robbins, M.B. & R.S. Ridgely, 1991. *Sipia rosenbergi* (Formicariidae) is a synonym of *Myrmeciza* [*laemosticta*] *nigricauda*, with comments on the validity of the genus *Sipia*. — *Bull. Brit. Orn. Cl.*, 111 (1): 11-18.
2082 Robbins, M.B. & R.S. Ridgely, 1992. Taxonomy and natural history of *Nyctiphrynus rosenbergi* (Caprimulgidae). — *Condor*, 94 (4): 984-987.
2083 Robbins, M.B. & R.S. Ridgely, 1993. A new name for *Myemeciza immaculata berlepschi* (Formicariidae). — *Bull. Brit. Orn. Cl.*, 113: 190.
2084 Robbins, M.B. *et al.*, 1994. A new species of cotinga (Cotingidae: *Doliornis*) from the Ecuadorian Andes, with comments on plumage sequences in *Doliornis* and *Ampelion*. — *Auk*, 111 (1): 1-7.
2085 Robbins, M.B. *et al.*, 1994. Voice, plumage and natural history of Anthony's Nightjar (*Caprimulgus anthonyi*). — *Condor*, 96 (1): 224-228.
2086 Robbins, M.B. & F.G. Stiles, 1999. A new species of Pygmy Owl (Strigidae : *Glaucidium*) from the Pacific slope of the northern Andes. — *Auk*, 116 (2): 305-315.
2087 Roberts, A., 1924. Synoptic Check-list of Birds of South Africa. — *Ann. Transvaal Mus.*, 10: 89-125.
2088 Roberts, A., 1933. Three new forms of South African Birds. — *Ann. Transvaal Mus.*, 15: 271-272.
2089 Roberts, A., 1937. Some results of the Barlow-Transvaal Mus. expedition to south-west Africa. — *Ostrich*, 8 (2): 84-111.
2090 Roberts, A., 1947. A new *Pternistis* from Salisbury, S. Rhodesia. — *Ostrich*, 18: 197.
2090A Roberts, T.J., 1992. *The Birds of Pakistan — Passeriformes*: 2. i-xxxvii, 1-617. — Oxford University Press, Karachi.
2091 Roberts, T.J. & B.F. King, 1986. Vocalizations of the owls of the genus *Otus* in Pakistan. — *Orn. Scand.*, 17: 299-305.
2092 Robertson, C.J.R. & G.B. Nunn, 1998. Towards a new taxonomy for Albatrosses. Pages 13-19. In *The Albatross, Biology and Conservation* (G. Robertson & R. Gales, eds.). — Beatty & Sons, Chipping Norton.
2093 Robertson, C.J.R. & J. Warham, 1992. Nomenclature of the New Zealand Wandering Albatross *Diomedea exulans*. — *Bull. Brit. Orn. Cl.*, 112 (2): 74-81.
2094 Robinson, H.C. & C.B. Kloss, 1930. A second collection of birds from Pulo Condore. — *J. Siam Soc., Nat. Hist. Suppl.*, 8 (2): 79-86.
2095 Robson, C., 2000. *A Field Guide to the Birds of South-east Asia*: 1-504. — New Holland, London.
2096 Rodner, C. *et al.*, 2000. *Checklist of the Birds of Northern South America*: [i-viii], 1-136. — Pica Press, Mountsfield, Sussex.
2097 Rogers, C.H., 1939. A new swift from the United States. — *Auk*, 56 (4): 465-468.
2098 Rohwer, S. *et al.*, 2000. A critical evaluation of Kenyon's Shag (*Phalacrocorax* [*Stictocarbo*] *kenyoni*). — *Auk*, 117 (2): 308-320.
2098A Romero-Zambrano, H., 1977. Status taxonomica de *Catamenia oreophila* Todd. — *Lozania*, 23: 1-7.
2099 Romero-Zambrano, H., 1980. Una nueva subespecie colombiana de *Campylorhamphus pusillus* (Aves - Dendrocolaptidae). — *Lozania*, 31: 1-4.
2100 Romero-Zambrano, H. & J.I. Hernandez-Camacho, 1979. Una nueva subespecie colombiana de *Haplophaedia aureliae* (Aves: Trochilidae). — *Lozania*, 30: 1-6.
2101 Romero-Zambrano, H. & J.E. Morales-Sánchez, 1981. Descripcion de una nueva subespecie de *Leptotila verreauxi* Bonaparte, 1855 (Aves: Columbidae) del sureste de Colombia. — *Caldasia*, 13: 291-299.
2102 Roonwal, M.L. & B. Nath, 1949. Contributions to the fauna of Manipur state, Assam. 2. — *Rec. Indian Mus.*, 46: 127-181.
2103 Roselaar, C.S., 1992. A new species of Mountain finch *Leucosticte* from western Tibet. — *Bull. Brit. Orn. Cl.*, 112 (4): 225-231.
2104 Roselaar, C.S., 1993. Subspecies recognition in Knot *Calidris canutus* and occurrence of races in western Europe. — *Beaufortia*, 33: 97-109.
2105 Roselaar, C.S., 1993. New subspecies of Fan-tailed Raven and Greenfinch. — *Dutch Birding*, 15 (6): 258-262.
2105A Roselaar, C.S., 1994. Geographical variation within western populations of Clamorous Reed Warbler. — *Dutch Birding*, 16 (6): 237-239.
2106 Roselaar, C.S., 1994. Systematic notes on Megapodiidae (Aves, Galliformes), including the description of five new subspecies. — *Bull. Zool. Mus. Univ. Amsterdam*, 14 (2): 9-36.
2107 Roselaar, C.S., 1995. *Taxonomy, morphology and distribution of the songbirds of Turkey*: 1-240. — GMB, Haarlem.
2108 Roselaar, C.S. & T.G. Prins, 2000. List of type specimens of birds in the Zoological Museum of the University of Amsterdam (ZMA), including taxa described by ZMA staff but without types in the ZMA. — *Beaufortia*, 50 (5): 95-126.
2109 Rothschild, N., 1937. [Exhibition of coloured drawings of two new subspecies of cassowary.]. — *Bull. Brit. Orn. Cl.*, 57: 120-121.
2109A Rotthowe, K. & J.M. Starck, 1998. Evidence for a phylogenetic position of button quails (Turnicidae: Aves) among the gruiformes. – *J. Zool. Syst. Evol. Res.*, 36: 39-51.
2110 Round, P.D. & V.M. Loskot, 1995. A reappraisal of the taxonomy of the Spotted Bush-Warbler *Bradypterus thoracicus*. — *Forktail*, 10: 159-172.
2111 Round, P.D. & C. Robson, 2001. Provenance and affinities of the Cambodian Laughingthrush *Garrulax ferrarius*. — *Forktail*, 17: 41-44.
2112 Roux, J.P. *et al.*, 1983. Un nouvel albatros *Diomedea amsterdamensis* n. sp. découvert sur l'île Amsterdam (37°50'S, 77°35'E). — *L'Oiseau*, 53 (1): 1-11.
2113 Roux, J.P. *et al.*, 1986. Le Prion de MacGillivray. Données taxonomiques. — *L'Oiseau*, 56: 379-383.
2114 Rowley, I., 1970. The genus *Corvus* (Aves: Corvidae) in Australia. — *CSIRO Wildl. Res.*, 15: 27-71.
2115 Rowley, I. & E. Russell, 1997. *Fairy-wrens and grass-wrens. Maluridae*: i-xxii, 1-274. — Oxford Univ. Press, Oxford.
2116 Rowley, J.S., 1968. Geographic variation in four species of birds in Oaxaca, Mexico. — *Occ. Pap. Western Found. Vert. Zool.*, 1: 1-10.
2117 Rowley, J.S. & R.T. Orr, 1964. A new hummingbird from southern Mexico. — *Condor*, 66 (2): 81-84.
2118 Roy, M.S., 1999. Species accounts within the Family Trochilidae (Hummingbirds). Pages 537-547. In *Handbook of the Birds of the World*. Vol. 5 (J. del Hoyo, A. Elliott & J. Sargatal, eds.). — Lynx Edicions, Barcelona.

[4] No. 2078 was misplaced and has been renumbered 2090A.

871

2119   Roy, M.S. *et al.*, 1999. Molecular phylogeny and evolutionary history of the tit-tyrants (Aves : Tyrannidae). — *Molec. Phylogen. Evol.*, 11 (1): 67-76.

2120   Rozendaal, F.G., 1987. Description of a new species of bush warbler of the genus *Cettia* Bonaparte, 1834 (Aves: Sylviidae) from Yamdena, Tanimbar Islands, Indonesia. — *Zool. Meded., Leiden*, 61 (14): 177-202.

2121   Rozendaal, F.G., 1990. Vocalisations and taxonomic status of *Caprimulgus celebensis*. — *Dutch Birding*, 12: 79-81.

2122   Rozendaal, F.G., 1993. New subspecies of Blue-rumped Pitta from southern Indochina. — *Dutch Birding*, 15 (1): 17-22.

2123   Rozendaal, F.G., 1994. Species limits within [the] Garnet Pitta complex. — *Dutch Birding*, 16: 239-245.

2124   Rozendaal, F.G. & F.R. Lambert, 1999. The taxonomic and conservation status of *Pinarolestes sanghirensis* Oustalet, 1881. — *Forktail*, 15: 1-13.

2125   Rudebeck, G., 1970. A new race of the Spike-heeled Lark, *Chersomanes albofasciata*, from Angola. — *Orn. Scand.*, 1: 45-49.

2126   Rudge, D.W. & R.J. Raikow, 1992. The phylogenetic relationships of the *Margarornis* assemblage (Furnariidae). — *Condor*, 94 (3): 760-766.

2127   Ruschi, A., 1959. A trochilifauna de Brasilia, com a descrição de um novo representante de *Amazilia* (Aves). E o primeiro povoamento com essas aves ai realizado. — *Bol. Mus. Biol. Prof. Mello Leitao, Sér. Biol. [Zool.], Santa Teresa*, 22: 1-16.

2128   Ruschi, A., 1962. Um novo representante de Colibri (Trochilidae, Aves) da região de Andarai no Estado da Bahia. — *Bol. Mus. Biol. Prof. Mello Leitao, Sér. Biol. [Zool.], Santa Teresa*, 32: 1-7.

2129   Ruschi, A., 1963. A atual distribuição Geográfica das espécies do Gènero *Augastes*, com a descrição de uma nova sub-espécie: *Augastes scutatus soaresi* Ruschi e a chave artificial e analítica para o reconhecimento das mesmas. — *Bol. Mus. Biol. Prof. Mello Leitao, Sér. Divulgação, Santa Teresa*, 4: 1-4.

2130   Ruschi, A., 1963. Um novo representante de *Campylopterus*, da regiáo de Diamantina, no estado de Minas Gerais. (Trochilidae - Aves). — *Bol. Mus. Biol. Prof. Mello Leitao, Sér. Divulgação, Santa Teresa*, 39: 1-9.

2131   Ruschi, A., 1965. Um novo representante de *Phaethornis*, da regiao de Santa Barbara, no Estado de Minas Gerais (Trochilidae - Aves). —*Bol. Mus. Biol. Prof. Mello Leitao, Sér. Divulgação, Santa Teresa*, 22 (?24): 1-15.

2132   Ruschi, A., 1972. Uma nova espécie de beija-flor do E. E. Santo. — *Bol. Mus. Biol. Prof. Mello Leitao, Sér. Divulgação, Santa Teresa*, 35: 1-5.

2133   Ruschi, A., 1973. Uma nova espécie de beija-flor do E. E. Santo. — *Bol. Mus. Biol. Prof. Mello Leitao, Sér. Divulgação, Santa Teresa*, 36: 1-3.

2134   Ruschi, A., 1973. Uma nova espécie de *Threnetes* (Aves, Trochilidae) *Threnetes grzimeki* sp. n. — *Bol. Mus. Biol. Prof. Mello Leitao, Sér. Divulgação, Santa Teresa*, 37: 1-7.

2135   Ruschi, A., 1973. Uma nova subespécie de beija-flor: *Glaucis hirsuta abrawayae* n. s. sp. — *Bol. Mus. Biol. Prof. Mello Leitao, Sér. Divulgação, Santa Teresa*, 43: 1-3.

2136   Ruschi, A., 1975. *Phaethornis pretrei schwarti* n. s. sp. — *Bol. Mus. Biol. Prof. Mello Leitao, Sér. Divulgação, Santa Teresa*, 82: 1-4.

2137   Ruschi, A., 1975. *Threnetes cristinae* n. sp. — *Bol. Mus. Biol. Prof. Mello Leitao, Sér. Divulgação, Santa Teresa*, 83: 1-3.

2138   Ruschi, A., 1976. Beija-flores do Amapé com a descricao de uma nova subespécie *Threnetes niger freirei* n. subsp. — *Bol. Mus. Biol. Prof. Mello Leitao, Sér. Divulgação, Santa Teresa*, 84: 1-4.

2139   Ruschi, A., 1978. *Ramphodon naevius freitasi* n. subsp. — *Bol. Mus. Biol. Prof. Mello Leitao, Sér. Divulgação, Santa Teresa*, 93: 1-6.

2140   Ruschi, A., 1982. Uma nova especie de Beija-flor do Brasil : *Amazilia rondoniae* n.sp. e a chave para determinar as especies de Amazilia que ocorrem no Brasil. — *Bol. Mus. Biol. Prof. Mello Leitao, Sér. Divulgação, Santa Teresa*, 100: 1-2.

2141   Russell, S.M., 1963. A new race of *Centurus aurifons* (Aves: Picidae) from British Honduras. — *Occ. Pap. Mus. Zool., Louisiana St. Univ.*, 25: 1-3.

2142   Rustamov, A.K., 1948. Novaya forma kamennoi kuropatki s Ust?-Urta. [A new form of the Rock Partridge from Ust?-Urt.]. — *Okhrana prirodi*, 5: 106-108.

2143   Ryan, P.G., 1998. The taxonomic and conservation status of the Spectacled Petrel *Procellaria conspicillata*. — *Bird Conserv.* Int., 8: 223-235.

2144   Ryan, P.G. & P. Bloomer, 1999. The Long-billed Lark complex: A species mosaic in southwestern Africa. — *Auk*, 116 (1): 194-208.

2145   Ryan, P.G. *et al.*, 1998. Barlow's Lark: a new species in the Karoo Lark *Certhilauda albescens* complex of southwest Africa. — *Ibis*, 140 (4): 605-619.

2146   Saetre, G.-P. *et al.*, 2001. A new bird species? The taxonomic status of "the Atlas Flycatcher" assessed from DNA sequence analysis. — *Ibis*, 143: 494-497.

2147   Safford, R.J. *et al.*, 1995. A new species of nightjar from Ethiopia. — *Ibis*, 137: 301-307.

2148   Salaman, P.G.W. & G. Stiles, 1996. A distinctive new species of vireo (Passeriformes : Vireonidae) from the Western Andes of Colombia. — *Ibis*, 138 (4): 610-619.

2149   Salomonsen, F., 1934. Four new birds and a new genus from Madagascar. — *Ibis*, 13 (4): 382-390.

2150   Salomonsen, F., 1936. Description of a new race of the Willow Grouse (*Lagopus lagopus variegatus*). — *Bull. Brit. Orn. Cl.*, 56: 99-100.

2151   Salomonsen, F., 1944. The Atlantic Alcidae. — *Goteborgs Kungl. Vetensk. Vitterhets-Samhalles Handl.*, 6, B, 3 (5): 1-138.

2152   Salomonsen, F., 1947. En ny Hjerpe (*Tetrastes bonasia* (L.)) fra Skandinavien. (A new hazel-grouse from Scandinavia). — *Dansk Orn. Foren. Tidsskr.*, 41 (3): 221-224.

2153   Salomonsen, F., 1950. En ny race af Fjaeldrype (*Lagopus mutus* (Montin)) fra Grønland. (A new race of the Rock-Ptarmigan (*Lagopus mutus* (Montin)) from Greenland. A collection of birds from North Atlantic). — *Dansk Orn. Foren. Tidsskr.*, 44: 219-226.

2154   Salomonsen, F., 1952. Systematic notes on some Philippine birds. — *Vidensk. Medd. fra Dansk naturh. Foren.*, 114: 341-364.

2155   Salomonsen, F., 1953. Miscellaneous notes on Philippine birds. — *Vidensk. Medd. fra Dansk naturh. Foren.*, 115: 205-281.

2156   Salomonsen, F., 1962. Whitehead's Swiftlet (*Collocalia whiteheadi* Ogilvie-Grant) in New Guinea and Melanesia. — *Vidensk. Medd. fra Dansk naturh. Foren.*, 125: 509-512.

2157   Salomonsen, F., 1964. Noona Dan Papers 9. Some remarkable new birds from Dyaul Is., Bismarck Archipelago, with zoogeographical notes. — *Biol. Skr. Dan. Vid. Selsk.*, 14 (1): 1-37.

2158   Salomonsen, F., 1967. Family Dicaeidae. Pages 166-208. In *Check-list of Birds of the World*. Vol. 12 (R.A. Paynter, Jr., ed.). — Mus. Comp. Zool., Cambridge, Mass.

2159   Salomonsen, F., 1967. Family Meliphagidae. Pages 338-450. In *Check-list of Birds of the World*. Vol. 12 (R.A. Paynter, Jr., ed.). — Mus. Comp. Zool., Cambridge, Mass.

2160   Salomonsen, F., 1972. New pigeons from the Bismarck Archipelago (Aves, Columbidae). — *Steenstrupia*, 2 (12): 183-189.

2161   Salomonsen, F., 1983. Revision of the Melanesian swiftlets (Apodes, Aves) and their conspecific forms in the Indo-Australian and Polynesian region. — *Biol. Skr. Dan. Vid. Selsk.*, 23 (5): 4-112.

2161A   Salvadori, T., 1874. Catalogo sistematico degli uccelli di Borneo. — *Ann. Mus. Civ. Stor. Nat. Genova*, 5: 1-429.

2161B   Salvadori, T., 1891. *A Catalogue of the Birds in the British Museum*: XX: i-xvii, 1-659. — Trustees of the British Museum (Nat. Hist.), London.

2162   Salvadori, T., 1896. Uccelli raccolti da Don Eugenio dei Principi Ruspoli durante l'ultimo suo Viaggio nelle regioni dei Somalie e dei Galla. — *Ann. Mus. Civ. Stor. Nat. Genova*, 36: 43-46.

2163   Sánchez Osés, C., 1999. Species accounts within the Family Trochilidae (Hummingbirds). Pages 549-680. In *Handbook of the Birds of the World*. Vol. 5. (J. del Hoyo, A. Elliott & J. Sargatal, eds.). — Lynx Edicions, Barcelona.

2164   Sanft, K., 1954. *Tockus nasutus dorsalis* subsp. nova. — *J. f. Orn.*, 95: 416.

2165   Sanft, K., 1960. Aves, Upupidae: Bucerotidae. — *Das Tierreich*, 76: 1-174.

2165A   Sangster, G. *et al.*, 1999. Dutch avifaunal list: species concepts, taxonomic instability, and taxonomic changes in 1977-1998. — *Ardea*, 87 (1): 139-166.

2166   Sangster, G. *et al.*, 2001. The taxonomic status of Green-winged Teal *Anas carolinensis*. — *Brit. Birds*, 94: 218-226.

2167   Sangster, G. *et al.*, 2002. Taxonomic recommendations for European birds. — *Ibis*, 144 (1): 153-159.

2168   Saunders, G.B., 1951. A new White-winged Dove from Guatemala. — *Proc. Biol. Soc. Wash.*, 64: 83-87.

2169   Saunders, G.B., 1968. Seven new White-winged Doves from Mexico, Central America and the southwestern United States. — *North Amer. Fauna*, (65): 1-30.

2170   Scarlett, R.J., 1966. A pelican in New Zealand. — *Notornis*, 13: 204-217.

2171   Schäfer, E. & R. Meyer de Schauensee, 1939. Zoological results of the second Dolan Expedition to Western China and Eastern Tibet, 1934-1936. Part II. Birds. — *Proc. Acad. Nat. Sci. Philad.*, 90: 185-260 (1938).

2172   Schiebel, G., 1934. *Alectoris graeca whitakeri* subsp. nova. — *Falco*, 30 (1): 2-3.

2173   Schifter, H., 1972. *Die Mausvögel (Coliidae)*: 459. 1-119. — A. Ziemsen Verlag, Wittenberg Lutherstadt.

2174   Schifter, H., 1975. Unterartgliederung und Verbreitung des Blaunackenmausvogels *Urocolius macrourus* (Coliiformes, Aves). — *Ann. Naturhist. Mus. Wien*, 79: 109-182.

2175   Schodde, R., 1975. *Interim list of Australian Songbirds*: Passerines. — R.A.O.U., Melbourne.

2176   Schodde, R., 1978. The identity of five type-specimens of New Guinean birds. — *Emu*, 78 (1): 1-6.

2177   Schodde, R., 1989. New subspecies of Australian birds. — *Canberra Bird Notes*, 13: 119-122 (1988).

2178   Schodde, R., 1992. Towards stabilizing the nomenclature of Australian birds: neotypification of *Myzomela sanguinolenta* (Latham, 1801), *Microeca fascinans* (Latham, 1801) and *Microeca leucophaea* (Latham, 1801). — *Bull. Brit. Orn. Cl.*, 112 (3): 185-190.

2179   Schodde, R., 1993. The bird fauna of western New South Wales: ecological patterns and conservation implications. Pages 107-121. In *The Future of Native Fauna in western New South Wales*. (D. Lunney, S. Hand, P. Reed & D. Butcher, eds.). — Royal Zool. Soc., Sydney.

2180   Schodde, R., 1993. Geographic forms of the Regent Parrot *Polytelis anthopeplus* (Lear), and their type localities. — *Bull. Brit. Orn. Cl.*, 113 (1): 44-47.

2181   Schodde, R. & L. Christidis, 1987. Genetic differentiation and subspeciation in the Grey Grasswren *Amytornis barbatus* (Maluridae). — *Emu*, 87: 188-192.

2182   Schodde, R. & R. de Naurois, 1982. Patterns of variation and dispersal in the Buff-banded Rail (*Gallirallus philippensis*) in the south-west Pacific, with description of a new subspecies. — *Notornis*, 29: 131-142.

2183   Schodde, R. & D.T. Holyoak, 1997. Application of *Halcyon ruficollaris* Holyoak and *Alcyone ruficollaris* Bankier. — *Bull. Brit. Orn. Cl.*, 72 (1): 32.

2184   Schodde, R. & I.J. Mason, 1976. A new subspecies of *Colluricincla megarhyncha* Quoy and Gaimard from the Northern Territory. — *Emu*, 76 (3): 109-114.

2185   Schodde, R. & I.J. Mason, 1976. Infra-specific variation in *Alcedo azurea* (Latham) (Alcedinidae). — *Emu*, 76: 161-166.

2186   Schodde, R. & I.J. Mason, 1980. *Nocturnal birds of Australia*: 1-136. — Lansdowne, Melbourne.

2187   Schodde, R. & I.J. Mason, 1991. Subspeciation in the Western Whipbird *Psophodes nigrogularis* and its zoogeographical significance, with descriptions of two new subspecies. — *Emu*, 91 (3): 133-144.

2188   Schodde, R. & I.J. Mason, 1997. Aves (Columbidae to Coraciidae). Pages i-xiii, 1-436. In *Zoological Catalogue of Australia*. Vol. 37. Pt. 2. (W.W.K. Houston & A. Wells, eds.). — C.S.I.R.O., Melbourne.

2189   Schodde, R. & I.J. Mason, 1999. *Directory of Australian Birds*. Passerines: i-x, 1-851. — CSIRO Publishing, Canberra.

2190   Schodde, R. *et al.*, 1979. A new subspecies of *Philemon buceroides* from Arnhem Land. — *Emu*, 79: 24-30.

2191   Schodde, R. *et al.*, 1980. Variation in the Striated Heron *Butorides striatus* in Australasia. — *Emu*, 80 (4): 203-212.

2192   Schodde, R. *et al.*, 1993. Geographical differentiation in the Glossy Black-Cockatoo *Calyptorhynchus lathami* (Temminck) and its history. — *Emu*, 93: 156-166.

2193   Schodde, R. & S.J. Mathews, 1977. Contributions to Papuasian Ornithology: V. Survey of the birds of Taam Island, Kai Group. — *CSIRO Wildl. Res. Tech. Pap.*, 33: 1-29.

2194   Schodde, R. & J.L. McKean, 1972. Distribution and taxonomic status of *Parotia lawesii helenae* De Vis. — *Emu*, 72: 113-114.

2195   Schodde, R. & J.L. McKean, 1973. The species of the genus *Parotia* (Paradisaeidae) and their relationships. — *Emu*, 73: 145-156.

2196   Schodde, R. & S.C. Tideman (eds), 1986. *Readers' Digest complete book of Australian birds*. 1-639. — Readers' Digest, Sydney.

2197   Schodde, R. & R.G. Weatherly, 1981. A new subspecies of the Southern Emu-wren *Stipiturus malachurus* from South Australia, with notes on its affinities. — *South Australian Orn.*, 28: 169-170.

2198   Schodde, R. & R.G. Weatherly, 1982. *The Fairy-wrens, A Monograph of the Maluridae*: 1-203. — Lansdowne Editions, Melbourne.

2199   Schouteden, H., 1952. Un Strigidé nouveau d'Afrique noire: *Phodilus Prigoginei* nov. sp. — *Rev. Zool. Bot. Afr.*, 46 (3/4): 423-428.

2199A   Schouteden, H., 1954. Faune du Congo-Belge et Ruanda, Urundi, 3. Non Passeres. — *Ann. Mus. Congo Belge*, 29: 1-434.

2200   Schouteden, H., 1954. Quelques oiseaux de la faune congolaise. — *Rev. Zool. Bot. Afr.*, 49 (3/4): 353-356.

2201   Schuchmann, K.-L., 1978. Allopatrische Artbildung bei der Kolibrigattung *Trochilus*. — *Ardea*, 66: 156-172.

2202   Schuchmann, K.-L., 1984. Two hummingbird species, one a new subspecies, new to Bolivia. — *Bull. Brit. Orn. Cl.*, 104: 5-7.

2203   Schuchmann, K.-L., 1999. Family Trochilidae (Hummingbirds). Pages 468-680. In *Handbook of the Birds of the World*. Vol. 5 (J. del Hoyo, A. Elliott & J. Sargatal, eds.). — Lynx Edicions, Barcelona.

2204   Schuchmann, K.-L. & H.E. Wolters, 1982. A new subspecies of *Serinus estherae* (Carduelidae) from Sulawesi. — *Bull. Brit. Orn. Cl.*, 102: 12-14.

2205   Schuchmann, K.-L. *et al.*, 2000. Biogeography and taxonomy of the Andean hummingbird genus *Haplophaedia* Simon (Aves : Trochilidae), with the description of a new subspecies from southern Ecuador. — *Orn. Anzeiger*, 39 (1): 17-42.

2206   Schuchmann, K.-L. *et al.*, 2001. Systematics and biogeography of the Andean genus *Eriocnemis* (Aves: Trochilidae). — *J. f. Orn.*, 142 (4): 433-482.

2207   Schuchmann, K.-L. & T. Zuchner, 1997. *Coeligena violifer albicaudata* (Aves : Trochilidae) : A new Hummingbird subspecies from the southern Peruvian Andes. — *Orn. Neotrop.*, 8: 247-253.

2208   Schulenberg, T.S., In Press. Vangas. In *The natural history of Madagascar* (S.M. Goodman & J.P. Benstead, eds.). — The University of Chicago Press, Chicago.

2209   Schulenberg, T.S. & L.C. Binford, 1985. A new species of tanager (Emberizidae: Thraupinae, *Tangara*) from southern Peru. — *Wilson Bull.*, 97 (4): 413-420.

2210   Schulenberg, T.S. & G.L. Graham, 1981. A new subspecies of *Anairetes agraphia* (Tyrannidae) from northern Peru. — *Bull. Brit. Orn. Cl.*, 101: 241-243.

2211   Schulenberg, T.S. & T.A. Parker, III., 1997. A new species of Tyrant-Flycatcher (Tyrannidae : *Tolmomyias*) from the western Amazon basin. — *Orn. Monogr.*, 48: 723-731.

2212   Schulenberg, T.S. & D.F. Stotz, 1991. The taxonomic status of *Myrmeciza stictothorax* (Todd). — *Auk*, 108: 731-733.

2213   Schulenberg, T.S. & M.D. Williams, 1982. A new species of antpitta (*Grallaria*) from northern Peru. — *Wilson Bull.*, 94 (2): 105-113.

2214   Schulenberg, T.S. *et al.*, 1984. Distributional records from the Cordillera Yanachaga, central Peru. — *Gerfaut*, 74: 57-70.

2215   Schulenberg, T.S. *et al.*, 1993. Genetic variation in two subspecies of *Nesillas typica* (Sylviidae) in south-eastern Madagascar. Pages 173-177. In *Proc. 8th. Pan African Orn. Congr., Bujumbura, Burundi*. 1992.

2216   Sclater, P.L., 1860. List of birds collected by Mr. Fraser in Ecuador, at Nanegas, Calacali, Perucho and Puellavo; with notes and description of a new species. — *Proc. Zool. Soc., Lond.*, 83-97.

2216A   Sclater, P.L., 1862. Catalogue of a collection of American birds belonging to P.L. Sclater: i-xvi, 1-368. – Trübner, London.

2217   Sclater, P.L., 1863. On some new and interesting animals recently acquired for the Society's Menagerie. — *Proc. Zool. Soc., Lond.*, 374-378.

2218   Sclater, P.L., 1866. Descriptions of eight new species of birds from Veragua. — *Proc. Zool. Soc., Lond.*, 67-76.

2219   Sclater, P.L., 1891. On a second collection of birds from the Province of Tarapacá, northern Chili. — *Proc. Zool. Soc., Lond.*, 395-404.

2219A   Sclater, P.L. & O. Salvin, 1876. Descriptions of new birds obtained by Mr. C. Buckley in Bolivia. — *Proc. Zool. Soc., Lond.*: 253-254.

2220   Scott, P., 1972. *A coloured key to the wildfowl of the world*: 1-96. Revised edition. — W.R. Royle & Son, London.

2221   Selander, R.K. & M. Alvarez del Toro, 1955. A new race of Booming Nighthawk from southern Mexico. — *Condor*, 57: 144-147.

2222   Serle, W., 1949. A new genus and species of babbler, and new races of a wood-hoopoe, swift, barbet, robin-chat, scrub-warblers and apalis. — *Bull. Brit. Orn. Cl.*, 69: 50-56.

2223   Serle, W., 1949. New races of a Warbler, a Flycatcher and an Owl from West Africa. — *Bull. Brit. Orn. Cl.*, 69: 74-76.

2224   Serle, W., 1952. The relationship of *Mesopicos johnstoni* (Shelley) and *Mesopicos ellioti* (Cassin). — *Bull. Brit. Orn. Cl.*, 72: 104-106.

2225   Serle, W., 1959. A new race of *Mirafra africana* Smith from British Cameroons. — *Bull. Brit. Orn. Cl.*, 79: 2-3.

2225A   Severtsov, N.A., 1875. Notes on some new Central-Asiatic Birds.—*Ibis* (3) 5: 487-494.

2226   Sharpe, R.B., 1875. Contribution to the ornithology of Madagascar. — *Proc. Zool. Soc., Lond.*, 70-78.

2227   Sharpe, R.B., 1876. Prof. Steere's expedition to the Philippines. — *Nature*, 14 (Aug 3.): 297-298.

2228   Sharpe, R.B., 1881. On a new genus of Timeliidae from Madagascar, with remarks on some other genera. — *Proc. Zool. Soc., Lond.*, 195-197.

2228A   Sharpe, R.B., 1882. *A Catalogue of the Birds in the British Museum*. VI: i-xiii, 1-421 (1881). — Trustees of the British Museum, London.

2229   Sharpe, R.B., 1885. *A Catalogue of the Birds in the British Museum*. X: i-xiii, 1-682. — Trustees of the British Museum (Nat. Hist.), London.

2230   Sharpe, R.B., 1887. Notes on a collection of birds made by Mr. John Whitehead on the Mountain of Kina Balu, in Northern Borneo, with descriptions of new species. — *Ibis*, (5) 5: 435-454.

2231   Sharpe, R.B., 1887. On a second collection of birds formed by Mr. L. Wray in the mountains of Perak, Malay Peninsula. — *Proc. Zool. Soc., Lond.*, 431-443.

2232   Sharpe, R.B., 1892. [An apparently new species of *Rhipidura* from the island of Dammar in the Banda Sea.]. — *Bull. Brit. Orn. Cl.*, 1: xviii-xix.

2233   Sharpe, R.B., 1896. *A Catalogue of the Birds in the British Museum*. Vol. XXIV. i-xii, 1-794. — Trustees of the British Museum (Nat. Hist.), London.

2234   Sheldon, F.H., 1985. The taxonomy and biogeography of the Thick-billed Flowerpecker complex in Borneo. — *Auk*, 102: 606-612.

2235   Sheldon, F.H. & F.B. Gill, 1996. A reconsideration of songbird phylogeny, with emphasis on the evolution of titmice and their sylvioid relatives. — *Syst. Biol.*, 45: 473-495.

2236   Sheldon, F.H. & D.W. Winkler, 1993. Intergeneric phylogentic relationships of swallows estimated by DNA-DNA hybridization. — *Auk*, 110: 798-824.

2237   Sheldon, F.H. *et al.*, 1995. Phylogentic relationships of the zigzag heron (*Zebrilus undulatus*) and white-crested bittern (*Tigriornis leucolophus*) estimated by DNA-DNA hybridization. — *Auk*, 112: 672-679.

2238   Sheldon, F.H. *et al.*, 1999. Comparison of cytochrome *b* and DNA hybridization data bearing on the phylogeny of swallows (Aves: Hirundinidae). — *Molec. Phylogen. Evol.*, 11: 320-331.

2239   Sheldon, F.H. *et al.*, 2000. Relative patterns and rates of evolution in heron nuclear and mitochrondrial DNA. — Mol. Biol. Evol., 17: 437-450.

2240   Shelley, G.E., 1906. *The Birds of Africa comprising all the species which occur in the Ethiopian region*. Vol. 5. (1). 1-163. — R. H. Portis, London.

2241   Sherborn, C.D., 1894. Dates of publication of Jardine and Selby's Illustrations of Ornithology. – *Ibis* (6) 6: 326.

2242   Sherborn, C.D., 1898. On the dates of Temminck's and Laugier's "Planches coloriées". – *Ibis* (7) 4: 485-488.

2242A   Sherborn, C.D. & B.B. Woodward, 1902. Notes on the dates of publication of the natural history portions of some French voyages. Part 2. Ferret and Galinier's 'Voyage en Abyssinie'; Lefebvre's 'Voyage en Abyssinie'; 'Exploration scientifique de l'Algérie'; Castelnau's 'Amérique du Sud'; Dumont d'Urville's 'Voyage de l'Astrolabe'; Laplace's 'Voyage sur la Favorite'; Jacquemont's 'Voyage dans l'Inde'; Tréhouart's 'Commission scientifique d'Islande'; Cailliaud, 'Voyage à Méroé'; 'Expédition scientifique de Morée'; Fabre, 'Commission scientifique du Nord'; Du Petit-Thouars, 'Voyage de la Vénus' and on the dates of the 'Faune Française'. — *Ann. Mag. Nat. Hist.*, (7) 8: 161-164.

2243   Sherman, P.T., 1996. Family Psophiidae (Trumpeters). Pages 96-107. In *Handbook of the Birds of the World*. Vol. 3 (J. del Hoyo, A. Elliott & J. Sargatal, eds.). — Lynx Edicions, Barcelona.

2244   Shirihai, H., 1988. A new subspecies of Arabian Warbler *Sylvia leucomelaena* from Israel. — *Bull. Brit. Orn. Cl.*, 108: 64-68.

2245   Shirihai, H. & D.A. Christie, 1996. A new taxon of small shearwater from the Indian Ocean. — *Bull. Brit. Orn. Cl.*, 116 (3): 180-186.

2246   Shirihai, H. & P.R. Colston, 1992. A new race of the Sand Martin *Riparia riparia* from Israel. — *Bull. Brit. Orn. Cl.*, 112 (2): 129-132.

2247   Shirihai, H. *et al.*, 2001. *Sylvia warblers. Identification, taxonomy and phylogeny of the genus* Sylvia: 1-576. — A. & C. Black, London.

2248   Short, L.L., Jr., 1969. A new species of blackbird (*Agelaius*) from Peru. — *Occas. Pap. Mus. Zool. Louisiana St. Univ.*, 36: 1-8.

2249   Short, L.L., Jr., 1972. Relationships among the four species of the superspecies *Celeus elegans* (Aves, Picidae). — *Amer. Mus. Novit.*, 2487: 1-26.

2250   Short, L.L., Jr., 1973. A new race of *Celeus spectabilis* from eastern Brazil. — *Wilson Bull.*, 85: 465-467.

2251   Short, L.L., Jr., 1974. Relationship of *Veniliornis* "*cassini*" *chocoensis* and V. "*cassini*" *caqetensis* with V. *affinis*. — *Auk*, 91: 631-634.

2252   Short, L.L., Jr., 1982. *Woodpeckers of the World*: i-xviii, 1-676. — Delaware Mus. Nat. Hist., Wilmington.

2253   Short, L.L., 1988. Various species accounts. In *The Birds of Africa*. Vol. 3: i-xvi, 1-611 (C.H. Fry, S. Keith & E.K. Urban, eds.). — Academic Press, London.

2254   Short, L.L. & J.F.M. Horne, 1988. Current speciation problems in Afrotropical Piiformes In *Internat. Ornith. Congress XIX.*, Ottawa. 1986: 2519-2527.

2255   Short, L.L. & J.F.M. Horne, 1988. Various species accounts In *The Birds of Africa*. Vol. 3: i-xiv, 1-611 (C.H. Fry, S. Keith & E.K. Urban, eds.). — Academic Press, London.

2255A   Short, L.L. & J.F.M. Horne, 2001. *Toucans, Barbets and Honeyguides. Rhampastidae, Capitonidae and Indicatoridae*. i-xxiii, 1-526. — O.U.P., Oxford.

2255B   Short, L.L. & J.F.M. Horne, 2002. Family Capitonidae (Barbets). Pages 140-219. In *Handbook of the Birds of the World*, 7 (J. del Hoyo, A. Elliott, and J. Sargatal, eds.). — Lynx Edicions, Barcelona.

2256   Short, L.L. *et al.*, 1987. *Indicator narokensis* Jackson is a synonym of *Indicator meliphilus* (Oberholser). — *Mitt. Zool. Mus. Berlin*, 63 (Suppl. Ann. Orn. 11): 161-168.

2257   Short, L.L. *et al.*, 1990. Annotated Check-list of the birds of East Africa. — *Proc. Western Found. Vert. Zool.*, 4 (3): 61-246.

2258   Sibley, C.G., 1955. The generic allocation of the Green-tailed Towhee. — *Auk*, 72: 420-423.

2259   Sibley, C.G. & J.E. Ahlquist, 1982. The relationships of the Australo-Papuan scrub-robins *Drymodes* as indicated by DNA-DNA hybridization. — *Emu*, 82: 101-105.

2260   Sibley, C.G. & J.E. Ahlquist, 1984. The relationships of the Papuan genus *Peltops*. — *Emu*, 84: 181-183.

2260A   Sibley, C.G. & J.E. Ahlquist, 1985. The phylogeny and classification of the Australo-Papuan passerine birds. – *Emu*, 85: 1-14.

2261   Sibley, C.G. & J.E. Ahlquist, 1990. *Phylogeny and classification of birds. A study in molecular evolution*: i-xxiii, 1-976. — Yale Univ. Press, New Haven, Conn.

2262   Sibley, C.G. & B.L. Monroe, Jr., 1990. *Distribution and taxonomy of birds of the world*: i-xxiv, 1-1111. — Yale University Press, New Haven, Conn.

2263   Sibley, C.G. & L.L. Short, Jr., 1959. Hybridization in some Indian bulbuls *Pycnonotus cafer* x *Pycnonotus leucogenys*. — *Ibis*, 101 (2): 177-182.

2264   Sibley, C.G. *et al.*, 1988. A classification of the living birds of the world based on DNA-DNA hybridization studies. — *Auk*, 105: 409-423.

2265   Sick, H., 1950. Eine neue Form von *Dendrocincla fuliginosa* vom Alto Xingu, Zentralbrasilien (*D. f. trumaii* subsp. nova). — *Orn. Ber., Darmstadt*, 3: 23-25.

2266   Sick, H., 1958. Resultados de uma excursão ornitológica do Museu Nacional a Brasilia, novo Distrito Federal Goiás, com a descrição de um novo representante de *Scytalopus* (Rhinocryptidae, Aves). — *Bol. Mus. Nac. Rio de Janeiro, Brasil, Nova Sér., Zool.*, 185: 1-41.

2267   Sick, H., 1959. Ein neuer Sittich aus Brasilien: *Aratinga cactorum paraënsis, subsp. nova.* — *J. f. Orn.*, 100 (4): 413-416.

2268   Sick, H., 1960. Zur Systematik und Biologie der Bürzelstelzer (Rhinocryptidae), speziell Brasiliens. — *J. f. Orn.*, 101 (1/2): 141-174.

2269   Sick, H., 1969. Über einige Töpfervögel (Furnariidae) aus Rio Grande do Sul, Brasilien, mit Beschreibung eines neuen *Cinclodes*. — *Beitr. Neotrop. Fauna*, 6 (2): 63-79.

2270   Sick, H., 1985. *Ornitologia Brasiliera, Uma Inrodução*: i-xxii, 1-827. — Universidade de Brasilia, Brasilia.

2271   Sick, H., 1991. Distribution and subspeciation of the Biscutate Swift *Streptoprocne biscutata*. — *Bull. Brit. Orn. Cl.*, 111 (1): 38-40.

2272   Siegel-Causey, D., 1988. Phylogeny of the Phalacrocoracidae. — *Condor*, 90: 885-905.

2273   Siegel-Causey, D., 1991. Systematics and biogeography of North Pacific shags, with a description of a new species. — *Occ. Pap. Mus. Nat. Hist., Univ. Kansas*, 140: 1-17.

2274   Simpson, S.F. & J. Cracraft, 1981. The phylogenetic relationships of the Piciformes (Class Aves). — *Auk*, 98: 481-494.

2275   Sims, R.W., 1954. A new race of Button-Quail (*Turnix maculosa*) from New Guinea. — *Bull. Brit. Orn. Cl.*, 74: 37-40.

2276   Sims, R.W. & R.M.L. Warren, 1955. The names of the races of Pompadour Pigeon, *Treron pompadora* (Gmelin), in Java and Celebes. — *Bull. Brit. Orn. Cl.*, 75: 96-97.

2277   Sinclair, I. & O. Langrand, 1998. *Birds of the Indian Ocean islands*: 1-184. — Struik Publishers (Pty.) Ltd., Cape Town.

2278   Slikas, B. *et al.*, 2000. Phylogenetic relationships of Micronesian White-eyes based on mitochondrial sequence data. — *Auk*, 117 (2): 355-365.

2278A   Slud, P., 1964. The birds of Costa Rica. Distribution and ecology. — *Bull. Amer. Mus. Nat. Hist.*, 128: 1-430.

2279   Smeenk, C., 1974. Comparative ecological studies of some East African birds of prey. — *Ardea*, 62 (1/2): 1-97.

2280   Smith, A., 1836. *Report of the Expedition for Exploring Central Africa from the Cape of Good Hope Jun 23, 1834*: 1-57. — Govt. Gazette.

2281   Smith, E.F.G. *et al.*, 1991. A new species of Shrike (Laniidae: *Laniarius*) from Somalia, verified by DNA sequence data from the only known individual. — *Ibis*, 133 (3): 227-235.

2282   Smith, T.B., 1990. Patterns of morphological and geographic variation in trophic bill morphs of the African finch *Pyrenestes*. — *Biol. J. Linn. Soc., Lond.*, 541: 381-414.

2283   Smithers, R.H.N., 1954. A new race of Nightjar from Northern Rhodesia. — *Bull. Brit. Orn. Cl.*, 74: 84.

2284   Smithers, R.H.N., 1954. A new race of Nightjar from the Caprivi Strip, South West Africa. — *Bull. Brit. Orn. Cl.*, 74: 83-84.

2285   Smythies, B.E., 1953. *The Birds of Burma*: i-xliii, 1-668. 2nd. edition. — Oliver and Boyd, Edinburgh.

2286   Smythies, B.E., 1957. An annotated checklist of the birds of Borneo. — *Sarawak Mus. J.*, 7: i-xv, 523-818.

2287   Smythies, B.E., 1960. *The Birds of Borneo*: i-xvi, 1-562. 1st edition. — Oliver and Boyd, Edinburgh.

2288   Smythies, B.E., 1981. *The Birds of Borneo*: i-xiv, 1-473. 3rd. edition. — The Sabah Society and the Malayan Nature Society, Kuala Lumpur.

2289   Smythies, B.E., 2000. *The Birds of Borneo*: i-xii, 1-710 (1999). 4th. edition. — Natural History Publications (Borneo), Kota Kinabalu.

2290   Snigirevskij, S.I., 1937. *Tetrao urogallus obsoletus* subsp. nov. In *Avifauna of the extra-Polar part of the northern Urals* (L.A. Portenko, ed.). 1-240. — Academy of Sciences, Leningrad.

2291   Snow, D.W., 1967. Family Aegithalidae. Pages 52-61. In *Check-list of Birds of the World*. Vol. 12 (R.A. Paynter, Jr., ed.). — Mus. Comp. Zool., Cambridge, Mass.

2292   Snow, D.W., 1967. Family Paridae. Pages 70-124. In *Check-list of Birds of the World*. Vol. 12 (R.A. Paynter, Jr., ed.). — Mus. Comp. Zool., Cambridge, Mass.

2293   Snow, D.W., 1979. Family Cotingidae. Pages 281-308. In *Check-list of Birds of the World*. Vol. 8 (M.A. Traylor, Jr., ed.). — Mus. Comp. Zool., Cambridge, Mass.

2294   Snow, D.W., 1980. A new species of cotinga from southeastern Brazil. — *Bull. Brit. Orn. Cl.*, 100: 213-215.

2294A   Snow, D.W., 1997. Should the biological be superseded by the phylogenetic species concept? — *Bull. Brit. Orn. Cl.*, 117: 110-121.

2295   Snow, D.W., 2001. Family Momotidae (Motmots). Pages 264-284. In *Handbook of the Birds of the World*. Vol. 6 (J. del Hoyo, A. Elliott & J. Sargatal, eds.). — Lynx Edicions, Barcelona.

2296   Snyder, L.L., 1961. On an unnamed population of the Great Horned Owl. — *Contr. R. Ontario Mus.*, 54: 1-7.

2297   Somadikarta, S., 1975. On the two new subspecies of Crested Tree Swift from Peleng Island, and Sula Islands (Aves: Hemiprocnidae). — *Treubia*, 28 (4): 119-127.

2298   Somadikarta, S., 1986. *Collocalia linchi* Horsfield & Moore - a revision. — *Bull. Brit. Orn. Cl.*, 106: 32-40.

2299   Somadikarta, S., 1994. The identity of the Marquesan Swiftlet *Collocalia ocista* Oberholser. — *Bull. Brit. Orn. Cl.*, 114 (4): 259-263.

2299A   Sorensen, M.D. & R.B. Payne, 2001. A single ancient origin of brood parasitism in African finches: implications for host parasite co-evolution. — *Evolution*, 55: 2550-2567.

2300   Sözer, R. & A.J.W.J. van der Heijden, 1997. An overview of the distribution, status and behavioural ecology of White-shouldered Ibis in East Kalimantan, Indonesia. — *Kukila*, 9: 126-140.

2301   Springer, M.S. *et al.*, 1995. Molecular evidence that the Bonin Islands "Honeyeater" is a white-eye. — *J. Yamashina Inst. Orn.*, 27 (94): 66-77.

2302   Stager, K.E., 1959. The Machris Brazilian Expedition. Ornithology: Two new birds from Central Goiás, Brasil. — *Contrib. Sci. Los Angeles Co. Mus.*, 33: 1-8.

2303   Stager, K.E., 1961. A new bird of the genus *Picumnus* from eastern Brazil. — *Contrib. Sci. Los Angeles Co. Mus.*, 46: 1-4.

2304   Stager, K.E., 1968. A new piculet from Amazonian Bolivia. — *Contrib. Sci. Los Angeles Co. Mus.*, 143: 1-14.

2305   Stager, K.E., 1968. A new piculet from southeastern Peru. — *Contrib. Sci. Los Angeles Co. Mus.*, 153: 1-4.

2306   Stanford, J.K., 1941. The Vernay Cutting Expedition to Northern Burma, Part V. With notes on the collection by Dr. Ernst Mayr. — *Ibis*: 479-518.

2307   Stanford, J.K. & C.B. Ticehurst, 1939. On the birds of northern Burma. — *Ibis*, (14)3: 1-45.

2307A   Statius Müller, P.L., 1776. *Linne's Natursystems: Supplements und Registerband*: 1-536. — Nürnberg.

2308   Stegmann, B., 1934. Eine neue Form von *Xema sabini* Sabine. — *Orn. Monatsber.*, 42: 25-26.

2309   Stegmann, B., 1934. Über die Formen der grossen Möwen "subgenus *Larus*" und ihre gegenseitigen Beziehungen. — *J. f. Orn.*, 82 (3): 340-380.

2310   Stegmann, B., 1934. Zur Kenntnis der sibirischen Moorschneehühner. — *Orn. Monatsber.*, 42 (5): 150-152.

2311   Stegmann, B., 1937. *Charadrius mongolus litoralis* subsp. n. — *Orn. Monatsber.*, 45 (1): 25-26.

2312   Stegmann, B., 1938. Eine neue Form von *Lerwa lerwa* (Hodgs.). — *Orn. Monatsber.*, 46 (2): 43-44.

2313   Stegmann, B., 1949. Geographical variation in the nightjar *Caprimulgus europaeus* L. [in Russian]. — *Nature Protection*, 6 (6): 103-114.

2314   Steinbacher, J., 1962. Beiträge zur Kenntnis der Vögel von Paraguay. — *Abhandl. Senckenbergischen Naturf. Gesellsch.*, (502): 1-106.

2315   Steinbacher, J., 1979. Order Ciconiiformes, Family Threskiornithidae. Pages 253-268. In *Check-list of Birds of the World*. Vol. 1. (2nd. Edit.) (E. Mayr & G.W. Cottrell, eds.). — Mus. Comp. Zool., Cambridge, Mass.

2316   Stepanyan, L., 1961. Geograficheskaya izmenchivost' saksaul'nogo vorob'ya (*Passer ammodendri* Gould.). [Geographical variation in the Saxaul Sparrow (*Passer ammodendri* Gould.)]. — *Arch. Zool. Mus. Moscow State Univ.*, 8: 217-222.

2317   Stepanyan, L.S., 1965. Geographical variability of the rock thrush *Monticola saxatilis*. — *Sbornik Trudy Zool. Muz. MGU* [Coll. Stud. Zool. Mus. Moscow Univ.], 9: 228-231 (1964).

2318 Stepanyan, L.S., 1972. A new subspecies *Synthliboramphus antiquus microrhynchos* subsp. nov. (Alcidae, Aves) from the Commander Is. — *Ornitologiya*, 10: 388-389.

2319 Stepanyan, L.S., 1974. A new subspecies *Parus lugubris talischensis* Stepanyan ssp. n. (Paridae, Aves) from Talish. — *Byull. Mosk. O-va Ispyt. Prir. (Otd. Biol.)*, 79: 143-145.

2320 Stepanyan, L.S., 1974. *Paradoxornis heudei polivanovi* Stepanyan ssp. n. (Paradoxornithidae, Aves) from the Khanka Lake basin. — *Zool. Zhurn.*, 53 (8): 1270-1272.

2321 Stepanyan, L.S., 1975. A new subspecies of lark from the Tuva ASSR. — *Ornitologiya*, 12: 246-247.

2322 Stepanyan, L.S., 1979. *Paradoxornis heudei mongolicus* Stepanyan ssp. nov. (Paradoxornithidae, Aves) from the eastern part of the Mongolian People's Republic. — *Bull. Moscow Soc. Nat., Biol. Ser.*, 84 (3): 53-55.

2323 Stepanyan, L.S., 1983. *Nadvidy i vidy-dvoiniki v avifauna SSSR*. (Superspecies and sibling species in the avifauna of the USSR): 296. — Akad. Nauk., Moscow.

2324 Stepanyan, L.S., 1985. *Aethopyga christinae sokolovi* Stepanyan ssp. n. (Nectariniidae, Aves) from southern Vietnam. — *Ornitologiya*, 20: 133-138.

2325 Stepanyan, L.S., 1988. *Dendrocopos atratus* (Picidae, Aves) - a new species of the Vietnamese fauna. — *Biol. Nauk.*, (7): 48-52.

2326 Stepanyan, L.S., 1990. *Conspectus of the ornithological fauna of the USSR*: 1-727. — Moscow Nauka, Moscow.

2327 Stepanyan, L.S., 1992. On the problem of the taxonomic structure of the genus *Spilornis* G. R. Gray, 1840 (Accipitridae, Aves). [In Russian]. Pages 205-223. In *Zoologicheskie Issledovaniya vo V'etname* (= Zoological studies in Vietnam) (V.E. Sokolov, ed.) —, Moscow.

2328 Stepanyan, L.S., 1993. New data on the taxonomy of the genus *Spilornis* (Accipitridae, Aves). — *Zool. Zhurn.*, 72 (10): 132-145.

2329 Stepanyan, L.S., 1996. Notes on taxonomic composition of the genus *Myophonus* (Aves, Muscicapidae). — *Zool. Zhurn.*, 75 (12): 1815-1827.

2330 Stepanyan, L.S., 1998. A new subspecies of *Garrulax lineatus* (Aves, Timaliidae) from Badakhshan Mountains (the Western Pamir). — *Zool. Zhurn.*, 77 (5): 615-618.

2331 Stepanyan, L.S., 1998. On independent species status of *Paradoxornis polivanovi* (Paradoxornithidae, Aves). — *Zool. Zhurn.*, 77 (10): 1158-1161.

2332 Stepanyan, L.S., 1998. *Regulus ignicapillus caucasicus* Stepanyan subsp. n. (Regulidae, Aves) from the Western Caucasus. — *Zool. Zhurn.*, 77: 1077-1079.

2333 Stepanyan, L.S., 2000. The new subspecies *Certhia brachydactyla* (Aves, Certhiidae) from the northwest Caucasus. — *Zool. Zhurn.*, 79 (3): 333-337.

2334 Stephan, B., 1999. Zur Taxonomie mediterraner Sperlinge der Gattung *Passer* - Probleme weiterhin aktuell: Hybridisation, *italiae*, *tingitanus*. — *Mitt. Mus. Naturkd. Berlin, Zool. Reihe*, 75: 3-9.

2335 Stephan, B., 2000. Die Arten der Familie Passeridae (Gattungen *Montifringilla, Petronia, Passer*) und ihre phylogenetischen Beziehungen. — *Bonn. Zool. Beitr.*, 49: 39-70.

2336 Stevenson, H.M., 1973. An undescribed insular race of the Carolina Wren. — *Auk*, 90: 35-38.

2337 Stevenson, H.M., 1978. The populations of Boat-tailed Grackles in the southeastern United States. — *Proc. Biol. Soc. Wash.*, 91: 27-51.

2338 Stiles, F.G., 1985. Geographic variation in the Fiery-throated Hummingbird, *Panterpe insignis*. — *Orn. Monogr.*, 36: 22-30.

2339 Stiles, F.G., 1992. A new species of Antpitta (Formicariidae : *Grallaria*) from the eastern Andes of Colombia. — *Wilson Bull.*, 104 (3): 389-399.

2340 Stiles, F.G., 1995. Dos nuevas subespecies de aves de la Serrania del Chiribiquete, Departamento del Caqueta, Colombia. — *Lozania*, 66: 1-16.

2341 Stiles, F.G., 1996. A new species of Emerald Hummingbird (Trochilidae, *Chlorostilbon*) from the Sierra de Chiribiquete, southeastern Colombia, with a review of the *C. mellisugus* complex. — *Wilson Bull.*, 108 (1): 1-27.

2342 Stiles, F.G., 1996. Dos nuevas subespecies de aves de la Serrania del Chiribiquete, Departamento del Caqueta, Colombia. — *Lozania*, 66: 1-16.

2343 Stiles, F.G., 1996. When black plus white equals gray: the nature of variation in the Variable Seedeater complex (Emberizinae: *Sporophila*). — *Orn. Neotrop.*, 7: 75-107.

2344 Stiles, F.G., 1999. Species accounts within the Family Trochilidae (Hummingbirds). Pages 549-680. In *Handbook of the Birds of the World*. Vol. 5 (J. del Hoyo, A. Elliott & J. Sargatal, eds.). — Lynx Edicions, Barcelona.

2345 Stone, W., 1931. Three new birds from Honduras. — *Proc. Acad. Nat. Sci. Philad.*, 83: 1-3.

2346 Stonehouse, B., 1970. Geographic variation in Gentoo Penguins *Pygoscelis papua*. — *Ibis*, 112 (1): 52-57.

2347 Storer, R.W., 1950. Geographic variation in the pigeon guillemots of North America. — *Condor*, 52 (1): 28-31.

2348 Storer, R.W., 1970. Family Emberizidae, Subfamily Thraupinae. Pages 246-408. In *Check-list of Birds of the World*. Vol. 13 (R.A. Paynter, Jr., ed.). — Mus. Comp. Zool., Cambridge, Mass.

2349 Storer, R.W., 1979. Order Gaviiformes. Pages 135-139. In *Check-list of Birds of the World*. Vol. 1. (2ⁿᵈ. Edit.) (E. Mayr & G.W. Cottrell, eds.). — Mus. Comp. Zool., Cambridge, Mass.

2350 Storer, R.W., 1979. Order Podicipediformes. Pages 140-155. In *Check-list of Birds of the World*. Vol. 1. (2ⁿᵈ. Edit.) (E. Mayr & G.W. Cottrell, eds.). — Mus. Comp. Zool., Cambridge, Mass.

2351 Storer, R.W., 1988. *Type specimens of birds in the collections of the University of Michigan Museum of Zoology*. — Misc. Publ. Mus. Zool. Univ. Michigan, 174: i-iv, 1-69.

2352 Storer, R.W. & T. Getty, 1985. Geographic variation in the Least Grebe (*Tachybaptus dominicus*). — *Orn. Monogr.*, 36: 31-39.

2353 Storr, G.M., 1980. The western subspecies of the Cape Barren Goose *Cereopsis novaehollandiae grisea* (Vieillot). — *West. Austral. Nat.*, 14: 202-203.

2354 Stotz, D.F., 1990. The taxonomic status of *Phyllornis reiseri*. — *Bull. Brit. Orn. Cl.*, 110 (4): 184-187.

2355 Stotz, D.F., 1992. A new subspecies of *Aramides cajanea* from Brazil. — *Bull. Brit. Orn. Cl.*, 112 (4): 231-234.

2356 Stotz, D.F., 1992. Specific status and nomenclature of *Hemitriccus minimus* and *Hemitriccus aenigma*. — *Auk*, 109 (4): 916-917.

2357 Straneck, R. & F. Vidoz, 1995. Sobre el estado taxonomico de *Strix rufipes* (King) y de *Strix chocoensis* (Cherrie & Reichenberger). — *Notulas Faunisticas*, 74: 1-5.

2357A Straube, F.C., 1994. On the validity of *Anumbius annumbi machrisi* Stager, 1959 (Furnariidae, Aves). — *Bull. Brit. Orn. Cl.*, 114: 46-47.

2358 Straube, F.C. & M.R. Bornschein, 1991. Revisão das subespécies de *Baryphthengus ruficapillus* (Coraciiformes : Momotidae). — *Ararajuba*, 2: 65-67.

2359 Stresemann, E., 1936. A nominal list of the birds of the Celebes. — *Ibis*, (13) 6: 356-369.

2360 Stresemann, E., 1937. *Francolinus coqui hoeschianus* subsp. nova. — *Orn. Monatsber.*, 45 (2): 66-67.

2361 Stresemann, E., 1938. *Turnix sylvatica arenaria* subsp. nova. — *Orn. Monatsber.*, 46: 26.

2362 Stresemann, E., 1939. Zwei neue Vogelrassen aus Südwest-Afrika. — *Orn. Monatsber.*, 47: 61-62.

2363 Stresemann, E., 1940. Die Vogel von Celebes. III. Systematik und Biologie. — *J. f. Orn.*, 88 (3): 389-487.

2364 Stresemann, E., 1941. Die Vogel von Celebes. III. Systematik und Biologie. — *J. f. Orn.*, 89 (1): 2-102.

2365 Stresemann, E., 1950. Birds collected during Capt. James Cook's last expedition (1776-1780). — *Auk*, 67: 66-88.

2366 Stresemann, E., 1952. On the birds collected by Pierre Poivre in Canton, Manila, India and Madagascar (1751-1756). — *Ibis*, 94 (3): 499-523.

2367 Stresemann, E., 1953. Vögel, gesammelt von Labilliardiere während der 'Voyage à la recherche de la Perouse' 1791-1794. — *Mitt. Zool. Mus. Berlin*, 29 (1): 75-106.

2368   Stresemann, E. & D. Amadon, 1979. Order Falconiformes. Pages 271-425. In *Check-list of Birds of the World*. Vol. 1. (2nd. Edit.) (E. Mayr & G.W. Cottrell, eds.). — Mus. Comp. Zool., Cambridge, Mass.

2369   Stresemann, E. *et al.*, 1934. Vorläufiges über die ornithologischen Ergebnisse der Expedition Stein 1931-1932. II. Zur Ornithologie des Weyland-Gebirges in Niederländisch Neuguinea. — *Orn. Monatsber.*, 42: 43-46.

2370   Stresemann, E. & V. Stresemann, 1972. Die postnuptiale und die praenuptiale Vollmauser von *Pericrocotus divaricatus* Raffles. — *J. f. Orn.*, 113: 435-439.

2371   Stuart, S.N., 1992. Various species accounts. In *The Birds of Africa*. Vol. 4: i-xv, 1-609 (S. Keith, E.K. Urban & C.H. Fry, eds.). — Academic Press Inc., San Diego.

2372   Stuart, S.N. & N.J. Collar, 1985. Subspeciation in the Karamoja Apalis *Apalis karamojae*. — *Bull. Brit. Orn. Cl.*, 105: 86-89.

2373   Sudilovskaya, A.M., 1934. Zametki o nekotorikh tsentralno-asiatskikh formakh ptits. [Notes on some central Asian forms of birds.]. — *Sbornik Trudy Zool. Muz. MGU* [Coll. Stud. Zool. Mus. Moscow Univ.], 1: 109-111.

2374   Summers-Smith, D., 1984. The Rufous Sparrows of the Cape Verde Islands. — *Bull. Brit. Orn. Cl.*, 104 (4): 138-142.

2375   Sutter, E., 1955. Uber die mauser einiger laufhuhnchen und die rassen von *Turnix maculosa* und *sylvatica* im Indo-Australischen Gebiet. — *Verh. Naturf. Ges. Basel*, 66 (1): 85-139.

2376   Sutton, G.M., 1941. A new race of *Chaetura vauxi* from Tamaulipas. — *Wilson Bull.*, 53 (4): 231-233.

2377   Sutton, G.M., 1955. A new race of Olivaceous Woodcreeper from Mexico. — *Wilson Bull.*, 67 (3): 209-211.

2378   Sutton, G.M. & T.D. Burleigh, 1939. A new screech owl from Nuevo Leon. — *Auk*, 56: 174-175.

2379   Svensson, L., 2001. The correct name of the Iberian Chiffchaff *Phylloscopus ibericus* Ticehurst, 1937, its identification and new evidence of its winter grounds. — *Bull. Brit. Orn. Cl.*, 121 (4): 281-296.

2380   Swierczewski, E.V. & R.J. Raikow, 1981. Hind limb morphology, phylogeny, and classification of the Piciformes. — *Auk*, 98: 466-480.

2381   Swinhoe, R., 1859. Notes on some new species of birds found on the island of Formosa. — *J. North China Br. Roy. Asiat. Soc.*, 1: 225-230.

2382   Swinhoe, R., 1864. Descriptions of four new species of Formosan birds. – *Ibis* (1) 6: 361-363.

2382A   Swinhoe, R., 1870. On the ornithology of Hainan. – *Ibis*, (2) 6: 77-97, 230-256, 342-367.

2383   Taka-Tsukasa, N., 1944. *Studies on the Galli of Nippon*: 1-1000⁵. Privately published, Tokyo.

2384   Tan, Yao-kuang & Wu, Zhikang, 1981. A new subspecies of the Silver Pheasant from Guizhou, China - *Lophura nyctemera [sic] rongjiangensis*. — *Zool. Res.* [China], 2 (4): 301-305.

2385   Tamplin, J.W. *et al.*, 1993. Biochemical and morphometric relationships among some members of the Cardinalinae. — *Wilson Bull.*, 105: 93-113.

2386   Taylor, P.B., 1992. Various species accounts. In *The Birds of Africa*. Vol. 4: i-xv, 1-609 (S. Keith, E.K. Urban & C.H. Fry, eds.). — Academic Press Inc., San Diego.

2387   Taylor, P.B., 1996. Family Rallidae (Rails, Gallinules and Coots). Pages 108-209. In *Handbook of the Birds of the World*. Vol. 3 (J. del Hoyo, A. Elliott & J. Sargatal, eds.). — Lynx Edicions, Barcelona.

2388   Teixeira, D.M., 1987. A new tyrannulet (*Phylloscartes*) from northwestern Brazil. — *Bull. Brit. Orn. Cl.*, 107: 37-41.

2389   Teixeira, D.M. & N. Carnevalli, 1989. Nova especie de *Scytalopus* Gould, 1837, do nordeste do Brasil (Passeriformes, Rhinocryptidae). — *Bol. Mus. Nac. Rio de Janeiro, Brasil, Nova Sér., Zool.*, 331: 1-11.

2390   Teixeira, D.M. & L.P. Gonzaga, 1983. A new antwren from northeastern Brazil. — *Bull. Brit. Orn. Cl.*, 103: 133-135.

2391   Teixeira, D.M. & L.P. Gonzaga, 1983. Um novo Furnariidae do nordeste do Brasil: *Philydor novaesi* sp. nov. (Aves, Passeriformes). — *Bol. Mus. Para. Emílio Goeldi, Nov. Ser., Zool.*, 124: 1-22.

2392   Teixeira, D.M. & L.P. Gonzaga, 1985. Uma nova subespecie de *Myrmotherula unicolor* (Ménétriès, 1835) (Passeriformes, Formicariidae) do nordeste do Brasil. — *Bol. Mus. Nac. Rio de Janeiro, Brasil, Nova Sér., Zool.*, 310: 1-16.

2393   Teixeira, D.M. *et al.*, 1987. Notes on some birds of northeastern Brazil (2). — *Bull. Brit. Orn. Cl.*, 107 (4): 151-157.

2394   Teixeira, D.M. *et al.*, 1989. Notes on some birds of north-eastern Brazil (4). — *Bull. Brit. Orn. Cl.*, 109 (3): 152-157.

2395   Temminck, C.J. & M. Laugier, 1820-1839. *Nouveaux receuil de planches coloriées d'oiseaux, pour servir de suite et de complement aux planches enluminées de Buffon*: 102 livraisons (600 pls.). — Levrault; Dufour et d'Ocagne, Paris.

2396   Themido, A., 1937. Un nouveau "*Francolinus*" de l'Angola. — *12e Congr. Int. Zool., Lisboa, Comptes rendus in Arquivos do Mus. Bocage*, 6a, 3 (9): 1833-1834.

2397   Thiede, W., 1963. *The distribution of the Redshank (*Tringa totanus L.).*, University of Bonn, PhD. thesis, Bonn.

2398   Thiollay, J.M., 1994. Family Accipitridae. In *Handbook of the Birds of the World*. Vol. 2: 1-638 (J. del Hoyo, A. Elliott & J. Sargatal, eds.). — Lynx Edicions, Barcelona.

2399   Thomas, B.T., 1996. Family Opisthocomidae (Hoatzin). Pages 24-32. In *Handbook of the Birds of the World*. Vol. 3 (J. del Hoyo, A. Elliott & J. Sargatal, eds.). — Lynx Edicions, Barcelona.

2400   Thompson, M.C. & P. Temple, 1964. Geographic variations in the Coot in New Guinea. — *Proc. Biol. Soc. Wash.*, 77: 251-252.

2401   Thonglongya, K., 1968. A new martin of the genus *Pseudochelidon* from Thailand. — *Thai Nat. Sci. Pap., Fauna Ser.*, 1: 3-10.

2402   Thumser, N.N. & J.D. Karron, 1994. Patterns of genetic polymorphism in five species of penguins. — *Auk*, 111: 1018-1022.

2403   Ticehurst, C.B., 1936. A note on Razorbills. — *Ibis*, 13 (6): 381-383.

2404   Tien, D.V., 1961. Recherches zoologiques dans la région de Thai-Nguyen (Nord-Vietnam). — *Zool. Anz. Leipzig*, 166: 298-308.

2405   Todd, W.E.C., 1937. The Pigeons of the *Columba plumbea* group. — *Proc. Biol. Soc. Wash.*, 50: 185-190.

2406   Todd, W.E.C., 1940. Eastern races of the Ruffed Grouse. — *Auk*, 57: 390-397.

2407   Todd, W.E.C., 1947. A new name for *Bonasa umbellus canescens*. — *Auk*, 64 (2): 326.

2408   Todd, W.E.C., 1947. New South American Parrots. — *Ann. Carnegie Mus.*, 30: 331-338.

2409   Todd, W.E.C., 1947. The Venezuelan races of *Piaya cayana*. — *Proc. Biol. Soc. Wash.*, 60: 59-60.

2410   Todd, W.E.C., 1947. Two new South American pigeons. — *Proc. Biol. Soc. Wash.*, 60: 67-68.

2411   Todd, W.E.C., 1947. Two new owls from Bolivia. — *Proc. Biol. Soc. Wash.*, 60: 95-96.

2412   Todd, W.E.C., 1948. Critical remarks on the Wood-hewers. — *Ann. Carnegie Mus.*, 31 (2): 5-18.

2413   Todd, W.E.C., 1950. Two apparently new oven-birds from Colombia. — *Proc. Biol. Soc. Wash.*, 63: 85-88.

2414   Todd, W.E.C., 1953. A taxonomic study of the American Dunlin (*Erolia alpina* sub-spp.). — *J. Wash. Acad. Sci.*, 43: 85-88.

2415   Todd, W.E.C., 1954. A new gallinule from Bolivia. — *Proc. Biol. Soc. Wash.*, 67: 85-86.

2416   Tomkovich, P.S., 1986. Geographical variability of the Dunlin in the Far East. — *Byull. Mosk. O-va Ispyt. Prir. (Otd. Biol.)*, 91 (5): 3-15.

2417   Tomkovich, P.S., 1990. Analysis of geographical variability in Knot *Calidris canutus* (L.). — *Byull. Mosk. O-va Ispyt. Prir. (Otd. Biol.)*, 95 (5): 59-72.

2418   Tomkovich, P.S., 2001. A new subspecies of Red Knot *Calidris canutus* from the New Siberian Islands. — *Bull. Brit. Orn. Cl.*, 121 (4): 257-263.

2419   Tomkovich, P.S. & L. Serra, 1999. Morphometrics and prediction of breeding origin in some Holarctic waders. — *Ardea*, 87: 289-300.

2420   Toschi, A., 1958. Una nuova forma di Francolino dall' Abissinia. — *Ric. Zool. appl. Lab Zool. Univ. Bologna, Caccia, Suppl.*, 2 (9): 285-291.

2421   Traylor, M.A., Jr., 1948. New birds from Peru and Ecuador. — *Fieldiana, Zool.*, 31 (24): 195-200.

2422   Traylor, M.A., Jr., 1951. Notes on some Peruvian birds. — *Fieldiana, Zool.*, 31: 613-621.

2423   Traylor, M.A., Jr., 1952. A new race of *Otus ingens* (Salvin) from Colombia. — *Nat. Hist. Misc., Chicago Acad. Sci.*, 99: 1-4.

2424   Traylor, M.A., Jr., 1962. New birds from Barotseland. — *Fieldiana, Zool.*, 44: 113-115.

⁵ Copy not located and pagination unknown.

2425   Traylor, M.A., Jr., 1962. A new pipit from Angola. — *Bull. Brit. Orn. Cl.*, 82: 76-77.
2426   Traylor, M.A., Jr., 1964. Three new birds from Africa. — *Bull. Brit. Orn. Cl.*, 84: 81-84.
2427   Traylor, M.A., Jr., 1967. A collection of birds from Szechwan. — *Fieldiana, Zool.*, 53 (1): 1-67.
2428   Traylor, M.A., Jr., 1968. Family Ploceidae, subfamily Viduinae. Pages 390-397. In *Check-list of Birds of the World*. Vol. 14 (R.A. Paynter, Jr., ed.). — Mus. Comp. Zool., Cambridge, Mass.
2429   Traylor, M.A., Jr., 1970. Two new birds from the Ivory Coast. — *Bull. Brit. Orn. Cl.*, 90: 78-80.
2430   Traylor, M.A., Jr., 1970. A new race of *Serinus citrinelloides*. — *Bull. Brit. Orn. Cl.*, 90: 83-86.
2431   Traylor, M.A., Jr., 1979. Family Tyrannidae, Subfamily Elaeniinae. Pages 3-112. In *Check-list of Birds of the World*. Vol. 8 (M.A. Traylor, Jr., ed.). — Mus. Comp. Zool., Cambridge, Mass.
2432   Traylor, M.A., Jr., 1979. Family Tyrannidae, Subfamily Fluvicolinae. Pages 112-186. In *Check-list of Birds of the World*. Vol. 8 (M.A. Traylor, Jr., ed.). — Mus. Comp. Zool., Cambridge, Mass.
2433   Traylor, M.A., Jr., 1979. Family Tyrannidae, Subfamily Tyranninae. Pages 186-229. In *Check-list of Birds of the World*. Vol. 8 (M.A. Traylor, Jr., ed.). — Mus. Comp. Zool., Cambridge, Mass.
2434   Traylor, M.A., Jr., 1982. Notes on tyrant flycatchers (Aves: Tyrannidae). — *Fieldiana, Zool.*, N.S. 13: 1-22.
2435   Traylor, M.A., Jr., 1988. Geographic variation and evolution in South American *Cistothorus platensis* (Aves: Troglodytidae). — *Fieldiana, Zool.*, 48: i-iii, 1-35.
2436   Tréca, B. & C. Érard, 2000. A new subspecies of the Red-billed Hornbill, *Tockus erythrorhynchus* from West Africa. — *Ostrich*, 71: 363-366.
2437   Trewick, S.A., 1997. Flightlessness and phylogeny amongst the endemic rails (Aves: Rallidae) of the New Zealand region. — *Phil. Trans. Roy. Soc. London*, B. 352: 429-446.
2438   Triggs, S.J. & C.H. Daugherty, 1996. Conservation and genetics of New Zealand parakeets. — *Bird Conserv. Int.*, 6: 89-101.
2438A Trimen, R., 1882. On an apparently undescribed Sun-Bird from Tropical South-Western Africa. — *Proc. Zool. Soc., Lond.*,: 451-452.
2439   Trischitta, A., 1939. *Alcune nuove forme di Uccelli Italiani*: 1-5. — Arti Grafiche Soluto, Bagheria.
2440   Trischitta, A., 1939. *Altre nuove forme di Uccelli Italiani*: 1-4. — Arti Grafiche Soluto, Bagheria.
2441   Turner, D.A., 1997. Family Musophagidae (Turacos). Pages 480-506. In *Handbook of the Birds of the World*. Vol. 4 (J. del Hoyo, A. Elliott & J. Sargatal, eds.). — Lynx Edicions, Barcelona.
2442   Turner, D.A. *et al.*, 1991. Taxonomic notes on some East African birds. Pt. 1. Non-passerines. — *Scopus*, 14 (2): 84-91.
2442A Tweeddale, A., Marquis of, 1877. Descriptions of three new species of birds from the Indian region. — *Proc. Zool. Soc., Lond.*,: 366-367.
2443   Twomey, A.C., 1950. A new race of paroquet of the species *Aratinga astec* from the Republic of Honduras. — *Ann. Carnegie Mus.*, 31: 297-298.
2444   Tye, A., 1989. The systematic position of the Buff-streaked Chat (*Oenanthe/Saxicola bifasciata*). — *Bull. Brit. Orn. Cl.*, 109 (1): 53-58.
2445   Tye, A., 1991. A new subspecies of Forest Scrub-Robin *Cercotrichas leucosticta* from West Africa. — *Malimbus*, 13 (2): 74-77.
2446   Tye, A., 1992. A new subspecies of *Cisticola bulliens* from northern Angola. — *Bull. Brit. Orn. Cl.*, 112 (1): 55-56.
2447   Tye, A., 1992. Various species accounts. In *The Birds of Africa*. Vol. 4: i-xv, 1-609 (S. Keith, E.K. Urban & C.H. Fry, eds.). — Academic Press Inc., San Diego.
2448   Tye, A., 1997. Various species accounts. In *The Birds of Africa*. Vol. 5: i-xix, 1-669 (E.K. Urban, C.H. Fry & S. Keith, eds.). — Academic Press, London.
2449   Unitt, P. & A.M. Rea, 1997. Taxonomy of the Brown Creeper in California. Pages 177-185. In *The Era of Allan R. Phillips: A Festschrift* (R.W. Dickerman, ed.) — Albuquerque.
2450   Unitt, P. *et al.*, 1996. Taxonomy of the Marsh Wren in southern California. — *Proc. San Diego Soc. Nat. Hist.*, 31: 1-20.
2451   Urban, E.K., 1986. Various species accounts. In *The Birds of Africa*. Vol. 2: i-xvi, 1-552 (E.K. Urban, C.H. Fry & S. Keith, eds.). — Academic Press, London.
2452   Urban, E.K., 1997. Various species accounts. In *The Birds of Africa*. Vol. 5: i-xix, 1-669 (E.K. Urban, C.H. Fry & S. Keith, eds.). — Academic Press, London.
2453   Urban, E.K., 2000. Various species accounts. In *The Birds of Africa*. Vol. 6: i-xvii, 1-724 (C.H. Fry, S. Keith & E.K. Urban, eds.). — Academic Press, London.
2454   Urban, E.K. & S. Keith, 1992. Various species accounts. In *The Birds of Africa*. Vol. 4: i-xv, 1-609 (S. Keith, E.K. Urban & C.H. Fry, eds.). — Academic Press Inc., San Diego.
2455   Urban, E.K. *et al.* (eds) 1997. *The Birds of Africa*. Vol. 5: i-xix, 1-669. — Academic Press, London.
2456   Uttal, L.J., 1939. Subspecies of the Spruce Grouse. — *Auk*, 56 (4): 460-464.
2457   Valqui, T. & J. Fjeldså, 1999. New brush-finch *Atlapetes* from Peru. — *Ibis*, 141 (2): 194-198.
2457A Valqui, T. & J. Fjeldså, 2002. *Atlapetes melanopsis* nom. nov. for the Black-faced Brush-Finch. — *Ibis*, 144 (2): 347.
2458   van Balen, S., 1993. The identification of tit-babblers and red sunbirds on Java. — *OBC Bull.*, 18: 26-28.
2459   van Balen, S., 1998. A hybrid munia? — *Bull. Brit. Orn. Cl.*, 118: 118-119.
2460   van Bemmel, A.C.V., 1940. Ornithologische Notizen I-III. — *Treubia*, 17: 333-335.
2461   van Bemmel, A.C.V. & A. Hoogerwerf, 1940. The birds of Goenoeng Api. — *Treubia*, 17 (5): 421-472.
2462   van Bemmel, A.C.V. & K.H. Voous, 1951. The birds of the islands of Muna and Buton, S. E. Celebes. — *Treubia*, 21: 27-104.
2463   van den Elzen, R., 1985. Systematics and evolution of African canaries and seedeaters. In *Proc. Int. Symp. Afr. Vertebr. Mus. A. Koenig, Bonn* (ed. K.-L. Schuchmann). 1984: 435-451.
2463A van den Elzen, R. & C. König, 1983. Vögel des (Süd-)Sudan: taxonomische und tiergeographische Bemerkungen. – *Bonn Zool. Beitr.*, 34: 149-196.
2464   van den Hoek Ostende, L.W. *et al.*, 1997. Type-specimens of birds in the National Museum of Natural History, Leiden. — *NNM Tech. Bull.*, 1: 1-248.
2465   van Franeker, J.A., 1995. Kleurfasen van de Noordse Stormvogel *Fulmarus glacialis* in de Noordatlantische Oceaan. Colour phases of the Fulmar *Fulmarus glacialis* in the North Atlantic. — *Sula*, 9: 93-106.
2466   van Franeker, J.A. & J. Wattel, 1982. Geographical variation of the fulmar in the North Atlantic. — *Ardea*, 70: 31-44.
2467   van Gils, J. & P. Wiersma, 1996. Species accounts for the family Scolopacidae (Sandpipers, Snipes and Phalaropes). Pages 489-533. In *Handbook of the Birds of the World*. Vol. 3 (J. del Hoyo, A. Elliott & J. Sargatal, eds.). — Lynx Edicions, Barcelona.
2468   van Marle, J.G. & K.H. Voous, 1988. The birds of Sumatra, an annotated checklist. — *BOU Check-list Ser. No.* 10: 1-265.
2469   van Rossem, A.J., 1934. Critical notes on Middle American birds. A. Notes on some species and subspecies of Guatemala. — *Bull. Mus. Comp. Zool.*, 77: 387-405.
2470   van Rossem, A.J., 1934. Critical notes on Middle American Birds. C. A systematic report on the Brewster collection of Mexican birds. — *Bull. Mus. Comp. Zool.*, 77 (7): 424-490.
2471   van Rossem, A.J., 1934. A race of *Porzana flaviventer* from Central America. — *Condor*, 36 (6): 243-244.
2472   van Rossem, A.J., 1934. Two new races of the Black Chachalaca from Central America. — *Trans. San Diego Soc. Nat. Hist.*, 7: 363-365.
2473   van Rossem, A.J., 1937. A review of the races of the mountain quail. — *Condor*, 39: 20-24.
2474   van Rossem, A.J., 1938. Descriptions of three new birds from western Mexico. — *Trans. San Diego Soc. Nat. Hist.*, 9: 9-12.
2475   van Rossem, A.J., 1939. An overlooked race of the California Quail. — *Auk*, 56 (1): 68-69.
2476   van Rossem, A.J., 1939. Some new races of birds from Mexico. — *Ann. Mag. Nat. Hist.*, 11 (4): 439-443.
2477   van Rossem, A.J., 1941. A race of the Poor-will from Sonora. — *Condor*, 43: 247.

2478   van Rossem, A.J., 1942. Notes on some Mexican and Californian birds, with descriptions of six undescribed races. — *Trans. San Diego Soc. Nat. Hist.*, 9: 377-384.

2479   van Rossem, A.J., 1946. The California Quail of Central Baja California. — *Condor*, 48: 265-267.

2480   van Rossem, A.J., 1947. Comments on certain birds of Baja California, including descriptions of three new races. — *Proc. Biol. Soc. Wash.*, 60: 51-58.

2481   van Rossem, A.J. & M. Hachisuka, 1937. A further report on birds from Sonora, Mexico, with descriptions of two new races. — *Trans. San Diego Soc. Nat. Hist.*, 8 (23): 321-336.

2482   van Rossem, A.J. & M. Hachisuka, 1937. A race of Verreaux's Dove from Sonora. — *Proc. Biol. Soc. Wash.*, 50: 199-200.

2483   van Rossem, A.J. & M. Hachisuka, 1939. A race of the Military Macaw from Sonora. — *Proc. Biol. Soc. Wash.*, 52: 13-14.

2484   van Saceghem, R., 1942. Une nouvelle variété de francolin. *Francolinus camerunensis* var. *Ruandae*. — Zooleo (Bull. Soc. Bot. Zool. Cong.), 5: 18-19.

2485   van Someren, V.G.L., 1938. A new race of Grey-wing Francolin from Kenya Colony. — *Bull. Brit. Orn. Cl.*, 59: 7.

2486   van Someren, V.G.L., 1939. Reports on the Corydon Museum Expedition to the Chyulu Hills. 2. The birds of the Chyulu Hills. — *J. East Afr. & Uganda Nat. Hist. Soc.*, 14: 15-129.

2487   van Tets, G.F., 1965. A comparative study of some social communication patterns in the Pelecaniformes. — *Orn. Monogr.*, 2.

2488   van Tuinen, M. *et al.*, 2001. Convergence and divergence in the evolution of aquatic birds. — *Proc. Roy. Soc., Lond.*, 268B: 1-6.

2489   van Tuinen, M. *et al.*, 2000. The early history of modern birds inferred from DNA sequences of nuclear and mitochondrial ribosomal genes. — *Mol. Biol. Evol.*, 17: 451-457.

2490   van Tyne, J. & W.N. Koelz, 1936. Seven new birds from the Punjab. — *Occ. Pap. Mus. Zool., Univ. Michigan*, 334: 1-6.

2491   van Tyne, J. & M.B. Trautman, 1941. New Birds from Yucatán. — *Occ. Pap. Mus. Zool., Univ. Michigan*, 439: 1-11.

2492   Vaurie, C., 1951. Notes on some Asiatic swallows. — *Amer. Mus. Novit.*, 1529: 1-47.

2493   Vaurie, C., 1954. Systematic Notes on Palearctic Birds. No. 6. Timaliinae and Paradoxornithinae. — *Amer. Mus. Novit.*, 1669: 1-12.

2494   Vaurie, C., 1954. Systematic Notes on Palearctic Birds. No. 10. Sylviinae: the genera *Cettia, Hippolais* and *Locustella*. — *Amer. Mus. Novit.*, 1691: 1-9.

2495   Vaurie, C., 1954. Systematic Notes on Palearctic Birds. No. 12. Muscicapinae, Hirundinidae, and Sturnidae. — *Amer. Mus. Novit.*, 1694: 1-18.

2496   Vaurie, C., 1955. Systematic Notes on Palearctic Birds. No. 16. Troglodytiinae, Cinclidae, and Prunellidae. — *Amer. Mus. Novit.*, 1751: 1-25.

2497   Vaurie, C., 1955. Systematic Notes on Palearctic Birds. No. 18. Supplementary Notes on Corvidae, Timaliinae, Alaudidae, Sylviinae, Hirundinidae, and Turdinae. — *Amer. Mus. Novit.*, 1753: 1-19.

2498   Vaurie, C., 1956. Systematic Notes on Palearctic Birds. No. 20. Fringillidae: the Genera *leucosticte, Rhodopechys, Carpodacus, Pinicola, Loxia, Uragus, Urocynchramus* and *Propyrrhula*. — *Amer. Mus. Novit.*, 1786: 1-37.

2499   Vaurie, C., 1957. Systematic Notes on Palearctic Birds. No. 27. Paridae: the Genera *Parus* and *Sylviparus*. — *Amer. Mus. Novit.*, 1852: 1-43.

2500   Vaurie, C., 1958. Systematic Notes on Palearctic Birds. No. 32. Oriolidae, Dicruridae, Bombycillidae, Pycnonotidae, Nectariniidae and Zosteropidae. — *Amer. Mus. Novit.*, 1869: 1-28.

2501   Vaurie, C., 1958. Remarks on some corvidae of Indo-Malaya and the Australian region. — *Amer. Mus. Novit.*, 1915: 1-13.

2502   Vaurie, C., 1959. *The birds of the Palearctic Fauna. Order Passeriformes*: i-xiii, 1-762. — H. F. & G. Witherby, London.

2503   Vaurie, C., 1959. Systematic notes on Palearctic birds. No. 35. Picidae. The genus *Dendrocopos* (part 1). — *Amer. Mus. Novit.*, 1946: 1-29.

2504   Vaurie, C., 1959. Systematic Notes on Palearctic Birds. No. 37. Picidae: The Subfamilies Jynginae and Picumninae. — *Amer. Mus. Novit.*, 1963: 1-16.

2505   Vaurie, C., 1960. Systematic notes on Palearctic birds. No. 39. Caprimulgidae: A new species of *Caprimulgus*. — *Amer. Mus. Novit.*, 1985: 1-10.

2506   Vaurie, C., 1961. A new subspecies of the Nubian Bustard. — *Bull. Brit. Orn. Cl.*, 81: 26-27.

2507   Vaurie, C., 1961. Systematic Notes on Palearctic Birds. No. 44. Falconidae: The Genus *Falco* (Part 1. *Falco peregrinus* and *Falco peregrinoides*). — *Amer. Mus. Novit.*, 2035: 1-19.

2508   Vaurie, C., 1962. Family Dicruridae. Pages 137-157. In *Check-list of Birds of the World*. Vol. 15 (E. Mayr & J.C. Greenway, Jr., eds.). — Mus. Comp. Zool., Cambridge, Mass.

2509   Vaurie, C., 1963. Systematic Notes on Palearctic Birds. No. 51. A review of *Burhinus oedicnemus*. — *Amer. Mus. Novit.*, 2131: 1-13.

2510   Vaurie, C., 1965. *The birds of the Palearctic Fauna. Non Passeriformes*: i-xx, 1-763. — H. F. & G. Witherby, London.

2511   Vaurie, C., 1965. Distribution regionale et altitudinale des genres *Garrulax* et *Babax* et notes sur leur systematique. — *L'Oiseau*, 35 (Special): 141-152.

2512   Vaurie, C., 1965. Systematic Notes on the bird family Cracidae. No. 3. *Ortalis guttata, Ortalis superciliaris* and *Ortalis motmot*. — *Amer. Mus. Novit.*, 2232: 1-21.

2513   Vaurie, C., 1966. Systematic Notes on the bird family Cracidae. No. 5. *Penelope purpurascens, Penelope jacquacu* and *Penelope obscura*. — *Amer. Mus. Novit.*, 2250: 1-23.

2514   Vaurie, C., 1966. Systematic Notes on the bird family Cracidae. No. 6. Reviews of 9 species of Penelope. — *Amer. Mus. Novit.*, 2251: 1-30.

2515   Vaurie, C., 1971. Systematic status of *Synallaxis demissa* and *S. poliophrys*. — *Ibis*, 113 (4): 520-521.

2516   Vaurie, C., 1980. Taxonomy and geographical distribution of the Furnariidae (Aves, Passeriformes). — *Bull. Amer. Mus. Nat. Hist.*, 166: 1-357.

2517   Vaurie, C. *et al.*, 1960. Family Motacillidae. Pages 129-167. In *Check-list of Birds of the World*. Vol. 9 (E. Mayr & J.C. Greenway, Jr., eds.). — Mus. Comp. Zool., Cambridge, Mass.

2518   Vaurie, C. *et al.*, 1972. Taxonomy of *Schizoeaca fuliginosa* (Furnariidae), with description of two new subspecies. — *Bull. Brit. Orn. Cl.*, 92: 142-144.

2519   Verheyen, R., 1941. Etude des formes géographiques de la faune ornithologiques belge. — *Bull. Mus. Royal Hist. Nat. Belg.*, 17 (33): 1-32.

2520   Verheyen, R., 1946. Notes sur la faune ornithologique de l'Afrique centrale. VI. Description d'un nouveau Strigidé du congo belge. — *Bull. Mus. Hist. nat. Belge*, 22 (21): 1-2.

2521   Verheyen, R., 1947. Notes sur la faune ornithologique de l'Afrique centrale. VII. Liste d'une collection d'oiseaux rares réunie à Albertville et description d'un nouveau Touraco du Congo belge. — *Bull. Mus. Hist. nat. Belge*, 23 (9): 1-4.

2522   Verheyen, R., 1951. Notes sur la faune ornithologique de l'Afrique centrale. IX. Description de trois oiseaux nouveaux du Katanga (Congo belge). — *Bull. Mus. Hist. Nat. Belge*, 27 (50): 1-2.

2523   Veron, G. & B.J. Winney, 2000. Phylogenetic relationships within the turacos (Musophagidae). — *Ibis*, 142 (3): 446-456.

2524   Vieillot, L.P., 1818. *Nouveau Dictionnaire d'Histoire Naturelle*. 23. NIL-ORC. 1-612. — Paris.

2525   Vieillot, L.P., 1819. *Nouveau Dictionnaire d'Histoire Naturelle*. 32. SPH-TAZ. 1-595. — Paris.

2526   Vielliard, J., 1976. La Sitelle Kabyle. — *Alauda*, 44 (3): 351-352.

2527   Vielliard, J., 1989. Uma nova espécie de *Glaucidium* (Aves, Strigidae) da Amazonia. — *Rev. Bras. Zool.*, 6: 685-693.

2528   Vielliard, J., 1990. Uma nova especie de *Asthenes* da serra do Cipo, Minas Gerais, Brasil. — *Ararajuba*, 1: 121-122.

2529   Vigors, N.A., 1832. Observations on a collection of birds from the Himalayan Mountains, with characters of new genera and species. — *Proc. Comm. Zool. Soc., Lond.*, 1: 170-176 (1830/1831).

2530   Vincent, J., 1949. Systematic notes (Part IV). — *Ostrich*, 20: 145-152.

2531   Vo Quy, 1975. *Birds of Vietnam*. Vol. 1 [in Vietnamese: Chim Viêt Nam - hình thái và phân loai]: 1-649. — Science and Technology, Hanoi.

2532   Vo Quy, 1981. *Birds of Vietnam*. Vol. 2 [in Vietnamese: Chim Viêt Nam - hình thái và phân loai]: 1-394. — Science and Technology, Hanoi.

2533   Voelker, G., 1999. Dispersal, vicariance, and clocks: historical biogeography and speciation in a cosmopolitan passerine genus (Anthus: Motacillidae). — *Evolution*, 53: 1536-1552.

2534   Voelker, G., 1999. Molecular evolutionary relationships in the avian genus *Anthus* (Pipits: Motacillidae). — *Molec. Phylogen. Evol.*, 11: 84-94.

2535   Voisin, J.-F., 1971. Description de *Sterna virgata mercuri* de l'Archipel Crozet. — *Terres Aust. Antarct. Franc.*, 54: 44-49.

2536   von Berlepsch, H., 1889. Systematisches Verzeichniss der von Herrn Gustav Garlepp in Brasilien und Nord-Peru, im Gebiete des oberen Amazonas, gesammelten Vogelbälge. — *J. f. Orn.*, 37: 97-101; 289-321.

2536A   von Berlepsch, H., 1895. Description of two new species of the genera *Phoenicophaes* and *Spilornis* with a note on *Oriolus consobrinus* Rams. — *Novit. Zool.*, 2: 70-75.

2537   von Berlepsch, H. & J. Stolzmann, 1901. Descriptions d'oiseaux nouveaux du Pérou central recueillis par le voyageur Polonais Jean Kalinowski. — *Ornis*, 11: 191-195.

2538   von Ihering, H., 1908. *Catalogos da Fauna Brazileira*: 1. 1-485. — Museu Paulista, Sao Paulo.

2539   von Jordans, A., 1940. Ein Beitrag zur Kenntnis der Vogelwelt Bulgariens. — *Bull. Inst. Roy. Hist. Nat. Sofia*, 13: 49-152.

2540   von Jordans, A., 1970. Die westpalaearktischen Rassen des Formenkreises *Parus major* (Aves, Paridae). — *Zool. Abh. Staatl. Mus. Tierk. Dresden*, 31: 205-225.

2541   von Jordans, A. & G. Schiebel, 1944. *Tetrao bonasia styriacus* form. nov. — *Falco*, 40 (1): 1.

2542   von Jordans, A. & J. Steinbacher, 1942. Beiträge zur Avifauna der Iberischen Halbinsel. — *Ann. Naturhist. Mus. Wien*, 52 (1941): 200-244.

2543   von Pelzeln, A., 1859. Über neue Arten der Gattungen *Synallaxis, Anabates* und *Xenops* in der Kais. ornithologischen Sammlung nebst Auszügen aus Johann Natterer's nachgelassenen Notizen über die von ihm in Brasilien gesammelten Arten der Subfamilien: Furnarinae und Synallaxiniae. — *Sitzber. math. naturw. Cl. kaiserl. Akad. Wiss*, 34: 99-134.

2544   von Sneidern, K., 1955. Notas ornitologicas sobre la coleccion de Museo de Historia natural de la Universidad del Cauca. — *Noved. Colomb.*, 2: 35-44.

2545   Voous, K.H., 1961. Birds collected by Carl Lumholtz in eastern and central Borneo. — *Nytt Mag. Zool.*, 10: 127-181.

2546   Voous, K.H., 1964. Wood Owls of the genera *Strix* and *Ciccaba*. — *Zool. Meded., Leiden*, 39: 471-478.

2547   Voous, K.H., 1973. List of recent holarctic bird species. — *Ibis*, 115: 612-638.

2548   Voous, K.H., 1977. List of recent holarctic bird species. — *Ibis*, 119: 223-250, 376-406.

2549   Vorob'ev, K.A., 1951. Novaya forma indiiskoi kukushki *Cuculus micropterus ognevi* subsp. nov. — *Dokl. Akad. Nauk. SSSR*, 77 (3): 511-512.

2550   Vuilleumier, F., 1969. Systematics and evolution in *Diglossa* (Aves, Coeribidae). — *Amer. Mus. Novit.*, 2381: 1-44.

2551   Vuilleumier, F. & E. Mayr, 1987. New species of birds described from 1976 to 1980. — *J. f. Orn.*, 128: 137-150.

2552   Vuilleumier, F. *et al.*, 1992. New species of birds described from 1981 to 1990. — *Bull. Brit. Orn. Cl.*, 112A: 267-309.

2553   Walkinshaw, L.H., 1965. A new sandhill crane from Central Canada. — *Can. Field-Nat.*, 79: 181-184.

2554   Walters, M., 1993. On the status of the Christmas Island Sandpiper *Aechmorhynchus cancellatus*. — *Bull. Brit. Orn. Cl.*, 113 (2): 97-102.

2555   Walters, M., 1996. The correct citation of the Blue-chinned Sapphire *Chlorestes notatus*. — *Bull. Brit. Orn. Cl.*, 116 (4): 270-271.

2556   Walters, M., 1997. On the identity of *Lophornis melaniae* Floericke (Trochilidae). — *Bull. Brit. Orn. Cl.*, 117 (3): 235-236.

2557   Walters, M., 1998. What is *Psittacus borneus* Linnaeus? — *Forktail*, 13: 124.

2558   Walters, M., 2001. The correct scientific name of the White-bellied Heron. — *Bull. Brit. Orn. Cl.*, 121 (4): 234-236.

2559   Warner, D.W., 1951. A new race of the cuckoo, *Chalcites lucidus*, from the New Hebrides Islands. — *Auk*, 68: 106-107.

2560   Warner, D.W. & B.E. Harrell, 1955. Una nueva raza de *Dactylortyx thoracicus* de Quintana Roo, Mexico. — *Rev. Soc. mex. Hist. nat.*, 14 (1953): 205-206.

2561   Warner, D.W. & B.E. Harrell, 1957. The systematics and biology of the singing quail, *Dactylortyx thoracicus*. — *Wilson Bull.*, 69 (2): 123-148.

2562   Watling, D., 1982. *Birds of Fiji, Tonga and Samoa*: 1-176. — Millwood Press, Wellington, NZ.

2563   Watson, G.E., 1962. Three sibling species of *Alectoris* Partridge. — *Ibis*, 104: 353-367.

2564   Watson, G.E. *et al.*, 1986. Family Sylviidae. Pages 3-294. In *Check-list of Birds of the World*. Vol. 11 (E. Mayr & G.W. Cottrell, eds.). — Mus. Comp. Zoology, Cambridge, Mass.

2565   Watson, G.E. *et al.*, 1986. Family Muscicapidae (*sensu stricto*). Pages 295-375. In *Check-list of Birds of the World*. Vol. 11 (E. Mayr & G.W. Cottrell, eds.). — Mus. Comp. Zoology, Cambridge, Mass.

2566   Watson, G.E. *et al.*, 1986. Family Monarchidae, Subfamily Monarchinae. Pages 464-526. In *Check-list of Birds of the World*. Vol. 11 (E. Mayr & G.W. Cottrell, eds.). — Mus. Comp. Zoology, Cambridge, Mass.

2567   Watson, G.E. *et al.*, 1991. A new subspecies of the Double-crested Cormorant, *Phalacrocorax auritus*, from San Salvador, Bahama Islands. — *Proc. Biol. Soc. Wash.*, 104 (2): 356 - 364.

2568   Webster, J.D., 1983. A new subspecies of Fox Sparrow from Alaska. — *Proc. Biol. Soc. Wash.*, 96: 664-668.

2569   Wee, M.B. *et al.*, 2001. The Norfolk Island Green Parrot and the New Caledonian Red-crowned Parrakeet are distinct species. — *Emu*, 101: 113-121.

2570   Weick, F., 2001. Zur Taxonomie der Amerikanischen Uhus (*Bubo* spp). Unter Berücksichtigung eines größtenteils parallel variierenden Polymorphismus innerhalb der Subspecies. — *Ökol. Vögel*, 21: 363-387 (1999).

2571   Welch, G.R. & H.J. Welch, 1988. A new subspecies of *Pytilia melba* from Djibouti, East Africa. — *Bull. Brit. Orn. Cl.*, 108: 68-70.

2572   Weller, A.A., 1999. On types of trochilids in The Natural History Museum, Tring II. Re-evaluation of *Erythronota* (?) *elegans* Gould 1860: a presumed extinct species of the genus *Chlorostilbon*. — *Bull. Brit. Orn. Cl.*, 119 (3): 197-202.

2573   Weller, A.A., 1999. Species accounts within the Family Trochilidae (Hummingbirds). Pages 549-680. In *Handbook of the Birds of the World*. Vol. 5 (J. del Hoyo, A. Elliott & J. Sargatal, eds.). — Lynx Edicions, Barcelona.

2574   Weller, A.A., 2000. Biogeography, geographic variation and habitat preference in the Amazilia Hummingbird, *Amazilia amazilia* Lesson (Aves: Trochilidae), with notes on the status of *Amazilia alticola* Gould. — *J. f. Orn.*, 141 (1): 93-101.

2575   Weller, A.A., 2000. A new hummingbird subspecies from southern Bolivar, Venezuela, with notes on biogeography and taxonomy of the *Saucerottia viridigaster - cupreicauda* species group. — *Orn. Neotrop.*, 11 (2): 143-154.

2576   Weller, A.A., 2001. On types of trochilids in the Natural History Museum, Tring. III. *Amazilia alfaroana* Underwood (1896), with notes on biogeography and geographical variation in *Saucerottia saucerottei* superspecies. — *Bull. Brit. Orn. Cl.*, 121 (2): 98-107.

2576A   Weller, A.A. & S.C. Renner, 2001. A new subspecies of *Heliodoxa xanthogonys* (Aves, Trochilidae) from the southern Pantepui highlands, with biogeographical and taxonomic notes. — *Ararajuba*, 9: 1-5.

2577   Weller, A.A. & K.-L. Schuchmann, 1999. Geographical variation in the southern distributional range of the Rufous-tailed Hummingbird, *Amazilia tzacatl* De la Llave, 1832: a new subspecies from Narino, southwestern Colombia. — *J. f. Orn.*, 140 (4): 457-466.

2578   Wells, D.R., 1975. The moss-nest swiftlet *Collocalia vanikorensis* Quoy and Gaimard in Sumatra. — *Ardea*, 63: 148-151.

2579   Wells, D.R., 1999. The Birds of the Thai-Malay Peninsula: 1. i-liii, 1-648. — Academic Press, London.

2580   Wells, D.R., 1999. Family Hemiprocnidae (Treeswifts). Pages 458-466. In *Handbook of the Birds of the World*. Vol. 5 (J. del Hoyo, A. Elliott & J. Sargatal, eds.). — Lynx Edicions, Barcelona.

2581   Wells, D.R. & J.H. Becking, 1975. Vocalizations and status of Little and Himalayan Cuckoos *Cuculus poliocephalus* and *C. saturatus* in Southeast Asia. — *Ibis*, 117 (3): 366-371.

2582   Wells, D.R. *et al.*, 2001. Systematic notes on Asian birds. 21. Babbler jungle: a re-evaluation of the '*pyrrogenys*' group of Asian pellorneines (Timaliidae). — *Zool. Verhand., Leiden*, 335: 235-252.

2583   Wells, R.W. & R. Wellington, 1992. A classification of the cockatoos and parrots (Aves: Psittaciformes) of Australia. — *Sydney Basin Nat.*, 1: 107-169.

2584   Weske, J.S., 1985. A new subspecies of Collared Inca Hummingbird (*Coeligena torquata*) from Peru. — *Orn. Monogr.*, 36: 40-45.
2585   Weske, J.S. & J.W. Terborgh, 1971. A new subspecies of curassow of the genus *Pauxi* from Peru. — *Auk*, 88 (2): 233-238.
2586   Weske, J.S. & J.W. Terborgh, 1974. *Hemispingus parodii*, a new species of tanager from Peru. — *Wilson Bull.*, 86 (2): 97-103.
2587   Weske, J.S. & J.W. Terborgh, 1977. *Phaethornis koepckeae*, a new species of hummingbird from Peru. — *Condor*, 79 (2): 143-147.
2588   Weske, J.S. & J.W. Terborgh, 1981. *Otus marshalli*, a new species of screech-owl from Peru. — *Auk*, 98 (1): 1-7.
2589   Wetherbee, D.K., 1985. The extinct Cuban and Hispaniolan Macaws (*Ara*, Psittacidae), and description of a new species, *Ara cubensis*. — *Carib. J. Sci.*, 21: 169-175.
2590   Wetmore, A., 1941. New forms of birds from Mexico and Columbia. — *Proc. Biol. Soc. Wash.*, 54: 203-210.
2591   Wetmore, A., 1946. New Birds from Colombia. — *Smithsonian Misc. Coll.*, 106 (16): 1-14.
2592   Wetmore, A., 1946. New forms of birds from Panamá and Columbia. — *Proc. Biol. Soc. Wash.*, 59: 49-54.
2593   Wetmore, A., 1950. Additional forms of birds from the Republics of Panamá and Colombia. — *Proc. Biol. Soc. Wash.*, 63: 171-174.
2594   Wetmore, A., 1951. Additional forms of birds from Colombia and Panamá. — *Smithsonian Misc. Coll.*, 117 (2): 1-11.
2595   Wetmore, A., 1952. The birds of the islands of Taboga, Taboguilla and Urava, Panamá. — *Smithsonian Misc. Coll.*, 121 (2): 16-17.
2596   Wetmore, A., 1953. Further additions to the birds of Panamá and Colombia. — *Smithsonian Misc. Coll.*, 122 (8): 1-12.
2597   Wetmore, A., 1956. Additional forms of birds from Panamá and Columbia. — *Proc. Biol. Soc. Wash.*, 69: 123-126.
2598   Wetmore, A., 1957. The birds of Isla Coiba, Panamá. — *Smithsonian Misc. Coll.*, 134 (9): 1-105.
2599   Wetmore, A., 1958. Additional subspecies of birds from Colombia. — *Proc. Biol. Soc. Wash.*, 71: 1-4.
2599A   Wetmore, A., 1959. The birds of Isla Escudo de Veraguas, Panamá. – *Smithsonian Misc. Coll.*, 138 (4): 1-24.
2600   Wetmore, A., 1962. Systematic Notes concerned with the Avifauna of Panamá. — *Smithsonian Misc. Coll.*, 145 (1): 1-14.
2601   Wetmore, A., 1963. Additions to records of birds known from the Republic of Panamá. — *Smithsonian Misc. Coll.*, 145 (6): 1-11.
2602   Wetmore, A., 1965. Additions to the list of birds of the Republic of Colombia. — *L'Oiseau*, 35 (No. Spec.): 156-162.
2603   Wetmore, A., 1965. The birds of the Republic of Panamá. 1. Tinamidae (Tinamous) to Rynchopidae (Skimmers). — *Smithsonian Misc. Coll.*, 150: i-iv, 1-483.
2604   Wetmore, A., 1967. Further systematic notes on the avifauna of Panamá. — *Proc. Biol. Soc. Wash.*, 80: 229-242.
2605   Wetmore, A., 1968. The birds of the Republic of Panamá. 2. Columbidae (Pigeons) to Picidae (Woodpeckers). — *Smithsonian Misc. Coll.*, 150 (2): i-v, 1-605.
2606   Wetmore, A., 1970. Descriptions of additional forms of birds from Panamá and Colombia. — *Proc. Biol. Soc. Wash.*, 82: 767-776.
2608[6] Wetmore, A., 1972. The birds of the Republic of Panamá. 3. Passeriformes: Dendrocolaptidae (Woodcreepers) to Oxyruncidae (Sharpbills). — *Smithsonian Misc. Coll.*, 150 (3): i-iv, 1-631.
2609   Wetmore, A. & J.I. Borrero, 1964. Description of a race of the Double-striped Thick-knee (Aves, family Burhinidae). — *Auk*, 81: 231-233.
2610   Wetmore, A. & K.C. Parkes, 1962. A new subspecies of Ivory-billed Woodhewer from Mexico. — *Proc. Biol. Soc. Wash.*, 75: 57-60.
2611   Wetmore, A. & W.H. Phelps, 1943. Description of a third form of curassow of the genus *Pauxi*. — *J. Wash. Acad. Sci.*, 33 (5): 142-146.
2612   Wetmore, A. & W.H. Phelps, Jr., 1952. A new form of hummingbird from the Perijá Mountains of Venezuela and Columbia. — *Proc. Biol. Soc. Wash.*, 65: 135-136.
2613   Wetmore, A. & W.H. Phelps, 1953. Notes on the Rufous Goatsuckers of Venezuela. — *Proc. Biol. Soc. Wash.*, 66: 15-20.
2614   Wetmore, A. & W.H. Phelps, Jr., 1956. Further additions to the list of birds of Venezuela. — *Proc. Biol. Soc. Wash.*, 69: 1-12.
2615   Wheeler, A., 1998. Dates of publication of J. E. Gray's Illustrations of Indian Zoology (1830-1835). — *Arch. Nat. Hist.*, 25 (3): 345-354.
2616   Whistler, H., 1937. [Description of a new subspecies.]. — *Bull. Brit. Orn. Cl.*, 58: 9.
2617   Whistler, H., 1939. A new race of the Indian Bush-Quail. — *Bull. Brit. Orn. Cl.*, 59: 76.
2618   Whistler, H., 1941. Recognition of new subspecies of Birds in Ceylon. — *Ibis*, (14) 5: 319-320.
2619   Whistler, H., 1944. The avifaunal survey of Ceylon conducted jointly by the British and Colombo Museums. — *Spolia Zeylanica*, 23 (3/4): 117-321.
2620   Whistler, H. & N.B. Kinnear, 1936. The Vernay Scientific Survey of the Eastern Ghats. XIV. — *J. Bombay Nat. Hist. Soc.*, 38: 672-698.
2621   Whistler, H. & N.B. Kinnear, 1937. The Vernay Scientific Survey of the Eastern Ghats, 15. — *J. Bombay Nat. Hist. Soc.*, 39 (2): 246-263.
2622   White, C.M.N., 1937. [A new form of Chukor from Crete.]. — *Bull. Brit. Orn. Cl.*, 57: 65-66.
2623   White, C.M.N., 1938. A new form of *Talegalla fuscirostris* Salvadori (Ann. Mus. Civ. Genova, ix. 1877, p. 332). – *Ibis*, (14) 2: 763-764.
2624   White, C.M.N., 1943. A new race of Green Pigeon from Northern Rhodesia. — *Bull. Brit. Orn. Cl.*, 63: 63-64.
2625   White, C.M.N., 1943. Three new races from Northern Rhodesia. — *Bull. Brit. Orn. Cl.*, 64: 19-22.
2626   White, C.M.N., 1944. A new race of Scrub Robin and a new race of Red-winged Francolin from Northern Rhodesia. — *Bull. Brit. Orn. Cl.*, 64: 49-50.
2627   White, C.M.N., 1944. A new race of Francolin from Northern Rhodesia. — *Bull. Brit. Orn. Cl.*, 65: 7-8.
2628   White, C.M.N., 1945. On Francolins from Angola and Northern Rhodesia. — *Bull. Brit. Orn. Cl.*, 65: 38-40.
2629   White, C.M.N., 1945. A new race of Bustard from Northern Rhodesia. — *Bull. Brit. Orn. Cl.*, 65: 47-48.
2630   White, C.M.N., 1945. The ornithology of the Kaonde-Lunda province, Northern Rhodesia. Part III. Systematic List. — *Ibis*, 87: 309-345.
2631   White, C.M.N., 1947. Two new races of Francolins from Northern Rhodesia and some records from Lundazi. — *Bull. Brit. Orn. Cl.*, 67: 72-73.
2632   White, C.M.N., 1947. Notes on Central African birds. — *Bull. Brit. Orn. Cl.*, 68: 34-36.
2633   White, C.M.N., 1948. A new race of Lemon Dove from Northern Rhodesia. — *Bull. Brit. Orn. Cl.*, 69: 20-21.
2634   White, C.M.N., 1952. On the genus *Pternistis*. — *Ibis*, 94: 306-309.
2635   White, C.M.N., 1956. A new race of Swallow from Somaliland. — *Bull. Brit. Orn. Cl.*, 76 (9): 160.
2636   White, C.M.N., 1958. The names of some francolins. — *Bull. Brit. Orn. Cl.*, 78: 76.
2637   White, C.M.N., 1958. A new lark from Northern Rhodesia. — *Bull. Brit. Orn. Cl.*, 78: 163-164.
2638   White, C.M.N., 1959. A new Finch Lark from South West Africa. — *Bull. Brit. Orn. Cl.*, 79: 51-52.
2639   White, C.M.N., 1960. A Check List of the Ethiopian Muscicapidae (Sylviinae). Part I. — *Occ. Pap. Nat. Mus. Sthn. Rhodesia*, 3 (24B): 399-430.
2640   White, C.M.N., 1960. The Ethiopian and allied forms of *Calandrella cinerea* (Gmelin). — *Bull. Brit. Orn. Cl.*, 80: 24-25.
2641   White, C.M.N., 1960. A new lark from Nigeria. — *Bull. Brit. Orn. Cl.*, 80: 11.
2642   White, C.M.N., 1960. A note on *Certhilauda curvirostris*. — *Bull. Brit. Orn. Cl.*, 80: 25.
2643   White, C.M.N., 1960. The Somali forms of *Calandrella rufescens*. — *Bull. Brit. Orn. Cl.*, 80: 132.
2644   White, C.M.N., 1961. A new form of Spike-heeled Lark from Bechuanaland. — *Bull. Brit. Orn. Cl.*, 81: 33.
2645   White, C.M.N., 1961. *A Revised Check List of African Broadbills, Pittas, Larks, Swallows, Wagtails and Pipits*: i-iv, 1-84. — Govt. printer, Lusaka.
2646   White, C.M.N., 1962. *A Revised Check List of African Shrikes, Orioles, Drongos, Starlings, Crows, Waxwings, Cuckoo-Shrikes, Bulbuls, Accentors, Thrushes and Babblers*: i-vi, 1-176. — Govt. printer, Lusaka.
2647   White, C.M.N., 1962. A Check List of the Ethiopian Muscicapidae (Sylviinae). Part II. — *Occ. Pap. Nat. Mus. Sthn. Rhodesia*, 3 (26B): 653-738.
2648   White, C.M.N., 1963. *A Revised Check List of African Flycatchers, Tits, Tree Creepers, Sunbirds, White-eyes, Honey Eaters, Buntings, Finches, Weavers and Waxbills*: i-vi, 1-218. — Govt. printer, Lusaka.
2649   White, C.M.N., 1965. *A Revised Check List of African Non-Passerine Birds*: i-v, 1-299. — Govt. printer, Lusaka.
2650   White, C.M.N., 1967. A recently described African Swift. — *Bull. Brit. Orn. Cl.*, 87: 63.

---

[6] No. 2607 has been deleted.

2651   White, C.M.N., 1974. Some questionable records of Celebes birds. — *Bull. Brit. Orn. Cl.*, 94: 144-145.

2652   White, C.M.N. & M.D. Bruce, 1986. *The birds of Wallacea*: 1-524. B.O.U. Check-list No. 7 — British Ornithologists' Union, London.

2653   Whitney, B.M., 1994. A new *Scytalopus* Tapaculo (Rhinocryptidae) from Bolivia, with notes on other Bolivian members of the genus and the *magellanicus* complex. — *Wilson Bull.*, 106 (4): 585-614.

2654   Whitney, B.M. & J.A. Alonso, 1998. A new *Herpsilochmus* antwren (Aves: Thamnophilidae) from northern Amazonian Peru and adjacent Ecuador: The role of edaphic heterogeneity of Terra Firme forest. — *Auk*, 115 (3): 559-576.

2655   Whitney, B.M. & J.F. Pacheco, 1997. Behavior, vocalizations, and relationships of some *Myrmotherula* antwrens (Thamnophilidae) in eastern Brazil, with comments on the "Plain-winged" group. — *Orn. Monogr.*, 48: 809-819.

2656   Whitney, B.M. & J.F. Pacheco, 1999. The valid name for Blue-winged Parrotlet and designation of the lectotype of *Psittaculus xanthopterygius* Spix, 1824. — *Bull. Brit. Orn. Cl.*, 119 (4): 211-214.

2657   Whitney, B.M. *et al.*, 1995. *Hylopezus nattereri* (Pinto, 1937) is a valid species (Passeriformes : Formicariidae). — *Ararajuba*, 3: 37-42.

2658   Whitney, B.M. *et al.*, 1995. Two species of *Neopelma* in southeastern Brazil and diversification within the *Neopelma/Tyranneutes* complex: implications of the subspecies concept for conservation (Passeriformes: Tyrannidae). — *Ararajuba*, 3: 43-53.

2659   Whitney, B.M. *et al.*, 2000. Systematic revision and biogeography of the *Herpsilochmus pileatus* complex, with description of a new species from northeastern Brazil. — *Auk*, 117 (4): 869-891.

2660   Wickler, W., 1973. Artunterschiede im Duettgesang zwischen *Trachyphonus darnaudii usambiro* und den anderen Unterarten von *T.darnaudii*. — *J. f. Orn.*, 114: 123-128.

2661   Wiedenfeld, D.A., 1994. A new subspecies of Scarlet Macaw and its status and conservation. – *Orn. Neotrop.*, 5: 99-104.

2662   Wiersma, P., 1996. Species accounts for the family Charadriidae (Plovers). Pages 411-442. In *Handbook of the Birds of the World*. Vol. 3 (J. del Hoyo, A. Elliott & J. Sargatal, eds.). — Lynx Edicions, Barcelona.

2663   Wiggins, D., 2000. Various species accounts. In *The Birds of Africa*. Vol 6: i-xvii, 1-724 (C.H. Fry, S. Keith & E.K. Urban, eds.). — Academic Press, London.

2664   Wilkinson, R., 2000. Various species accounts. In *The Birds of Africa*. Vol. 6: i-xvii, 1-724 (C.H. Fry, S. Keith & E.K. Urban, eds.). — Academic Press, London.

2665   Williams, J.G., 1966. A new species of swallow from Kenya. — *Bull. Brit. Orn. Cl.*, 86: 40.

2666   Williams, J.G., 1966. A new race of *Lybius torquatus* from Tanzania. — *Bull. Brit. Orn. Cl.*, 86: 47-48.

2667   Willis, E.O., 1983. Trans-Andean *Xiphorhynchus* (Aves, Dendrocolaptidae) as army ant followers. — *Rev. Bras. Biol.*, 43: 125-131.

2668   Willis, E.O., 1991. Sibling species of Greenlets (Vireonidae) in southern Brazil. — *Wilson Bull.*, 103 (4): 559-567.

2669   Willis, E.O., 1992. Comportamento e ecologia do Arapaçu-barrado *Dendrocolaptes certhia* (Aves, Dendrocolaptidae). — *Bol. Mus. Para. Emilio Goeldi, Zool.*, 8: 151-216.

2670   Willis, E.O., 1992. Three *Chamaeza* Antthrushes in eastern Brazil (Formicariidae). — *Condor*, 94: 110-116.

2671   Willis, E.O. & Y. Oniki, 1992. A new *Phylloscartes* (Tyrannidae) from southeastern Brazil. — *Bull. Brit. Orn. Cl.*, 112 (3): 158-165.

2672   Wilson, S.B. & A.H. Evans, 1890-1899. *Aves Hawaiienses; the birds of the Sandwich Islands*: 1-256. — R.H. Porter, London.

2673   Winker, K., 1995. *Xiphorhynchus striatigularis* (Dendrocolaptidae): Nomen monstrositatum. — *Auk*, 112: 1066-1070.

2674   Winker, K., 1997. A new form of *Anabacerthia variegaticeps* (Furnariidae) from western Mexico. Pages 203-208. In *The Era of Allan R. Phillips: A Festschrift* (R.W. Dickerman, ed.) — Albuquerque.

2675   Winker, K., 2000. A new subspecies of Toucanet (*Aulacorhynchus prasinus*) from Veracruz, Mexico. — *Orn. Neotrop.*, 11: 253-257.

2676   Winkler, H. *et al.*, 1995. *Woodpeckers. A guide to the woodpeckers, piculets and wrynecks of the world*: [i-x], 1-406. — Pica Press, Mountsfield, Sussex.

2677   Winterbottom, J.M., 1956. Systematic notes on birds of the Cape Province. Pt. I. *Mirafra apiata* (Vieillot). — *Ostrich*, 27: 156-158.

2678   Winterbottom, J.M., 1958. Review of the races of the Spike-heeled Lark *Certhilauda albofasciata* Lafresnaye. — *Ann. S. Afr. Mus.*, 44: 53-67.

2679   Winterbottom, J.M., 1963. The South African subspecies of the Buffy Pipit, *Anthus vaalensis* Shelley. — *Ann. S. Afr. Mus.*, 46 (13): 341-350.

2680   Winterbottom, J.M., 1964. Results of the Percy FitzPatrick Institute-Windhoek State Museum joint ornithological expeditions: report on the birds of Game Reserve no. 2. — *Cimbebasia*, 9: 1-75.

2681   Winterbottom, J.M., 1966. Results of the Percy FitzPatrick Institute-Windhoek State Museum joint ornithological expeditions: 5. The birds of the Kaokoveld and Kunene River. — *Cimbebasia*, 19: 1-71.

2682   Wolters, H.E., 1969. Die geographische Variation von *Amandava subflava* (Aves: Estrildidae). — *Bonn. Zool. Beitr.*, 20 (1-3): 60-68.

2683   Wolters, H.E., 1972. Aus der ornithologischen Sammlung des Museums Alexander Koenig. II. — *Bonn. Zool. Beitr.*, 23: 87-94.

2684   Wolters, H.E., 1974. Aus der ornithologischen Sammlung des Museums Alexander Koenig. III. Ein neuer *Malimbus* (Ploceidae, Aves) von der Elfenbeinküste. — *Bonn. Zool. Beitr.*, 25 (4): 283-291.

2684A   Wolters, H.E., 1976. *Die Vogelarten der Erde. Eine systematische Liste mit Verbreitungsangaben sowie deutschen und englischen Namen*. Pt. 2.: 81-160. — Paul Parey, Hamburg.

2685   Wolters, H.E., 1979. *Die Vogelarten der Erde. Eine systematische Liste mit Verbreitungsangaben sowie deutschen und englischen Namen*. Pt. 4: 241-320. — Paul Parey, Hamburg.

2686   Wolters, H.E., 1983. *Die Vögel Europas im System der Vögel. Eine Übersicht*: 1-70. — Biotropic Verlag, Baden-Baden.

2687   Wolters, H.E. & P.A. Clancey, 1969. A new race of Green-headed Oriole from Southern Moçambique. — *Bull. Brit. Orn. Cl.*, 89: 108-109.

2688   Won, H.G., 1962. On the occurence of *Dendrocopus minor kemaensis* and *Dendrocopus minor nojidaensis* in the Kaema Highland. — *Kwahakwon Tongbo* [Bull. Acad. Sci. P'yongyang], 1962 (2): 31-32.

2689   Woodall, P.F., 2001. Family Alcedinidae (KIngfishers). Pages 130-249. In *Handbook of the Birds of the World*. Vol. 6 (J. del Hoyo, A. Elliott & J. Sargatal, eds.). — Lynx Edicions, Barcelona.

2690   Woods, R.W., 1993. Cobb's Wren *Troglodytes* (*aedon*) *cobbi* of the Falkland Islands. — *Bull. Brit. Orn. Cl.*, 113 (4): 195-207.

2691   Wotzkow, C., 1991. New subspecies of Gundlach's Hawk, *Accipiter gundlachi* (Lawrence). — *Birds of Prey Bull.*, 4: 271-292.

2692   Yamagishi, S. *et al.*, 2001. Extreme endemic radiation of the Malagasy Vangas (Aves: Passeriformes). — *J. Mol. Evol.*, 53: 39-46.

2693   Yamashina, Y., 1942. On a new subspecies of *Micropus pacificus* residing in Formosa and Botel Tobago. — *Bull. Biogeog. Soc. Japan*, 12 (2): 79-80.

2694   Yamashina, Y., 1943. On a new subspecies of *Chloroceryle americana* from Ecuador [in Japanese]. — *Bull. Biogeog. Soc. Japan*, 13 (19): 145-146.

2695   Yamashina, Y., 1944. On a new subspecies of *Goura scheepmakeri* from South New Guinea. — *Bull. Biogeog. Soc. Japan*, 14 (1): 1-2.

2696   Yamashina, Y. & T. Mano, 1981. A new species of rail from Okinawa Island. — *J. Yamashina Inst. Orn.*, 13 (3): 147-152.

2697   Yang, Lan & Li, Gui-yuan, 1989. A new subspecies of the *Athene brama* (Spotted Little Owl) *Athene brama poikila* (Belly-mottled Little Owl) (Strigiformes: Strigidae). — *Zool. Res.* [China], 10: 303-308.

2698   Yang, Lan & Xu, Yan-gong, 1987. A new subspecies of the Tibetan Snowcock - *Tetraogallus tibetanus yunnanensis* (Galliformes: Phasianidae). — *Acta Zootax. Sinica*, 12 (1): 104-109.

2699   Yang, Lan *et al.*, 1994. On the taxonomy of Blood Pheasant (*Ithaginis*). — *Zool. Res.* [China], 15: 28-30.

2700   Yen, Kwok-Yung & Chong, Ling-ting, 1937. Notes additionnelles sur l'avifaune du Kwangsi. — *L'Oiseau*, 7 (4): 546-553.

2701   Yosef, R. *et al.*, A new subspecies of the Booted Eagle from southern Africa, inferred from biometrics and mitochondrial DNA. Pages 43-49 In: *Raptors at risk* (R.D. Chancellor & B.-U. Meyburg, eds). Proc. 5th. World Conf. Birds of Prey and Owls. World Working Group on Birds of Prey and Owls. 2000: 1-895.

2702   Young, J.R. *et al.*, 2000. A new species of Sage Grouse (Phasianidae: *Centrocercus*) from southwestern Colorado, USA. — *Wilson Bull.*, 112: 445-453.

2703   Zalataev, V.S., 1962. The Caspian Eagle-Owl (*Bubo bubo gladkovi* subsp. nov.) [In Russian.]. — *Ornitologia*, 4: 190-193.

2705[7] Zhang, Yin-sun *et al.*, 1989. A new subspecies of the Chukor Partridge - *Alectoris chukar ordoscensis* (Galliformes: Phasianidae). — *Acta Zootax. Sinica*, 14 (4): 496-499.

2706   Zheng, Guangmei *et al.*, 2000. A new species of flycatcher (*Ficedula*) from China (Aves: Passeriformes: Muscicapidae). — *J. Beijing Normal Univ., (Nat. Sci.)*, 36 (3): 405- 409.

2707   Zimmer, J.T., 1926. Catalogue of the Edward E. Ayer Ornithological Library. — *Field Mus. Nat. Hist., Zool.*, 16 (1): i-x, 1-364.

2707A   Zimmer, J.T., 1932. Studies of Peruvian birds. VI. The formicarian genera *Myrmoborus* and *Myrmeciza* in Peru. — *Amer. Mus. Novit.*, 545: 1-24.

2708   Zimmer, J.T., 1941. Studies of Peruvian birds No. XXXVII. The genera *Sublegatus, Phaeomyias, Camptostoma, Xanthomyias, Phyllomyias* and *Tyrannus*. — *Amer. Mus. Novit.*, 1109: 1-25.

2709   Zimmer, J.T., 1945. A new swift from Central and South America. — *Auk*, 62: 586-592.

2710   Zimmer, J.T., 1947. New birds from Pernambuco, Brazil. — *Proc. Biol. Soc. Wash.*, 60: 99-104.

2711   Zimmer, J.T., 1948. Studies of Peruvian birds. No. 53. The family Trogonidae. — *Amer. Mus. Novit.*, 1380: 1-56.

2712   Zimmer, J.T., 1948. Two new Peruvian hummingbirds of the genus *Coeligena*. — *Auk*, 65: 410-416.

2713   Zimmer, J.T., 1950. Studies of Peruvian birds. No. 55. The humming bird genera *Doryfera, Glaucis, Threnetes* and *Phaethornis*. — *Amer. Mus. Novit.*, 1449: 1-51.

2714   Zimmer, J.T., 1950. Studies of Peruvian birds. No. 56. The genera *Eutorxeres, Campylopterus, Eupetomena* and *Florisuga*. — *Amer. Mus. Novit.*, 1450: 1-14.

2715   Zimmer, J.T., 1950. Studies of Peruvian birds. No. 57. The genera *Colibri, Anthracothorax, Klais, Lophornis* and *Chlorestes*. — *Amer. Mus. Novit.*, 1463: 1-18.

2716   Zimmer, J.T., 1950. Studies of Peruvian birds. No. 58. The genera *Chlorostilbon, Thalurania, Hylocharis* and *Chrysuronia*. — *Amer. Mus. Novit.*, 1474: 1-31.

2717   Zimmer, J.T., 1951. Studies of Peruvian birds. No. 60. The genera *Heliodoxa, Phlogophilus, Urosticte, Polyplancta, Adelomyia, Coeligena, Ensifera, Oreotrochilus* and *Topaza*. — *Amer. Mus. Novit.*, 1513: 1-45.

2718   Zimmer, J.T., 1951. Studies of Peruvian birds. No. 61. The genera *Aglaeactes, Lafresnaya, Pterophanes, Boissonneaua, Heliangelus, Eriocnemis, Haplophaedia, Ocreatus* and *Lesbia*. — *Amer. Mus. Novit.*, 1540: 1-55.

2719   Zimmer, J.T., 1953. Studies of Peruvian birds. No. 63. The hummingbird genera *Oreonympha, Schistes, Heliothryx, Loddigesia, Heliomaster, Rhodopsis, Thaumastura, Calliphlox, Myrtis, Acestrura*. — *Amer. Mus. Novit.*, 1604: 1-26.

2720   Zimmer, J.T., 1953. Studies of Peruvian birds. No. 64. The swifts: family Apodidae. — *Amer. Mus. Novit.*, 1609: 1-20.

2721   Zimmer, J.T., 1954. A new subspecies of *Upucerthia dumetaria* (Family Furnariidae) from Peru. — *Proc. Biol. Soc. Wash.*, 67: 189-193.

2721A   Zimmer, J.T. & E. Mayr, 1943. New species of birds described from 1938 to 1941. — *Auk*, 60: 249-262.

2722   Zimmer, J.T. & W.H. Phelps, 1944. New species and subspecies of birds from Venezuela. 1. — *Amer. Mus. Novit.*, 1270: 1-9.

2723   Zimmer, J.T. & W.H. Phelps, 1946. Twenty-three new subspecies of birds from Venezuela and Brazil. — *Amer. Mus. Novit.*, 1312: 1-22.

2724   Zimmer, J.T. & W.H. Phelps, 1947. Seven new subspecies of birds from Venezuela and Brazil. — *Amer. Mus. Novit.*, 1338: 1-7.

2725   Zimmer, J.T. & W.H. Phelps, 1949. Four new subspecies of birds from Venezuela. — *Amer. Mus. Novit.*, 1395: 1-9.

2726   Zimmer, J.T. & W.H. Phelps, 1950. Three new Venezuelan birds. — *Amer. Mus. Novit.*, 1455: 1-7.

2727   Zimmer, J.T. & W.H. Phelps, 1951. New subspecies of birds from Surinam and Venezuela. — *Amer. Mus. Novit.*, 1511: 1-10.

2728   Zimmer, J.T. & W.H. Phelps, 1952. New birds from Venezuela. — *Amer. Mus. Novit.*, 1544: 1-7.

2729   Zimmer, J.T. & W.H. Phelps, 1955. Three new subspecies of birds from Venezuela. — *Amer. Mus. Novit.*, 1709: 1-6.

2730   Zimmer, K.J., 1997. Species limits in *Cranioleuca vulpina*. — *Orn. Monogr.*, 48: 849-864.

2731   Zimmer, K.J., 1999. Behaviour and vocalizations of the Caura and Yapacana Antbirds. — *Wilson Bull.*, 111: 195-209.

2732   Zimmer, K.J. & A. Whittaker, 2000. The Rufous Cachalote (Furnariidae: *Pseudoseisura*) is two species. — *Condor*, 102 (2): 409-422.

2733   Zimmer, K.J. & A. Whittaker, 2000. Species limits in Pale-tipped Tyrannulets (*Inezia* : Tyrannidae). — *Wilson Bull.*, 112 (1): 51-66.

2734   Zimmer, K.J. *et al.*, 2001. A cryptic new species of flycatcher (Tyrannidae: *Suiriri*) from the Cerrado region of central South America. — *Auk*, 118: 56-78.

2735   Zimmerman, D.A. *et al.*, 1996. *Birds of Kenya and northern Tanzania*: 1-752. — Christoper Helm, London.

2736   Zink, R.M., 1986. Patterns and evolutionary significance of geographic variation in the *schistacea* group of the Fox Sparrow (*Passerella iliaca*). — *Orn. Monogr.*, 40: i-viii, 1-119.

2737   Züchner, T., 1999. Species accounts within the Family Trochilidae (Hummingbirds). Pages 549-680. In *Handbook of the Birds of the World*. Vol. 5 (J. del Hoyo, A. Elliott & J. Sargatal, eds.). — Lynx Edicions, Barcelona.

2738   Zusi, R.L. & J.R. Jehl, 1970. The systematic position of *Aechmorhynchus, Prosobonia* and *Phlegornis* (Charadriiformes: Charadrii). — *Auk*, 87: 760-780.

2739   Zusi, R. & J.T. Marshall, Jr., 1970. A comparison of Asiatic and North American Sapsuckers. — *Nat. Hist. Bull., Siam Soc.*, 23 (3): 393-407.

---

[7] No. 2704 has been renumbered.

Index of Scientific Names

andina, Recurvirostra 133
andinus, Chaetocercus jourdanii 277
andinus, Cyanoliseus patagonus 198
andinus, Hemitriccus granadensis 360
andinus, Hymenops perspicillatus 370
andinus, Phoenicoparrus 81
andinus, Podiceps 80
andinus, Polytmus guainumbi 264
andinus, Sclerurus mexicanus 418
andium, Anas flavirostris 67
andrei, Chaetura vauxi 251
andrei, Crypturellus soui 31
andrei, Dysithamnus mentalis 383
andrei, Taeniotriccus 362
andrewi, Hirundo 532
andrewsi, Chrysocolaptes lucidus 329
andrewsi, Fregata 89
andrewsi, Gallirallus philippensis 120
andria, Saurothera merlini 213
ANDRODON 257
andromedae, Zoothera 661
ANDROPADUS 568
ANDROPHOBUS 452
anerythra, Pitta 339
anerythra, Pyrrhura lepida 198
anga, Aplonis opaca 652
angarensis, Motacilla flava 739
angelae, Dendroica 762
angelica, Leptotila verreauxi 168
angelinae, Otus 220
angelorum, Ploceus philippinus 723
anggiensis, Fulica atra 127
angkanensis, Aethopyga nipalensis 713
anglicus, Dendrocopos major 320
anglorum, Regulus regulus 632
angolae, Pluvianus aegyptius 145
angolensis, Anthreptes longuemarei 704
angolensis, Apalis rufogularis 561
angolensis, Buphagus erythrorhynchus 659
angolensis, Cercomela familiaris 686
angolensis, Cisticola robustus 554
angolensis, Dryoscopus 459
angolensis, Eremomela scotops 595
angolensis, Estrilda astrild 730
angolensis, Euplectes capensis 725
angolensis, Francolinus coqui 50
angolensis, Gallinago nigripennis 139
angolensis, Gypohierax 101
angolensis, Hirundo 535
angolensis, Indicator meliphilus 310
angolensis, Mirafra 541
angolensis, Monticola 688
angolensis, Muscicapa adusta 692
angolensis, Nigrita canicapillus 727
angolensis, Oriolus larvatus 488
angolensis, Oryzoborus 793
angolensis, Phoeniculus purpureus 297
angolensis, Pitta 339
angolensis, Ploceus 723
angolensis, Pogoniulus coryphaeus 307
angolensis, Tricholaema hirsuta 308
angolensis, Turdoides melanops 610
angolensis, Uraeginthus 729
angolensis, Urolestes melanoleucus 479
angolicus, Prionops plumatus 456
angoniensis, Stactolaema whytii 307
anguitimens, Eurocephalus 479
angulata, Gallinula 126
angulensis, Vireo pallens 483
angusta, Neocichla gutturalis 659
anguste, Dinopium shorii 329
angusticauda, Cisticola 555
angustifasciata, Ochthoeca cinnamomeiventris 373
angustifrons, Coracina papuensis 467
angustifrons, Melanerpes formicivorus 314
angustifrons, Psarocolius 768
angustipluma, Chaetoptila 431
angustirostris, Lepidocolaptes 425

angustirostris, Marmaronetta 68
angustirostris, Myrmoborus leucophrys 390
angustirostris, Phylloscartes ventralis 356
angustirostris, Phytotoma rutila 346
angustirostris, Sayornis nigricans 366
angustirostris, Todus 293
ANHIMA 61
ANHINGA 92
anhinga, Anhinga 93
ani, Crotophaga 217
ANISOGNATHUS 808
anjouanensis, Zosterops maderaspatanus 630
anjuanensis, Foudia eminentissima 724
ankafanae, Xanthomixis zosterops 588
ankoberensis, Serinus 748
ankole, Cisticola brachypterus 554
anna, Calypte 277
anna, Ciridops 759
annabelae, Sialia mexicana 663
annabellae, Myzomela boiei 442
annae, Ammomanes deserti 545
annae, Anthus cinnamomeus 742
annae, Artamella viridis 462
annae, Cettia 579
annae, Dicaeum 700
annae, Ithaginis cruentus 57
annae, Ocreatus underwoodii 273
annalisa, Ficedula hyperythra 694
annamarulae, Melaenornis 689
annamense, Pellorneum tickelli 600
annamensis, Aegithalos concinnus 539
annamensis, Aethopyga gouldiae 712
annamensis, Alcippe peracensis 620
annamensis, Arborophila rufogularis 55
annamensis, Blythipicus pyrrhotis 330
annamensis, Celeus brachyurus 325
annamensis, Criniger pallidus 572
annamensis, Ficedula hyperythra 694
annamensis, Garrulax merulinus 614
annamensis, Harpactes erythrocephalus 280
annamensis, Hypopicus hyperythrus 317
annamensis, Lophura nycthemera 58
annamensis, Megalaima oorti 305
annamensis, Napothera crispifrons 604
annamensis, Pericrocotus ethologus 472
annamensis, Picus chlorolophus 327
annamensis, Pnoepyga pusilla 605
annamensis, Pomatorhinus schisticeps 603
annamensis, Pteruthius flaviscapis 617
annamensis, Seicercus castaniceps 593
annamensis, Treron phoenicopterus 173
anneae, Euphonia 822
annectans, Dicrurus 492
annectans, Heterophasia 621
annectens, Cyanocitta stelleri 507
annectens, Iduna rama 586
annectens, Nothura maculosa 34
annectens, Passerella iliaca 781
annectens, Pseudotriccus pelzelni 355
annectens, Tangara punctata 811
annectens, Tephrodornis virgatus 456
annectens, Thraupis cyanocephala 807
annectens, Todirostrum maculatum 362
annulosa, Taeniopygia bichenovii 733
annumbi, Anumbius 412
ANODORHYNCHUS 195
anolaimae, Columba subvinacea 160
anomala, Aethopyga saturata 713
anomala, Brachycope 724
anomala, Cossypha 677
ANOMALOSPIZA 737
anomalus, Eleothreptus 245
anomalus, Eubucco bourcierii 304
anomalus, Rhinopomastus aterrimus 297
anomalus, Zosterops 627
anonymus, Centropus sinensis 216
anonymus, Cisticola 551
ANOPETIA 256
ANORRHINUS 298

ANOUS 152
anoxanthus, Loxipasser 795
anselli, Centropus 216
anselli, Cisticola textrix 555
ANSER 62
anser, Anser 62
ANSERANAS 61
ansorgeanus, Criniger barbatus 572
ansorgei, Acrocephalus rufescens 584
ansorgei, Andropadus 569
ansorgei, Anthoscopus caroli 530
ansorgei, Anthus leucophrys 743
ansorgei, Chlorophoneus bocagei 458
ansorgei, Cisticola rufilatus 553
ansorgei, Dyaphorophyia concreta 454
ansorgei, Euplectes gierowii 725
ansorgei, Indicator willcocksi 310
ansorgei, Lagonosticta rhodopareia 729
ansorgei, Mirafra sabota 542
ansorgei, Nesocharis 727
ansorgei, Ortygospiza atricollis 731
ansorgei, Parisoma subcaeruleum 598
ansorgei, Parmoptila woodhousei 726
ansorgei, Plocepasser mahali 715
ansorgei, Prinia flavicans 559
ansorgei, Pseudoalcippe abyssinica 602
ansorgei, Pterocles bicinctus 157
ansorgei, Sylvietta rufescens 595
ansorgei, Tchagra australis 458
ansorgei, Treron calvus 173
ansorgei, Xenocopsychus 678
ansorgii, Tricholaema hirsuta 308
antarctica, Geositta 402
antarctica, Thalassoica 73
antarcticus, Anthus 745
antarcticus, Cinclodes 404
antarcticus, Lopholaimus 180
antarcticus, Podilymbus podiceps 79
antarcticus, Pygoscelis 70
antarcticus, Rallus 121
antarcticus, Stercorarius 153
antelius, Larus fuscus 148
antelius, Mimus gilvus 649
antelius, Pionus sordidus 202
anteocularis, Polioptila plumbea 644
ANTHOBAPHES 705
ANTHOCEPHALA 267
ANTHOCHAERA 437
anthoides, Asthenes 407
anthoides, Corythopis torquata 355
anthoides, Pericrocotus brevirostris 472
anthonyi, Butorides virescens 86
anthonyi, Campylorhynchus brunneicapillus 634
anthonyi, Caprimulgus 242
anthonyi, Cinclus mexicanus 698
anthonyi, Dicaeum 701
anthonyi, Geotrygon albifacies 169
anthonyi, Lanius ludovicianus 480
anthonyi, Troglodytes bewickii 636
anthopeplus, Polytelis 193
anthophilus, Phaethornis 256
ANTHORNIS 435
ANTHOSCOPUS 530
anthracina, Columba vitiensis 159
anthracina, Pipra pipra 342
anthracinus, Buteogallus 109
anthracinus, Knipolegus aterrimus 369
anthracinus, Turdus chiguanco 670
ANTHRACOCEROS 299
ANTHRACOTHORAX 259
ANTHREPTES 704
ANTHROPOIDES 128
ANTHUS 741
Antibyx 133
Antichromus 458
antigone, Grus 128
antillarum, Buteo platypterus 111
antillarum, Columbina passerina 167
antillarum, Mimus gilvus 648

antillarum, Myiarchus 378
antillarum, Podilymbus podiceps 79
antillarum, Sterna 151
antillarum, Zonotrichia capensis 782
ANTILOPHIA 341
antinorii, Cisticola lateralis 551
antinorii, Psalidoprocne pristoptera 531
antioproctum, Dicaeum trigonostigma 701
antioquensis, Picumnus granadensis 313
antioquiae, Ampelion rufaxilla 346
antioquiae, Anisognathus somptuosus 808
antioquiae, Machaeropterus regulus 340
antioquiae, Piprites chloris 349
antioquiae, Zenaida auriculata 166
antioxantha, Culicicapa ceylonensis 698
antipodensis, Diomedea exulans 72
antipodes, Megadyptes 71
antiquus, Synthliboramphus 155
antisianus, Pharomachrus 282
antisiensis, Cranioleuca 411
antofagastae, Zonotrichia capensis 783
antonensis, Aimophila rufescens 787
antoniae, Carpodectes 348
antonii, Mirafra angolensis 541
antonii, Rhipidura dahli 496
antonii, Sarothrura affinis 117
anulus, Passerculus sandwichensis 784
ANUMBIUS 412
ANUROLIMNAS 118
ANUROPHASIS 55
anxia, Anabacerthia striaticollis 414
aolae, Alcedo pusilla 292
aolae, Micropsitta finschii 182
apache, Accipiter gentilis 108
apache, Regulus satrapa 633
APALIS 560
APALODERMA 279
APALOPTERON 631
apatetes, Pachycephala olivacea 475
apega, Stachyris erythroptera 608
apertus, Buarremon brunneinuchus 799
apertus, Caprimulgus cayennensis 241
apetzii, Calandrella rufescens 546
aphaea, Zonotrichia leucophrys 783
aphanes, Chaetura vauxi 251
APHANOTRICCUS 366
APHANTOCHROA 258
APHELOCEPHALA 449
APHELOCOMA 507
aphrasta, Eremophila alpestris 550
APHRASTURA 404
APHRIZA 143
aphrodite, Parus major 524
APHRODROMA 74
apiaster, Eumomota superciliosa 294
apiaster, Merops 296
apiata, Mirafra 542
apicalis, Acanthiza 449
apicalis, Amazilia fimbriata 266
apicalis, Loriculus philippensis 181
apicalis, Moho 431
apicalis, Myiarchus 377
apicalis, Phaethornis guy 256
apicauda, Treron 173
apivorus, Dicrurus adsimilis 491
apivorus, Pernis 99
APLONIS 651
aplonotus, Parus xanthogenys 525
Aplopelia 160
apo, Dicaeum ignipectum 703
apo, Pyrrhula leucogenis 756
apo, Rhipidura superciliaris 493
apo, Sitta oenochlamys 646
apoda, Paradisaea 518
apoensis, Aerodramus mearnsi 248
apoensis, Pachycephala philippinensis 475
apolinari, Cistothorus 635
apolis, Cinnyris souimanga 712
appalachiensis, Catharus ustulatus 665

appalachiensis, Sphyrapicus varius 315
apperti, Xanthomixis 588
approximans, Cercomacra nigrescens 389
approximans, Circus 103
approximans, Euplectes capensis 725
approximans, Leptotila verreauxi 168
approximans, Malaconotus blanchoti 457
approximans, Pachycephalopsis poliosoma 520
approximans, Vireo pallens 483
apricaria, Pluvialis 134
apricus, Ramphocelus sanguinolentus 806
APROSMICTUS 192
apsleyi, Coracina papuensis 467
apsleyi, Entomyzon cyanotis 435
apsleyi, Ramsayornis fasciatus 440
APTENODYTES 70
APTERYX 35
apuliae, Galerida cristata 547
apurensis, Athene cunicularia 233
apurensis, Cranioleuca vulpina 410
apurensis, Emberizoides herbicola 791
apurensis, Myospiza aurifrons 785
apurensis, Picumnus squamulatus 312
APUS 253
apus, Apus 253
aquaemontis, Melaenornis pallidus 689
aquatica, Muscicapa 692
aquaticus, Cinclus cinclus 698
aquaticus, Rallus 121
aquaticus, Sayornis nigricans 366
AQUILA 112
aquila, Eutoxeres 255
aquila, Fregata 89
aquilonalis, Turdus fumigatus 671
aquilonia, Bombycilla cedrorum 522
aquilonifer, Tetraogallus tibetanus 48
aquilonis, Acrocephalus caffer 585
aquilonis, Chondrohierax uncinatus 99
aquilonis, Iole propinqua 573
aquilonius, Charadrius obscurus 135
aquitanicus, Tetrao urogallus 45
ARA 195
arabica, Coturnix delegorguei 54
arabica, Emberiza tahapisi 778
arabica, Ptyonoprogne obsoleta 536
arabica, Streptopelia roseogrisea 161
arabicus, Anthus similis 742
arabicus, Cinnyricinclus leucogaster 657
arabicus, Milvus migrans 100
arabicus, Pterocles lichtensteinii 156
arabistanicus, Francolinus francolinus 50
arabs, Ardeotis 114
arabs, Zosterops abyssinicus 629
aracari, Pteroglossus 302
ARACHNOTHERA 714
arada, Cyphorhinus 642
araea, Falco 95
aragonica, Pterocles orientalis 156
araguaiae, Furnarius leucopus 404
araguayae, Sakesphorus luctuosus 380
ARAMIDES 122
ARAMIDOPSIS 122
ARAMUS 129
aranea, Sterna nilotica 149
araneipes, Rhea americana 35
ararauna, Ara 195
ARATINGA 196
araucana, Columba 159
araucuan, Ortalis guttata 37
arausiaca, Amazona 204
arborea, Dendrocygna 61
arborea, Lullula 549
arborea, Spizella 785
arboricola, Progne subis 533
ARBOROPHILA 55
arcaei, Bangsia 807
arcana, Haplospiza rustica 789
Arcanator 625
arcanum, Apaloderma narina 279

arcanus, Cisticola erythrops 551
arcanus, Ptilinopus 177
archboldi, Aegotheles 246
archboldi, Dacelo tyro 285
archboldi, Eurostopodus 240
archboldi, Newtonia 462
archboldi, Petroica 521
ARCHBOLDIA 427
archeri, Buphagus erythrorhynchus 659
archeri, Buteo augur 112
archeri, Cossypha 677
archeri, Falco tinnunculus 95
archeri, Francolinus levaillantoides 51
archeri, Heteromirafra 543
archeri, Platalea leucorodia 83
archeri, Tachymarptis melba 253
archibaldi, Acanthiza pusilla 448
ARCHILOCHUS 277
archipelagicus, Indicator 311
archipelagicus, Macronous gularis 608
arcosi, Eriocnemis vestita 272
arctica, Calidris alpina 144
arctica, Fratercula 155
arctica, Gavia 71
arctica, Sitta europaea 644
arcticincta, Hirundo angolensis 535
arcticola, Calidris alpina 144
arcticola, Eremophila alpestris 550
arcticus, Cepphus grylle 154
arcticus, Picoides 323
arcticus, Pipilo erythrophthalmus 797
arctitorquis, Pachycephala monacha 478
arctoa, Leucosticte 752
arctolegus, Agelaius phoeniceus 773
arctus, Pycnonotus striatus 565
arcuata, Dendrocygna 62
arcuata, Emberiza fucata 778
arcuata, Pipreola 346
arcus, Perisoreus canadensis 505
ARDEA 87
ardeleo, Ramphocaenus melanurus 643
ardens, Arborophila 56
ardens, Cossypha caffra 677
ardens, Euplectes 726
ardens, Harpactes 279
ardens, Oriolus traillii 489
ardens, Piranga leucoptera 819
ardens, Promerops gurneyi 715
ardens, Pyrocephalus rubinus 369
ardens, Selasphorus 278
ardens, Sericulus aureus 427
ARDEOLA 86
ardeola, Dromas 132
ARDEOTIS 114
ardescens, Corydon sumatranus 336
ardesiaca, Conopophaga 394
ardesiaca, Egretta 88
ardesiaca, Fulica 127
ardesiaca, Sporophila 792
ardesiacus, Cinclus mexicanus 698
ardesiacus, Melaenornis 689
ardesiacus, Myrmoborus myotherinus 390
ardesiacus, Platylophus galericulatus 504
ardesiacus, Rhopornis 389
ardesiacus, Thamnomanes 384
ardosiaceus, Falco 95
ardosiaceus, Turdus plumbeus 672
ardosiaceus, Contopus fumigatus 367
aremoricus, Aegithalos caudatus 538
arenacea, Spizella pusilla 785
arenaceus, Charadrius marginatus 135
arenaceus, Mimus saturninus 649
arenalensis, Zonotrichia capensis 783
ARENARIA 142
arenaria, Chersomanes albofasciata 544
arenarius, Lanius isabellinus 479
arenarius, Pterocles orientalis 156
arenarius, Turnix sylvatica 129
arenarum, Percnostola 391

bensoni, *Phyllastrephus terrestris* 570
bensoni, *Poeoptera kenricki* 658
bensoni, *Pseudocossyphus sharpei* 688
bensoni, *Sheppardia gunningi* 675
bensonorum, *Carduelis flavirostris* 751
bentet, *Lanius schach* 480
benti, *Turdus assimilis* 671
berard, *Pelecanoides urinatrix* 78
berauensis, *Trichoglossus haematodus* 185
BERENICORNIS 300
berezowskii, *Ithaginis cruentus* 57
berezowskii, *Parus flavipectus* 526
bergii, *Phasianus colchicus* 60
bergii, *Sterna* 150
berigora, *Falco* 97
beringiae, *Limosa fedoa* 141
berlai, *Baryphthengus ruficapillus* 294
berlandieri, *Thryothorus ludovicianus* 638
berlepschi, *Aglaiocercus kingi* 275
berlepschi, *Anthocephala floriceps* 267
berlepschi, *Asthenes* 406
berlepschi, *Campylopterus cuvierii* 258
berlepschi, *Chaetocercus* 277
berlepschi, *Chamaeza campanisona* 398
berlepschi, *Chlorostilbon aureoventris* 261
berlepschi, *Colius striatus* 278
berlepschi, *Columba subvinacea* 160
berlepschi, *Conopias trivirgatus* 375
berlepschi, *Crypturellus* 31
berlepschi, *Dacnis* 813
berlepschi, *Hemispingus melanotis* 804
berlepschi, *Hylopezus* 401
berlepschi, *Leptasthenura aegithaloides* 405
berlepschi, *Lesbia victoriae* 273
berlepschi, *Merganetta armata* 64
berlepschi, *Mitrephanes phaeocercus* 367
berlepschi, *Myrmeciza* 392
berlepschi, *Myrmeciza immaculata* 392
berlepschi, *Myrmoborus lugubris* 390
berlepschi, *Myrmotherula menetriesii* 386
berlepschi, *Ochthoeca fumicolor* 373
berlepschi, *Parotia carolae* 516
berlepschi, *Phimosus infuscatus* 83
berlepschi, *Picoides mixtus* 321
berlepschi, *Pipra erythrocephala* 343
berlepschi, *Polioptila dumicola* 644
berlepschi, *Pseudotriccus pelzelni* 355
berlepschi, *Pyrrhura melanura* 199
berlepschi, *Rhegmatorhina* 393
berlepschi, *Tangara nigroviridis* 812
berlepschi, *Thamnophilus tenuepunctatus* 381
berlepschi, *Thraupis episcopus* 807
berlepschi, *Thripophaga* 412
berlepschi, *Tolmomyias sulphurescens* 363
berlepschi, *Turdus albicollis* 672
berlepschi, *Xiphocolaptes promeropirhynchus* 422
BERLEPSCHIA 414
berlepschii, *Myiarchus oberi* 378
berliozi, *Apus* 253
berliozi, *Canirallus kioloides* 117
berliozi, *Heterospingus xanthopygius* 814
berliozi, *Lophura nycthemera* 58
bermudensis, *Sialia sialis* 663
bermudianus, *Vireo griseus* 482
bernali, *Haplophaedia aureliae* 273
bernardi, *Sakesphorus* 380
bernicla, *Branta* 63
bernieri, *Anas* 67
bernieri, *Malurus lamberti* 429
bernieri, *Oriolia* 461
bernieri, *Threskiornis aethiopicus* 82
BERNIERIA 588
bernsteini, *Centropus* 215
bernsteini, *Chalcopsitta atra* 184
bernsteini, *Monarcha trivirgatus* 501
bernsteini, *Pitta erythrogaster* 338
bernsteini, *Sterna* 149
bernsteinii, *Megapodius* 36
bernsteinii, *Ptilinopus* 175

berthelotii, *Anthus* 744
berthemyi, *Garrulax poecilorhynchus* 614
bertrandi, *Chlorophoneus olivaceus* 457
bertrandi, *Ploceus* 720
beryllina, *Amazilia* 267
beryllinus, *Loriculus* 181
besra, *Accipiter virgatus* 107
bessereri, *Basileuterus tristriatus* 767
bessophilus, *Lampornis clemenciae* 268
besti, *Dicaeum trigonostigma* 701
bestiarum, *Buphagus erythrorhynchus* 659
besuki, *Pellorneum pyrrogenys* 600
bethelae, *Garrulax affinis* 615
bethelae, *Pucrasia macrolopha* 57
bethunei, *Sterna vittata* 150
bewickii, *Cygnus columbianus* 63
bewickii, *Thryomanes* 636
bewsheri, *Turdus* 667
bhamoensis, *Pomatorhinus ruficollis* 603
bhamoensis, *Stachyris ruficeps* 606
biaki, *Eclectus roratus* 192
bianchii, *Certhia familiaris* 647
bianchii, *Lanius excubitor* 480
bianchii, *Phasianus colchicus* 60
biarcuata, *Melozone* 798
biarmicus, *Falco* 97
biarmicus, *Panurus* 622
BIAS 454
BIATAS 380
biblicus, *Passer domesticus* 716
bicalcarata, *Galloperdix* 57
bicalcaratum, *Polyplectron* 60
bicalcaratus, *Francolinus* 52
bichenovii, *Taeniopygia* 733
bicincta, *Cariama cristata* 116
bicinctus, *Charadrius* 136
bicinctus, *Hypnelus ruficollis* 333
bicinctus, *Pterocles* 157
bicinctus, *Treron* 172
bicinia, *Coracina montana* 469
bicknelli, *Catharus* 665
bicolor, *Accipiter* 108
bicolor, *Campylorhynchus griseus* 633
bicolor, *Conirostrum* 815
bicolor, *Coracina* 467
bicolor, *Cyanophaia* 262
bicolor, *Dendrocygna* 61
bicolor, *Dicaeum* 701
bicolor, *Ducula* 180
bicolor, *Garrulax leucolophus* 612
bicolor, *Gymnopithys leucaspis* 393
bicolor, *Laniarius* 460
bicolor, *Lonchura* 734
bicolor, *Microrhopias quixensis* 387
bicolor, *Nigrita* 726
bicolor, *Parus* 529
bicolor, *Perisoreus canadensis* 505
bicolor, *Ploceus* 723
bicolor, *Porzana* 123
bicolor, *Rhyacornis* 682
bicolor, *Saxicola caprata* 684
bicolor, *Speculipastor* 658
bicolor, *Sporophila leucoptera* 793
bicolor, *Spreo* 658
bicolor, *Stachyris erythroptera* 608
bicolor, *Tachycineta* 533
bicolor, *Tiaris* 795
bicolor, *Trichastoma* 600
bicolor, *Turdoides* 611
bicornis, *Buceros* 299
bicornis, *Eremophila alpestris* 551
biddulphi, *Podoces* 511
biddulphi, *Pucrasia macrolopha* 57
biddulphi, *Strix aluco* 228
bidentata, *Piranga* 818
bidentatus, *Harpagus* 100
bidentatus, *Lybius* 309
bido, *Spilornis cheela* 103
bidoupensis, *Alcippe ruficapilla* 619

biedermanni, *Picus canus* 328
biesenbachi, *Turdus poliocephalus* 668
bieti, *Alcippe vinipectus* 619
bieti, *Garrulax* 614
bifasciata, *Certhidea olivacea* 796
bifasciata, *Loxia leucoptera* 756
bifasciatus, *Campicoloides* 684
bifasciatus, *Caprimulgus longirostris* 241
bifasciatus, *Cinnyris* 710
bifasciatus, *Platyrinchus mystaceus* 364
bifasciatus, *Psarocolius* 769
bifrontatus, *Charadrius tricollaris* 135
bigilalae, *Lonchura tristissima* 735
bihe, *Prinia flavicans* 559
bilbali, *Anthus australis* 741
bilineata, *Amphispiza* 786
bilineata, *Polioptila plumbea* 644
bilineatus, *Pogoniulus* 308
bilkevitchi, *Garrulax lineatus* 615
bilkevitschi, *Carduelis chloris* 749
billitonis, *Chloropsis cochinchinensis* 699
billitonis, *Eurylaimus javanicus* 336
billitonis, *Megalaima rafflesii* 305
billitonis, *Pycnonotus plumosus* 568
billonis, *Francolinus francolinus* 50
billypayni, *Francolinus francolinus* 50
bilopha, *Eremophila* 551
bilophus, *Heliactin* 275
bimaculata, *Gallicolumba tristigmata* 170
bimaculata, *Melanocorypha* 544
bimaculata, *Peneothello* 519
bimaculatus, *Ammodramus savannarum* 784
bimaculatus, *Caprimulgus macrurus* 243
bimaculatus, *Cnemotriccus fuscatus* 366
bimaculatus, *Monarcha trivirgatus* 501
bimaculatus, *Pycnonotus* 567
bimaensis, *Dicrurus hottentottus* 492
bineschensis, *Cinnyris chloropygius* 708
binfordi, *Grallaria guatimalensis* 399
binghami, *Stachyris chrysaea* 606
binotata, *Apalis* 561
binotatus, *Veles* 240
bipartita, *Eumomota superciliosa* 294
bipartitus, *Phrygilus alaudinus* 788
birchalli, *Catharus aurantiirostris* 664
birgitae, *Anthreptes malacensis* 704
birmanus, *Merops orientalis* 296
birostris, *Ocyceros* 299
birulae, *Larus argentatus* 147
birulai, *Lagopus lagopus* 47
birwae, *Ptyonoprogne fuligula* 536
biscutata, *Streptoprocne* 247
bishaensis, *Streptopelia lugens* 160
bishopi, *Catharopeza* 762
bishopi, *Moho* 431
bisignatus, *Rhinoptilus africanus* 145
bismarckii, *Aviceda subcristata* 98
bispecularis, *Garrulus glandarius* 509
bistictus, *Ptilinopus solomonensis* 176
bistriatus, *Anthus australis* 741
bistriatus, *Burhinus* 131
bistrigiceps, *Acrocephalus* 585
bitorquata, *Melanopareia maximiliani* 395
bitorquata, *Streptopelia* 161
bitorquatus, *Pteroglossus* 302
bitorquatus, *Rhinoptilus* 145
bivittata, *Buettikoferella* 577
bivittata, *Petroica* 521
bivittatus, *Basileuterus* 766
bivittatus, *Lepidocolaptes angustirostris* 425
bivittatus, *Trochocercus cyanomelas* 498
BIZIURA 70
blaauwi, *Geophaps smithii* 165
blainvillii, *Peltops* 464
blakei, *Catharus dryas* 665
blakei, *Grallaria* 400
blakei, *Melanerpes superciliaris* 315
blakistoni, *Anthus spinoletta* 744
blakistoni, *Bubo* 227
blakistoni, *Turnix suscitator* 130

blanchoti, Malaconotus 457
blanci, Aethopyga nipalensis 713
blancoi, Contopus latirostris 368
blancoui, Numida meleagris 40
blandi, Tricholaema melanocephala 308
blandini, Houbaropsis bengalensis 116
blandus, Cacomantis variolosus 209
blanfordi, Calandrella 546
blanfordi, Dendrocopos mahrattensis 319
blanfordi, Parus major 524
blanfordi, Prinia inornata 559
blanfordi, Psalidoprocne pristoptera 531
blanfordi, Pycnonotus 568
blanfordi, Pyrgilauda 719
blanfordi, Sylvia leucomelaena 597
blanfordii, Turnix tanki 129
blasii, Hypothymis azurea 498
blasii, Lonchura punctulata 735
blasii, Myzomela 441
blatteus, Pyrocephalus rubinus 369
blayneyi, Anthus caffer 744
BLEDA 572
blewitti, Heteroglaux 233
blewitti, Perdicula erythrorhyncha 55
blicki, Cinnyris venustus 710
blighi, Bubo nipalensis 226
blighi, Myophonus 659
blissetti, Dyaphorophyia 454
blissi, Crateroscelis nigrorufa 445
bloodi, Epimachus meyeri 517
bloxami, Cyanoliseus patagonus 198
blumenbachii, Crax 40
blythi, Carpodacus thura 754
blythi, Cyornis rufigastra 697
blythi, Prinia inornata 559
blythii, Garrulax affinis 615
blythii, Onychognathus 658
blythii, Psittaculirostris desmarestii 205
blythii, Sturnus malabaricus 655
blythii, Tragopan 57
BLYTHIPICUS 330
boanensis, Monarcha 501
boanensis, Tanysiptera galatea 284
boavistae, Alaemon alaudipes 544
bocagei, Anthus cinnamomeus 742
bocagei, Bostrychia olivacea 82
bocagei, Chlorophoneus 458
bocagei, Estrilda melanotis 730
bocagei, Euplectes axillaris 725
bocagei, Illadopsis rufipennis 601
bocagei, Lybius torquatus 309
bocagei, Sheppardia 675
bocagei, Tchagra australis 458
bocagei, Turdus pelios 666
BOCAGIA 458
bocagii, Amaurocichla 588
bocagii, Nectarinia 708
bochaiensis, Luscinia cyane 676
bocki, Cuculus sparverioides 207
bodalyae, Numida meleagris 40
bodessa, Cisticola 552
bodini, Amazona festiva 203
boehmeri, Afrotis afra 116
boehmi, Dinemellia dinemelli 719
boehmi, Francolinus afer 53
boehmi, Lanius excubitoroides 481
boehmi, Merops 295
boehmi, Muscicapa 692
boehmi, Neafrapus 250
boehmi, Parisoma 598
boehmi, Sarothrura 117
boehmi, Trachyphonus darnaudii 310
boehmii, Anthus campestris 742
bogdanovi, Francolinus francolinus 50
bogolubovi, Cursorius cursor 145
bogotense, Camptostoma obsoletum 352
bogotensis, Anthus 745
bogotensis, Asio flammeus 236
bogotensis, Chrysomus icterocephalus 772

bogotensis, Cinnycerthia olivascens 635
bogotensis, Colinus cristatus 42
bogotensis, Columba subvinacea 160
bogotensis, Contopus cinereus 367
bogotensis, Gallinula melanops 126
bogotensis, Ixobrychus exilis 85
bogotensis, Myiopagis gaimardii 350
bogotensis, Polystictus pectoralis 354
bogotensis, Sicalis luteola 791
bogotensis, Sporophila intermedia 792
bogotensis, Tangara guttata 811
bogotensis, Thryothorus leucotis 638
bohndorffi, Anthus leucophrys 743
bohndorffi, Cyanomitra verticalis 706
bohndorffi, Phyllanthus atripennis 620
bohndorffi, Ploceus cucullatus 721
boholensis, Coracina striata 467
boholensis, Otus megalotis 221
boholensis, Pachycephala philippinensis 475
boholensis, Rhinomyias ruficauda 690
boholensis, Stachyris nigrocapitata 606
boholensis, Zosterops everetti 626
boiei, Myzomela 442
boineti, Amaurornis 124
BOISSONNEAUA 269
boissonneautii, Pseudocolaptes 414
bojeri, Ploceus 721
bokermanni, Antilophia 341
bokharensis, Parus 525
bolivari, Coereba flaveola 776
bolivari, Pipra pipra 342
bolivari, Synallaxis cinnamomea 408
boliviana, Athene cunicularia 233
boliviana, Carduelis magellanica 750
boliviana, Chamaeza campanisona 398
boliviana, Chiroxiphia 341
boliviana, Coeligena coeligena 270
boliviana, Eupetomena macroura 258
boliviana, Grallaricula flavirostris 401
boliviana, Hemithraupis guira 814
boliviana, Henicorhina leucophrys 641
boliviana, Lesbia nuna 273
boliviana, Monasa morphoeus 334
boliviana, Nothura darwinii 34
boliviana, Phaeothlypis rivularis 768
boliviana, Phibalura flavirostris 346
boliviana, Piaya cayana 214
boliviana, Piprites chloris 349
boliviana, Poospiza 790
boliviana, Pulsatrix perspicillata 230
boliviana, Tangara mexicana 810
boliviana, Thalurania furcata 263
boliviana, Thraupis sayaca 807
bolivianum, Camptostoma obsoletum 352
bolivianum, Glaucidium 230
bolivianum, Philydor rufum 416
bolivianum, Ramphomicron microrhynchum 274
bolivianum, Ramphotrigon megacephalum 378
bolivianus, Agelaioides badius 773
bolivianus, Anairetes alpinus 353
bolivianus, Ara militaris 195
bolivianus, Attila 379
bolivianus, Capito auratus 304
bolivianus, Chlorornis riefferii 809
bolivianus, Chlorospingus ophthalmicus 818
bolivianus, Iridosornis jelskii 809
bolivianus, Lathrotriccus euleri 366
bolivianus, Lepidocolaptes lacrymiger 425
bolivianus, Myioborus melanocephalus 765
bolivianus, Oreopsar 773
bolivianus, Oreotrochilus estella 270
bolivianus, Otus guatemalae 224
bolivianus, Phaethornis malaris 256
bolivianus, Pitangus sulphuratus 374
bolivianus, Polioxolmis rufipennis 372
bolivianus, Scytalopus 396
bolivianus, Thinocorus rumicivorus 138

bolivianus, Thryothorus genibarbis 637
bolivianus, Trogon curucui 280
bolivianus, Vireolanius leucotis 482
bolivianus, Zimmerius 355
boliviensis, Lampropsar tanagrinus 772
bollei, Phoeniculus 297
bollii, Columba 158
bolovenensis, Pitta oatesi 337
boltoni, Aethopyga 712
bombus, Chaetocercus 277
BOMBYCILLA 522
bonairensis, Coereba flaveola 776
bonairensis, Margarops fuscatus 651
bonairensis, Vireo altiloquus 485
bonana, Icterus 770
bonapartei, Coeligena 271
bonapartei, Gymnobucco 306
bonapartei, Loriculus philippensis 181
bonapartei, Nothocercus 31
bonapartei, Turdus viscivorus 669
bonapartii, Malurus cyanocephalus 429
bonapartii, Sarothrura rufa 117
bonariae, Troglodytes aedon 640
bonariensis, Molothrus 773
bonariensis, Thraupis 807
BONASA 44
bonasia, Tetrastes 45
bondi, Dicrurus leucophaeus 492
bondi, Turdus fumigatus 671
bondi, Tyto alba 219
bonelli, Phylloscopus 590
bonensis, Hylocitrea 474
boneratensis, Oriolus chinensis 488
bonga, Dicaeum ignipectum 703
bongaoensis, Parus elegans 525
bonita, Aethopyga bella 712
bonnetiana, Zonotrichia capensis 782
bonthaina, Ficedula 694
bonthaina, Hylocitrea bonensis 474
bonvaloti, Aegithalos 539
bonvaloti, Babax lanceolatus 612
boobook, Ninox 234
boodang, Petroica 521
boothi, Petrophassa albipennis 165
boranensis, Turdoides aylmeri 610
boraquira, Nothura 34
borbae, Picumnus aurifrons 312
borbae, Pipra aureola 342
borbae, Skutchia 394
borbae, Taraba major 380
borbonica, Phedina 532
borbonicus, Hypsipetes 575
borbonicus, Zosterops 630
borealis, Basilinna leucotis 268
borealis, Buteo jamaicensis 111
borealis, Calonectris diomedea 75
borealis, Campylorhamphus pusillus 426
borealis, Cettia diphone 579
borealis, Contopus 367
borealis, Cypseloides niger 246
borealis, Lanius excubitor 480
borealis, Numenius 141
borealis, Otus flammeolus 222
borealis, Parus montanus 527
borealis, Phylloscopus 591
borealis, Picoides 321
borealis, Psittacula krameri 193
borealis, Somateria mollissima 69
borealis, Troglodytes troglodytes 639
borealoides, Phylloscopus 591
borelli, Buarremon torquatus 799
borelliana, Strix virgata 229
boreonesioticus, Sericornis magnirostra 446
boreus, Corvus tasmanicus 514
borin, Sylvia 596
borinquensis, Ammodramus savannarum 785
borkumensis, Perdix perdix 53
bornea, Eos 185
borneana, Chalcoparia singalensis 704

borneanum, Dicaeum concolor 702
borneanus, Actenoides concretus 284
borneense, Glaucidium brodiei 230
borneensis, Aviceda jerdoni 98
borneensis, Caloperdix oculeus 56
borneensis, Dicrurus hottentottus 492
borneensis, Enicurus leschenaulti 683
borneensis, Erythrura hyperythra 733
borneensis, Eupetes macrocerus 453
borneensis, Macropygia emiliana 163
borneensis, Megalaima rafflesii 305
borneensis, Melanoperdix niger 54
borneensis, Myophonus glaucinus 659
borneensis, Ninox scutulata 234
borneensis, Psarisomus dalhousiae 336
borneensis, Pycnonotus squamatus 566
borneensis, Rhopodytes diardi 212
borneensis, Stachyris nigriceps 607
borneensis, Terpsiphone paradisi 499
borneensis, Zanclostomus curvirostris 213
bornensis, Anthreptes malacensis 704
bornensis, Macronous gularis 608
bornensis, Pomatorhinus montanus 603
borneoensis, Buceros rhinoceros 299
borneoensis, Orthotomus ruficeps 564
borneonense, Dinopium javanense 329
borodinonis, Microscelis amaurotis 575
bororensis, Camaroptera brachyura 562
bororensis, Gymnoris superciliaris 718
borosi, Ammomanes deserti 545
borreroi, Anas cyanoptera 67
borreroi, Xiphorhynchus picus 423
borrisowi, Bubo bubo 226
boscaweni, Oenanthe lugubris 685
boschmai, Lonchura castaneothorax 736
BOSTRYCHIA 82
BOTAURUS 84
botelensis, Otus elegans 222
bottae, Oenanthe 684
bottanensis, Pica pica 511
bottegi, Francolinus castaneicollis 52
botterii, Aimophila 786
bottomeana, Chloroceryle americana 292
boucardi, Aimophila ruficeps 787
boucardi, Amazilia 266
boucardi, Crypturellus 33
boucardi, Granatellus sallaei 768
boucardi, Microrhopias quixensis 387
boucardi, Myrmeciza longipes 391
bouchellii, Streptoprocne zonaris 247
bougainvillei, Accipiter novaehollandiae 106
bougainvillei, Actenoides 284
bougainvillei, Alcedo pusilla 292
bougainvillei, Coracina caledonica 466
bougainvillei, Pachycephala orioloides 477
bougainvillei, Phylloscopus poliocephalus 593
bougainvillei, Stresemannia 439
bougainvillei, Turdus poliocephalus 668
bougainvillii, Phalacrocorax 92
bougueri, Urochroa 269
boulboul, Turdus 667
boultoni, Bradypterus lopesi 581
boultoni, Margarornis rubiginosus 413
bourbonnensis, Terpsiphone 499
bourcierci, Chlorochrysa calliparaea 809
bourcieri, Geotrygon frenata 169
bourcieri, Phaethornis 256
bourcierii, Eubucco 304
bourdellei, Hemixos flavala 575
bourdilloni, Eurostopodus macrotis 240
bourdilloni, Rhopocichla atriceps 608
bourdilloni, Turdus merula 668
bourkii, Neopsephotus 189
bournei, Ardea purpurea 87
bourneorum, Calamanthus fuliginosus 444
bournsi, Loriculus philippensis 181
bouroensis, Oriolus 487
bourquii, Francolinus 50
bourreti, Chaetura caudacuta 251

bouruensis, Otus magicus 222
bouruensis, Rhipidura rufiventris 494
bouvieri, Cinnyris 710
bouvieri, Scotopelia 227
bouvreuil, Sporophila 793
bouvronides, Sporophila 792
bowdleri, Alcedo leucogaster 291
bowdleri, Melaenornis pallidus 689
bowdleri, Turdoides rubiginosa 610
boweni, Chersomanes albofasciata 544
boweni, Glareola pratincola 146
bowensis, Pardalotus striatus 444
boweri, Colluricincla 489
boweri, Epthianura crocea 443
bowmani, Phlegopsis nigromaculata 394
boyciana, Ciconia 81
boydi, Dryoscopus angolensis 459
boydi, Puffinus assimilis 76
boyeri, Coracina 467
boylei, Diglossa gloriosissima 816
braba, Glyciphila melanops 440
braccata, Amazilia saucerrottei 266
braccatus, Moho 431
braccatus, Trogon violaceus 280
bracei, Chlorostilbon 261
brachidactylus, Geothlypis trichas 763
brachipus, Lewinia pectoralis 122
BRACHYCOPE 724
brachydactyla, Calandrella 545
brachydactyla, Carpospiza 717
brachydactyla, Certhia 648
BRACHYGALBA 331
brachylophus, Lophornis 260
brachyphorus, Dicrurus paradiseus 493
brachyptera, Athene cunicularia 233
brachyptera, Elaenia chiriquensis 351
brachyptera, Tachycineta thalassina 533
BRACHYPTERACIAS 283
brachypterus, Buteo 112
brachypterus, Cisticola 554
brachypterus, Cypsiurus parvus 252
brachypterus, Dasyornis 443
brachypterus, Donacobius atricapilla 642
brachypterus, Ploceus nigricollis 720
brachypterus, Podargus strigoides 237
brachypterus, Quiscalus niger 775
brachypterus, Tachybaptus dominicus 79
brachypterus, Tachyeres 64
BRACHYPTERYX 673
brachypus, Pyrrhocorax pyrrhocorax 512
BRACHYRAMPHUS 155
brachyrhyncha, Egretta intermedia 87
brachyrhyncha, Megalaima henricii 306
brachyrhyncha, Rhipidura 496
brachyrhynchos, Corvus 513
brachyrhynchus, Anser 62
brachyrhynchus, Contopus fumigatus 367
brachyrhynchus, Larus canus 146
brachyrhynchus, Oriolus 488
brachytarsus, Contopus cinereus 367
brachyura, Camaroptera 562
brachyura, Chaetura 251
brachyura, Galerida cristata 548
brachyura, Myrmotherula 384
brachyura, Pitta 339
brachyura, Poecilodryas 518
brachyura, Sylvietta 595
brachyura, Synallaxis 409
brachyura, Uropsila leucogastra 641
brachyurum, Apaloderma narina 279
brachyurus, Accipiter 107
brachyurus, Anthus 744
brachyurus, Buteo 111
brachyurus, Celeus 325
brachyurus, Graydidascalus 202
brachyurus, Idiopsar 789
brachyurus, Myiarchus tyrannulus 378
brachyurus, Ramphocinclus 650
brachyurus, Surniculus lugubris 211

brachyurus, Thamnophilus multistriatus 381
brachyurus, Todiramphus chloris 288
bracki, Phrygilus alaudinus 788
bracteatus, Dicrurus 493
bracteatus, Nyctibius 238
bradfieldi, Apus 253
bradfieldi, Emberiza capensis 778
bradfieldi, Francolinus hartlaubi 52
bradfieldi, Mirafra sabota 542
bradfieldi, Motacilla capensis 740
bradfieldi, Ortygospiza atricollis 731
bradfieldi, Tockus 298
Bradornis 690
bradshawi, Certhilauda subcoronata 543
BRADYPTERUS 580
brama, Athene 232
brandti, Eremophila alpestris 551
brandti, Leucosticte 752
brandtii, Garrulus glandarius 508
brandtii, Parus palustris 527
brangeri, Cymbilaimus lineatus 379
branickii, Heliodoxa 269
branickii, Leptosittaca 198
branickii, Nothoprocta ornata 33
branickii, Odontorchilus 634
branickii, Tangara vassorii 812
branickii, Theristicus melanopis 83
bransfieldensis, Phalacrocorax atriceps 92
BRANTA 63
brasherii, Basileuterus culicivorus 766
brasiliana, Cercomacra 389
brasiliana, Hydropsalis 245
brasiliana, Leptocoma sperata 707
brasilianum, Glaucidium 231
brasilianus, Anthus hellmayri 745
brasilianus, Phalacrocorax 91
brasiliensis, Amazona 203
brasiliensis, Amazonetta 65
brasiliensis, Caryothraustes canadensis 824
brasiliensis, Cathartes 93
brasiliensis, Chelidoptera tenebrosa 334
brasiliensis, Leptotila verreauxi 168
brasiliensis, Sicalis flaveola 791
brasiliensis, Tangara mexicana 810
brassi, Melidectes belfordi 438
brassi, Philemon 436
brauni, Apalis rufogularis 561
brauni, Laniarius luehderi 460
braunianus, Dicrurus aeneus 492
braziliensis, Tityra cayana 343
brazzae, Pachycoccyx audeberti 207
brazzae, Phedina 532
bredoi, Ruwenzorornis johnstoni 206
brehmeri, Turtur 164
brehmi, Amazilia tzacatl 265
brehmi, Ptilopachus petrosus 50
brehmii, Monarcha 502
brehmii, Phylloscopus collybita 589
brehmii, Psittacella 190
brehmorum, Apus pallidus 253
brenchleyi, Ducula 179
brenchleyi, Rhipidura albiscapa 495
bres, Criniger 573
bresilius, Ramphocelus 806
brevibarba, Pithys albifrons 392
brevicarinatus, Ramphastos sulfuratus 303
brevicauda, Amalocichla incerta 522
brevicauda, Aplonis magna 651
brevicauda, Muscicapa caerulescens 691
brevicauda, Muscigralla 373
brevicauda, Paradigalla 516
brevicauda, Paramythia montium 452
brevicauda, Passerella iliaca 781
brevicauda, Phalacrocorax melanoleucos 91
brevicauda, Prinia socialis 558
brevicauda, Zosterops flavifrons 628
brevicaudata, Camaroptera brachyura 562
brevicaudata, Napothera 604
brevicaudata, Nesillas 582

chadensis, Batis orientalis 455
chadensis, Bradypterus baboecala 580
chadensis, Mirafra cantillans 540
CHAETOCERCUS 277
CHAETOPS 624
CHAETOPTILA 431
CHAETORHYNCHUS 491
CHAETORNIS 578
CHAETURA 251
chagwensis, Andropadus gracilirostris 569
CHAIMARRORNIS 682
chakei, Cinnycerthia unirufa 635
chalcauchenia, Leptotila verreauxi 168
chalceus, Cinnyris cupreus 711
chalceus, Lamprotornis mevesii 657
chalcocephala, Galbula albirostris 331
chalcolophus, Tauraco livingstonii 205
chalcomelas, Cinnyris 710
CHALCOMITRA 706
chalconota, Amazilia yucatanensis 265
chalconota, Ducula 179
chalconota, Gallicolumba jobiensis 170
chalconotus, Phalacrocorax 92
CHALCOPARIA 703
CHALCOPHAPS 164
CHALCOPSITTA 184
chalcoptera, Phaps 164
chalcopterus, Pionus 202
chalcopterus, Rhinoptilus 145
chalcospilos, Turtur 164
Chalcostetha 708
CHALCOSTIGMA 274
chalcothorax, Galbula 331
chalcothorax, Parotia carolae 516
chalcurum, Polyplectron 60
chalcurus, Lamprotornis 656
chalcurus, Ptilinopus 176
challengeri, Eos histrio 184
chalybaeus, Lamprotornis 656
chalybaeus, Surniculus velutinus 211
chalybatus, Manucodia 516
chalybea, Dyaphorophyia 454
chalybea, Euphonia 821
chalybea, Progne 533
chalybea, Psalidoprocne pristoptera 531
chalybeata, Vidua 737
chalybeocephala, Myiagra alecto 504
chalybeus, Centropus 215
chalybeus, Cinnyris 709
chalybeus, Lophornis 260
CHALYBURA 267
CHAMAEA 609
CHAMAEPETES 39
CHAMAEZA 398
chamberlaini, Lagopus muta 47
chamelum, Pellorneum ruficeps 599
chamnongi, Centropus toulou 216
chanco, Phalacrocorax olivaceus 91
chandleri, Acanthiza lineata 449
changamwensis, Bias musicus 454
changamwensis, Cyanomitra olivacea 706
chapadense, Philydor rufum 416
chapadensis, Otus choliba 223
chapadensis, Xenops rutilans 419
chapalensis, Geothlypis trichas 763
chaparensis, Piaya minuta 214
chapini, Apalis 561
chapini, Cinnyris stuhlmanni 709
chapini, Corvinella corvina 478
chapini, Dryoscopus cubla 459
chapini, Francolinus nobilis 52
chapini, Guttera edouardi 40
chapini, Kupeornis 620
chapini, Mirafra africana 541
chapini, Motacilla clara 740
chapini, Schoutedenapus myoptilus 250
chapini, Sheppardia bocagei 675
chapini, Sylvietta leucophrys 596
chaplinae, Grallaria haplonota 399

chaplini, Lybius 309
chapmani, Amazona farinosa 204
chapmani, Basileuterus coronatus 766
chapmani, Chaetura 251
chapmani, Chordeiles minor 239
chapmani, Chubbia jamesoni 140
chapmani, Columba plumbea 160
chapmani, Conopophaga castaneiceps 394
chapmani, Formicivora rufa 388
chapmani, Herpetotheres cachinnans 94
chapmani, Hirundo fulva 538
chapmani, Mecocerculus leucophrys 353
chapmani, Micropygia schomburgkii 118
chapmani, Myiodynastes maculatus 375
chapmani, Oceanodroma leucorhoa 78
chapmani, Phylloscartes 356
chapmani, Pulsatrix perspicillata 230
chapmani, Pyrrhura melanura 199
chapmani, Schistes geoffroyi 275
chapmani, Sicalis luteola 791
chapmani, Synallaxis brachyura 409
CHARADRIUS 135
charienturus, Thryomanes bewickii 636
chariessa, Apalis 561
CHARITOSPIZA 801
charlottae, Iole olivacea 573
charltonii, Arborophila 56
CHARMOSYNA 187
charmosyna, Estrilda 731
chaseni, Batrachostomus javensis 237
chaseni, Charadrius peronii 136
chaseni, Enicurus leschenaulti 683
chaseni, Harpactes erythrocephalus 280
chaseni, Strix leptogrammica 228
chaseni, Treron curvirostra 172
CHASIEMPIS 499
chathamensis, Anthus novaeseelandiae 741
chathamensis, Cyanoramphus novaezelandiae 188
chathamensis, Eudyptula minor 71
chathamensis, Haematopus unicolor 132
chathamensis, Hemiphaga novaeseelandiae 180
chathamensis, Pelecanoides urinatrix 78
chathamensis, Petroica macrocephala 522
chathamensis, Prosthemadera novaeseelandiae 435
chauleti, Cissa hypoleuca 510
CHAUNA 61
CHAUNOPROCTUS 755
chaupensis, Tangara labradorides 812
chavaria, Chauna 61
chavezi, Cyanocorax melanocyanea 506
chayulensis, Paradoxornis fulvifrons 623
cheela, Spilornis 102
cheeputi, Attagis malouinus 138
cheesmani, Ammomanes deserti 545
cheimomnestes, Oceanodroma leucorhoa 78
cheleensis, Calandrella 546
CHELICTINIA 100
chelicuti, Halcyon 287
chelidonia, Glyciphila melanops 440
CHELIDOPTERA 334
Chelidorynx 493
Chen 62
chendoola, Galerida cristata 548
cheniana, Mirafra 540
CHENONETTA 65
CHERAMOECA 532
cherina, Cisticola 555
cheriway, Caracara 93
chermesina, Myzomela cardinalis 442
cherriei, Cypseloides 246
cherriei, Elaenia fallax 352
cherriei, Myospiza aurifrons 785
cherriei, Myrmotherula 384
cherriei, Synallaxis 409
cherriei, Thripophaga 411
cherriei, Tolmomyias sulphurescens 363
cherrug, Falco 97

CHERSOMANES 543
chersonesophilus, Macronous gularis 608
chersonesus, Chrysocolaptes lucidus 329
chersonesus, Megalaima asiatica 305
chersophila, Streptopelia semitorquata 161
CHERSOPHILUS 547
chersophilus, Anthus nyassae 743
chiapensis, Arremonops rufivirgatus 798
chiapensis, Campylorhynchus 633
chiapensis, Caprimulgus vociferus 241
chiapensis, Catharus frantzii 665
chiapensis, Dactylortyx thoracicus 43
chiapensis, Hylomanes momotula 293
chiapensis, Otus cooperi 223
chiapensis, Passerina caerulea 826
chiapensis, Pipilo maculatus 797
chicquera, Falco 96
chienfengensis, Sitta solangiae 646
chiguanco, Turdus 670
chiguancoides, Turdus pelios 666
chihi, Plegadis 83
chihuahuae, Sitta pygmaea 645
chilcatensis, Passerella iliaca 781
chilensis, Accipiter bicolor 108
chilensis, Anthus correndera 745
chilensis, Cinclodes patagonicus 404
chilensis, Elaenia albiceps 351
chilensis, Phoenicopterus 80
chilensis, Rollandia rolland 79
chilensis, Stercorarius 153
chilensis, Tangara 810
chilensis, Troglodytes aedon 640
chilensis, Vanellus 134
chilensis, Zonotrichia capensis 783
CHILIA 403
chillagoensis, Aerodramus terraereginae 248
chiloensis, Diuca diuca 789
chimachima, Milvago 93
chimaera, Uratelornis 283
chimango, Milvago 94
chimantae, Diglossa major 817
chimantae, Lochmias nematura 418
chimborazo, Oreotrochilus 270
chinchipensis, Synallaxis stictothorax 410
chinchorrensis, Elaenia martinica 351
chinensis, Amaurornis phoenicurus 123
chinensis, Cissa 510
chinensis, Coturnix 54
chinensis, Eudynamys scolopaceus 211
chinensis, Garrulax 613
chinensis, Jynx torquilla 311
chinensis, Oriolus 487
chinensis, Picumnus innominatus 312
chinensis, Riparia paludicola 532
chinensis, Streptopelia 162
chinensis, Tyto capensis 219
chinganicus, Garrulax davidi 613
chiniana, Cisticola 552
chio, Pternistis afer 53
CHIONIS 131
chionogaster, Accipiter striatus 108
chionogaster, Amazilia 264
chionogaster, Chamaeza turdina 398
chionogenys, Myza sarasinorum 439
chionopectus, Amazilia brevirostris 265
chionoptera, Diomedea exulans 72
chionura, Elvira 262
chionurus, Trogon viridis 280
chiribiquetensis, Hemitriccus margaritaceiventer 360
chirindensis, Apalis 561
chiripepe, Pyrrhura frontalis 198
chiriquensis, Elaenia 351
chiriquensis, Geothlypis aequinoctialis 764
chiriquensis, Geotrygon 169
chiriquensis, Scytalopus argentifrons 396
chiriquensis, Vireo leucophrys 484
chiriri, Brotogeris 200
Chirocylla 348

chrysorrhoides, Pycnonotus aurigaster 567
chrysorrhos, Eopsaltria australis 519
chrysorrhous, Ptilinopus melanospilus 177
chrysosema, Brotogeris chrysoptera 201
chrysostoma, Neophema 189
chrysostoma, Thalassarche 72
CHRYSOTHLYPIS 815
chrysotis, Alcippe 618
chrysotis, Tangara 811
chrysura, Campethera abingoni 316
chrysura, Hylocharis 264
CHRYSURONIA 264
chthonia, Grallaria 399
Chthonicola 445
chthonium, Pellorneum ruficeps 599
chuana, Certhilauda 543
chuancheica, Sylvia curruca 597
chubbi, Cichlopsis leucogenys 664
chubbi, Cisticola 552
chubbi, Mandingoa nitidula 727
chubbi, Sylvietta ruficapilla 596
chubbii, Tyrannus 376
chugurensis, Atlapetes latinuchus 801
chuka, Zoothera gurneyi 661
chukar, Alectoris 49
chumbiensis, Alcippe vinipectus 619
chunchotambo, Xiphorhynchus ocellatus 423
CHUNGA 116
chyuluensis, Anthus similis 742
chyuluensis, Francolinus squamatus 51
cia, Emberiza 777
CICCABA 229
ciceliae, Cuculus varius 207
CICHLADUSA 678
CICHLHERMINIA 672
CICHLOCOLAPTES 416
CICHLOPSIS 664
Cichlornis 577
CICINNURUS 517
CICONIA 81
ciconia, Ciconia 81
cienagae, Dendroica petechia 761
cinchoneti, Conopias 375
cincinatus, Phalacrocorax auritus 91
CINCLIDIUM 682
CINCLOCERTHIA 651
CINCLODES 403
CINCLORAMPHUS 577
cinclorhynchus, Monticola 688
CINCLOSOMA 453
CINCLUS 698
cinclus, Cinclus 698
cincta, Dichrozona 386
cincta, Notiomystis 431
cincta, Poephila 733
cincta, Riparia 533
cinctura, Ammomanes 545
cinctus, Erythrogonys 134
cinctus, Parus 528
cinctus, Ptilinopus 174
cinctus, Rhinoptilus 145
cinctus, Rhynchortyx 44
cinctus, Saltator 825
cinderella, Urolais epichlorus 560
cineracea, Ducula 180
cineracea, Emberiza 778
cineracea, Myzomela 441
cineracea, Nonnula rubecula 334
cineracea, Pyrrhula pyrrhula 757
cineraceus, Contopus fumigatus 367
cineraceus, Garrulax 614
cineraceus, Orthotomus ruficeps 564
cineraceus, Otus kennicotti 223
cineraceus, Parus inornatus 529
cineraceus, Sturnus 655
cinerascens, Acanthiza apicalis 449
cinerascens, Aplonis 653
cinerascens, Bubo 226
cinerascens, Camptostoma obsoletum 352

cinerascens, Cercomacra 388
cinerascens, Certhidea olivacea 796
cinerascens, Circaetus 102
cinerascens, Colius striatus 278
cinerascens, Dendrocitta 510
cinerascens, Fraseria 689
cinerascens, Gerygone olivacea 448
cinerascens, Monarcha 501
cinerascens, Myiarchus 378
cinerascens, Myiodynastes chrysocephalus 375
cinerascens, Nothoprocta 33
cinerascens, Oedistoma iliolophum 451
cinerascens, Pachycephala griseonota 478
cinerascens, Parisoma subcaeruleum 598
cinerascens, Parus 524
cinerascens, Prinia burnesii 557
cinerascens, Rhipidura albicollis 494
cinerascens, Rynchops niger 153
cinerascens, Synallaxis 408
cinerea, Alcippe 618
cinerea, Amphispiza belli 786
cinerea, Apalis 562
cinerea, Ardea 87
cinerea, Batara 379
cinerea, Butorides striata 86
cinerea, Calandrella 546
cinerea, Cercotrichas coryphaeus 679
cinerea, Coracina 468
cinerea, Creatophora 656
cinerea, Gallicrex 125
cinerea, Gerygone 448
cinerea, Glareola 146
cinerea, Motacilla 740
cinerea, Mycteria 81
cinerea, Myiagra ferrocyanea 503
cinerea, Myiopagis caniceps 350
cinerea, Piezorhina 789
cinerea, Poospiza 790
cinerea, Porzana 124
cinerea, Procellaria 75
cinerea, Rhipidura rufiventris 494
cinerea, Serpophaga 354
cinerea, Struthidea 515
cinerea, Synallaxis 408
cinerea, Turdoides plebejus 611
cinerea, Zoothera 660
cinereicapilla, Zimmerius 355
cinereicapillus, Colaptes rupicola 324
cinereicapillus, Spizixos semitorques 565
cinereicauda, Lampornis castaneoventris 268
cinereiceps, Alcippe 619
cinereiceps, Garrulax cineraceus 614
cinereiceps, Gerygone chloronota 448
cinereiceps, Grallaria varia 399
cinereiceps, Gymnobucco bonapartei 306
cinereiceps, Ixos siquijorensis 574
cinereiceps, Laterallus albigularis 119
cinereiceps, Macropygia amboinensis 162
cinereiceps, Malacocincla 601
cinereiceps, Ortalis 37
cinereiceps, Orthotomus 564
cinereiceps, Peneoenanthe pulverulenta 519
cinereiceps, Phapitreron 172
cinereiceps, Phyllomyias 350
cinereiceps, Polihierax insignis 94
cinereiceps, Thamnophilus amazonicus 382
cinereiceps, Tolmomyias sulphurescens 363
cinereiceps, Xanthomixis 588
cinereicollis, Arachnothera longirostra 714
cinereicollis, Basileuterus 766
cinereicollis, Orthotomus cucullatus 563
cinereifrons, Garrulax 612
cinereifrons, Meliphaga gracilis 434
cinereifrons, Poecilodryas albispecularis 518
cinereifrons, Pseudelaenia leucospodia 355
cinereifrons, Pycnonotus plumosus 568
cinereifrons, Pycnopygius ixoides 436
cinereigenae, Minla strigula 618
cinereigula, Dendrocopos moluccensis 318

cinereigula, Pericrocotus solaris 472
cinereigulare, Dicaeum trigonostigma 701
cinereigulare, Oncostoma 361
cinereipectus, Poecilotriccus plumbeiceps 361
cinereiventris, Chaetura 251
cinereiventris, Grallaria albigula 400
cinereiventris, Hellmayrea gularis 410
cinereiventris, Microbates 643
cinereiventris, Myrmotherula menetriesii 386
cinereiventris, Pachyramphus polychopterus 345
cinereocapilla, Malacopteron magnirostre 601
cinereocapilla, Motacilla flava 739
cinereocapilla, Prinia 557
cinereocauda, Chaetura brachyura 252
cinereocephalus, Chlorospingus ophthalmicus 818
cinereogenys, Oriolus steerii 487
cinereola, Muscicapa caerulescens 691
cinereola, Sporophila leucoptera 793
cinereolus, Cisticola 552
cinereoniger, Thamnophilus nigrocinereus 381
cinereovinacea, Euschistospiza 728
cinerescens, Elaenia martinica 351
cinereum, Conirostrum 815
cinereum, Malacopteron 601
cinereum, Todirostrum 362
cinereum, Toxostoma 650
cinereus, Acridotheres 654
cinereus, Artamus 464
cinereus, Calamonastes undosus 563
cinereus, Callaeas 452
cinereus, Circaetus 102
cinereus, Circus 104
cinereus, Coccyzus 213
cinereus, Contopus 367
cinereus, Cracticus torquatus 463
cinereus, Crypturellus 31
cinereus, Hemixos flavala 575
cinereus, Muscisaxicola 370
cinereus, Myadestes occidentalis 663
cinereus, Odontorchilus 634
cinereus, Parus major 525
cinereus, Ptilogonys 523
cinereus, Pycnopygius 436
cinereus, Seiurus aurocapilla 763
cinereus, Thryothorus sinaloa 638
cinereus, Vanellus 134
cinereus, Xenus 142
cinereus, Xolmis 371
cinereus, Zosterops 629
cinericia, Polioptila plumbea 644
cineritius, Empidonax difficilis 368
cinnamomea, Emberiza capensis 778
cinnamomea, Macropygia emiliana 163
cinnamomea, Neopipo 365
cinnamomea, Pachycephala orioloides 477
cinnamomea, Sporophila 793
cinnamomea, Synallaxis 408
cinnamomea, Terpsiphone 499
cinnamomea, Tringa solitaria 142
cinnamomeigula, Automolus rubiginosus 417
cinnamomeipectus, Hemitriccus 360
cinnamomeiventris, Ochthoeca 373
cinnamomeiventris, Thamnolaea 687
cinnamomeoventris, Iole propinqua 573
cinnamomeum, Cinclosoma 453
cinnamomeum, Pellorneum albiventre 599
cinnamomeus, Acrocephalus baeticatus 586
cinnamomeus, Ailuroedus buccoides 426
cinnamomeus, Anthus 742
cinnamomeus, Attila 378
cinnamomeus, Bradypterus 581
cinnamomeus, Certhiaxis 411
cinnamomeus, Cisticola 556
cinnamomeus, Crypturellus 32
cinnamomeus, Furnarius leucopus 404
cinnamomeus, Hypocryptadius 632
cinnamomeus, Ixobrychus 85
cinnamomeus, Pachyramphus 345
cinnamomeus, Passer rutilans 716

crissalis, Habia rubica 819
crissalis, Loxigilla noctis 795
crissalis, Pachycephala albiventris 475
crissalis, Pheucticus aureoventris 823
crissalis, Pipilo 797
crissalis, Sterna fuscata 151
crissalis, Strepera graculina 463
crissalis, Trogon violaceus 280
crissalis, Vermivora 760
crissoleucus, Picoides tridactylus 322
cristata, Alcedo 291
cristata, Calyptura 349
cristata, Cariama 116
cristata, Corythaeola 205
cristata, Coua 215
cristata, Cyanocitta 507
cristata, Elaenia 351
cristata, Eucometis penicillata 805
cristata, Fulica 126
cristata, Galerida 547
cristata, Goura 171
cristata, Gubernatrix 802
cristata, Habia 819
cristata, Lophostrix 229
cristata, Lophotibis 83
cristata, Pseudoseisura 414
cristata, Rhegmatorhina 393
cristata, Sterna bergii 150
cristata, Tadorna 65
cristatella, Aethia 155
cristatellus, Acridotheres 654
cristatellus, Cyanocorax 506
cristatellus, Tachyphonus cristatus 805
cristatus, Aegotheles 246
cristatus, Colinus 42
cristatus, Furnarius 404
cristatus, Lanius 479
cristatus, Orthorhyncus 260
cristatus, Oxyruncus 347
cristatus, Pandion haliaetus 98
cristatus, Parus 529
cristatus, Pavo 61
cristatus, Phyllomyias griseiceps 350
cristatus, Pitohui 490
cristatus, Podiceps 80
cristatus, Prionops plumatus 456
cristatus, Psophodes 452
cristatus, Sakesphorus 380
cristatus, Tachyphonus 805
cristinae, Threnetes 255
crocatus, Ploceus ocularis 720
crocea, Epthianura 443
croceus, Macronyx 741
CROCIAS 621
crockettorum, Meliphaga flavirictus 434
croconotus, Icterus icterus 770
croconotus, Loriculus stigmatus 181
crocopygius, Serinus albogularis 748
crocotius, Paradoxornis nipalensis 623
croizati, Accipiter tachiro 105
croizati, Catherpes mexicanus 635
croizati, Minla cyanouroptera 618
croizati, Turdus assimilis 672
crookshanki, Zosterops fuscicapilla 627
crossensis, Chalcomitra rubescens 706
crossini, Vireolanius melitophrys 482
crossleyi, Atelornis 283
crossleyi, Mystacornis 462
crossleyi, Zoothera 661
CROSSLEYIA 588
CROSSOPTILON 59
crossoptilon, Crossoptilon 59
croteta, Chloroceryle americana 292
crotopezus, Turdus albicollis 672
CROTOPHAGA 217
crozettensis, Chionis minor 131
cruciana, Dendroica petechia 761
crucigerus, Otus choliba 223
crudigularis, Arborophila 55

cruentata, Chalcomitra senegalensis 707
cruentata, Myzomela 441
cruentata, Pyrrhura 198
cruentatum, Dicaeum 703
cruentatus, Malurus melanocephalus 430
cruentatus, Melanerpes 314
cruentipectus, Dendrocopos cathpharius 319
cruentus, Ithaginis 57
cruentus, Malaconotus 457
cruentus, Oriolus 488
cruentus, Rhodophoneus 458
cruentus, Rhodospingus 802
crumeniferus, Leptoptilos 81
cruralis, Brachypteryx montana 673
cruralis, Cincloramphus 577
cruzi, Francolinus 52
cruziana, Columbina 167
crymophilus, Catharus guttatus 665
cryphius, Telmatodytes palustris 635
CRYPSIRINA 510
crypta, Ficedula 694
crypta, Iole olivacea 573
crypta, Spizocorys conirostris 547
crypta, Sporophila bouvreuil 793
crypta, Tyto alba 219
cryptanthus, Pomatorhinus schisticeps 603
crypterythrus, Myiophobus fasciatus 359
cryptica, Brachypterx 599
crypticus, Francolinus hartlaubi 52
cryptoleuca, Myrmecocichla aethiops 687
cryptoleuca, Peneothello 519
cryptoleuca, Platysteira peltata 455
cryptoleuca, Progne 533
cryptoleucus, Corvus 515
cryptoleucus, Notharchus macrorhynchos 332
cryptoleucus, Smithornis capensis 335
cryptoleucus, Thamnophilus 381
cryptolophus, Snowornis 348
CRYPTOPHAPS 180
cryptorhynchus, Cicinnurus regius 517
CRYPTOSPIZA 727
CRYPTOSYLVICOLA 588
cryptoxanthus, Myiophobus 359
cryptoxanthus, Poicephalus 195
CRYPTURELLUS 31
cryptus, Cypseloides 246
cryptus, Thryomanes bewickii 636
crystallochresta, Cryptospiza salvadorii 728
cuandocubangensis, Macronyx grimwoodi 741
cuanhamae, Corythaixoides concolor 206
cubanensis, Buteo platypterus 110
cubanensis, Caprimulgus 241
cubanensis, Colinus virginianus 42
cubensis, Ara 196
cubensis, Tyrannus 376
cubensis, Vireo crassirostris 483
cubla, Dryoscopus 459
cuchacanchae, Asthenes sclateri 407
cuchivera, Pyrrhura picta 198
cuchiverus, Philydor rufum 416
cucullata, Andigena 303
cucullata, Buthraupis montana 808
cucullata, Carduelis 750
cucullata, Carpornis 346
cucullata, Cecropis 537
cucullata, Coracina cinerea 468
cucullata, Crypsirina 510
cucullata, Cyanolyca 505
cucullata, Grallaricula 401
cucullata, Lonchura 734
cucullata, Melanodryas 520
cucullata, Pachycephala caledonica 477
cucullata, Pitta sordida 338
cucullata, Tangara 811
cucullata, Urocissa flavirostris 509
cucullatus, Coryphospingus 802
cucullatus, Icterus 770
cucullatus, Lophodytes 69
cucullatus, Orthotomus 563

cucullatus, Ploceus 721
cucullatus, Raphus 157
cucullatus, Sphecotheres vieilloti 486
cucullatus, Tchagra senegalus 459
cuculloides, Aviceda 98
cuculloides, Glaucidium 231
CUCULUS 207
cui, Alcippe dubia 619
cuicui, Microeca flavovirescens 520
cujubi, Pipile 39
CULICICAPA 698
CULICIVORA 359
culicivorus, Basileuterus 766
culik, Selenidera 302
culionensis, Prionochilus plateni 700
culminans, Turdus olivaceus 667
culminata, Coracina fimbriata 470
culminator, Lamprotornis nitens 656
culminatus, Corvus macrorhynchos 514
culminatus, Ramphastos vitellinus 303
cumanensis, Diopsittaca nobilis 196
cumanensis, Grallaricula nana 402
cumanensis, Pipile 39
cumanensis, Thryothorus rufalbus 638
cumanensis, Turdus serranus 670
cumanensis, Zimmerius chrysops 356
cumatilis, Cyanoptila cyanomelaena 695
cumatilis, Thraupis episcopus 807
cumbreanus, Dysithamnus mentalis 383
cumingi, Lepidogrammus 213
cumingii, Megapodius 36
cunctata, Macropygia amboinensis 162
cuneata, Geopelia 165
cuneata, Zoothera lunulata 662
cuneatus, Glyphorynchus spirurus 421
cuneicauda, Thinocorus rumicivorus 138
cunenensis, Acrocephalus gracilirostris 584
cunenensis, Scleroptila jugularis 51
cunhaci, Leiothrix argentauris 616
cunicularia, Athene 233
cunicularia, Geositta 402
cupido, Tympanuchus 46
cuprea, Ducula badia 180
cupreicauda, Amazilia 267
cupreicauda, Trogon rufus 281
cupreicaudus, Centropus 216
cupreiceps, Elvira 262
cupreipennis, Bostrychia olivacea 82
cupreocauda, Lamprotornis 657
cupreonitens, Nectarinia famosa 708
cupreoventris, Eriocnemis 272
cupreus, Chrysococcyx 210
cupreus, Cinnyris 711
cupripennis, Aglaeactis 270
CURAEUS 772
curaeus, Curaeus 772
curasoensis, Icterus nigrogularis 771
curiosus, Phaethornis augusti 257
curonicus, Charadrius dubius 135
curruca, Sylvia 596
currucoides, Sialia 663
cursitans, Cisticola juncidis 555
cursitans, Crypturellus erythropus 32
cursor, Coua 215
cursor, Cursorius 145
CURSORIUS 145
curtata, Cranioleuca 411
curtata, Polioptila melanura 643
curtatus, Pipilo maculatus 797
curucui, Trogon 280
curvipennis, Campylopterus 258
curvirostra, Loxia 756
curvirostra, Treron 172
curvirostre, Toxostoma 650
curvirostris, Andropadus 569
curvirostris, Campylorhynchus zonatus 633
curvirostris, Certhilauda 543
curvirostris, Limnornis 412
curvirostris, Nothoprocta 34

delphinae, Colibri 259
deltae, Prinia gracilis 558
deltana, Dendrocincla fuliginosa 420
deltanus, Celeus elegans 325
deltanus, Piculus rubiginosus 323
deltanus, Xiphorhynchus picus 423
deltarhyncha, Passerina caerulea 826
DELTARHYNCHUS 378
deluzae, Formicivora grisea 387
dementievi, Alauda arvensis 549
dementievi, Alectoris graeca 49
dementievi, Caprimulgus europaeus 242
dementjevi, Accipiter nisus 107
demersus, Spheniscus 71
demeryi, Zosterops senegalensis 629
deminuta, Chalcomitra amethystina 707
deminutus, Anthracoceros malayanus 299
demissa, Cranioleuca 411
demissus, Vanellus coronatus 133
demonstratus, Xiphorhynchus susurrans 424
DENDRAGAPUS 46
DENDREXETASTES 421
dendrobates, Ceuthmochares aereus 212
DENDROCINCLA 419
DENDROCITTA 510
DENDROCOLAPTES 422
dendrocolaptoides, Clibanornis 412
DENDROCOPOS 317
DENDROCYGNA 61
DENDROICA 760
DENDRONANTHUS 739
dendrophilus, Phyllastrephus flavostriatus 571
DENDROPICOS 316
DENDRORTYX 41
denhami, Neotis 115
deningeri, Lichmera 439
deningeri, Turdus poliocephalus 668
deningeri, Zanclostomus curvirostris 213
denisei, Atlapetes semirufus 800
dennisi, Petroica multicolor 521
dennistouni, Stachyris 606
denotata, Niltava sundara 697
densirostris, Margarops fuscatus 651
densus, Dicrurus hottentottus 492
dentata, Gymnoris 718
dentatus, Creurgops 803
denti, Pholidornis rushiae 531
denti, Sylvietta 596
dentirostris, Scenopoeetes 427
deplanchii, Trichoglossus haematodus 185
deppei, Polioptila caerulea 643
derbiana, Psittacula 193
derbianus, Aulacorhynchus 301
derbianus, Oreophasis 39
derbianus, Orthotomus 564
derbianus, Pitangus sulphuratus 374
derbyanus, Nyctidromus albicollis 240
derbyi, Artamus minor 465
derbyi, Cacatua pastinator 183
derbyi, Eriocnemis 272
derjugini, Parus ater 526
deroepstorffi, Tyto alba 219
derooi, Cyanomitra alinae 706
DEROPTYUS 204
deschauenseei, Coracina coerulescens 468
deschauenseei, Ortalis vetula 37
deserti, Ammomanes 545
deserti, Mirafra apiata 542
deserti, Oenanthe 685
deserti, Serinus atrogularis 747
deserti, Sylvia nana 597
deserticola, Amphispiza bilineata 786
deserticola, Apus bradfieldi 253
deserticola, Cistothorus palustris 636
deserticola, Megarynchus pitangua 375
deserticola, Meliphaga fusca 433
deserticola, Sylvia 597
deserticolor, Carpodacus thura 755
deserticolor, Geositta cunicularia 402

deserticolor, Otus bakkamoena 220
deserticolor, Rallus aquaticus 121
desertorum, Alaemon alaudipes 544
desertorum, Caprimulgus ruficollis 242
desgodinsi, Heterophasia melanoleuca 621
desiderata, Collocalia esculenta 248
desidiosa, Piranga flava 818
desiradensis, Loxigilla noctis 795
desmaresti, Tangara 810
desmarestii, Phalacrocorax aristotelis 92
desmarestii, Psittaculirostris 205
desmursi, Dendrocopos darjellensis 319
desmursii, Sylviorthorhynchus 404
desolata, Pachyptila 73
desolata, Terpsiphone bourbonnensis 499
desolatus, Aerodramus spodiopygius 248
despecta, Colluricincla megarhyncha 489
despotes, Tyrannus melancholicus 376
destructa, Lonchura grandis 736
destructus, Formicarius nigricapillus 398
detersus, Calorhamphus fuliginosus 306
detorta, Tyto alba 218
deuteronymus, Copsychus saularis 680
deva, Galerida 548
devia, Eudromia elegans 34
devillei, Amazilia beryllina 267
devillei, Dendrexetastes rufigula 421
devillei, Drymophila 388
devillei, Pyrrhura 198
devillii, Arremon flavirostris 798
devisi, Rhipidura brachyrhyncha 496
devittatus, Lorius hypoinochrous 186
devius, Campylorhamphus trochilirostris 426
devius, Dactylortyx thoracicus 43
devronis, Sporophila peruviana 793
dewittei, Anthus similis 742
dewittei, Francolinus albogularis 50
dexter, Cisticola fulvicapilla 554
dextra, Piranga flava 818
dextralis, Psophia viridis 127
dharmakumarsinhjii, Alauda gulgula 549
diabolicus, Carpodacus rubicilla 755
diabolicus, Eurostopodus 240
diabolicus, Melaenornis pammelaina 689
diabolicus, Nigrita canicapillus 727
diadema, Amazona autumnalis 203
diadema, Catamblyrhynchus 822
diadema, Charmosyna 187
diadema, Ochthoeca 373
diademata, Alethe 673
diademata, Cyanocitta stelleri 507
diademata, Tricholaema 308
diademata, Yuhina 622
diadematus, Euplectes 725
diadematus, Indicator minor 311
diadematus, Monarcha trivirgatus 501
diadematus, Stephanophorus 809
dialeucos, Odontophorus 43
dialilaemus, Cyornis rubeculoides 696
diamantina, Amytornis barbatus 430
diamantinensis, Campylopterus largipennis 258
diamesus, Anous minutus 152
diamondi, Phonygammus keraudrenii 516
diana, Myiomela 682
diaoluoensis, Cyornis unicolor 696
diaphora, Eremophila alpestris 550
diaphora, Sclateria naevia 391
diardi, Criniger phaeocephalus 573
diardi, Garrulax leucolophus 612
diardi, Lophura 59
diardi, Rhopodytes 212
diardii, Harpactes 279
diatropus, Dendrocopos darjellensis 319
diazi, Anas platyrhynchos 66
dibapha, Certhia 442
DICAEUM 700
dichroa, Aplonis 652
dichroa, Cossypha 678
dichroides, Parus dichrous 529

dichromata, Myzomela rubratra 442
dichrorhyncha, Pelargopsis melanorhyncha 286
dichrorhynchus, Myophonus caeruleus 660
dichroura, Coeligena violifer 271
dichrous, Cyphorhinus thoracicus 642
dichrous, Parus 529
dichrous, Pitohui 490
dichrous, Puffinus lherminieri 76
DICHROZONA 386
dickermani, Cinclus mexicanus 698
dickermani, Hylophilus decurtatus 486
dickermani, Icterus pustulatus 771
dickeyae, Icterus graduacauda 771
dickeyae, Passerina versicolor 826
dickeyi, Colinus leucopogon 42
dickeyi, Cyanocorax 506
dickeyi, Egretta rufescens 88
dickeyi, Penelopina nigra 39
dickeyi, Phalaenoptilus nuttallii 240
dickinsoni, Falco 96
dickinsoni, Pomatorhinus ferruginosus 603
dicolorus, Ramphastos 303
dicrocephalus, Ploceus 722
dicruriformis, Dicrurus andamanensis 493
dicruroides, Surniculus lugubris 211
DICRURUS 491
dictator, Timalia pileata 609
didimus, Accipiter fasciatus 106
didimus, Glossopsitta concinna 186
dido, Acrocephalus caffer 585
DIDUNCULUS 171
didymus, Pteroglossus inscriptus 302
dieffenbachii, Gallirallus 120
diemenensis, Acanthiza pusilla 448
diemenensis, Alcedo azurea 291
diemenensis, Calamanthus fuliginosus 445
diemenensis, Dromaius novaehollandiae 35
diemenensis, Philemon 437
diemenensis, Platycercus eximius 189
diemenianus, Dromaius 35
diesingii, Cyanocorax chrysops 506
differens, Anthus trivialis 743
differens, Turdus plebejus 670
difficilis, Coracina striata 467
difficilis, Empidonax 368
difficilis, Geospiza 795
difficilis, Myiodynastes maculatus 375
difficilis, Phylloscartes 357
difficilis, Pomatorhinus schisticeps 603
difficilis, Thamnophilus doliatus 381
difficilis, Zosterops montanus 626
diffusus, Oriolus chinensis 487
diffusus, Passer 717
digglesi, Pitta erythrogaster 338
digitatus, Pyrrhocorax graculus 512
DIGLOSSA 816
Diglossopsis 817
dignissima, Grallaria 399
dignus, Veniliornis 322
digressa, Ortygospiza atricollis 731
dilectissimus, Touit 201
diligens, Parus rufiventris 524
dilloni, Pellorneum ruficeps 599
dillonii, Bubo capensis 226
dillonripleyi, Rallus longirostris 121
dillwyni, Megapodius cumingii 36
diloloensis, Jynx ruficollis 311
diluta, Amazilia rutila 265
diluta, Napothera epilepidota 605
diluta, Rhipidura 494
diluta, Riparia 532
dilutescens, Ploceus cucullatus 722
dilutior, Arachnothera longirostra 714
dilutior, Euphonia xanthogaster 822
dilutior, Illadopsis fulvescens 602
dilutum, Dinopium benghalense 329
dilutum, Ornithion brunneicapillus 352
dilutus, Atlapetes pileatus 800
dilutus, Ceyx madagascariensis 290

enertera, Eremophila alpestris 550
enganensis, Aplonis panayensis 652
enganensis, Coracina striata 467
enganensis, Gracula religiosa 654
enganensis, Otus umbra 222
engelbachi, Heterophasia melanoleuca 621
engelbachi, Lophura nycthemera 58
engelbachi, Macropygia ruficeps 163
ENICOGNATHUS 199
enicura, Doricha 276
ENICURUS 682
enigma, Ninox connivens 234
enigma, Sterna bergii 150
enigma, Todiramphus 289
ennosiphyllus, Sclerurus guatemalensis 418
enochrus, Sittasomus griseicapillus 420
ENODES 653
ENSIFERA 271
ensifera, Cercomela sinuata 686
ensifera, Ensifera 271
ensipennis, Campylopterus 258
entebbe, Cisticola ayresii 556
enthymia, Eremophila alpestris 550
ENTOMODESTES 666
ENTOMYZON 435
enucleator, Pinicola 755
enunciator, Anthus leucophrys 743
eoa, Dendroica petechia 761
EOLOPHUS 183
EOPHONA 757
EOPSALTRIA 519
EOS 184
eos, Carpodacus 754
eos, Coeligena bonapartei 271
eous, Garrulax albogularis 612
eous, Pycnonotus finlaysoni 568
epauletta, Pyrrhoplectes 758
ephemeralis, Aechmophorus occidentalis 80
EPHIPPIORHYNCHUS 81
Ephthianura see Epthianura 443
epia, Ptilinopus subgularis 174
epichlorus, Urolais 560
epiensis, Aerodramus spodiopygius 248
epilepidota, Napothera 605
EPIMACHUS 517
epirhinus, Tockus nasutus 298
episcopalis, Rhipidura leucothorax 495
episcopus, Ciconia 81
episcopus, Phaethornis ruber 257
episcopus, Thraupis 807
epomophora, Diomedea 72
epops, Upupa 297
EPTHIANURA 443
epulata, Muscicapa 692
eques, Euplectes albonotatus 726
eques, Myzomela 441
equicaudatus, Cisticola exilis 556
equifasciatus, Veniliornis nigriceps 322
erckelii, Francolinus 53
erebus, Calyptorhynchus lathami 183
EREMALAUDA 547
eremica, Calandrella blanfordi 546
eremica, Emberiza impetuani 778
eremicus, Cisticola aridulus 555
eremicus, Picoides scalaris 321
EREMIORNIS 577
eremita, Alcippe peracensis 620
eremita, Geronticus 82
eremita, Megapodius 37
eremita, Nesocichla 672
eremita, Thalassarche cauta 72
eremnus, Philetairus socius 715
eremnus, Thamnophilus doliatus 380
EREMOBIUS 413
eremobius, Ploceus baglafecht 720
eremodites, Eremalauda dunni 547
eremoeca, Aimophila ruficeps 787
eremogiton, Halcyon chelicuti 287
EREMOMELA 594

eremonoma, Phaeomyias murina 354
EREMOPHILA 550
eremophilus, Oreortyx pictus 41
eremophilus, Pipilo crissalis 797
eremophilus, Thryomanes bewickii 636
EREMOPTERIX 549
eremus, Acrocephalus caffer 585
ERGATICUS 765
ericrypta, Melospiza georgiana 782
erikssoni, Chersomanes albofasciata 544
erikssoni, Cinnyris ludovicensis 709
erimacrus, Treron curvirostra 172
erimelas, Copsychus saularis 680
ERIOCNEMIS 272
eriphaea, Alcippe brunneicauda 620
eriphile, Thalurania furcata 263
erithaca, Ceyx 290
erithachorides, Dendroica petechia 761
ERITHACUS 675
erithacus, Liosceles thoracicus 395
erithacus, Psittacus 194
eritreae, Francolinus levaillantoides 51
erlangeri, Batis minor 455
erlangeri, Bostrychia hagedash 82
erlangeri, Calandrella blanfordi 546
erlangeri, Camaroptera brachyura 562
erlangeri, Cercotrichas quadrivirgata 679
erlangeri, Cinnyris nectarinioides 710
erlangeri, Colius striatus 278
erlangeri, Coturnix coturnix 54
erlangeri, Eupodotis senegalensis 115
erlangeri, Falco biarmicus 97
erlangeri, Galerida theklae 548
erlangeri, Glareola pratincola 146
erlangeri, Gyps rueppelli 102
erlangeri, Halcyon albiventris 287
erlangeri, Laniarius aethiopicus 460
erlangeri, Melaenornis pallidus 689
erlangeri, Prinia somalica 559
erlangeri, Pterocles exustus 156
erlangeri, Rhinoptilus africanus 145
erlangeri, Riparia cincta 533
erlangeri, Salpornis spilonotus 648
erlangeri, Scops 225
erlangeri, Tyto alba 218
erlangshanicus, Paradoxornis zappeyi 623
ermanni, Streptopelia senegalensis 162
ernesti, Cercomela sordida 687
ernesti, Falco peregrinus 97
ernesti, Lonchura grandis 736
ernesti, Oenanthe leucopyga 686
ernesti, Pachycephalopsis hattamensis 520
ernesti, Tyto alba 218
ernsti, Hirundapus cochinchinensis 251
ernsti, Parus palustris 527
ernstmayri, Myzomela pammelaena 442
ERPORNIS 622
erratilis, Gymnocichla nudiceps 390
erro, Prinia hodgsonii 557
errolius, Andropadus importunus 569
erromangae, Todiramphus chloris 288
erwini, Collocalia esculenta 247
erwini, Dryoscopus gambensis 459
erythaca, Pyrrhula 756
erythopthalmus, Coccyzus 213
erythrauchen, Accipiter 107
erythreae, Dryoscopus gambensis 459
erythrina, Myzomela cruentata 441
erythrinus, Carpodacus 753
erythrocampe, Otus bakkamoena 220
erythrocephala, Amadina 727
erythrocephala, Mirafra 542
erythrocephala, Myzomela 441
erythrocephala, Pipra 343
erythrocephala, Piranga 819
erythrocephala, Pyrrhula 756
erythrocephala, Streptopelia orientalis 161
erythrocephalus, Chrysocolaptes lucidus 329
erythrocephalus, Cisticola exilis 556

erythrocephalus, Garrulax 615
erythrocephalus, Harpactes 280
erythrocephalus, Hylocryptus 418
erythrocephalus, Melanerpes 314
erythrocephalus, Trachyphonus 310
erythrocercum, Philydor 416
ERYTHROCERCUS 576
erythrocercus, Cinnyris 710
erythrochlamys, Certhilauda 543
erythrochroa, Ammomanes deserti 545
erythrocnemis, Pomatorhinus erythrogenys 602
erythrogaster, Hirundo rustica 535
erythrogaster, Laniarius 460
erythrogaster, Malimbus 724
erythrogaster, Pitta 338
erythrogastrus, Phoenicurus 682
erythrogenys, Aratinga 197
erythrogenys, Emberiza citrinella 777
erythrogenys, Microhierax 94
erythrogenys, Pomatorhinus 602
erythrogenys, Tityra inquisitor 343
erythrognathus, Aulacorhynchus sulcatus 301
erythrognathus, Zanclostomus curvirostris 213
ERYTHROGONYS 134
erythrogyna, Brachypteryx montana 673
erytholaema, Habia fuscicauda 819
erytholaemus, Garrulax erythrocephalus 615
erytholeuca, Grallaria 400
erytholophus, Tauraco 206
erythromelas, Ixobrychus exilis 85
erythromelas, Myzomela 443
erythromelas, Periporphyrus 824
erythronemius, Accipiter striatus 108
erythronota, Rhipidura spilodera 496
erythronota, Zoothera 660
erythronotos, Amazilia tobaci 267
erythronotos, Estrilda 731
erythronotos, Formicivora 387
erythronotum, Philydor fuscipenne 415
erythronotus, Lanius schach 480
erythronotus, Phoenicurus 681
erythronotus, Phrygilus 788
erythronotus, Pseudocossyphus sharpei 688
erythronotus, Stachyris nigricollis 607
erythropareia, Geotrygon frenata 169
erythrophris, Enodes 653
erythrophrys, Myrmoborus leucophrys 390
erythrophrys, Poospiza 790
erythrophthalma, Batis capensis 454
erythrophthalma, Lophura 59
erythrophthalma, Netta 68
erythrophthalmus, Phacellodomus 412
erythrophthalmus, Pipilo 797
erythrophthalmus, Pycnonotus 568
erythropis, Piculus flavigula 323
erythropleura, Prinia atrogularis 557
erythropleura, Ptiloprora 438
erythropleurus, Turdus poliocephalus 668
erythropleurus, Zosterops 625
erythrops, Cisticola 551
erythrops, Climacteris 428
erythrops, Cranioleuca 411
erythrops, Dryocopus lineatus 326
erythrops, Myiagra oceanica 503
erythrops, Neocrex 124
erythrops, Odontophorus 43
erythrops, Quelea 724
erythroptera, Gallicolumba 170
erythroptera, Mirafra 543
erythroptera, Ortalis 37
erythroptera, Phlegopsis 394
erythroptera, Stachyris 608
erythropterum, Philydor 416
erythropterus, Andropadus virens 569
erythropterus, Aprosmictus 192
erythropterus, Heliolais 559
erythropterus, Myiozetetes cayanensis 374
erythropus, Accipiter 106
erythropus, Anser 62

flavicans, Prinia 559
flavicans, Prioninurus 191
flavicans, Trichoglossus haematodus 185
flavicapilla, Xenopipo 342
flavicaudata, Pytilia melba 727
flavicaudus, Thapsinillas affinis 574
flaviceps, Atlapetes 800
flaviceps, Auriparus 531
flaviceps, Campochaera sloetii 470
flaviceps, Dendroica petechia 760
flaviceps, Ploceus manyar 722
flaviceps, Rhodacanthis 758
flaviceps, Zosterops lateralis 629
flaviclunis, Dicaeum trigonostigma 701
flavicollis, Chlorocichla 570
flavicollis, Dupetor 85
flavicollis, Hemithraupis 814
flavicollis, Lichenostomus 432
flavicollis, Macronus 609
flavicollis, Macronyx 741
flavicollis, Pipra aureola 342
flavicollis, Ptilinopus regina 176
flavicollis, Yuhina 621
flavicrissalis, Eremomela 594
flavicrissus, Cacicus cela 769
flavicrissus, Saltator atriceps 824
flavida, Apalis 560
flavida, Dendroica petechia 760
flavida, Gerygone palpebrosa 448
flavida, Meliphaga analoga 433
flavidicollis, Saltator striatipectus 825
flavidifrons, Zimmerius chrysops 356
flavidior, Myiarchus nuttingi 378
flavidior, Trogon aurantiiventris 281
flavidiventris, Amandava amandava 731
flavidorsalis, Hemispingus frontalis 804
flavidus, Acrocephalus caffer 585
flavifrons, Amblyornis 427
flavifrons, Anthoscopus 530
flavifrons, Euphonia musica 821
flavifrons, Megalaima 305
flavifrons, Melanerpes 314
flavifrons, Pachycephala 477
flavifrons, Picumnus aurifrons 312
flavifrons, Poicephalus 195
flavifrons, Vireo 483
flavifrons, Zosterops 628
flavigaster, Arachnothera 714
flavigaster, Hyliota 587
flavigaster, Microeca 521
flavigaster, Xiphorhynchus 424
flavigastra, Cinnyris jugularis 711
flavigula, Chlorocichla flavicollis 570
flavigula, Gymnoris superciliaris 718
flavigula, Manorina 434
flavigula, Melanerpes formicivorus 314
flavigula, Piculus 323
flavigula, Serinus 747
flavigularis, Apalis thoracica 560
flavigularis, Chlorospingus 818
flavigularis, Chrysococcyx 210
flavigularis, Machetornis rixosa 374
flavigularis, Platyrinchus 364
flavilateralis, Zosterops abyssinicus 629
flavimentalis, Abroscopus schisticeps 594
flavimentum, Phyllomyias nigrocapillus 350
flavimentum, Pogoniulus subsulphureus 307
flavinucha, Muscisaxicola 370
flavinucha, Picus 327
flavinucha, Sitta pygmaea 645
flavinuchus, Anisognathus somptuosus 808
flavinuchus, Melanerpes chrysogenys 314
flavipectus, Aethopyga bella 712
flavipectus, Cyclarhis gujanensis 482
flavipectus, Oedistoma pygmaeum 451
flavipectus, Parus 526
flavipectus, Philydor ruficaudatum 415
flavipennis, Chloropsis 699
flavipes, Hylophilus 486

flavipes, Ketupa 227
flavipes, Notiochelidon 534
flavipes, Platalea 83
flavipes, Platycichla 666
flavipes, Ploceus 723
flavipes, Tringa 142
flaviprymna, Lonchura 736
flavipunctata, Tricholaema hirsuta 308
flavirictus, Meliphaga 434
flavirostra, Amaurornis 123
flavirostris, Amblycercus holosericeus 769
flavirostris, Anairetes 353
flavirostris, Anas 67
flavirostris, Anser albifrons 62
flavirostris, Arremon 798
flavirostris, Calonectris 75
flavirostris, Carduelis 751
flavirostris, Ceuthmochares aereus 212
flavirostris, Chlorophonia 822
flavirostris, Columba 159
flavirostris, Grallaricula 401
flavirostris, Humblotia 693
flavirostris, Monasa 334
flavirostris, Myophonus caeruleus 660
flavirostris, Paradoxornis 623
flavirostris, Phibalura 346
flavirostris, Picus squamatus 328
flavirostris, Porphyrio 126
flavirostris, Prinia inornata 559
flavirostris, Pteroglossus azara 302
flavirostris, Rynchops 153
flavirostris, Syma torotoro 290
flavirostris, Tockus 298
flavirostris, Urocissa 509
flaviscapis, Pteruthius 617
flaviserinus, Serinus canarius 746
flavisquamatus, Pogoniulus scolopaceus 307
flavissima, Microeca flavigaster 521
flavissima, Motacilla flava 739
flavissima, Sicalis luteola 791
flavissimus, Forpus xanthopterygius 200
flavissimus, Zosterops chloris 626
flaviventer, Dacnis 813
flaviventer, Machaerirhynchus 461
flaviventer, Porzana 124
flaviventer, Xanthotis 431
flaviventre, Camptostoma obsoletum 352
flaviventris, Abroscopus superciliaris 594
flaviventris, Apalis thoracica 560
flaviventris, Cettia vulcania 579
flaviventris, Chlorocichla 570
flaviventris, Cyclarhis gujanensis 481
flaviventris, Emberiza 779
flaviventris, Emberiza flaviventris 779
flaviventris, Empidonax 368
flaviventris, Eopsaltria 520
flaviventris, Hylophilus hypoxanthus 486
flaviventris, Lathrotriccus euleri 366
flaviventris, Leptocoma zeylonica 707
flaviventris, Motacilla 740
flaviventris, Pezoporus wallicus 190
flaviventris, Phylloscartes 356
flaviventris, Prinia 558
flaviventris, Pseudocolopteryx 354
flaviventris, Pycnonotus melanicterus 565
flaviventris, Serinus 747
flaviventris, Sphecotheres vieilloti 486
flaviventris, Sylvietta virens 596
flaviventris, Tachyphonus luctuosus 805
flaviventris, Tolmomyias 364
flaviventris, Toxorhamphus novaeguineae 452
flavivertex, Heterocercus 342
flavivertex, Myioborus 765
flavivertex, Myiopagis 350
flavivertex, Serinus canicollis 746
flaviviridis, Hemitriccus zosterops 360
flavocincta, Apalis flavida 561
flavocincta, Chloropsis cochinchinensis 699
flavocinctus, Oriolus 487

flavocinerea, Stigmatura budytoides 355
flavocristata, Melanochlora sultanea 529
flavodorsalis, Prodotiscus insignis 310
flavogaster, Elaenia 351
flavogriseum, Pachycare 474
flavogularis, Asthenes pyrrholeuca 406
flavogularis, Seicercus xanthoschistos 593
flavolateralis, Gerygone 447
flavolivacea, Cettia 579
flavoolivaceus, Tolmomyias sulphurescens 363
flavopalliatus, Lorius garrulus 186
flavopectus, Chlorospingus ophthalmicus 817
flavoptera, Pyrrhura molinae 198
flavostriata, Aethopyga siparaja 713
flavostriatus, Phyllastrephus 571
flavostriatus, Phylloscopus trivirgatus 592
flavotectus, Tolmomyias assimilis 363
flavotectus, Trichoglossus capistratus 185
flavotincta, Grallaria 400
flavotincta, Lichmera incana 439
flavotinctus, Oriolus flavocinctus 487
flavotinctus, Picumnus olivaceus 313
flavovelata, Geothlypis 763
flavovirens, Basileuterus signatus 766
flavovirens, Chlorospingus 818
flavovirens, Phylloscartes 356
flavovirescens, Microeca 520
flavoviridis, Hartertula 588
flavoviridis, Merops orientalis 296
flavoviridis, Trichoglossus 185
flavoviridis, Vireo 485
flavum, Malacopteron magnirostre 601
flavum, Oedistoma iliolophum 451
flavus, Celeus 325
flavus, Hemignathus 758
flavus, Lichenostomus 432
flavus, Rhynchocyclus olivaceus 363
flavus, Xanthopsar 772
flavus, Zosterops 626
fleayi, Aquila audax 113
flecki, Centropus senegalensis 216
flecki, Sylvietta rufescens 595
flemingorum, Emberiza cia 777
flemmingi, Dysithamnus puncticeps 383
fleurieuensis, Platycercus elegans 189
flexipes, Geranospiza caerulescens 109
flindersi, Sericornis frontalis 446
floccosus, Pycnoptilus 444
floccus, Haplophaedia aureliae 272
florensis, Corvus 513
florensis, Ninox scutulata 234
florentiae, Ptilopachus petrosus 50
florentinoi, Centurus superciliaris 315
florentinoi, Contopus caribaeus 367
floresae, Emberizoides herbicola 791
floresiana, Alcedo atthis 292
floresiana, Pelargopsis capensis 286
floresianus, Geoffroyus geoffroyi 191
floresianus, Turnix maculosus 129
floriceps, Anthocephala 267
florida, Dendroica pinus 762
florida, Tangara 810
floridae, Ninox jacquinoti 235
floridana, Athene cunicularia 233
floridana, Rhipidura cockerelli 495
floridanus, Agelaius phoeniceus 774
floridanus, Ammodramus savannarum 784
floridanus, Cardinalis cardinalis 823
floridanus, Colinus virginianus 42
floridanus, Otus asio 223
floridanus, Phalacrocorax auritus 91
floridanus, Zosterops metcalfii 628
floris, Brachypteryx montana 673
floris, Coracina personata 466
floris, Phylloscopus presbytes 592
floris, Seicercus montis 593
floris, Spizaetus cirrhatus 114
floris, Terpsiphone paradisi 499
floris, Treron 173

ginginianus, Acridotheres 654
ginginianus, Neophron percnopterus 101
giraudi, Eremophila alpestris 550
giraudi, Peucedramus taeniatus 739
giraudii, Icterus chrysater 771
girensis, Brachypternus benghalensis 329
gironieri, Ptilinopus leclancheri 175
girrenera, Haliastur indus 101
githagineus, Bucanetes 753
giulianetti, Meliphaga flaviventer 431
giulianettii, Phylloscopus poliocephalus 592
glaber, Sublegatus arenarum 358
glabricollis, Cephalopterus 349
glabripes, Otus bakkamoena 220
glabrirostris, Melanoptila 648
glacialis, Fulmarus 73
glacialis, Picoides pubescens 321
glacialoides, Fulmarus 73
gladiator, Malaconotus 457
gladkovi, Bubo bubo 226
glandarius, Clamator 207
glandarius, Garrulus 508
glanvillei, Apus barbatus 253
GLAREOLA 146
glareola, Tringa 142
glaszneri, Garrulus glandarius 508
glauca, Diglossa 817
glaucescens, Cyanocorax yncas 506
glaucescens, Larus 147
glaucicomans, Cyornis rubeculoides 696
GLAUCIDIUM 230
glaucinus, Myophonus 659
GLAUCIS 255
glaucocaerulea, Cyanoloxia 825
glaucocauda, Ducula aenea 178
glaucocolpa, Thraupis 807
glaucogularis, Aegithalos caudatus 538
glaucogularis, Ara 195
glaucogularis, Dacnis cayana 813
glaucogularis, Eubucco versicolor 304
glaucoides, Larus 147
glaucopis, Thalurania 263
glaucopoides, Eriocnemis 272
glaucops, Tyto 219
glaucura, Pachycephala pectoralis 476
glaucurus, Eurystomus 282
glaucus, Anodorhynchus 195
glaucus, Thamnomanes caesius 384
glaucus, Treron calvus 173
glaux, Athene noctua 232
globulosa, Crax 40
gloriae, Xiphidiopicus percussus 315
gloriosa, Diglossa 816
gloriosissima, Diglossa 816
GLOSSOPSITTA 186
Glychichaera 443
GLYCIFOHIA 439
GLYCIPHILA 440
GLYPHORYNCHUS 421
gnatho, Euphonia hirundinacea 821
gnoma, Glaucidium 230
GNORIMOPSAR 772
godeffroyi, Monarcha 502
godeffroyi, Todiramphus 289
godefrida, Claravis 167
godfreii, Ixoreus naevius 662
godfreyi, Bradypterus barratti 581
godini, Eriocnemis 272
godlewskii, Anthus 742
godlewskii, Emberiza 777
godmani, Basileuterus culicivorus 766
godmani, Colinus virginianus 42
godmani, Euphonia affinis 820
godmani, Psittaculirostris desmarestii 205
godwini, Garrulax erythrocephalus 615
godwini, Pomatorhinus ruficollis 603
goeldii, Myrmeciza 392
goeldii, Sicalis columbiana 791
goeringi, Brachygalba 331

goeringi, Hemispingus 804
goertae, Dendropicos 317
GOETHALSIA 262
gofanus, Francolinus castaneicollis 52
goffini, Cacatua 183
goffinii, Trachylaemus purpuratus 309
goiavier, Pycnonotus 568
goisagi, Gorsachius 85
golandi, Ploceus 722
goldiei, Cinclosoma ajax 454
goldiei, Macropygia amboinensis 162
goldiei, Psitteuteles 186
goldii, Ninox theomacha 235
goldmani, Aimophila botterii 786
goldmani, Coturnicops noveboracensis 118
goldmani, Crypturellus cinnamomeus 32
goldmani, Dendroica coronata 761
goldmani, Geothlypis beldingi 763
goldmani, Geotrygon 169
goldmani, Melospiza melodia 782
goldmani, Momotus momota 294
goldmani, Trogon elegans 281
GOLDMANIA 262
goliath, Ardea 87
goliath, Centropus 215
goliath, Ducula 179
goliath, Probosciger aterrimus 182
goliathi, Melipotes fumigatus 437
goliathina, Charmosyna papou 187
golzi, Apalis flavida 561
gomesi, Mirafra africana 541
gommaensis, Serinus mozambicus 747
gongonensis, Passer griseus 717
gongshanensis, Paradoxornis guttaticollis 623
gonzalesi, Pericrocotus flammeus 473
goodalli, Asthenes humicola 406
goodenovii, Petroica 521
goodfellowi, Brachypteryx montana 673
goodfellowi, Ceyx lepidus 290
goodfellowi, Eos 185
goodfellowi, Lophozosterops 631
goodfellowi, Pitta sordida 339
goodfellowi, Regulus 632
goodfellowi, Rhinomyias 691
goodfellowi, Turdus ignobilis 671
goodi, Pycnonotus barbatus 567
goodsoni, Anthus leucophrys 743
goodsoni, Colluricincla megarhyncha 489
goodsoni, Columba 160
goodsoni, Coracina melas 469
goodsoni, Gallirallus philippensis 120
goodsoni, Macropygia mackinlayi 163
goodsoni, Pachycephala citreogaster 477
goodsoni, Phylloscopus reguloides 591
goodsoni, Stachyris ruficeps 606
goodsoni, Tangara arthus 810
goodwini, Ptilinopus rarotongensis 175
goramensis, Myiagra galeata 503
gordoni, Cecropis semirufa 537
gordoni, Nesocichla eremita 672
gordoni, Philemon buceroides 436
gorgonae, Coereba flaveola 776
gorgonae, Thamnophilus atrinucha 382
GORSACHIUS 85
goslingi, Apalis 562
goslingi, Emberiza tahapisi 778
gossii, Porzana flaviventer 124
goudoti, Lepidopyga 263
goudotii, Chamaepetes 39
goughensis, Rowettia 789
goulburni, Megalurus gramineus 577
gouldi, Acrocephalus australis 584
gouldi, Ardetta 85
gouldi, Euphonia 821
gouldi, Pelargopsis capensis 286
gouldi, Pterodroma macroptera 74
gouldi, Turdus rubrocanus 669
gouldi, Zosterops lateralis 629
gouldiae, Aethopyga 712

gouldiae, Erythrura 734
gouldii, Anthus leucophrys 743
gouldii, Colluricincla megarhyncha 489
gouldii, Lesbia nuna 273
gouldii, Lophornis 260
gouldii, Melospiza melodia 782
gouldii, Monarcha trivirgatus 502
gouldii, Phonygammus keraudrenii 516
gouldii, Phylidonyris niger 440
gouldii, Selenidera 302
gounellei, Anopetia 256
GOURA 171
gourdini, Pycnonotus goiavier 568
govinda, Milvus migrans 100
goyana, Cranioleuca semicinerea 411
goyderi, Amytornis 430
grabae, Fratercula arctica 155
graberi, Cistothorus platensis 635
graceannae, Icterus 770
graciae, Dendroica 762
gracileus, Sittasomus griseicapillus 420
gracilipes, Amazona vittata 203
gracilipes, Zimmerius 356
gracilirostris, Acrocephalus 584
gracilirostris, Andropadus 569
gracilirostris, Anthracothorax prevostii 259
gracilirostris, Caridonax fulgidus 285
gracilirostris, Catharus 664
gracilirostris, Chloropeta 587
gracilirostris, Tyto alba 218
gracilirostris, Vireo 485
gracilirostris, Xiphorhynchus guttatus 424
gracilis, Anas 67
gracilis, Andropadus 569
gracilis, Aphelocoma ultramarina 507
gracilis, Buteo magnirostris 110
gracilis, Coracina lineata 467
gracilis, Cypsiurus parvus 252
gracilis, Eutoxeres condamini 255
gracilis, Geranospiza caerulescens 109
gracilis, Heterophasia 621
gracilis, Leptosomus discolor 283
gracilis, Lesbia nuna 273
gracilis, Lophozosterops goodfellowi 631
gracilis, Meliphaga 434
gracilis, Melospiza lincolnii 782
gracilis, Mimus gilvus 648
gracilis, Oceanites 77
gracilis, Piaya minuta 214
gracilis, Polioptila caerulea 643
gracilis, Prinia 558
gracilis, Rhinoptilus africanus 145
gracilis, Sterna dougallii 150
gracilis, Stigmatura budytoides 355
GRACULA 654
graculina, Strepera 463
graculinus, Prionops retzii 457
graculus, Pyrrhocorax 512
gradaria, Columba leuconota 157
graduacauda, Icterus 771
graeca, Alectoris 48
graecus, Garrulus glandarius 508
graeffi, Ptilinopus porphyraceus 175
graeffii, Pachycephala vitiensis 477
graellsii, Larus fuscus 148
GRAFISIA 658
grahami, Garrulax maesi 613
graingeri, Lichenostomus plumulus 432
GRALLARIA 398
grallaria, Athene cunicularia 233
grallaria, Fregetta 77
GRALLARICULA 401
grallarius, Burhinus 131
GRALLINA 503
graminea, Melospiza melodia 782
gramineus, Megalurus 577
gramineus, Pooecetes 785
gramineus, Tanygnathus 192
GRAMINICOLA 578

gularis, *Halcyon smyrnensis* 286
gularis, *Heliodoxa* 269
gularis, *Hellmayrea* 410
gularis, *Icterus* 771
gularis, *Illadopsis fulvescens* 601
gularis, *Macronous* 608
gularis, *Mecocerculus leucophrys* 353
gularis, *Melithreptus* 435
gularis, *Merops* 295
gularis, *Monticola* 688
gularis, *Myrmotherula* 385
gularis, *Nicator* 576
gularis, *Paradoxornis* 624
gularis, *Paroaria* 802
gularis, *Piculus rubiginosus* 323
gularis, *Pycnonotus melanicterus* 565
gularis, *Rhinomyias* 690
gularis, *Rhipidura rufiventris* 494
gularis, *Sericornis frontalis* 446
gularis, *Serinus* 748
gularis, *Sylvia atricapilla* 596
gularis, *Tephrodornis* 456
gularis, *Thamnistes anabatinus* 383
gularis, *Turdoides* 610
gularis, *Yuhina* 622
gulgula, *Alauda* 549
gulielmi, *Poicephalus* 194
gulielmitertii, *Cyclopsitta* 204
gulmergi, *Muscicapa sibirica* 691
gumia, *Platyrinchus coronatus* 364
gunax, *Puffinus lherminieri* 76
gundlachi, *Colaptes auratus* 324
gundlachi, *Dendroica petechia* 760
gundlachii, *Accipiter* 108
gundlachii, *Buteogallus anthracinus* 109
gundlachii, *Chordeiles* 239
gundlachii, *Mimus* 649
gundlachii, *Quiscalus niger* 775
gundlachii, *Vireo* 483
gunni, *Larus novaehollandiae* 148
gunningi, *Sheppardia* 675
gurgaoni, *Caprimulgus asiaticus* 244
gurneyi, *Anaplectes melanotis* 724
gurneyi, *Aquila* 113
gurneyi, *Aviceda subcristata* 98
gurneyi, *Mimizuku* 225
gurneyi, *Pitta* 338
gurneyi, *Podiceps nigricollis* 80
gurneyi, *Promerops* 715
gurneyi, *Zoothera* 661
gurue, *Cossypha anomala* 677
gustavi, *Anthus* 743
gustavi, *Brotogeris cyanoptera* 201
gusti, *Aplonis panayensis* 652
guttacristatus, *Chrysocolaptes lucidus* 329
guttata, *Arborophila rufogularis* 55
guttata, *Certhilauda albescens* 543
guttata, *Chlamydera* 428
guttata, *Cichladusa* 678
guttata, *Dendrocygna* 61
guttata, *Myrmotherula* 385
guttata, *Ortalis* 37
guttata, *Stachyris striolata* 607
guttata, *Stagonopleura* 732
guttata, *Taeniopygia* 733
guttata, *Tangara* 811
guttata, *Tyto alba* 218
guttata, *Zoothera* 661
guttaticollis, *Ailuroedus crassirostris* 427
guttaticollis, *Alcippe cinereiceps* 619
guttaticollis, *Napothera epilepidota* 605
guttaticollis, *Paradoxornis* 623
guttatoides, *Xiphorhynchus guttatus* 424
guttatum, *Todirostrum chrysocrotaphum* 362
guttatum, *Toxostoma* 649
guttatus, *Campylorhynchus brunneicapillus* 634
guttatus, *Catharus* 665
guttatus, *Colaptes puncticula* 324
guttatus, *Enicurus maculatus* 683

guttatus, *Eurostopodus* 240
guttatus, *Hypoedaleus* 379
guttatus, *Ixonotus* 570
guttatus, *Laniarius bicolor* 460
guttatus, *Odontophorus* 43
guttatus, *Passerculus sandwichensis* 784
guttatus, *Psilorhamphus* 395
guttatus, *Salpinctes obsoletus* 634
guttatus, *Sericornis spilodera* 445
guttatus, *Tinamus* 31
guttatus, *Xiphorhynchus* 424
GUTTERA 40
guttifer, *Accipiter bicolor* 108
guttifer, *Caprimulgus ruwenzorii* 244
guttifer, *Picumnus albosquamatus* 313
guttifer, *Pogonocichla stellata* 674
guttifer, *Tringa* 142
guttistriatus, *Campylorhamphus trochilirostris* 426
guttula, *Monarcha* 501
guttulata, *Megaceryle lugubris* 293
guttulata, *Syndactyla* 415
guttulatus, *Icterus pectoralis* 770
guttuligera, *Premnornis* 413
gutturalis, *Anthus* 745
gutturalis, *Aramides cajanea* 122
gutturalis, *Atlapetes albinucha* 800
gutturalis, *Chalcomitra senegalensis* 707
gutturalis, *Cinclocerthia* 651
gutturalis, *Corapipo* 340
gutturalis, *Criniger bres* 573
gutturalis, *Francolinus levaillantoides* 51
gutturalis, *Habia* 819
gutturalis, *Hirundo rustica* 535
gutturalis, *Hylophylax poecilinotus* 394
gutturalis, *Irania* 677
gutturalis, *Myrmotherula* 385
gutturalis, *Neocichla* 659
gutturalis, *Oreoica* 490
gutturalis, *Oreoscopus* 445
gutturalis, *Parula* 760
gutturalis, *Pseudoseisura* 414
gutturalis, *Pterocles* 156
gutturalis, *Saxicola* 680
gutturata, *Cranioleuca* 411
gutturosus, *Manacus manacus* 341
guy, *Phaethornis* 256
guyanensis, *Otus choliba* 223
guzuratus, *Orthotomus sutorius* 564
gwendolenae, *Pomatostomus superciliosus* 450
gwendolenae, *Sterna bergii* 150
GYALOPHYLAX 410
GYGIS 153
gymnocephala, *Pityriasis* 465
gymnocephalus, *Picathartes* 522
GYMNOCICHLA 390
GYMNOCREX 123
gymnocycla, *Columba livia* 157
GYMNODERUS 349
gymnogenys, *Turdoides* 611
GYMNOGLAUX 225
GYMNOGYPS 93
GYMNOMYSTAX 771
GYMNOMYZA 434
GYMNOPHAPS 180
gymnophthalmus, *Dendrocopos moluccensis* 318
gymnopis, *Cacatua sanguinea* 183
GYMNOPITHYS 393
gymnops, *Melipotes* 437
gymnops, *Metriopelia ceciliae* 168
gymnops, *Rhegmatorhina* 393
GYMNORHINA 463
GYMNORHINUS 508
gymnorhynchus, *Cicinnurus regius* 517
GYMNORIS 718
gymnostoma, *Jacana spinosa* 138
GYPAETUS 101
GYPOHIERAX 101

GYPS 102
gyrator, *Apus affinis* 254
gyrola, *Tangara* 811

haastii, *Apteryx* 35
habeli, *Camarhynchus psittacula* 796
habenichti, *Pitta erythrogaster* 338
habessinicus, *Cinnyris* 710
habessinicus, *Tchagra senegalus* 459
HABIA 819
HABROPTILA 125
habroptila, *Strigops* 181
hachisuka, *Lanius validirostris* 480
hachisukae, *Pycnonotus plumosus* 568
hachisukai, *Chloroceryle americana* 292
hades, *Myzomela pammelaena* 442
hades, *Turdus poliocephalus* 669
hadii, *Zoothera piaggiae* 661
hadropus, *Buteo jamaicensis* 111
haemacephala, *Megalaima* 306
haemachalanus, *Gypaetus barbatus* 101
haemalea, *Piranga flava* 818
haemastica, *Limosa* 140
haematina, *Spermophaga* 728
haematocephala, *Cisticola galactotes* 553
haematocephala, *Lagonosticta rubricata* 729
HAEMATODERUS 349
haematodus, *Trichoglossus* 185
haematogaster, *Campephilus* 326
haematogaster, *Northiella* 189
haematonota, *Myrmotherula* 385
haematonotus, *Psephotus* 189
haematophagus, *Buphagus africanus* 659
HAEMATOPUS 131
haematopus, *Himantornis* 117
haematopygia, *Leucosticte brandti* 752
haematopygus, *Aulacorhynchus* 301
haematorrhous, *Northiella haematogaster* 189
HAEMATORTYX 56
HAEMATOSPIZA 755
haematostictum, *Dicaeum* 701
haematotis, *Pionopsitta* 202
haematribon, *Chrysocolaptes lucidus* 329
haematuropygia, *Cacatua* 183
haemorrhous, *Aratinga acuticaudata* 196
haemorrhous, *Cacicus* 769
haemorrhous, *Hypopicus hyperythrus* 317
haemorrhousus, *Pycnonotus cafer* 567
haesitatus, *Cisticola* 555
hafizi, *Luscinia megarhynchos* 677
hagedash, *Bostrychia* 82
hagenbecki, *Phasianus colchicus* 60
hagenensis, *Oreostruthus fuliginosus* 732
hagenensis, *Peneothello sigillata* 519
hahasima, *Apalopteron familiare* 631
hainana, *Napothera epilepidota* 605
hainana, *Streptopelia chinensis* 162
hainanensis, *Pycnonotus jocosus* 566
hainanus, *Blythipicus pyrrhotis* 330
hainanus, *Caprimulgus macrurus* 243
hainanus, *Corvus macrorhynchos* 514
hainanus, *Cyornis* 695
hainanus, *Dendrocopos major* 320
hainanus, *Harpactes erythrocephalus* 280
hainanus, *Paradoxornis gularis* 624
hainanus, *Parus major* 524
hainanus, *Phylloscopus* 592
hainanus, *Pomatorhinus hypoleucos* 602
hainanus, *Pycnonotus sinensis* 566
hainanus, *Rhopodytes tristis* 212
hainanus, *Tephrodornis virgatus* 456
hainanus, *Treron curvirostra* 172
hainanus, *Zosterops japonicus* 625
halconensis, *Zosterops montanus* 626
HALCYON 286
halfae, *Galerida cristata* 548
HALIAEETUS 101
haliaetus, *Pandion* 98
HALIASTUR 101

horsfieldii, Pomatorhinus 602
hortensis, Sylvia 597
hortorum, Dendrocopos minor 318
hortulana, Emberiza 778
hortulorum, Turdus 667
horus, Apus 254
horvathi, Colinus cristatus 42
hosei, Pitta caerulea 337
hosii, Calyptomena 335
hosii, Dicaeum cruentatum 703
hosii, Oriolus 488
hoskinsii, Glaucidium gnoma 230
hottentota, Anas 68
hottentottus, Dicrurus 492
hottentottus, Turnix 129
HOUBAROPSIS 116
houghtoni, Sheppardia cyornithopsis 675
housei, Amytornis 430
houstonensis, Ammodramus henslowii 784
houyi, Turdoides reinwardtii 611
hova, Asio capensis 236
hova, Mirafra 540
howardi, Dendragapus obscurus 46
howei, Sericornis magnirostra 446
howei, Strepera versicolor 464
howelli, Chordeiles minor 239
howelli, Stactolaema olivacea 307
hoya, Spilornis cheela 103
hoyi, Carduelis magellanica 750
hoyi, Geositta rufipennis 402
hoyi, Otus 224
hoyi, Pycnonotus sinensis 566
hoyti, Eremophila alpestris 550
hrota, Branta bernicla 63
huachamacarii, Chamaeza campanisona 398
huallagae, Asthenes urubambensis 407
huallagae, Aulacorhynchus 302
huallagae, Chlorospingus parvirostris 818
huallagae, Phaethornis syrmatophorus 256
huallagae, Synallaxis gujanensis 409
huallagae, Thamnomanes saturninus 384
huallagae, Thamnophilus punctatus 382
huambina, Hemitriaupis guira 814
huambo, Cisticola natalensis 554
huancabambae, Anairetes flavirostris 353
huancabambae, Zonotrichia capensis 783
huancavelicae, Asthenes dorbignyi 406
huancavelicae, Upucerthia serrana 403
huarandosae, Tachyphonus cristatus 805
hubbardi, Francolinus coqui 50
huberi, Tangara cayana 811
huberi, Thamnophilus nigrocinereus 381
hudsoni, Asthenes 407
hudsoni, Knipolegus 369
hudsonia, Calidris alpina 144
hudsonia, Pica 511
hudsonicus, Numenius phaeopus 141
hudsonicus, Parus 528
hudsonius, Circus cyaneus 104
huegeli, Coenocorypha aucklandica 139
huei, Galerida theklae 548
hueskeri, Macropygia amboinensis 162
hueti, Alcippe morrisonia 620
huetii, Touit 201
hueyi, Dendroica petechia 761
hueyi, Phalaenoptilus nuttallii 240
hueyi, Tympanuchus phasianellus 46
hufufae, Passer domesticus 716
hufufae, Prinia gracilis 558
hughlandi, Melanerpes aurifrons 315
hugonis, Abroscopus albogularis 594
huhula, Ciccaba 229
huilae, Calamonastes undosus 563
huilae, Malacoptila fulvogularis 333
huilae, Tiaris bicolor 795
huilensis, Cisticola chiniana 552
huillensis, Serinus canicollis 746
hulli, Nesomimus parvulus 649
hulliana, Aplonis fusca 653

hullianus, Puffinus carneipes 77
humaythae, Schistocichla leucostigma 391
humayuni, Pycnonotus cafer 566
humbloti, Ardea 87
humbloti, Cinnyris 712
HUMBLOTIA 693
humboldti, Phrygilus alaudinus 788
humboldti, Pteroglossus inscriptus 302
humboldti, Spheniscus 71
humboldtii, Francolinus afer 53
humboldtii, Hylocharis grayi 264
humei, Aegithina tiphia 465
humei, Artamus leucorynchus 464
humei, Celeus brachyurus 325
humei, Dendrocopos macei 319
humei, Phylloscopus 591
humei, Sphenocichla 606
humeralis, Agelaius 774
humeralis, Aimophila 786
humeralis, Cossypha 678
humeralis, Diglossa 816
humeralis, Euplectes hartlaubi 726
humeralis, Geopelia 165
humeralis, Lanius collaris 481
humeralis, Myospiza 785
humeralis, Ninox rufa 233
humeralis, Parkerthraustes 824
humeralis, Ptilinopus iozonus 177
humeralis, Terenura 388
humiae, Syrmaticus 59
humicola, Asthenes 406
humii, Carpodacus puniceus 755
humii, Coccothraustes coccothraustes 757
humii, Hypsipetes leucocephalus 575
humii, Megalaima mystacophanos 305
humii, Paradoxornis nipalensis 623
humii, Picus mentalis 327
humii, Pycnonotus leucogenys 566
humii, Sturnus vulgaris 656
humii, Todiramphus chloris 288
humilis, Asthenes 406
humilis, Campylorhynchus rufinucha 634
humilis, Cisticola chiniana 552
humilis, Euphonia minuta 822
humilis, Eupodotis 115
humilis, Ichthyophaga 101
humilis, Pomatorhinus schisticeps 603
humilis, Pseudopodoces 511
humilis, Sericornis frontalis 446
humilis, Serinus gularis 748
humilis, Streptopelia tranquebarica 161
humilis, Yuhina flavicollis 622
humphreysi, Orthotomus atrogularis 564
hunanensis, Pomatorhinus ruficollis 603
hungarica, Calandrella brachydactyla 545
hunsteini, Diphyllodes magnificus 517
hunsteini, Lonchura 736
hunsteini, Pachycephalopsis poliosoma 520
hunsteini, Phonygammus keraudrenii 516
hunteri, Chalcomitra 707
hunteri, Cisticola 552
huonensis, Meliphaga montana 433
huonensis, Psittacella madaraszi 190
huonensis, Ptilinopus coronulatus 176
huriensis, Galerida theklae 548
huszari, Otus senegalensis 221
hutchinsii, Branta canadensis 63
hutchinsoni, Rhipidura nigrocinnamomea 496
hutsoni, Indicator willcocksi 310
huttoni, Otus spilocephalus 220
huttoni, Phoebetria palpebrata 72
huttoni, Prunella atrogularis 738
huttoni, Ptilinopus 176
huttoni, Puffinus 76
huttoni, Turdoides caudata 610
huttoni, Vireo 484
hutzi, Pitta elegans 339
hutzi, Pycnonotus plumosus 568
hyacinthina, Halcyon leucocephala 287

hyacinthinus, Anodorhynchus 195
hyacinthinus, Cyornis 697
hyacinthinus, Geoffroyus heteroclitus 191
hyalinus, Phaethornis anthophilus 256
hybrida, Chlidonias 152
hybrida, Chloephaga 64
hybrida, Eudynamys scolopaceus 211
hybrida, Hemiparra crassirostris 133
hybrida, Pteruthius xanthochlorus 617
hybridus, Colluricincla megarhyncha 489
hybridus, Tanygnathus lucionensis 191
HYDROBATES 77
hydrocharis, Tanysiptera 285
HYDROCHOUS 247
hydrocorax, Buceros 299
HYDROPHASIANUS 138
Hydroprogne 149
HYDROPSALIS 245
hyemalis, Clangula 69
hyemalis, Junco 783
HYETORNIS 214
hygrica, Pogonocichla stellata 674
hygricus, Macronyx croceus 741
hygrophila, Pytilia melba 727
hygroscopus, Turdus poliocephalus 668
Hylacola 444
hylebata, Phylloscopus borealis 591
hyleorus, Dendrocolaptes certhia 422
HYLEXETASTES 421
HYLIA 588
HYLIOTA 587
hylobius, Philydor 417
HYLOCHARIS 263
HYLOCICHLA 666
HYLOCITREA 474
HYLOCRYPTUS 418
HYLOCTISTES 415
hylodroma, Grallaria gigantea 399
hylodromus, Xiphorhynchus triangularis 424
hylodytes, Pogoniulus simplex 307
HYLOMANES 293
HYLONYMPHA 269
HYLOPEZUS 400
hylophila, Halcyon albiventris 287
hylophila, Strix 228
HYLOPHILUS 485
hylophilus, Monticola angolensis 688
hylophona, Cossypha natalensis 678
hylophona, Cryptolybia woodwardi 307
HYLOPHYLAX 393
Hylopsar 656
HYLORCHILUS 635
hyloscopus, Picoides villosus 321
hymenaicus, Ploceus hypoxanthus 722
HYMENOLAIMUS 63
HYMENOPS 370
hyogastrus, Ptilinopus 177
hypaethrus, Aimophila rufescens 787
HYPARGOS 728
hypenema, Allenia fusca 650
hyperborea, Lagopus muta 47
hyperborea, Uria aalge 154
hyperboreus, Larus 147
hyperboreus, Plectrophenax 781
hyperemnus, Pycnonotus atriceps 565
HYPERGERUS 562
hypermelaenus, Parus 527
hypermetra, Mirafra 541
hypermetra, Tyto alba 218
hypernephela, Cercomela sinuata 686
hyperorius, Theristicus caudatus 83
hyperriphaeus, Parus cyanus 526
hyperrynchus, Notharchus macrorhynchos 332
hyperytha, Turdoides subrufa 610
hyperythra, Arborophila 56
hyperythra, Brachypteryx 673
hyperythra, Cecropis daurica 537
hyperythra, Diglossa sittoides 816
hyperythra, Dumetia 608

kikuyuensis, Cinnyris reichenowi 709
kikuyuensis, Eremomela scotops 595
kikuyuensis, Francolinus levaillantii 51
kikuyuensis, Lagonosticta senegala 729
kikuyuensis, Turdoides jardineii 611
kikuyuensis, Zosterops poliogastrus 630
kilgoris, Spermophaga ruficapilla 728
kilianensis, Carpodacus puniceus 755
kilimense, Glaucidium brodiei 230
kilimensis, Cryptospiza salvadorii 728
kilimensis, Dendropicos griseocephalus 317
kilimensis, Estrilda quartinia 730
kilimensis, Linurgus olivaceus 749
kilimensis, Nectarinia 708
kilimensis, Serinus burtoni 748
kilimensis, Stactolaema leucotis 307
kilimensis, Zoothera piaggiae 661
kimberli, Tyto novaehollandiae 218
kimbutui, Cossypha archeri 677
kinabalu, Rhipidura albicollis 494
kinabaluensis, Phylloscopus trivirgatus 592
kinabaluensis, Spilornis 103
kingi, Acrocephalus familiaris 584
kingi, Aglaiocercus 275
kingi, Anthochaera paradoxa 437
kingi, Heterophasia melanoleuca 621
kingi, Melanodryas vittata 520
kingi, Melithreptus validirostris 435
kingi, Oenanthe chrysopygia 685
kingi, Oriolus flavocinctus 487
kinneari, Chloropsis cochinchinensis 699
kinneari, Cinnyris habessinicus 710
kinneari, Dendrocitta vagabunda 510
kinneari, Macronous gularis 608
kinneari, Melidectes belfordi 438
kinneari, Sasia ochracea 313
kinneari, Sitta carolinensis 646
kinneari, Spelaeornis chocolatinus 606
kinneari, Turnix maculosus 129
kinnisii, Turdus merula 668
kioloides, Canirallus 117
kirghizorum, Carduelis flavirostris 751
kirhizica, Petronia petronia 718
kirhocephalus, Pitohui 490
kirki, Prionops scopifrons 457
kirki, Zosterops maderaspatanus 630
kirkii, Chalcomitra amethystina 707
kirkii, Turdoides jardineii 611
kirkii, Veniliornis 322
kirmanensis, Parus lugubris 527
kirmanensis, Streptopelia senegalensis 162
kirmanica, Columba palumbus 158
kirtlandii, Dendroica 762
kiskensis, Troglodytes troglodytes 639
kismayensis, Laniarius ruficeps 460
kisserensis, Philemon citreogularis 436
kistchinskii, Calidris alpina 144
kitsutsuki, Picoides major 320
kittlitzi, Carduelis sinica 749
kiusiuensis, Aegithalos caudatus 539
kivuense, Glaucidium tephronotum 231
kivuensis, Cinnyris regius 709
kivuensis, Cossypha caffra 677
kivuensis, Pseudoalcippe abyssinica 602
kivuensis, Ruwenzorornis johnstoni 206
kivuensis, Terpsiphone viridis 499
kiwanukae, Estrilda charmosyna 731
kiwuensis, Colius striatus 278
kizuki, Dendrocopos 318
kizuki, Dendrocopos kizuki 318
klaas, Chrysococcyx 210
klagesi, Drymophila caudata 388
klagesi, Lophornis chalybeus 260
klagesi, Myrmotherula 384
klagesi, Taeniotriccus andrei 362
klagesi, Thripadectes virgaticeps 416
klagesi, Tolmomyias poliocephalus 364
KLAIS 260
klei, Tinamus tao 31

kleini, Alectoris chukar 49
kleinschmidti, Anthus petrosus 744
kleinschmidti, Columba palumbus 158
kleinschmidti, Erythrura 734
kleinschmidti, Galerida cristata 547
kleinschmidti, Halcyon erythrorhampha 284
kleinschmidti, Lamprolia victoriae 504
kleinschmidti, Parus montanus 527
kleinschmidti, Petroica multicolor 521
klinikowskii, Margarops fuscatus 651
klossi, Alcippe castaneceps 619
klossi, Cinnyris jugularis 711
klossi, Cissa chinensis 510
klossi, Cyornis rubeculoides 696
klossi, Harpactes erythrocephalus 280
klossi, Pachycephala soror 476
klossi, Phylloscopus davisoni 592
klossi, Pomatorhinus schisticeps 603
klossi, Prinia atrogularis 557
klossi, Pycnonotus aurigaster 567
klossi, Rallina rubra 118
klossi, Spilornis 103
klossi, Sylviparus modestus 529
klossii, Bubo coromandus 227
KNIPOLEGUS 369
knoxi, Crypturellus obsoletus 32
knudseni, Himantopus himantopus 133
kobayashii, Myzomela rubratra 442
kobdensis, Carpodacus rubicilla 755
kobdensis, Luscinia svecica 676
kobylini, Lanius collurio 479
kochi, Pitta 338
kochii, Coracina striata 467
kodonophonos, Pitohui cristatus 490
koelzi, Aethopyga nipalensis 713
koelzi, Dendrocopos mahrattensis 319
koelzi, Sitta castanea 645
koenigi, Alectoris barbara 49
koenigi, Aratinga acuticaudata 196
koenigi, Dendropicos goertae 317
koenigi, Lanius meridionalis 480
koenigi, Sicalis flaveola 791
koenigi, Sylvia atricapilla 596
koenigi, Troglodytes troglodytes 639
koenigorum, Myrmoborus leucophrys 390
koeniswaldiana, Pulsatrix 230
koepckeae, Cacicus 769
koepckeae, Otus 224
koepckeae, Pauxi unicornis 40
koepckeae, Phaethornis 256
koepckeae, Rhodopis vesper 276
koesteri, Caprimulgus ruwenzorii 244
kogo, Picus canus 328
koktalensis, Parus cyanus 526
kolichisi, Cettia vulcania 579
kolichisi, Colluricincla harmonica 489
koliensis, Serinus 746
kollari, Synallaxis 407
kollmannspergeri, Bubo africanus 226
kollmanspergeri, Ammomanes deserti 545
kolymensis, Charadrius hiaticulus 135
kolymensis, Riparia riparia 532
kolymensis, Tetrao parvirostris 46
kolymensis, Tetrastes bonasia 45
komadori, Luscinia 676
kombok, Amblyornis macgregoriae 427
komensis, Lagopus muta 47
kona, Chloridops 758
konigseggi, Francolinus clappertoni 52
konkakinhensis, Garrulax 614
koratensis, Chalcoparia singalensis 703
kordensis, Monarcha chrysomela 502
kordensis, Rhipidura rufiventris 495
kordoana, Charmosyna rubronotata 187
kordofanensis, Corvus capensis 513
kordofanensis, Thamnolaea cinnamomeiventris 687
koreensis, Cyanopica cyanus 509
korejevi, Carduelis flavirostris 751

korejewi, Emberiza schoeniclus 780
korejewi, Otis tarda 114
korejewi, Passer ammodendri 715
korejewi, Rallus aquaticus 121
koreni, Lagopus lagopus 47
kori, Ardeotis 115
korinchi, Picus flavinucha 328
koroana, Pachycephala vitiensis 477
koroviakovi, Alectoris chukar 49
korthalsi, Treron sphenurus 174
korustes, Sterna dougallii 150
koslowi, Babax 612
koslowi, Emberiza 779
koslowi, Prunella 738
koslowi, Tetraogallus himalayensis 48
kosswigi, Panurus biarmicus 622
kotataki, Dendrocopos kizuki 318
kotschubeii, Carpodacus rhodochlamys 755
kowaldi, Ifrita 515
kozlowae, Lagopus lagopus 47
KOZLOWIA 755
krakari, Charmosyna rubrigularis 187
krakari, Macropygia mackinlayi 163
krameri, Psittacula 193
krascheninnikowi, Lagopus muta 47
kreffti, Mino 653
kretschmeri, Macrosphenus 587
kreyenborgi, Falco 97
kriderii, Buteo jamaicensis 111
kriegi, Pyrrhura frontalis 198
krishnarajui, Malacocincla abbotti 600
krishnarkumarsinhji, Calandrella raytal 546
kronei, Phylloscartes 356
kruensis, Anthreptes seimundi 705
krueperi, Sitta 645
krugii, Gymnoglaux 224
krynicki, Garrulus glandarius 508
kuatunensis, Emberiza fucata 779
kuatunensis, Parus ater 526
kubanensis, Carpodacus erythrinus 753
kubaryi, Corvus 513
kubaryi, Gallicolumba 170
kubaryi, Rhipidura rufifrons 497
kuboriensis, Daphoenositta miranda 474
kubtchecki, Amazilia versicolor 265
kubuna, Rhipidura rufidorsa 496
kuehni, Dicaeum celebicum 703
kuehni, Dicrurus hottentottus 493
kuehni, Gallirallus torquatus 119
kuehni, Gerygone dorsalis 447
kuehni, Myzomela 441
kuehni, Pachycephala griseonota 478
kuehni, Pitta erythrogaster 338
kuehni, Zosterops 627
kuenzeli, Streptoprocne zonaris 247
kuhli, Eolophus roseicapilla 183
kuhlii, Leucopternis 109
kuhlii, Vini 186
kuhni, Cyornis hyacinthinus 697
kuiperi, Turnix suscitator 130
kukenamensis, Grallaricula nana 402
kukunoorensis, Calandrella cheleensis 546
kulalensis, Zosterops poliogastrus 630
kulambangrae, Coracina caledonica 466
kulambangrae, Petroica multicolor 521
kulambangrae, Turdus poliocephalus 668
kulambangrae, Zosterops 628
kulczynskii, Thamnophilus nigrocinereus 381
kumaiensis, Leiothrix lutea 616
kumaonensis, Picus flavinucha 327
kumawa, Melipotes fumigatus 437
kumbaensis, Dyaphorophyia concreta 454
kumboensis, Cecropis daurica 537
kumboensis, Muscicapa adusta 692
kumerloevei, Eremophila alpestris 551
kumlieni, Larus glaucoides 147
kumusi, Meliphaga flaviventer 431
kumusi, Rhipidura rufidorsa 496
kumusii, Lonchura caniceps 736

leucopyga, Nyctiprogne 239
leucopyga, Oenanthe 686
leucopyga, Tachycineta meyeni 533
leucopygia, Coracina 467
leucopygia, Iodopleura pipra 344
leucopygia, Lepidothrix isidorei 340
leucopygia, Phaeothlypis fulvicauda 767
leucopygia, Turdoides 611
leucopygialis, Artamus leucorynchus 464
leucopygialis, Dicrurus caerulescens 492
leucopygialis, Lalage nigra 470
leucopygialis, Rhaphidura 250
leucopygius, Aerodramus spodiopygius 248
leucopygius, Serinus 746
leucopygius, Todiramphus 288
leucopygos, Lanius meridionalis 480
leucopyrrhus, Laterallus 119
leucoramphus, Cacicus chrysonotus 769
leucorhoa, Oceanodroma 78
leucorhoa, Oenanthe oenanthe 685
leucorhynchus, Laniarius 459
leucorhynchus, Pitohui ferrugineus 490
leucorodia, Platalea 83
leucorrhoa, Corapipo 340
leucorrhoa, Tachycineta 533
leucorrhous, Buteo 111
leucorrhous, Polytmus theresiae 264
leucorynchus, Artamus 464
leucoryphus, Haliaeetus 101
leucoryphus, Platyrinchus 364
LEUCOSARCIA 165
leucoscepus, Francolinus 53
leucosoma, Hirundo 536
leucosomus, Accipiter novaehollandiae 105
leucospila, Rallina 118
leucospilus, Otus magicus 222
leucospodia, Pseudelaenia 355
leucostephes, Melidectes 438
leucosterna, Cheramoeca 532
leucosternos, Tachybaptus novaehollandiae 79
leucosticta, Cercotrichas 679
leucosticta, Certhia americana 647
leucosticta, Henicorhina 641
leucosticta, Lonchura tristissima 735
leucosticta, Napothera brevicaudata 605
leucosticta, Ptilorrhoa 453
LEUCOSTICTE 752
leucostictus, Bubo 227
leucostictus, Dysithamnus plumbeus 384
leucostigma, Anas sparsa 66
leucostigma, Rhagologus 474
leucostigma, Schistochlia 391
leucothorax, Gerygone chrysogaster 448
leucothorax, Hypsipetes leucocephalus 575
leucothorax, Lanio 806
leucothorax, Rhipidura 495
leucotis, Basilinna 268
leucotis, Colinus cristatus 42
leucotis, Colius striatus 278
leucotis, Entomodestes 666
leucotis, Eremopterix 549
leucotis, Galbalcyrhynchus 331
leucotis, Garrulus glandarius 509
leucotis, Lichenostomus 432
leucotis, Melozone 798
leucotis, Monarcha 501
leucotis, Phapitreron 171
leucotis, Poephila personata 732
leucotis, Ptilopsis 225
leucotis, Pycnonotus leucogenys 566
leucotis, Pyrrhura 199
leucotis, Stachyris 607
leucotis, Stactolaema 307
leucotis, Tangara ruficervix 812
leucotis, Turaco 206
leucotis, Thryothorus 638
leucotis, Vireolanius 482
leucotos, Dendrocopos 319
leucura, Lagopus 47

leucura, Montifringilla nivalis 718
leucura, Myiomela 682
leucura, Oenanthe 686
leucura, Peneoenanthe pulverulenta 519
leucura, Pinicola enucleator 755
leucura, Tanysiptera sylvia 285
leucura, Tityra 343
leucura, Urochroa bougueri 269
leucurus, Agriornis montanus 371
leucurus, Baeopogon indicator 570
leucurus, Elanus 100
leucurus, Monarcha 502
leucurus, Picoides pubescens 321
leucurus, Saxicola 684
leucurus, Threnetes niger 255
leucurus, Vanellus 134
leuphotes, Aviceda 98
levaillantii, Clamator 207
levaillantii, Corvus macrorhynchos 514
levaillantii, Francolinus 51
levaillantoides, Francolinus 51
levantina, Acrocephalus stentoreus 584
levantina, Sitta europaea 644
leverianus, Cissopis 803
levigaster, Gerygone 447
levipes, Rallus longirostris 120
levis, Rostrhamus sociabilis 100
levis, Sittasomus griseicapillus 420
levraudi, Laterallus 119
levyi, Catharus fuscescens 665
LEWINIA 121
lewinii, Meliphaga 434
lewis, Melanerpes 314
lewisi, Actinodura egertoni 617
lewisi, Lophura nycthemera 58
lewisii, Ptilinopus viridis 177
leyboldi, Sephanoides fernandensis 271
leylandi, Colinus leucopogon 42
leymebambae, Grallaricula ferrugineipectus 401
leytensis, Dendrocopos maculatus 318
leytensis, Gallicolumba crinigera 170
leytensis, Micromacronus 609
leytensis, Pericrocotus flammeus 473
leytensis, Rhabdornis inornatus 651
lhamarum, Alauda gulgula 549
lherminieri, Cichlherminia 672
lherminieri, Puffinus 76
lhuysii, Lophophorus 58
liangshanensis, Garrulax lunulatus 614
libanotica, Oenanthe oenanthe 685
liberalis, Malacocincla sepiaria 600
liberatus, Laniarius 460
liberatus, Tyto capensis 219
liberiae, Glareola nuchalis 146
liberiae, Pseudhirundo griseopyga 532
liberiensis, Turtur afer 164
libertatis, Tachuris rubrigastra 359
libonyanus, Turdus 666
libs, Coracopsis nigra 194
LICHENOSTOMUS 431
lichenya, Anthus cinnamomeus 742
lichiangense, Crossoptilon crossoptilon 59
LICHMERA 439
lichtensteini, Amazilia beryllina 267
lichtensteini, Philydor 416
lichtensteinii, Pterocles 156
lictor, Pitangus 374
licua, Glaucidium perlatum 230
lidthi, Garrulus 509
lifuensis, Coracina caledonica 466
lifuensis, Gerygone flavolateralis 447
lifuensis, Myzomela cardinalis 442
ligea, Sitta magna 647
lightoni, Apalis melanocephala 561
lignarius, Picoides 317
lignator, Centropus bengalensis 216
lignicida, Lepidocolaptes affinis 425
ligus, Cyornis banyumas 696
lihirensis, Accipiter novaehollandiae 105

lihirensis, Aerodramus vanikorensis 249
lilacea, Sitta oenochlamys 646
lilacina, Amazona autumnalis 203
lilae, Podargus strigoides 237
lilfordi, Dendrocopos leucotos 319
lilfordi, Grus grus 128
lilianae, Agapornis 194
lilianae, Sturnella magna 775
lilith, Athene noctua 232
lilliae, Lepidopyga 263
lilloi, Asthenes sclateri 407
limae, Picumnus 313
limarius, Gallirallus torquatus 119
limatus, Phaethon aethereus 88
limatus, Ramphocelus dimidiatus 806
limbata, Glareola pratincola 146
limbata, Lichmera indistincta 439
limbata, Myrmecocichla albifrons 687
limbata, Turdoides leucopygia 611
limborgi, Chrysococcyx 210
limes, Certhia himalayana 648
LIMICOLA 144
limicola, Rallus 121
limitans, Picus vittatus 328
limnaeetus, Spizaetus cirrhatus 114
limnatis, Geothlypis speciosa 764
limnetis, Rallus longirostris 121
Limnoctites 412
LIMNODROMUS 140
LIMNORNIS 412
LIMNOTHLYPIS 763
limoncochae, Ixobrychus exilis 85
LIMOSA 140
limosa, Limosa 140
limpopoensis, Emberiza capensis 778
limpopoensis, Laniarius aethiopicus 460
linae, Halcyon coromanda 286
linae, Harpactes ardens 279
linaraborae, Aethopyga 712
linchi, Collocalia 248
lincolnii, Melospiza 782
lindenii, Oxypogon guerinii 274
lindsayi, Actenoides 284
linearis, Chiroxiphia 341
linearis, Geotrygon 169
lineata, Acanthiza 449
lineata, Conopophaga 394
lineata, Coracina 467
lineata, Coturnix chinensis 54
lineata, Dacnis 813
lineata, Lophura nycthemera 58
lineata, Megalaima 305
lineata, Pytilia 727
lineata, Syndactyla subalaris 415
lineata, Thalassidroma 77
lineaticeps, Lepidocolaptes souleyetii 425
lineatocapilla, Xiphorhynchus ocellatus 423
lineatocephalus, Xiphocolaptes
    promeropirhynchus 422
lineatula, Coturnix chinensis 54
lineatum, Tigrisoma 84
lineatus, Buteo 110
lineatus, Cymbilaimus 379
lineatus, Dryocopus 326
lineatus, Garrulax 615
lineatus, Melanerpes formicivorus 314
lineatus, Milvus migrans 100
lineifrons, Grallaricula 402
lineiventris, Anthus 744
lineocapilla, Cisticola exilis 556
lineola, Bolborhynchus 200
lineola, Sporophila 792
lineolata, Pachycephala griseonota 478
lineolatus, Dactylortyx thoracicus 43
lingshuiensis, Pteruthius flaviscapis 617
linnaei, Turdus grayi 671
linteatus, Heterocercus 342
lintoni, Myiophobus 359
LINURGUS 749

lorti, Francolinus levaillantoides 51
lory, Lorius 186
lotenius, Cinnyris 712
loudoni, Carduelis carduelis 751
louisae, Rhynchostruthus socotranus 749
louisiadensis, Coracina papuensis 467
louisiadensis, Cracticus 463
louisiadensis, Myzomela nigrita 441
louisiadensis, Rhipidura rufifrons 497
lovejoyi, Picumnus squamulatus 312
lovensis, Ashbyia 443
loveridgei, Campethera cailliautii 316
loveridgei, Cinnyris 709
loveridgei, Pterocles decoratus 156
loveridgei, Serinus sulphuratus 747
loveridgei, Turdoides aylmeri 610
lowei, Cyanomitra olivacea 706
lowei, Fregata magnificens 89
lowei, Sheppardia 675
lowei, Stellula calliope 278
lowei, Tachymarptis aequatorialis 253
lowei, Treron apicauda 173
loweryi, Quiscalus mexicanus 775
loweryi, Xenoglaux 232
lowi, Aerodramus maximus 249
lowi, Jacana spinosa 138
lowii, Coereba flaveola 776
lowii, Sarcops calvus 653
LOXIA 755
LOXIGILLA 795
LOXIOIDES 758
LOXIPASSER 795
LOXOPS 759
loyca, Sturnella 775
loyei, Molothrus aeneus 773
loyemilleri, Puffinus lherminieri 76
lozanoana, Tangara nigroviridis 812
lozanoi, Gallinula chloropus 126
lualabae, Muscicapa aquatica 692
lualabae, Urocolius indicus 278
luangwae, Melocichla mentalis 582
luangwae, Passer diffusus 717
luapula, Cisticola galactotes 553
lubomirskii, Pipreola 347
lubricus, Trogon elegans 281
lucaris, Emberizoides herbicola 791
lucasanus, Picoides scalaris 321
lucasanus, Vireo cassinii 484
lucasi, Thapsinillas affinis 574
lucayana, Tyto alba 219
lucaysiensis, Myiarchus sagrae 378
lucens, Cinnyris mariquensis 709
luciae, Amazilia 266
luciae, Otus spilocephalus 220
luciae, Vermivora 760
luciana, Turnix nana 129
luciani, Eriocnemis 272
lucianii, Pyrrhura picta 199
lucida, Cyanocompsa parellina 826
lucida, Gallinula chloropus 126
lucida, Hirundo 535
lucida, Myiagra alecto 504
lucida, Perdix perdix 53
lucida, Polioptila melanura 643
lucida, Strix occidentalis 228
lucidipectus, Cinnyris pulchellus 709
lucidiventris, Lybius torquatus 309
lucidus, Chrysococcyx 210
lucidus, Chrysocolaptes 329
lucidus, Cistothorus platensis 635
lucidus, Cyanerpes 813
lucidus, Hemignathus 758
lucidus, Phalacrocorax carbo 91
lucifer, Calothorax 276
lucifer, Carpodacus rubicilloides 755
lucifer, Melidectes ochromelas 438
lucilleae, Napothera epilepidota 605
lucionensis, Tanygnathus 191
lucioniensis, Lanius cristatus 479

luconensis, Prioniturus 191
luctisonus, Otus choliba 223
luctuosa, Ducula bicolor 180
luctuosa, Myrmotherula axillaris 385
luctuosa, Sporophila 792
luctuosa, Sterna fuscata 151
luctuosus, Sakesphorus 380
luctuosus, Saxicola gutturalis 684
luctuosus, Tachyphonus 805
ludibunda, Amazilia edward 266
ludlowi, Alcippe 619
ludlowi, Athene noctua 232
ludlowi, Dendrocopos cathpharius 319
ludlowi, Phylloscopus trochiloides 591
ludovicae, Doryfera 257
ludovicensis, Cinnyris 709
ludoviciae, Turdus olivaceus 667
ludoviciana, Conuropsis carolinensis 198
ludoviciana, Piranga 819
ludovicianus, Caprimulgus 244
ludovicianus, Lanius 480
ludovicianus, Pheucticus 823
ludovicianus, Thryothorus 638
ludwigii, Dicrurus 491
ludwigii, Neotis 115
luebberti, Ploceus intermedius 721
luehderi, Laniarius 460
lufira, Cisticola woosnami 551
lugens, Conirostrum albifrons 816
lugens, Haplophaedia 273
lugens, Motacilla alba 740
lugens, Oenanthe 685
lugens, Parisoma 598
lugens, Parus lugubris 527
lugens, Sarothrura 117
lugens, Streptopelia 160
lugens, Tetrao urogallus 45
Lugensa 74
lugentoides, Oenanthe lugubris 685
lugubris, Aerodramus vanikorensis 249
lugubris, Anthus richardi 741
lugubris, Brachygalba 331
lugubris, Celeus 325
lugubris, Cisticola galactotes 553
lugubris, Contopus 367
lugubris, Dendropicos gabonensis 317
lugubris, Fulica atra 127
lugubris, Garrulax 613
lugubris, Megaceryle 293
lugubris, Melaenornis edolioides 689
lugubris, Melampitta 515
lugubris, Myiotheretes fumigatus 372
lugubris, Myrmoborus 390
lugubris, Ninox scutulata 234
lugubris, Oenanthe 685
lugubris, Parus 527
lugubris, Phalacrocorax carbo 91
lugubris, Poeoptera 658
lugubris, Qusicalus 774
lugubris, Speirops 632
lugubris, Surniculus 211
lugubris, Vanellus 133
luhderi, Cinnyris chloropygius 708
luizae, Asthenes 407
lukolelae, Glaucidium brasilianum 231
LULLULA 549
lulu, Tyto alba 219
luluae, Poecilotriccus 361
lumachella, Augastes 275
lumholtzi, Daphoenositta chrysoptera 473
luminosa, Lepidopyga goudoti 263
luminosus, Qusicalus lugubris 774
lumsdeni, Babax waddelli 612
lunata, Sterna 151
lunatipectus, Microcerculus ustulatus 642
lunatus, Melithreptus 435
lunatus, Serilophus 336
Lunda 155
lundazi, Francolinus swainsonii 53

lungae, Colius striatus 278
lungchowensis, Sphenocercus sphenurus 174
lunigera, Tangara parzudakii 811
lunulata, Anthochaera 437
lunulata, Galloperdix 57
lunulata, Zoothera 662
lunulatus, Anisognathus igniventris 808
lunulatus, Bolbopsittacus 205
lunulatus, Garrulax 614
lunulatus, Gymnopithys 393
lurida, Ninox boobook 234
luridus, Pardirallus sanguinolentus 125
lurio, Cisticola aberrans 552
luristanica, Luscinia svecica 676
LUROCALIS 238
luschi, Myiopsitta monachus 199
LUSCINIA 675
luscinia, Luscinia 677
luscinia, Microcerculus marginatus 642
luscinia, Upcerthia certhioides 403
luscinioides, Locustella 583
luscinius, Acrocephalus 584
lusitanicus, Garrulus glandarius 508
lusitanius, Larus argentatus 147
lusituensis, Vidua funerea 737
lutea, Leiothrix 616
lutea, Manorina flavigula 434
lutea, Motacilla flava 739
lutea, Piranga flava 818
lutea, Sicalis 790
luteicapilla, Euphonia 821
luteifrons, Hylophilus ochraceiceps 486
luteifrons, Nigrita 726
luteirostris, Zosterops 628
luteiventris, Lophotriccus pileatus 361
luteiventris, Myiodynastes 375
luteiventris, Myiozetetes 374
luteiventris, Sicalis luteola 791
luteocephala, Sicalis 790
luteola, Certhidea olivacea 796
luteola, Coereba flaveola 776
luteola, Leiothrix lutea 616
luteola, Sicalis 791
luteolus, Ploceus 720
luteolus, Pycnonotus 568
luteoschistaceus, Accipiter 106
luteoventris, Bradypterus 581
luteovirens, Ptilinopus 177
luteoviridis, Basileuterus 766
luteoviridis, Pselliophorus 801
lutescens, Anthus 745
lutescens, Chlorothraupis carmioli 820
lutescens, Hylophilus ochraceiceps 486
lutescens, Macronous gularis 608
lutescens, Mionectes oleagineus 357
lutescens, Trogon mexicanus 280
lutescens, Vermivora celata 759
lutetiae, Coeligena 271
luteus, Colaptes auratus 324
luteus, Orthotomus sutorius 564
luteus, Passer 717
luteus, Zosterops 628
lutleyi, Tangara cyanotis 812
lutosa, Caracara 93
luxuosus, Cyanocorax yncas 506
luzonensis, Arachnothera clarae 714
luzonensis, Phylloscopus cebuensis 592
luzonica, Anas 66
luzonica, Gallicolumba 170
luzonica, Grus antigone 128
luzonicus, Zosterops nigrorum 626
luzoniense, Dicaeum ignipectum 703
luzoniensis, Copsychus 681
luzoniensis, Ficedula hyperythra 694
luzoniensis, Harpactes ardens 279
luzoniensis, Loxia curvirostra 756
lwenarum, Anthus hoeschi 742
lwenarum, Mirafra rufocinnamomea 541
lwenarum, Serinus atrogularis 747

maculifrons, Hemispingus superciliaris 804
maculifrons, Veniliornis 323
maculipectus, Garrulax ocellatus 614
maculipectus, Phacellodomus striaticollis 412
maculipectus, Phapitreron amethystinus 172
maculipectus, Rhipidura 495
maculipectus, Thryothorus 637
maculipennis, Larus 148
maculipennis, Phylloscopus 590
maculipennis, Pygiptila stellaris 384
maculirostris, Cinclodes antarcticus 404
maculirostris, Muscisaxicola 370
maculirostris, Selenidera 303
maculirostris, Turdus nudigenis 671
maculosa, Anas fulvigula 66
maculosa, Campethera 316
maculosa, Columba 159
maculosa, Lalage 471
maculosa, Nothura 34
maculosa, Prinia 559
maculosa, Tyto capensis 219
maculosus, Ailuroedus crassirostris 427
maculosus, Burhinus capensis 131
maculosus, Caprimulgus 242
maculosus, Nyctibius 238
maculosus, Psarocolius decumanus 768
maculosus, Turnix 129
mada, Gymnophaps 180
mada, Prioniturus 191
madagarensis, Margaroperdix 53
madagascariensis, Accipiter 107
madagascariensis, Alectroenas 178
madagascariensis, Anastomus lamelligerus 81
madagascariensis, Ardea purpurea 87
madagascariensis, Asio 236
madagascariensis, Aviceda 98
madagascariensis, Bernieria 588
madagascariensis, Calicalicus 461
madagascariensis, Caprimulgus 244
madagascariensis, Ceyx 290
madagascariensis, Foudia 724
madagascariensis, Hypsipetes 575
madagascariensis, Numenius 141
madagascariensis, Otus 222
madagascariensis, Oxylabes 588
madagascariensis, Phedina borbonica 532
madagascariensis, Porphyrio porphyrio 125
madagascariensis, Rallus 121
madagascarinus, Cyanolanius 462
MADANGA 631
madaraszi, Carduelis chloris 749
madaraszi, Colluricincla megarhyncha 489
madaraszi, Eremopterix leucotis 550
madaraszi, Neafrapus boehmi 250
madaraszi, Psittacella 190
madaraszi, Xanthotis flaviventer 431
madeira, Pterodroma 74
madeirae, Lepidocolaptes albolineatus 425
madeirae, Tachyphonus cristatus 805
madeirensis, Anthus berthelotii 744
madeirensis, Regulus ignicapilla 632
madens, Falco peregrinus 97
maderaspatanus, Oriolus xanthornus 488
maderaspatanus, Zosterops 630
maderaspatensis, Motacilla 741
maderaspatensis, Pomatorhinus horsfieldii 603
maderensis, Columba palumbus 158
maderensis, Fringilla coelebs 746
madoci, Monticola solitarius 688
madrensis, Accipiter striatus 108
madrensis, Columba flavirostris 159
madzoedi, Terpsiphone paradisi 499
maeandrina, Colluricincla megarhyncha 489
maesi, Garrulax 613
mafalu, Malurus alboscapulatus 429
maforense, Dicaeum geelvinkianum 702
maforensis, Coracina lineata 467
maforensis, Leptocoma sericea 708
maforensis, Macropygia amboinensis 162

maforensis, Phylloscopus poliocephalus 592
mafulu, Coturnix ypsilophora 54
magdalenae, Malacoptila panamensis 333
magdalenae, Microbates cinereiventris 643
magdalenae, Pachyramphus cinnamomeus 345
magdalenae, Passerculus sandwichensis 784
magdalenae, Rallus longirostris 120
magdalenae, Thamnophilus nigriceps 381
magdalenae, Thripadectes virgaticeps 416
magdalenae, Thryothorus felix 637
magdalenensis, Thryomanes bewickii 636
magellani, Pelecanoides 78
magellanica, Carduelis 750
magellanica, Gallinago paraguaiae 140
magellanicus, Bubo virginianus 225
magellanicus, Campephilus 327
magellanicus, Phalacrocorax 92
magellanicus, Scytalopus 397
magellanicus, Spheniscus 71
magellanicus, Turdus falcklandii 670
magentae, Pterodroma 74
magicus, Cynanthus latirostris 262
magicus, Otus 222
magister, Myiarchus tyrannulus 378
magister, Vireo 485
magna, Acanthornis 444
magna, Alectoris 49
magna, Amazona ochrocephala 204
magna, Aplonis 651
magna, Arachnothera 714
magna, Galerida cristata 548
magna, Luscinia svecica 676
magna, Macropygia 162
magna, Sitta 647
magna, Sturnella 775
magnifica, Aethopyga siparaja 713
magnifica, Megalaima virens 304
magnificens, Fregata 89
magnificus, Calyptorhynchus 183
magnificus, Diphyllodes 517
magnificus, Gorsachius 85
magnificus, Lamprotornis regius 657
magnificus, Lophornis 260
magnificus, Ptilinopus 175
magnificus, Ptiloris 517
magniplumis, Buteo magnirostris 110
magnirostra, Sericornis 446
magnirostrata, Cyanomitra cyanolaema 706
magnirostre, Malacopteron 601
magnirostris, Amaurornis olivacea 123
magnirostris, Aviceda jerdoni 98
magnirostris, Burhinus 131
magnirostris, Buteo 110
magnirostris, Capsiempis flaveola 354
magnirostris, Cardinalis cardinalis 823
magnirostris, Chlorothraupis carmioli 820
magnirostris, Coereba flaveola 776
magnirostris, Coracina atriceps 466
magnirostris, Cuculus solitarius 208
magnirostris, Cyornis banyumas 696
magnirostris, Daphoenositta chrysoptera 473
magnirostris, Diuca speculifera 789
magnirostris, Eophona personata 757
magnirostris, Eopsaltria australis 519
magnirostris, Esacus 131
magnirostris, Galerida 548
magnirostris, Geospiza 795
magnirostris, Gerygone 448
magnirostris, Hippolais languida 587
magnirostris, Melithreptus brevirostris 435
magnirostris, Microscelis amaurotis 575
magnirostris, Mimus gilvus 649
magnirostris, Myiarchus 378
magnirostris, Oriolus sagittatus 487
magnirostris, Oryzoborus maximiliani 793
magnirostris, Paradoxornis ruficeps 624
magnirostris, Phylloscopus 591
magnirostris, Pipilo maculatus 797
magnirostris, Psittacula eupatria 193

magnirostris, Ramphocelus carbo 806
magnirostris, Sasia abnormis 313
magnirostris, Strepera graculina 463
magnirostris, Treron capellei 173
magnirostris, Urocissa erythrorhyncha 509
magnirostris, Zosterops novaeguineae 628
magnistriata, Eudromia elegans 34
magnoides, Saltator maximus 824
magnolia, Dendroica 761
magnum, Malacopteron 601
magnus, Aegithalos caudatus 539
magnus, Aegolius funereus 233
magnus, Gampsonyx swainsonii 99
magnus, Monticola solitarius 688
magnus, Piculus flavigula 323
magnus, Vireo gundlachii 483
magrathi, Troglodytes troglodytes 639
maguari, Ciconia 81
maguirei, Myioborus brunniceps 765
mahali, Plocepasser 715
maharao, Francolinus coqui 50
mahendrae, Prinia sylvatica 558
mahrattarum, Parus major 524
mahrattensis, Acridotheres fuscus 654
mahrattensis, Caprimulgus 243
mahrattensis, Dendrocopos 319
maillardi, Circus 103
mailliardi, Melospiza melodia 782
mailliardorum, Agelaius phoeniceus 774
maingayi, Strix leptogrammica 228
maior, Lagopus lagopus 47
maior, Polioptila plumbea 644
maja, Lonchura 736
major, Aegithalos caudatus 538
major, Aegotheles cristatus 246
major, Alcippe rufogularis 619
major, Brachypteryx 673
major, Bradypterus 581
major, Campylorhamphus trochilirostris 426
major, Carduelis carduelis 751
major, Centropus viridis 216
major, Cettia 579
major, Chloropeta natalensis 587
major, Cissopis leverianus 803
major, Cisticola textrix 555
major, Crotophaga 217
major, Cypseloides 246
major, Dendrocopos 320
major, Diglossa 817
major, Eurypyga helias 117
major, Garrulax mitratus 614
major, Halcyon coromanda 286
major, Heliolais erythropterus 560
major, Heliothryx barroti 275
major, Hemiprocne comata 254
major, Laniarius aethiopicus 460
major, Leptopoecile sophiae 539
major, Lerwa lerwa 48
major, Myzomela rubratra 442
major, Nectarinia famosa 708
major, Neopsittacus musschenbroekii 187
major, Nothura maculosa 34
major, Numida meleagris 40
major, Oreopsittacus arfaki 187
major, Pachycephala homeyeri 475
major, Pachyramphus 345
major, Parus 524
major, Penelope superciliaris 38
major, Phaethornis bourcieri 256
major, Platalea leucorodia 83
major, Podiceps 80
major, Psittacella madaraszi 190
major, Psittacula alexandri 193
major, Ptilopachus petrosus 50
major, Pyrocephalus rubinus 369
major, Quiscalus 775
major, Rostrhamus sociabilis 100
major, Scapaneus leucopogon 327
major, Schiffornis 343

melanotos, Calidris 143
melanotos, Pica pica 511
melanotos, Sarkidiornis 64
Melanotrochilus 258
melanotus, Climacteris picumnus 428
melanotus, Malurus splendens 429
melanotus, Porphyrio porphyrio 125
melanotus, Sarcops calvus 653
melanotus, Turnix maculosus 129
melanoxantha, Hemithraupis flavicollis 815
melanoxantha, Phainoptila 523
melanoxanthum, Dicaeum 700
melanoxanthus, Ploceus nigricollis 720
melanozanthos, Mycerobas 757
melanterus, Psarocolius decumanus 768
melanura, Anthornis 435
melanura, Cercomela 687
melanura, Chilia 403
melanura, Euphonia laniirostris 821
melanura, Myiagra caledonica 503
melanura, Myrmeciza atrothorax 392
melanura, Pachycephala 477
melanura, Polioptila 643
melanura, Pyrrhura 199
melanuroides, Limosa limosa 140
melanurus, Centropus phasianinus 215
melanurus, Ceyx 290
melanurus, Cisticola 555
melanurus, Climacteris 428
melanurus, Copsychus malabaricus 680
melanurus, Himantopus himantopus 133
melanurus, Myophonus 659
melanurus, Myrmoborus 390
melanurus, Passer 717
melanurus, Pomatorhinus horsfieldii 602
melanurus, Psaltriparus minimus 539
melanurus, Ramphocaenus 643
melanurus, Taraba major 380
melanurus, Trogon 281
melanurus, Zosterops palpebrosus 625
melas, Capito maculicoronatus 303
melas, Coracina 469
melaschistos, Accipiter nisus 107
melaschistos, Coracina 470
melasmenus, Orthonyx spaldingii 450
melba, Pytilia 727
melba, Tachymarptis 253
melbina, Pseudhirundo griseopyga 532
meleagrides, Agelastes 40
MELEAGRIS 44
meleagris, Numida 40
MELIARCHUS 438
MELICHNEUTES 311
melicus, Saltator striatipectus 825
MELIDECTES 438
MELIDORA 284
MELIERAX 104
meligerus, Troglodytes troglodytes 639
MELIGNOMON 310
MELILESTES 439
melindae, Anthus 744
MELIPHAGA 433
meliphilus, Indicator 310
meliphonus, Catherpes mexicanus 635
MELIPOTES 437
MELITHREPTUS 435
MELITOGRAIS 436
melitophrys, Vireolanius 482
mellea, Zonotrichia capensis 783
melleri, Anas 66
melleus, Hylophilus flavipes 486
melli, Garrulax monileger 612
melli, Tyto capensis 219
melli, Zoothera citrina 660
melliana, Chloropsis hardwickei 699
mellianus, Oriolus 489
MELLISUGA 276
mellisugus, Chlorostilbon 261
mellitus, Campylopterus hemileucurus 258

mellivora, Florisuga 258
melloni, Aratinga astec 197
mellori, Corvus 514
mellori, Gallirallus philippensis 120
mellori, Meliphaga penicillata 433
mellori, Sericornis frontalis 446
MELOCICHLA 582
meloda, Zenaida 166
melodia, Melospiza 781
melodus, Charadrius 135
melodus, Dactylortyx thoracicus 43
melophilus, Erithacus rubecula 675
MELOPHUS 776
melopogenys, Treron fulvicollis 172
MELOPSITTACUS 190
MELOPYRRHA 794
meloryphus, Euscarthmus 355
MELOSPIZA 781
MELOZONE 798
melpoda, Estrilda 730
melpomene, Catharus aurantiirostris 664
meltoni, Lichenostomus melanops 432
melvillensis, Carbo melanoleucos 91
melvillensis, Chalcophaps indica 164
melvillensis, Chlamydera nuchalis 428
melvillensis, Coracina tenuirostris 469
melvillensis, Ducula bicolor 180
melvillensis, Lichenostomus flavescens 433
melvillensis, Lichmera indistincta 439
melvillensis, Manorina flavigula 434
melvillensis, Melanodryas cucullata 520
melvillensis, Mirafra javanica 540
melvillensis, Pandion haliaetus 98
melvillensis, Pardalotus striatus 444
melvillensis, Philemon argenticeps 437
melvillensis, Tyto novaehollandiae 218
membranaceus, Malacorhynchus 65
memnon, Caprimulgus indicus 242
menachensis, Serinus 748
menachensis, Turdus 667
menagei, Batrachostomus septimus 237
menagei, Dendrocopos maculatus 318
menagei, Dicrurus hottentottus 492
menagei, Gallicolumba 170
menaiensis, Zosterops maderaspatanus 630
menawa, Ptilorrhoa leucosticta 453
menbeki, Centropus 215
menckei, Monarcha 502
mendanae, Acrocephalus caffer 585
mendanae, Rhyticeros plicatus 300
mendeni, Columba vitiensis 159
mendeni, Hemiprocne longipennis 254
mendeni, Napothera epilepidota 605
mendeni, Otus magicus 222
mendeni, Stachyris melanothorax 608
mendeni, Zoothera erythronota 660
mendiculus, Spheniscus 71
mendosus, Serinus gularis 748
mendozae, Nothoprocta pentlandii 33
mendozae, Pomarea 500
mendozae, Sicalis olivascens 790
meneliki, Oriolus monacha 488
menetriesi, Buteo buteo 111
menetriesii, Myrmotherula 386
mengeli, Caprimulgus sericocaudatus 241
mengliensis, Garrulax canorus 614
mengtszensis, Corvus macrorhynchos 514
meninting, Alcedo 292
mennelli, Serinus 748
menstruus, Pionus 202
mentalis, Artamus leucorynchus 464
mentalis, Catharus fuscater 664
mentalis, Celeus loricatus 325
mentalis, Certhidea olivacea 796
mentalis, Cracticus 463
mentalis, Diglossa caerulescens 817
mentalis, Dysithamnus 383
mentalis, Galbula tombacea 331
mentalis, Melocichla 582

mentalis, Merops muelleri 295
mentalis, Muscisaxicola maclovianus 370
mentalis, Pachycephala 476
mentalis, Picus 327
mentalis, Pipra 343
mentalis, Platysteira peltata 455
mentalis, Ploceus bicolor 723
mentalis, Prionops rufiventris 456
mentalis, Syndactyla subalaris 415
mentalis, Turdoides aylmeri 610
mentalis, Xiphorhynchus flavigaster 424
mentawi, Oriolus xanthonotus 487
mentawi, Otus 220
mentor, Andropadus importunus 570
mentoris, Zosterops chloris 626
MENURA 426
menzbieri, Anthus gustavi 743
menzbieri, Buteo lagopus 112
menzbieri, Limosa lapponica 140
menzbieri, Phylloscopus collybita 589
menzbieri, Remiz pendulinus 530
mercedesfosterae, Mionectes macconnelli 357
mercenaria, Amazona 204
mercierii, Ptilinopus 176
mercuri, Sterna virgata 150
MERGANETTA 64
merganser, Mergus 70
MERGELLUS 69
MERGUS 70
meridae, Acestrura heliodor 277
meridae, Anthus bogotensis 745
meridae, Atlapetes albofrenatus 800
meridae, Cistothorus 635
meridae, Eriocnemis luciani 272
meridae, Mionectes olivaceus 357
meridae, Piculus rivolii 324
meridae, Pseudocolaptes boissonneautii 414
meridana, Cyanolyca armillata 505
meridana, Henicorhina leucophrys 641
meridana, Ochthoeca diadema 373
meridana, Synallaxis unirufa 407
meridanus, Basileuterus tristriatus 767
meridanus, Scytalopus 396
meridensis, Notiochelidon murina 534
meridensis, Otus albogularis 224
meridensis, Pheucticus aureoventris 823
meridensis, Piculus rubiginosus 323
meridensis, Thlypopsis fulviceps 804
meridianus, Dumetella carolinensis 648
meridionale, Delichon urbicum 536
meridionalis, Acrocephalus stentoreus 584
meridionalis, Amadina fasciata 727
meridionalis, Bubo bubo 225
meridionalis, Buteogallus 109
meridionalis, Caprimulgus europaeus 242
meridionalis, Certhia discolor 648
meridionalis, Chaetura 351
meridionalis, Cormobates placens 428
meridionalis, Dendrocincla homochroa 420
meridionalis, Dendropicos goertae 317
meridionalis, Eumyias panayensis 695
meridionalis, Eurypyga helias 117
meridionalis, Galerida cristata 547
meridionalis, Gallinula chloropus 126
meridionalis, Garrulax jerdoni 615
meridionalis, Gypaetus barbatus 101
meridionalis, Kaupifalco monogrammicus 108
meridionalis, Lanius 480
meridionalis, Loxia curvirostra 756
meridionalis, Merops pusillus 295
meridionalis, Microhierax erythrogenys 94
meridionalis, Myrmotherula ornata 385
meridionalis, Myza celebensis 439
meridionalis, Nestor 181
meridionalis, Otus flammeolus 222
meridionalis, Pachycephala sulfuriventer 475
meridionalis, Pachyramphus versicolor 344
meridionalis, Pitohui kirhocephalus 490
meridionalis, Ptilinopus fischeri 174

minuscula, Turdus albicollis 672
minussensis, Sterna hirundo 150
minuta, Aethopyga bella 712
minuta, Bocagia 458
minuta, Calidris 143
minuta, Collocalia esculenta 247
minuta, Columbina 167
minuta, Euphonia 822
minuta, Ficedula tricolor 695
minuta, Piaya 214
minuta, Ptilocichla mindanensis 604
minuta, Sporophila 793
minutilla, Calidris 143
minutillus, Chrysococcyx 210
minutissimum, Glaucidium 231
minutissimus, Picumnus 312
minutus, Anous 152
minutus, Anthoscopus 530
minutus, Chlorostilbon aureoventris 261
minutus, Corvus palmarum 513
minutus, Ixobrychus 85
minutus, Larus 148
minutus, Numenius 141
minutus, Xenops 419
minutus, Zosterops 628
minyanyae, Mirafra angolensis 541
minythomelas, Pericrocotus flammeus 473
miombensis, Nilaus afer 461
MIONECTES 357
miosnomensis, Pachycephala simplex 476
miquelii, Ptilinopus rivoli 176
mira, Pomarea mendozae 500
mira, Porzana pusilla 123
mira, Scolopax 139
mirabilis, Ammodramus maritimus 784
mirabilis, Catharus fuscater 664
mirabilis, Copsychus malabaricus 680
mirabilis, Cyanolyca 505
mirabilis, Dicrurus balicassius 492
mirabilis, Eriocnemis 272
mirabilis, Loddigesia 275
MIRAFRA 540
miranda, Daphoenositta 474
mirandae, Hemitriccus 360
mirandae, Philydor guttulatus 415
mirandae, Vireo leucophrys 484
mirandollei, Micrastur 94
mirandus, Basilornis 653
mirei, Ammomanes deserti 545
mirifica, Lewinia 121
mirus, Chondrohierax uncinatus 99
mirus, Otus 221
mirus, Rhynchocyclus olivaceus 363
miserabilis, Myiornis ecaudatus 361
misimae, Collocalia esculenta 248
misoliensis, Aepypodius arfakianus 36
misoliensis, Ailuroedus crassirostris 427
misoliensis, Colluricincla megarhyncha 489
misoriense, Dicaeum geelvinkianum 702
misoriensis, Accipiter novaehollandiae 105
misoriensis, Chrysococcyx minutillus 210
misoriensis, Micropsitta geelvinkiana 182
misoriensis, Phylloscopus poliocephalus 592
mississippiensis, Ictinia 100
mista, Ducula aenea 178
misulae, Accipiter novaehollandiae 105
mitchelli, Dendrocopos canicapillus 318
mitchellii, Calliphlox 276
mitchellii, Phegornis 137
mitchellii, Trichoglossus forsteni 185
mitrata, Aratinga 197
mitrata, Cyanolyca cucullata 505
mitratus, Garrulax 614
mitratus, Numida meleagris 40
mitratus, Parus cristatus 529
MITREPHANES 366
MITROSPINGUS 803
MITU 39
mitu, Mitu 39

mituensis, Poecilotriccus latirostris 362
mixta, Batis 454
mixta, Cisticola exilis 556
mixta, Heterophasia annectans 621
mixta, Lalage maculosa 471
mixta, Meliphaga notata 434
mixtus, Alcedo azurea 291
mixtus, Batrachostomus poliolophus 237
mixtus, Melidectes torquatus 438
mixtus, Picoides 320
mixtus, Poecilotriccus latirostris 362
mixtus, Tolmomyias sulphurescens 363
miyakoensis, Todiramphus 289
miza, Treron vernans 172
mizorhina, Halcyon coromanda 286
mjobergi, Anthreptes malacensis 704
mjobergi, Ficedula hyperythra 694
mlokosiewiczi, Lyrurus 46
mnionophilus, Premnoplex brunnescens 413
MNIOTILTA 762
moabiticus, Passer 716
moae, Ninox boobook 234
mochae, Scelorchilus rubecula 395
mocinno, Pharomachrus 282
moco, Anthus similis 742
mocoa, Aglaiocercus kingi 275
mocquerysi, Colinus cristatus 42
moderatus, Automolus rubiginosus 417
moderatus, Serilophus lunatus 336
moderatus, Thripadectes holostictus 416
modesta, Acanthiza nana 449
modesta, Arachnothera 714
modesta, Ardea alba 87
modesta, Asthenes 406
modesta, Cercomela familiaris 686
modesta, Cittura cyanotis 285
modesta, Elaenia albiceps 351
modesta, Erythrura trichroa 733
modesta, Galerida 548
modesta, Geothlypis trichas 763
modesta, Gerygone igata 447
modesta, Lalage maculosa 471
modesta, Myrmothera campanisona 401
modesta, Neochmia 732
modesta, Pachycephala 475
modesta, Progne 533
modesta, Psittacula 190
modesta, Psittacula longicauda 194
modesta, Sarothrura lugens 117
modesta, Turacoena 163
modestum, Dicaeum agile 700
modestus, Amytornis textilis 430
modestus, Charadrius 137
modestus, Cisticola lateralis 551
modestus, Crypturellus soui 31
modestus, Dicrurus 491
modestus, Gallirallus 120
modestus, Larus 146
modestus, Melaenornis pallidus 689
modestus, Myiobius atricaudus 365
modestus, Otus sunia 222
modestus, Ramsayornis 440
modestus, Sublegatus 358
modestus, Sylviparus 529
modestus, Thryothorus 638
modestus, Treron seimundi 174
modestus, Veniliornis passerinus 322
modestus, Vireo 483
modestus, Zosterops 630
modicus, Onychognathus neumanni 658
modiglianii, Macropygia emiliana 163
modiglianii, Pericrocotus flammeus 473
modularis, Prunella 738
modulator, Cyphorhinus arada 642
modulator, Mimus saturninus 649
MODULATRIX 625
moebii, Cinnyris notatus 712
moesta, Amaurospiza 794
moesta, Lalage atrovirens 470

moesta, Oenanthe 684
moesta, Synallaxis 409
moestissima, Fregetta 77
moffitti, Branta canadensis 63
moffitti, Lophura leucomelanos 58
mogenseni, Piaya cayana 214
mohavensis, Parus inornatus 529
mohelicus, Cinnyris humbloti 712
moheliensis, Coracina cinerea 468
moheliensis, Hypsipetes parvirostris 575
moheliensis, Nesillas typica 582
moheliensis, Otus 222
moheliensis, Turdus bewsheri 667
MOHO 431
MOHOUA 450
mohun, Mulleripicus pulverulentus 331
molesworthi, Tragopan blythii 57
molinae, Pyrrhura 198
molinensis, Certhia americana 647
molitor, Batis 455
molleri, Prinia 559
mollis, Cacatua leadbeateri 184
mollis, Lanius excubitor 480
mollis, Pterodroma 74
mollissima, Chamaea 398
mollissima, Somateria 69
mollissima, Zoothera 661
molochinus, Myioborus miniatus 765
molochinus, Ramphocelus dimidiatus 806
moloneyana, Illadopsis fulvescens 601
MOLOTHRUS 773
moltchanovi, Parus ater 526
moltonii, Streptopelia turtur 160
moltonii, Sylvia cantillans 598
molucca, Lonchura 734
molucca, Threskiornis 82
moluccana, Amaurornis olivacea 123
moluccanus, Trichoglossus haematodus 185
moluccarum, Aerodramus vanikorensis 249
moluccarum, Butorides striata 86
moluccensis, Cacatua 184
moluccensis, Chloropsis cochinchinensis 699
moluccensis, Dendrocopos 318
moluccensis, Falco 95
moluccensis, Philemon 436
moluccensis, Pitta 339
molunduensis, Francolinus bicalcaratus 52
molybdophanes, Corythaixoides concolor 206
molybdophanes, Ptilogonys cinereus 523
molybdophanes, Struthio camelus 34
mombassica, Campethera 316
mombassicus, Colius striatus 278
momboloensis, Francolinus levaillantii 51
momiyamae, Strix uralensis 229
momota, Momotus 294
momotula, Hylomanes 293
MOMOTUS 293
momus, Sylvia melanocephala 598
monacha, Coracina tenuirostris 468
monacha, Crateroscelis murina 445
monacha, Ducula oceanica 179
monacha, Grus 128
monacha, Oenanthe 686
monacha, Oriolus 488
monacha, Pachycephala 478
monacha, Ploceus pelzelni 720
monacha, Ptilinopus 176
MONACHELLA 520
monachus, Actenoides 284
monachus, Aegypius 102
monachus, Artamus 464
monachus, Centropus 216
monachus, Garrulax chinensis 613
monachus, Leptodon cayanensis 99
monachus, Myiopsitta 199
monachus, Necrosyrtes 101
monachus, Philemon corniculatus 437
monachus, Pseudoalcippe abyssinica 602
monachus, Tyrannus savana 376

nigrorum, Tanygnathus lucionensis 192
nigrorum, Todiramphus winchelli 287
nigrorum, Turdus poliocephalus 668
nigrorum, Turnix sylvaticus 129
nigrorum, Zosterops 626
nigrosquamatus, Francolinus clappertoni 52
nigrostellatus, Pomatorhinus ruficollis 603
nigrostriatus, Cisticola juncidis 555
nigrotecta, Poephila cincta 733
nigrotectus, Monarcha browni 502
nigroventris, Euplectes 725
nigroviridis, Colaptes melanochloros 324
nigroviridis, Tangara 812
nihonensis, Charadrius alexandrinus 136
nijoi, Acrocephalus luscinius 585
nikolskii, Bubo bubo 225
nikolskii, Strix uralensis 229
NILAUS 461
nilesi, Orthotomus derbianus 564
nilghiriensis, Anthus 744
nilgiriensis, Saxicola caprata 684
nilgiriensis, Zosterops palpebrosus 625
nilotica, Bostrychia hagedash 82
nilotica, Sterna 149
niloticus, Cisticola erythrops 551
niloticus, Lanius senator 481
niloticus, Passer domesticus 716
niloticus, Phoeniculus purpureus 297
NILTAVA 697
nimbae, Muscicapa olivascens 692
nimia, Anas crecca 68
NINOX 233
nipalense, Delichon 536
nipalensis, Aceros 300
nipalensis, Actinodura 618
nipalensis, Aethopyga 713
nipalensis, Alcippe 620
nipalensis, Apus 254
nipalensis, Aquila 113
nipalensis, Brachypteryx leucophrys 673
nipalensis, Bubo 226
nipalensis, Carpodacus 753
nipalensis, Cecropis daurica 537
nipalensis, Certhia 648
nipalensis, Coracina macei 465
nipalensis, Cutia 616
nipalensis, Paradoxornis 623
nipalensis, Parus major 524
nipalensis, Pitta 337
nipalensis, Prunella collaris 738
nipalensis, Psittacula eupatria 193
nipalensis, Pucrasia macrolopha 57
nipalensis, Pyrrhula 756
nipalensis, Spizaetus 114
nipalensis, Treron curvirostra 172
nipalensis, Troglodytes troglodytes 639
nipalensis, Turdoides 609
nippon, Dendrocopos kizuki 318
nippon, Nipponia 82
nipponensis, Lymnocryptes minimus 139
NIPPONIA 82
nisicolor, Cuculus fugax 207
nisoides, Accipiter gularis 107
nisoria, Coracina tenuirostris 469
nisoria, Lonchura punctulata 735
nisoria, Sylvia 596
nisorius, Thryothorus pleurostictus 638
nisosimilis, Accipiter nisus 107
nisus, Accipiter 107
nitens, Chlorostilbon mellisugus 261
nitens, Collocalia esculenta 247
nitens, Columba janthina 158
nitens, Lamprotornis 656
nitens, Malimbus 723
nitens, Phainopepla 523
nitens, Psalidoprocne 531
nitens, Trochocercus 498
nitida, Aplonis metallica 651
nitida, Asturina 110

nitida, Euphonia chrysopasta 822
nitidifrons, Amazilia versicolor 265
nitidior, Basileuterus trifasciatus 767
nitidior, Capito auratus 304
nitidissima, Alectroenas 178
nitidissima, Chlorochrysa 809
nitidissimus, Tachyphonus luctuosus 805
nitidula, Lagonosticta 728
nitidula, Mandingoa 727
nitidum, Dicaeum 702
nitidus, Carpodectes 348
nitidus, Cyanerpes 813
nitidus, Monarcha chrysomela 502
nitidus, Orthotomus atrogularis 564
nitidus, Phylloscopus 591
nitidus, Vireo hypochryseus 484
nivalis, Montifringilla 718
nivalis, Phalacrocorax atriceps 92
nivalis, Plectrophenax 781
nivarius, Phrygilus unicolor 788
nivea, Chionarchus minor 131
nivea, Oenanthe oenanthe 685
nivea, Pagodroma 73
niveicapilla, Cossypha 678
niveicauda, Tyto alba 219
niveiceps, Colonia colonus 373
niveiceps, Monticola brevipes 688
niveiceps, Turdus poliocephalus 668
niveifrons, Aleadryas rufinucha 474
niveigularis, Platyrinchus mystaceus 364
niveigularis, Tyrannus 376
niveiventris, Rhipidura rufiventris 495
nivenorum, Emberiza tahapisi 778
niveogularis, Aegithalos 535
niveoguttatus, Hypargos 728
niveoventer, Amazilia edward 266
nivescens, Anthus similis 742
niveus, Eurocephalus anguitimens 479
niveus, Xolmis irupero 371
nivicola, Strix aluco 228
nivosa, Campethera 316
nivosus, Charadrius alexandrinus 136
nivosus, Otus scops 221
njikena, Aviceda subcristata 98
njombe, Cisticola 552
noanamae, Bucco 332
nobilior, Sturnus vulgaris 656
nobilis, Chamaeza 398
nobilis, Diopsittaca 196
nobilis, Francolinus 52
nobilis, Gallinago 140
nobilis, Lampornis amethystinus 268
nobilis, Lophura ignita 59
nobilis, Moho 431
nobilis, Myiodynastes maculatus 375
nobilis, Oreonympha 274
nobilis, Otidiphaps 171
nobilis, Rhea americana 35
nocticolor, Diglossa humeralis 816
noctipetens, Asio stygius 236
noctis, Loxigilla 795
noctitherus, Caprimulgus 241
noctivagus, Crypturellus 32
noctivagus, Lurocalis semitorquatus 238
noctivigulus, Caprimulgus rufus 241
noctua, Athene 232
noctuvigilis, Caprimulgus macrurus 243
noguchii, Sapheopipo 330
NOMONYX 70
NONNULA 333
nonnula, Estrilda 731
noomei, Andropadus importunus 569
noonaedanae, Aerodramus spodiopygius 248
nordmanni, Glareola 146
nordmanni, Lamprotornis chalybaeus 656
norfolkensis, Gallirallus philippensis 120
norkolkiensis, Todiramphus sanctus 289
normani, Artamus cinereus 464
normani, Cisticola juncidis 555

normani, Colluricincla megarhyncha 489
normani, Coracina larvata 466
normantoni, Acanthiza chrysorrhoa 449
normantoni, Cacatua sanguinea 183
normantoni, Mirafra javanica 540
normantoniensis, Melithreptus laetior 435
noronha, Zenaida auriculata 166
norrisae, Sylvia melanocephala 598
NORTHIELLA 189
northropi, Icterus dominicensis 770
nortoni, Siptornis striaticollis 413
notabilis, Anisognathus 809
notabilis, Campylorhamphus trochilirostris 426
notabilis, Glycifohia 439
notabilis, Lichmera 440
notabilis, Nestor 181
notaea, Hypocnemis cantator 390
notata, Campethera 316
notata, Carduelis 750
notata, Cercomacra nigrescens 389
notata, Gerygone chrysogaster 448
notata, Meliphaga 434
notatus, Bleda eximius 572
notatus, Chlorostilbon 261
notatus, Cinnyris 712
notatus, Coturnicops 118
notatus, Elanus 100
notatus, Francolinus afer 53
notatus, Mecocerculus leucophrys 353
notatus, Oriolus auratus 487
notatus, Rhinopomastus aterrimus 297
notatus, Xiphorhynchus obsoletus 423
NOTHARCHUS 332
NOTHOCERCUS 31
NOTHOCRAX 39
NOTHOPROCTA 33
NOTHURA 34
nothus, Tchagra senegalus 459
notia, Leptotila plumbeiceps 168
NOTIOCHELIDON 534
NOTIOMYSTIS 431
notius, Basileuterus coronatus 766
notius, Vireo plumbeus 483
notophila, Lissotis melanogaster 116
notosticta, Aimophila 787
nouhuysi, Melidectes 438
nouhuysi, Sericornis 446
novaanglica, Corvus tasmanicus 514
novacapitalis, Scytalopus 397
novaecaledoniae, Turnix varius 130
novaeguineae, Coturnix chinensis 54
novaeguineae, Dacelo 285
novaeguineae, Falco berigora 97
novaeguineae, Fulica atra 127
novaeguineae, Harpyopsis 112
novaeguineae, Mearnsia 250
novaeguineae, Orthonyx 450
novaeguineae, Paradisaea apoda 518
novaeguineae, Philemon buceroides 436
novaeguineae, Pitta sordida 339
novaeguineae, Toxorhamphus 452
novaeguineae, Zosterops 625
novaehiberniae, Todiramphus chloris 288
novaehibernicae, Pitta erythrogaster 338
novaehollandiae, Accipiter 105
novaehollandiae, Anhinga melanogaster 92
novaehollandiae, Cereopsis 62
novaehollandiae, Coracina 466
novaehollandiae, Dromaius 35
novaehollandiae, Egretta 88
novaehollandiae, Eudyptula minor 71
novaehollandiae, Larus 148
novaehollandiae, Menura 426
novaehollandiae, Phalacrocorax carbo 91
novaehollandiae, Phylidonyris 440
novaehollandiae, Recurvirostra 133
novaehollandiae, Scythrops 212
novaehollandiae, Tachybaptus 79
novaehollandiae, Tyto 218

novaehollandiae, Vanellus miles 134
novaenorciae, Lichenostomus leucotis 432
novaeseelandiae, Anthus 741
novaeseelandiae, Aythya 68
novaeseelandiae, Finschia 450
novaeseelandiae, Hemiphaga 180
novaeseelandiae, Ninox 234
novaeseelandiae, Prosthemadera 435
novaeseelandiae, Thinornis 137
novaesi, Chordeiles pusillus 239
novaesi, Philydor 416
novaesi, Zonotrichia capensis 782
novaezealandiae, Pelecanus conspicillatus 90
novaezelandiae, Falco 96
novaezelandiae, Apteryx australis 35
novaezelandiae, Coturnix 54
novaezelandiae, Cyanoramphus 188
novaezelandiae, Himantopus 133
novaezelandiae, Ixobrychus 85
novaolindae, Capito auratus 304
noveboracensis, Coturnicops 118
noveboracensis, Seiurus 763
noveboracensis, Vireo griseus 482
novicius, Chlorospingus ophthalmicus 817
novus, Machaerirhynchus flaviventer 461
novus, Pelecanoides georgicus 78
novus, Pericrocotus flammeus 473
novus, Rhagologus leucostigma 474
nuba, Neotis 115
nubica, Campethera 315
nubicoides, Merops nubicus 296
nubicola, Amblyornis macgregoriae 427
nubicola, Chaetura 247
nubicola, Glaucidium 230
nubicolus, Parus major 525
nubicus, Caprimulgus 243
nubicus, Lanius 481
nubicus, Merops 296
nubicus, Torgus tracheliotus 102
nubifuga, Tachymarptis melba 253
nubila, Collocalia esculenta 247
nubilosa, Prinia flavicans 559
nubilosa, Sterna fuscata 151
nubivagus, Lampornis viridipallens 268
nuchale, Hypogramma hypogrammicum 705
nuchalis, Campylorhynchus 633
nuchalis, Chlamydera 428
nuchalis, Cisticola robustus 554
nuchalis, Ducula aenea 178
nuchalis, Garrulax 613
nuchalis, Glareola 146
nuchalis, Grallaria 400
nuchalis, Megalaima oorti 305
nuchalis, Melidectes torquatus 438
nuchalis, Parus 524
nuchalis, Pomatorhinus schisticeps 603
nuchalis, Pteroglossus torquatus 302
nuchalis, Sphyrapicus 315
nuchalis, Strix woodfordi 229
NUCIFRAGA 511
nudiceps, Gymnocichla 390
nudiceps, Gyps tenuirostris 102
nudicollis, Procnias 348
nudifrons, Phimosus infuscatus 83
nudigenis, Turdus 671
nudigula, Pachycephala 478
nudipes, Hirundapus caudacutus 251
nudipes, Otus 224
nudirostris, Treron calvus 173
nuditarsus, Aerodramus 249
nugator, Myiarchus 378
nugax, Puffinus lherminieri 76
nukuhivae, Pomarea mendozae 500
NUMENIUS 141
numforana, Coracina tenuirostris 468
numforensis, Collocalia esculenta 247
NUMIDA 40
numida, Eudromia elegans 34
numidus, Dendrocopos major 320

nuna, Lesbia 273
nuntius, Orthotomus sericeus 564
nupera, Tangara gyrola 811
nupta, Lichmera indistincta 439
nuristani, Garrulax variegatus 613
nusae, Todiramphus chloris 288
nuttalli, Pica 511
nuttalli, Zonotrichia leucophrys 783
nuttallii, Phalaenoptilus 240
nuttallii, Picoides 321
nuttingi, Leptotila verreauxi 168
nuttingi, Myiarchus 378
nuttingi, Oryzoborus 793
nyansae, Campethera cailliautii 316
nyansae, Cisticola galactotes 553
nyansae, Estrilda astrild 730
nyansae, Platysteira cyanea 455
nyanzae, Francolinus afer 53
nyaritensis, Agelaius phoeniceus 774
nyasa, Cisticola erythrops 551
nyasae, Illadopsis pyrrhoptera 602
nyassae, Anthreptes longuemarei 704
nyassae, Anthus 743
nyassae, Bradypterus cinnamomeus 581
NYCTANASSA 86
NYCTEA 225
nycthemera, Lophura 58
NYCTIBIUS 238
NYCTICORAX 85
nycticorax, Nycticorax 85
NYCTICRYPHES 137
NYCTIDROMUS 240
nyctilampis, Pellorneum capistratum 600
nyctiphasma, Strix leptogrammica 228
NYCTIPHRYNUS 240
NYCTIPROGNE 239
NYCTYORNIS 295
nyeanus, Melanerpes superciliaris 315
nyika, Cisticola aberrans 552
nyikae, Mirafra africana 541
nyikae, Ploceus baglafecht 720
nyikae, Turdus olivaceus 667
nyikensis, Melaenornis fischeri 689
nyikensis, Nectarinia johnstoni 708
nymani, Myzomela eques 441
nympha, Pitta 339
nympha, Tanysiptera 285
NYMPHICUS 184
nyombensis, Kupeornis chapini 620
nyroca, Aythya 68
NYSTALUS 332

oahuensis, Arenaria interpres 142
oahuensis, Sterna fuscata 151
oatesi, Garrulus glandarius 509
oatesi, Lophura leucomelanos 58
oatesi, Niltava vivida 697
oatesi, Paradoxornis atrosuperciliaris 624
oatesi, Pitta 337
oatesi, Spelaeornis chocolatinus 606
oatleyi, Cercotrichas signata 679
oaxacae, Aphelocoma unicolor 507
oaxacae, Caprimulgus vociferus 241
oaxacae, Dendrortyx macroura 41
oaxacae, Eremophila alpestris 551
oaxacae, Pipilo maculatus 797
oaxacae, Thryothorus pleurostictus 638
oaxacae, Turdus assimilis 671
obbiensis, Spizocorys 547
obcura, Porzana pusilla 123
oberholseri, Collocalia esculenta 247
oberholseri, Empidonax 368
oberholseri, Eurystomus orientalis 283
oberholseri, Glaucidium palmarum 231
oberholseri, Hypothymis azurea 497
oberholseri, Myadestes occidentalis 663
oberholseri, Pseudocolaptes boissonneautii 414
oberholseri, Sitta carolinensis 646
oberholseri, Thryothorus ludovicianus 638

oberholseri, Toxostoma curvirostre 650
oberholseri, Treron fulvicollis 172
oberi, Dysithamnus mentalis 383
oberi, Icterus 770
oberi, Myiarchus 378
oberi, Pseudoscops clamator 236
oberlaenderi, Zoothera 661
oberon, Psophodes leucogaster 452
obfuscata, Ochthoeca rufipectoralis 373
obidensis, Dendrocincla merula 420
obidensis, Hemithraupis flavicollis 815
obidensis, Thamnomanes ardesiacus 384
obidensis, Thripophaga fusciceps 412
obiensis, Accipiter novaehollandiae 105
obiensis, Cacomantis variolosus 209
obiensis, Caprimulgus macrurus 243
obiensis, Coracina tenuirostris 468
obiensis, Ducula basilica 179
obiensis, Eos squamata 184
obiensis, Eumyias panayensis 695
obiensis, Geoffroyus geoffroyi 190
obiensis, Lycocorax pyrrhopterus 515
obiensis, Pachycephala mentalis 476
obiensis, Pitta erythrogaster 338
obiensis, Rhipidura rufiventris 494
obiensis, Tanysiptera galatea 284
obiensis, Turnix maculosus 129
obira, Otus magicus 222
objurgans, Prinia rufescens 557
objurgatus, Falco tinnunculus 95
oblectans, Garrulax sannio 615
oblita, Cettia flavolivacea 580
oblita, Coereba flaveola 776
oblita, Pitta erythrogaster 338
oblita, Rynchops nigra 153
oblitus, Myiophobus pulcher 359
oblitus, Passerculus sandwichensis 784
oblitus, Prionochilus maculatus 700
oblitus, Pycnonotus simplex 568
oblitus, Ramphastos tucanus 303
oblitus, Thryothorus pleurostictus 638
oblitus, Treron sieboldii 174
oblitus, Turdus assimilis 671
oblitus, Zosterops ugiensis 628
obrieni, Celeus spectabilis 326
obrieni, Strix nivicola 228
obscura, Aimophila ruficeps 787
obscura, Anthornis melanura 435
obscura, Aphelocoma californica 508
obscura, Aviceda subcristata 98
obscura, Bonasa umbellus 44
obscura, Chamaeza campanisona 398
obscura, Coeligena coeligena 270
obscura, Coereba flaveola 776
obscura, Colluricincla megarhyncha 489
obscura, Cyanomitra olivacea 706
obscura, Dendroica petechia 761
obscura, Diglossa cyanea 817
obscura, Elaenia 352
obscura, Grallaria rufula 400
obscura, Inezia subflava 358
obscura, Leptopoecile sophiae 539
obscura, Lichenostomus 431
obscura, Lonchura atricapilla 735
obscura, Luscinia 676
obscura, Manorina flavigula 435
obscura, Muscicapa adusta 692
obscura, Myrmotherula 384
obscura, Myzomela 441
obscura, Nesillas typica 582
obscura, Ninox scutulata 234
obscura, Penelope 38
obscura, Piaya cayana 214
obscura, Polioptila caerulea 643
obscura, Prinia bairdii 559
obscura, Prunella modularis 738
obscura, Psalidoprocne 532
obscura, Psophia viridis 127
obscura, Pyrrhura egregia 199

olivascens, Muscicapa 692
olivascens, Myiarchus tuberculifer 377
olivascens, Neocrex erythrops 124
olivascens, Oedistoma pygmaeum 451
olivascens, Pterocles exustus 156
olivascens, Saltator coerulescens 825
olivascens, Sclerurus caudacutus 418
olivascens, Sicalis 790
olivascens, Tinamus major 31
olivascentior, Amalocichla incerta 522
olivater, Turdus 670
olivea, Tesia 578
oliveri, Phalacrocorax punctatus 92
olivetorum, Hippolais 587
oliviae, Columba 158
olivicyanea, Thraupis cyanocephala 807
olivieri, Amaurornis 123
olivii, Turnix 130
olivii, Turnix castanotus 130
olivina, Elaenia pallatangae 352
olivina, Schistochlamys melanopis 802
olivinus, Cercococcyx 208
olivinus, Veniliornis passerinus 322
olmecorum, Euphonia affinis 820
olor, Cygnus 63
olrogi, Cinclodes 403
olrogi, Micrastur ruficollis 94
olsoni, Vireo pallens 483
olympicus, Cinclus cinclus 698
omalurus, Cisticola juncidis 555
omaruru, Mirafra africanoides 542
ombriosa, Coracina lineata 467
ombriosa, Fringilla coelebs 746
ombriosus, Parus caeruleus 526
omeiensis, Liocichla 616
omeiensis, Lophura nycthemera 58
omeiensis, Seicercus 593
omeiensis, Yuhina gularis 622
omiltemensis, Catharus frantzii 664
omiltemensis, Xiphocolaptes promeropirhynchus 421
omissa, Coeligena torquata 271
omissa, Emberiza godlewskii 777
omissa, Euphonia fulvicrissa 821
omissa, Foudia 724
omissa, Myrmotherula menetriesii 386
omissa, Synallaxis rutilans 409
omissus, Bubo bubo 226
omissus, Campylorhamphus trochilirostris 426
omissus, Copsychus malabaricus 681
omissus, Cyornis rufigastra 697
omissus, Dendrocopos canicapillus 318
omissus, Larus argentatus 147
omissus, Tiaris bicolor 795
omo, Cisticola robustus 554
omoensis, Anthus leucophrys 743
omoensis, Cercomela familiaris 686
omoensis, Cossypha albicapilla 678
omoensis, Emberiza affinis 779
omoensis, Mirafra rufocinnamomea 541
omoensis, Phylloscopus umbrovirens 589
omoensis, Turdoides leucopygia 611
omoensis, Zosterops abyssinicus 629
ONCOSTOMA 361
oneho, Anthornis melanura 435
oneilli, Nephelornis 805
onerosa, Gerygone magnirostris 448
onguati, Streptopelia capicola 161
onocrotalus, Pelecanus 89
onslowi, Phalacrocorax 92
onusta, Meleagris gallopavo 44
ONYCHOGNATHUS 658
ONYCHORHYNCHUS 364
ONYCHOSTRUTHUS 718
oocleptica, Aphelocoma californica 508
oorti, Clytomyias insignis 430
oorti, Megalaima 305
oorti, Sericornis nouhuysi 446
opaca, Aplonis 652

opaca, Iduna pallida 586
opaca, Strix albitarsis 229
opacus, Scytalopus canus 397
opertaneus, Catharus fuscater 664
ophryophanes, Falco rufigularis 96
OPHRYSIA 55
ophthalmica, Cacatua 184
ophthalmica, Sporophila corvina 792
ophthalmicus, Chlorospingus 817
ophthalmicus, Phylloscartes 356
opicus, Perisoreus infaustus 505
opistherythra, Rhipidura 496
opisthisus, Copsychus malabaricus 680
OPISTHOCOMUS 205
opisthocyanea, Hypothymis azurea 497
opistholeuca, Oenanthe picata 685
opisthomelas, Puffinus 76
opisthopelus, Copsychus malabaricus 680
OPISTHOPRORA 275
Opopsitta 204
OPORORNIS 763
oppenheimi, Sturnus vulgaris 656
optata, Pachycephala vitiensis 477
optatus, Cuculus saturatus 208
opthalmicus, Haematopus fuliginosus 132
opuntia, Amphispiza bilineata 786
orangensis, Serinus albogularis 748
oraria, Dendroica petechia 761
oratrix, Amazona 203
orbignyianus, Thinocorus 138
orbitalis, Phylloscartes 356
orbitalis, Sylvia conspicillata 598
orbitatus, Hemitriccus 360
orbygnesius, Bolborhynchus 200
orcesi, Pyrrhura 199
ORCHESTICUS 802
ordii, Notharchus 332
ordoscensis, Alectoris chukar 49
oreas, Geotrygon goldmani 169
oreas, Lessonia 369
oreas, Picathartes 522
oreganus, Junco hyemalis 783
oregonus, Pipilo maculatus 797
oreinus, Dendragapus obscurus 46
orenocensis, Certhiaxis cinnamomeus 411
orenocensis, Formicivora grisea 387
orenocensis, Knipolegus 369
orenocensis, Pyroderus scutatus 349
orenocensis, Saltator 825
orenocensis, Thalurania furcata 263
orenocensis, Veniliornis affinis 323
orenocensis, Xiphocolaptes promeropirhynchus 422
oreobates, Apus barbatus 253
oreobates, Cinclodes fuscus 403
oreobates, Cisticola lais 553
oreobates, Merops 295
oreobates, Saxicola torquatus 684
OREOCHARIS 452
oreodytes, Cisticola lais 553
OREOICA 490
OREOMANES 816
OREOMYSTIS 759
oreonesus, Pogoniulus bilineatus 308
OREONYMPHA 274
OREOPHASIS 39
oreophila, Catamenia homochroa 794
oreophila, Cettia vulcania 579
oreophila, Oenanthe deserti 685
OREOPHILAIS 559
oreophilos, Pluvialis apricaria 134
oreophilus, Buteo 112
oreophilus, Cacomantis variolosus 209
oreophilus, Cisticola tinniens 553
oreophilus, Zosterops novaeguineae 628
OREOPHOLUS 137
OREOPHYLAX 406
oreopola, Metallura tyrianthina 274
OREOPSAR 773

oreopsar, Agelaioides 773
OREOPSITTACUS 187
OREORNIS 433
OREORTYX 41
OREOSCOPTES 649
OREOSCOPUS 445
OREOSTRUTHUS 732
OREOTHRAUPIS 799
OREOTROCHILUS 270
oresbia, Tangara icterocephala 810
oreskios, Harpactes 279
oressinoma, Euphonia xanthogaster 822
orestera, Vermivora celata 759
orestera, Zonotrichia capensis 782
orestes, Lafresnaya lafresnayi 270
orestes, Myrmecocichla formicivora 687
oreum, Pellorneum ruficeps 599
oriantha, Zonotrichia leucophrys 783
oriens, Cercotrichas paena 679
orientale, Cinclidium frontale 682
orientale, Glaucidium passerinum 230
orientalis, Acrocephalus 584
orientalis, Aerodramus 249
orientalis, Ammomanes deserti 545
orientalis, Anthreptes 704
orientalis, Aquila nipalensis 113
orientalis, Arborophila 56
orientalis, Athene noctua 232
orientalis, Basileuterus coronatus 766
orientalis, Batis 455
orientalis, Branta bernicla 63
orientalis, Calandrella brachydactyla 546
orientalis, Certhia familiaris 647
orientalis, Cettia cetti 580
orientalis, Chlamydera nuchalis 428
orientalis, Corvus corone 513
orientalis, Corydon sumatranus 336
orientalis, Emberiza cabanisi 779
orientalis, Eubucco bourcierii 304
orientalis, Eudynamys scolopaceus 211
orientalis, Eurystomus 283
orientalis, Gallicolumba rufigula 170
orientalis, Gallinula chloropus 126
orientalis, Geoffroyus geoffroyi 191
orientalis, Halcyon albiventris 287
orientalis, Hypocnemoides maculicauda 390
orientalis, Ketupa zeylonensis 227
orientalis, Lonchura cantans 734
orientalis, Lophophorus sclateri 58
orientalis, Macropygia ruficeps 163
orientalis, Mayrornis lessoni 500
orientalis, Megalaima australis 306
orientalis, Melanocharis longicauda 451
orientalis, Meliphaga 433
orientalis, Melocichla mentalis 582
orientalis, Mergus merganser 70
orientalis, Merops 296
orientalis, Minla cyanouroptera 618
orientalis, Mino anais 653
orientalis, Myadestes occidentalis 663
orientalis, Myioparus plumbeus 693
orientalis, Neomixis tenella 563
orientalis, Numenius arquata 141
orientalis, Parus caeruleus 525
orientalis, Pernis ptilorhynchus 99
orientalis, Phylloscopus 590
orientalis, Pipilo maculatus 797
orientalis, Pogonocichla stellata 674
orientalis, Pomatorhinus ferruginosus 603
orientalis, Psalidoprocne pristoptera 531
orientalis, Psephotus varius 189
orientalis, Pseudocolaptes boissonneautii 414
orientalis, Pterocles 156
orientalis, Pterodroma cookii 75
orientalis, Pyrrhurus scandens 570
orientalis, Reinarda squamata 252
orientalis, Serinus estherae 748
orientalis, Spizaetus nipalensis 114
orientalis, Stachyris thoracica 608

orientalis, Streptopelia 161
orientalis, Tchagra senegalus 459
orientalis, Treron calvus 173
orientalis, Turdoides striata 610
orientalis, Upupa epops 297
orientalis, Vidua 737
orientalis, Vireo gundlachii 483
orienticola, Amazilia brevirostris 265
orienticola, Electron platyrhynchum 294
orienticola, Penelope jacquacu 38
orientis, Tephrodornis pondicerianus 456
orientis, Zoothera citrina 660
origenis, Aerodramus whiteheadi 249
ORIGMA 444
orii, Aplonis opaca 652
orii, Dendrocopos kizuki 318
orii, Garrulus glandarius 509
orii, Parus varius 527
orii, Pycnonotus sinensis 566
orii, Streptopelia orientalis 161
orii, Todiramphus chloris 288
orii, Turdus chrysolaus 669
orina, Coeligena bonapartei 271
orinocensis, Leptogogon amaurocephalus 358
orinocensis, Picumnus spilogaster 312
orinocensis, Sublegatus arenarum 358
orinocensis, Tachyphonus cristatus 805
orinocensis, Thlypopsis sordida 804
orinocensis, Turdus fumigatus 671
orinocoensis, Pyrrhura picta 198
orinomus, Anas cyanoptera 67
orinomus, Cnemarchus erythropygius 372
orinus, Acrocephalus 586
ORIOLIA 461
oriolinus, Lobotos 471
orioloides, Pachycephala 477
ORIOLUS 486
oriolus, Oriolus 487
oriomo, Coracina papuensis 467
orissae, Rhipidura albicollis 494
orissae, Turdoides striata 610
oritis, Cyanomitra 706
ORITURUS 787
orius, Picoides villosus 321
orix, Euplectes 725
orlandoi, Alectoris graeca 48
ornata, Charmosyna placentis 187
ornata, Emberiza aureola 779
ornata, Hemithraupis flavicollis 814
ornata, Myrmotherula 385
ornata, Nothoprocta 33
ornata, Pachycephala vitiensis 477
ornata, Poospiza 790
ornata, Thlypopsis 804
ornata, Thraupis 807
ornata, Urocissa 509
ornatus, Auriparus flaviceps 531
ornatus, Calcarius 781
ornatus, Cephalopterus 349
ornatus, Cinnyris jugularis 711
ornatus, Lamprotornis 657
ornatus, Lichenostomus 432
ornatus, Lophornis 260
ornatus, Merops 296
ornatus, Myioborus 765
ornatus, Myiotriccus 359
ornatus, Orthorhyncus cristatus 260
ornatus, Pachyramphus albogriseus 345
ornatus, Pardalotus striatus 444
ornatus, Ptilinopus 175
ornatus, Spizaetus 114
ornatus, Todiramphus chloris 288
ornatus, Trichoglossus 185
ornatus, Xiphorhynchus elegans 424
Ornismya 265
ORNITHION 352
OROAETUS 114
orosae, Capito auratus 304
orostruthus, Modulatrix 625

oroyae, Haplochelidon andecola 534
orphea, Cossypha heuglini 678
orpheanum, Parisoma subcaeruleum 598
orpheus, Mimus polyglottos 648
orpheus, Pachycephala 476
orphna, Anas discors 67
orphnum, Camptostoma obsoletum 352
orphnus, Parus cinerascens 524
orphogaster, Cinnyris chloropygius 708
orquillensis, Quiscalus lugubris 774
orrhophaeus, Harpactes 279
orru, Corvus 514
ORTALIS 37
ORTHOGONYS 803
ORTHONYX 450
orthonyx, Acropternis 397
ORTHOPSITTACA 196
ORTHORHYNCUS 260
ORTHOTOMUS 563
ortiva, Chlorocichla flaviventris 570
ortizi, Incaspiza 789
ortleppi, Prinia flavicans 559
ortoni, Penelope 38
Ortygocichla 577
ORTYGOSPIZA 731
ORTYXELOS 130
oryzivora, Lonchura 736
oryzivorus, Dolichonyx 776
oryzivorus, Molothrus 773
ORYZOBORUS 793
osai, Corvus macrorhynchos 514
osberti, Chlorostilbon canivetii 261
osburni, Vireo 483
osceola, Meleagris gallopavo 44
oscillans, Rhinomyias 690
oscitans, Anastomus 81
osculans, Chrysococcyx 209
osculans, Coeligena violifer 271
osculans, Haematopus ostralegus 132
osculans, Sericornis maculatus 446
osea, Cinnyris 710
oseryi, Psarocolius 768
osgoodi, Aulacorhynchus derbianus 301
osgoodi, Canchites canadensis 45
osgoodi, Catharus guttatus 665
osgoodi, Momotus momota 294
osgoodi, Tinamus 31
oshiroi, Buteo buteo 112
osiris, Cinnyris mariquensis 709
osmastoni, Cettia pallidipes 579
osmastoni, Pelargopsis capensis 286
osmastoni, Pericrocotus cinnamomeus 472
ossifragus, Corvus 513
ostenta, Coracina 469
ostjakorum, Perisoreus infaustus 505
ostralegus, Haematopus 132
ostrinus, Pyrenestes 728
otero, Heliodoxa leadbeateri 269
OTIDIPHAPS 171
otiosus, Caprimulgus rufus 241
otiosus, Dicrurus paradiseus 493
OTIS 114
otofuscus, Ptilogonys cinereus 523
otomitra, Zoothera gurneyi 661
Otophanes 240
ottolanderi, Pomatorhinus montanus 603
ottomeyeri, Lalage leucomela 470
ottomeyeri, Pachycephala citreogaster 477
ottonis, Asthenes 406
ottonis, Phylloscartes ophthalmicus 356
ottonis, Ptilinopus cinctus 174
ottowi, Geositta rufipennis 402
OTUS 220
otus, Asio 236
oustaleti, Cinclodes 404
oustaleti, Cinnyris 710
oustaleti, Garrulax affinis 615
oustaleti, Megapodius freycinet 36
oustaleti, Nothoprocta pentlandii 33

oustaleti, Phylloscartes 357
ovamboensis, Cercotrichas leucophrys 679
ovamboensis, Cinnyris mariquensis 709
ovampensis, Accipiter 107
ovandensis, Catharus dryas 665
ovandensis, Lampornis viridipallens 268
oweni, Amytornis striatus 430
oweni, Chlamydera nuchalis 428
oweni, Megalurus timoriensis 577
owenii, Apteryx 35
owstoni, Aethopyga siparaja 713
owstoni, Cymochorea 78
owstoni, Dendrocopos leucotos 320
owstoni, Ficedula narcissina 693
owstoni, Gallirallus 120
owstoni, Garrulax canorus 614
owstoni, Nucifraga caryocatactes 511
owstoni, Parus varius 527
owstoni, Pyrrhula erythaca 756
owstoni, Todiramphus chloris 288
owstoni, Zosterops semperi 626
oxycerca, Cercibis 83
OXYLABES 588
Oxylophus 207
OXYPOGON 274
oxyptera, Anas flavirostris 67
OXYRUNCUS 347
OXYURA 70
oxyurus, Treron 174

pabsti, Cinclodes 403
pacaraimae, Amazilia cupreicauda 267
pacaraimae, Myrmothera simplex 401
pacatus, Pinicola enucleator 755
pachistorhina, Aplonis panayensis 652
pachiteae, Chaetura cinereiventris 251
PACHYCARE 474
PACHYCEPHALA 475
pachycephaloides, Clytorhynchus 500
PACHYCEPHALOPSIS 520
PACHYCOCCYX 207
Pachyphantes 722
PACHYPTILA 73
pachyrampha, Aplonis tabuensis 652
PACHYRAMPHUS 344
pachyrhyncha, Myrmecocichla albifrons 687
pachyrhyncha, Rhynchopsitta 196
pachyrhynchos, Cacicus haemorrhous 769
pachyrhynchus, Eudyptes 71
pachyrhynchus, Indicator exilis 310
pachyrhynchus, Odontophorus gujanensis 43
pacifica, Amazilia candida 265
pacifica, Amphispiza bilineata 786
pacifica, Ardea 87
pacifica, Calidris alpina 144
pacifica, Coereba flaveola 776
pacifica, Drepanis 759
pacifica, Ducula 178
pacifica, Gallinula 126
pacifica, Gavia 71
pacifica, Malacoptila mystacalis 333
pacifica, Myiopagis viridicata 351
pacifica, Myrmotherula 384
pacifica, Parula pitiayumi 760
pacifica, Pyriglena leuconota 389
pacifica, Pyrrhura melanura 199
pacifica, Synallaxis erythrothorax 409
pacifica, Uropsila leucogastra 641
pacificus, Anthus rubescens 744
pacificus, Apus 254
pacificus, Aramides cajanea 122
pacificus, Bubo virginianus 225
pacificus, Cacicus uropygialis 769
pacificus, Cyanerpes cyaneus 814
pacificus, Eurystomus orientalis 283
pacificus, Falco columbarius 96
pacificus, Gallirallus 120
pacificus, Histrionicus histrionicus 69
pacificus, Hylophilus ochraceiceps 486

paradoxus, Paradoxornis 622
paradoxus, Syrrhaptes 156
paradoxus, Zosterops kulambangrae 628
paraensis, Aratinga pertinax 197
paraensis, Automolus infuscatus 417
paraensis, Dendrexetastes rufigula 421
paraensis, Formicarius analis 398
paraensis, Glyphorynchus spirurus 421
paraensis, Granatellus pelzelni 768
paraensis, Hylopezus macularius 400
paraensis, Iodopleura isabellae 344
paraensis, Myrmotherula longipennis 386
paraensis, Notharchus macrorhynchos 332
paraensis, Phlegopsis nigromaculata 394
paraensis, Piculus chrysochloros 323
paraensis, Polioptila guianensis 644
paraensis, Thamnophilus amazonicus 382
paraensis, Tolmomyias assimilis 363
paraensis, Xiphocolaptes promeropirhynchus 422
paraguae, Anthreptes malacensis 704
paraguaiae, Gallinago 140
paraguanae, Dendroica petechia 761
paraguanae, Melanerpes rubricapillus 314
paraguanae, Sakesphorus canadensis 380
paraguanae, Xiphorhynchus picus 423
paraguayae, Furnarius rufus 404
paraguayensis, Dacnis cayana 813
paraguayensis, Phaethornis eurynome 256
paraguayensis, Thamnophilus caerulescens 382
paraguayensis, Turdus albicollis 672
paraguena, Eudynamys scolopaceus 211
paraguensis, Microbates collaris 643
paraheinei, Zoothera dauma 662
paralios, Sturnella magna 775
parallelus, Stercorarius parasiticus 153
parambae, Attila spadiceus 379
parambae, Grallaria haplonota 399
parambae, Myiopagis caniceps 350
parambae, Odontophorus erythrops 43
parambanus, Turdus obsoletus 671
paramillo, Eriocnemis vestita 272
paramo, Cinclodes fuscus 403
PARAMYTHIA 452
paranensis, Leptasthenura fuliginiceps 405
paraquensis, Atlapetes personatus 800
paraquensis, Automolus roraimae 417
paraquensis, Knipolegus poecilurus 369
paraquensis, Piculus rubiginosus 323
paraquensis, Turdus olivater 670
parasiticus, Stercorarius 153
parasitus, Milvus migrans 100
paratermus, Gallirallus striatus 120
parcus, Mionectes oleagineus 357
pardalotum, Glaucidium brodiei 230
PARDALOTUS 444
pardalotus, Xiphorhynchus 424
PARDIRALLUS 125
pardus, Campylorhynchus nuchalis 633
pareensis, Apalis thoracica 560
pareensis, Lybius leucocephalus 309
parelii, Myioparus griseigularis 692
parellina, Cyanocompsa 826
parens, Cettia 579
parensis, Momotus momota 294
pareola, Chiroxiphia 341
pariae, Basileuterus tristriatus 767
pariae, Grallaria haplonota 399
pariae, Grallaricula nana 402
pariae, Leptopogon superciliaris 358
pariae, Myioborus 765
pariae, Pipreola formosa 347
pariae, Premnoplex tatei 413
pariae, Pyrrhomyias cinnamomeus 365
pariae, Synallaxis cinnamomea 408
parimeda, Stipiturus malachurus 430
parina, Xenodacnis 815
parishi, Loxigilla violacea 795
parisii, Coturnix coturnix 54
PARISOMA 598

parisorum, Icterus 770
parkerae, Francolinus francolinus 50
parkeri, Acanthagenys rufogularis 437
parkeri, Calamanthus pyrrhopygius 444
parkeri, Cercomacra 389
parkeri, Glaucidium 231
parkeri, Herpsilochmus 386
parkeri, Metallura theresiae 274
parkeri, Phainoptila melanoxantha 523
parkeri, Phylloscartes 357
parkeri, Scytalopus 396
parkeri, Thamnophilus stictocephalus 382
PARKERTHRAUSTES 824
parkesi, Buarremon brunneinuchus 799
parkesi, Chrysococcyx flavigularis 210
parkesi, Dendroica petechia 760
parkesi, Ixos philippinus 574
parkesi, Mulleripicus funebris 330
parkesi, Oryzoborus maximiliani 793
parkesi, Saltator aurantiirostris 825
parkesi, Vireo huttoni 484
parkesi, Zosterops montanus 626
parkini, Passer domesticus 716
parkinsoni, Procellaria 75
parkmanii, Troglodytes aedon 640
PARMOPTILA 726
parnaguae, Megaxenops 419
PAROARIA 802
parodii, Hemispingus 803
paropanisi, Carduelis carduelis 751
PAROPHASMA 620
paroptus, Ploceus cucullatus 721
PAROREOMYZA 759
PAROTIA 516
paroticalis, Pycnonotus cyaniventris 566
parroti, Centropus sinensis 216
parroti, Dendrocopos major 320
parryi, Egretta novaehollandiae 88
parryi, Pardalotus rubricatus 444
pars, Arachnothera affinis 714
parsonsi, Hirundo neoxena 535
parsonsi, Prionochilus olivaceus 699
particeps, Lonchura punctulata 735
partridgei, Athene cunicularia 233
partridgei, Platyrinchus mystaceus 364
parui, Atlapetes personatus 800
parui, Diglossa duidae 817
parui, Mecocerculus leucophrys 353
parui, Pachyramphus castaneus 345
PARULA 760
parulus, Anairetes 353
parumstriata, Prinia crinigera 557
PARUS 523
parus, Melithreptus laetior 435
parva, Carduelis carduelis 751
parva, Columba fasciata 159
parva, Cyclarhis gujanensis 482
parva, Diglossa baritula 816
parva, Eupodotis canicollis 115
parva, Ficedula 693
parva, Microdynamis 211
parva, Mirafra javanica 540
parva, Parula superciliosa 760
parva, Porzana 123
parva, Rallina forbesi 118
parva, Sporophila minuta 793
parva, Tangara gyrola 811
parvicristatus, Colinus cristatus 42
parvimaculata, Nothoprocta cinerascens 33
parvimaculatus, Copsychus luzoniensis 681
parvior, Strepera fuliginosa 464
parvior, Tockus flavirostris 298
parvipes, Amazona auropalliata 204
parvipes, Branta canadensis 63
parvirostris, Aechmorhynchus 142
parvirostris, Ammomanes deserti 545
parvirostris, Atlapetes albinucha 800
parvirostris, Attila spadiceus 379
parvirostris, Chloropsis sonnerati 699

parvirostris, Chlorospingus 818
parvirostris, Colorhamphus 373
parvirostris, Crypturellus 33
parvirostris, Dendrocopos pubescens 321
parvirostris, Elaenia 351
parvirostris, Emberiza schoeniclus 780
parvirostris, Hypsipetes 575
parvirostris, Jabouilleia danjoui 604
parvirostris, Microchera albocoronata 267
parvirostris, Nystalus maculatus 333
parvirostris, Parus griseiventris 524
parvirostris, Phylloscopus trivirgatus 592
parvirostris, Pipilo albicollis 797
parvirostris, Polioptila plumbea 644
parvirostris, Prinia crinigera 557
parvirostris, Psittacula krameri 193
parvirostris, Scytalopus 396
parvirostris, Tetrao 46
parvissima, Colluricincla megarhyncha 489
parvistriatus, Picumnus sclateri 312
parvula, Aglaeactis cupripennis 270
parvula, Cacatua sulphurea 184
parvula, Colluricincla megarhyncha 489
parvula, Columbina passerina 167
parvula, Coracina 467
parvula, Dendrocitta vagabunda 510
parvula, Heliodoxa leadbeateri 269
parvula, Riparia cincta 533
parvulus, Aeronautes andecolus 252
parvulus, Anthoscopus 530
parvulus, Camarhynchus 796
parvulus, Caprimulgus 242
parvulus, Nesomimus 649
parvus, Acrocephalus gracilirostris 584
parvus, Anthus lutescens 745
parvus, Conopias albovittatus 375
parvus, Cypsiurus 252
parvus, Dryocopus javensis 326
parvus, Harpactes fasciatus 279
parvus, Hemignathus 758
parvus, Icterus galbula 771
parvus, Lanio versicolor 806
parvus, Melaenornis pallidus 689
parvus, Numenius americanus 141
parvus, Phodilus badius 219
parvus, Picoides scalaris 321
parvus, Treron vernans 172
parvus, Xiphorhynchus erythropygius 424
parvus, Zimmerius vilissimus 355
parzudakii, Tangara 811
paschae, Pterodroma heraldica 74
pascuus, Corvus brachyrhynchos 513
pasiphae, Sylvia melanocephala 598
pasquieri, Garrulax monileger 612
pasquieri, Rimator malacoptilus 604
passekii, Aegithalos caudatus 538
PASSER 715
PASSERCULUS 783
PASSERELLA 781
PASSERINA 826
passerina, Columbina 166
passerina, Emberiza schoeniclus 780
passerina, Mirafra 540
passerina, Spizella 785
passerinii, Ramphocelus 806
passerinum, Glaucidium 230
passerinus, Cacomantis 209
passerinus, Forpus 200
passerinus, Veniliornis 322
pastazae, Galbula 331
pastinator, Cacatua 183
pastinator, Corvus frugilegus 513
Pastor 655
patachonica, Oidemia 64
patachonicus, Tachyeres 64
PATAGONA 271
patagonica, Asthenes 406
patagonica, Eudromia elegans 34
patagonica, Pygochelidon cyanoleuca 534

percnerpes, Meiglyptes tukki 330
percnopterus, Neophron 101
PERCNOSTOLA 391
perconfusa, Calandrella somalica 546
percussus, Prionochilus 700
percussus, Xiphidiopicus 315
perdicaria, Nothoprocta 33
PERDICULA 55
perdita, Hypositta 462
perdita, Petrochelidon 537
PERDIX 53
perdix, Brachyramphus 155
perdix, Perdix 53
peregrina, Eremophila alpestris 551
peregrina, Vermivora 759
peregrinator, Falco peregrinus 97
peregrinus, Falco 97
peregrinus, Pericrocotus cinnamomeus 472
perenensis, Tangara punctata 811
perennius, Cisticola juncidis 555
pererrata, Coracina tenuirostris 468
perezchinchillorum, Zonotrichia capensis 782
PERICROCOTUS 471
perijana, Anabacerthia striaticollis 414
perijana, Asthenes wyatti 407
perijana, Grallaria ruficapilla 399
perijana, Habia rubica 819
perijana, Piprites chloris 349
perijana, Schizoeaca 405
perijanus, Buarremon torquatus 799
perijanus, Dromococcyx pavoninus 217
perijanus, Myiophobus flavicans 358
perijanus, Picumnus cinnamomeus 313
perijanus, Picumnus olivaceus 313
perijanus, Platyrinchus mystaceus 364
perijanus, Sittasomus griseicapillus 420
perijanus, Xenops rutilans 419
perimacha, Eremomela icteropygialis 594
perionca, Psittacula alexandri 193
periophthalmica, Terpsiphone atrocaudata 499
periophthalmicus, Dicrurus leucophaeus 492
periophthalmicus, Hylopezus perspicillatus 400
periophthalmicus, Monarcha frater 501
peripheris, Turdus libonyanus 666
PERIPORPHYRUS 824
periporphyrus, Icterus cayanensis 770
PERISOREUS 505
PERISSOCEPHALUS 349
perissuria, Spizella pusilla 785
peristephes, Actenoides concretus 284
peritum, Glaucidium brodiei 230
perkeo, Batis 455
perlata, Pyrrhura 198
perlata, Rhipidura 494
perlatum, Glaucidium 230
perlatus, Margarornis squamiger 414
perlatus, Microbates collaris 643
perlatus, Ptilinopus 175
perlonga, Hemiprocne longipennis 254
perlutus, Picus miniaceus 327
permagnus, Treron formosae 174
permista, Campethera cailliautii 316
permistus, Poicephalus gulielmi 194
permixtus, Turdus migratorius 672
permutatus, Dendrocopos kizuki 318
pernambucensis, Picumnus exilis 312
pernambucensis, Pyriglena leuconota 389
pernambucensis, Thamnophilus caerulescens 382
pernambucensis, Tinamus solitarius 31
perneglecta, Collocalia esculenta 247
perneglecta, Pachycephala simplex 476
perneglecta, Rhipidura rufiventris 494
perniger, Hypsipetes leucocephalus 575
perniger, Ictinaetus malayensis 112
pernigra, Certhia americana 647
PERNIS 99
pernix, Myiotheretes 371
pernotus, Phylloscopus cantator 592
pernyii, Dendrocopos cathpharius 319

perobscurus, Accipiter striatus 107
perolivaceus, Rhinomyias olivaceus 690
peromissus, Cyornis rufigastra 697
peronii, Charadrius 136
peronii, Zoothera 660
perousii, Ptilinopus 175
perpallida, Coracina papuensis 467
perpallida, Ptyonoprogne obsoleta 536
perpallida, Synallaxis albescens 408
perpallidus, Ammodramus savannarum 784
perpallidus, Falco tinnunculus 95
perpallidus, Monarcha cinerascens 501
perpallidus, Pipilo fuscus 797
perplexa, Carduelis pinus 750
perplexa, Iole olivacea 573
perplexa, Polioptila caerulea 643
perplexa, Vireo flavoviridis 485
perplexus, Aerodramus fuciphagus 250
perplexus, Cisticola aridulus 555
perplexus, Corvus coronoides 514
perplexus, Phylloscopus armandii 590
perplexus, Pycnonotus simplex 568
perplexus, Spilornis cheela 103
perplexus, Xiphorhynchus ocellatus 423
perplexus, Zosterops flavifrons 628
perpulchra, Halcyon smyrnensis 286
perpullus, Cisticola tinniens 553
perquisitor, Vireo griseus 482
perreini, Estrilda 730
perrotii, Hylexetastes 421
perrufus, Megapodius nicobariensis 36
perrygoi, Cyclarhis gujanensis 482
persa, Tauraco 205
persaturatus, Garrulus glandarius 509
persaturatus, Picumnus cinnamomeus 313
persica, Calandrella cheleensis 546
persica, Certhia familiaris 647
persica, Motacilla alba 740
persica, Oenanthe lugens 685
persica, Sitta europaea 644
persiccus, Colinus nigrogularis 42
persicus, Cinclus cinclus 698
persicus, Merops 296
persicus, Parus caeruleus 526
persicus, Passer domesticus 716
persicus, Phasianus colchicus 60
persicus, Puffinus lherminieri 76
persimile, Glaucidium cuculoides 231
persimilis, Chlorospingus ophthalmicus 817
persimilis, Cyanocorax yncas 506
persimilis, Thamnomanes caesius 384
personata, Apalis 561
personata, Coracina 466
personata, Emberiza spodocephala 779
personata, Eophona 757
personata, Gerygone palpebrosa 448
personata, Hypothymis helenae 498
personata, Incaspiza 789
personata, Motacilla alba 740
personata, Poephila 732
personata, Prosopeia 188
personata, Rhipidura 496
personata, Spizocorys 547
personata, Sula dactylatra 90
personata, Tityra semifasciata 343
personatus, Agapornis 194
personatus, Artamus 464
personatus, Atlapetes 800
personatus, Corythaixoides 206
personatus, Heliopais 127
personatus, Nesomimus parvulus 649
personatus, Psaltriparus minimus 539
personatus, Pterocles 156
personatus, Trogon 281
personus, Turdus fumigatus 671
perspicax, Penelope 38
perspicillaris, Accipiter rufiventris 107
perspicillata, Conopophaga melanops 394
perspicillata, Ducula 178

perspicillata, Malacocincla 600
perspicillata, Melanitta 69
perspicillata, Pulsatrix 230
perspicillata, Streptopelia decipiens 161
perspicillatus, Artamus cinereus 464
perspicillatus, Garrulax 612
perspicillatus, Hylopezus 400
perspicillatus, Hymenops 370
perspicillatus, Malaconotus monteiri 457
perspicillatus, Phalacrocorax 91
perspicillatus, Sericornis 446
perstriata, Alcippe vinipectus 619
perstriata, Ptiloprora 439
perstriatus, Saltator striatipectus 825
persuasus, Oriolus xanthonotus 487
perthi, Artamus cyanopterus 464
pertinax, Aratinga 197
pertinax, Contopus 367
pertinax, Myiarchus cinerascens 378
peruana, Carduelis magellanica 750
peruana, Cinnycerthia 635
peruana, Hemithraupis flavicollis 814
peruana, Monasa morphoeus 334
peruana, Upucerthia dumetaria 403
peruana, Uropsalis lyra 245
peruanum, Glaucidium 231
peruanum, Todirostrum cinereum 362
peruanus, Chlorostilbon mellisugus 261
peruanus, Gymnopithys leucaspis 393
peruanus, Knipolegus poecilurus 369
peruanus, Mionectes macconnelli 357
peruanus, Ocreatus underwoodii 273
peruanus, Pachyramphus viridis 344
peruanus, Thryothorus leucotis 638
peruviana, Conopophaga 394
peruviana, Delothraupis castaneoventris 809
peruviana, Dendroica petechia 761
peruviana, Fulica americana 127
peruviana, Geositta 402
peruviana, Geothlypis aequinoctialis 764
peruviana, Grallaricula 401
peruviana, Habia rubica 819
peruviana, Hapalopsittaca melanotis 202
peruviana, Hypocnemis cantator 390
peruviana, Jacana jacana 138
peruviana, Leptasthenura andicola 405
peruviana, Metallura tyrianthina 274
peruviana, Nothoprocta curvirostris 34
peruviana, Nothura darwinii 34
peruviana, Patagona gigas 271
peruviana, Pygochelidon cyanoleuca 534
peruviana, Schizoeaca fuliginosa 405
peruviana, Siptornis gutturata 411
peruviana, Sporophila 793
peruviana, Tangara 811
peruviana, Terenura callinota 388
peruviana, Vini 186
peruvianus, Aeronautes andecolus 252
peruvianus, Akletos 392
peruvianus, Anthus lutescens 745
peruvianus, Cacicus chrysonotus 769
peruvianus, Campephilus pollens 326
peruvianus, Celeus flavus 325
peruvianus, Charadrius vociferus 135
peruvianus, Chlorospingus ophthalmicus 818
peruvianus, Cichlopsis leucogenys 664
peruvianus, Colaptes atricollis 324
peruvianus, Crypturellus tataupa 33
peruvianus, Falco sparverius 95
peruvianus, Hylophylax naevius 393
peruvianus, Ixobrychus exilis 85
peruvianus, Lanio fulvus 806
peruvianus, Leptogogon amaurocephalus 358
peruvianus, Margarornis squamiger 414
peruvianus, Masius chrysopterus 340
peruvianus, Microbates cinereiventris 643
peruvianus, Myiobius villosus 365
peruvianus, Phacellodomus rufifrons 412
peruvianus, Phrygilus fruticeti 788

poliophrys, Alethe 674
poliophrys, Buarremon torquatus 799
poliophrys, Synallaxis 408
poliopis, Malacoptila panamensis 333
poliopleura, Emberiza 779
poliopsa, Coracina schisticeps 469
poliopsis, Accipiter badius 105
polioptera, Coracina 469
polioptera, Cossypha 678
poliopterus, Melierax 104
poliopterus, Toxorhamphus 452
POLIOPTILA 643
poliosoma, Pachycephalopsis 520
poliothorax, Kakamega 602
poliothorax, Pseudoalcippe abyssinica 602
poliotis, Actinodura waldeni 618
poliotis, Lichmera incana 439
poliotis, Paradoxornis nipalensis 623
polioxantha, Eremomela icteropygialis 594
POLIOXOLMIS 372
poliurus, Ptilinopus magnificus 175
polius, Turdus olivaceus 667
polivanovi, Paradoxornis 624
polleni, Xenopirostris 461
pollenii, Columba 158
pollenorum, Francolinus jacksoni 52
pollens, Agriornis albicauda 371
pollens, Campephilus 326
pollens, Coracina personata 466
pollicaris, Rissa tridactyla 149
pollux, Cercomela schlegelii 686
poltaratskyi, Sturnus vulgaris 656
POLYBOROIDES 104
Polyborus 93
polychloros, Eclectus roratus 192
polychopterus, Pachyramphus 345
polychopterus, Pachyramphus polychopterus 345
polychroa, Prinia 557
polyclystus, Cinnyris jugularis 711
polycryptus, Accipiter fasciatus 106
Polyerata 264, 266
polyglotta, Hippolais 587
polyglottos, Mimus 648
polyglottus, Cistothorus platensis 635
polygrammica, Lalage leucomela 470
polygrammus, Melanerpes aurifrons 315
polygrammus, Xanthotis 431
polymorpha, Petroica multicolor 521
polynesiae, Puffinus lherminieri 76
POLYONYMUS 273
polyosoma, Buteo 111
polyphonus, Melidectes torquatus 438
POLYPLECTRON 60
POLYSTICTA 69
polysticta, Asthenes humicola 406
POLYSTICTUS 354
polystictus, Ocreatus underwoodii 273
polystictus, Xiphorhynchus guttatus 424
POLYTELIS 192
POLYTMUS 264
polytmus, Trochilus 261
polyzonoides, Accipiter badius 105
polyzonus, Dendrocolaptes certhia 422
polyzonus, Piculus chrysochloros 323
POMAREA 500
pomarea, Pomarea nigra 500
pomarina, Aquila 112
pomarinus, Stercorarius 153
POMATORHINUS 602
POMATOSTOMUS 450
pompadora, Treron 172
pompalis, Tyrannus crassirostris 376
pompata, Calocitta formosa 507
ponapensis, Aerodramus inquietus 249
ponapensis, Aplonis opaca 652
ponapensis, Asio flammeus 236
ponapensis, Ptilinopus porphyraceus 175
ponapensis, Zosterops cinereus 629
pondalowiensis, Psophodes leucogaster 452

pondicerianus, Francolinus 50
pondicerianus, Tephrodornis 456
pondoensis, Bradypterus sylvaticus 580
pondoensis, Laniarius ferrugineus 460
pondoensis, Prinia subflava 558
pondoensis, Turdus olivaceus 667
pons, Charadrius marginatus 135
ponsi, Chlorospingus ophthalmicus 817
ponsi, Pionus sordidus 202
ponticus, Larus argentatus 147
pontifex, Clytorhynchus vitiensis 500
pontifex, Dicaeum hypoleucum 701
pontifex, Sericornis magnirostra 446
pontilis, Brachyramphus hypoleucus 155
pontilis, Geothlypis poliocephala 764
pontilis, Polioptila melanura 643
pontilus, Junco hyemalis 783
pontius, Psittinus cyanurus 190
POOECETES 785
poortmani, Chlorostilbon 262
POOSPIZA 789
Popelairia 260
popelairii, Discosura 260
porculae, Phaethornis griseogularis 257
porculae, Picumnus sclateri 312
porini, Turdus olivaceus 667
porphyraceus, Ptilinopus 175
porphyreolophus, Tauraco 206
porphyreus, Blythipicus pyrrhotis 330
porphyreus, Ptilinopus 174
porphyreus, Pycnonotus plumosus 568
PORPHYRIO 125
porphyrio, Porphyrio 125
porphyrocephala, Glossopsitta 186
porphyrocephalus, Iridosornis 809
PORPHYROLAEMA 348
porphyrolaema, Apalis 561
porphyrolaema, Leptocoma sericea 707
porphyrolaema, Porphyrolaema 348
porphyronotus, Sturnus vulgaris 656
PORPHYROSPIZA 787
Porphyrula 126
portenkoi, Carpodacus roseus 754
portoricensis, Asio flammeus 236
portoricensis, Coereba flaveola 776
portoricensis, Columbina passerina 167
portoricensis, Icterus dominicensis 770
portoricensis, Loxigilla 795
portoricensis, Melanerpes 314
portoricensis, Otus choliba 223
portoricensis, Spindalis 820
portovelae, Myiobius atricaudus 365
PORZANA 123
porzana, Porzana 123
postocularis, Chlorospingus ophthalmicus 817
postrema, Ducula pistrinaria 179
postremus, Acrocephalus caffer 585
potanini, Alectoris chukar 49
potanini, Pyrgilauda davidiana 718
potior, Dicrurus forficatus 491
potosina, Aphelocoma ultramarina 507
potosinus, Carpodacus mexicanus 754
potosinus, Cistothorus platensis 635
potosinus, Pipilo fuscus 797
powelli, Clytorhynchus vitiensis 500
powelli, Turnix suscitator 130
practicus, Parus atricapillus 528
praecognita, Stachyris ruficeps 606
praecox, Icterus prosthemelas 770
praecox, Thamnophilus 381
praedatus, Lepidocolaptes angustirostris 425
praedi, Malimbus rubricollis 724
praedicta, Ptiloprora perstriata 439
praedo, Gallirallus philippensis 120
praepectoralis, Neocossyphus poensis 659
praepes, Crypturellus cinnamomeus 32
praetermissa, Euphonia gouldi 821
praetermissa, Galerida theklae 548
praetermissa, Megalaima faiostricta 305

praevelox, Chaetura brachyura 251
prageri, Emberiza cia 777
prasina, Erythrura 733
prasina, Hylia 588
prasina, Melanocharis striativentris 451
prasinonota, Aleadryas rufinucha 474
prasinorrhous, Ptilinopus rivoli 176
prasinus, Aulacorhynchus 301
pratensis, Ammodramus savannarum 784
pratensis, Anthus 743
pratensis, Grus canadensis 128
prateri, Sitta castanea 645
praticola, Eremophila alpestris 550
praticola, Sturnella magna 775
pratincola, Glareola 146
pratincola, Tyto alba 219
pratti, Crateroscelis robusta 445
prattii, Haematopus palliatus 132
preciosa, Tangara 811
preissi, Rhipidura albiscapa 495
PREMNOPLEX 413
PREMNORNIS 413
prenticei, Halcyon malimbica 287
prentissgrayi, Halcyon albiventris 287
prepositus, Francolinus pondicerianus 50
presaharica, Ptyonoprogne obsoleta 536
presbytes, Phylloscopus 592
pressa, Modulatrix stictigula 625
pretiosa, Claravis 167
pretiosa, Terpsiphone mutata 499
pretoriae, Monticola brevipes 688
pretoriae, Ptyonoprogne fuligula 536
pretrei, Amazona 203
pretrei, Phaethornis 257
pretrei, Spindalis zena 820
preussi, Cinnyris reichenowi 709
preussi, Coracina caesia 468
preussi, Onychognathus walleri 658
preussi, Petrochelidon 537
preussi, Ploceus 723
prevostii, Anthracothorax 259
prevostii, Euryceros 462
pridii, Chloropsis aurifrons 699
priesti, Bradypterus barratti 581
prigoginei, Caprimulgus 244
prigoginei, Chlorocichla 570
prigoginei, Cinnyris stuhlmanni 709
prigoginei, Linurgus olivaceus 749
prigoginei, Parisoma lugens 598
prigoginei, Phodilus 219
prigoginei, Turdus cameronensis 661
prillwitzi, Arachnothera longirostra 714
prillwitzi, Macronus flavicollis 609
prillwitzi, Pycnonotus simplex 568
primarius, Anthus similis 742
primigenia, Aethopyga 712
primitiva, Myzomela eques 441
primolina, Metallura williami 274
PRIMOLIUS 196
primrosei, Pycnonotus cafer 566
primulus, Myiozetetes similis 374
princei, Zoothera 661
princeps, Accipiter 106
princeps, Actenoides 284
princeps, Emberiza flaviventris 779
princeps, Grallaria guatimalensis 399
princeps, Leucopternis 109
princeps, Melidectes 438
princeps, Passerculus sandwichensis 784
princeps, Ploceus 721
princeps, Psittacus erithacus 194
princetoniana, Melanodera melanodera 788
principalis, Campephilus 327
principalis, Columba larvata 160
principalis, Corvus corax 514
principalis, Phasianus colchicus 60
principum, Glaucidium radiatum 232
pringlii, Dryoscopus 459
PRINIA 557

roborowskii, Kozlowia 755
robsoni, Alcippe chrysotis 618
robusta, Arachnothera 714
robusta, Asthenes humilis 406
robusta, Aviceda subcristata 98
robusta, Coracina papuensis 467
robusta, Crateroscelis 445
robusta, Gracula religiosa 654
robusta, Motacilla cinerea 740
robusta, Pachycephala melanura 477
robusta, Perdix perdix 53
robusta, Tringa totanus 141
robustipes, Cettia fortipes 579
robustipes, Phasianus versicolor 60
robustirostris, Acanthiza 449
robustirostris, Cinnyris jugularis 711
robustus, Asio stygius 236
robustus, Campephilus 327
robustus, Cisticola 553
robustus, Eudyptes 71
robustus, Melichneutes 311
robustus, Phylloscopus fuscatus 590
robustus, Poicephalus 194
robustus, Tinamus major 31
roccattii, Anthoscopus caroli 530
rochei, Mirafra somalica 541
rochii, Cuculus 208
rochussenii, Scolopax 139
rockefelleri, Cinnyris 709
rocki, Ithaginis cruentus 57
rocki, Paradoxornis conspicillatus 623
rocki, Prinia polychroa 557
rocki, Spelaeornis troglodytoides 605
rodericanus, Acrocephalus 584
rodericanus, Necropsar 651
rodgeri, Picus chlorolophus 327
rodgersi, Swynnertonia swynnertoni 674
rodgersii, Fulmarus glacialis 73
rodinogaster, Petroica 521
rodochroa, Carpodacus 754
rodolphei, Stachyris 606
rodolphei, Synallaxis albigularis 408
rodolphi, Rhopodytes sumatranus 212
rodopeplus, Carpodacus 754
rodwayi, Euphonia violacea 821
roehli, Andropadus masukuensis 568
roehli, Apus barbatus 253
roehli, Picumnus squamulatus 312
roehli, Turdus olivaceus 667
rogersi, Aerodramus brevirostris 248
rogersi, Anas superciliosa 66
rogersi, Anthus australis 741
rogersi, Atlantisia 122
rogersi, Butorides striata 86
rogersi, Calidris canutus 143
rogersi, Chalcophaps indica 164
rogersi, Cyornis rubeculoides 696
rogersi, Gerygone olivacea 448
rogersi, Malurus lamberti 429
rogersi, Podargus papuensis 237
rogersi, Sterna anaethetus 151
rogersi, Yuhina flavicollis 622
rogosowi, Perisoreus infaustus 505
rolland, Rollandia 79
ROLLANDIA 79
rolleti, Lybius 309
rolleti, Oriolus larvatus 488
rolli, Arborophila 56
ROLLULUS 56
romblonis, Otus mantananensis 222
romeroana, Grallaria rufocinerea 400
rondoniae, Amazilia 265
rongjiangensis, Lophura nycthemera 58
rooki, Coracina tenuirostris 468
rooki, Myzomela cineracea 441
rookmakeri, Leiothrix argentauris 616
roosevelti, Uraeginthus ianthinogaster 730
roquettei, Phylloscartes 357
roraima, Columba fasciata e 159

roraimae, Automolus 417
roraimae, Basileuterus bivittatus 766
roraimae, Caprimulgus longirostris 241
roraimae, Chlorophonia cyanea 822
roraimae, Coereba flaveola 776
roraimae, Grallaria guatimalensis 399
roraimae, Hemithraupis guira 814
roraimae, Herpsilochmus 387
roraimae, Mecocerculus leucophrys 353
roraimae, Mionectes macconnelli 357
roraimae, Myiophobus 359
roraimae, Otus vermiculatus 224
roraimae, Parula pitiayumi 760
roraimae, Pteroglossus aracari 302
roraimae, Trogon personatus 281
roraimae, Turdus olivater 670
roraimae, Zonotrichia capensis 782
RORAIMIA 413
roratus, Eclectus 192
rosa, Harpactes erythrocephalus 280
rosacea, Ducula 179
rosacea, Piranga flava 818
rosacea, Pyrrhula pyrrhula 757
rosaceus, Aegithalos caudatus 538
rosae, Chaetocercus jourdanii 277
rosea, Habia rubica 819
rosea, Megalaima haemacephala 306
rosea, Micropsitta bruijnii 182
rosea, Petroica 521
rosea, Rhodinocichla 820
rosea, Rhodostethia 149
roseata, Psittacula 193
roseatus, Anthus 744
roseatus, Carpodacus erythrinus 753
roseicapilla, Eolophus 183
roseicapilla, Ptilinopus 176
roseicollis, Agapornis 194
roseicrissa, Estrilda paludicola 730
roseifrons, Pyrrhura picta 199
roseigaster, Priotelus 280
roseilia, Sitta europaea 645
roseipectus, Carpodacus mexicanus 754
roseipileum, Ptilinopus regina 176
roselaari, Calidris canutus 143
rosenbergi, Amazilia 266
rosenbergi, Nyctiphrynus 241
rosenbergi, Pittasoma rufopileatum 398
rosenbergi, Rhytipterna holerythra 376
rosenbergi, Schiffornis turdina 343
rosenbergi, Sipia 392
rosenbergi, Xiphorhynchus susurrans 424
rosenbergii, Gymnocrex 123
rosenbergii, Myzomela 443
rosenbergii, Pitta sordida 339
rosenbergii, Scolopax 139
rosenbergii, Trichoglossus haematodus 185
rosenbergii, Tyto 218
roseoaxillaris, Ninox jacquinoti 235
roseogrisea, Streptopelia 161
roseogularis, Piranga 819
roseotinctus, Phaethon rubricauda 88
roseus, Carpodacus 754
roseus, Pericrocotus 471
roseus, Phoenicopterus ruber 80
roseus, Sturnus 655
rosinae, Acanthiza iredalei 449
rosinae, Eopsaltria griseogularis 520
rosinae, Sericornis frontalis 446
rositae, Passerina 826
rossae, Musophaga 206
rosseli, Dicaeum nitidum 702
rosseliana, Gerygone magnirostris 448
rosseliana, Ninox theomacha 235
rosseliana, Pachycephala citreogaster 477
rosseliana, Tanysiptera galatea 285
rosselianus, Accipiter cirrhocephalus 107
rosselianus, Lorius hypoinochrous 186
rosselianus, Monarcha cinerascens 501
rossica, Lagopus lagopus 47

rossicus, Anser fabalis 62
rossii, Anser 62
rossikowi, Pyrrhula pyrrhula 757
rossocaucasica, Certhia brachydactyla 648
rossorum, Saxicola caprata 684
rostrata, Asthenes modesta 406
rostrata, Athene cunicularia 233
rostrata, Carduelis flammea 751
rostrata, Coracina tenuirostris 469
rostrata, Geositta peruviana 402
rostrata, Geothlypis 764
rostrata, Hylocharis cyanus 264
rostrata, Melampitta lugubris 515
rostrata, Nothoprocta ornata 33
rostrata, Pterodroma 75
rostrata, Streptopelia picturata 161
rostrata, Urosticte benjamini 273
ROSTRATULA 137
rostratum, Trichastoma 600
rostratus, Aegolius ridgwayi 233
rostratus, Colibri coruscans 259
rostratus, Coryphospingus pileatus 802
rostratus, Hypsipetes madagascariensis 575
rostratus, Mimus gilvus 649
rostratus, Mirafra africana 541
rostratus, Passerculus sandwichensis 784
rostratus, Pheucticus ludovicianus 823
rostratus, Premnoplex brunnescens 413
rostratus, Turnix suscitator 130
rostratus, Xiphocolaptes promeropirhynchus 422
ROSTRHAMUS 100
rostrifera, Thalurania colombica 263
rotensis, Zosterops 625
rothschildi, Amazona barbadensis 203
rothschildi, Arachnothera longirostra 714
rothschildi, Astrapia 516
rothschildi, Bangsia 808
rothschildi, Bostrychia olivacea 82
rothschildi, Buteo buteo 111
rothschildi, Charmosyna pulchella 187
rothschildi, Cypseloides 246
rothschildi, Daphoenositta chrysoptera 473
rothschildi, Dromaius novaehollandiae 35
rothschildi, Fregata minor 89
rothschildi, Geospiza scandens 796
rothschildi, Geotrygon saphirina 169
rothschildi, Hirundo lucida 535
rothschildi, Leucopsar 655
rothschildi, Megalaima lagrandieri 305
rothschildi, Muscicapa sibirica 691
rothschildi, Nucifraga caryocatactes 511
rothschildi, Phasianus colchicus 60
rothschildi, Poecilodryas albispecularis 518
rothschildi, Pyrenestes ostrinus 728
rothschildi, Scaeophaethon rubricauda 89
rothschildi, Sericulus chrysocephalus 427
rothschildi, Serilophus lunatus 336
rothschildi, Serinus 747
rothschildi, Uraeginthus ianthinogaster 730
rothschildi, Xiphirhynchus superciliaris 604
rothschildi, Zosterops mayeri 627
rothschildii, Cyanocompsa cyanoides 825
rotiensis, Ninox boobook 234
rotumae, Aplonis tabuensis 652
rotumae, Lalage maculosa 471
rougeoti, Jynx ruficollis 311
rougetii, Rougetius 122
ROUGETIUS 122
rouloul, Rollulus 56
roundi, Heterophasia annectans 621
rourei, Nemosia 803
rouxi, Gerygone flavolateralis 447
rouxi, Yuhina flavicollis 622
rovianae, Gallirallus 120
rovuma, Francolinus sephaena 51
rovumae, Cercotrichas quadrivirgata 679
rowani, Grus canadensis 128
rowei, Zoothera piaggiae 661
ROWETTIA 789

ruficapilla, Synallaxis 408
ruficapilla, Vermivora 760
ruficapillus, Baryphthengus 294
ruficapillus, Charadrius 136
ruficapillus, Chrysomus 772
ruficapillus, Enicurus 683
ruficapillus, Schistochlamys 802
ruficapillus, Thamnophilus 382
ruficauda, Aglaeactis cupripennis 270
ruficauda, Aimophila 786
ruficauda, Chamaeza 398
ruficauda, Cichladusa 678
ruficauda, Cinclocerthia 651
ruficauda, Galbula 331
ruficauda, Gerygone 448
ruficauda, Histurgops 715
ruficauda, Muscicapa 691
ruficauda, Myrmeciza 391
ruficauda, Neochmia 732
ruficauda, Ortalis 37
ruficauda, Ramphotrigon 378
ruficauda, Rhinomyias 690
ruficauda, Touit surda 201
ruficauda, Zenaida auriculata 166
ruficaudatum, Philydor 415
ruficaudatus, Thryothorus genibarbis 637
ruficaudus, Upucerthia 403
ruficaudus, Xenops minutus 419
ruficeps, Aimophila 787
ruficeps, Chalcostigma 274
ruficeps, Cisticola 554
ruficeps, Coua 215
ruficeps, Dendrocincla homochroa 420
ruficeps, Elaenia 351
ruficeps, Euphonia xanthogaster 822
ruficeps, Formicarius colma 397
ruficeps, Garrulax albogularis 612
ruficeps, Grallaria nuchalis 400
ruficeps, Gymnopithys leucaspis 393
ruficeps, Laniarius 460
ruficeps, Luscinia 676
ruficeps, Macropygia 163
ruficeps, Ortalis motmot 38
ruficeps, Orthotomus 564
ruficeps, Paradoxornis 624
ruficeps, Pellorneum 599
ruficeps, Poecilotriccus 361
ruficeps, Pomatostomus 450
ruficeps, Pseudotriccus 355
ruficeps, Pyrrhocoma 805
ruficeps, Stachyris 606
ruficeps, Stipiturus 430
ruficeps, Thlypopsis 805
ruficeps, Turdus poliocephalus 669
ruficeps, Veniliornis affinis 323
ruficervix, Caprimulgus longirostris 241
ruficervix, Tangara 812
ruficollaris, Alcedo azurea 291
ruficollaris, Todiramphus tutus 289
ruficollis, Branta 63
ruficollis, Calidris 143
ruficollis, Caprimulgus 242
ruficollis, Cathartes aura 93
ruficollis, Chrysococcyx 210
ruficollis, Corvus 514
ruficollis, Egretta tricolor 88
ruficollis, Falco chicquera 96
ruficollis, Garrulax 614
ruficollis, Gerygone 447
ruficollis, Hypnelus 333
ruficollis, Jynx 311
ruficollis, Loxigilla violacea 795
ruficollis, Madanga 631
ruficollis, Micrastur 94
ruficollis, Myiagra 504
ruficollis, Oreopholus 137
ruficollis, Pernis ptilorhynchus 99
ruficollis, Pomatorhinus 603
ruficollis, Pucrasia macrolopha 57

ruficollis, Pyrgilauda 719
ruficollis, Rhyticeros plicatus 300
ruficollis, Sporophila 793
ruficollis, Stelgidopteryx 534
ruficollis, Syndactyla 415
ruficollis, Tachybaptus 79
ruficollis, Turdus 669
ruficolor, Galerida theklae 548
ruficoronatus, Myioborus melanocephalus 765
ruficrissa, Amaurornis olivacea 123
ruficrissa, Eumyias indigo 695
ruficrissa, Ortalis ruficauda 37
ruficrissa, Rhinomyias ruficauda 690
ruficrissa, Syndactyla subalaris 415
ruficrissa, Urosticte benjamini 273
ruficrissus, Criniger ochraceus 572
ruficrista, Lophotis 116
rufidorsa, Rhipidura 496
rufidorsalis, Arremon aurantiirostris 799
rufidorsalis, Passer domesticus 716
rufidorsalis, Urorhipis rufifrons 562
rufifacies, Sceloglaux albifacies 236
rufifacies, Schistocichla leucostigma 391
rufifrons, Acanthiza ewingii 449
rufifrons, Basileuterus 767
rufifrons, Cyornis caerulatus 696
rufifrons, Formicarius 398
rufifrons, Fulica 127
rufifrons, Garrulax 612
rufifrons, Malacopteron cinereum 601
rufifrons, Percnostola 391
rufifrons, Phacellodomus 412
rufifrons, Rhipidura 496
rufifrons, Stachyris 606
rufifrons, Urorhipis 562
rufigaster, Ducula 179
rufigaster, Ninox novaeseelandiae 234
rufigastra, Alcedo meninting 292
rufigastra, Cyornis 696
rufigastra, Threnetes niger 255
rufigena, Caprimulgus 242
rufigenis, Atlapetes 801
rufigenis, Cranioleuca erythrops 411
rufigenis, Poecilotriccus ruficeps 361
rufigenis, Sylvietta ruficapilla 596
rufigenis, Tangara 812
rufigenis, Yuhina castaniceps 621
rufigula, Dendrexetastes 421
rufigula, Ducula rubricera 179
rufigula, Ficedula 694
rufigula, Gallicolumba 170
rufigula, Gymnopithys 393
rufigula, Petrochelidon 537
rufigula, Ptyonoprogne fuligula 536
rufigula, Tangara 811
rufigularis, Falco 96
rufigularis, Glyphorynchus spirurus 421
rufigularis, Hemitriccus 360
rufigularis, Hyetornis 214
rufigularis, Ixos 574
rufigularis, Sclerurus 418
rufilata, Luscinia cyanura 677
rufilata, Prunella collaris 738
rufilateralis, Rhipidura spilodera 496
rufilatus, Cisticola 553
rufilatus, Turnix suscitator 130
rufiloris, Zanclostomus calyorhynchus 213
rufimarginatus, Herpsilochmus 387
rufina, Melospiza melodia 781
rufina, Netta 68
rufinucha, Aleadryas 474
rufinucha, Atlapetes 801
rufinucha, Campylorhynchus 634
rufinucha, Charadrius wilsonia 135
rufinucha, Leptotila cassini 169
rufinuchalis, Laniarius ruficeps 460
rufinus, Buteo 112
rufinus, Empidonomus varius 375
rufipectoralis, Ochthoeca 373

rufipectus, Arborophila 56
rufipectus, Automolus rubiginosus 417
rufipectus, Formicarius 398
rufipectus, Leptopogon 358
rufipectus, Napothera 604
rufipectus, Nonnula ruficapilla 334
rufipectus, Parus ater 526
rufipectus, Spilornis 103
rufipectus, Stachyris ruficeps 606
rufipennis, Aplonis zelandica 652
rufipennis, Butastur 109
rufipennis, Cichladusa guttata 678
rufipennis, Cinnyris 710
rufipennis, Columbina talpacoti 167
rufipennis, Cranioleuca pyrrhophia 410
rufipennis, Geositta 402
rufipennis, Illadopsis 601
rufipennis, Macropygia 163
rufipennis, Myiophobus roraimae 359
rufipennis, Myiozetetes cayanensis 374
rufipennis, Neomorphus 218
rufipennis, Otus sunia 221
rufipennis, Pachycephala simplex 475
rufipennis, Petrophassa 165
rufipennis, Pitangus sulphuratus 374
rufipennis, Polioxolmis 372
rufipes, Hemitriccus margaritaceiventer 360
rufipes, Lophura nycthemera 58
rufipes, Strix 228
rufipilea, Mirafra apiata 542
rufipileatus, Automolus 418
rufipileus, Colaptes auratus 324
rufirostris, Tockus erythrorhynchus 298
rufitergum, Garrulus glandarius 508
rufitergum, Gymnoris superciliaris 718
rufitergum, Tadorna radjah 65
rufitinctus, Garrulax rufogularis 614
rufitorques, Accipiter 106
rufitorques, Turdus 672
rufiventer, Eudynamys scolopaceus 211
rufiventer, Napothera brevicaudata 604
rufiventer, Pteruthius 617
rufiventer, Tachyphonus 805
rufiventer, Terpsiphone 498
rufiventris, Accipiter 107
rufiventris, Ardeola 86
rufiventris, Cercomacra tyrannina 389
rufiventris, Chamaepetes goudotii 39
rufiventris, Cinclus cinclus 698
rufiventris, Colluricincla harmonica 490
rufiventris, Euphonia 822
rufiventris, Formicivora grisea 387
rufiventris, Hellmayrea gularis 410
rufiventris, Lalage leucomela 471
rufiventris, Lamprotornis pulcher 657
rufiventris, Lurocalis 238
rufiventris, Malacocincla sepiaria 600
rufiventris, Mionectes 357
rufiventris, Monticola 688
rufiventris, Myiagra vanikorensis 504
rufiventris, Neoxolmis 372
rufiventris, Numenius phaeopus 141
rufiventris, Pachycephala 478
rufiventris, Parus 524
rufiventris, Phoenicurus ochruros 681
rufiventris, Picumnus 313
rufiventris, Pitta erythrogaster 338
rufiventris, Poicephalus 195
rufiventris, Prionops 456
rufiventris, Ramphocaenus melanurus 643
rufiventris, Rhipidura 494
rufiventris, Saltator 825
rufiventris, Thryothorus leucotis 638
rufiventris, Turdus 670
rufivertex, Dendroica petechia 760
rufivertex, Euphonia anneae 822
rufivertex, Iridosornis 809
rufivertex, Muscisaxicola 370
rufivirgatus, Arremonops 798

rufoaxillaris, Molothrus 773
rufobrunneus, Serinus 748
rufobrunneus, Thripadectes 416
rufocarpalis, Calicalicus 461
rufociliatus, Troglodytes 640
rufocinctus, Kupeornis 620
rufocinctus, Passer motitensis 716
rufocinerea, Grallaria 400
rufocinerea, Terpsiphone 498
rufocinereus, Monticola 688
rufocinnamomea, Mirafra 541
rufocollaris, Petrochelidon 538
rufocrissalis, Melidectes 438
rufodorsalis, Minla cyanouroptera 618
rufofronta, Rhipidura rufifrons 497
rufofuscum, Trichastoma celebense 600
rufofuscus, Buteo 112
rufofuscus, Passerculus sandwichensis 784
rufofuscus, Tchagra senegalus 459
rufogaster, Colluricincla megarhyncha 489
rufogularis, Acanthagenys 437
rufogularis, Alcippe 619
rufogularis, Anthus cervinus 744
rufogularis, Apalis 561
rufogularis, Arborophila 55
rufogularis, Conopophila 440
rufogularis, Garrulax 614
rufogularis, Odontophorus gujanensis 43
rufogularis, Pachycephala 475
rufolateralis, Myiagra alecto 504
rufolateralis, Smithornis 335
rufolavatus, Tachybaptus 79
rufomarginatus, Euscarthmus 355
rufomarginatus, Mecocerculus leucophrys 353
rufomerus, Chrysococcyx minutillus 210
rufonuchalis, Parus 526
rufoolivacea, Dendrocincla fuliginosa 420
rufopalliatus, Turdus 672
rufopectoralis, Buccanodon leucotis 307
rufopectus, Ochthoeca rufipectoralis 373
rufopectus, Poliocephalus 79
rufopicta, Lagonosticta 728
rufopictus, Francolinus 53
rufopileata, Dendroica petechia 761
rufopileatum, Pittasoma 398
rufopunctatus, Chrysocolaptes lucidus 329
rufoscapulatus, Plocepasser 715
rufoschistaceus, Accipiter novaehollandiae 106
rufostrigata, Carduelis flavirostris 751
rufostrigata, Ninox connivens 234
rufosuperciliaris, Hemispingus 804
rufosuperciliata, Syndactyla 415
rufotibialis, Accipiter virgatus 107
rufovelatus, Malimbus rubricollis 724
rufoviridis, Galbula ruficauda 331
rufuensis, Turdoides hypoleuca 611
rufula, Cecropis daurica 537
rufula, Chamaea fasciata 609
rufula, Grallaria 400
rufula, Prinia hodgsonii 557
rufuloides, Anthus cinnamomeus 742
rufulus, Anthus richardi 741
rufulus, Gampsorhynchus 617
rufulus, Troglodytes 640
rufum, Campylorhynchus rufinucha 634
rufum, Conirostrum 815
rufum, Philydor 416
rufum, Toxostoma 649
rufus, Attila 379
rufus, Bathmocercus 582
rufus, Campylopterus 258
rufus, Caprimulgus 241
rufus, Casiornis 377
rufus, Cinclodes fuscus 403
rufus, Cisticola 554
rufus, Climacteris 428
rufus, Cursorius 145
rufus, Furnarius 404
rufus, Megalurulus 578

rufus, Neocossyphus 659
rufus, Otus guatemalae 224
rufus, Pachyramphus 344
rufus, Selasphorus 277
rufus, Tachyphonus 806
rufus, Trogon 281
rufusater, Philesturnus carunculatus 452
rugensis, Metabolus 501
rugosus, Aceros corrugatus 300
rukensis, Aerodramus inquietus 249
rukensis, Anas superciliosa 66
ruki, Rukia 631
RUKIA 630
rukwensis, Mirafra albicauda 540
rumicivorus, Thinocorus 138
rungweensis, Stactolaema olivacea 307
rupatensis, Cyornis turcosus 696
rupchandi, Apus acuticaudus 254
rupchandi, Arborophila atrogularis 55
rupchandi, Batrachostomus hodgsoni 237
rupchandi, Chaetura cochinchinensis 251
rupchandi, Otus spilocephalus 220
rupchandi, Tragopan blythii 57
rupestris, Chordeiles 239
rupestris, Columba 157
rupestris, Lagopus muta 47
rupestris, Monticola 688
rupestris, Ptyonoprogne 536
rupestris, Tetrastes bonasia 45
RUPICOLA 347
rupicola, Aimophila ruficeps 787
rupicola, Colaptes 324
rupicola, Pyrrhura 199
rupicola, Rupicola 347
rupicola, Sitta neumayer 646
rupicolaeformis, Falco tinnunculus 95
rupicoloides, Falco 95
rupicolus, Falco tinnunculus 95
ruppelli, Anas undulata 66
rupurumii, Phaethornis 257
rushiae, Pholidornis 531
ruspolii, Tauraco 206
russata, Haplophaedia aureliae 273
russata, Rhipidura rufifrons 497
russatus, Catharus aurantiirostris 664
russatus, Chlorostilbon 262
russatus, Chrysococcyx minutillus 210
russatus, Poecilotriccus 361
russelli, Cistothorus platensis 635
russelli, Euphonia hirundinacea 821
russelli, Oreortyx pictus 41
russeolus, Certhiaxis cinnamomeus 411
russeus, Thryothorus sinaloa 638
russicus, Panurus biarmicus 622
rustica, Emberiza 779
rustica, Haplospiza 788
rustica, Hirundo 535
rusticola, Scolopax 139
rusticolus, Falco 97
rusticus, Cisticola exilis 556
rutenbergi, Butorides striata 86
rutha, Myadestes lanaiensis 663
ruthae, Anthoscopus flavifrons 530
ruthae, Estrilda paludicola 730
ruthae, Fraseria cinerascens 689
ruthenus, Bubo bubo 225
ruthenus, Perisoreus infaustus 505
rutherfordi, Spilornis cheela 103
ruticilla, Setophaga 762
rutila, Amazilia 265
rutila, Emberiza 779
rutila, Phytotoma 346
rutila, Streptoprocne 246
rutilans, Lanius senator 481
rutilans, Passer 716
rutilans, Synallaxis 409
rutilans, Xenops 419
rutilus, Caprimulgus rufus 241
rutilus, Otus 222

rutilus, Thryothorus 637
ruwenzori, Cryptospiza salvadorii 728
ruwenzori, Dendropicos griseocephalus 317
ruwenzori, Psalidoprocne pristoptera 531
ruwenzoria, Mirafra africana 540
ruwenzorii, Apalis 560
ruwenzorii, Caprimulgus 244
ruwenzorii, Phyllastrephus flavostriatus 571
ruwenzorii, Pogonocichla stellata 674
ruwenzorii, Zoothera piaggiae 661
RUWENZORORNIS 206
ruweti, Ploceus 721
RYNCHOPS 153

saani, Erpornis zantholeuca 622
saba, Malacopteron magnum 601
sabini, Bonasa umbellus 44
sabini, Dryoscopus 459
sabini, Rhaphidura 250
sabini, Xema 149
sabinoi, Cercomacra laeta 389
sabota, Mirafra 542
sabotoides, Mirafra sabota 542
sabrina, Tanysiptera galatea 284
sabyi, Numida meleagris 40
sacculatus, Criniger ochraceus 572
sacer, Todiramphus chloris 289
sacerdos, Pericrocotus cinnamomeus 472
sacerdotis, Ceyx lepidus 290
sacerdotum, Monarcha 501
sachalinensis, Parus montanus 527
sacra, Egretta 88
sadai, Thryomanes bewickii 636
sadiecoatsae, Myiophobus roraimae 359
saffordi, Myzomela rubratra 442
sagitta, Heliodoxa leadbeateri 269
SAGITTARIUS 98
sagittatus, Oriolus 487
sagittatus, Otus 220
sagittatus, Pyrrholaemus 445
sagittirostris, Hemignathus 758
sagrae, Myiarchus 378
saharae, Athene noctua 232
saharae, Burhinus oedicnemus 130
saharae, Caprimulgus aegyptius 243
saharae, Passer simplex 717
saharae, Scotocerca inquieta 556
sahari, Emberiza striolata 778
saipanensis, Rhipidura rufifrons 496
saira, Piranga flava 818
saisseti, Cyanoramphus 188
sakaiorum, Abroscopus superciliaris 594
sakalava, Ploceus 722
sakeratensis, Dendrocitta vagabunda 510
SAKESPHORUS 380
sakhalina, Calidris alpina 144
sakhalinensis, Cettia diphone 579
sakhalinensis, Perisoreus infaustus 505
sakhalinensis, Pinicola enucleator 755
sala, Alauda gulgula 549
salamonis, Gallicolumba 170
salangana, Aerodramus 249
salangensis, Dicrurus leucophaeus 491
salax, Saxicola torquatus 684
salebrosus, Troglodytes troglodytes 639
salentina, Athene noctua 232
salicamans, Carduelis tristis 750
salicaria, Passerina caerulea 826
salicarius, Parus montanus 527
salicicola, Catharus fuscescens 665
saliens, Cinnyris afer 709
saliens, Rhopodytes tristis 212
salimali, Perdicula argoondah 55
salimalii, Apus pacificus 254
salimalii, Carpodacus synoicus 754
salimalii, Chalcophaps indica 164
salimalii, Cisticola juncidis 555
salimalii, Ploceus megarhynchus 723
salimalii, Pomatorhinus schisticeps 603

salimalii, *Zosterops palpebrosus* 625
salinarum, *Xolmis rubetra* 371
salinasi, *Laterallus jamaicensis* 119
salinicola, *Certhilauda albofasciata* 544
salita, *Calliphlox evelynae* 276
sallaei, *Crypturellus cinnamomeus* 32
sallaei, *Granatellus* 768
sallei, *Cyrtonyx montezumae* 44
salmoni, *Brachygalba* 331
salmoni, *Chrysothlypis* 815
salmoni, *Myrmotherula fulviventris* 385
salmoni, *Psarocolius angustifrons* 768
salmoni, *Tigrisoma fasciatum* 84
salomonensis, *Alcedo atthis* 292
salomonis, *Coracina* 469
salomonis, *Turnix maculosus* 129
salomonseni, *Dicaeum pygmaeum* 702
salomonseni, *Pseudocossyphus sharpei* 688
SALPINCTES 634
SALPORNIS 648
saltarius, *Caprimulgus rufus* 241
SALTATOR 824
SALTATRICULA 801
saltonis, *Melospiza melodia* 781
saltuarius, *Crypturellus erythropus* 32
saltuarius, *Xiphorhynchus flavigaster* 424
saltuensis, *Amazona albifrons* 203
saltuensis, *Grallaria rufula* 400
saltuensis, *Thryothorus genibarbis* 636
salvadoranus, *Dactylortyx thoracicus* 44
salvadorensis, *Attila spadiceus* 379
salvadori, *Caprimulgus macrurus* 243
salvadori, *Loriculus* 181
salvadori, *Lybius undatus* 309
salvadori, *Pitohui kirhocephalus* 490
salvadori, *Salpornis spilonotus* 648
salvadoriana, *Tanysiptera sylvia* 285
salvadorii, *Aegotheles albertisi* 246
salvadorii, *Chrysococcyx crassirostris* 210
salvadorii, *Coracina morio* 469
salvadorii, *Cryptospiza* 728
salvadorii, *Dicaeum maugei* 702
salvadorii, *Ducula pinon* 179
salvadorii, *Eremomela icteropygialis* 594
salvadorii, *Eudynamys scolopaceus* 211
salvadorii, *Leptocoma sericea* 708
salvadorii, *Lichenostomus subfrenatus* 431
salvadorii, *Lorius lory* 186
salvadorii, *Merops philippinus* 296
salvadorii, *Nectarinia johnstoni* 708
salvadorii, *Nothura darwinii* 34
salvadorii, *Onychognathus* 658
salvadorii, *Pachycephala hyperythra* 475
salvadorii, *Paradisaea raggiana* 518
salvadorii, *Psittaculirostris* 205
salvadorii, *Ptilinopus viridis* 177
salvadorii, *Pycnonotus erythrophthalmus* 568
salvadorii, *Sphecotheres vieilloti* 486
salvadorii, *Tanygnathus lucionensis* 192
salvadorii, *Treron calvus* 173
salvadorii, *Turdoides caudata* 610
salvadorii, *Zenaida aurita* 166
salvadorii, *Zosterops* 626
SALVADORINA 65
salvini, *Amazona autumnalis* 203
salvini, *Basileuterus rufifrons* 767
salvini, *Camarhynchus parvulus* 796
salvini, *Caprimulgus* 241
salvini, *Chlorostilbon canivetii* 261
salvini, *Claravis mondetoura* 167
salvini, *Colinus virginianus* 42
salvini, *Cyphorhinus arada* 642
salvini, *Empidonax flavescens* 368
salvini, *Eubucco bourcierii* 304
salvini, *Eutoxeres aquila* 255
salvini, *Gymnopithys* 393
salvini, *Habia fuscicauda* 819
salvini, *Knipolegus poecilurus* 369
salvini, *Lampornis amethystinus* 268

salvini, *Mitu* 39
salvini, *Neomorphus geoffroyi* 217
salvini, *Oryzoborus funereus* 793
salvini, *Pachyptila* 73
salvini, *Pachyramphus albogriseus* 345
salvini, *Picumnus exilis* 312
salvini, *Sclerurus guatemalensis* 418
salvini, *Sicalis olivascens* 790
salvini, *Spindalis zena* 820
salvini, *Thalassarche cauta* 72
salvini, *Tumbezia* 372
salvini, *Vireo pallens* 483
salvinii, *Icterus mesomelas* 770
samaipatae, *Synallaxis azarae* 408
samaliyae, *Columba larvata* 160
samaliyae, *Serinus mozambicus* 747
samamisicus, *Phoenicurus phoenicurus* 681
samarensis, *Ceyx melanurus* 290
samarensis, *Corvus enca* 512
samarensis, *Dicrurus hottentottus* 492
samarensis, *Dryocopus javensis* 326
samarensis, *Ficedula basilanica* 694
samarensis, *Ixos everetti* 574
samarensis, *Oriolus steerii* 487
samarensis, *Orthotomus* 565
samarensis, *Penelopides panini* 300
samarensis, *Prionochilus olivaceus* 699
samarensis, *Pycnonotus goiavier* 568
samarensis, *Rhinomyias ruficauda* 690
samarensis, *Rhipidura superciliaris* 493
samarensis, *Sarcophanops steerii* 336
samharensis, *Ammomanes deserti* 545
samios, *Garrulus glandarius* 508
samoensis, *Gymnomyza* 434
samoensis, *Porphyrio porphyrio* 125
samoensis, *Turdus poliocephalus* 669
samoensis, *Zosterops* 628
samueli, *Calyptorhynchus banksii* 183
samueli, *Cinclosoma cinnamomeum* 453
samueli, *Lichenostomus chrysops* 432
samueli, *Malurus cyaneus* 429
samueli, *Stagonopleura bella* 732
samuelis, *Melospiza melodia* 782
samveasnae, *Motacilla* 740
sana, *Lophortyx gambelii* 41
sanaka, *Melospiza melodia* 782
sanblasiana, *Cyanocorax* 506
sanborni, *Nothoprocta perdicaria* 33
sanborni, *Strix rufipes* 228
sanborni, *Zonotrichia capensis* 783
sanchezi, *Glaucidium* 231
sanctaecatarinae, *Otus* 224
sanctaecrucis, *Clytorhynchus nigrogularis* 500
sanctaecrucis, *Gallicolumba* 170
sanctaecrucis, *Myzomela cardinalis* 442
sanctaecrucis, *Zosterops* 628
sanctaehelenae, *Charadrius* 135
sanctaeluciae, *Cichlherminia lherminieri* 672
sanctaeluciae, *Lophotriccus pileatus* 361
sanctaeluciae, *Myadestes genibarbis* 663
sanctaeluciae, *Myiarchus oberi* 378
sanctaeluciae, *Ramphocinclus brachyurus* 650
sanctaemariae, *Cymbilaimus* 379
sanctaemariae, *Regulus regulus* 632
sanctaemartae, *Asthenes wyatti* 407
sanctaemartae, *Catharus fuscater* 664
sanctaemartae, *Lepidocolaptes lacrymiger* 425
sanctaemartae, *Myioborus miniatus* 765
sanctaemartae, *Myrmotherula schisticolor* 385
sanctaemartae, *Scytalopus* 396
sanctaemartae, *Trogon personatus* 281
sanctaemartae, *Turdus olivater* 670
sanctaemartae, *Tyrannus savana* 376
sanctaemartae, *Xiphocolaptes*
    promeropirhynchus 422
sanctaemarthae, *Chamaepetes goudotii* 39
sanctaemarthae, *Ramphocaenus melanurus* 643
sanctaethomae, *Ploceus* 723
sanctamartae, *Gymnocichla nudiceps* 390

sanctihieronymi, *Panyptila* 252
sanctijohannis, *Buteo lagopus* 112
sanctijohannis, *Dendrocopos medius* 319
sanctijohannis, *Phaethornis yaruqui* 256
sanctinicolai, *Strix aluco* 228
sanctithomae, *Brotogeris* 201
sanctithomae, *Coereba flaveola* 776
sanctithomae, *Dendrocolaptes* 422
sanctithomae, *Treron* 173
sanctorum, *Aimophila ruficeps* 787
sanctorum, *Passerculus sandwichensis* 784
sanctorum, *Picoides villosus* 321
sanctus, *Todiramphus* 289
sanderi, *Apalis rufogularis* 561
sandgroundi, *Chlorophoneus nigrifrons* 457
sandiae, *Pyrrhura rupicola* 199
sandiegensis, *Campylorhynchus brunneicapillus*
    634
sandlandi, *Acanthiza chrysorrhoa* 449
sandlandi, *Pycnoptilus floccosus* 444
sandvicensis, *Branta* 63
sandvicensis, *Gallinula chloropus* 126
sandvicensis, *Sterna* 149
sandwichensis, *Asio flammeus* 236
sandwichensis, *Chalcophaps indica* 164
sandwichensis, *Chasiempis* 499
sandwichensis, *Passerculus* 783
sandwichensis, *Porzana* 124
sandwichensis, *Pterodroma phaeopygia* 74
sanfelipensis, *Centurus superciliaris* 315
sanfelipensis, *Vireo gundlachii* 483
sanfordi, *Archboldia papuensis* 427
sanfordi, *Asio flammeus* 236
sanfordi, *Crateroscelis robusta* 445
sanfordi, *Cyornis* 697
sanfordi, *Diomedea epomophora* 72
sanfordi, *Erythrura trichroa* 733
sanfordi, *Gallirallus torquatus* 119
sanfordi, *Haliaeetus* 101
sanfordi, *Lagopus muta* 47
sanfordi, *Micrathene whitneyi* 232
sanfordi, *Mino kreffti* 653
sanfordi, *Myzomela cardinalis* 442
sanfordi, *Pachycephala orioloides* 477
sanfordi, *Penelopides exarhatus* 299
sanfordi, *Rhamphomantis megarhynchus* 211
sanghae, *Emberiza striolata* 778
sanghensis, *Stiphrornis* 674
sanghirana, *Pitta sordida* 339
sanghirense, *Dicaeum celebicum* 703
sanghirensis, *Aplonis panayensis* 652
sanghirensis, *Cittura cyanotis* 285
sanghirensis, *Colluricincla* 489
sanghirensis, *Macropygia amboinensis* 162
sanghirensis, *Megapodius cumingii* 36
sanghirensis, *Oriolus chinensis* 488
sangirensis, *Ceyx fallax* 290
sangirensis, *Leptocoma sericea* 707
sangirensis, *Tanygnathus sumatranus* 192
sangirensis, *Treron griseicauda* 172
sanguinea, *Cacatua* 183
sanguinea, *Himatione* 759
sanguineus, *Cnemophilus macgregorii* 451
sanguineus, *Pteroglossus* 302
sanguineus, *Pyrenestes ostrinus* 728
sanguineus, *Rhodopechys* 753
sanguineus, *Veniliornis* 322
sanguiniceps, *Haematortyx* 56
sanguiniceps, *Picus canus* 328
sanguinipectus, *Aethopyga saturata* 713
sanguinodorsalis, *Lagonosticta* 729
sanguinolenta, *Myzomela* 442
sanguinolenta, *Piranga bidentata* 818
sanguinolentum, *Dicaeum* 703
sanguinolentus, *Pardirallus* 125
sanguinolentus, *Ramphocelus* 806
sanguinolentus, *Rupicola peruvianus* 347
sanguinolentus, *Uragus sibiricus* 753
sanguinolentus, *Veniliornis fumigatus* 322

TRICHIXOS 681
Trichocichla 578
TRICHODERE 440
TRICHOGLOSSUS 185
TRICHOLAEMA 308
TRICHOLESTES 573
trichopsis, Otus 223
trichorrhos, Macronous ptilosus 609
TRICHOTHRAUPIS 805
trichroa, Erythrura 733
TRICLARIA 204
tricollaris, Charadrius 135
tricolor, Acridotheres melanopterus 655
tricolor, Agelaius 774
tricolor, Alectrurus 372
tricolor, Ara 196
tricolor, Atlapetes 800
tricolor, Coereba flaveola 776
tricolor, Copsychus malabaricus 680
tricolor, Coracina holopolia 469
tricolor, Egretta 88
tricolor, Epthianura 443
tricolor, Erythrura 733
tricolor, Ficedula 695
tricolor, Furnarius leucopus 404
tricolor, Lalage 470
tricolor, Lanius schach 480
tricolor, Perissocephalus 349
tricolor, Phalaropus 144
tricolor, Phyllastrephus icterinus 571
tricolor, Ploceus 722
tricolor, Poecilotriccus 361
tricolor, Prionops retzii 457
tricolor, Pycnonotus barbatus 567
tricolor, Rallina 118
tricolor, Tachybaptus ruficollis 79
tricolor, Terpsiphone rufiventer 498
tricolor, Vanellus 134
tridactyla, Jacamaralcyon 331
tridactyla, Rissa 149
tridactylus, Picoides 322
trifasciatus, Basileuterus 767
trifasciatus, Carpodacus 754
trifasciatus, Hemispingus 804
trifasciatus, Nesomimus 649
trigeminus, Ptilinopus coronulatus 176
TRIGONOCEPS 102
trigonostigma, Dicaeum 701
TRINGA 141
trinitae, Chlorocharis emiliae 631
trinitatis, Euphonia 821
trinitatis, Fregata ariel 89
trinitatis, Geotrygon linearis 169
trinitatis, Icterus nigrogularis 771
trinitatis, Manacus manacus 341
trinitatis, Myiopagis gaimardii 350
trinitatis, Piculus rubiginosus 323
trinitatis, Pitangus sulphuratus 374
trinitatis, Pulsatrix perspicillata 230
trinitatis, Ramphocaenus melanurus 643
trinitatis, Sakesphorus canadensis 380
trinitatis, Synallaxis albescens 408
trinitatis, Tangara guttata 811
trinitatis, Toxostoma crissale 650
trinkutensis, Coturnix chinensis 54
trinotatus, Accipiter 105
tristanensis, Sterna vittata 150
tristigma, Caprimulgus 244
tristigmata, Gallicolumba 170
tristis, Acridotheres 655
tristis, Carduelis 750
tristis, Corvus 513
tristis, Meiglyptes 330
tristis, Pachyramphus polychopterus 345
tristis, Phylloscopus collybita 589
tristis, Regulus regulus 633
tristis, Rhopodytes 212
tristissima, Lonchura 735
tristoides, Maina 655

tristrami, Dicaeum 702
tristrami, Emberiza 778
tristrami, Micropsitta finschii 182
tristrami, Myzomela 442
tristrami, Oceanodroma 78
tristrami, Ptilinopus mercierii 176
tristrami, Todiramphus chloris 288
tristramii, Onychognathus 658
tristrani, Pterodroma externa 74
tristriatus, Basileuterus 767
tristriatus, Serinus 748
triton, Cacatua galerita 184
triurus, Mimus 649
trivialis, Anthus 743
trivialis, Philemon buceroides 436
trivirgatus, Accipiter 104
trivirgatus, Accipiter trivirgatus 104
trivirgatus, Aegithalos caudatus 539
trivirgatus, Conopias 375
trivirgatus, Monarcha 501
trivirgatus, Phylloscopus 592
trivirgatus, Pomatostomus temporalis 450
trizonatus, Buteo oreophilus 112
trobriandi, Colluricincla megarhyncha 489
trobriandi, Lalage leucomela 470
trobriandi, Manucodia comrii 516
trocaz, Columba 158
trochila, Columbina passerina 167
trochileum, Dicaeum 703
trochilirostris, Campylorhamphus 425
trochiloides, Acanthorhynchus tenuirostris 440
trochiloides, Icterus cucullatus 770
trochiloides, Phylloscopus 591
TROCHILUS 261
trochilus, Leptocoma sperata 707
trochilus, Phylloscopus 589
TROCHOCERCUS 498
TROGLODYTES 639
troglodytes, Athene cunicularia 233
troglodytes, Caprimulgus ridgwayi 241
troglodytes, Cisticola 554
troglodytes, Collocalia 248
troglodytes, Estrilda 730
troglodytes, Icterus gularis 771
troglodytes, Troglodytes 639
troglodytoides, Spelaeornis 605
TROGON 280
trophis, Pericrocotus igneus 472
tropica, Fregetta 77
tropica, Streptopelia capicola 161
tropicalis, Accipiter minullus 106
tropicalis, Falco sparverius 95
tropicalis, Icteria virens 768
tropicalis, Mirafra africana 540
tropicalis, Thryothorus ludovicianus 638
tropicalis, Turdus libonyanus 666
Tropicoperdix 55
TROPICRANUS 299
tropicus, Anas cyanoptera 67
tropicus, Euplectes ardens 726
tropicus, Zosterops lateralis 629
trothae, Francolinus shelleyi 51
trothae, Ploceus rubiginosus 722
trouessarti, Pseudobulweria rostrata 75
troughtoni, Macropygia rufa 163
trudeaui, Sterna 150
trudiae, Aegithina tiphia 465
trudis, Chlorospingus ophthalmicus 817
TRUGON 165
trumaii, Dendrocincla fuliginosa 420
TRYNGITES 144
tsaidamensis, Lanius isabellinus 479
tsaidamensis, Rallus aquaticus 121
tsanae, Acrocephalus gracilirostris 584
tsaratananae, Pseudocossyphus sharpei 688
tsavoensis, Cinnyris 710
tschadensis, Estrilda melpoda 730
tschebaiewi, Luscinia pectoralis 676
tschimenensis, Tetraogallus tibetanus 48

tschitscherini, Sitta neumayer 646
tschudii, Ampelioides 347
tschudii, Chamaepetes goudotii 39
tschudii, Pardirallus sanguinolentus 125
tschudii, Piprites chloris 349
tschudii, Thamnophilus nigrocinereus 381
tschuktschorum, Calidris ptilocnemis 144
tschuktschorum, Xema sabini 149
tschusii, Carduelis carduelis 751
tschusii, Emberiza schoeniclus 780
tschusii, Jynx torquilla 311
tschusii, Lyrurus tetrix 46
tschutschensis, Motacilla flava 739
tsubame, Aerodramus vanikorensis 249
tuberculifer, Myiarchus 377
tuberosum, Mitu 39
tucanus, Ramphastos 303
tucinkae, Eubucco 304
tucopiae, Aplonis tabuensis 652
tucopiae, Myzomela cardinalis 442
tucumana, Amazona 203
tucumana, Carduelis magellanica 750
tucumana, Ochthoeca leucophrys 373
tucumanum, Glaucidium brasilianum 231
tucumanus, Cistothorus platensis 635
tucumanus, Cyanocorax chrysops 506
tucumanus, Phrygilus unicolor 788
tucumanus, Phylloscartes ventralis 356
tucumanus, Piculus rubiginosus 323
tucumanus, Picumnus cirratus 312
tucuyensis, Dysithamnus plumbeus 384
tuerosi, Laterallus jamaicensis 119
tuftsi, Asio otus 236
tuidara, Tyto alba 219
tuipara, Brotogeris chrysoptera 201
tukki, Meiglyptes 330
tullbergi, Campethera 316
tumbezana, Phaeomyias murina 354
TUMBEZIA 372
tumultuosus, Pionus 202
tumulus, Megapodius reinwardt 37
tundrae, Charadrius hiaticula 135
tundrius, Falco peregrinus 97
tuneti, Tachymarptis melba 253
tunneyi, Epthianura crocea 443
tunneyi, Puffinus assimilis 76
TURACOENA 163
turanicus, Otus scops 221
turati, Picoides pubescens 321
turatii, Laniarius 460
turcestanicus, Phasianus colchicus 60
turcmenica, Sylvia mystacea 598
turcomana, Acrocephalus palustris 586
turcomanus, Bubo bubo 226
turcosa, Cyanolyca 505
turcosa, Irena puella 632
turcosus, Cyornis 696
turdina, Chamaeza 398
turdina, Dendrocincla fuliginosa 420
turdina, Schiffornis 343
turdinus, Automolus ochrolaemus 417
turdinus, Campylorhynchus 634
turdinus, Ptyrticus 602
TURDOIDES 609
turdoides, Cataponera 662
TURDUS 666
turensis, Tetrao urogalloides 46
turgidus, Euplectes orix 725
turipavae, Megalurulus whitneyi 578
turipavae, Zoothera talaseae 662
turkana, Cercomela scotocerca 686
turkanae, Cinnyris habessinicus 710
turkestanica, Carduelis chloris 749
turkestanica, Columba rupestris 157
turkestanicus, Parus bokharensis 525
TURNAGRA 474
turnagra, Turnagra capensis 474
turneffensis, Melanerpes aurifrons 315
turneri, Colius leucocephalus 278

UPUPA 297
URAEGINTHUS 729
URAGUS 753
uralensis, Cinclus cinclus 698
uralensis, Dendrocopos leucotos 319
uralensis, Parus montanus 527
uralensis, Strix 228
uralensis, Tetrao urogallus 45
uraniae, Rhipidura rufifrons 496
URATELORNIS 283
urbanoi, Cyphorhinus arada 642
urbicum, Delichon 536
URIA 154
urichi, Phyllomyias 350
urile, Phalacrocorax 92
urinator, Pelecanus occidentalis 90
urinatrix, Pelecanoides 78
UROCHROA 269
urochrysia, Chalybura 267
UROCISSA 509
UROCOLIUS 278
UROCYNCHRAMUS 753
URODYNAMIS 212
urogalloides, Tetrao 46
urogallus, Tetrao 45
UROGLAUX 235
UROLAIS 560
UROLESTES 478
Uromyias 353
UROPELIA 168
urophasianus, Centrocercus 46
UROPSALIS 245
UROPSILA 641
uropygialis, Acanthiza 449
uropygialis, Cacicus 769
uropygialis, Carduelis 750
uropygialis, Ceyx lepidus 290
uropygialis, Cisticola juncidis 555
uropygialis, Coereba flaveola 776
uropygialis, Collocalia esculenta 248
uropygialis, Ducula rufigaster 179
uropygialis, Lipaugus 348
uropygialis, Lonchura castaneothorax 736
uropygialis, Melanerpes 314
uropygialis, Nigrita fusconotus 726
uropygialis, Pachyramphus major 345
uropygialis, Pardalotus striatus 444
uropygialis, Pheucticus aureoventris 823
uropygialis, Phyllomyias 350
uropygialis, Pitohui kirhocephalus 490
uropygialis, Pogoniulus pusillus 308
uropygialis, Ptilorrhoa castanonota 453
uropygialis, Sicalis 790
uropygialis, Stelgidopteryx ruficollis 534
uropygialis, Zosterops 627
uropygiatus, Attila spadiceus 379
URORHIPIS 562
UROSPHENA 578
urosticta, Myrmotherula 386
UROSTICTE 273
urostictus, Pycnonotus 567
UROTHRAUPIS 801
UROTRIORCHIS 108
urschi, Lophotibis cristata 83
ursulae, Cinnyris 711
urubambae, Elaenia albiceps 351
urubambae, Formicivora rufa 388
urubambae, Hemispingus superciliaris 804
urubambae, Ochthoeca leucophrys 373
urubambae, Scytalopus 397
urubambae, Synallaxis azarae 408
urubambae, Tangara parzudakii 811
urubambae, Thraupis episcopus 807
urubambensis, Asthenes 407
urubambensis, Carduelis magellanica 750
urubitinga, Buteogallus 109
uruguaiensis, Otus choliba 223
urumutum, Nothocrax 39
urutani, Automolus roraimae 417

usambara, Hyliota australis 587
usambarae, Andropadus nigriceps 569
usambarae, Bradypterus lopesi 580
usambarae, Francolinus squamatus 51
usambarae, Sheppardia sharpei 675
usambaricus, Cinnyris mediocris 709
usambaricus, Laniarius fuelleborni 460
usambiro, Trachyphonus darnaudii 310
usheri, Asthenes dorbignyi 406
usheri, Eremialector bicinctus 157
ussheri, Indicator conirostris 311
ussheri, Muscicapa 692
ussheri, Pholidornis rushiae 531
ussheri, Pitta granatina 338
ussheri, Scotopelia 227
ussheri, Tchagra australis 458
ussheri, Telacanthura 250
ussurianus, Urosphena squameiceps 578
ussuriensis, Anthus richardi 741
ussuriensis, Bubo bubo 226
ussuriensis, Carduelis sinica 749
ussuriensis, Coturnix japonica 54
ussuriensis, Lyrurus tetrix 46
ussuriensis, Ninox scutulata 234
ussuriensis, Otus bakkamoena 220
ussuriensis, Tringa totanus 141
ussuriensis, Uragus sibiricus 753
usta, Otus watsonii 224
usticollis, Eremomela 595
ustulata, Phlegopsis erythroptera 394
ustulatus, Catharus 665
ustulatus, Microcerculus 642
utahensis, Eremophila alpestris 550
utakwensis, Lichenostomus subfrenatus 431
uthaii, Yuhina gularis 622
utilensis, Buteogallus anthracinus 109
utupuae, Pachycephala vitiensis 477
utupuae, Rhipidura rufifrons 497
utupuae, Todiramphus chloris 288
uvaeensis, Eunymphicus cornutus 188
uzungwensis, Francolinus squamatus 51
uzungwensis, Phyllastrephus flavostriatus 571

v-nigra, Somateria mollissima 69
vaalensis, Anthus 743
vacillans, Streptopelia chinensis 162
vafer, Cettia major 579
vaga, Strix leptogrammica 228
vagabunda, Dendrocitta 510
vagans, Cuculus 207
vagans, Melilestes megarhynchus 439
vagans, Todiramphus sanctus 289
vagilans, Calandrella cinerea 546
vaillantii, Picus 328
vaillantii, Trachyphonus 309
valdizani, Veniliornis dignus 322
valencianus, Certhiaxis cinnamomeus 411
valenciobuenoi, Icterus cayanensis 770
valens, Erithacus rubecula 675
valens, Indicator minor 311
valentinae, Alcippe vinipectus 619
valentini, Seicercus 593
valida, Prinia sylvatica 558
valida, Sicalis flaveola 791
validior, Cercomela scotocerca 686
validirostris, Dendrocopos maculatus 317
validirostris, Lanius 480
validirostris, Melithreptus 435
validirostris, Pteruthius flaviscapis 617
validirostris, Upucerthia 403
validus, Corvus 513
validus, Dendrocolaptes picumnus 423
validus, Myiarchus 378
validus, Pachycoccyx audeberti 207
validus, Pachyramphus 345
validus, Reinwardtipicus 330
valisineria, Aythya 68
vallicola, Ortalis vetula 37
valuensis, Zosterops lateralis 629

vana, Lonchura 736
vanbemmelli, Accipiter virgatus 107
vanderbilti, Collocalia esculenta 247
vanderbilti, Malacocincla sepiaria 600
vanderbilti, Serinus estherae 748
vandevenderi, Parus wollweberi 529
vandewateri, Otus spilocephalus 220
vandeweghei, Alethe poliocephala 674
VANELLUS 133
vanellus, Vanellus 133
VANGA 461
vanheurni, Spizaetus cirrhatus 114
vanheysti, Picus chlorolophus 327
vanikorensis, Aerodramus 249
vanikorensis, Lalage maculosa 471
vanikorensis, Myiagra 504
vanikorensis, Pachycephala caledonica 477
vanikorensis, Turdus poliocephalus 668
vanmarli, Carduelis chloris 749
vanrossemi, Aramides cajanea 122
vanrossemi, Eumomota superciliosa 294
vanrossemi, Momotus mexicanus 293
vanrossemi, Polioptila albiloris 644
vanrossemi, Sterna nilotica 149
vanrossemi, Thryothorus modestus 638
vanrossemi, Vireosylva flavoviridis 485
vansoni, Serinus mozambicus 747
vantynei, Aimophila botterii 787
vantynei, Pycnonotus melanicterus 566
vantynei, Taccocua leschenaultii 212
vanwyckii, Ducula pistrinaria 179
varennei, Garrulax maesi 613
varia, Grallaria 399
varia, Mniotilta 762
varia, Strix 228
varia, Tangara 811
variabilis, Amaurornis phoenicurus 123
variabilis, Emberiza 780
variabilis, Eudyptula minor 71
variabilis, Xenicus longipes 335
variegata, Anas rhynchotis 67
variegata, Lagopus lagopus 47
variegata, Ninox 235
variegata, Sula 90
variegata, Tadorna 65
variegaticeps, Alcippe 618
variegaticeps, Anabacerthia 414
variegatus, Certhionyx 441
variegatus, Crypturellus 33
variegatus, Garrulax 613
variegatus, Indicator 311
variegatus, Legatus leucophaius 374
variegatus, Merops 295
variegatus, Mesitornis 116
variegatus, Numenius phaeopus 141
variegatus, Odontophorus atrifrons 43
variegatus, Saxicola torquatus 683
variolosus, Cacomantis 209
varius, Corvus corax 514
varius, Cuculus 207
varius, Empidonomus 375
varius, Fregilupus 656
varius, Gallus 58
varius, Parus 527
varius, Phalacrocorax 91
varius, Psephotus 189
varius, Sphyrapicus 315
varius, Turnix 130
varnak, Perisoreus infaustus 505
varonai, Torreornis inexpectata 787
varzeae, Picumnus 312
vasa, Coracopsis 194
vassali, Garrulax 613
vassorii, Tangara 812
vasta, Microligea palustris 764
vastitas, Pterocles coronatus 156
vatensis, Zosterops lateralis 629
vatuanus, Clytorhynchus vitiensis 500
vauana, Lalage maculosa 471

Index of English Names

Laughing-thrush Ashy-headed 612, Barred 614, Black 613, Black-faced 615, Black-hooded 613, Black-throated 613, Blue-winged 615, Brown-capped 615, Brown-cheeked 615, Chestnut-backed 613, Chestnut-capped 614, Chestnut-crowned 615, Chestnut-eared 614, Collared 616, Elliot's 615, Giant 614, Greater Necklaced 612, Grey 613, Grey-breasted 615, Grey-sided 614, Lesser Necklaced 612, Masked 612, Melodious 614, Moustached 614, Nilgiri 615, Plain 613, Red-tailed 616, Red-winged 616, Rufous-chinned 614, Rufous-fronted 612, Rufous-necked 614, Rufous-vented 613, Rusty 614, Scaly 615, Snowy-cheeked 613, Spot-breasted 614, Spotted 614, Streaked 615, Striated 613, Striped 615, Sunda 612, Variegated 613, White-browed 615, White-cheeked 613, White-crested 612, White-necked 613, White-speckled 614, White-throated 612, White-whiskered 615, Yellow-throated 613
Leaf-love 570
Leafbird Blue-masked 699, Blue-winged 699, Golden-fronted 699, Greater Green 699, Lesser Green 699, Orange-bellied 699, Philippine 699, Yellow-throated 699
Leaftosser Black-tailed 418, Grey-throated 418, Rufous-breasted 418, Scaly-throated 418, Short-billed 418, Tawny-throated 418
Leiothrix Red-billed 616
Limpkin 129
Linnet Common 752, Warsangli 752, Yemeni 752
Liocichla Emei Shan 616, Red-faced 616, Steere's 616
Logrunner Australian 450, Papuan 450
Longbill Bocage's 588, Grey 587, Kemp's 587, Kretschmer's 587, Plumed 451, Pulitzer's 587, Pygmy 451, Slaty-chinned 452, Yellow 587, Yellow-bellied 452
Longclaw Abyssinian 741, Cape 741, Fülleborn's 741, Grimwood's 741, Pangani 741, Rosy-breasted 741, Sharpe's 741, Yellow-throated 741
Longspur Chestnut-collared 781, Lapland 780, McCown's 780, Smith's 780
Longtail Green 560
Loon 71, Arctic 71, Common 71, Pacific 71, Yellow-billed 71
Lorikeet Blue 186, Blue-crowned 186, Blue-fronted 187, Duchess 187, Fairy 187, Flores 185, Goldie's 186, Henderson 186, Iris 186, Josephine's 187, Kuhl's 186, Little 186, Marigold 185, Meek's 187, Mindanao 185, Musk 186, New Caledonian 187, Olive-headed 185, Orange-billed 187, Ornate 185, Palm 187, Papuan 187, Plum-faced 187, Pohnpei 185, Purple-crowned 186, Pygmy 187, Rainbow 185, Red-chinned 187, Red-collared 185, Red-flanked 187, Red-fronted 187, Red-throated 187, Scaly-breasted 185, Scarlet-breasted 185, Striated 187, Ultramarine 186, Varied 186, Yellow-and-green 185, Yellow-billed 187
Lory Black 184, Black-capped 186, Black-winged 185, Blue-eared 185, Blue-

streaked 185, Brown 184, Cardinal 184, Chattering 186, Collared 186, Dusky 185, Purple-bellied 186, Purple-naped 186, Red 185, Red-and-blue 184, Violet-necked 184, White-naped 186, Yellow-bibbed 186, Yellow-streaked 184
Lovebird Black-cheeked 194, Black-collared 194, Black-winged 194, Fischer's 194, Grey-headed 194, Nyasa 194, Red-headed 194, Rosy-faced 194, Yellow-collared 194
Lyrebird Prince Albert's 426, Superb 426

Macaw Blue-and-yellow 195, Blue-headed 196, Blue-throated 195, Blue-winged 196, Chestnut-fronted 196, Glaucous 195, Great Green 195, Hispaniolan 196, Hyacinth 195, Indigo 195, Military 195, Red-and-green 196, Red-bellied 196, Red-fronted 196, Red-shouldered 196, Scarlet 195, Spix's 195, Yellow-collared 196
Magpie Australian 463, Azure-winged 509, Black 505, Black-billed 511, Common 511, Common Green 510, Indochinese Green 510, Red-billed Blue 509, Short-tailed Green 510, Sri Lanka Blue 509, Taiwan Blue 509, White-winged 509, Yellow-billed 511, Yellow-billed Blue 509
Magpie-Goose 61
Magpie-Jay Black-throated 507, White-throated 507
Magpie-lark 503
Magpie-Mannikin 734
Magpie-Robin Madagascan 680, Oriental 680, Seychelles 680
Magpie-Shrike 478
Magpie-Starling 658
Magpie-Tanager 803
Maleo 36
Malia 624
Malimbe Black-throated 723, Blue-billed 723, Crested 724, Ibadan 723, Rachel's 723, Red-bellied 724, Red-crowned 723, Red-headed 724, Red-vented 723, Tai 723
Malkoha Black-bellied 212, Blue-faced 212, Chestnut-bellied 212, Chestnut-breasted 213, Green-billed 212, Raffles's 212, Red-billed 212, Red-faced 213, Rough-crested 213, Scale-feathered 213, Sirkeer 212, Yellow-billed 213
Mallard 66
Malleefowl 36
Mamo Black 759, Hawaii 759
Manakin Araripe 341, Band-tailed 342, Bearded 341, Black 342, Blue 341, Blue-backed 341, Blue-crowned 340, Blue-rumped 340, Cerulean-capped 341, Club-winged 340, Crimson-hooded 342, Fiery-capped 340, Flame-crowned 342, Golden-crowned 341, Golden-headed 343, Golden-winged 340, Green 342, Helmeted 341, Jet 342, Lance-tailed 341, Long-tailed 341, Olive 342, Opal-crowned 341, Orange-bellied 341, Orange-crowned 342, Pin-tailed 340, Red-capped 343, Red-headed 343, Round-tailed 343, Scarlet-horned 342, Snow-capped 341, Striped 340, White-bibbed 340, White-crowned 342, White-fronted 341, White-ruffed 340, White-throated 340, Wire-tailed 342, Yellow-crowned 342, Yellow-headed

342, Yungas 341
Mango Antillean 259, Black-throated 259, Green 259, Green-breasted 259, Green-throated 259, Jamaican 259, Veraguas 259
Mannikin Alpine 736, Bismarck 736, Black 736, Black-and-white 734, Black-breasted 736, Bronze 734, Chestnut-breasted 736, Grand 736, Grey-banded 736, Grey-crowned 736, Grey-headed 736, Hooded 736, New Ireland 736, Snow Mountain 736, Streak-headed 735, White-cowled 736, Yellow-rumped 736
Manucode Crinkle-collared 516, Curl-crested 516, Glossy-mantled 515, Jobi 516, Trumpet 516
Marabou 81
Marshbird Brown-and-yellow 773, Yellow-rumped 772
Martin African River 531, Asian House 536, Banded 533, Brazza's 532, Brown-chested 534, Caribbean 533, Collared Sand 532, Congo Sand 532, Cuban 533, Dusky Crag 536, Eurasian Crag 536, Fairy 538, Galapagos 533, Grey-breasted 533, Mascarene 532, Nepal House 536, Northern House 536, Pale Crag 536, Pale Sand 532, Peruvian 533, Plain 532, Purple 533, Rock 536, Sinaloa 533, Southern 533, Tree 538, White-eyed River 531
Meadowlark Eastern 775, Long-tailed 775, Pampas 775, Peruvian 775, Western 775
Megapode Biak 36, Dusky 36, Forster's 37, Melanesian 37, Micronesian 36, Moluccan 37, New Guinea 37, Nicobar 36, Philippine 36, Polynesian 36, Sula 36, Tanimbar 36, Vanuatu 37
Melampitta Greater 515, Lesser 515
Merganser Auckland Island 70, Brazilian 70, Common 70, Hooded 69, Red-breasted 70, Scaly-sided 70
Merlin 96
Mesia Silver-eared 616
Mesite Brown 116, Subdesert 116, White-breasted 116
Metaltail Black 274, Coppery 274, Fire-throated 274, Neblina 274, Perija 274, Scaled 274, Tyrian 274, Violet-throated 274, Viridian 274
Millerbird 584
Miner Bell 434, Campo 402, Coastal 402, Common 402, Creamy-rumped 402, Dark-winged 402, Greyish 402, Noisy 434, Puna 402, Rufous-banded 402, Short-billed 402, Slender-billed 402, Thick-billed 402, Yellow-throated 434
Minivet Ashy 472, Fiery 472, Flores 472, Grey-chinned 472, Long-tailed 472, Rosy 471, Scarlet 473, Short-billed 472, Small 472, Sunda 472, Swinhoe's 472, White-bellied 472
Minla Blue-winged 618, Chestnut-tailed 618, Red-tailed 618
Mistletoebird 703
Mockingbird Bahama 649, Blue 650, Blue-and-white 650, Brown-backed 649, Chalk-browed 649, Charles 649, Chatham 649, Chilean 649, Galapagos 649, Hood 649, Long-tailed 649, Northern 648, Patagonian 649, Socorro 649, Tropical 648, White-banded 649
Monal Chinese 58, Himalayan 58,